Instrumental Data for Drug Analysis

Volume 6

CRC SERIES IN FORENSIC AND POLICE SCIENCE

BARRY A. J. FISHER, Series Editor
L.A. County Sheriff's Department

TECHNIQUES OF CRIME SCENE INVESTIGATION
Fifth Edition
Barry A. J. Fisher

SCIENTIFIC EXAMINATION OF QUESTIONED DOCUMENTS
Revised Edition
Ordway Hilton

ADVANCES IN FINGERPRINT TECHNOLOGY
Henry C. Lee
R. E. Gaensslen

INSTRUMENTAL DATA FOR DRUG ANALYSIS
Second Edition, Volumes 1–4
Terry Mills III
J. Conrad Roberson

INSTRUMENTAL DATA FOR DRUG ANALYSIS
Second Edition, Volume 5
Terry Mills III
J. Conrad Roberson
H. Horton McCurdy
William H. Wall

INSTRUMENTAL DATA FOR DRUG ANALYSIS
Volume 6-7
Terry Mills III
J. Conrad Roberson
William H. Wall
Kevin L. Lothridge
William D. McDougall
Michael W. Gilbert

Instrumental Data for Drug Analysis

Volume 6

Terry Mills III, J. Conrad Roberson,
William H. Wall, Kevin L. Lothridge,
William D. McDougall, and Michael W. Gilbert

CRC Press
Boca Raton New York London Tokyo

Publisher/Life Sciences: Robert B. Stern
Editorial Assistant: Jean Jarboe
Project Editor: Carrie L. Unger
Marketing Manager: Greg Daurelle
Direct Marketing Manager: Arline Massey
Cover design: Denise Craig
PrePress: Kevin Luong
Manufacturing: Sheri Schwartz

Library of Congress Cataloging-in-Publication Data

Mills, Terry
Instrumental data for drug analysis / by Terry Mills III, J. Conrad Roberson, William H. Wall
v. < 7 >; cm. — (CRC series on forensic and police science)
Includes index
Includes bibliographical references and index.
ISBN 0-8493-8115-0 (alk. paper)
1. Drugs—Analysis—Standards. 2. Drugs—Sepctra—Standards. 3. Pesticides—Analysis—Standards. 4. Pesticides—Spectra—Standards. 5. Instrumental analysis—Standards—Indexes. I. Roberson, J. Conrad. II. Wall, William H. III. Title IV. Series
RA1228.M54 1996
615′.1901—dc20

96-31352

CIP

Direct all inquiries to CRC Press, Inc., 2000 Corporate Blvd., N.W., Boca Raton, Florida 33431.

International Standard Book Number 0-8493-8115-0
Library of Congress Card Number 96-31352
Printed in the United States of America 1 2 3 4 5 6 7 8 9 0
Printed on acid-free paper

CONTENTS

PREFACE

Almost everyone engaged in the analysis of drug-related compounds, whether in the forensic, clinical, or university laboratory, has an accumulation of analytical data and thus has acquired a data base for the analysis of these compounds. Some of the information contained in Volume 6 of *Instrumental Data for Drug Analysis* is available in the literature; however, there is no single source that contains timely, quality data of this type presented in a large, easily usable format.

As in our previous volumes, we have included the six currently most popular analytical techniques: ultraviolet (UV), infrared (IR) spectrometry, proton nuclear magnetic resonance (NMR) spectrometry, mass spectrometry (MS), gas chromatography (GC), and high pressure liquid chromatography (HPLC). As the quality of data presented was of paramount importance in a reference source, we generated all of our data in our laboratories under uniform, reproducible conditions using state-of-the-art technology and verified chemical standards.

In this Volume 6 of the second edition of *IDDA,* we have included a special section of analytical data especially designed for toxicologists. This text contains mass spectra of over 200 pesticides. In addition, there are over 600 GC/FTIR spectra of many of the drugs found in previous volumes of *IDDA*. Six new tabulations of mass spectra data on these drugs and pesticides are included in the Cumulative Index Volume 7. In the Cumulative Index all indices have been expanded to include all data from volumes 1 through 6 with the corresponding page and volume number given after each entry.

Terry Mills III
J. Conrad Roberson
William H. Wall
Kevin L. Lothridge
William D. McDougall
Michael Gilbert

INTRODUCTION

This reference book consists of chromatographic and spectral data on over one hundred and thirty selected drug compounds as well as over 600 GC/FTIR drug spectra and over 200 pesticides mass spectra. Each monograph is accompanied by UV, NMR, IR, and MS spectra and tabulation of GC and HPLC data where available. The information on a specific compound should be located by using the alphabetical index, the GC, IR, and MS tables or the UV maxima indexes found in Volume 7 of *IDDA*.

CHEMICAL STANDARDS

Every effort was made to secure chemical standards of the highest purity available. Where possible, the data presented in this book were obtained from samples secured as "pure" drug standards from the Drug Enforcement Administration (DEA), Applied Sciences Laboratories, Sigma Chemical Company, or various pharmaceutical companies. The pesticides were obtained from Chem Service or the former EPA-Pesticides and Industrial Chemicals Repository and were in excess of 95% purity. When necessary, samples were purified by extraction methods followed by recrystallization to constant literature melting points and verified by thin layer chromatography, In almost all cases, the data presented on each compound were obtained from one sample. The sample purity was usually greater than 95% and, in many cases, greater than 99%. Where available, each spectrum generated was confirmed by previously published data.

DRUG MONOGRAPHS

The monograph chiefly consists of the chemical title, molecular formula and weights, synonyms and trade names, usage, and structure. In most cases the chemical title, which appears above each spectrum, either is listed in the Federal Drug Code, Title 21, or is the most common name. The molecular weights are based on the current acceptable IUPAC convention to the nearest hundredth of a decimal place. The value in parentheses represents the weight using the most abundant naturally occurring isotope of each element.

Generally, the first name listed as the synonym is the uninverted form of the Chemical Abstracts' name. Other alternate names such as common chemical or trivial names follow the Chemical Abstracts listing. The trade names include those that are currently available as listed in the 1996 Physician's Desk Reference (PDR). The major therapeutic actions of the drugs are listed in the use section. The structure presented on each compound is, in most cases, not intended to represent spatial configuration. The structure were drawn by using Chemwindow program from SoftShell International.

GAS CHROMATOGRAPHY

The gas chromatography data are presented in the monographs in the form of Kovats indices calculated by the following formula:

$$I = 100\ [2\ (\log T_D / T_X) / (\log T_Y / T_X) + X]$$

where

I = Kovats index

T_D = Retention time of the drug

T_x = Retention time of an even numbered normal hydrocarbon whose carbon number is X

T_y = Retention time of an even numbered normal hydrocarbon whose carbon number is Y where $Y = X+2$ and $T_x \leq T_D \leq T_Y$

For compounds eluting between dotriacontane and hexatriacontane, the "2" in the above equation was changed to a "4" to calculate indices between 3200 and 3600. The C34 hydrocarbon standard was not available . The chromatograph used was a Hewlett-Packard (HP) 5890 Series II model attached to an HP 5988 mass spectrometer, which was used to verify the identity of the spectral peaks used to produce the data presented here. The column used was an HP1 capillary column with an inside diameter of 0.2 mm and a 0.33 micron film thickness of methyl silicone stationary phase. The carrier gas was helium at a flow rate of 0.9 ml/mint All measurements were made with isothermal temperature programs. The data are presented as Kovats indices followed by the column temperatures, in degrees Centigrade, used to produce the data. To reproduce the results presented in this book, the same temperature, stationary phase, carrier gas, gas flow rate, and column dimensions should be used. The Kovats indices presented here cannot be compared with indices measured with a packed column. A table of retention indices appears in Appendix E in Volume 7 of *IDDA*.

MASS SPECTROMETRY

The mass spectra were acquired on a Hewlett-Packard 5970 GC/MS operating in the electron impact (EI) mode with an electron energy of 70 ev. Unless otherwise noted, samples were introduced via a methyl silicone column into the MS source, which was maintained at 200° C.

Every effort was made to standardize the sample size and mass spectrometer tuning to ensure consistent spectra throughout this collection. When it was necessary to manipulate the spectra, e.g., to remove traces of the injection solvent or column background, due care was exercised to avoid distorting the data. Mass calibration was checked several times a day, and the inertness of the interface was demonstrated daily by the analysis of cholesterol, which produced a 386/368 ion ratio of greater than 2:1.

The mass spectra were plotted on a HP 7550A graphics plotter. Although prominent ions in each spectrum are labeled, the user should be aware that these masses were selected on the basis of abundance and may not indicate the most significant fragments for each compound.

Several cumulative indices of the mass spectra, sorted by base peak, by exact molecular mass, and by alphabetical order for both drugs and pesticides are included as Appendix D of Volume 7 of *IDDA*.

The abbreviation "OA" within a name on the pesticide mass spectra refers to the "oxygen-analog" of the organophosphate pesticide.

HIGH PRESSURE LIQUID CHROMATOGRAPHY

High pressure liquid chromatography was carried out on an HP 1090 Series II liquid chromatograph with the Chemstation and diode array detector. On-the-fly ultraviolet spectra were taken to verify the identity of the peaks used to generate the retention time data presented here. A Hewlett-Packard 4.6 by 100 mm column containing 5 micron ODS Hypersil was used with the corresponding Brownlee Newguard precolumn. The flow rate was 0.5/min with a column "dead Volume" of 1.5 ml. The following solvents were used:
Fisher Optima grade methanol and ethanol
Data is presented with solvent used followed by retention times in minutes.

ULTRAVIOLET SPECTROPHOTOMETRY

The absorption spectra were obtained with a Hewlett-Packard 8451A diode array spectrophotometer and plotted as a wavelength verus transmittance from 220 nm to 340 nm. Sample solutions were prepared by dissolving an appropriate amount of chemical into the proper solvent. In the printed spectrum, the solvent solutions are represented by a solid line for the 0.2N H2SO4 solution and by a dashed line for strongly basic solution. A dotted line represents ethanolic solutions. Fisher-brand Suprasil ultraviolet cells were used for the sample solutions. The solutions were made basic by the addition of several drops of concentrated sodium hydroxide solution.

A listing of compounds with their respective UV maxima values can be found in Appendix B in Volume 7 of *IDDA*.

NUCLEAR MAGNETIC RESONANCE SPECTROMETRY

The proton nuclear magnetic spectra were recorded on a General Electric QE-300 Superconducting F1NMR spectrometer operating at 300 MHz. This NMR spectrometer is equipped with a 70.5 KG NB-Ti superconducting magnet, a magnet bore of 44 mm and a dual ^{1}H/ ^{13}C 5-mm probe.
Samples were prepared by dissolving the compound in the appropriate solvent (Aldrich deuterochloroform 99.5% containing 0.039% TMS unless otherwise noted on the NMR spectrum). Where possible, the sample concentration was maintained at a level judged to be the best compromise between solute interactions and instrument response. In some instances, low solubility or low sample concentration (less than 5 mg of compound was available in some instances) has resulted in spectra showing high noise level or trace contamination. All spectra were observed at a constant thermostated probe temperature. Sample solutions were equilibrated to the probe temperature before the spectra were recorded. Most spectra were recorded at a spinning rate of 15 to 25 rps, 32 acquisitions with quadrature phase detection, observed frequency of 300.151851 MHz, spectral width of 6024 Hz, 32768 data size resolution, no line broading, and a pulse width of 2.33 sec (30°). Most spectra include 0.03% TMS as a reference. Precautions were taken in handling all NMR solvents to minimize contamination with atmospheric moisture.

Spectra of the various Aldrich NMR solvents used can be found in Appendix A in Volume 7 of *IDDA*.

INFRARED SPECTROPHOTOMETRY

The infrared spectra were produced using a Nicolet 170SX Fourier transform infrared spectrometer. This infrared spectrometer is equipped with a laser-referenced Michelson interferometer with an absolute wavenumber accuracy specified better than $\pm$ 0.01 cm^{-1}.

The constant spectral resolution was kept at 4 cm^{-1} by collecting 64 one-second scans (4096 data points/scan). Because the FTIR spectrometer is a single beam instrument, some of the spectra may have small absorption bands due to CO_2 present in the sample chamber when the data was collected. This doublet can be found at 2360 cm^{-1} and 2340 cm^{-1}.

Unless otherwise stated, the compounds were prepared for spectral analysis by using potassium bromide (KBr) pellets. The KBr powder was oven dried and then kept in a desiccator. Every effort was made to remove water, however, many times both the sample as well as the KBr were hygroscopic, and water bands may have appeared in some of the spectra. It must be noted that spectra reproducibility may be difficult to regulate without careful weighing of both the KBr and the sample. In addition, there may be KBr interactions with the sample, especially amine compounds. Although these complications exist, KBr pellets generally give much better resolution than other techniques. In most cases, each spectrum was expanded to give full scale presentation of the data. Representative peaks as listed on each spectrum are intended to aid the user. They were determined by a Peak-Picker Program and should be used only for approximate values. A cumulative index of the infrared spectra, sorted by prominent peaks identified by a computer program

GAS CHROMATOGRAPHY/INFRARED SPECTROSCOPY

All gas chromatography/fourier transform infrared spectra (GC/FTIR) were obtained using a Hewlett-Packard 5965B IRD in combination with a Hewlett-Packard 5970 MSD. Samples were introduced into the system via a Hewlett-Packard 5890 gas chromatograph using He as a carrier gas. Temperature programming was used ranging from 80° C to 320° C. The column was a DB5MS 0.32 mm column in the IRD and a DB5MS 0.2 mm column in the MSD. All GC/FTIR spectra were verified by their corresponding mass spectra. GC/FTIR spectra were collected by the Pinellas County Forensic Laboratory in Largo, Florida, and by the Denver Police Department Crime Laboratory Bureau in Denver, Colorado.

ACKNOWLEDGMENTS

The authors wish to express appreciation to the many individuals who contributed to this reference book by offering helpful suggestions, information, assistance, and reviews.

We are grateful to Dr. Byron Dawson, Director and to Mr. Milton Nix, Director of the Georgia Bureau of Investigation, for making the facilities of the Georgia crime laboratory available for this work.

We are grateful to Denver Chief of Police David L. Michaud and Crime Laboratory Bureau Commander Captain Steve C. Metros and Forensic Supervisor Sergeant Mark R. Olin for making the facilities of the Denver crime laboratory available for this work.

We are grateful for Kevin Lothridge, forensic Laboratory Director for making the facilities of the Pinellas County crime laboratory available for this work.

We appreciate the assistance of Patricia Price for helping with the mass spectral data, and the assistance of Mathew Philip and Len Brown for helping with the computer word-processing. We also appreciate the help from William Pearson, Jeff Martinez, Adel Ashcroft, Adam Dryden, Devona Vix, and members of the crime laboratories for their help. We are indebted to Bradley Mills for drawing most of the chemical structures.

We appreciate the program ChemWindow from SoftShell International.

DRUG DATA

ACLARUBICIN

$C_{42}H_{53}NO_{15}$

Molecular Weight: 811.88 (811.34)

Synonyms: 2-Ethyl-1,2,3,4,6,11-hexahydro-2,5,7-trihydroxy-6,11-dioxo-4-[[2,3,6-trideoxy-4-O-[2,6-dideoxy-4-O-[(2R-trans)-tetrahydro-6-methyl-5-oxo2H-pyran-2-yl]-α-L-lyxo-hexopyranosyl]-3-(dimethylamino)-α-L-lyxo-hexopyranosyl]oxy]-1-naphthacenecarboxylic acid methyl ester; aclacinomycin A; Jaclacin
Trade Names: Aclarubicin, Aclacinon, Aclaplastin
Use: Antineoplastic

HPLC:
GC:

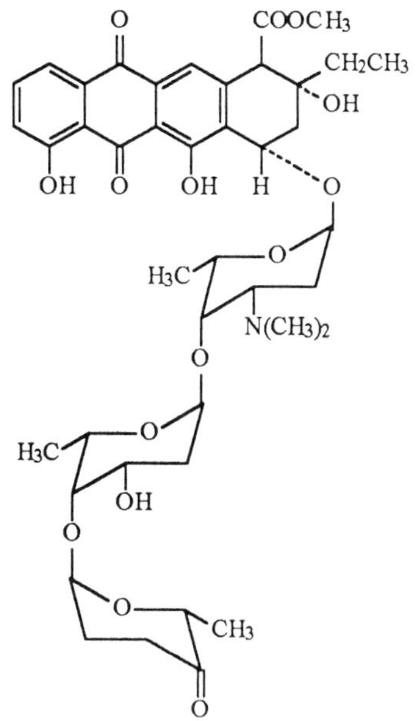

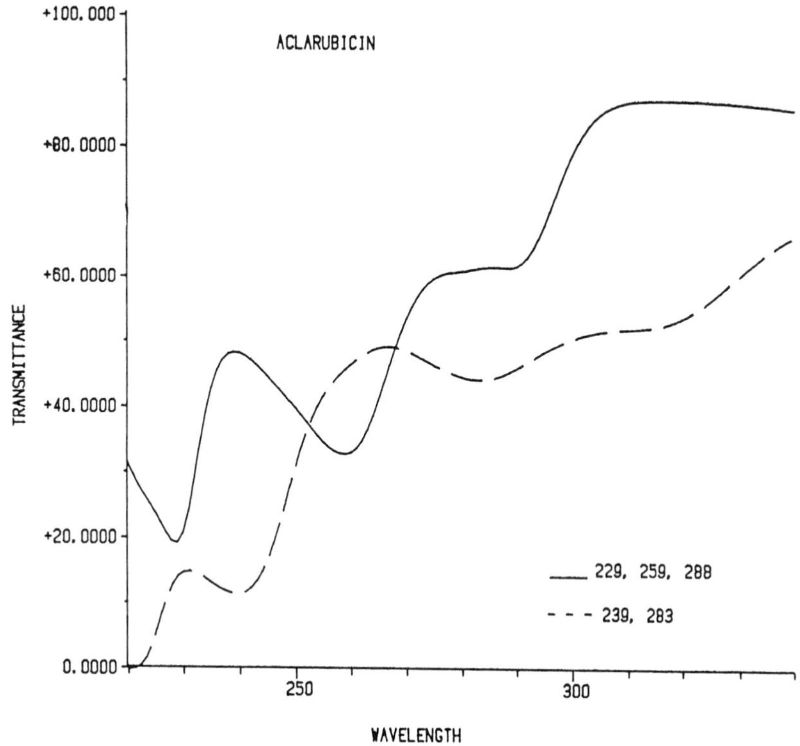

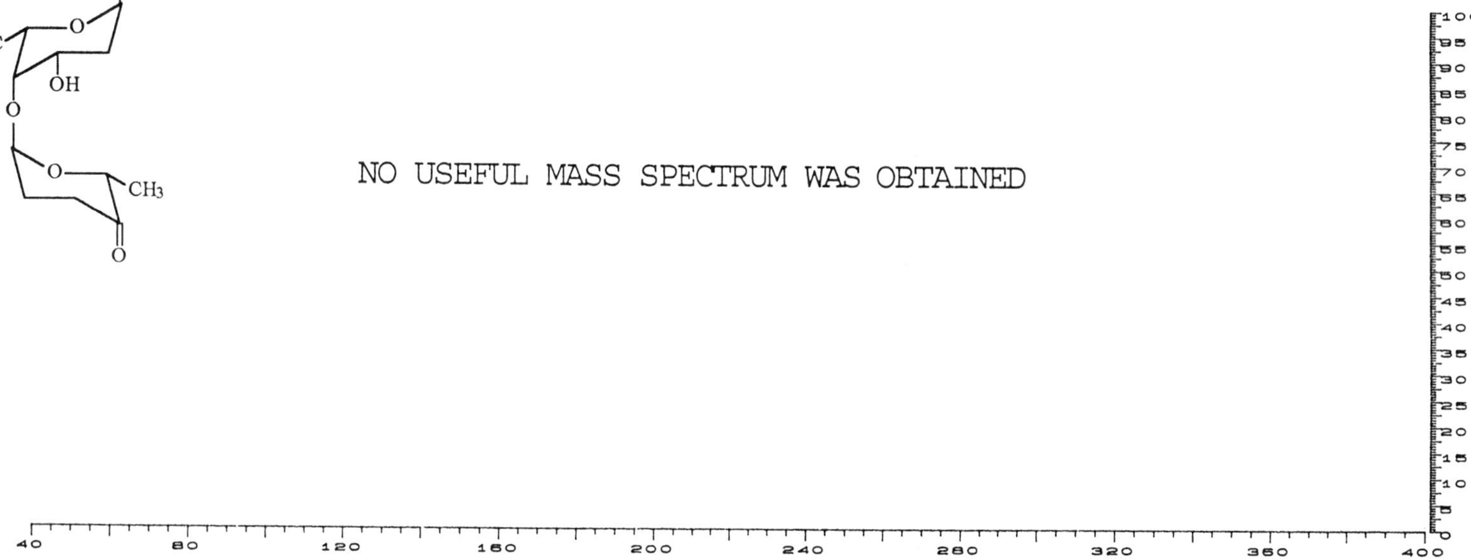

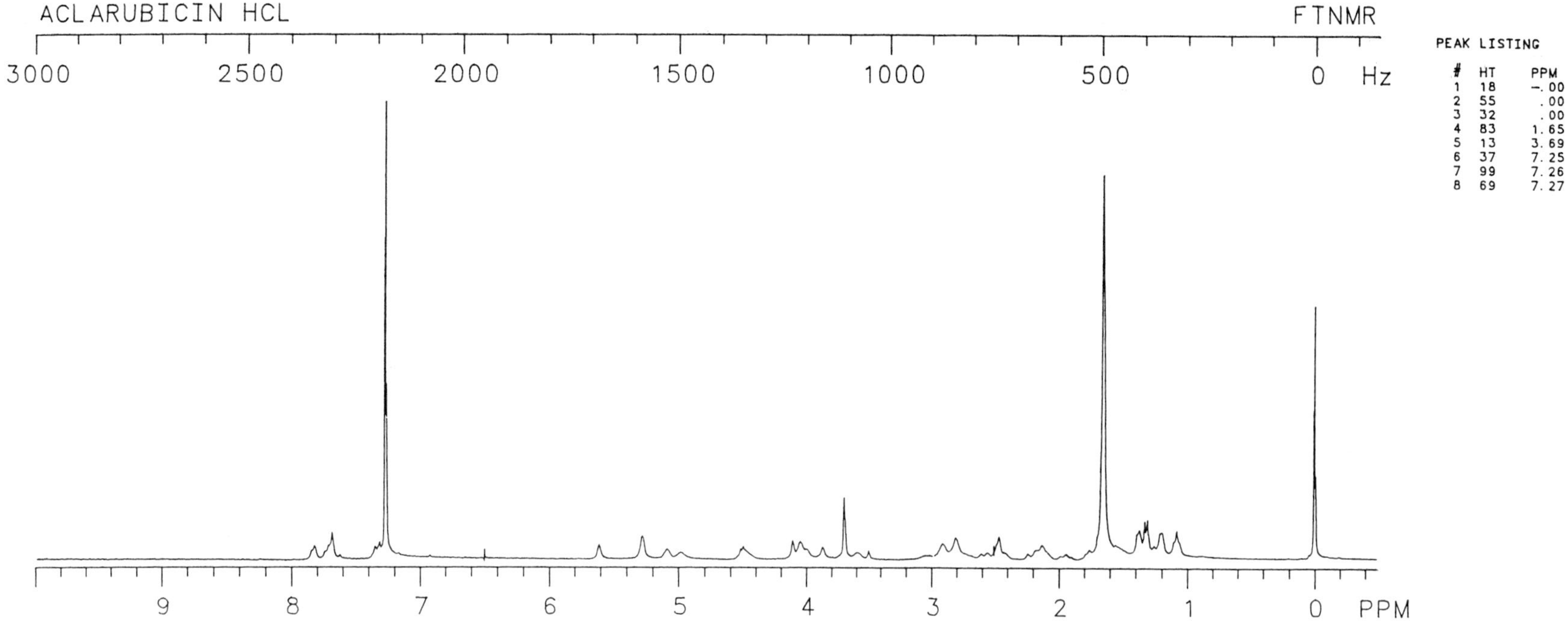

ACLARUBICIN HCL
FTNMR
3000 2500 2000 1500 1000 500 0 Hz
9 8 7 6 5 4 3 2 1 0 PPM
PEAK LISTING
HT PPM
1 18 -.00
2 55 .00
3 32 .00
4 83 1.65
5 13 3.69
6 37 7.25
7 99 7.26
8 69 7.27

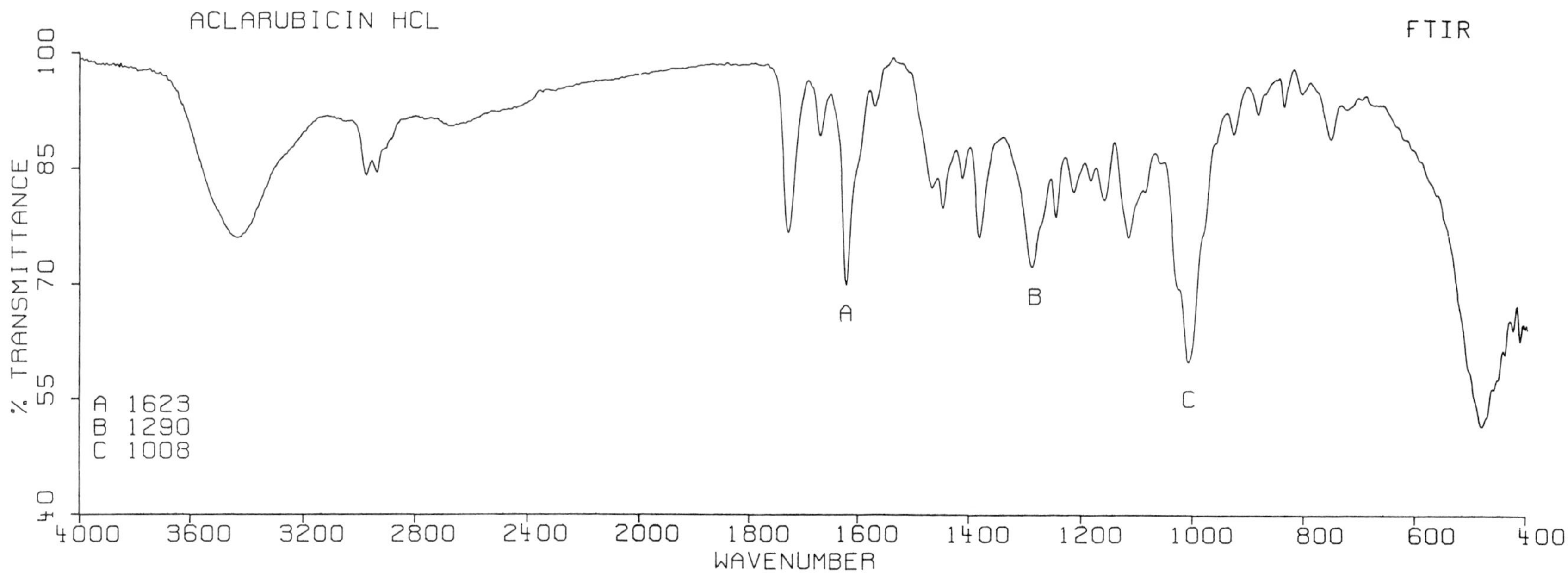

ACLARUBICIN HCL
FTIR
% TRANSMITTANCE
100 85 70 55 40
A 1623
B 1290
C 1008
A
B
C
4000 3600 3200 2800 2400 2000 1800 1600 1400 1200 1000 800 600 400
WAVENUMBER

ALTRETAMINE

$C_9H_{18}N_6$

Molecular Weight: 210.28 (210.06)

Synonyms: N,N,N',N',N",N"-Hexamethyl-1,3,5-triazine-2,3,6-triamine; hemel; hexamethylmelamine
Trade Names: Hexalen: Hexastat;
Use: Antineoplastic

HPLC: Methanol: 3.2
GC:

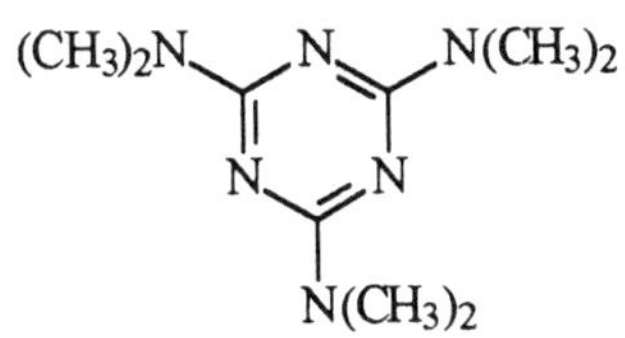

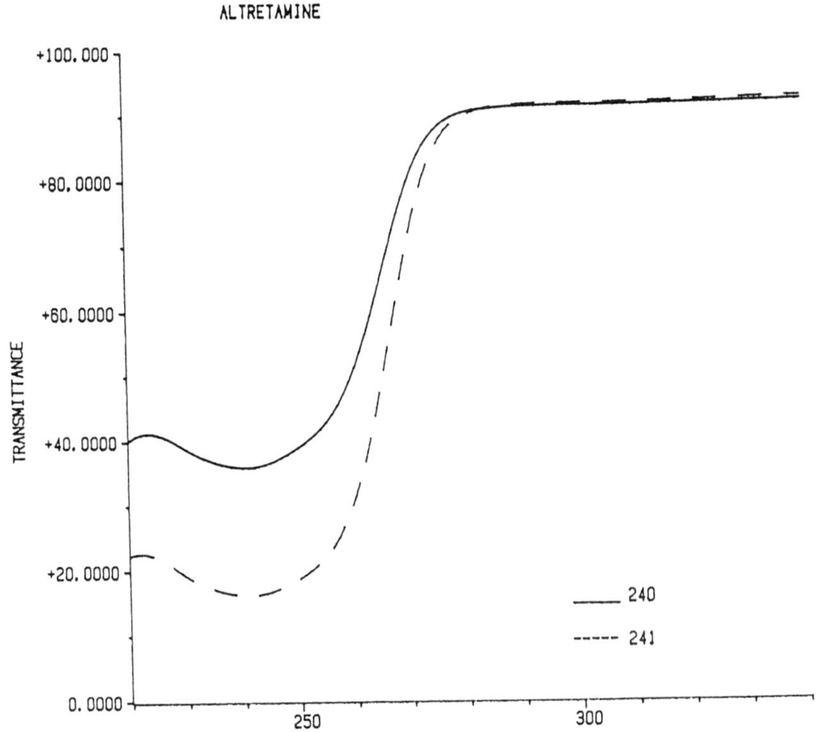

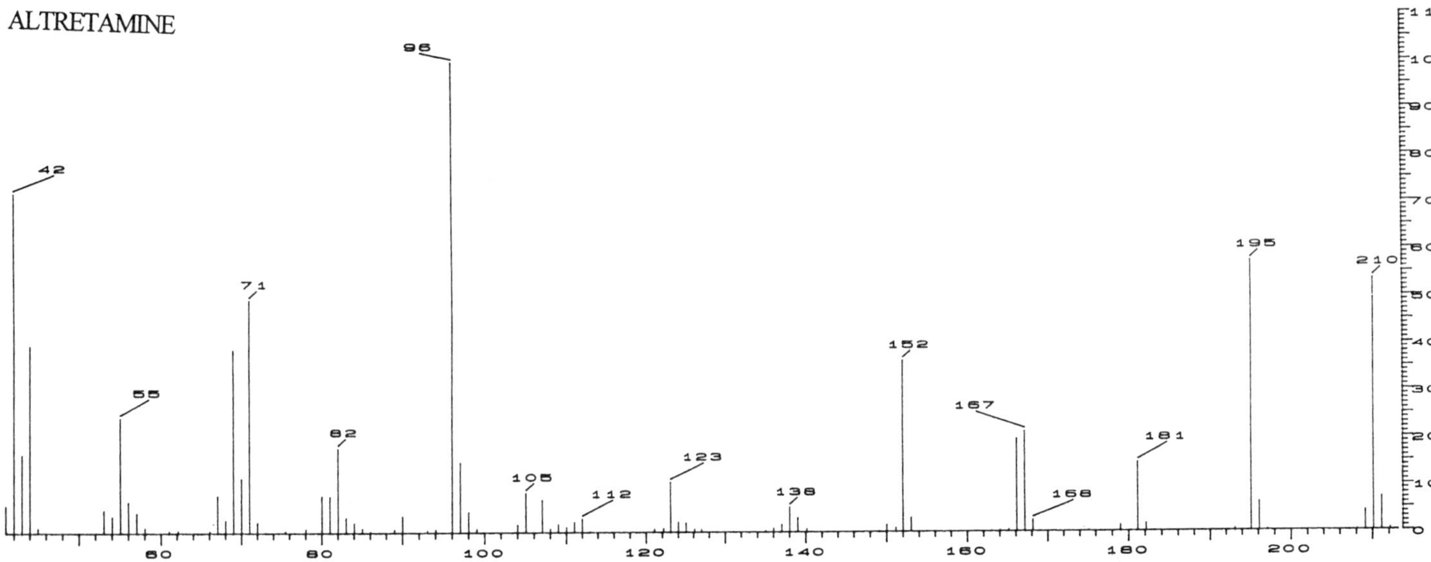

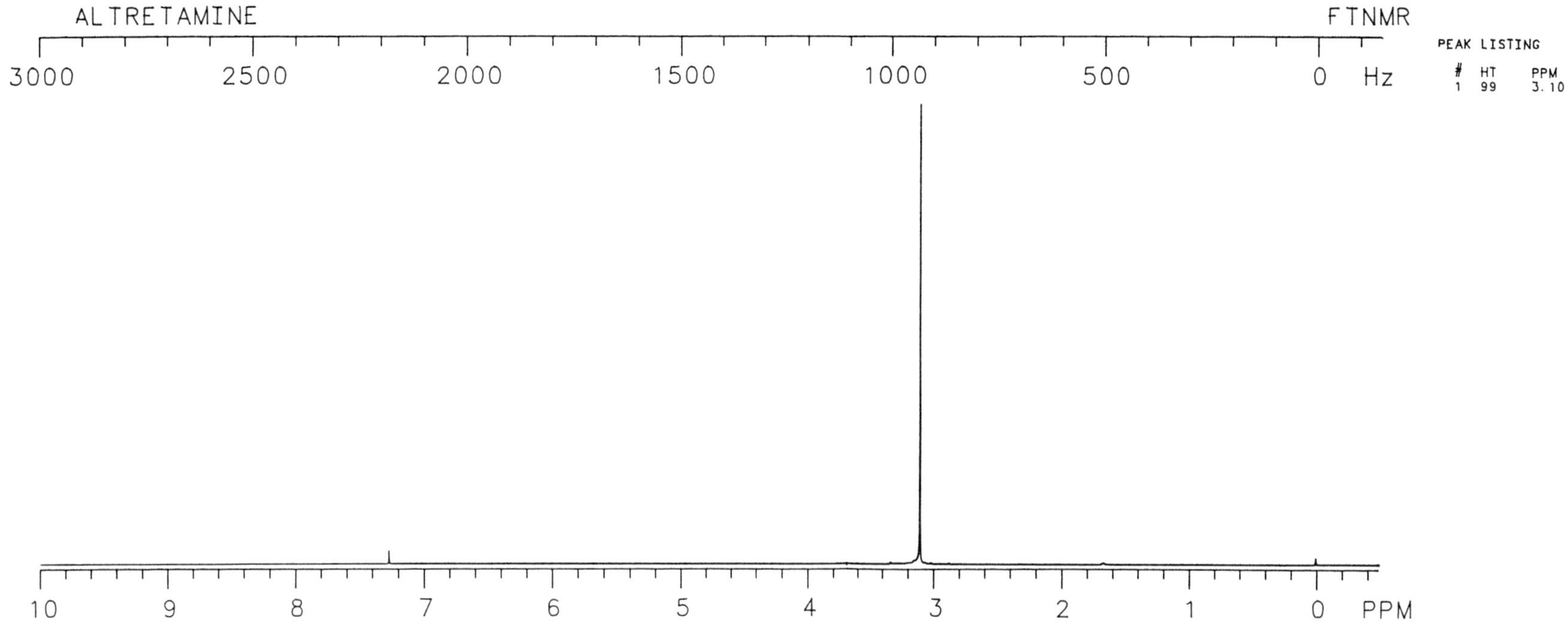
ALTRETAMINE
FTNMR
3000
2500
2000
1500
1000
500
0 Hz
PEAK LISTING
HT PPM
1 99 3.10
10
9
8
7
6
5
4
3
2
1
0 PPM

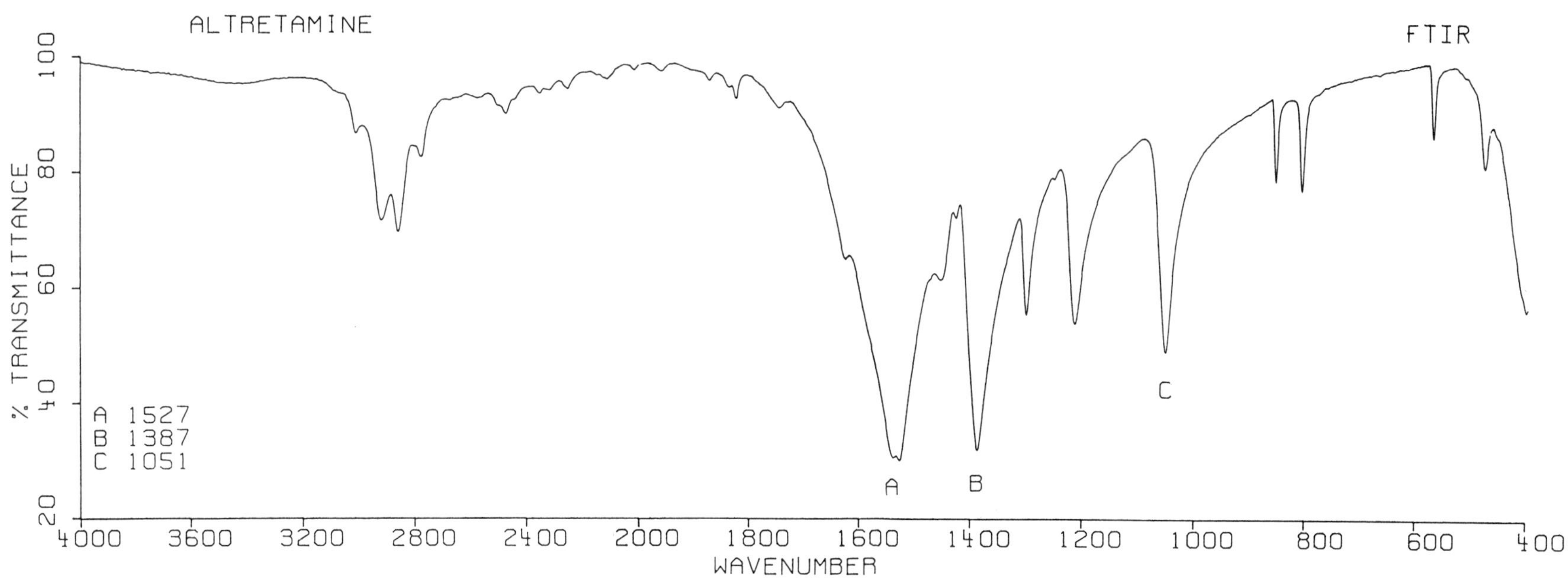
ALTRETAMINE
FTIR
% TRANSMITTANCE
100
80
60
40
20
A 1527
B 1387
C 1051
A
B
C
4000
3600
3200
2800
2400
2000
1800
1600
1400
1200
1000
800
600
400
WAVENUMBER

AMINOREX

$C_9H_{10}N_2O$

Molecular Weight: 162.19 (162.08)

Synonyms: 4,5-Dihydro-5-phenyl-2-oxazolamine; 2-amino-5-phenyl-2-oxazoline; aminoxafen; aminoxaphen
Trade Names: Menocil, Apiquel
Use: Anorexic

HPLC:
GC: 1512; 140°

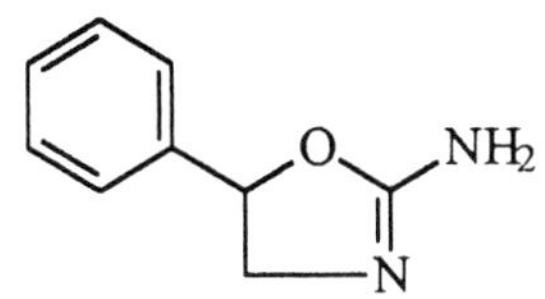

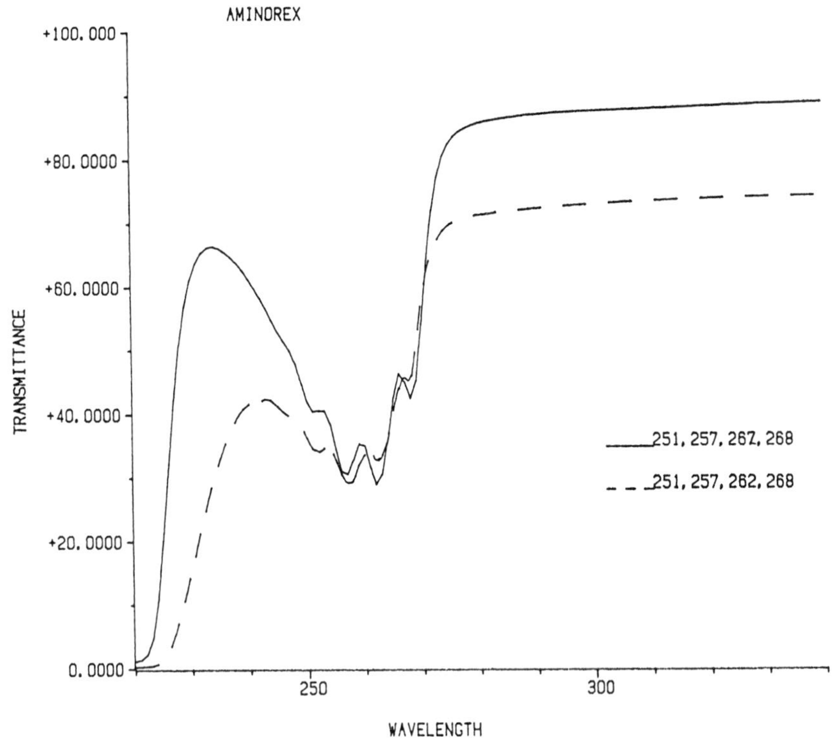

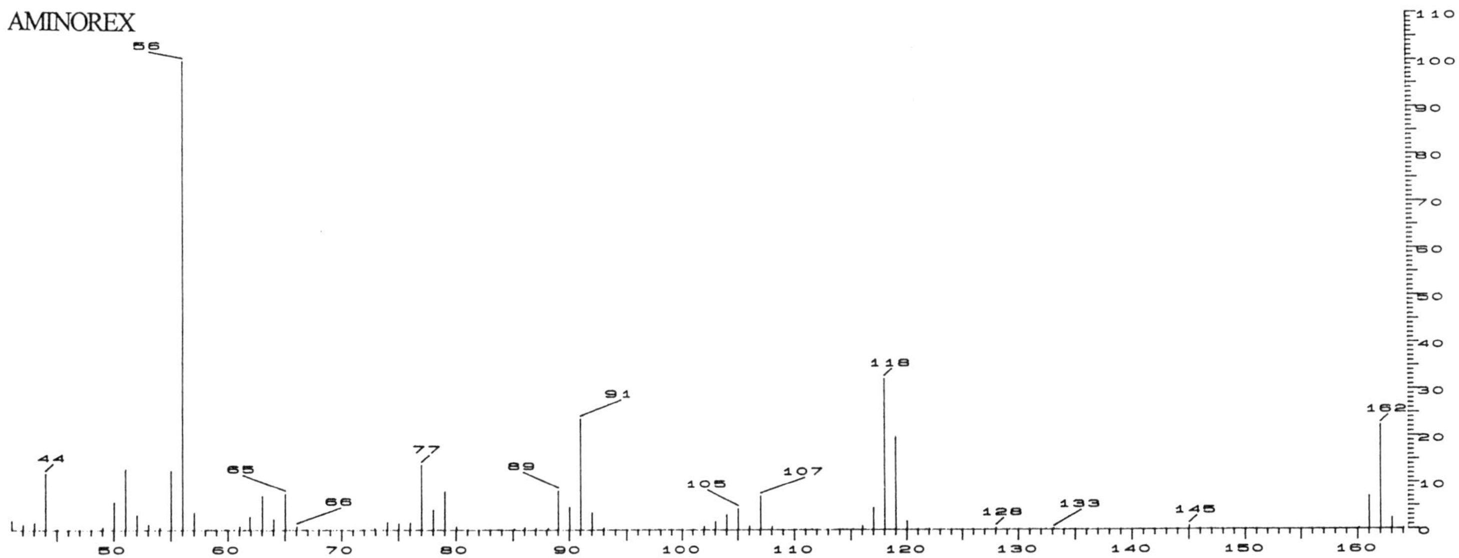

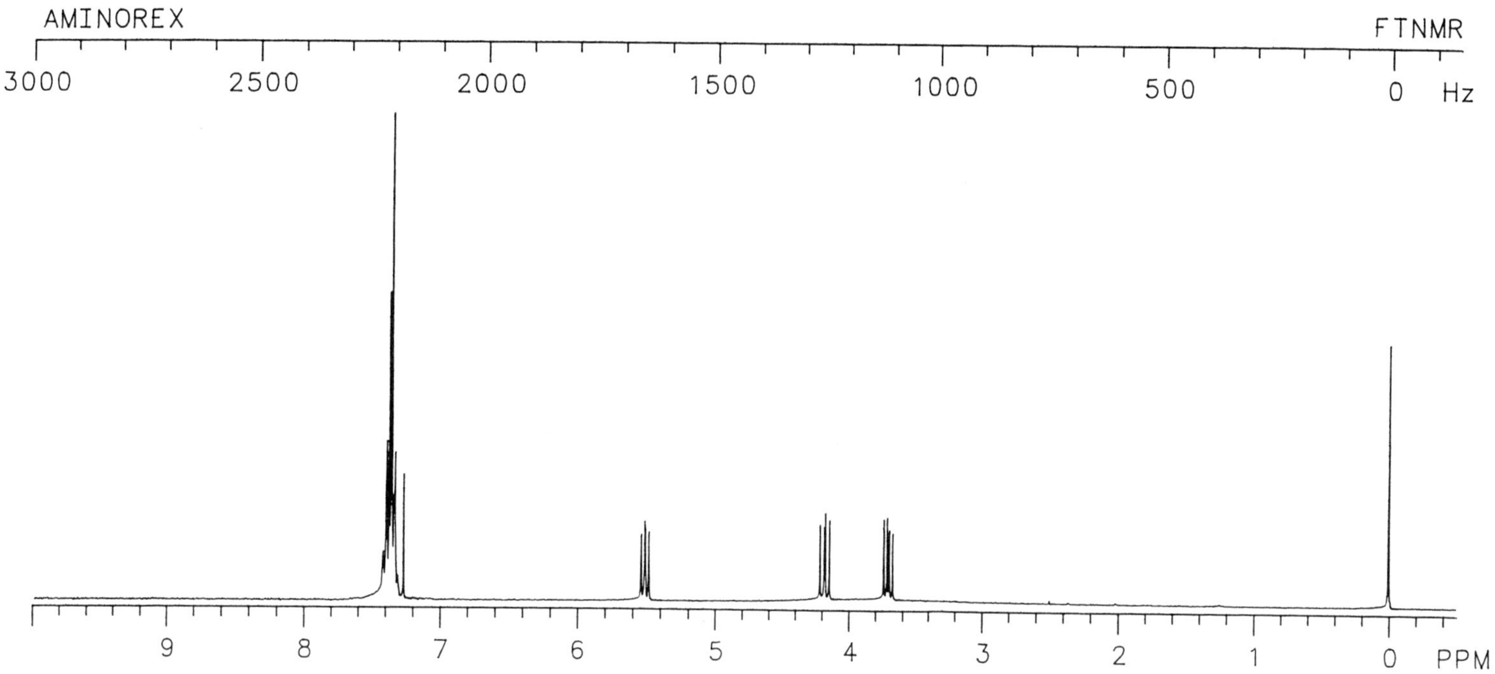

PEAK LISTING

#	HT	PPM
1	53	.00
2	14	3.66
3	14	3.69
4	17	3.71
5	17	3.73
6	16	4.13
7	18	4.16
8	15	4.18
9	15	4.21
10	14	5.47
11	15	5.50
12	16	5.50
13	13	5.53
14	25	7.26
15	30	7.32
16	21	7.33
17	20	7.33
18	99	7.34
19	62	7.36
20	15	7.38
21	32	7.38

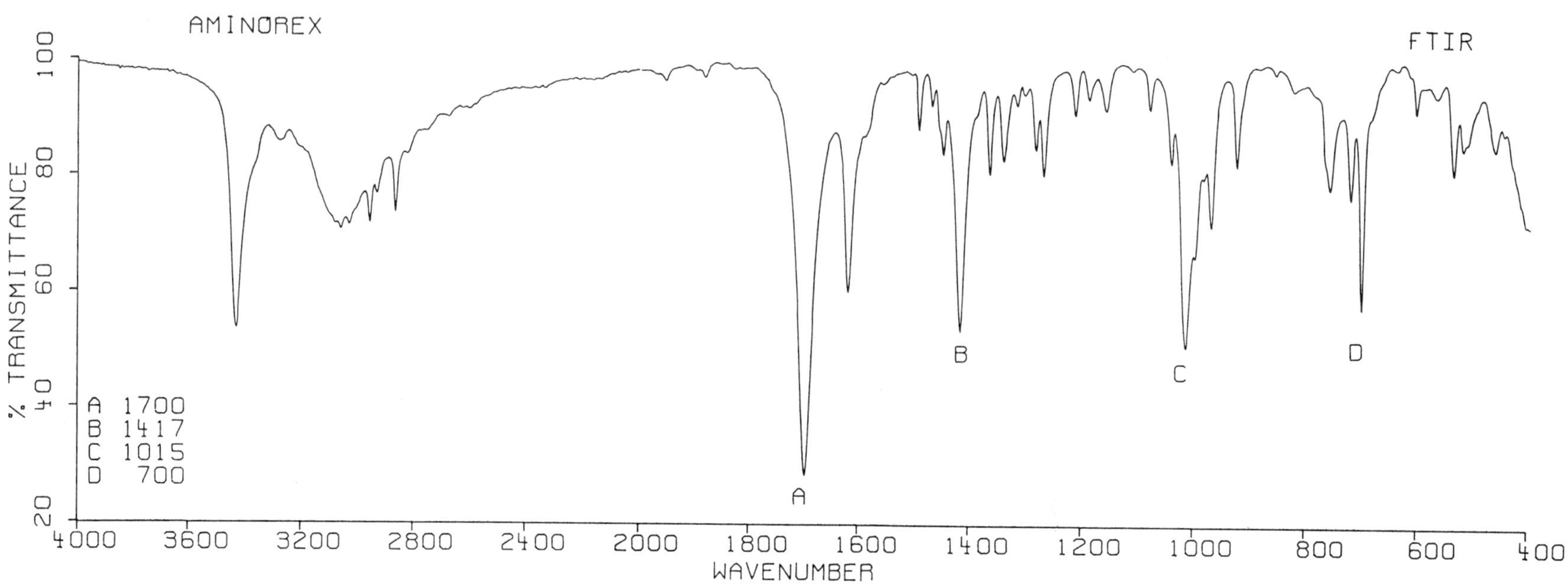

AMIODARONE

$C_{25}H_{29}I_2NO_3$

Molecular Weight: 645.32 (645.02)

Synonyms: (2-Butyl-3-benzofuranyl)[4-[2-(2-(diethylamino)ethoxy]-3,5-diiodophenyl]methanone
Trade Names: Amiodar, Ancoron, Angiodarona, Atlansil, Cordarex, Cordarone,Cordarone X, Miocard, Miodaron, Ortacrone, Ritmocardyl, Rythmarone, Trangorex
Use: Anti-arrhythmic; anti-anginal

HPLC:
GC:

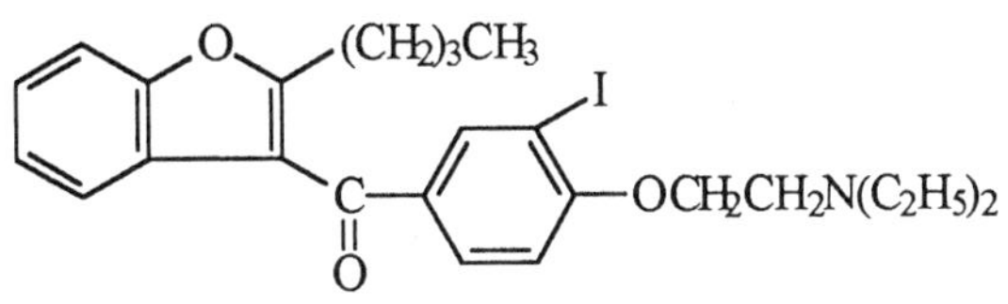

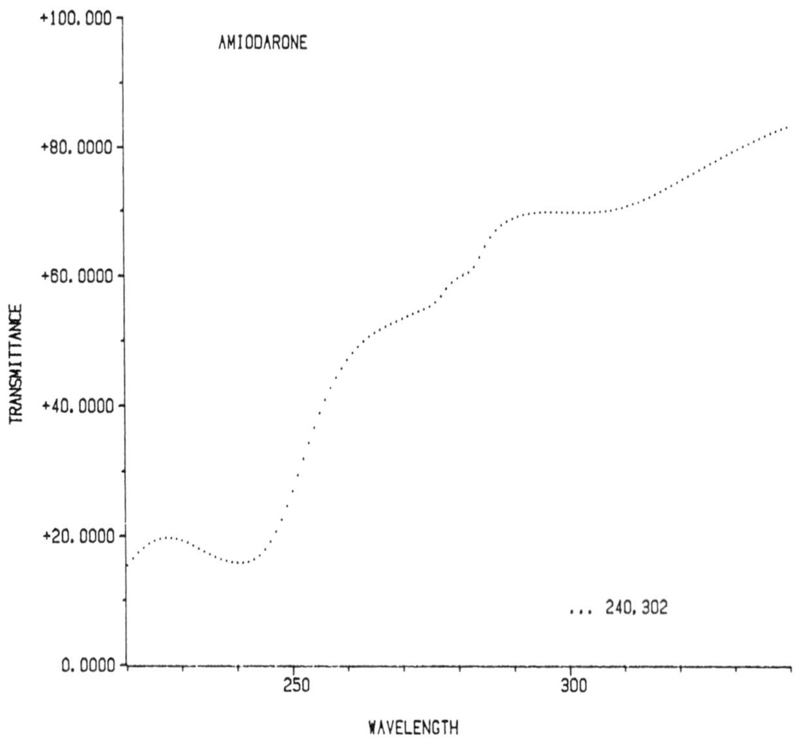

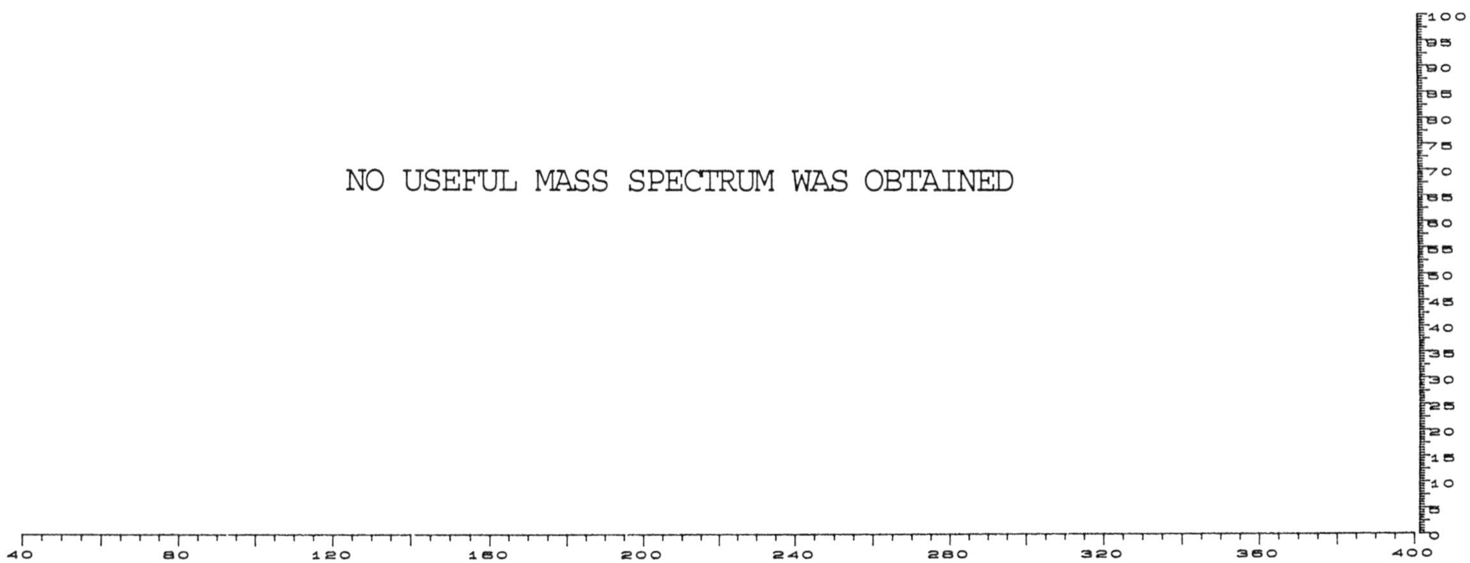

NO USEFUL MASS SPECTRUM WAS OBTAINED

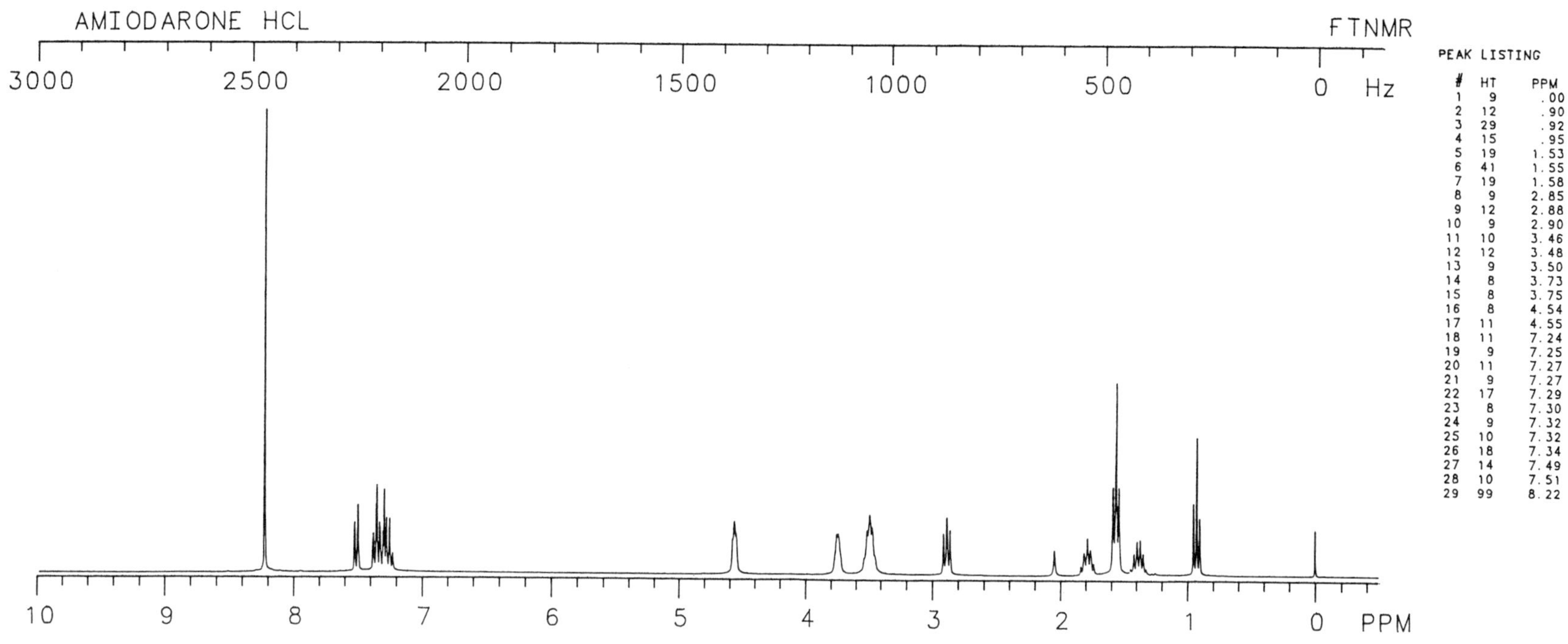

AMIODARONE HCL
FTNMR
3000
2500
2000
1500
1000
500
0 Hz
PEAK LISTING
HT PPM
1 9 .00
2 12 .90
3 29 .92
4 15 .95
5 19 1.53
6 41 1.55
7 19 1.58
8 9 2.85
9 12 2.88
10 9 2.90
11 10 3.46
12 12 3.48
13 9 3.50
14 8 3.73
15 8 3.75
16 8 4.54
17 11 4.55
18 11 7.24
19 9 7.25
20 11 7.27
21 9 7.27
22 17 7.29
23 8 7.30
24 9 7.32
25 10 7.32
26 18 7.34
27 14 7.49
28 10 7.51
29 99 8.22
10
9
8
7
6
5
4
3
2
1
0 PPM

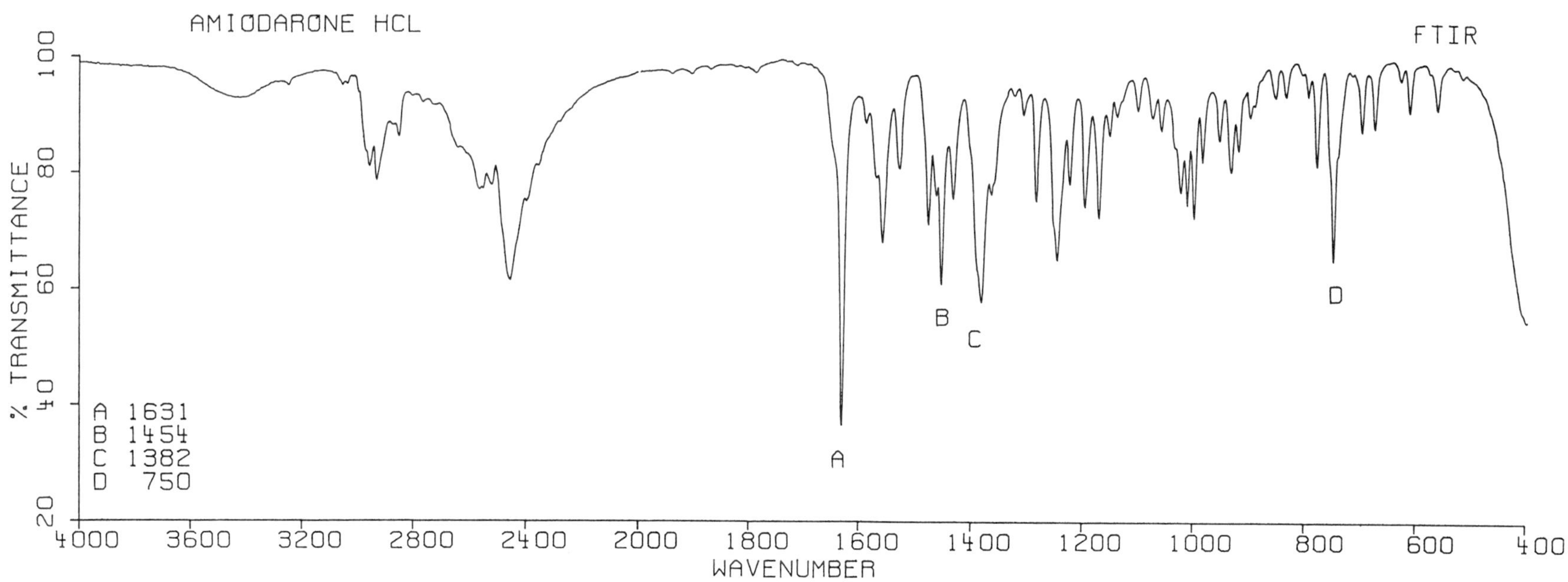

AMIODARONE HCL
FTIR
% TRANSMITTANCE
100
80
60
40
20
A 1631
B 1454
C 1382
D 750
A
B
C
D
4000
3600
3200
2800
2400
2000
1800
1600
1400
1200
1000
800
600
400
WAVENUMBER

AMIPRILOSE

$C_{14}H_{27}NO_6$

Molecular Weight: 305.37 (305.18)

Synonyms: 3-O-[3-(dimethylamino)propyl]-1,2-O-(1-methylethylidene)-α-D-glucofuranose
Trade Names: Therafectin
Use: Immunomodulator

HPLC:
GC:

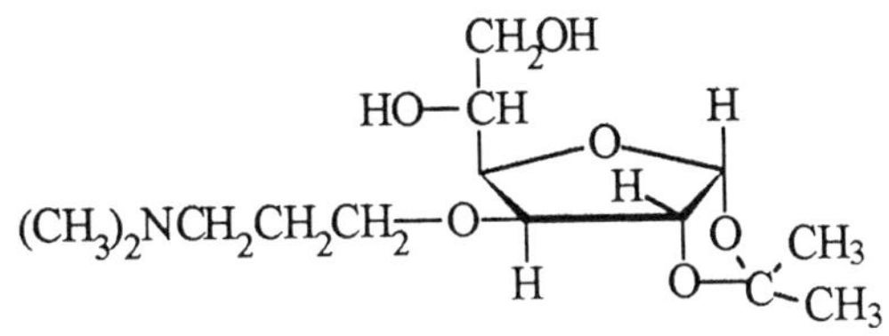

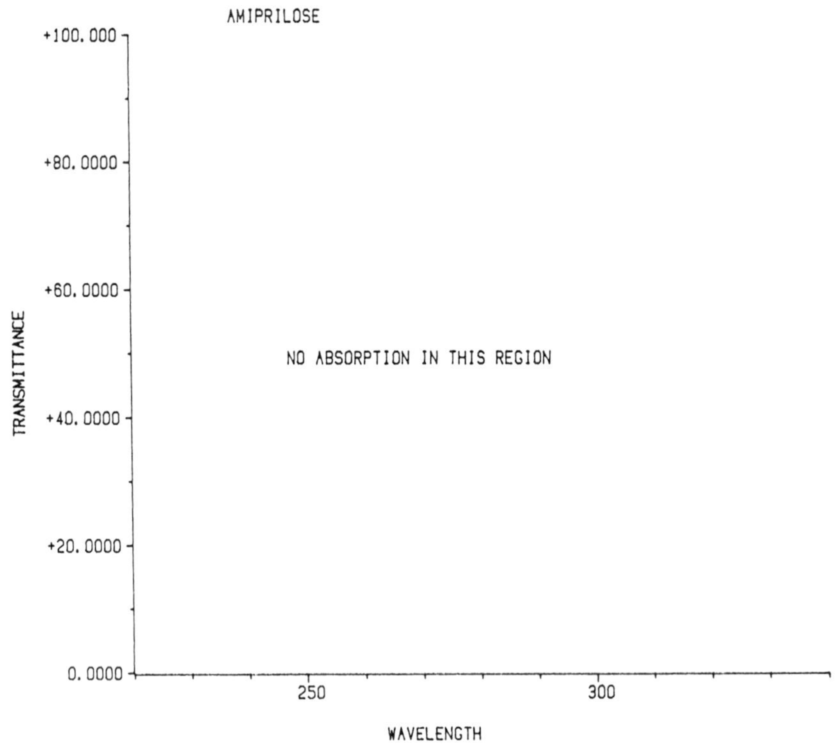

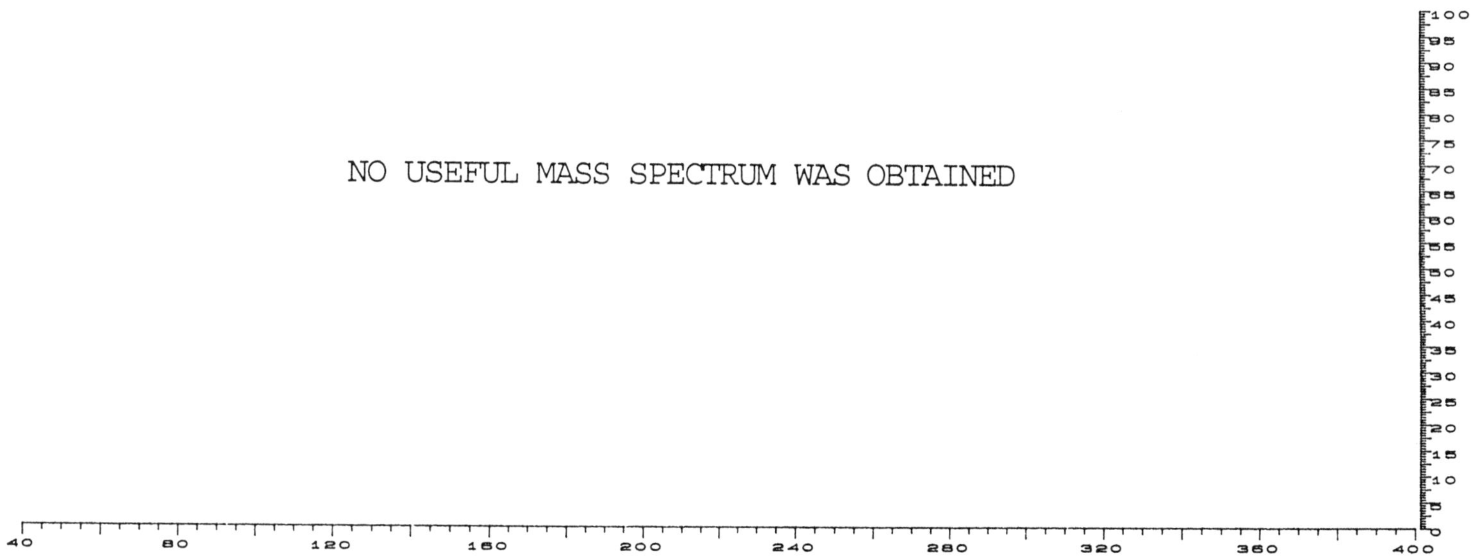

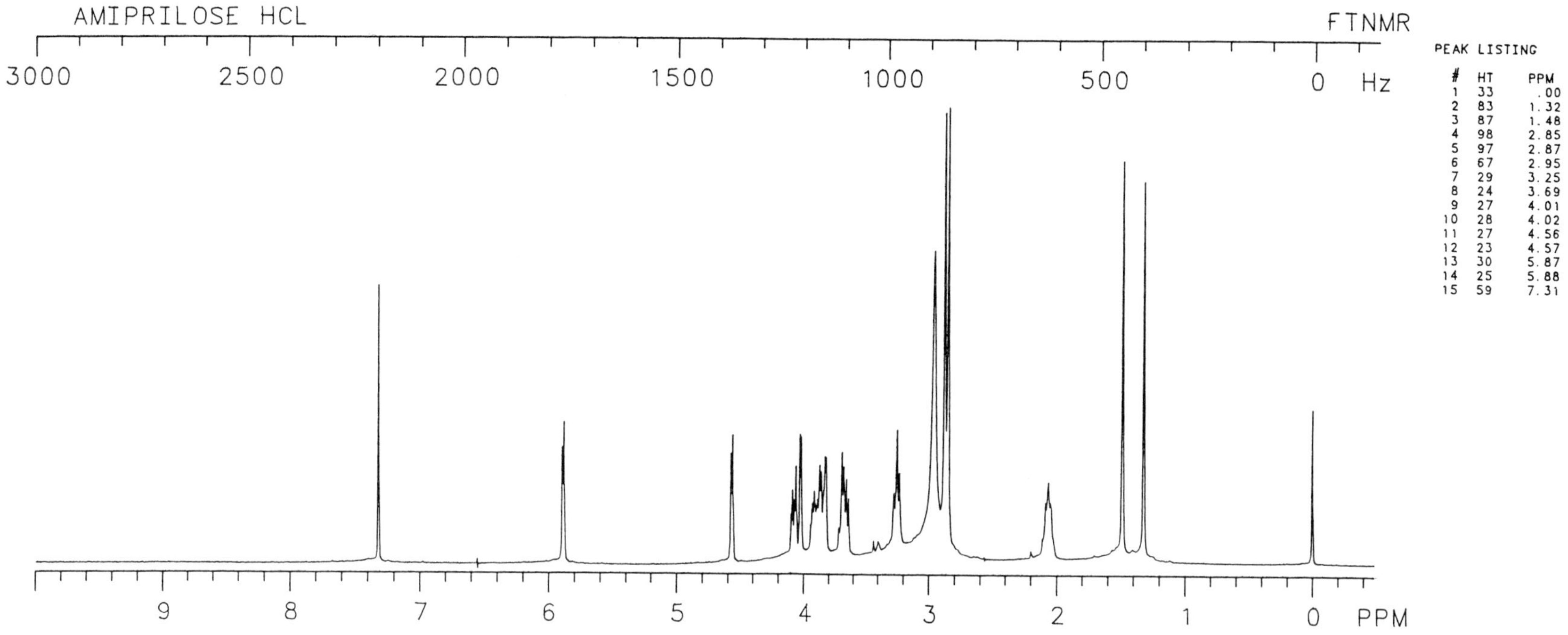
AMIPRILOSE HCL
FTNMR
3000 2500 2000 1500 1000 500 0 Hz
9 8 7 6 5 4 3 2 1 0 PPM
PEAK LISTING
HT PPM
1 33 .00
2 83 1.32
3 87 1.48
4 98 2.85
5 97 2.87
6 67 2.95
7 29 3.25
8 24 3.69
9 27 4.01
10 28 4.02
11 27 4.56
12 23 4.57
13 30 5.87
14 25 5.88
15 59 7.31

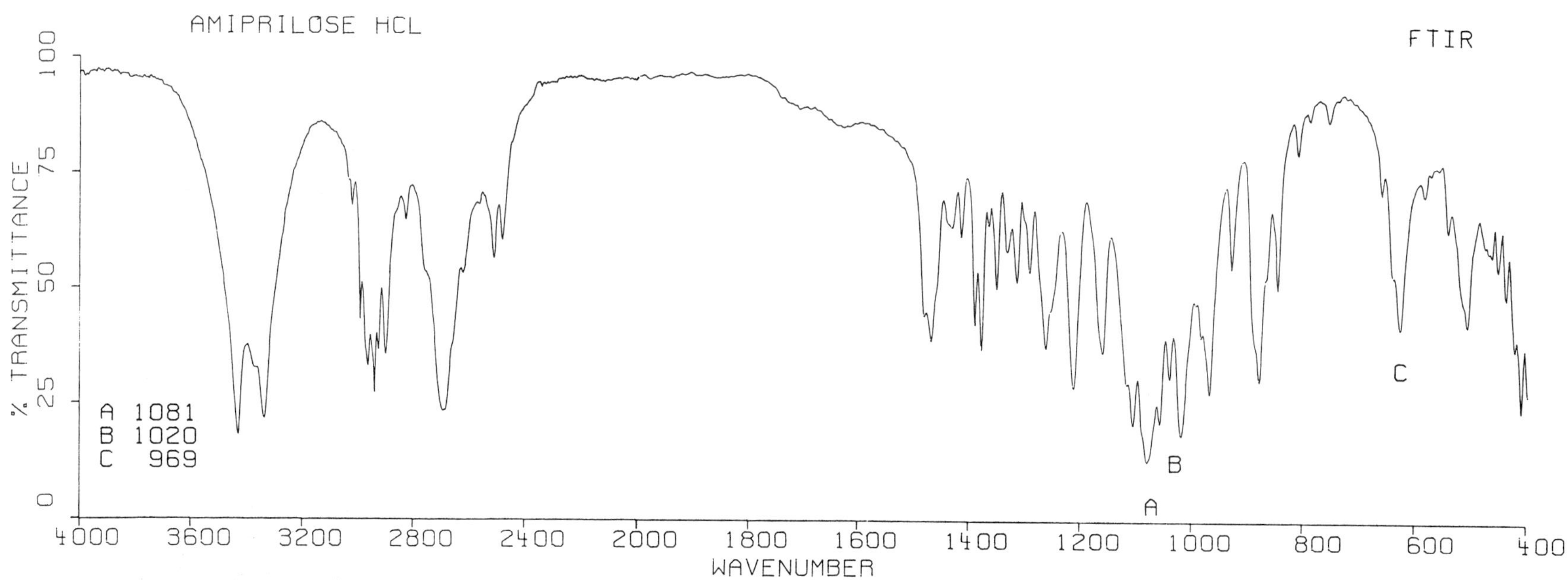
AMIPRILOSE HCL
FTIR
% TRANSMITTANCE
0 25 50 75 100
A 1081
B 1020
C 969
A
B
C
4000 3600 3200 2800 2400 2000 1800 1600 1400 1200 1000 800 600 400
WAVENUMBER

AMLODIPINE BESYLATE

$C_{20}H_{25}ClN_2O_5 \cdot C_6H_6O_3S$

Molecular Weight: 567.05 (566.15)

Synonyms: (R,S)-3-Ethyl-5-methyl-2-(2-aminoethoxymethyl)-4-(2-chlorophenyl)-1,4-dihydro-6-methyl-3,5-pyridinedicarboxylate benzenesulphonate
Trade Names: Norvasc
Use:

HPLC:
GC:

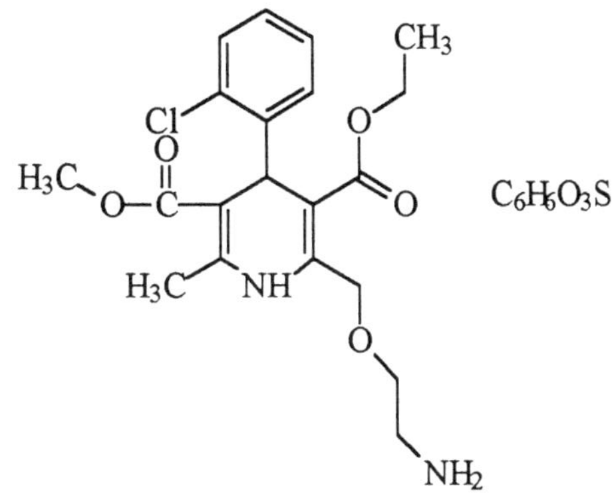

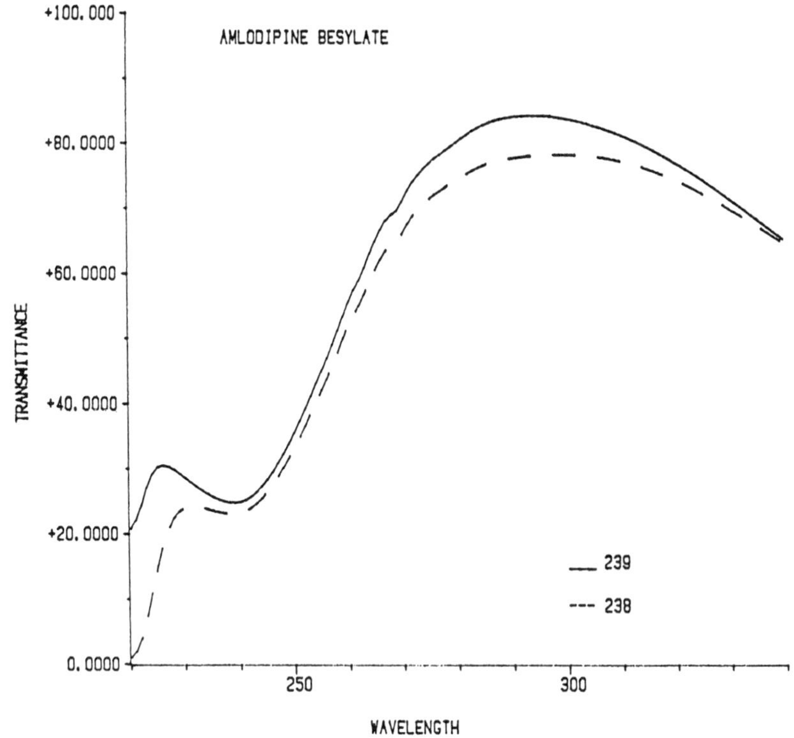

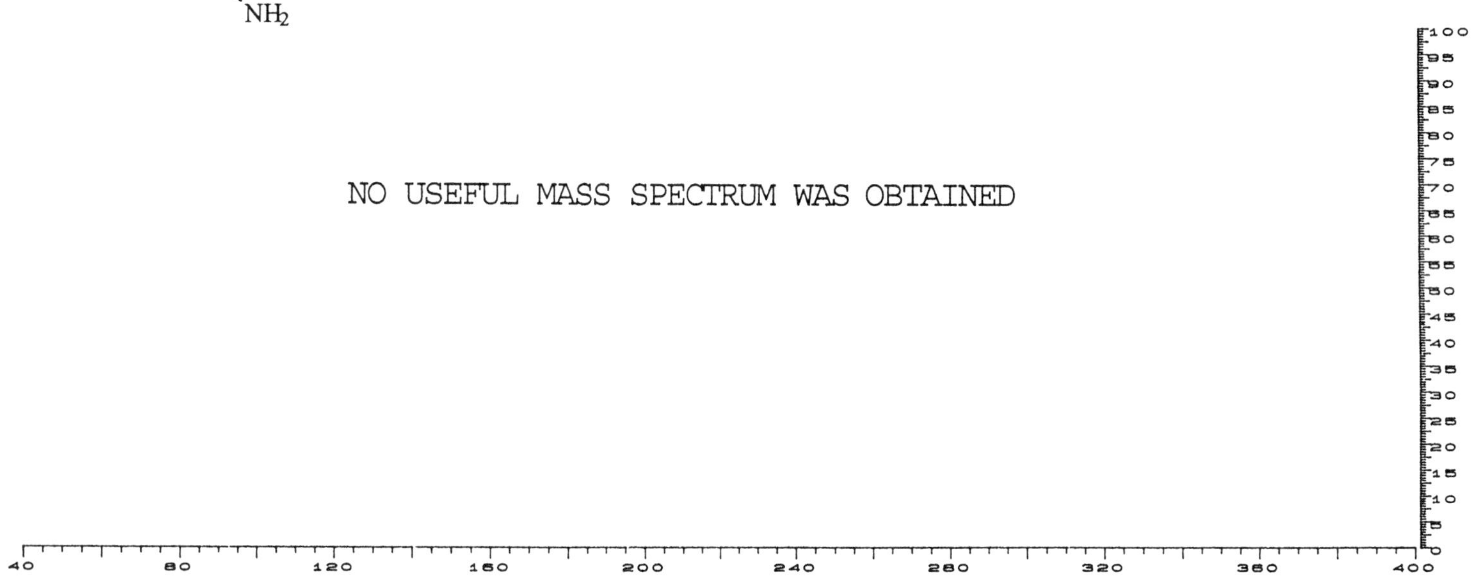

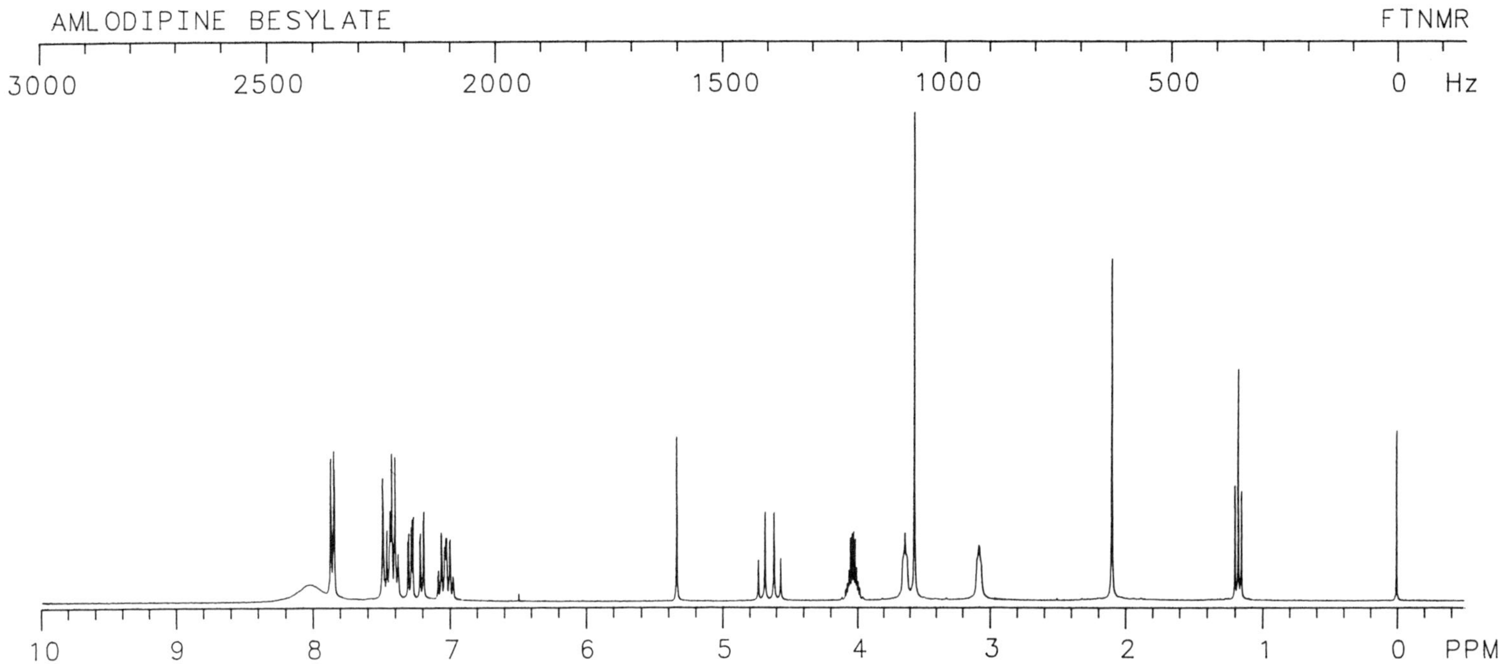

PEAK LISTING

#	HT	PPM
1	34	.00
2	22	1.15
3	47	1.17
4	23	1.19
5	69	2.09
6	99	3.55
7	33	5.33
8	18	7.18
9	29	7.39
10	30	7.42
11	18	7.43
12	25	7.48
13	26	7.83
14	30	7.84
15	29	7.86

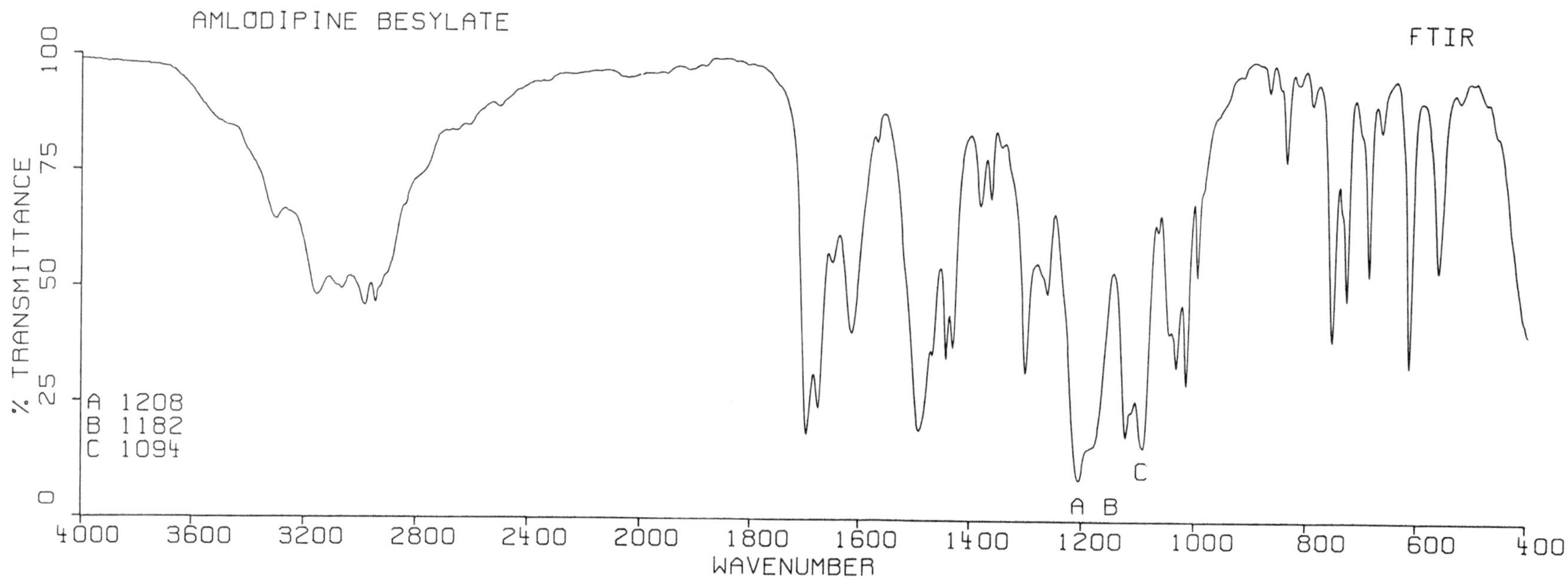

ANDROSTANOLONE BENZOATE

$C_{26}H_{34}O_3$

Molecular Weight: 394.55 (394.25)
Synonyms: 5α-Androstan-17β-ol-3-one benzoate; dihydrotestosterone benzoate; stanolone benzoate
Trade Names:
Use: Anabolic

HPLC: Methanol: 4.2
GC: 3298; 280°

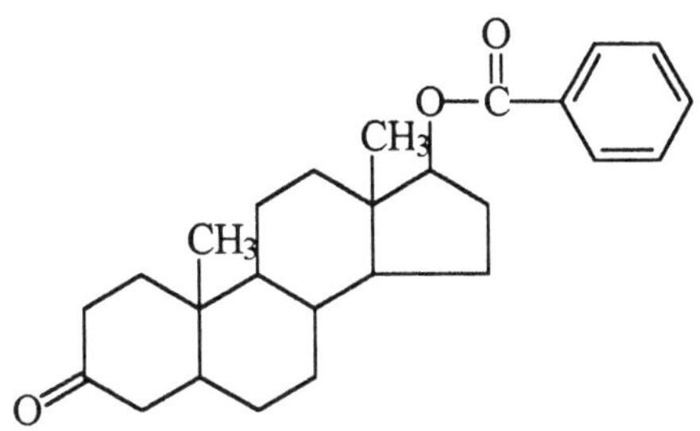

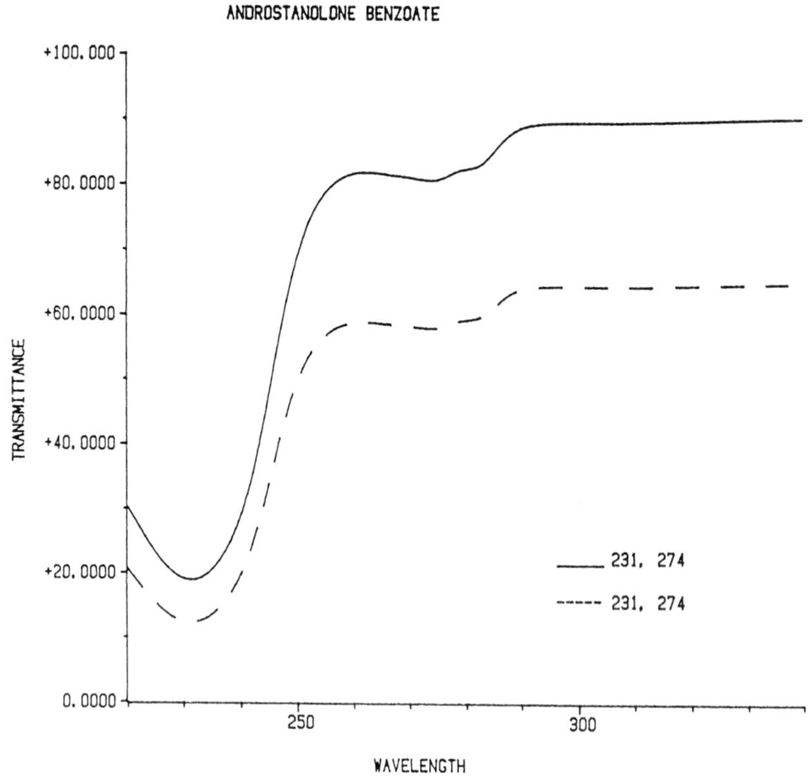

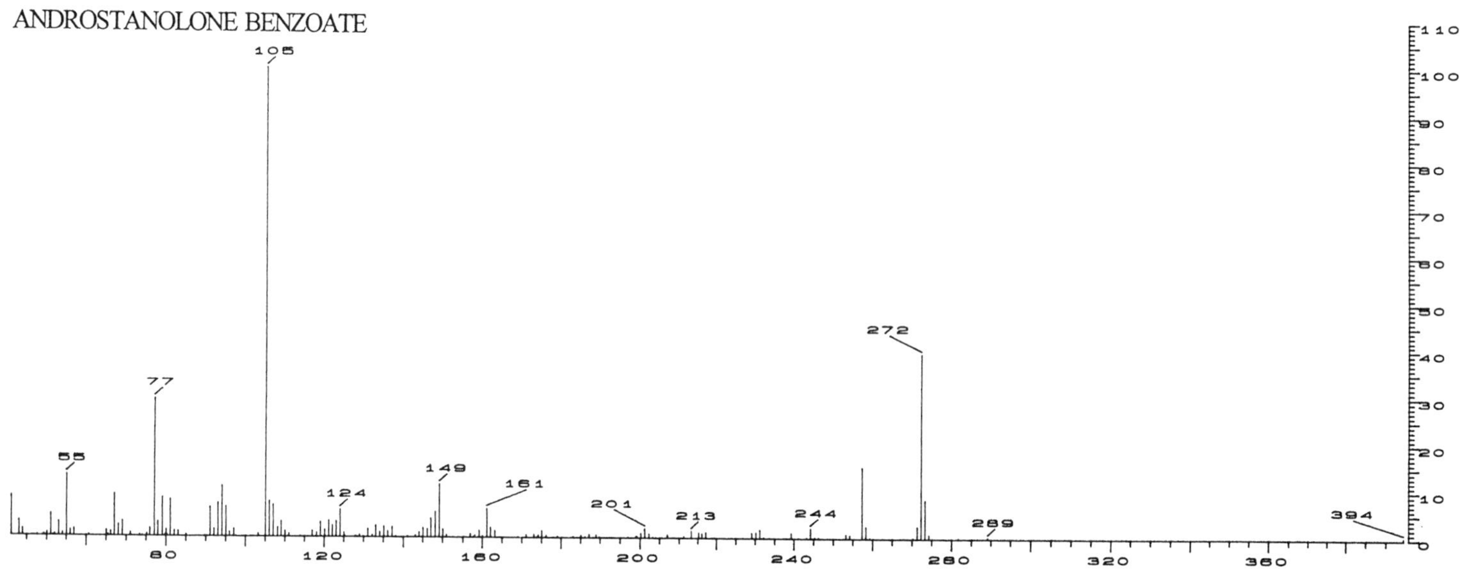

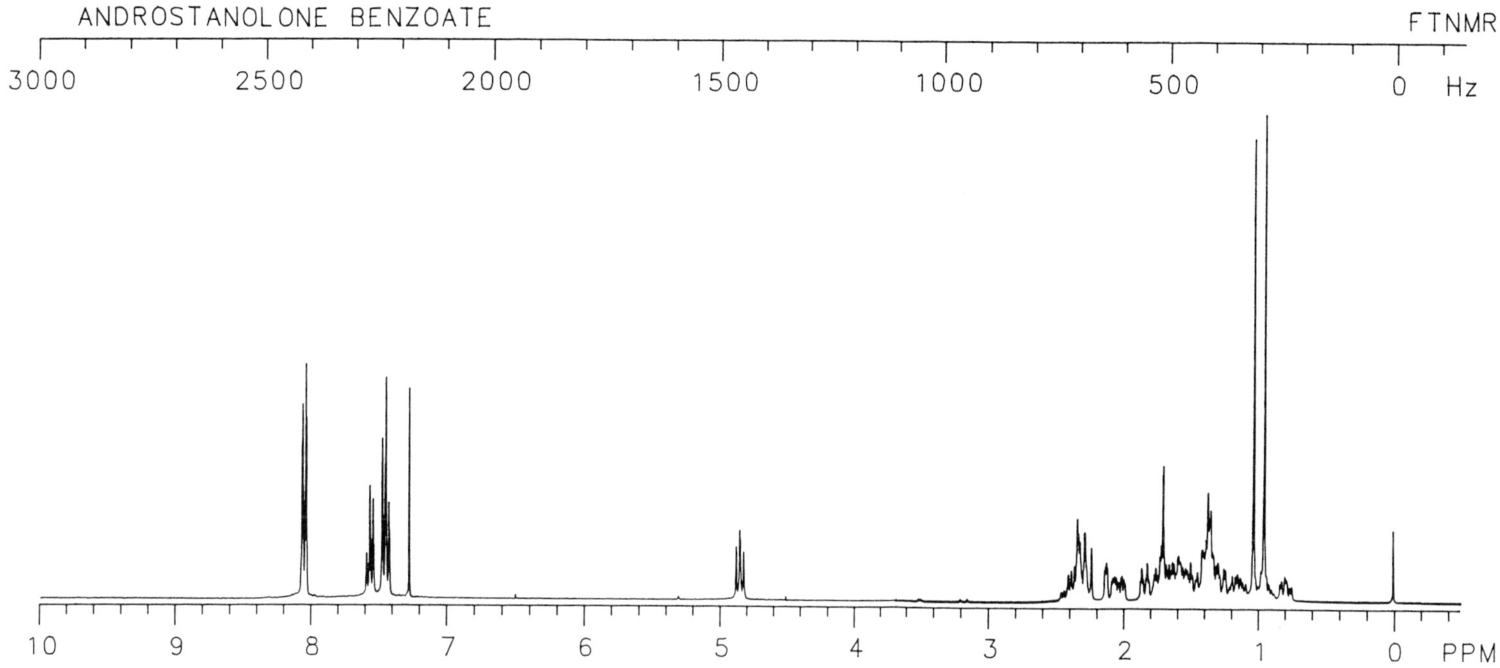

PEAK LISTING

#	HT	PPM
1	14	.00
2	99	.95
3	94	1.03
4	18	1.35
5	15	1.36
6	22	1.37
7	12	1.38
8	27	1.70
9	26	1.70
10	14	2.27
11	13	2.28
12	12	2.31
13	14	2.32
14	17	2.33
15	12	4.84
16	14	4.84
17	42	7.26
18	14	7.41
19	19	7.41
20	45	7.44
21	26	7.44
22	32	7.46
23	20	7.53
24	23	7.55
25	36	8.02
26	47	8.03
27	13	8.03
28	39	8.05
29	31	8.05

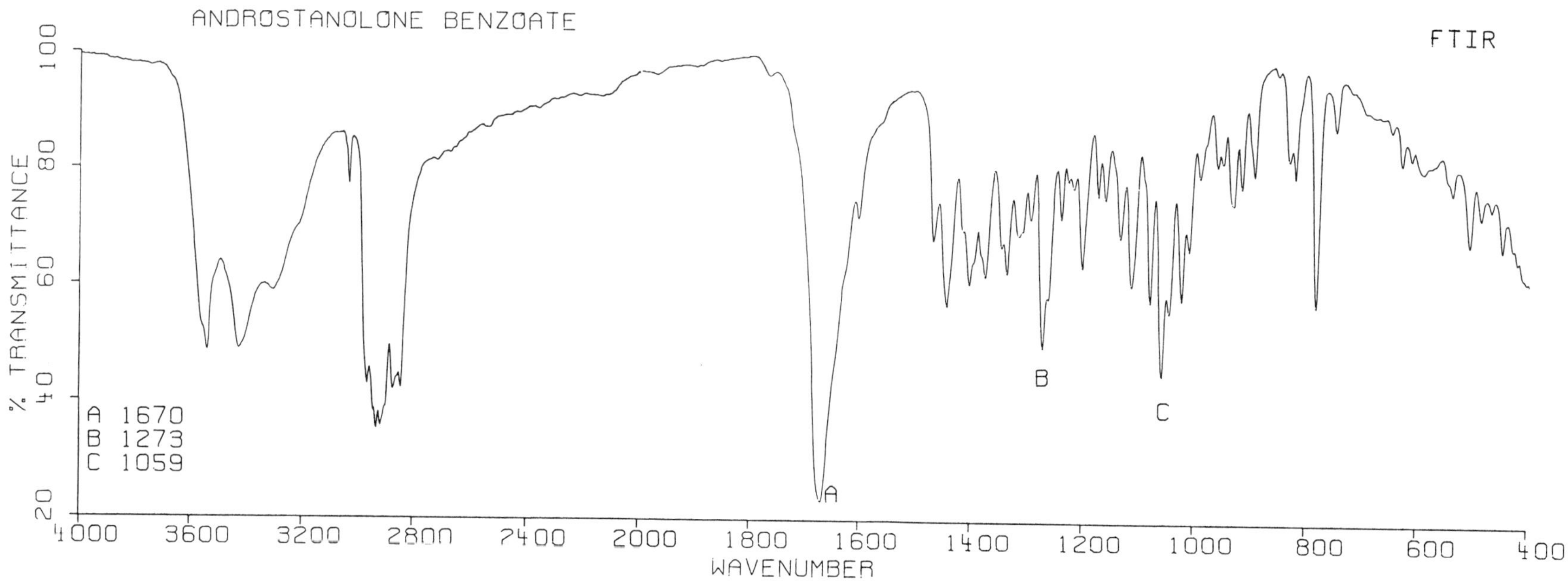

15

APOCODEINE

$C_{18}H_{19}NO_2$

Molecular Weight: 281.35 (281.14)

Synonyms: (R)-5,6,6a,7-Tetrahydro-10-methoxy-6-methyl-4H-dibenzo[de,g]quinolin-11-ol; R(-)-10-methoxy-11-hydroxyaporphine
Trade Names:
Use: Dopamine agonist, apomorphine metabolite

HPLC:
GC: 2621; 250°

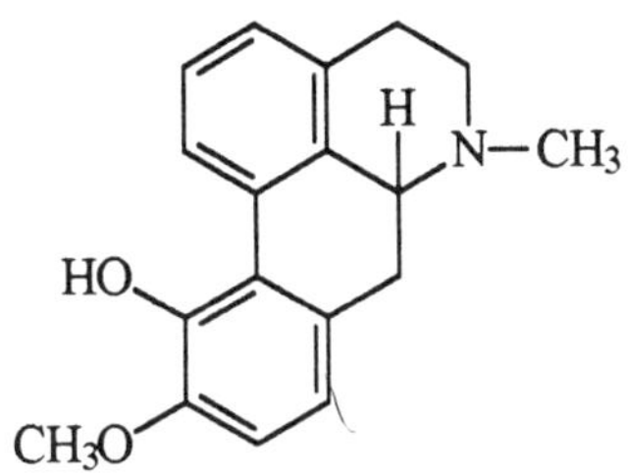

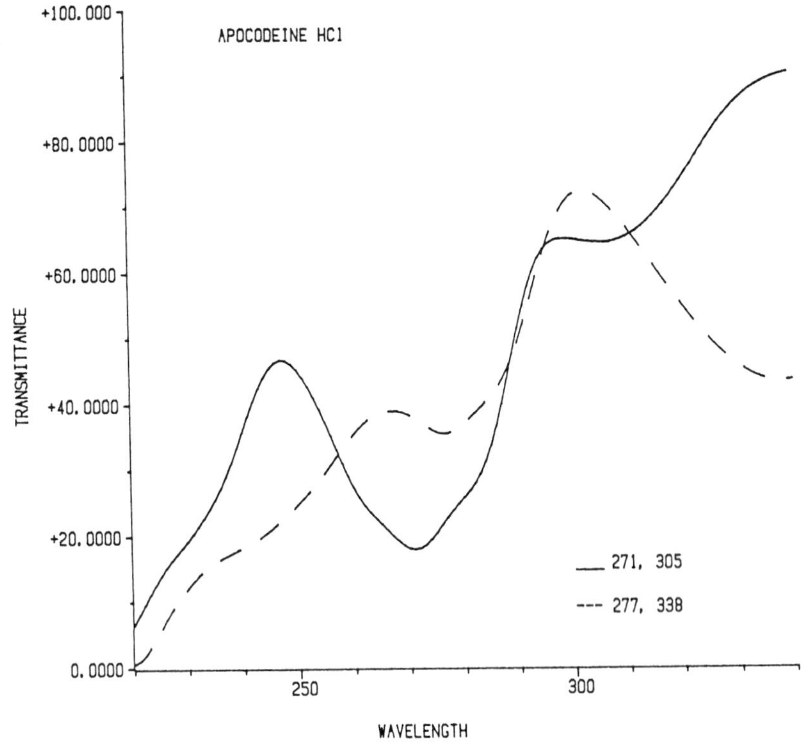

APOCODEINE

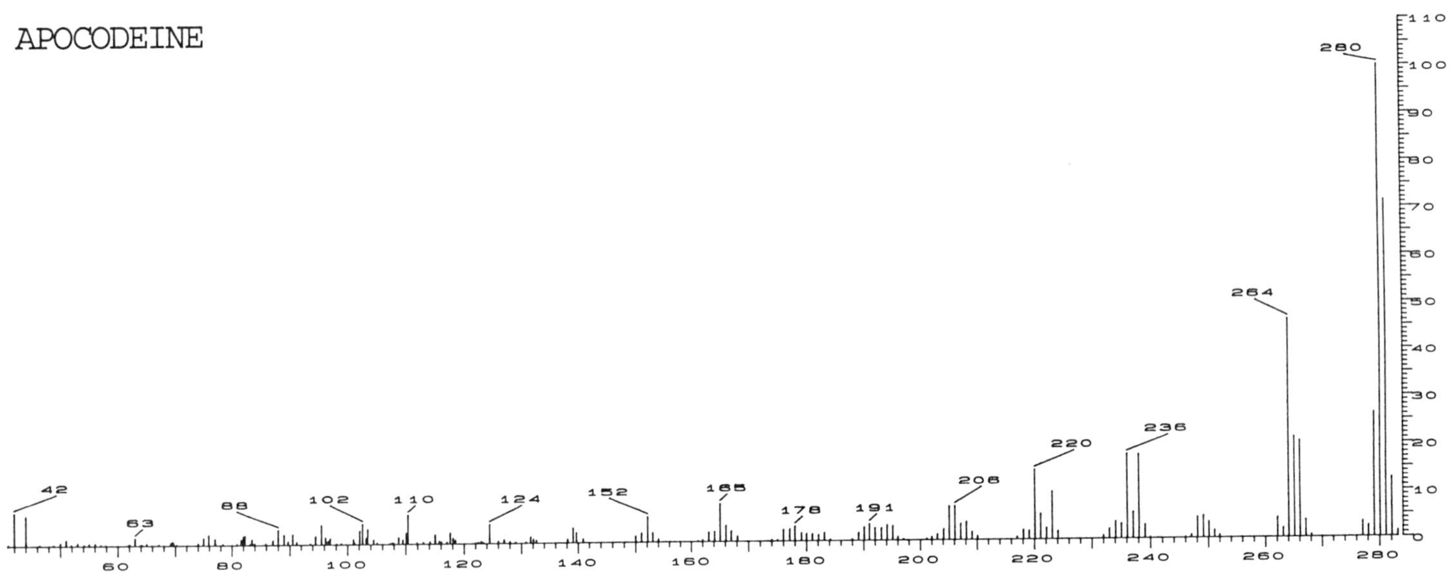

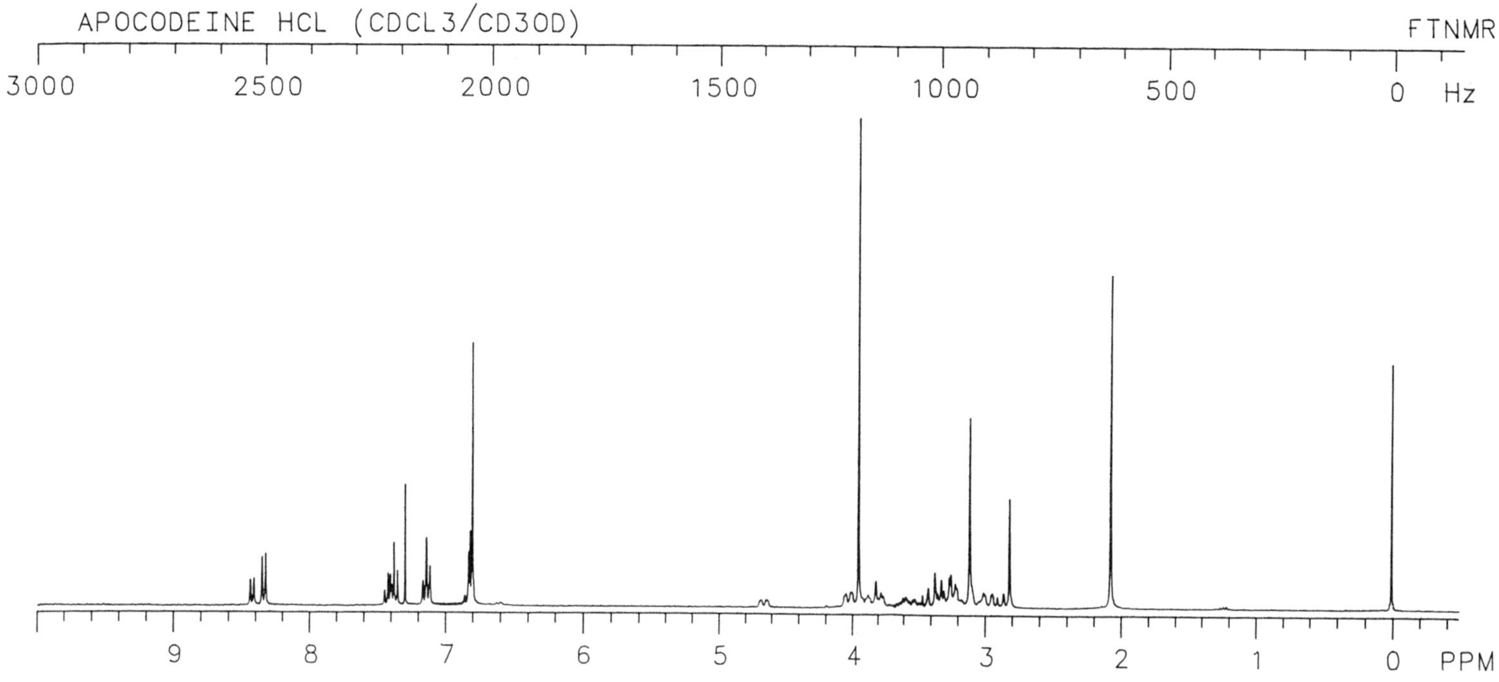

PEAK LISTING

#	HT	PPM
1	50	.00
2	67	2.06
3	22	2.81
4	38	3.10
5	99	3.94
6	53	6.79
7	15	6.81
8	24	7.29

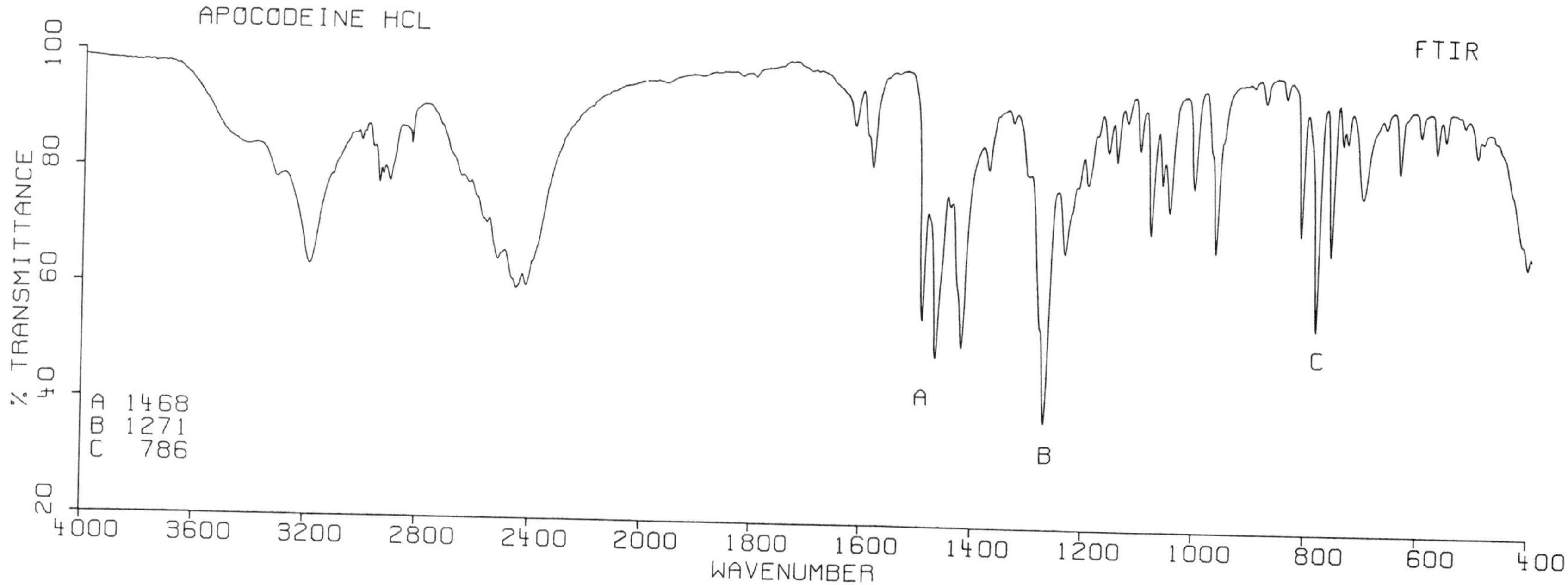

ATOVAQUONE

$C_{22}H_{19}ClO_3$

Molecular Weight: 366.84 (366.10)

Synonyms: Trans-2-[4-(-Chlorophenyl)cyclohexyl]-3-hydroxy-1,4-naphthalenedione
Trade Names: Mepron
Use: Antiprotozoal

HPLC: Methanol: 2.8
GC:

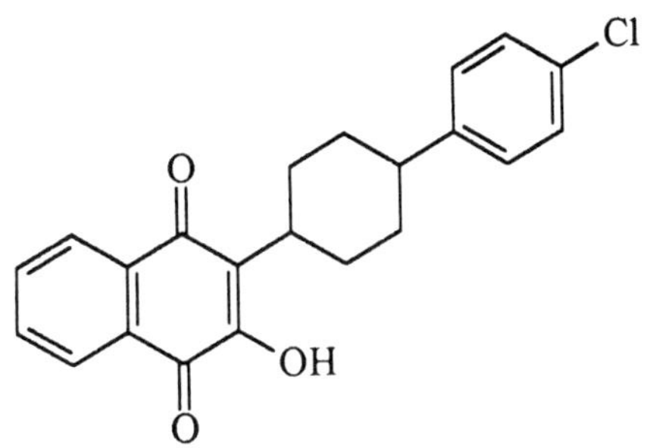

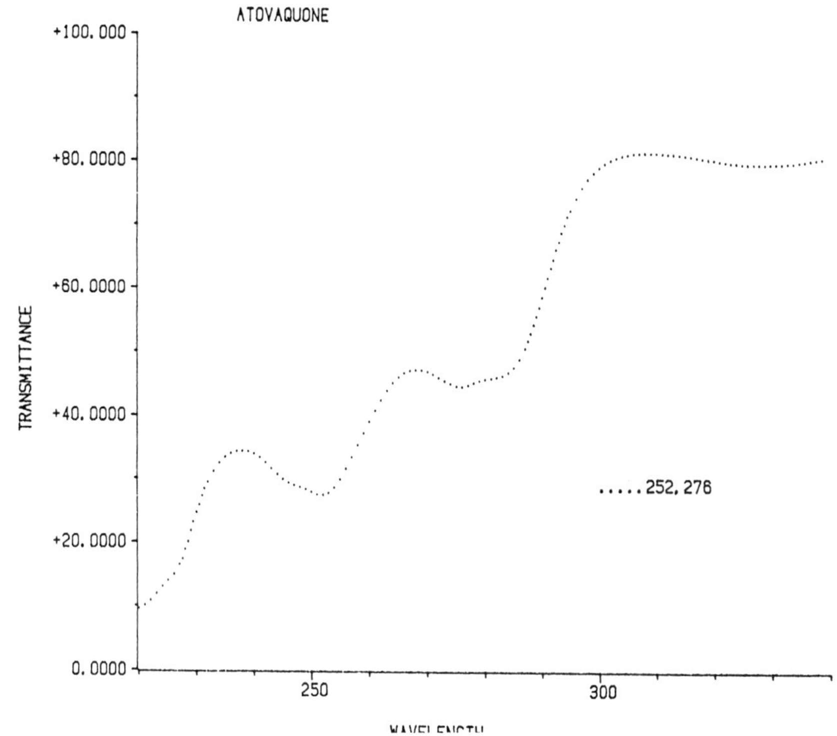

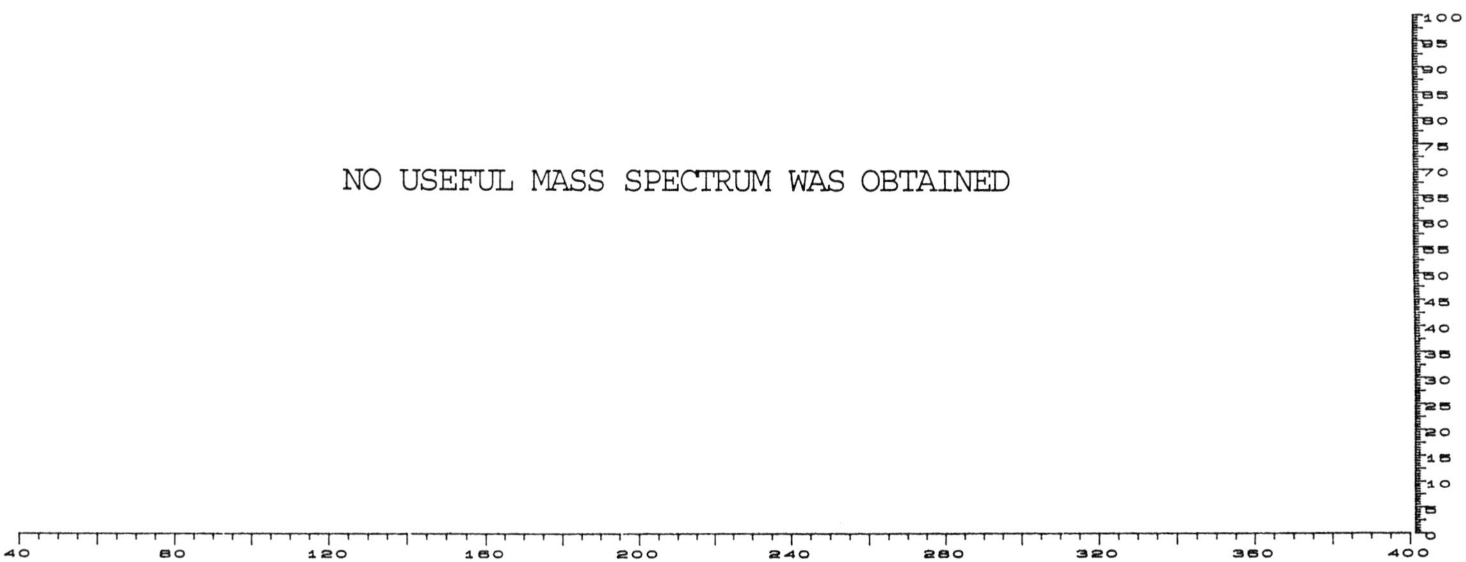

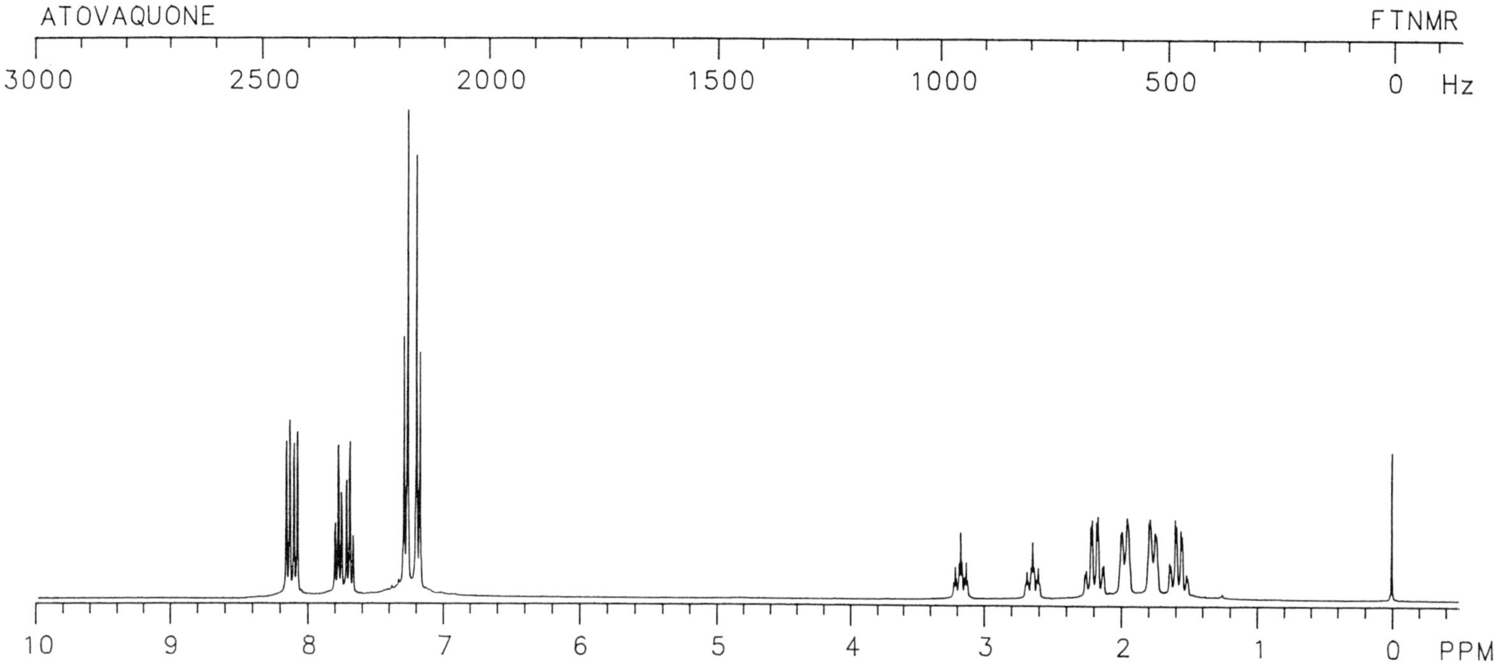

PEAK LISTING

#	HT	PPM
1	30	.00
2	14	1.56
3	15	1.59
4	16	1.60
5	16	1.78
6	16	1.79
7	16	1.95
8	14	1.98
9	17	2.16
10	16	2.17
11	16	2.20
12	15	2.21
13	50	7.16
14	90	7.19
15	99	7.25
16	44	7.26
17	53	7.28
18	32	7.68
19	24	7.70
20	23	7.71
21	20	7.74
22	21	7.74
23	31	7.76
24	31	7.77
25	15	7.79
26	33	8.06
27	33	8.06
28	31	8.09
29	36	8.12
30	34	8.12
31	32	8.14

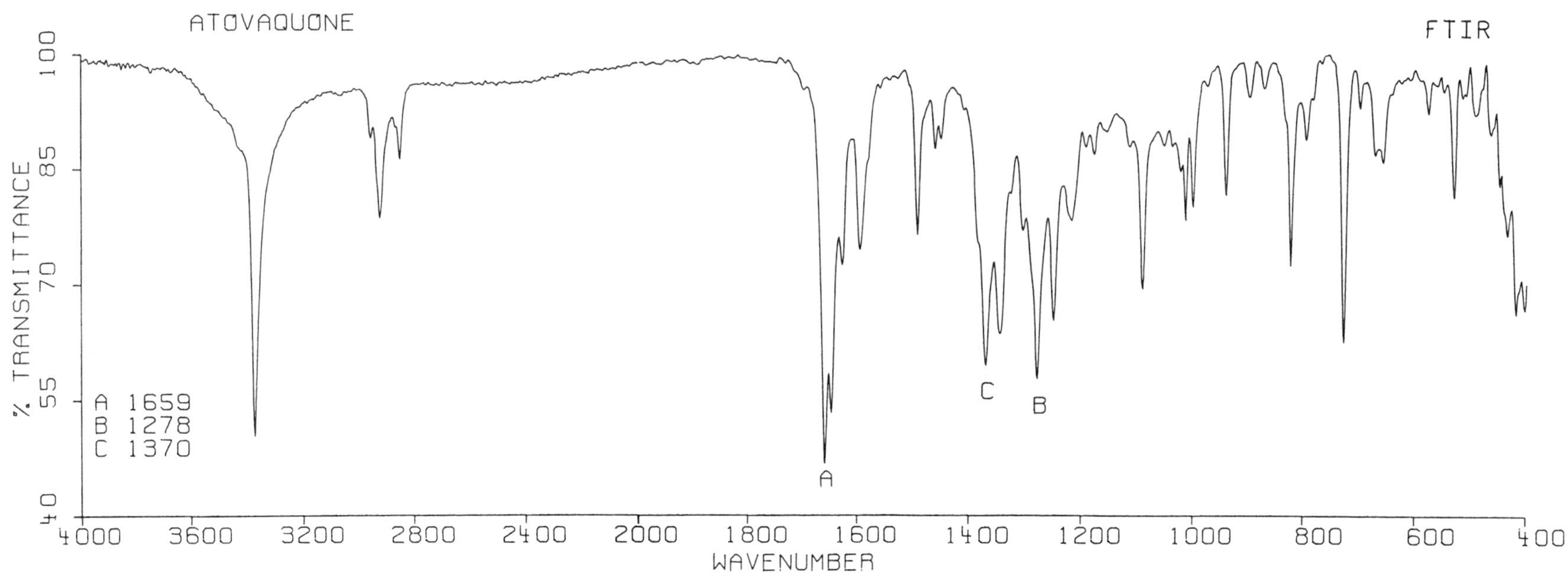

AURANOFIN

$C_{20}H_{34}AuO_9PS$

Molecular Weight: 678.48 (678.13)

Synonyms: (1-Thio-β-D-glucopyranase-2,3,4,6-tetraacetato-S)(triethylphosphine)gold
Trade Names: Aktil, Crisinar, Crisofin, Ridaura, Ridauran
Use: Antirheumatic

HPLC: Methanol: 2.5
GC:

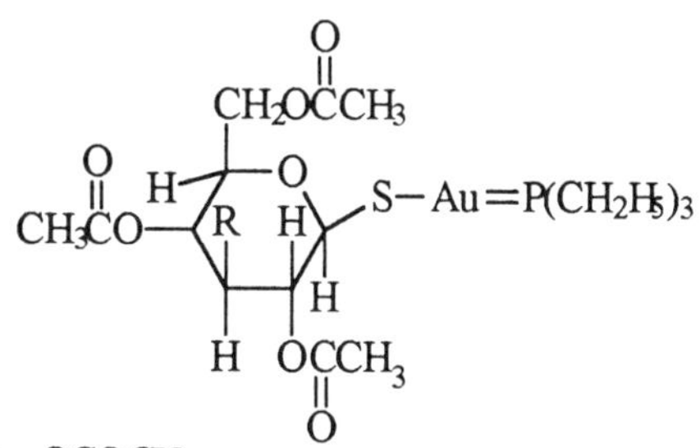

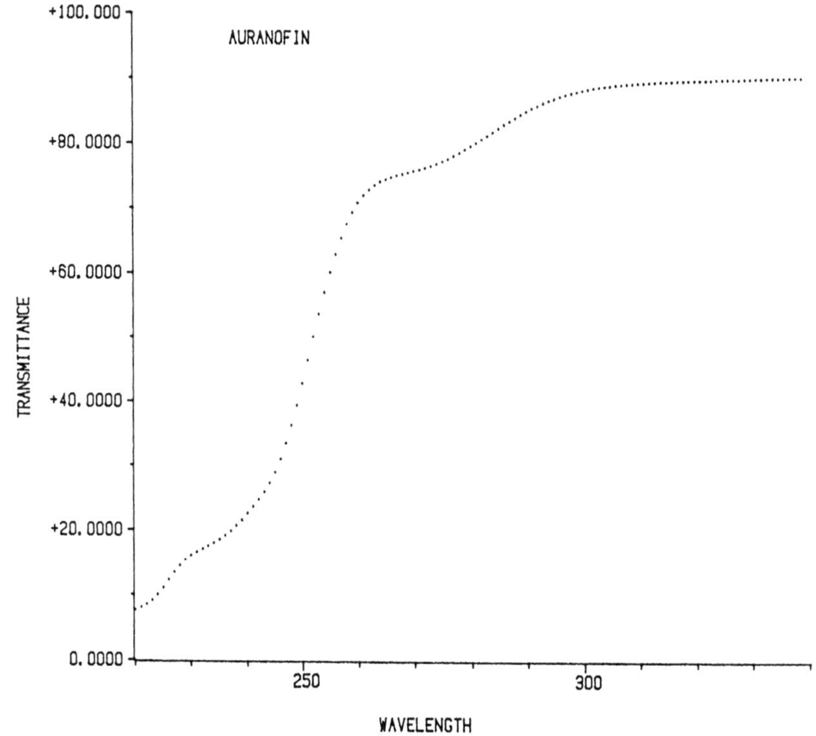

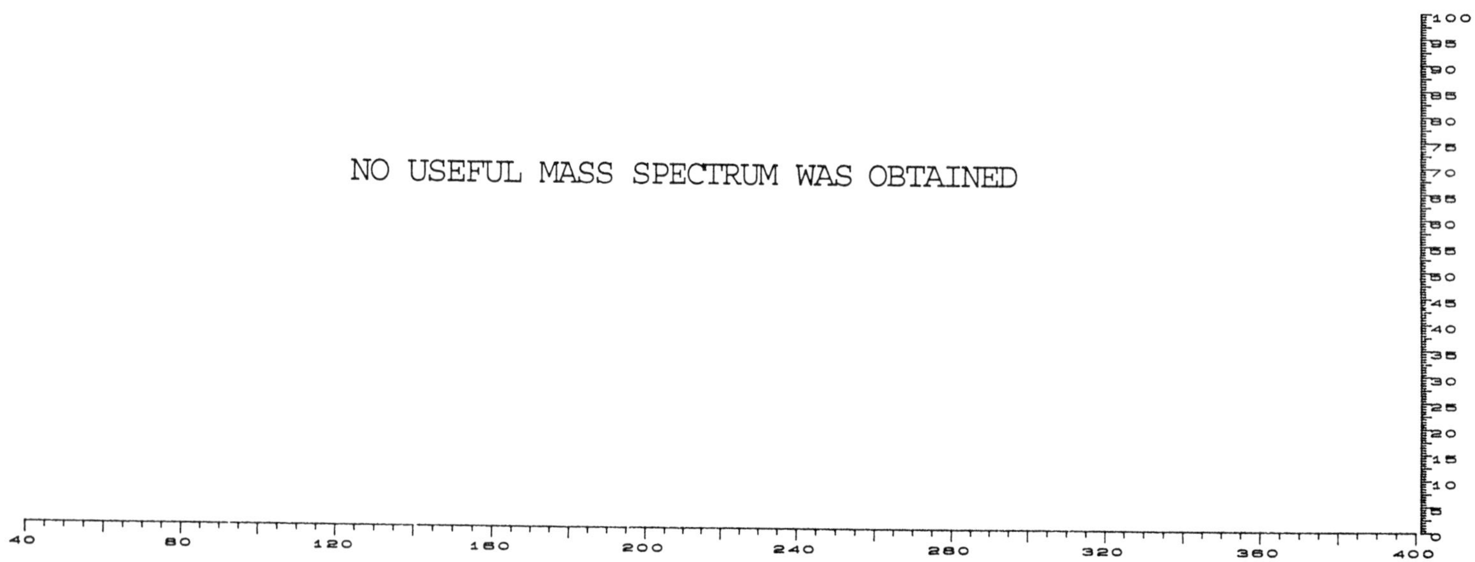

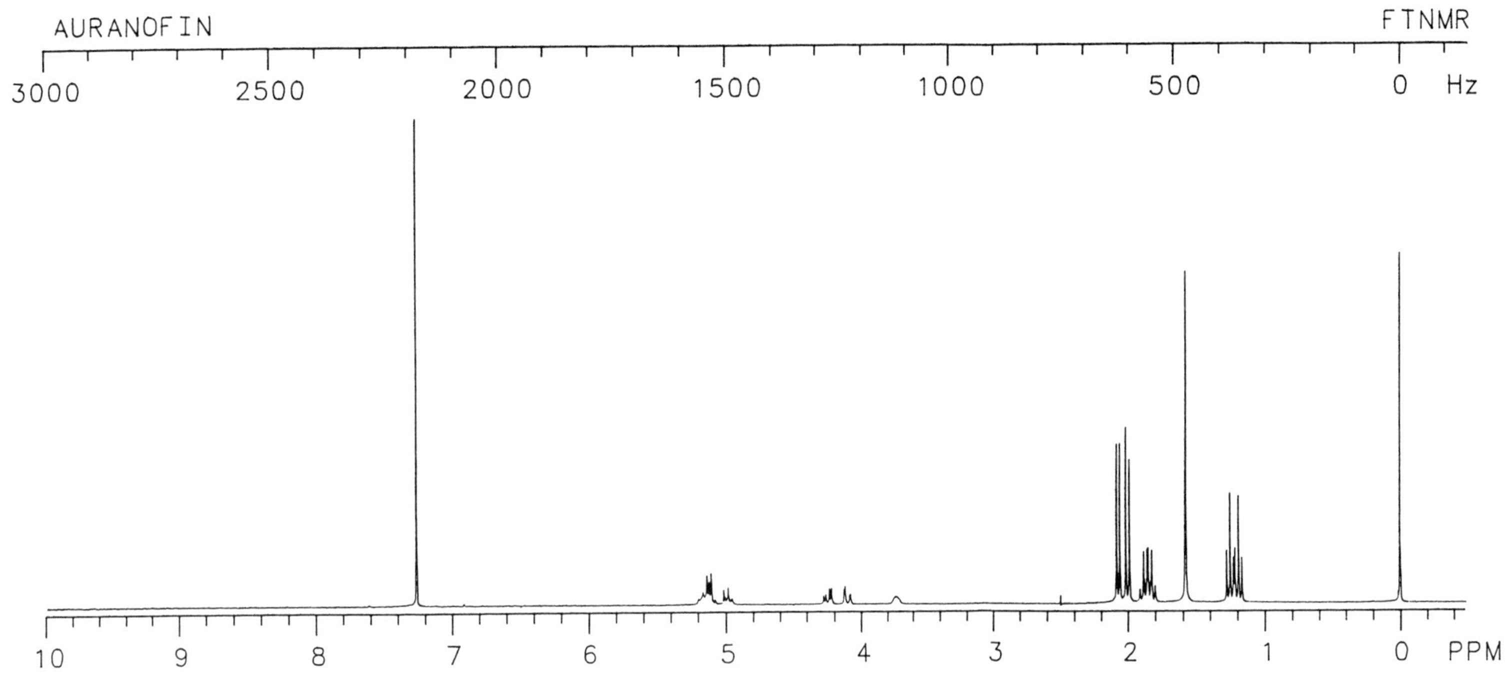

PEAK LISTING

#	HT	PPM
1	71	.00
2	21	1.19
3	22	1.25
4	67	1.58
5	29	1.98
6	35	2.01
7	32	2.05
8	32	2.08
9	99	7.26

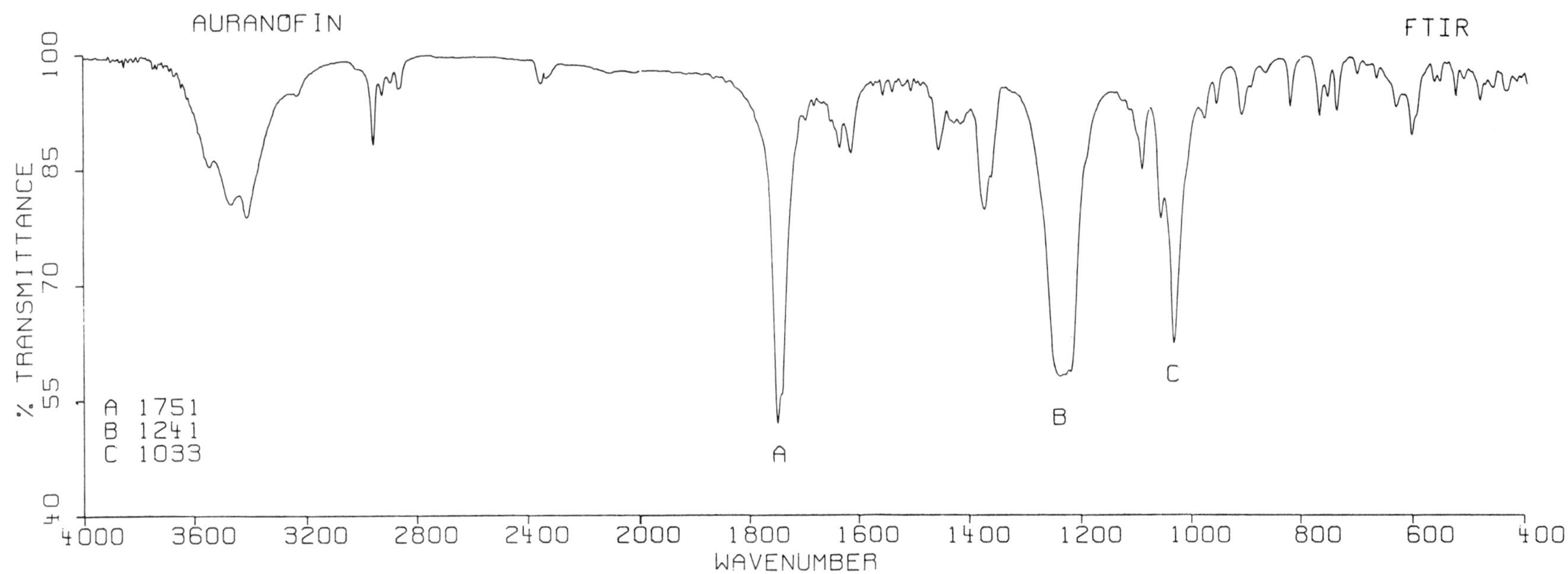

AZITHROMYCIN

$C_{38}H_{72}N_2O_{12}$

Molecular Weight: 748.99 (748.51)

Synonyms: [2R-(2R*,3S*,4R*,5R*,8R*,10R*,11R*,12S*,13S*,14R*)]-13-[(2,6-Dideoxy-3-C-methyl-3-O-methyl-α-L-ribo-hexopyranosyl)oxy]-2-ethyl-3,4,10-trihydroxy-3,5,6,8,10,12,14-heptamethyl-11-[[3,4,6-trideoxy-3-(dimethylamino)-β-D-xylo-hexopyranosyl]oxy]-1-oxa-6-azocyclopentadecan-15-one
Trade Names: Sumamed, Zithromax
Use: Antibacterial

HPLC:
GC:

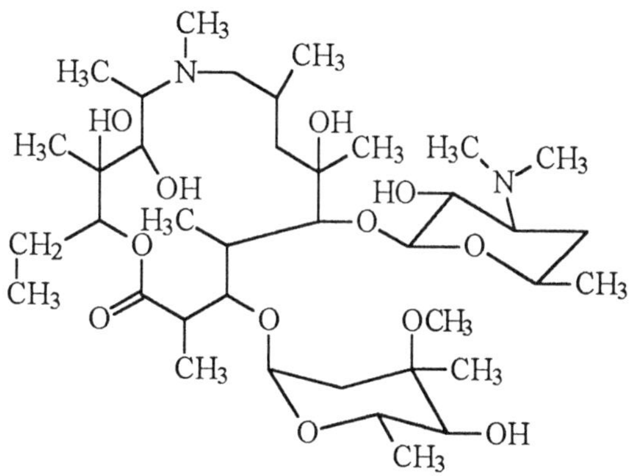

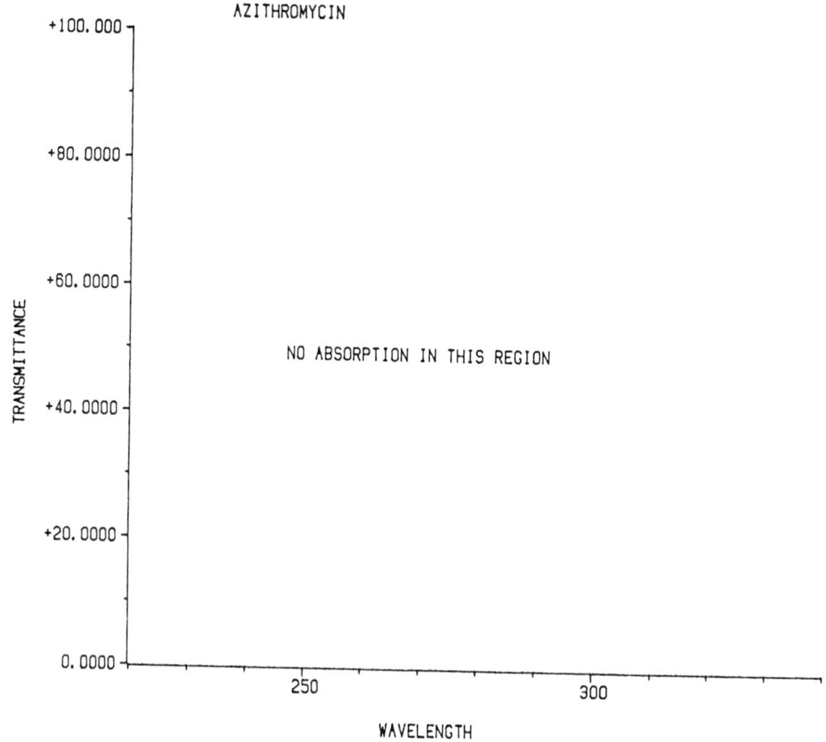

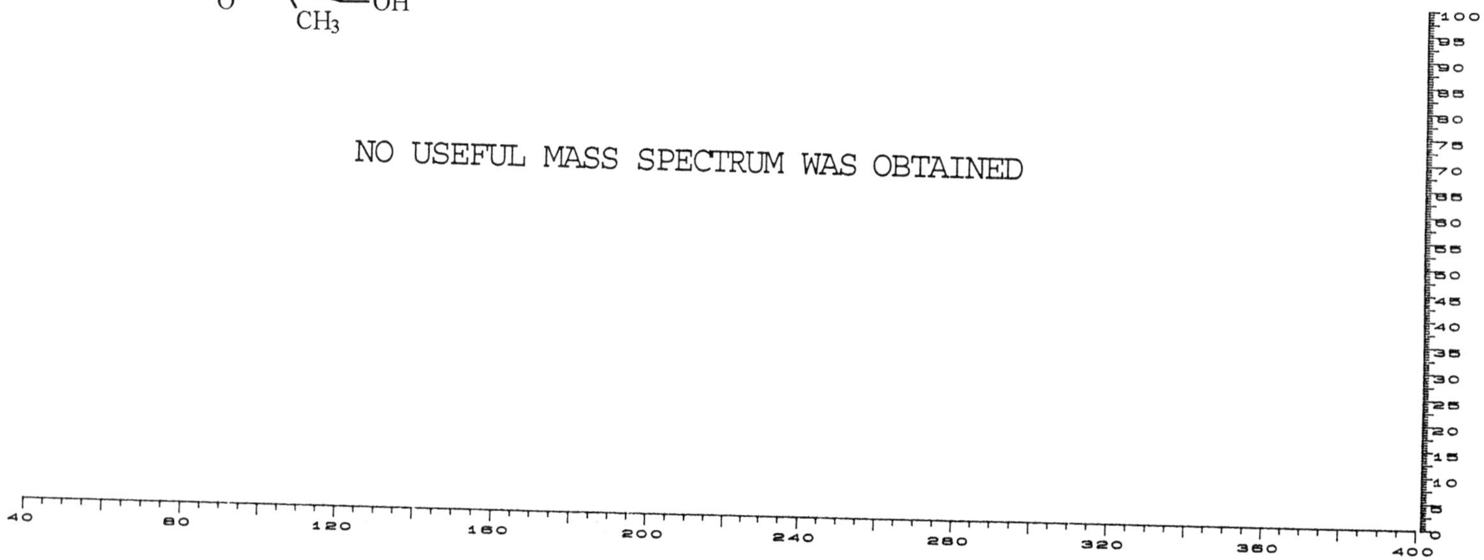

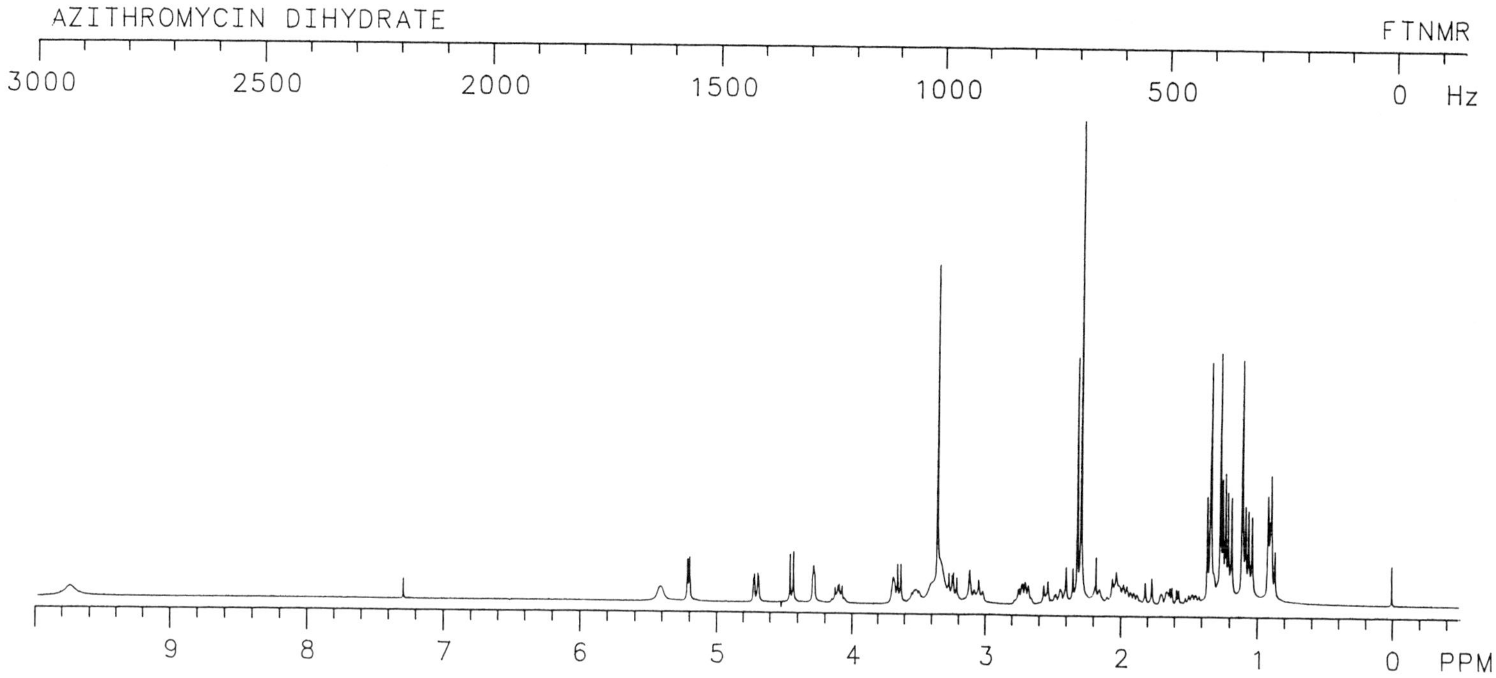

PEAK LISTING

#	HT	PPM
1	26	.88
2	17	.89
3	22	.91
4	18	1.03
5	19	1.05
6	20	1.07
7	50	1.10
8	22	1.17
9	23	1.20
10	27	1.21
11	25	1.23
12	51	1.25
13	49	1.32
14	22	1.35
15	98	2.28
16	50	2.31
17	69	3.35

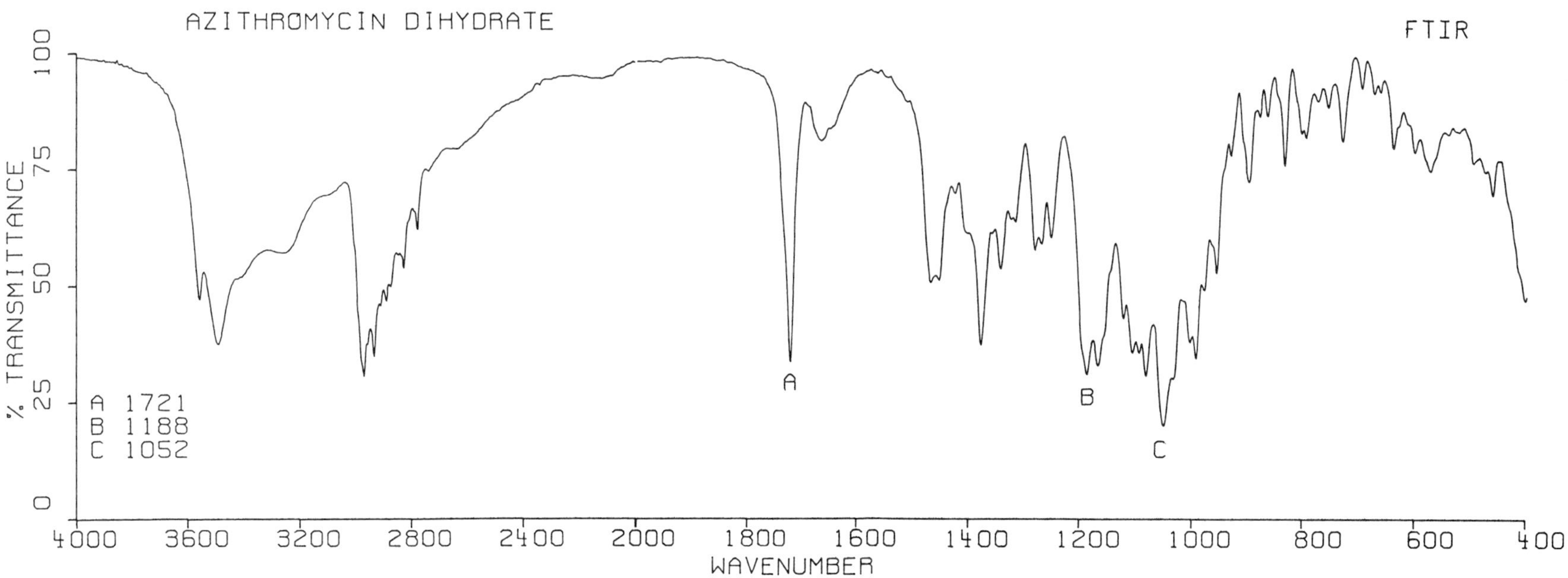

BACAMPICILLIN

$C_{21}H_{27}N_3O_7S$

Molecular Weight: 465.52 (465.16)

Synonyms: 6-[(Aminophenylacetyl)amino]-3,3-dimethyl-7-oxo-4-thia-1-azabicyclo[3.2.0]heptane-2-carboxylic acid 1-[(ethoxycarbonyl)oxy]ethyl ester
Trade Names: Ambacamp, Ambaxin, Bacacil, Bacampicine, Penglobe, Spectrobid,
Use: Antibacterial

HPLC:
GC:

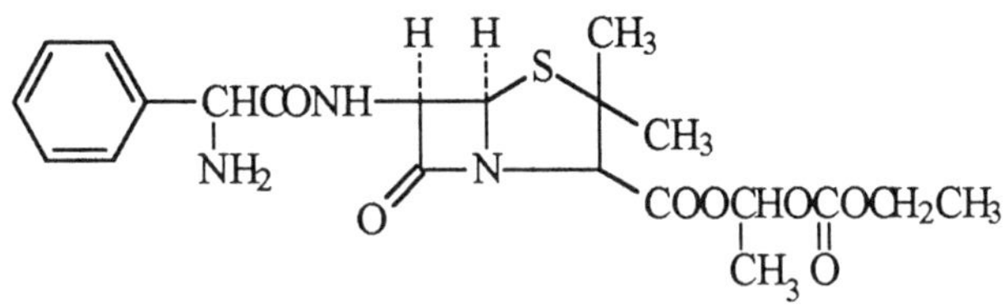

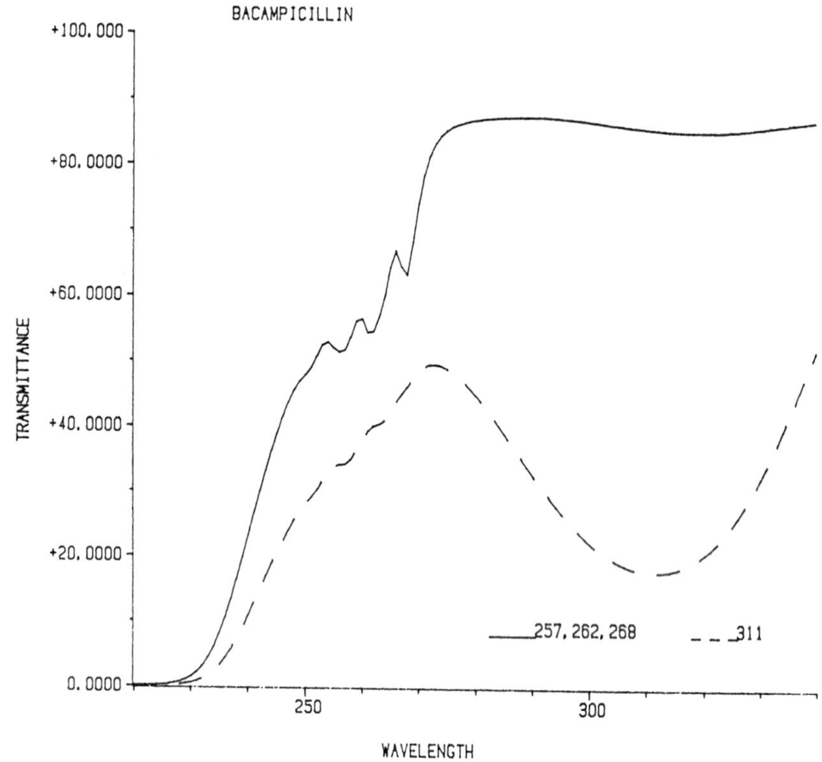

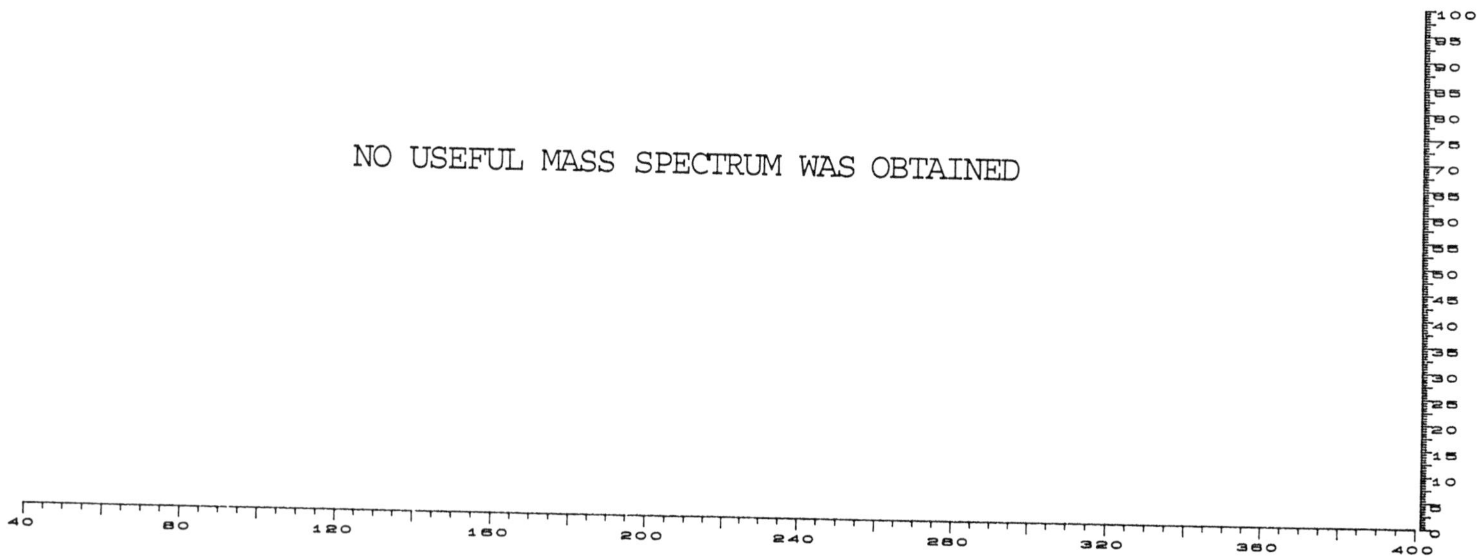

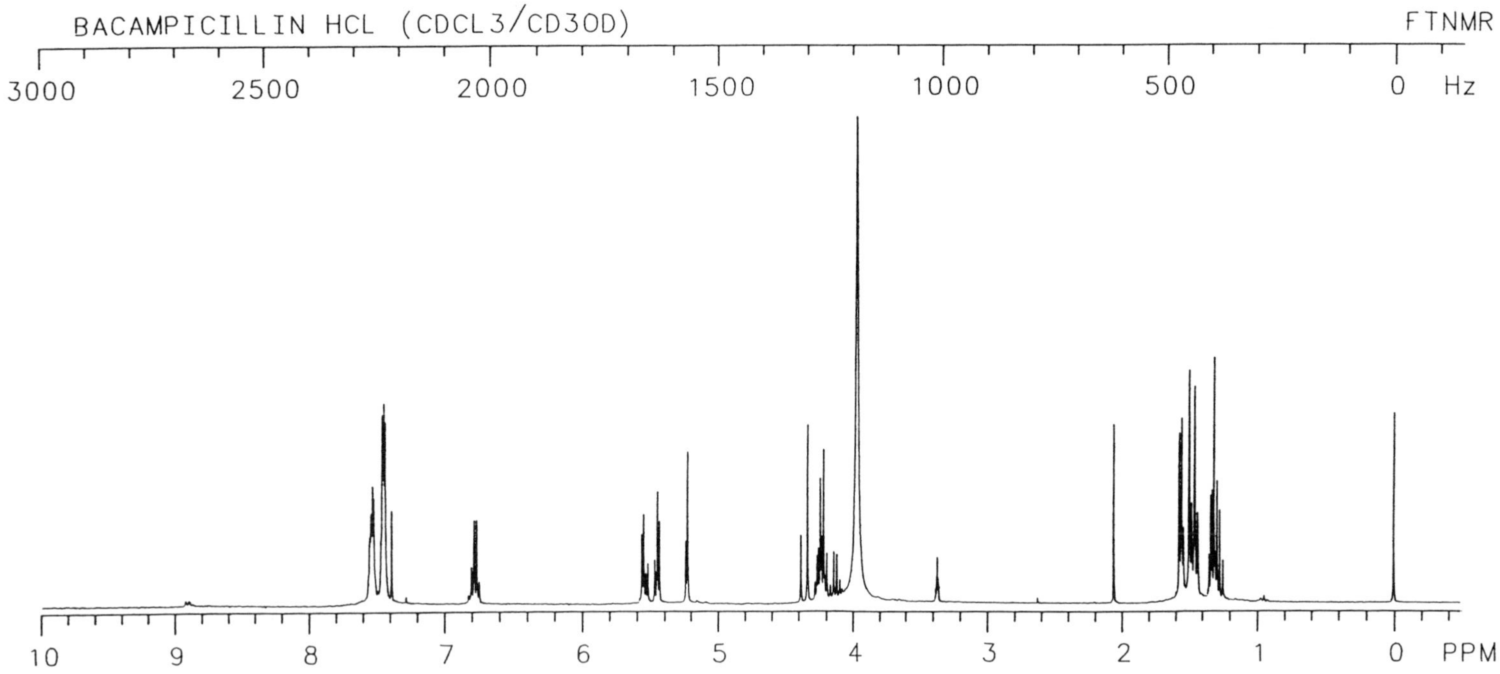

PEAK LISTING

#	HT	PPM
1	38	.00
2	25	1.28
3	50	1.31
4	44	1.46
5	47	1.50
6	37	1.55
7	34	1.57
8	36	2.05
9	99	3.96
10	31	4.21
11	25	4.23
12	36	4.32
13	31	5.21
14	37	7.43
15	41	7.44
16	38	7.45

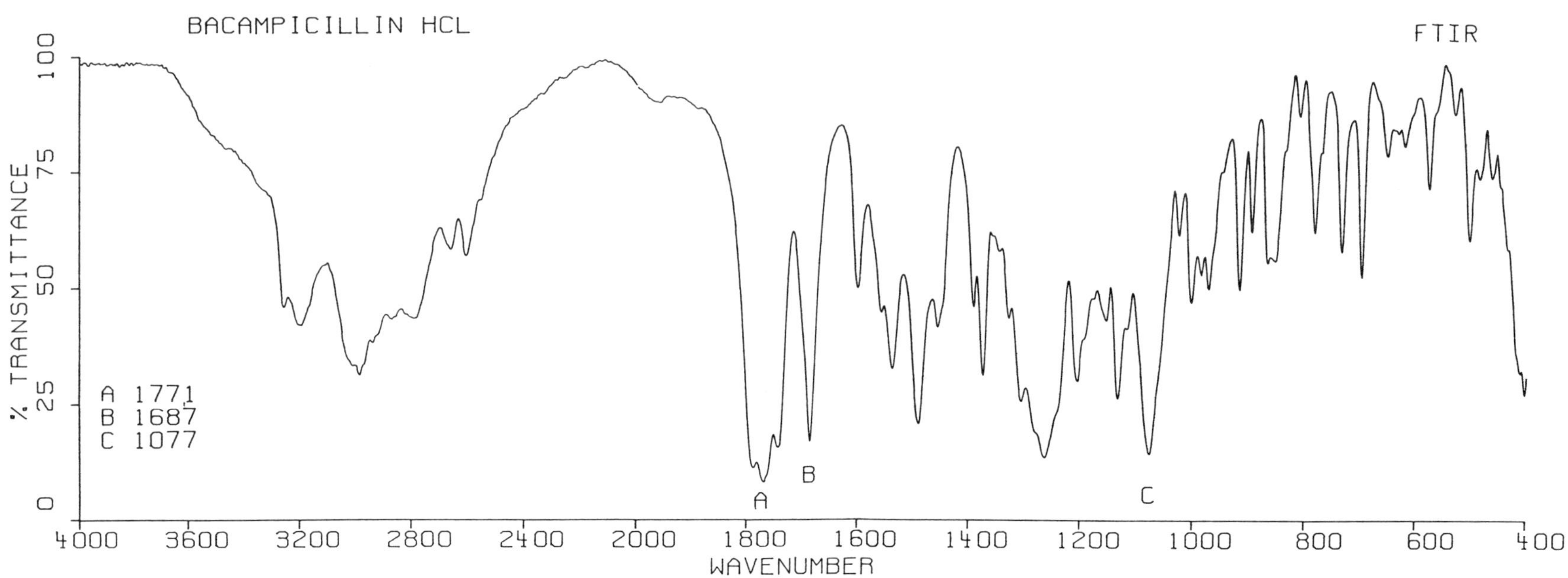

BENAZEPRIL

$C_{24}H_{28}N_2O_5$

Molecular Weight: 424.49 (424.20)

Synonyms: 3-[[1-(Ethoxycarbonyl)-3-phenyl-(1S)-propyl]amino]2,3,4,5-tetrahydro-2-oxo-1H-(3S)-benzazepine-1-acetic acid
Trade Names: Lotensin, Lotensin HCT, Lotrel
Use: Enzyme inhibitor

HPLC:
GC:

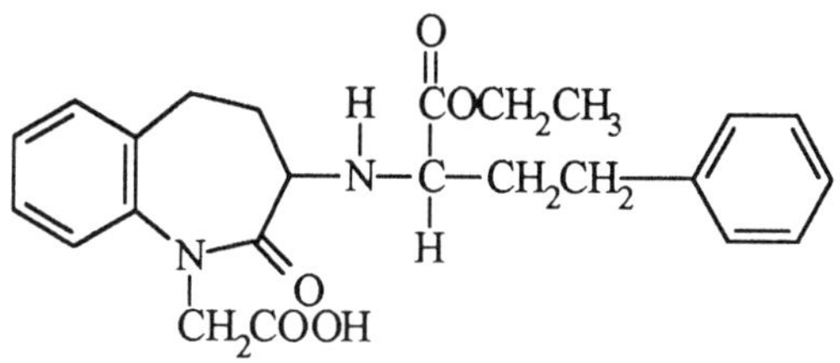

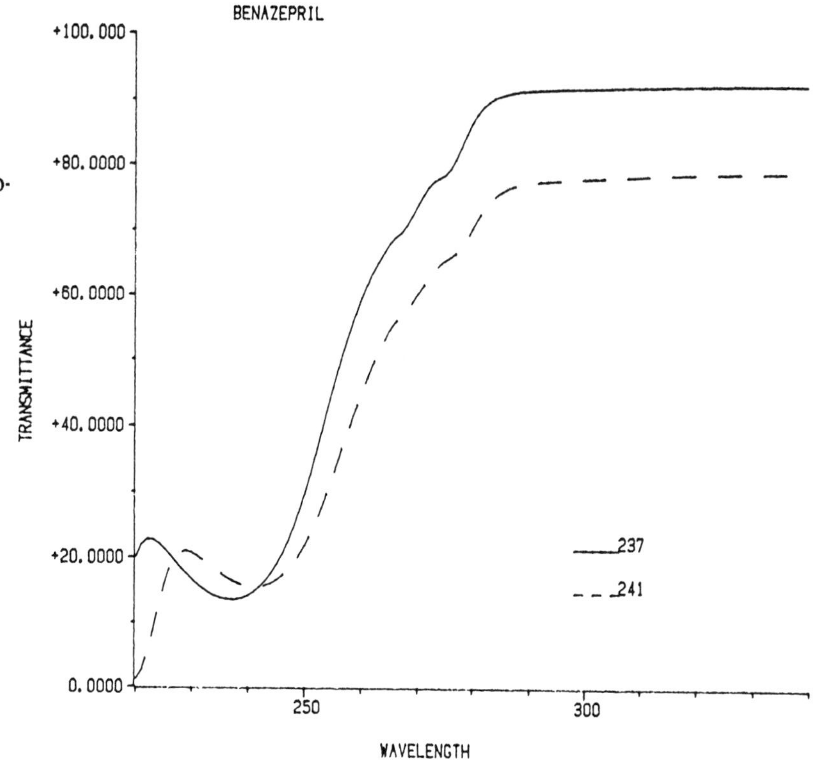

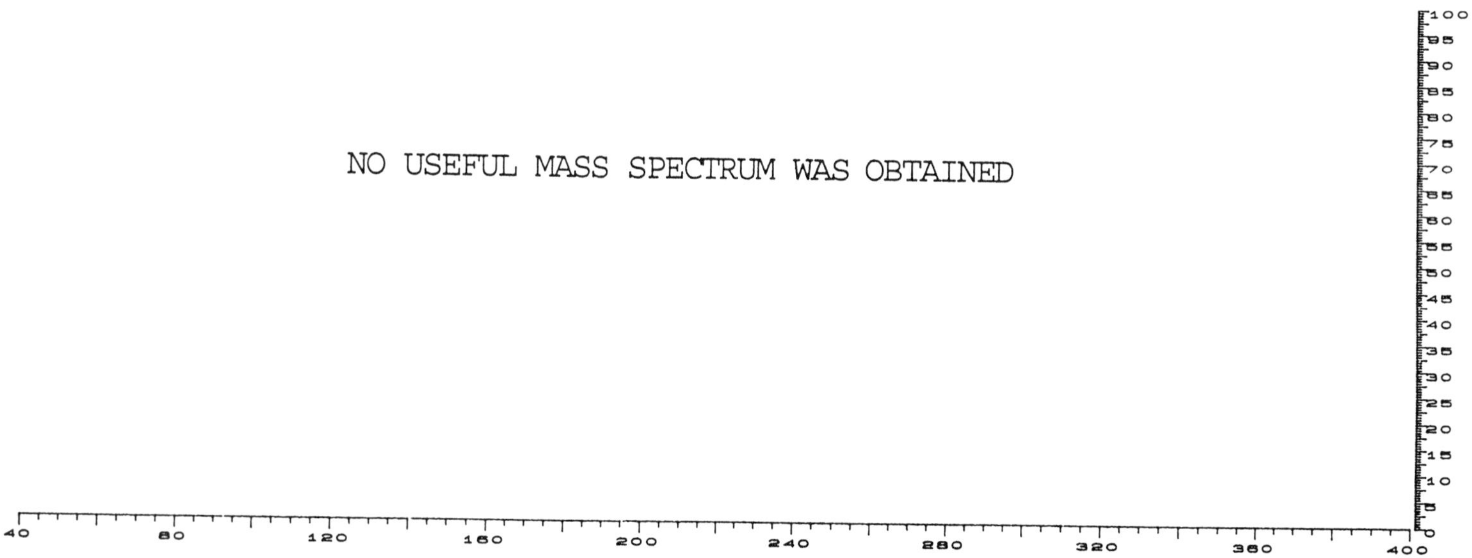

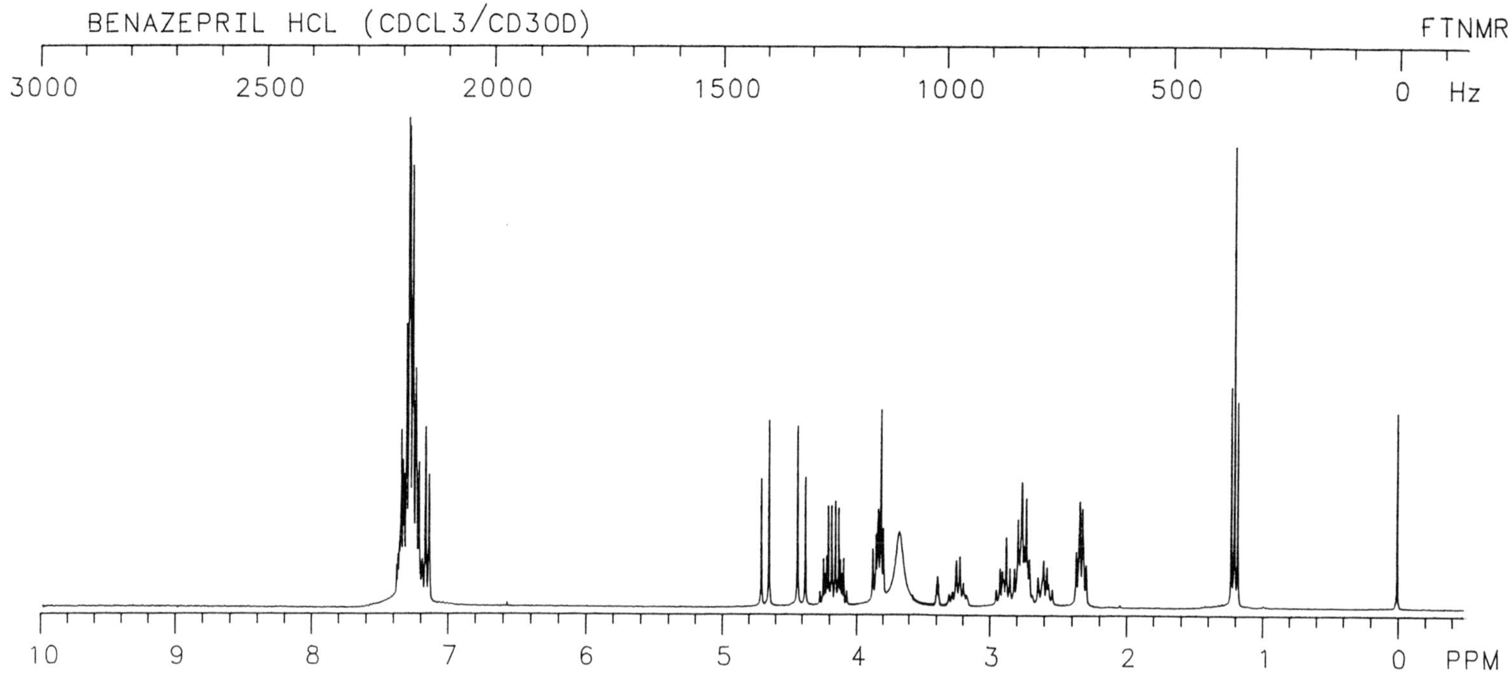

PEAK LISTING

#	HT	PPM
1	39	.00
2	42	1.18
3	93	1.20
4	45	1.22
5	25	2.75
6	40	3.81
7	26	4.36
8	36	4.42
9	38	4.64
10	26	4.70
11	26	7.13
12	36	7.15
13	29	7.20
14	48	7.22
15	89	7.25
16	97	7.27
17	99	7.28
18	57	7.29
19	27	7.31
20	29	7.32
21	35	7.33

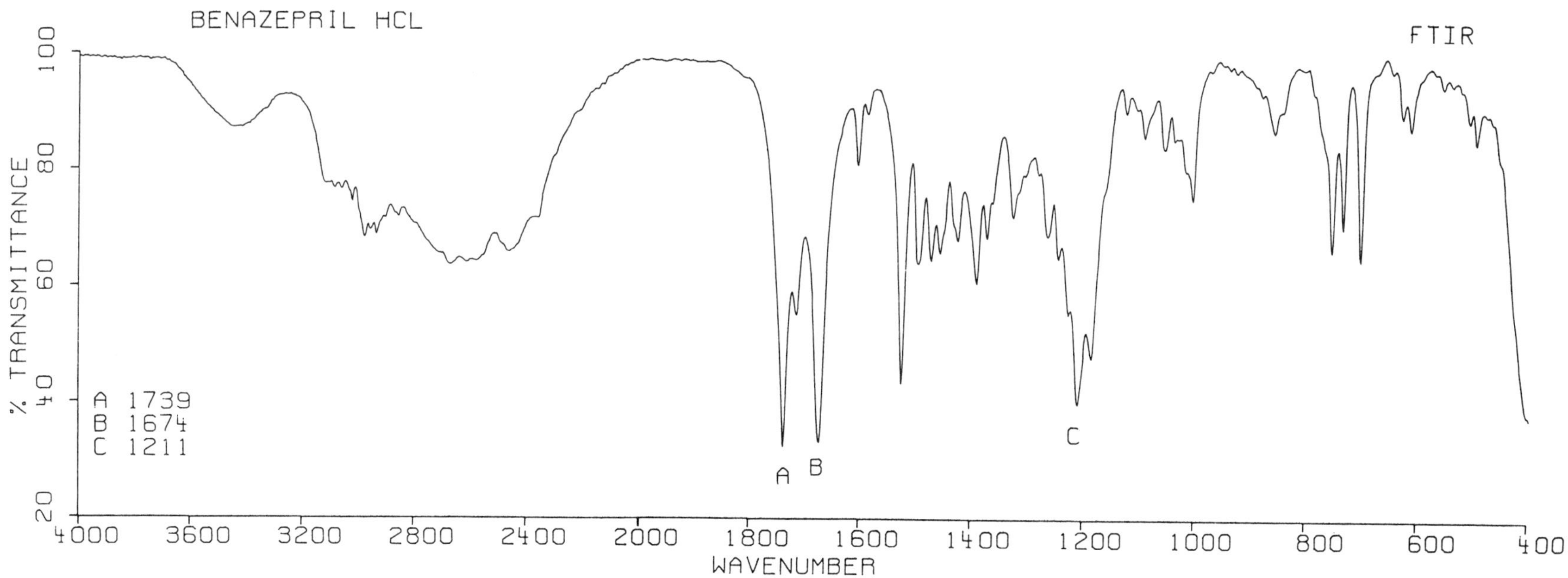

BENZAMIL

$C_{13}H_{14}ClN_7O$

Molecular Weight: 319.75 (319.09)

Synonyms: 3,5-Diamino-N-[amino(benzylamino)methylene]-6-chloropyrazinecarboxamide; benzylamiloride
Trade Names:
Use: Sodium inhibitor

HPLC: Methanol: 2.4
GC:

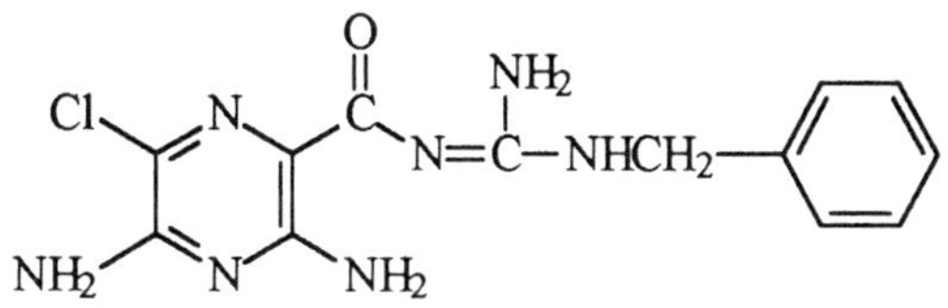

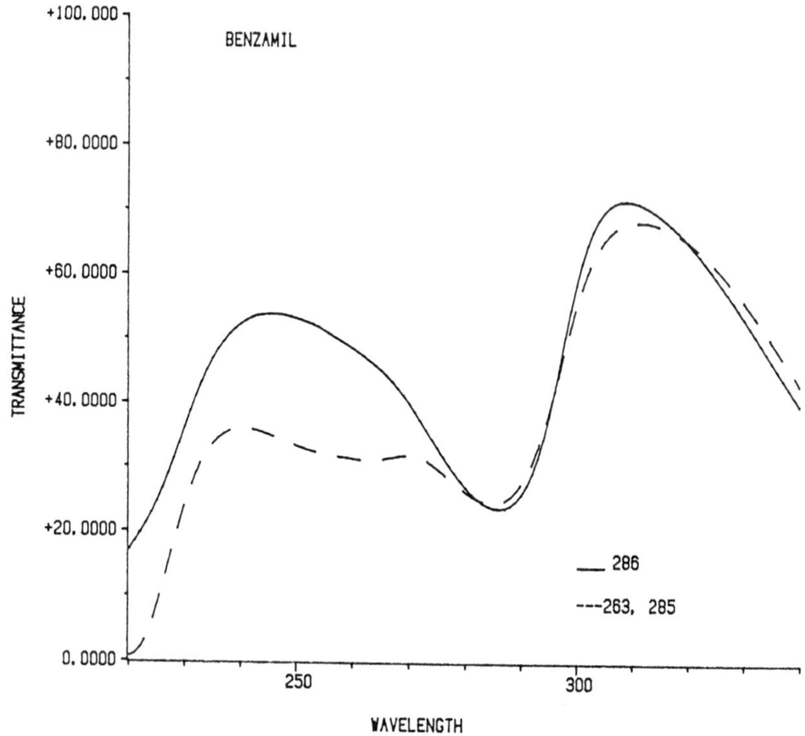

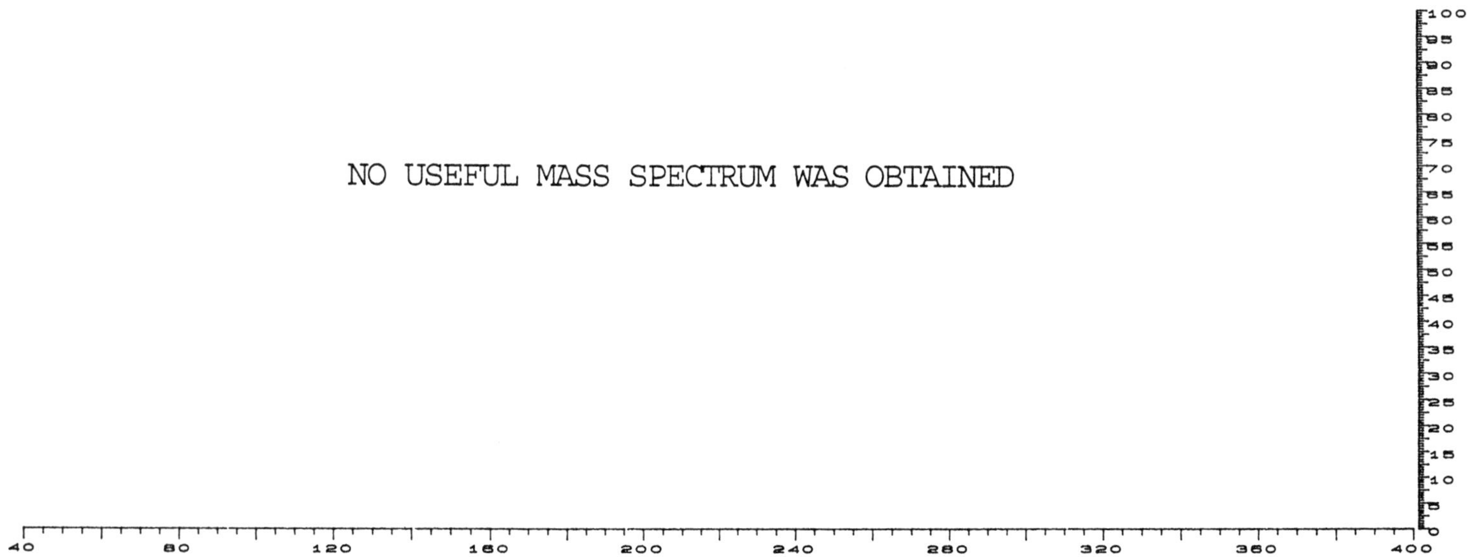

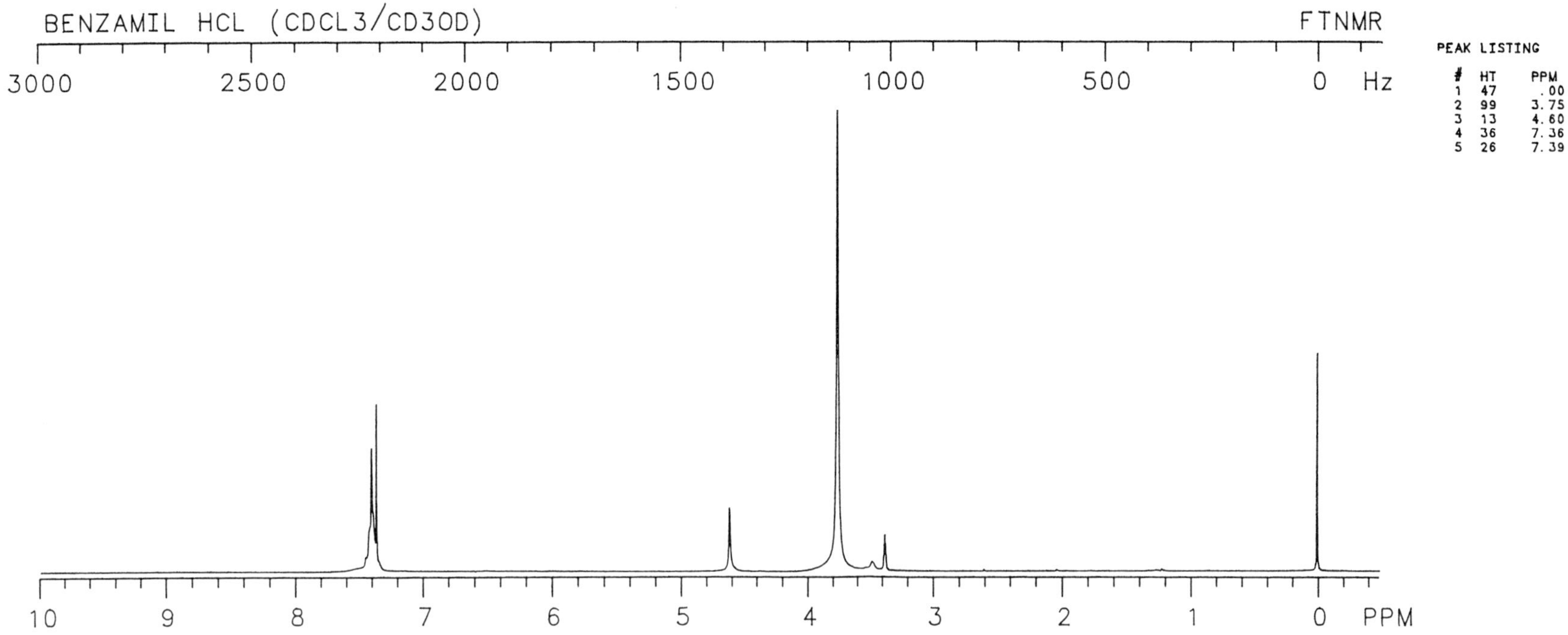
BENZAMIL HCL (CDCL3/CD3OD)
FTNMR
3000 2500 2000 1500 1000 500 0 Hz
10 9 8 7 6 5 4 3 2 1 0 PPM
PEAK LISTING
HT PPM
1 47 .00
2 99 3.75
3 13 4.60
4 36 7.36
5 26 7.39

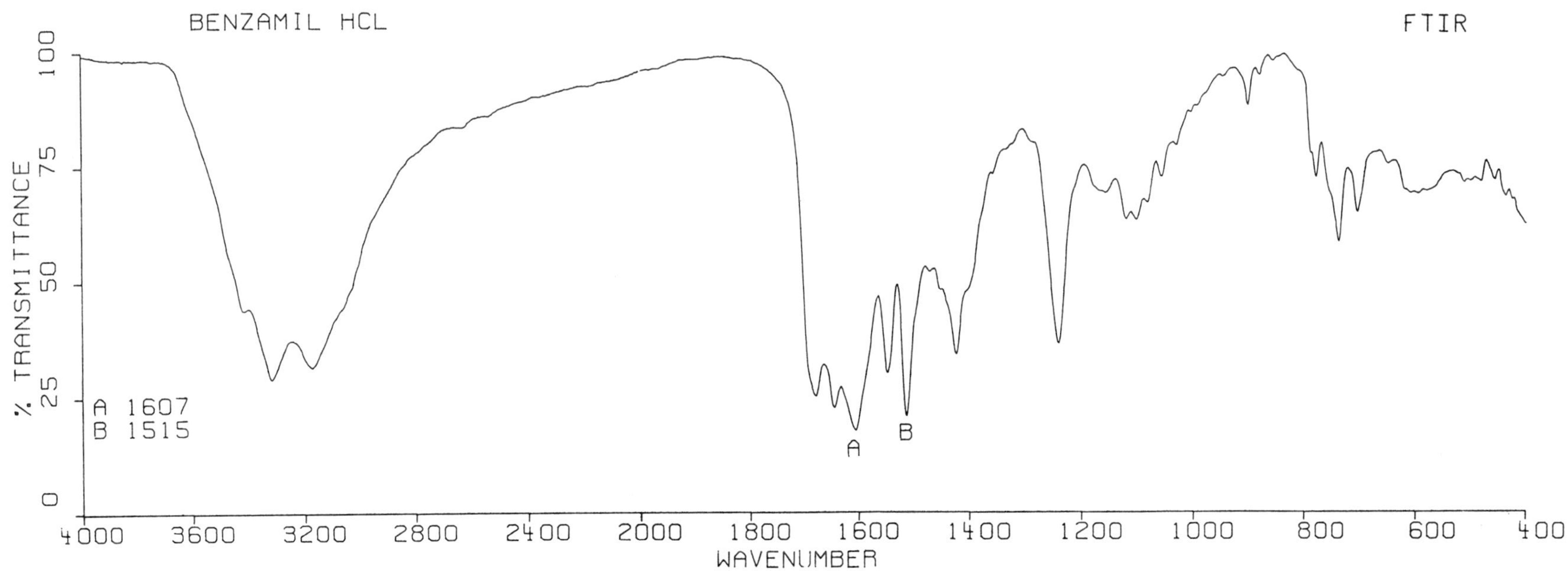
BENZAMIL HCL
FTIR
% TRANSMITTANCE
0 25 50 75 100
A 1607
B 1515
A
B
4000 3600 3200 2800 2400 2000 1800 1600 1400 1200 1000 800 600 400
WAVENUMBER

BENZOYLNORECGONINE

$C_{15}H_{17}NO_4$

Molecular Weight: 275.30 (275.12)

Synonyms: [1r-(exo,exo)]-3-(Benzoyloxy)-8-azabicyclo[3.2.1]octane-2β-carboxylic acid.
Trade Names:
Use: Cocaine metabolite

HPLC: Methanol: 2.6
GC:

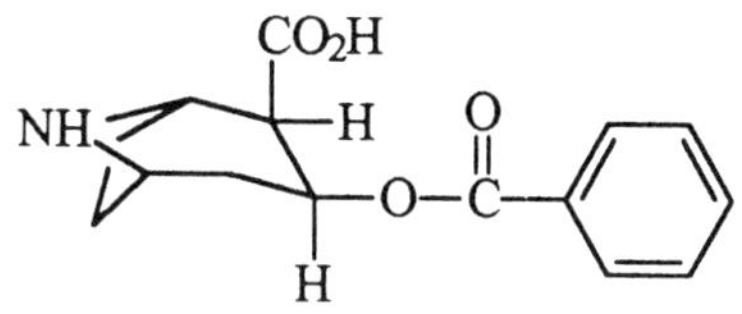

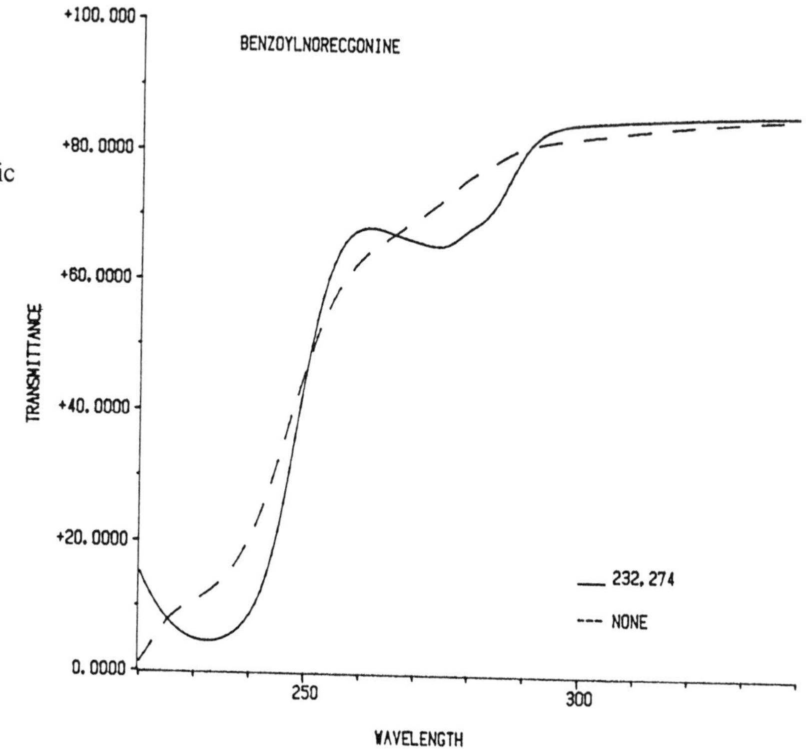

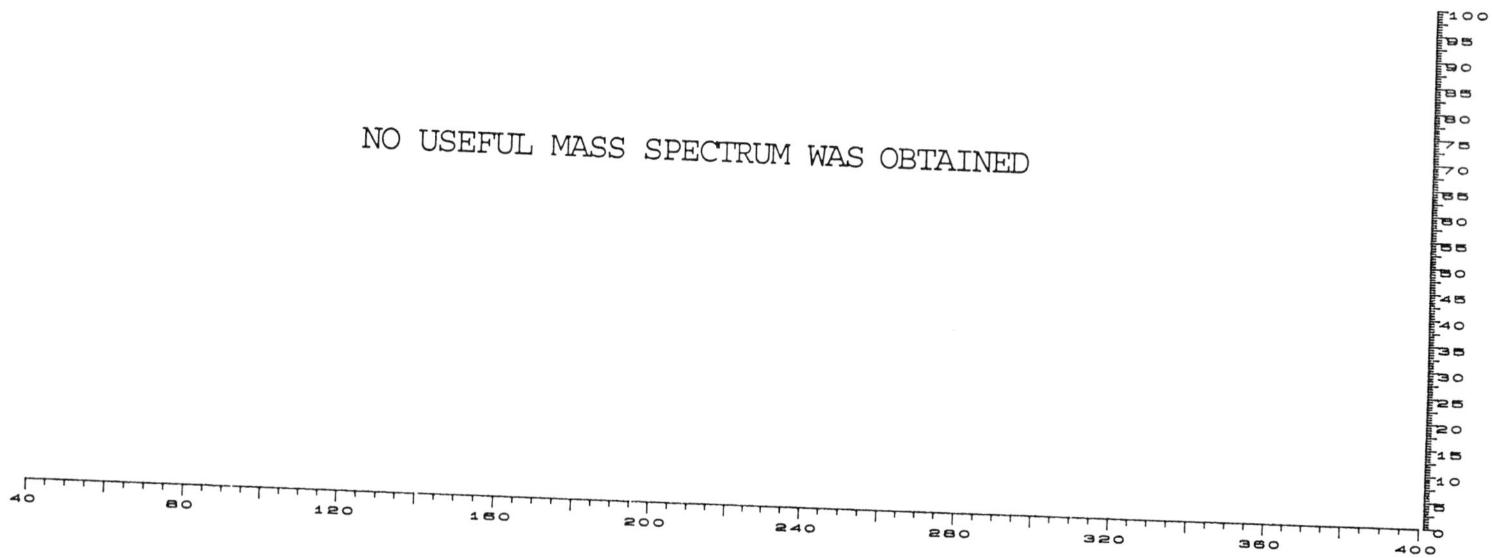

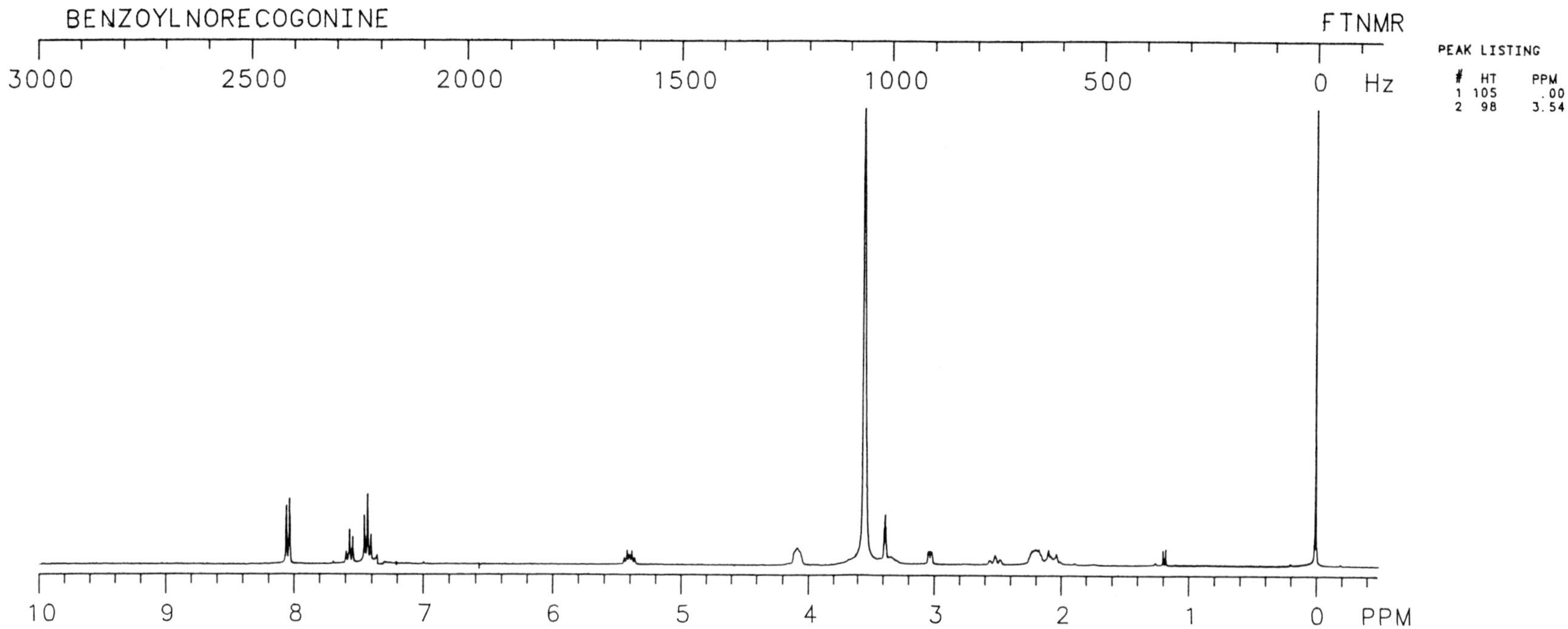
BENZOYLNORECOGONINE
FTNMR
3000 2500 2000 1500 1000 500 0 Hz
PEAK LISTING
HT PPM
1 105 .00
2 98 3.54
10 9 8 7 6 5 4 3 2 1 0 PPM

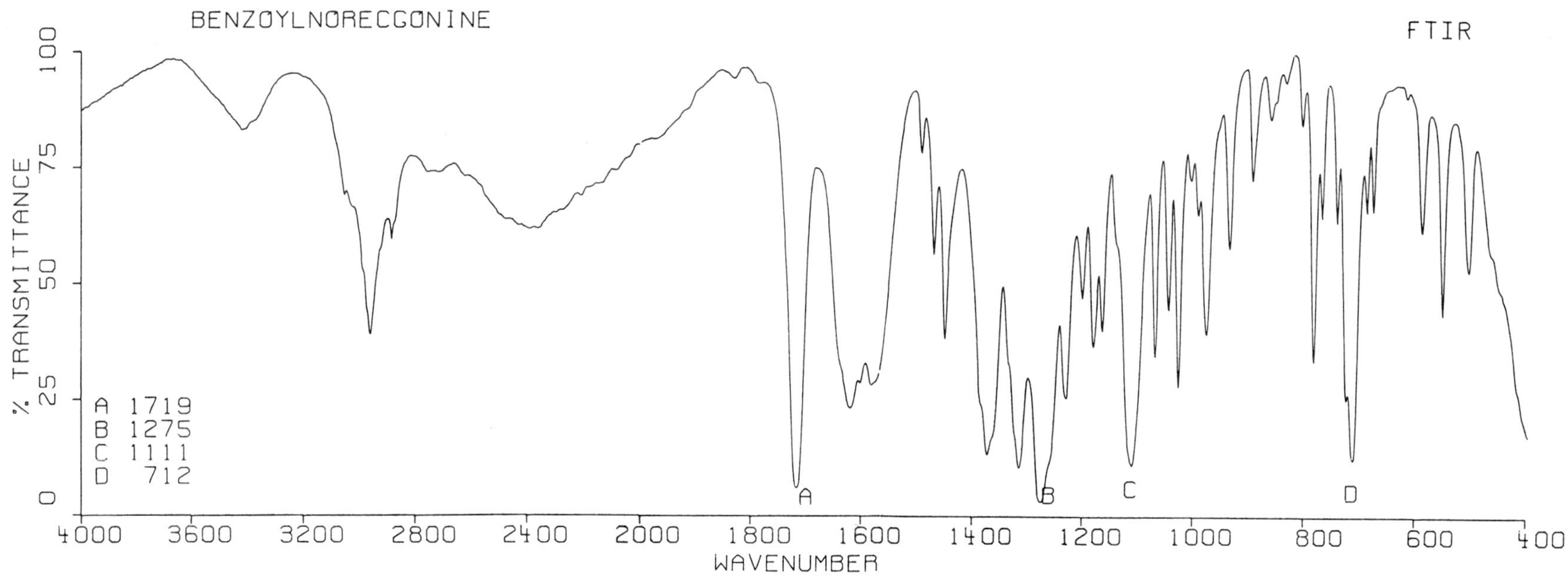
BENZOYLNORECGONINE
FTIR
% TRANSMITTANCE
100 75 50 25 0
A 1719
B 1275
C 1111
D 712
A B C D
4000 3600 3200 2800 2400 2000 1800 1600 1400 1200 1000 800 600 400
WAVENUMBER

BEPRIDIL

$C_{24}H_{34}N_2O$

Molecular Weight: 366.55 (366.27)

Synonyms: β-[(2-Methylpropoxy)methyl]-N-phenyl-N-(phenylmethyl)-1-pyrrolidineethanamine; 3-isobutoxy-2-pyrrolidino-N-phenyl-N-benzylpropylamine
Trade Names: Angopril, Bepadin, Cordium, Vascor
Use: Antianginal

HPLC: Methanol: 2.0
GC: 2594; 280°

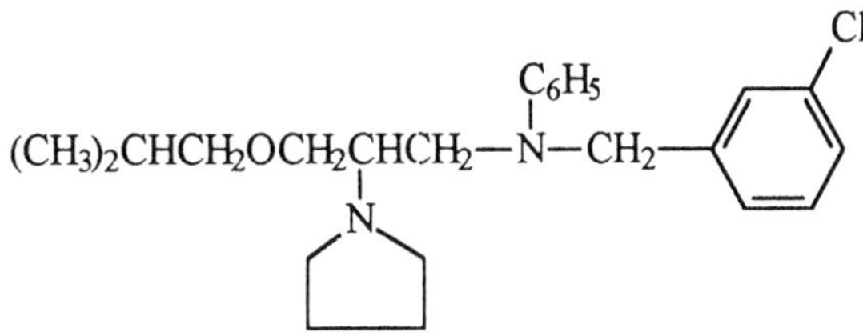

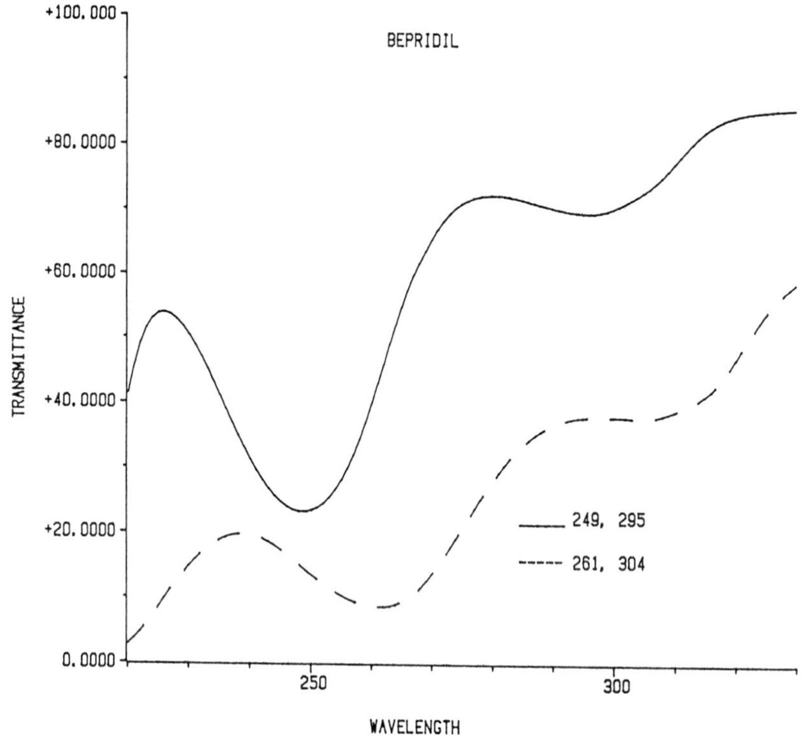

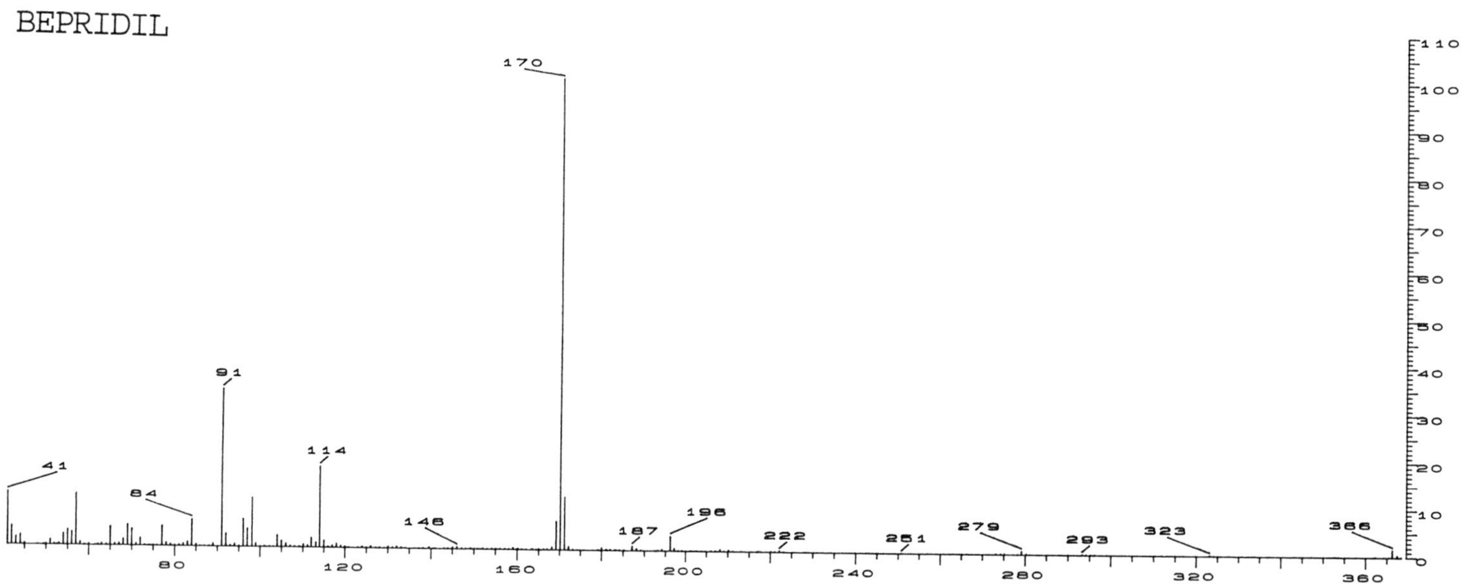

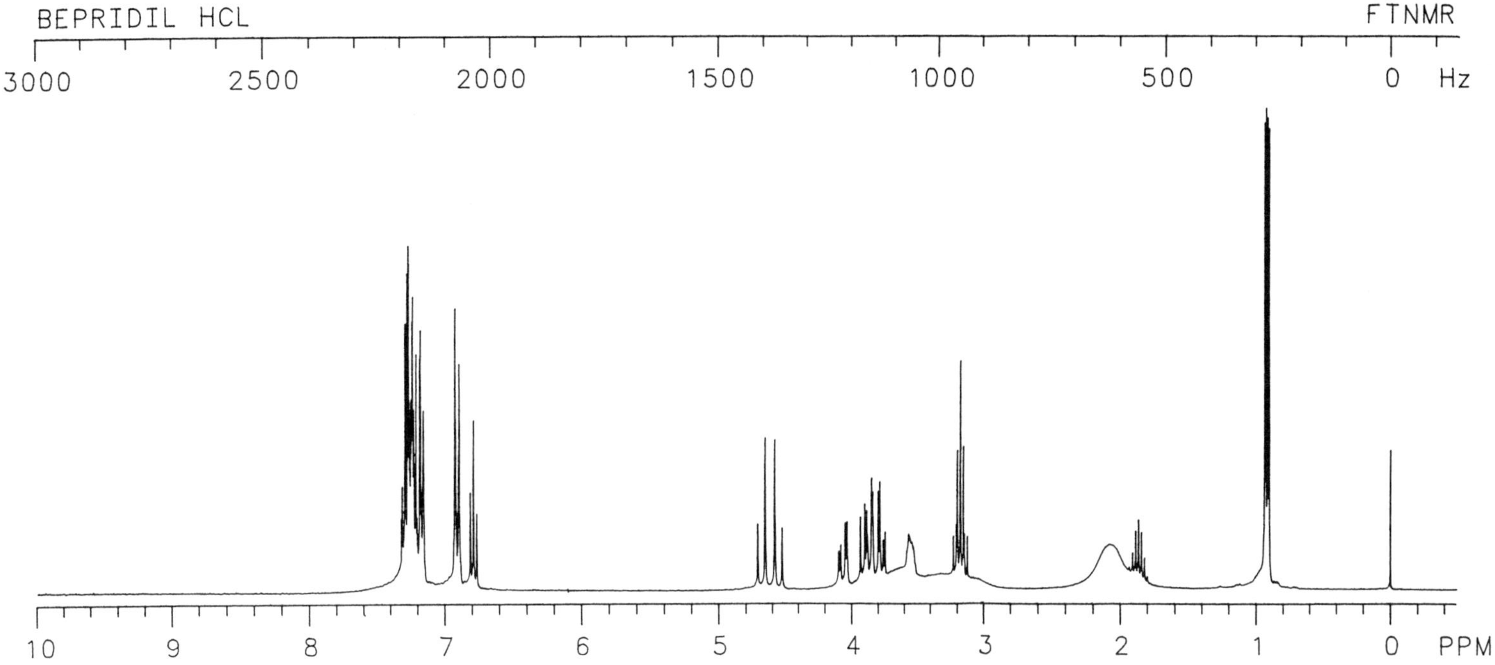

PEAK LISTING

#	HT	PPM
1	28	.00
2	94	.90
3	96	.90
4	98	.92
5	95	.93
6	29	3.14
7	46	3.17
8	28	3.19
9	30	4.57
10	31	4.64
11	34	6.78
12	40	6.89
13	46	6.89
14	57	6.92
15	36	7.15
16	53	7.17
17	47	7.18
18	48	7.21
19	36	7.22
20	59	7.23
21	54	7.24
22	38	7.25
23	27	7.25
24	66	7.26
25	70	7.26
26	33	7.27
27	64	7.27
28	33	7.28
29	54	7.29

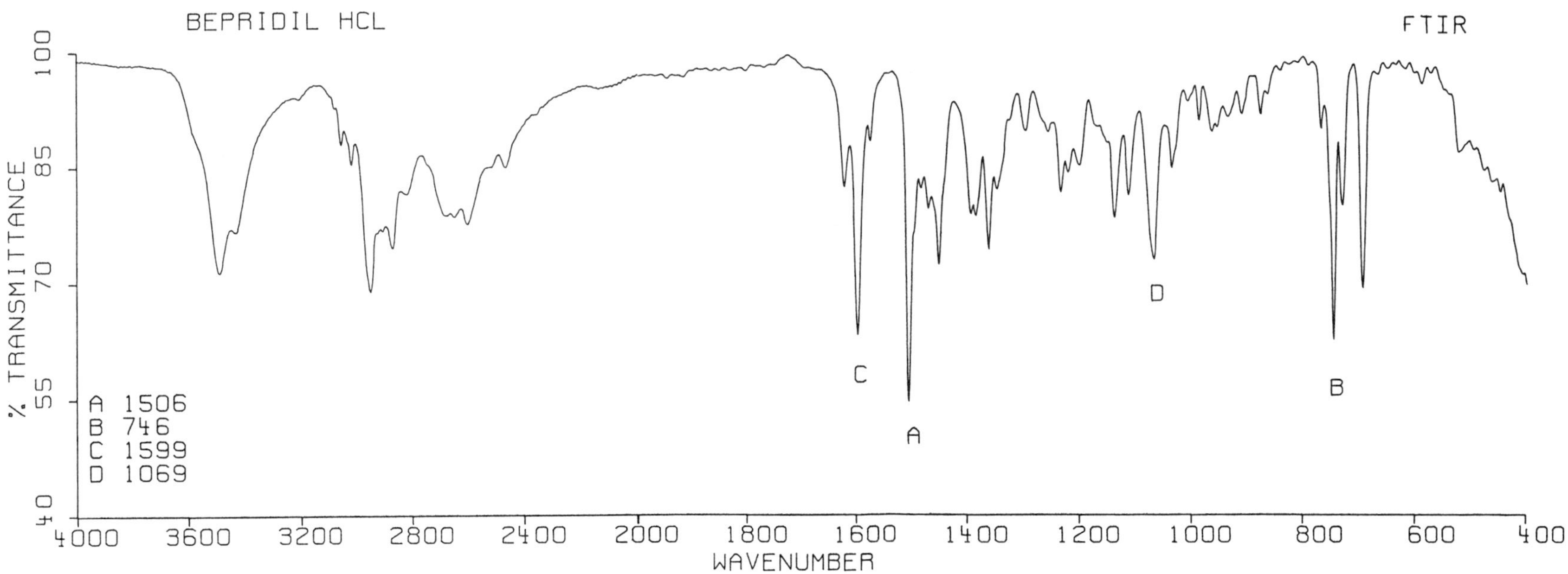

BETAMETHASONE-21-ACETATE

$C_{24}H_{31}FO_6$

Molecular Weight: 434.50 (434.21)

Synonyms: (9α-Fluoro-16β-methyl-11β,17α,21-trihydroxy-1,4-pregnadiene-3,20-dione-21-acetate
Trade Names: Betafluorene
Use: Glucocorticoid

HPLC: Methanol: 2.7
GC:

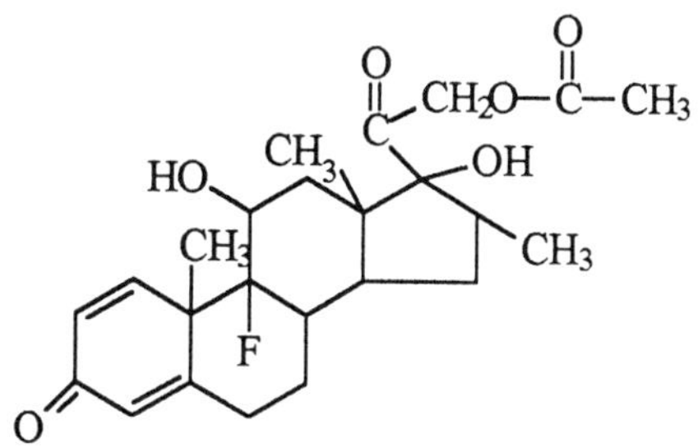

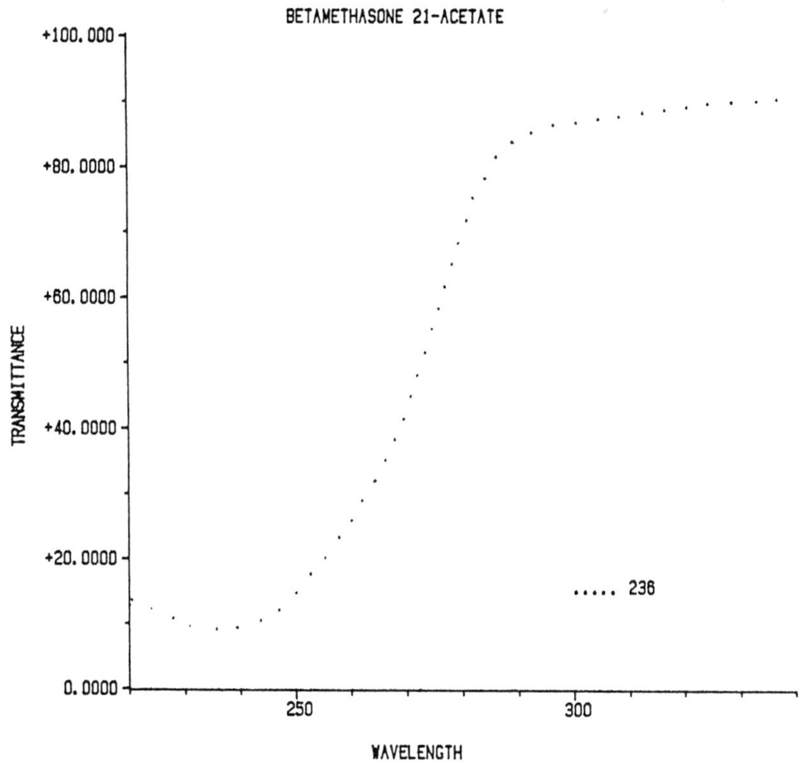

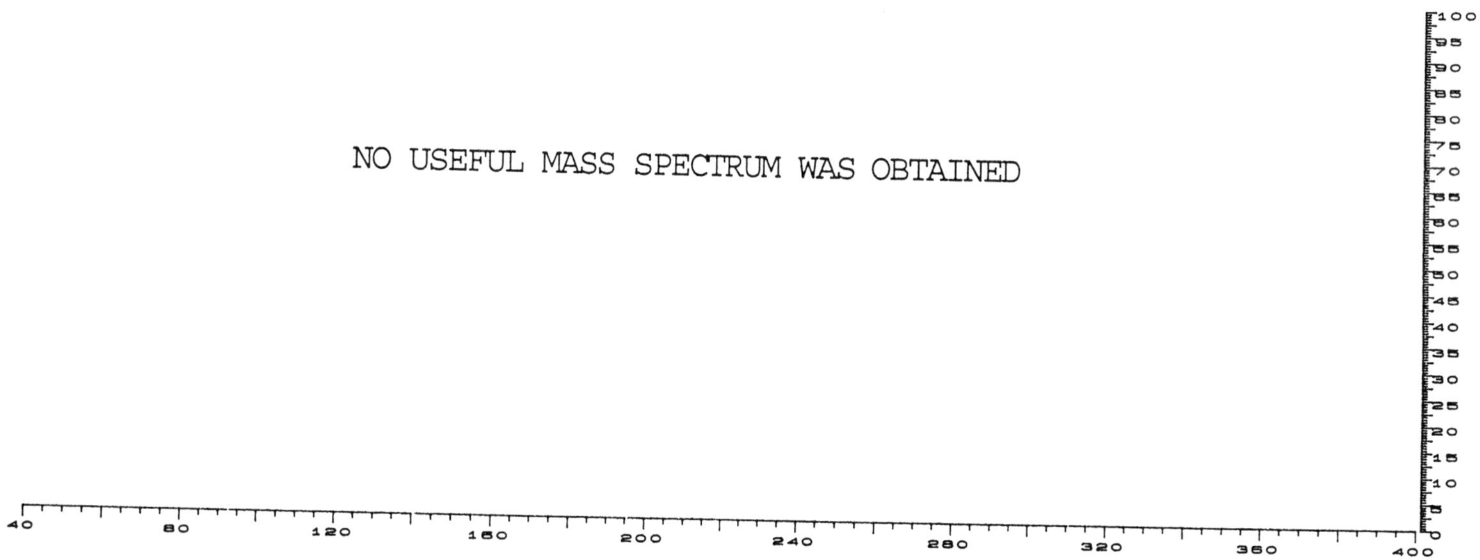

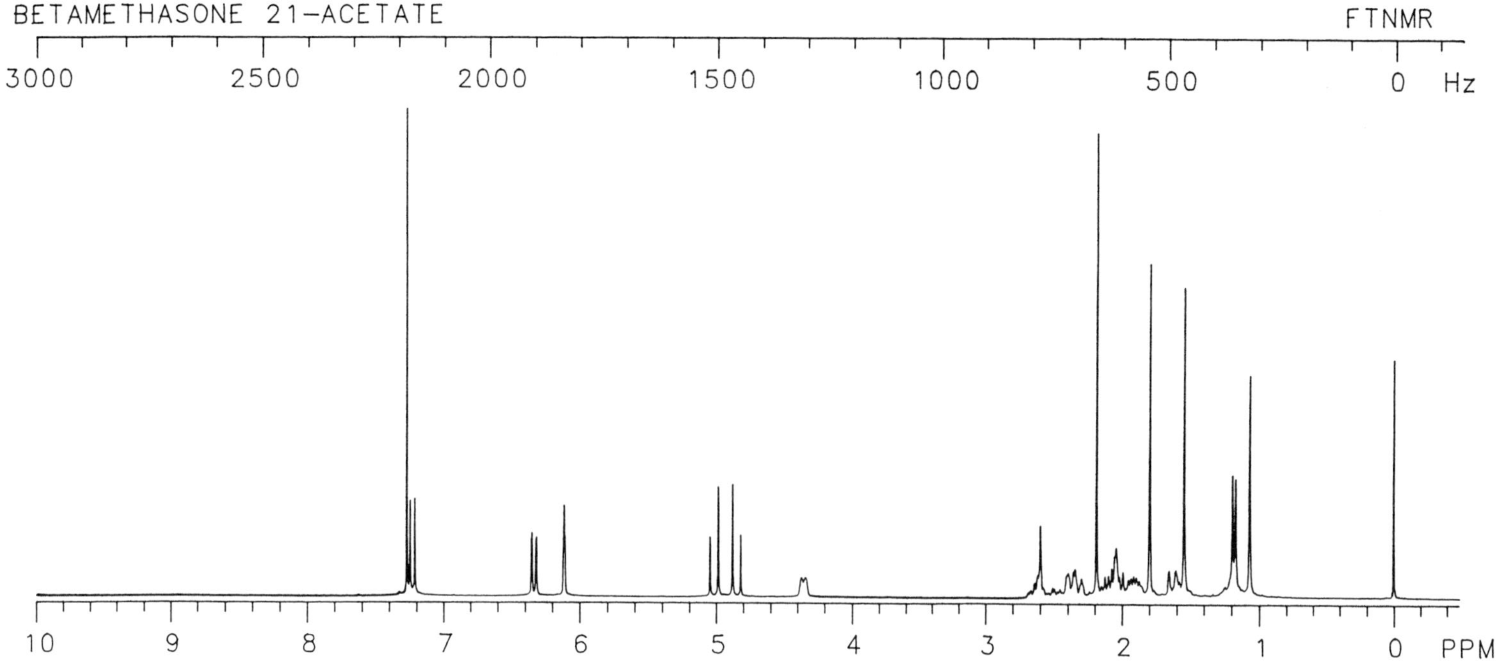

PEAK LISTING

#	HT	PPM
1	48	.00
2	45	1.06
3	24	1.16
4	25	1.18
5	63	1.54
6	68	1.79
7	95	2.18
8	14	2.59
9	12	4.81
10	23	4.87
11	22	4.97
12	12	5.03
13	18	6.10
14	12	8.31
15	13	6.34
16	11	6.35
17	19	7.20
18	19	7.24
19	99	7.26

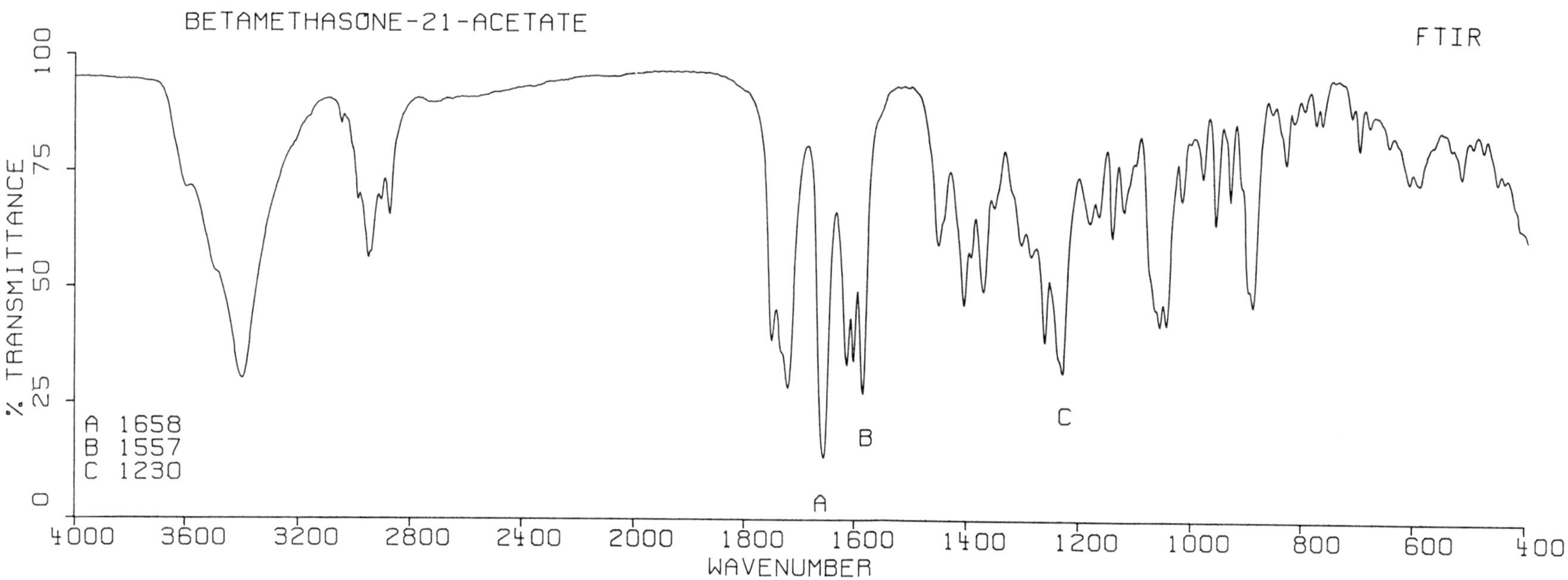

BETAMETHASONE-17,21-DIPROPIONATE

$C_{28}H_{37}FO_7$

Molecular Weight: 504.60 (504.25)

Synonyms: (9α-Fluoro-16β-methyl-11β,17α21-trihydroxy-1,4-pregnadiene-3,20-dione-17,21-dipropionate
Trade Names: Diproderm, Diprophos, Diprosis, Diprosone, Maxivate, Rinderon-DP
Use: Glucocorticoid

HPLC: Methanol: 2.7
GC:

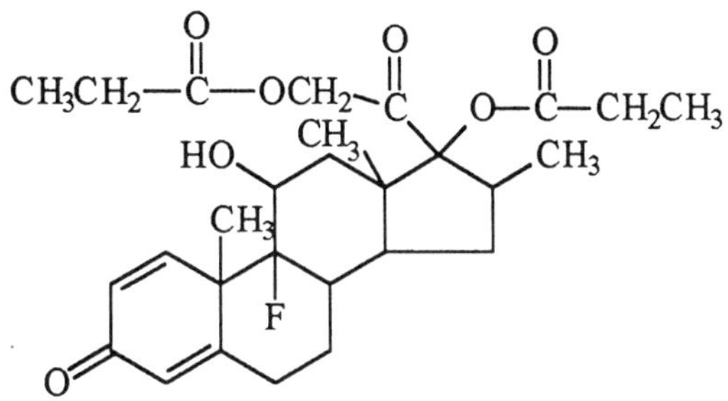

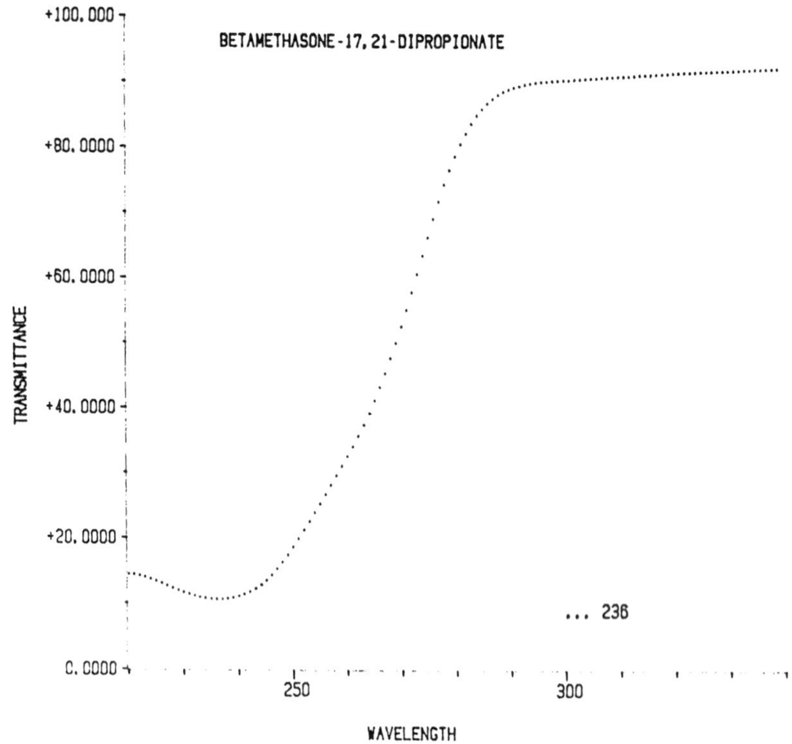

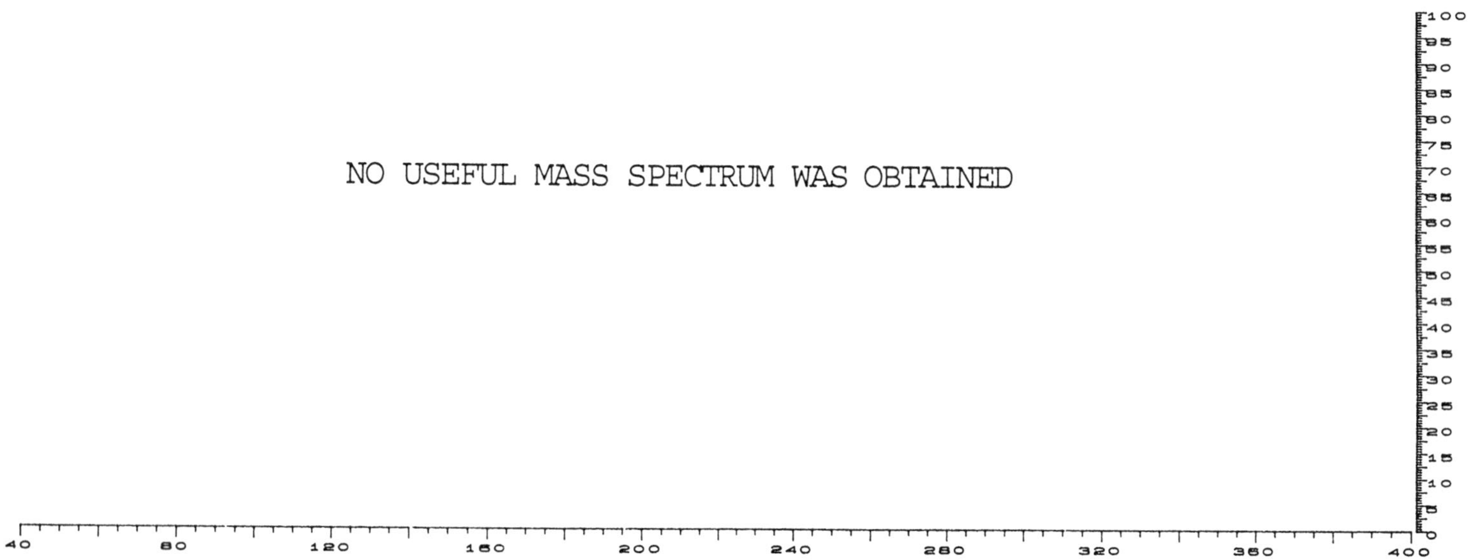

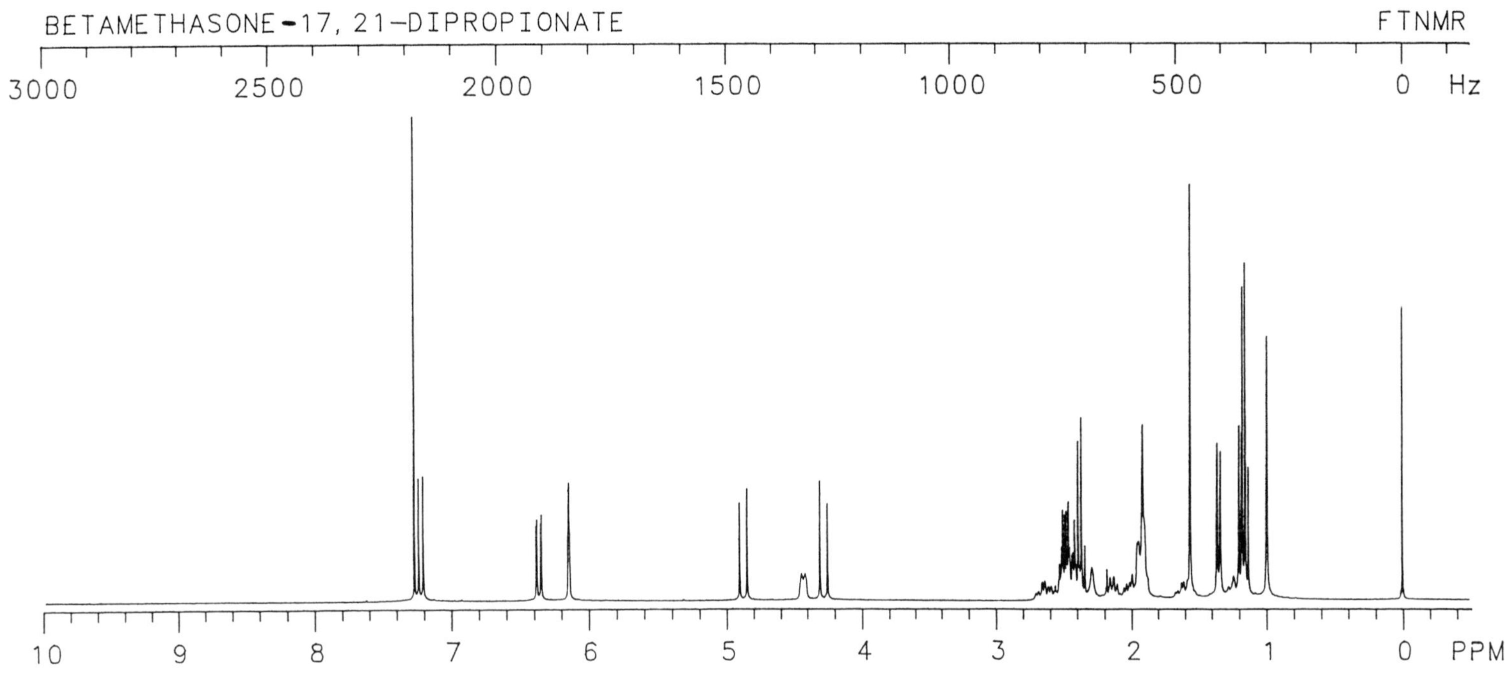

PEAK LISTING

#	HT	PPM
1	59	.00
2	54	.99
3	27	1.12
4	37	1.14
5	69	1.15
6	64	1.17
7	34	1.17
8	35	1.19
9	30	1.33
10	32	1.35
11	85	1.55
12	17	1.90
13	35	1.91
14	37	2.36
15	32	2.39
16	16	2.41
17	20	2.46
18	18	2.47
19	17	2.48
20	18	2.50
21	19	4.24
22	24	4.30
23	23	4.84
24	20	4.89
25	24	6.13
26	17	6.33
27	16	6.34
28	16	6.37
29	15	6.37
30	25	7.20
31	25	7.23
32	99	7.26

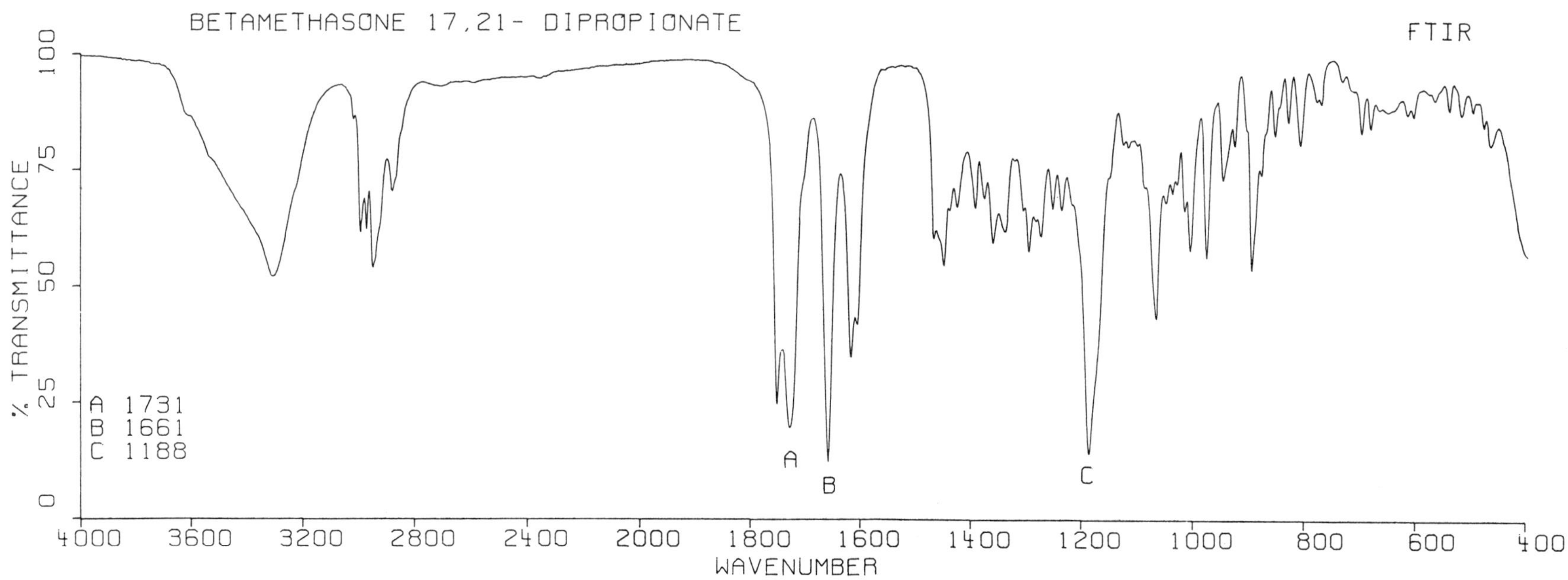

BETAMETHASONE-17-VALERATE

$C_{27}H_{37}FO_6$

Molecular Weight: 476.59 (476.26)

Synonyms: 9α-fluoro-16β-methyl-11β,17α,21-trihydroxy-1,4-pregnadiene-3,20-dione-17-valerate; 9α-fuoro-16β-methylprednisolone-17-valerate;
Trade Names: Bedermin, Betnesol-V, Betneval, Betnovate, Bextasol, Celestan-V, Celestoderm-V, Dermosol, Dermovaleas, Ecoval-70, Hormezon, Tokuderm, Valisone
Use: Glucocorticoid

HPLC: Methanol: 2.6
GC:

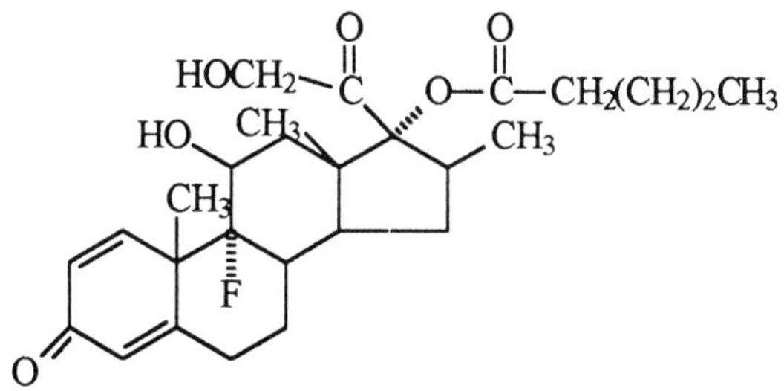

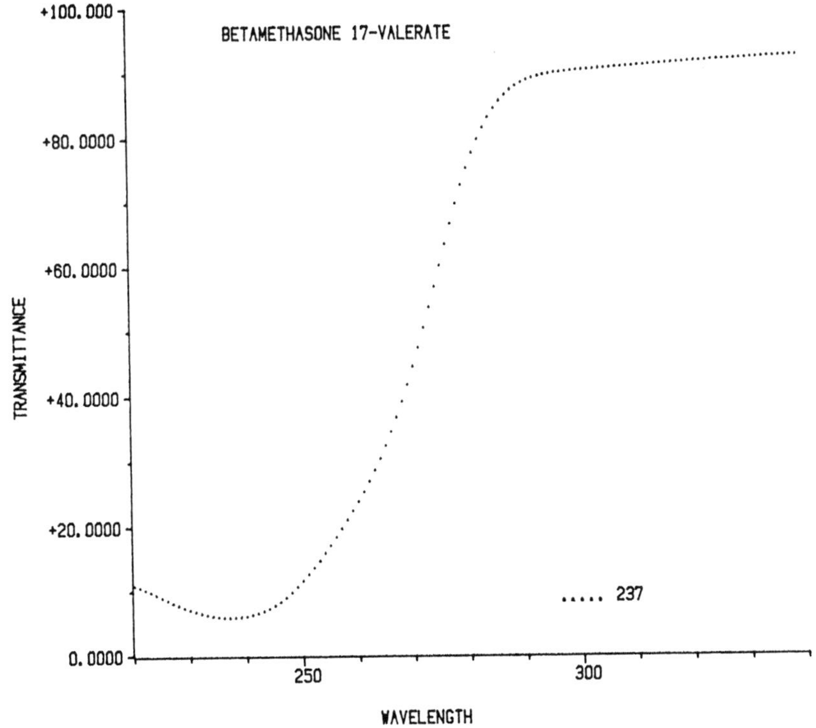

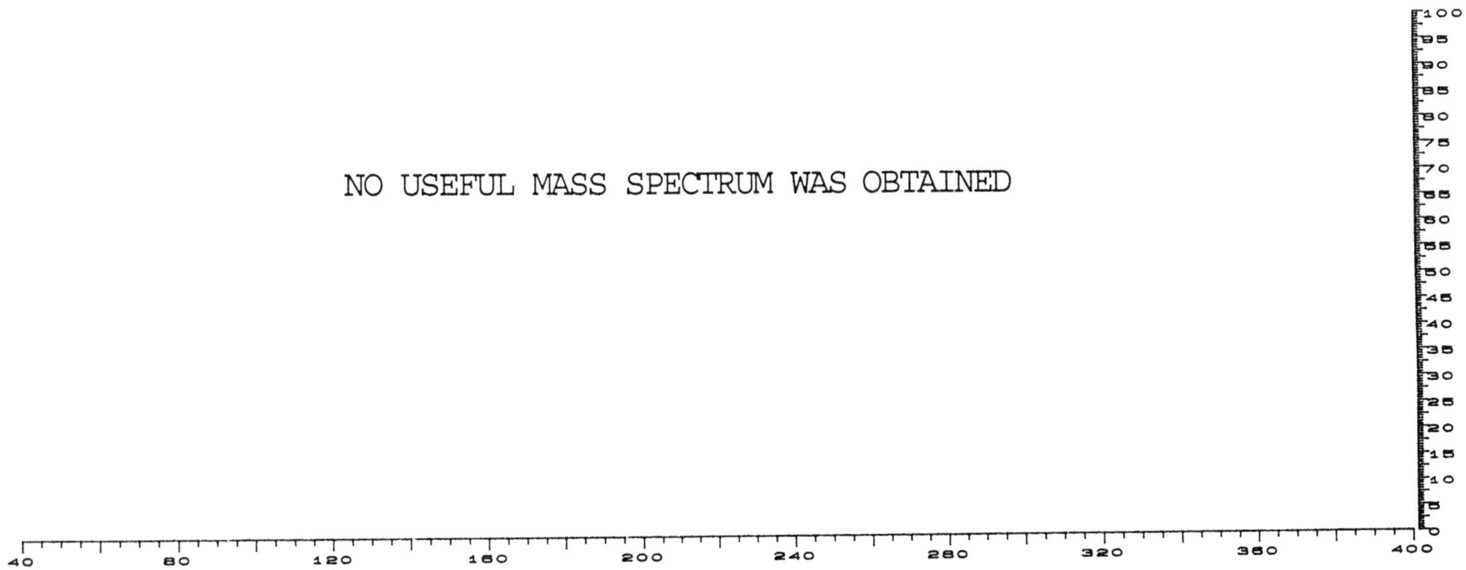

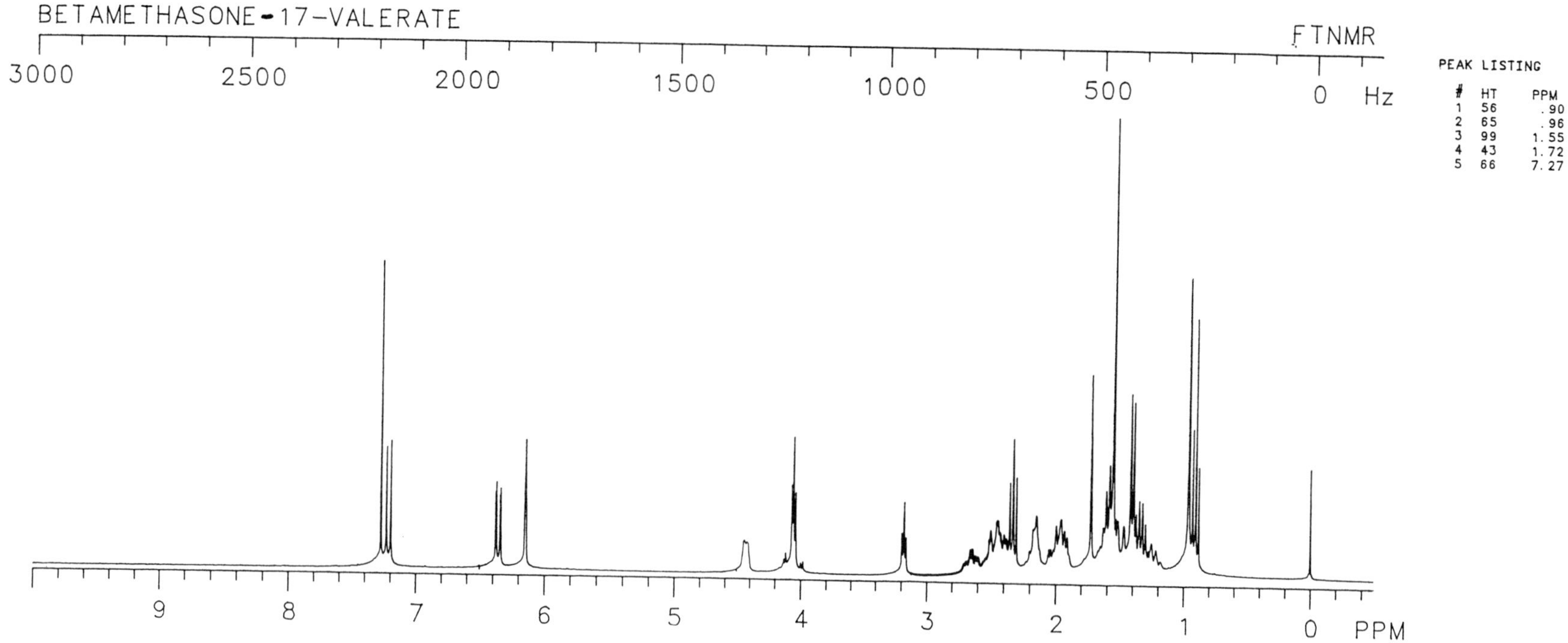
BETAMETHASONE-17-VALERATE
FTNMR
3000 2500 2000 1500 1000 500 0 Hz
9 8 7 6 5 4 3 2 1 0 PPM
PEAK LISTING
HT PPM
1 56 .90
2 65 .96
3 99 1.55
4 43 1.72
5 66 7.27

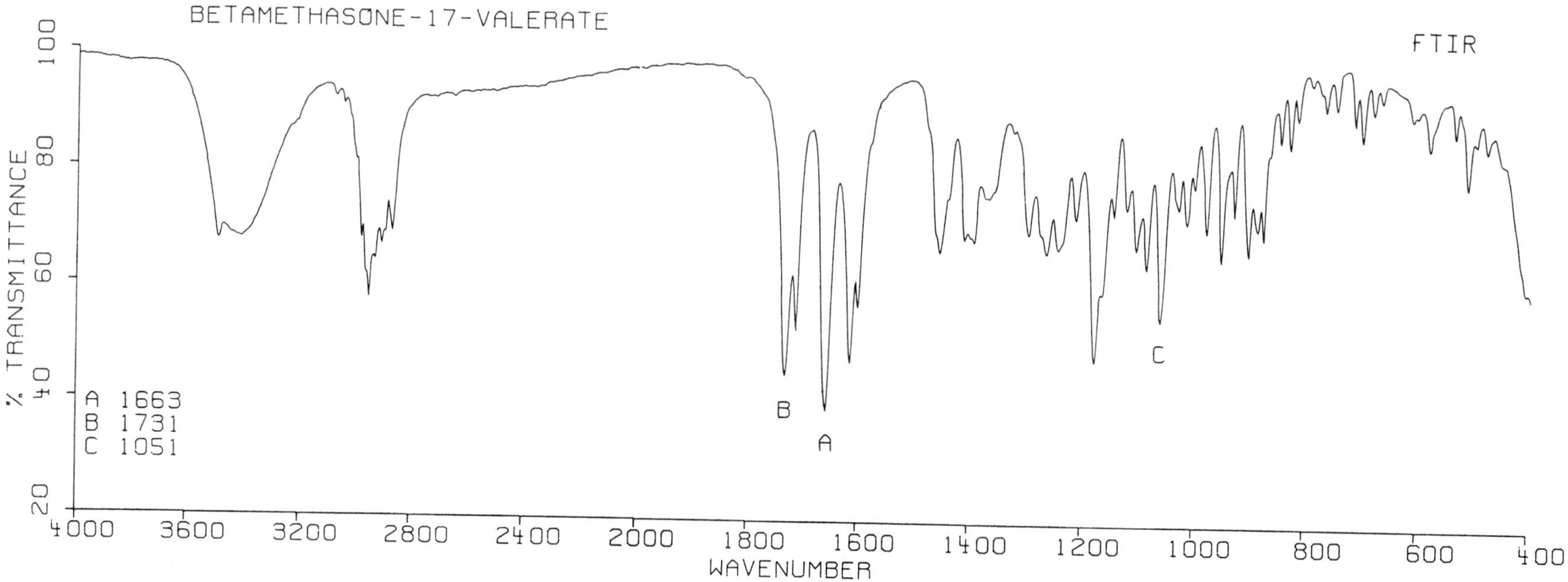
BETAMETHASONE-17-VALERATE
FTIR
% TRANSMITTANCE
100 80 60 40 20
A 1663
B 1731
C 1051
B
A
C
4000 3600 3200 2800 2400 2000 1800 1600 1400 1200 1000 800 600 400
WAVENUMBER

BUTYROLACTONE

$C_4H_6O_2$

Molecular Weight: 86.09 (86.04)

Synonyms: Dihydro-2(3H)-furanone; 1,2-butanolide; 1,4-butanolide; 3-hydroxybutyric acie lactone; γ-hydroxybutyrolactone; γ-hydoxybutyric acid lactone
Trade Names:
Use: Synthesis

HPLC:
GC: 992; 80°

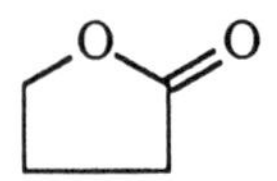

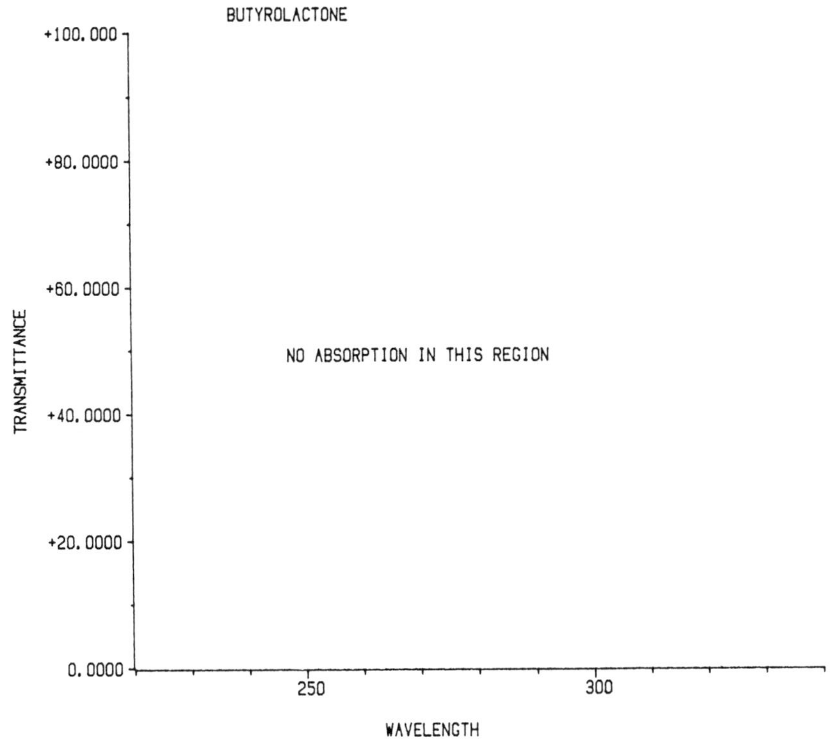

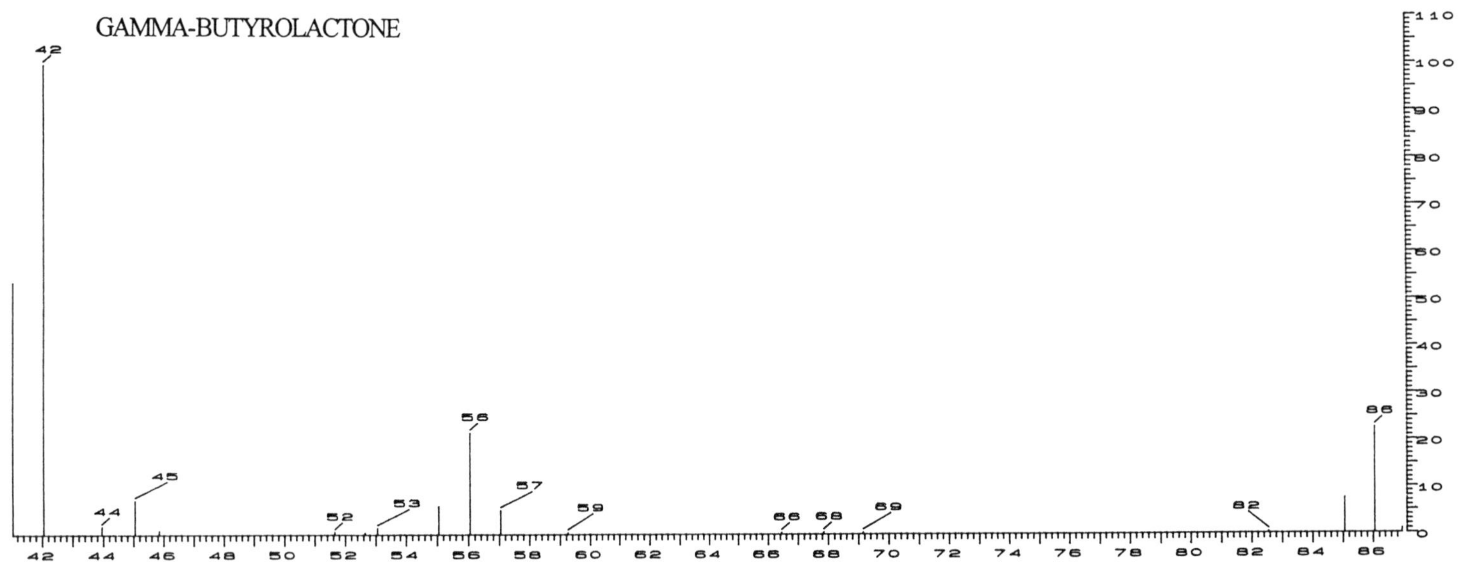

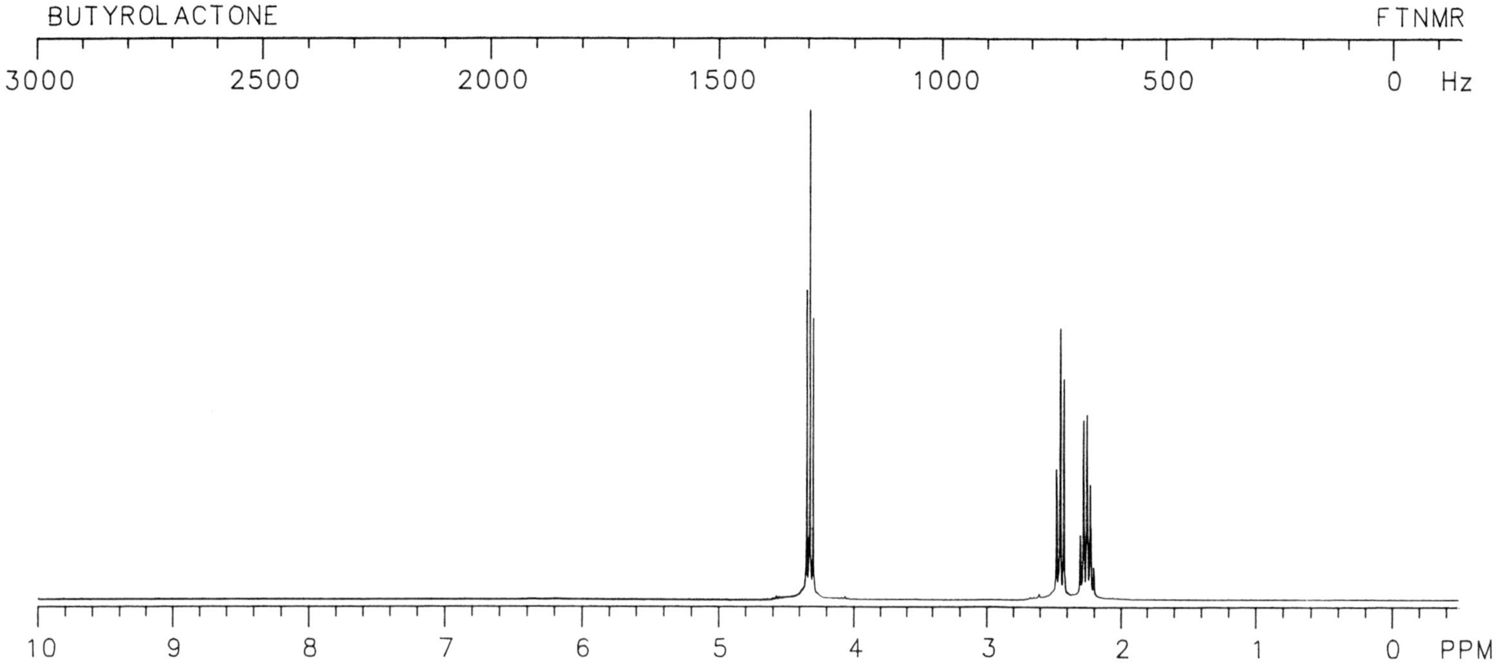

PEAK LISTING

#	HT	PPM
1	23	2.22
2	37	2.25
3	36	2.27
4	45	2.42
5	55	2.44
6	25	2.47
7	26	2.47
8	57	4.29
9	99	4.31
10	63	4.33

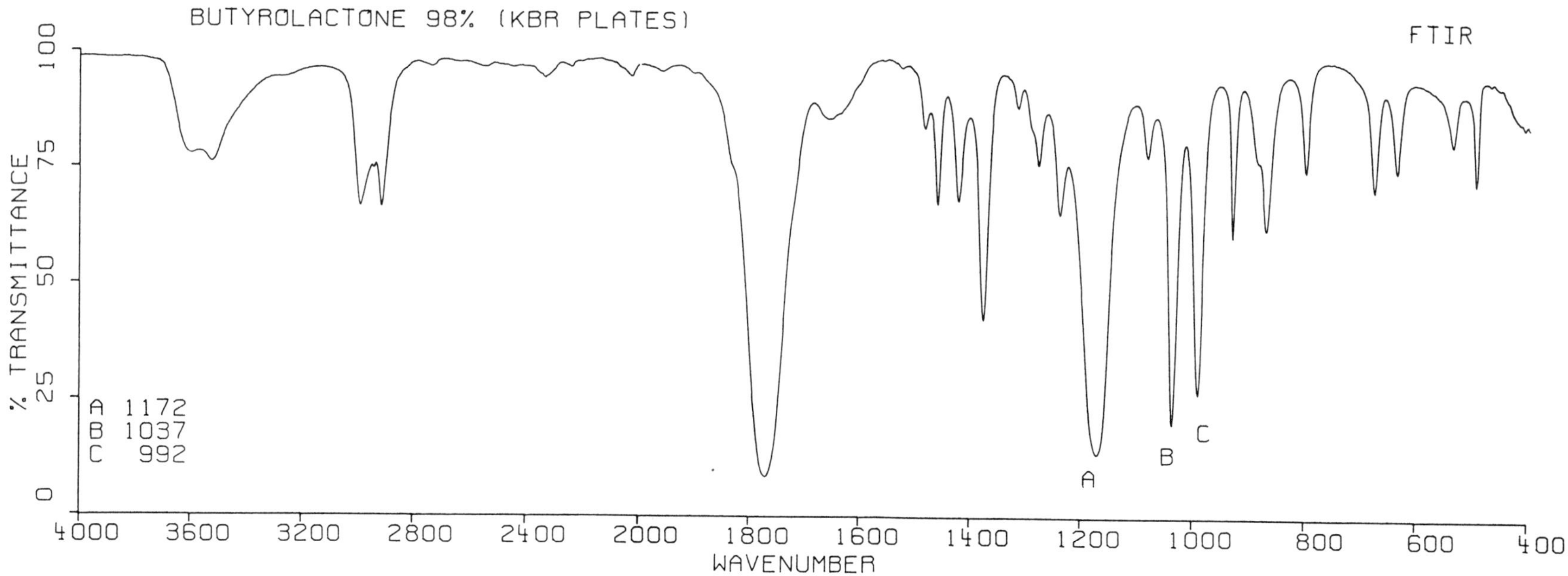

CATHINONE

$C_9H_{11}NO$

Molecular Weight: 149.19 (149.08)

Synonyms: (S)-2-Amino-1-phenyl-1-propanone; α-aminopropiophenone
Trade Names:
Use: Alkaloid from *Catha edulis*

HPLC: Methanol: 2.1
GC: 1278; 140°

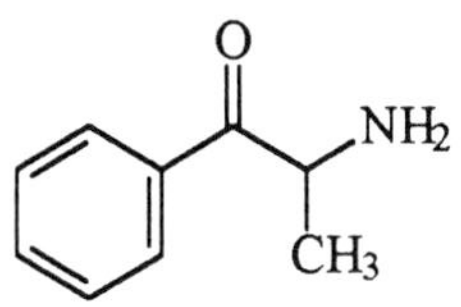

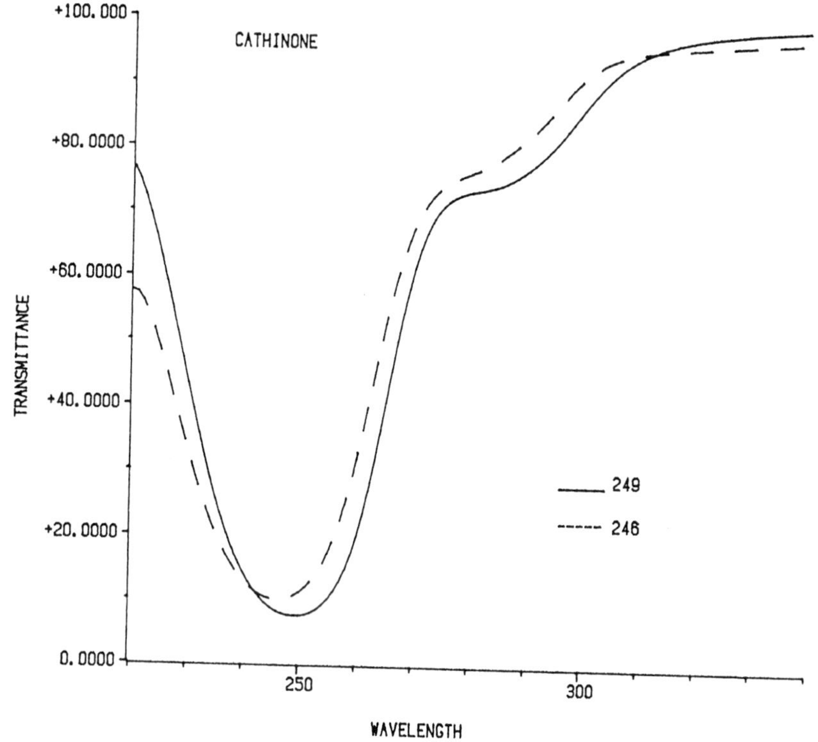

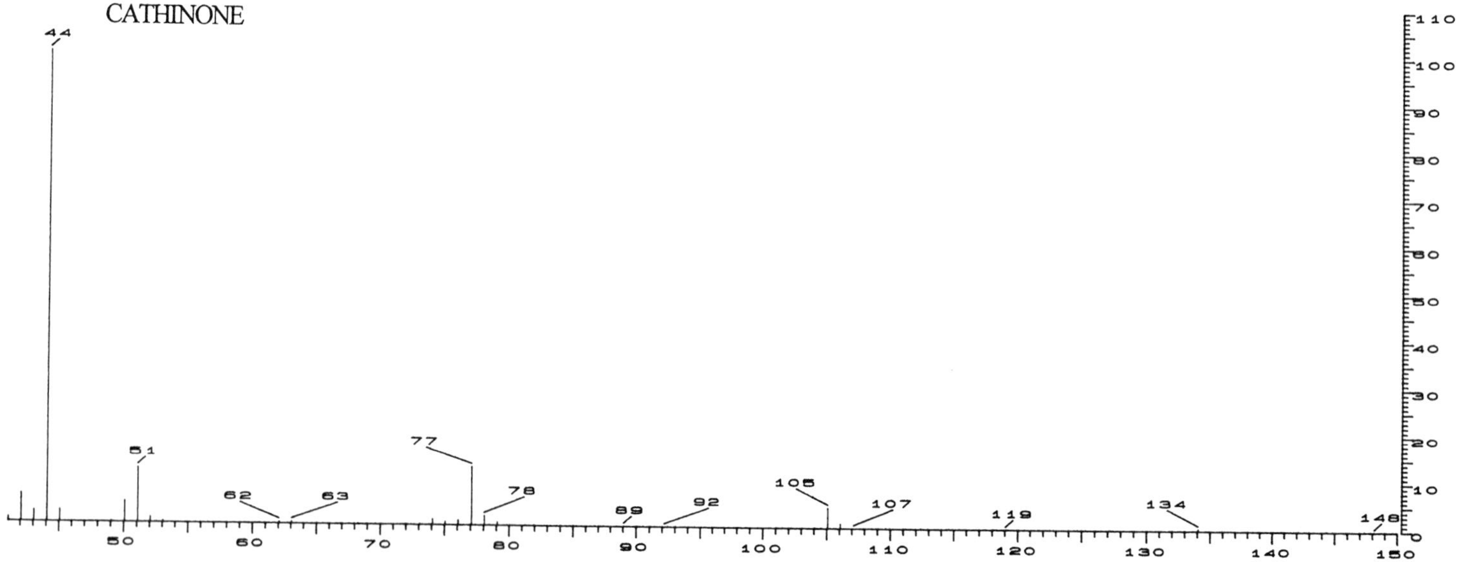

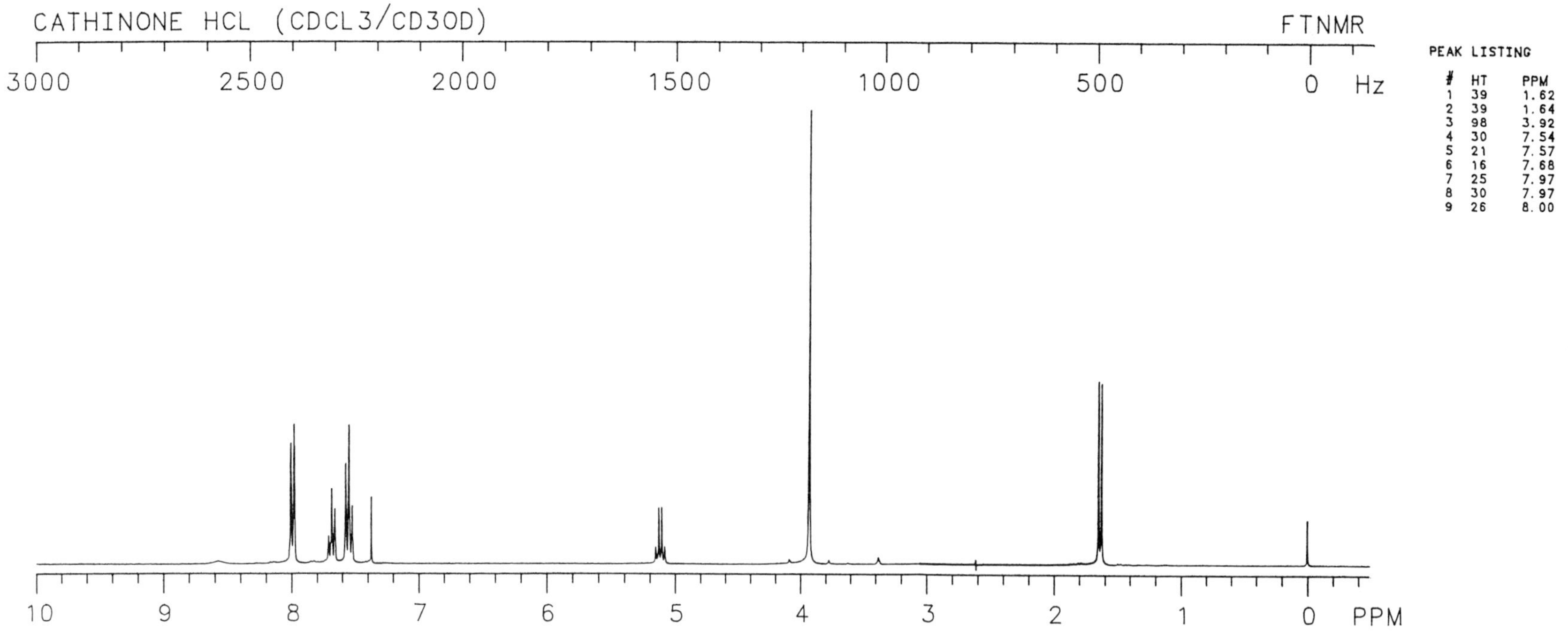
CATHINONE HCL (CDCL3/CD3OD)
FTNMR
3000
2500
2000
1500
1000
500
0 Hz
10
9
8
7
6
5
4
3
2
1
0 PPM
PEAK LISTING
HT PPM
1 39 1.62
2 39 1.64
3 98 3.92
4 30 7.54
5 21 7.57
6 16 7.68
7 25 7.97
8 30 7.97
9 26 8.00

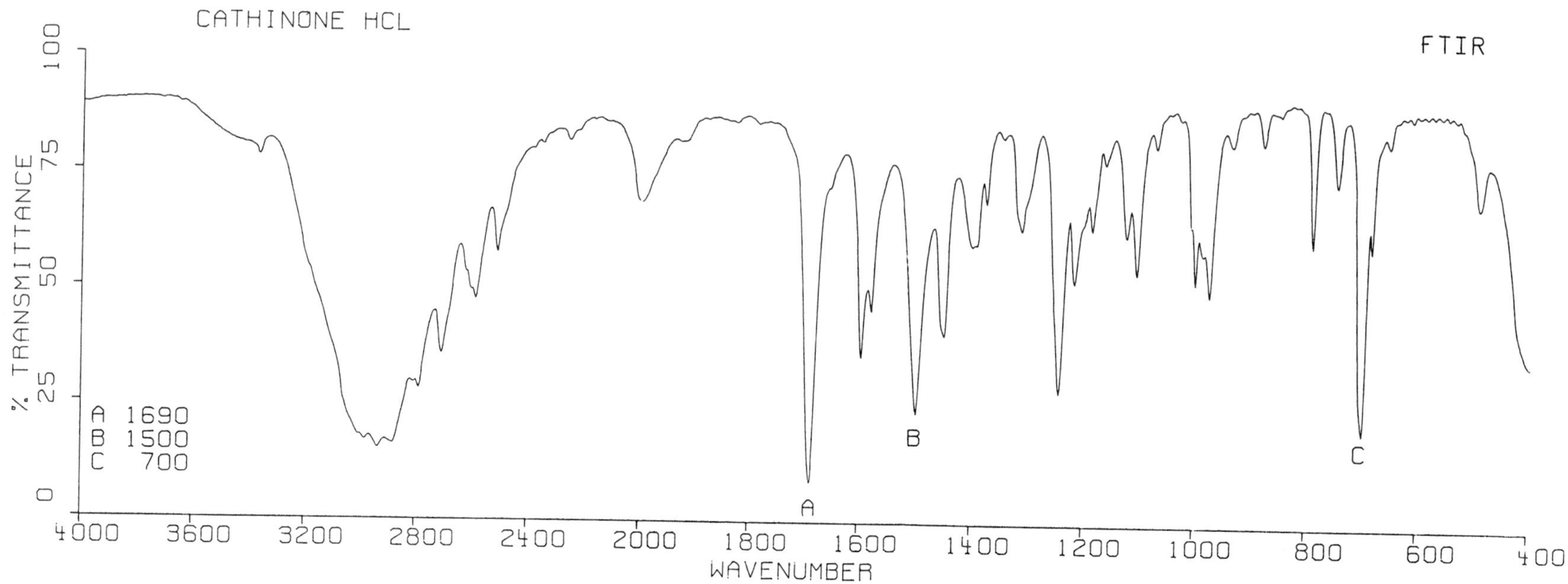
CATHINONE HCL
FTIR
% TRANSMITTANCE
100
75
50
25
0
A 1690
B 1500
C 700
A
B
C
4000
3600
3200
2800
2400
2000
1800
1600
1400
1200
1000
800
600
400
WAVENUMBER

CEFIXIME

$C_{16}H_{15}N_5O_7S_2$

Molecular Weight: 453.44 (453.04)

Synonyms: [6R-[6α,7β(Z)]]-7-[[(2-Amino-4-thiazolyl)[(carboxymethoxy)imino]acetyl]amino]-3-ethenyl-8-oxo-5-thia-1-azabicyclo[4.20]oct-2-ene-2-carboxylic acid
Trade Names: Cefspan, Oroken, Suprax
Use: Antibacterial

HPLC: Methanol: 2.9 (Tails)
GC:

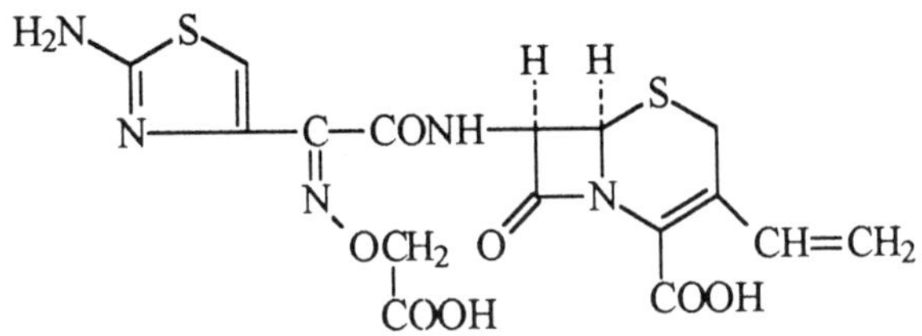

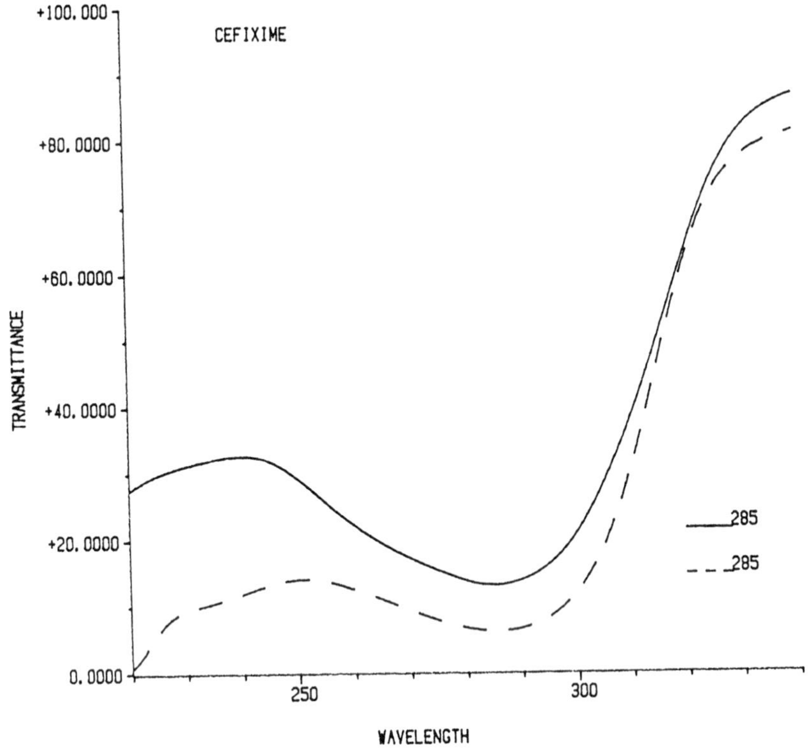

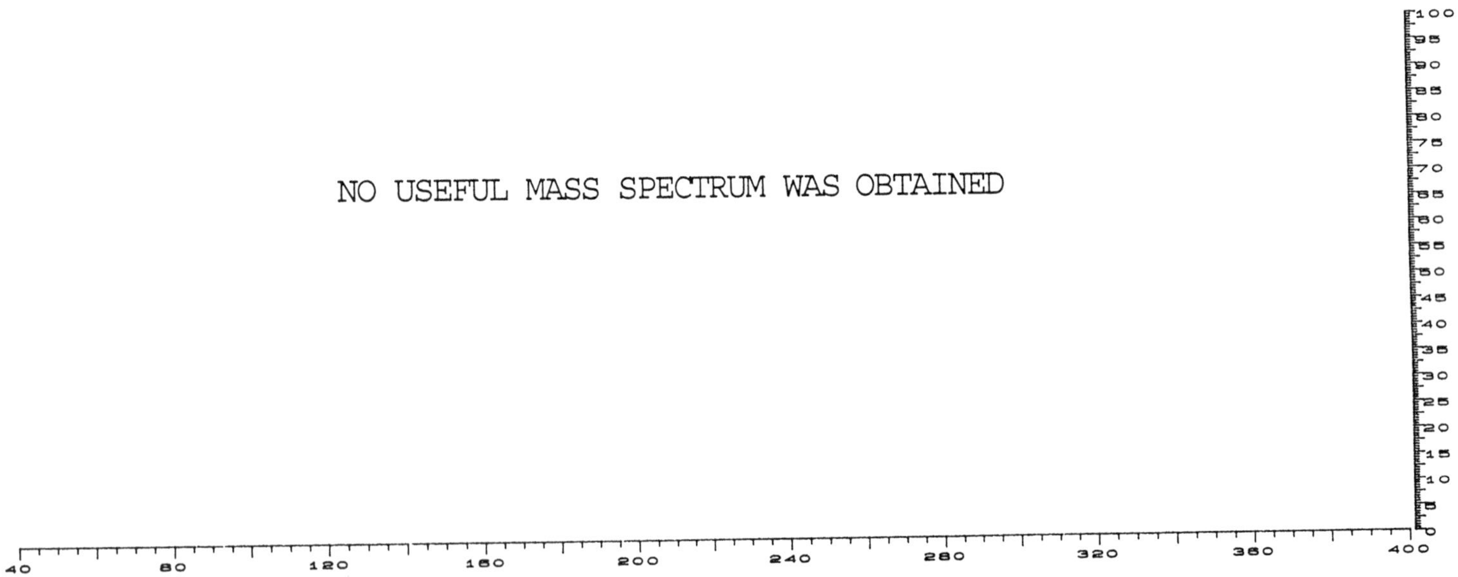

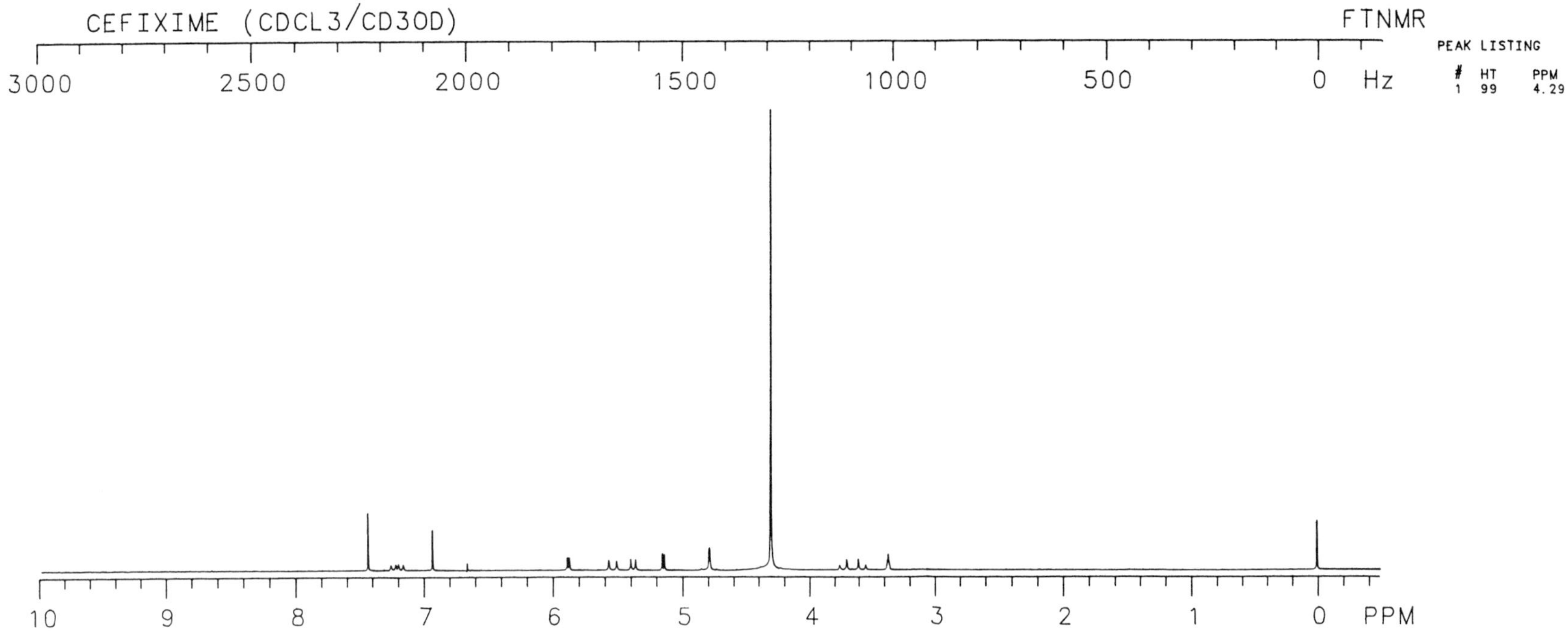

CEFIXIME (CDCL3/CD3OD)
FTNMR
PEAK LISTING
HT PPM
1 99 4.29
3000 2500 2000 1500 1000 500 0 Hz
10 9 8 7 6 5 4 3 2 1 0 PPM

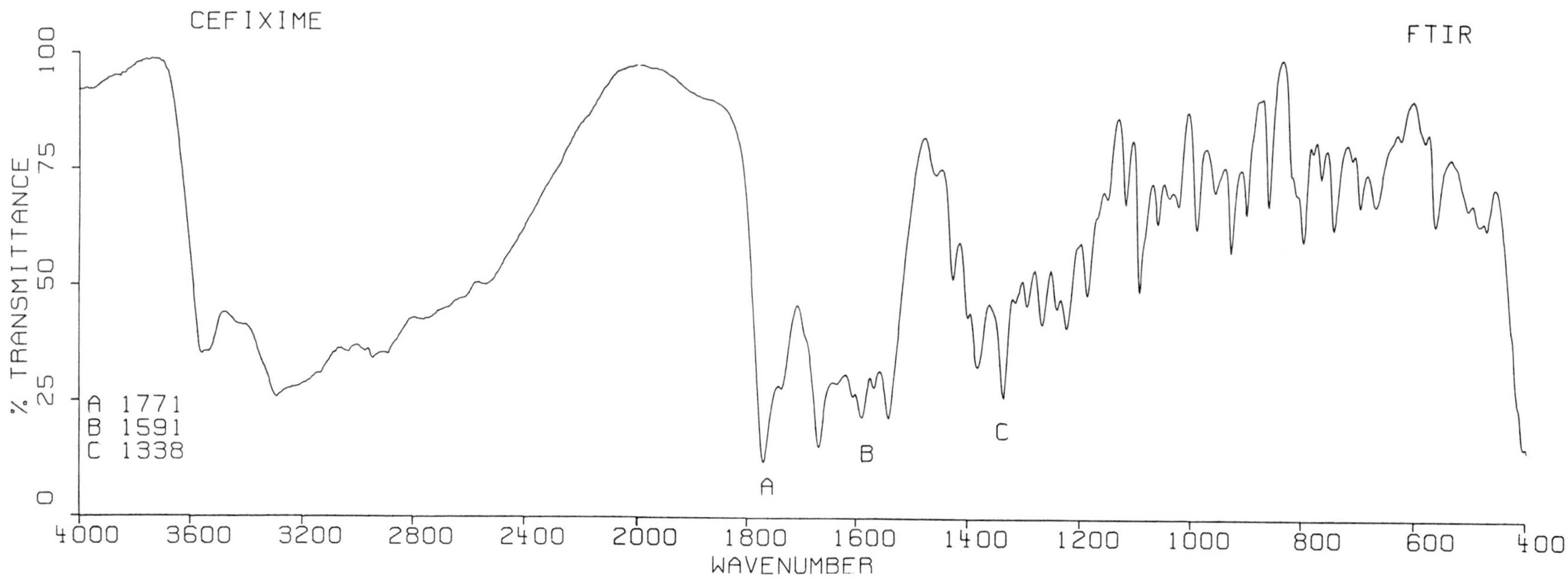

CEFIXIME
FTIR
% TRANSMITTANCE
0 25 50 75 100
A 1771
B 1591
C 1338
A
B
C
4000 3600 3200 2800 2400 2000 1800 1600 1400 1200 1000 800 600 400
WAVENUMBER

CEFMETAZOLE

$C_{15}H_{17}N_5O_7S_3$

Molecular Weight: 471.52 (471.05)
Synonyms: 7-[[[(Cyanomethyl)thio]acetyl]amino]-7-methoxy-3-[[(1-methyl-1H-tetrazol-5-yl)thio]methyl]-8-oxo-5-thia-1-azabicyclo[4.2.0]oct-2-ene-2-carboxylic acid
Trade Names: Cefmetazon, Zefazone
Use: Antibacterial

HPLC:
GC:

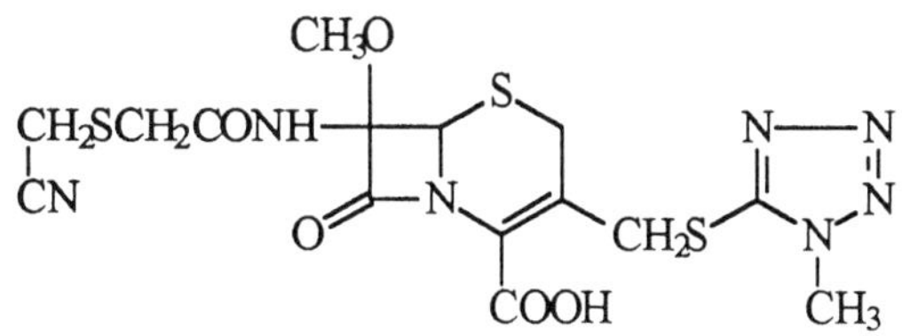

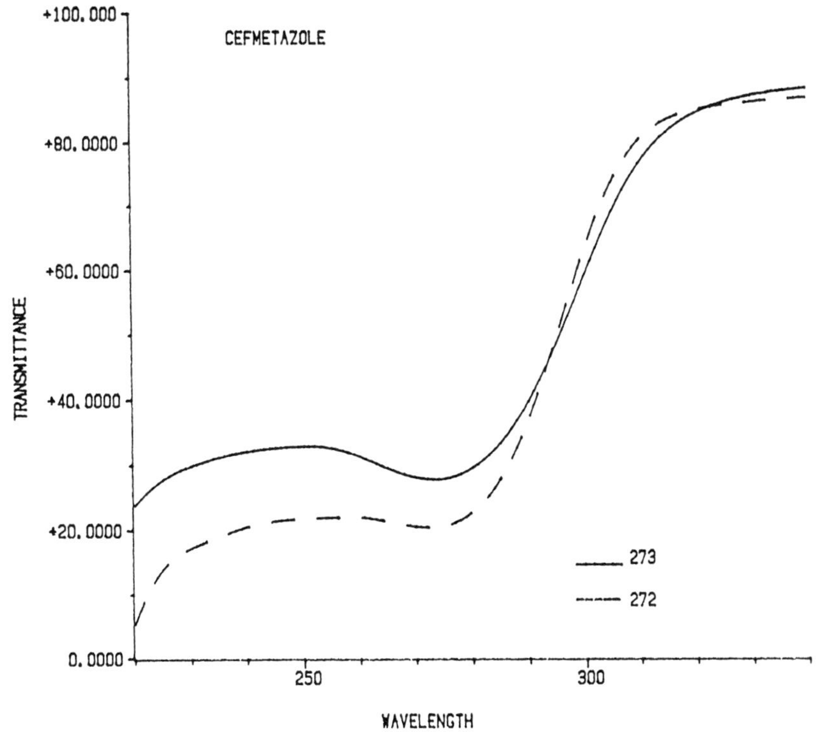

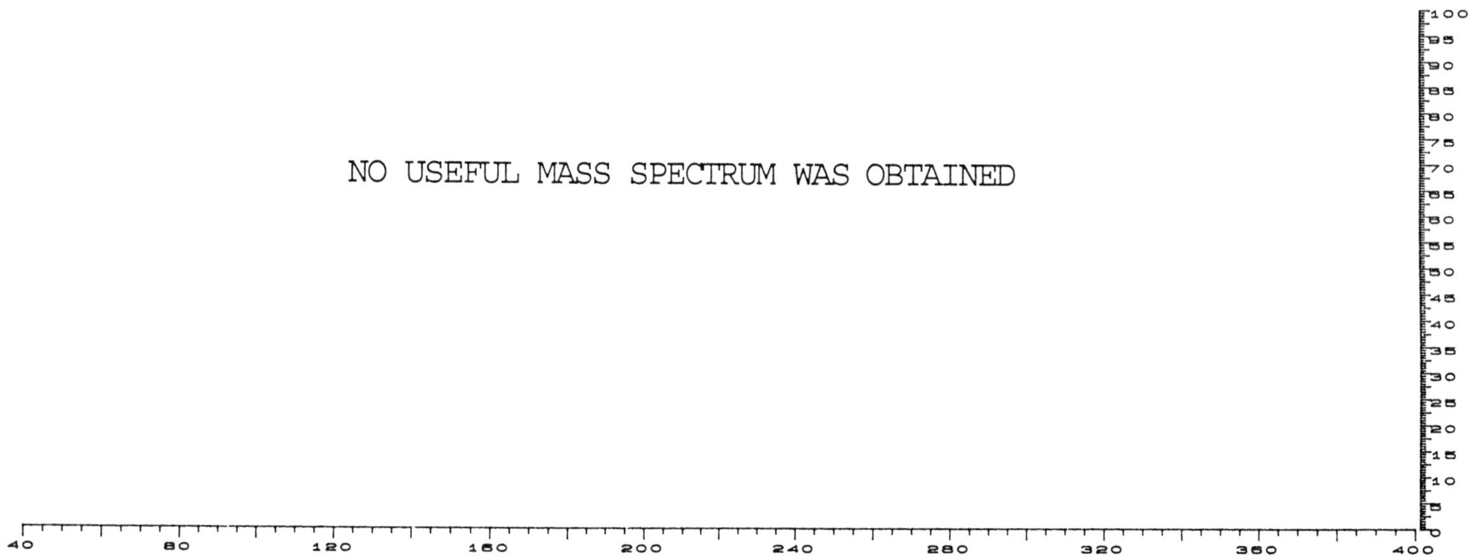

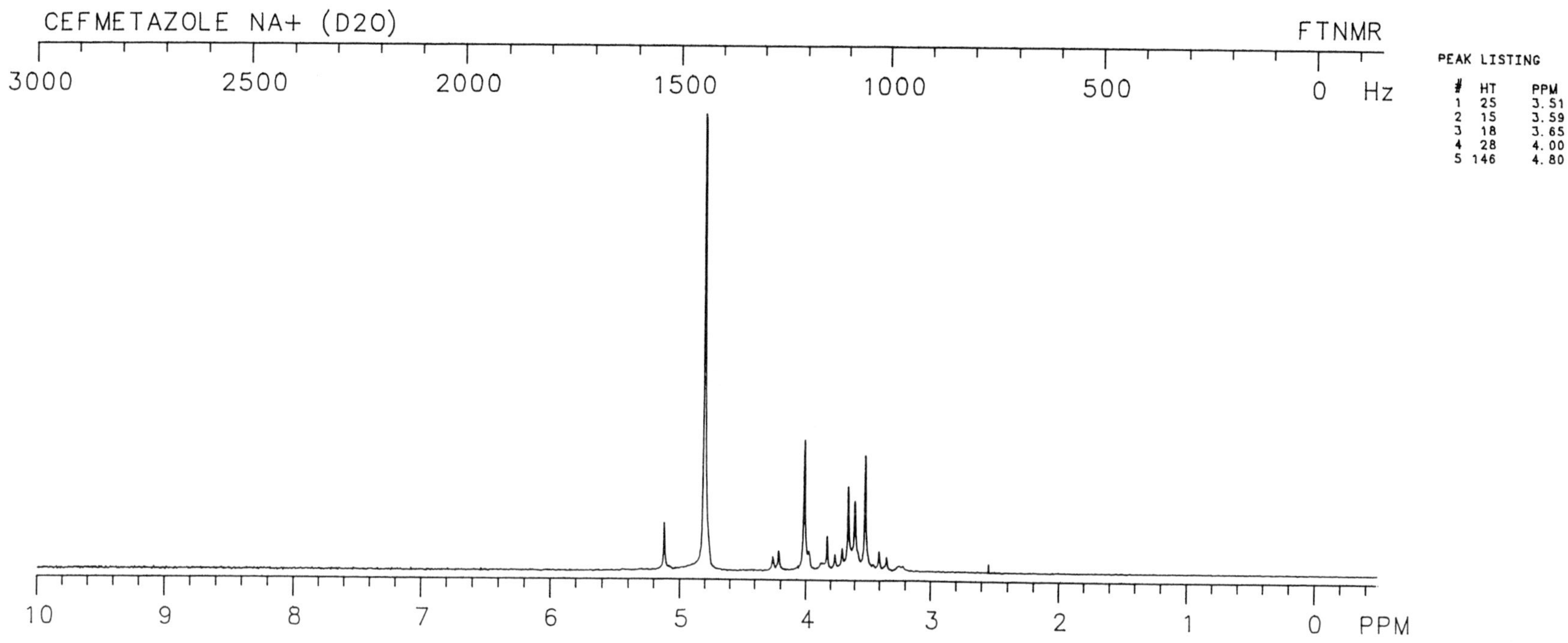
CEFMETAZOLE NA+ (D2O)
FTNMR
3000 2500 2000 1500 1000 500 0 Hz
PEAK LISTING
HT PPM
1 25 3.51
2 15 3.59
3 18 3.65
4 28 4.00
5 146 4.80
10 9 8 7 6 5 4 3 2 1 0 PPM

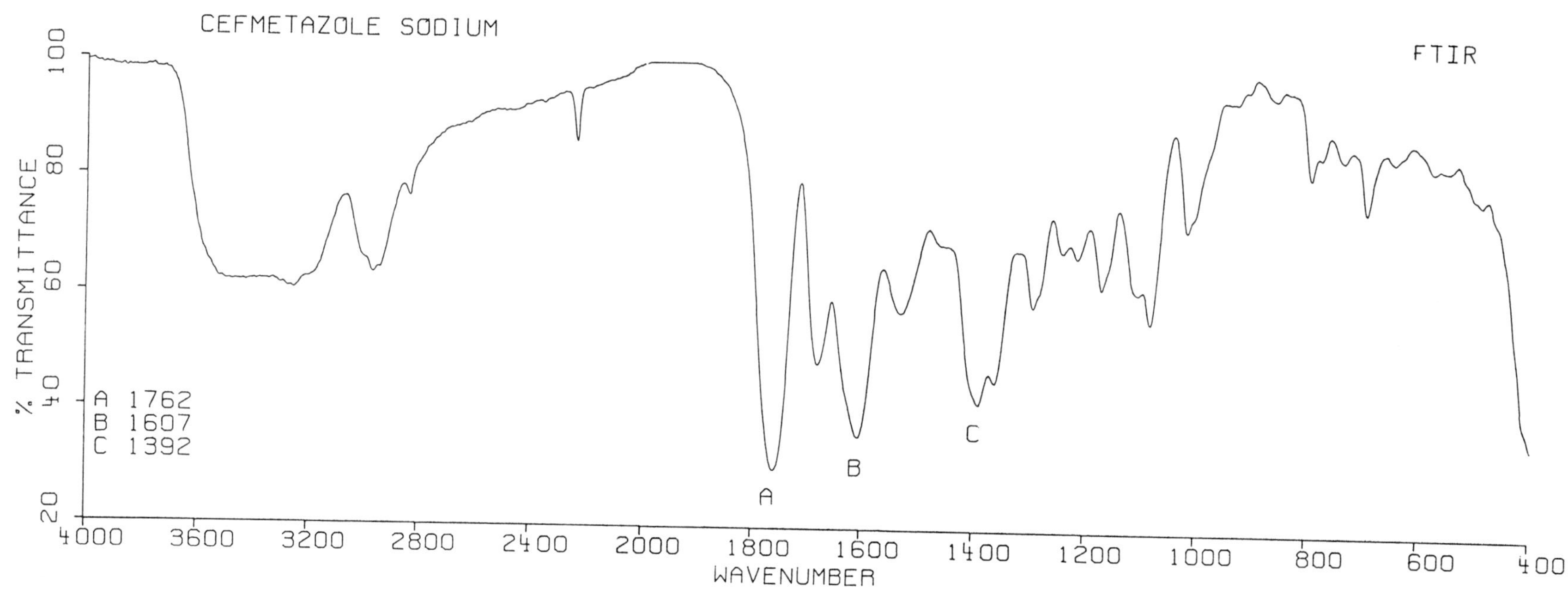
CEFMETAZOLE SODIUM
FTIR
% TRANSMITTANCE
100 80 60 40 20
A 1762
B 1607
C 1392
A
B
C
4000 3600 3200 2800 2400 2000 1800 1600 1400 1200 1000 800 600 400
WAVENUMBER

CEFOTETAN

$C_{17}H_{17}N_7O_8S_4$

Molecular Weight: 575.60 (575.00)

Synonyms: [6R-(6α,7α)]-7-[[[4-(2-Amino-1-carboxy-2-oxoethylidene)-1,3-dithietan-2-yl]carbonyl]-amino]-7-methoxy-3-[[(1-methyl-1H-tetrazol-5-yl)thio]methyl]-8-oxo-5-thia-1-azabicyclo[4.2.0]oct-2-ene-2-carboxylic acid
Trade Names: Apatef, Cefotan, Ceftenon, Cepan, Darvilen, Yamatetan
Use: Antibacterial

HPLC:
GC:

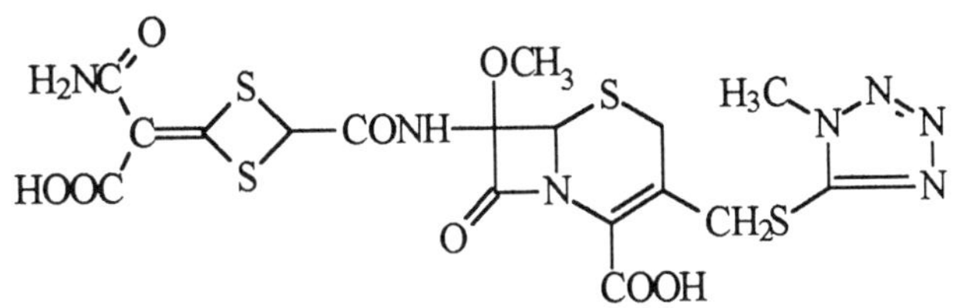

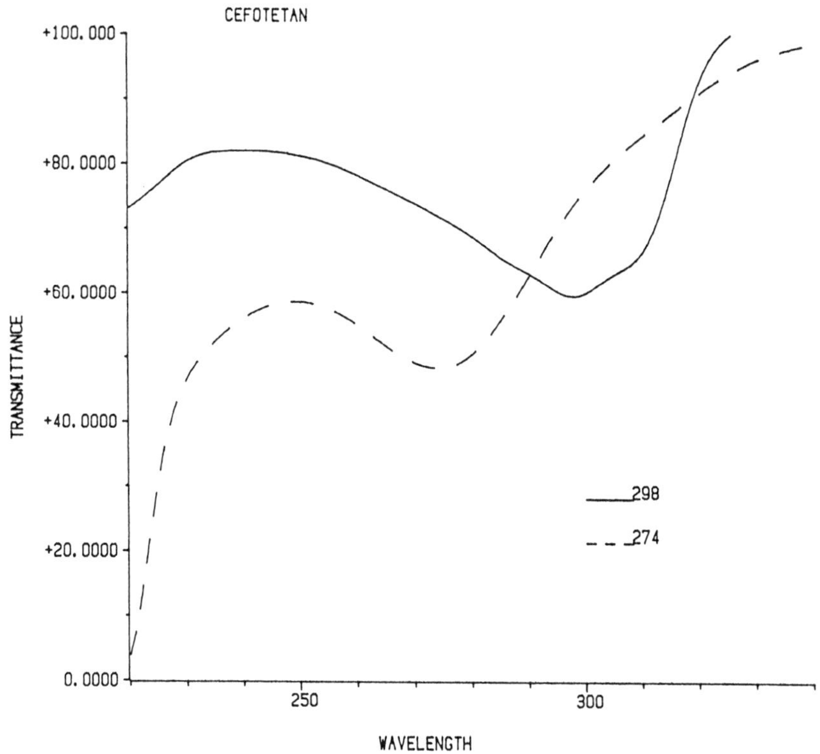

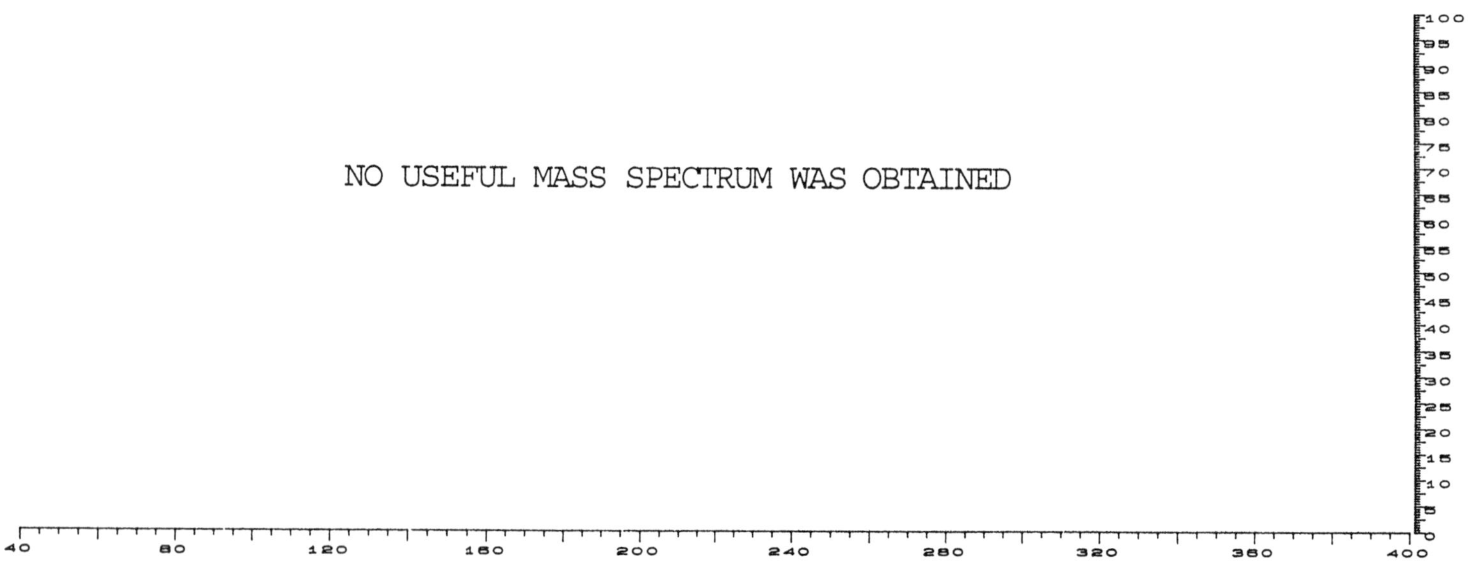

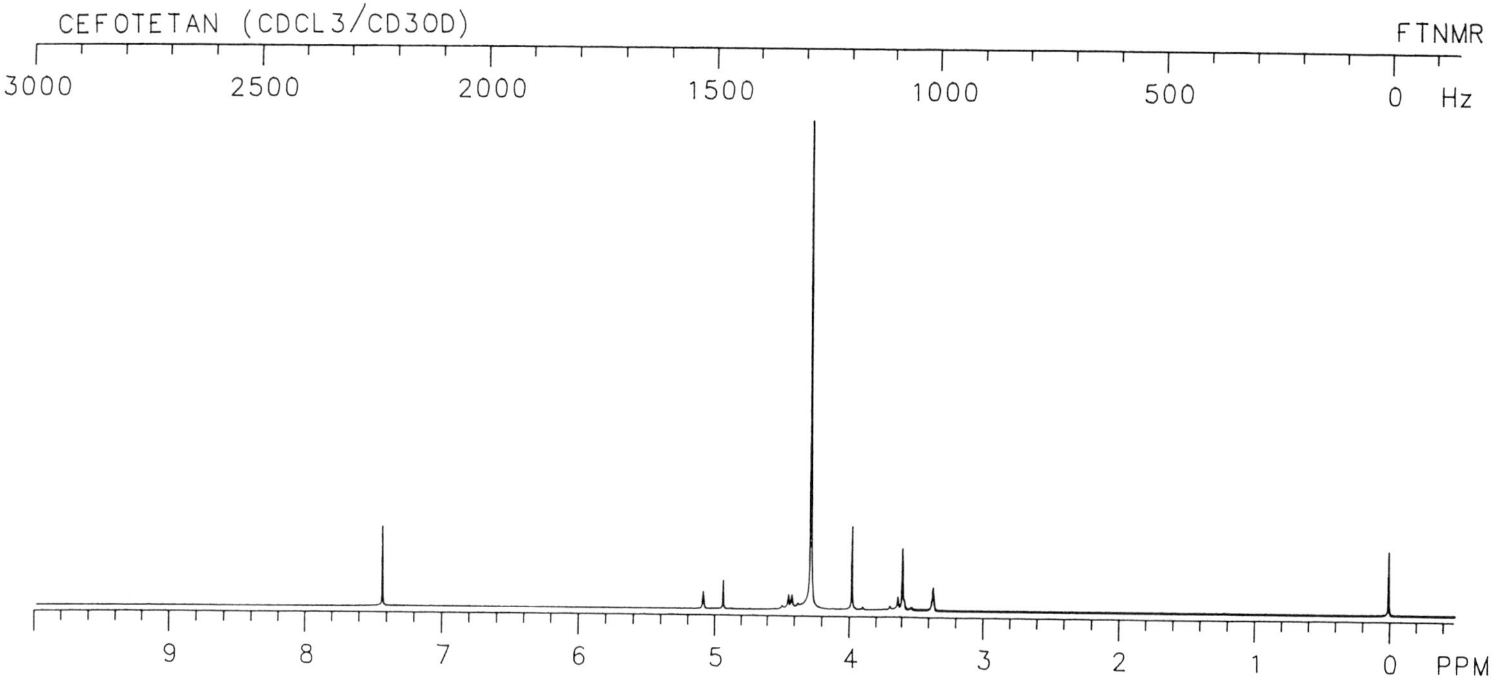

PEAK LISTING

#	HT	PPM
1	12	.00
2	11	3.58
3	12	3.59
4	17	3.97
5	99	4.27
6	16	7.42

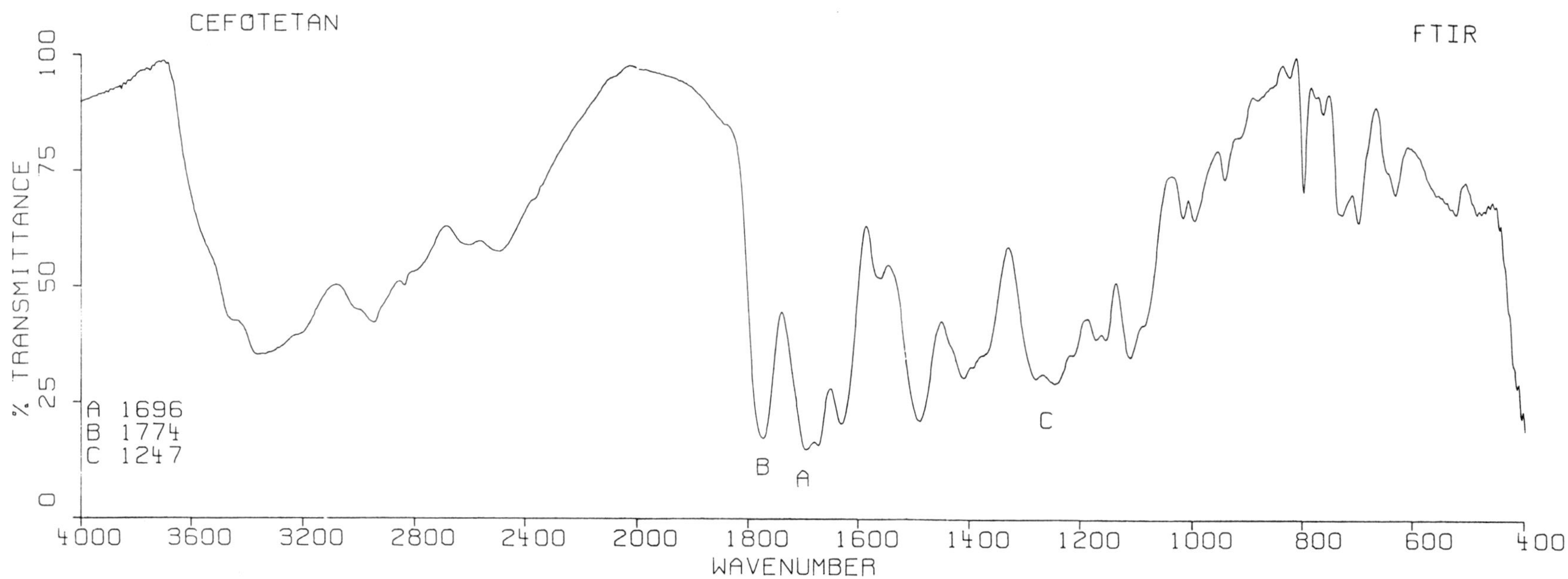

CEFTAZIDIME

$C_{22}H_{22}N_6O_7S_2$

Molecular Weight: 546.57 (546.10)

Synonyms: [6R-[6α,7β(Z)]]-1-[[7-[[(2-Amino-4-thiazolyl)[(1-carboxy-1-methylethoxy)imino]acetyl]amino]-2-carboxy-8-oxo-5-thia-1-azabicyclo[4.2.0]oct-2-en-3-yl]methyl]pyridinium hydroxide, inner salt; Fortam; Ketamin; Modacin

Trade Names: Ceftim, Ceptaz, Fortaz, Fortum, Glazidim, Kefazim, Panzid, Spectru,. Tazicef, Tazidime,

Use: Antibacterial

HPLC:
GC:

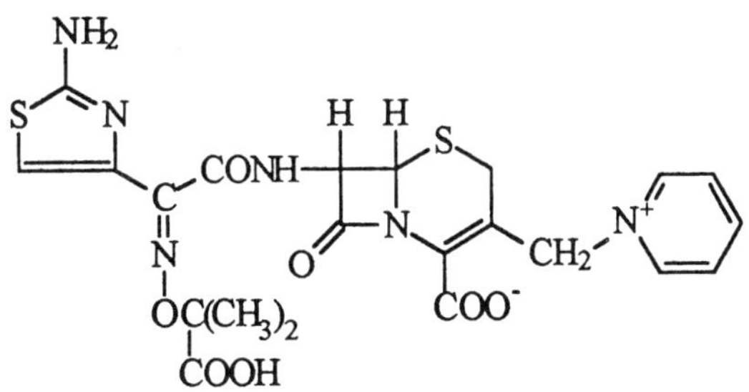

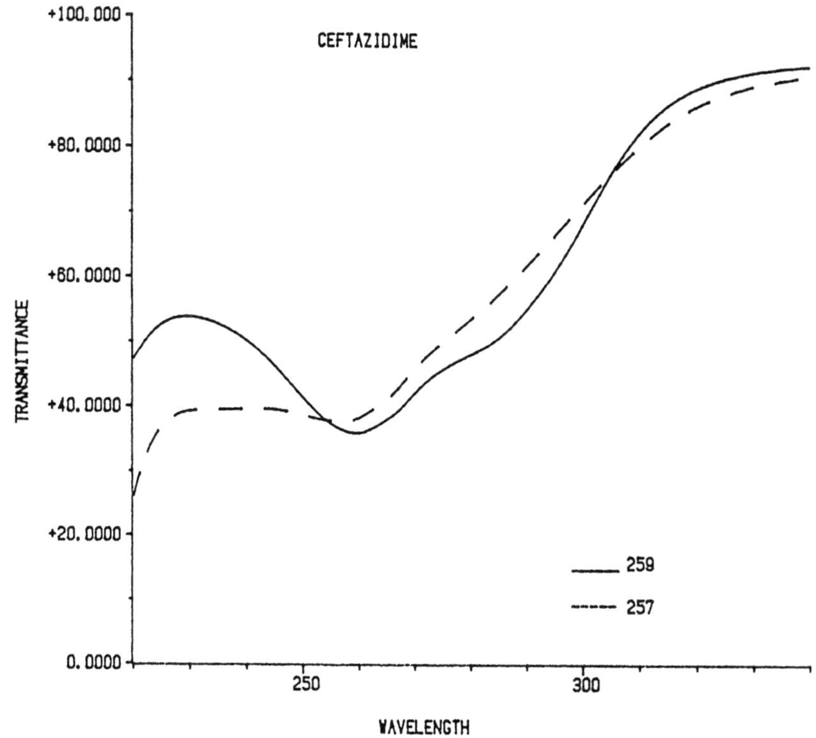

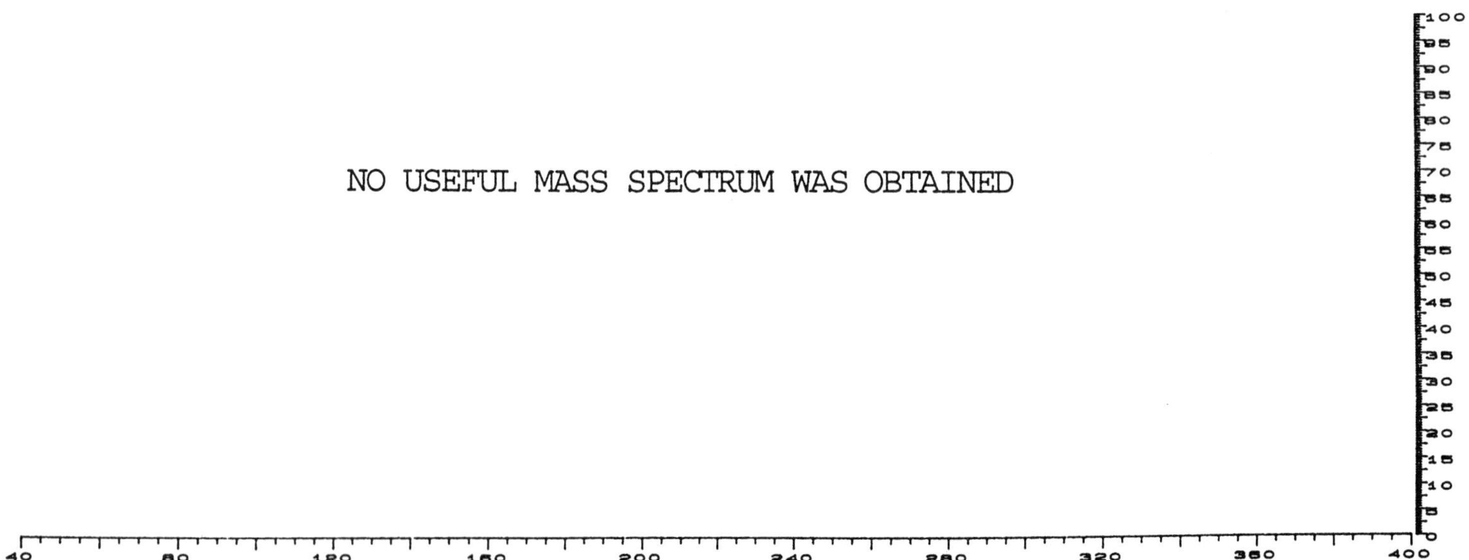

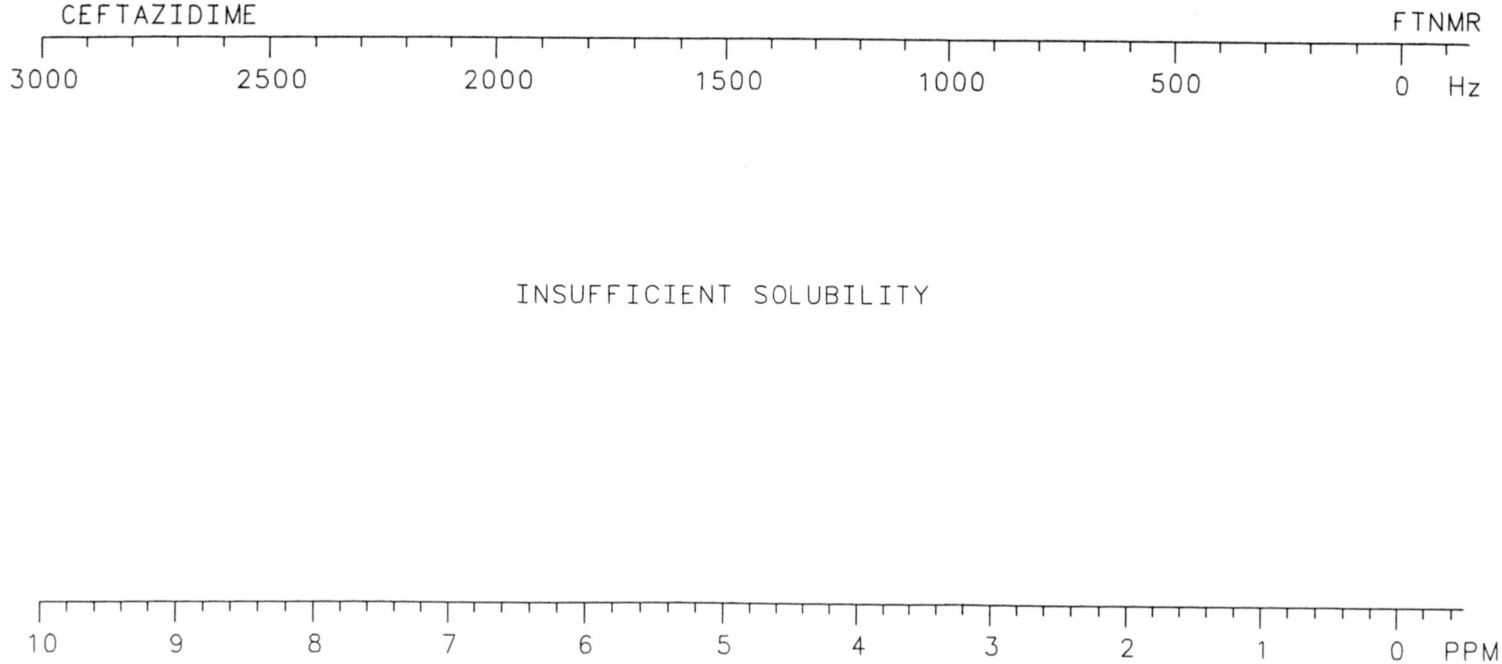
CEFTAZIDIME
FTNMR
3000 2500 2000 1500 1000 500 0 Hz
INSUFFICIENT SOLUBILITY
10 9 8 7 6 5 4 3 2 1 0 PPM

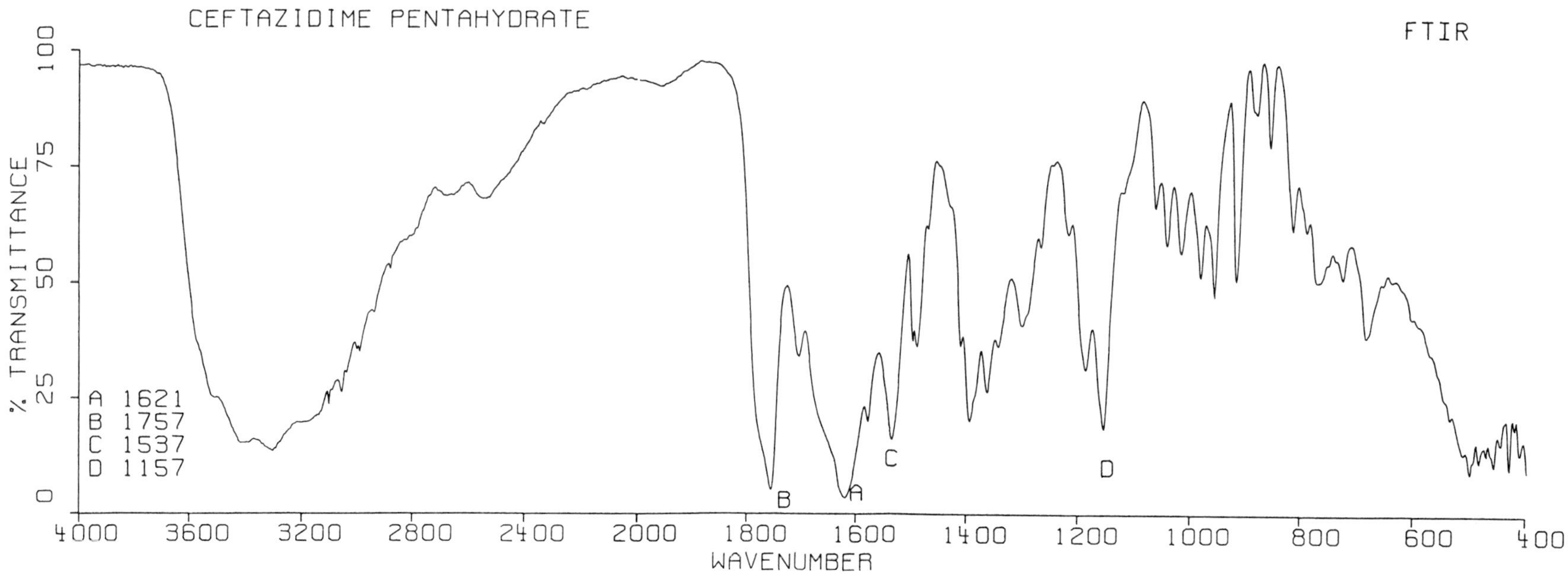
CEFTAZIDIME PENTAHYDRATE
FTIR
% TRANSMITTANCE
0 25 50 75 100
A 1621
B 1757
C 1537
D 1157
A B C D
4000 3600 3200 2800 2400 2000 1800 1600 1400 1200 1000 800 600 400
WAVENUMBER

CLARITHROMYCIN

$C_{38}H_{69}NO_{13}$

Molecular Weight: 747.96 (747.48)

Synonyms: 6-O-Methylerythromycin; clathromycin
Trade Names: Biaxin
Use: Antibacterial

HPLC:
GC:

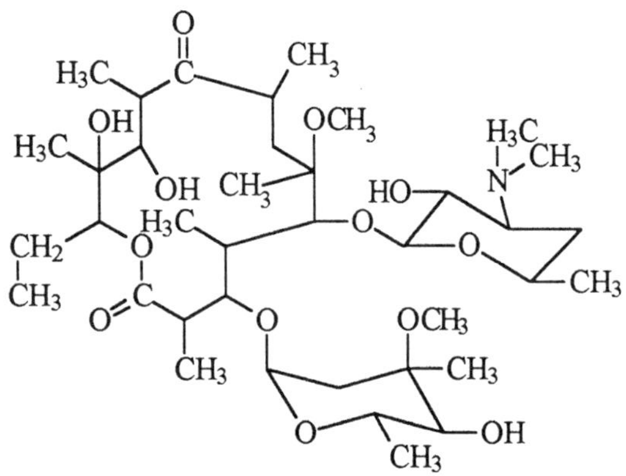

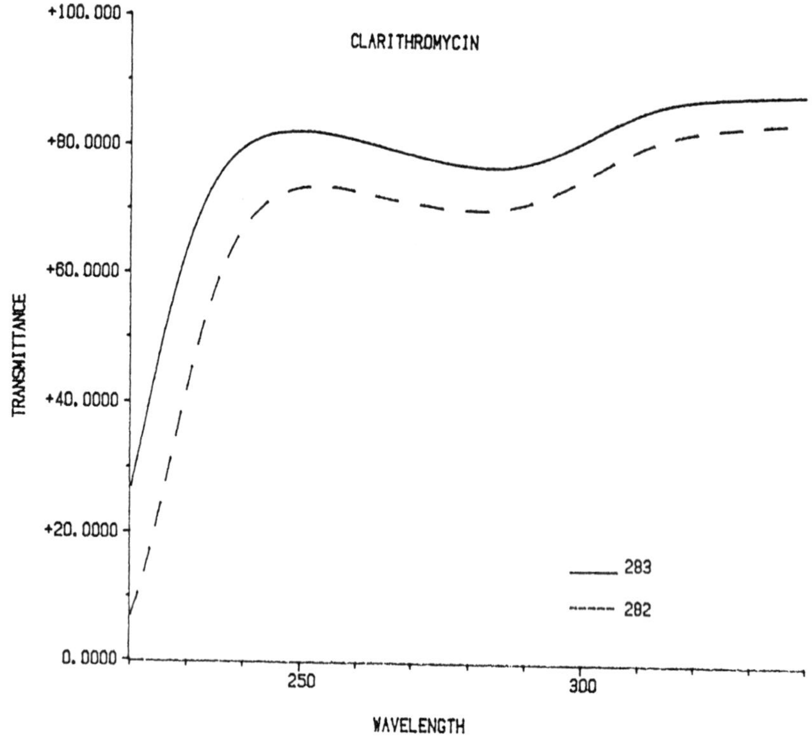

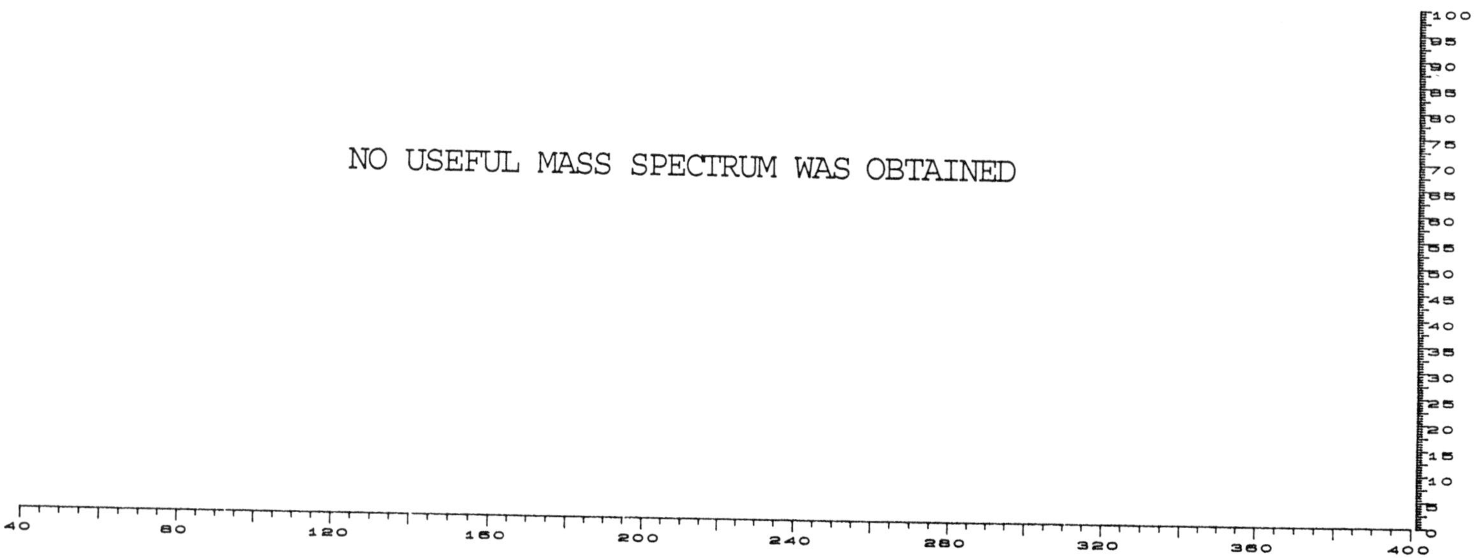

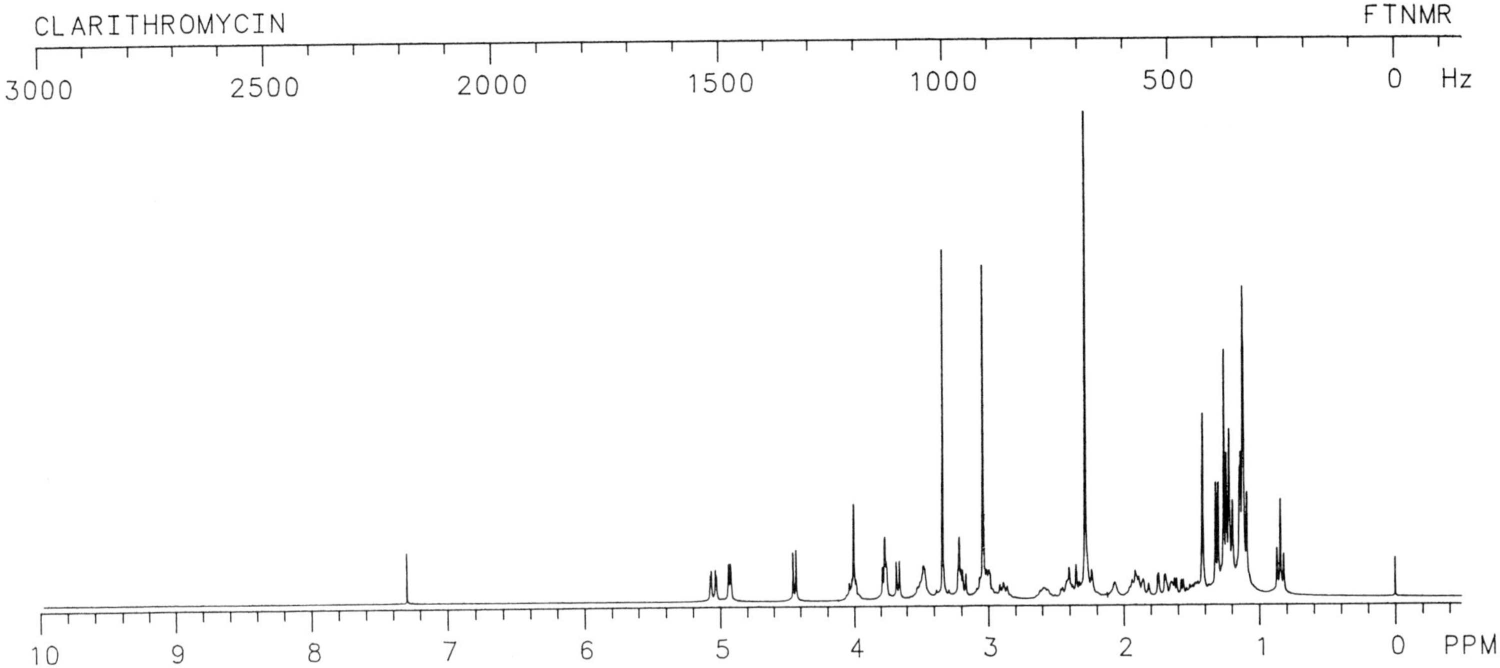

PEAK LISTING

#	HT	PPM
1	20	.84
2	21	1.08
3	53	1.11
4	63	1.11
5	30	1.13
6	27	1.14
7	20	1.19
8	34	1.21
9	29	1.23
10	50	1.25
11	23	1.29
12	23	1.31
13	37	1.41
14	99	2.28
15	68	3.03
16	71	3.33
17	19	4.00

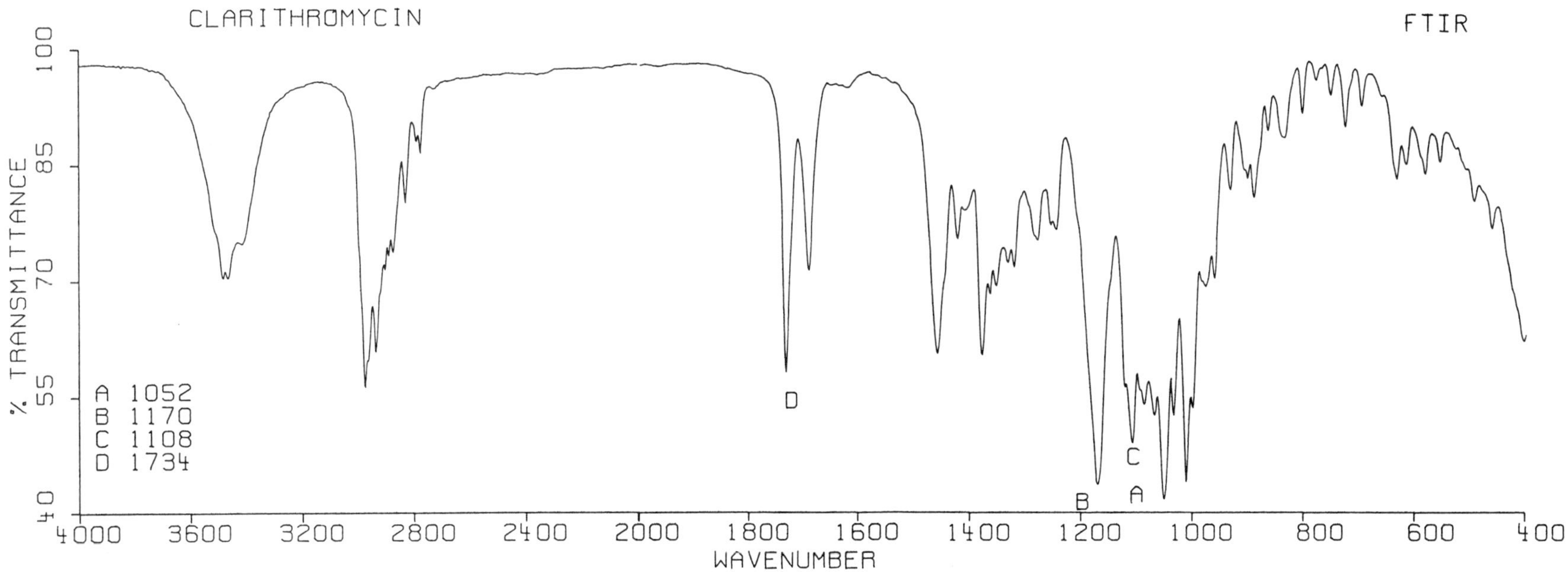

CLOBETASONE BUTYRATE

$C_{26}H_{32}ClFO_5$

Molecular Weight: 478.99 (478.19)

Synonyms: 21-Chloro-9-fluoro-17-hydroxy-16-methyl-pregna-1,4-diene-3,11,20-trione-17-butyrate; 21-chloro-11-dehydrobetamethasone-17-butyrate
Trade Names: Emovate, Eumovate, Molivate
Use: Anti-inflammatory

HPLC: Methanol: 2.9
GC:

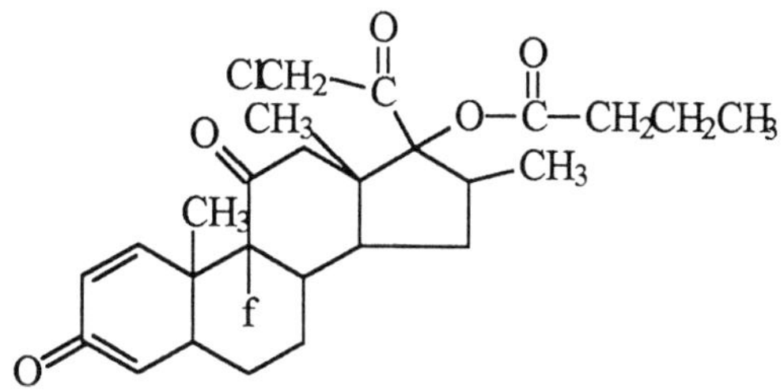

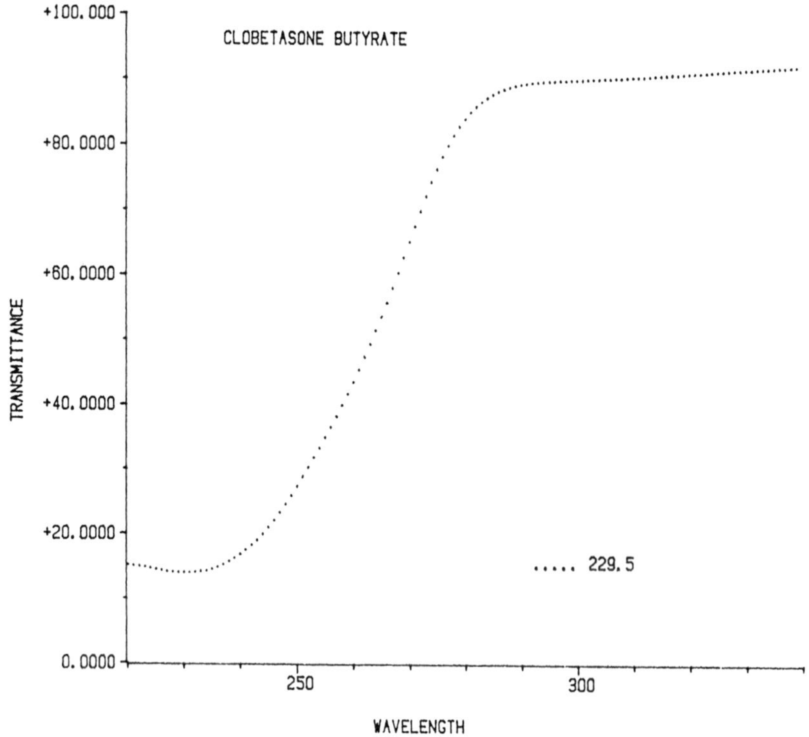

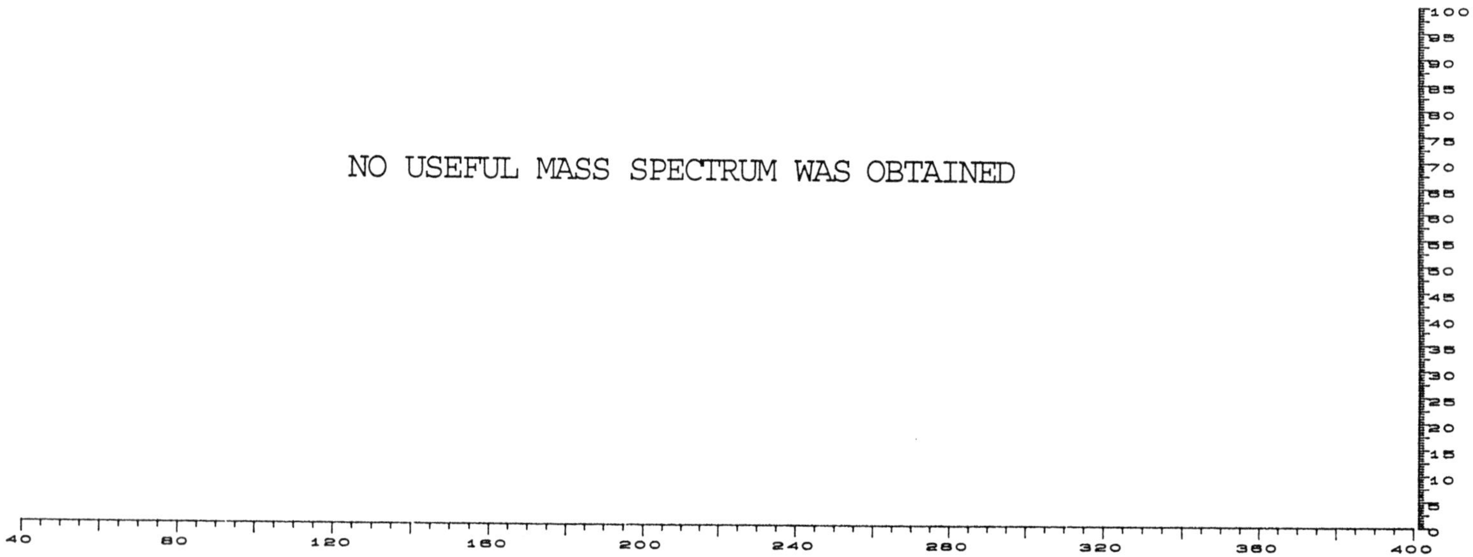

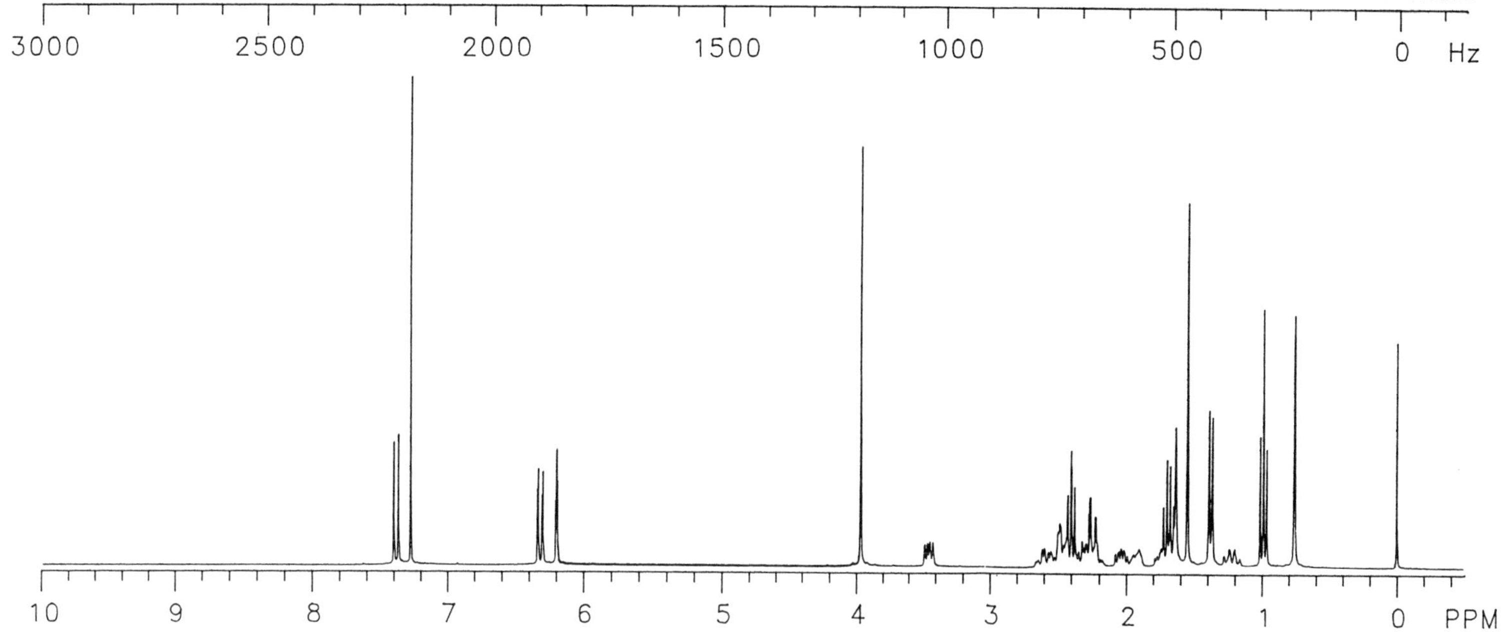

PEAK LISTING

#	HT	PPM
1	46	.00
2	51	.75
3	52	.98
4	30	1.36
5	32	1.38
6	74	1.55
7	28	1.63
8	85	3.96
9	99	7.27

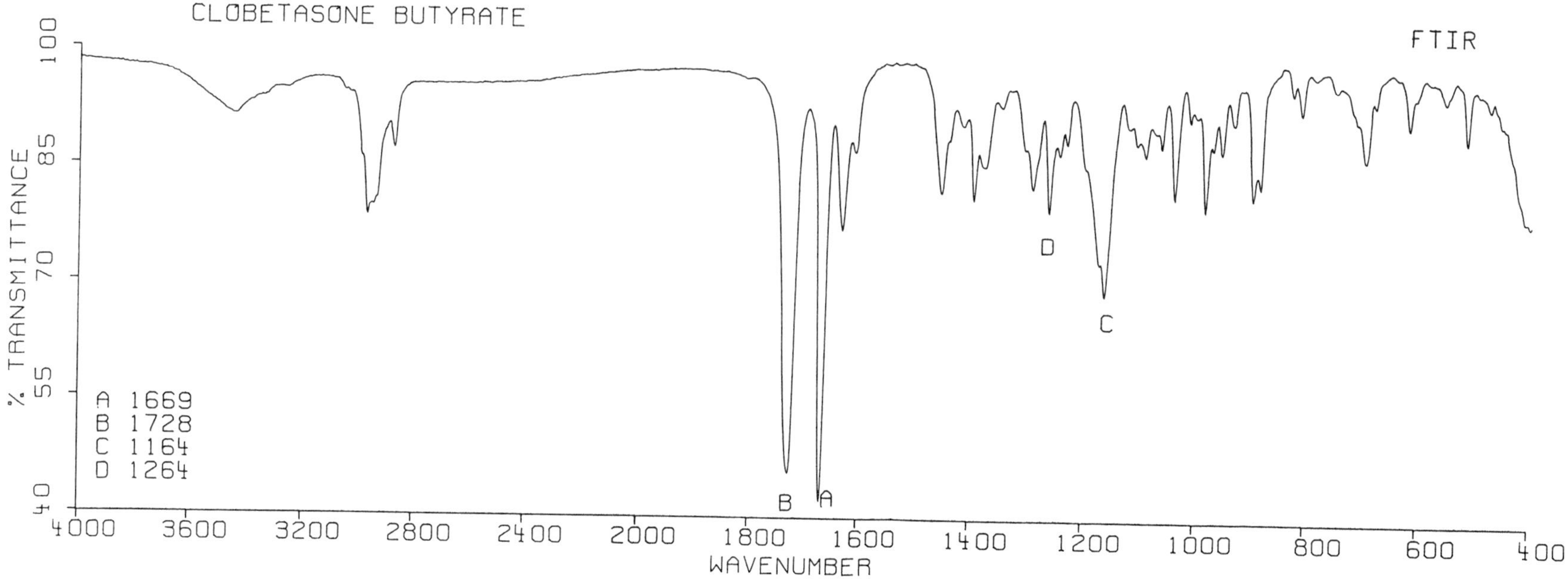

CLOFAZIMINE

$C_{27}H_{22}Cl_2N_4$

Molecular Weight: 473.41 (472.12)

Synonyms: N,5-Bis-(4-chlorophenyl)-3,5-dihydro-3-[(1-methylethyl)imino]-2-phenazinamine;
Trade Names: Lamprene
Use: Antibacterial

HPLC: Methanol: 4.0
GC:

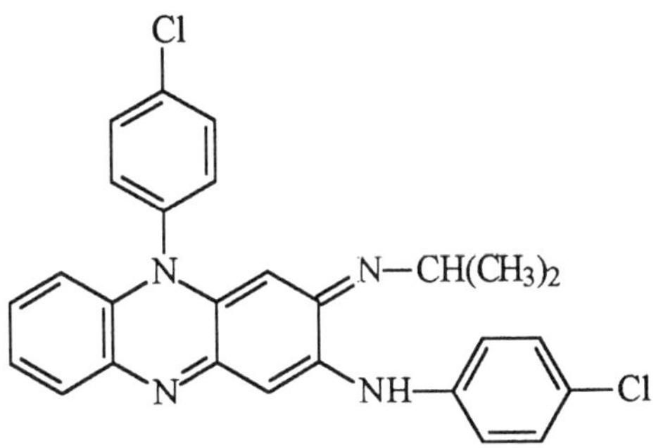

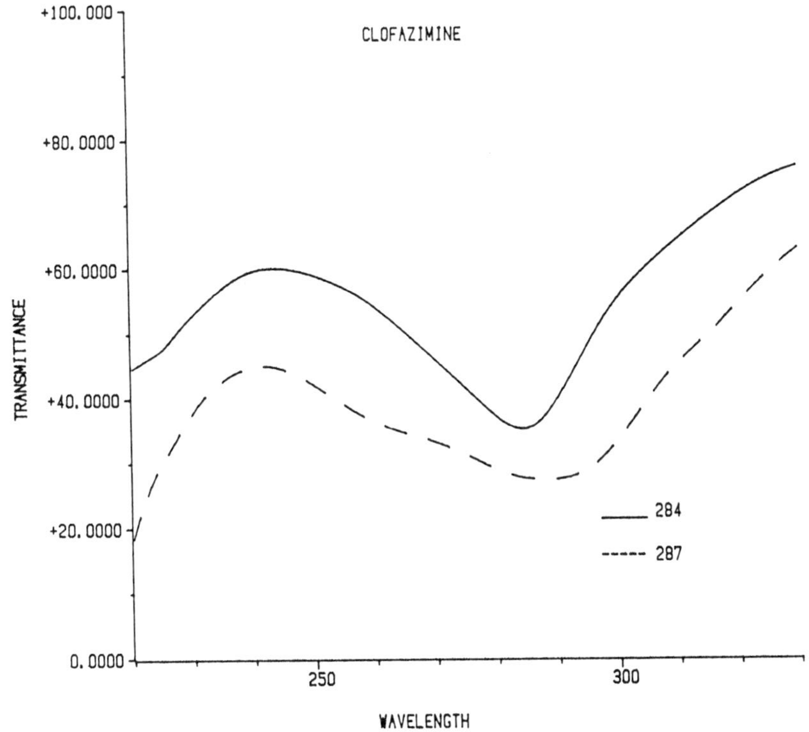

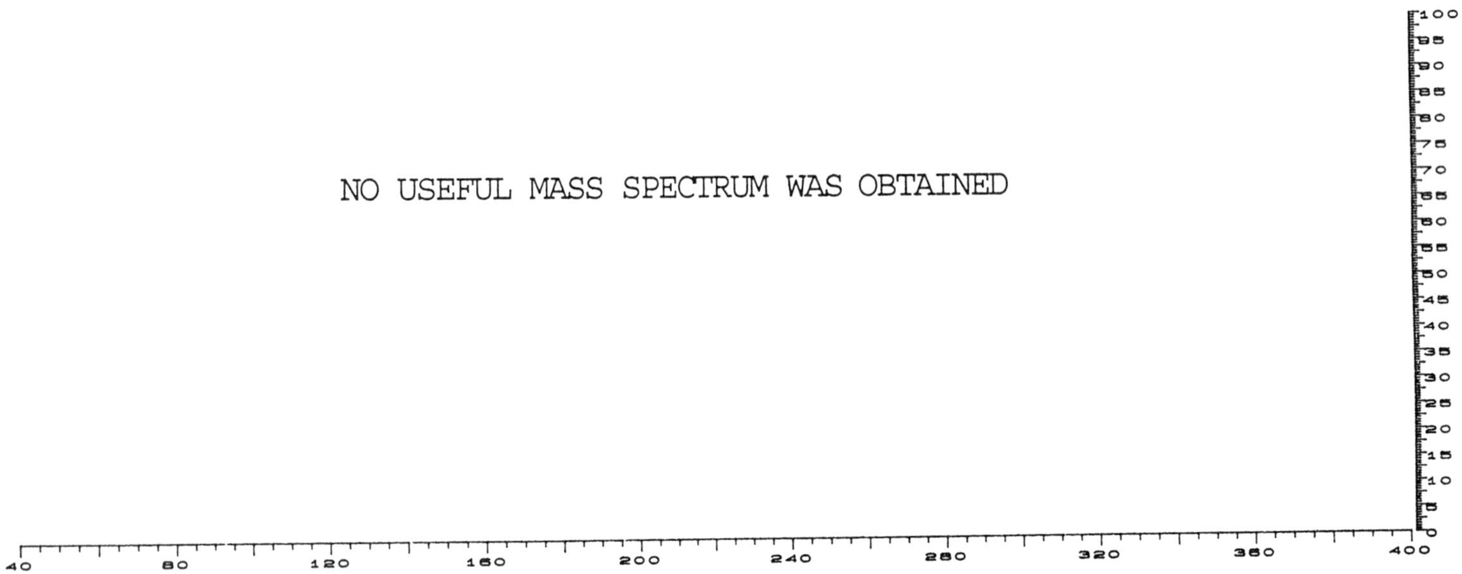

NO USEFUL MASS SPECTRUM WAS OBTAINED

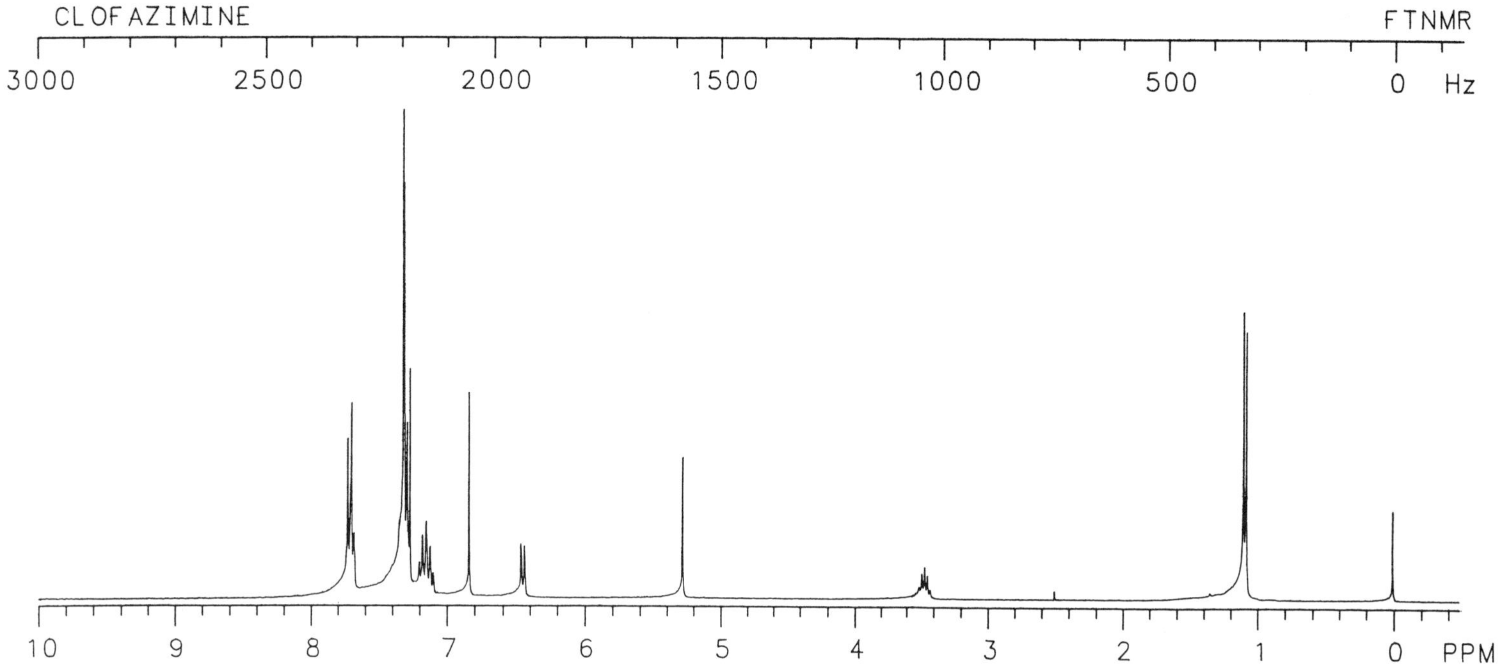

PEAK LISTING

#	HT	PPM
1	18	.00
2	54	1.07
3	58	1.09
4	28	5.27
5	41	6.83
6	13	7.13
7	15	7.14
8	13	7.14
9	12	7.16
10	13	7.17
11	46	7.26
12	13	7.27
13	35	7.28
14	18	7.28
15	84	7.30
16	99	7.30
17	23	7.31
18	17	7.32
19	16	7.33
20	13	7.66
21	13	7.67
22	39	7.68
23	16	7.71
24	32	7.71

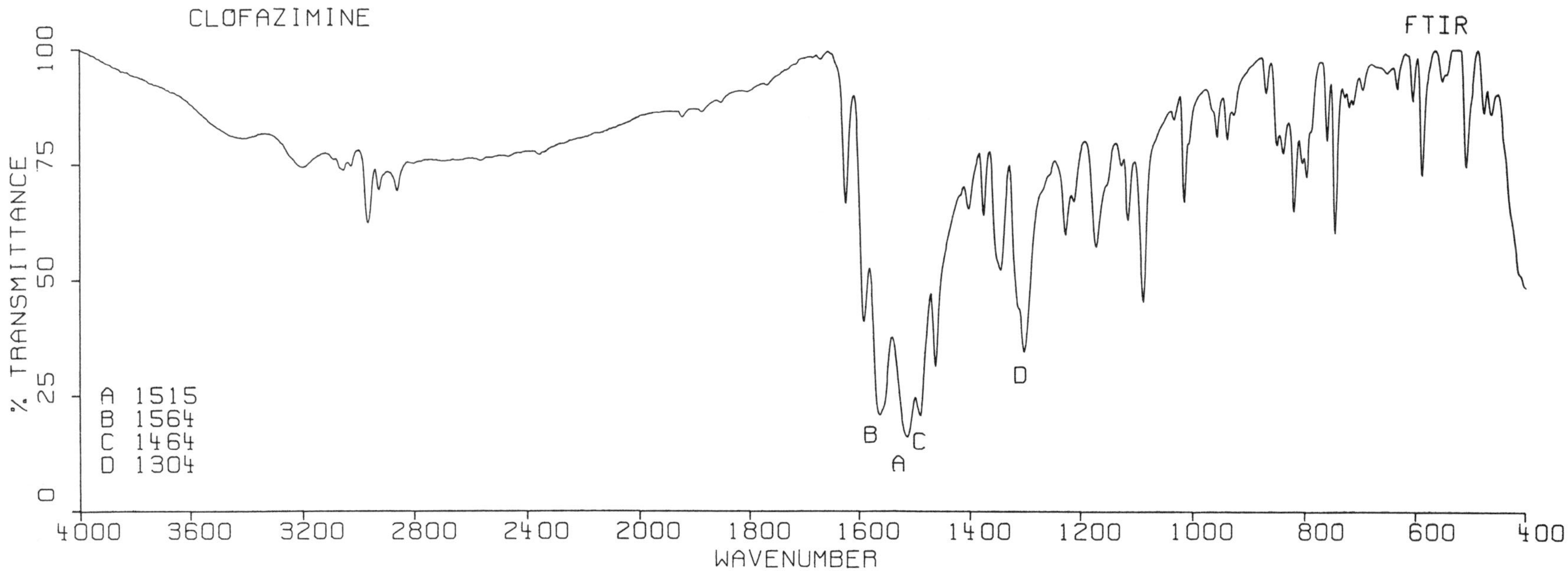

CLOZAPINE

$C_{18}H_{19}ClN_4$

Molecular Weight: 326.83 (326.13)

Synonyms: 8-Chloro-11-(4-methyl-1-piperazinyl)-5H-dibenzo[b,e][1,4]diazepine
Trade Names: Clozaril
Use: Antipsychotic

HPLC: Methanol: 13.8 (Tails)
GC: 2911; 250°

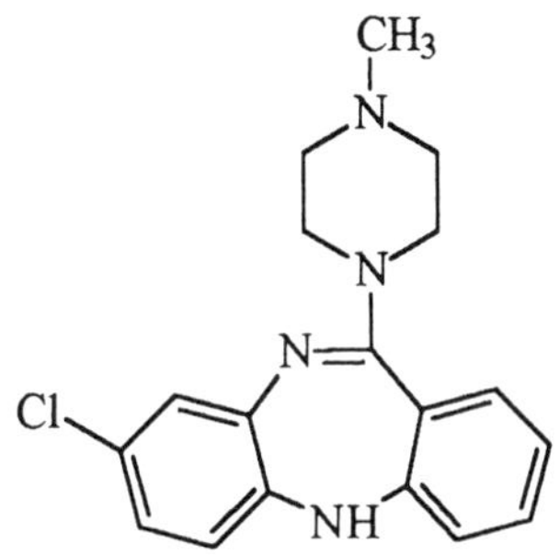

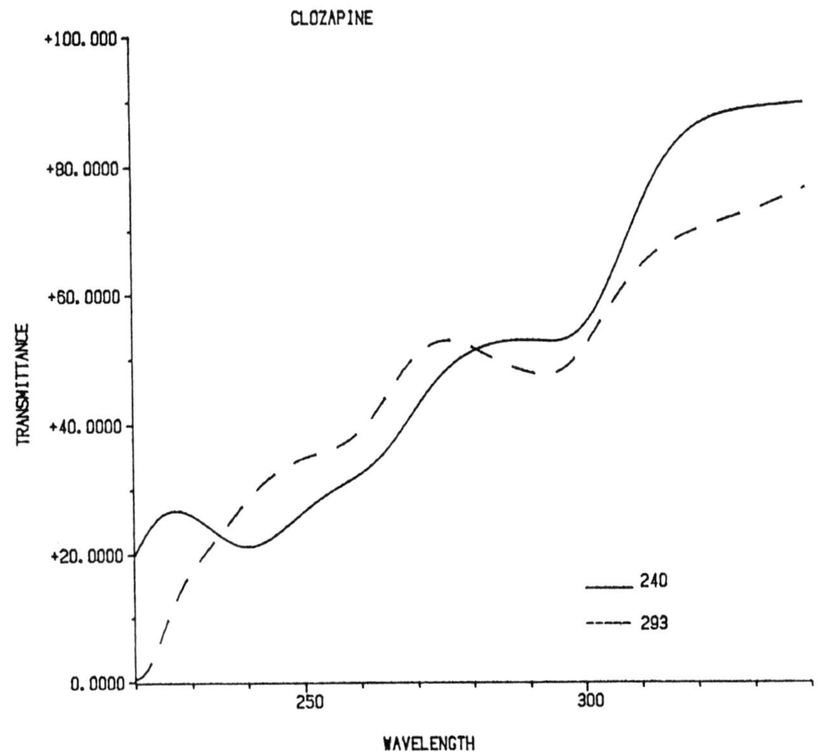

CLOZAPINE

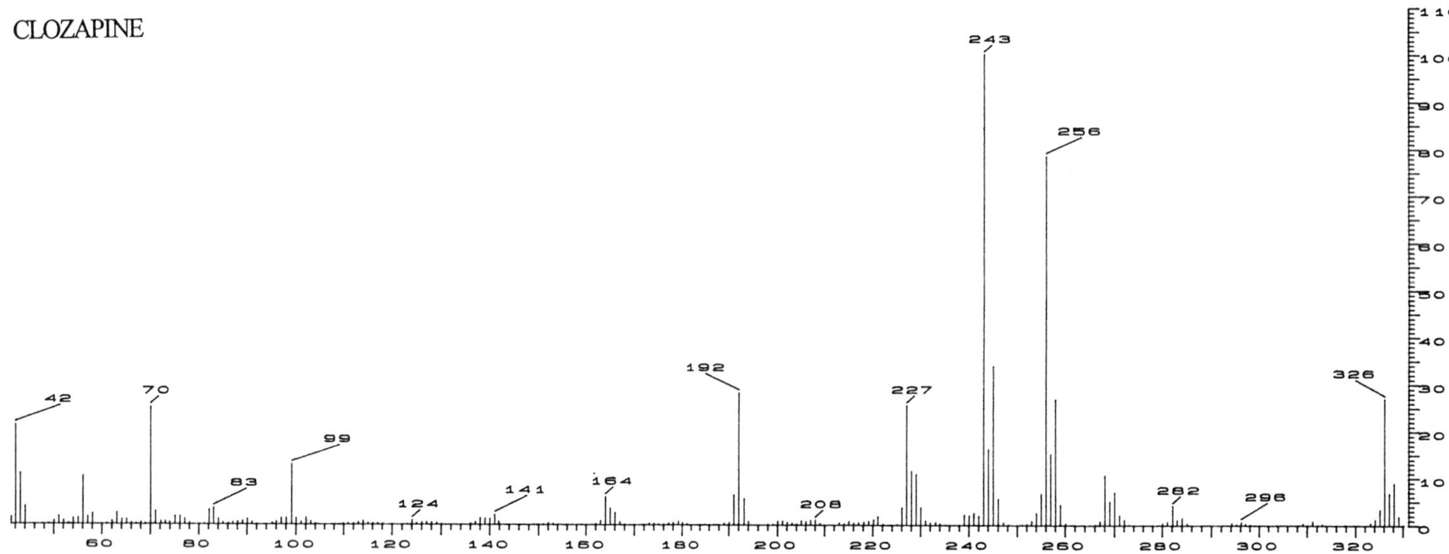

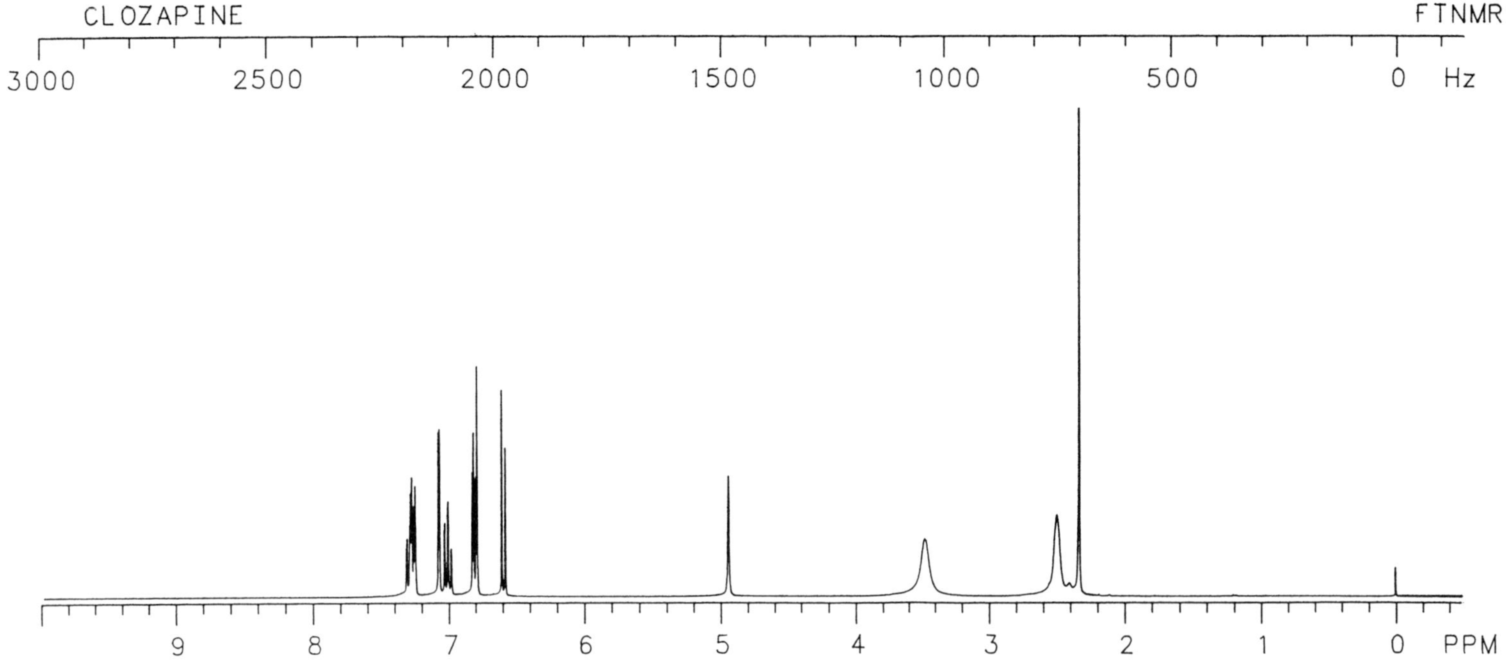

PEAK LISTING

#	HT	PPM
1	99	2.33
2	16	2.49
3	11	3.46
4	24	4.93
5	30	6.57
6	42	6.60
7	46	6.78
8	24	6.79
9	33	6.81
10	25	6.82
11	19	7.00
12	18	7.00
13	15	7.02
14	13	7.02
15	34	7.06
16	33	7.07
17	22	7.24
18	18	7.25
19	15	7.26
20	24	7.26
21	21	7.27
22	14	7.28
23	11	7.30

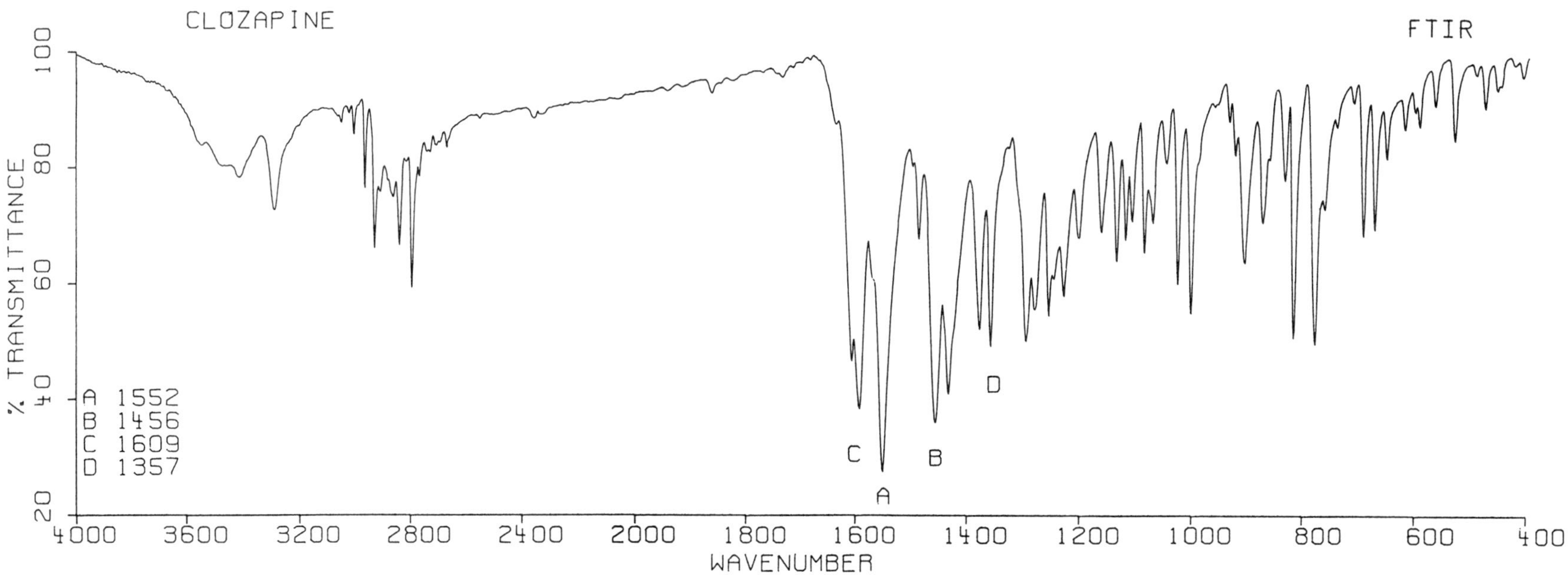

COCAETHYLENE

$C_{18}H_{23}NO_4$

Molecular Weight: 317.38 (317.16)

Synonyms: [1R-(exo,exo)]-3-(Benzoyloxy)-8-methyl-8-azabicyclo[3.2.1]octane-2-carboxylic acid ethyl ester; ecgonine ethyl ester benzoate; ethylbenzoylecgonine; homocaine; O-benzoyl-1-ecgonine ethyl ester
Trade Names:
Use: Local anesthetic

HPLC: Methanol: 2.8
GC: 2281; 200°

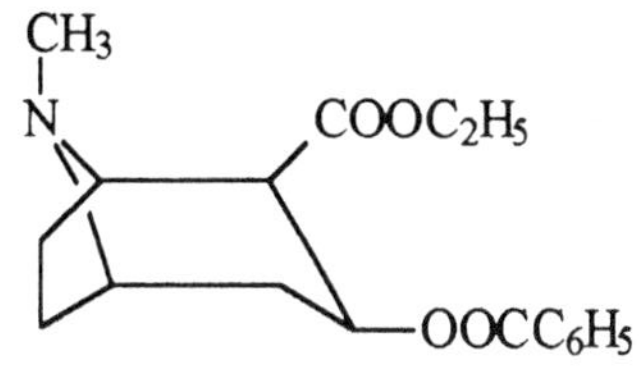

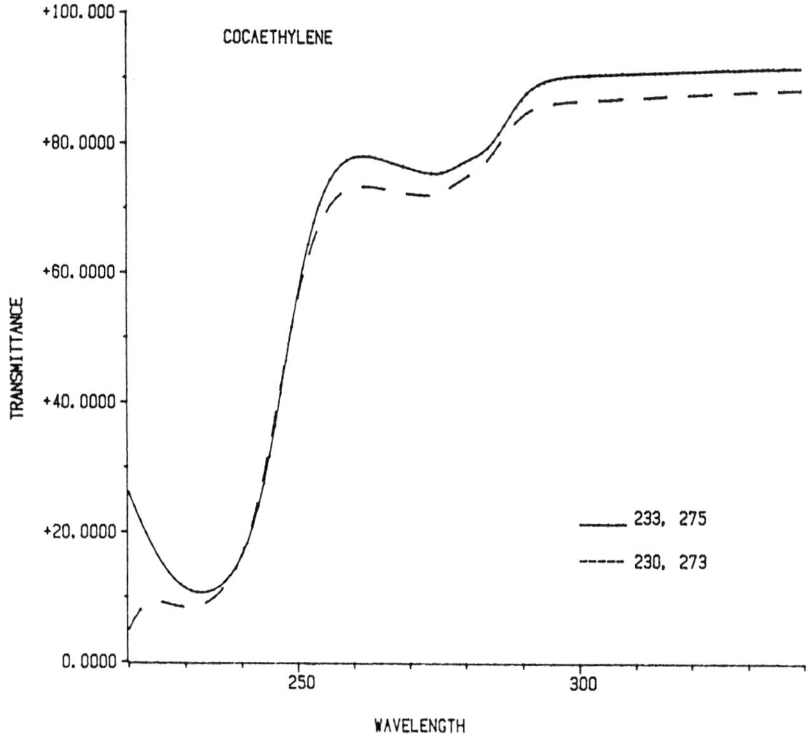

COCAETHYLENE

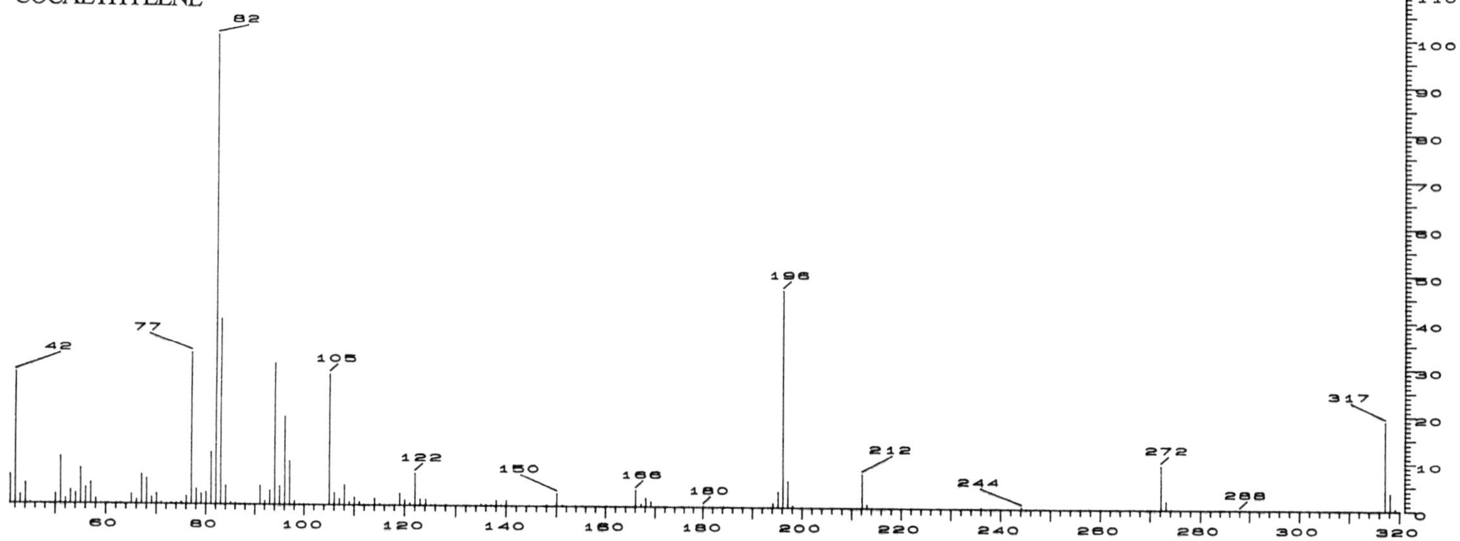

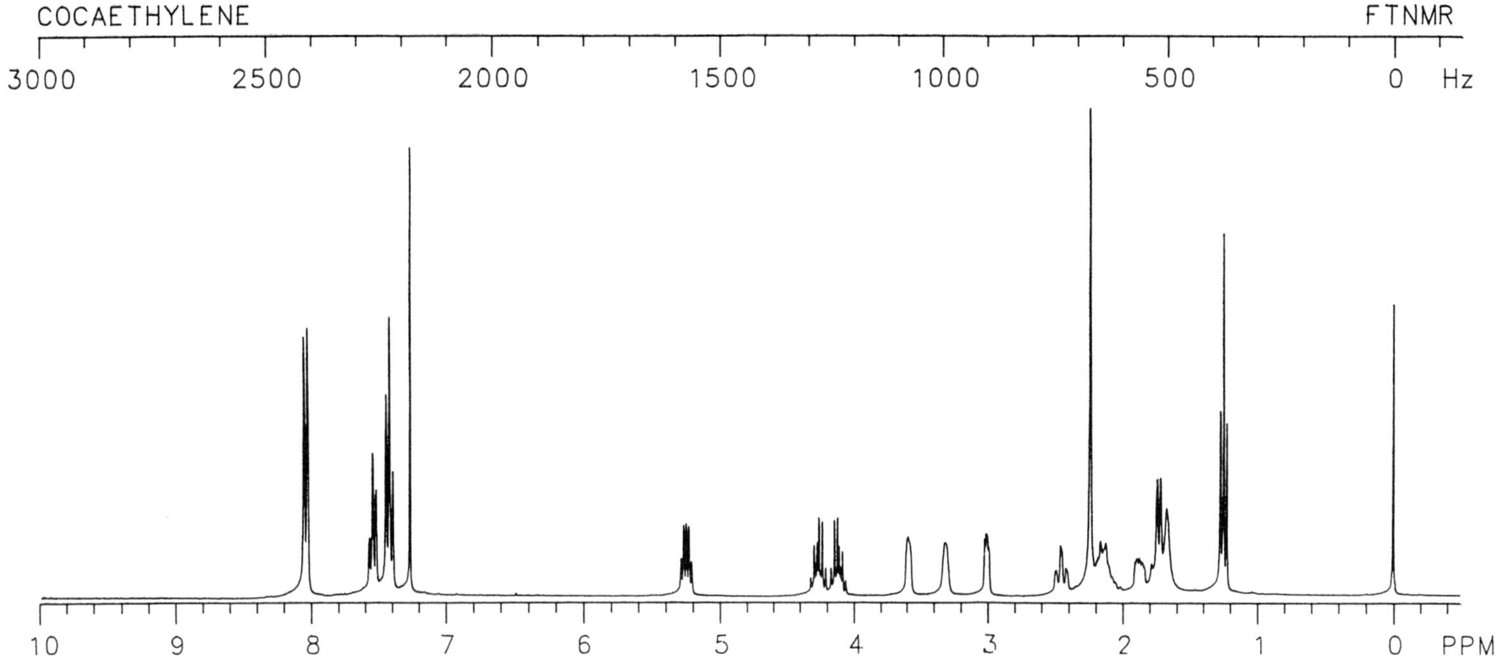

PEAK LISTING

#	HT	PPM
1	59	.00
2	48	.00
3	35	1.21
4	74	1.24
5	37	1.26
6	24	1.71
7	24	1.73
8	99	2.23
9	91	7.26
10	25	7.39
11	56	7.41
12	41	7.44
13	21	7.51
14	20	7.51
15	29	7.54
16	54	8.02
17	53	8.04

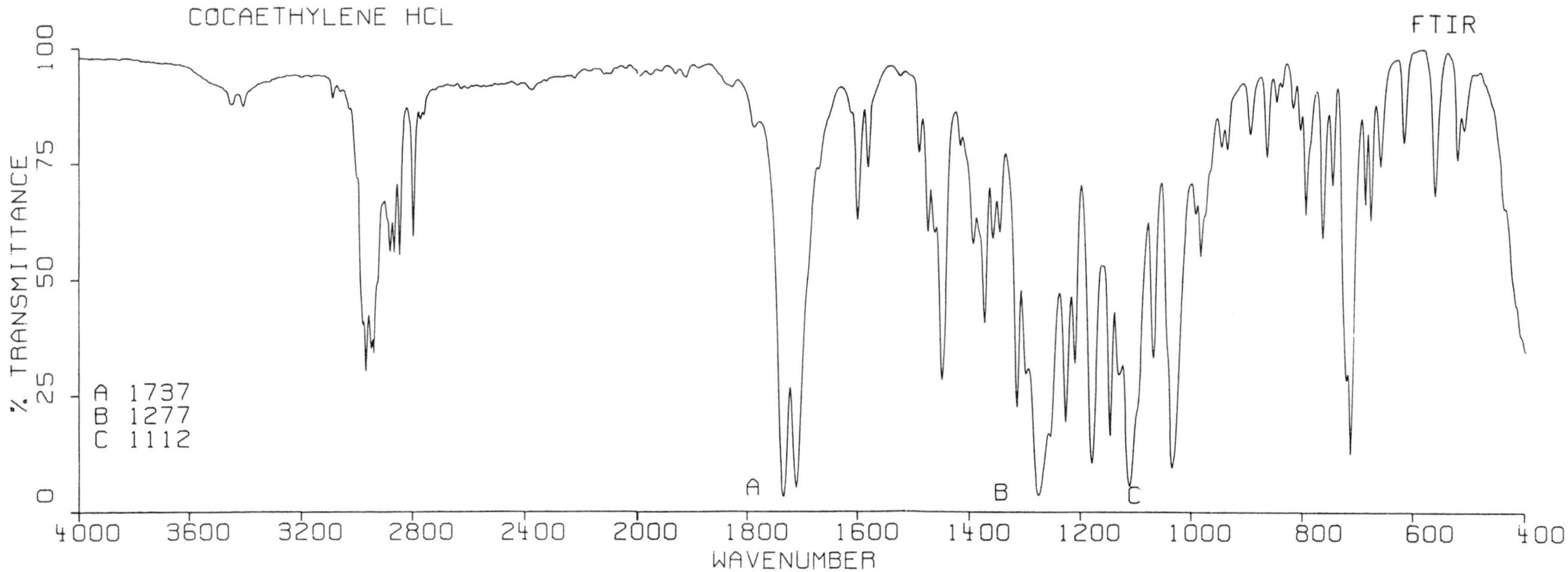

DEHYDRO-17-METHYLTESTOSTERONE

$C_{20}H_{28}O_2$

Molecular Weight: 300.44 (300.21)

Synonyms: 17β-Hydroxy-17α-methyl-4,9[11]-androstadien-3-one
Trade Names:
Use:

HPLC: Methanol: 2.9
GC: 2643; 280°

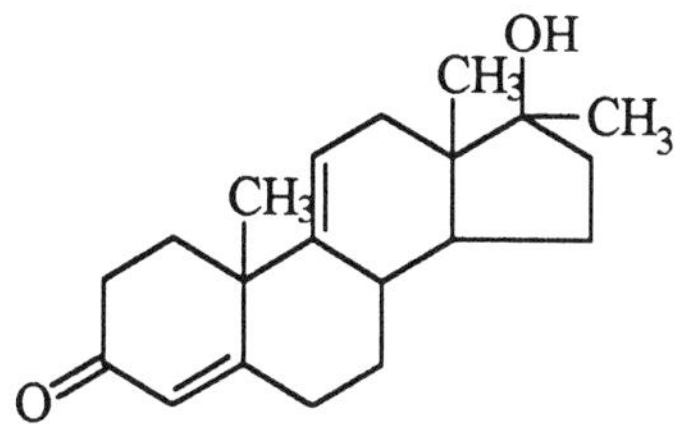

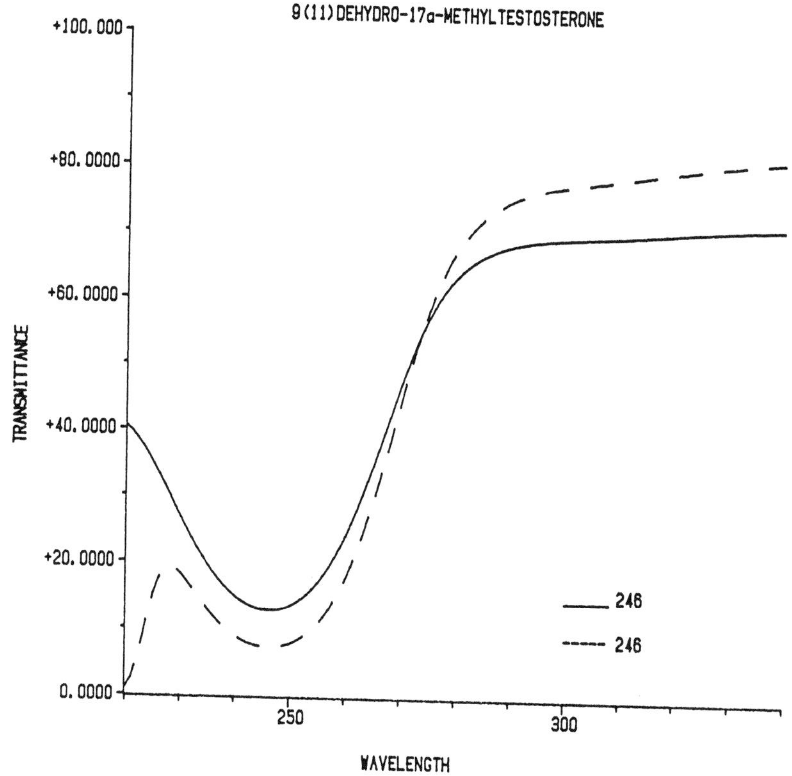

DEHYDROMETHYLTESTOSTERONE

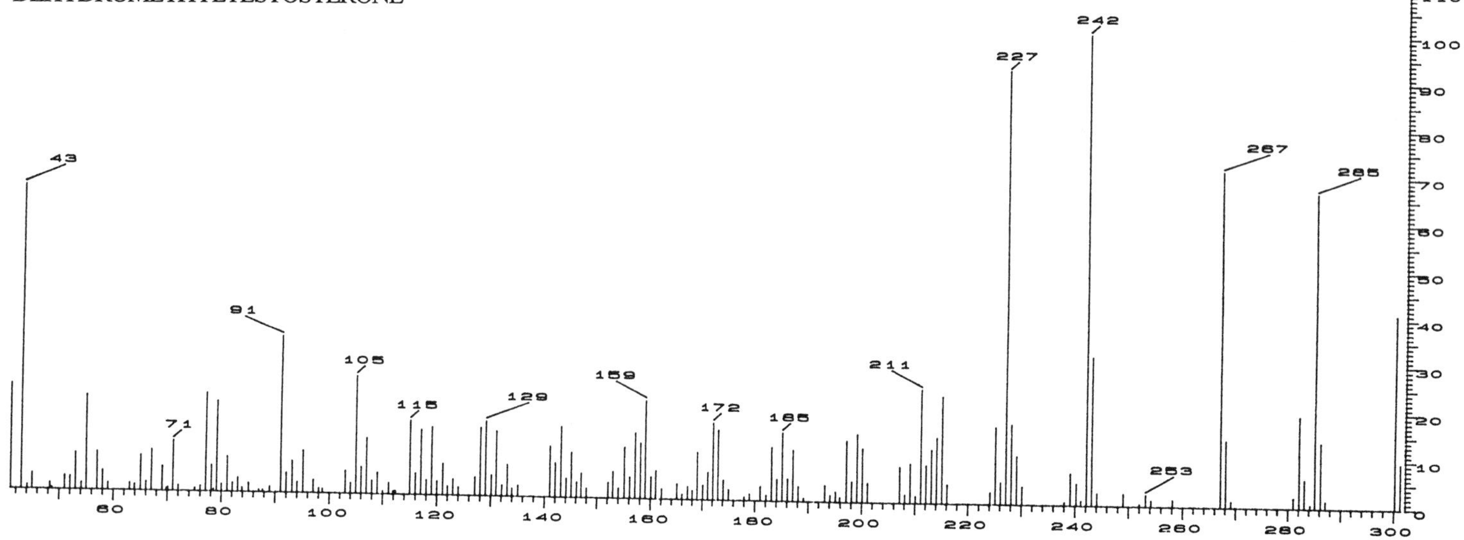

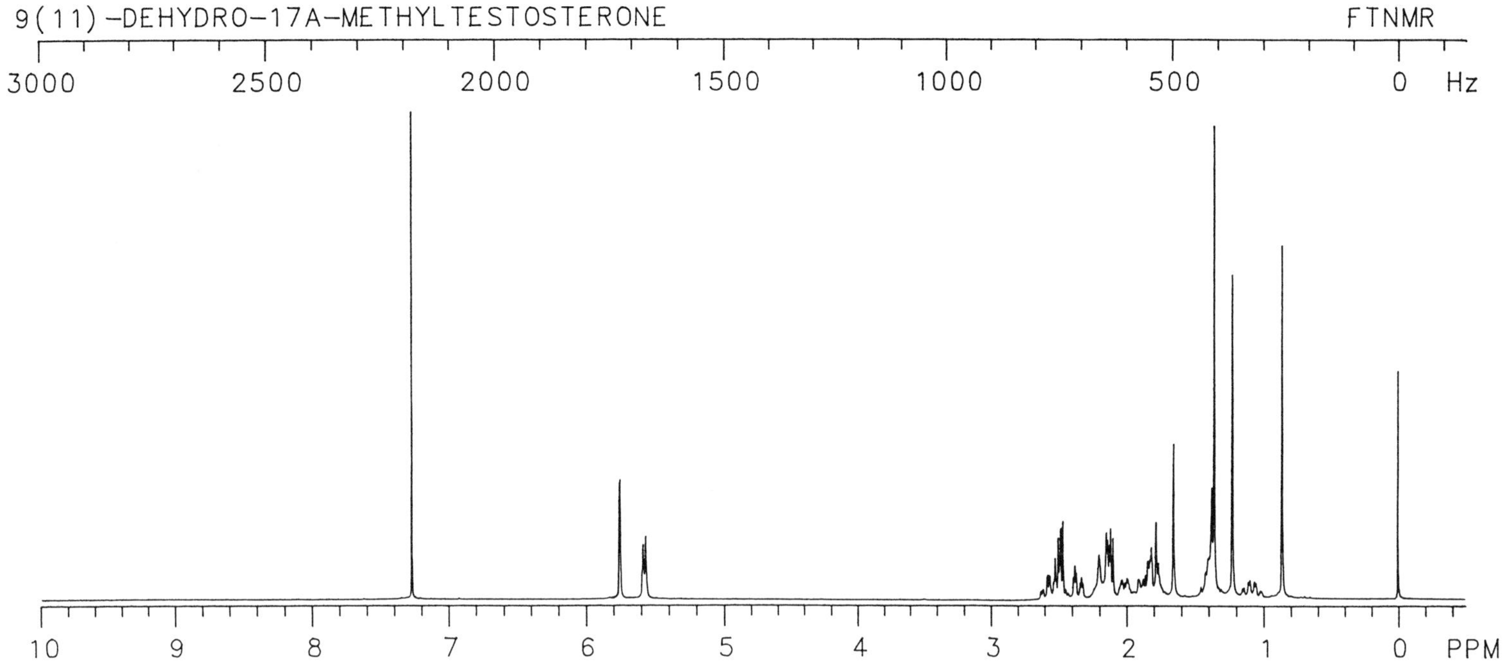

PEAK LISTING

#	HT	PPM
1	46	.00
2	16	.00
3	72	.86
4	66	1.22
5	96	1.35
6	23	1.37
7	31	1.65
8	16	1.78
9	14	2.11
10	16	2.46
11	14	2.48
12	24	5.74
13	23	5.75
14	99	7.26

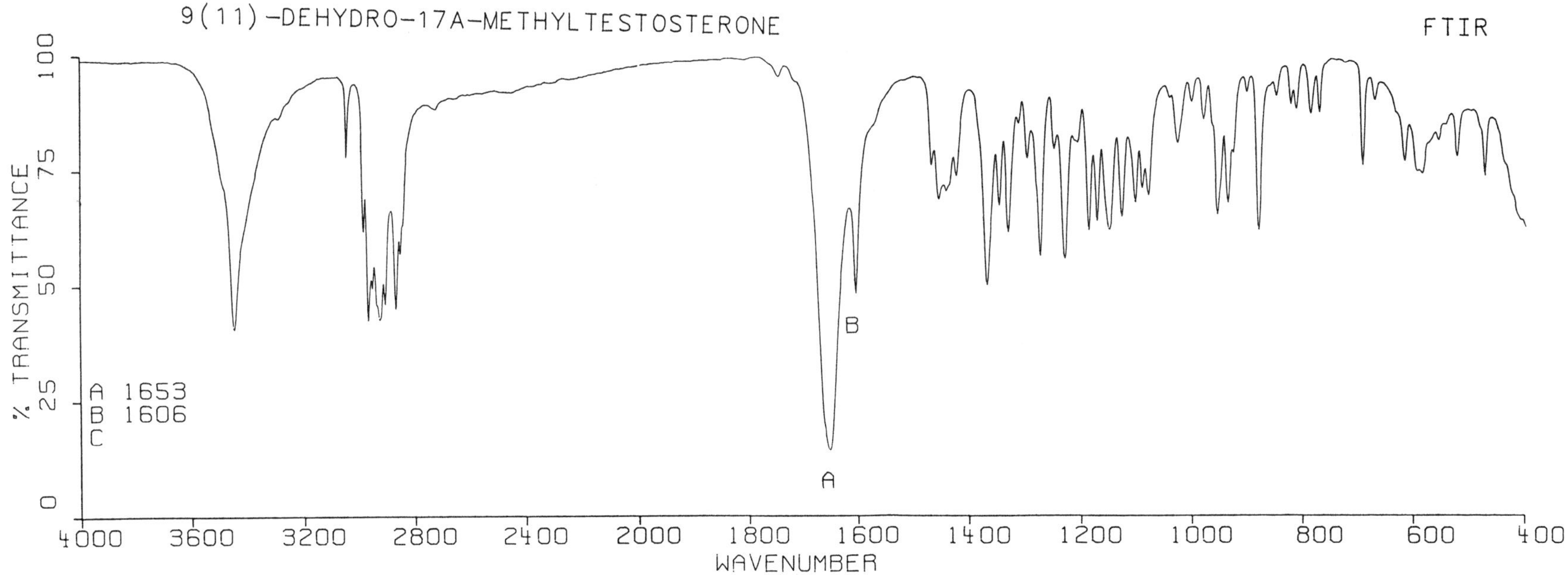

DEPRENYL

$C_{13}H_{17}N$

Molecular Weight: 187.28 (187.14)

Synonyms: N,α-dimethyl-N-2-propynylbenzeneethanamine; phenylisopropylmethylpropynylamine; deprenalin; deprenil
Trade Names: Eldeprine, Movergan, Selegiline
Use: Antiparkinsonian

HPLC:
GC:

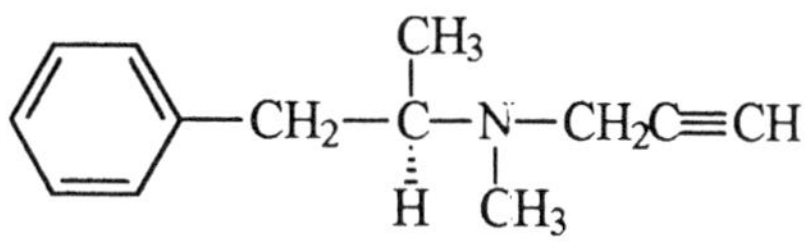

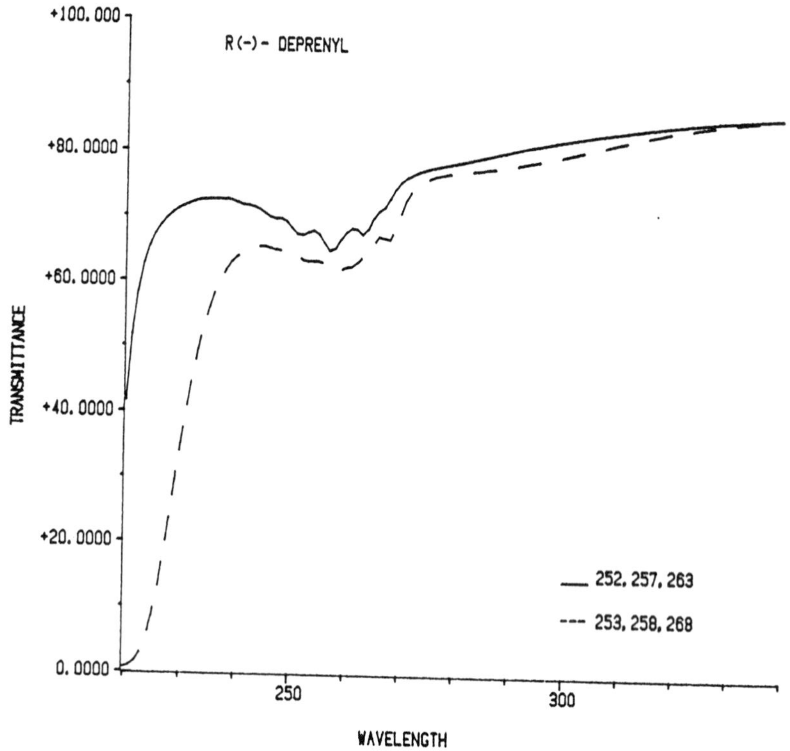

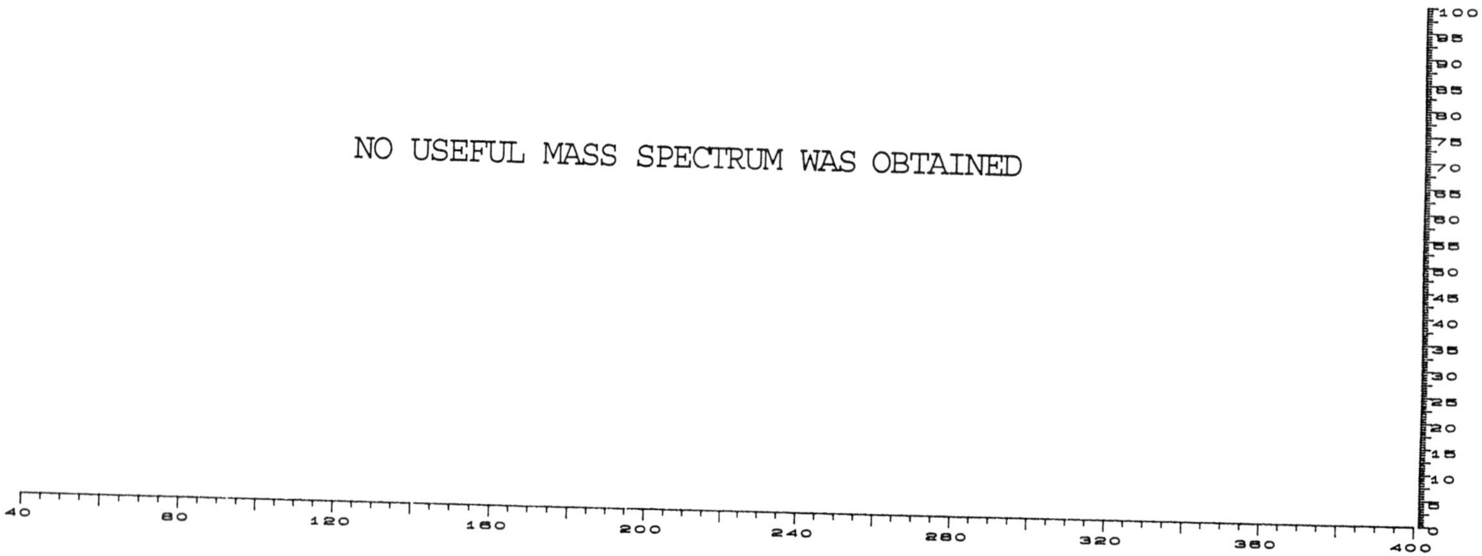

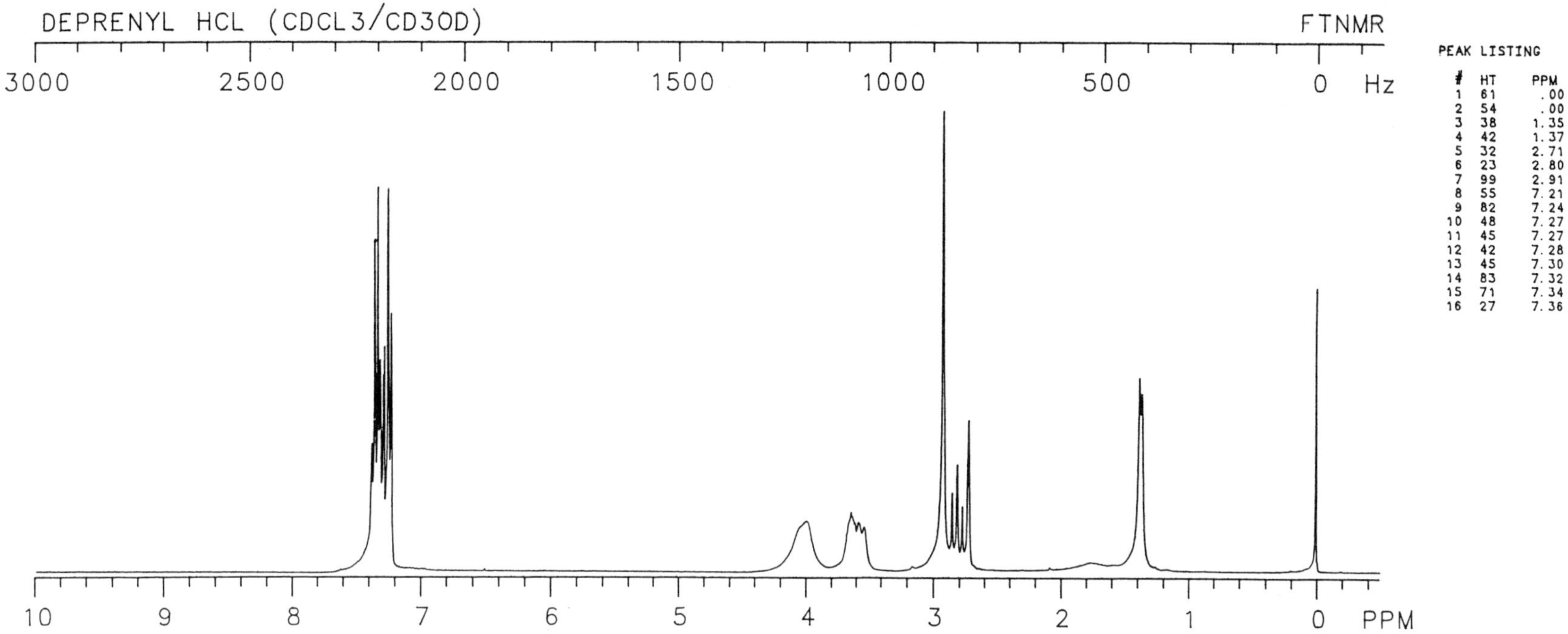
DEPRENYL HCL (CDCL3/CD3OD)
FTNMR
3000 2500 2000 1500 1000 500 0 Hz
10 9 8 7 6 5 4 3 2 1 0 PPM
PEAK LISTING
HT PPM
1 61 .00
2 54 .00
3 38 1.35
4 42 1.37
5 32 2.71
6 23 2.80
7 99 2.91
8 55 7.21
9 82 7.24
10 48 7.27
11 45 7.27
12 42 7.28
13 45 7.30
14 83 7.32
15 71 7.34
16 27 7.36

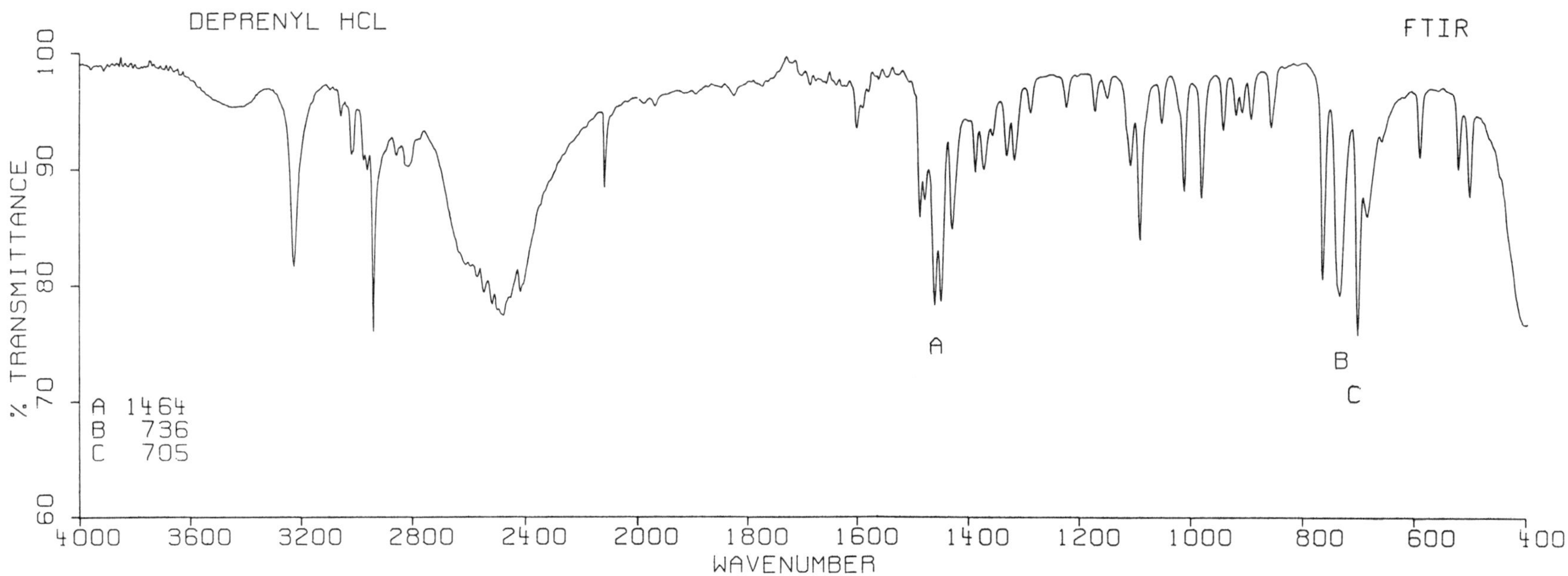
DEPRENYL HCL
FTIR
% TRANSMITTANCE
100 90 80 70 60
A 1464
B 736
C 705
A
B
C
4000 3600 3200 2800 2400 2000 1800 1600 1400 1200 1000 800 600 400
WAVENUMBER

DESOXIMETASONE

$C_{22}H_{29}FO_4$

Molecular Weight: 376.47 (376.20)

Synonyms: 9-Fluoro-11,21-dihydroxy-16-methylpregna-1,4-diene-3,20-dione; desoxymethasone
Trade Names: Esperson, Stiedex, Topicort, Topiderm, Topisolon
Use: Anti-inflammatory

HPLC: Methanol: 2.9
GC:

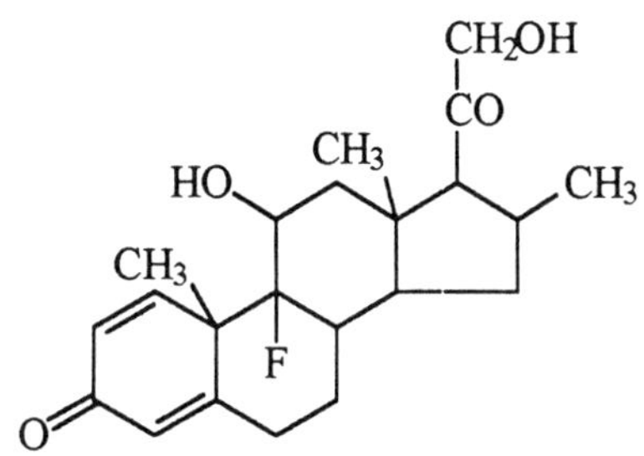

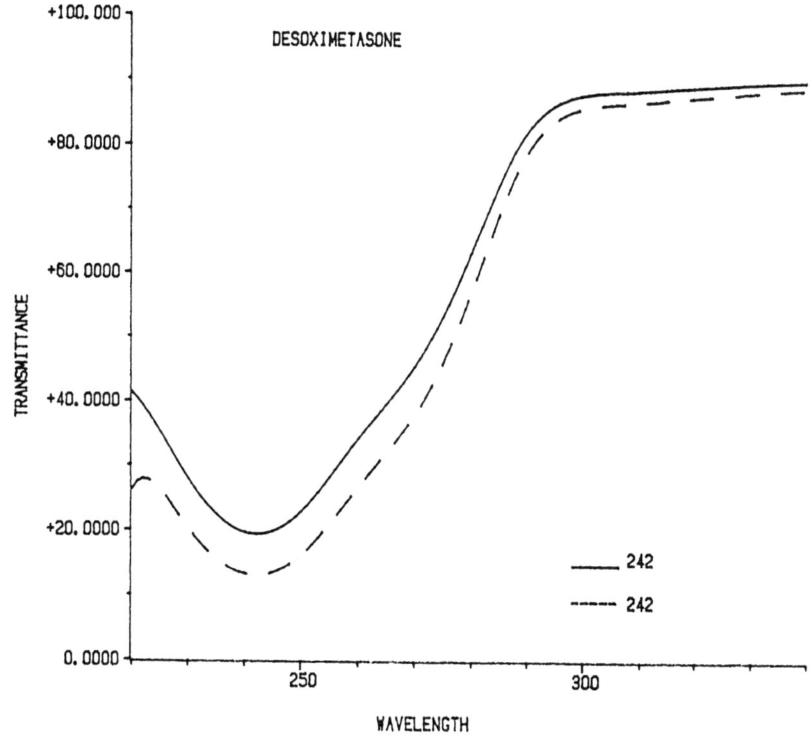

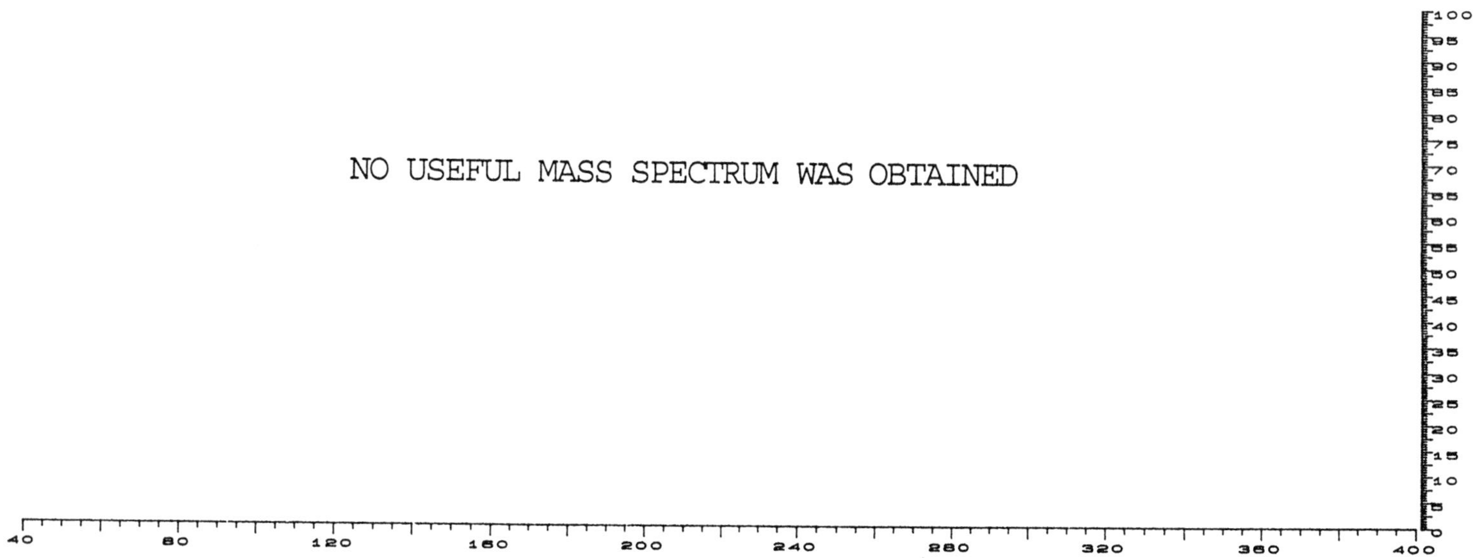

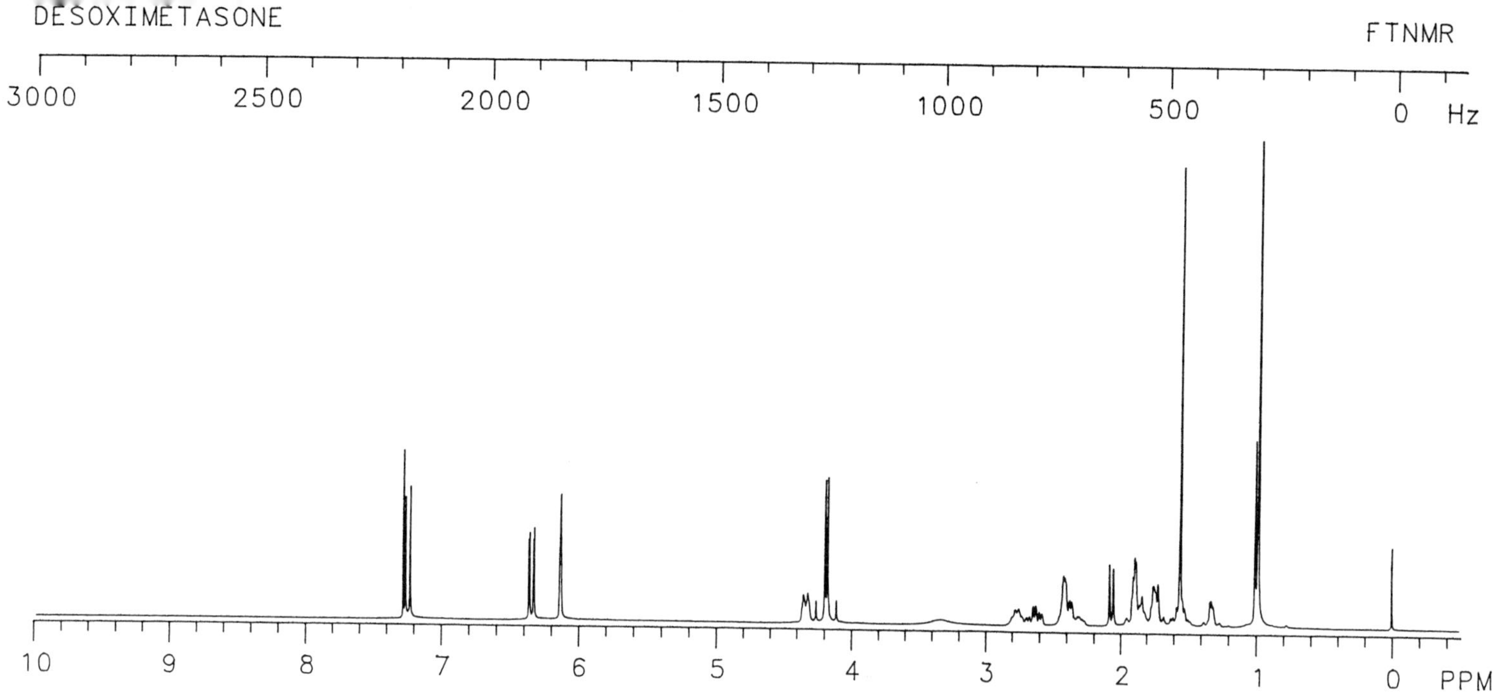

PEAK LISTING

#	HT	PPM
1	16	.00
2	99	.98
3	38	1.00
4	94	1.55
5	13	1.88
6	14	1.89
7	29	4.17
8	29	4.18
9	17	6.12
10	25	6.12
11	18	6.31
12	16	6.32
13	17	6.35
14	16	6.35
15	27	7.22
16	25	7.25
17	34	7.27

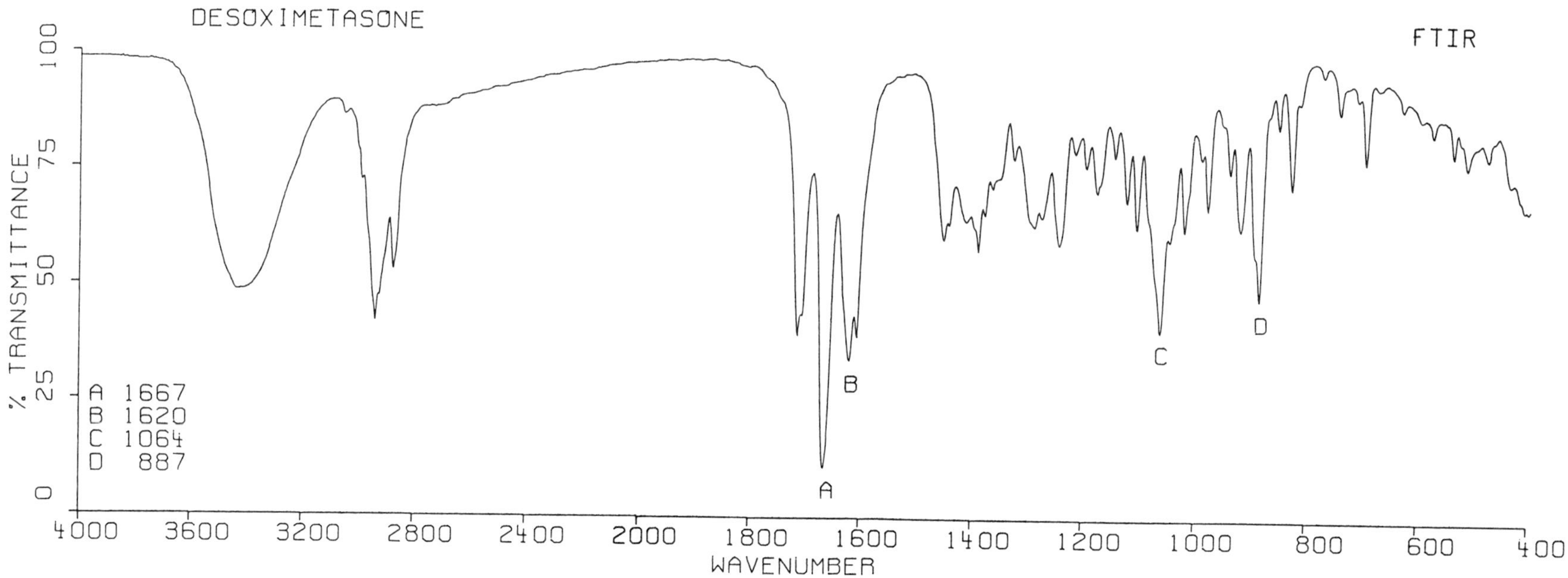

DIMETHYLAMILORIDE

$C_8H_{12}ClN_7O$

Molecular Weight: 257.68 (257.08)

Synonyms: 3-Amino-5-(N,N-dimethyl)amino-N-(aminoiminomethyl-6-chloropyrazinecarboxamide; 5-(N,N-dimethyl)amiloride
Trade Names:
Use: Diuretic

HPLC: Ethanol: 2.7
GC:

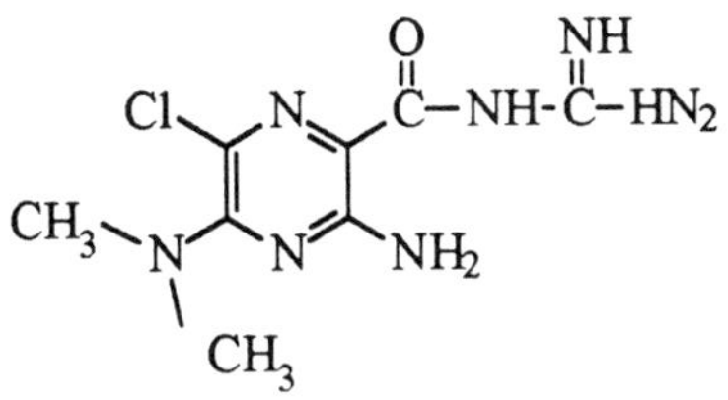

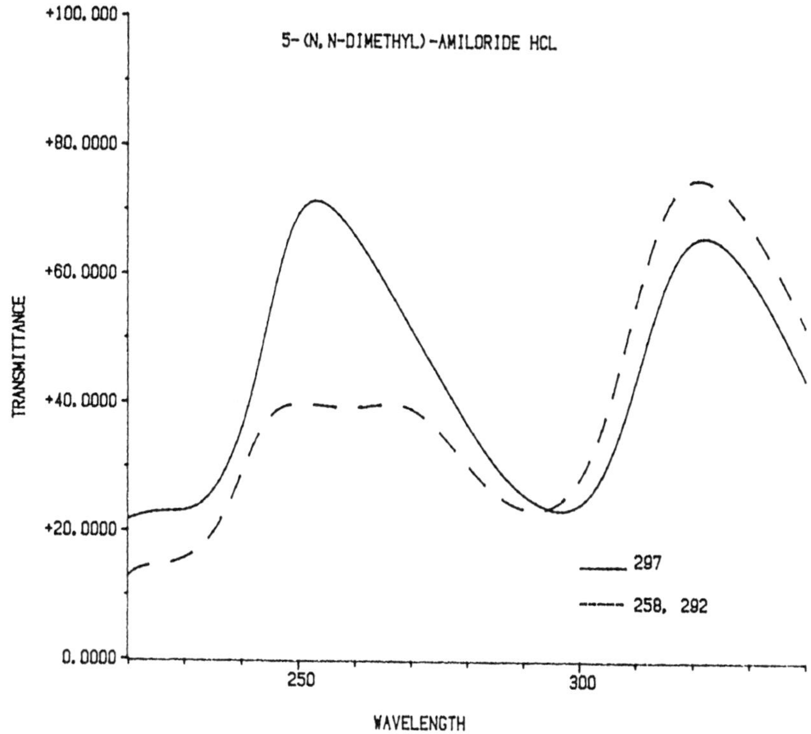

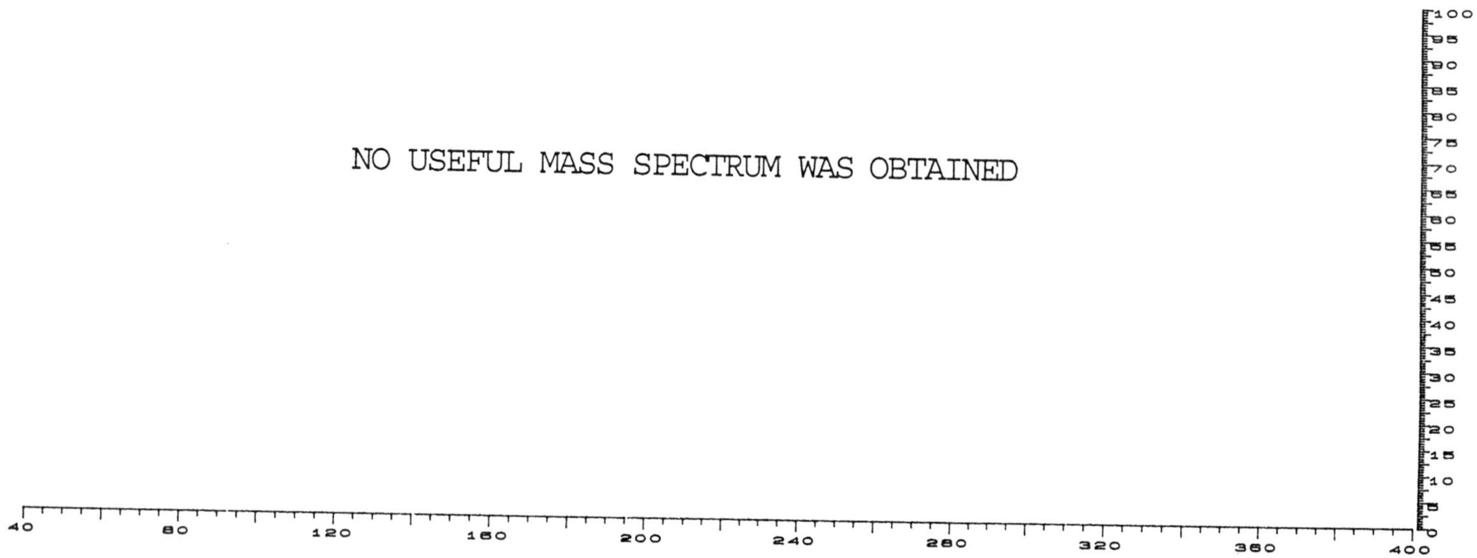

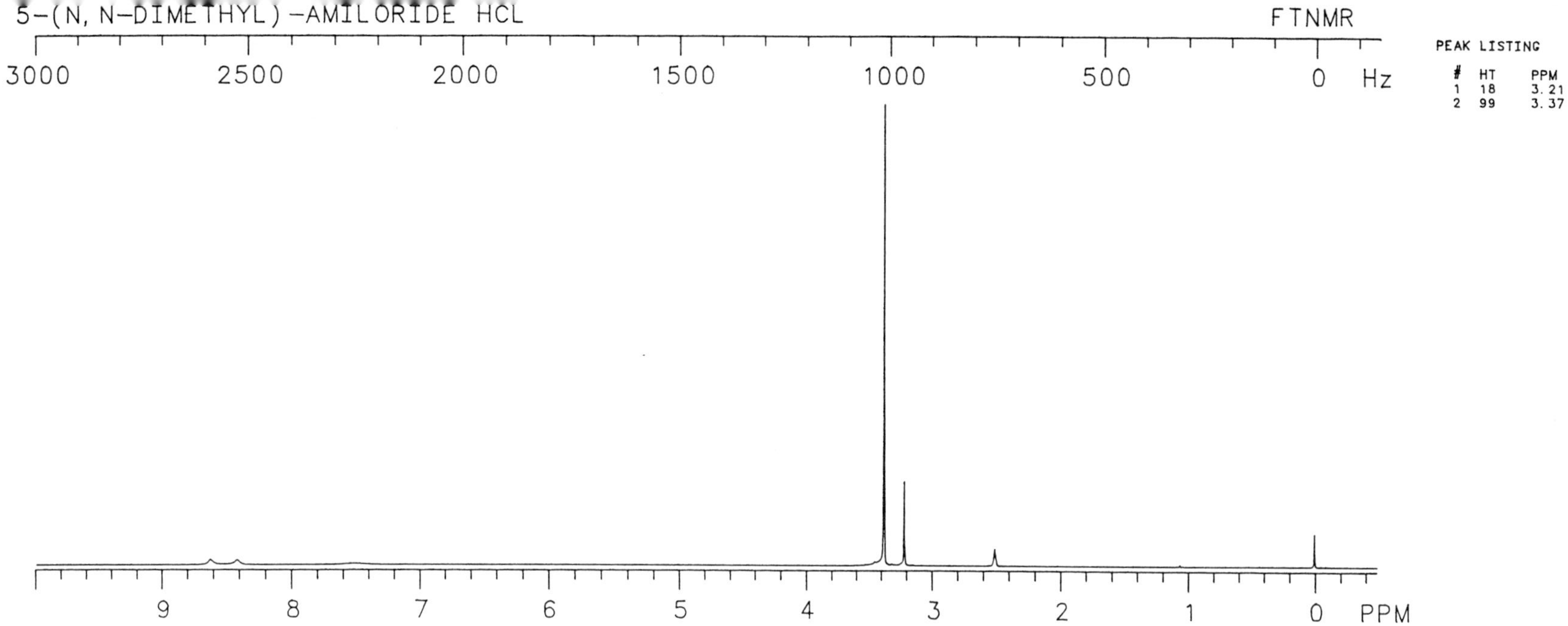
5-(N, N-DIMETHYL)-AMILORIDE HCL
FTNMR
3000
2500
2000
1500
1000
500
0 Hz
PEAK LISTING
HT PPM
1 18 3.21
2 99 3.37
9
8
7
6
5
4
3
2
1
0 PPM

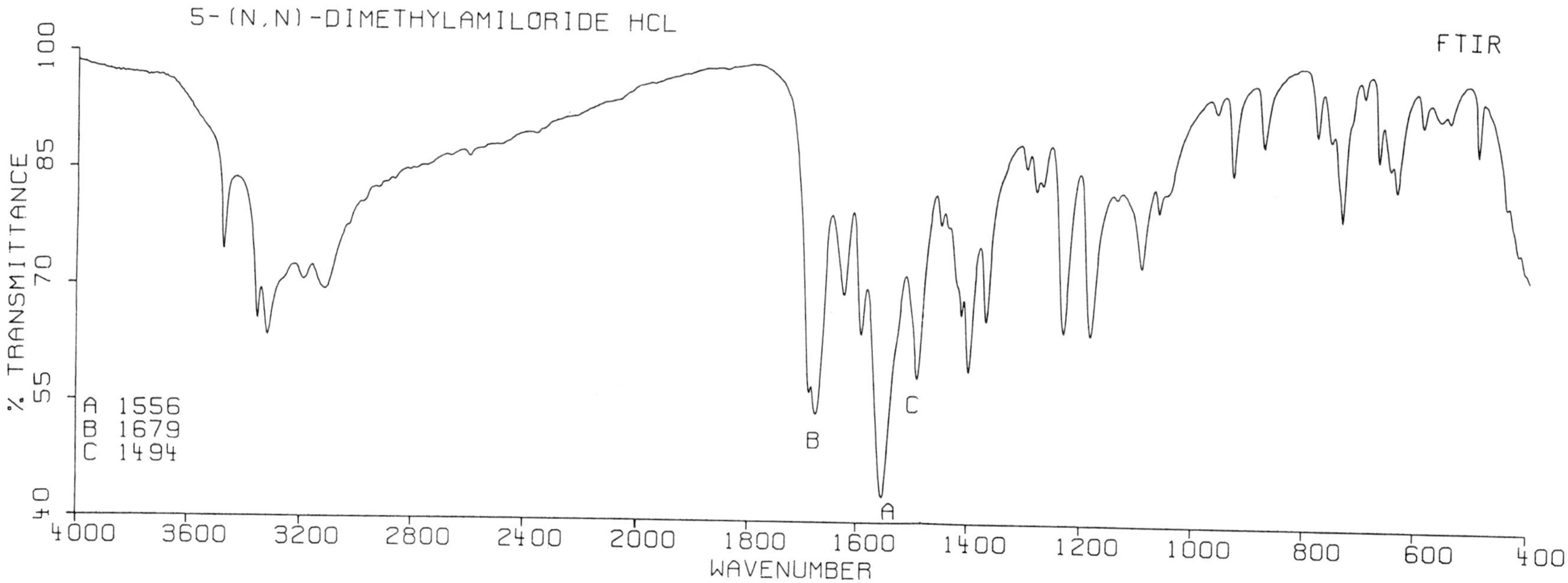
5-(N,N)-DIMETHYLAMILORIDE HCL
FTIR
% TRANSMITTANCE
100
85
70
55
40
A 1556
B 1679
C 1494
B
A
C
4000
3600
3200
2800
2400
2000
1800
1600
1400
1200
1000
800
600
400
WAVENUMBER

DIOCTYL SULFOSUCCINATE SODIUM

$C_{20}H_{37}O_7SNa$

Molecular Weight: 444.56 (444.22)

Synonyms: Sulfobutanedioic acid 1,4-bis-(2-ethylhexyl) ester sodium salt; sulfosuccinic acid 1,4-bis(2-ethylhexyl) ester sodium salt; docusate sodium
Trade Names: Aerosol OT, Alphasol OT, Colace, Comfolax, Complemix, Coprol, Dialose, Diocto, dioctylal, Diotilan, Diovac, Disonate, Docusate Sodium, Doxinate, Doxol, Dulsivac, Feet Sof-Lax, Hemaspan, Modane, Molatoc, Molcer, Molofac, Nevax, Peri-Colace, Regutol, Senokot-S, Softil, Soliwax, Solusol, Sulfimel, Valsol, Velmol, Waxsol, Yal
Use: Laxative

HPLC:
GC:

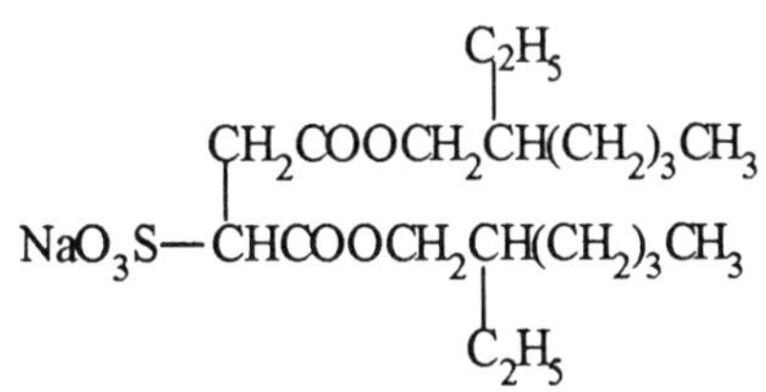

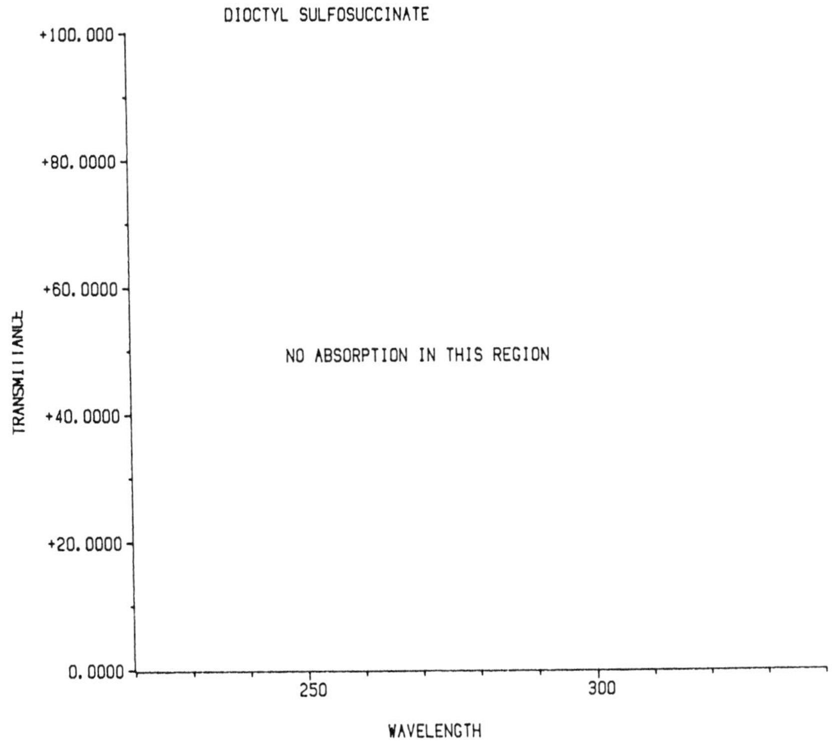

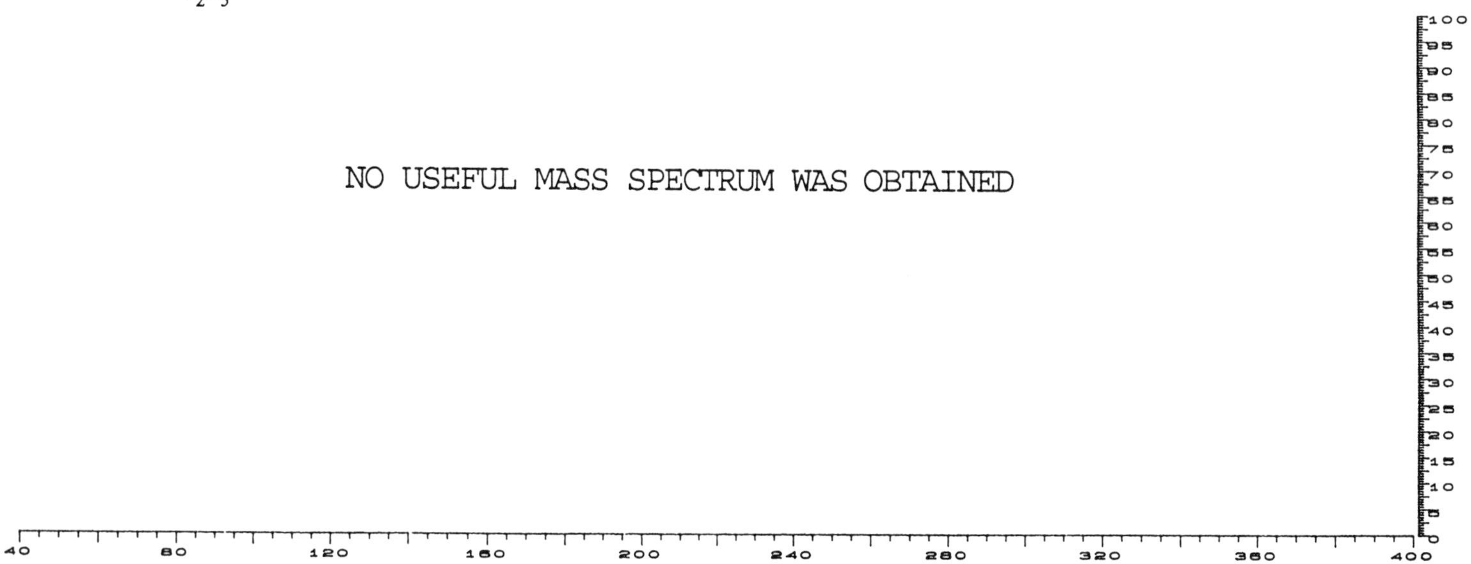

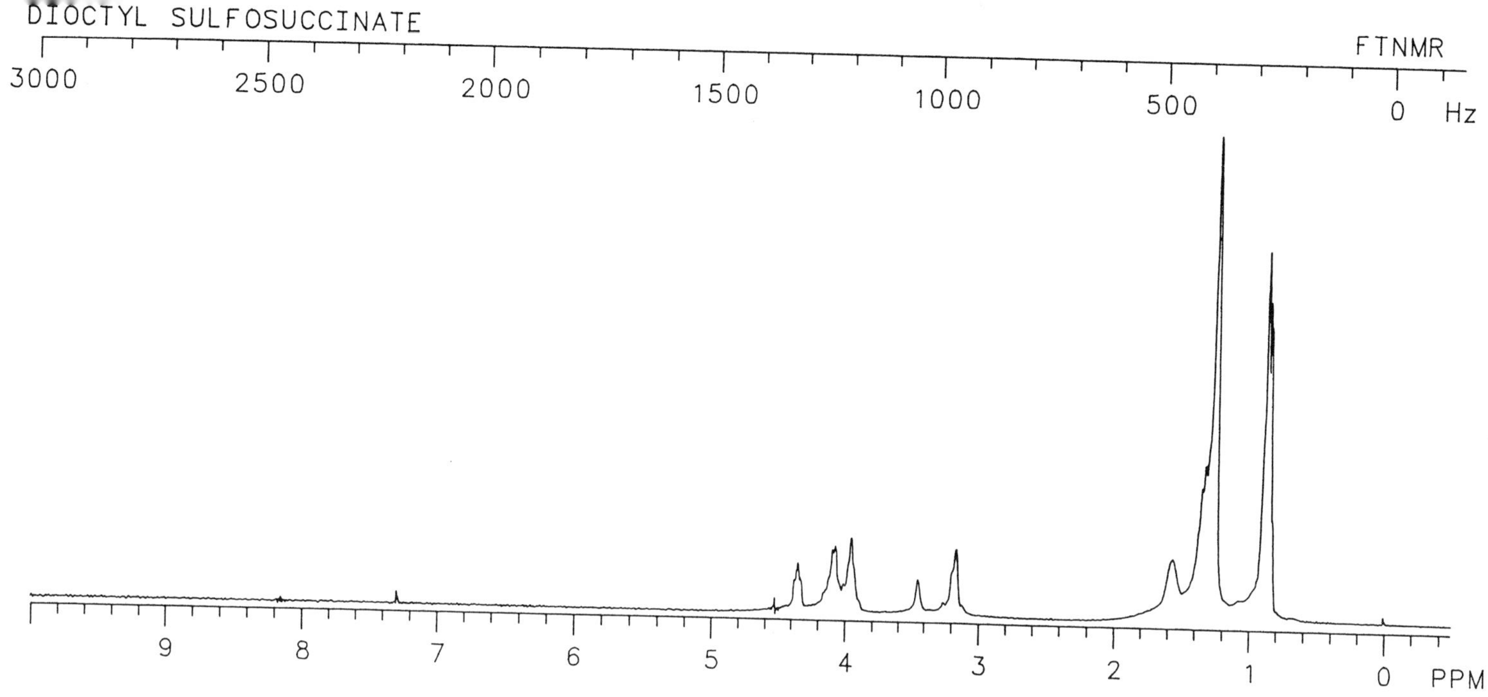

PEAK LISTING

#	HT	PPM
1	21	.83
2	60	.85
3	65	.86
4	75	.88
5	67	.89
6	99	1.26
7	31	1.32
8	27	1.34
9	15	3.94

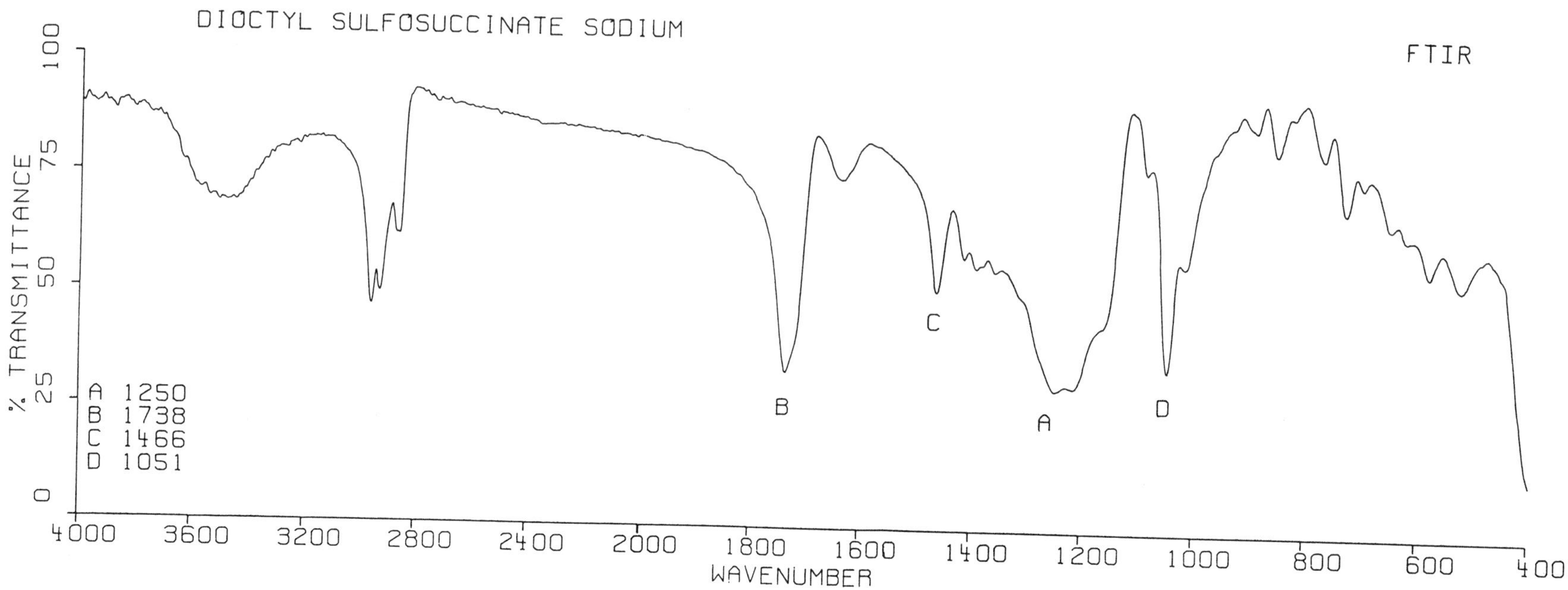

ECHINOMYCIN

$C_{51}H_{64}N_{12}O_{12}S_2$

Molecular Weight: 1101.26 (1100.42)

Synonyms: Quinomycin A
Trade Names:
Use: Antibiotic

HPLC:
GC:

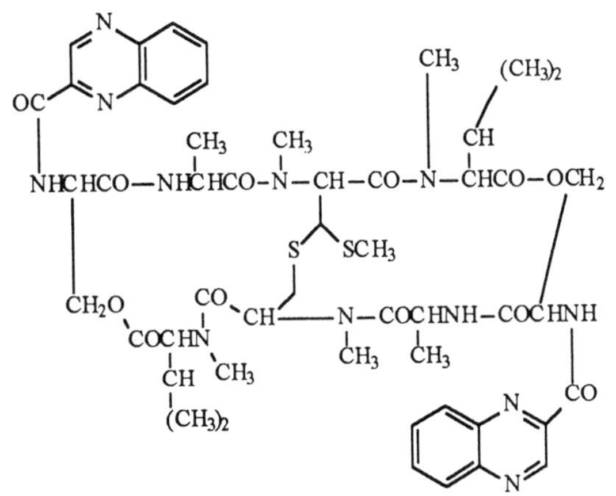

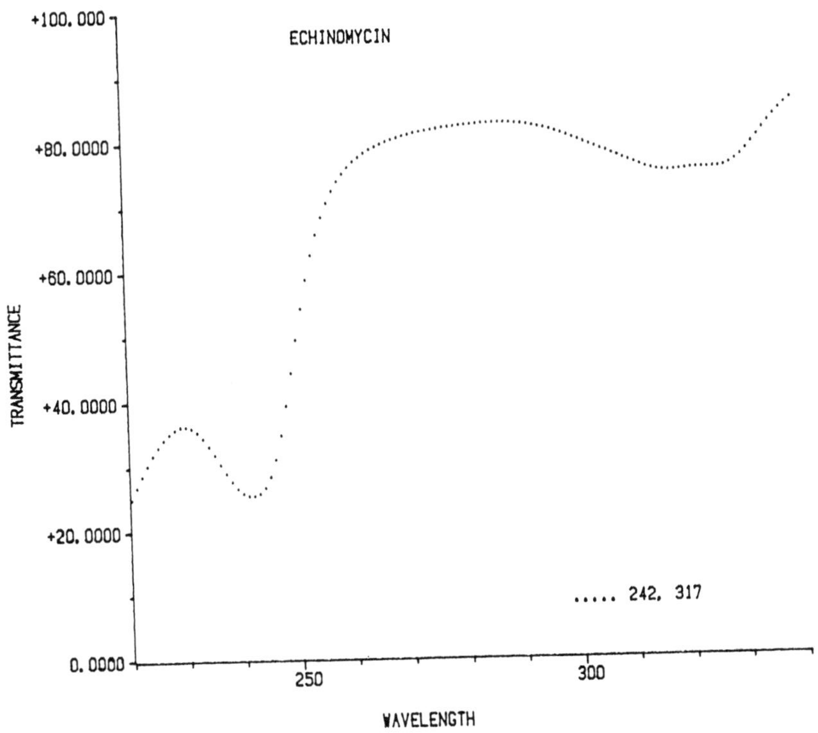

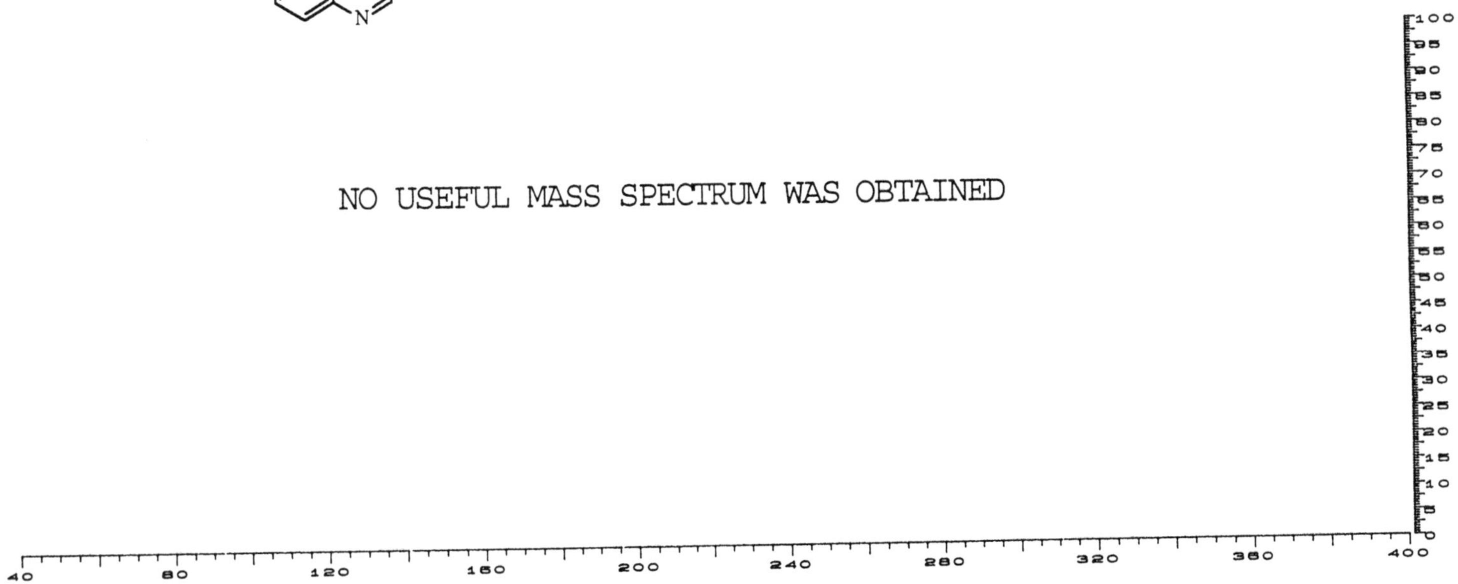

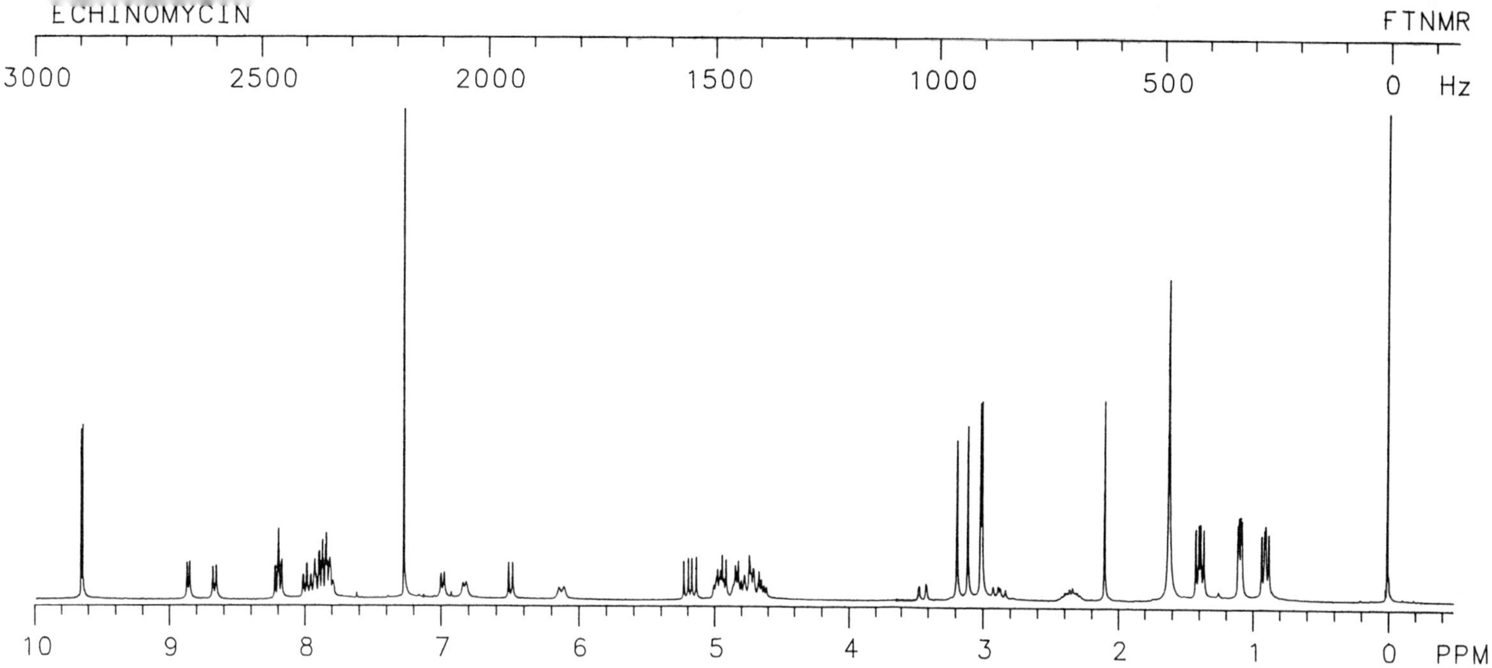

PEAK LISTING

#	HT	PPM
1	111	.00
2	15	.90
3	16	1.07
4	17	1.08
5	17	1.09
6	15	1.10
7	15	1.38
8	15	1.39
9	65	1.61
10	40	2.09
11	40	2.99
12	40	3.01
13	35	3.10
14	32	3.18
15	166	7.26
16	35	9.63
17	34	9.64

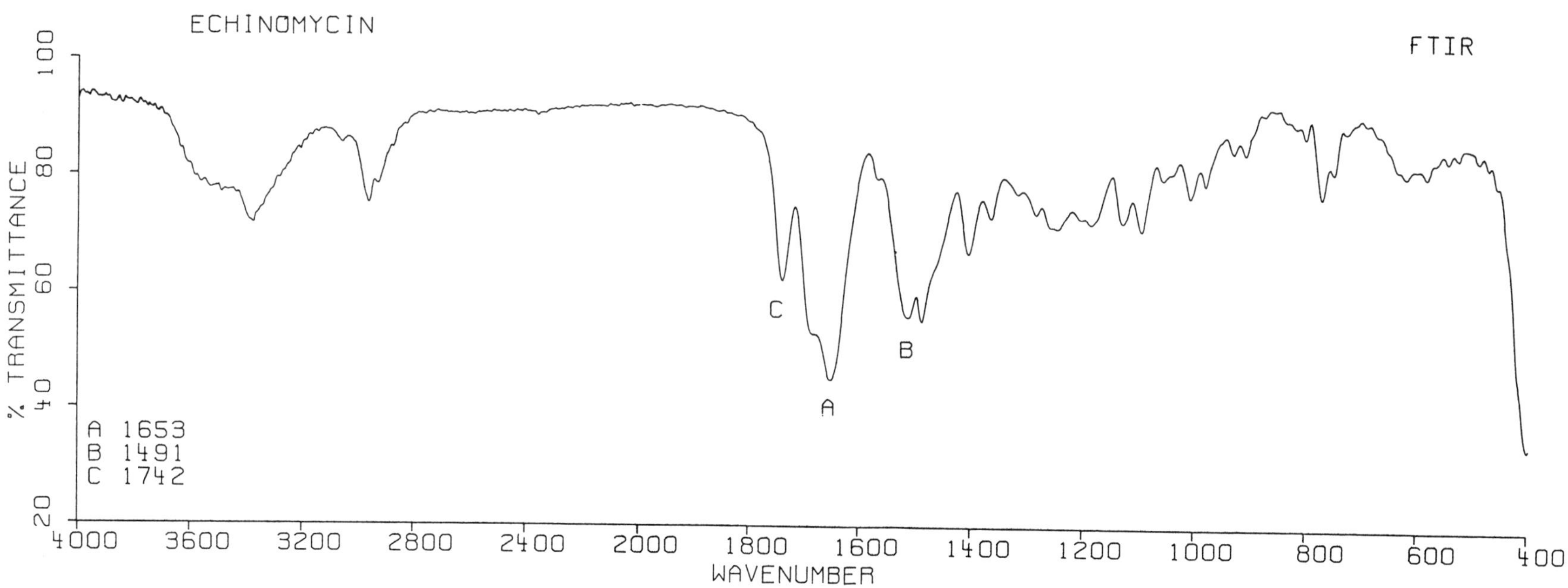

ENALAPRIL MALEATE

$C_{20}H_{28}N_2O_5 \cdot C_4H_4O_4$

Molecular Weight: 492.53 (492.21)

Synonyms: (S)-1-[N-[1-(Ethoxycarbonyl)-3-phenylpropyl]-L-alanyl]-L-proline, (Z)-2-butenedionate salt (1:1)
Trade Names: Vaseretic, Vasotec
Use: Enzyme inhibitor

HPLC:
GC:

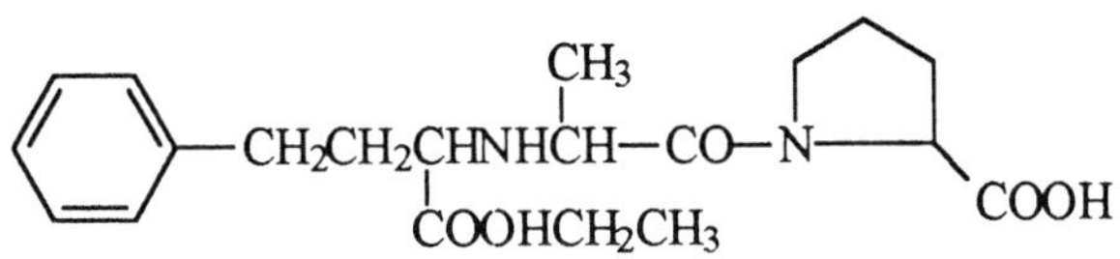

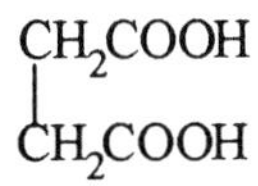

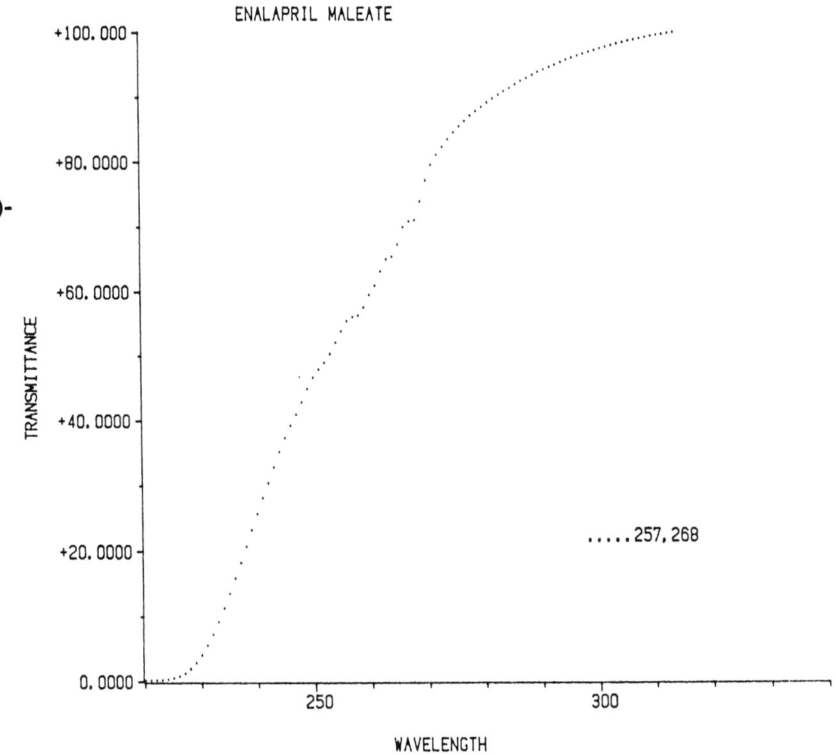

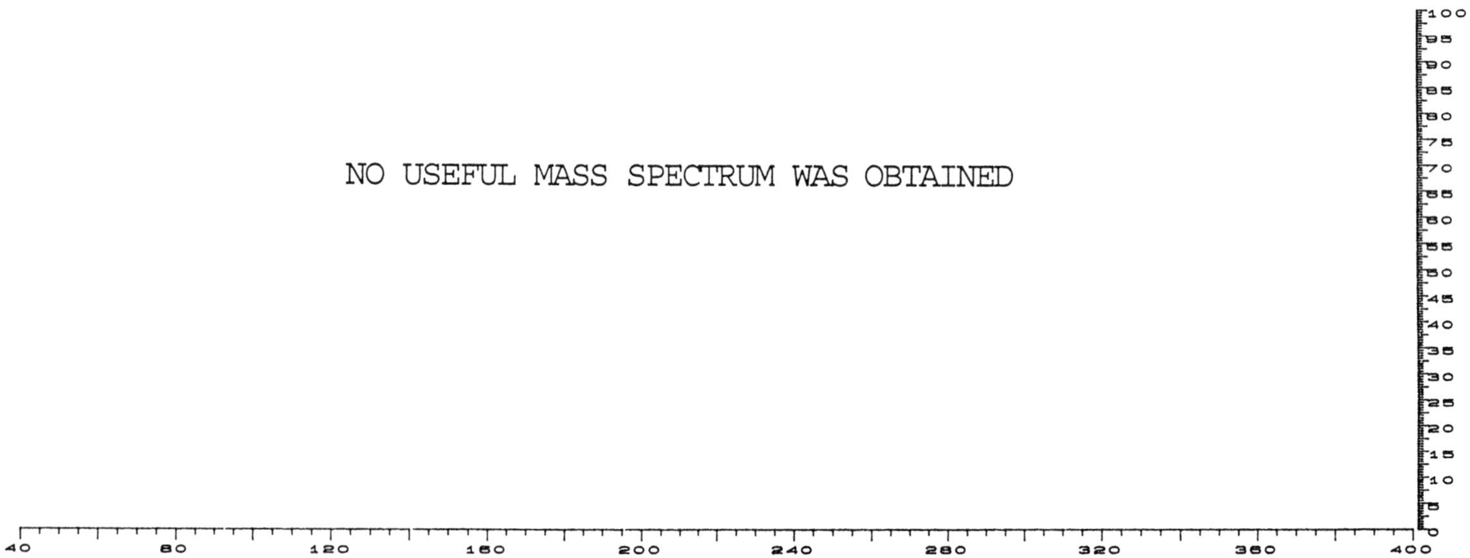

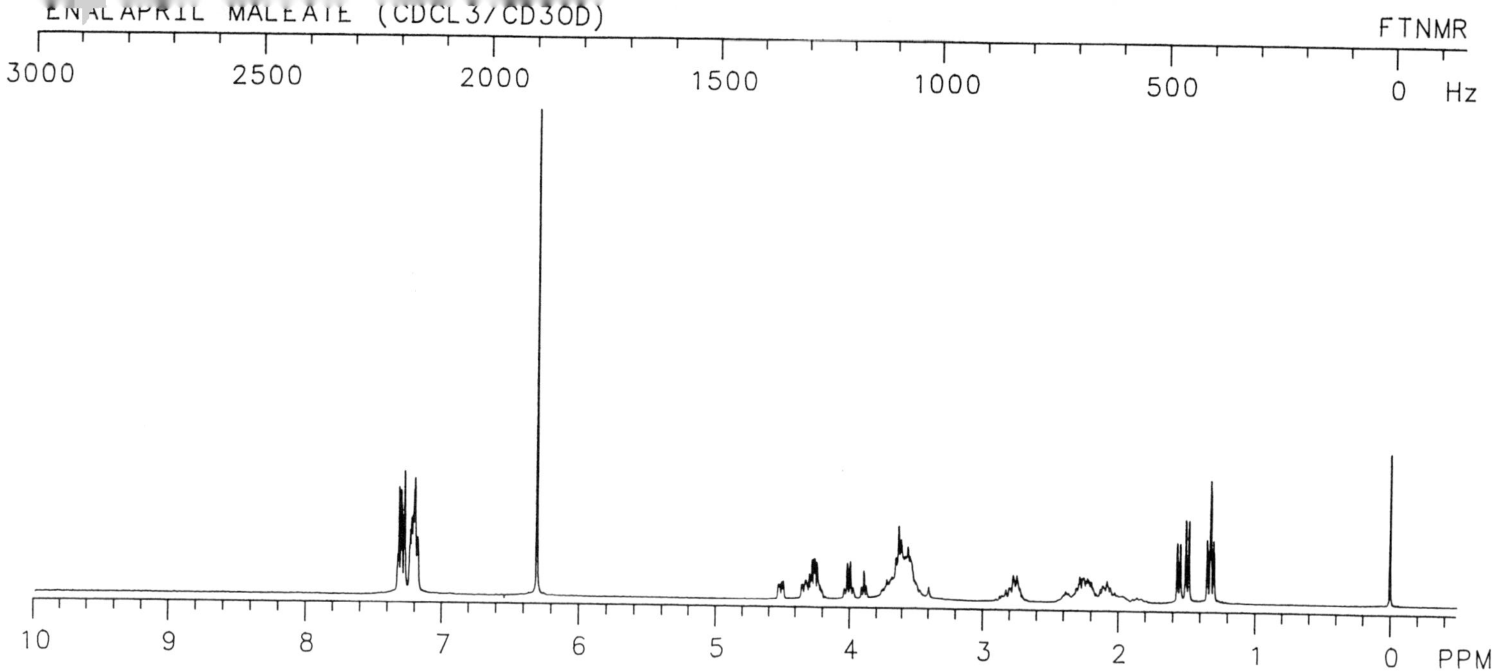

PEAK LISTING

#	HT	PPM
1	31	.00
2	23	1.31
3	25	1.31
4	17	1.47
5	17	1.49
6	15	3.62
7	99	6.30
8	23	7.19
9	21	7.19
10	16	7.20
11	15	7.21
12	25	7.27
13	21	7.29
14	21	7.31

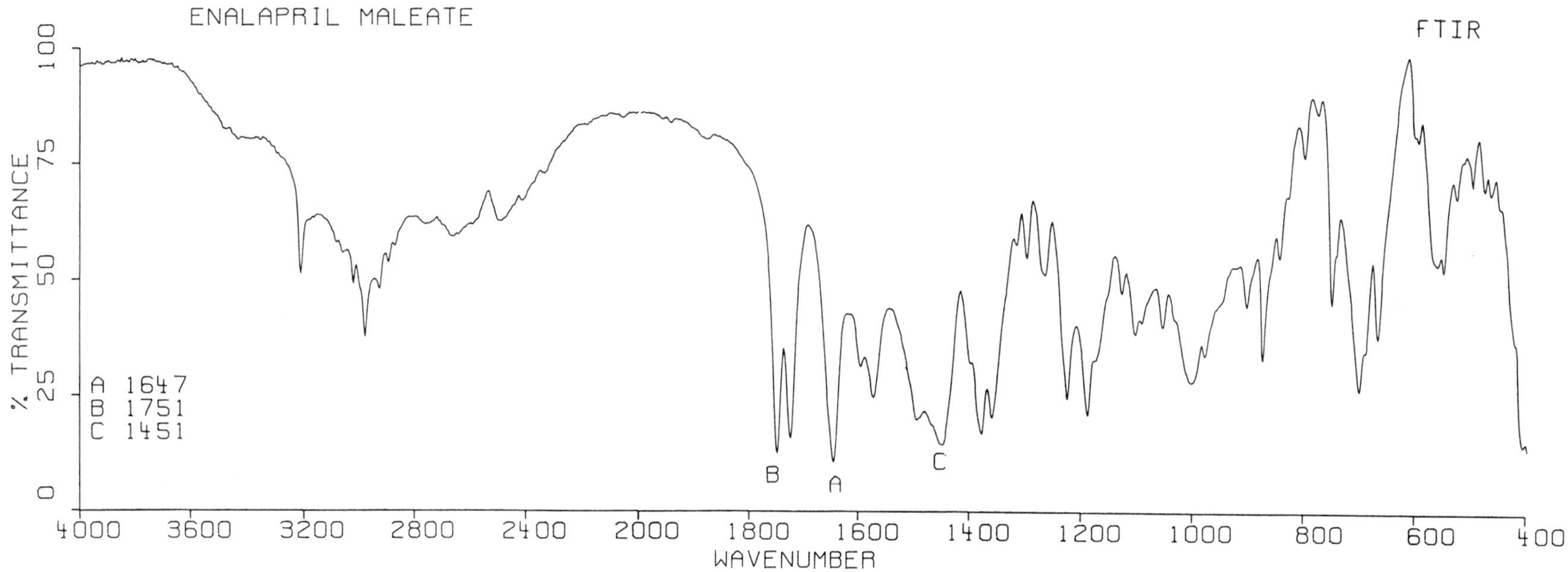

ENOXACIN

$C_{15}H_{17}FN_4O_3$

Molecular Weight: 320.32 (320.13)

Synonyms: 1-Ethyl-6-fluoro-1,4-dihydro-4-oxo-7-(1-piperazinyl)-1,8-naphthyridine-3-carboxylic acid
Trade Names: Bactidan, Comprecin, Enoram, Enoxen, Flumark, Gyramid, Lentor, Penetrex
Use: Antibacterial

HPLC:
GC:

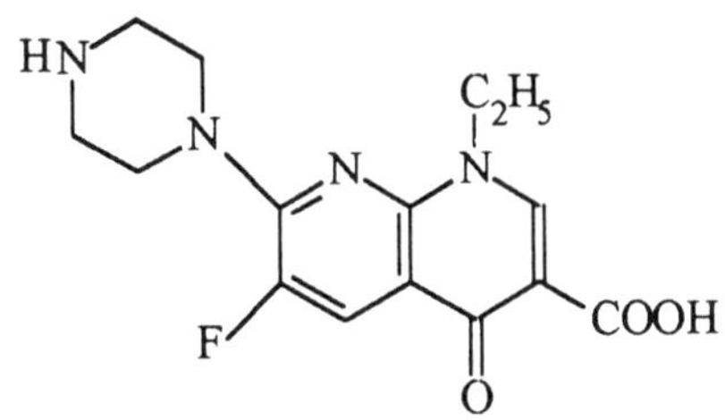

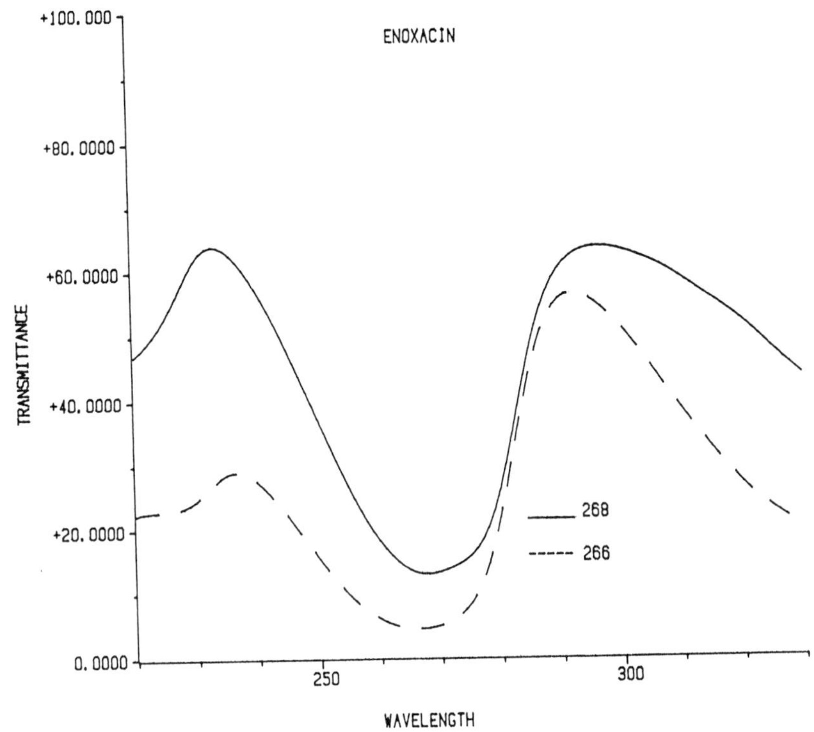

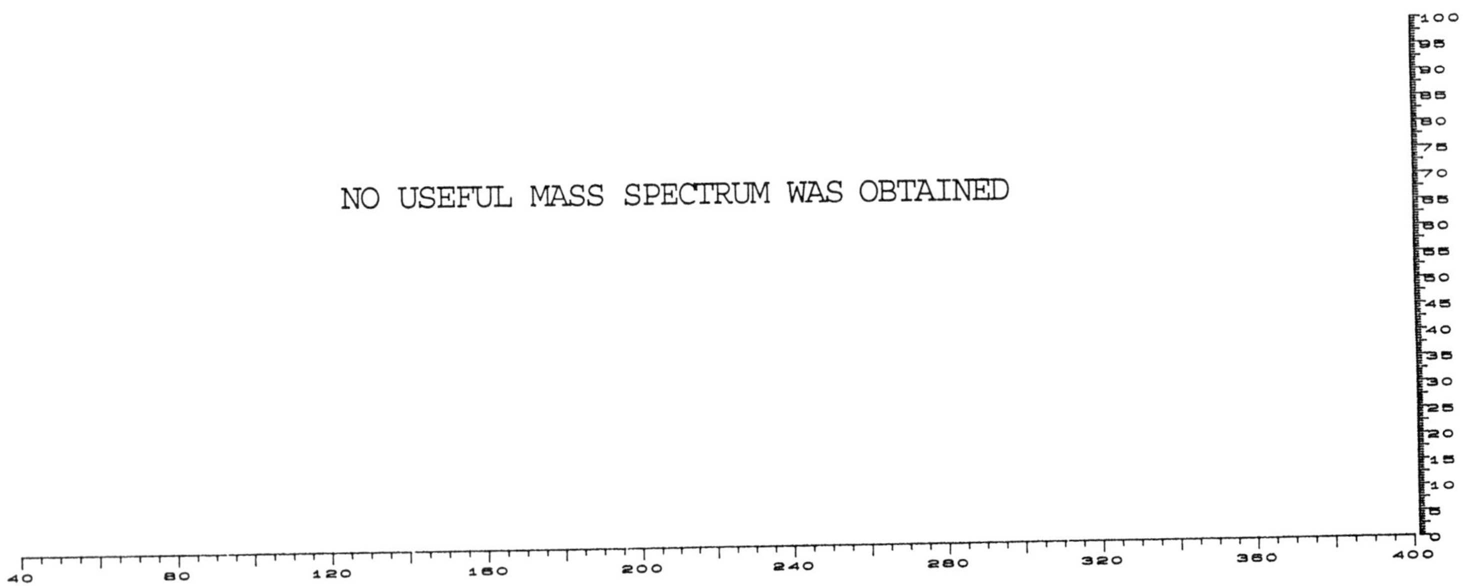

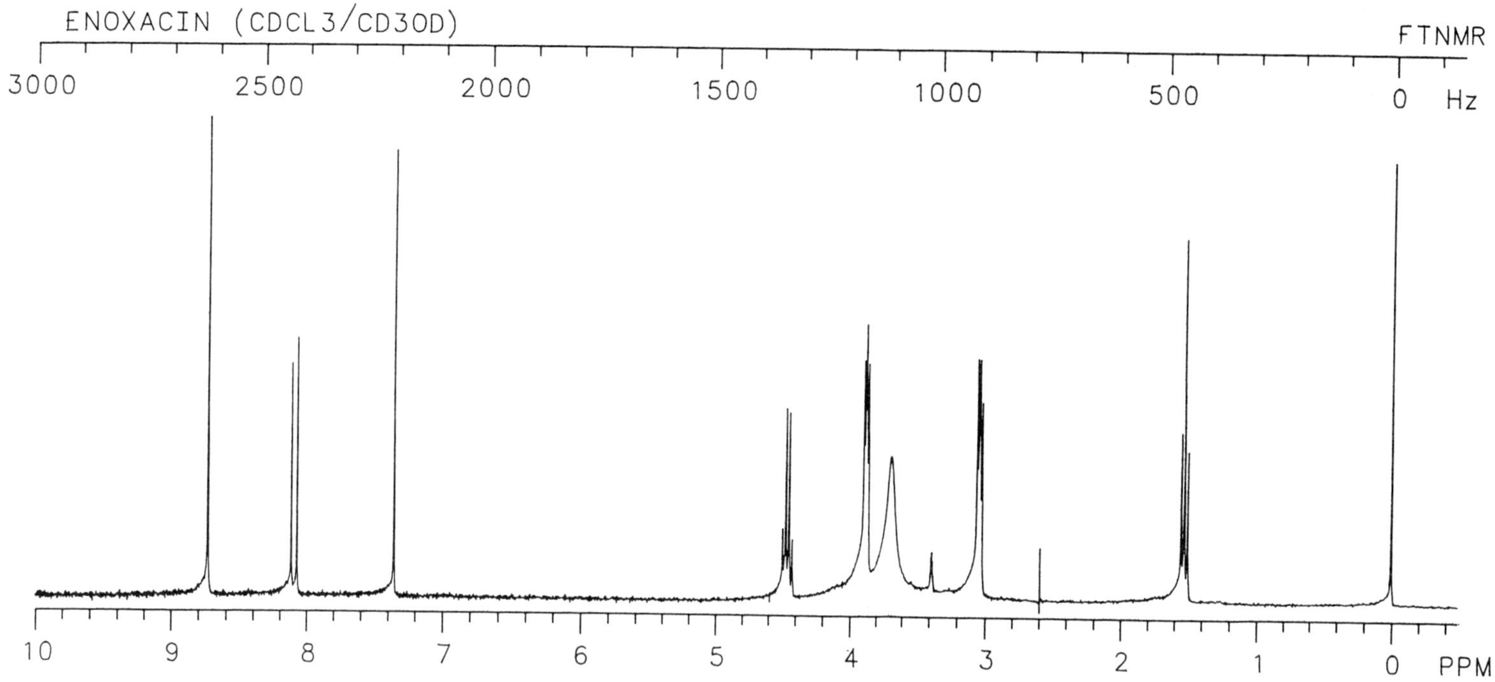

PEAK LISTING

#	HT	PPM
1	90	.00
2	74	1.52
3	34	1.54
4	40	3.02
5	49	3.04
6	49	3.05
7	48	3.86
8	56	3.88
9	49	3.89
10	38	4.44
11	39	4.46
12	91	7.35
13	52	8.07
14	47	8.11
15	97	8.72

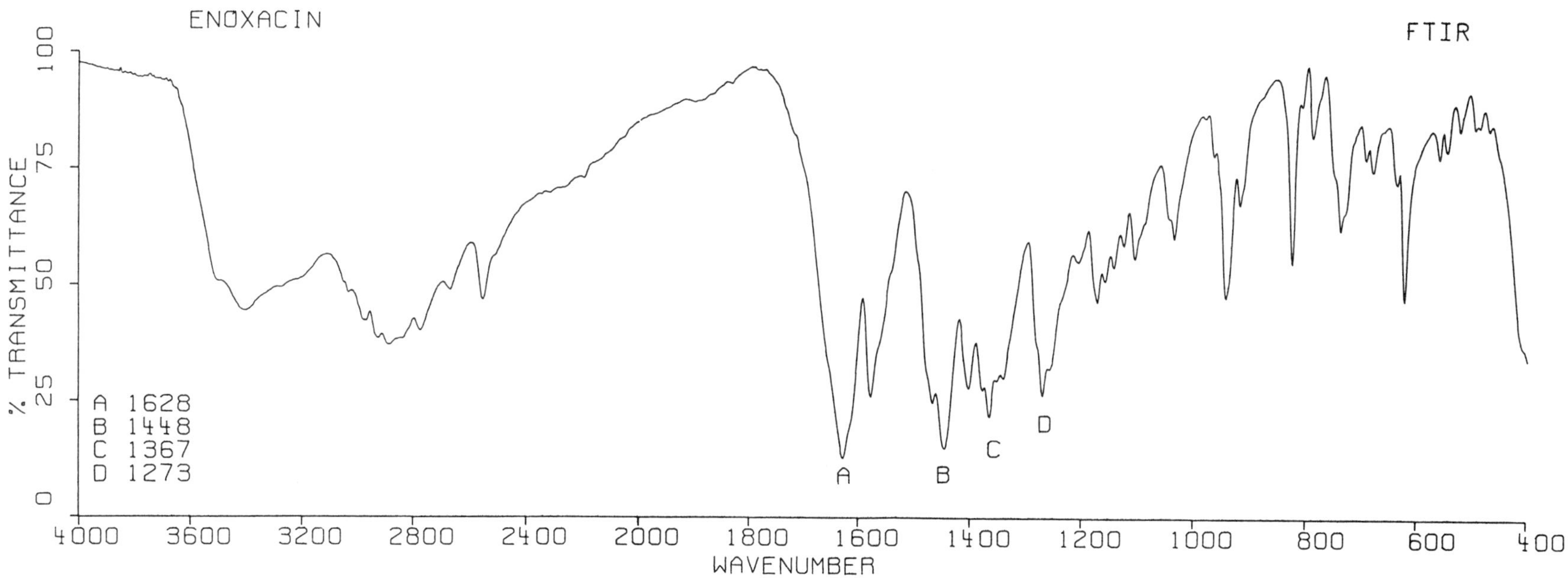

EPINEPHRINE METHYL ETHER

$C_{10}H_{15}NO_3$

Molecular Weight: 197.23 (197.11)

Synonyms: 4-[1-Methoxy-2-(methylamino)ethyl]-1,2-benzenediol
Trade Names:
Use:

HPLC:
GC:

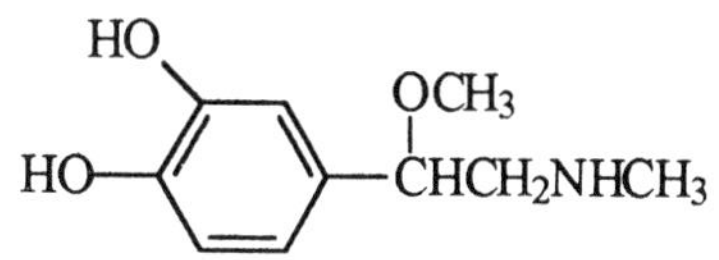

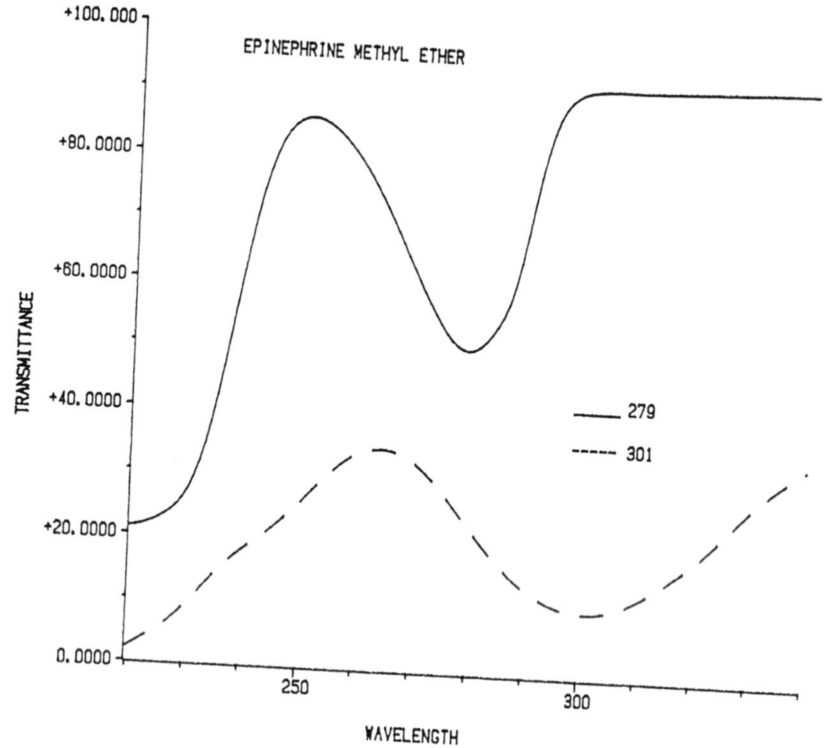

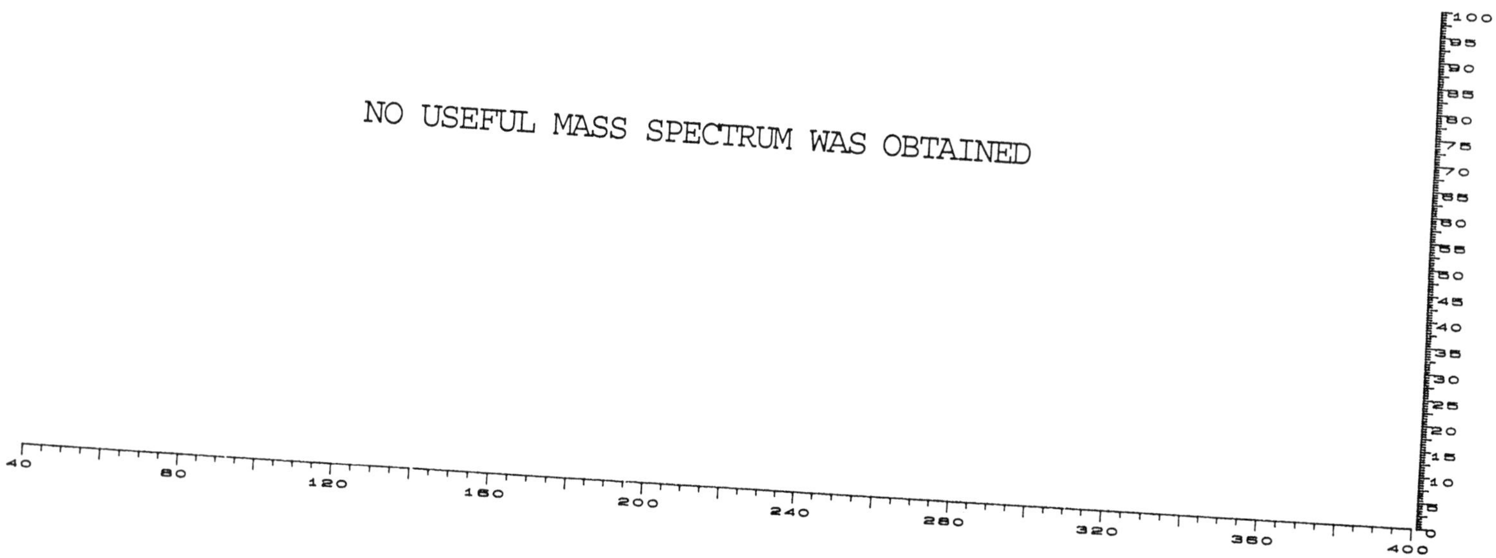

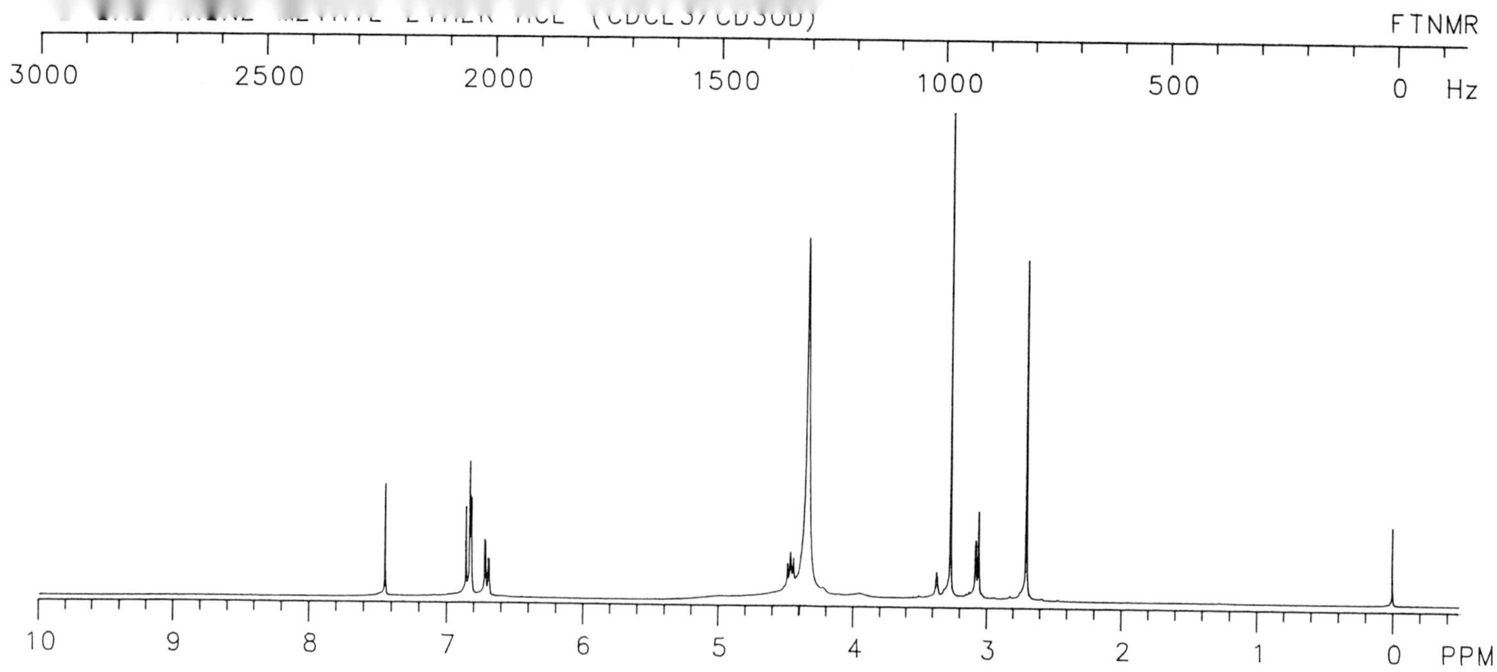

PEAK LISTING

#	HT	PPM
1	16	.00
2	69	2.69
3	18	3.05
4	12	3.07
5	11	3.07
6	99	3.26
7	74	4.33
8	8	4.43
9	9	4.45
10	10	4.46
11	11	6.70
12	11	6.71
13	20	6.80
14	20	6.81
15	27	6.82
16	18	6.84
17	23	7.44

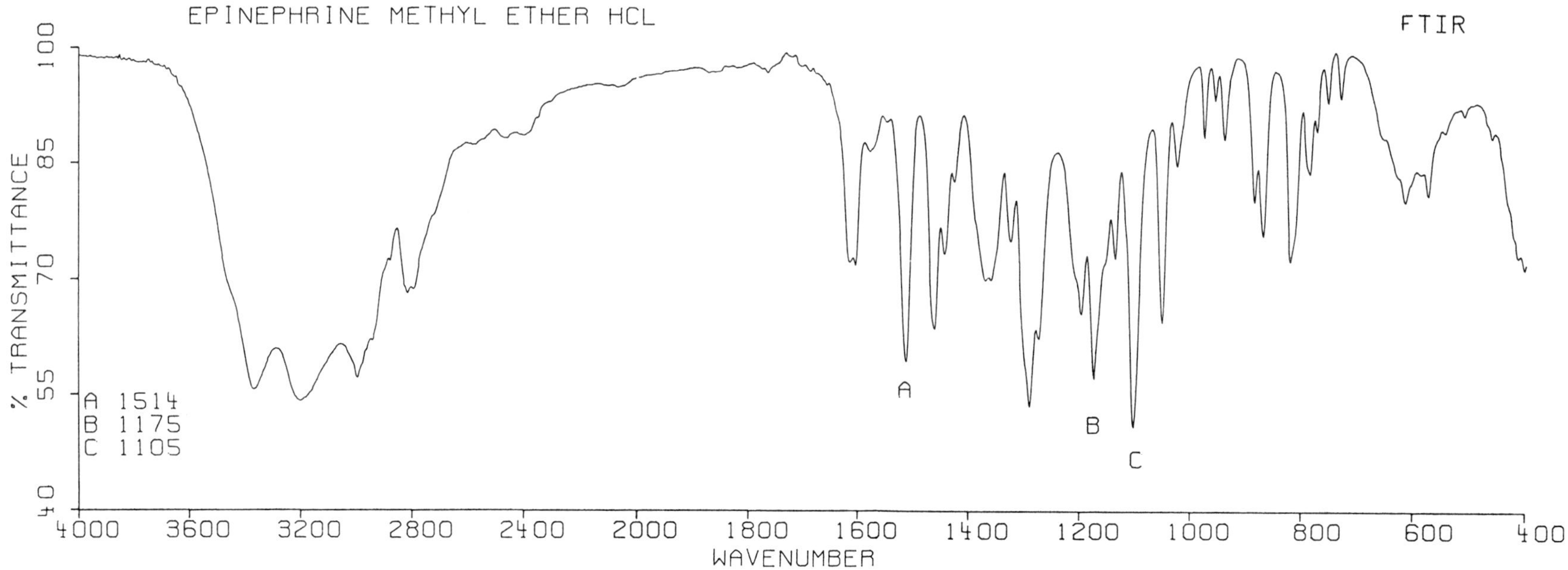

ETODOLAC

$C_{17}H_{21}NO_3$

Molecular Weight: 287.36 (287.15)

Synonyms: 1,8-Diethyl-1,3,4,9-tetrahydropyano-[3,4-b]indole-1-acetic acid, etodolic acid
Trade Names: Edolan, Lodine, Ramodar, Ultradol, Zedolac
Use: Anti-inflammatory

HPLC: Methanol: 2.8
GC: 2333; 250°

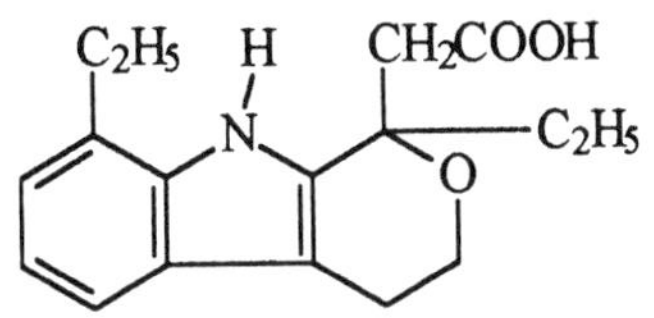

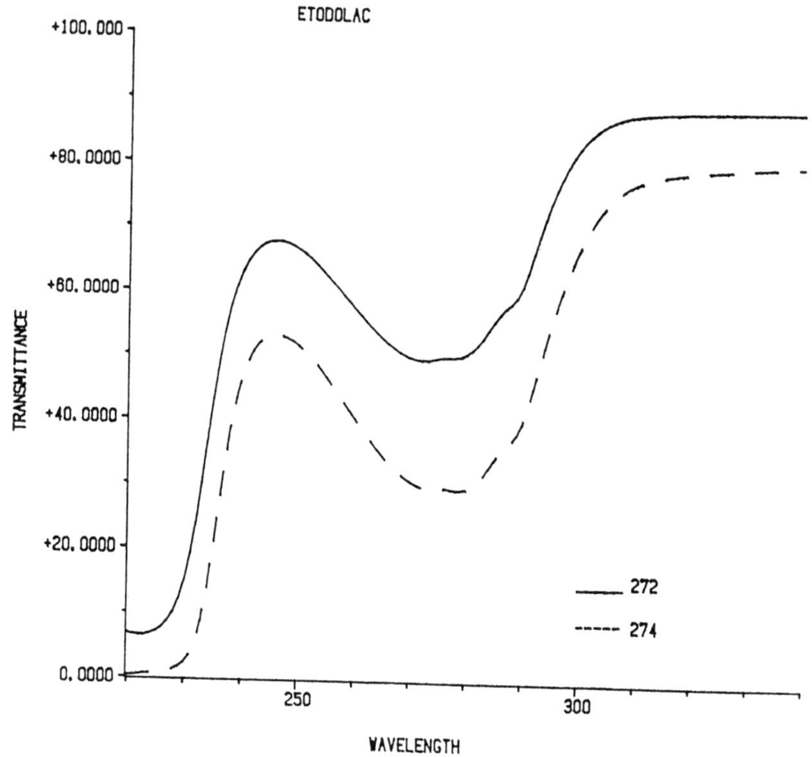

ETODOLAC

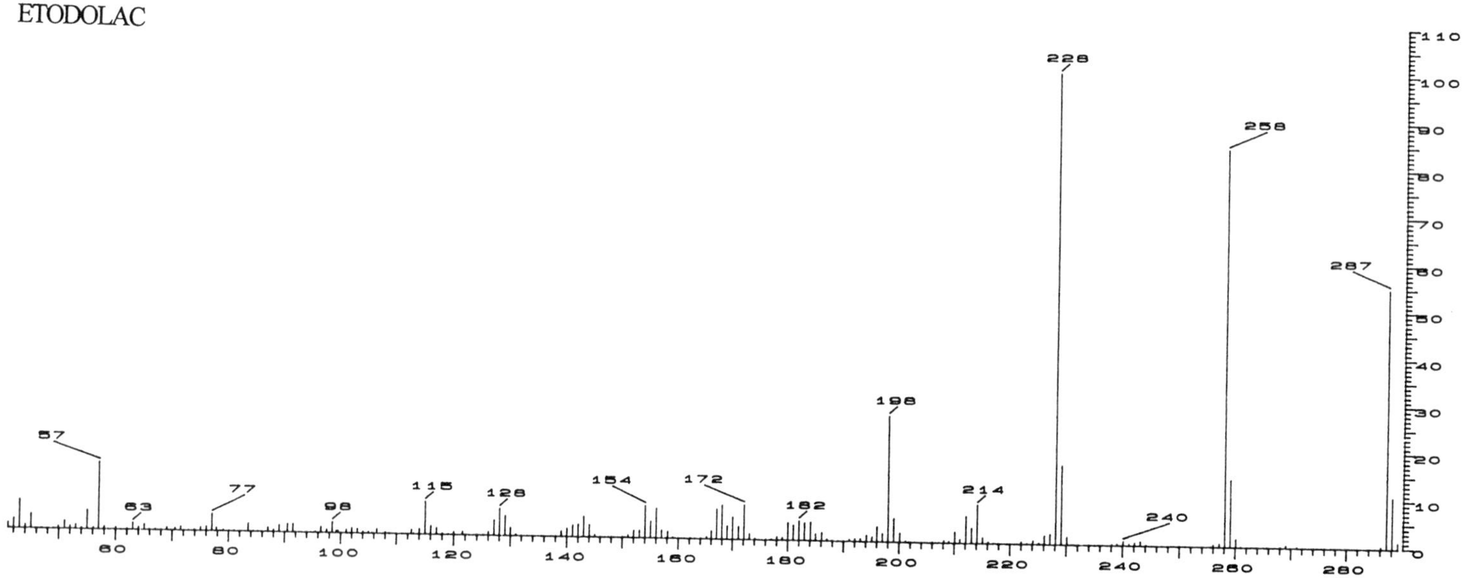

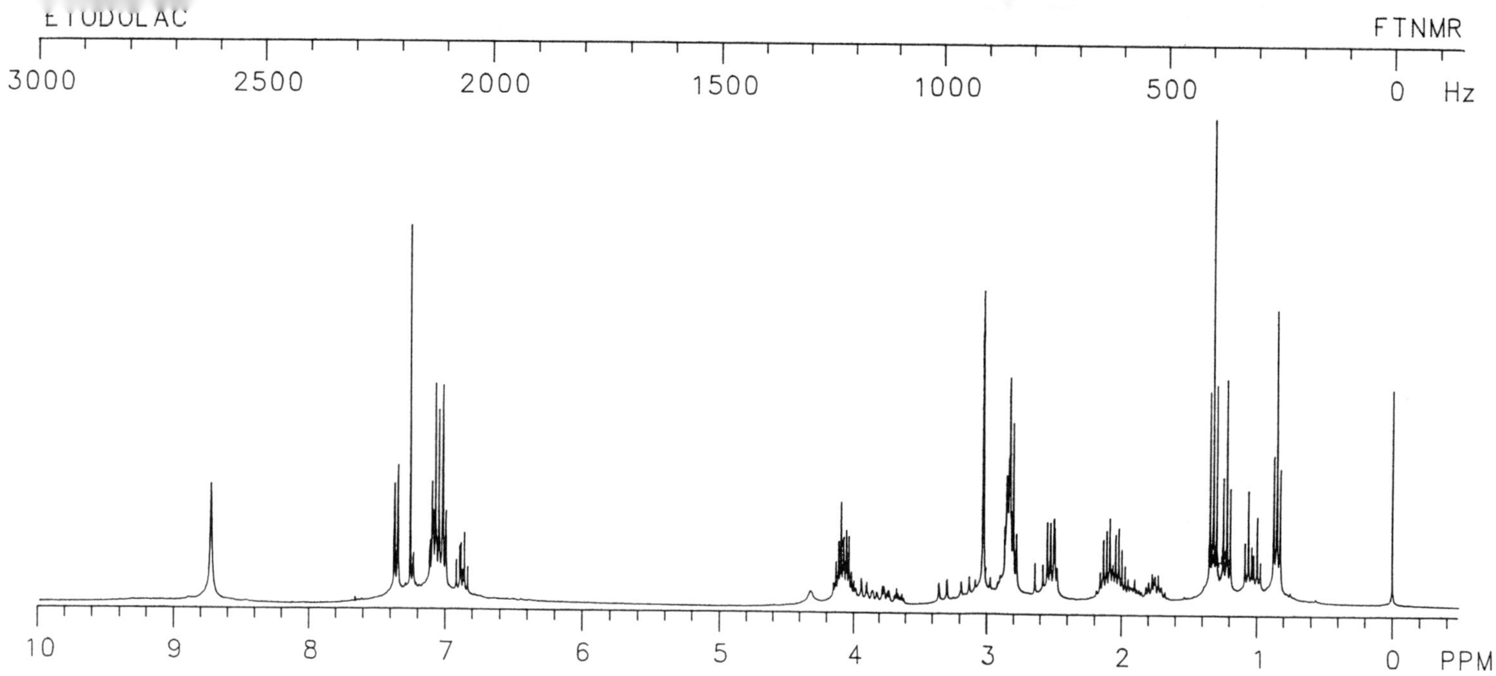

PEAK LISTING

#	HT	PPM
1	44	.00
2	28	.82
3	60	.85
4	31	.87
5	24	1.06
6	24	1.19
7	46	1.21
8	26	1.24
9	45	1.29
10	99	1.31
11	44	1.34
12	37	2.80
13	46	2.82
14	30	2.83
15	25	2.84
16	24	2.84
17	26	2.85
18	64	3.02
19	55	3.02
20	44	7.01
21	39	7.04
22	44	7.07
23	24	7.09
24	76	7.25
25	27	7.34
26	24	7.37
27	24	8.71

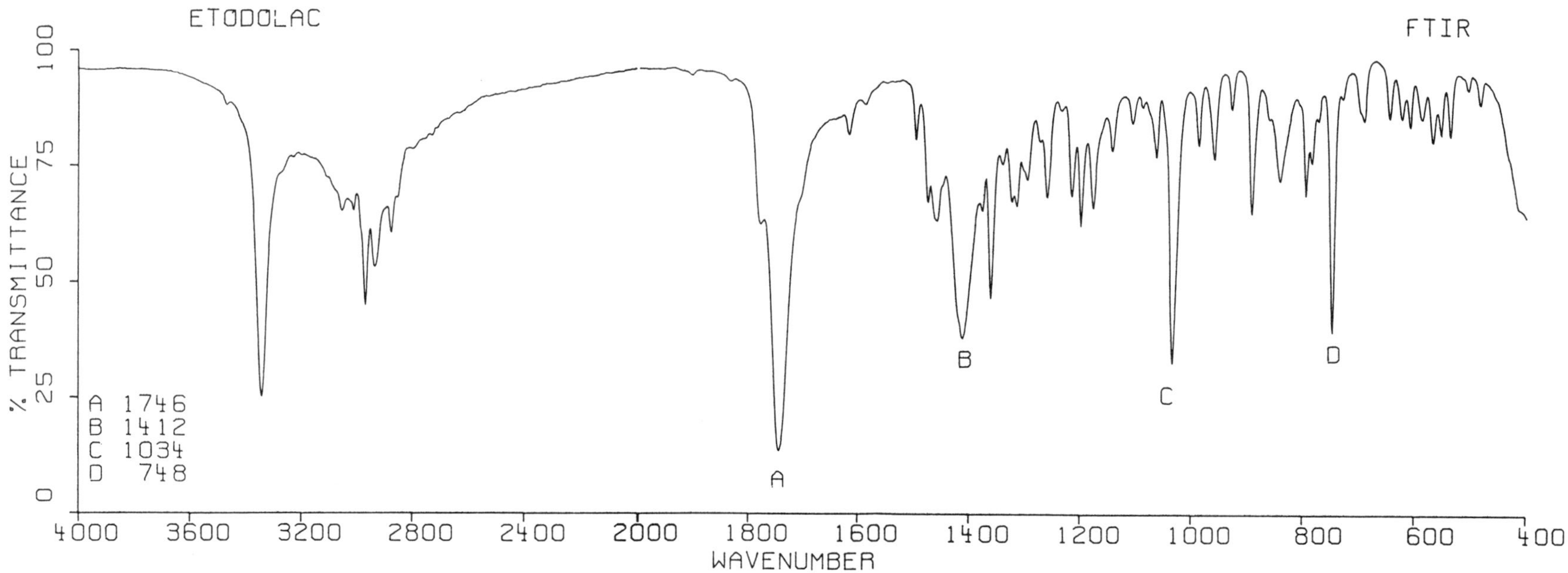

FAMCICLOVIR

$C_{14}H_{19}N_5O_4$

Molecular Weight: 321.36 (321.14)

Synonyms: 2-[2-(2-Amino-9H-purin-9-yl)ethyl]-1,3-propanediol diacetate
Trade Names: Famvir
Use: Antiviral agent

HPLC:
GC: 2552; 280°

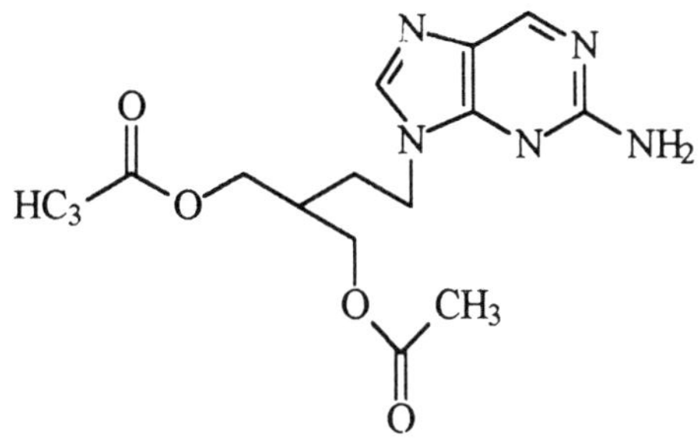

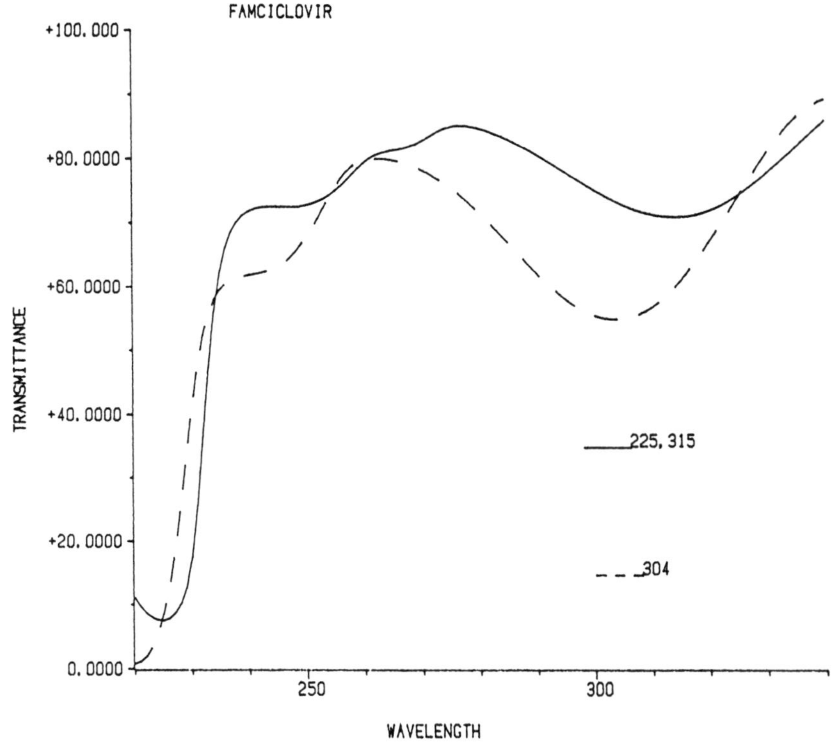

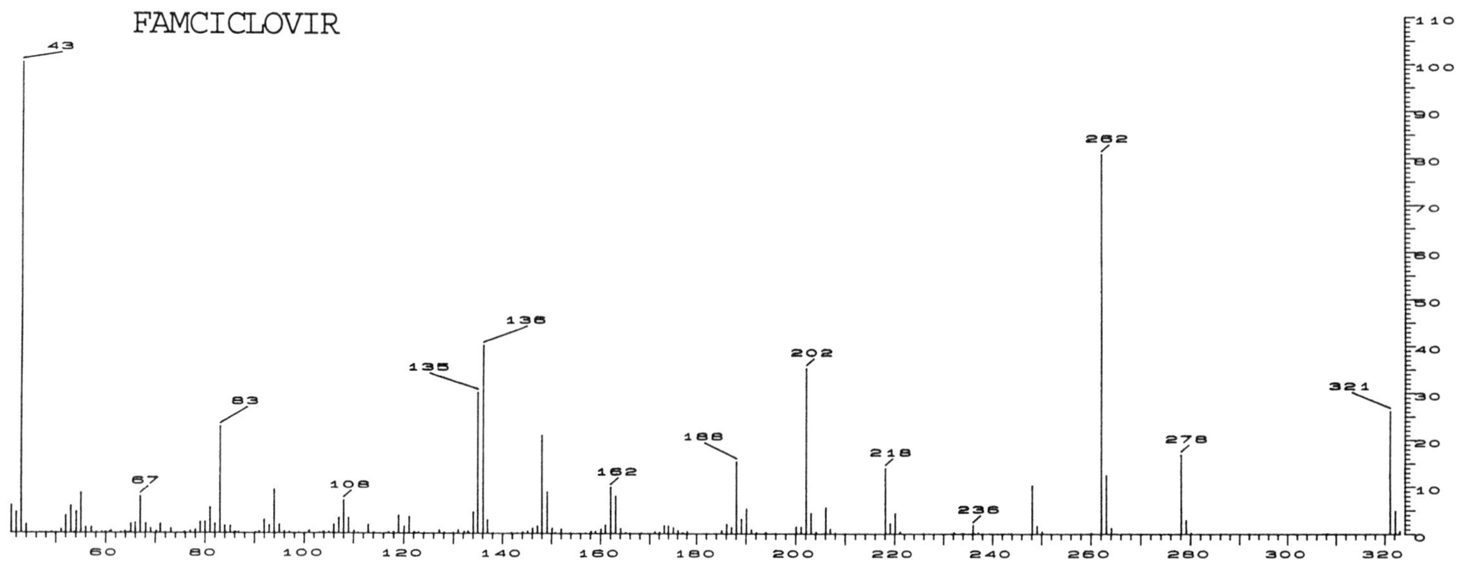

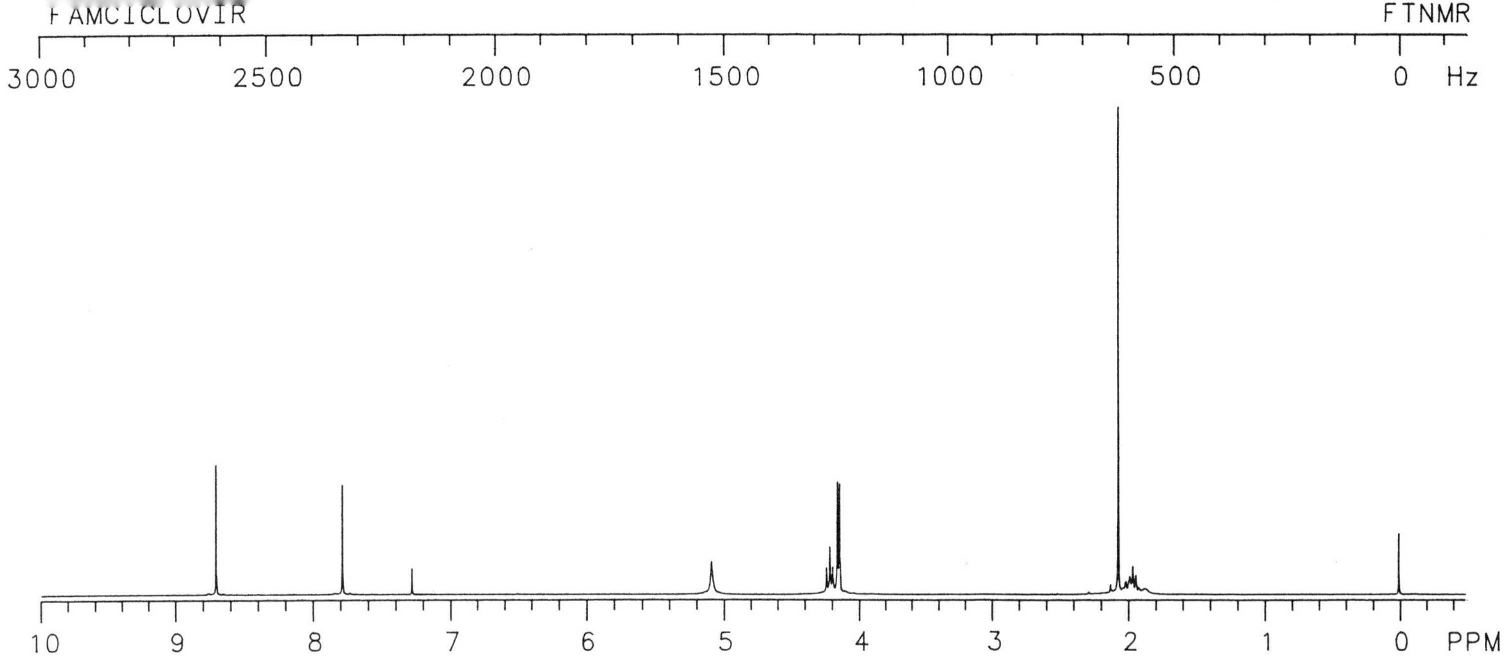

PEAK LISTING

#	HT	PPM
1	99	2.07
2	22	4.13
3	22	4.15
4	22	7.77
5	26	8.70

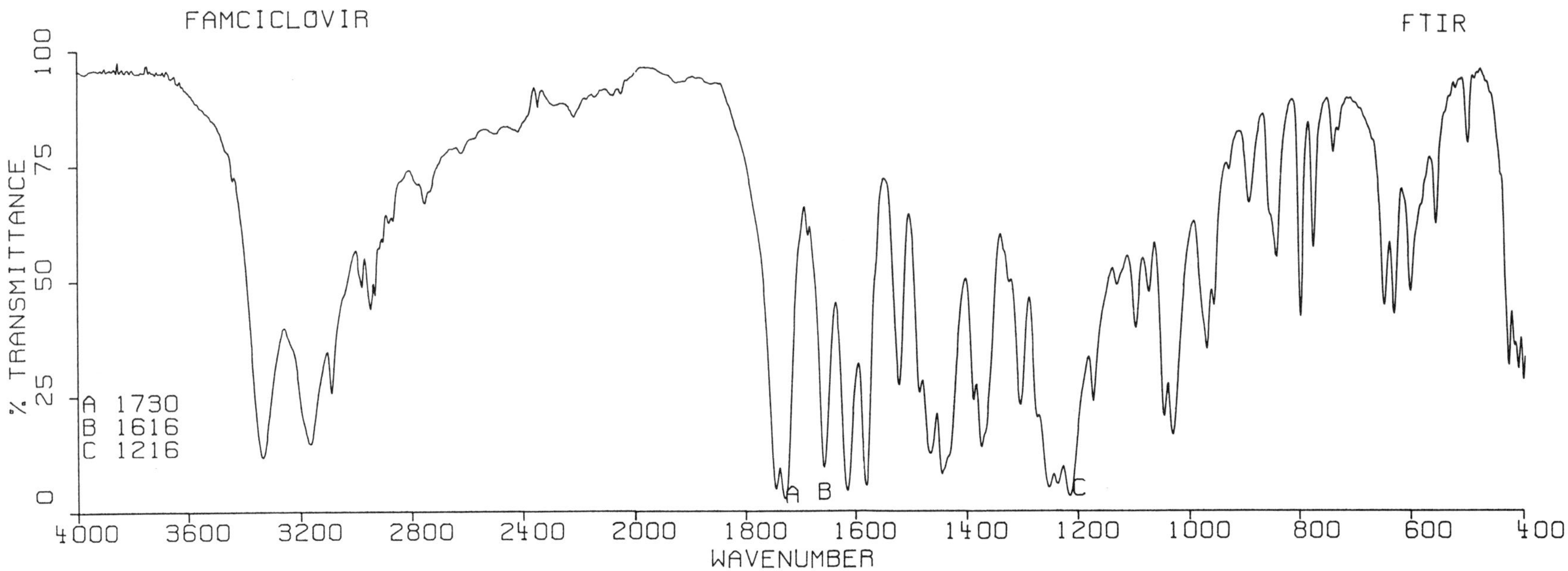

FELBAMATE

$C_{11}H_{14}N_2O_4$

Molecular Weight: 238.24 (238.10)

Synonyms: 2-Phenyl-1,3-propanediol dicarbamate
Trade Names: Felbatol
Use: Antiepiletic

HPLC:
GC:

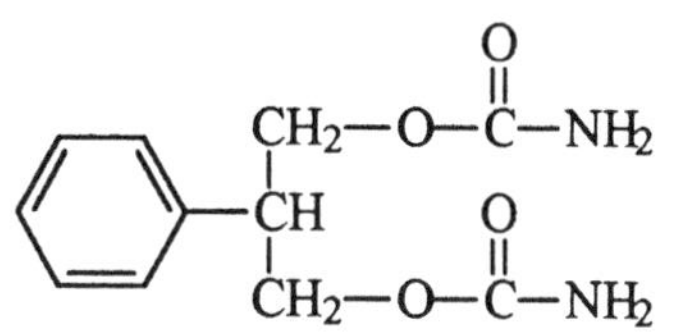

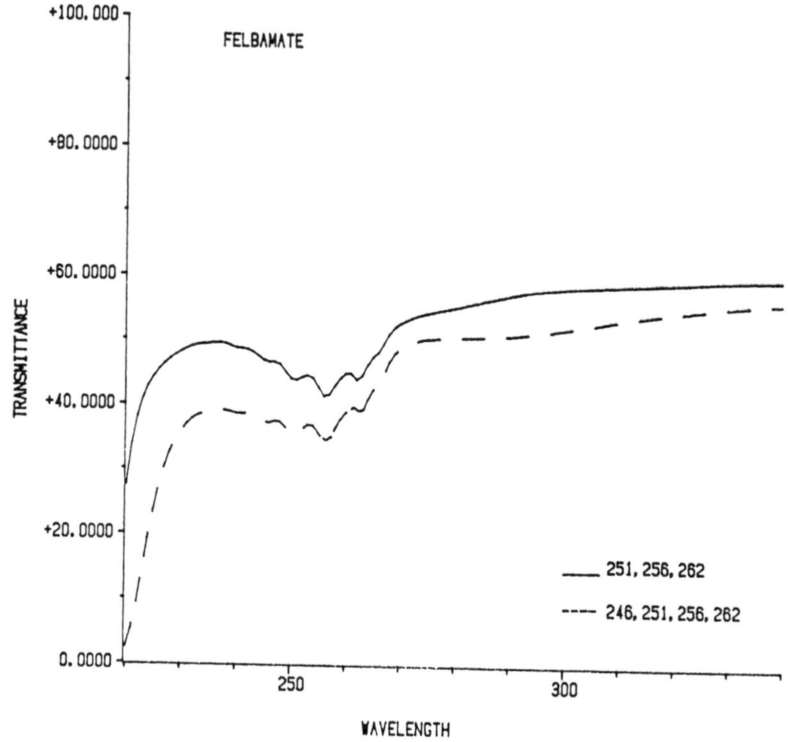

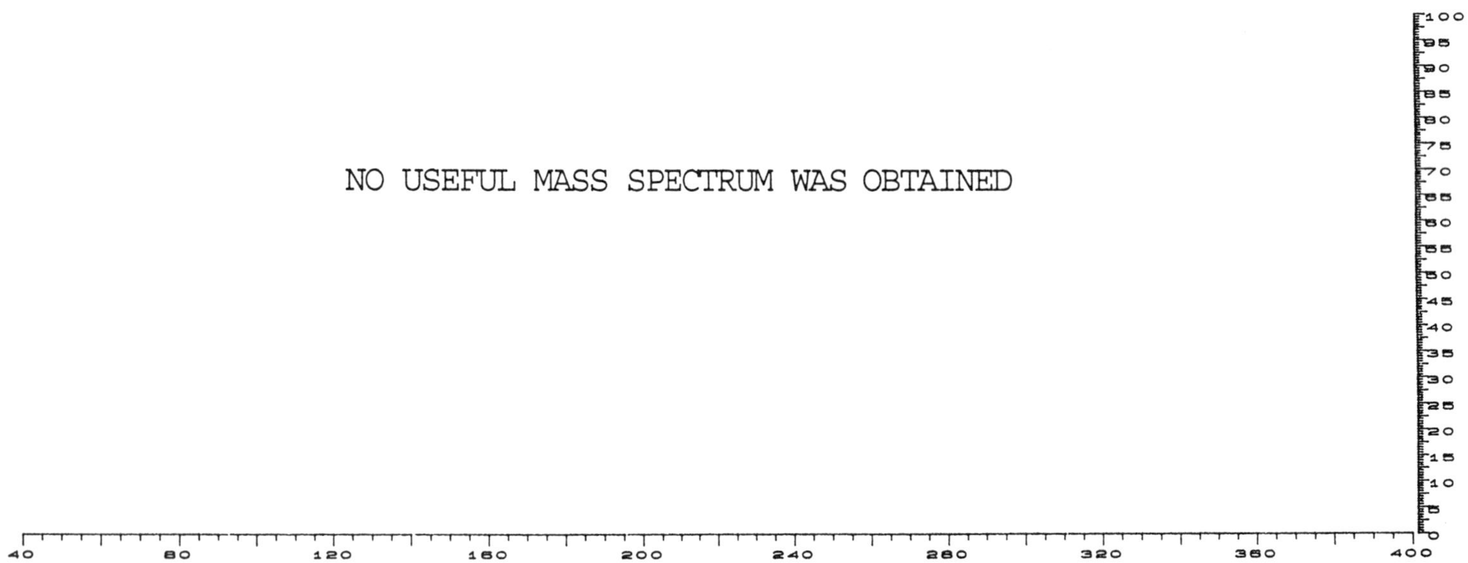

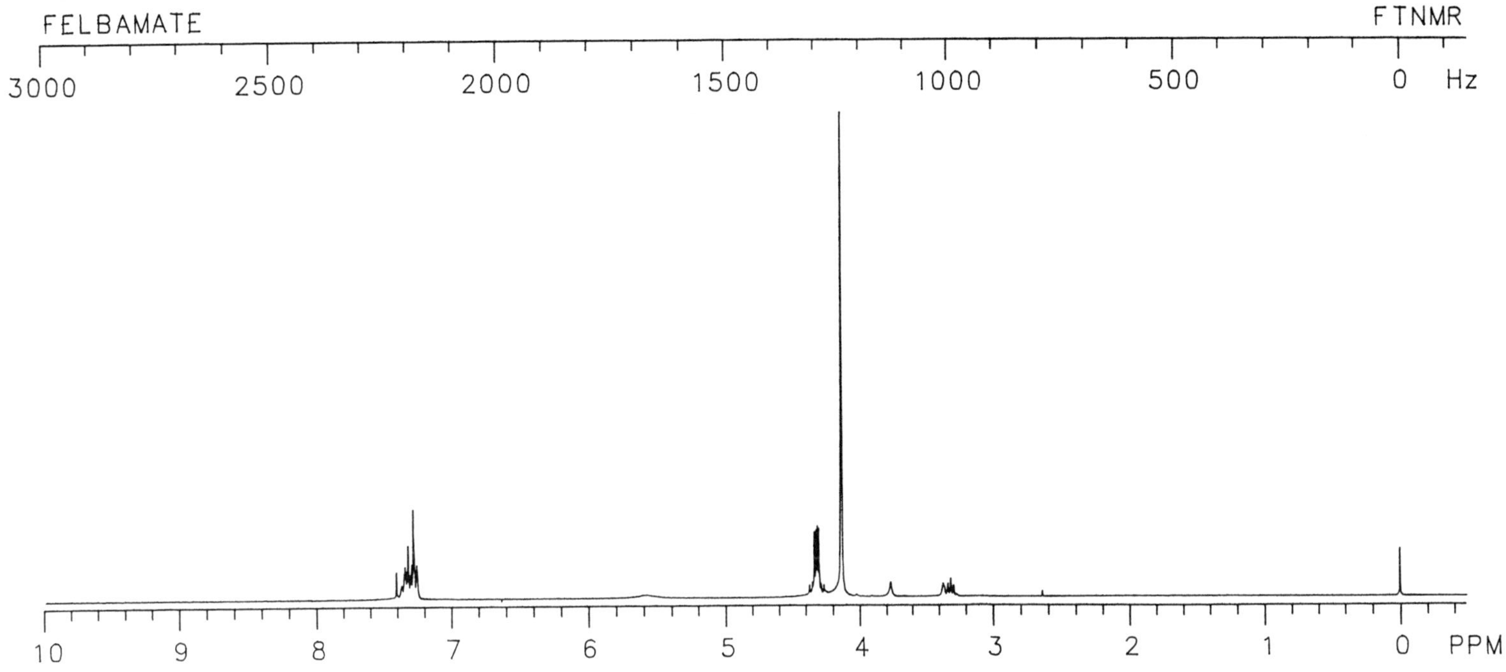

PEAK LISTING

#	HT	PPM
1	99	4.12
2	14	4.29
3	14	4.30
4	13	4.31
5	13	4.32
6	18	7.27

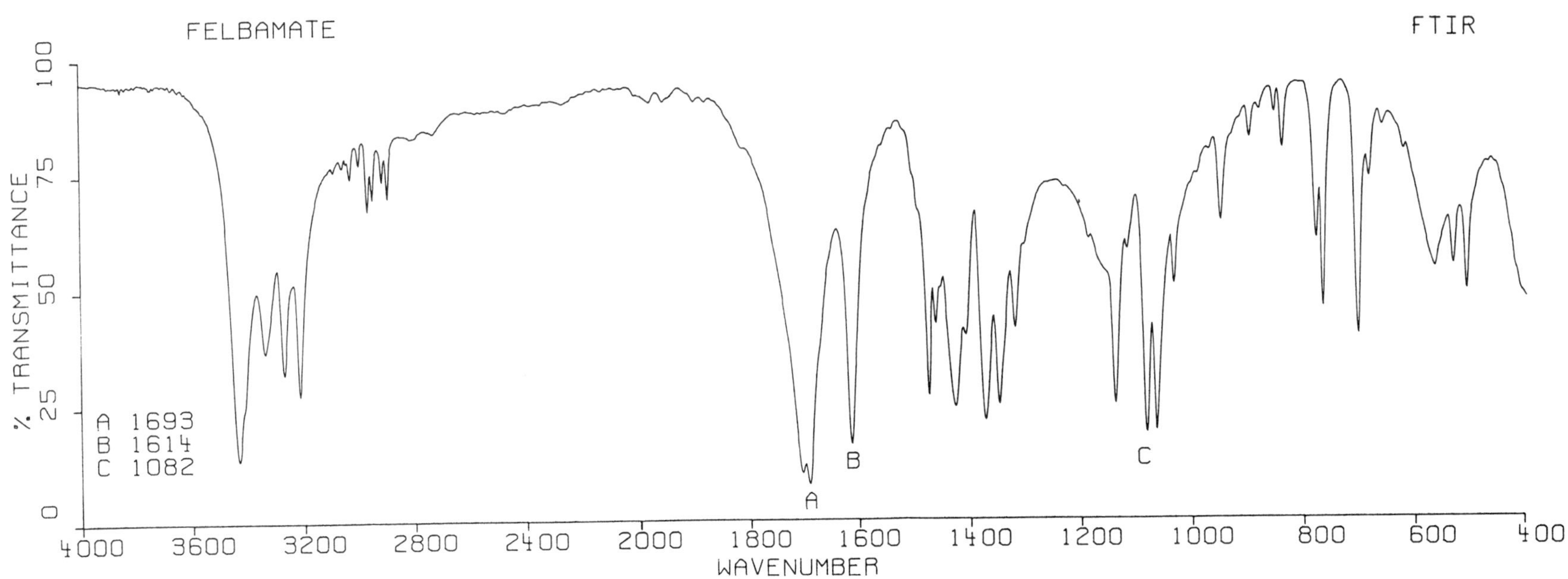

FELODIPINE

$C_{18}H_{19}Cl_2NO_4$

Molecular Weight: 384.26 (383.07)

Synonyms: 4-(2,3-Dichlorophenyl)-1,4-dihydro-2,6-dimethyl-3,5-pyridinedicarboxylic acid ethyl methyl ester
Trade Names: Agon, Hydac, Munobal, Plendil, Splendil
Use: Antihypertensive

HPLC: Methanol: 3.5
GC: 2647; 280°

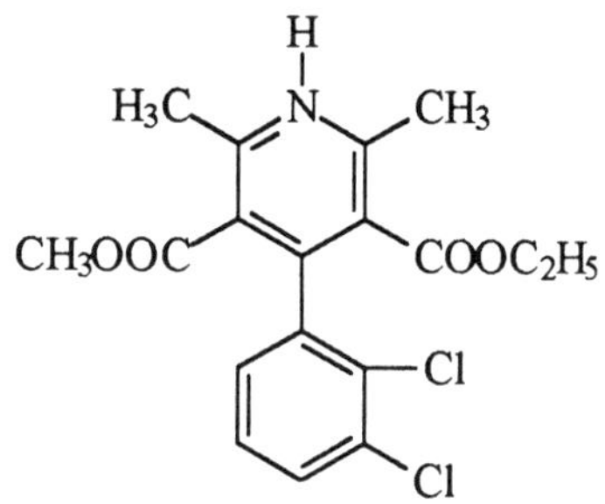

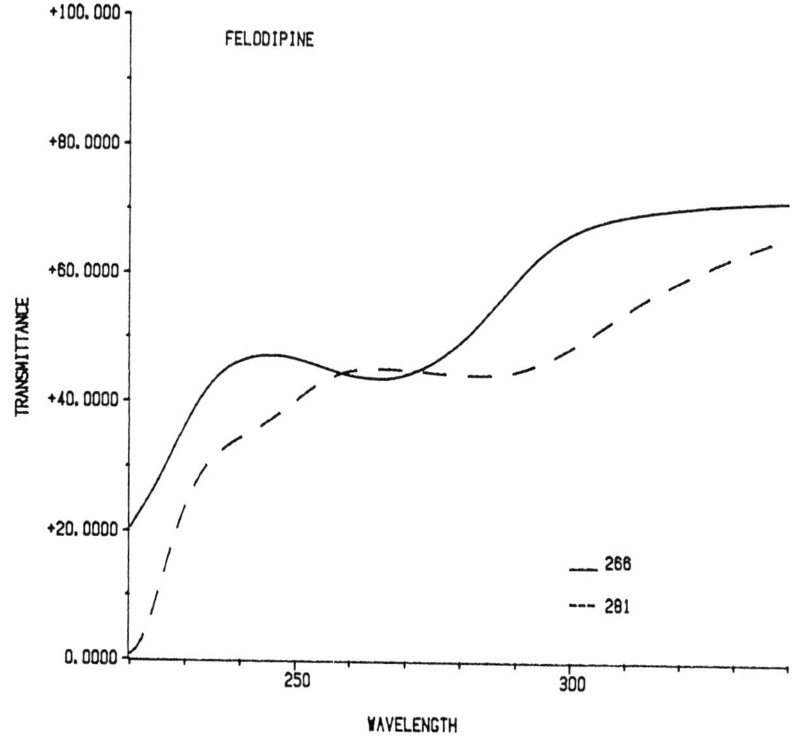

FELODIPINE

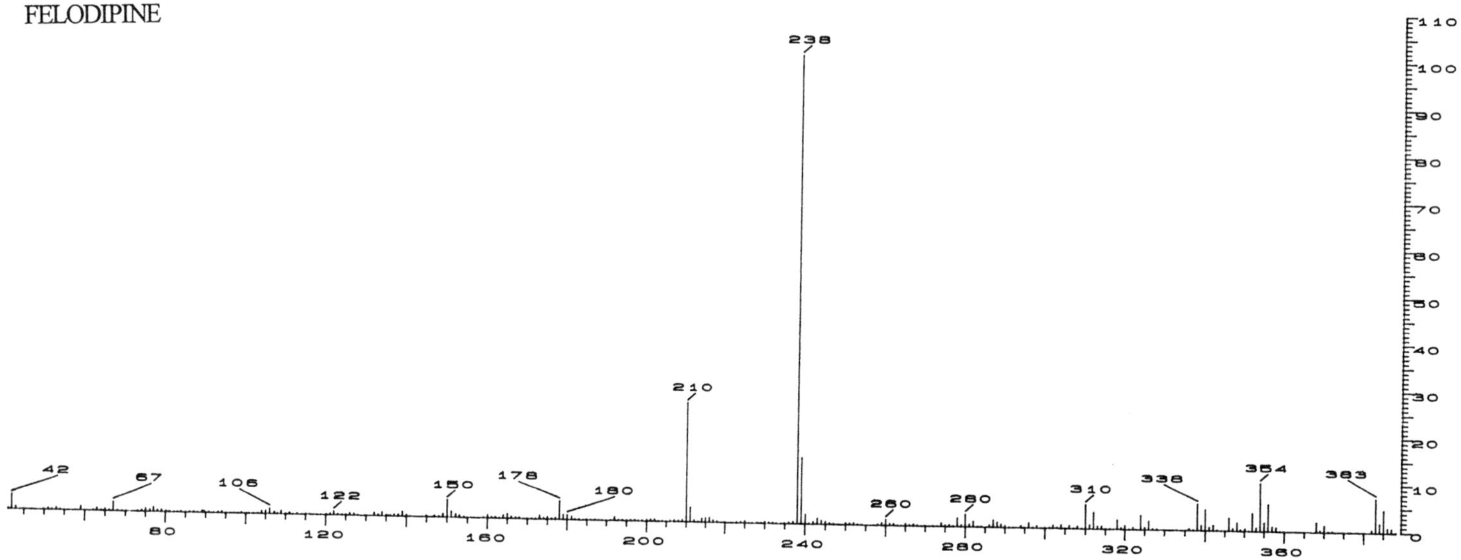

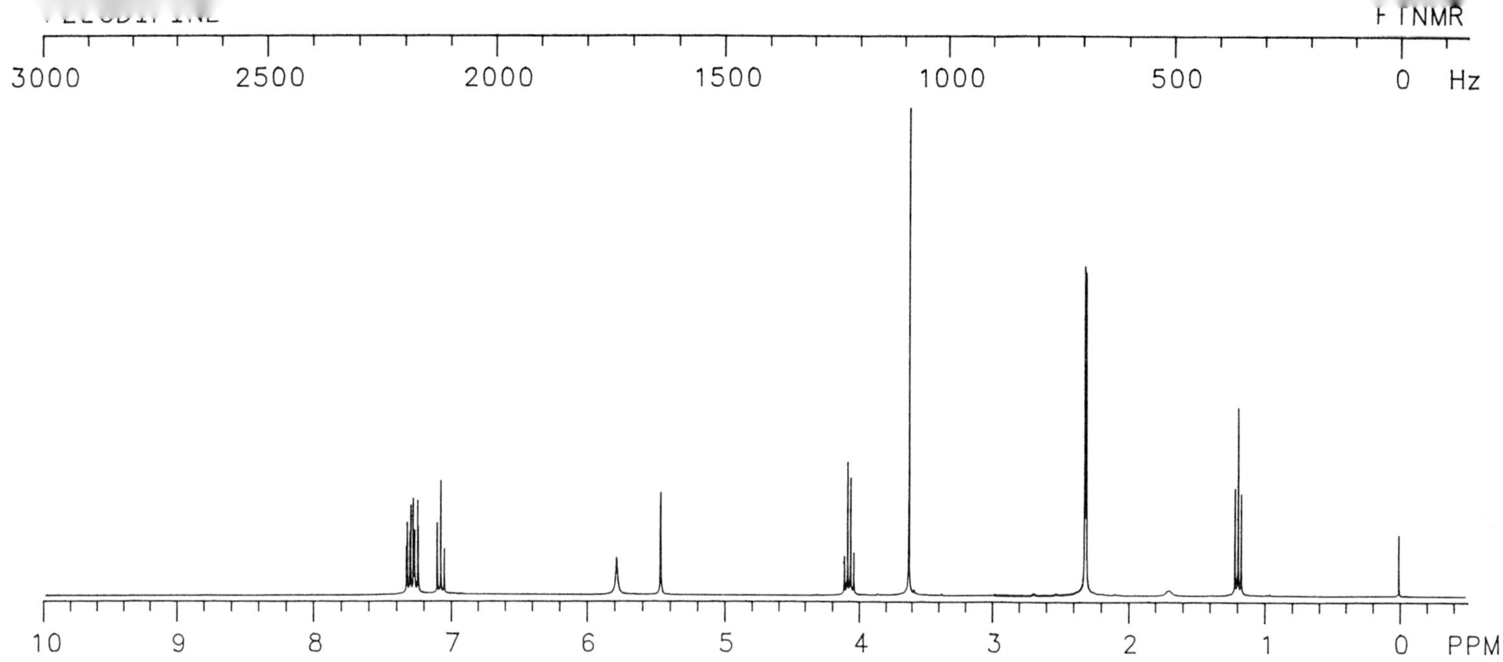

PEAK LISTING

#	HT	PPM
1	12	.00
2	21	1.16
3	38	1.18
4	22	1.20
5	66	2.30
6	67	2.31
7	99	3.61
8	8	4.03
9	24	4.05
10	27	4.08
11	8	4.10
12	21	5.46
13	9	7.04
14	23	7.06
15	14	7.09
16	14	7.23
17	19	7.23
18	13	7.25
19	12	7.26
20	19	7.26
21	18	7.28
22	13	7.29
23	14	7.31
24	10	7.31

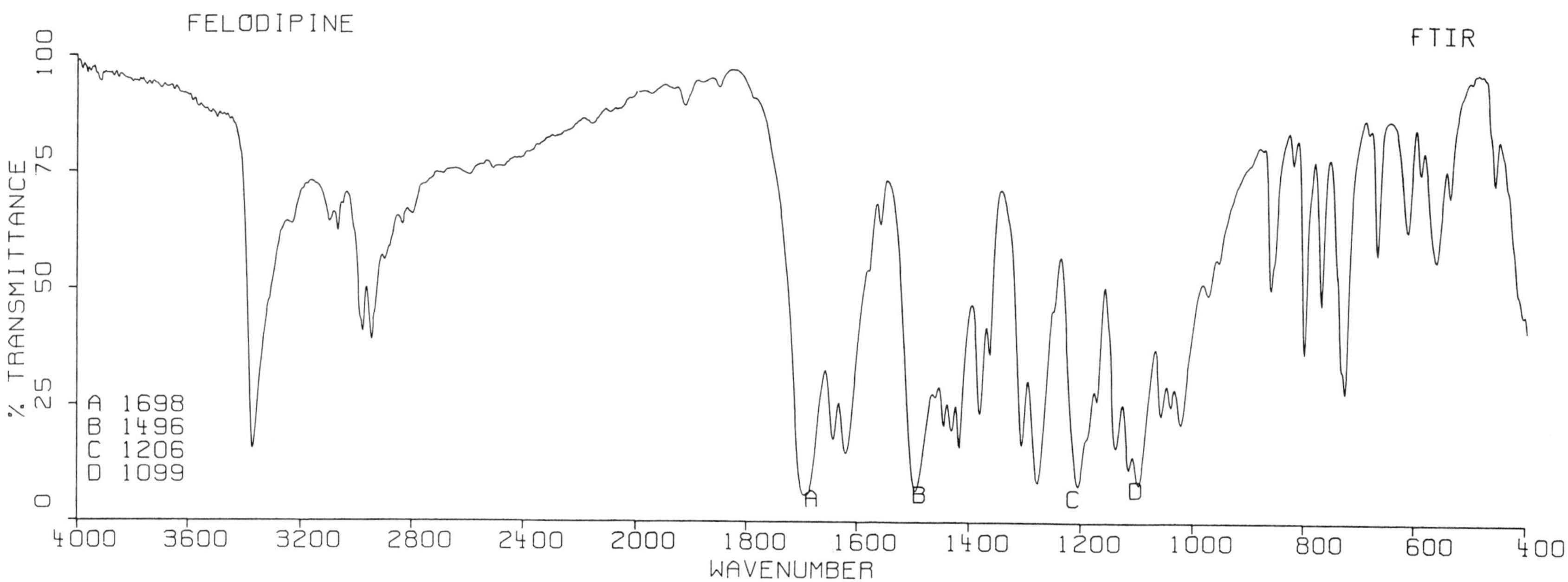

FLUCONAZOLE

$C_{13}H_{12}F_2N_6O$

Molecular Weight: 306.27 (306.10)

Synonyms: α-(2,4-Difluorophenyl)-α-(1H-1,2,4-driazol-1-ylmethyl)-1H-1,2,4-triazole-1-ethanol
Trade Names: Diflucan, Triblucan

Use: Antifungal

HPLC: Methanol: 2.4
GC: 2137; 250°

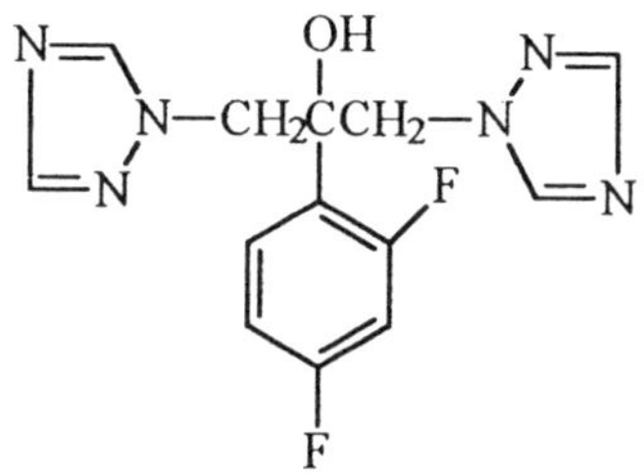

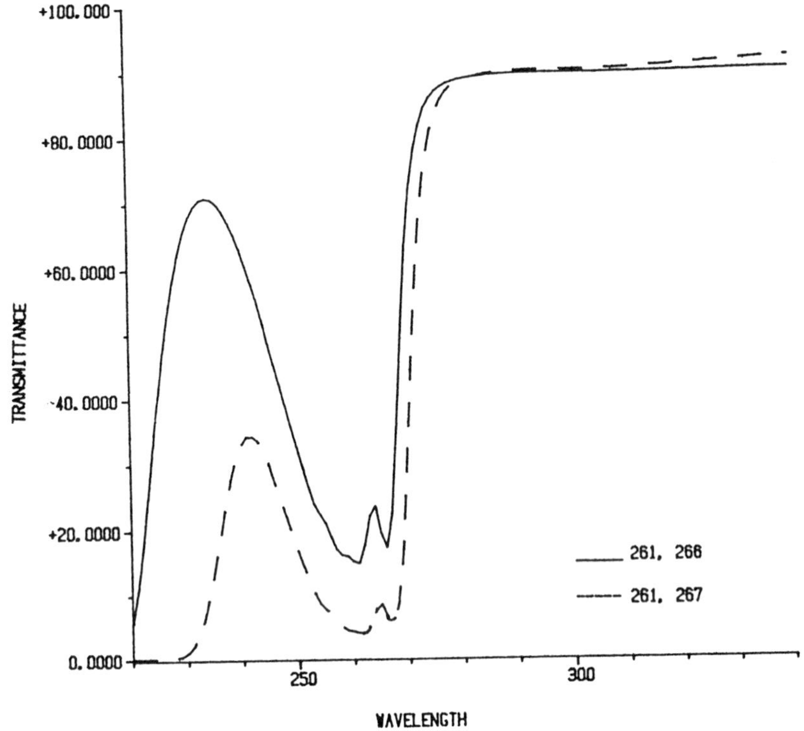

FLUCONAZOLE

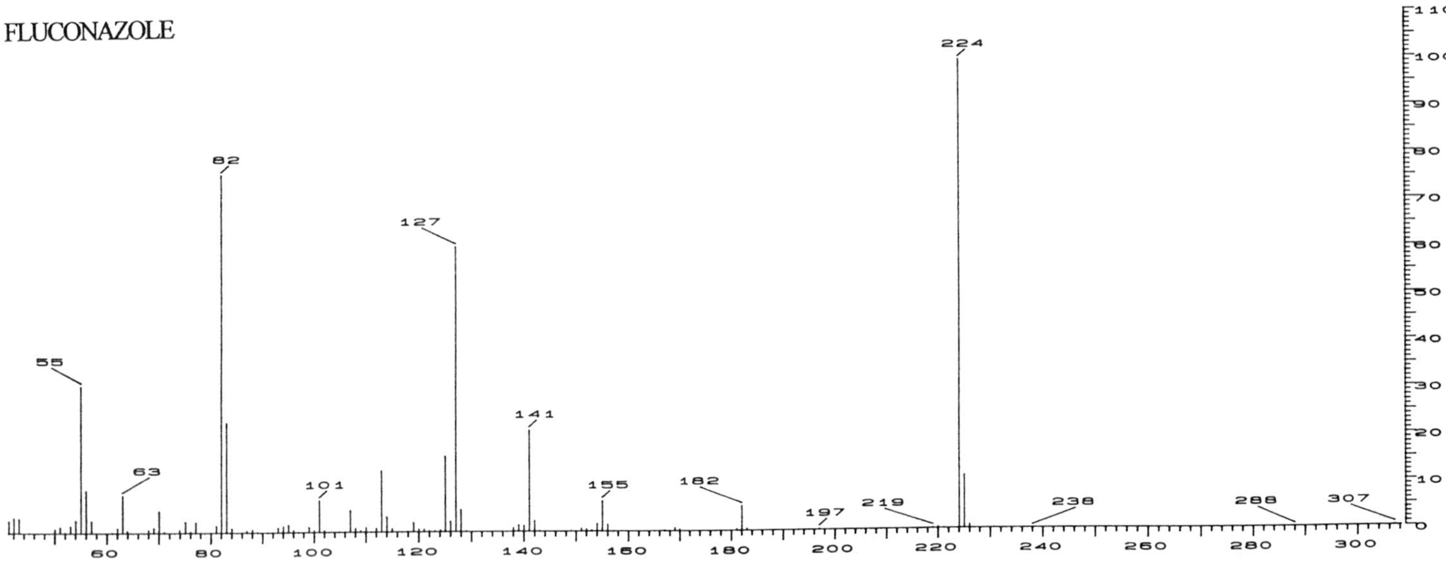

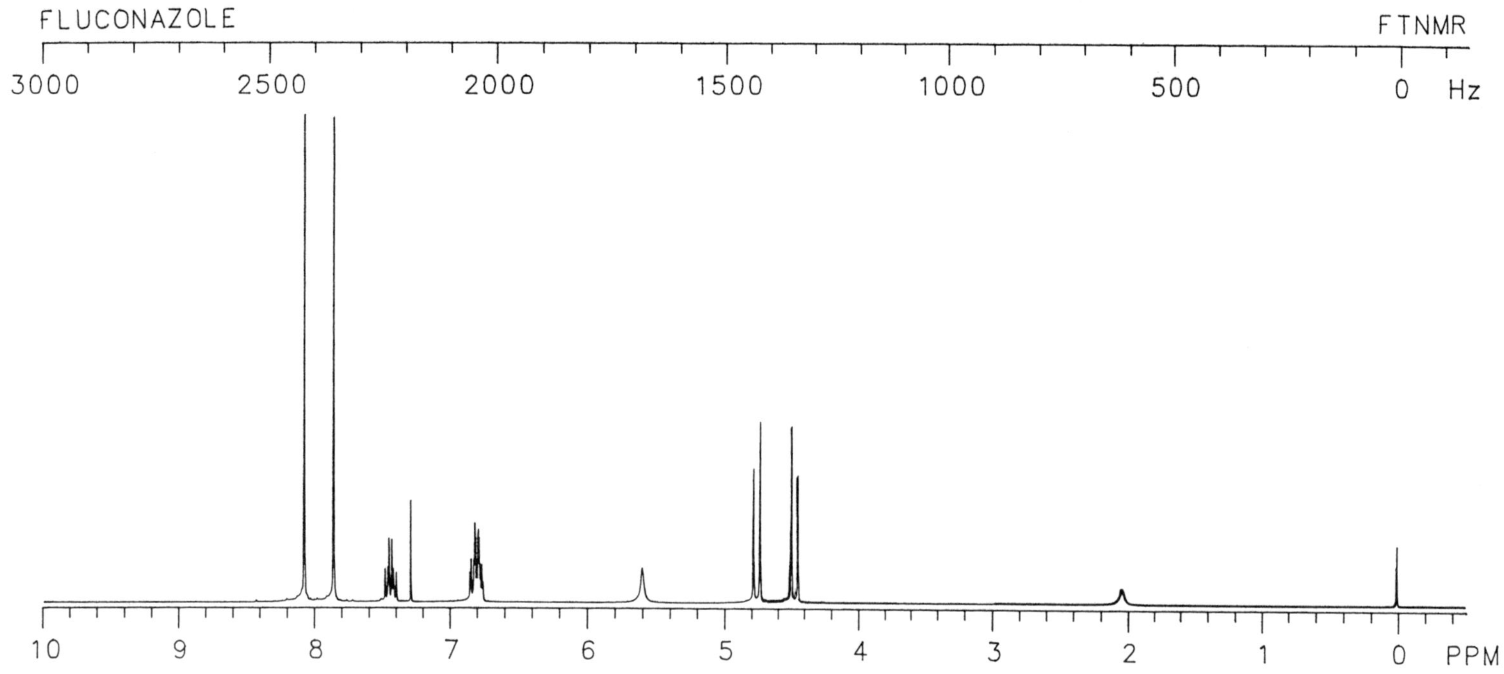

PEAK LISTING

#	HT	PPM
1	12	.00
2	26	4.44
3	36	4.49
4	37	4.73
5	27	4.77
6	13	6.77
7	15	6.78
8	13	6.80
9	16	6.81
10	20	7.28
11	13	7.42
12	13	7.44
13	99	7.85
14	99	8.06

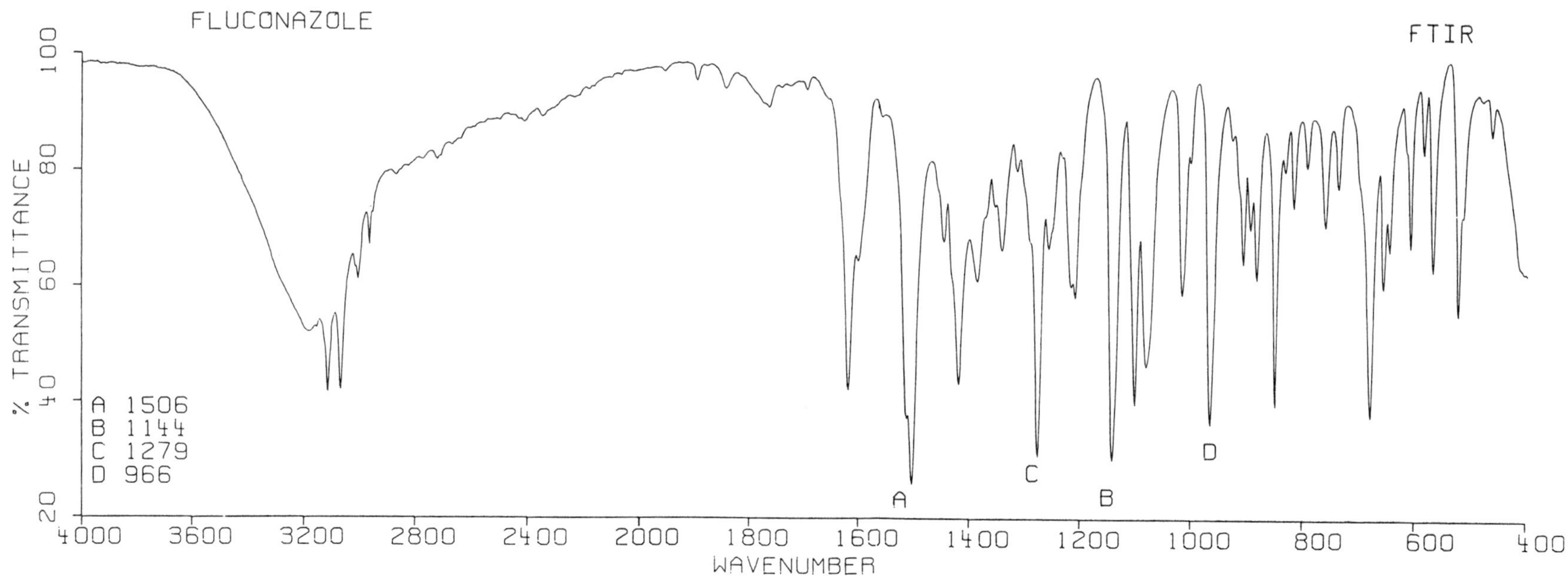

FLUMAZENIL

$C_{15}H_{14}FN_3O_3$

Molecular Weight: 303.29 (303.10)

Synonyms: 8-Fluoro-5,6-dihydro-5-methyl-6-oxo-4H-imidazo[1,5-a][1,4]benzodiazepine-3-carboxylic acid ethyl ester; flumazepil
Trade Names: Anexate, Lanexat
Use: Benzodiazepine antagonist

HPLC: Methanol: 3.4
GC: 2514; 250°

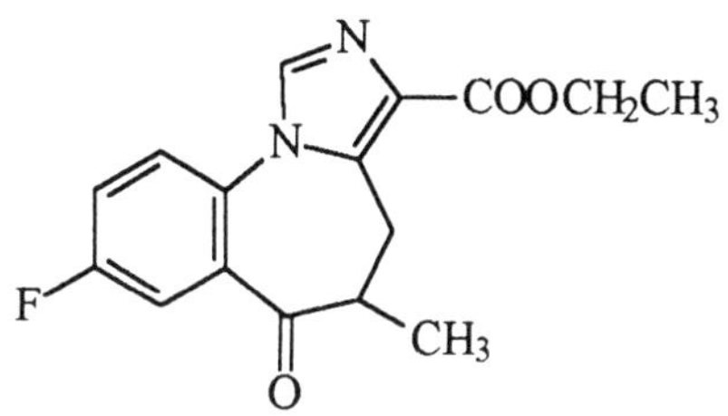

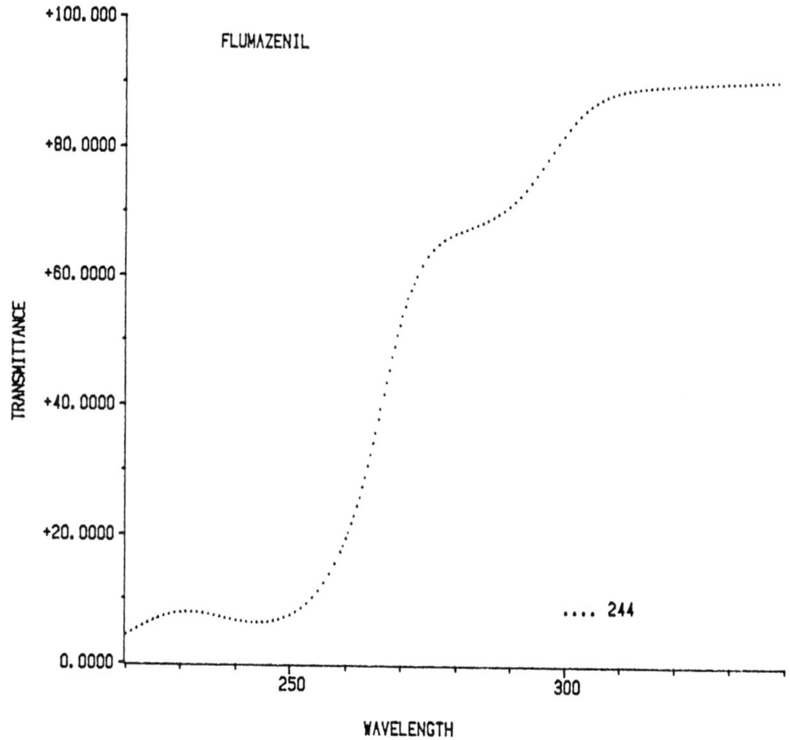

FLUMAZENIL

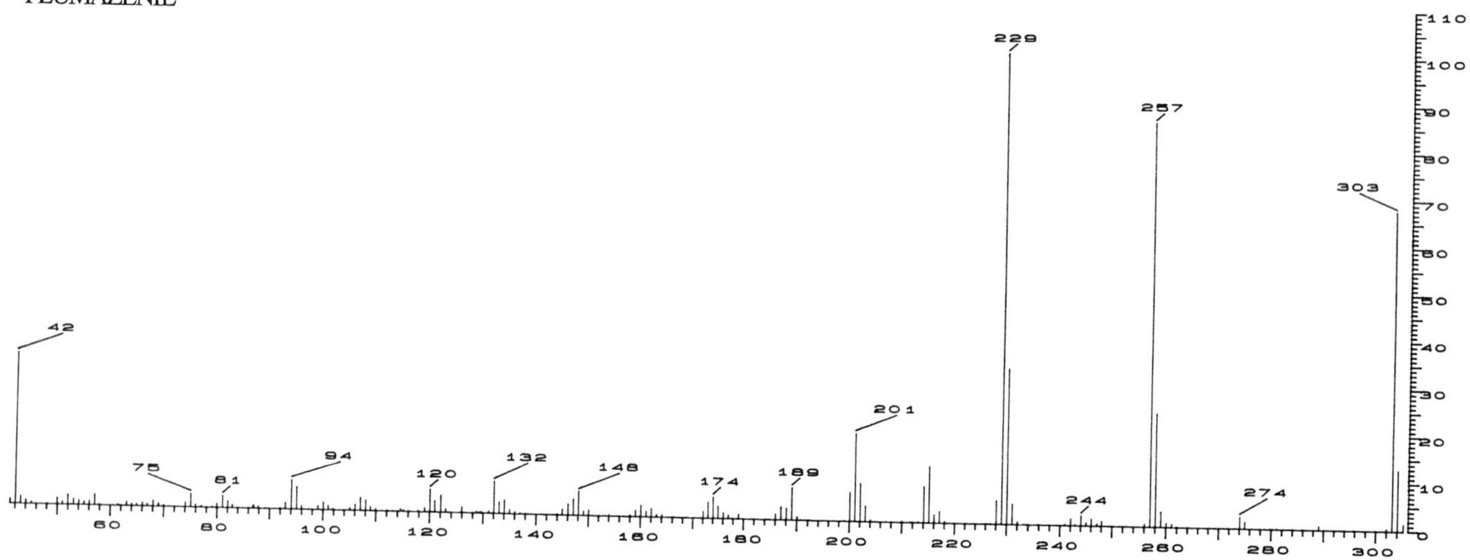

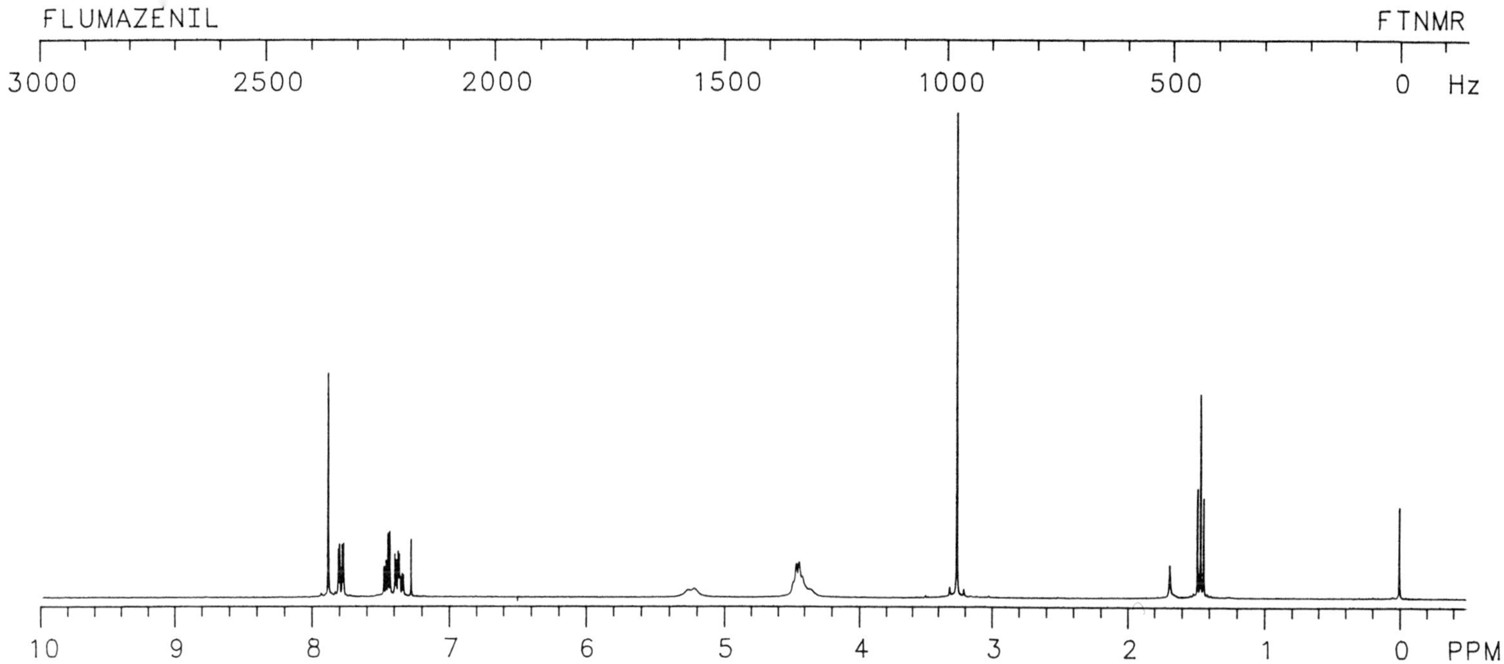

PEAK LISTING

#	HT	PPM
1	18	.00
2	20	1.43
3	41	1.45
4	22	1.47
5	99	3.25
6	45	7.87

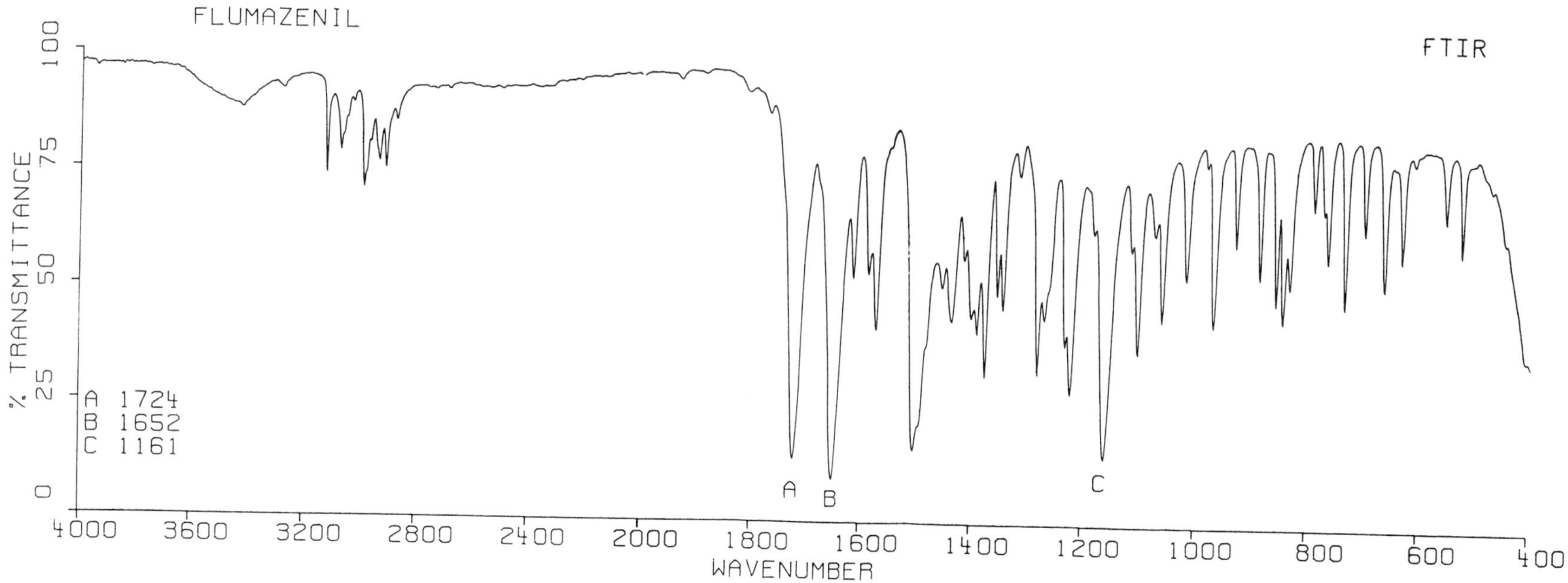

FLUNISOLIDE

$C_{24}H_{31}FO_6$

Molecular Weight: 434.50 (434.21)

Synonyms: 6-Fluoro-11,21-dihydroxy-16,17-[(1-methylethylidene)bis(oxy)]pregna-1,4-diene-3,20-dione
Trade Names: AeroBid, Lunis, Nasalide, Nasarel, Rhinalar, Syntaris, Yafrol
Use: Antiasthmatic

HPLC: Methanol: 2.8
GC:

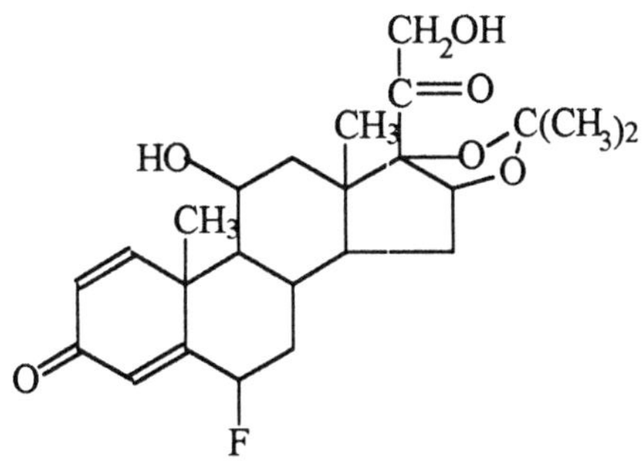

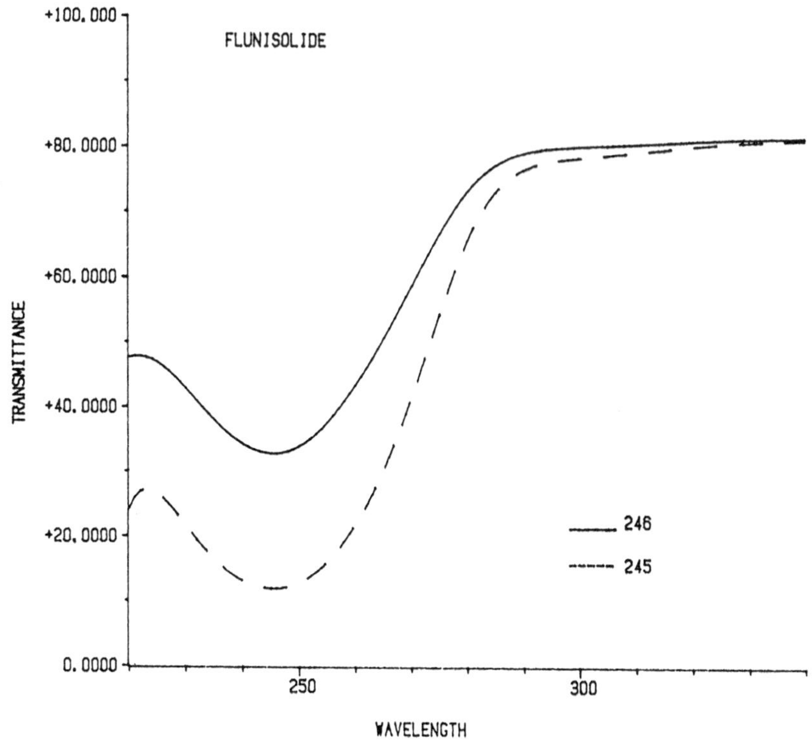

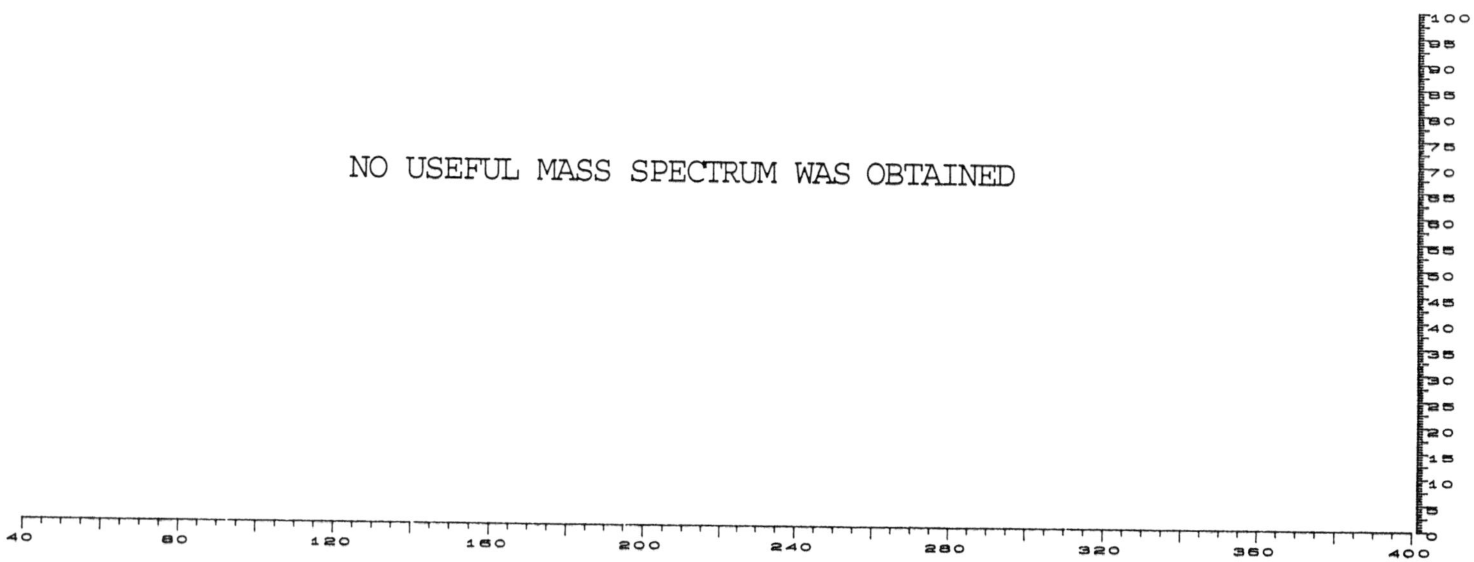

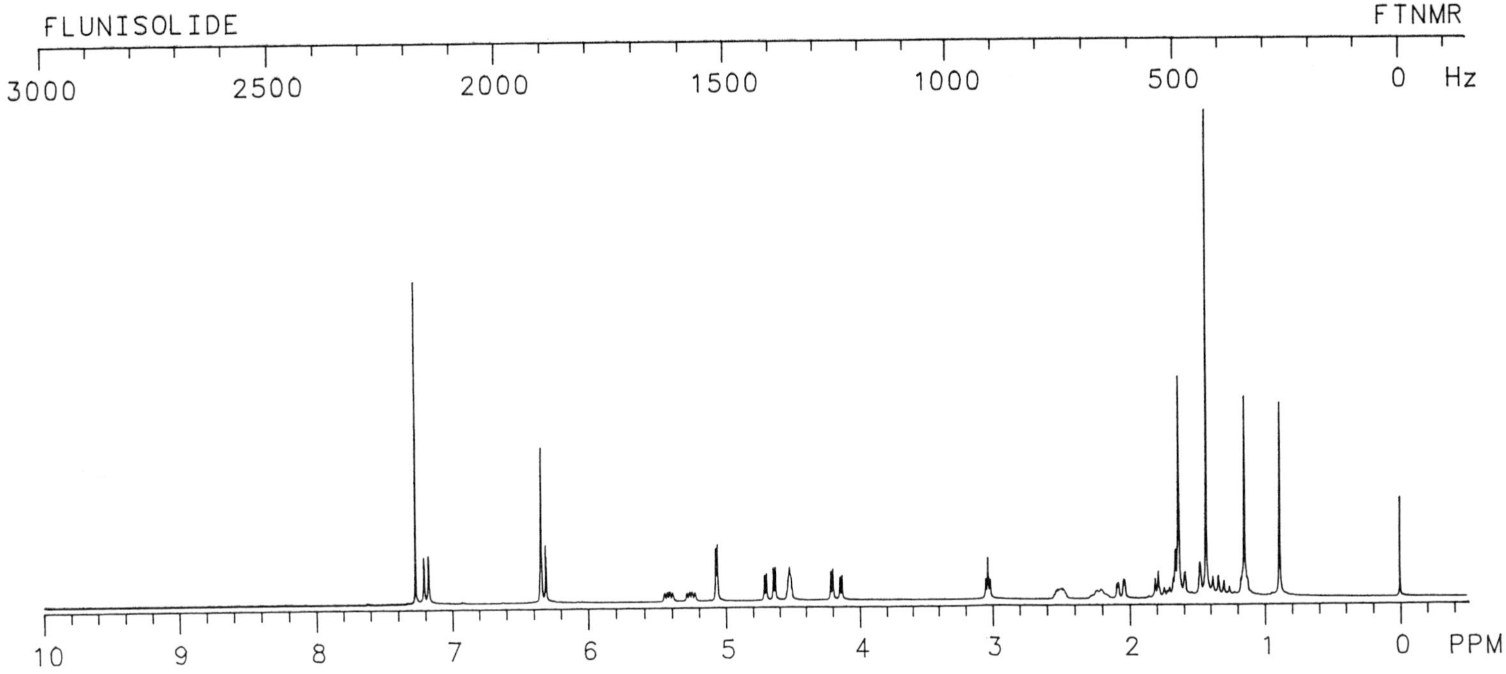

PEAK LISTING

#	HT	PPM
1	20	.00
2	39	.88
3	41	1.14
4	99	1.43
5	45	1.63
6	11	5.05
7	11	6.30
8	31	6.33
9	65	7.26

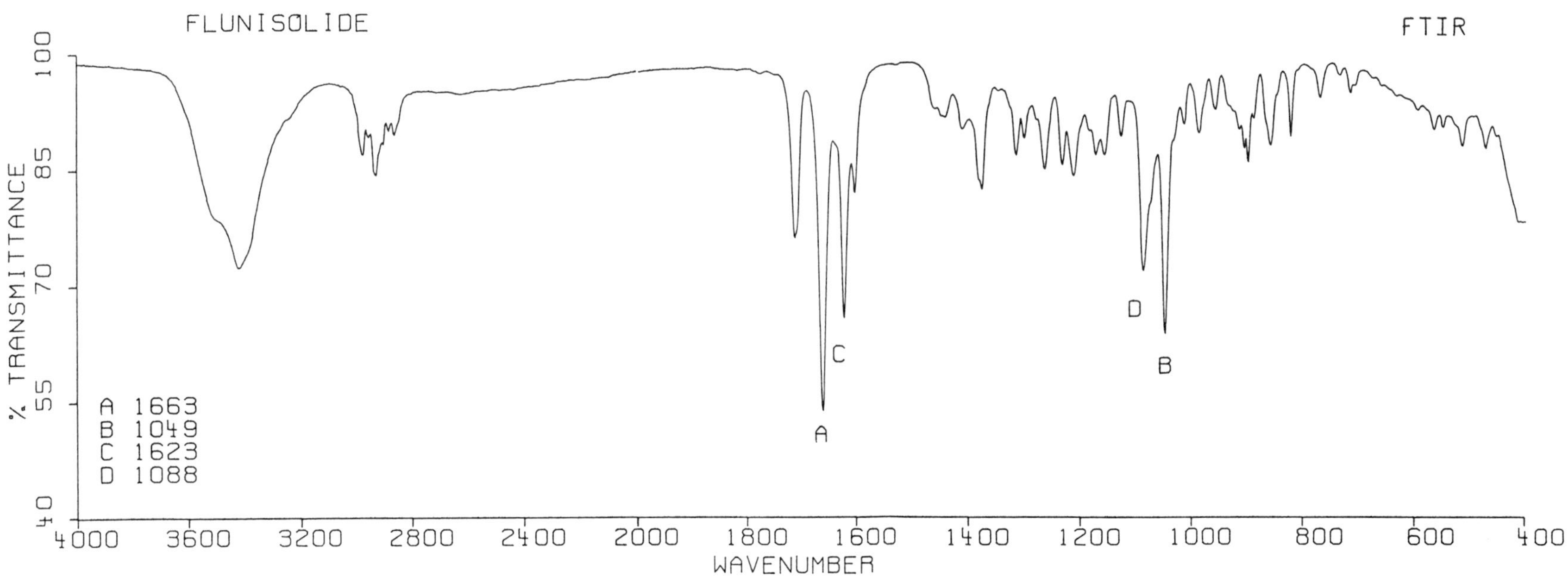

FLUOCINONIDE

$C_{26}H_{32}F_2O_7$

Molecular Weight: 494.52 (494.21)

Synonyms: 6α,9α-Difluoro-16α-hydroxyoredbusikibe-16,17-acetonide-21-acetate; fluocinolone acetonide-21-acetate
Trade Names: Lidex, Dermacin, Fluocinonide E
Use: Anti-inflammatory

HPLC: Ethanol: 2.5
GC:

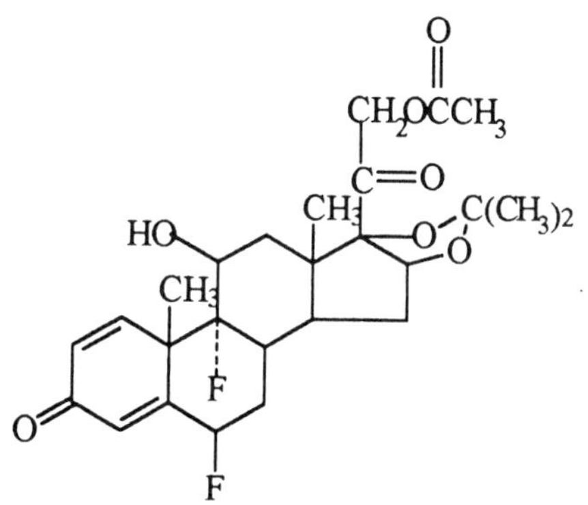

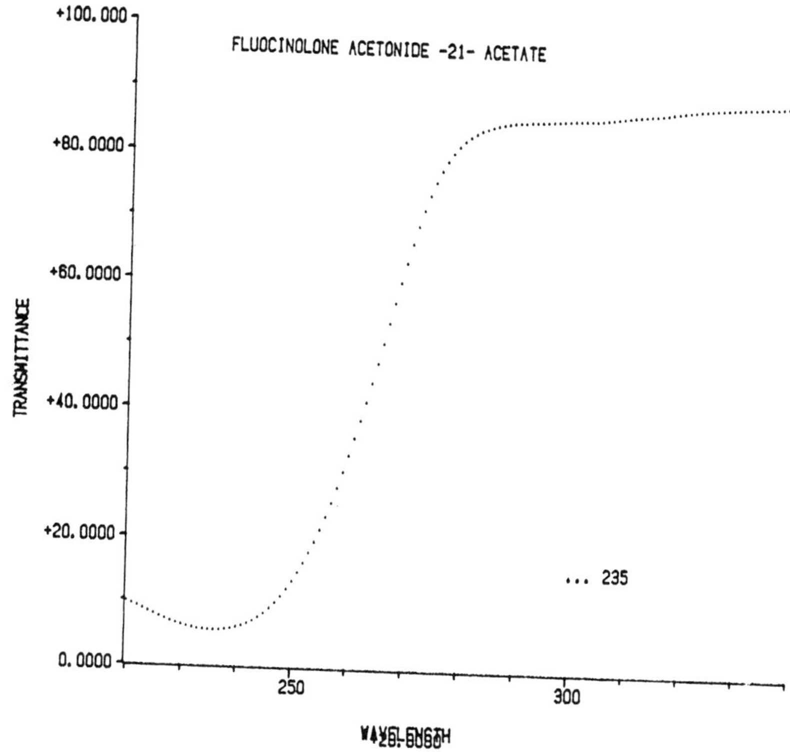

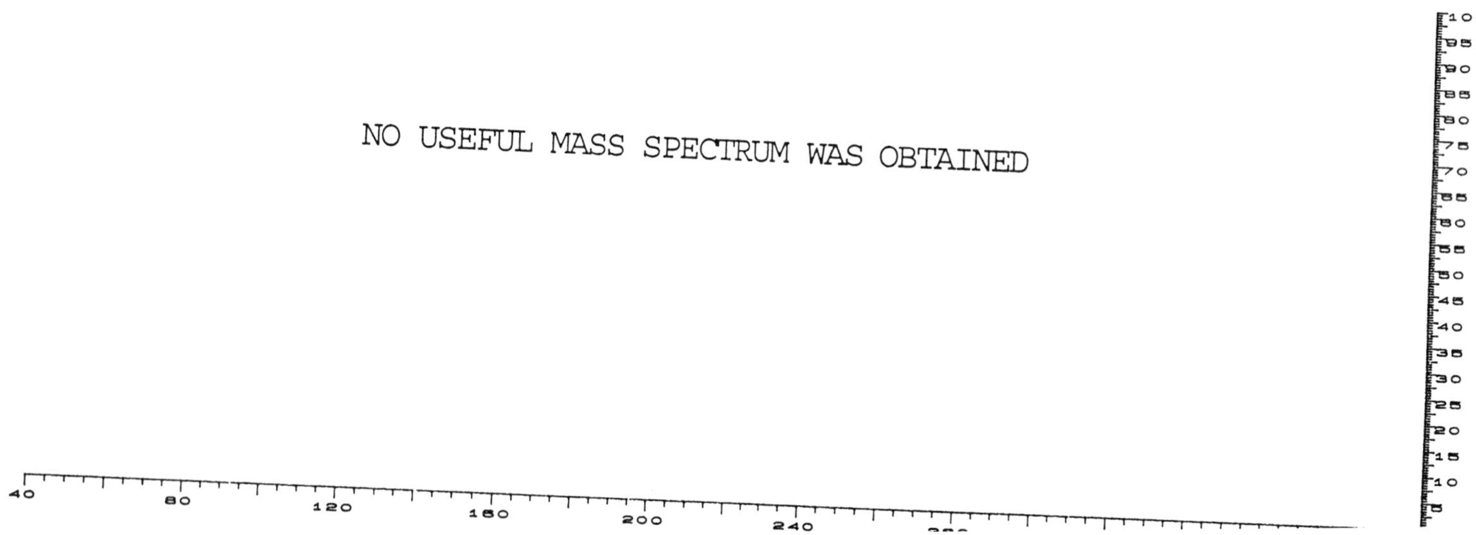

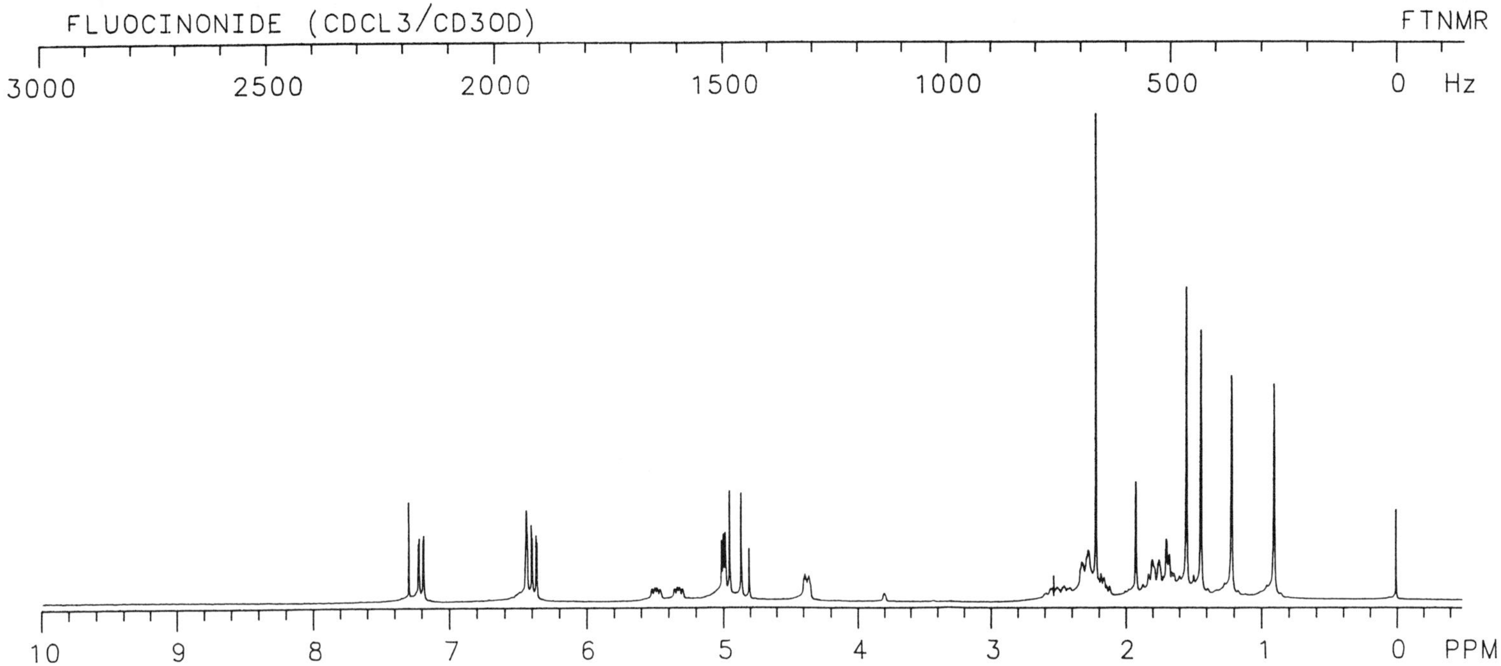

PEAK LISTING

#	HT	PPM
1	18	.00
2	44	.90
3	46	1.21
4	55	1.43
5	64	1.54
6	9	1.66
7	12	1.68
8	8	1.74
9	8	1.79
10	8	1.79
11	24	1.91
12	99	2.21
13	10	2.26
14	10	2.27
15	9	2.28
16	10	4.80
17	22	4.86
18	22	4.94
19	14	4.97
20	13	4.99
21	12	5.00
22	12	6.35
23	13	6.36
24	13	6.38
25	15	6.39
26	18	6.43
27	13	7.18
28	12	7.18
29	13	7.21
30	11	7.21
31	20	7.29

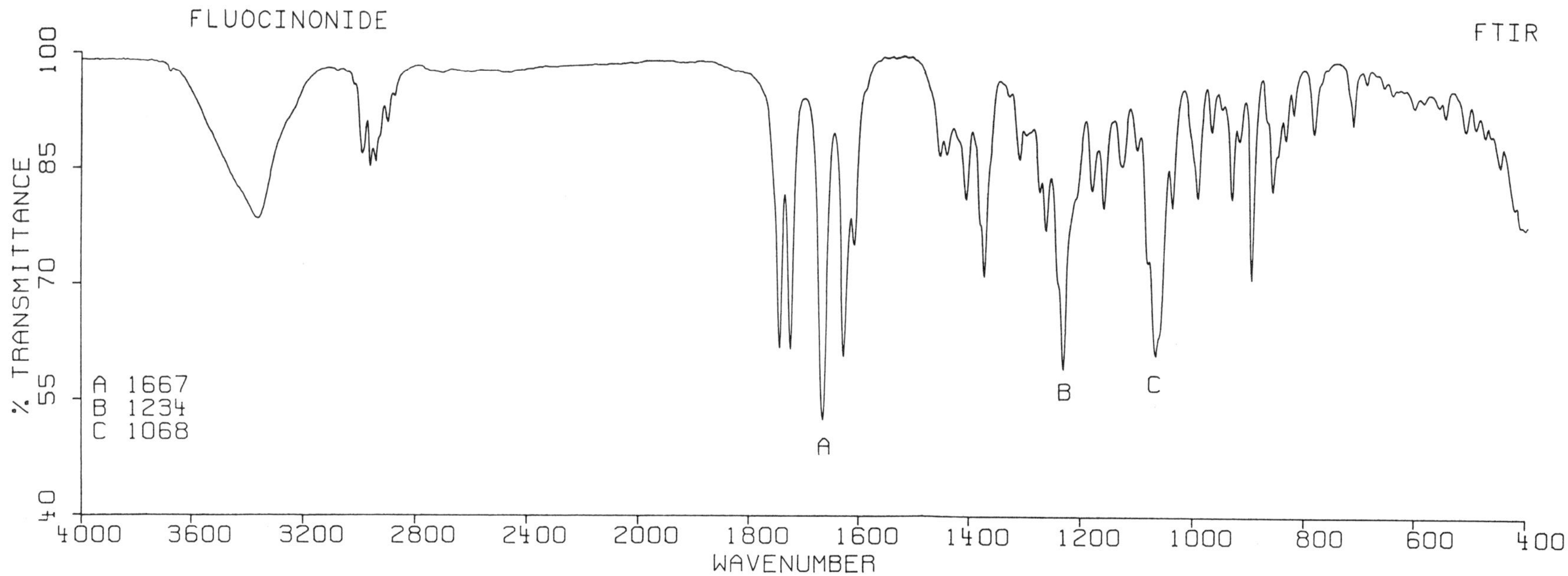

FLUPENTIXOL

$C_{23}H_{25}F_3N_2OS$

Molecular Weight: 434.54 (434.16)

Synonyms: 4-[3-[2-(Trifluoromethyl)-9H-thioxanthen-9-ylidene]propyl]-1 piperazineethanol; flupenthixol
Trade Names: Depixol, Emergil, Fluanxol, Viscoleo
Use: Antipsychotic

HPLC:
GC:

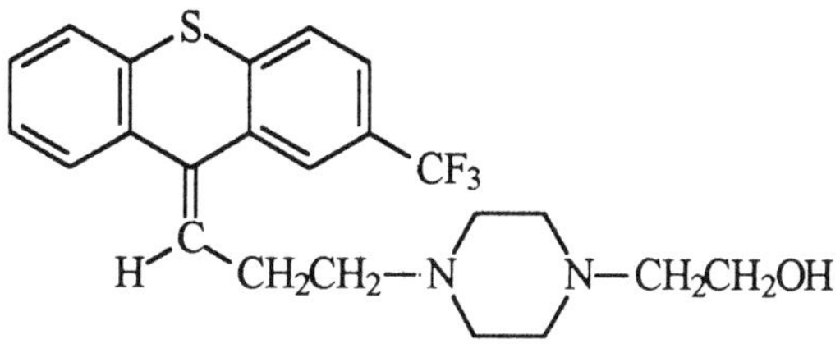

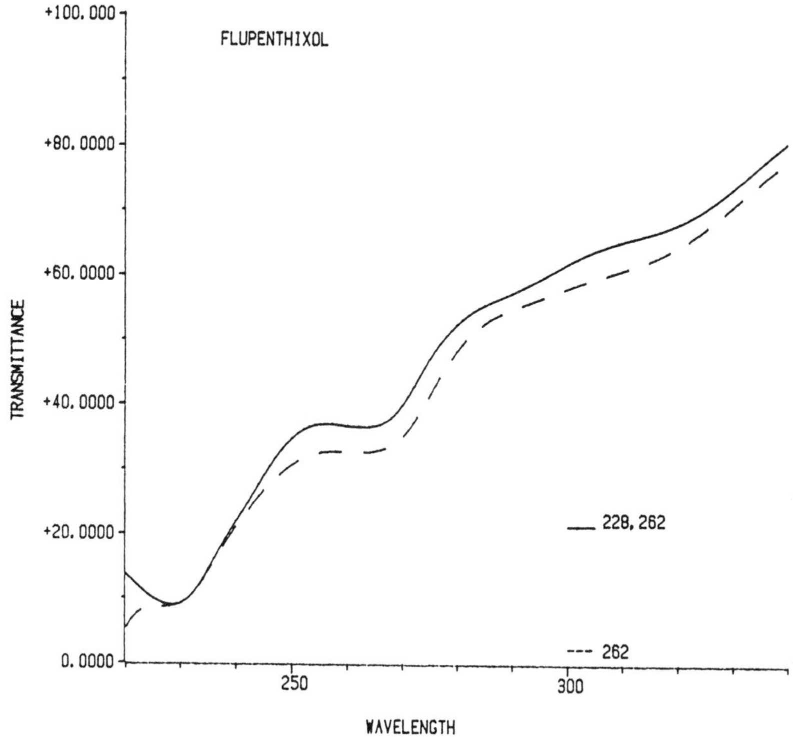

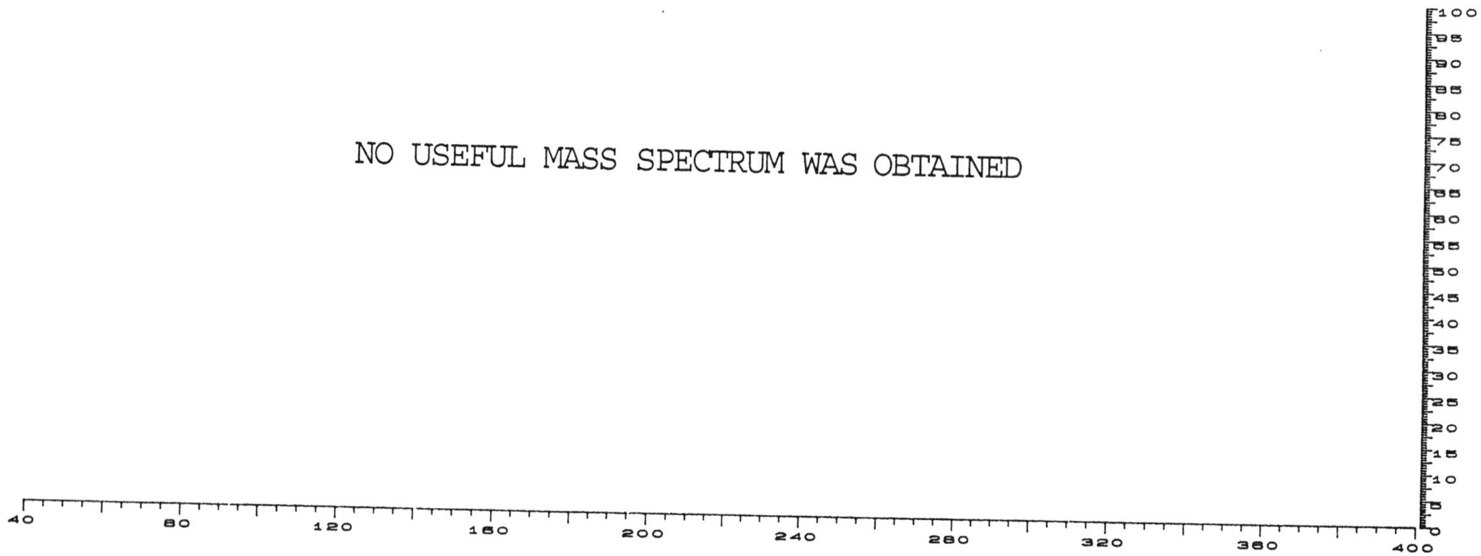

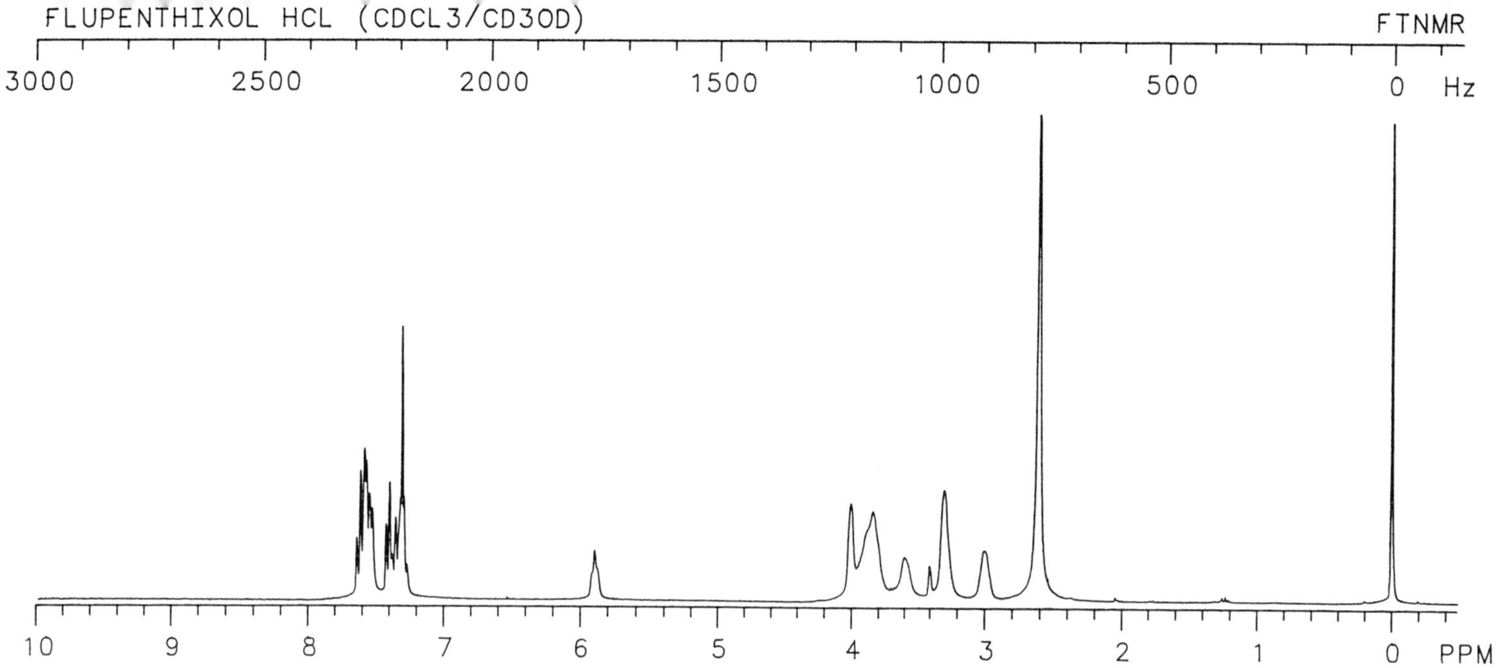

PEAK LISTING

#	HT	PPM
1	98	.00
2	99	2.61
3	23	3.29
4	18	3.83
5	18	3.83
6	15	3.86
7	15	3.86
8	14	3.87
9	14	3.88
10	20	3.99
11	21	7.28
12	49	7.30
13	55	7.30
14	20	7.31
15	16	7.34
16	23	7.39
17	15	7.41
18	18	7.51
19	21	7.53
20	28	7.56
21	30	7.57
22	26	7.60

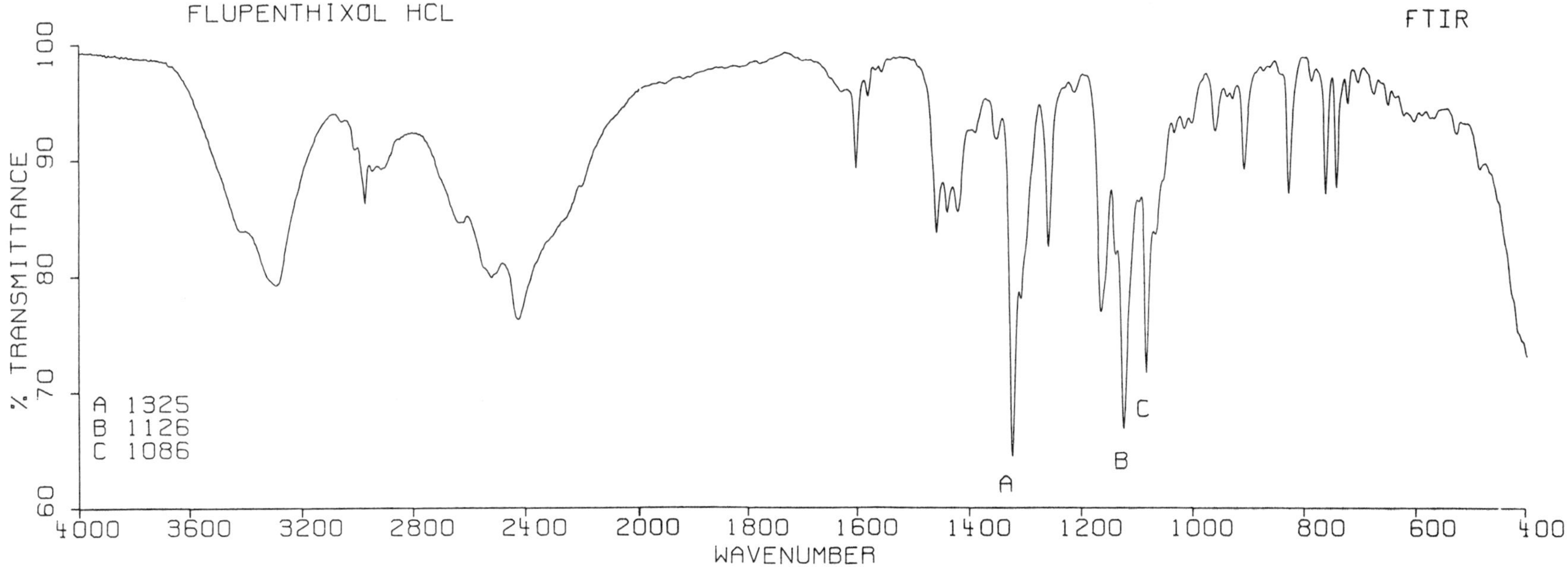

FOSINOPRIL

$C_{30}H_{46}NO_7P$

Molecular Weight: 563.67 (563.30)

Synonyms: (2α,4β)-4-Cyclohexyl-1-[[[2-methyl-1-(1-oxopropoxy)propoxy](4-phenylbutyl)phosphinyl]acetyl]-L-proline; fosenopril
Trade Names: Monopril
Use: Antihypertensive

HPLC: Methanol: 3.5
GC:

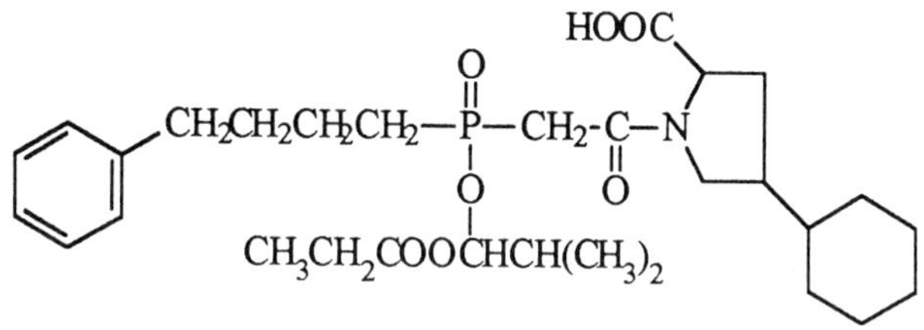

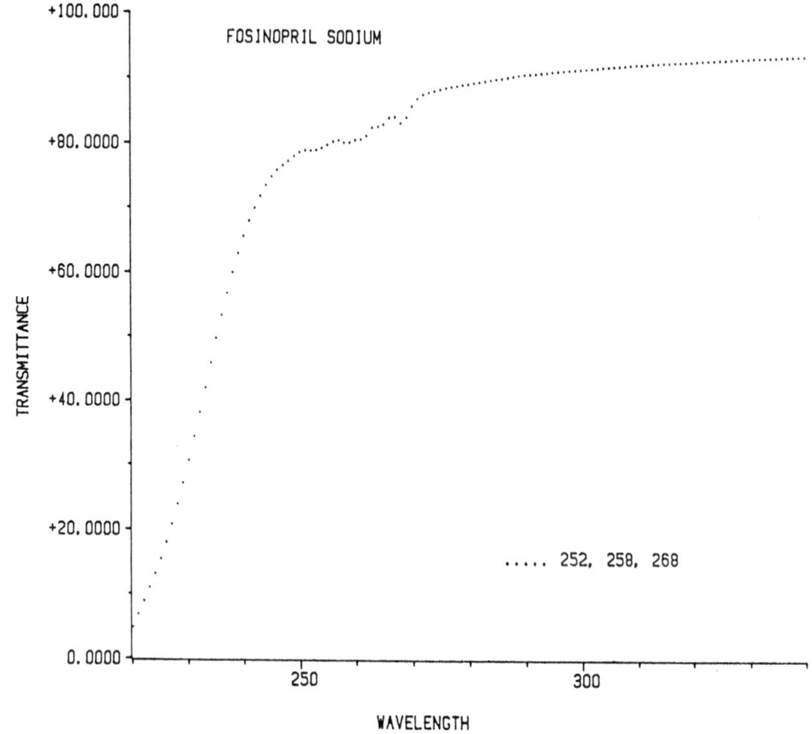

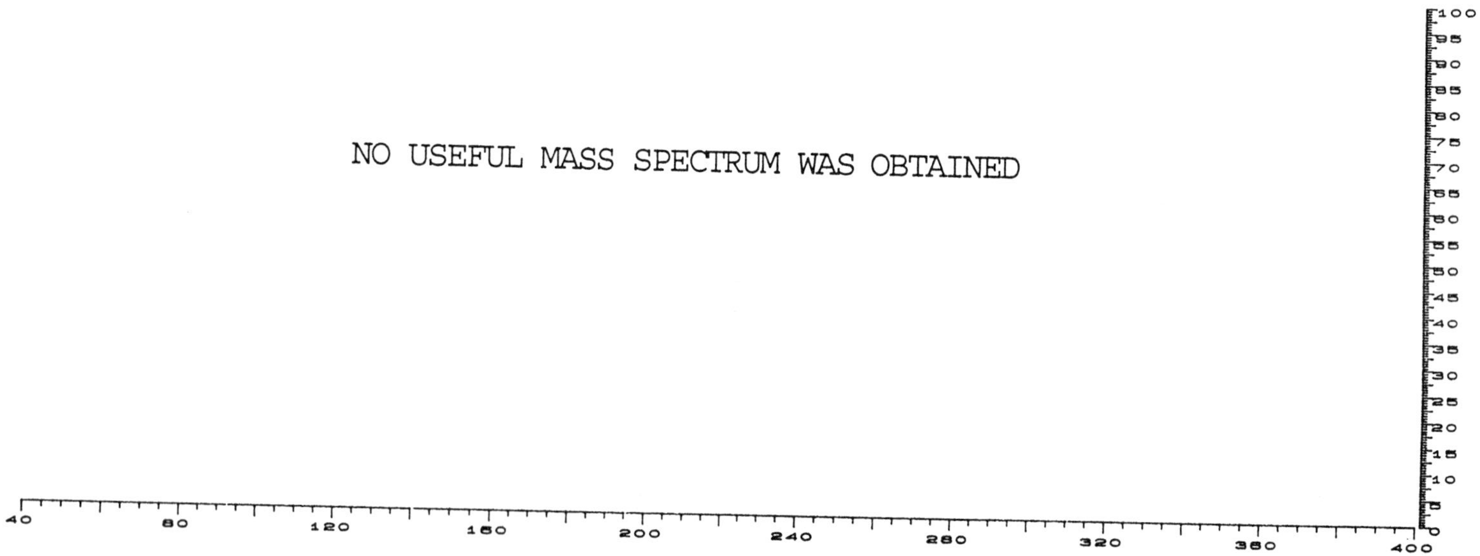

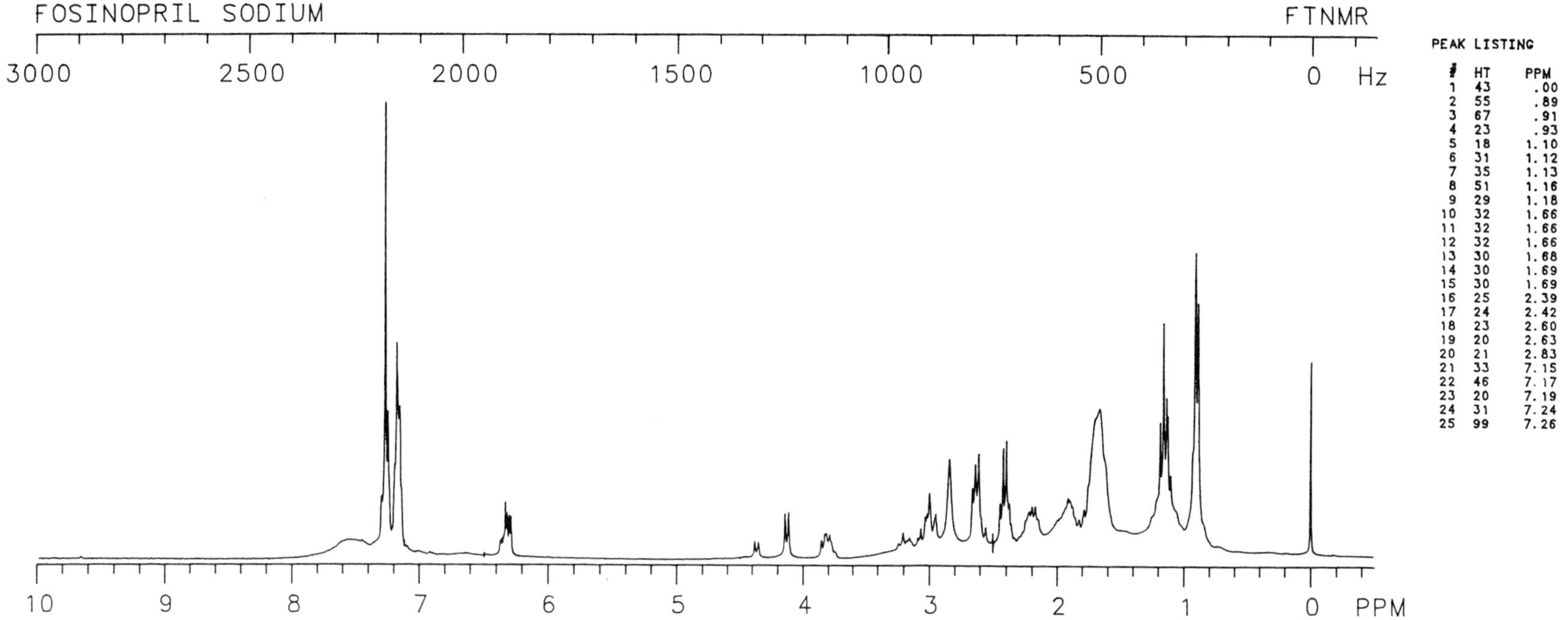

FOSINOPRIL SODIUM
FTNMR
3000 2500 2000 1500 1000 500 0 Hz
10 9 8 7 6 5 4 3 2 1 0 PPM
PEAK LISTING
HT PPM
1 43 .00
2 55 .89
3 67 .91
4 23 .93
5 18 1.10
6 31 1.12
7 35 1.13
8 51 1.16
9 29 1.18
10 32 1.66
11 32 1.66
12 32 1.66
13 30 1.68
14 30 1.69
15 30 1.69
16 25 2.39
17 24 2.42
18 23 2.60
19 20 2.63
20 21 2.83
21 33 7.15
22 46 7.17
23 20 7.19
24 31 7.24
25 99 7.26

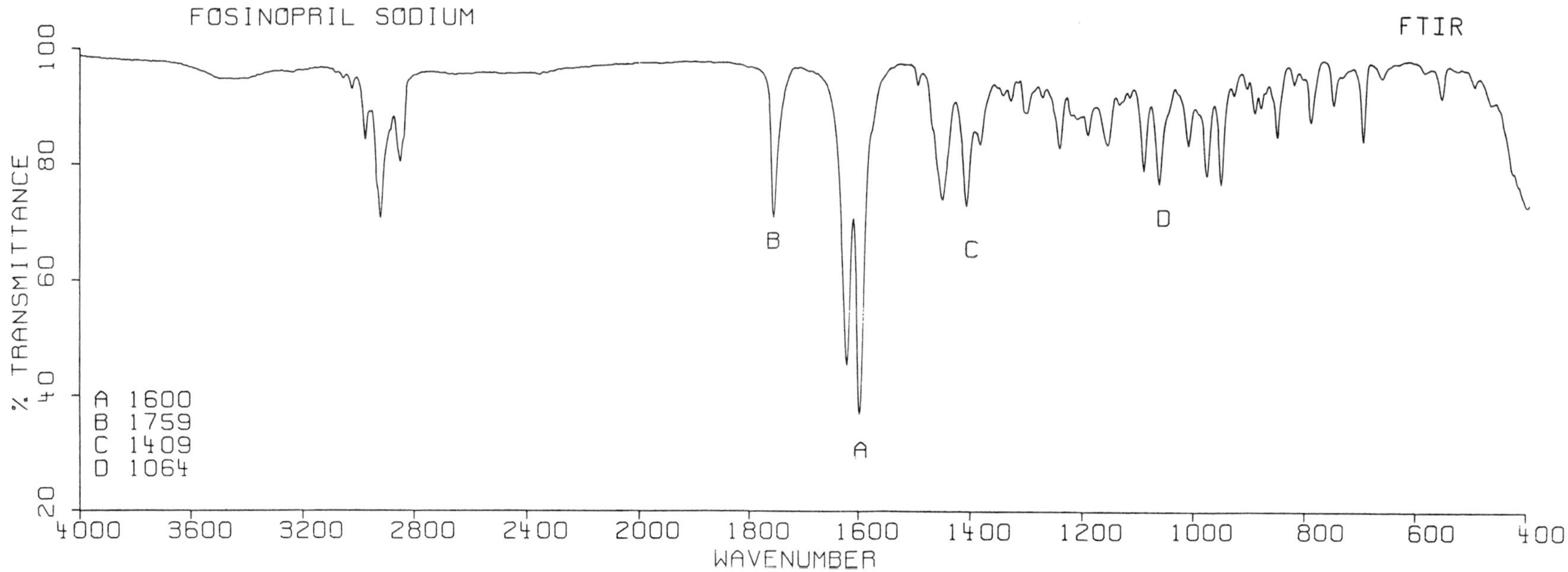

FOSINOPRIL SODIUM
FTIR
% TRANSMITTANCE
100 80 60 40 20
A 1600
B 1759
C 1409
D 1064
B
A
C
D
4000 3600 3200 2800 2400 2000 1800 1600 1400 1200 1000 800 600 400
WAVENUMBER

GLUCONIC ACID POTASSIUM

$C_6H_{12}O_7$

Molecular Weight: 196.16 (196.06)

Synonyms: D-Gluconic acid potassium salt; dextronic acid potassium; maltonic acid potassium; glyconic acid potassium; glycogenic acid potassium; pentahydroxycaproic acid potassium
Trade Names: Almora, Glucomag , Rubozinc, Ultra-Mg
Use: Antispasmodic

HPLC:
GC:

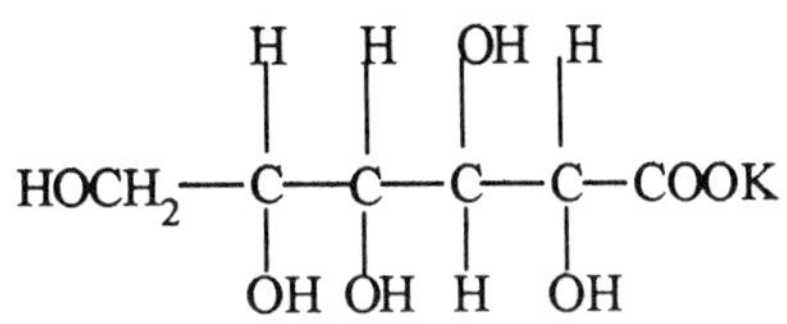

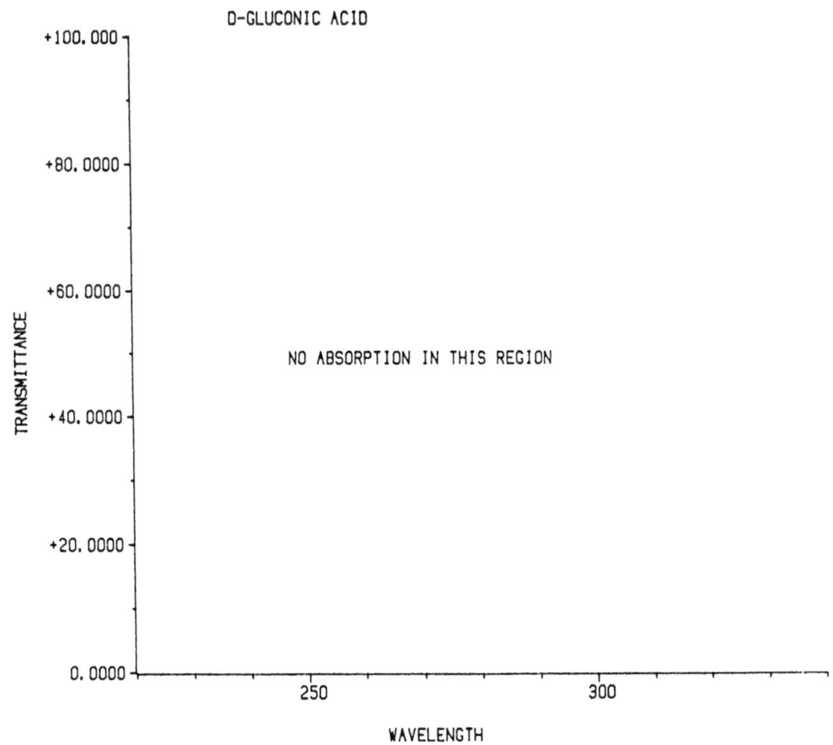

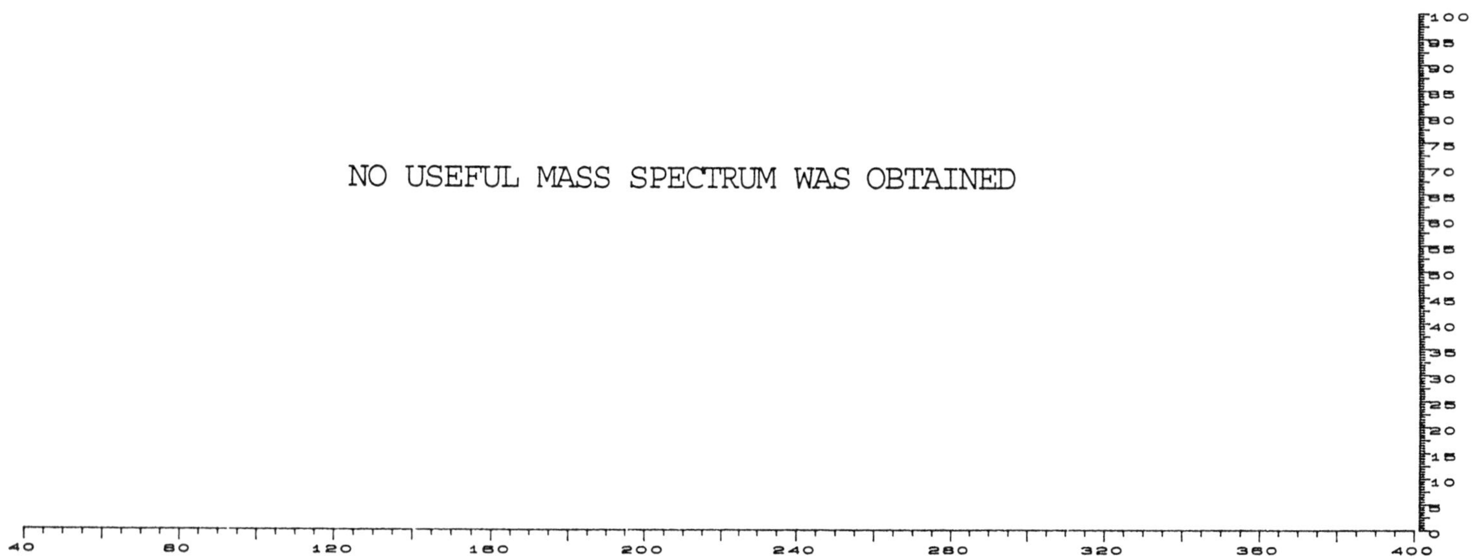

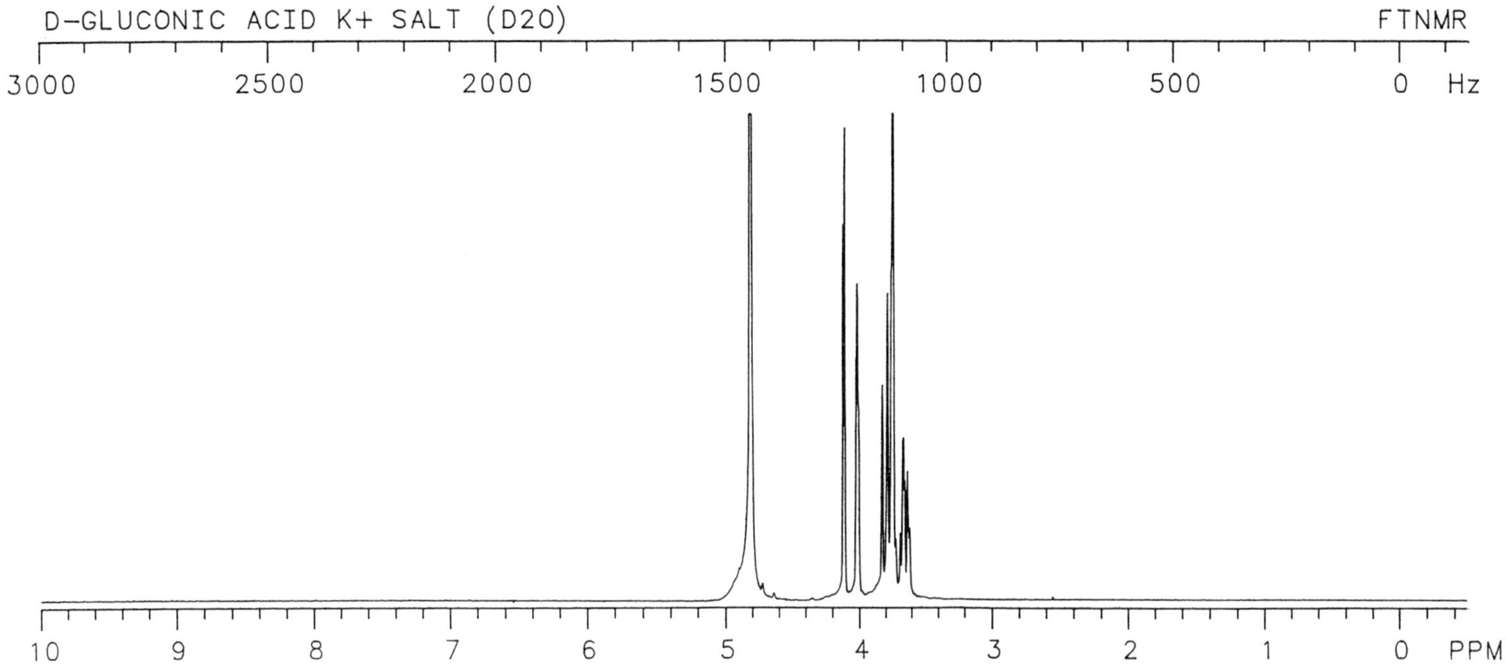

PEAK LISTING

#	HT	PPM
1	131	3.73
2	74	3.74
3	62	3.77
4	64	4.00
5	96	4.09
6	89	4.09
7	76	4.10
8	560	4.80

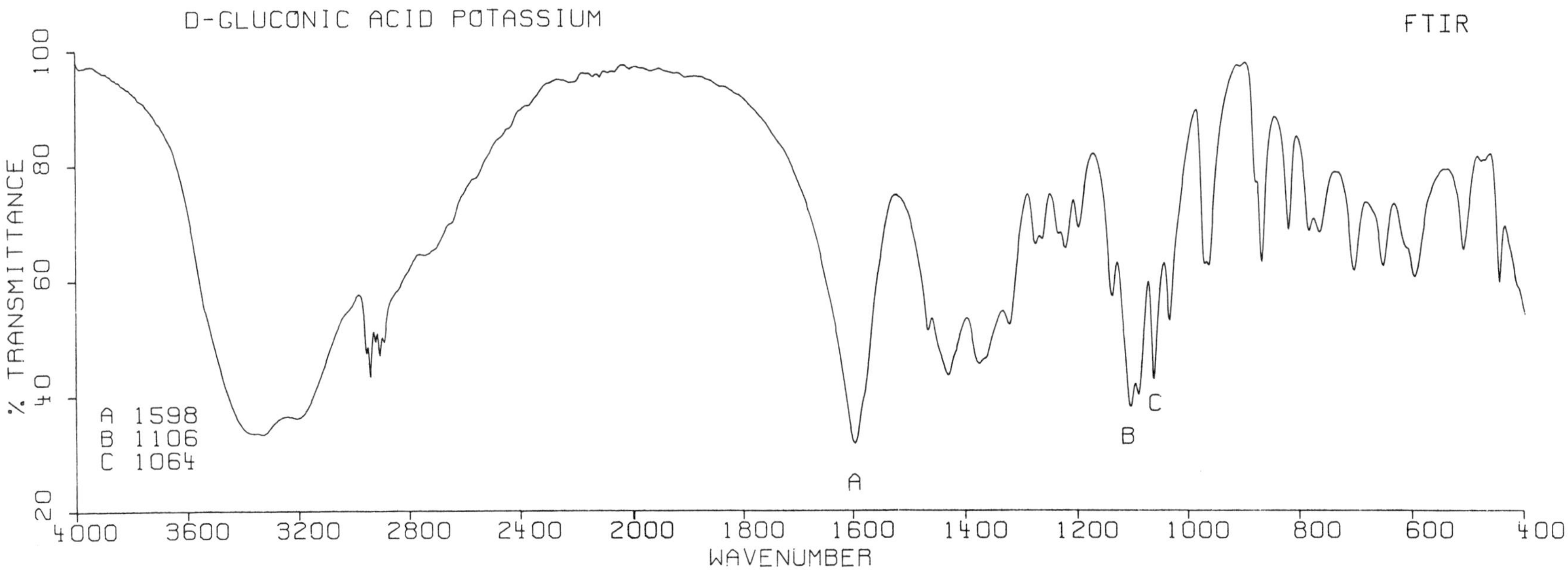

HEXAMETHYLENE AMILORIDE

$C_{12}H_{18}ClN_7O$

Molecular Weight: 311.77 (311.13)

Synonyms: 3-Amino-5-(N,N-hexamethylene)amino-N-(aminoiminomethyl)-6-chloropyrazinecarboxamid; 5-(N,N-hexamethylene)-amiloride
Trade Names:
Use: Sodium inhibitor

HPLC: Methanol: 3.0
GC:

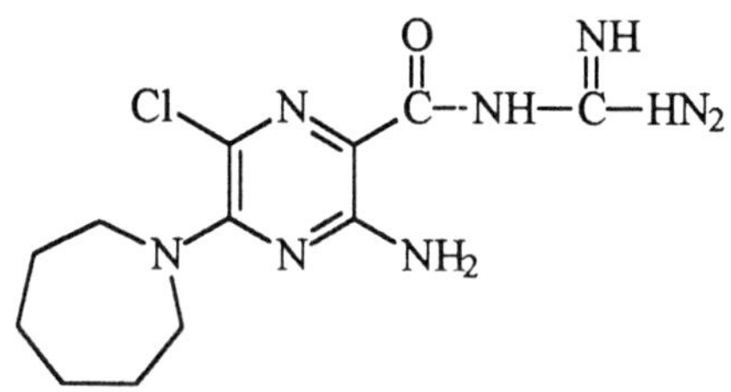

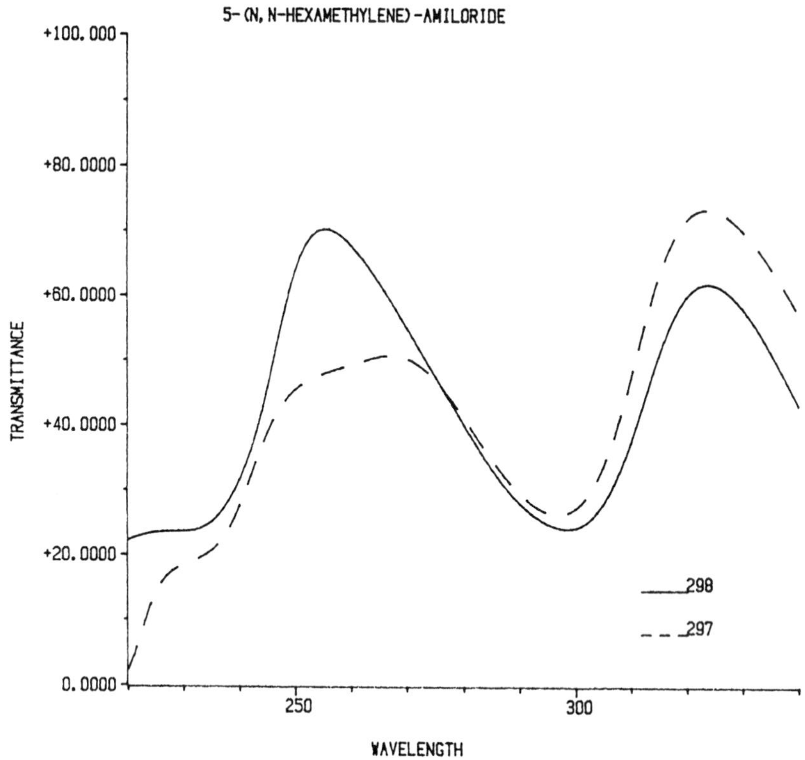

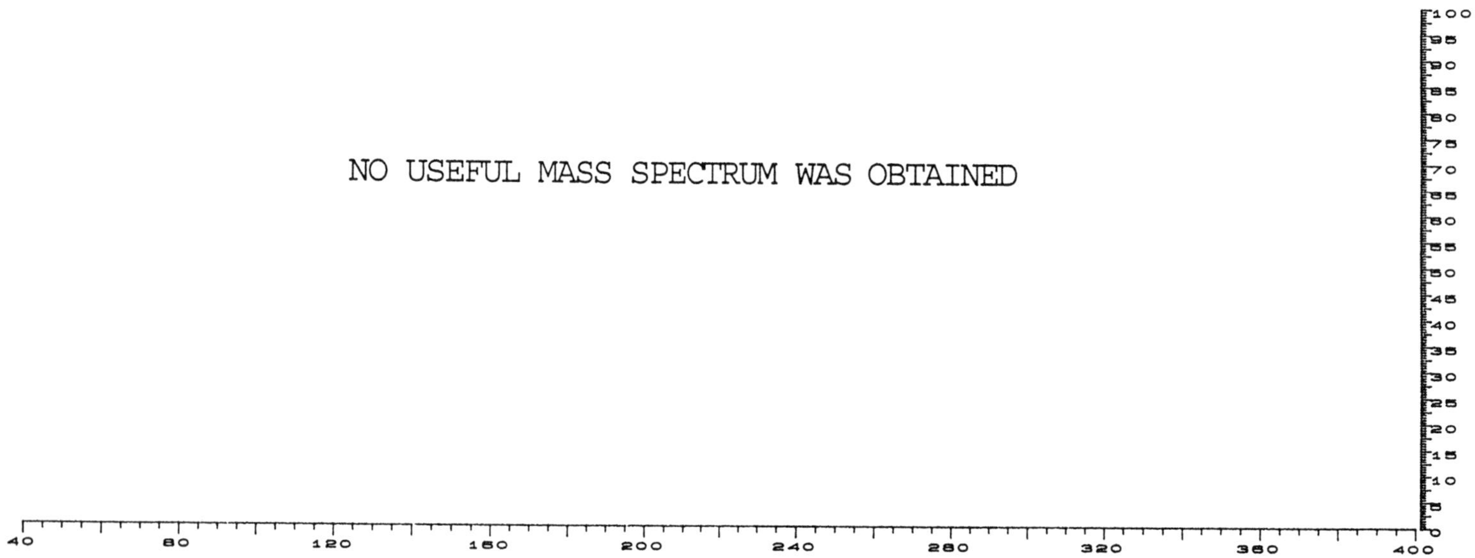

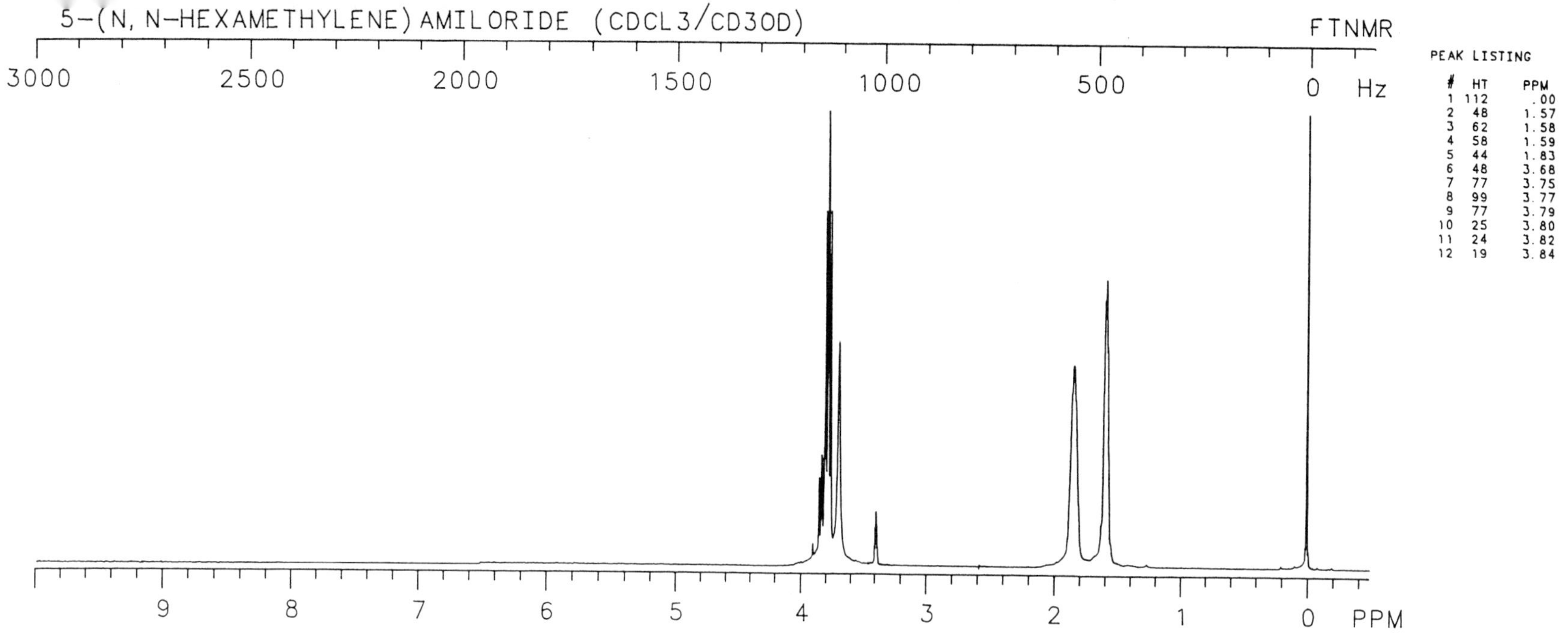
5-(N, N-HEXAMETHYLENE) AMILORIDE (CDCL3/CD3OD)
FTNMR
3000 2500 2000 1500 1000 500 0 Hz
9 8 7 6 5 4 3 2 1 0 PPM
PEAK LISTING
HT PPM
1 112 .00
2 48 1.57
3 62 1.58
4 58 1.59
5 44 1.83
6 48 3.68
7 77 3.75
8 99 3.77
9 77 3.79
10 25 3.80
11 24 3.82
12 19 3.84

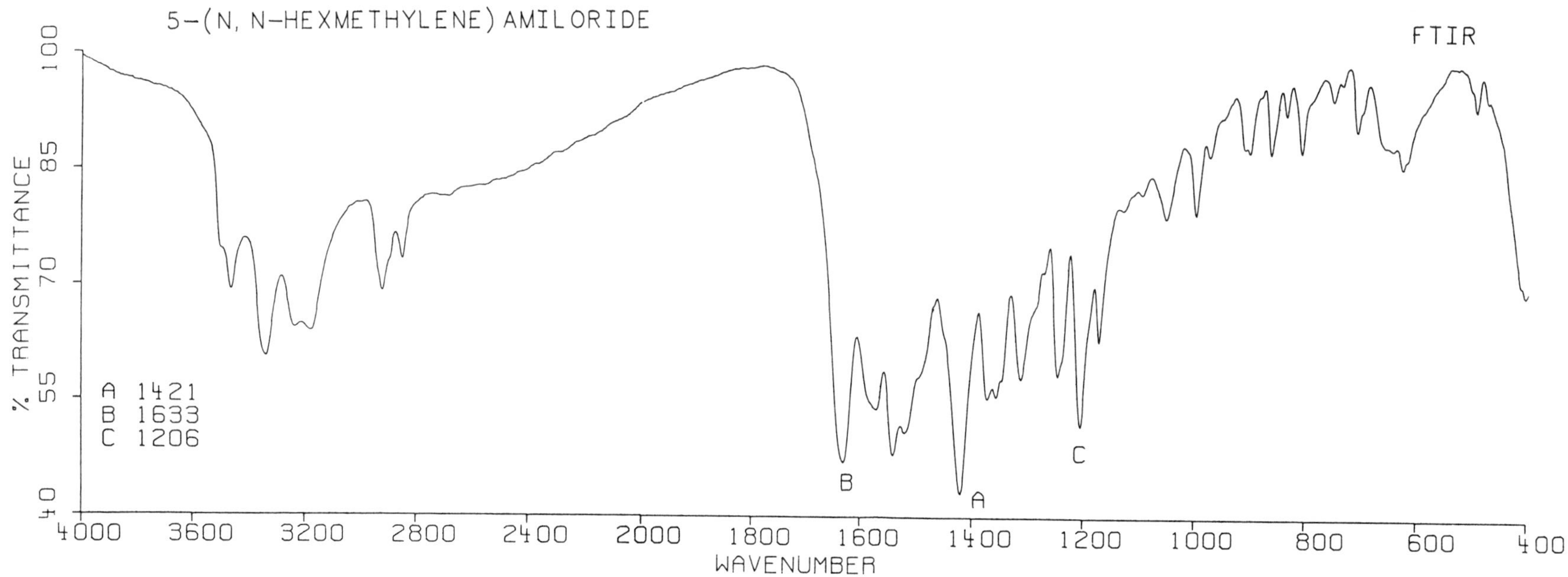
5-(N, N-HEXMETHYLENE) AMILORIDE
FTIR
% TRANSMITTANCE
100 85 70 55 40
A 1421
B 1633
C 1206
B
A
C
4000 3600 3200 2800 2400 2000 1800 1600 1400 1200 1000 800 600 400
WAVENUMBER

HYDROCORTISONE-17-BUTYRATE

$C_{25}H_{36}O_6$

Molecular Weight: 432.56 (432.25)

Synonyms: 11β,17,21-Trihydroxypregn-4-ene-3,20-dione-17-butyrate; cortisol-17-butyrate
Trade Names: Alfason, Locoid, Plancol
Use: Anti-inflammatory

HPLC: Methanol: 3.0
GC:

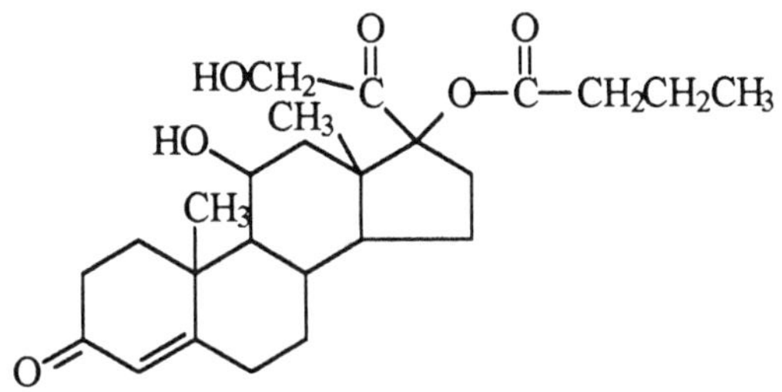

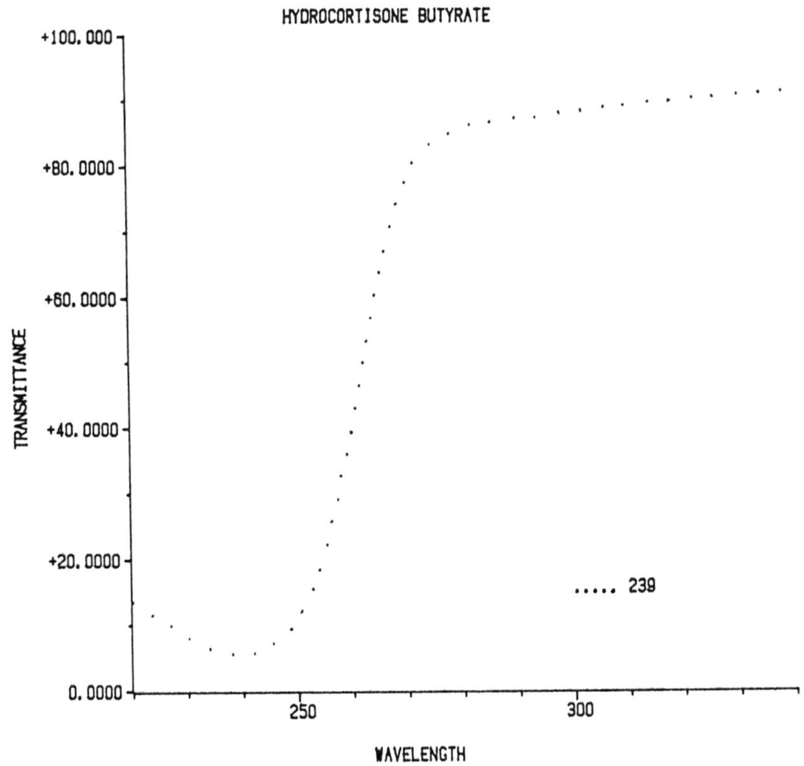

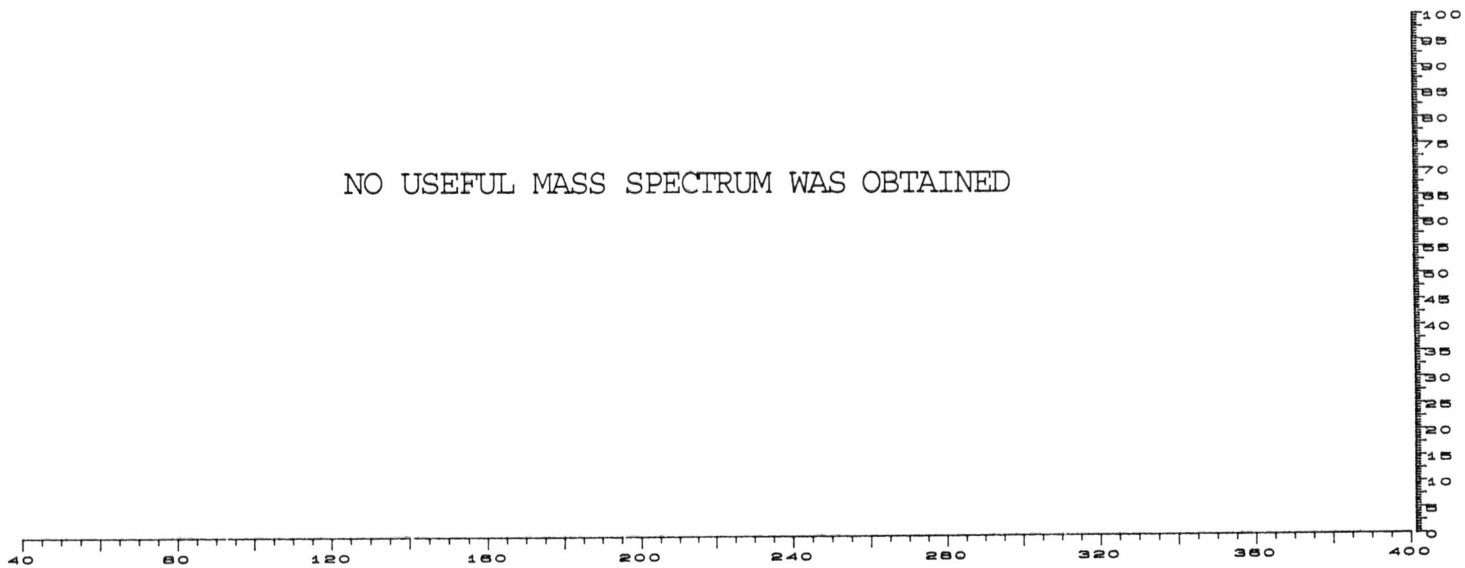

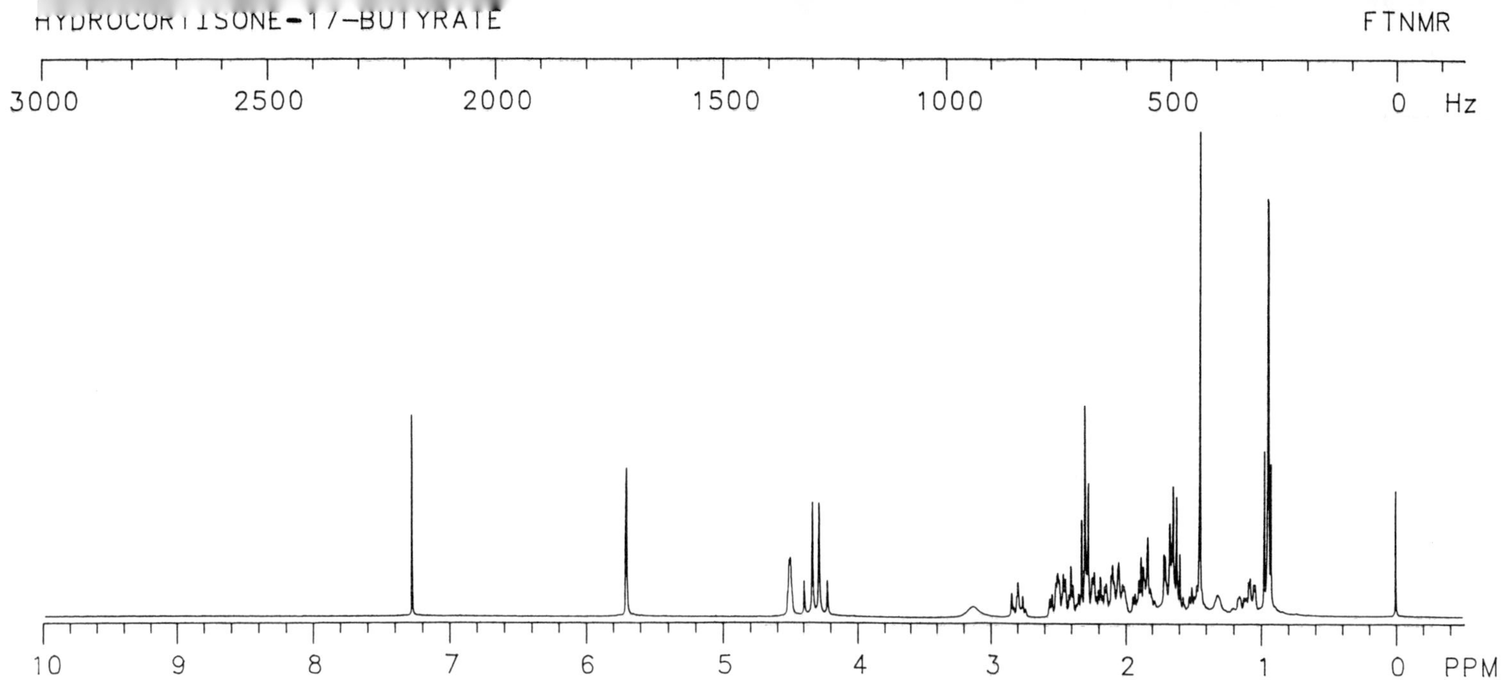

PEAK LISTING

#	HT	PPM
1	26	.00
2	31	.92
3	85	.94
4	85	.95
5	34	.97
6	99	1.44
7	13	1.59
8	24	1.61
9	27	1.64
10	13	1.65
11	19	1.66
12	12	1.69
13	13	1.70
14	16	1.83
15	27	2.27
16	43	2.29
17	20	2.32
18	23	4.27
19	23	4.32
20	30	5.69
21	41	7.27

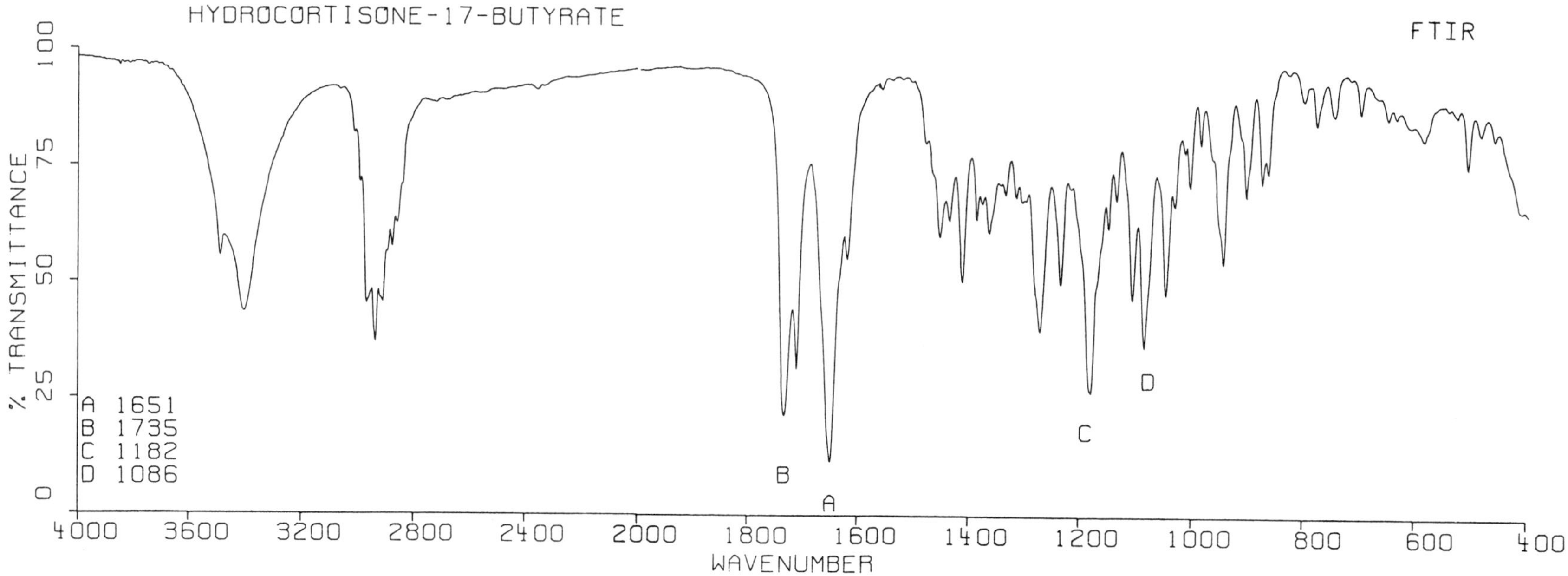

HYDROCORTISONE-21-CYPIONATE

$C_{29}H_{42}O_6$

Molecular Weight: 486.65 (486.30)

Synonyms: 11β,17,21-Trihydroxypregn-4-ene-3,20-dione-21-cypionate; cortisol-21-cyclopentanepropionate
Trade Names:
Use: Anti-inflammatory

HPLC: Methanol: 3.5
GC:

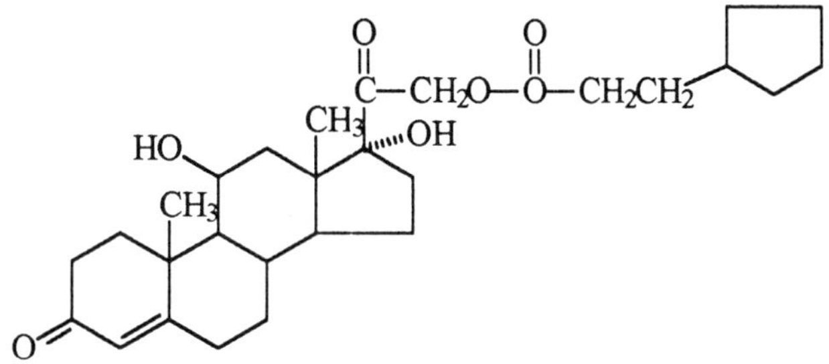

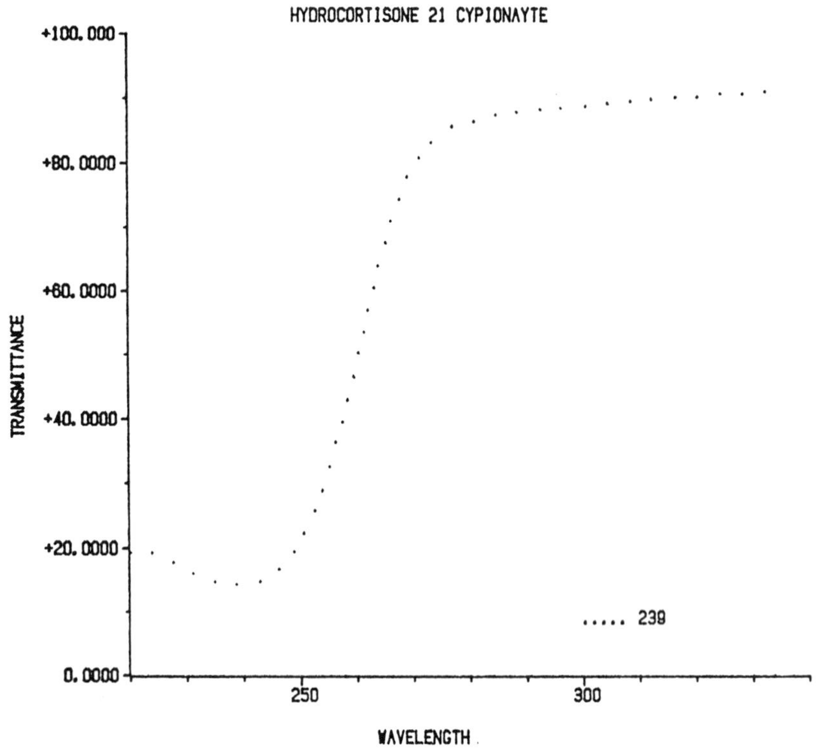

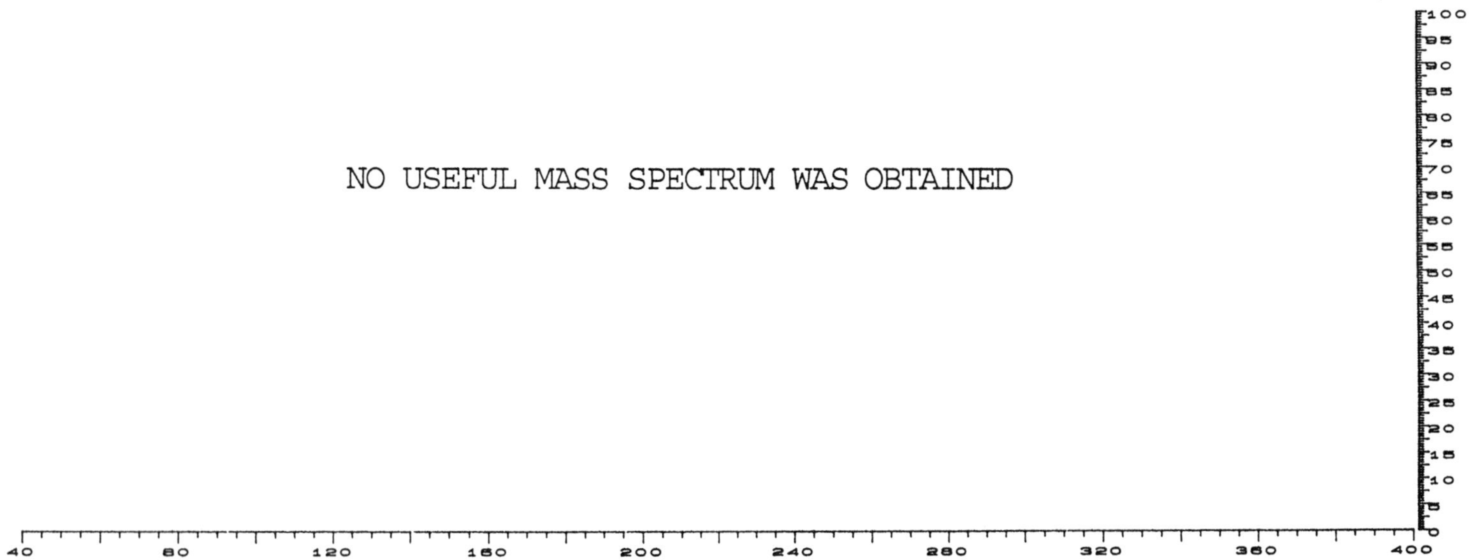

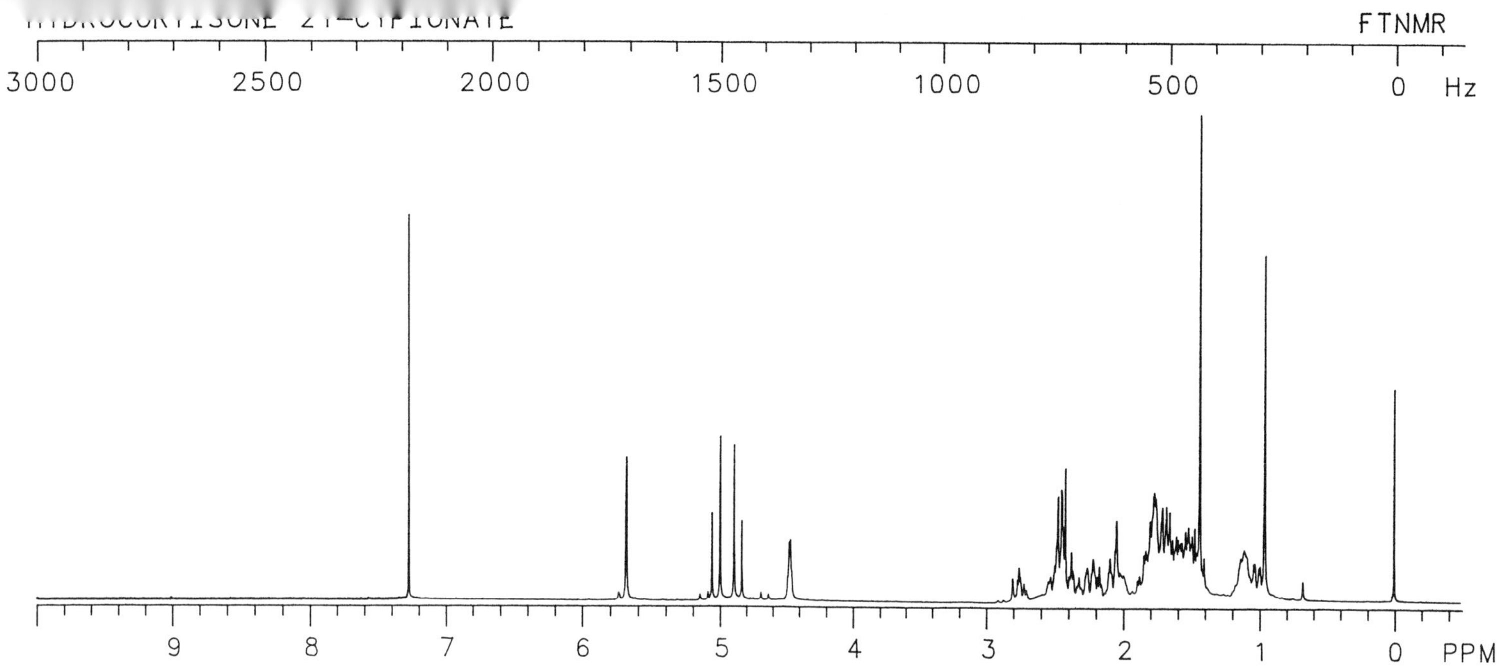

PEAK LISTING

#	HT	PPM
1	43	.00
2	70	.95
3	99	1.43
4	31	4.89
5	33	4.99
6	29	5.67
7	78	7.27

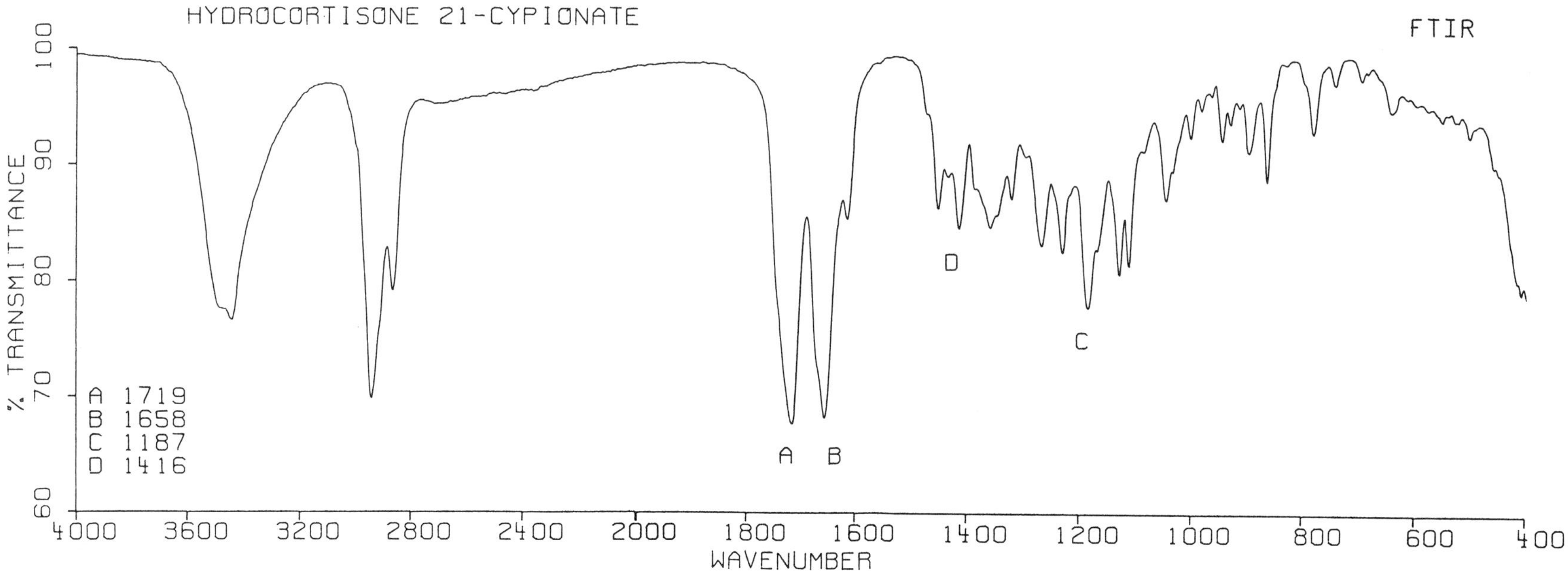

HYDROCORTISONE-17-VALERATE

$C_{26}H_{38}O_6$

Molecular Weight: 446.58 (446.27)

Synonyms: 11β,17,21-Trihydroxypregn-4-ene-3,20-dione-17-valerate; cortisol-17-valerate
Trade Names: Westcort
Use: Anti-inflammatory

HPLC: Methanol: 3.2
GC:

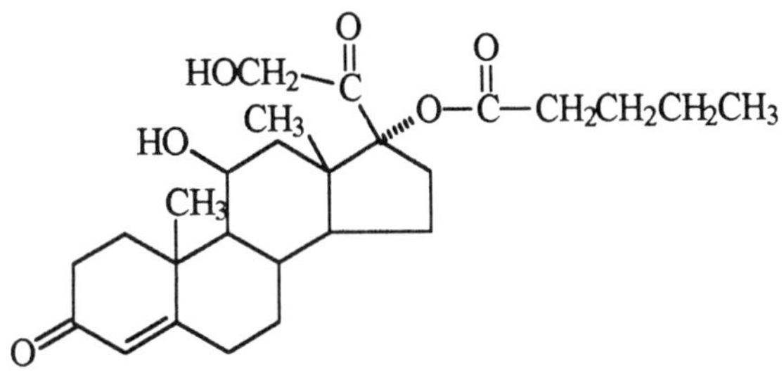

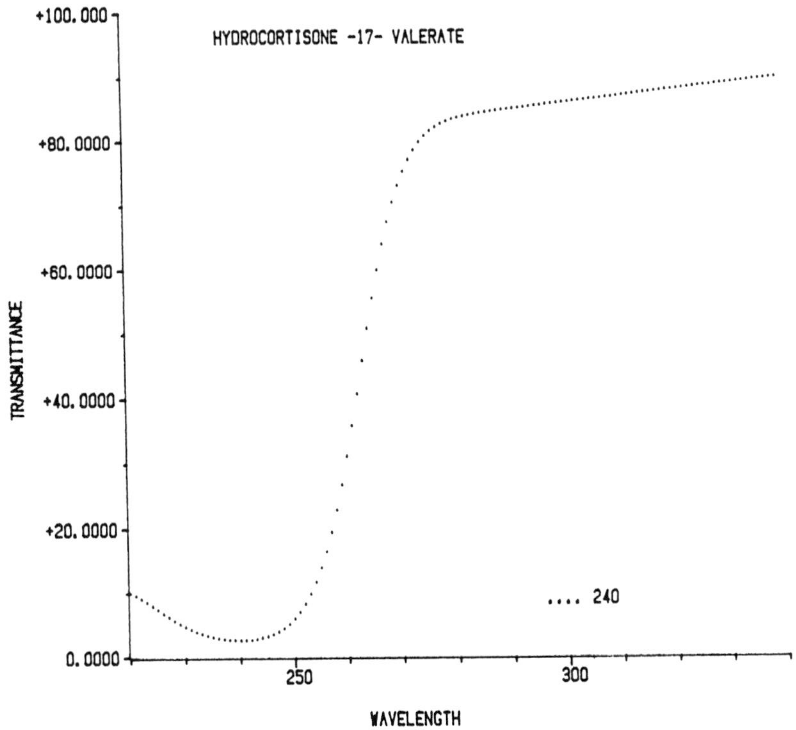

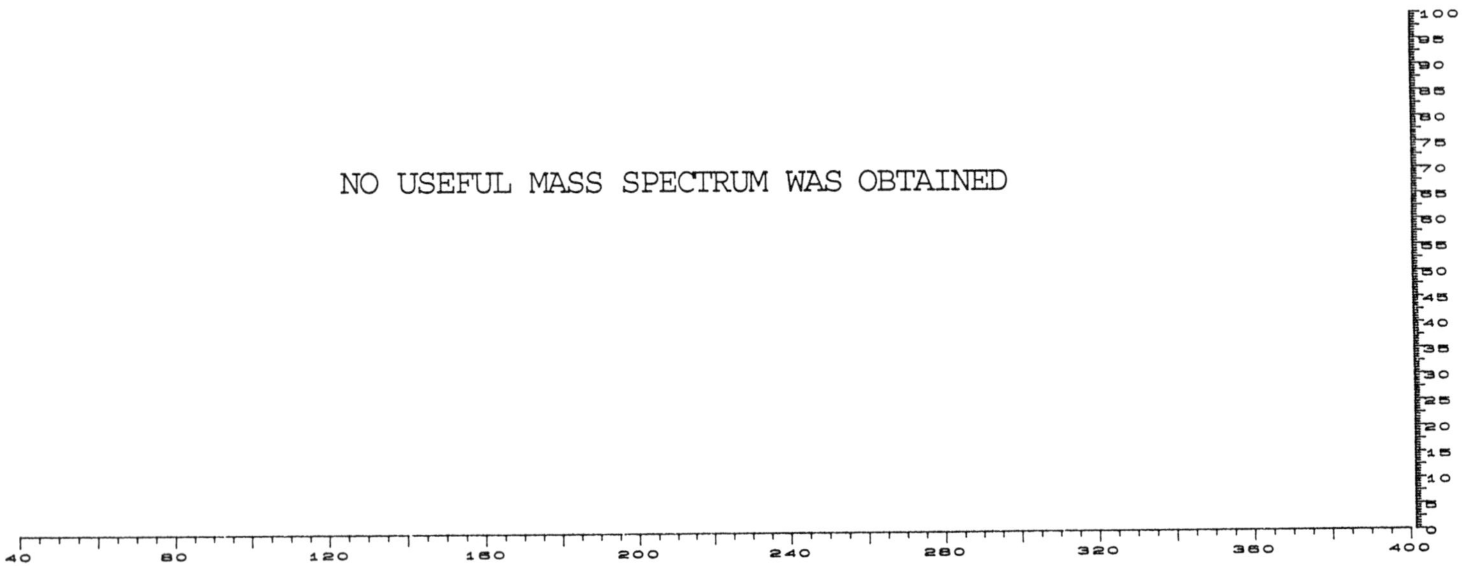

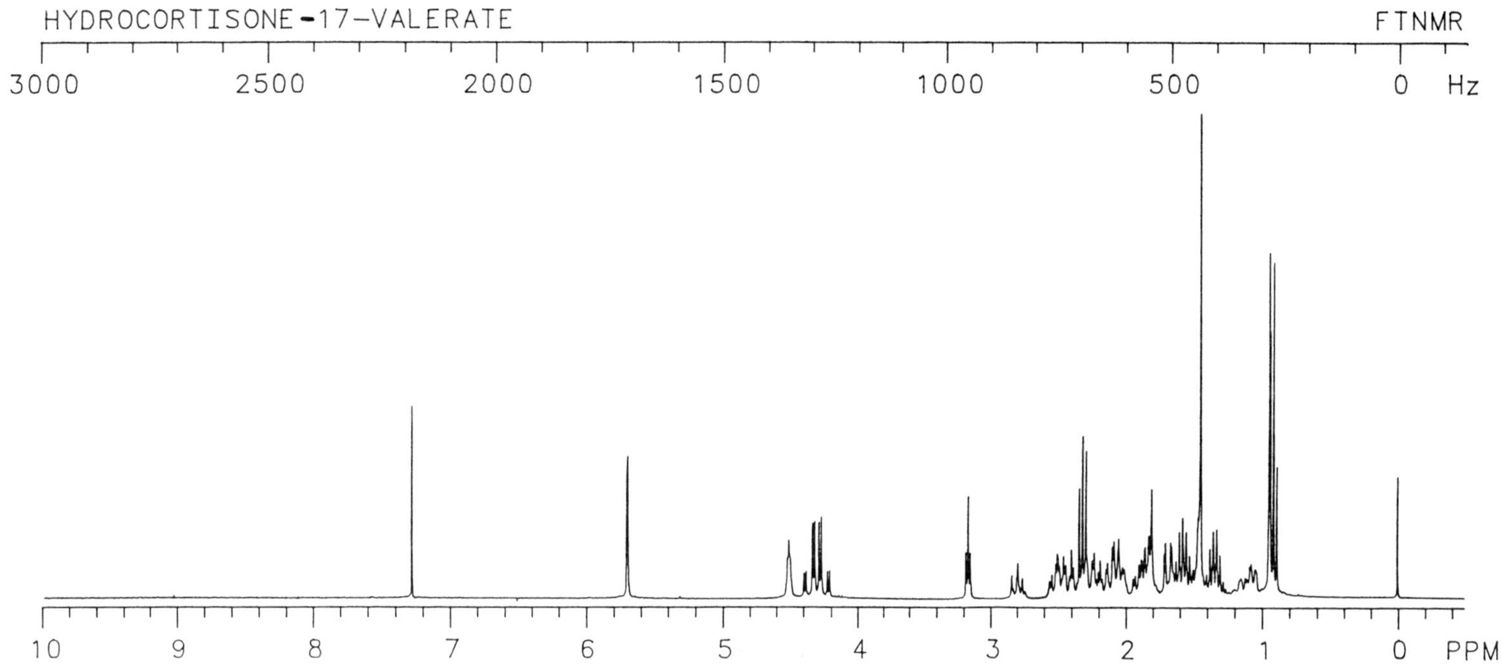

PEAK LISTING

#	HT	PPM
1	24	.00
2	27	.88
3	68	.91
4	54	.93
5	70	.94
6	14	1.32
7	13	1.35
8	99	1.44
9	16	1.46
10	13	1.55
11	16	1.57
12	13	1.60
13	22	1.80
14	12	1.82
15	30	2.29
16	33	2.31
17	22	2.34
18	20	3.16
19	16	4.26
20	15	4.28
21	15	4.31
22	15	4.32
23	29	5.69
24	39	7.27

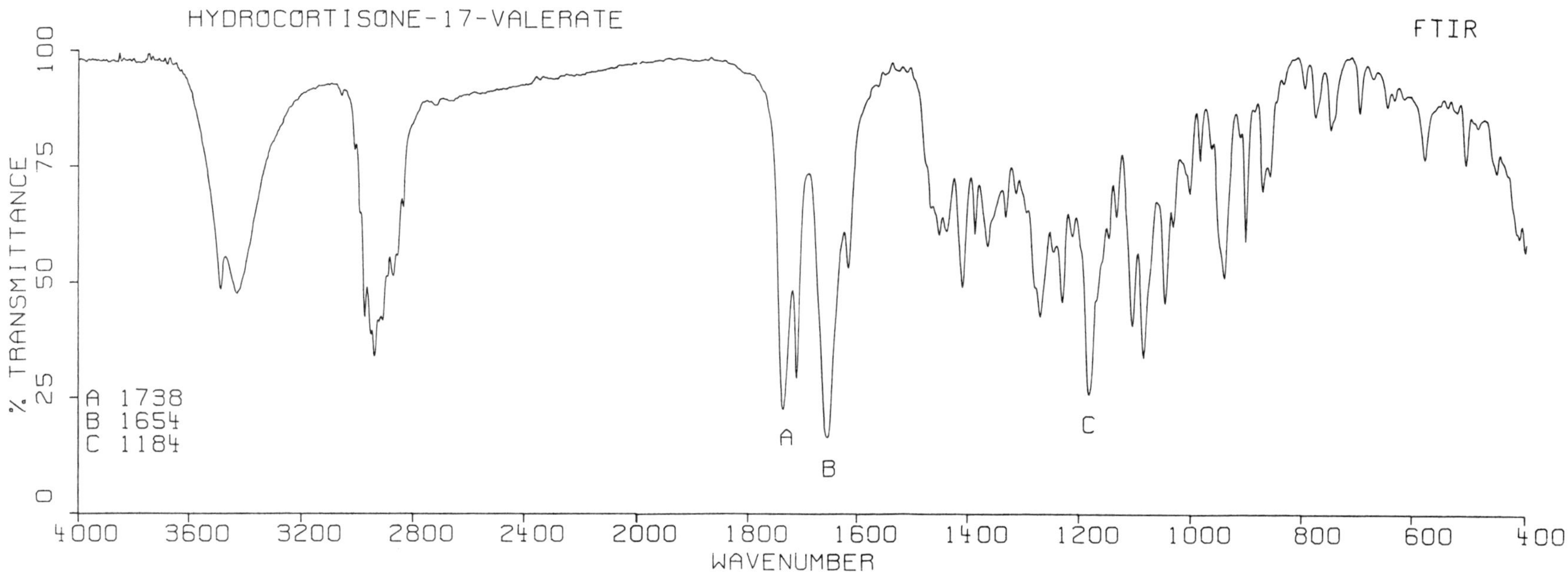

γ-HYDROXYBUTYRIC ACID SODIUM

$C_4H_7O_3Na$

Molecular Weight: 126.09 (126.03)

Synonyms: 3-Hydroxybutanoic acid sodium, GHB sodium
Trade Names:
Use:

HPLC:
GC:

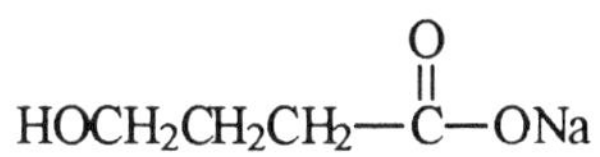

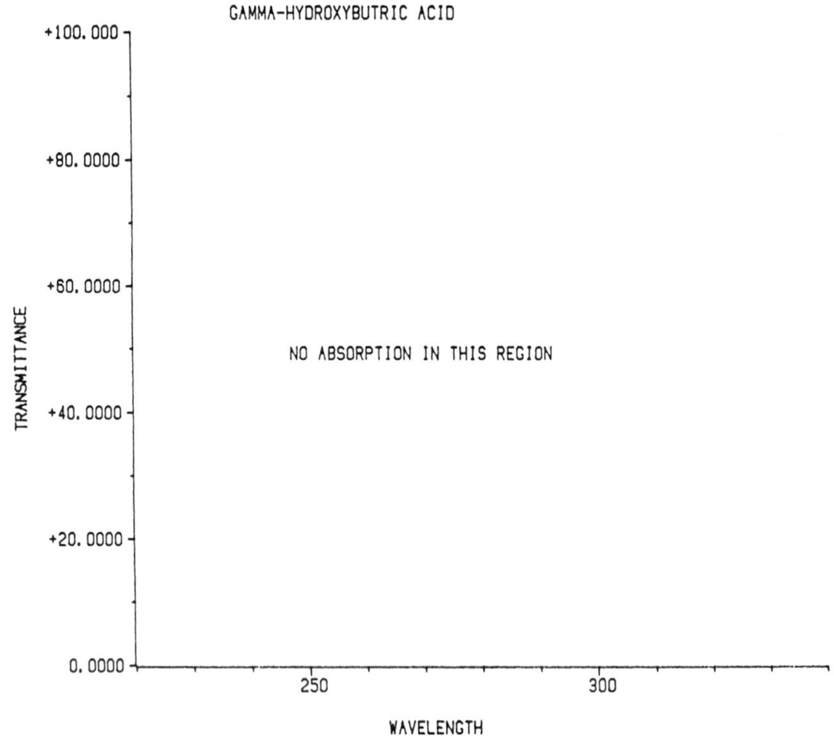

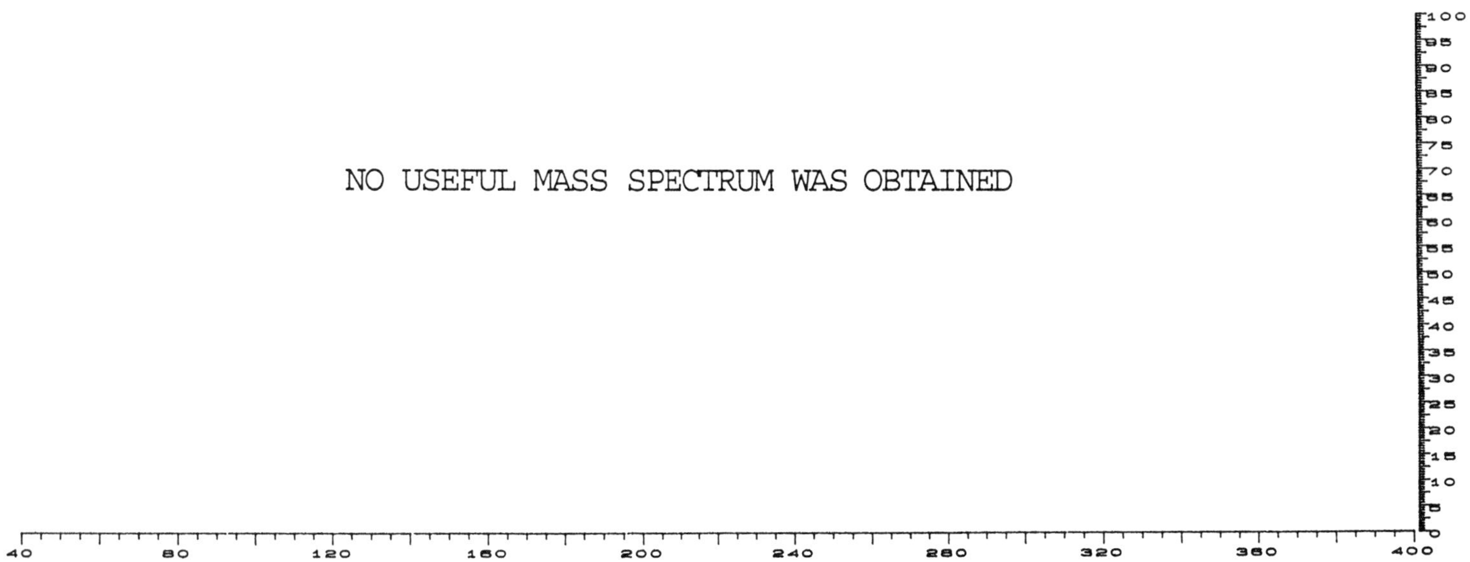

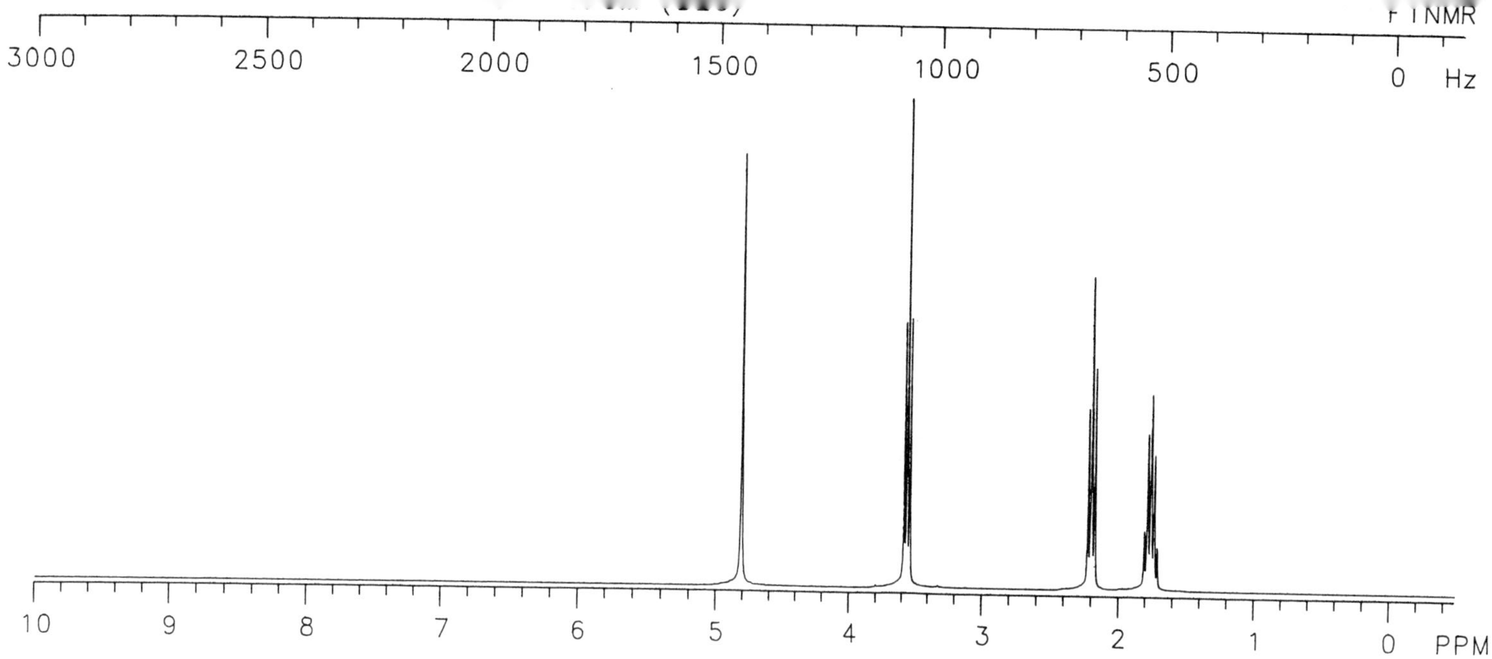

PEAK LISTING

#	HT	PPM
1	27	1.72
2	40	1.75
3	33	1.75
4	25	1.77
5	32	1.77
6	12	1.79
7	45	2.16
8	37	2.18
9	64	2.19
10	37	2.21
11	54	3.53
12	99	3.55
13	54	3.57
14	88	4.80

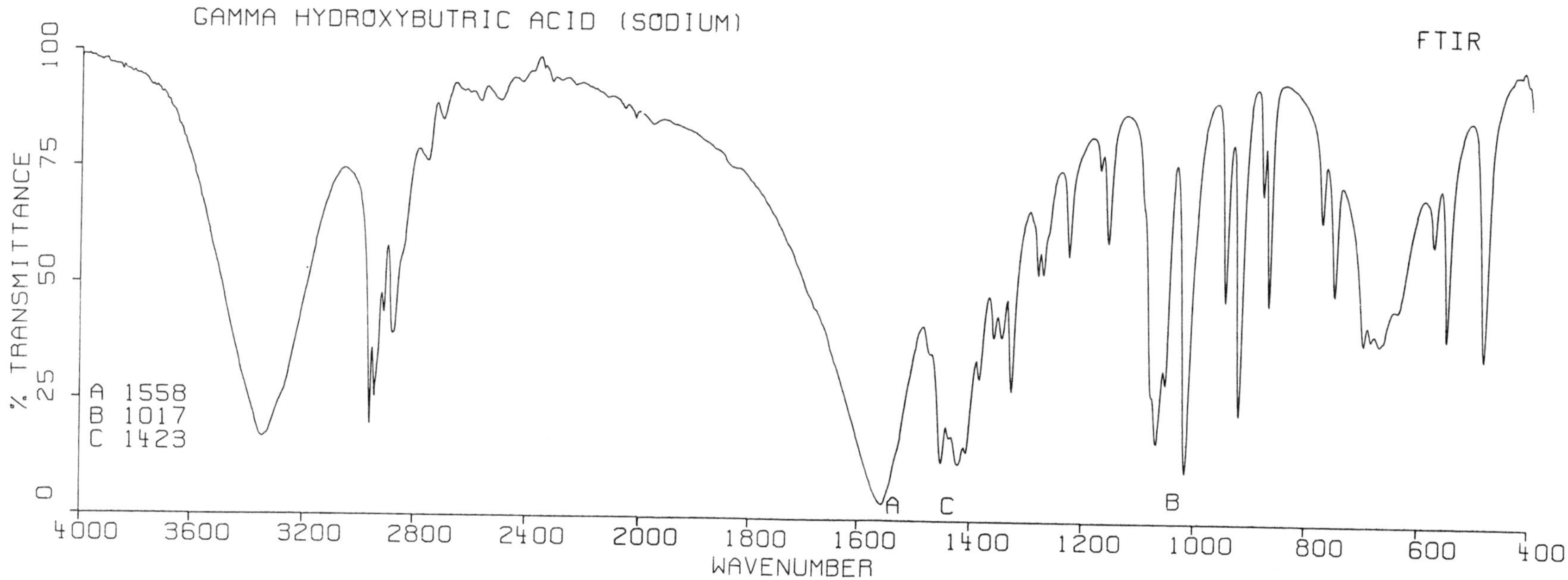

HYDROXYMETHAMPHETAMINE

$C_{10}H_{15}NO$

Molecular Weight: 165.24 (165.12)

Synonyms: 4-Hydroxy-N,α-dimethylbenzeneethanamine; p-hydroxymethamphetamine; pholderin
Trade Names:
Use: Central stimulant; metabolite of methamphetamine

HPLC:
GC:

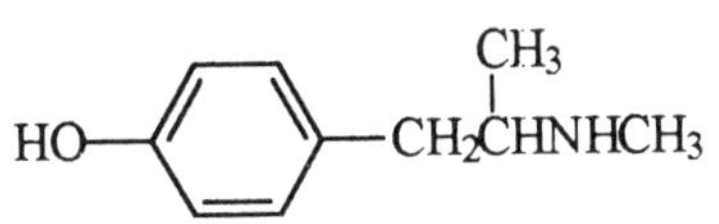

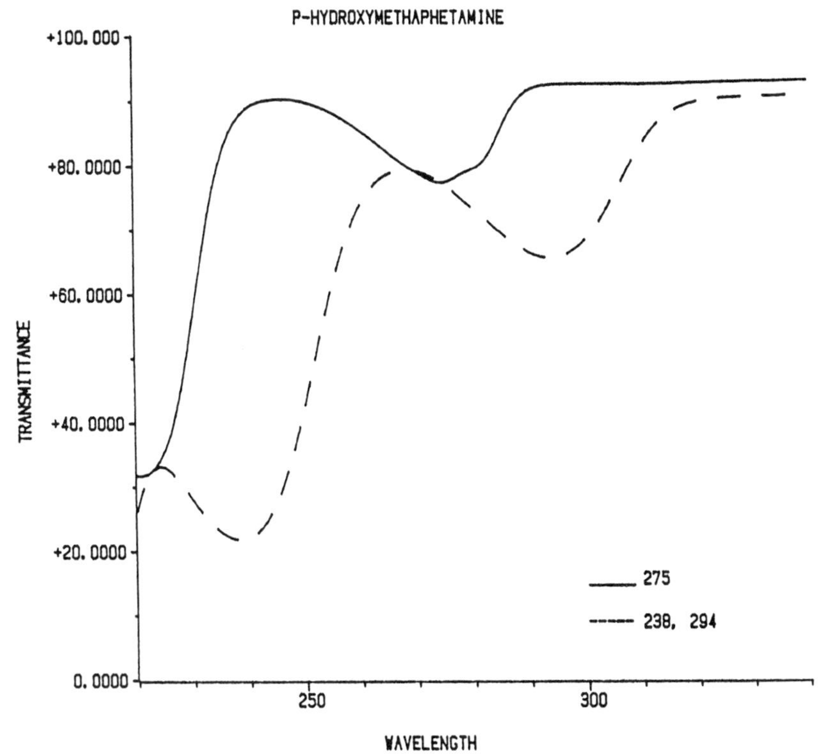

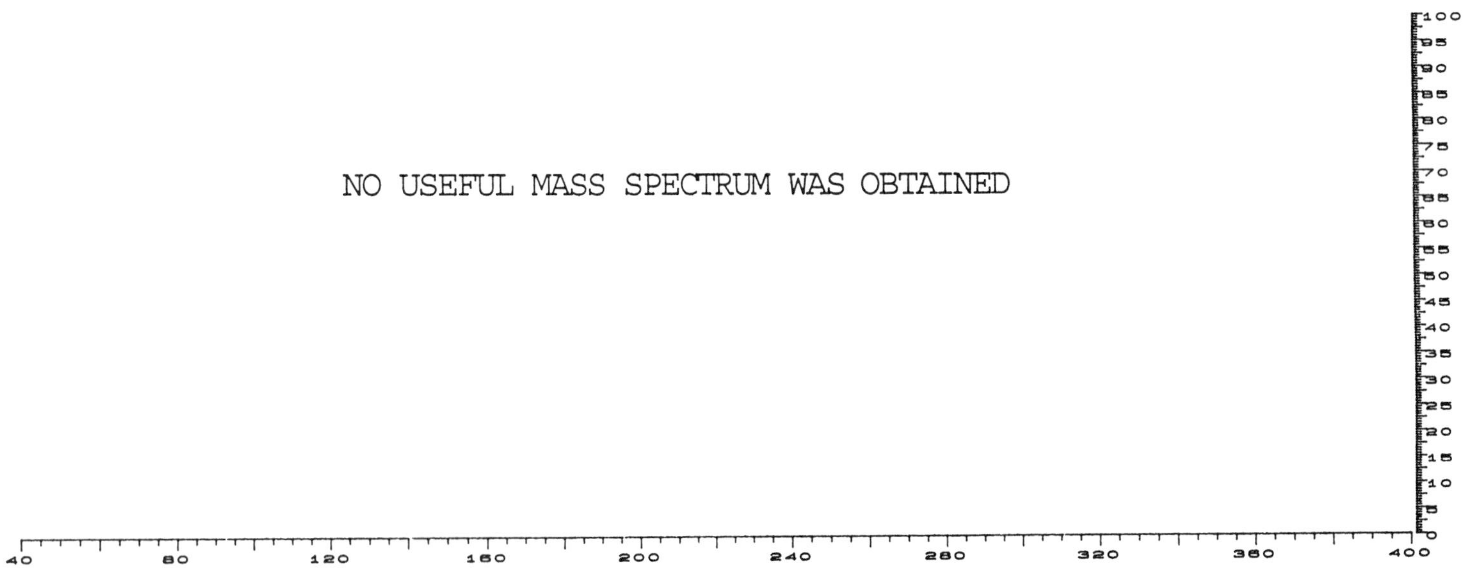

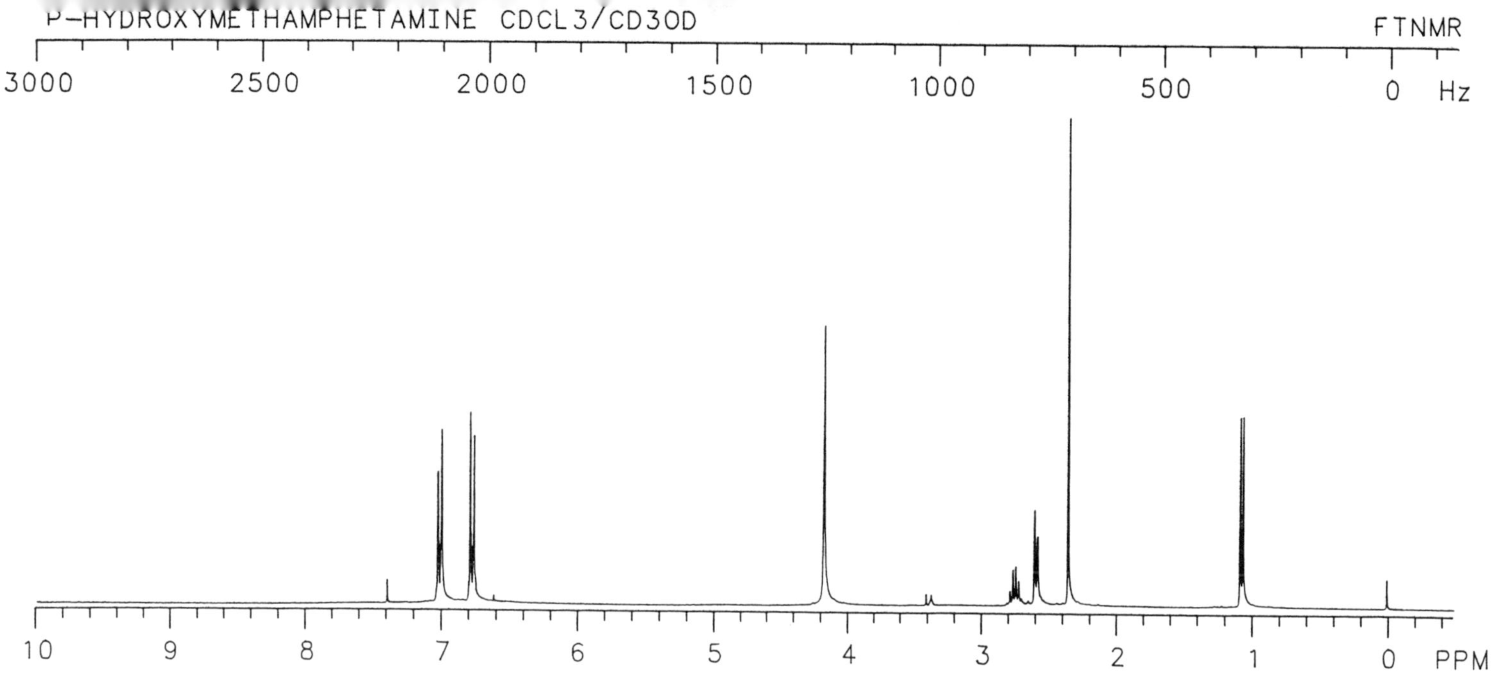

PEAK LISTING

#	HT	PPM
1	39	1.06
2	38	1.08
3	99	2.35
4	19	2.60
5	57	4.17
6	34	6.75
7	38	6.78
8	35	6.99
9	26	7.02

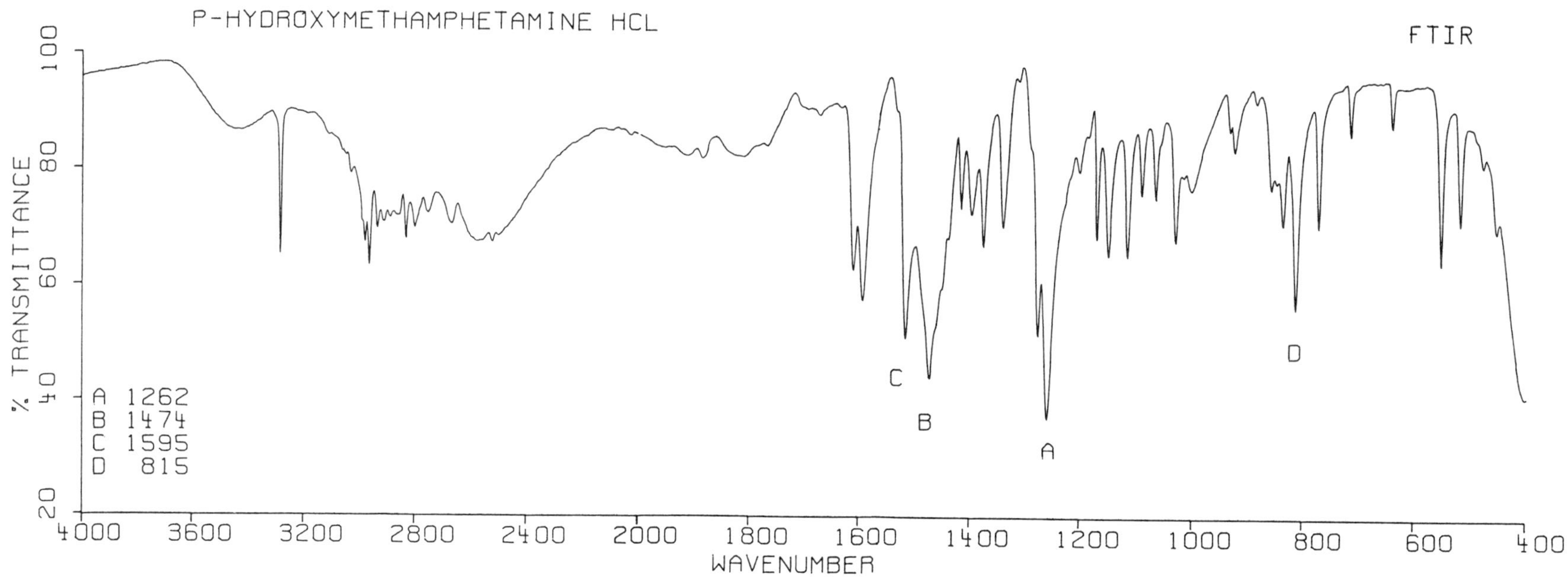

HYDROXYMETHYLETHYLPHENETHYLAMINE

$C_{11}H_{17}NO$

Molecular Weight: 179.26 (179.13)

Synonyms: 4-Methyl-α-ethyl-m-tyramine; 3-hydroxy-4-methyl-α-ethylphenethylamine; α-ethyl-3-hydroxy-4-methylphenethylamine
Trade Names:
Use:

HPLC: Ethanol: 2.0
GC: 1600; 200°

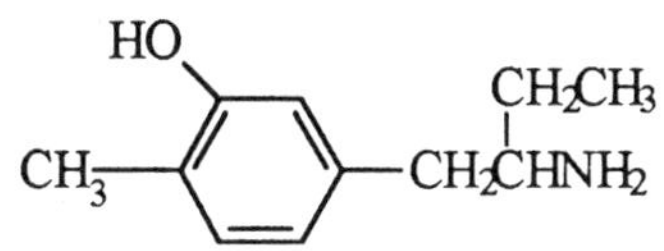

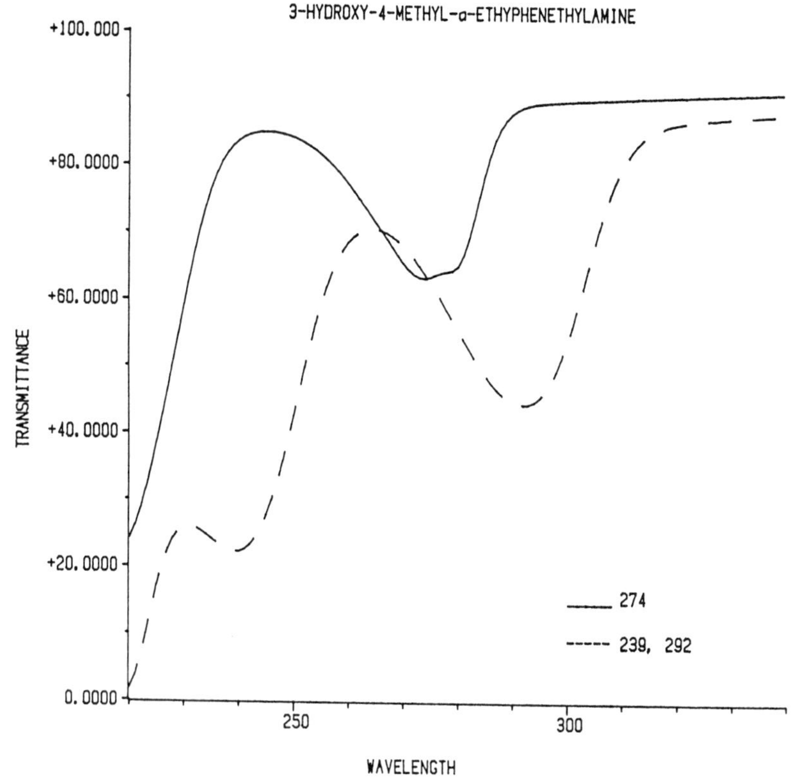

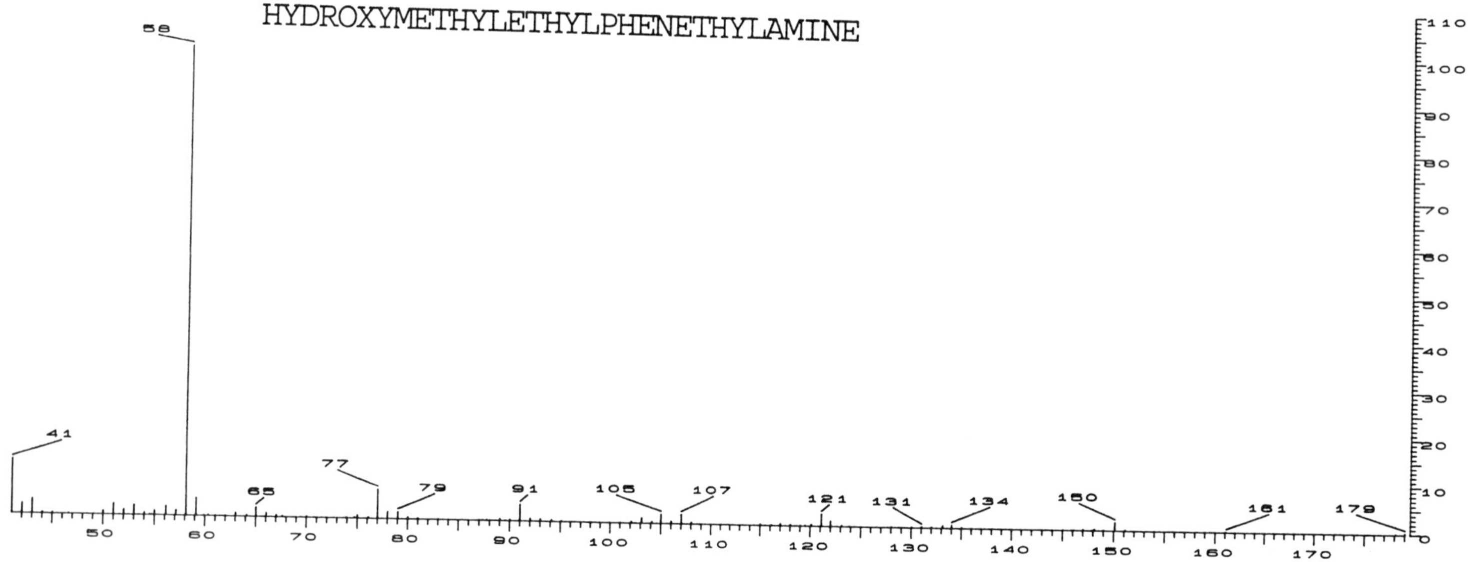

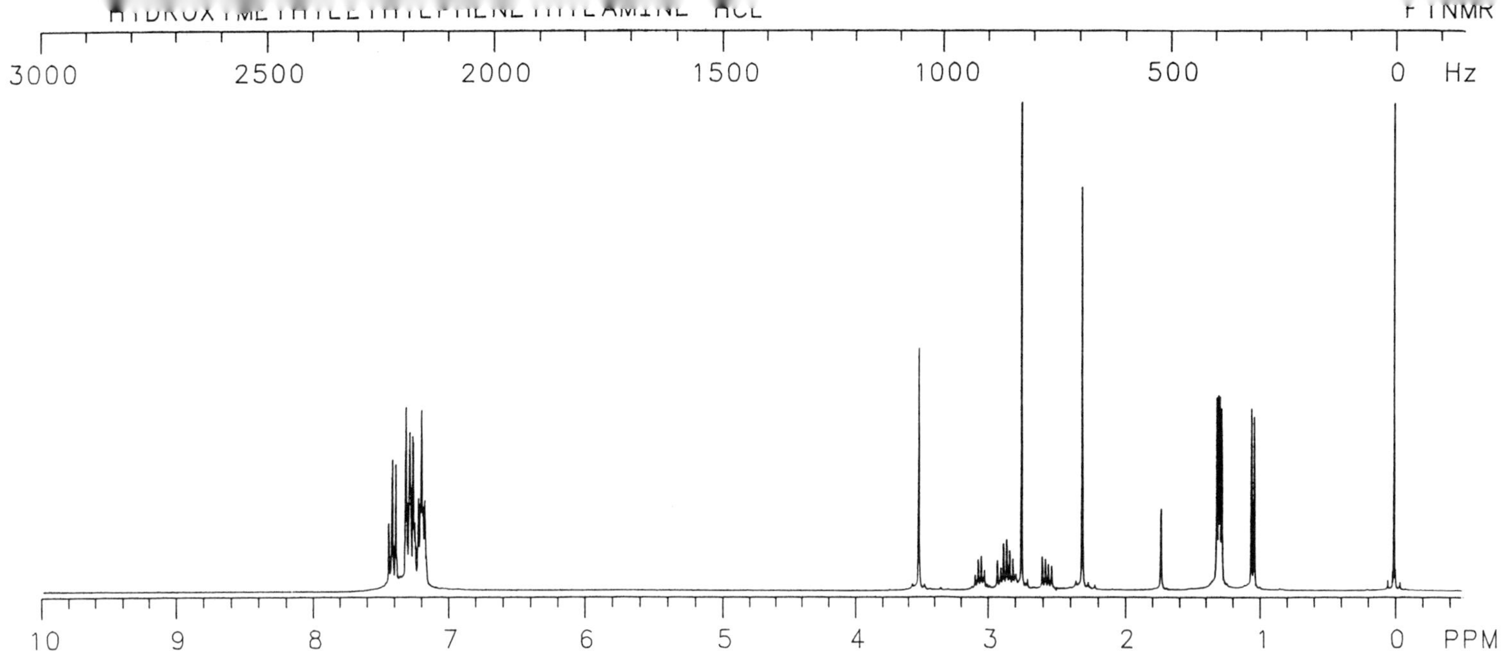

PEAK LISTING

#	HT	PPM
1	128	.00
2	35	1.03
3	37	1.05
4	37	1.27
5	39	1.28
6	39	1.29
7	39	1.31
8	16	1.72
9	82	2.30
10	99	2.74
11	49	3.51
12	18	7.16
13	17	7.18
14	36	7.19
15	17	7.20
16	18	7.21
17	14	7.24
18	30	7.25
19	31	7.25
20	32	7.28
21	18	7.30
22	37	7.30
23	28	7.31
24	25	7.38
25	26	7.40
26	14	7.43

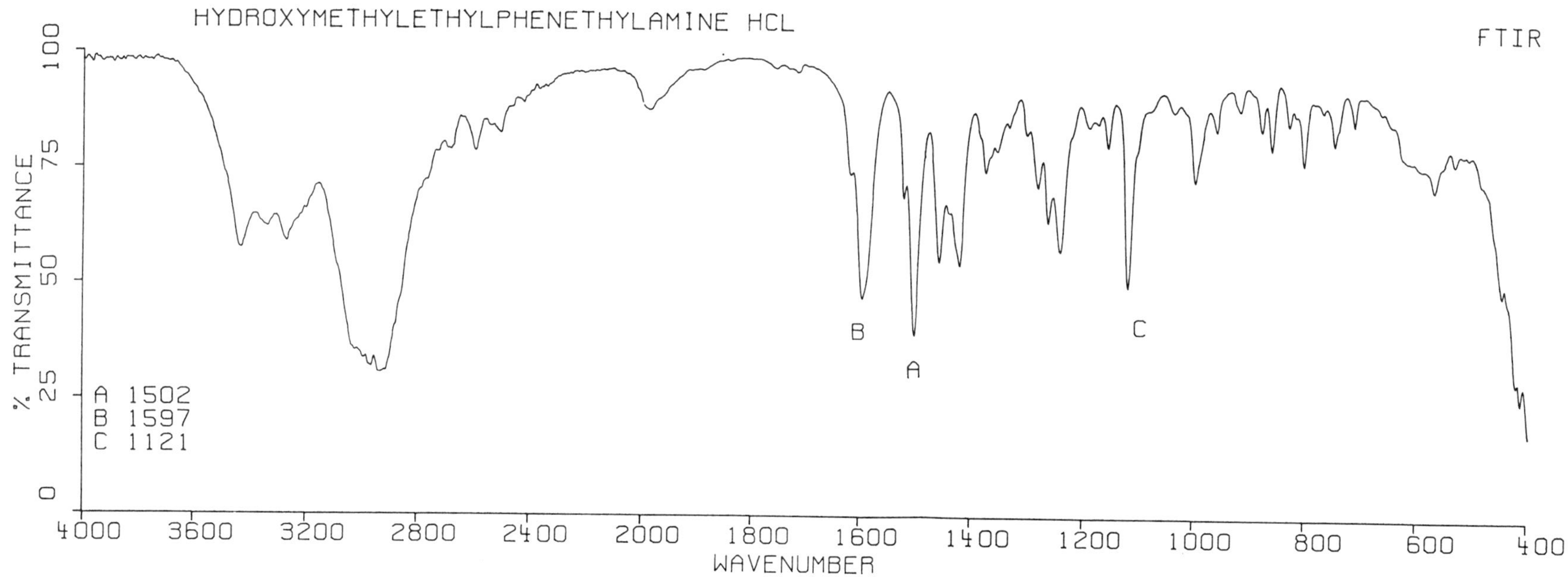

2α-HYDROXYTESTOSTERONE

$C_{19}H_{28}O_3$

Molecular Weight: 304.43 (304.20)

Synonyms: 4-Androstene-2α,17β-diol-3-one
Trade Names:
Use: Testosterone metabolite

HPLC:
GC: 2781; 280°

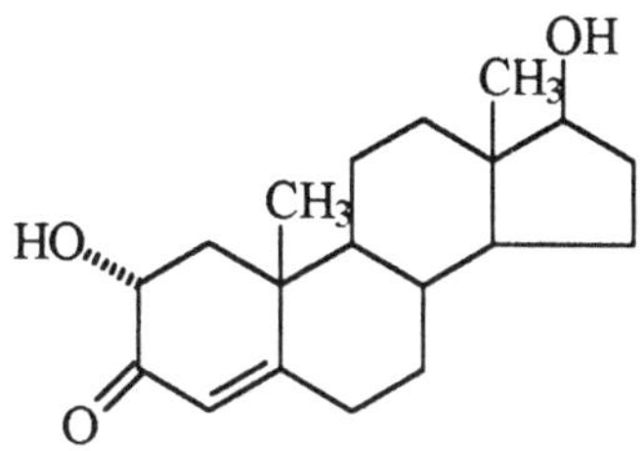

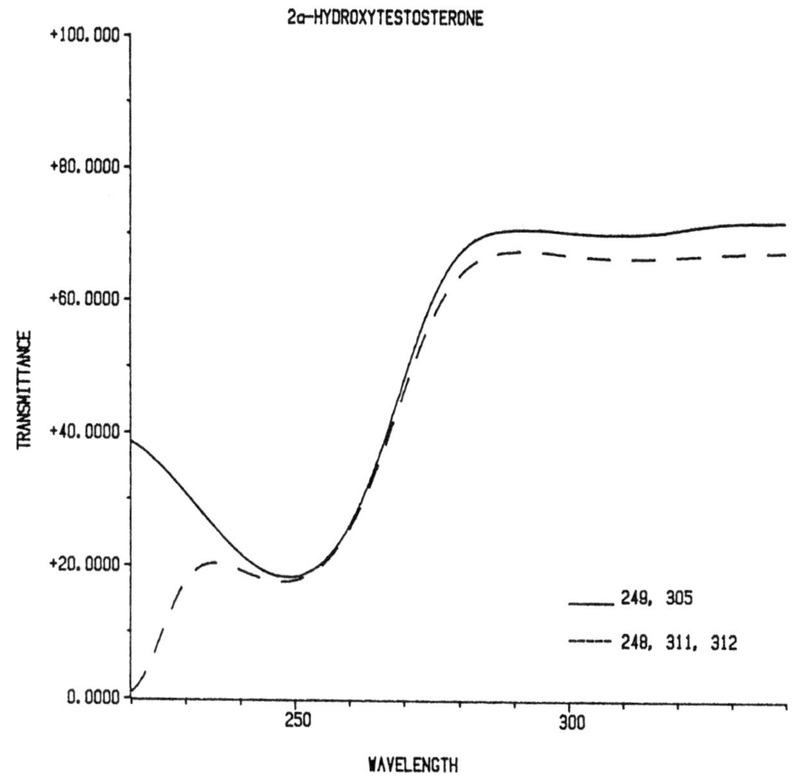

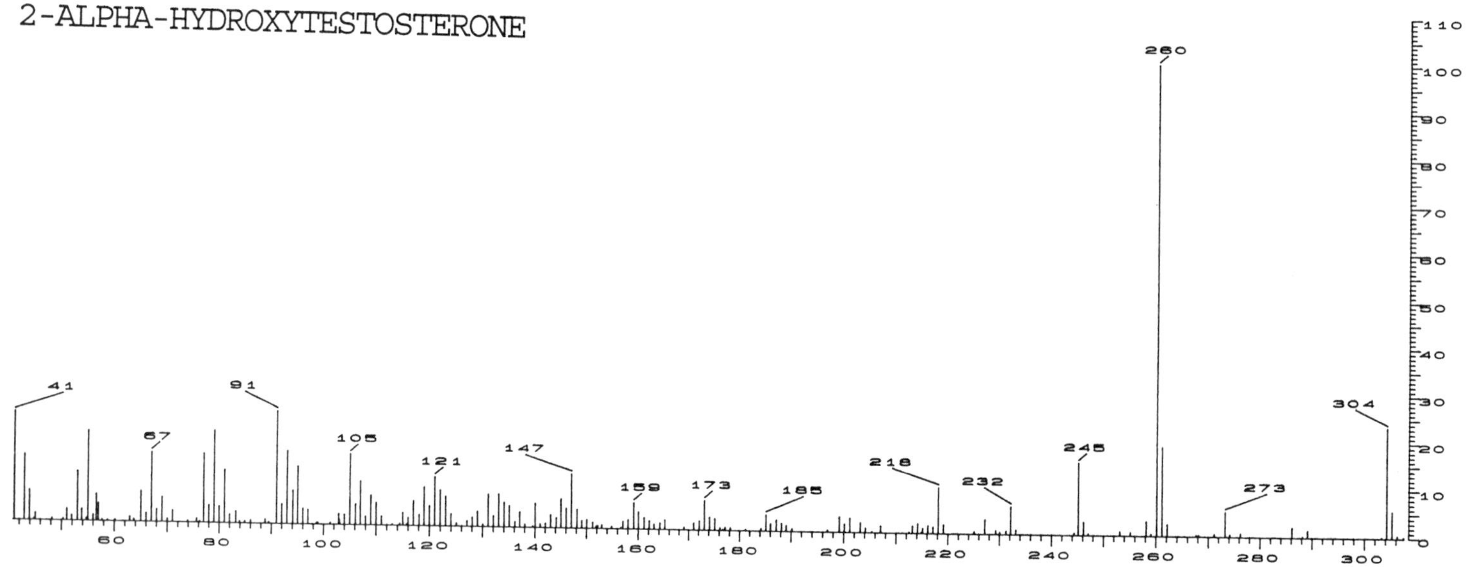

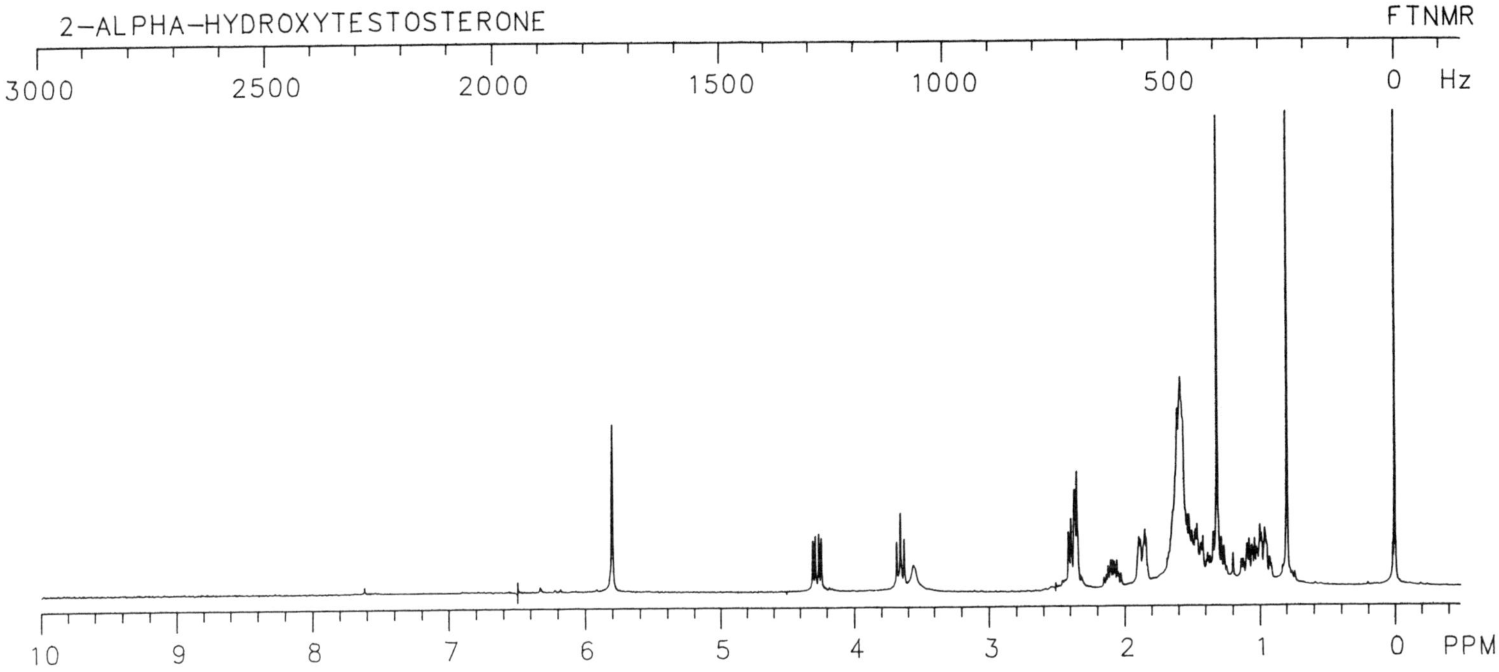

PEAK LISTING

#	HT	PPM
1	176	.00
2	97	.79
3	96	1.30
4	35	1.56
5	38	1.56
6	43	1.57
7	36	1.60
8	23	1.61
9	24	2.34
10	34	5.79
11	28	5.79

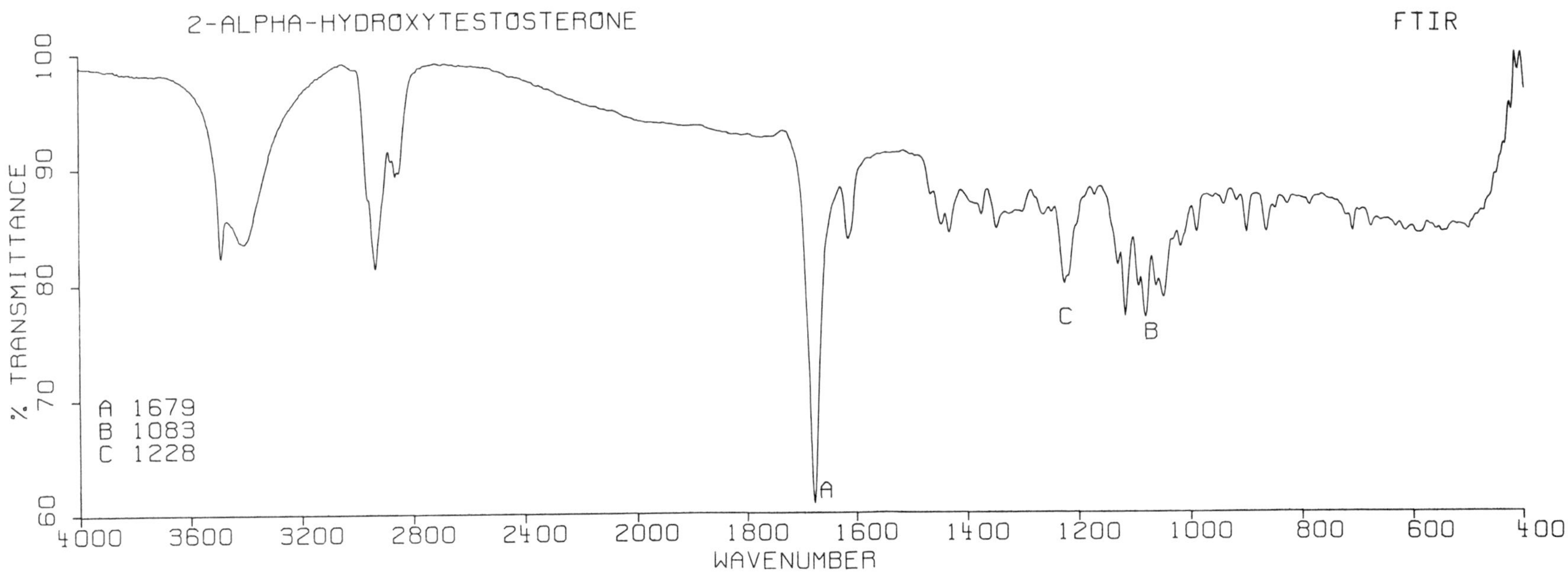

6β-HYDROXYTESTOSTERONE

$C_{19}H_{28}O_3$

Molecular Weight: 304.43 (304.20)

Synonyms: 4-Androstene-6β,17β-diol-3-one
Trade Names:
Use: Testosterone metabolite

HPLC: Ethanol: 2.5
GC: 2745; 250°

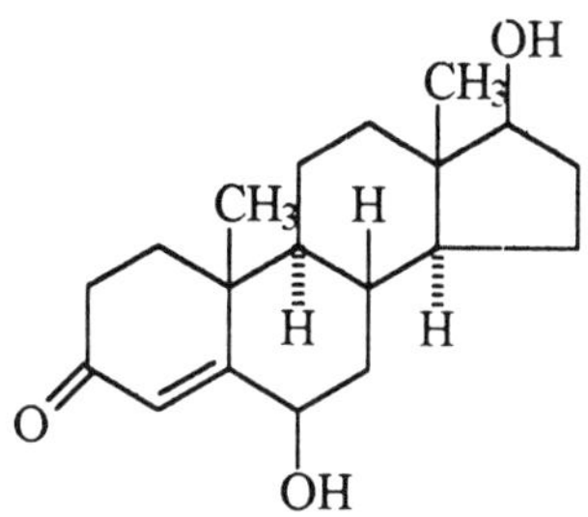

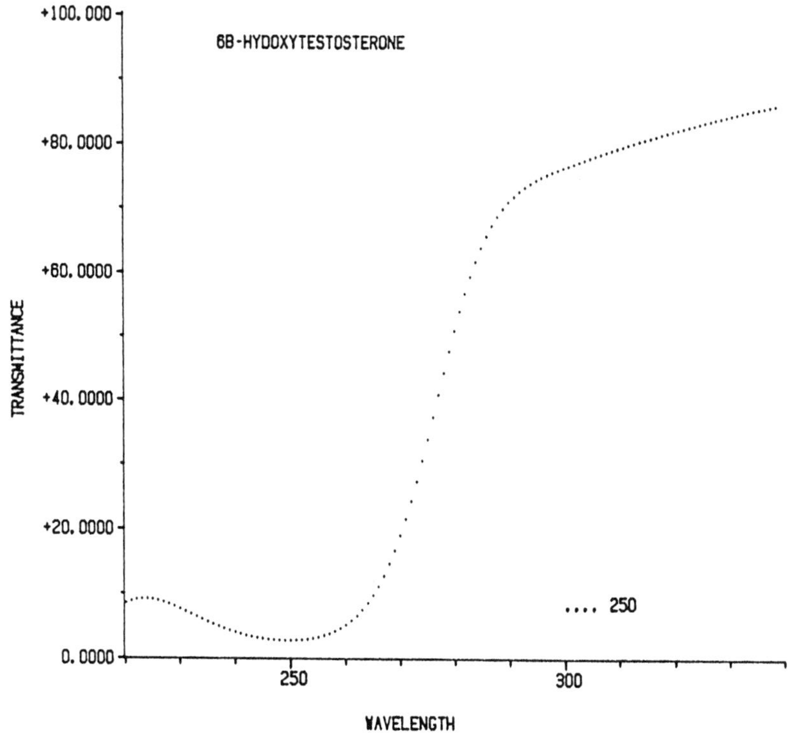

6-BETA-HYDROXYTESTOSTERONE

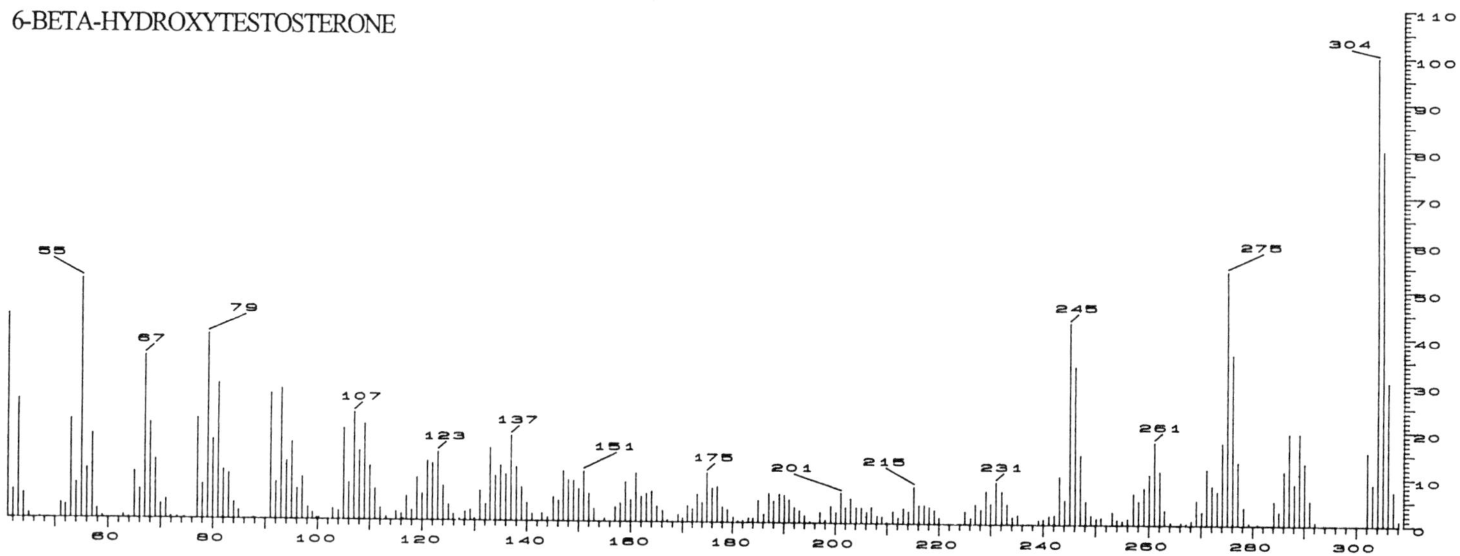

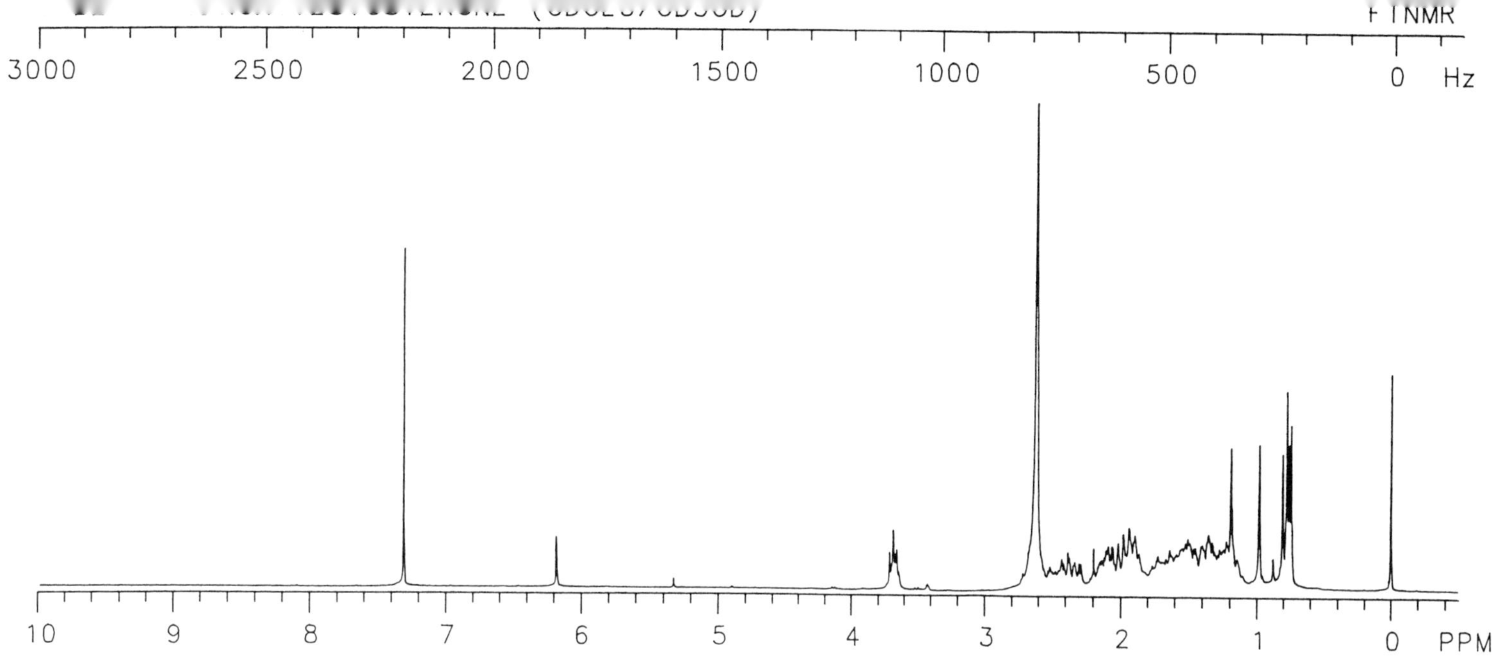

PEAK LISTING

#	HT	PPM
1	44	.00
2	33	.74
3	29	.76
4	40	.77
5	28	.80
6	30	.98
7	29	1.19
8	99	2.62
9	69	7.31

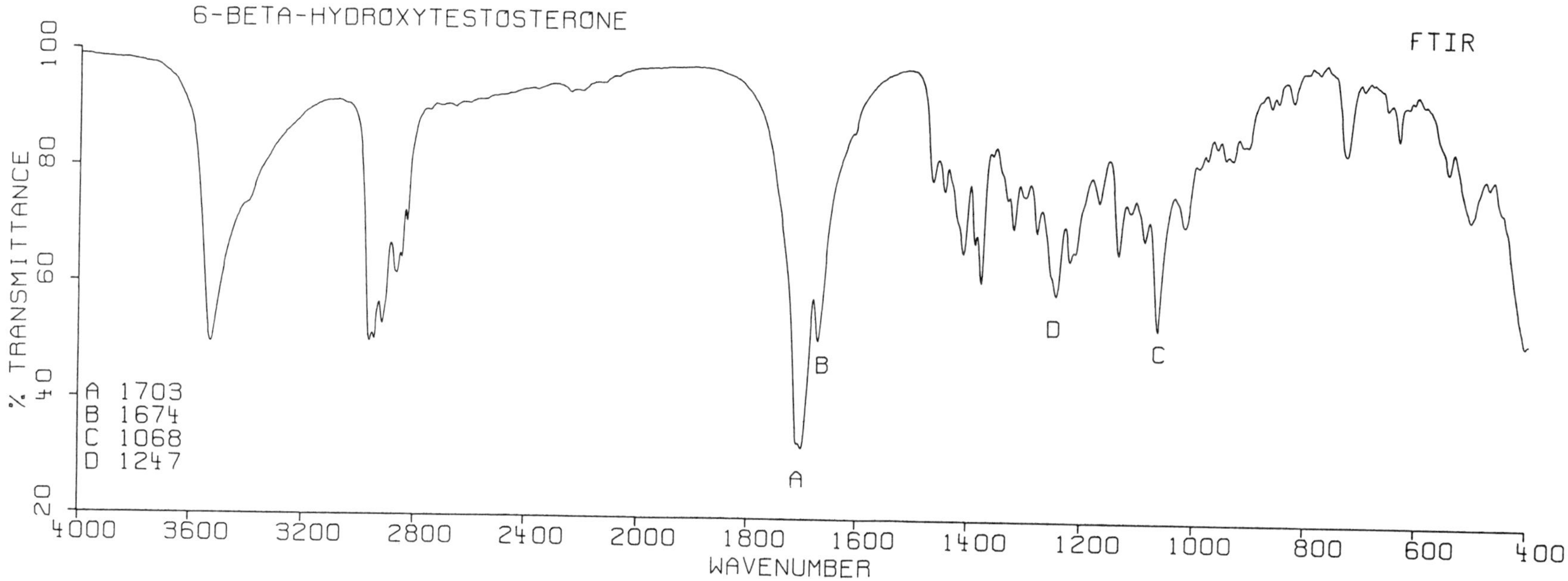

IFENPRODIL

$C_{21}H_{27}NO_2$

Molecular Weight: 325.45 (325.20)

Synonyms: α-(4-Hydroxyphenyl)-β-methyl-4-(phenylmethyl)-1-piperidineethanol
Trade Names: Cerocral, Dilvax, Vadilex, Validex
Use: Vasodilator

HPLC: Methanol: 2.3
GC:

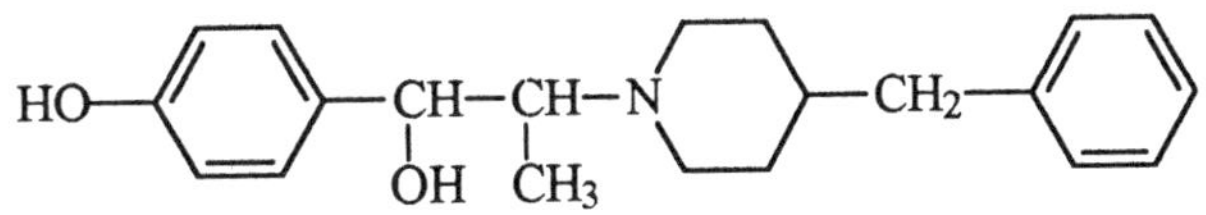

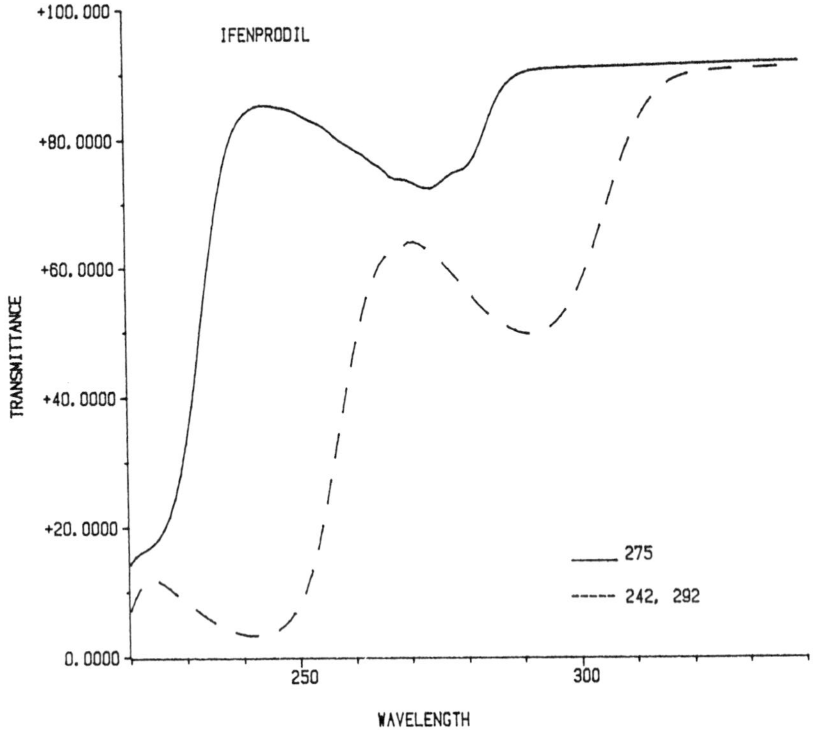

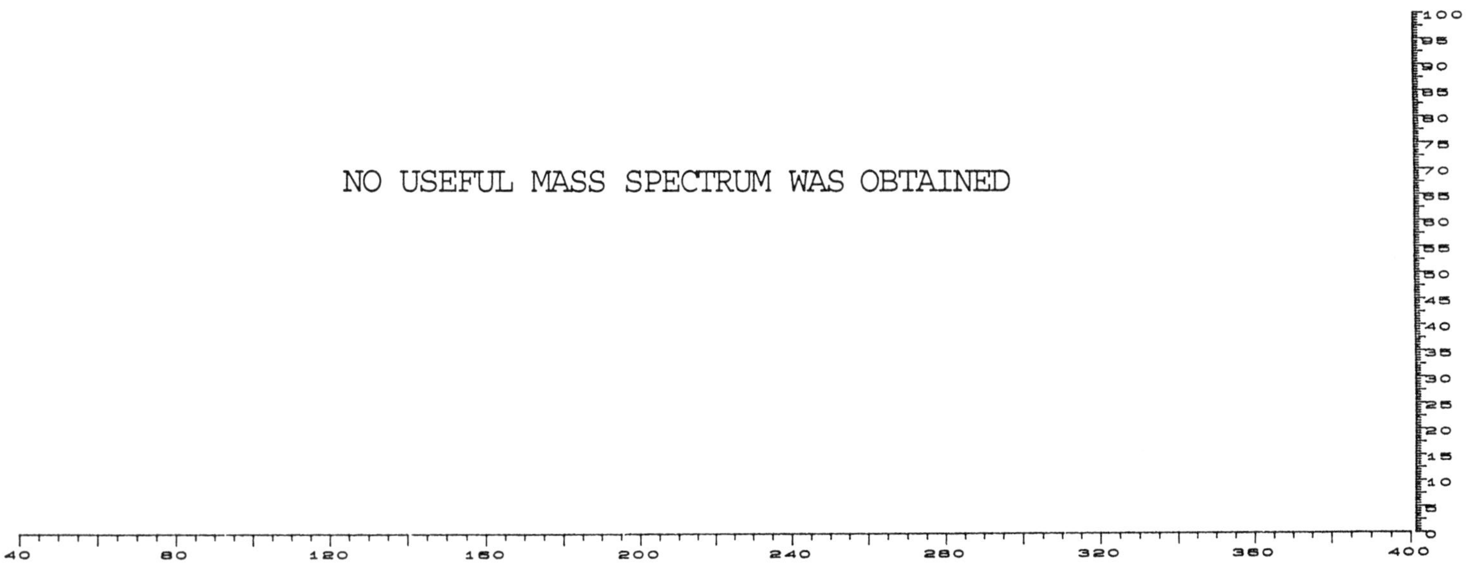

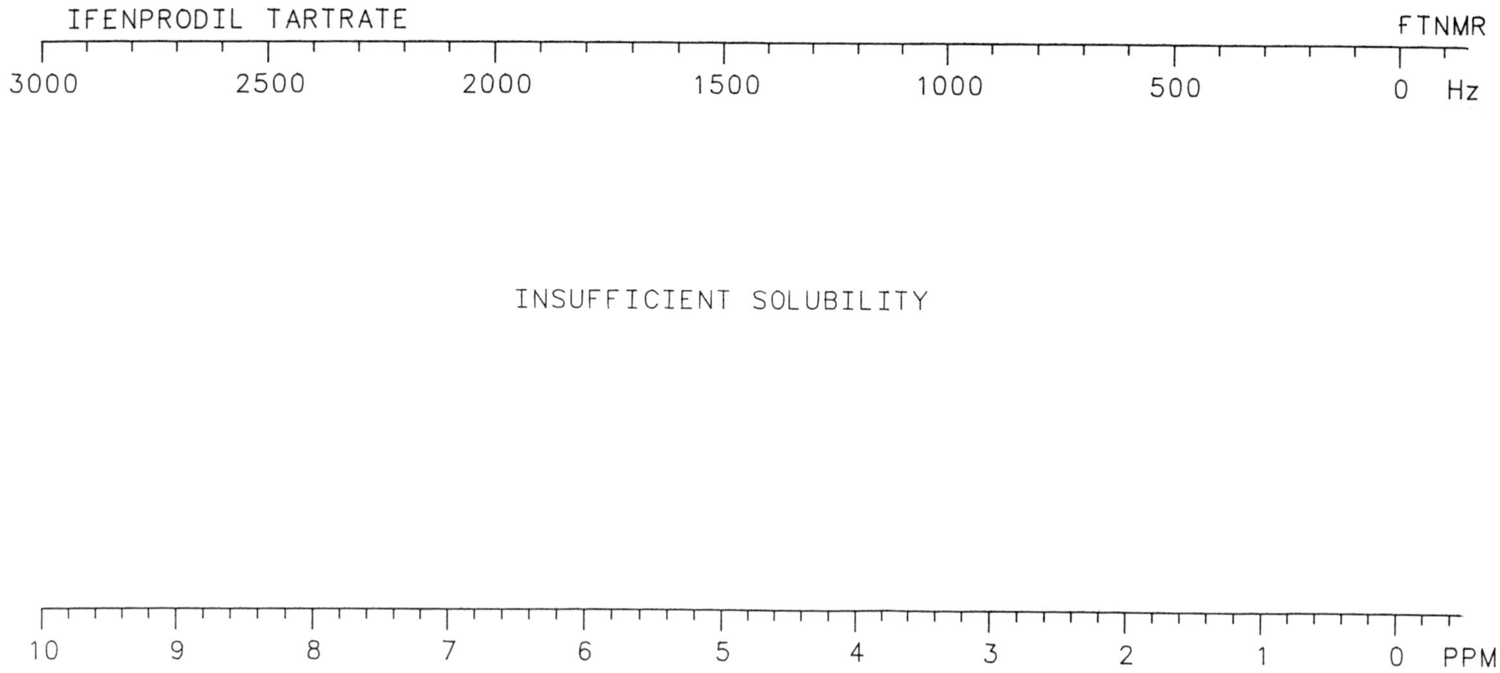
IFENPRODIL TARTRATE
FTNMR
3000
2500
2000
1500
1000
500
0 Hz
INSUFFICIENT SOLUBILITY
10
9
8
7
6
5
4
3
2
1
0 PPM

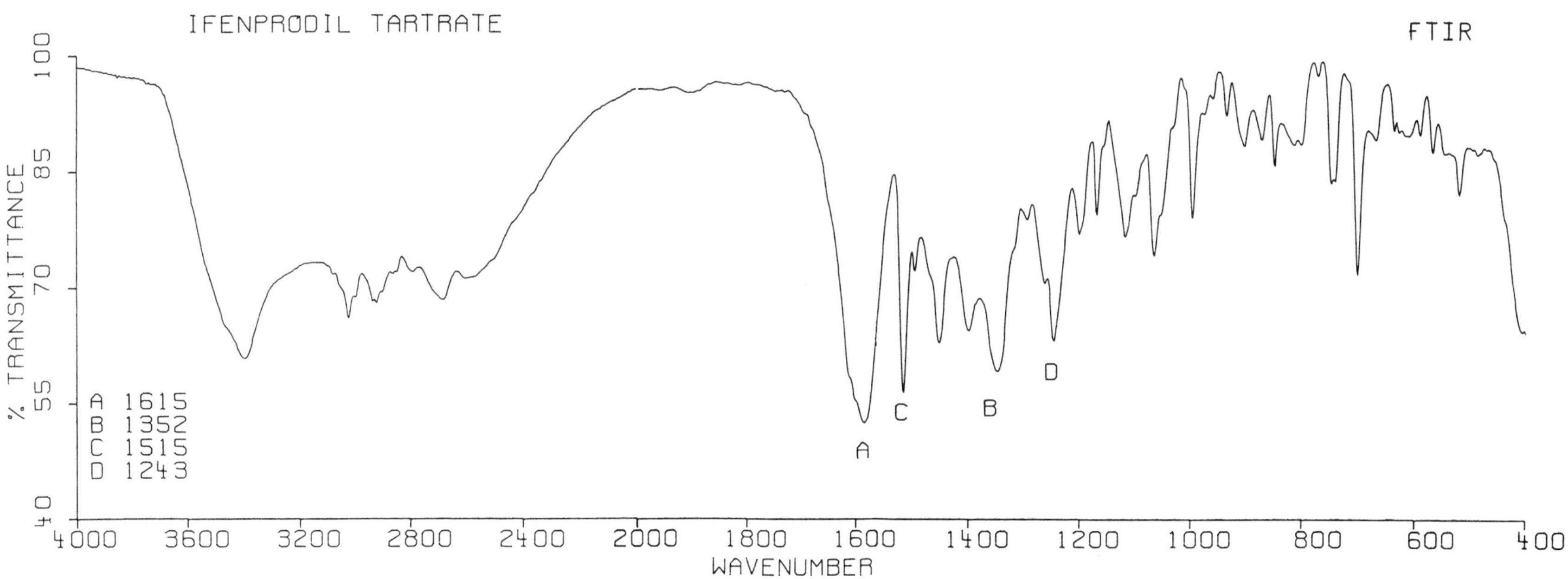
IFENPRODIL TARTRATE
FTIR
% TRANSMITTANCE
100
85
70
55
40
A 1615
B 1352
C 1515
D 1243
A
B
C
D
4000
3600
3200
2800
2400
2000
1800
1600
1400
1200
1000
800
600
400
WAVENUMBER

IOPHENOXIC ACID

$C_{11}H_{11}I_3O_3$

Molecular Weight: 571.92 (571.78)

Synonyms: α-Ethyl-3-hydroxy-2,4,6-triiodobenzenepropanoic acid; triiodoethionic acid
Trade Names: Teridax
Use: Diagnostic aid

HPLC: Methanol: 2.6
GC:

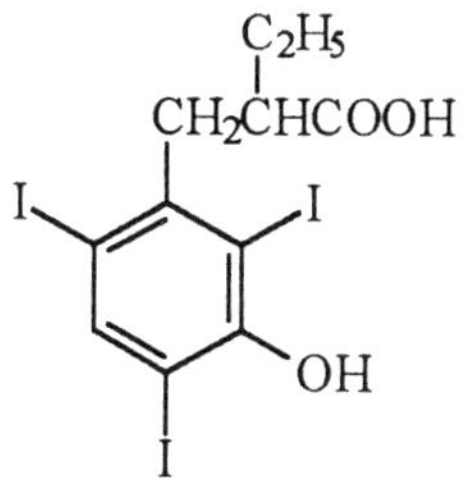

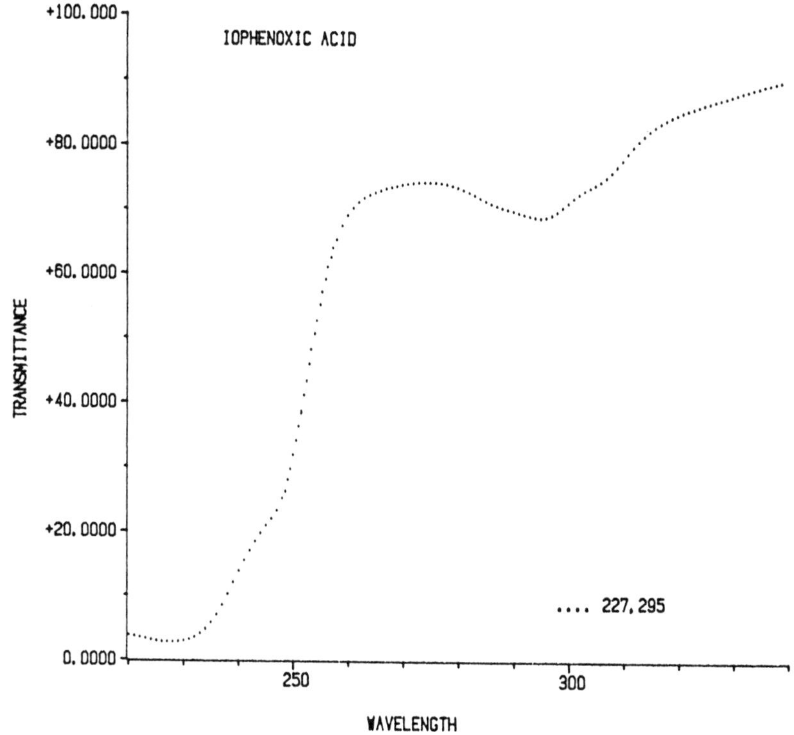

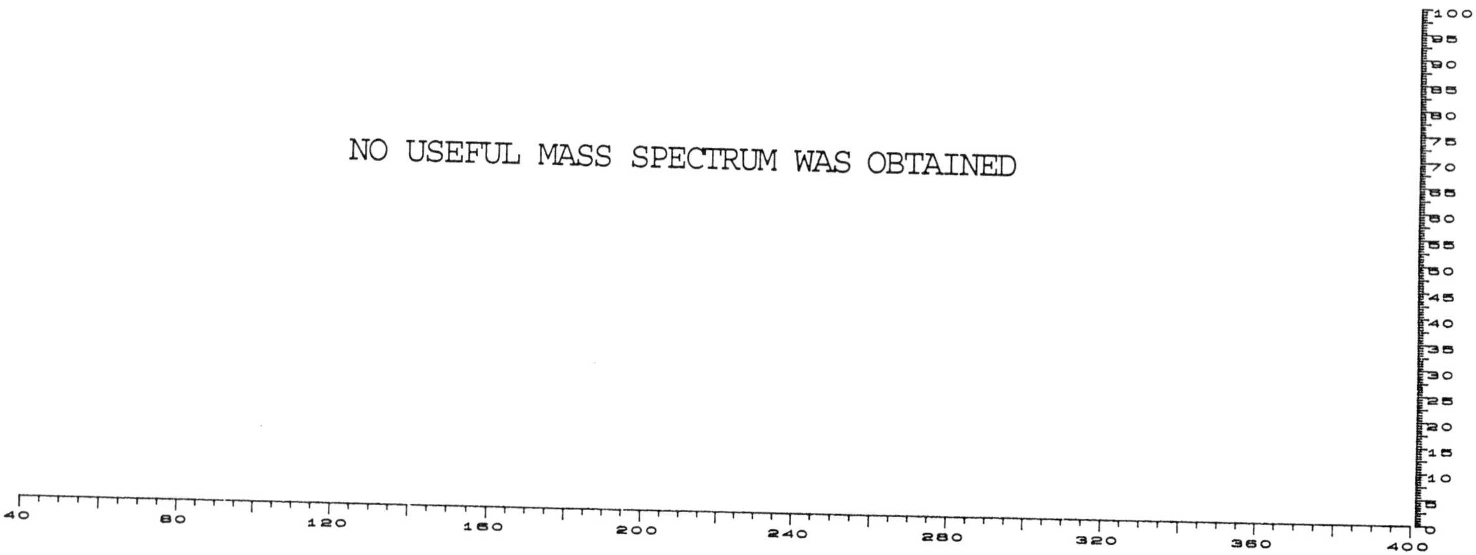

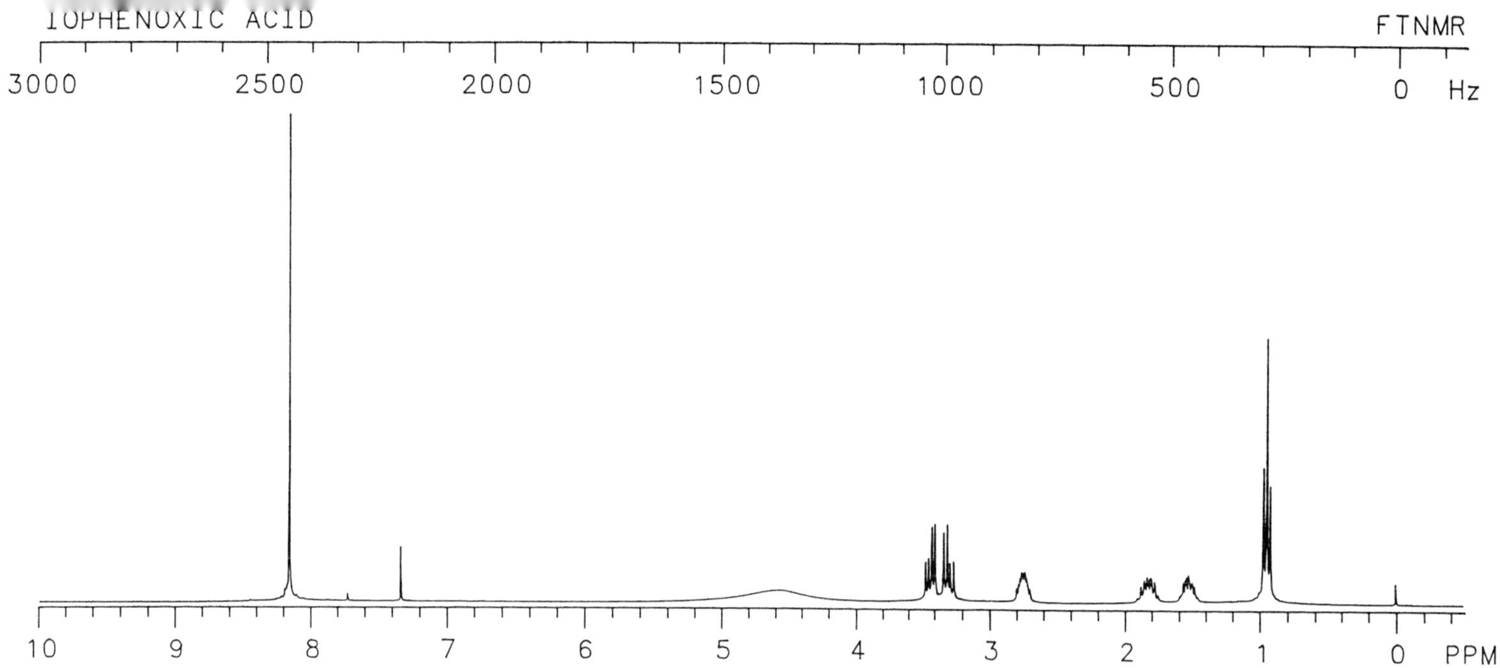

PEAK LISTING

#	HT	PPM
1	24	.91
2	54	.94
3	28	.96
4	16	3.30
5	14	3.32
6	16	3.39
7	15	3.41
8	9	3.43
9	11	7.33
10	99	8.15

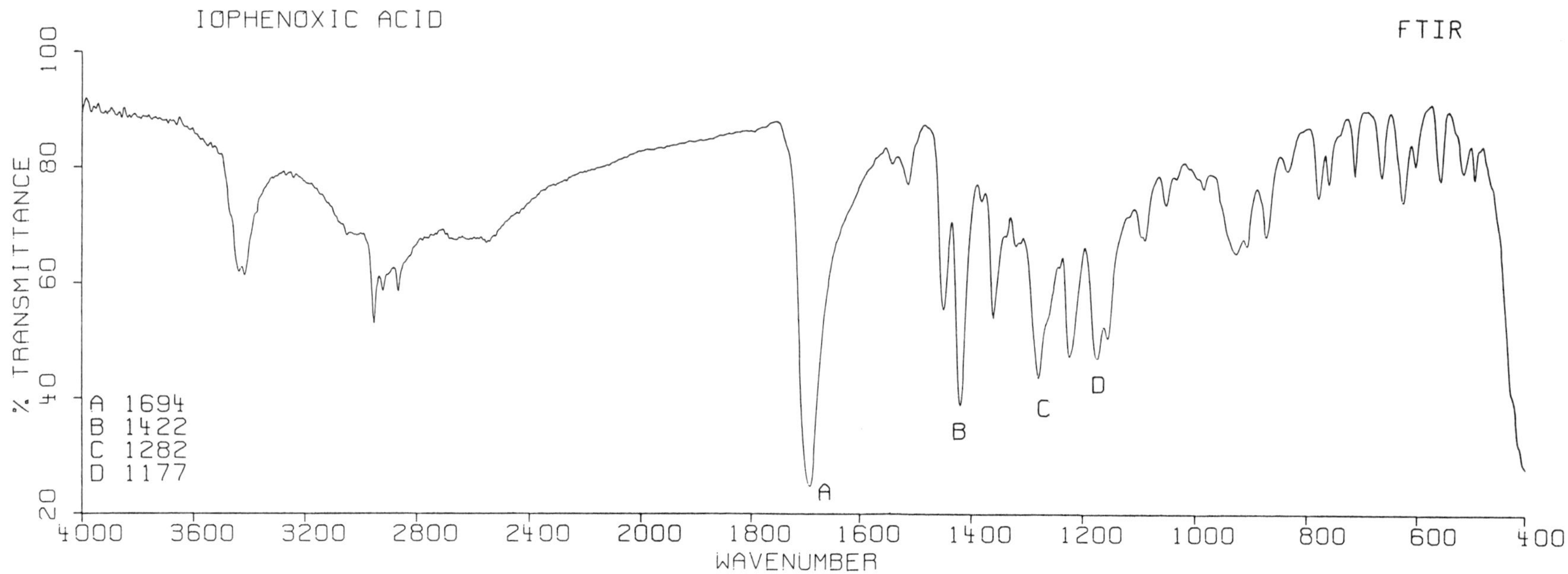

IPRATROPIUM BROMIDE

$C_{20}H_{30}BrNO_3$

Molecular Weight: 412.37 (411.14)

Synonyms: 3-(3-Hydroxy-1-oxo-2-phenylpropoxy)-8-methyl-8-(1-methylethyl)-8-azoniabicyclo[3.2.1]octane bromide; 8-isopropylnoratropine methobromide; atropine isopropylbromide
Trade Names: Atem, Atrovent, Bitrop, Itrop, Rinatec
Use: Bronchodilator

HPLC:
GC:

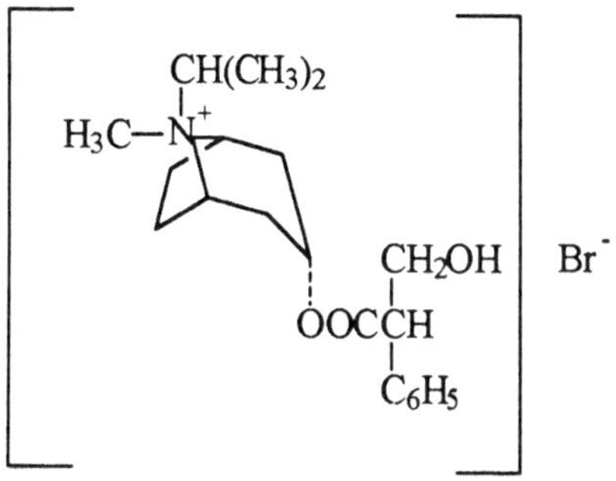

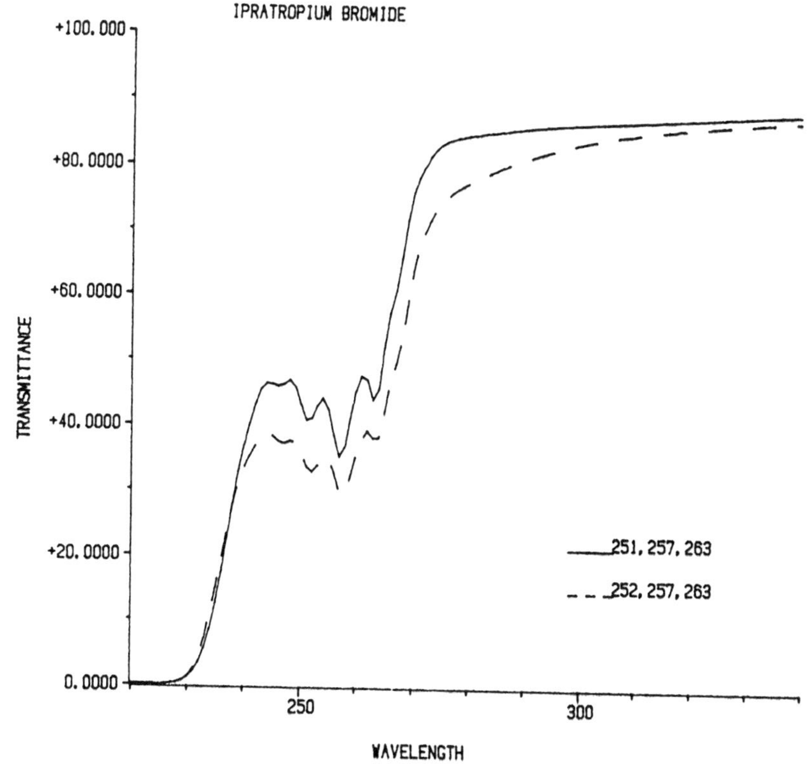

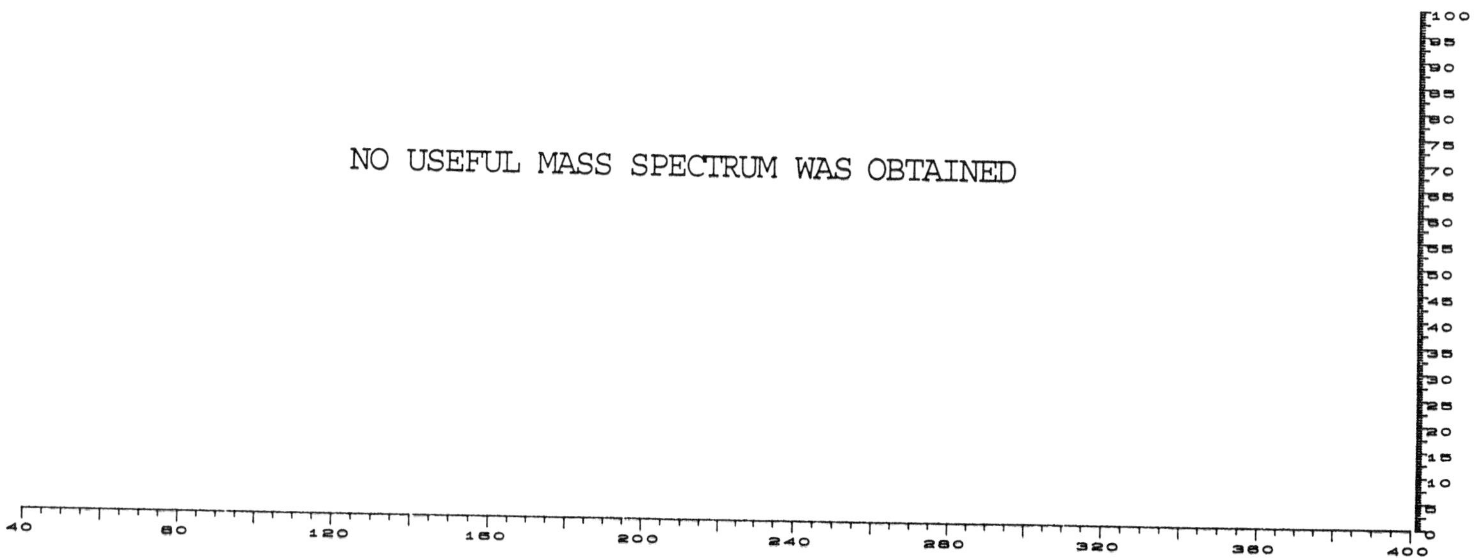

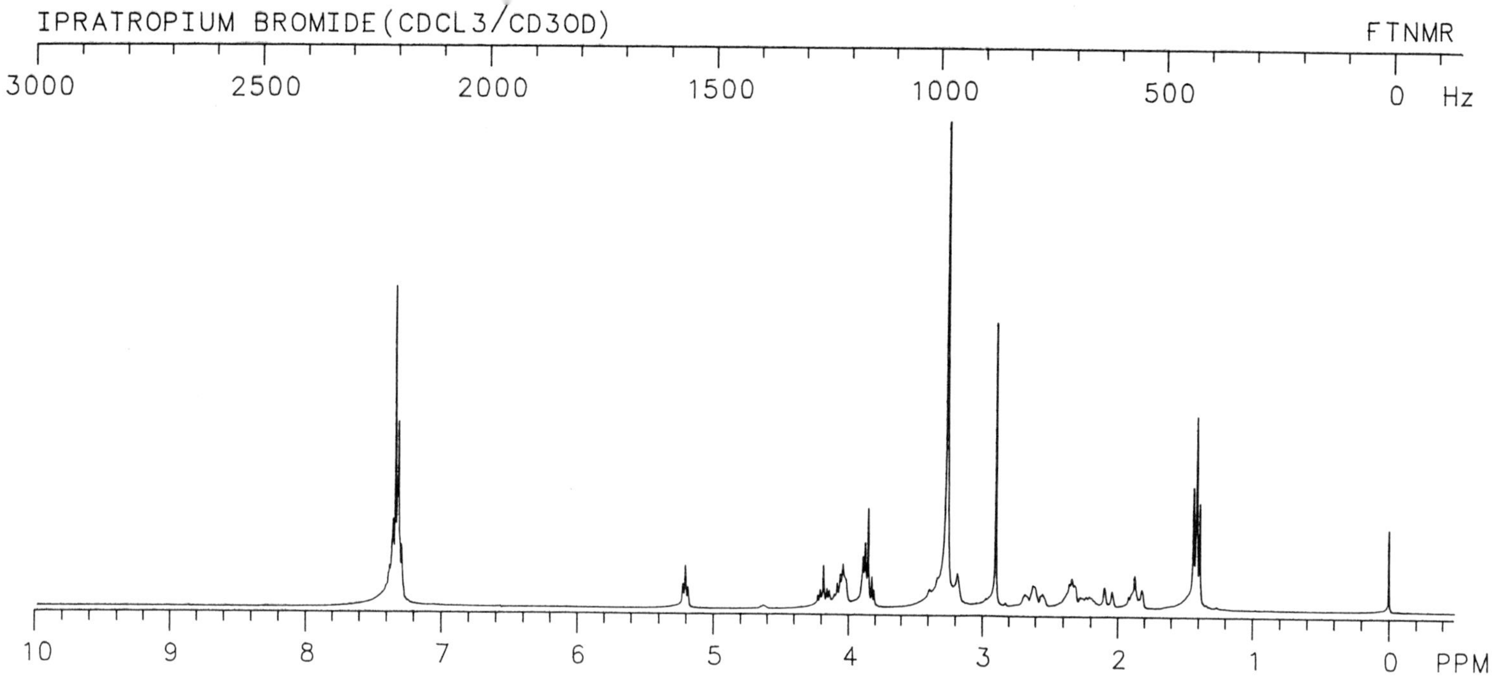

PEAK LISTING

#	HT	PPM
1	17	.00
2	22	1.38
3	39	1.40
4	25	1.43
5	58	2.89
6	99	3.25
7	20	3.84
8	37	7.30
9	22	7.31
10	65	7.32
11	18	7.34

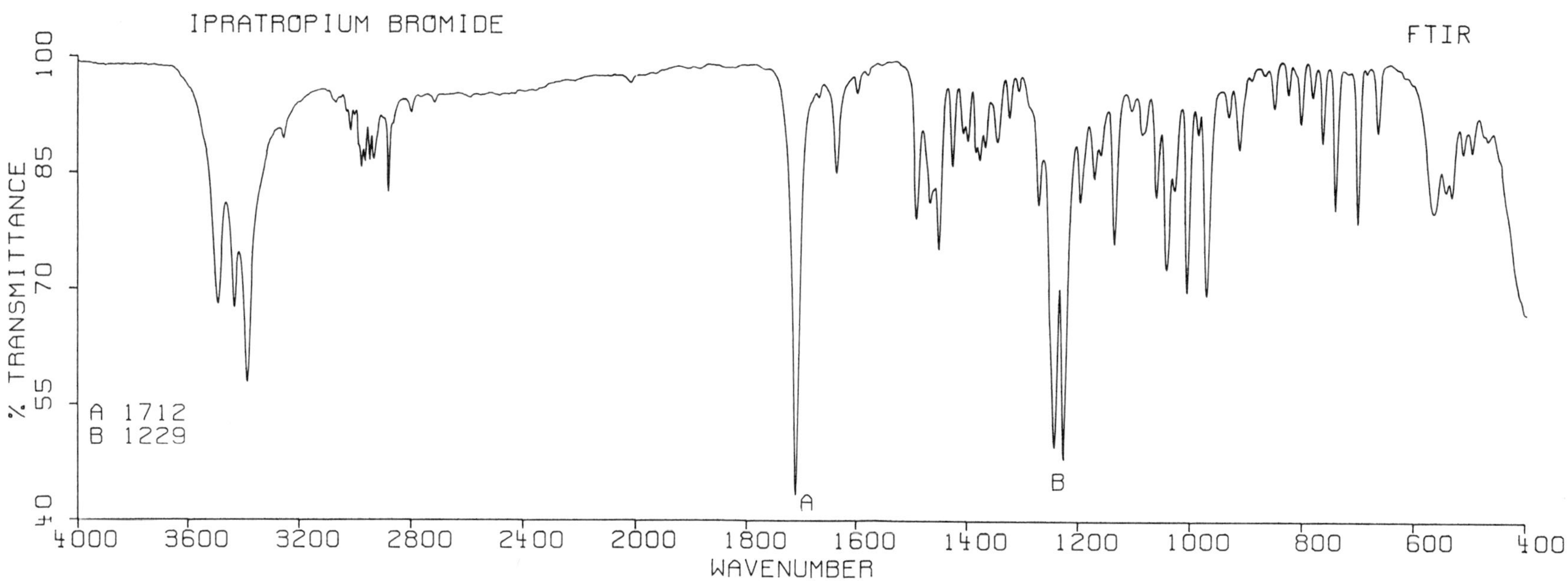

ISRADIPINE

$C_{19}H_{21}N_3O_5$

Molecular Weight: 371.39 (371.15)

Synonyms: 4-(4-Benzofurazanyl)-1,4-dihydro-2,6-dimethyl-3,5-pyridinedicarboxylic acid methyl 1-methylethyl ester; isrodipine
Trade Names: DynaCirc, Dynacrine, Lomir, Prescal, Rebriden
Use: Antihypertensive

HPLC: Methanol: 3.7
GC: 2619; 280°

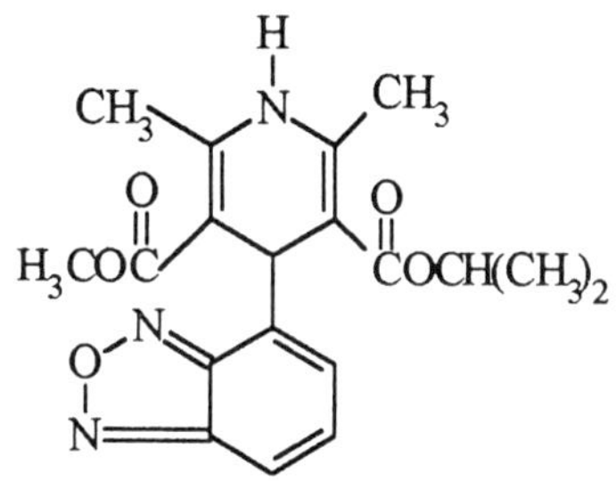

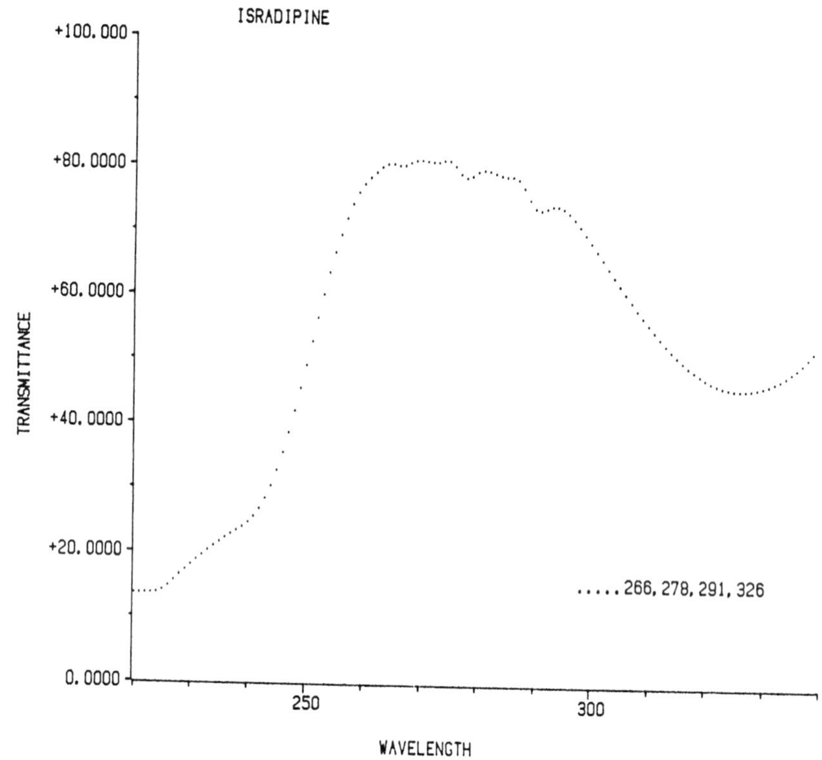

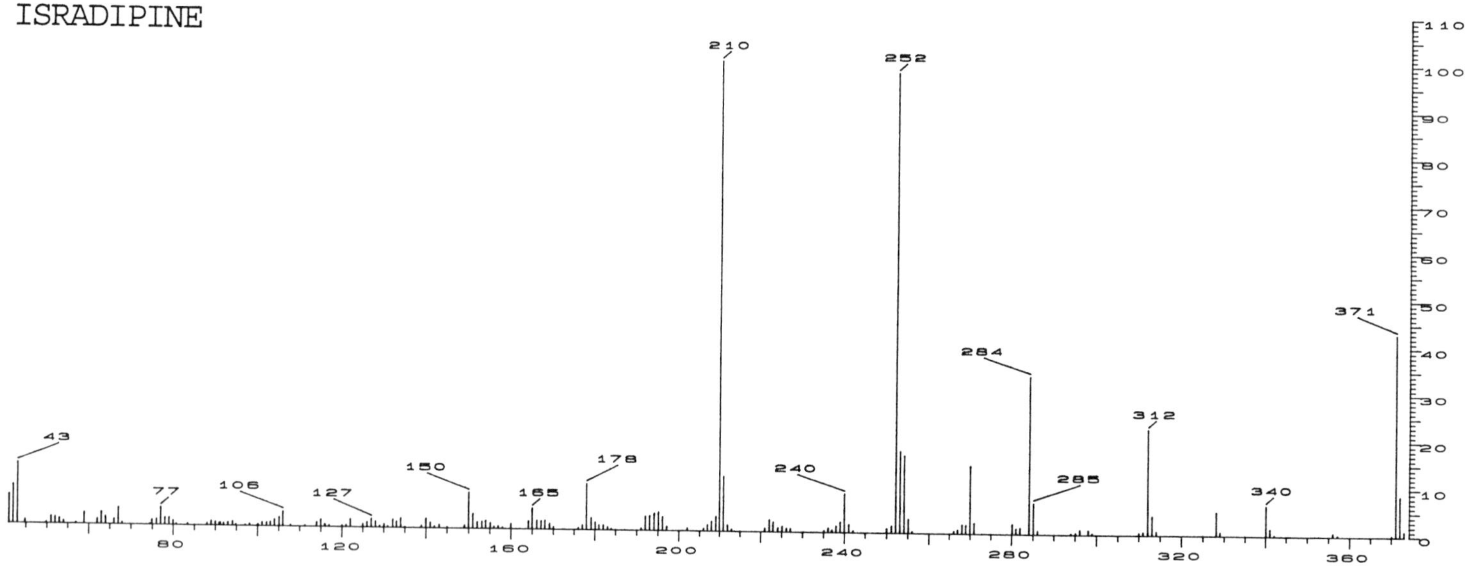

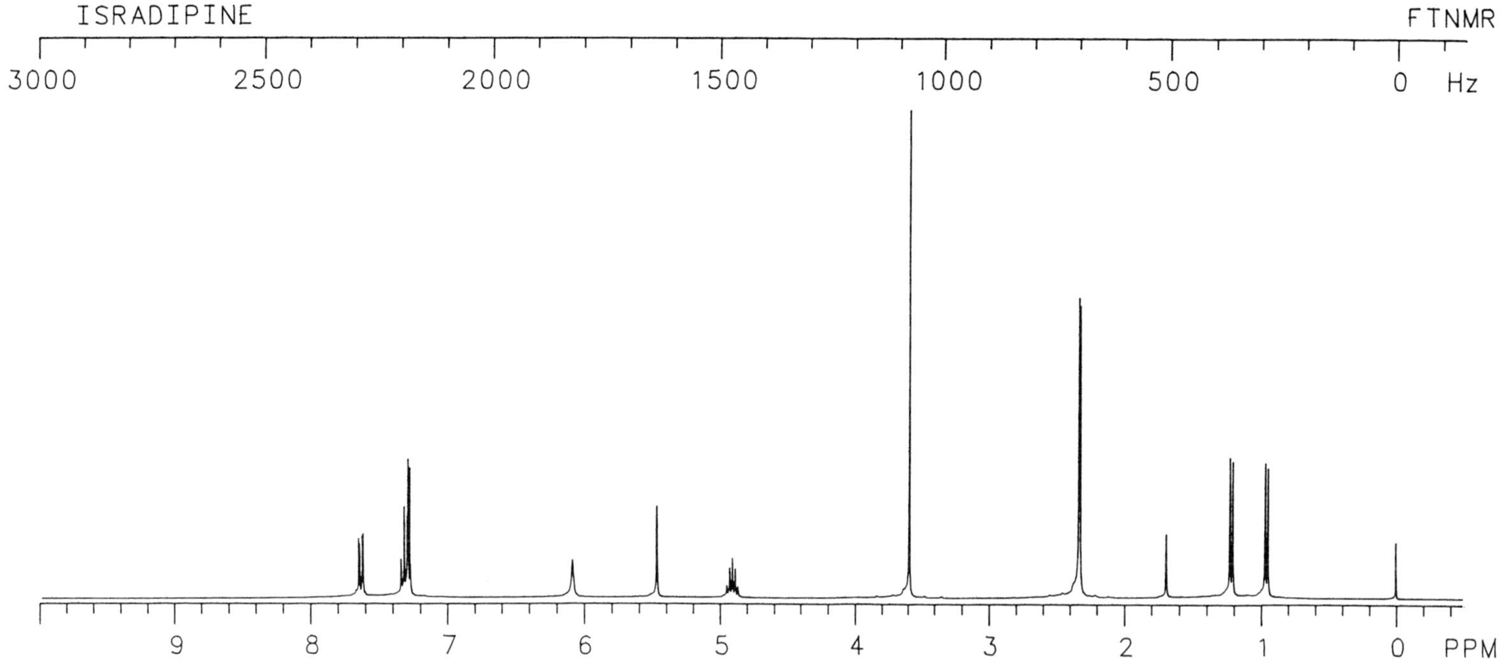

PEAK LISTING

#	HT	PPM
1	26	.94
2	27	.96
3	28	1.20
4	28	1.22
5	59	2.32
6	61	2.33
7	99	3.58
8	26	7.27
9	28	7.28

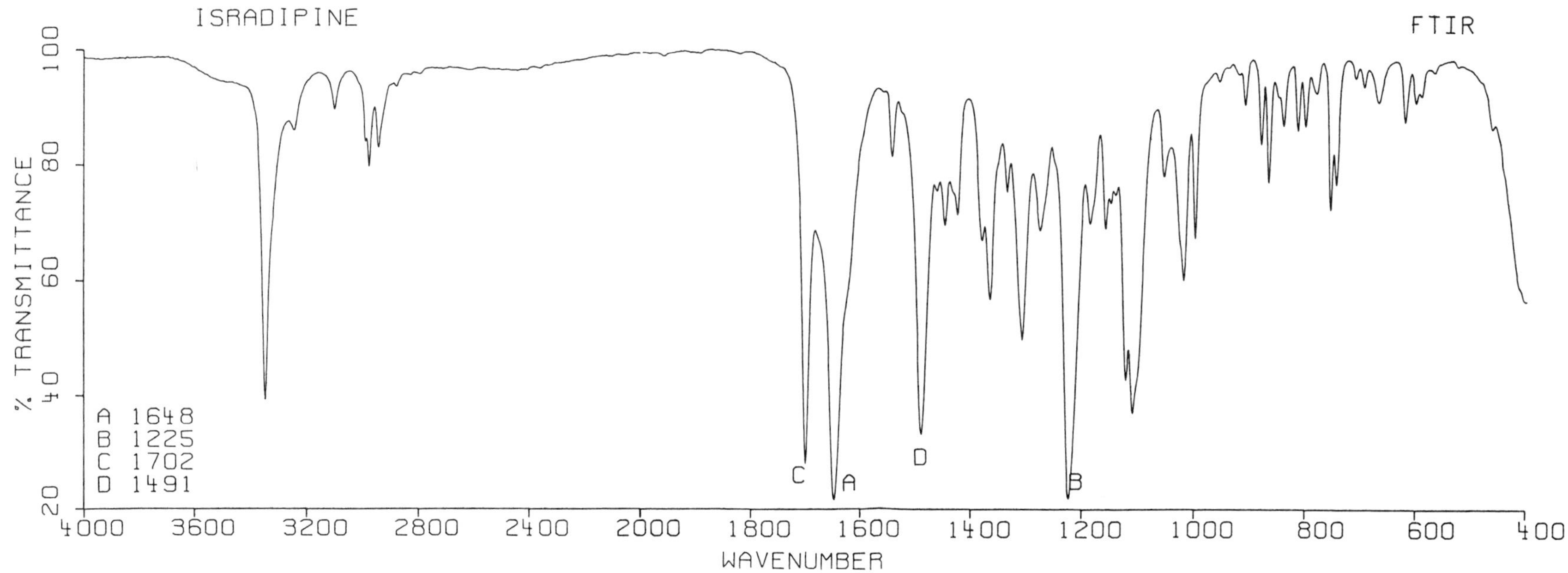

KETOTESTOSTERONE

$C_{19}H_{26}O_3$

Molecular Weight: 302.41 (302.19)

Synonyms: 4-Androsten-17β-ol-3,11-dione; 17β-hydroxy-4-androstene-3,11-dione; 11-ketotestosterone
Trade Names:
Use:

HPLC: Methanol: 3.4
GC:

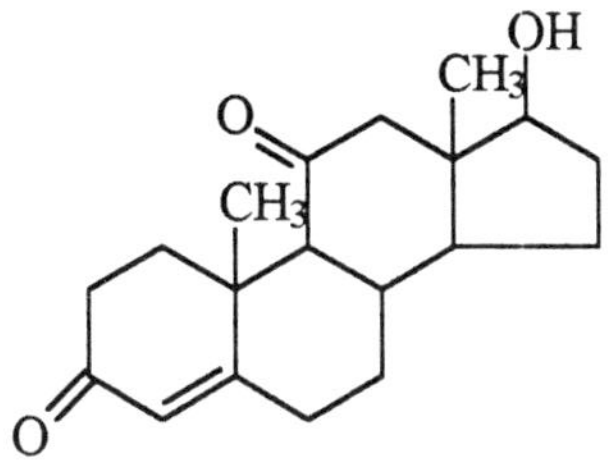

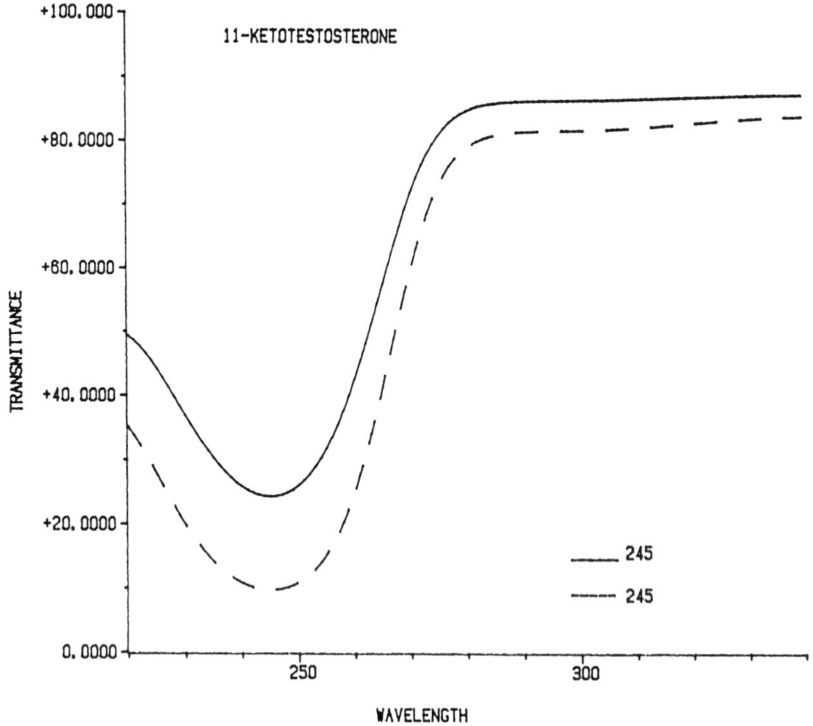

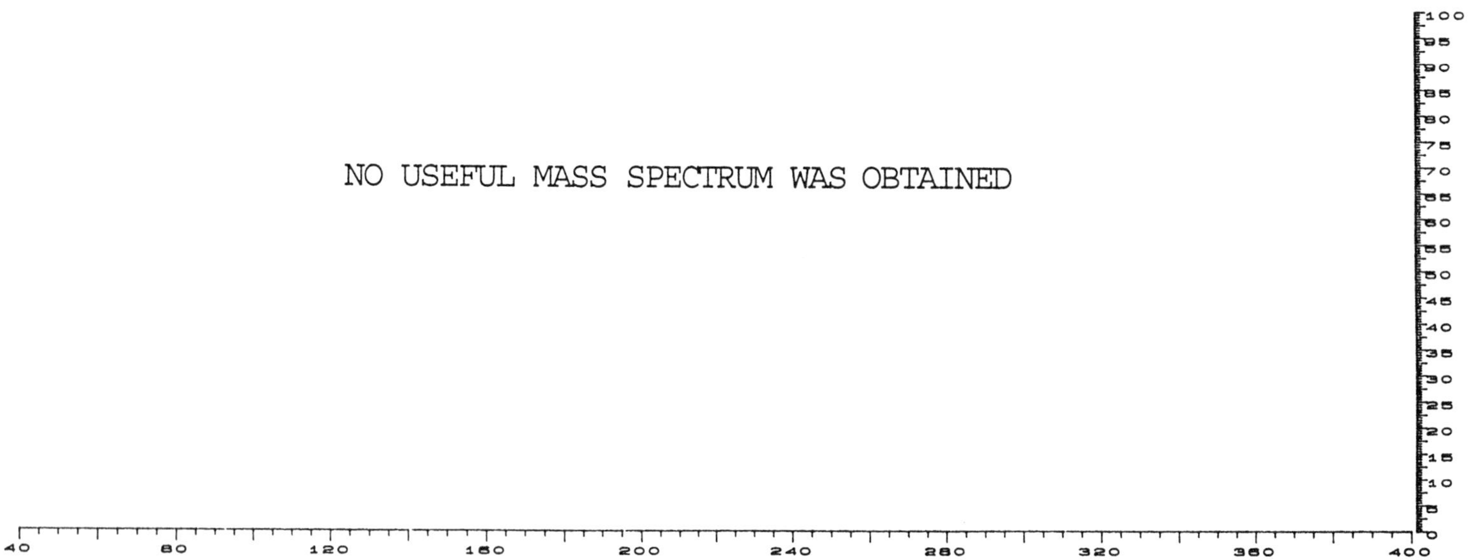

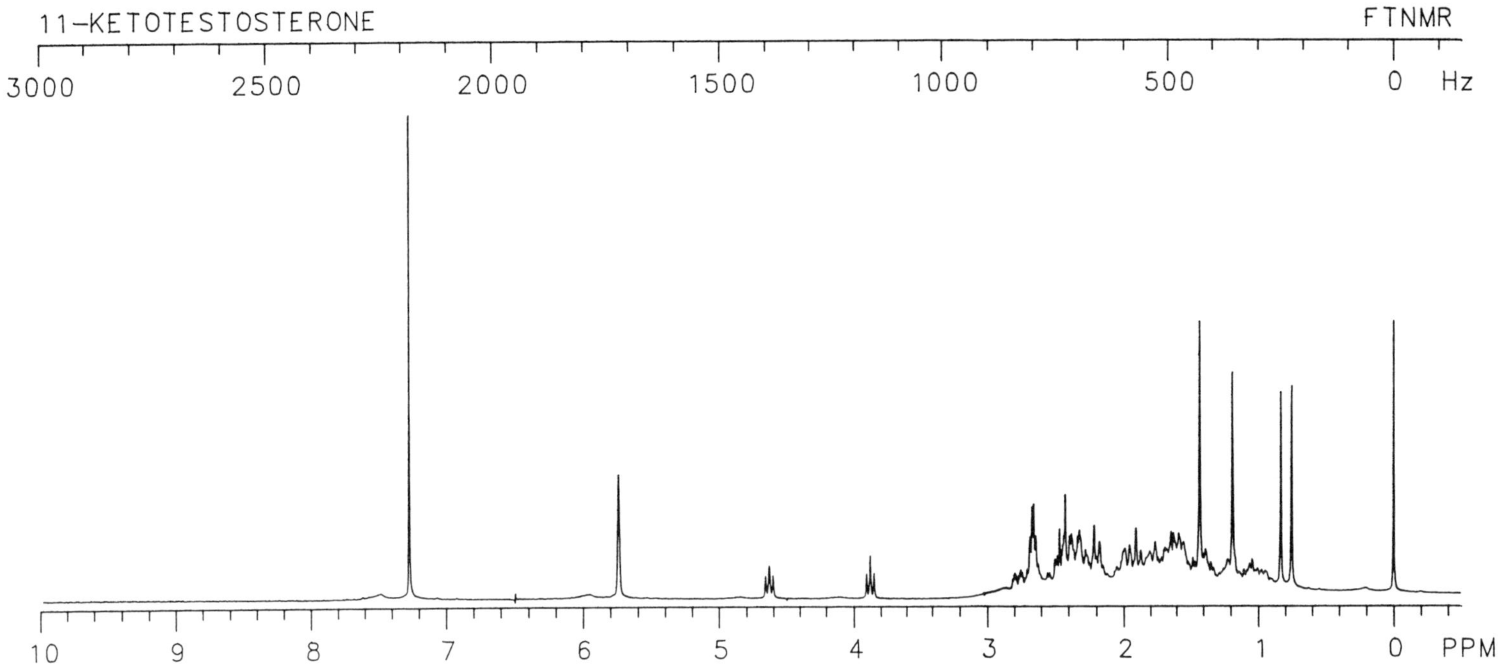

PEAK LISTING

#	HT	PPM
1	56	.00
2	42	.75
3	41	.83
4	45	1.19
5	56	1.42
6	12	1.57
7	13	1.58
8	13	1.62
9	13	1.63
10	14	1.90
11	14	2.21
12	13	2.31
13	13	2.32
14	12	2.33
15	13	2.37
16	12	2.38
17	20	2.42
18	12	2.42
19	13	2.46
20	12	2.63
21	19	2.65
22	18	2.66
23	12	2.67
24	25	5.73
25	99	7.26

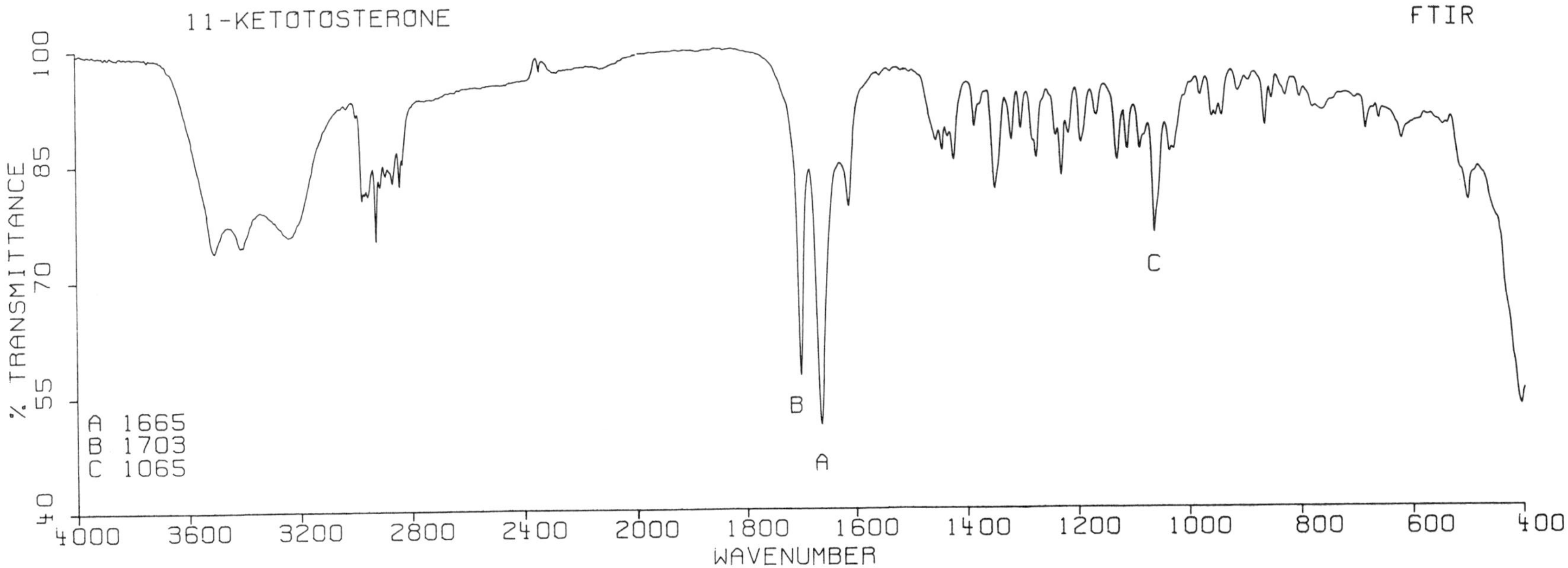

LISINOPRIL

$C_{21}H_{31}N_3O_3$

Molecular Weight: 405.49 (405.23)

Synonyms: (S)-1-[N^2-(1-Carboxy-3-phenylpropyl)-L-lysyl]-L-proline dihydrate
Trade Names: Acerbon, Carace, Novatec, Prinivil, Prinzide, Renacor, Vivatec, Zestoretic, Zestril
Use: Antihypertensive

HPLC:
GC:

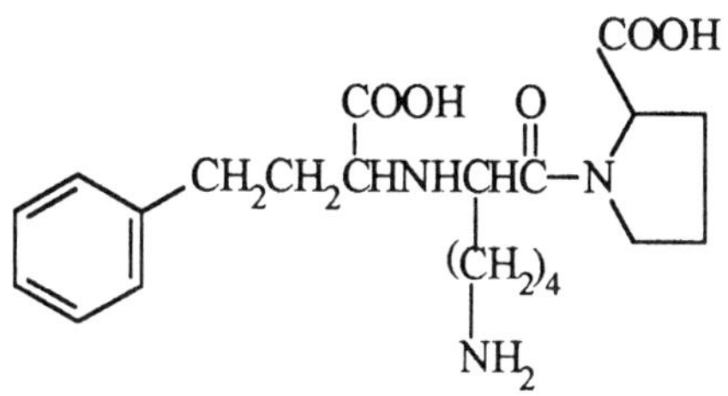

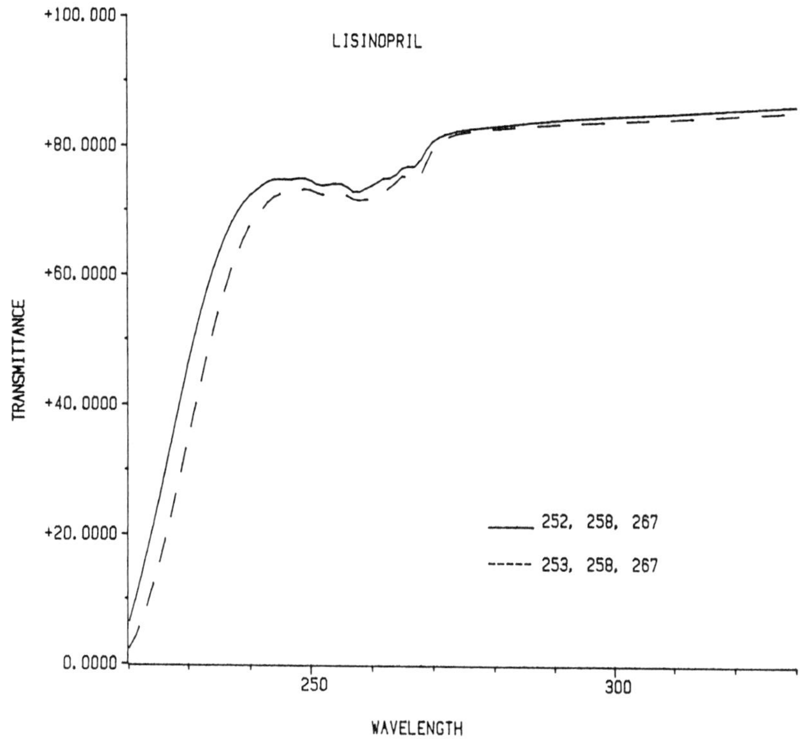

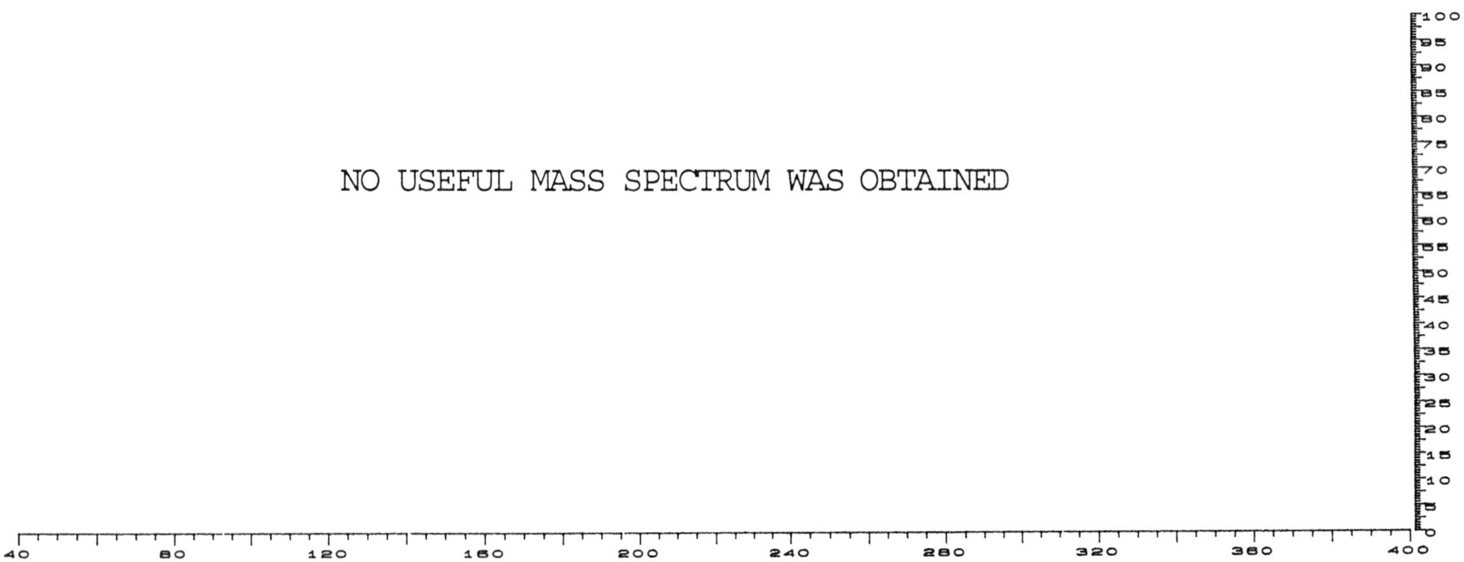

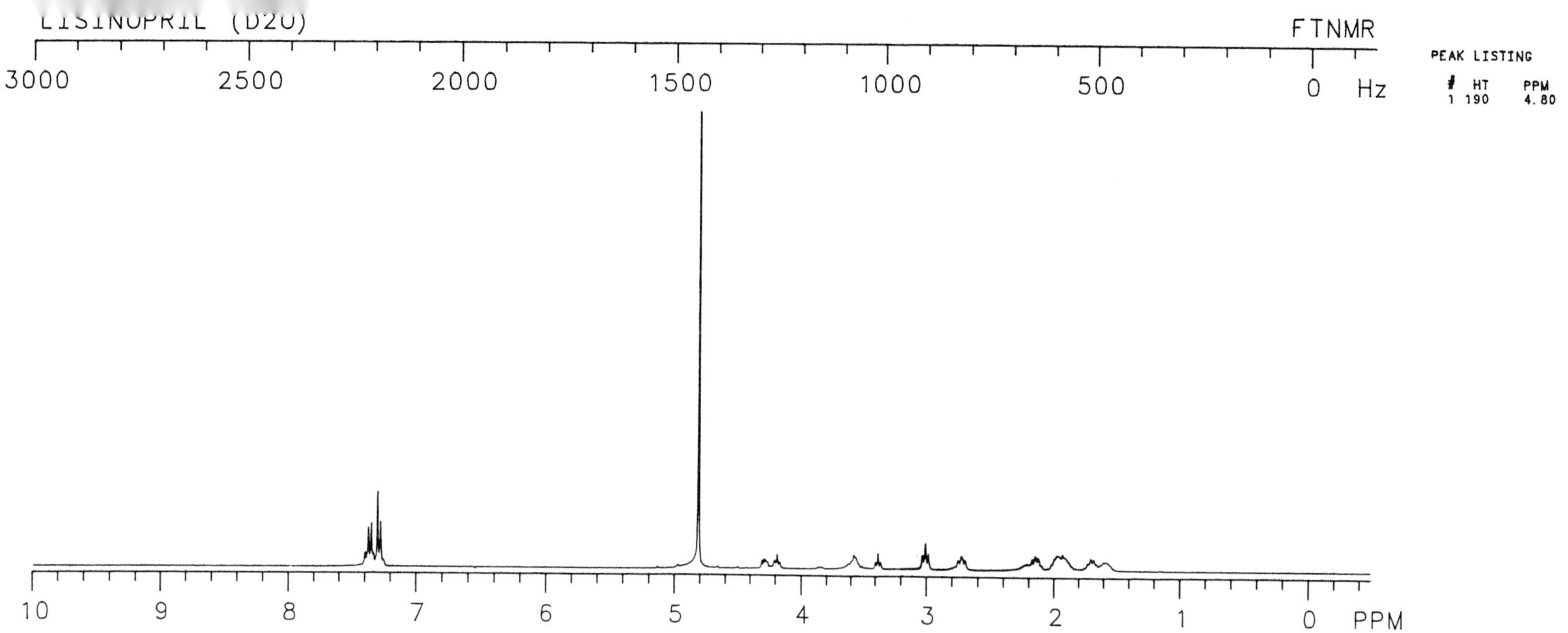

LISINOPRIL (D2O)
FTNMR
3000 2500 2000 1500 1000 500 0 Hz
10 9 8 7 6 5 4 3 2 1 0 PPM
PEAK LISTING
HT PPM
1 190 4.80

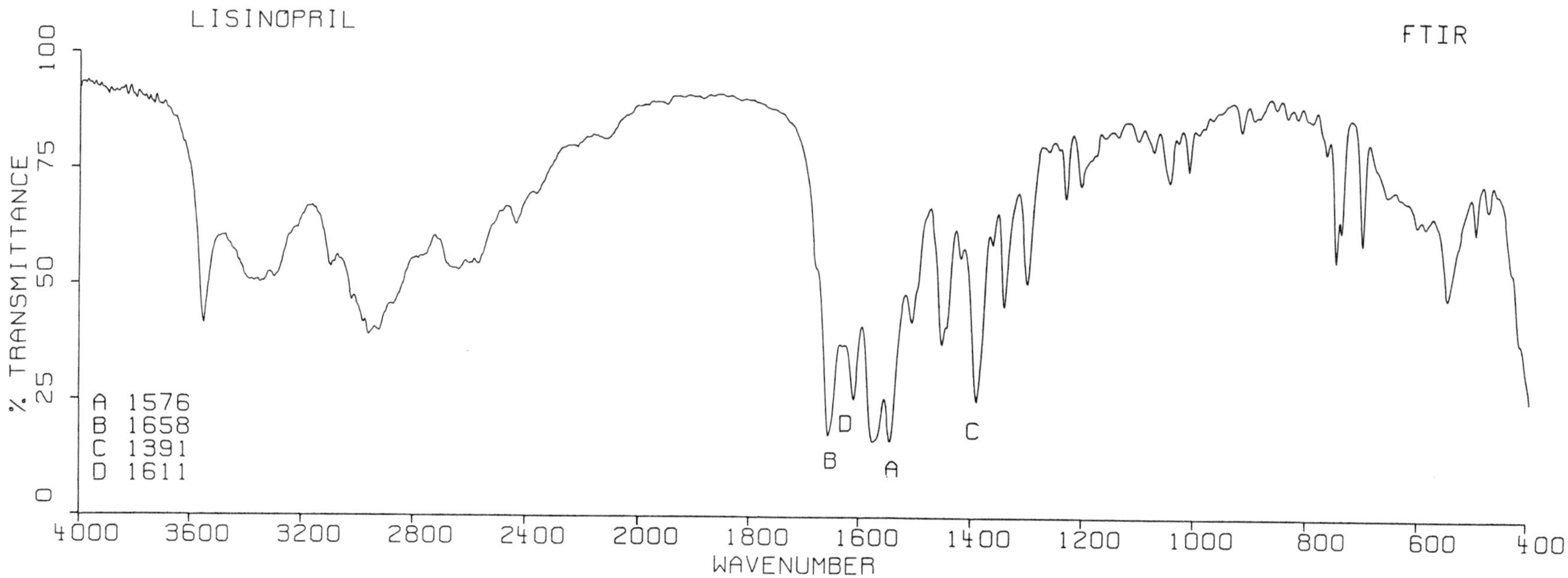

LISINOPRIL
FTIR
% TRANSMITTANCE
0 25 50 75 100
A 1576
B 1658
C 1391
D 1611
4000 3600 3200 2800 2400 2000 1800 1600 1400 1200 1000 800 600 400
WAVENUMBER

LOMEFLOXACIN

$C_{17}H_{19}F_2N_3O_3$

Molecular Weight: 351.35 (351.14)

Synonyms: 1-Ethyl-6,8-difluoro-1,4-dihydro-7-(3-methyl-1-piperazinyl)-4-oxo-3-quinolinecarboxylic acid
Trade Names: Bareon, Maxaquin
Use: Antibacterial

HPLC:
GC:

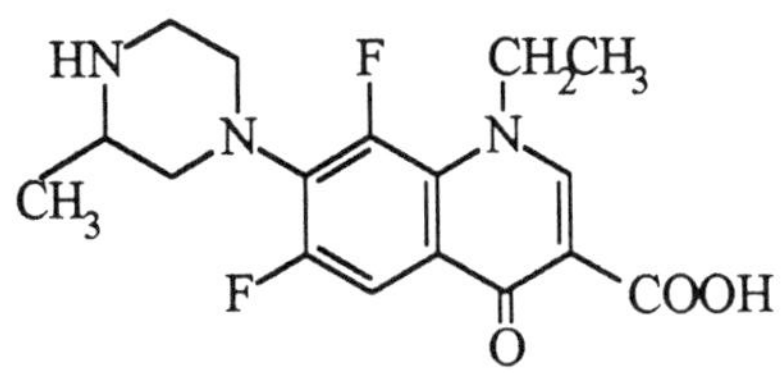

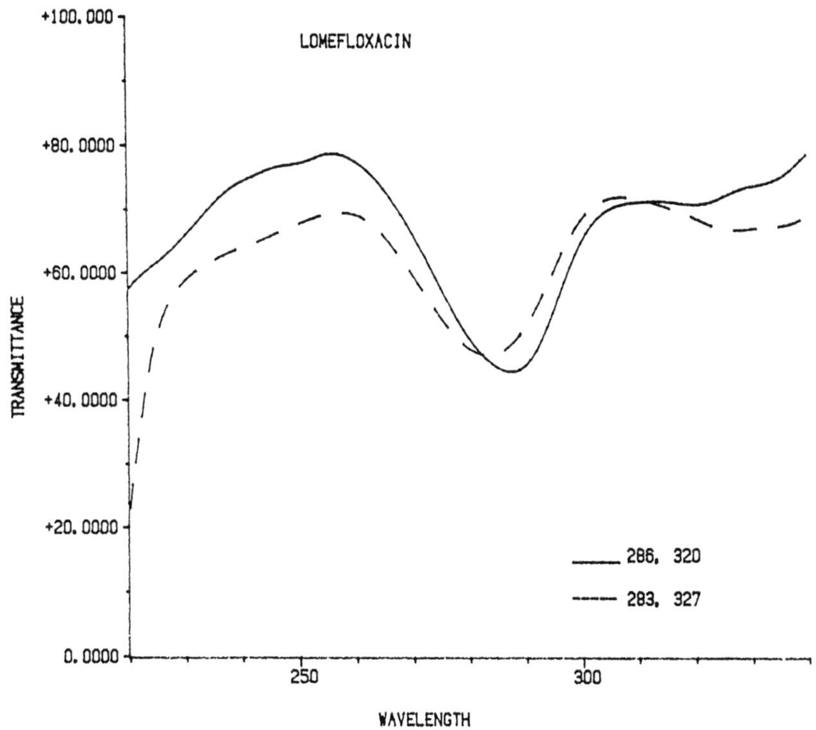

NO USEFUL MASS SPECTRUM WAS OBTAINED

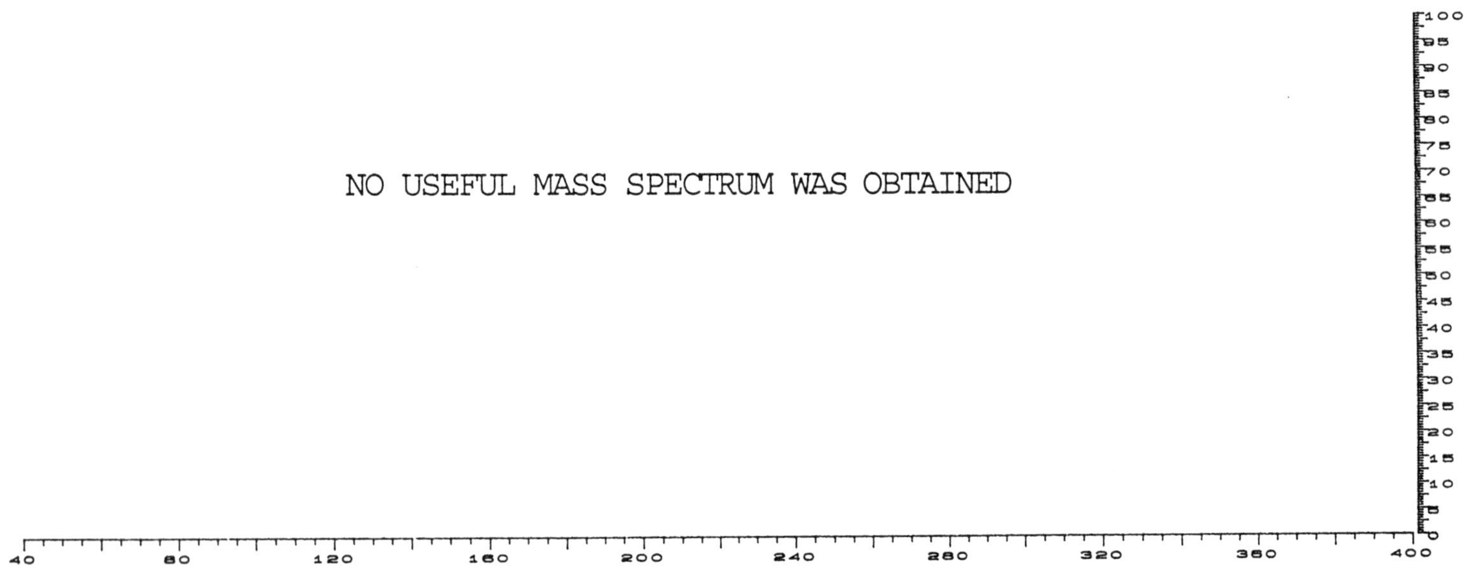

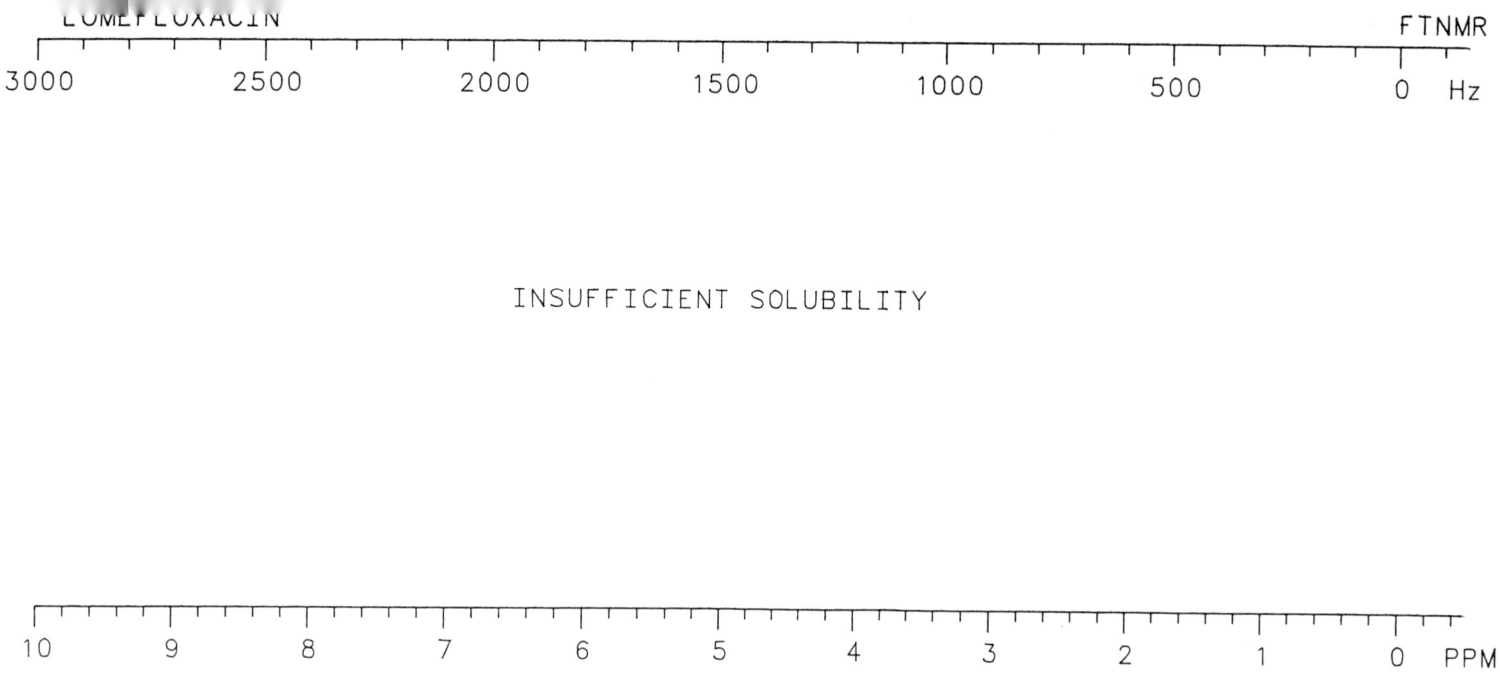
LOMEFLOXACIN
FTNMR
3000
2500
2000
1500
1000
500
0 Hz
INSUFFICIENT SOLUBILITY
10
9
8
7
6
5
4
3
2
1
0 PPM

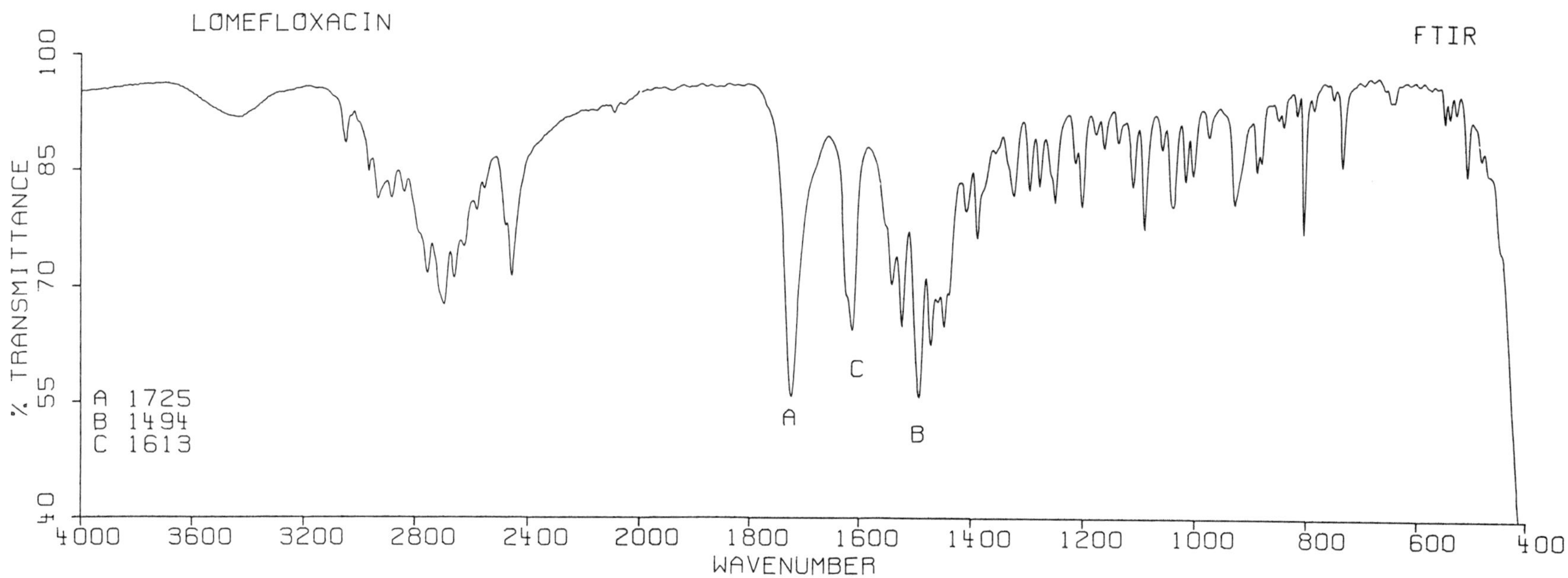
LOMEFLOXACIN
FTIR
% TRANSMITTANCE
100
85
70
55
40
A 1725
B 1494
C 1613
A
B
C
4000
3600
3200
2800
2400
2000
1800
1600
1400
1200
1000
800
600
400
WAVENUMBER

LORATADINE

$C_{22}H_{23}ClN_2O_2$

Molecular Weight: 382.88 (381.14)

Synonyms: 4-(8-Chloro-5,6-dihydro-11H-benzo[5.6]cyclohepta[1.2-b]pyridin-11-ylidene)-1-piperidinecarboxylic acid ethyl ester
Trade Names: Clantin, Claritin, Clarityne, Lisino
Use: Antihistaminic

HPLC: Methanol: 4.2
GC: 3098; 280°

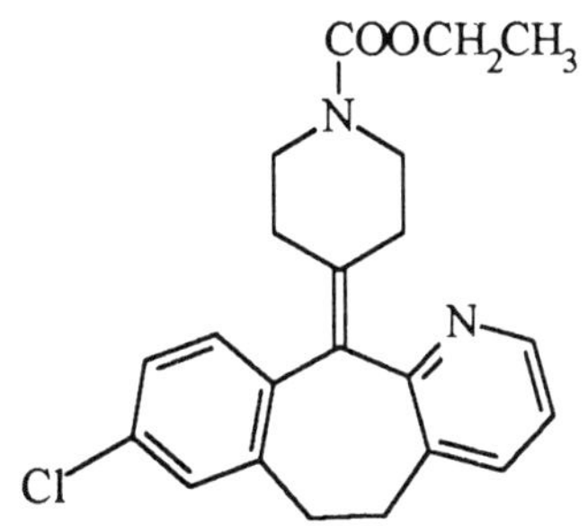

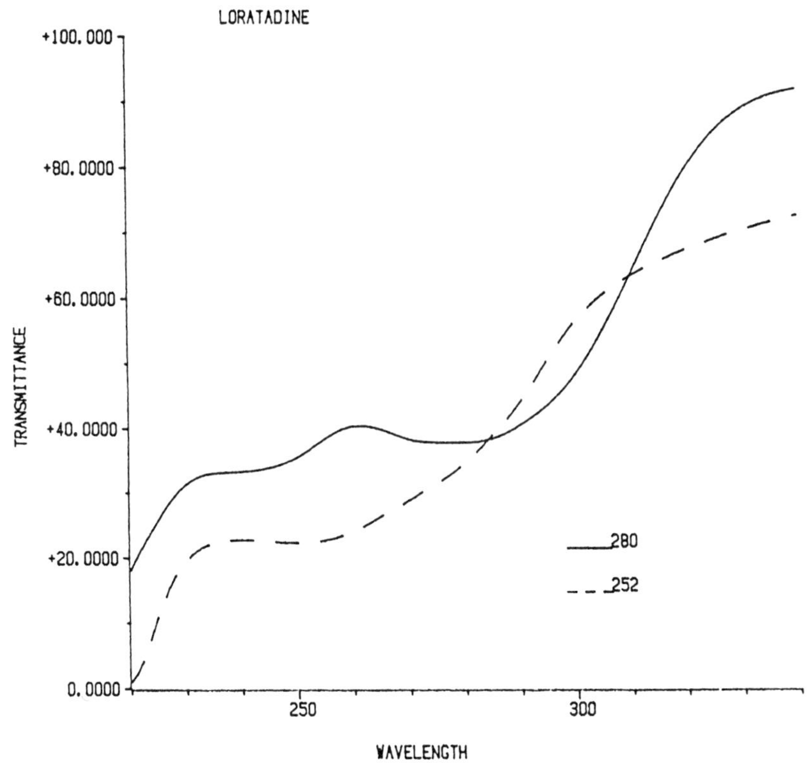

LORATADINE

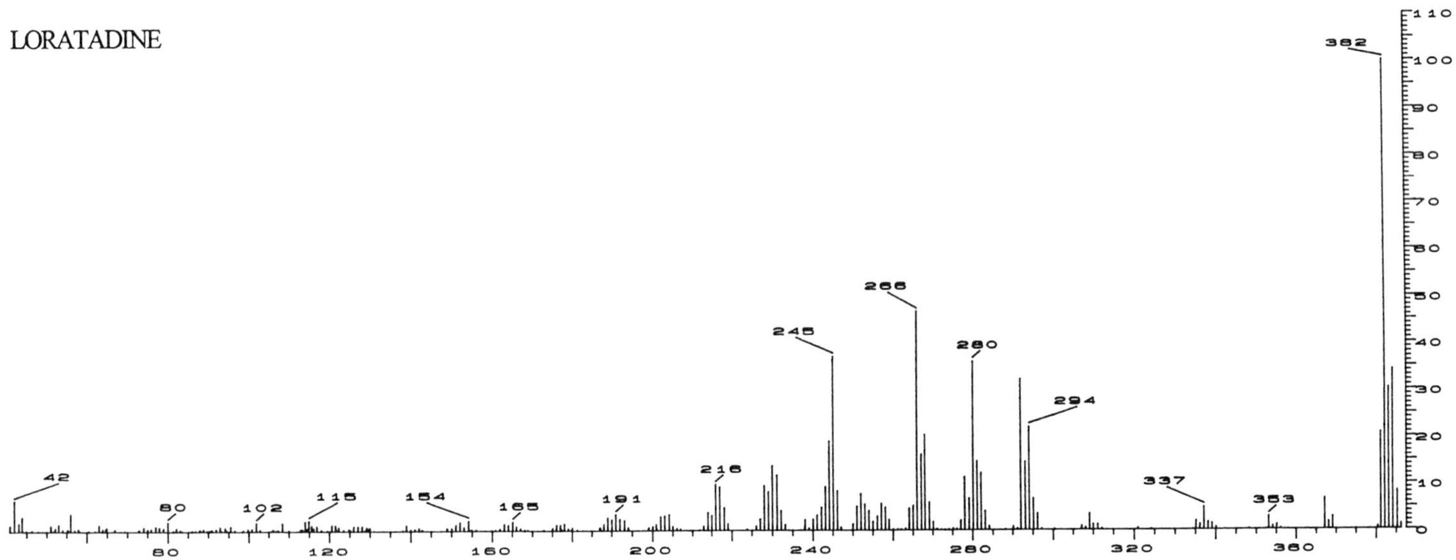

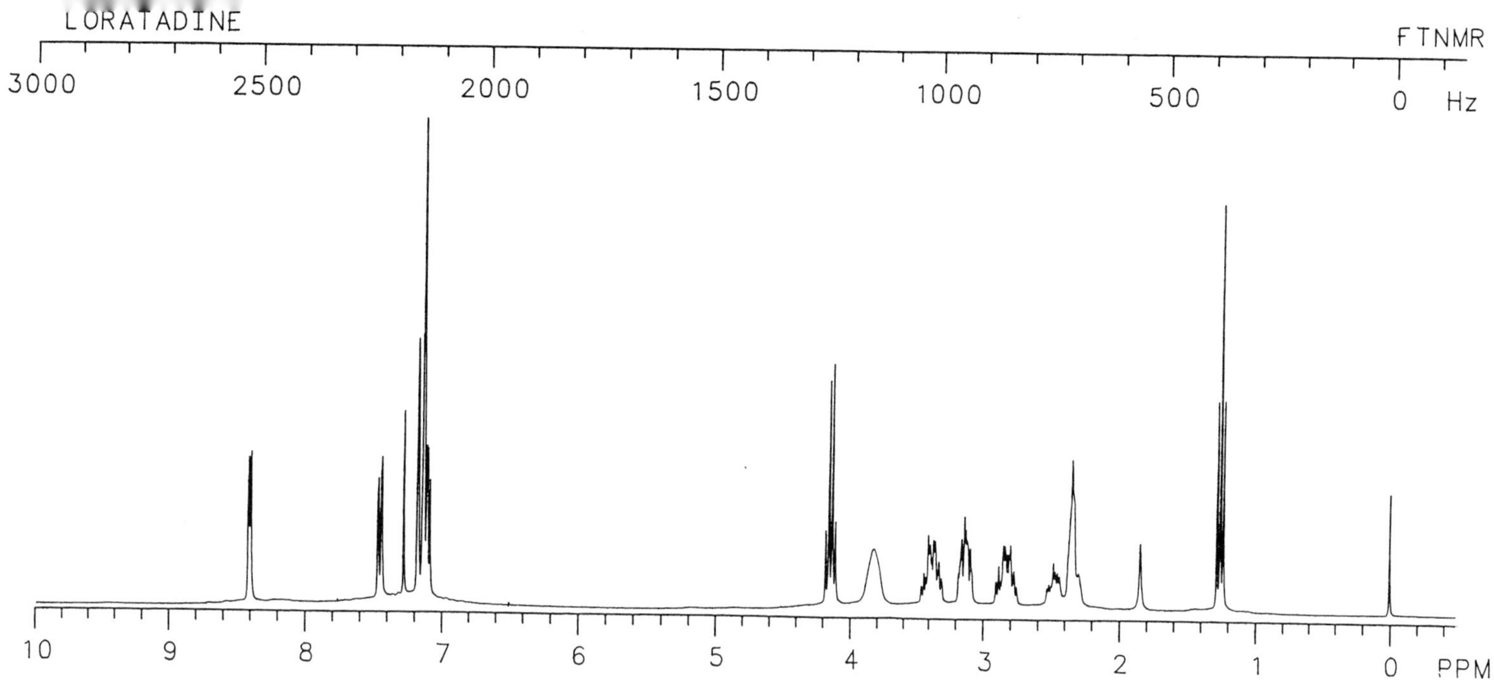

PEAK LISTING

#	HT	PPM
1	25	.00
2	43	1.22
3	83	1.25
4	43	1.27
5	24	2.33
6	31	2.34
7	50	4.12
8	46	4.14
9	25	7.08
10	32	7.09
11	32	7.10
12	99	7.13
13	55	7.13
14	54	7.16
15	39	7.27
16	30	7.43
17	26	7.45
18	31	8.39
19	28	8.39
20	29	8.40

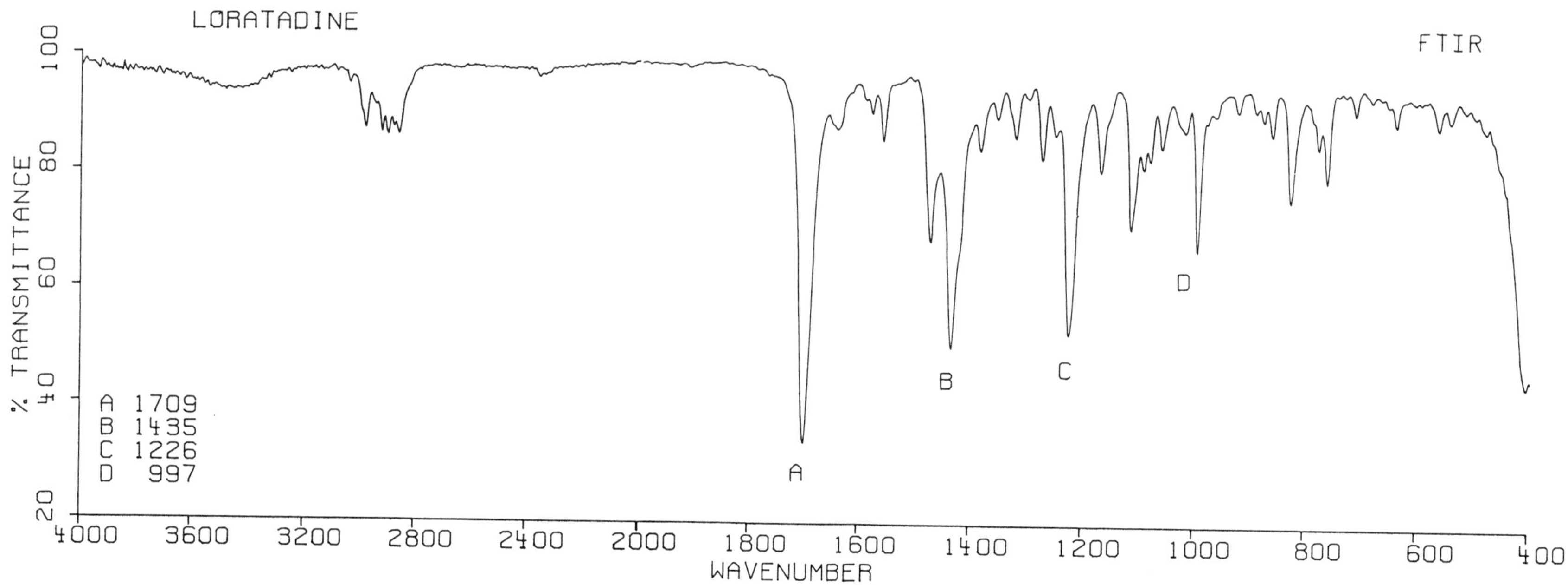

MEDROXYPROGESTERONE ACETATE

$C_{24}H_{34}O_4$

Molecular Weight: 386.53 (386.25)

Synonyms: (6α)-17-Hydroxy-6α-methylpregn-4-ene-3,20-dione-17-acetate
Trade Names: Agestal, Amen, Clinovir, Curretab, Cycrin, Depo-Clinovir, Depo-Provera, Deporone, Farlutal, Gestapuran, Gestapuron, Nidaxin, Oragest, Perlutex, Prodasone, Provera, Veramix
Use: Progestogen

HPLC: Methanol: 2.8
GC: 3097; 280°

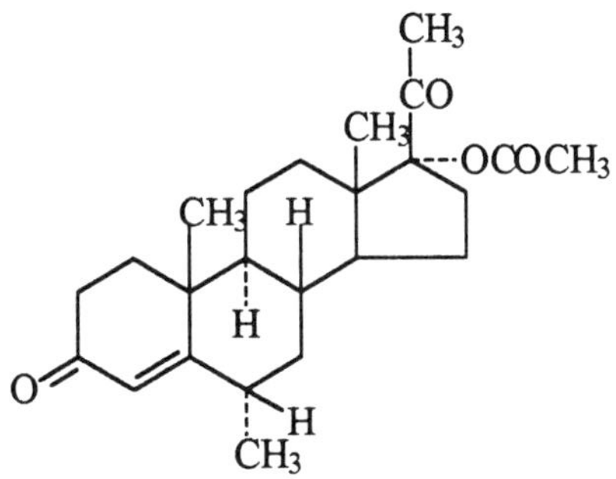

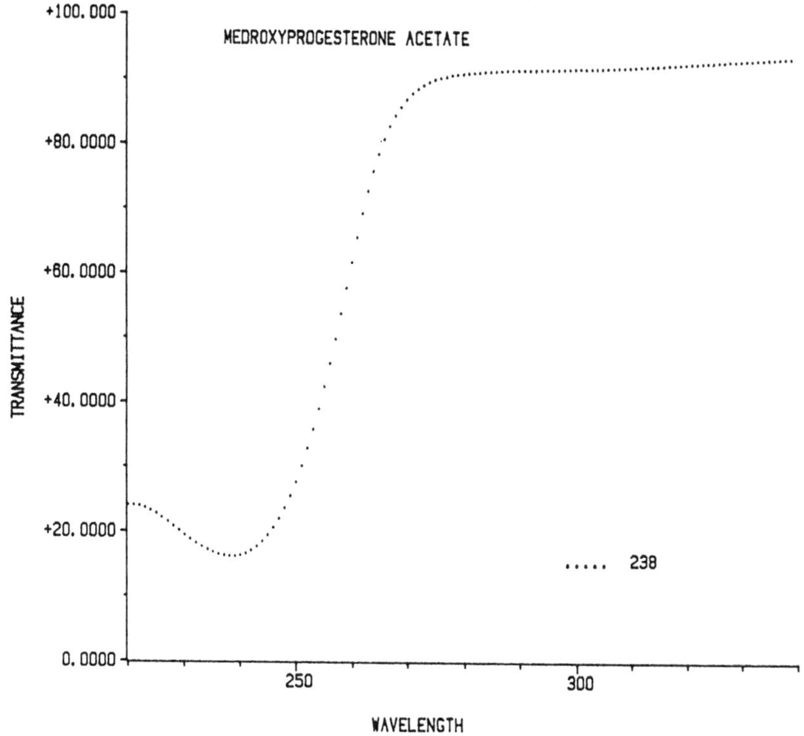

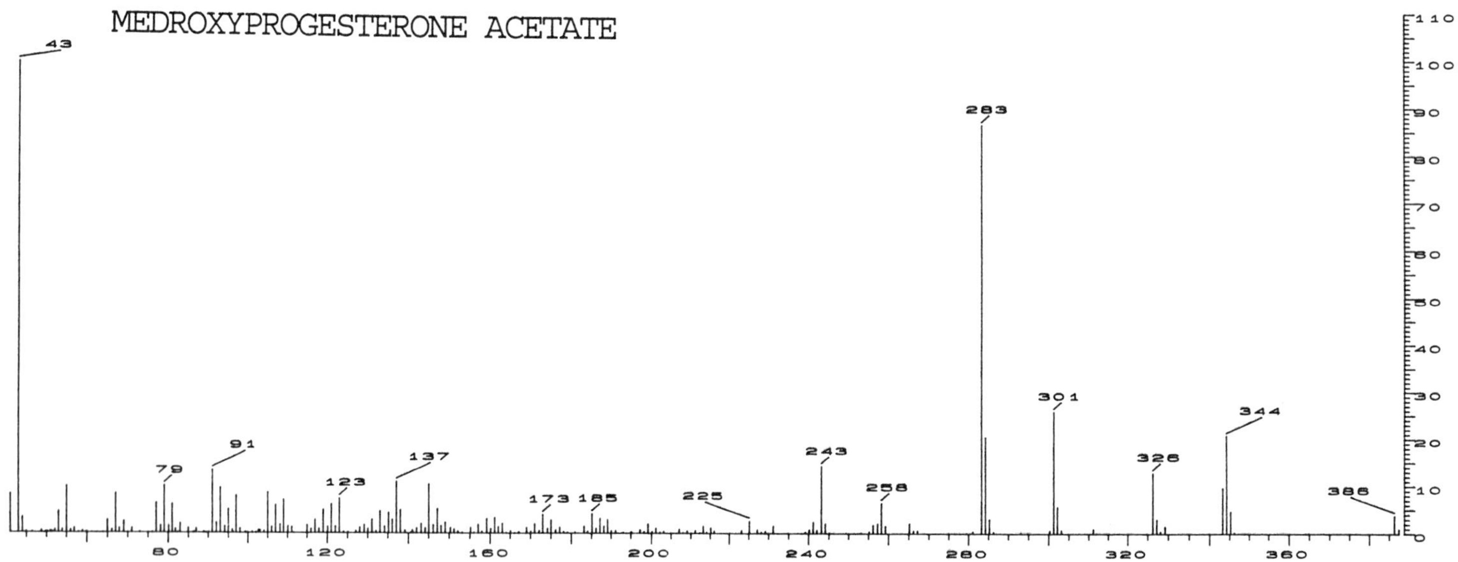

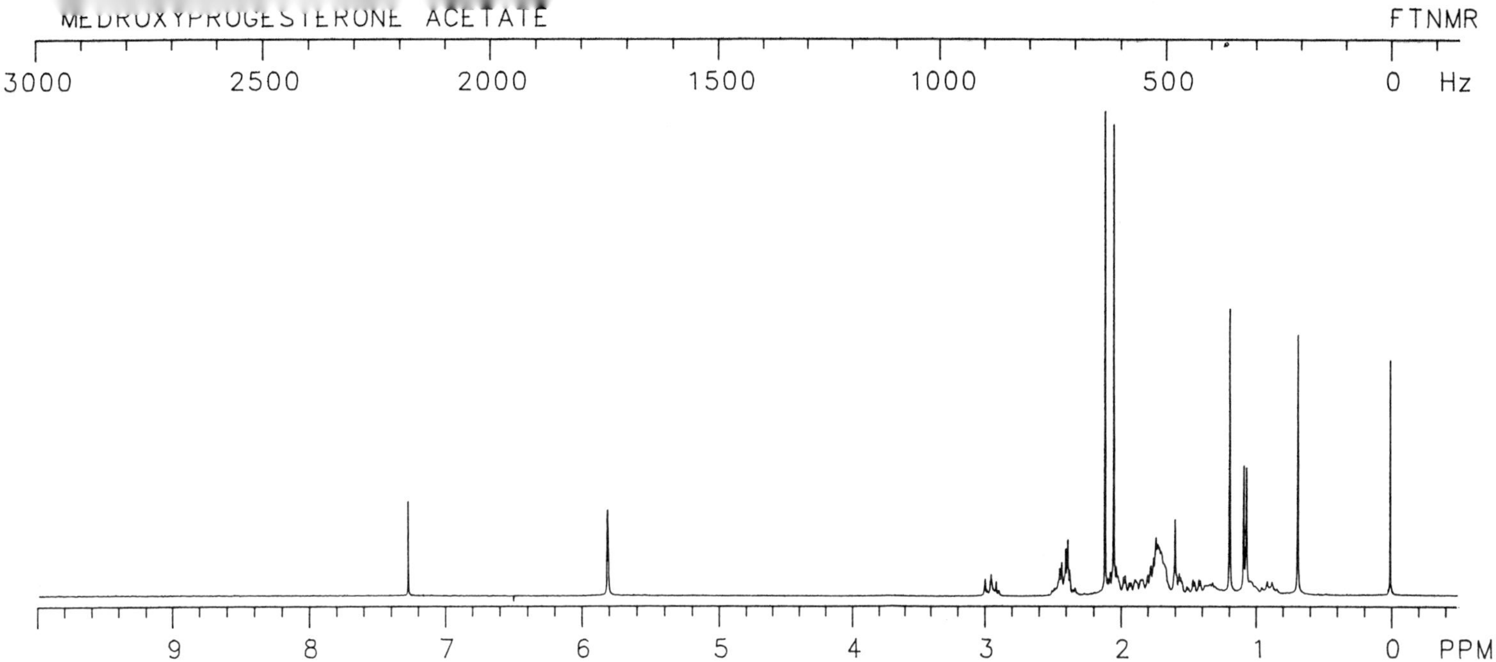

PEAK LISTING

#	HT	PPM
1	47	.00
2	53	.68
3	25	1.06
4	26	1.09
5	58	1.19
6	15	1.59
7	9	1.69
8	9	1.69
9	10	1.70
10	10	1.71
11	11	1.73
12	96	2.04
13	98	2.10
14	11	2.37
15	9	2.39
16	17	5.80
17	16	5.80
18	19	7.26

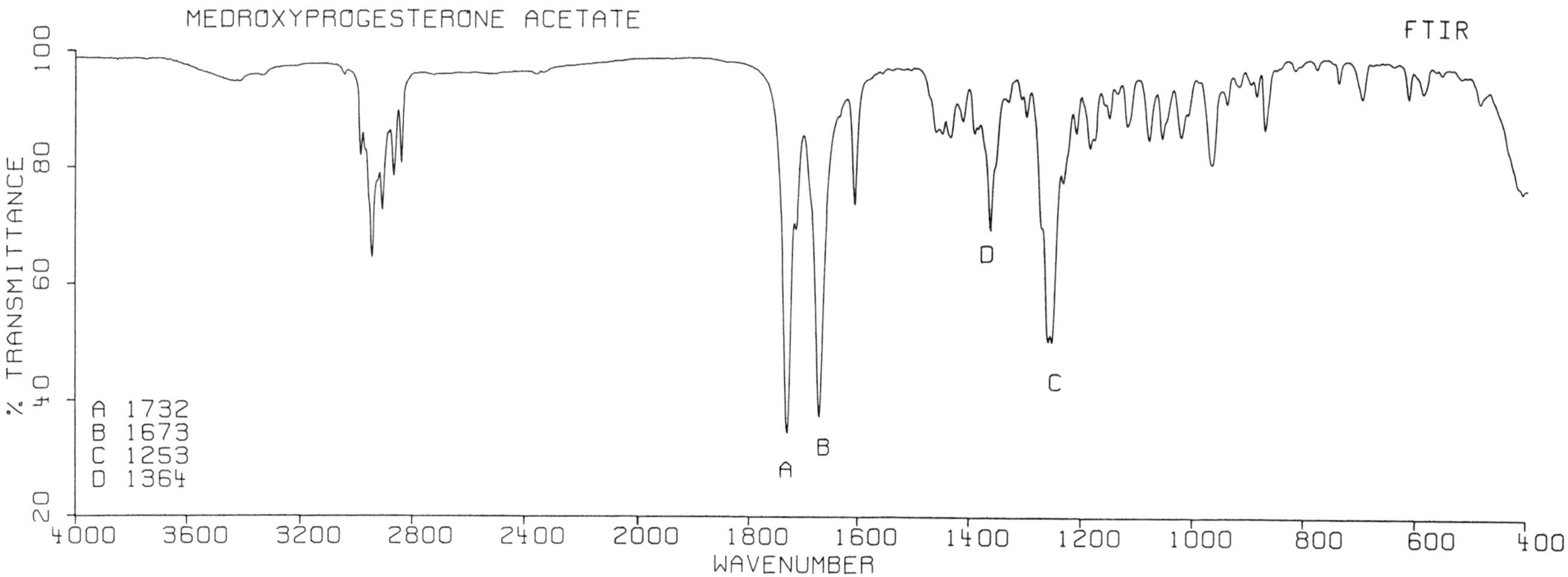

MEFLOQUINE

$C_{17}H_{16}F_6N_2O$

Molecular Weight: 378.32 (378.12)

Synonyms: 2-Piperidinyl-2,8-bis(trifluoromethyl)-4-quinolinemethanol
Trade Names: Lariam
Use: Antimalarial

HPLC: Methanol: 2.3
GC:

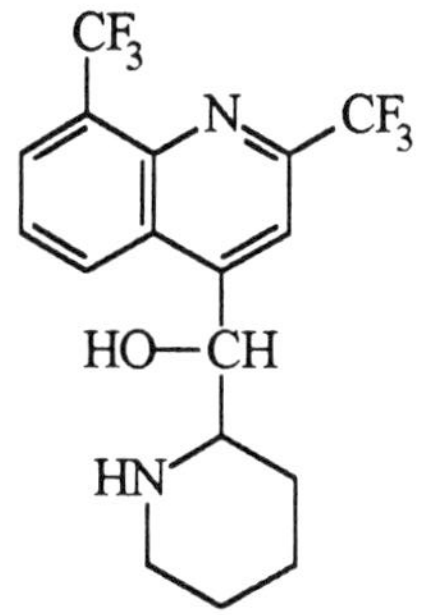

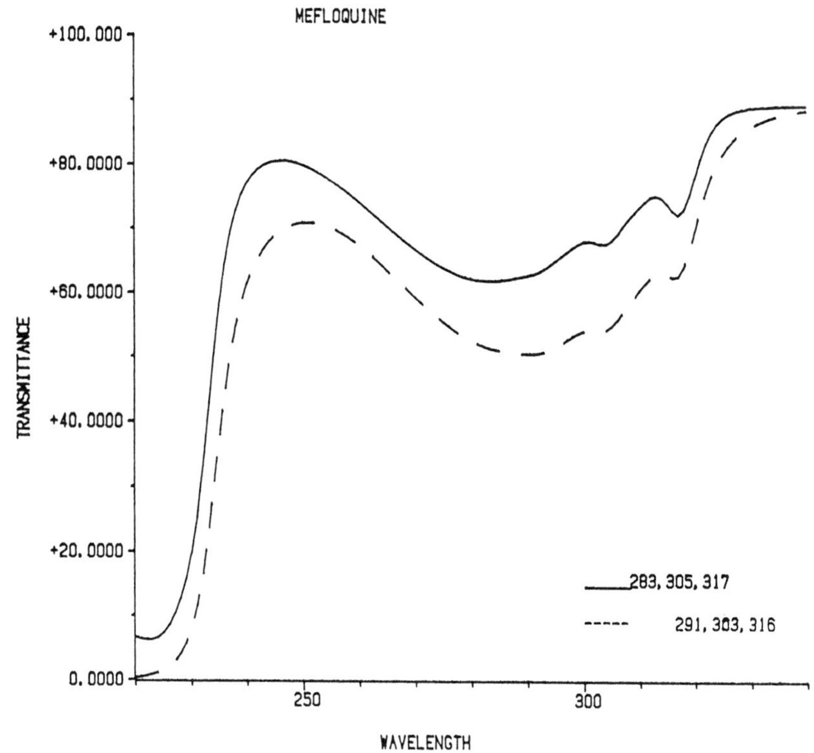

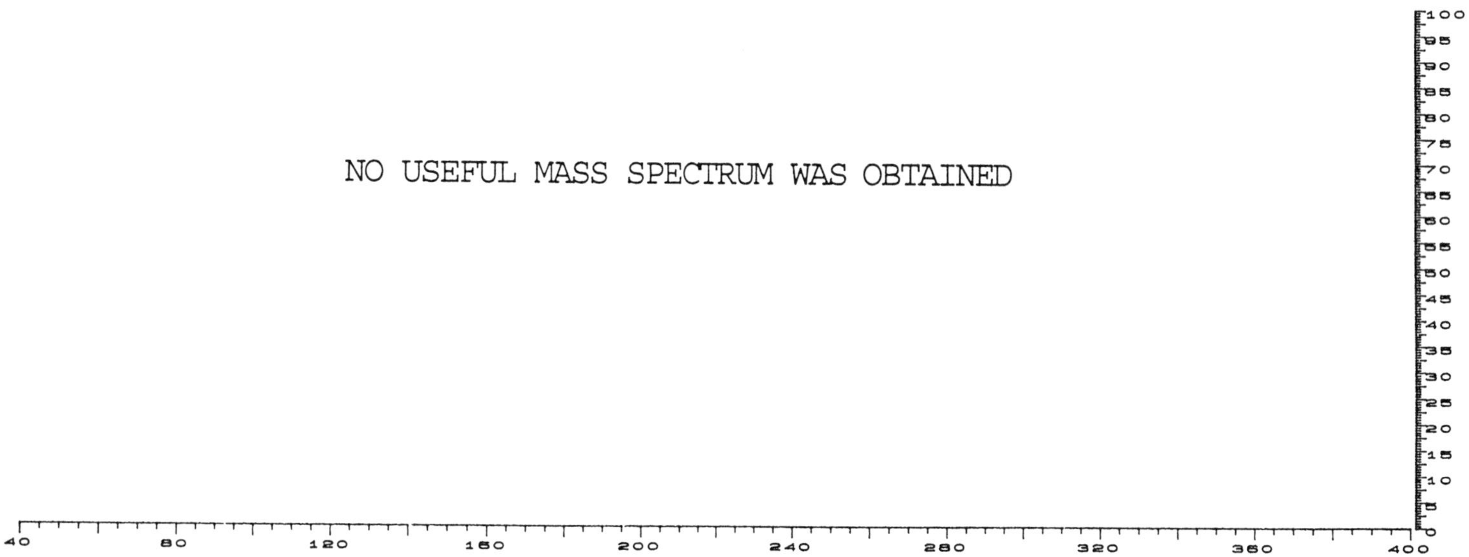

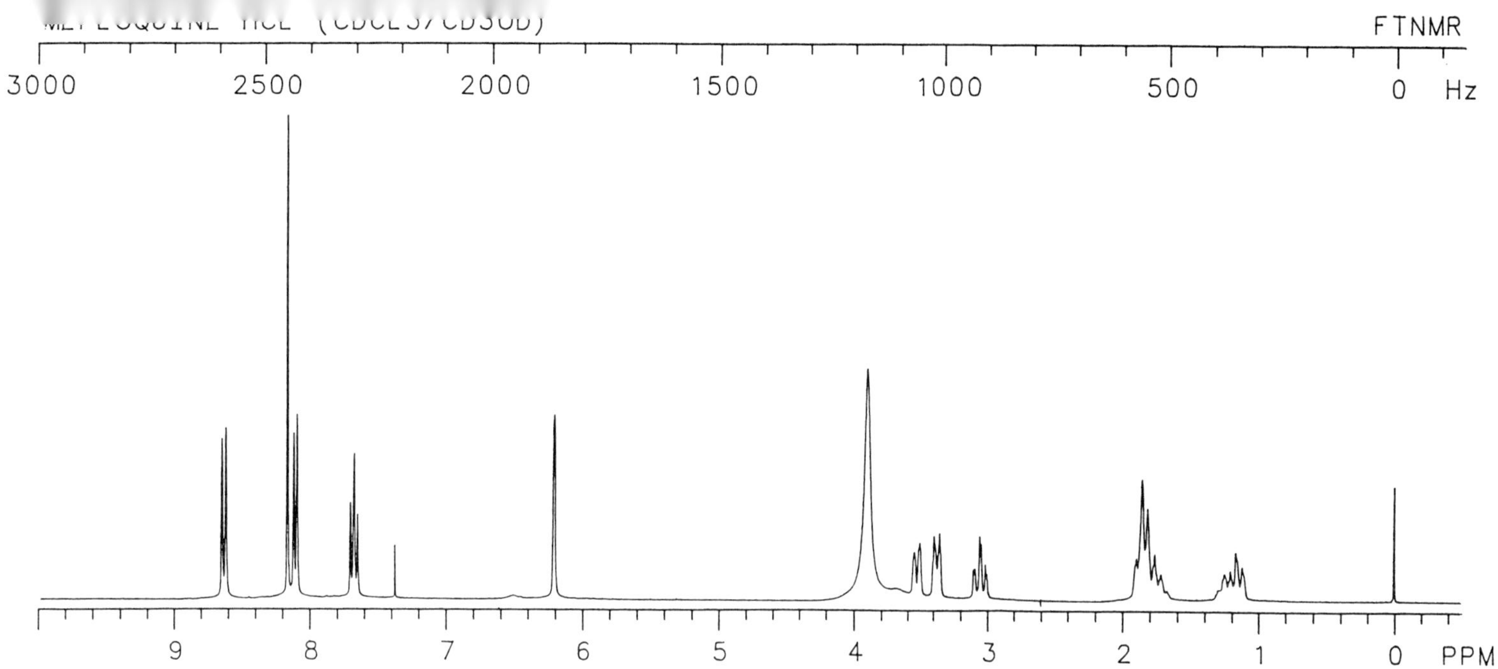

PEAK LISTING

#	HT	PPM
1	23	.00
2	19	1.81
3	25	1.85
4	47	3.90
5	37	6.20
6	36	6.21
7	29	7.67
8	19	7.70
9	37	8.09
10	33	8.11
11	99	8.16
12	34	8.61
13	32	8.64

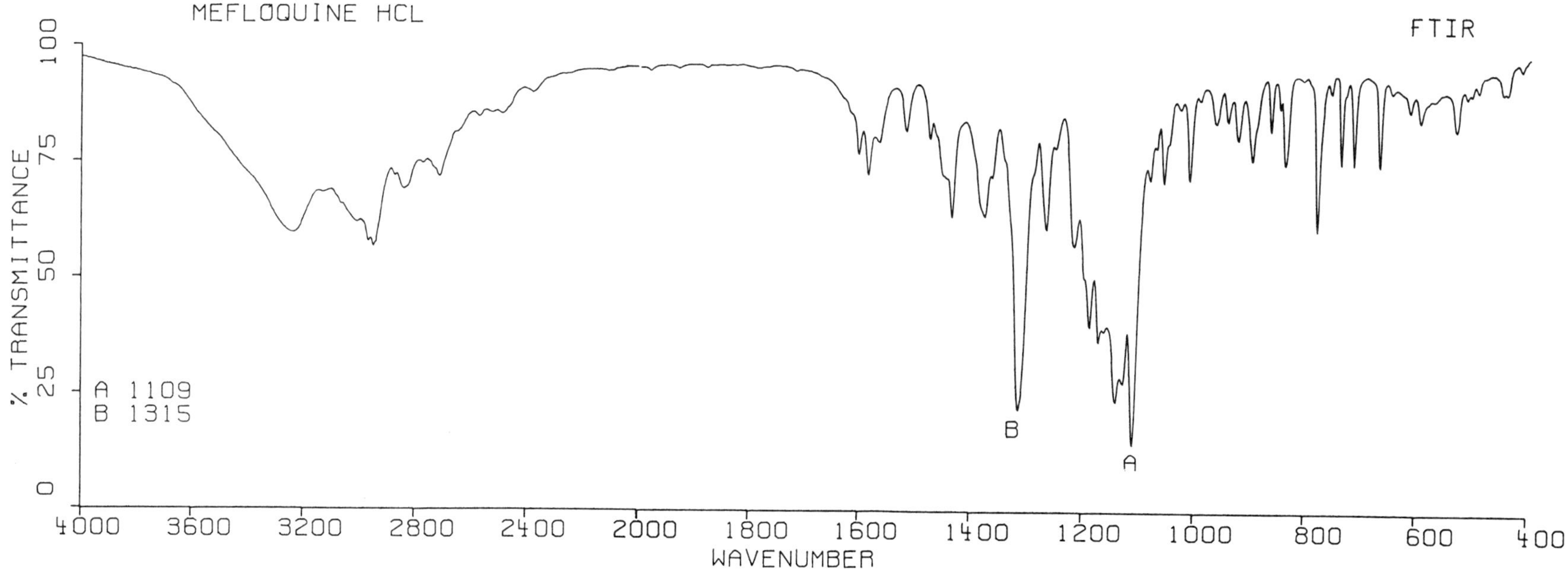

MESTEROLONE

$C_{20}H_{32}O_2$

Molecular Weight: 304.47 (304.24)

Synonyms: 17-Hydroxy-1-methylandrostan-3-one
Trade Names: Androviron, Mestoranum, Proviron
Use: Androgen

HPLC:
GC: 2626; 250°

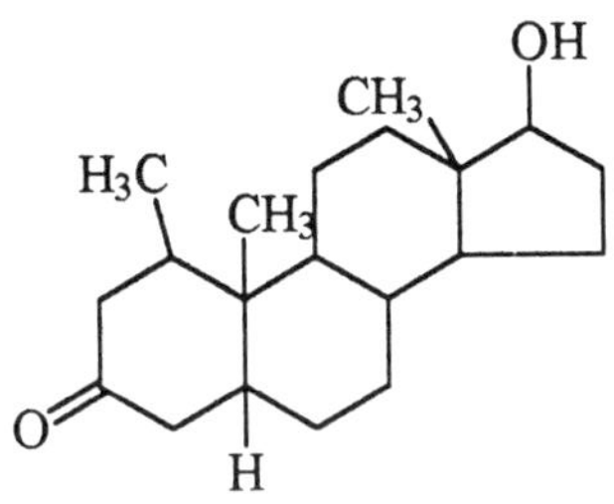

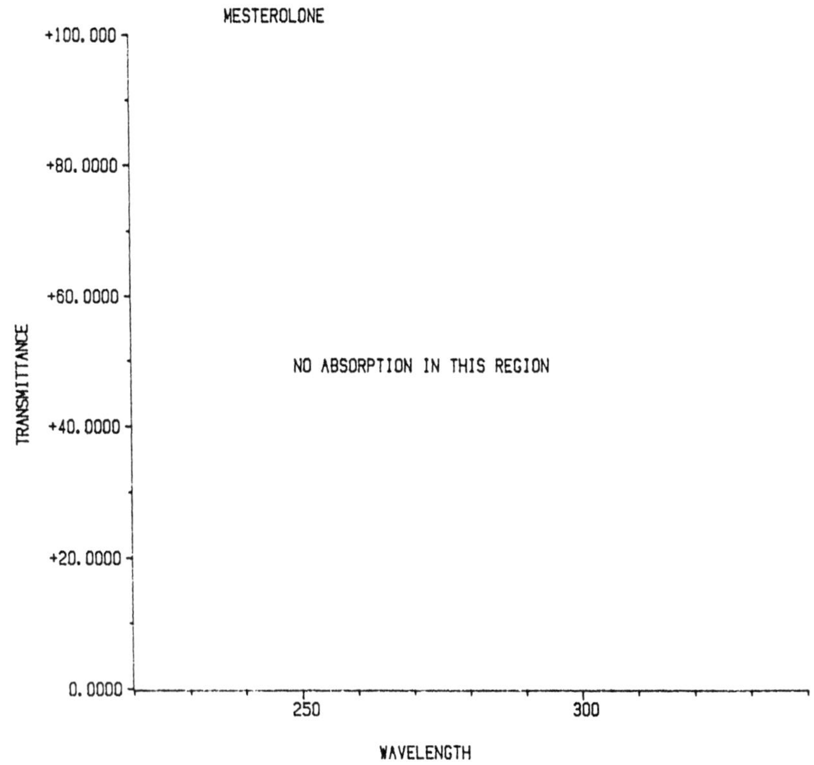

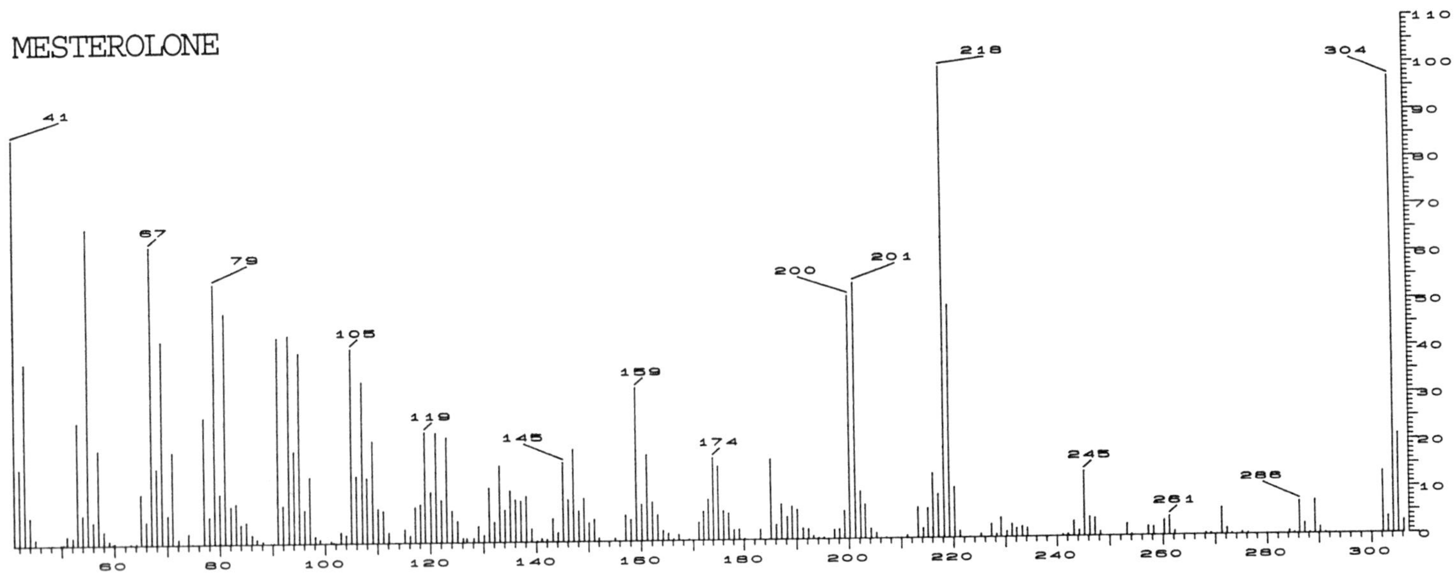

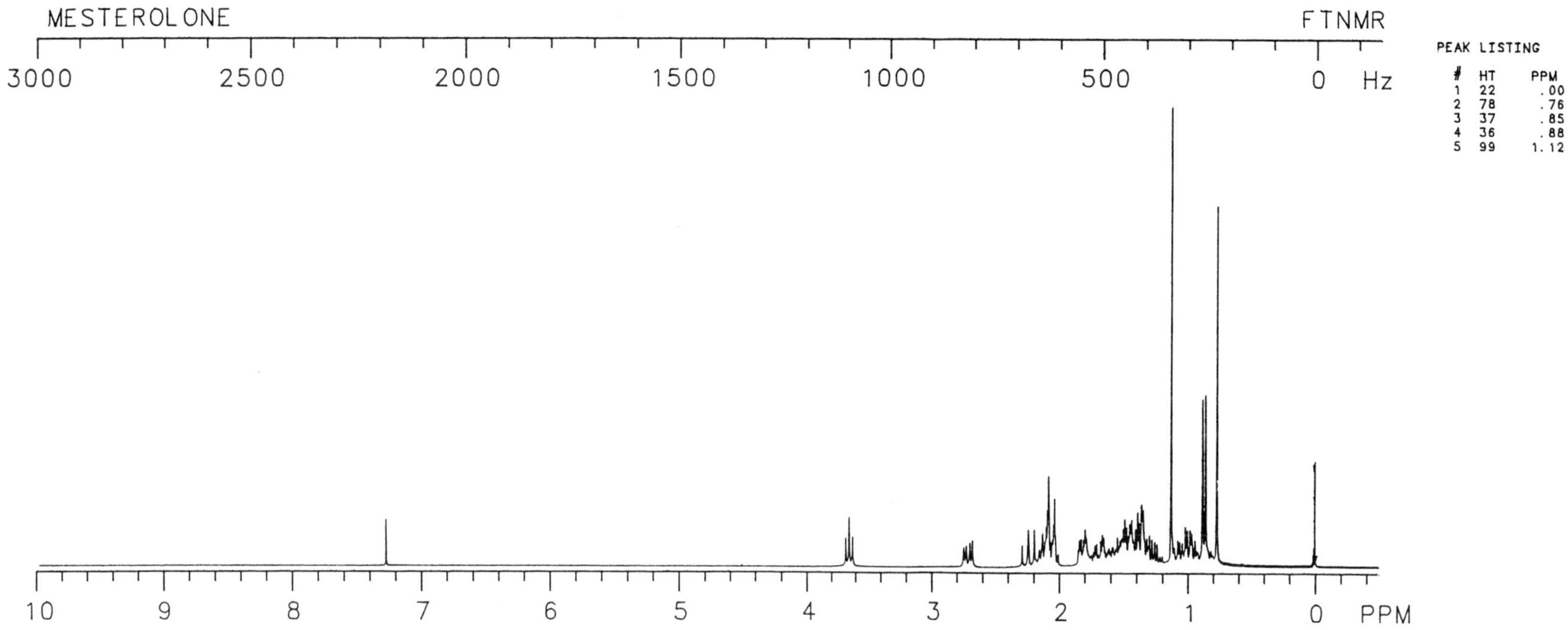
MESTEROLONE
FTNMR
3000 2500 2000 1500 1000 500 0 Hz
PEAK LISTING
HT PPM
1 22 .00
2 78 .76
3 37 .85
4 36 .88
5 99 1.12
10 9 8 7 6 5 4 3 2 1 0 PPM

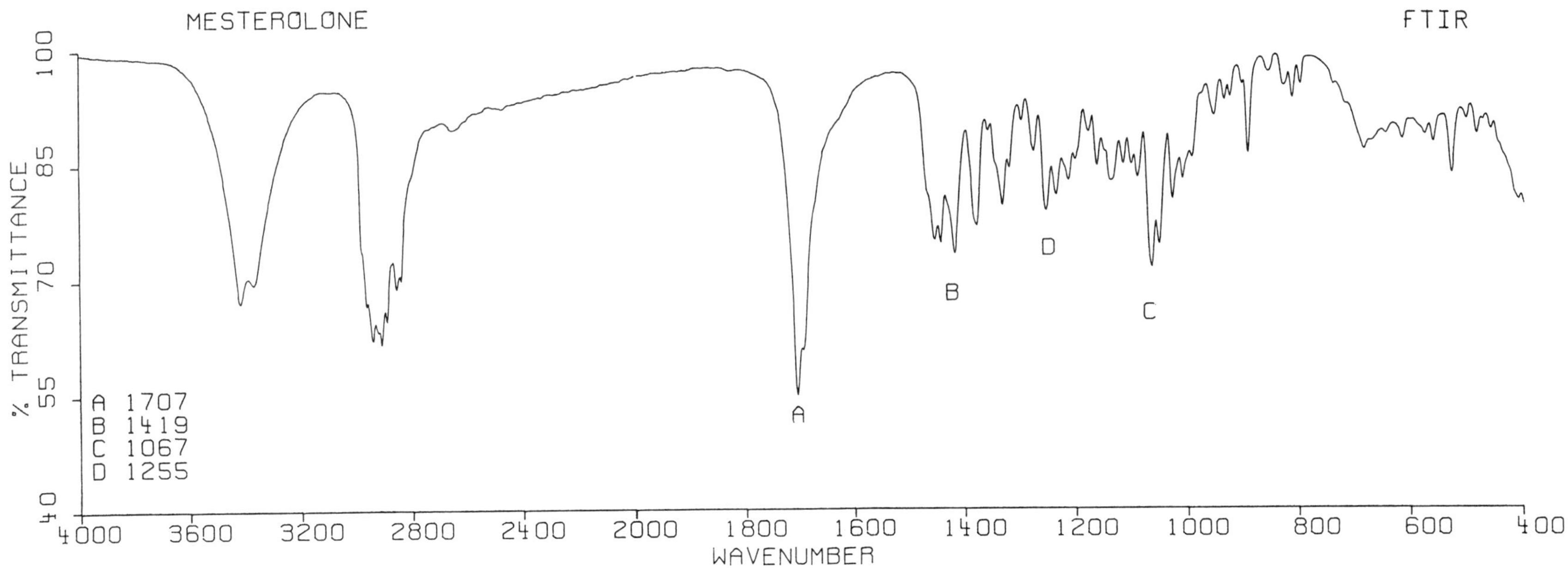
MESTEROLONE
FTIR
% TRANSMITTANCE
100 85 70 55 40
A 1707
B 1419
C 1067
D 1255
A
B
C
D
4000 3600 3200 2800 2400 2000 1800 1600 1400 1200 1000 800 600 400
WAVENUMBER

METHCATHINONE

$C_{10}H_{13}NO$

Molecular Weight: 163.22 (163.10)

Synonyms: 2-(Methylamino)-propiophenone
Trade Names:
Use: Central stimulant

HPLC:
GC: 1323; 140°

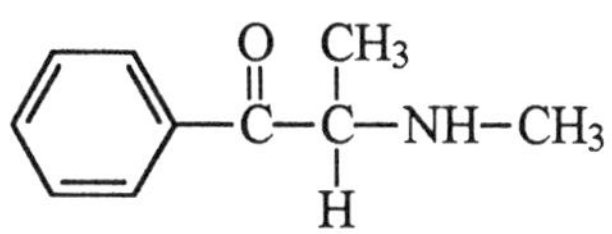

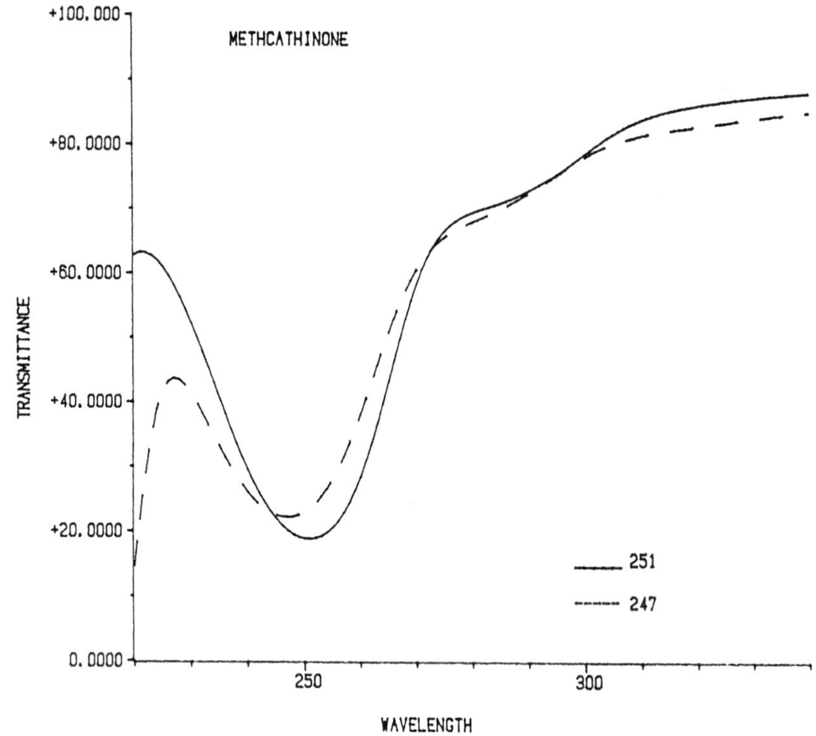

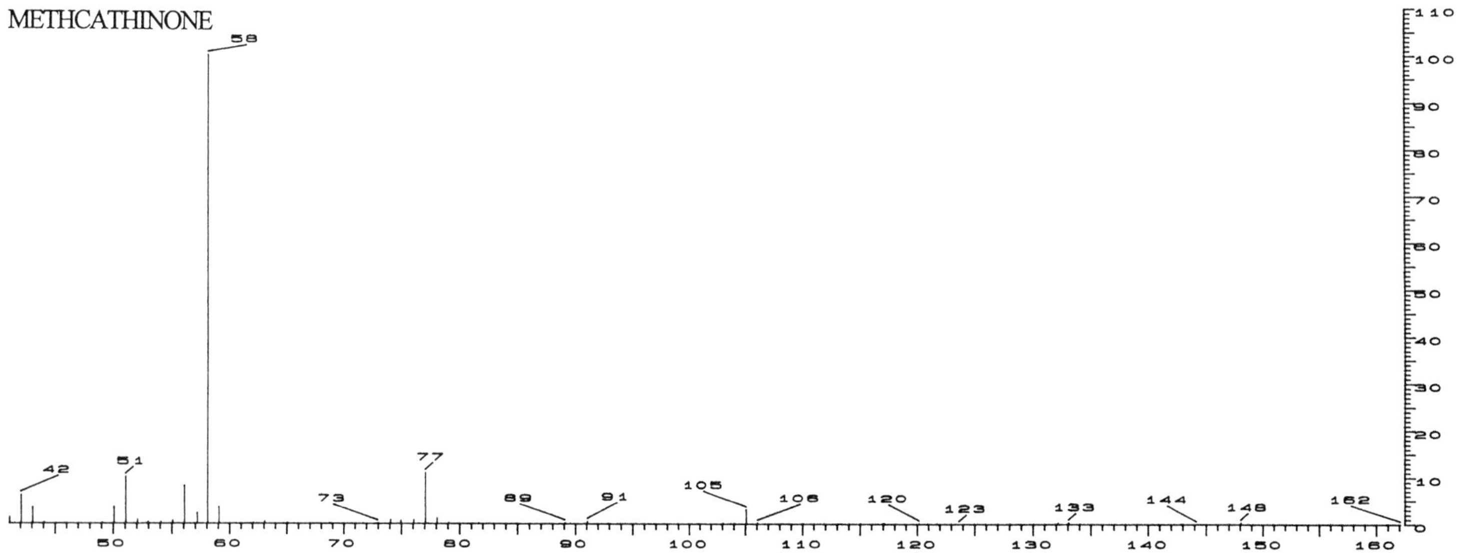

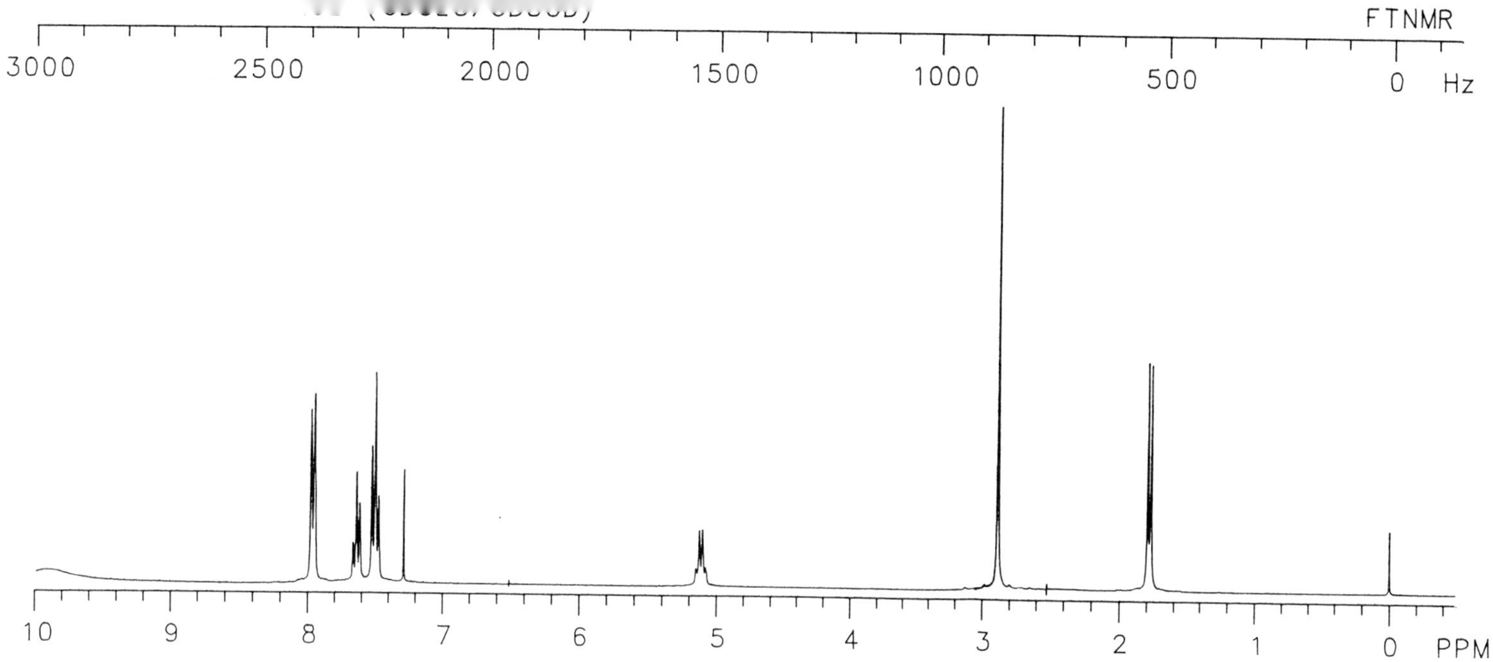

PEAK LISTING

#	HT	PPM
1	46	1.76
2	46	1.78
3	99	2.88
4	23	7.28
5	43	7.49
6	28	7.52
7	22	7.63
8	38	7.94
9	37	7.94
10	35	7.97

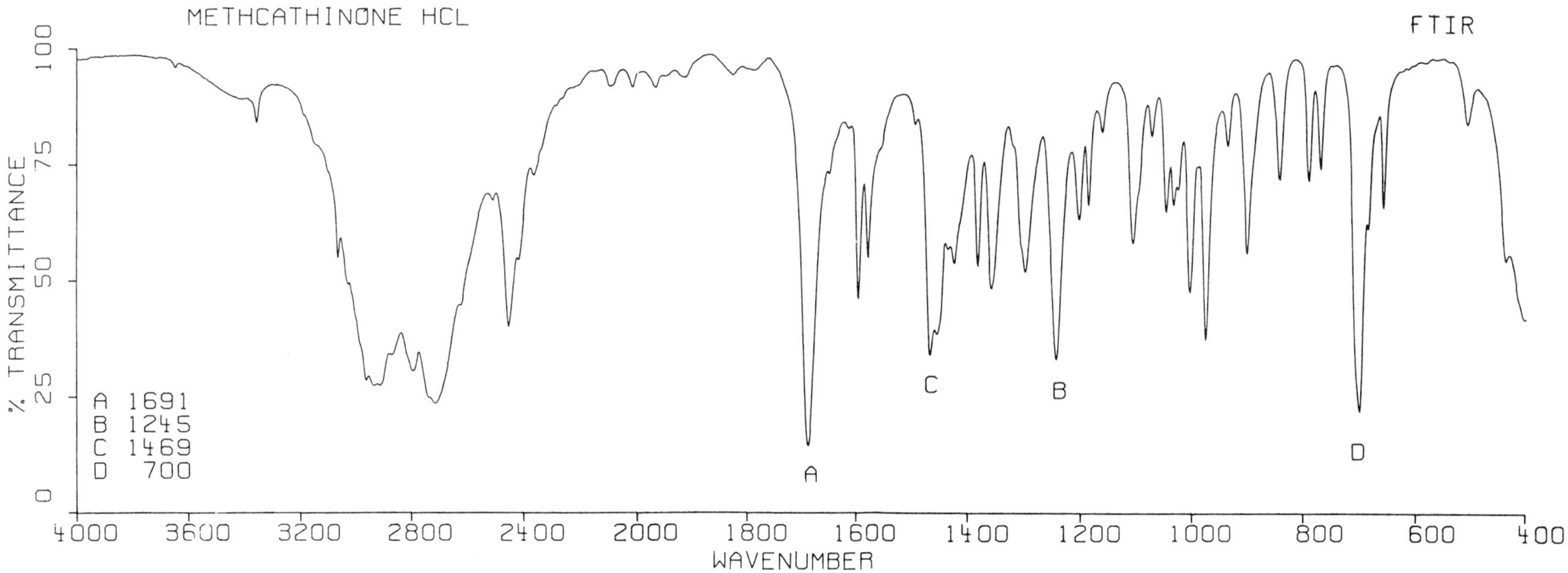

METHOXAMINE

$C_{11}H_{17}NO_3$

Molecular Weight: 211.26 (211,12)

Synonyms: α-(1-Aminoethyl)-2,5-dimethoxybenzenemethanol; 2,5-dimethoxynorephedrine
Trade Names: Pressomin, Vasoxine, Vasoxyl, Vasylox
Use: Adrenergic

HPLC: Ethanol: 2.1
GC:

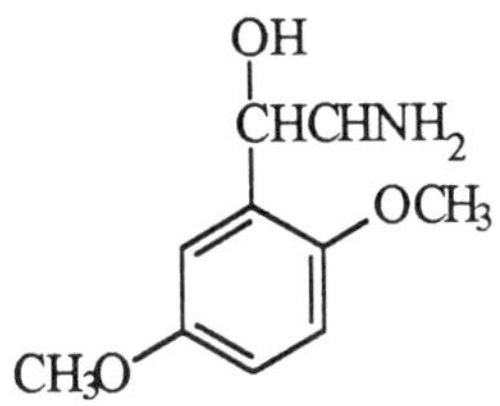

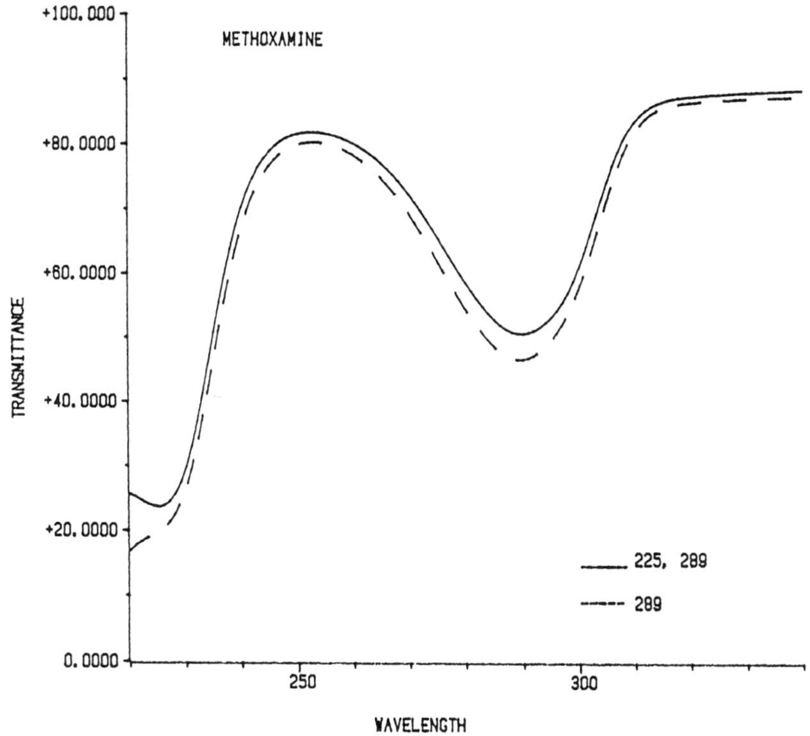

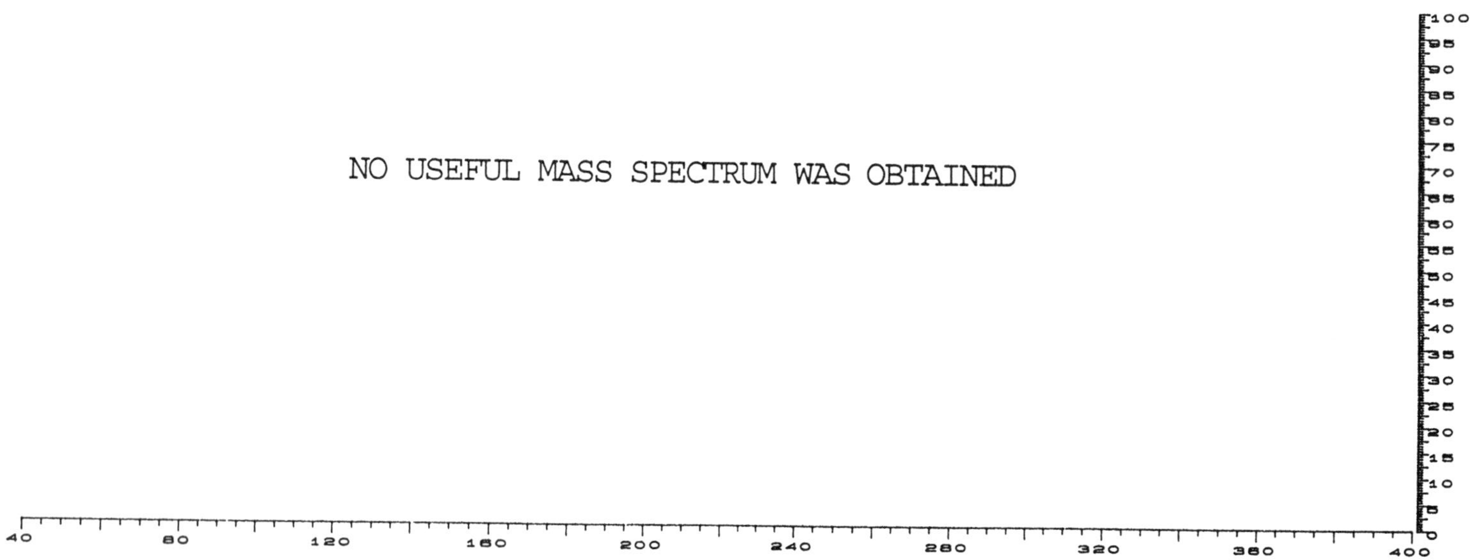

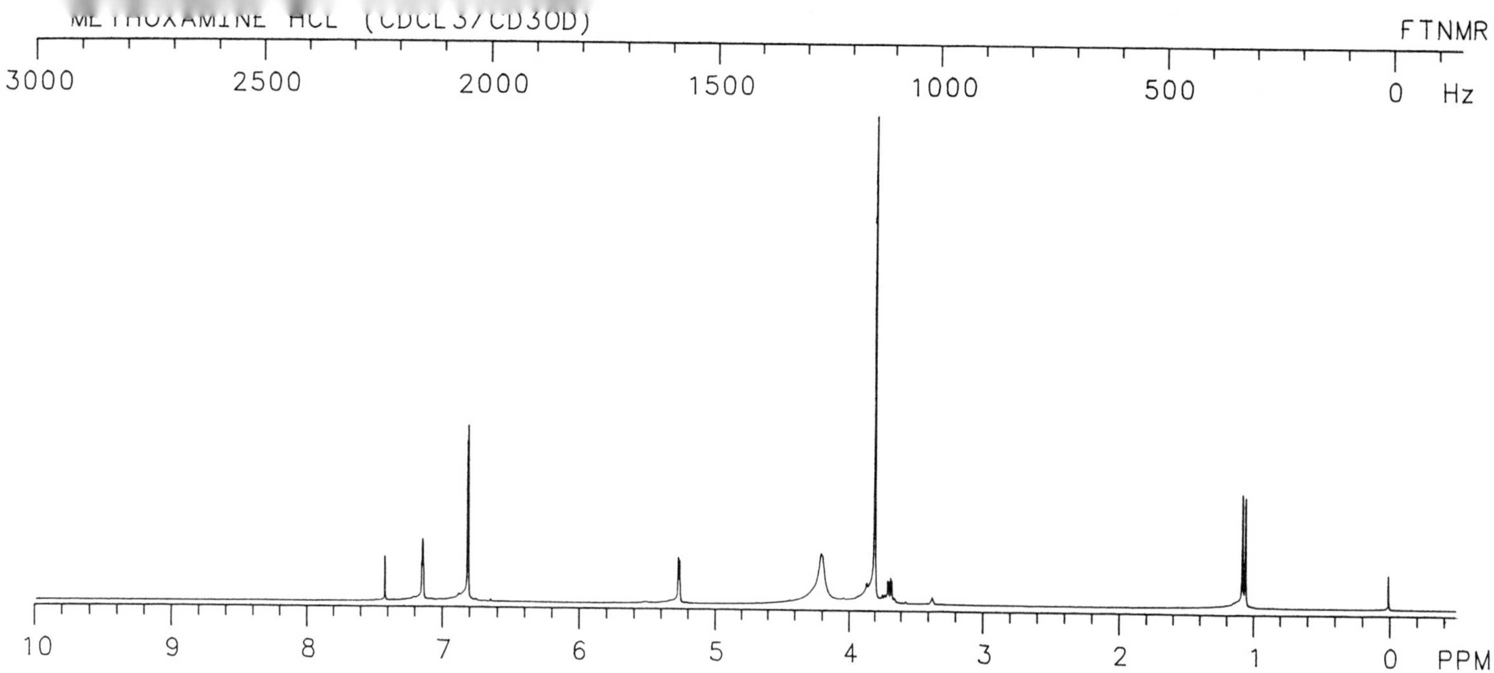

PEAK LISTING

#	HT	PPM
1	22	1.05
2	23	1.07
3	99	3.79
4	88	3.79
5	10	4.19
6	10	4.20
7	9	5.25
8	35	6.80
9	32	6.80
10	12	7.13

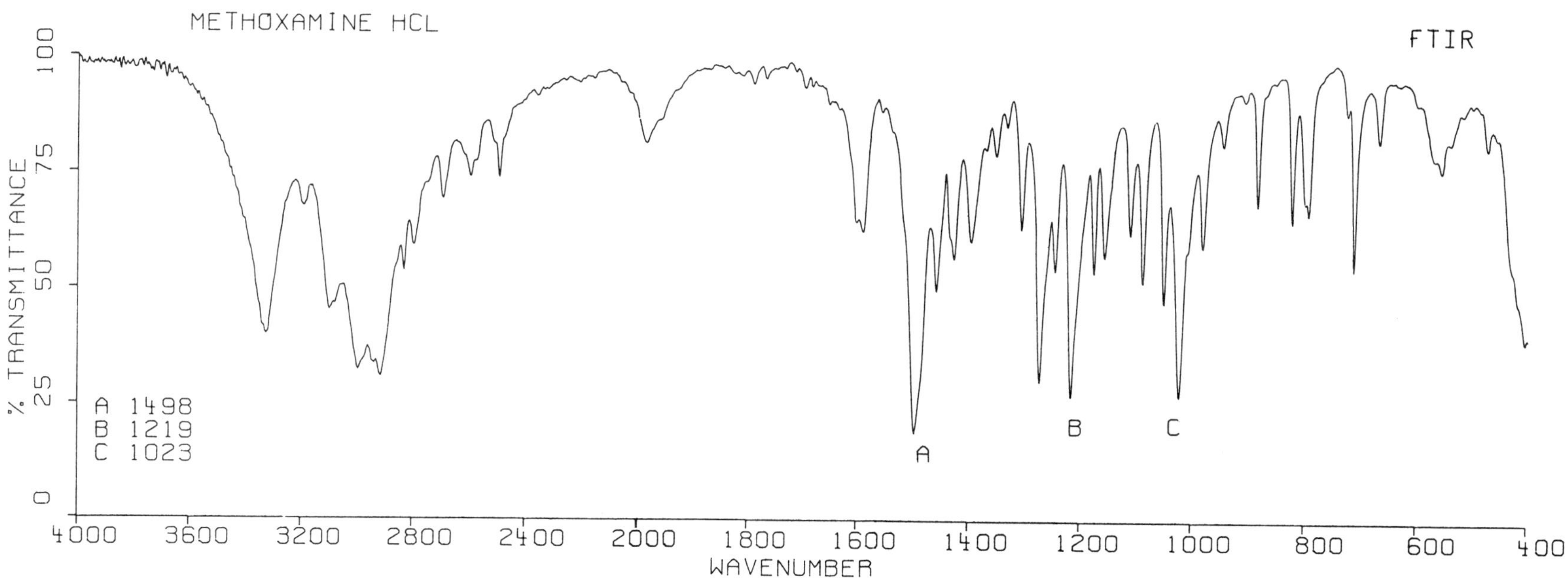

3-METHOXYTYRAMINE

$C_9H_{13}NO_2$

Molecular Weight: 167.21 (167.09)
Synonyms: 3-Methoxy-4-hydroxy-phenethylamine
Trade Names:
Use:

HPLC:
GC:

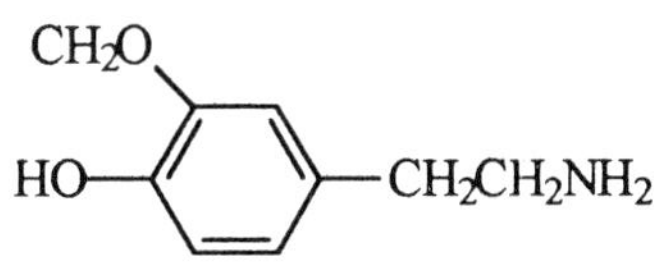

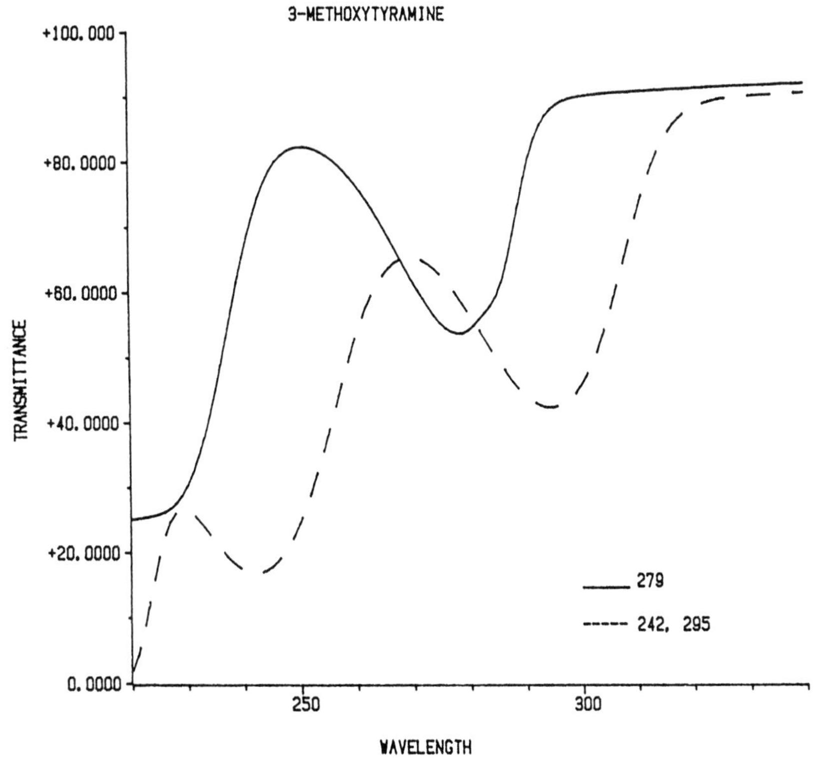

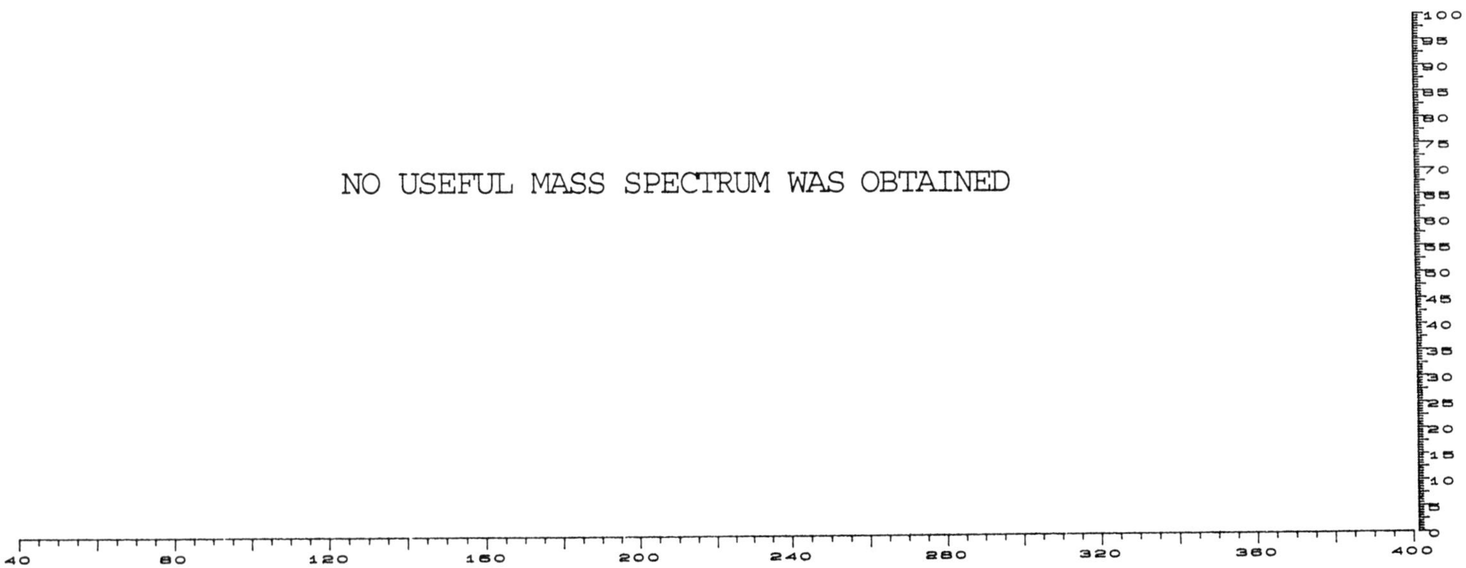

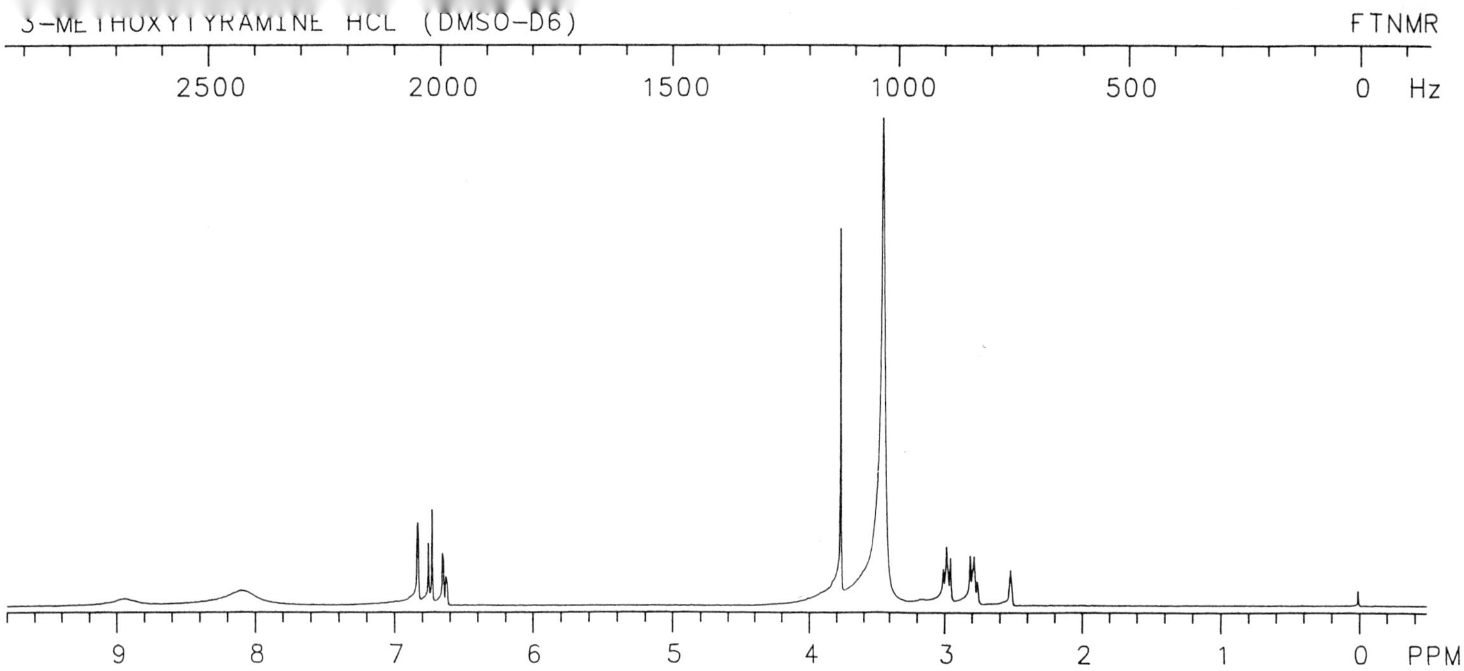

PEAK LISTING

#	HT	PPM
1	12	2.97
2	99	3.44
3	77	3.76
4	19	6.71
5	12	6.74
6	17	6.81
7	16	6.82

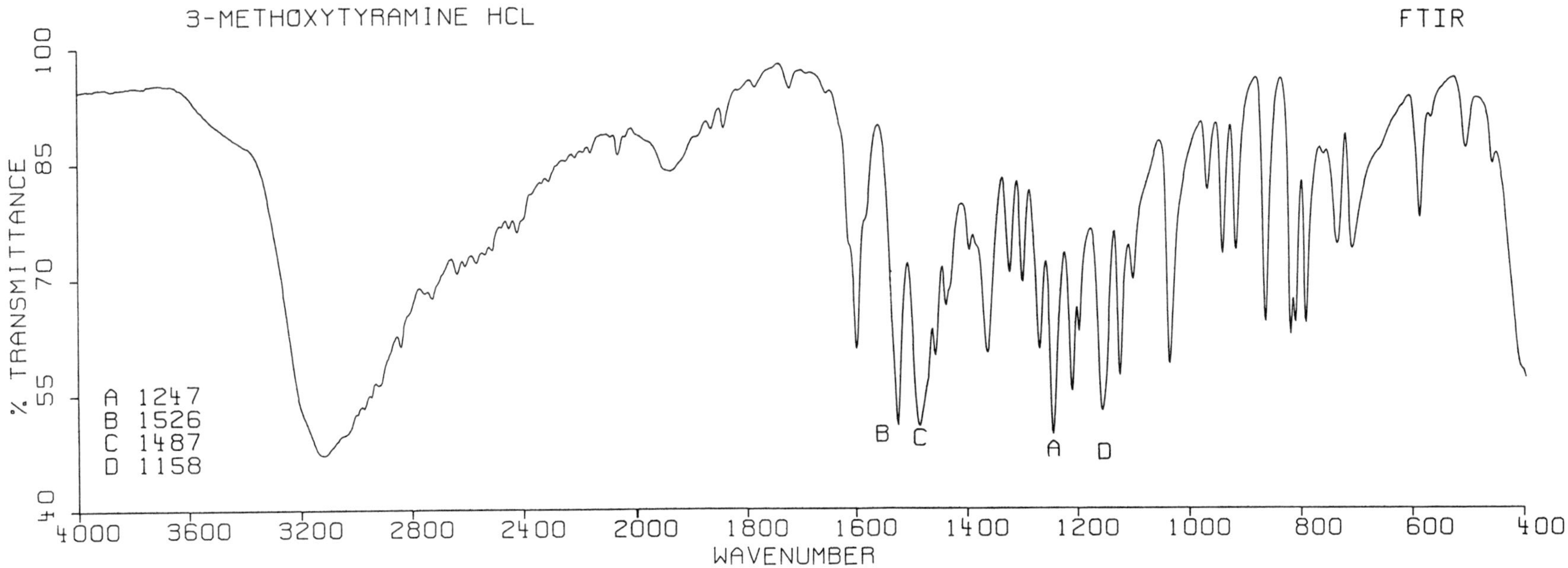

METHYLEPHEDRINE

$C_{11}H_{17}NO$

Molecular Weight: 179.26 (179.13)

Synonyms: α-[1-(Dimethylamino)ethyl]-benzenemethanol; 2-dimethylamino-1-phenylpropanol; N,N-dimethylnorephedrine; N-methylephedrine
Trade Names: Tybraine
Use: Analeptic

HPLC:
GC: 1400; 140°

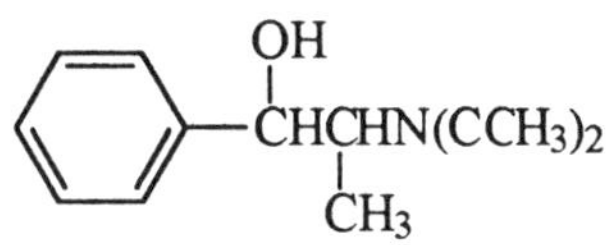

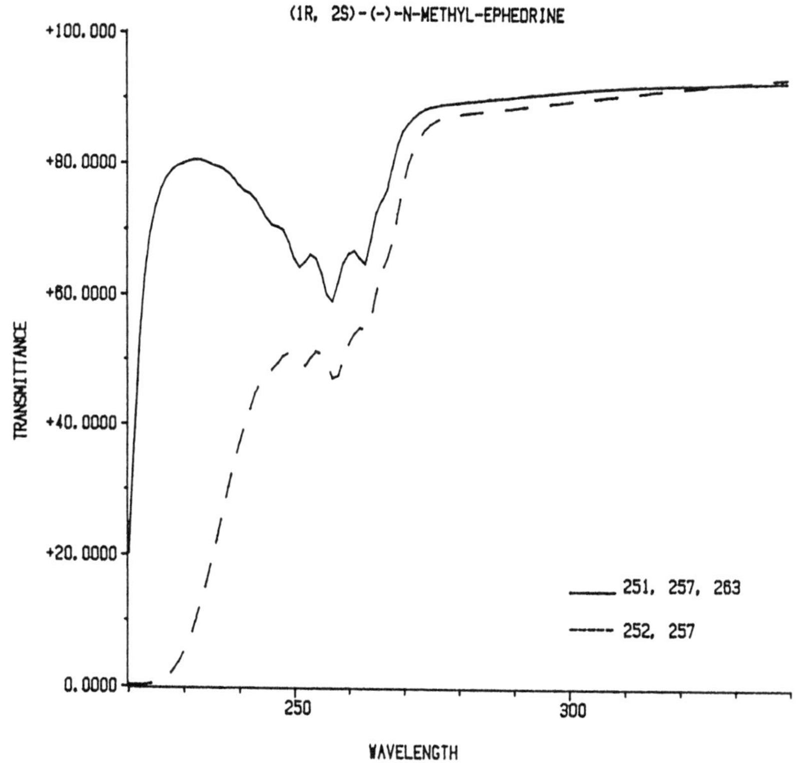

METHYLEPHEDRINE

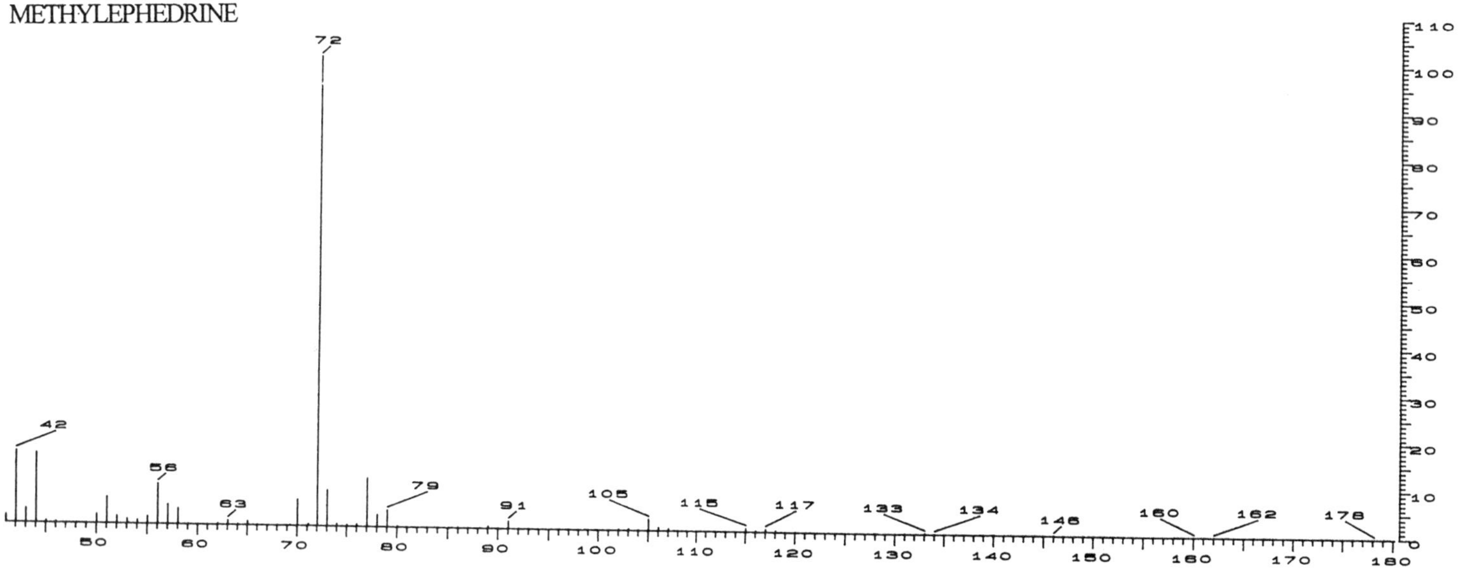

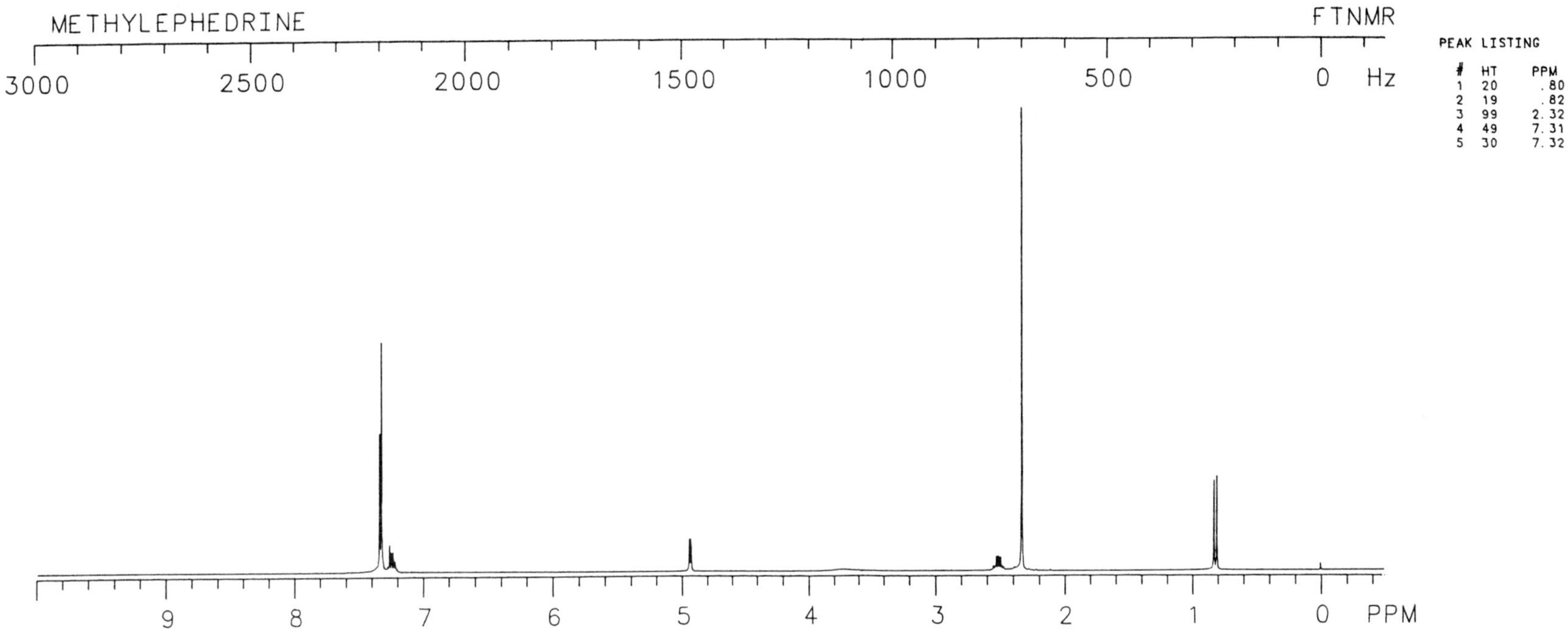
METHYLEPHEDRINE
FTNMR
3000
2500
2000
1500
1000
500
0 Hz
PEAK LISTING
HT PPM
1 20 .80
2 19 .82
3 99 2.32
4 49 7.31
5 30 7.32
9
8
7
6
5
4
3
2
1
0 PPM

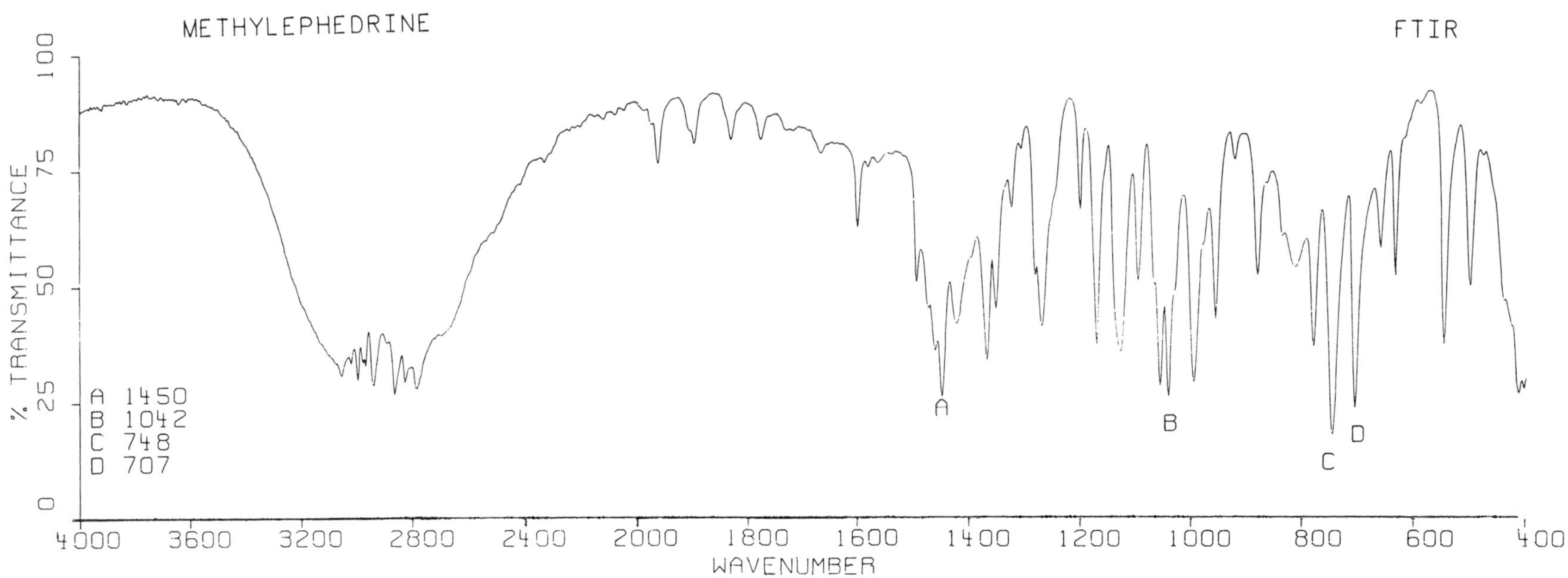
METHYLEPHEDRINE
FTIR
% TRANSMITTANCE
100
75
50
25
0
A 1450
B 1042
C 748
D 707
A
B
C
D
4000
3600
3200
2800
2400
2000
1800
1600
1400
1200
1000
800
600
400
WAVENUMBER

METHYLPREDNISOLONE-21-HEMISUCCINATE

$C_{26}H_{33}O_8Na$

Molecular Weight: 496.53 (496.21)

Synonyms: 6α-Methyl-11β,17α,21-trihydroxy-1,4-diene-3,20-dione-21-hemisuccinate
Trade Names: Depo-Medrol, Solu-Medrol, Urbason
Use: Glucocorticoid

HPLC: Ethanol: 2.5
GC:

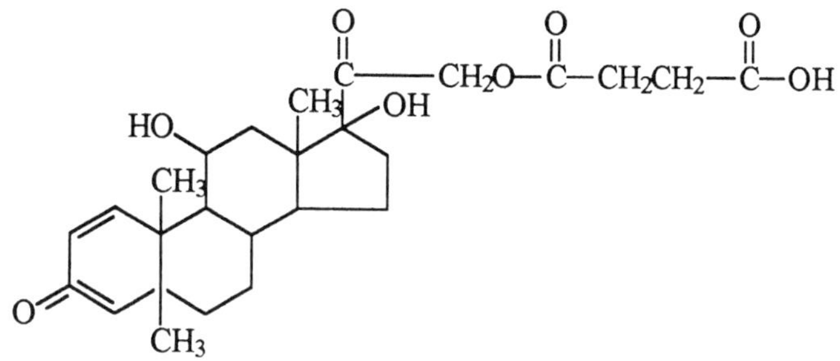

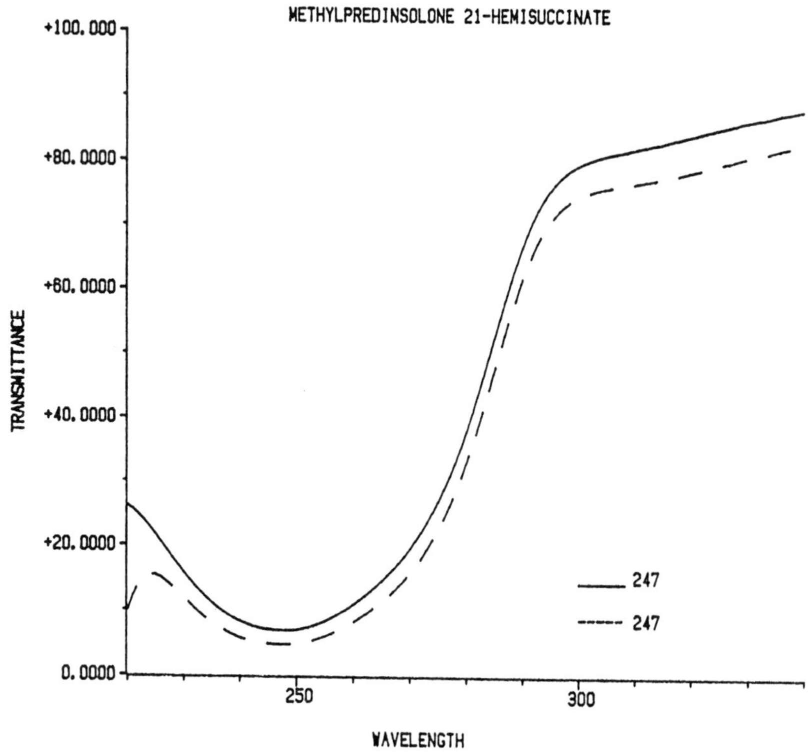

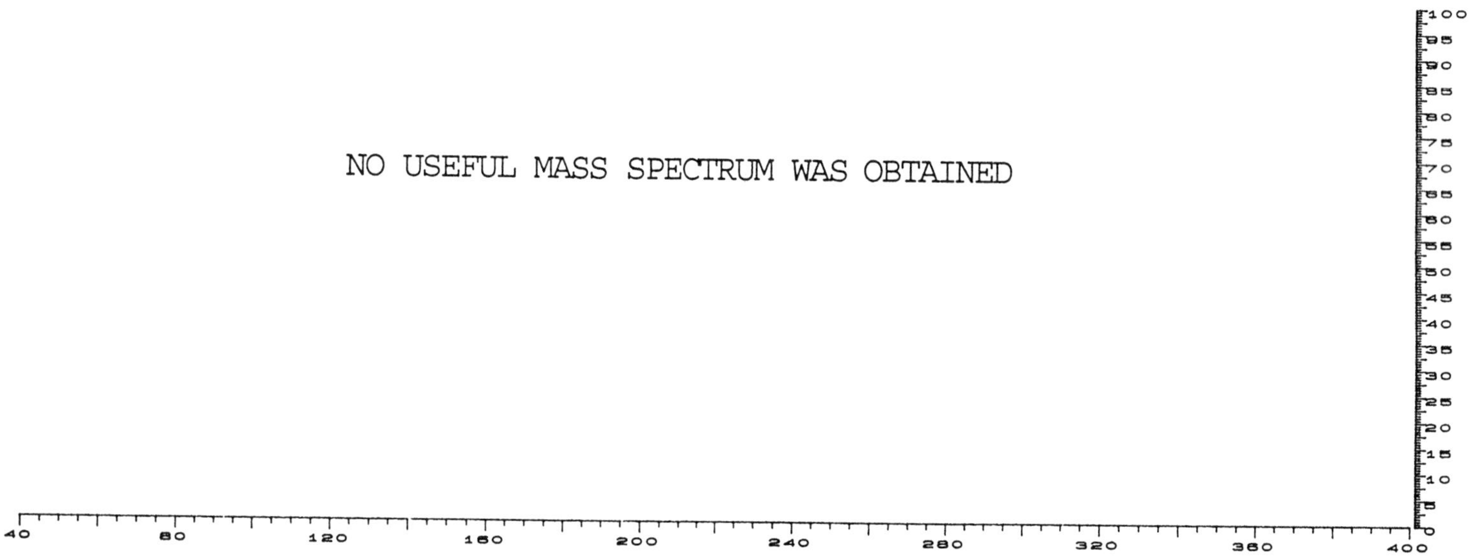

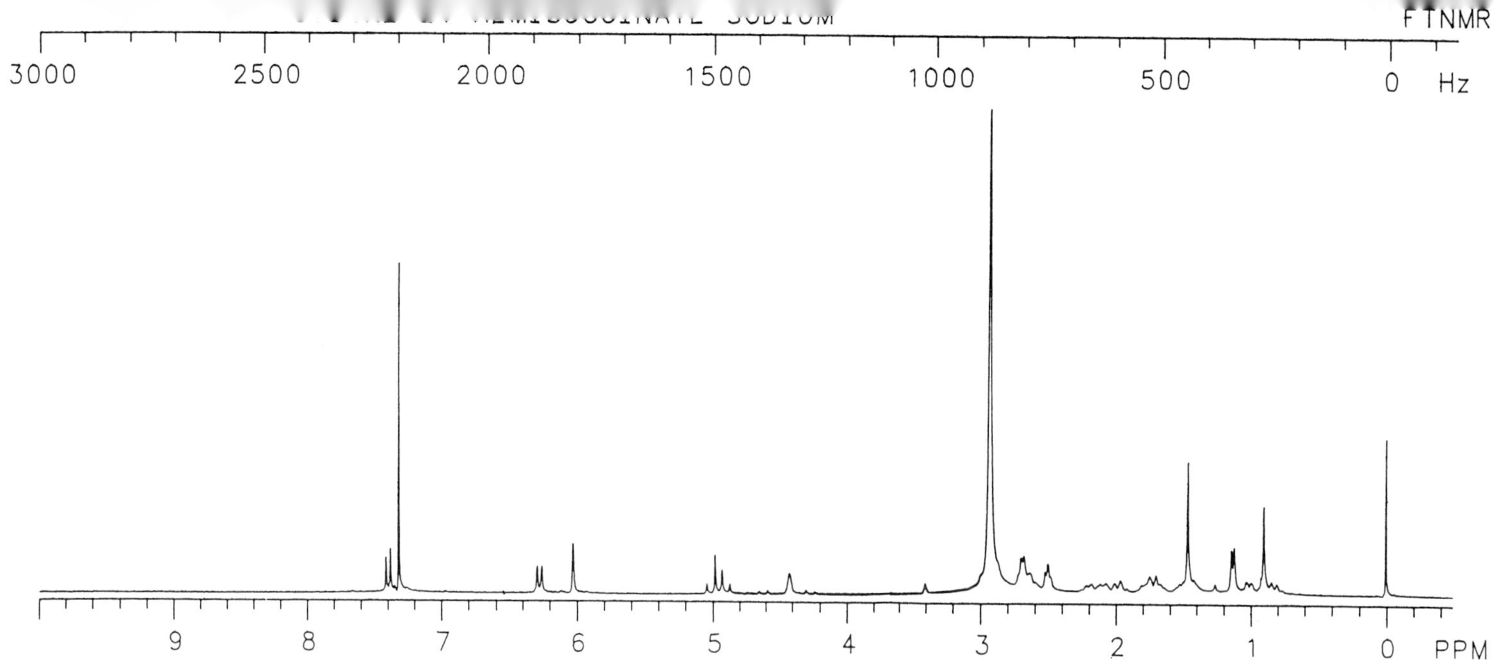

PEAK LISTING

#	HT	PPM
1	32	.00
2	18	.89
3	27	1.45
4	99	2.92
5	67	7.31

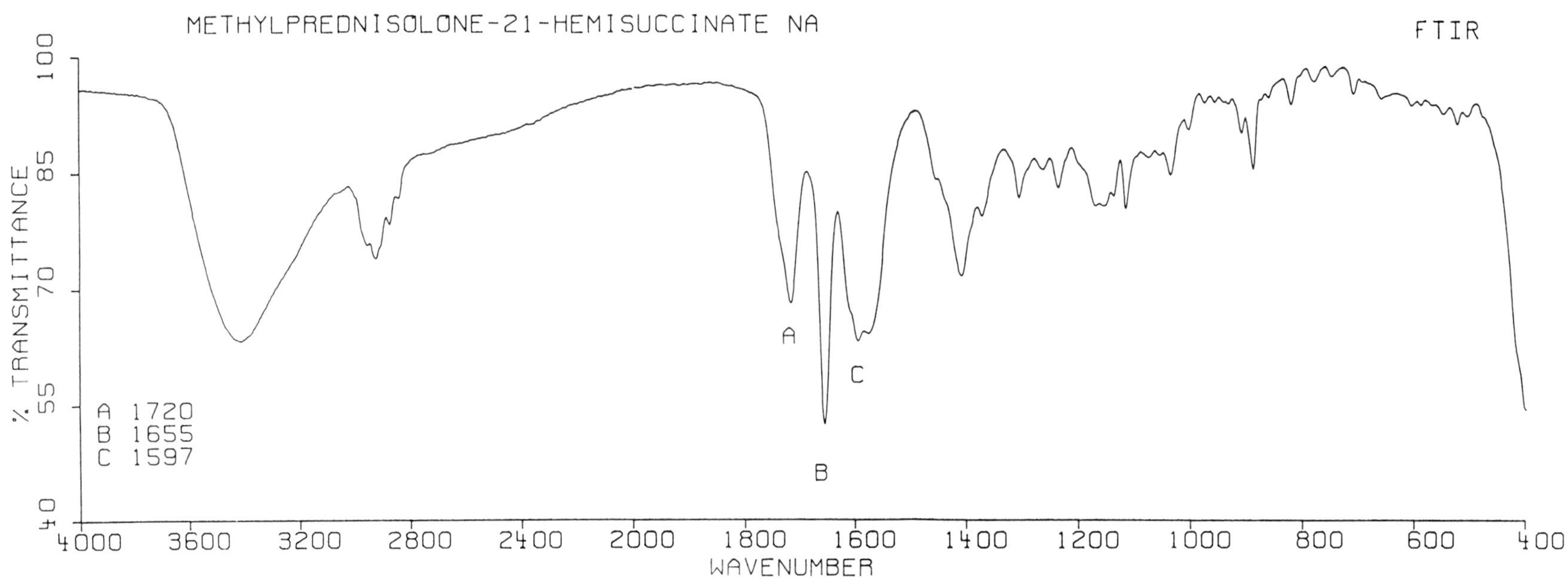

MIANSERIN

$C_{18}H_{20}N_2$

Molecular Weight: 264.37 (264.16)

Synonyms: 1,2,3,4,10,14b-Hexahydro-2-methyldibenzo[c,f]pyrazino[1,2-a]azepine
Trade Names: Athymil, Bolvidon, Lantanon, Norval, Tetramide, Tolvin, Tolvon
Use: Antidepressant

HPLC:
GC: 2248; 250°

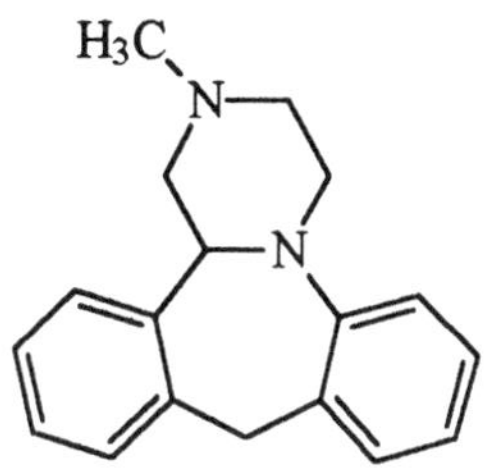

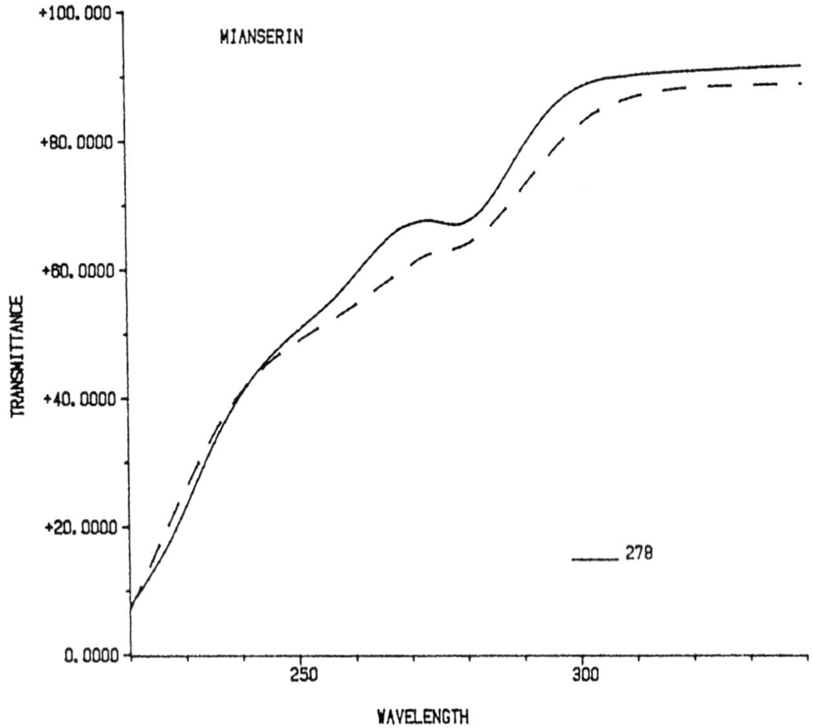

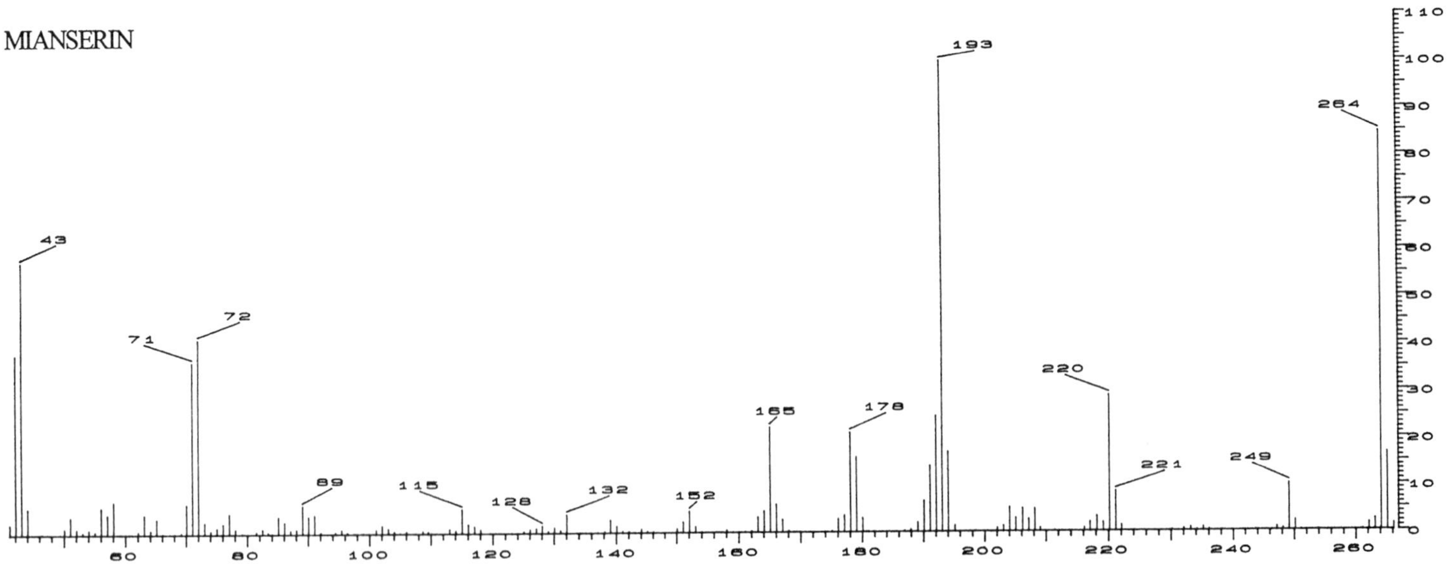

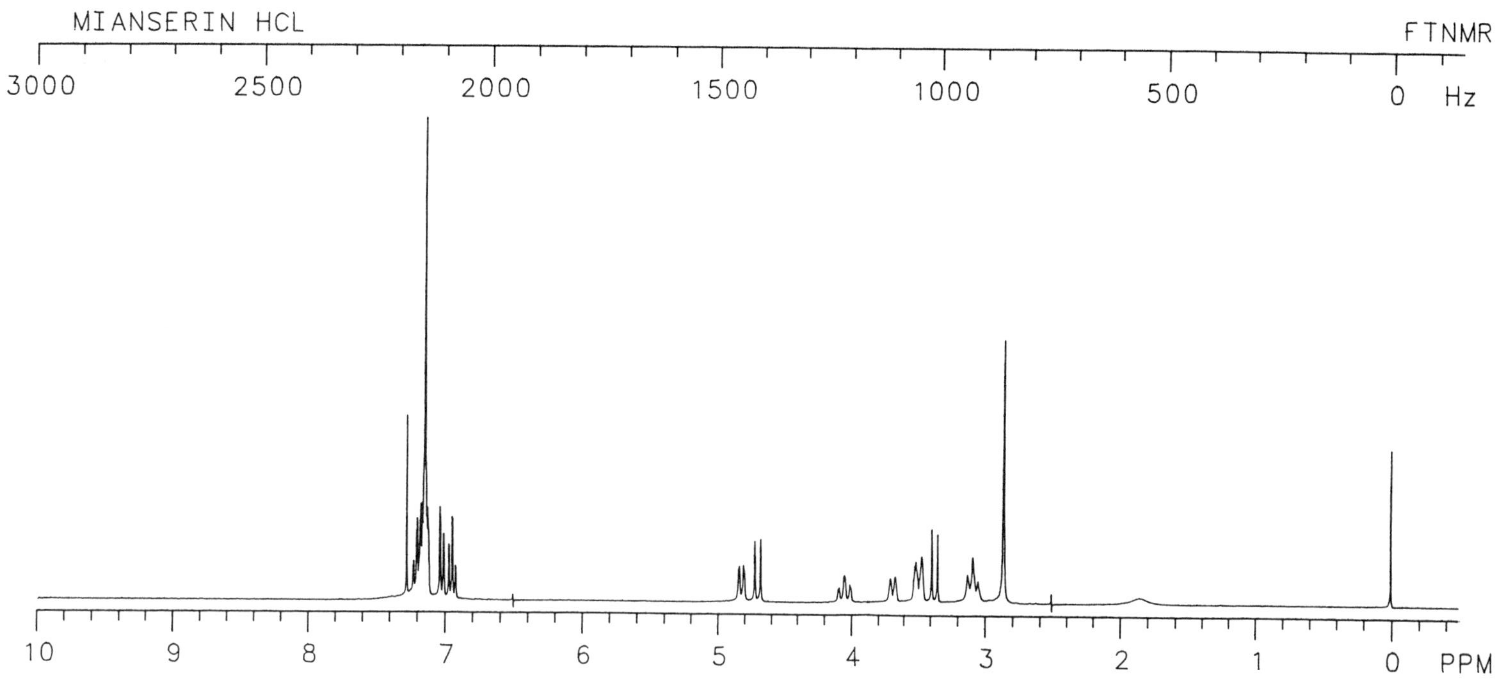

PEAK LISTING

#	HT	PPM
1	31	.00
2	53	2.86
3	9	3.08
4	14	3.34
5	15	3.38
6	9	3.45
7	9	3.46
8	12	4.67
9	12	4.71
10	16	6.93
11	17	6.94
12	11	6.96
13	11	6.96
14	13	7.00
15	13	7.00
16	19	7.02
17	15	7.11
18	18	7.11
19	18	7.12
20	27	7.12
21	98	7.13
22	19	7.15
23	19	7.16
24	16	7.17
25	16	7.19
26	14	7.19
27	37	7.27

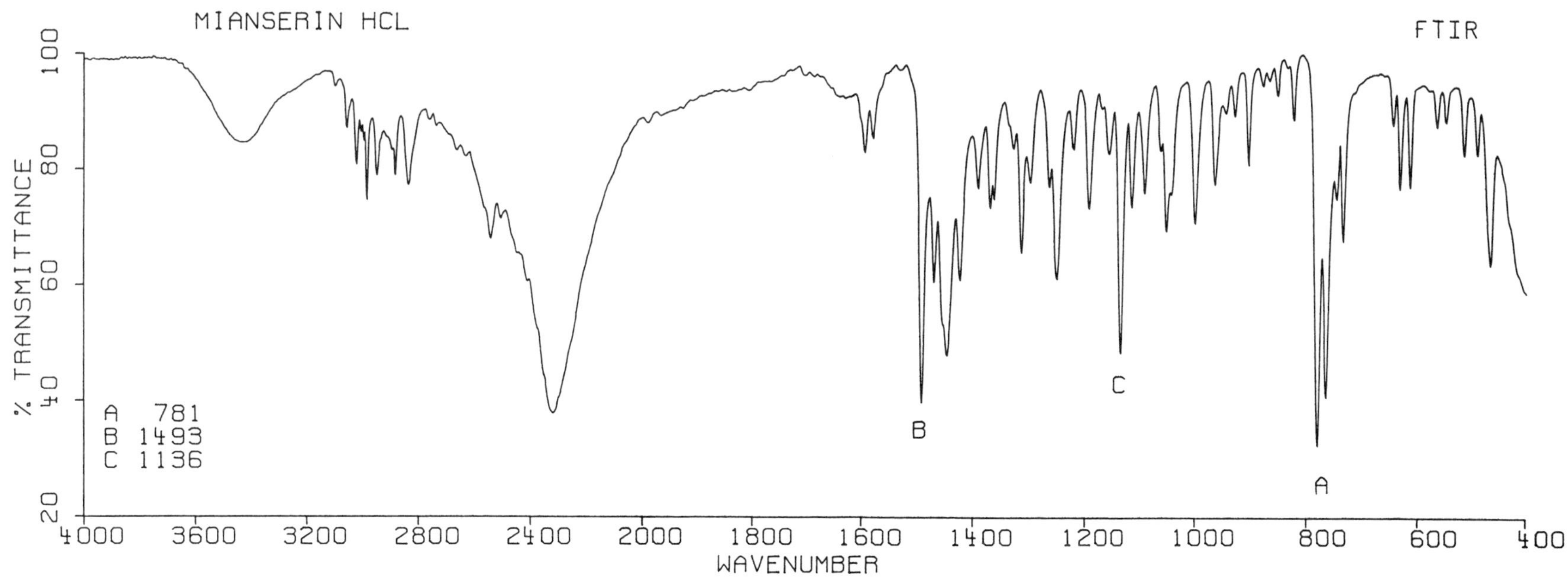

MILRINONE

$C_{12}H_9N_3O$

Molecular Weight: 211.22 (211.07)

Synonyms: 1,6-Dihydro-2-methyl-6-oxo-(3,4-bipyridine)-5-carbonitrile
Trade Names: Corotrope, Primacor
Use: Cardiotonic

HPLC:
GC:

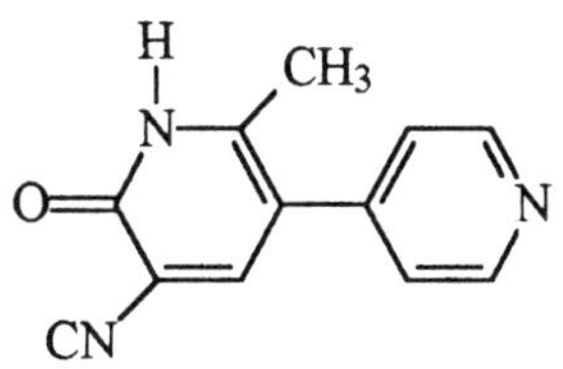

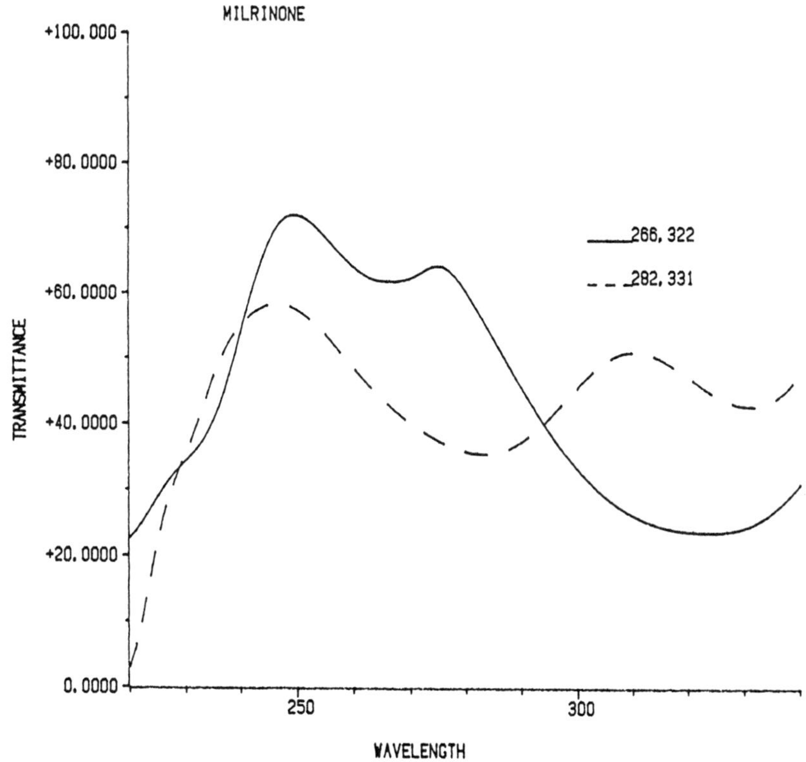

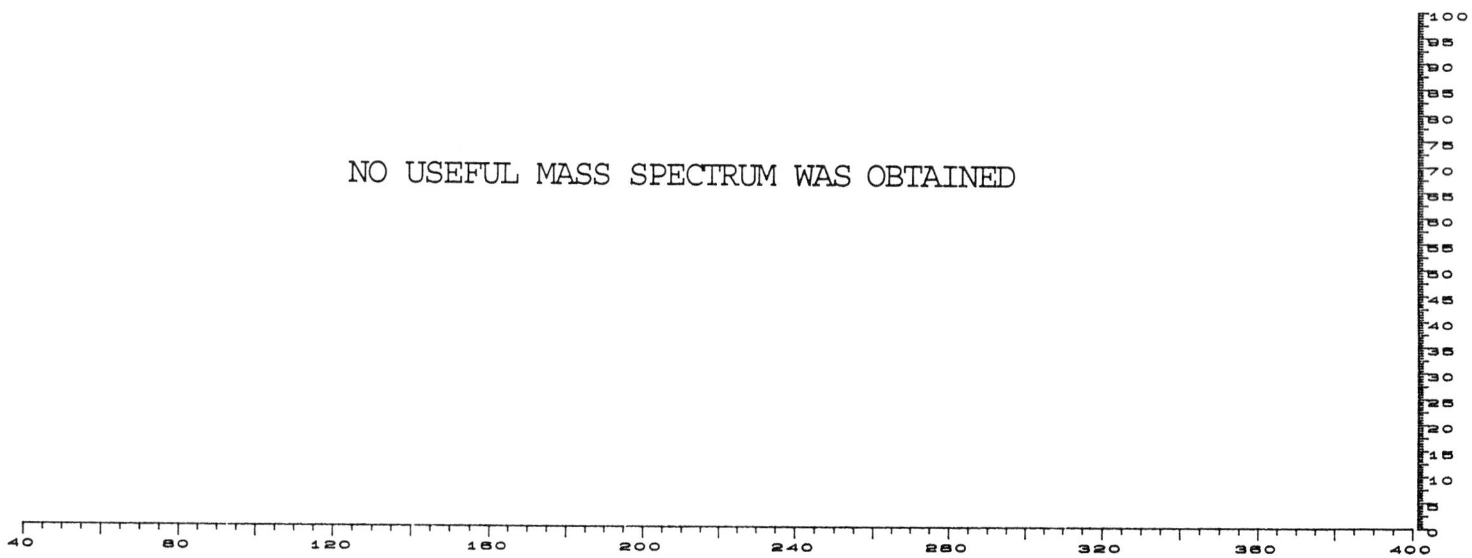

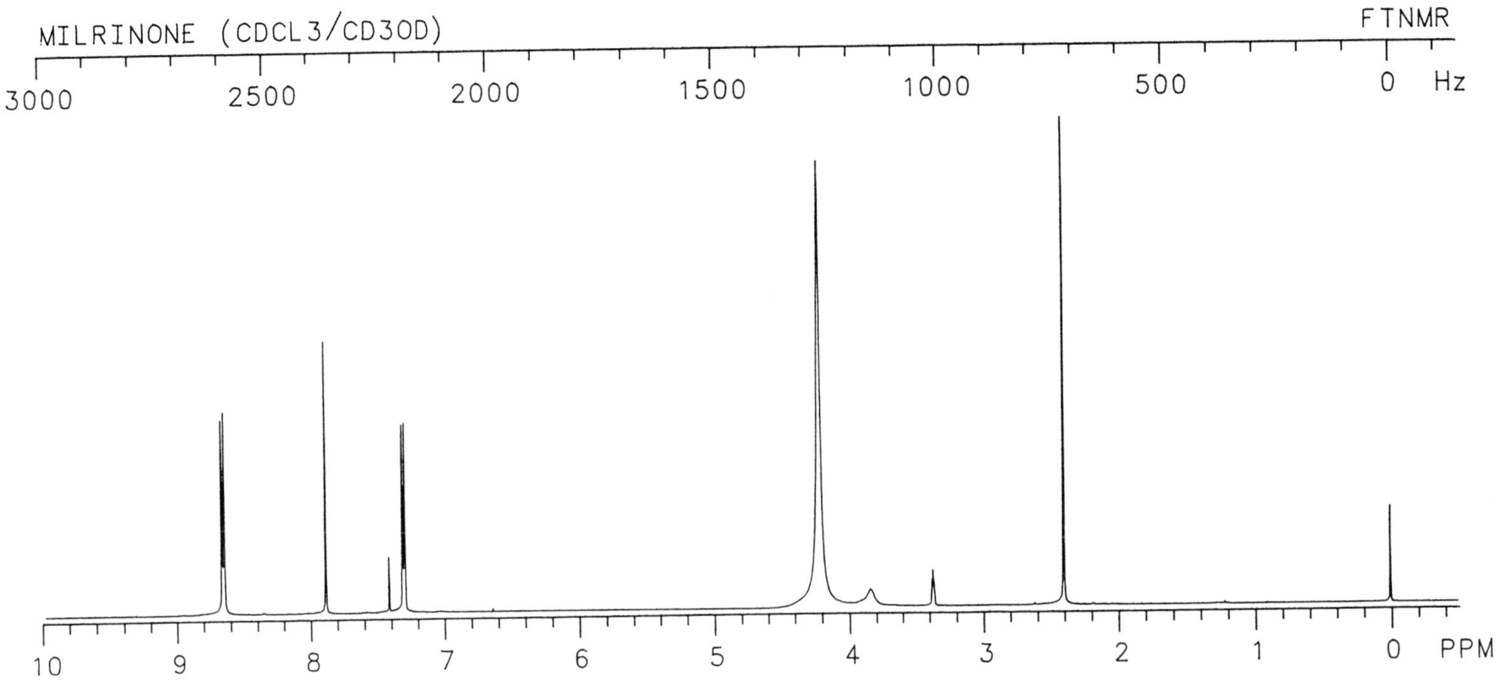

PEAK LISTING

#	HT	PPM
1	19	.00
2	99	2.40
3	91	4.21
4	38	7.29
5	23	7.29
6	26	7.30
7	38	7.31
8	55	7.88
9	41	8.64
10	25	8.64
11	39	8.66

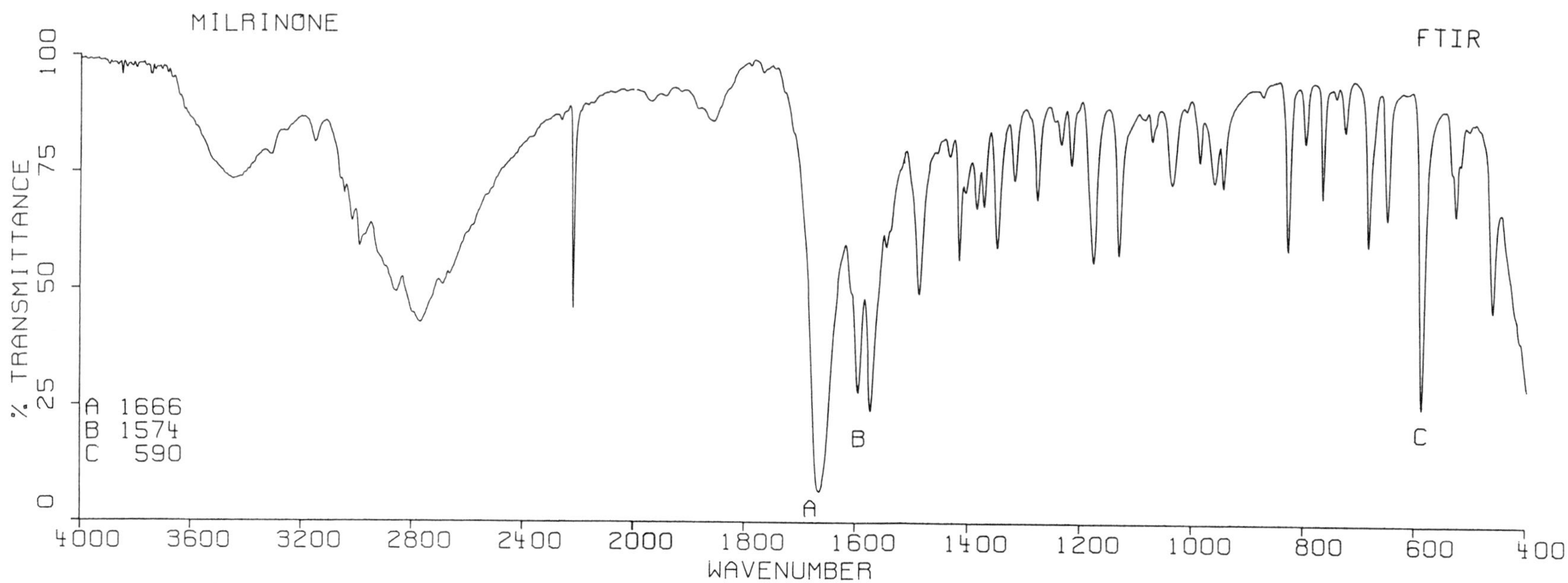

MIVACURIUM CHLORIDE

$C_{58}H_{80}Cl_2N_2O_{14}$

Molecular Weight: 1100.18 (1098.50)

Synonyms: [R,-[R*,R*-(E)]]-2,2'-[(1,8-dioxo-4-octene-1,8-diyl)bis(oxy-3,1-propanediyl)bis[1,2,3,4-tetrahydro-6,7-dimethoxy-2-methyl-1-[(3,4,5-trimethoxyphenyl)methyl]isoquinolinium]dichloride
Trade Names: Mivacron
Use: Muscle Relaxant

HPLC:
GC:

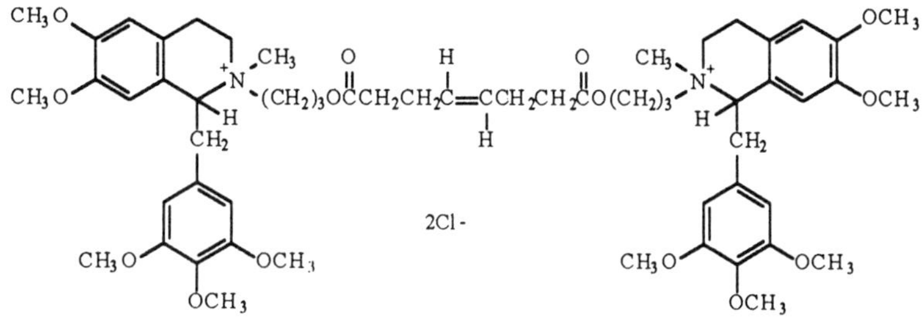

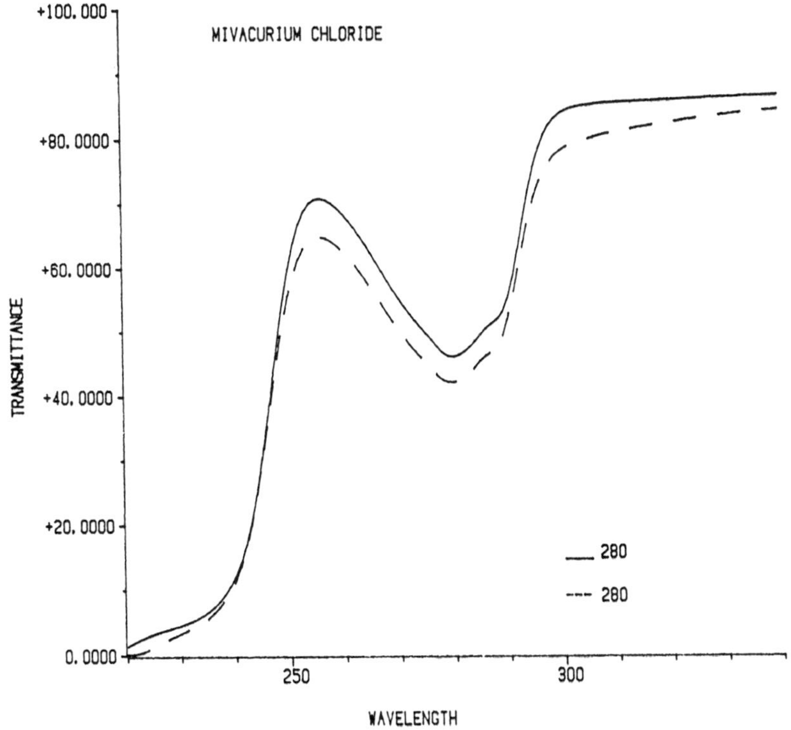

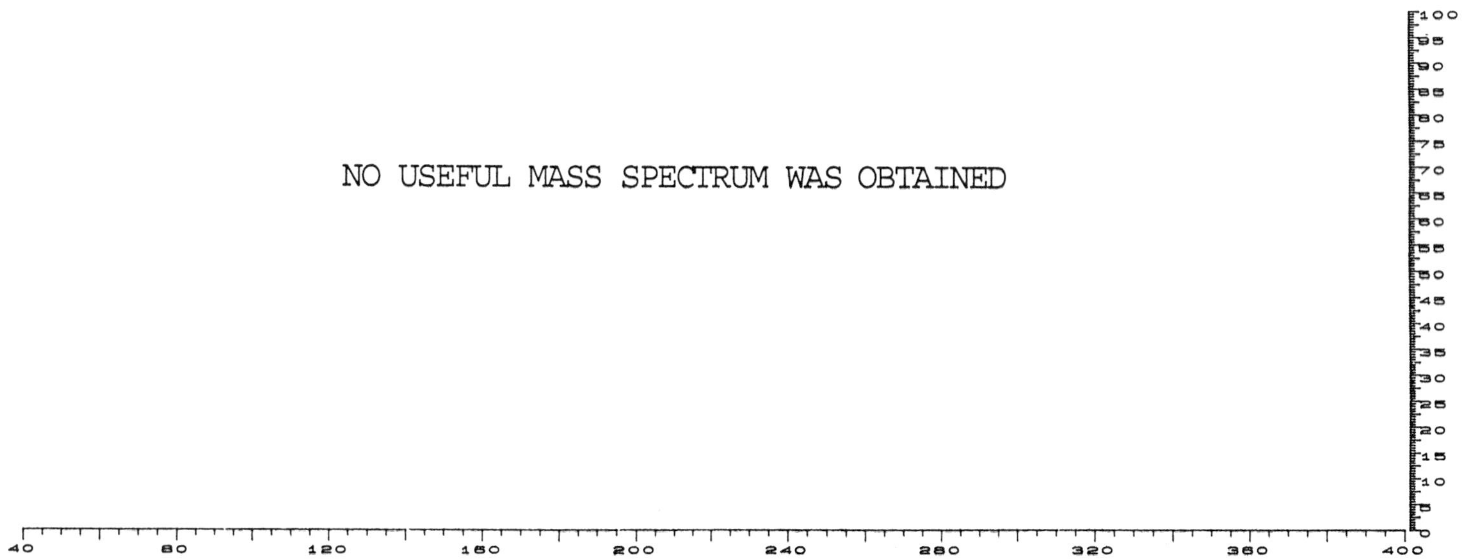

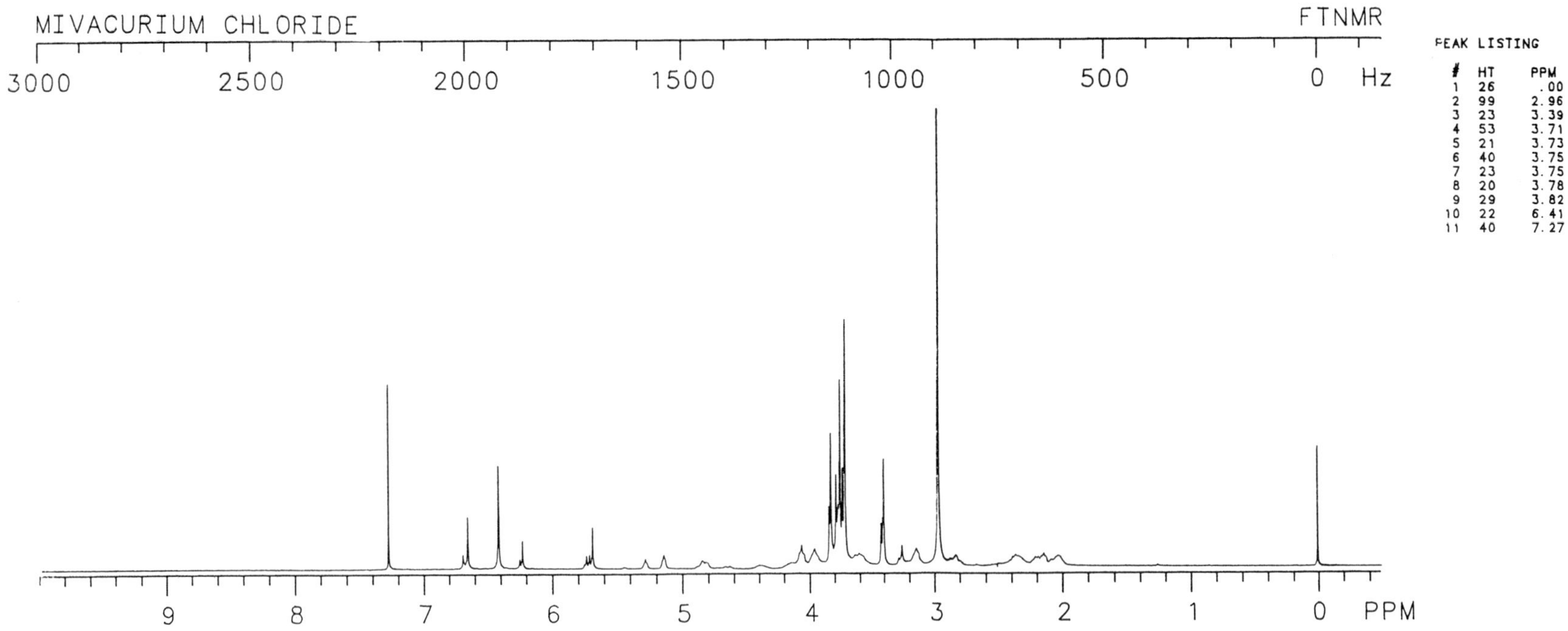
MIVACURIUM CHLORIDE
FTNMR
3000 2500 2000 1500 1000 500 0 Hz
9 8 7 6 5 4 3 2 1 0 PPM
PEAK LISTING
HT PPM
1 26 .00
2 99 2.96
3 23 3.39
4 53 3.71
5 21 3.73
6 40 3.75
7 23 3.75
8 20 3.78
9 29 3.82
10 22 6.41
11 40 7.27

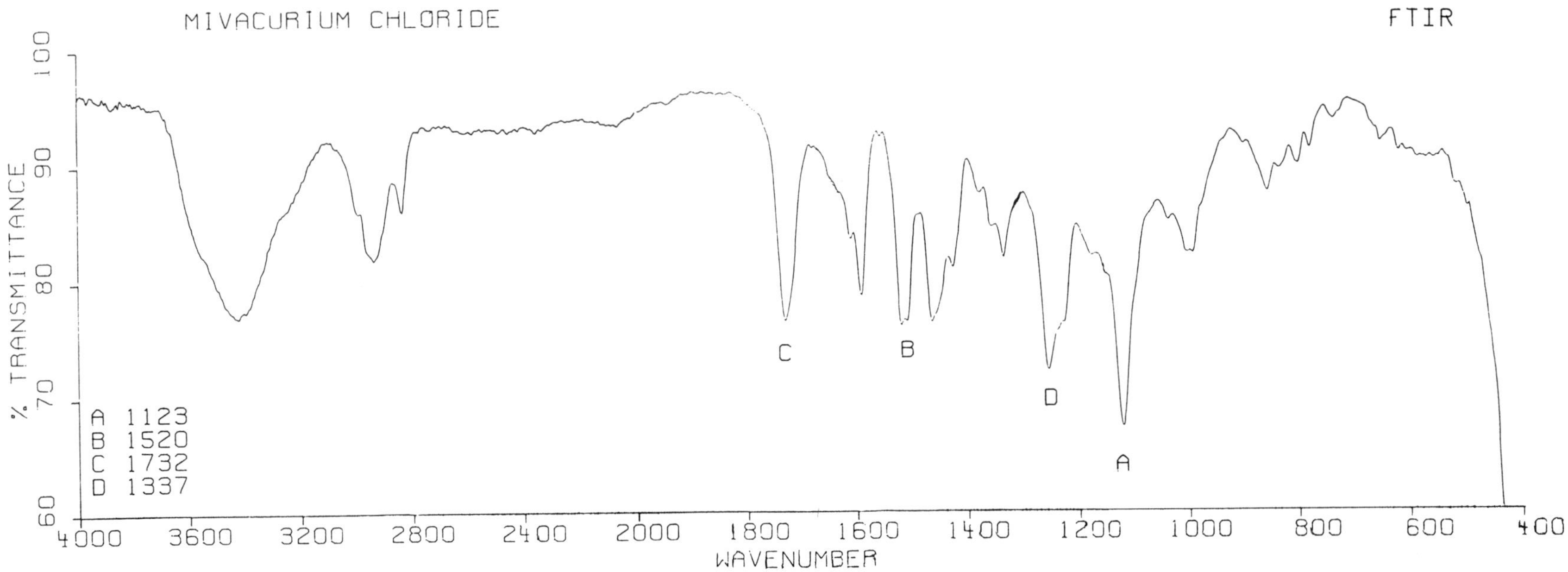
MIVACURIUM CHLORIDE
FTIR
% TRANSMITTANCE
100 90 80 70 60
A 1123
B 1520
C 1732
D 1337
C B D A
4000 3600 3200 2800 2400 2000 1800 1600 1400 1200 1000 800 600 400
WAVENUMBER

MORICIZINE

$C_{22}H_{25}N_3O_4S$

Molecular Weight: 427.52 (427.16)

Synonyms: [10-[3-(4-Morpholinyl)-oxopropyl]-10H-phenothiazin-2-yl]carbamic acid ethyl ester; ethmosine; ethmozin; etmozin; moracizine
Trade Names: Ethmozine, Etmozin
Use: Anti-arrhythmic

HPLC: Methanol: 5.6 (Tails)
GC: 3526; 280°

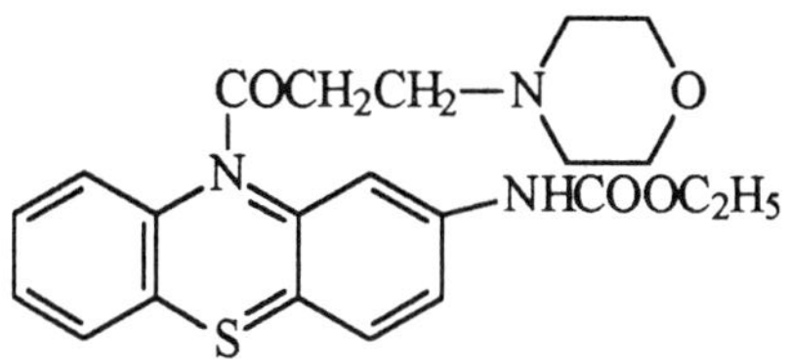

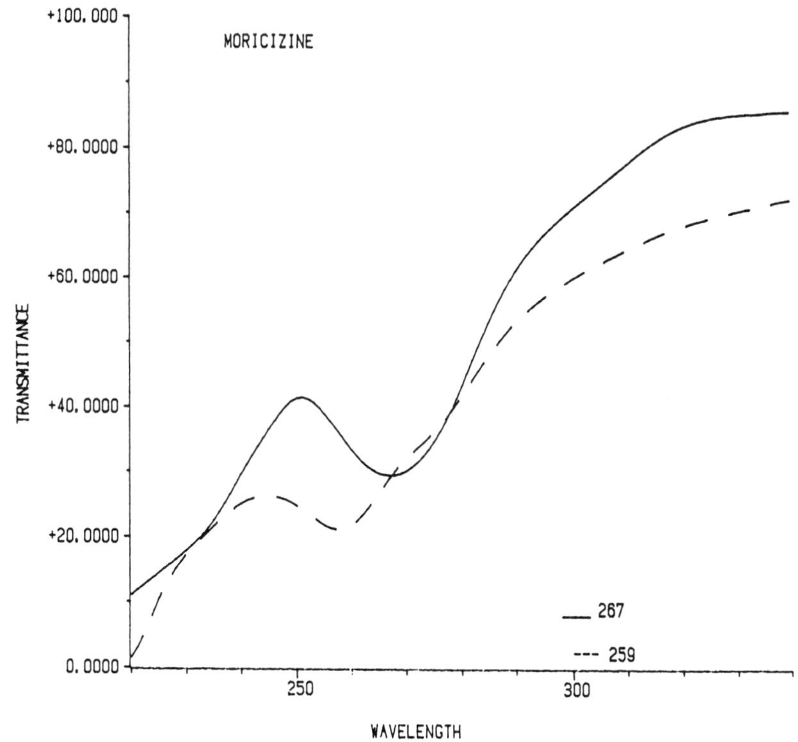

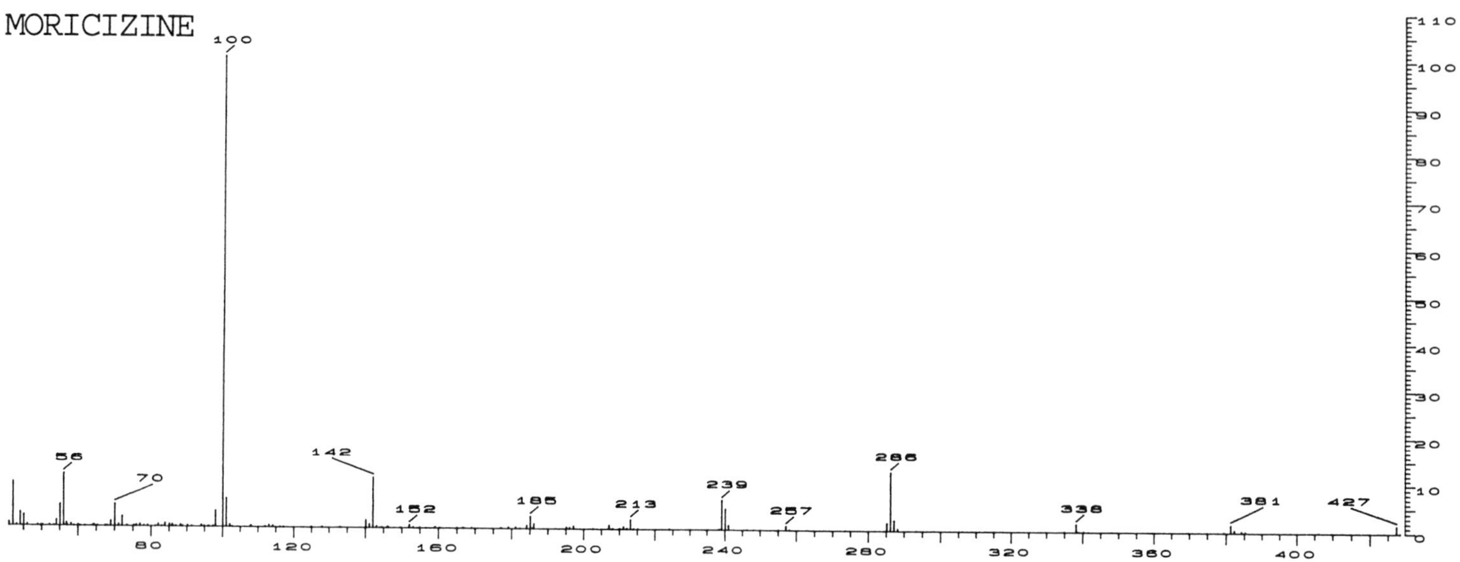

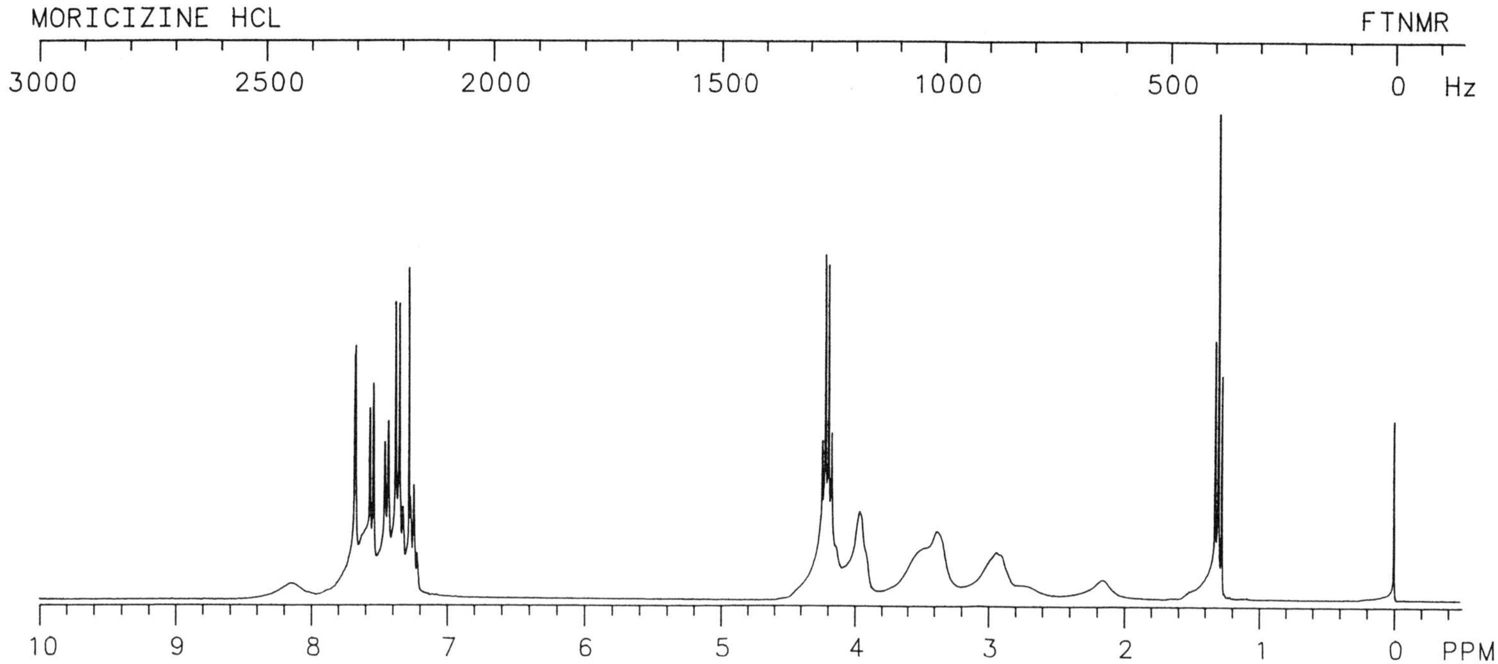

PEAK LISTING

#	HT	PPM
1	36	.00
2	45	1.27
3	99	1.29
4	53	1.32
5	34	4.16
6	68	4.18
7	70	4.21
8	32	4.23
9	23	7.23
10	20	7.26
11	67	7.27
12	60	7.34
13	60	7.37
14	38	7.42
15	32	7.45
16	44	7.53
17	39	7.56
18	51	7.66
19	49	7.67

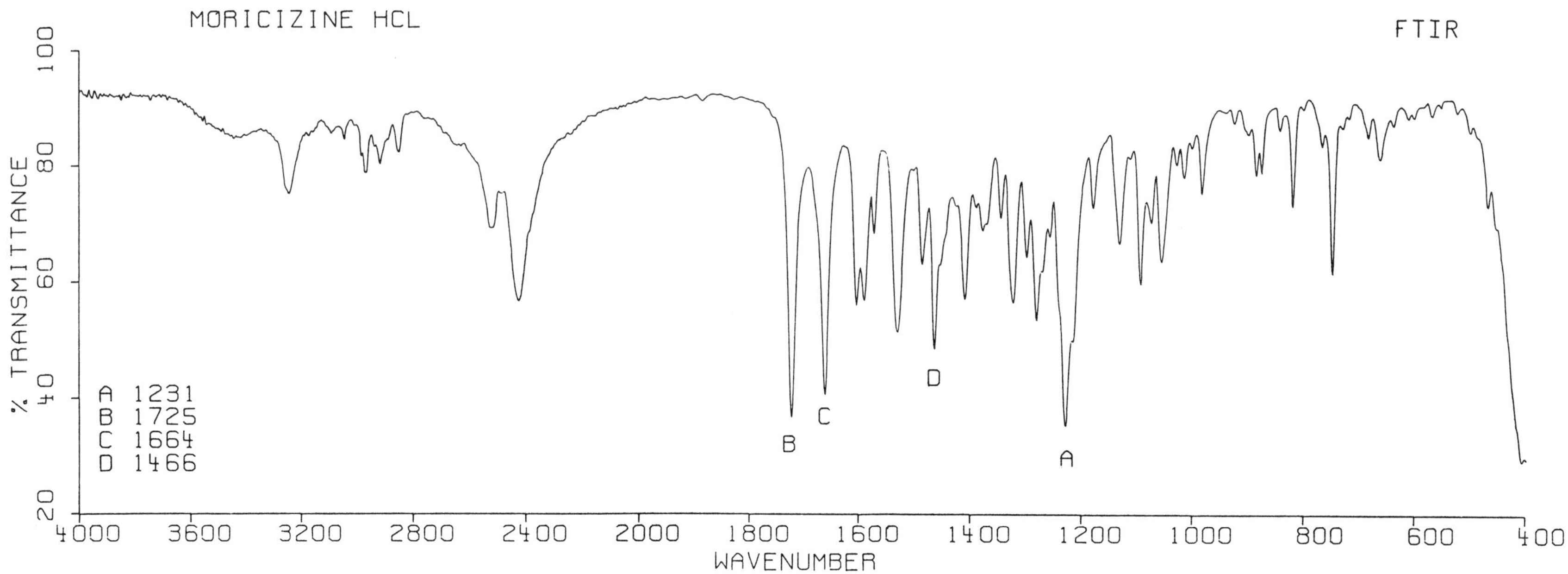

MUPIROCIN

$C_{26}H_{44}O_9$

Molecular Weight: 500.63 (500.30)

Synonyms: [2S-[2α(E),3β,4β,5α[2R*,3R*(1R*,2R*)]]]-9-3-Methyl-1-oxo-4-[tetrahydro-3,4-dihydroxy-5-[[3-(2-hydroxy-1-methylpropyl)oxiranyl]methyl]-2H-pyran-2-yl]-2-butenyl]oxy]nonanoic acid; pseudomonic acid A, trans-pseudomonic acid
Trade Names: Bactoderm, Bactroban, Eismycin
Use: Antibacterial

HPLC:
GC:

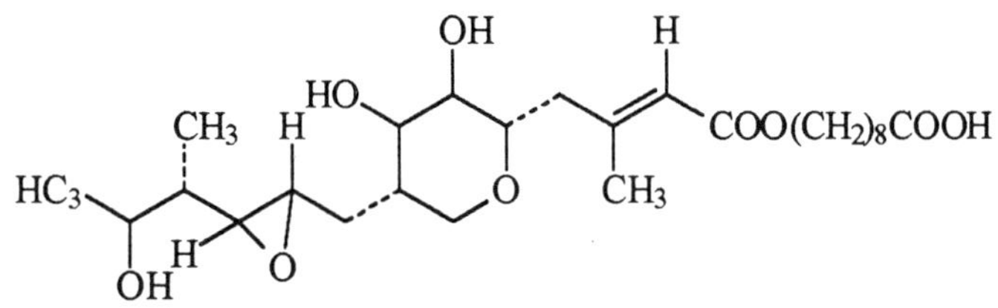

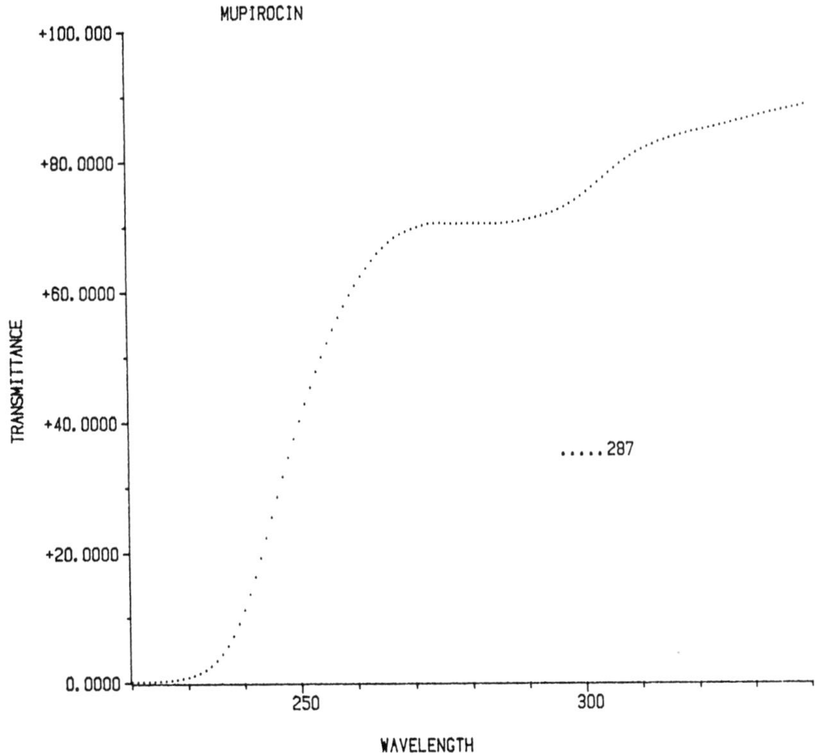

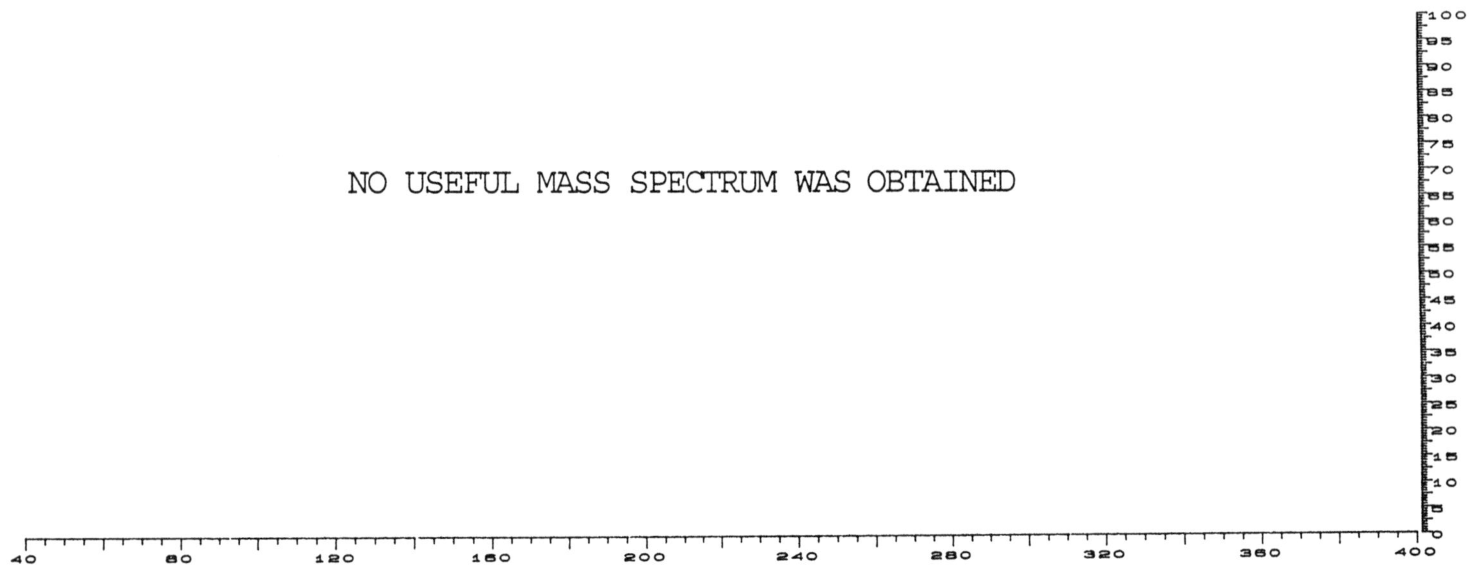

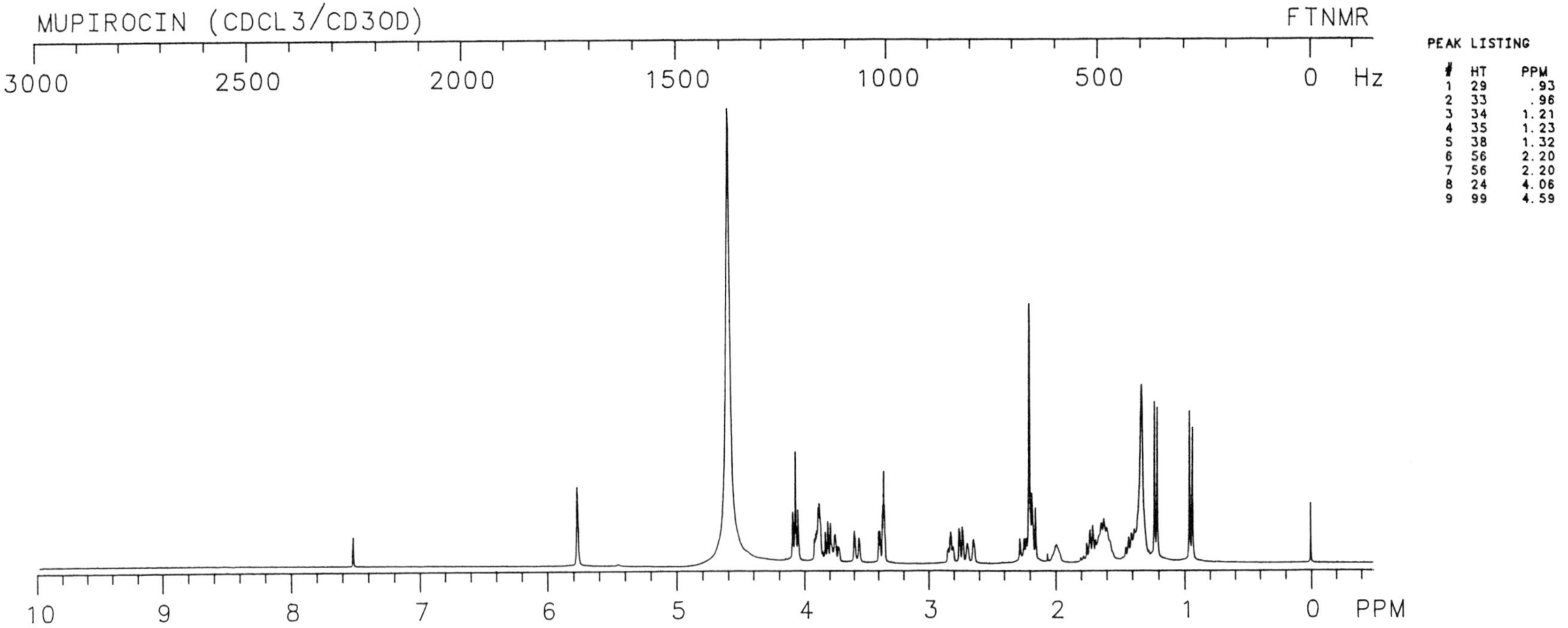
MUPIROCIN (CDCL3/CD3OD)
FTNMR
3000 2500 2000 1500 1000 500 0 Hz
10 9 8 7 6 5 4 3 2 1 0 PPM
PEAK LISTING
HT PPM
1 29 .93
2 33 .96
3 34 1.21
4 35 1.23
5 38 1.32
6 56 2.20
7 56 2.20
8 24 4.06
9 99 4.59

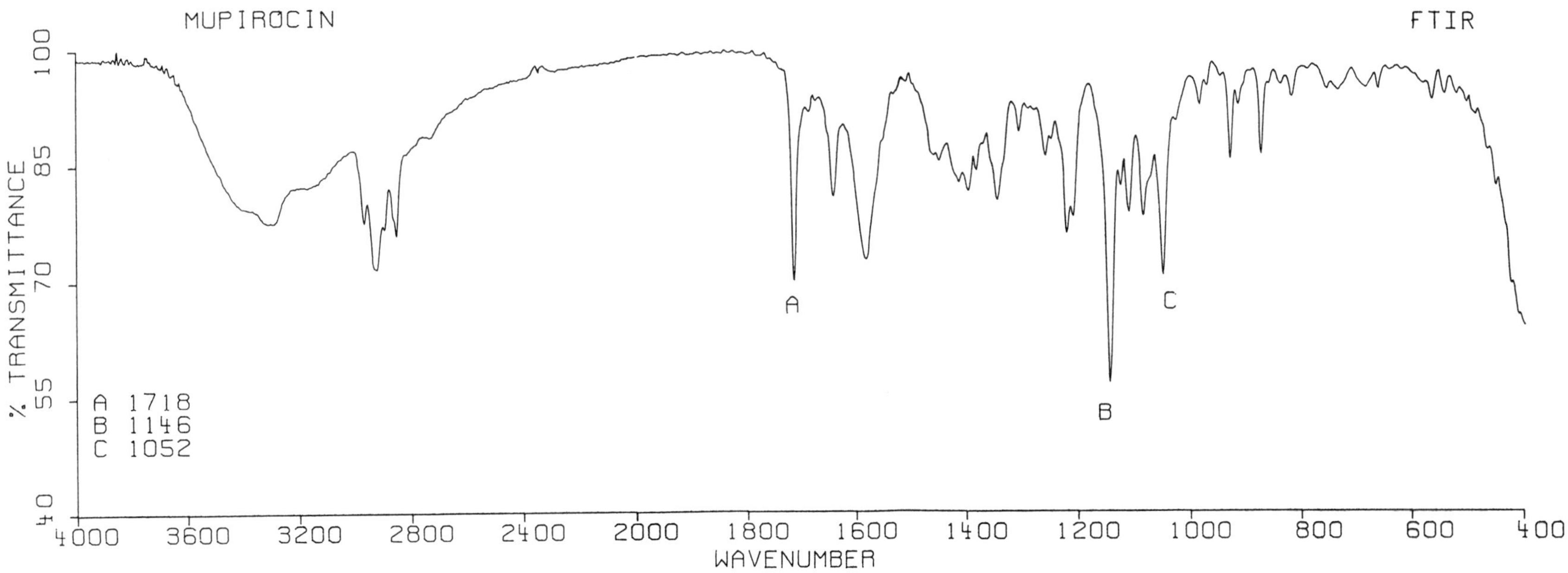
MUPIROCIN
FTIR
% TRANSMITTANCE
100 85 70 55 40
A 1718
B 1146
C 1052
A
B
C
4000 3600 3200 2800 2400 2000 1800 1600 1400 1200 1000 800 600 400
WAVENUMBER

NALMEFENE

$C_{21}H_{25}NO_3$

Molecular Weight: 339.43 (339.18)

Synonyms: (5α)-17-(Cyclopropylmethyl)-4,5-epoxy-6-methylenemorphinan-3,14-diol; nalmetrene
Trade Names: Revex
Use: Narcotic analgesic

HPLC:
GC: 2700; 250°

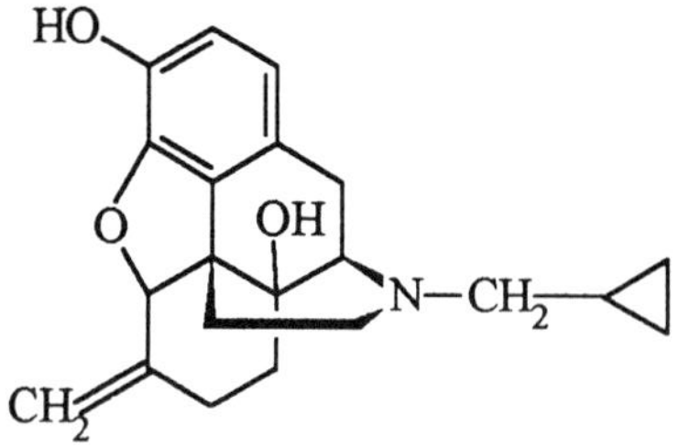

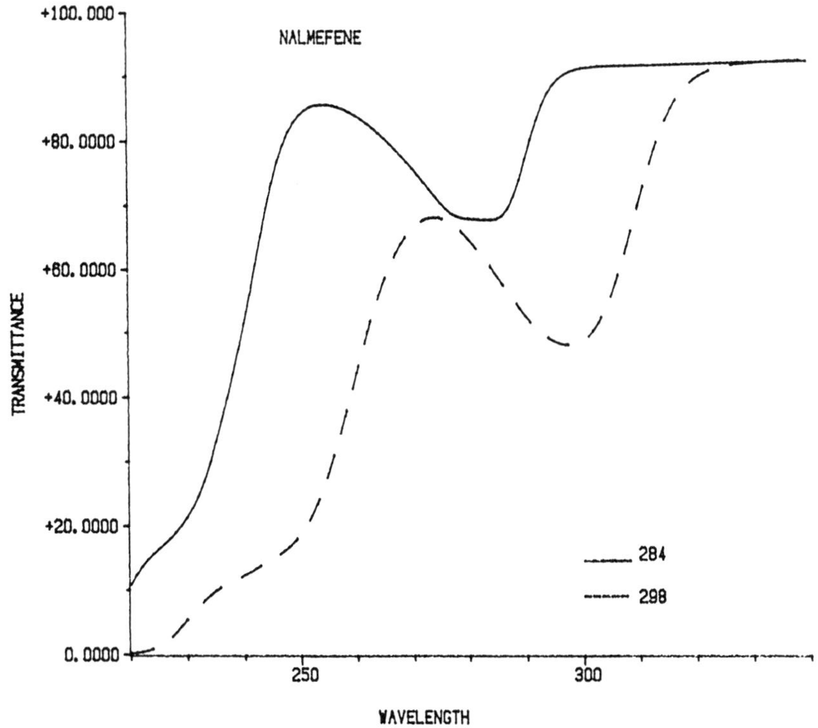

NALMEFENE

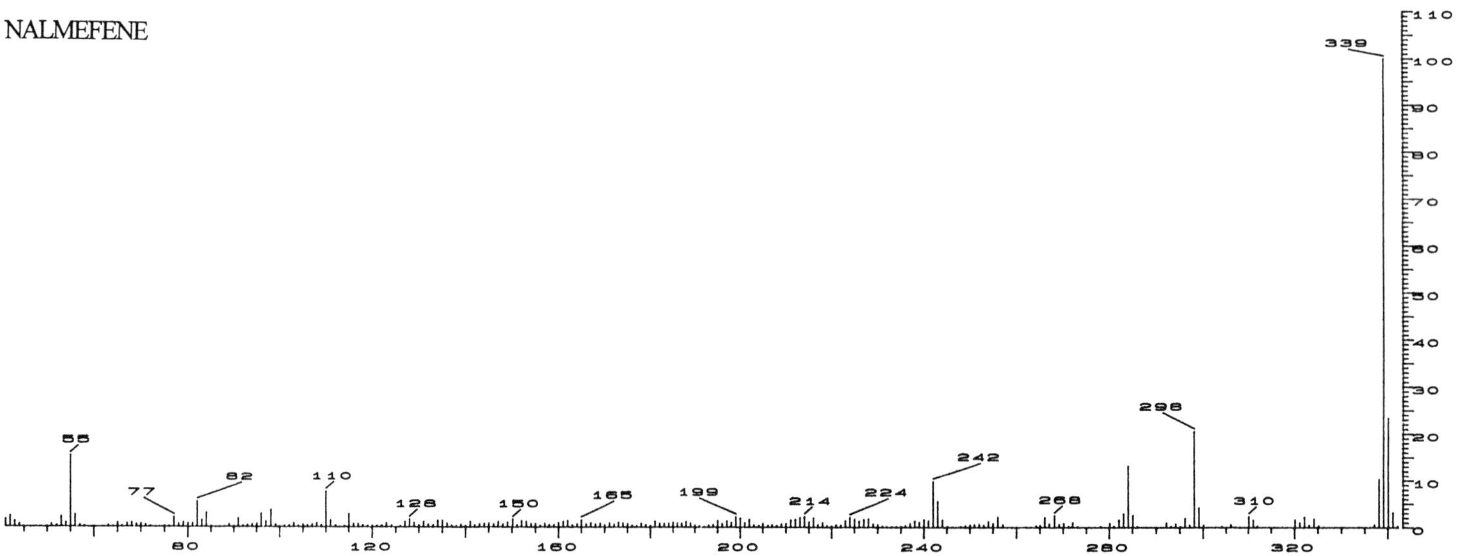

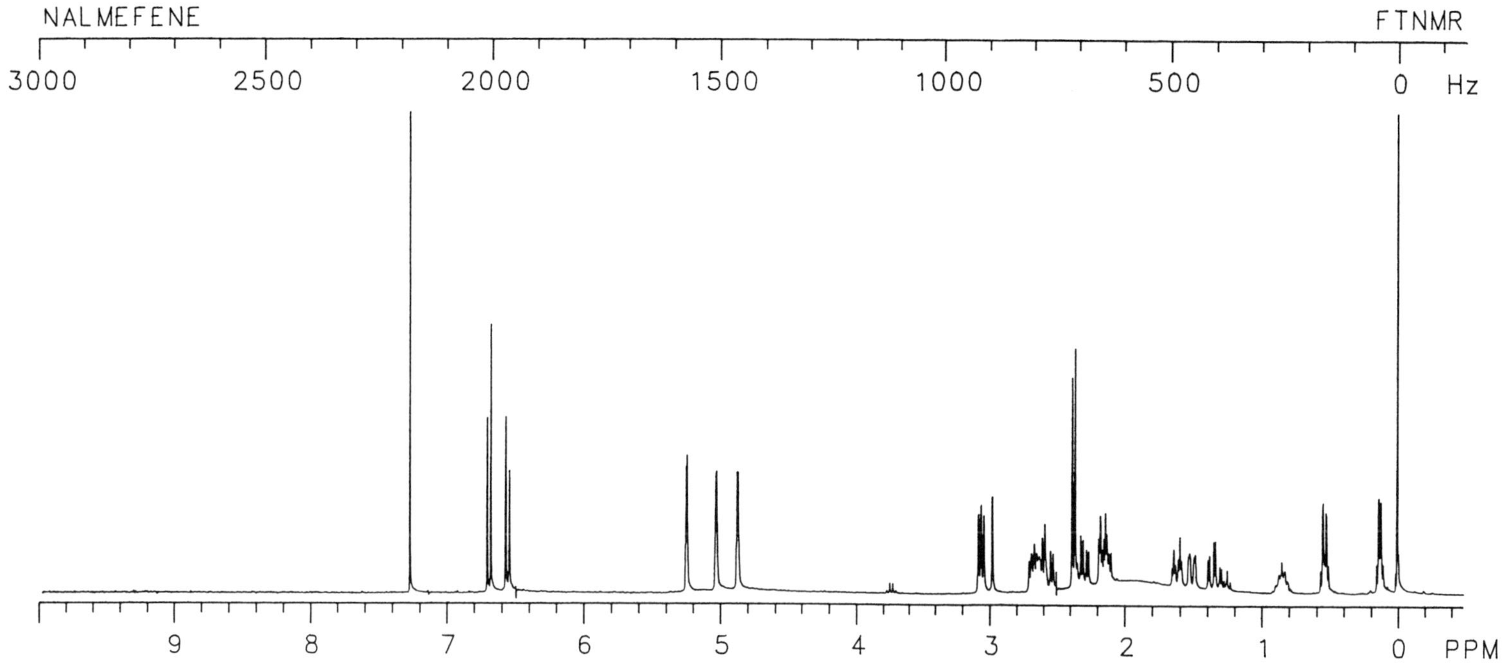

PEAK LISTING

#	HT	PPM
1	168	.00
2	50	2.35
3	44	2.38
4	28	5.23
5	35	6.56
6	54	6.66
7	35	6.69
8	98	7.26

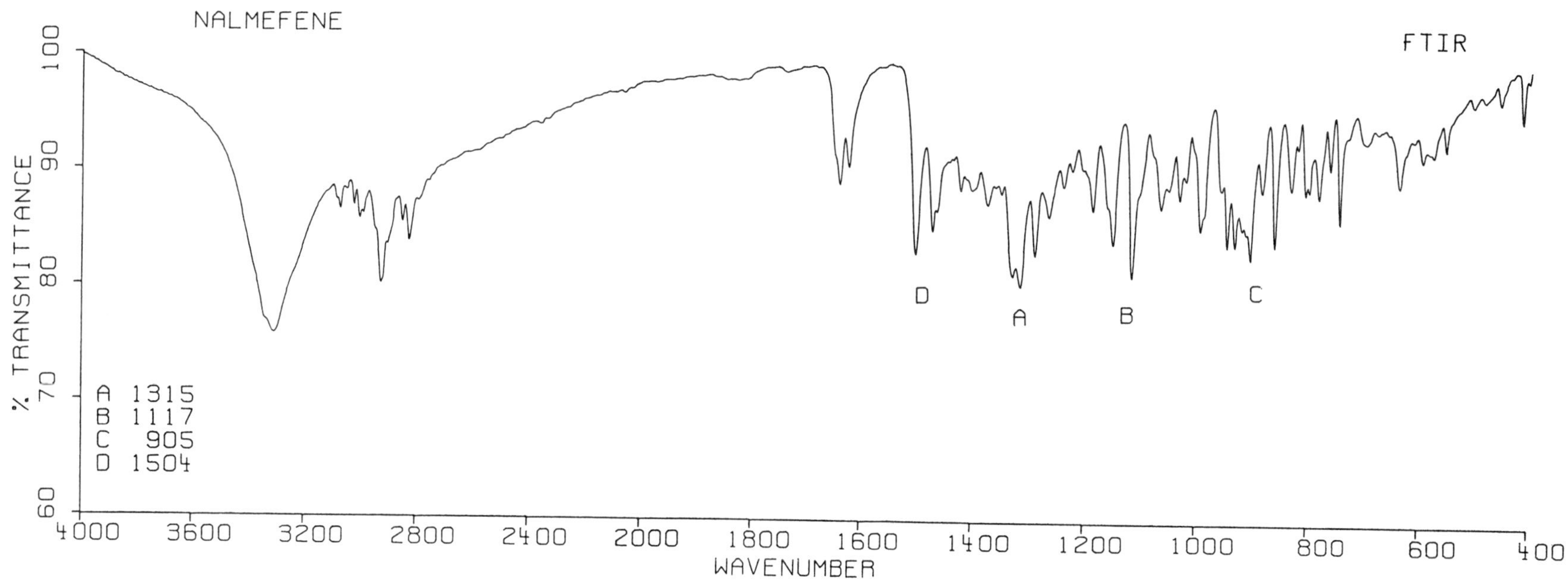

NANDROLONE DODECANOATE

$C_{30}H_{48}O_3$

Molecular Weight: 456.71 (456.36)

Synonyms: 4-Estren-17β-ol-3-one-17-dodecanoate; 17β-hydroxy-4-estren-3-one-17-dodecanoate; 17β-hydroxy-19-nor-4-androsten-3-one-17-dodecanoate; 19-nortestosterone-17-dodecanoate; 19-nortestosterone laurate
Trade Names:
Use: Steroid

HPLC: Methanol: 3.9
GC:

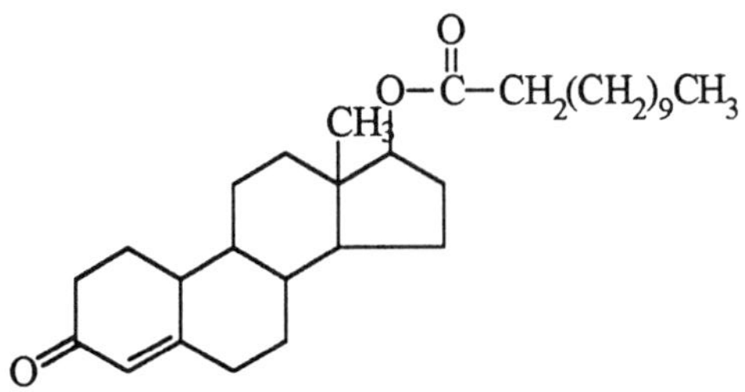

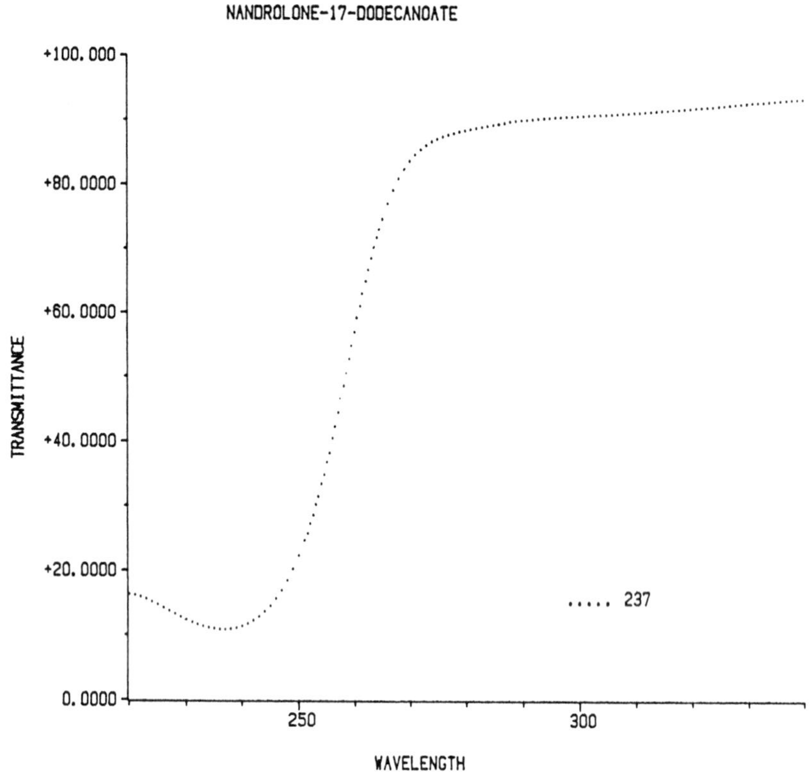

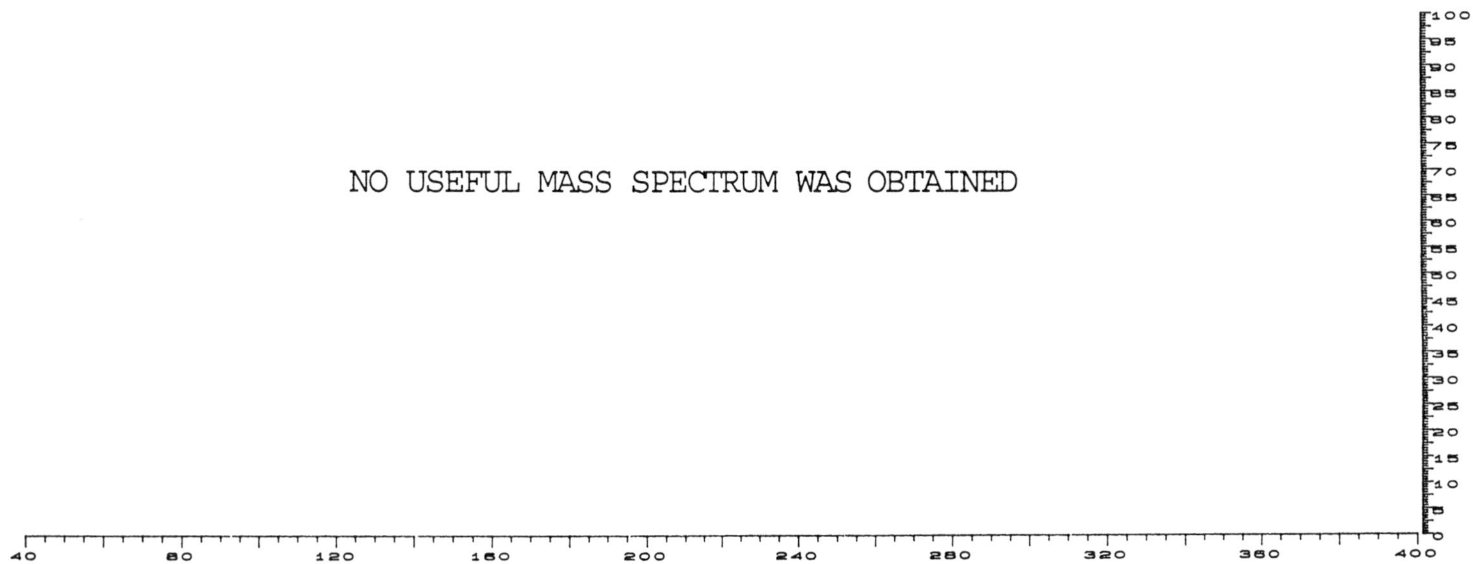

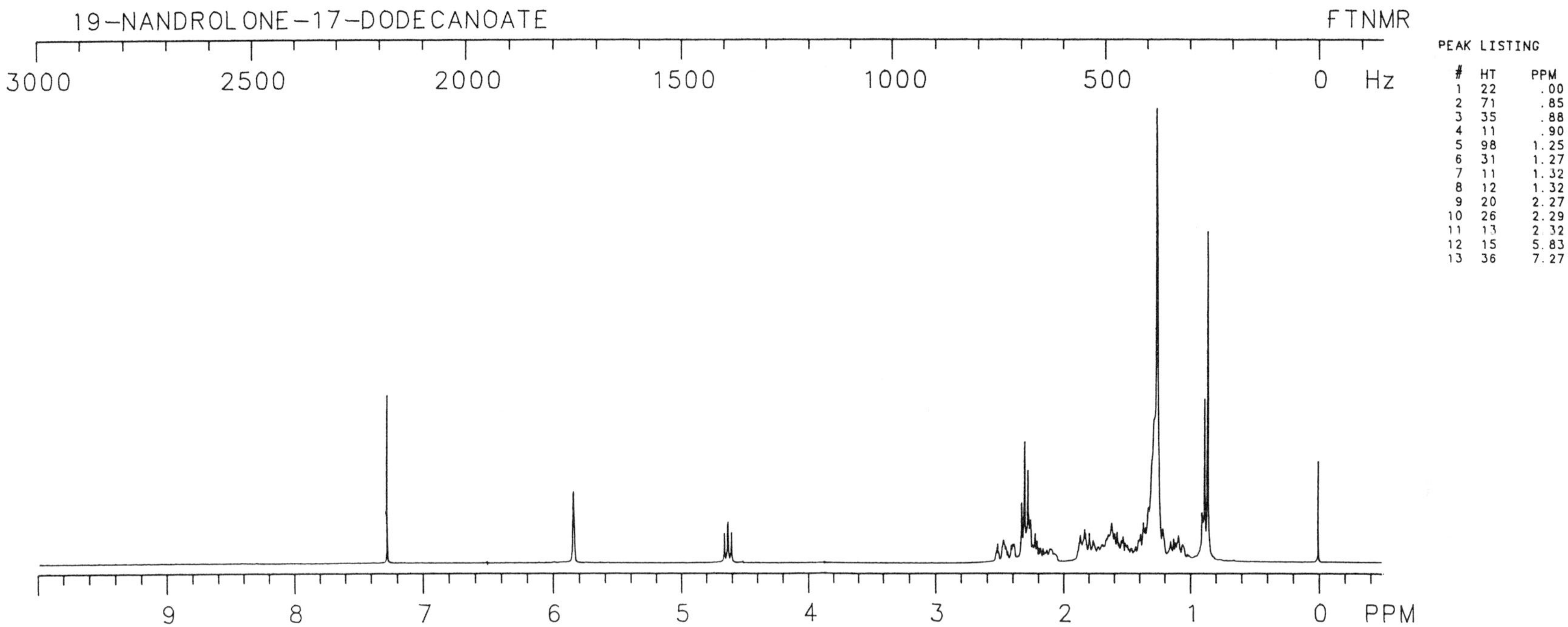
19-NANDROLONE-17-DODECANOATE
FTNMR
3000
2500
2000
1500
1000
500
0 Hz
9
8
7
6
5
4
3
2
1
0 PPM
PEAK LISTING
HT PPM
1 22 .00
2 71 .85
3 35 .88
4 11 .90
5 98 1.25
6 31 1.27
7 11 1.32
8 12 1.32
9 20 2.27
10 26 2.29
11 13 2.32
12 15 5.83
13 36 7.27

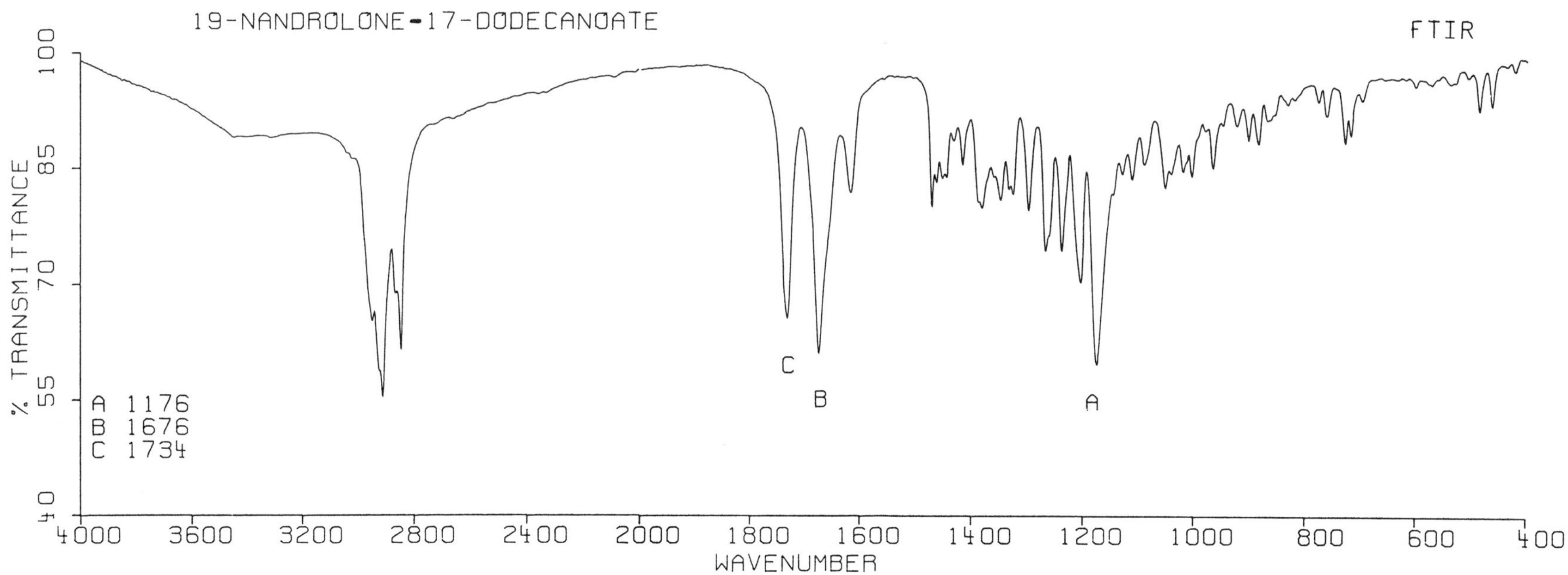
19-NANDROLONE-17-DODECANOATE
FTIR
% TRANSMITTANCE
100
85
70
55
40
A 1176
B 1676
C 1734
C
B
A
4000
3600
3200
2800
2400
2000
1800
1600
1400
1200
1000
800
600
400
WAVENUMBER

NARASIN

$C_{43}H_{72}O_{11}$

Molecular Weight: 765.04 (764.51)

Synonyms: (4S)-4-Methylsalinomycin;
Trade Names: Monteban
Use: Growth stimulant

HPLC:
GC:

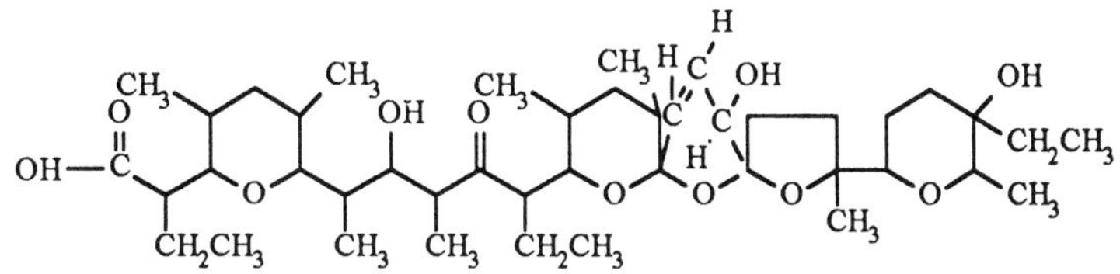

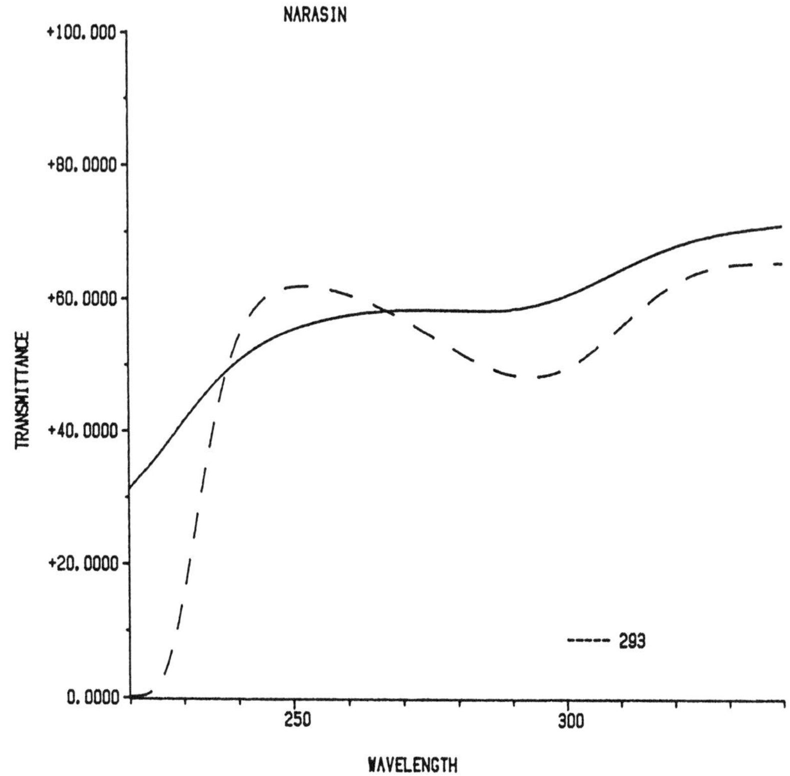

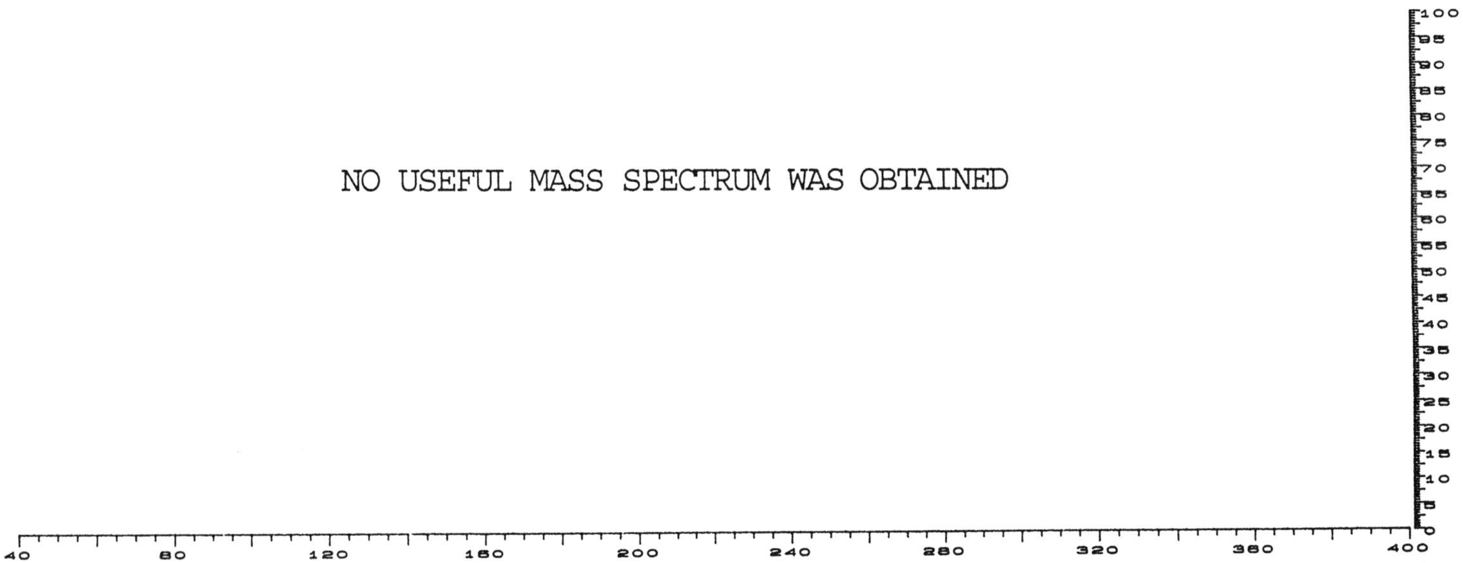

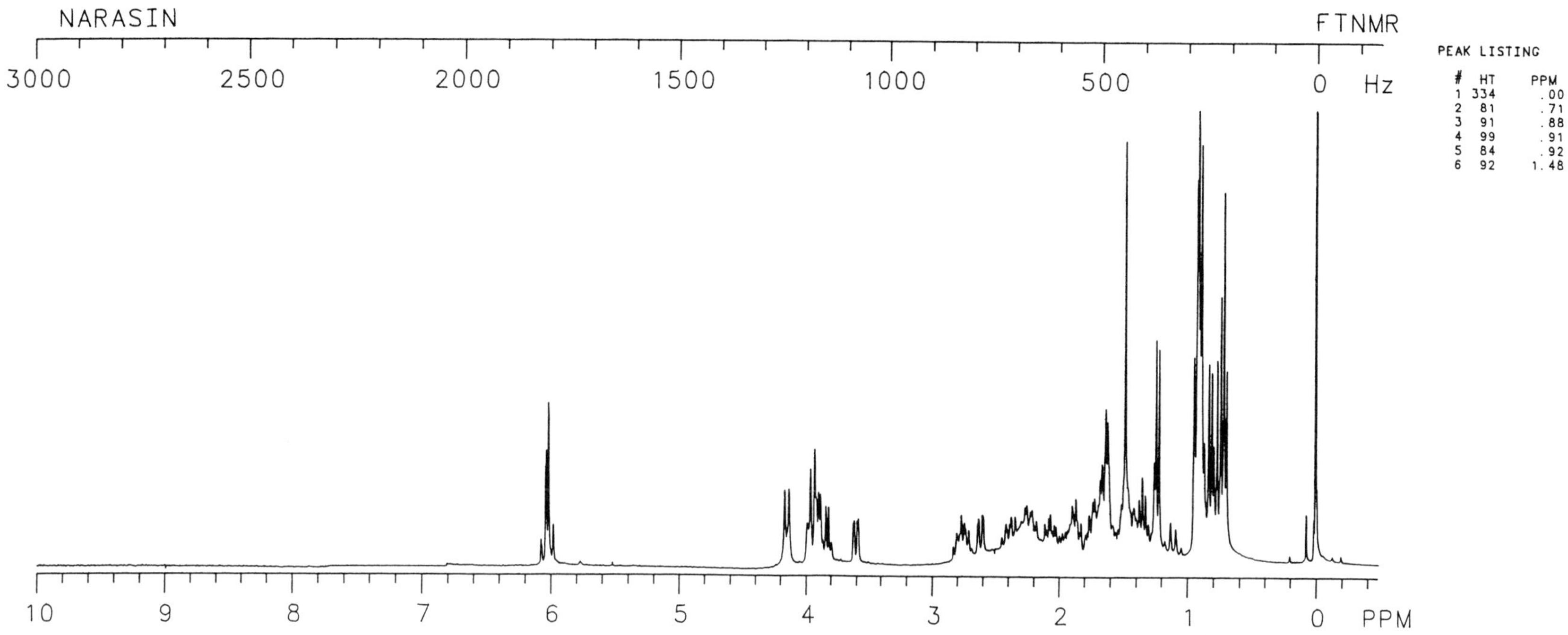

NARASIN
FTNMR
3000 2500 2000 1500 1000 500 0 Hz
PEAK LISTING
HT PPM
1 334 .00
2 81 .71
3 91 .88
4 99 .91
5 84 .92
6 92 1.48
10 9 8 7 6 5 4 3 2 1 0 PPM

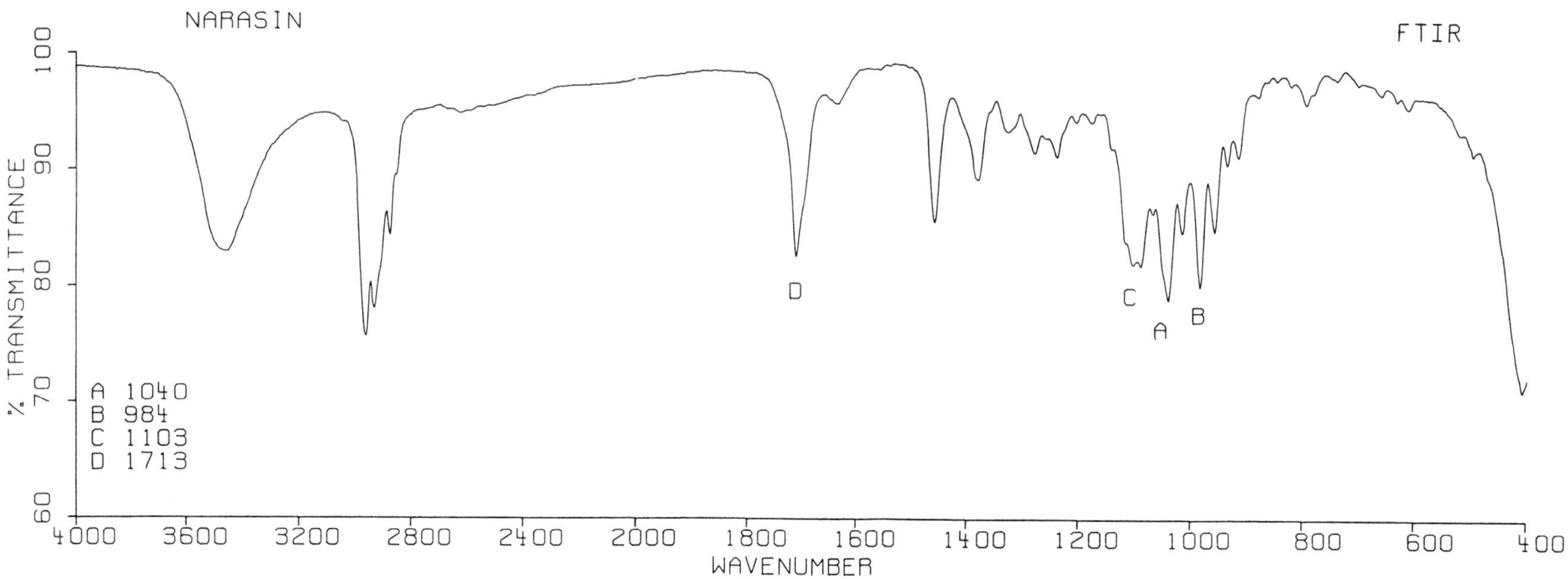

NARASIN
FTIR
% TRANSMITTANCE
100 90 80 70 60
D
C
A
B
A 1040
B 984
C 1103
D 1713
4000 3600 3200 2800 2400 2000 1800 1600 1400 1200 1000 800 600 400
WAVENUMBER

NEFAZODONE

$C_{25}H_{32}ClN_5O_2$

Molecular Weight: 433.55 (433.25)

Synonyms: 2-[3-[4-(3-Chlorophenyl)-1-piperazinyl]propyl]-5-ethyl-2,4-dihydro-4-(2-phenoxyethyl)-3H-1,2,4-triazol-3-one
Trade Names: Serzone
Use: Antidepressant

HPLC:
GC:

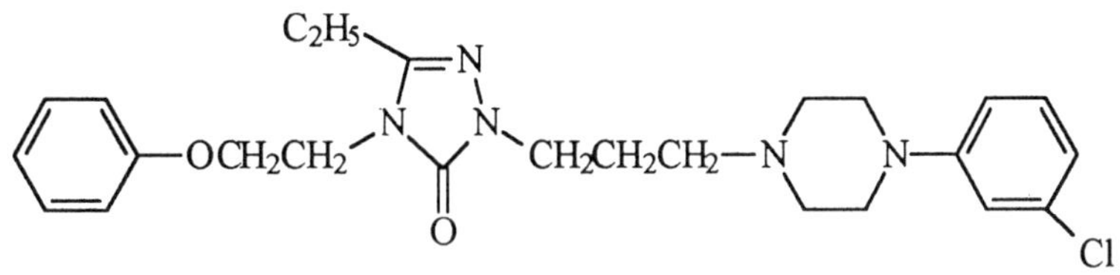

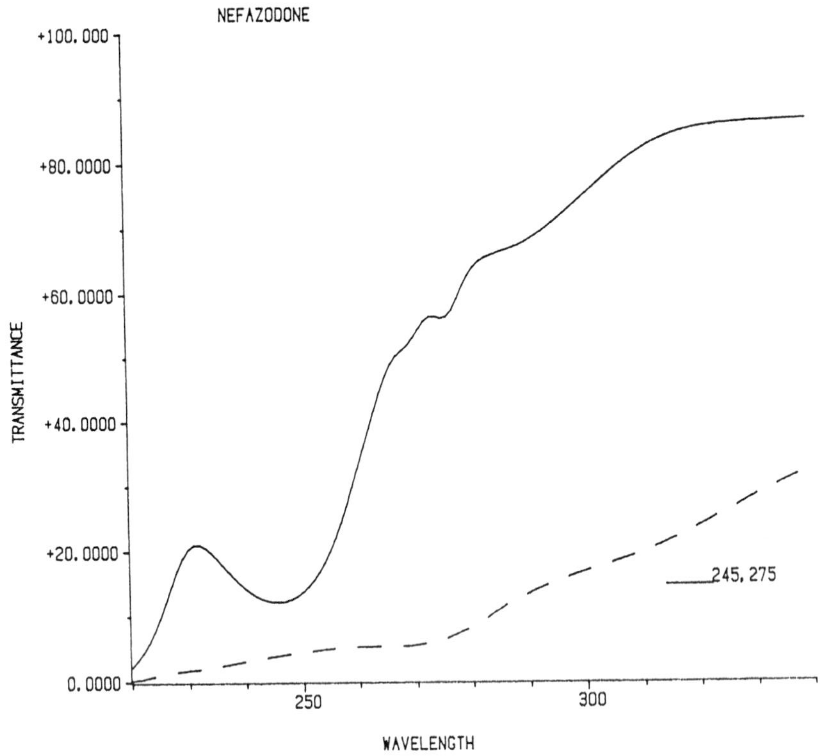

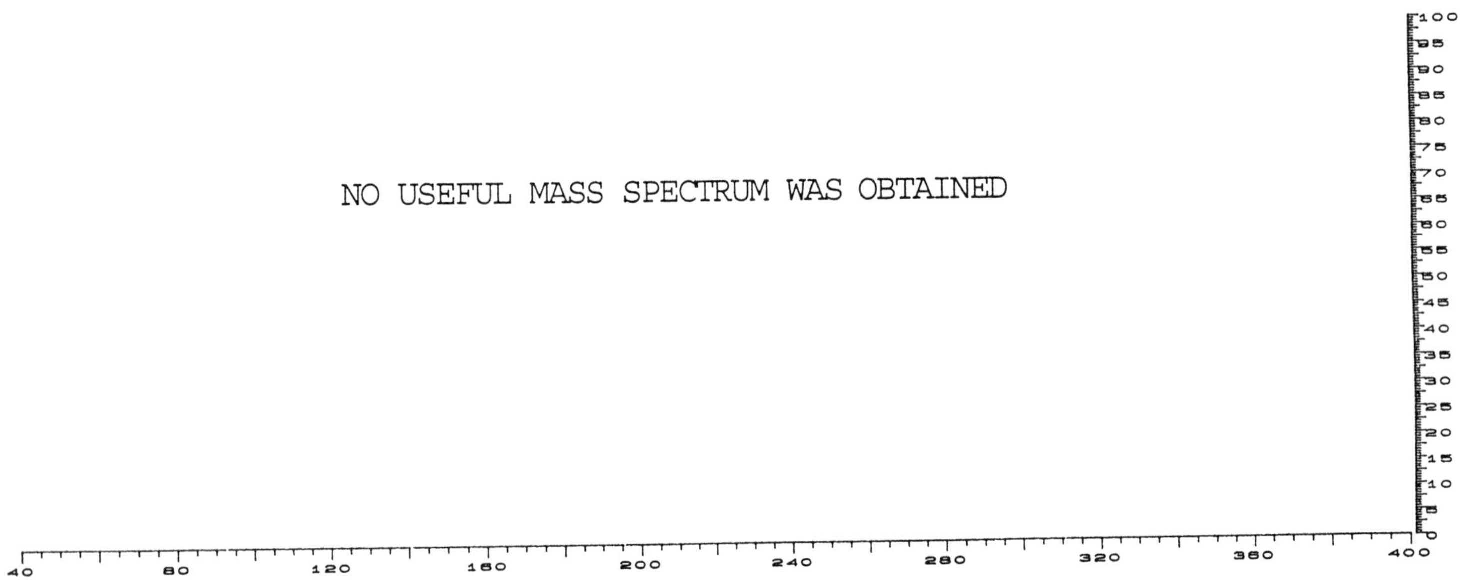

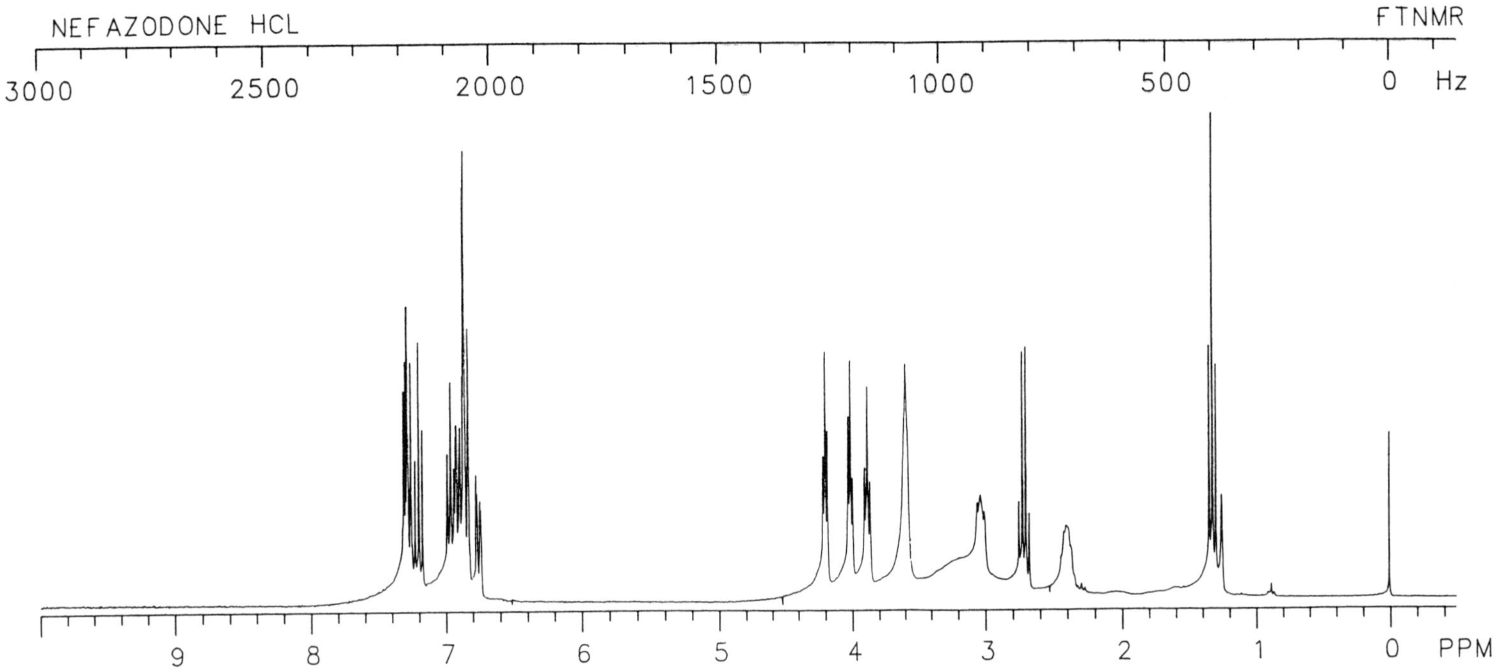

PEAK LISTING

#	HT	PPM
1	48	1.29
2	99	1.31
3	51	1.34
4	51	2.69
5	50	2.72
6	48	3.58
7	43	3.87
8	49	4.00
9	37	4.02
10	50	4.19
11	50	6.82
12	56	6.82
13	92	6.85
14	84	6.85
15	46	6.86
16	36	6.91
17	45	6.95
18	53	7.18
19	49	7.24
20	60	7.27
21	60	7.27
22	49	7.28
23	43	7.30

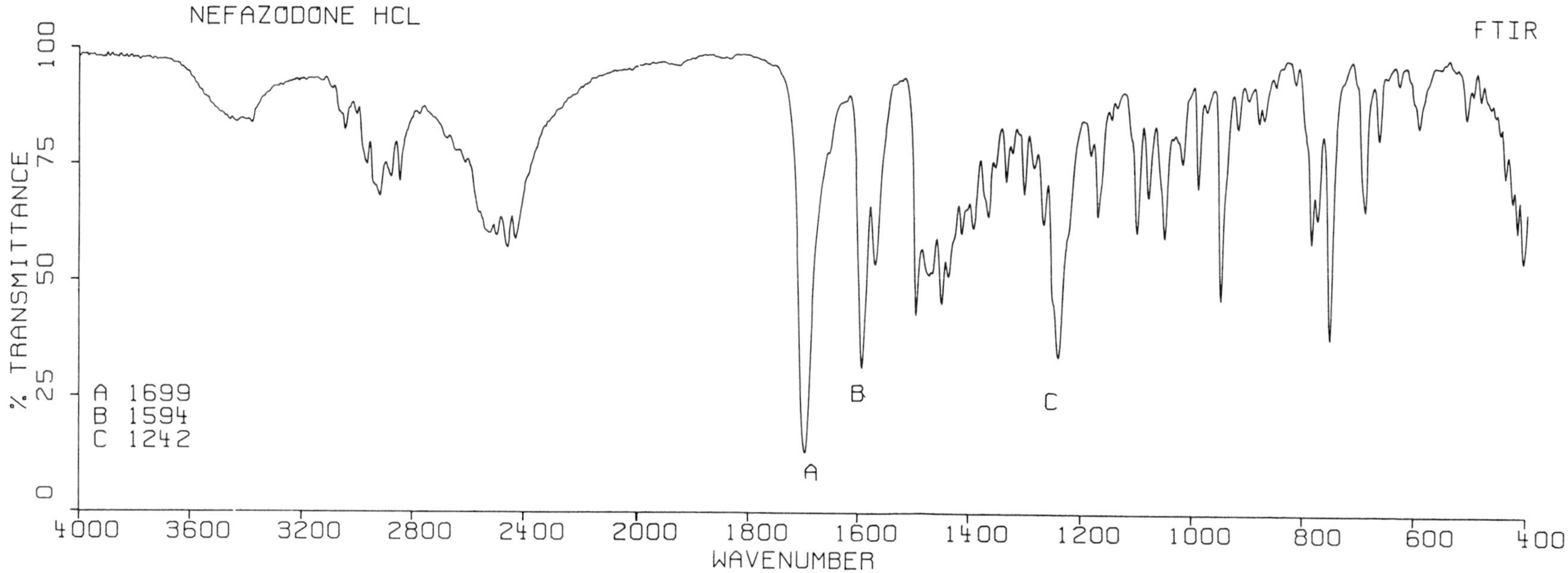

NIFENAZONE

$C_{17}H_{16}N_4O_2$

Molecular Weight: 308.34 (308.13)

Synonyms: N-(2,3-Dihydro-1,5-dimethyl-3-oxo-2-phenyl-1H-pyrazol-4-yl)-3-pyridinecarboxamide; N-nicotinoylaminoantipyrine
Trade Names: Dolongan, Nicopyron, Nikofezon, Niprazina, Phenicazone, Piralgo, Reupiron, Thylin
Use: Analgesic

HPLC: Methanol: 2.8
GC: 2982; 280°

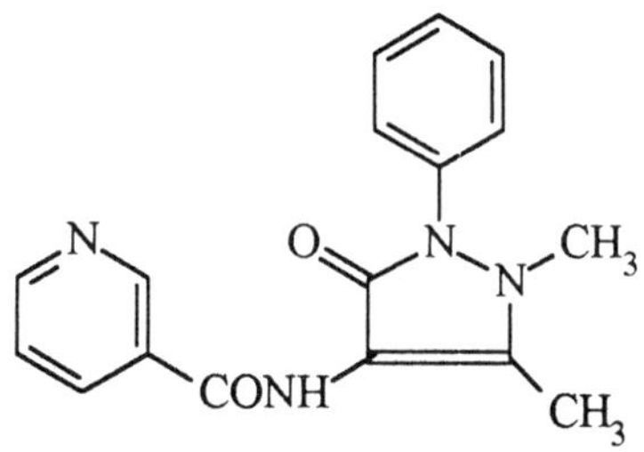

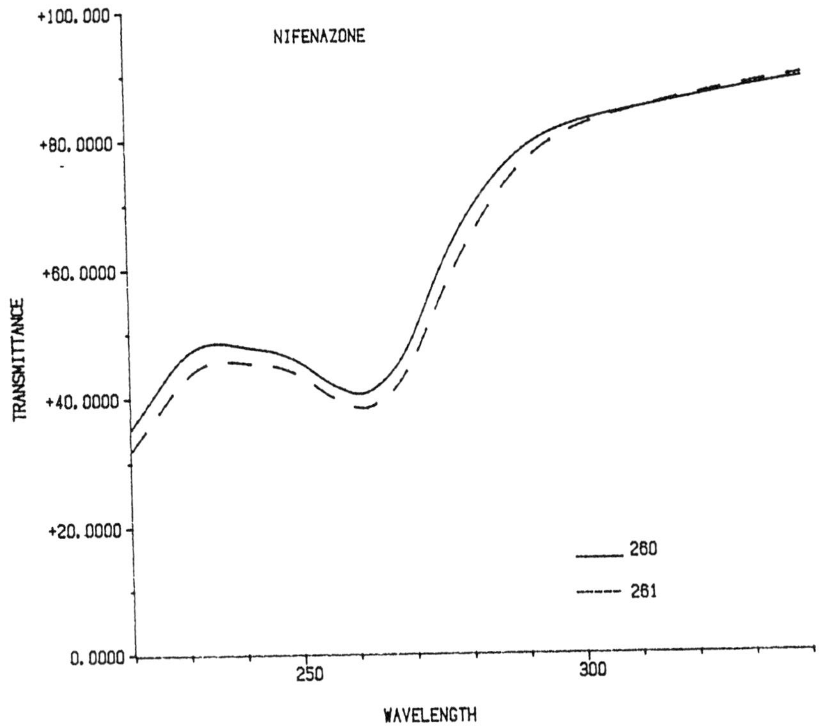

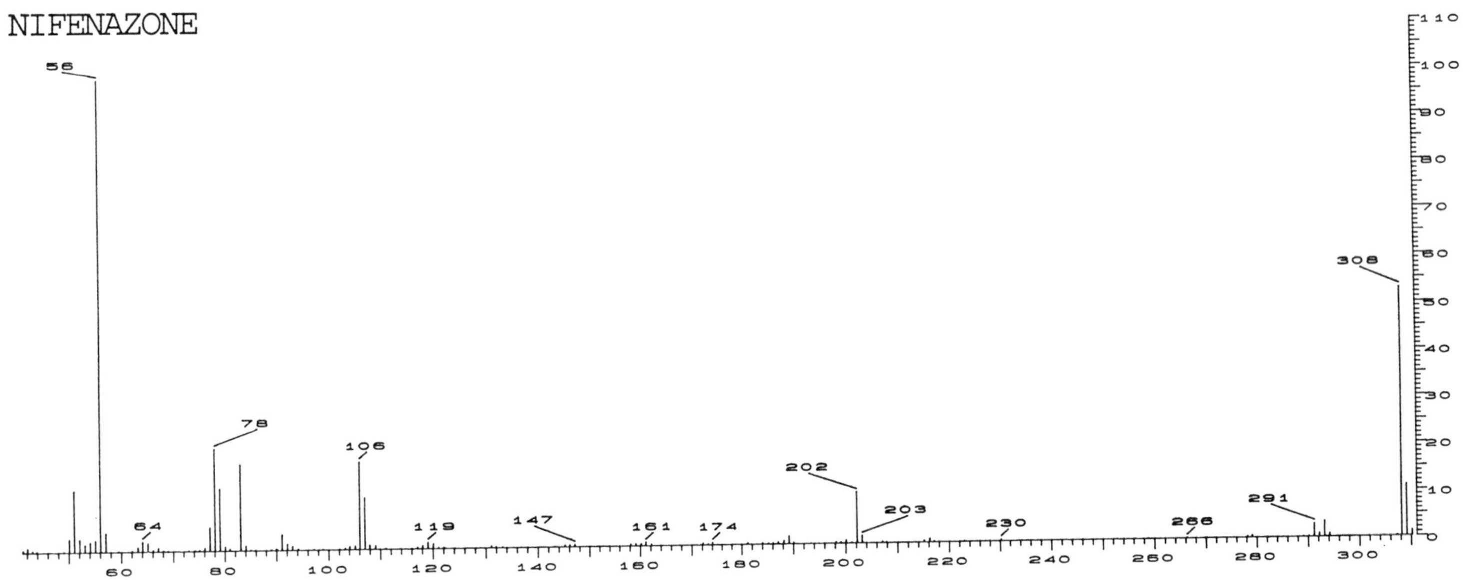

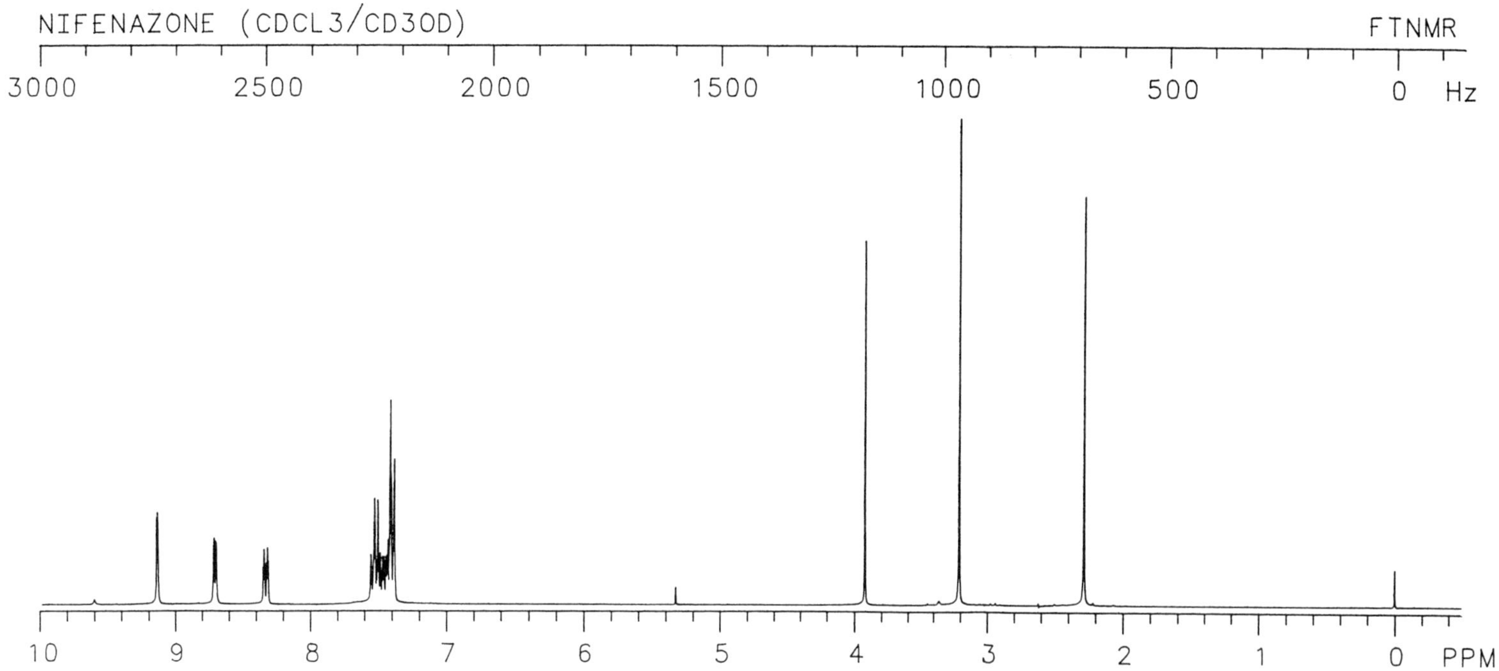

PEAK LISTING

#	HT	PPM
1	83	2.28
2	99	3.20
3	74	3.92
4	29	7.37
5	24	7.38
6	26	7.40
7	41	7.40
8	21	7.50
9	21	7.52
10	17	7.52
11	18	9.12
12	17	9.13

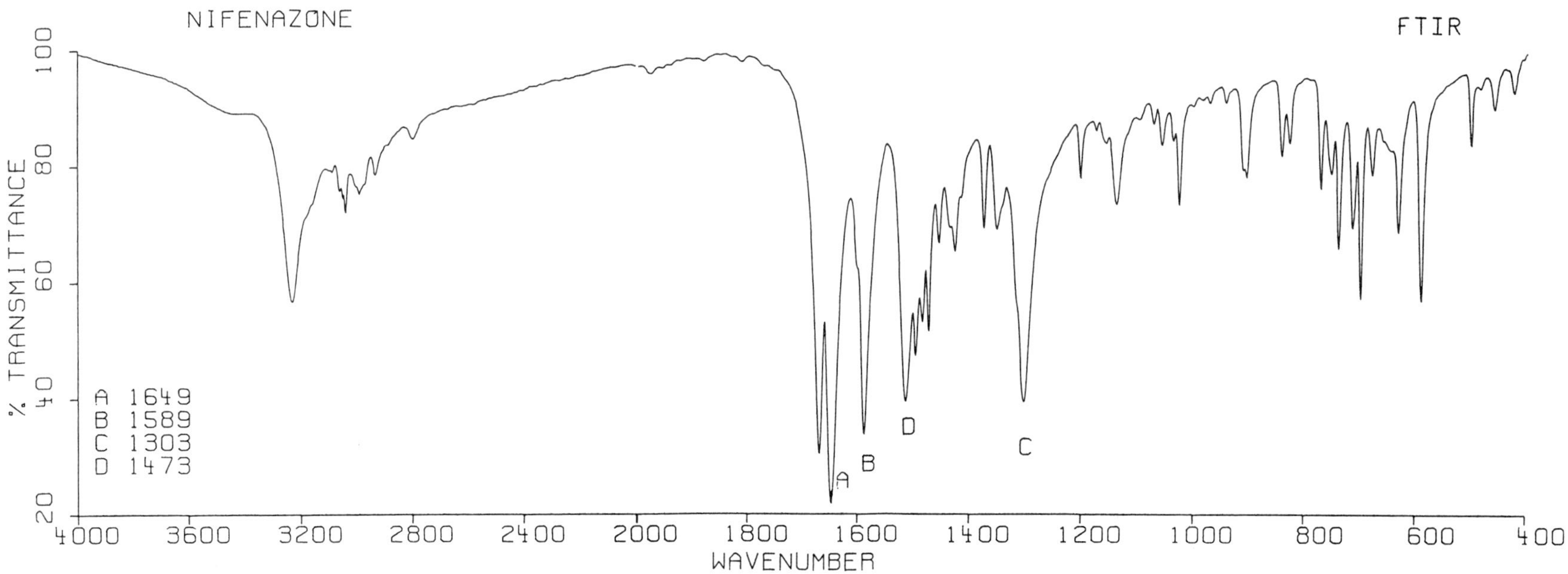

NIZATIDINE

$C_{12}H_{21}N_5O_2S_2$

Molecular Weight: 331.45 (331.11)

Synonyms: N-[2-[[[2-[(Dimethylamino)methyl]-4-thiazolyl]methyl]thio]ethyl]-N'-methyl-2-nitro-1,1-ethenediamine
Trade Names: Axid, Calmaxid, Cronizat, Gastrax, Naxidine, Nizax, Nizaxid, Zanizal
Use: Antiulcerative

HPLC:
GC:

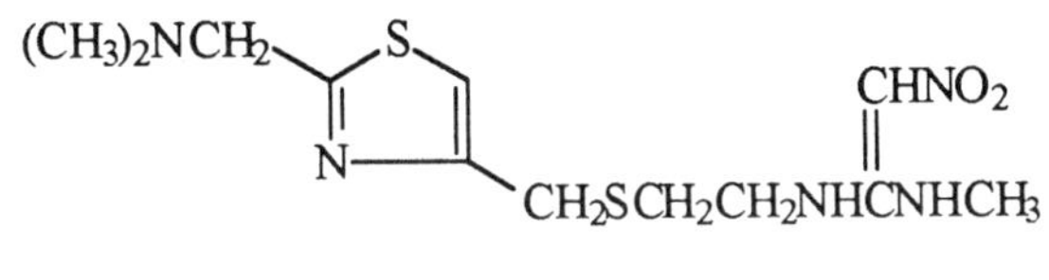

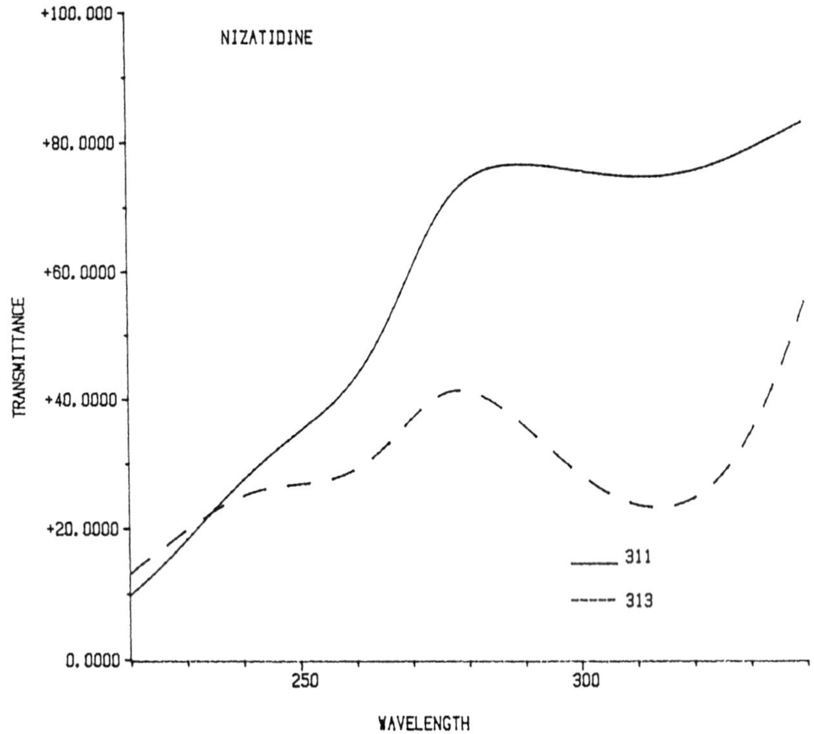

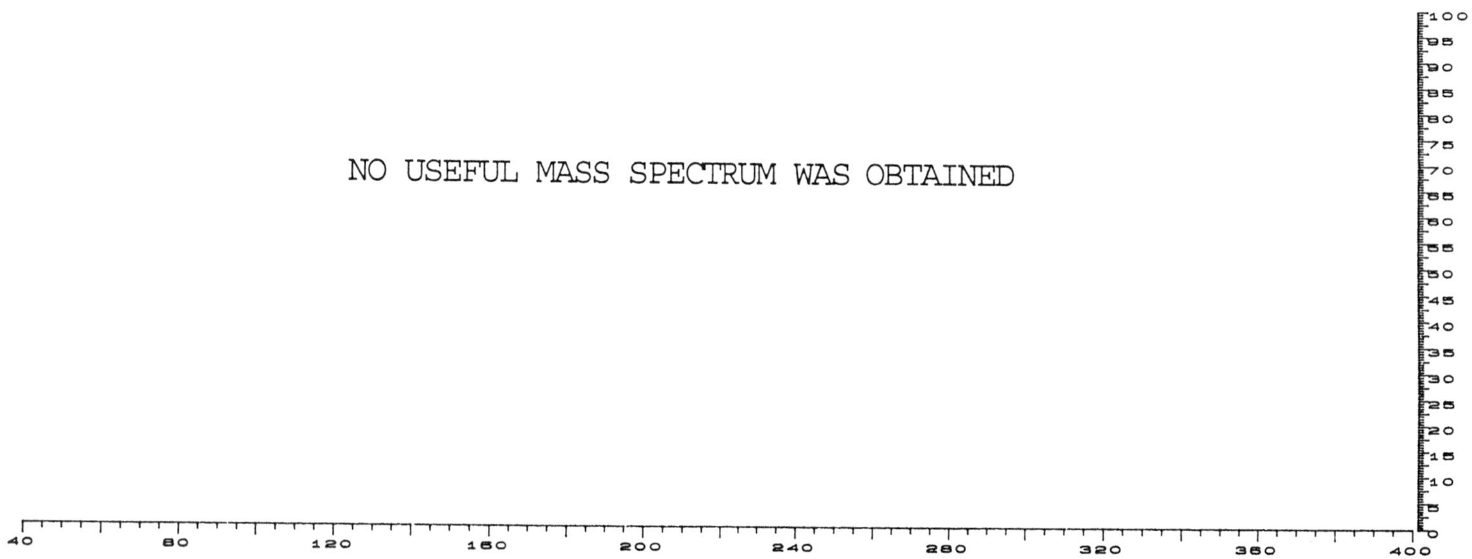

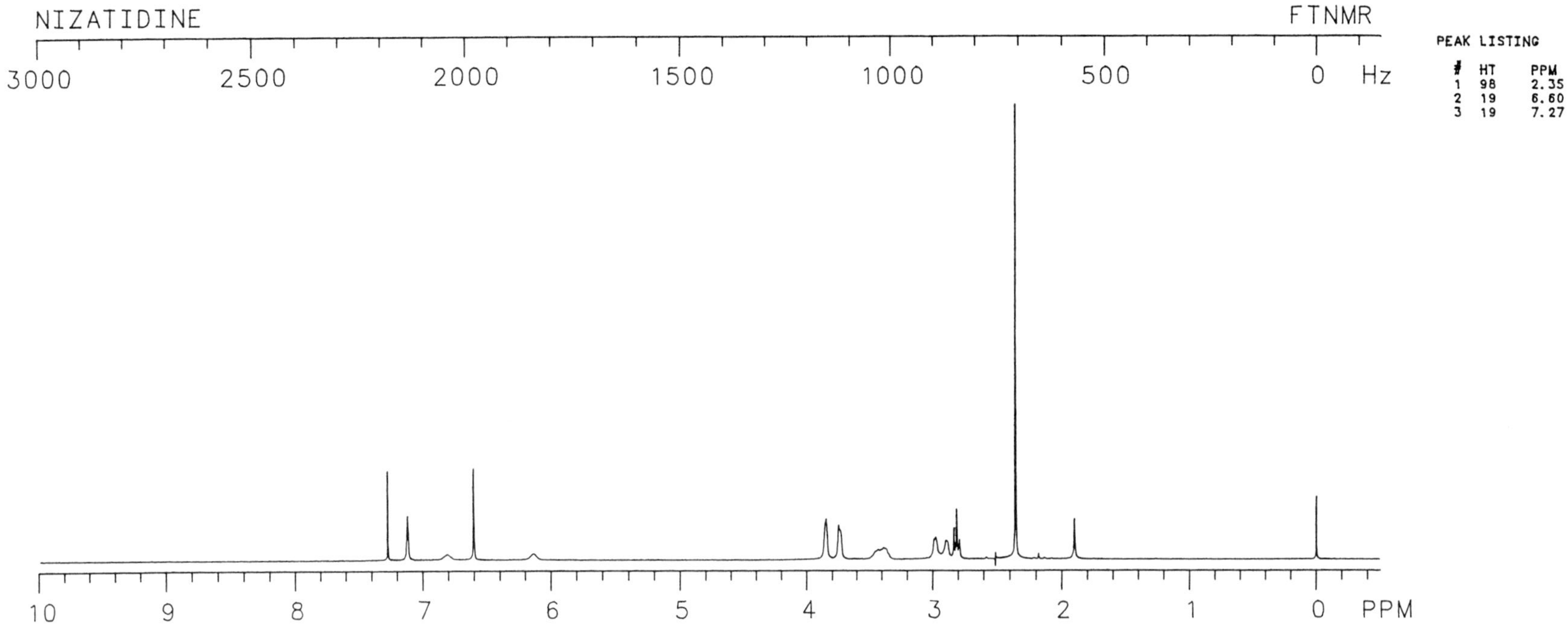
NIZATIDINE
FTNMR
3000
2500
2000
1500
1000
500
0
Hz
PEAK LISTING
HT PPM
1 98 2.35
2 19 6.60
3 19 7.27
10
9
8
7
6
5
4
3
2
1
0
PPM

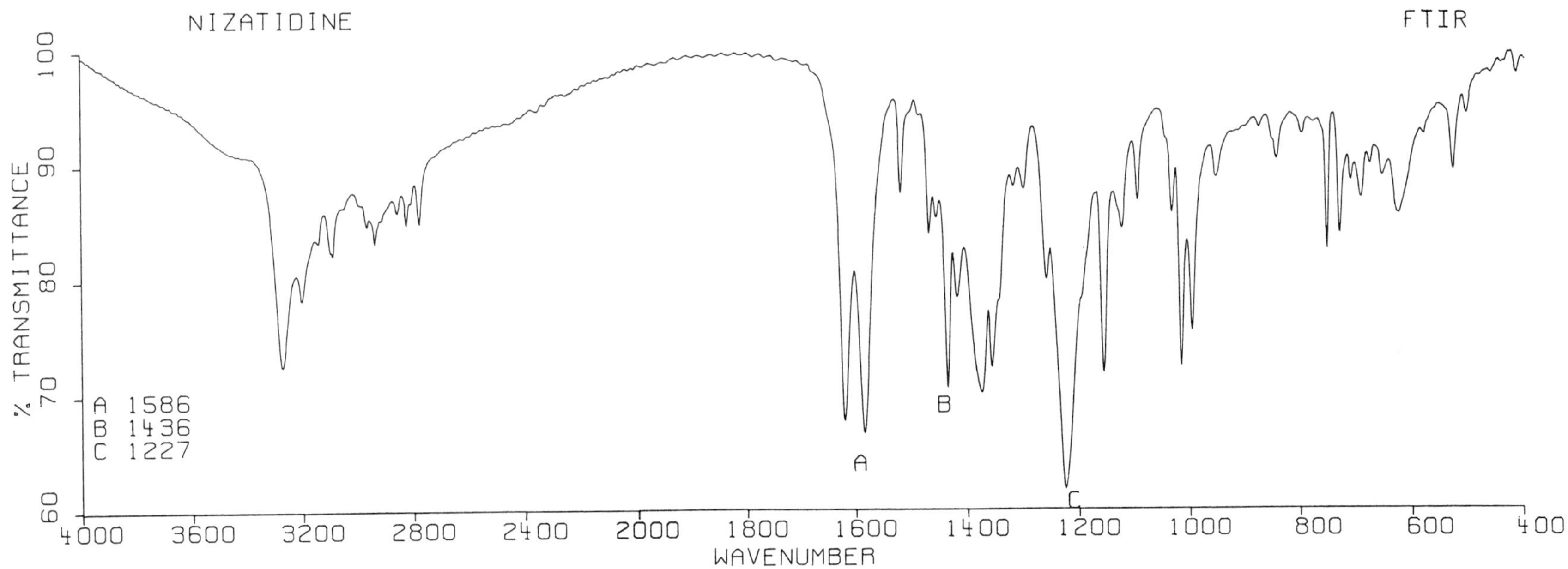
NIZATIDINE
FTIR
% TRANSMITTANCE
100
90
80
70
60
A 1586
B 1436
C 1227
A
B
C
4000
3600
3200
2800
2400
2000
1800
1600
1400
1200
1000
800
600
400
WAVENUMBER

NORCOCAINE

$C_{16}H_{19}NO_4$

Molecular Weight: 289.33 (289.13)

Synonyms: [1R-(exo,exo)]-3-(Benzoyloxy)-8-azabicyclo[3.2.1]octane-2-carboxylic acid methyl ester
Trade Names:
Use: Cocaine metabolite

HPLC: Methanol: 3.8
GC:

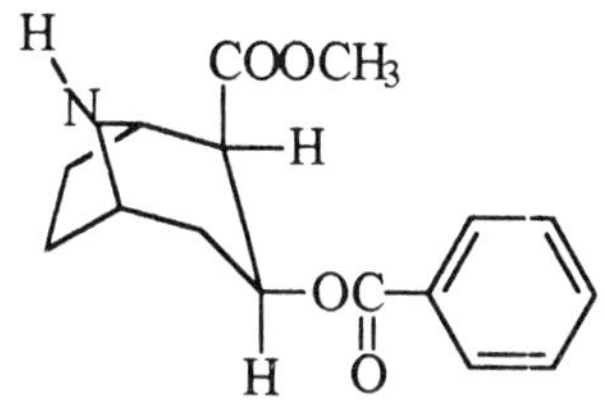

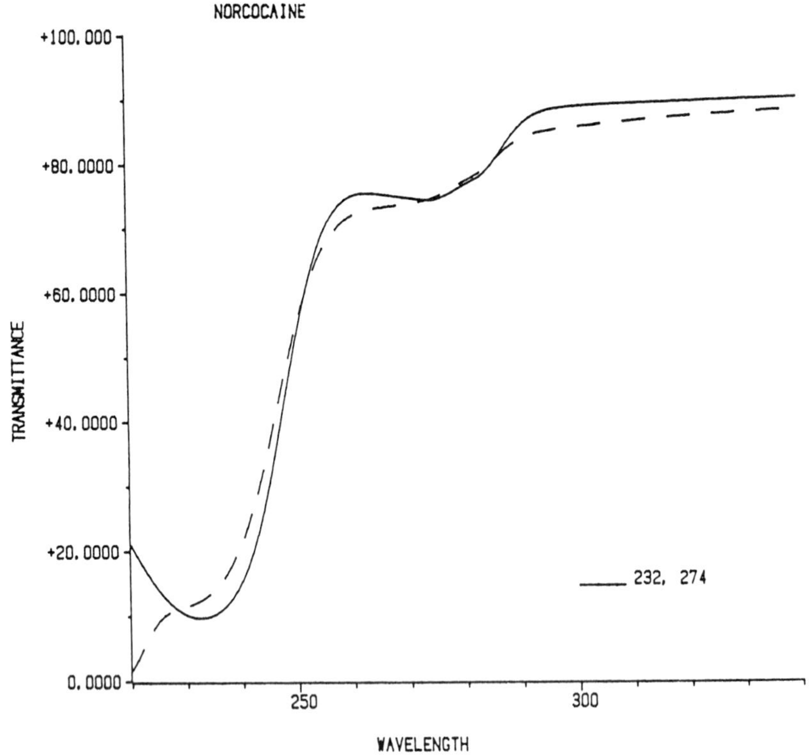

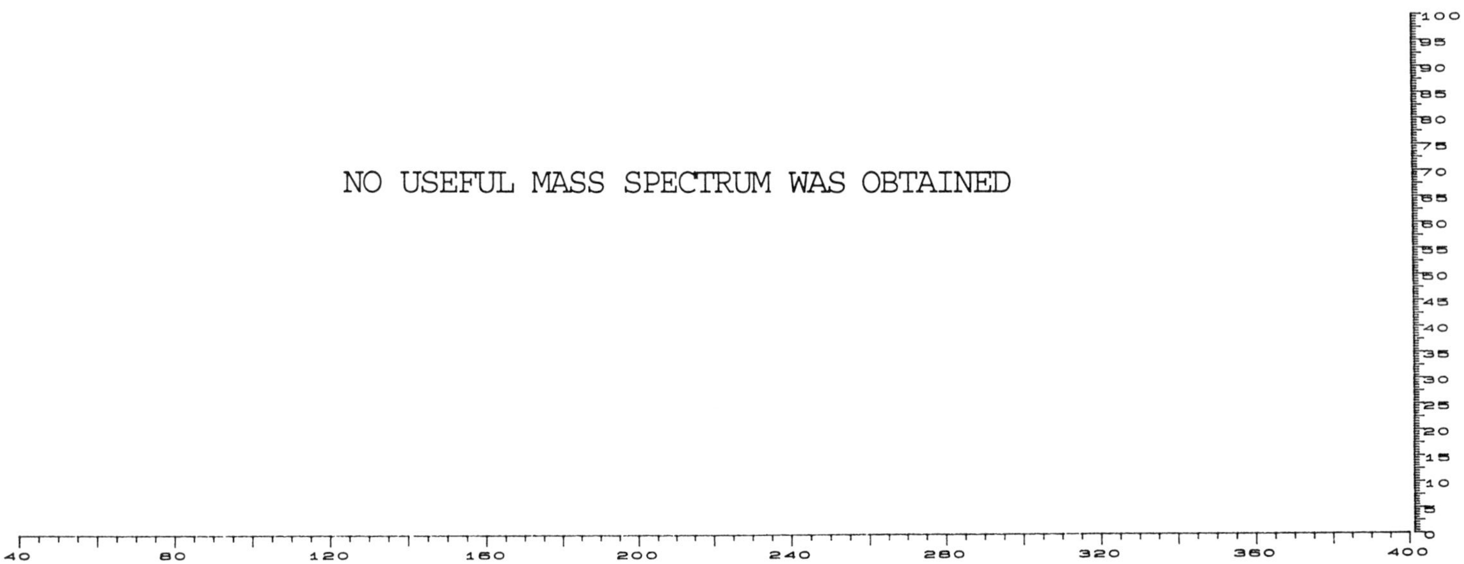

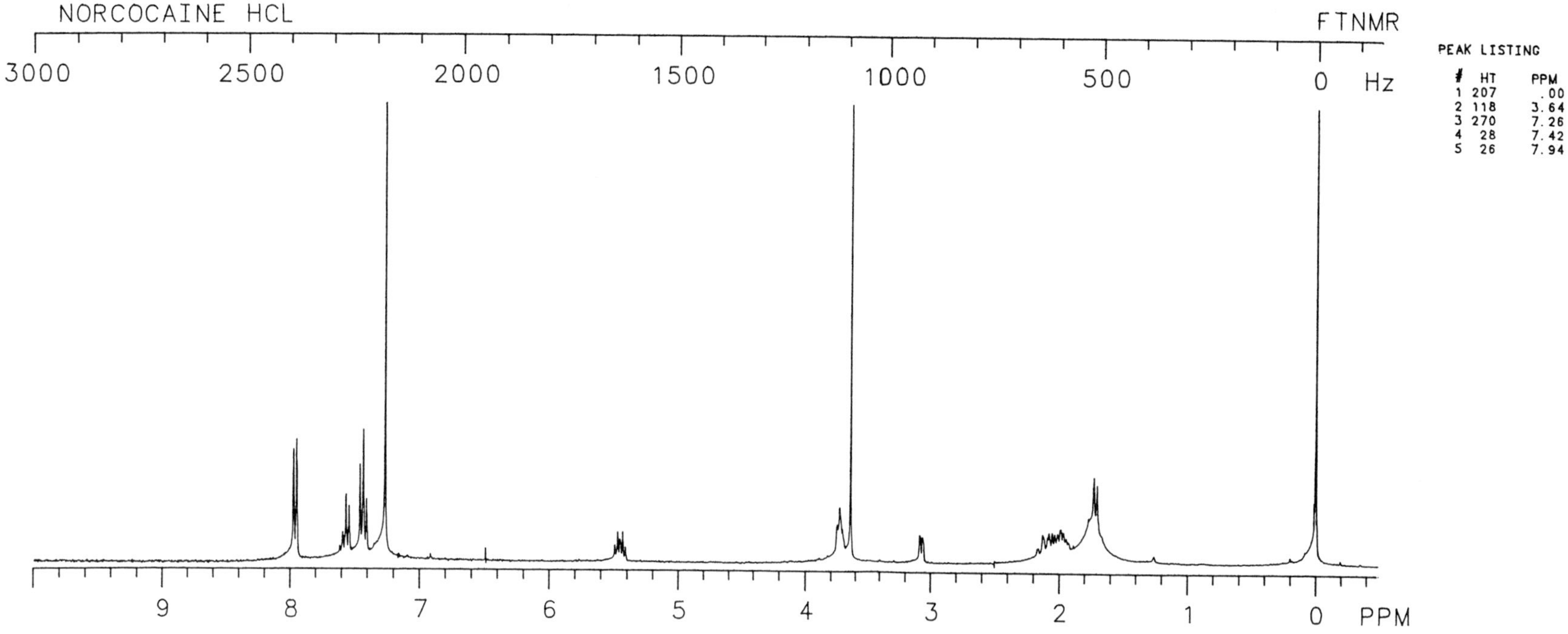
NORCOCAINE HCL
FTNMR
3000 2500 2000 1500 1000 500 0 Hz
PEAK LISTING
HT PPM
1 207 .00
2 118 3.64
3 270 7.26
4 28 7.42
5 26 7.94
9 8 7 6 5 4 3 2 1 0 PPM

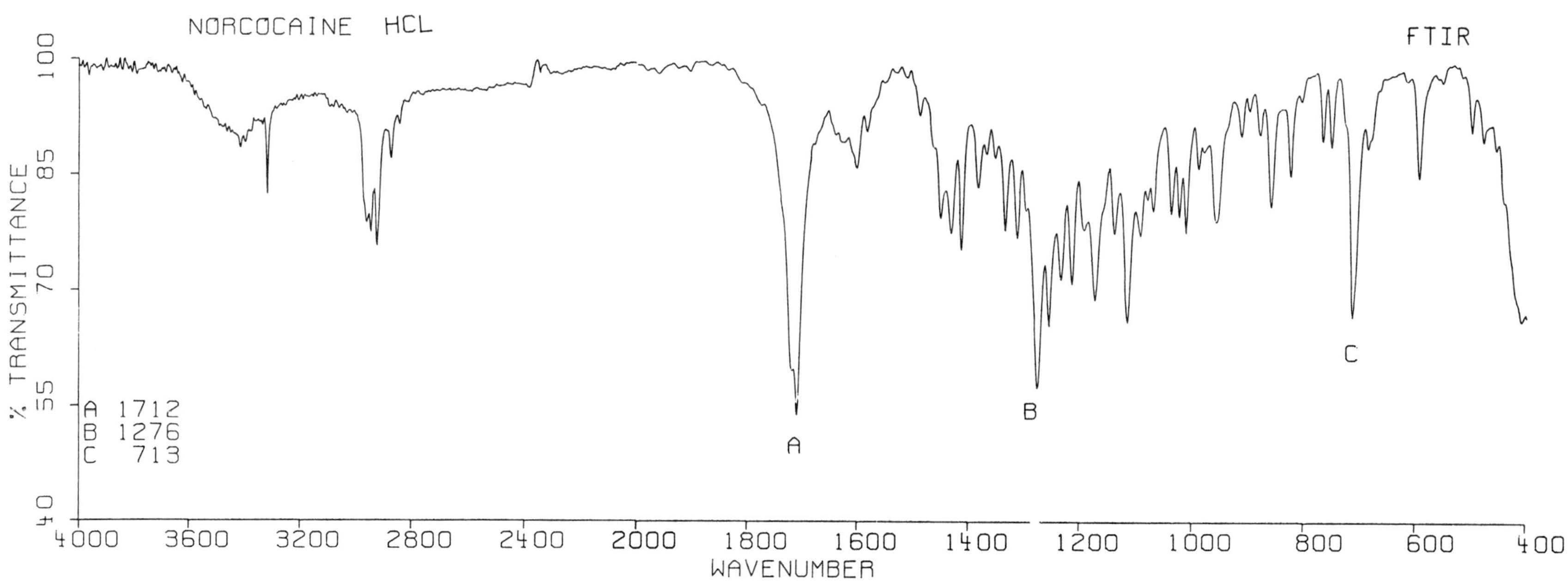
NORCOCAINE HCL
FTIR
% TRANSMITTANCE
100 85 70 55 40
A 1712
B 1276
C 713
A
B
C
4000 3600 3200 2800 2400 2000 1800 1600 1400 1200 1000 800 600 400
WAVENUMBER

NORDIHYDROGUAIARETIC ACID

$C_{18}H_{22}O_4$

Molecular Weight: 302.37 (302.15)

Synonyms: 4,4'-(2,3-Dimethyl-1,4-butanediyl)bis[1,2-benzenediol]
Trade Names:
Use: Antioxidant

HPLC:
GC:

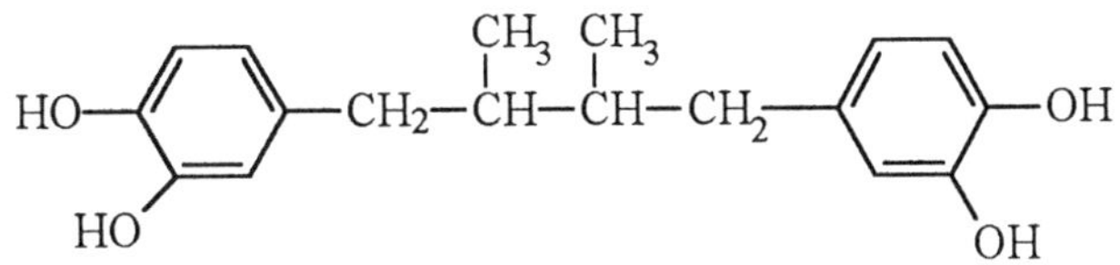

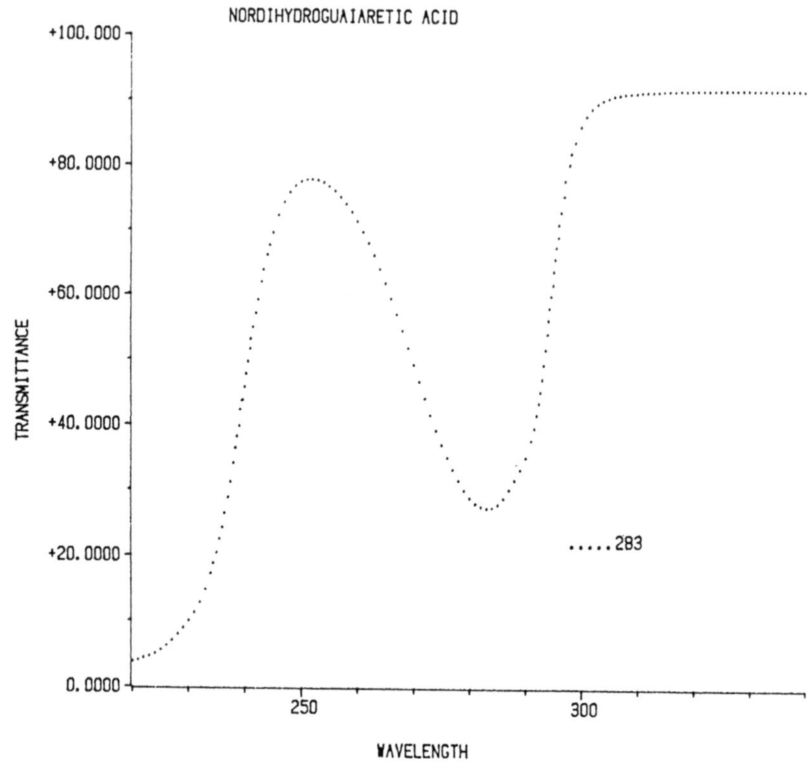

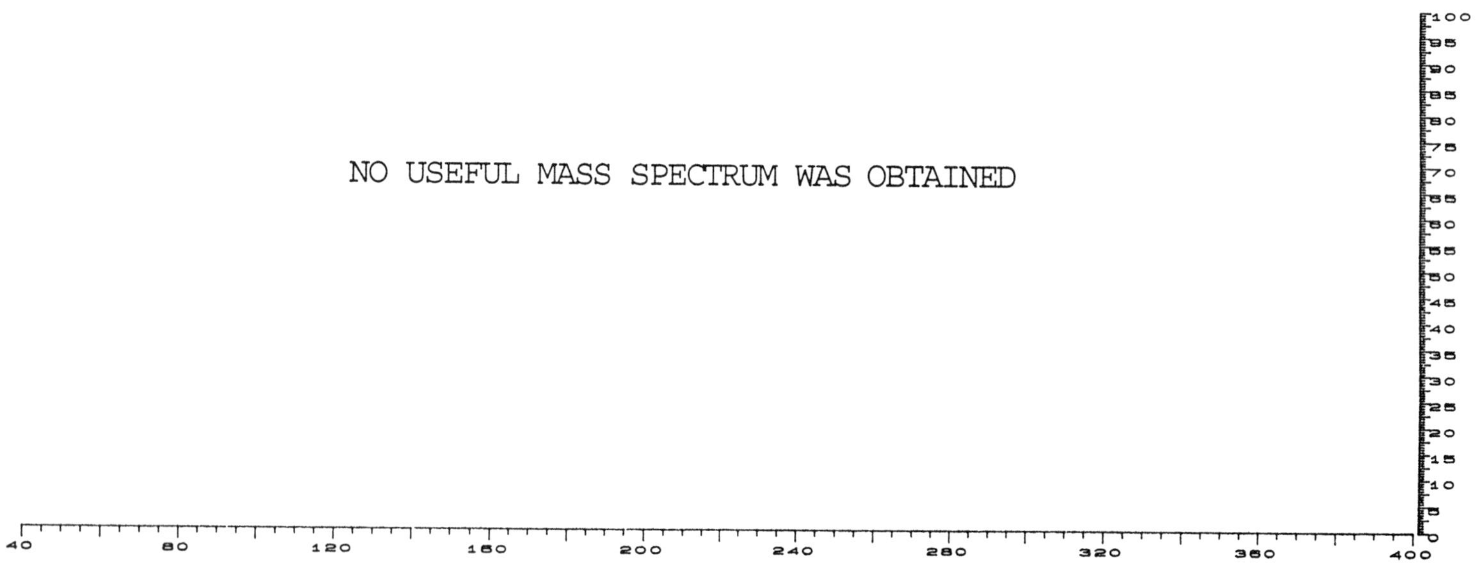

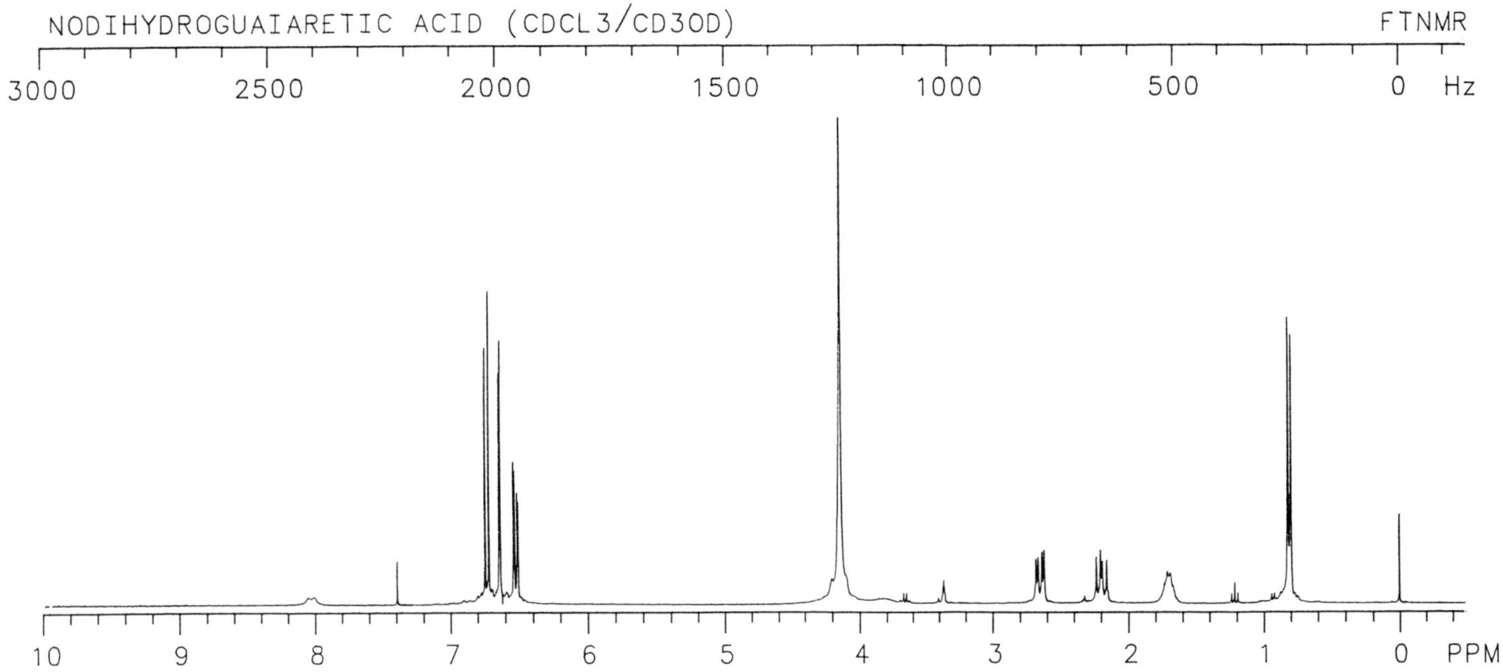

PEAK LISTING

#	HT	PPM
1	18	.00
2	54	.79
3	58	.81
4	99	4.14
5	20	6.51
6	22	6.51
7	27	6.53
8	29	6.54
9	53	6.63
10	47	6.64
11	63	6.72
12	52	6.74

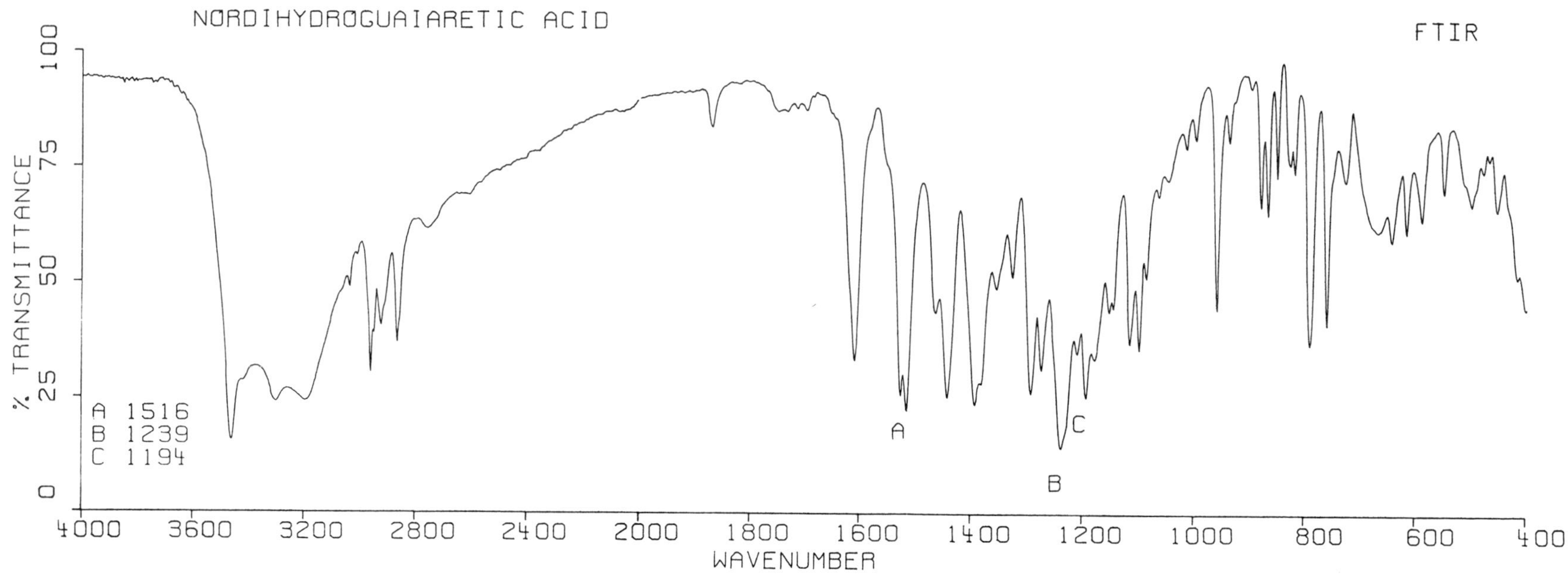

NORFLUOXETINE

$C_{16}H_{16}F_3NO$

Molecular Weight: 295.30 (295.12)

Synonyms: [4-(trifluoromethyl)phenoxy]benzenepropanamine
Trade Names:
Use: Fluoxetine metabolite

HPLC:
GC:

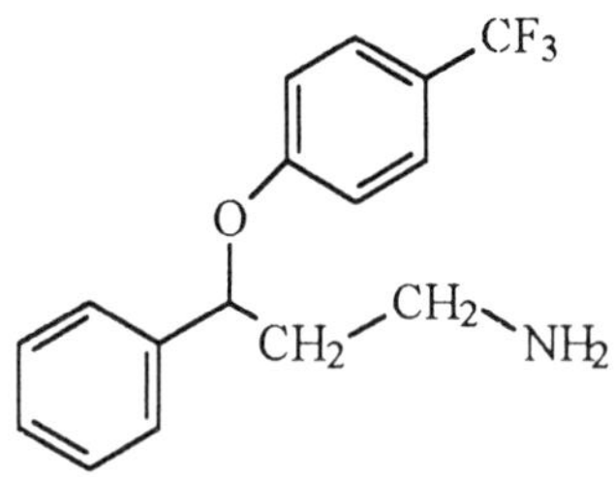

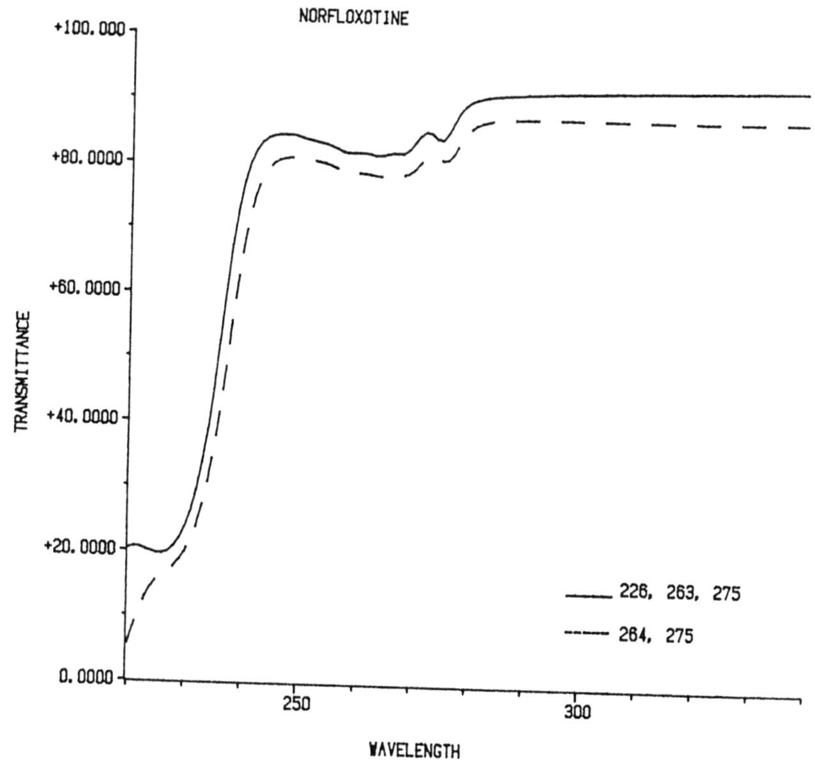

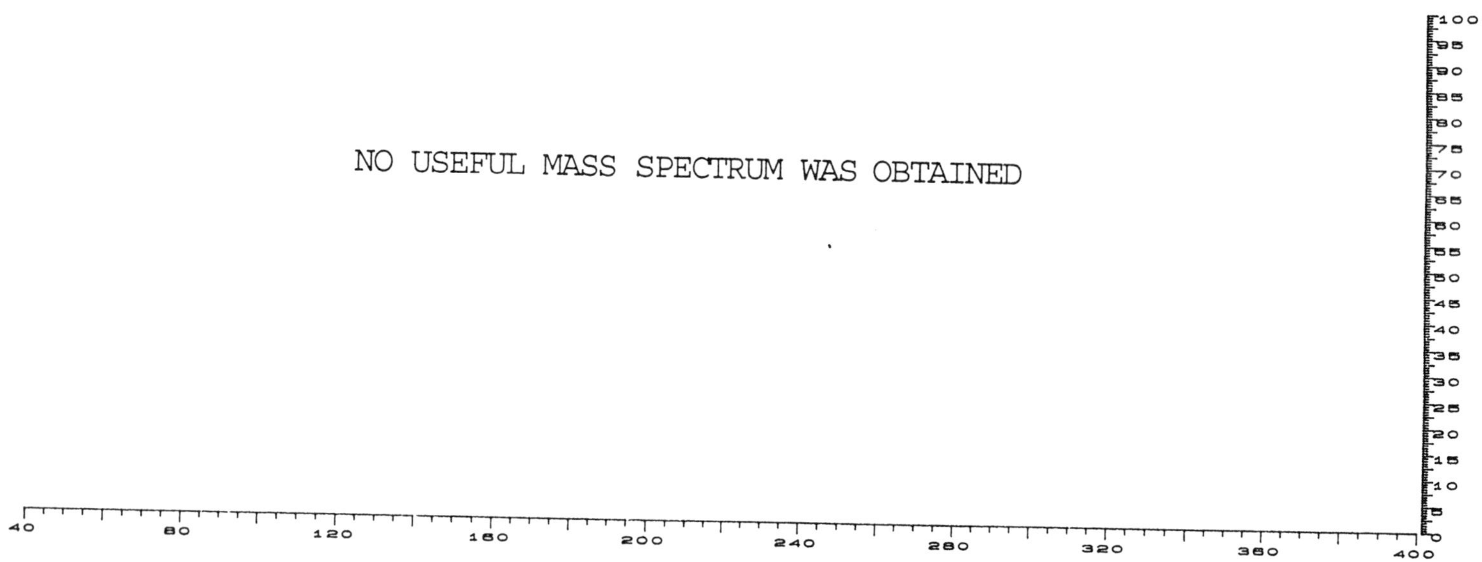

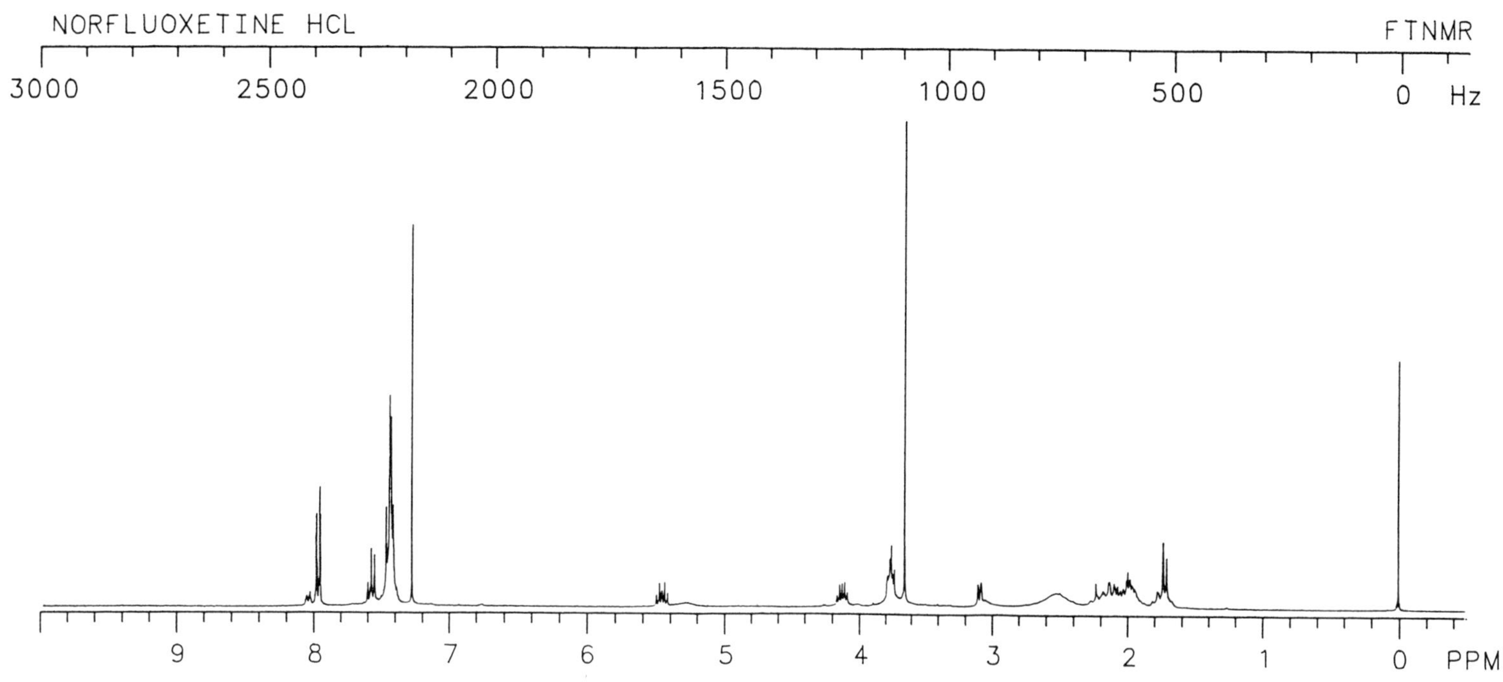

PEAK LISTING

#	HT	PPM
1	51	.00
2	99	3.64
3	78	7.26
4	18	7.40
5	20	7.40
6	38	7.41
7	43	7.42
8	20	7.45
9	18	7.94
10	24	7.94
11	18	7.96

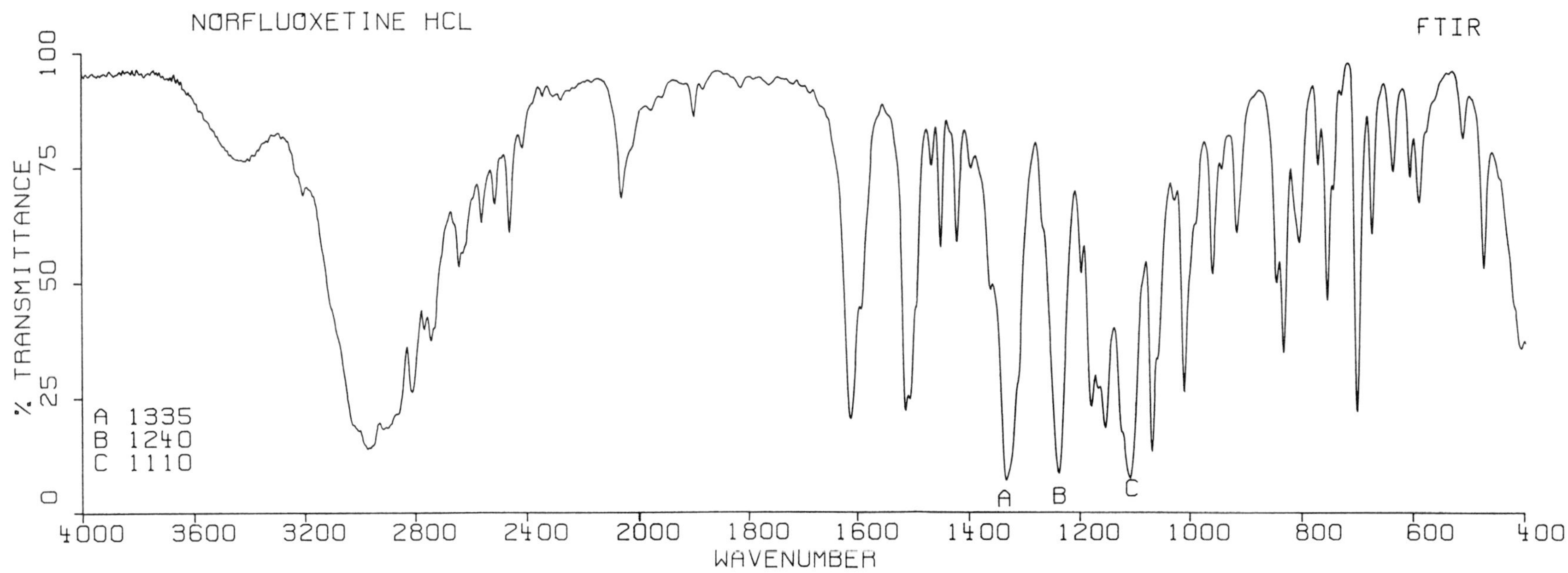

6-NORLYSERGIC ACID

$C_{15}H_{14}N_2O_2$

Molecular Weight: 254.29 (254.11)

Synonyms: 9,10-Didehydro-ergoline-8-carboxylic acid
Trade Names:
Use: Lysergic acid metabolite

HPLC:
GC:

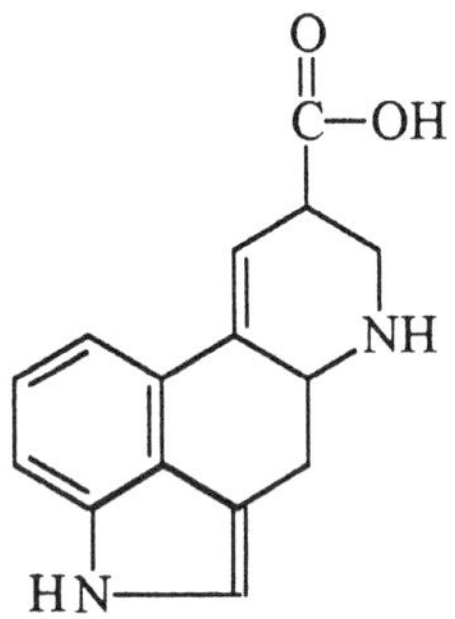

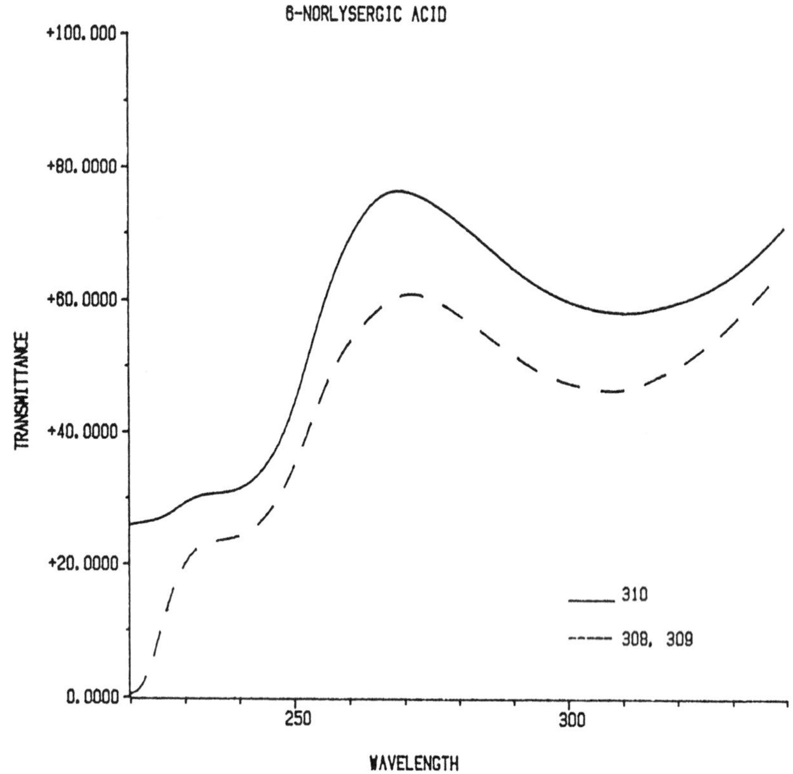

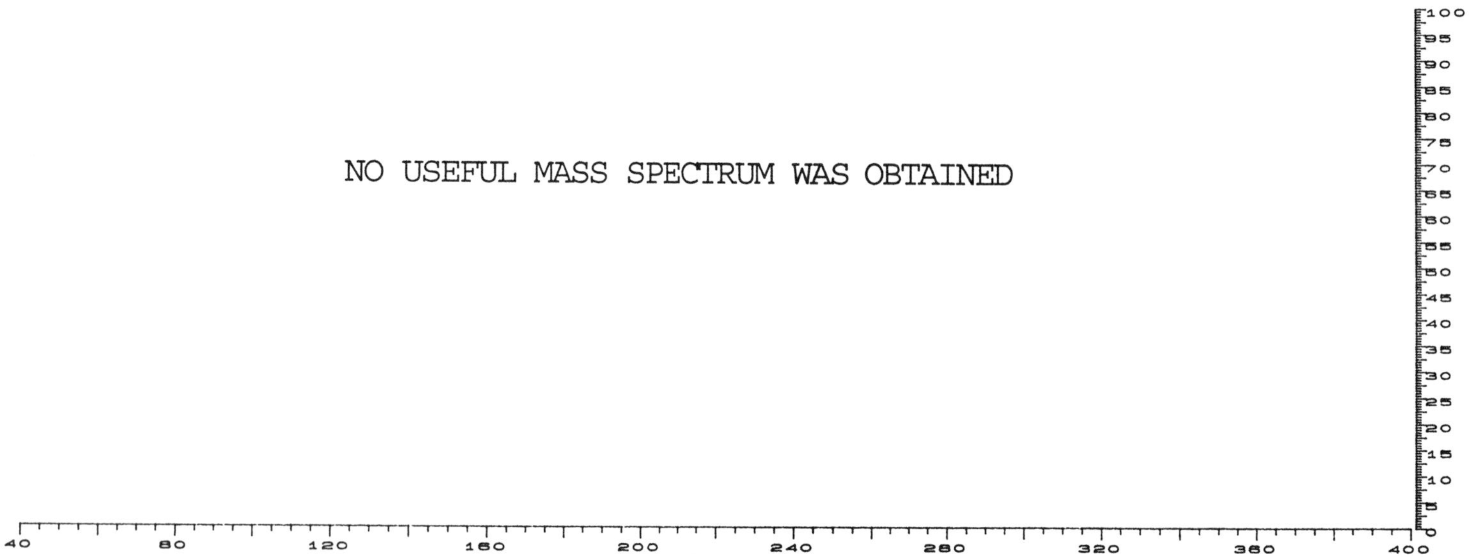

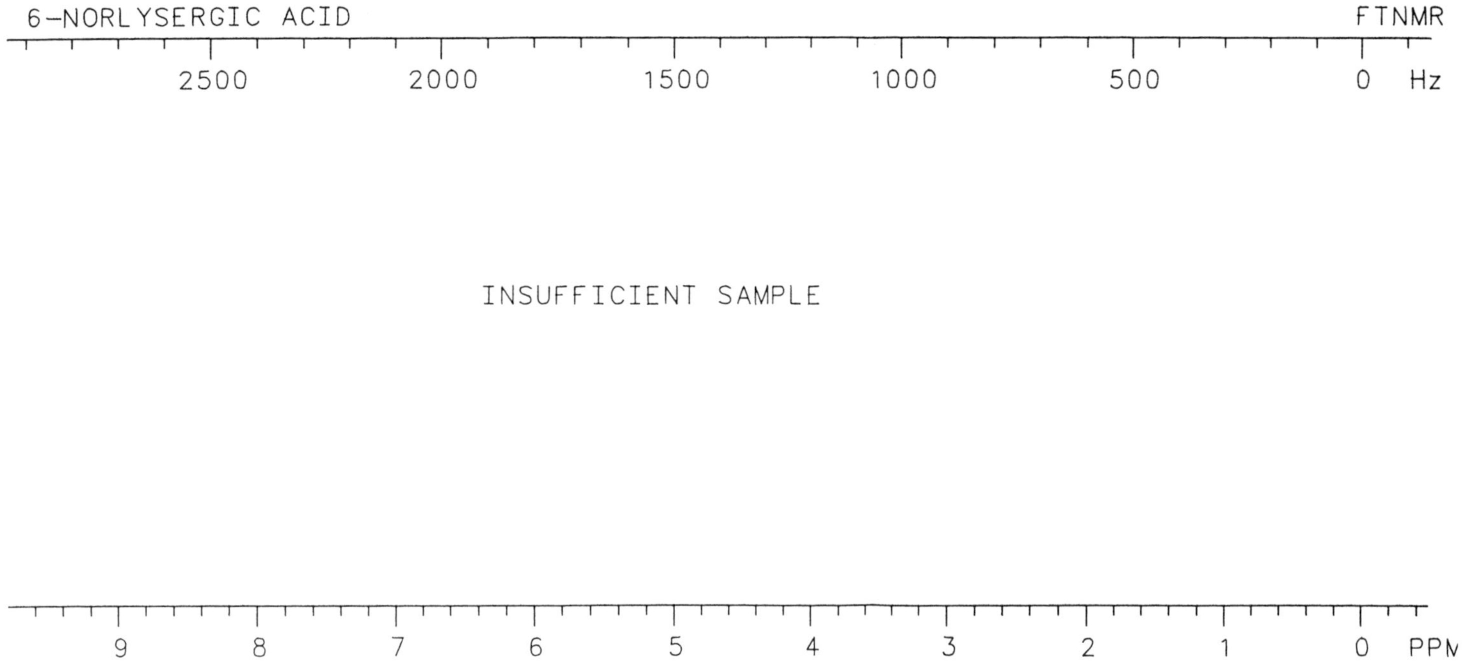
6-NORLYSERGIC ACID
FTNMR
2500
2000
1500
1000
500
0 Hz
INSUFFICIENT SAMPLE
9
8
7
6
5
4
3
2
1
0 PPM

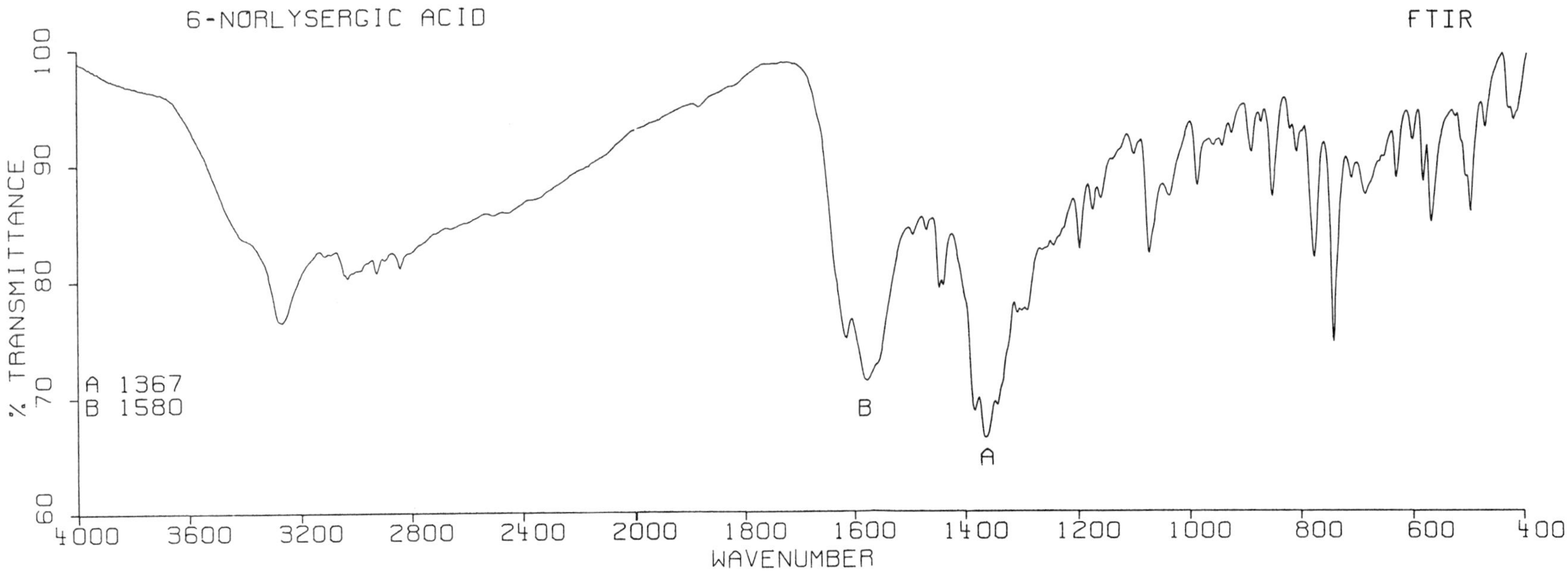
6-NORLYSERGIC ACID
FTIR
% TRANSMITTANCE
100
90
80
70
60
A 1367
B 1580
B
A
4000
3600
3200
2800
2400
2000
1800
1600
1400
1200
1000
800
600
400
WAVENUMBER

OFLOXACIN

$C_{18}H_{20}FN_3O_4$

Molecular Weight: 361.37 (361.14)

Synonyms: 9-Fluoro-2,3-dihydro-3-methyl-10-(4-methyl-1-puperazinyl)-7-oxo-7H-pyrido-[1,2,3-de]-1,4-benzoxazine-6-carboxylic acid
Trade Names: Floxin, Ocuflox
Use: Antibacterial

HPLC:
GC:

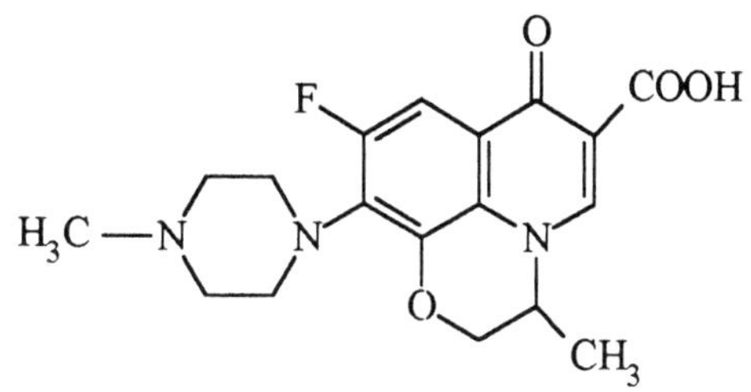

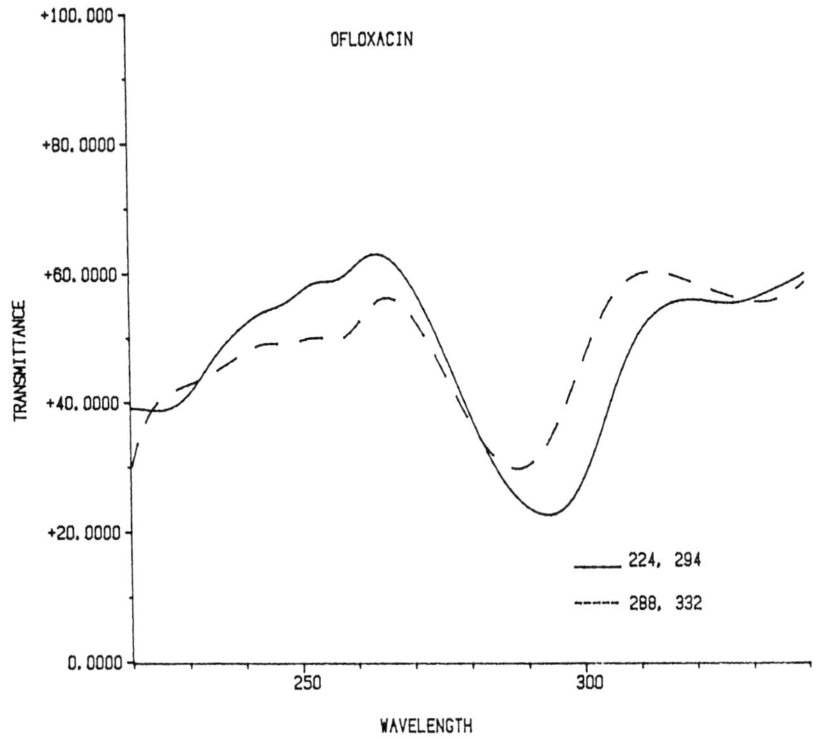

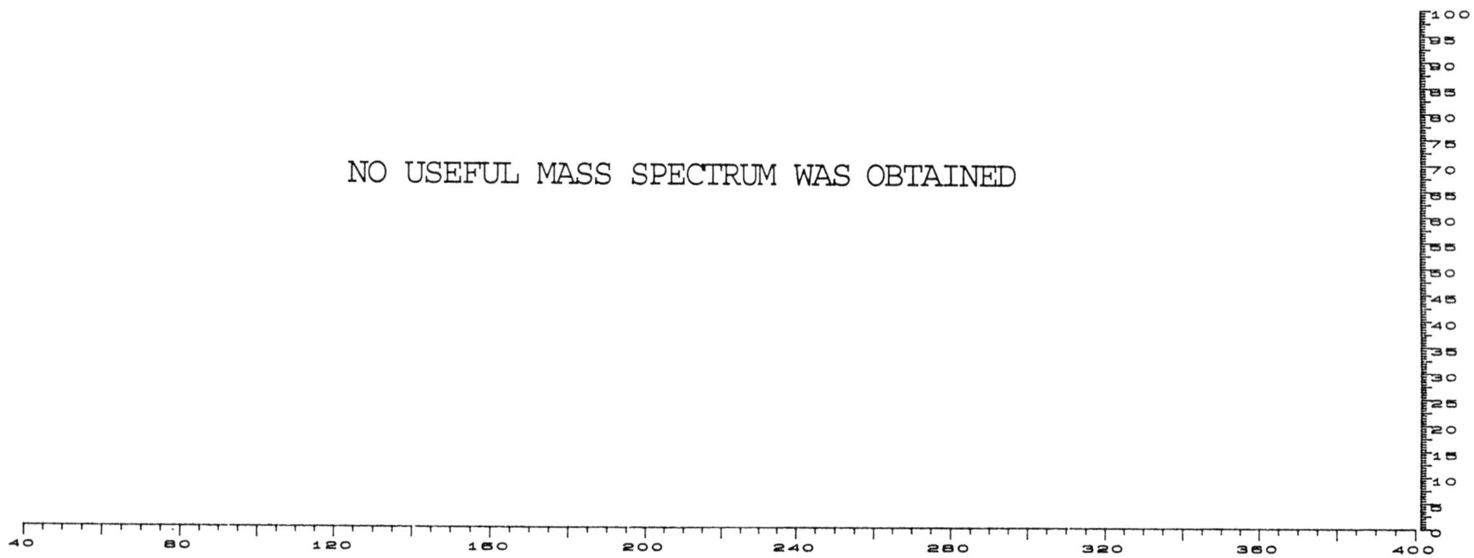

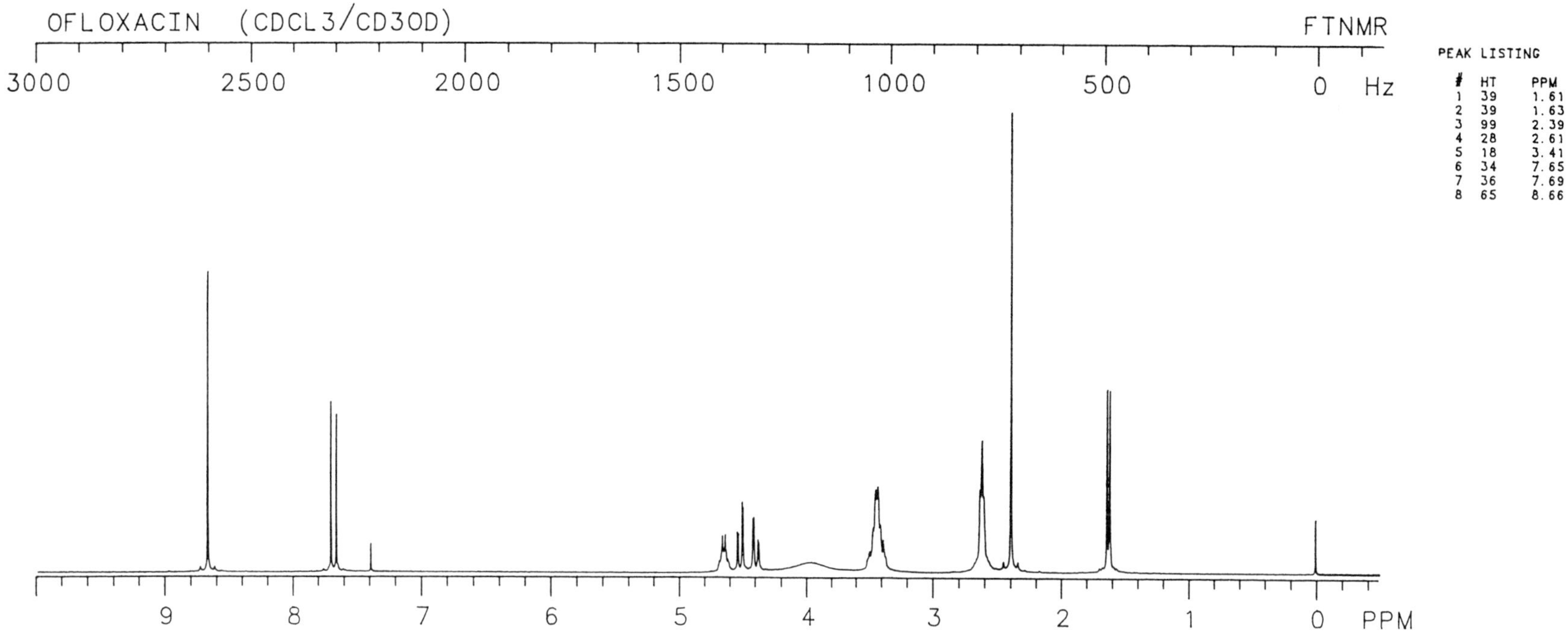

OFLOXACIN (CDCL3/CD3OD)
FTNMR
3000 2500 2000 1500 1000 500 0 Hz
PEAK LISTING
HT PPM
1 39 1.61
2 39 1.63
3 99 2.39
4 28 2.61
5 18 3.41
6 34 7.65
7 36 7.69
8 65 8.66
9 8 7 6 5 4 3 2 1 0 PPM

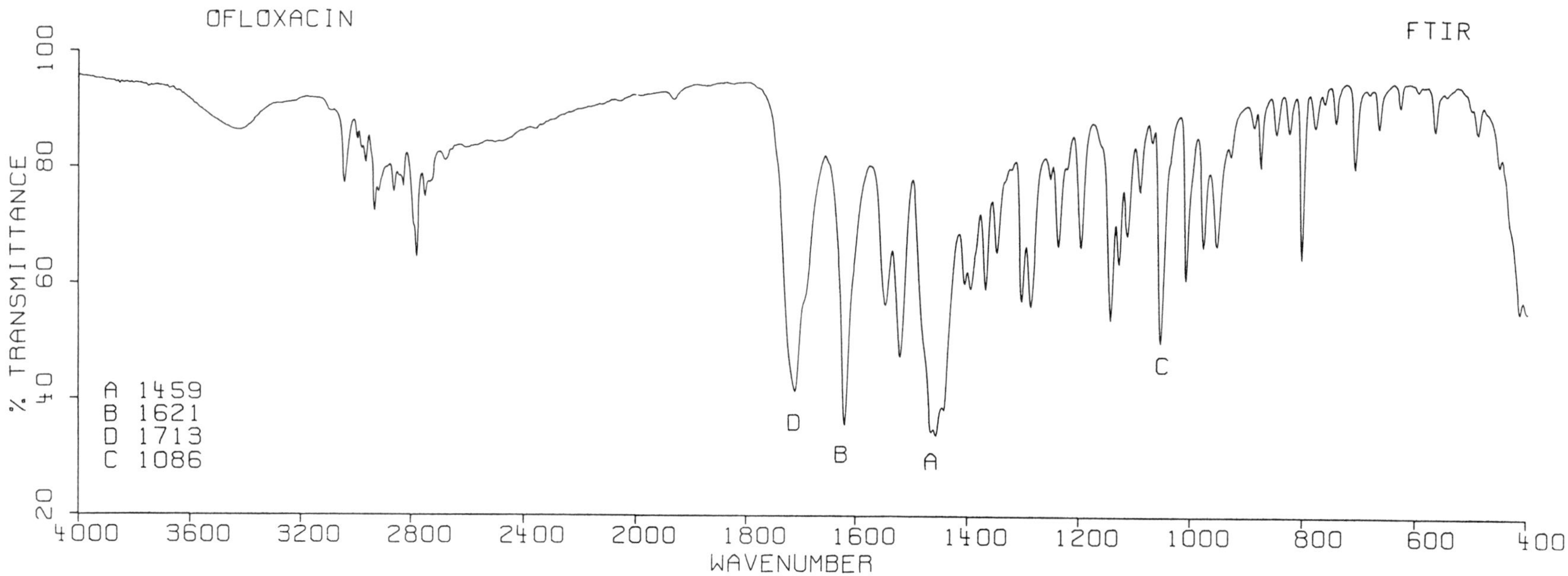

OFLOXACIN
FTIR
% TRANSMITTANCE
100 80 60 40 20
A 1459
B 1621
D 1713
C 1086
D B A C
4000 3600 3200 2800 2400 2000 1800 1600 1400 1200 1000 800 600 400
WAVENUMBER

OMEPRAZOLE

$C_{17}H_{19}N_3O_3S$

Molecular Weight: 345.42 (354.11)

Synonyms: 5-Methoxy-2-[[(4-methoxy-3,5-dimethyl-2-pyridinyl)methyl]sulfinyl]-1H-benzimidazole
Trade Names: Antra, Losec, Mopral, Omepral, Prilosec
Use: Anti-ulcerative

HPLC: Methanol: 2.7
GC:

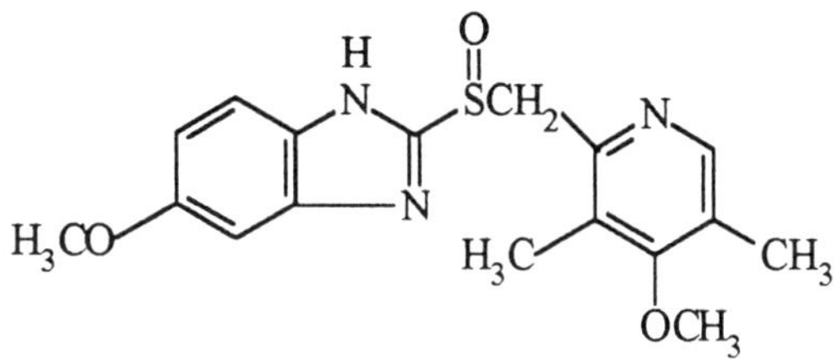

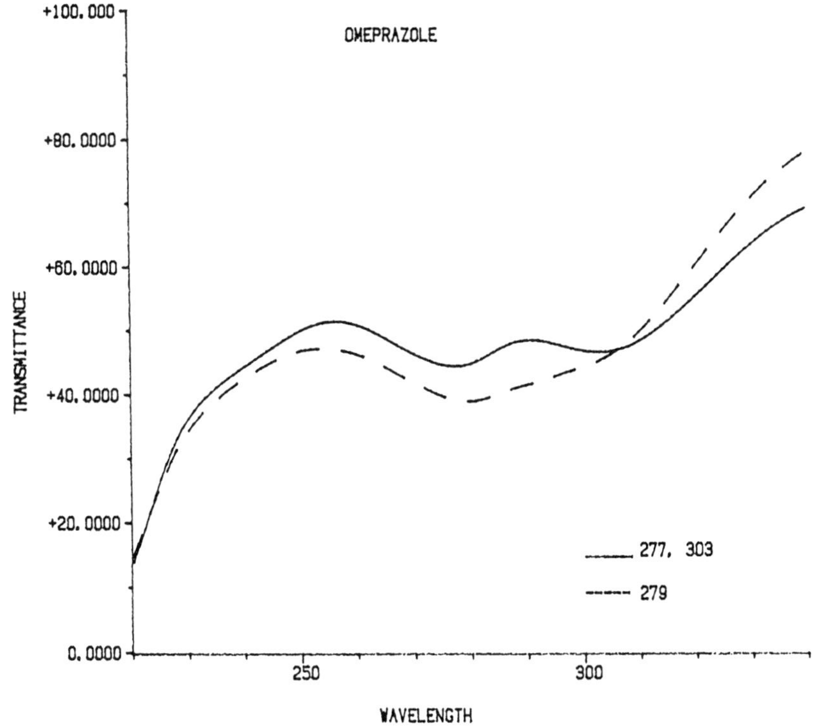

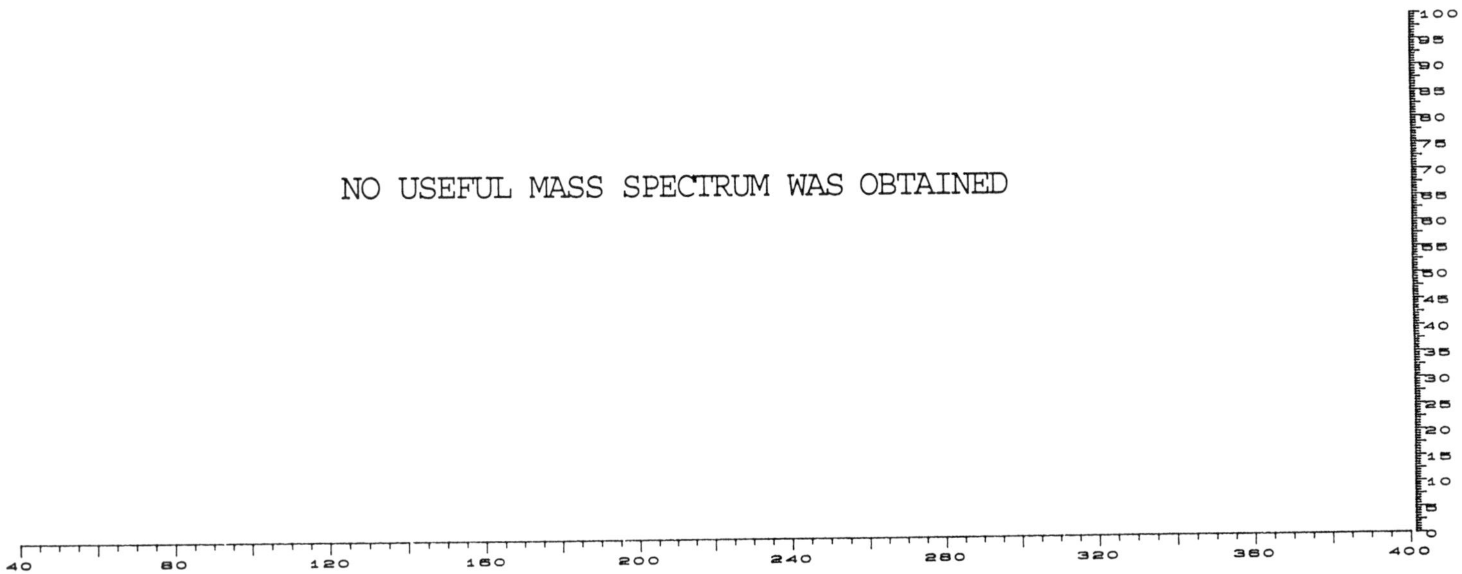

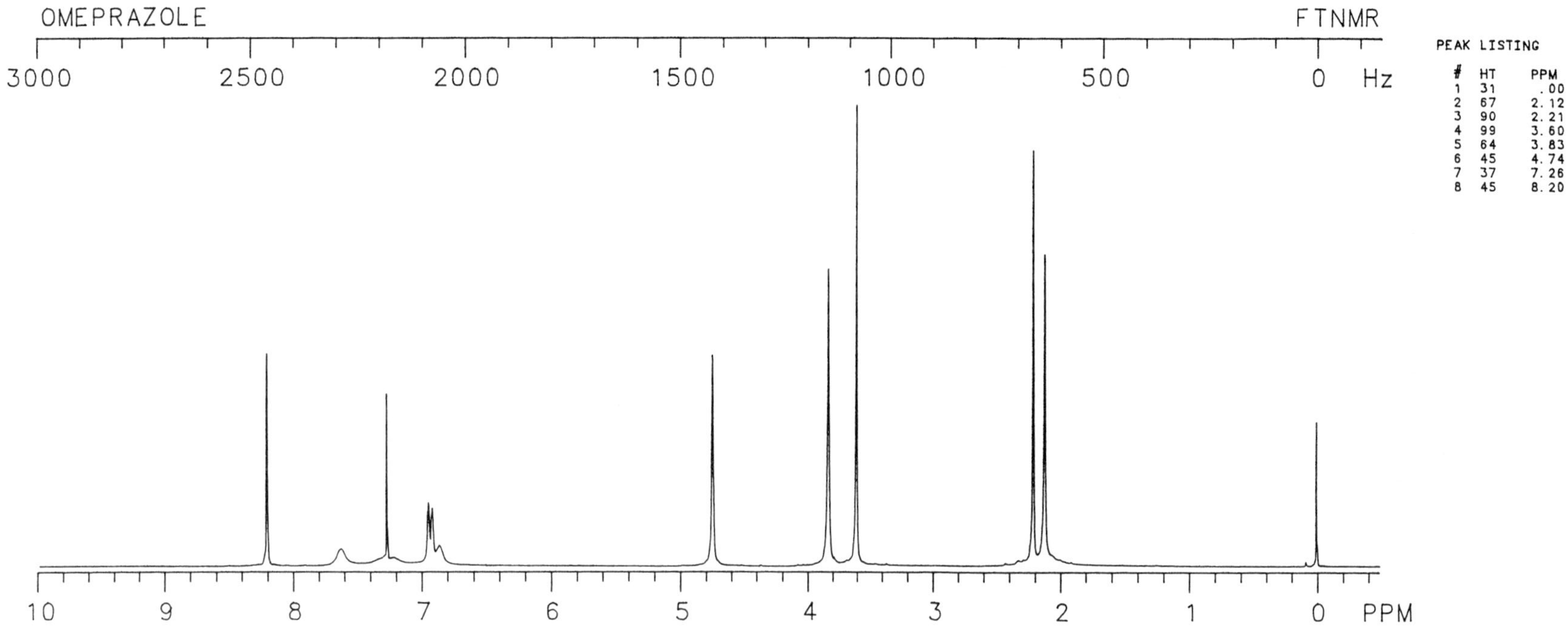
OMEPRAZOLE
FTNMR
3000
2500
2000
1500
1000
500
0 Hz
10
9
8
7
6
5
4
3
2
1
0 PPM
PEAK LISTING
HT PPM
1 31 .00
2 67 2.12
3 90 2.21
4 99 3.60
5 64 3.83
6 45 4.74
7 37 7.26
8 45 8.20

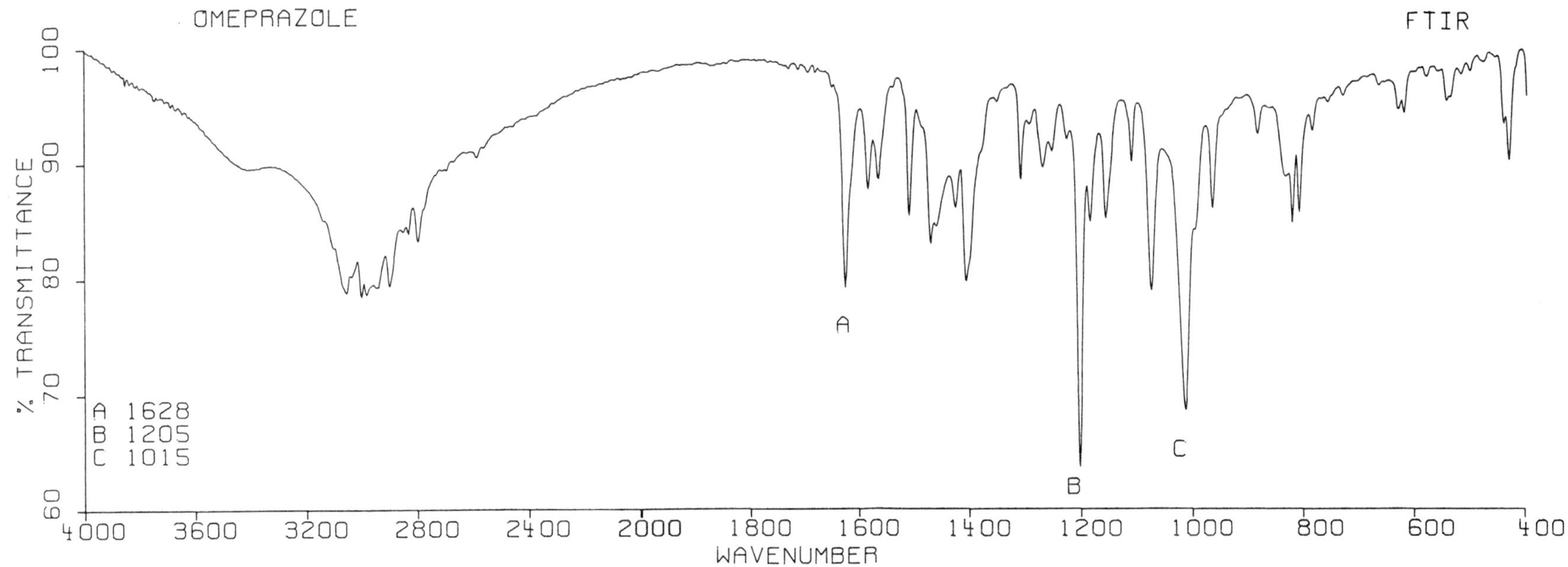
OMEPRAZOLE
FTIR
% TRANSMITTANCE
100
90
80
70
60
A
B
C
A 1628
B 1205
C 1015
4000
3600
3200
2800
2400
2000
1800
1600
1400
1200
1000
800
600
400
WAVENUMBER

OXAPROZIN

$C_{18}H_{15}NO_3$

Molecular Weight: 293.32 (293.11)

Synonyms: 4,5-Diphenyl-2-oxazolepropanoic acid
Trade Names: Alvo, Daypro, Durapro, Duraprost, Oxapro
Use: Anti-inflammatory

HPLC: Methanol: 2.9
GC:

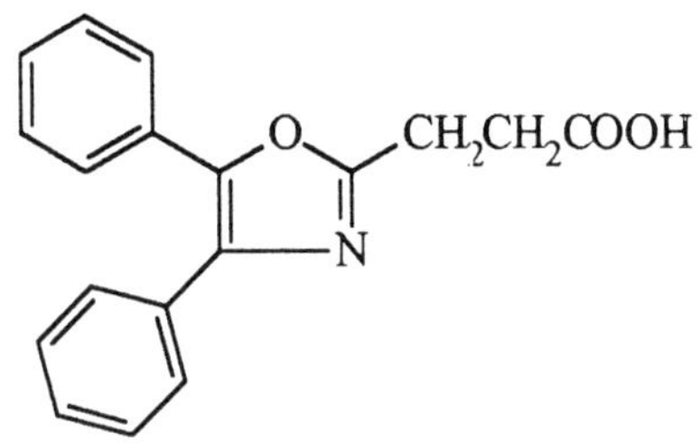

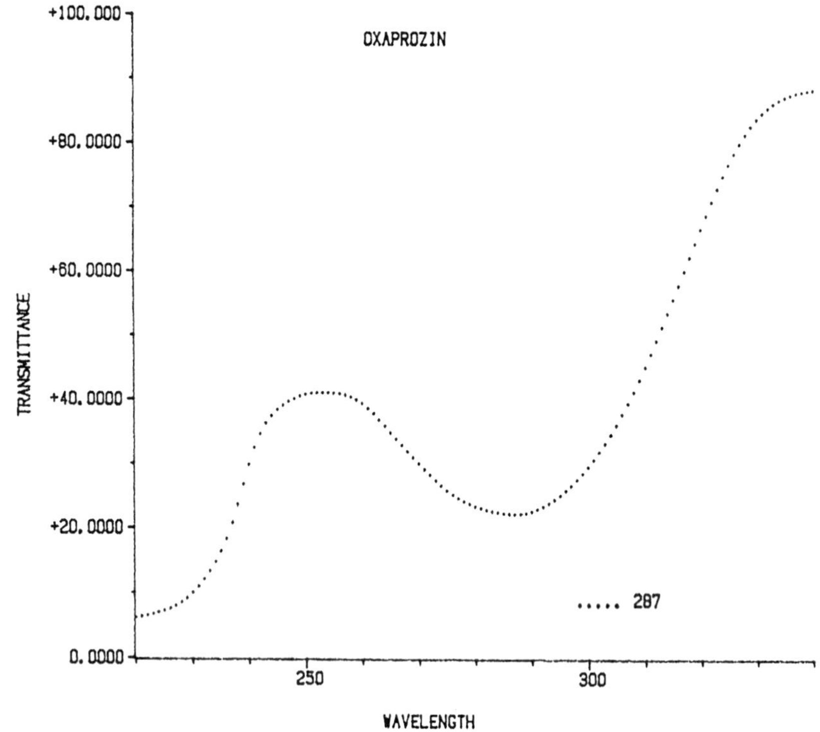

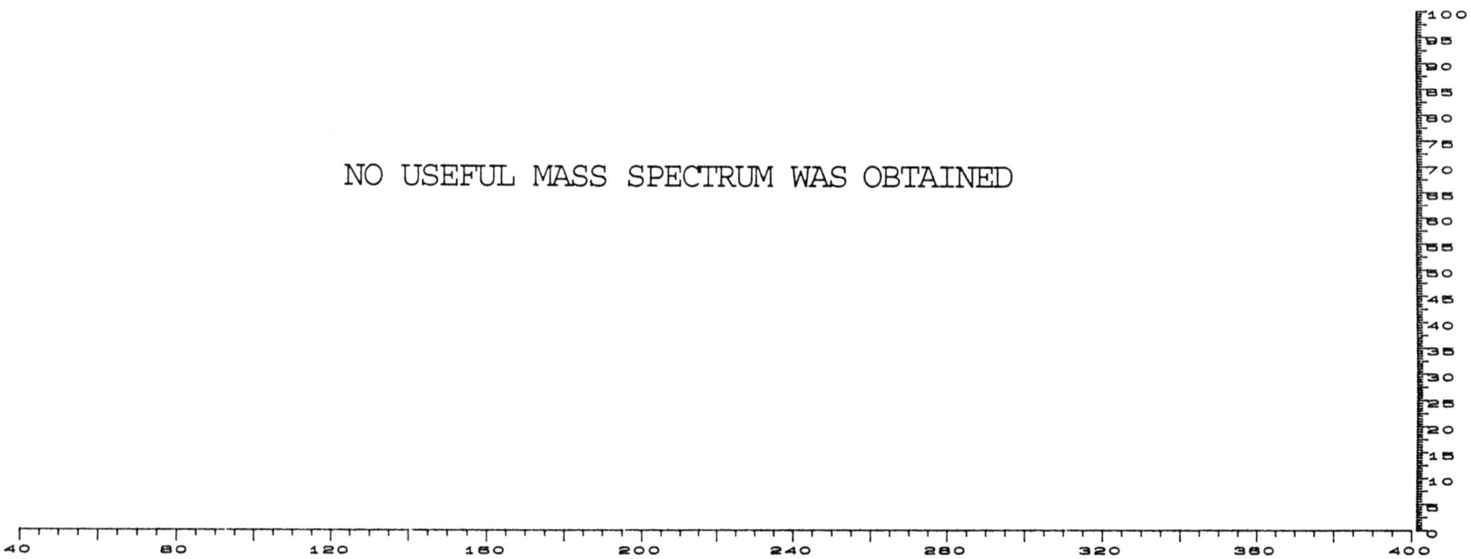

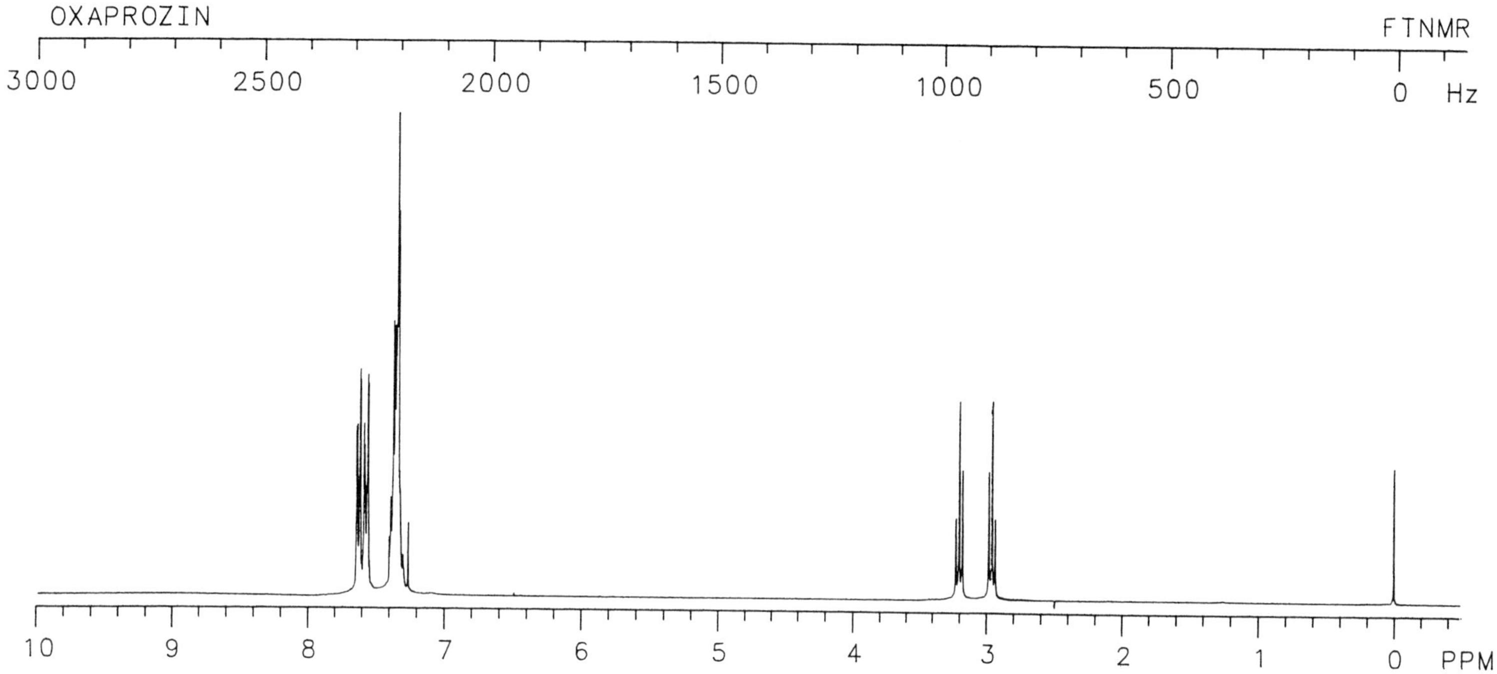

PEAK LISTING

#	HT	PPM
1	27	.00
2	16	2.93
3	40	2.95
4	26	2.98
5	26	3.17
6	40	3.19
7	16	3.22
8	21	7.31
9	25	7.32
10	78	7.33
11	98	7.34
12	54	7.34
13	55	7.35
14	56	7.36
15	22	7.37
16	20	7.38
17	43	7.54
18	45	7.55
19	22	7.55
20	30	7.57
21	35	7.57
22	43	7.60
23	46	7.60
24	15	7.62
25	34	7.62
26	34	7.63

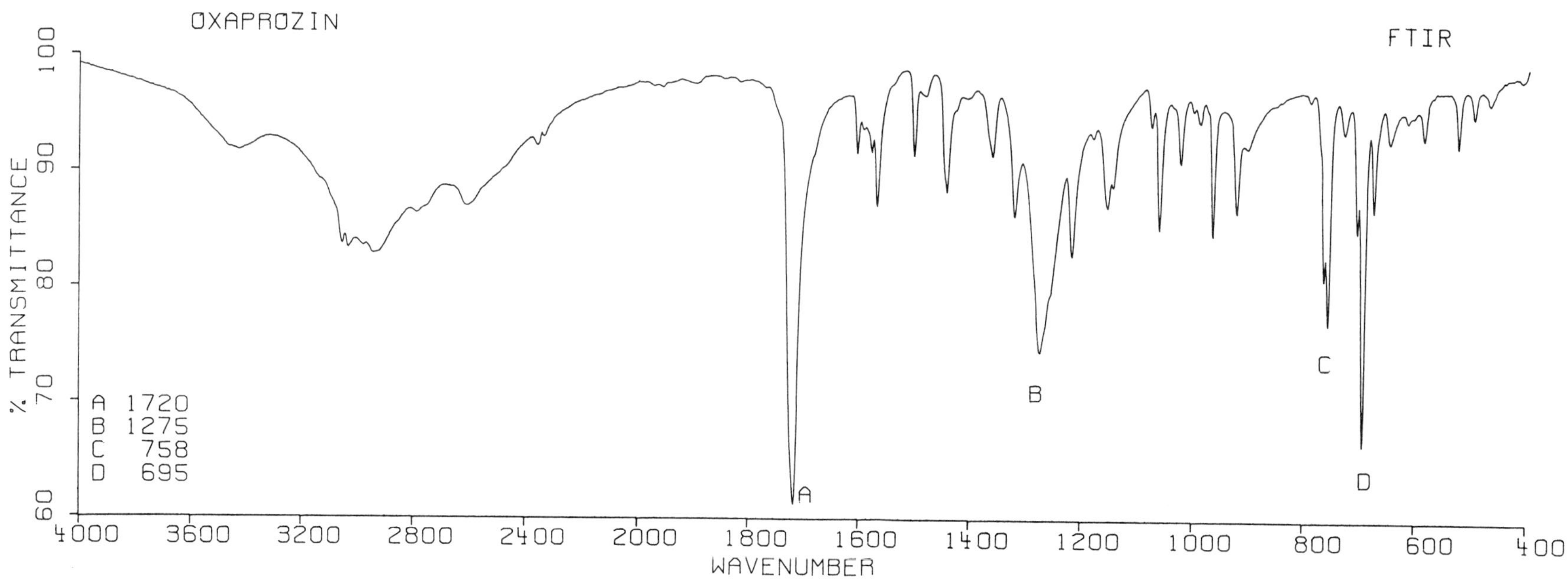

OXIBENDAZOLE

$C_{12}H_{15}N_3O_3$

Molecular Weight: 249.27 (249.11)

Synonyms: (5-Propoxy-1H-benzimidazol-2-yl)carbamic acid methyl ester
Trade Names: Anthelcide, Equitac
Use: Anthelmintic

HPLC: Ethanol: 3.0
GC:

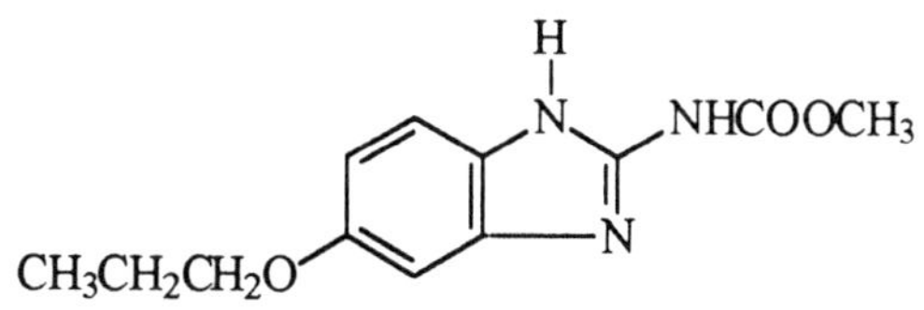

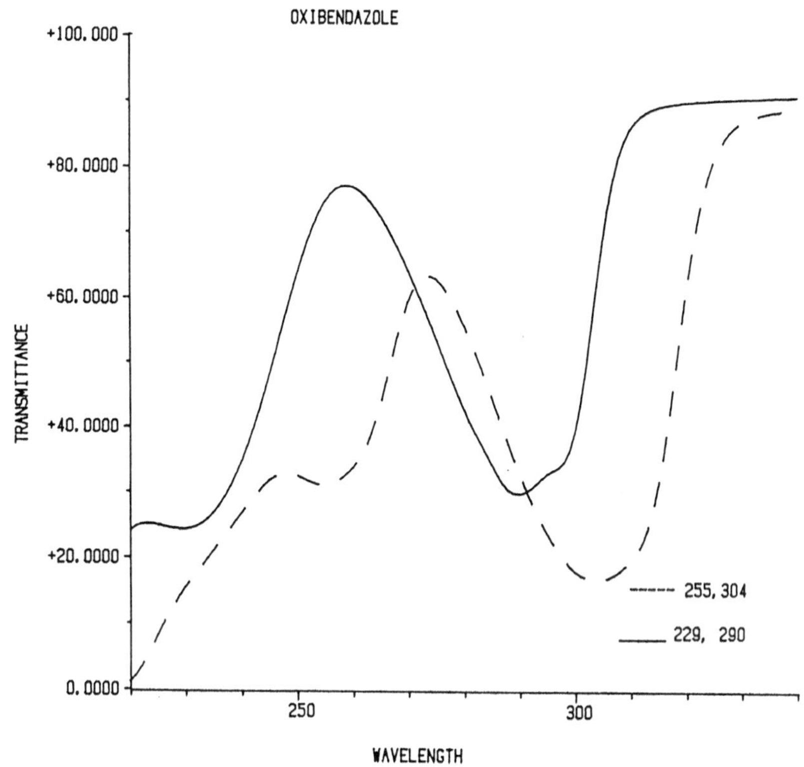

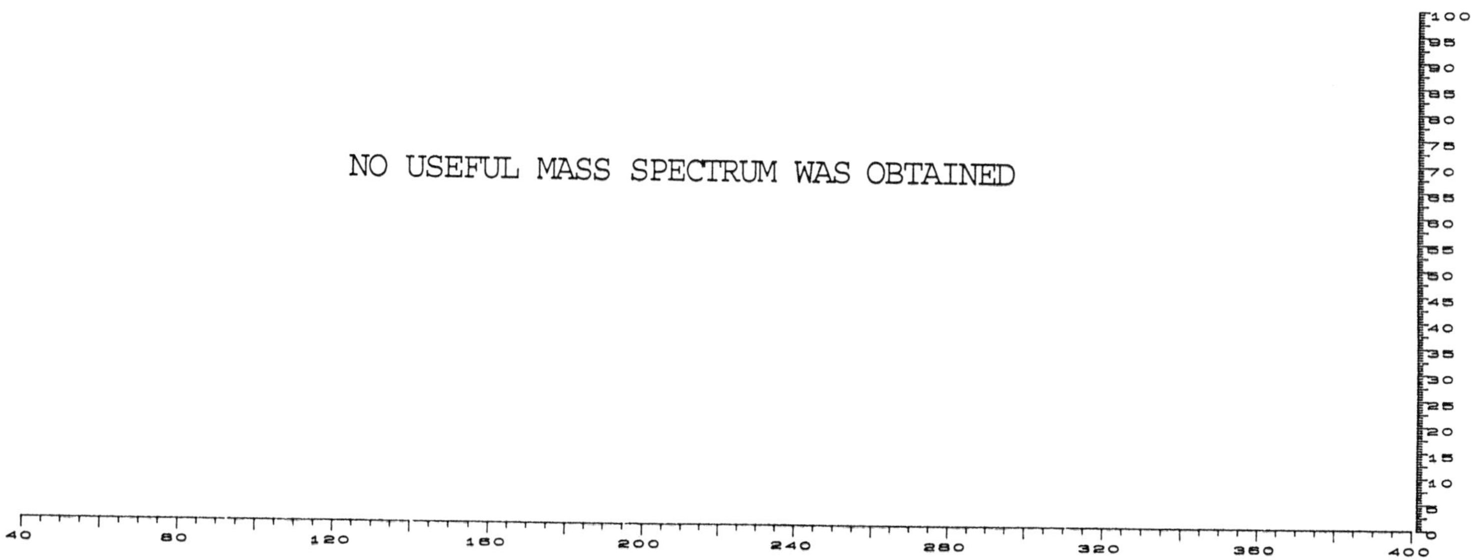

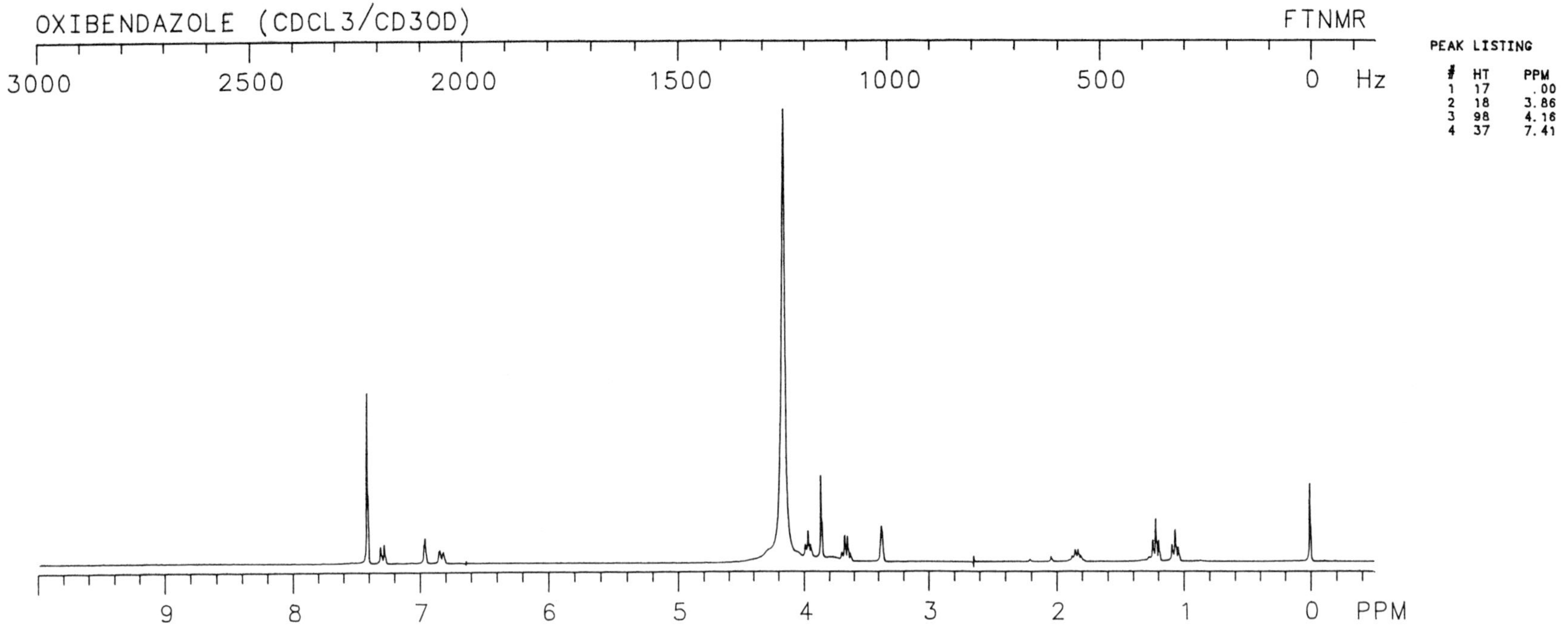
OXIBENDAZOLE (CDCL3/CD3OD)
FTNMR
3000 2500 2000 1500 1000 500 0 Hz
9 8 7 6 5 4 3 2 1 0 PPM
PEAK LISTING
HT PPM
1 17 .00
2 18 3.86
3 98 4.16
4 37 7.41

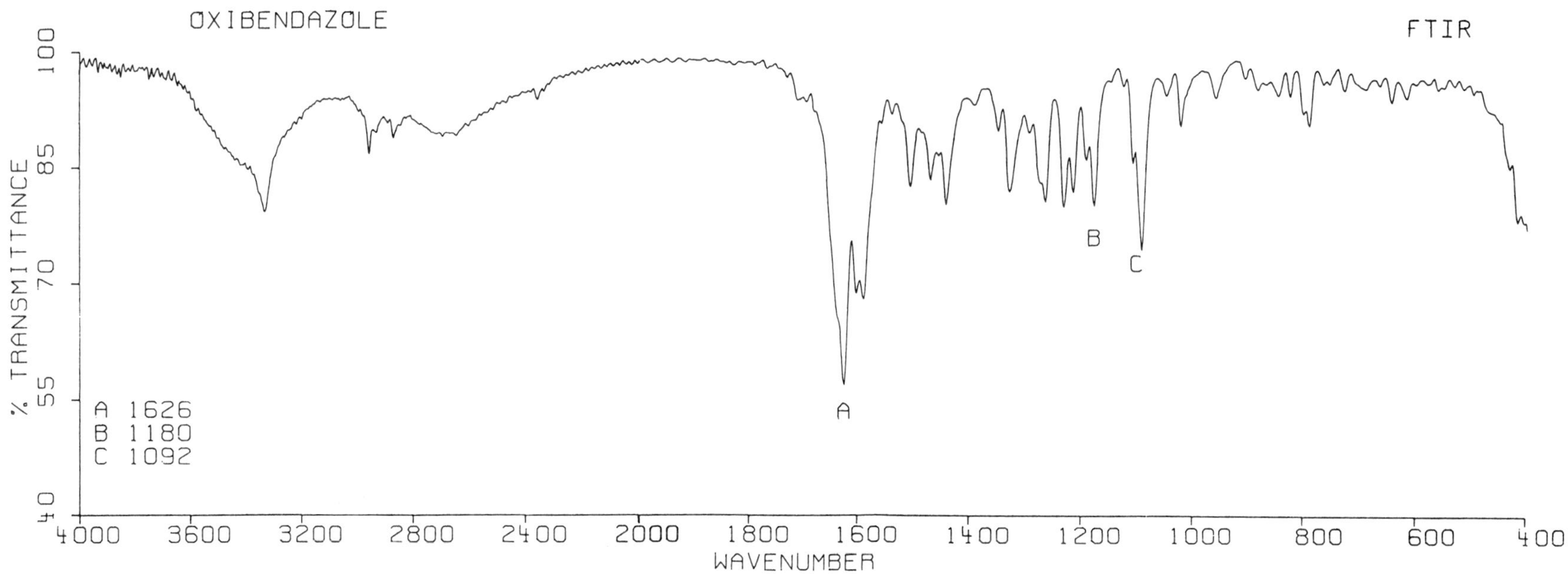
OXIBENDAZOLE
FTIR
% TRANSMITTANCE
100 85 70 55 40
A 1626
B 1180
C 1092
A
B
C
4000 3600 3200 2800 2400 2000 1800 1600 1400 1200 1000 800 600 400
WAVENUMBER

PAMIDRONATE DISODIUM

$C_3H_9NNa_2O_7P_2$

Molecular Weight: 279.03 (278.96)

Synonyms: (3-Amino-1-hydroxypropylidine)bisphosphonate disodium
Trade Names: Aminomux, Aredia
Use: Antipagetic

HPLC: Methanol: 3.4
GC:

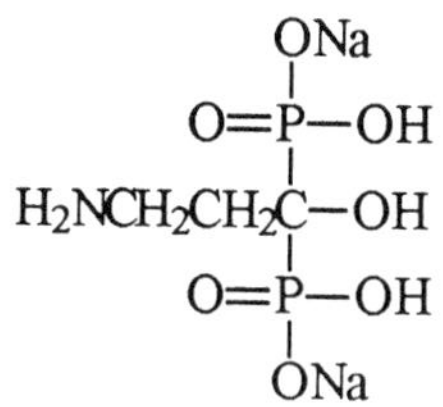

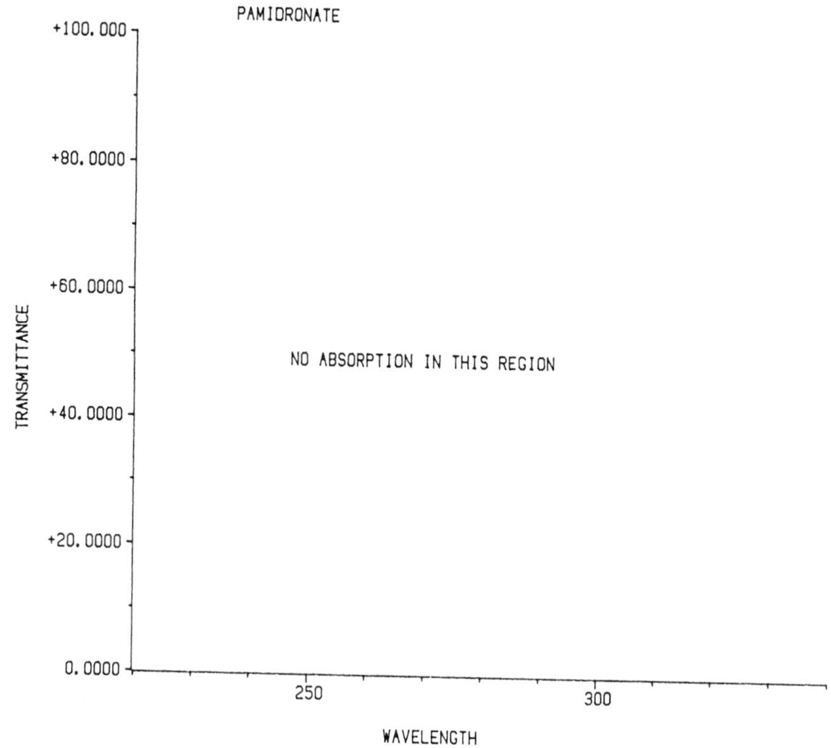

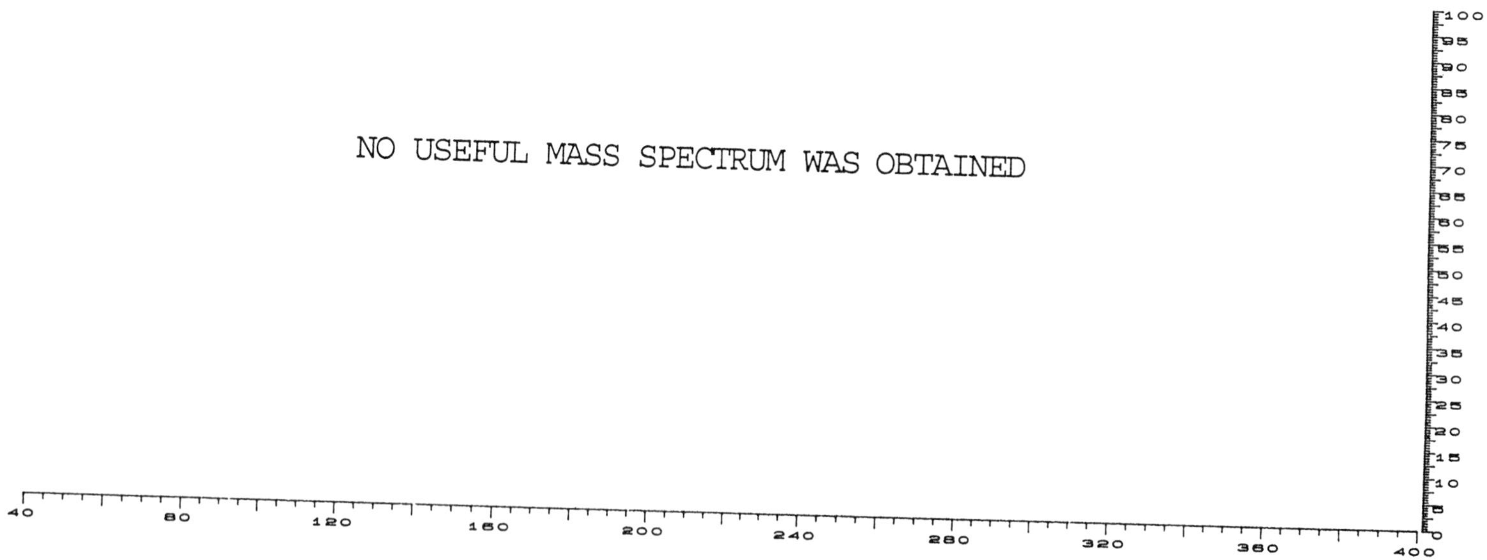

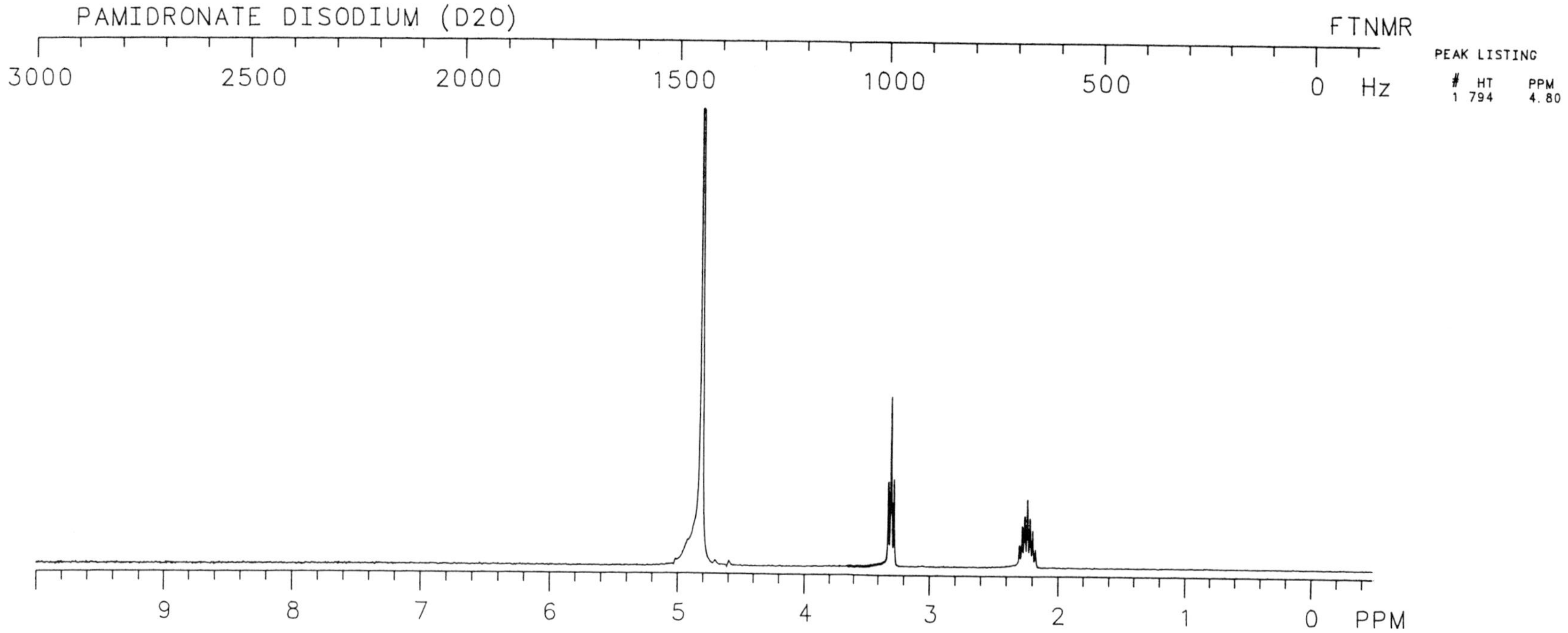
PAMIDRONATE DISODIUM (D2O)
FTNMR
PEAK LISTING
HT PPM
1 794 4.80
3000 2500 2000 1500 1000 500 0 Hz
9 8 7 6 5 4 3 2 1 0 PPM

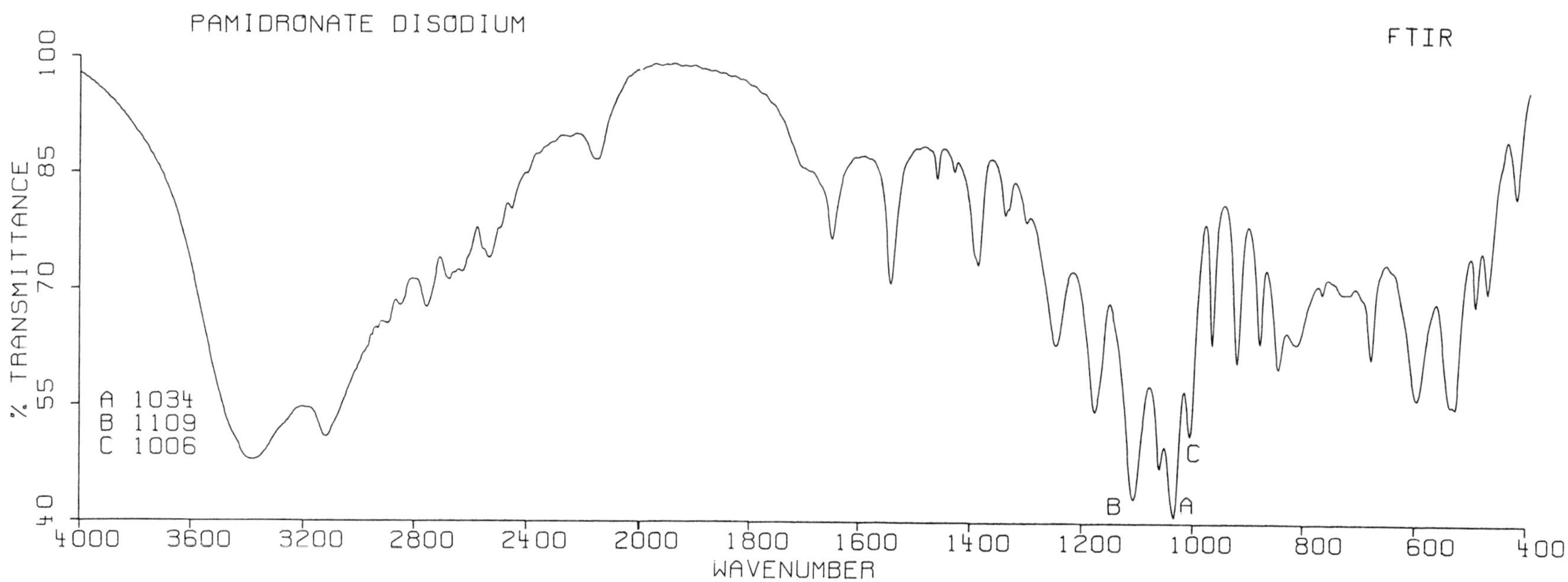
PAMIDRONATE DISODIUM
FTIR
% TRANSMITTANCE
100 85 70 55 40
A 1034
B 1109
C 1006
B A C
4000 3600 3200 2800 2400 2000 1800 1600 1400 1200 1000 800 600 400
WAVENUMBER

PAROXETINE

$C_{19}H_{20}FNO_3$

Molecular Weight: 329.37 (329.14)

Synonyms: 3-[(1,3-Benzodioxol-5-yloxy)methyl]-4-(4-fluorophenyl)piperidine
Trade Names: Aropax, Seroxat, Paxil
Use: Antidepressant

HPLC:
GC: 2492; 250°

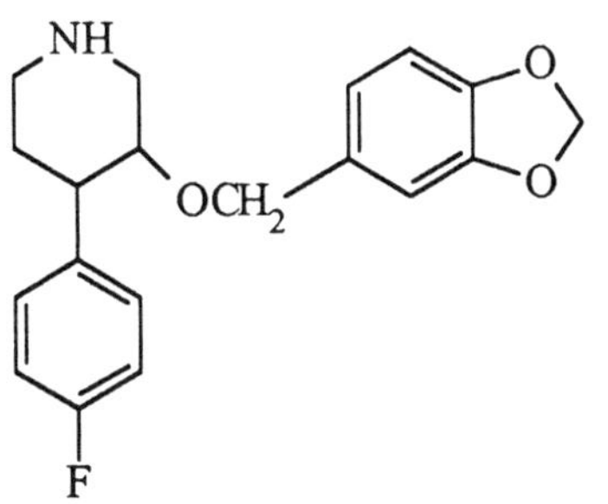

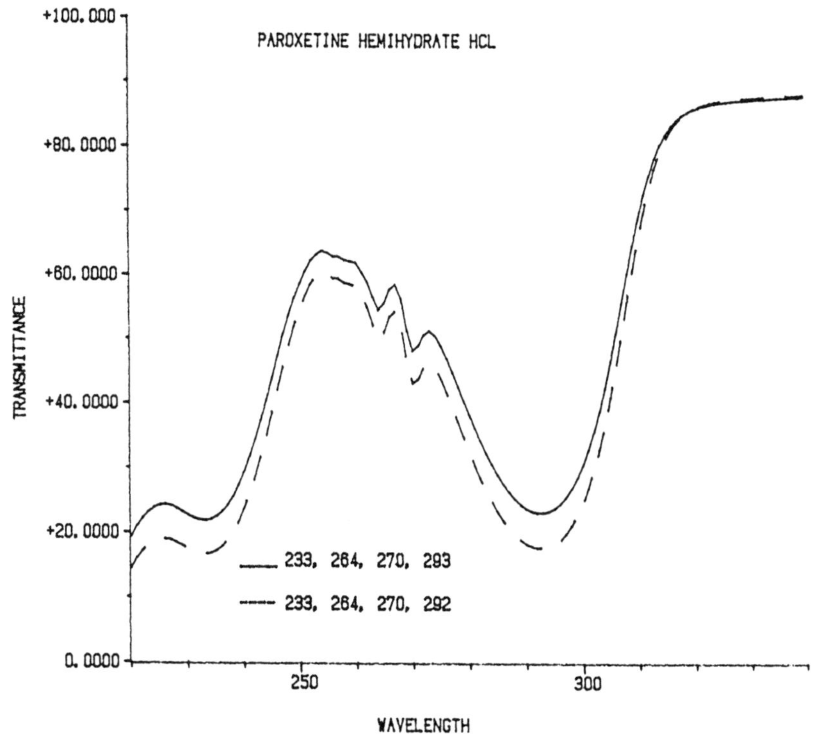

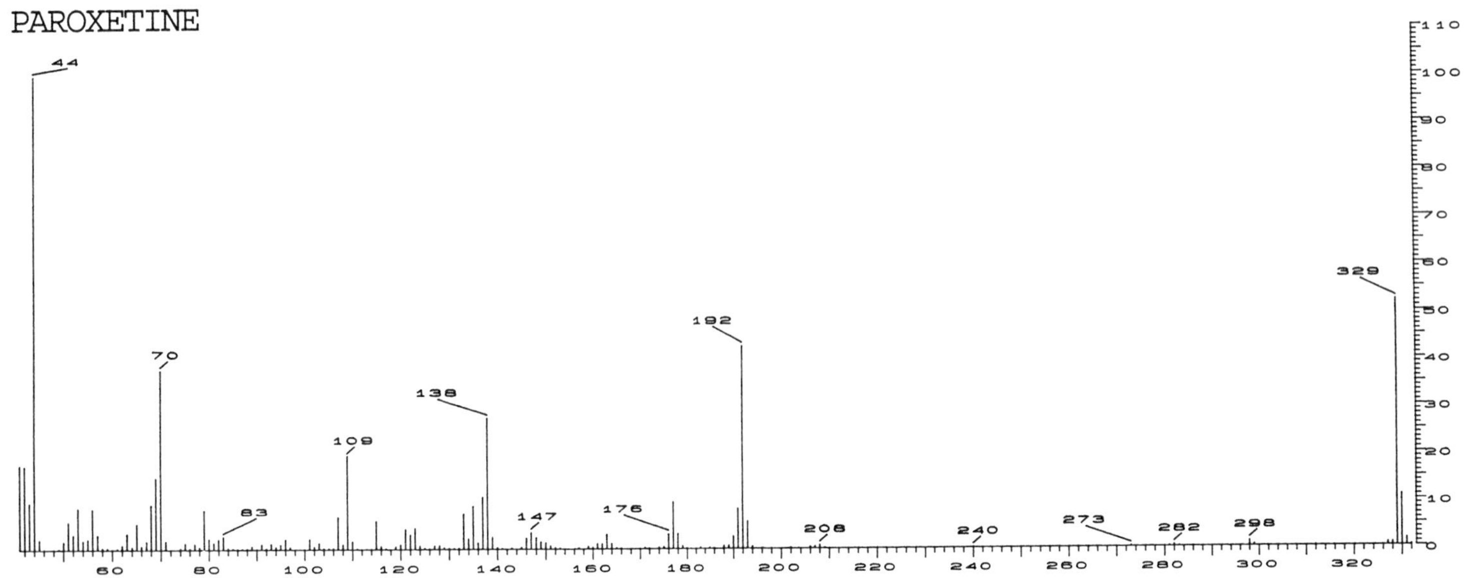

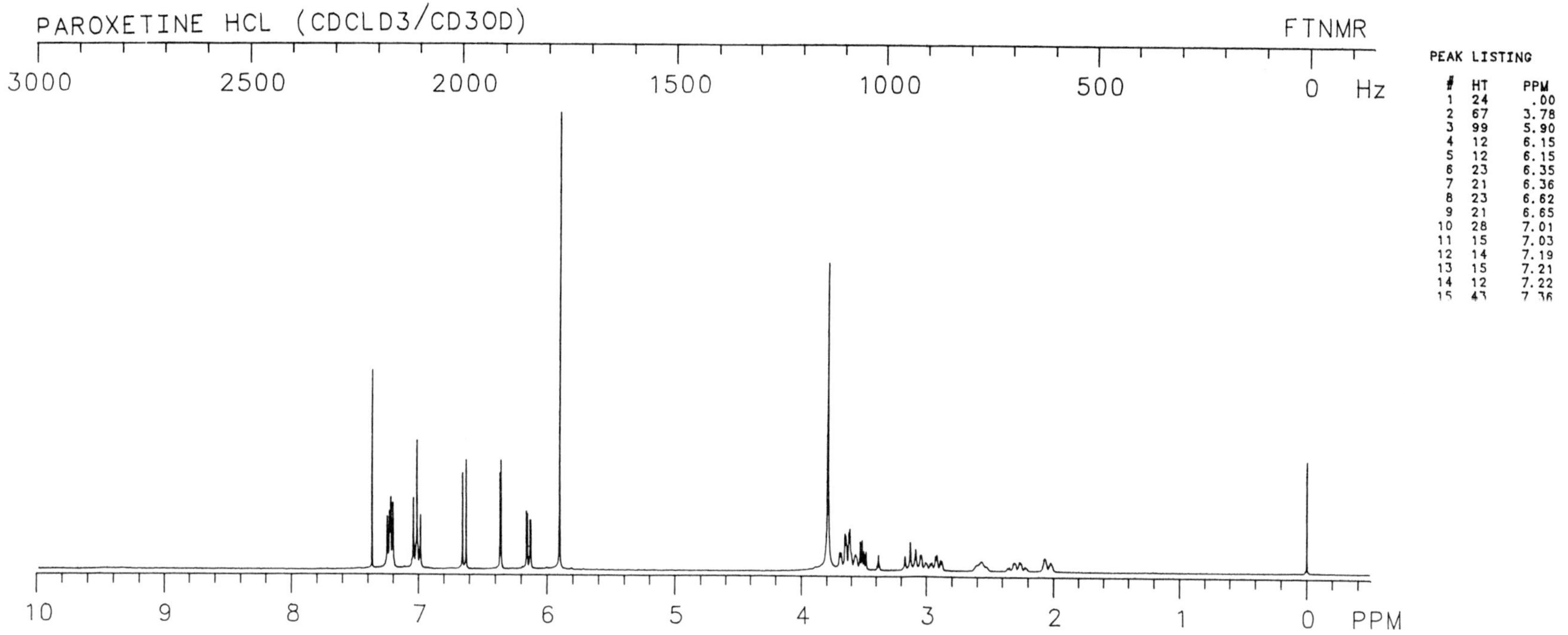
PAROXETINE HCL (CDCLD3/CD3OD)
FTNMR
3000
2500
2000
1500
1000
500
0 Hz
10
9
8
7
6
5
4
3
2
1
0 PPM
PEAK LISTING
HT PPM
1 24 .00
2 67 3.78
3 99 5.90
4 12 6.15
5 12 6.15
6 23 6.35
7 21 6.36
8 23 6.62
9 21 6.65
10 28 7.01
11 15 7.03
12 14 7.19
13 15 7.21
14 12 7.22

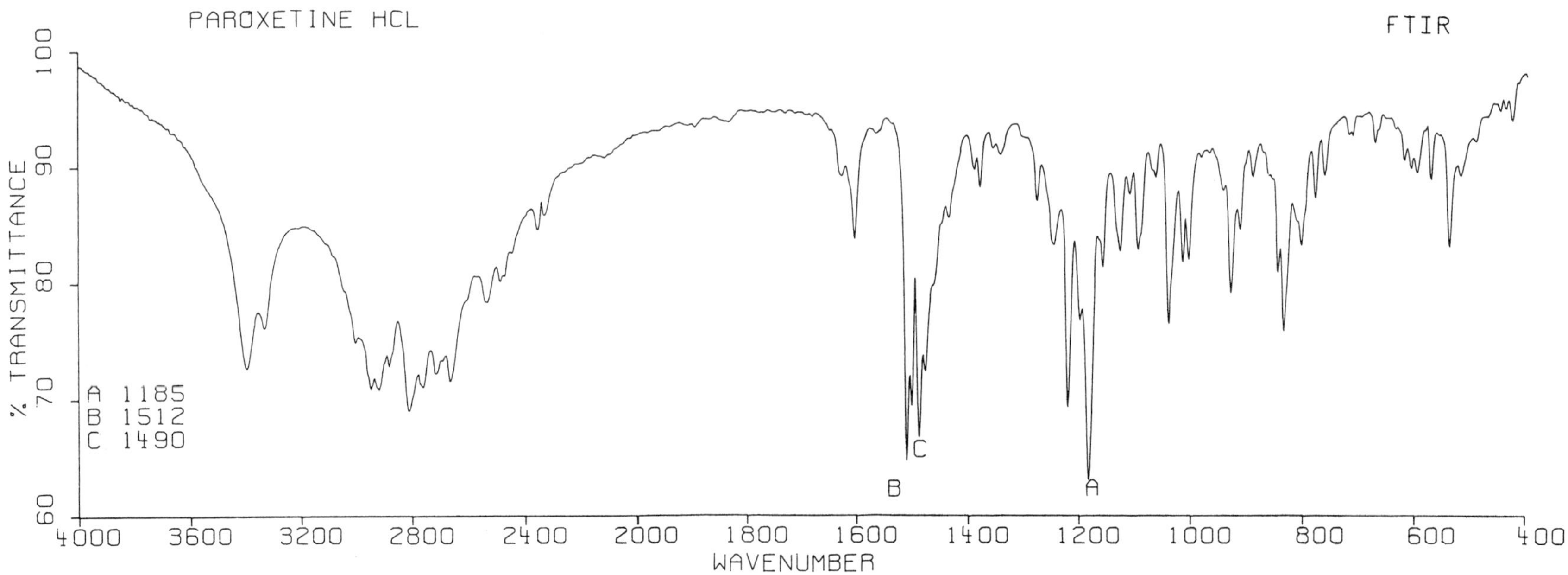
PAROXETINE HCL
FTIR
% TRANSMITTANCE
100
90
80
70
60
A 1185
B 1512
C 1490
B
C
A
4000
3600
3200
2800
2400
2000
1800
1600
1400
1200
1000
800
600
400
WAVENUMBER

PENTAMIDINE ISETHIONATE

$C_{23}H_{36}N_4O_{10}S_2$

Molecular Weight: 592.68 (592.19)

Synonyms: 4,4'-[1,5-Pentanediylbis(oxy)-benzenecarboximidamide isethionate
Trade Names: Aeropent, Benambax, NebuPent, Pentacarinat, Pentam
Use: Antiprotozoal

HPLC:
GC:

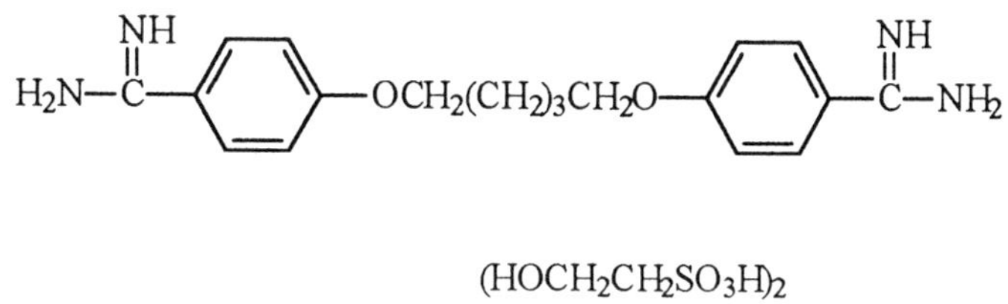

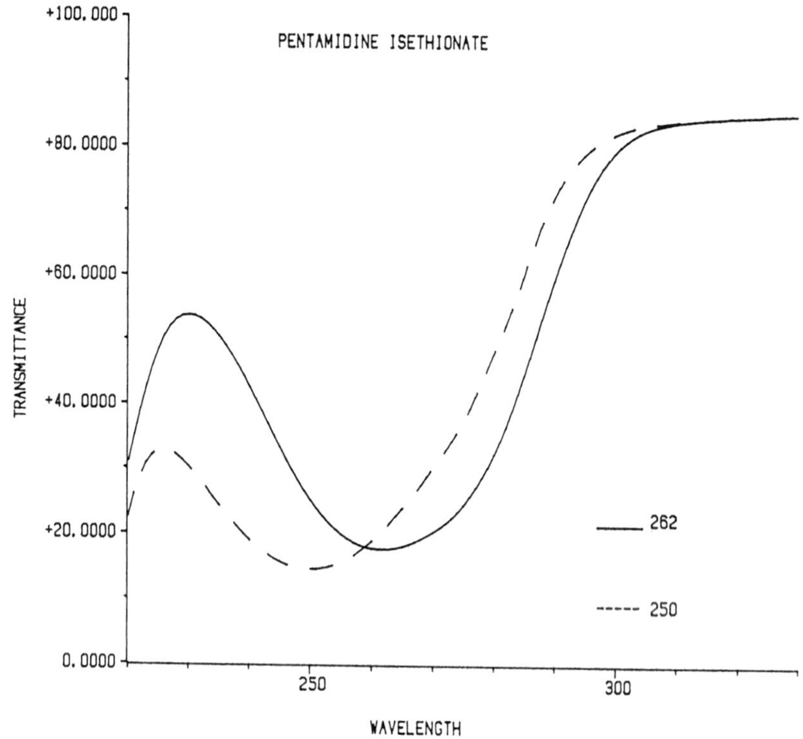

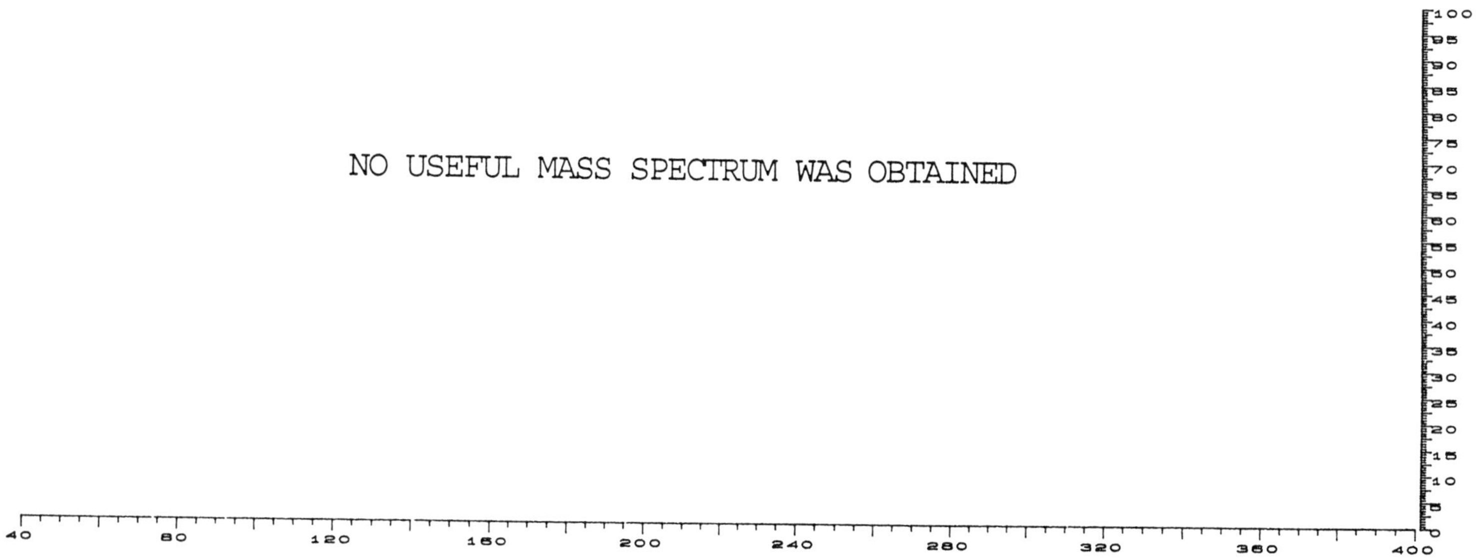

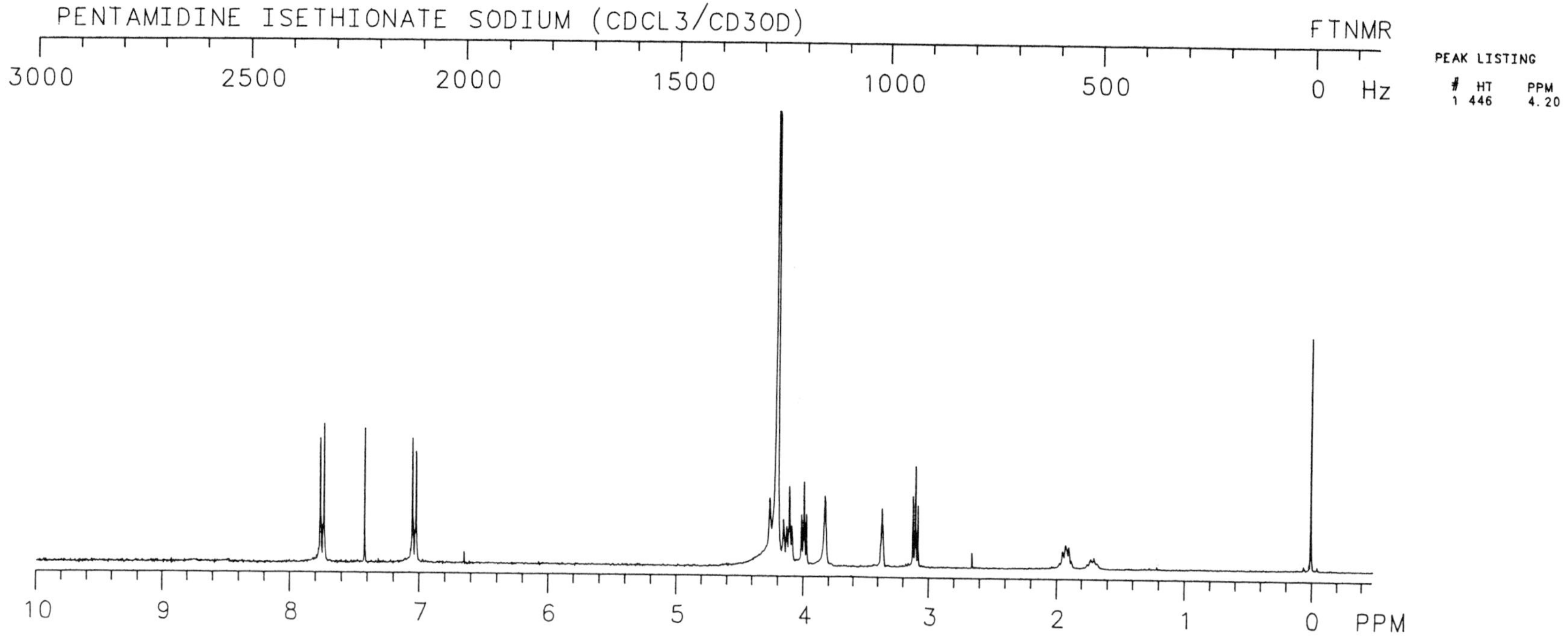
PENTAMIDINE ISETHIONATE SODIUM (CDCL3/CD3OD)
FTNMR
3000 2500 2000 1500 1000 500 0 Hz
PEAK LISTING
HT PPM
1 446 4.20
10 9 8 7 6 5 4 3 2 1 0 PPM

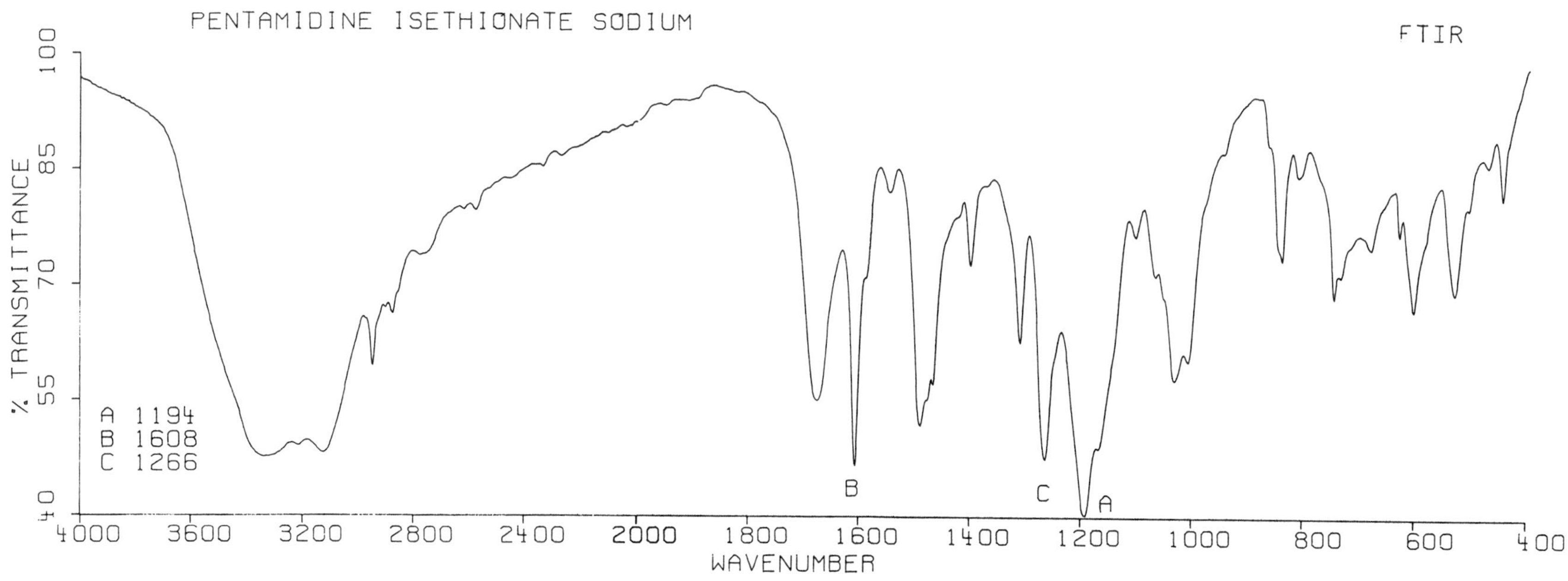
PENTAMIDINE ISETHIONATE SODIUM
FTIR
% TRANSMITTANCE
100 85 70 55 40
A 1194
B 1608
C 1266
B
C
A
4000 3600 3200 2800 2400 2000 1800 1600 1400 1200 1000 800 600 400
WAVENUMBER

PERGOLIDE MESYLATE

$C_{20}H_{30}N_2O_3S_2$

Molecular Weight: 410.59 (410.17)

Synonyms: 8-[(Methylthio)methyl]-6-propylergoline mesylate
Trade Names: Permax
Use: Antiparkinsonian

HPLC: Methanol: 3.6
GC:

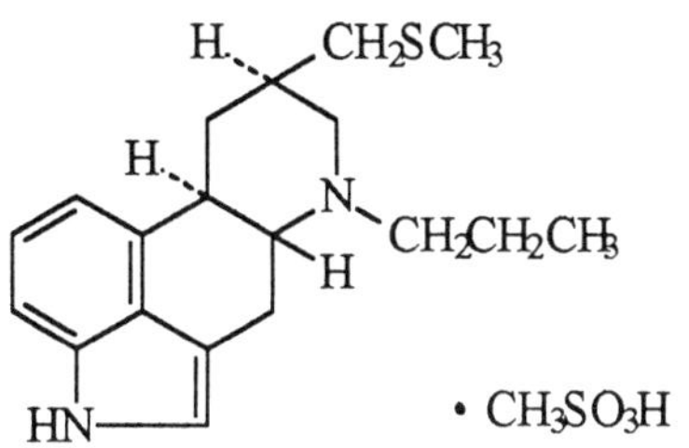

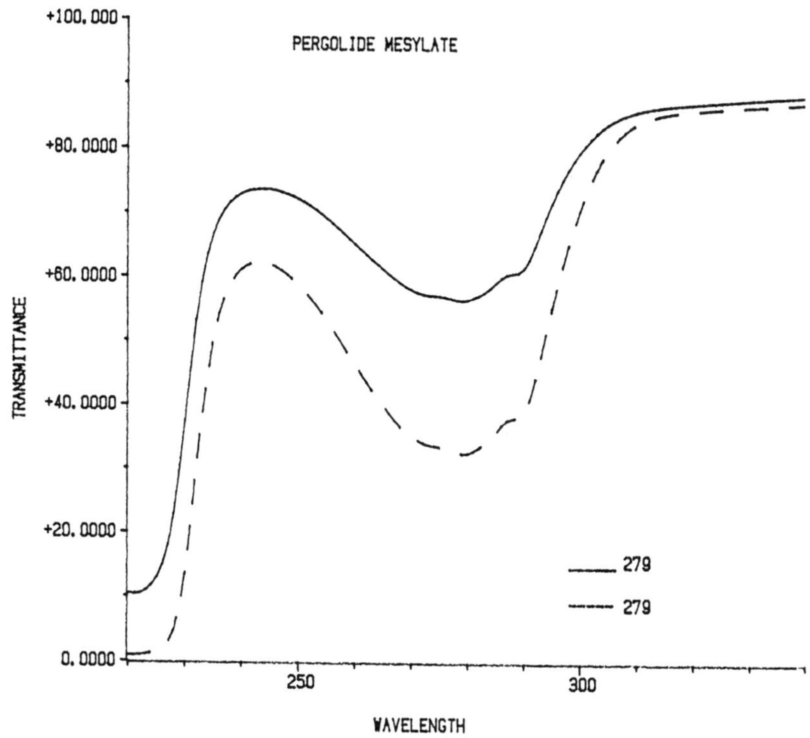

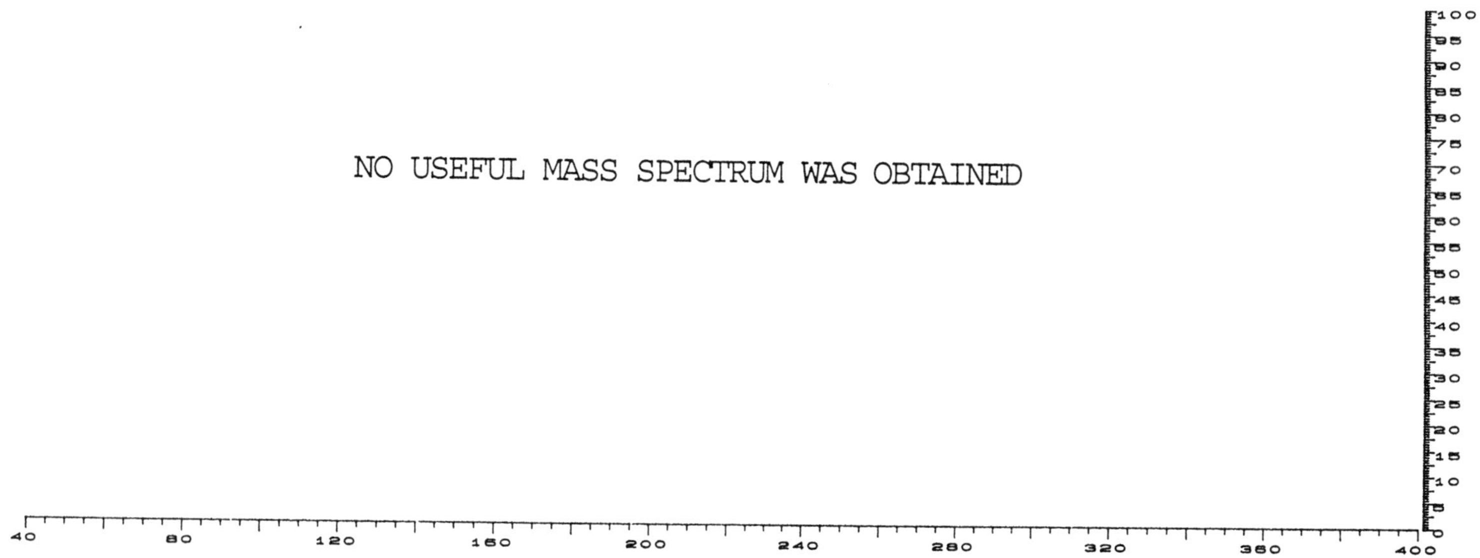

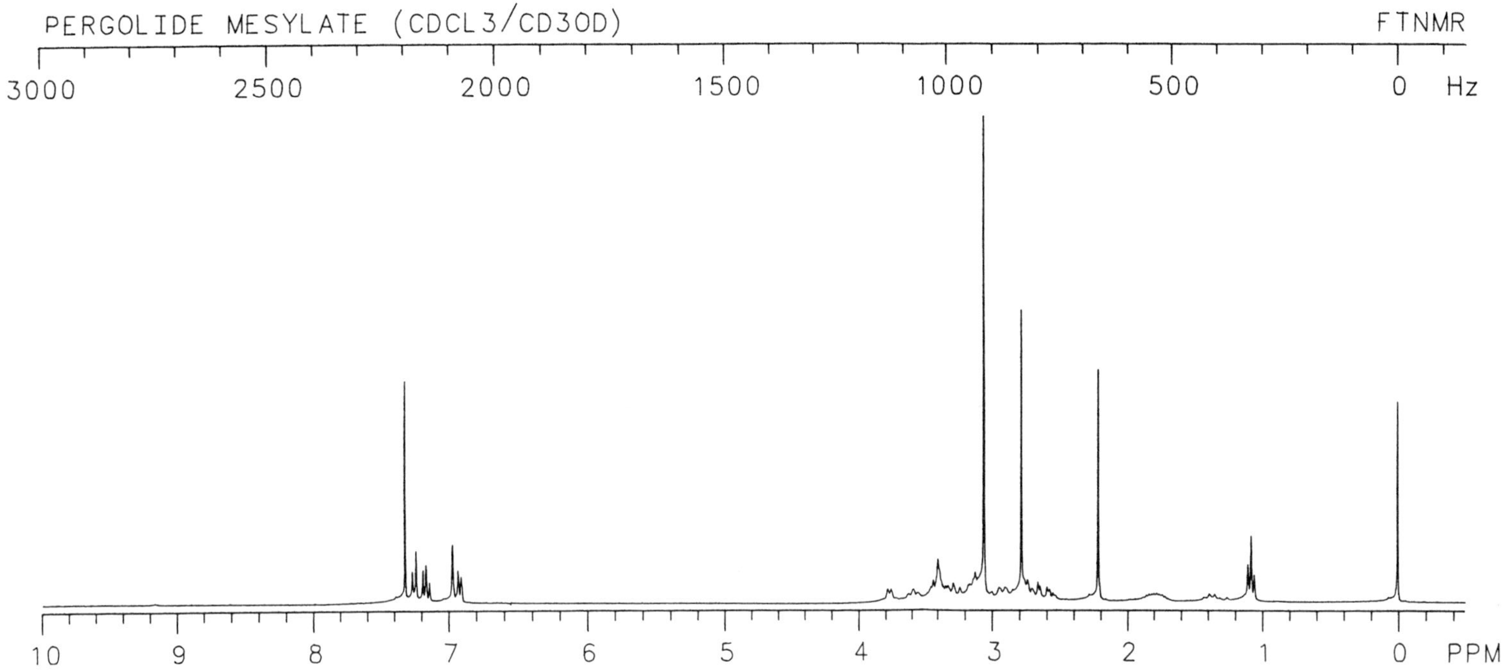

PEAK LISTING

#	HT	PPM
1	41	.00
2	13	1.08
3	7	1.10
4	47	2.21
5	59	2.78
6	99	3.05
7	7	3.38
8	8	3.39
9	12	6.96
10	7	7.16
11	10	7.23
12	45	7.31

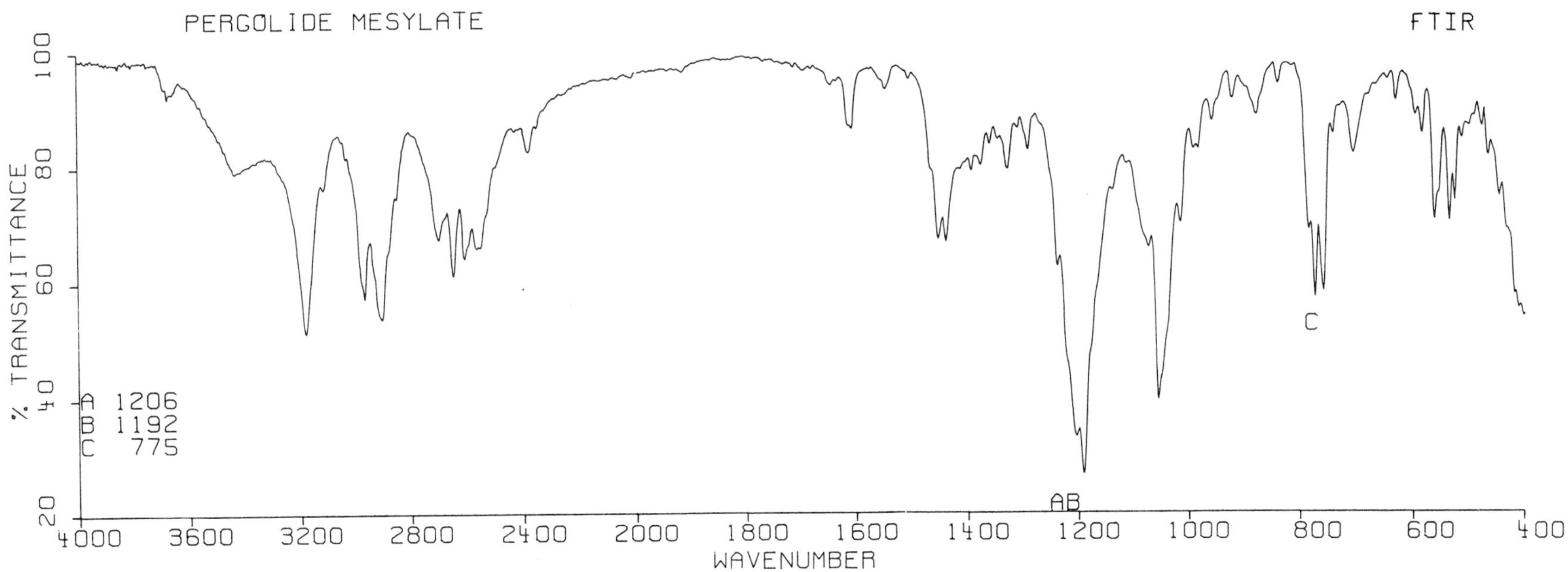

PHOSPHONOFORMIC ACID TRISODIUM

CO_5PNa_3

Molecular Weight: 191.95 (191.92)

Synonyms: Dihydroxyphosphinecarboxylic acid oxide trisodium salt; foscarnet; trisodium phosphonoformate
Trade Names: Foscavir
Use: Antiviral

HPLC:
GC:

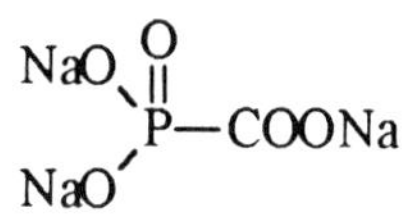

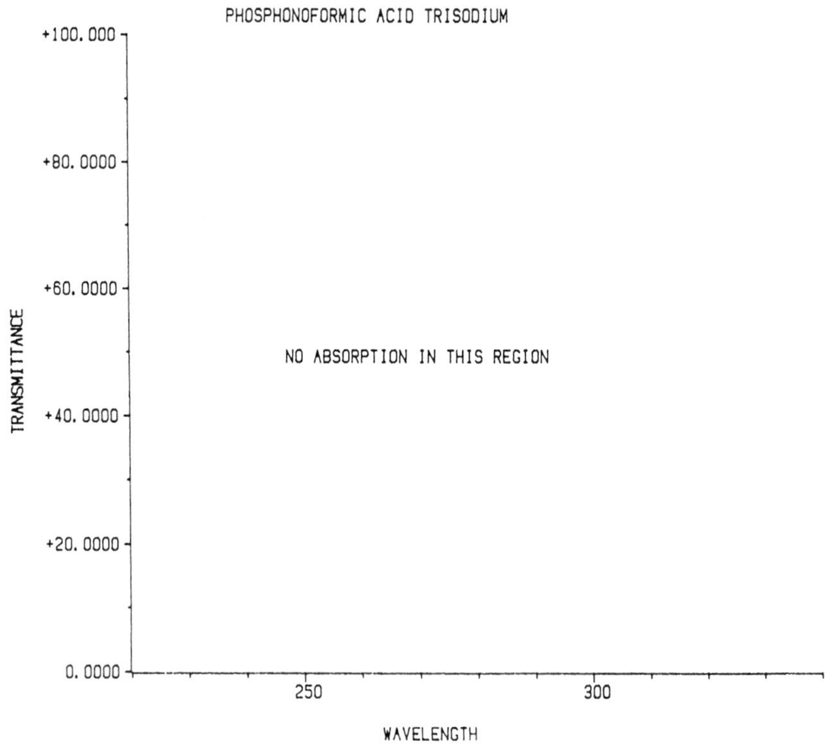

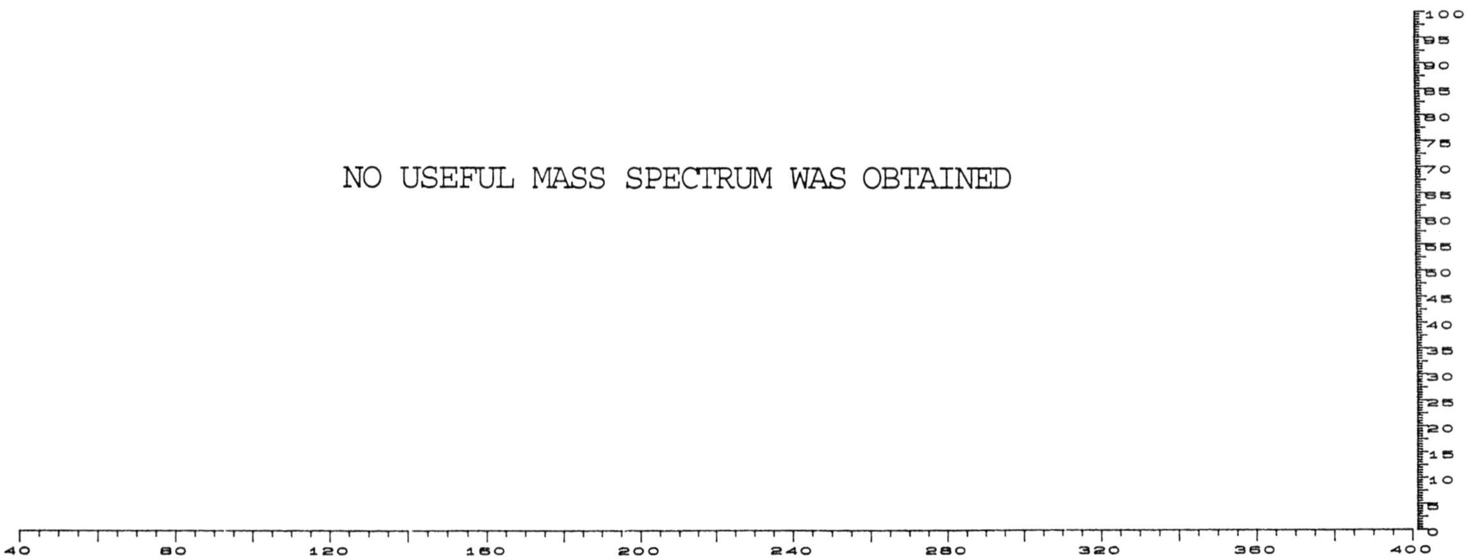

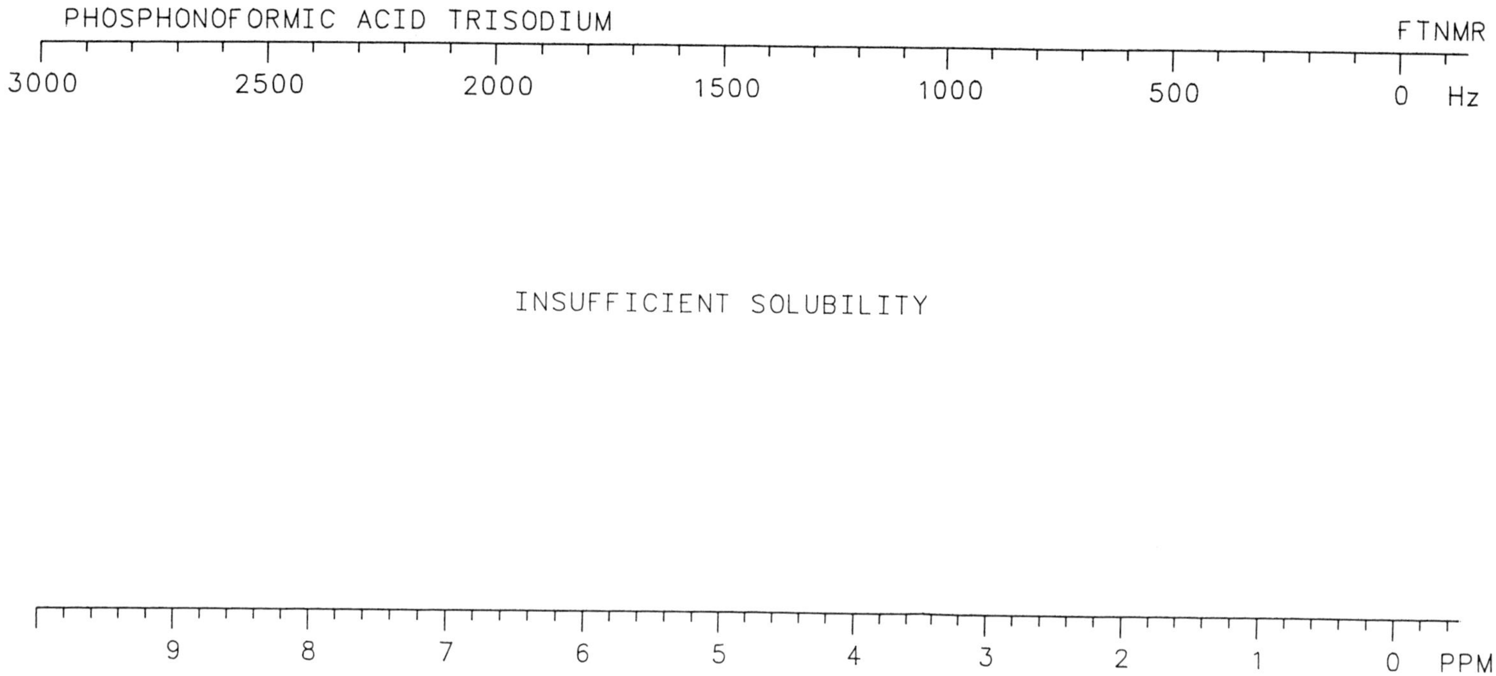
PHOSPHONOFORMIC ACID TRISODIUM
FTNMR
3000
2500
2000
1500
1000
500
0 Hz
INSUFFICIENT SOLUBILITY
9
8
7
6
5
4
3
2
1
0 PPM

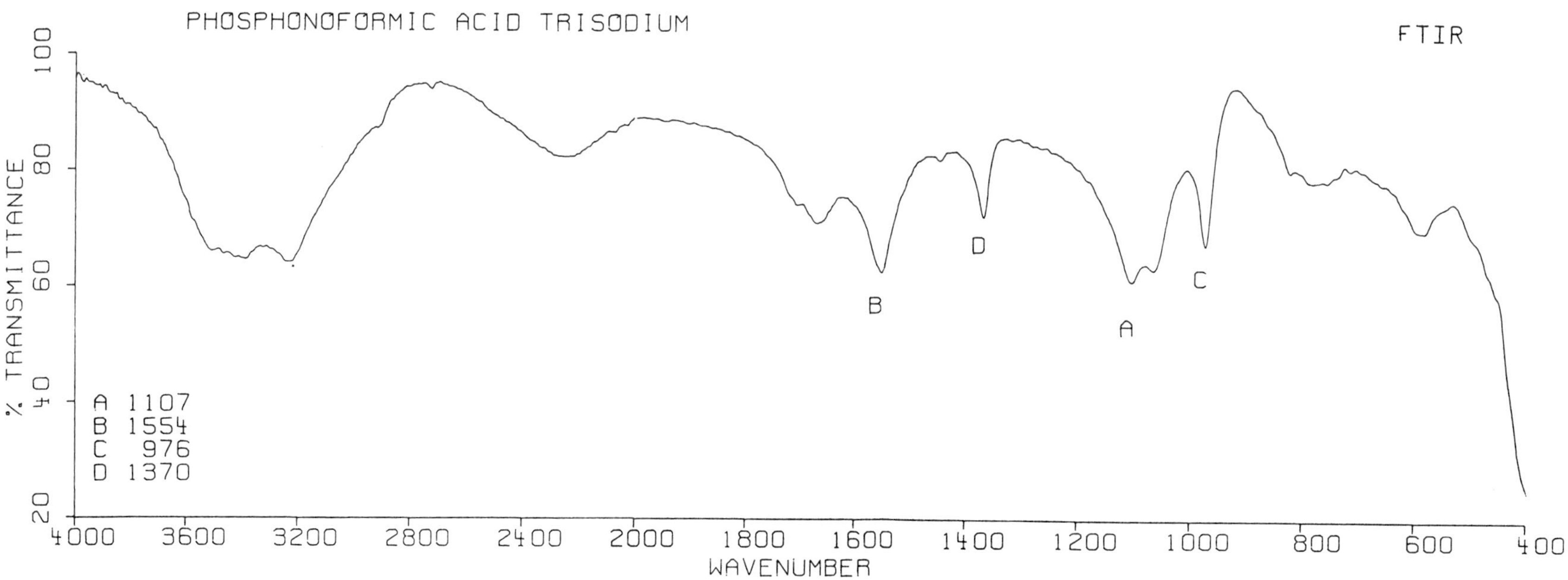
PHOSPHONOFORMIC ACID TRISODIUM
FTIR
% TRANSMITTANCE
100
80
60
40
20
B
D
A
C
A 1107
B 1554
C 976
D 1370
4000
3600
3200
2800
2400
2000
1800
1600
1400
1200
1000
800
600
400
WAVENUMBER

PIRBUTEROL ACETATE

$C_{14}H_{24}N_2O_5$

Molecular Weight: 300.35 (300.17)

Synonyms: α^6-[[(1,1-Dimethylethyl)amino]methyl]-3-hydroxy-2,6-pyridinedimethanol acetate
Trade Names: Maxair
Use: Bronchodilator

HPLC:
GC:

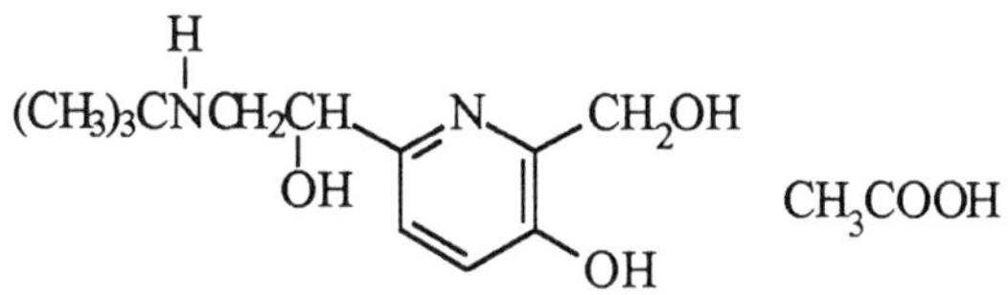

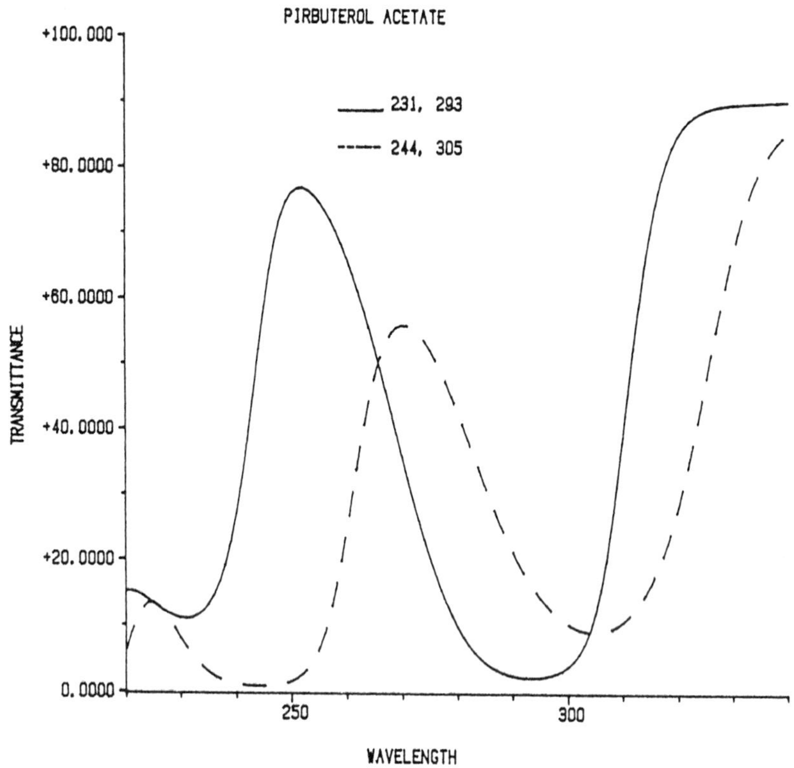

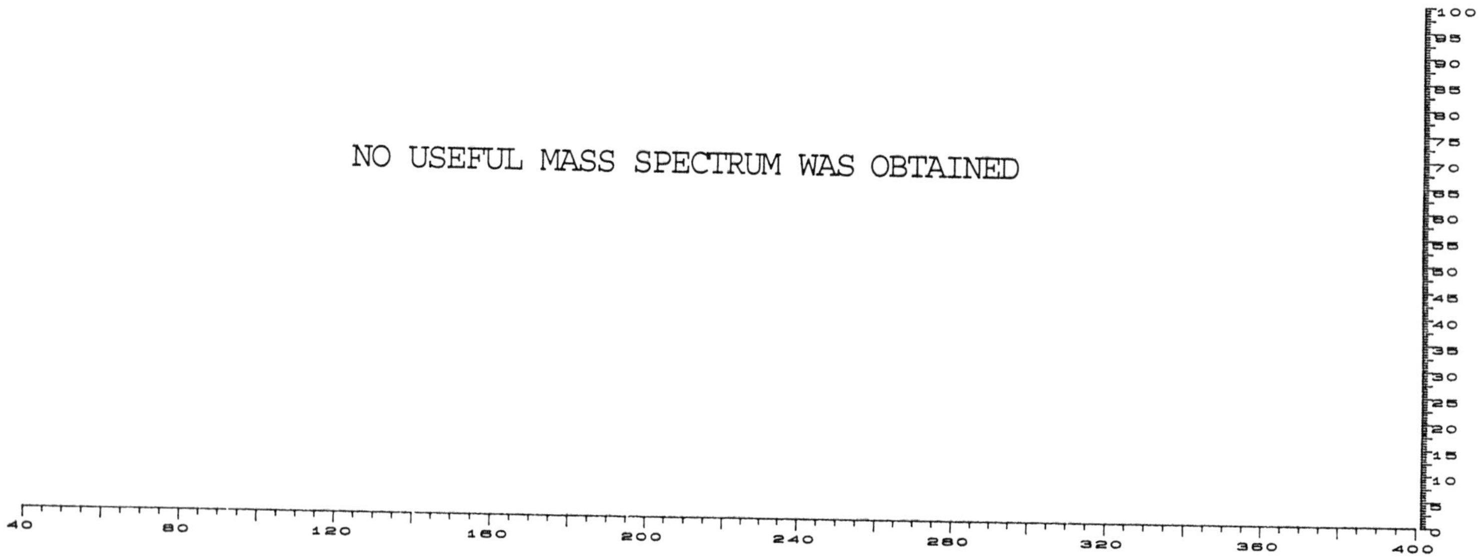

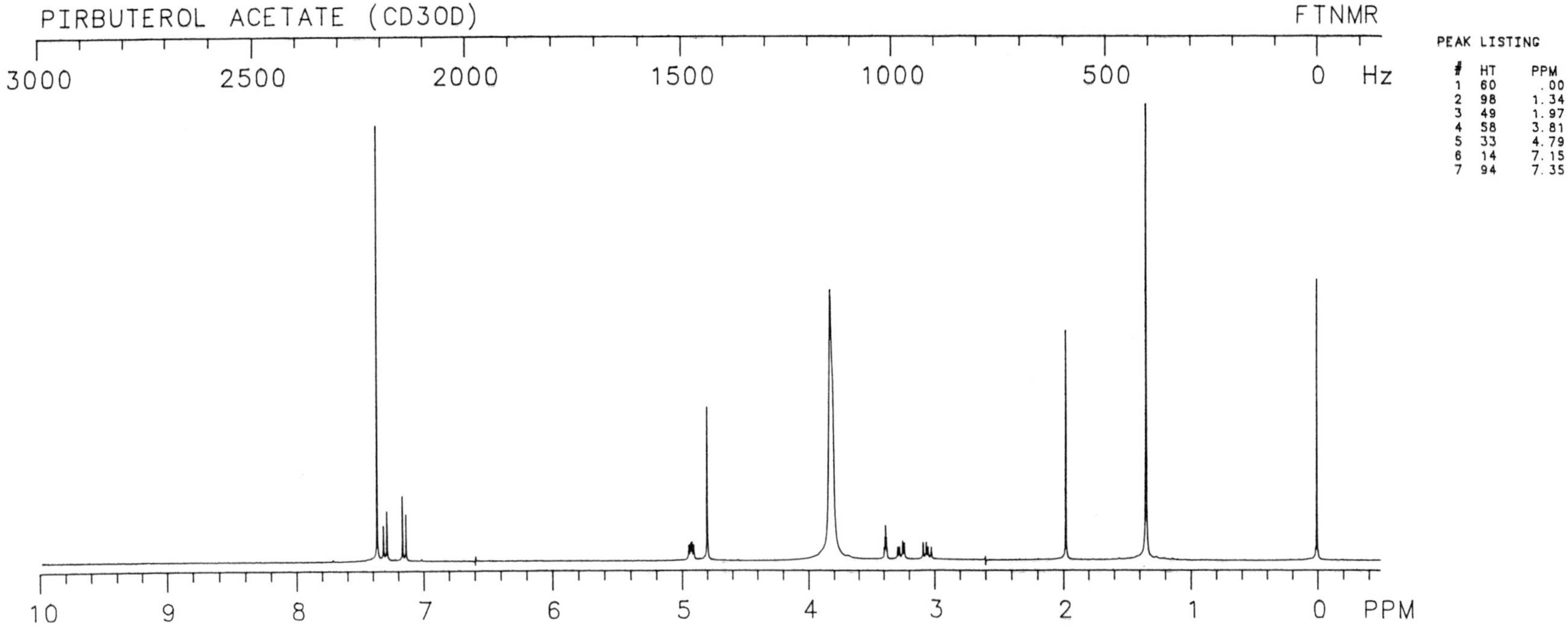
PIRBUTEROL ACETATE (CD3OD)
FTNMR
3000 2500 2000 1500 1000 500 0 Hz
PEAK LISTING
HT PPM
1 60 .00
2 98 1.34
3 49 1.97
4 58 3.81
5 33 4.79
6 14 7.15
7 94 7.35
10 9 8 7 6 5 4 3 2 1 0 PPM

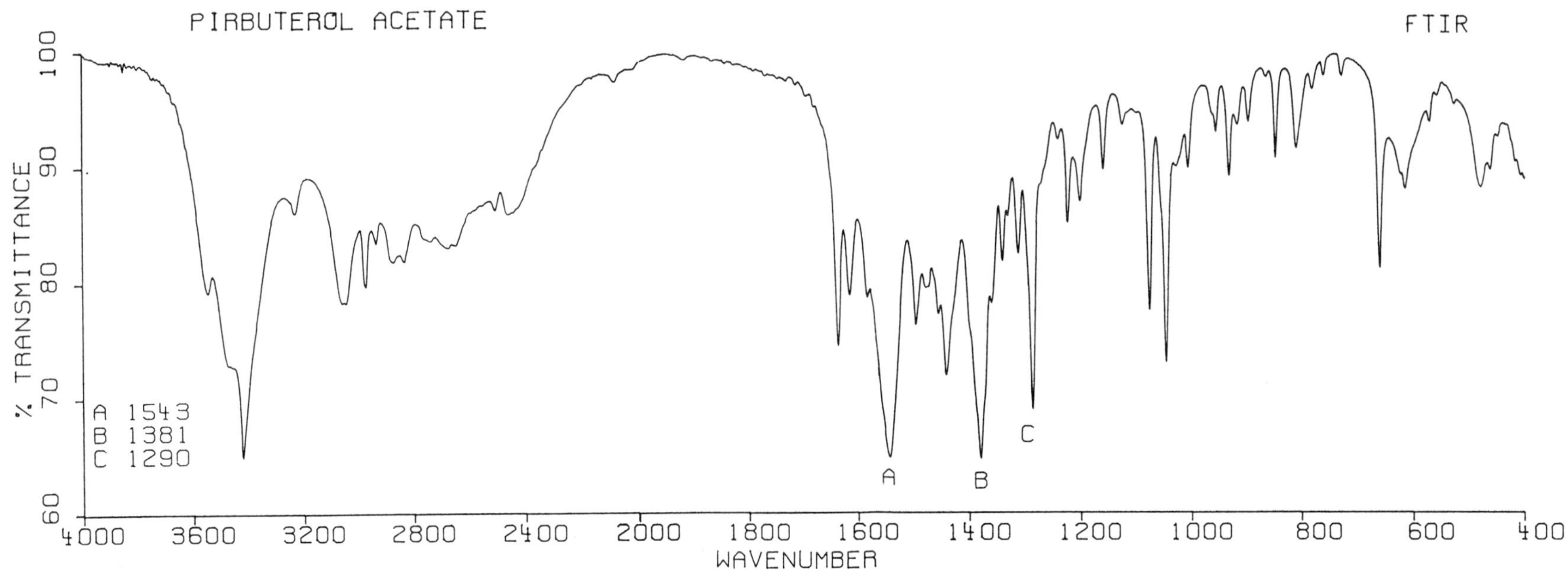
PIRBUTEROL ACETATE
FTIR
% TRANSMITTANCE
100 90 80 70 60
A 1543
B 1381
C 1290
A
B
C
4000 3600 3200 2800 2400 2000 1800 1600 1400 1200 1000 800 600 400
WAVENUMBER

PIRENPERONE

$C_{23}H_{24}FN_3O_2$

Molecular Weight: 393.46 (393.19)

Synonyms: 3-[2-[4-(4-Fluorobenzoyl)-1-piperidinyl]ethyl]-2-methyl-4H-pyridol[1,2-a]pyrimidin-4-one
Trade Names:
Use: Serotonin antagonist

HPLC:
GC: 3418; 280°

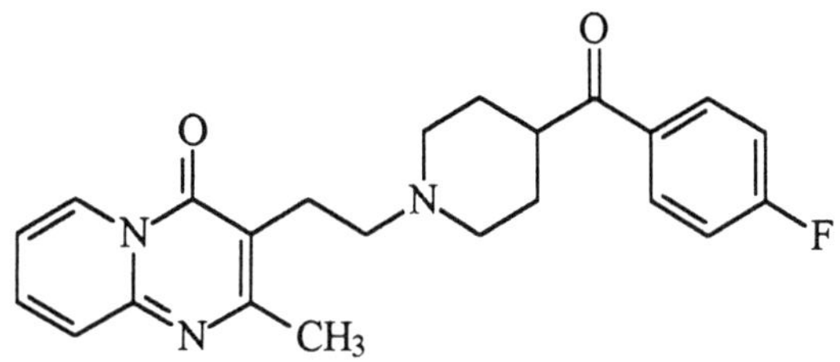

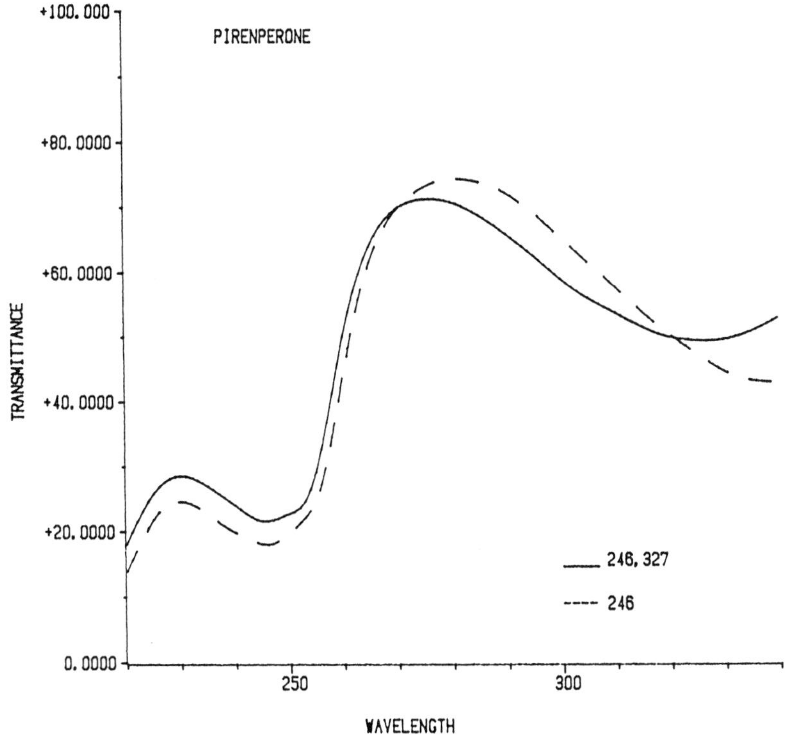

PIRENPERONE

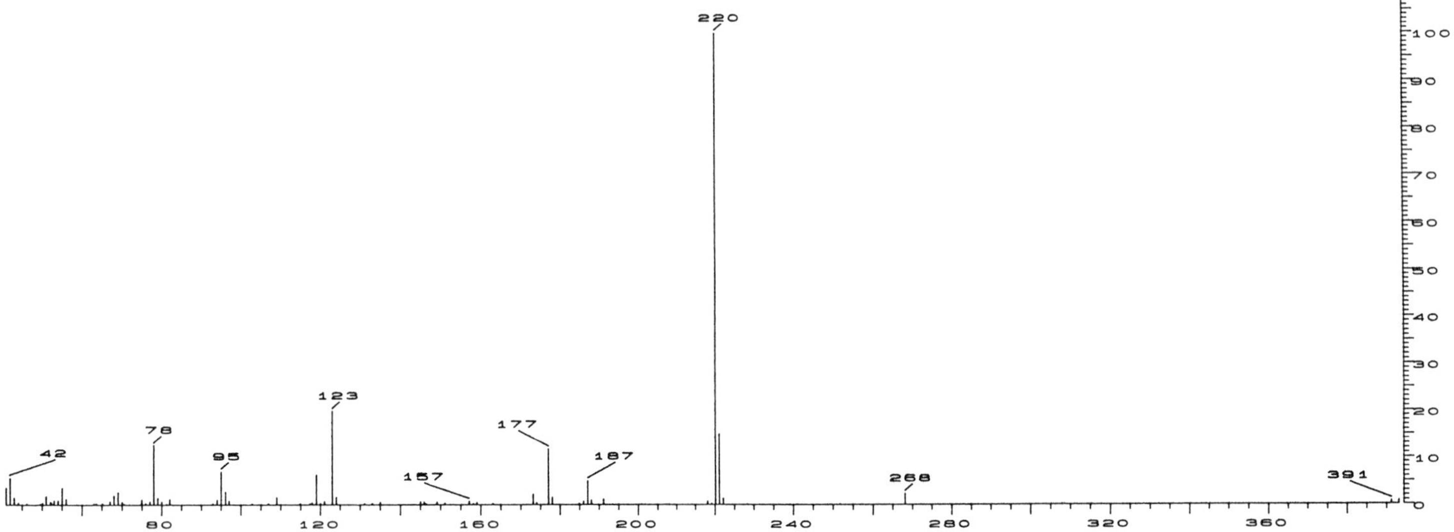

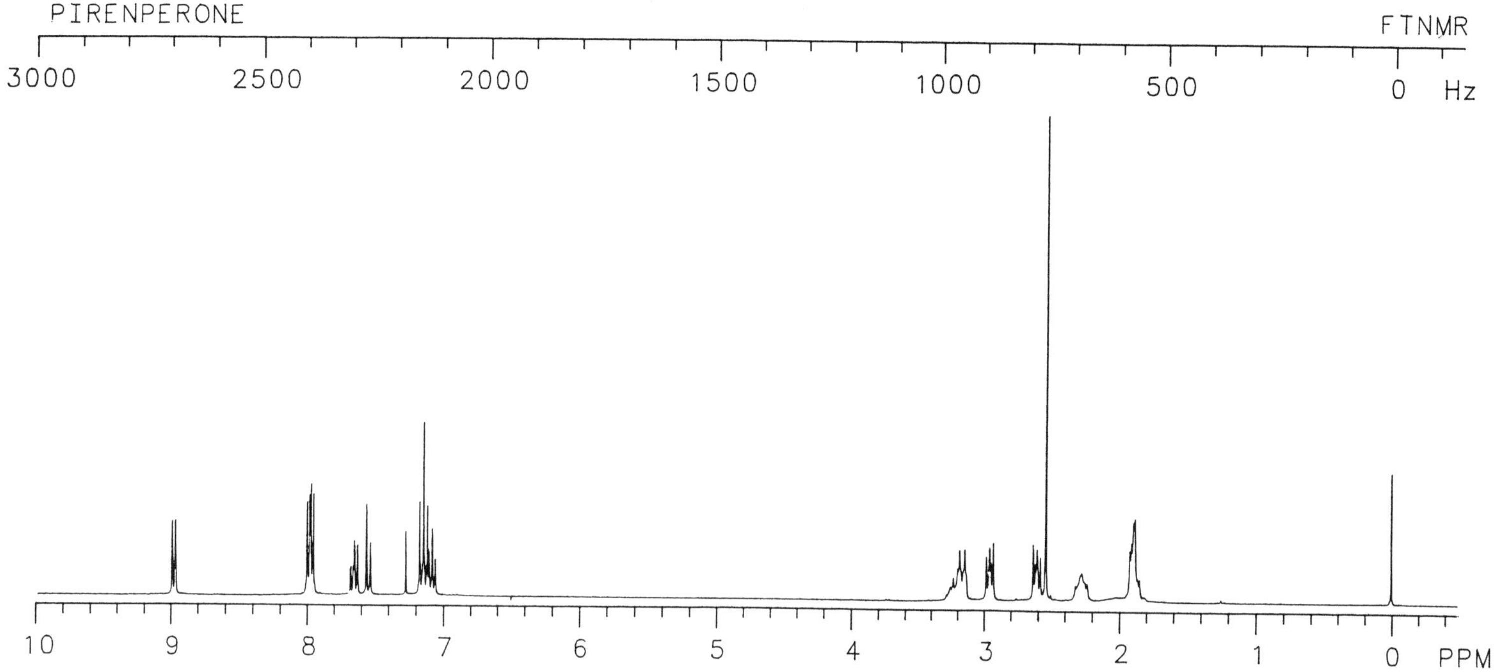

PEAK LISTING

#	HT	PPM
1	26	.00
2	17	1.88
3	99	2.54
4	18	7.11
5	35	7.14
6	19	7.17
7	18	7.55
8	20	7.95
9	22	7.97
10	20	7.98
11	19	7.99

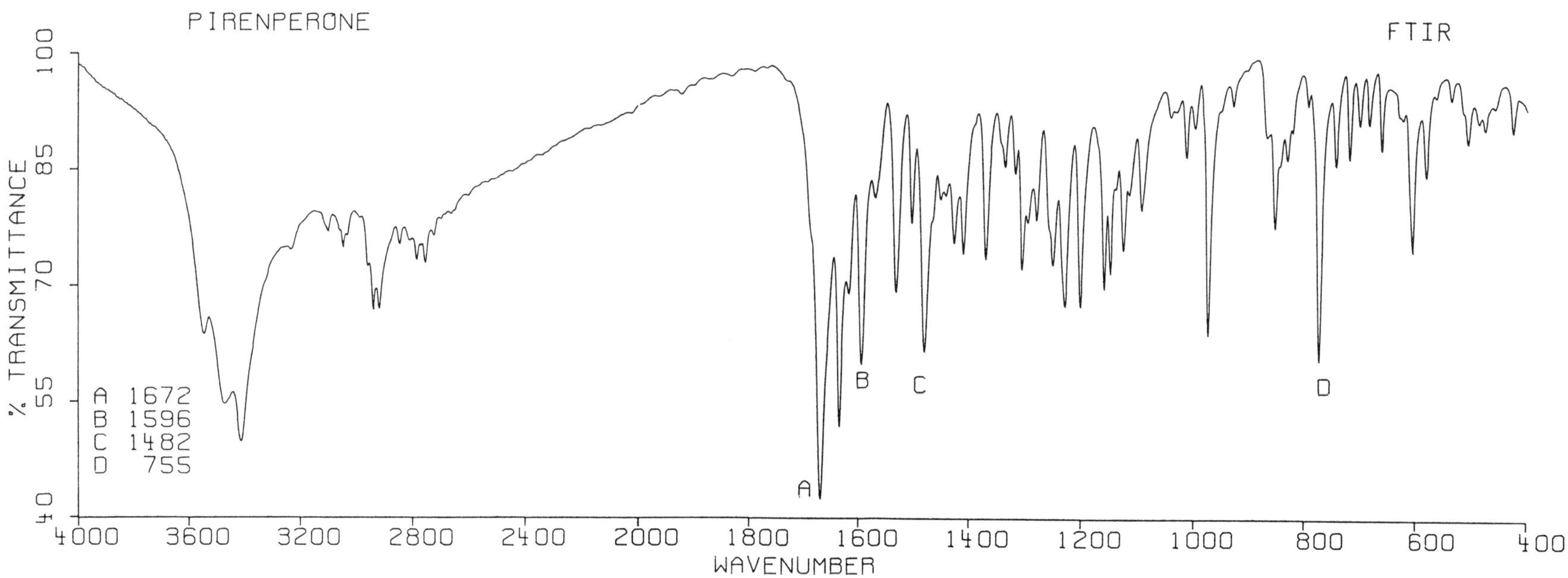

PODOPHYLLOTOXIN

$C_{22}H_{22}O_8$

Molecular Weight: 414.41 (414.13)

Synonyms: 5,8,8A,9-Tetrahydro-9-hydroxy-5-(3,4,5-trimethoxyphenyl)-furo[3',4':6,7]naphtho[2,3-d]-1,3-dioxol-6(5aH)-one; podophyllinic acid lactone; podofilox
Trade Names: Condyline, Condylox, Martec, Warticon
Use: Antiviral

HPLC: Methanol: 2.7
GC:

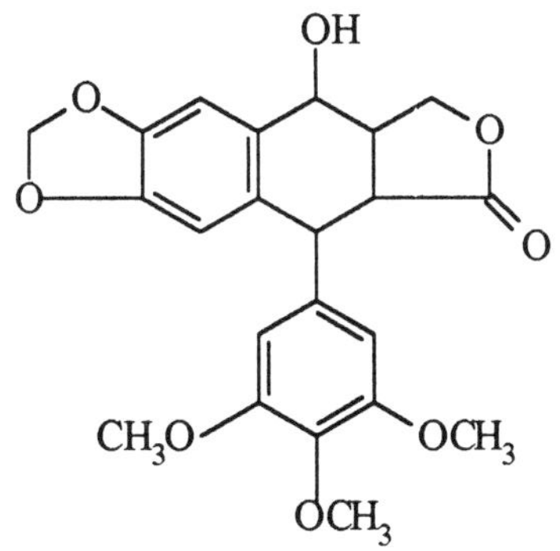

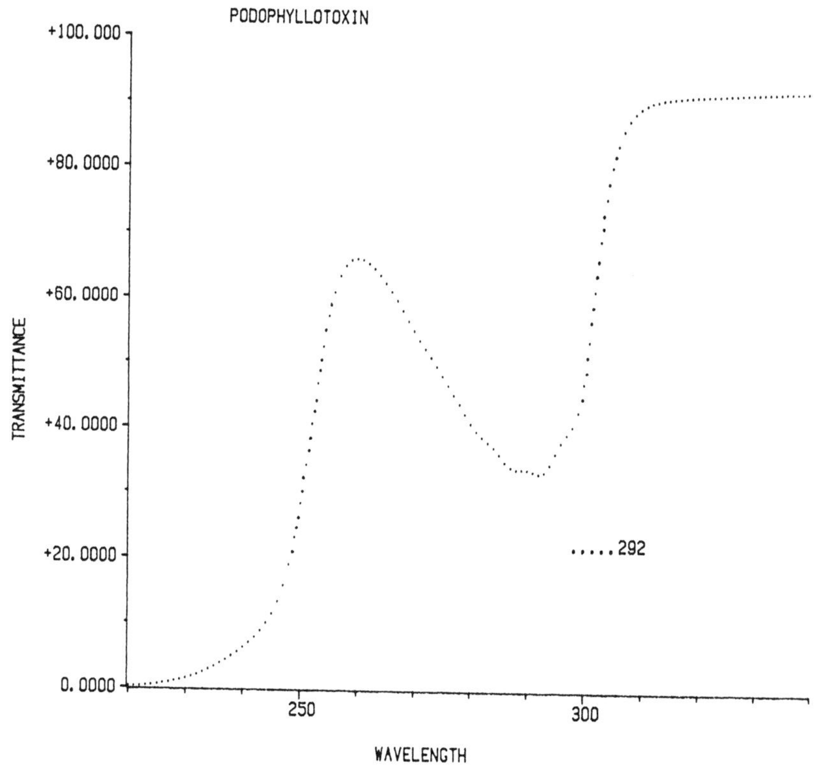

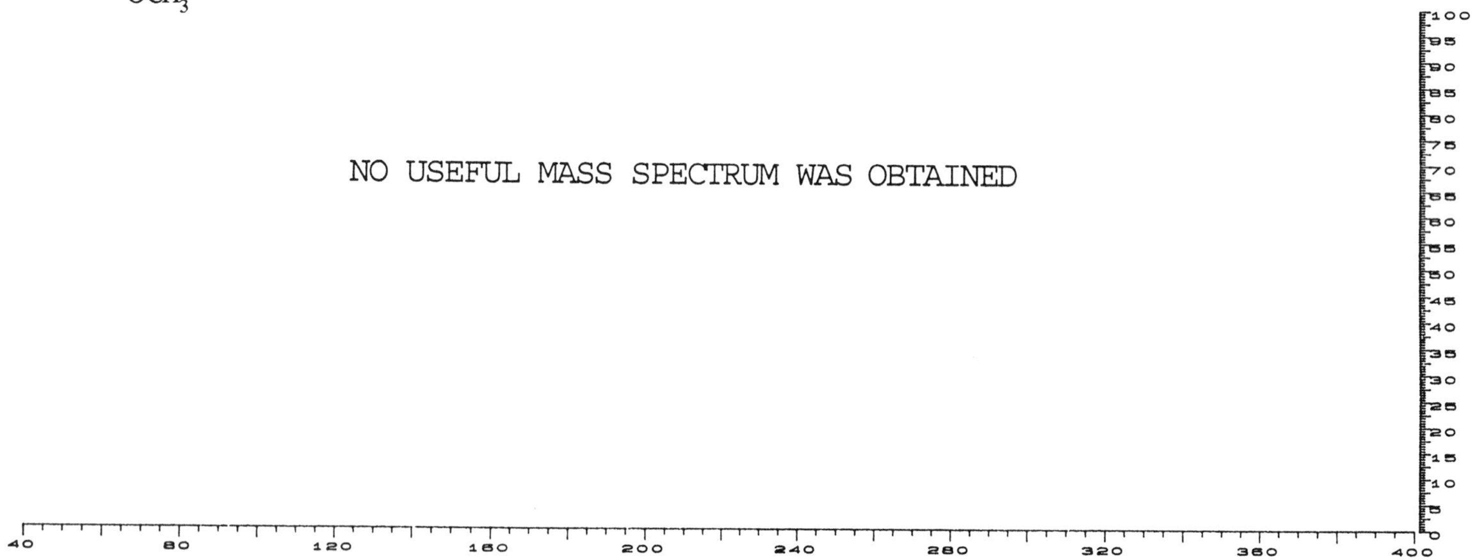

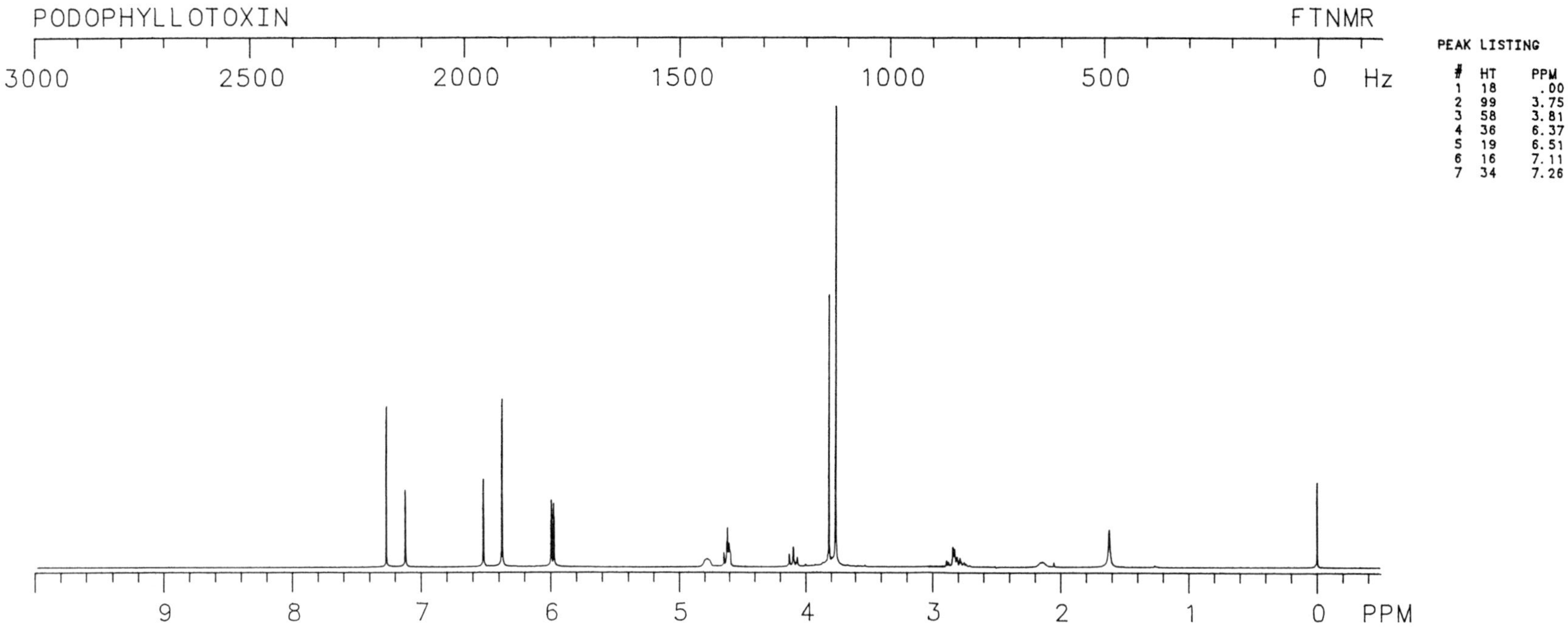
PODOPHYLLOTOXIN
FTNMR
3000 2500 2000 1500 1000 500 0 Hz
9 8 7 6 5 4 3 2 1 0 PPM
PEAK LISTING
HT PPM
1 18 .00
2 99 3.75
3 58 3.81
4 36 6.37
5 19 6.51
6 16 7.11
7 34 7.26

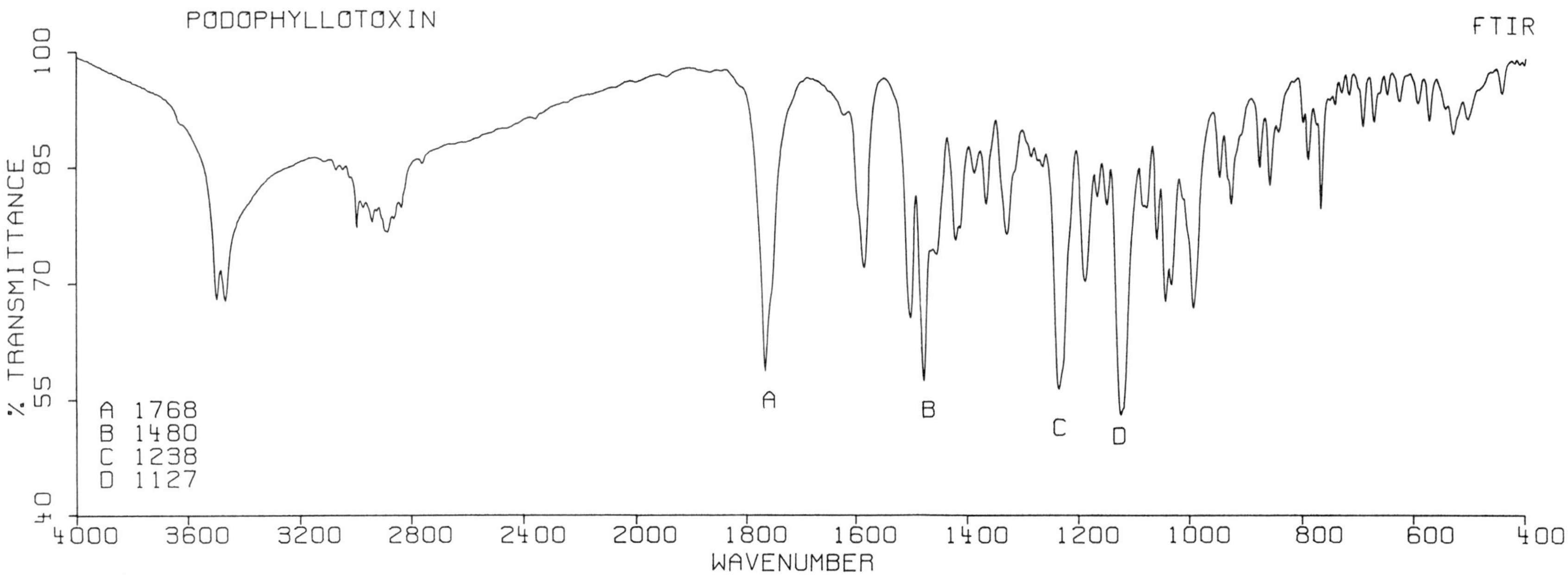
PODOPHYLLOTOXIN
FTIR
% TRANSMITTANCE
100 85 70 55 40
A 1768
B 1480
C 1238
D 1127
A B C D
4000 3600 3200 2800 2400 2000 1800 1600 1400 1200 1000 800 600 400
WAVENUMBER

PRAVASTATIN SODIUM

$C_{23}H_{35}NaO_7$

Molecular Weight: 446.52 (446.23)

Synonyms: [1S-[1α(βS*,δS*),2α,6α,8β(r*),8aα]]-1,2,6,7,8,8a-Hexahydro-β,δ,6-trihydroxy-2-methyl-8-(2-methyl-1-oxobutoxy)-1-naphthaleneheptanoic acid monosodium salt; 3β-hydroxycompactin sodium salt
Trade Names: Mevalotin, Pravachol
Use: Antihyperlipoproteinemic

HPLC: Methanol: 2.9
GC:

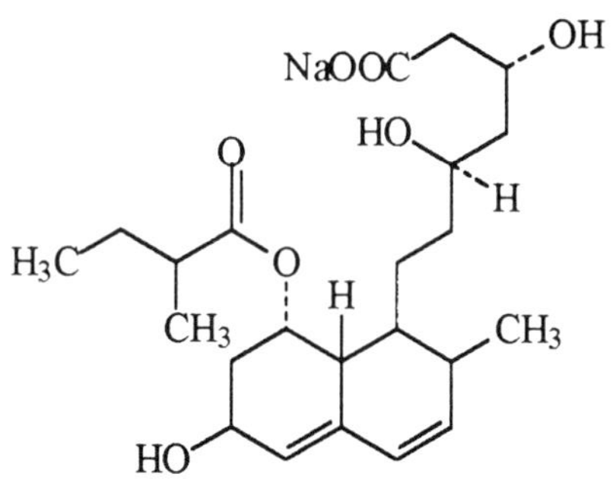

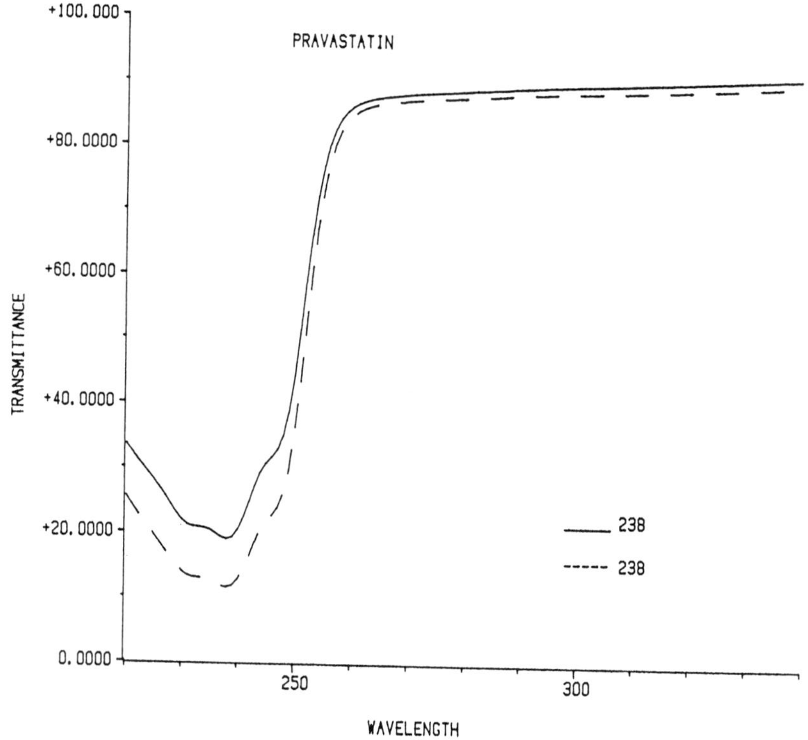

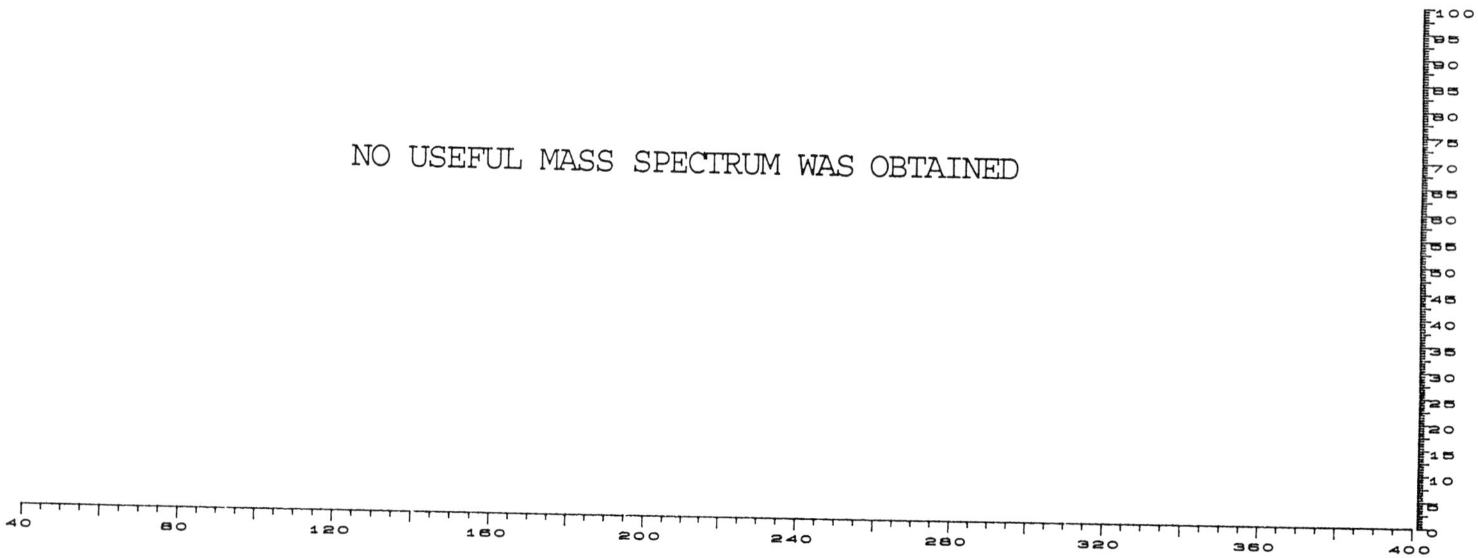

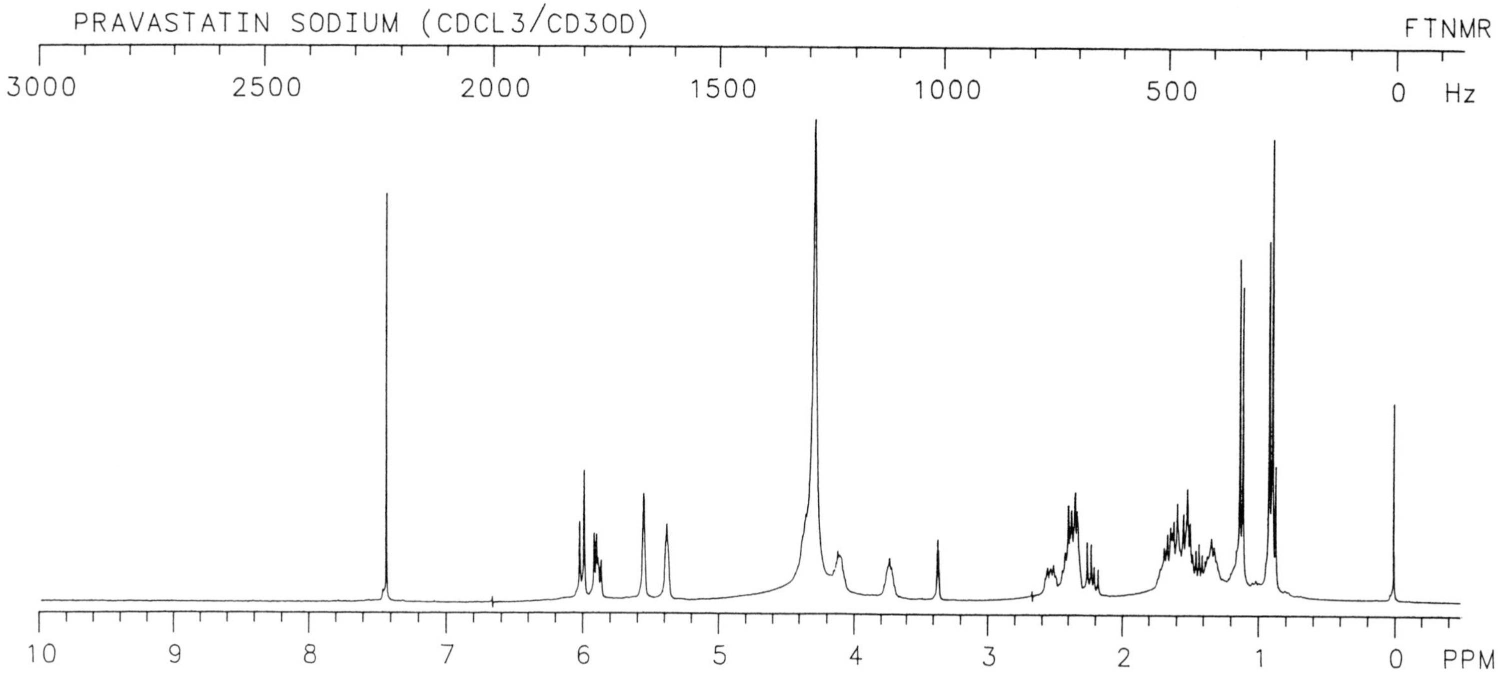

PEAK LISTING

#	HT	PPM
1	41	.00
2	28	.86
3	94	.89
4	74	.91
5	64	1.10
6	70	1.12
7	23	1.51
8	20	1.59
9	18	2.33
10	22	2.34
11	19	2.37
12	20	2.39
13	98	4.28
14	22	5.54
15	26	5.98
16	83	7.42

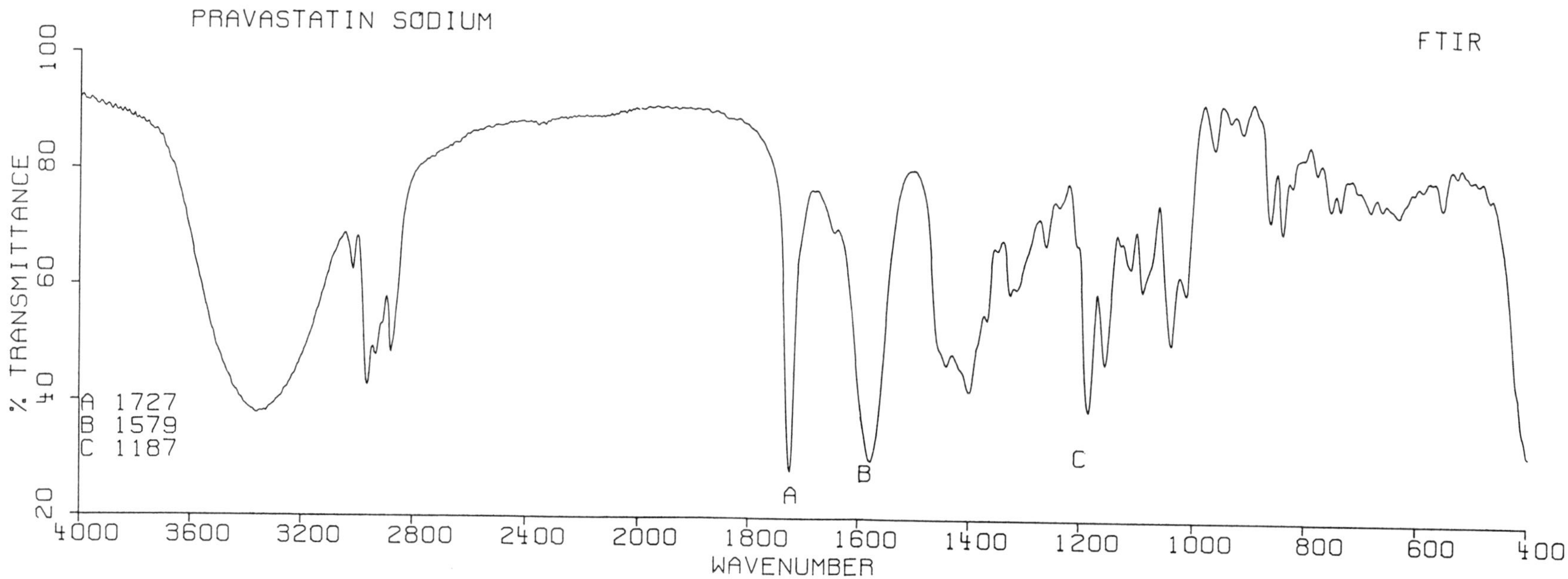

PROPENTOFYLLINE

$C_{15}H_{22}N_4O_3$

Molecular Weight: 306.36 (306.17)

Synonyms: 3,7-Dihydro-3-methyl-1-(5-oxohexyl)-7-propyl-1H-purine-2,6-dione
Trade Names: Hextol, Karsivan
Use: Vasodilator

HPLC: Ethanol: 2.6
GC:

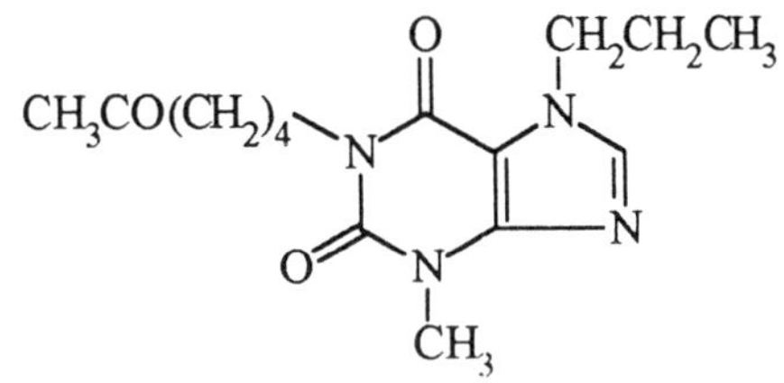

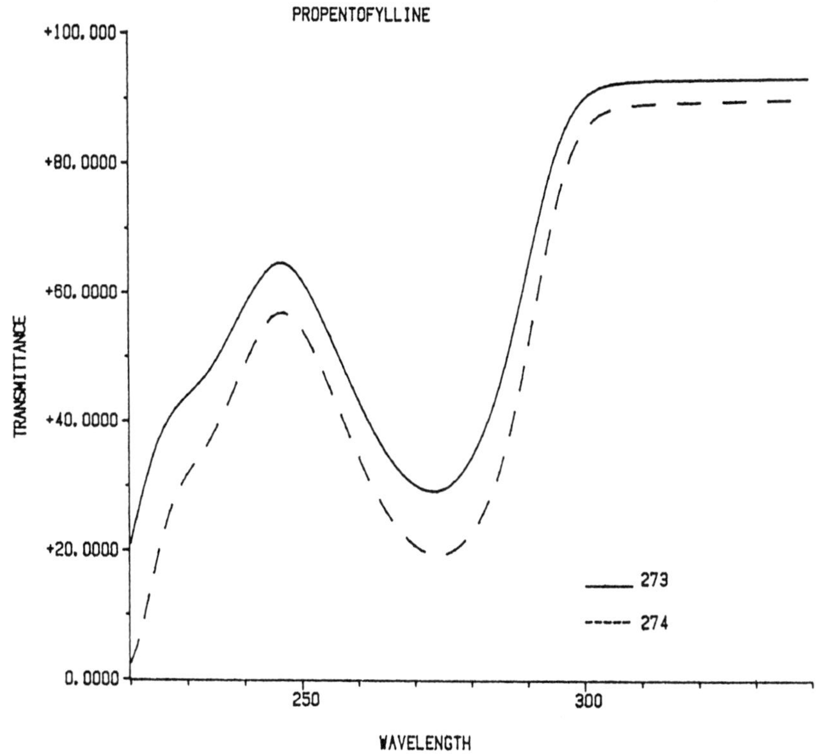

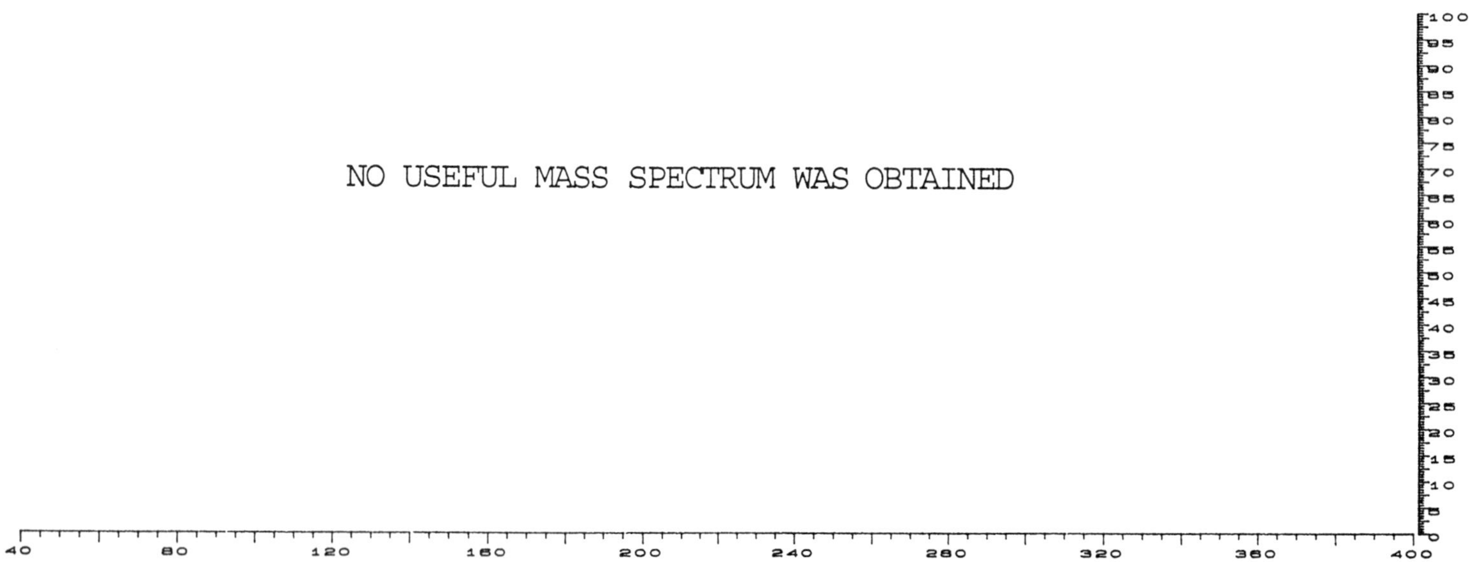

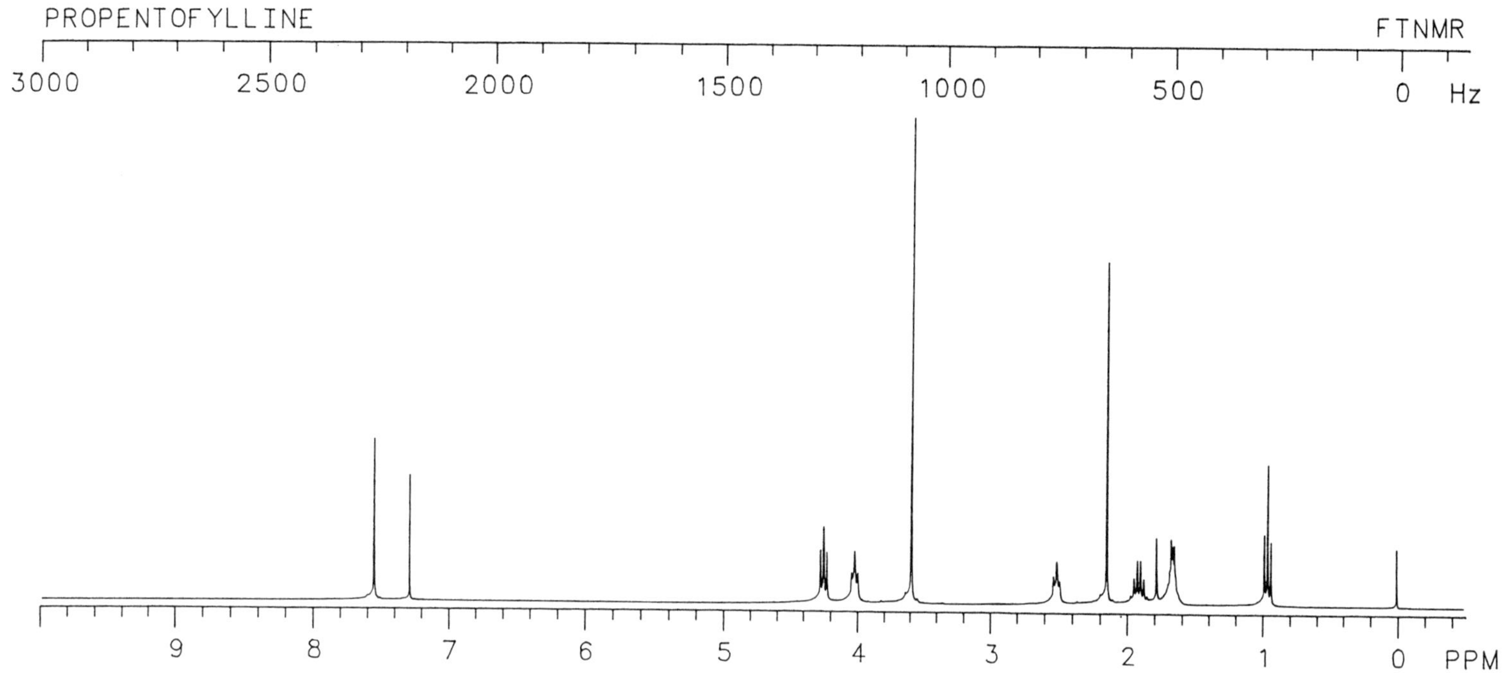

PEAK LISTING

#	HT	PPM
1	12	.00
2	13	.92
3	28	.95
4	14	.97
5	12	1.64
6	12	1.65
7	13	1.67
8	13	1.77
9	9	1.89
10	9	1.91
11	70	2.14
12	9	2.50
13	99	3.58
14	10	4.01
15	10	4.22
16	15	4.24
17	11	4.26
18	25	7.27
19	33	7.54

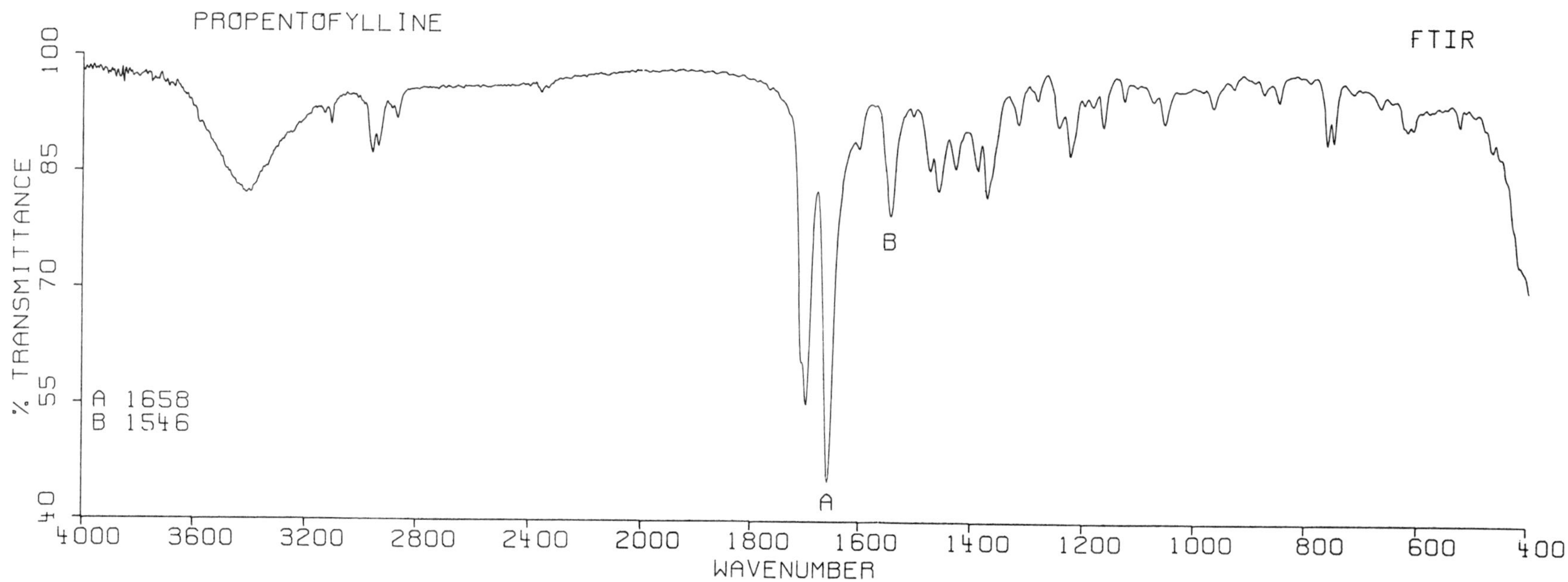

PROPIONYLPROMAZINE

$C_{20}H_{24}N_2O_5$

Molecular Weight: 340.48 (340.16)

Synonyms: 1-[10-[3-(Dimethylamino)propyl]-10H-phenothiazin-2-yl]-1-propanone; propiopromazine
Trade Names: Combelen, Tranvet, Tranvex
Use: Tranquilizer

HPLC: Ethanol: 2.0
GC: 2796; 250°

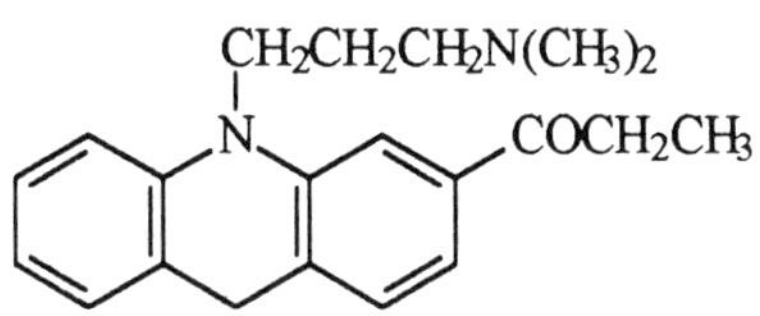

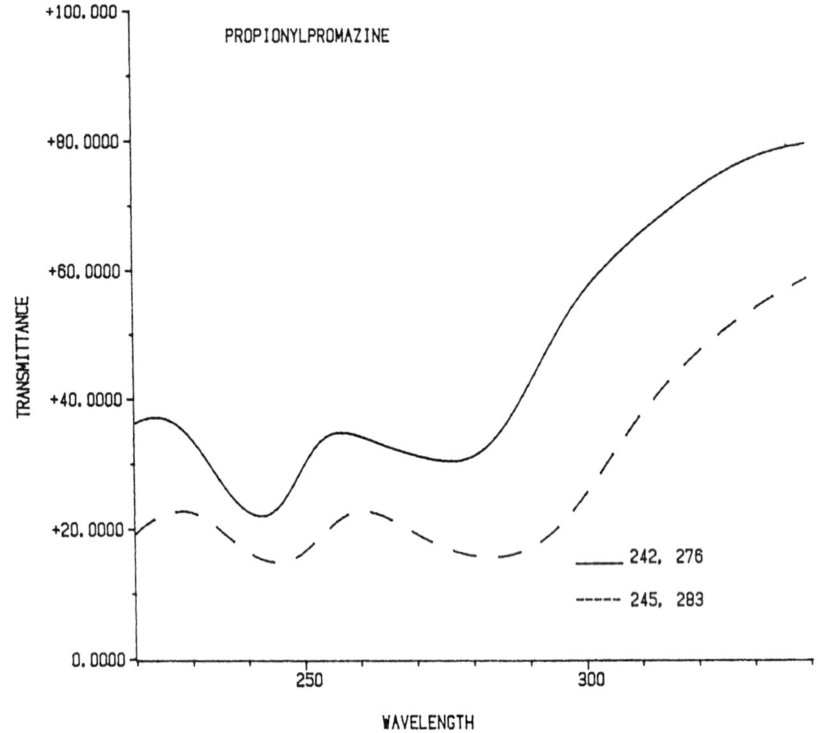

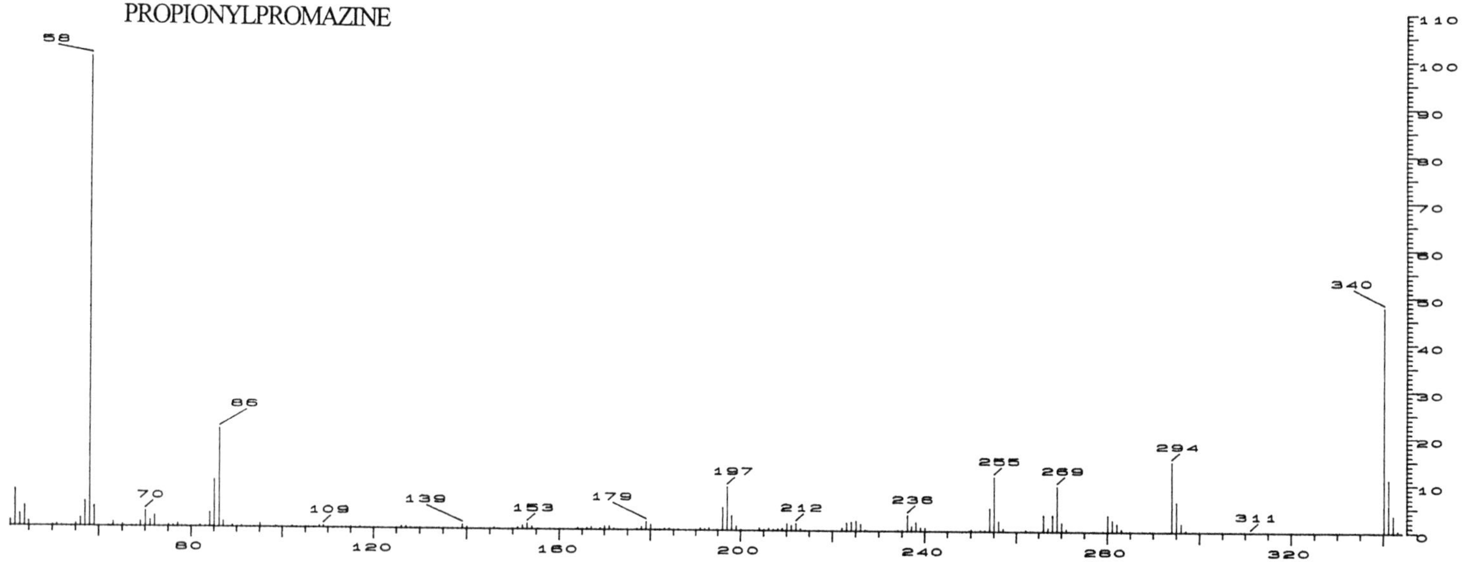

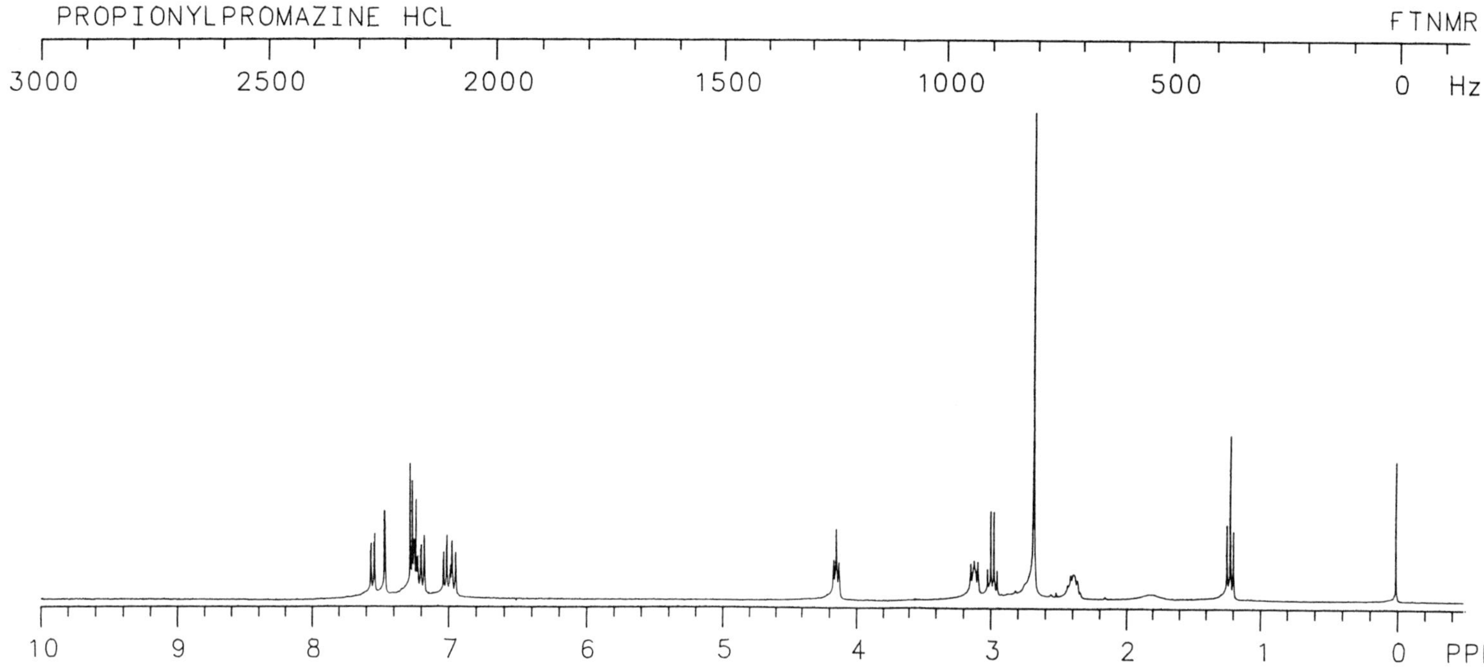

PEAK LISTING

#	HT	PPM
1	28	.00
2	14	1.19
3	33	1.21
4	15	1.24
5	99	2.67
6	18	2.96
7	18	2.99
8	14	4.13
9	11	6.96
10	13	7.00
11	12	7.16
12	12	7.17
13	20	7.23
14	12	7.24
15	24	7.25
10	27	7.27
17	18	7.45
18	18	7.46
19	13	7.53

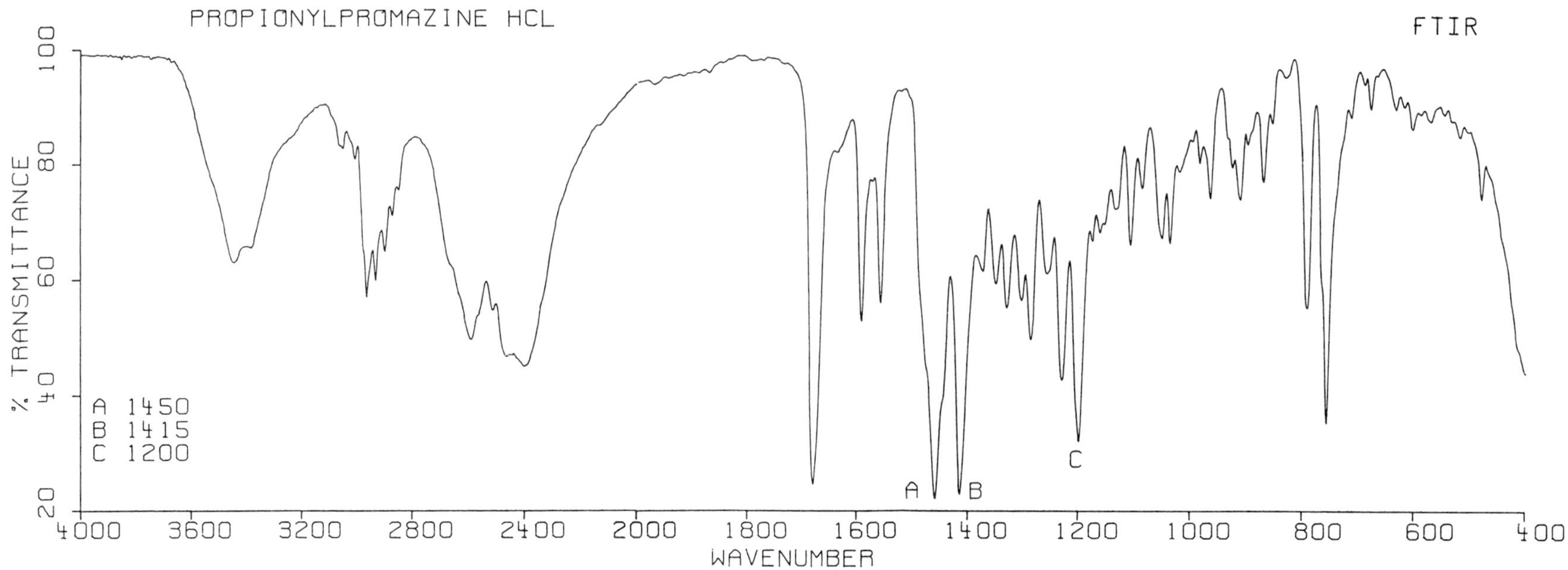

PROPOFOL

$C_{12}H_{18}O$

Molecular Weight: 178.27 (178.14)

Synonyms: 2,6-Bis(1-methylethyl)phenol; 2,6-diisopropylphenol; disoprofol
Trade Names: Diprivan, Disoprivan, Rapinovet
Use: Anesthetic

HPLC: Ethanol: 2.6
GC:

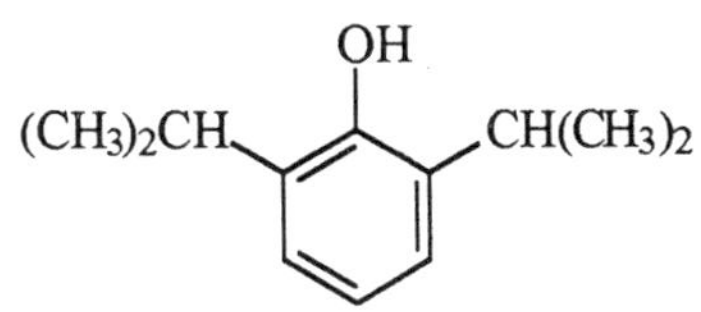

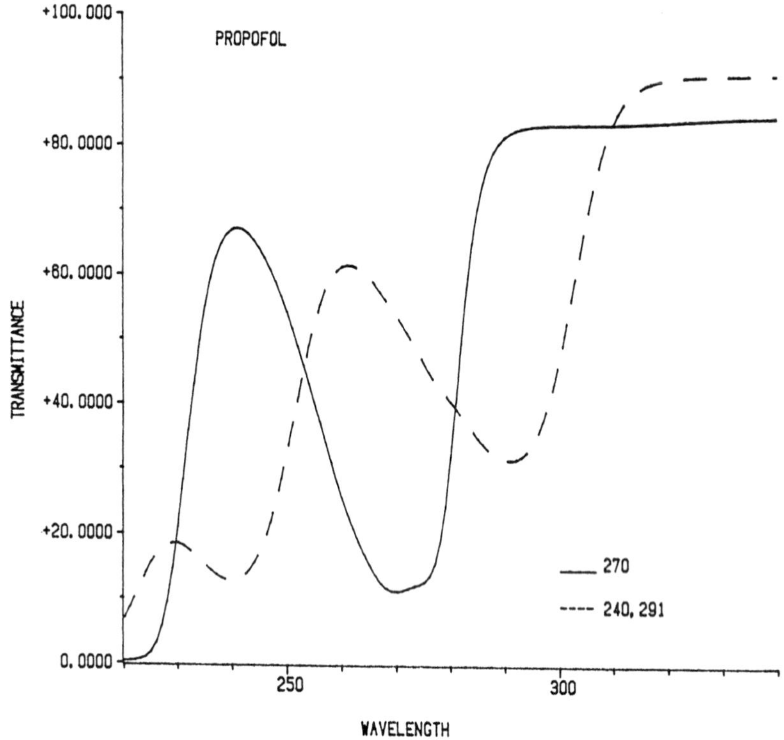

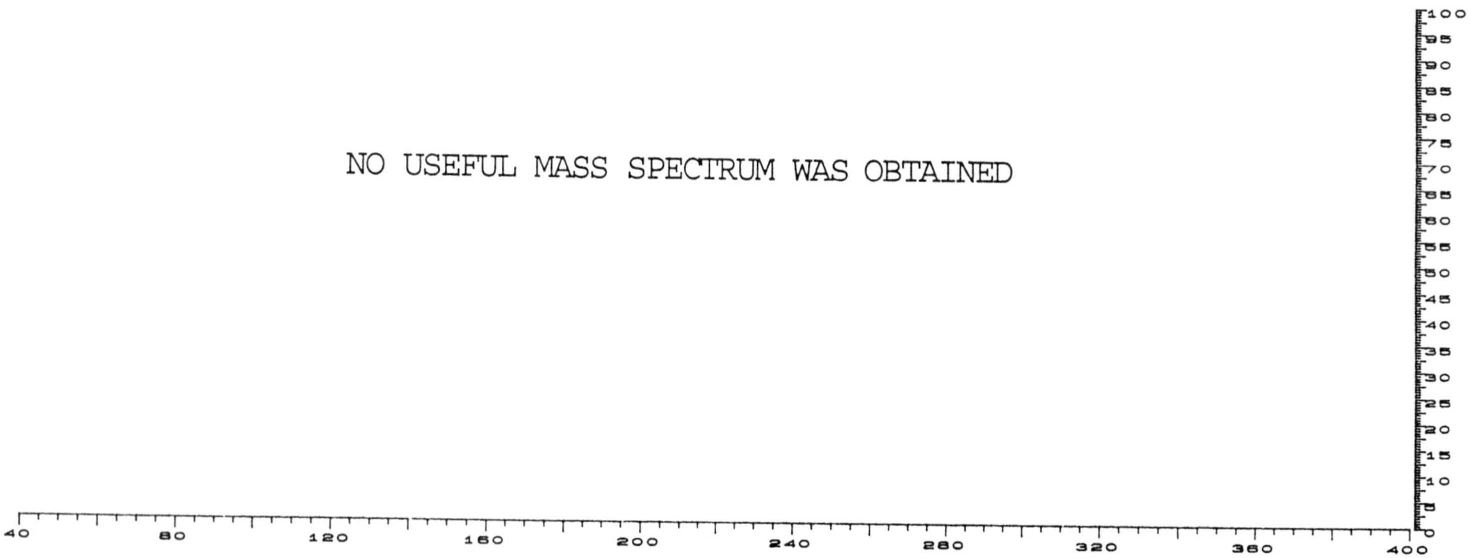

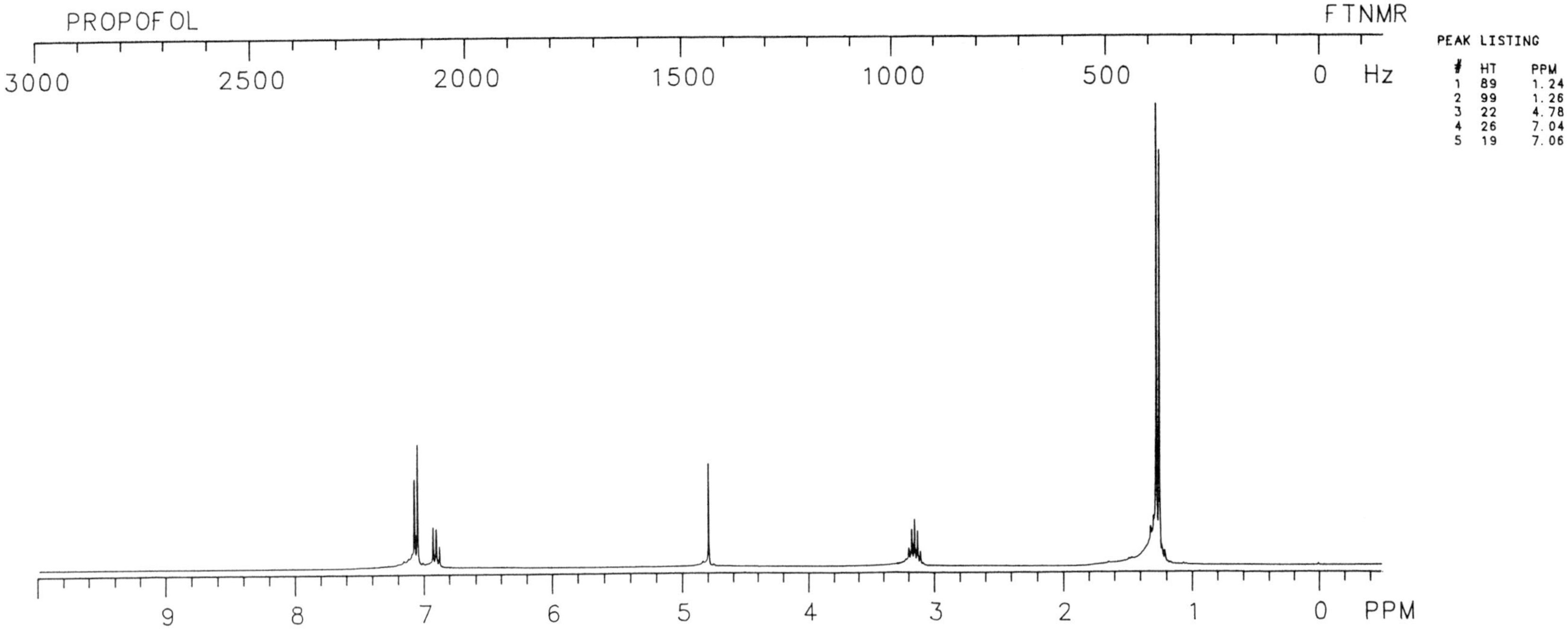
PROPOFOL
FTNMR
3000
2500
2000
1500
1000
500
0 Hz
PEAK LISTING
HT PPM
1 89 1.24
2 99 1.26
3 22 4.78
4 26 7.04
5 19 7.06
9
8
7
6
5
4
3
2
1
0 PPM

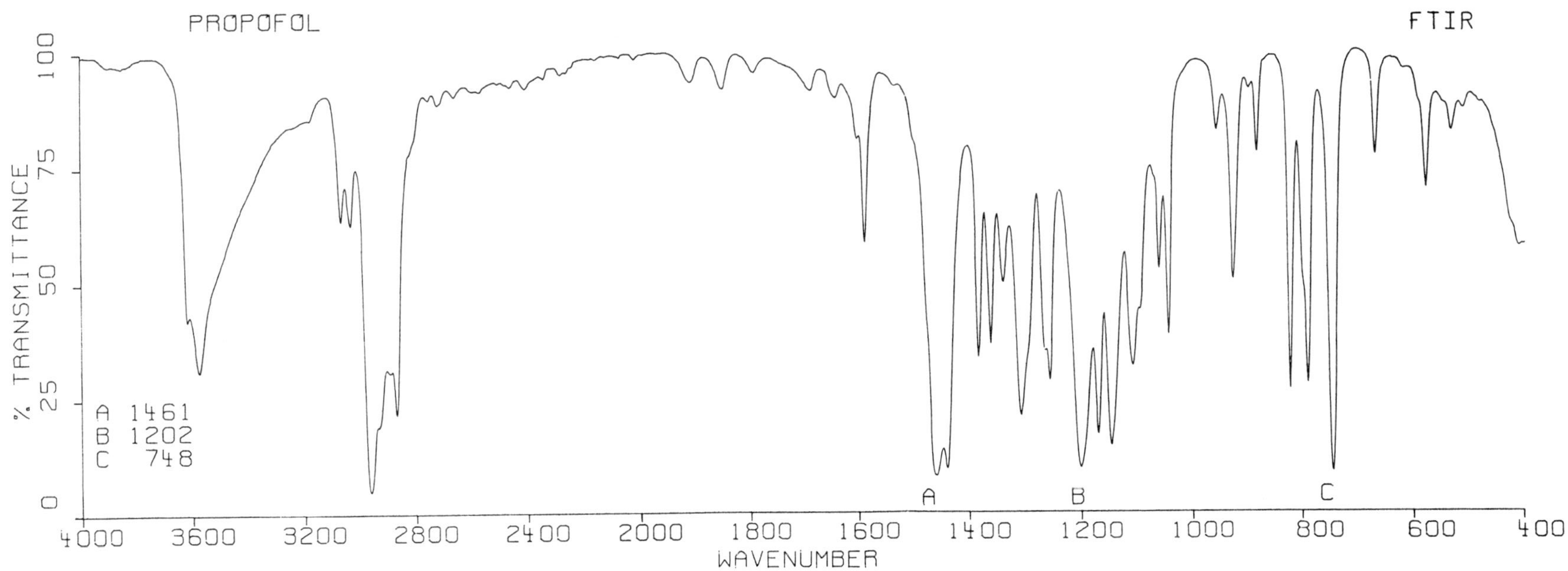
PROPOFOL
FTIR
% TRANSMITTANCE
100
75
50
25
0
A 1461
B 1202
C 748
A
B
C
4000
3600
3200
2800
2400
2000
1800
1600
1400
1200
1000
800
600
400
WAVENUMBER

QUAZEPAM

$C_{17}H_{11}ClF_4N_2S$

Molecular Weight: 386.79 (386.03)

Synonyms: 7-Chloro-5-(2-fluorophenyl)-1,3-dihydro-1-(2,2,2-trifluoroethyl)-2H-1,4-benzodiazepine-2-thione
Trade Names: Doral, Dormalin, Oniria, Prosedar, Quazium, Selepam
Use: Sedative, hypnotic

HPLC: Ethanol: 2.7
GC: 2475; 250°

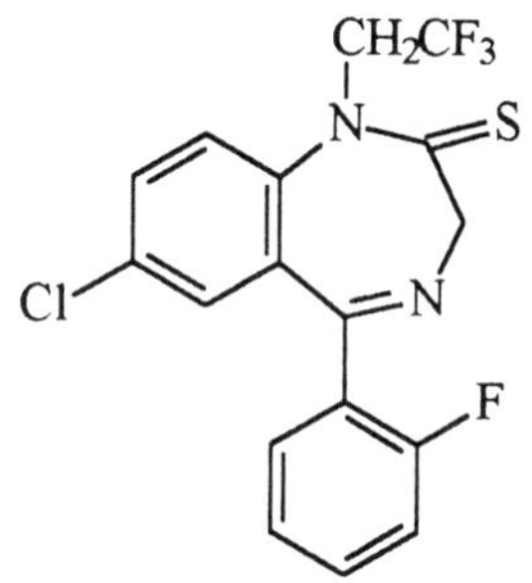

QUAZEPAM

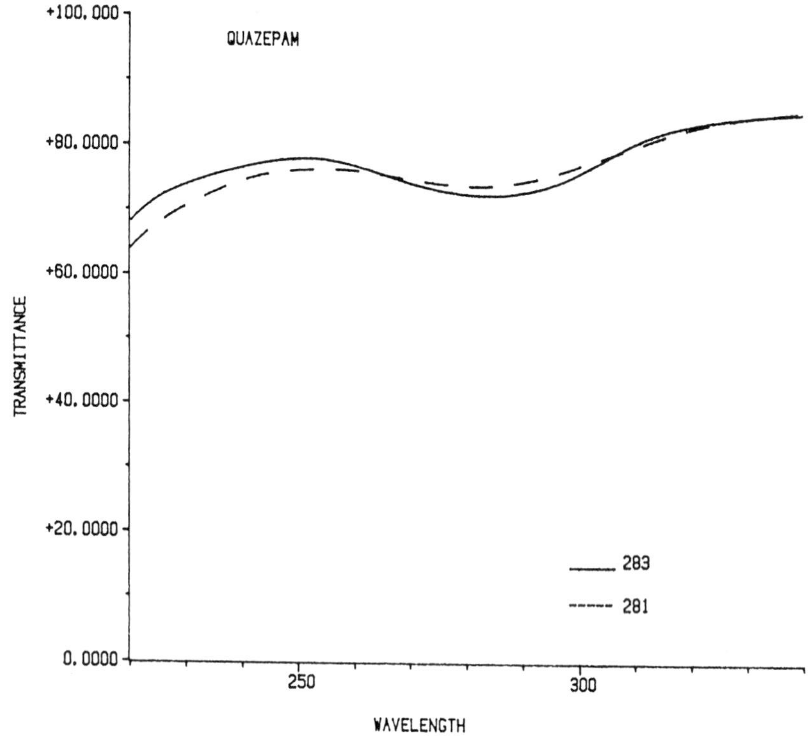

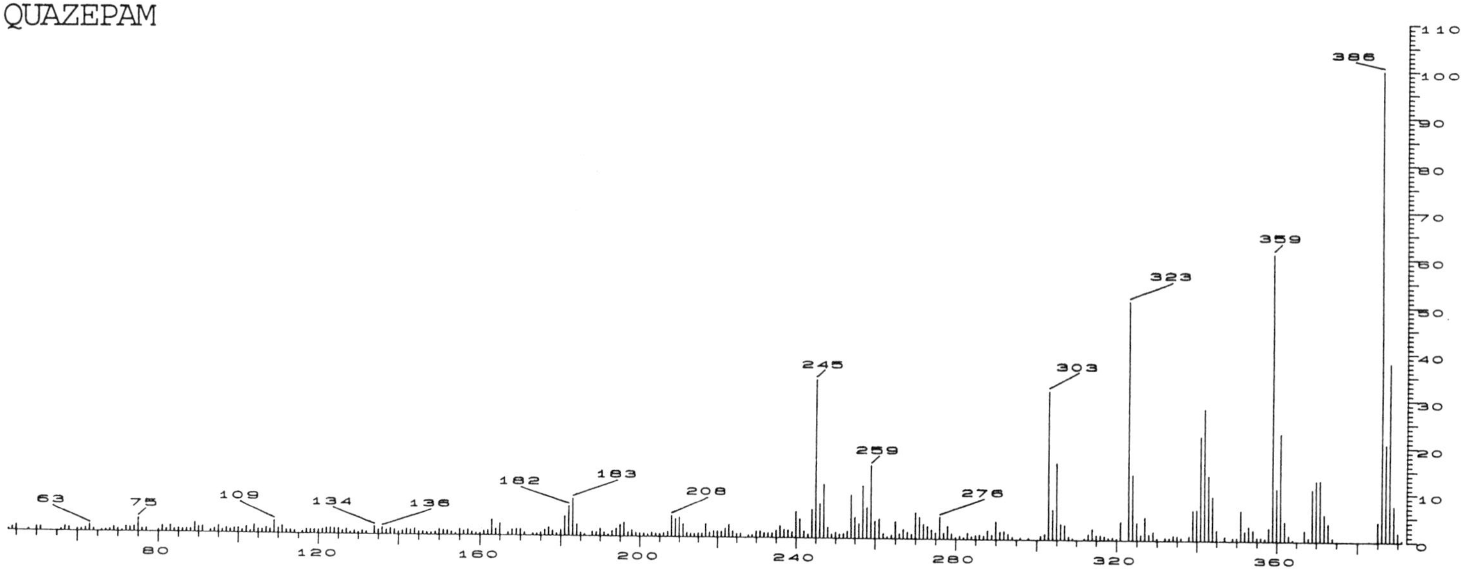

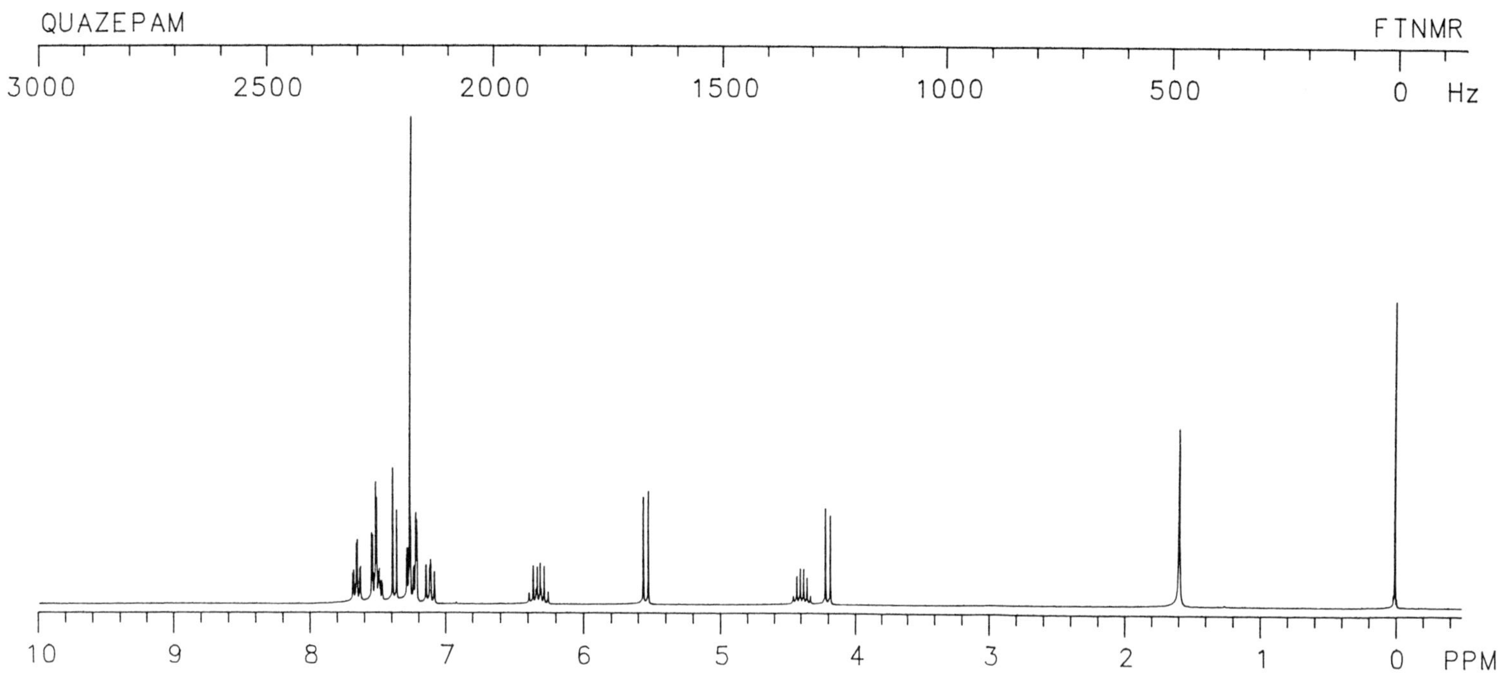

PEAK LISTING

#	HT	PPM
1	62	.00
2	36	1.59
3	18	4.17
4	19	4.21
5	23	5.52
6	22	5.55
7	17	7.20
8	15	7.21
9	18	7.21
10	17	7.25
11	99	7.26
12	19	7.35
13	27	7.38
14	21	7.50
15	24	7.51

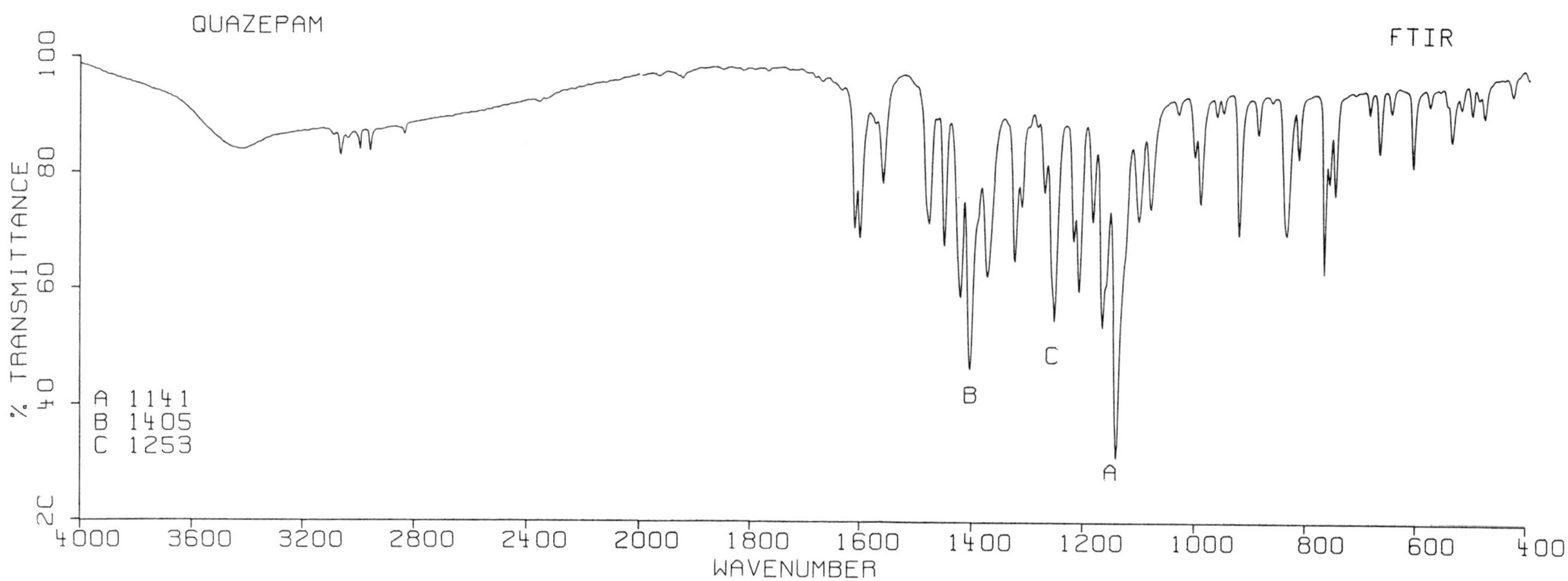

RAMIPRIL

$C_{23}H_{32}N_2O_5$

Molecular Weight: 416.52 (416.23)

Synonyms: [2s-[1[R*(R*)],2α,3aβ,6aβ]]-1-[2-[[1-(ethoxycarbonyl)-3-phenylpropyl]amino]-1-oxopropyl]octahydrocyclopenta[b]pyrrole-2-carboxylic acid
Trade Names: Altace, Cardace
Use: Antihypertensive

HPLC: Methanol: 2.7
GC:

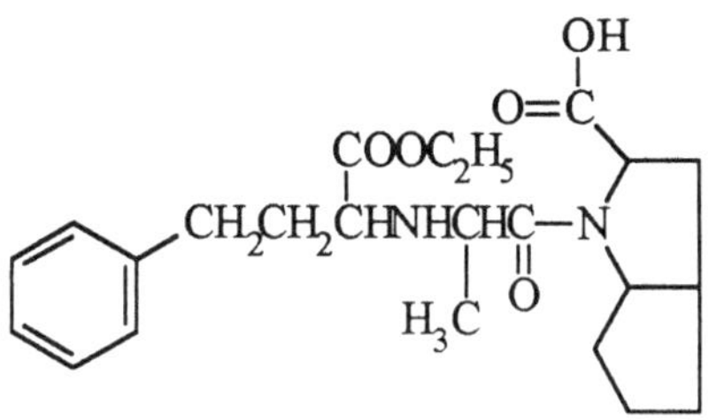

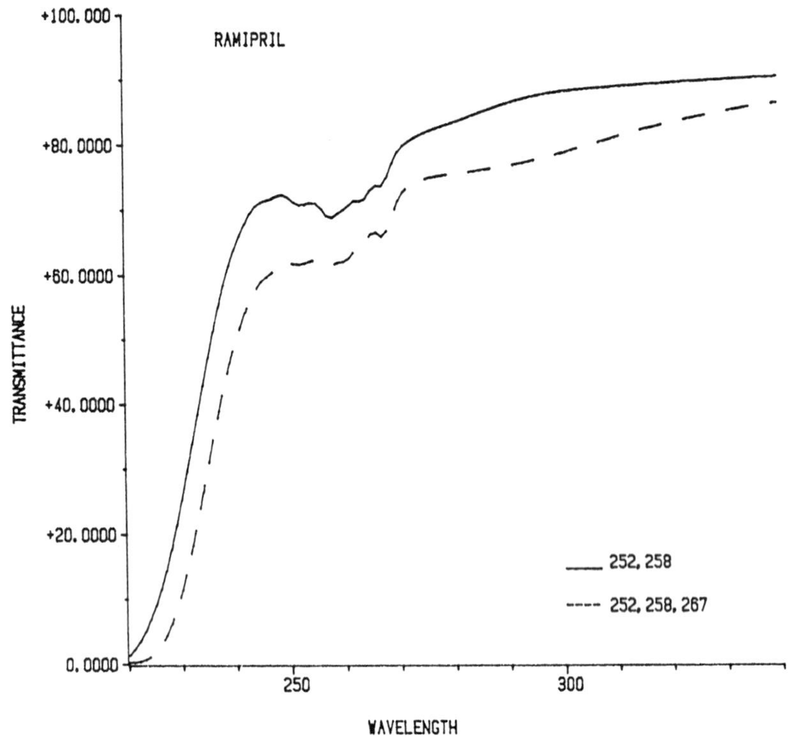

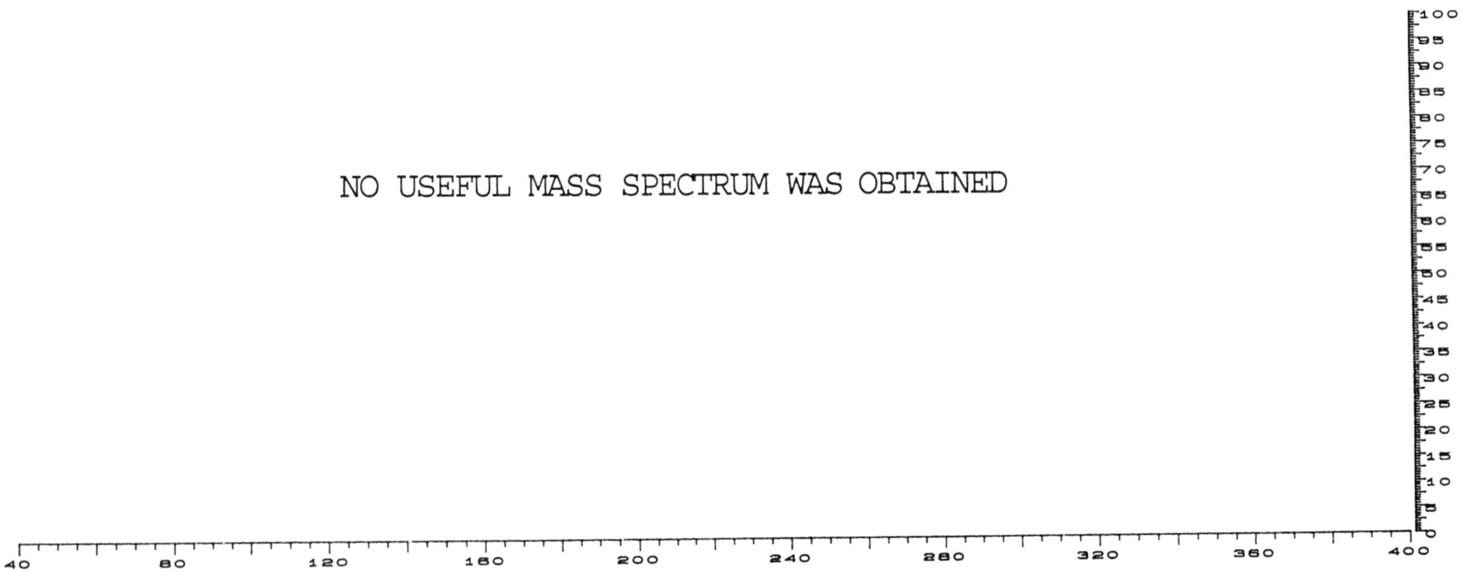

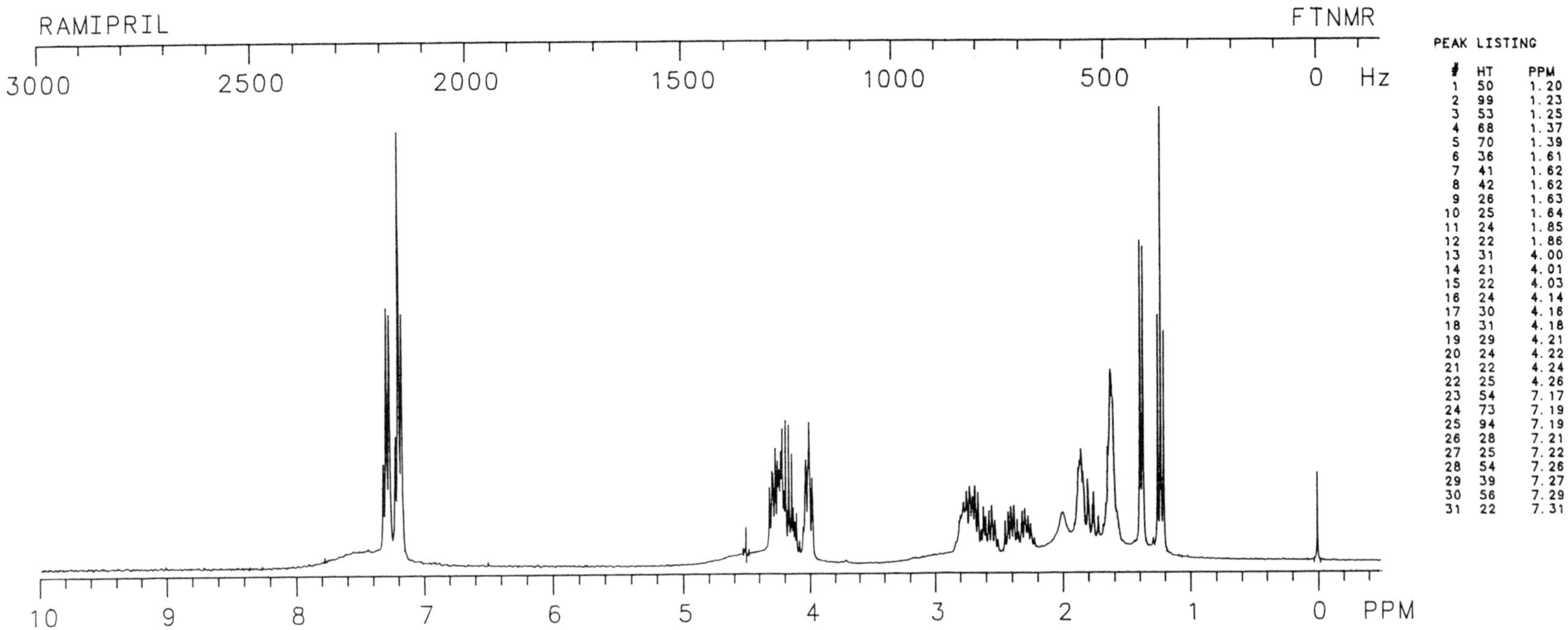

RAMIPRIL
FTNMR
3000 2500 2000 1500 1000 500 0 Hz
10 9 8 7 6 5 4 3 2 1 0 PPM
PEAK LISTING
HT PPM
1 50 1.20
2 99 1.23
3 53 1.25
4 68 1.37
5 70 1.39
6 36 1.61
7 41 1.62
8 42 1.62
9 26 1.63
10 25 1.64
11 24 1.85
12 22 1.86
13 31 4.00
14 21 4.01
15 22 4.03
16 24 4.14
17 30 4.16
18 31 4.18
19 29 4.21
20 24 4.22
21 22 4.24
22 25 4.26
23 54 7.17
24 73 7.19
25 94 7.19
26 28 7.21
27 25 7.22
28 54 7.26
29 39 7.27
30 56 7.29
31 22 7.31

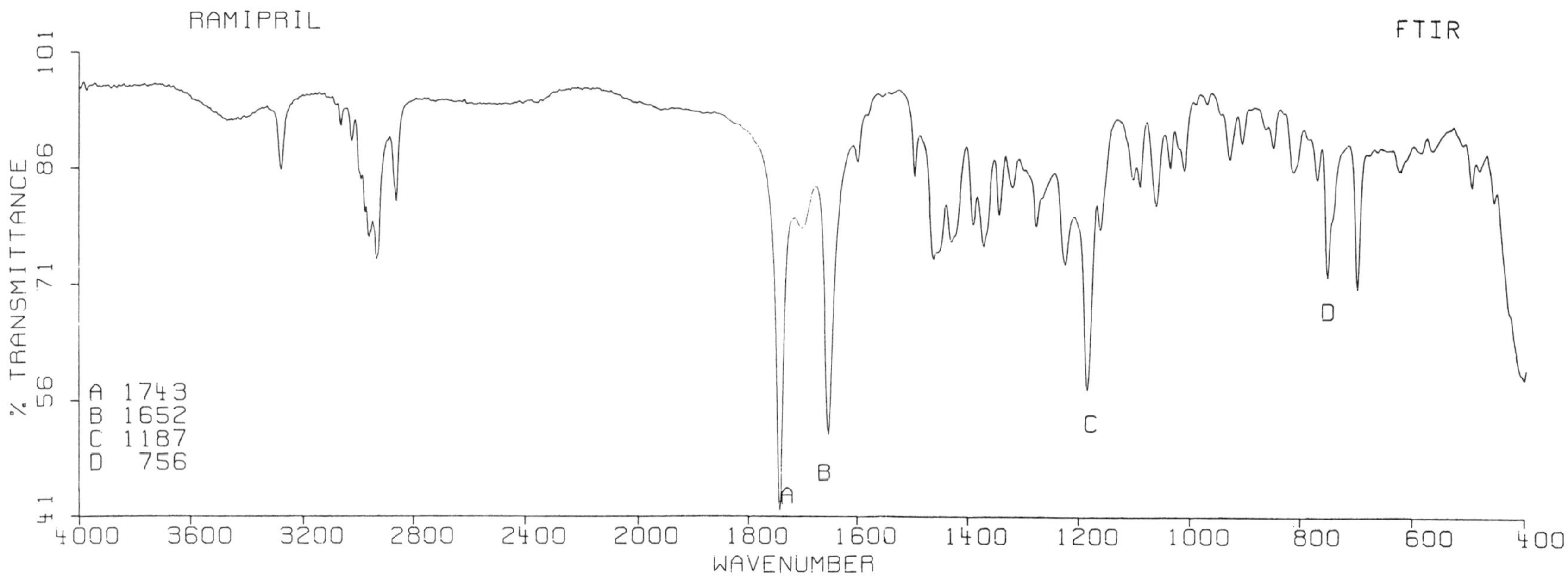

RAMIPRIL
FTIR
% TRANSMITTANCE
101 86 71 56 41
A 1743
B 1652
C 1187
D 756
A B C D
4000 3600 3200 2800 2400 2000 1800 1600 1400 1200 1000 800 600 400
WAVENUMBER

SALINOMYCIN

$C_{42}H_{70}O_{11}$

Molecular Weight: 751.01 (750.49)

Synonyms: Coxistac
Trade Names: Ovicox, Salocin
Use: Anticoccidial agent

HPLC:
GC:

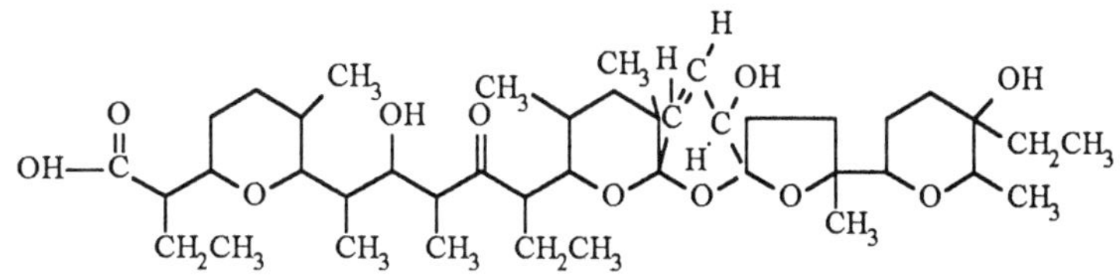

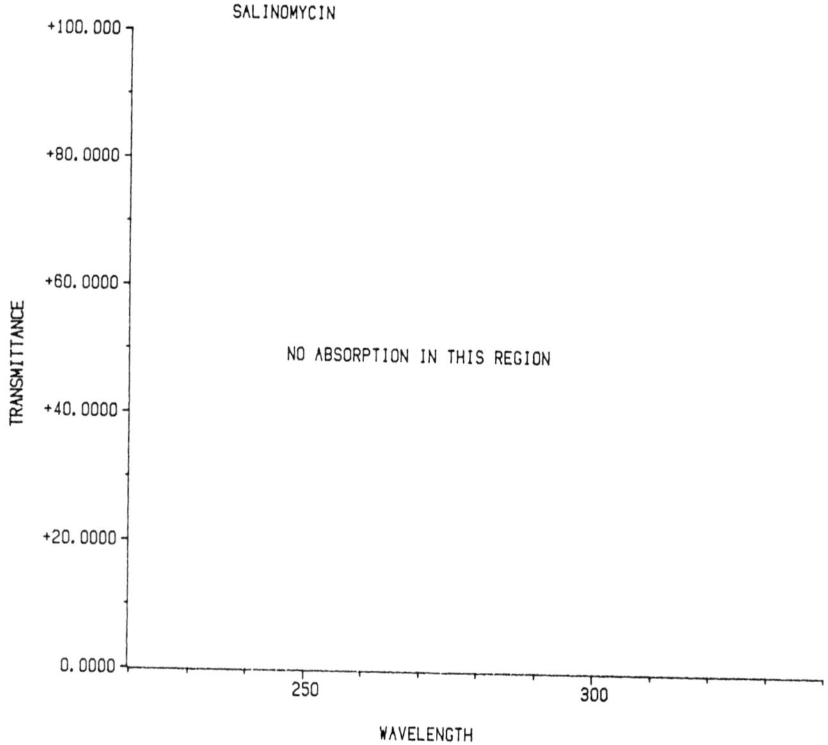

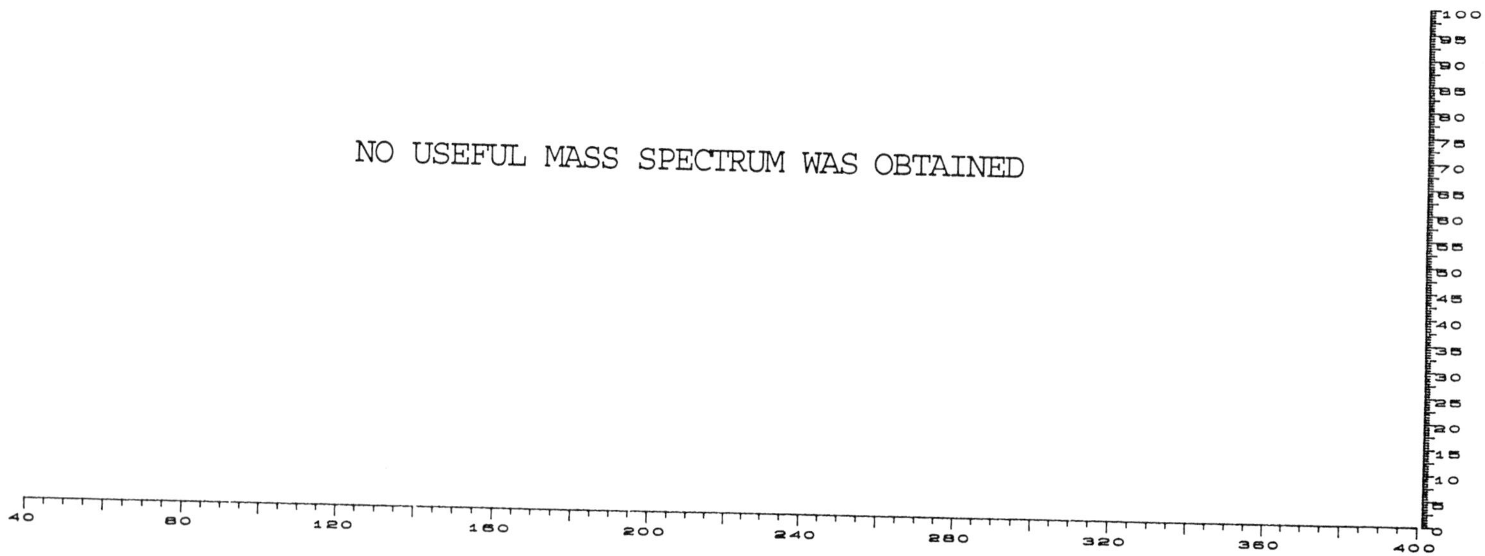

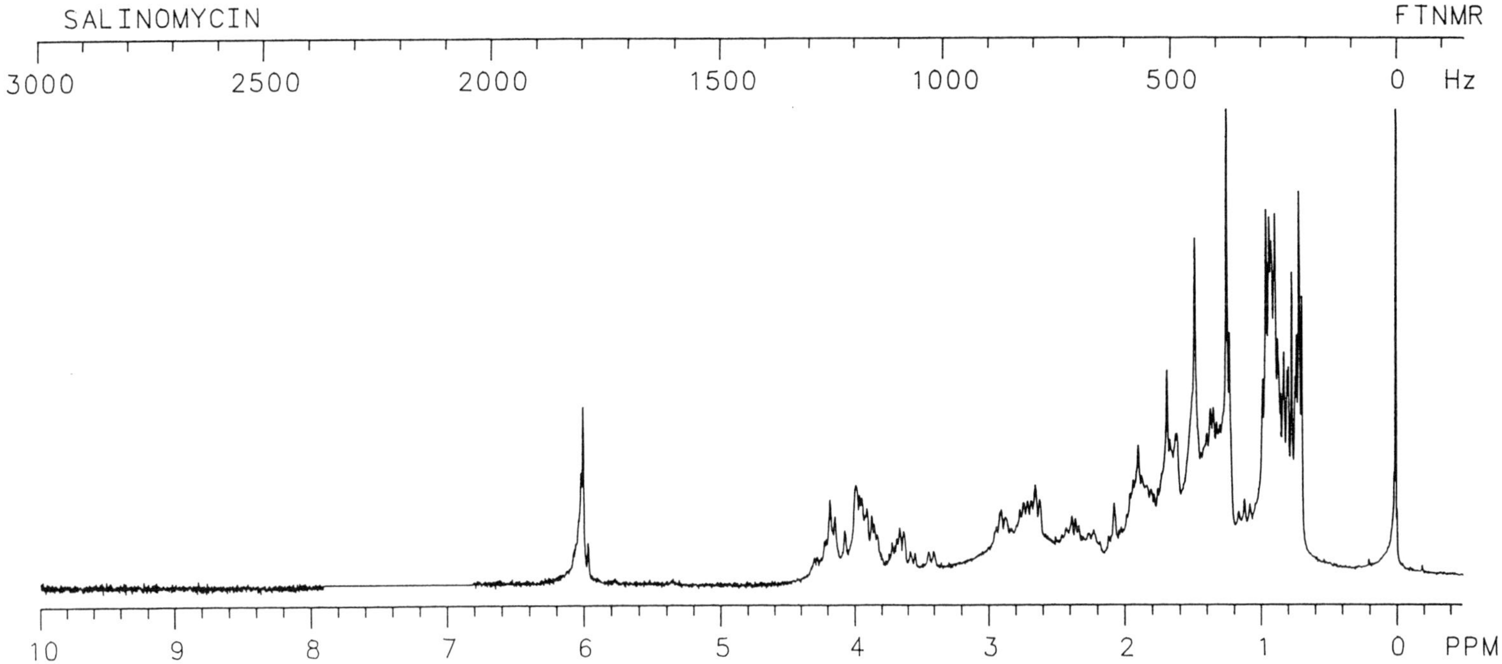

PEAK LISTING

#	HT	PPM
1	260	.00
2	58	.69
3	79	.71
4	63	.76
5	75	.89
6	63	.90
7	69	.91
8	69	.92
9	74	.93
10	76	.95
11	96	1.25
12	70	1.48

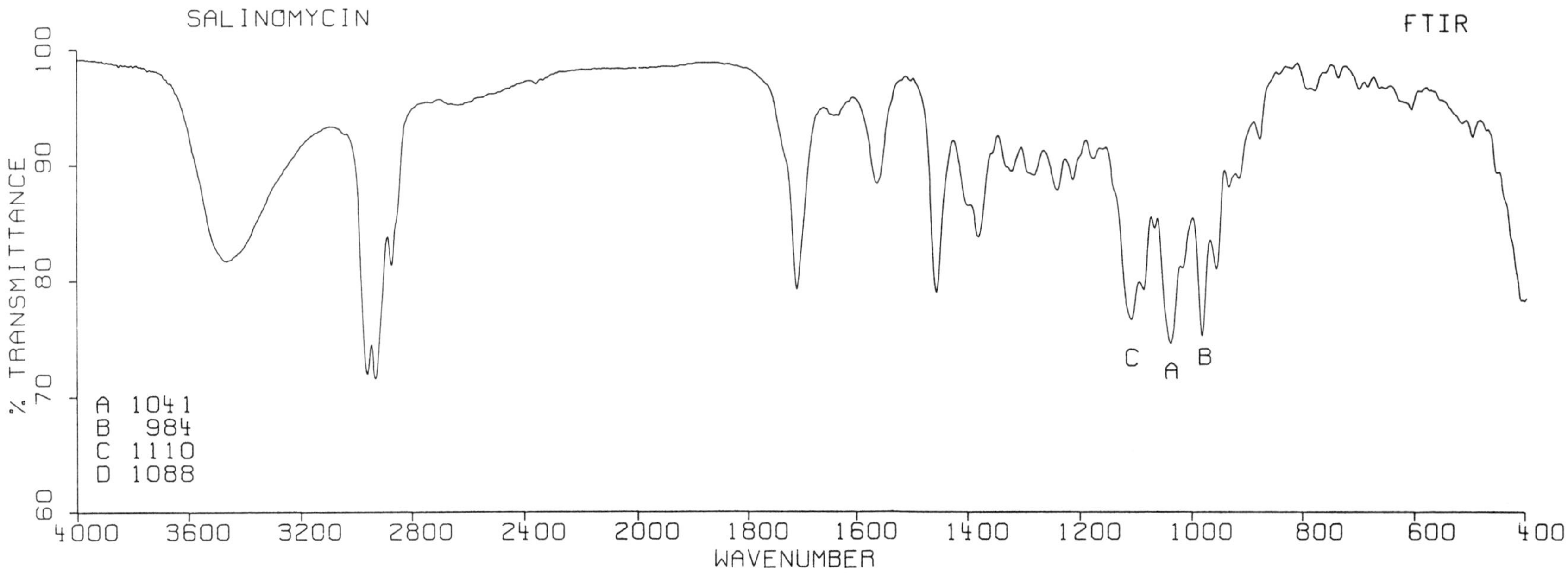

SANDOSTATIN

$C_{49}H_{66}N_{10}O_{10}S$

Molecular Weight: 987.18 (986.47)

Synonyms: D-Phenylalanyl-L-cysteinyl-L-phenylalanyl-D-tryptophyl-L-lysyl-L-threonyl-N-[2-hydroxy-1-(hydroxymethyl)propyl]-L-cysteinamide cyclic (2→7)-disulfide; octreotide
Trade Names: Sandostatin
Use: Gastric antisecretory agent

HPLC:
GC:

D—Phe—Cys—Phe—D—Trp—Lys—Thr—Cys—NHCH(CH_2OH)CHOHCH$_3$ (Cys—Cys bridged)

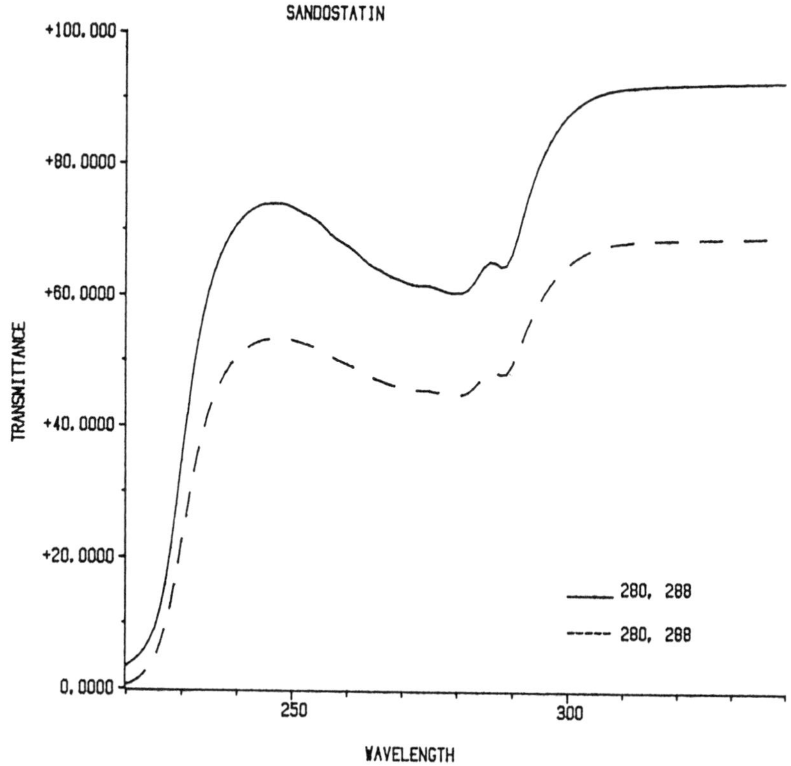

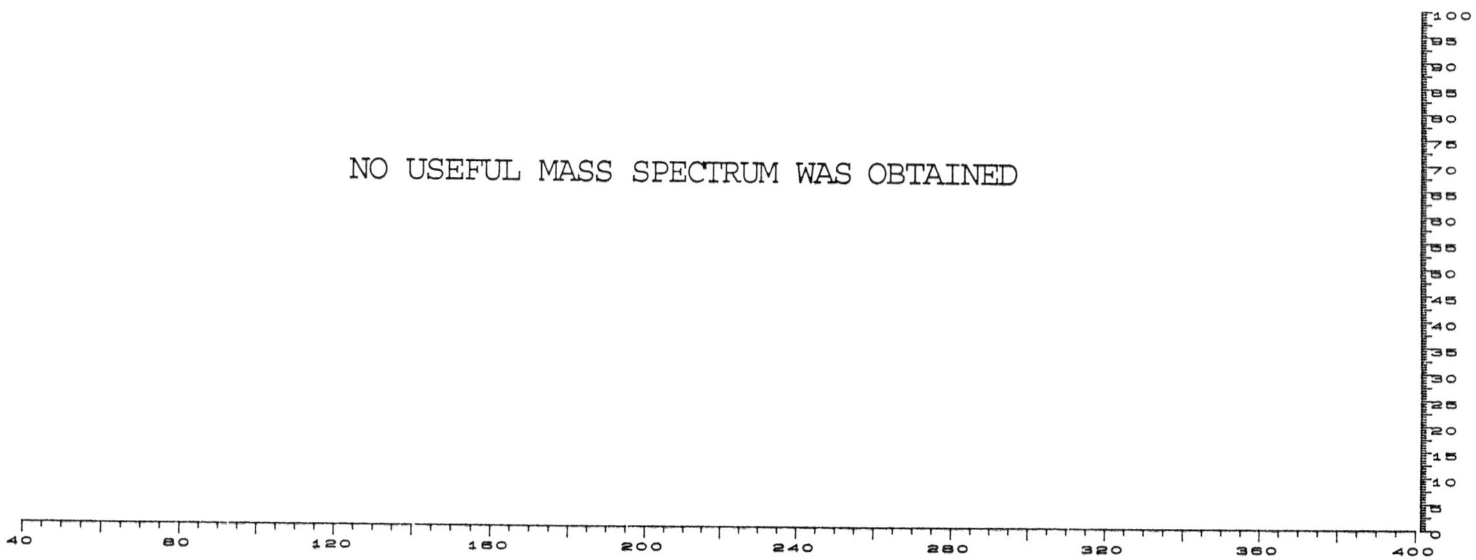

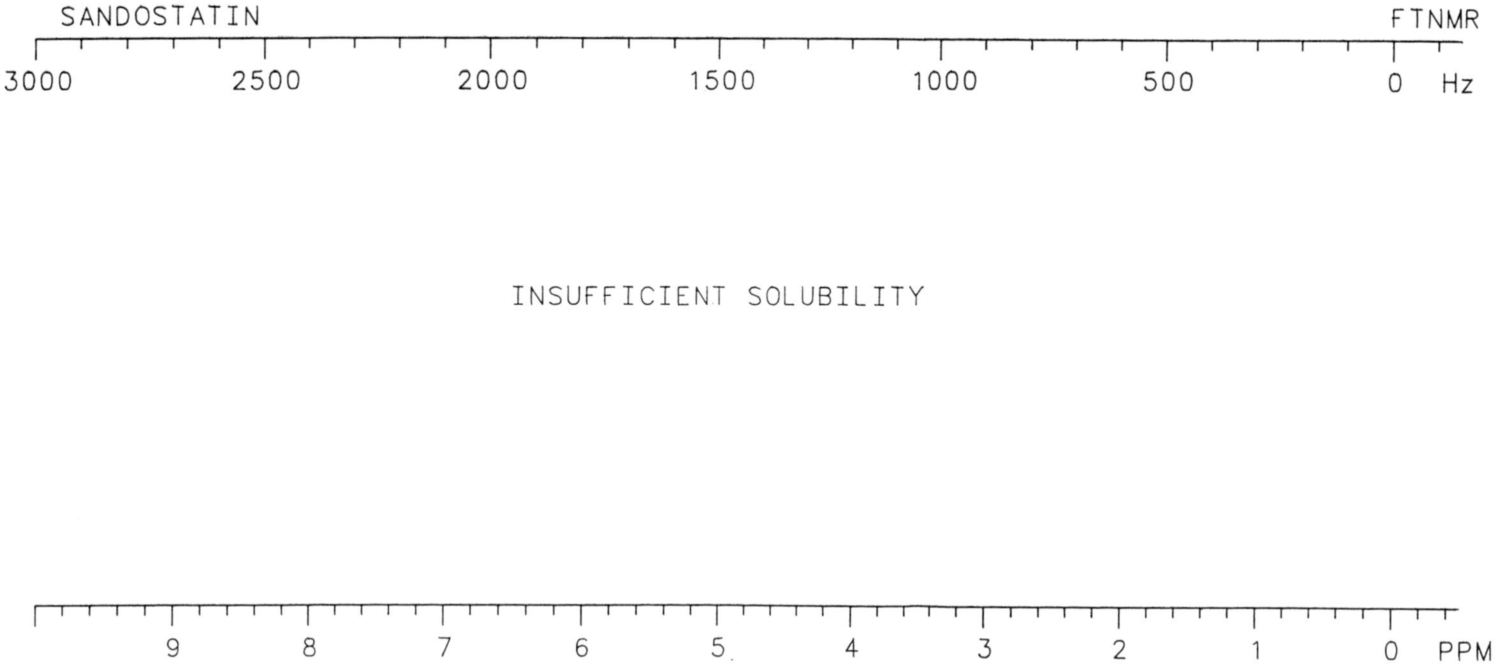
SANDOSTATIN
FTNMR
3000
2500
2000
1500
1000
500
0 Hz
INSUFFICIENT SOLUBILITY
9
8
7
6
5
4
3
2
1
0 PPM

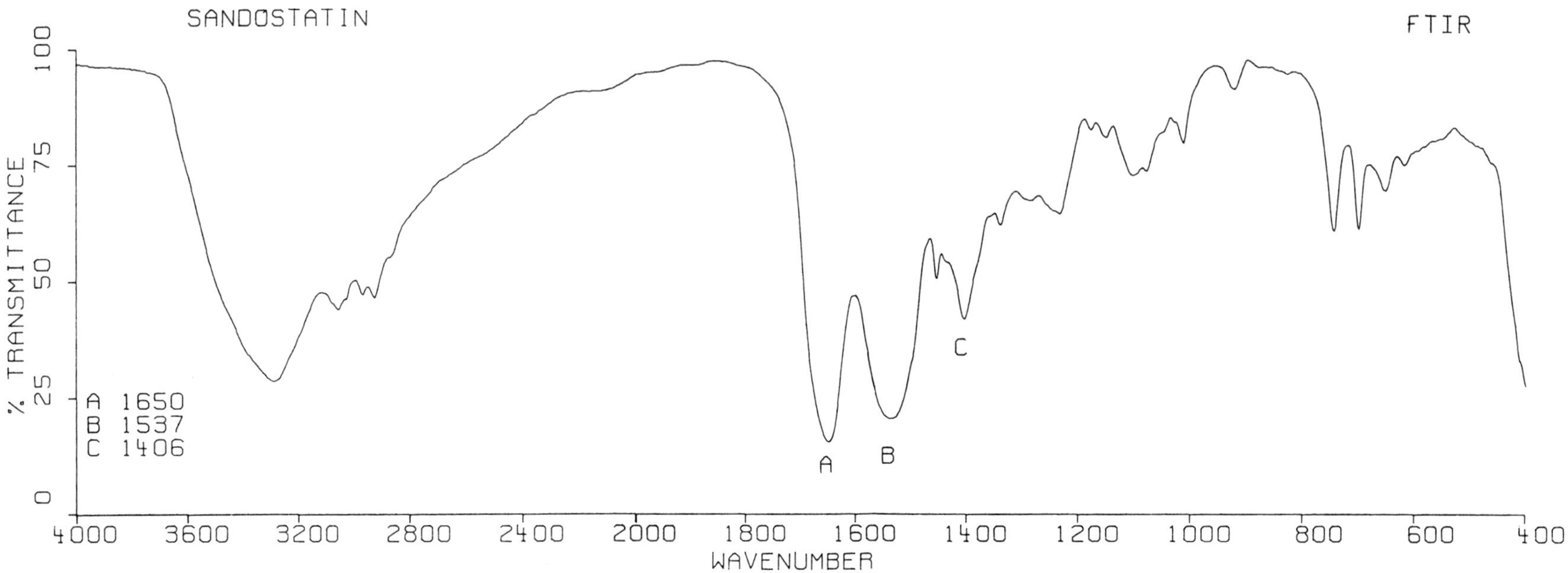
SANDOSTATIN
FTIR
% TRANSMITTANCE
100
75
50
25
0
A 1650
B 1537
C 1406
A
B
C
4000
3600
3200
2800
2400
2000
1800
1600
1400
1200
1000
800
600
400
WAVENUMBER

SERTRALINE

$C_{17}H_{17}Cl_2N$

Molecular Weight: 306.23 (305.07)

Synonyms: (1S-cis)-4-(3,4-Dichlorophenyl)-1,2,3,4-tetrahydro-N-methyl-1-naphthalenamine
Trade Names: Zoloft
Use: Antidepressant

HPLC:
GC: 2401; 250°

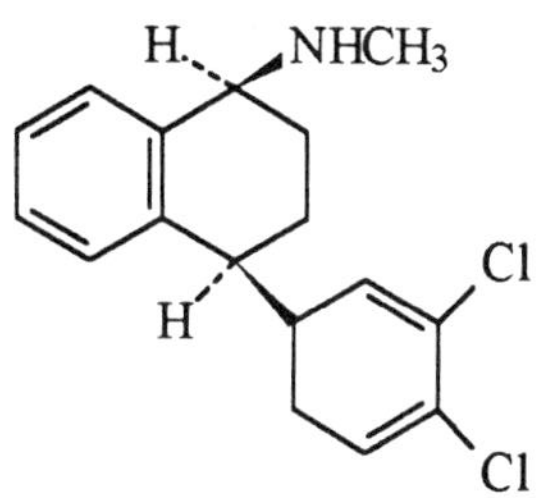

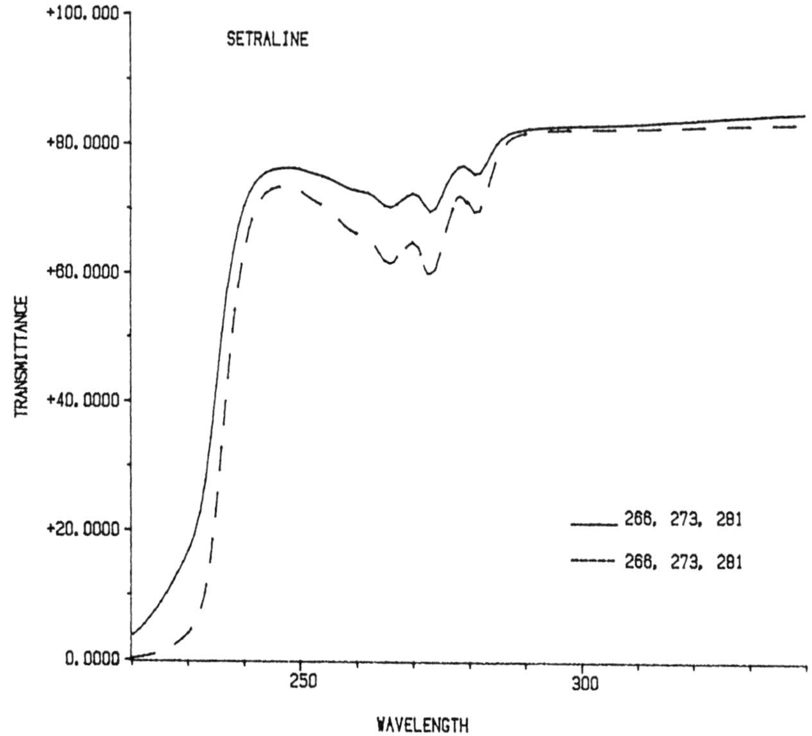

SERTRALINE

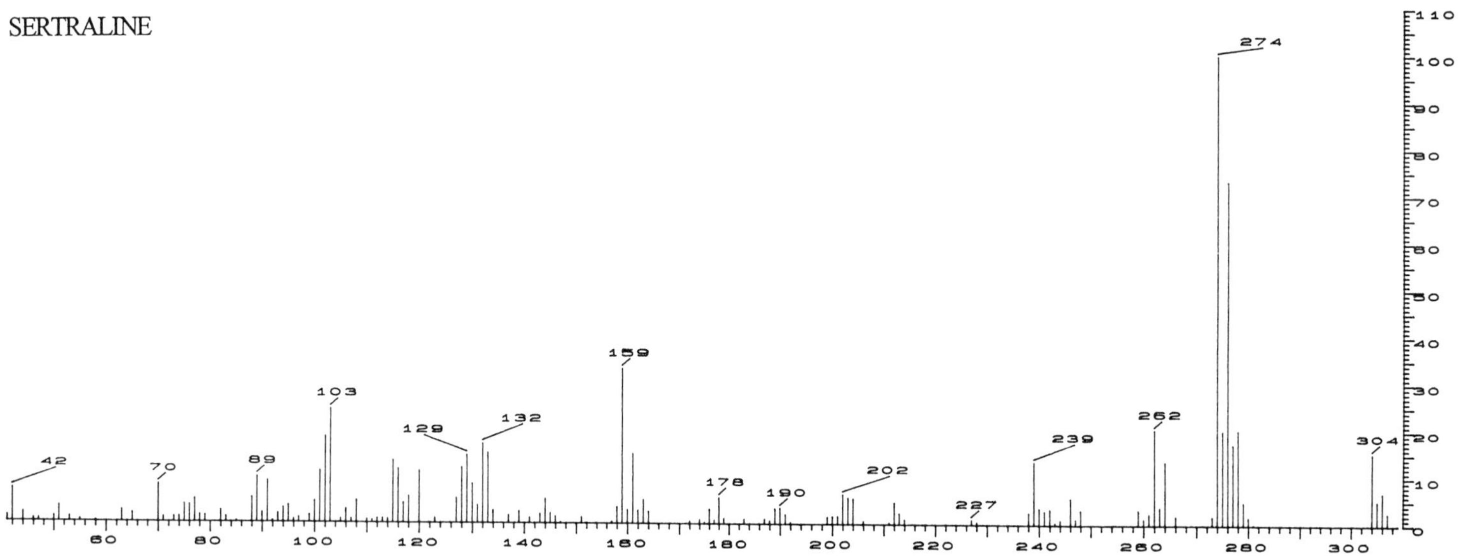

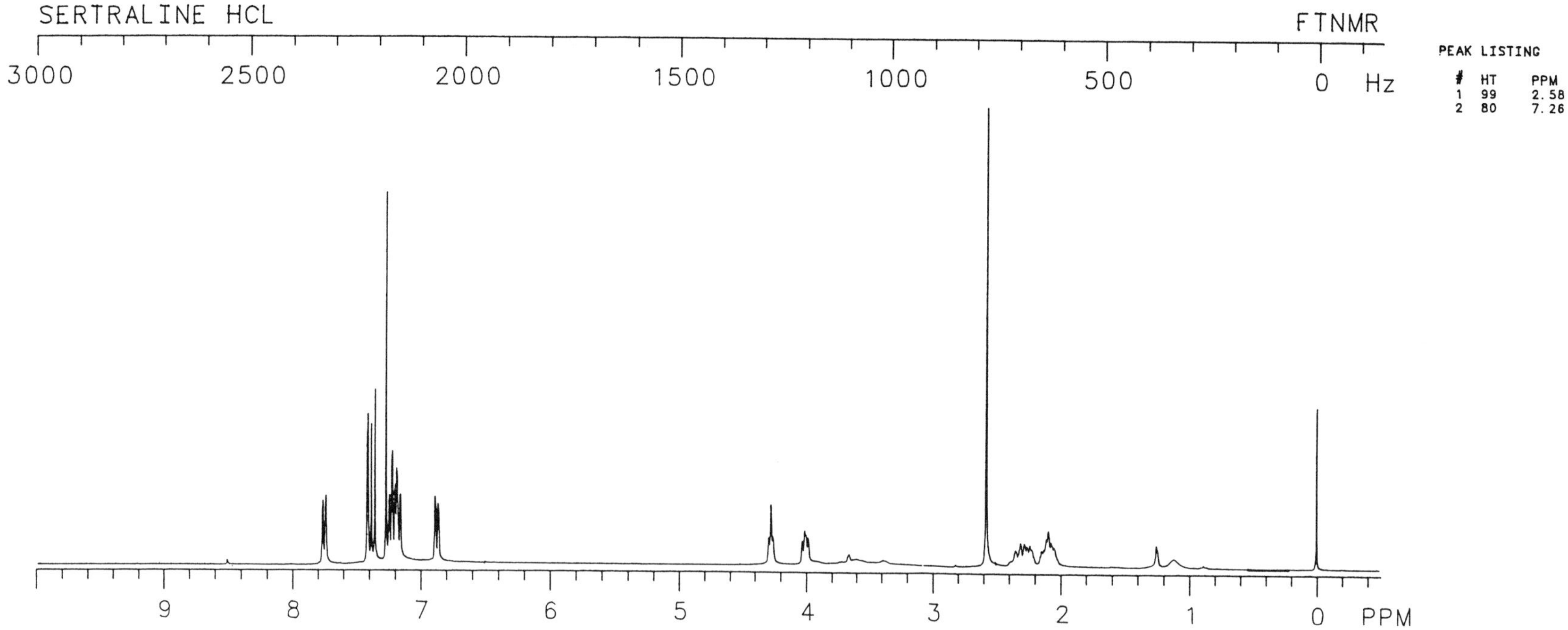
SERTRALINE HCL
FTNMR
3000 2500 2000 1500 1000 500 0 Hz
PEAK LISTING
HT PPM
1 99 2.58
2 80 7.26
9 8 7 6 5 4 3 2 1 0 PPM

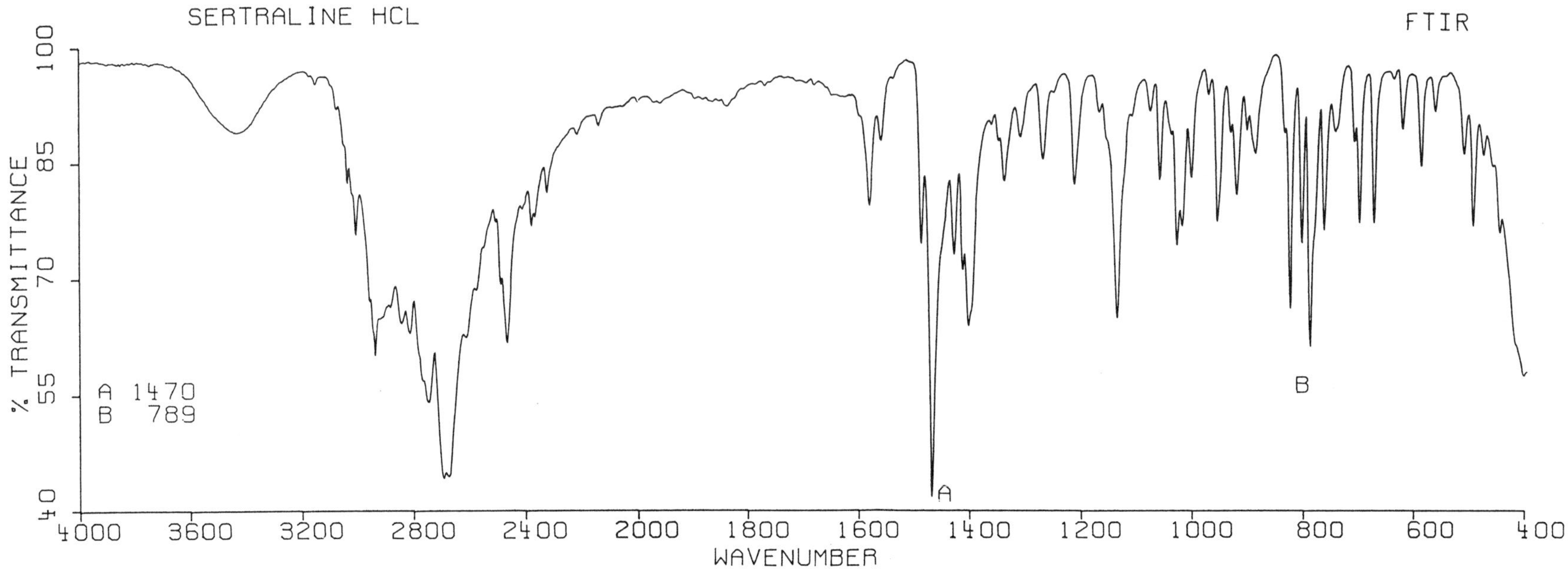
SERTRALINE HCL
FTIR
% TRANSMITTANCE
100 85 70 55 40
A 1470
B 789
A
B
4000 3600 3200 2800 2400 2000 1800 1600 1400 1200 1000 800 600 400
WAVENUMBER

SIMVASTATIN

$C_{25}H_{38}O_5$

Molecular Weight: 418.57 (418.27)

Synonyms: [1S-[1α,3α,7β,8β(2S*,4S*),8aβ]]-2,2-Dimethylbutanoic acid 1,2,3,7,8,8a-hexahydro-3,7-dimethyl-8-[2-(tetrahydro-4-hydroxy-6-oxo-2H-pyran-2-yl)ethyl]-1-naphthalenyl ester; synvinolin
Trade Names: Zocor, Zocord
Use: Antihyperlipidemic

HPLC: Methanol: 3.4
GC: 3158; 280°

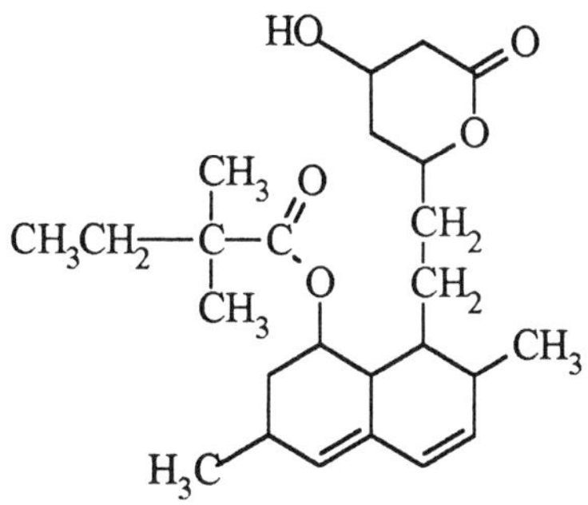

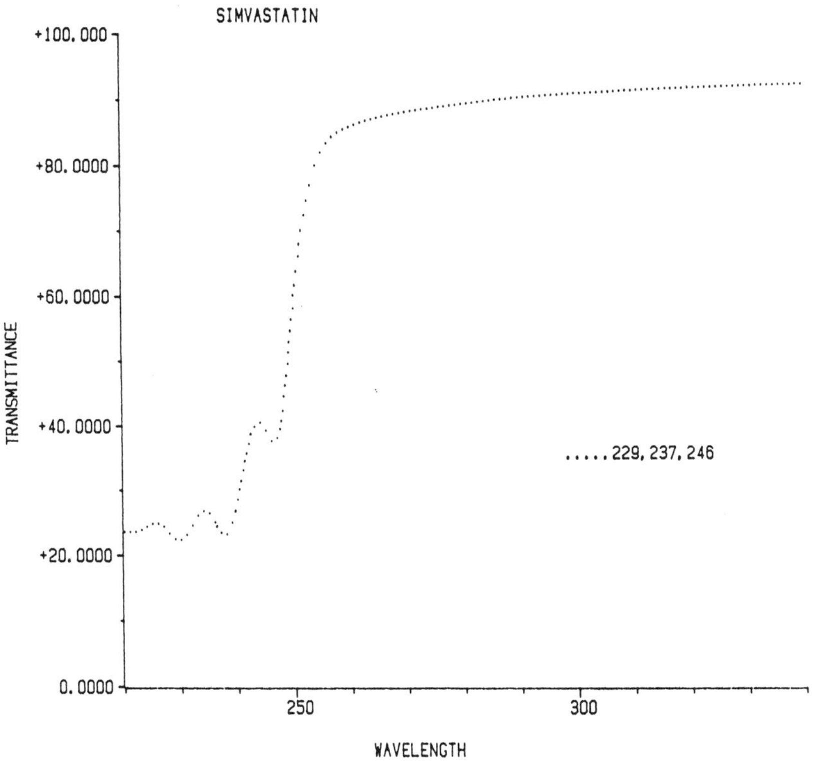

SIMVASTATIN

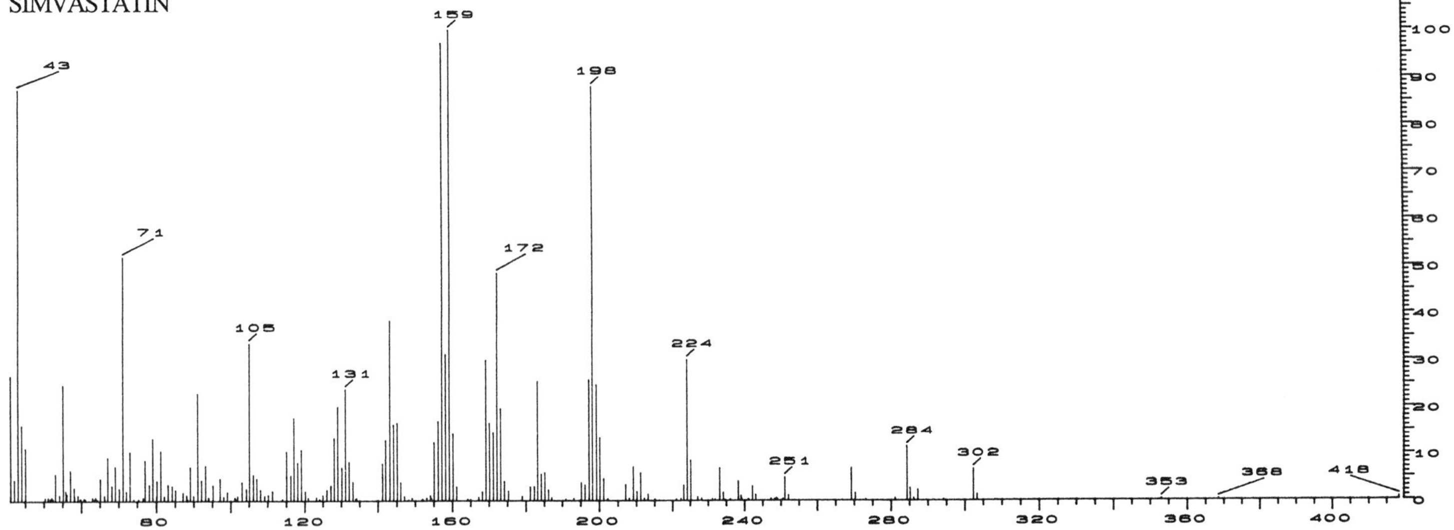

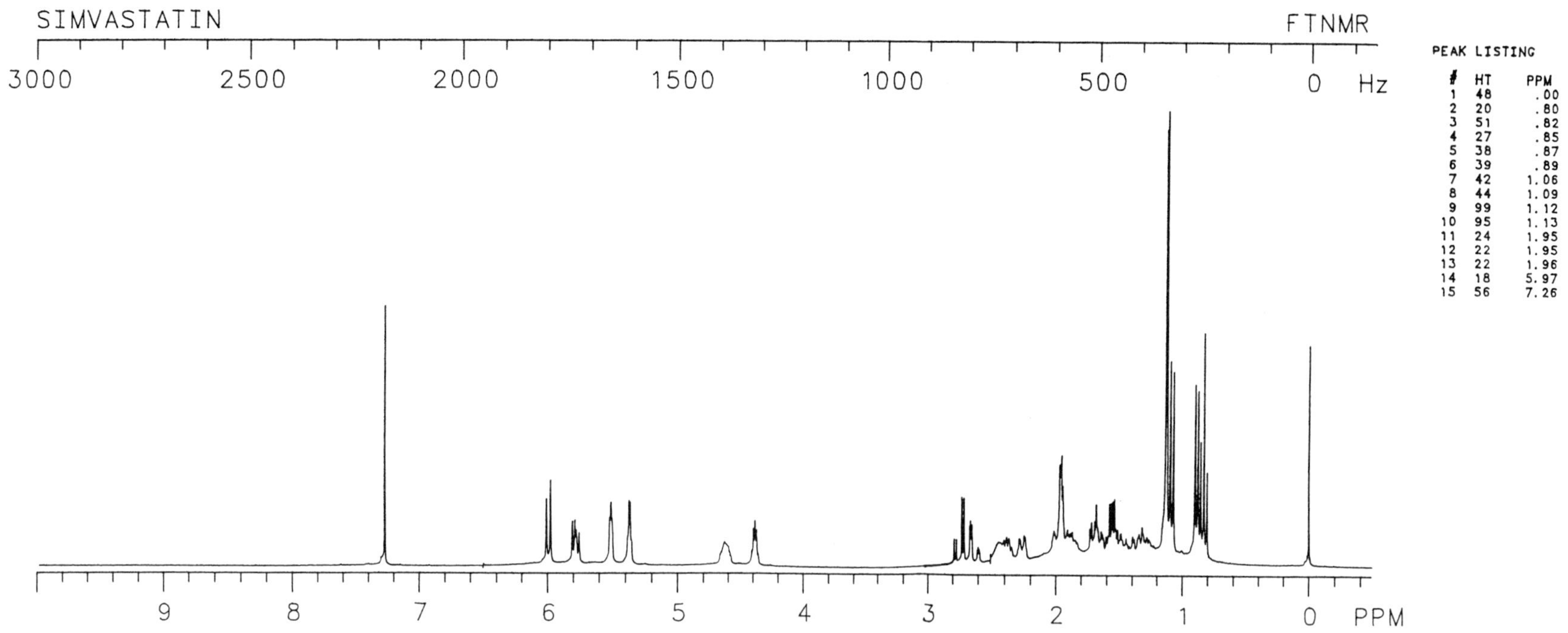
SIMVASTATIN
FTNMR
3000 2500 2000 1500 1000 500 0 Hz
9 8 7 6 5 4 3 2 1 0 PPM
PEAK LISTING
HT PPM
1 48 .00
2 20 .80
3 51 .82
4 27 .85
5 38 .87
6 39 .89
7 42 1.06
8 44 1.09
9 99 1.12
10 95 1.13
11 24 1.95
12 22 1.95
13 22 1.96
14 18 5.97
15 56 7.26

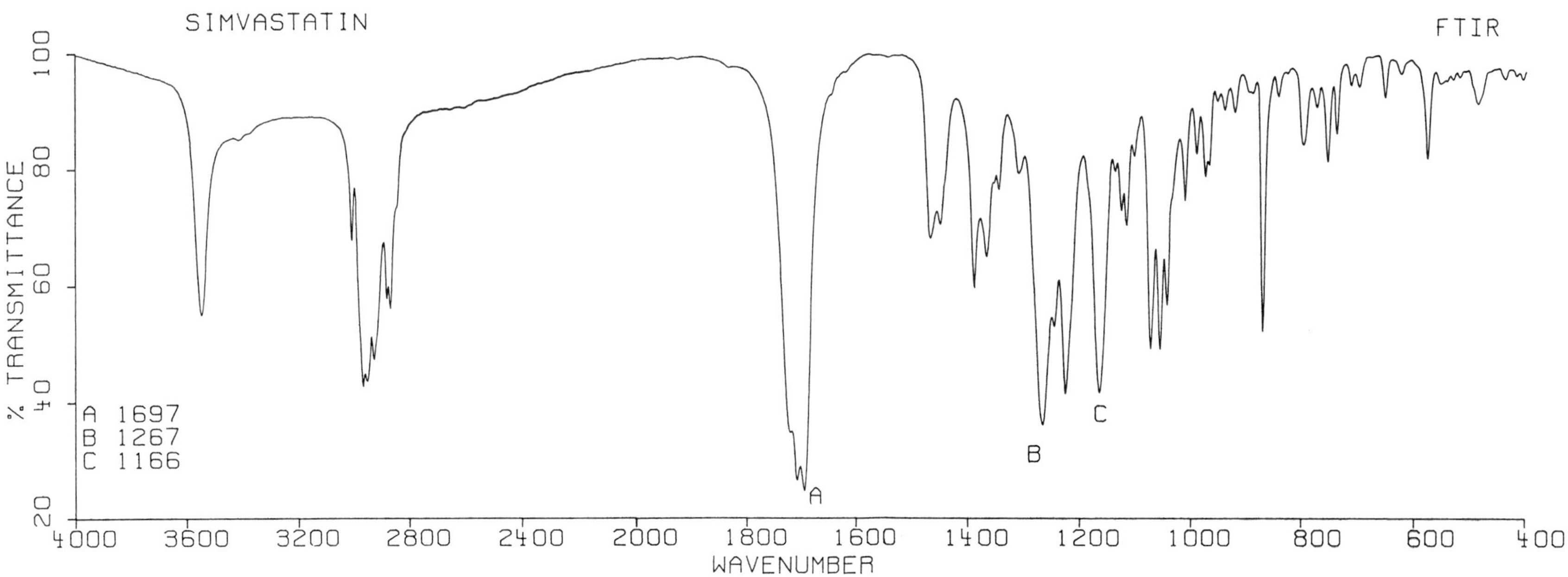
SIMVASTATIN
FTIR
% TRANSMITTANCE
100 80 60 40 20
A 1697
B 1267
C 1166
A
B
C
4000 3600 3200 2800 2400 2000 1800 1600 1400 1200 1000 800 600 400
WAVENUMBER

SOTALOL

$C_{12}H_{20}N_2O_3S$

Molecular Weight: 272.36 (272.12)

Synonyms: N-[4-[1-Hydroxy-2-[(1-methylethyl)amino]ethyl]phenyl]-methansulfonamide
Trade Names: Beta-Cardone, Betapace, Sotacor, Sotalex
Use: Antihypertensive

HPLC:
GC:

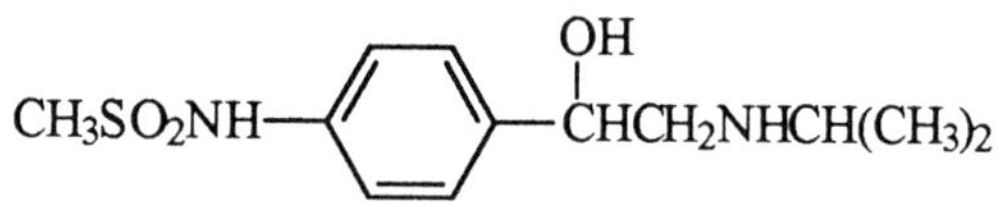

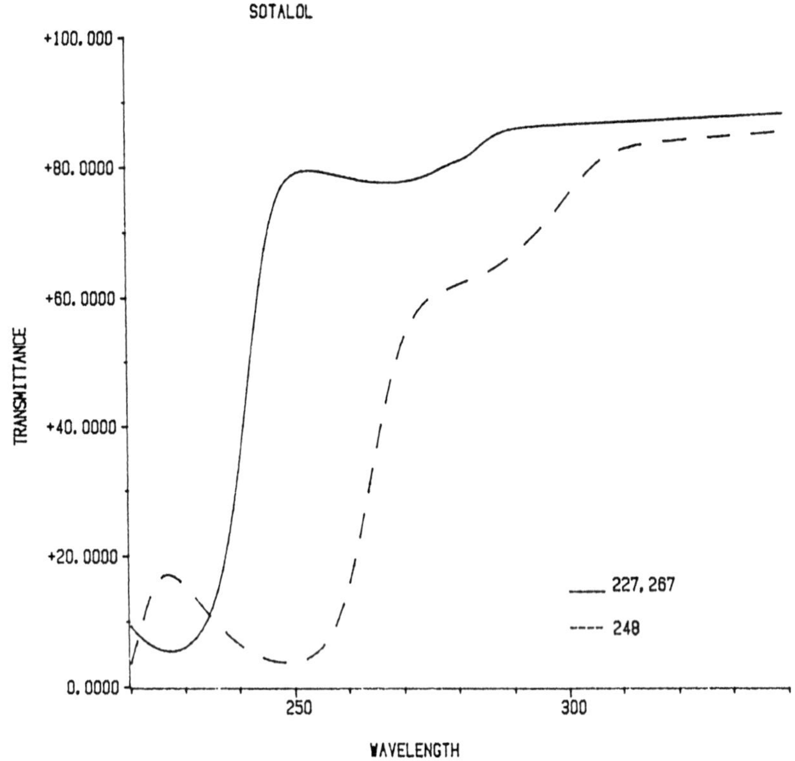

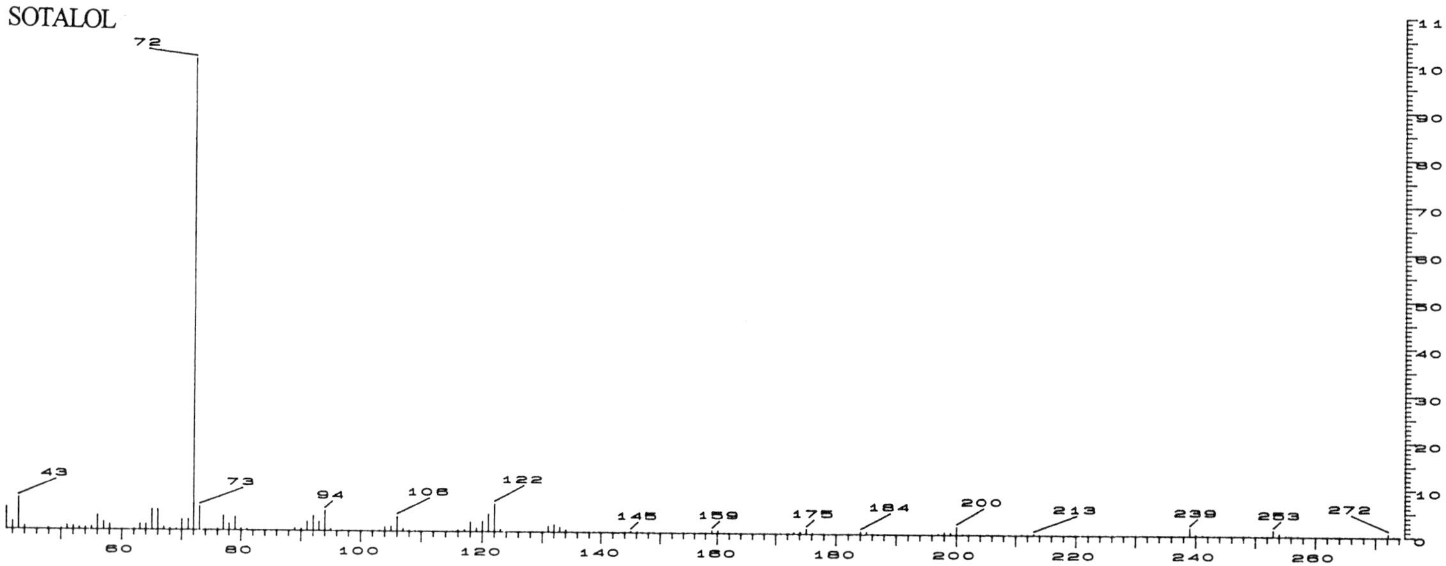

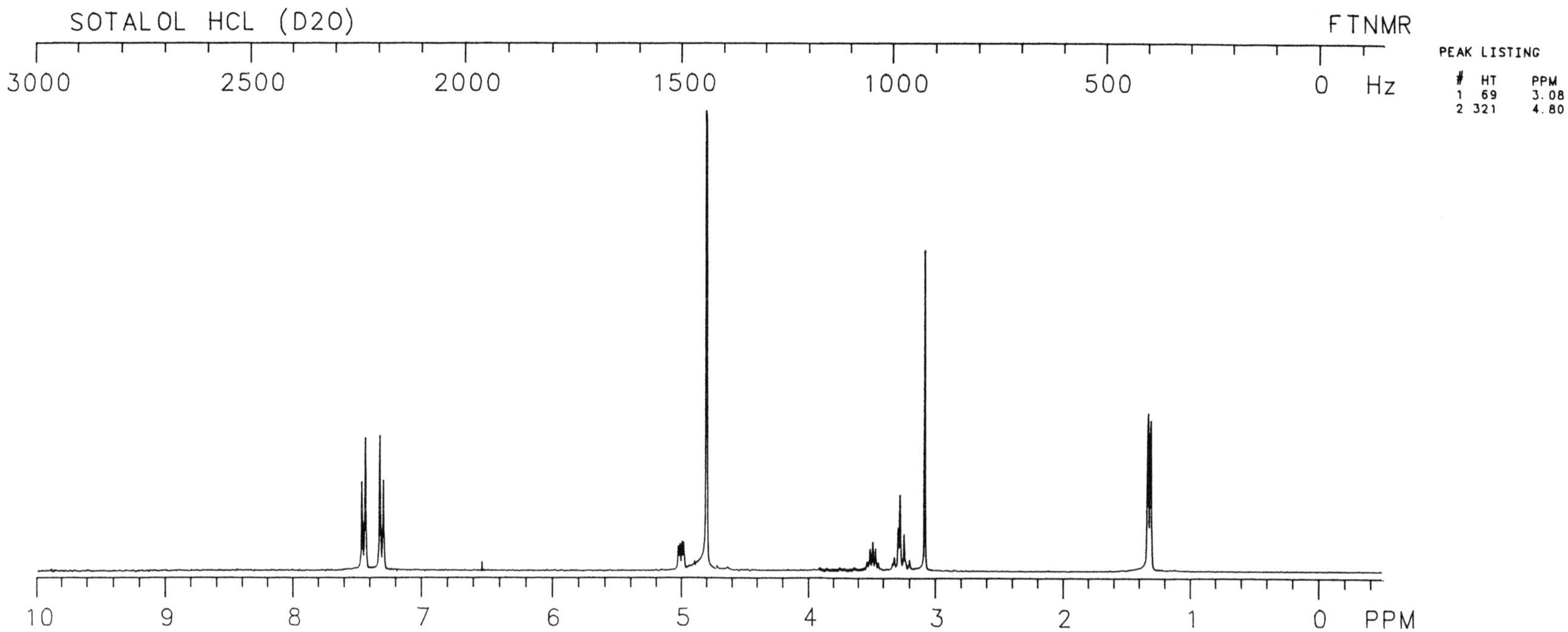
SOTALOL HCL (D2O)
FTNMR
3000 2500 2000 1500 1000 500 0 Hz
10 9 8 7 6 5 4 3 2 1 0 PPM
PEAK LISTING
HT PPM
1 69 3.08
2 321 4.80

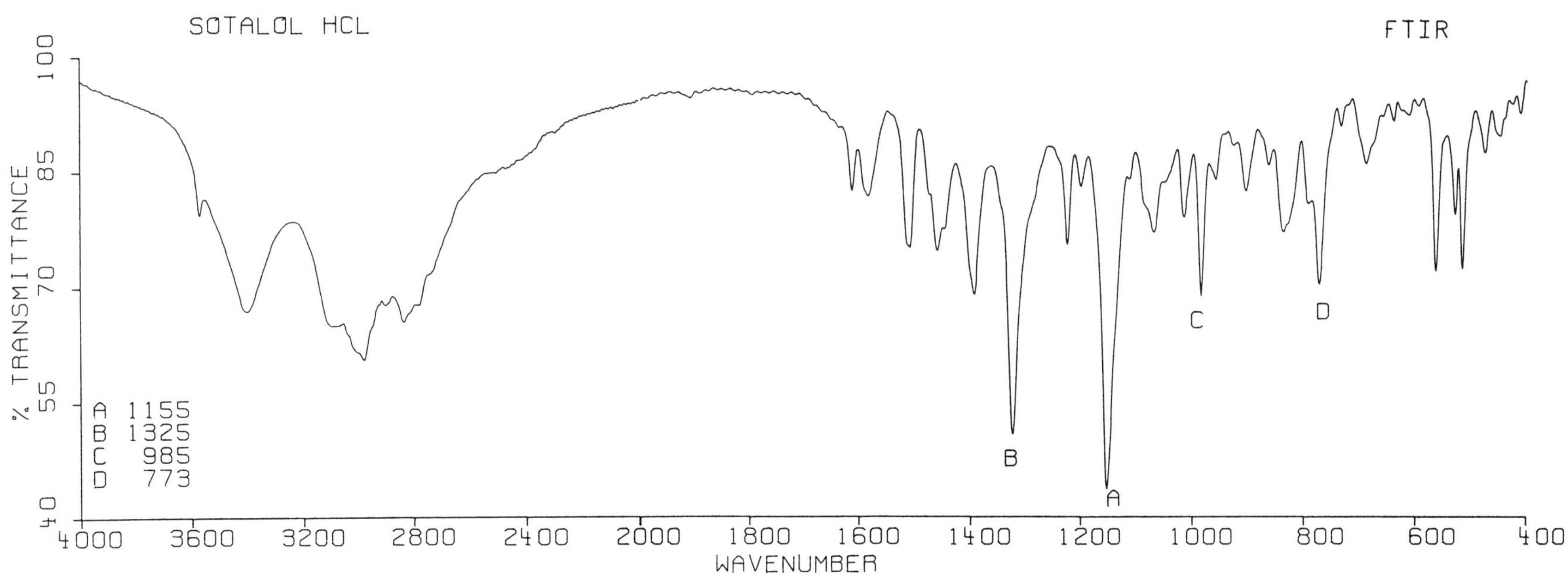
SOTALOL HCL
FTIR
% TRANSMITTANCE
100 85 70 55 40
4000 3600 3200 2800 2400 2000 1800 1600 1400 1200 1000 800 600 400
WAVENUMBER
A 1155
B 1325
C 985
D 773
A
B
C
D

STREPTOZOCIN

$C_8H_{15}N_3O_7$

Molecular Weight: 265.22 (265.09)

Synonyms: 2-Deoxy-2-[[(methylnitroamino)carbonyl]amino]-D-glucopyranose; streptozotocin
Trade Names: Zanosar
Use: Antineoplastic

HPLC: Methanol: 2.6
GC:

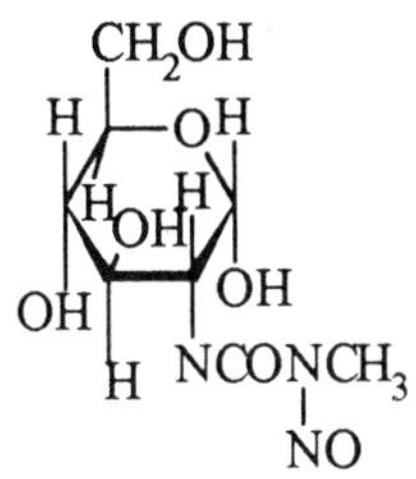

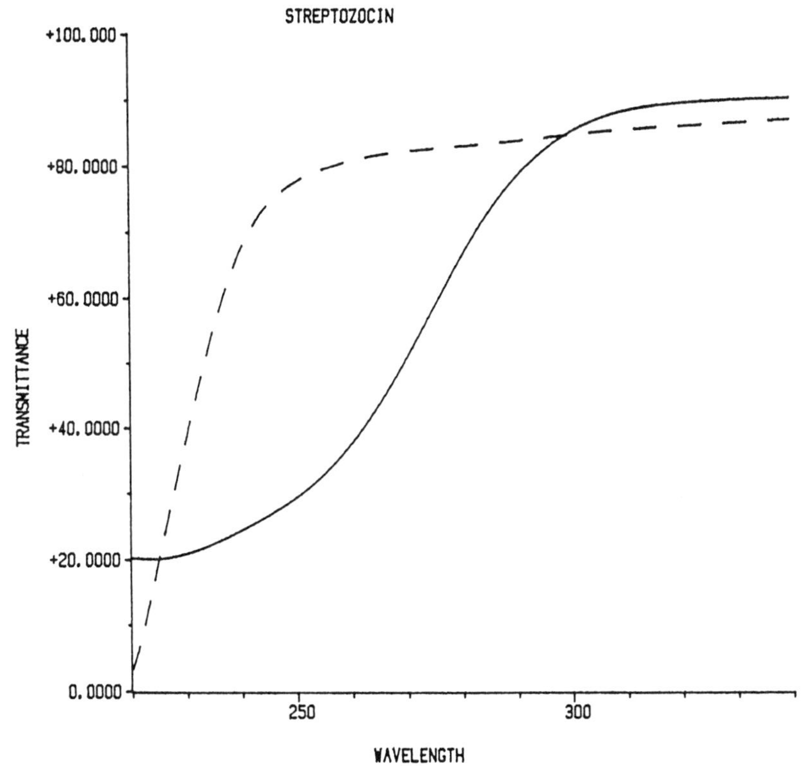

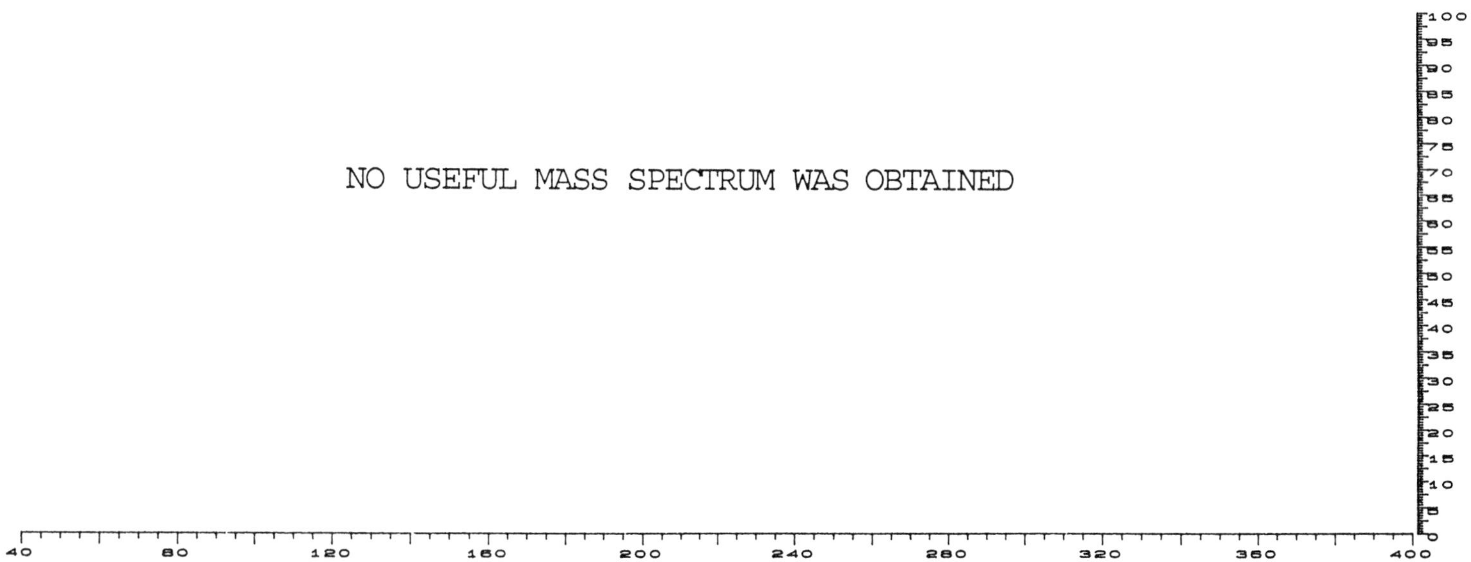

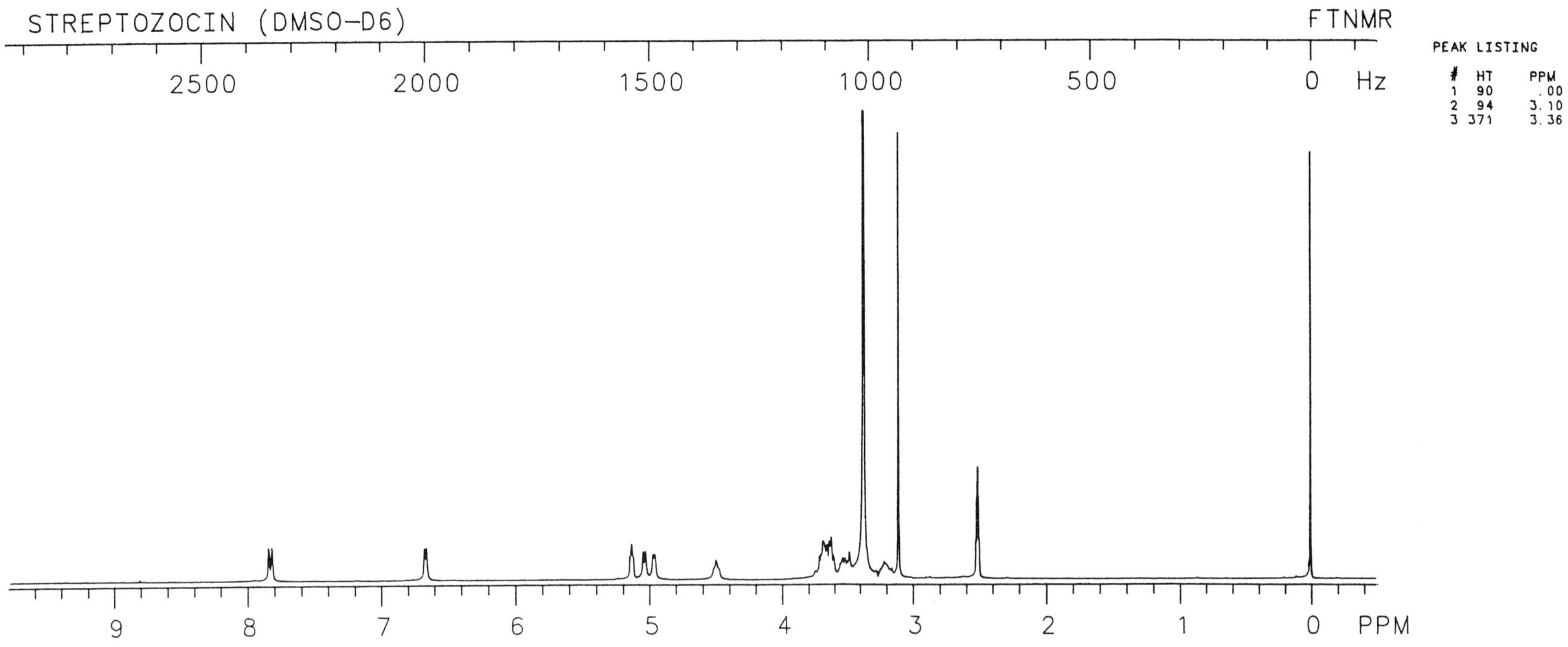

STREPTOZOCIN (DMSO-D6)
FTNMR
2500
2000
1500
1000
500
0 Hz
9
8
7
6
5
4
3
2
1
0 PPM
PEAK LISTING
HT PPM
1 90 .00
2 94 3.10
3 371 3.36

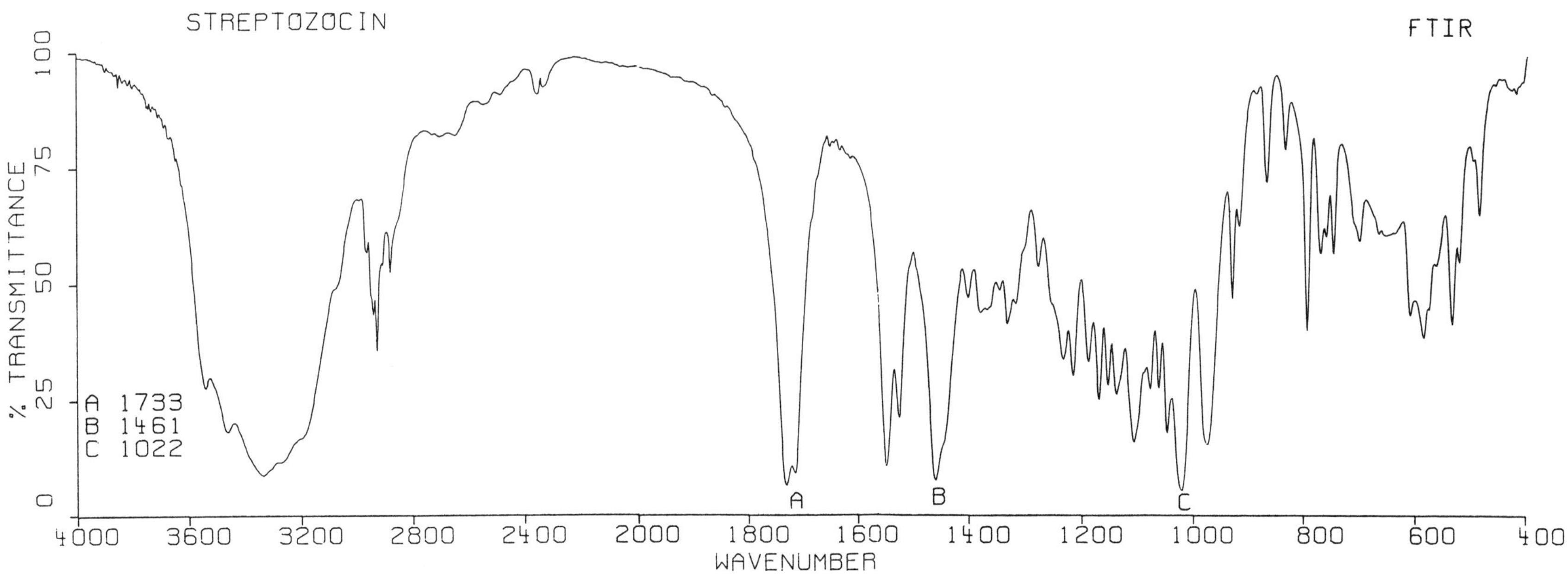

STREPTOZOCIN
FTIR
% TRANSMITTANCE
100
75
50
25
0
A 1733
B 1461
C 1022
A
B
C
4000
3600
3200
2800
2400
2000
1800
1600
1400
1200
1000
800
600
400
WAVENUMBER

SUCCIMER

$C_4H_6O_4S_2$

Molecular Weight: 182.21 (181.97)

Synonyms: (R*,S*)-2,3-Dimercaptobutanedioic; meso-2,3-dimercaptosuccinic acid
Trade Names: Chemet
Use: Chelating agent

HPLC:
GC:

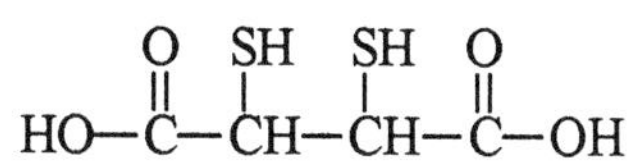

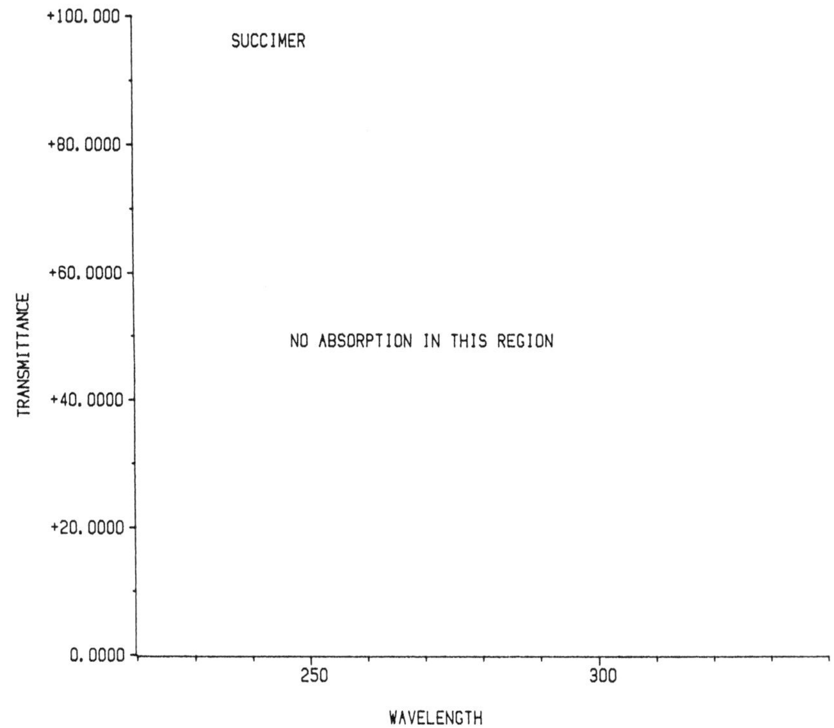

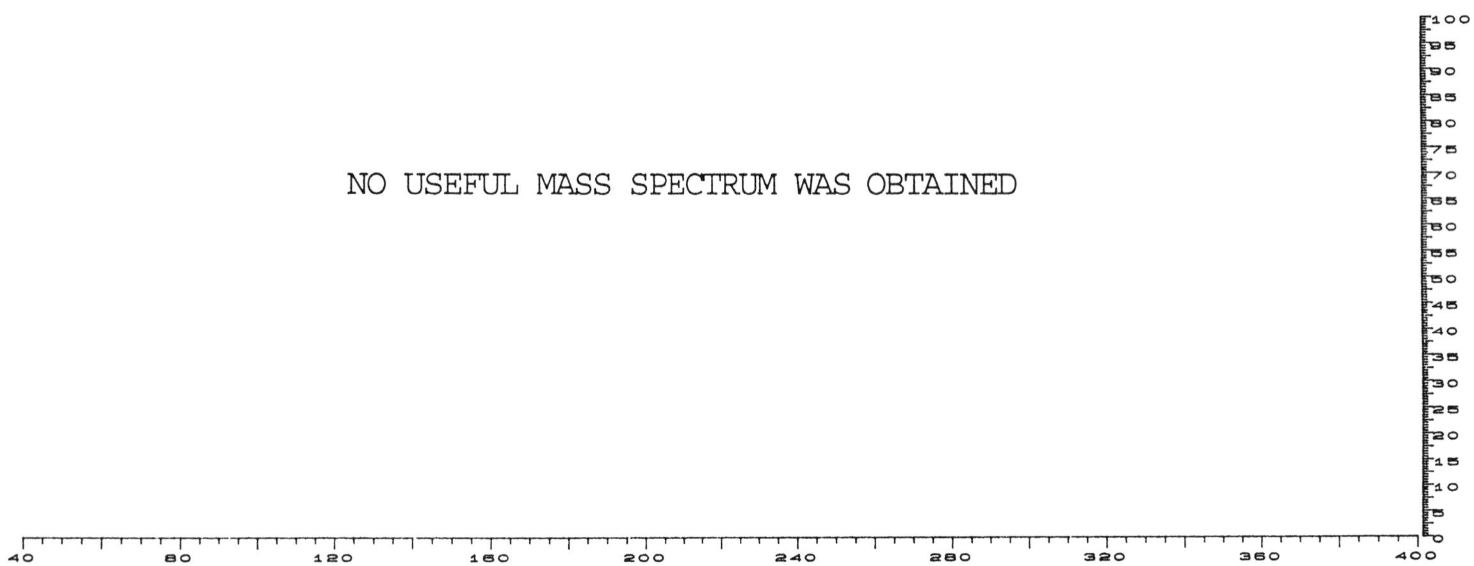

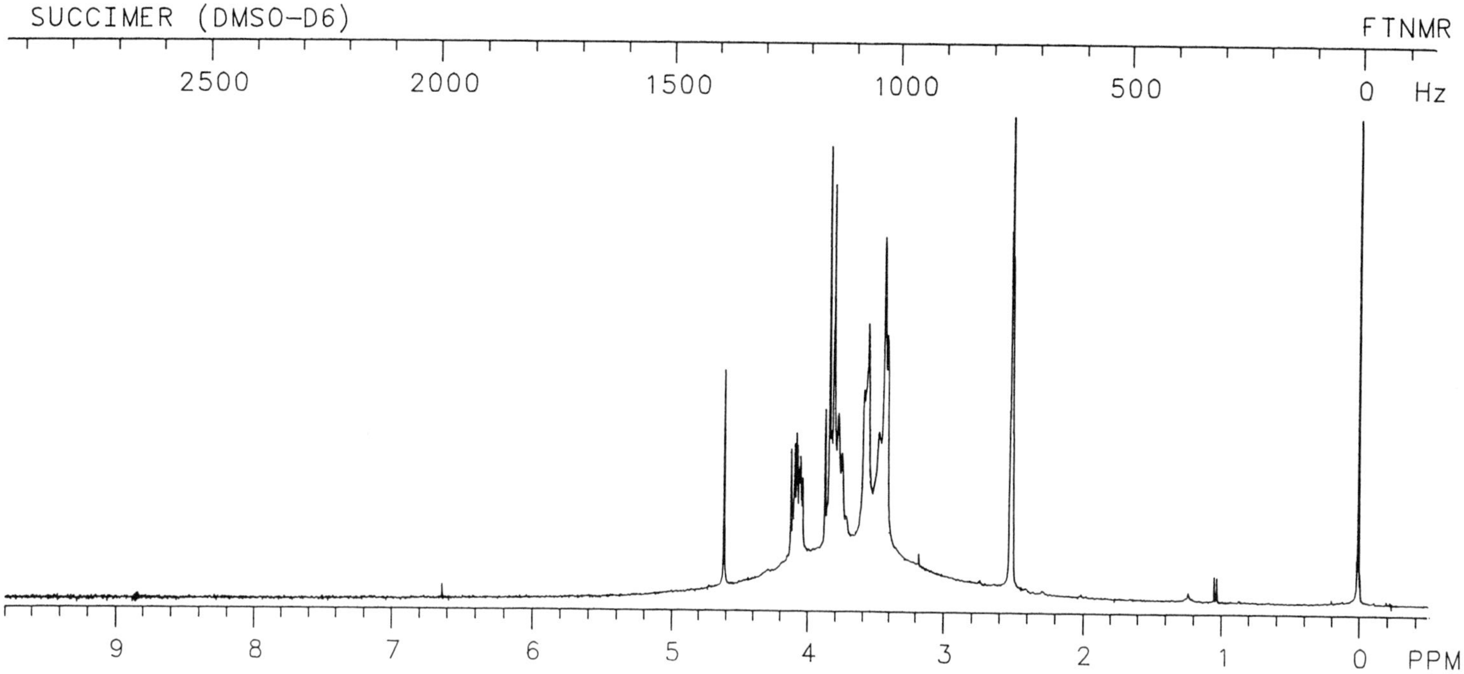

PEAK LISTING

#	HT	PPM
1	204	.00
2	70	2.50
3	98	2.50
4	75	2.51
5	54	3.41
6	74	3.43
7	56	3.55
8	84	3.80
9	55	3.81
10	92	3.83
11	46	4.60

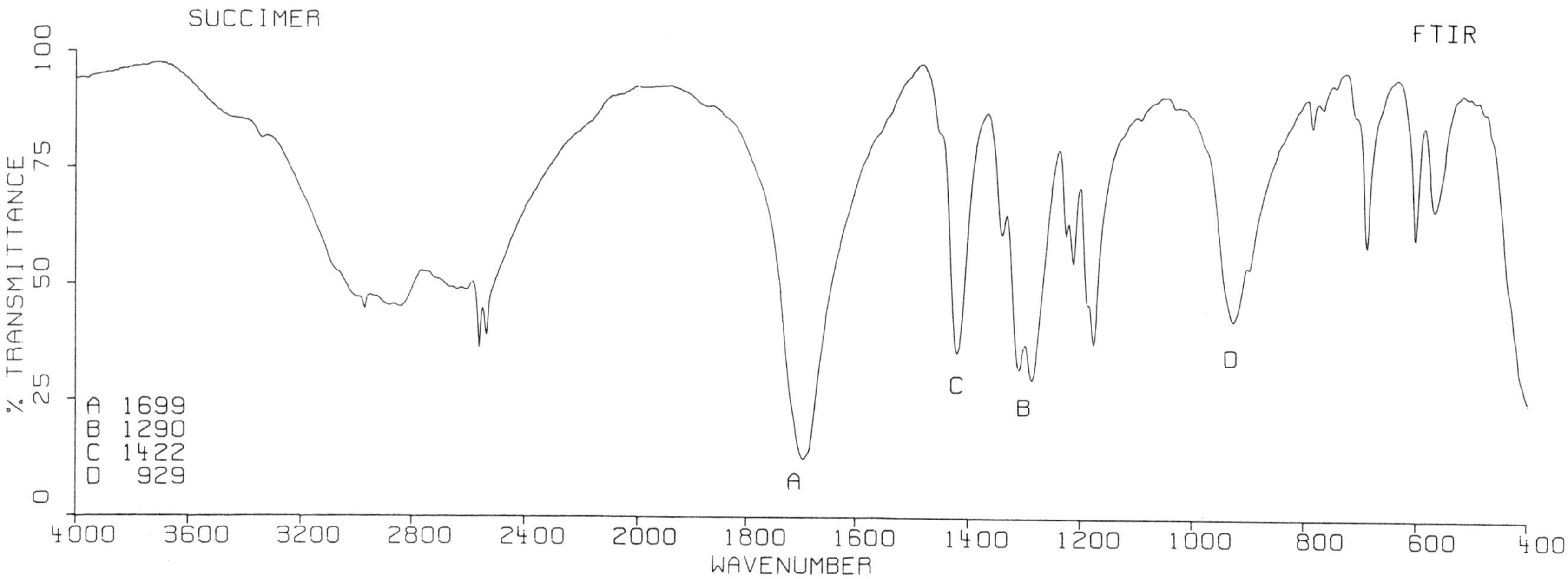

SULBACTAM SODIUM

$C_8H_{10}NNaO_5S$

Molecular Weight: 255.22 (255.02)

Synonyms: (2S-cis)-3,3-Dimethyl-7-oxo-4-thia-1-azabicyclo[3.2.0]heptane-2-carboxylic acid 4,4-dioxide sodium salt; penicillanic acid sulfone sodium salt
Trade Names: Loricin, Sulperazone, Unasyn, Unacid
Use: Antibacterial

HPLC:
GC:

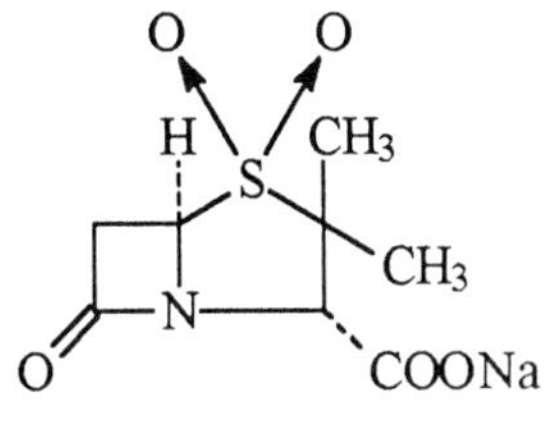

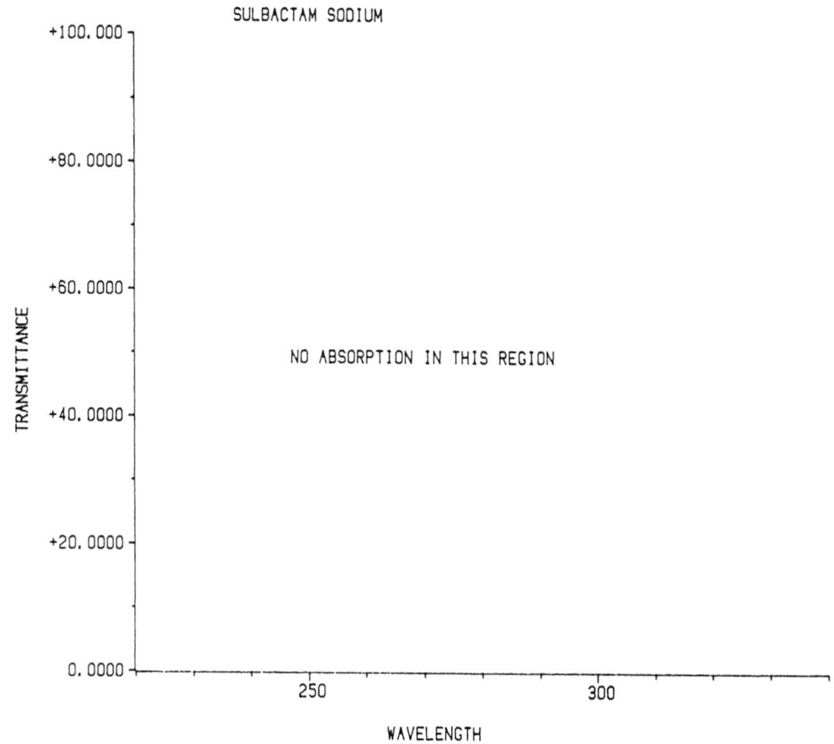

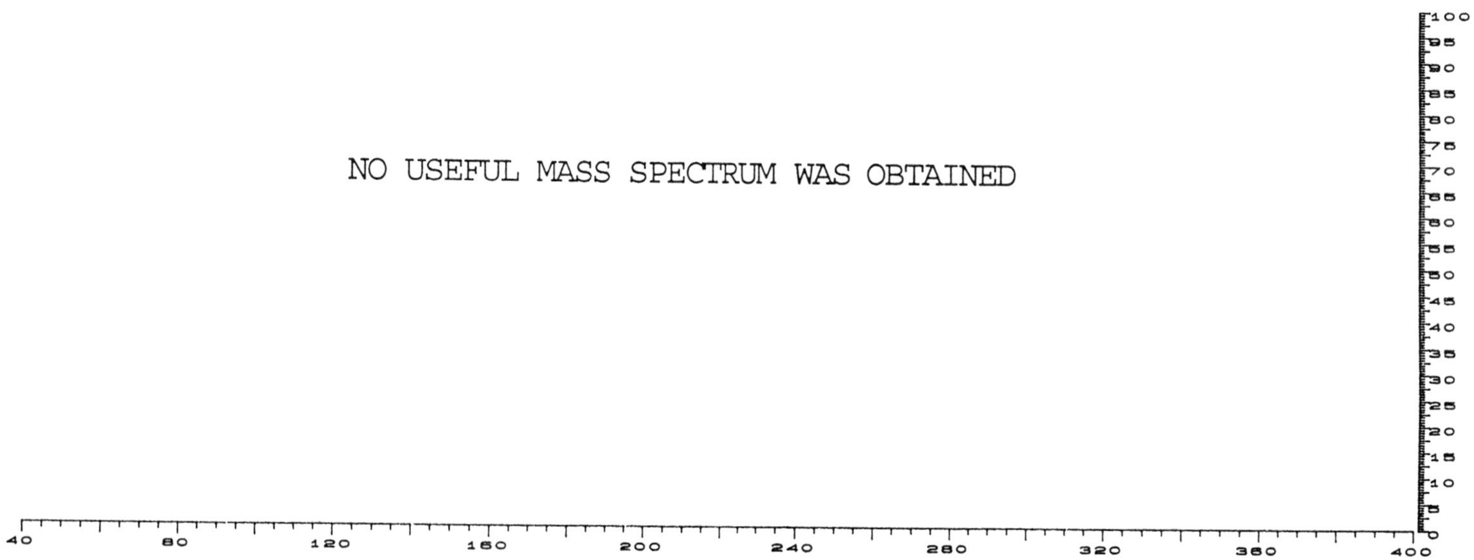

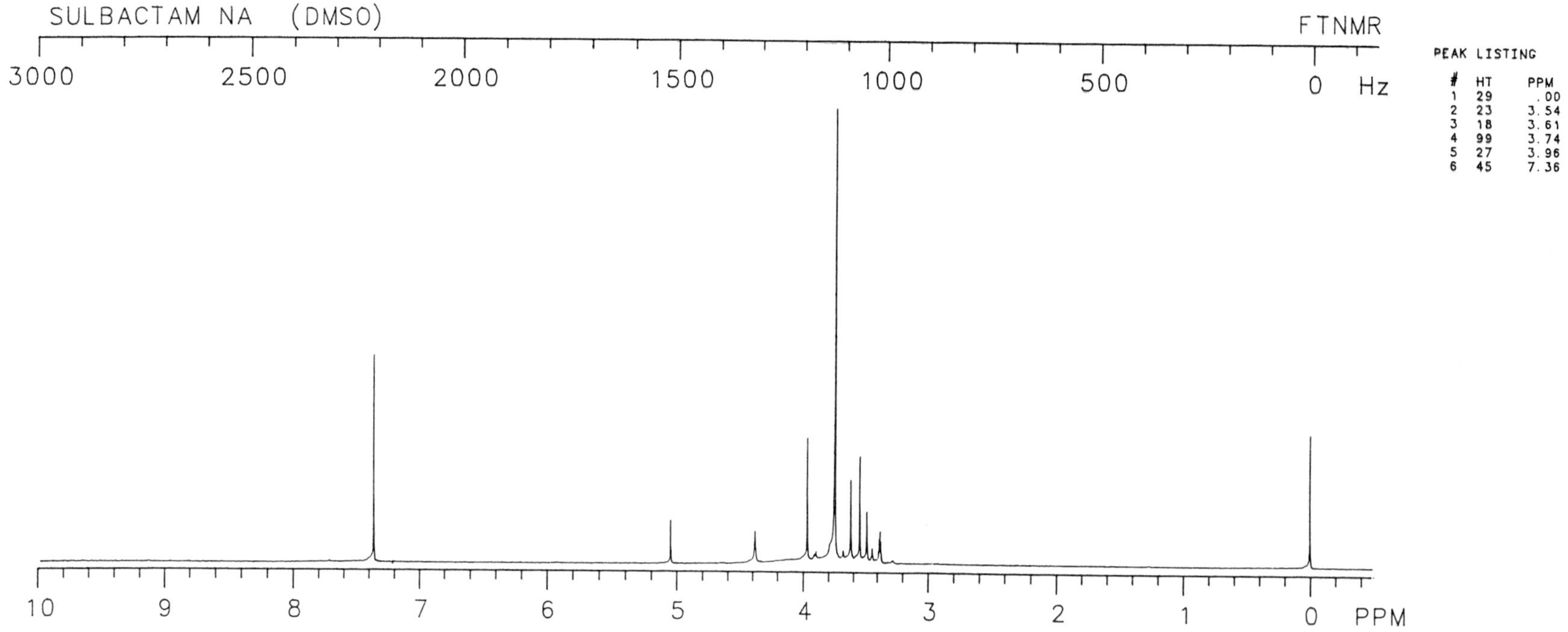
SULBACTAM NA (DMSO)
FTNMR
3000
2500
2000
1500
1000
500
0 Hz
PEAK LISTING
HT PPM
1 29 .00
2 23 3.54
3 18 3.61
4 99 3.74
5 27 3.96
6 45 7.36
10
9
8
7
6
5
4
3
2
1
0 PPM

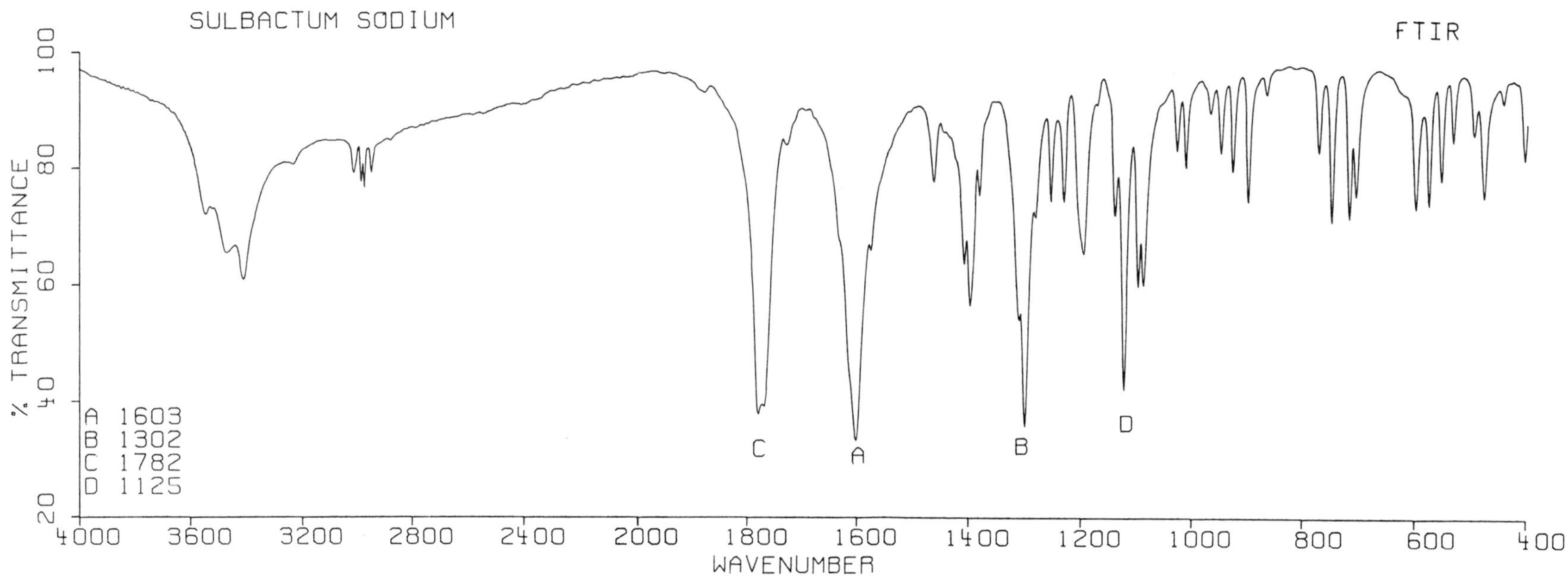
SULBACTUM SODIUM
FTIR
% TRANSMITTANCE
100
80
60
40
20
A 1603
B 1302
C 1782
D 1125
C
A
B
D
4000
3600
3200
2800
2400
2000
1800
1600
1400
1200
1000
800
600
400
WAVENUMBER

SULCONAZOLE

$C_{18}H_{15}Cl_3N_2S$

Molecular Weight: 397.75 (396.00)

Synonyms: 1-[2-[[(4-Chlorophenyl)methyl]thio]-2-(2,4-dichlorophenyl)ethyl]-1H-imidazole
Trade Names: Exelderm, Myk, Sulcosyn
Use: Antifungal

HPLC: Methanol: 3.6 (Tails)
GC:

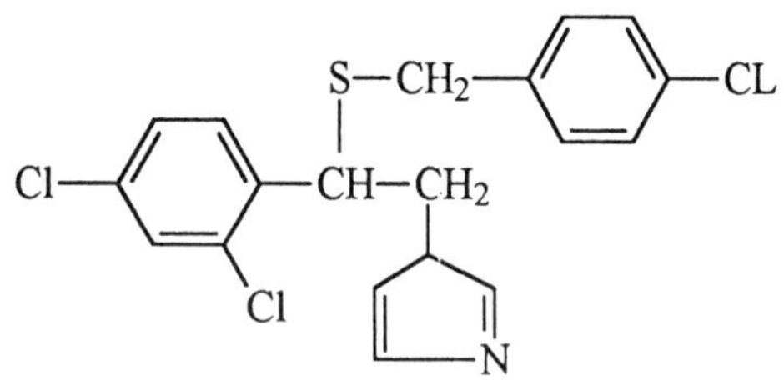

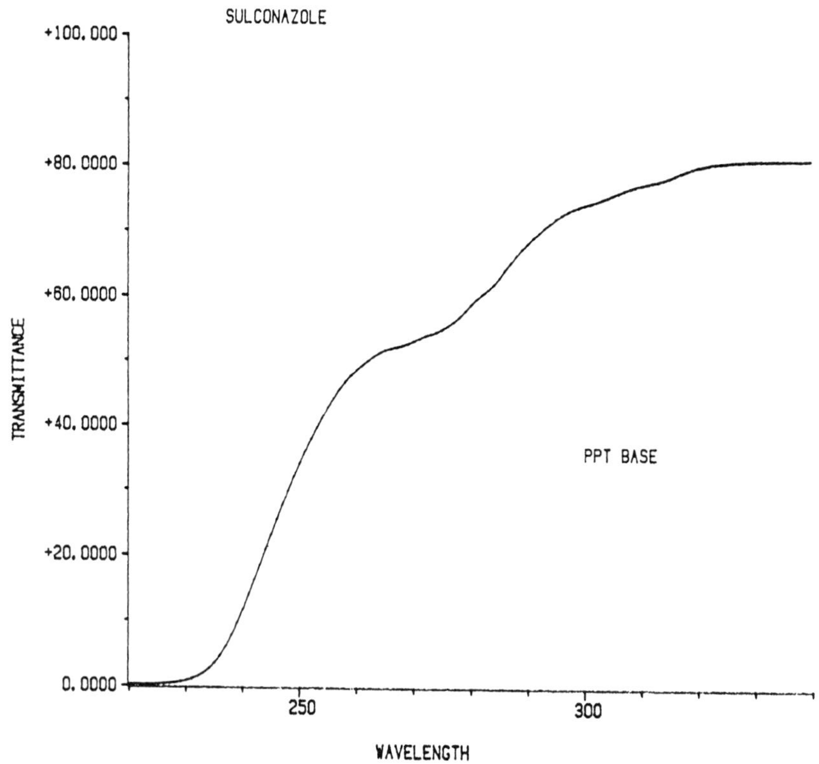

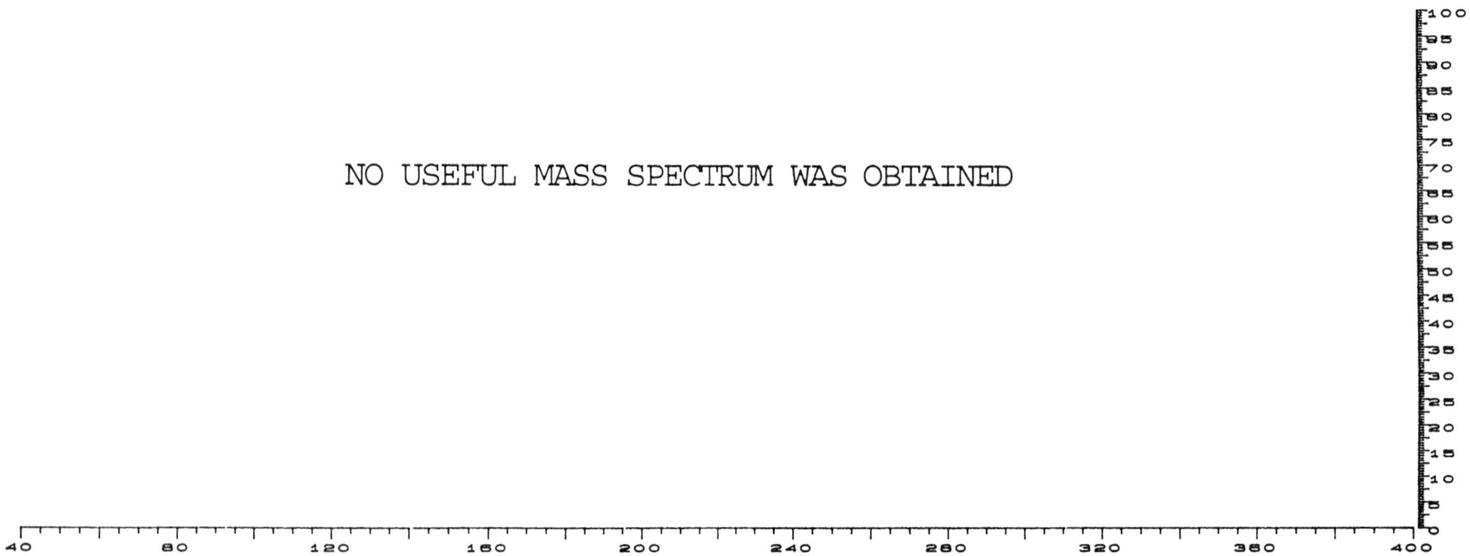

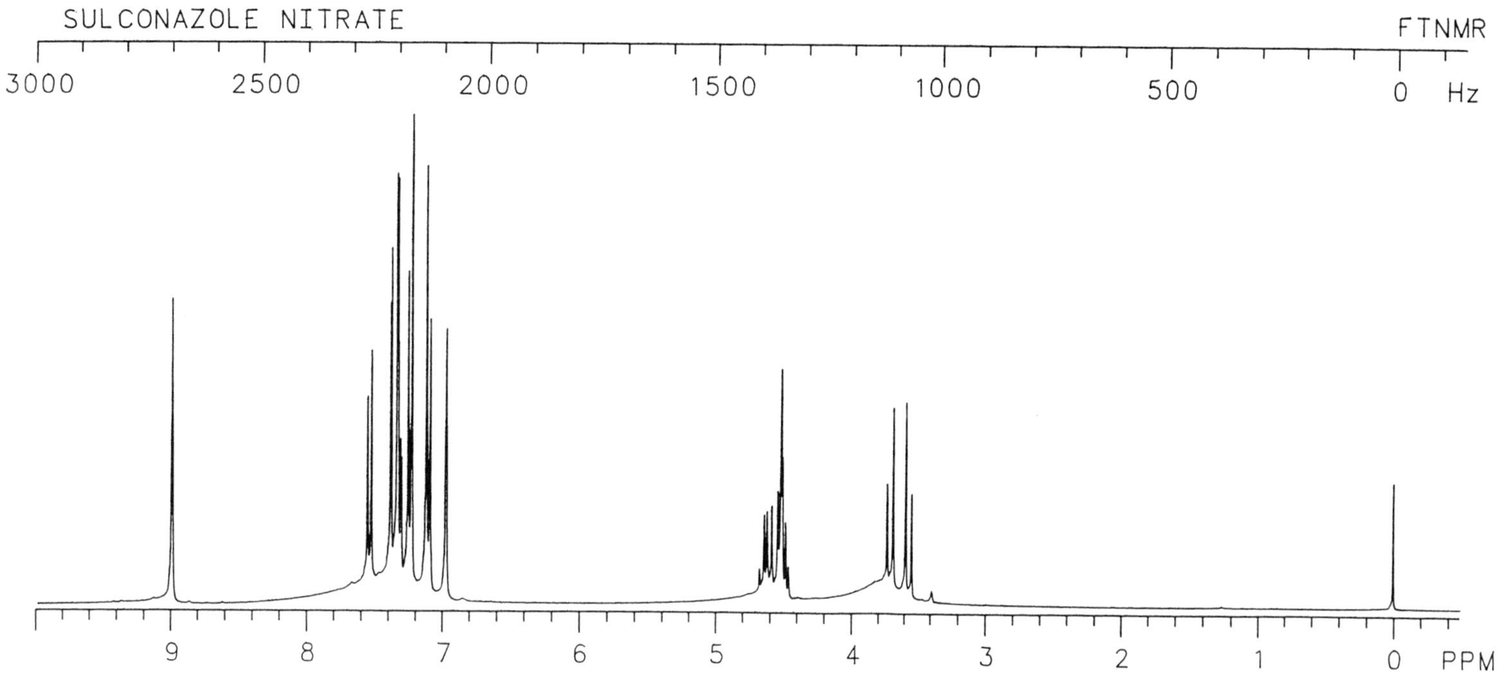

PEAK LISTING

#	HT	PPM
1	25	.00
2	23	3.53
3	41	3.58
4	40	3.68
5	25	3.72
6	17	4.48
7	30	4.50
8	48	4.51
9	23	4.53
10	20	4.58
11	19	4.61
12	18	4.63
13	36	6.96
14	56	6.97
15	36	6.97
16	58	7.08
17	89	7.11
18	99	7.22
19	35	7.22
20	35	7.24
21	67	7.25
22	16	7.25
23	29	7.29
24	33	7.30
25	53	7.32
26	86	7.32
27	87	7.33
28	72	7.37
29	61	7.37
30	51	7.51
31	42	7.54
32	62	8.98

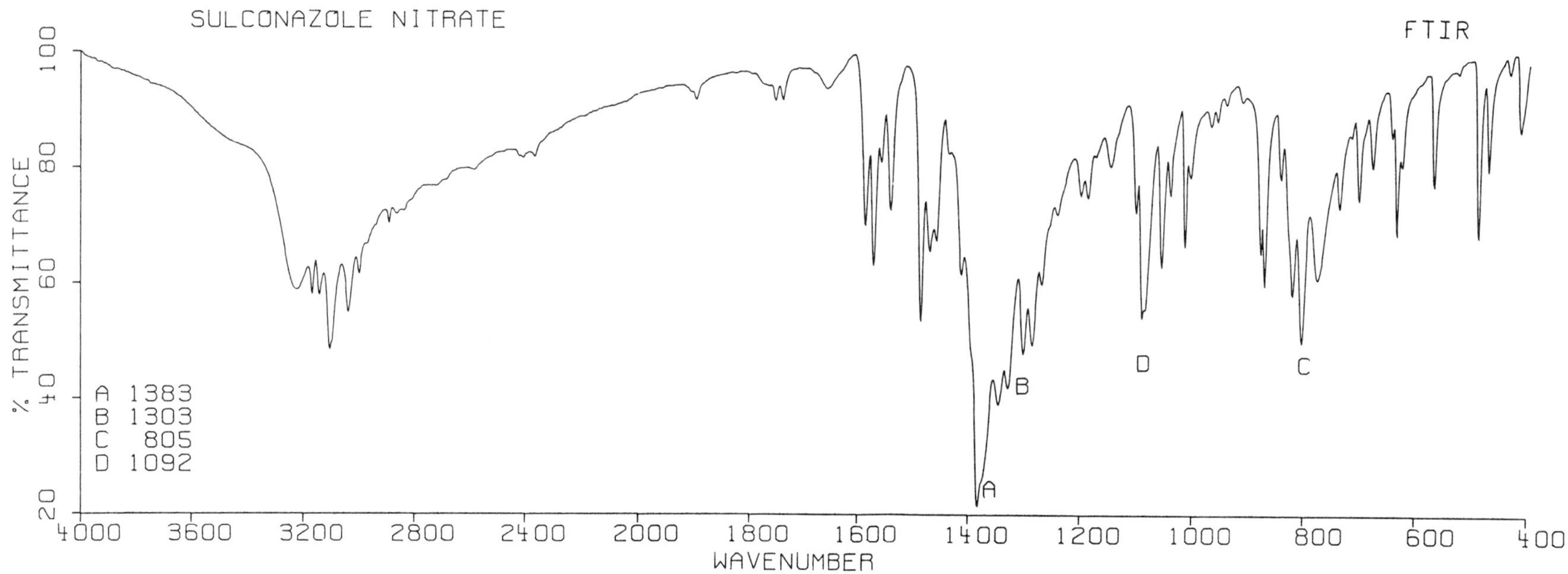

TAXOL

$C_{47}H_{51}NO_{14}$

Molecular Weight: 841.91 (841.23)

Synonyms: [2aR-[2aα,4β,4aβ,6β,9α(αR*,βS*),11α,12a,12aα,12bα]]-β-(Benzoylamino)-α-hydroxybenzenepropanoic acid 6,12b-bis(acetyloxy)-12-(benzoyloxy)-2a,3,4,4a,5,6,9,10,11,12,12a,12b-dodecahydro-4,11-dihydroxy-4a,8,13,13-tetramethyl-5-oxo-7,11-methano-1H-cyclodeca[3.4]benz[1,2-b]-oxet-9-yl ester
Trade Names: Taxol
Use: Antineoplastic

HPLC: Methanol: 2.6
GC:

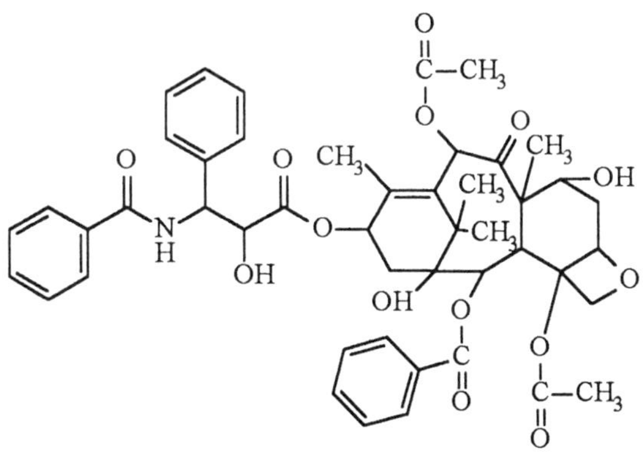

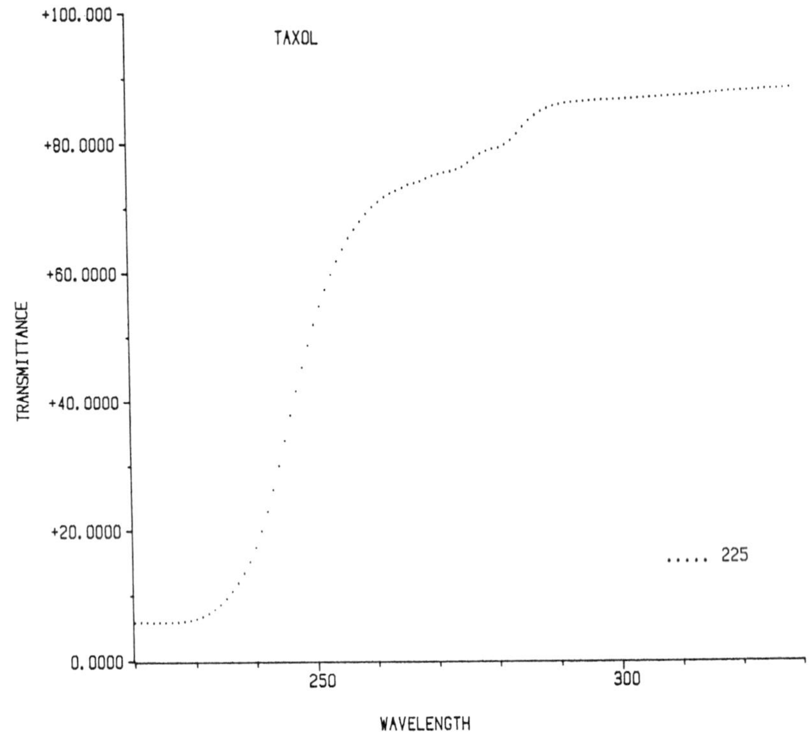

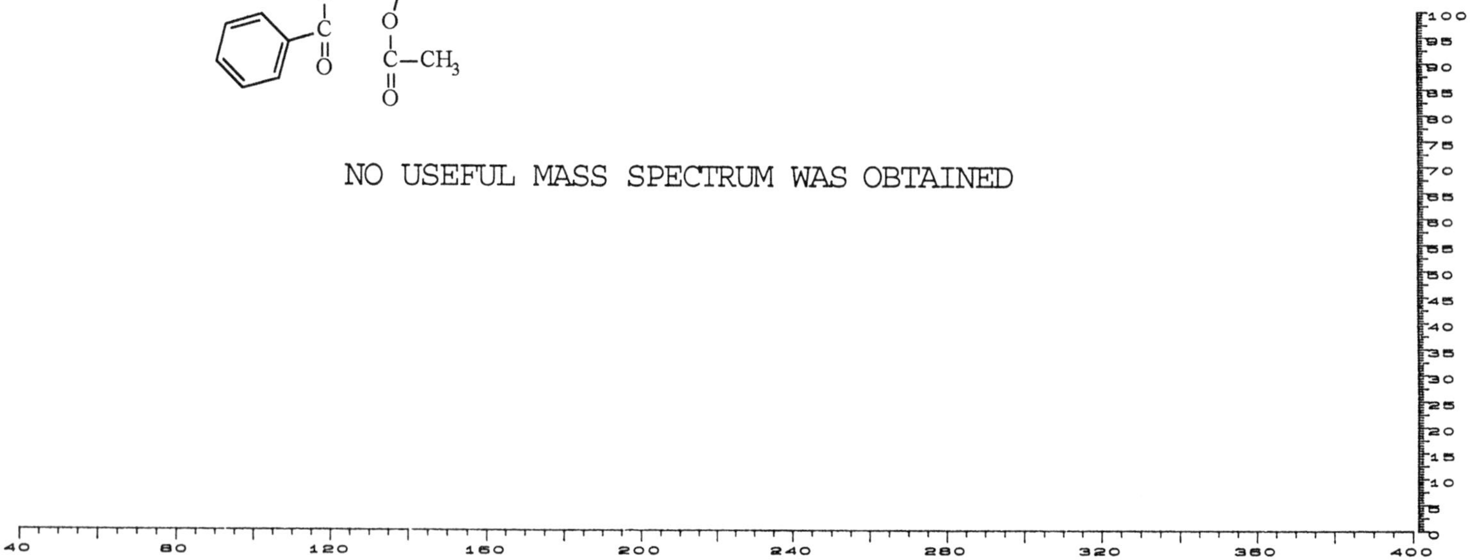

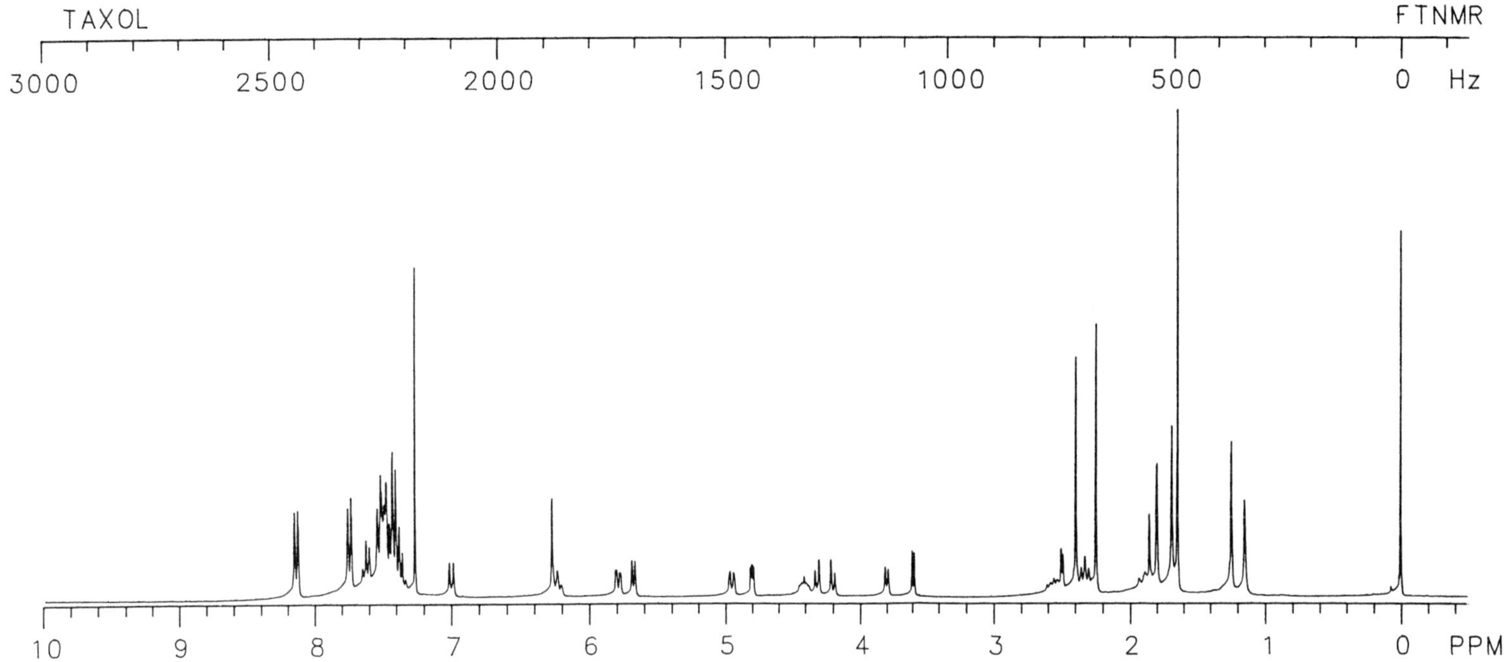

PEAK LISTING

#	HT	PPM
1	75	.00
2	20	1.14
3	31	1.24
4	99	1.64
5	35	1.68
6	27	1.79
7	17	1.84
8	55	2.24
9	49	2.38
10	20	6.26
11	67	7.26
12	14	7.37
13	26	7.40
14	30	7.42
15	15	7.44
16	12	7.45
17	23	7.47
18	23	7.47
19	19	7.47
20	19	7.48
21	19	7.48
22	17	7.49
23	16	7.50
24	22	7.50
25	25	7.51
26	18	7.53
27	10	7.59
28	11	7.62
29	17	7.72
30	20	7.73
31	18	7.75
32	14	8.11

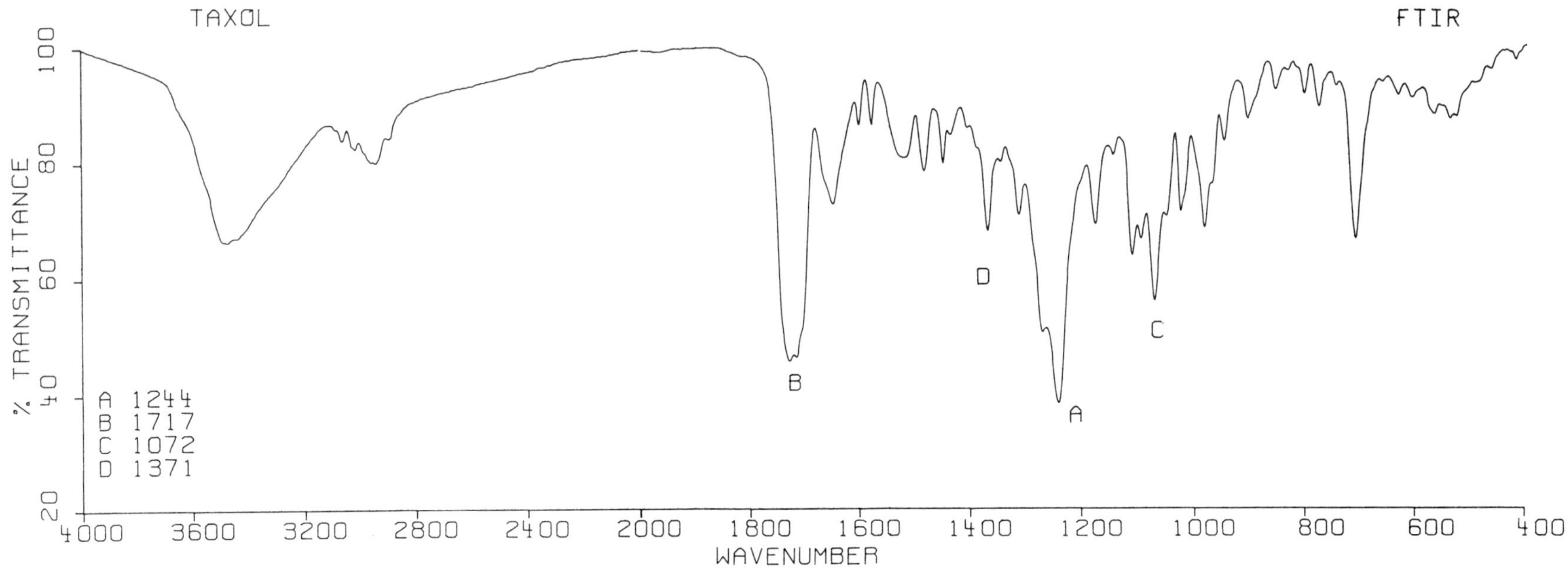

TENOXICAM

$C_{13}H_{11}N_3O_4S_2$

Molecular Weight: 337.37 (337.02)

Synonyms: 4-Hydroxy-2-methyl-N-2-pyridinyl-2H-thieno[2.3-e]-1,2-thiazine-3-carboxamide 1,1-dioxide
Trade Names: Liman, Mobiflex, Tilatil, Tilcotil
Use: Anti-inflammatory

HPLC: Methanol: 2.7
GC:

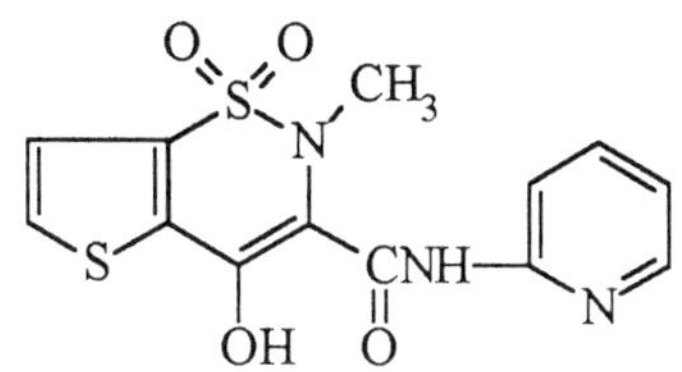

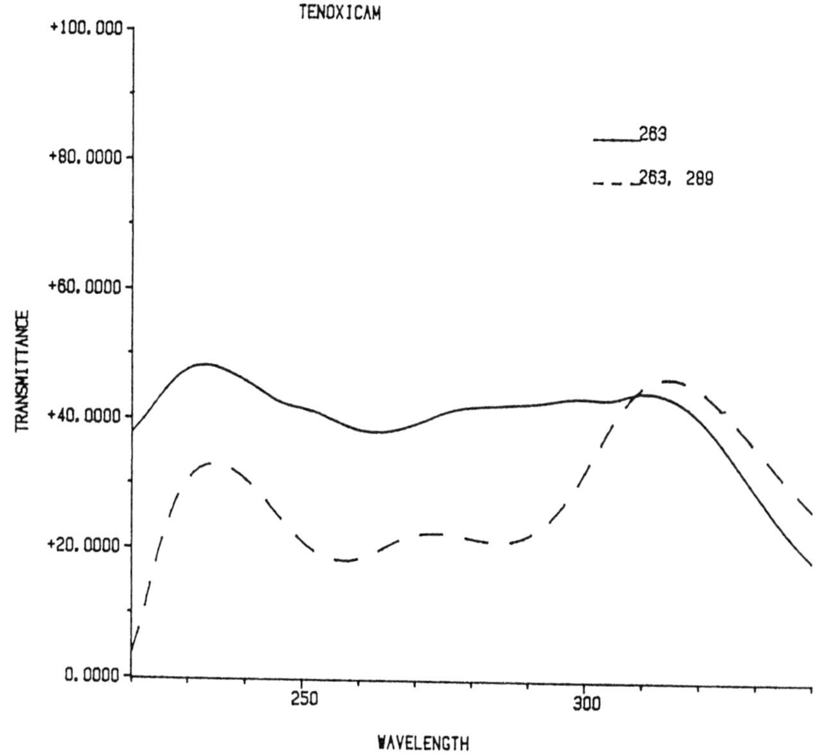

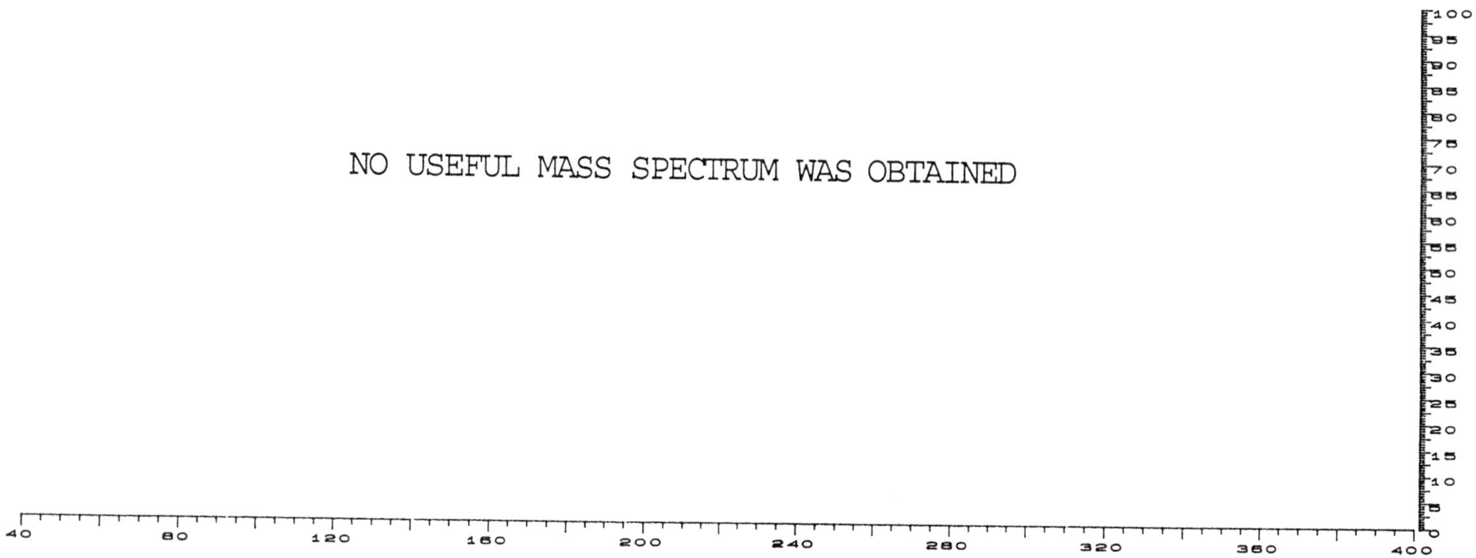

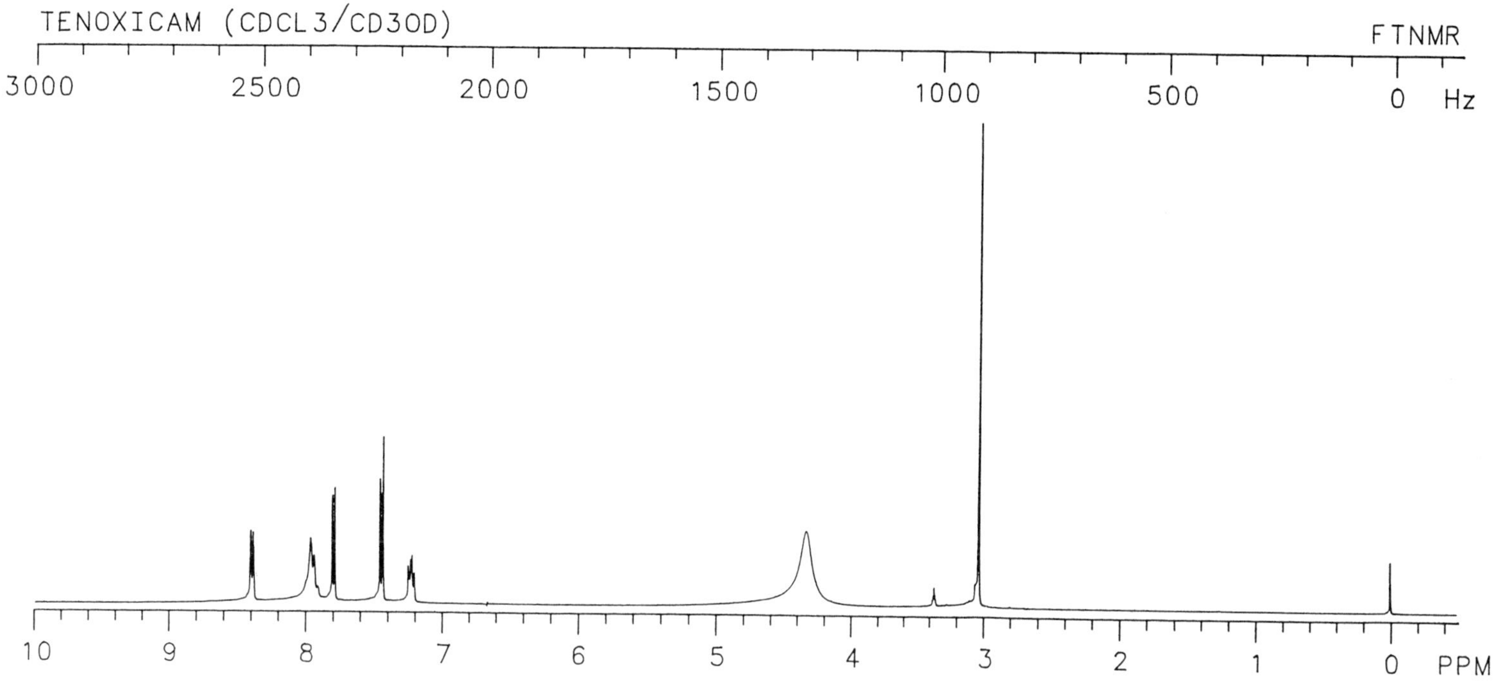

PEAK LISTING

#	HT	PPM
1	10	.00
2	99	3.03
3	15	4.32
4	15	4.32
5	9	7.21
6	34	7.43
7	22	7.43
8	25	7.45
9	23	7.78
10	22	7.80
11	10	7.92
12	13	7.95
13	11	8.37
14	14	8.37
15	12	8.39
16	15	8.39

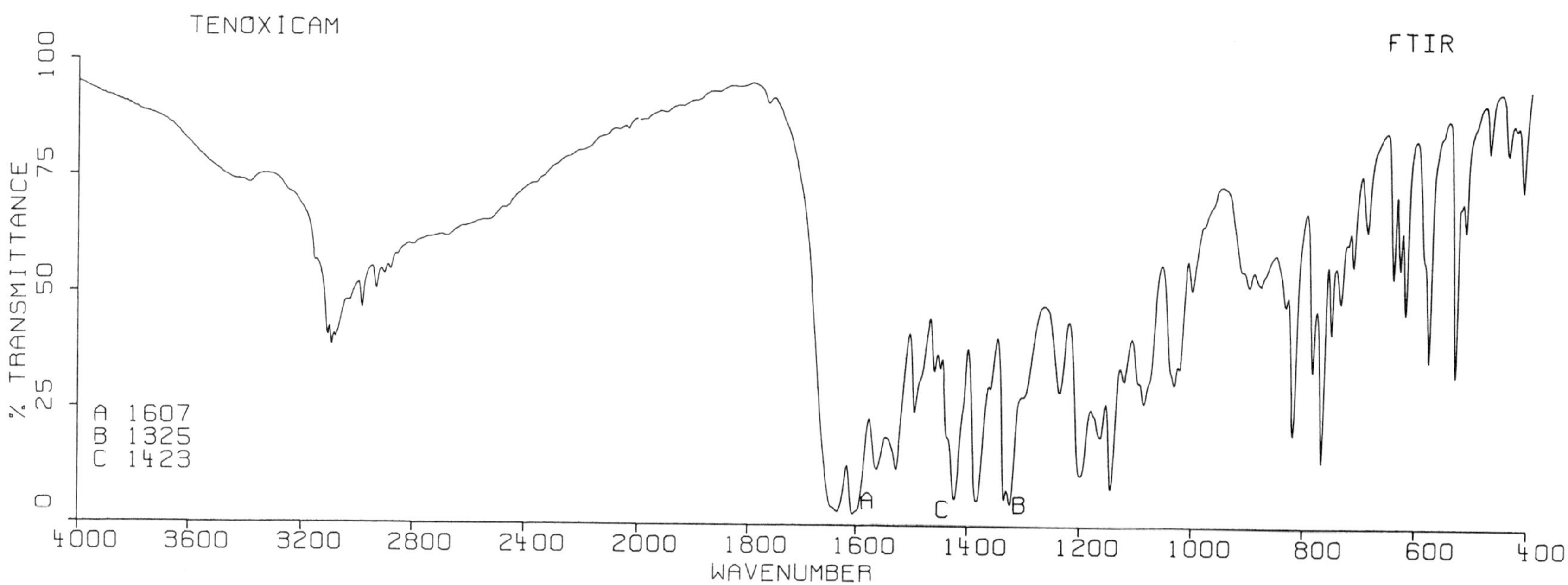

TERAZOSIN

$C_{19}H_{25}N_5O_4$

Molecular Weight: 387.44 (387.19)

Synonyms: 1-(4-Amino-6,7-dimethoxy-2-quinazolinyl)-4-[(tetrahydro-2-furanyl)carbonyl]piperazine
Trade Names: Heitrin, Hytracin, Hytrin, Hytrinex, Vasocard, Vasomet
Use: Antihypertensive

HPLC:
GC:

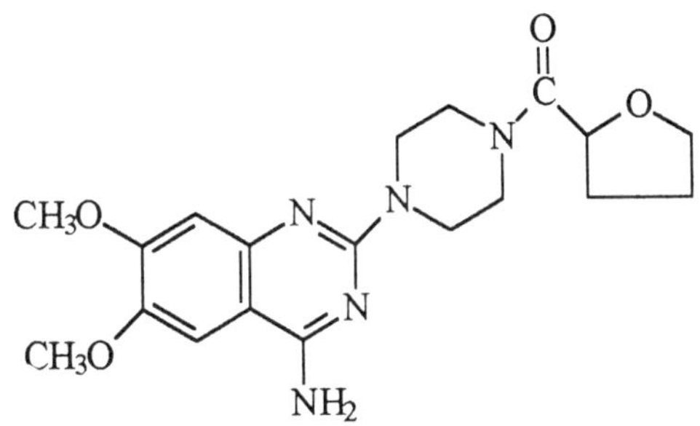

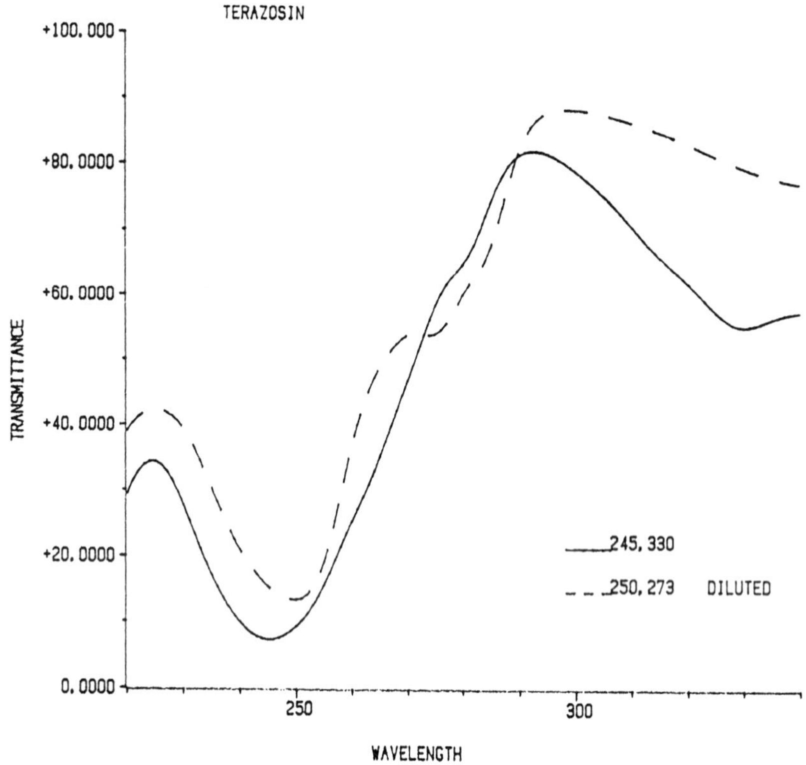

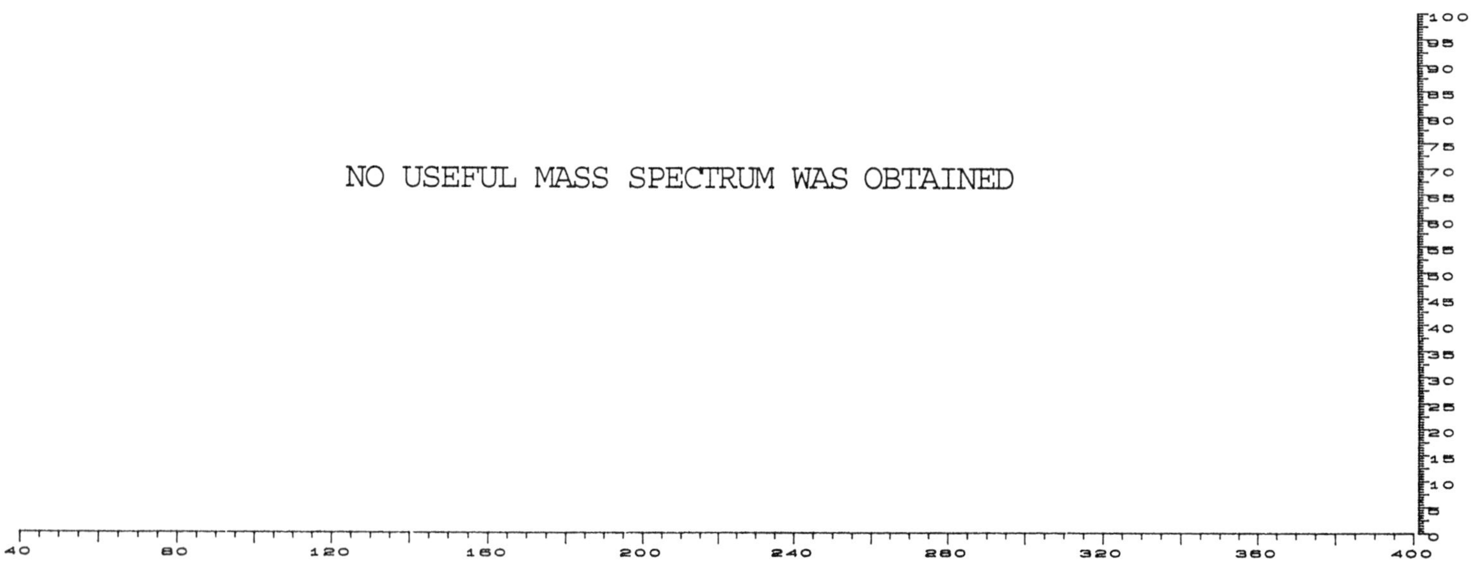

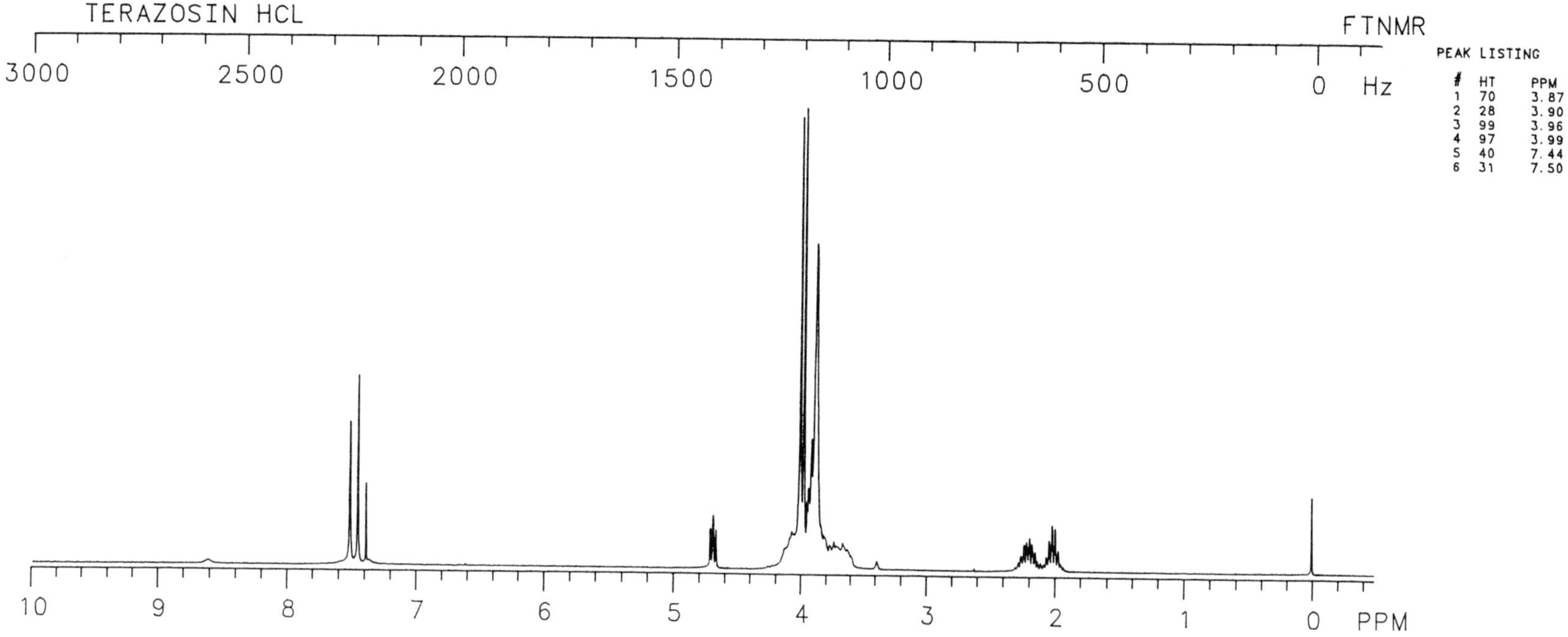
TERAZOSIN HCL
FTNMR
3000
2500
2000
1500
1000
500
0 Hz
PEAK LISTING
HT PPM
1 70 3.87
2 28 3.90
3 99 3.96
4 97 3.99
5 40 7.44
6 31 7.50
10
9
8
7
6
5
4
3
2
1
0 PPM

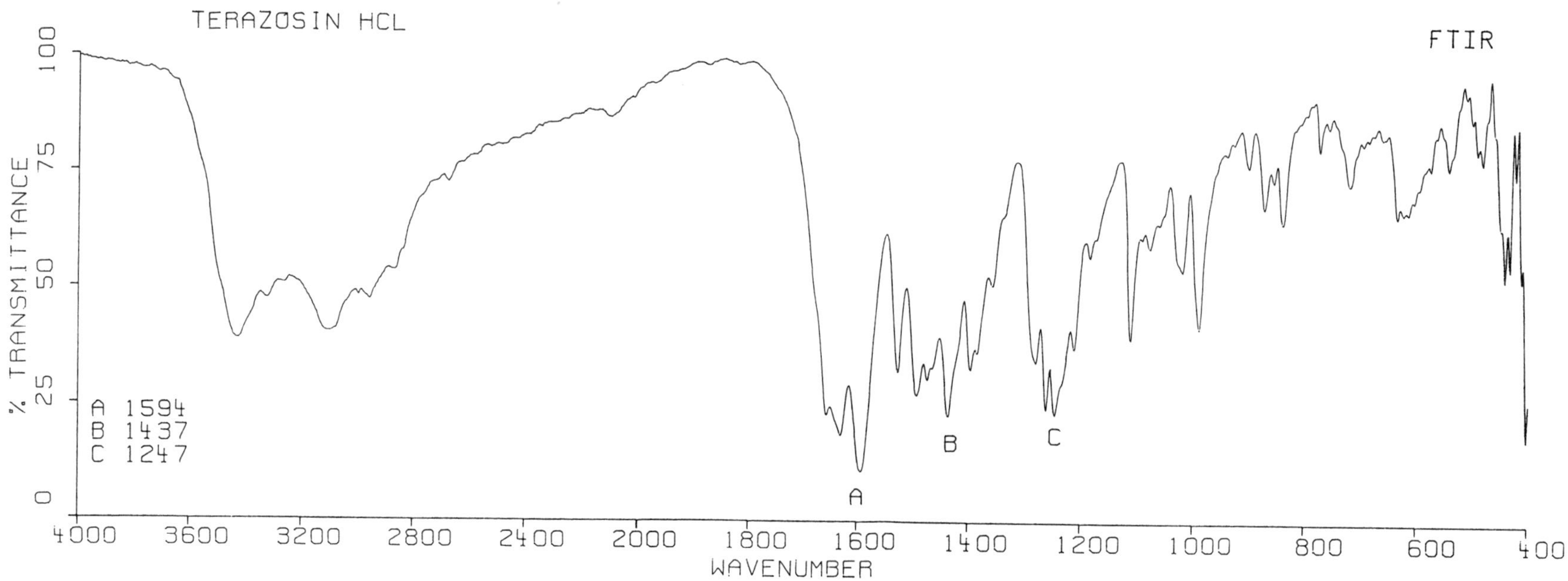
TERAZOSIN HCL
FTIR
% TRANSMITTANCE
100
75
50
25
0
A 1594
B 1437
C 1247
A
B
C
4000
3600
3200
2800
2400
2000
1800
1600
1400
1200
1000
800
600
400
WAVENUMBER

TERBINAFINE

$C_{21}H_{25}N$

Molecular Weight: 291.44 (291.20)

Synonyms: (E)-N-(6,6-Dimethyl-2-hepten-4-ynyl)-N-methyl-1-naphthalene methanamine
Trade Names: Lamisil
Use: Antifungal

HPLC:
GC:

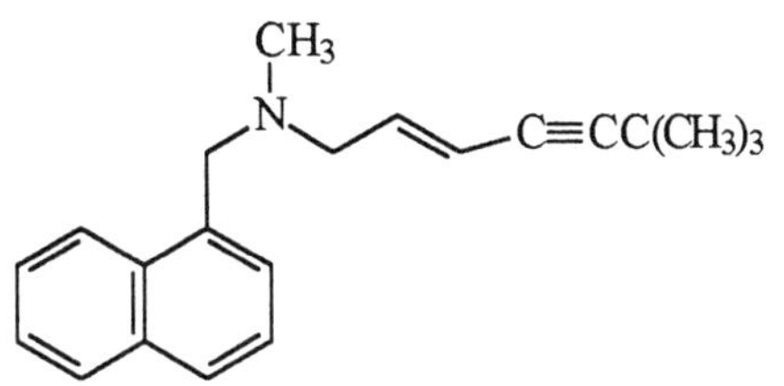

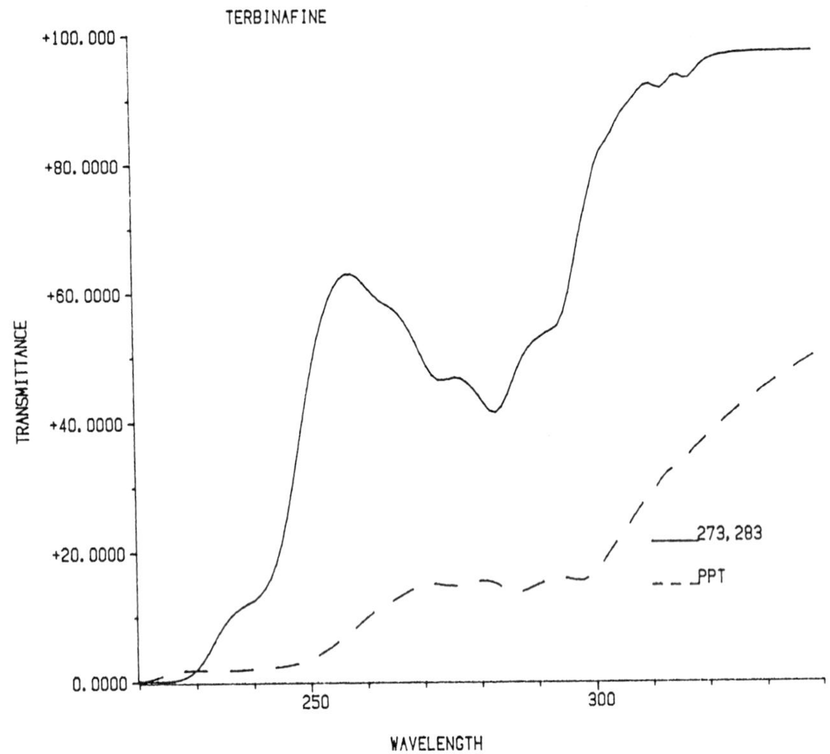

TERBINAFINE

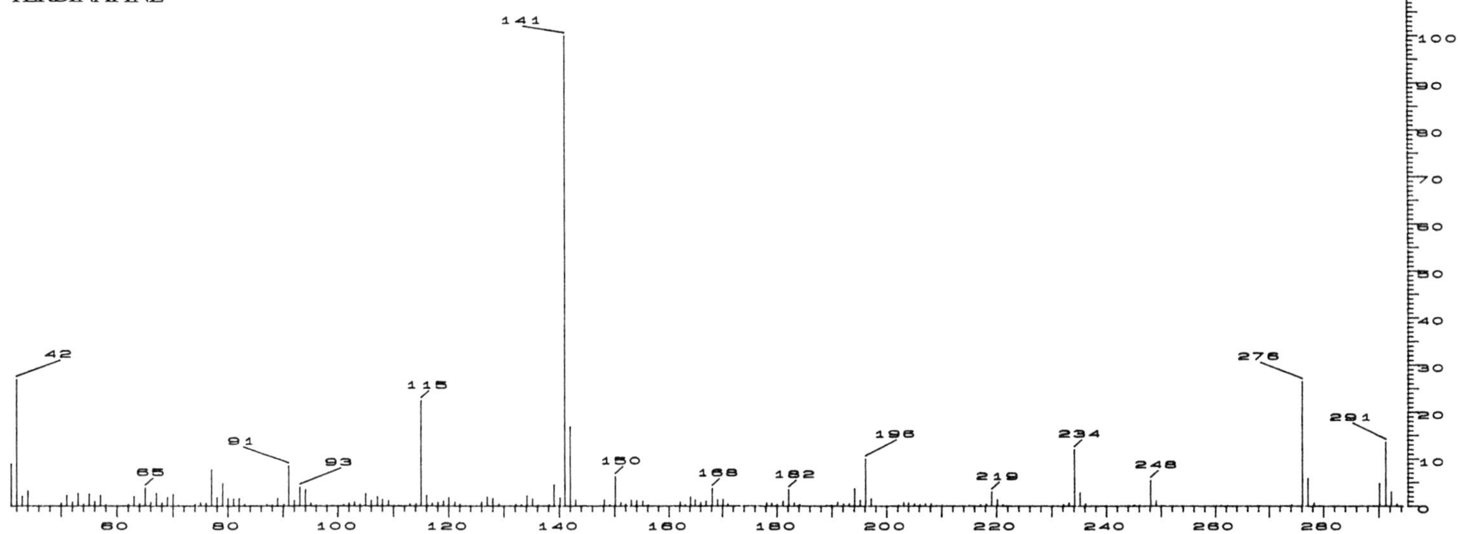

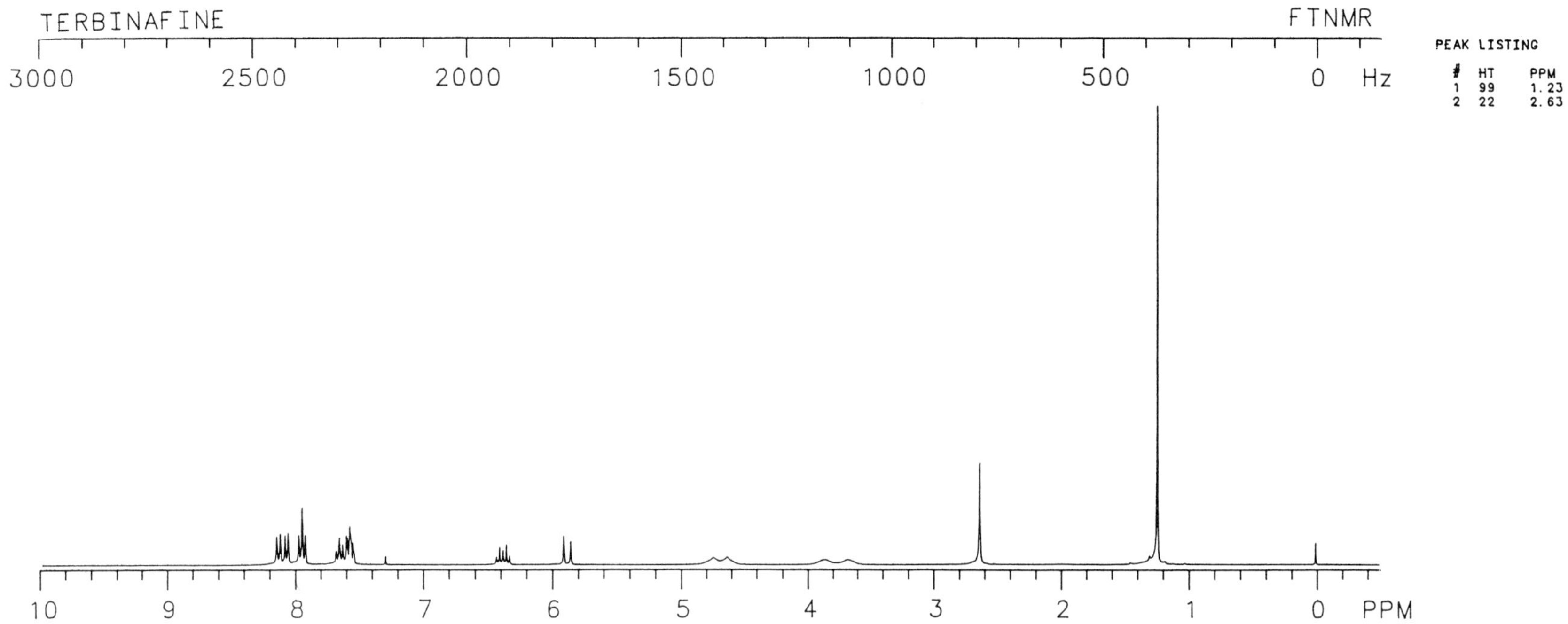
TERBINAFINE
FTNMR
3000 2500 2000 1500 1000 500 0 Hz
PEAK LISTING
HT PPM
1 99 1.23
2 22 2.63
10 9 8 7 6 5 4 3 2 1 0 PPM

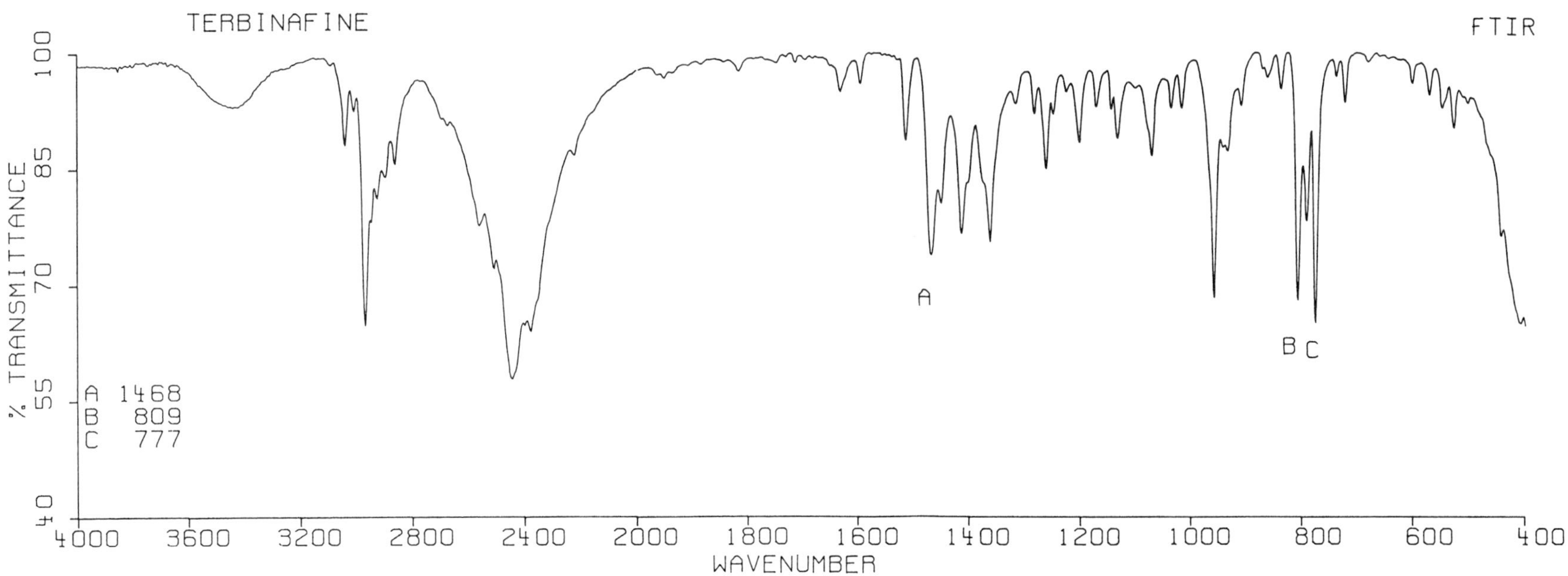
TERBINAFINE
FTIR
% TRANSMITTANCE
100 85 70 55 40
A 1468
B 809
C 777
A
B C
4000 3600 3200 2800 2400 2000 1800 1600 1400 1200 1000 800 600 400
WAVENUMBER

TESTOSTERONE DECANOATE

$C_{29}H_{46}O_3$

Molecular Weight: 442.68 (442.34)

Synonyms: 4-Androsten-17β-ol-one-17-caprate; 17β-hydroxy-4-androsten-3-one-17-decanoate
Trade Names:
Use: Androgen

HPLC: Methanol: 5.3
GC:

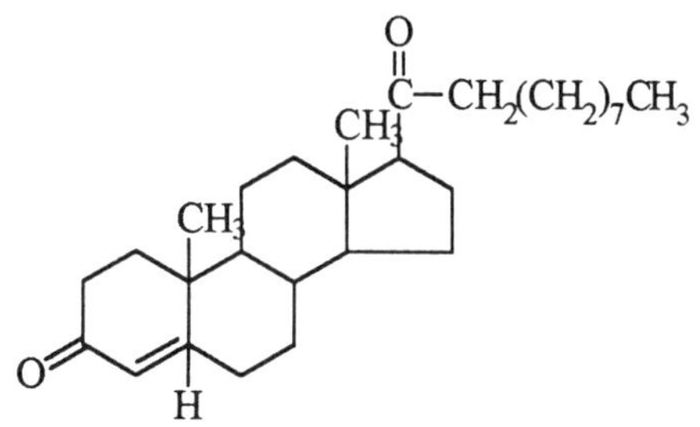

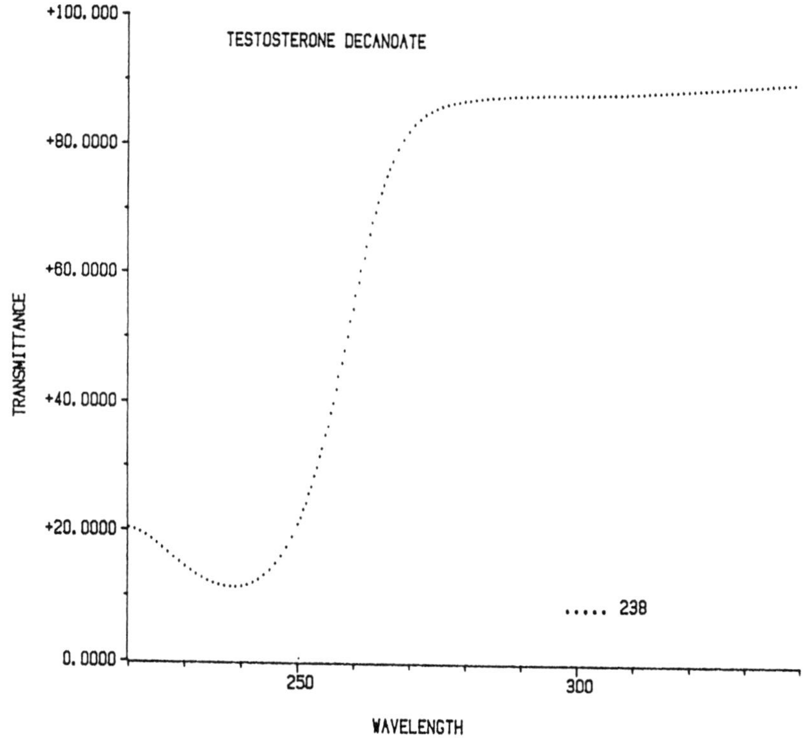

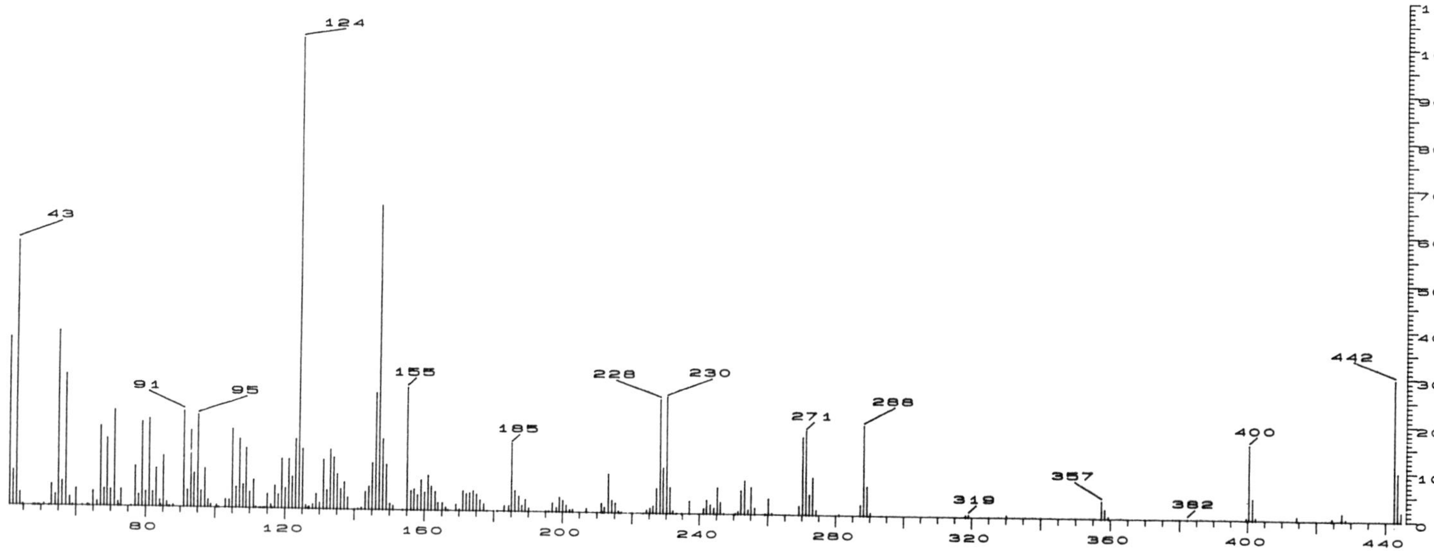

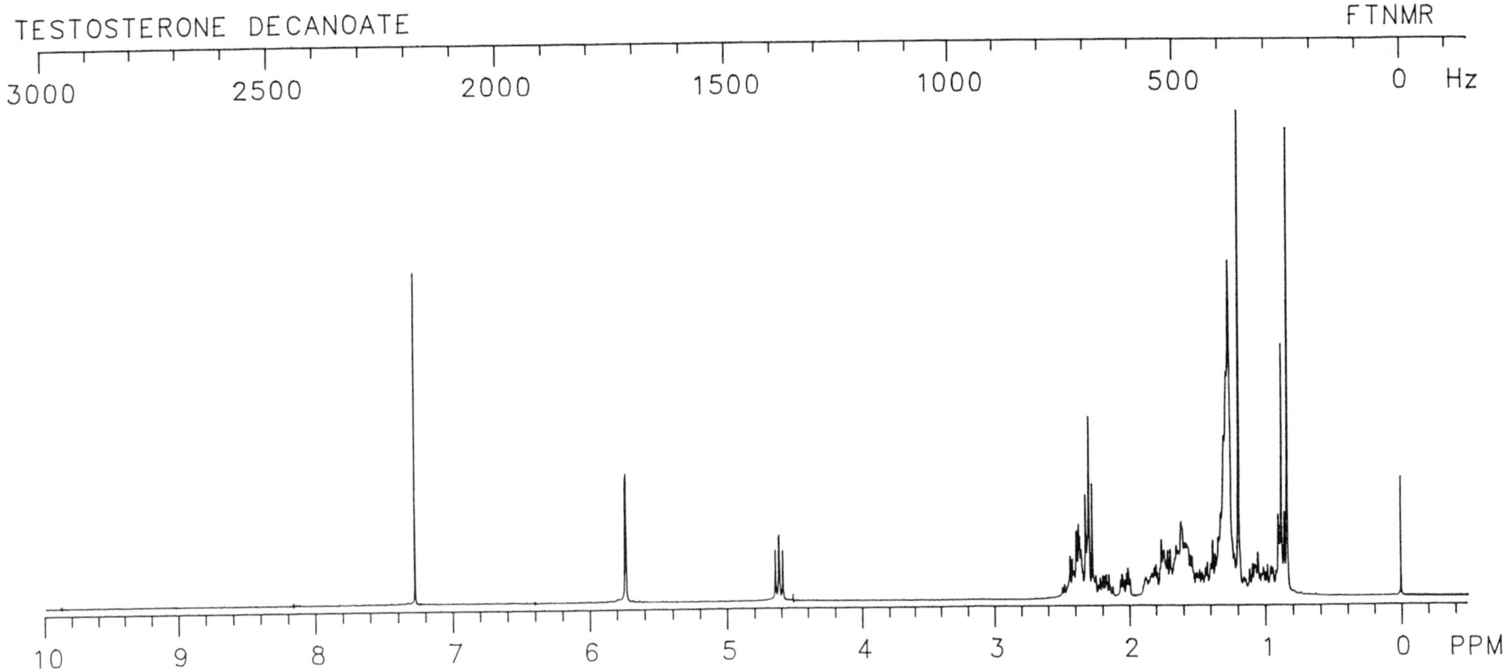

PEAK LISTING

#	HT	PPM
1	24	.00
2	95	.83
3	17	.85
4	51	.88
5	17	.90
6	99	1.19
7	68	1.26
8	46	1.28
9	33	1.29
10	17	1.32
11	14	1.60
12	15	1.61
13	23	2.27
14	37	2.29
15	21	2.32
16	15	2.37
17	13	2.38
18	13	4.60
19	26	5.73
20	25	5.73
21	67	7.27

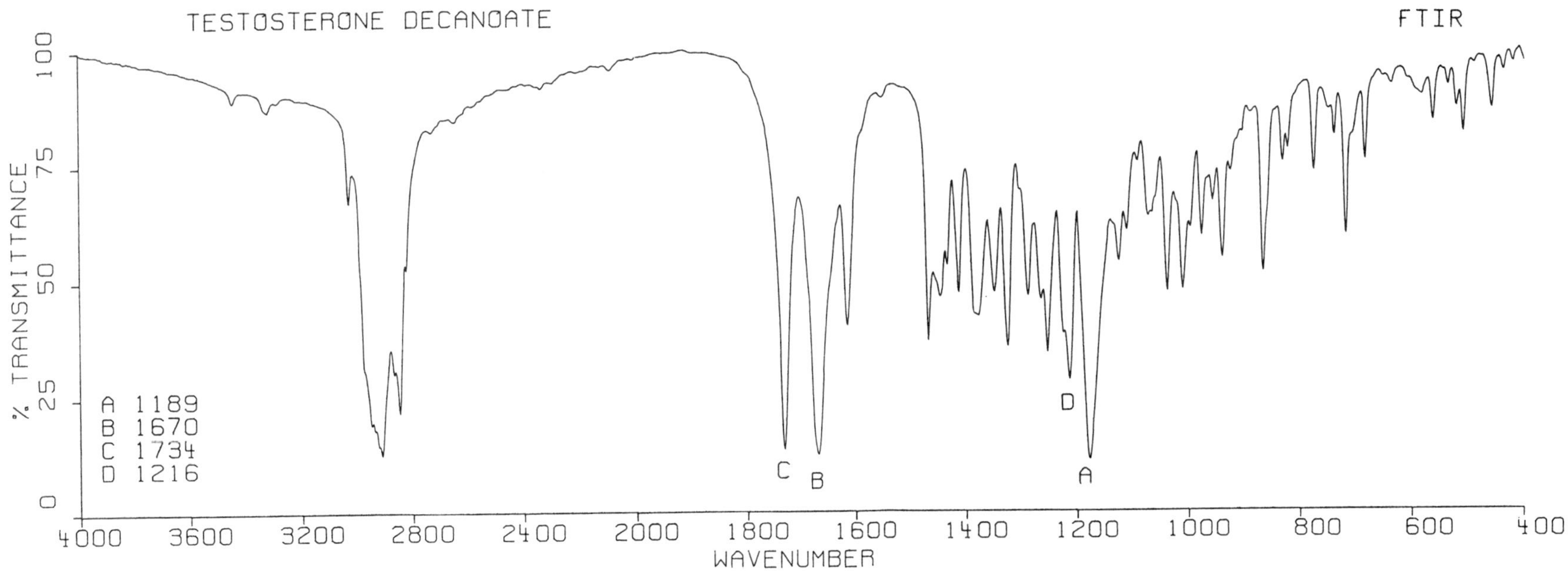

TESTOSTERONE-17β-HEMISUCCINATE

$C_{23}H_{32}O_5$

Molecular Weight: 388.50 (388.22)

Synonyms: 4-Androsten-17β-ol-3-one-17-hemisuccinate; 17β-hydroxy-4-androsten-3-one-17-hemisuccinate
Trade Names:
Use: androgen

HPLC: Methanol: 3.6
GC:

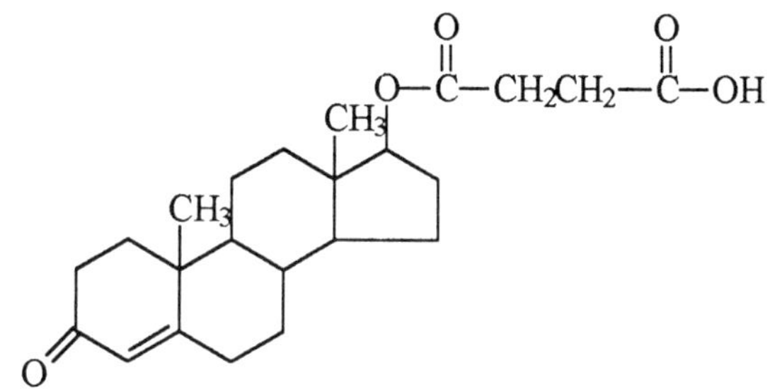

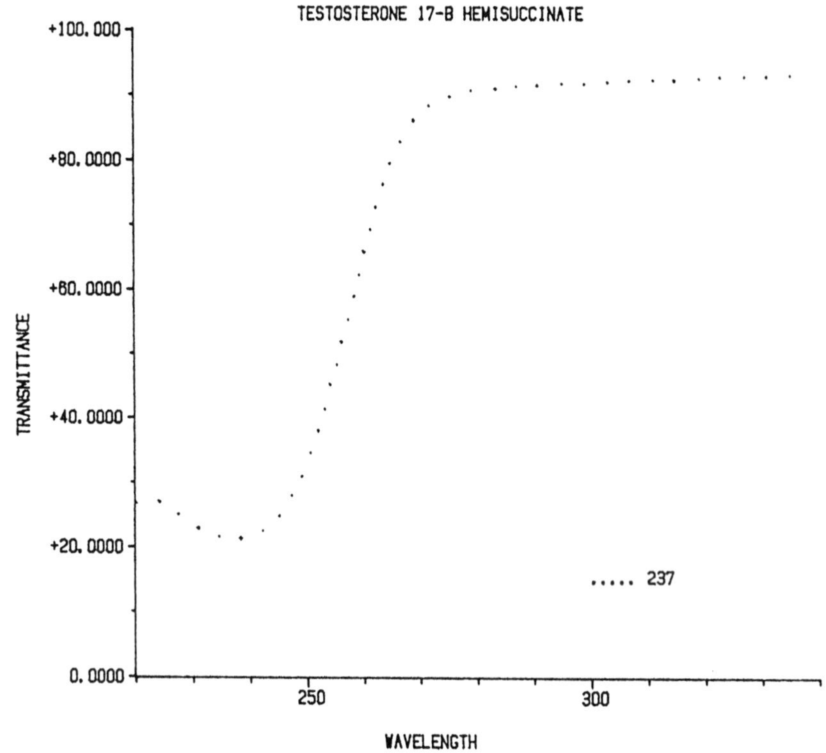

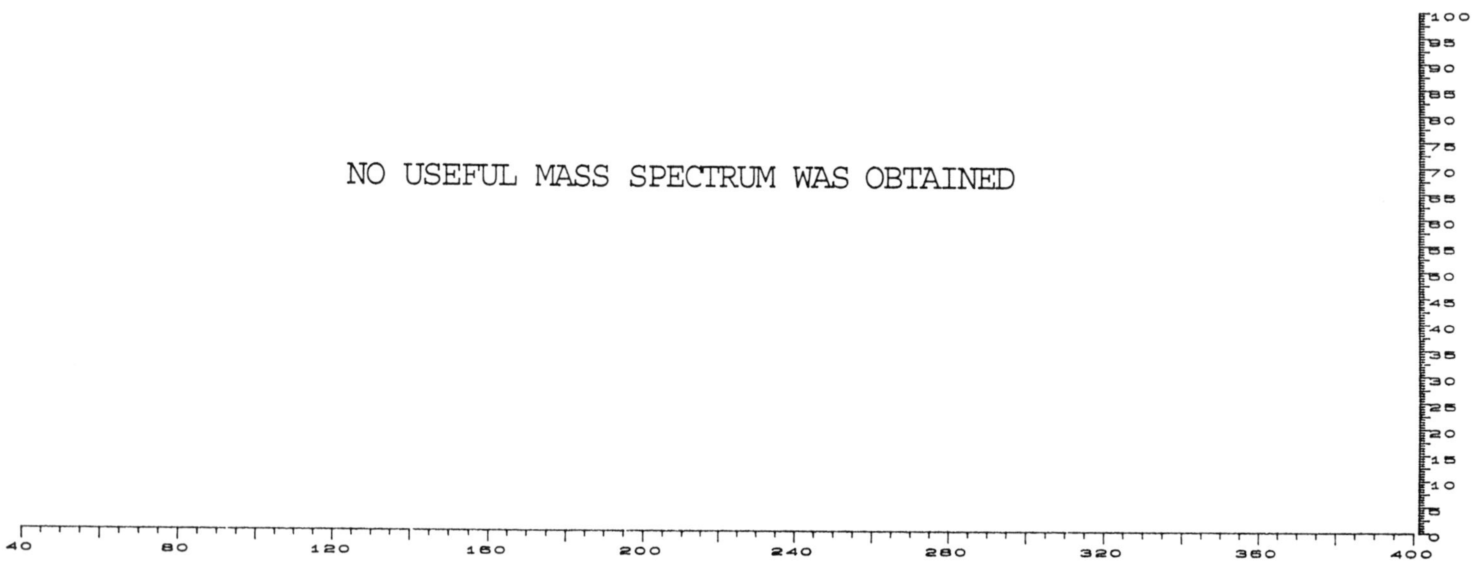

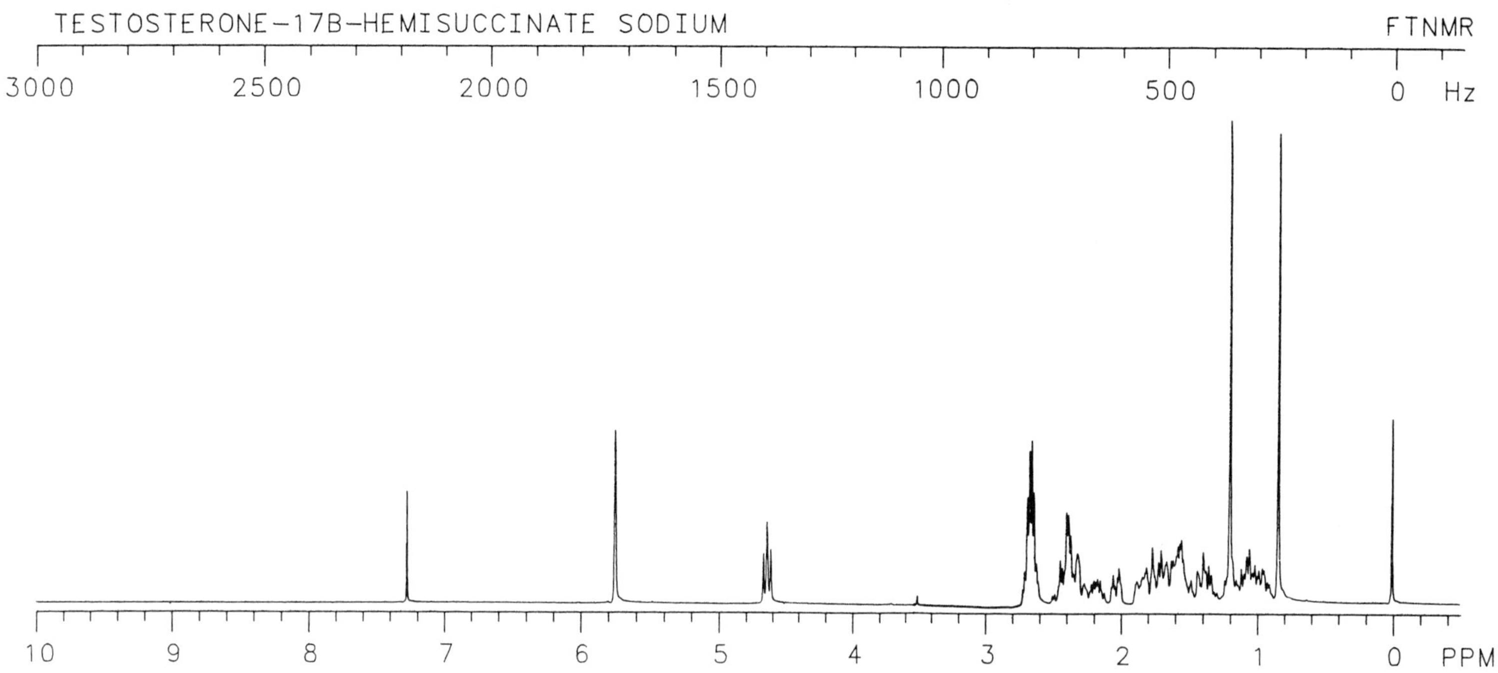

PEAK LISTING

#	HT	PPM
1	38	.00
2	21	.00
3	96	.83
4	11	1.04
5	99	1.19
6	13	1.54
7	12	1.55
8	12	1.56
9	11	1.75
10	14	2.36
11	18	2.37
12	18	2.39
13	22	2.63
14	33	2.65
15	31	2.66
16	22	2.68
17	16	4.63
18	35	5.74
19	22	7.26
20	12	7.27

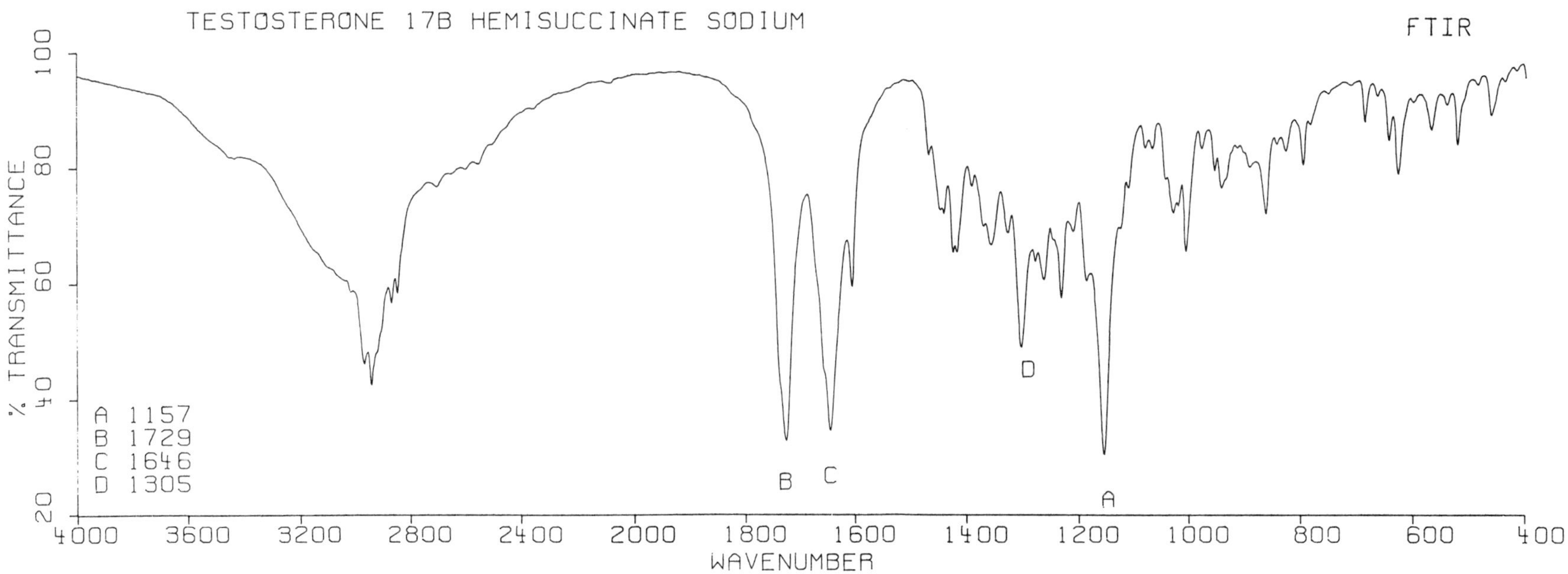

TESTOSTERONE ISOCAPROATE

$C_{25}H_{38}O_3$

Molecular Weight: 386.57 (386.28)

Synonyms: 4-Androsten-17β-ol-3-one-17-[4-methylpentanoate]; 17β-hydroxy-4-androsten-3-one-17-isocaprate
Trade Names:
Use: Androgen

HPLC:
GC:

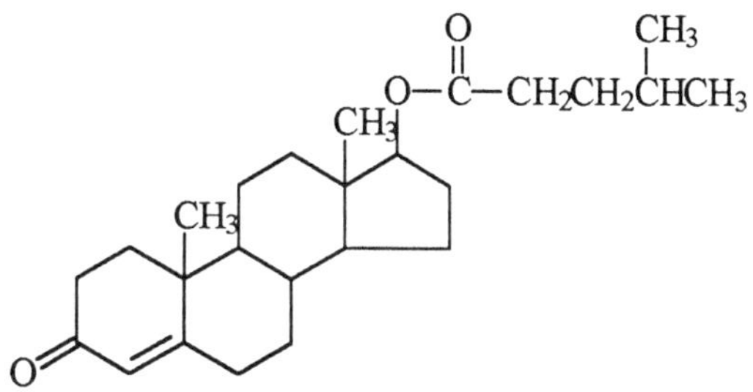

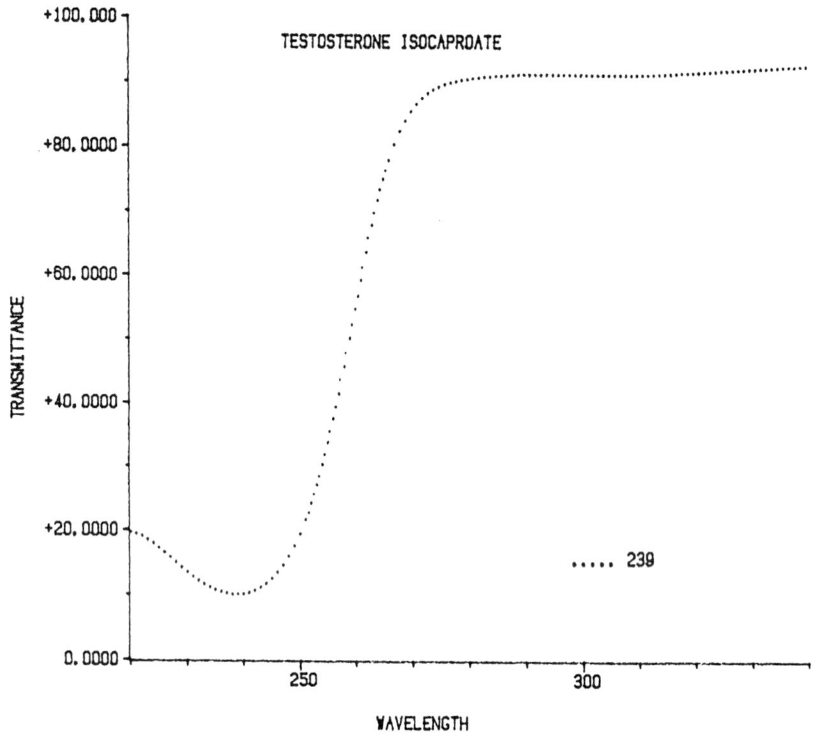

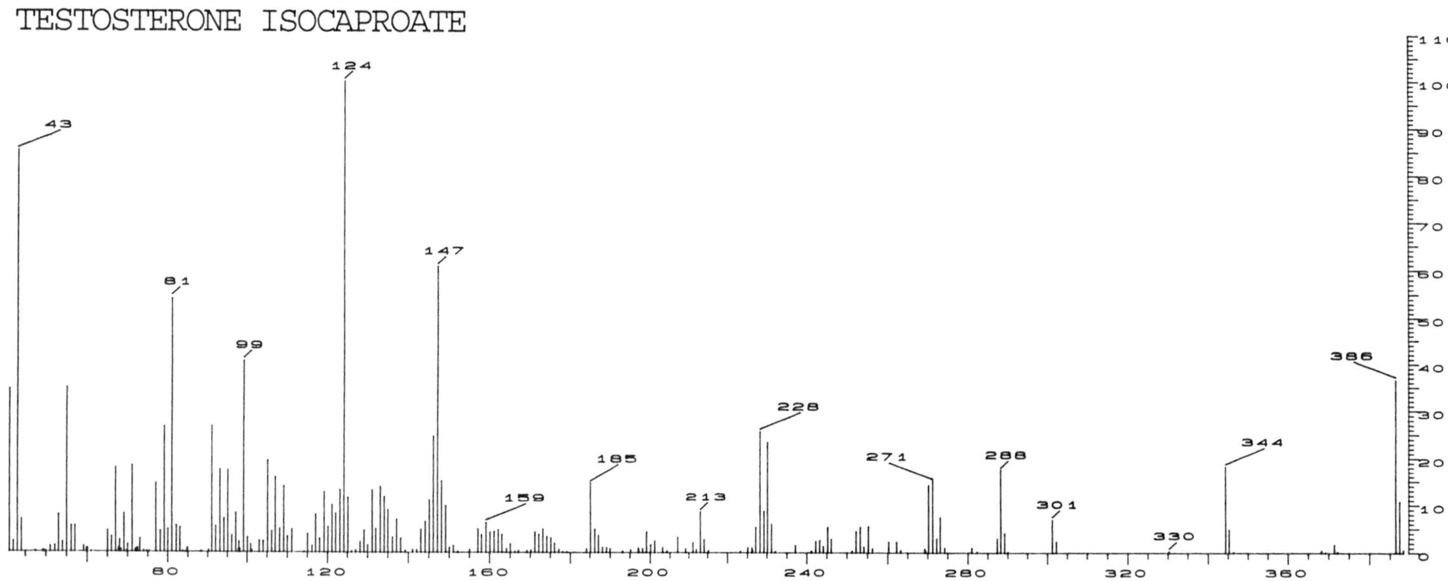

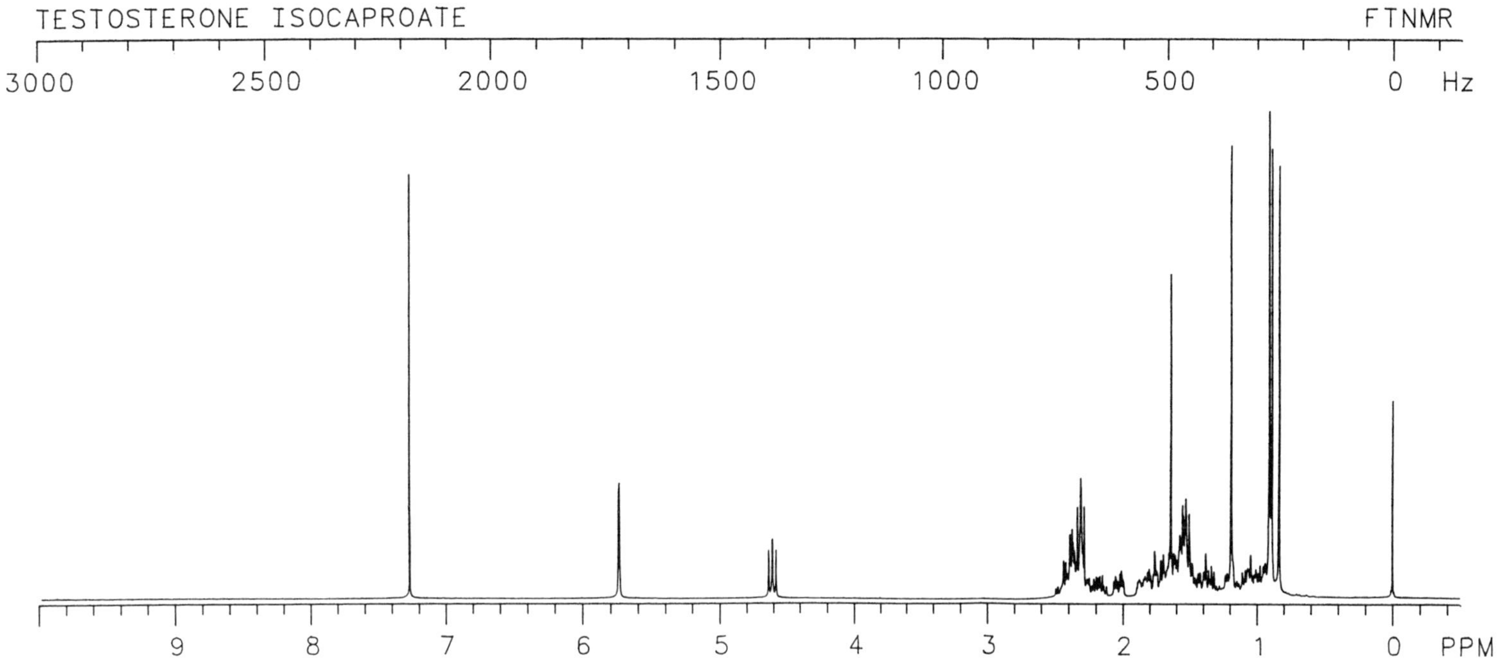

PEAK LISTING

#	HT	PPM
1	40	.00
2	88	.84
3	92	.89
4	17	.90
5	99	.91
6	92	1.19
7	17	1.50
8	20	1.52
9	15	1.53
10	17	1.54
11	19	1.55
12	13	1.57
13	66	1.63
14	18	2.28
15	24	2.30
16	18	2.33
17	14	2.37
18	13	2.38
19	23	5.73
20	22	5.73
21	86	7.26

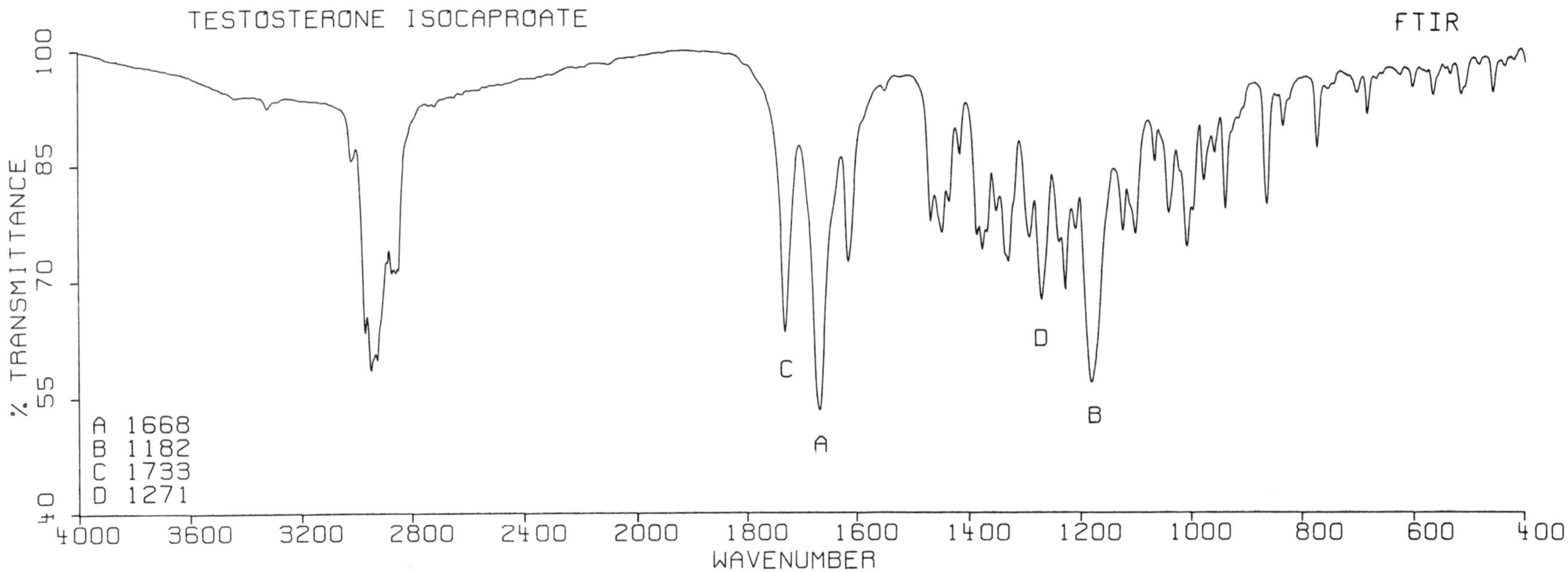

TESTOSTERONE-17-PHENYLPROPIONATE

$C_{28}H_{36}O_3$

Molecular Weight: 420.59 (420.27)

Synonyms: 4-Androsten-17β-ol-3-one-17-phenylpropionate; testosterone hydrocinnamate
Trade Names:
Use: Androgen

HPLC: Methanol: 3.9
GC:

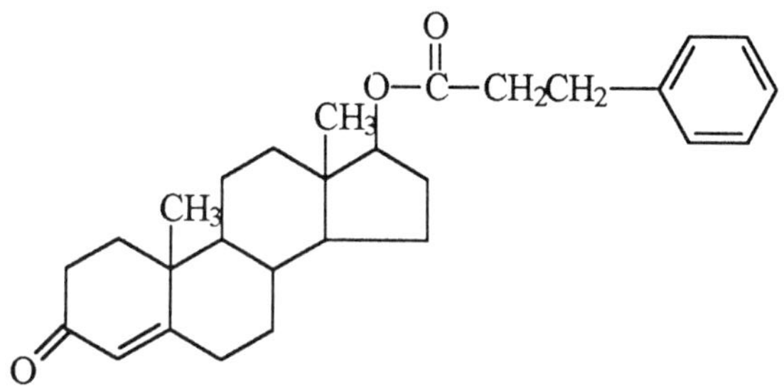

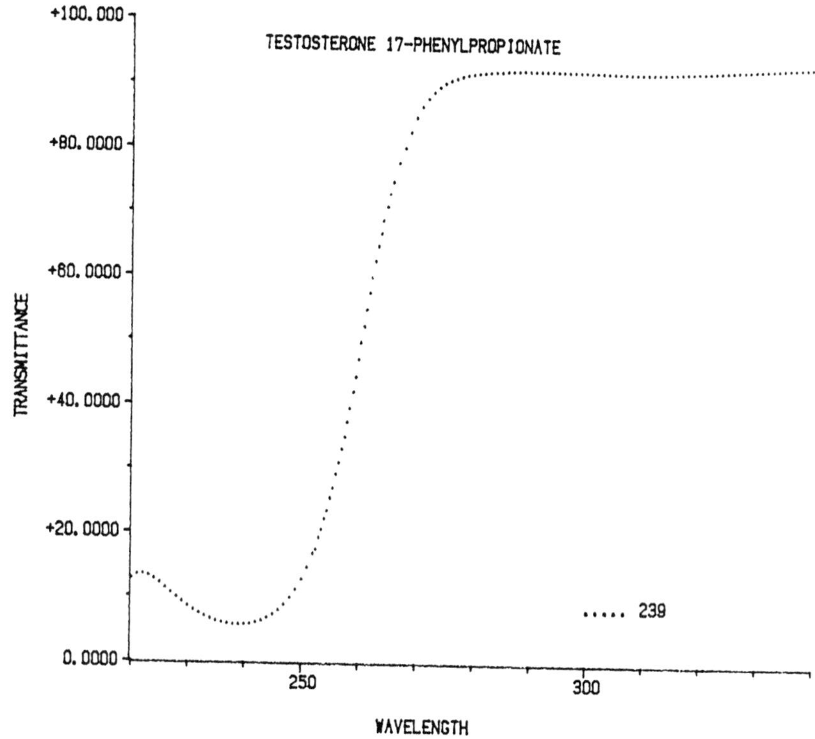

TESTOSTERONE PHENPROPIONATE

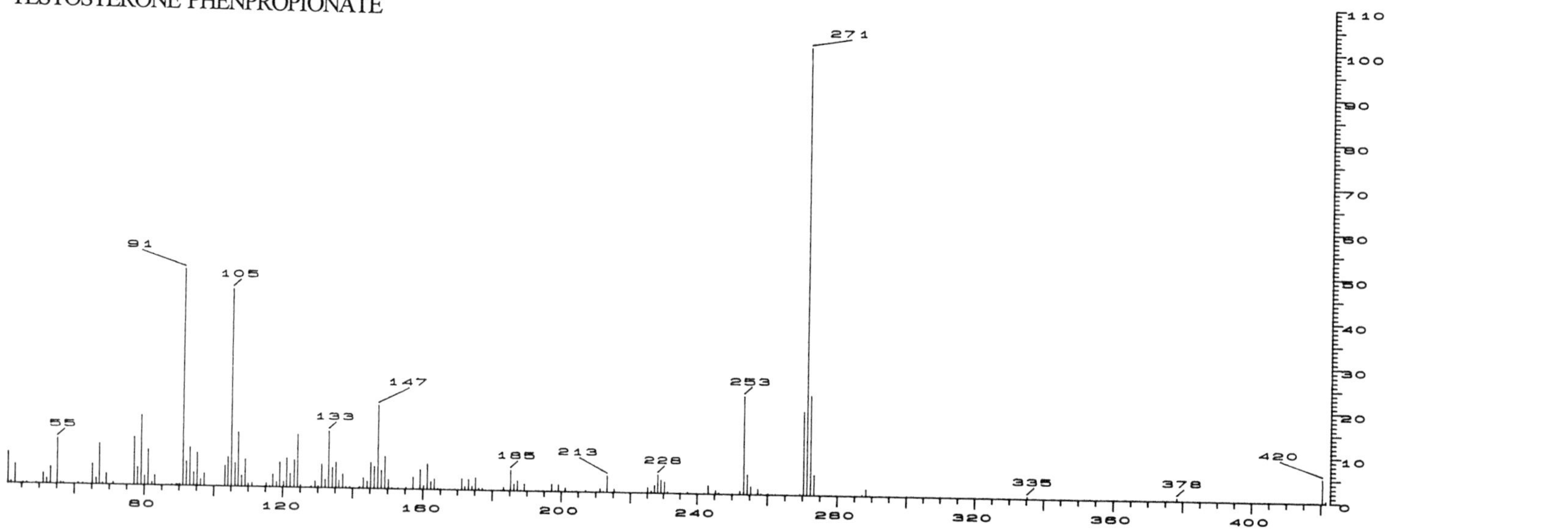

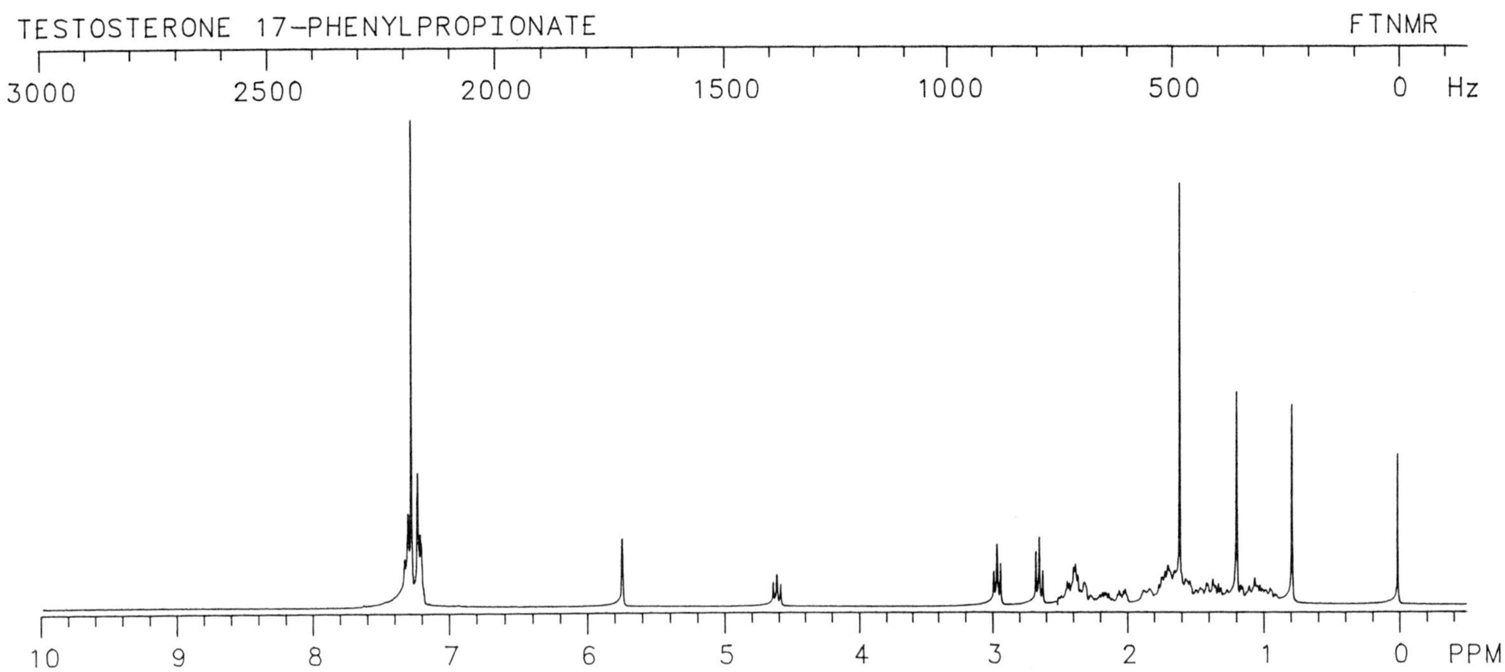

#	HT	PPM
1	30	.00
2	40	.78
3	43	1.19
4	86	1.61
5	14	2.64
6	11	2.67
7	12	2.96
8	13	5.74
9	13	7.20
10	15	7.20
11	27	7.22
12	99	7.27
13	19	7.29
14	9	7.32

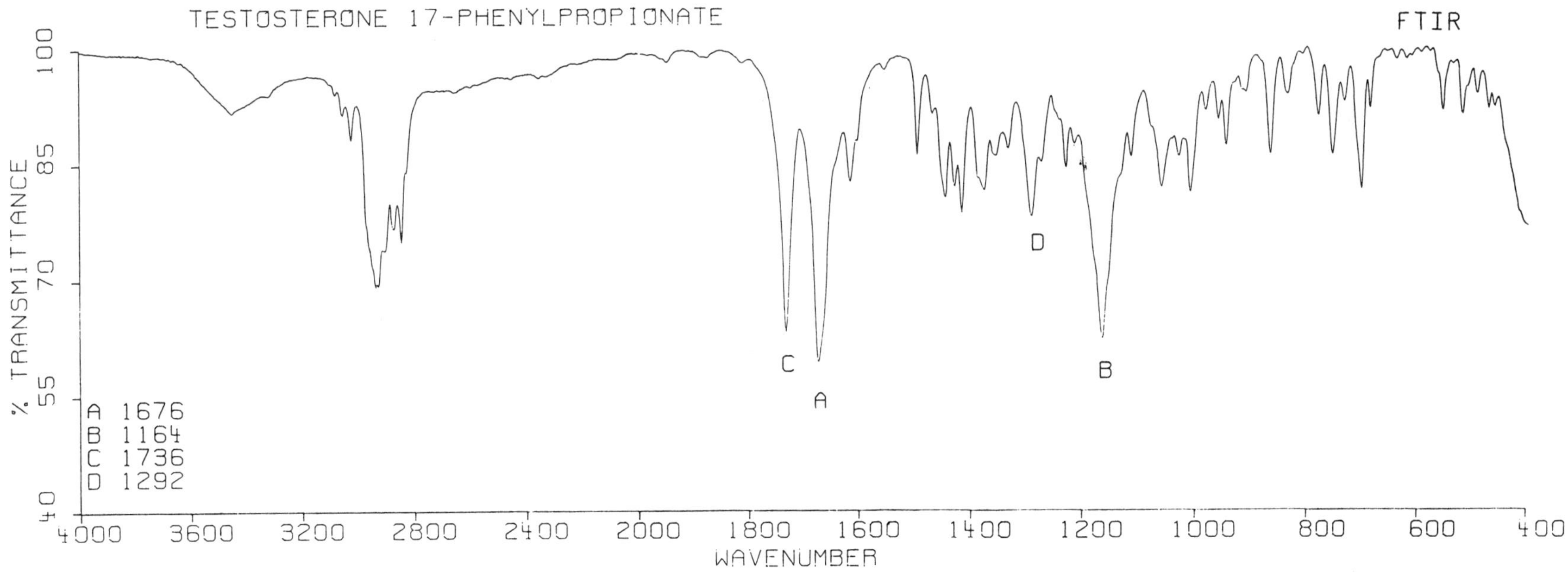

TETRAHYDROCORTISONE

$C_{21}H_{32}O_5$

Molecular Weight: 364.48 (364,22)

Synonyms: 3α,17α,21-Trihydroxy-5β-pregnane-11,20-dione; THE
Trade Names:
Use: Androgen

HPLC:
GC:

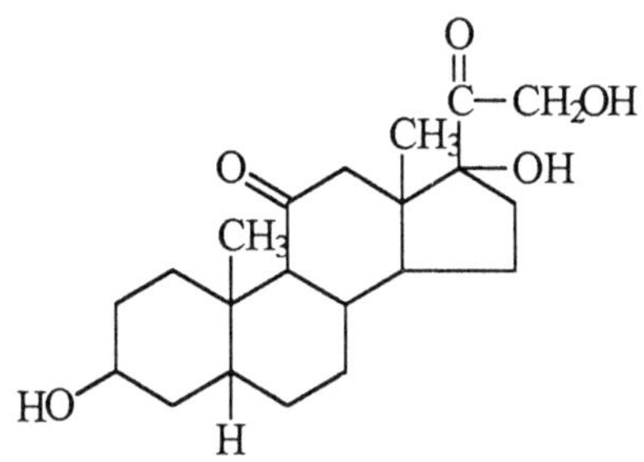

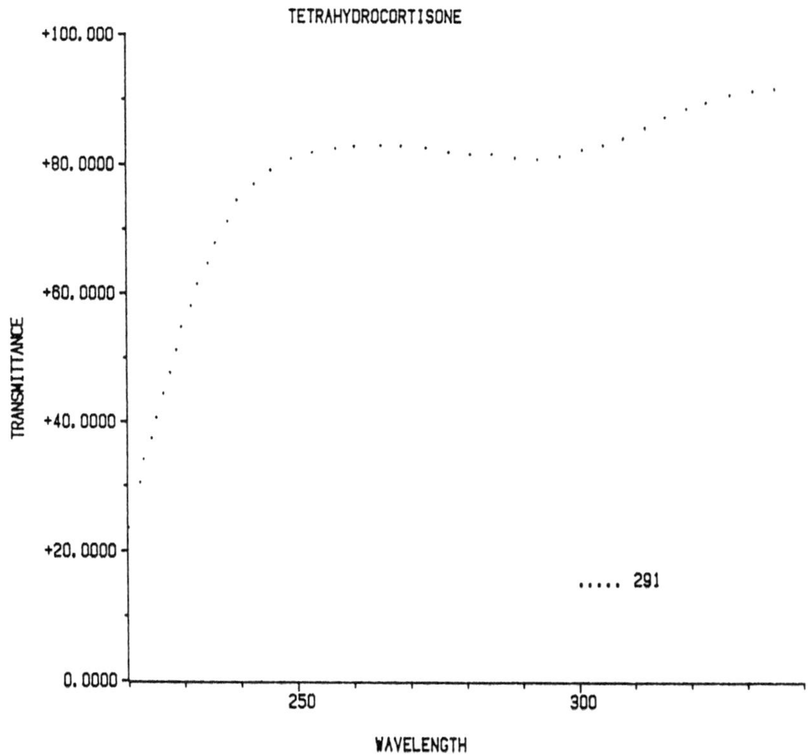

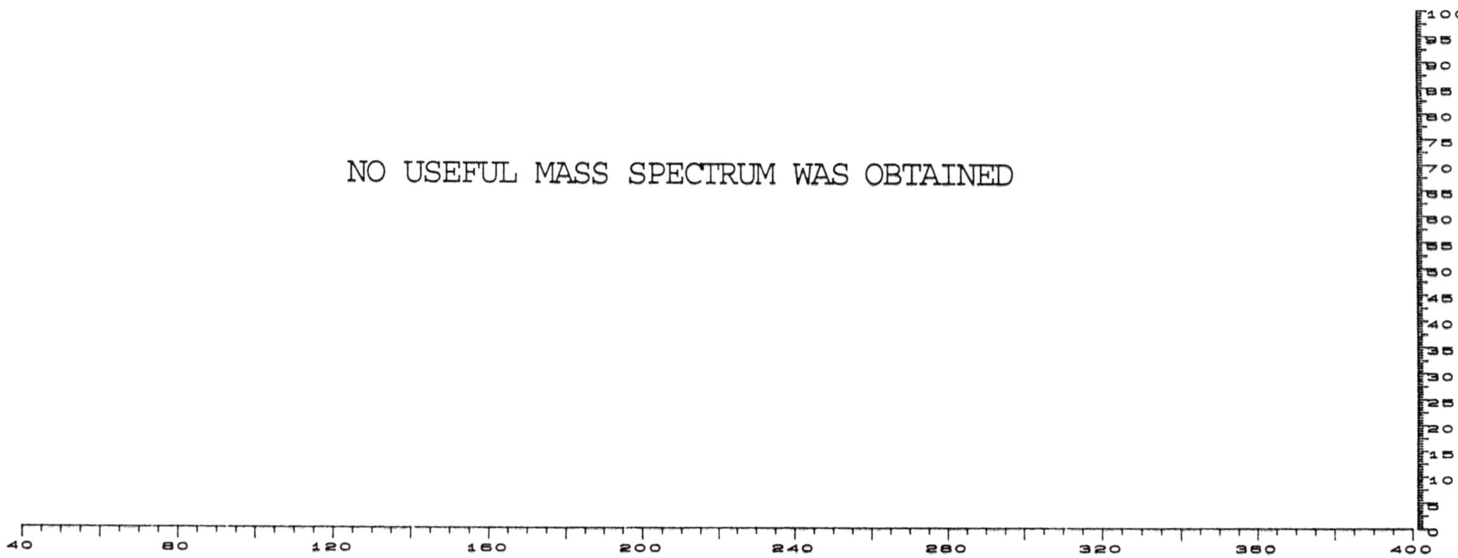

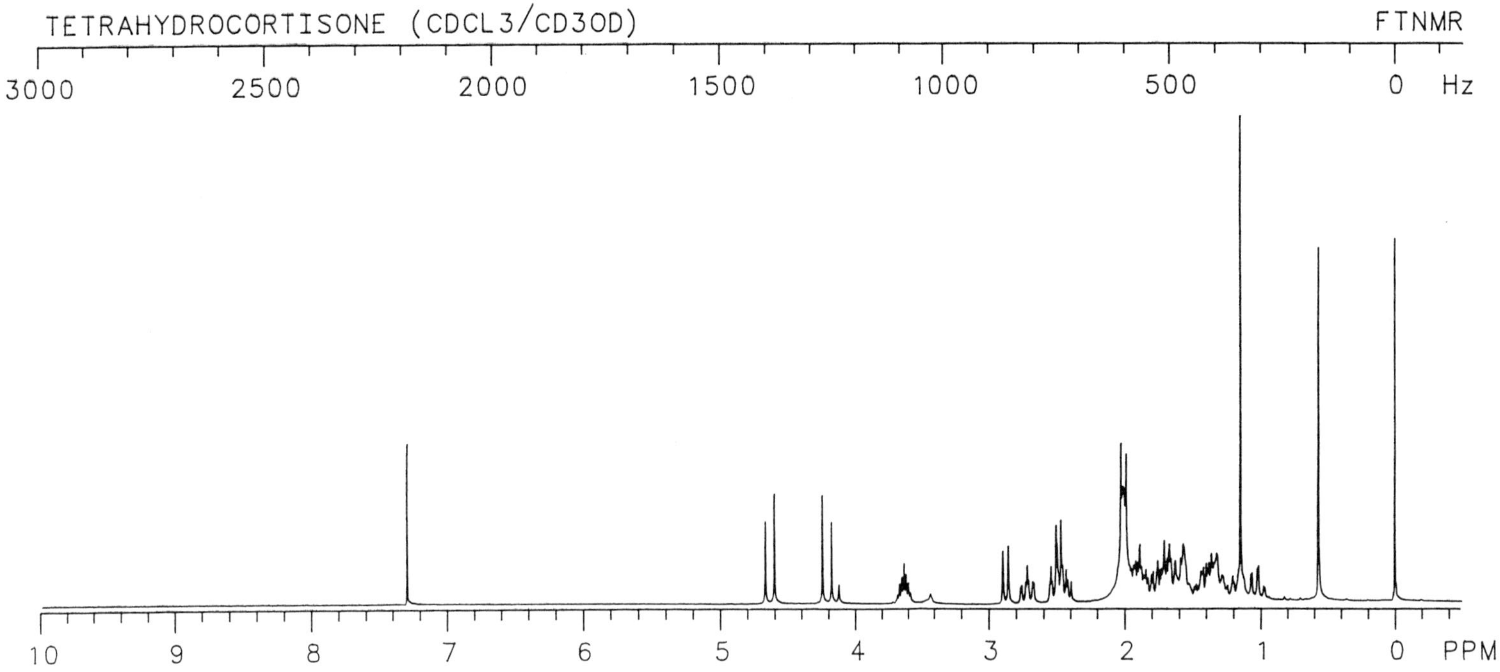

PEAK LISTING

#	HT	PPM
1	74	.00
2	72	.56
3	99	1.14
4	30	1.98
5	23	2.00
6	32	2.02
7	17	2.46
8	16	2.50
9	16	4.17
10	22	4.23
11	22	4.59
12	16	4.66
13	32	7.28

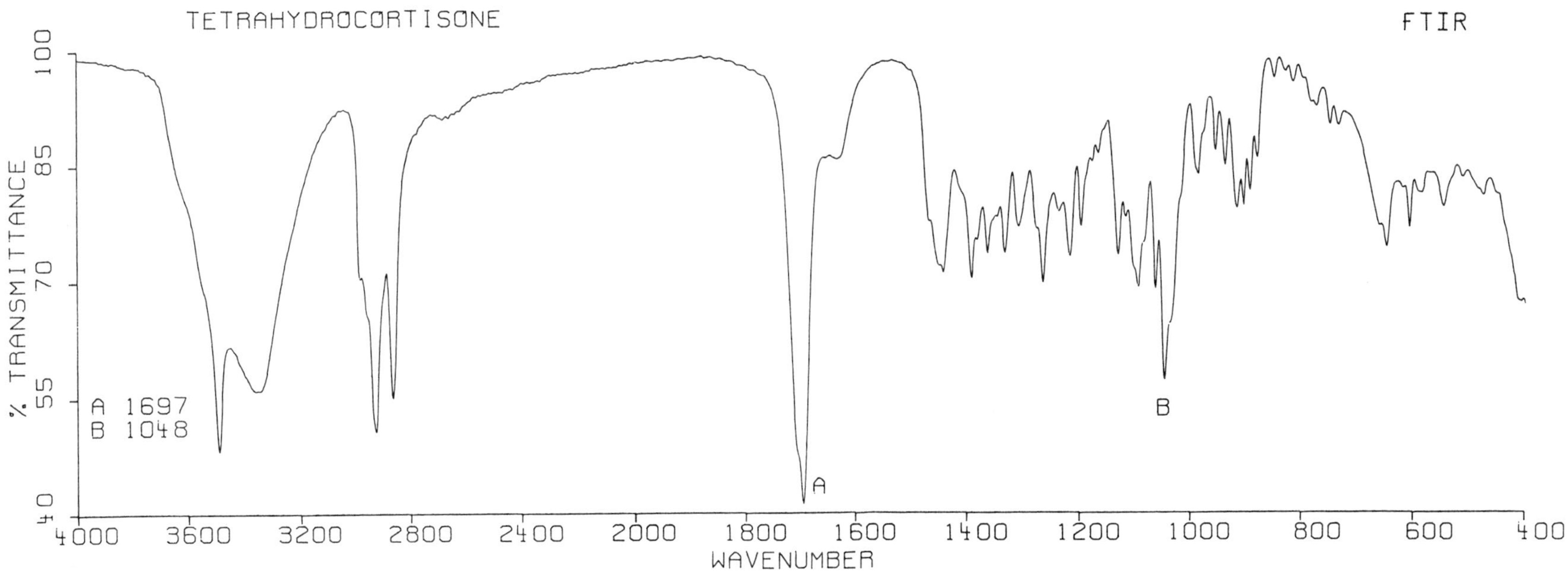

TETRAHYDROCORTISONE-3-ACETATE

$C_{23}H_{34}O_5$

Molecular Weight: 406.52 (406.24)

Synonyms: 3α,17α,21-Trihydroxy-5β-pregnane-11,20-dione-3-acetate
Trade Names:
Use: Androgen

HPLC:
GC:

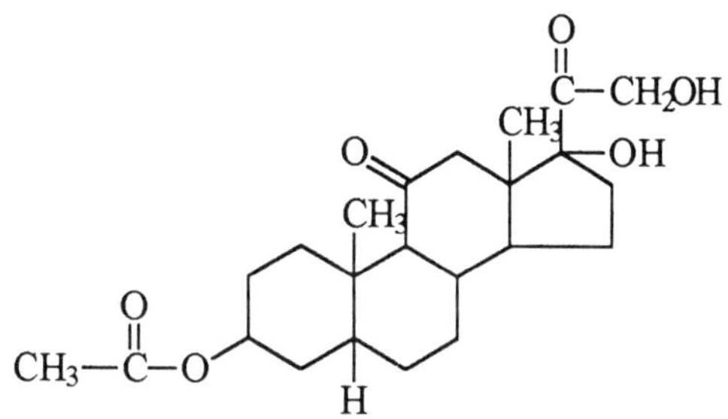

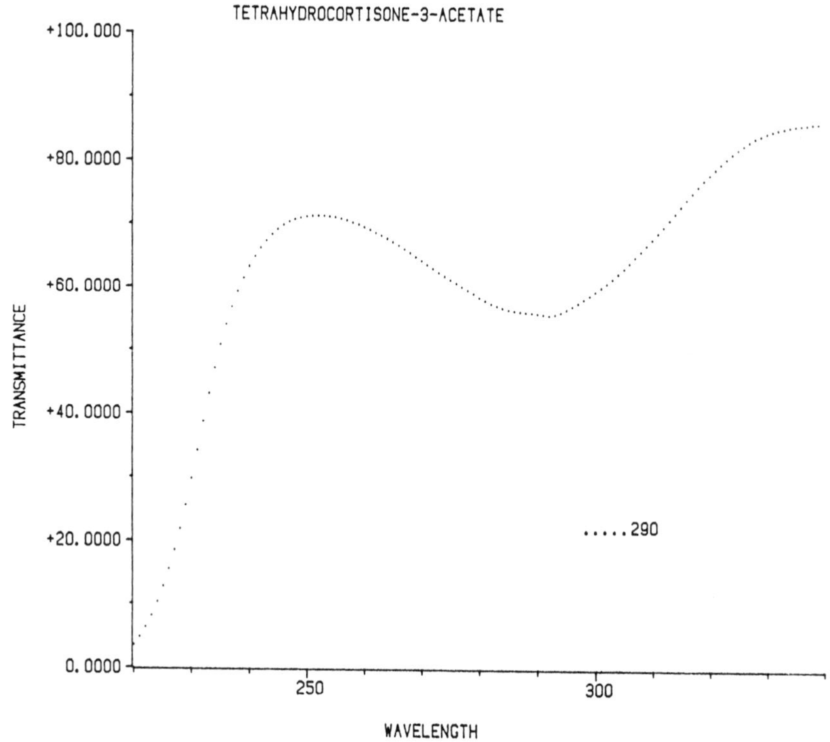

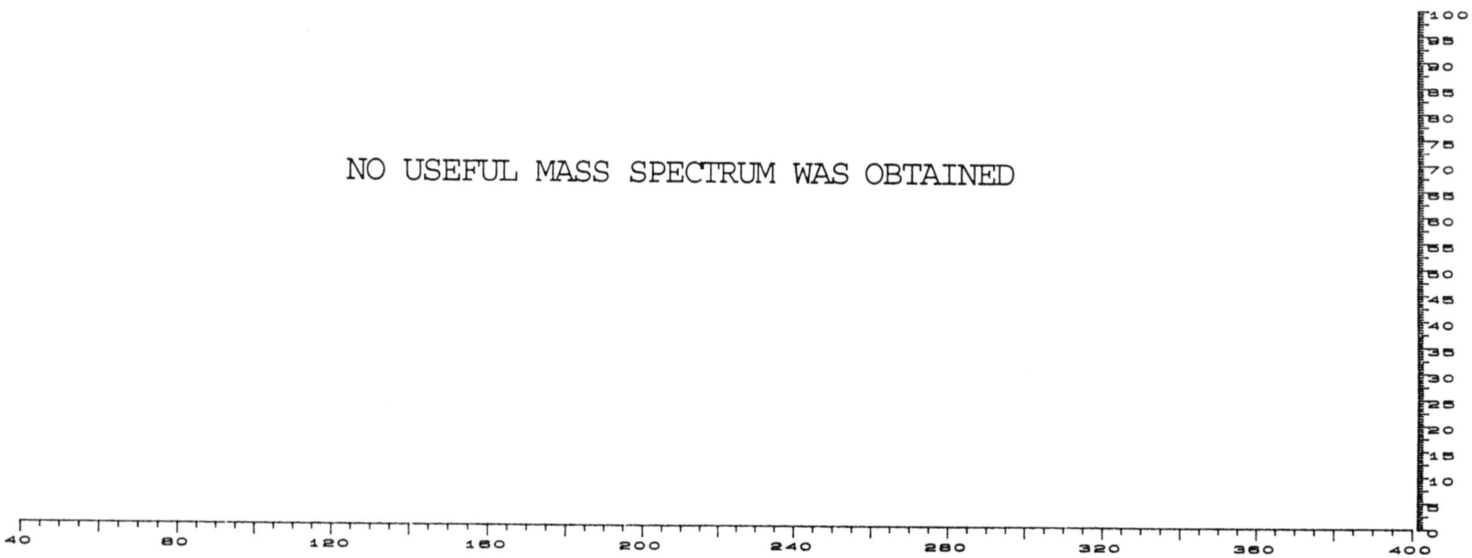

NO USEFUL MASS SPECTRUM WAS OBTAINED

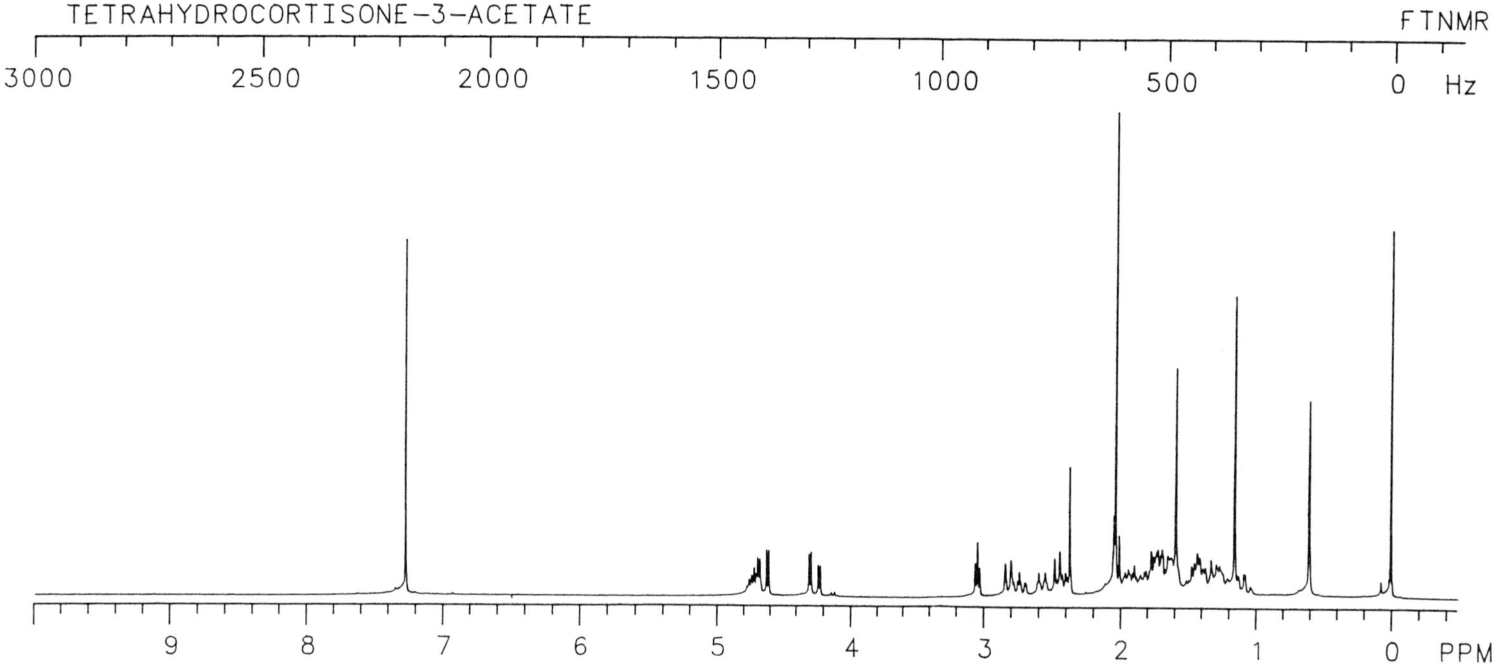

#	HT	PPM
1	75	.00
2	40	.59
3	61	1.15
4	46	1.58
5	12	2.00
6	99	2.03
7	16	2.04
8	26	2.36
9	11	3.03
10	72	7.26

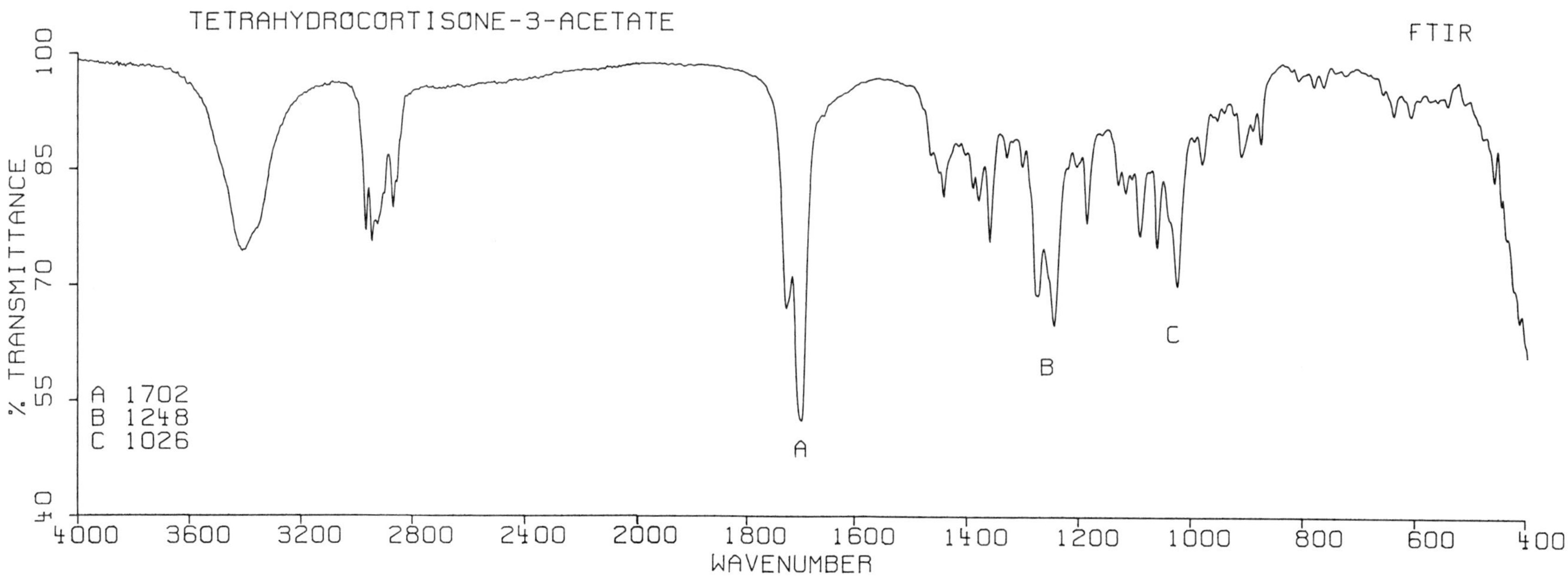

TRAMADOL

$C_{16}H_{25}NO_2$

Molecular Weight: 263.38 (263.19)

Synonyms: 2-[(dimethylamino)methyl]-1-(3-methoxyphenyl)cyclohexanol
Trade Names: Crispin, Tramal, Ultram
Use: Analgesic

HPLC:
GC: 1948; 200°

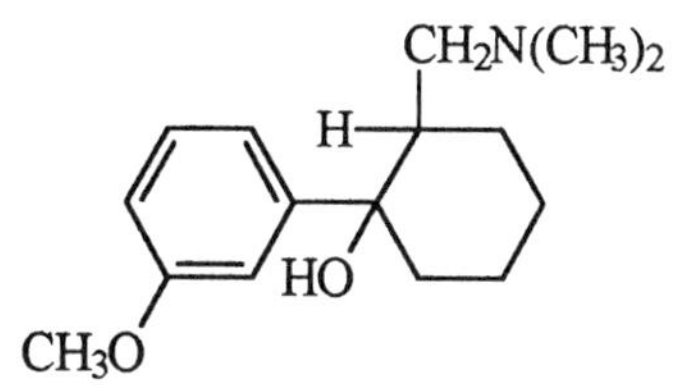

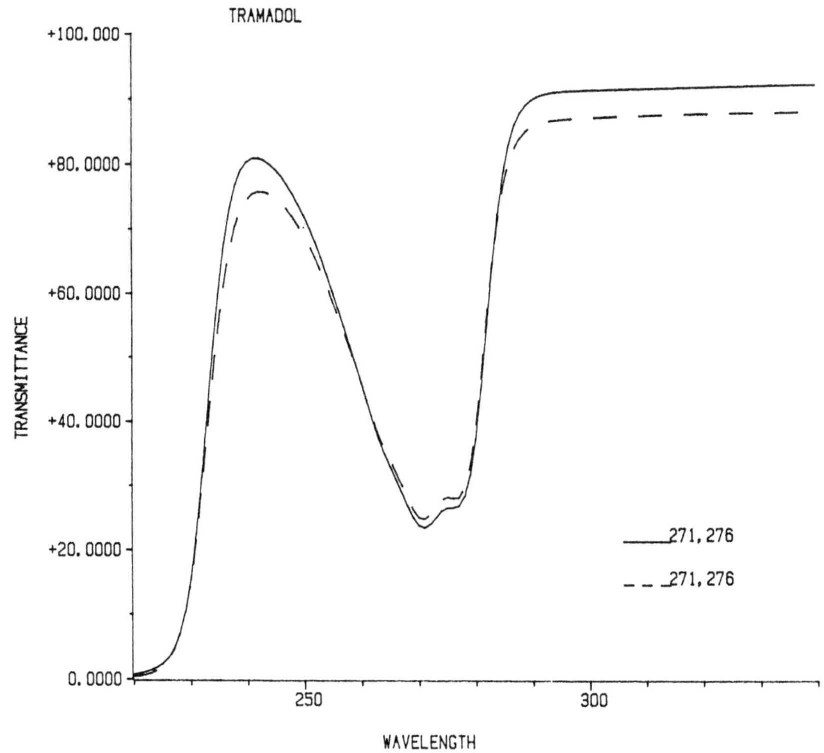

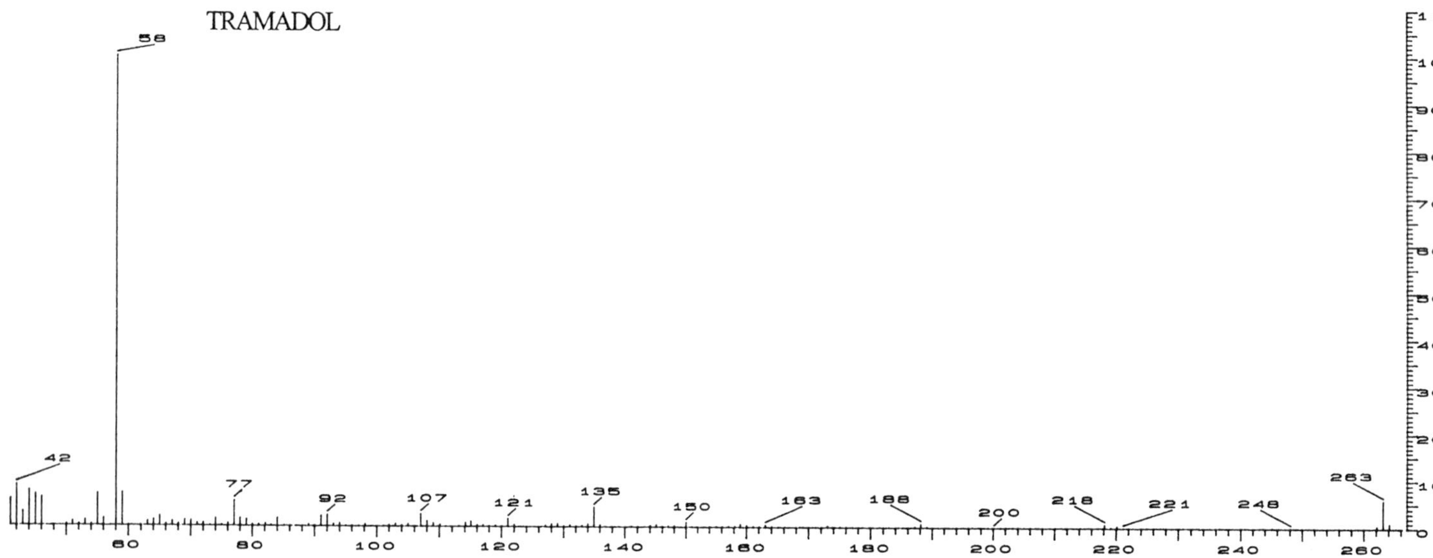

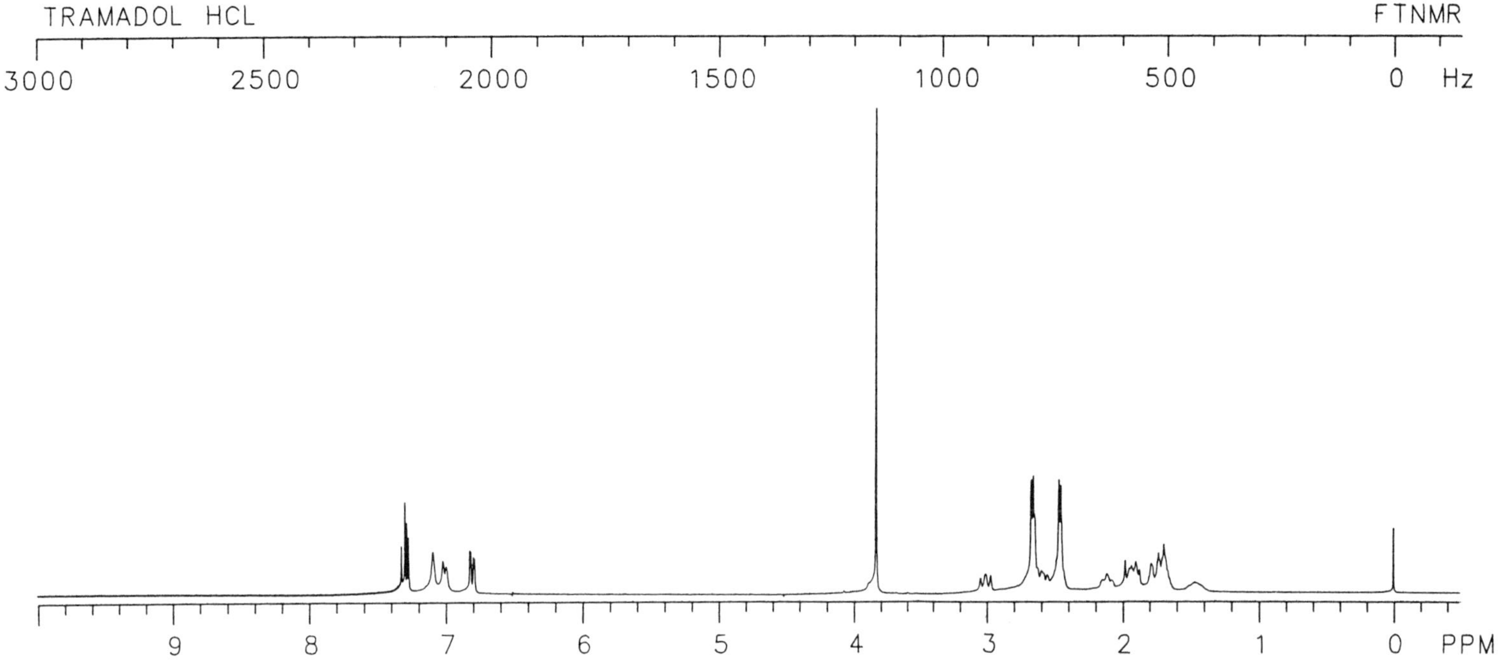

PEAK LISTING

#	HT	PPM
1	13	.00
2	10	1.69
3	8	1.73
4	22	2.45
5	23	2.46
6	16	2.64
7	24	2.65
8	23	2.67
9	99	3.83
10	7	6.79
11	8	6.81
12	9	6.82
13	8	7.09
14	11	7.27
15	14	7.28
16	18	7.29
17	9	7.32

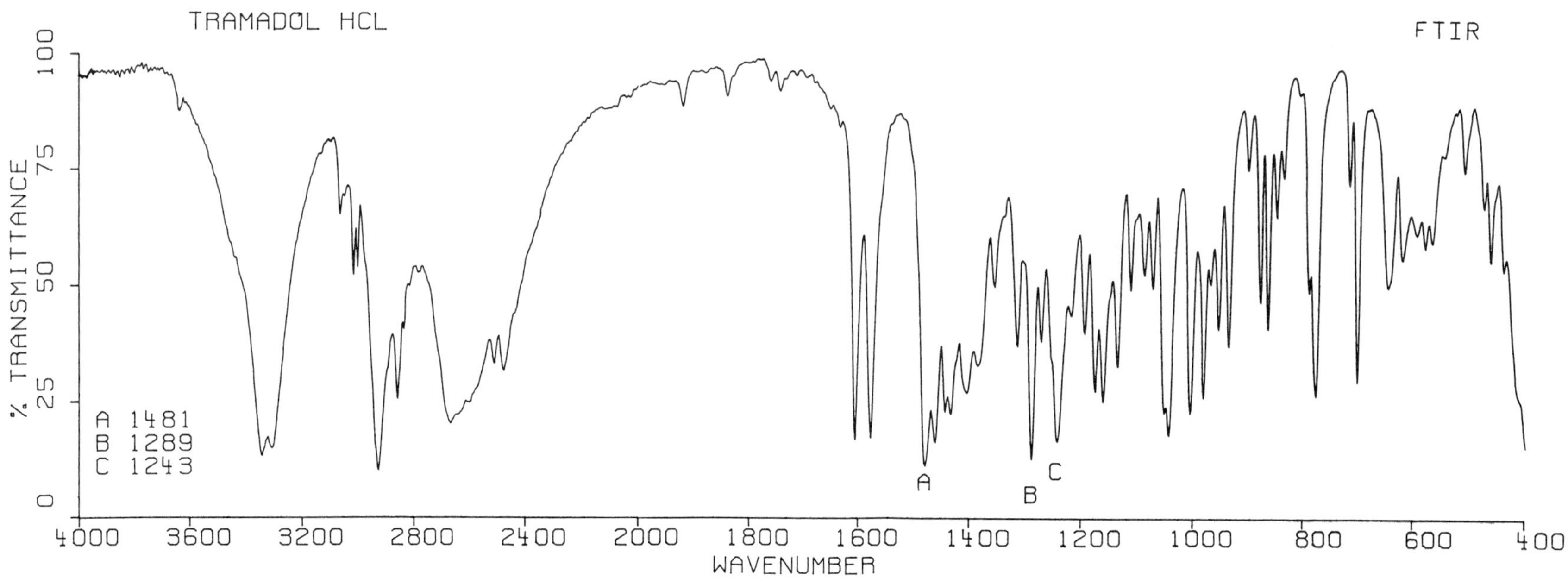

TRENBOLONE

$C_{18}H_{22}O_2$

Molecular Weight: 270.38 (270.16)

Synonyms: 17-Hydroxyestra-4,9,11-trien-3-one
Trade Names:
Use: Anabolic

HPLC:
GC: 2674; 250°

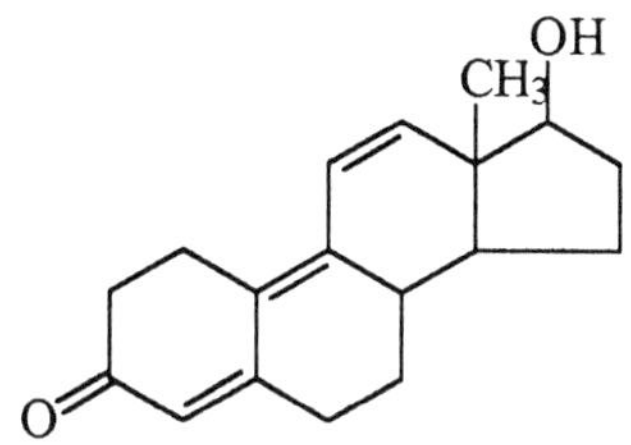

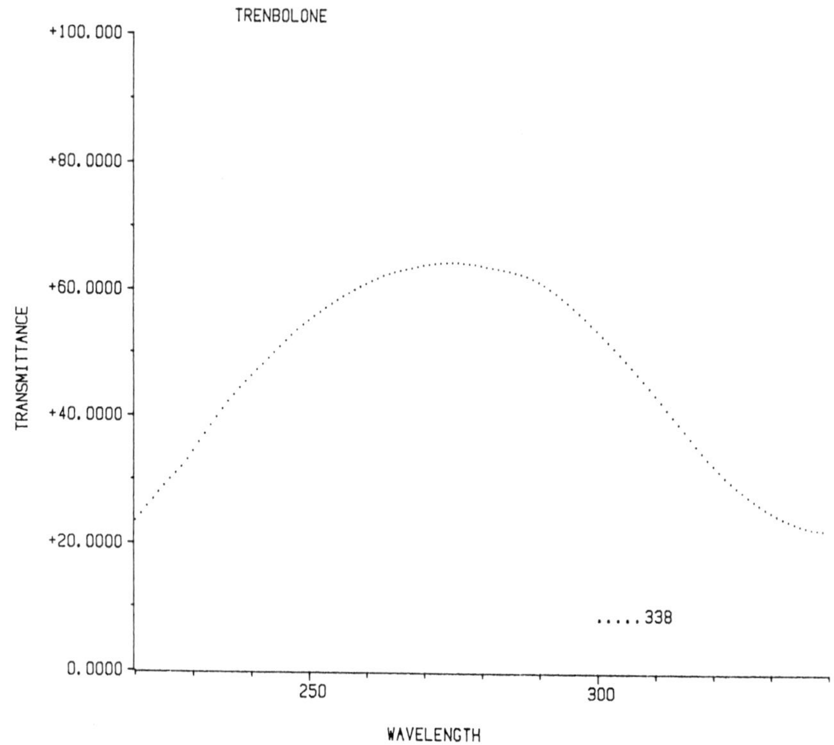

TRENBOLONE

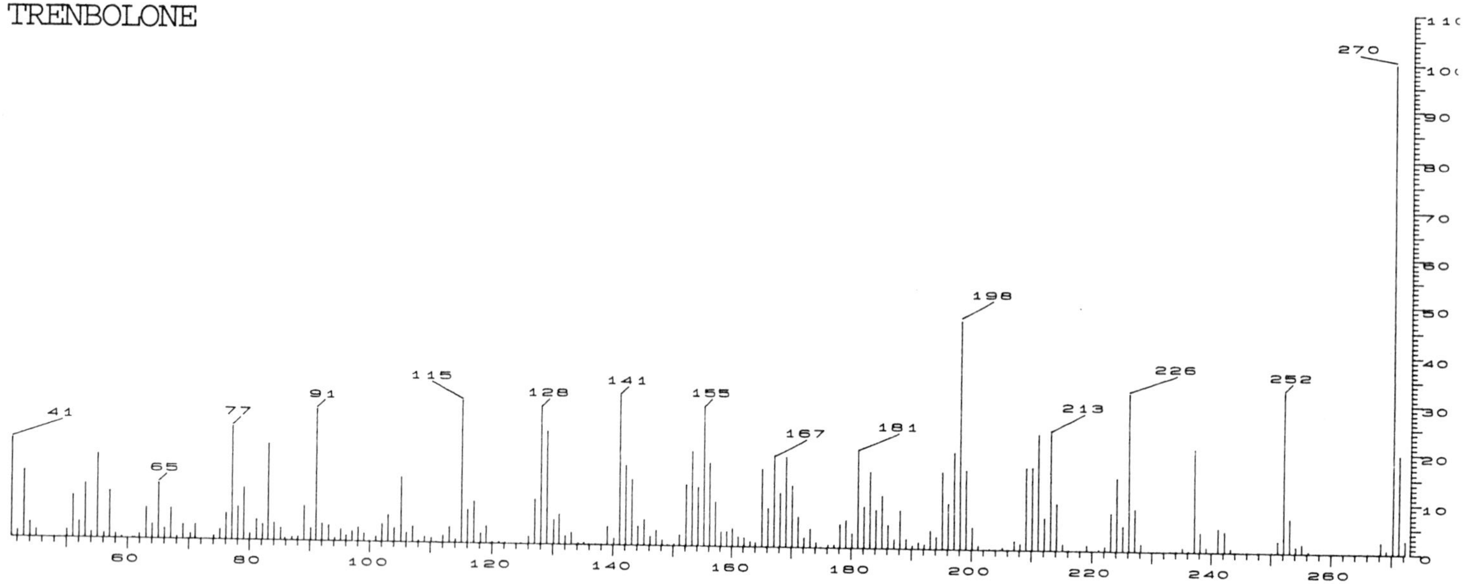

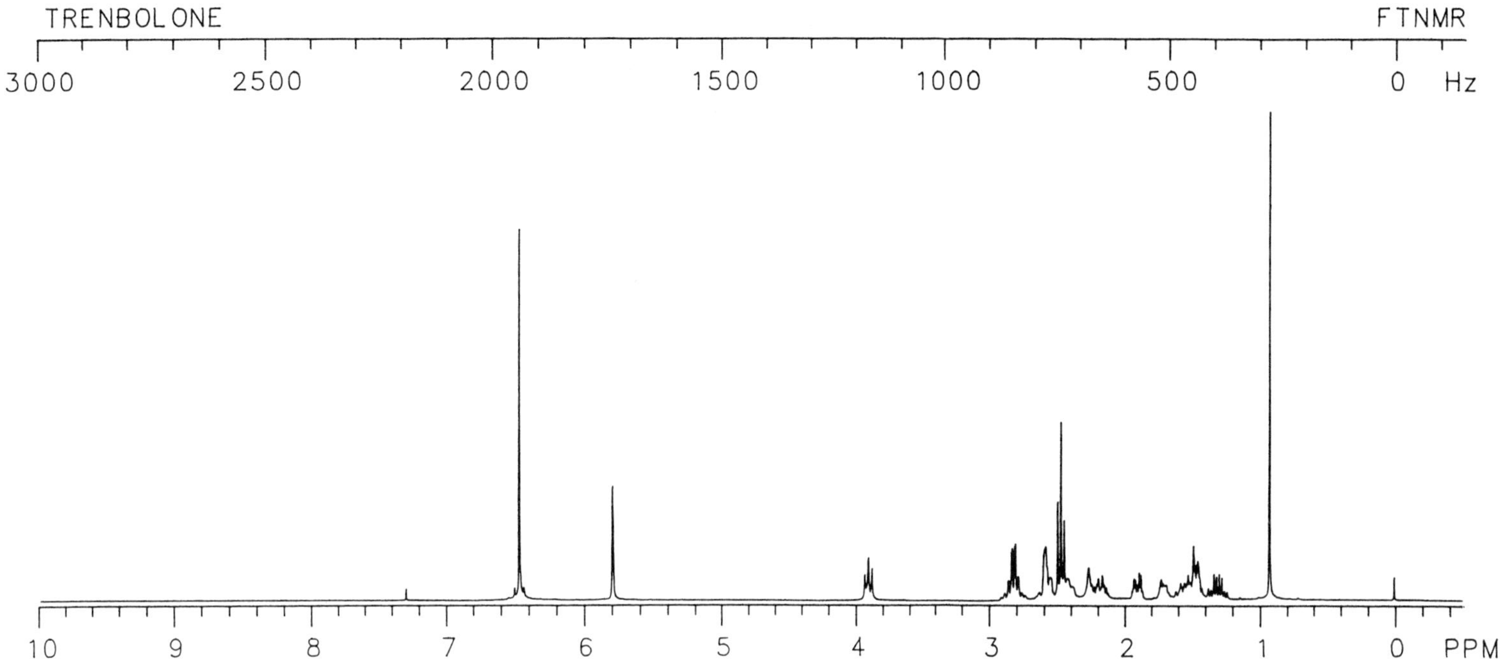

PEAK LISTING

#	HT	PPM
1	99	.92
2	16	2.44
3	36	2.46
4	20	2.49
5	23	5.78
6	75	6.46

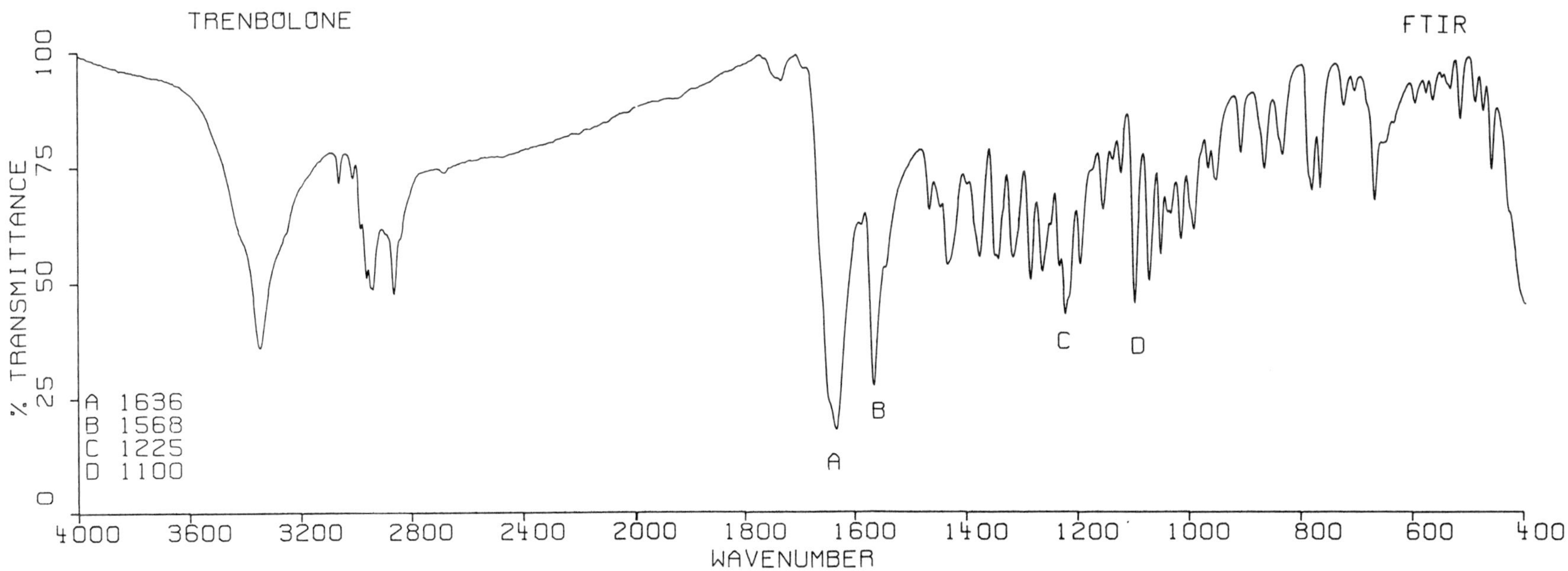

TRENBOLONE ACETATE

$C_{20}H_{24}O_3$

Molecular Weight: 312.41 (312.17)

Synonyms: 17-Hydroxyestra-4,9,11-trien-3-one acetate;
Trade Names: Finaplix
Use: Anabolic

HPLC:
GC: 2775; 250°

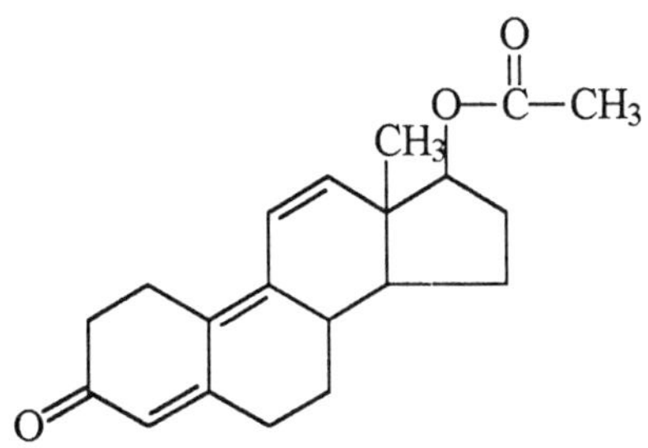

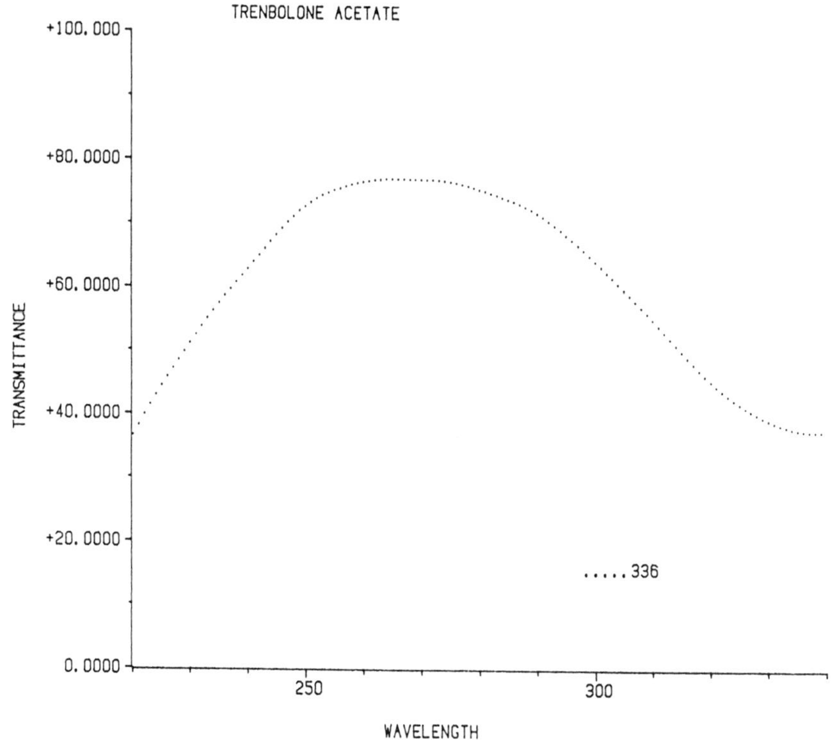

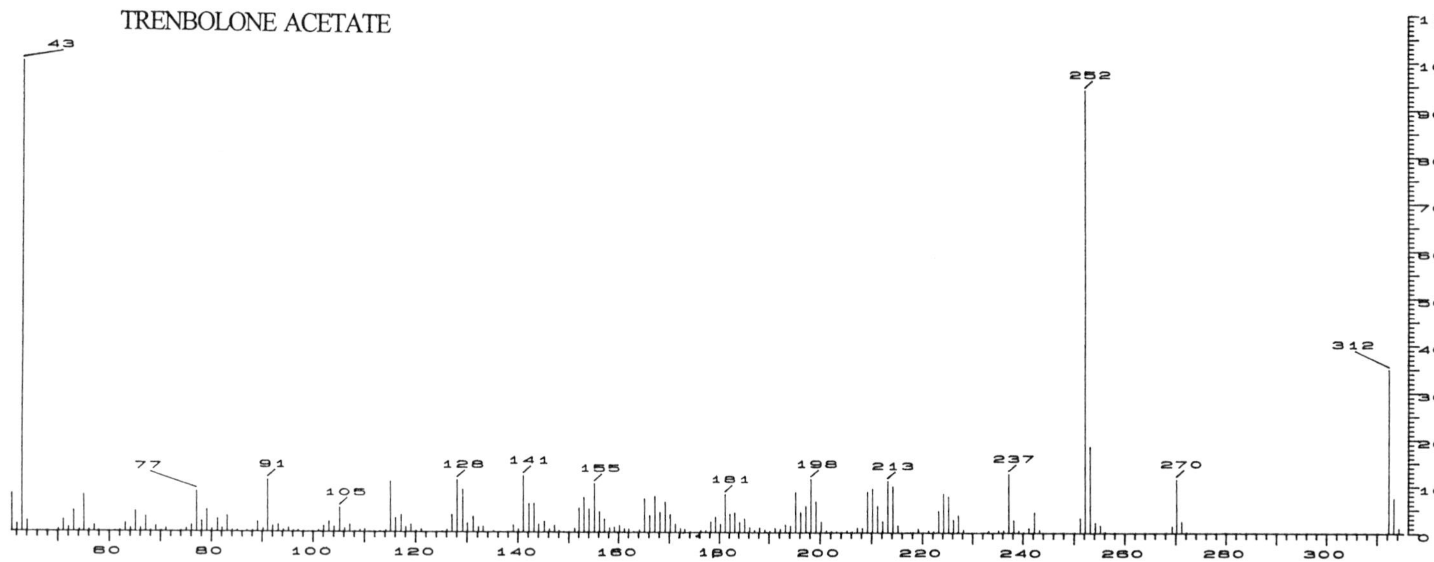

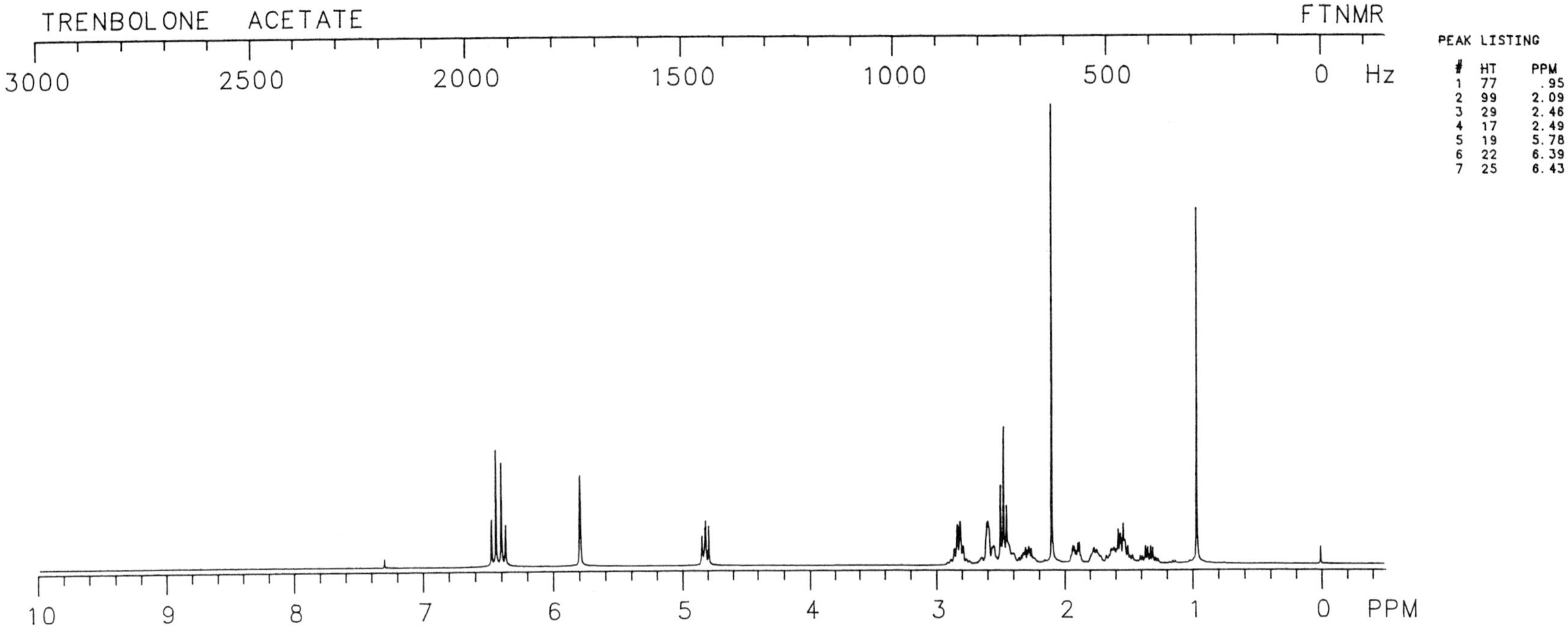
TRENBOLONE ACETATE
FTNMR
3000 2500 2000 1500 1000 500 0 Hz
PEAK LISTING
HT PPM
1 77 .95
2 99 2.09
3 29 2.46
4 17 2.49
5 19 5.78
6 22 6.39
7 25 6.43
10 9 8 7 6 5 4 3 2 1 0 PPM

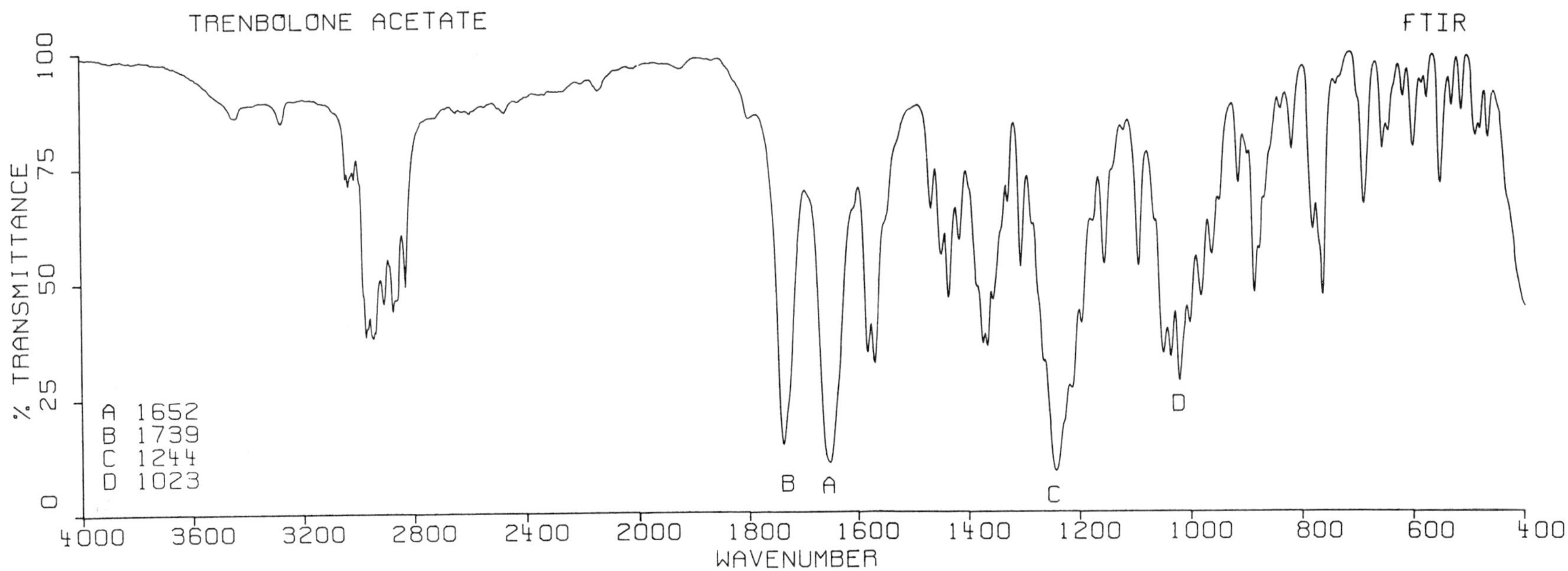
TRENBOLONE ACETATE
FTIR
% TRANSMITTANCE
0 25 50 75 100
A 1652
B 1739
C 1244
D 1023
B A C D
4000 3600 3200 2800 2400 2000 1800 1600 1400 1200 1000 800 600 400
WAVENUMBER

TROLEANDOMYCIN

$C_{41}H_{67}NO_{15}$

Molecular Weight: 813.98 (813.45)

Synonyms: Oleandomycin triacetate ester; triacetyloleandomycin
Trade Names: Cyclamin, Evramicina, Evramycin, Matromicina, Oleandocetine, Olicin, Spectrobact, Tao, Treilmicina, Wytrion,
Use: Antibacterial

HPLC:
GC:

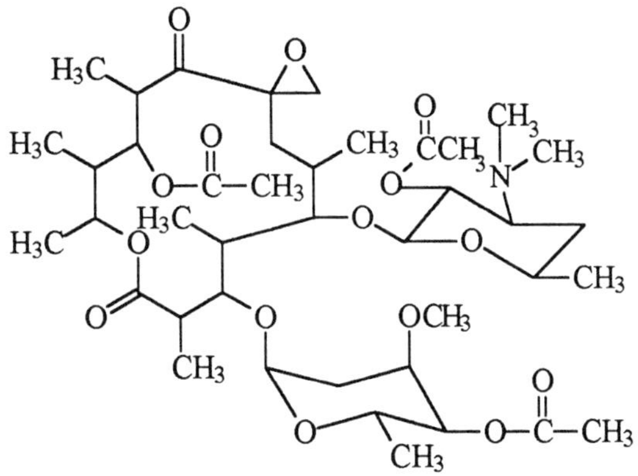

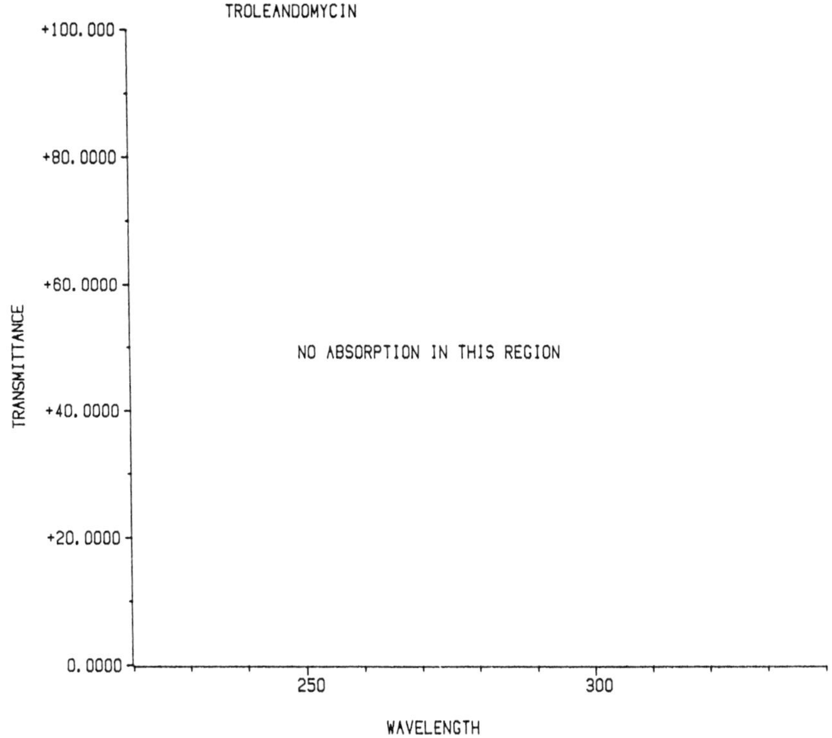

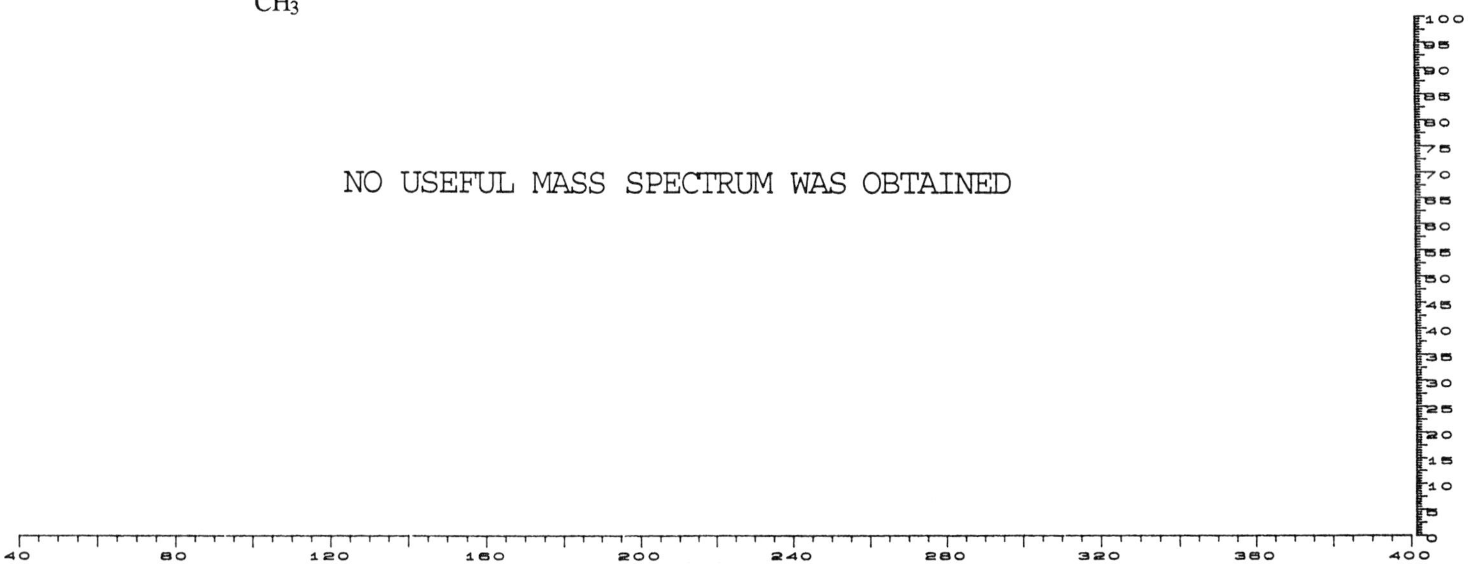

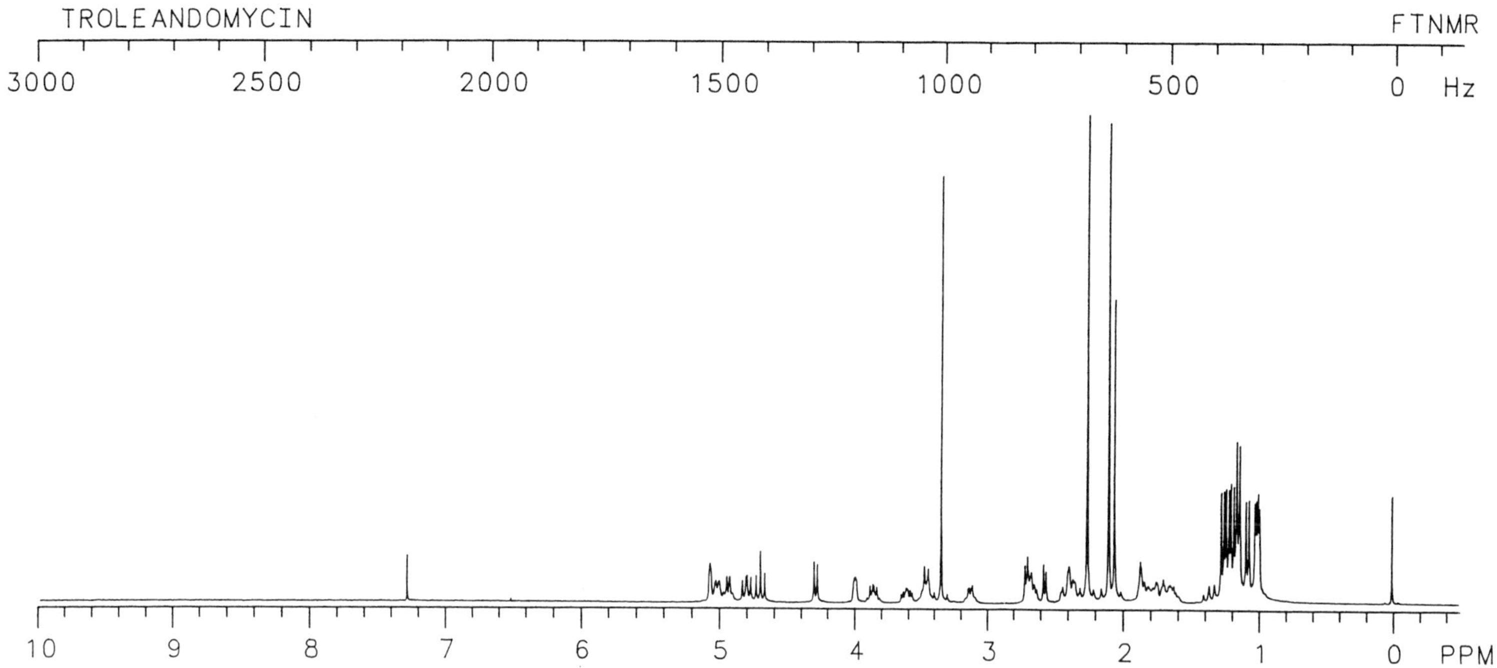

PEAK LISTING

#	HT	PPM
1	22	.00
2	19	.98
3	22	.99
4	21	1.01
5	20	1.02
6	21	1.06
7	20	1.08
8	32	1.13
9	33	1.15
10	23	1.17
11	24	1.19
12	23	1.21
13	23	1.23
14	23	1.24
15	22	1.26
16	62	2.06
17	97	2.10
18	91	2.10
19	99	2.25
20	87	3.34

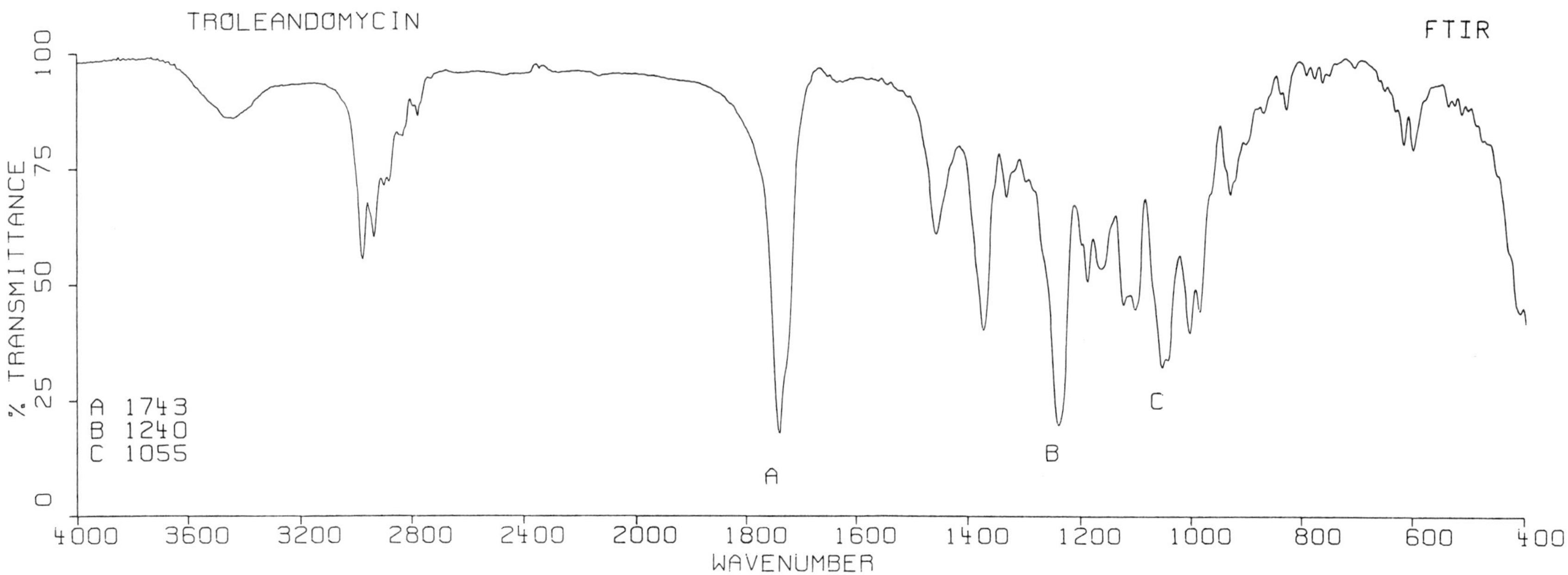

TROPINE-3,5-DICHLOROBENZOATE

$C_{15}H_{17}Cl_2NO_2$

Molecular Weight: 314.21 (313.06)

Synonyms: endo-8-Methyl-8-azabicyclo[3.2.1]octan-3-ol-3,5-dichlorobenzoate
Trade Names:
Use: Antagonist

HPLC:
GC: 2270; 250°

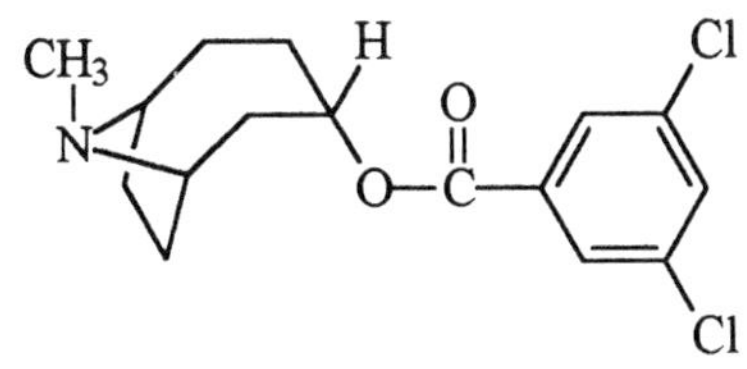

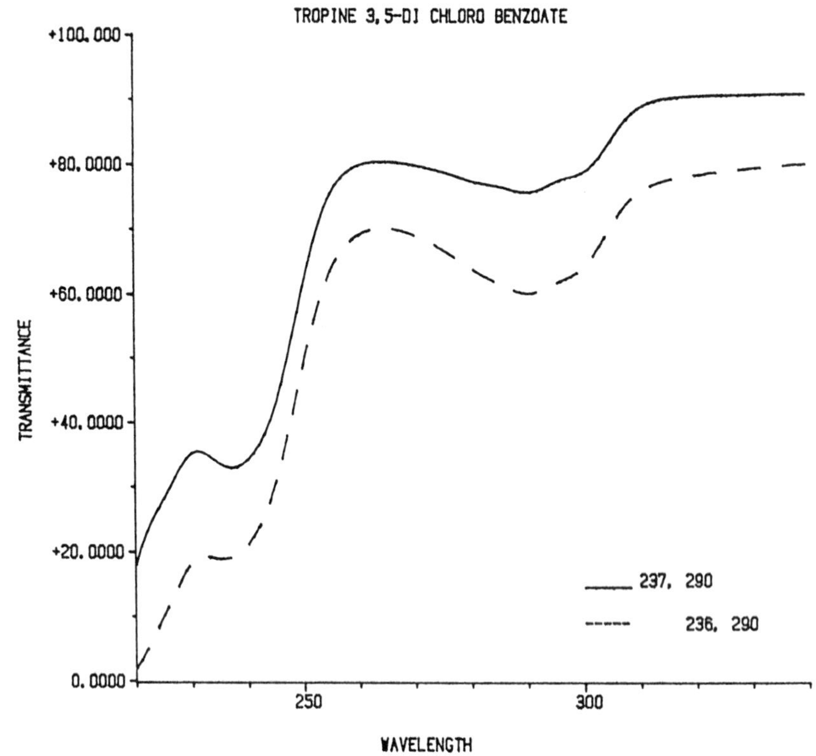

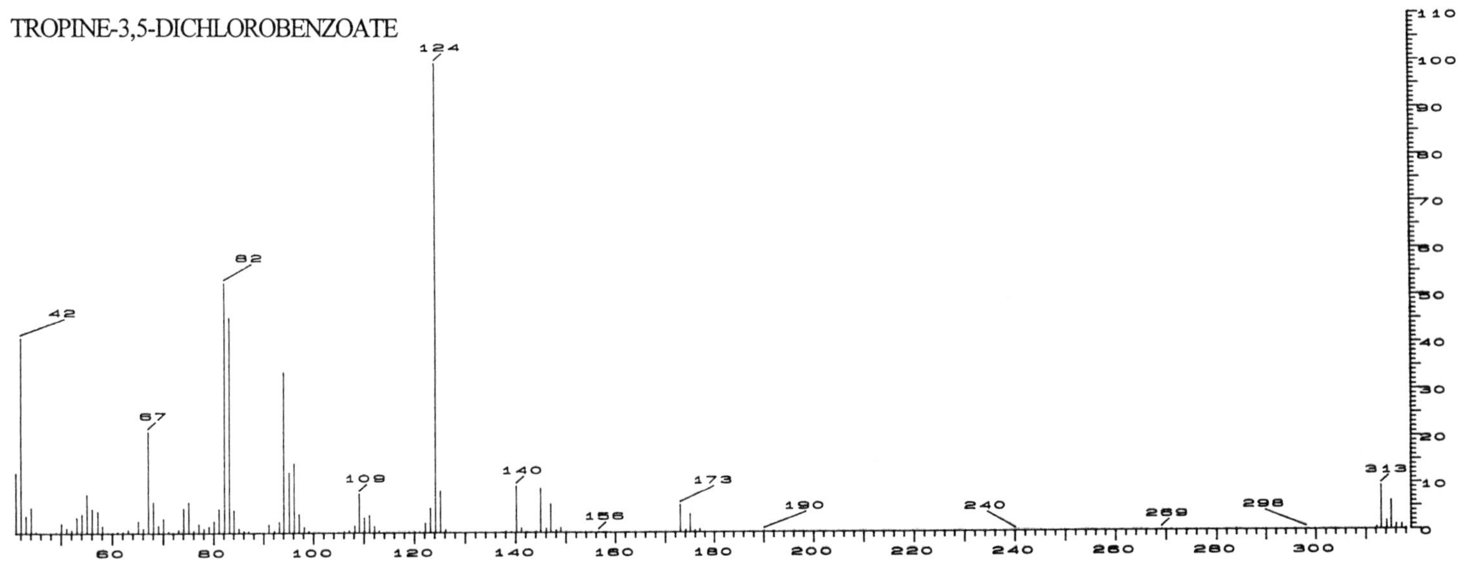

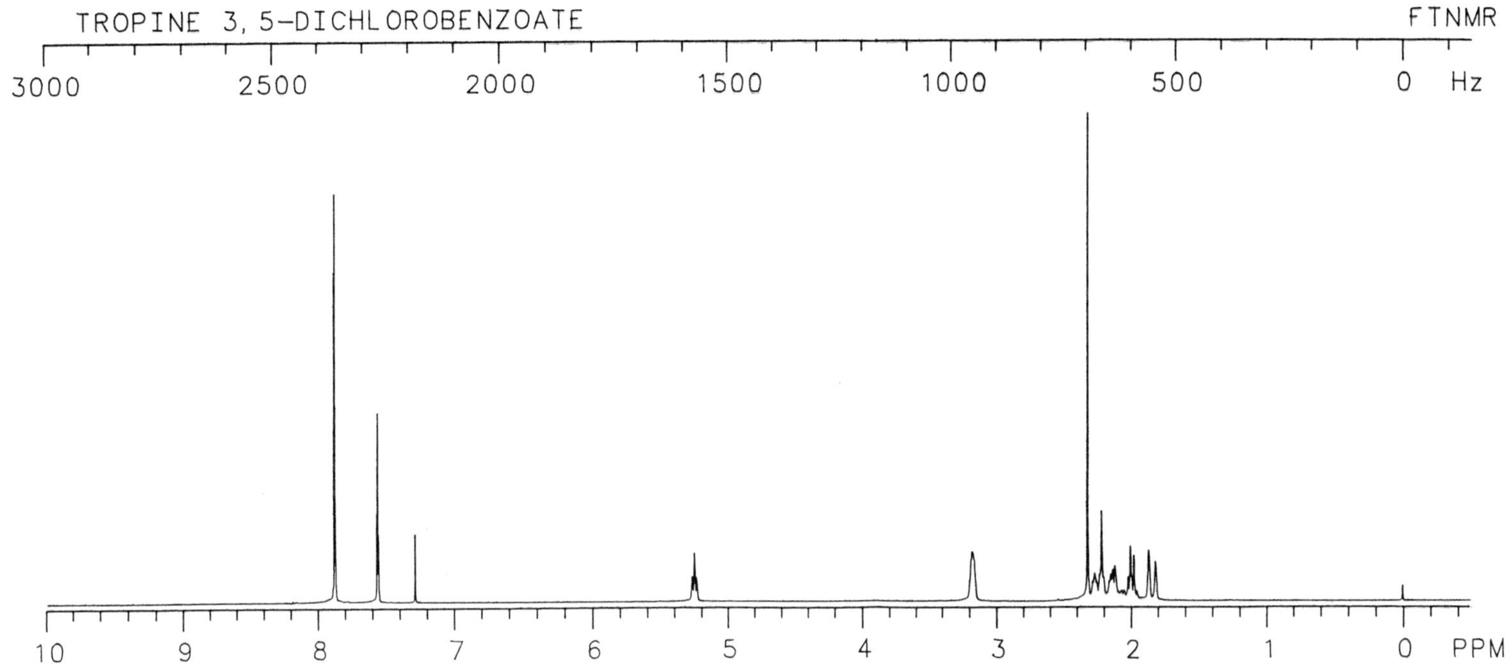

#	HT	PPM
1	11	2.00
2	18	2.21
3	99	2.31
4	13	7.28
5	13	7.54
6	38	7.55
7	18	7.55
8	83	7.86
9	67	7.87
10	36	7.87

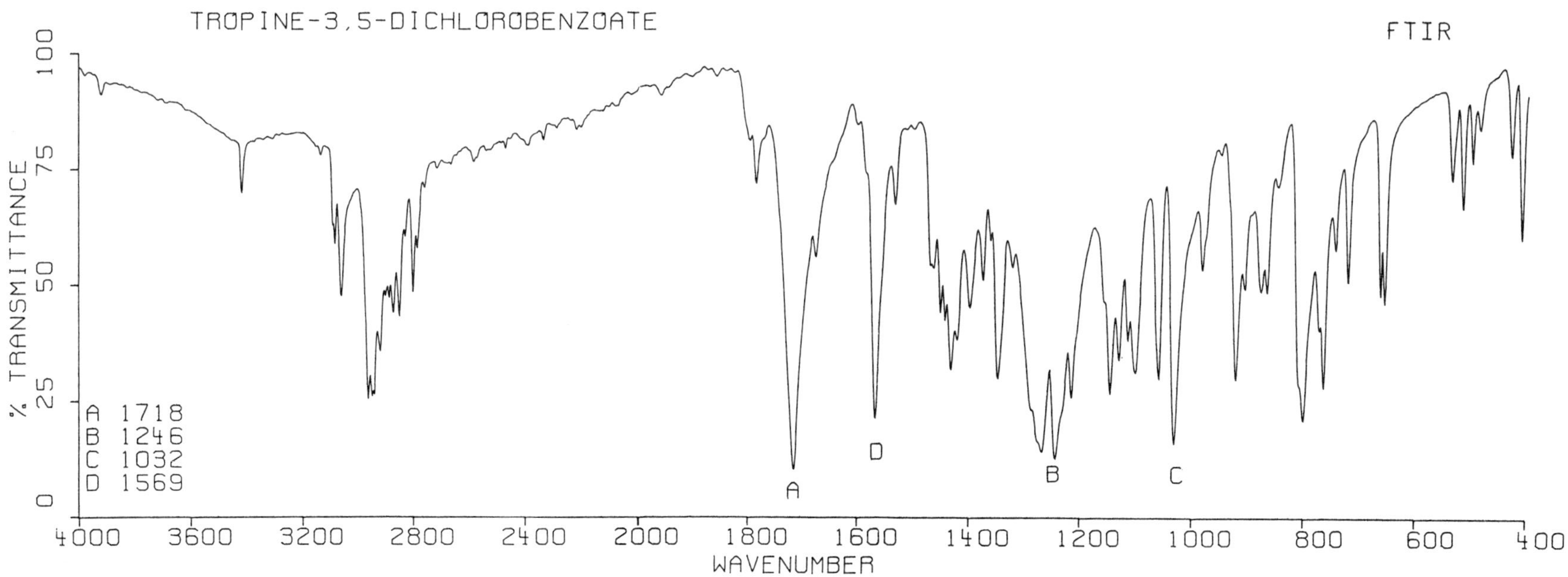

URAPIDIL

$C_{20}H_{29}N_5O_3$

Molecular Weight: 387.48 (387.23)

Synonyms: 6-[[3-[4-(2-Methoxyphenyl)-1-piperazinyl]propyl]amino]-1,3-dimethyl-2,4(1H,3H)-pyrimidinedione
Trade Names: Edrantil, Eupressyl, Uraprene
Use: Antihypertensive

HPLC: Ethanol: 7.9
GC:

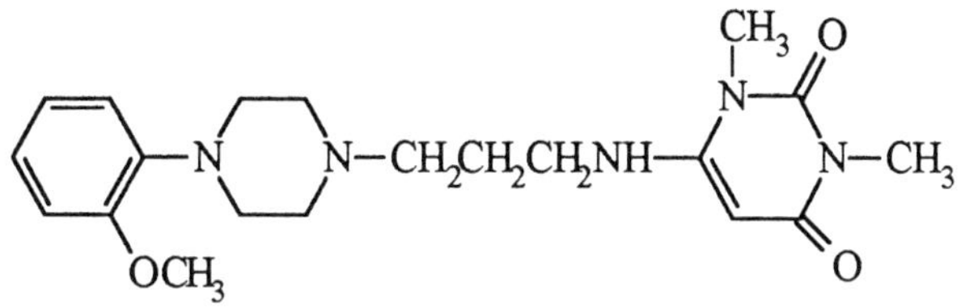

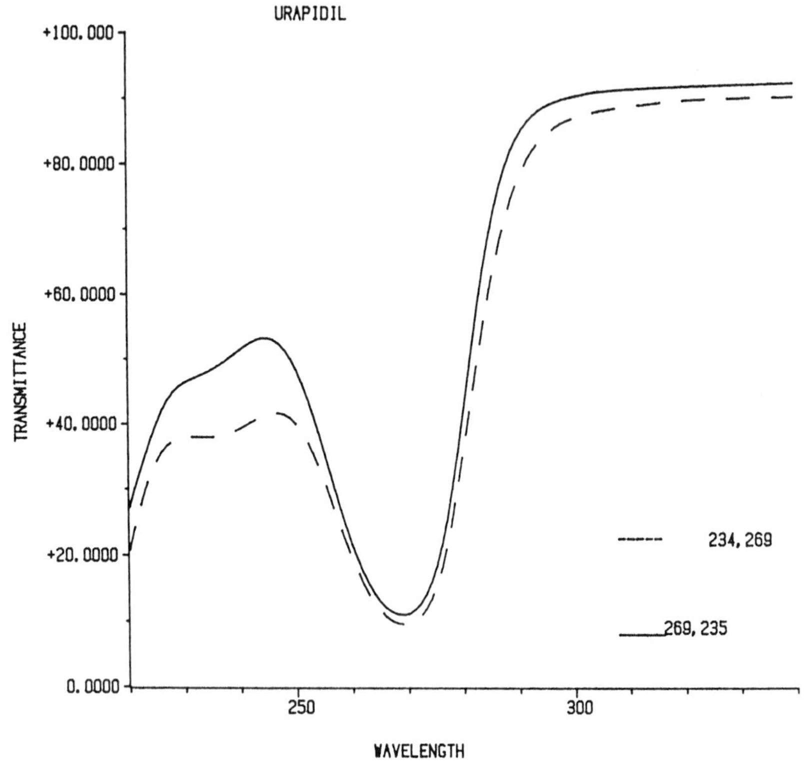

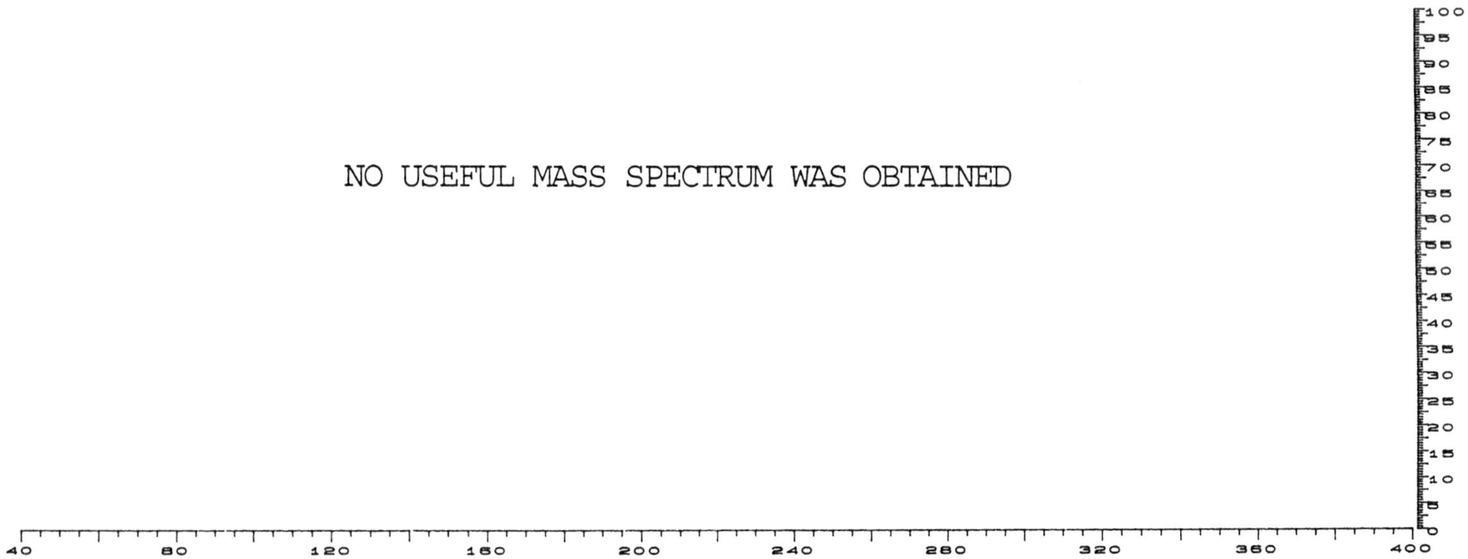

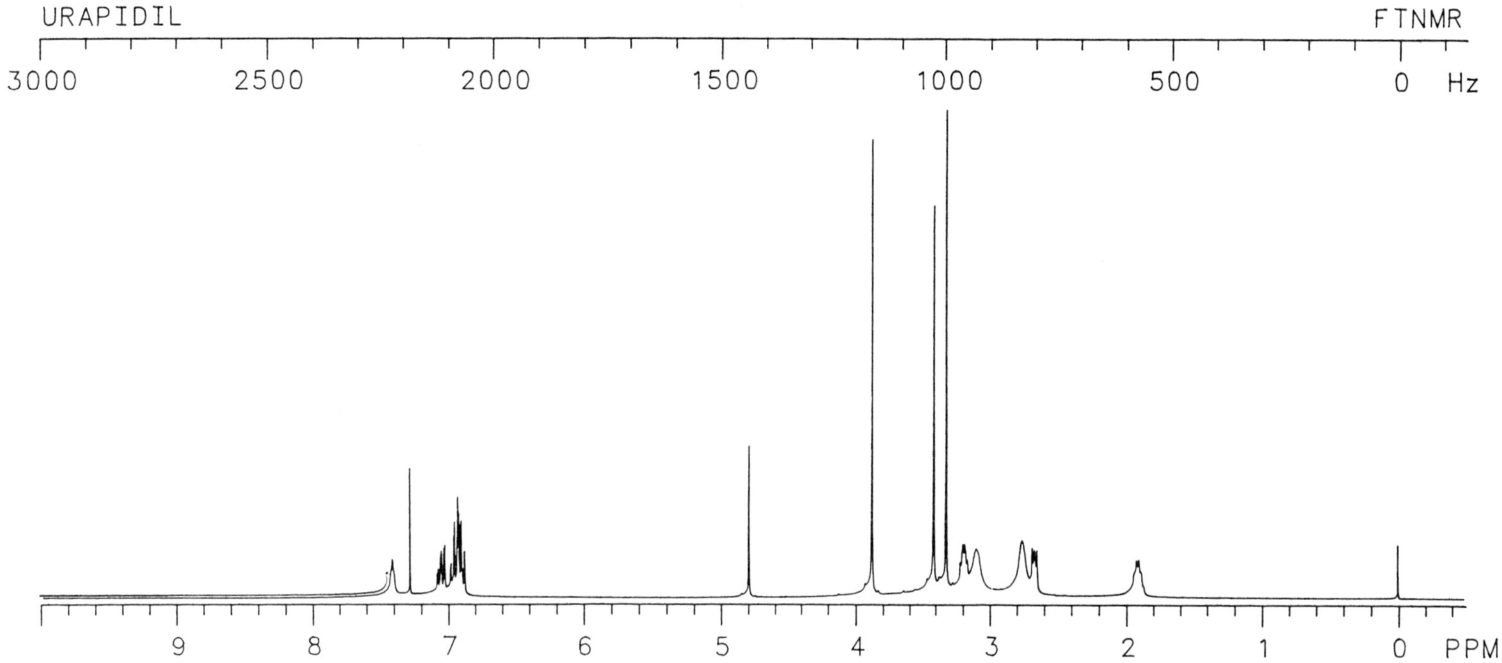

PEAK LISTING

#	HT	PPM
1	99	3.31
2	80	3.40
3	93	3.87
4	30	4.78
5	15	6.89
6	14	6.90
7	17	6.91
8	20	6.92
9	15	6.94
10	26	7.27

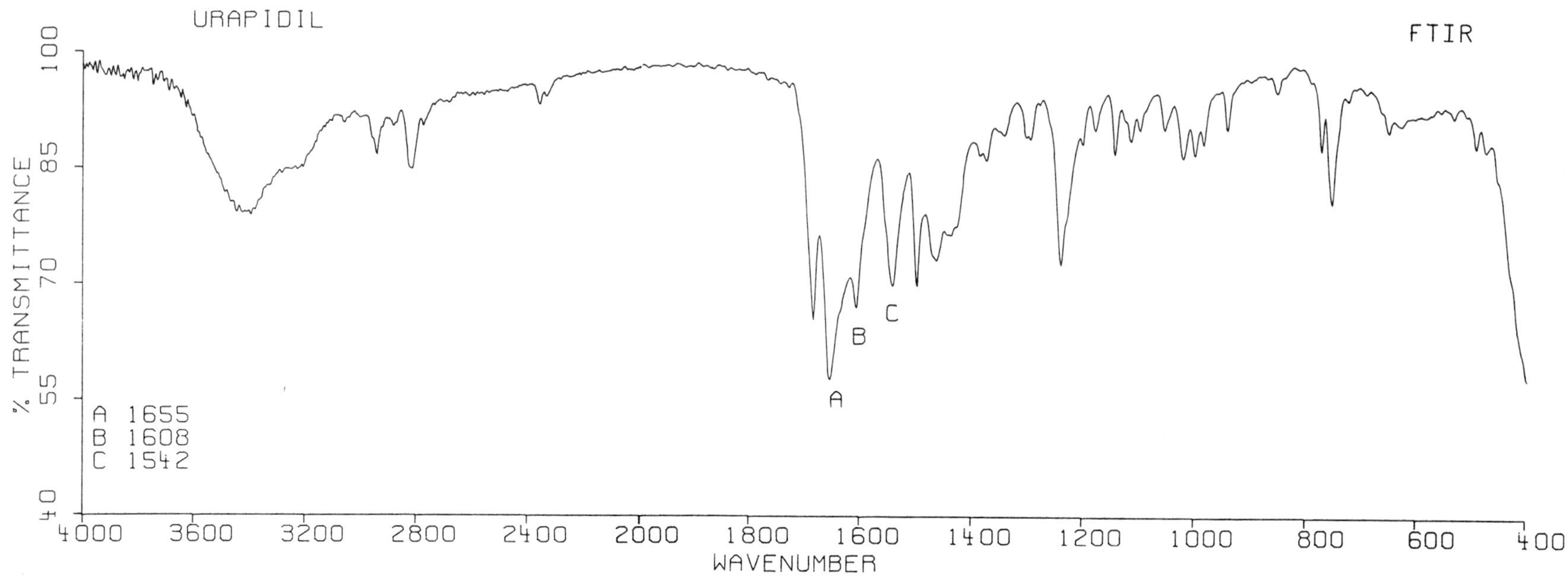

VENLAFAXINE

$C_{17}H_{27}NO_2$

Molecular Weight: 277.41 (277.20)

Synonyms: 1-[2-(Dimethylamino)-1-(4-methoxyphenyl)ethyl]cyclohexanol; 1-[α-[(dimethylamino)methyl]-p-methoxybenzyl]cyclohexanol
Trade Names: Effexor
Use: Antidepressant

HPLC:
GC:

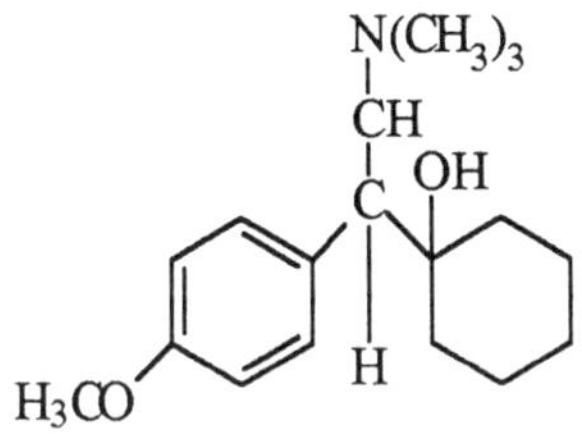

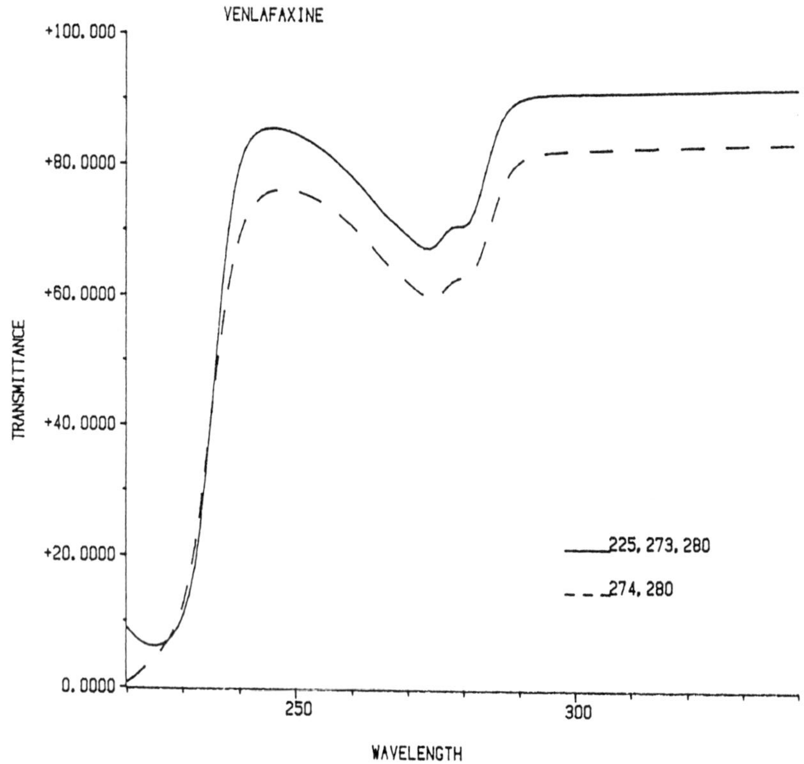

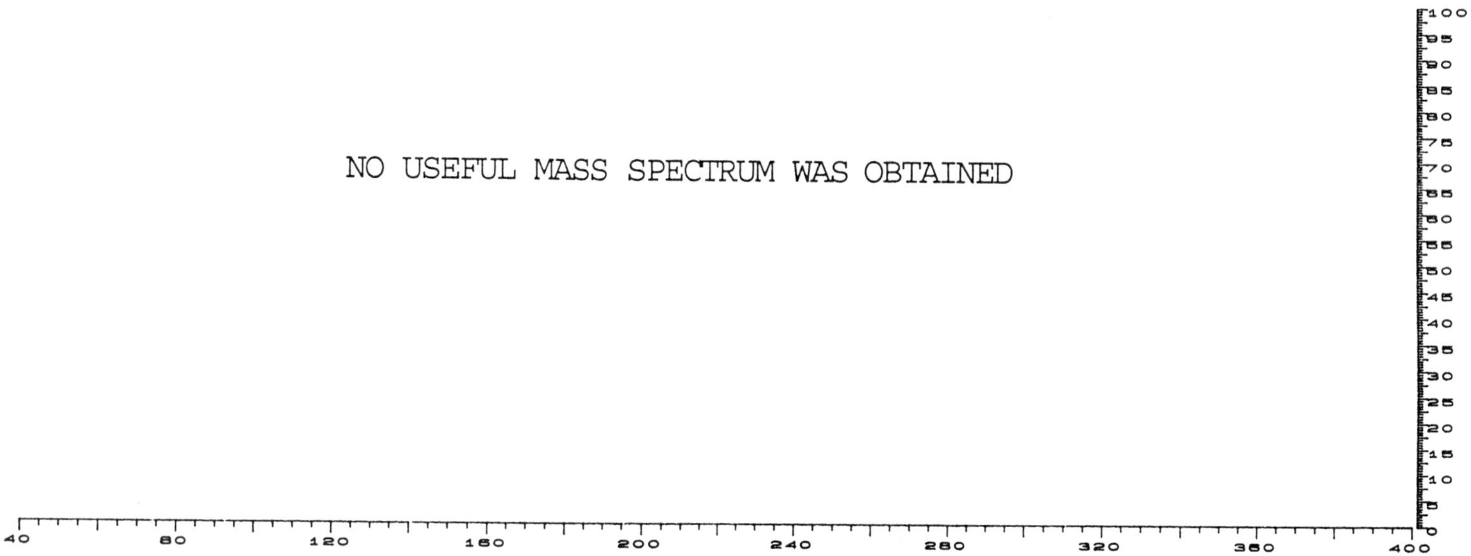

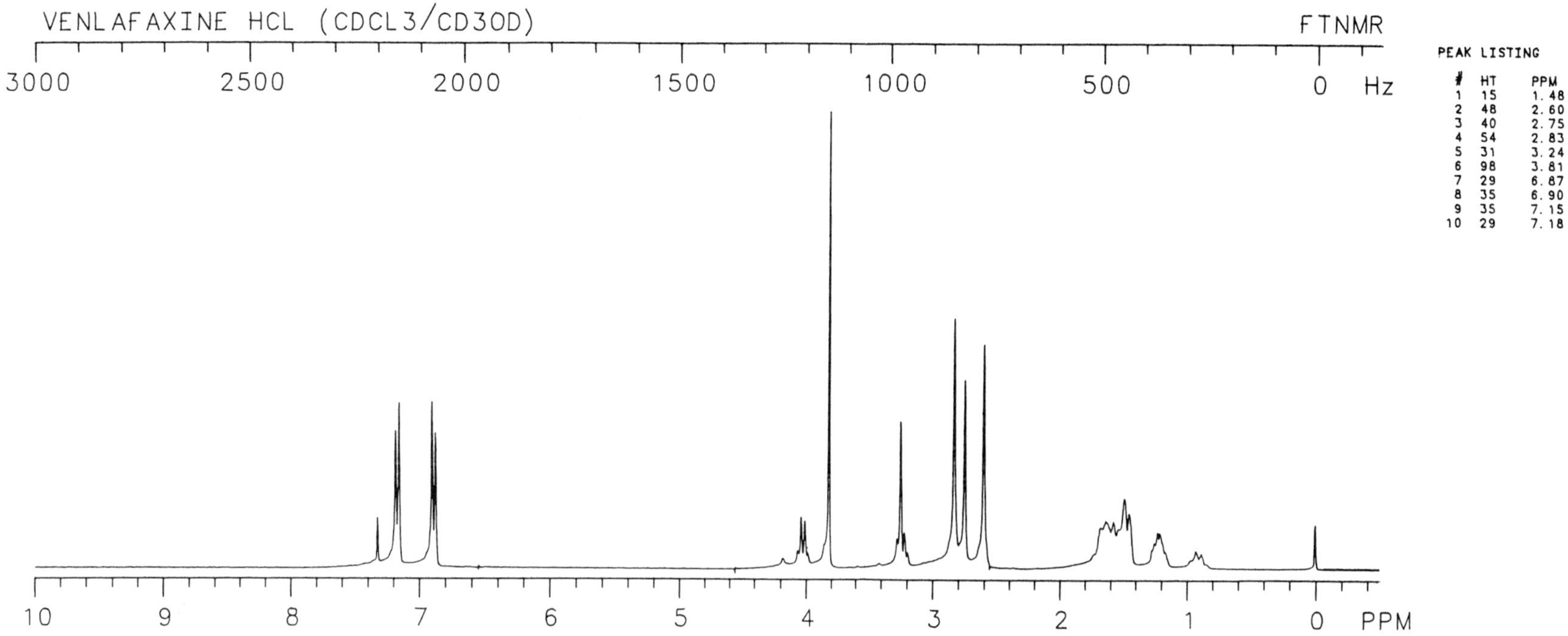
VENLAFAXINE HCL (CDCL3/CD3OD)
FTNMR
3000 2500 2000 1500 1000 500 0 Hz
10 9 8 7 6 5 4 3 2 1 0 PPM
PEAK LISTING
HT PPM
1 15 1.48
2 48 2.60
3 40 2.75
4 54 2.83
5 31 3.24
6 98 3.81
7 29 6.87
8 35 6.90
9 35 7.15
10 29 7.18

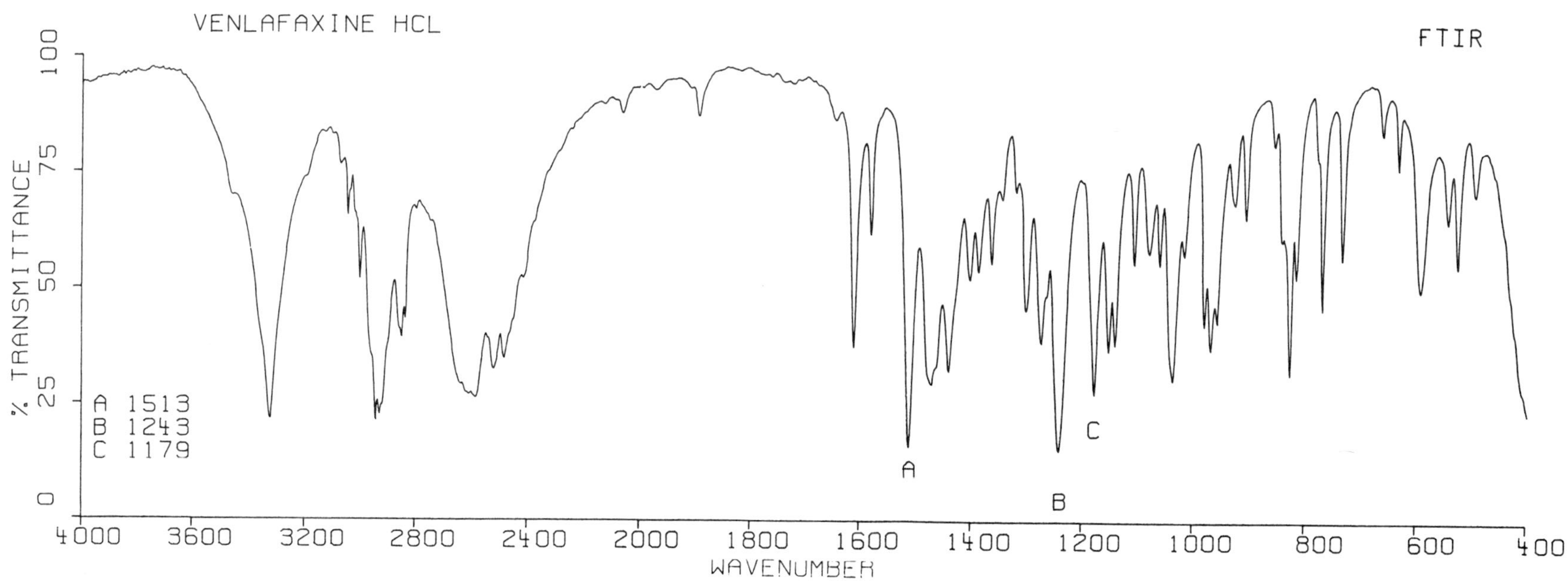
VENLAFAXINE HCL
FTIR
% TRANSMITTANCE
0 25 50 75 100
A 1513
B 1243
C 1179
A
B
C
4000 3600 3200 2800 2400 2000 1800 1600 1400 1200 1000 800 600 400
WAVENUMBER

ZALCITABINE

$C_9H_{13}N_3O_3$

Molecular Weight: 211.22 (211.10)

Synonyms: 2',3'-Dideoxycytidine; dideoxycytidine; ddCyd
Trade Names: Hivid
Use: Antiviral

HPLC: Methanol: 2.6
GC:

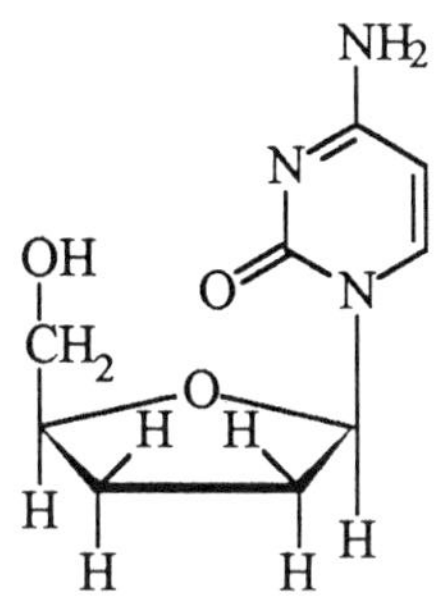

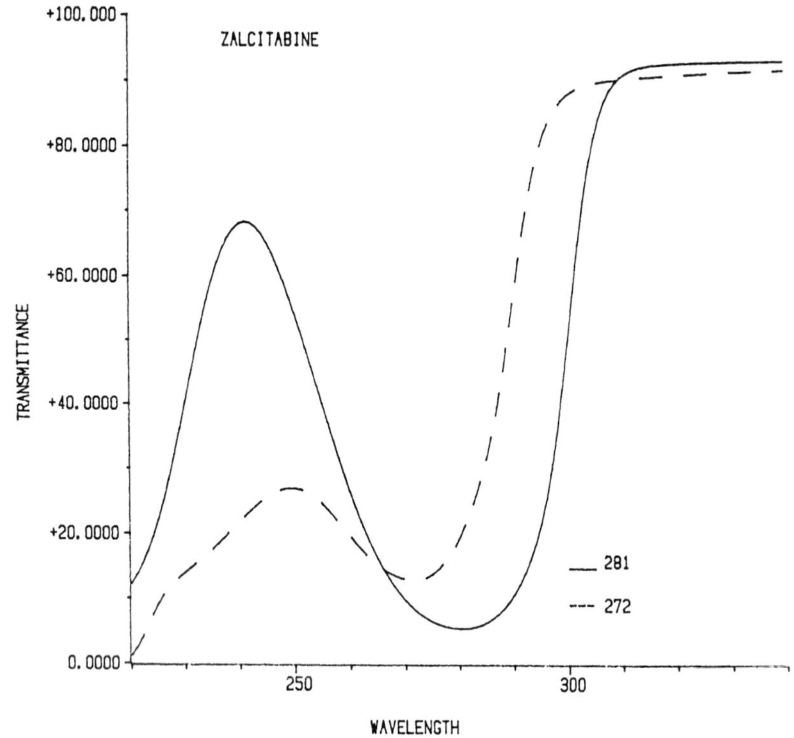

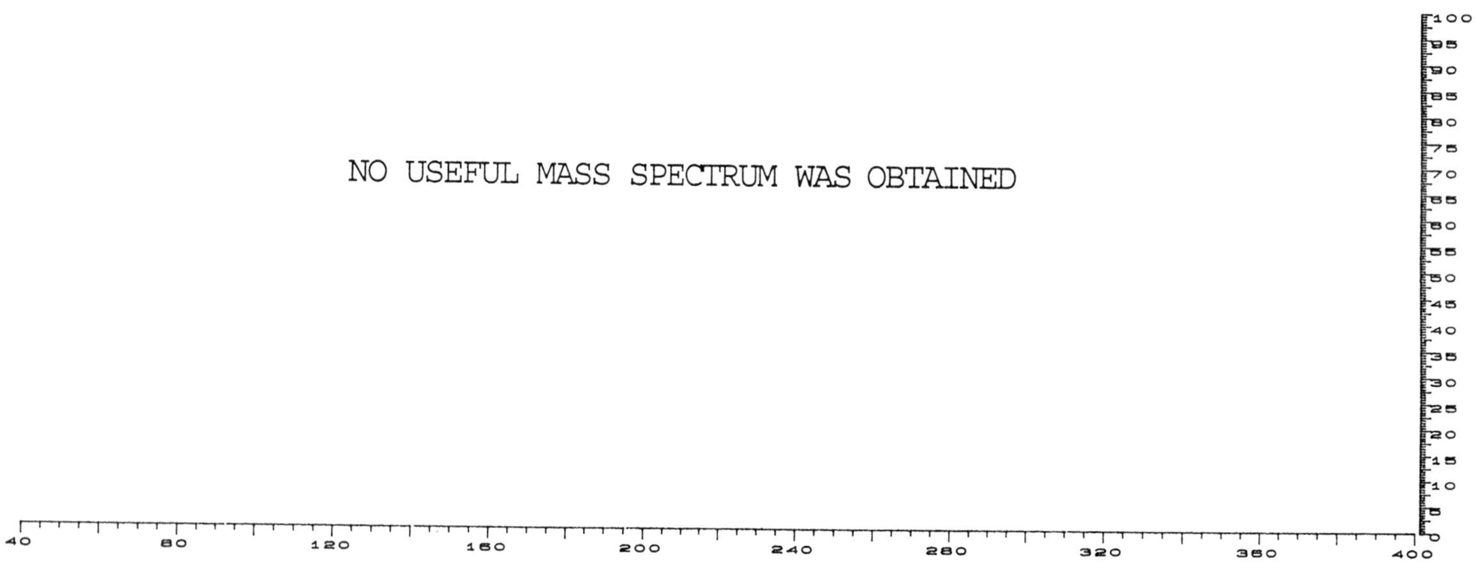

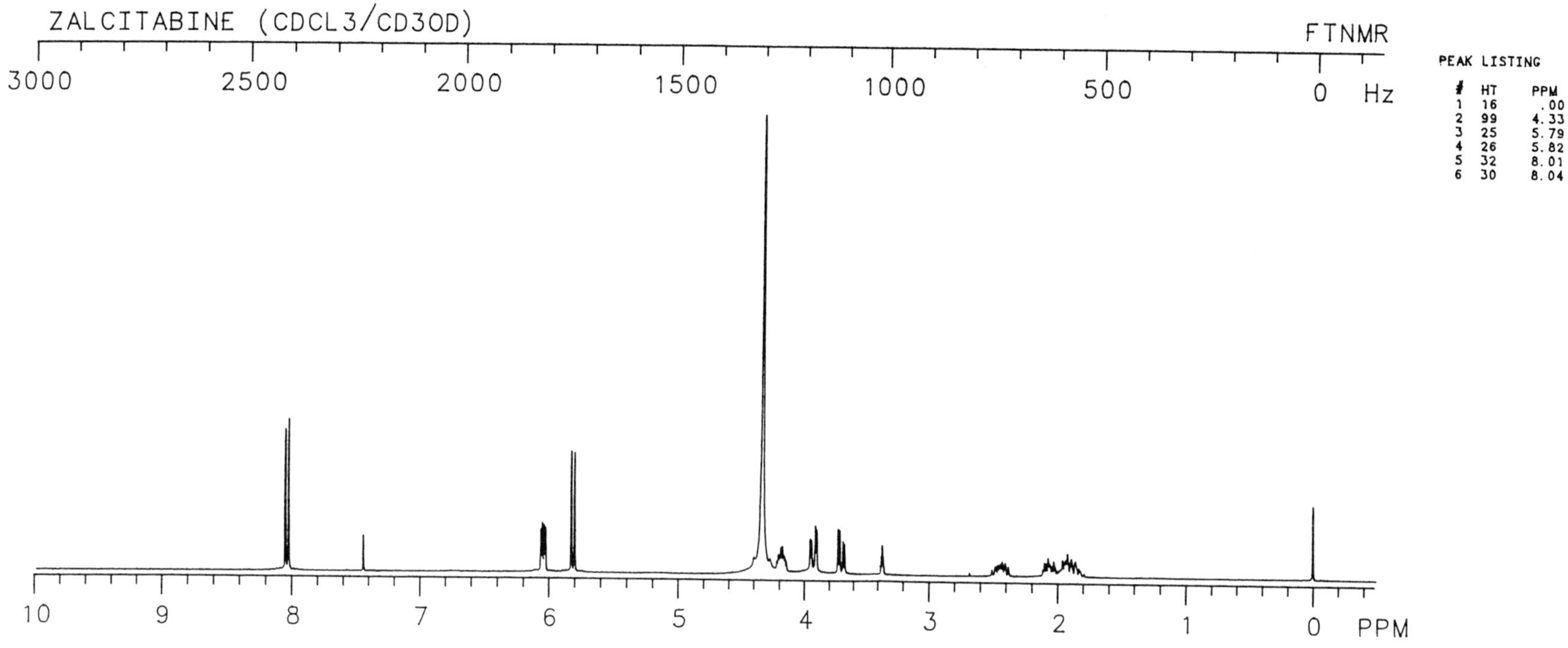
ZALCITABINE (CDCL3/CD3OD)
FTNMR
3000 2500 2000 1500 1000 500 0 Hz
10 9 8 7 6 5 4 3 2 1 0 PPM
PEAK LISTING
HT PPM
1 16 .00
2 99 4.33
3 25 5.79
4 26 5.82
5 32 8.01
6 30 8.04

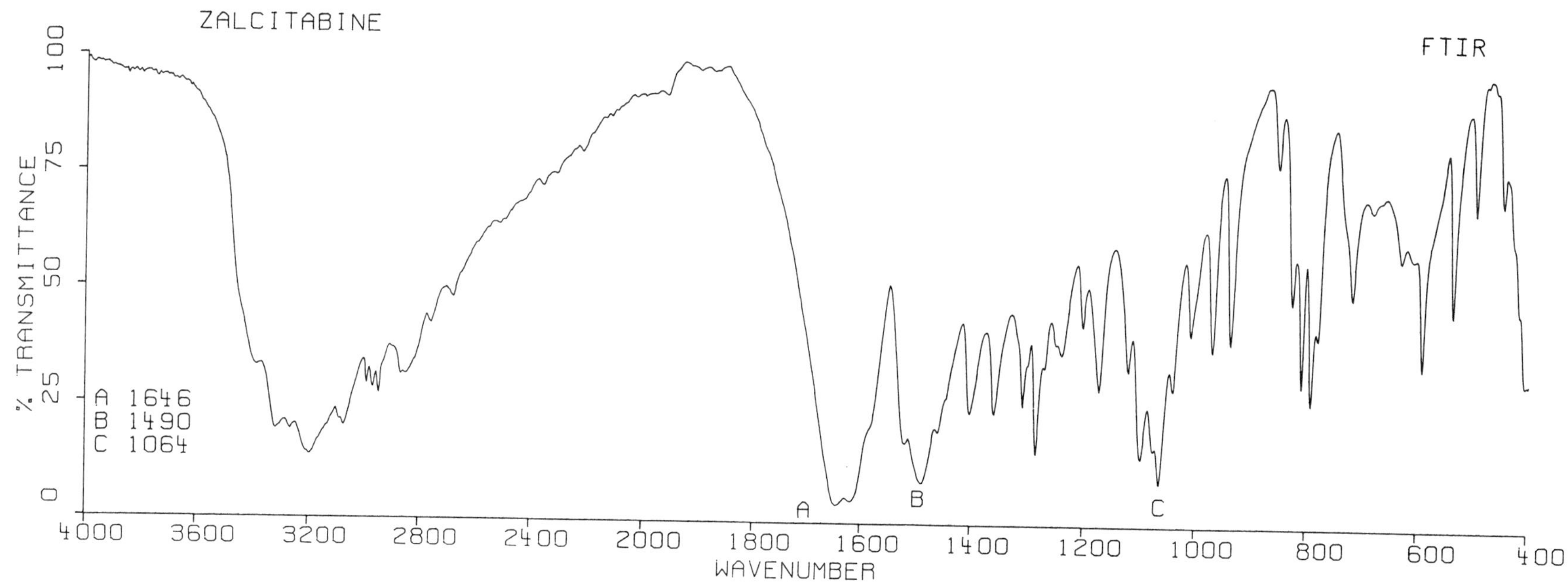
ZALCITABINE
FTIR
% TRANSMITTANCE
100
75
50
25
0
A 1646
B 1490
C 1064
A
B
C
4000 3600 3200 2800 2400 2000 1800 1600 1400 1200 1000 800 600 400
WAVENUMBER

ZEARALENONE

$C_{18}H_{22}O_5$

Molecular Weight: 318.37 (318.15)

Synonyms: 3,4,5,6,9,10-Hexahydro-14,16-dihydroxy-3-methyl-1H-2-benzoxacyclotetradecin-1,7(8H)-dione
Trade Names: FES
Use: Anabolic

HPLC: Ethanol: 2.6
GC:

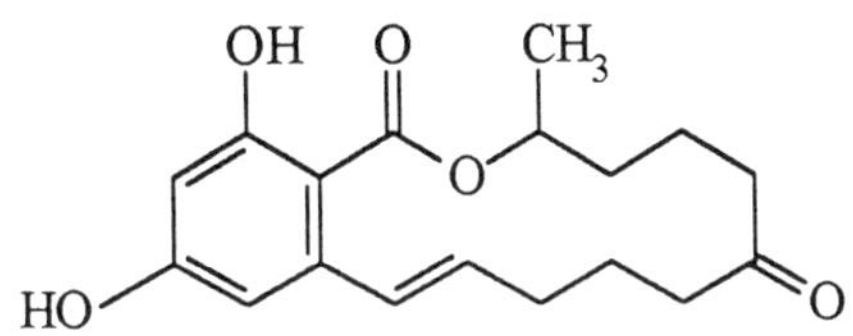

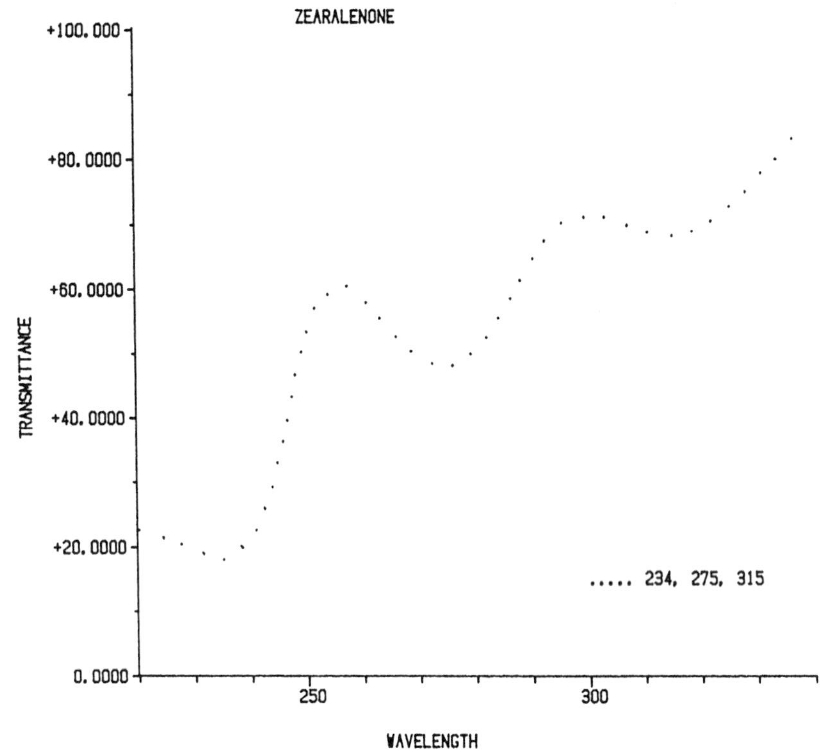

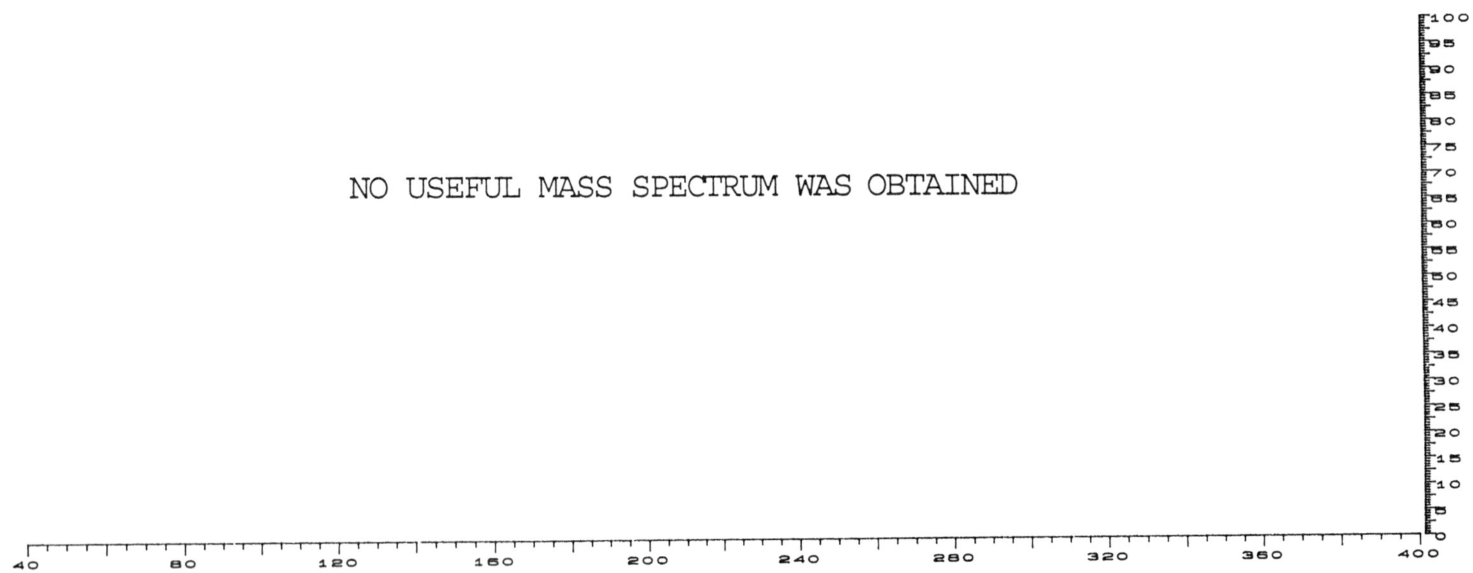

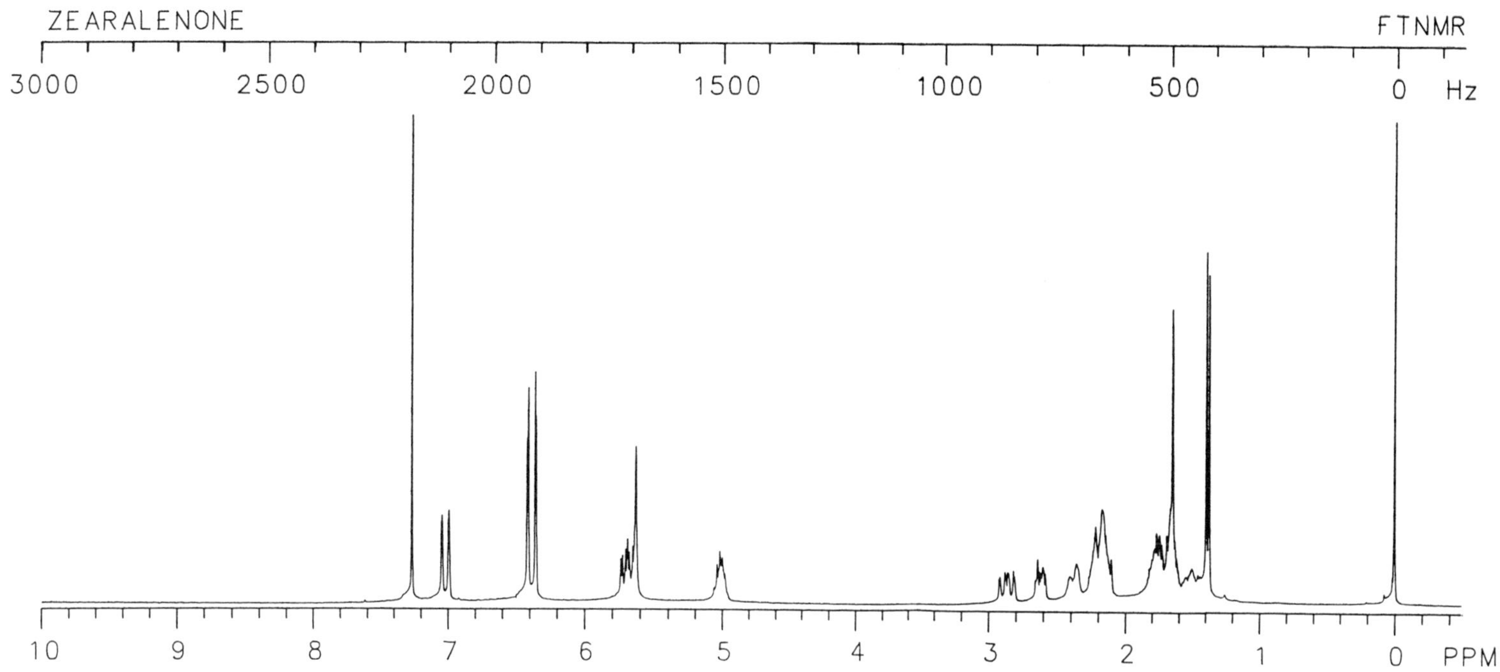

PEAK LISTING

#	HT	PPM
1	99	.00
2	67	1.37
3	72	1.39
4	60	1.64
5	20	1.66
6	14	1.68
7	14	1.73
8	14	1.76
9	14	2.14
10	19	2.16
11	19	2.17
12	14	2.18
13	16	2.21
14	32	5.62
15	38	6.34
16	47	6.35
17	44	6.40
18	33	6.41
19	18	6.98
20	17	6.99
21	18	7.03
22	16	7.04
23	99	7.26

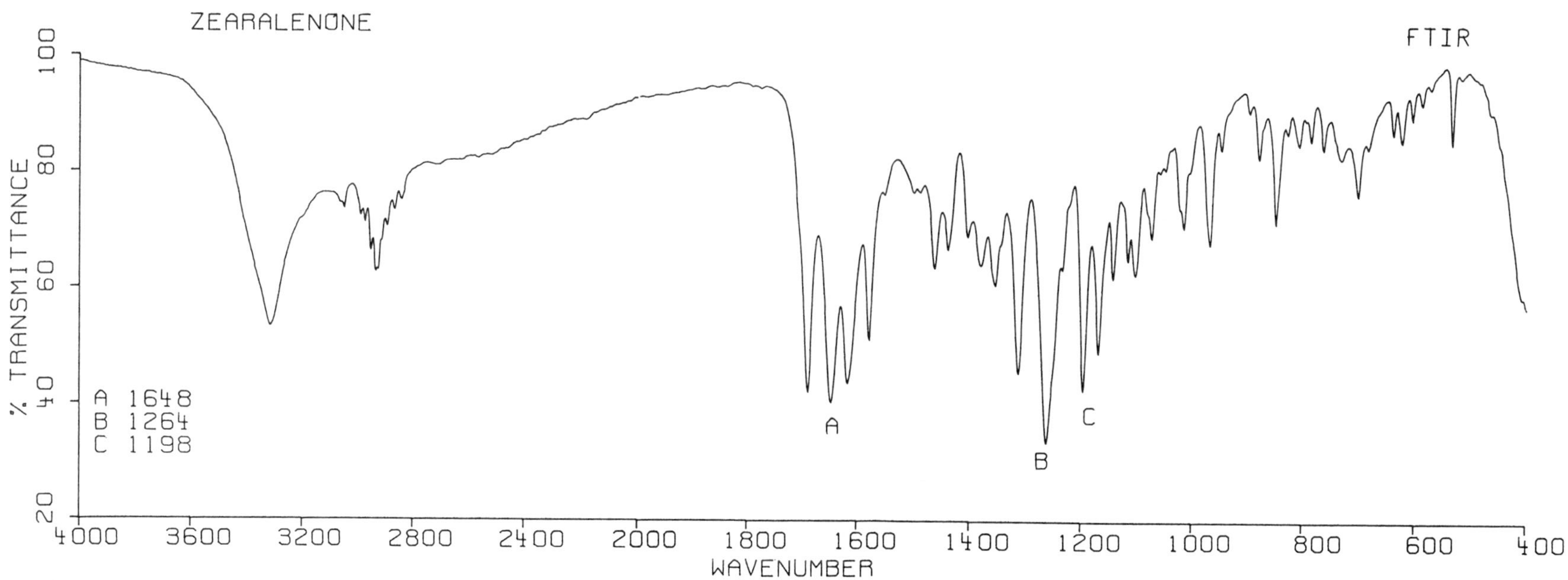

ZOLPIDEM

$C_{19}H_{21}N_3O$

Molecular Weight: 307.40 (307.17)

Synonyms: N,N,6-Trimethyl-2-(4-methylphenyl)-imidazo[1.2-a]pyridine-3-acetamide; N,N,6-trimetyl-2-p-tolylimidazo[1,2-a]pyridine-3-acetamide
Trade Names: Ambien, Stilnox
Use: Hypnotic

HPLC:
GC: 2781; 250°

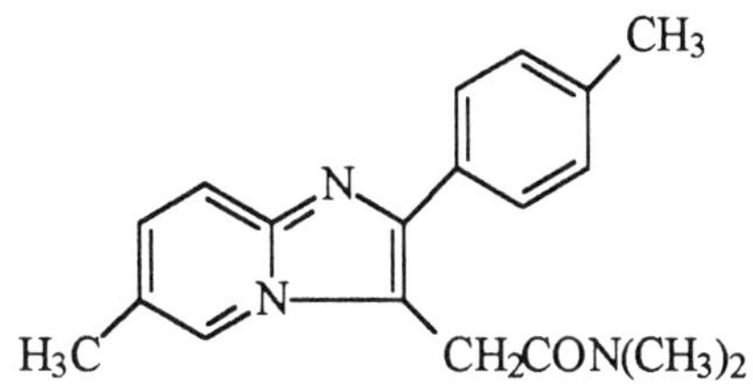

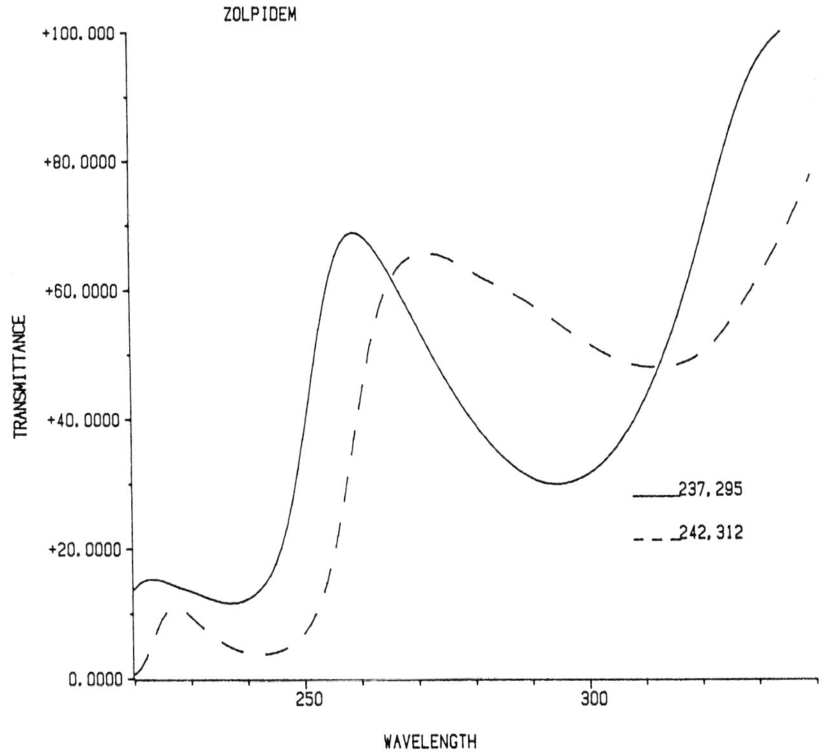

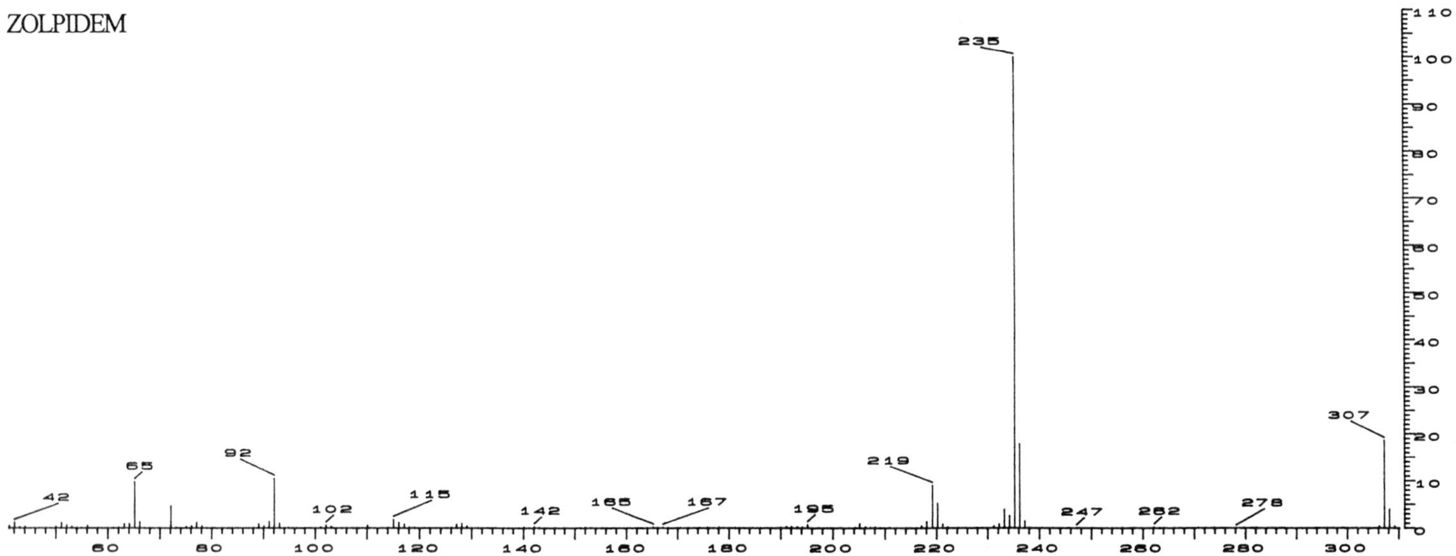

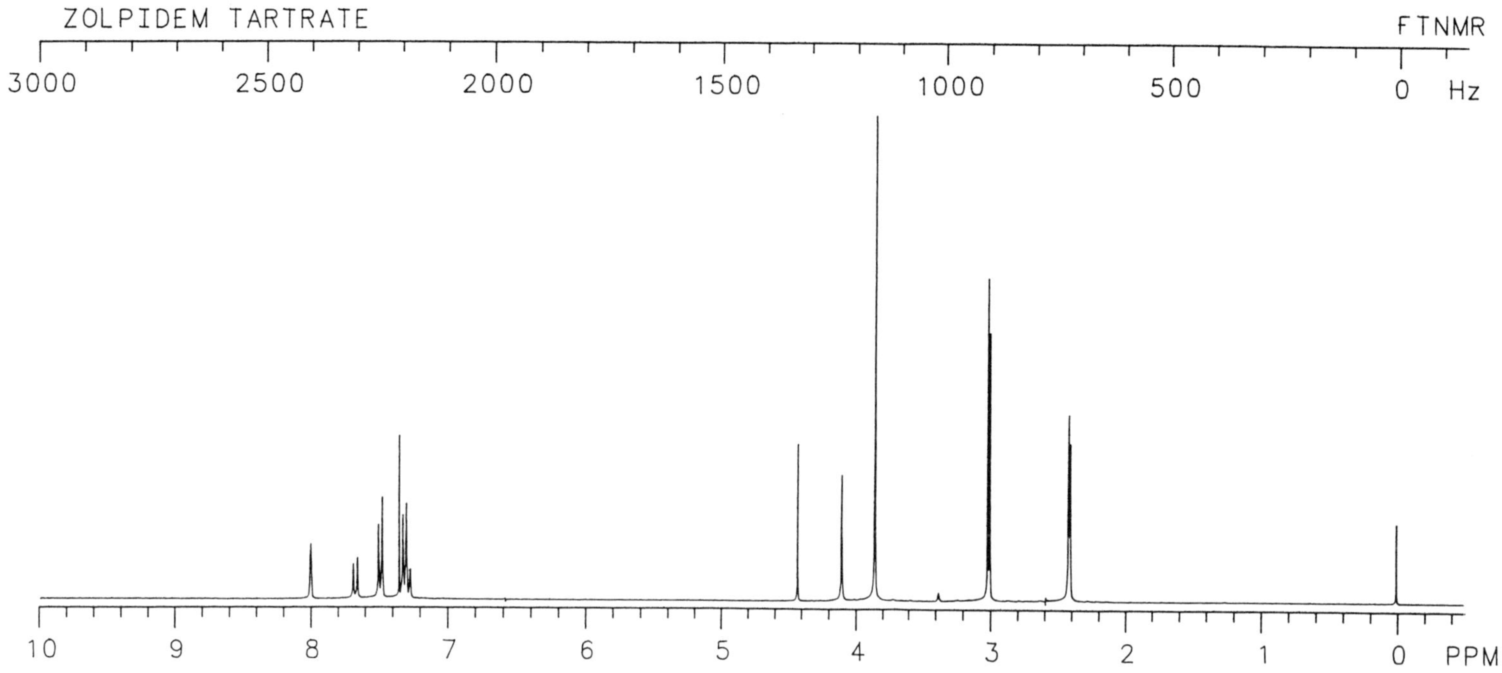

PEAK LISTING

#	HT	PPM
1	16	.00
2	32	2.40
3	32	2.40
4	38	2.41
5	54	2.99
6	65	3.01
7	99	3.85
8	25	4.10
9	32	4.42
10	19	7.29
11	17	7.32
12	33	7.34
13	20	7.47
14	15	7.49
15	11	7.99

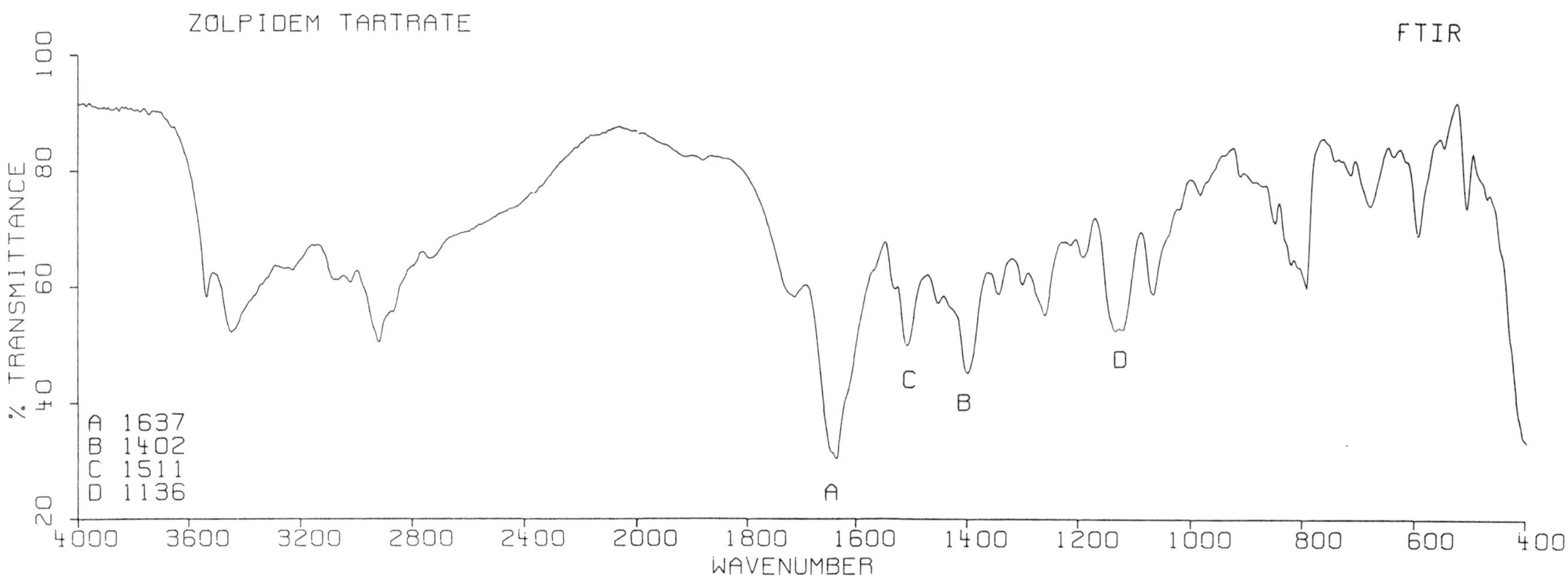

ZOPICLONE

$C_{17}H_{17}ClN_6O_3$

Molecular Weight: 388.81 (388.11)

Synonyms: 4-Methyl-1-piperazinecarboxylic acid-6-(5-chloro-2-pyridinyl)-6,7-dihydro-7-oxo-5H-pyrrolo[3,4-b]pyrazin-5-yl ester
Trade Names: Amoban, Amovane, Imovance, Imovane, Zimovane
Use: Hypnotic

HPLC:
GC: 3091; 280°

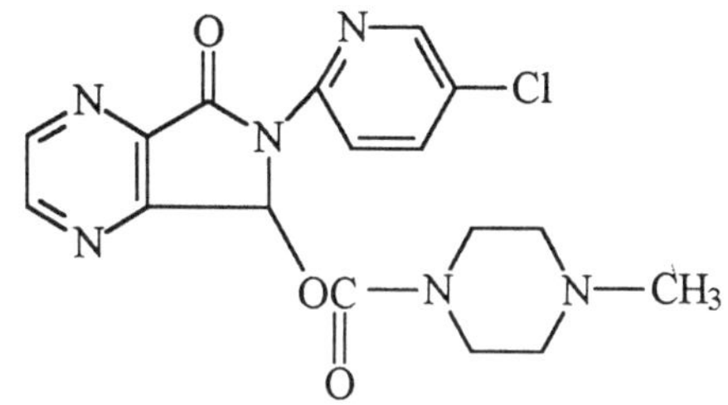

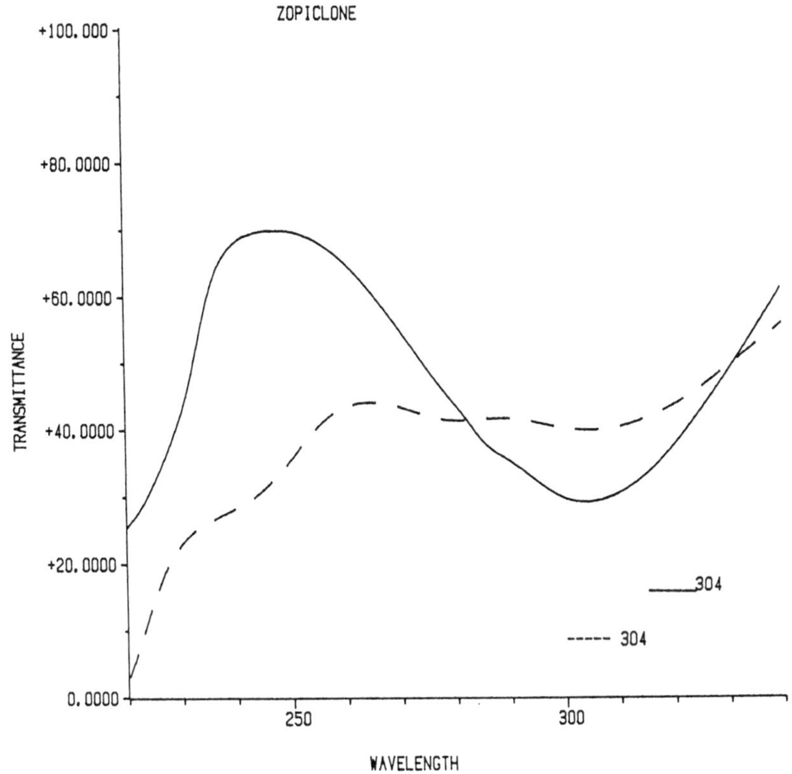

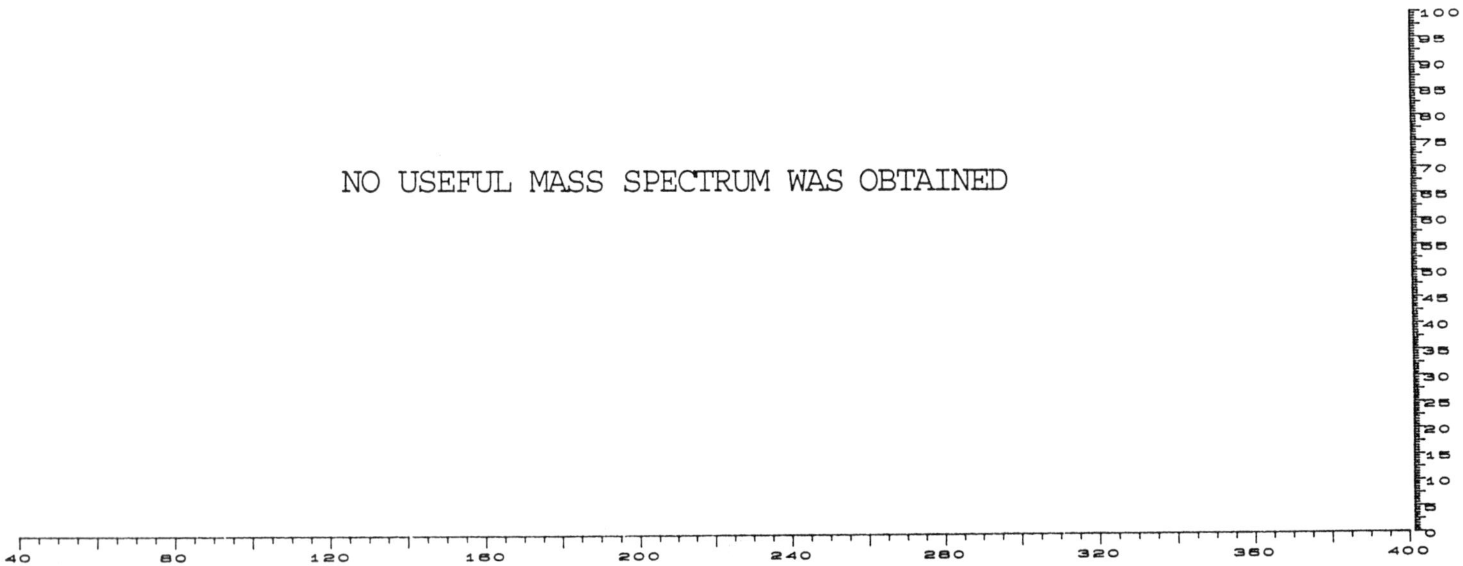

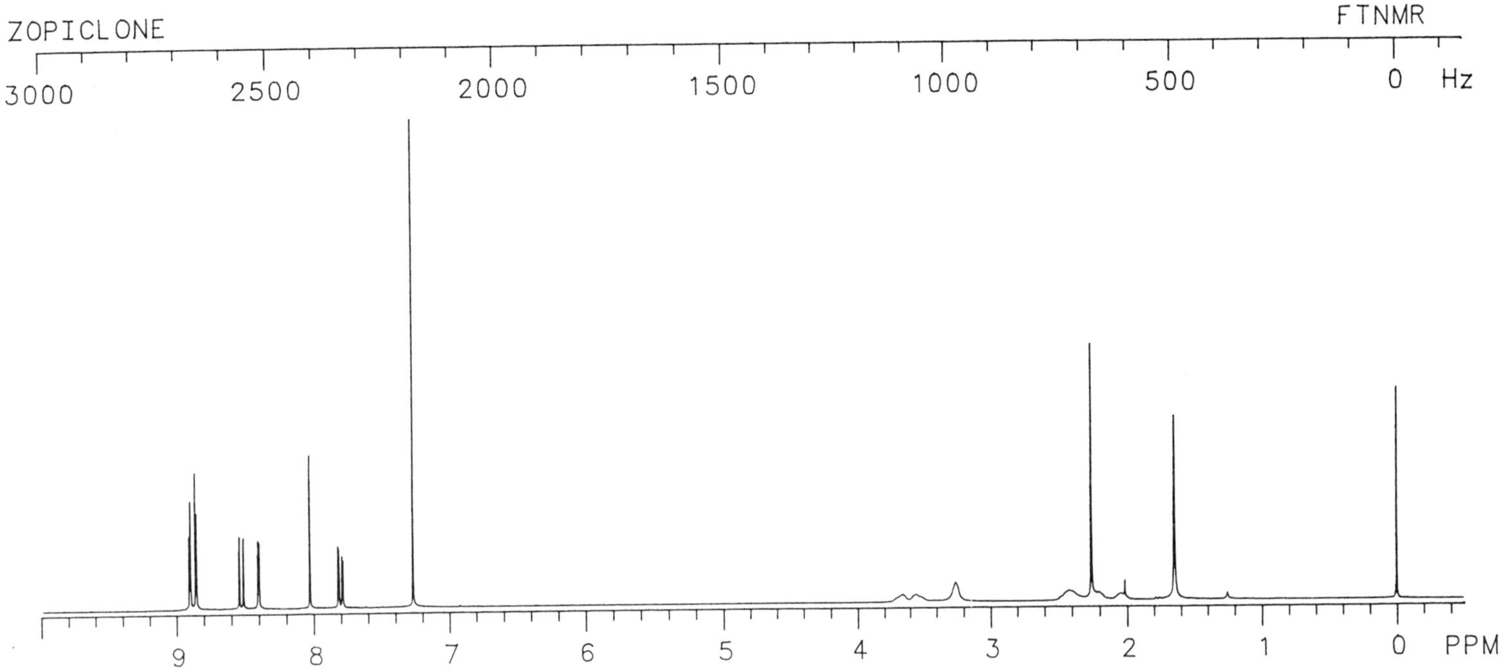

PEAK LISTING

#	HT	PPM
1	43	.00
2	37	1.64
3	52	2.25
4	99	7.26
5	10	7.78
6	12	7.81
7	12	7.81
8	31	8.02
9	13	8.39
10	14	8.40
11	11	8.40
12	14	8.50
13	14	8.53
14	11	8.53
15	19	8.84
16	27	8.85
17	22	8.89
18	15	8.90

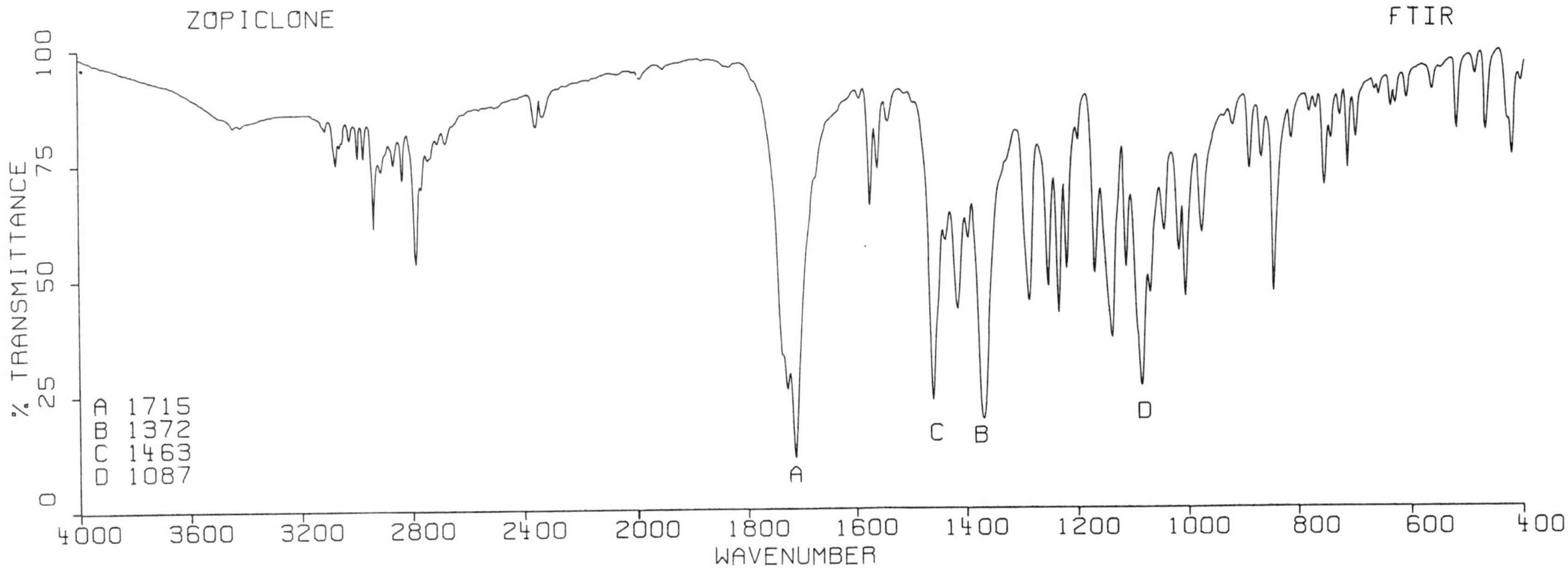

GAS CHROMATOGRAPHY/ INFRARED SPECTRA OF DRUGS

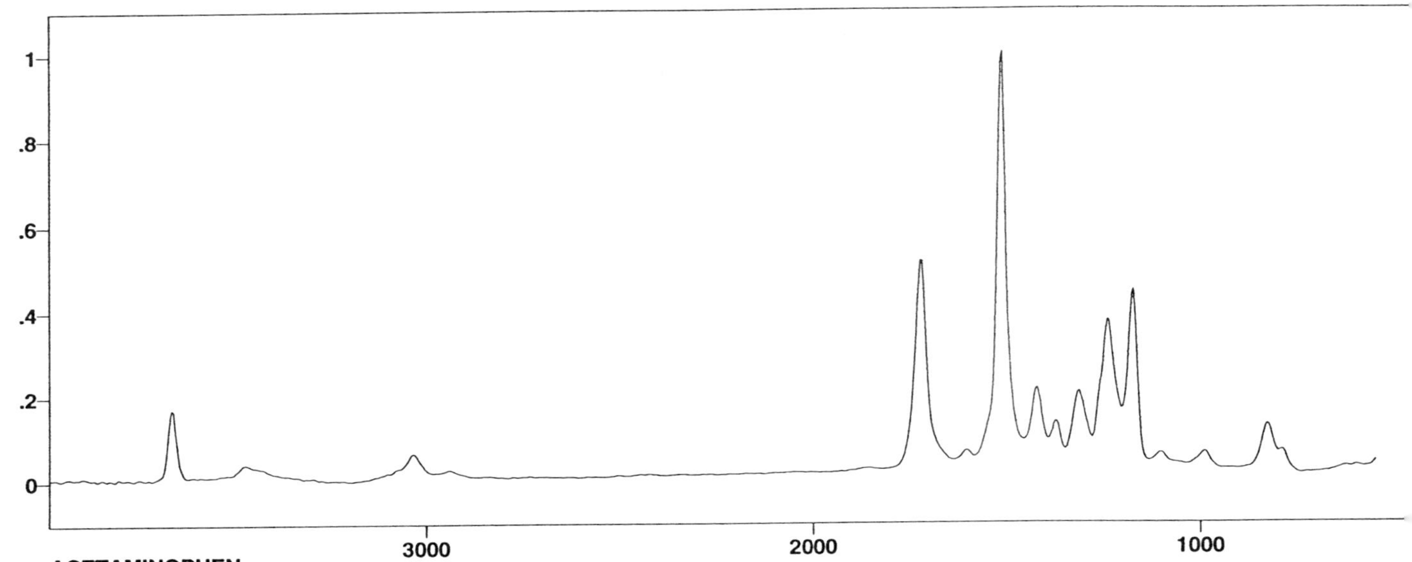

ACETAMINOPHEN

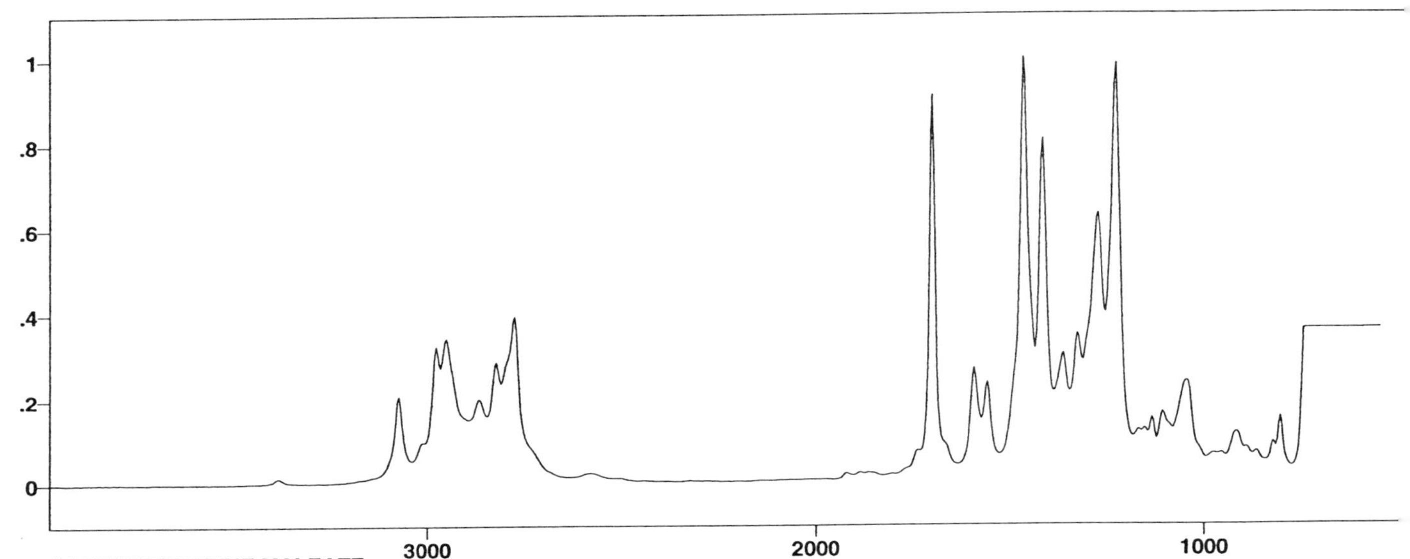

ACETOPROMAZINE MALEATE

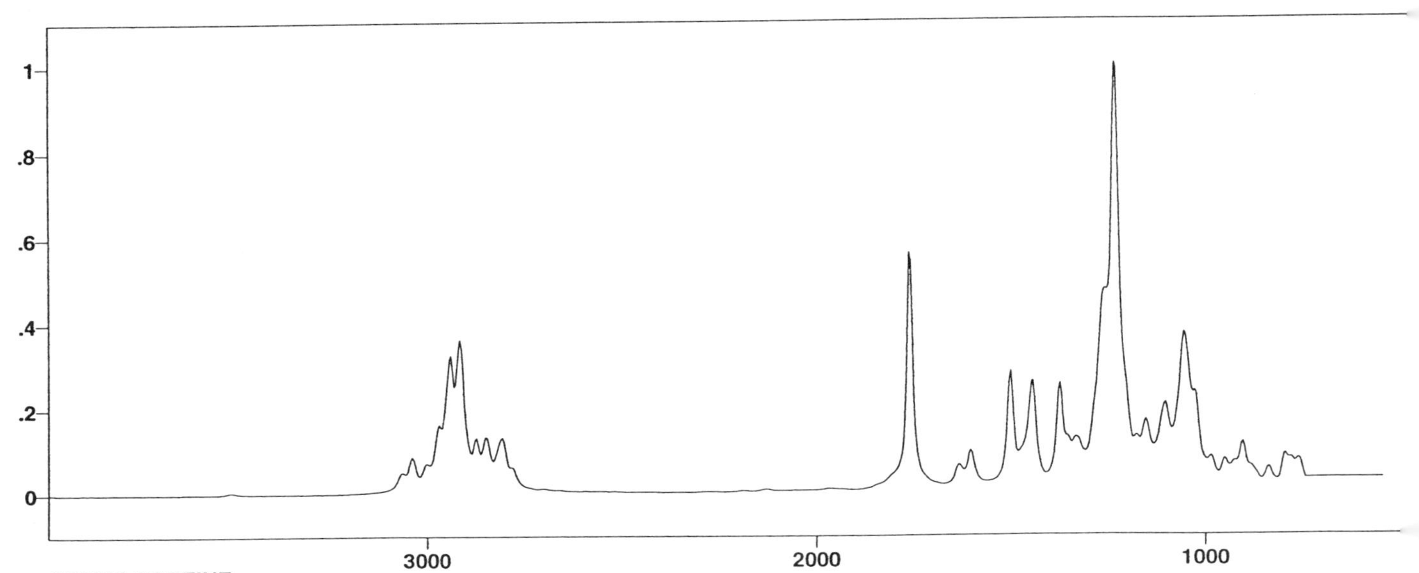

ACETYLCODEINE

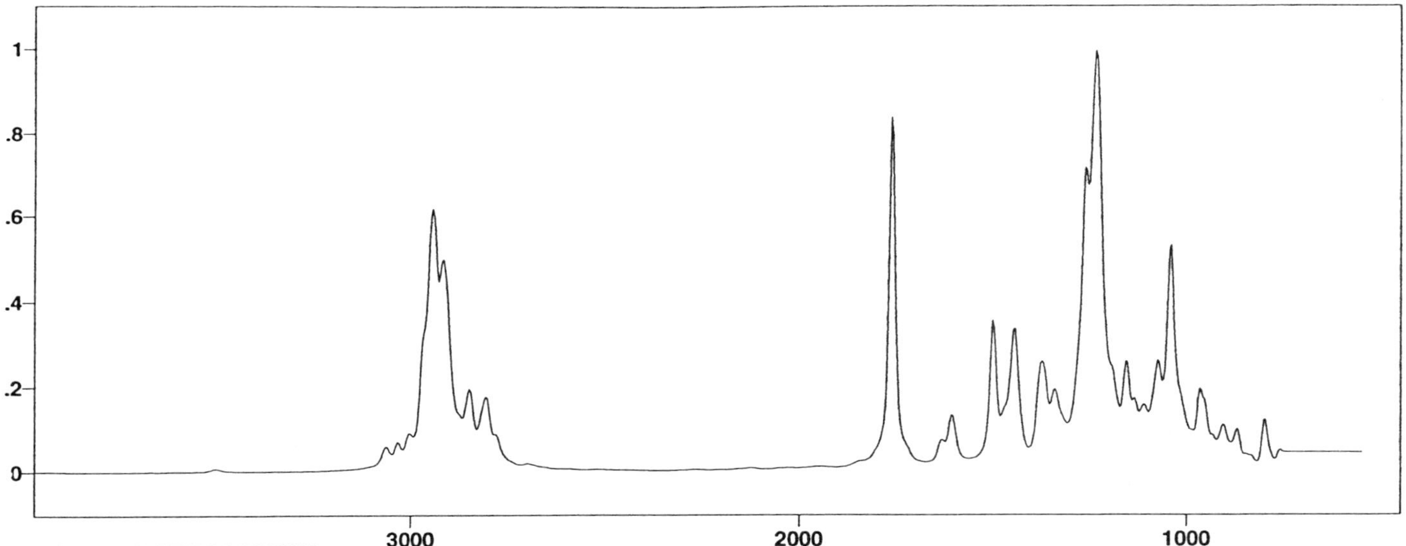

ACETYLDIHYDROCODEINE

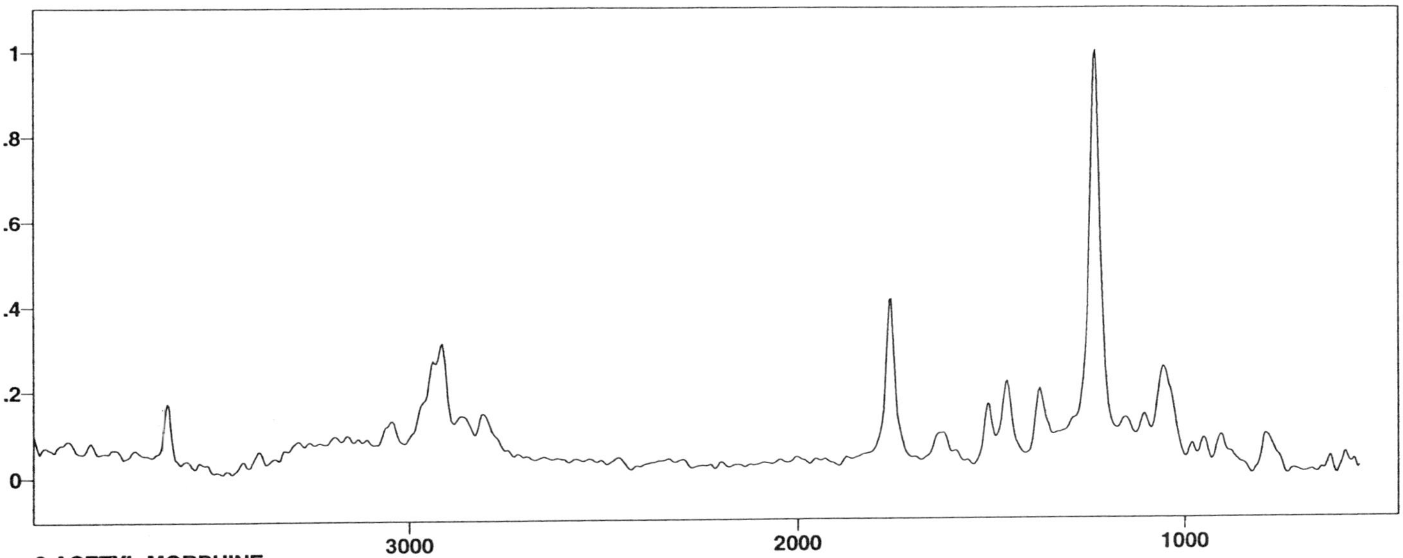

6-ACETYL-MORPHINE

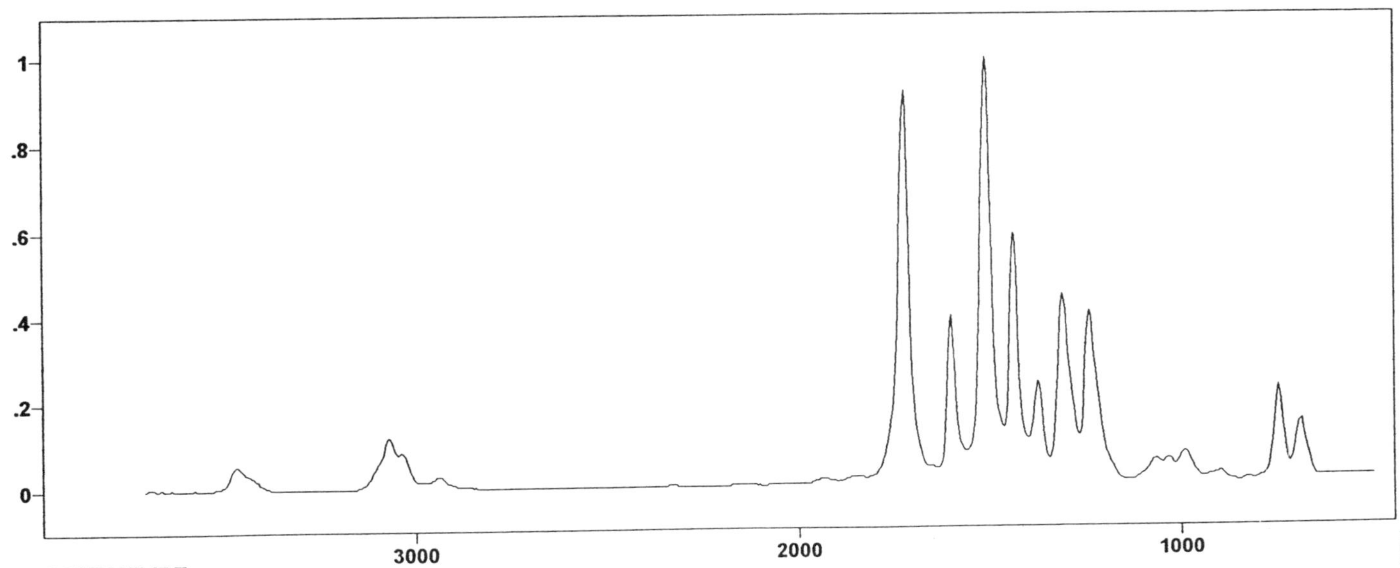

ACETANILIDE

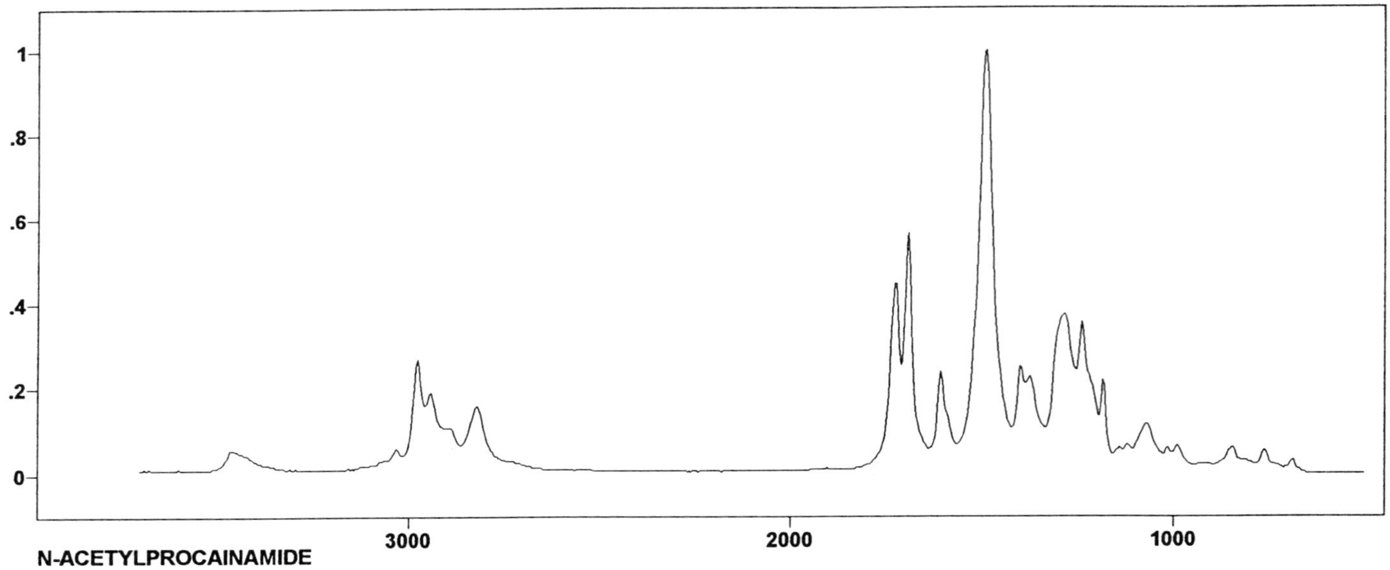

N-ACETYLPROCAINAMIDE

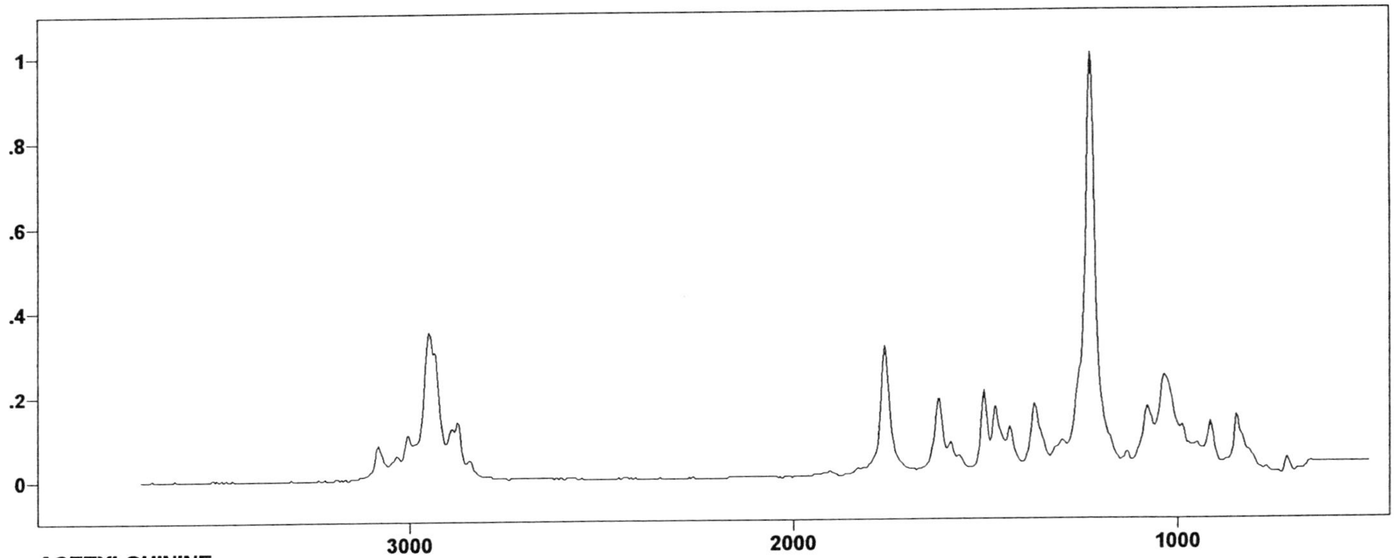

ACETYLQUININE

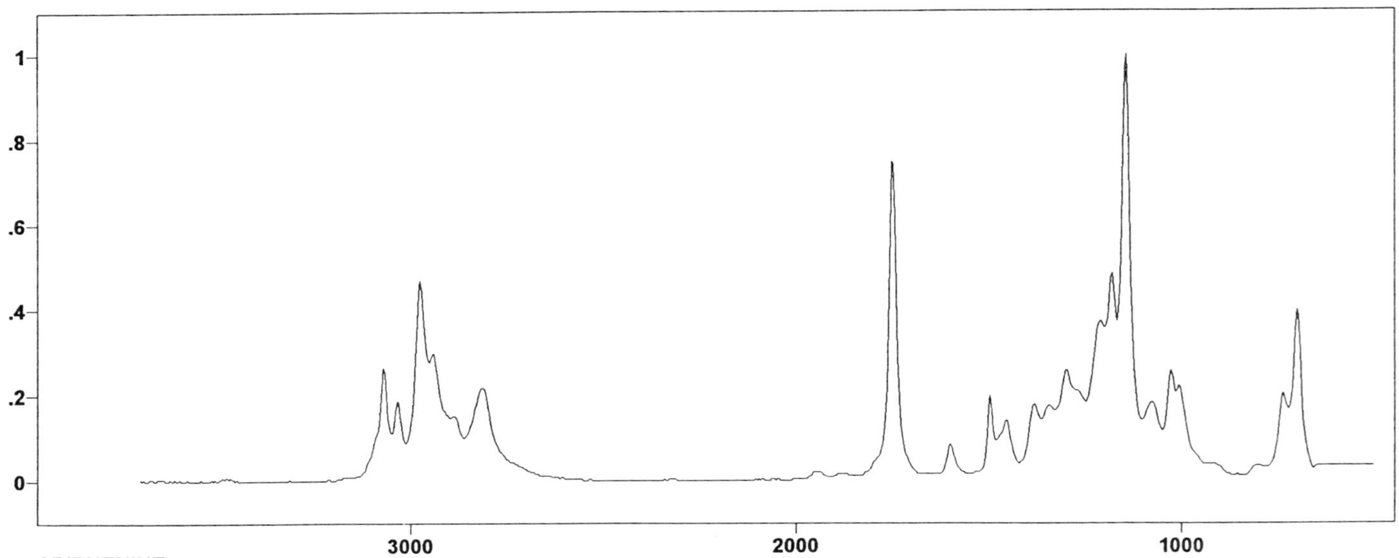

ADIPHENINE

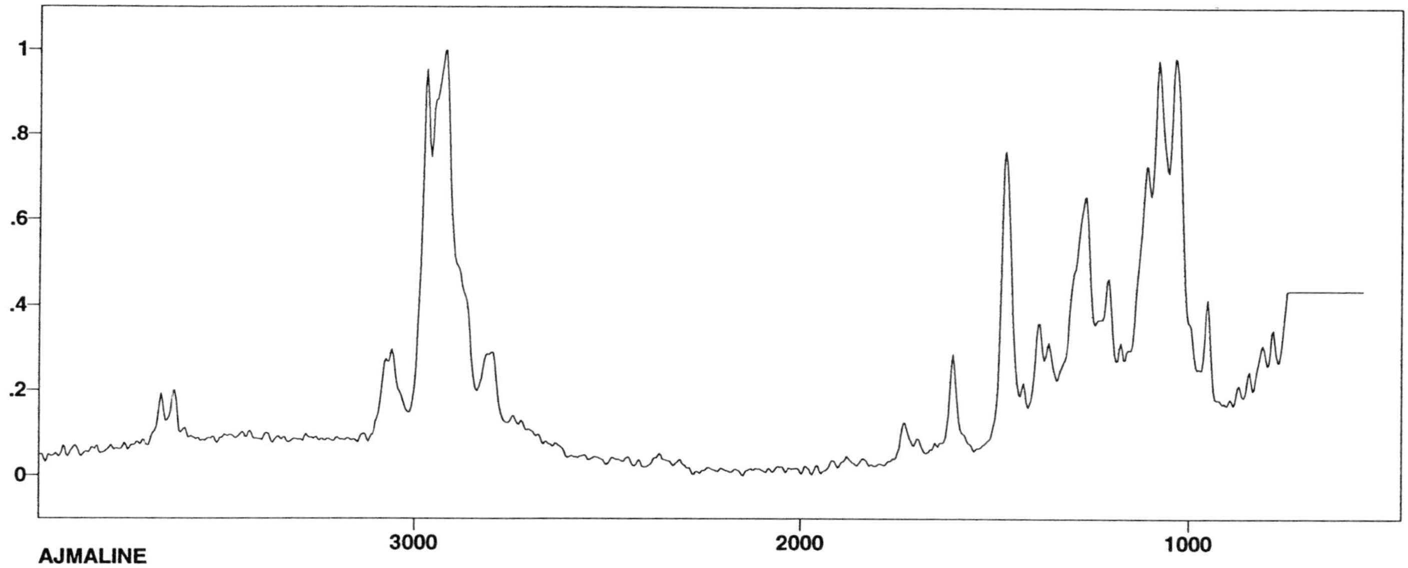

AJMALINE

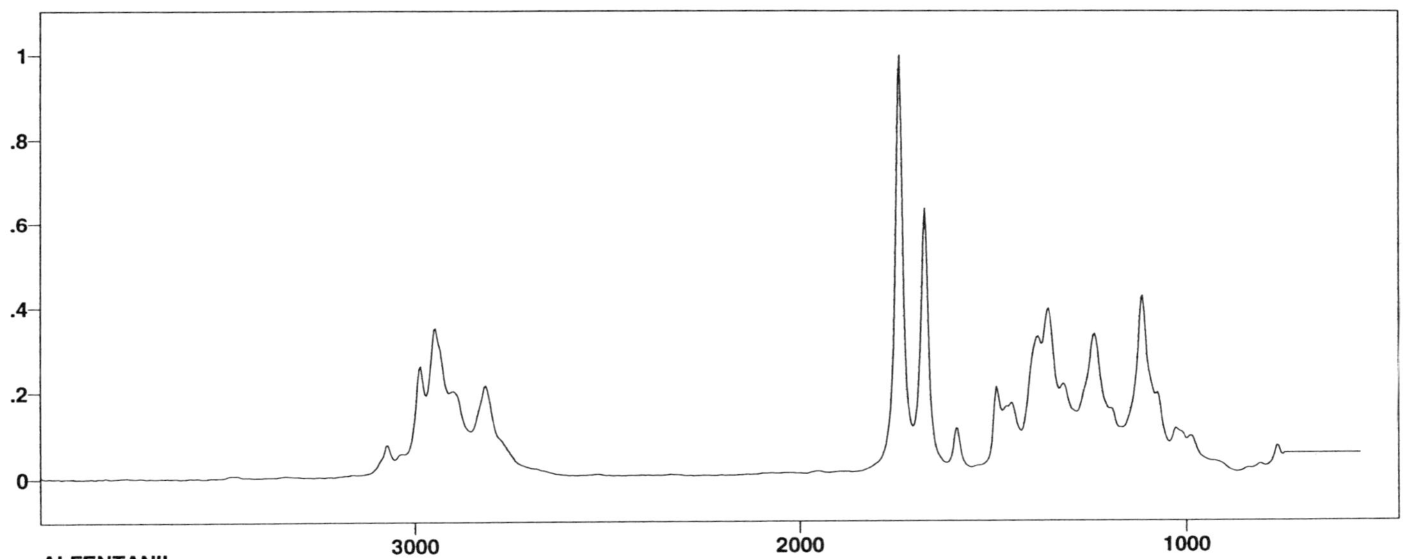

ALFENTANIL

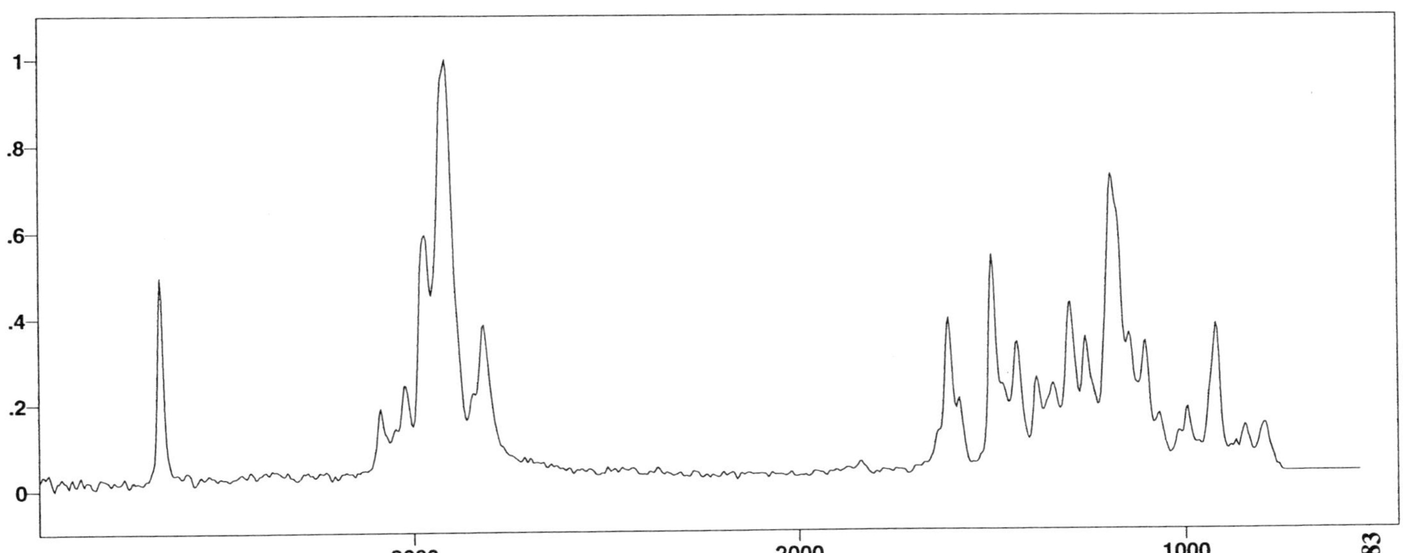

N-ALHYNORMETAZOCINE

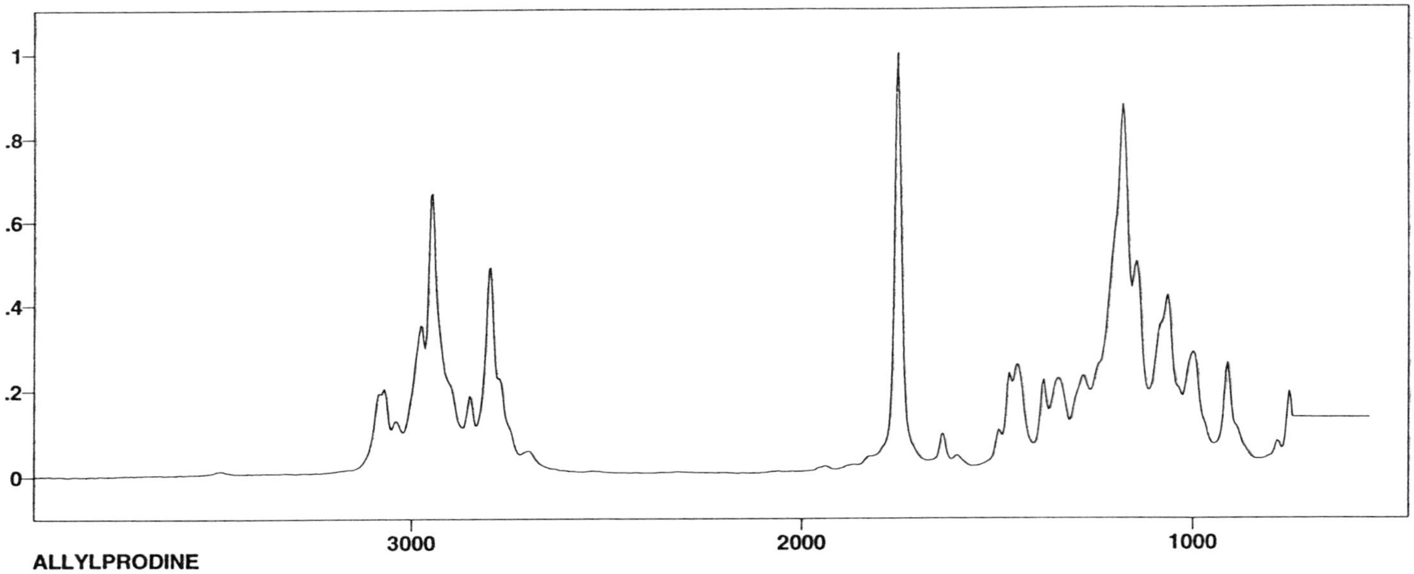

ALLYLPRODINE

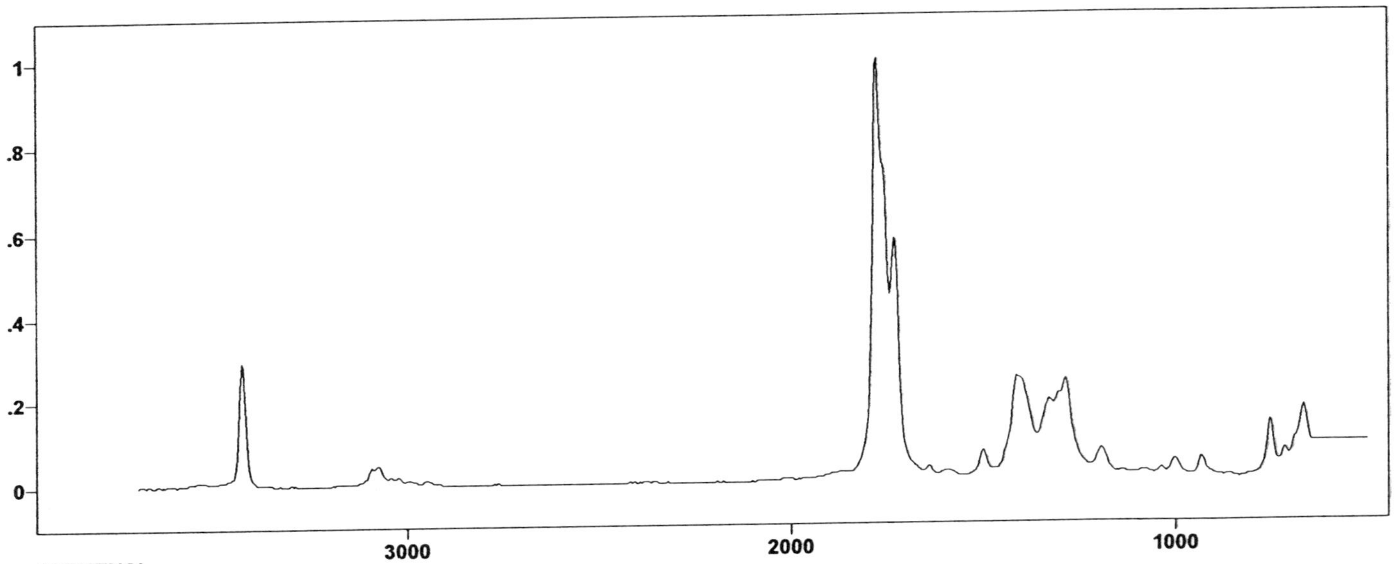

ALPHENAL

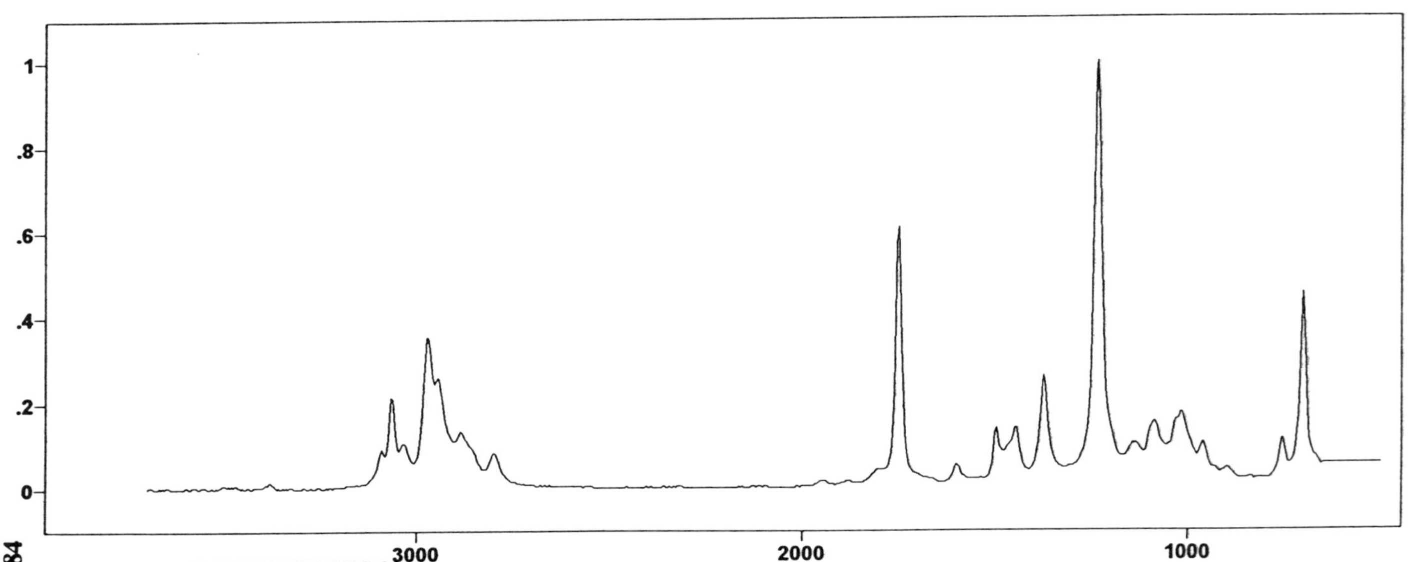

ALPHA-ACETYL-N-NORMETHADOL

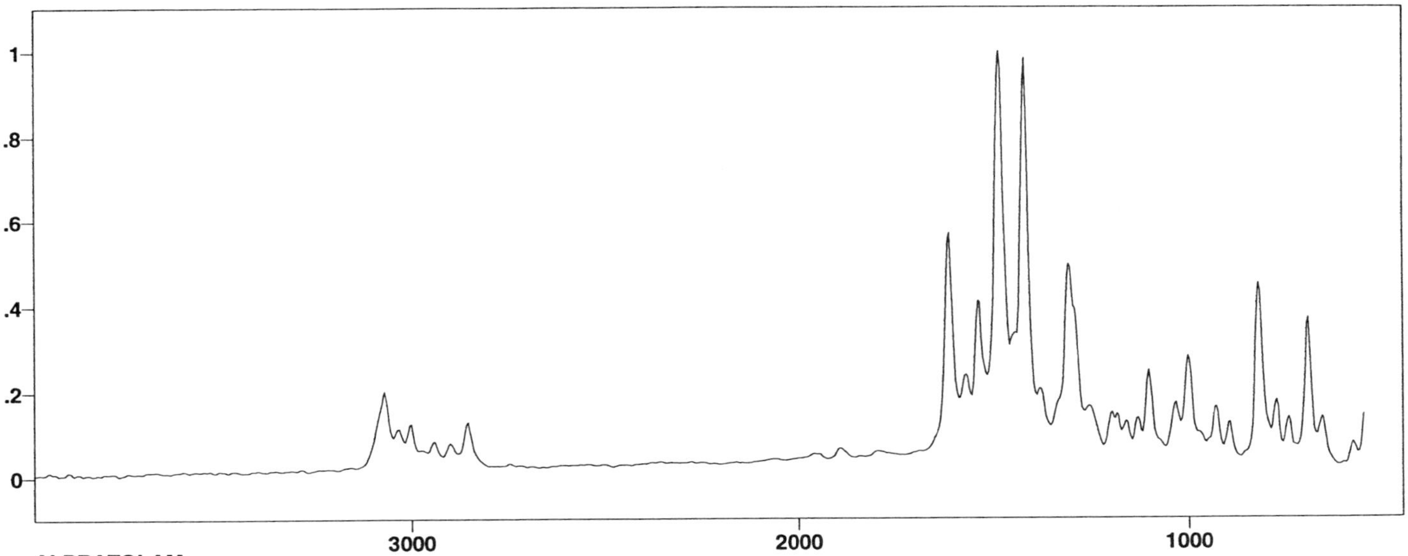

ALPRAZOLAM

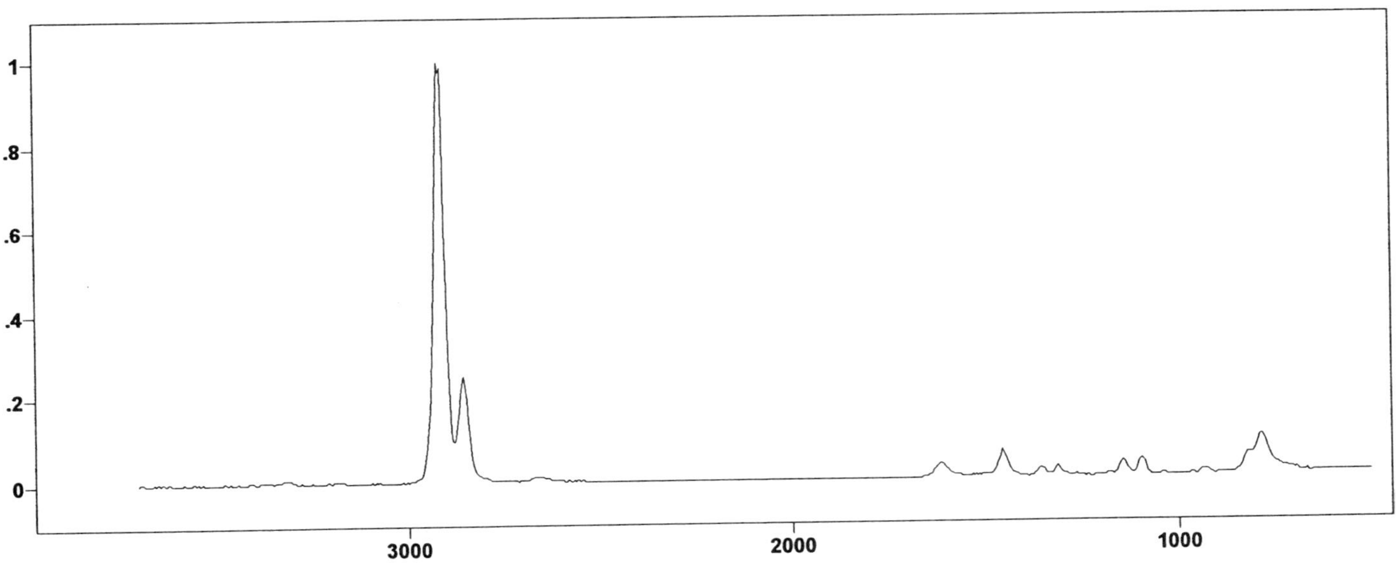

AMANTADINE

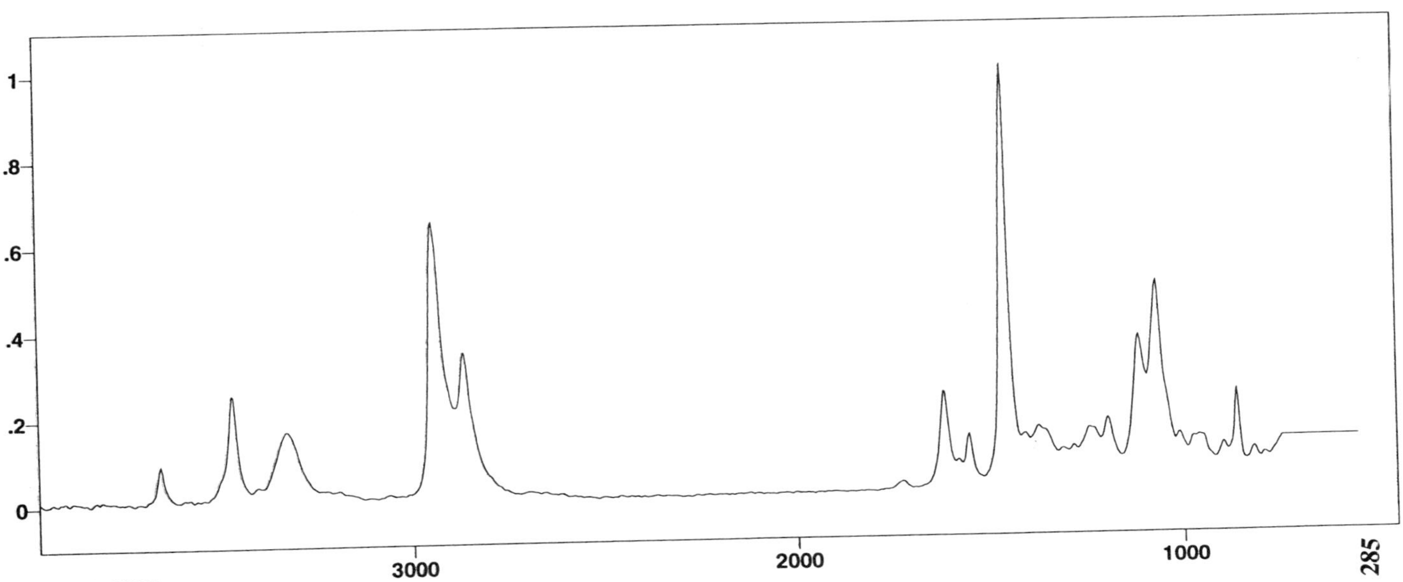

AMBROXOL

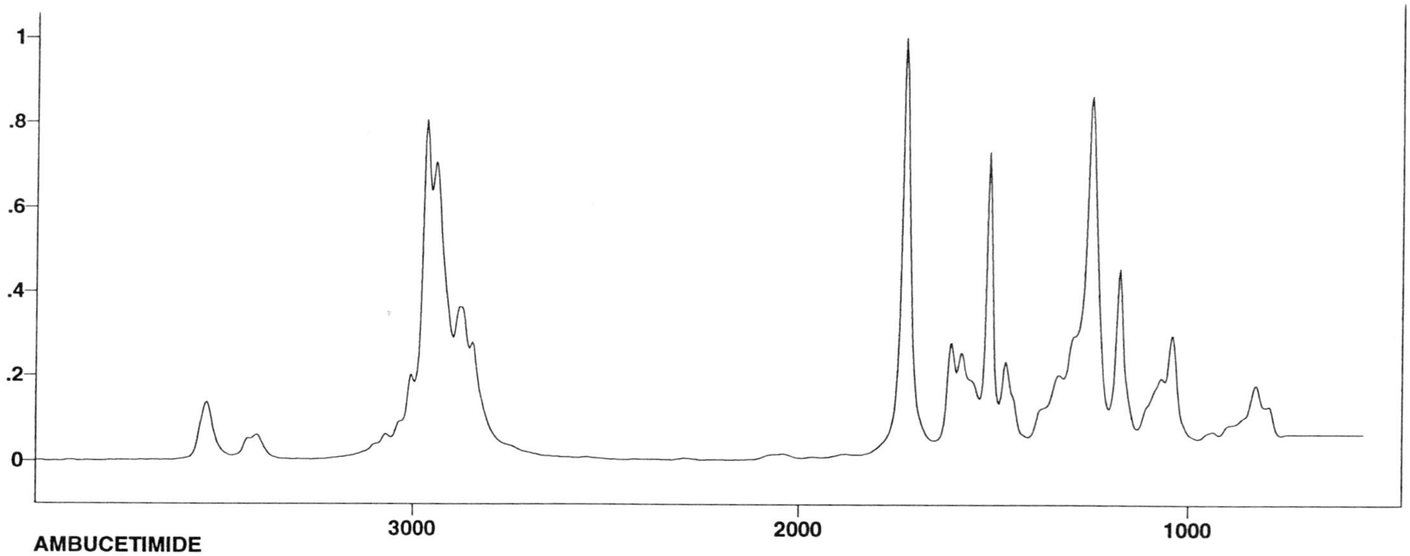

AMBUCETIMIDE

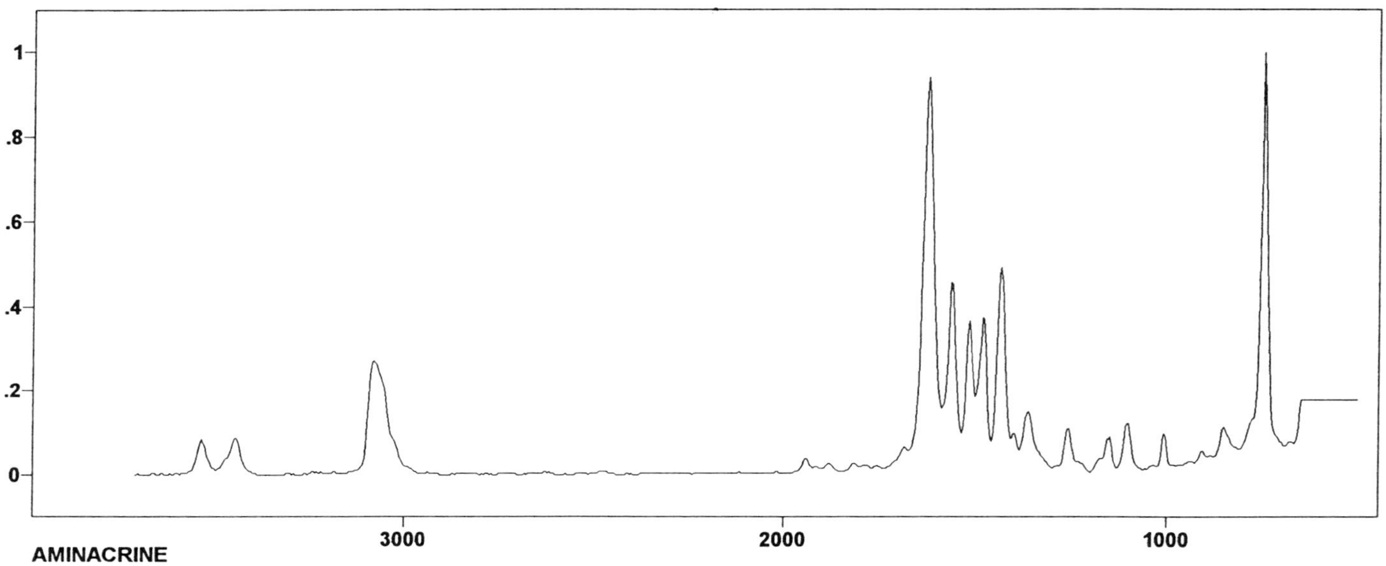

AMINACRINE

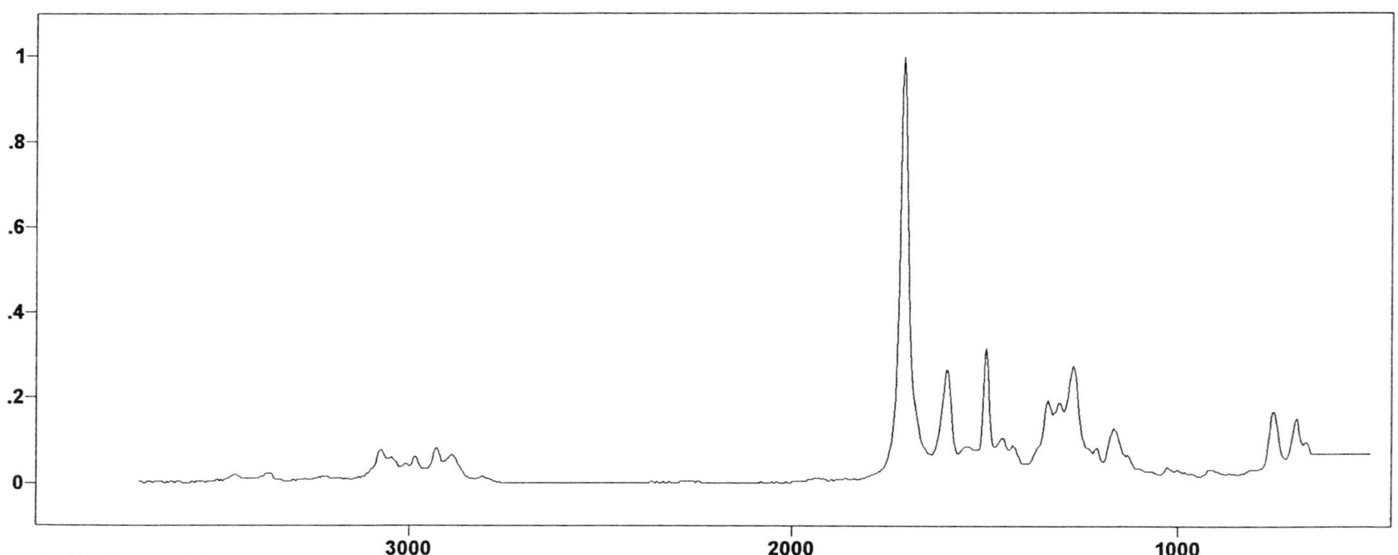

4-AMINOANTIPYRINE

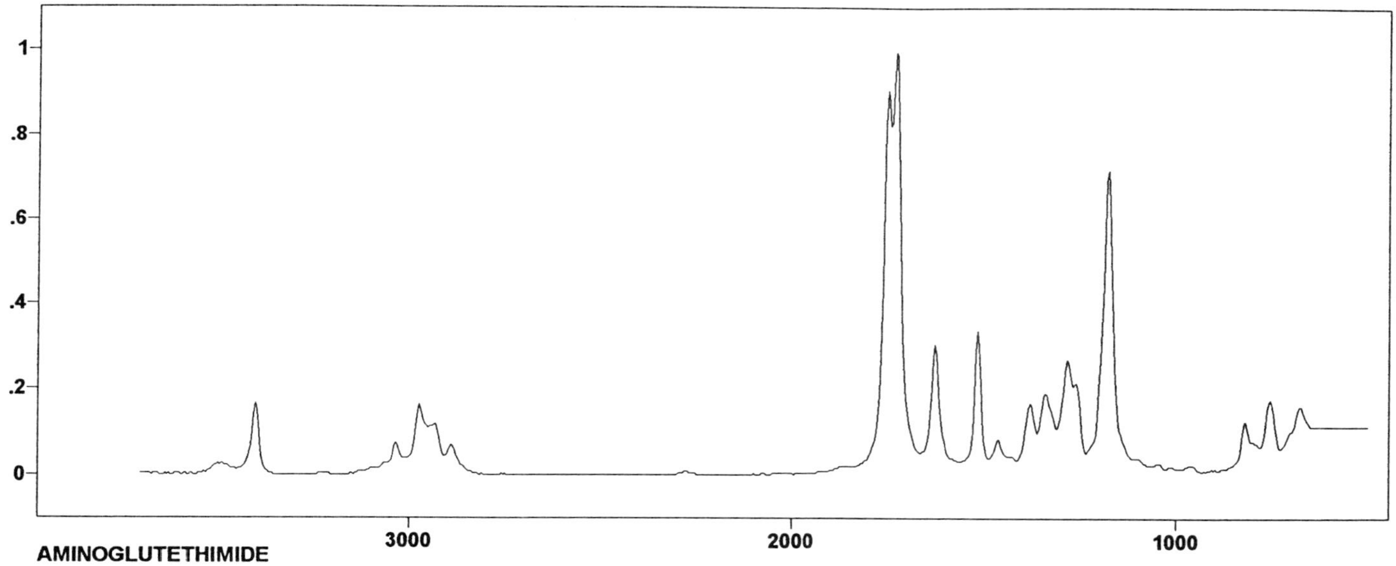

AMINOGLUTETHIMIDE

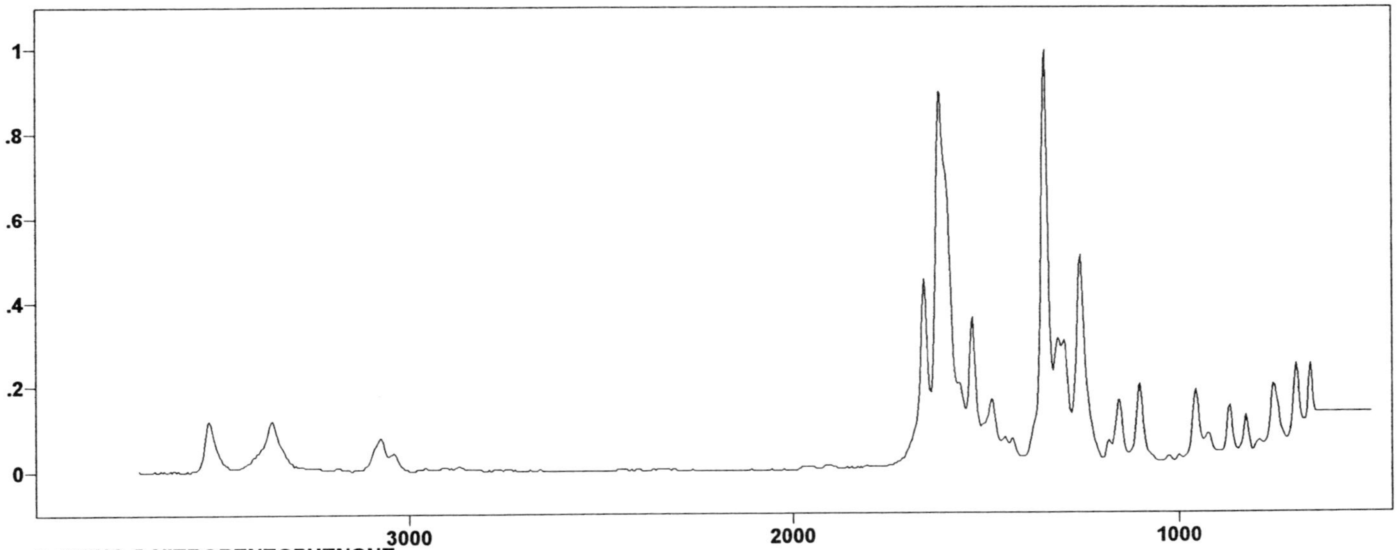

2-AMINO-5-NITROBENZOPHENONE

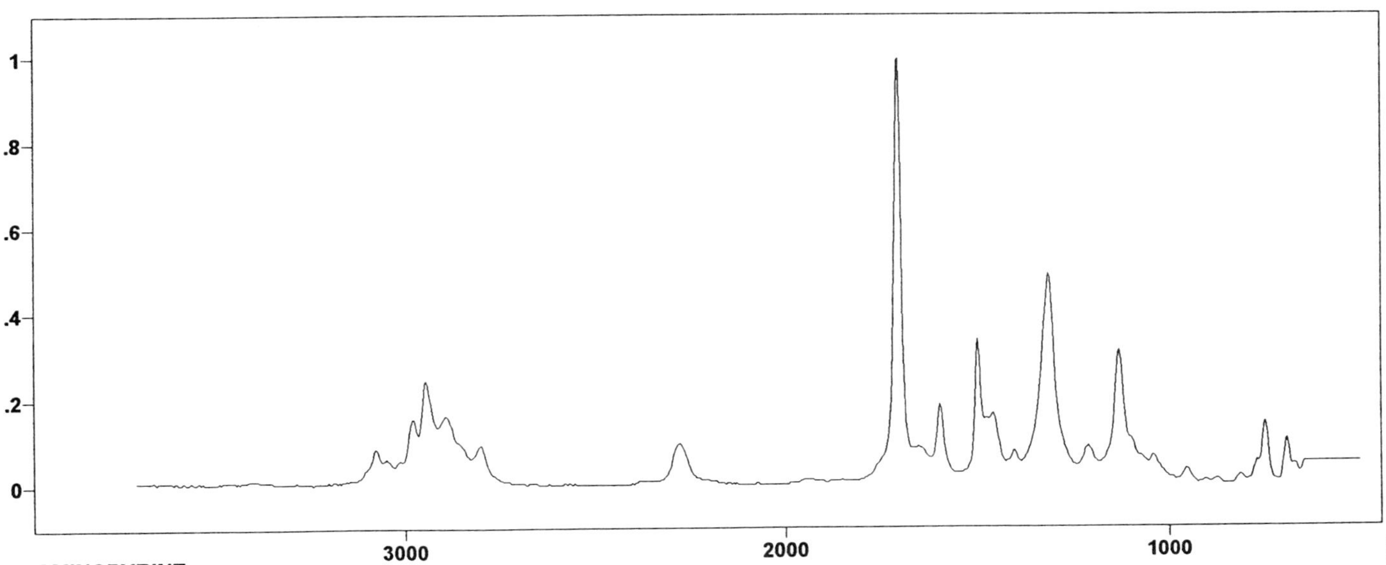

AMINOPYRINE

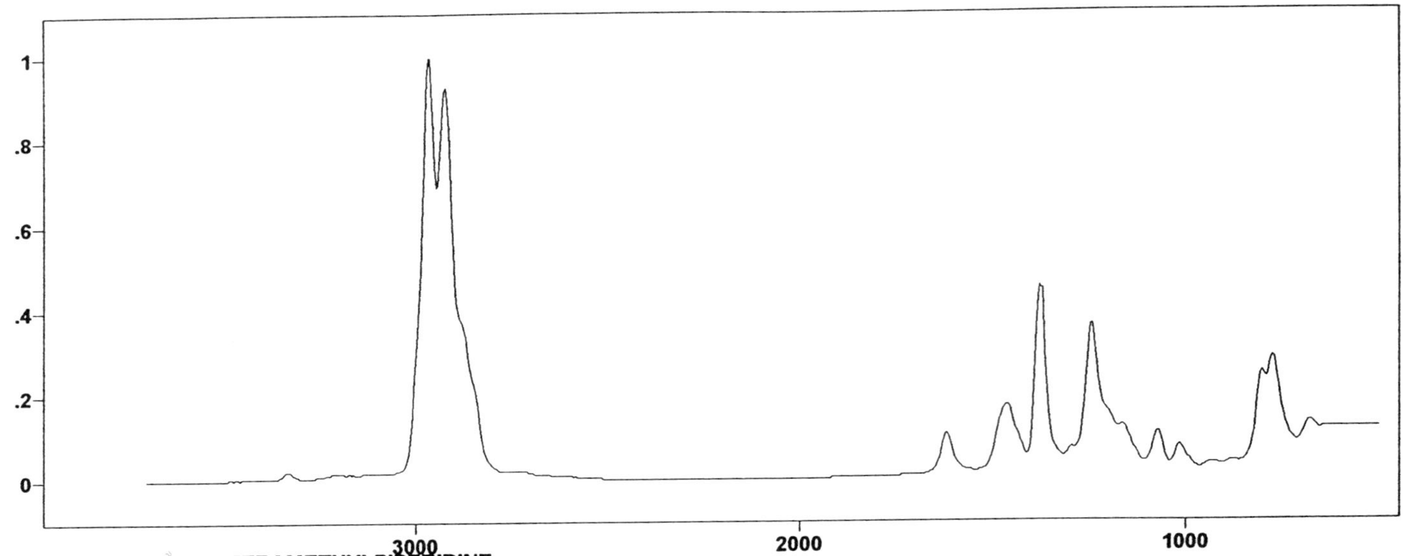

4-AMINO-2,2,6,6-TETRAMETHYLPIPERIDINE

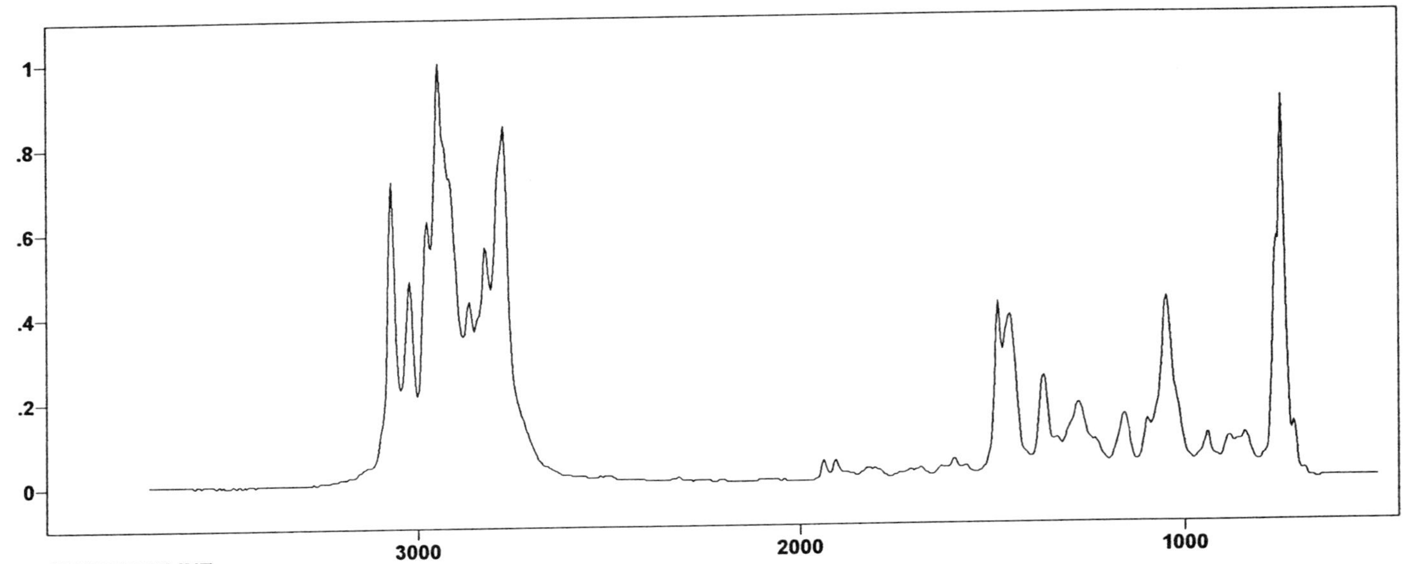

AMITRIPTYLINE

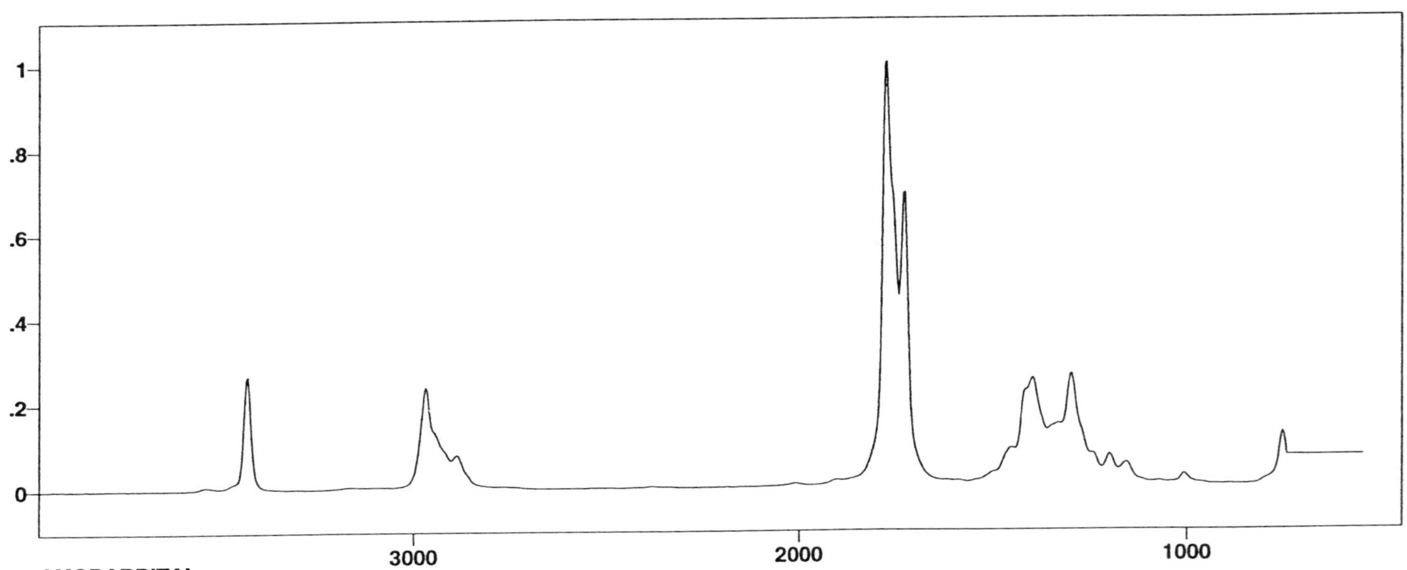

AMOBARBITAL

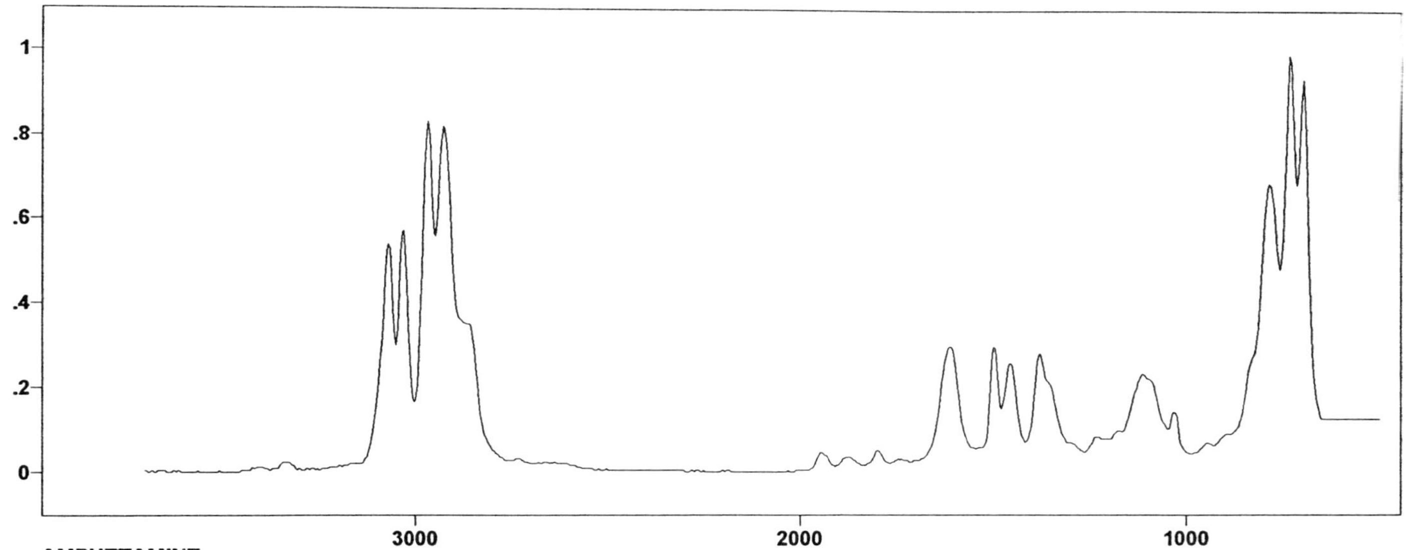

AMPHETAMINE

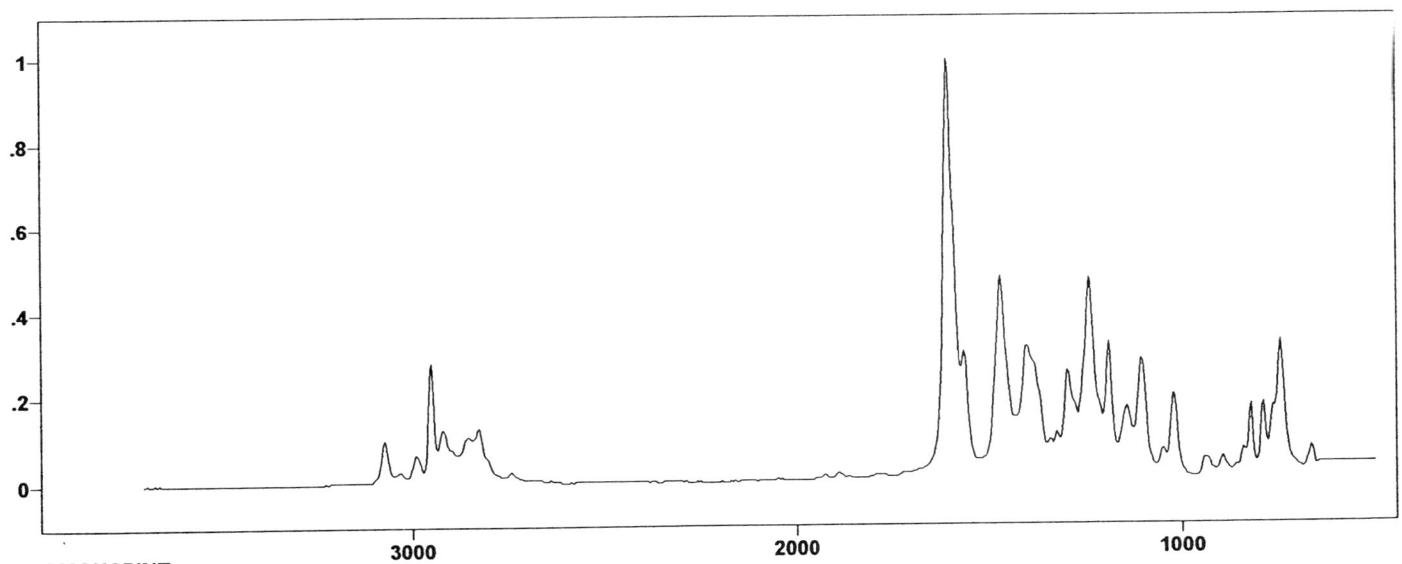

AMOXAPINE

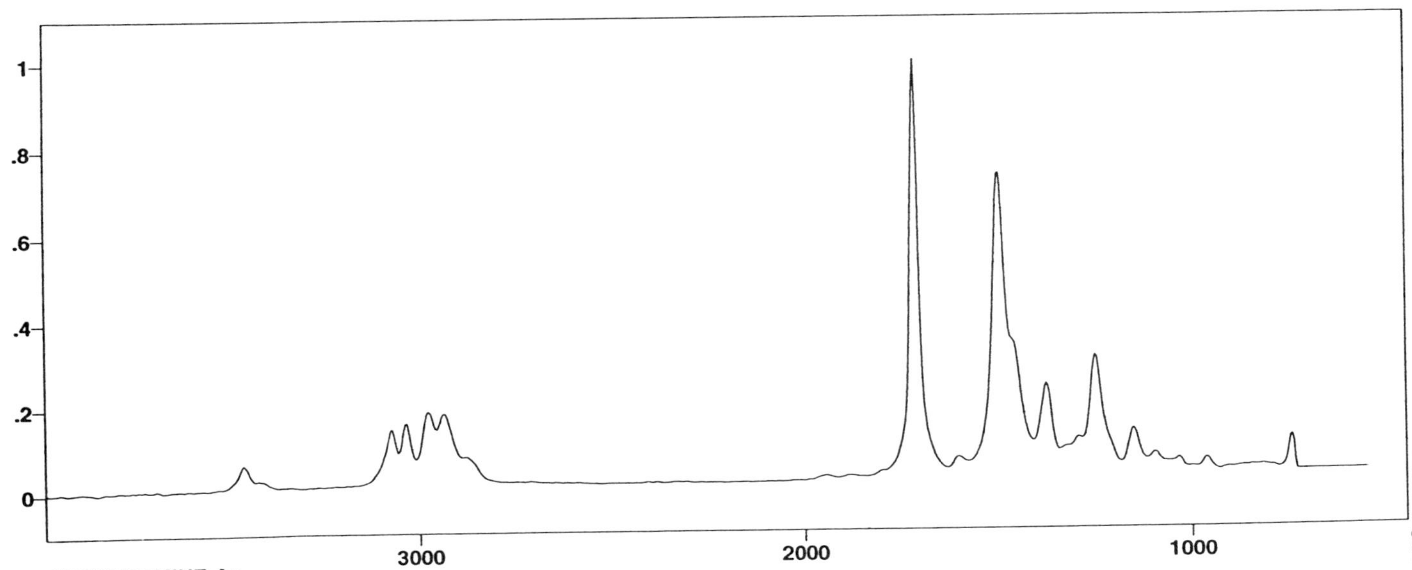

AMPHETAMINE-Ac

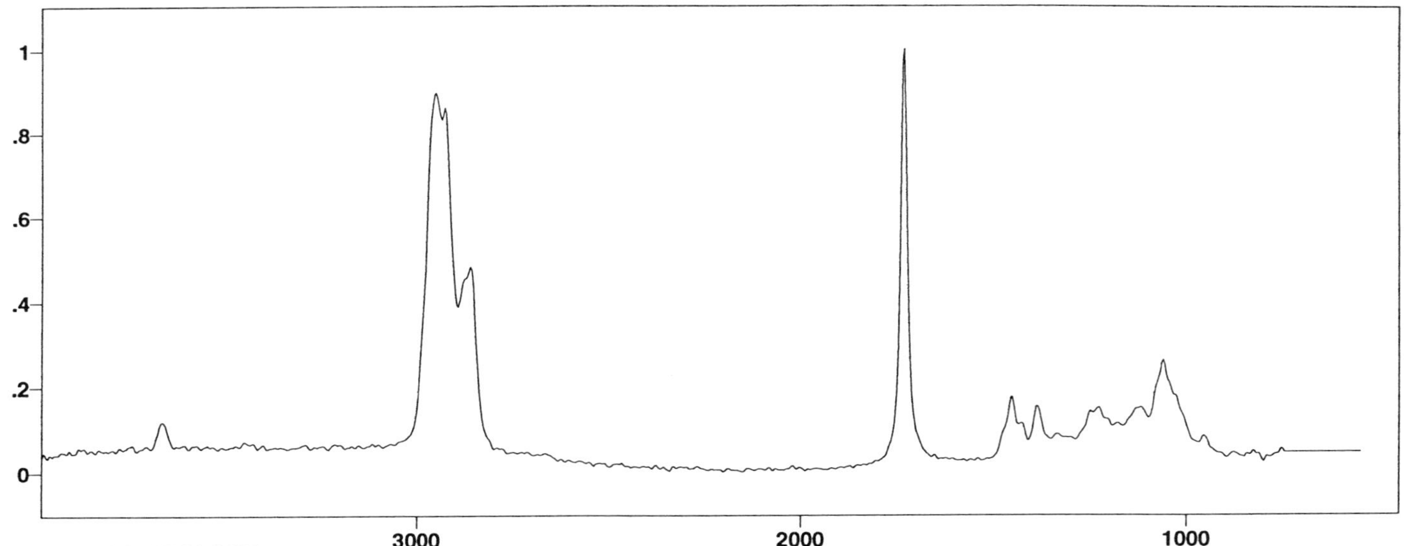

ANDROSTANOLONE

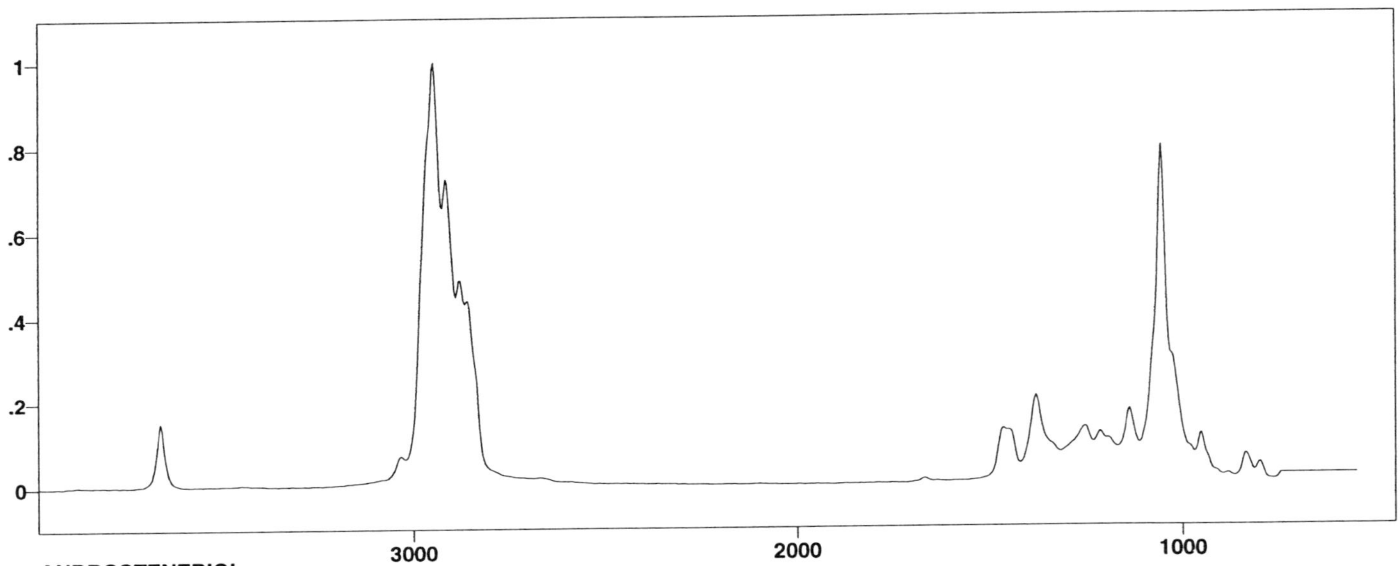

ANDROSTENEDIOL

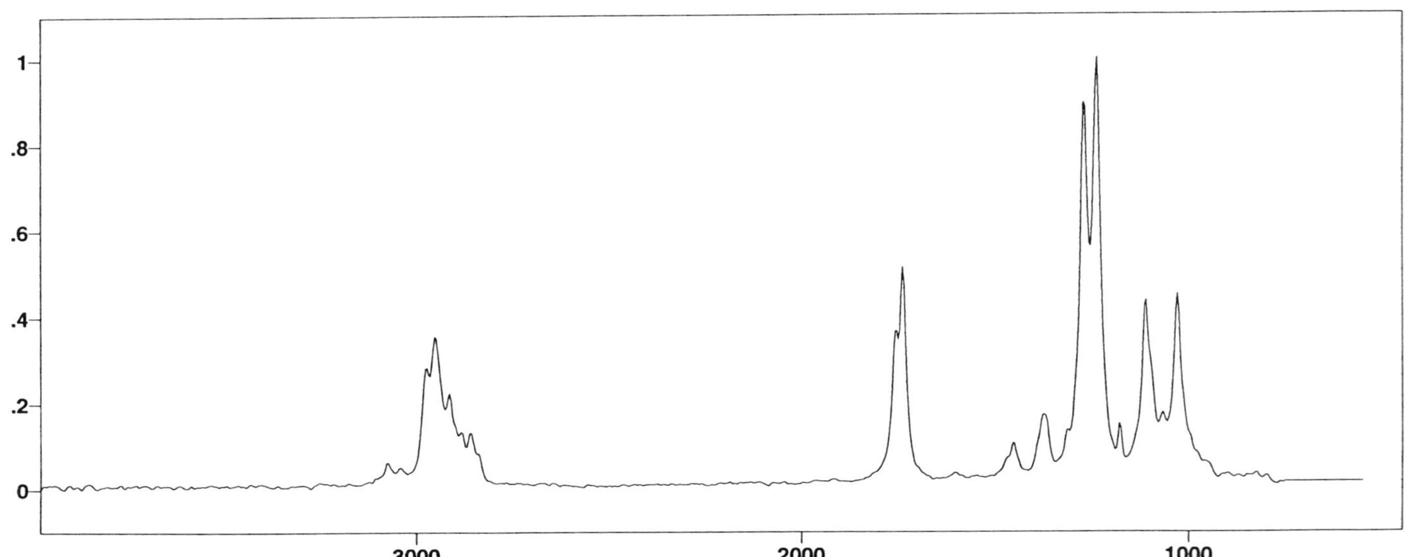

ANDROSTENEDIOL ACETATE BENZOATE

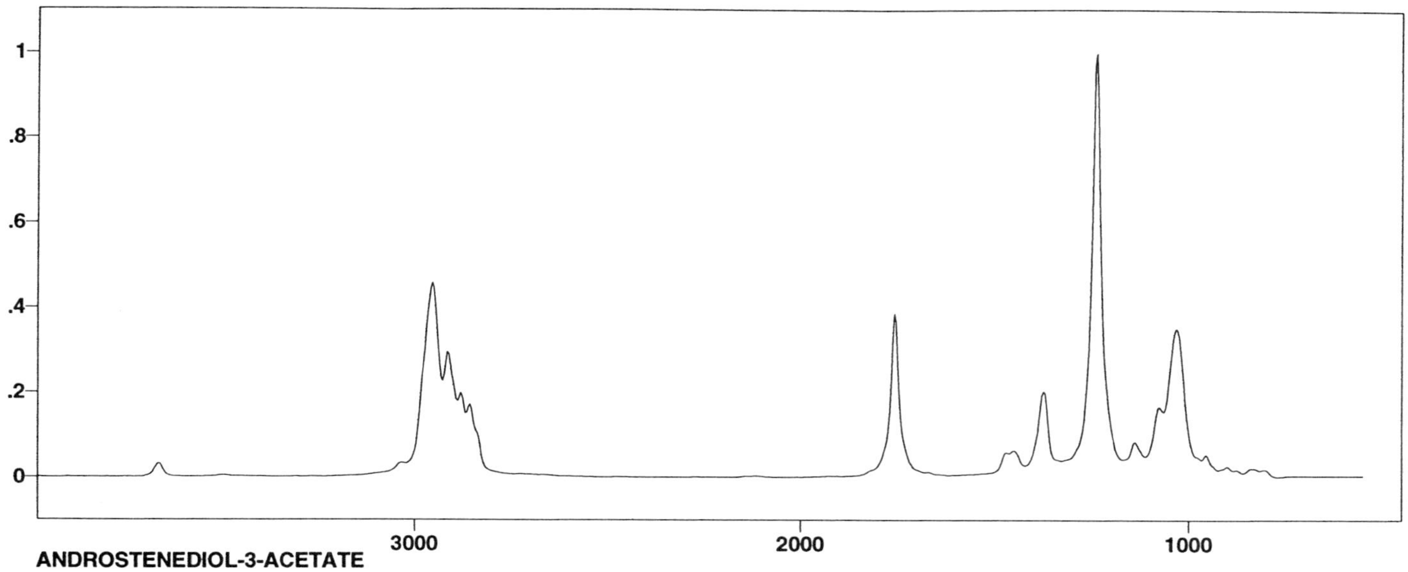

ANDROSTENEDIOL-3-ACETATE

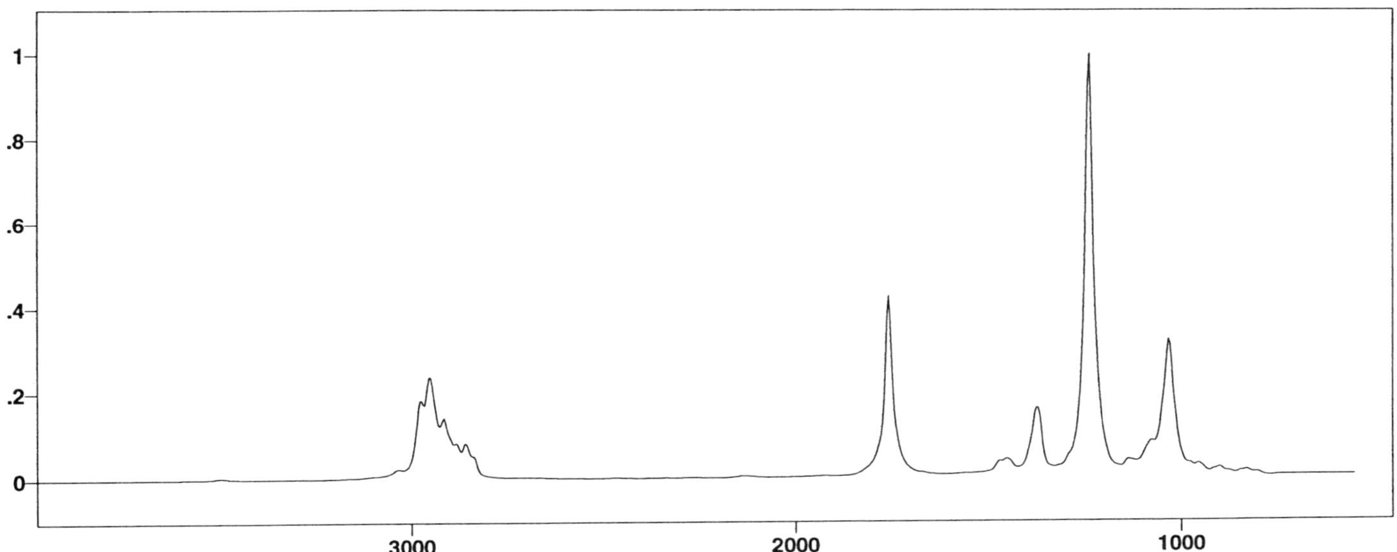

ANDROSTENEDIOL DIACETATE

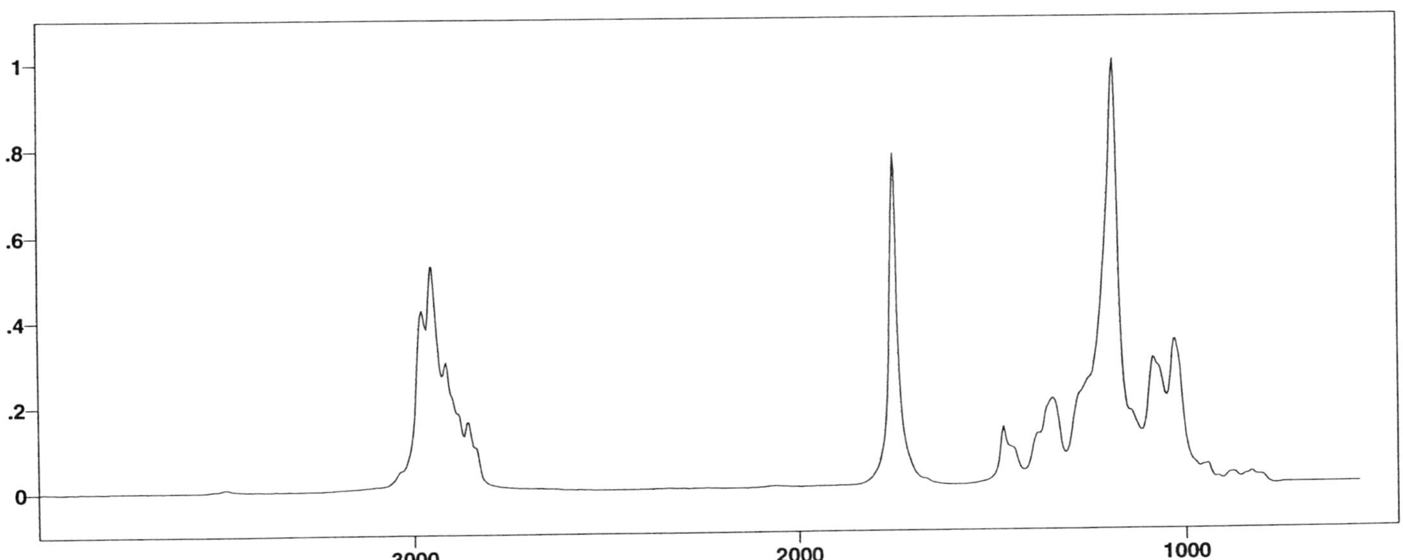

ANDROSTENEDIOL DIPROPIONATE

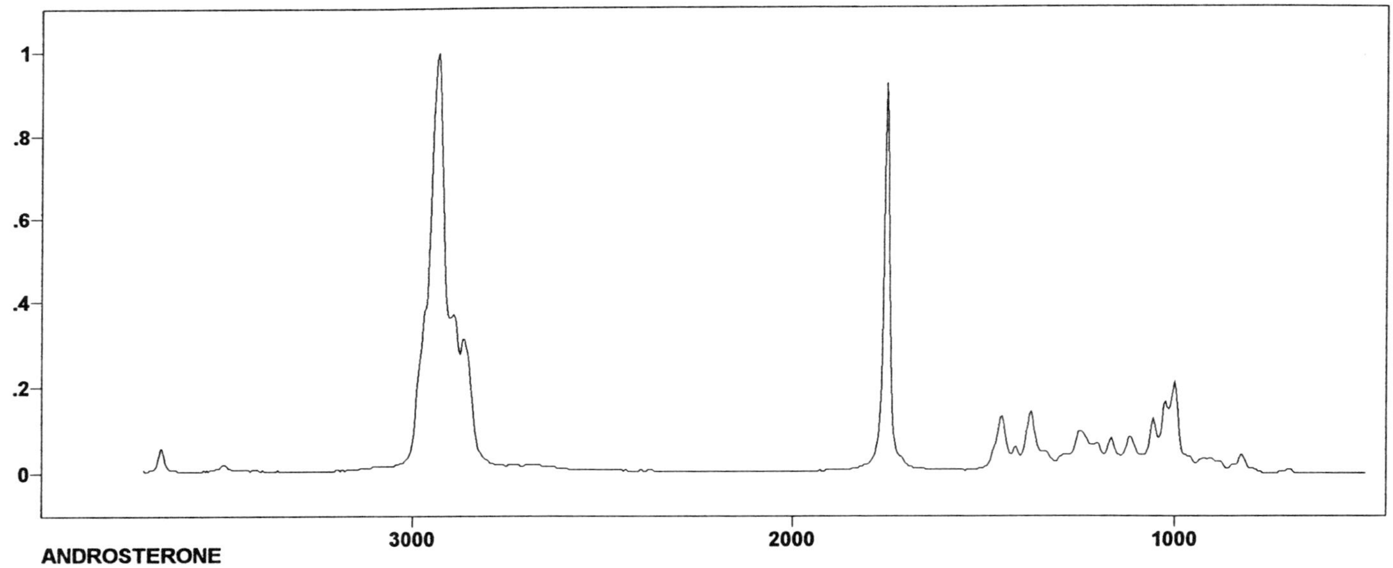

ANDROSTERONE

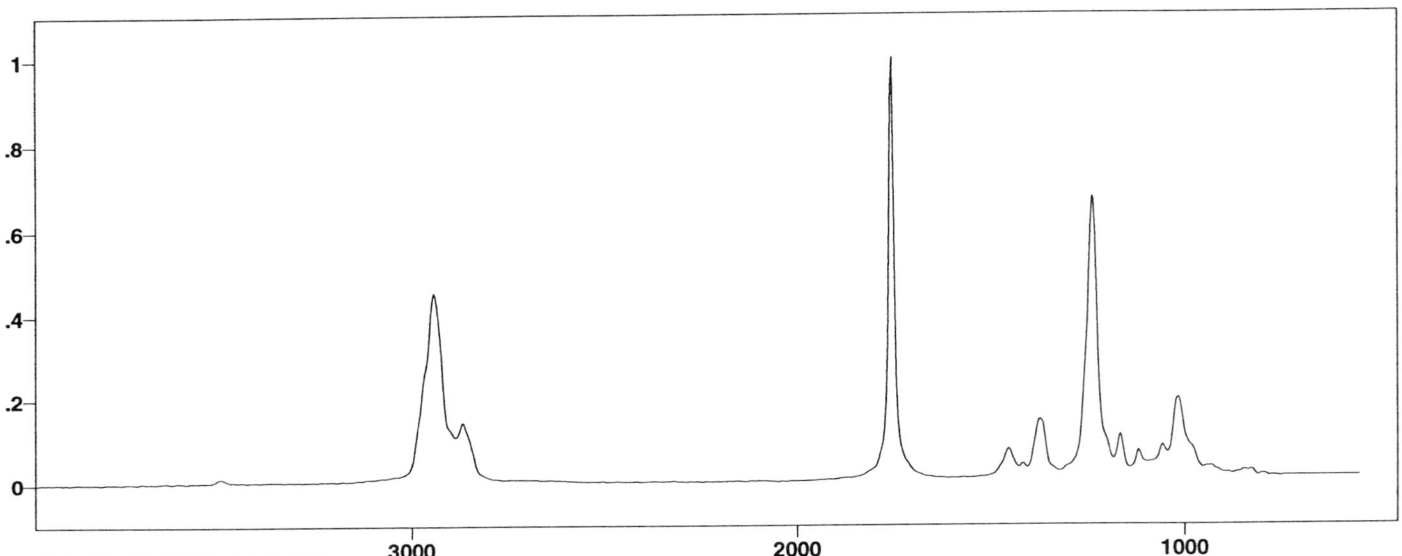

ANDROSTERONE ACETATE

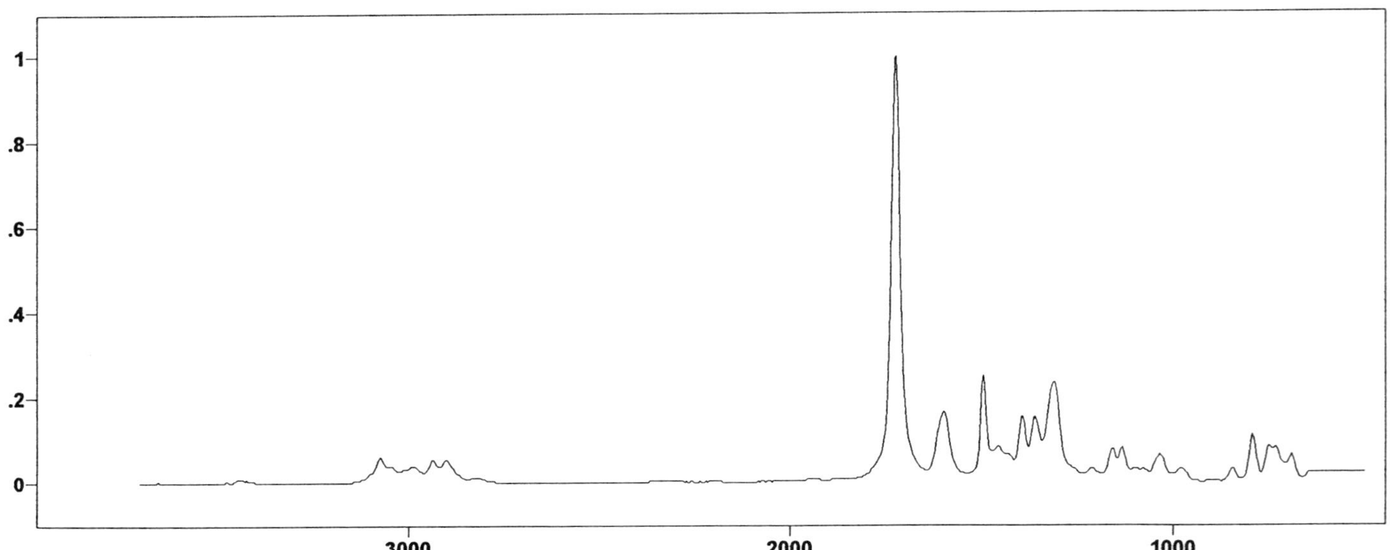

ANTIPYRINE

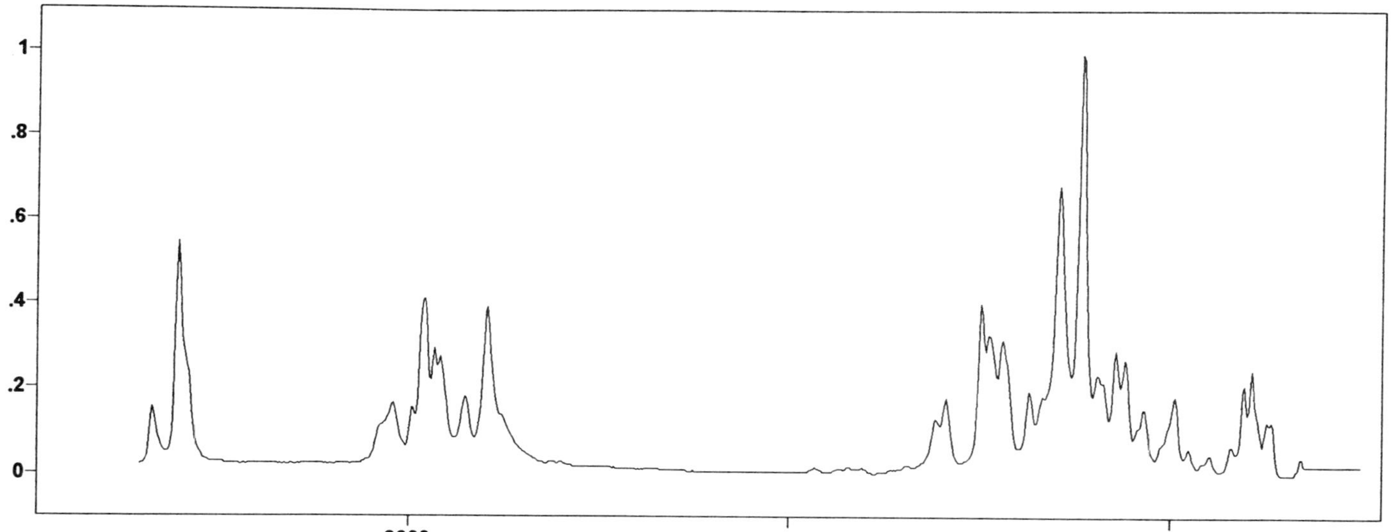

APOMORPHINE

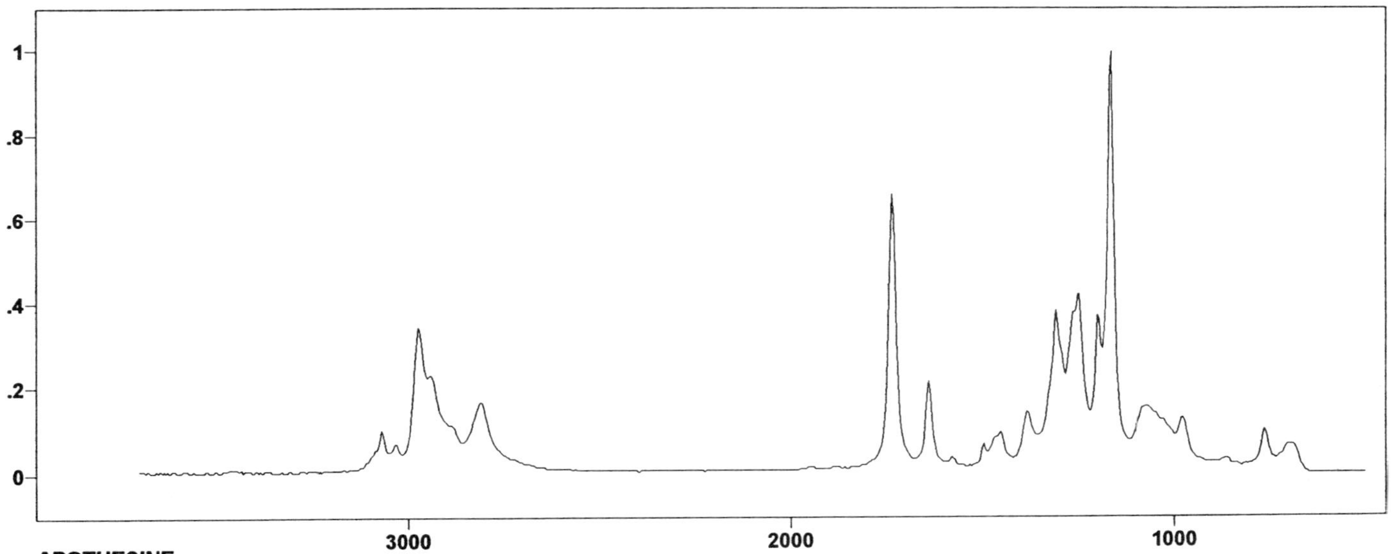

APOTHESINE

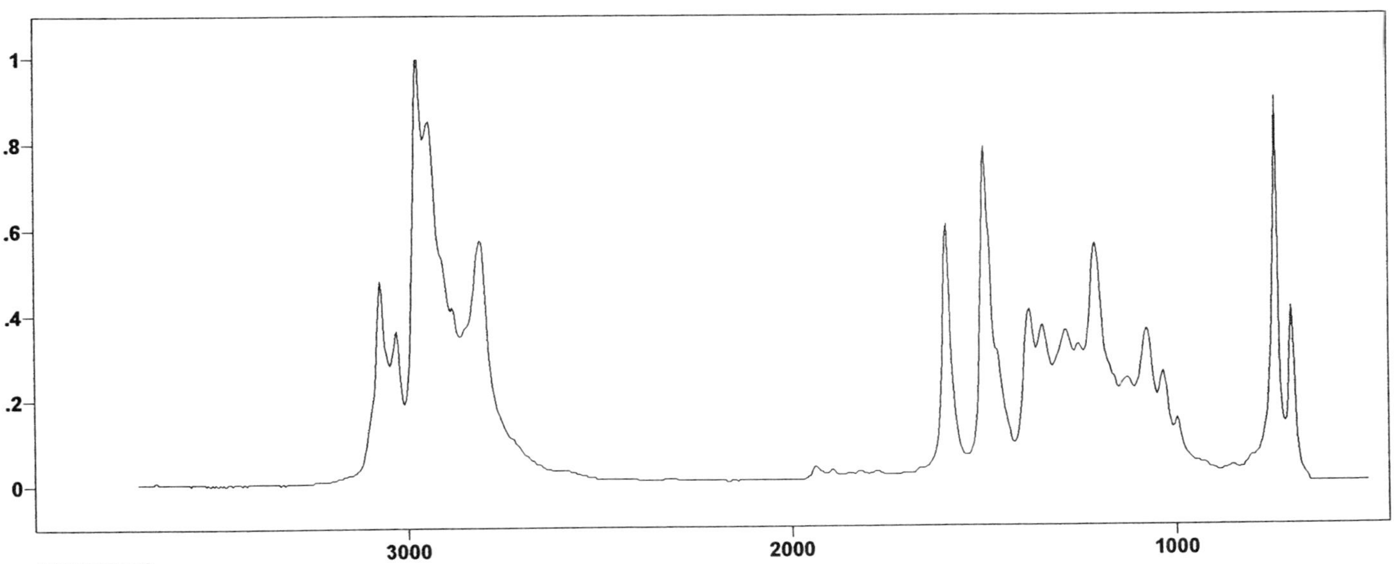

APRINDINE

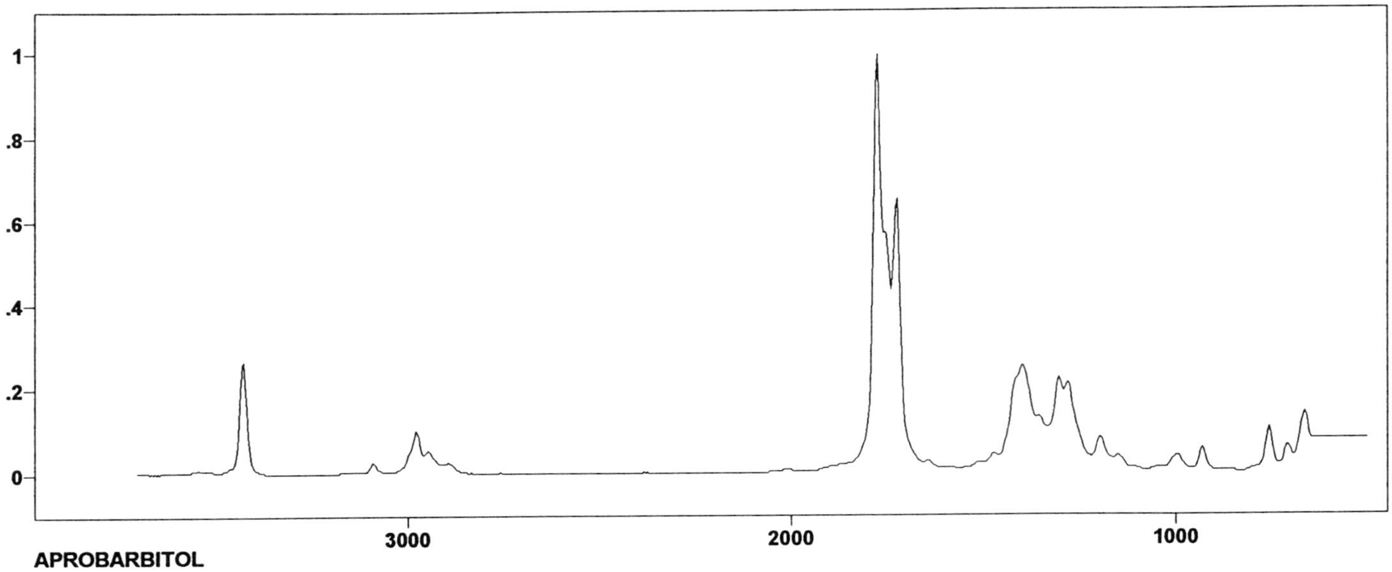

APROBARBITOL

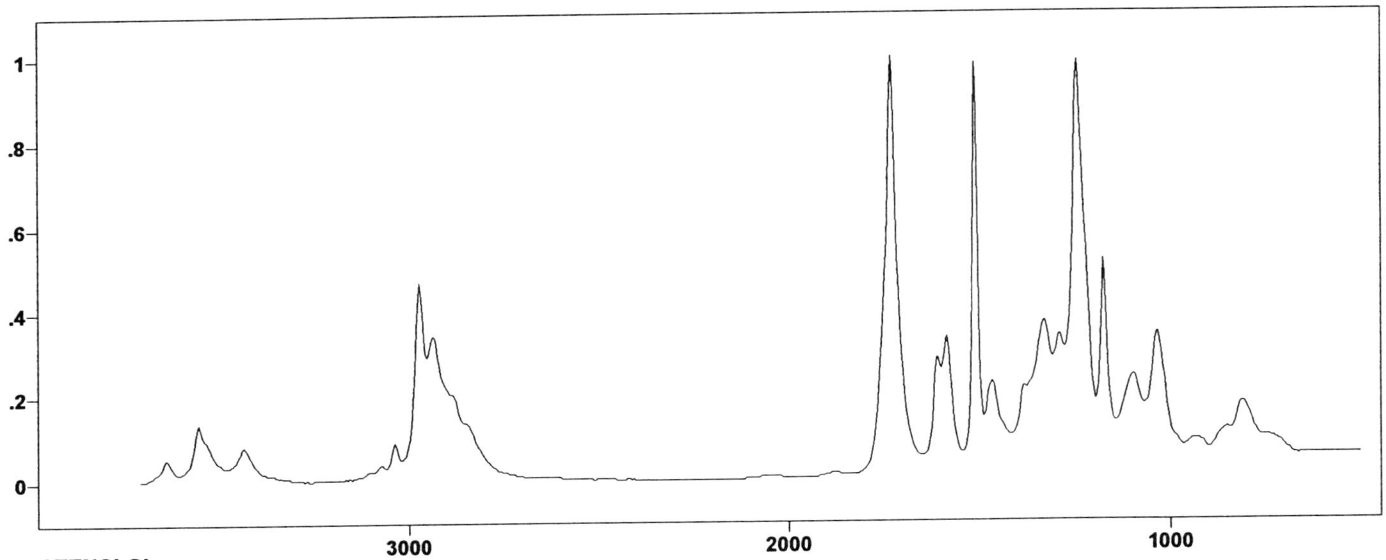

ATENOLOL

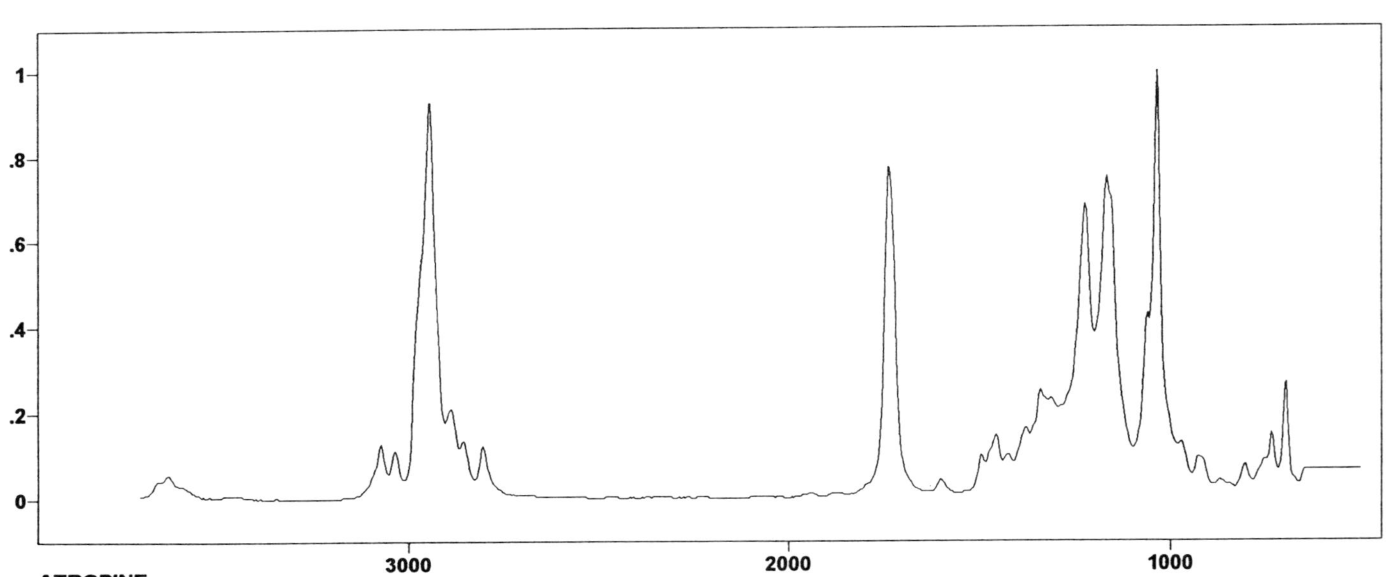

ATROPINE

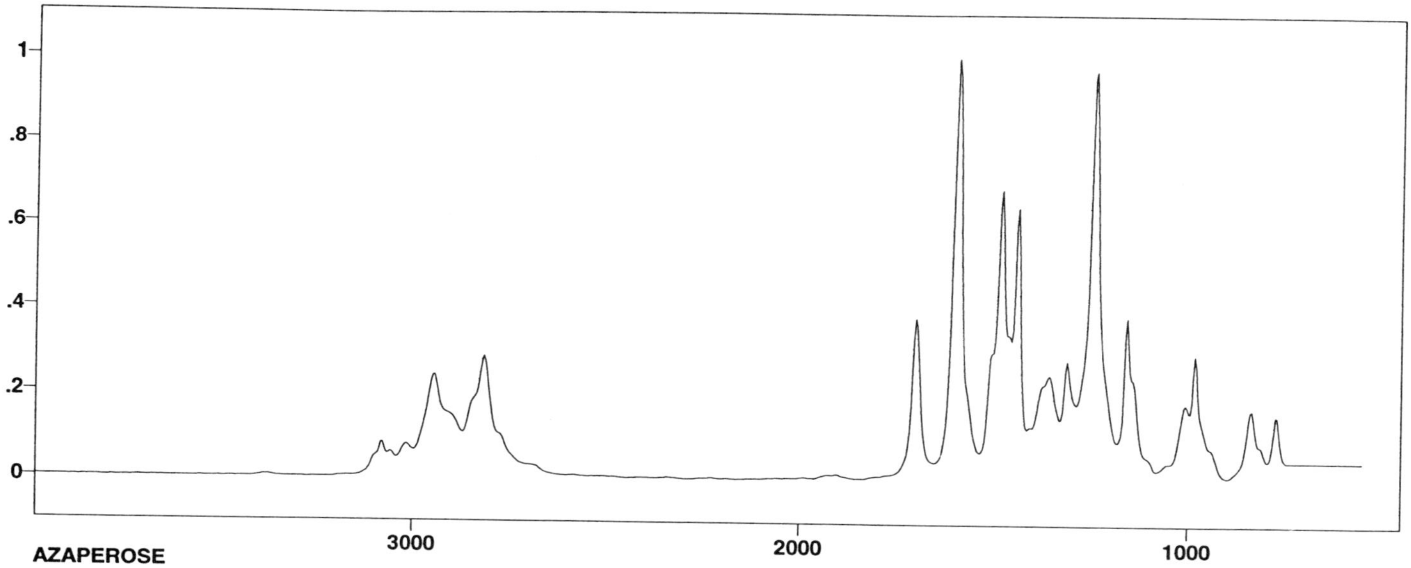

AZAPEROSE

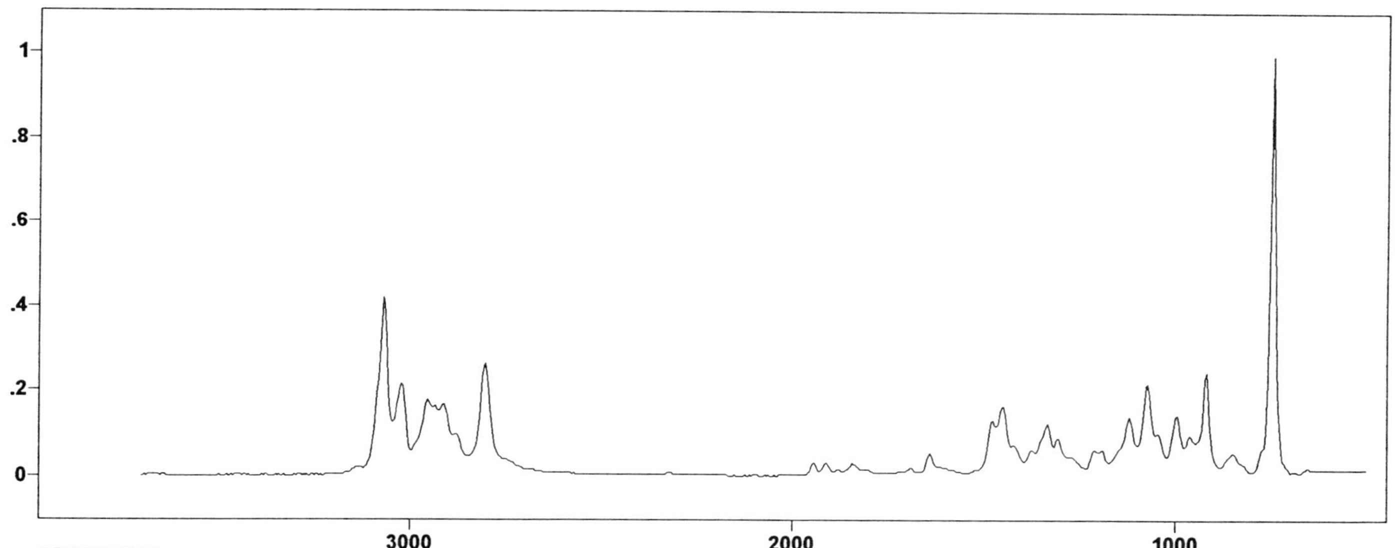

AZAPETINE

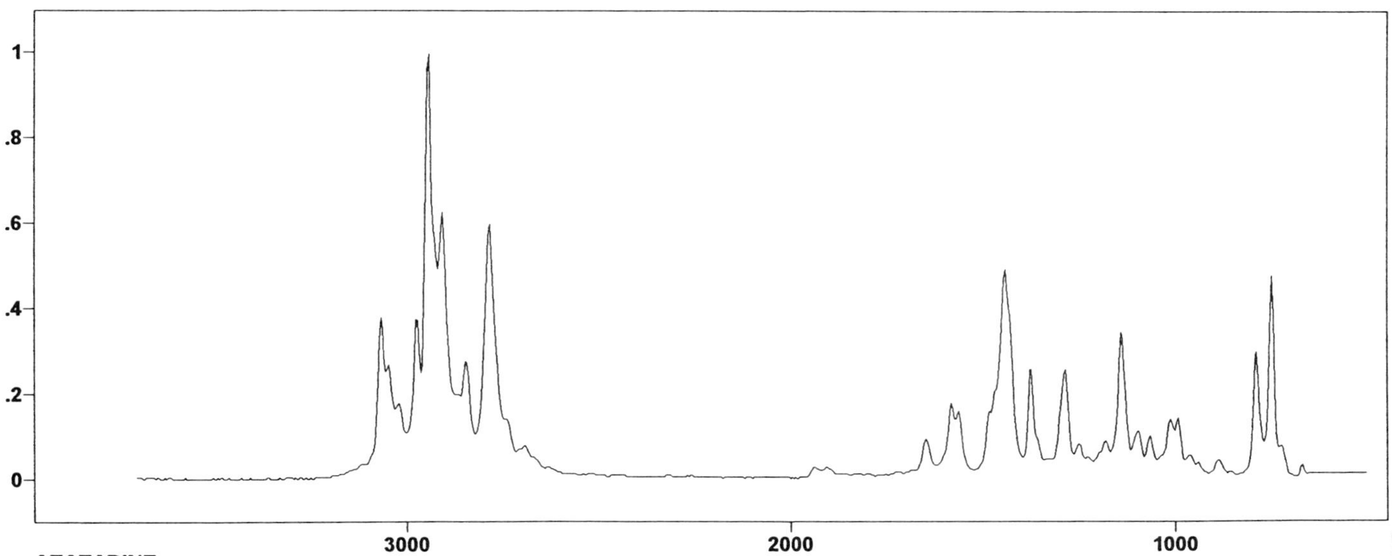

AZATADINE

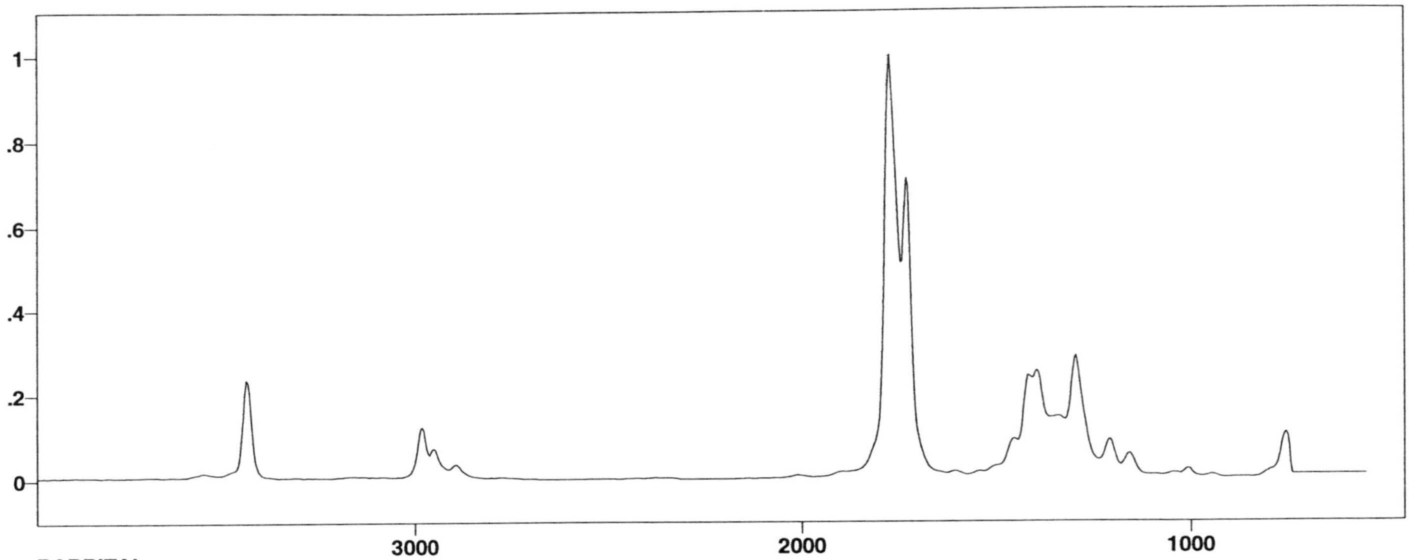

BARBITAL

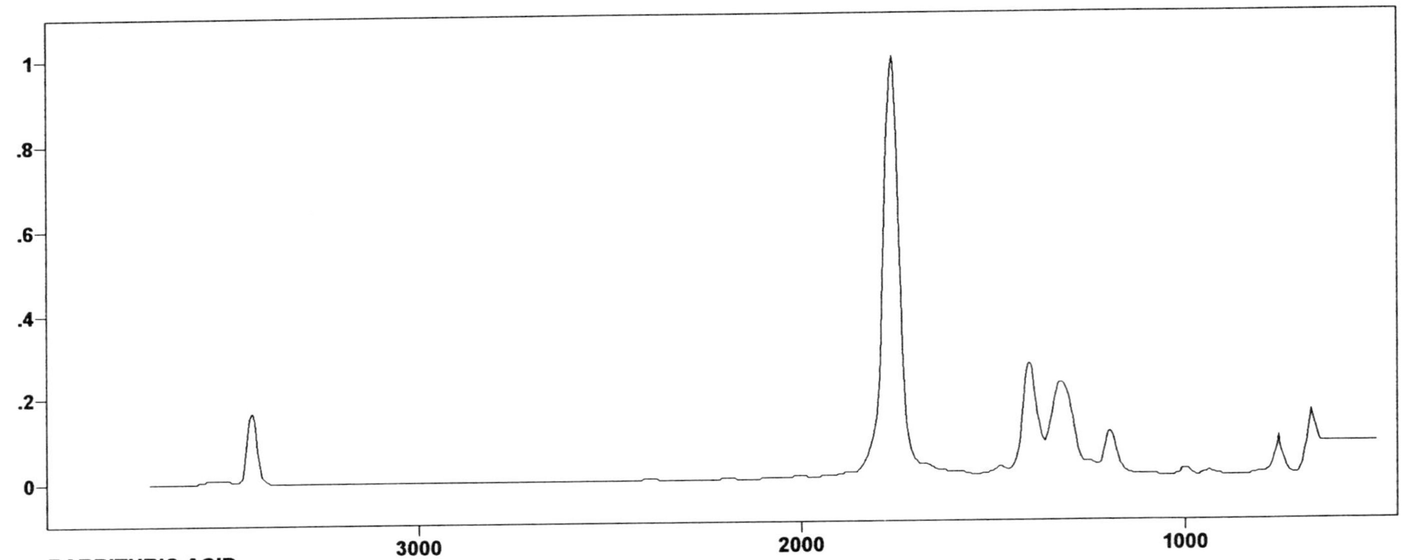

BARBITURIC ACID

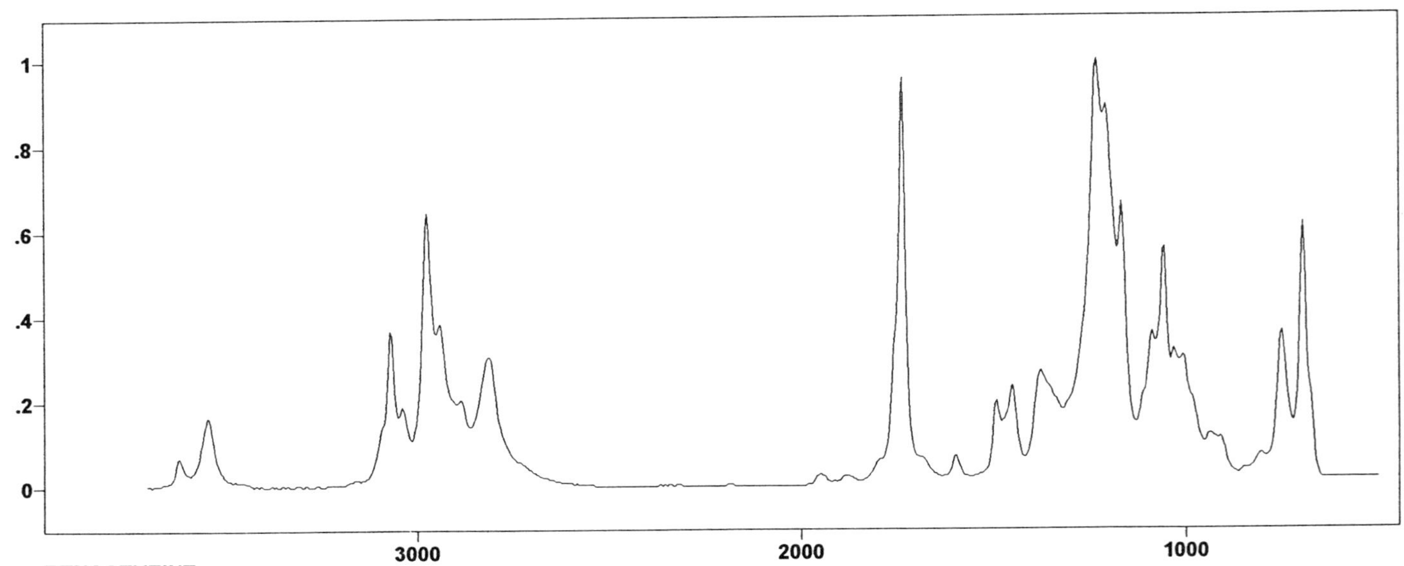

BENACTYZINE

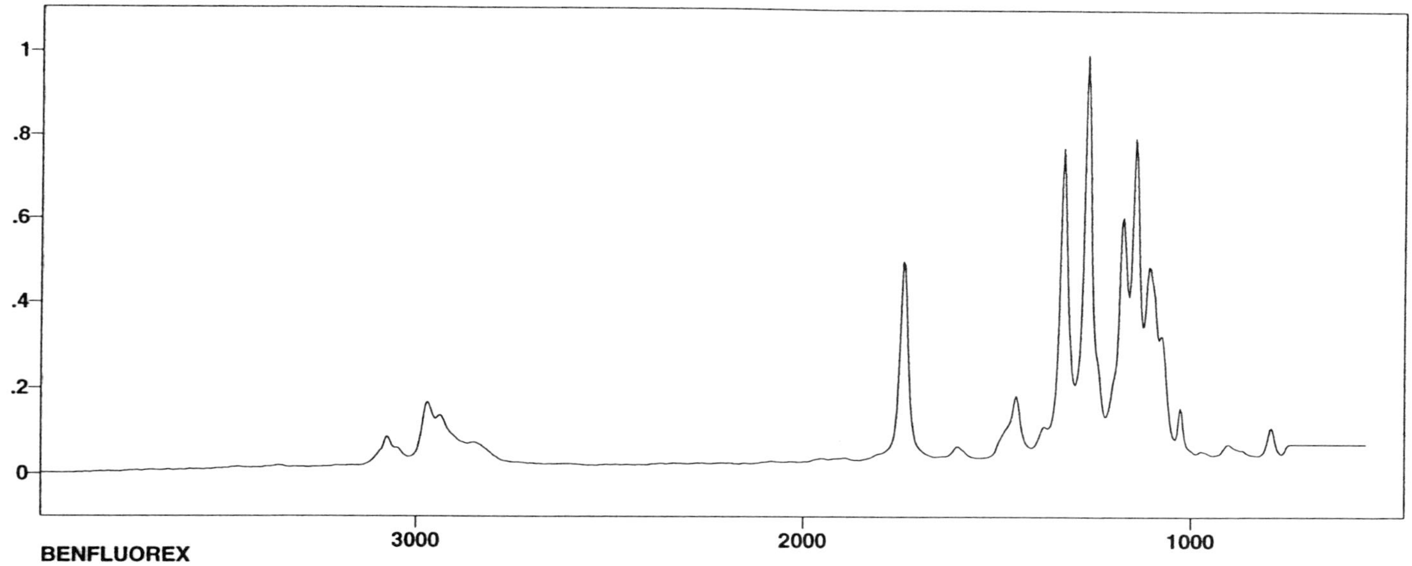

BENFLUOREX

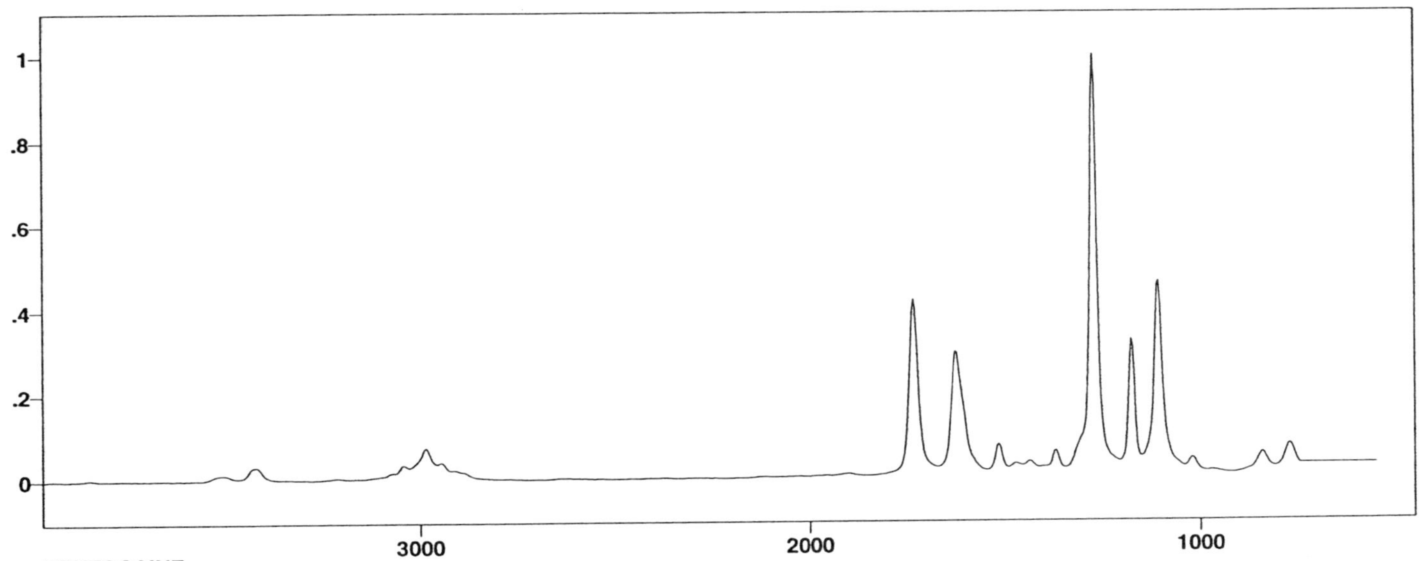

BENZOCAINE

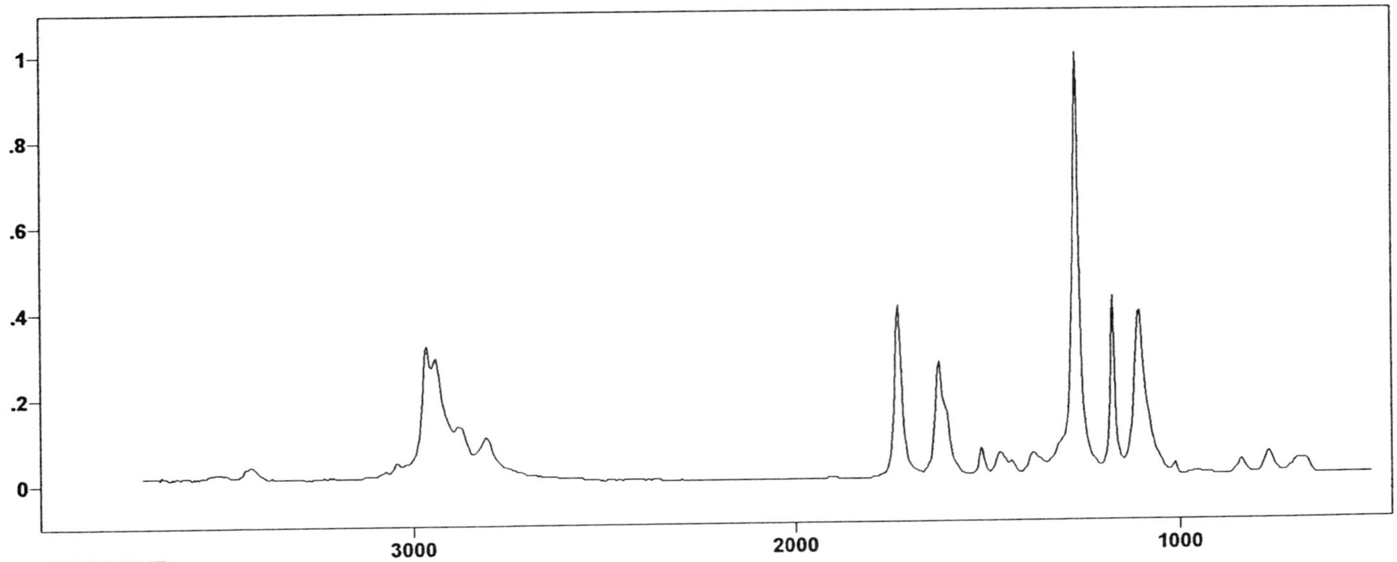

BUTACAINE

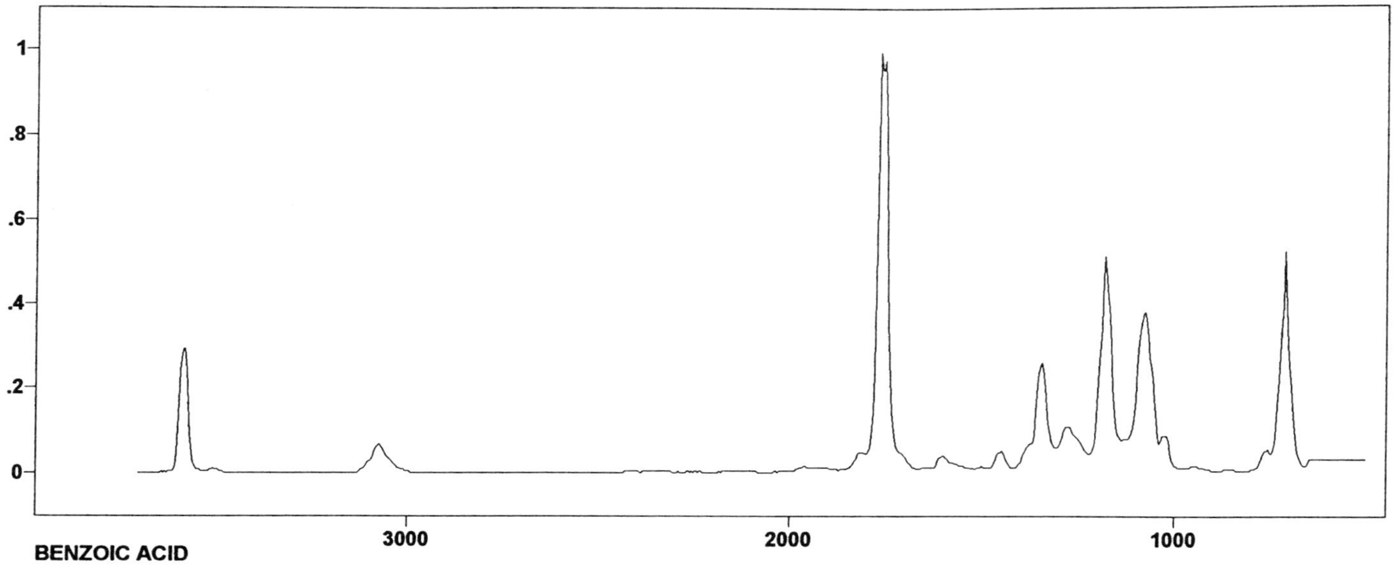

BENZOIC ACID

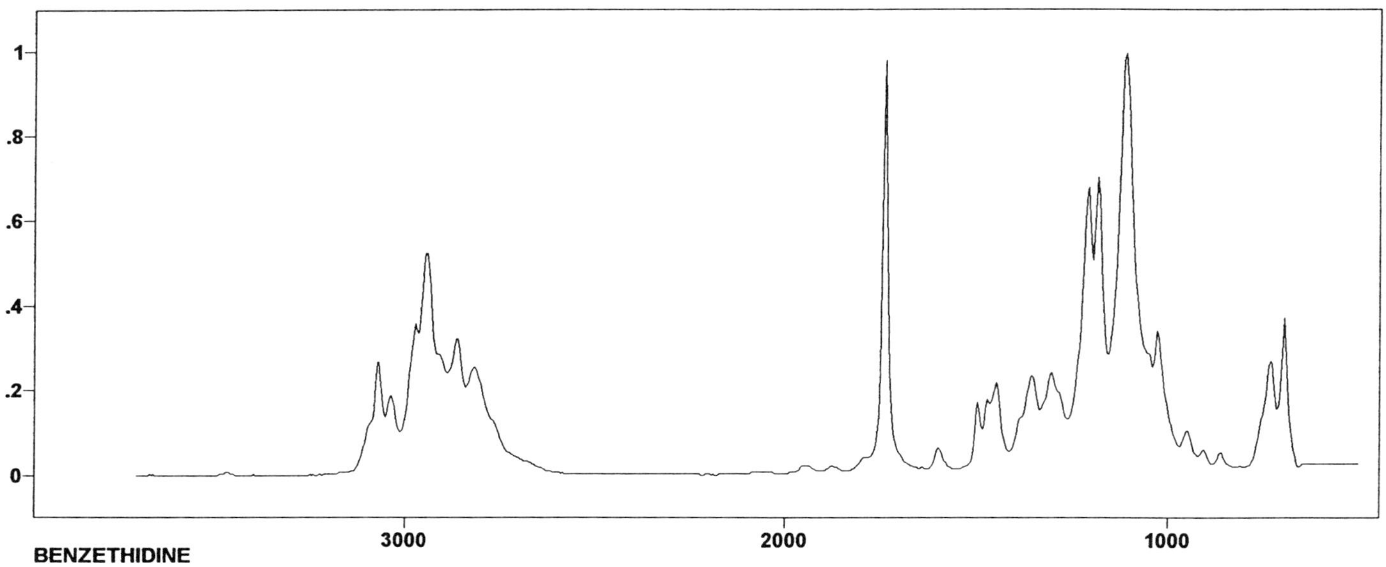

BENZETHIDINE

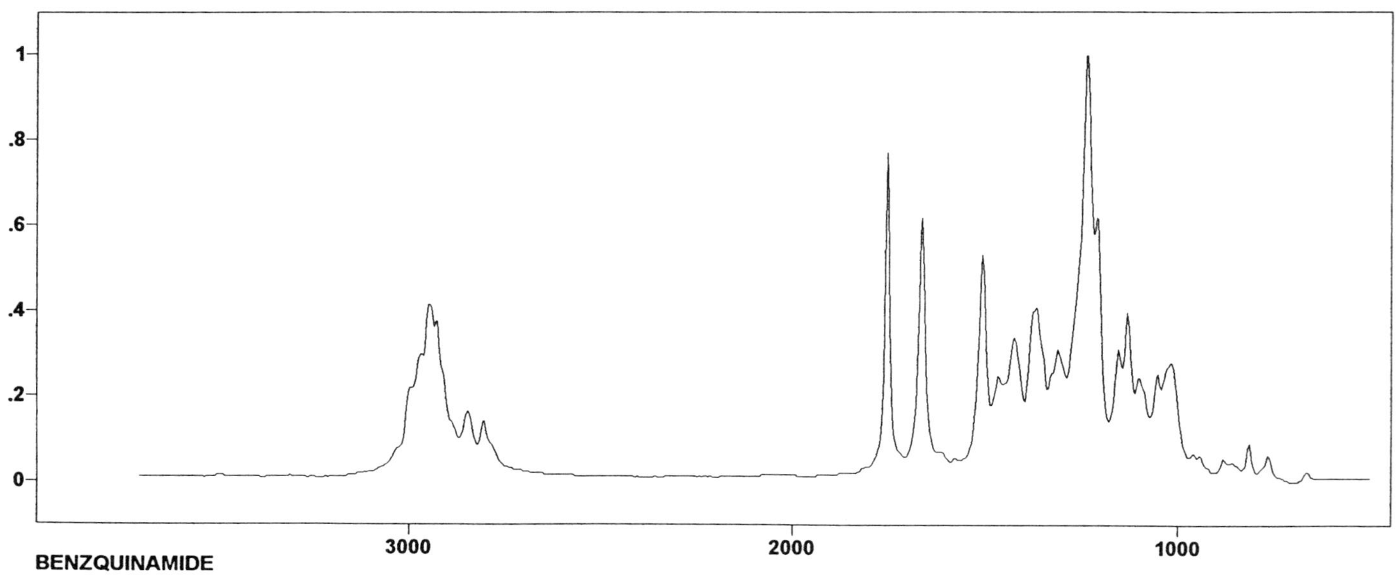

BENZQUINAMIDE

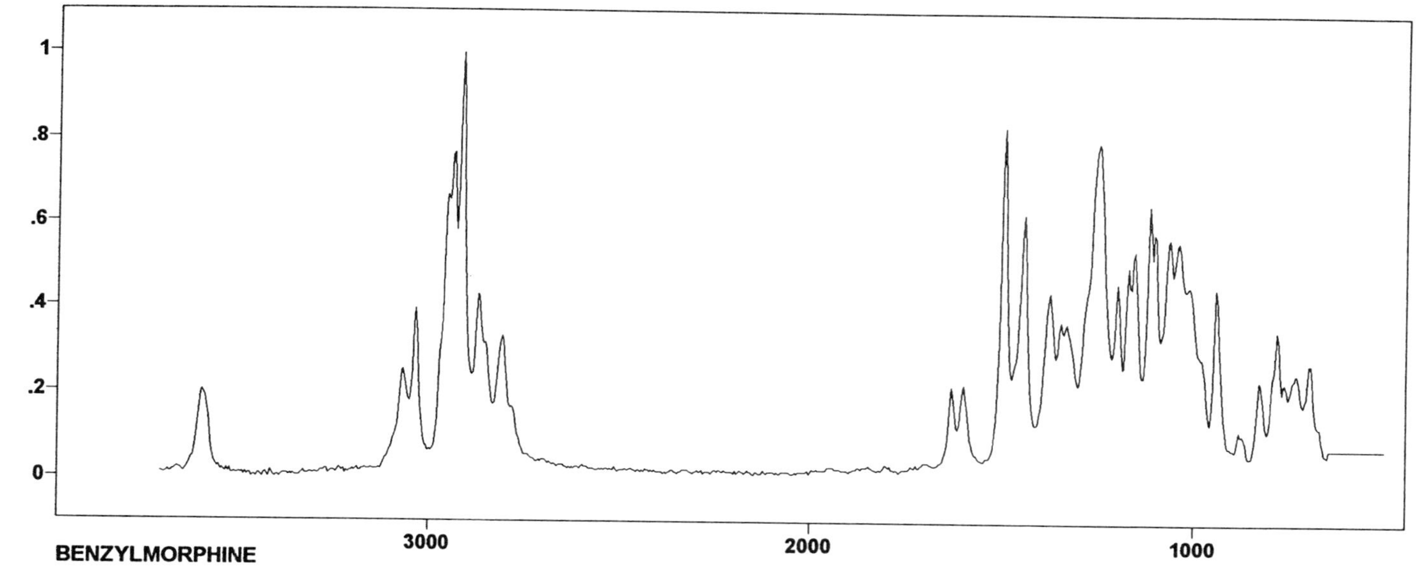

BENZYLMORPHINE

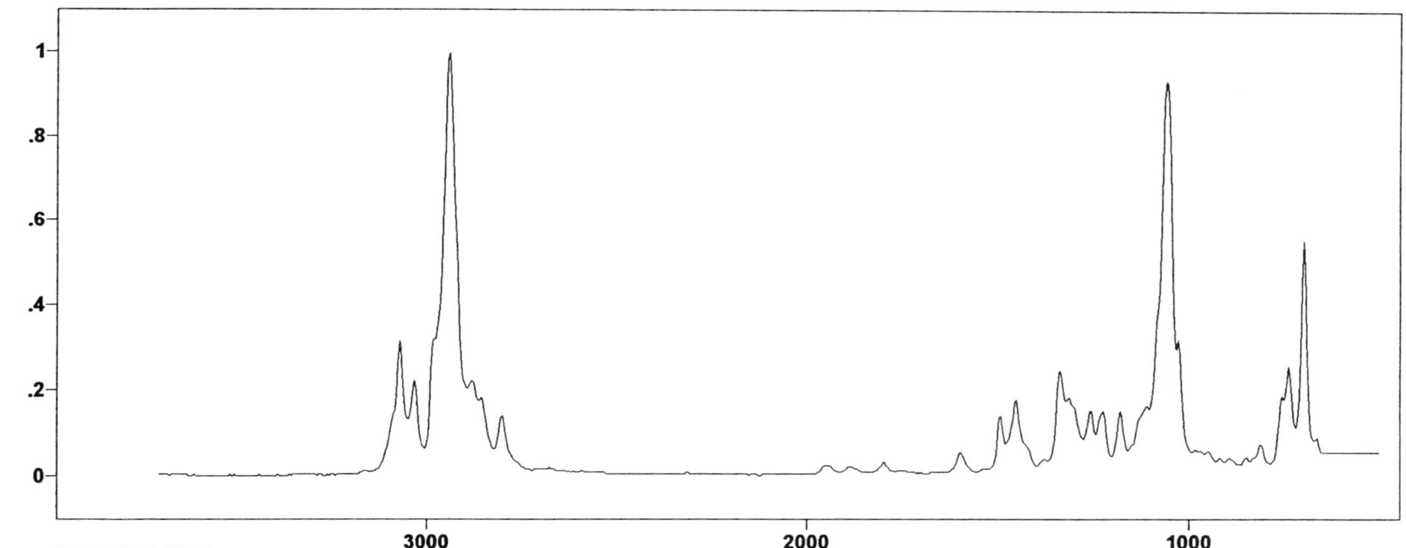

BENZTROPINE

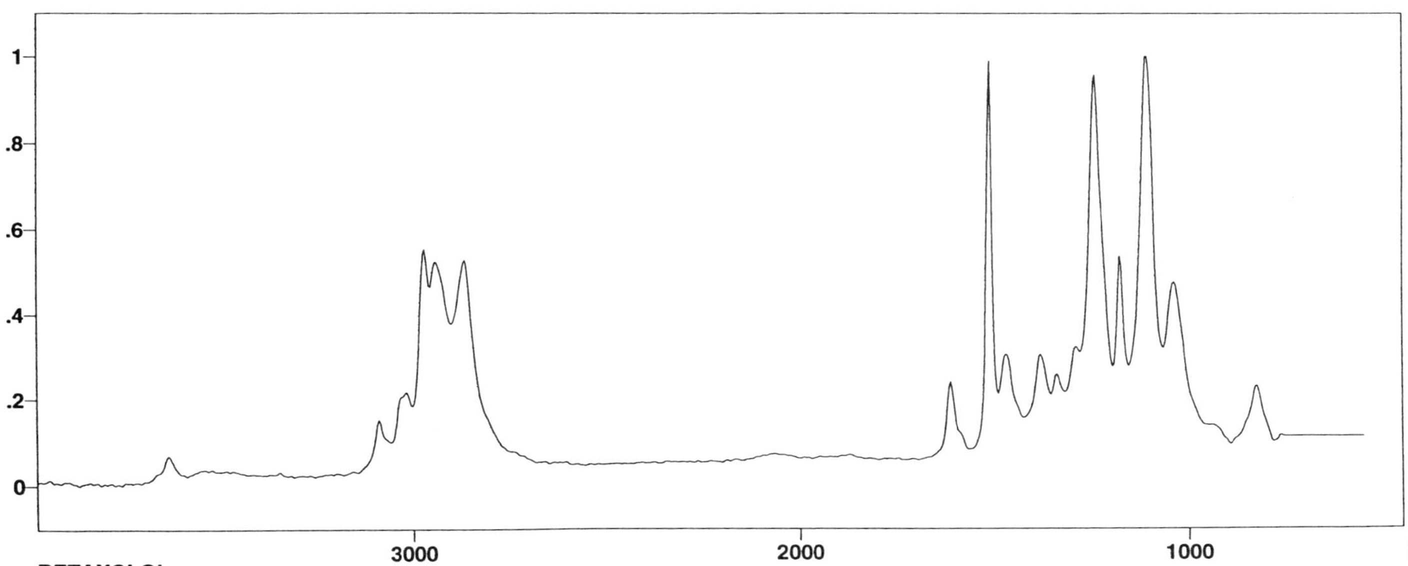

BETAXOLOL

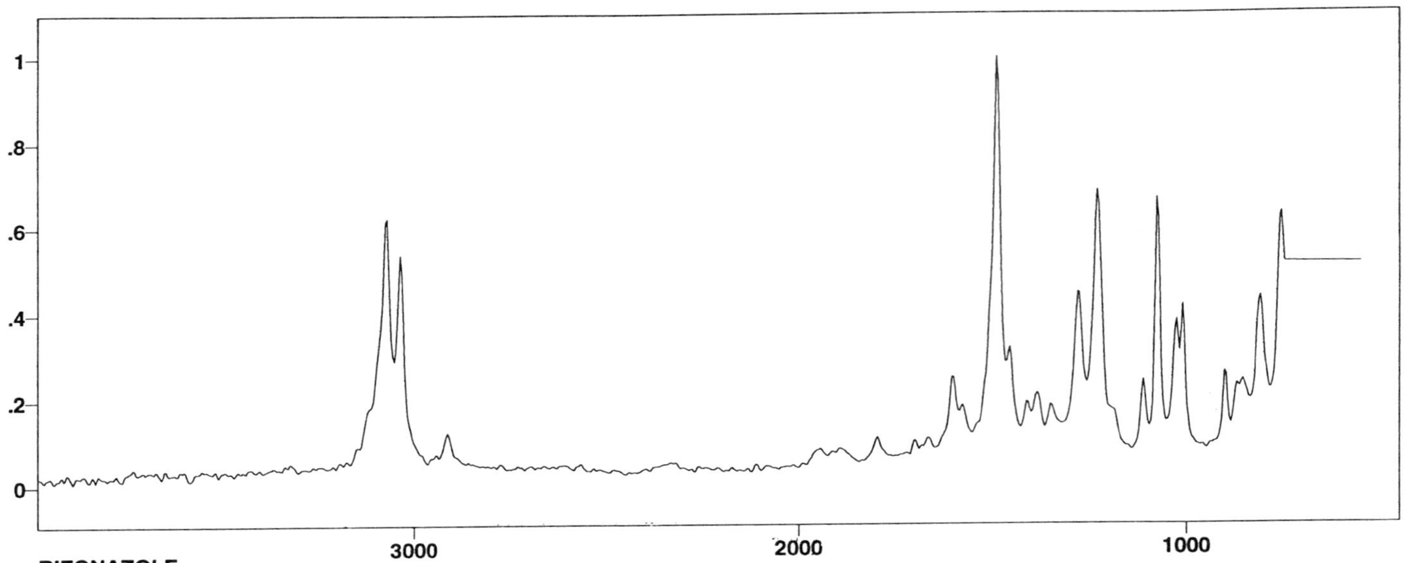

BIFONAZOLE

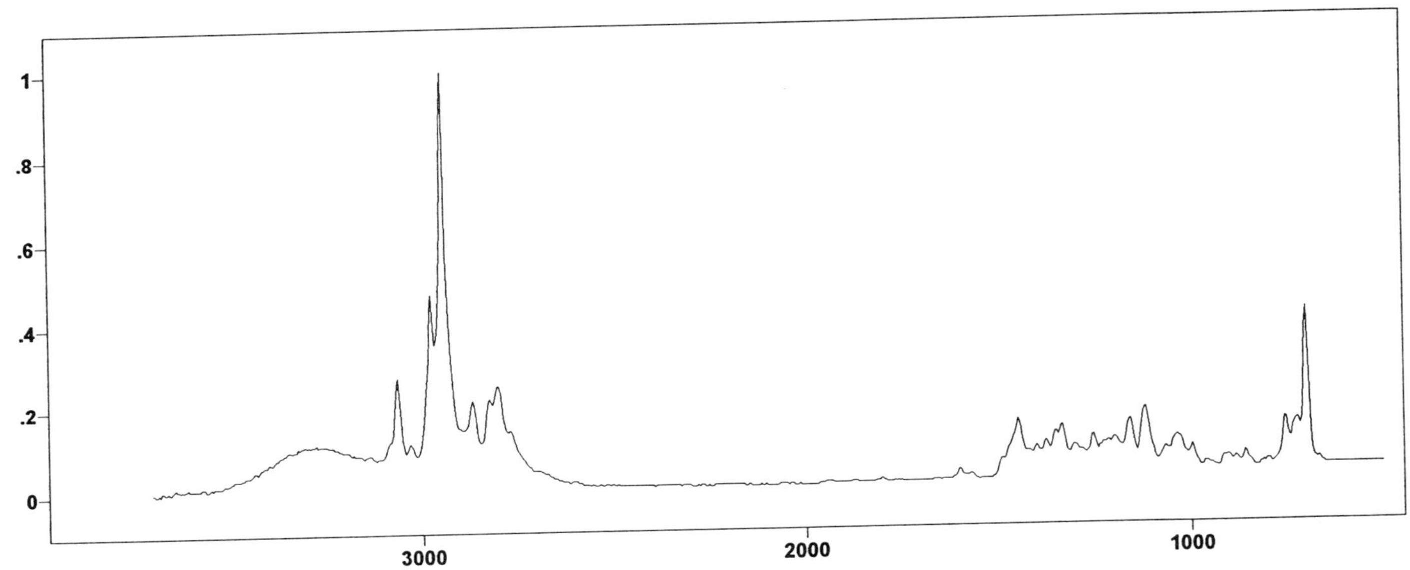

BIPERIDEN

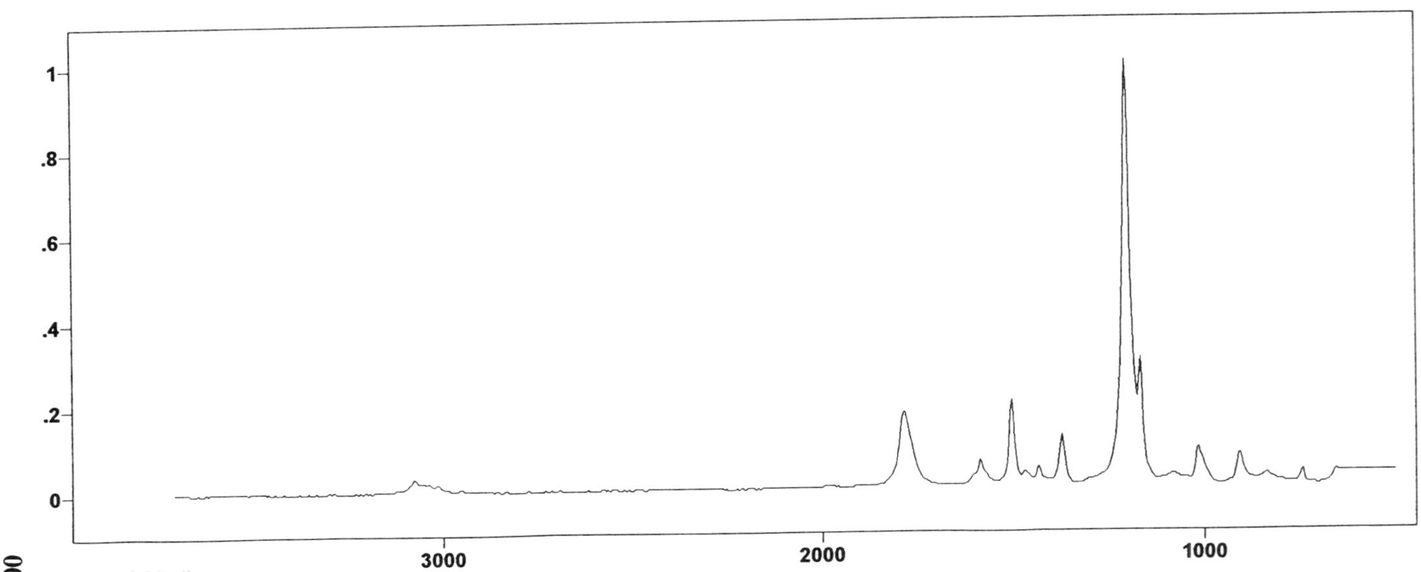

BISACODYL

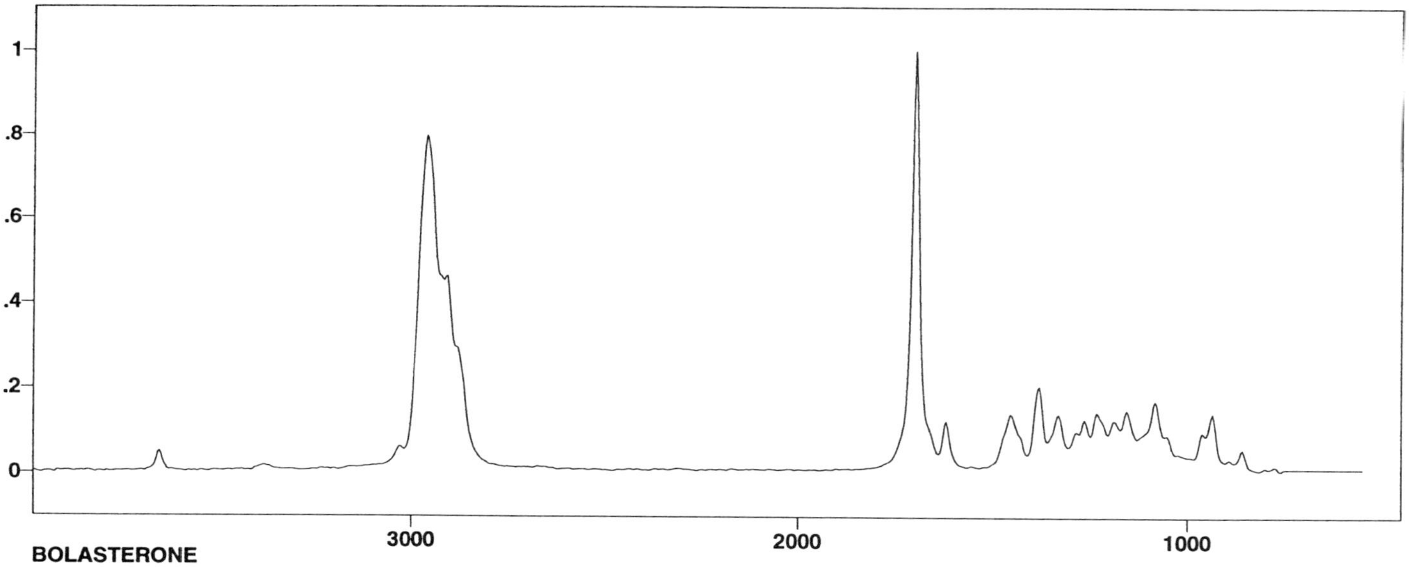

BOLASTERONE

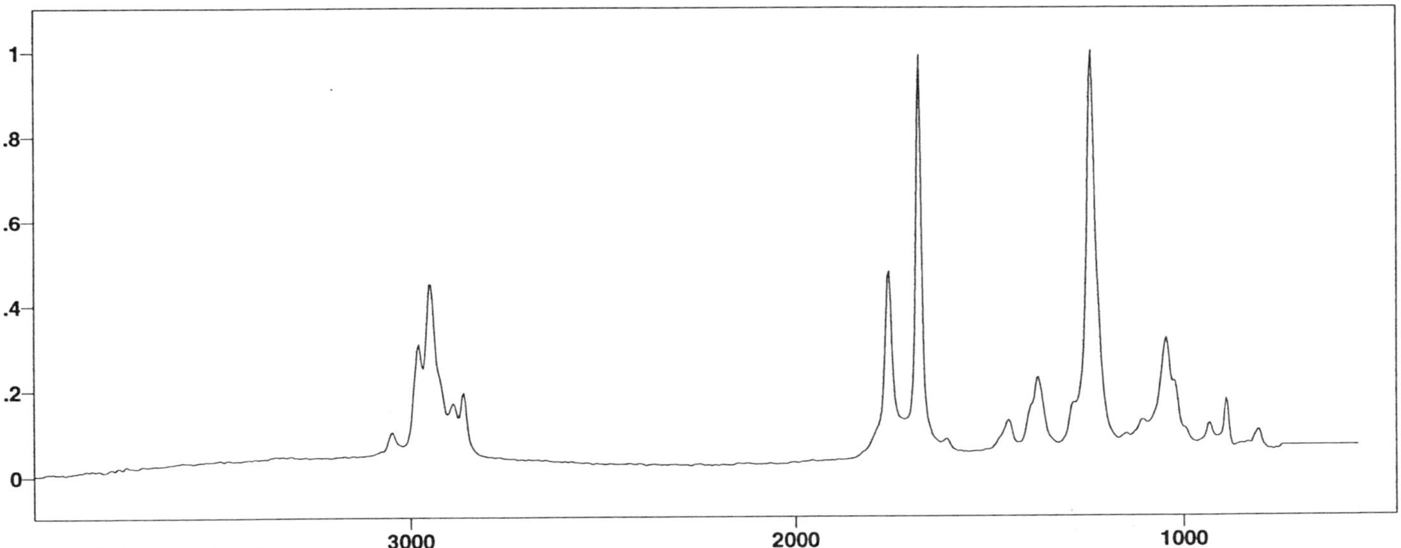

BOLDENONE ACETATE

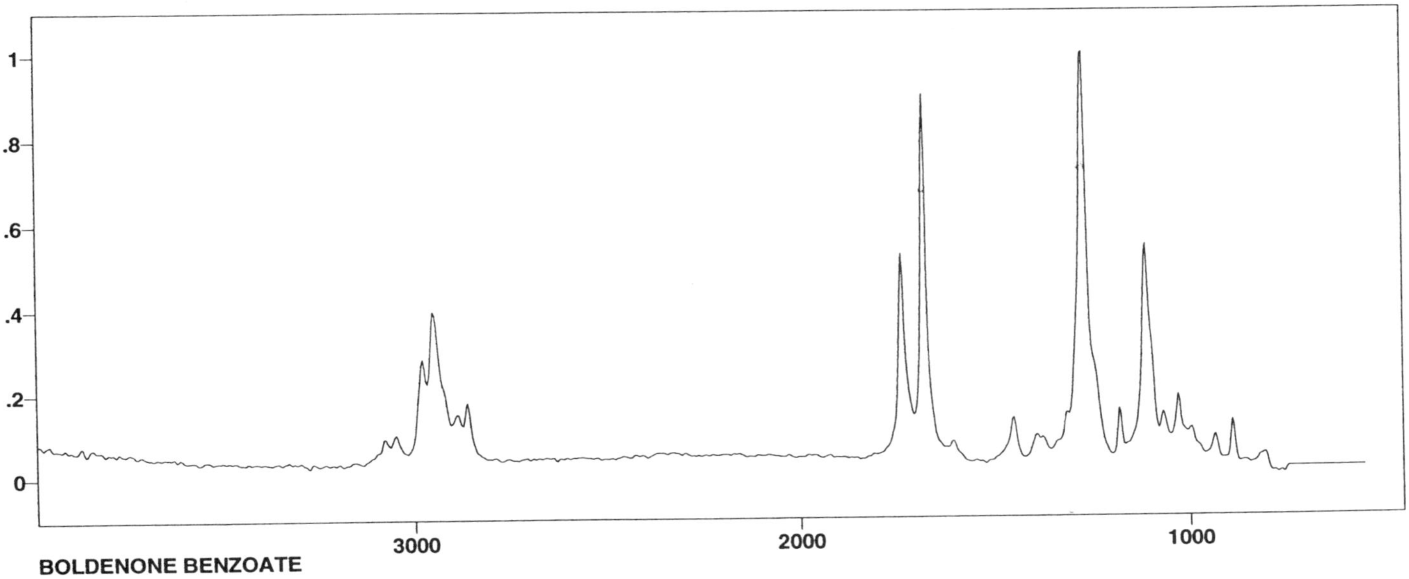

BOLDENONE BENZOATE

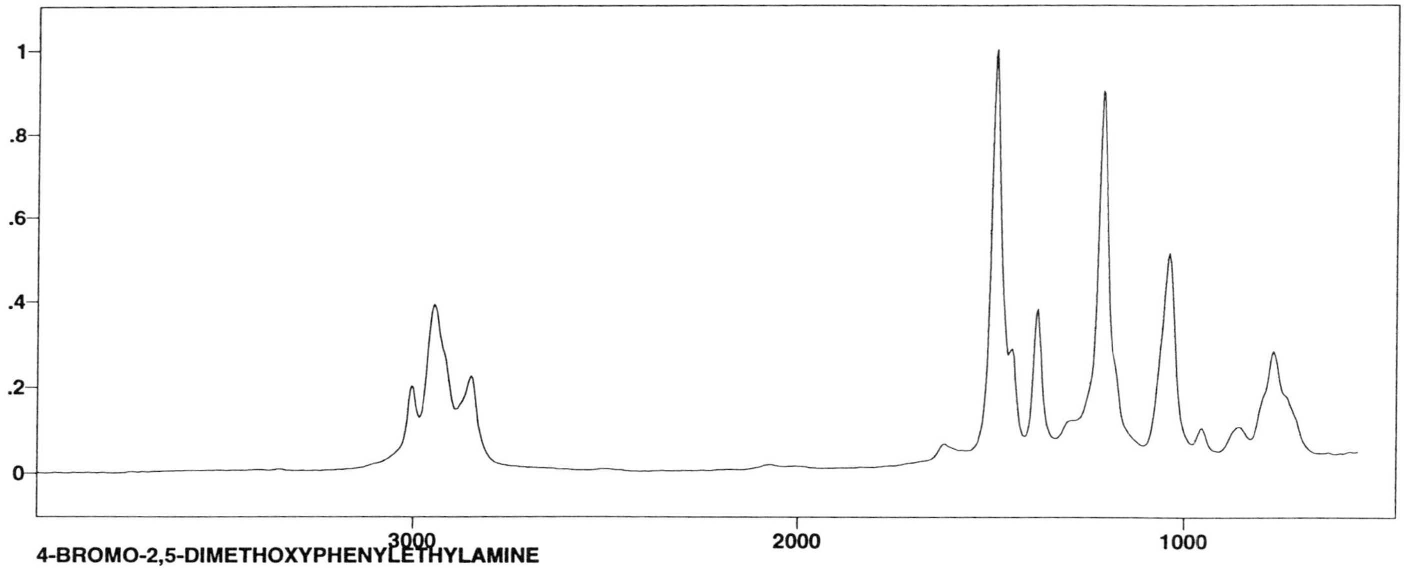

4-BROMO-2,5-DIMETHOXYPHENYLETHYLAMINE

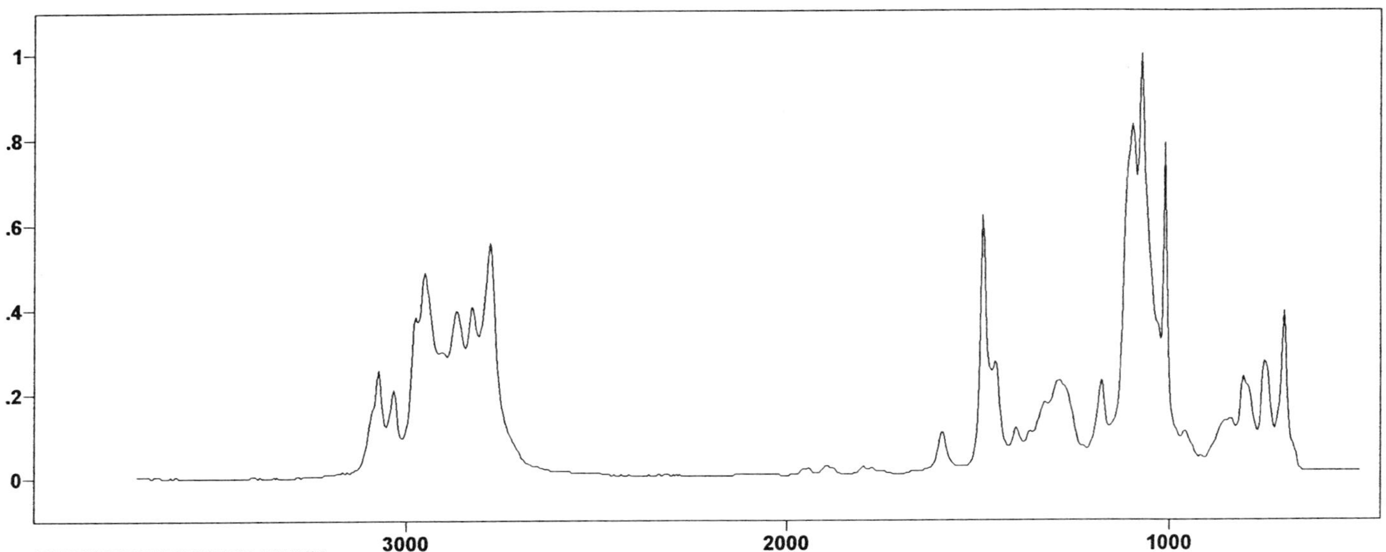

BROMODIPHENHYDRAMINE

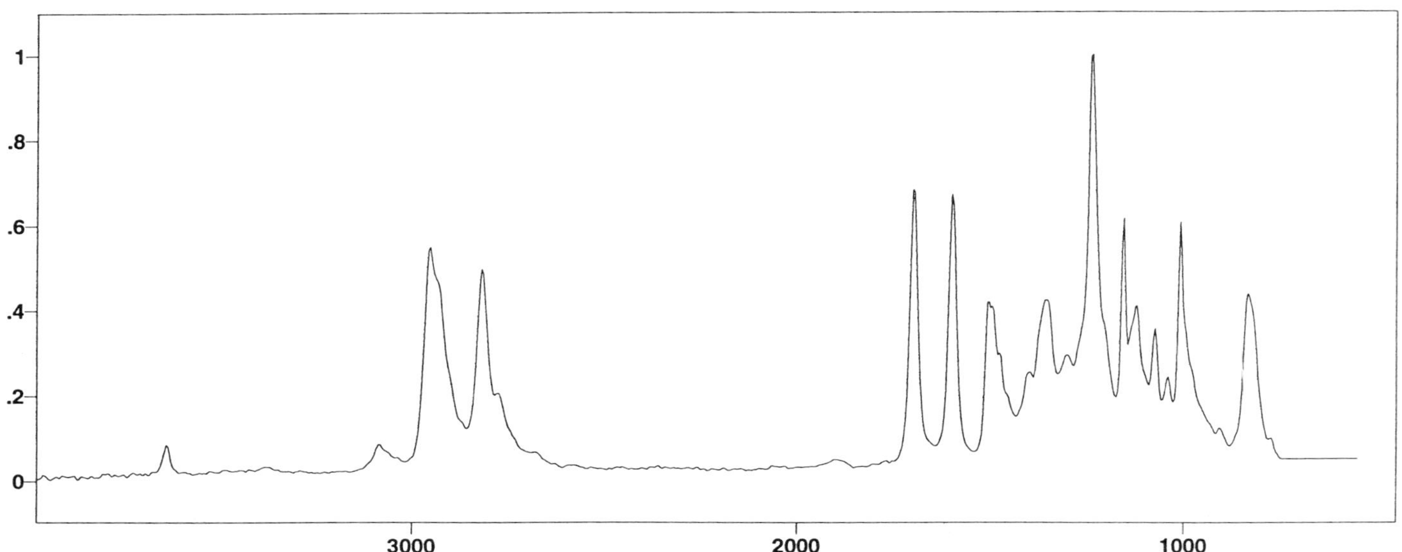

BROMPERIDOL

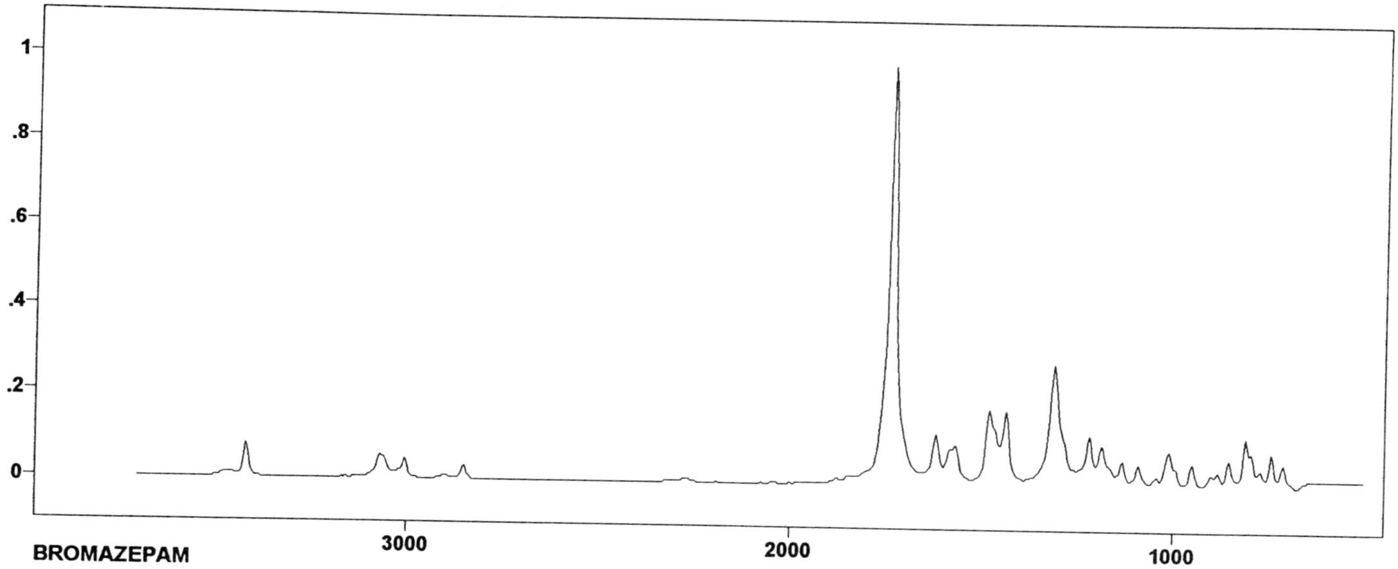

BROMAZEPAM

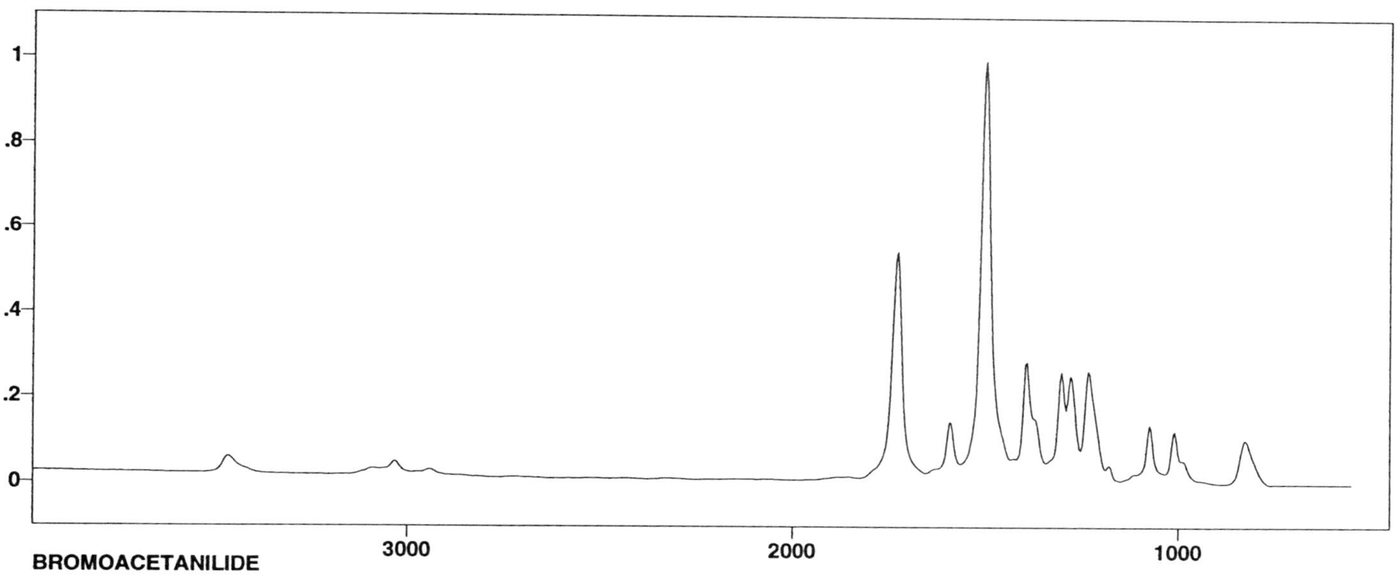

BROMOACETANILIDE

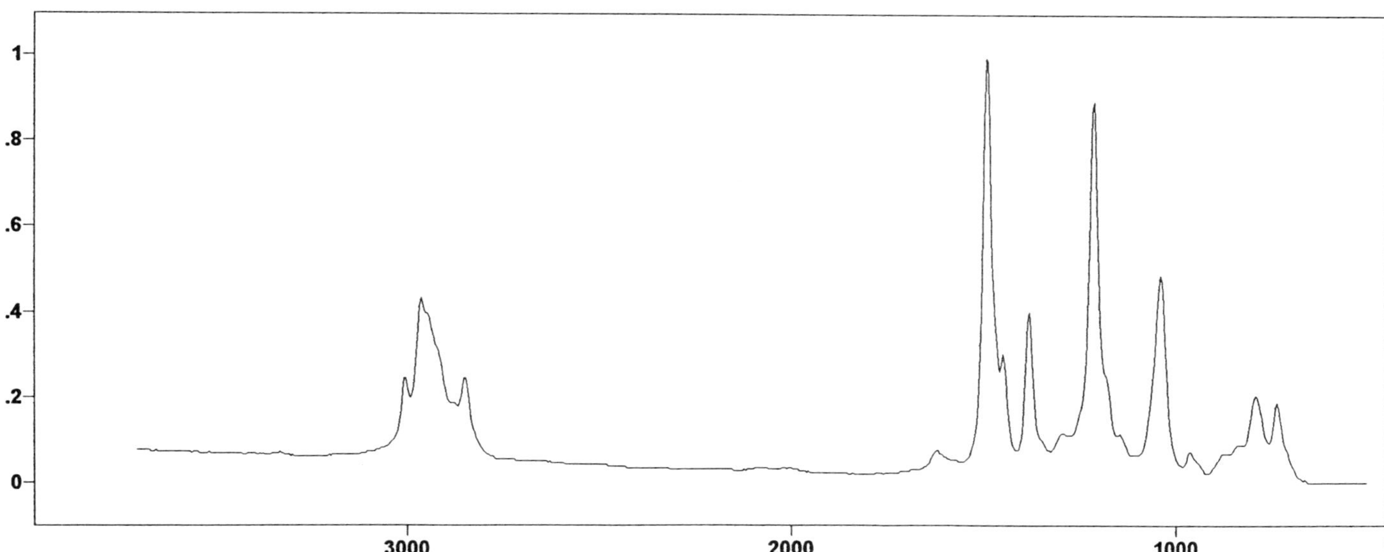

4-BROMO-2,5-DIMETHOXYAMPHETAMINE

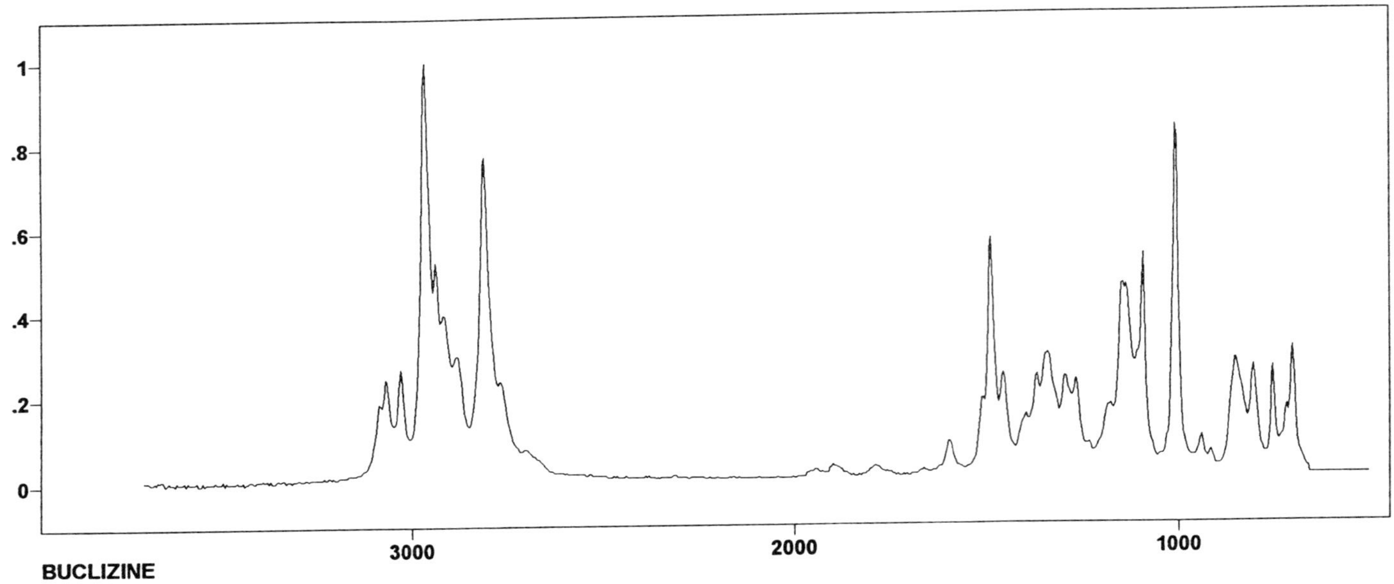

BUCLIZINE

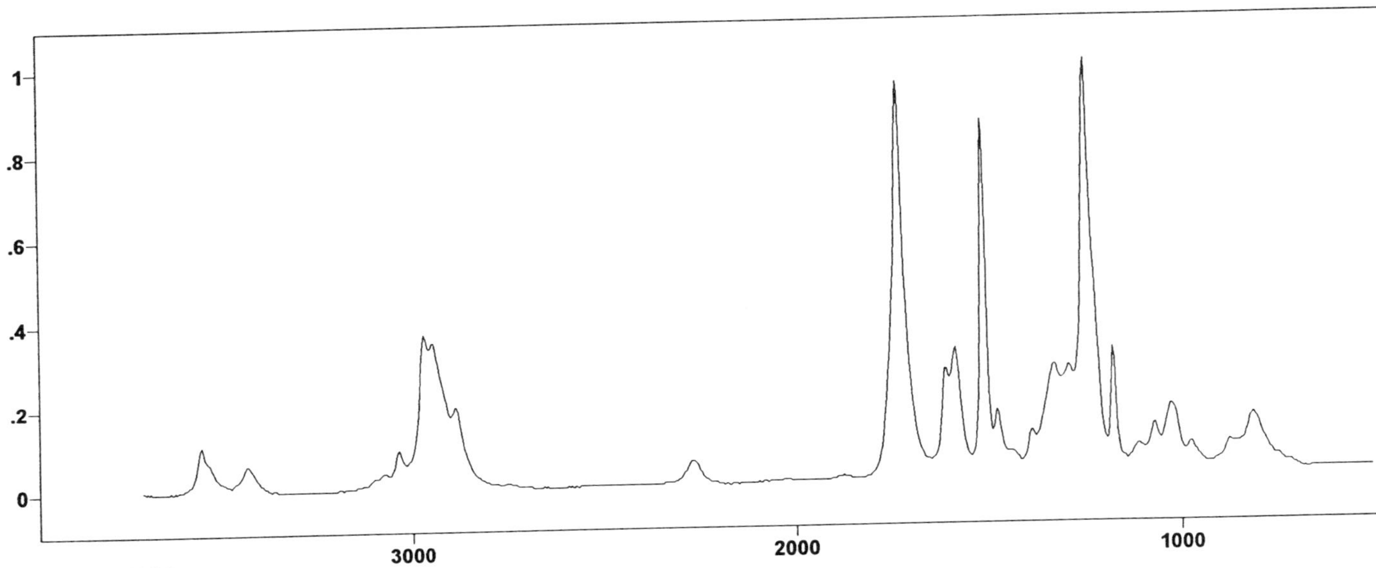

BUFEXAMAC

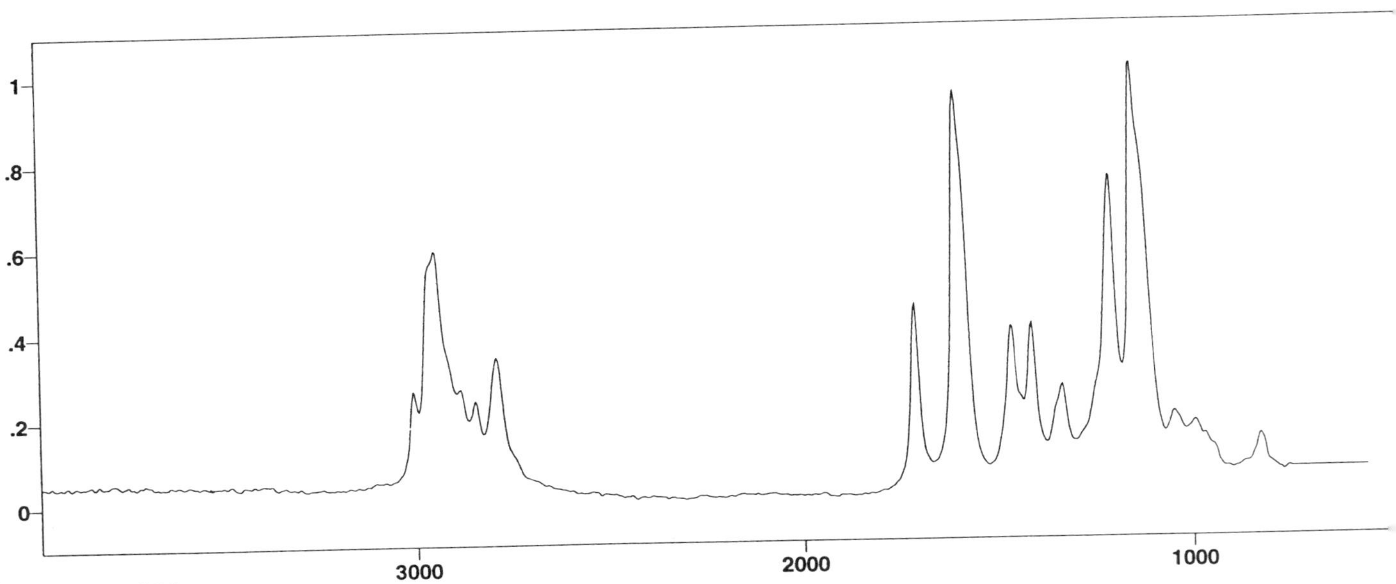

BUFLOMODIL

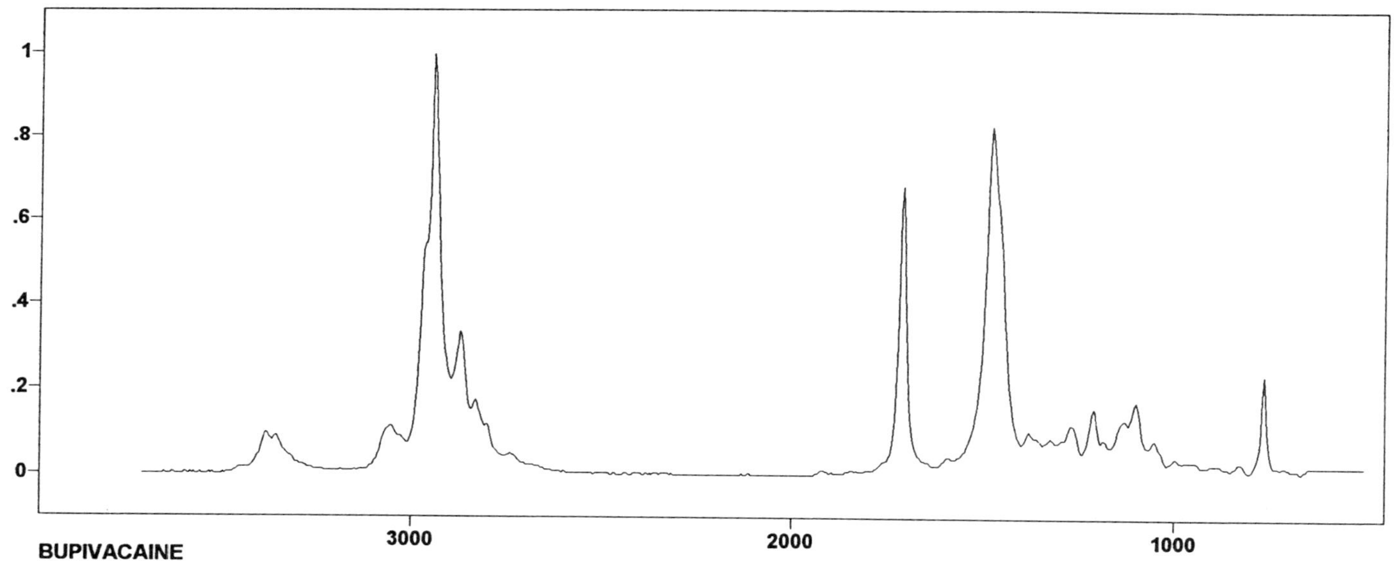

BUPIVACAINE

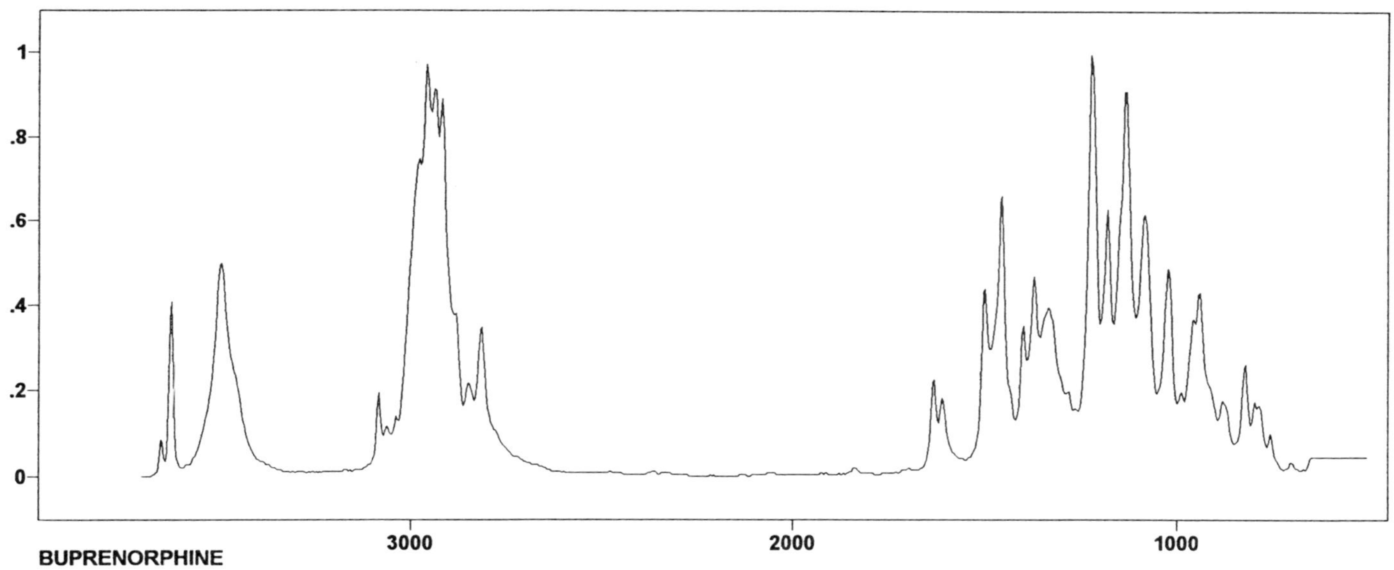

BUPRENORPHINE

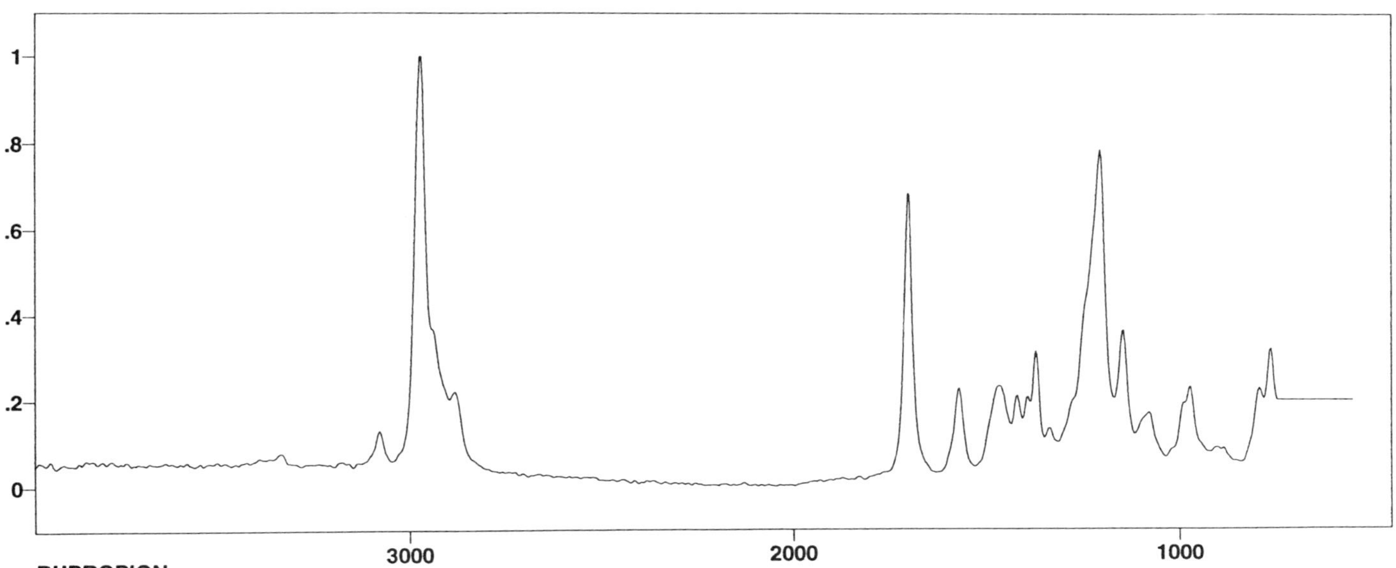

BUPROPION

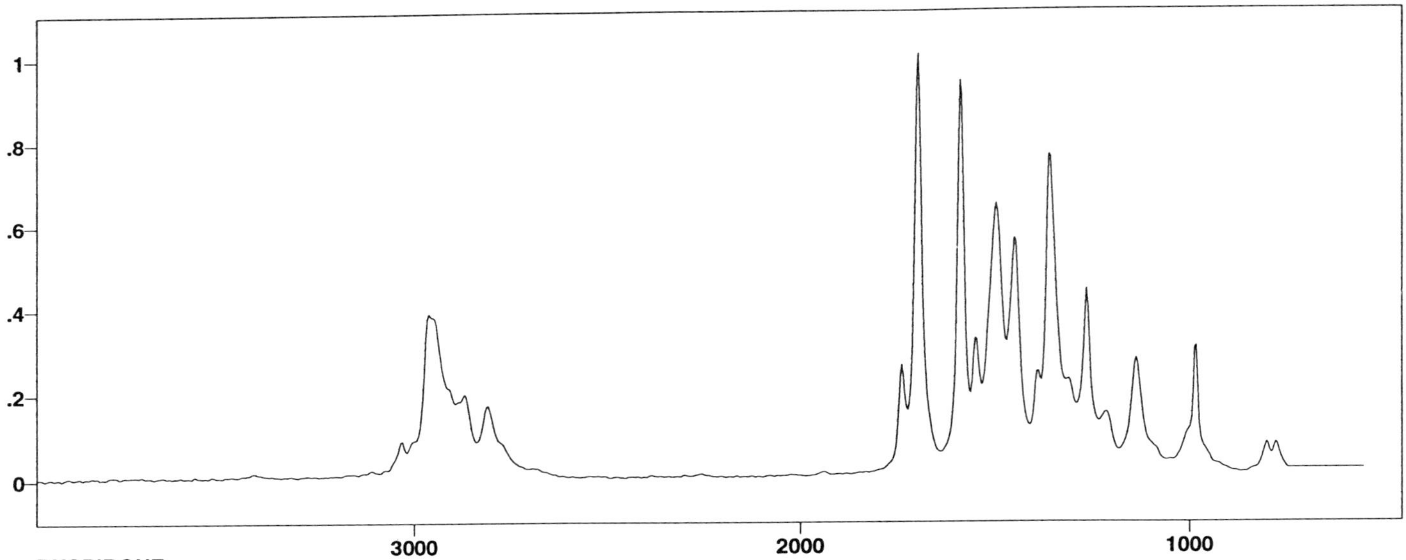

BUSPIRONE

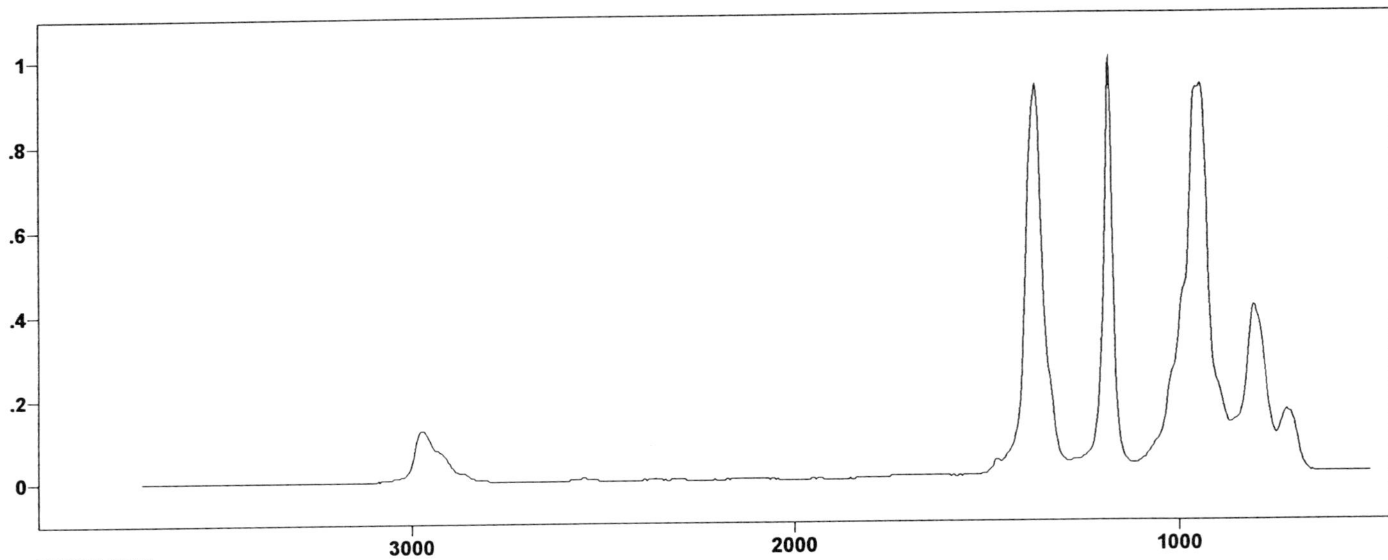

BUSULFAN

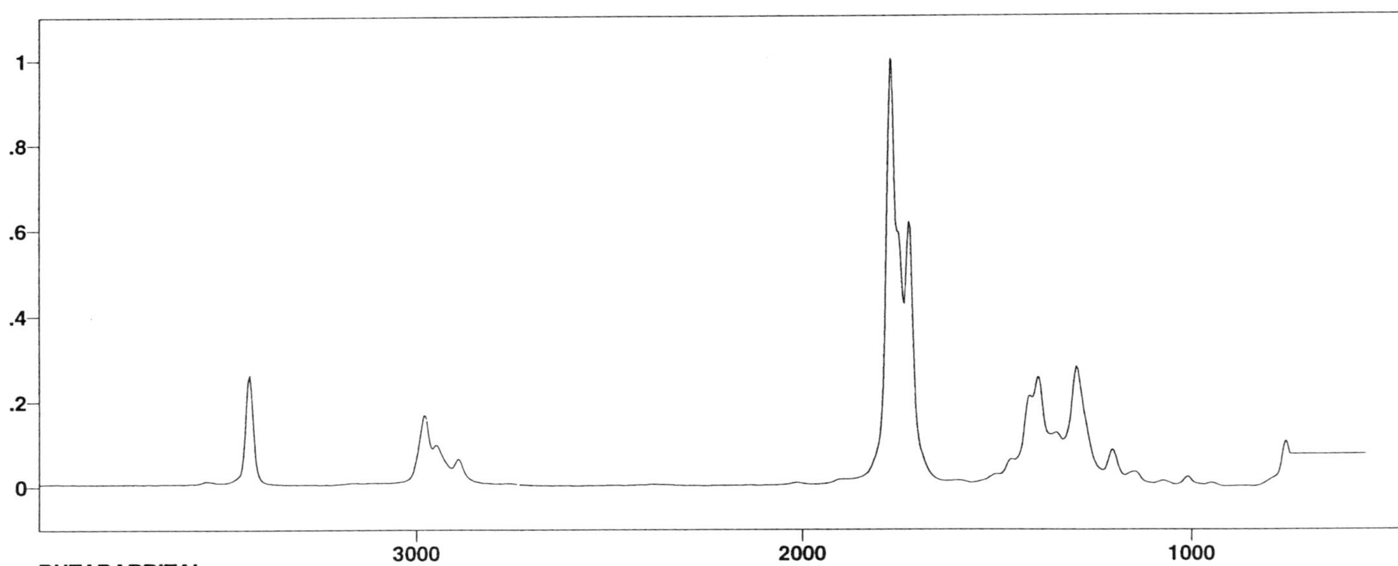

BUTABARBITAL

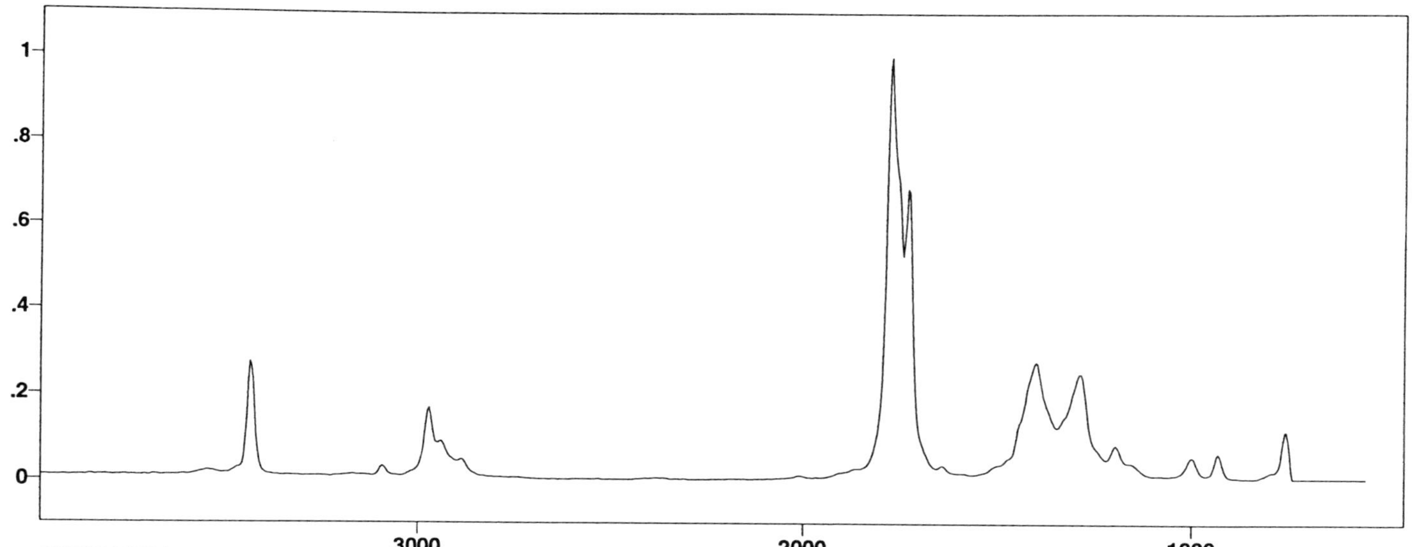

BUTALBITAL

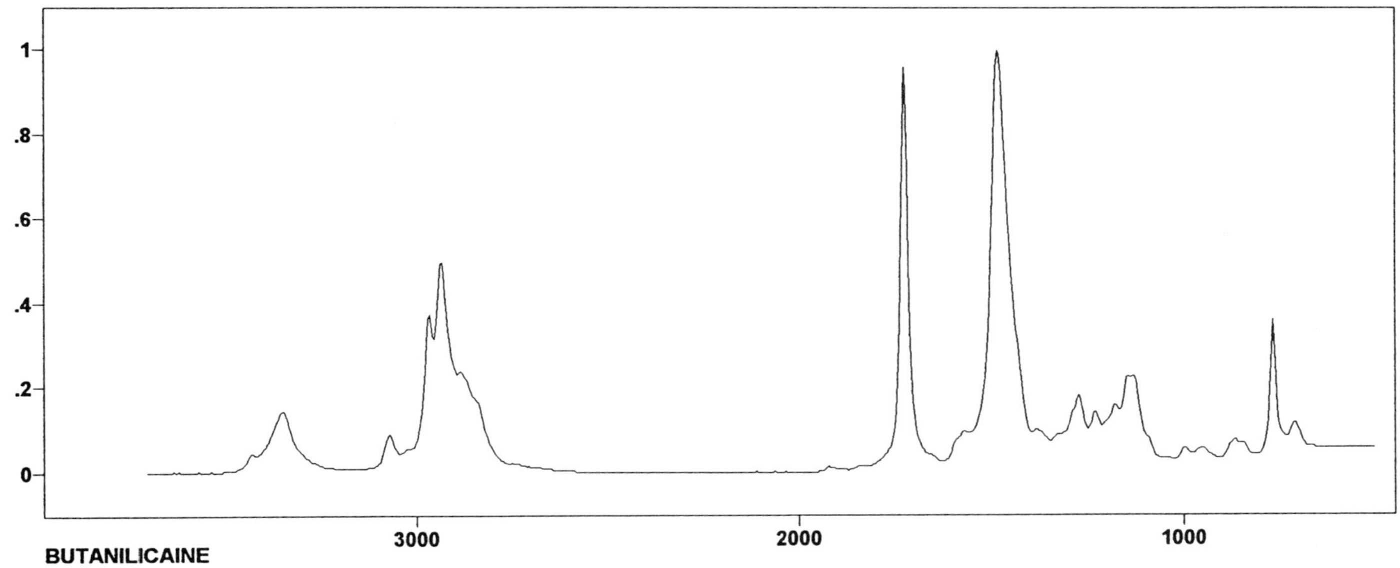

BUTANILICAINE

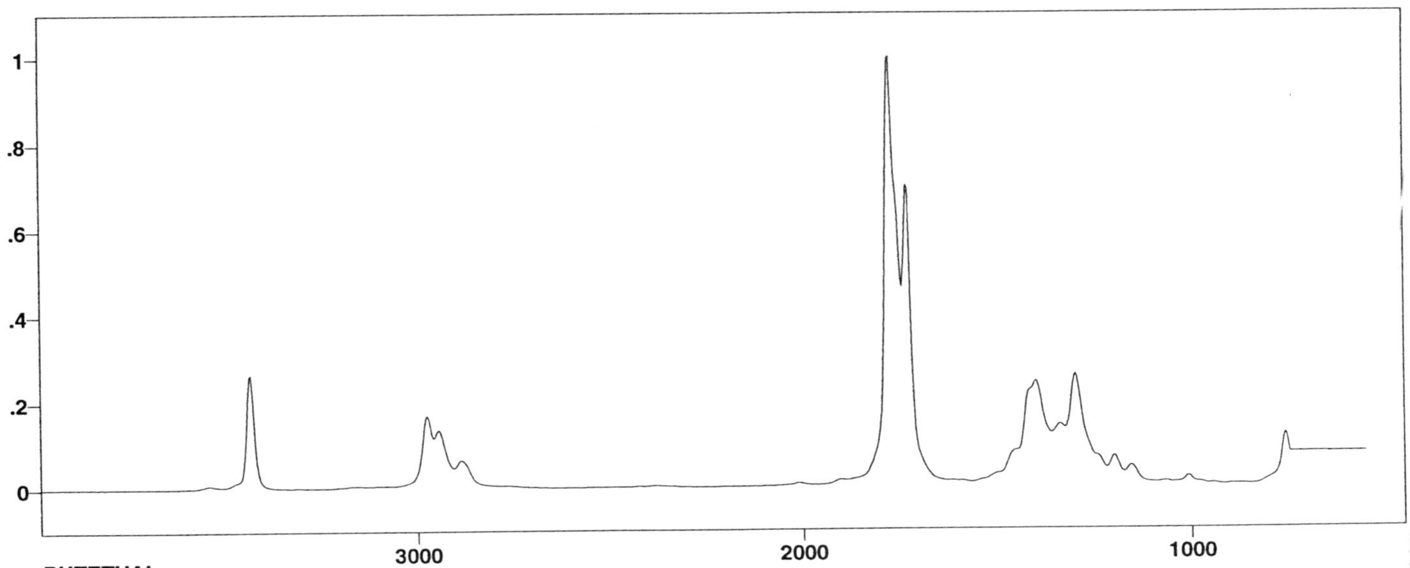

BUTETHAL

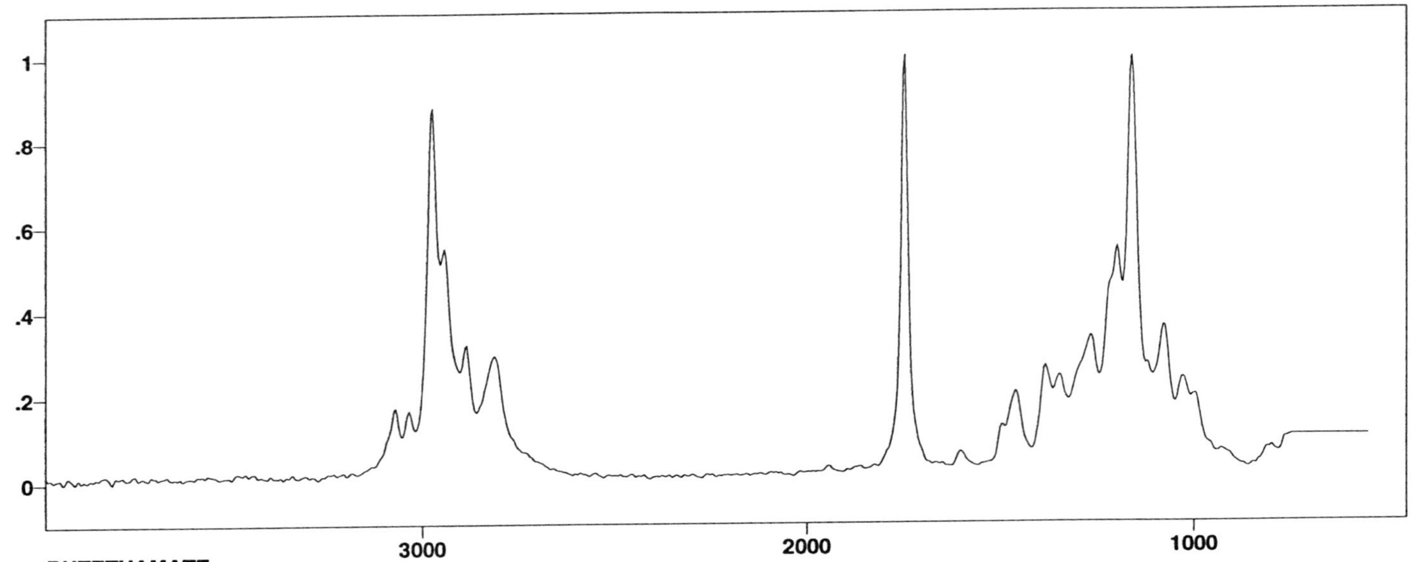

BUTETHAMATE

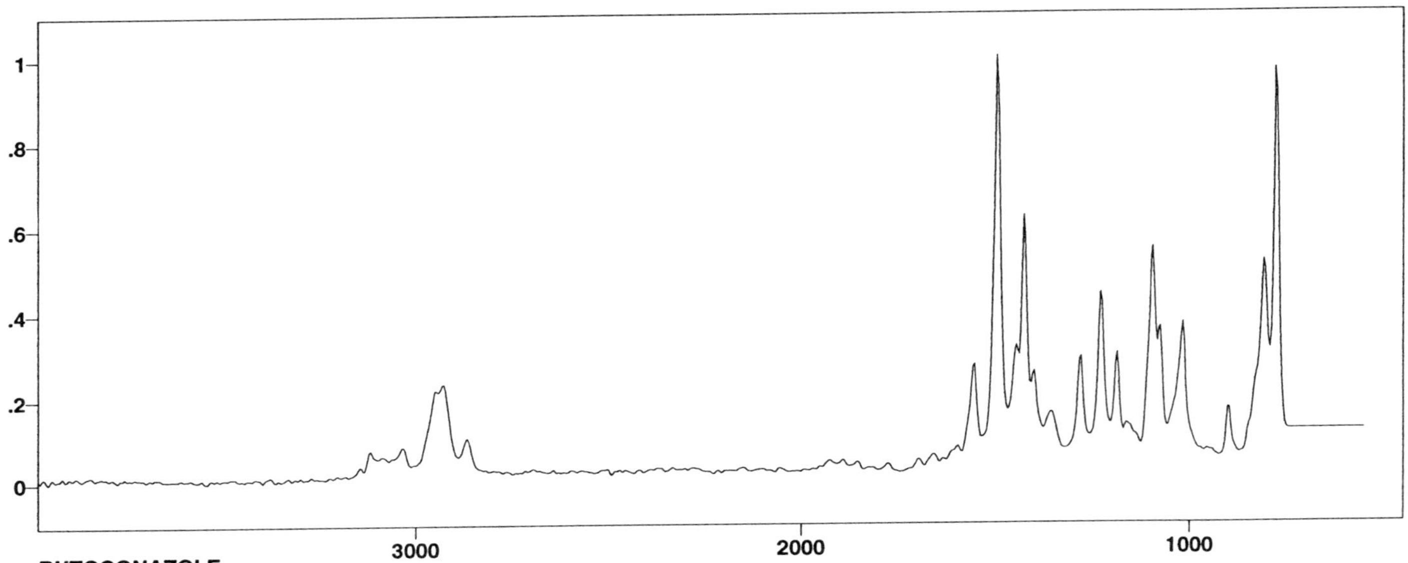

BUTOCONAZOLE

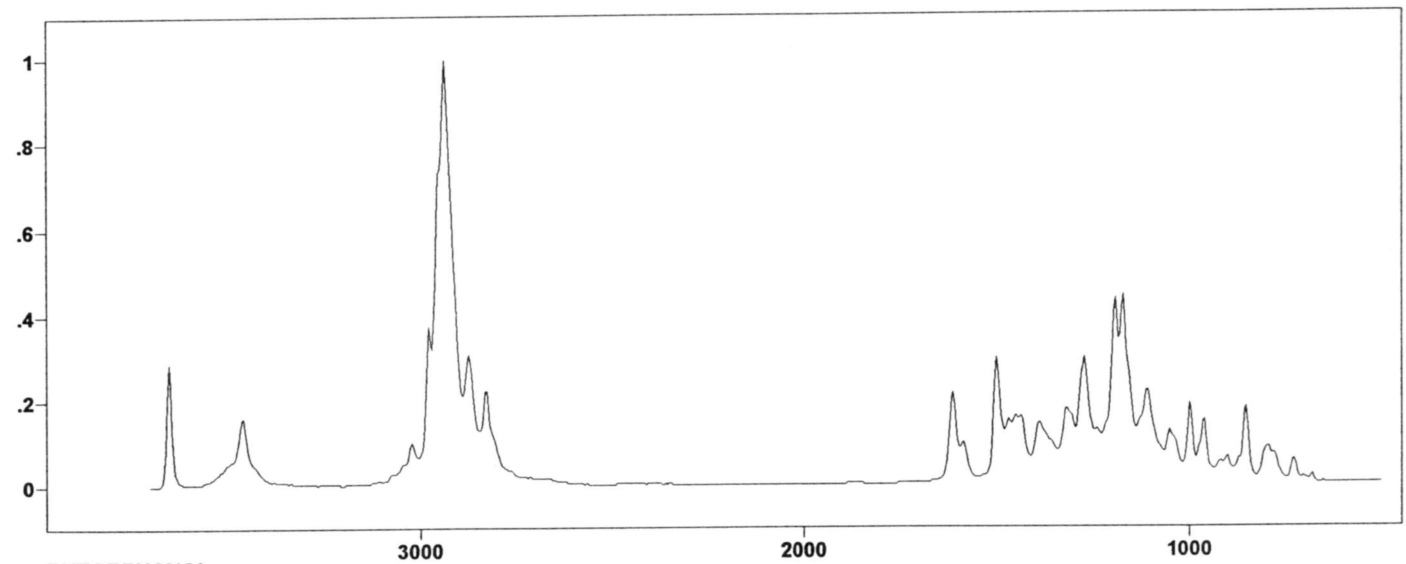

BUTORPHANOL

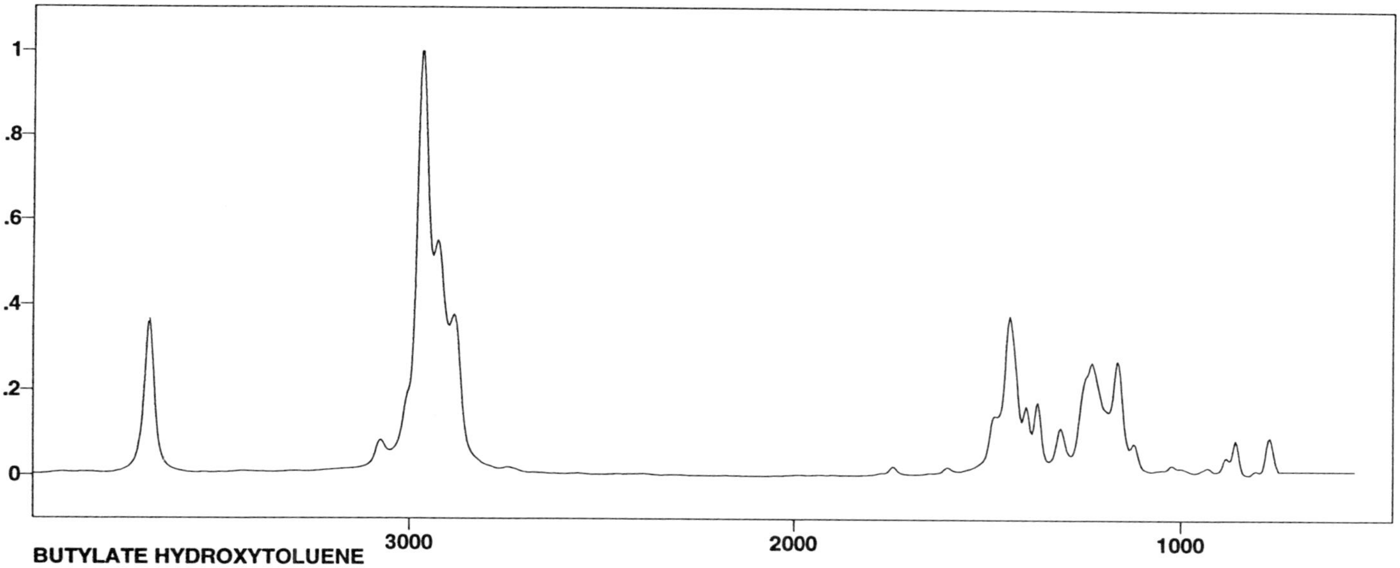

BUTYLATE HYDROXYTOLUENE

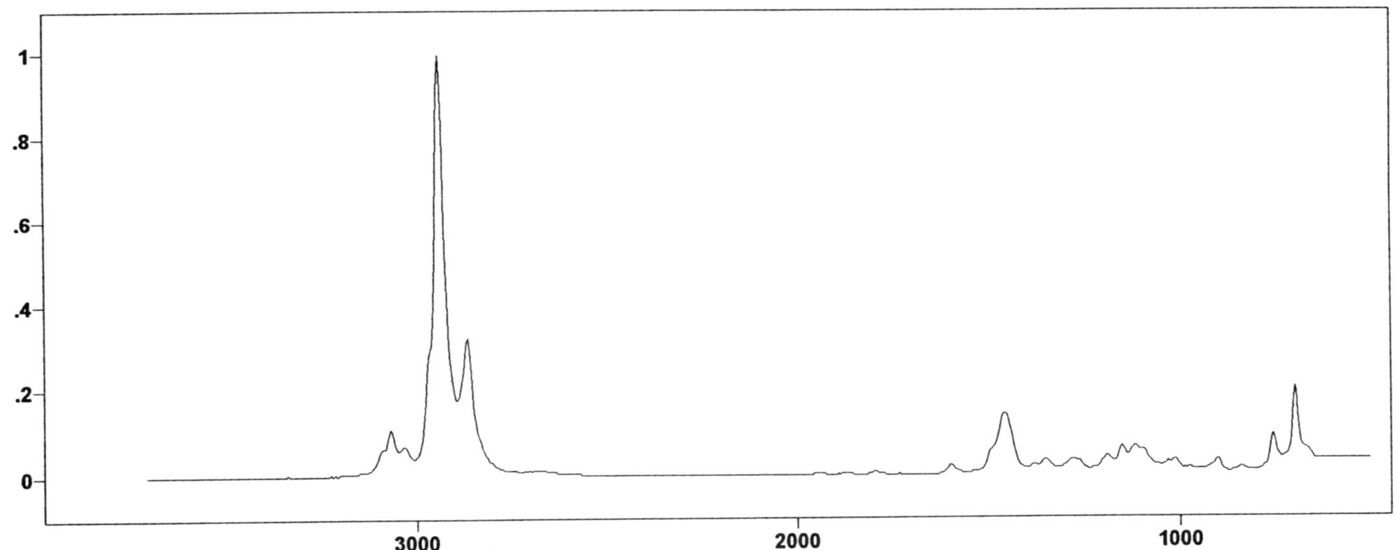

N-(n-BUTYL)-1-PHENYLCYCLOHEXYLAMINE

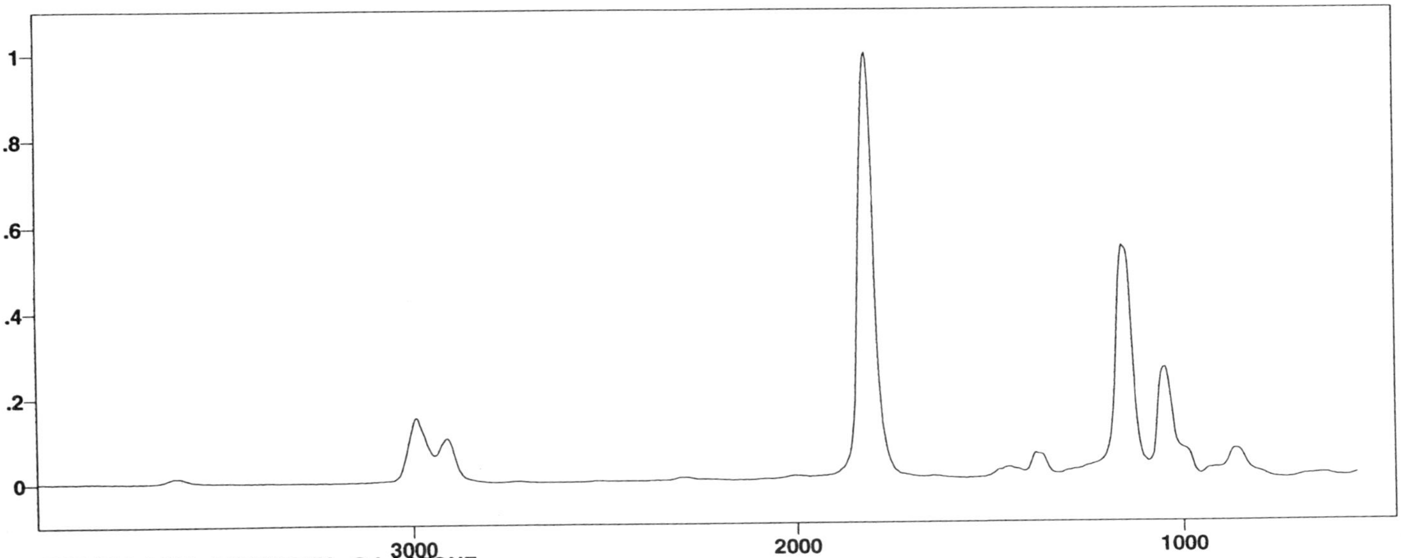

BUTYRIC ACID, 4-HYDROXY-, G-LACTONE

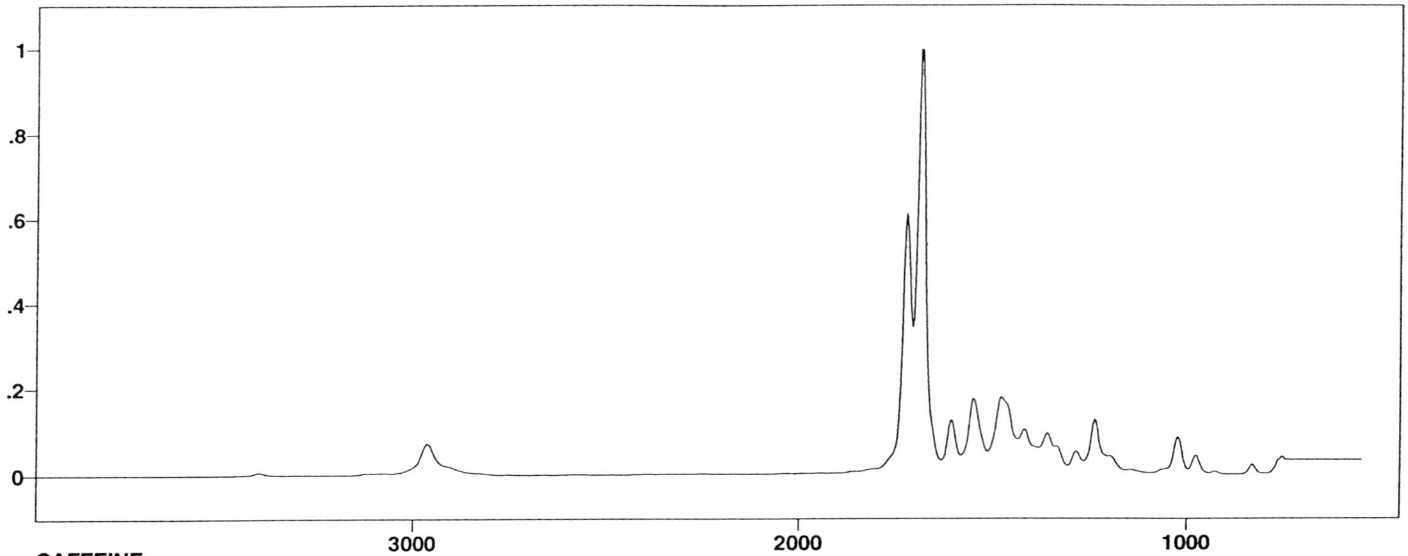

CAFFEINE

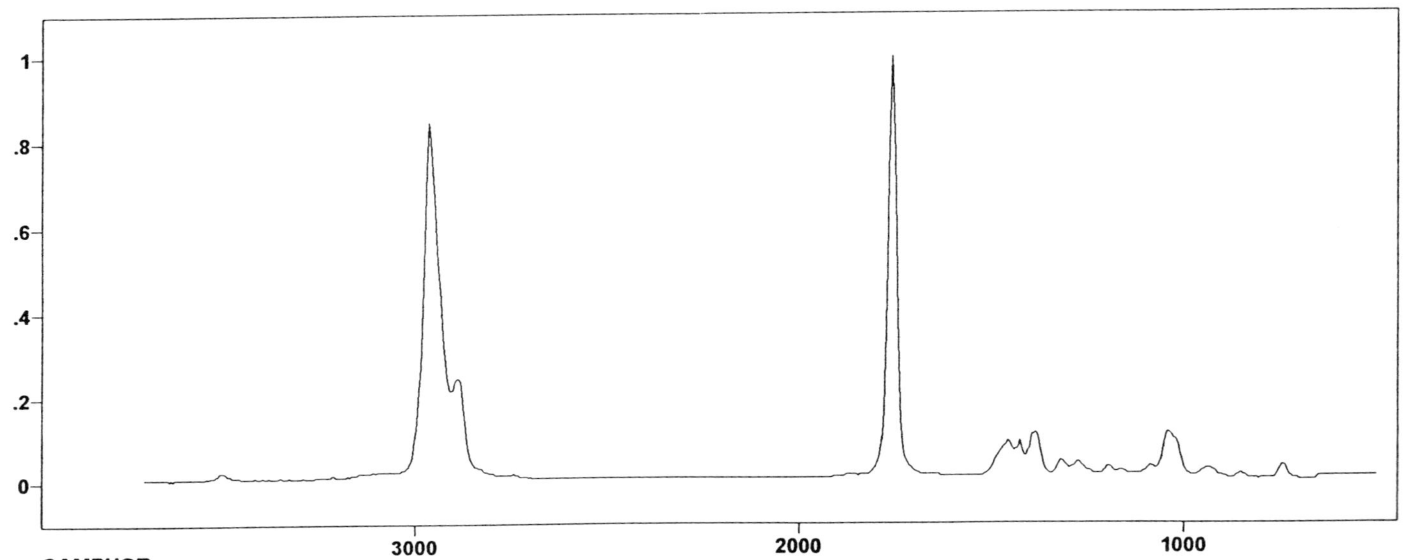

CAMPHOR

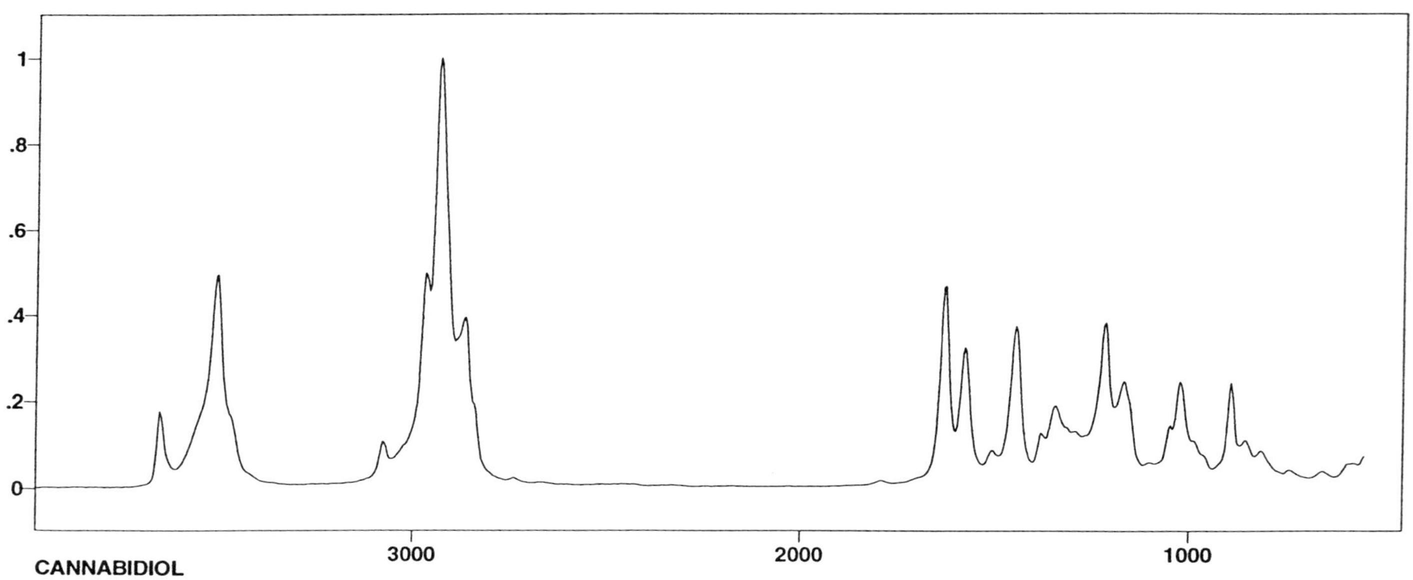

CANNABIDIOL

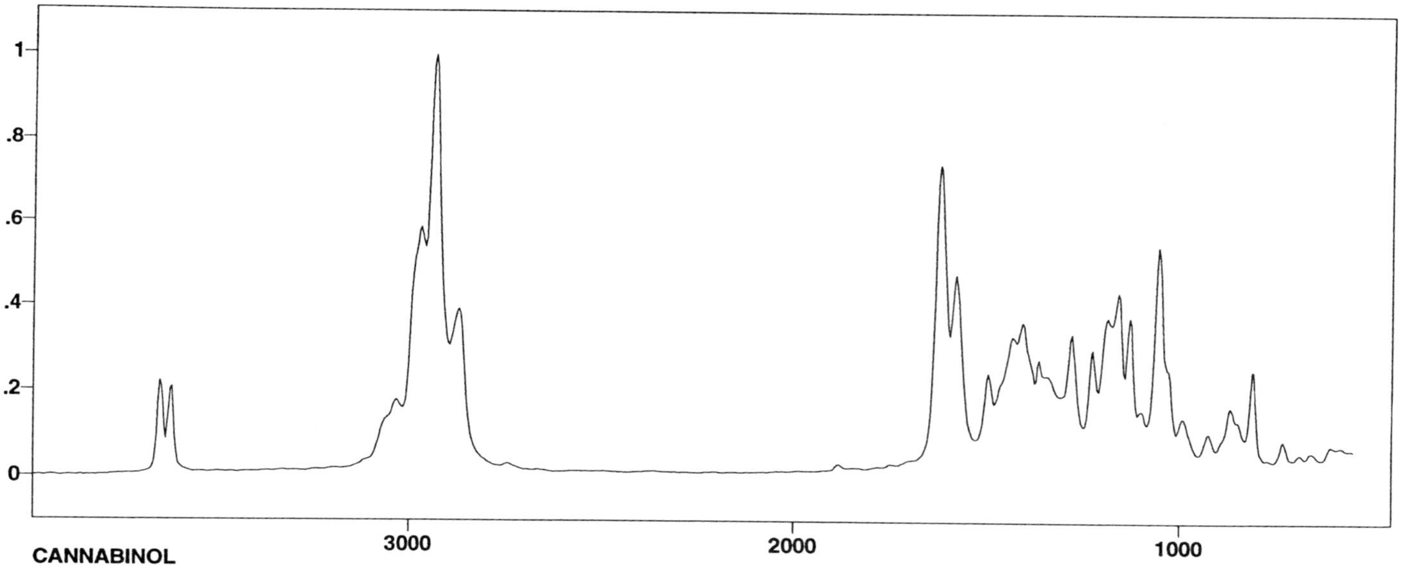

CANNABINOL

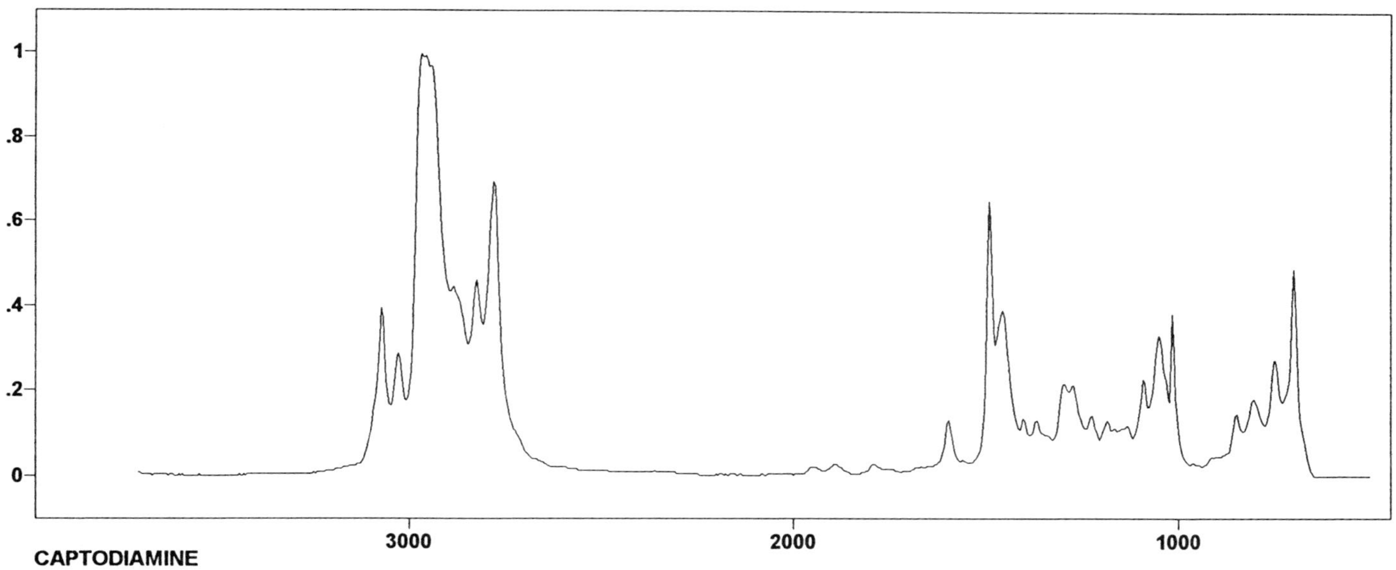

CAPTODIAMINE

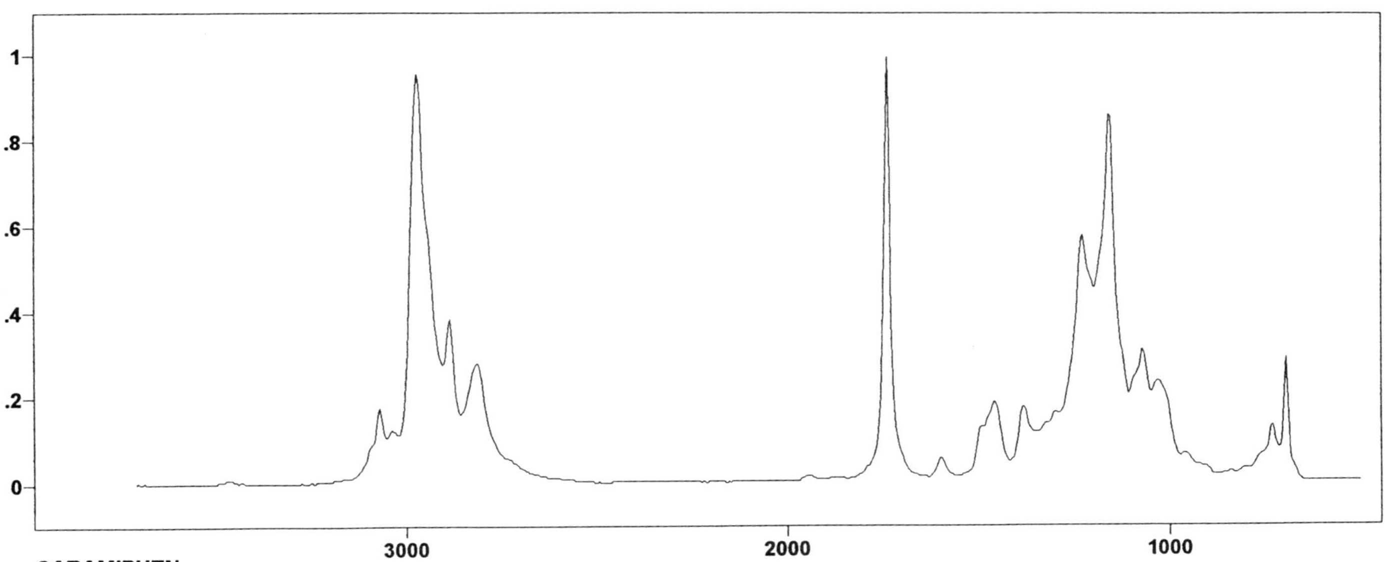

CARAMIPHEN

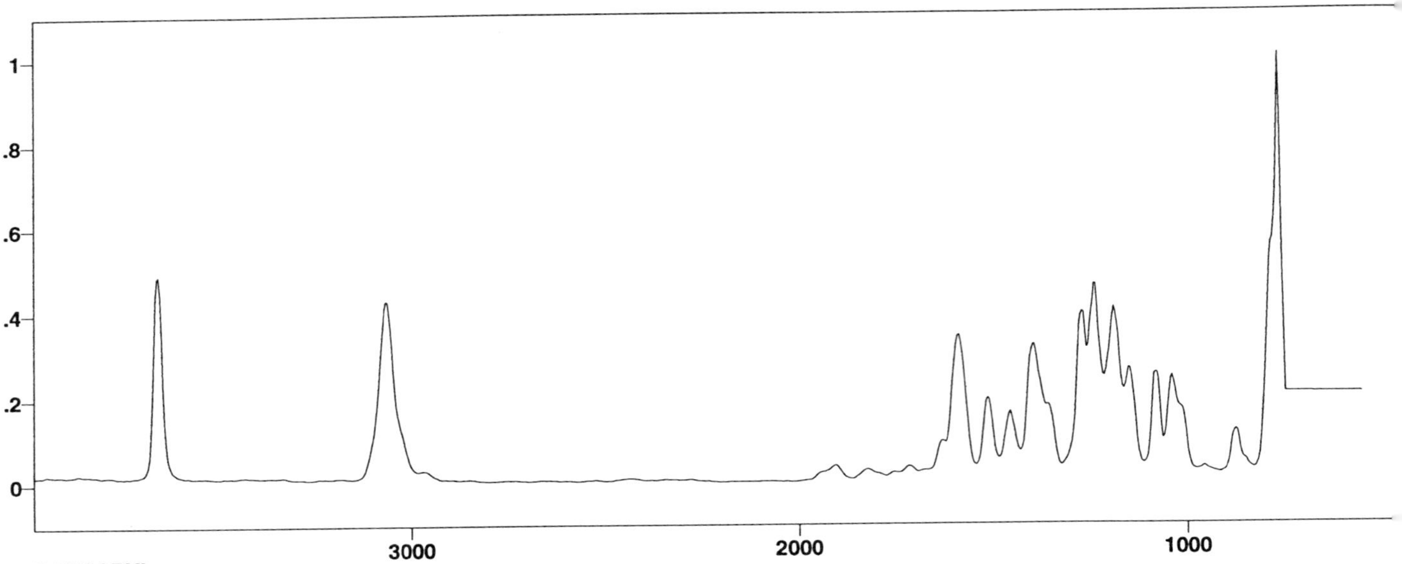

CARBARYL

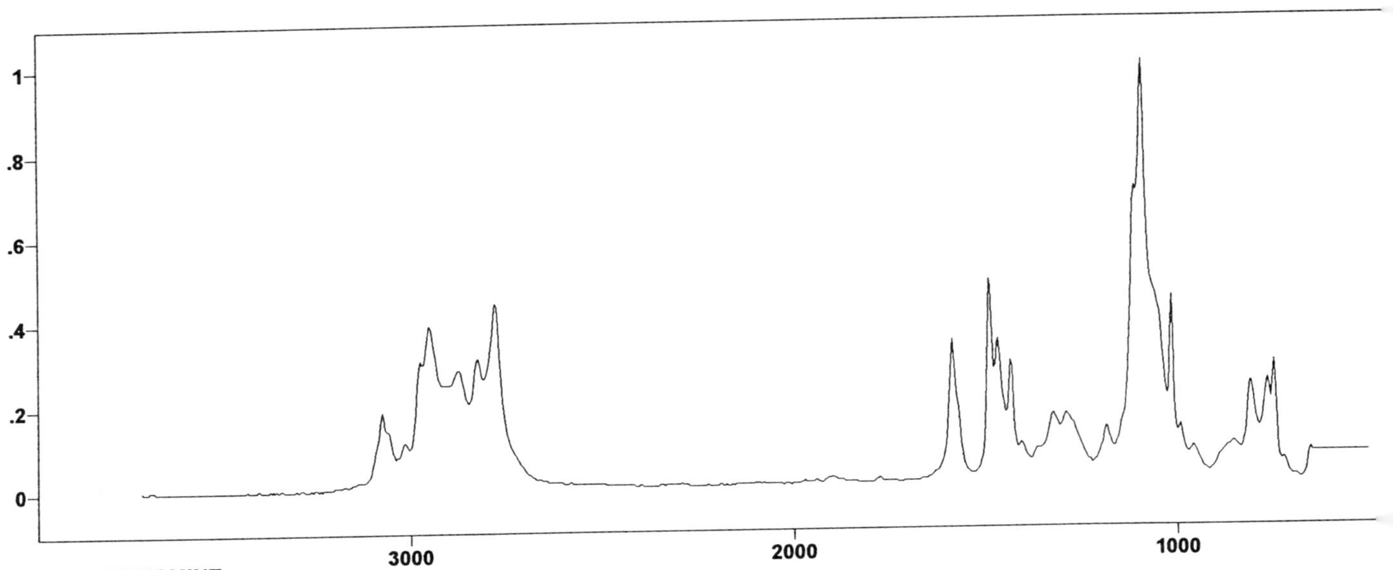

CARBINOXAMINE

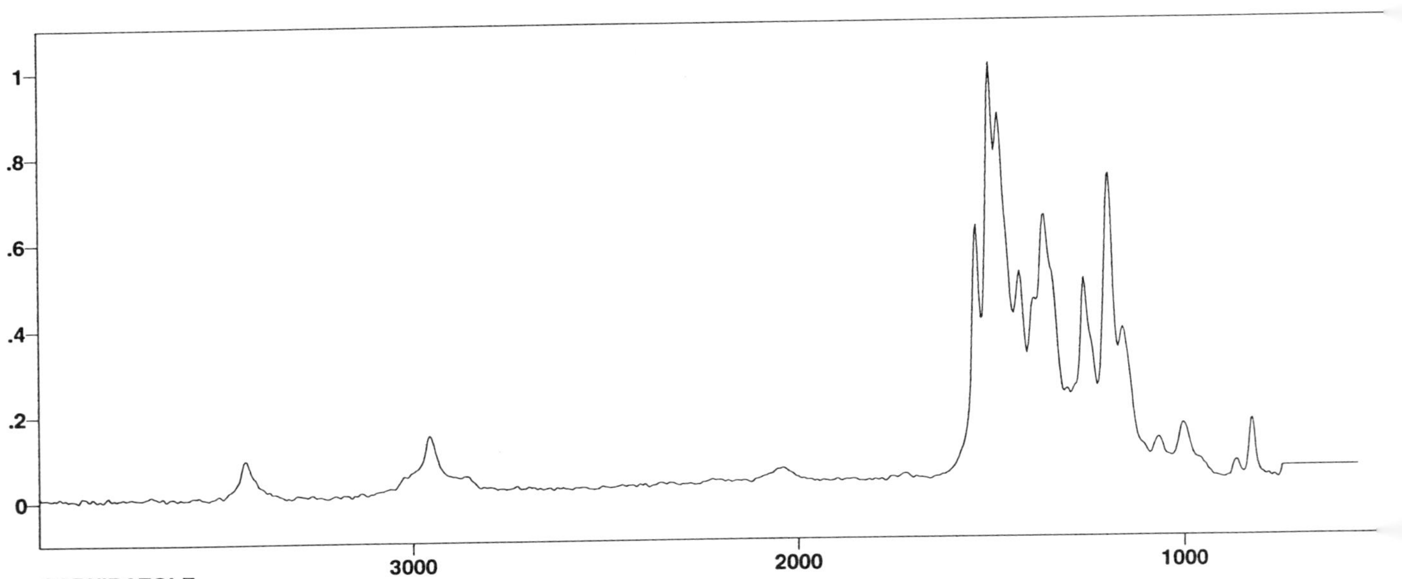

CARNIDAZOLE

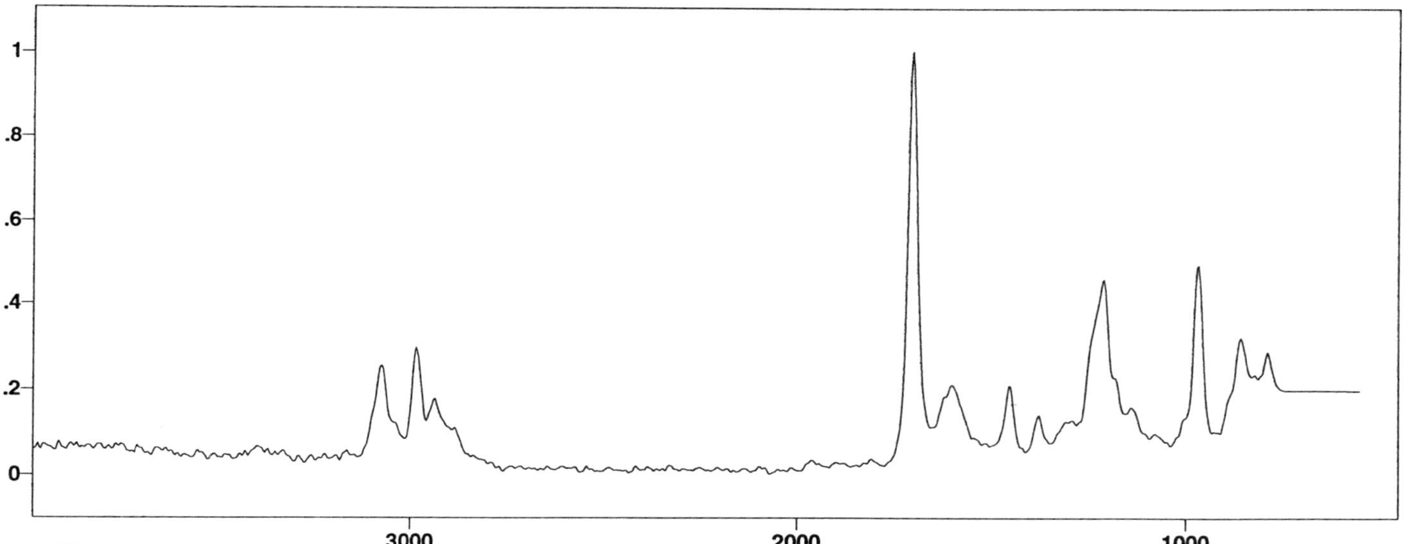

CATHINONE

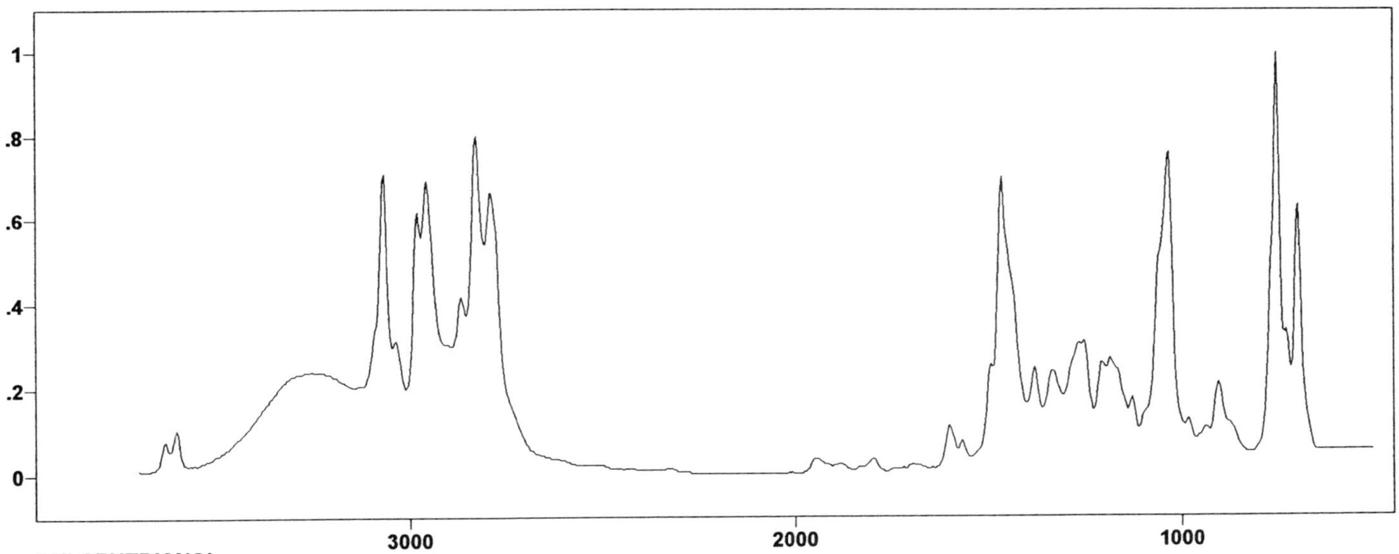

CHLOPHEDIANOL

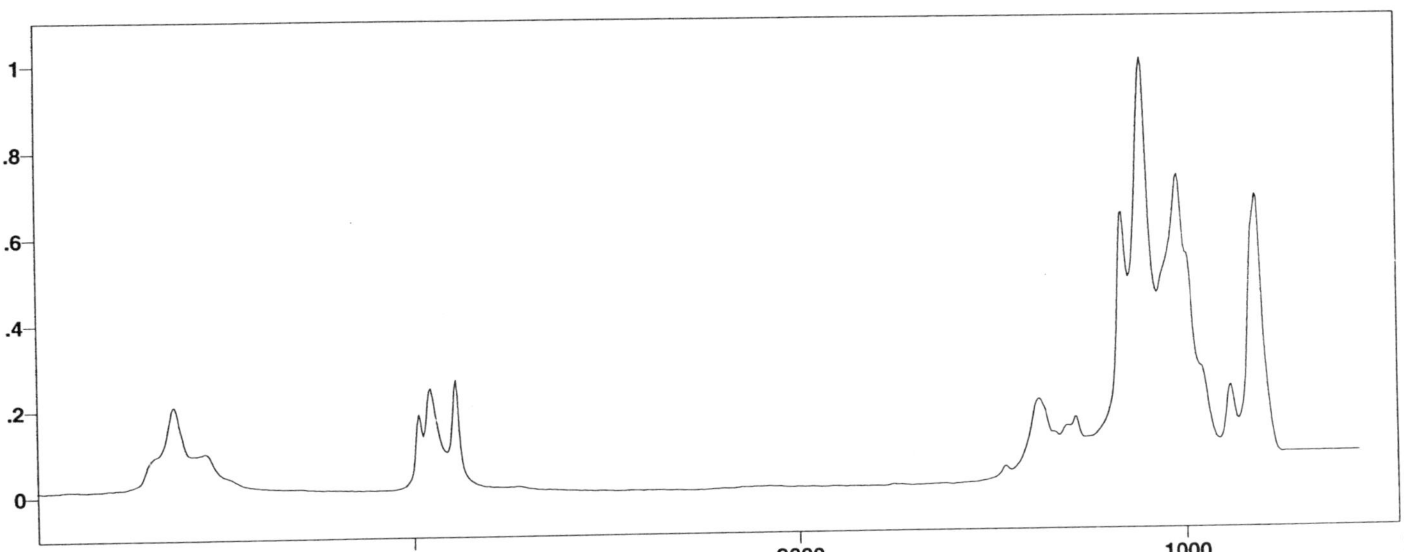

CHLORALOSE

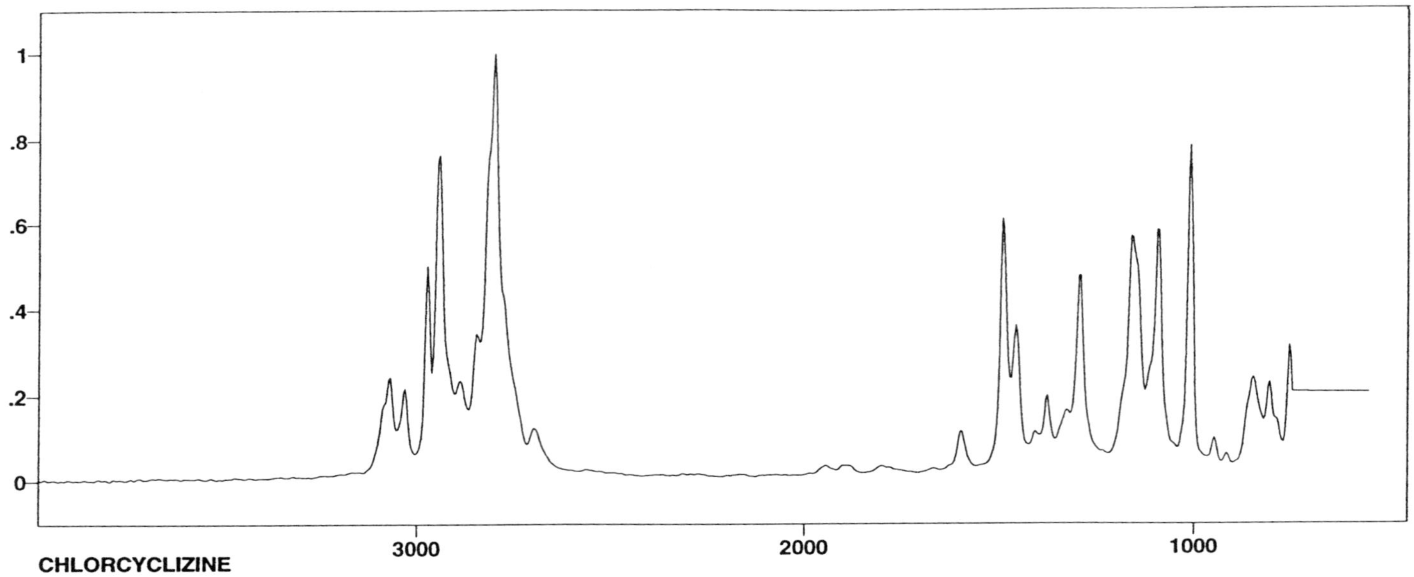

CHLORCYCLIZINE

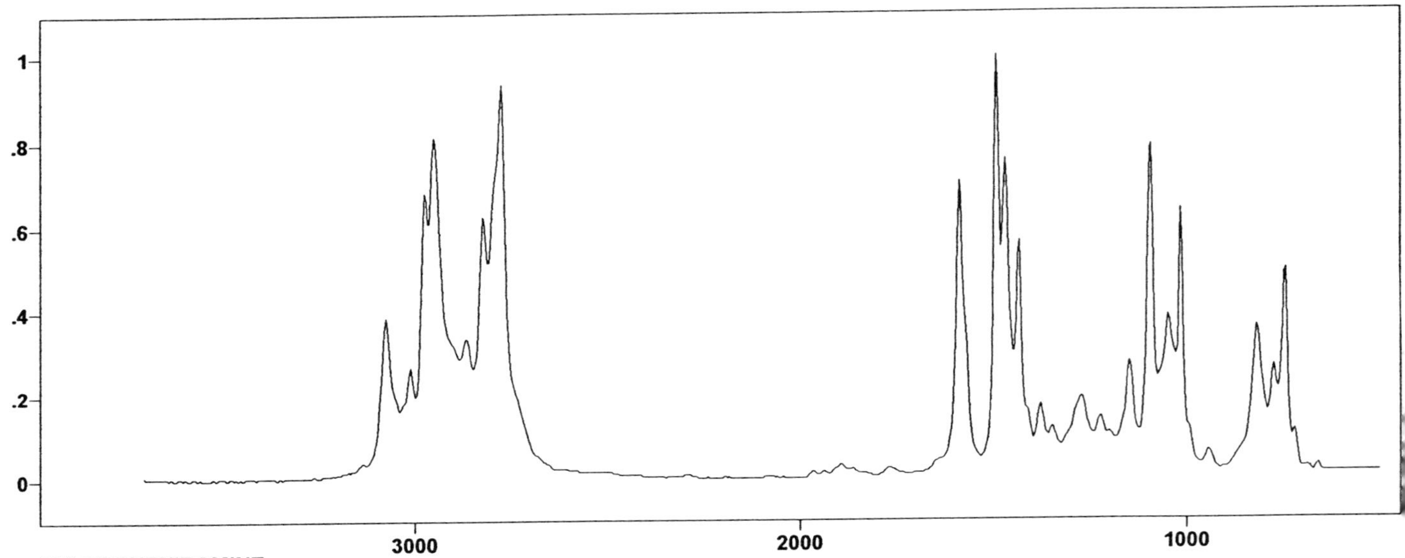

CHLORPHENIRAMINE

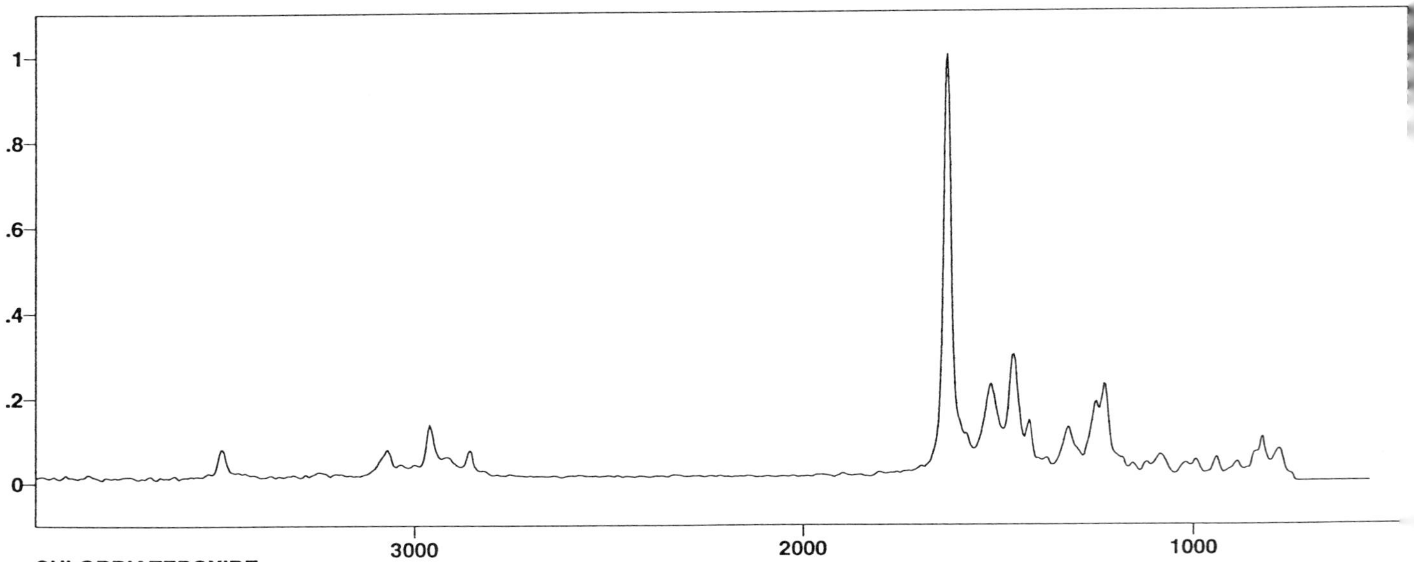

CHLORDIAZEPOXIDE

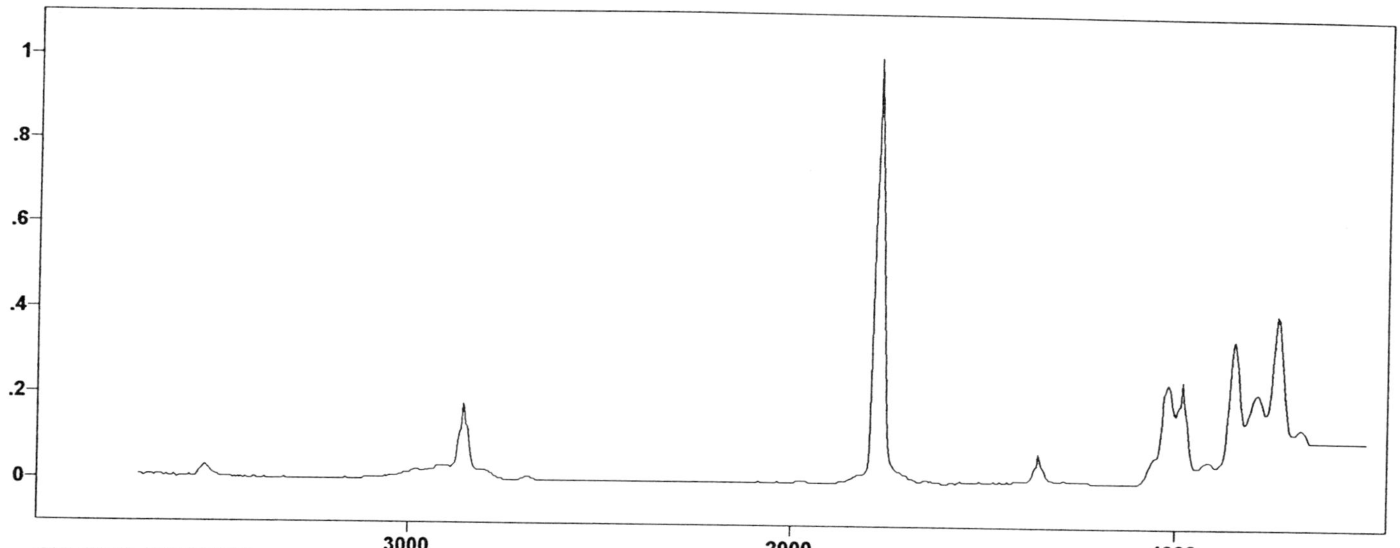

CHLORAL HYDRATE

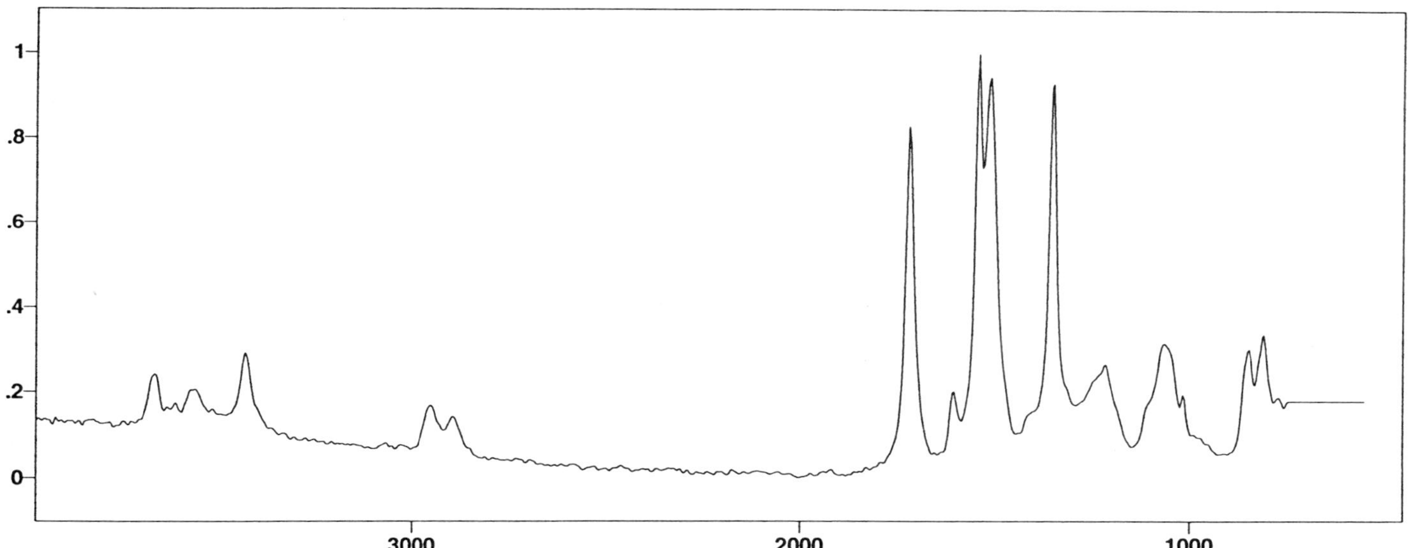

CHLORAMPHENICOL

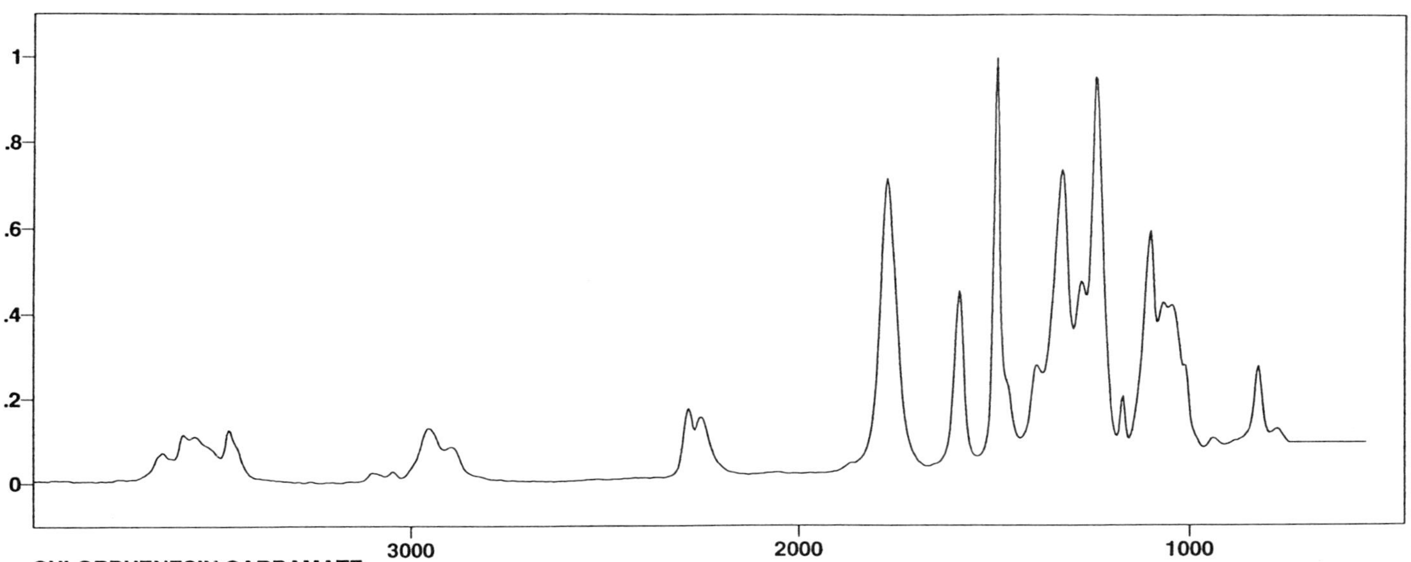

CHLORPHENESIN CARBAMATE

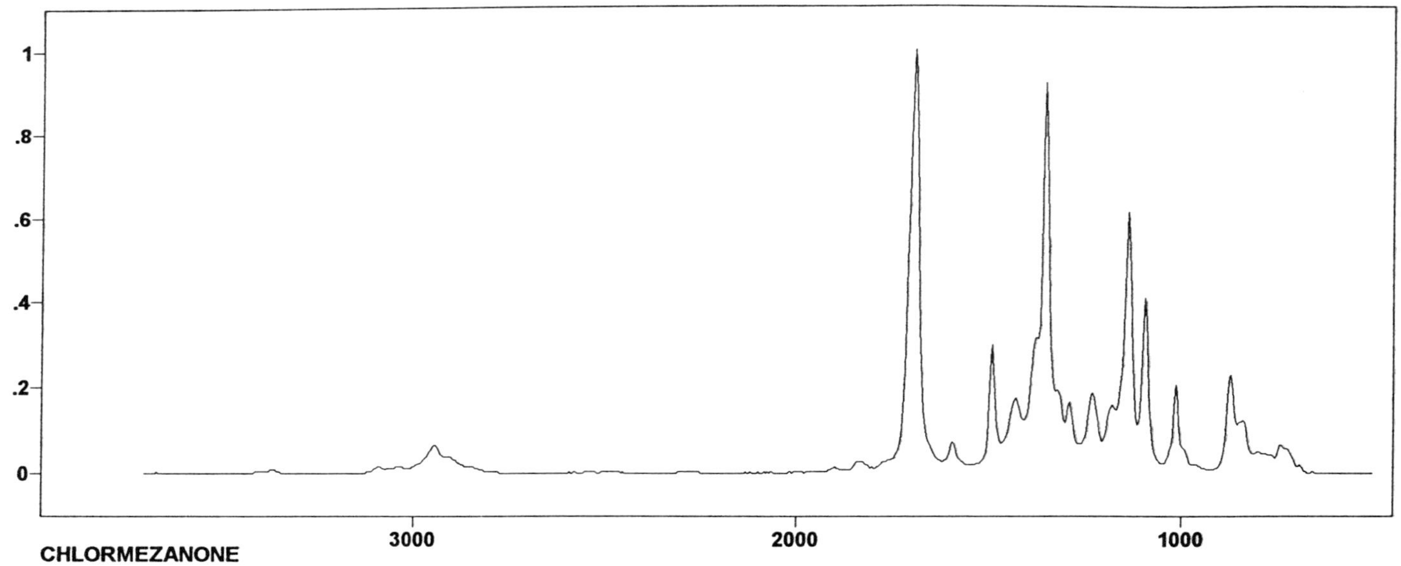

CHLORMEZANONE

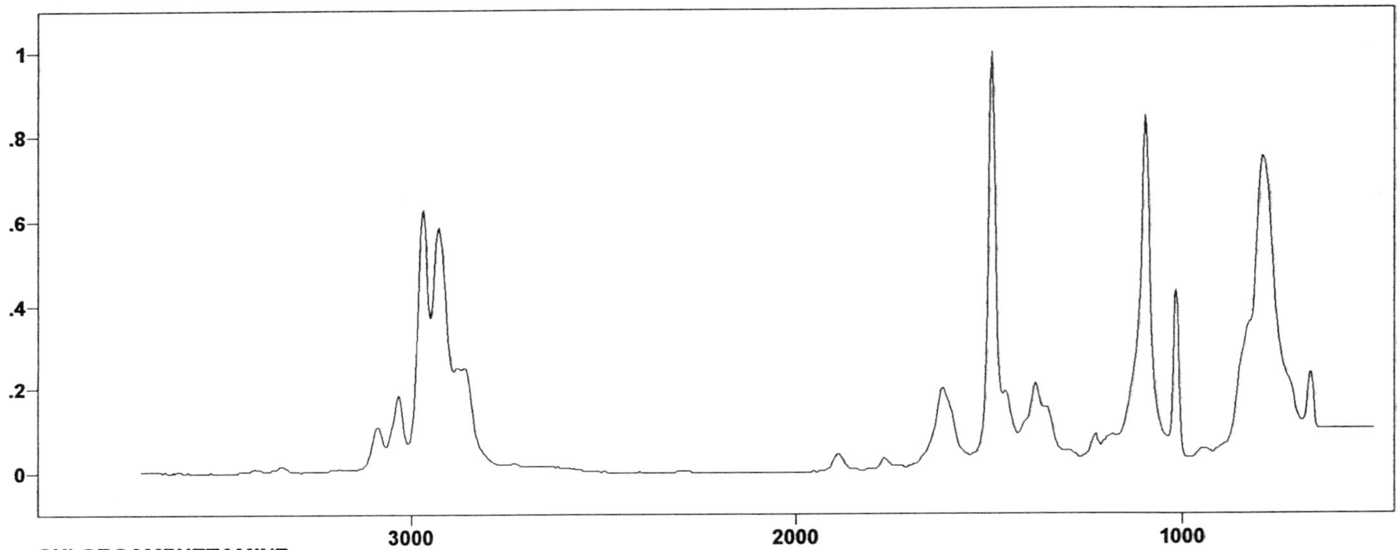

CHLOROAMPHETAMINE

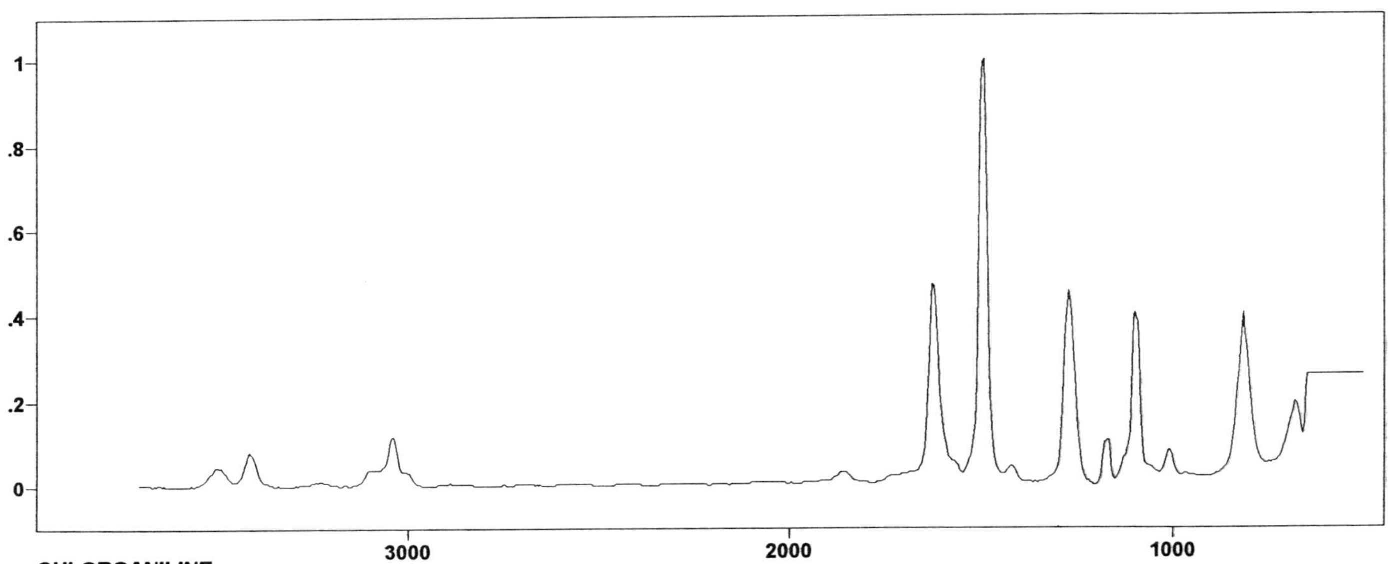

CHLOROANILINE

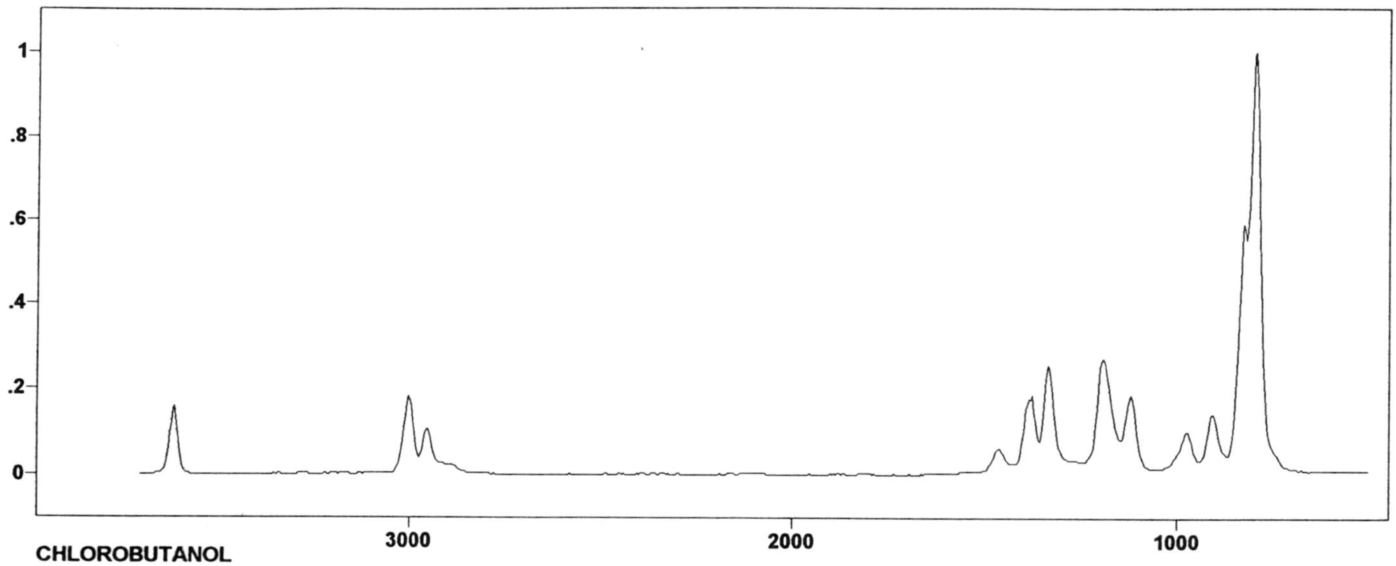

CHLOROBUTANOL

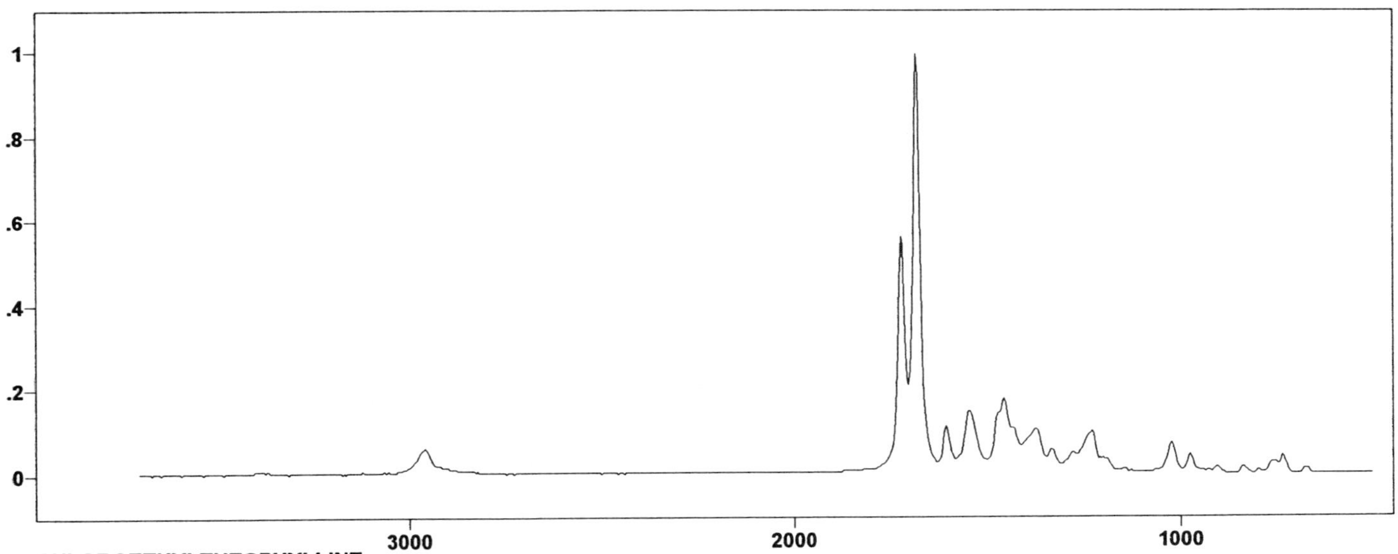

CHLOROETHYLTHEOPHYLLINE

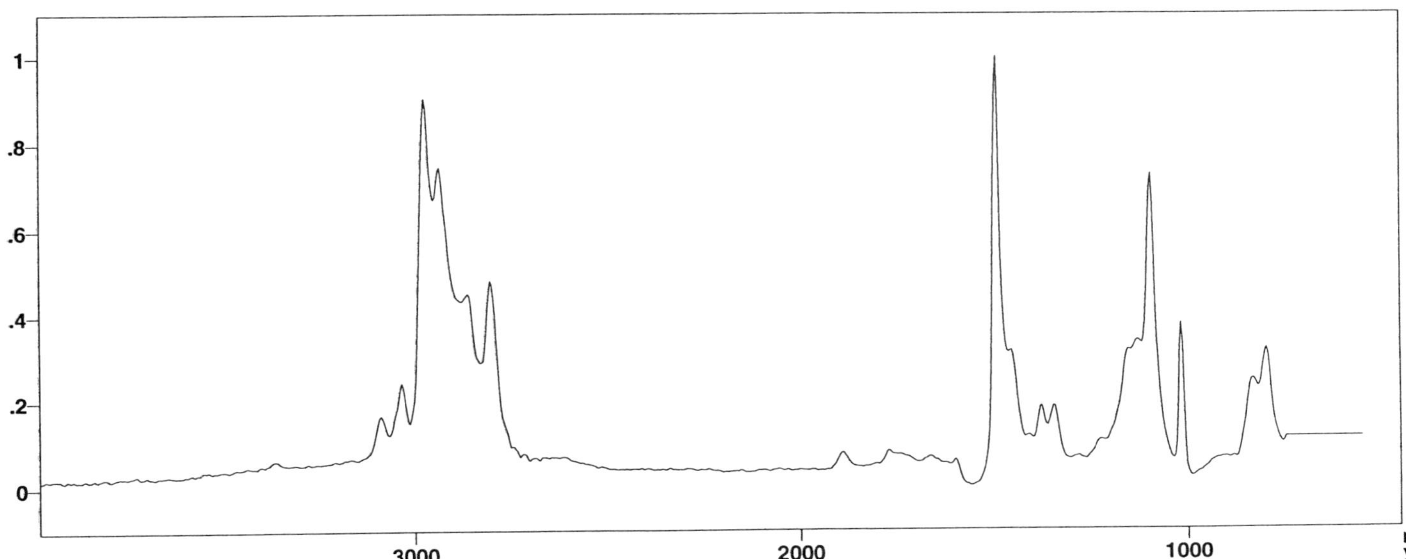

CHLOROMETHAMPHETAMINE

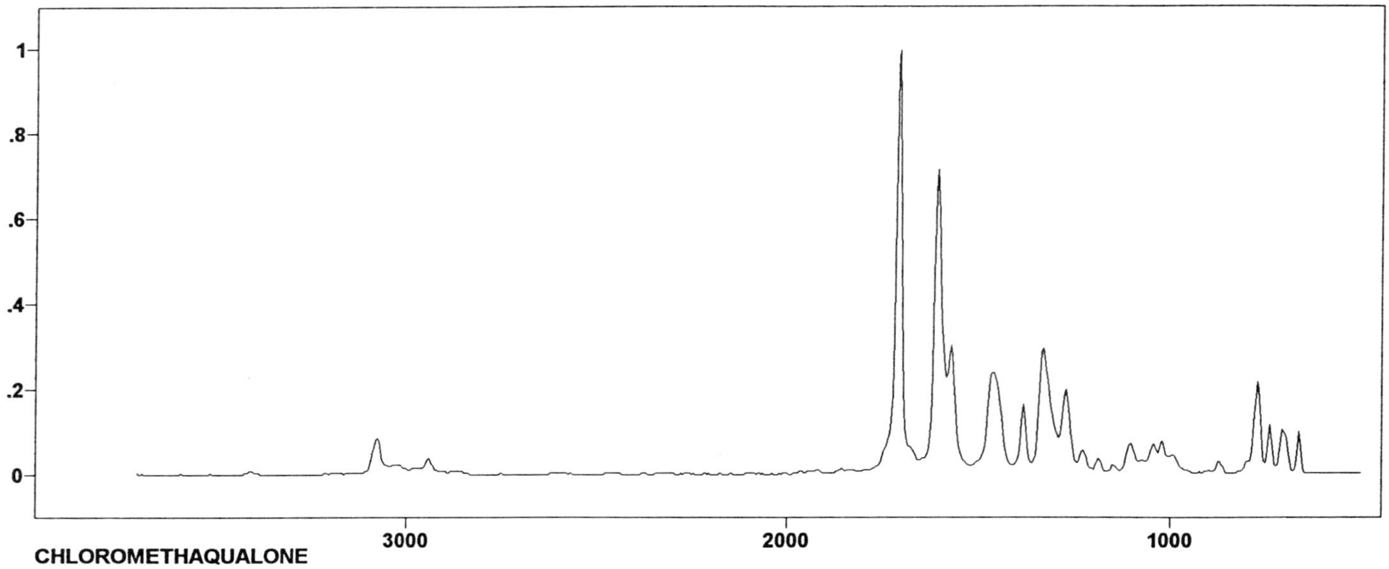

CHLOROMETHAQUALONE

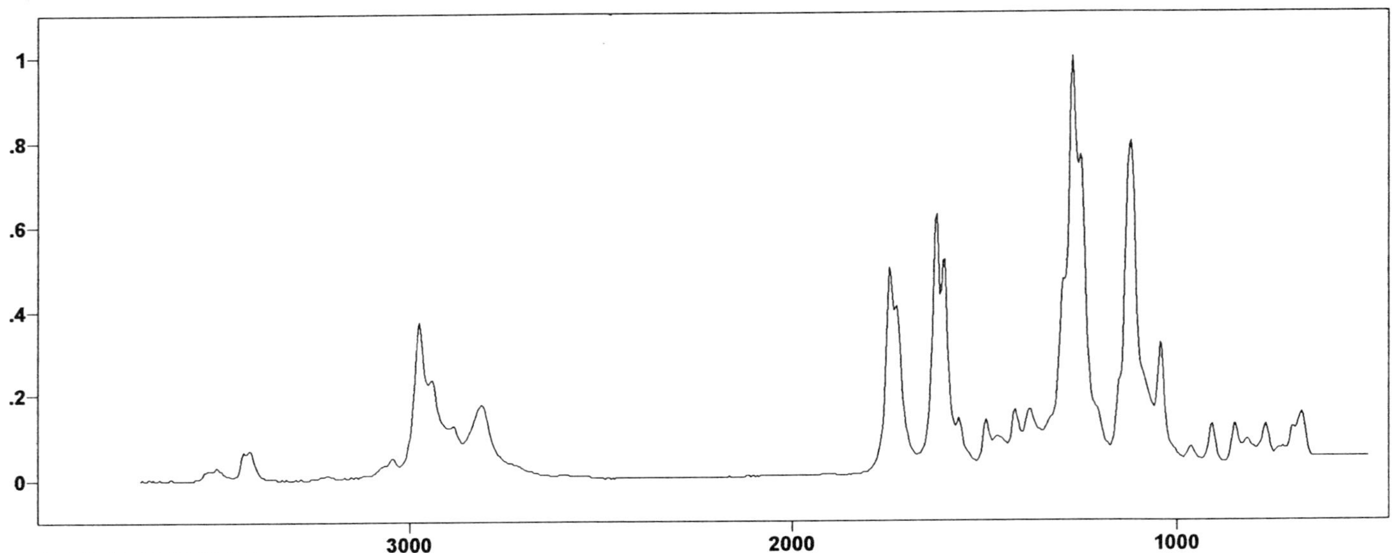

2-CHLOROPROCAINE

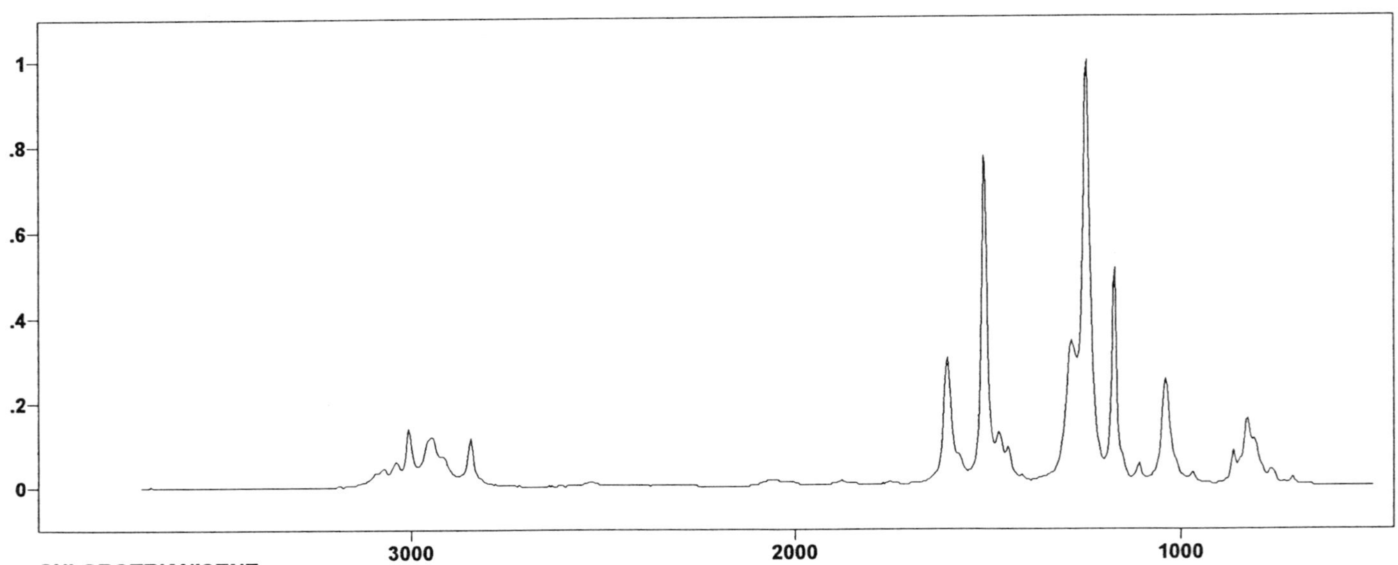

CHLOROTRIANISENE

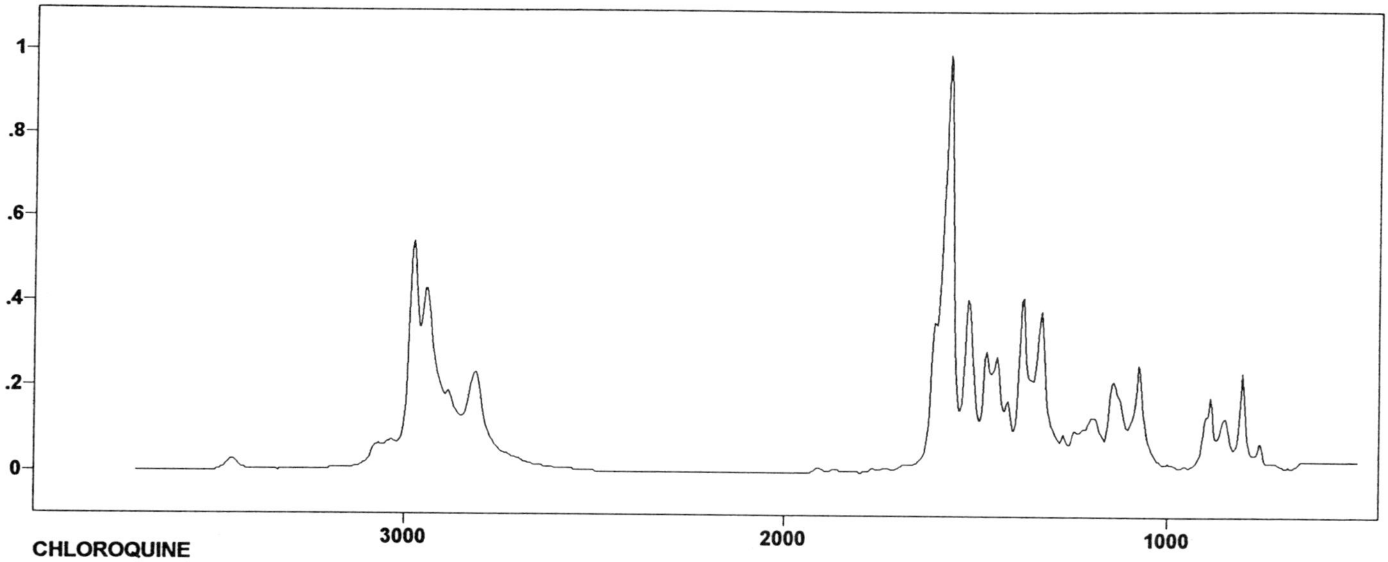

CHLOROQUINE

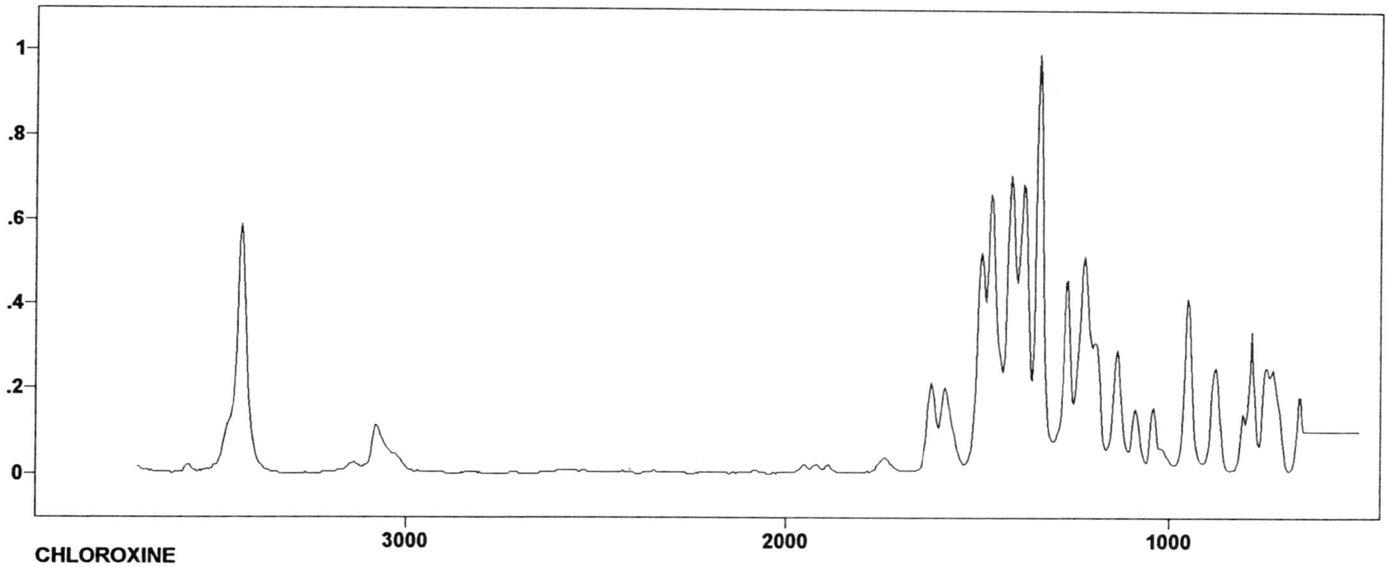

CHLOROXINE

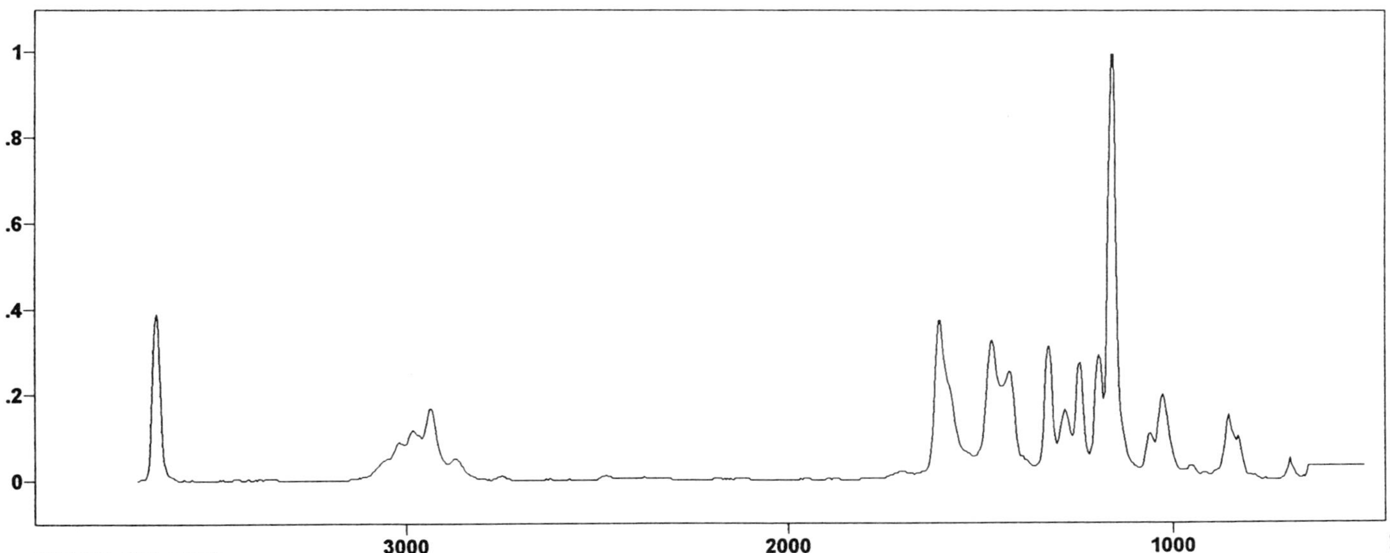

CHLOROXYLENOL

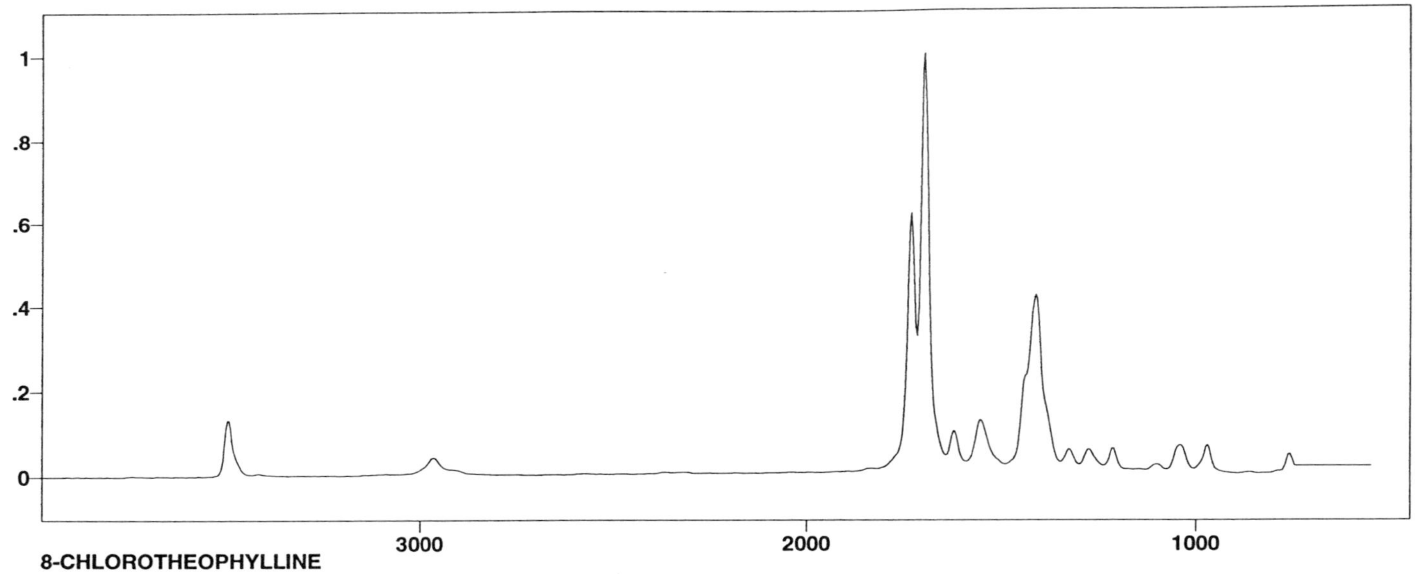

8-CHLOROTHEOPHYLLINE

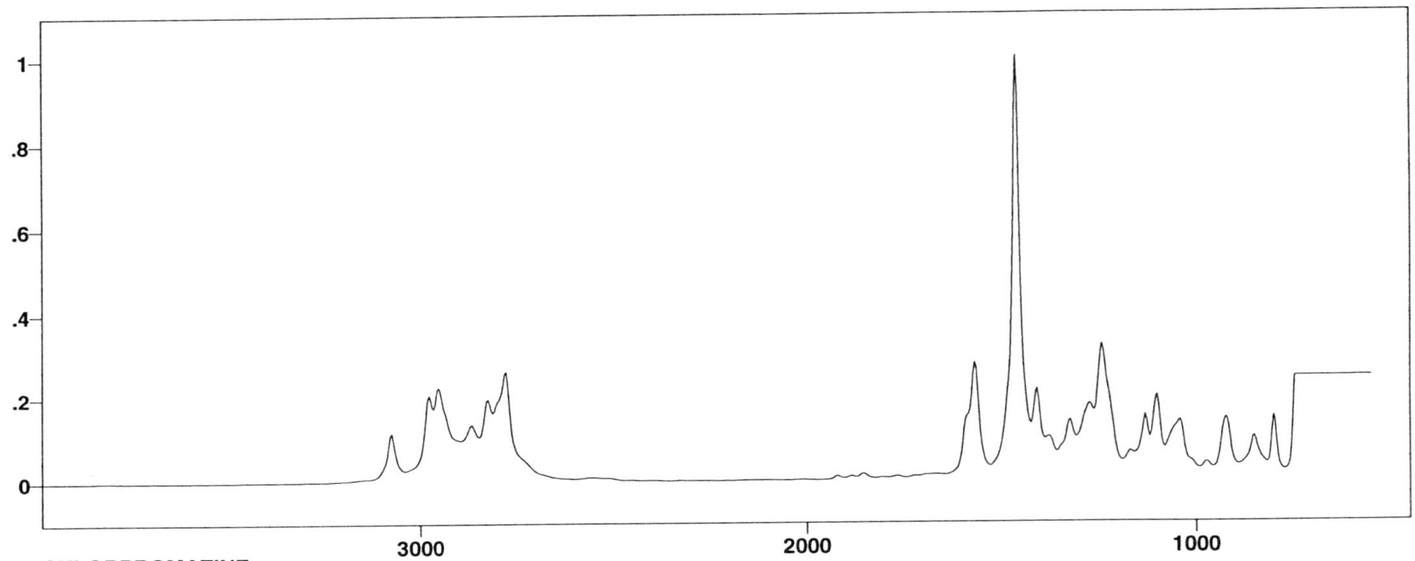

CHLORPROMAZINE

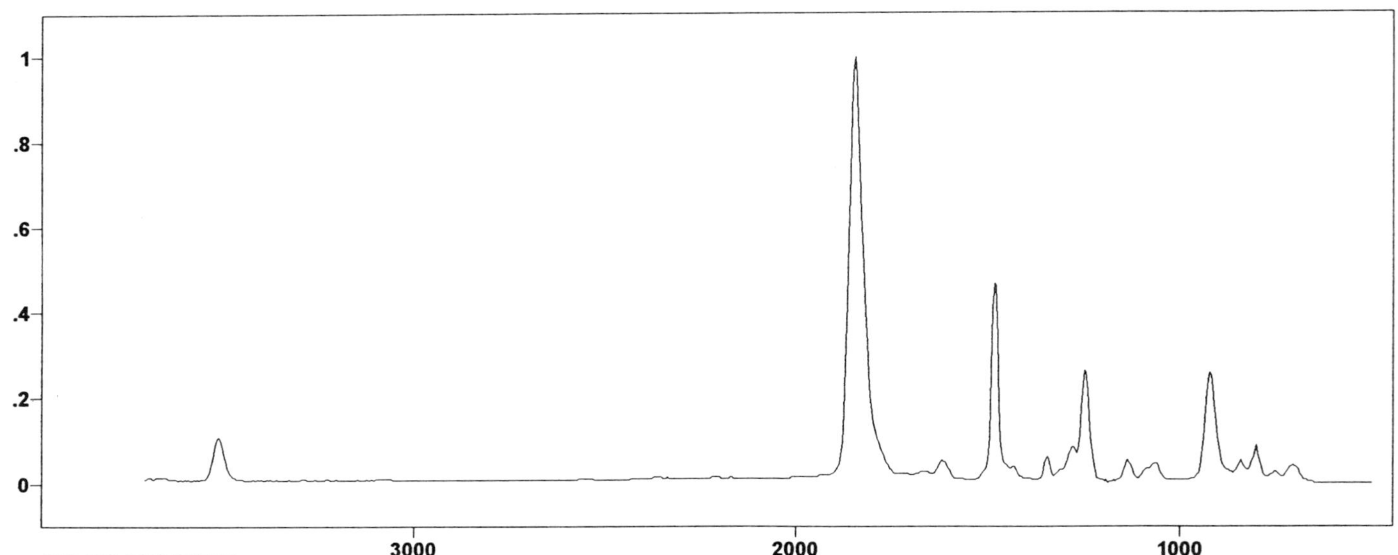

CHLORZOXAZONE

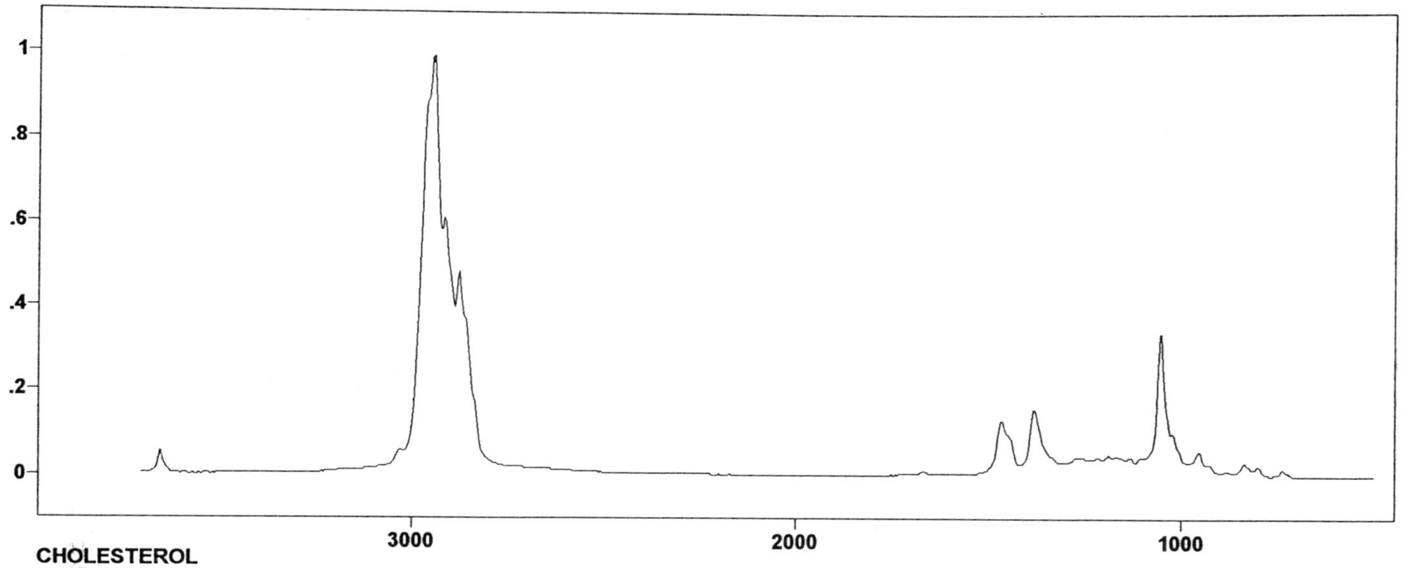

CHOLESTEROL

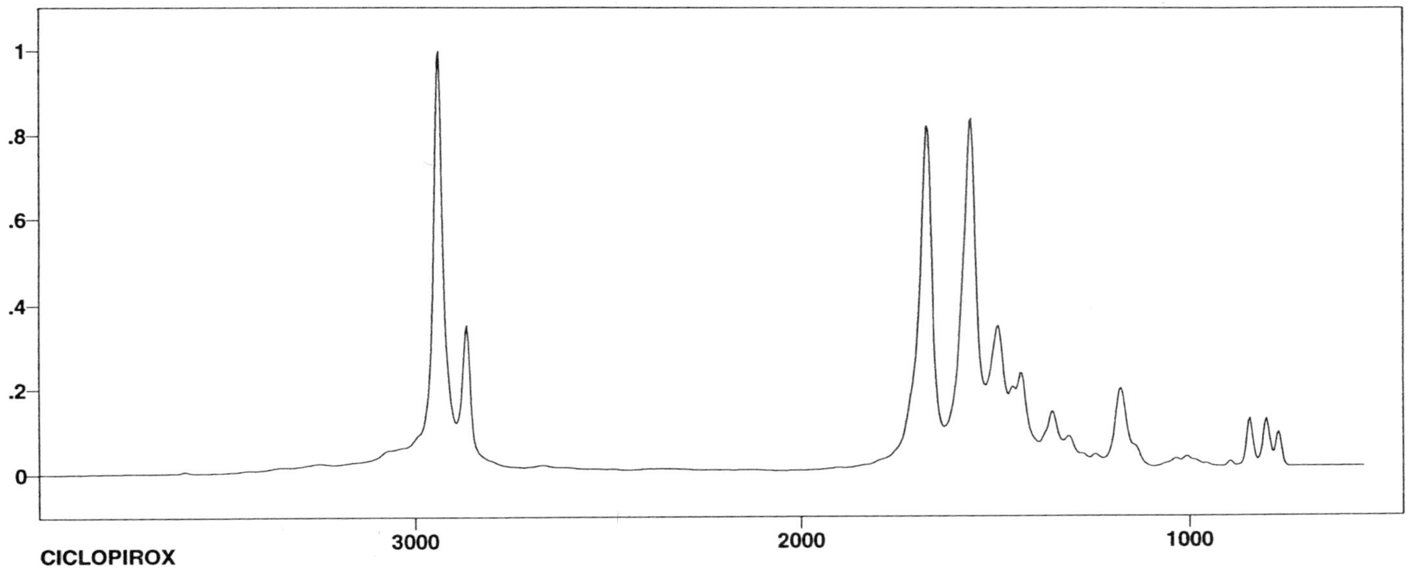

CICLOPIROX

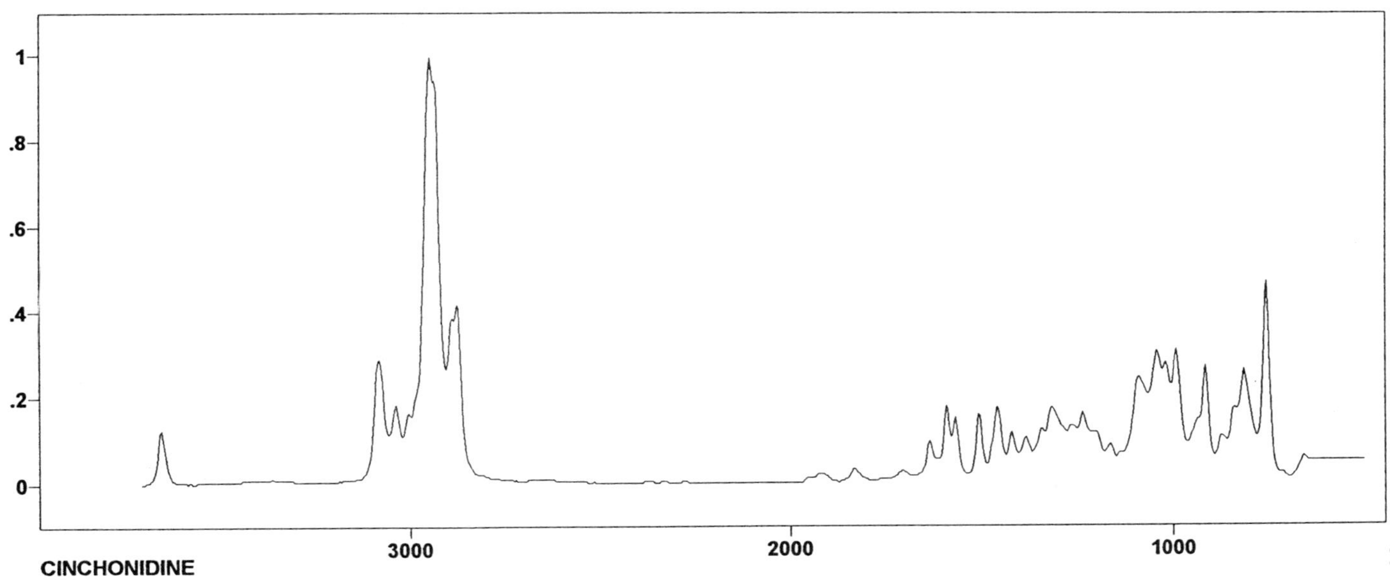

CINCHONIDINE

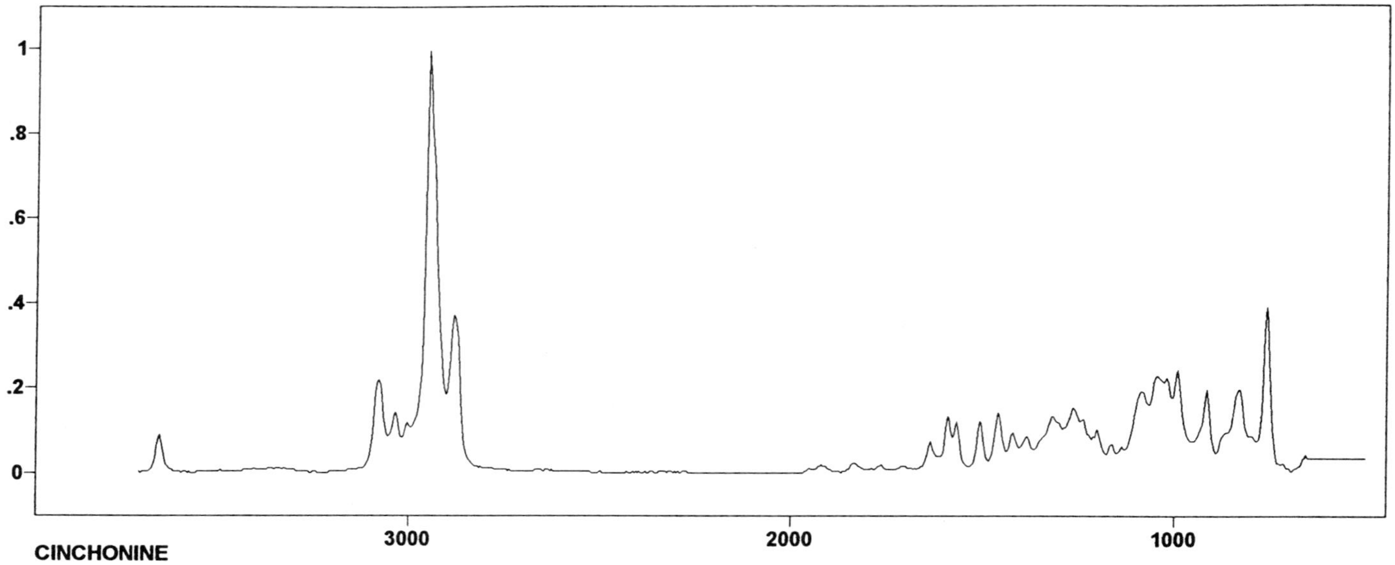

CINCHONINE

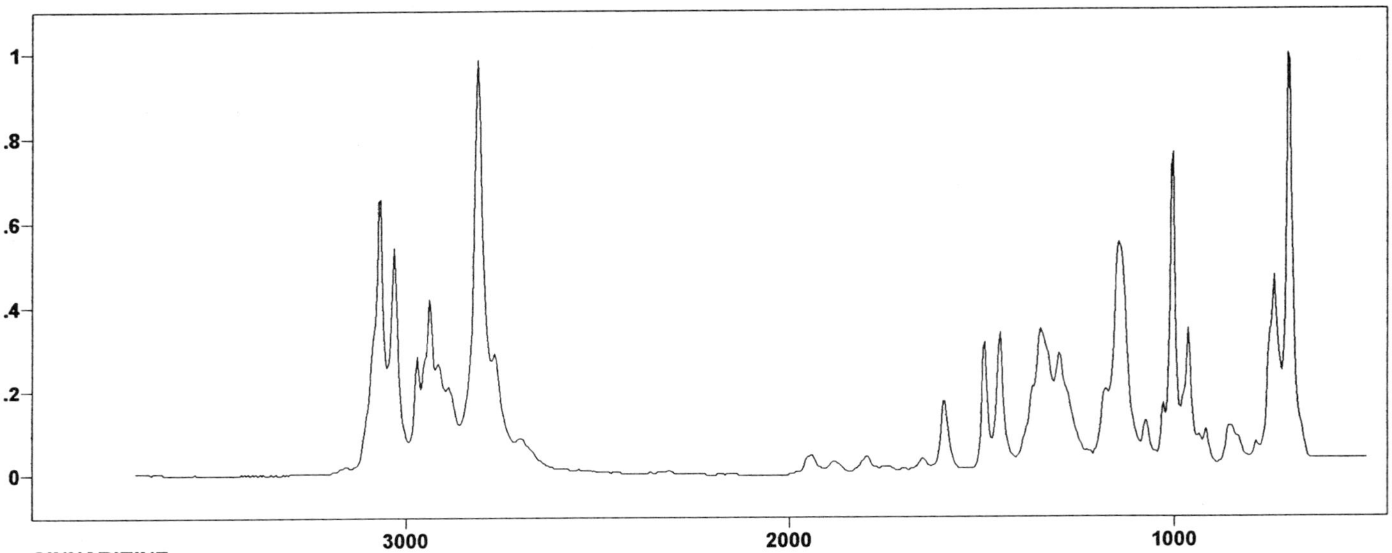

CINNARIZINE

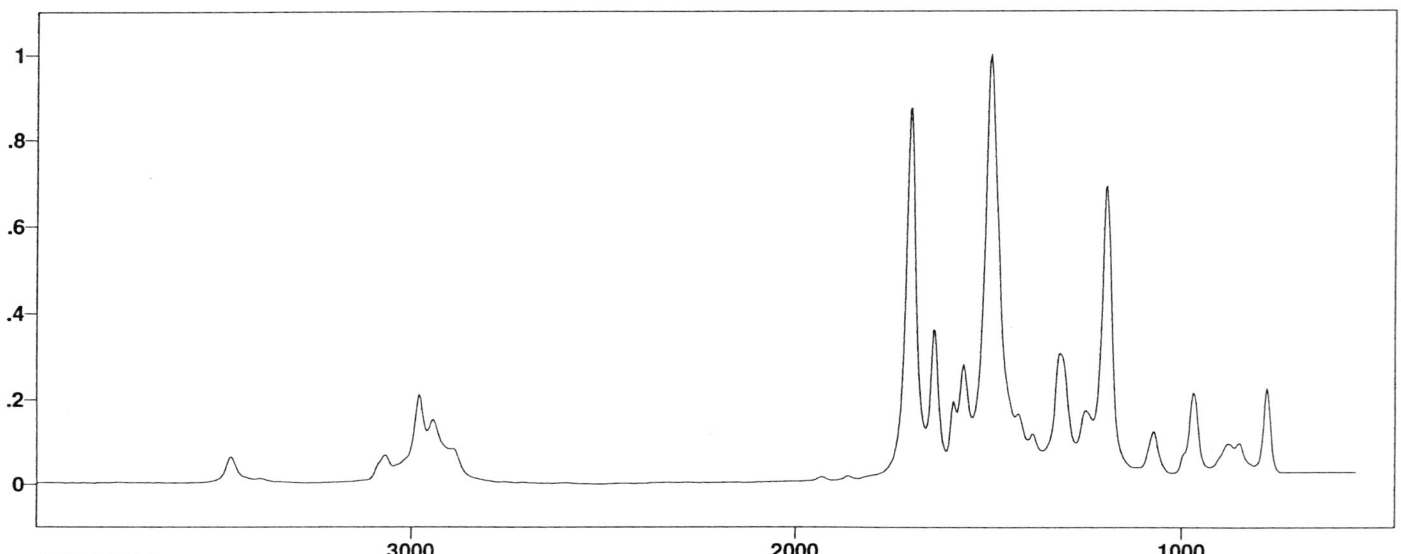

CINROMIDE

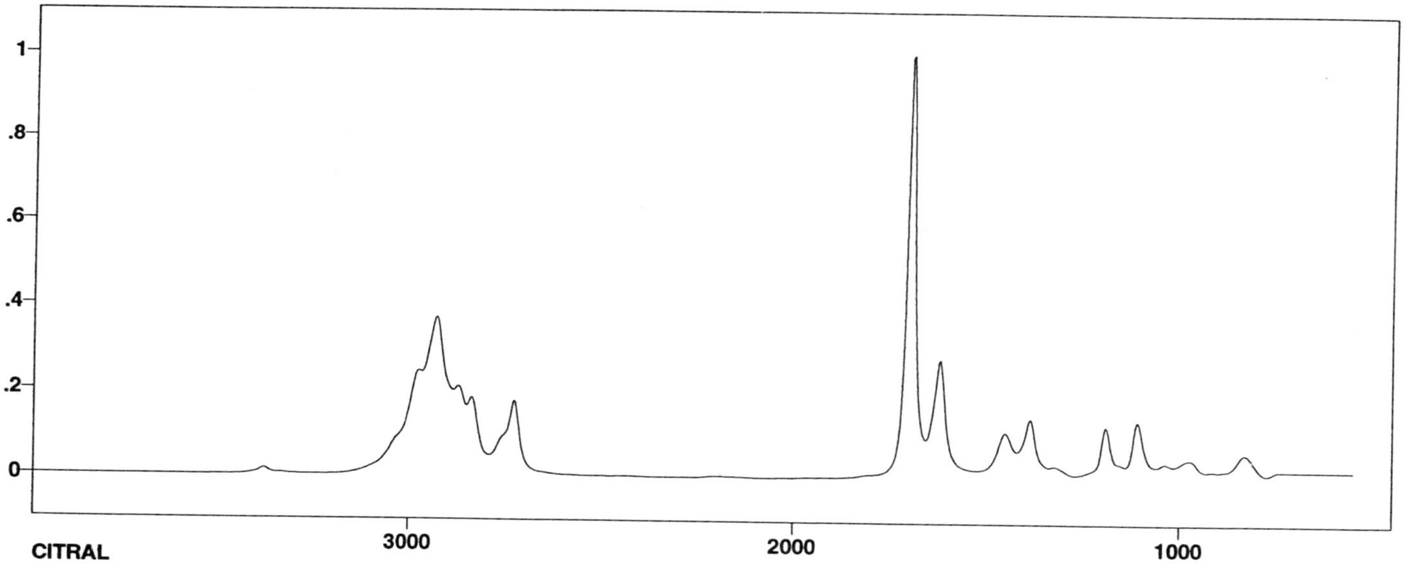

CITRAL

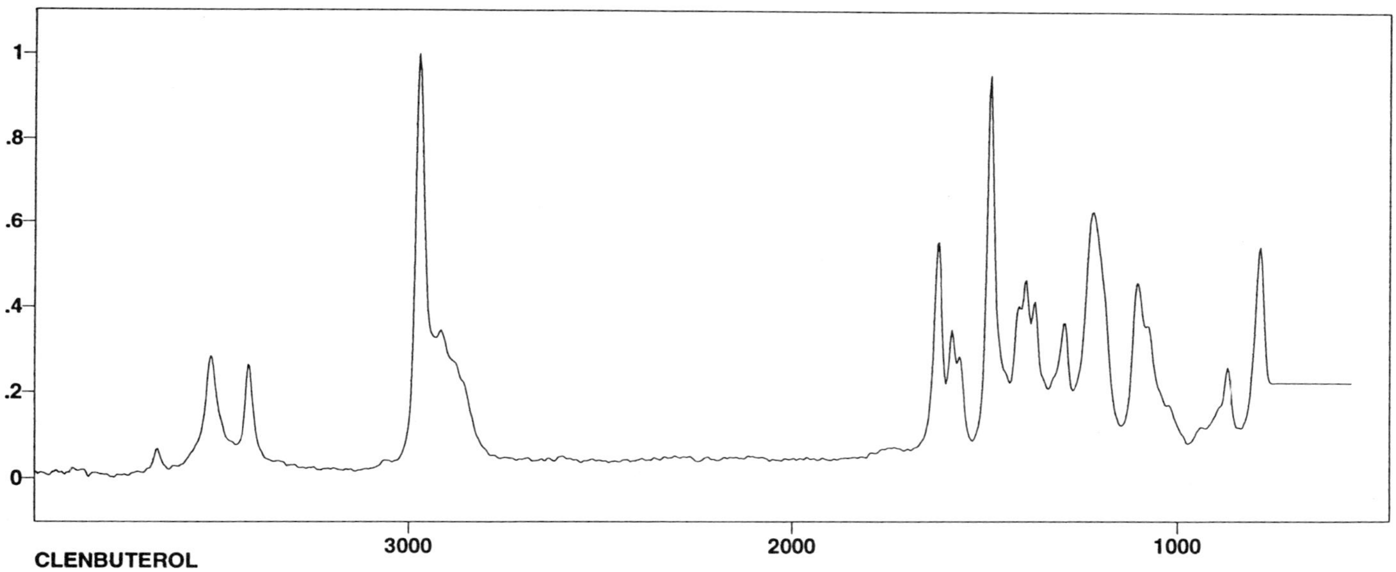

CLENBUTEROL

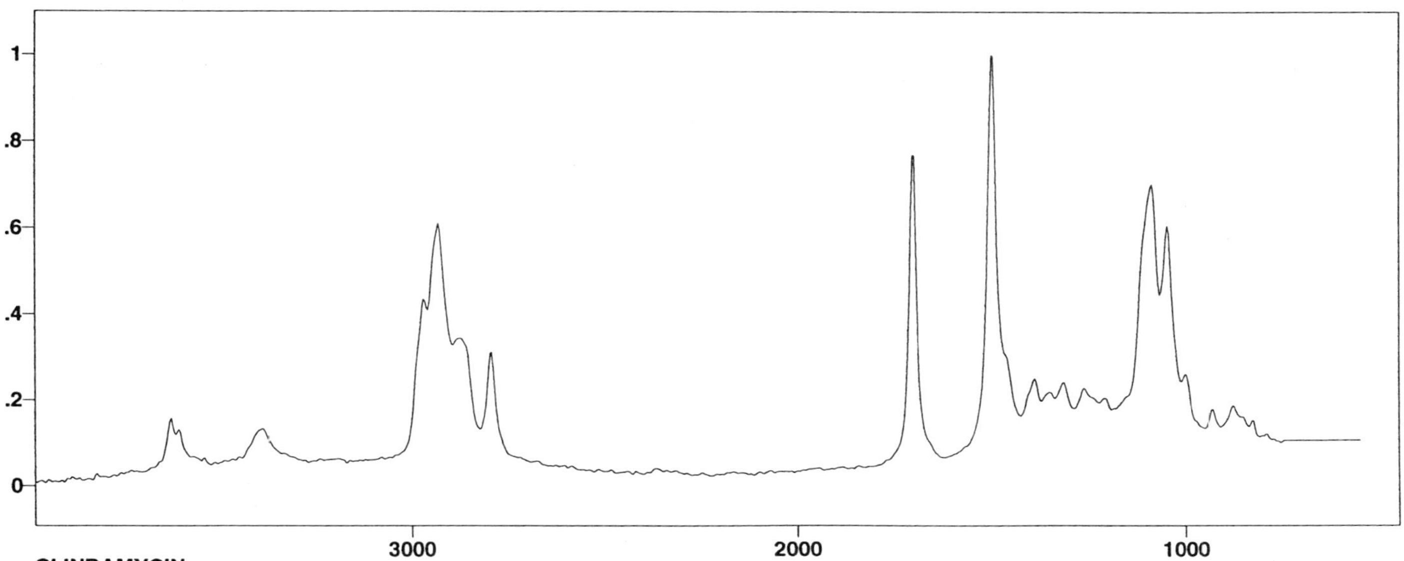

CLINDAMYCIN

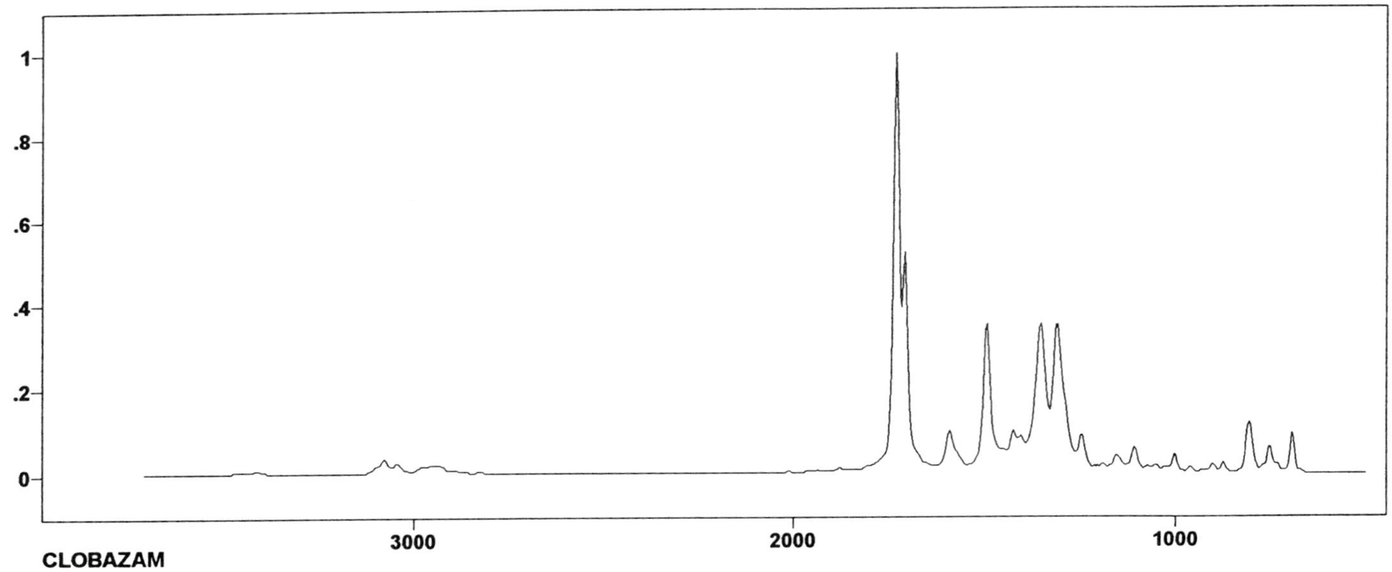

CLOBAZAM

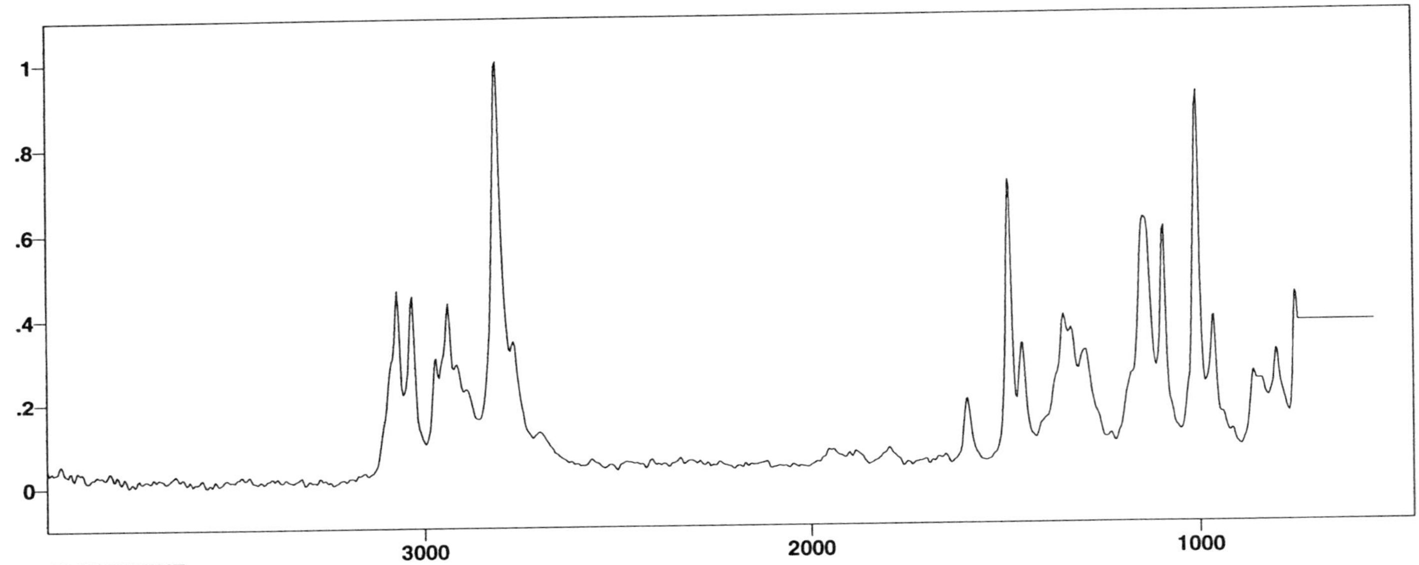

CLOCINIZINE

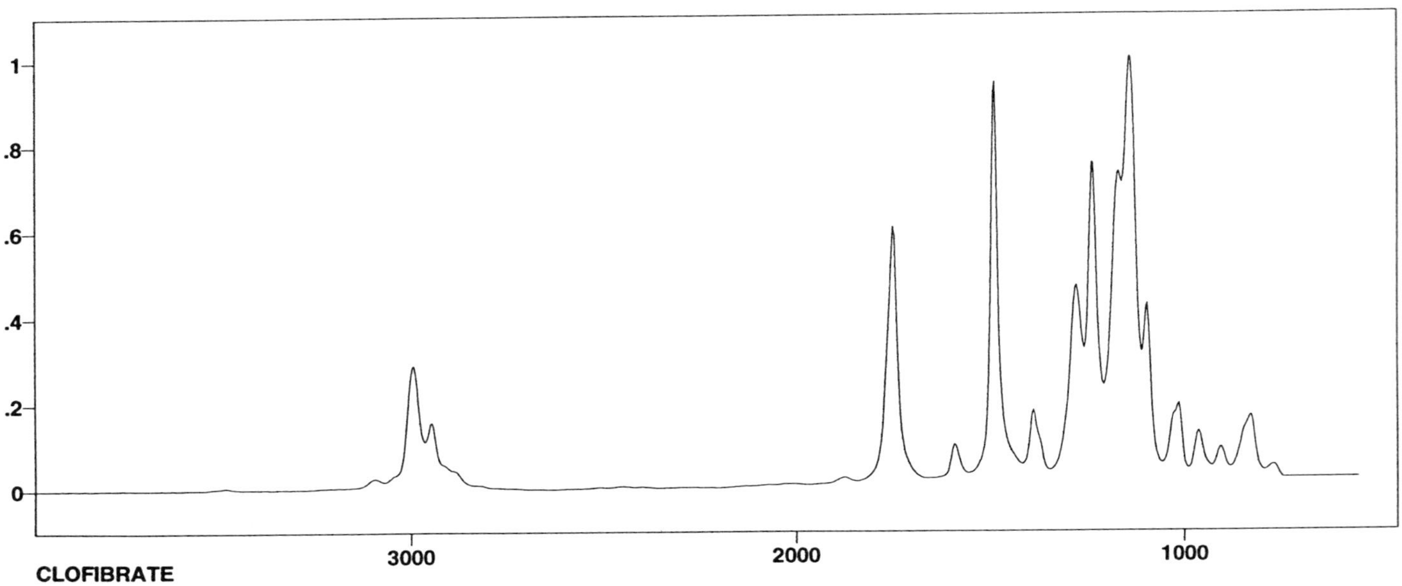

CLOFIBRATE

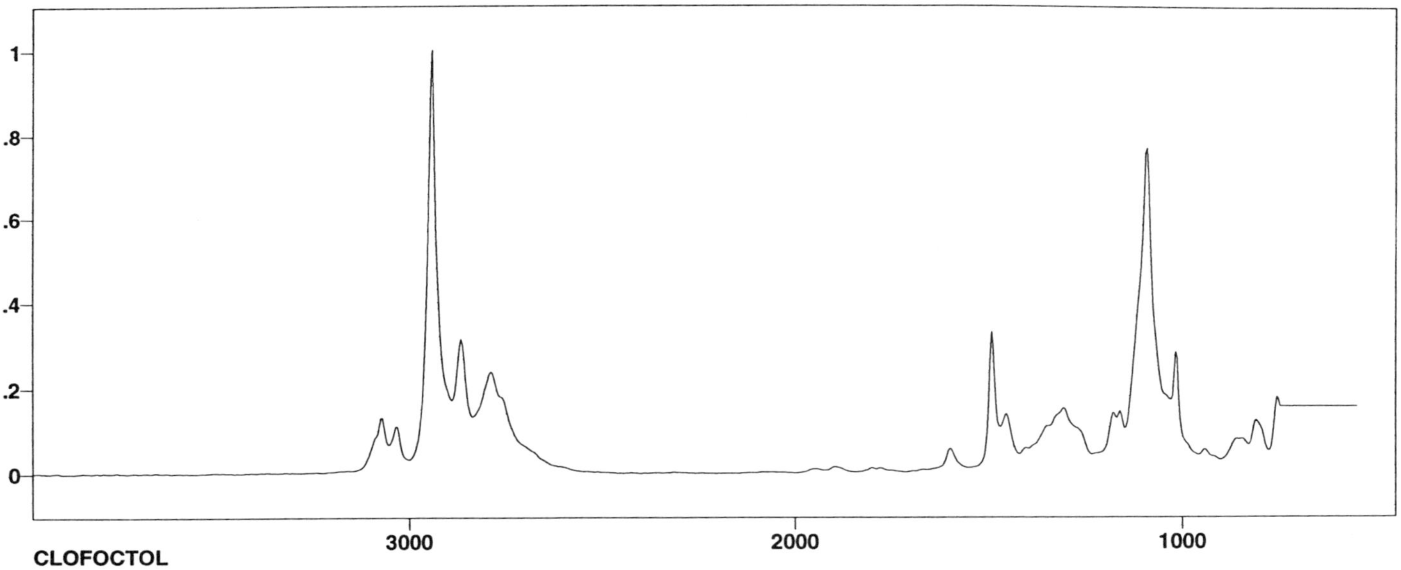

CLOFOCTOL

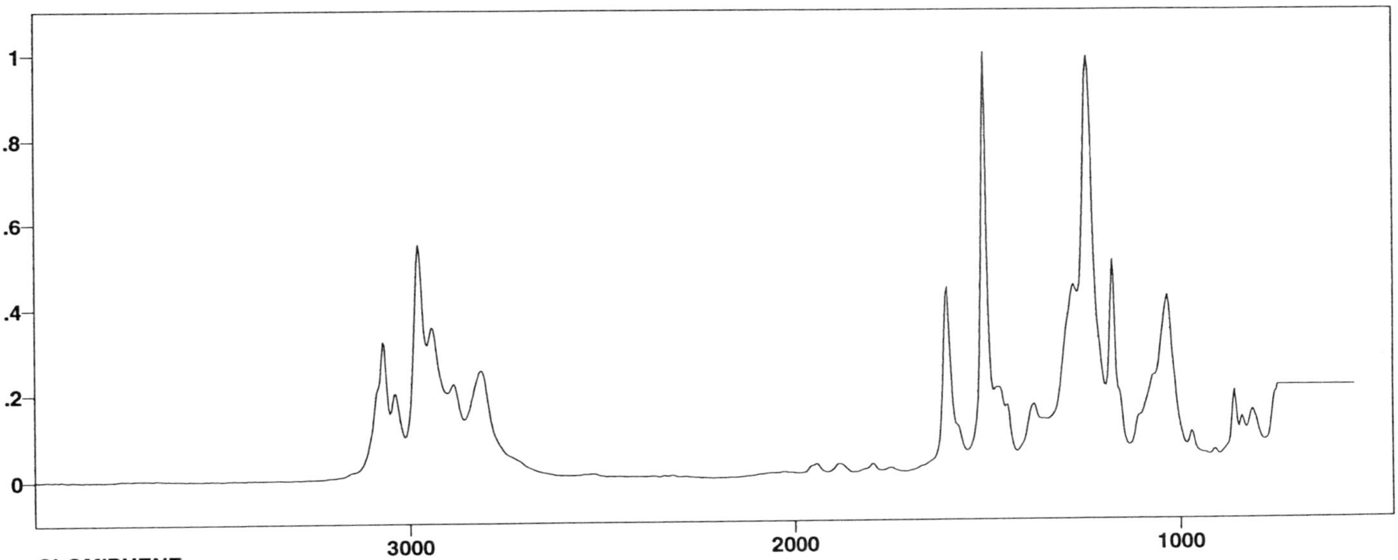

CLOMIPHENE

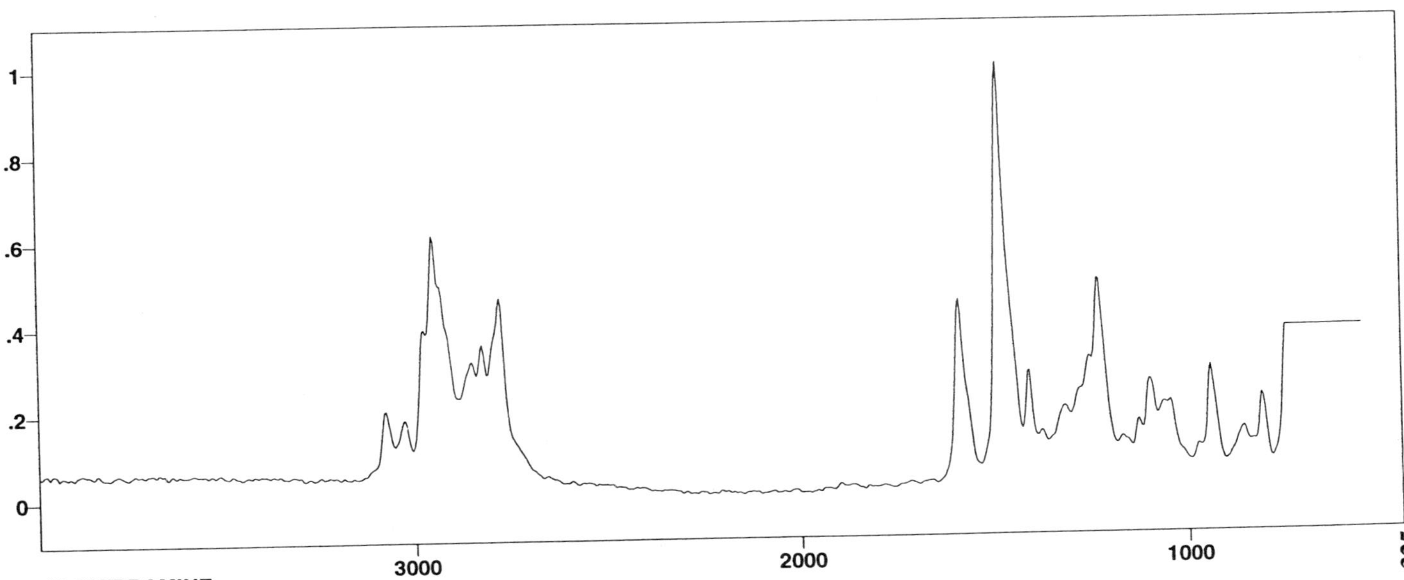

CLOMIPRAMINE

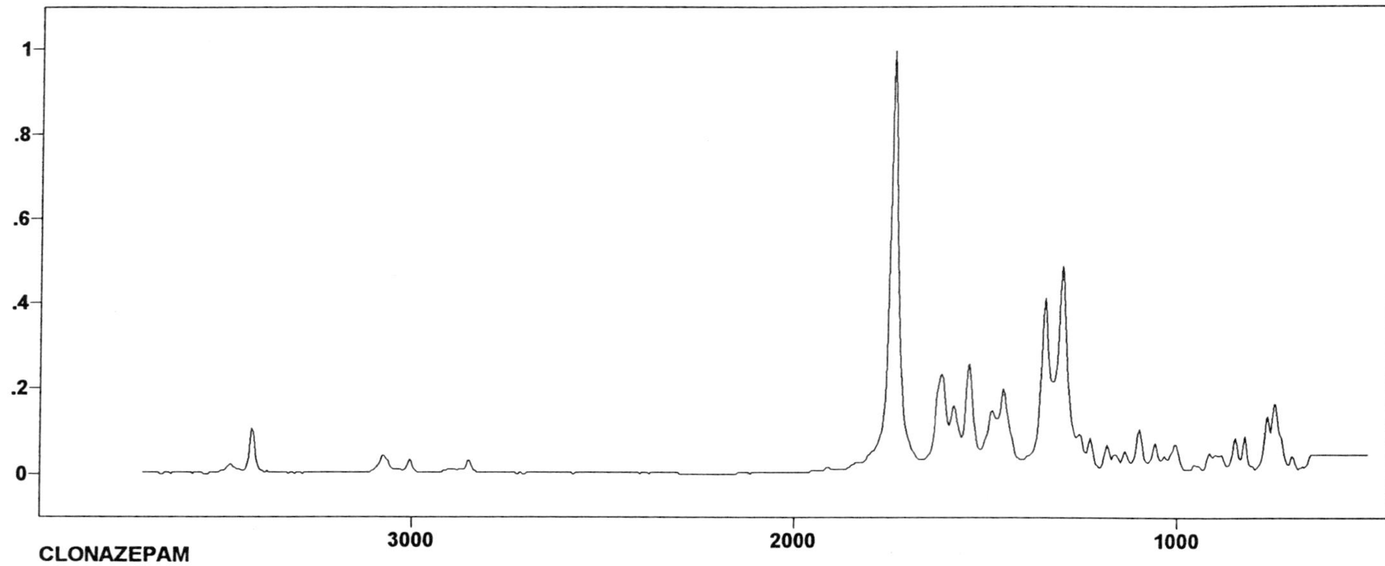

CLONAZEPAM

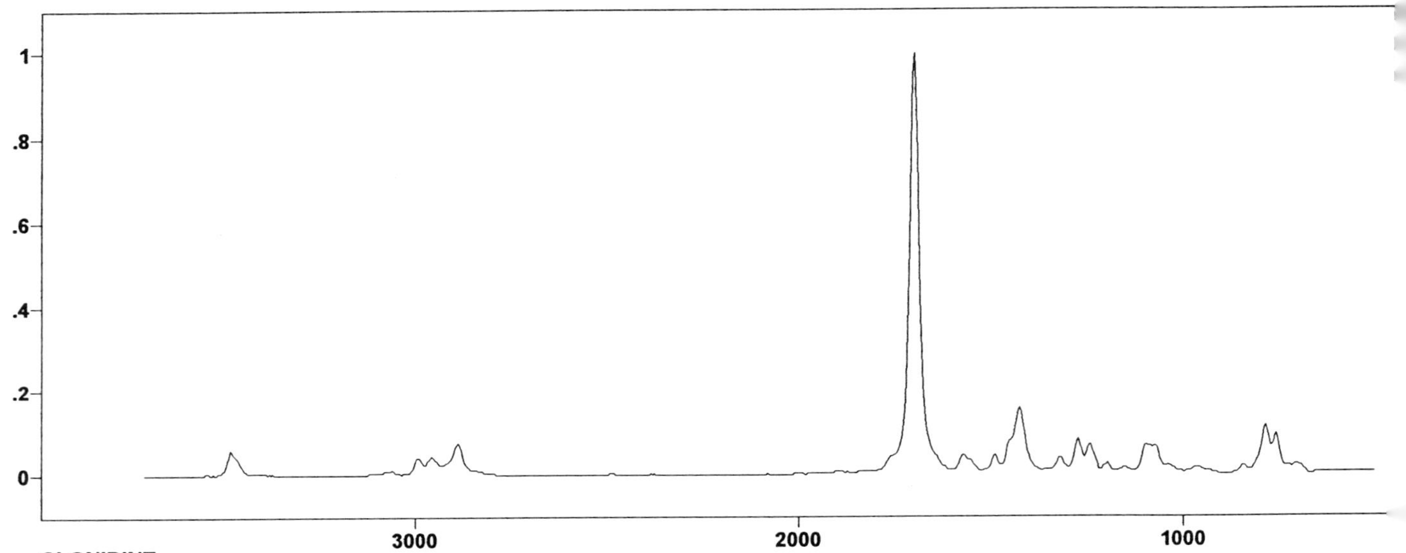

CLONIDINE

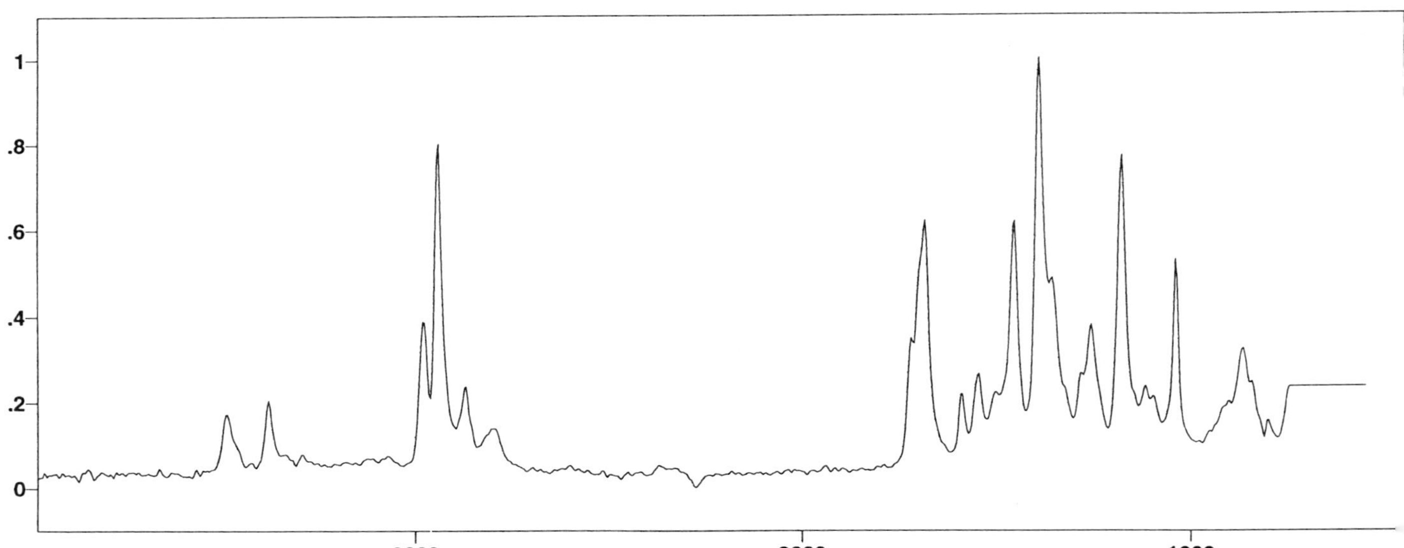

CLOPAMIDE

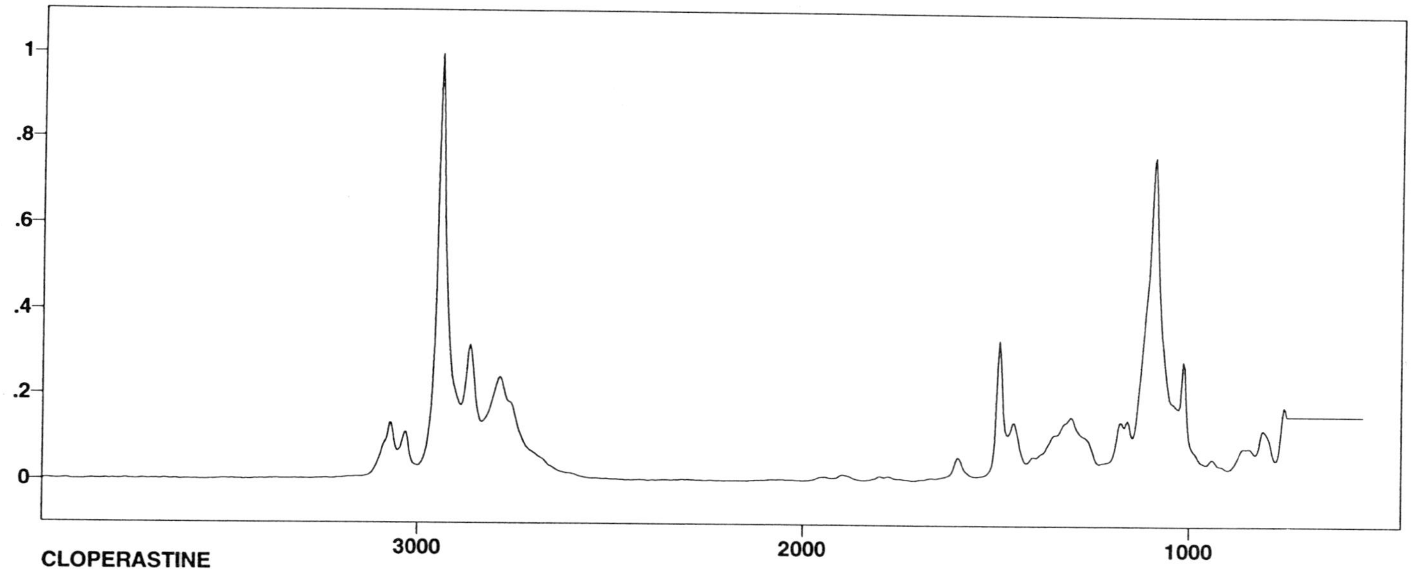

CLOPERASTINE

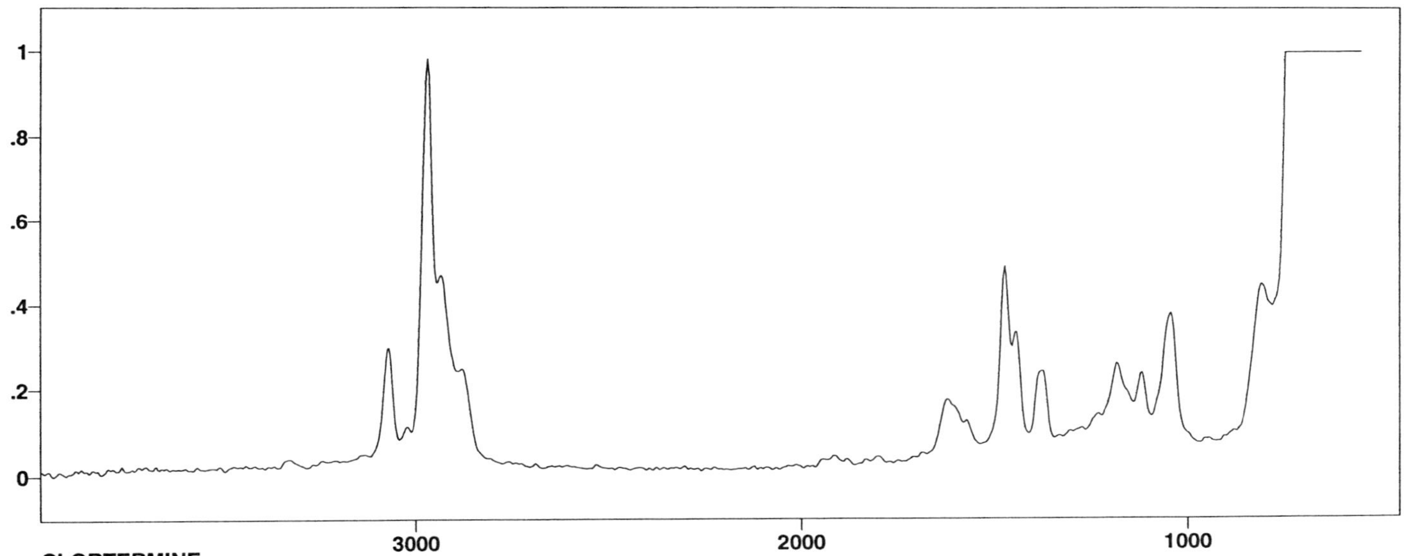

CLORTERMINE

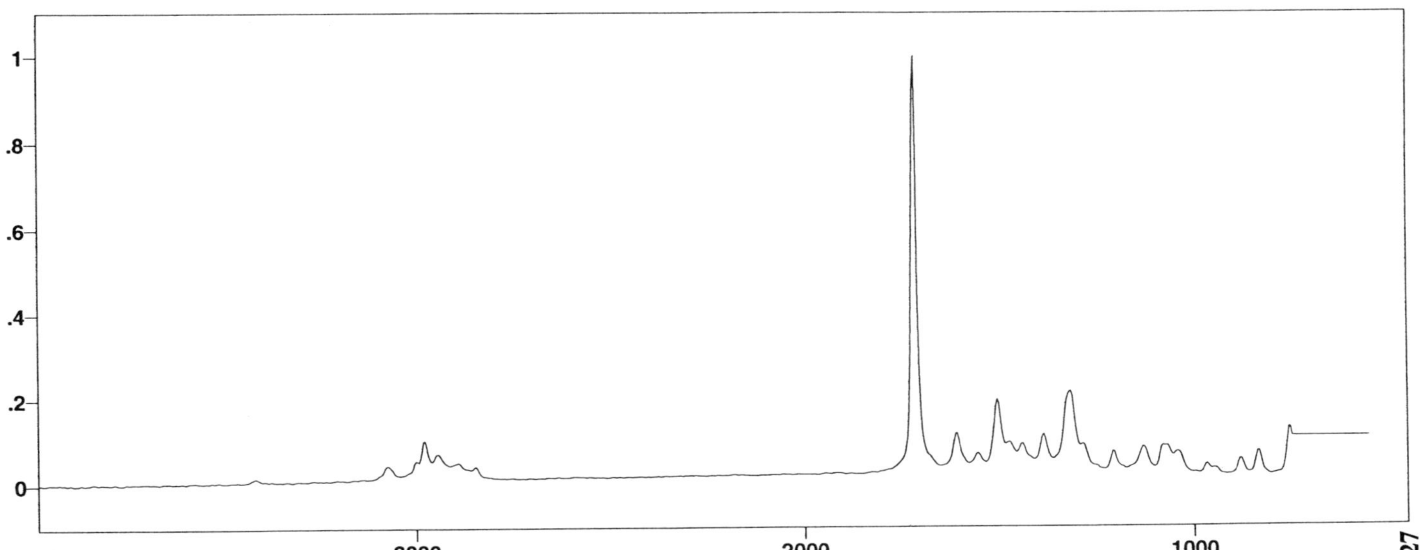

CLOTIAZEPAM

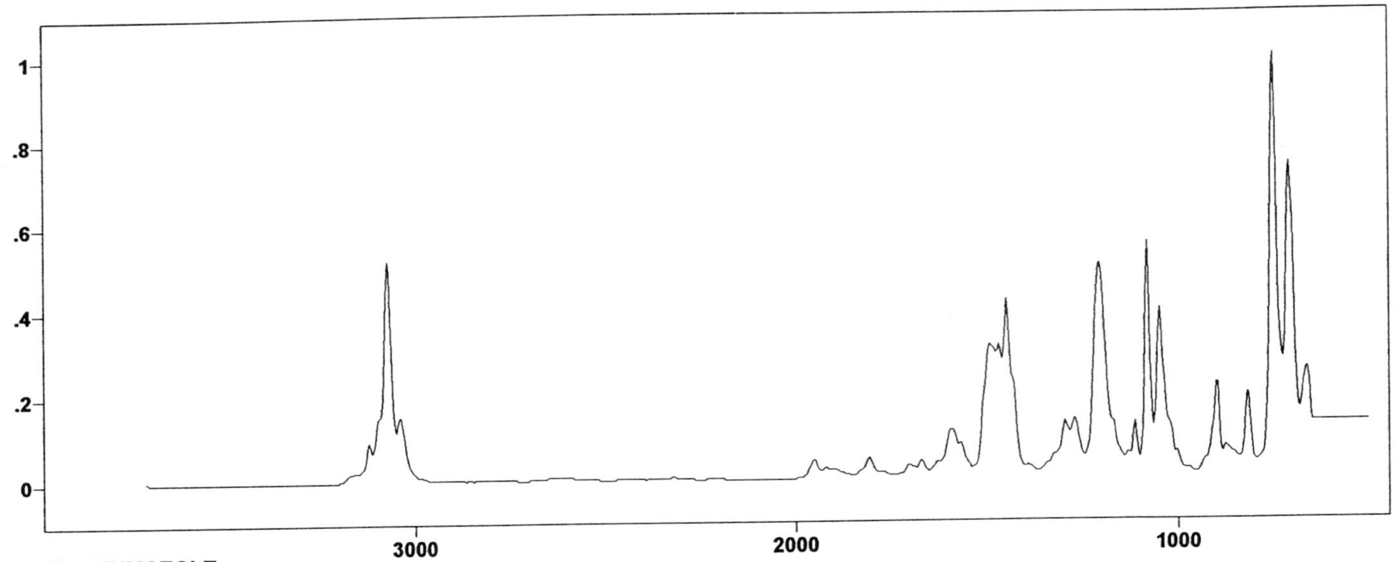

CLOTRIMAZOLE

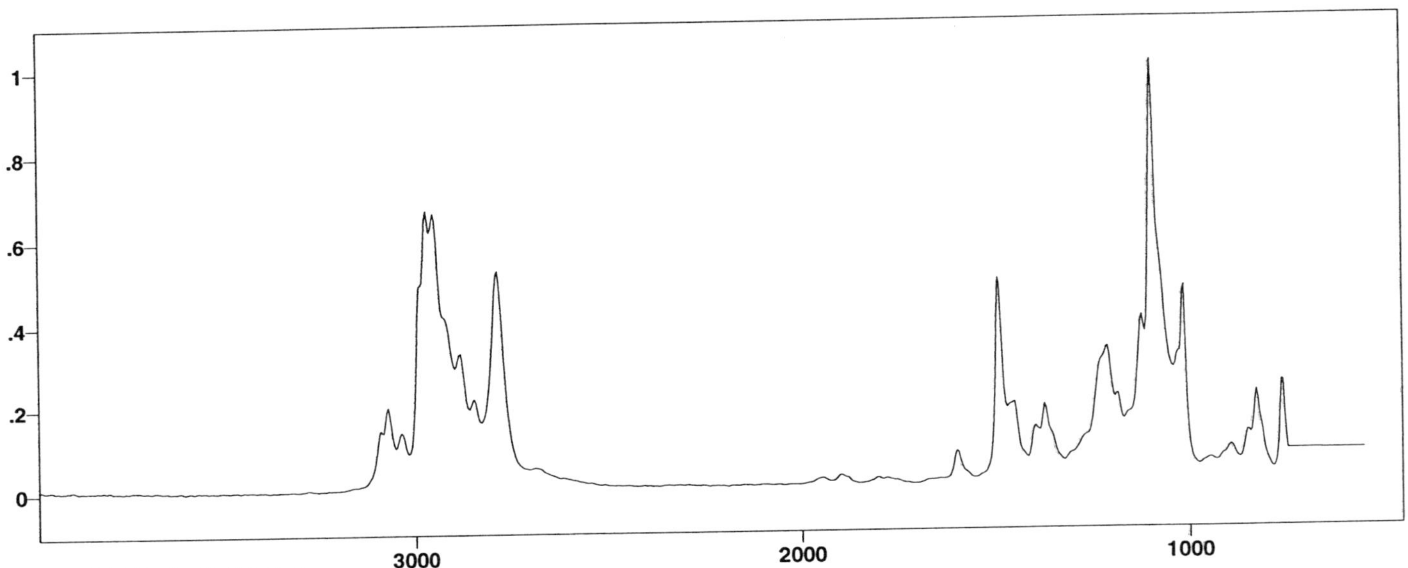

CLEMASTINE FUMARATE

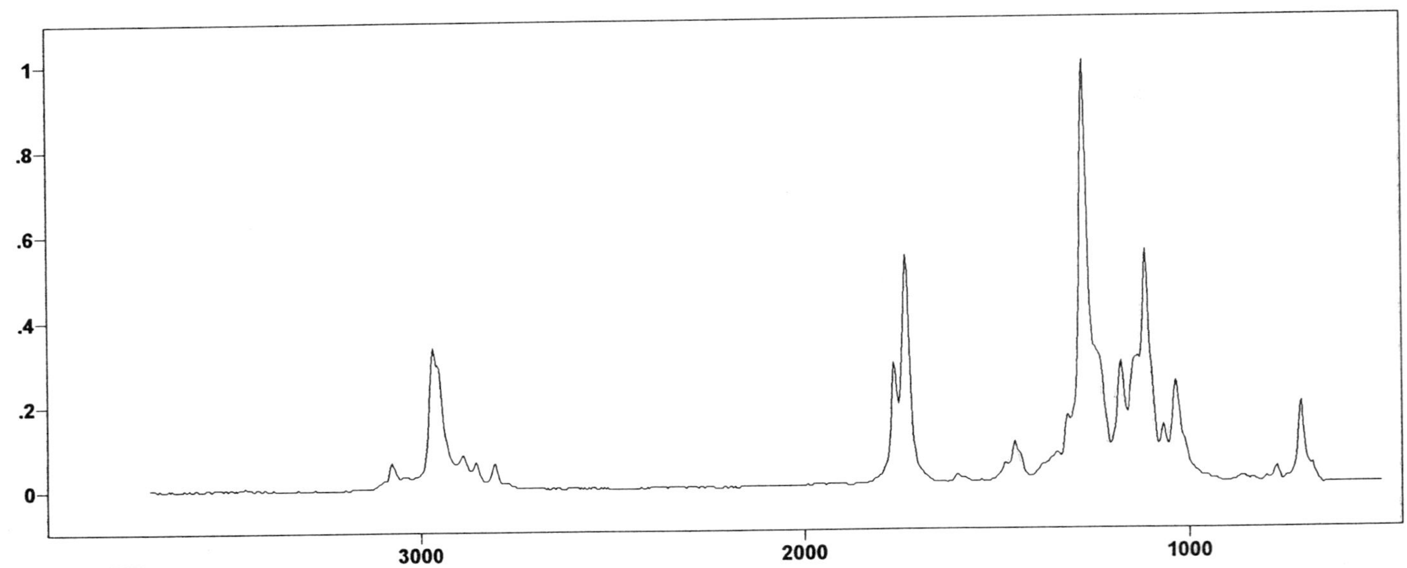

COCAINE

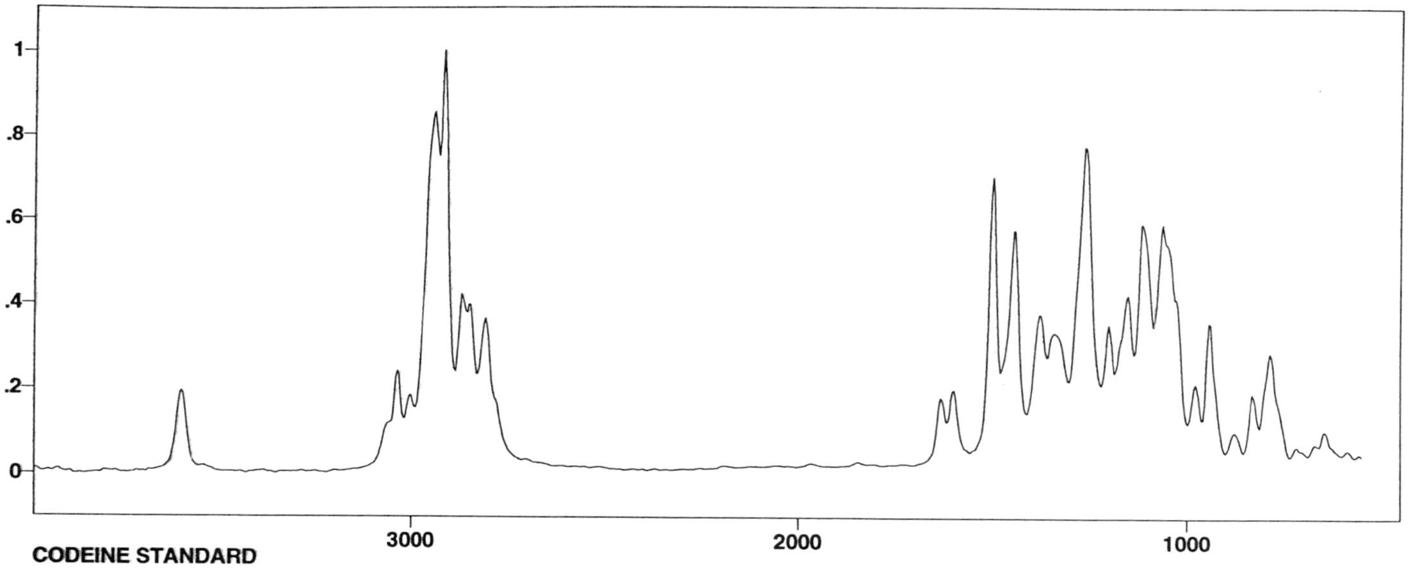

CODEINE STANDARD

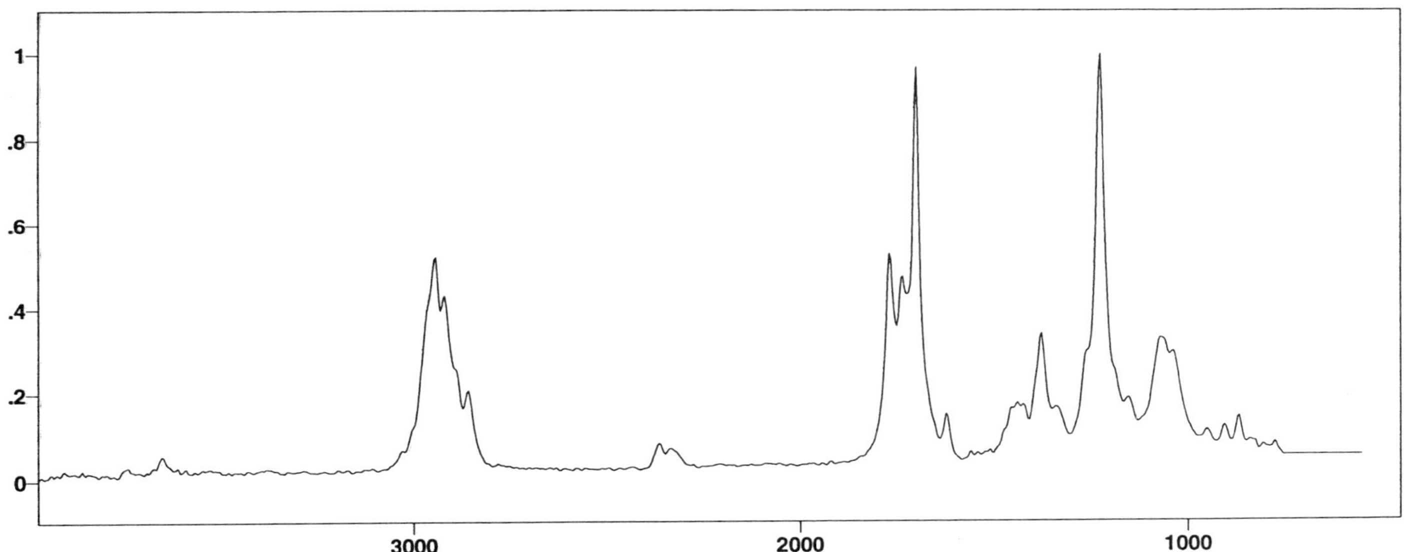

CORTICOSTERONE

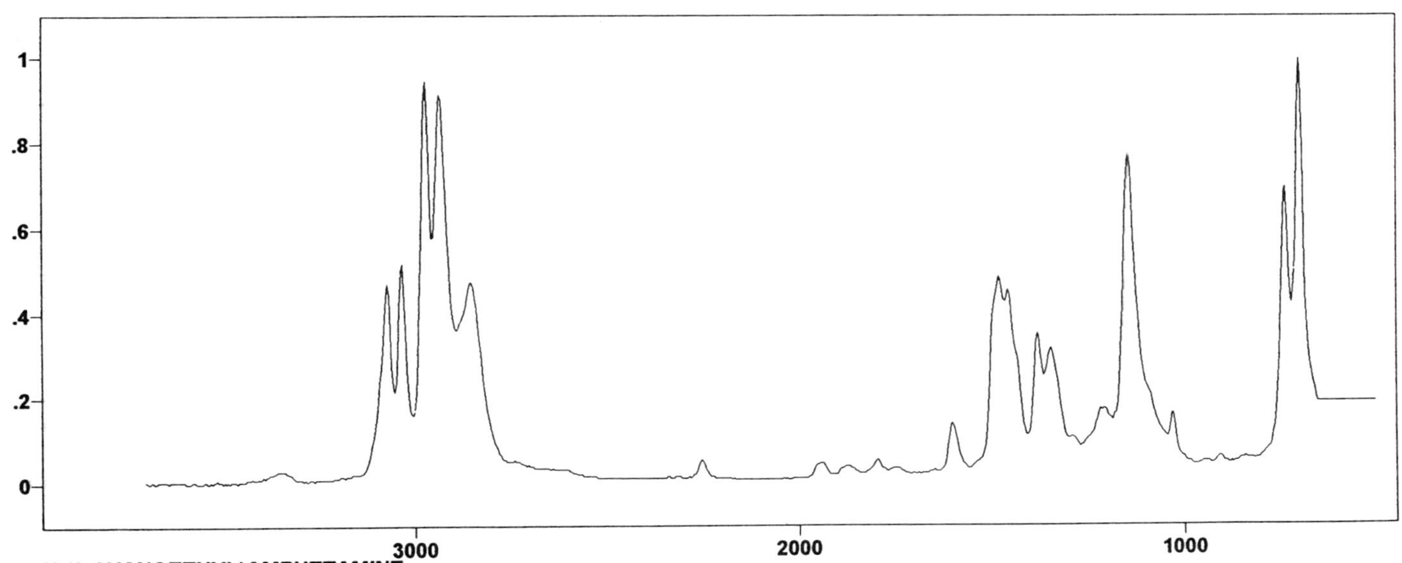

N-(2-CYANOETHYL)AMPHETAMINE

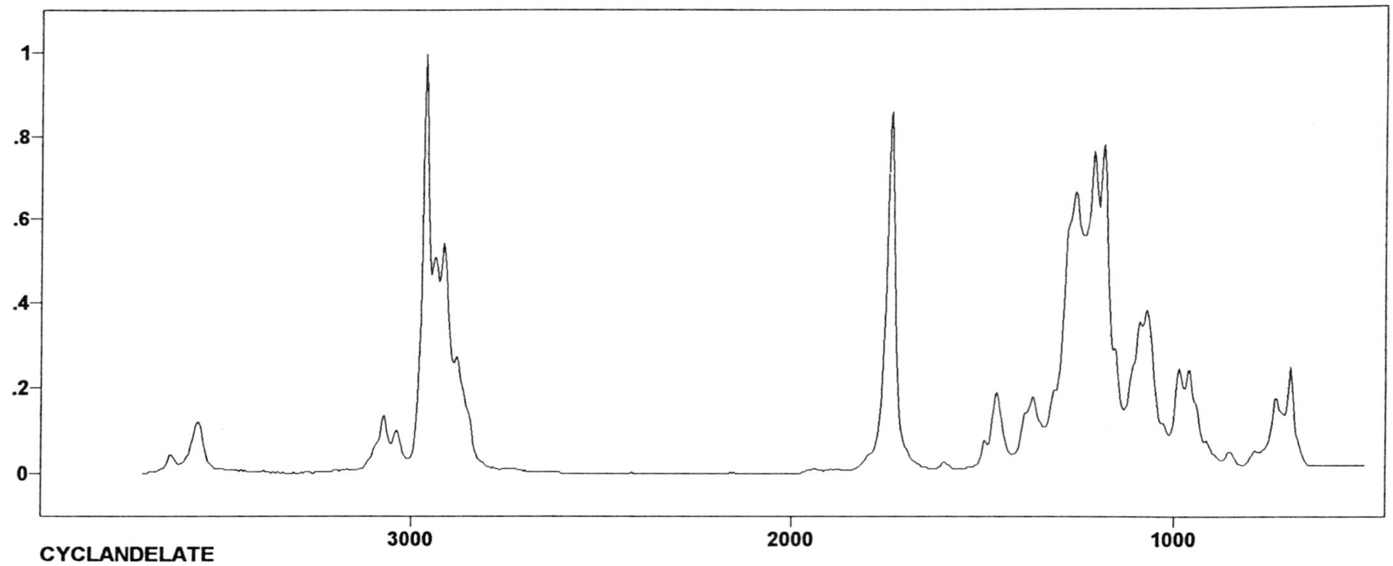

CYCLANDELATE

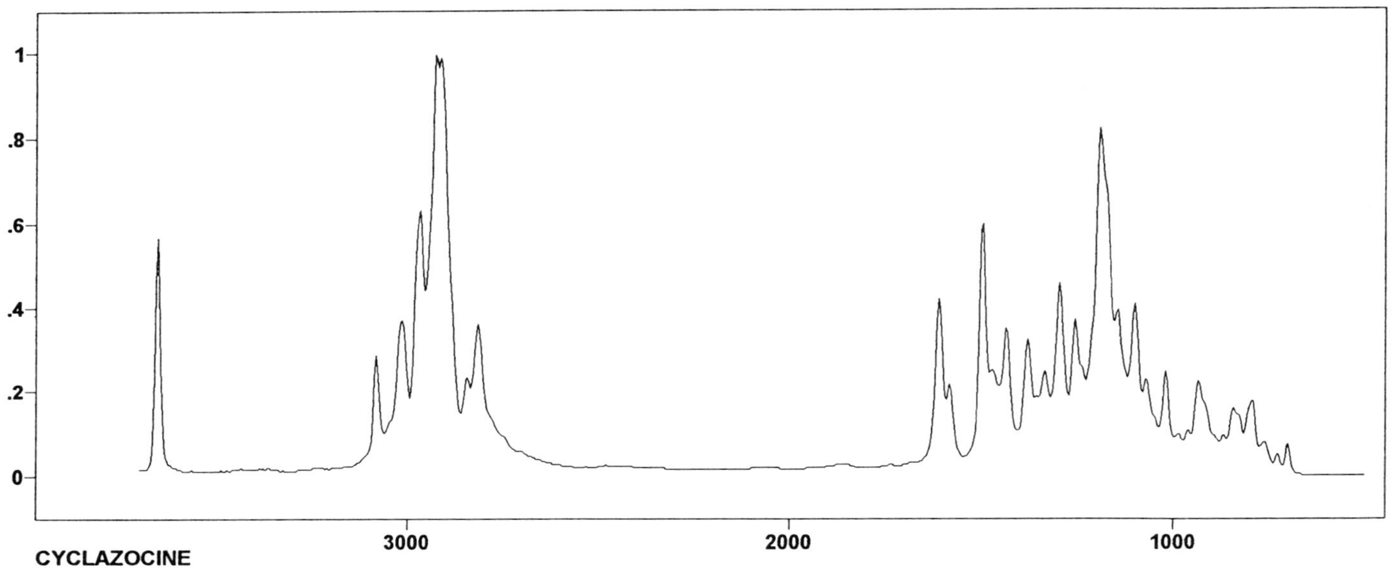

CYCLAZOCINE

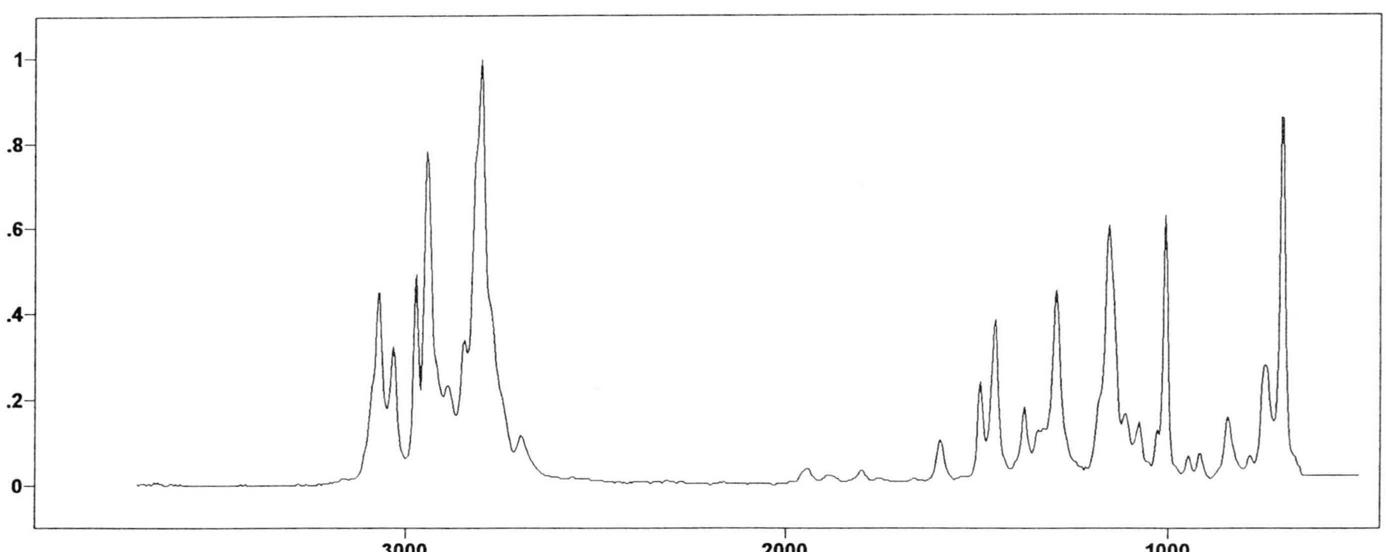

CYCLIZINE

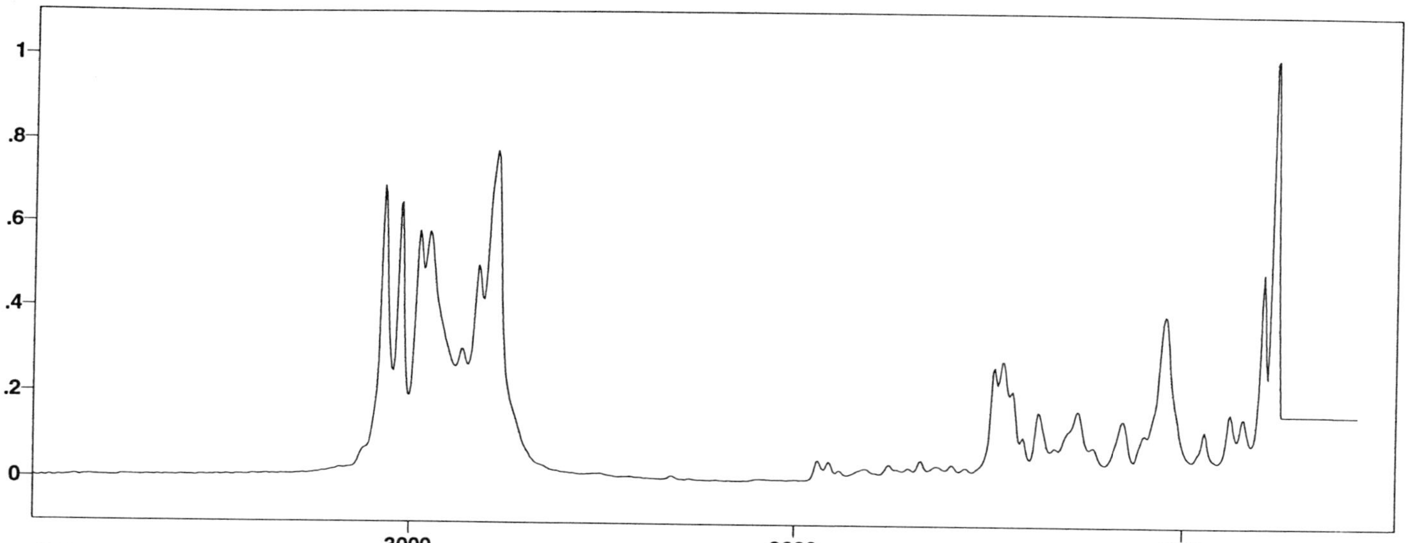

CYCLOBENZAPRINE

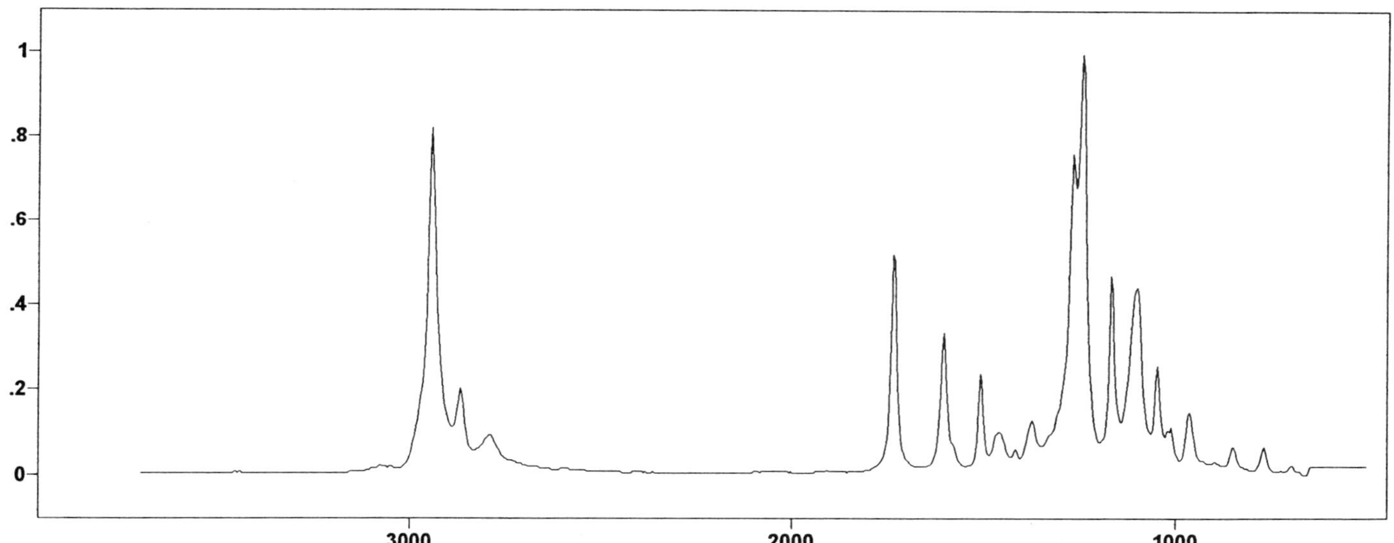

CYCLOMETHYCAINE

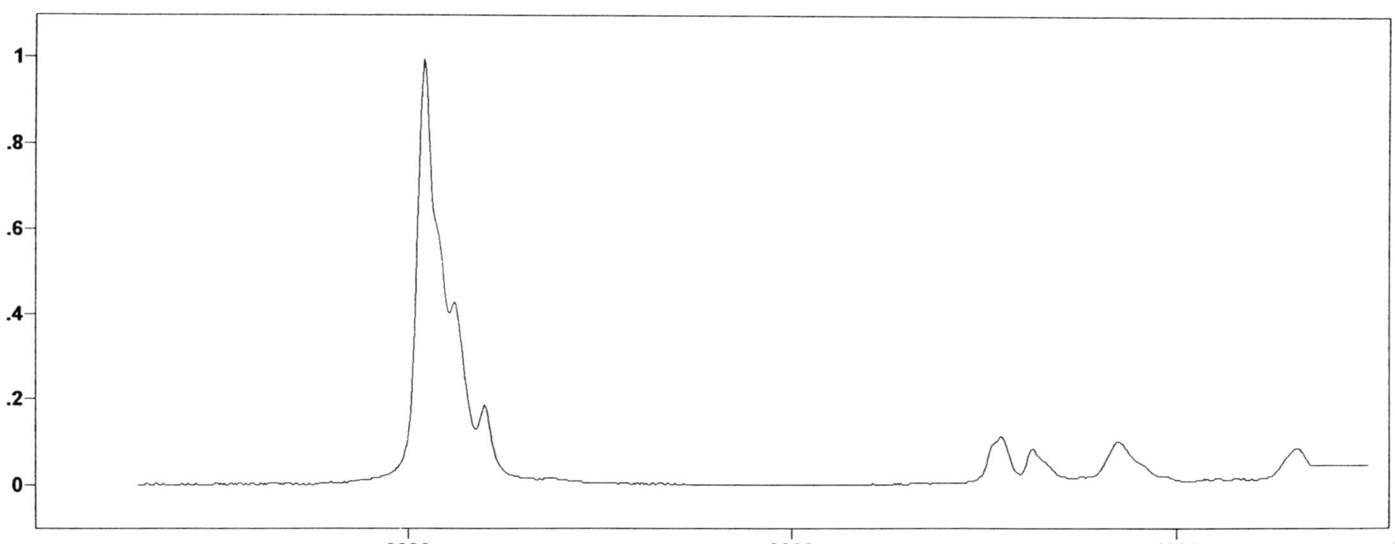

CYCLOPENTAMINE

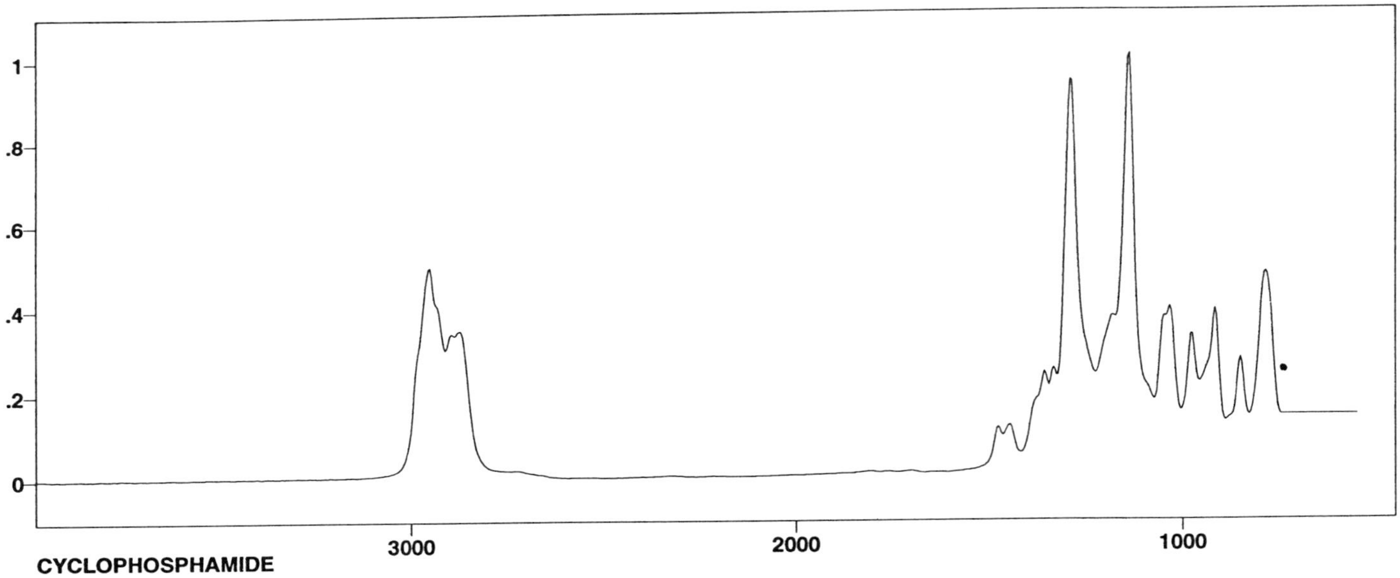

CYCLOPHOSPHAMIDE

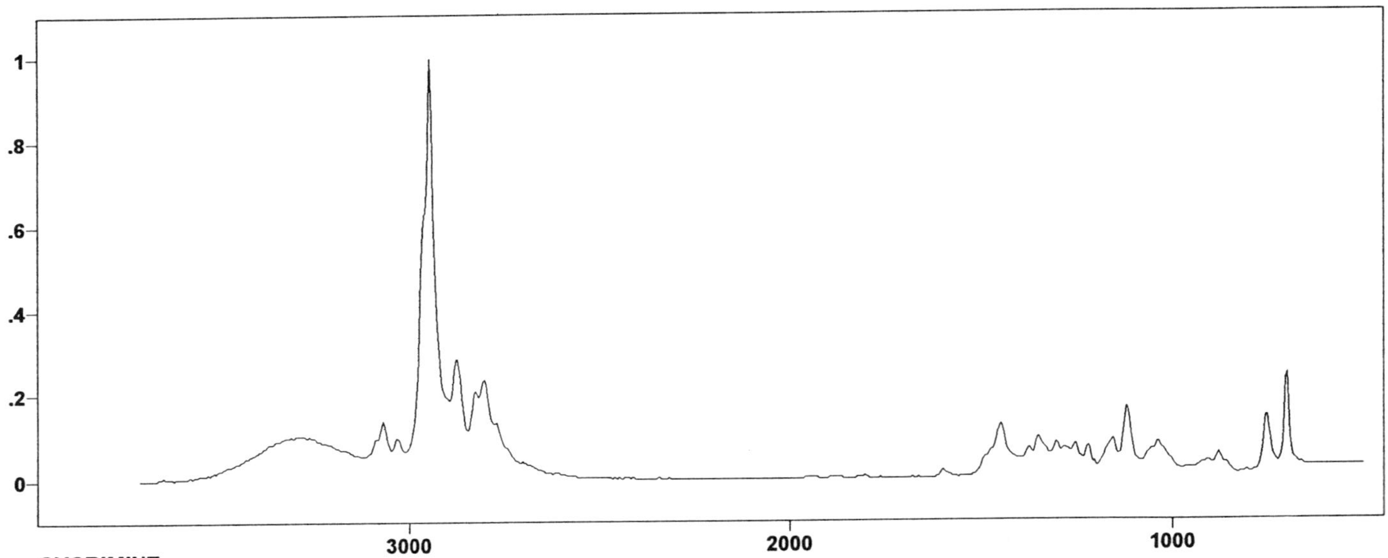

CYCRIMINE

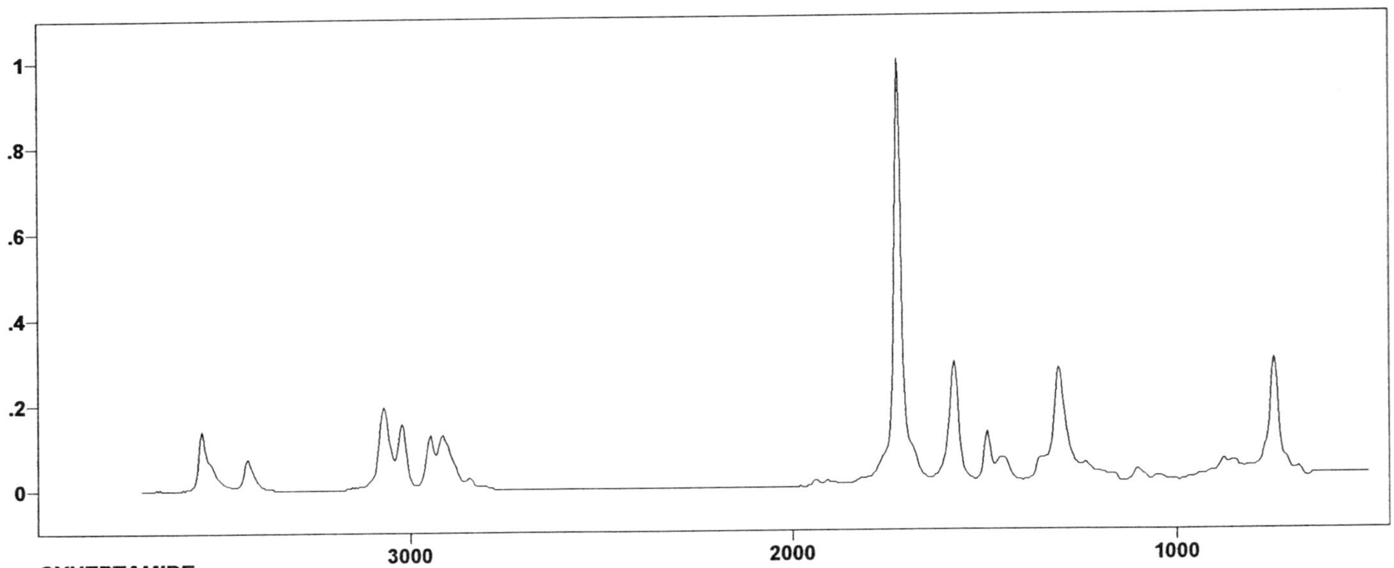

CYHEPTAMIDE

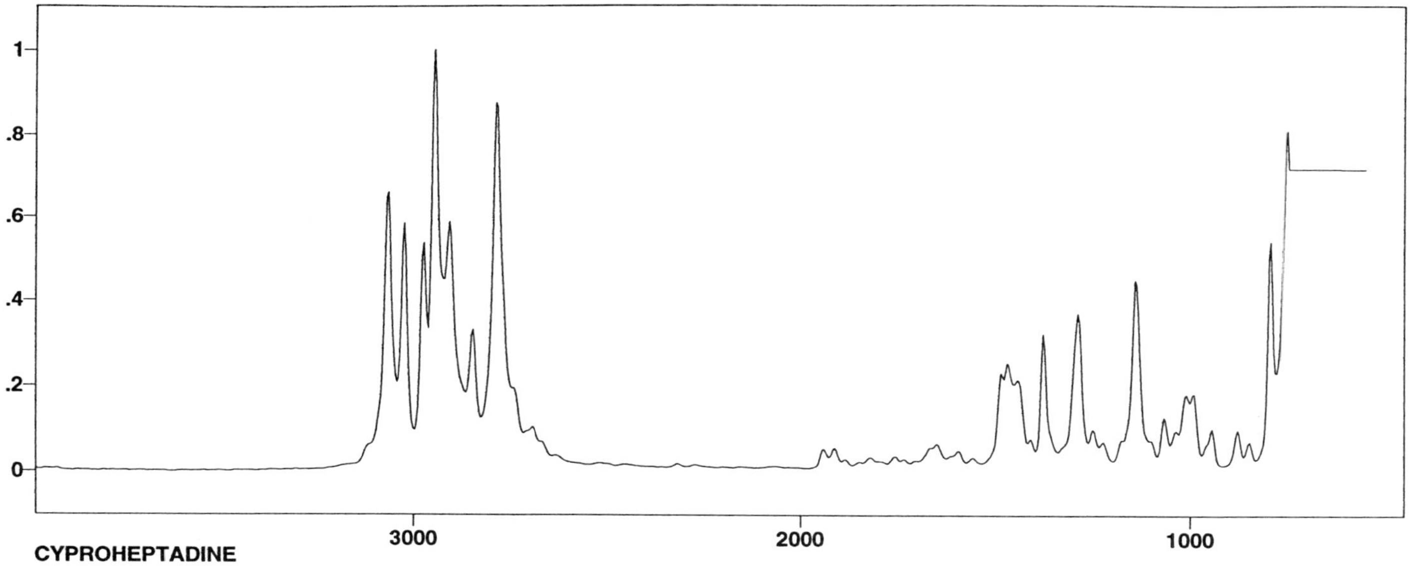

CYPROHEPTADINE

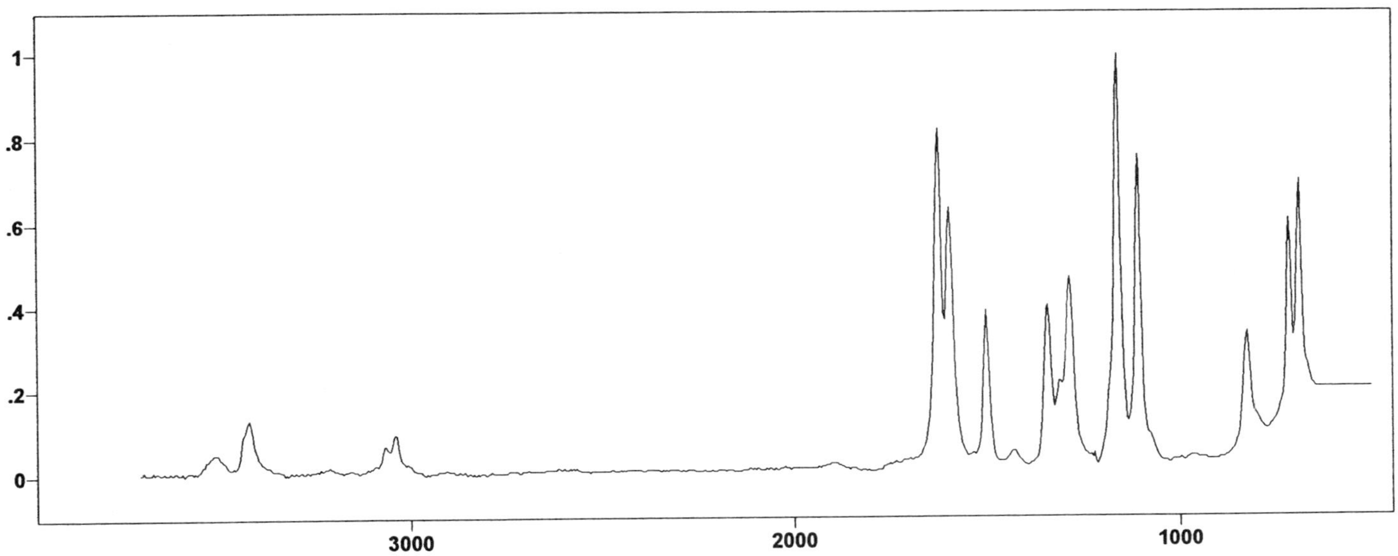

DAPSONE

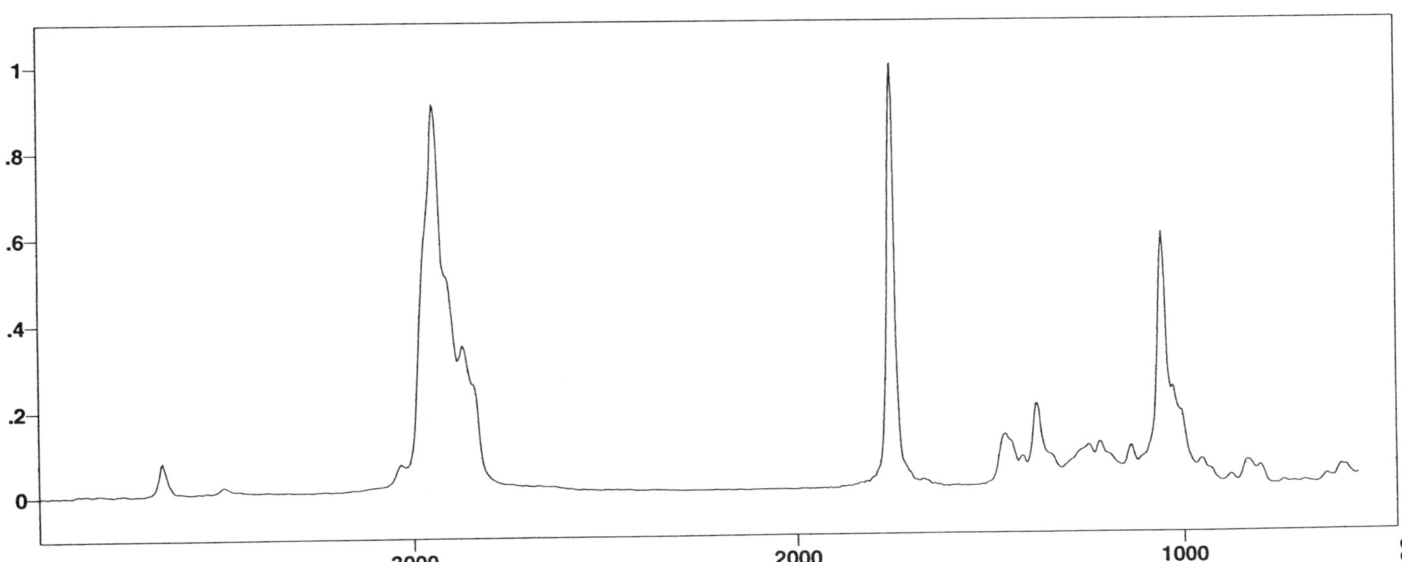

DEHYDROISOANDROSTERONE

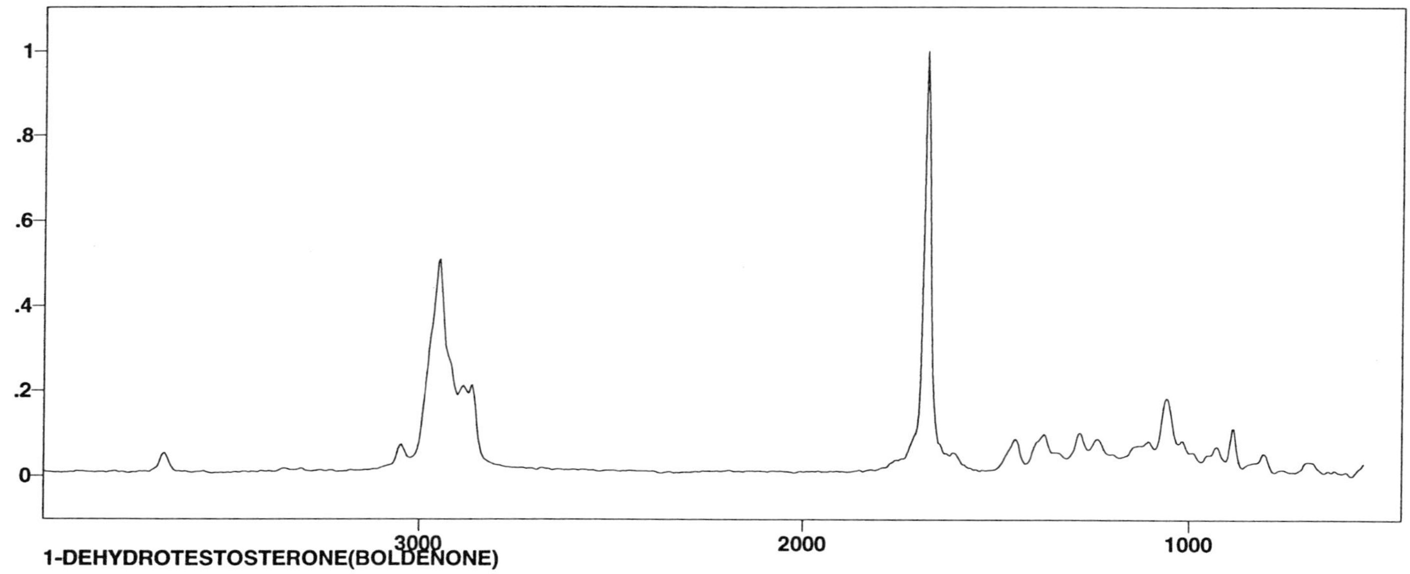

1-DEHYDROTESTOSTERONE(BOLDENONE)

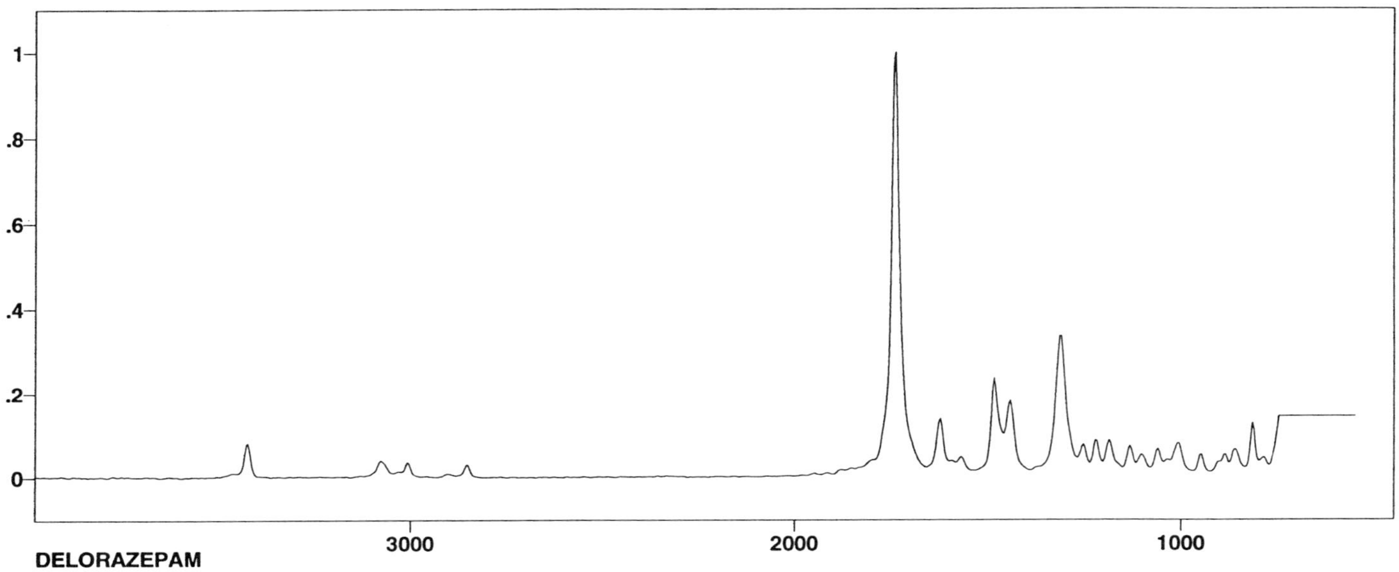

DELORAZEPAM

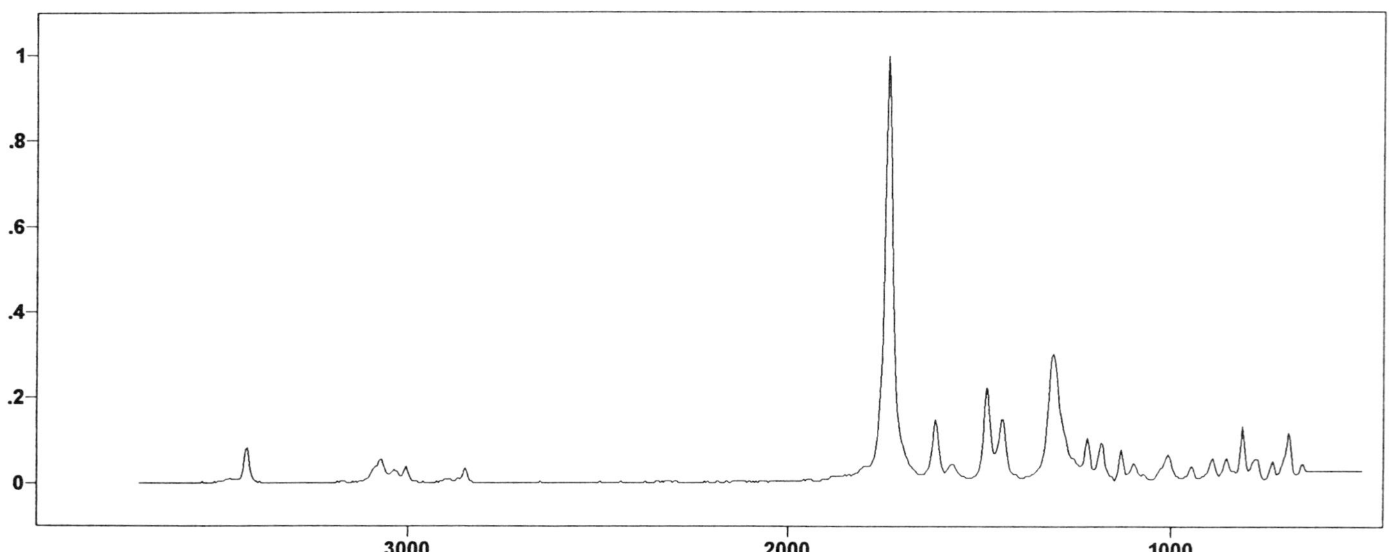

DEMOXEPAM

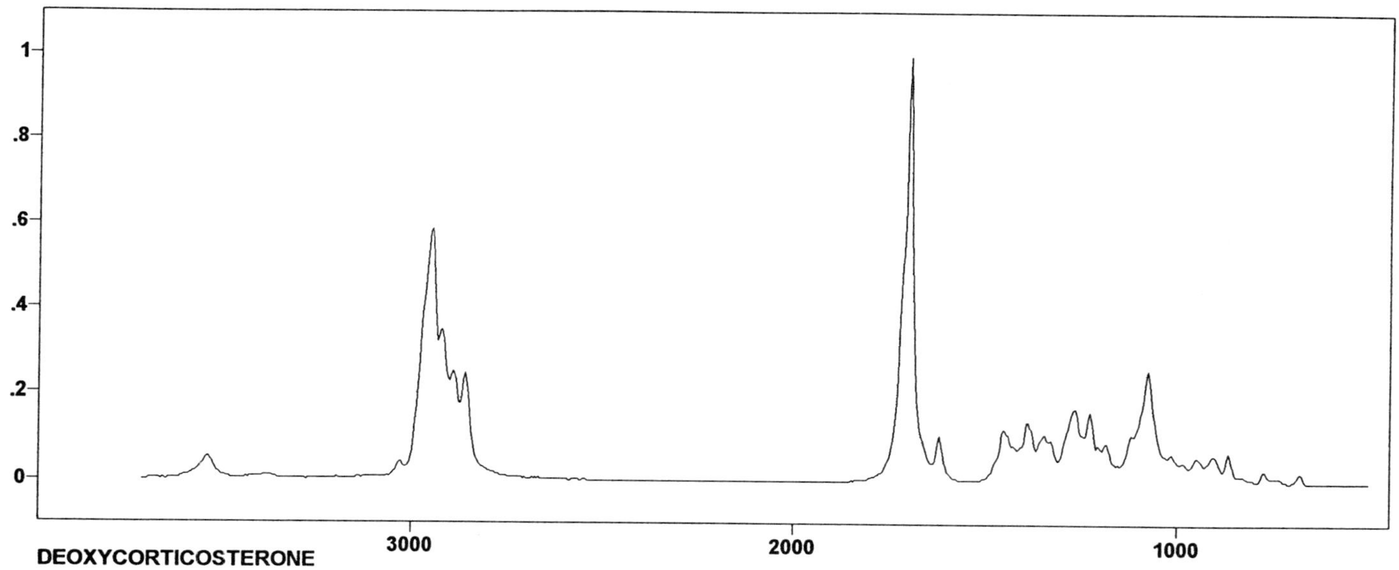

DEOXYCORTICOSTERONE

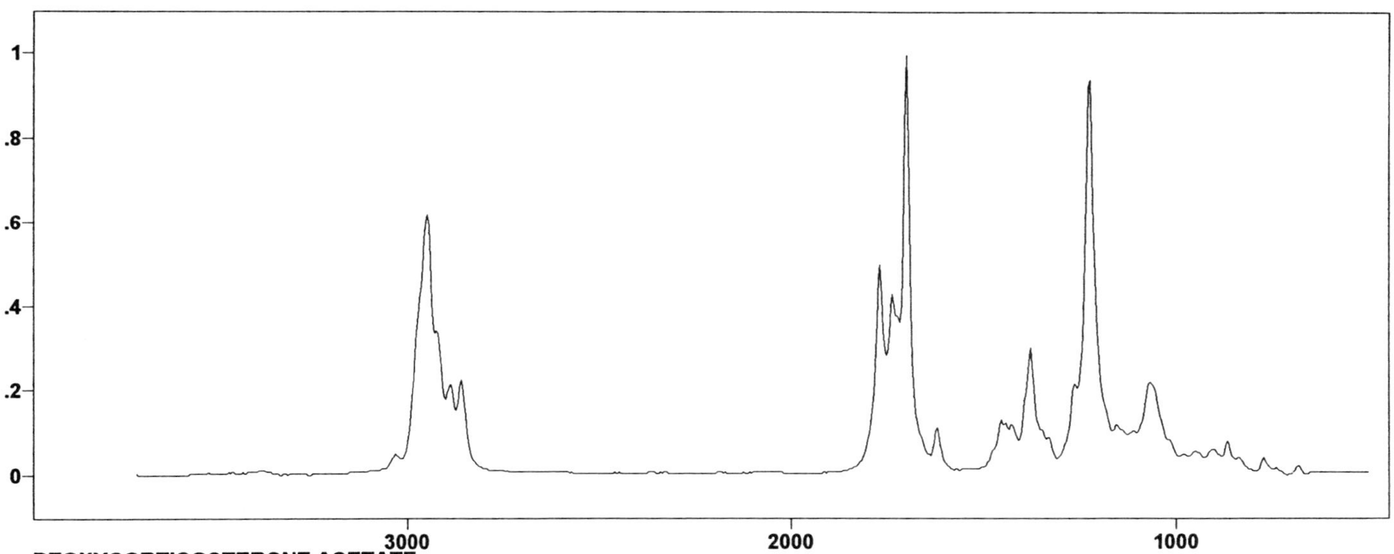

DEOXYCORTICOSTERONE ACETATE

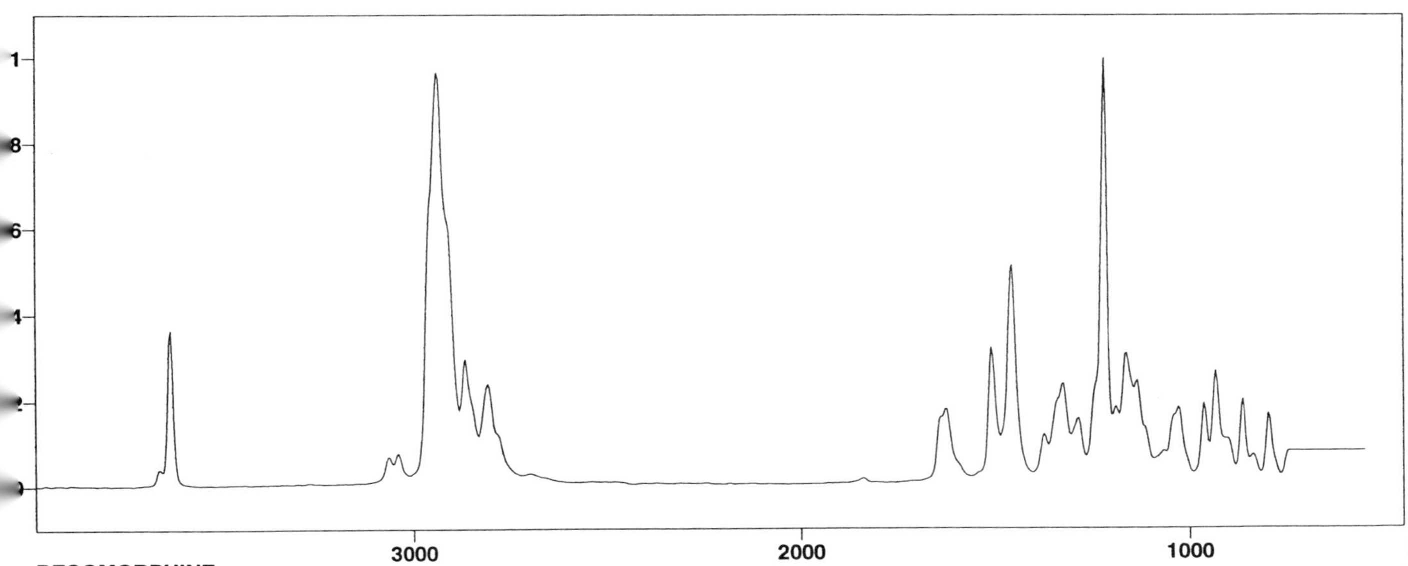

DESOMORPHINE

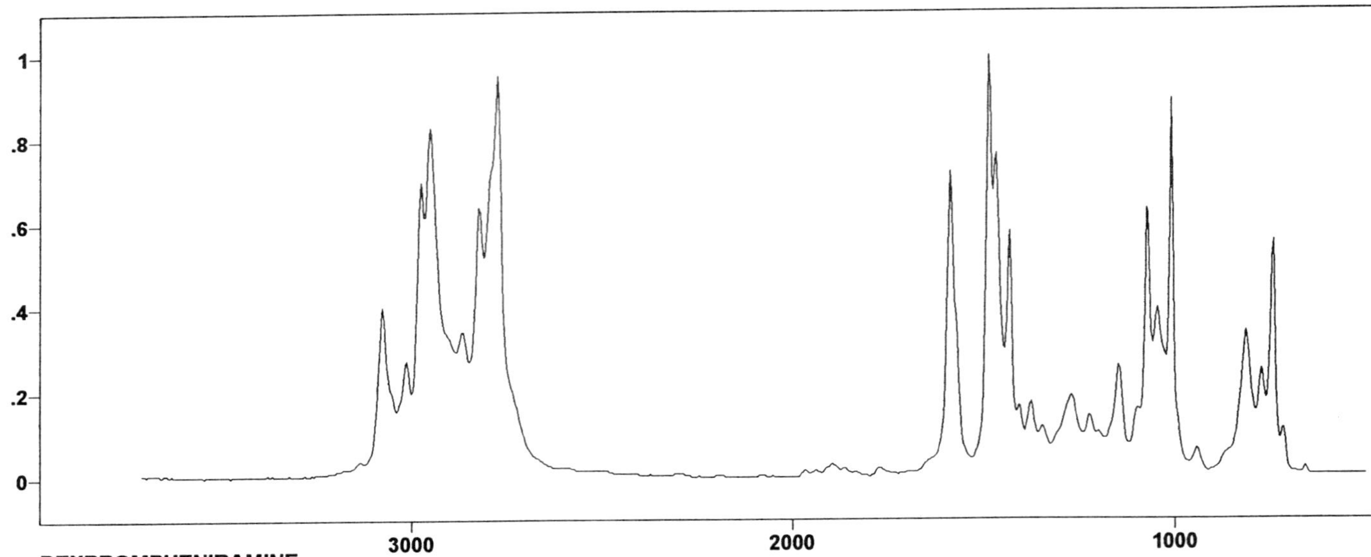

DEXBROMPHENIRAMINE

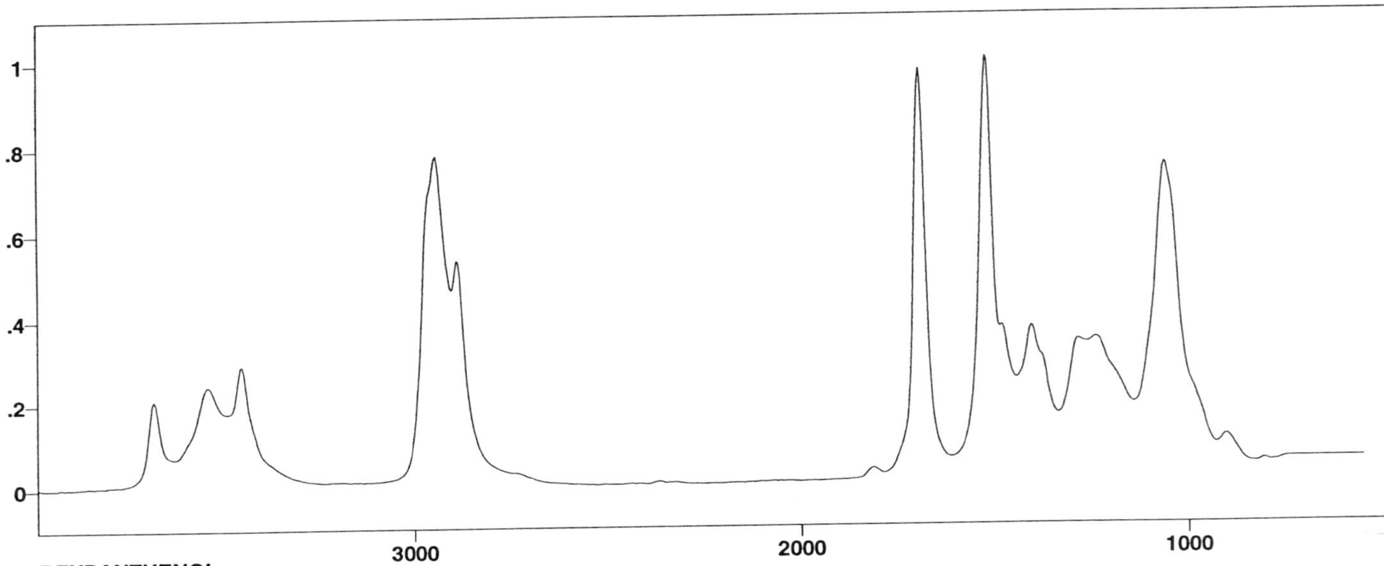

DEXPANTHENOL

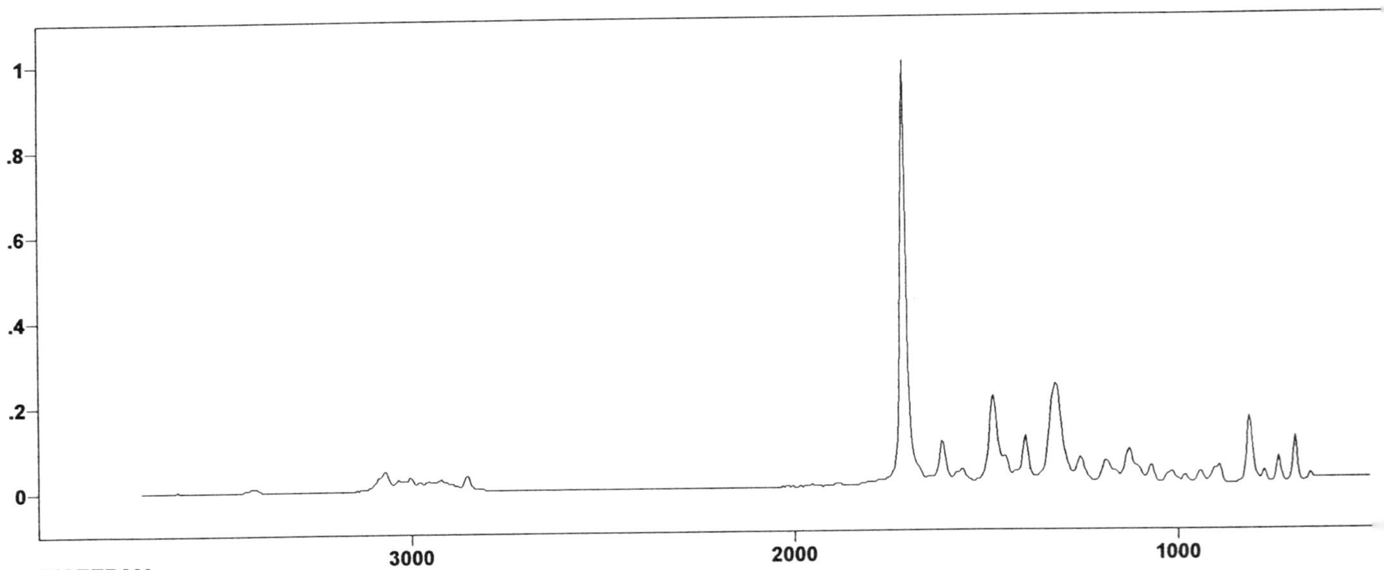

DIAZEPAM

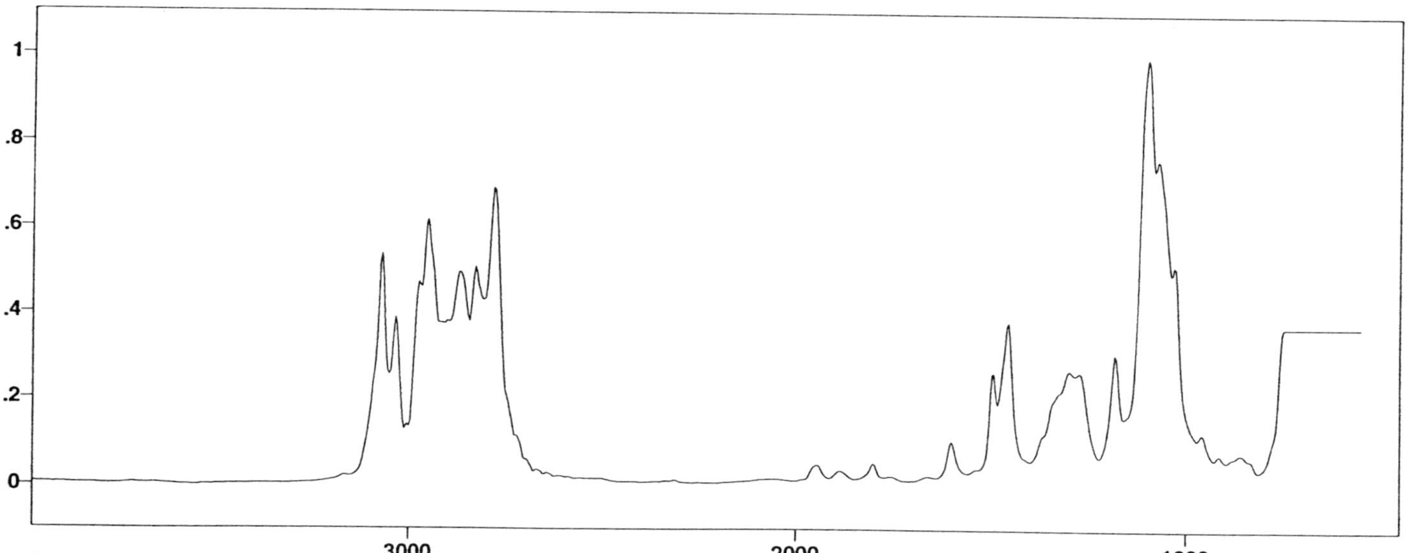

DIPHENHYDRAMINE

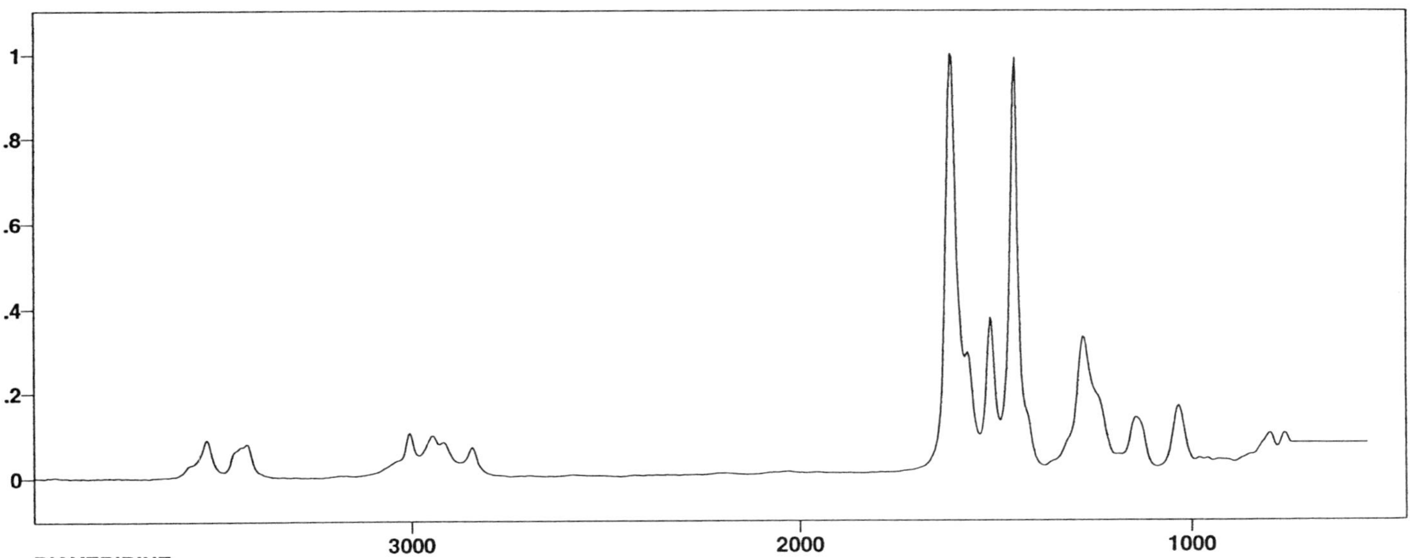

DIAVERIDINE

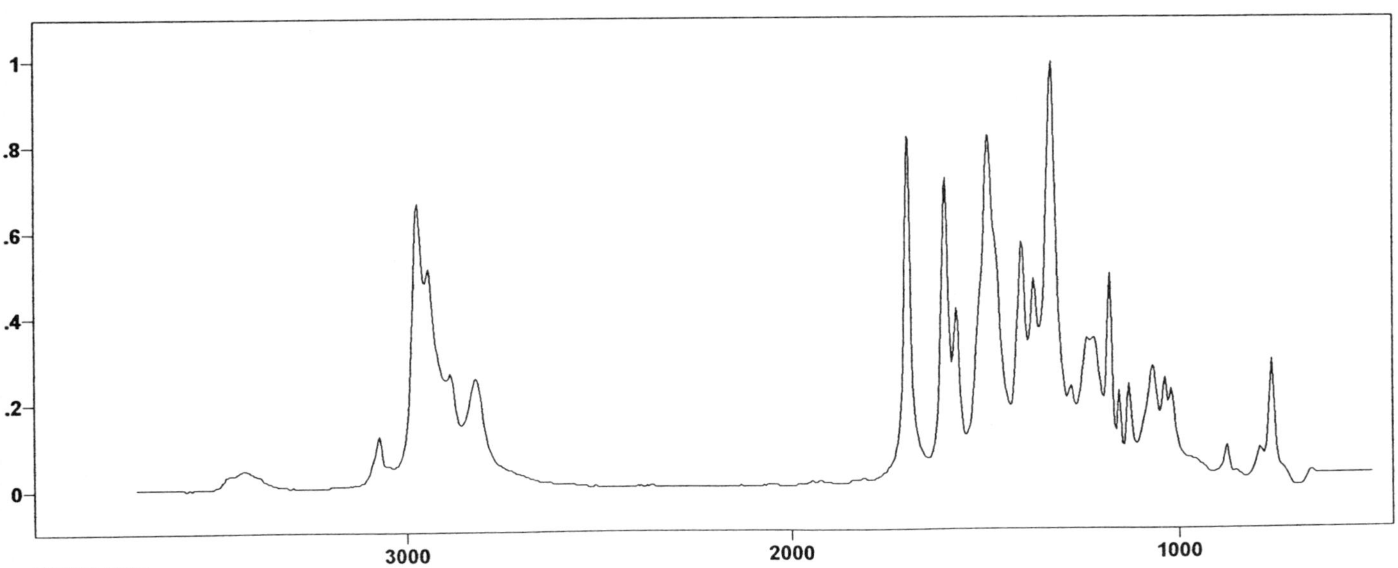

DIBUCAINE

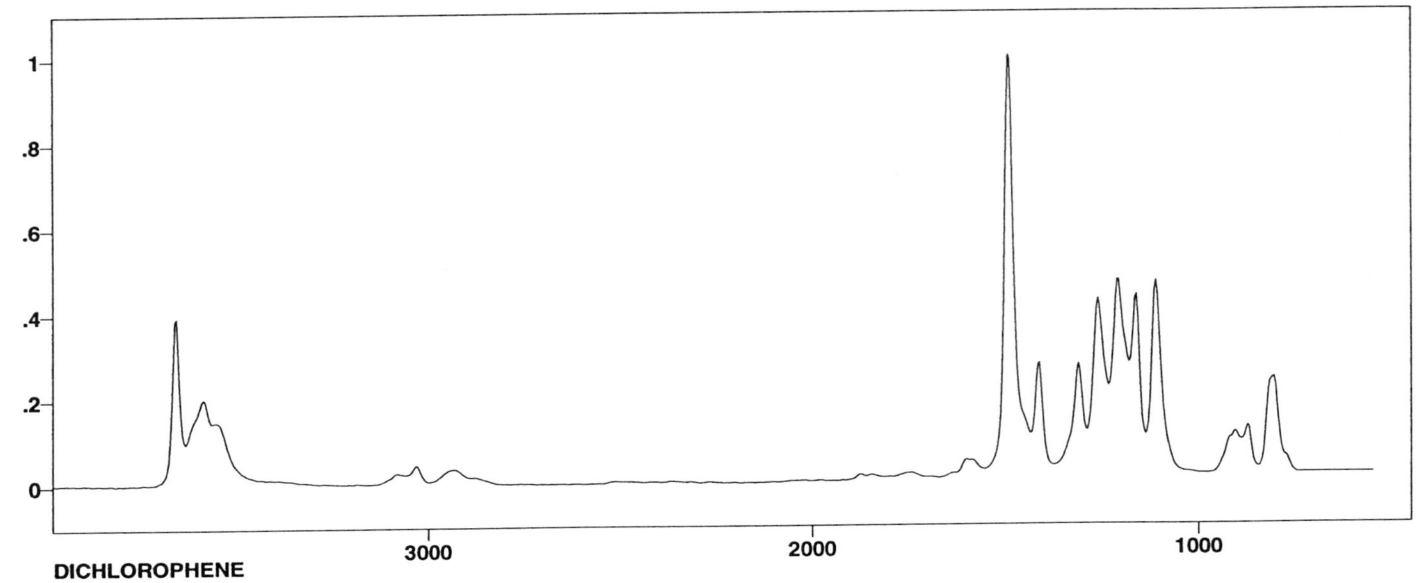

DICHLOROPHENE

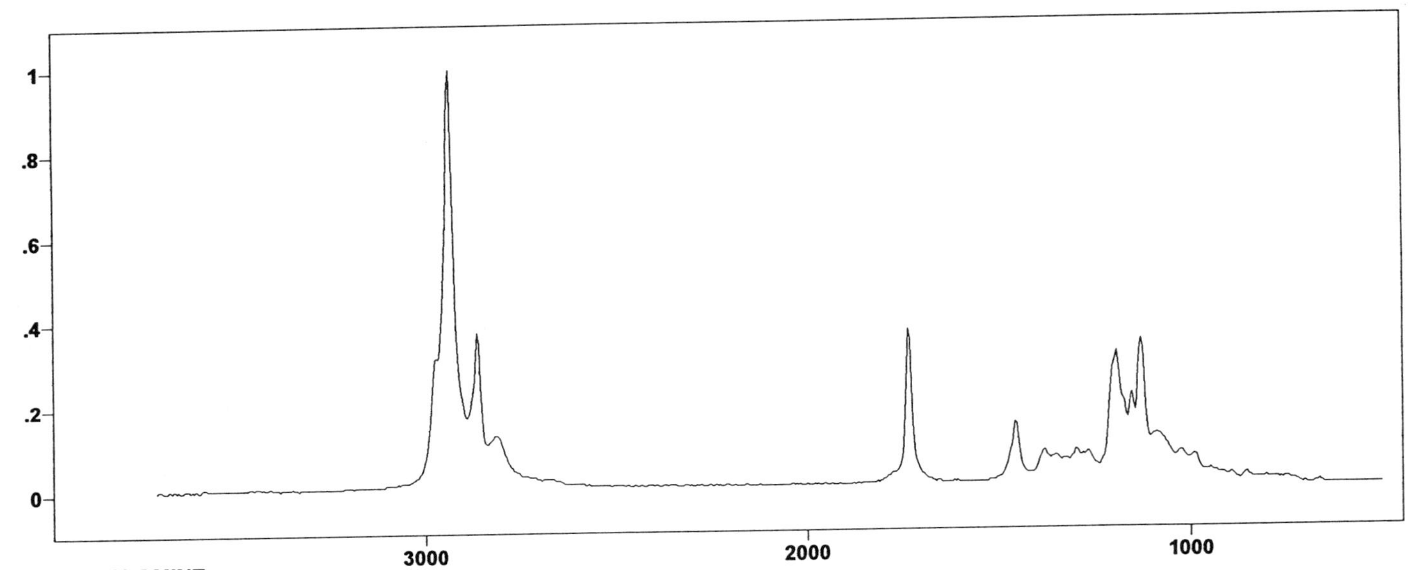

DICYCLOMINE

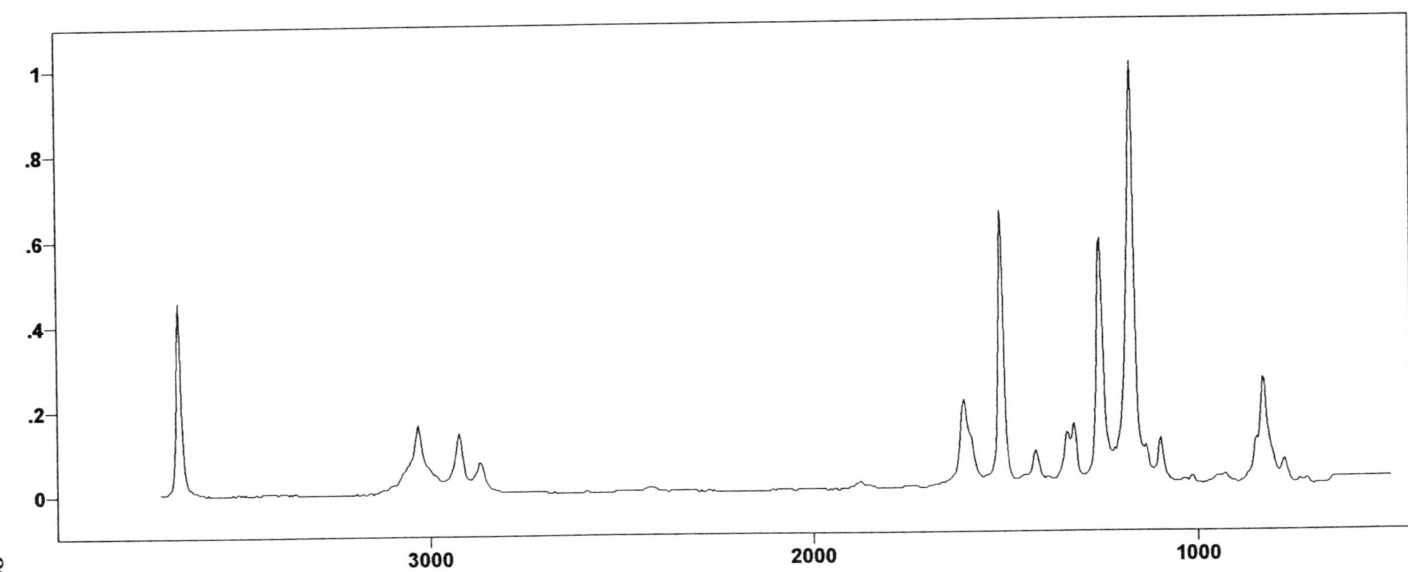

DIENESTROL

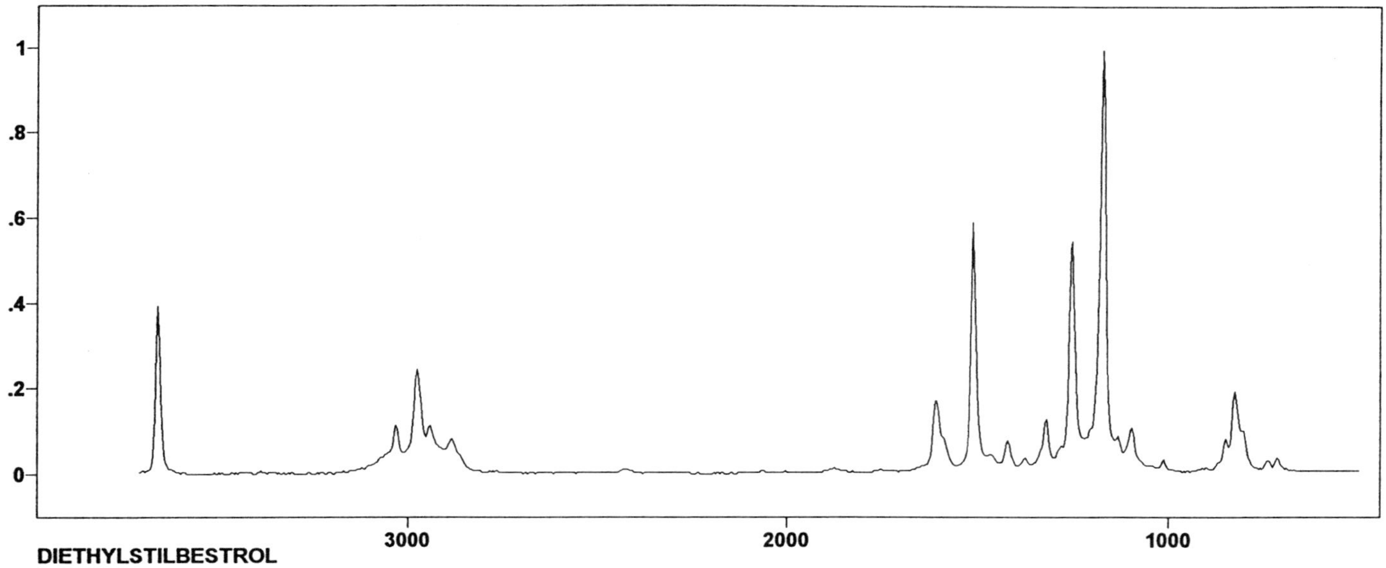

DIETHYLSTILBESTROL

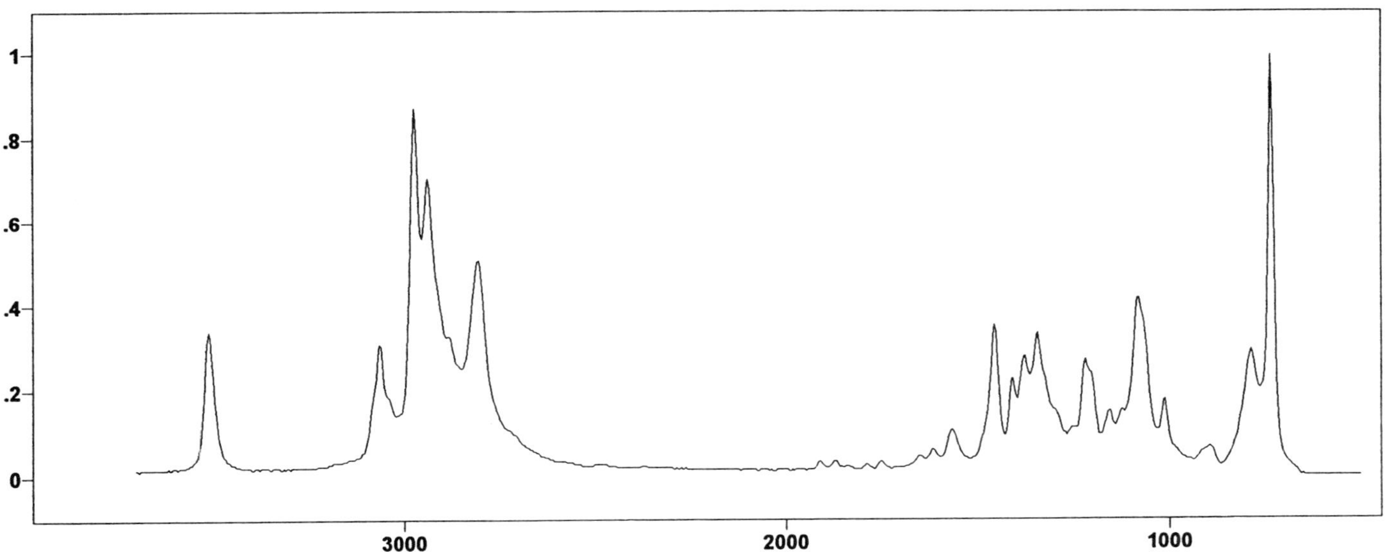

DIETHYLTRYPTAMINE

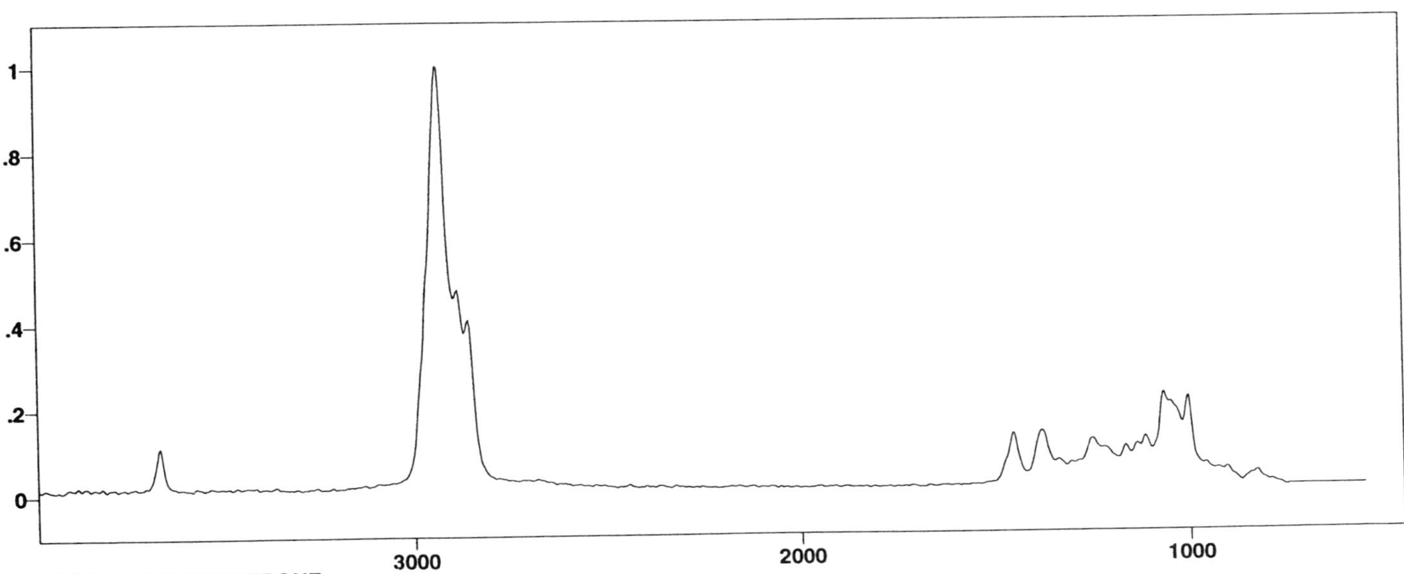

DIHYDROANDROSTERONE

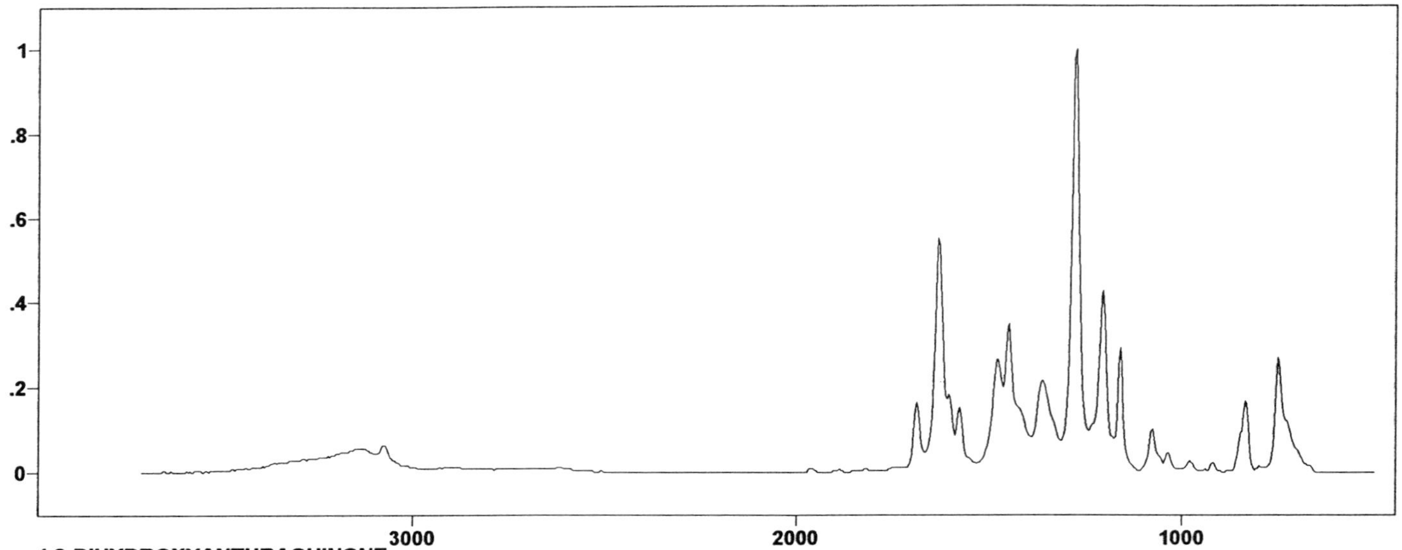

1,8-DIHYDROXYANTHRAQUINONE

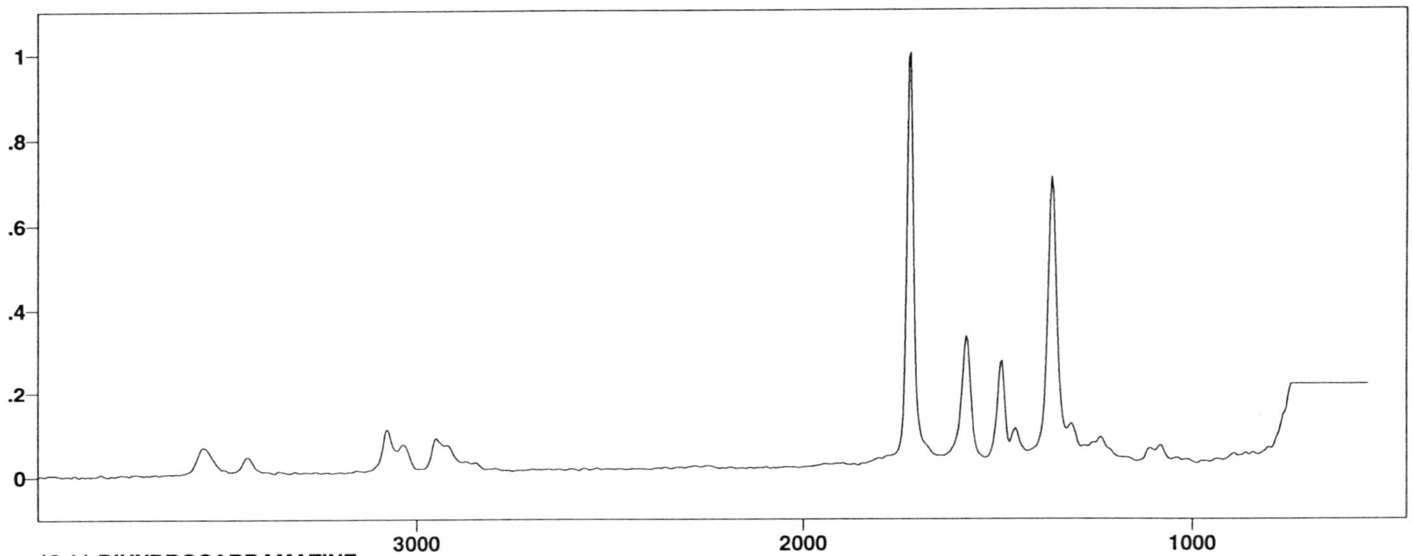

10,11-DIHYDROCARBAMAZINE

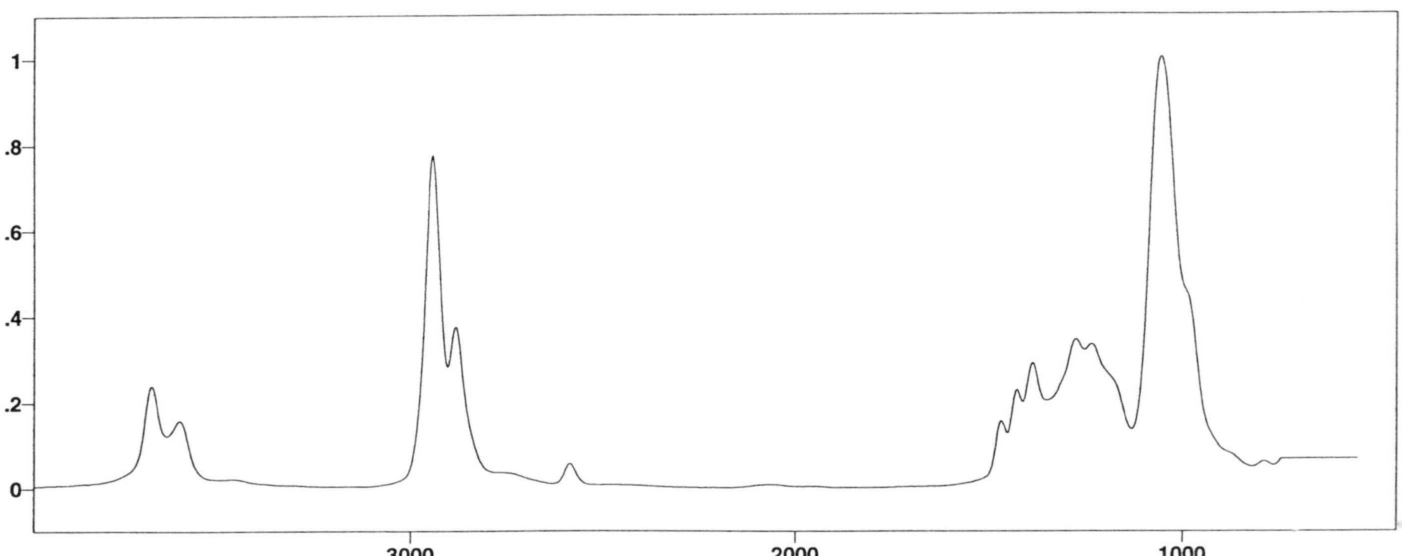

DIMERCAPROL

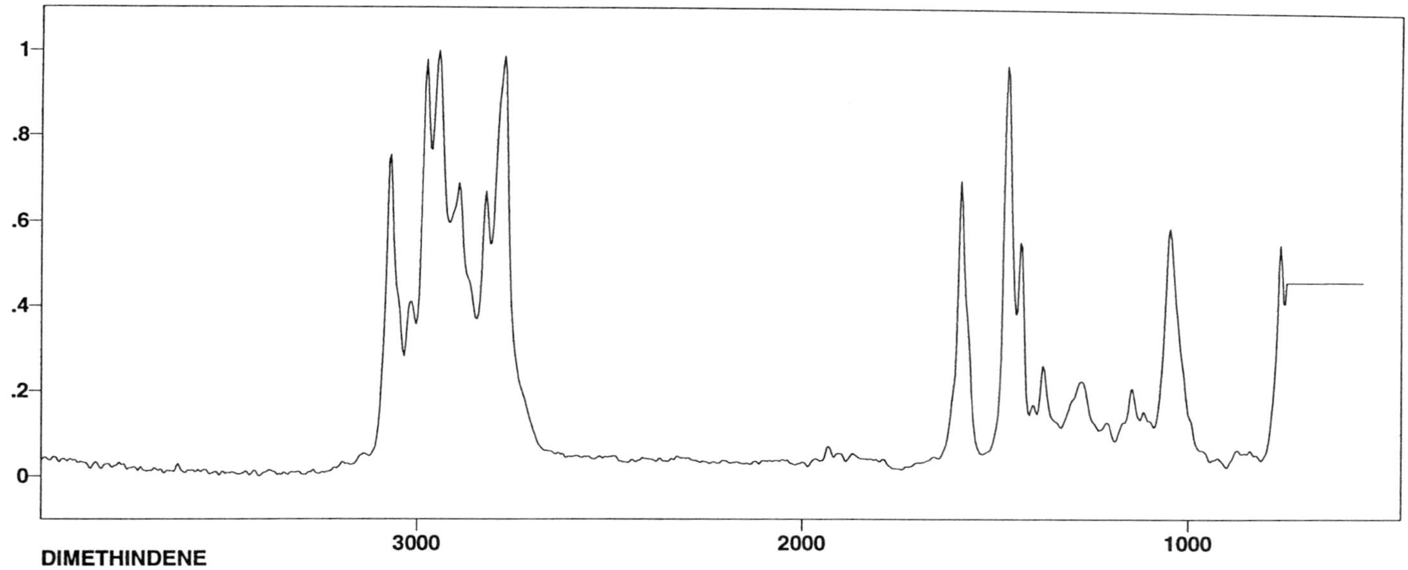

DIMETHINDENE

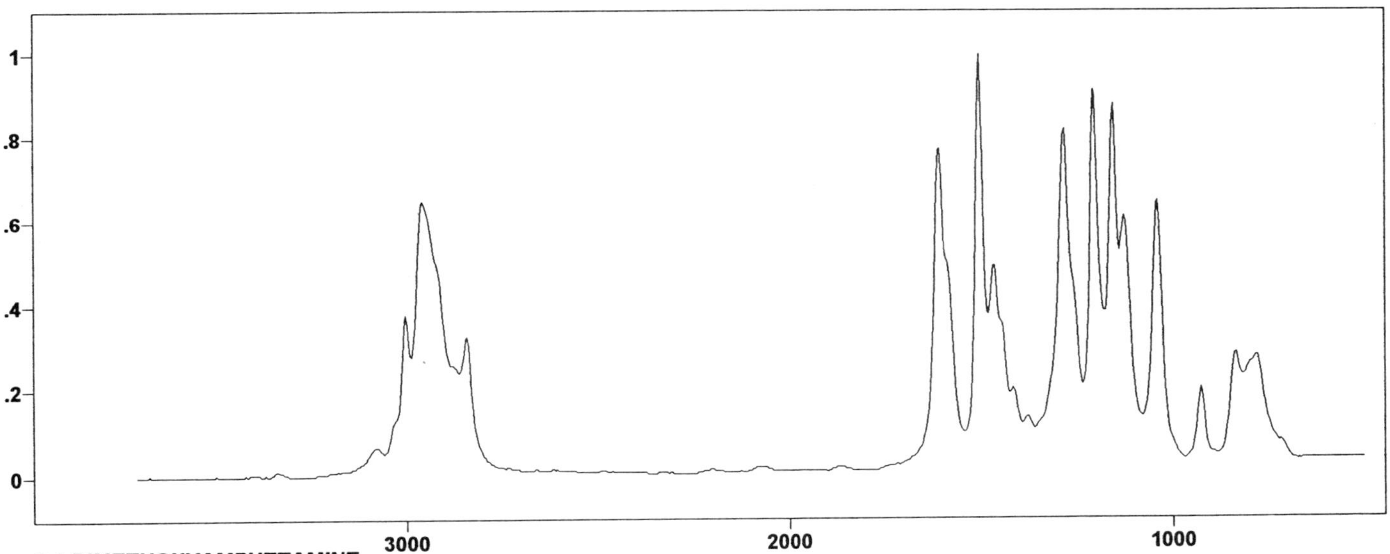

2,4-DIMETHOXYAMPHETAMINE

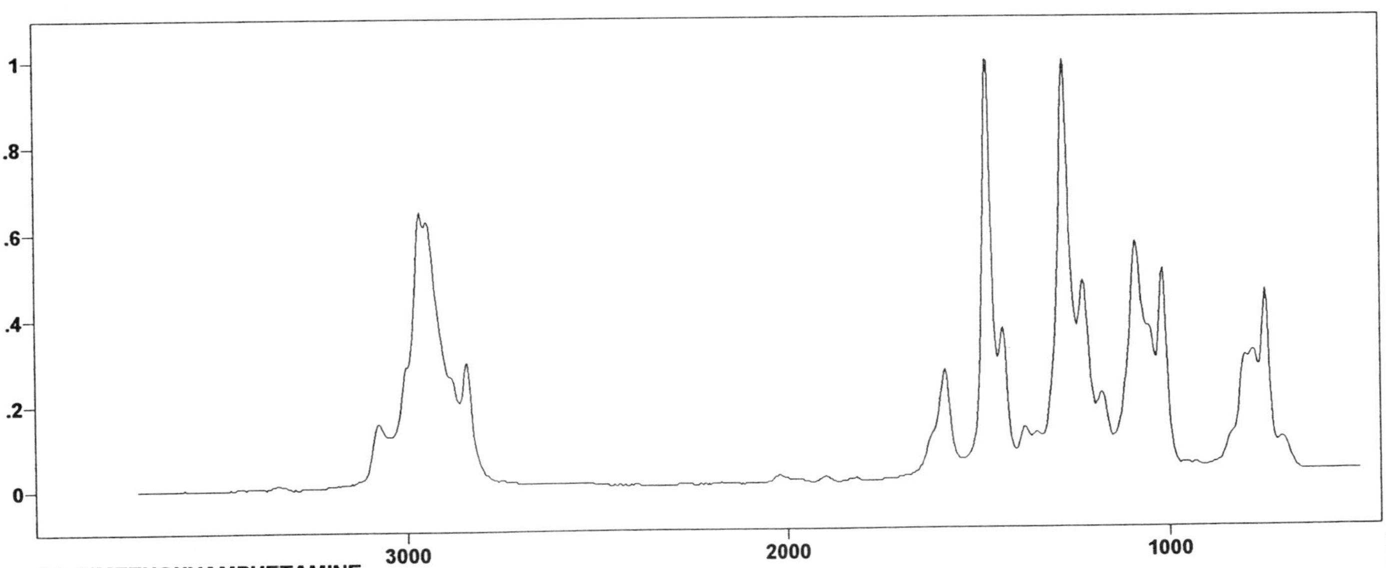

2,3-DIMETHOXYAMPHETAMINE

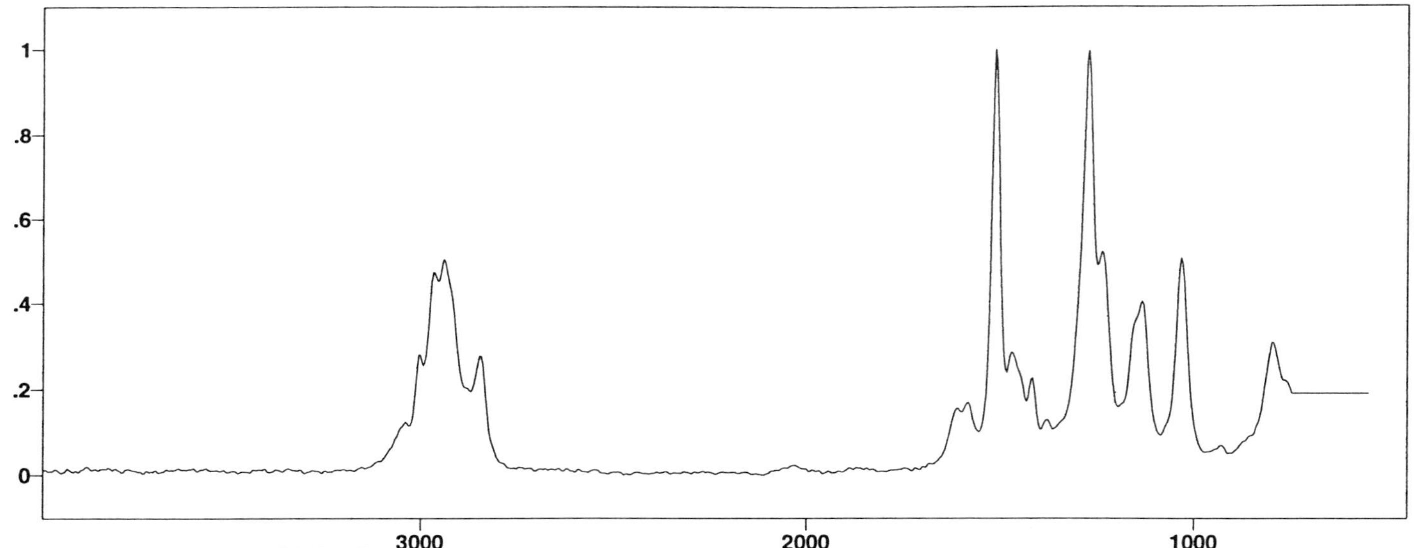

3,4-DIMETHOXYAMPHETAMINE

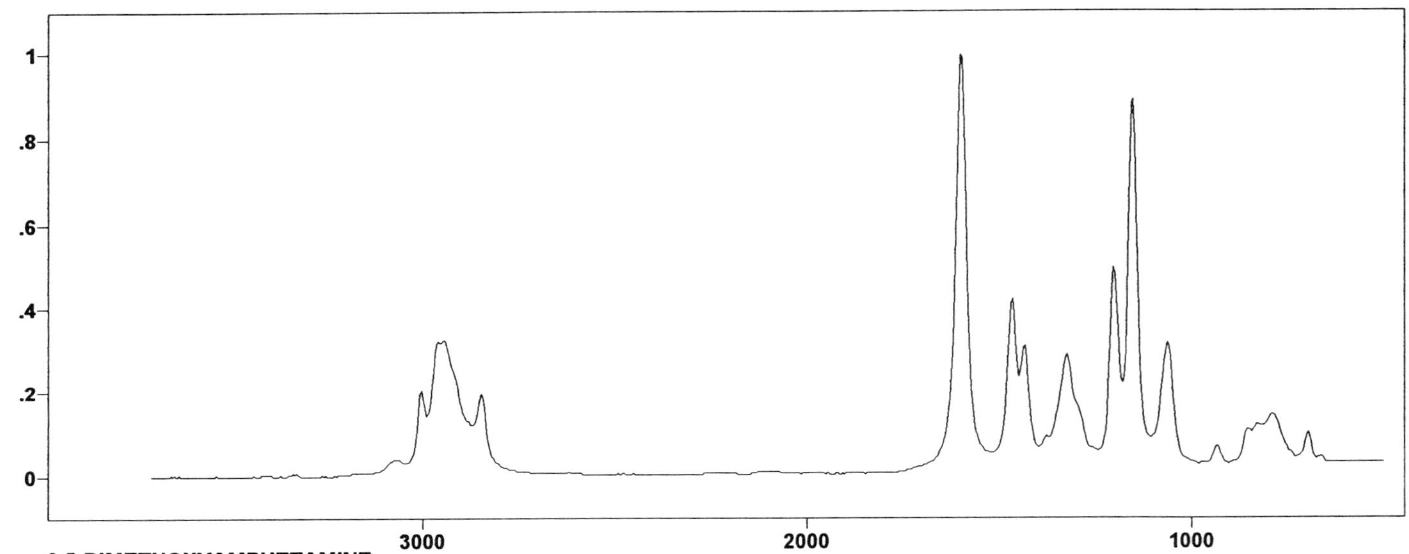

3,5-DIMETHOXYAMPHETAMINE

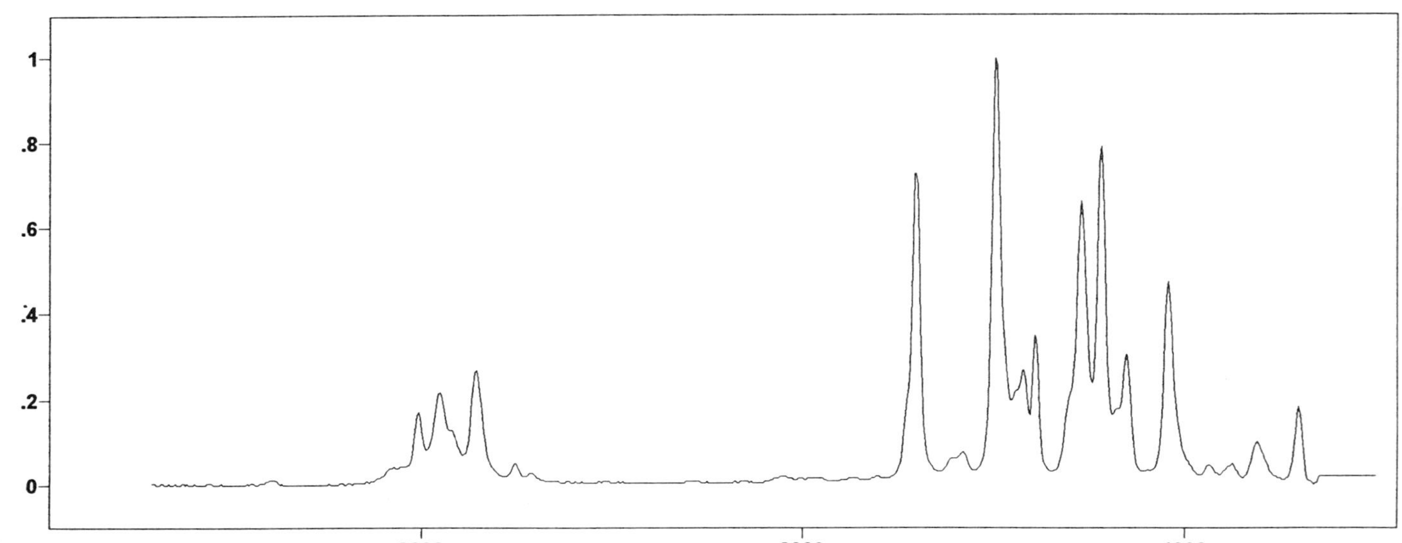

2,5-DIMETHOXYBENZALDEHYDE

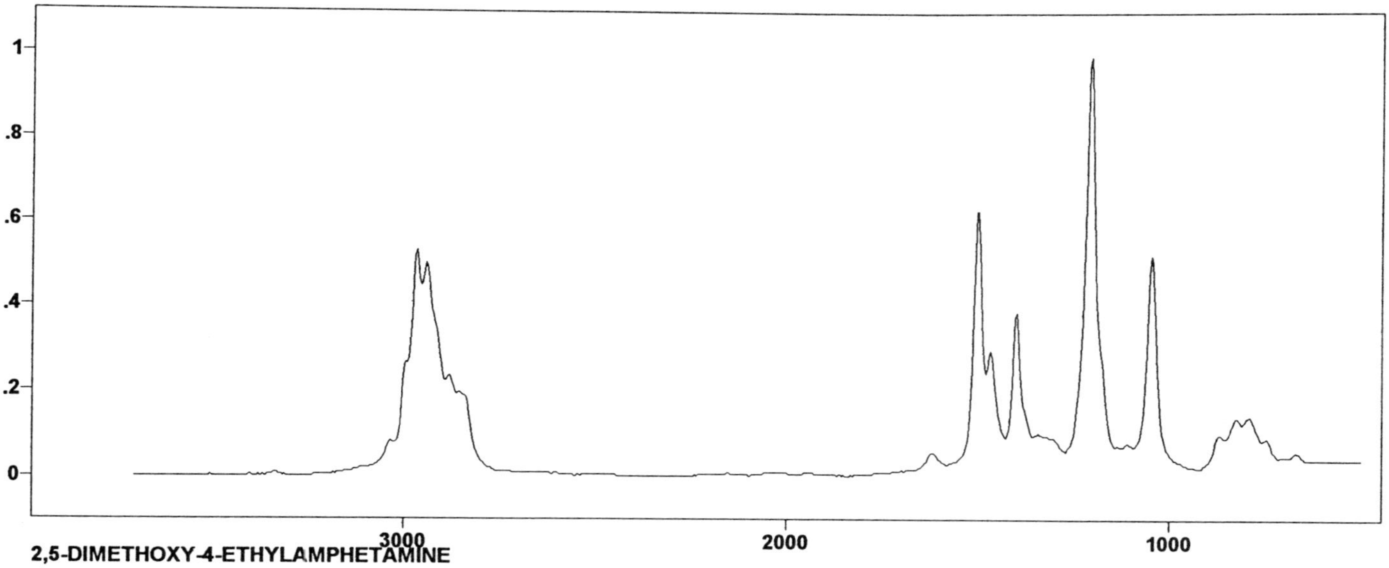

2,5-DIMETHOXY-4-ETHYLAMPHETAMINE

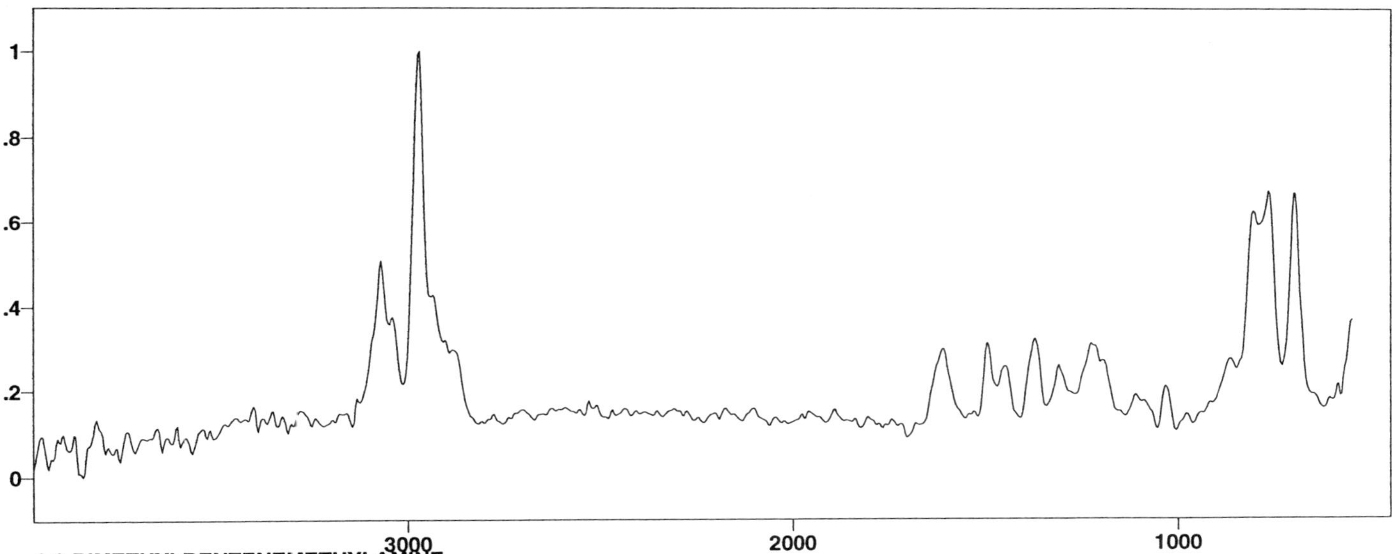

1,1-DIMETHYLBENZENEMETHYLAMINE

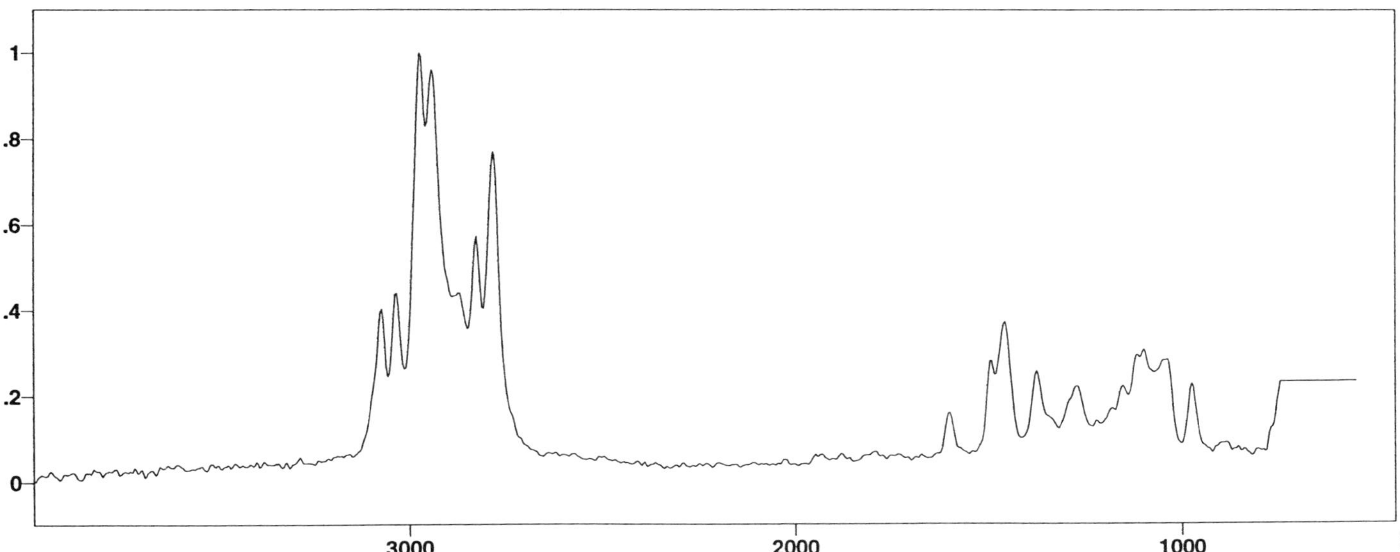

DIMETHYLAMPHETAMINE

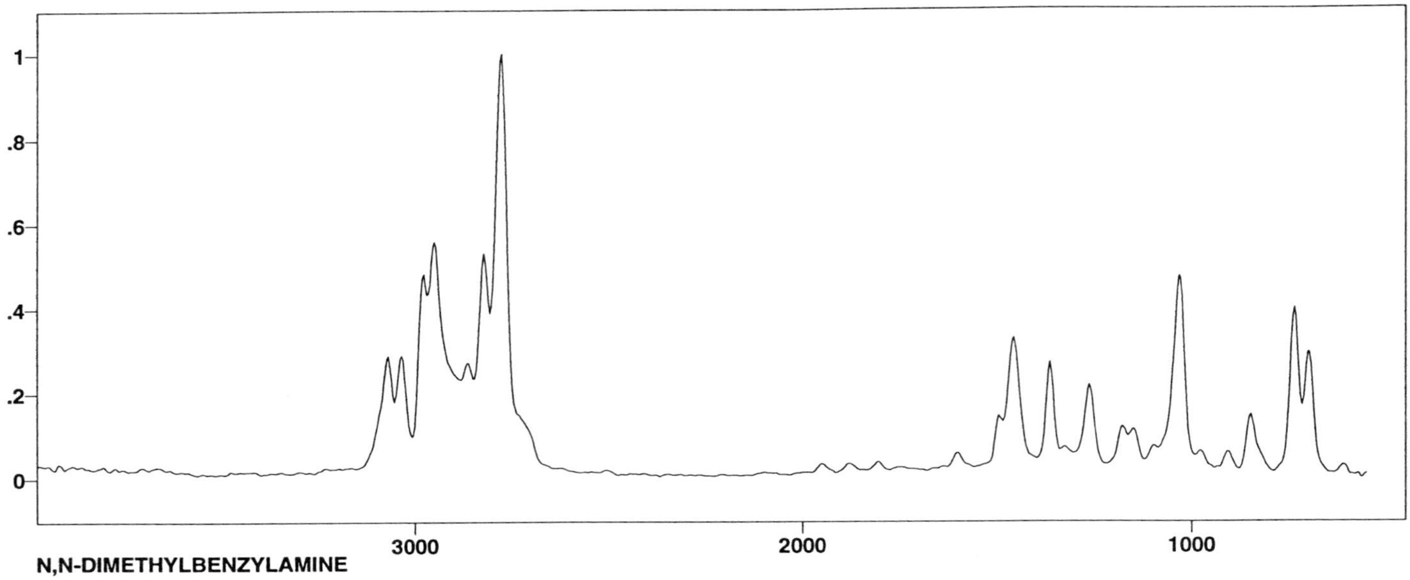

N,N-DIMETHYLBENZYLAMINE

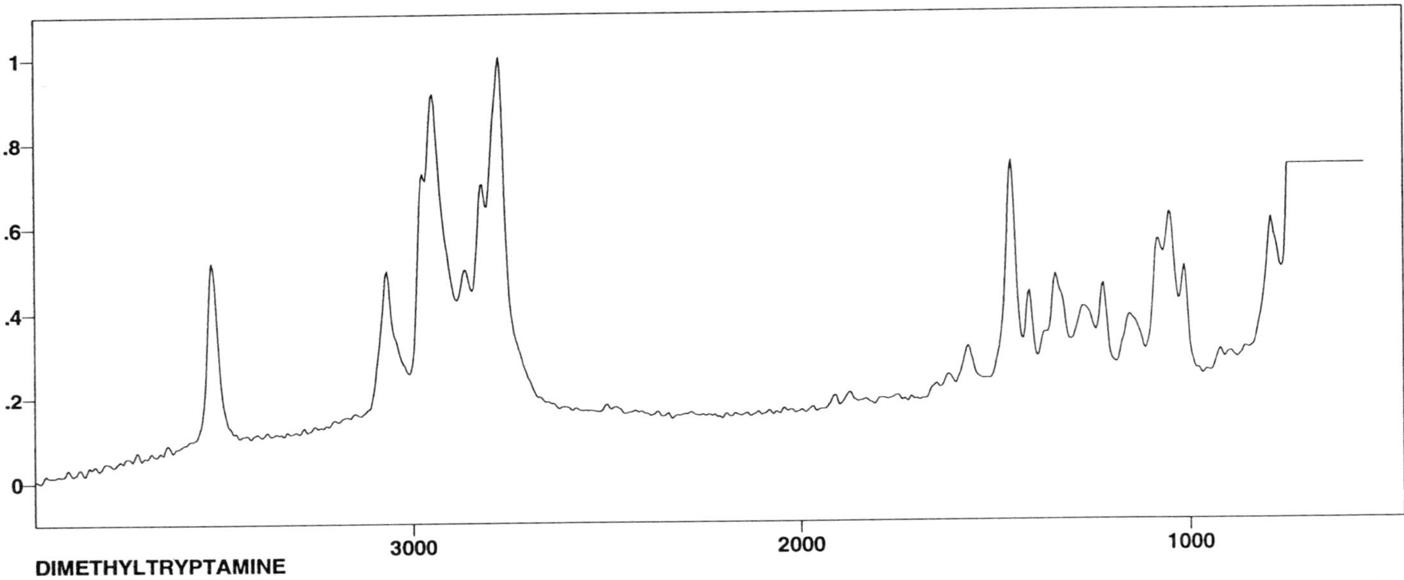

DIMETHYLTRYPTAMINE

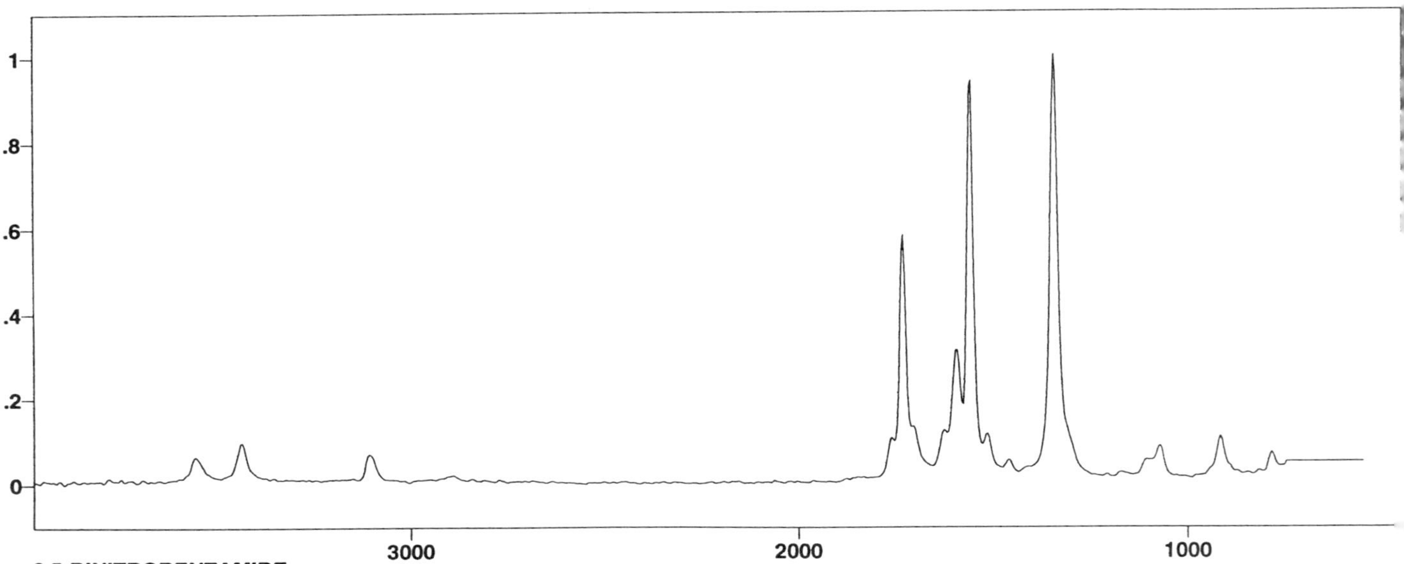

3,5-DINITROBENZAMIDE

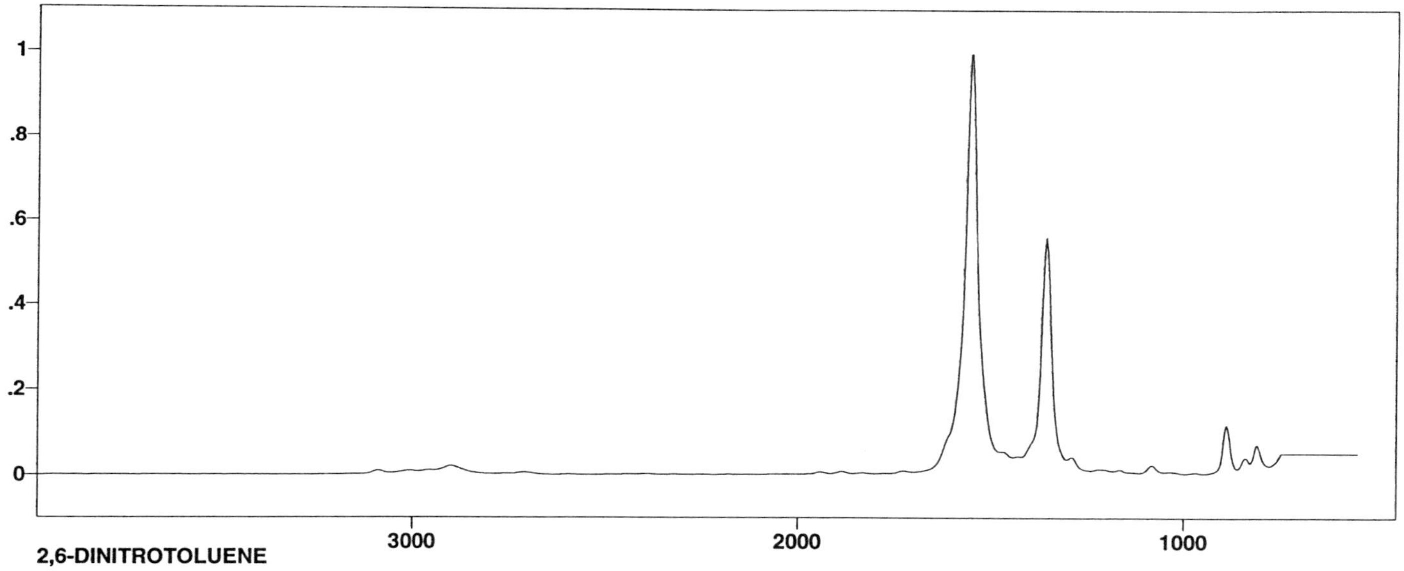

2,6-DINITROTOLUENE

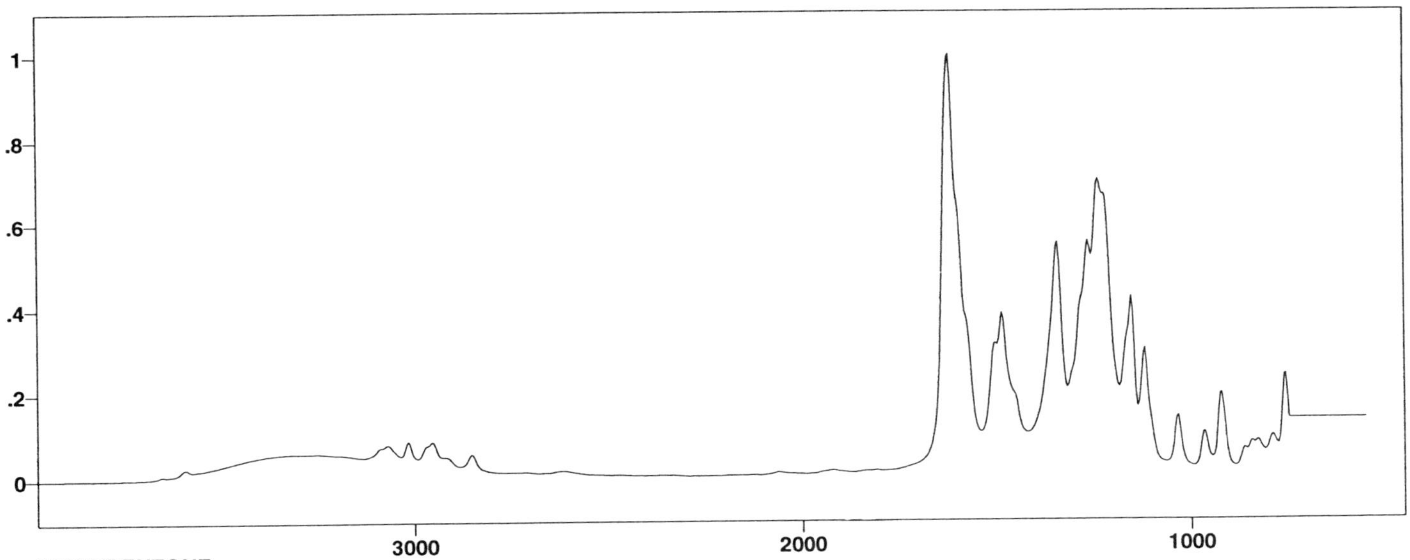

DIOXYBENZONE

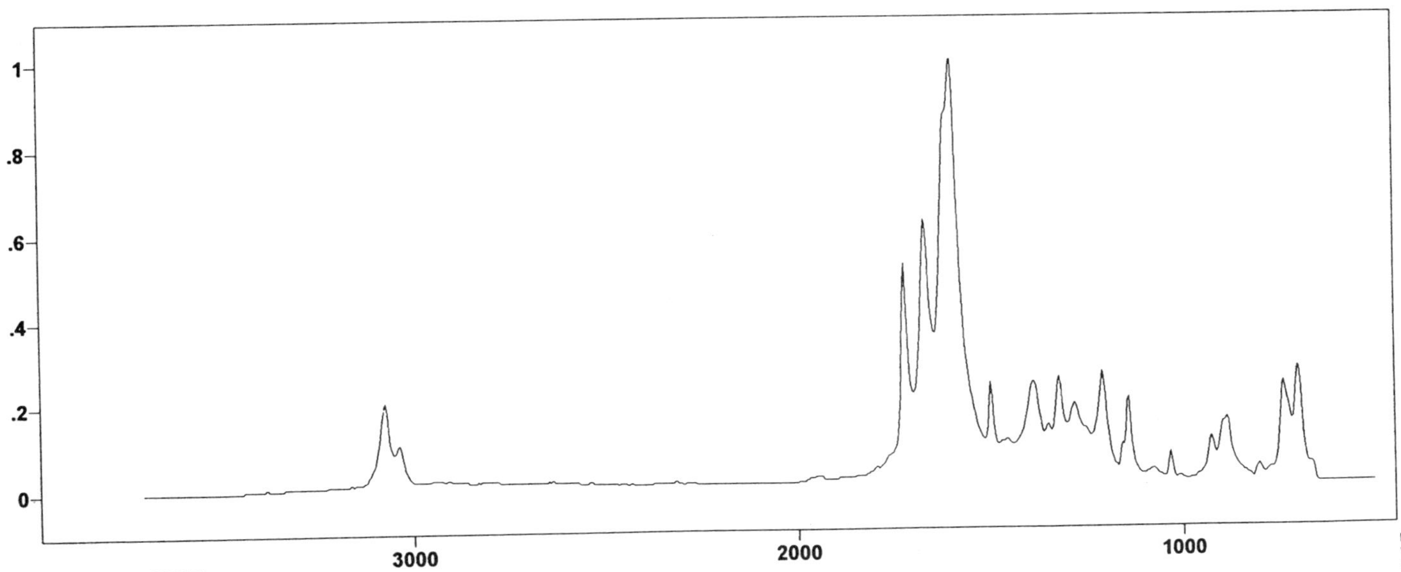

DIPHENADIONE

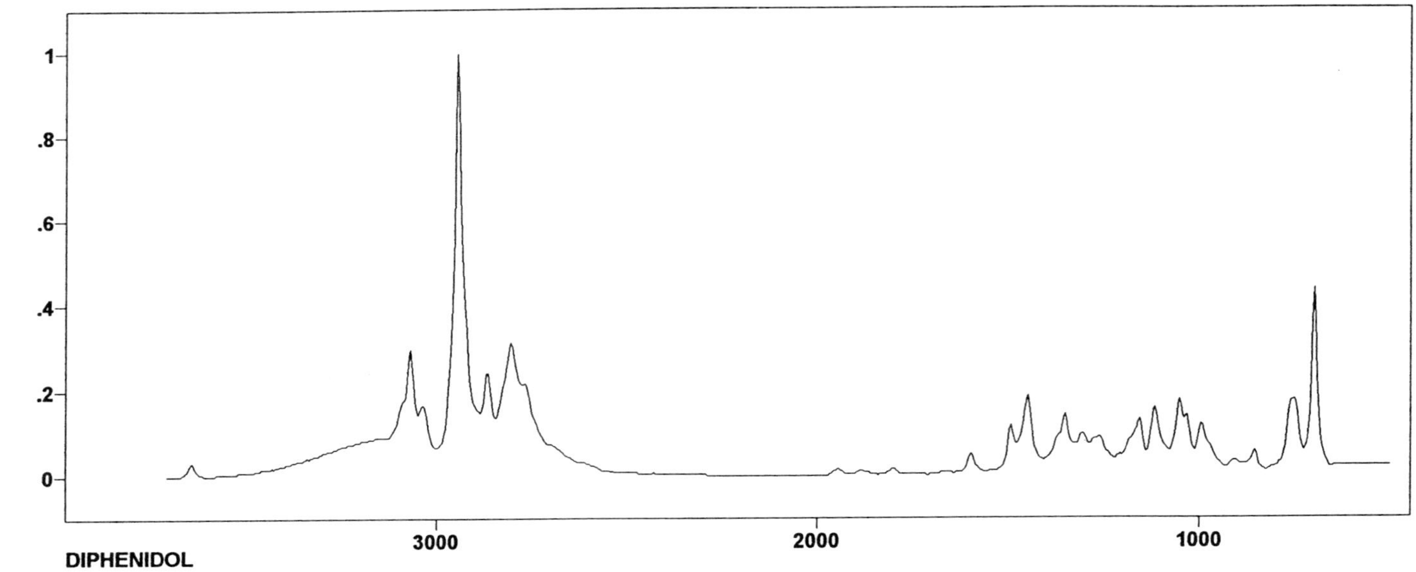

DIPHENIDOL

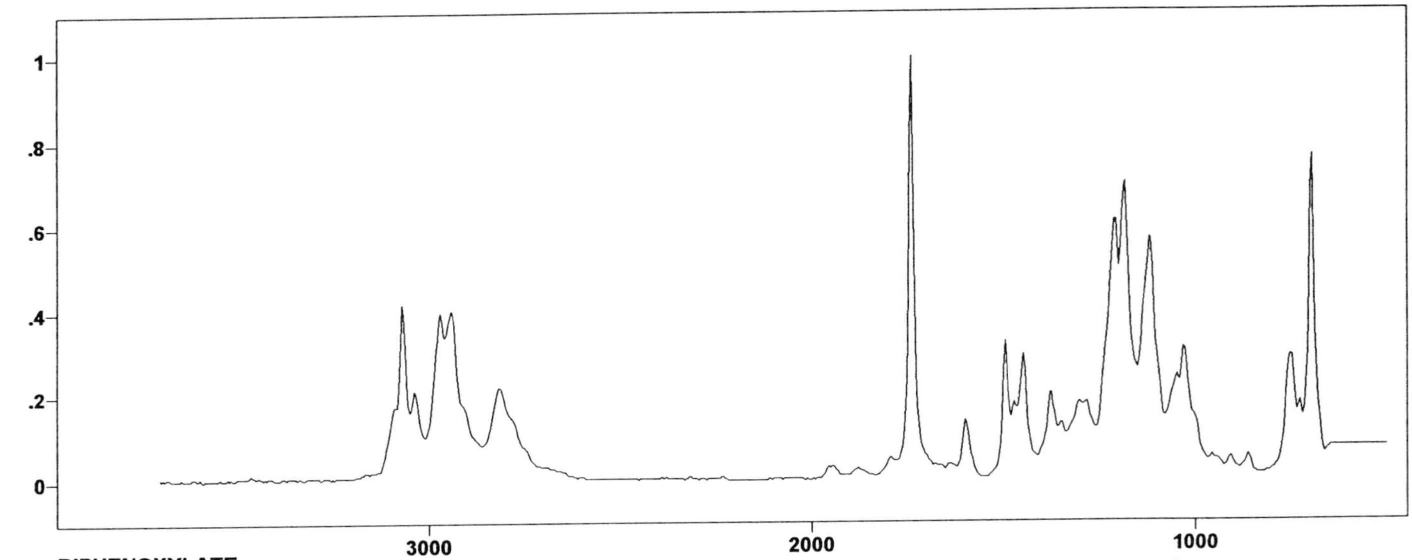

DIPHENOXYLATE

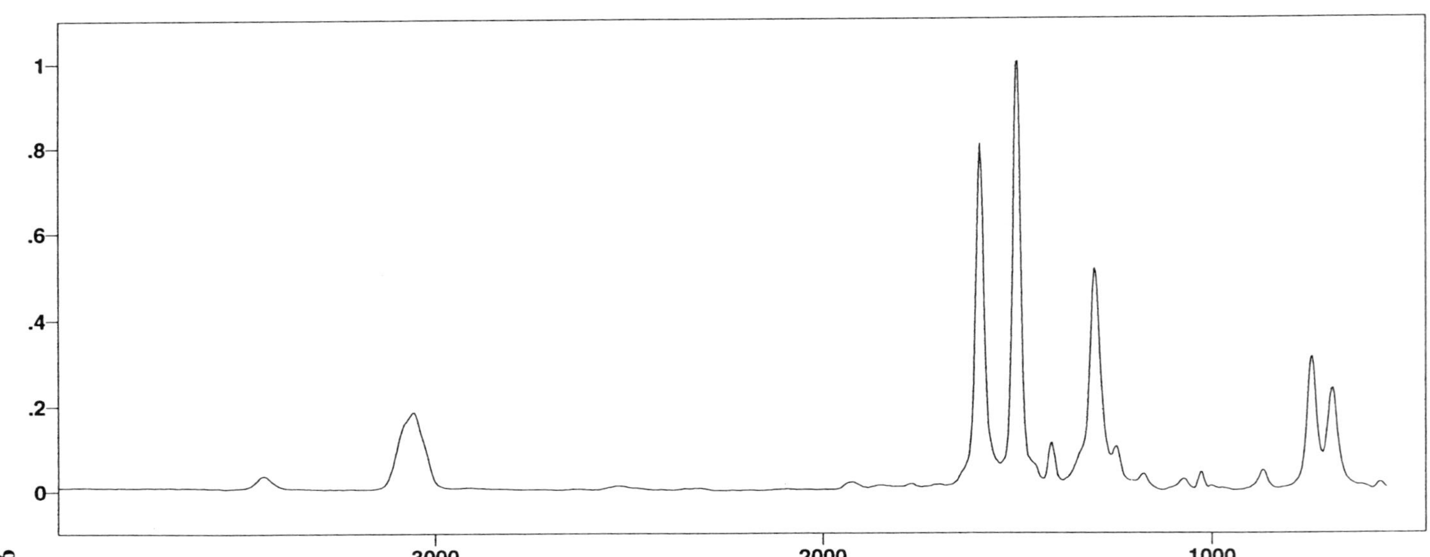

DIPHENYL AMINE

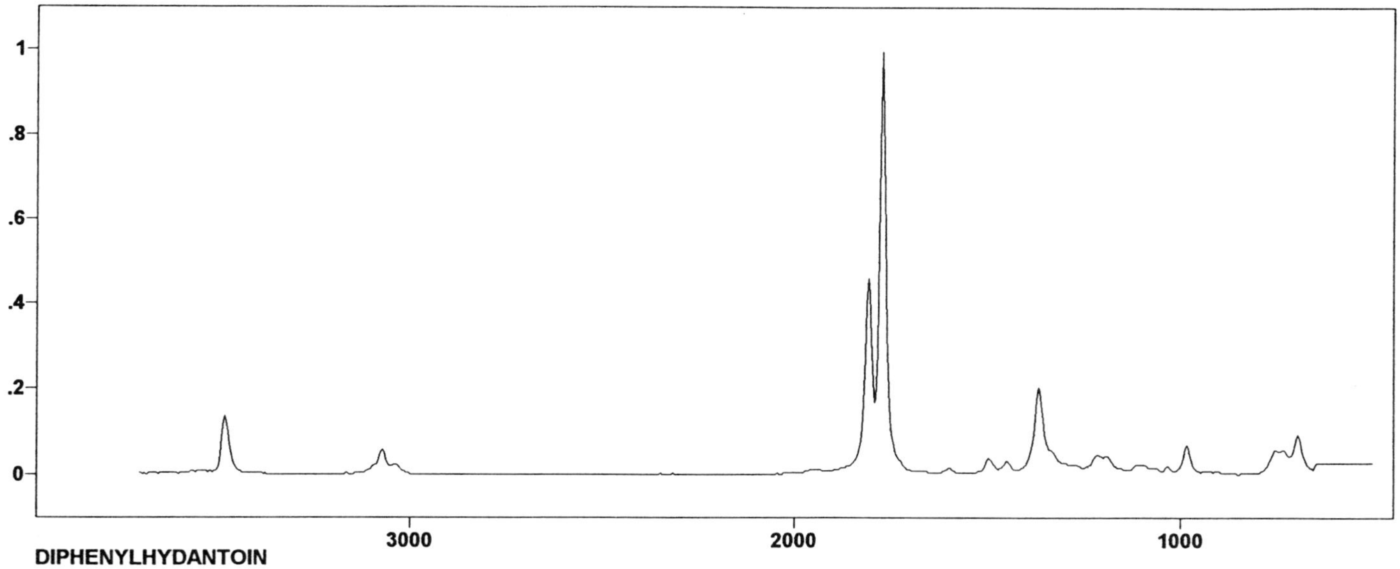

DIPHENYLHYDANTOIN

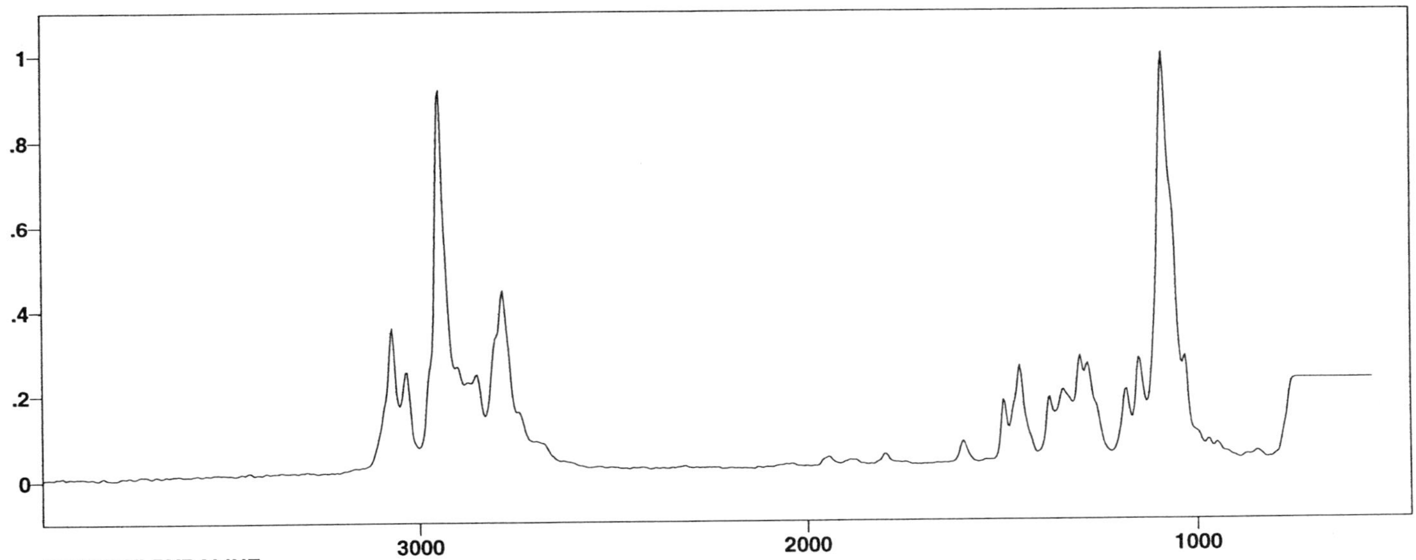

DIPHENYLPYRALINE

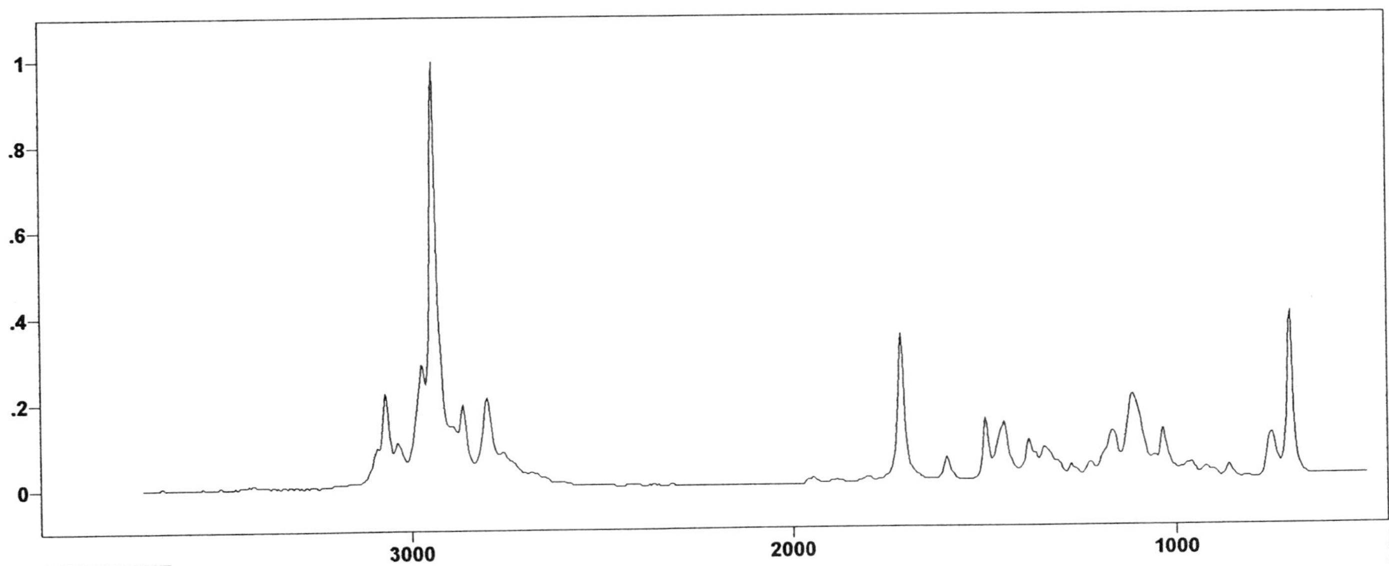

DIPIPANONE

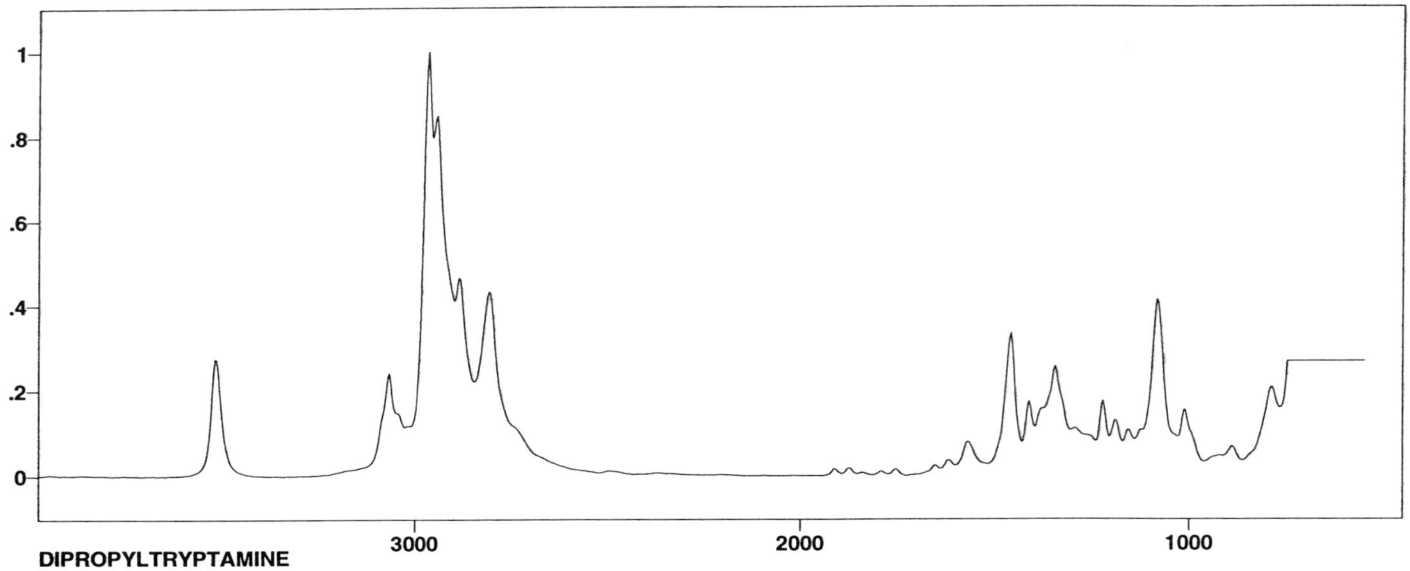

DIPROPYLTRYPTAMINE

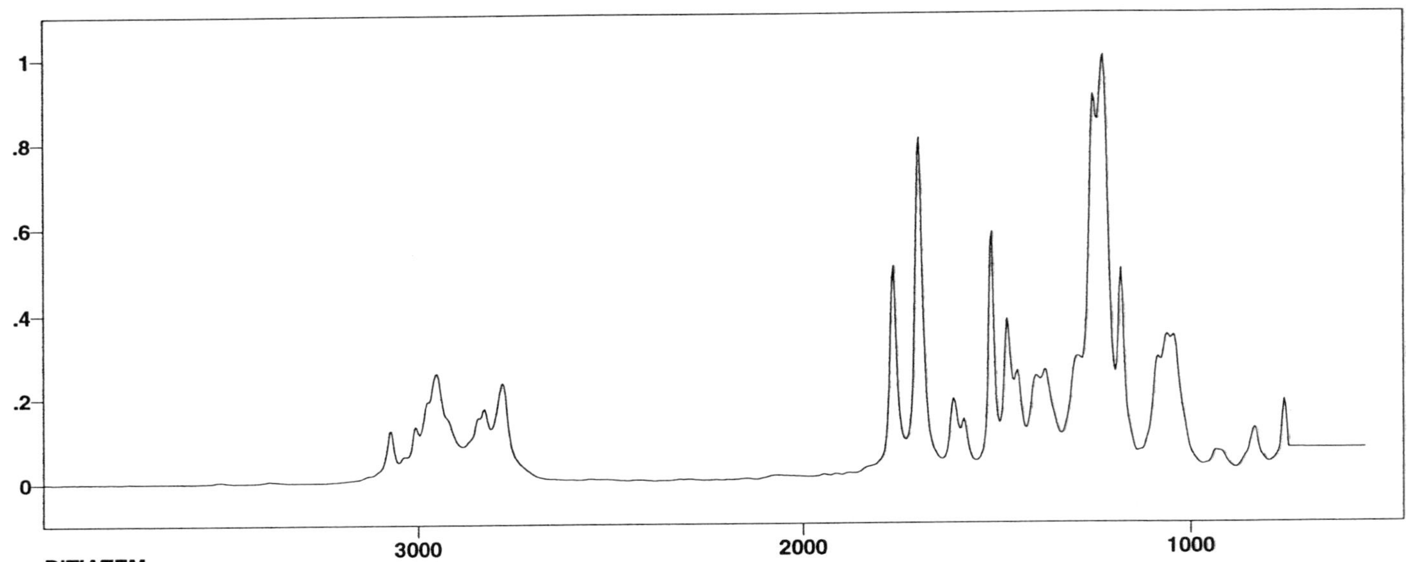

DITIAZEM

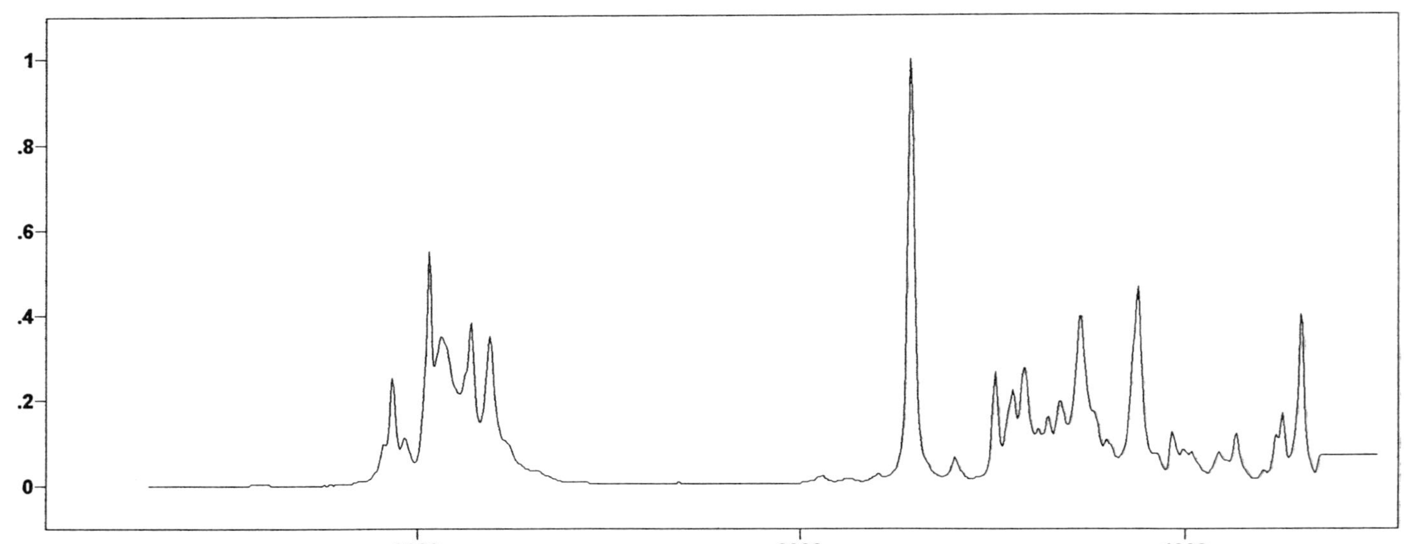

DOXAPRAM

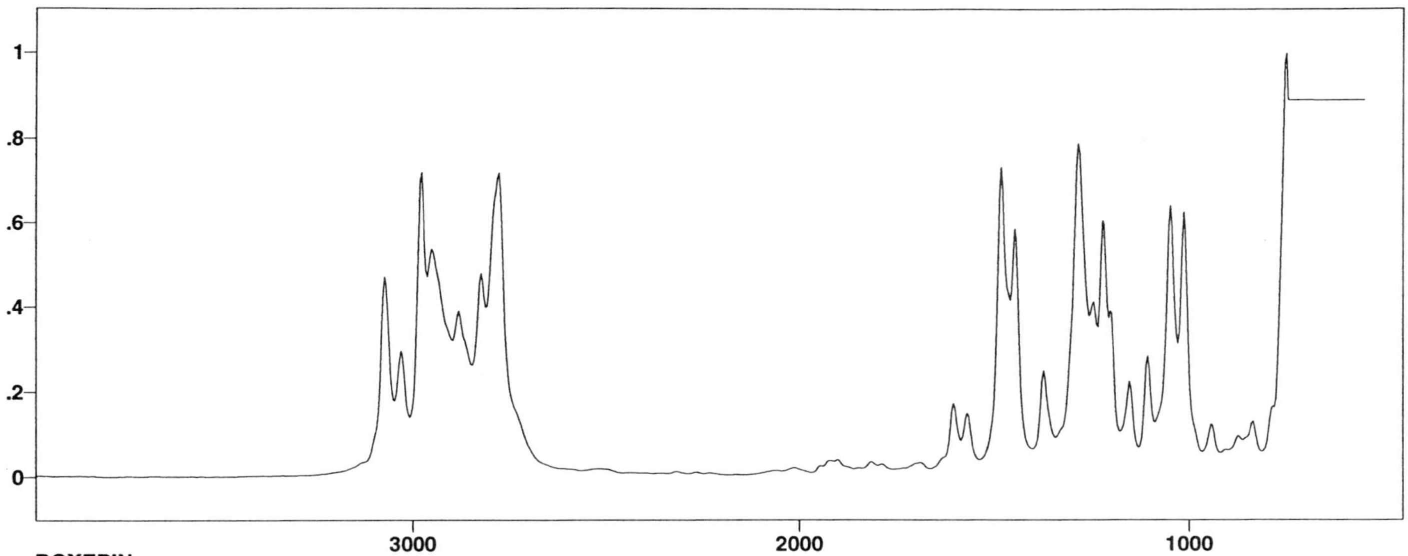

DOXEPIN

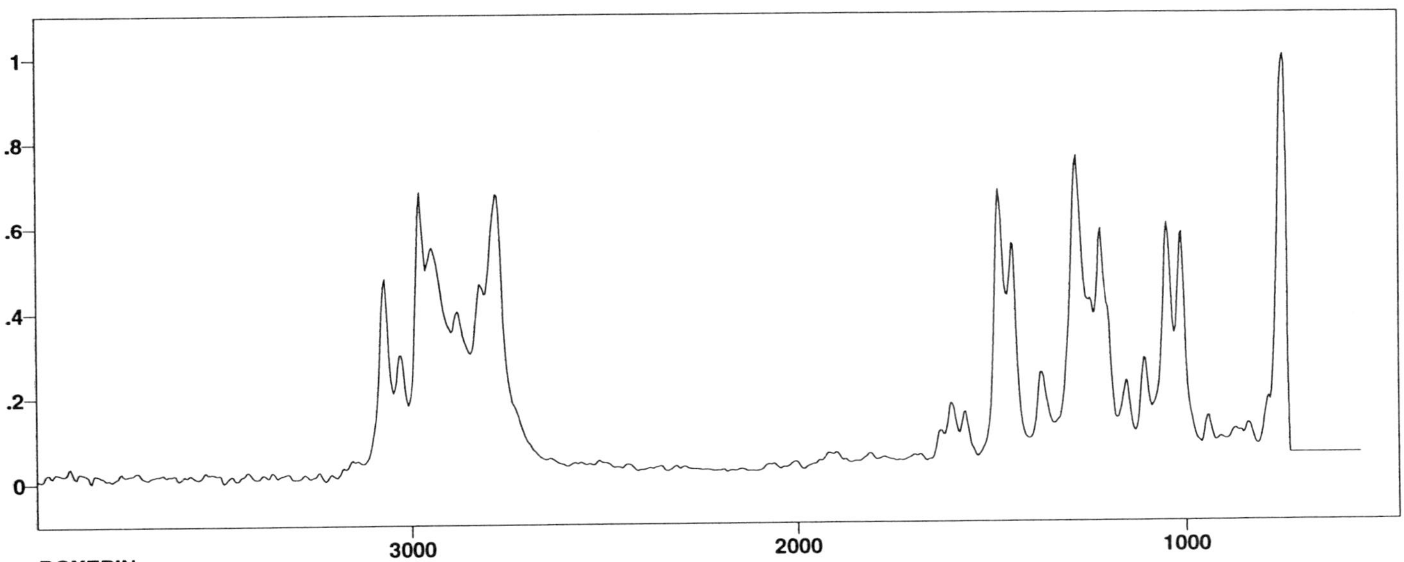

DOXEPIN

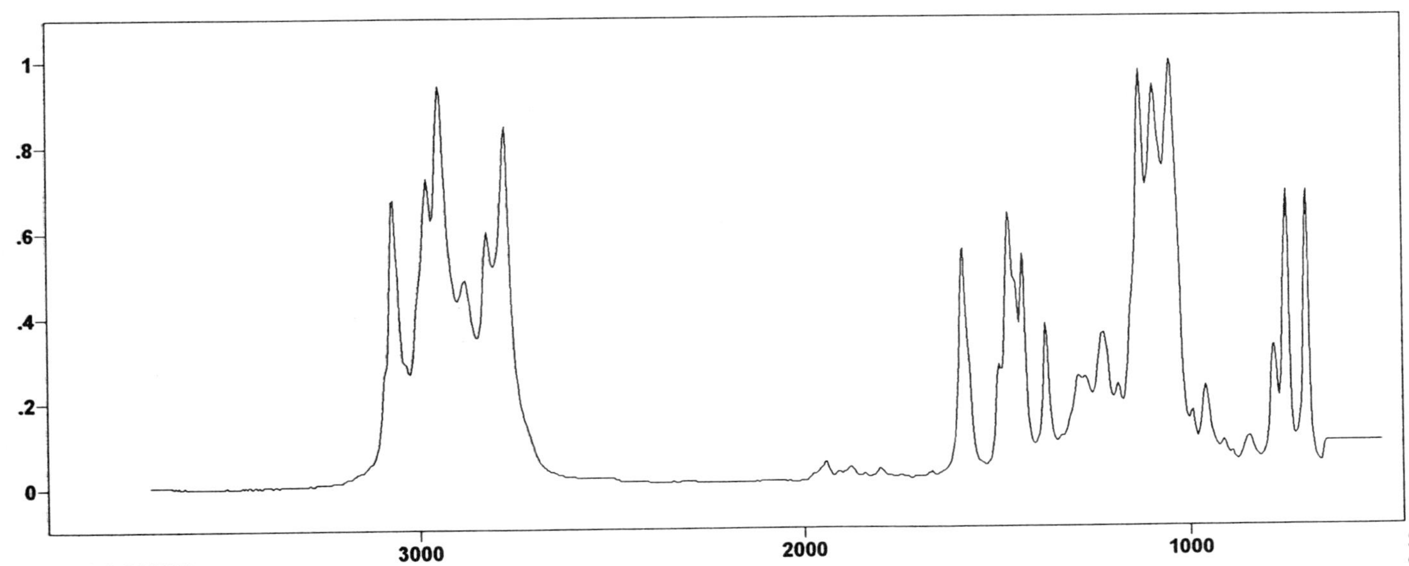

DOXYLAMINE

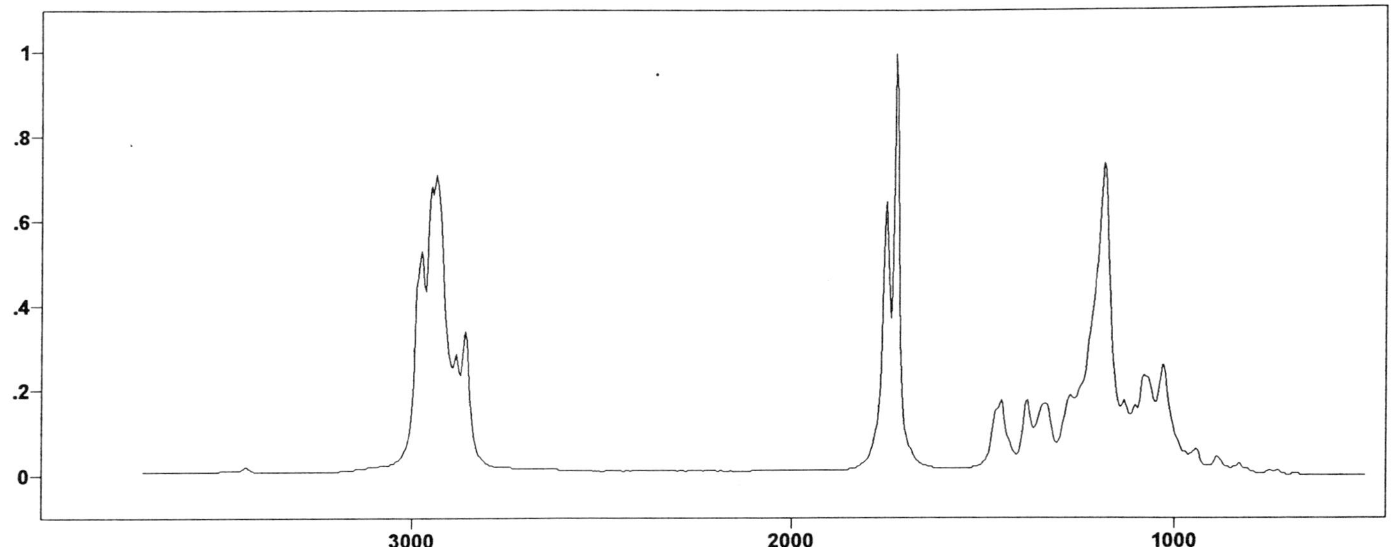

DROMOSTANOLONE PROPIONATE

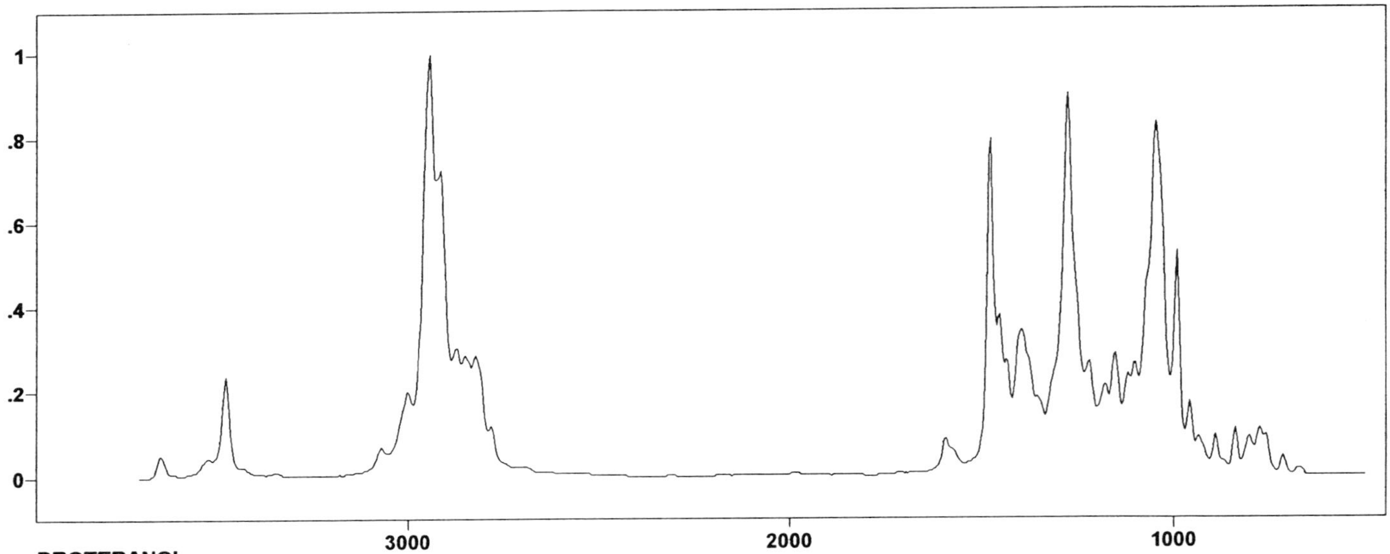

DROTEBANOL

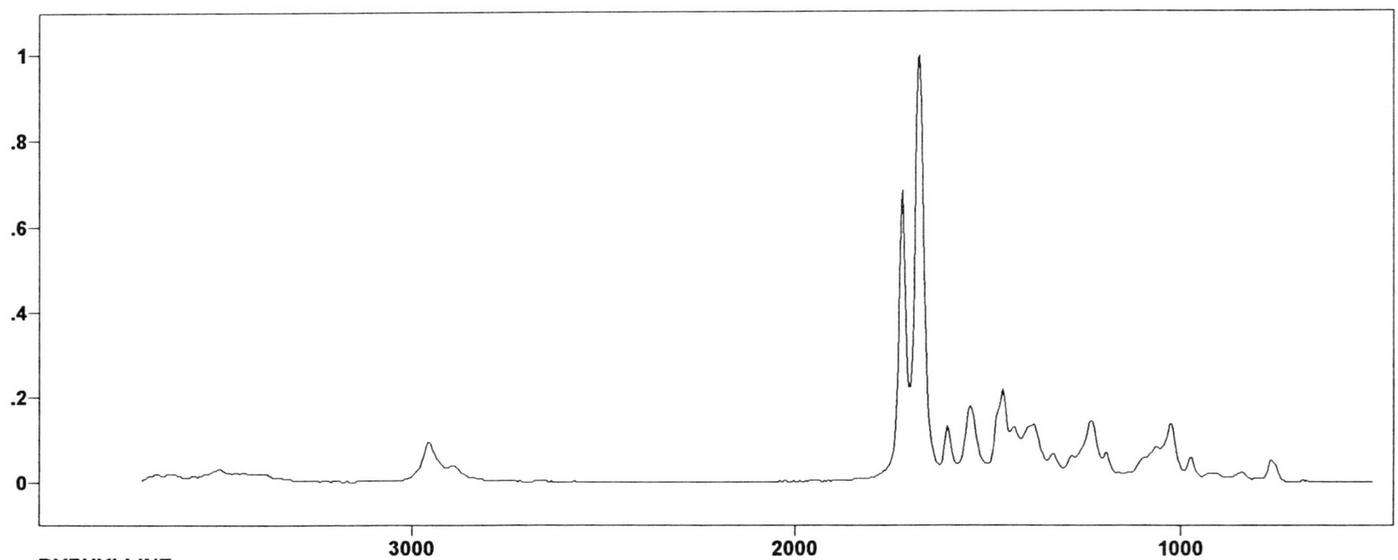

DYPHYLLINE

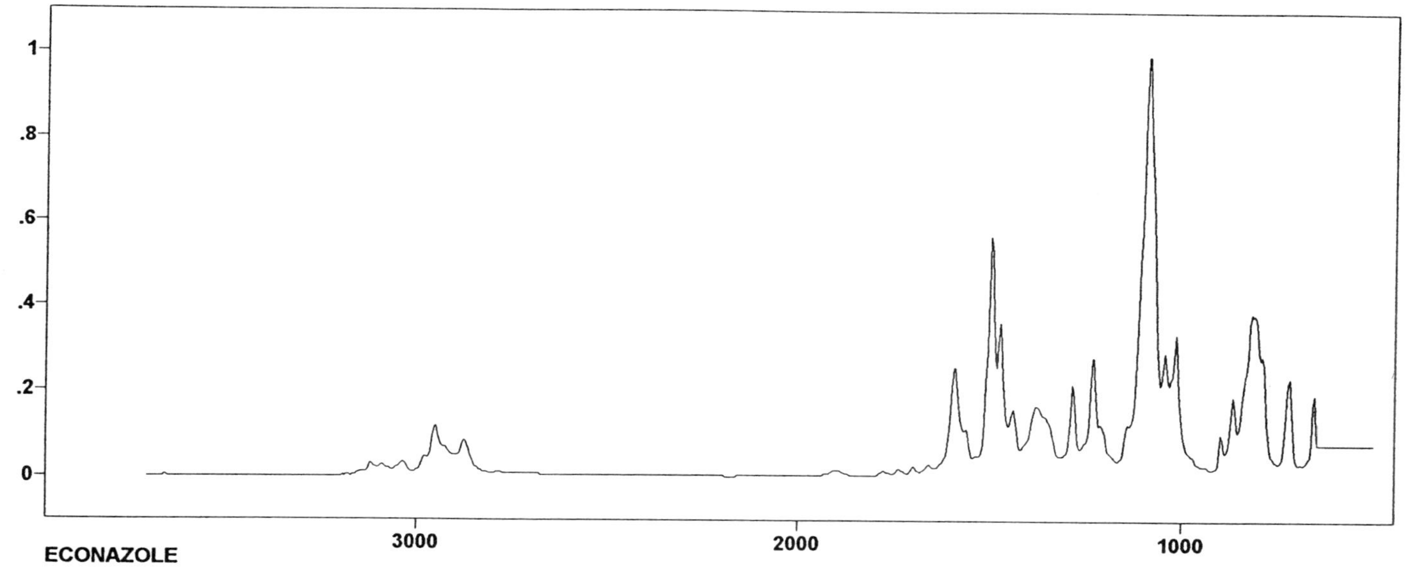

ECONAZOLE

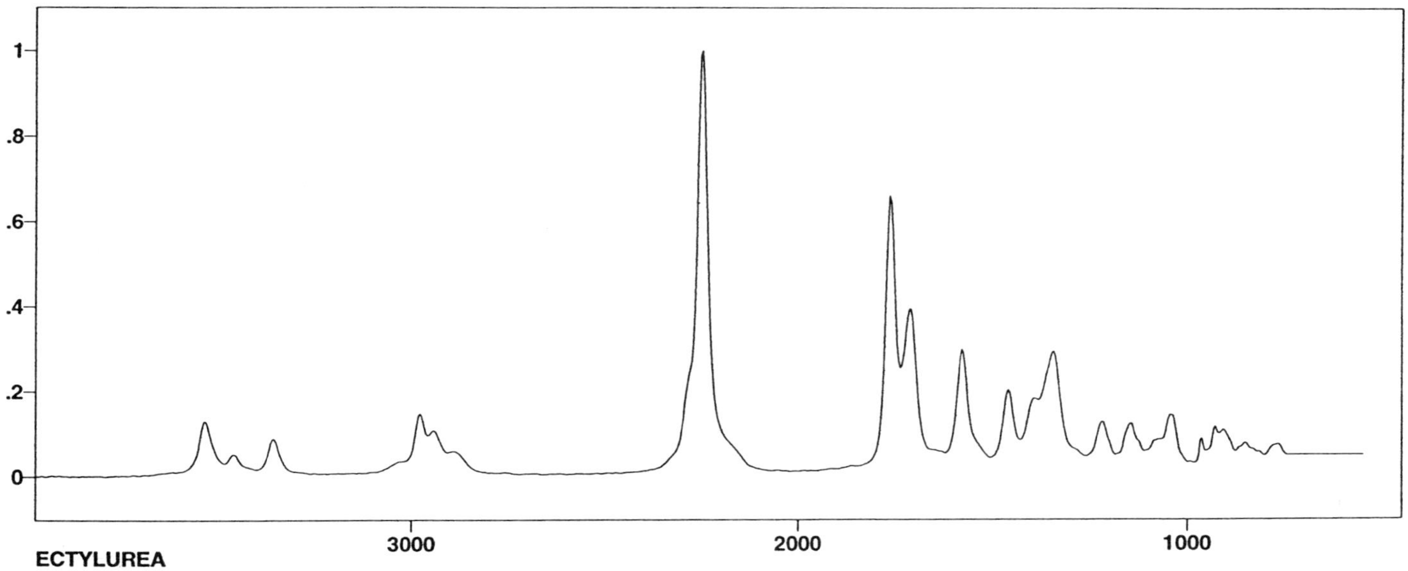

ECTYLUREA

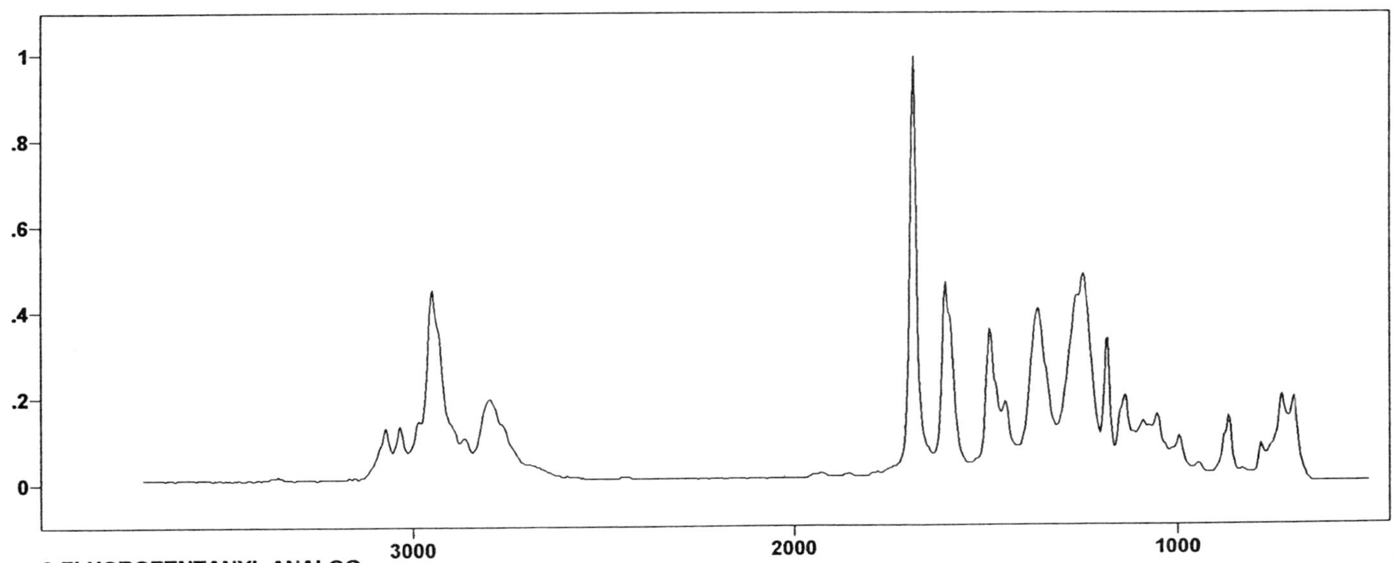

3-FLUOROFENTANYL ANALOG

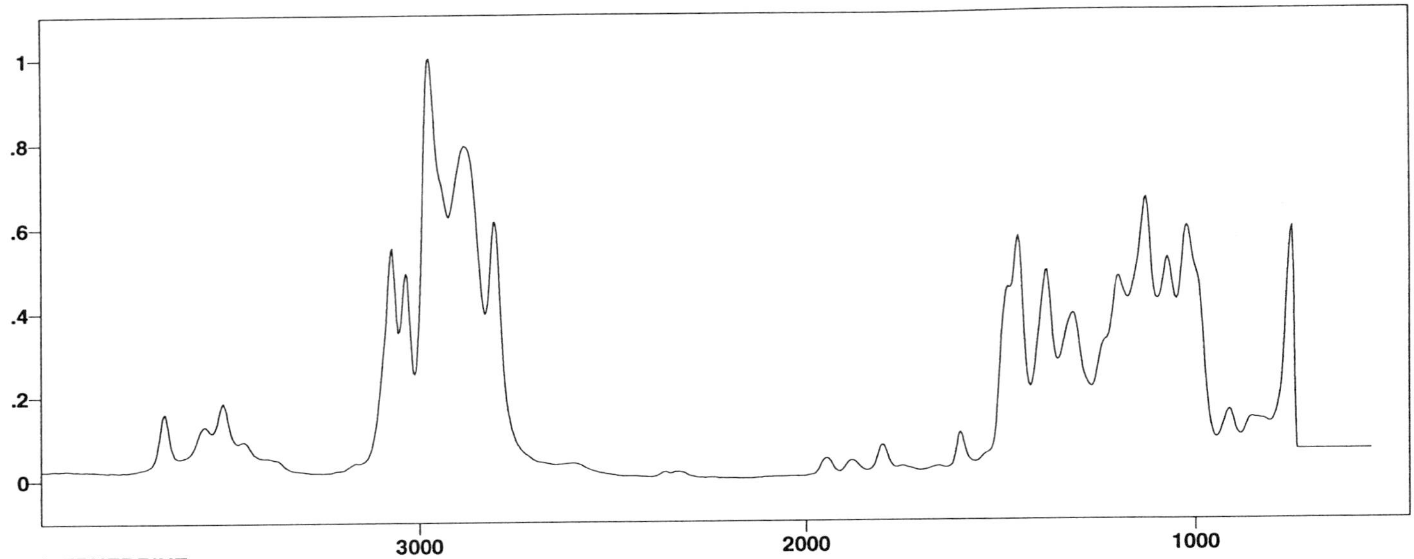

L-EPHEDRINE

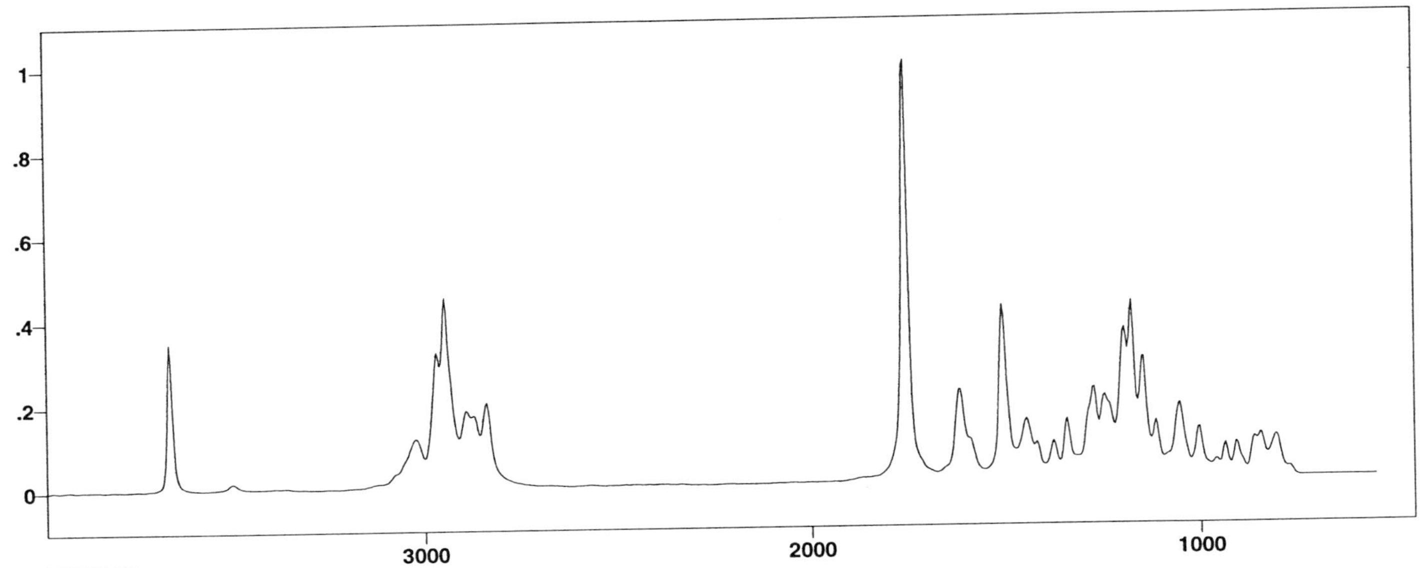

EQUILIN

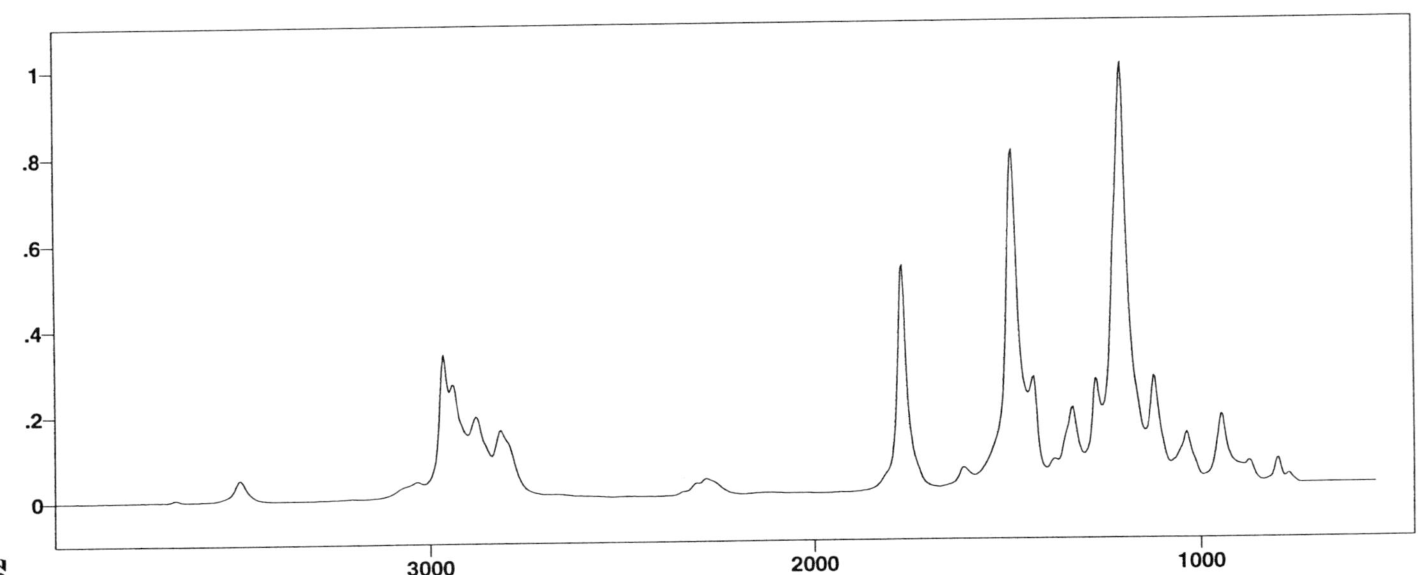

ESERINE

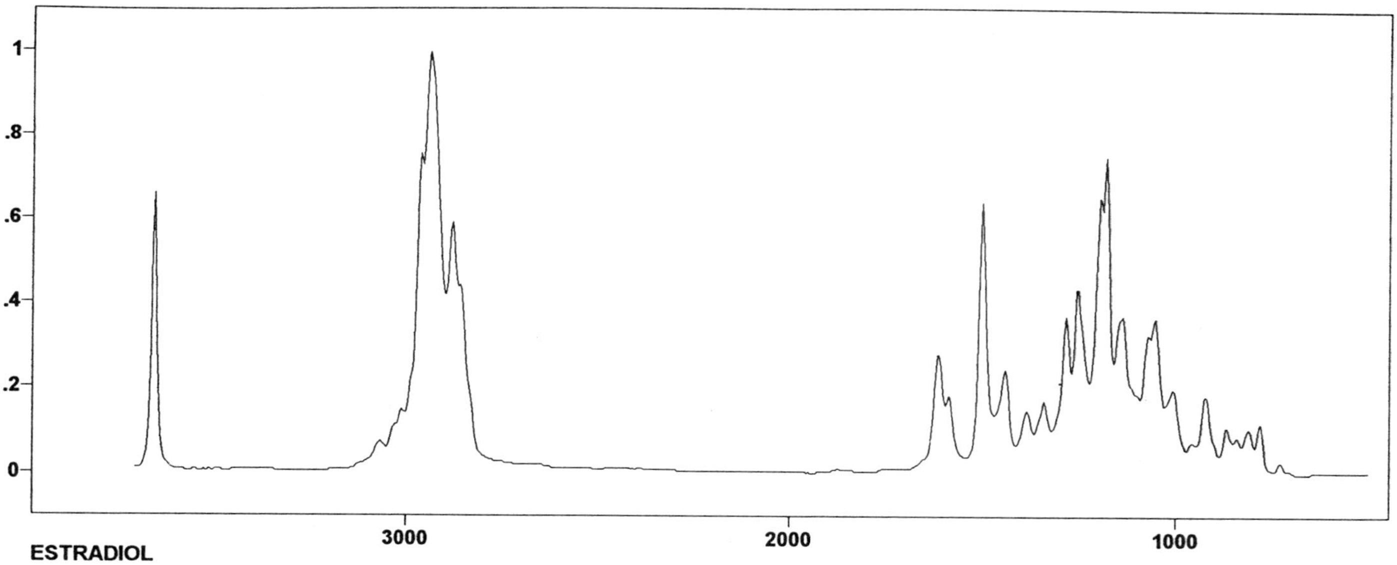

ESTRADIOL

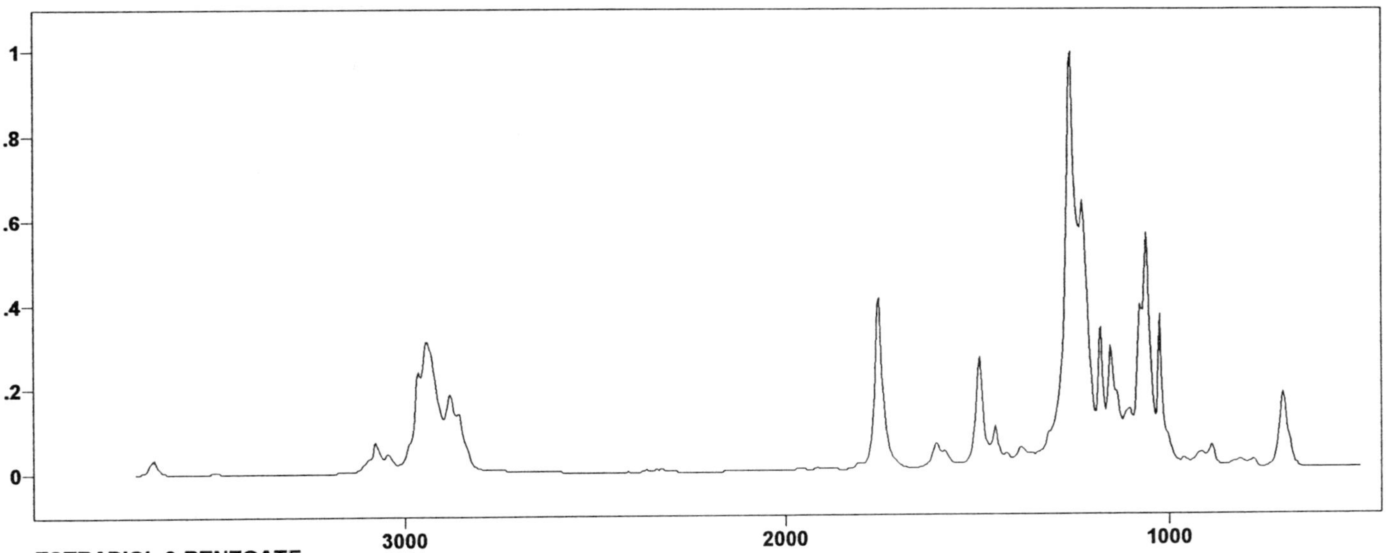

ESTRADIOL-3-BENZOATE

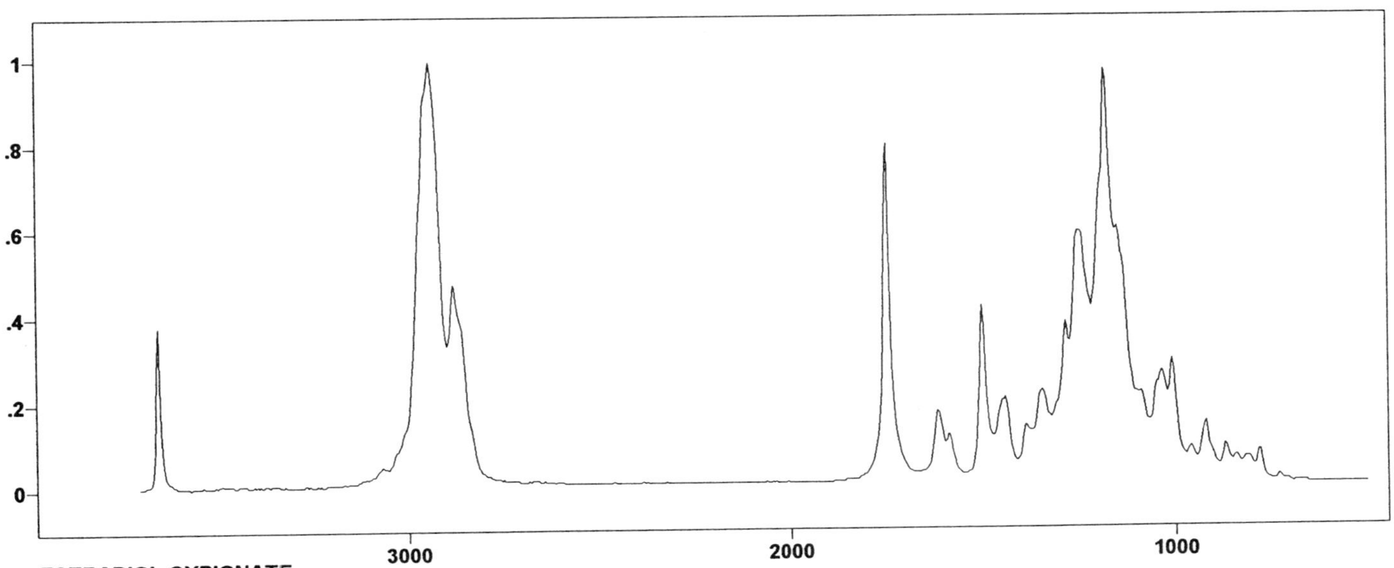

ESTRADIOL CYPIONATE

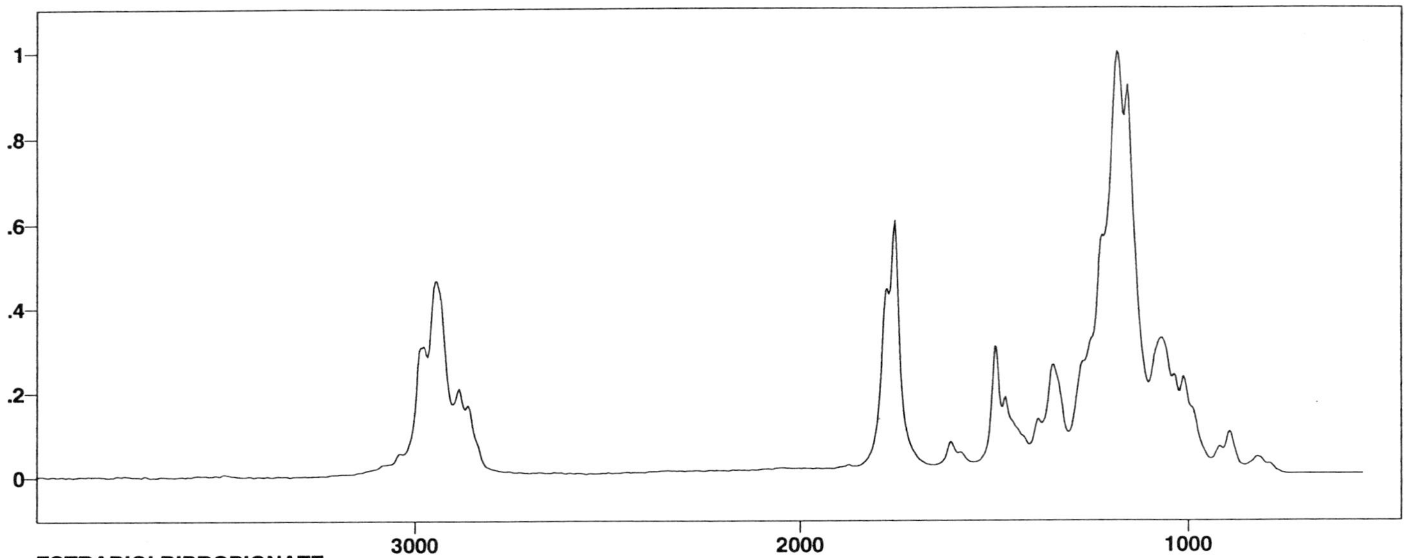

ESTRADIOLDIPROPIONATE

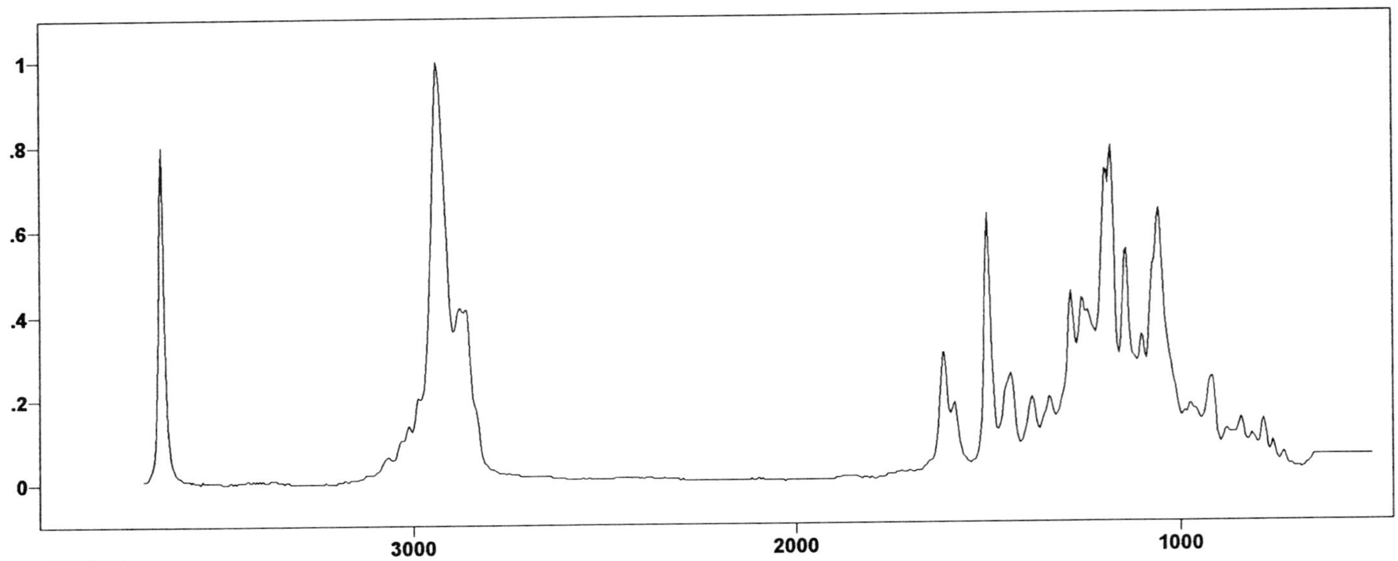

ESTRIOL

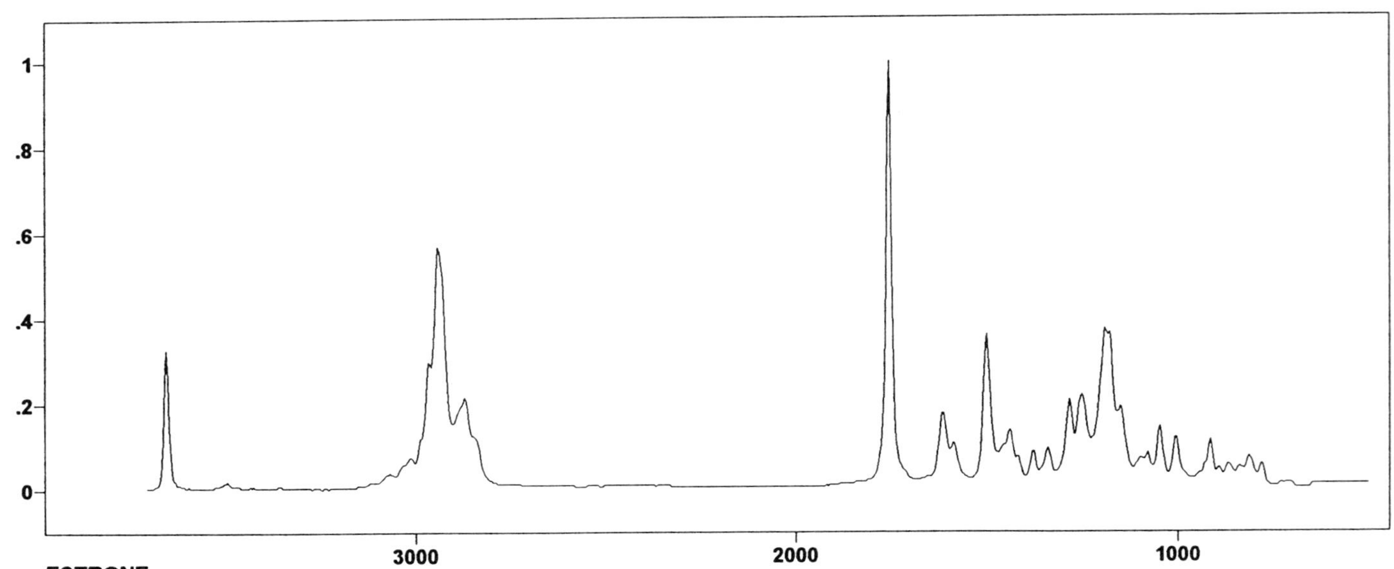

ESTRONE

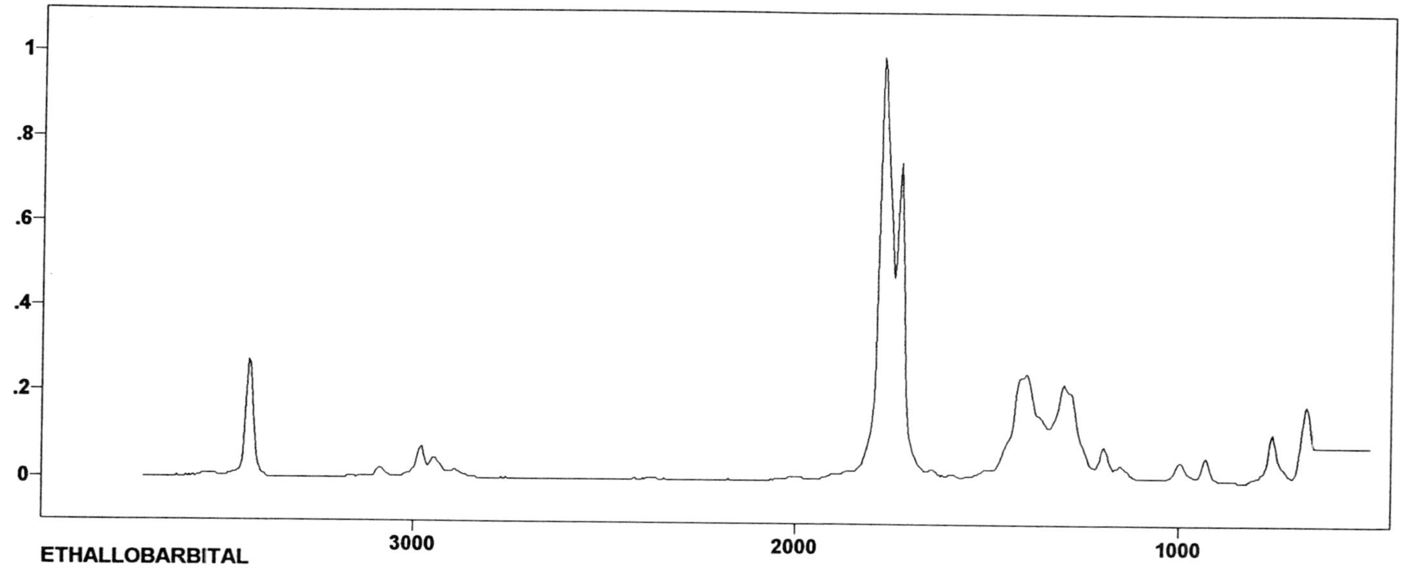

ETHALLOBARBITAL

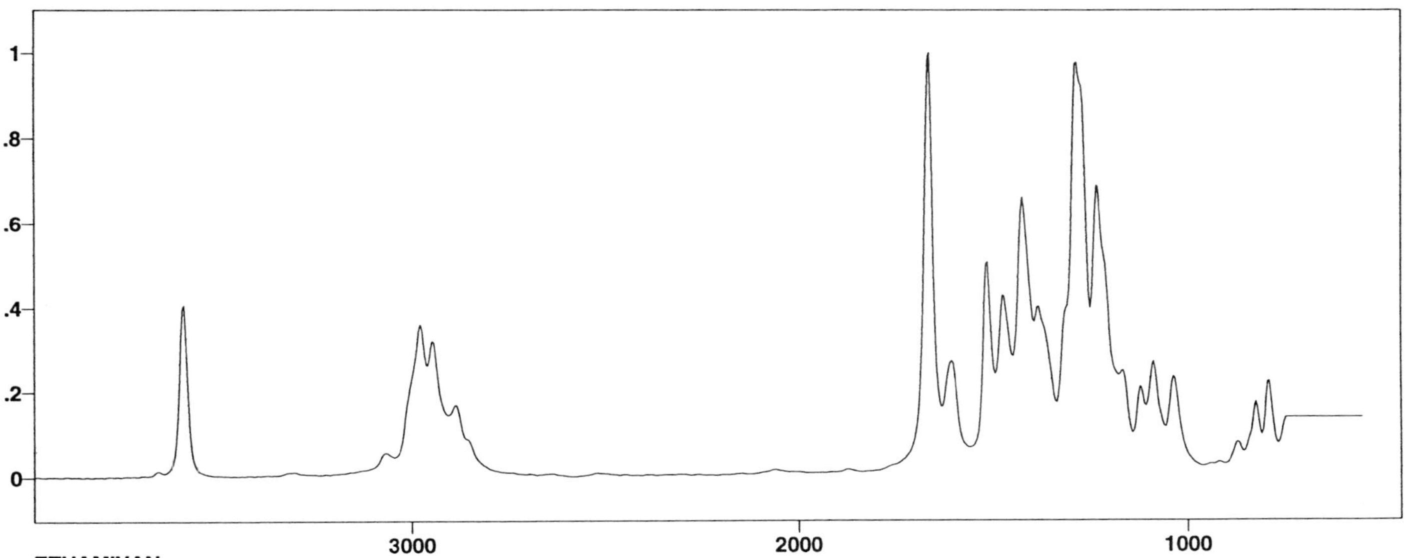

ETHAMIVAN

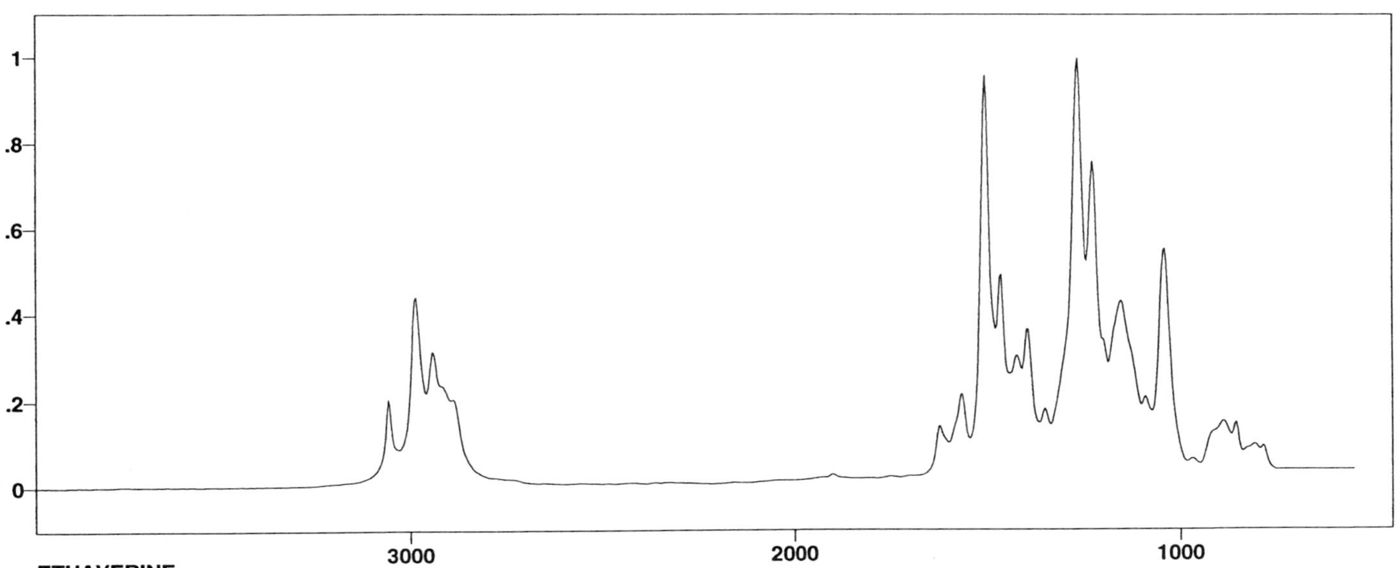

ETHAVERINE

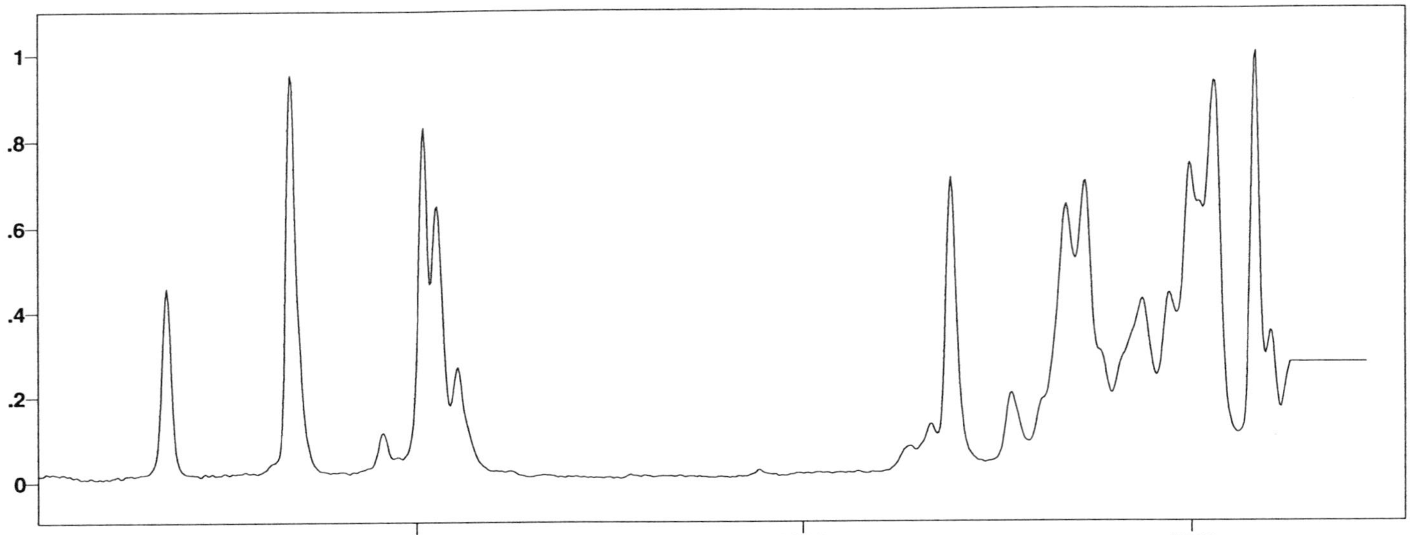

ETHCHLORVYNOL

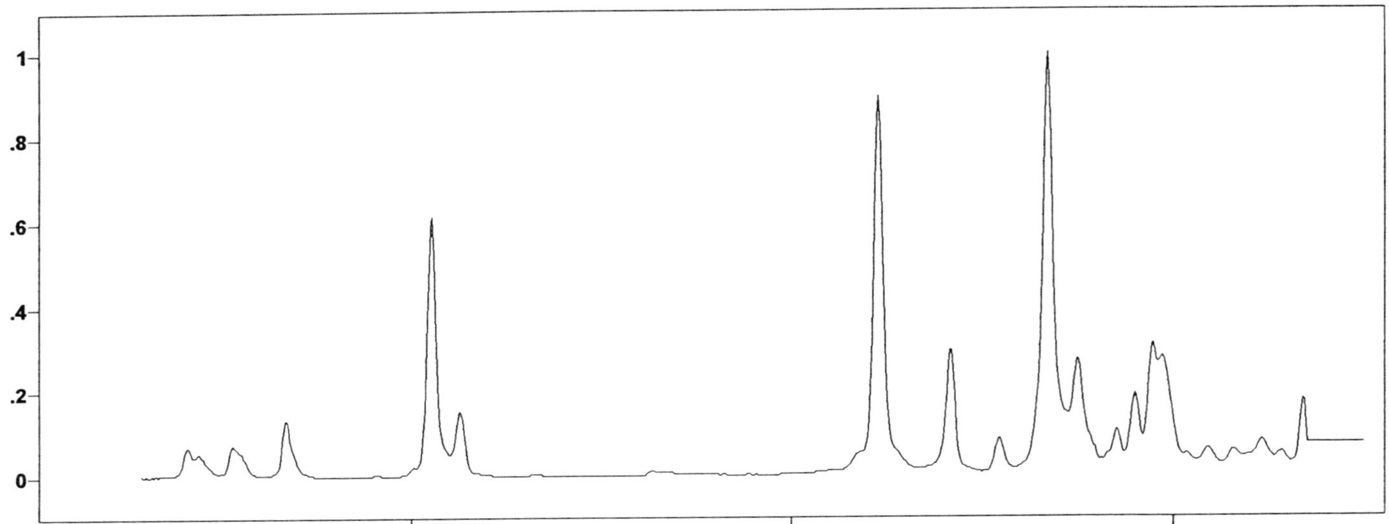

ETHINAMATE

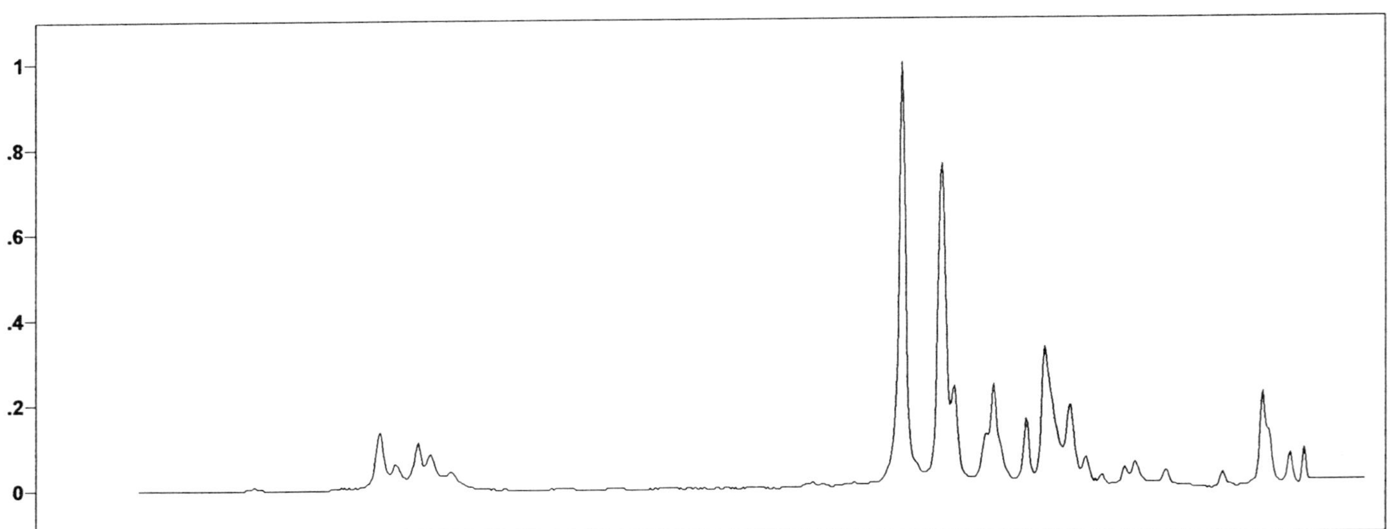

ETHINAZONE

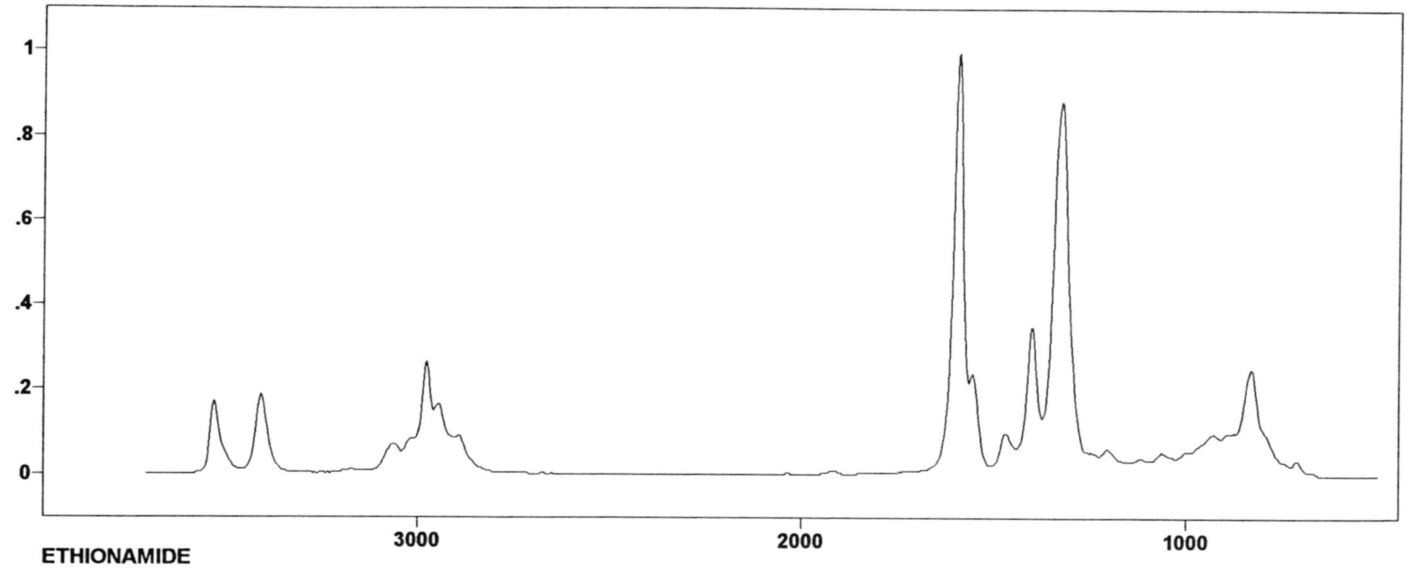

ETHIONAMIDE

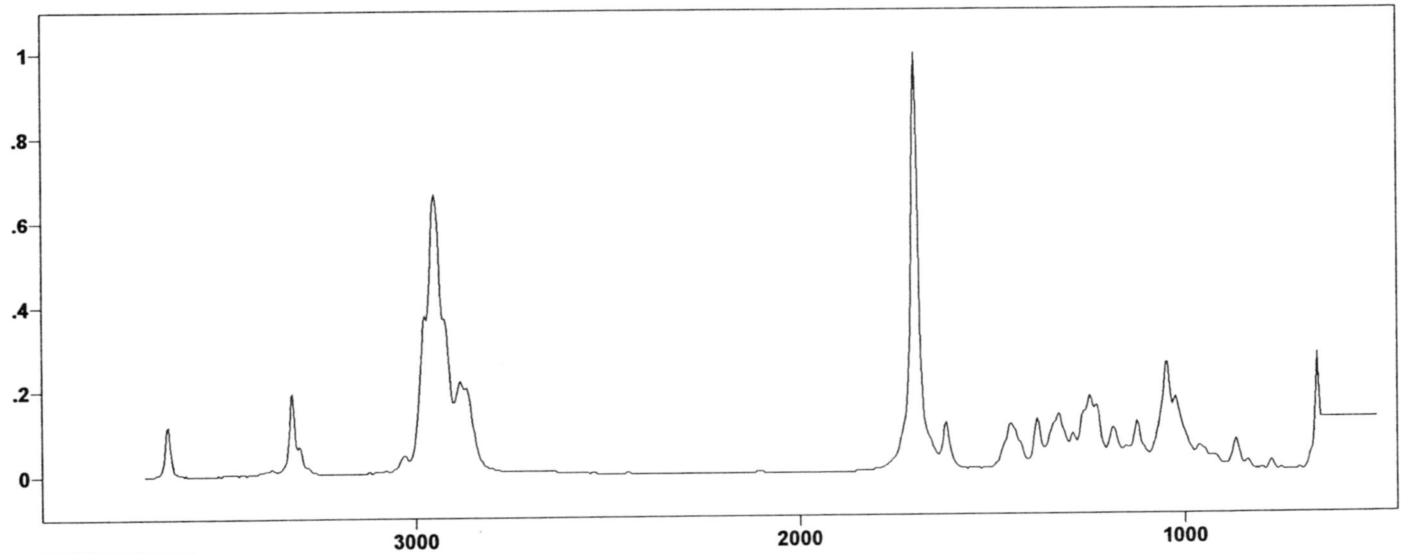

ETHISTERONE

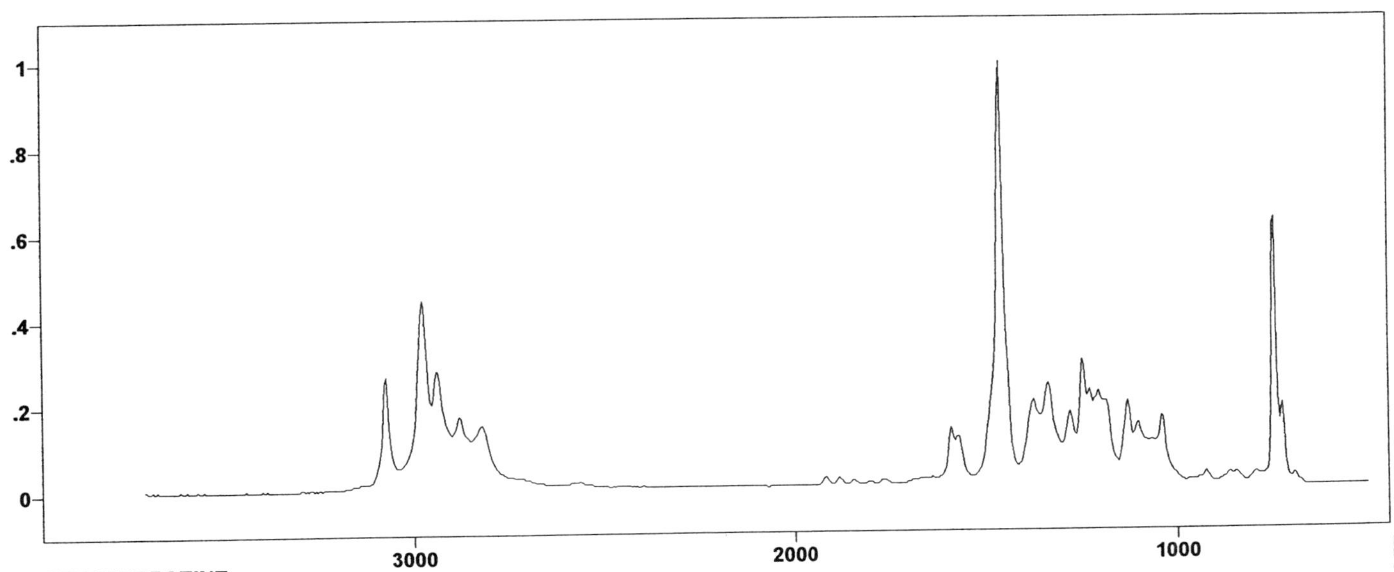

ETHOPROPAZINE

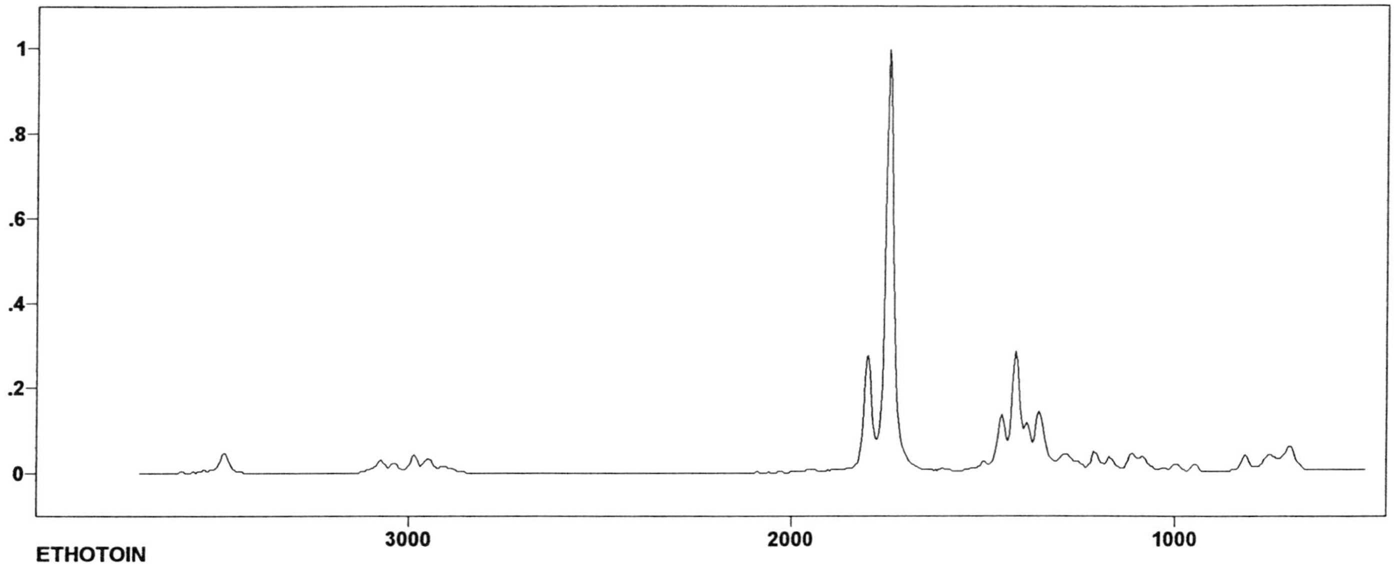

ETHOTOIN

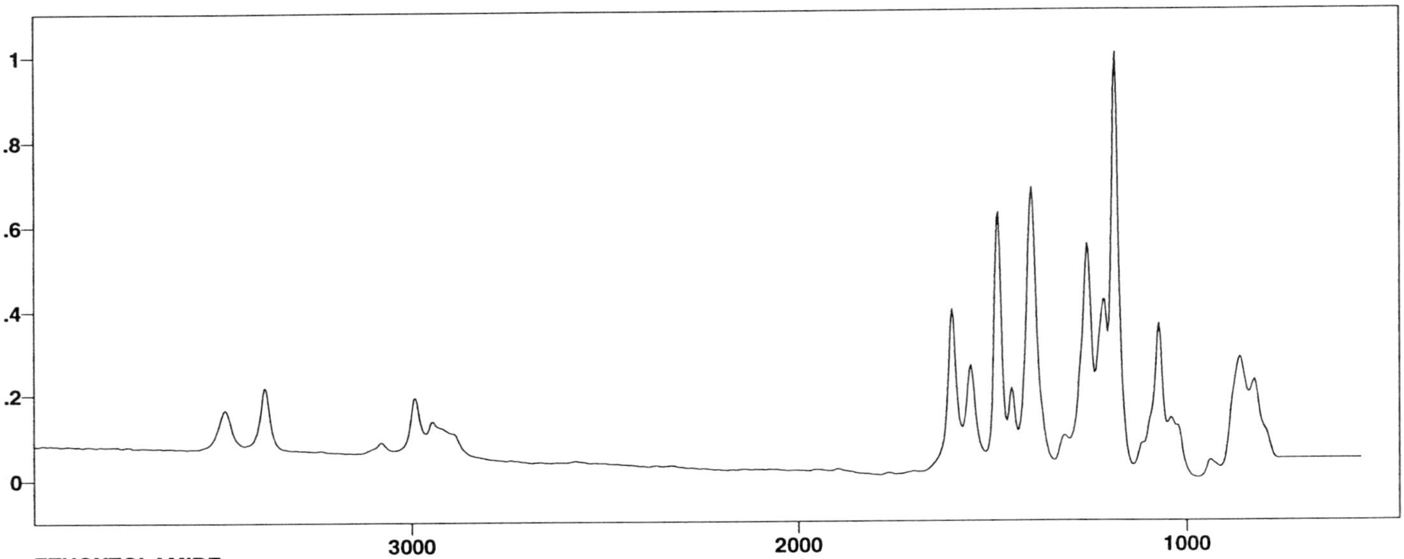

ETHOXZOLAMIDE

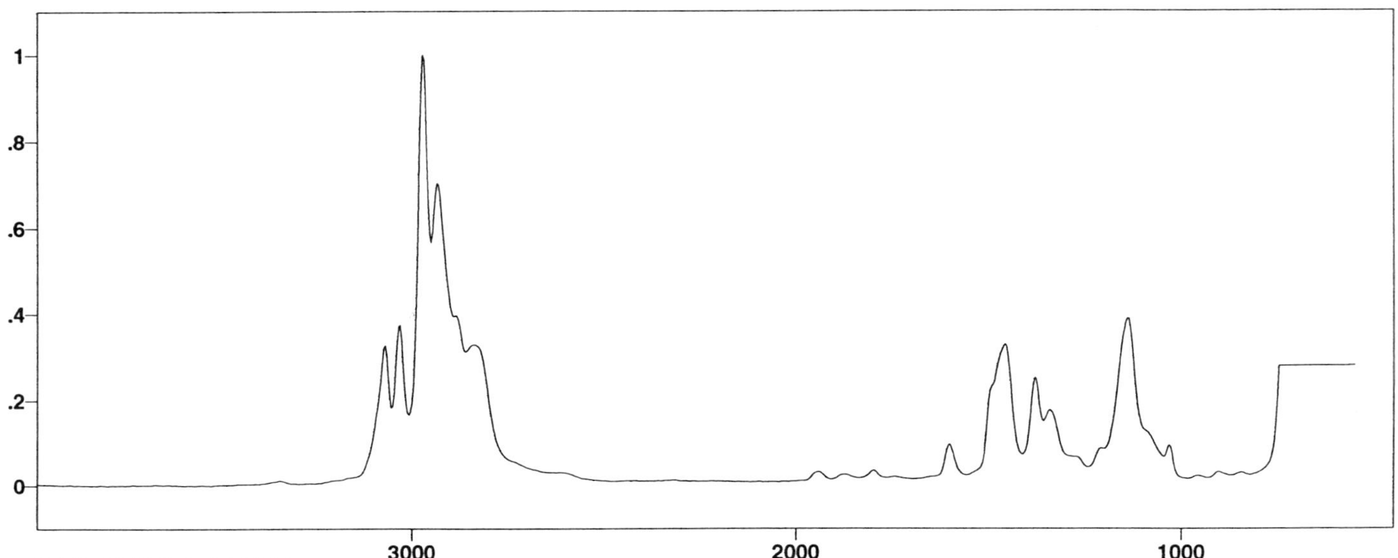

N-ETHYLAMPHETAMINE

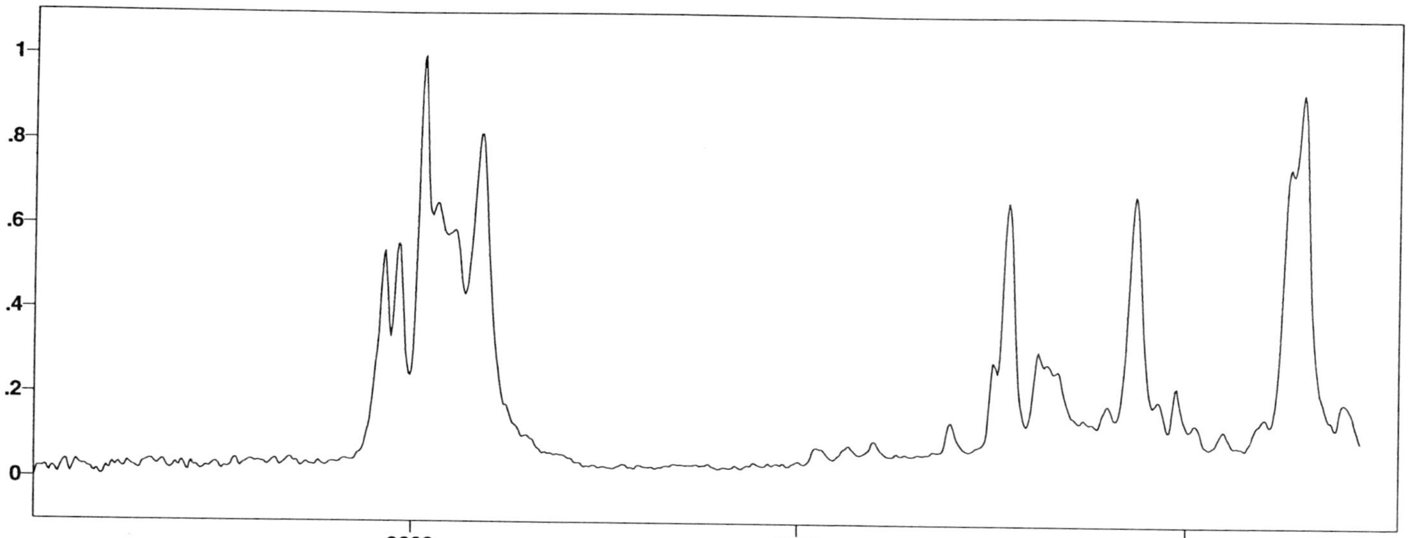

N-ETHYLBENZYLAMINE

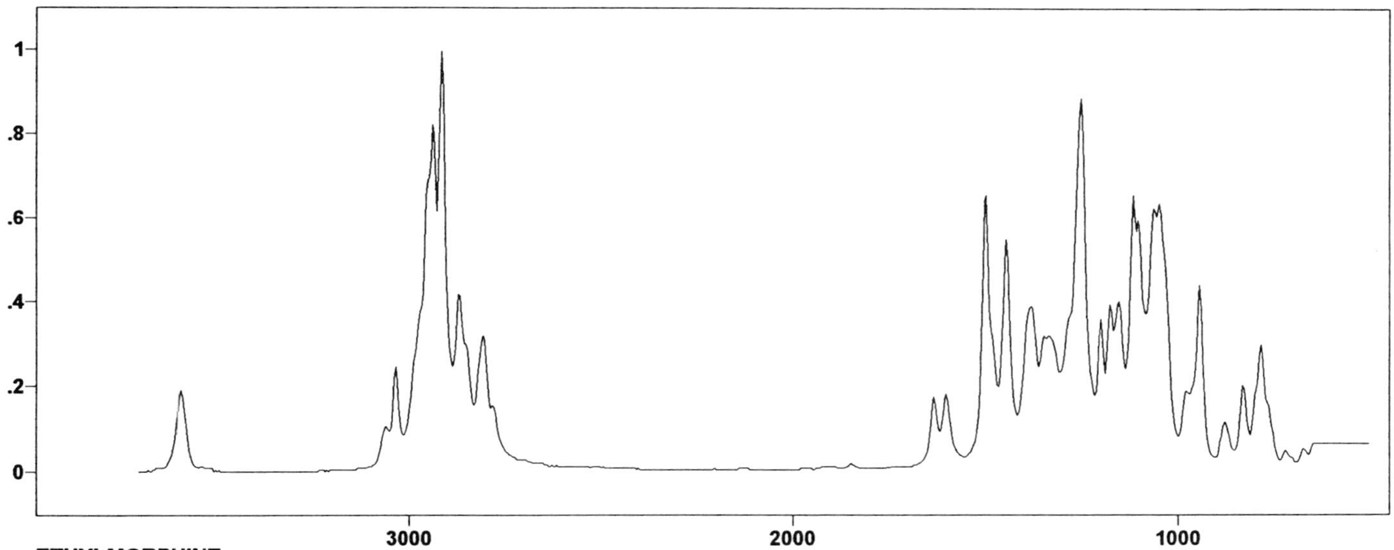

ETHYLMORPHINE

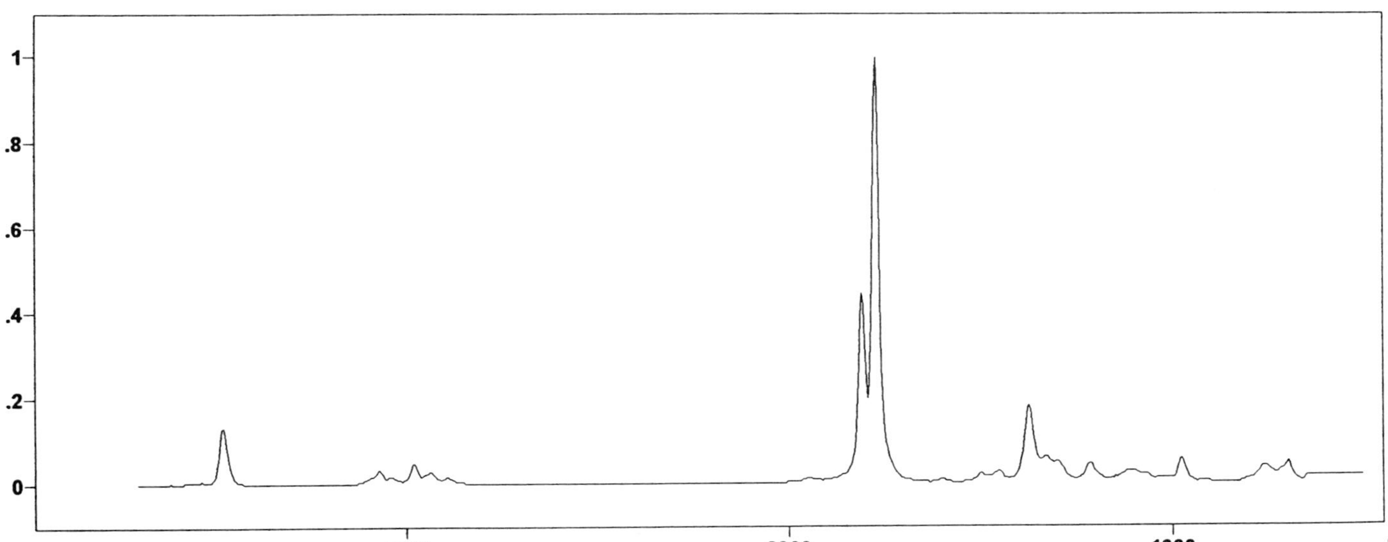

5-ETHYL-5-PHENYLHYDANTOIN

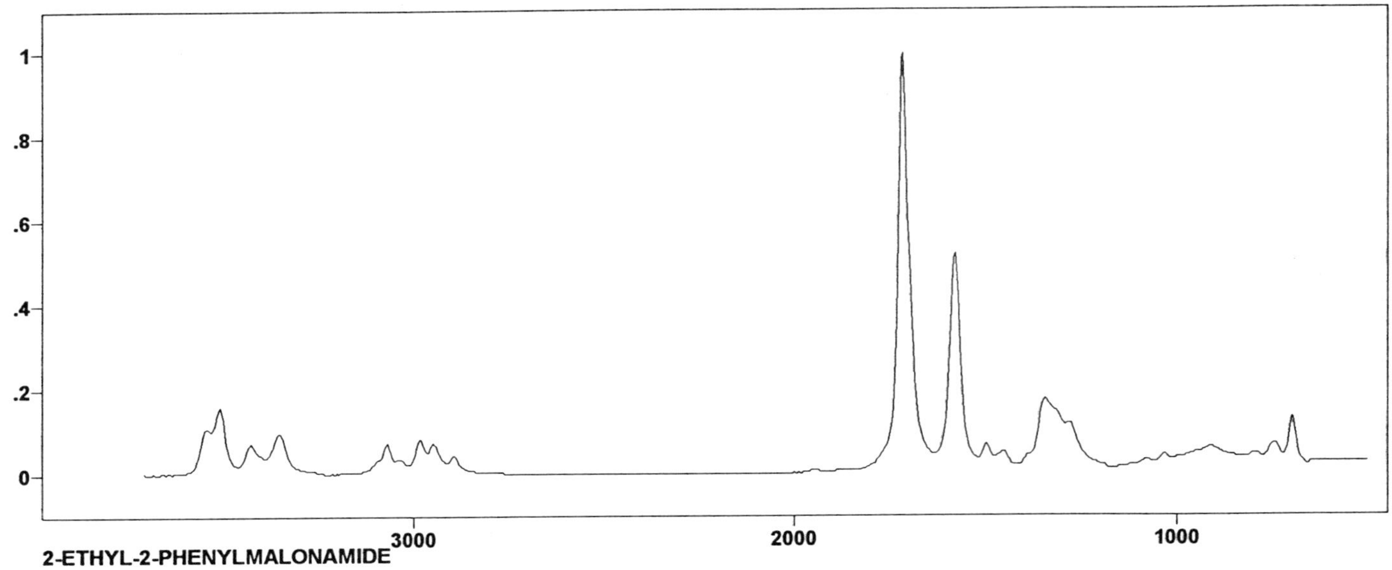

2-ETHYL-2-PHENYLMALONAMIDE

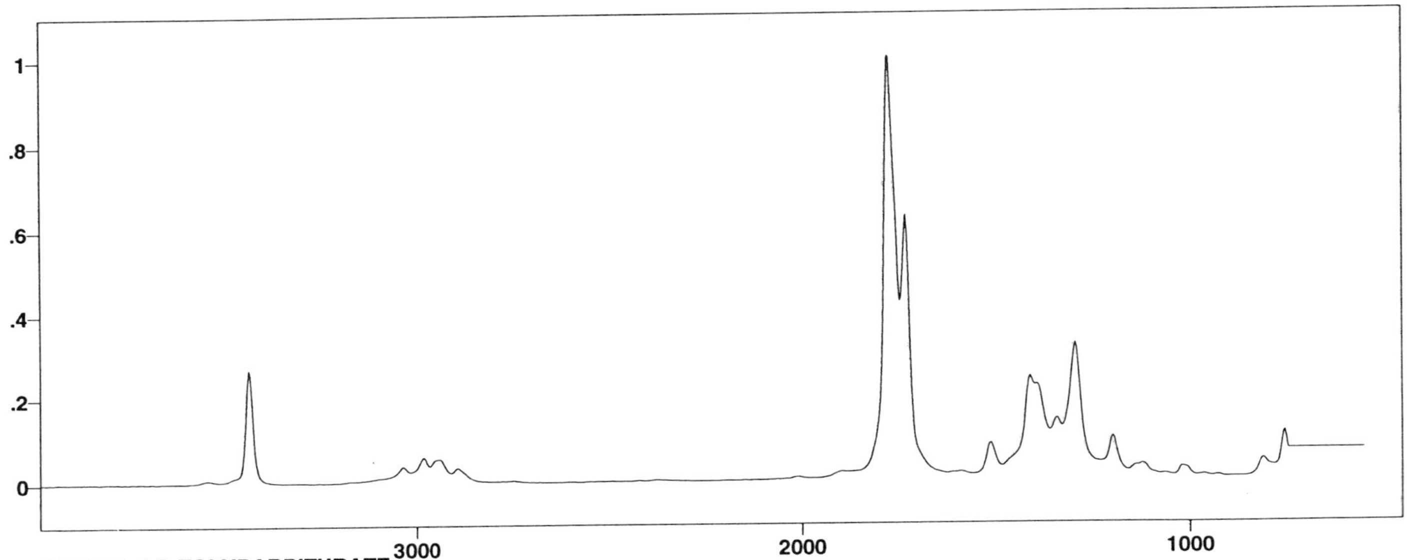

5-ETHYL-5-P-TOLYBARBITURATE

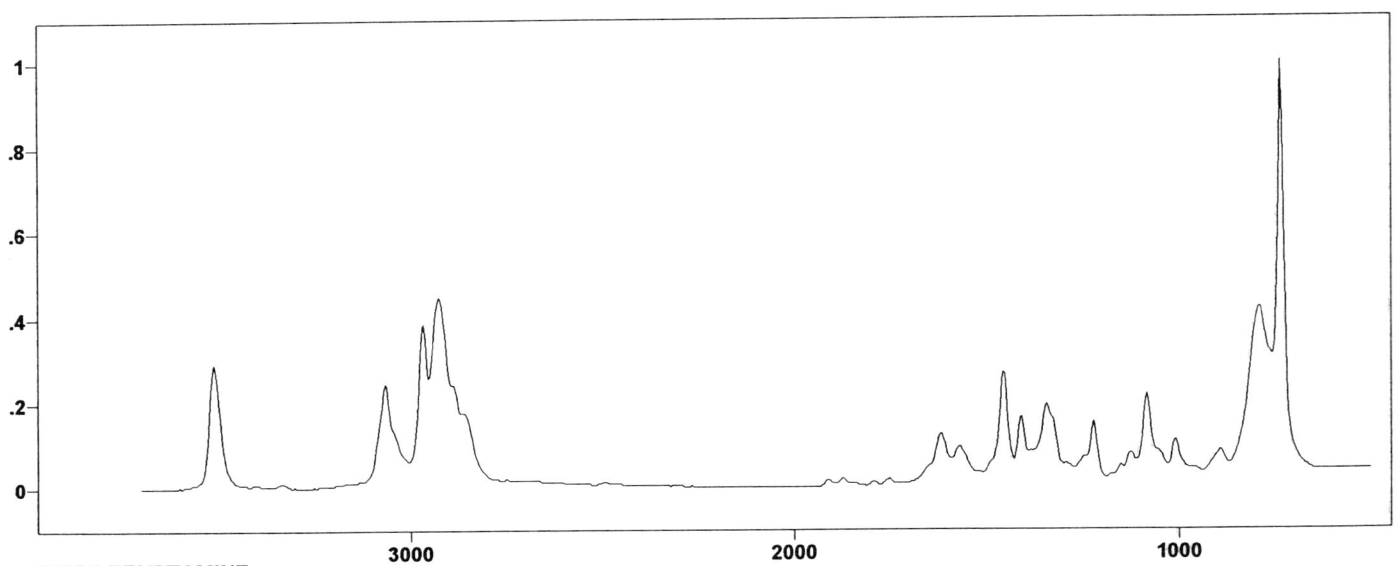

ETHYLTRYPTAMINE

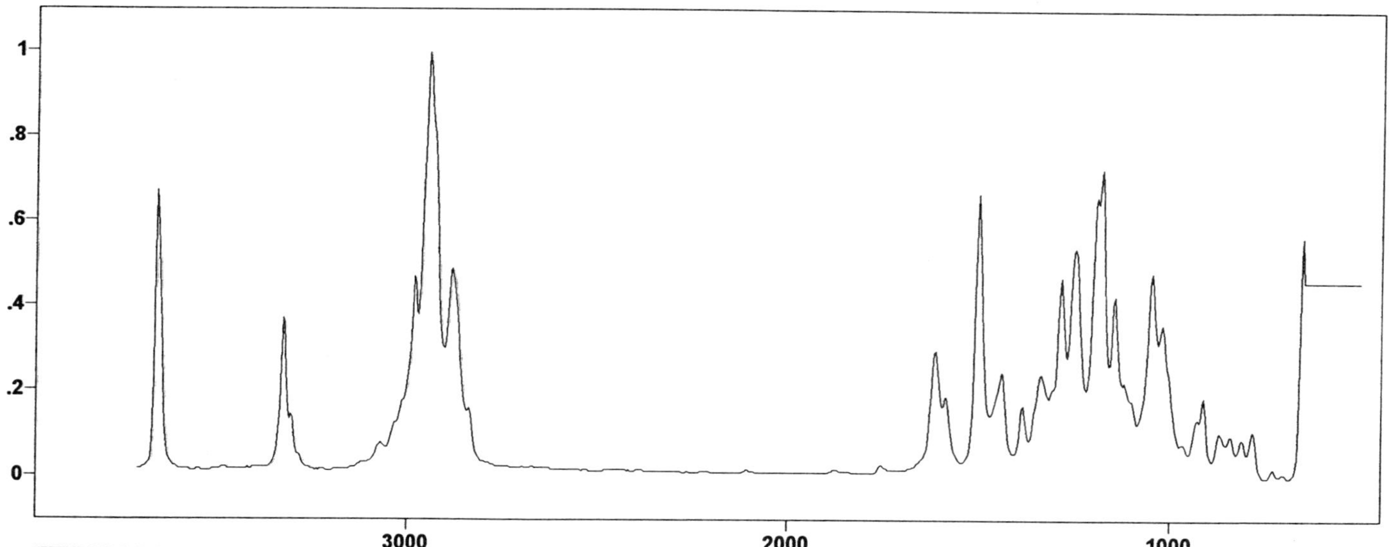

ETHYNYLESTRADIOL

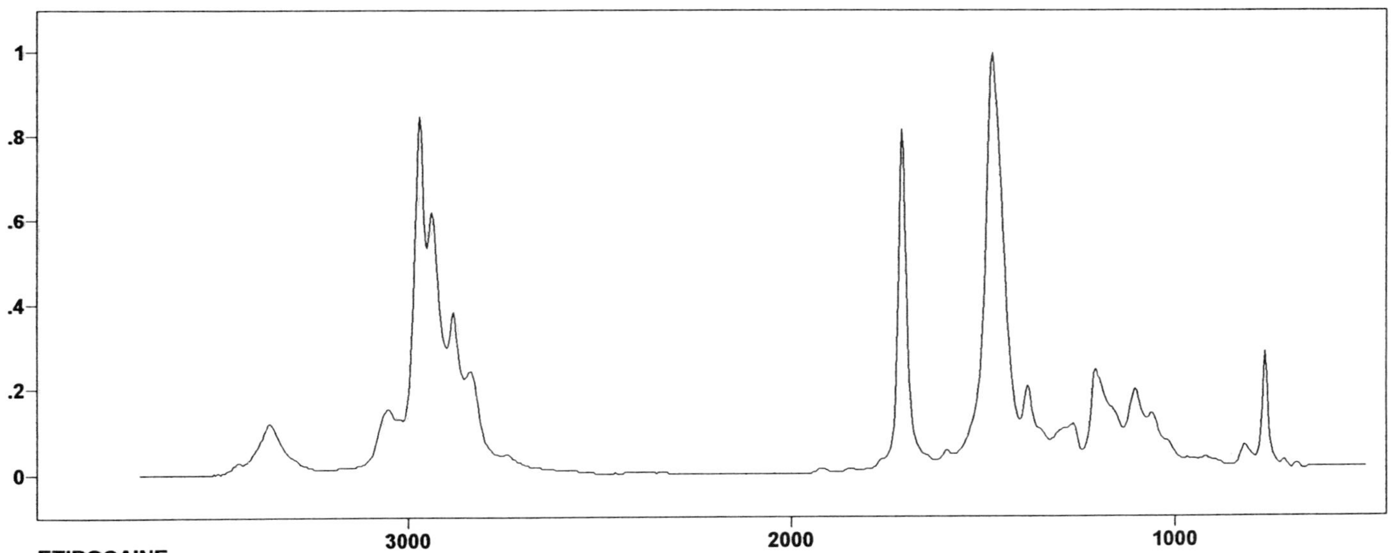

ETIDOCAINE

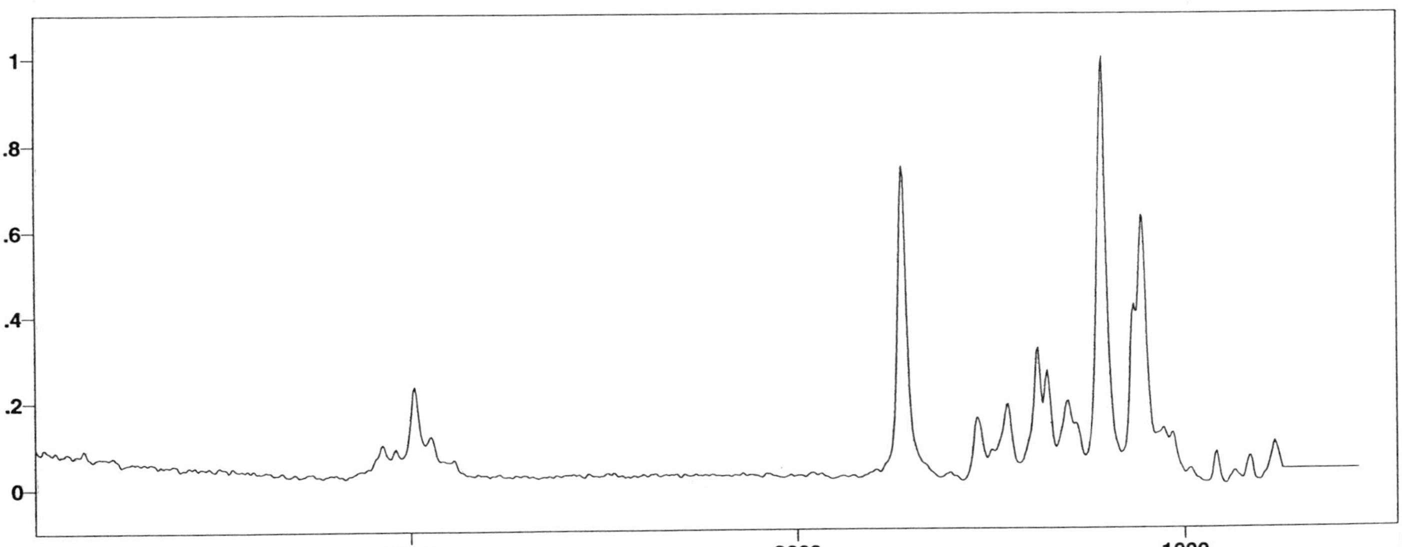

ETOMIDATE

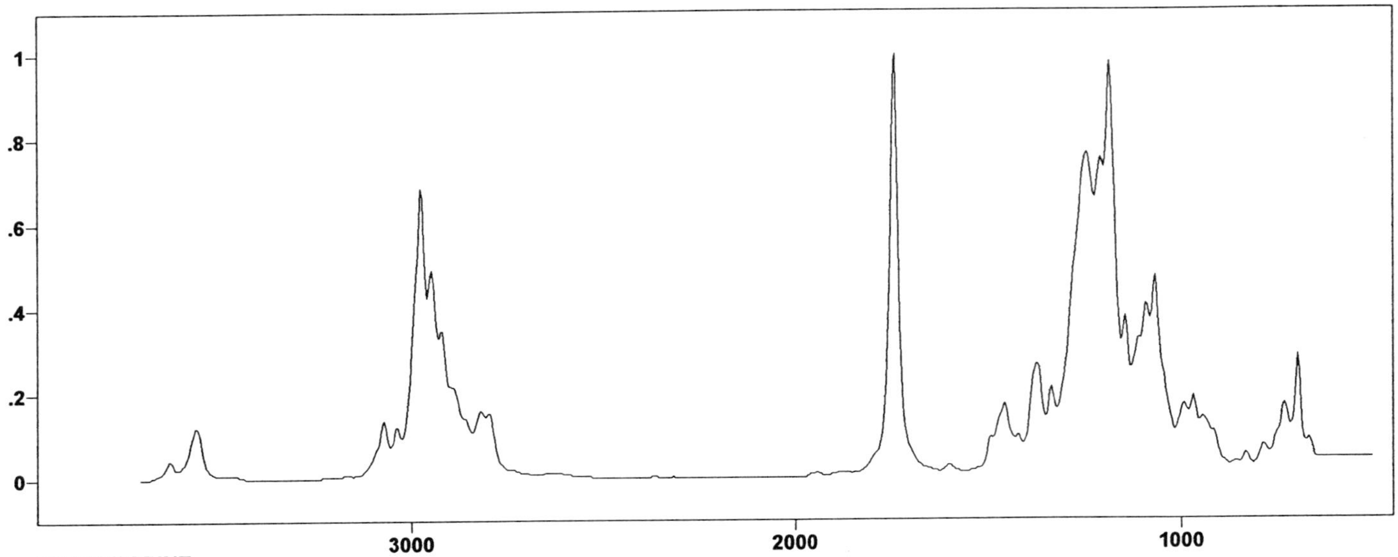

EUCATROPINE

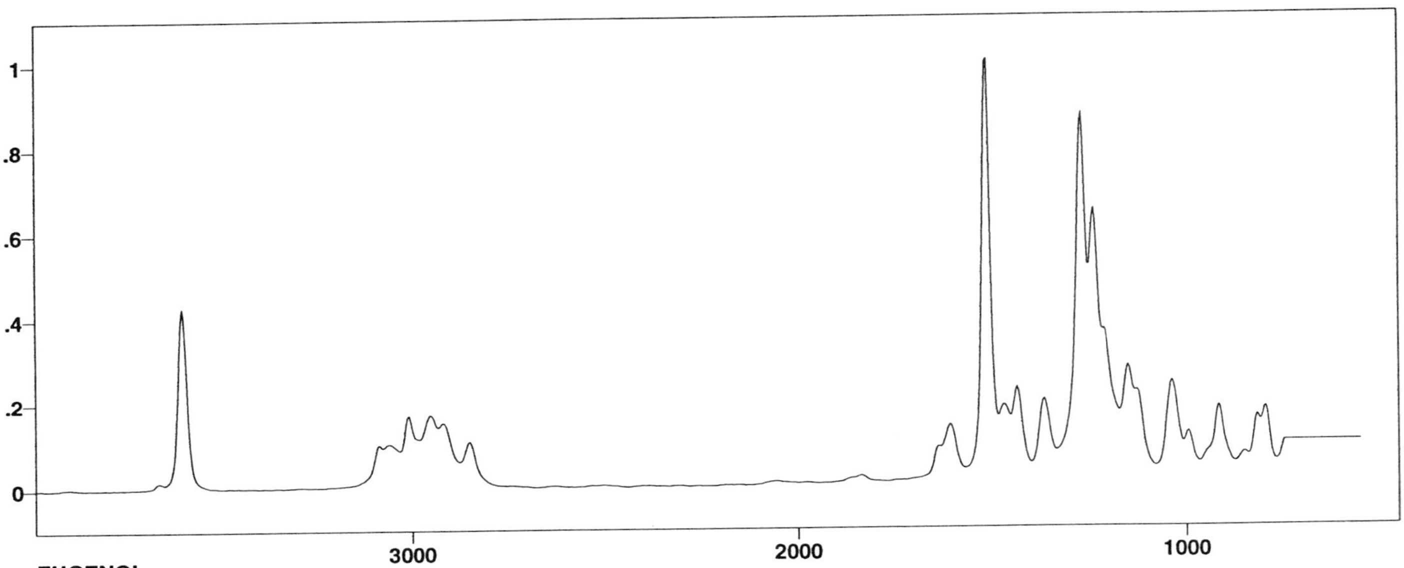

EUGENOL

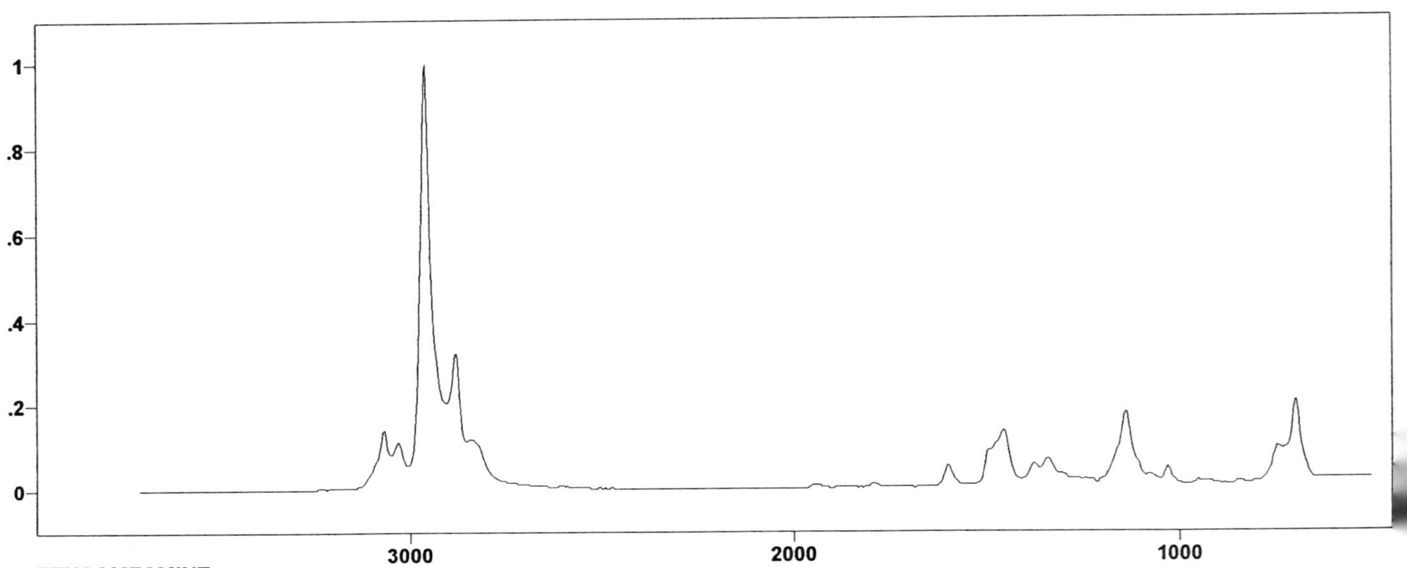

FENCAMFAMINE

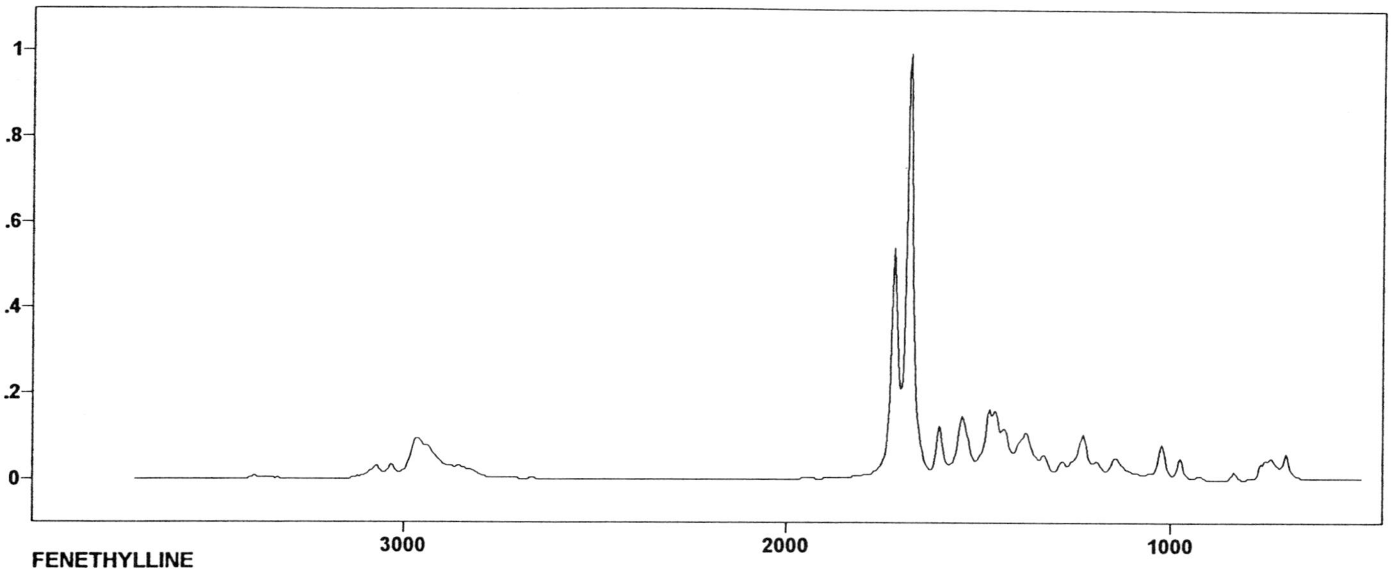

FENETHYLLINE

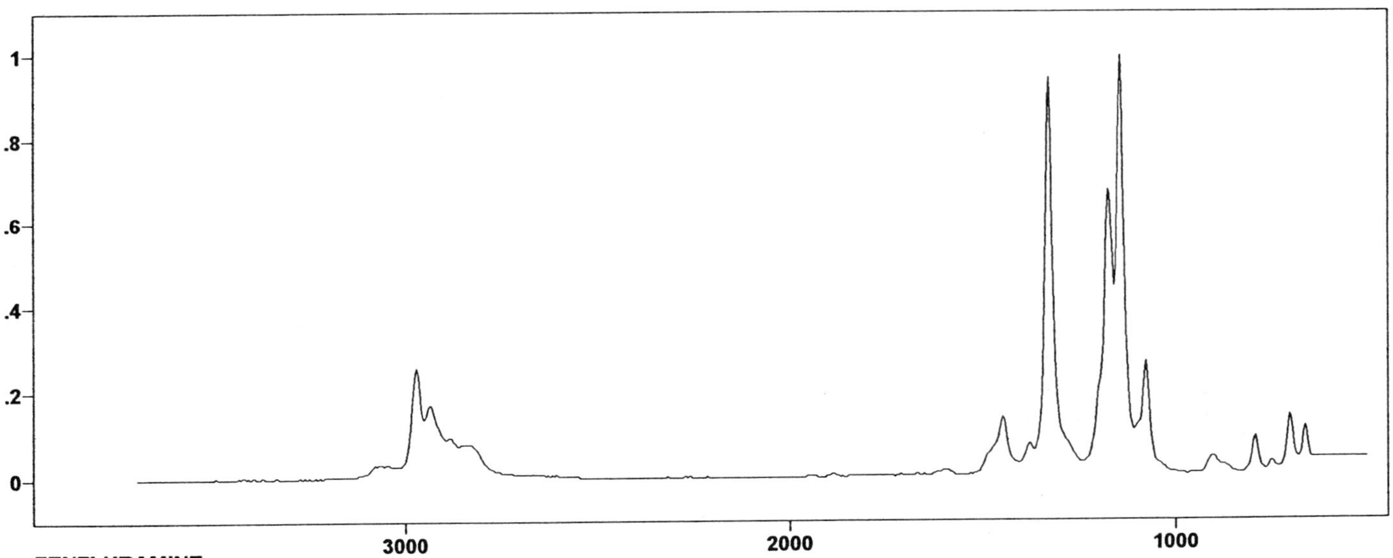

FENFLURAMINE

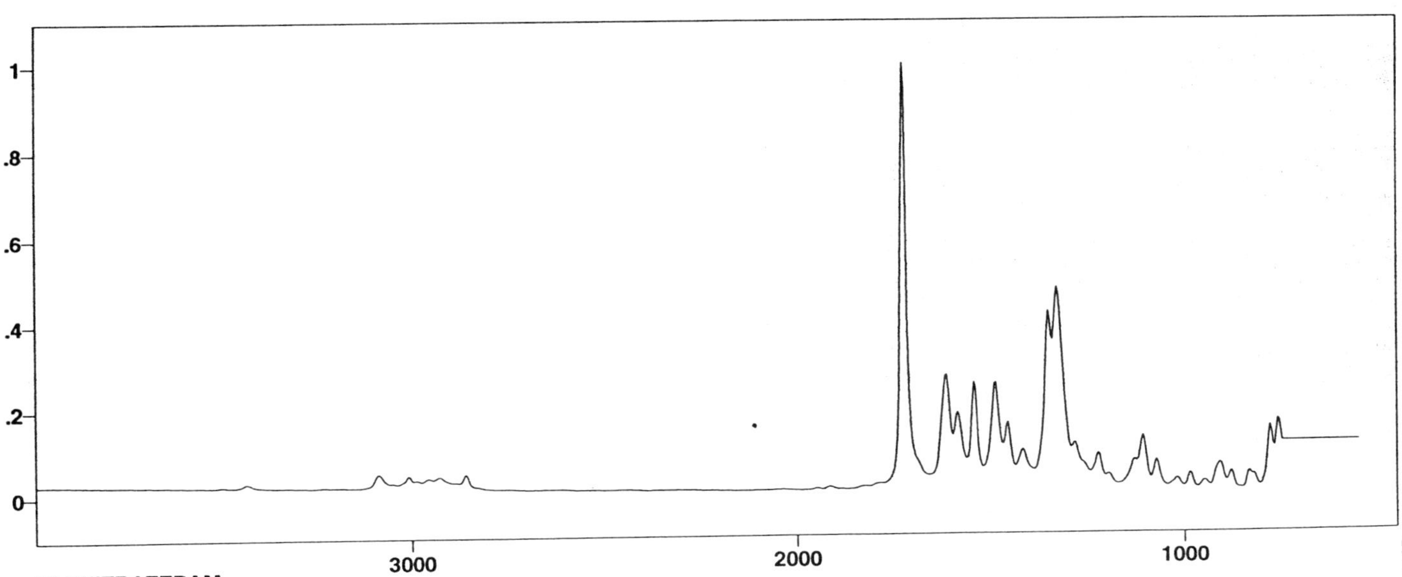

FLUNITRAZEPAM

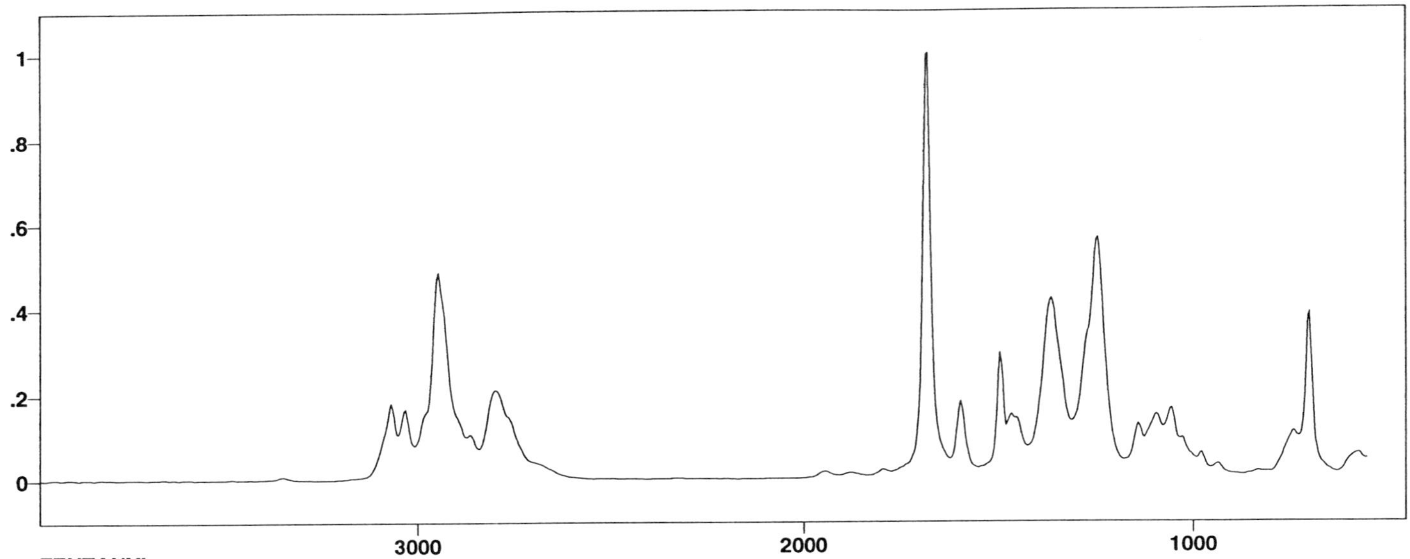

FENTANYL

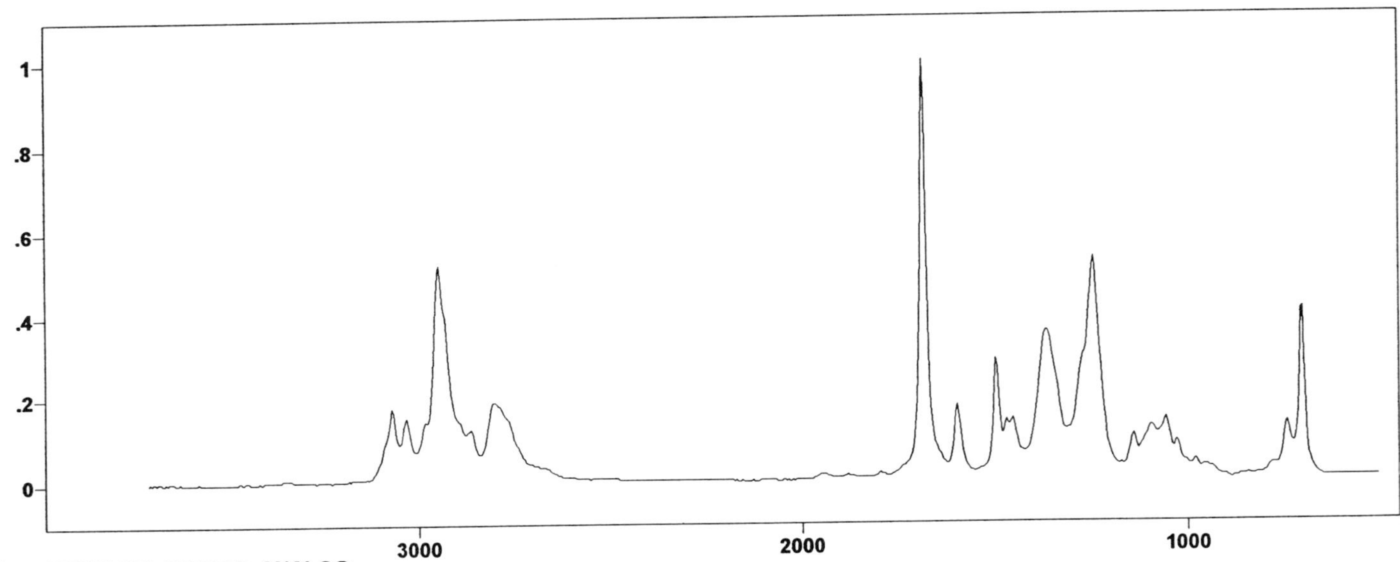

FENTANYL PROPYL ANALOG

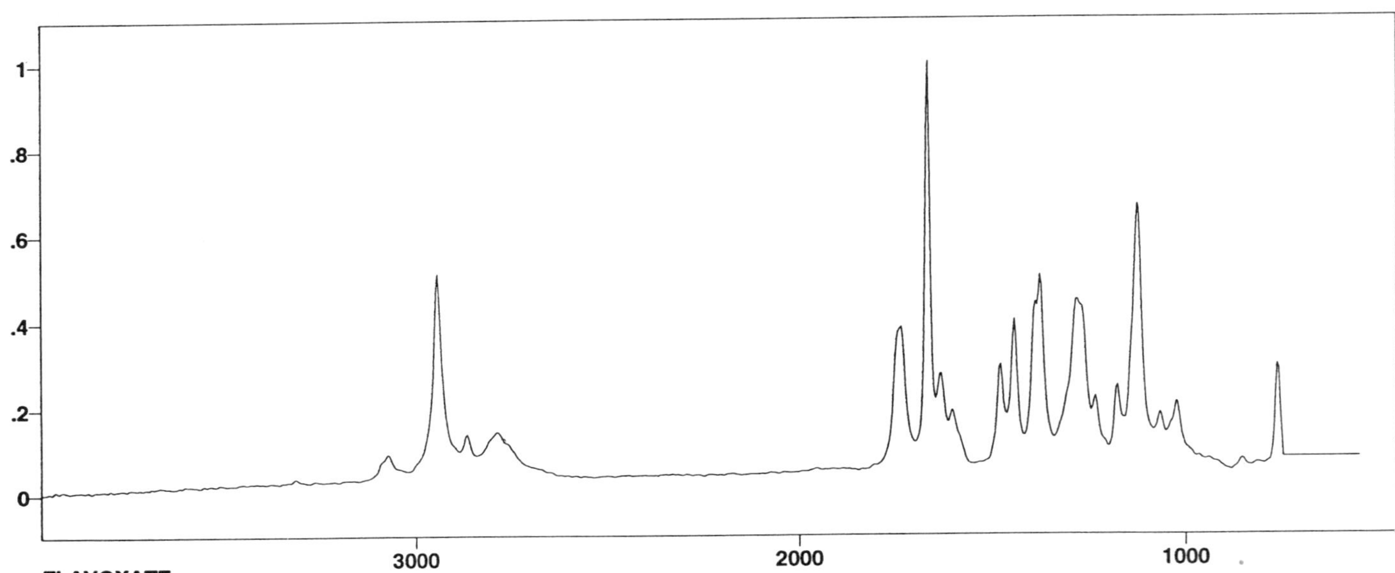

FLAVOXATE

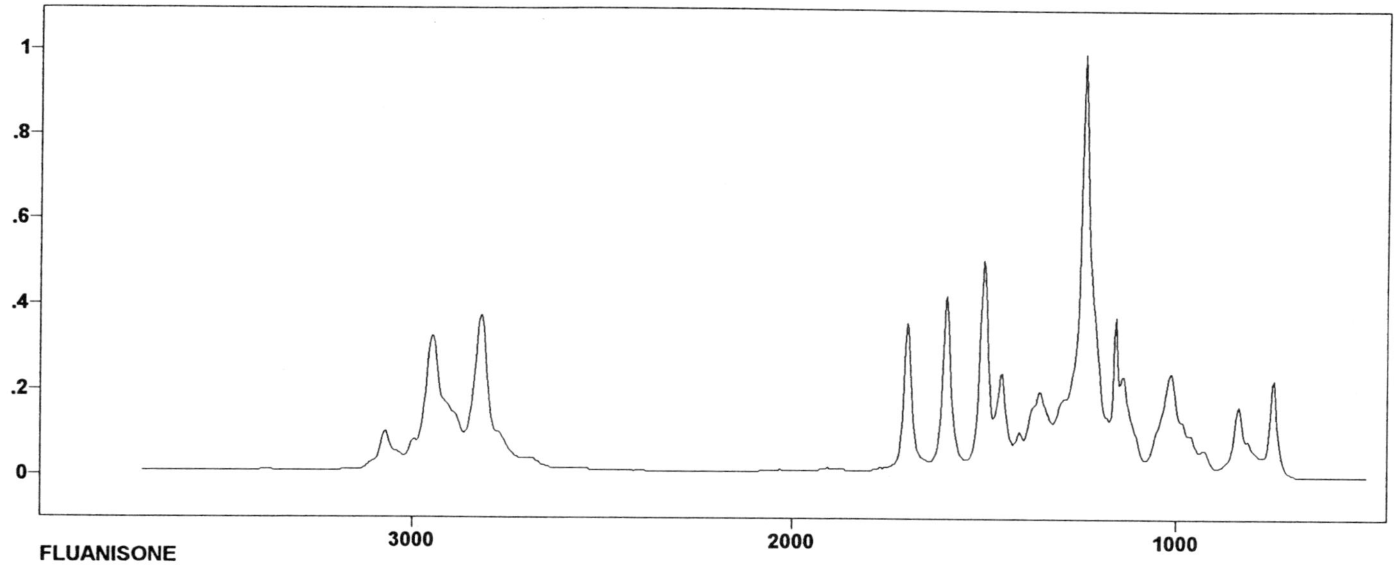

FLUANISONE

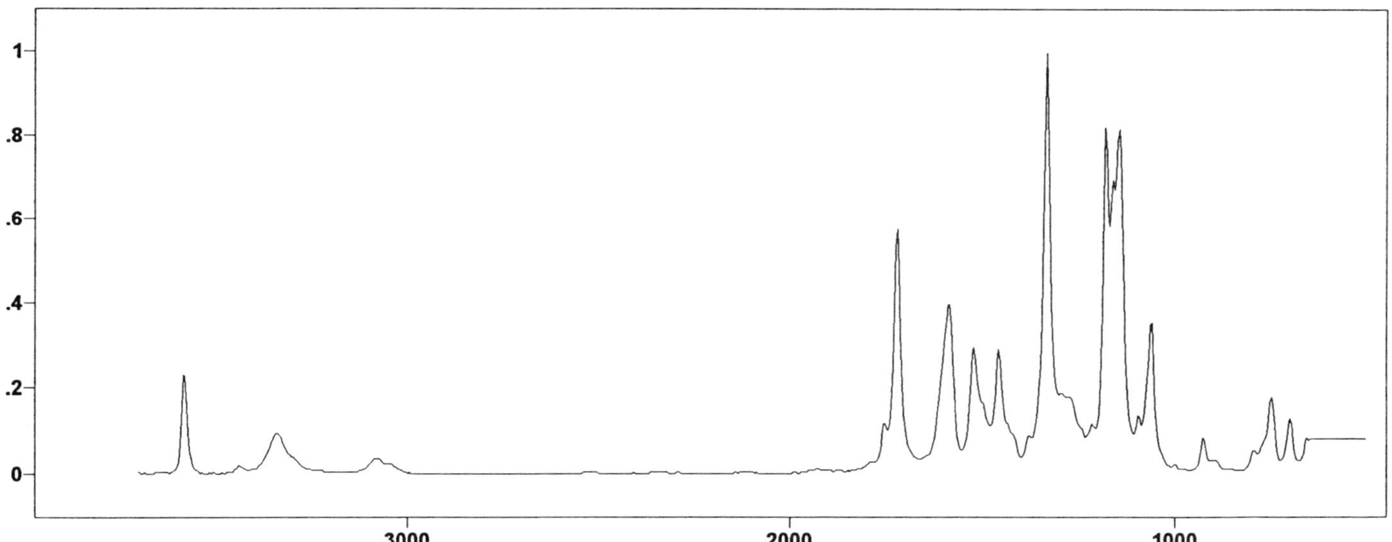

FLUFENAMIC ACID

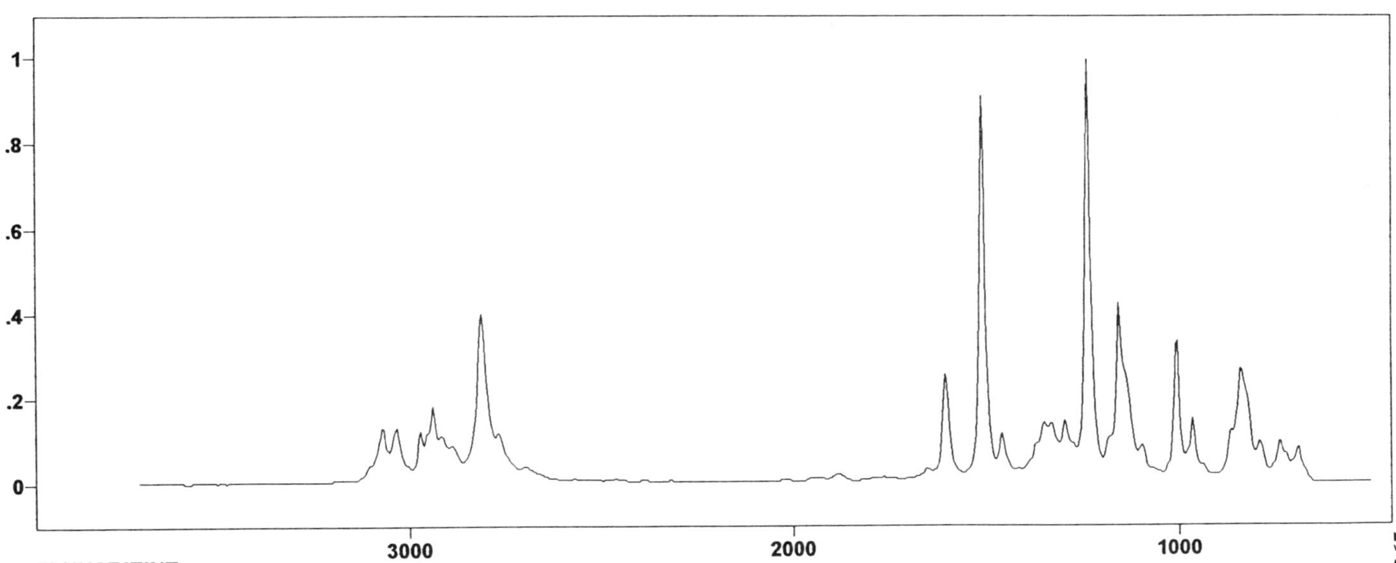

FLUNARIZINE

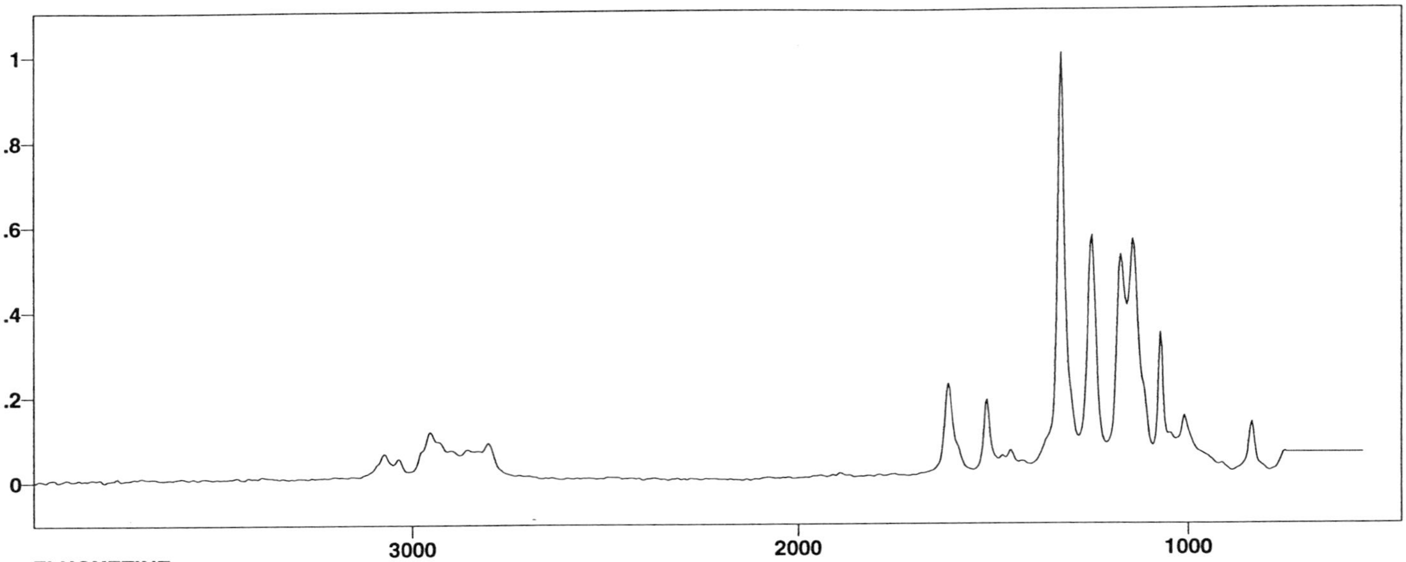

FLUOXETINE

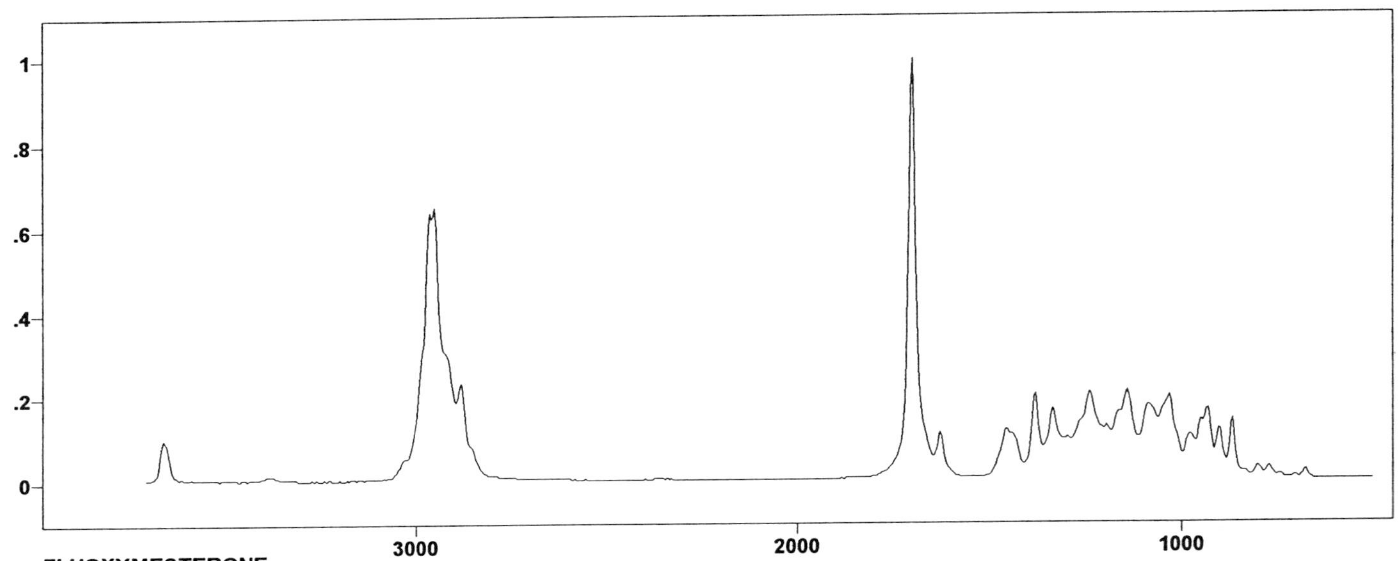

FLUOXYMESTERONE

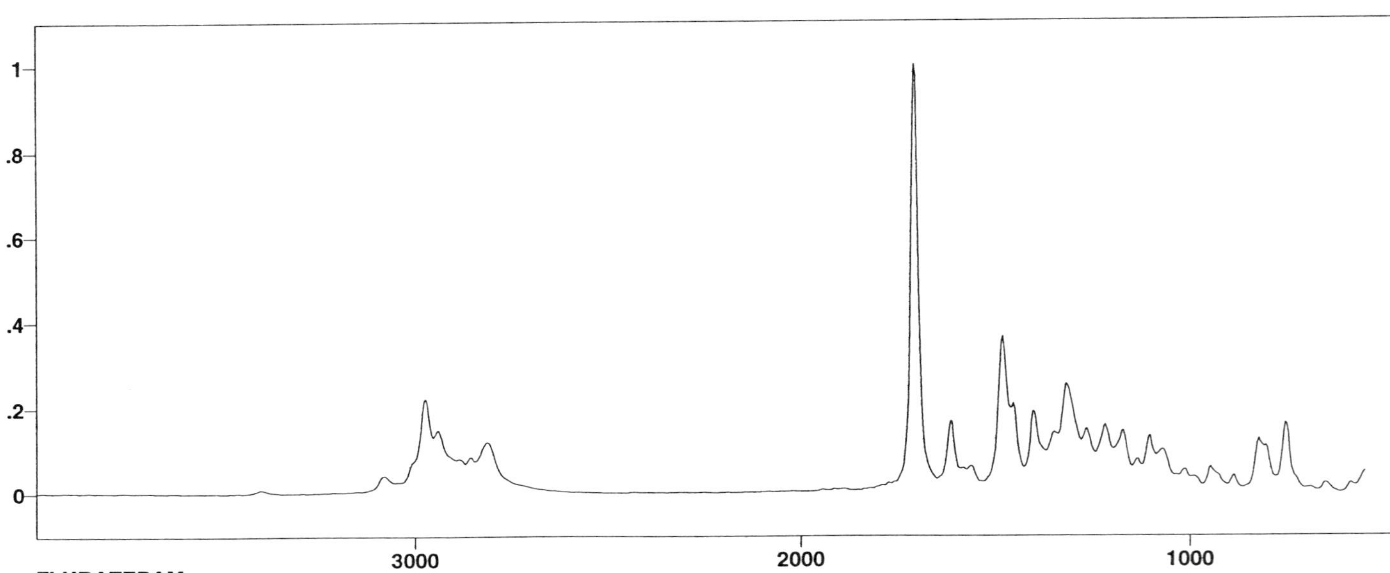

FLURAZEPAM

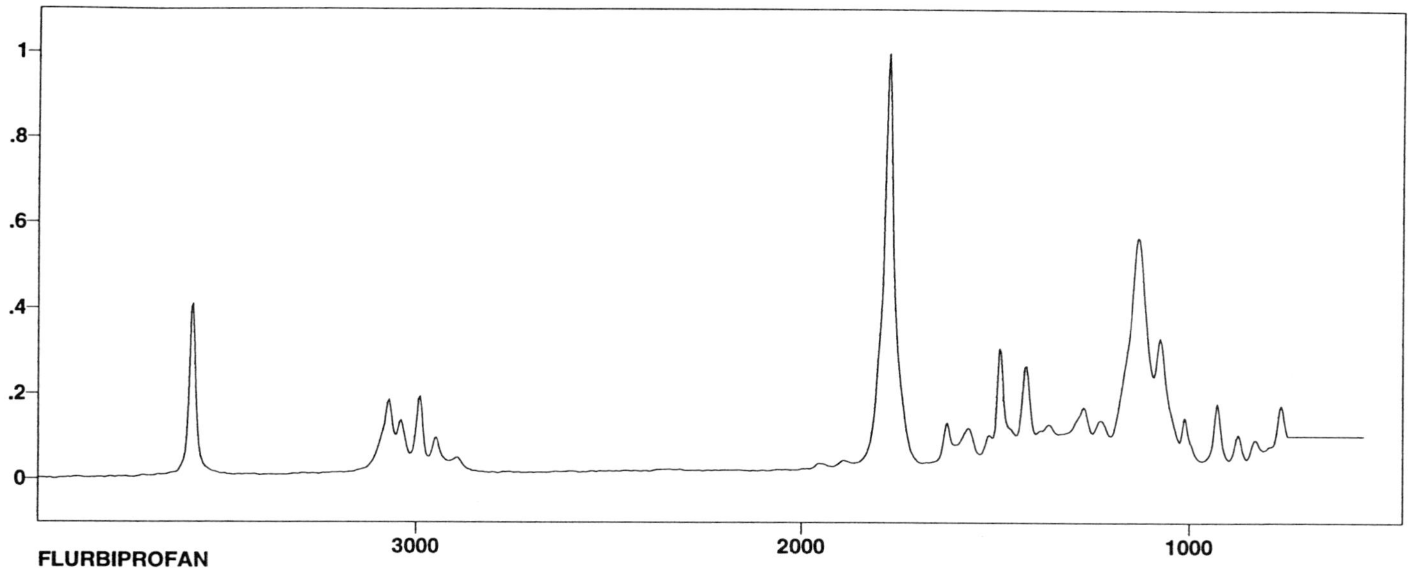

FLURBIPROFAN

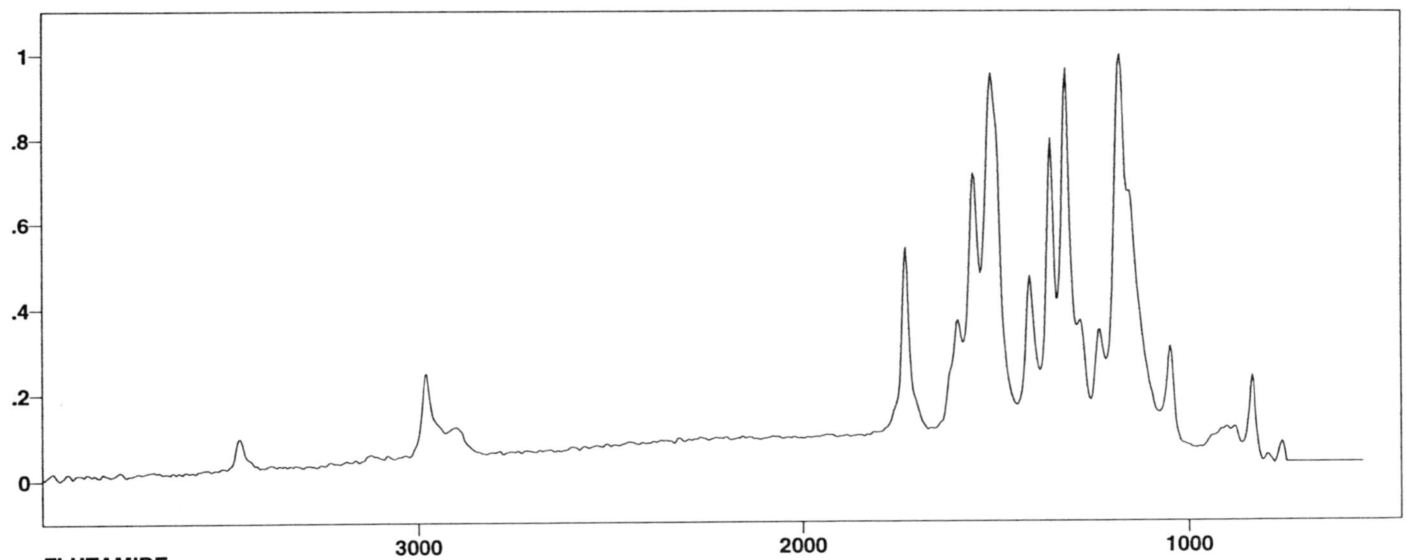

FLUTAMIDE

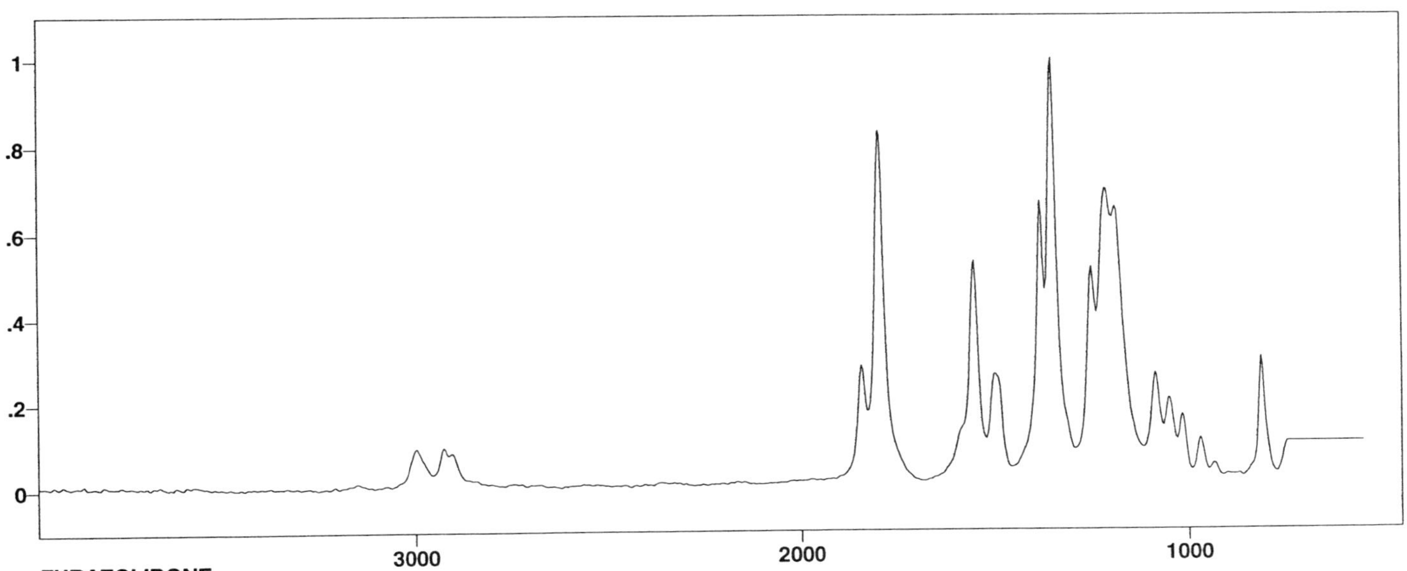

FURAZOLIDONE

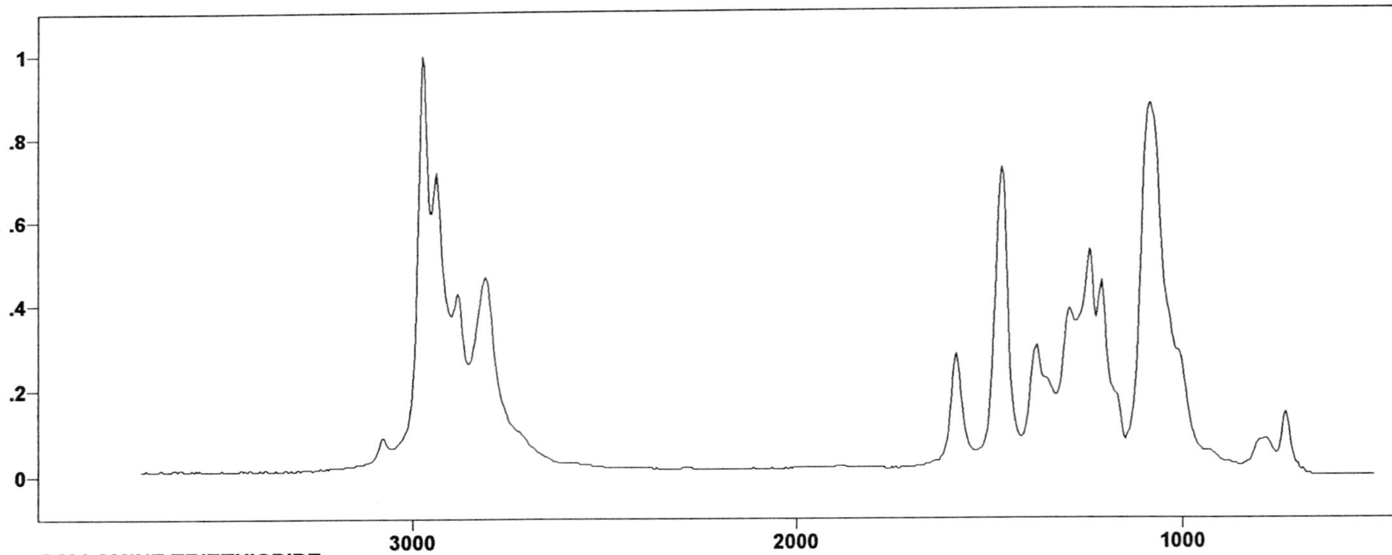

GALLAMINE TRIETHIODIDE

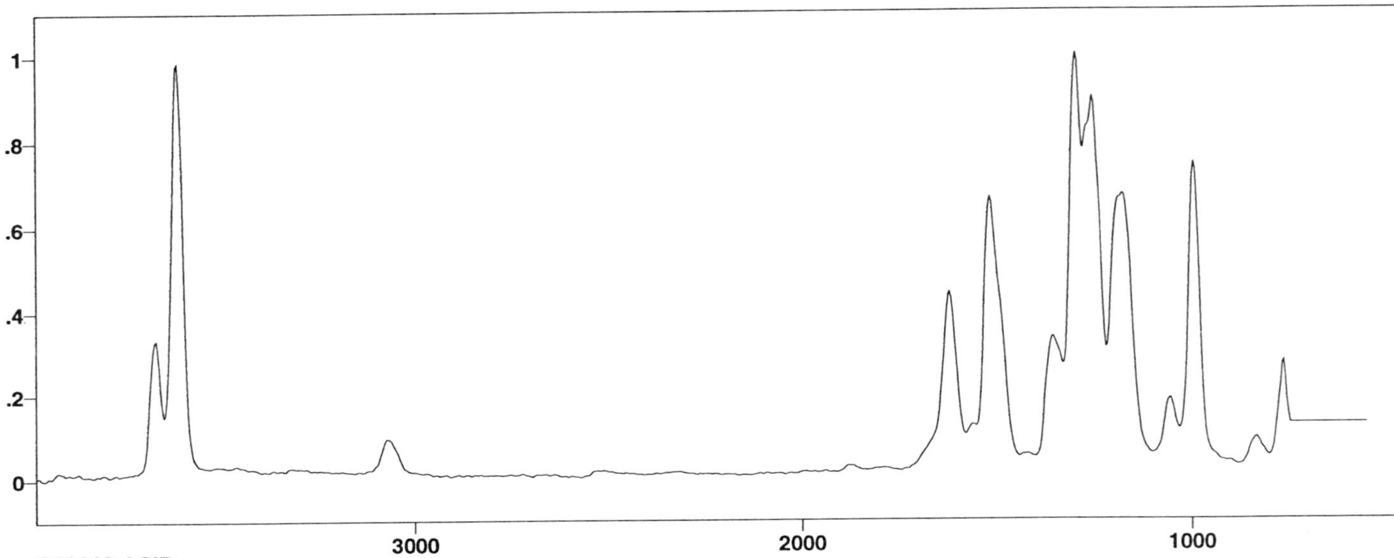

GALLIC ACID

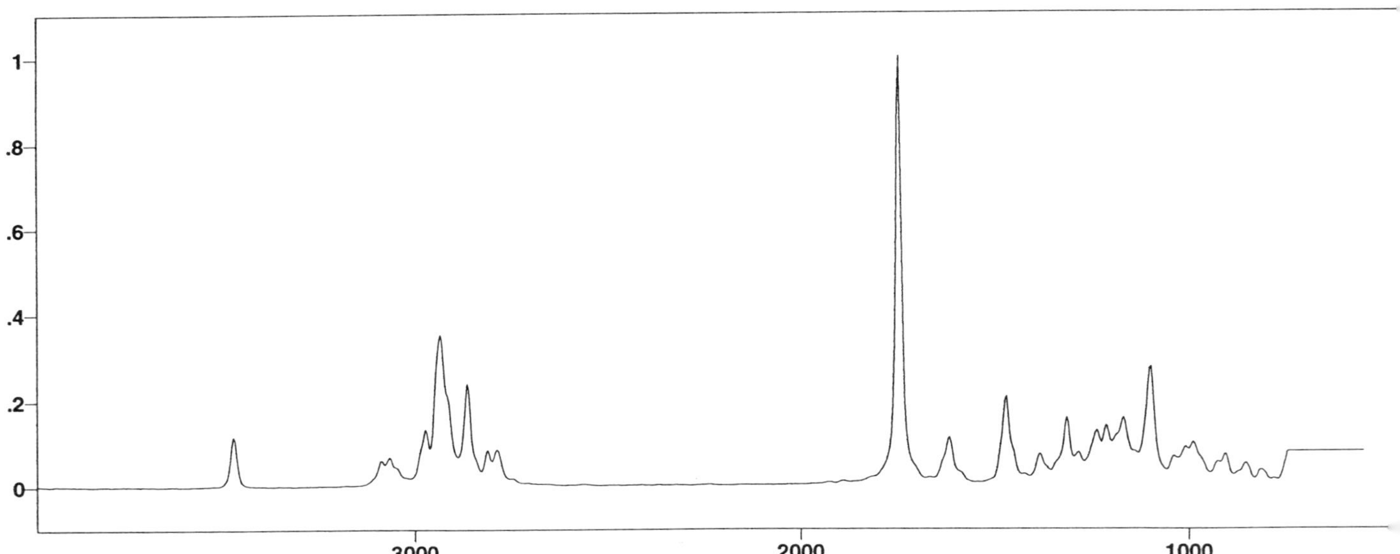

GELSEMINE

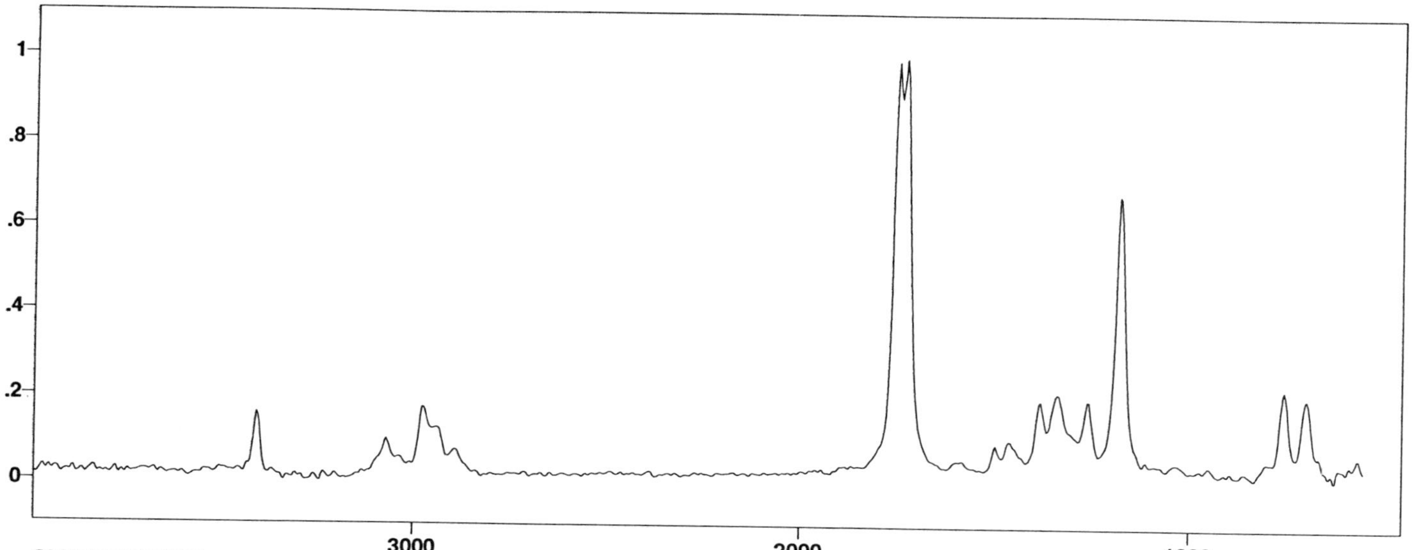

GLUTETHAMIDE

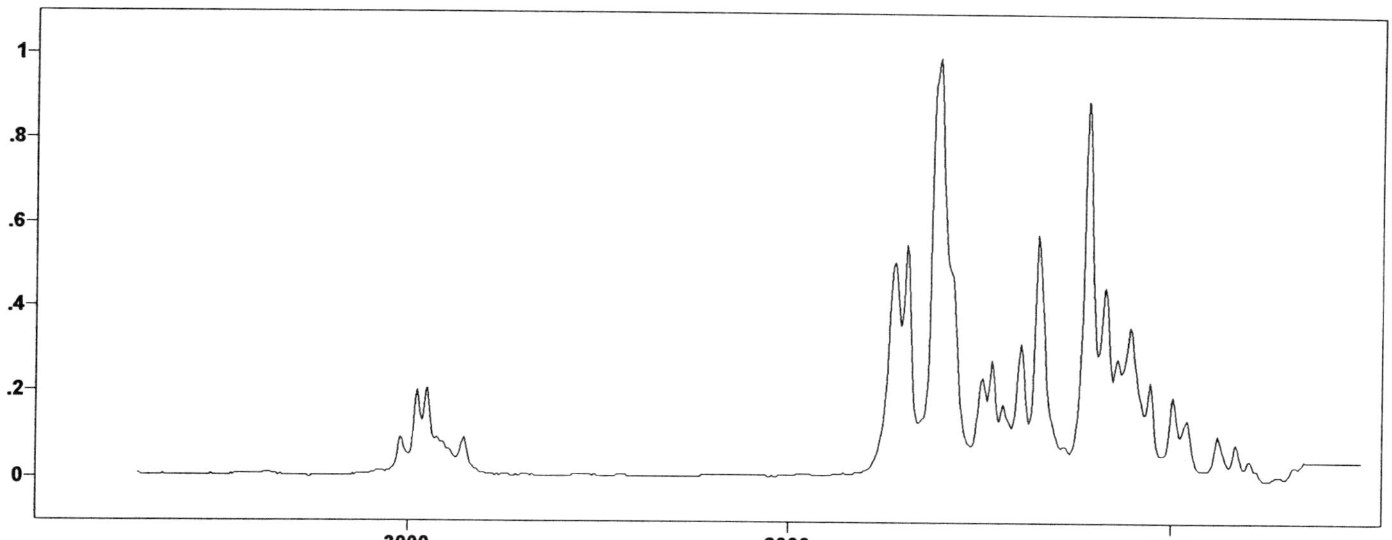

GRISEOFULVIN

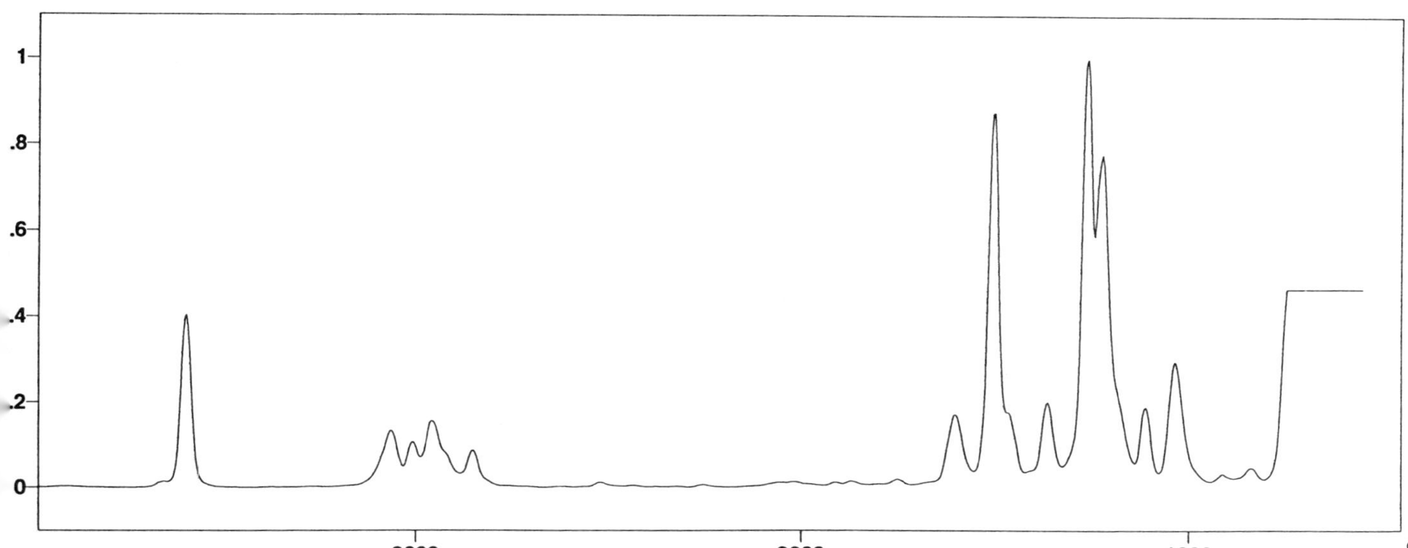

GUAIACOL

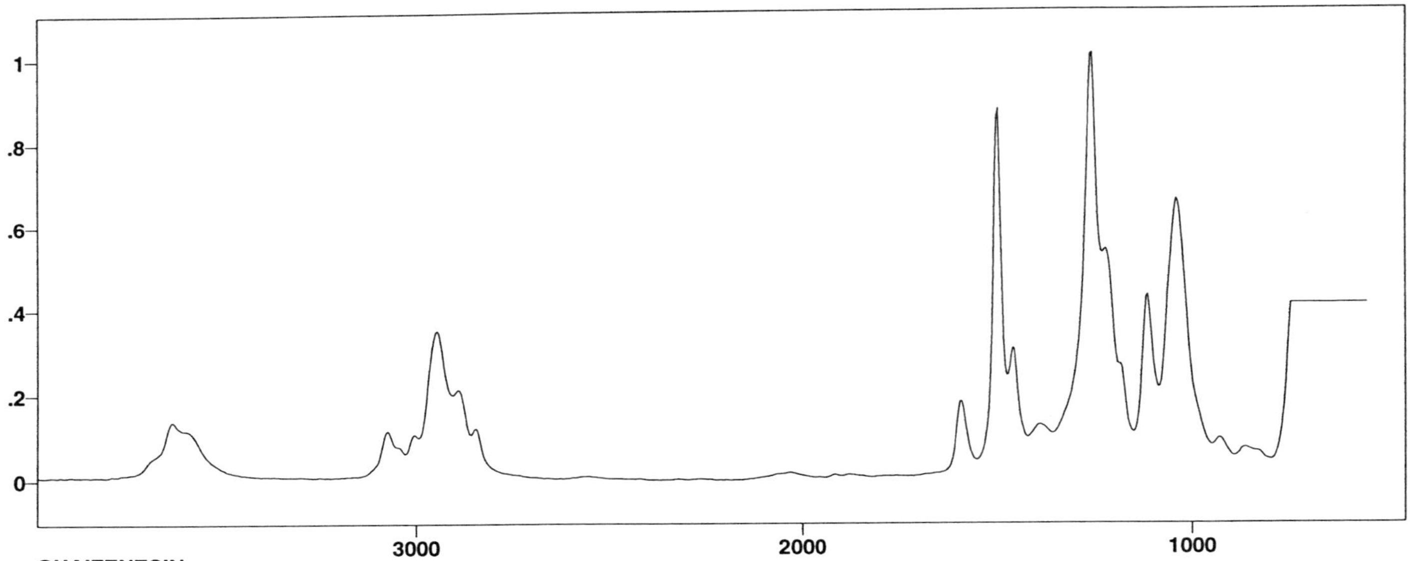

GUAIFENESIN

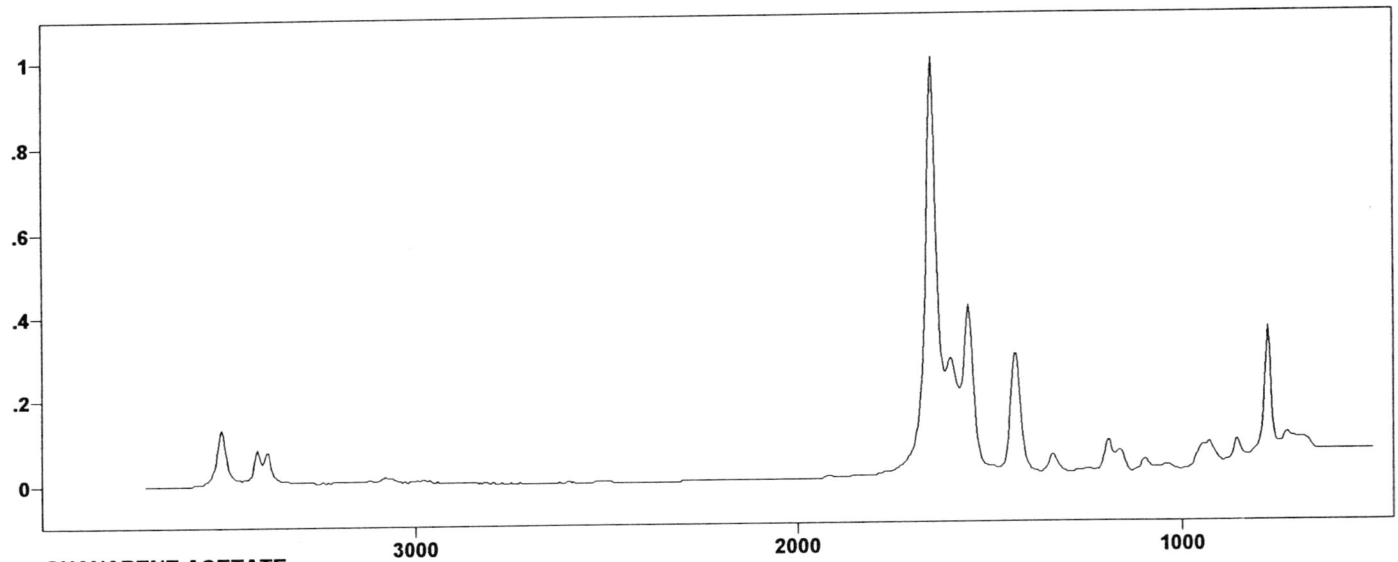

GUANABENZ ACETATE

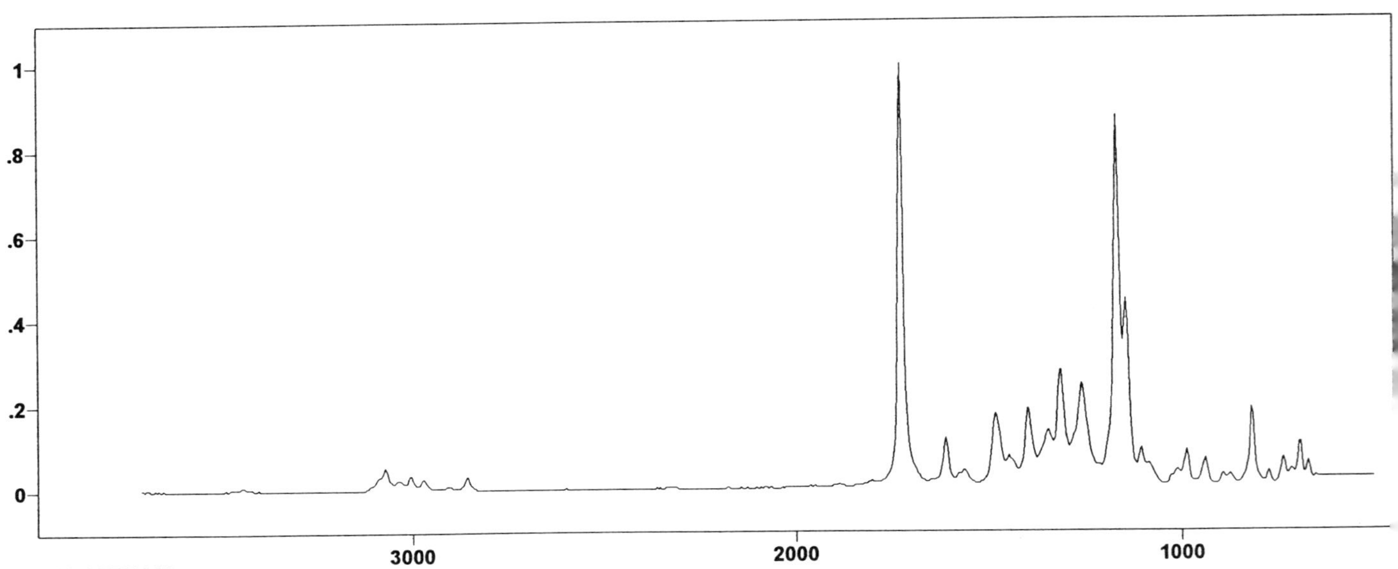

HALAZEPAM

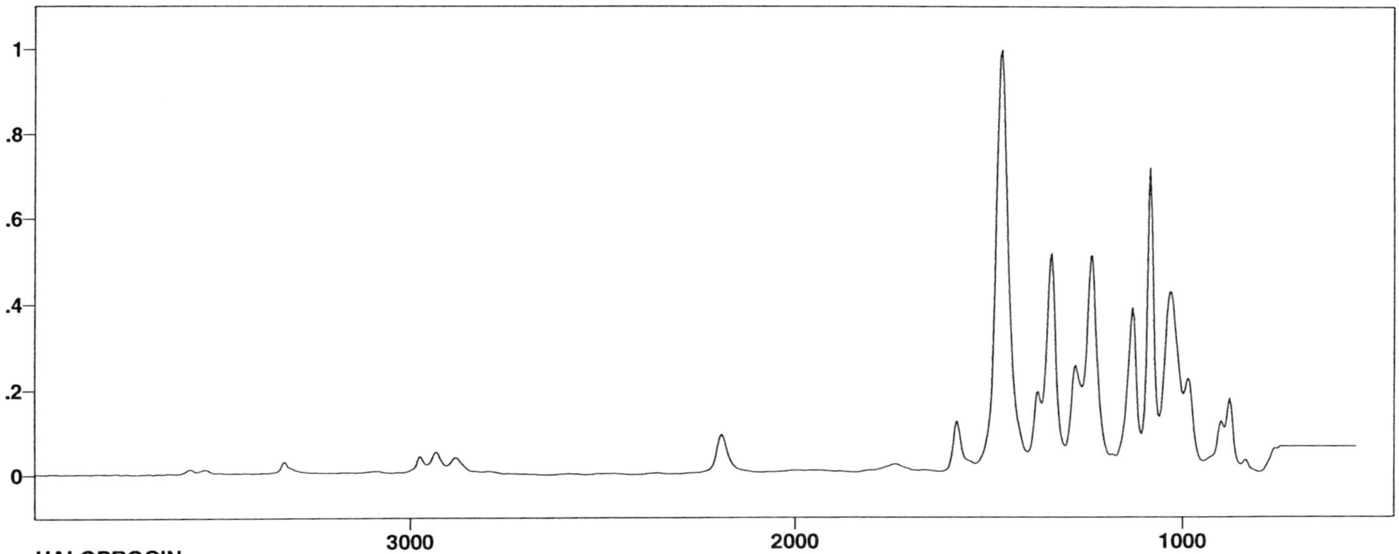

HALOPROGIN

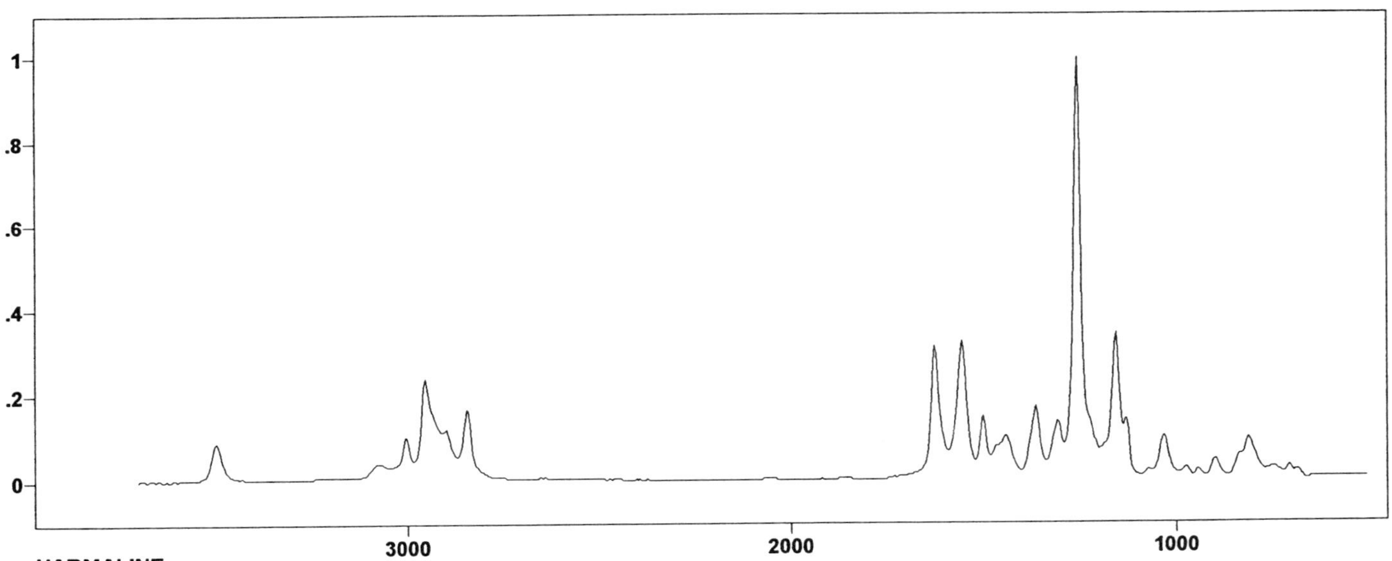

HARMALINE

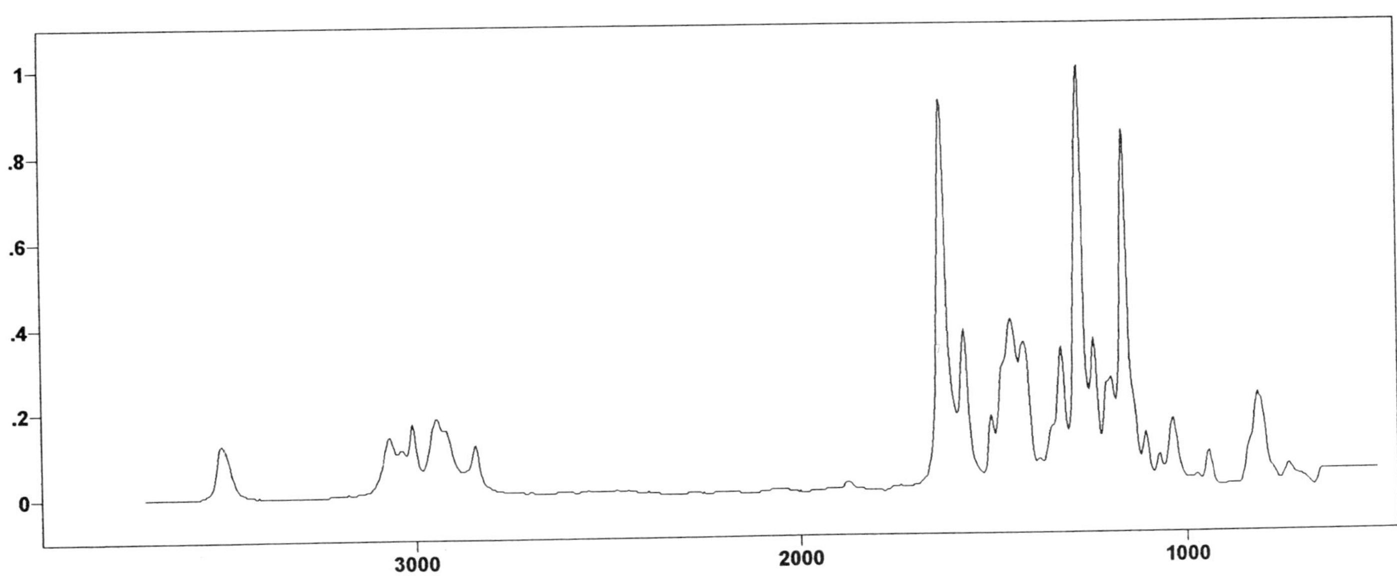

HARMINE

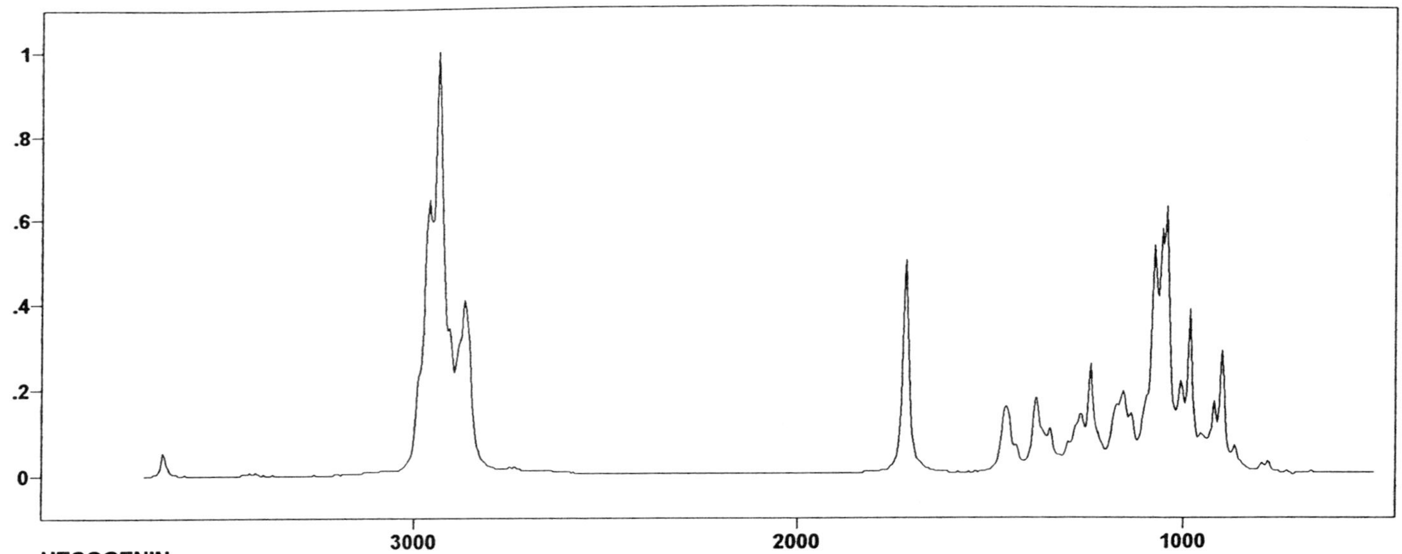

HECOGENIN

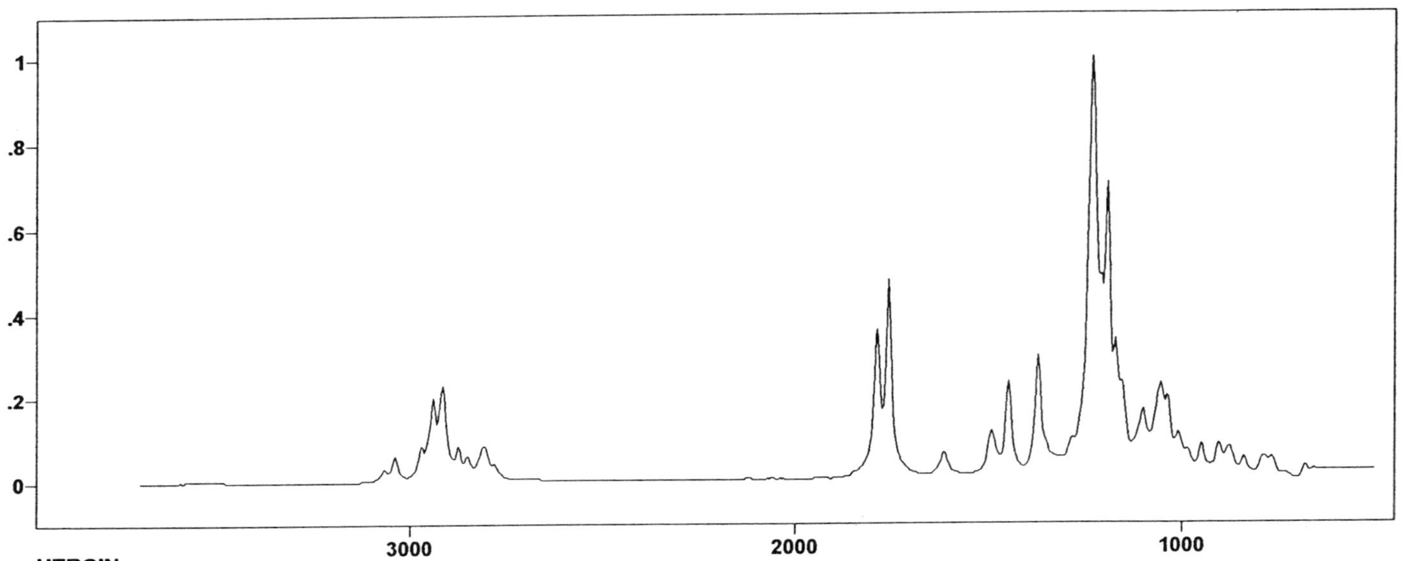

HEROIN

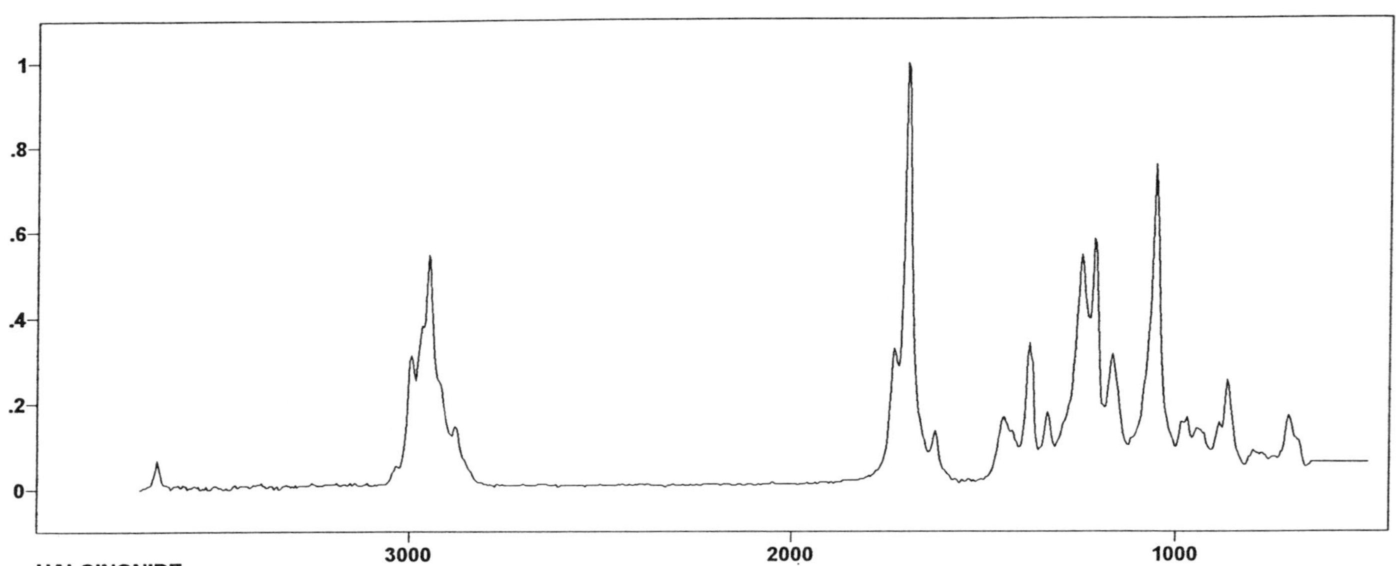

HALCINONIDE

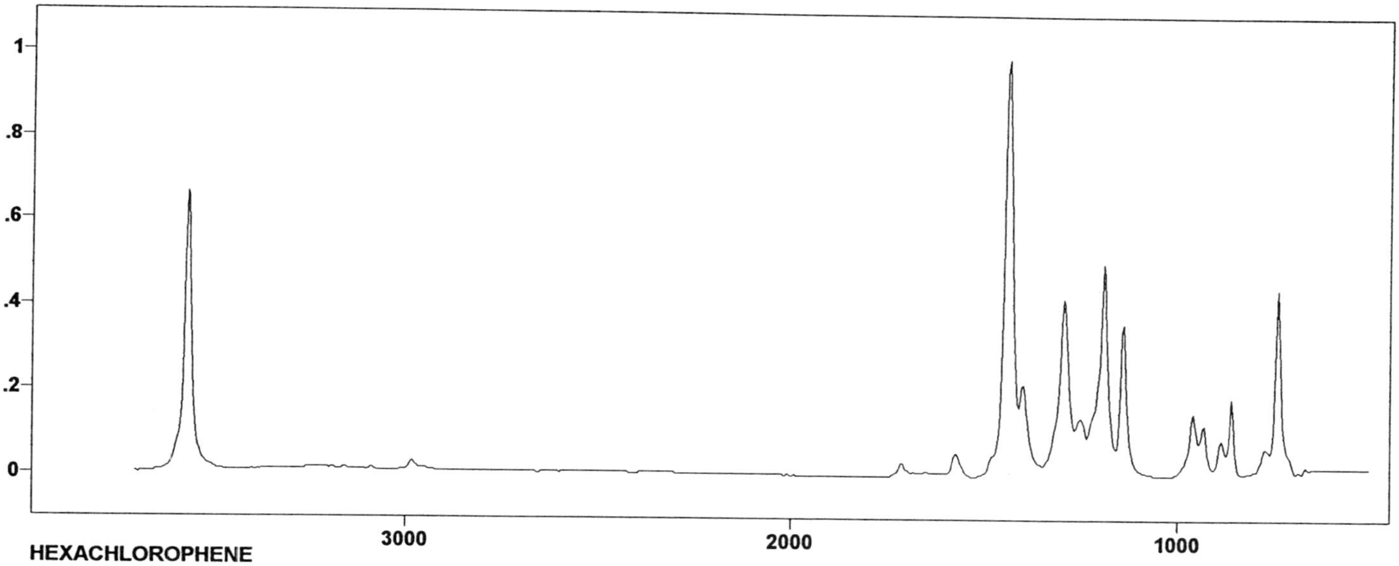

HEXACHLOROPHENE

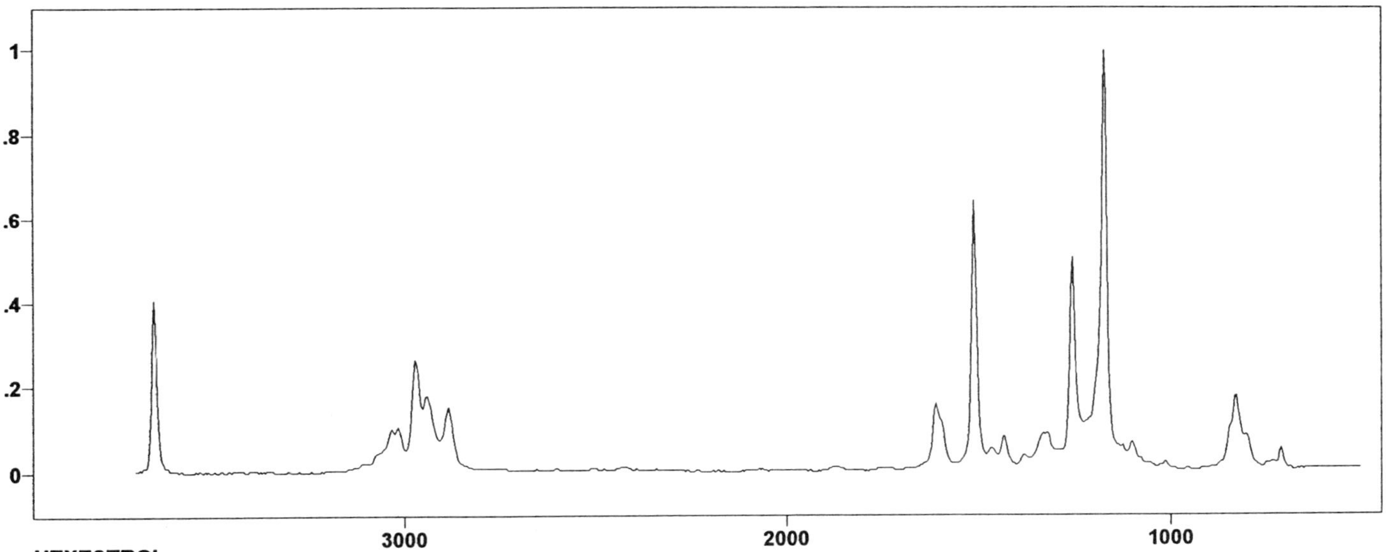

HEXESTROL

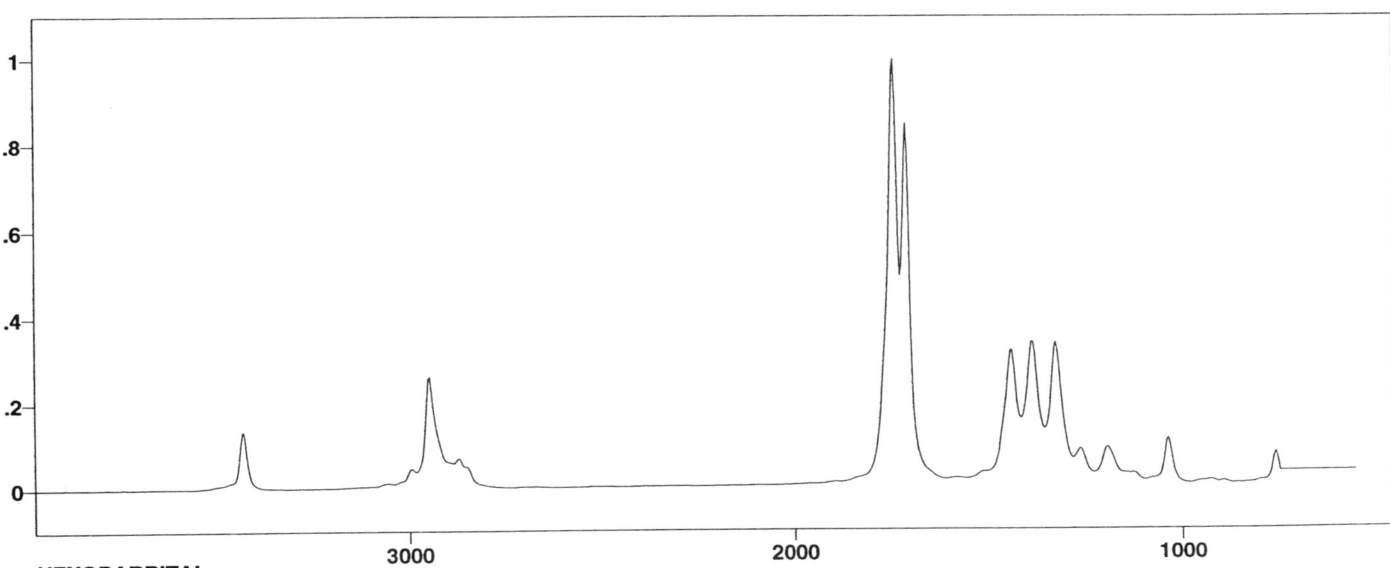

HEXOBARBITAL

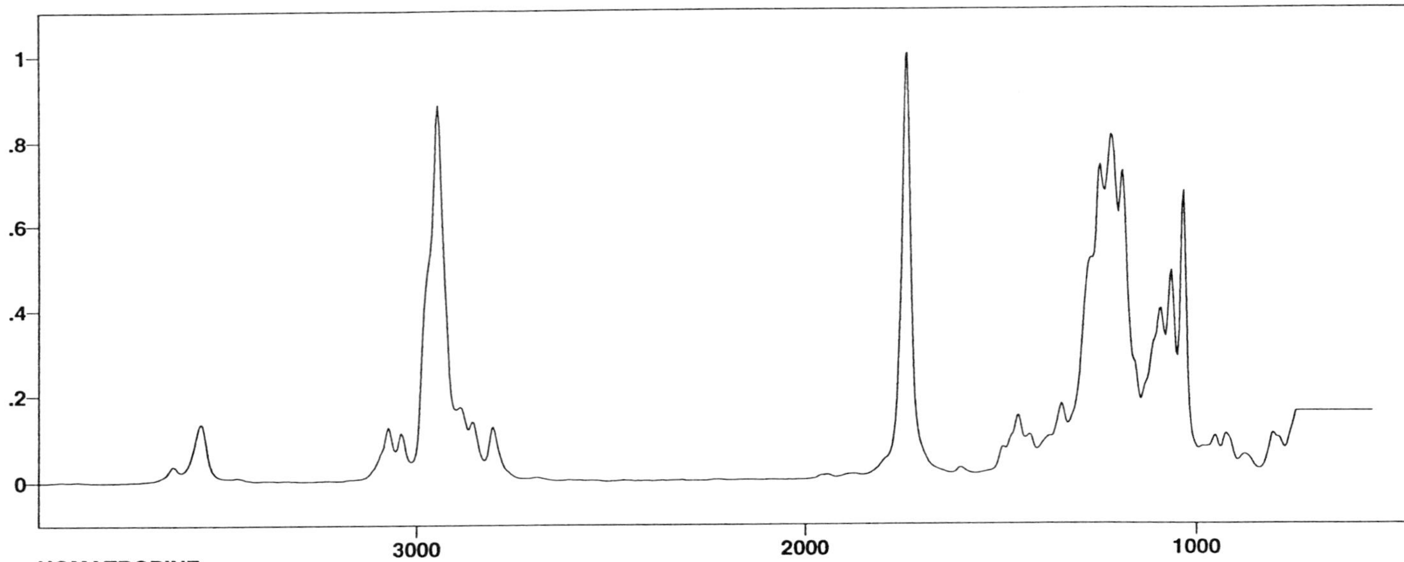

HOMATROPINE

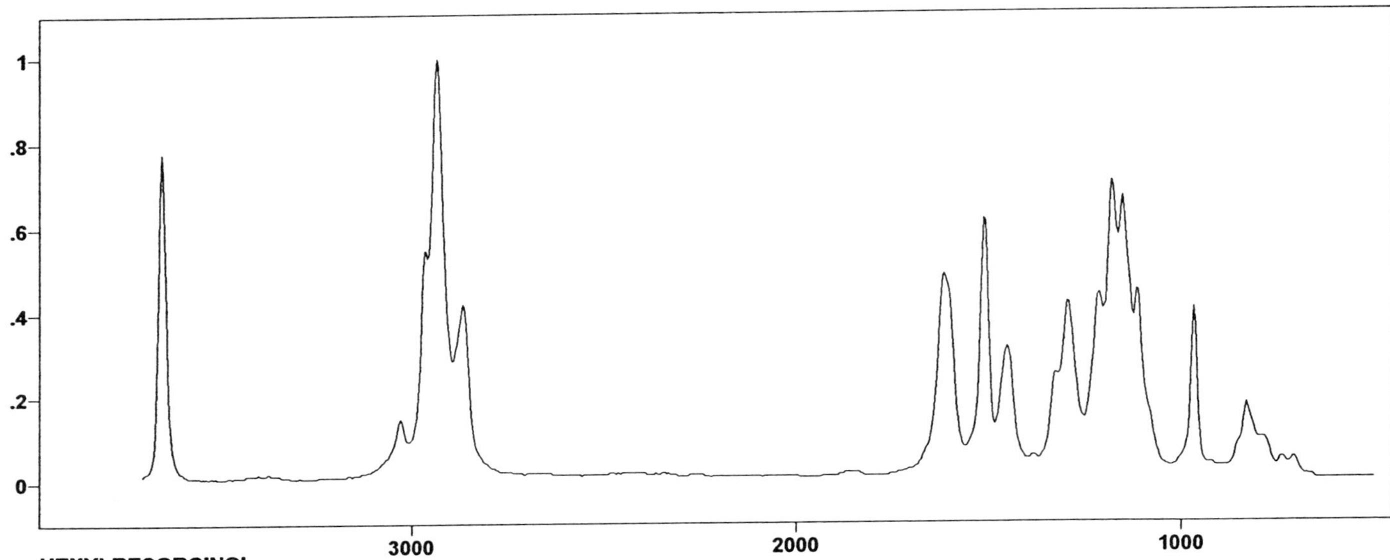

HEXYLRESORCINOL

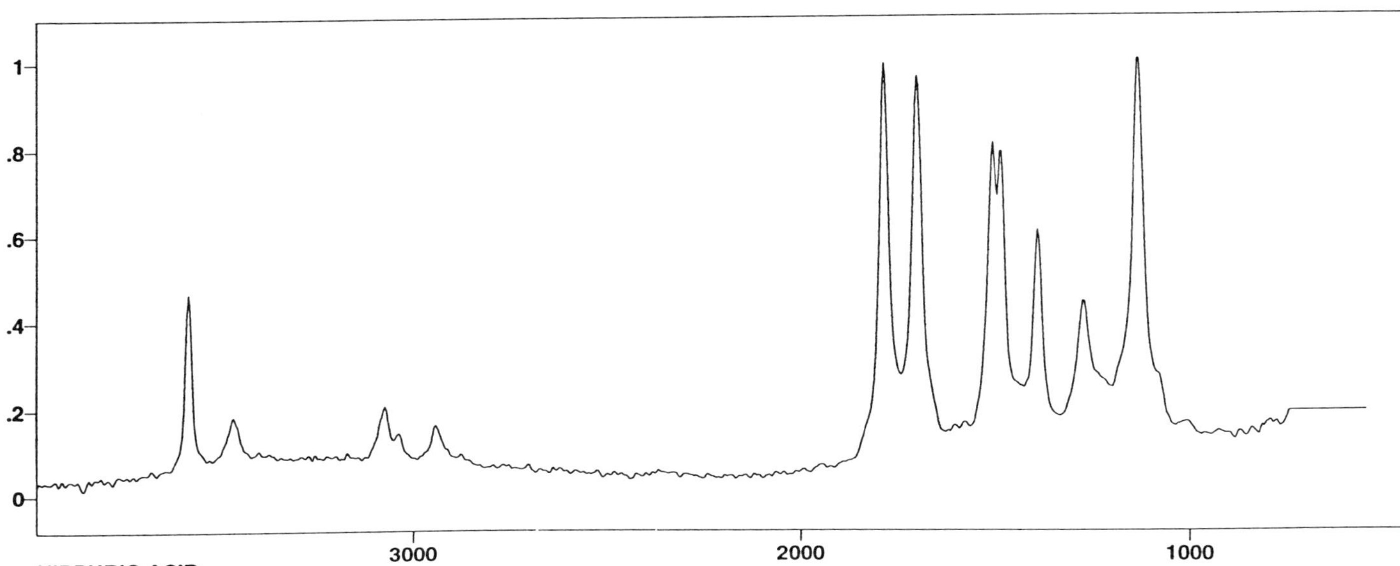

HIPPURIC ACID

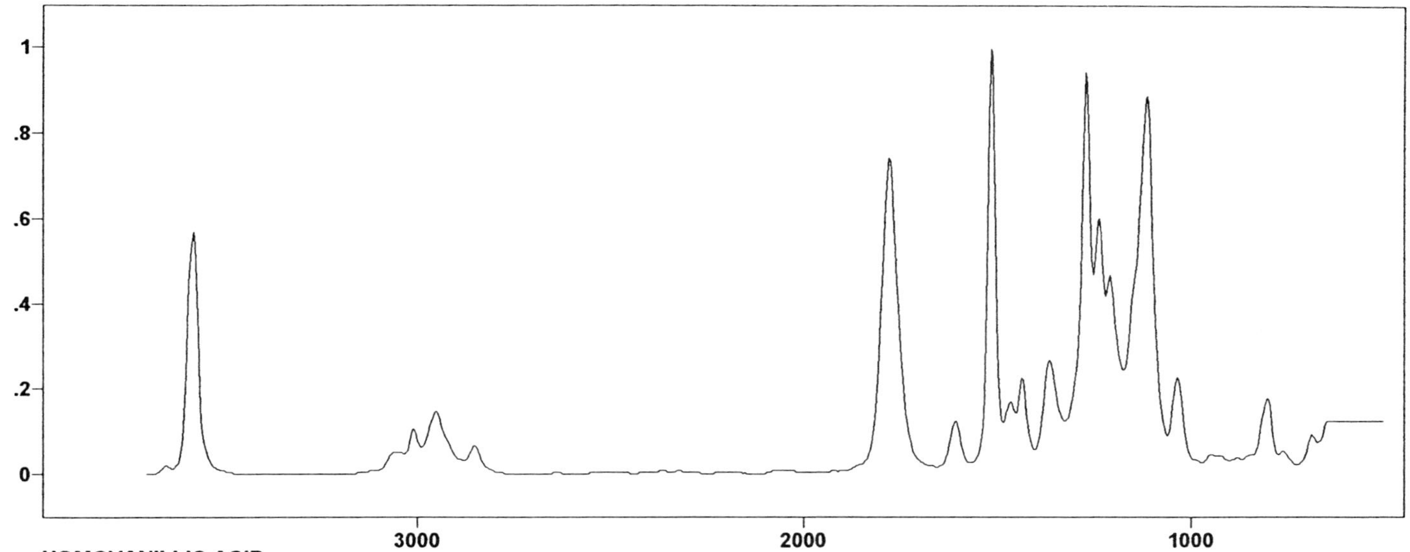

HOMOVANILLIC ACID

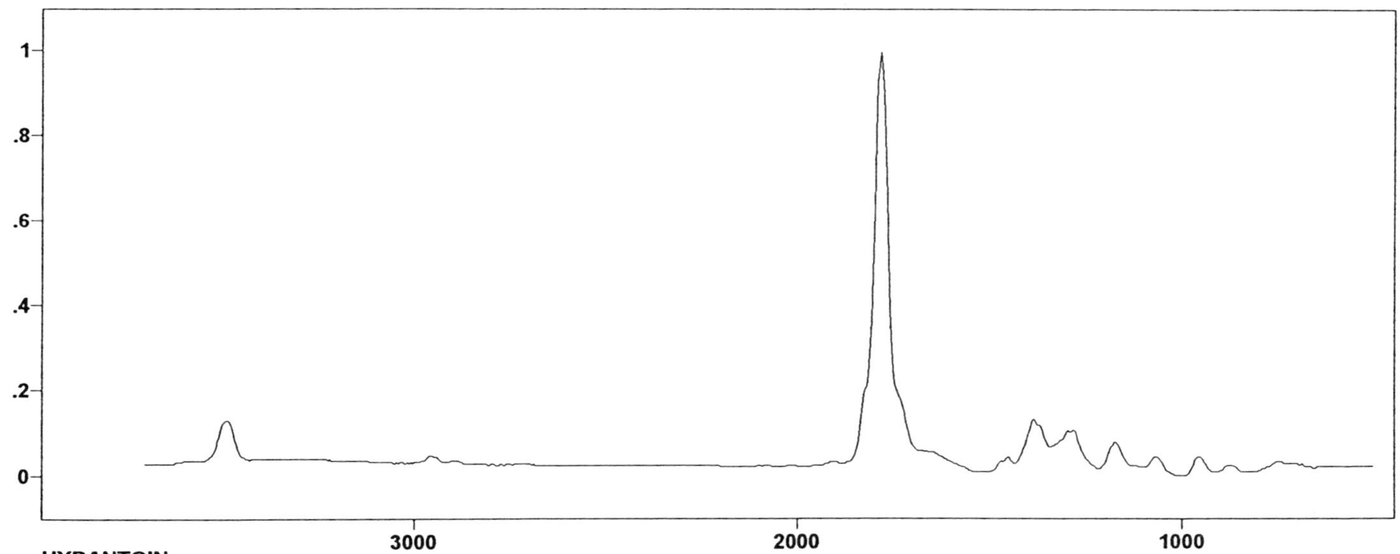

HYDANTOIN

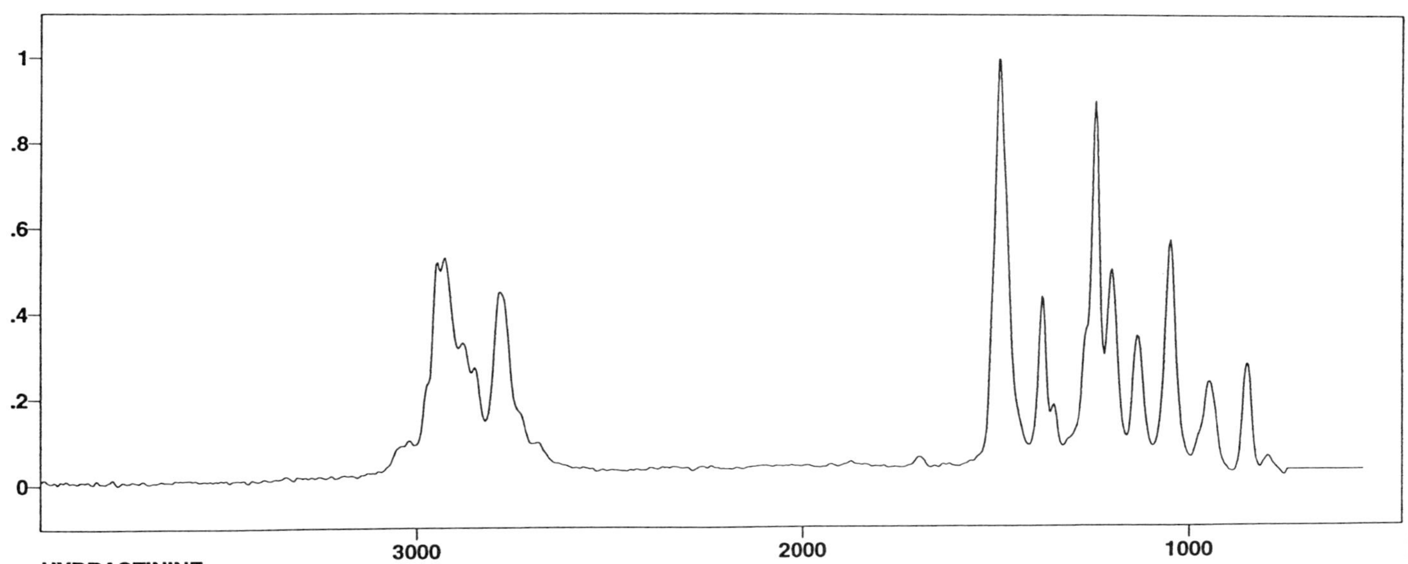

HYDRASTININE

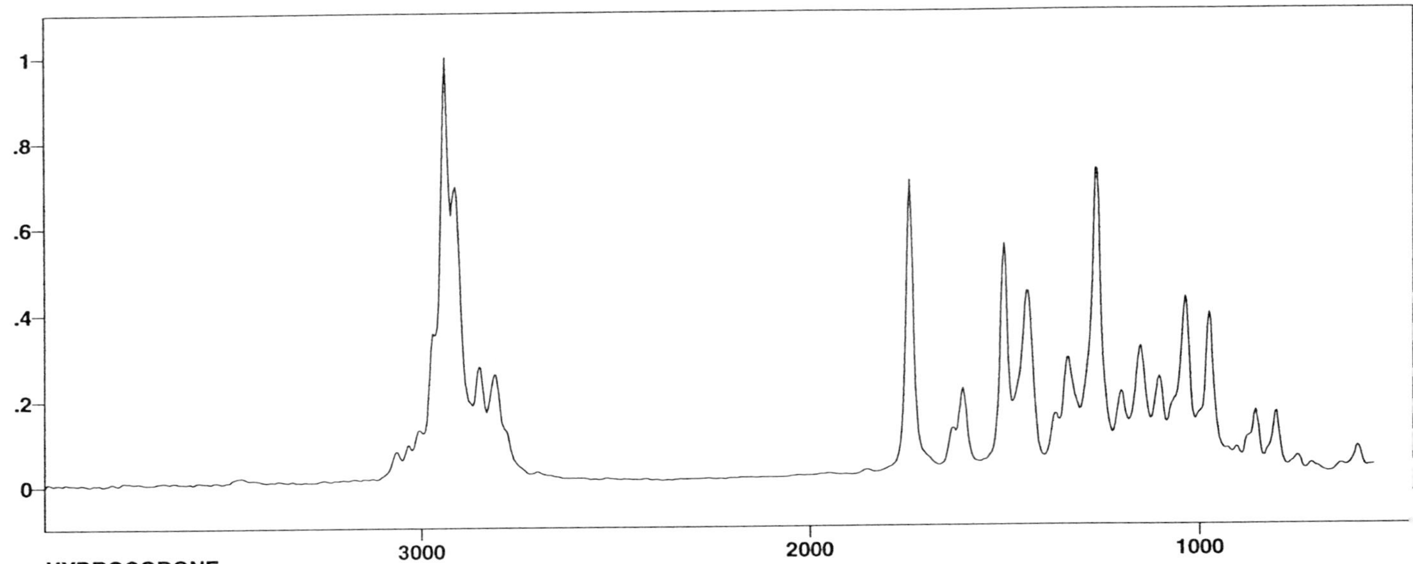

HYDROCODONE

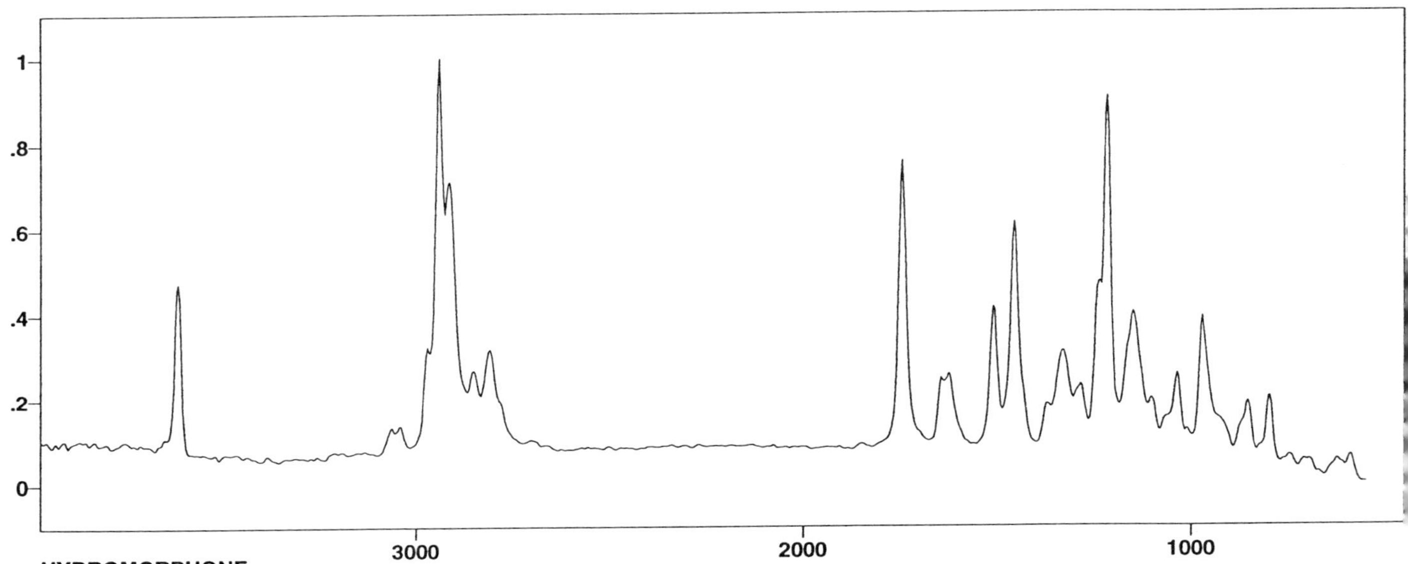

HYDROMORPHONE

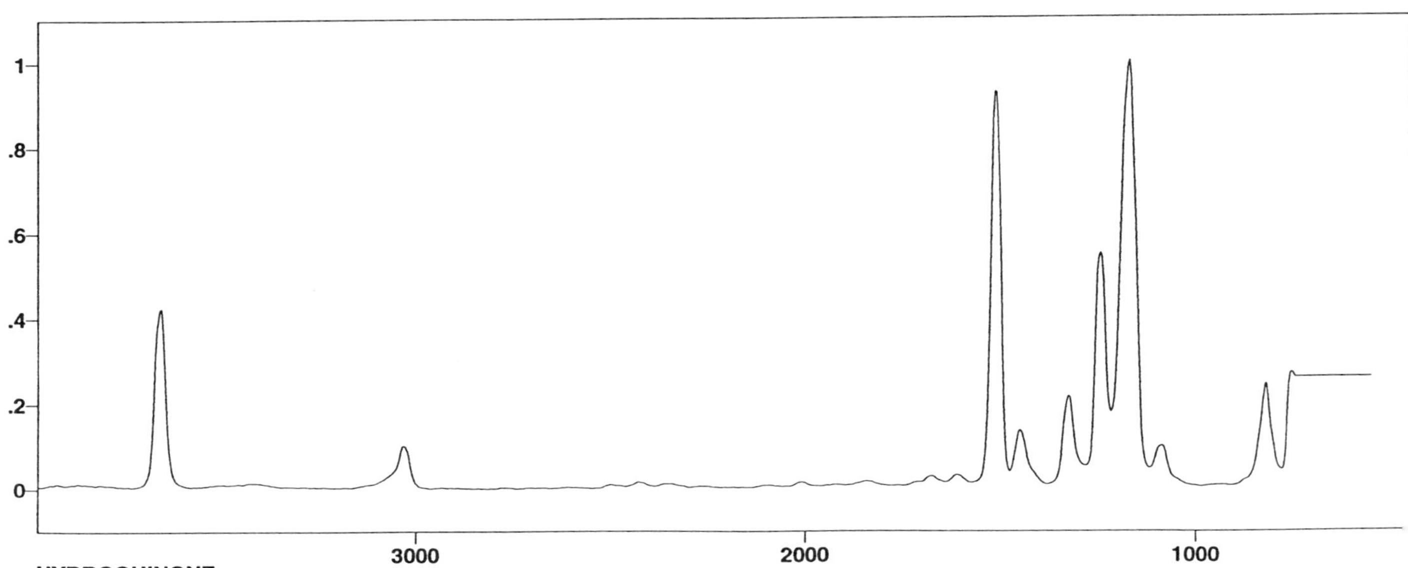

HYDROQUINONE

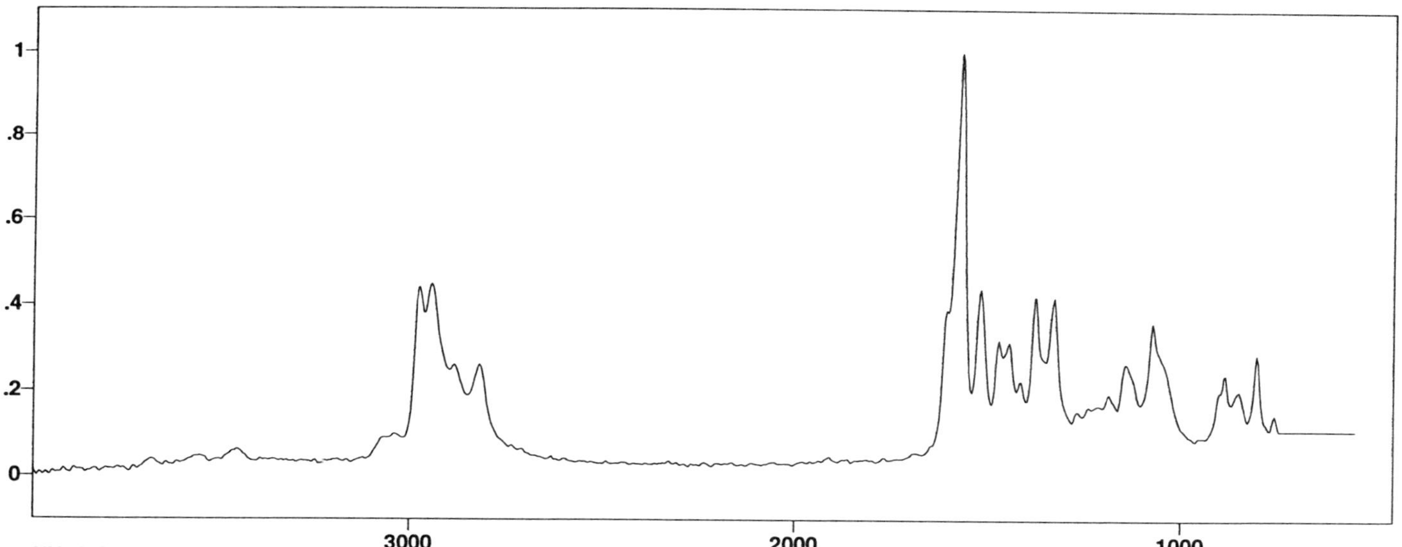

HYDROXYCHLOROQUINE

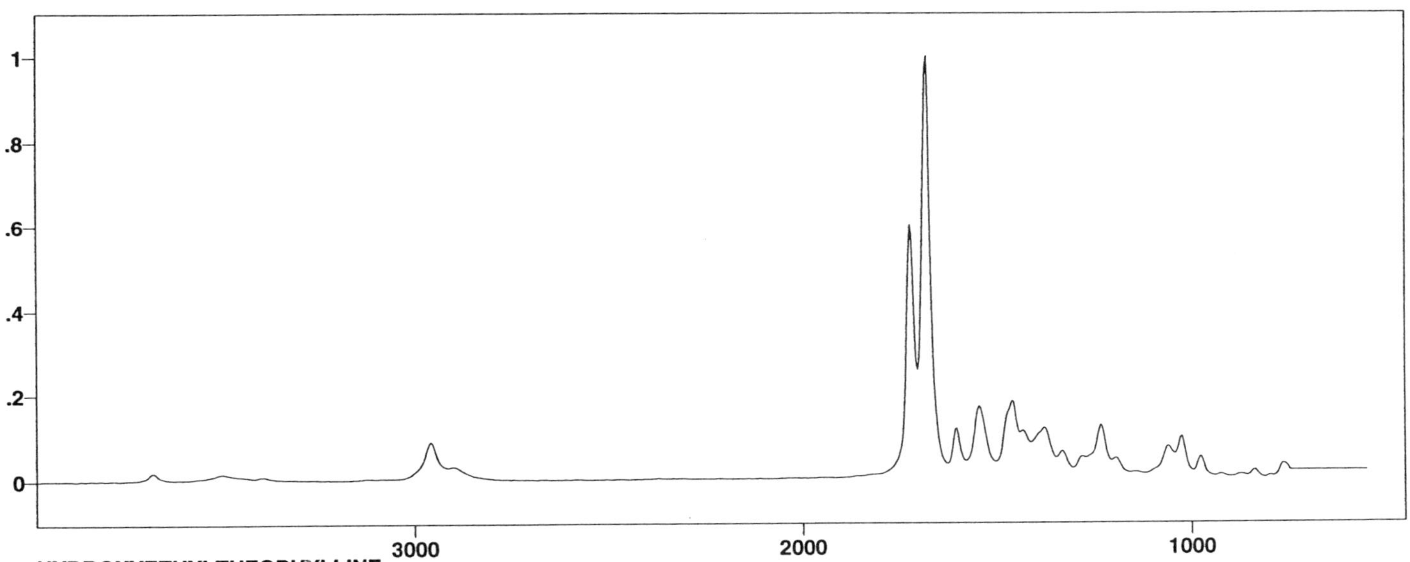

HYDROXYETHYLTHEOPHYLLINE

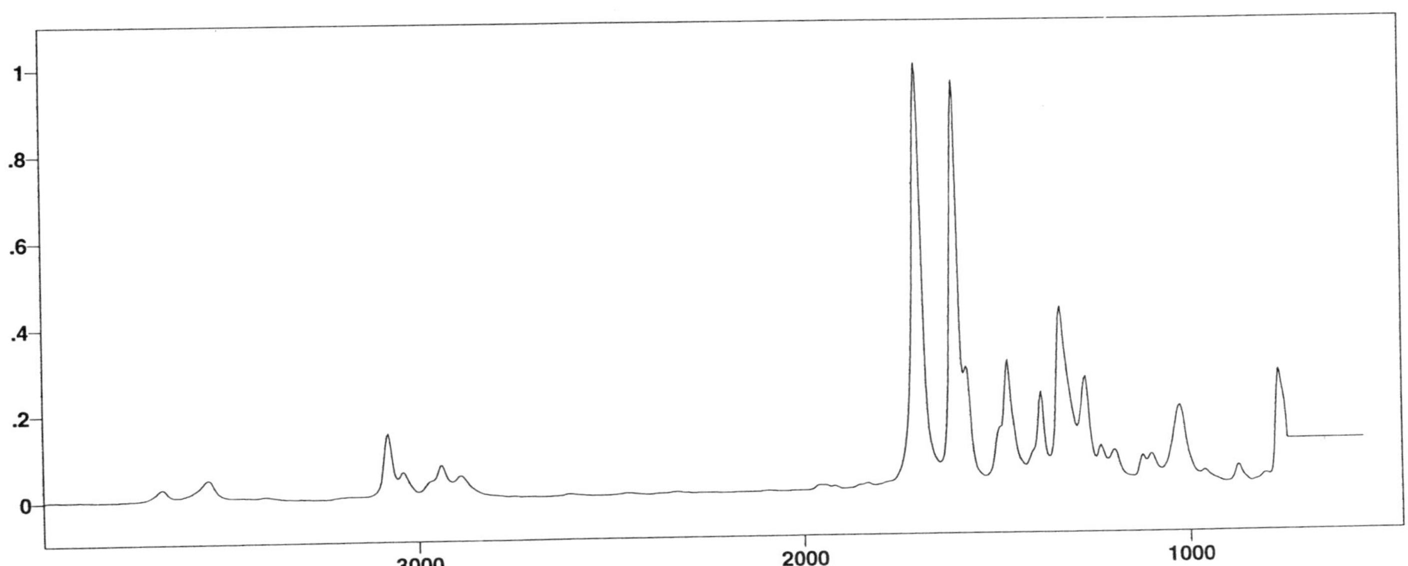

HYDROXYMETHYLMETHAQUALINE

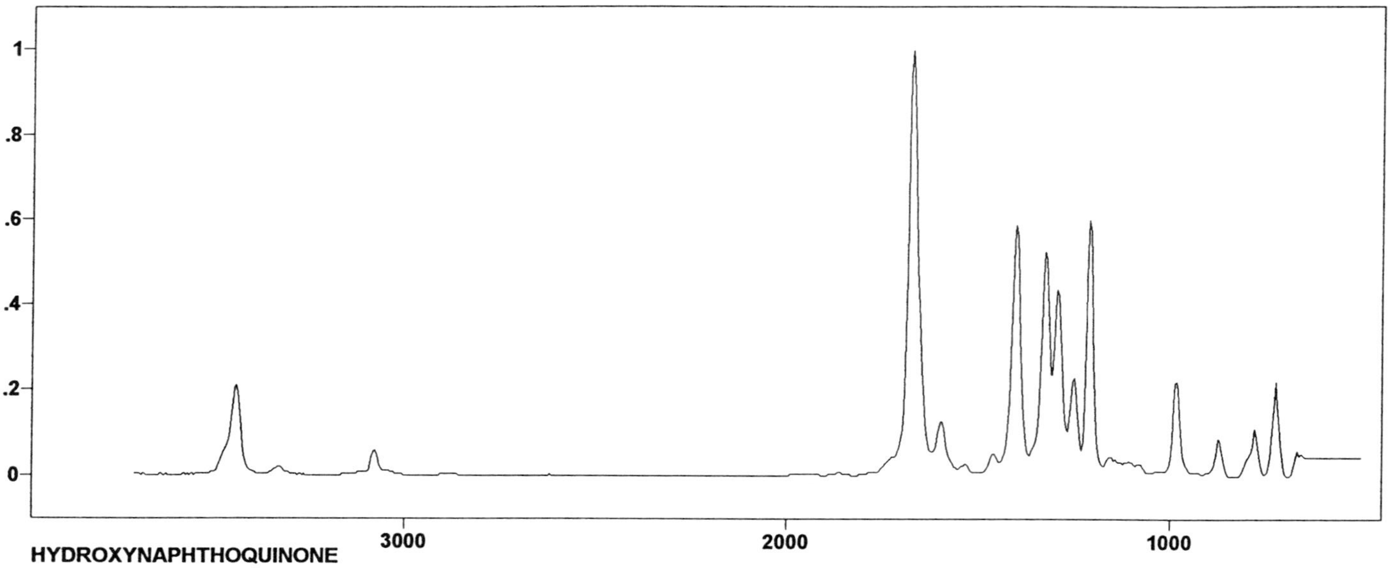

HYDROXYNAPHTHOQUINONE

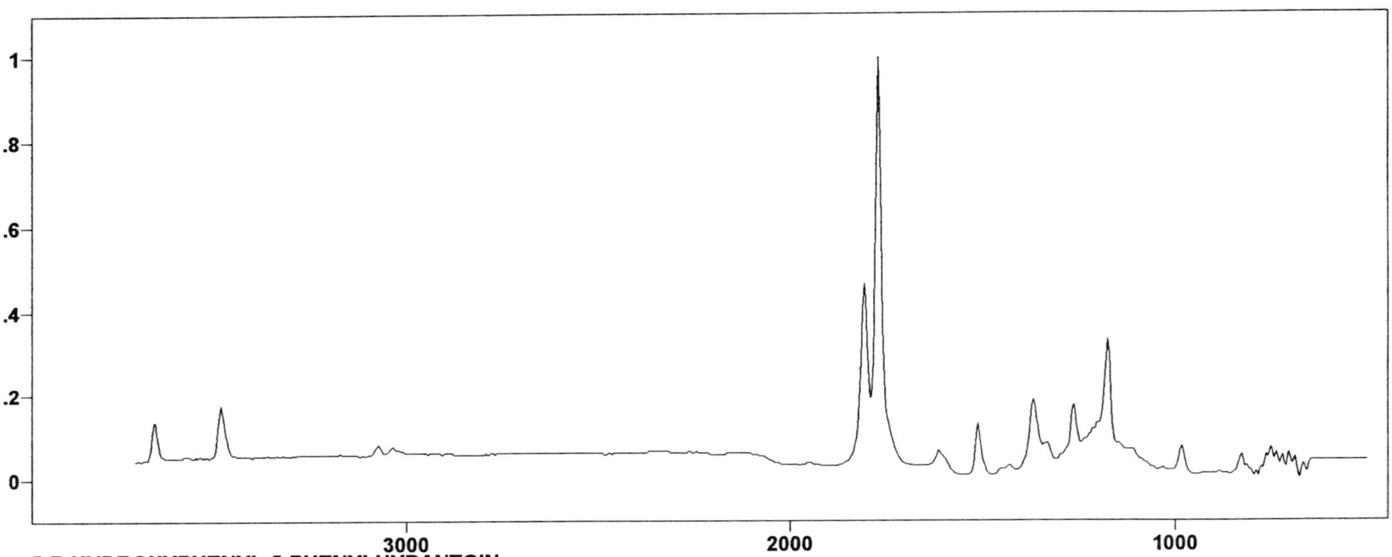

5-P-HYDROXYPHENYL-5-PHENYLHYDANTOIN

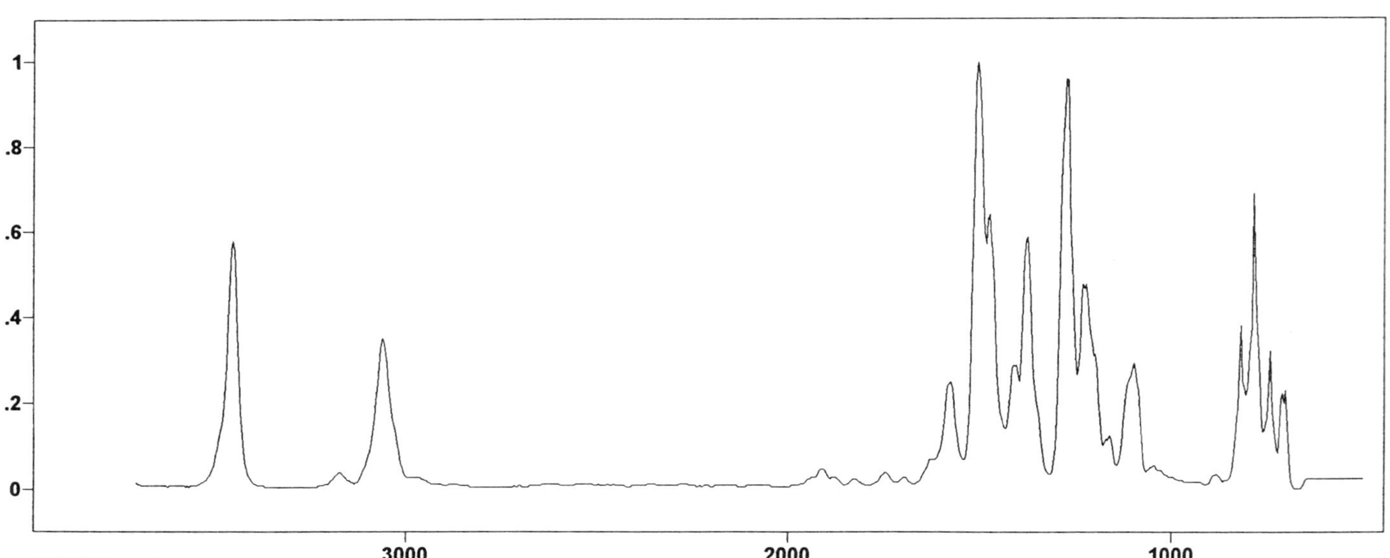

8-HYDROXYQUINOLINE

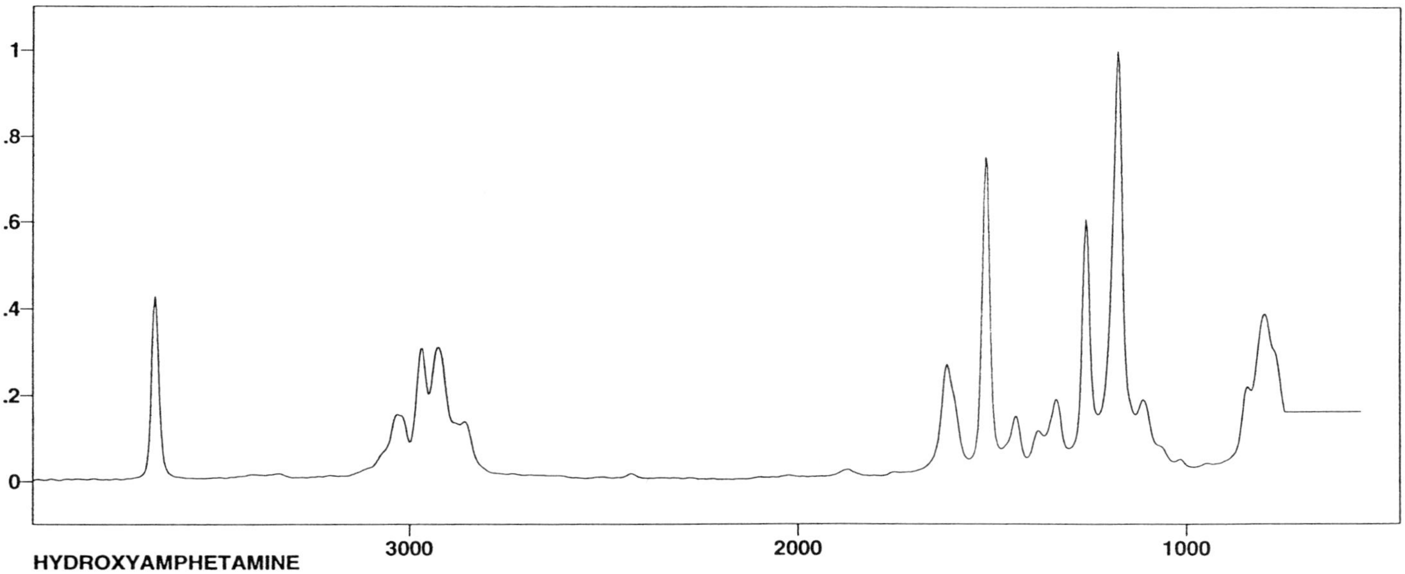

HYDROXYAMPHETAMINE

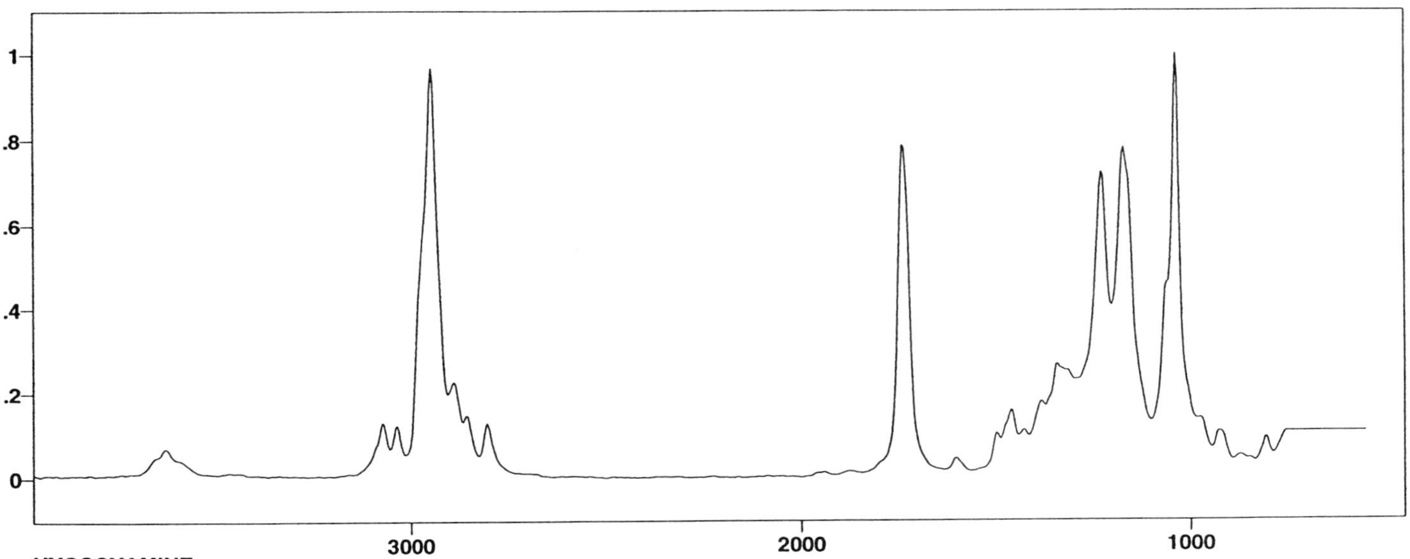

HYOSCYAMINE

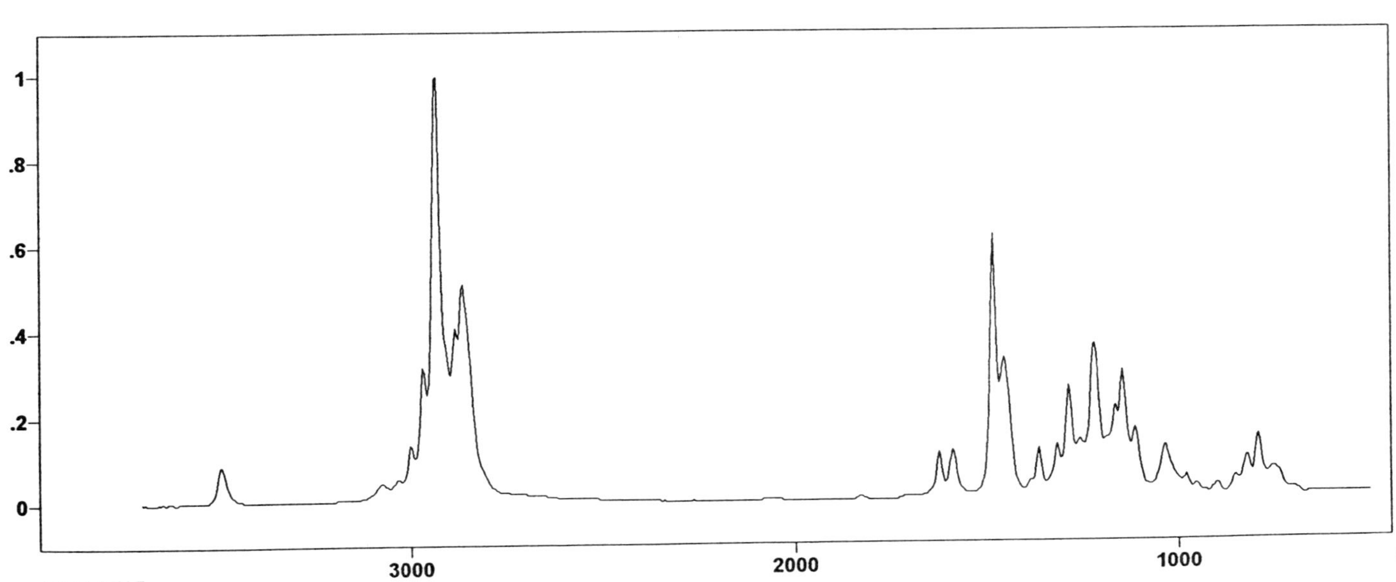

IBOGAINE

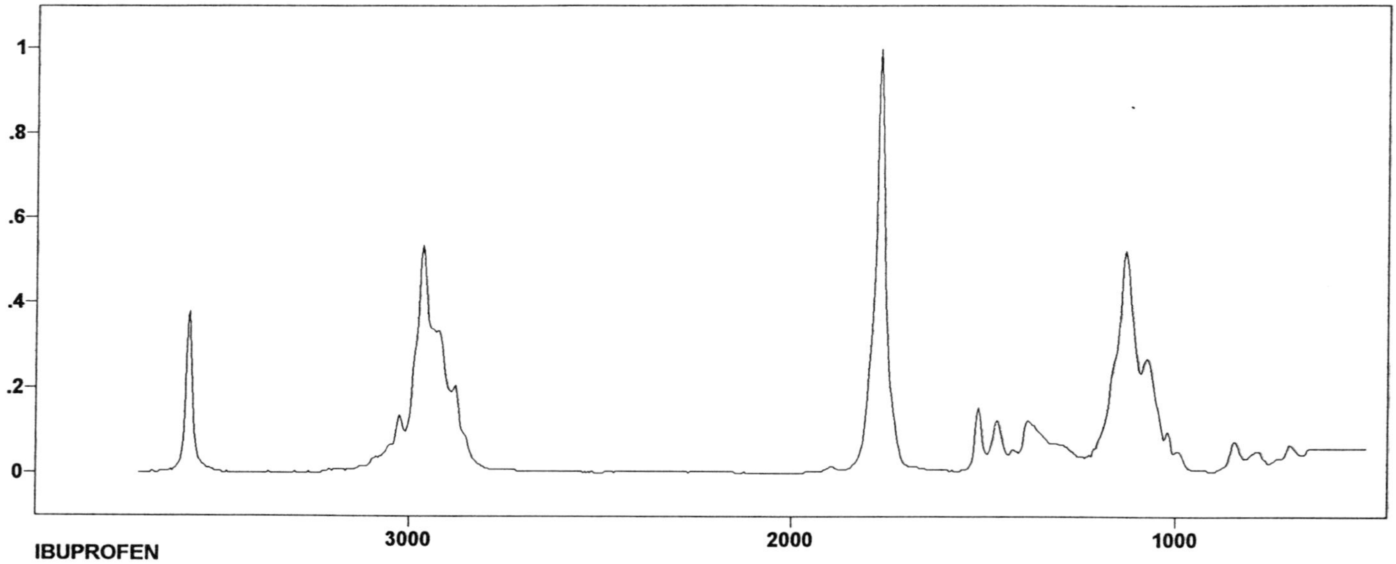

IBUPROFEN

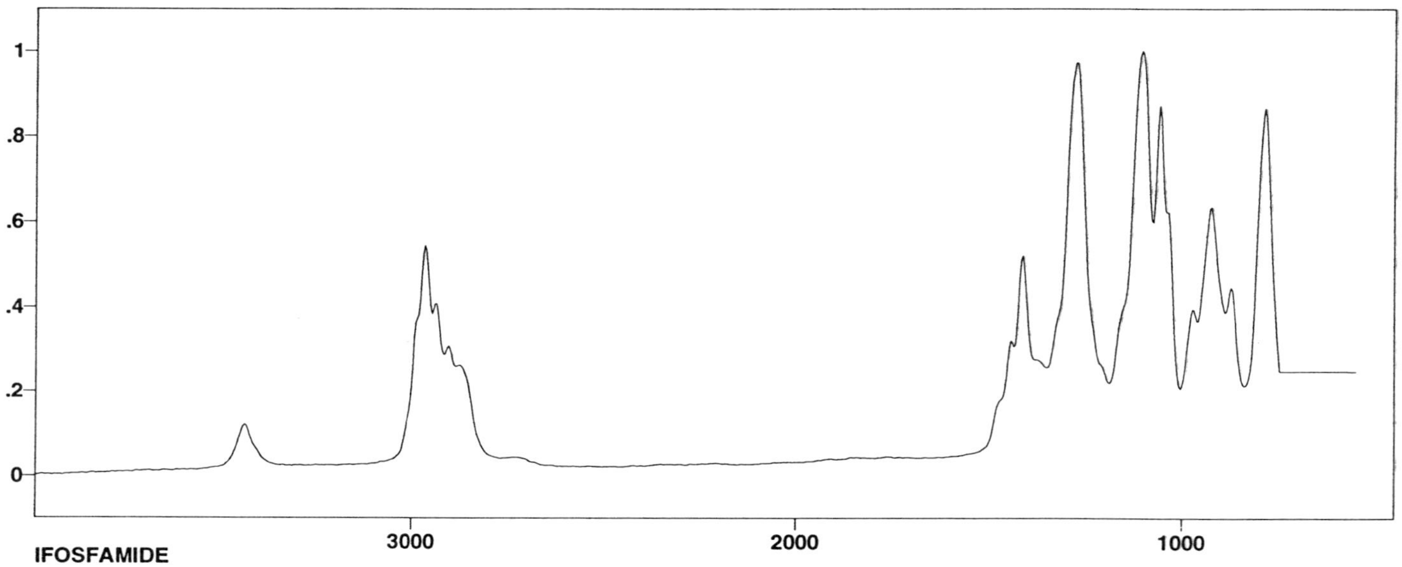

IFOSFAMIDE

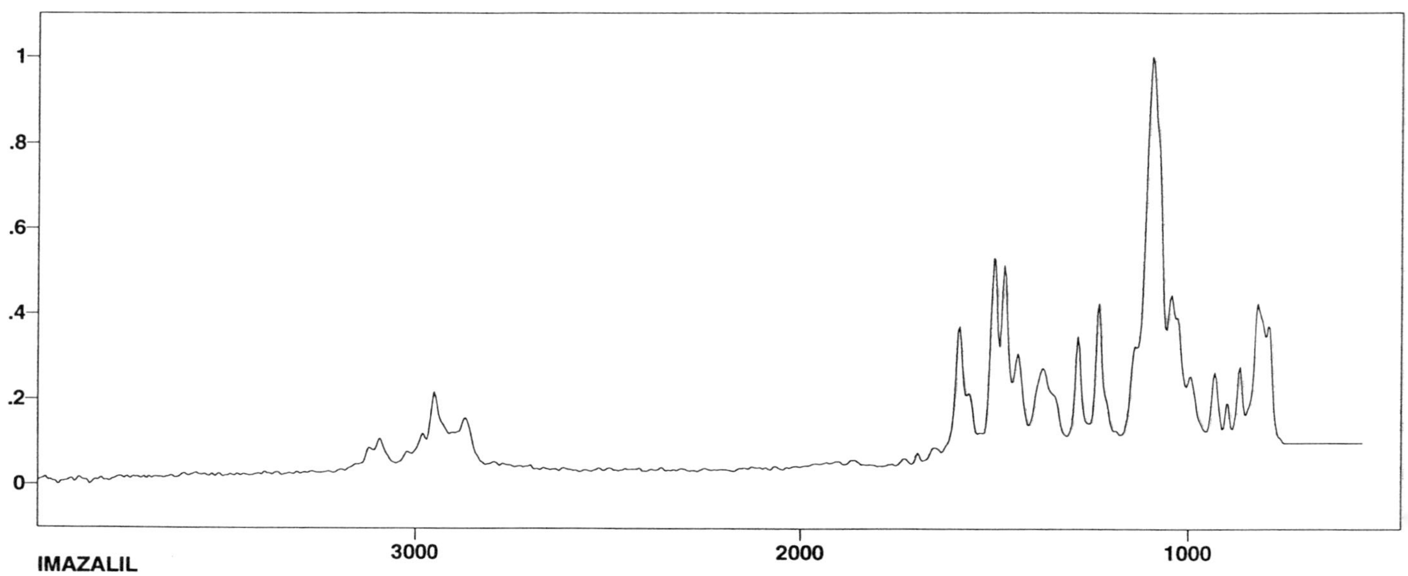

IMAZALIL

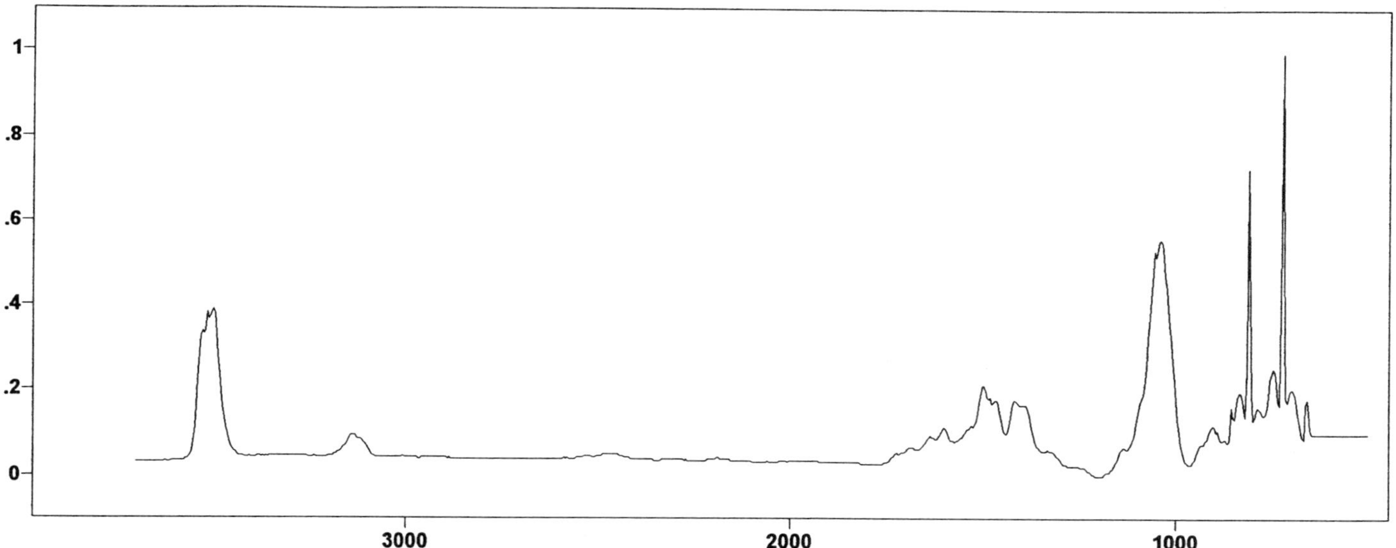

IMIDAZOLE

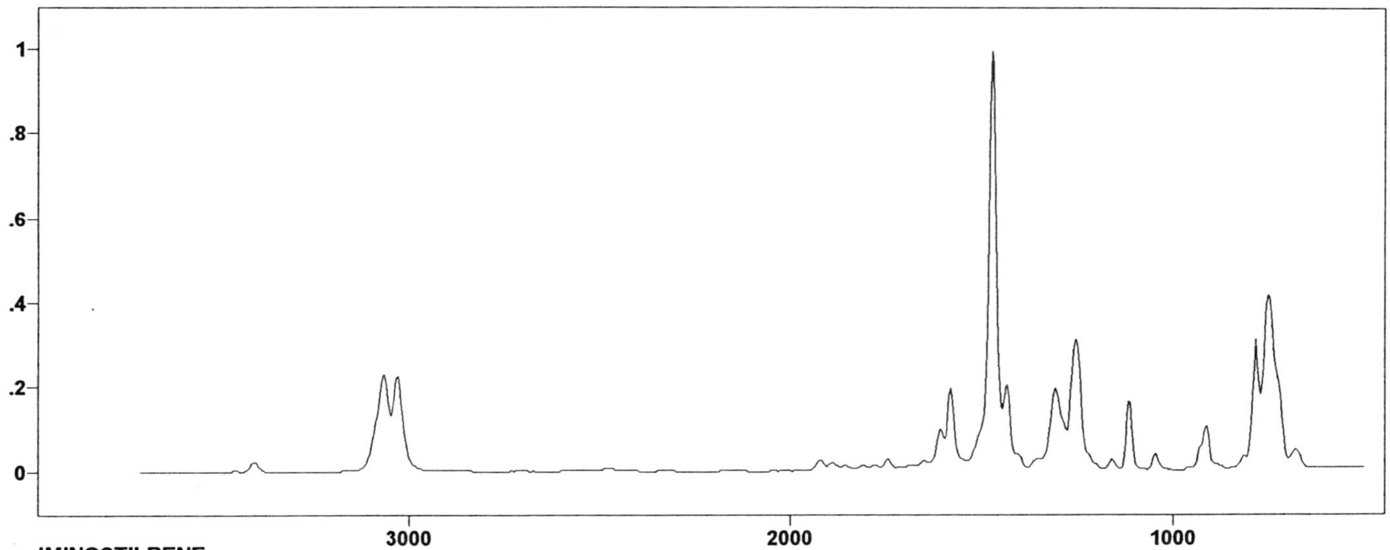

IMINOSTILBENE

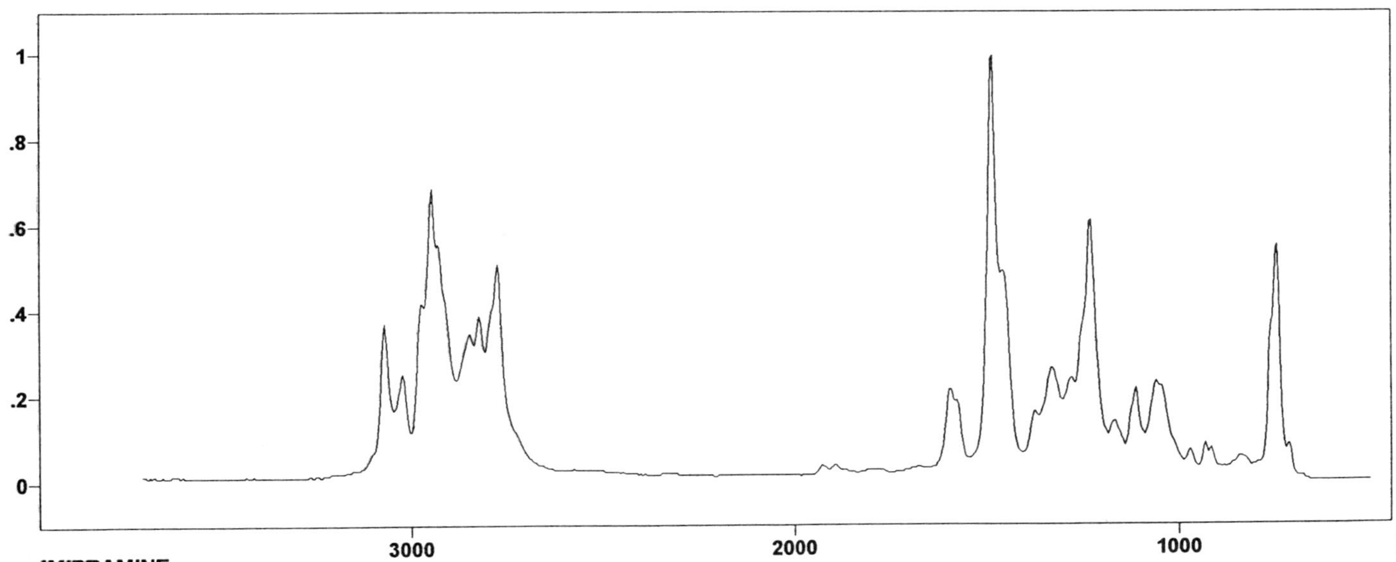

IMIPRAMINE

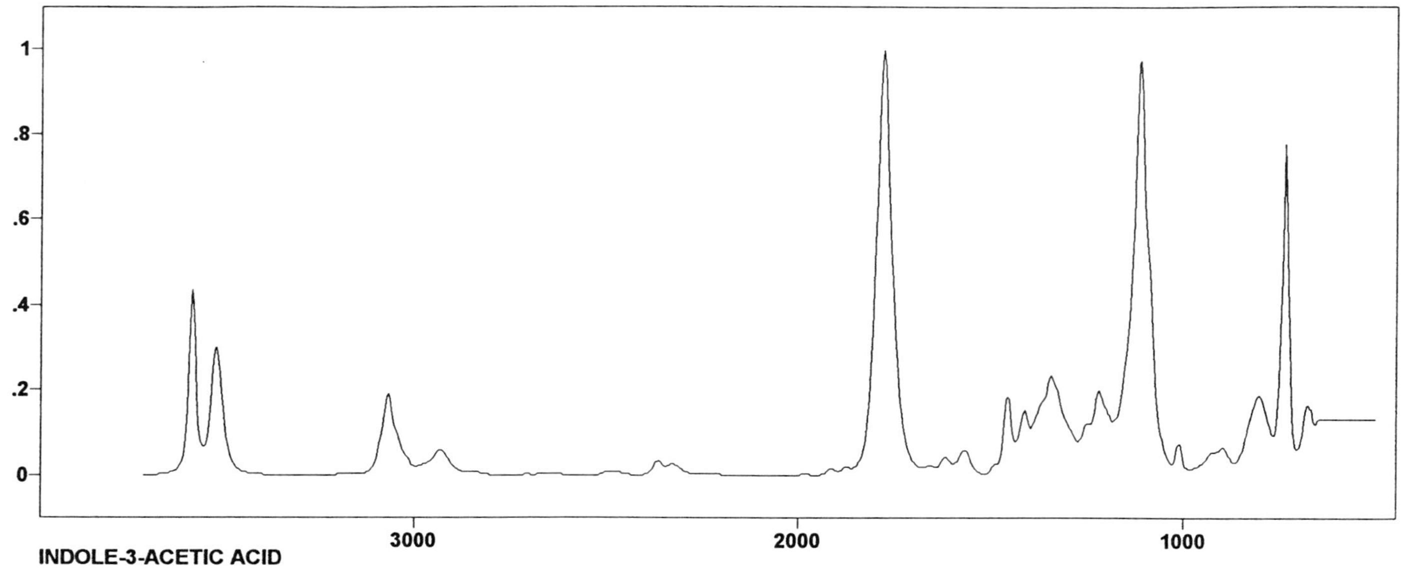

INDOLE-3-ACETIC ACID

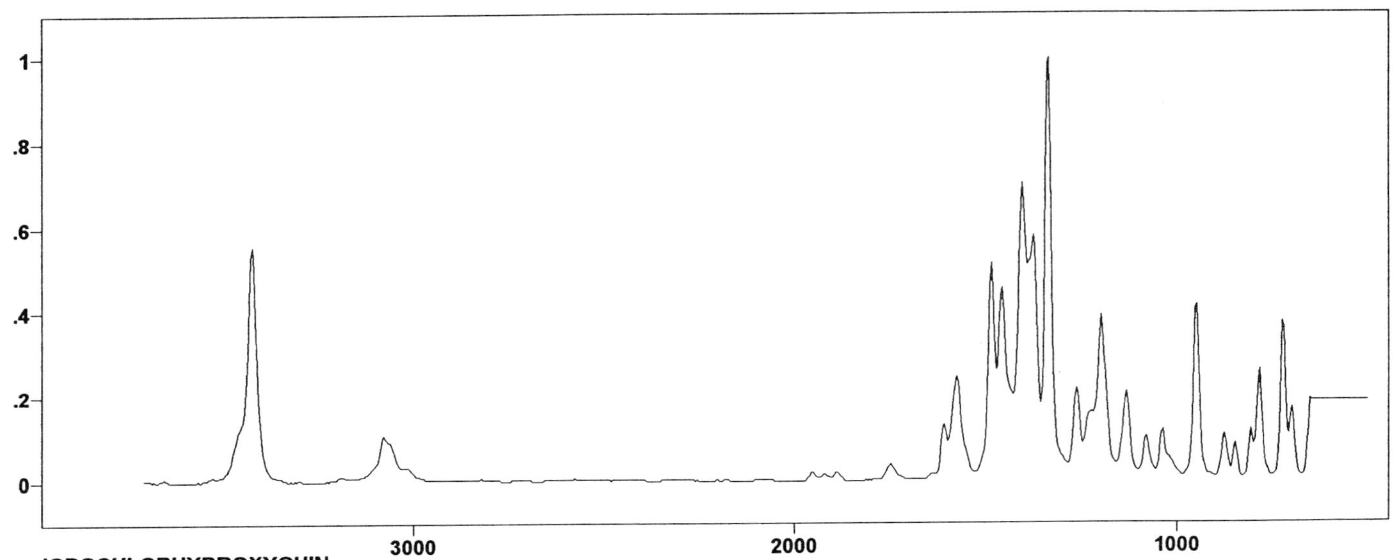

IODOCHLORHYDROXYQUIN

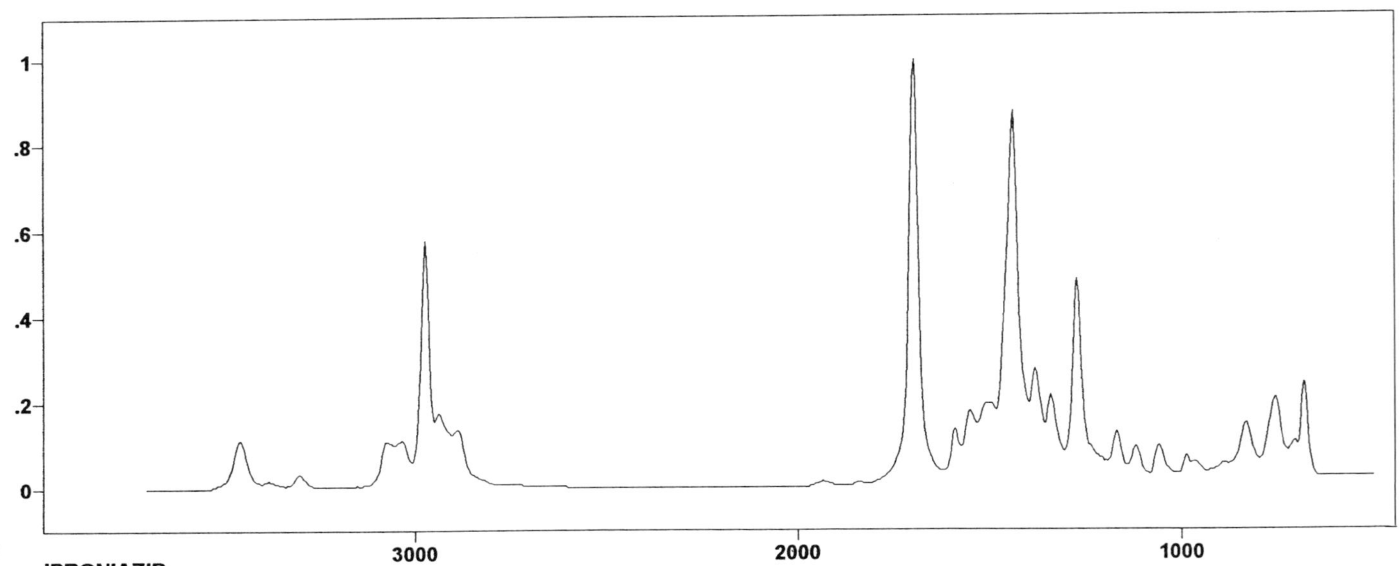

IPRONIAZID

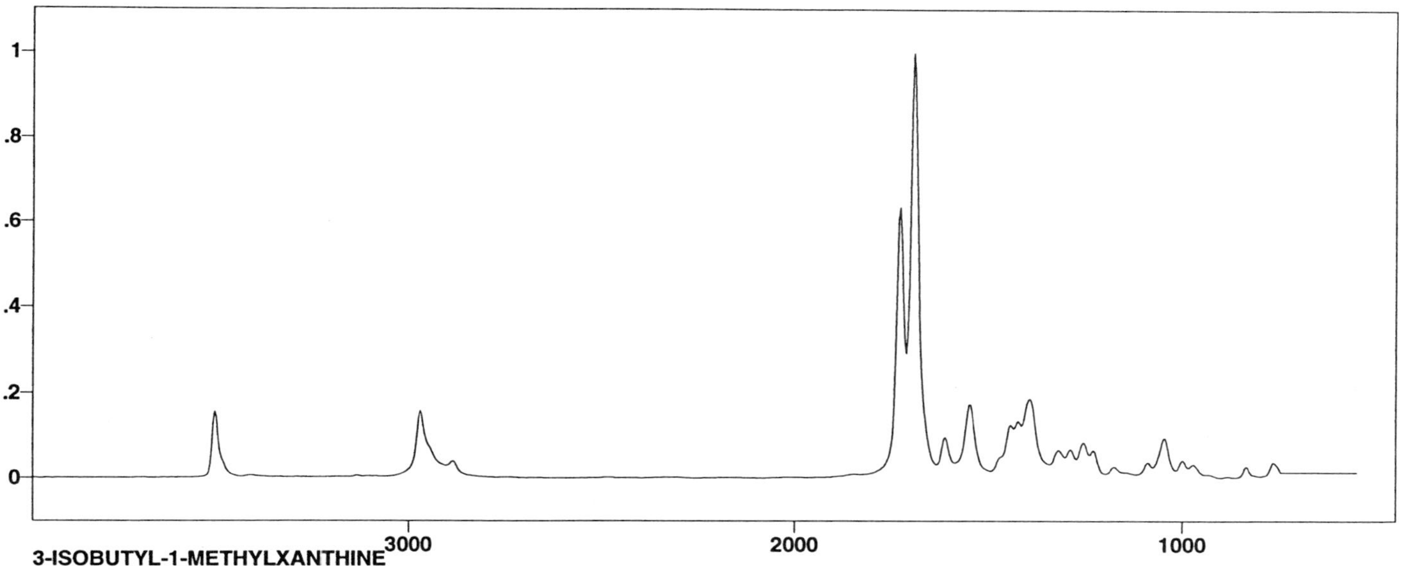

3-ISOBUTYL-1-METHYLXANTHINE

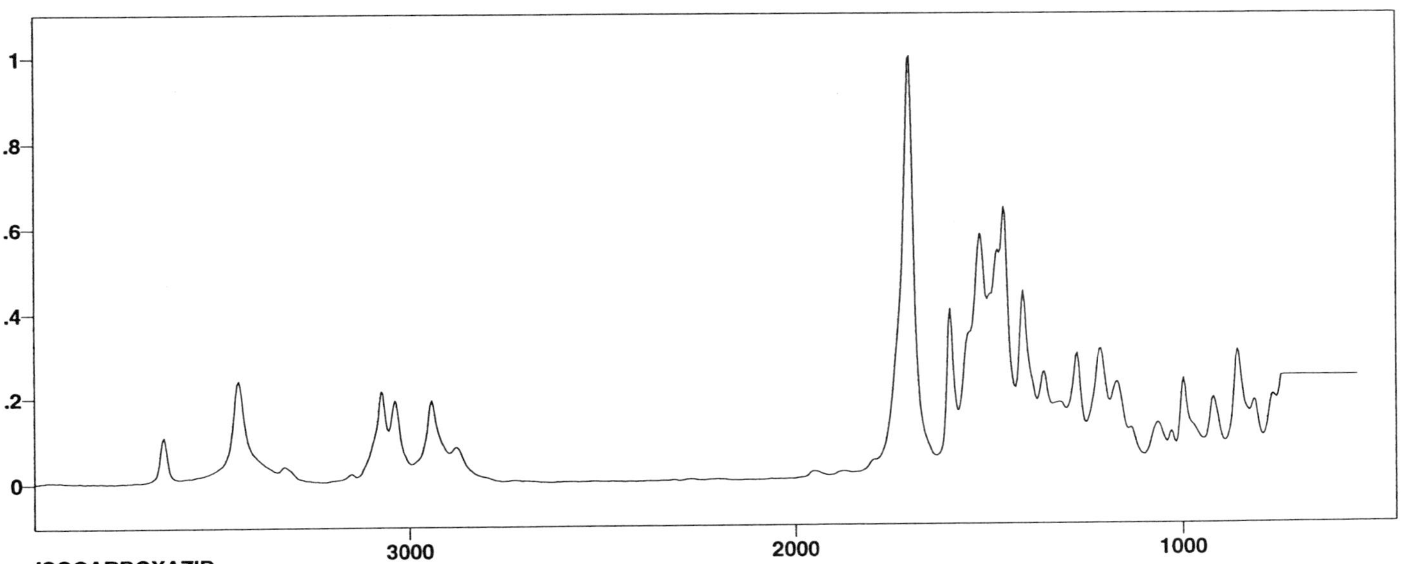

ISOCARBOXAZID

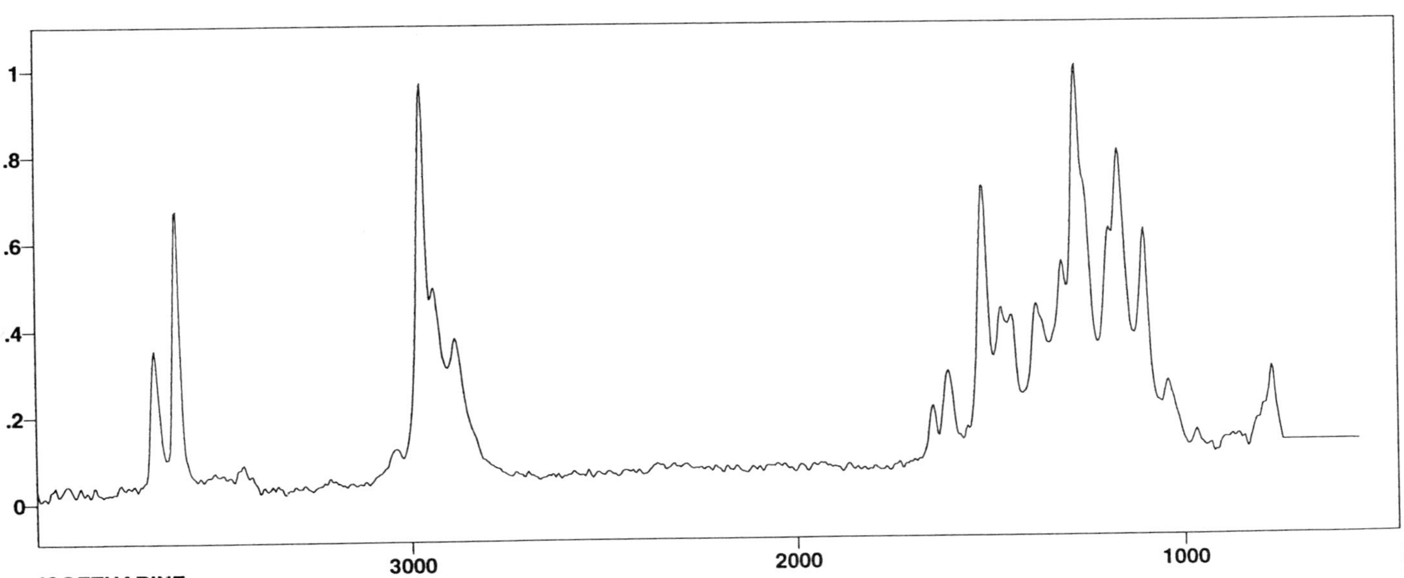

ISOETHARINE

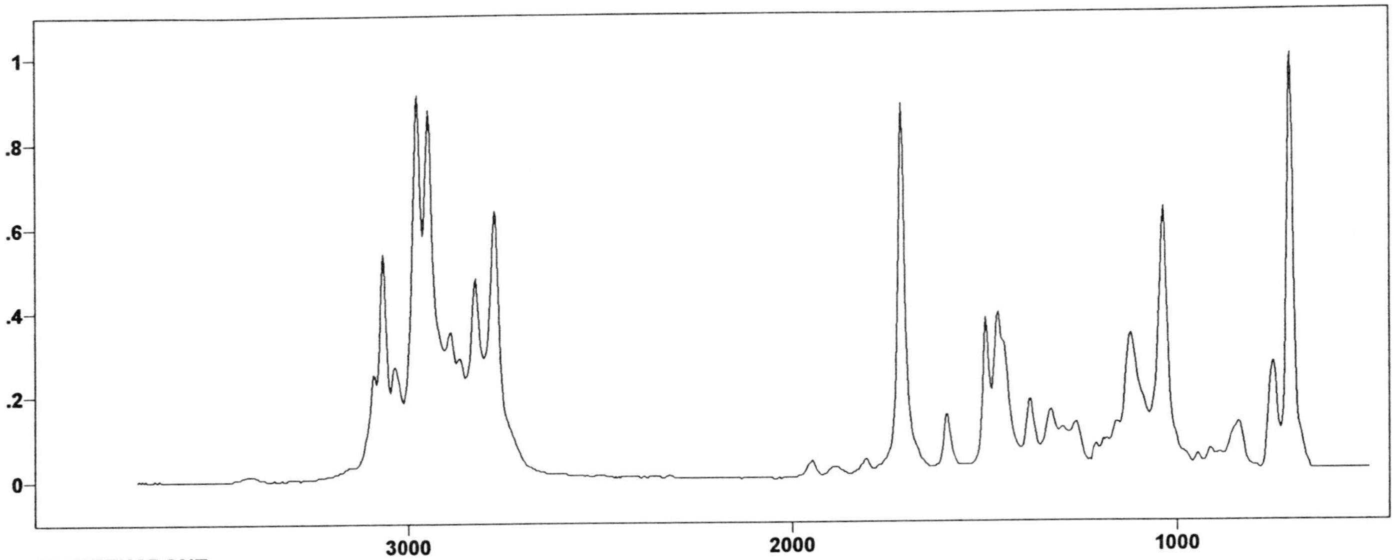

ISOMETHADONE

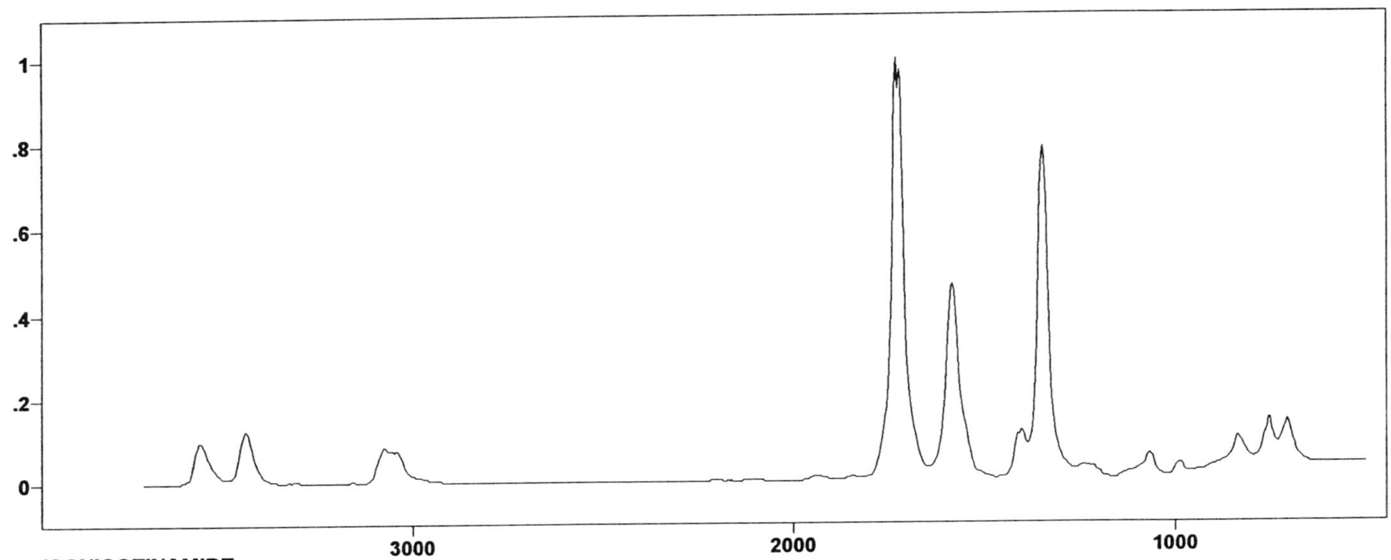

ISONICOTINAMIDE

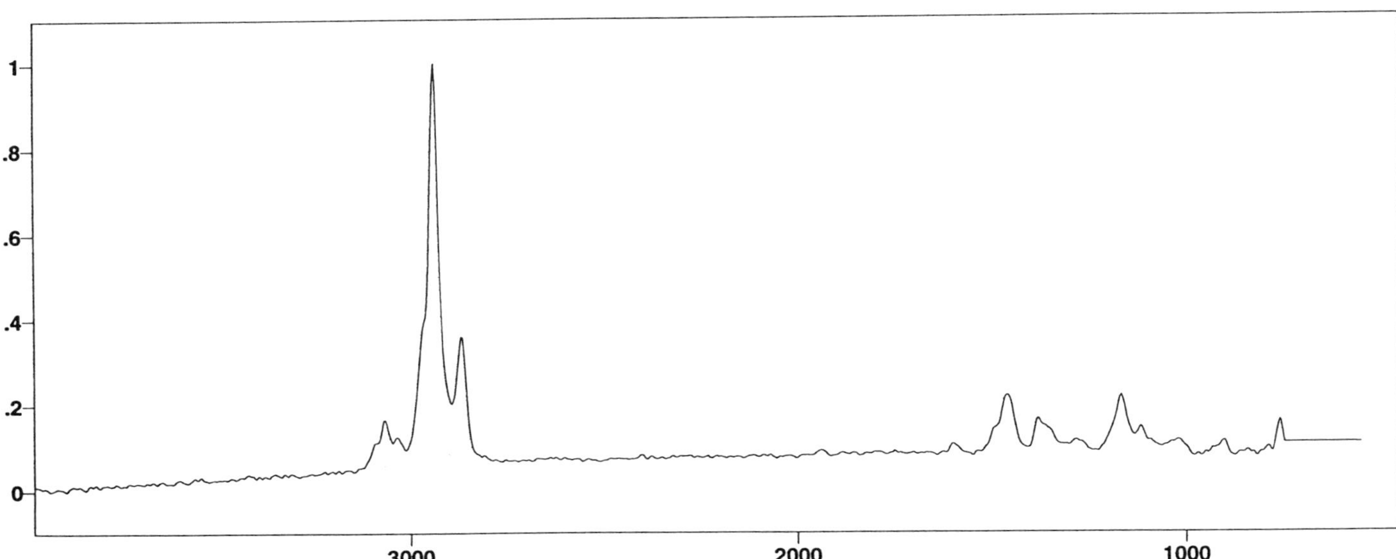

N-(ISOPROPYL)-1-PHENYLCYCLOHEXYLAMINE

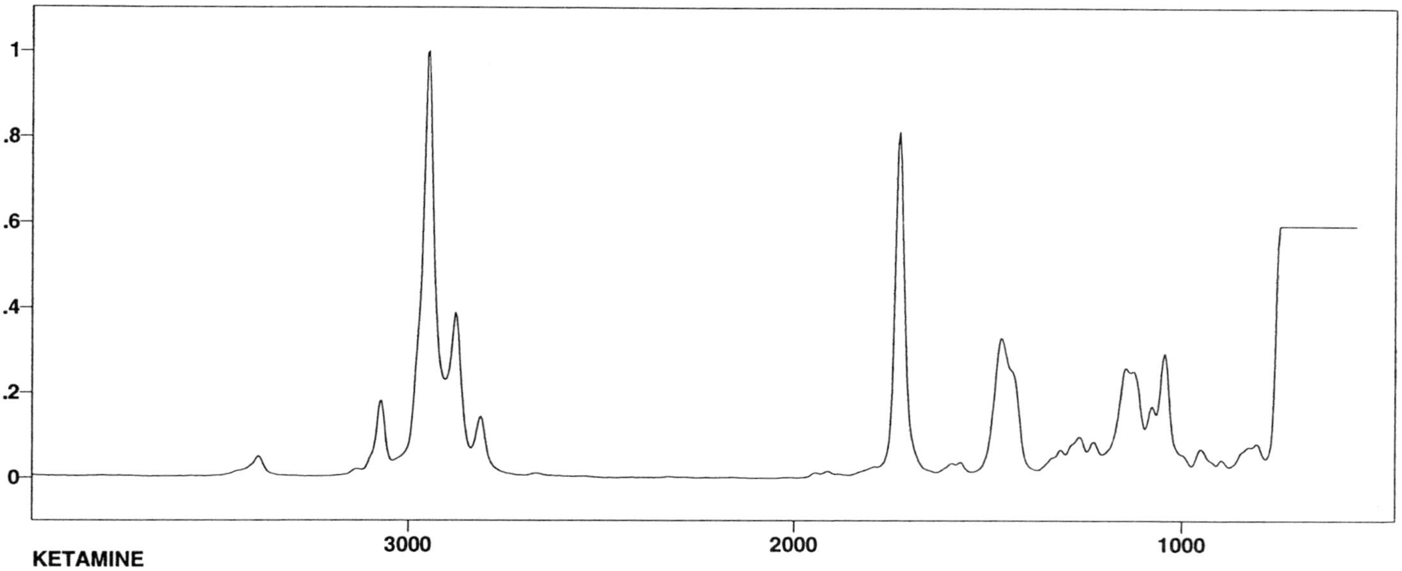

KETAMINE

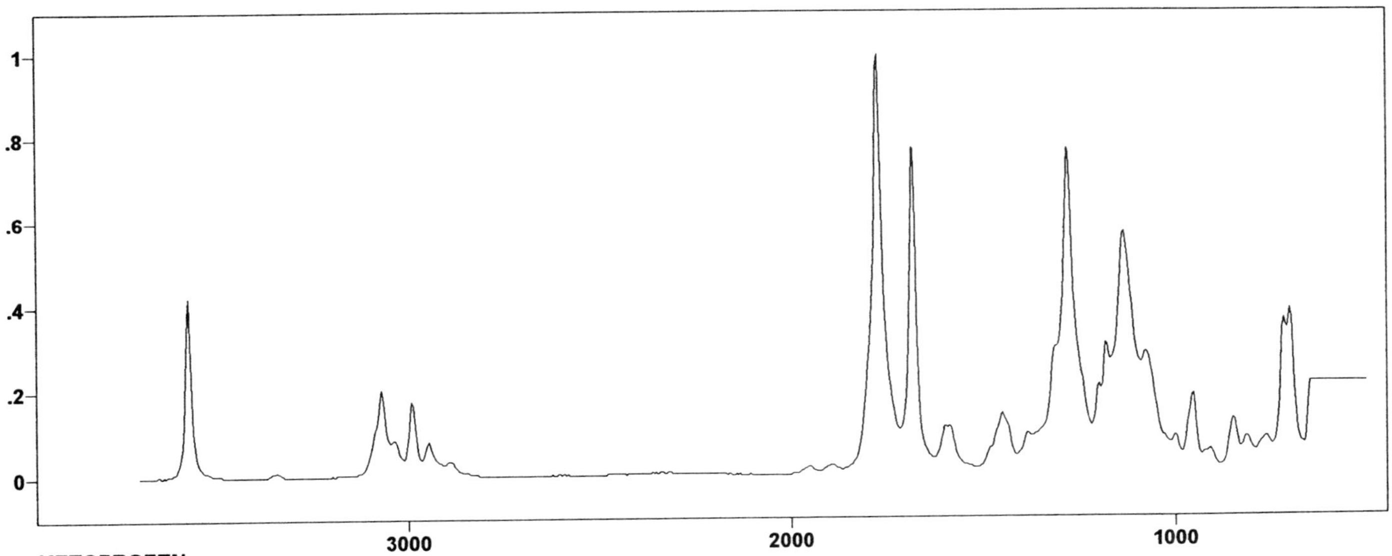

KETOPROFEN

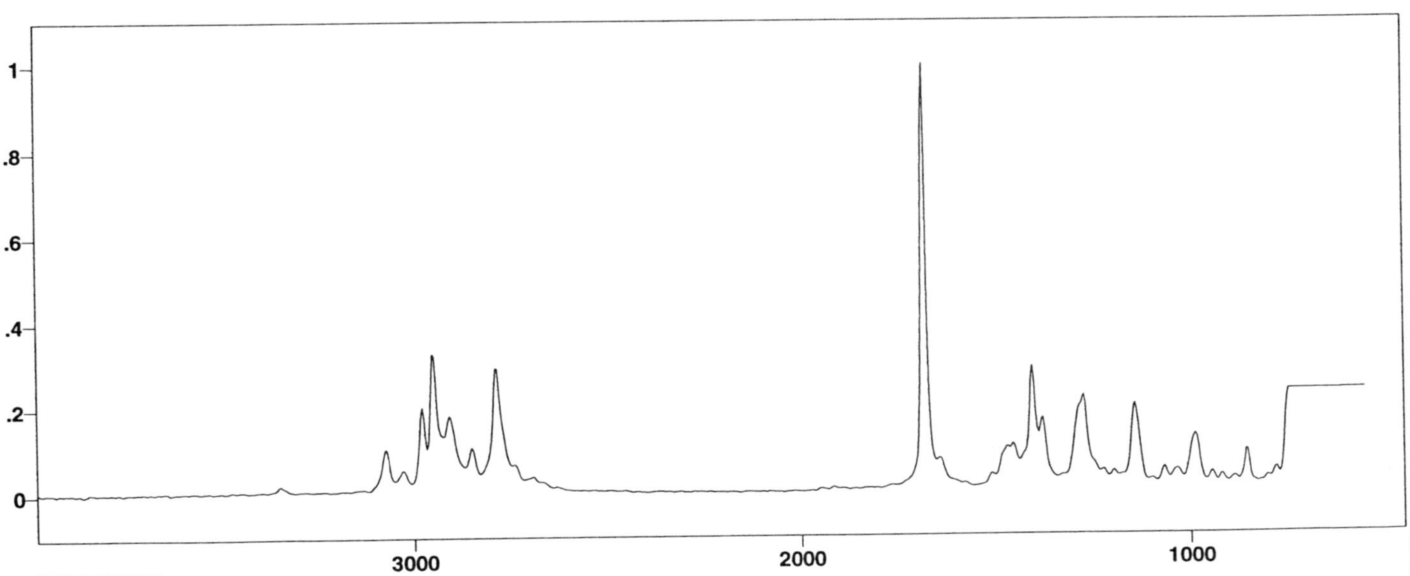

KETOTIFEN

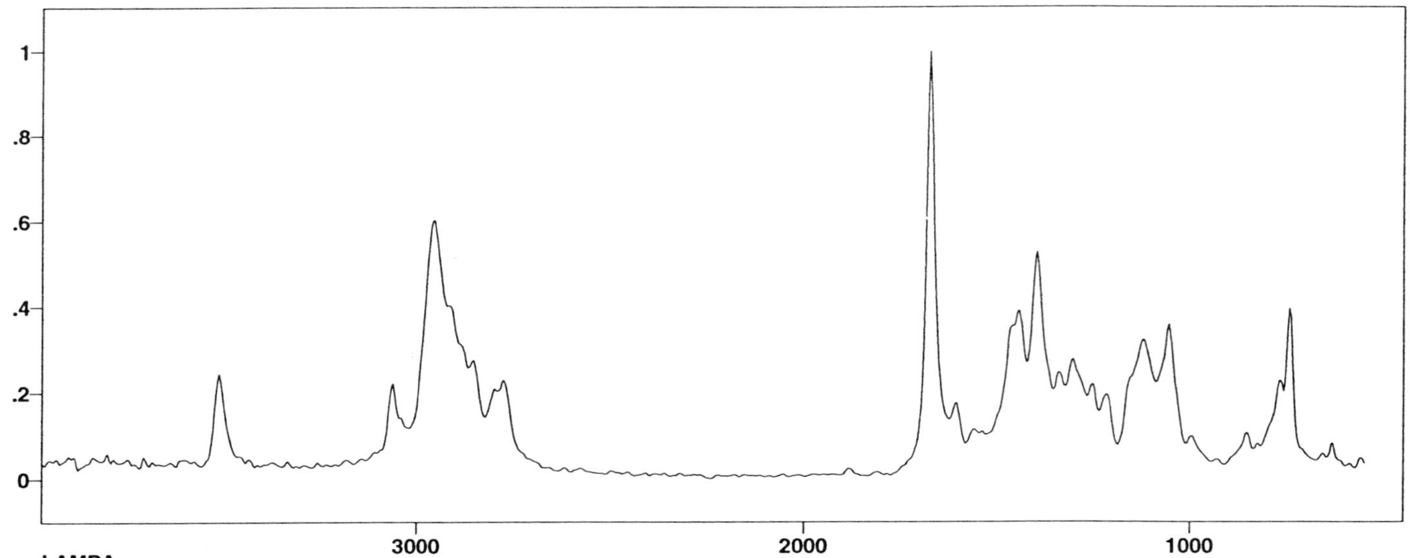

LAMPA

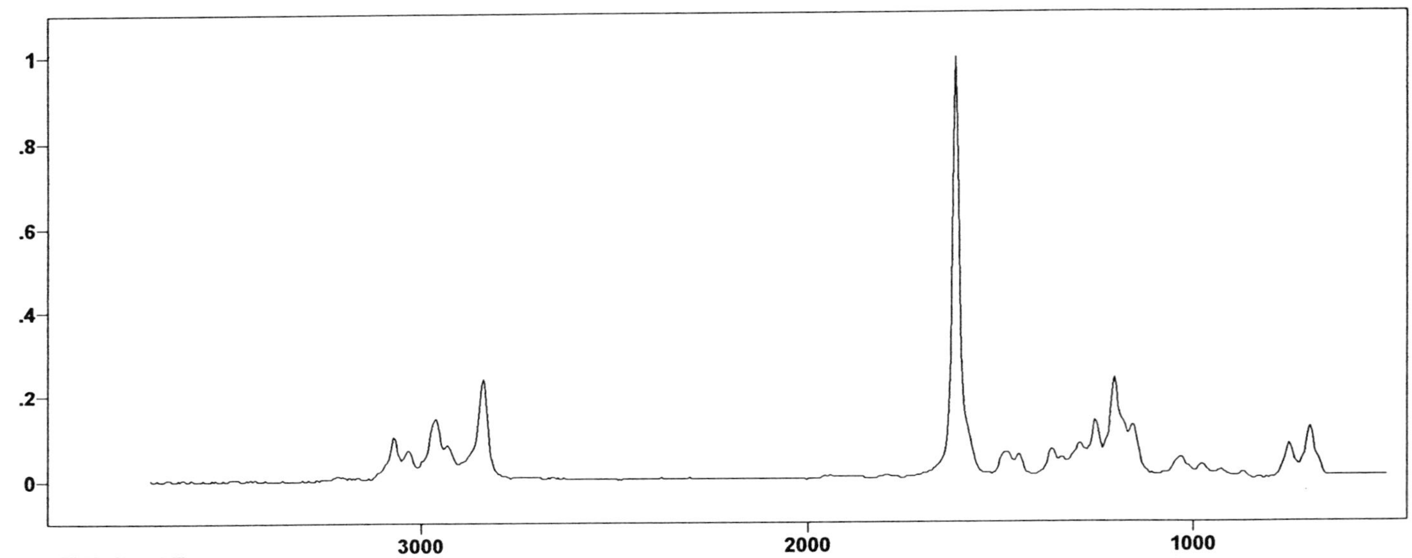

LEVAMISOLE

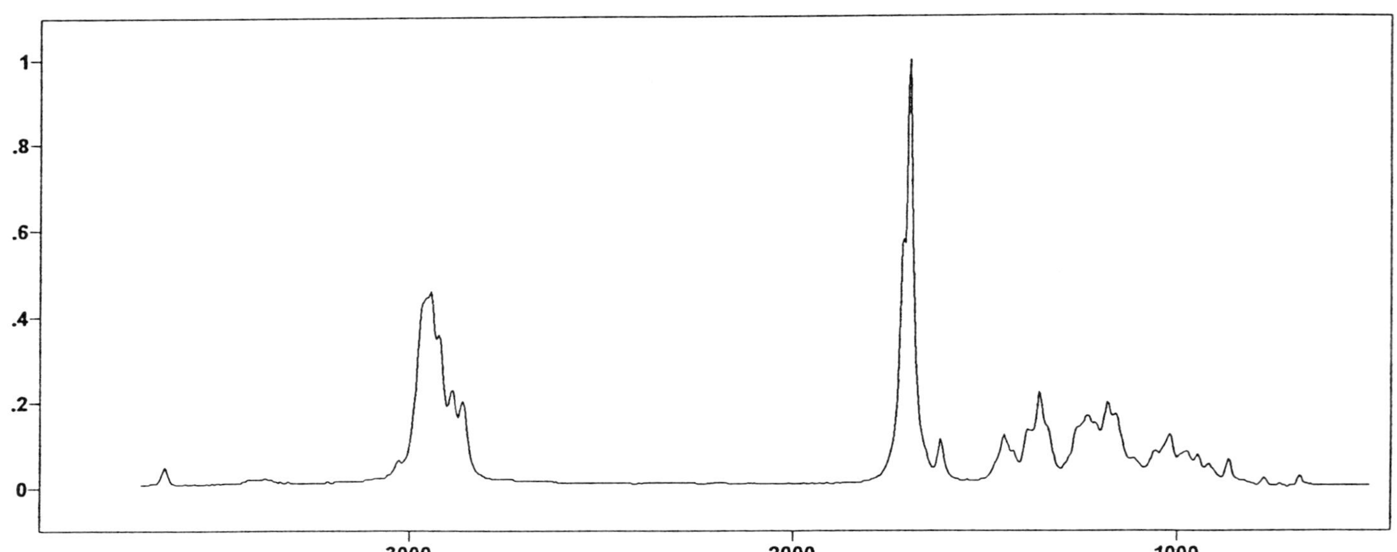

11-ALPHA-HYDROXYPROGESTERONE

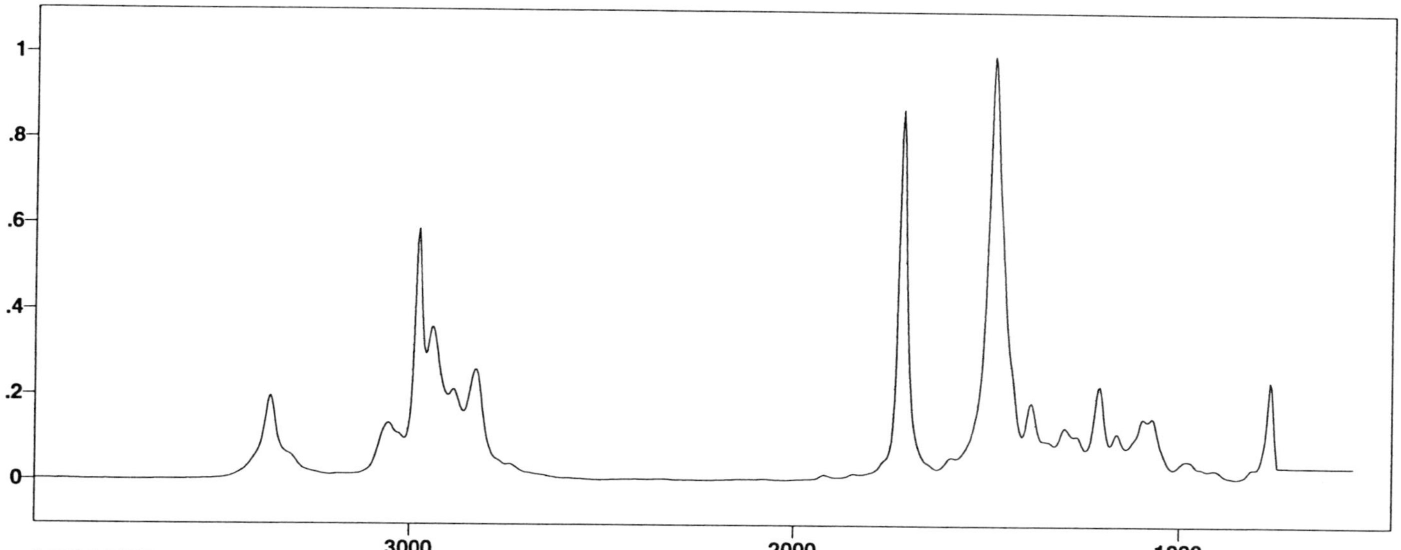

LIDOCAINE

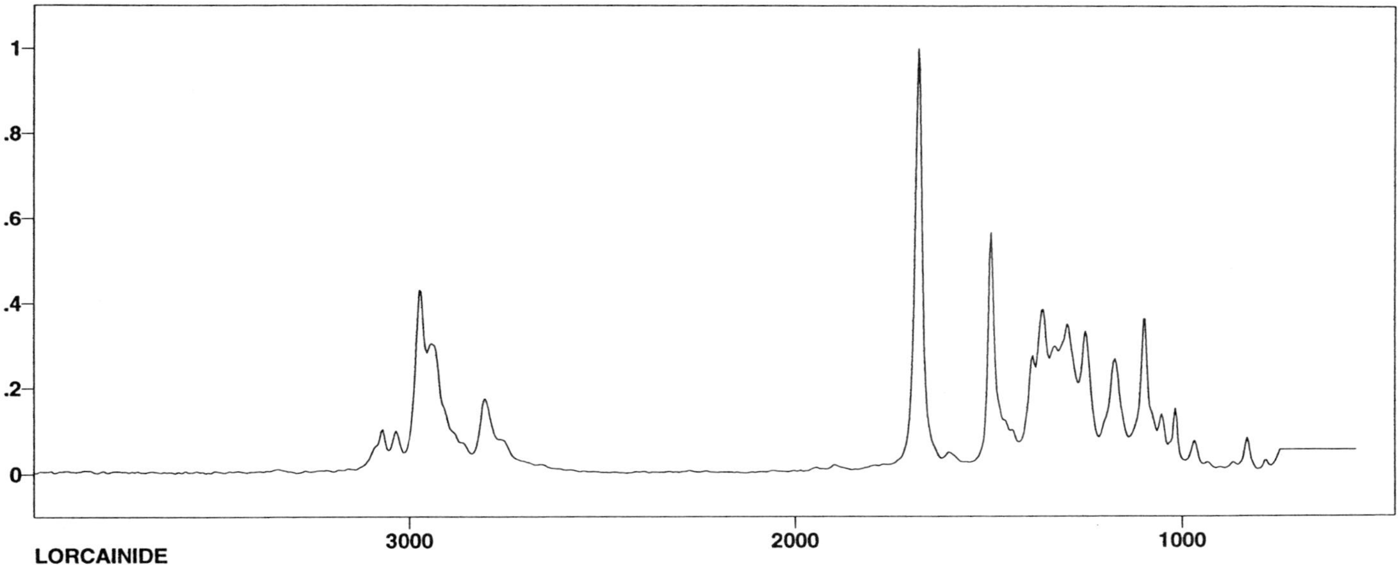

LORCAINIDE

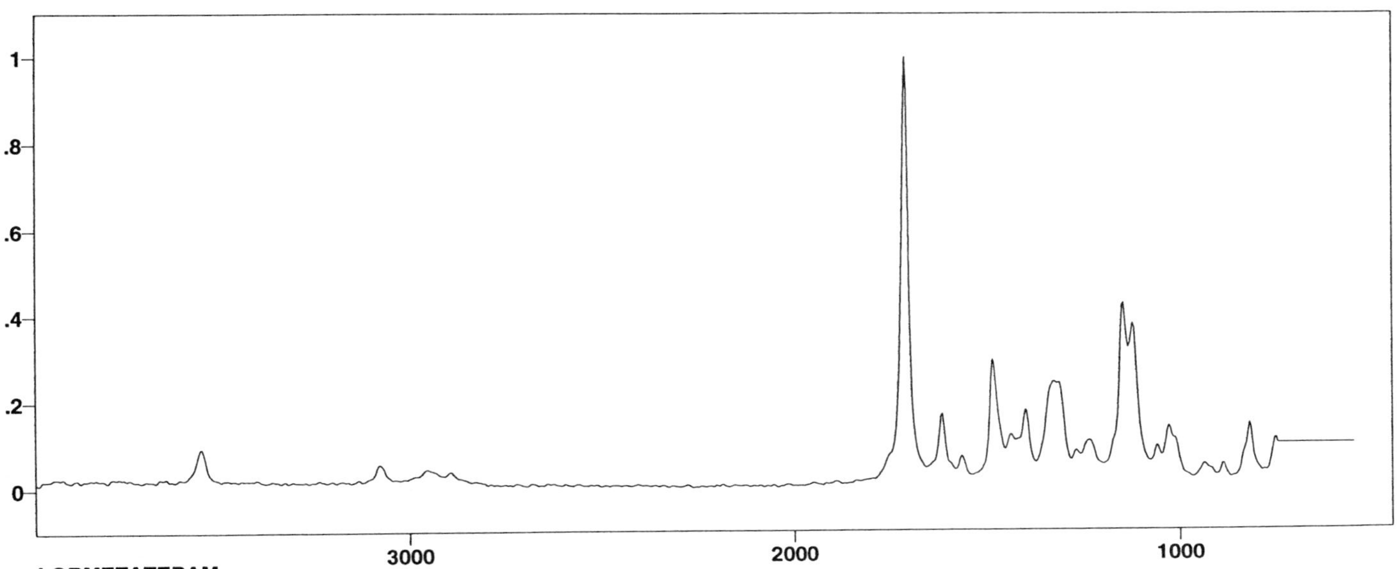

LORMETAZEPAM

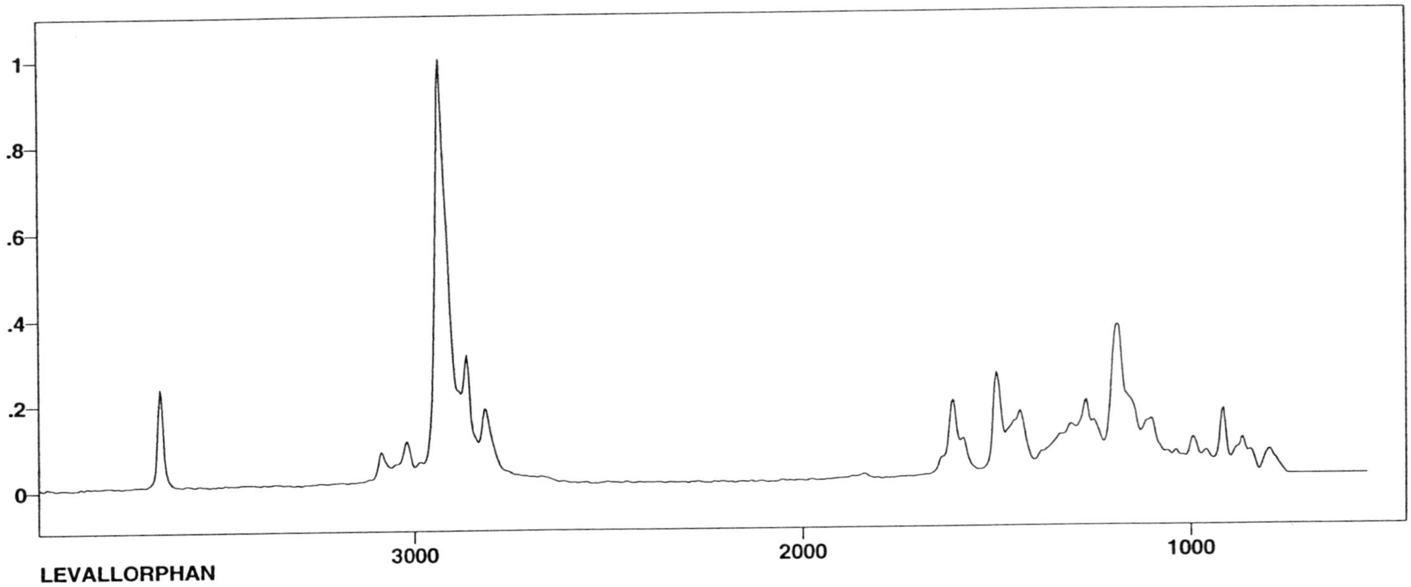

LEVALLORPHAN

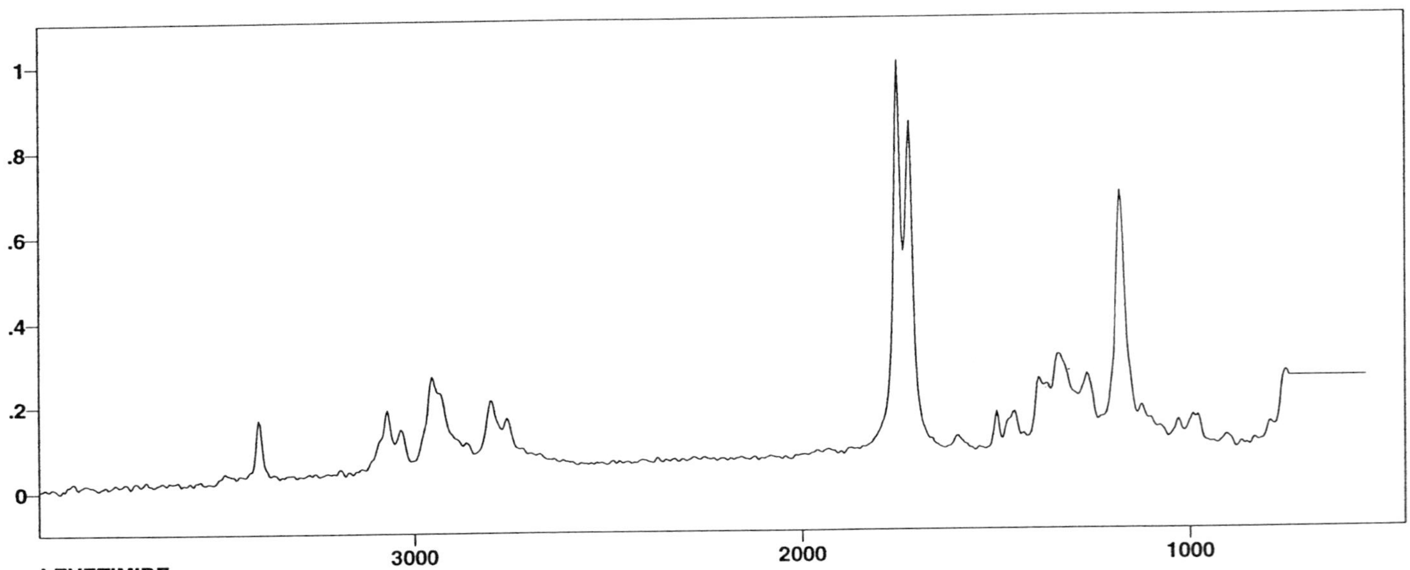

LEVETIMIDE

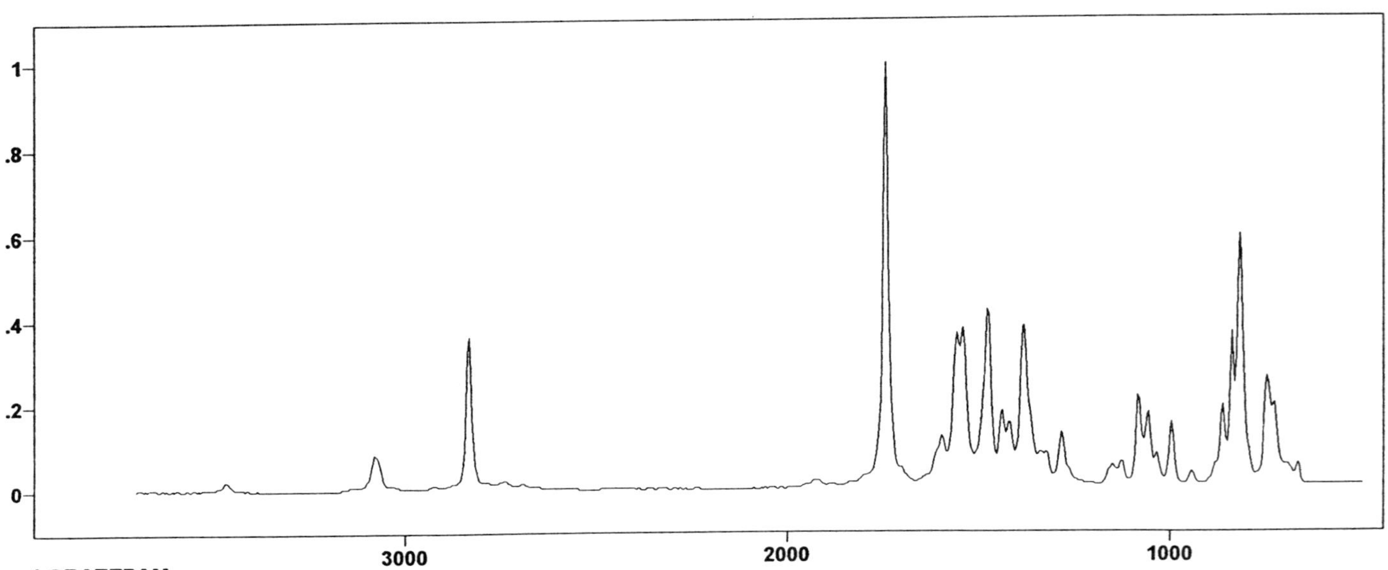

LORAZEPAM

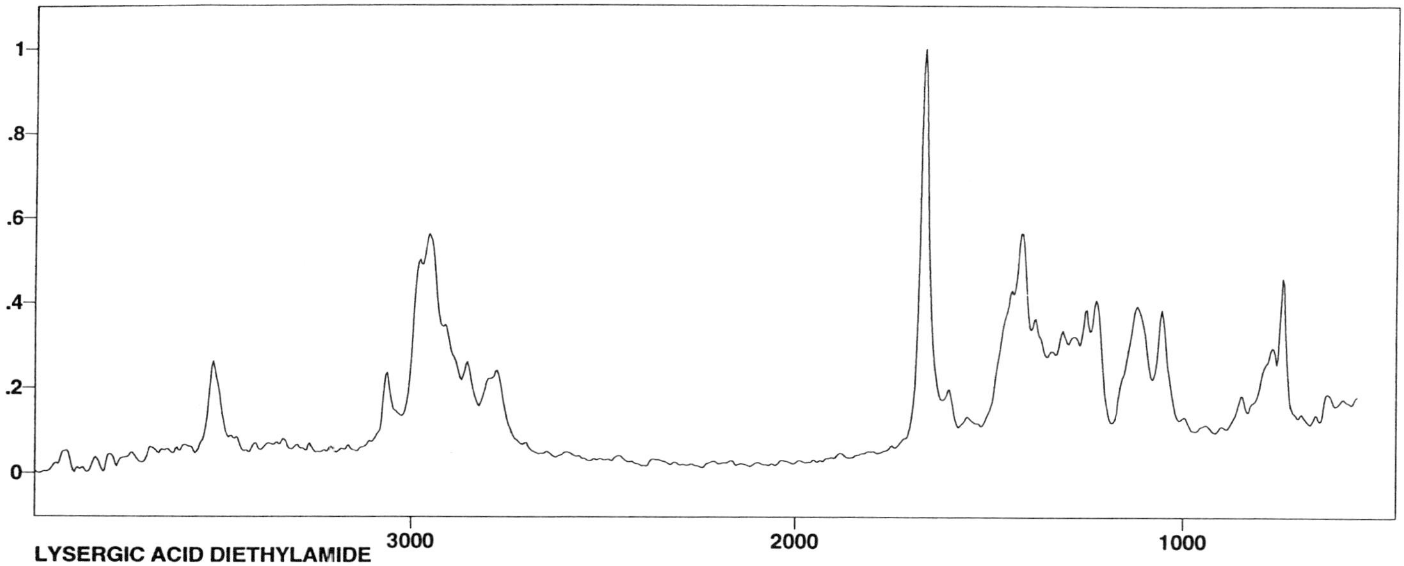

LYSERGIC ACID DIETHYLAMIDE

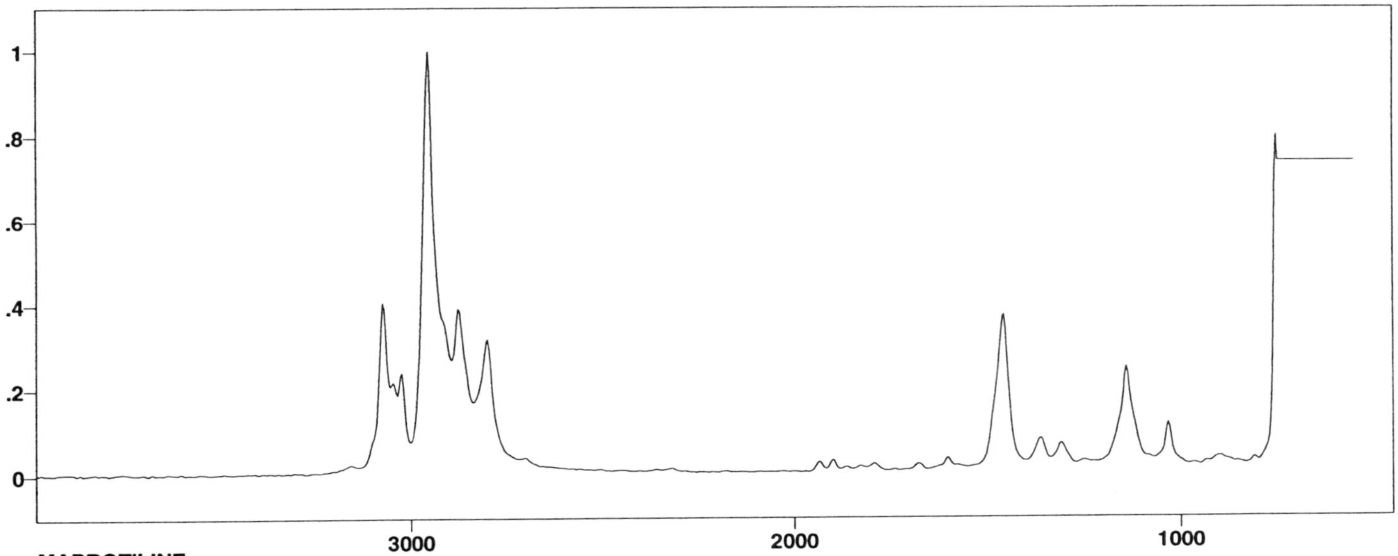

MAPROTILINE

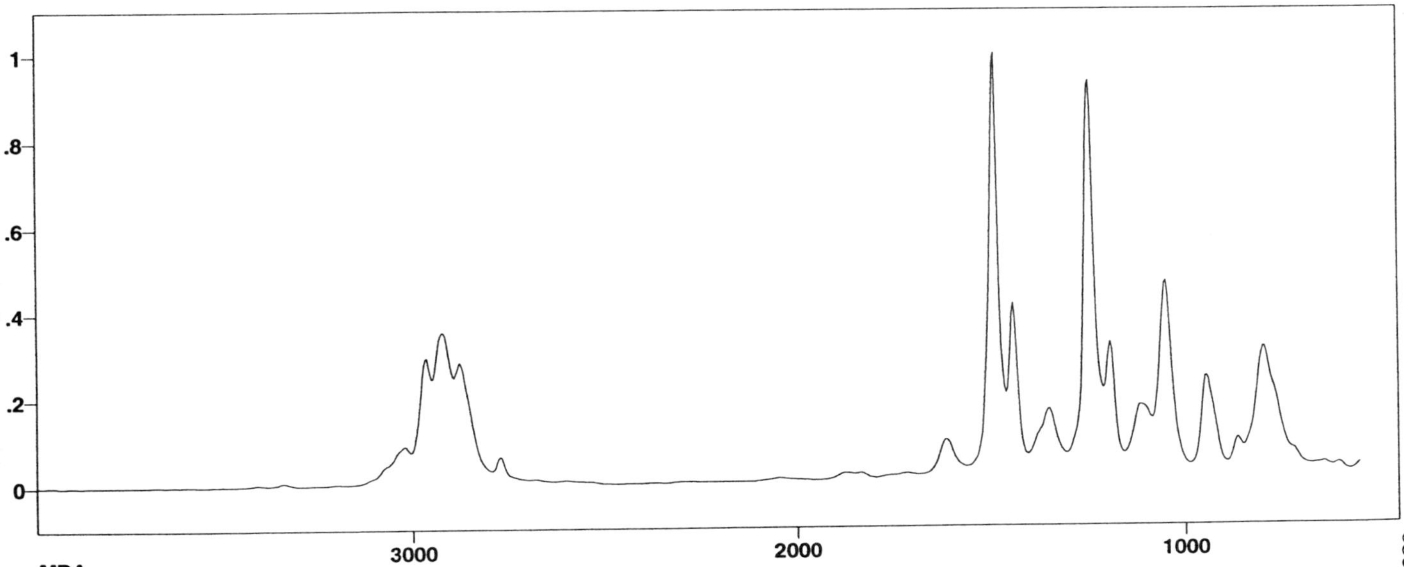

MDA

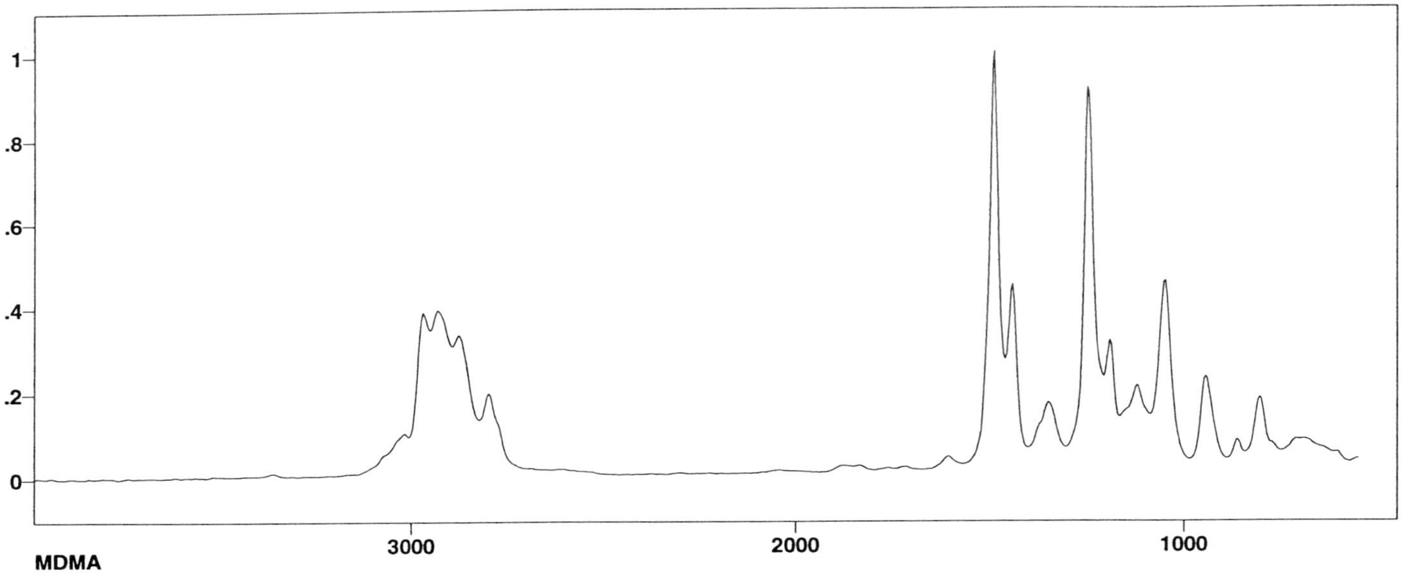

MDMA

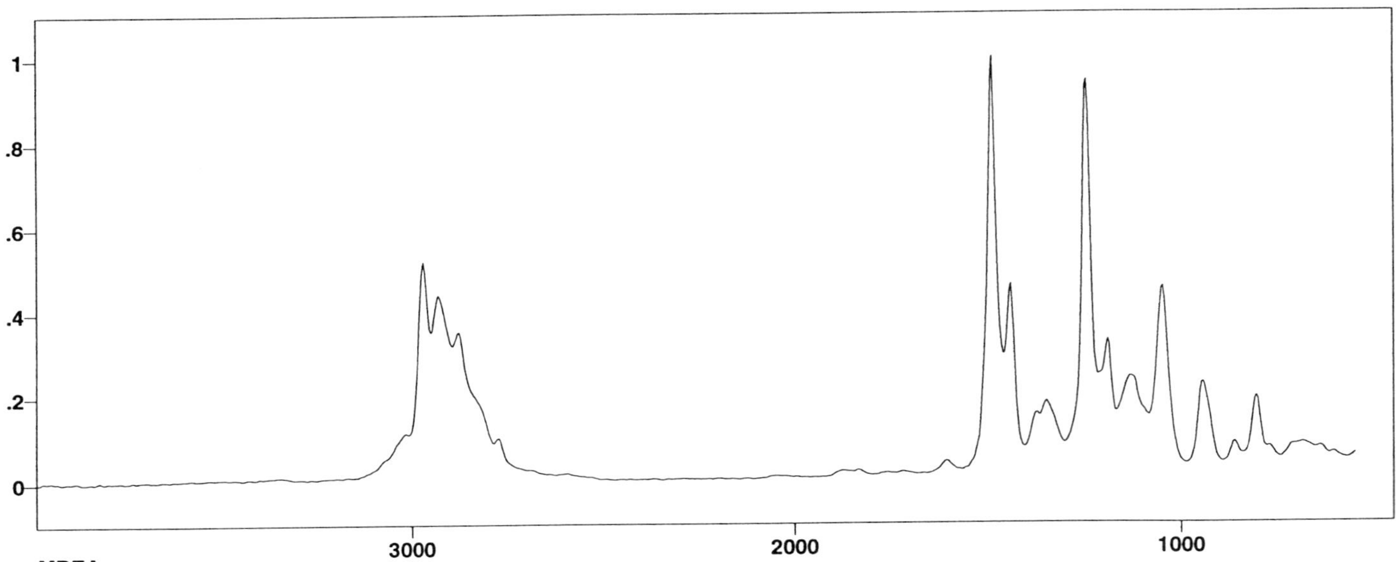

MDEA

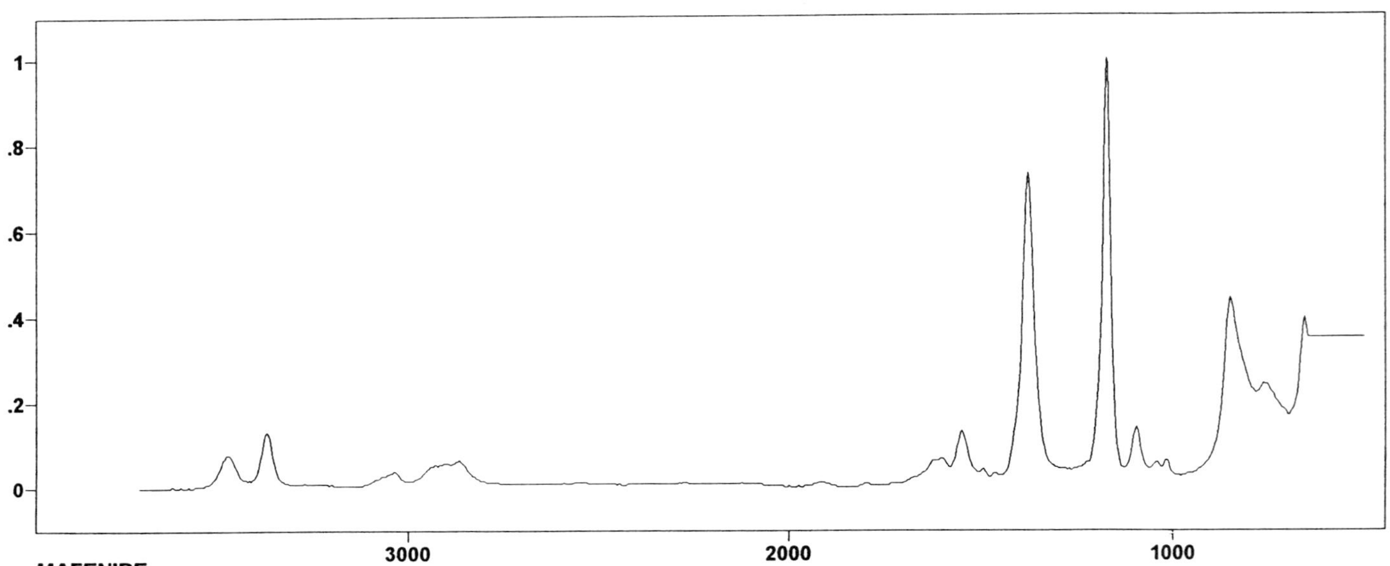

MAFENIDE

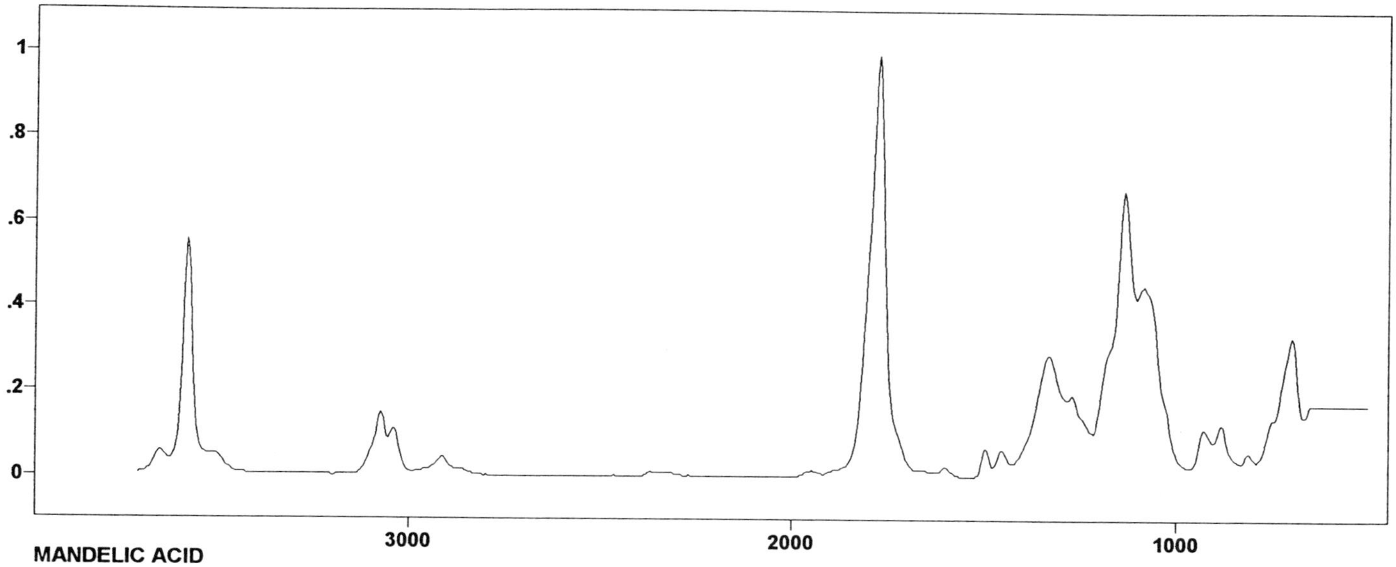

MANDELIC ACID

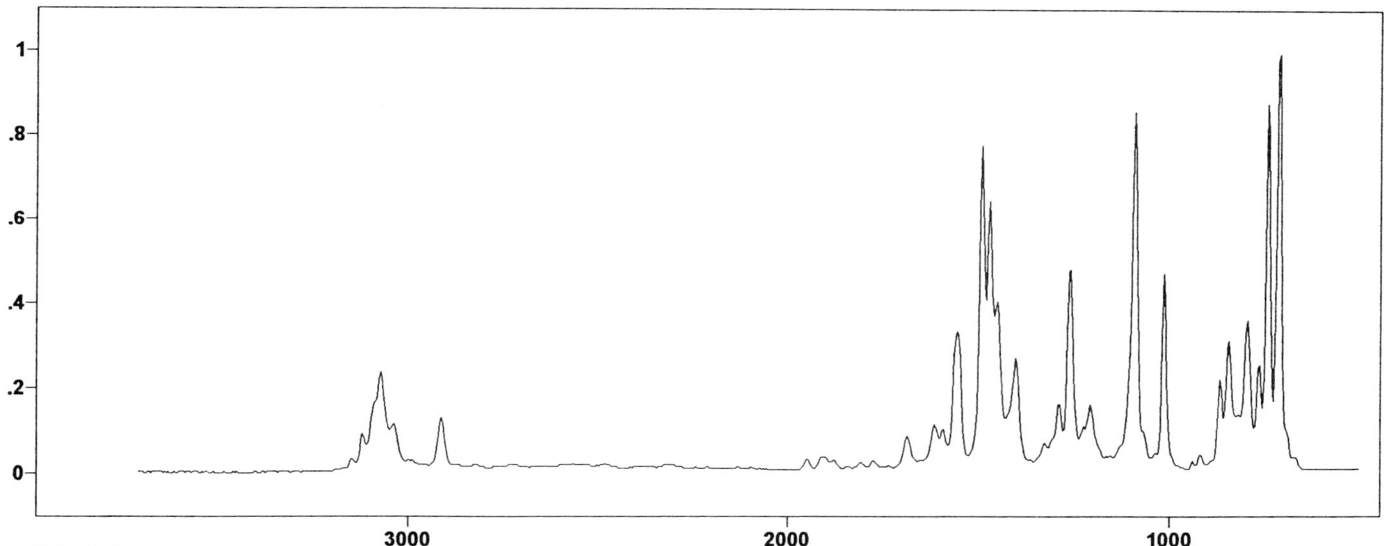

MAZINDOL

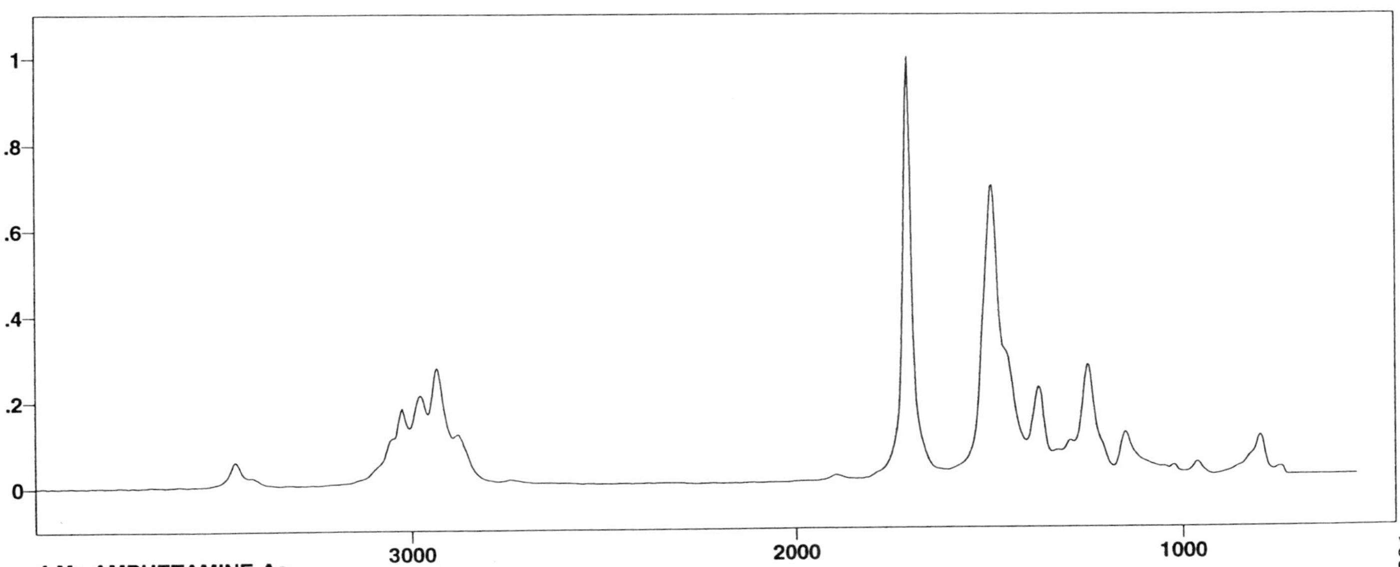

4-Me-AMPHETAMINE-Ac

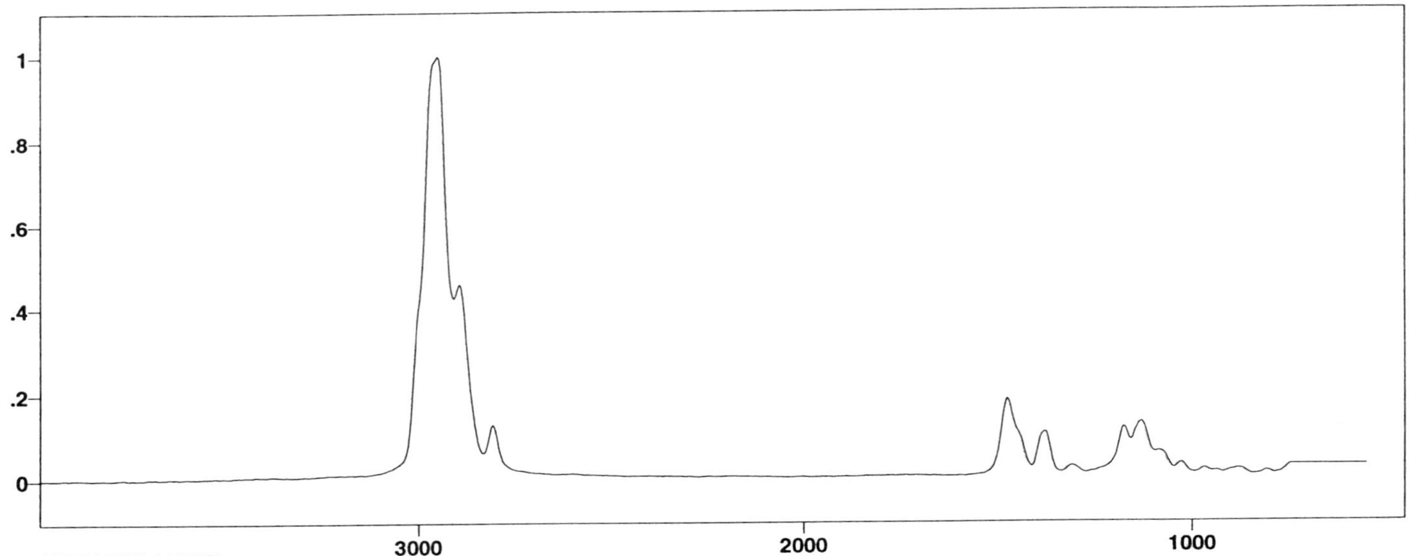

MECAMYLAMINE

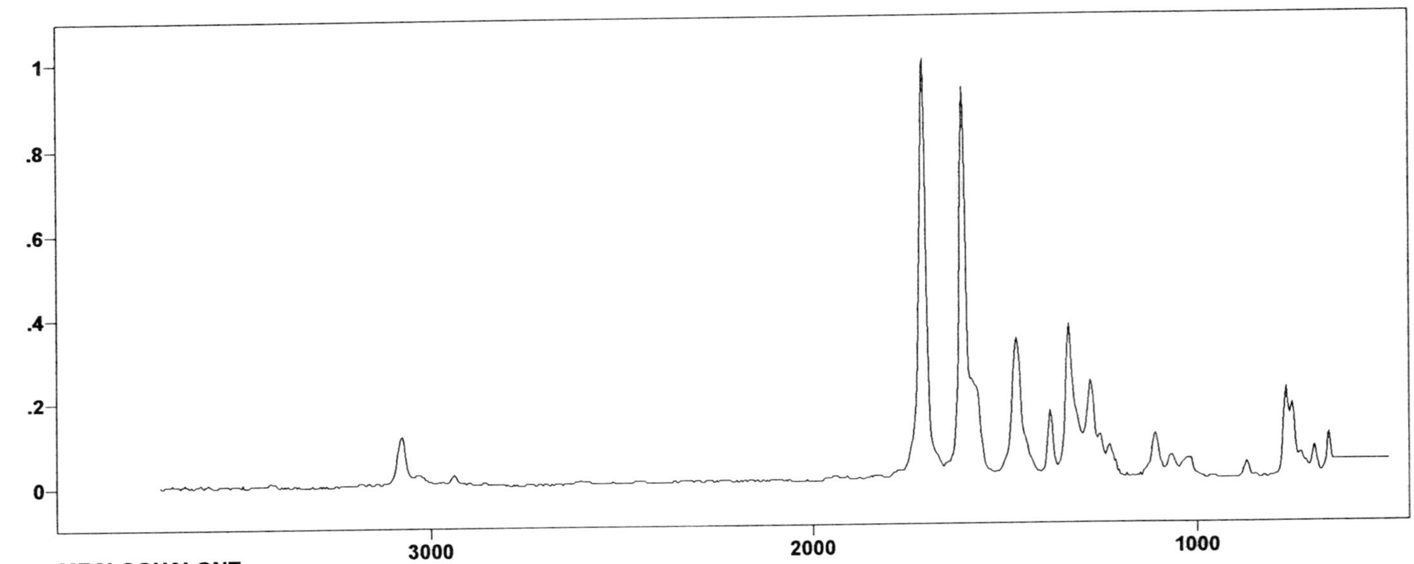

MECLOQUALONE

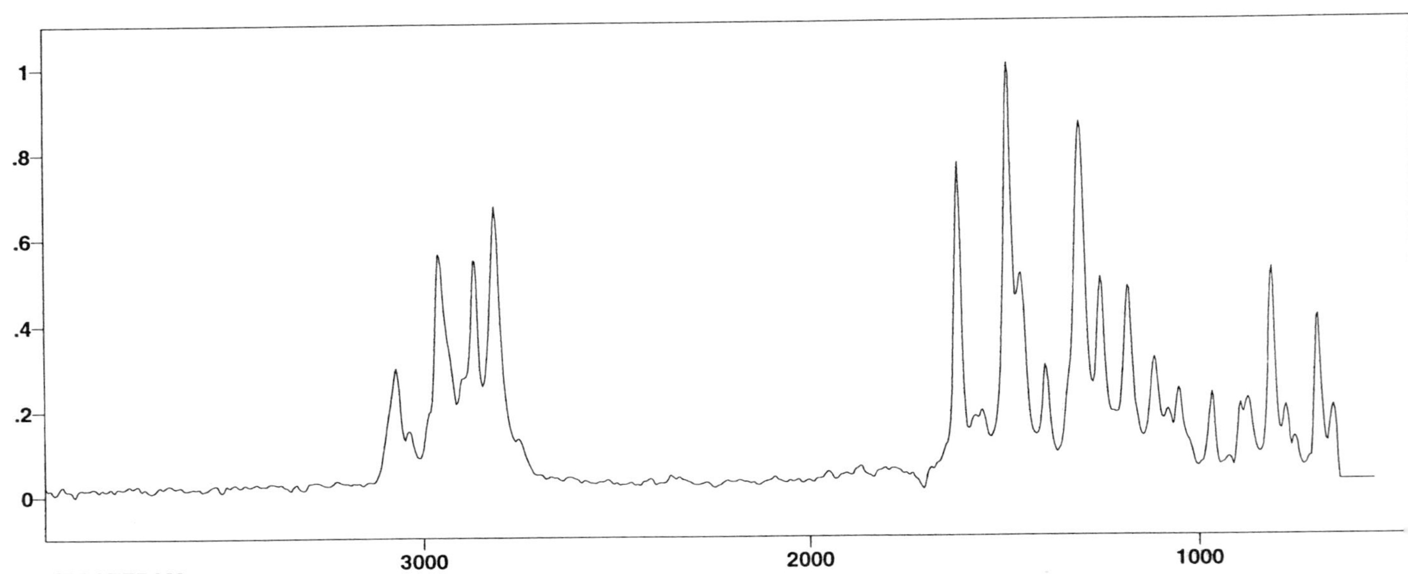

MEDAZEPAM

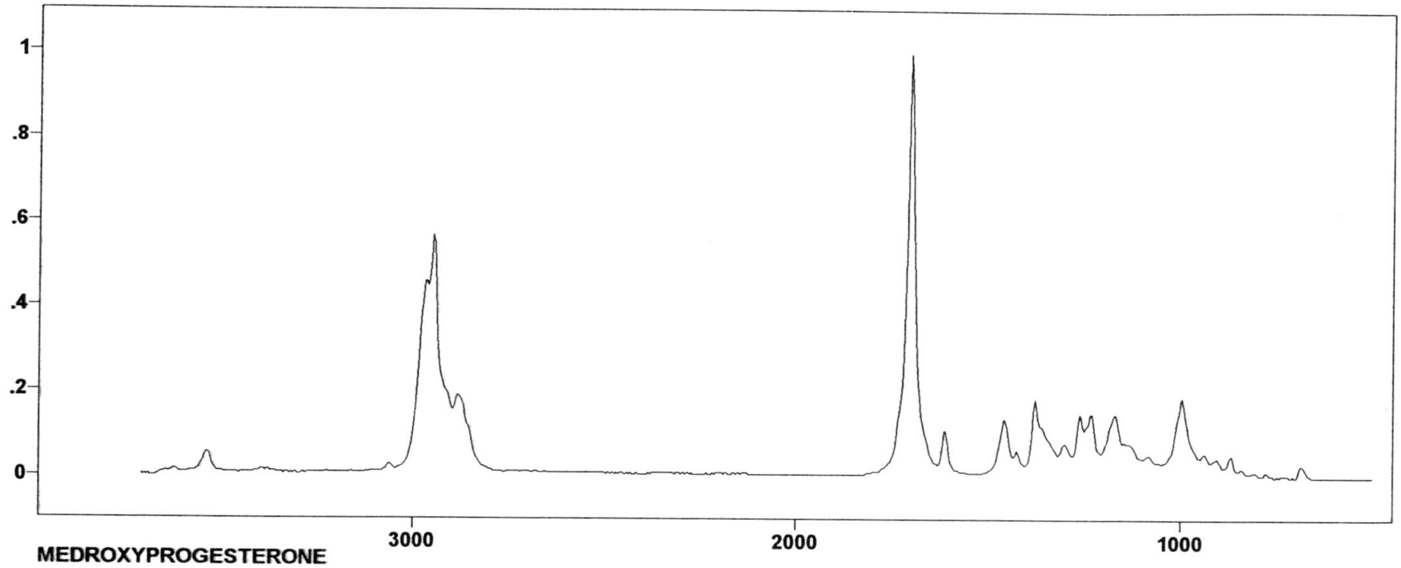

MEDROXYPROGESTERONE

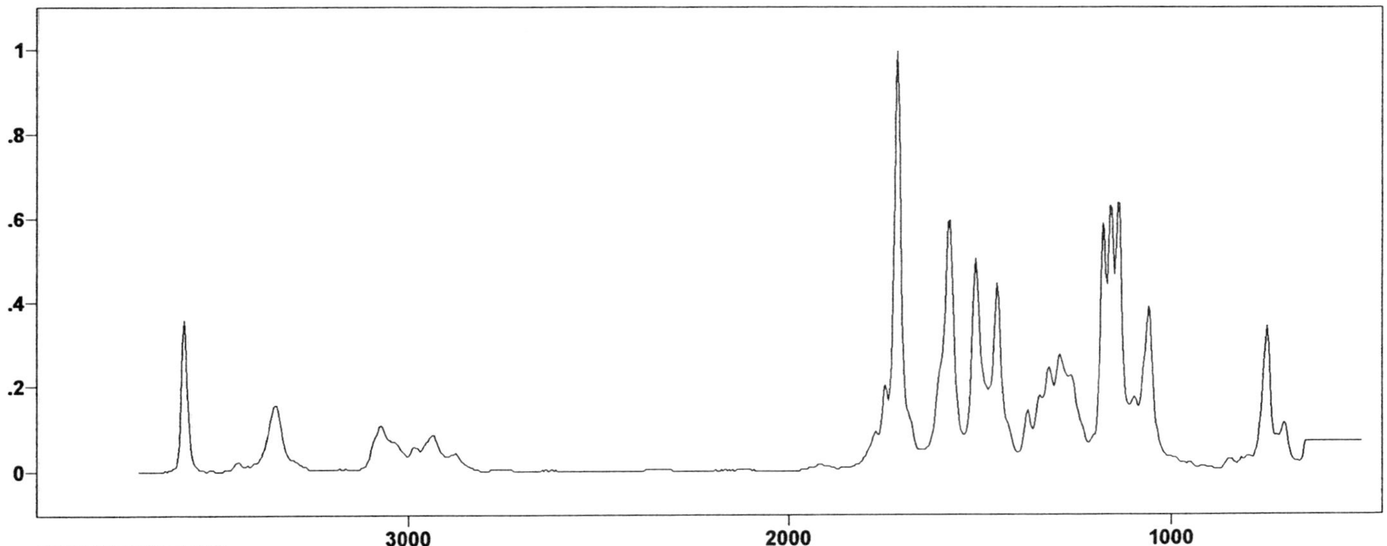

MEFENAMIC ACID

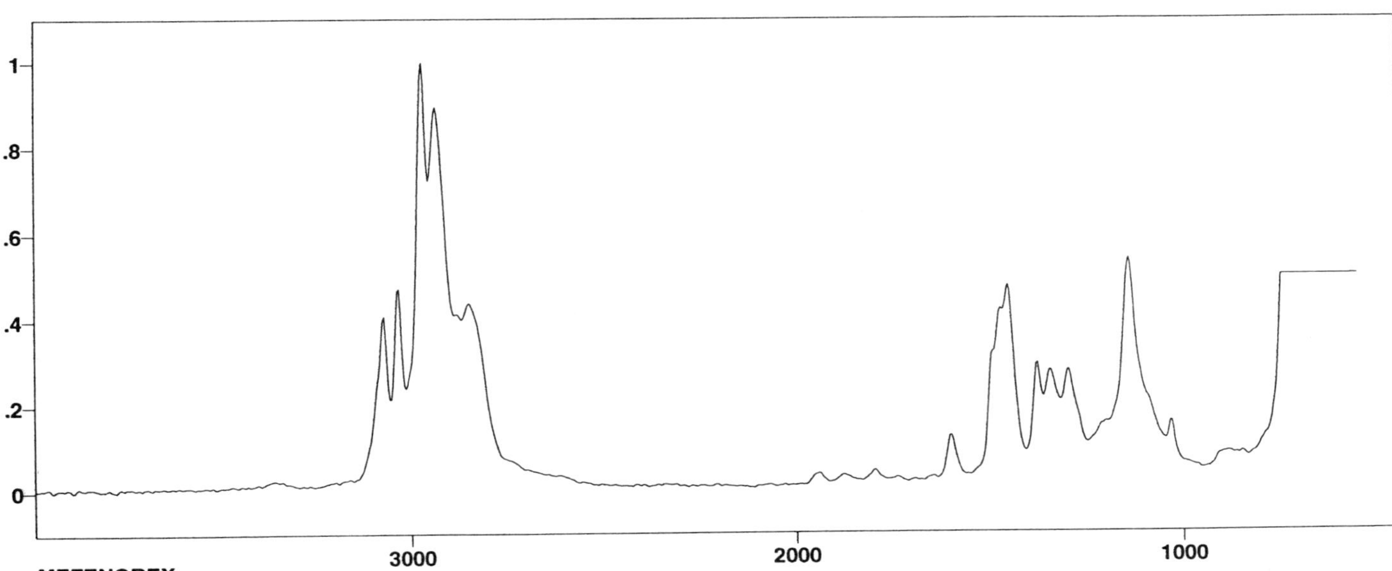

MEFENOREX

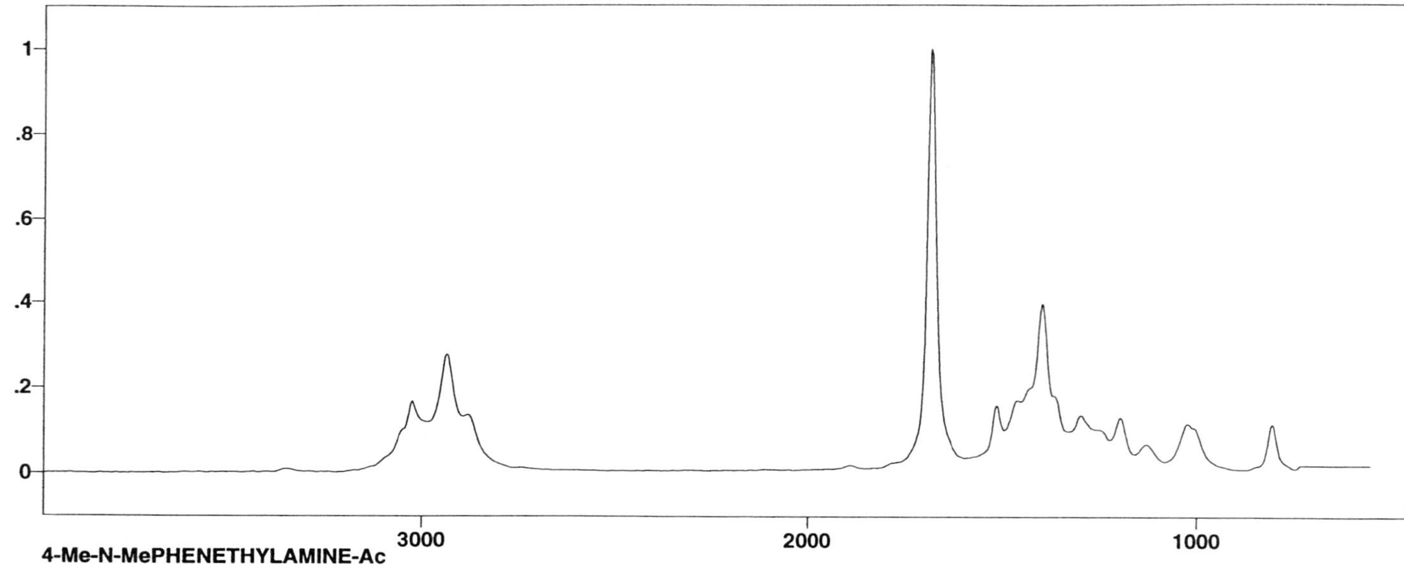

4-Me-N-MePHENETHYLAMINE-Ac

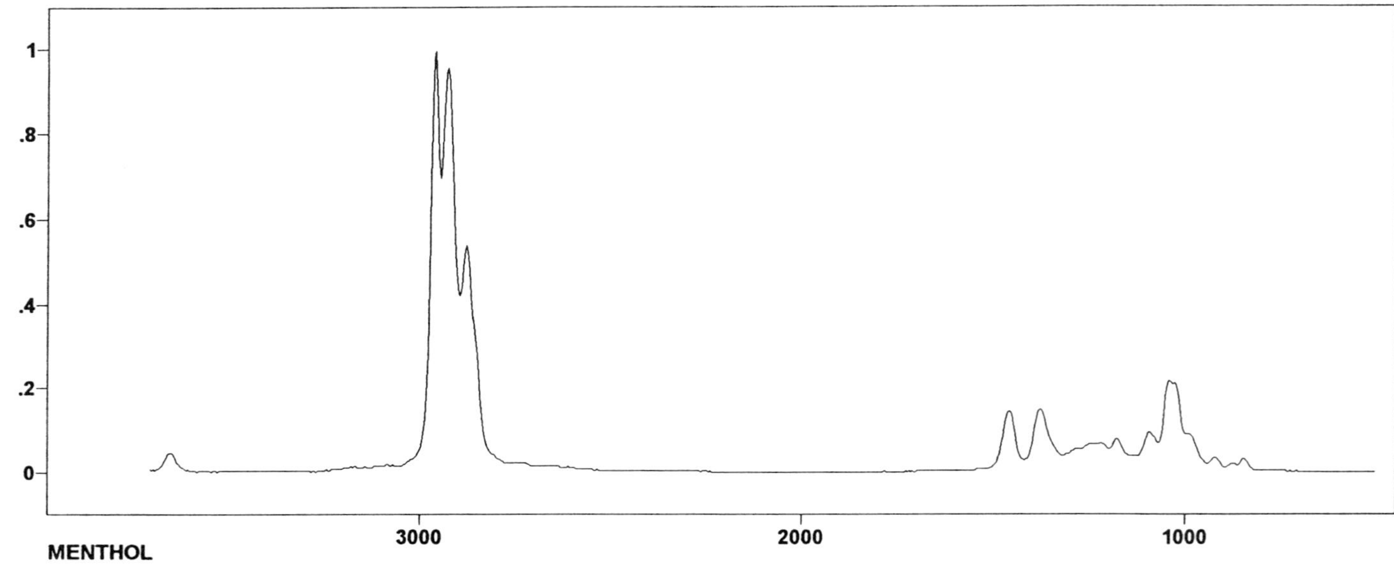

MENTHOL

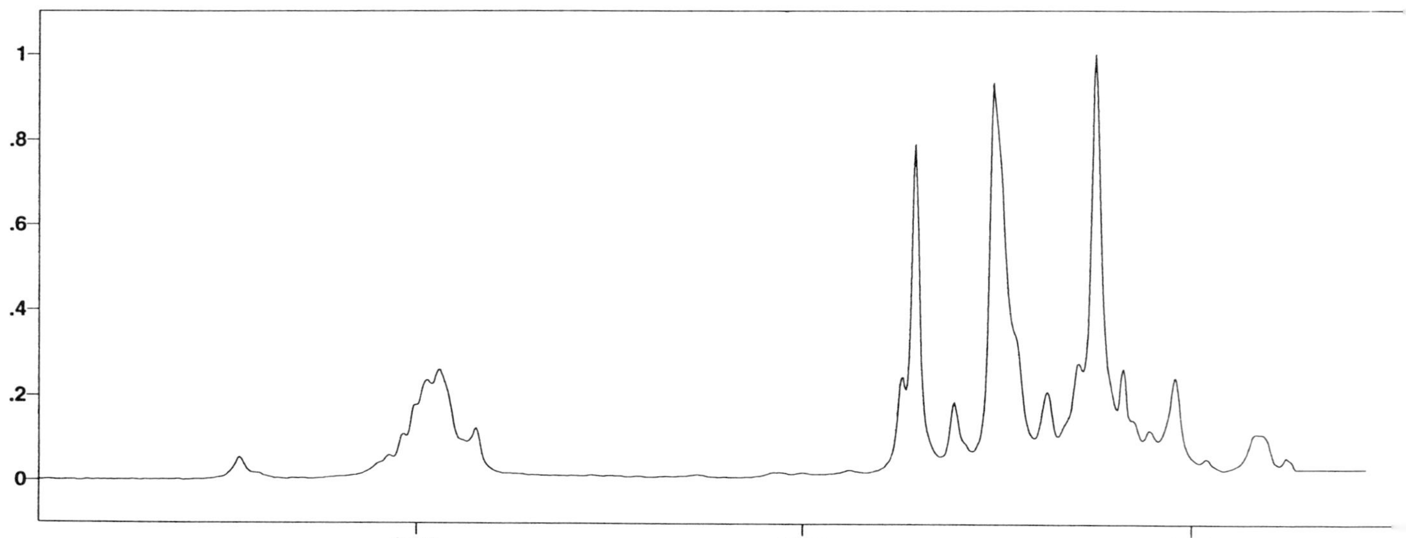

4-MeO-AMPHETAMINE-Ac

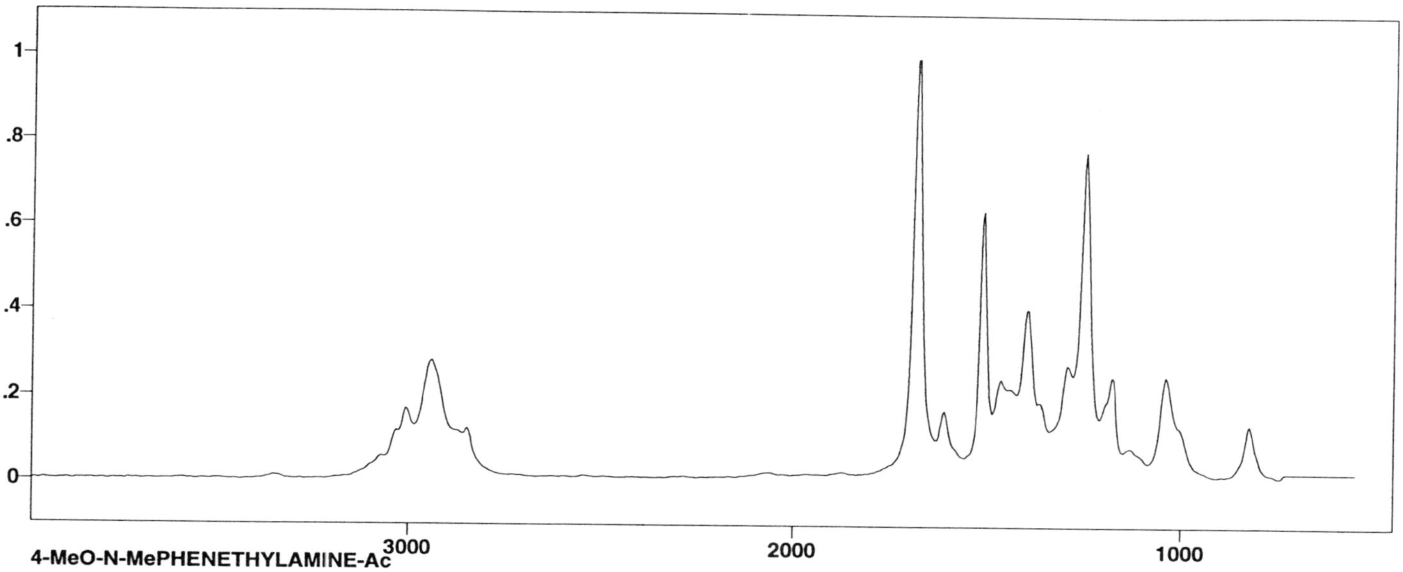

4-MeO-N-MePHENETHYLAMINE-Ac

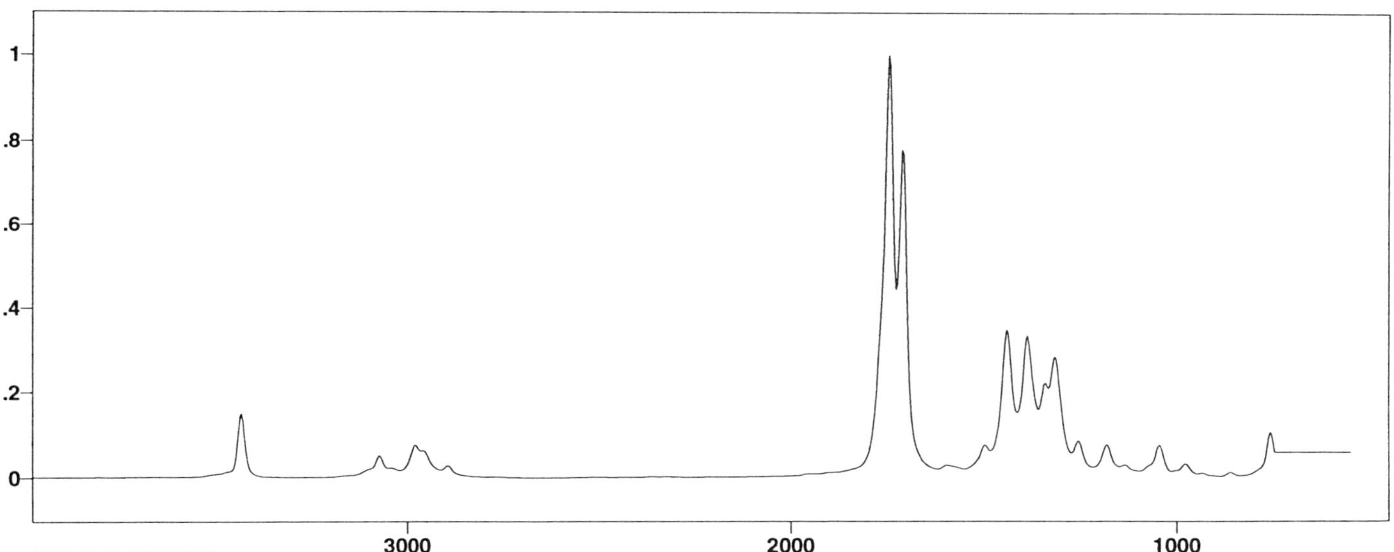

MEPHOBARBITAL

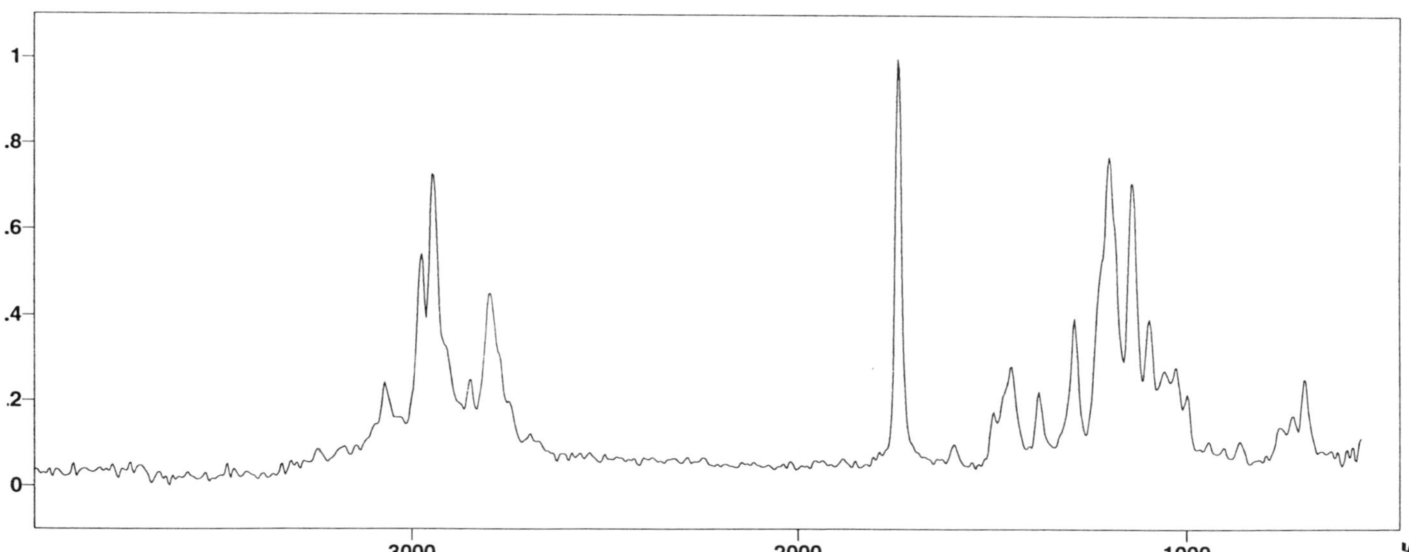

MEPERIDINE

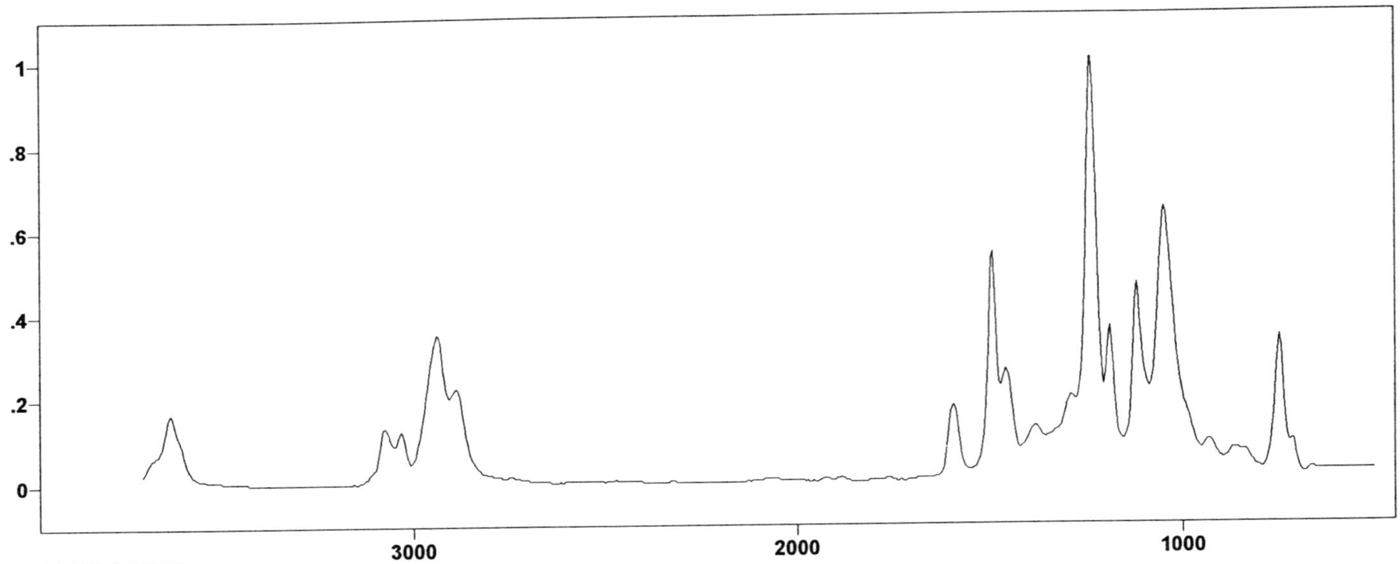

MEPHENESIN

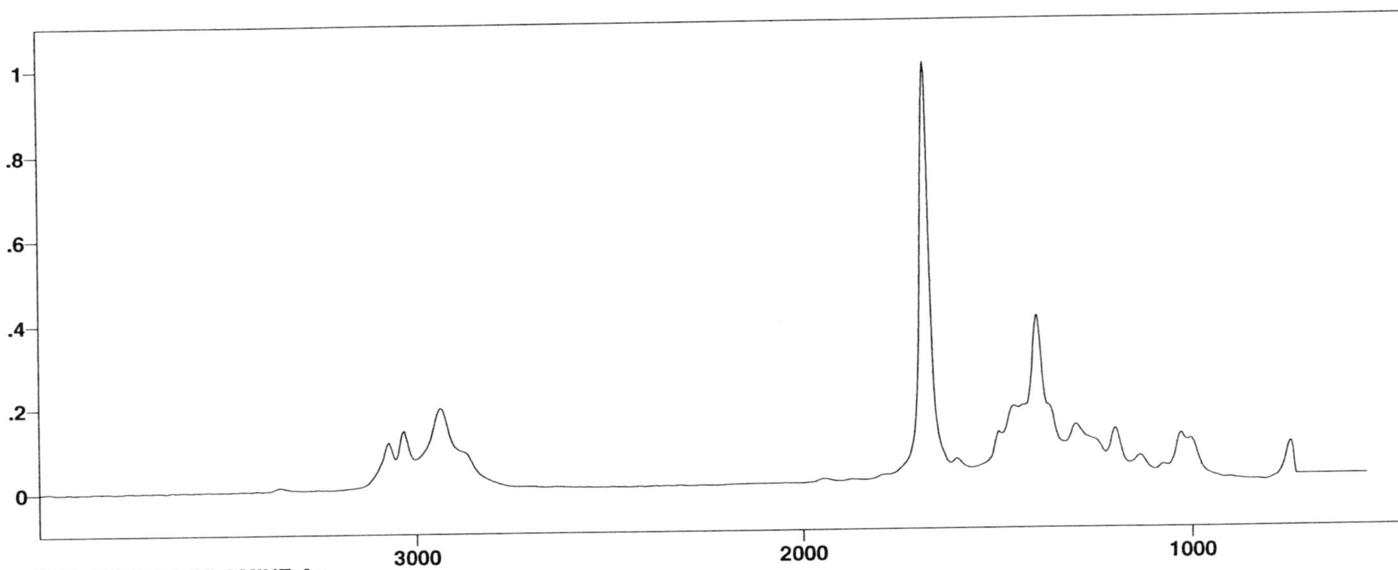

N-MePHENETHYLAMINE-Ac

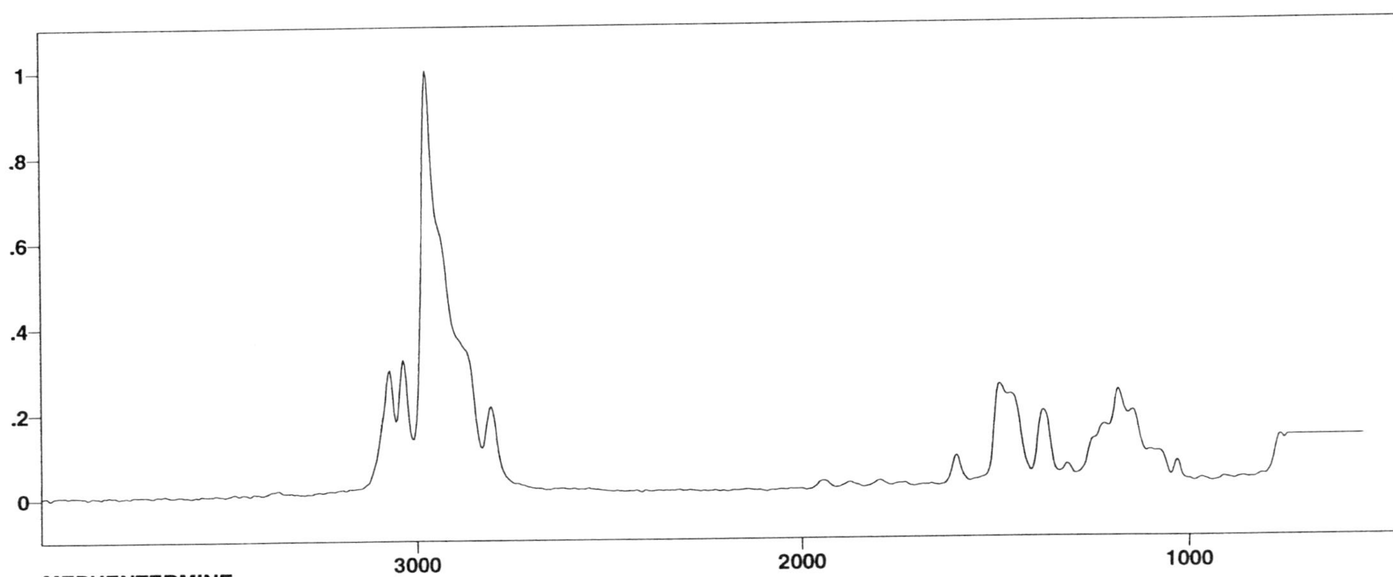

MEPHENTERMINE

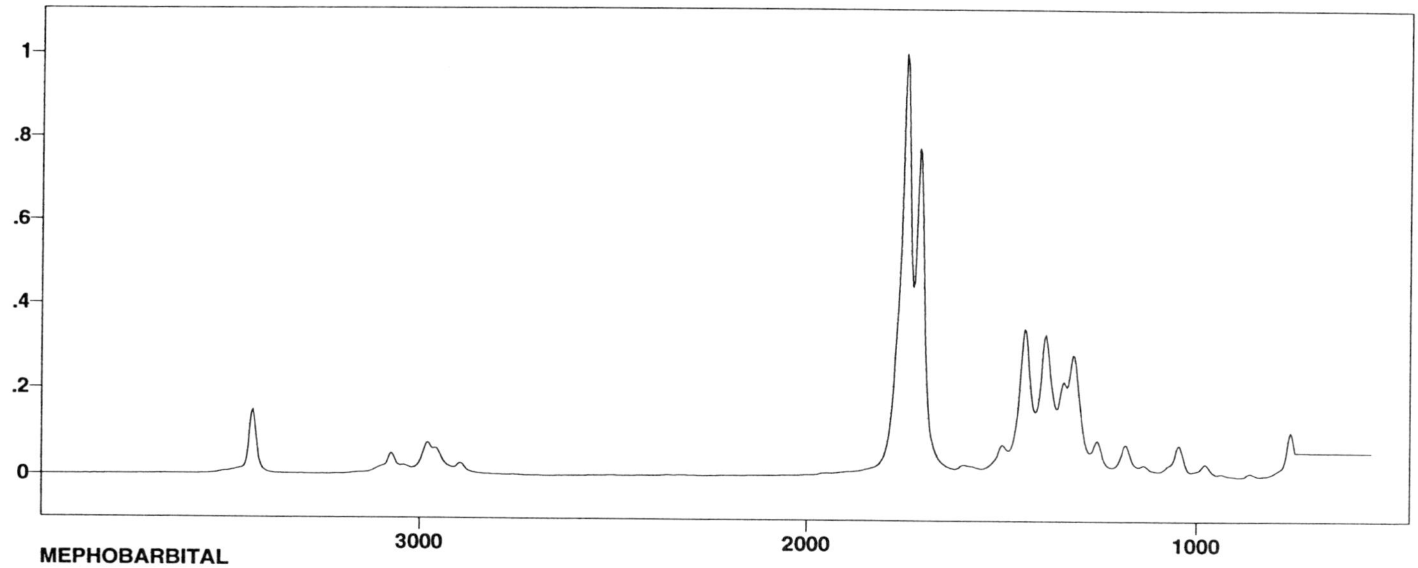

MEPHOBARBITAL

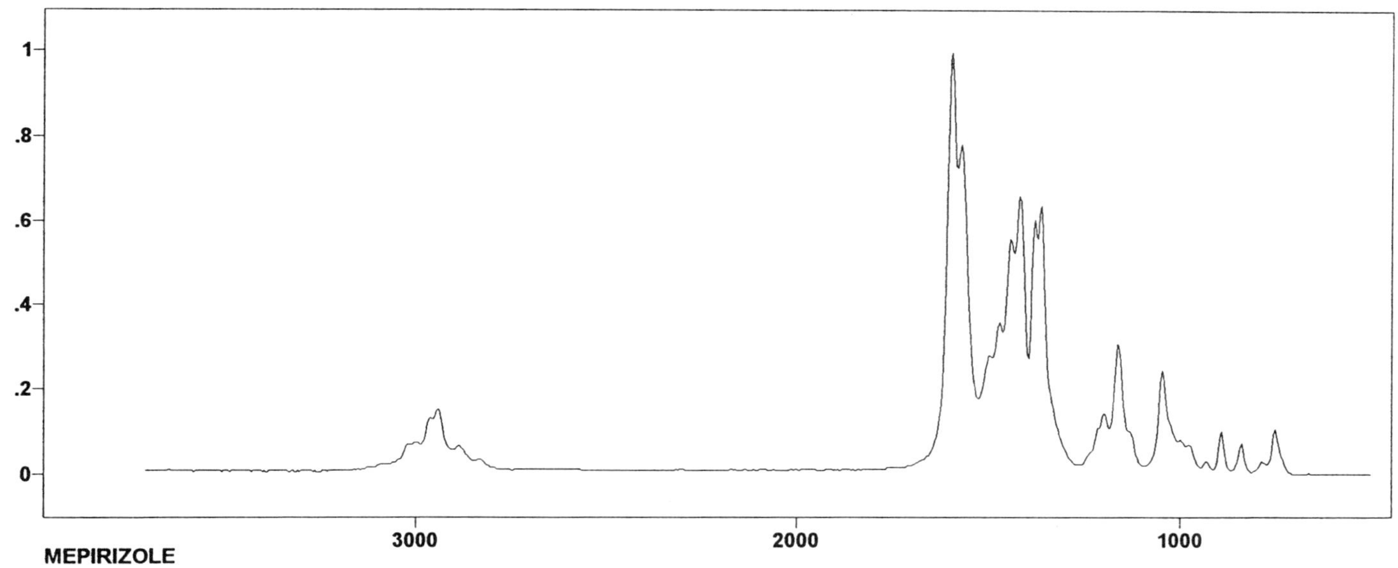

MEPIRIZOLE

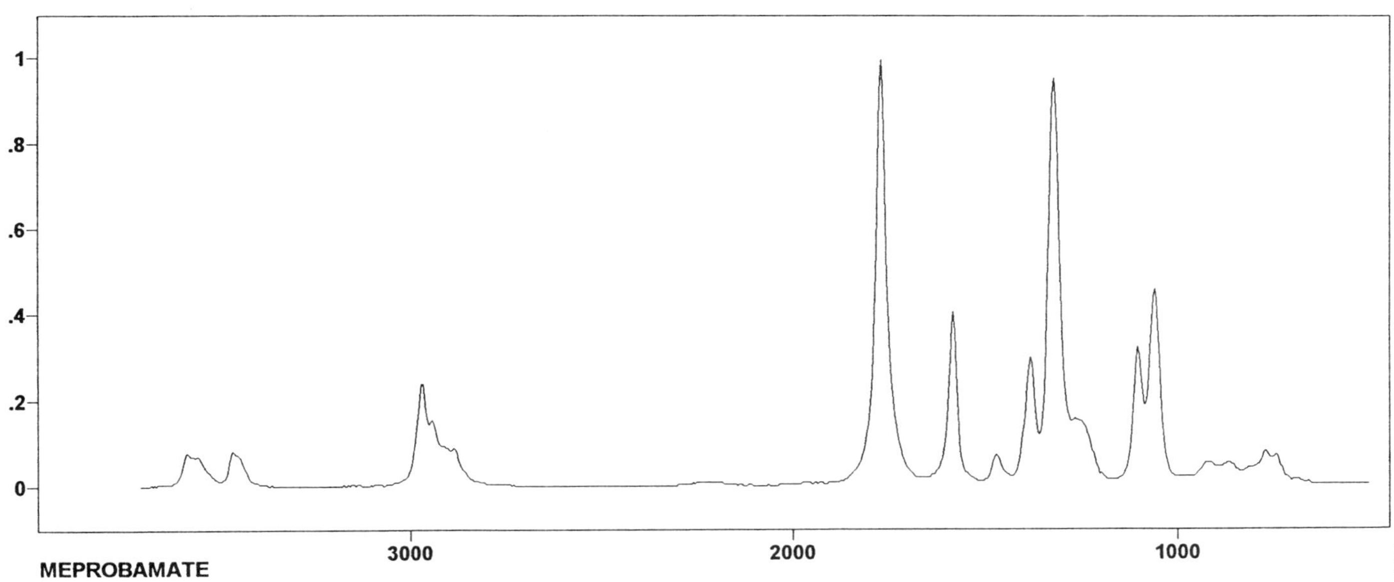

MEPROBAMATE

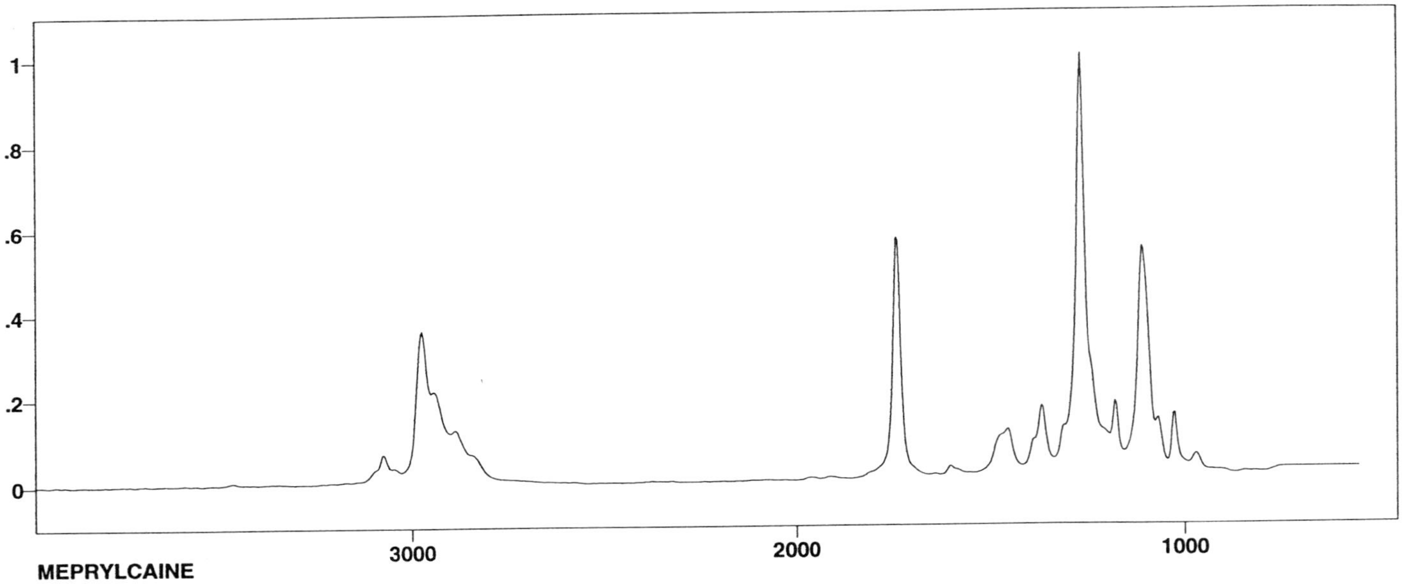

MEPRYLCAINE

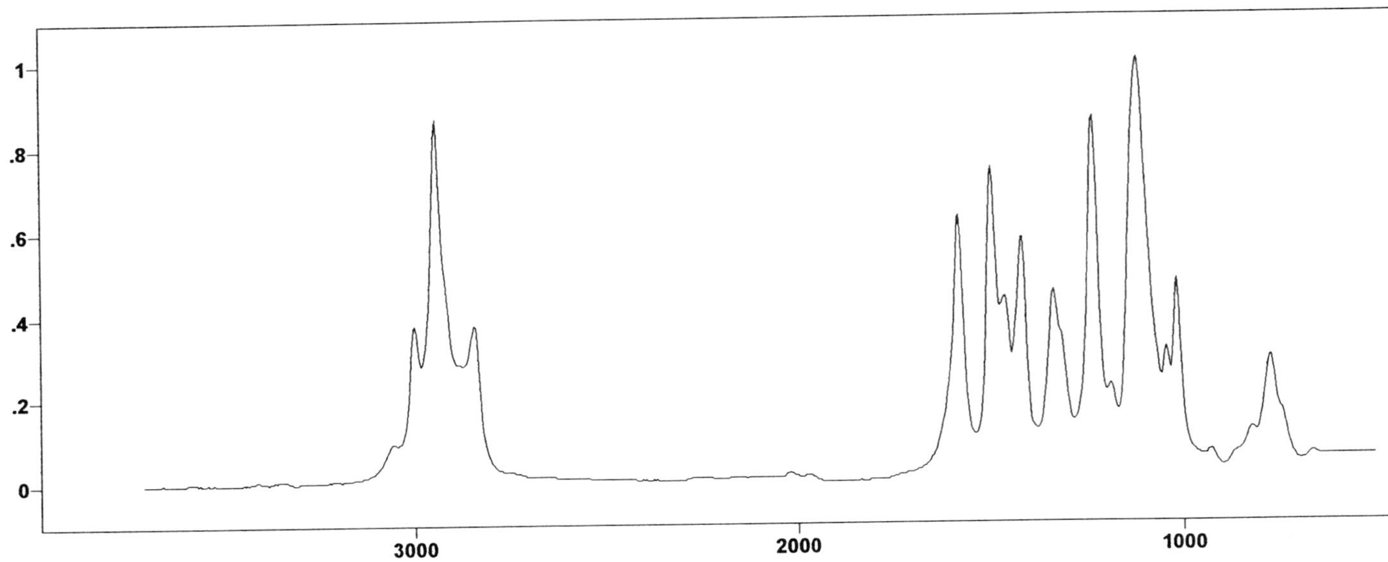

MESCALINE

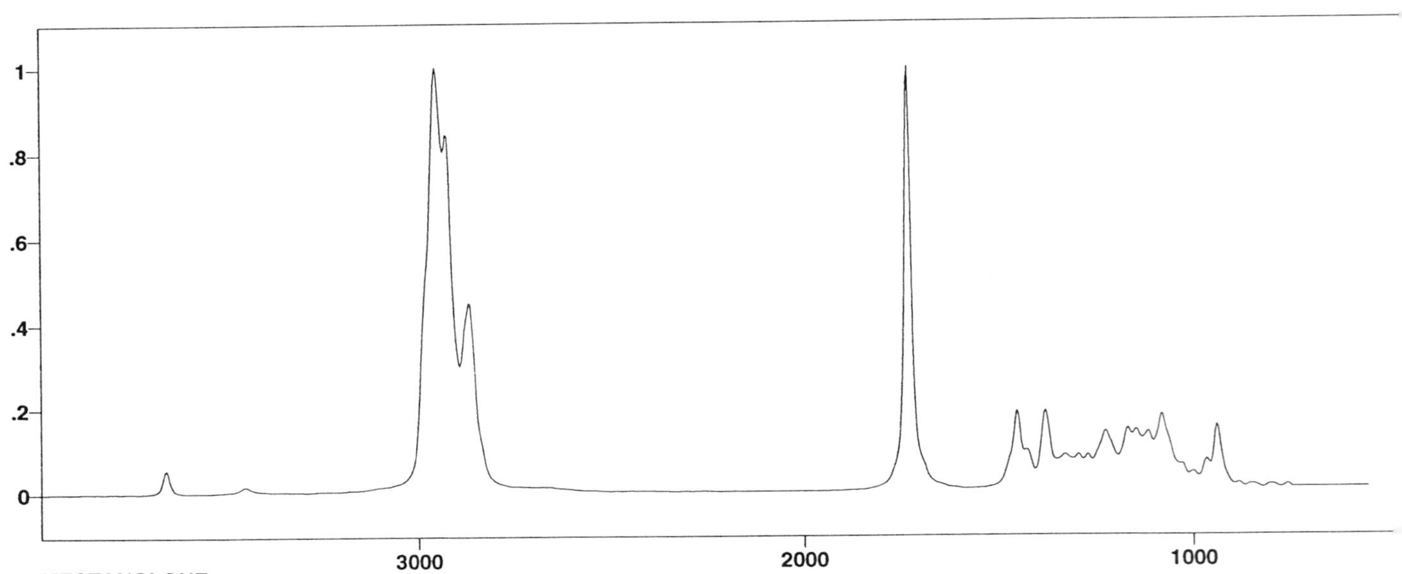

MESTANOLONE

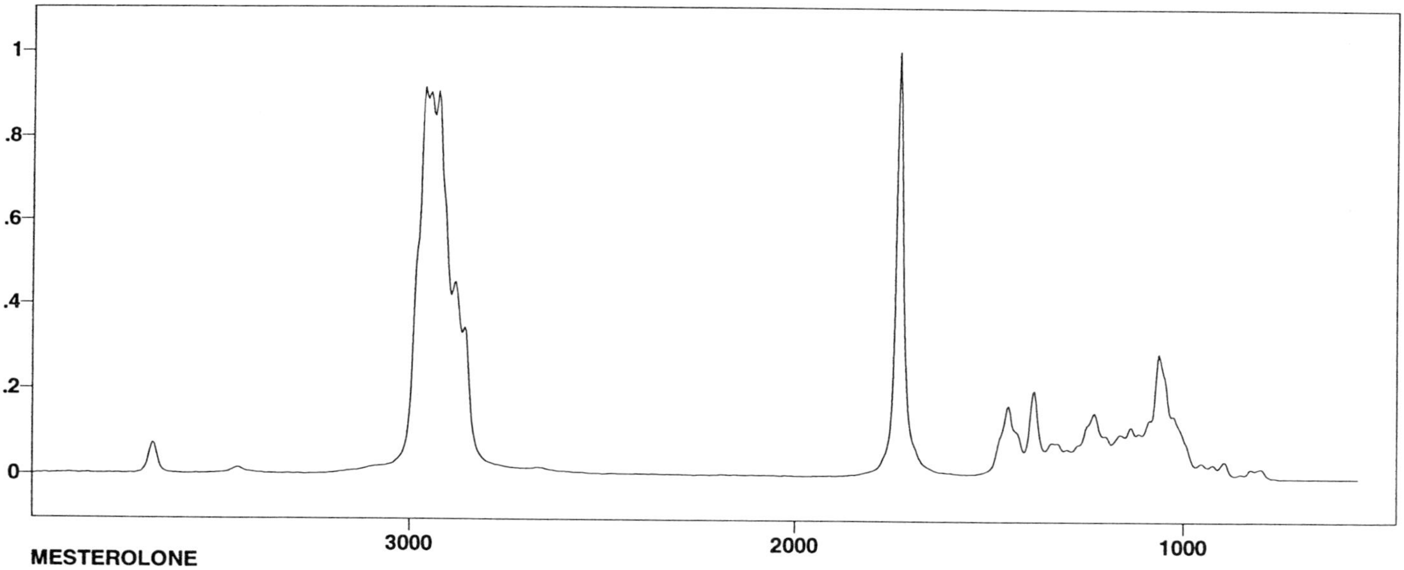

MESTEROLONE

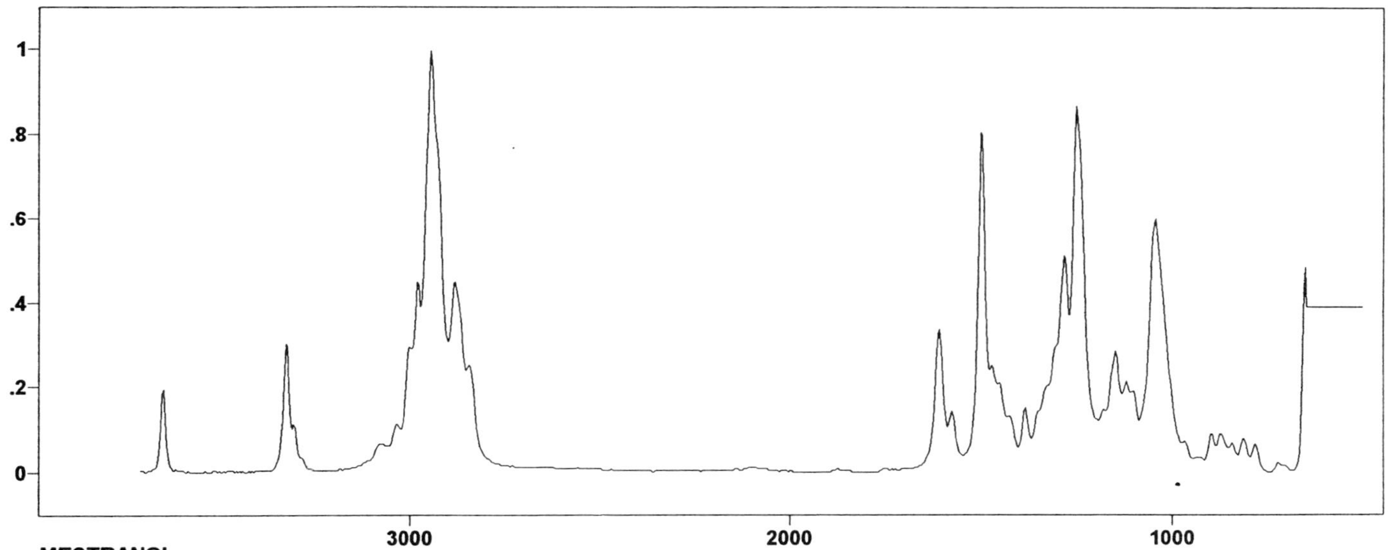

MESTRANOL

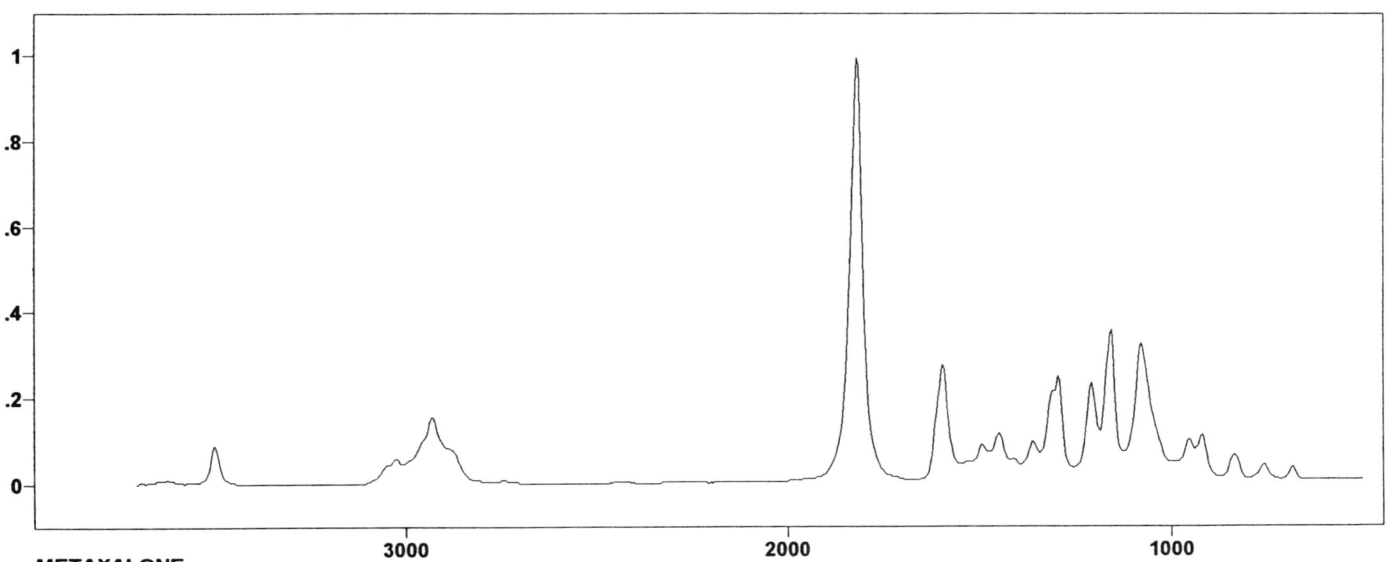

METAXALONE

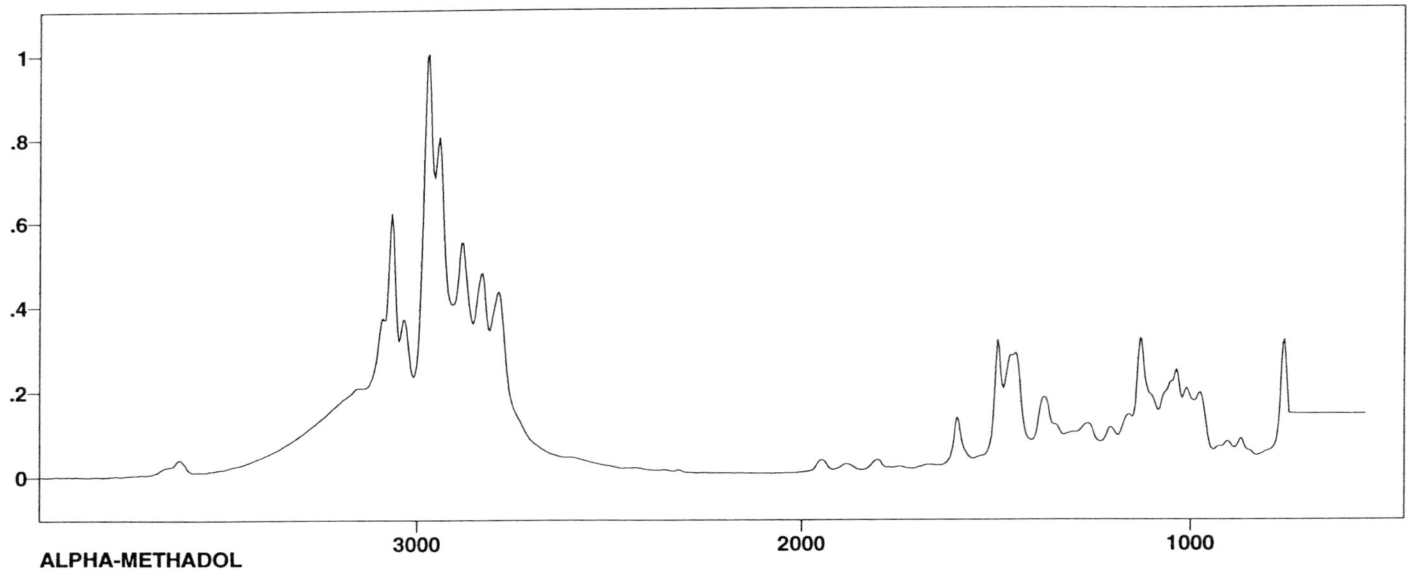

ALPHA-METHADOL

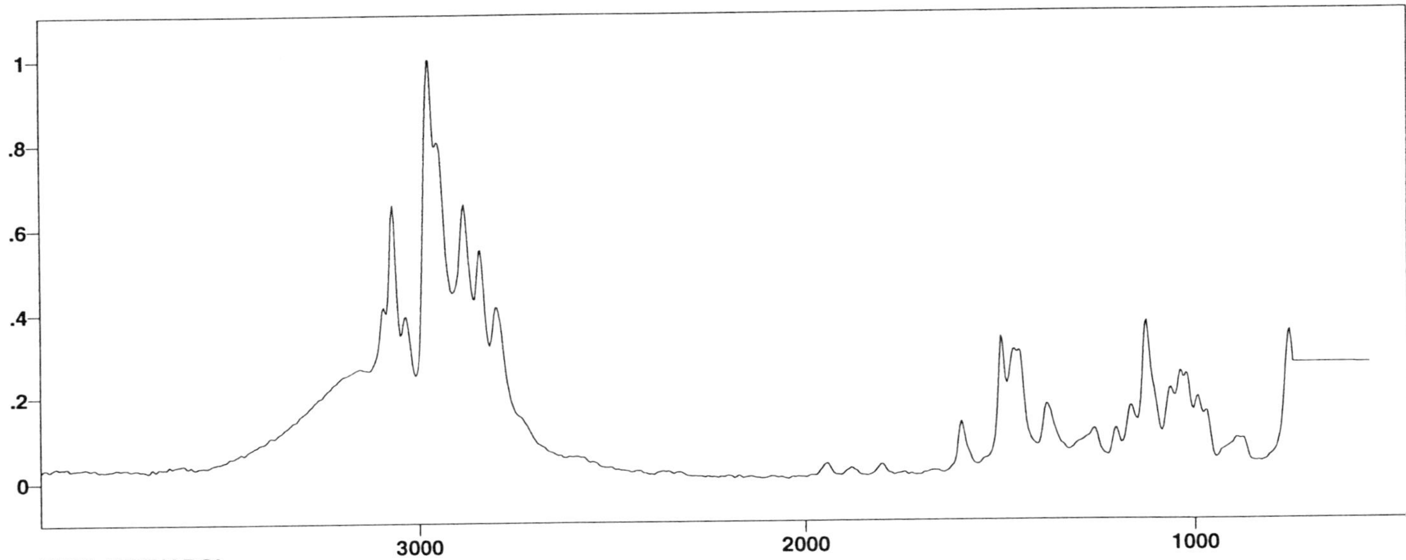

BETA-METHADOL

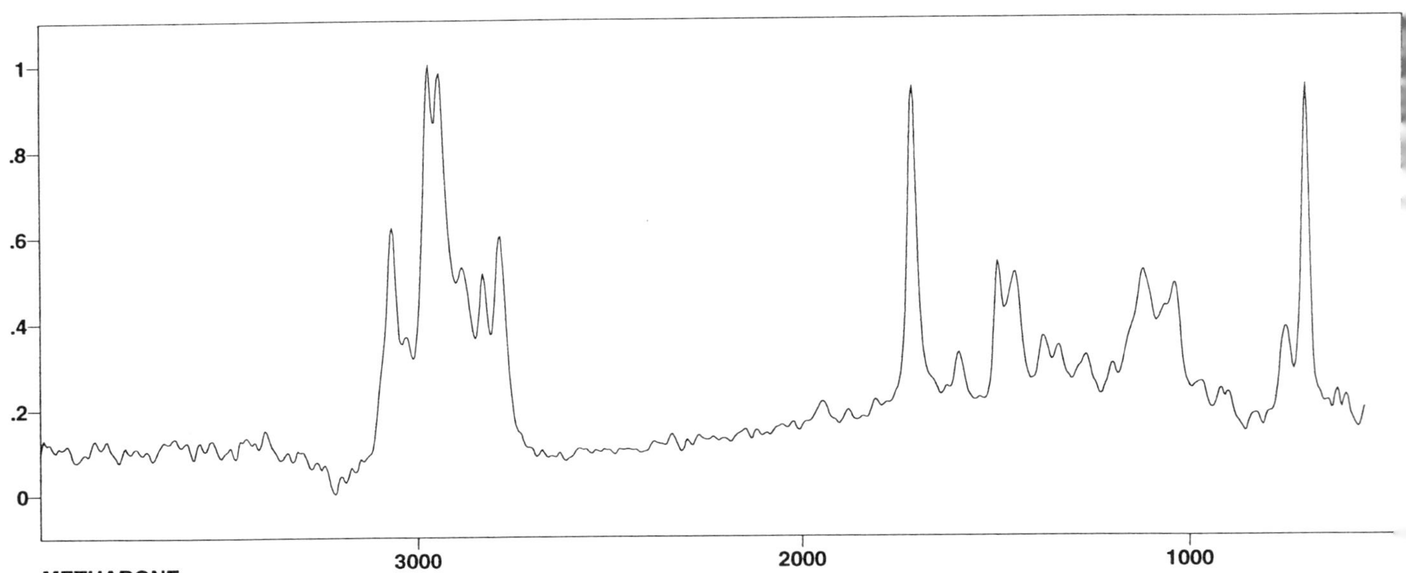

METHADONE

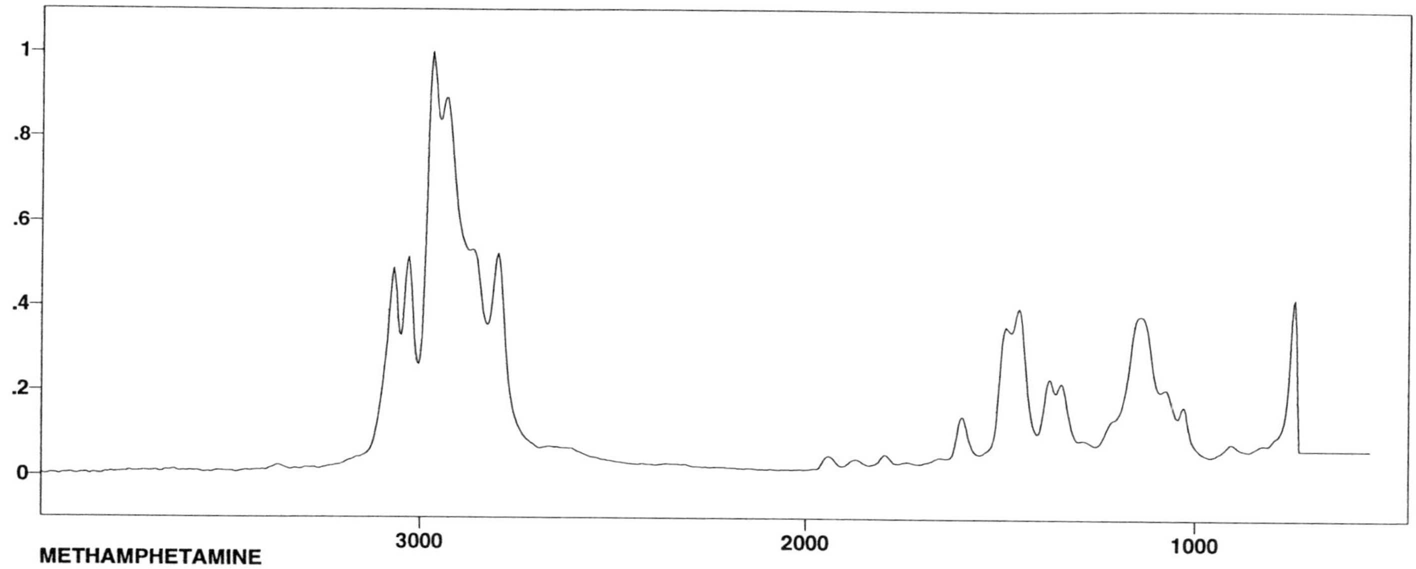

METHAMPHETAMINE

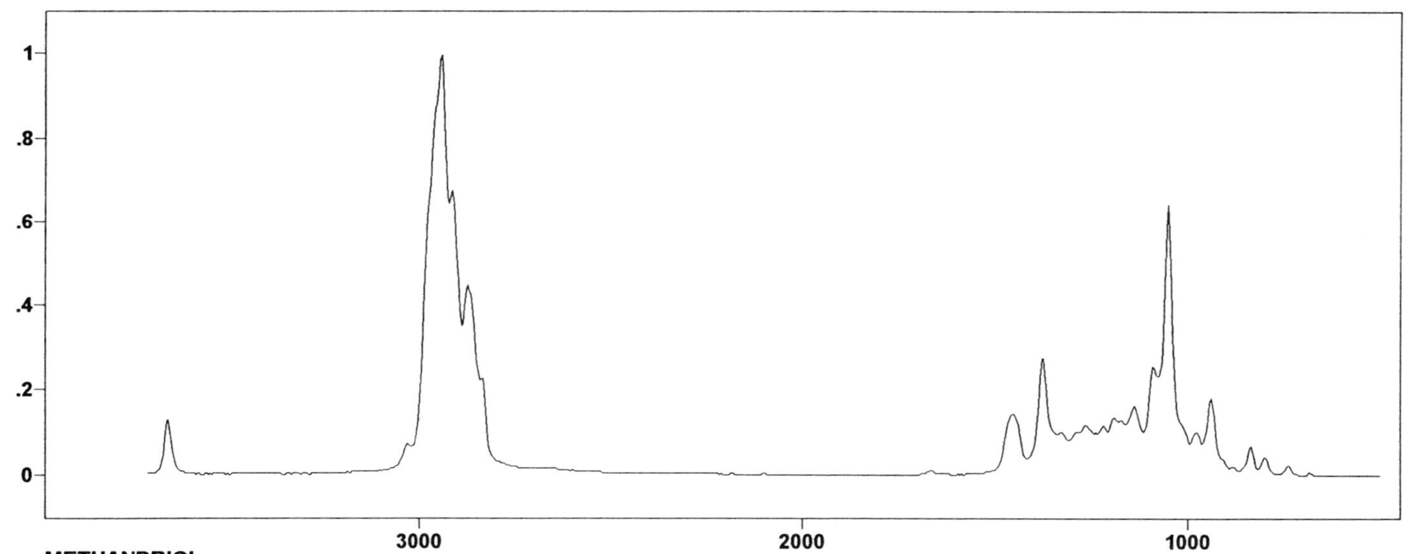

METHANDRIOL

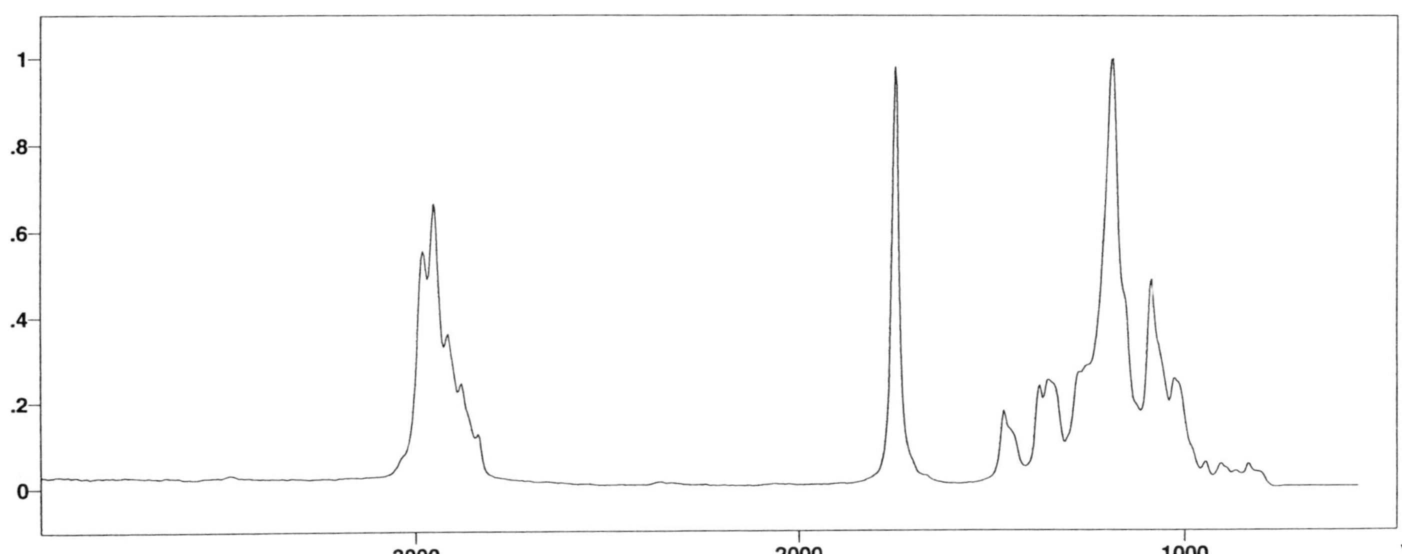

METHANDRIOL DIPROPIONATE

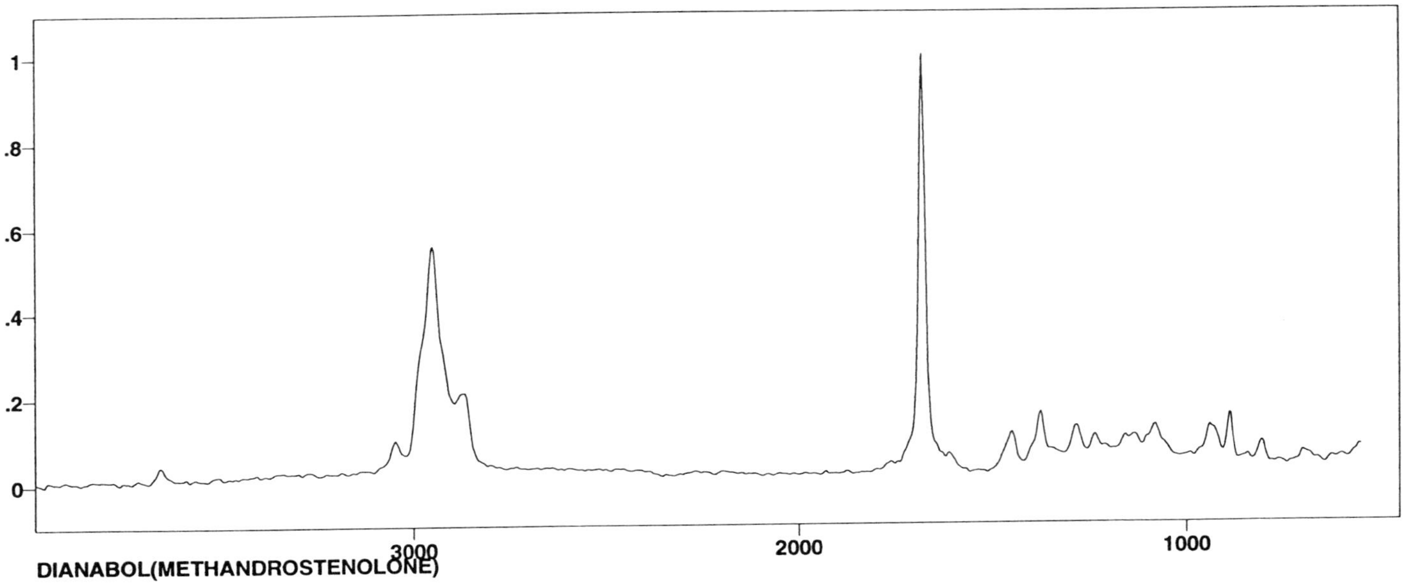

DIANABOL(METHANDROSTENOLONE)

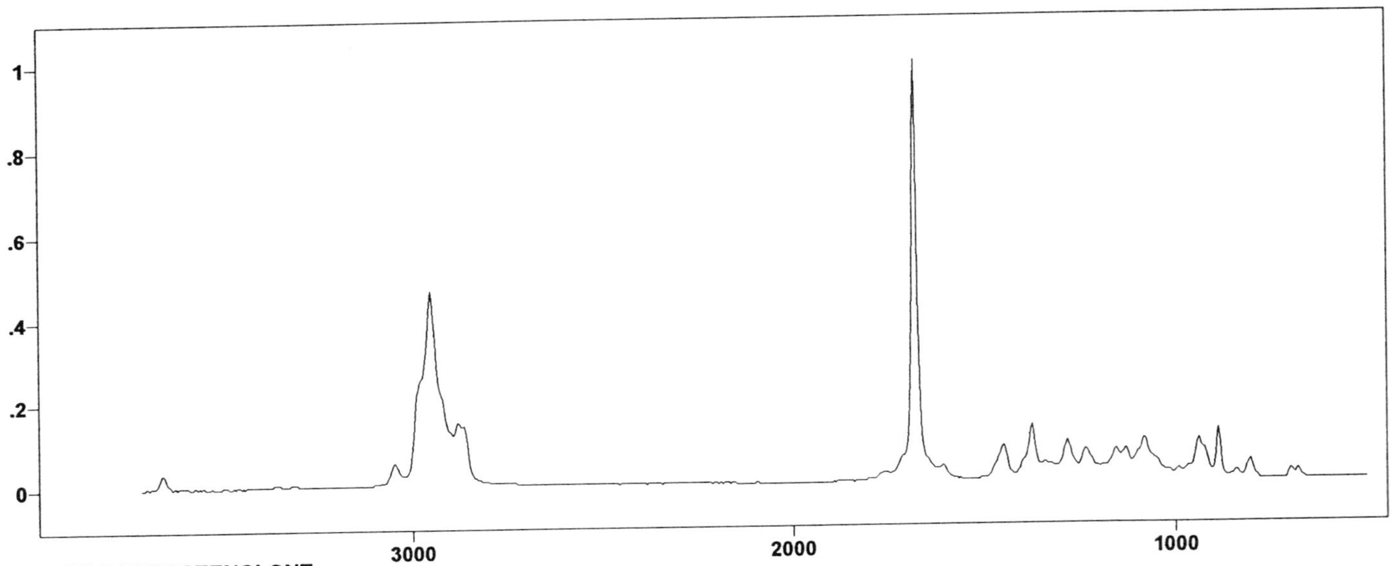

METHANDROSTENOLONE

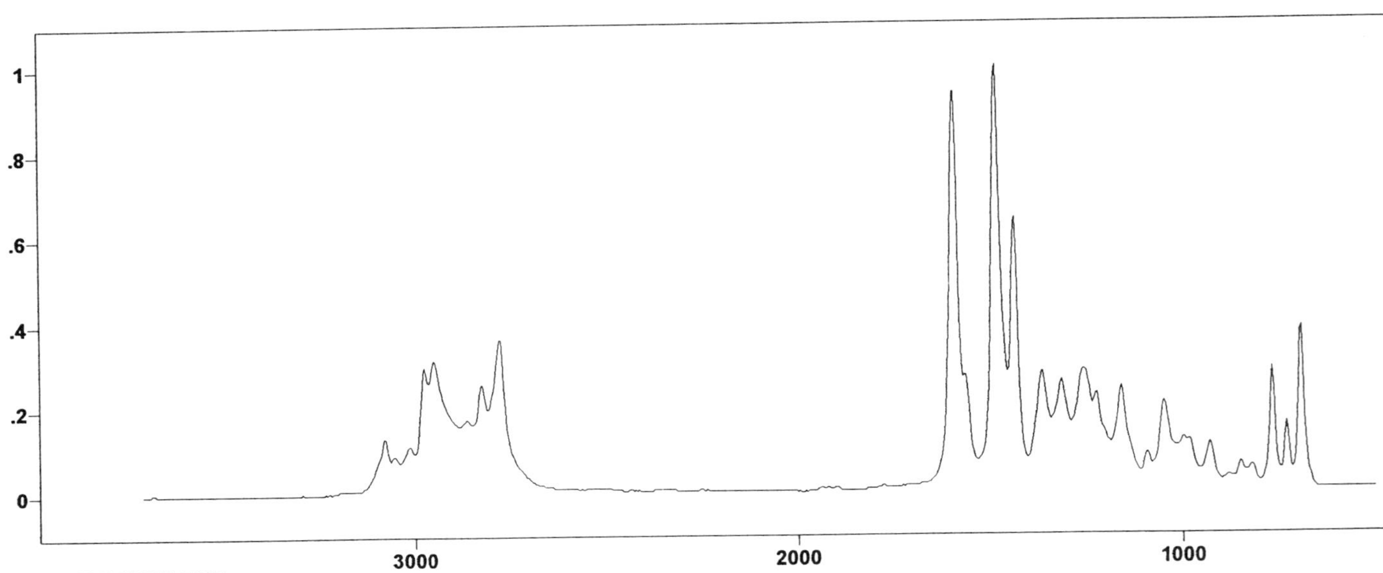

METHAPYRILENE

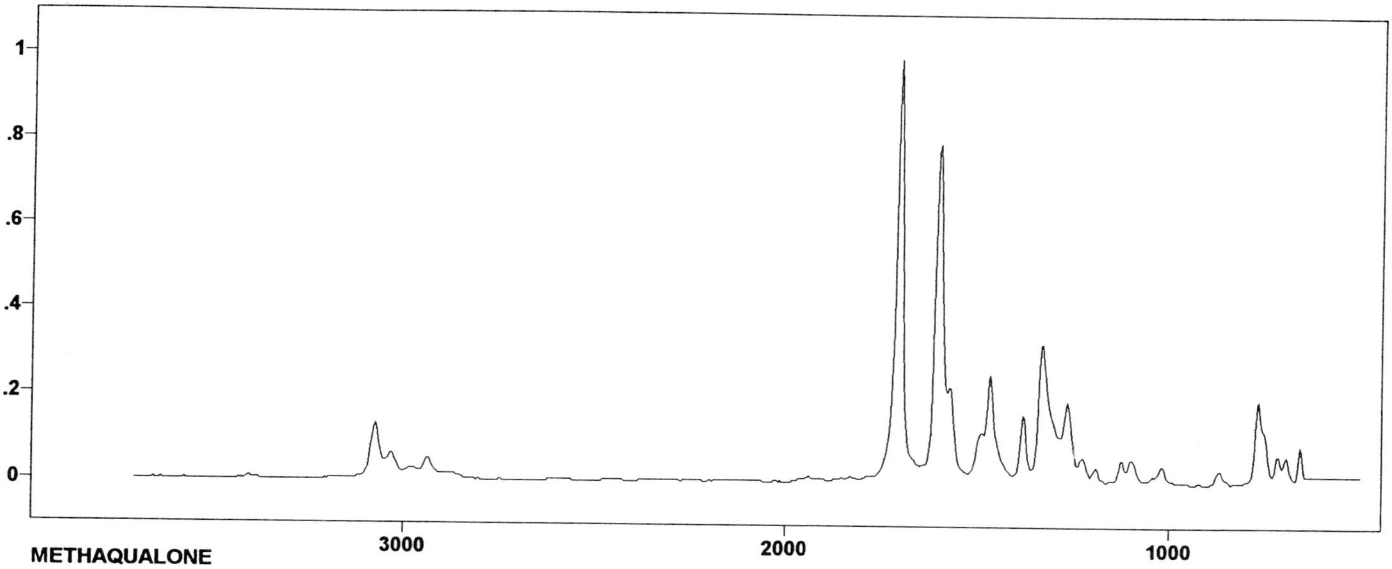

METHAQUALONE

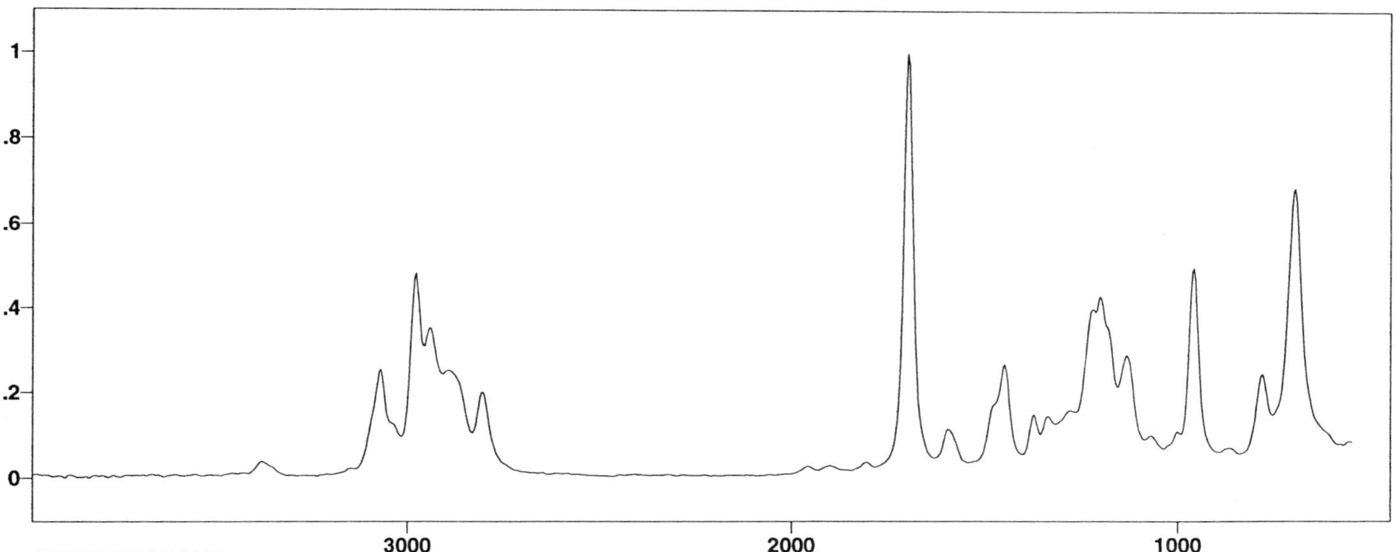

METHCATHINONE

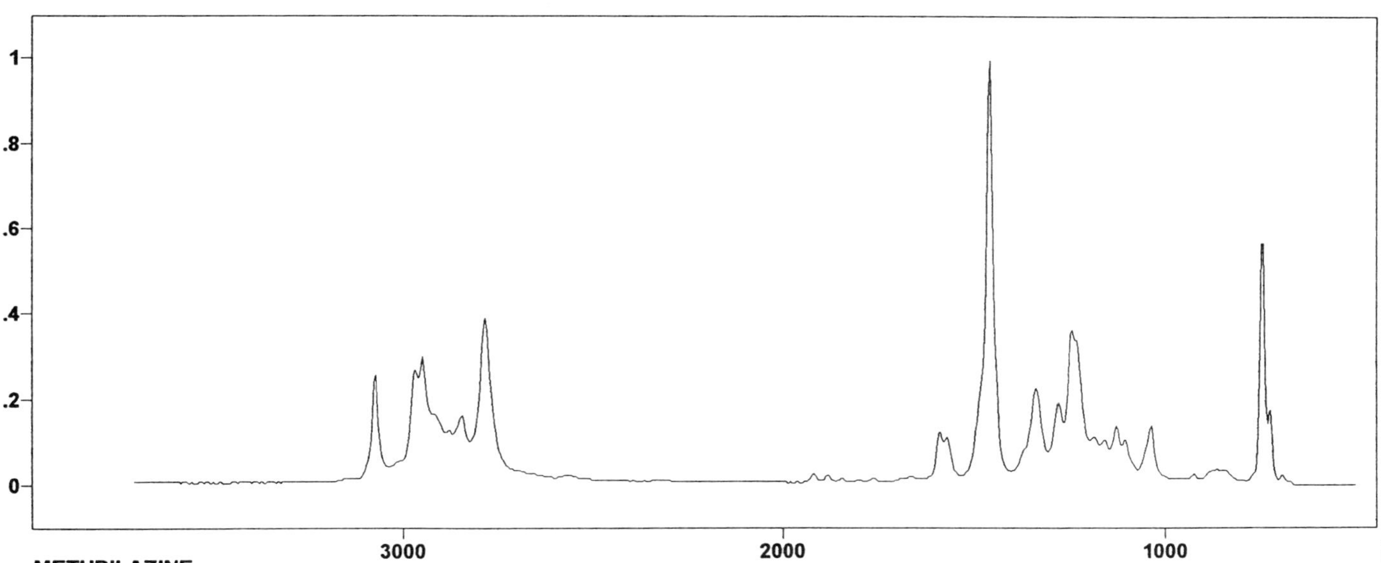

METHDILAZINE

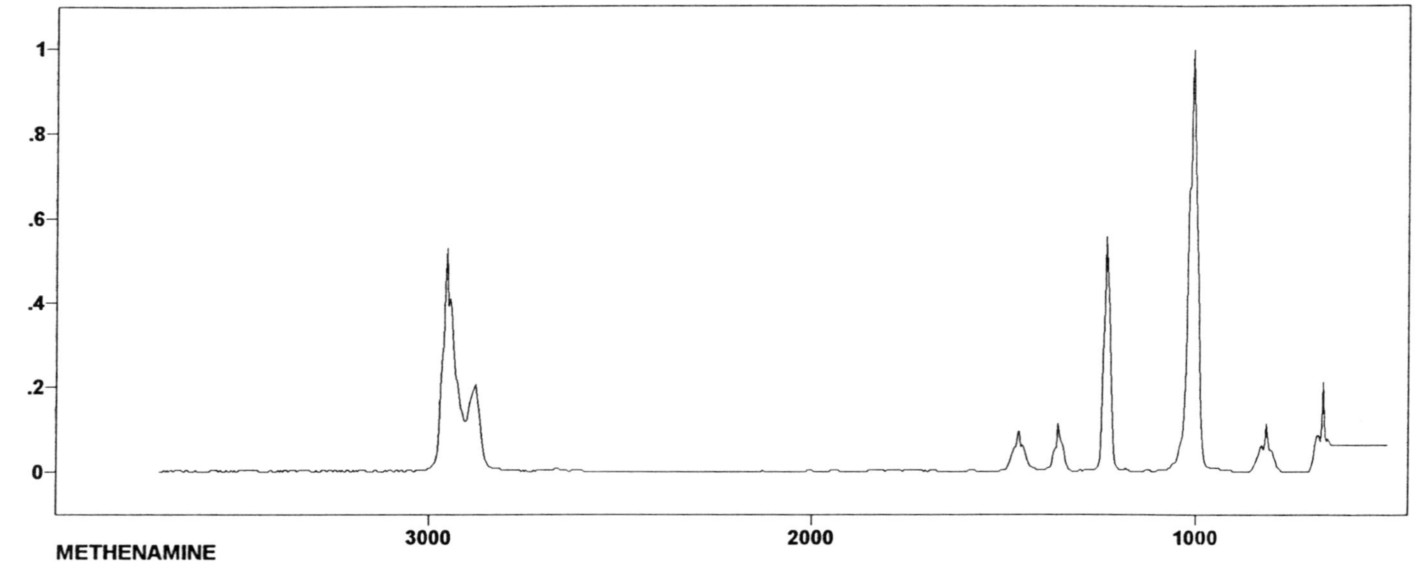

METHENAMINE

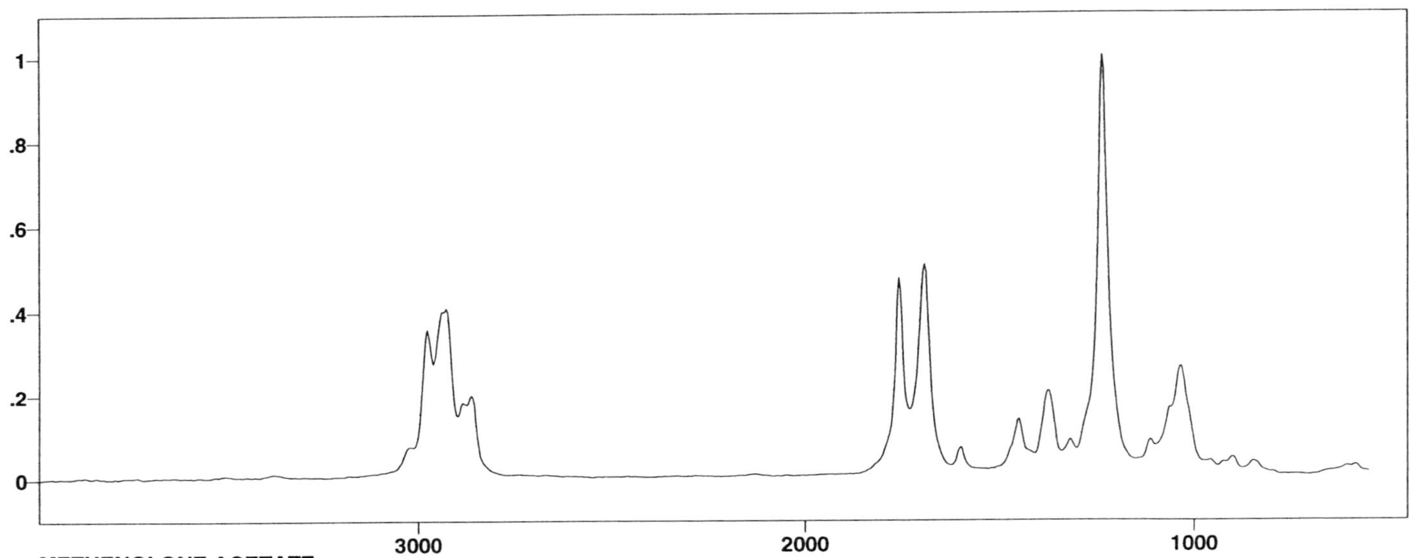

METHENOLONE ACETATE

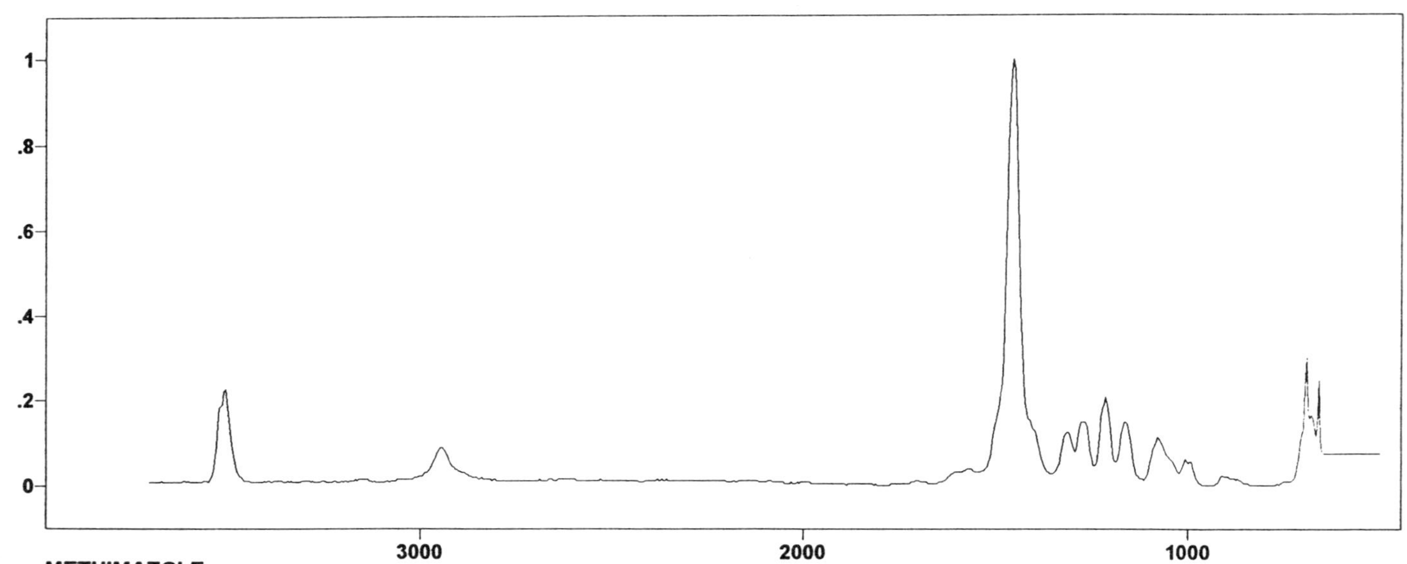

METHIMAZOLE

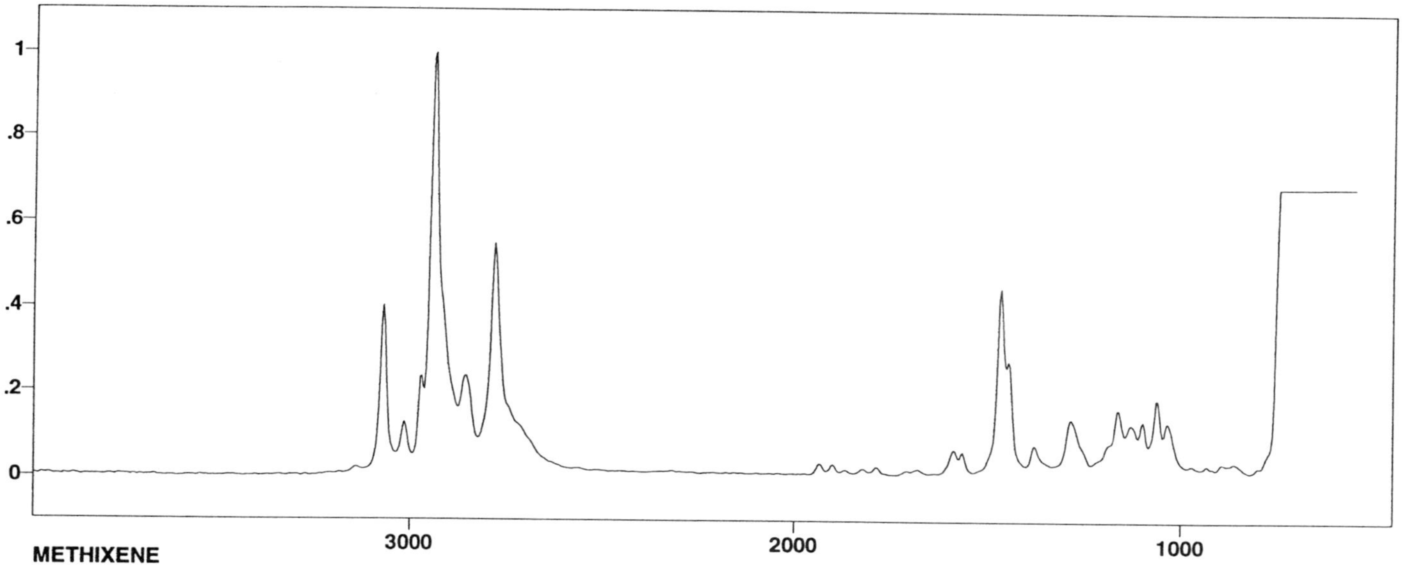

METHIXENE

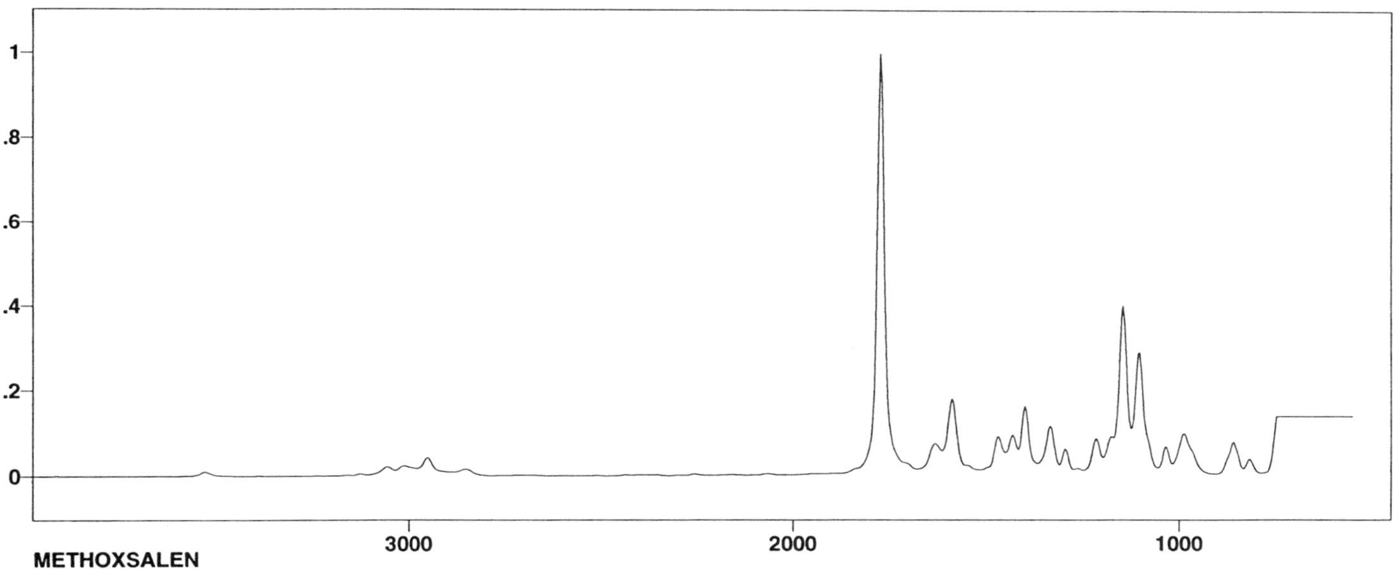

METHOXSALEN

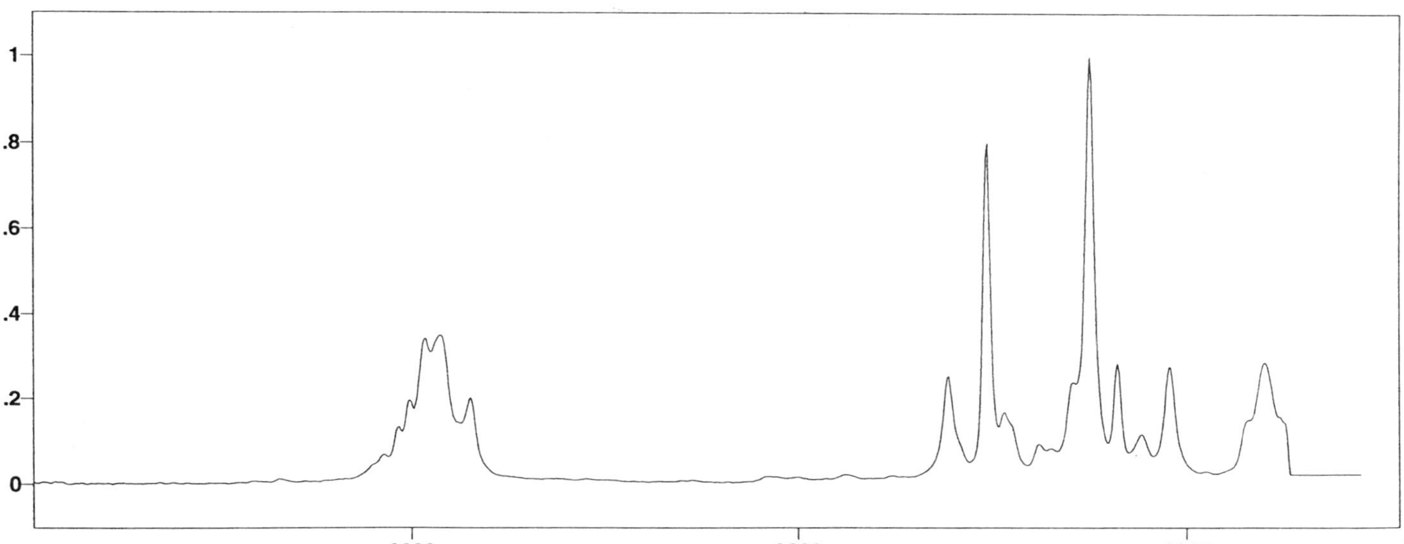

4-METHOXYAMPHETAMINE

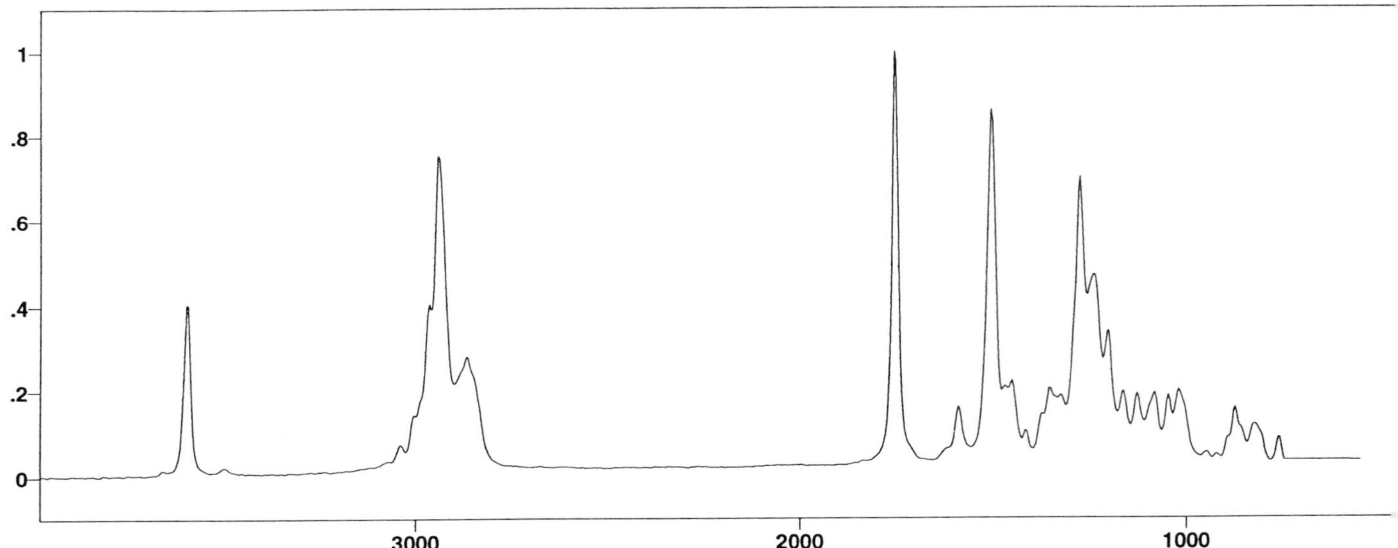

METHOXYESTRONE

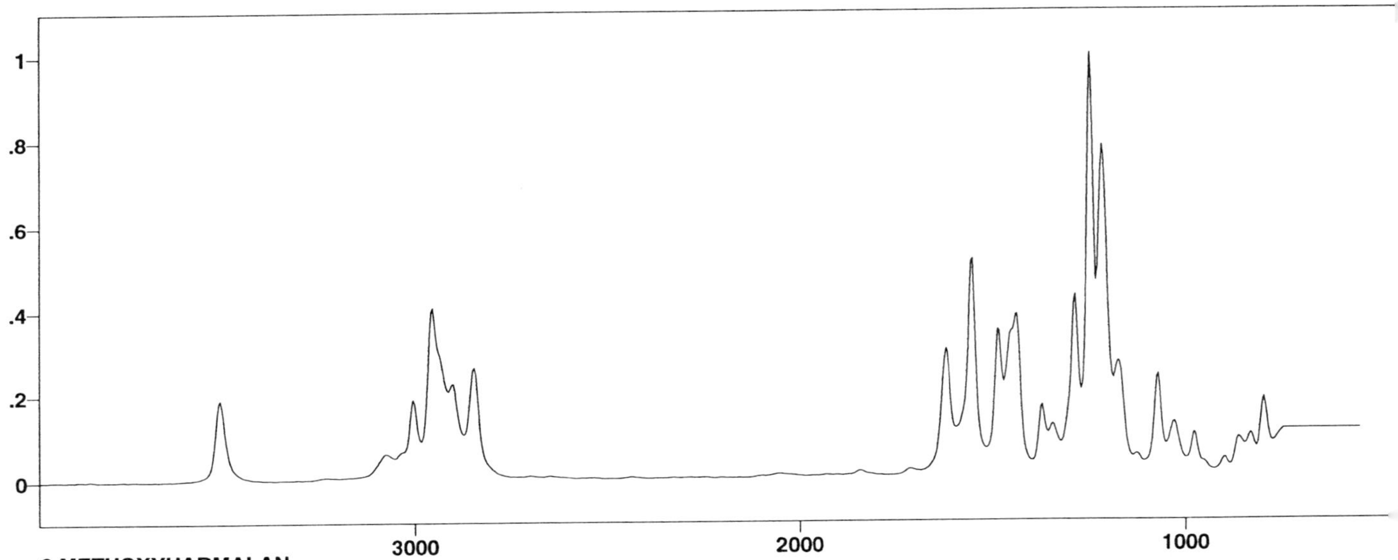

6-METHOXYHARMALAN

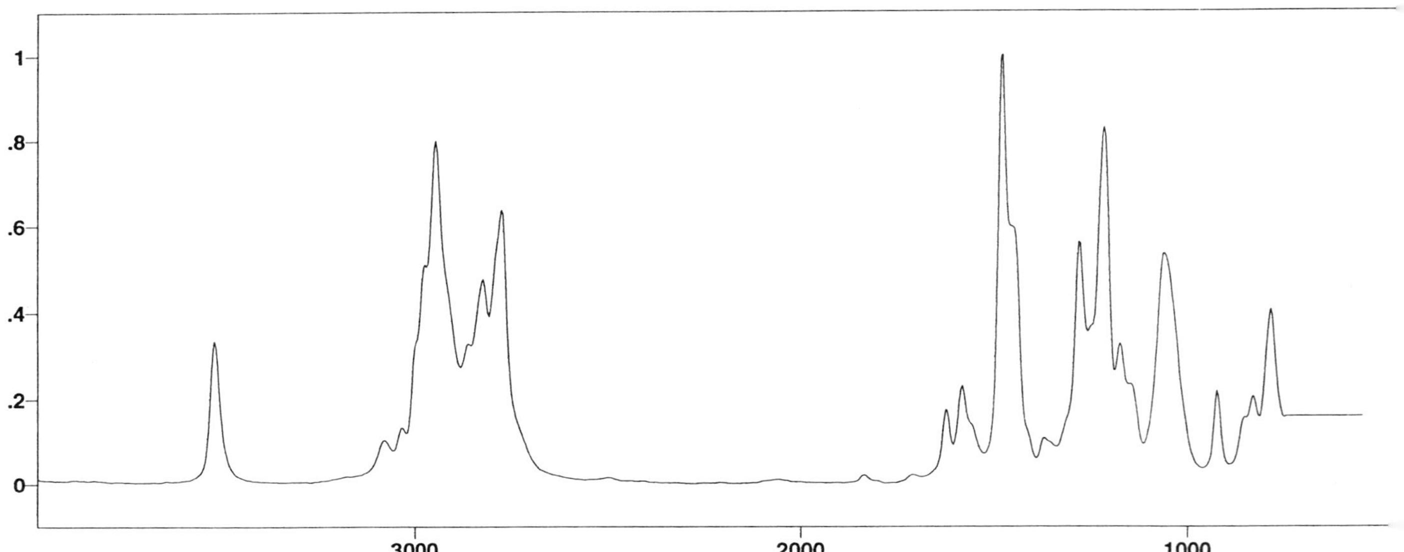

5-METHOXY-N,N-DIMETHYLTRYPTAMINE

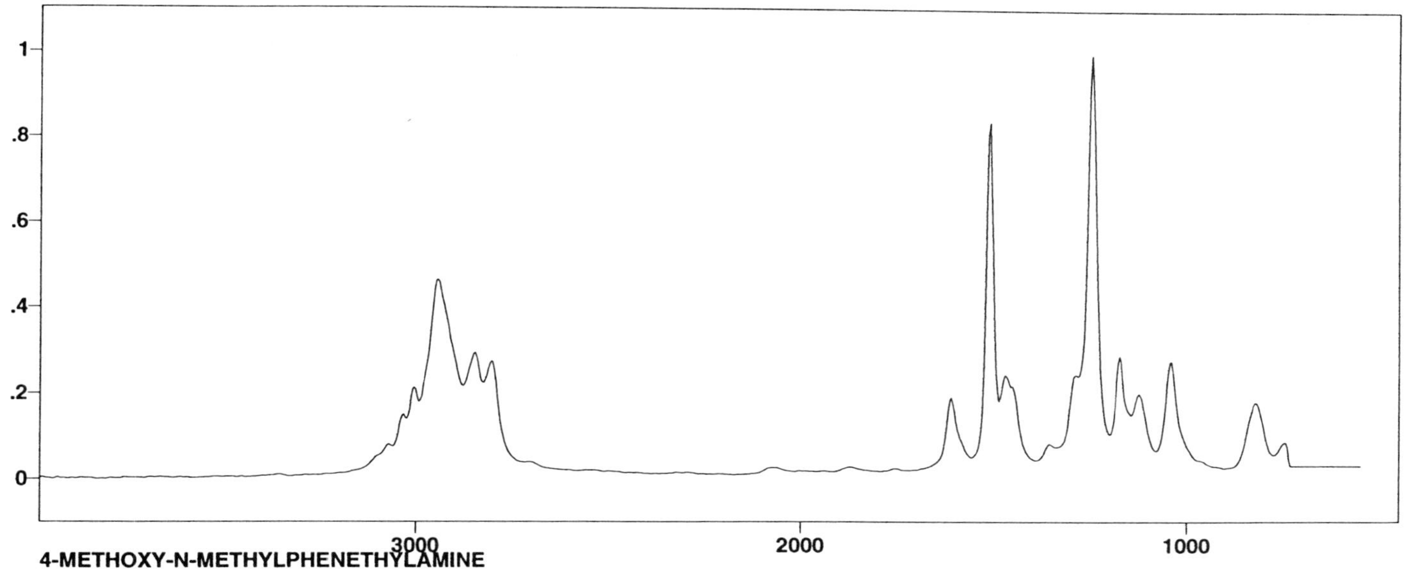

4-METHOXY-N-METHYLPHENETHYLAMINE

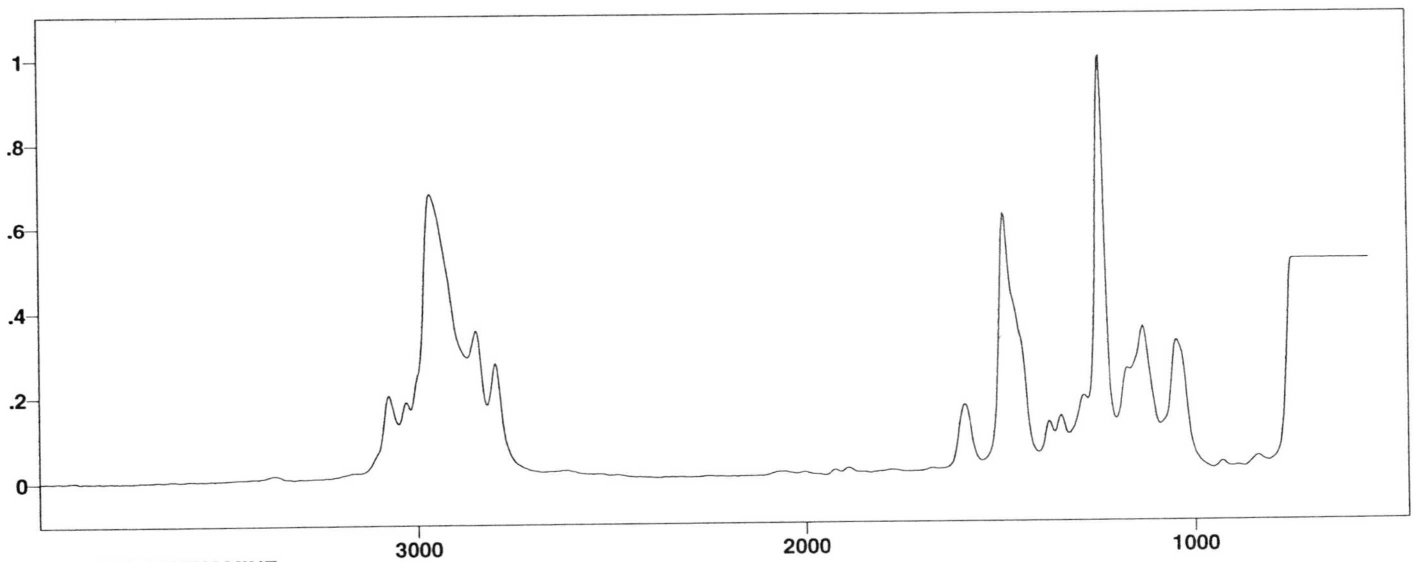

METHOXYPHENAMINE

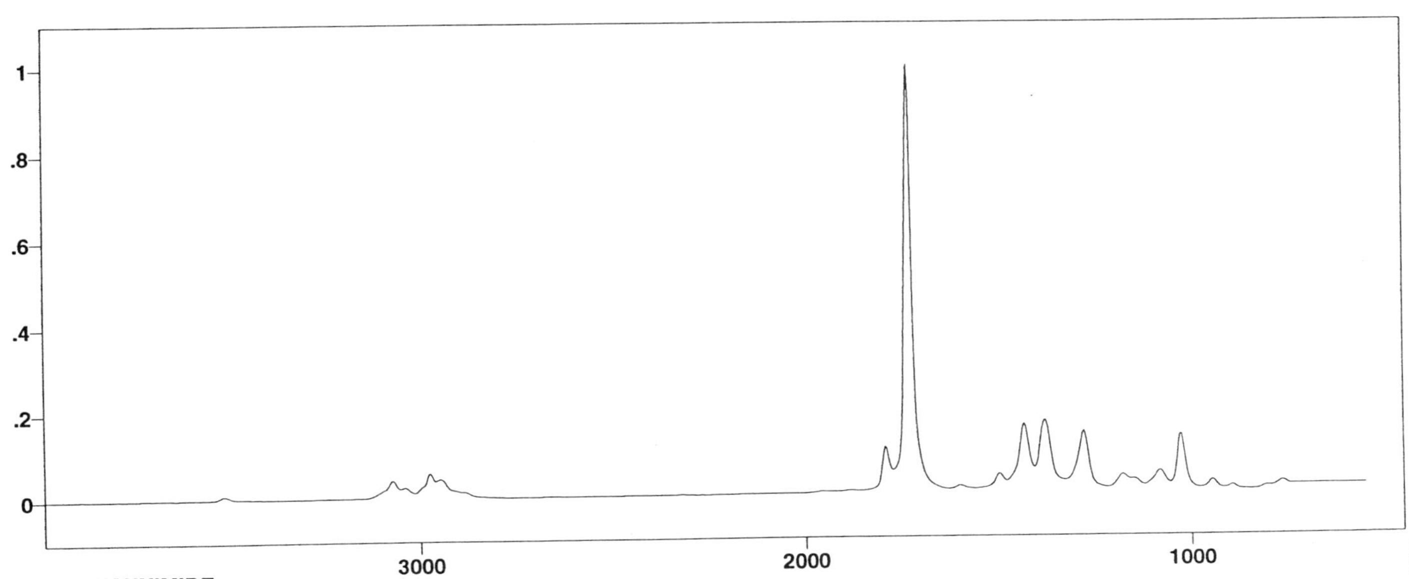

METHSUXIMIDE

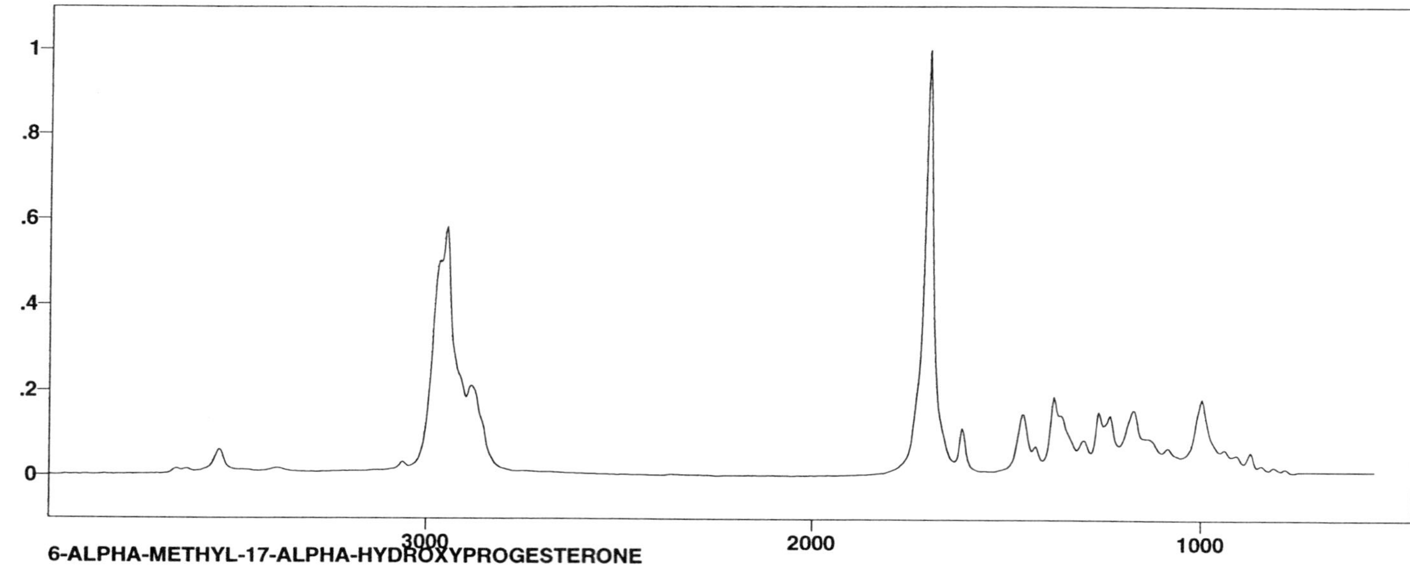

6-ALPHA-METHYL-17-ALPHA-HYDROXYPROGESTERONE

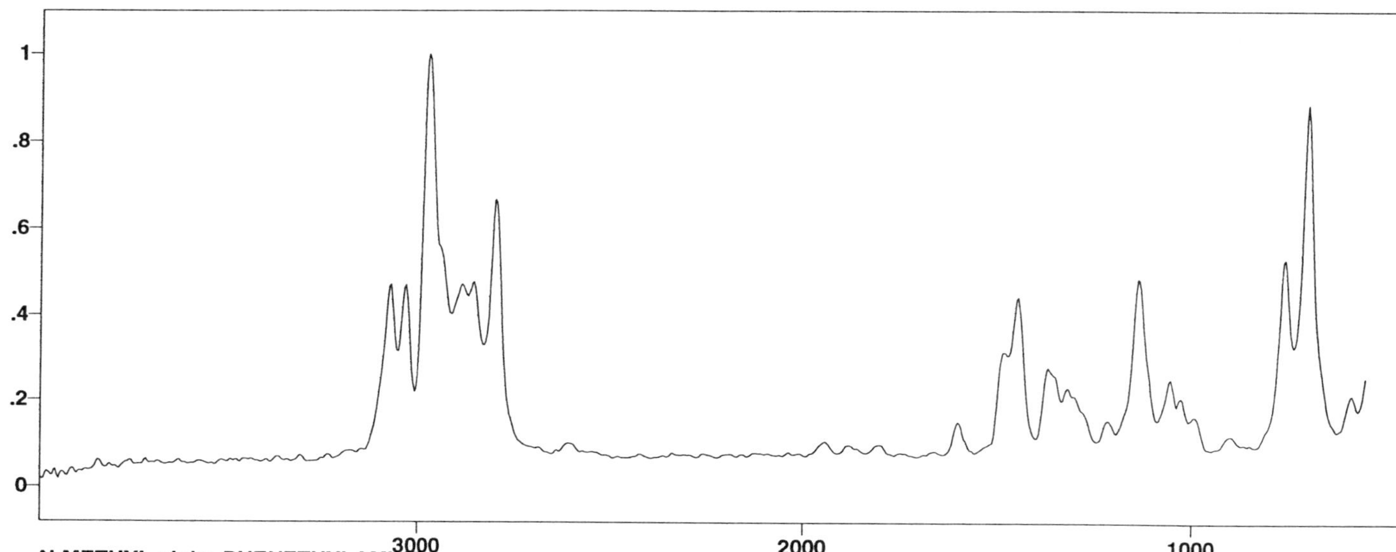

N-METHYL-alpha-PHENETHYLAMINE

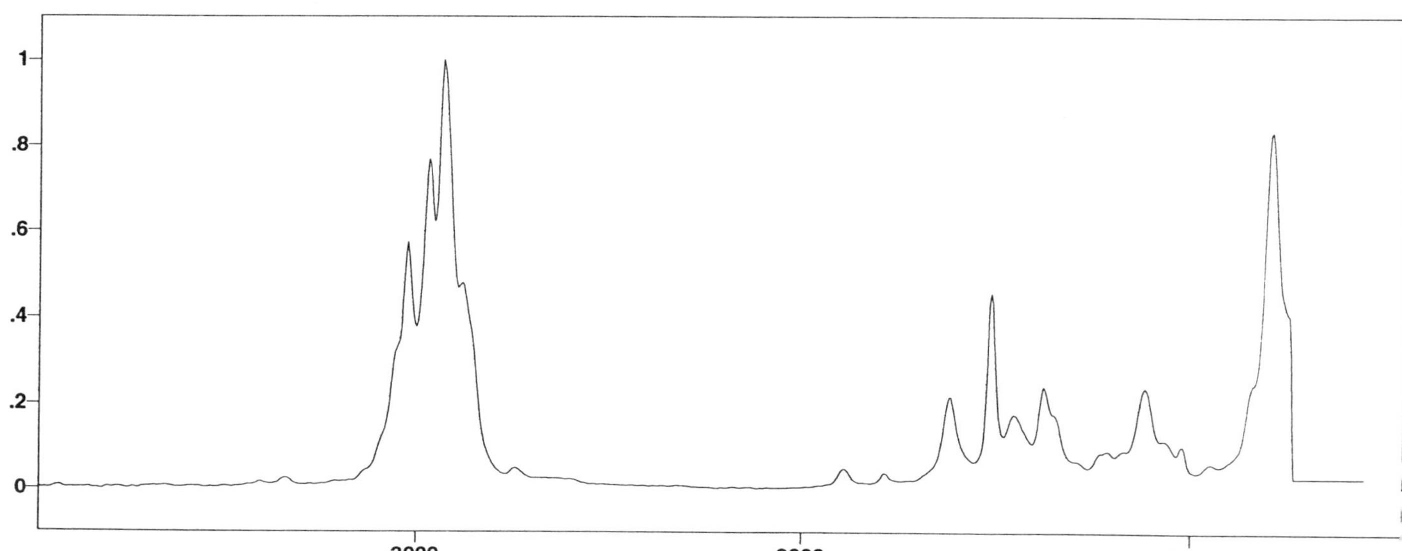

4-METHYLAMPHETAMINE

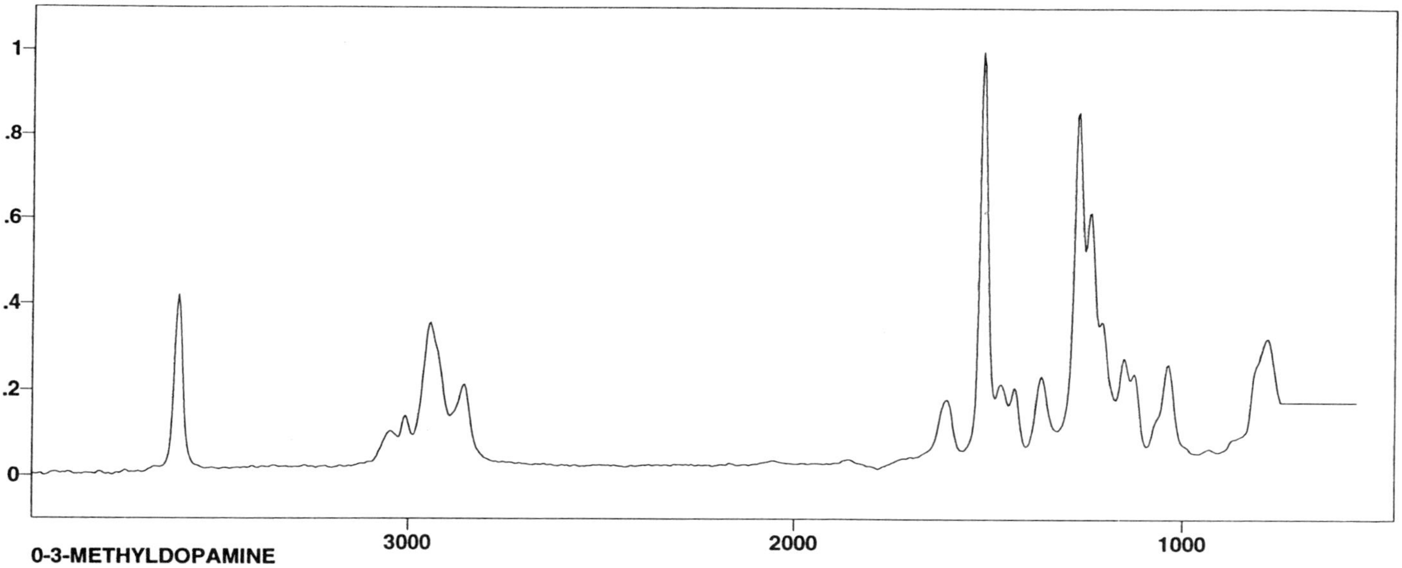

0-3-METHYLDOPAMINE

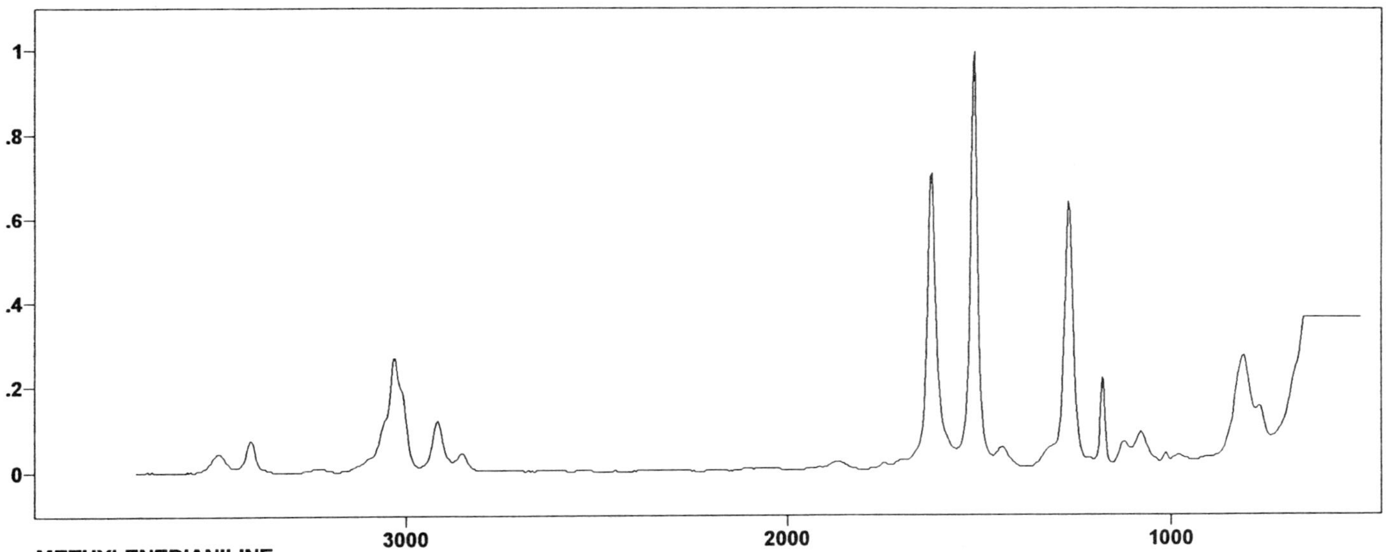

METHYLENEDIANILINE

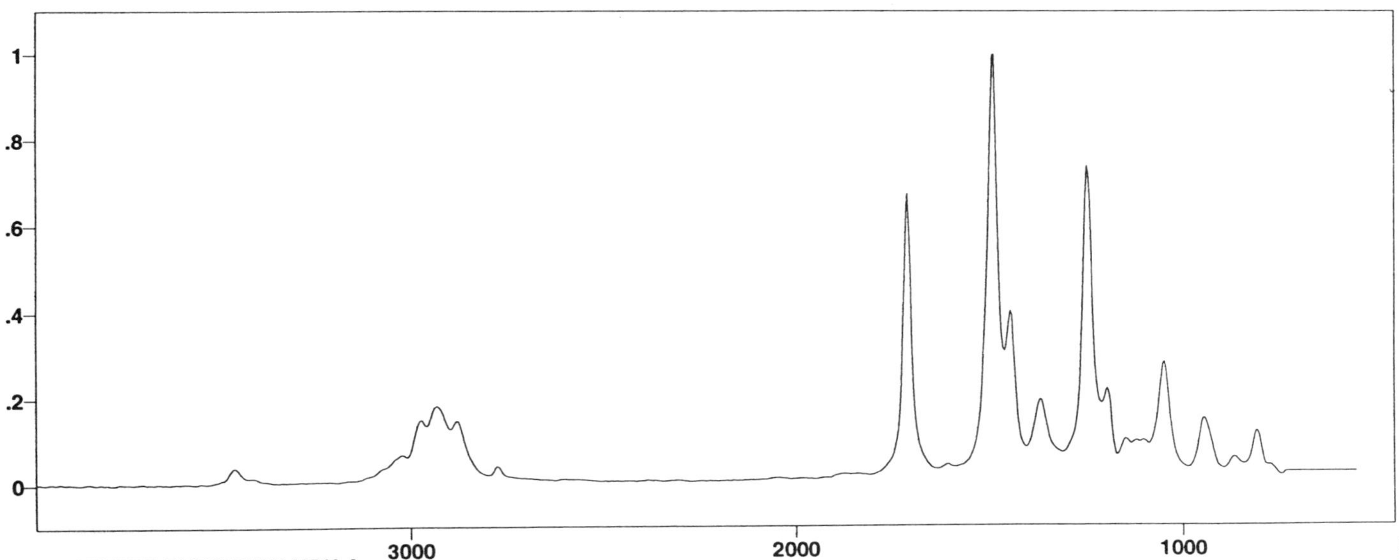

3,4-METHYLENEDIOXYAMPH-Ac

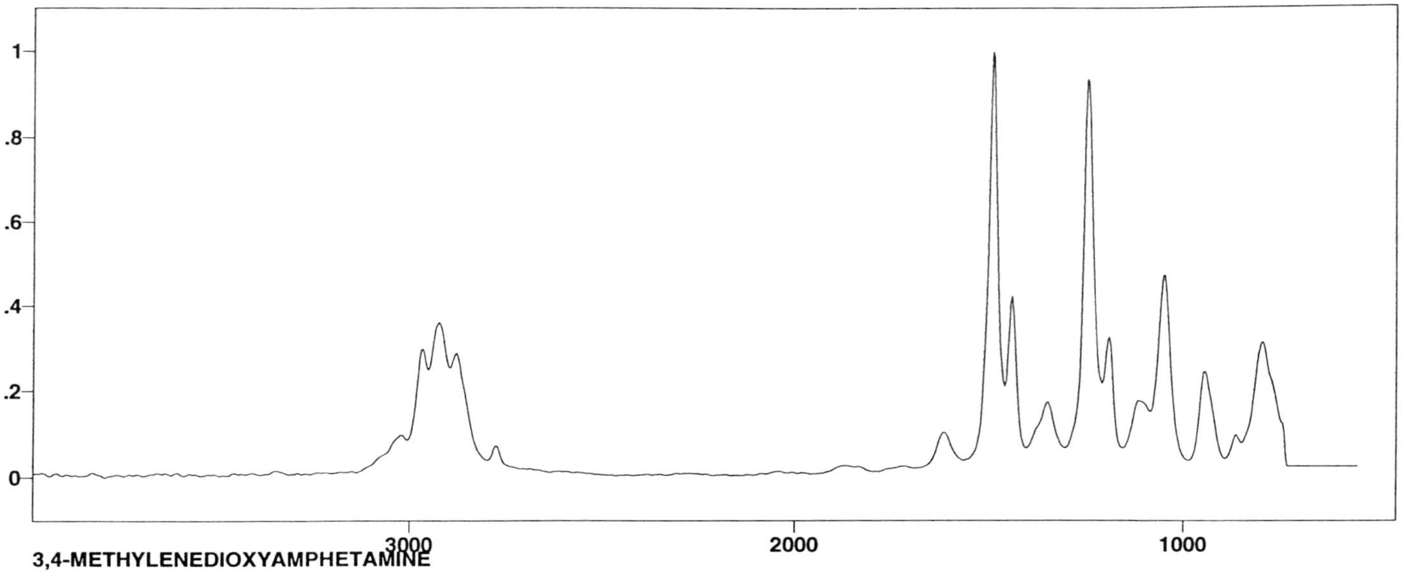

3,4-METHYLENEDIOXYAMPHETAMINE

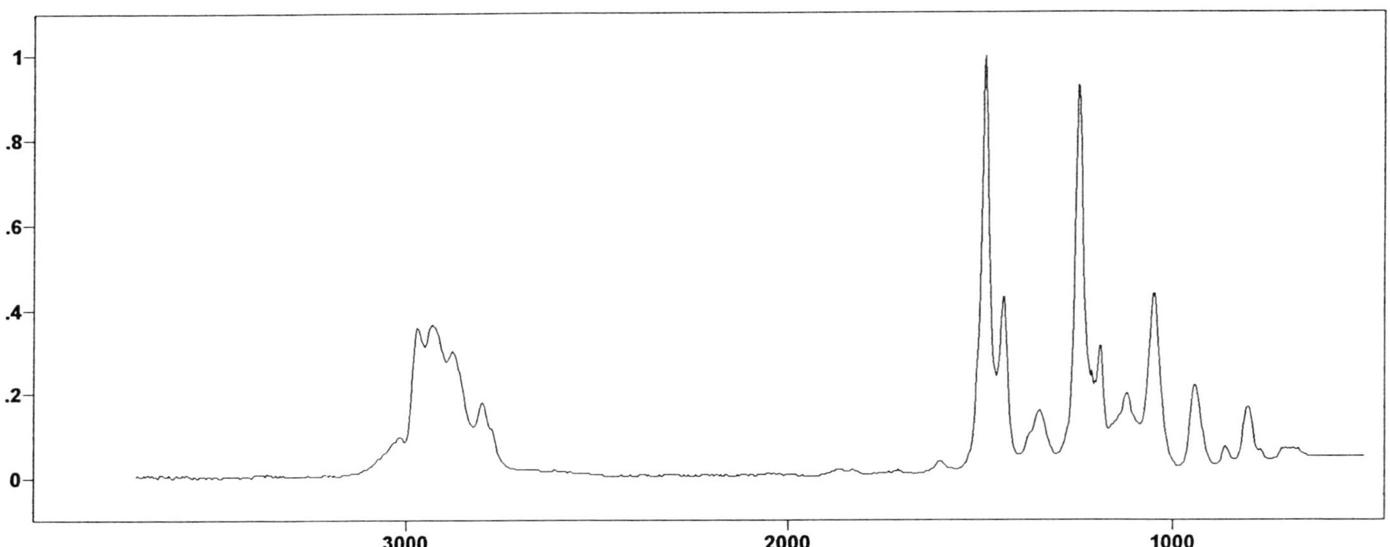

3,4-METHYLENEDIOXYMETHAMPHETAMINE

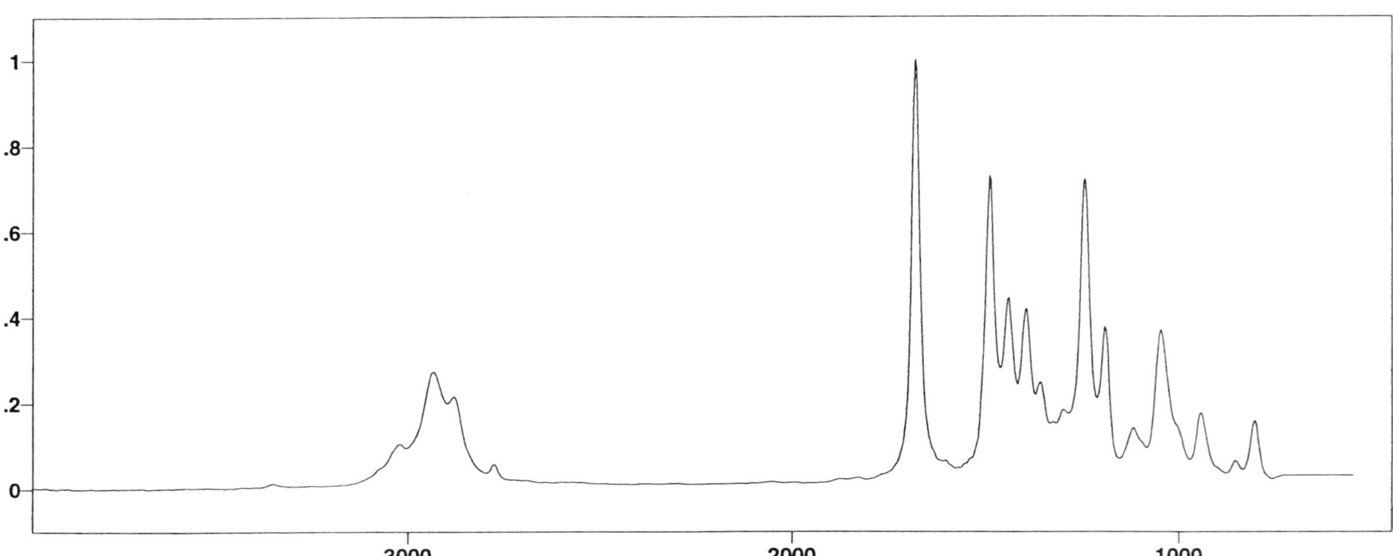

3,4-METHYLENEDIOXY-N-MePHENETHYLAM-Ac

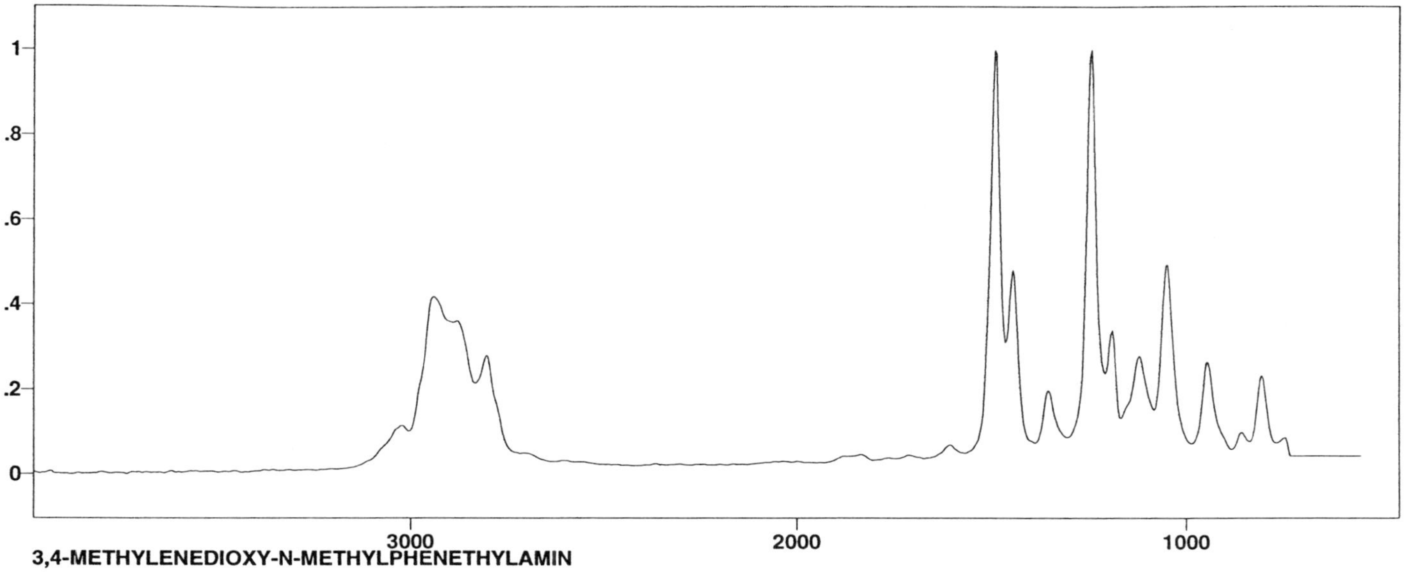

3,4-METHYLENEDIOXY-N-METHYLPHENETHYLAMIN

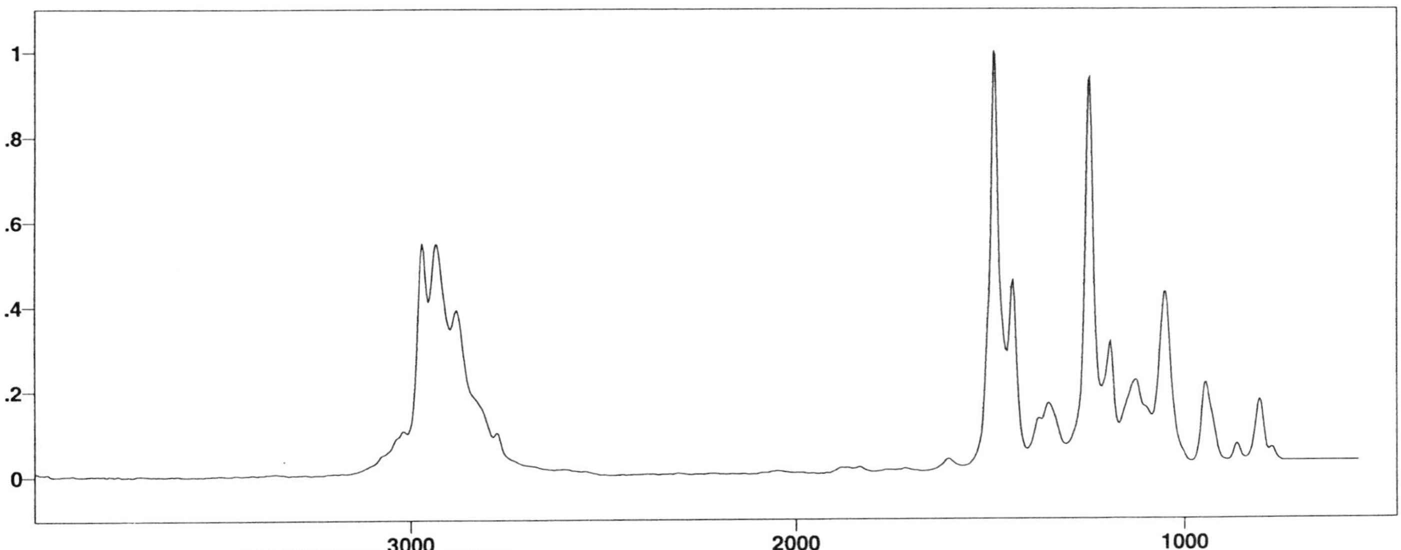

MDPA (3,4-METHYLENEDIOXYPROPYLAMINE)

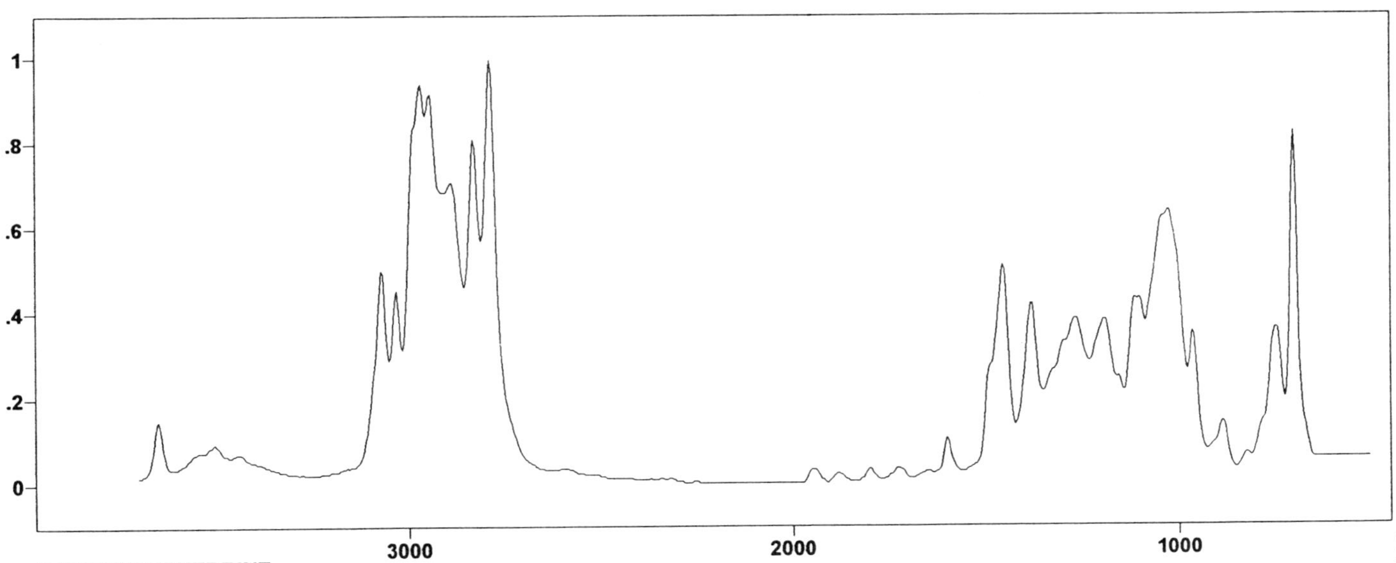

N-METHYLEPHEDRINE

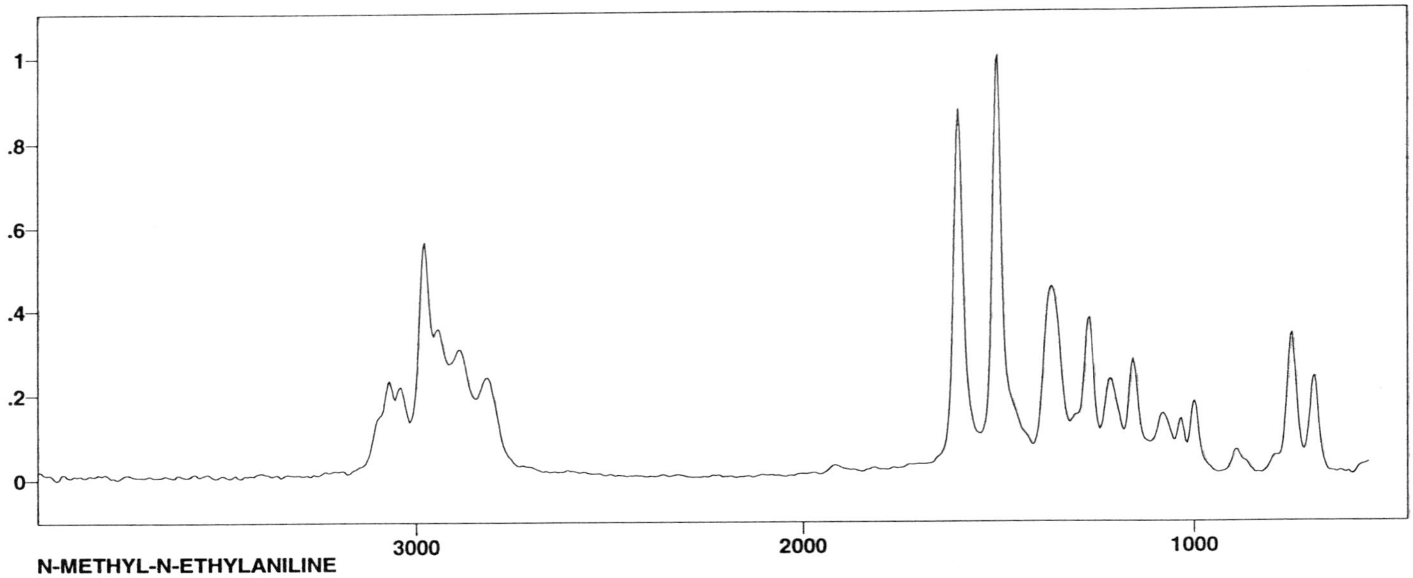

N-METHYL-N-ETHYLANILINE

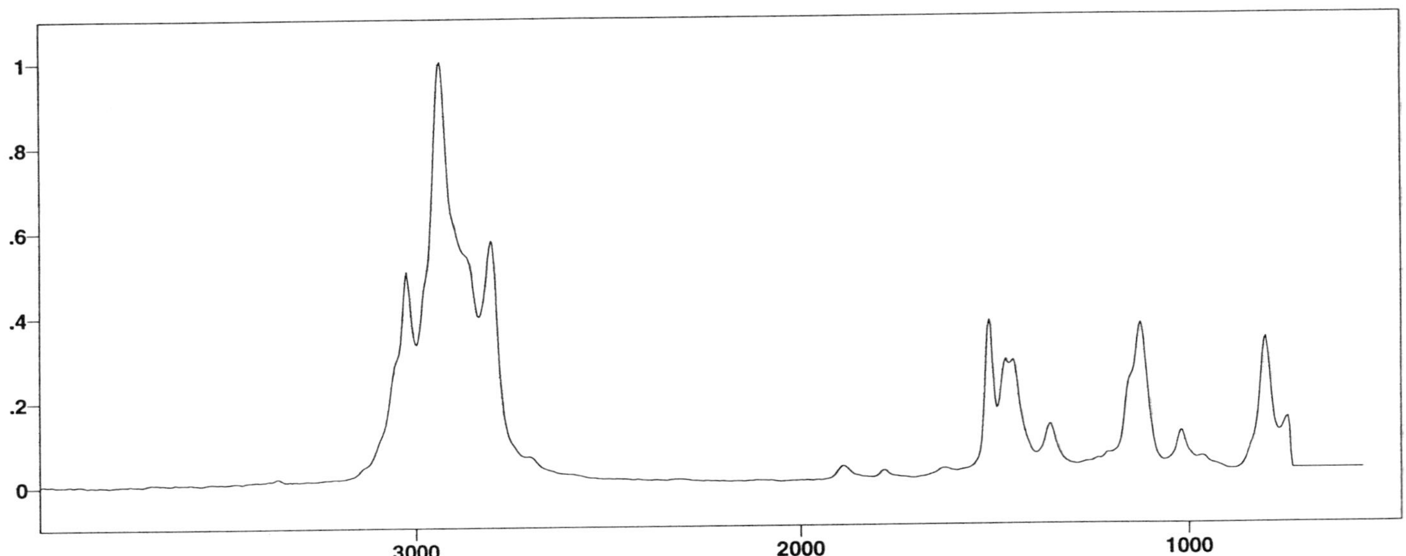

4-METHYL-N-METHYLPHENETHYLAMINE

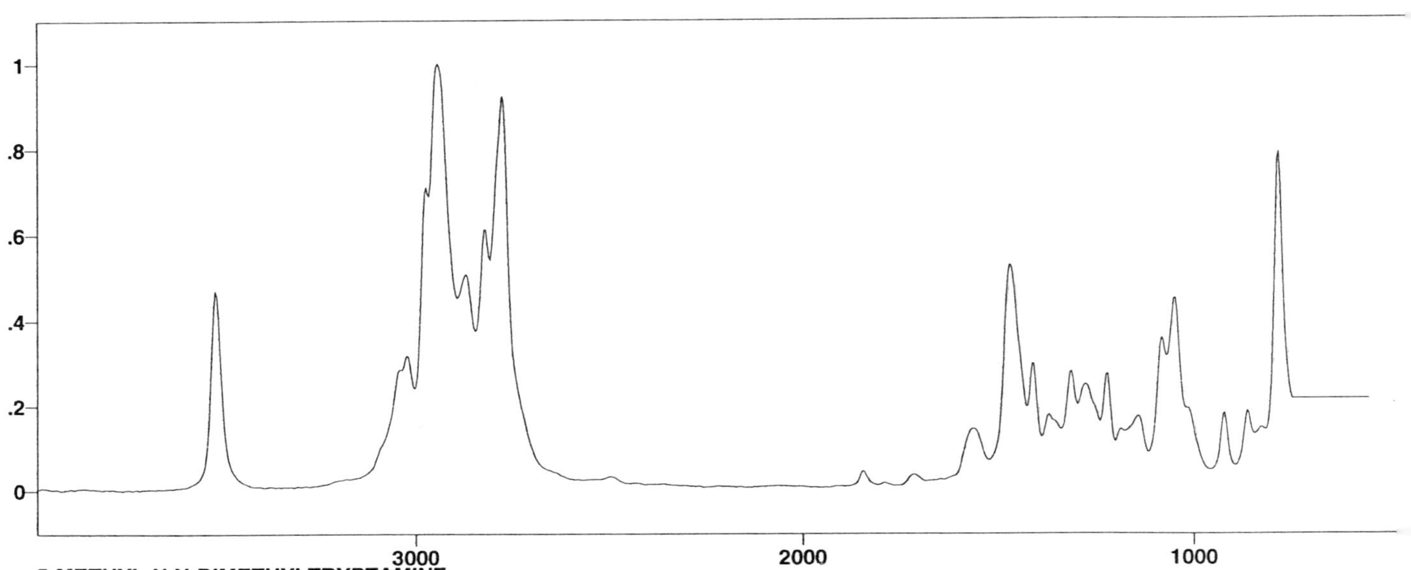

5-METHYL-N,N-DIMETHYLTRYPTAMINE

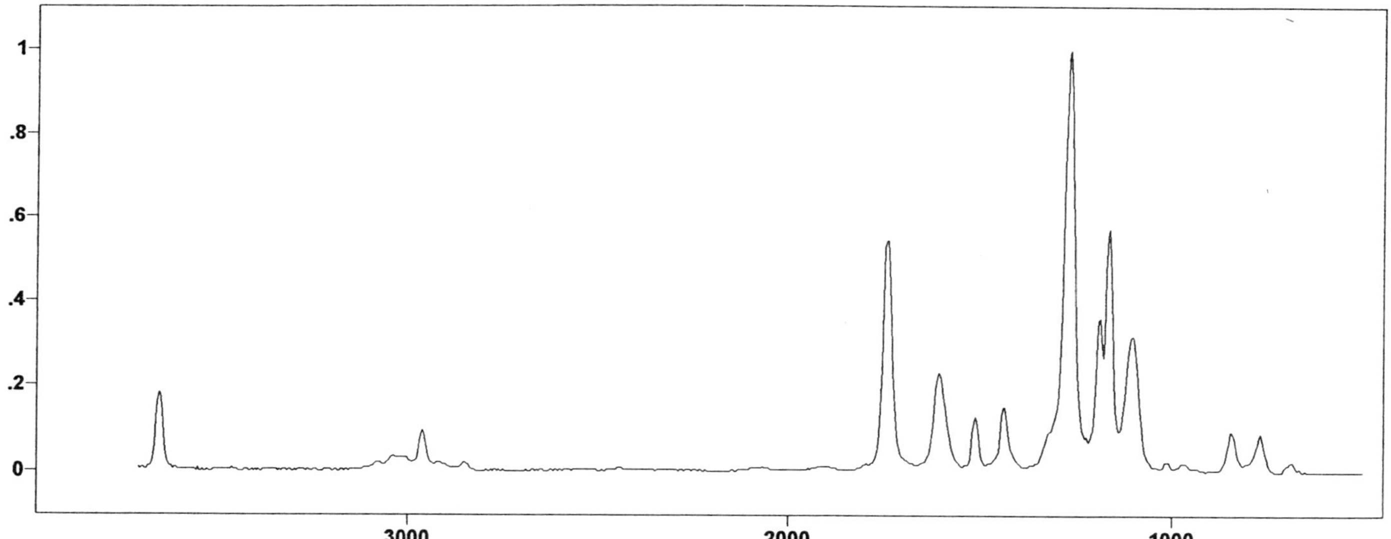

METHYLPARABEN

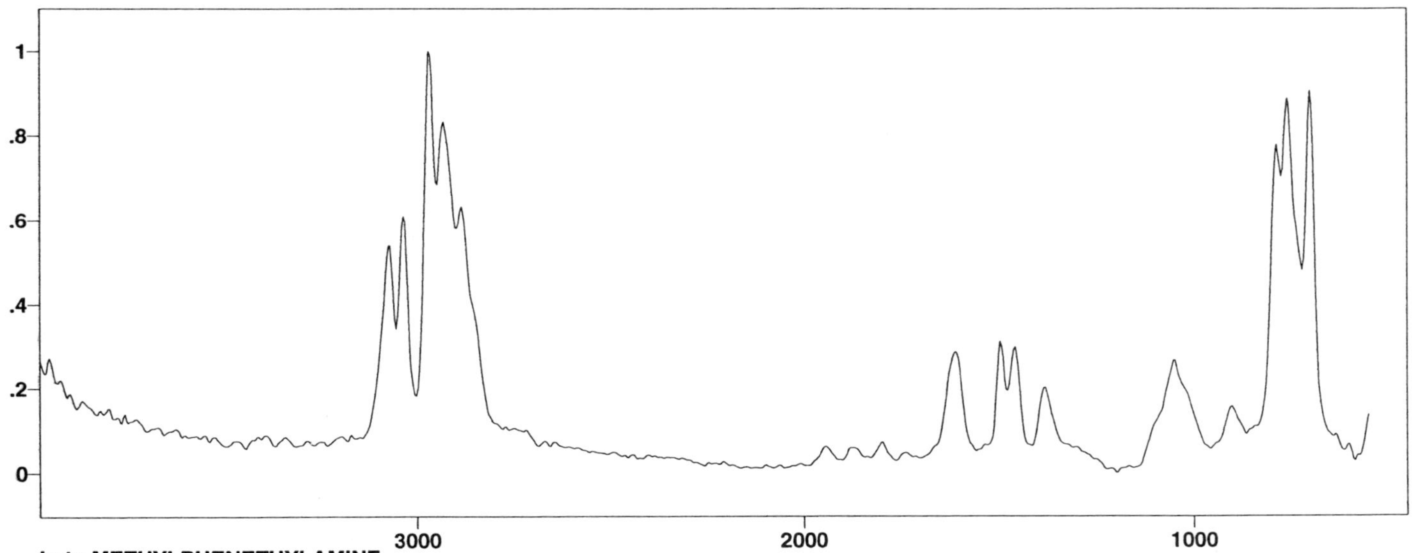

beta-METHYLPHENETHYLAMINE

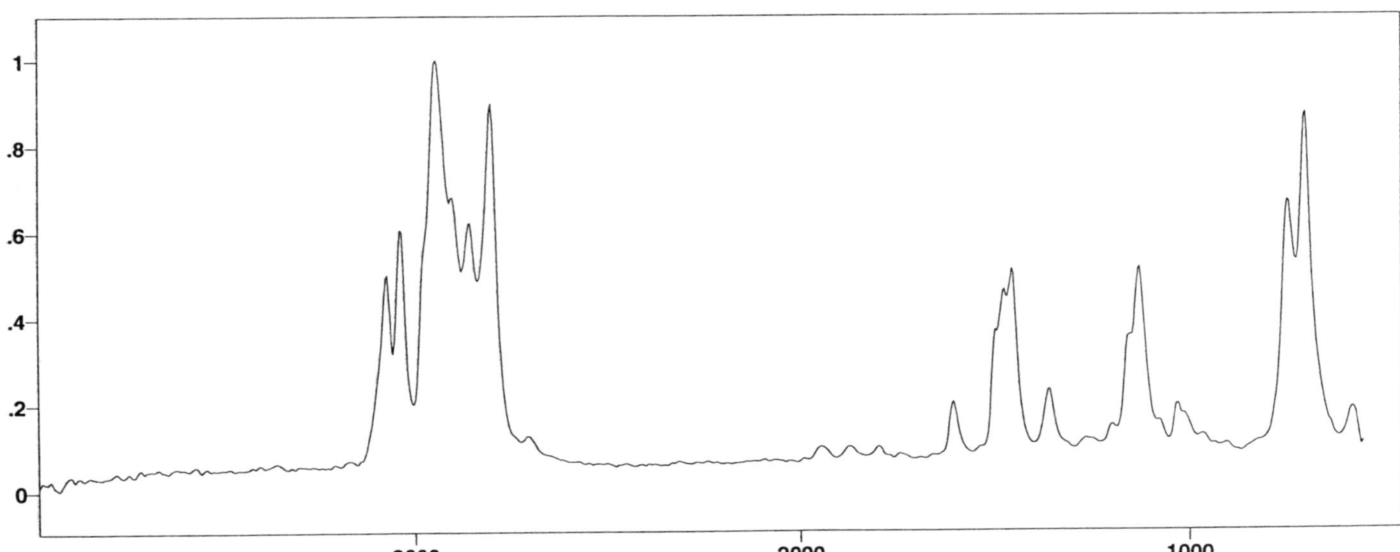

N-METHYLPHENETHYLAMINE

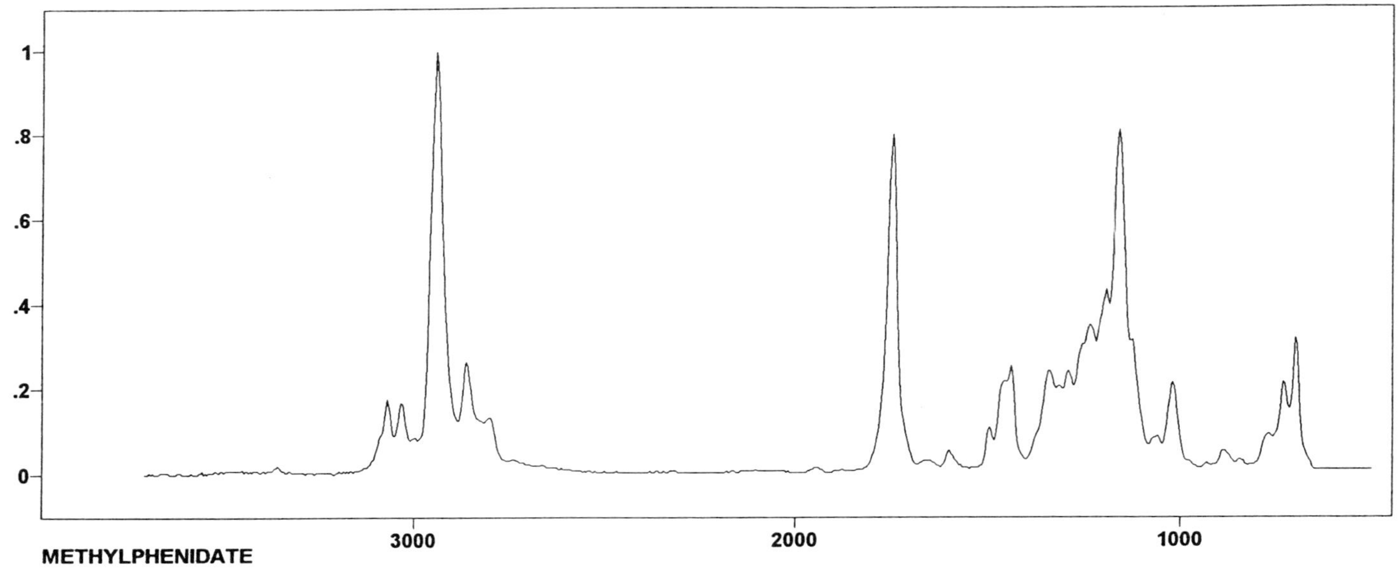

METHYLPHENIDATE

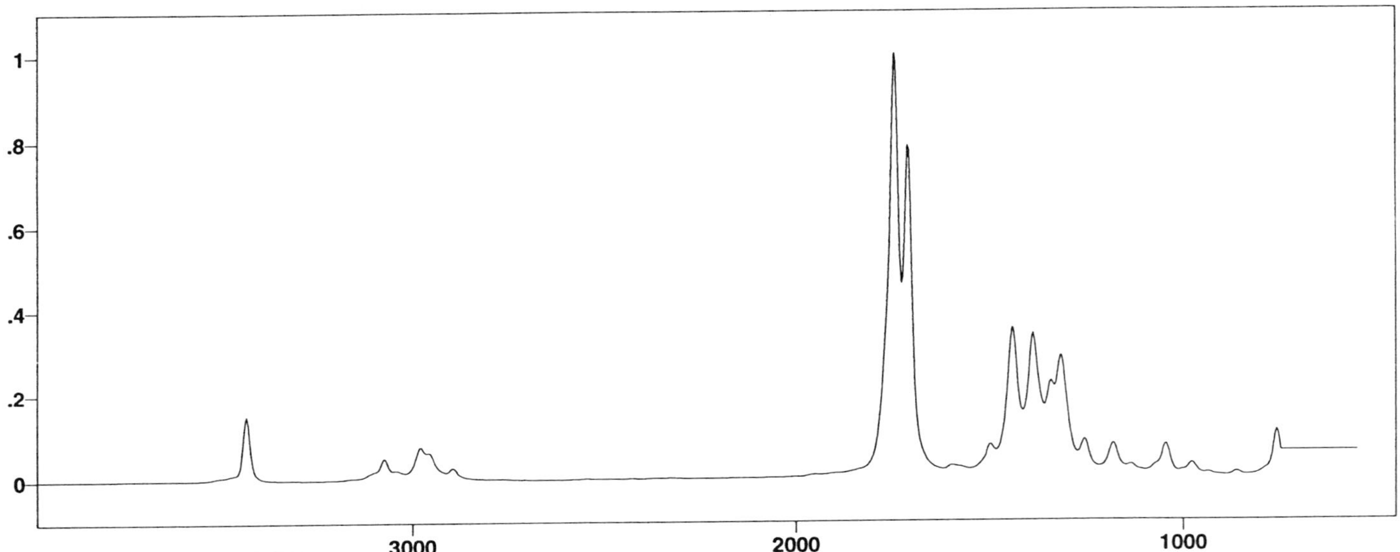

METHYLPHENOBARBITAL

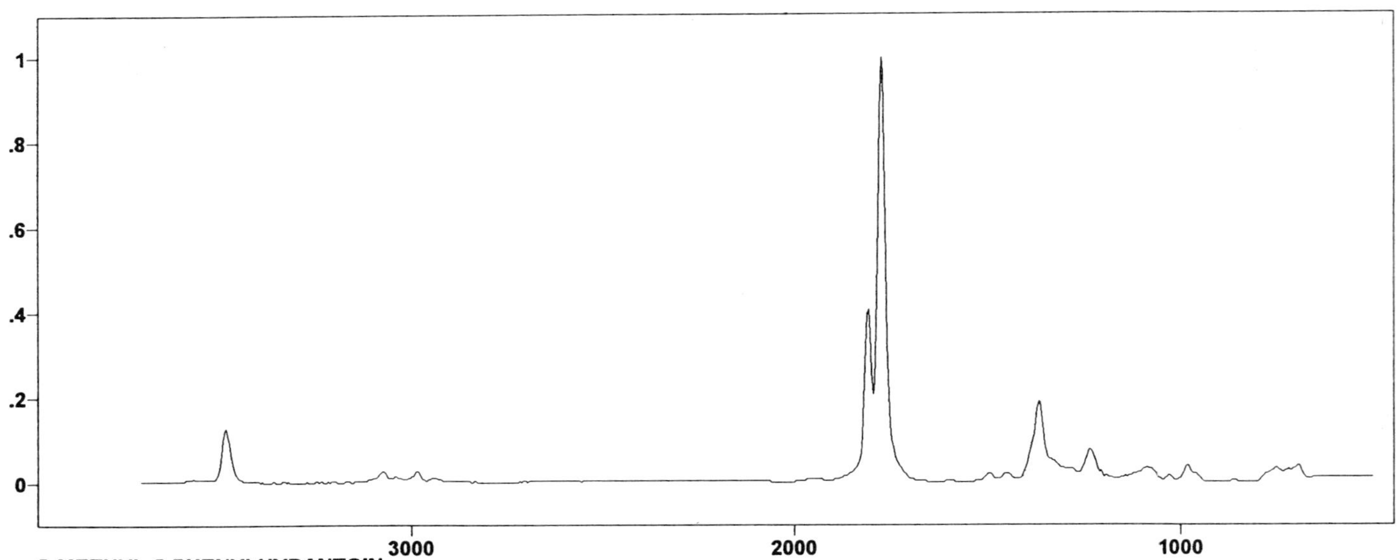

5-METHYL-5-PHENYLHYDANTOIN

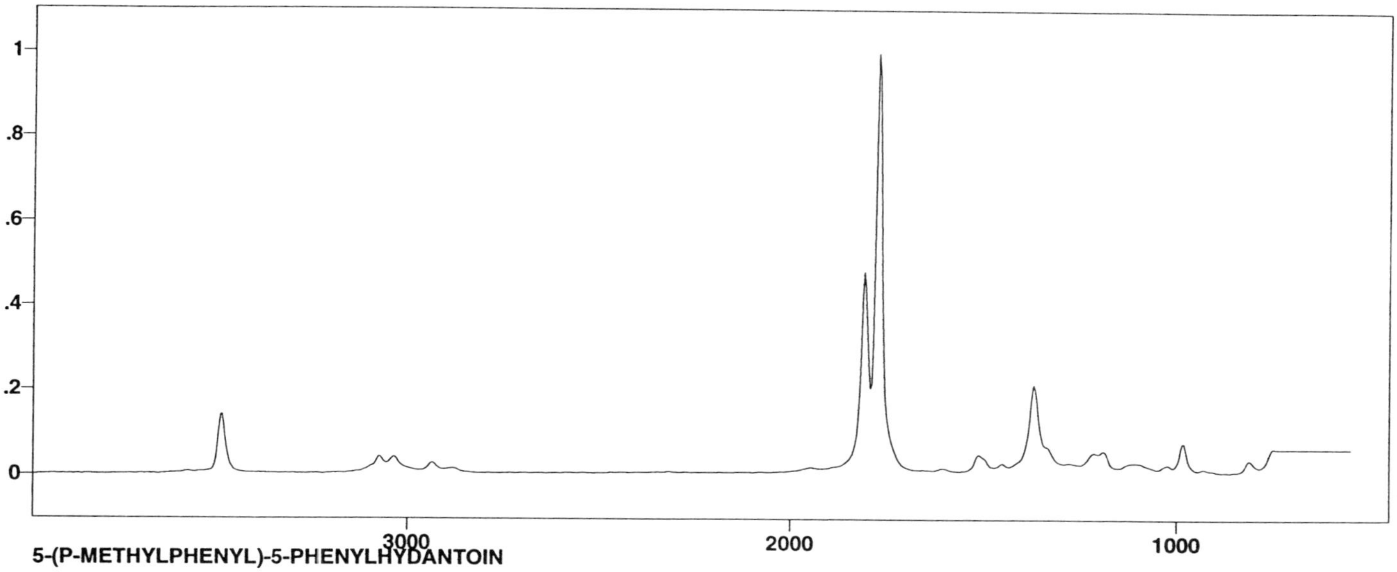

5-(P-METHYLPHENYL)-5-PHENYLHYDANTOIN

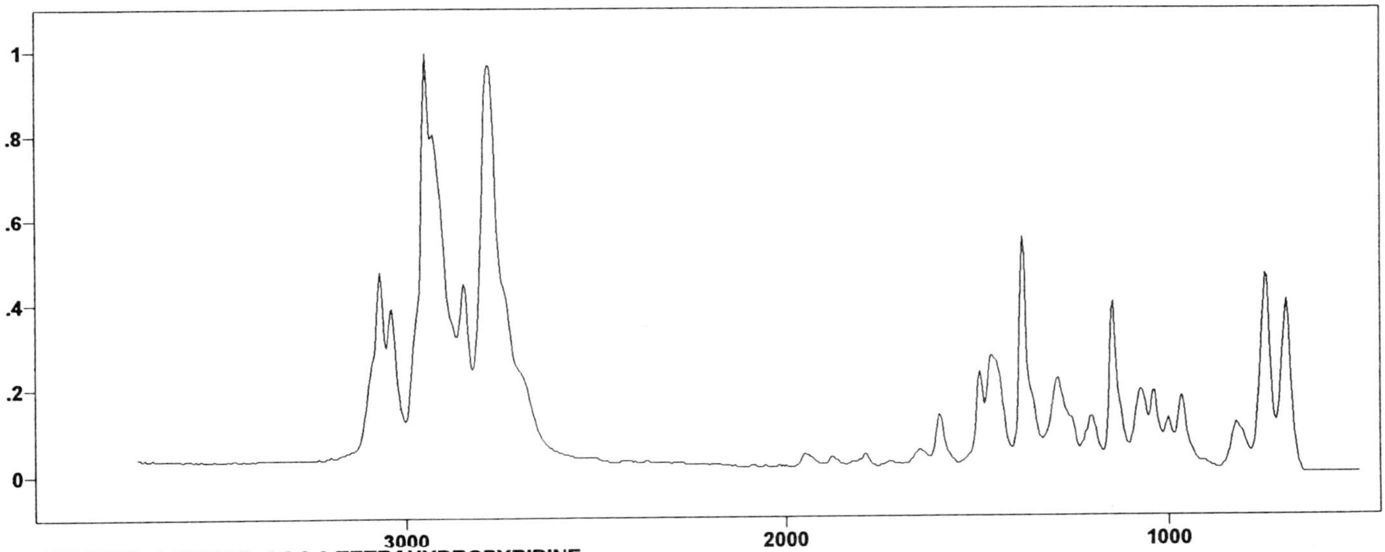

1-METHYL-4-PHENYL-1,2,3,6-TETRAHYDROPYRIDINE

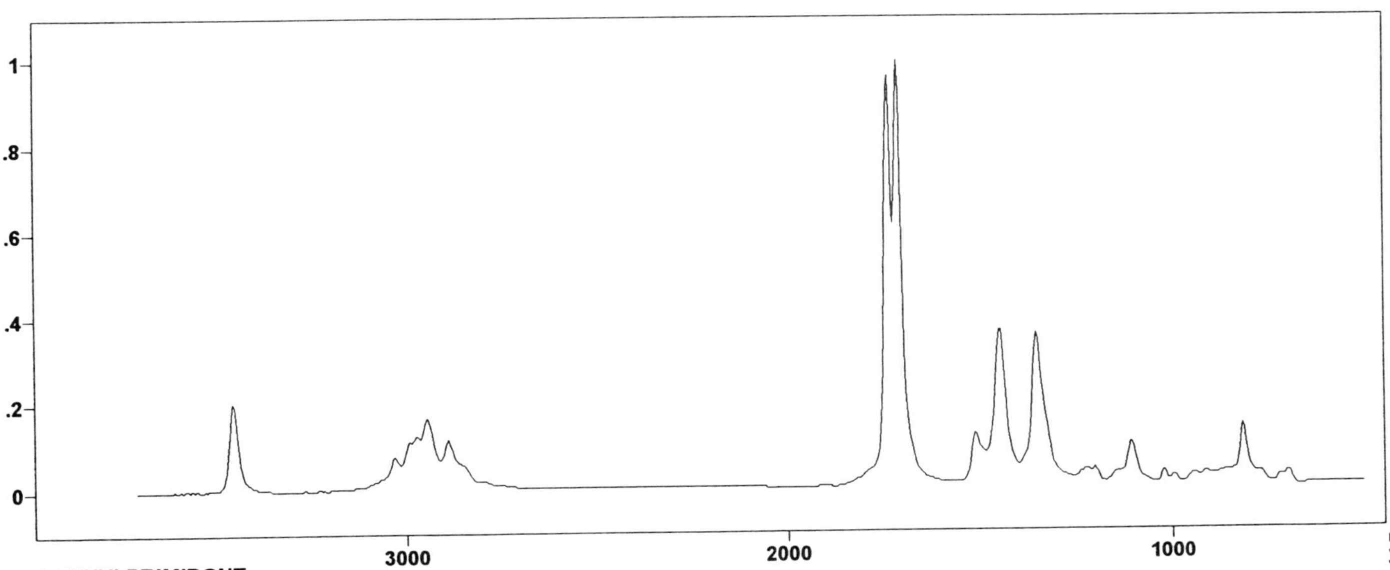

METHYLPRIMIDONE

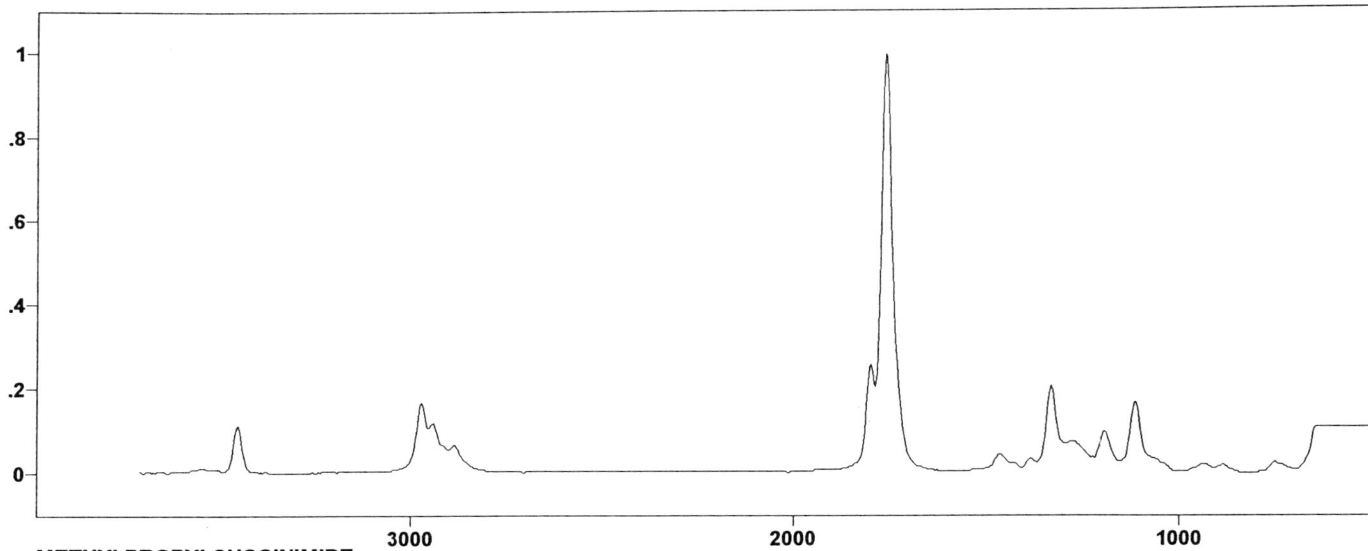

METHYLPROPYLSUCCINIMIDE

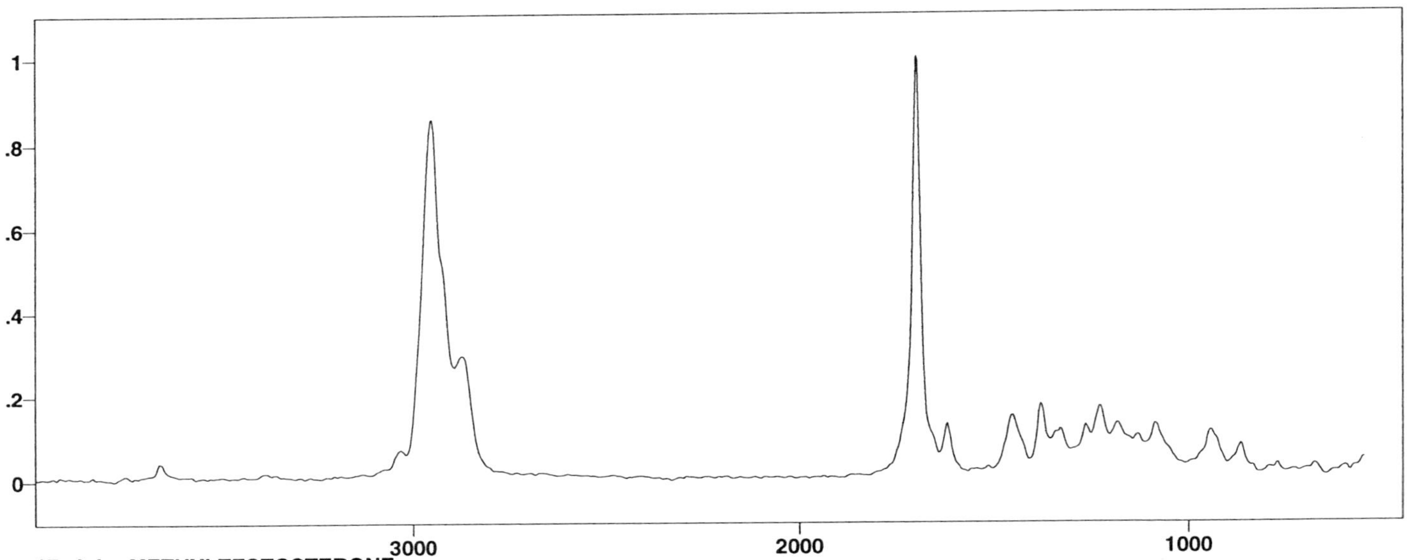

17 alpha-METHYLTESTOSTERONE

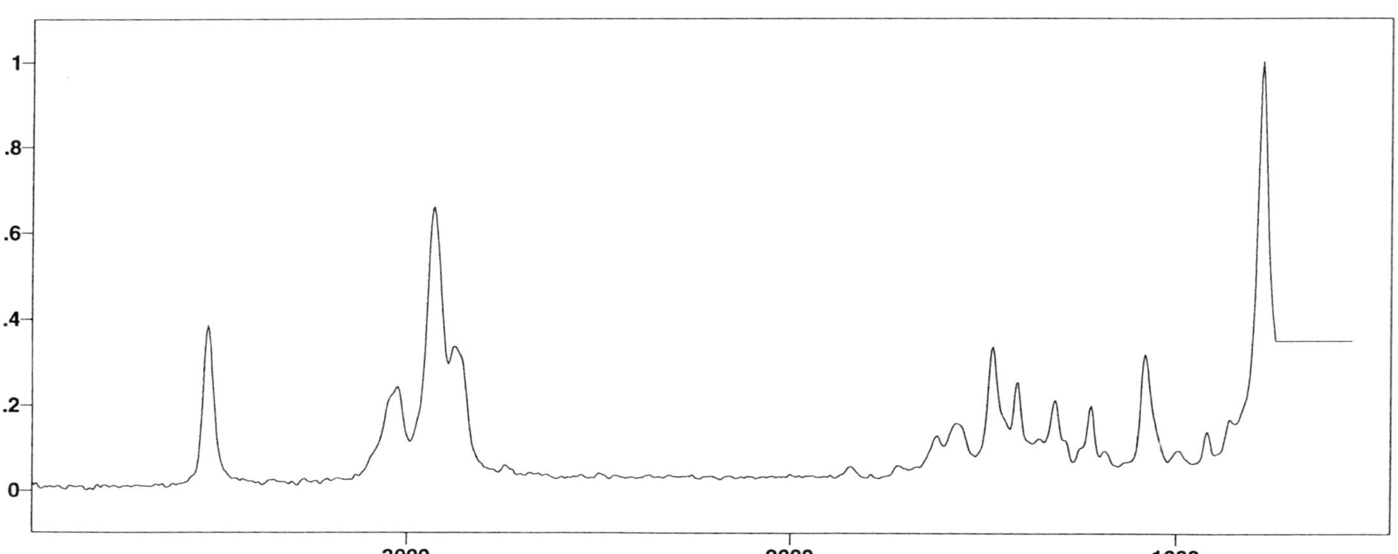

5-METHYLTRYPTAMINE

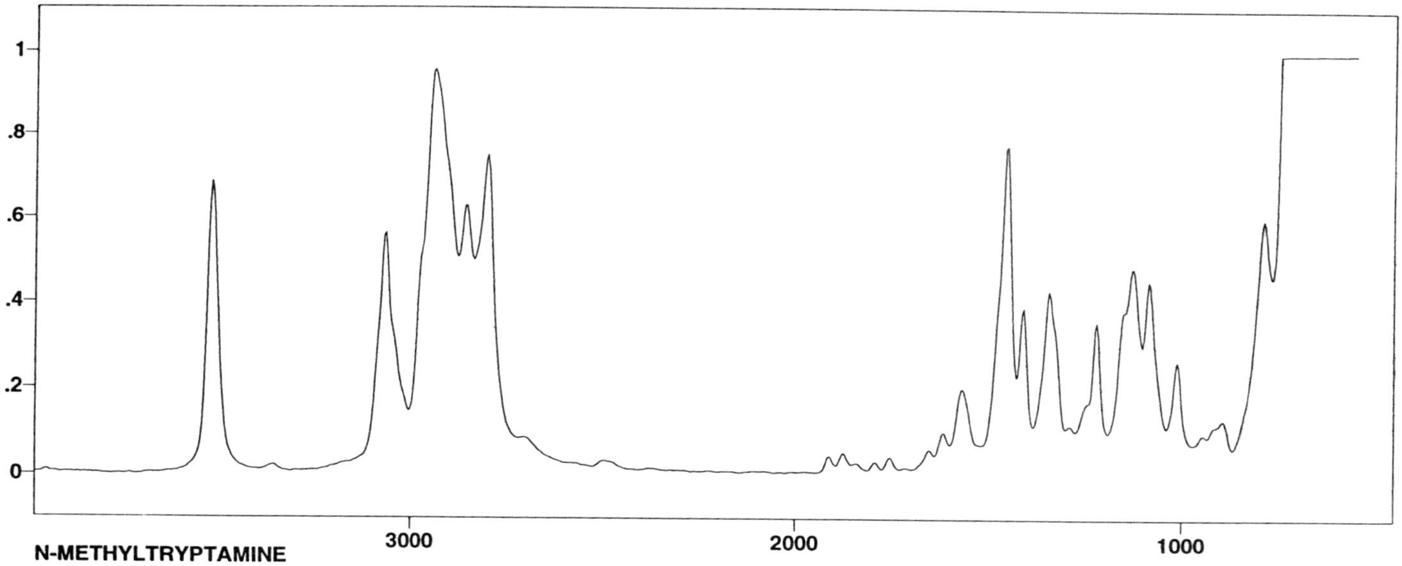

N-METHYLTRYPTAMINE

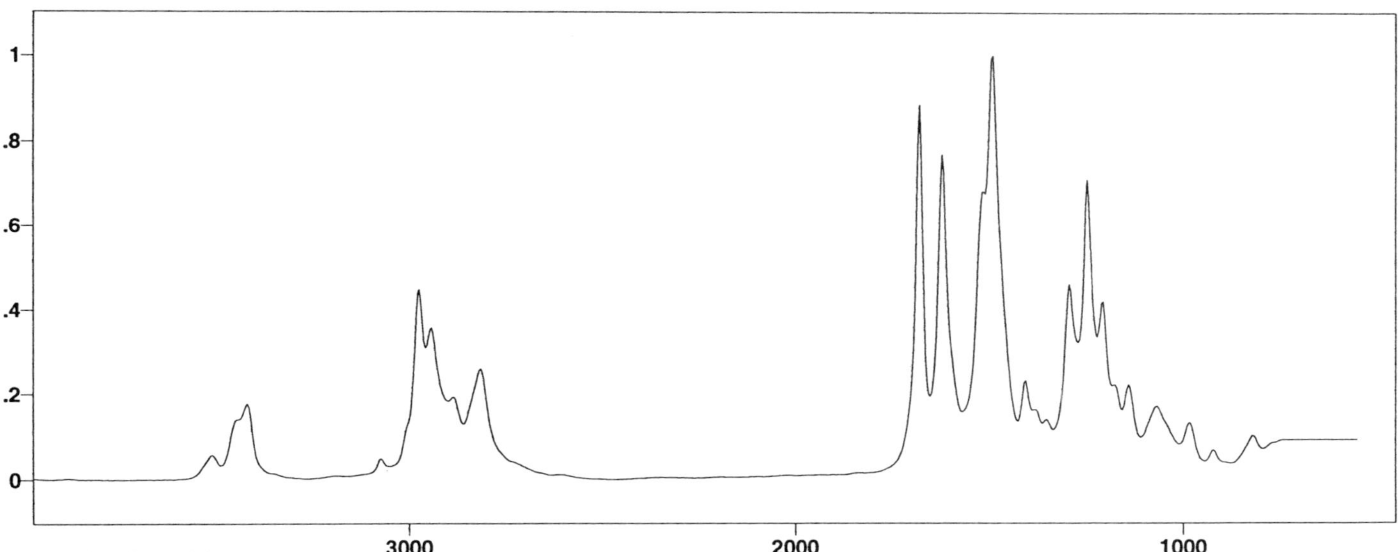

METOCLOPRAMIDE

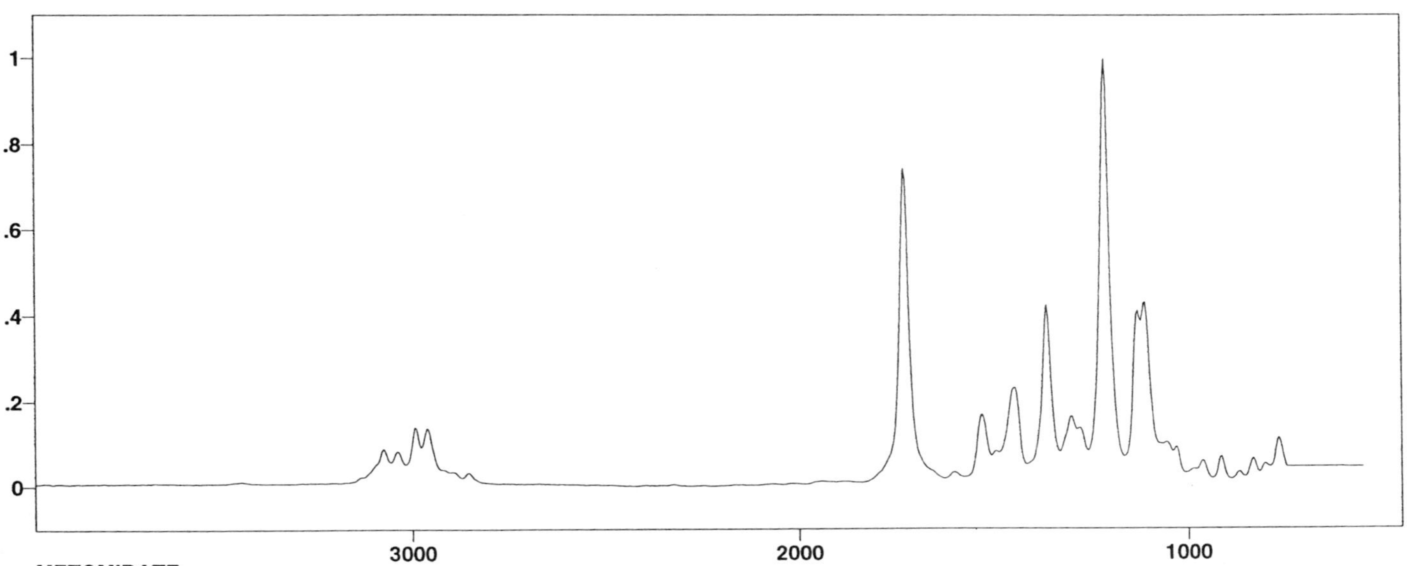

METOMIDATE

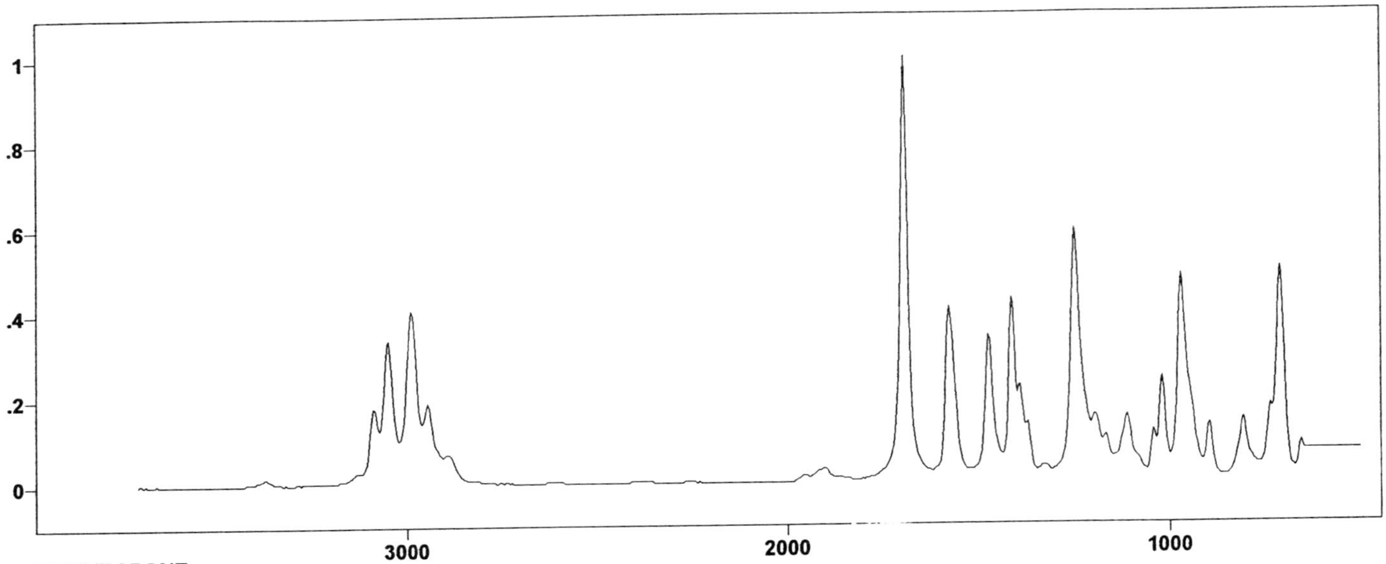

METYRAPONE

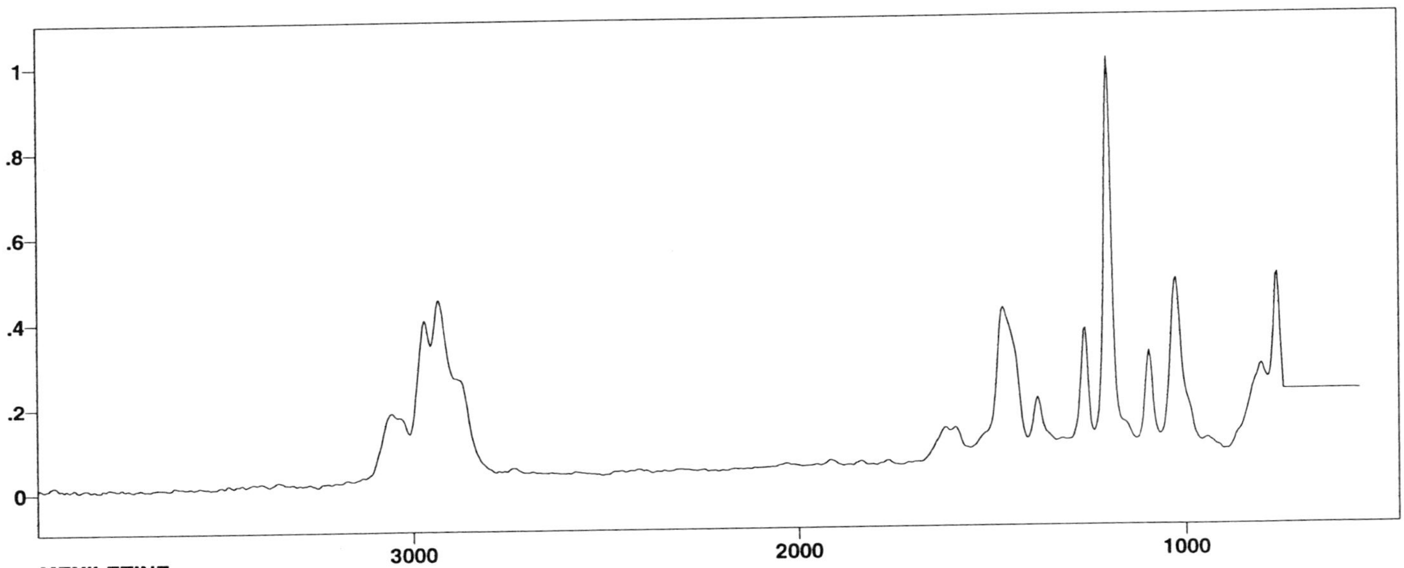

MEXILETINE

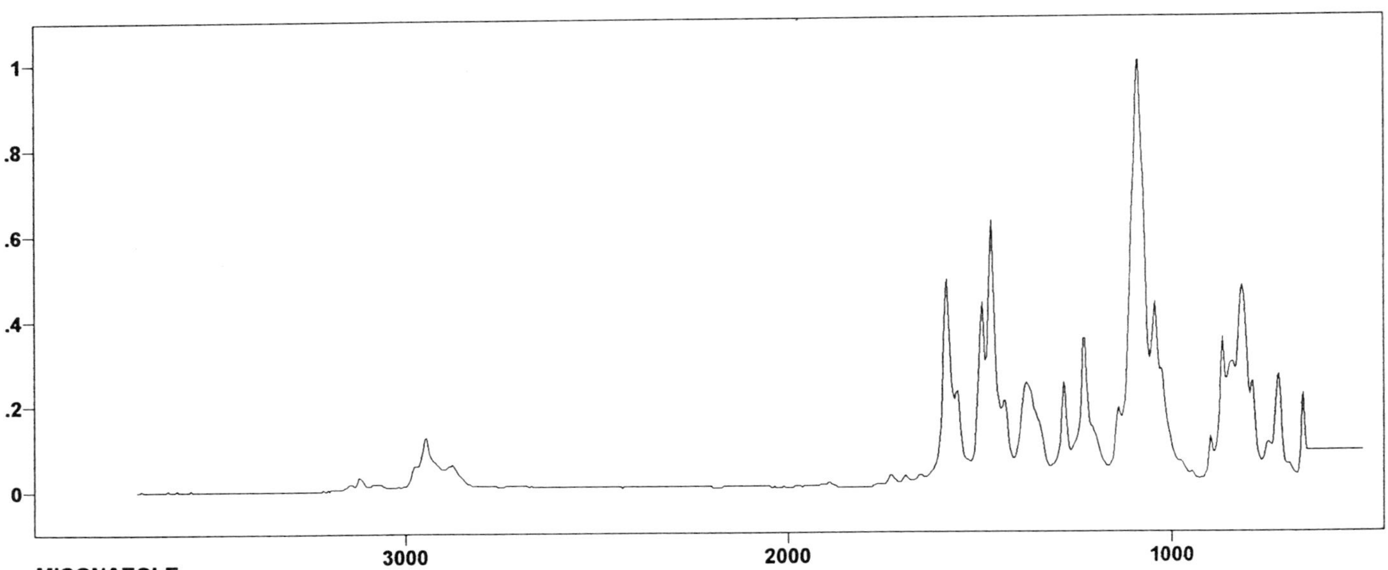

MICONAZOLE

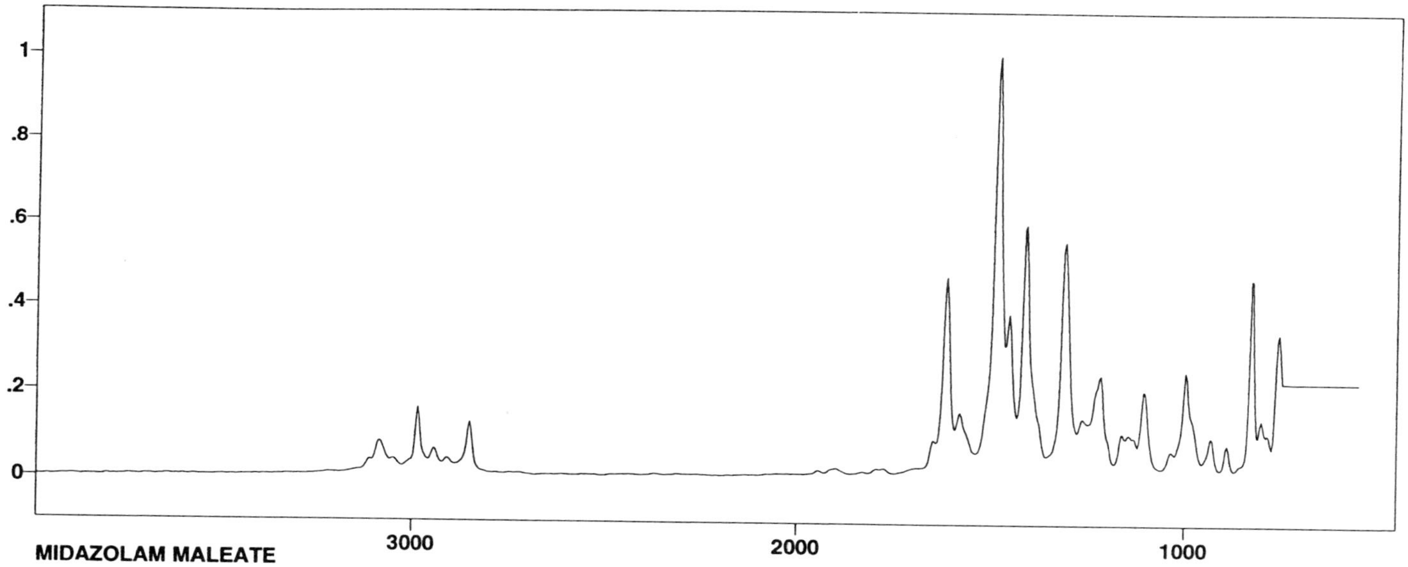

MIDAZOLAM MALEATE

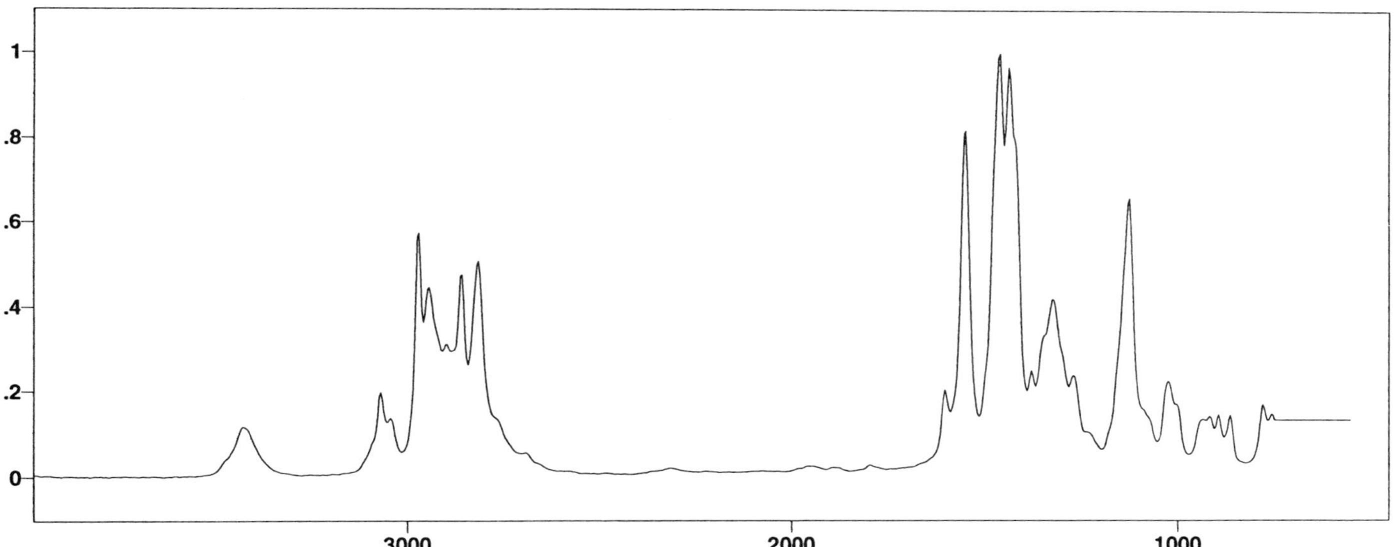

MINAPRINE

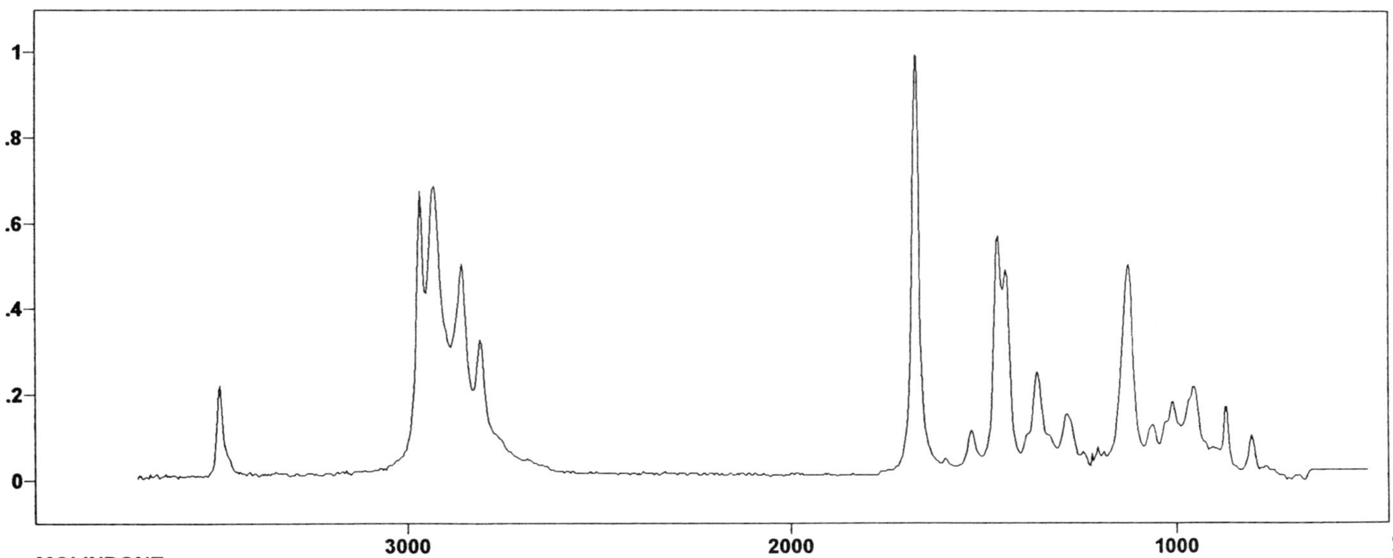

MOLINDONE

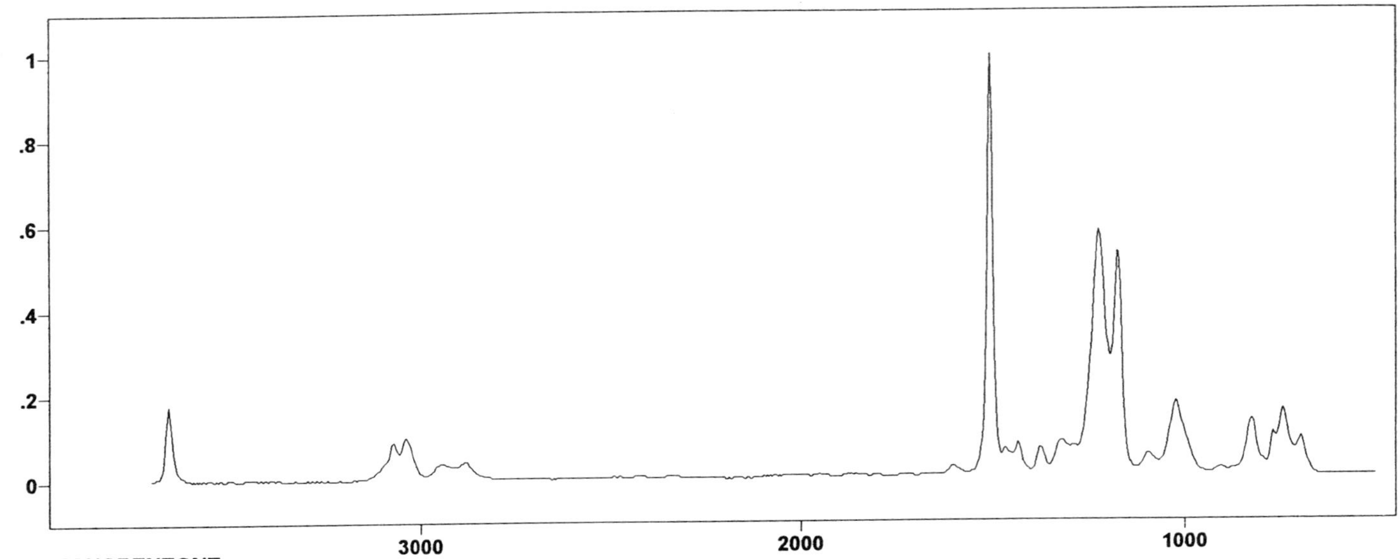

MONOBENZONE

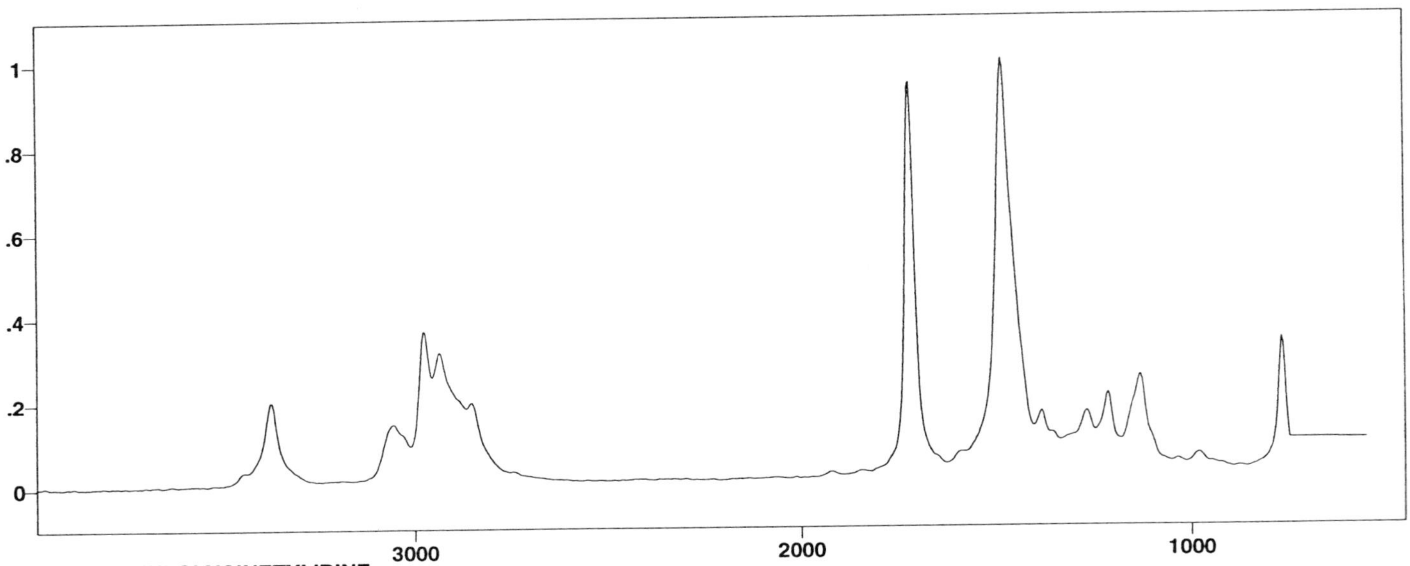

MONOETHYLGLYCINETYLIDINE

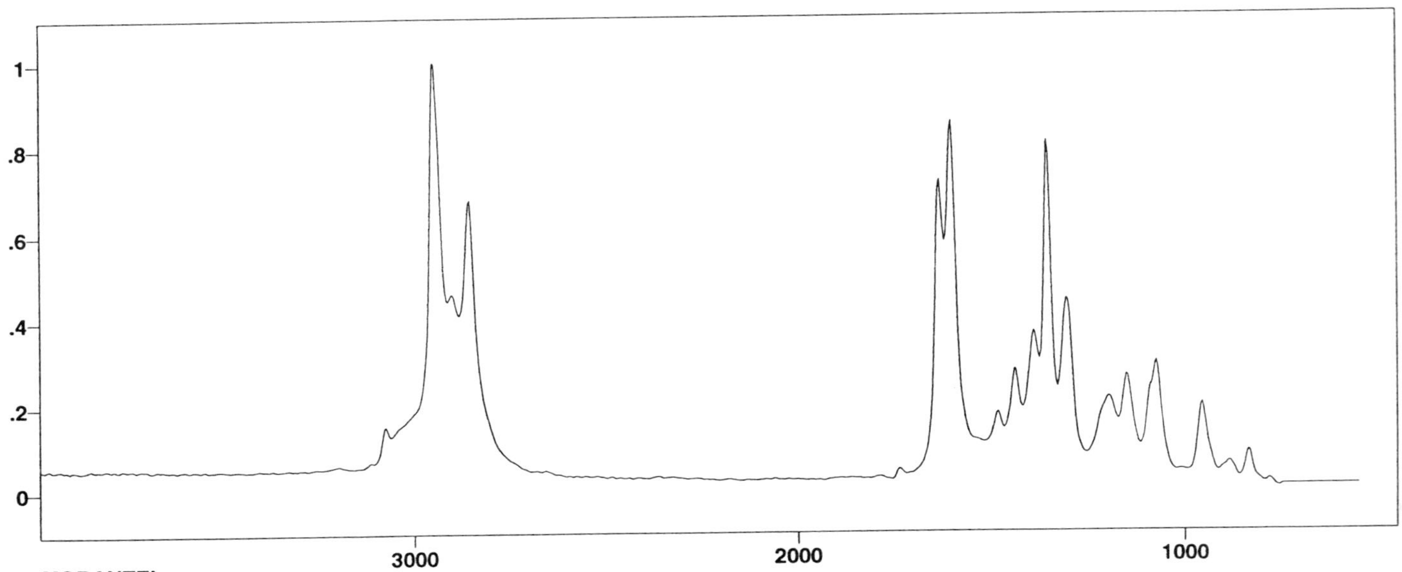

MORANTEL

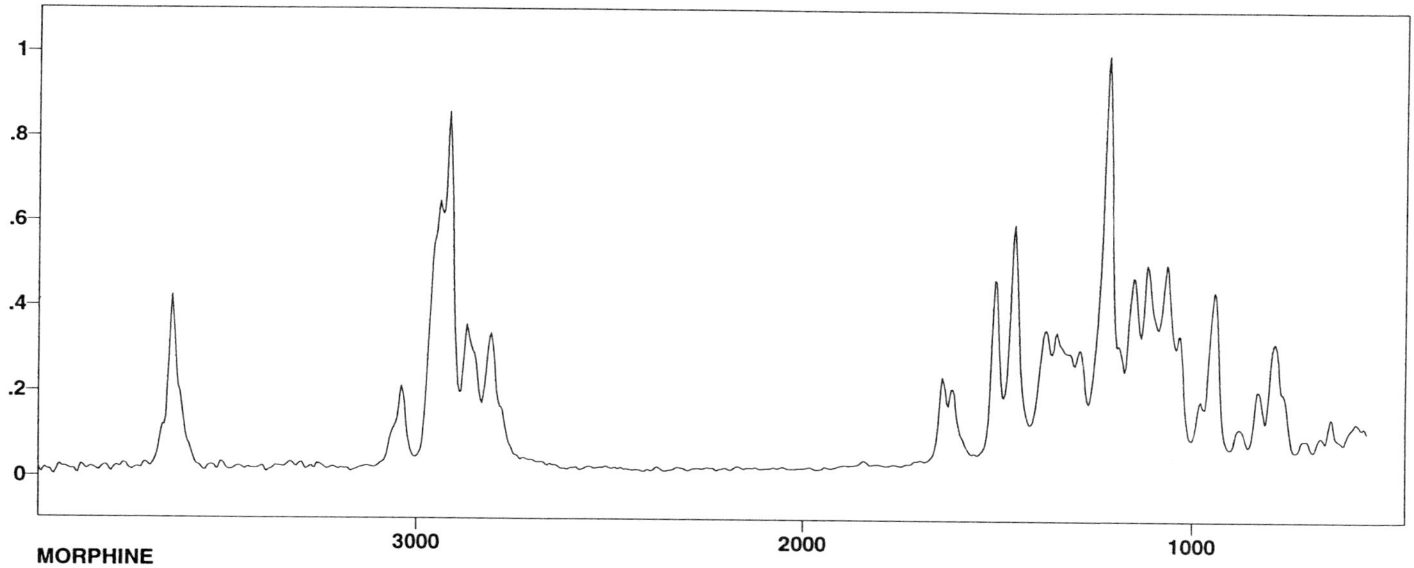

MORPHINE

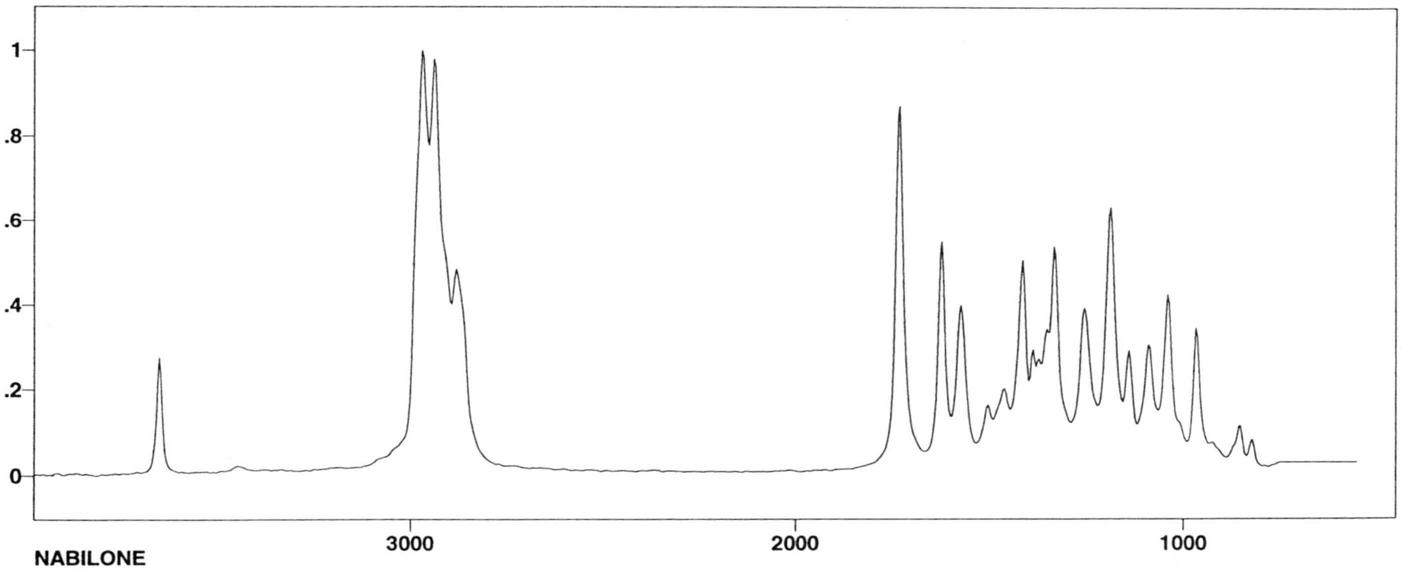

NABILONE

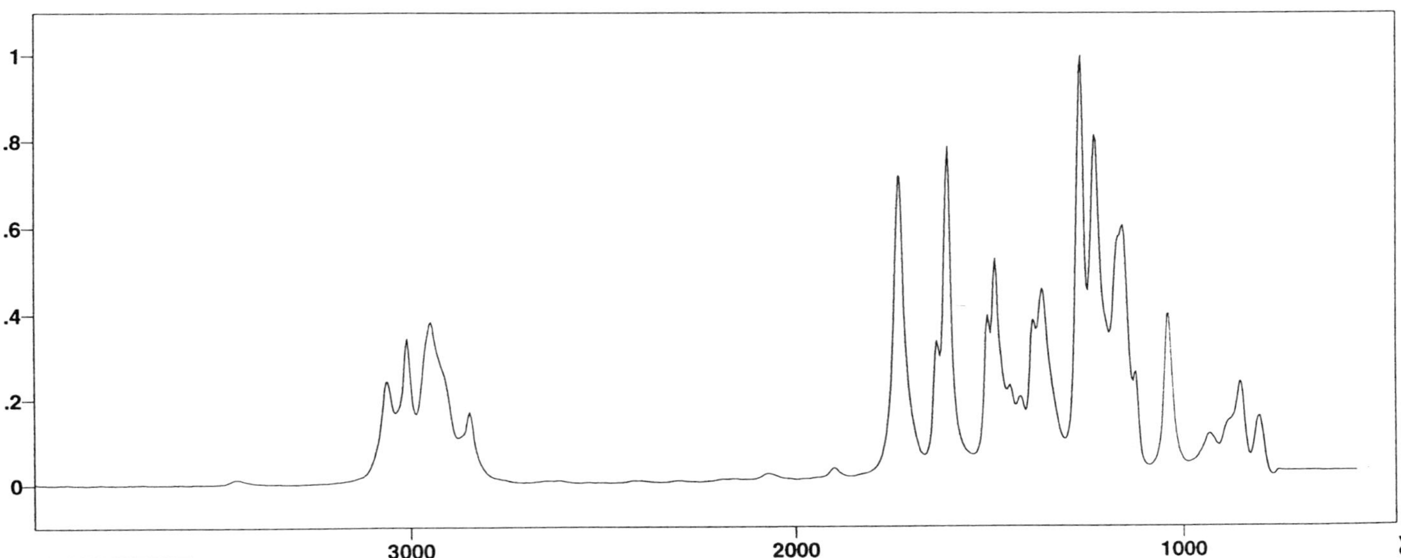

NABUMETONE

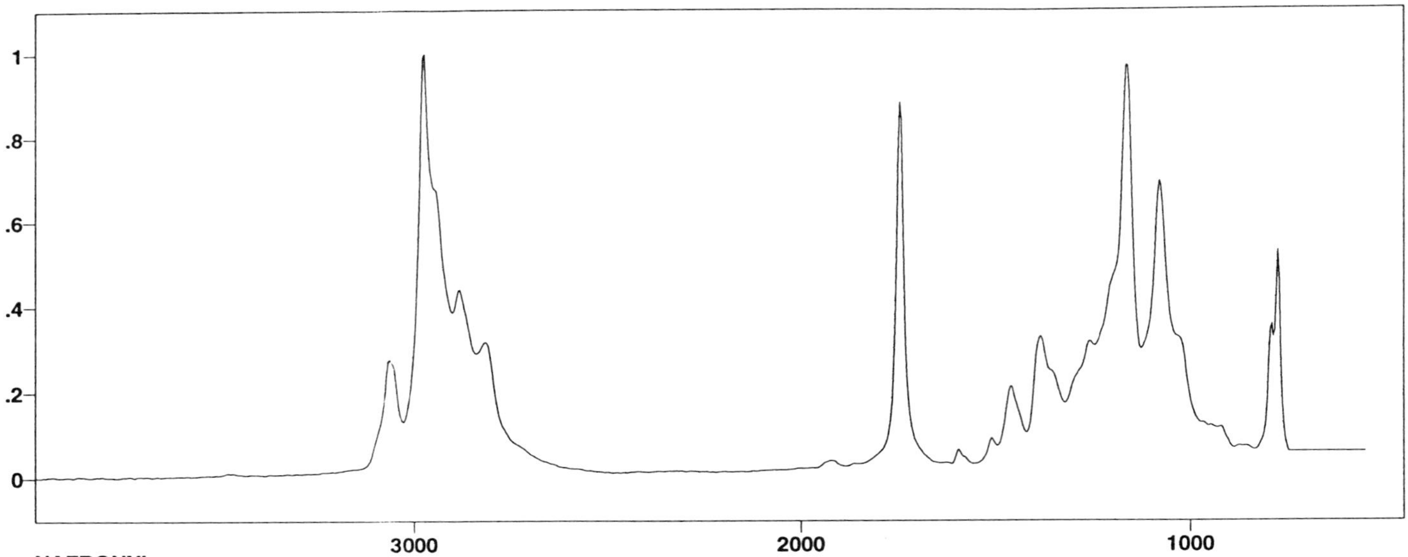

NAFRONYL

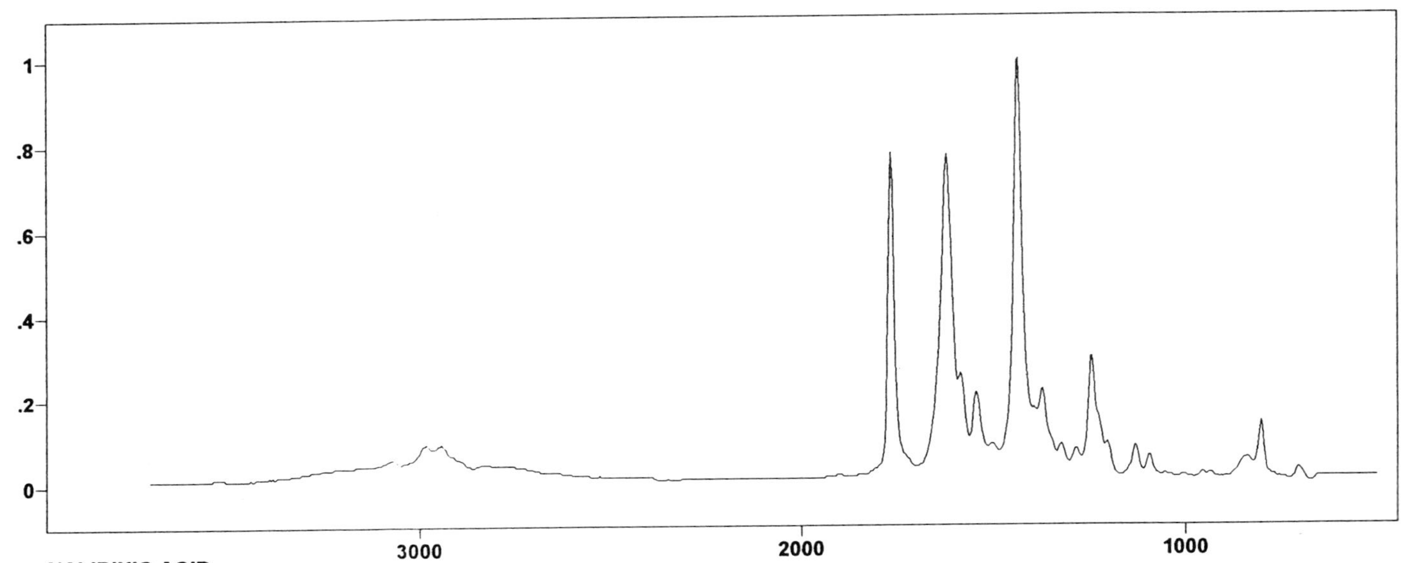

NALIDIXIC ACID

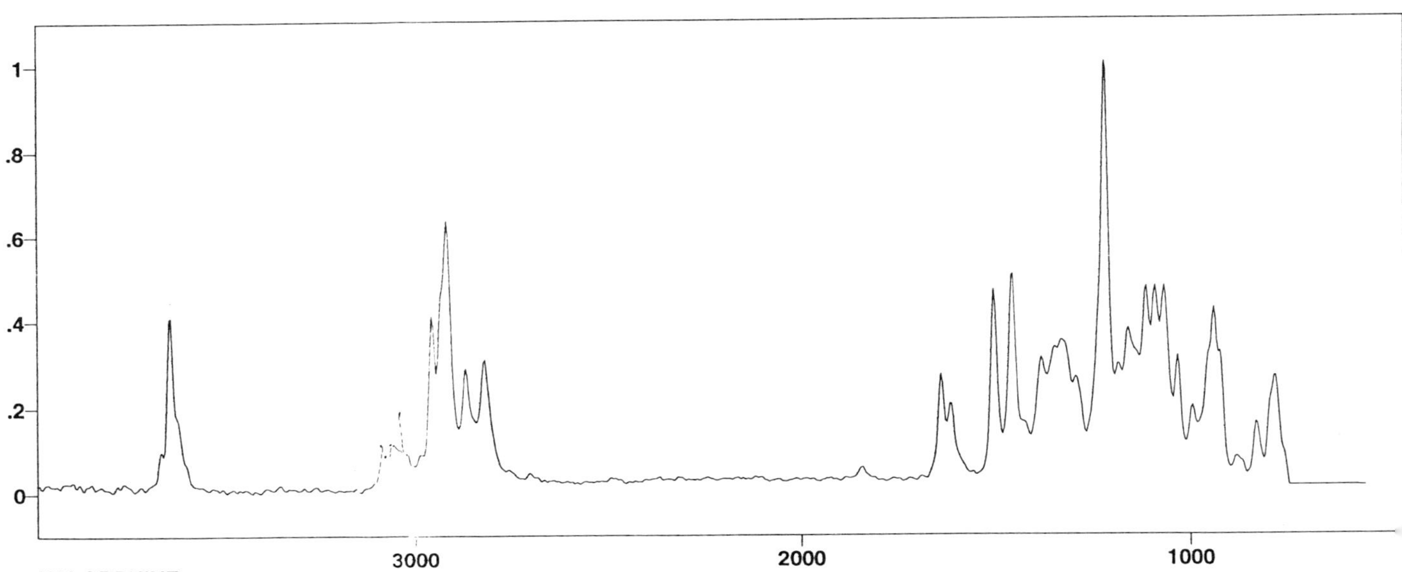

NALORPHINE

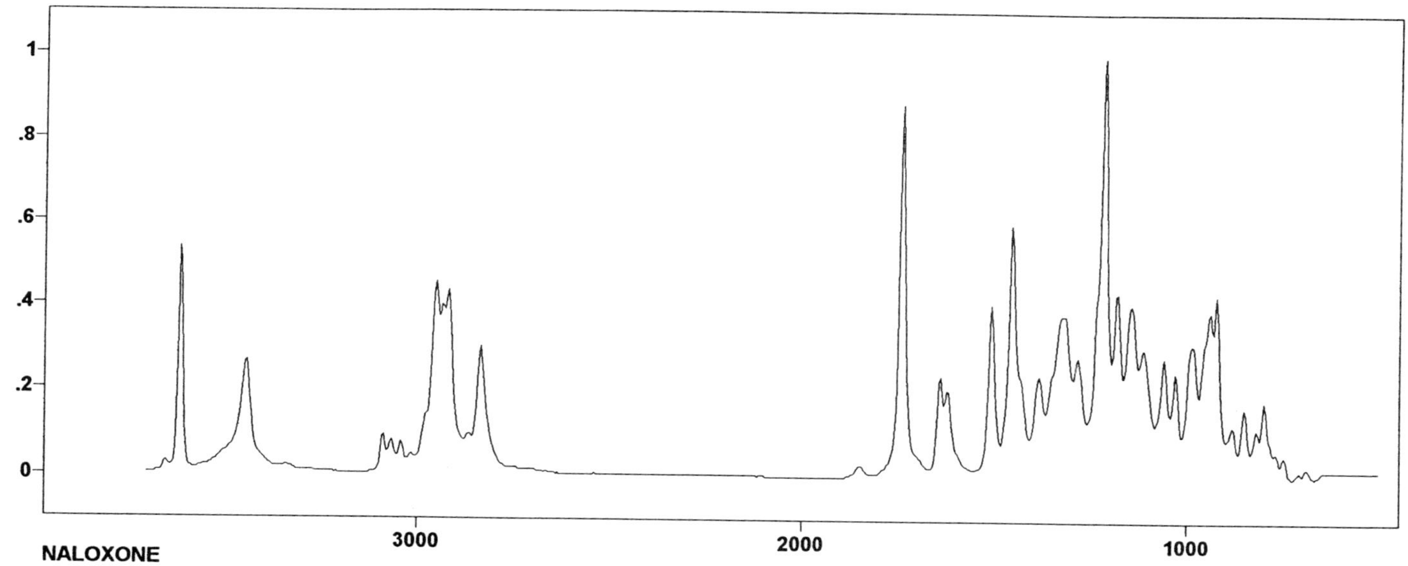

NALOXONE

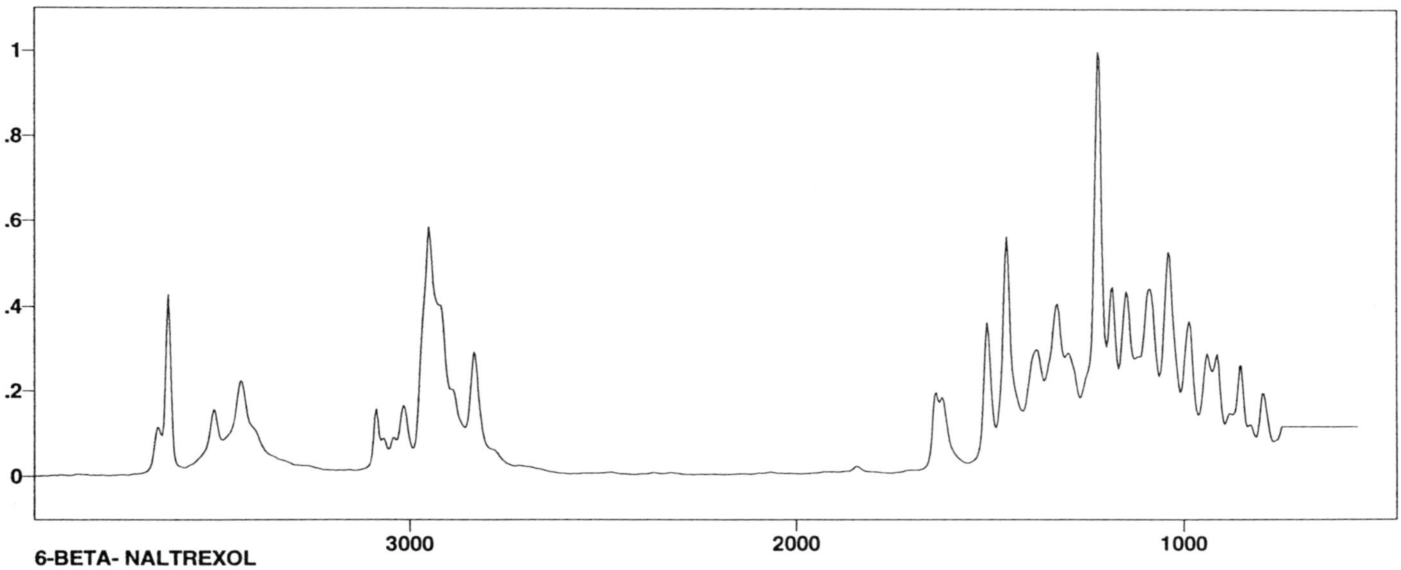

6-BETA- NALTREXOL

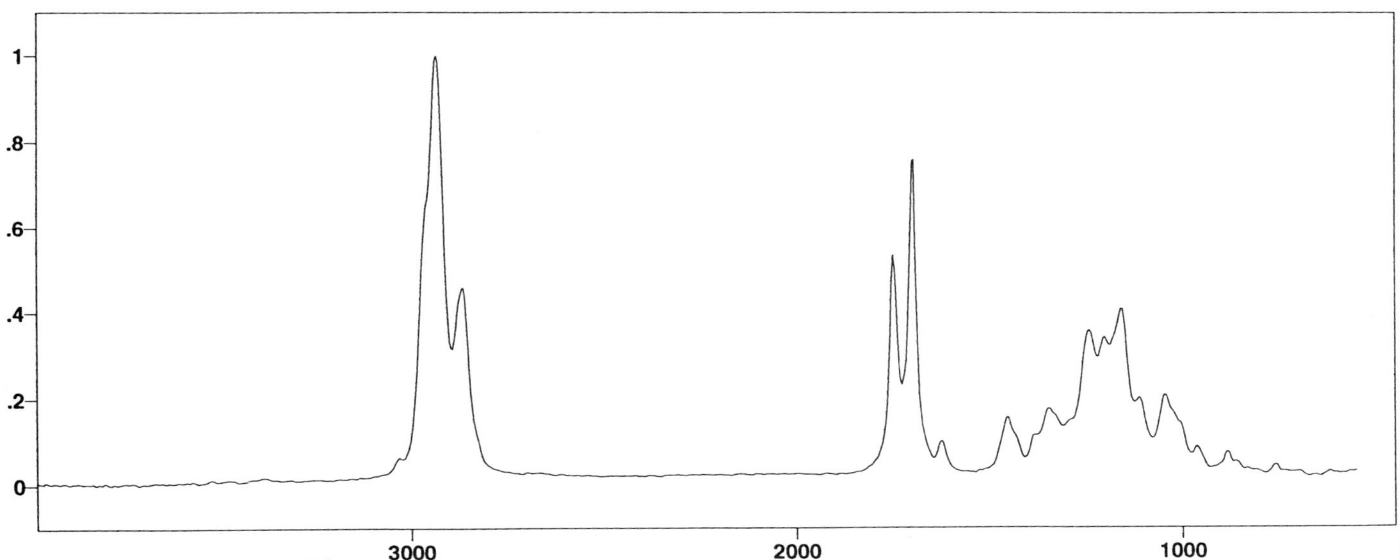

NANDROLONE DECANOATE

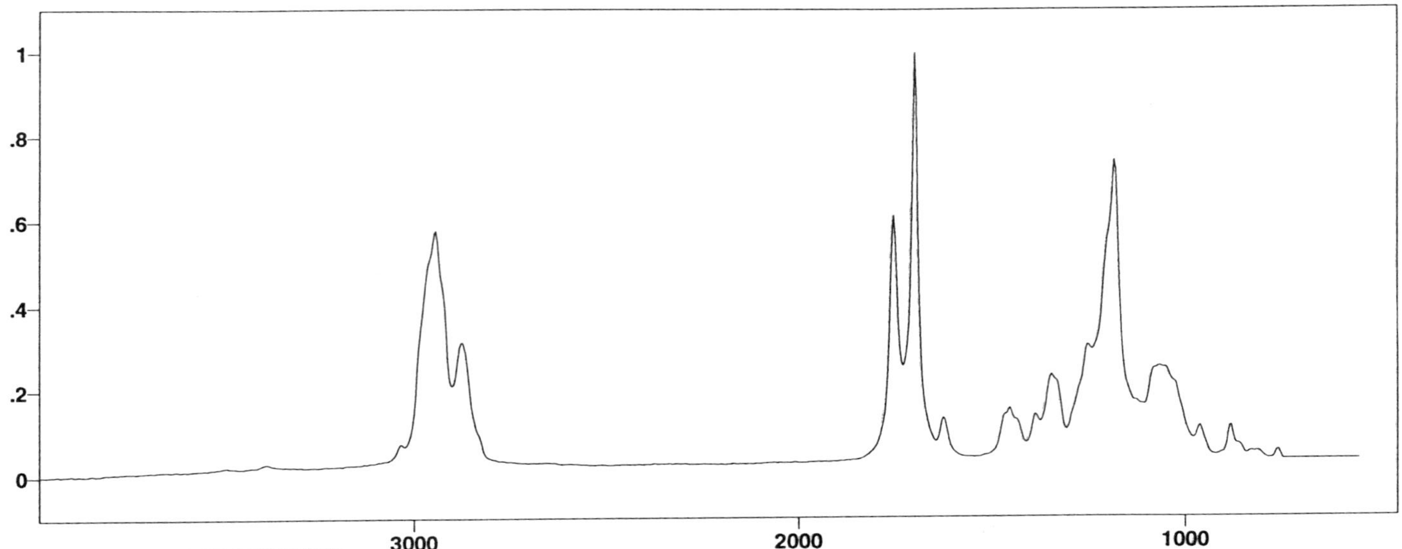

NANDROLONE PROPIONATE

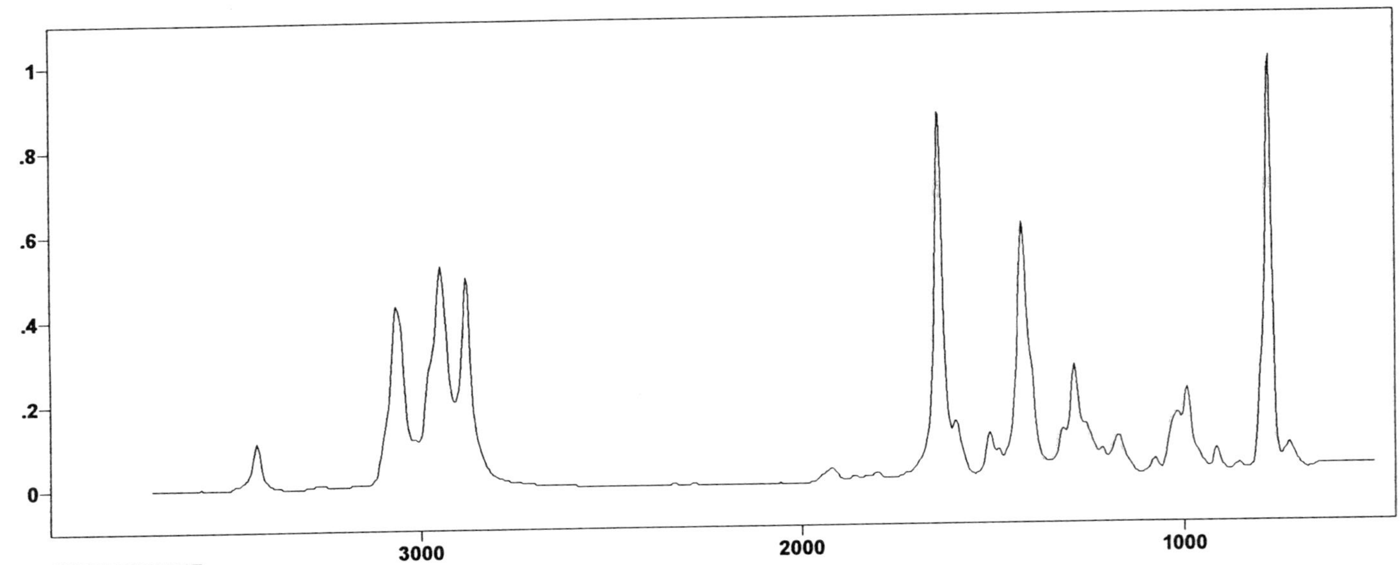

NAPHAZOLINE

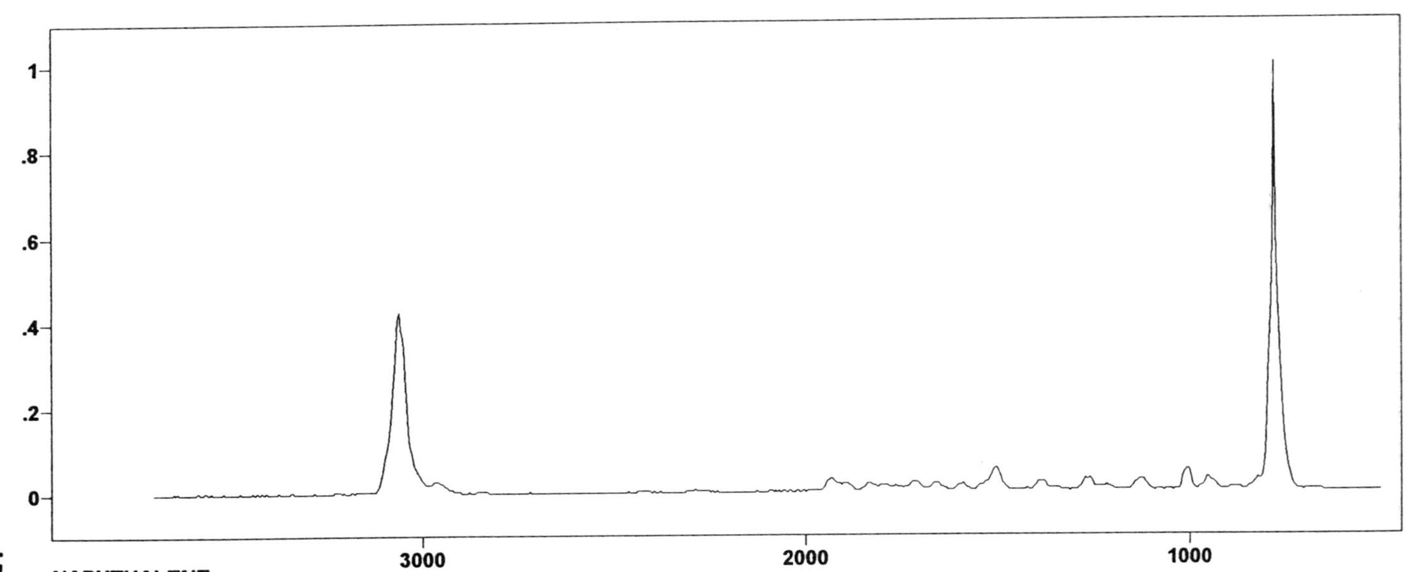

NAPHTHALENE

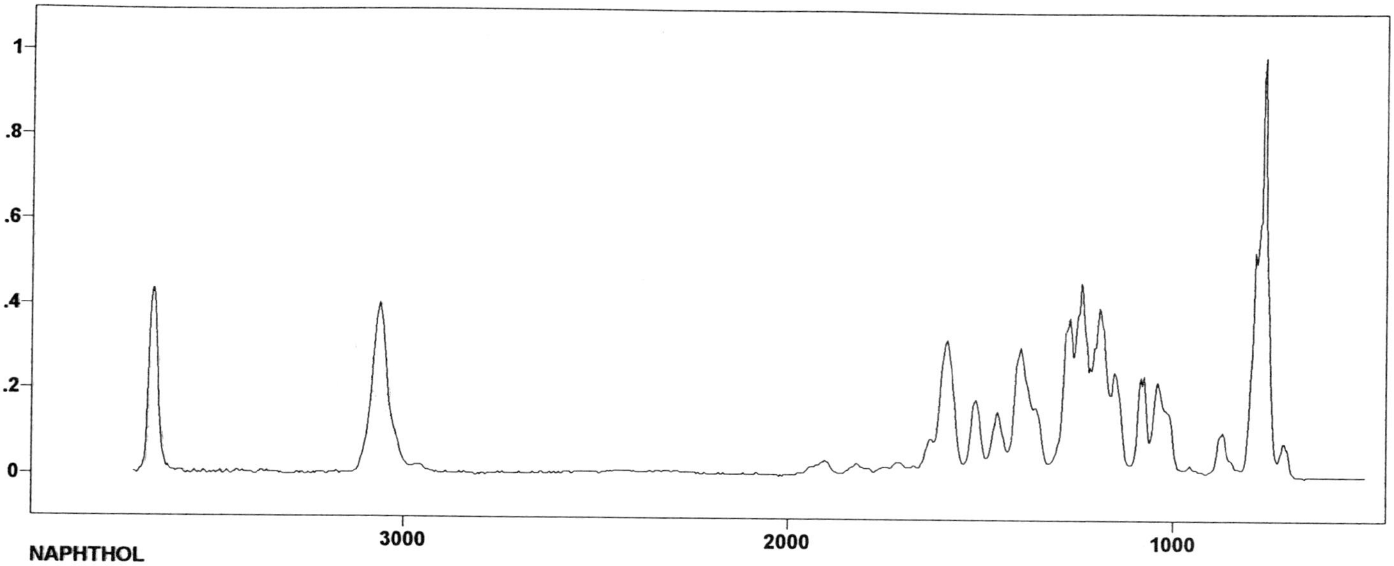

NAPHTHOL

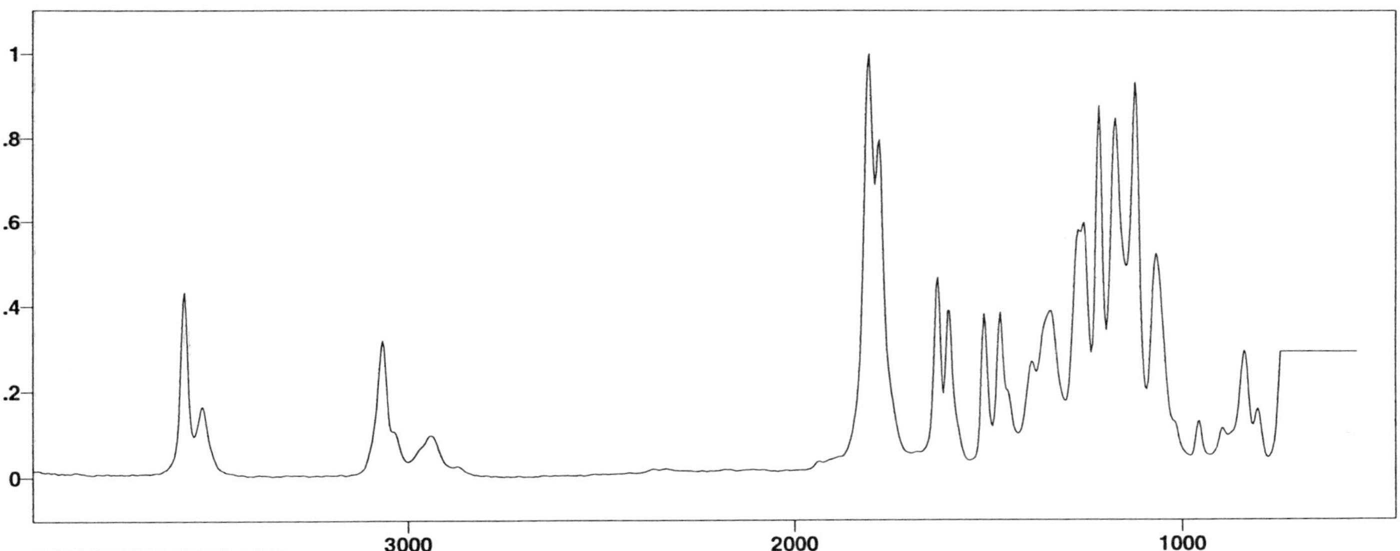

NAPHTOXYACETIC ACID

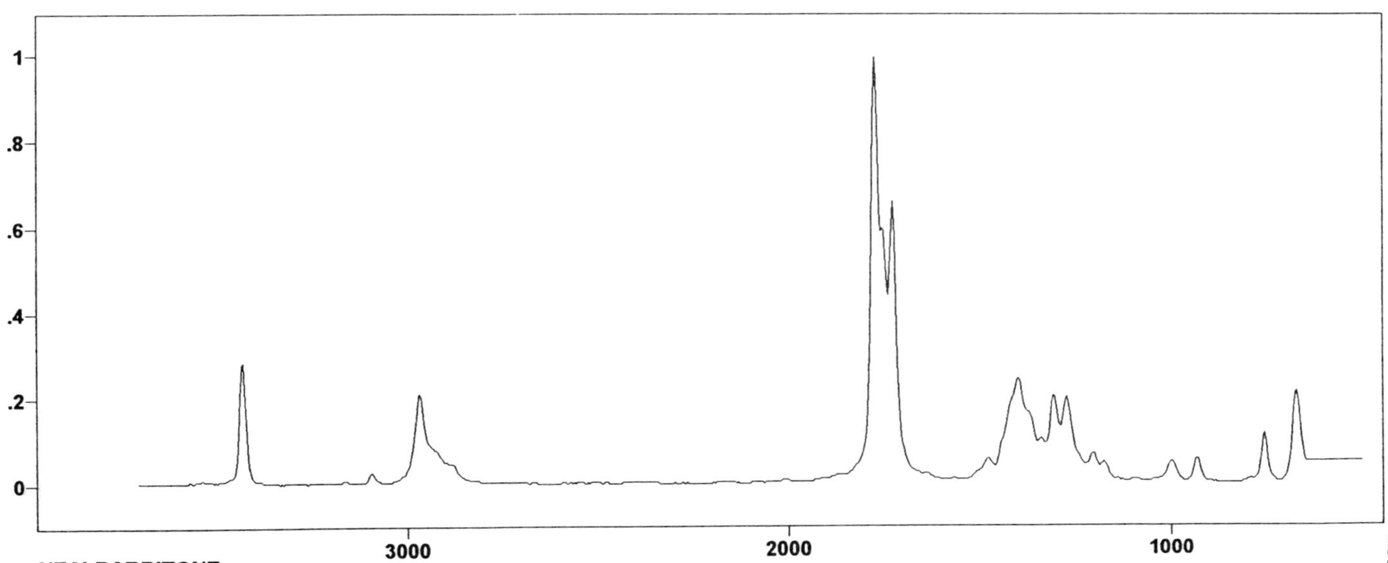

NEALBARBITONE

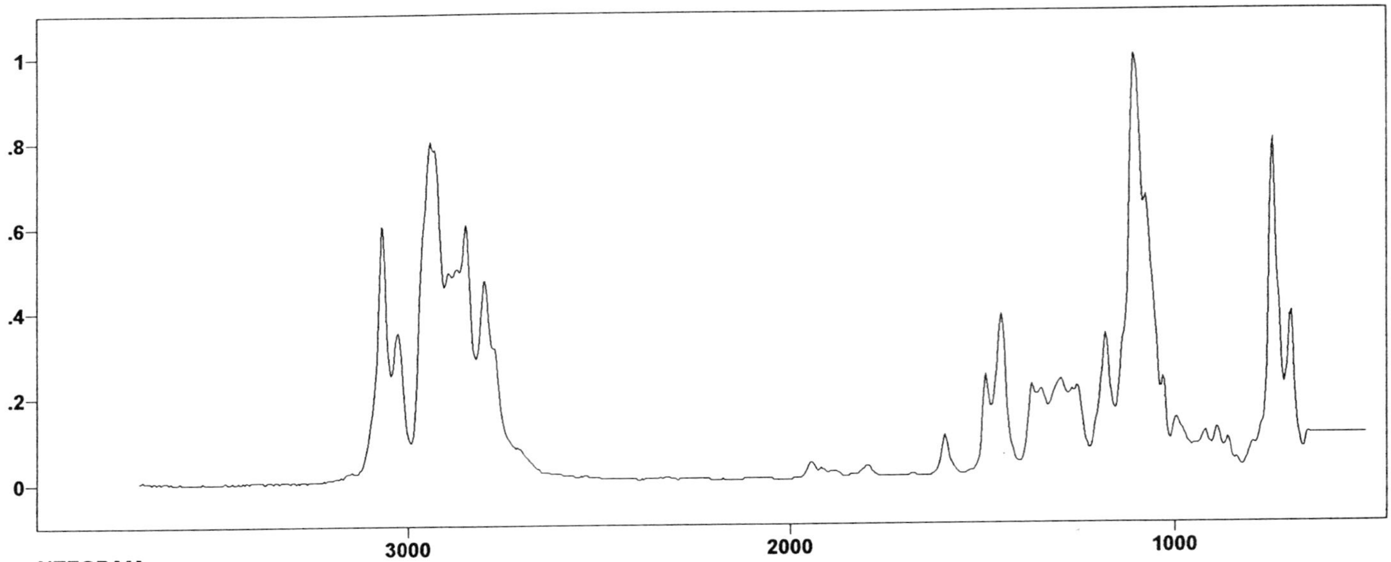

NEFOPAM

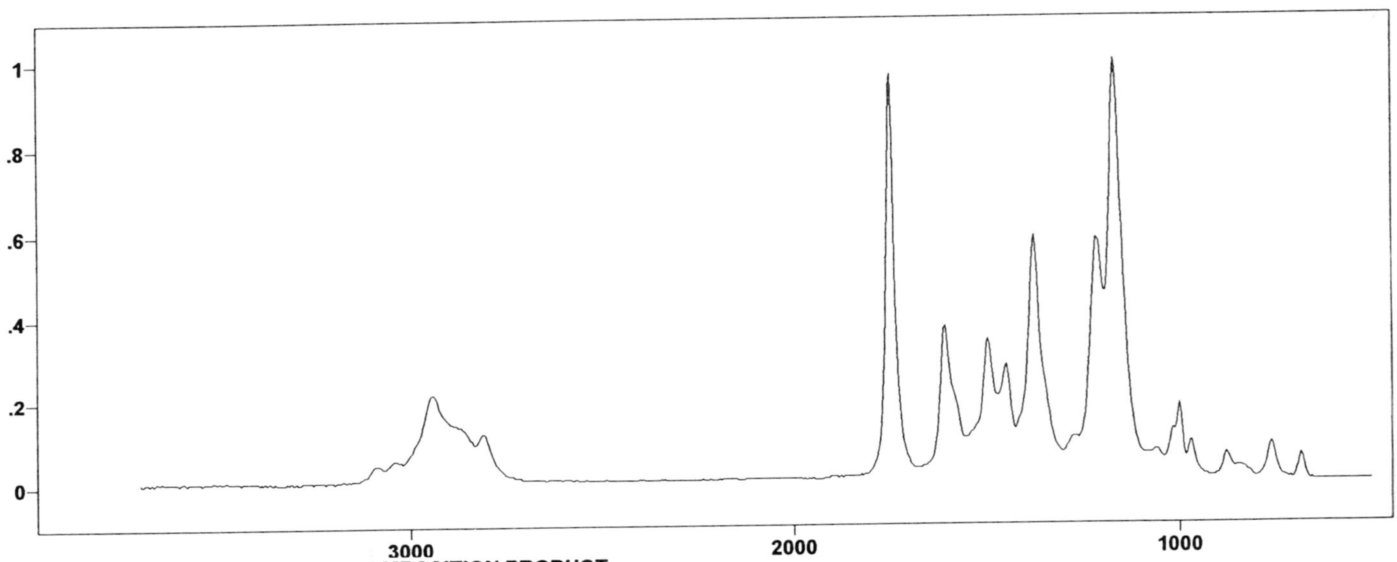

NEOSTIGMINE BROMIDE- - DECOMPOSITION PRODUCT

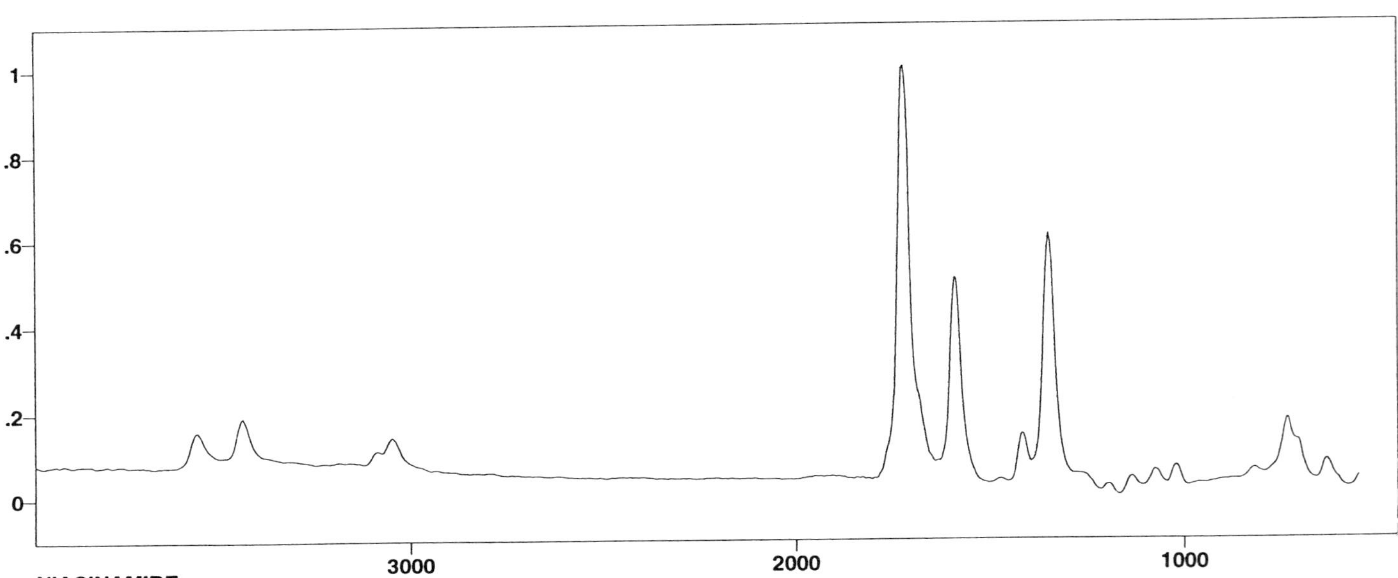

NIACINAMIDE

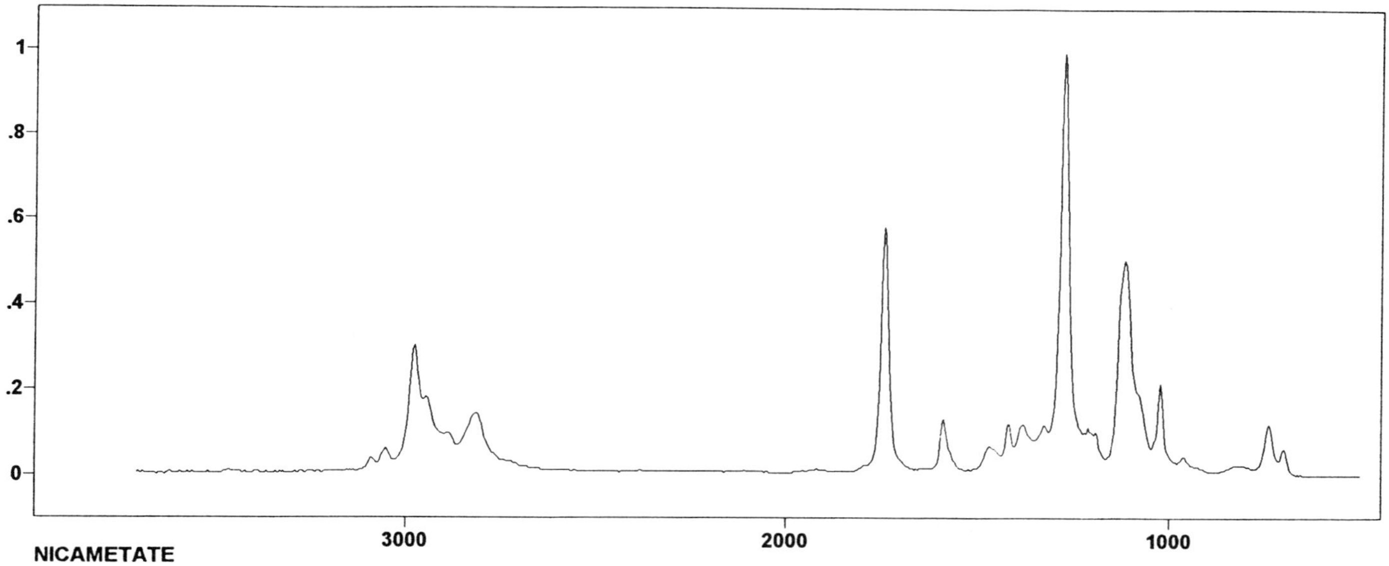

NICAMETATE

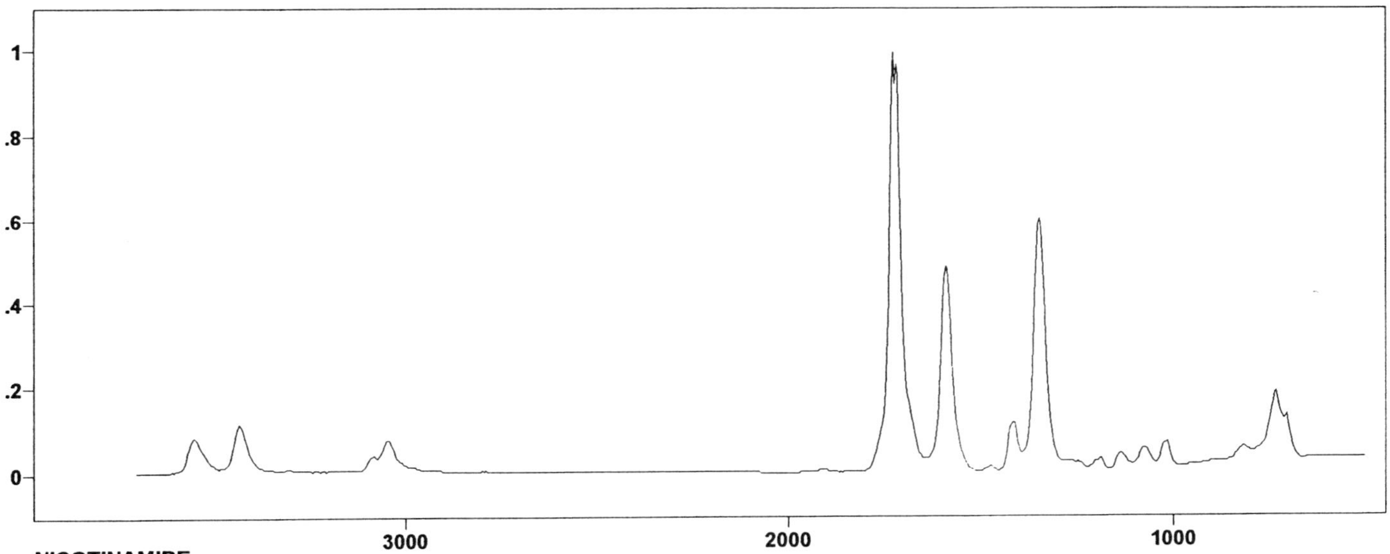

NICOTINAMIDE

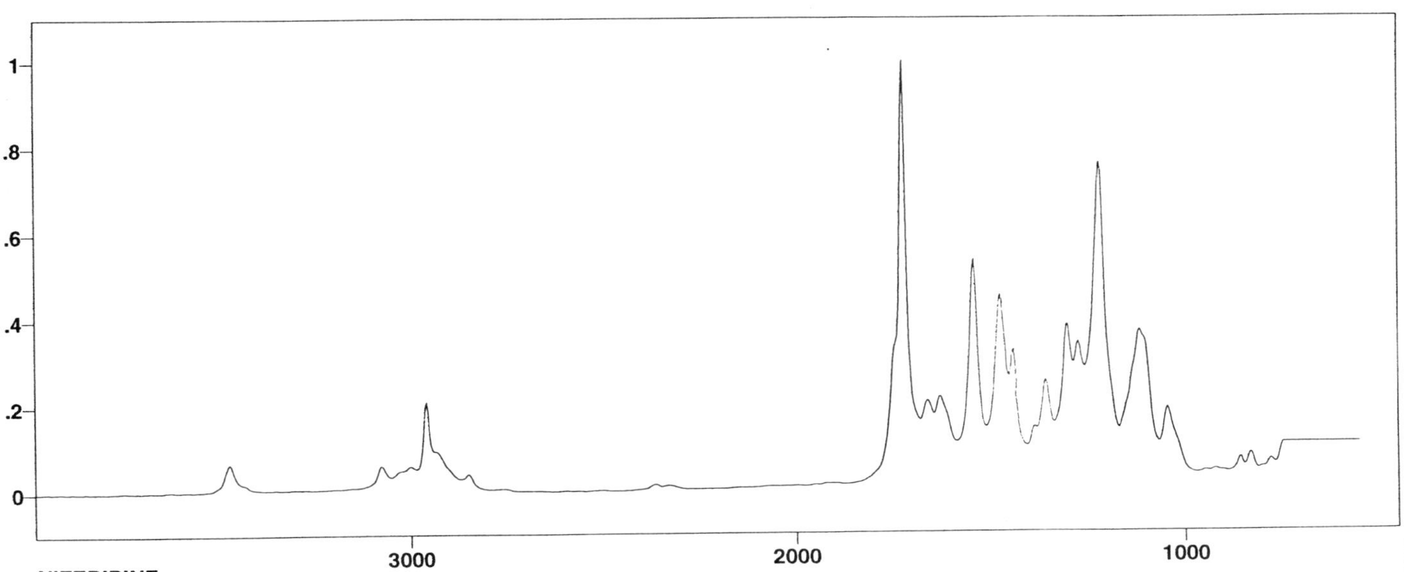

NIFEDIPINE

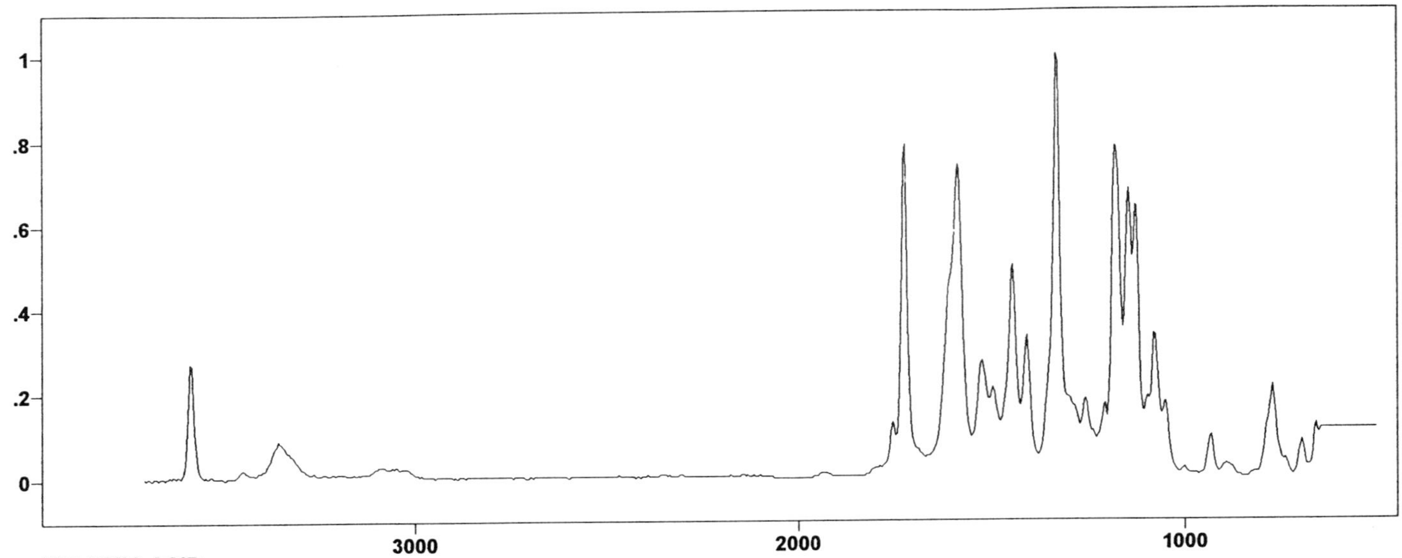

NIFLUMIC ACID

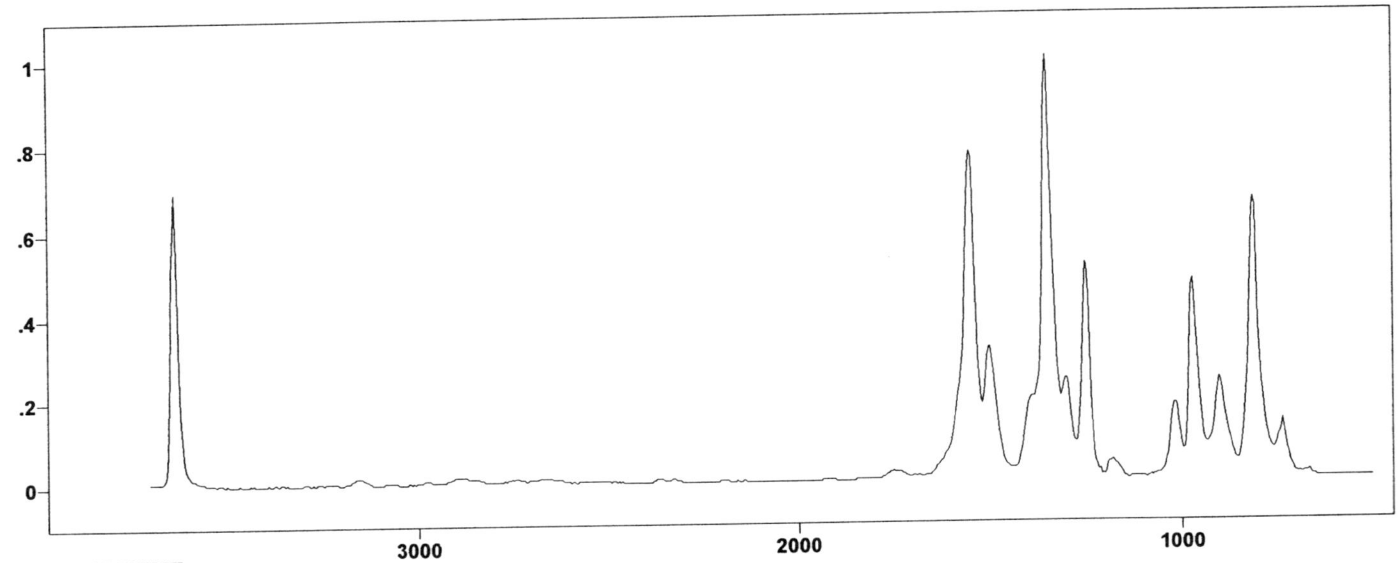

NIFUROXIME

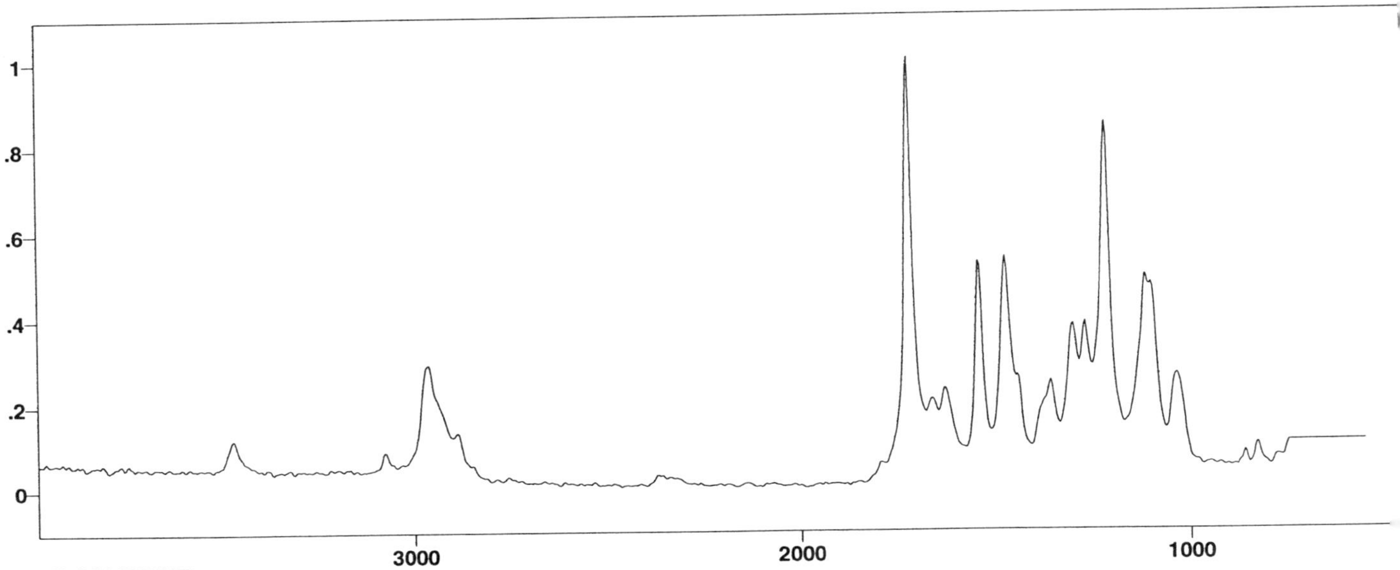

NILSOLDIPINE

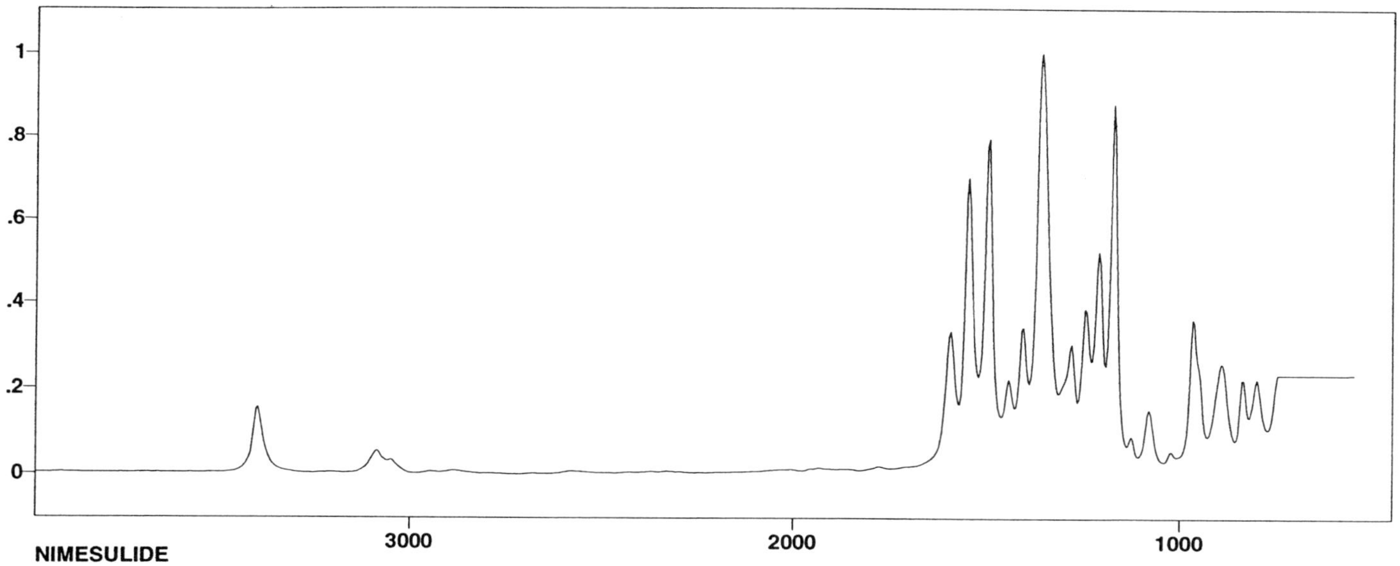

NIMESULIDE

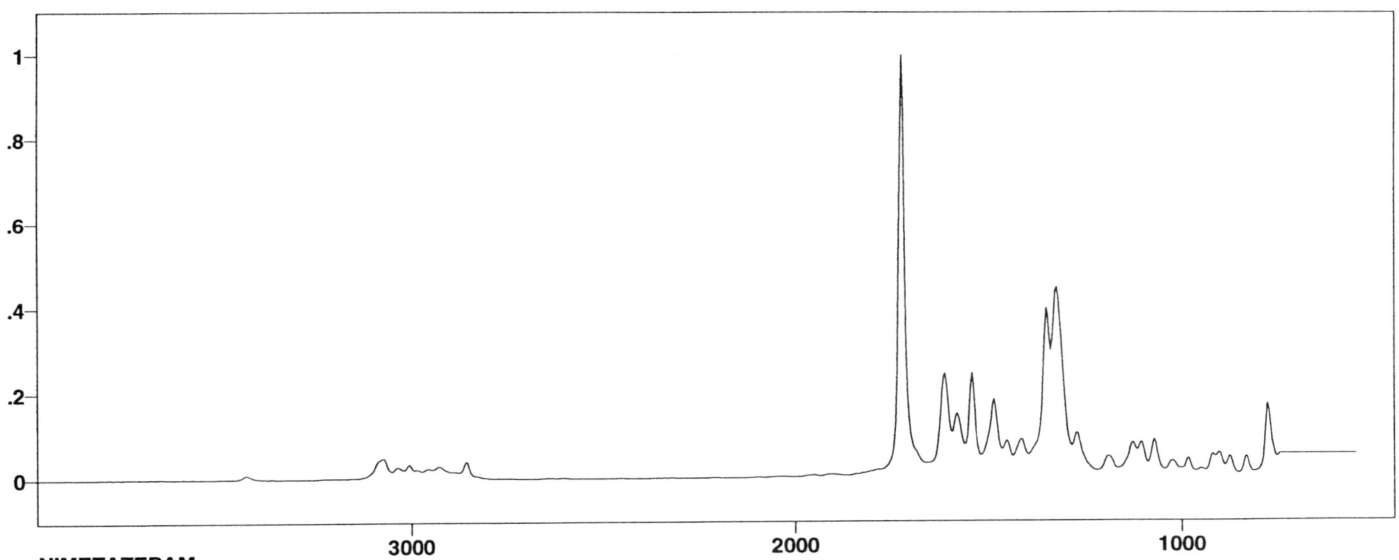

NIMETAZEPAM

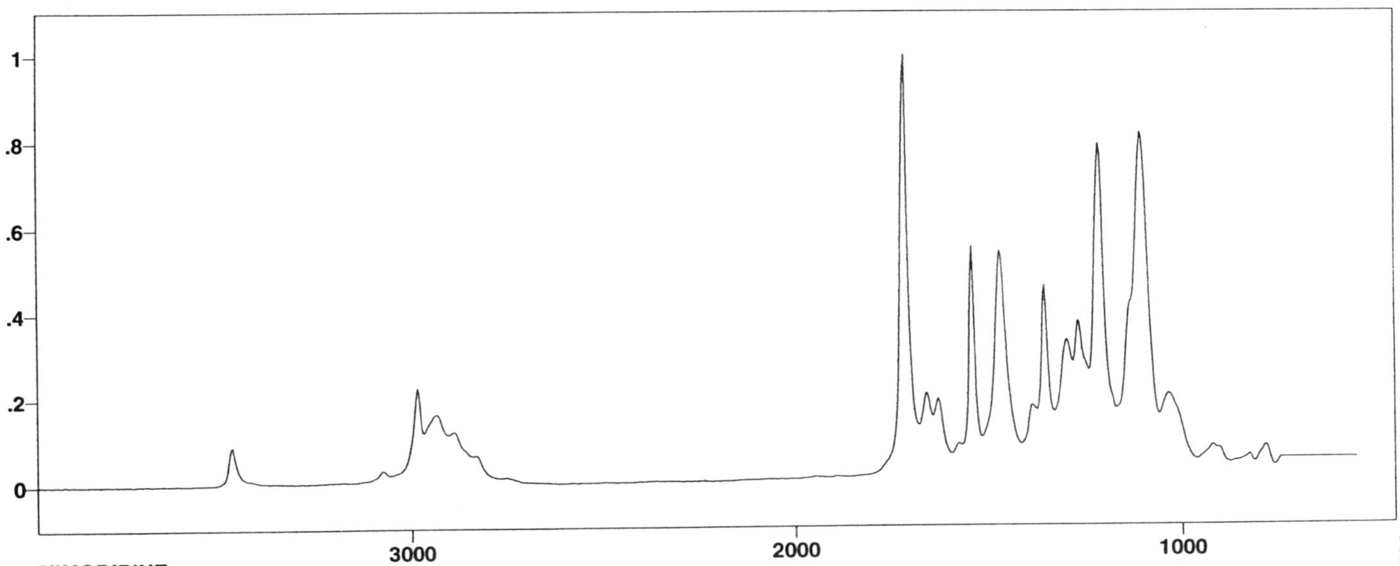

NIMODIPINE

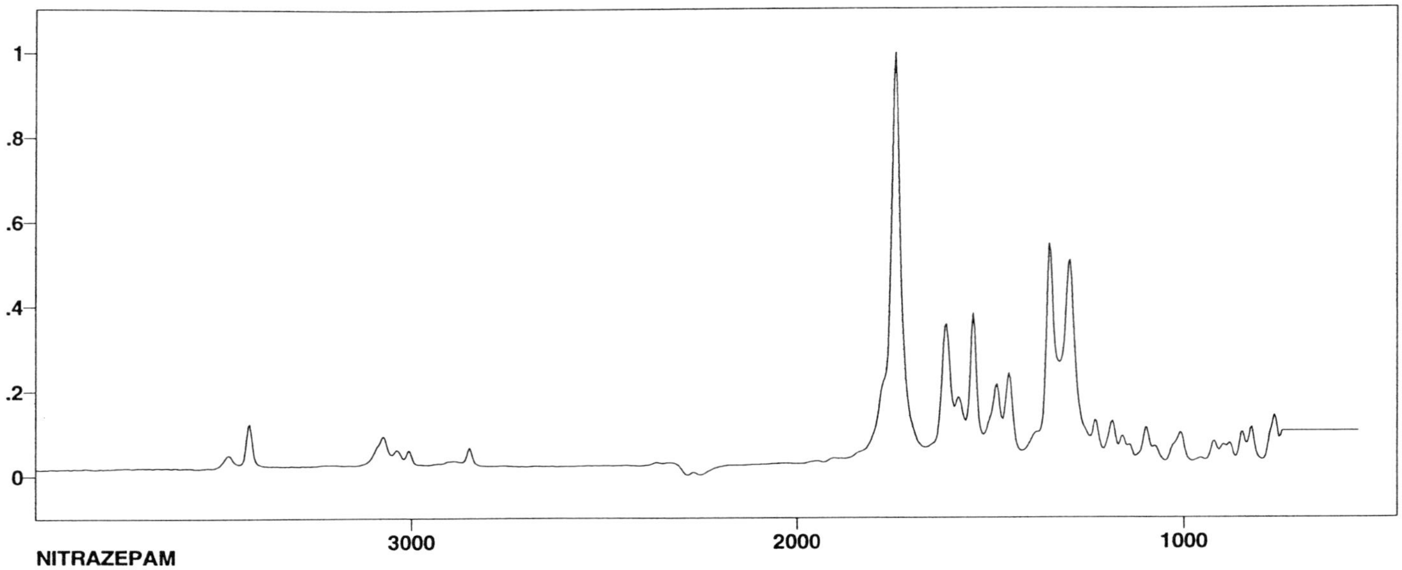

NITRAZEPAM

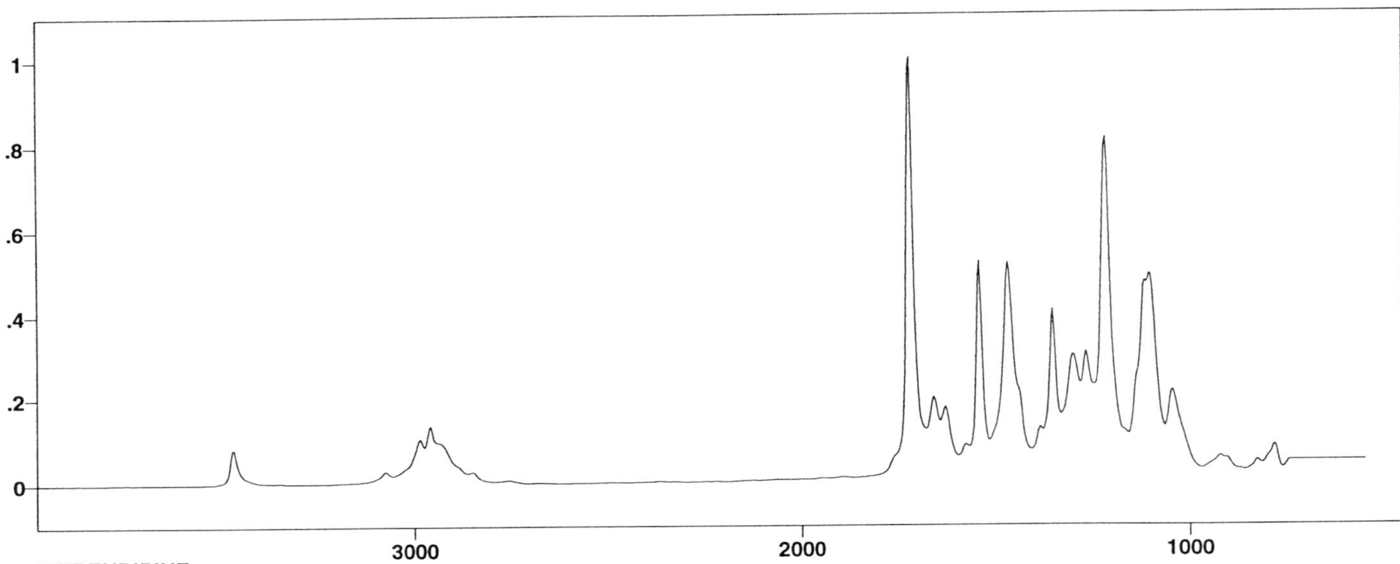

NITRENDIPINE

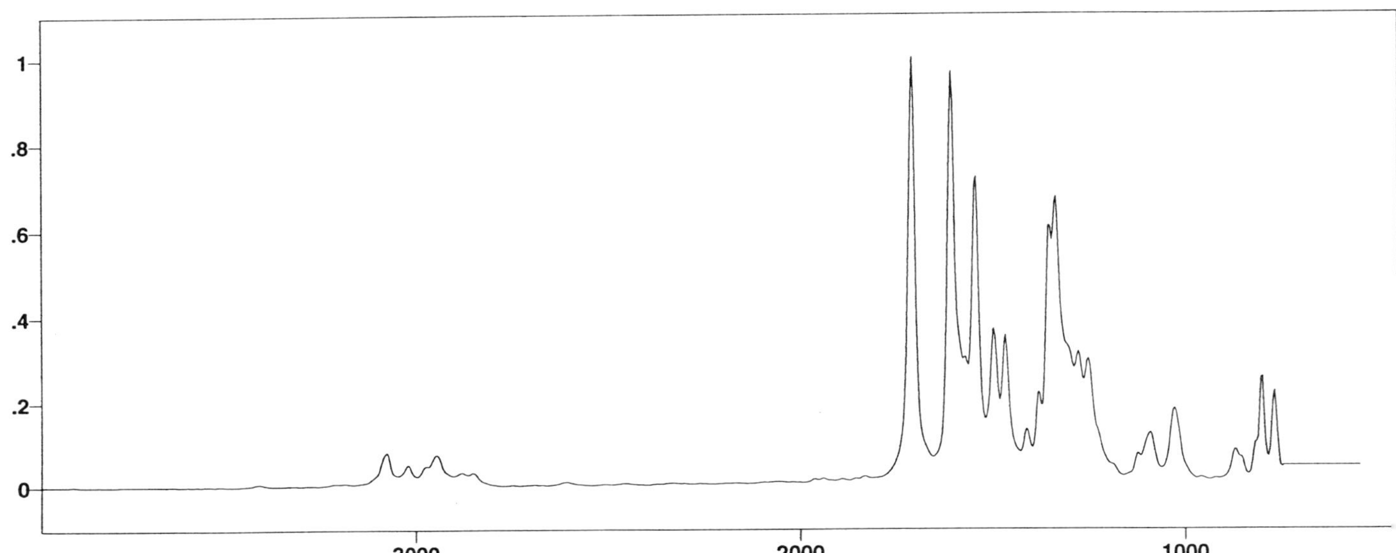

NITROMETHAQUALINE

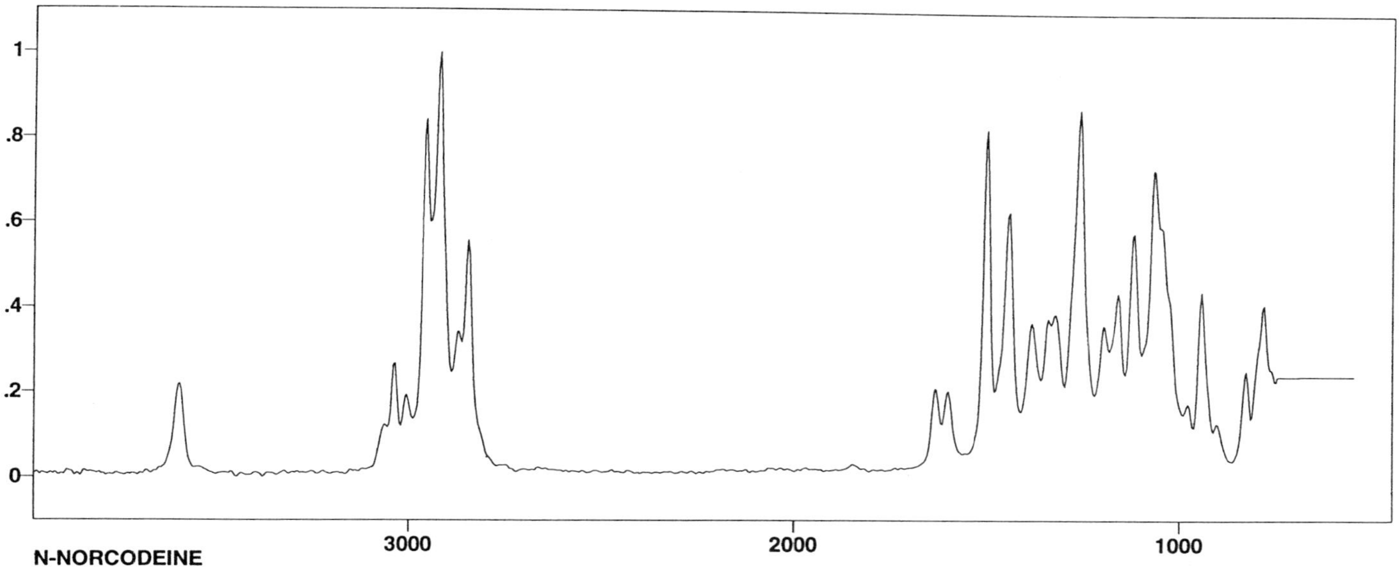

N-NORCODEINE

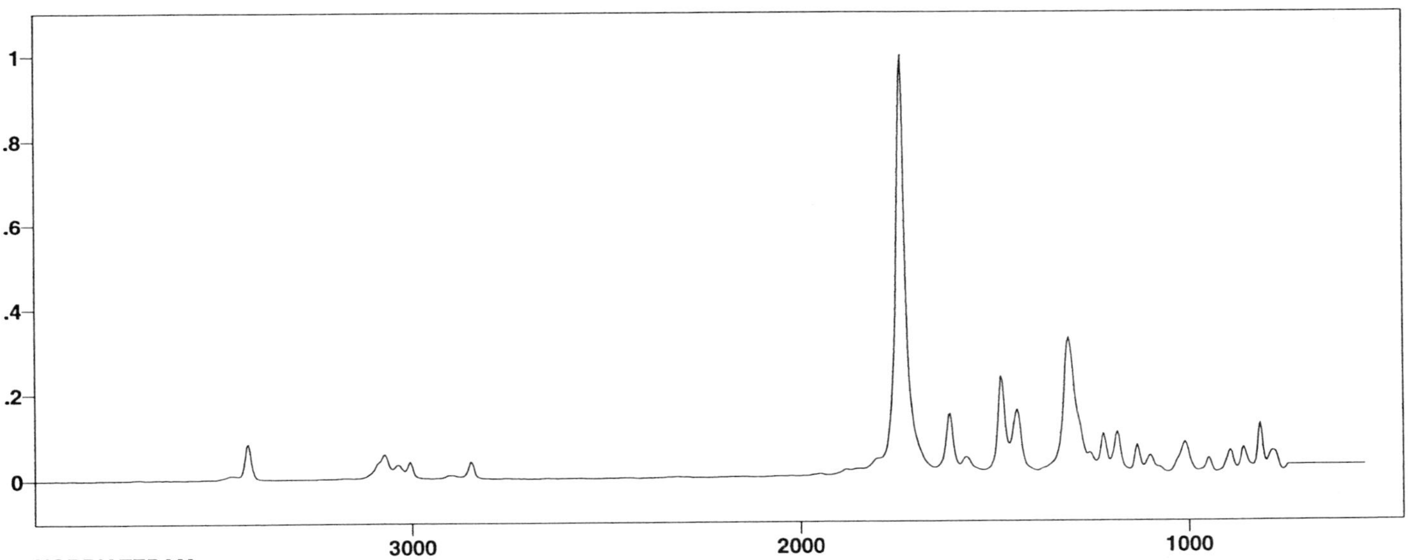

NORDIAZEPAM

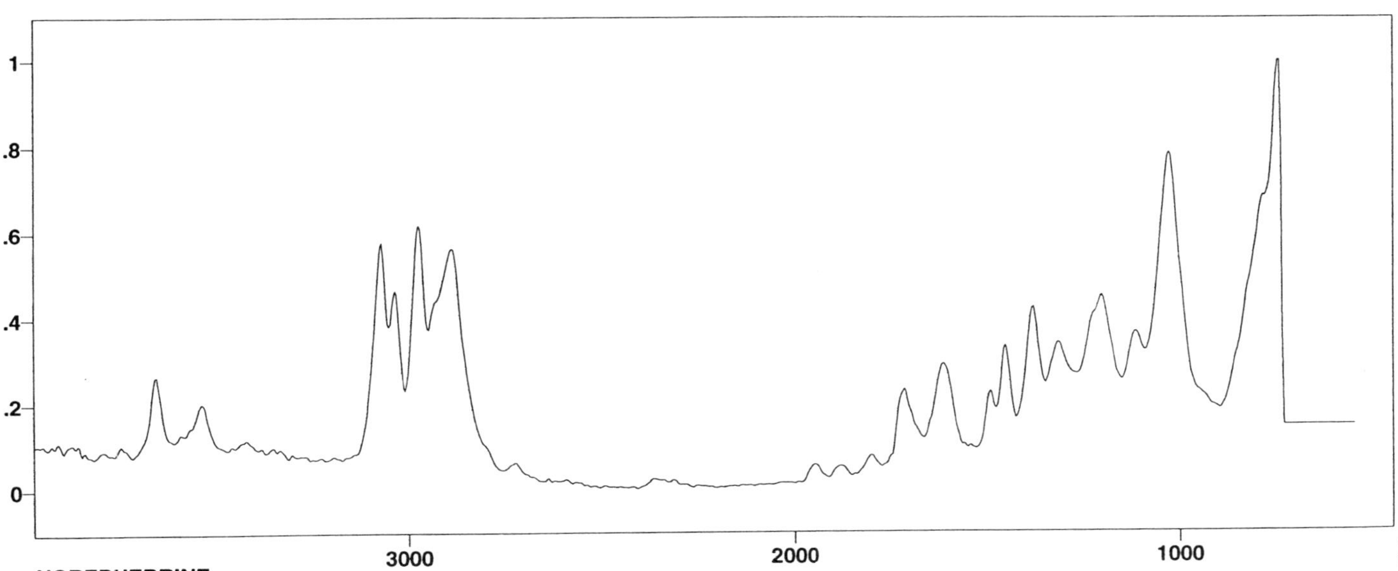

NOREPHEDRINE

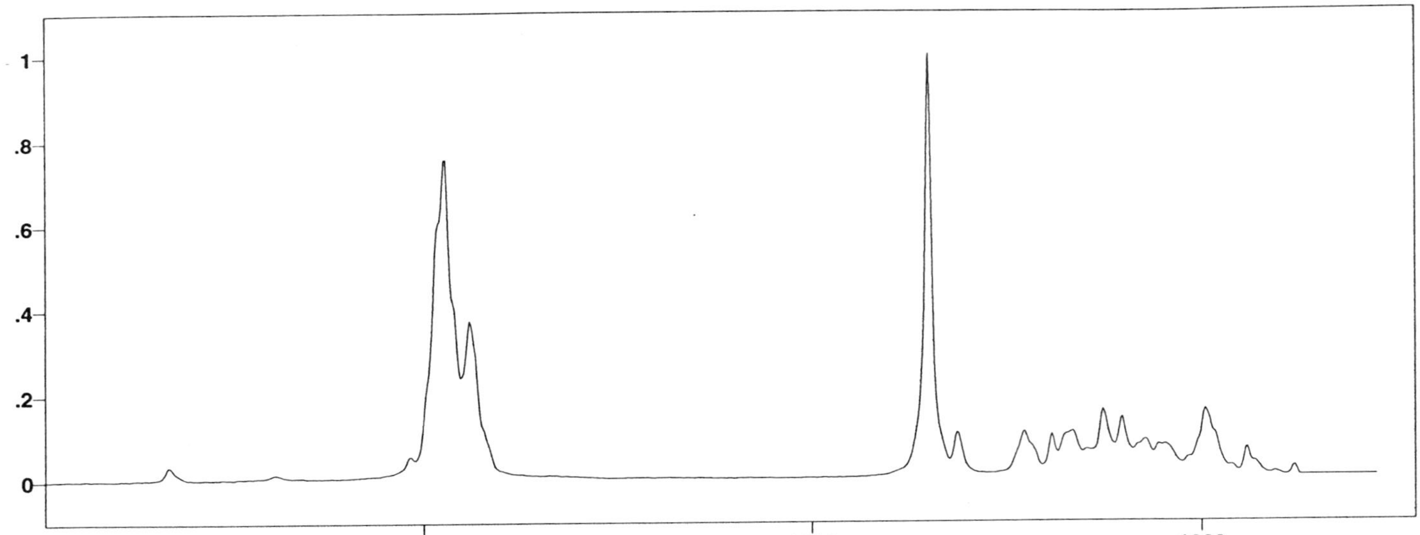

NORETHANDROLONE

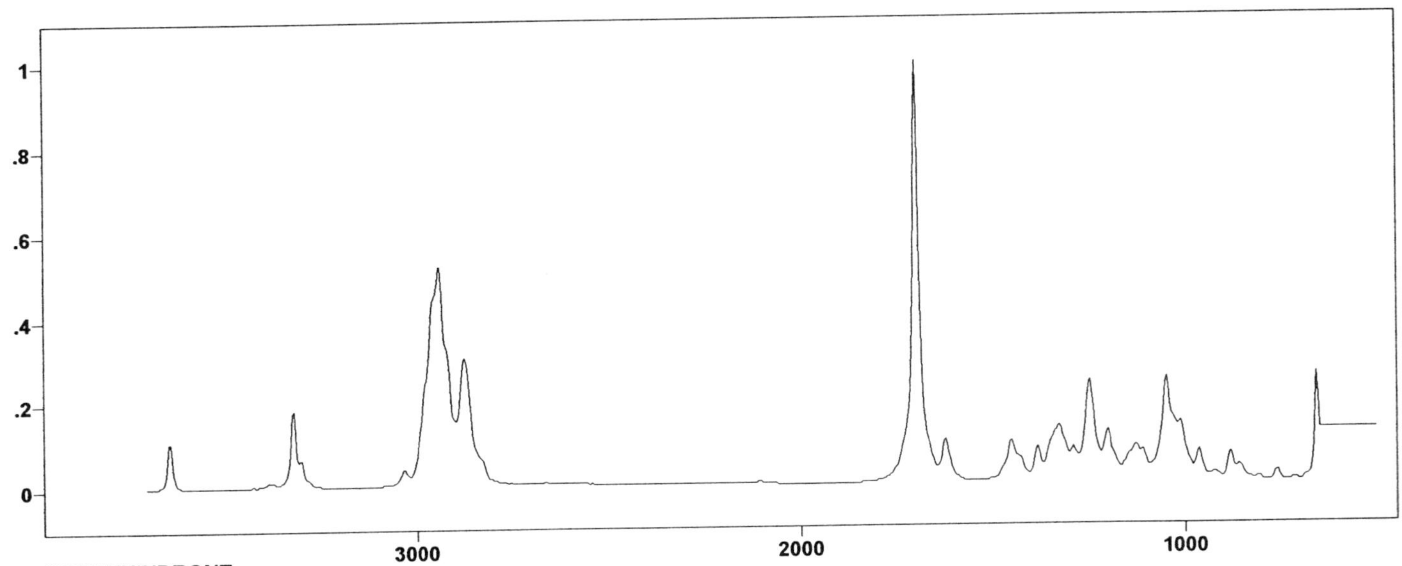

NORETHINDRONE

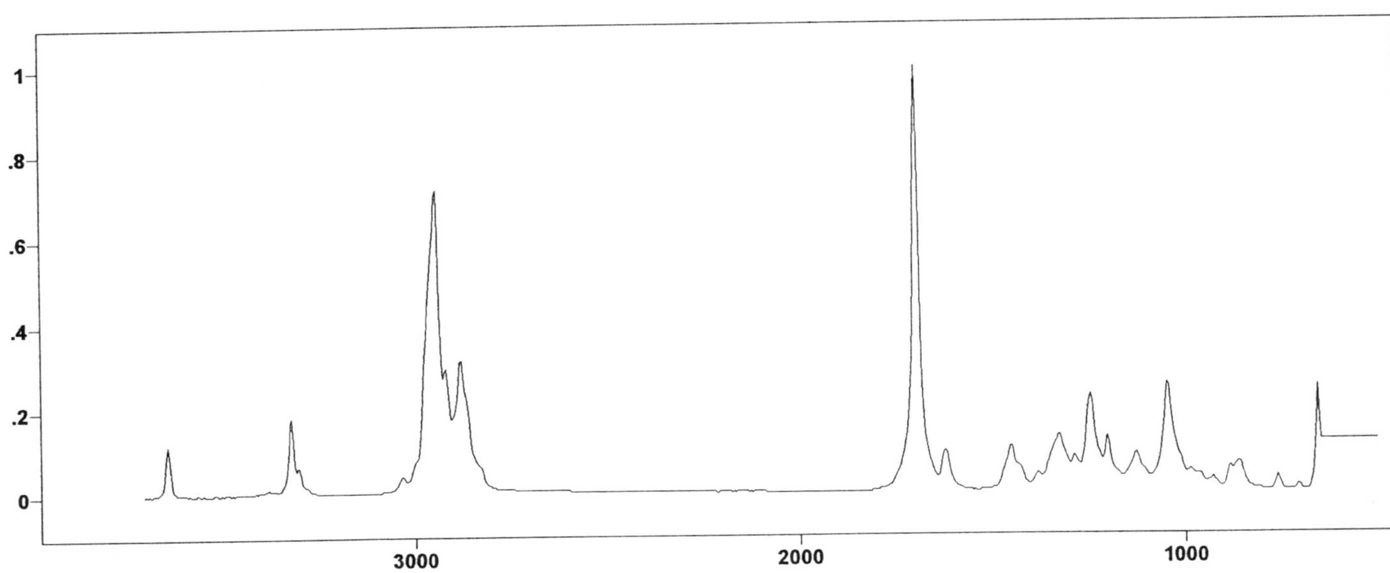

NORGESTREL

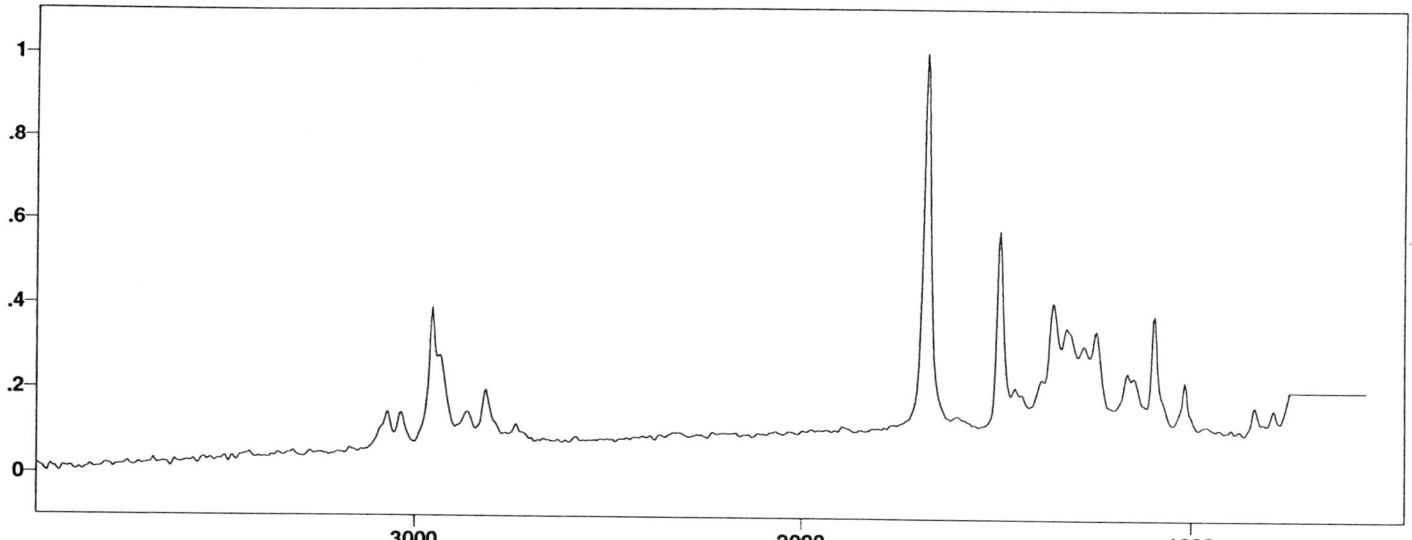

NORLORCAINIDE

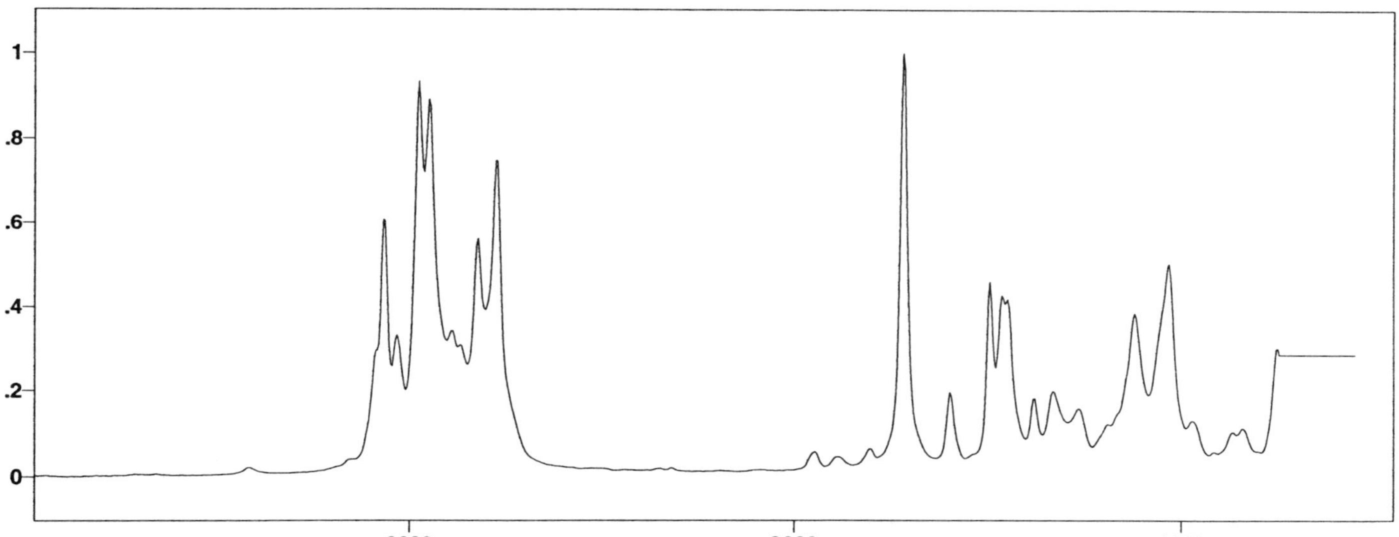

NORMETHADONE

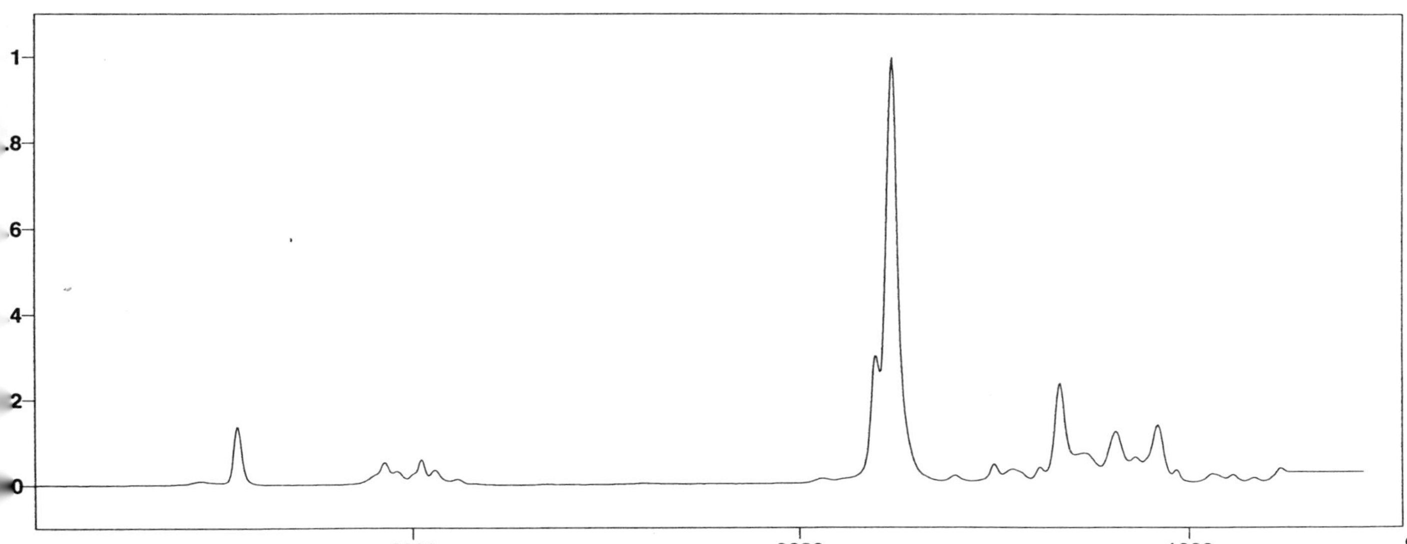

NORMETHSUXIMIDE

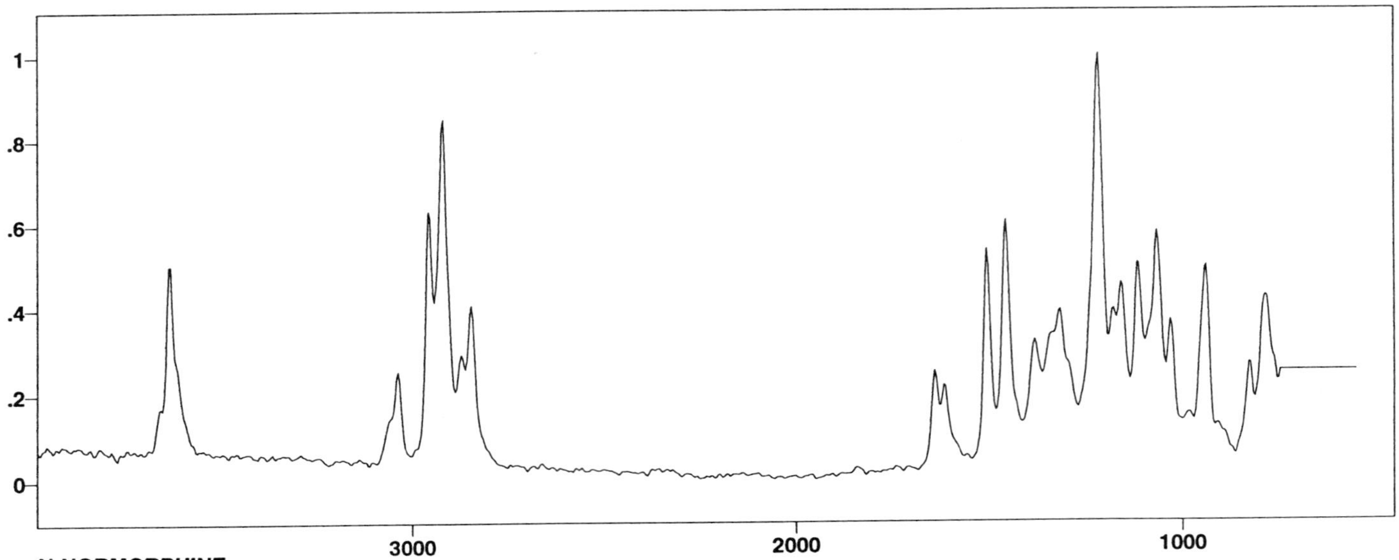

N-NORMORPHINE

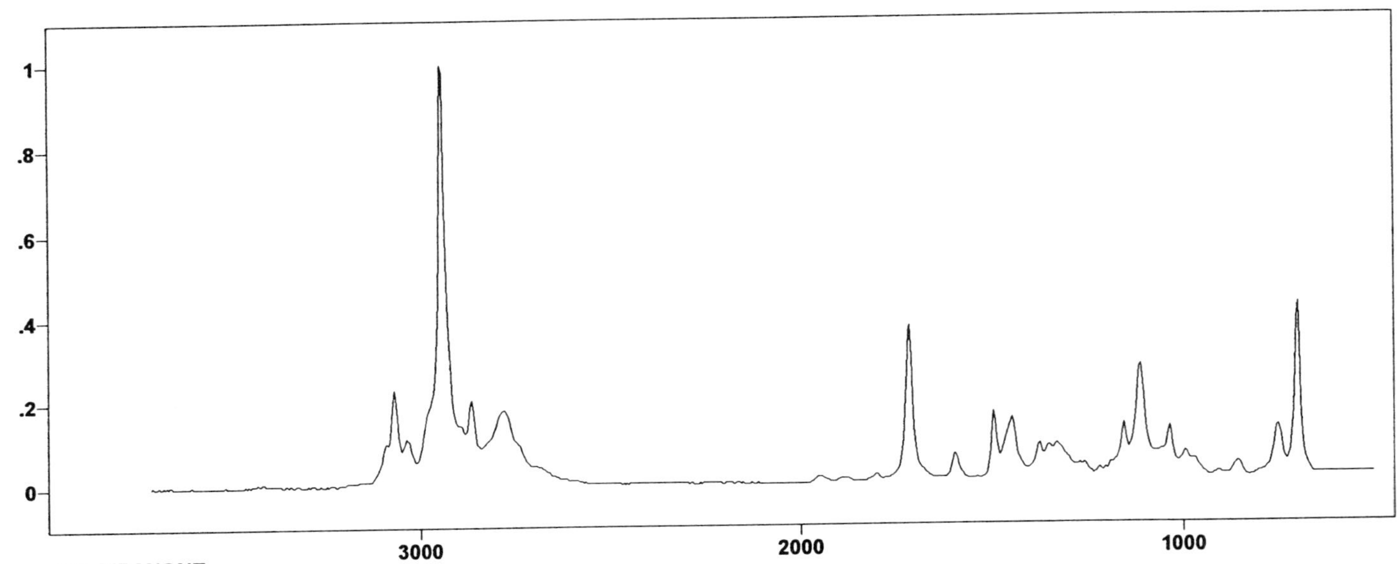

NORPIPANONE

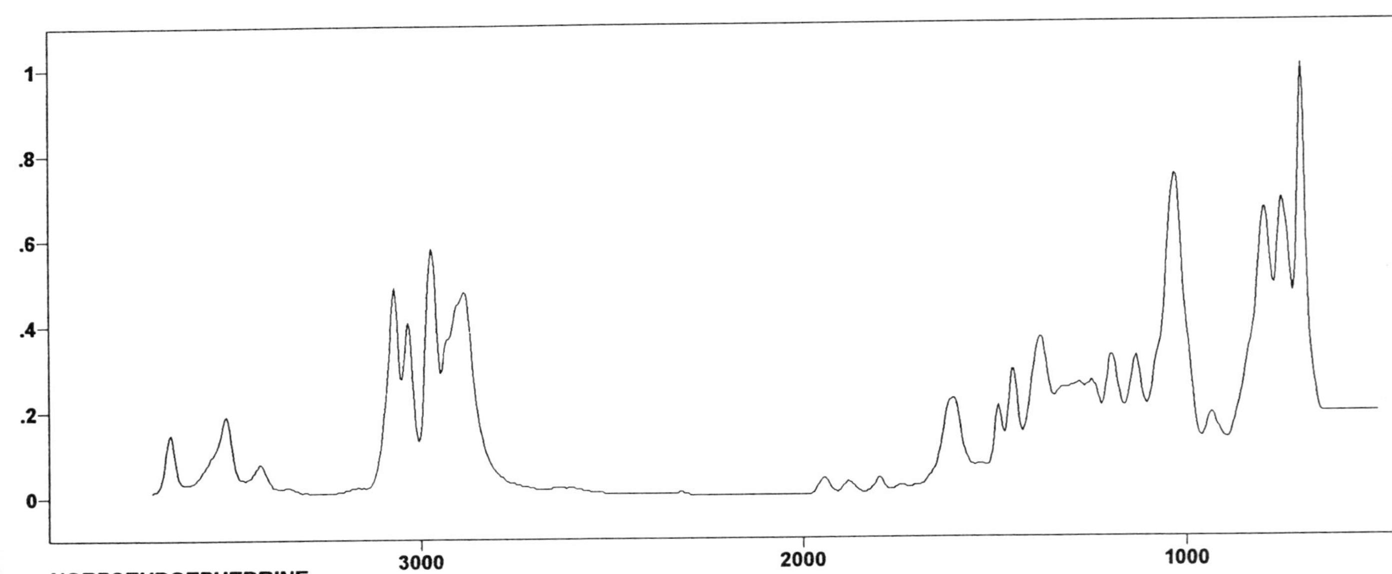

NORPSEUDOEPHEDRINE

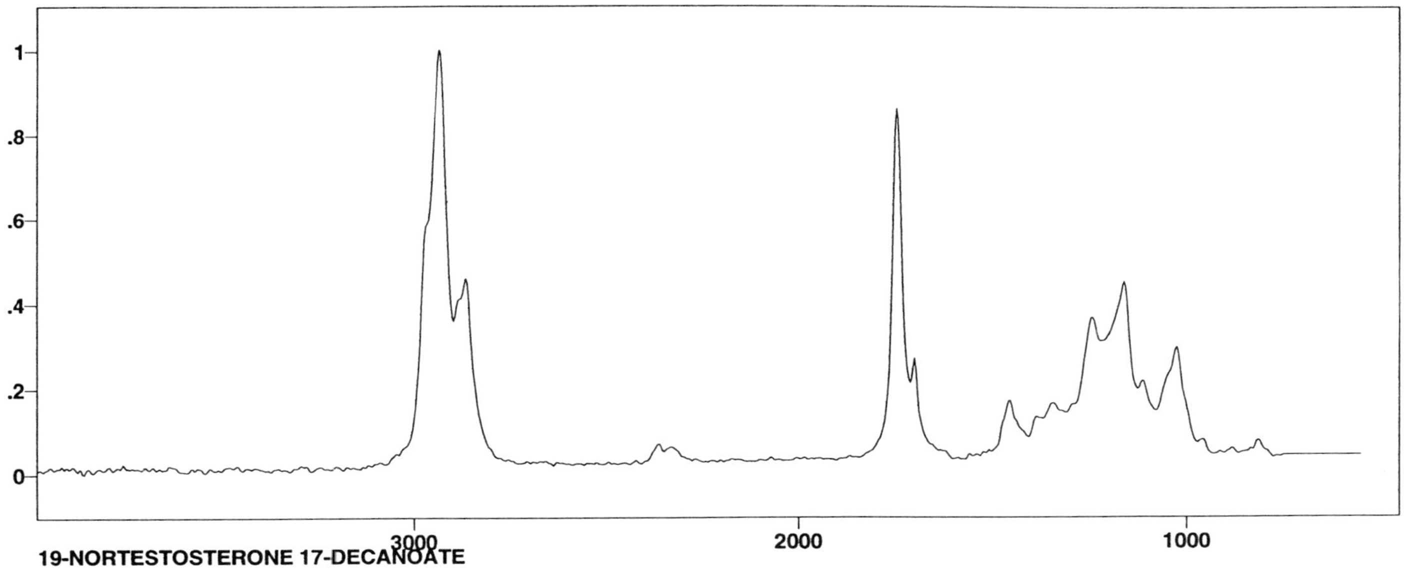

19-NORTESTOSTERONE 17-DECANOATE

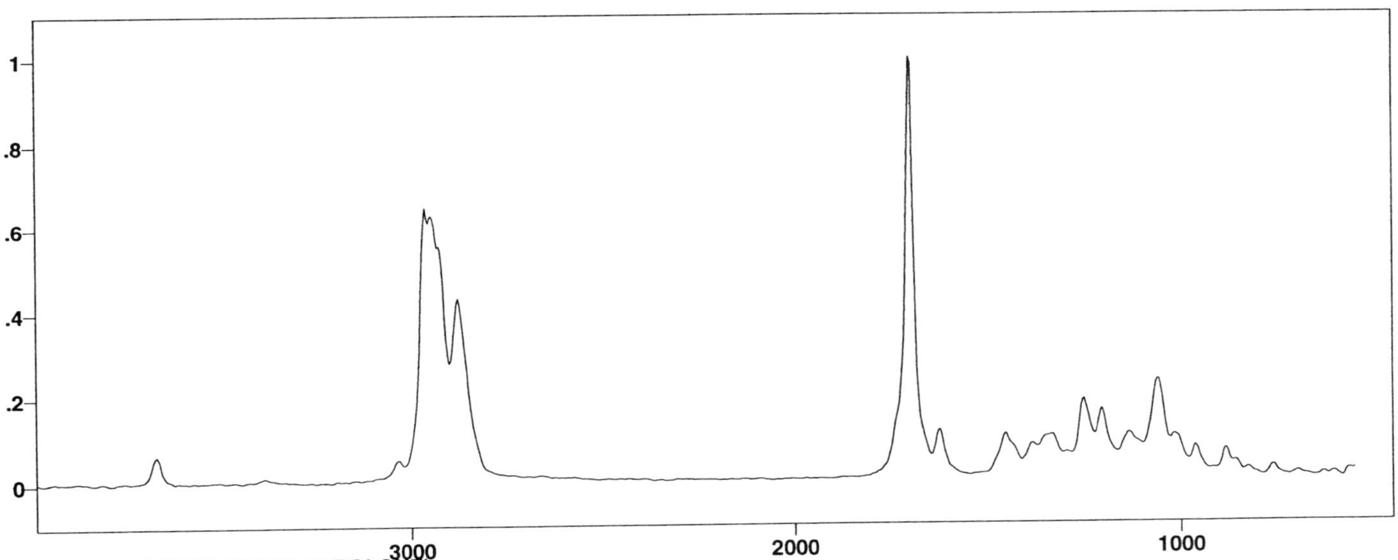

NORTESTOSTERONE(NANDROLONE)

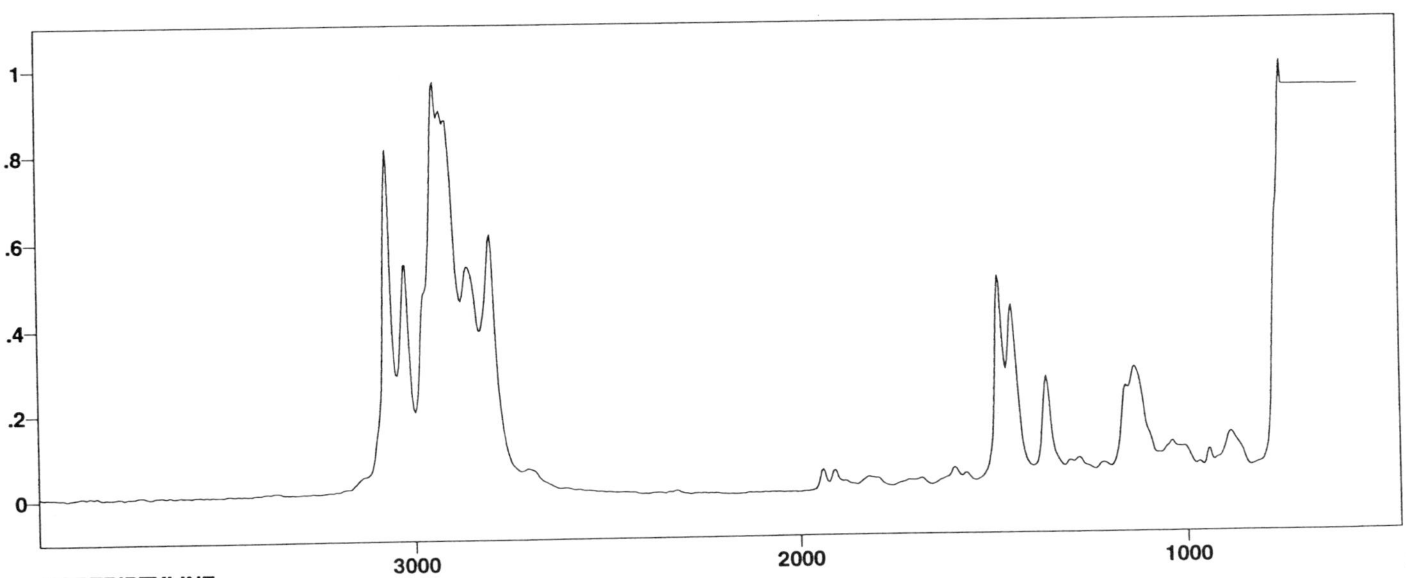

NORTRIPTYLINE

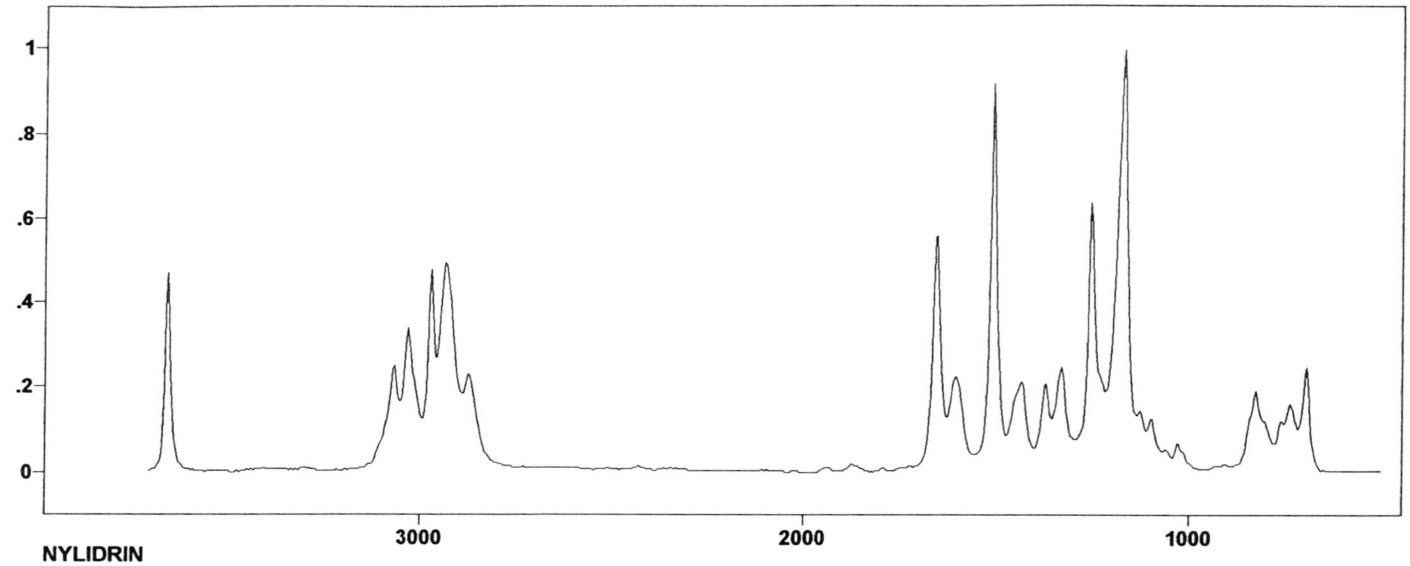

NYLIDRIN

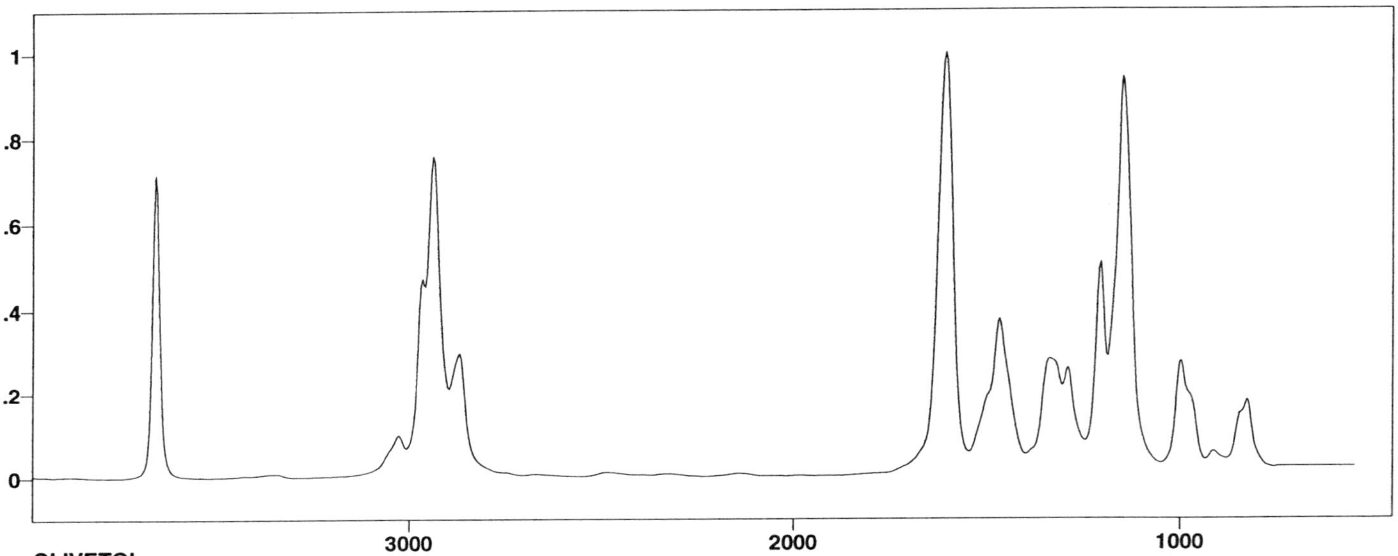

OLIVETOL

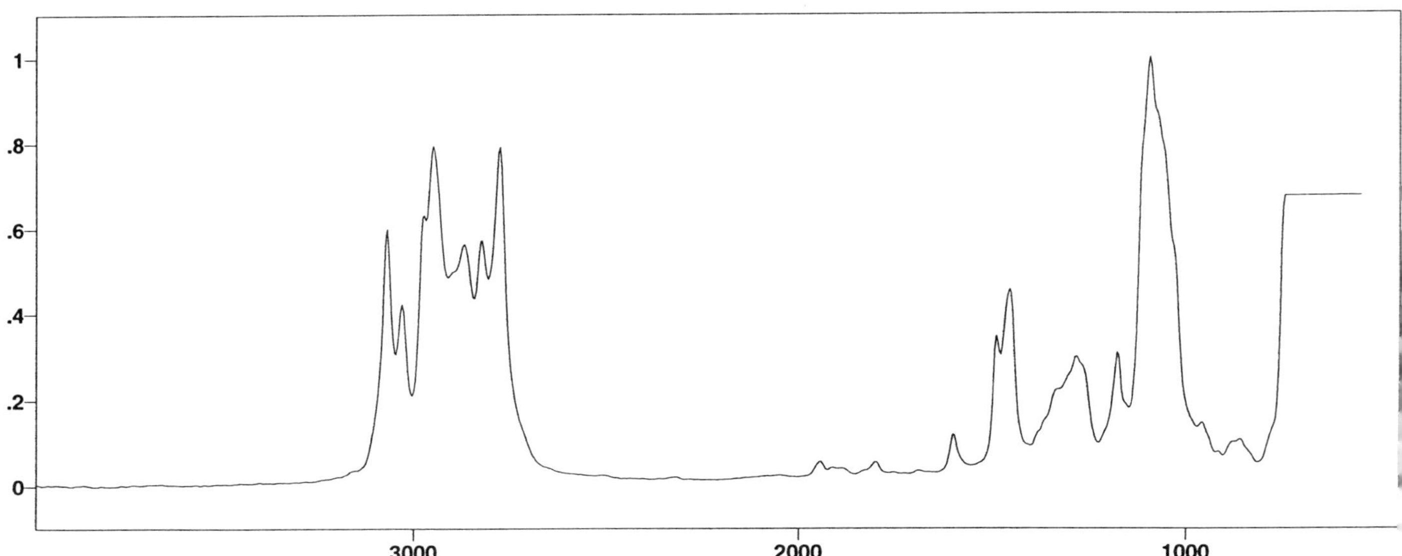

ORPHENADRINE

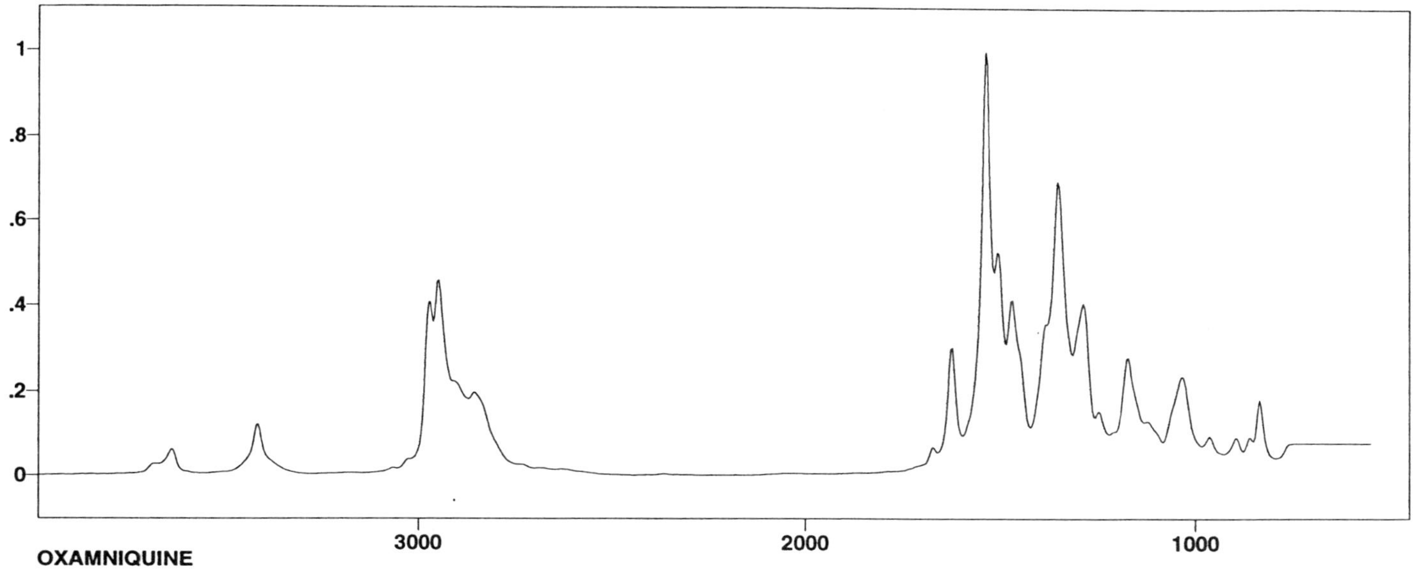

OXAMNIQUINE

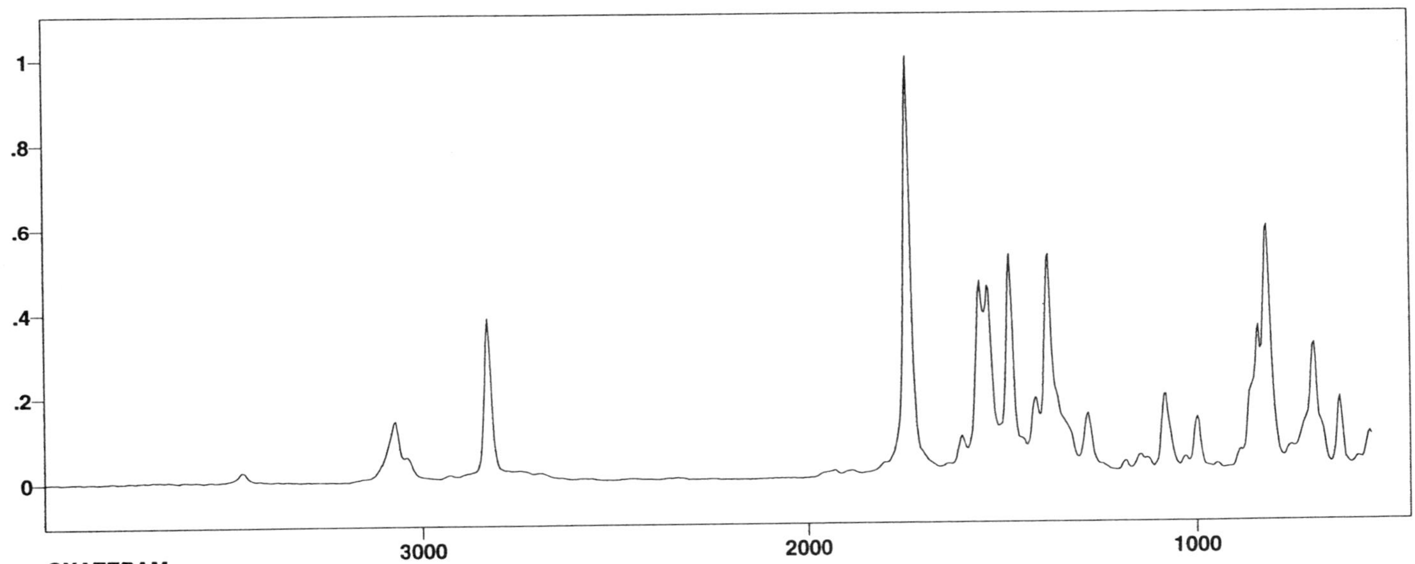

OXAZEPAM

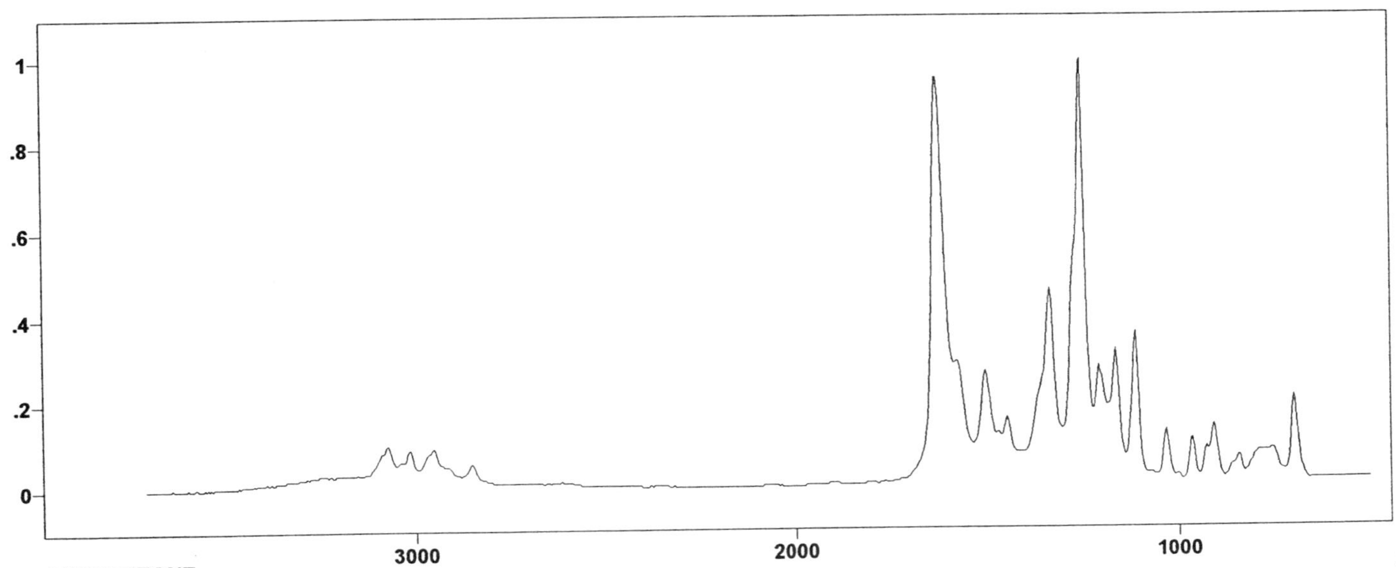

OXYBENZONE

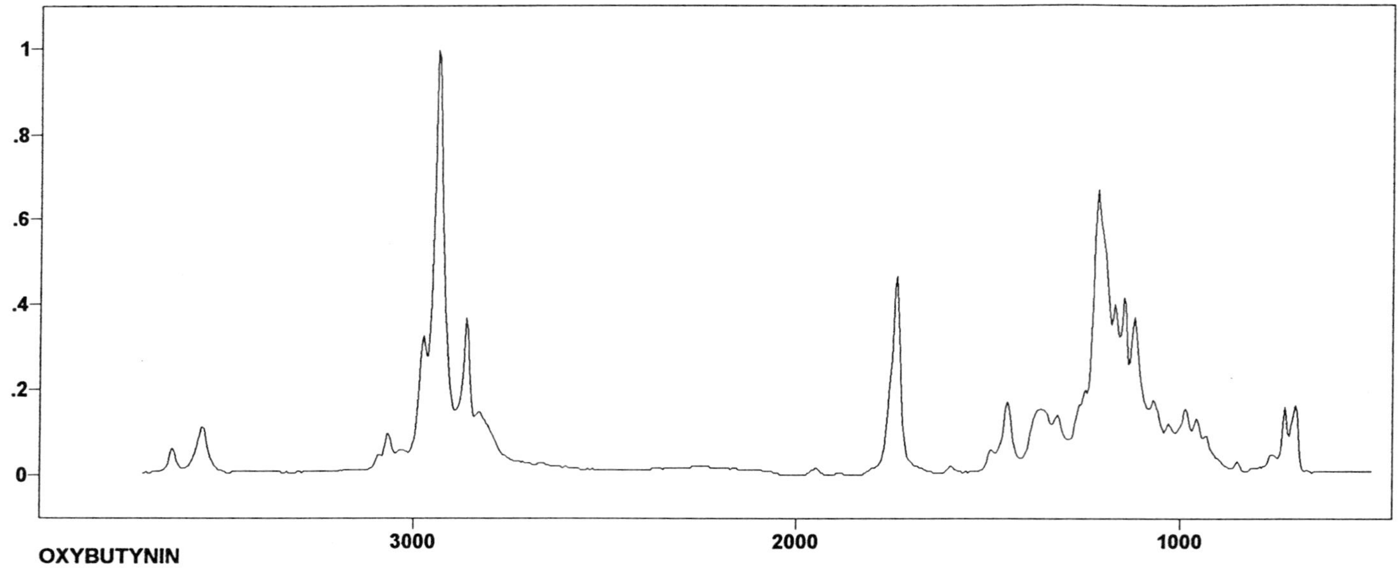

OXYBUTYNIN

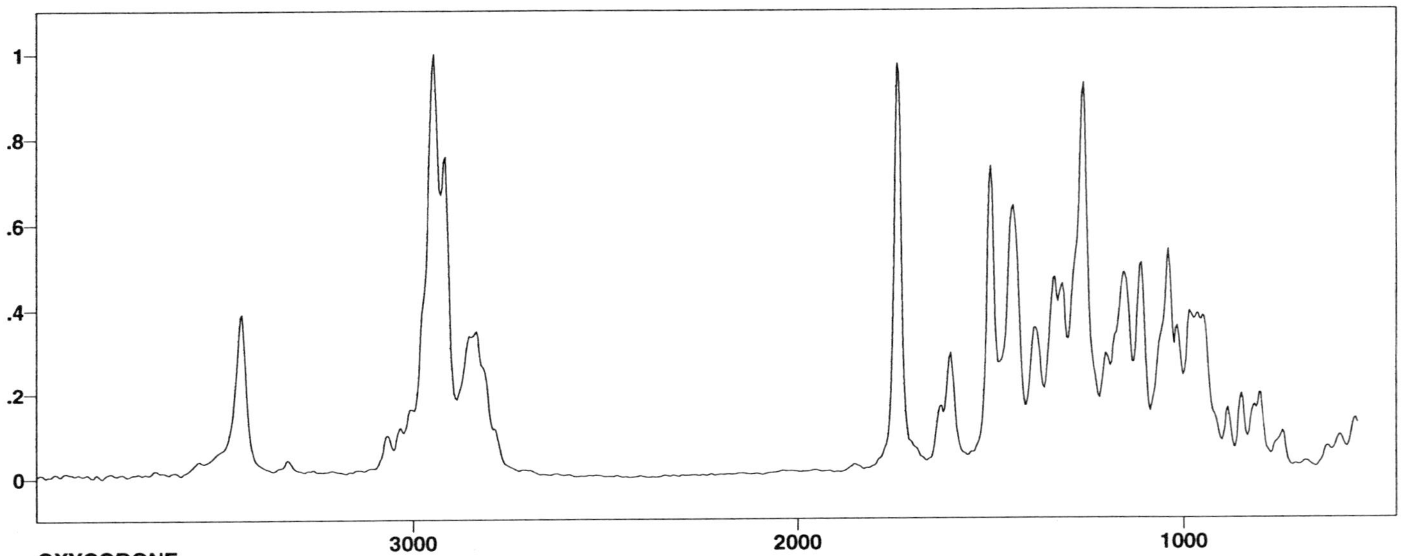

OXYCODONE

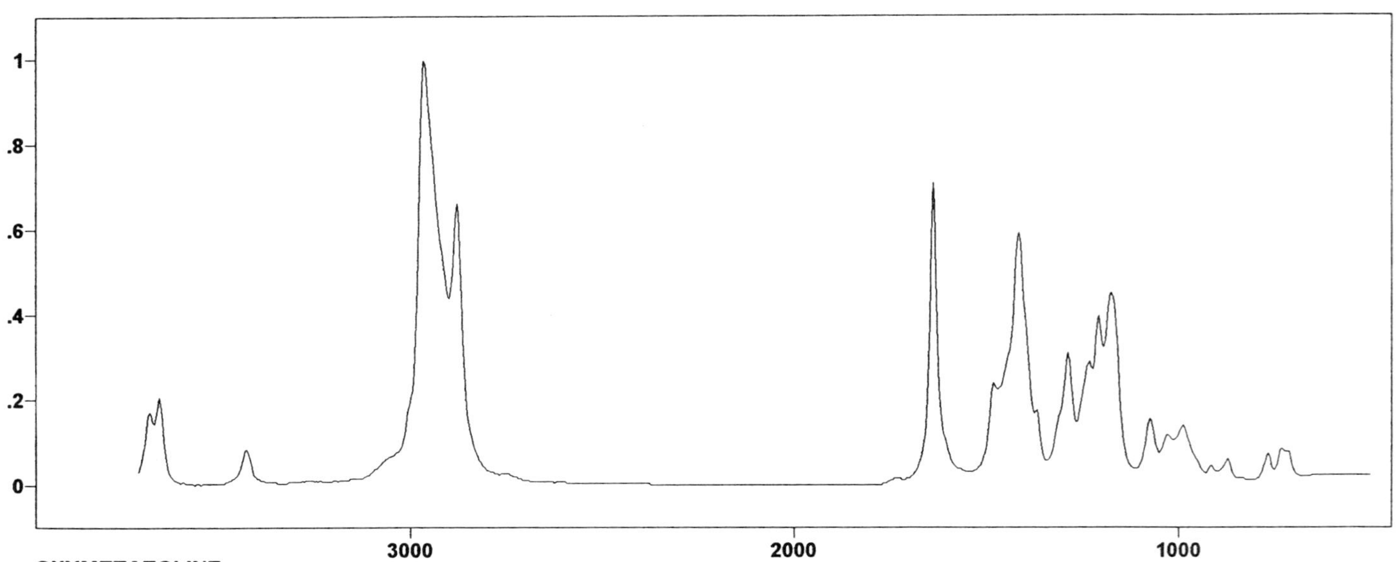

OXYMETAZOLINE

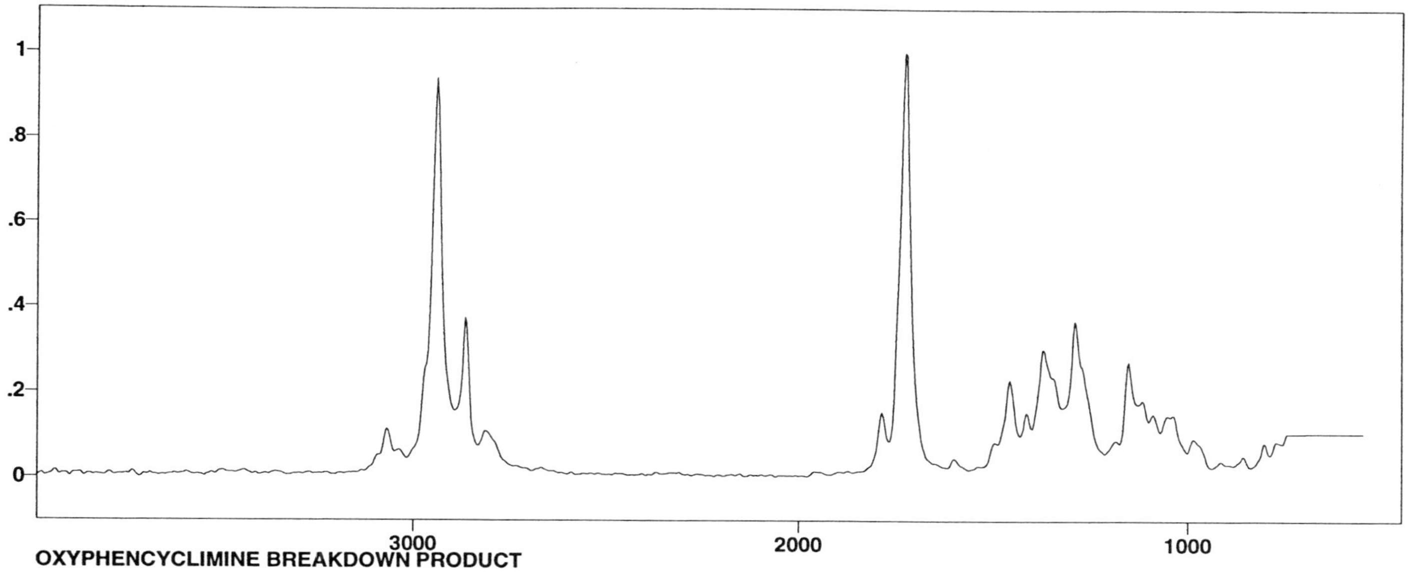

OXYPHENCYCLIMINE BREAKDOWN PRODUCT

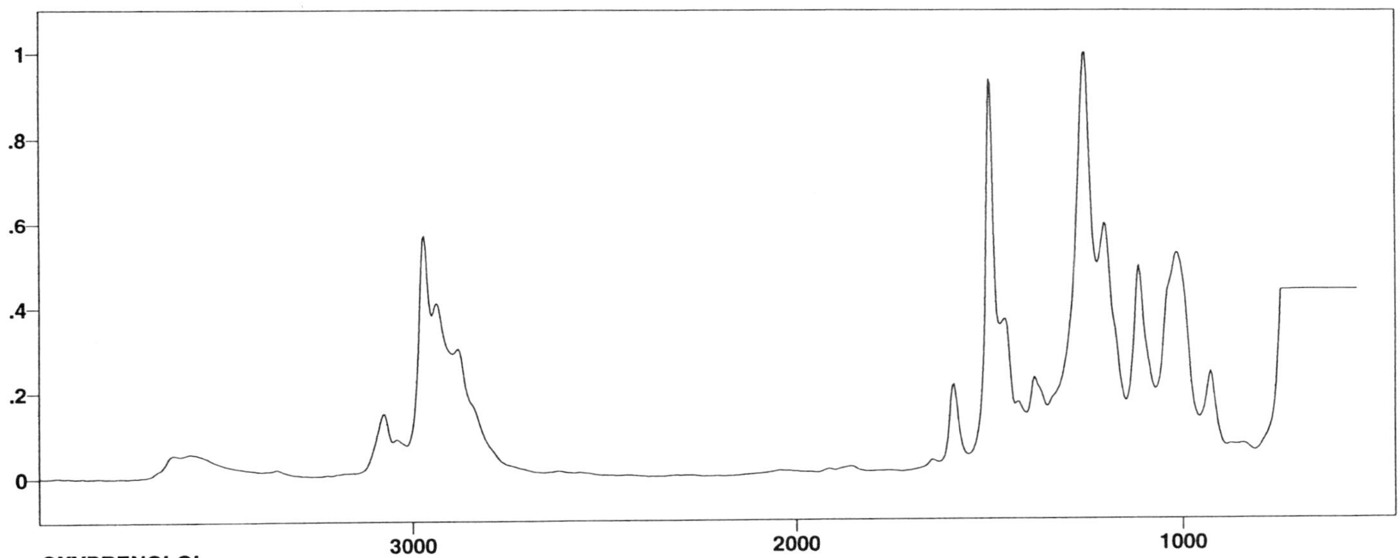

OXYPRENOLOL

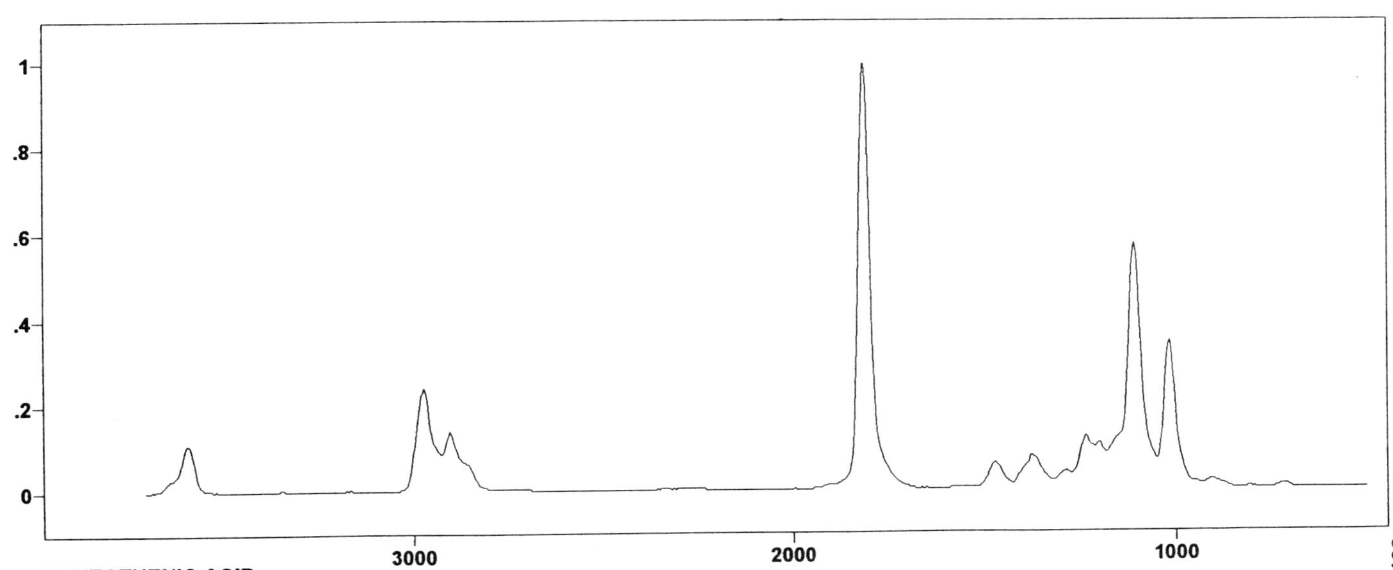

PANTOTHENIC ACID

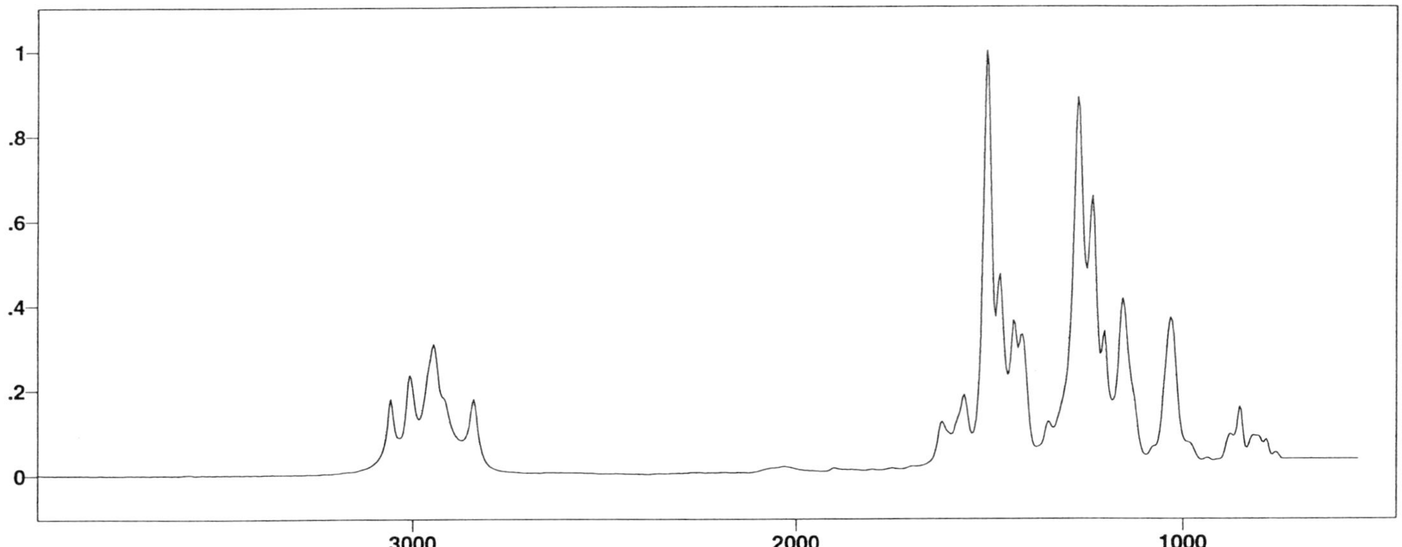

PAPAVERINE

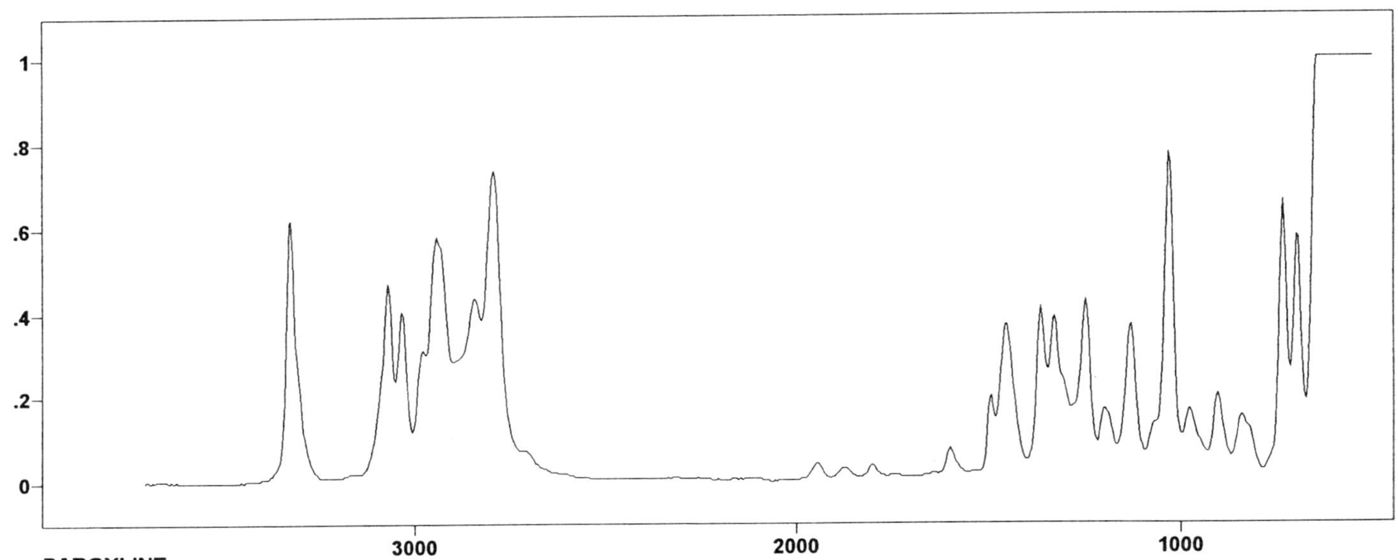

PARGYLINE

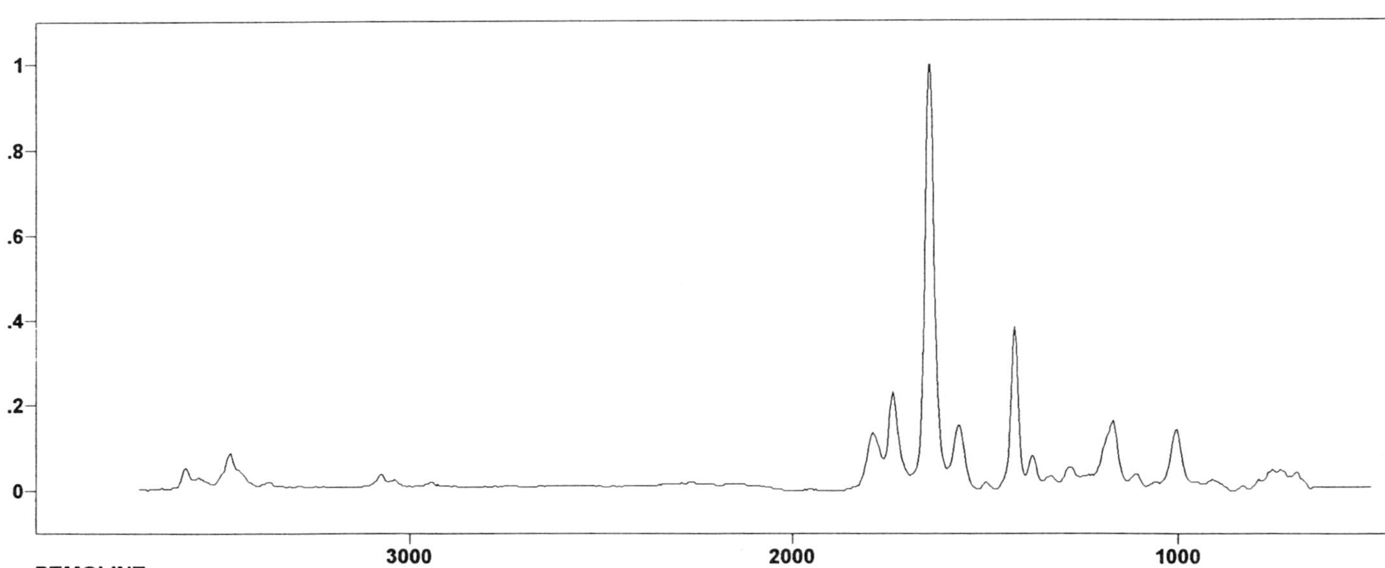

PEMOLINE

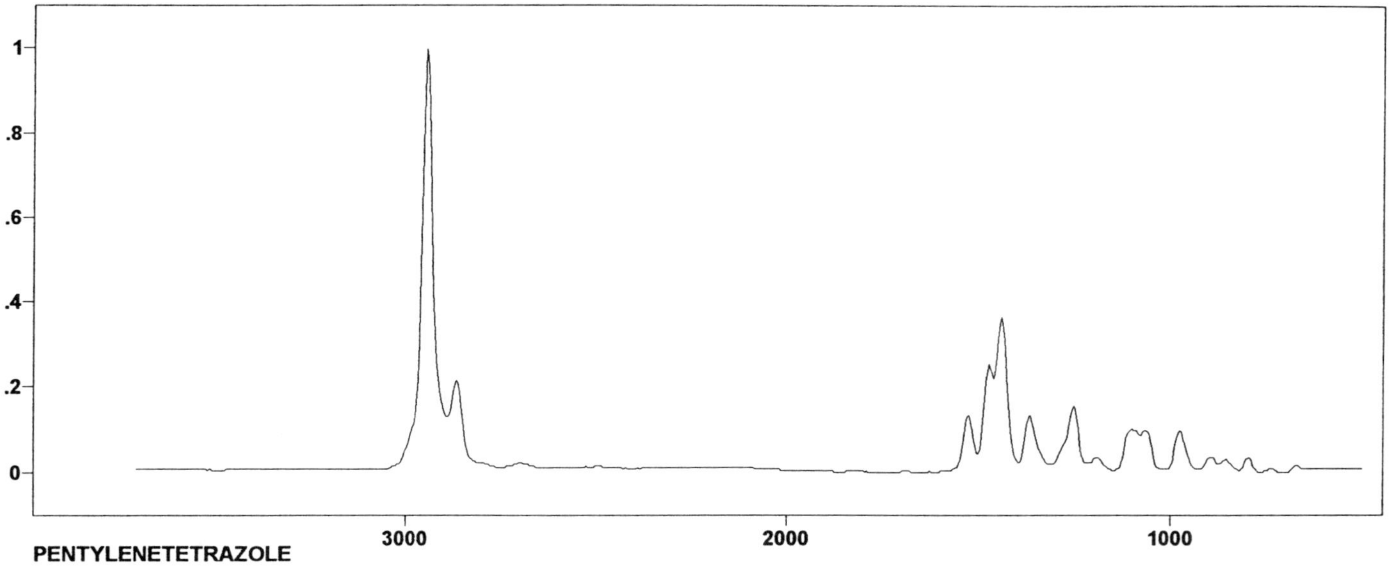

PENTYLENETETRAZOLE

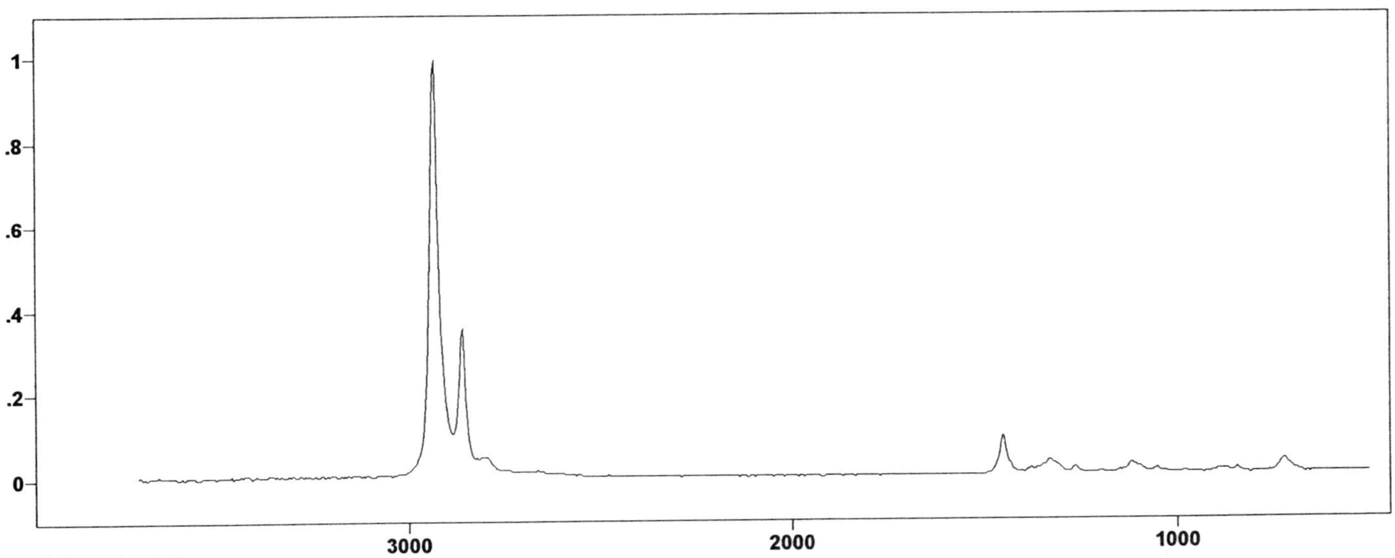

PERHEXILENE

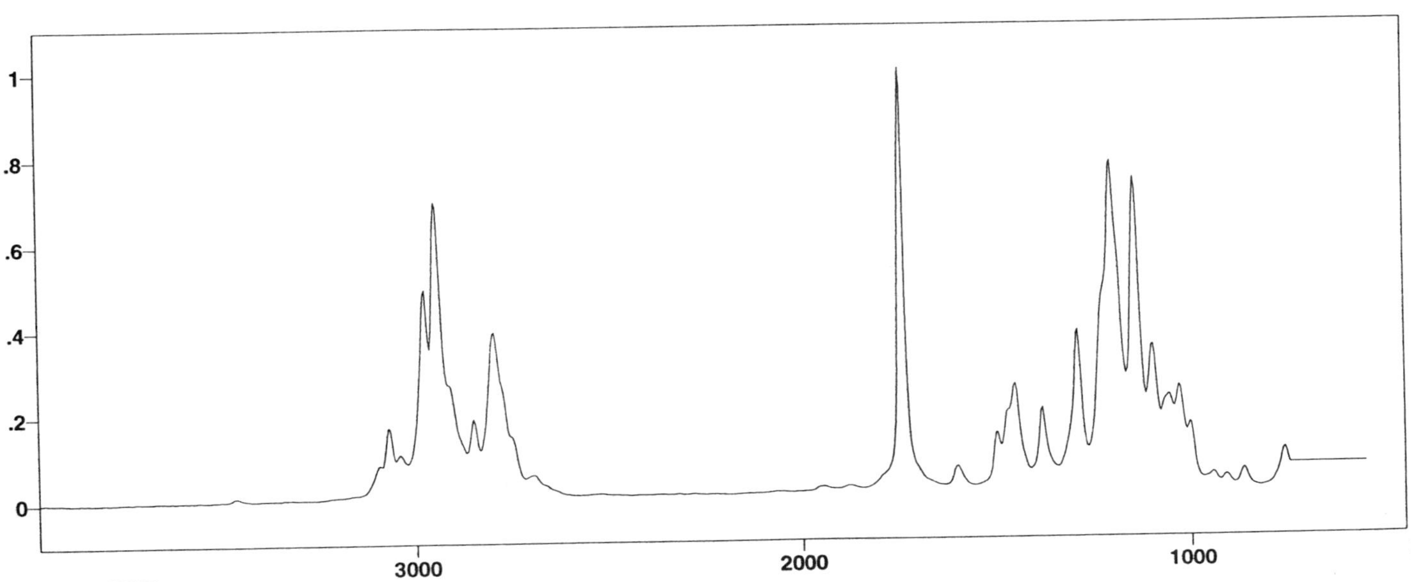

PETHIDINE

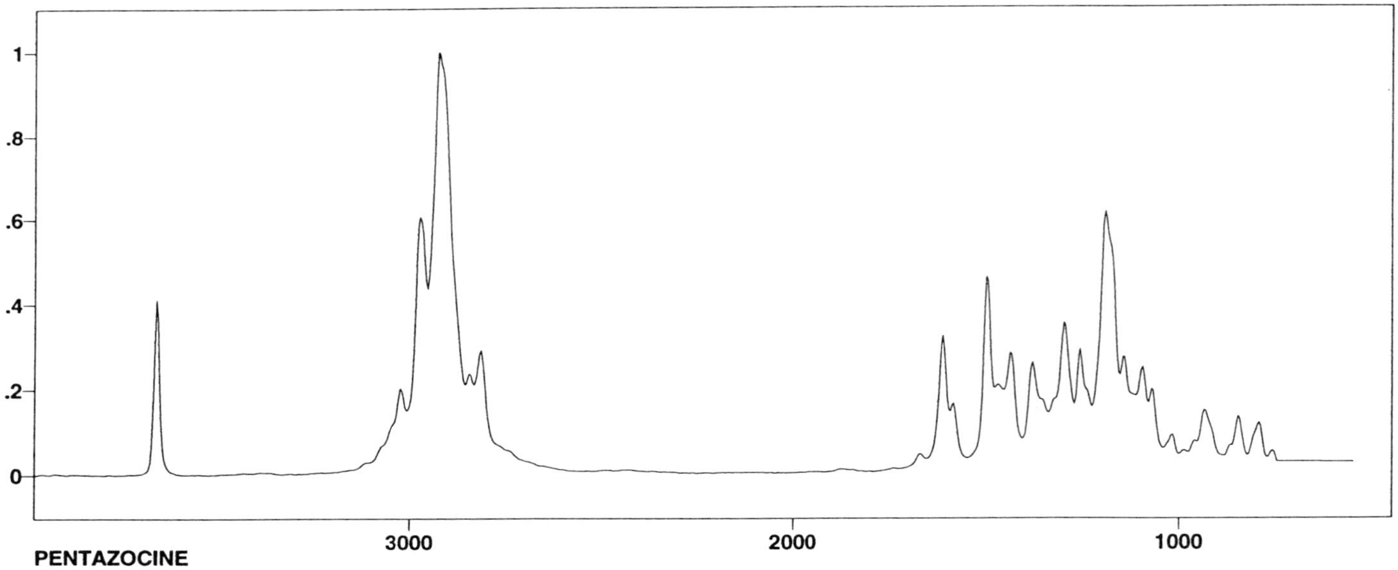

PENTAZOCINE

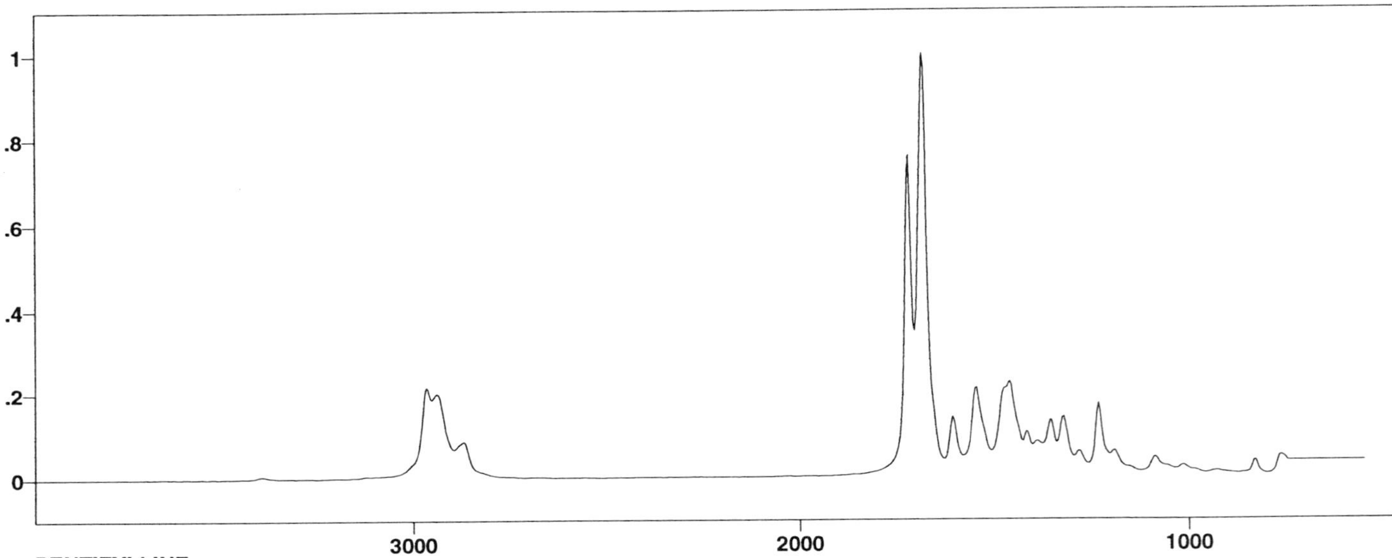

PENTIFYLLINE

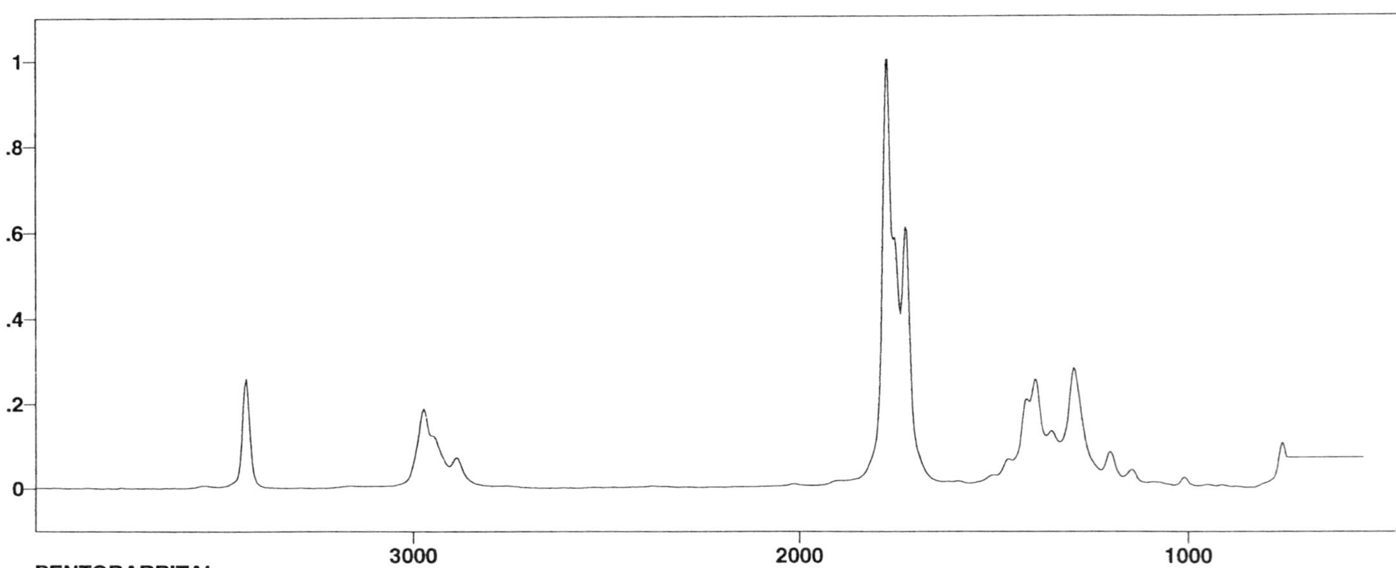

PENTOBARBITAL

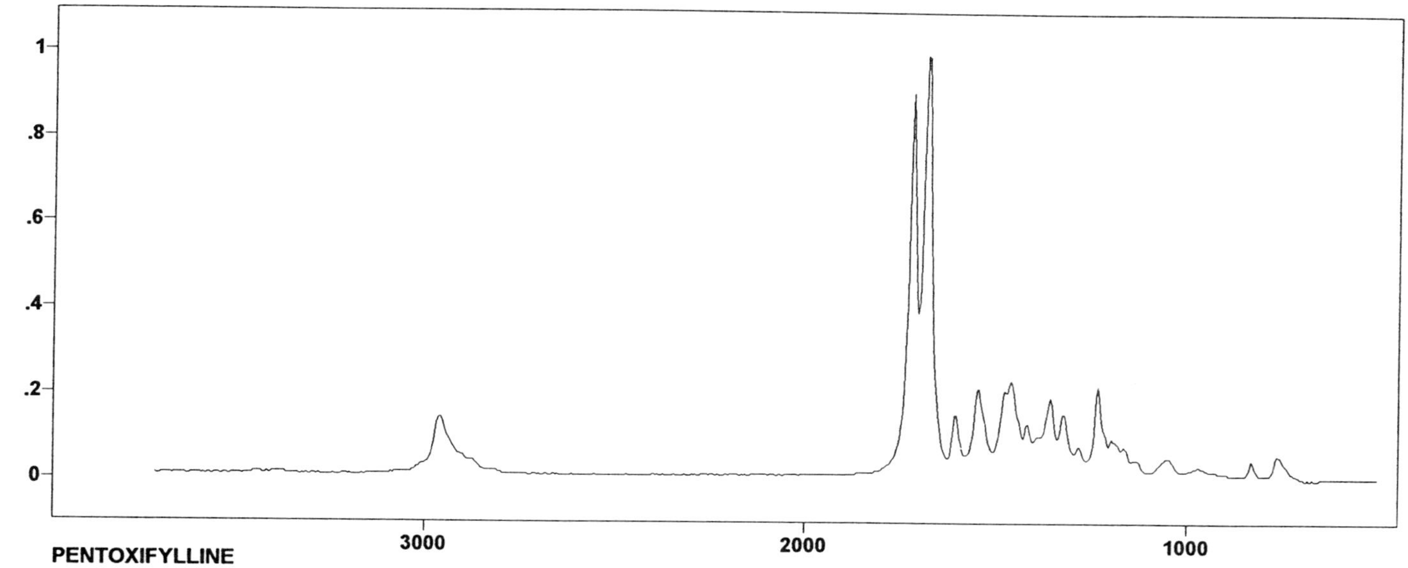

PENTOXIFYLLINE

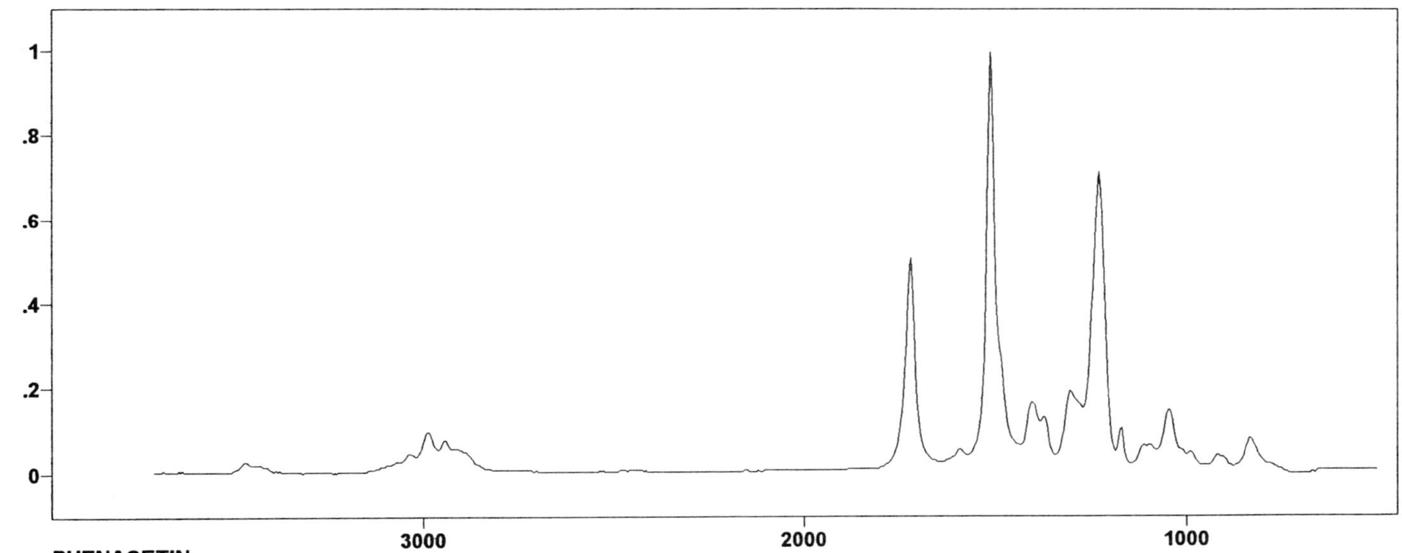

PHENACETIN

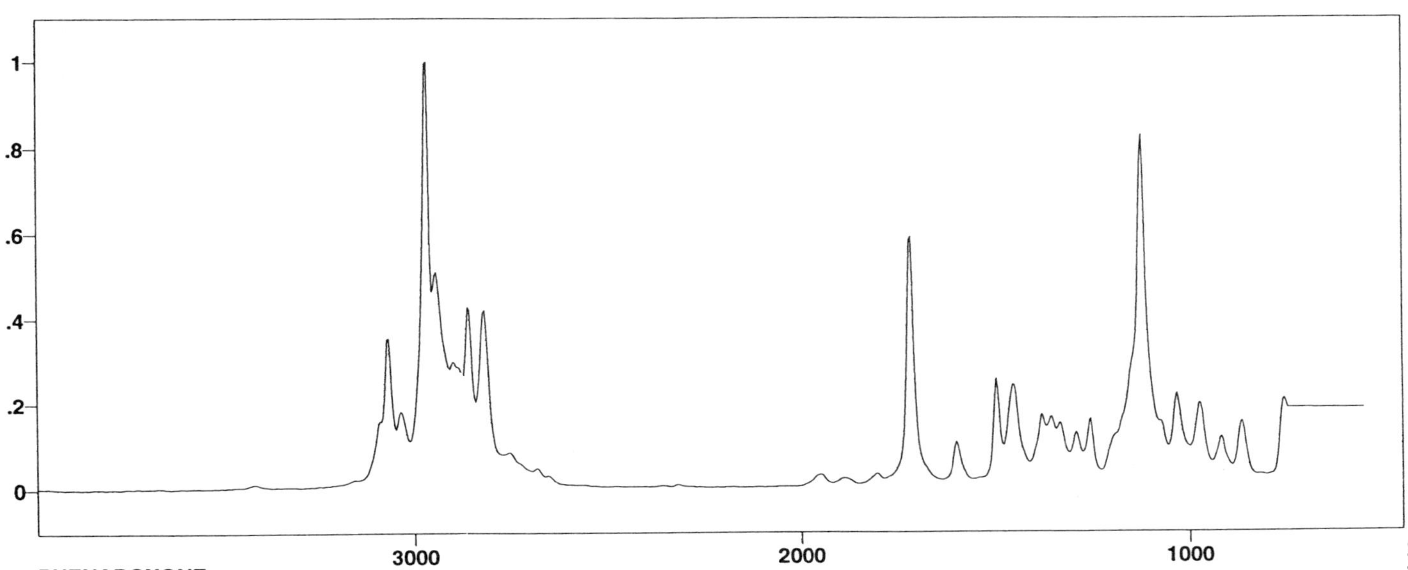

PHENADOXONE

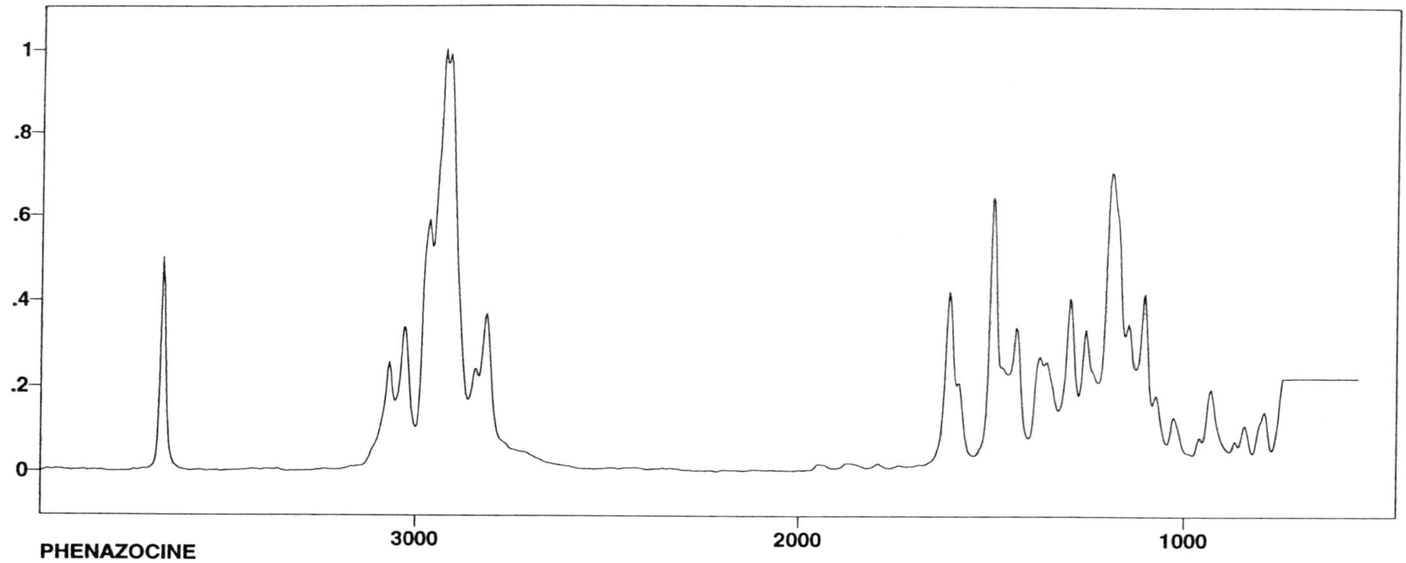

PHENAZOCINE

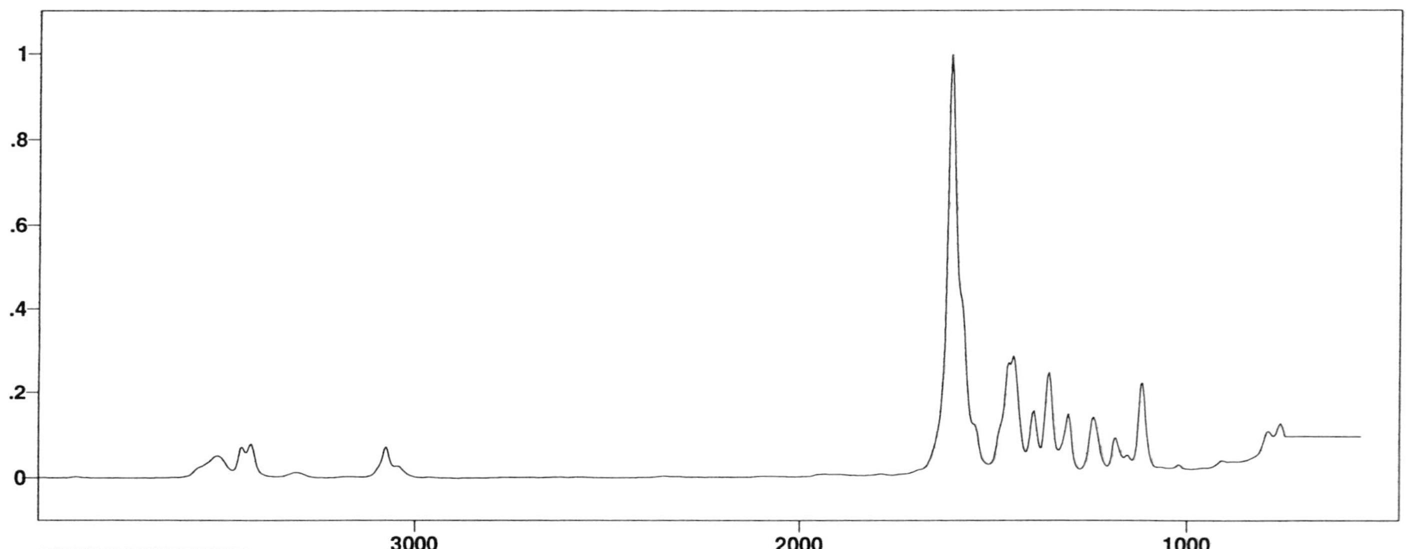

PHENAZOPYRIDINE

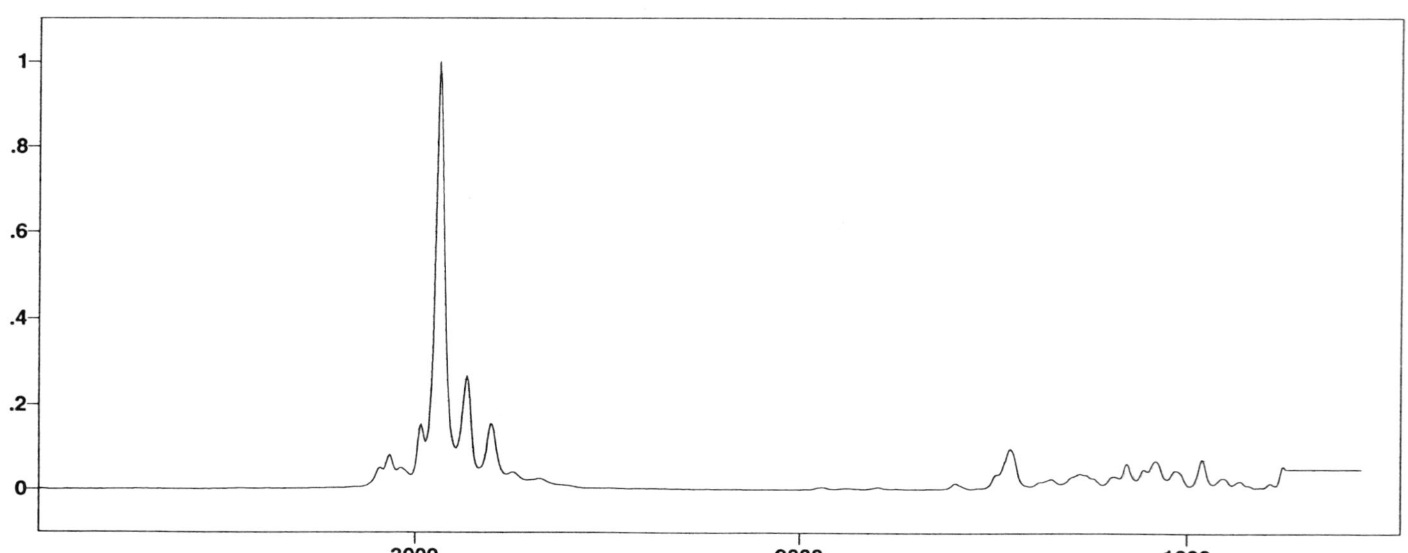

PHENCYCLIDINE

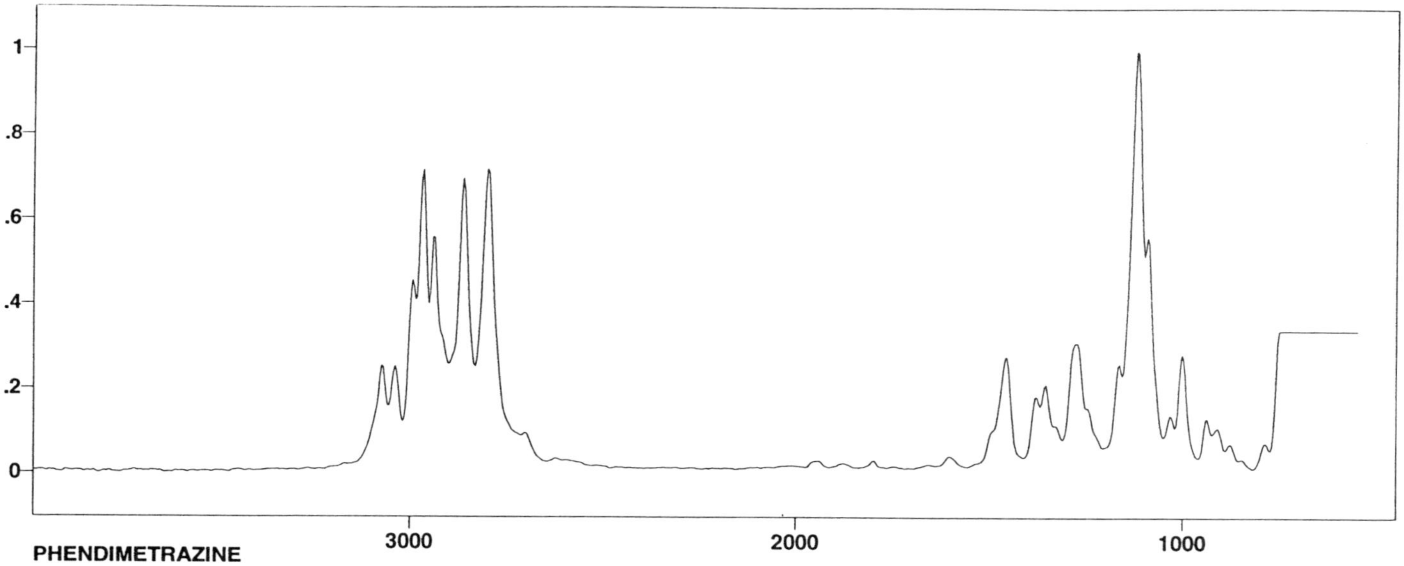

PHENDIMETRAZINE

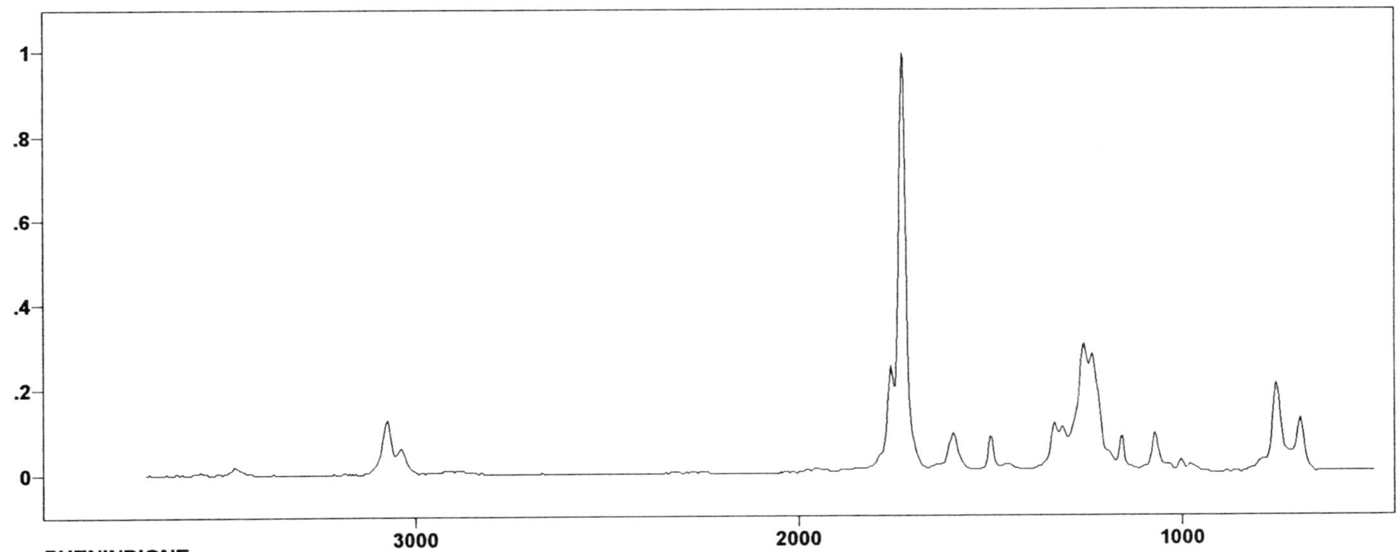

PHENINDIONE

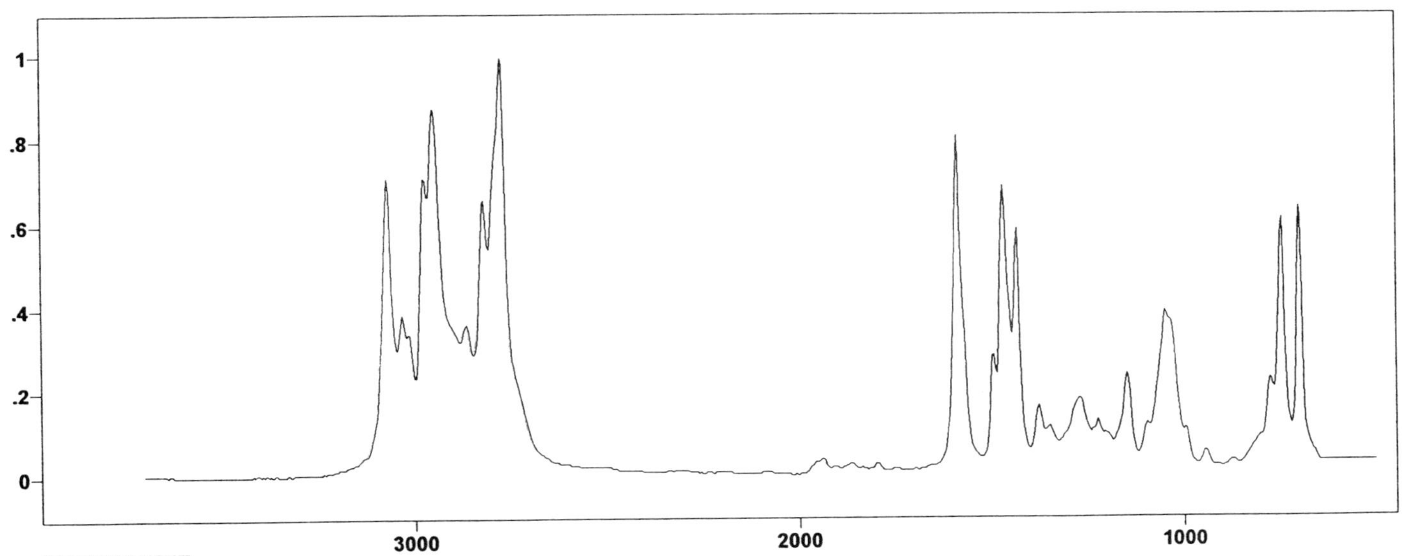

PHENIRAMINE

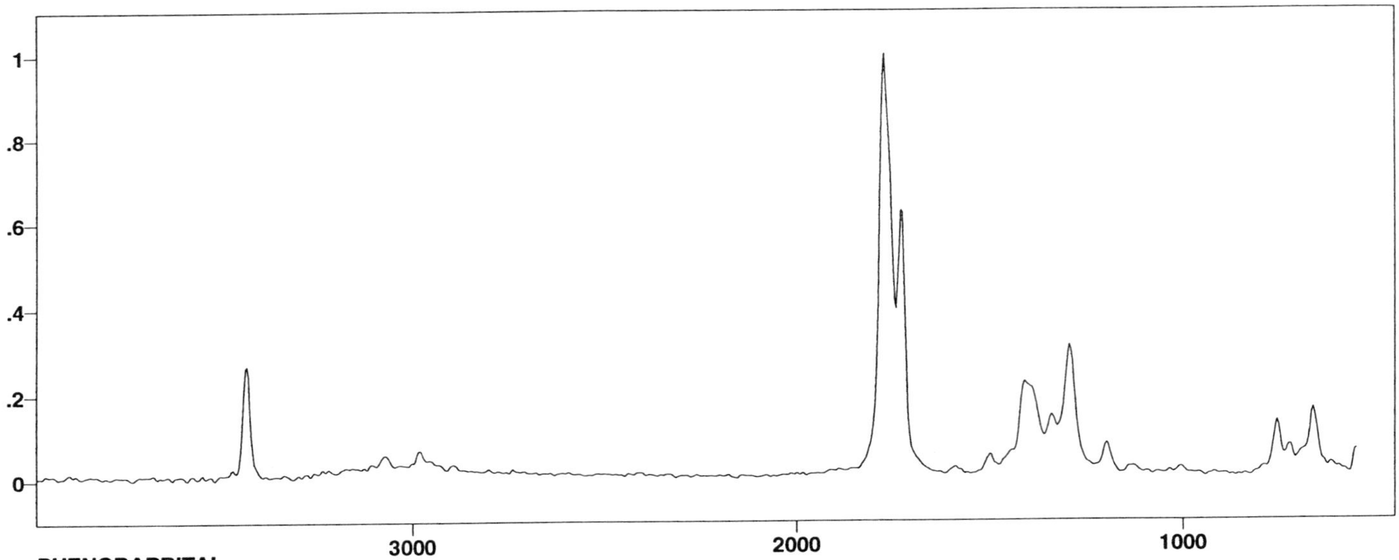

PHENOBARBITAL

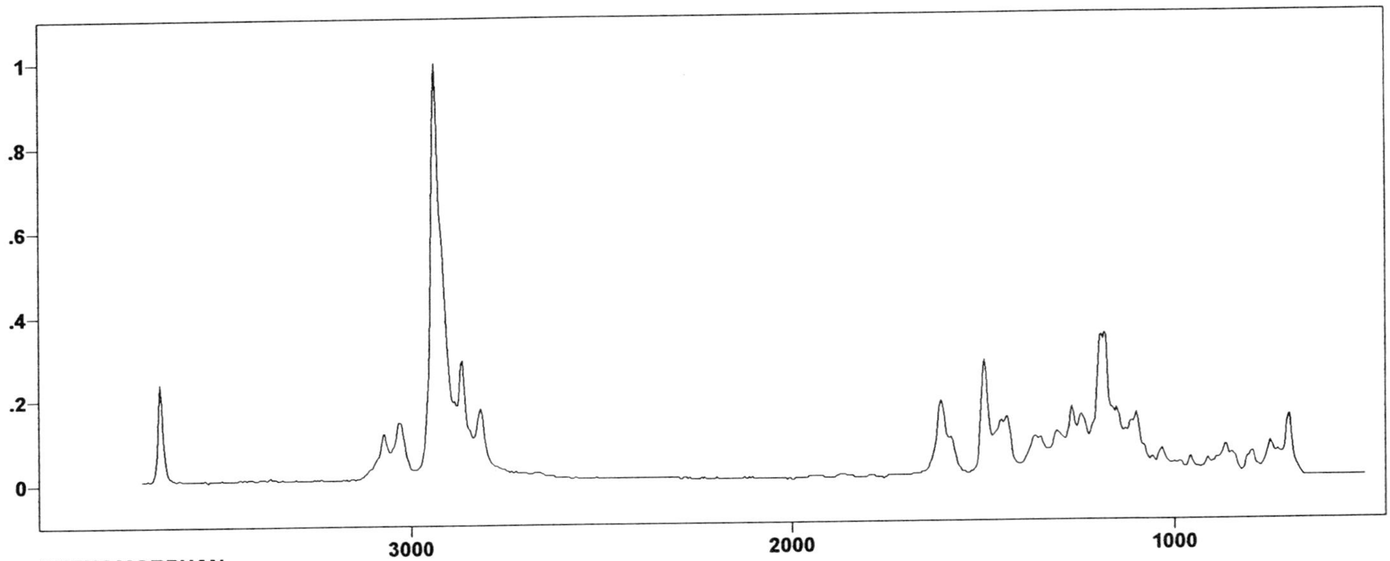

PHENOMORPHAN

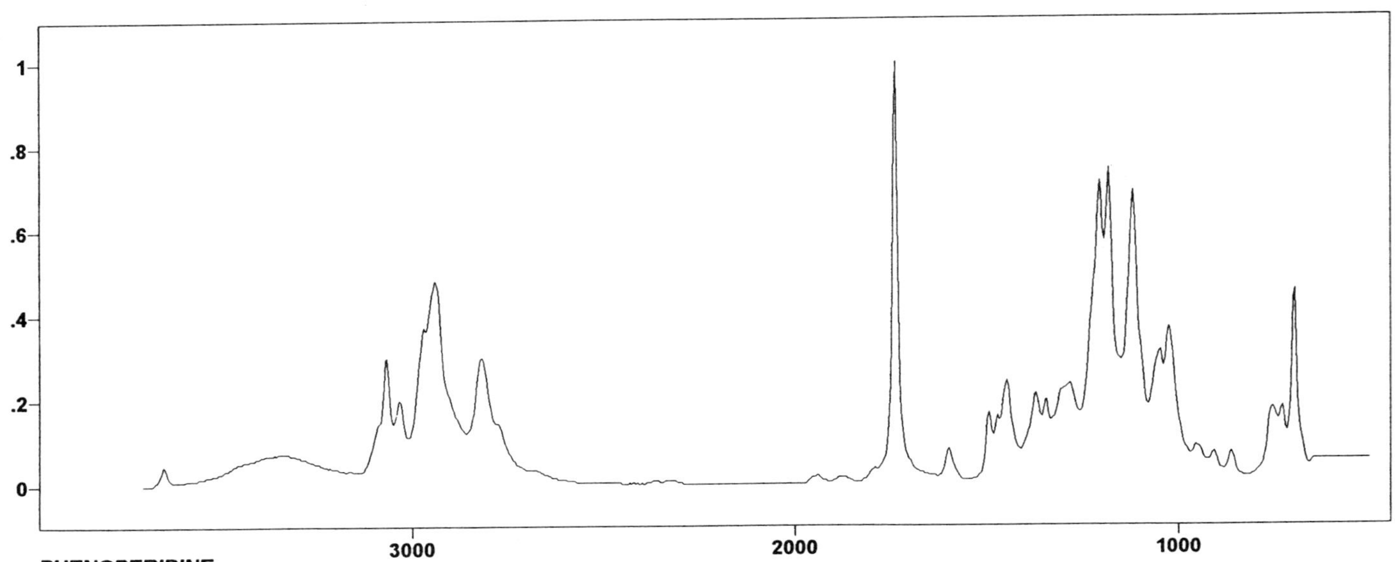

PHENOPERIDINE

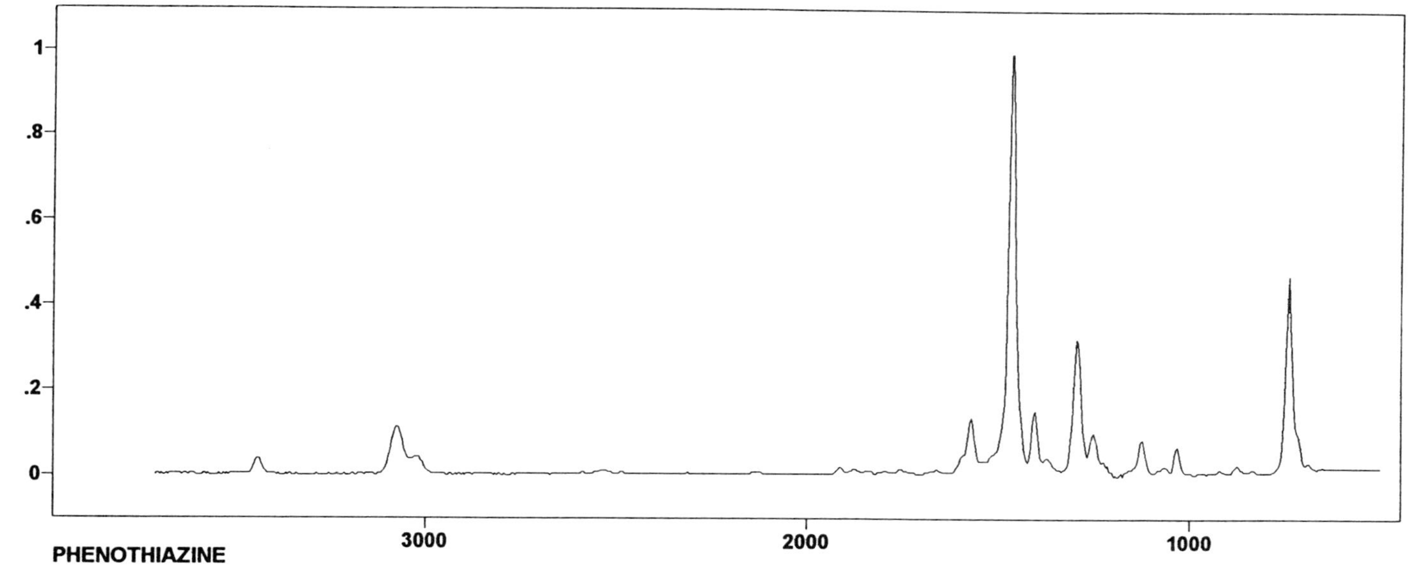

PHENOTHIAZINE

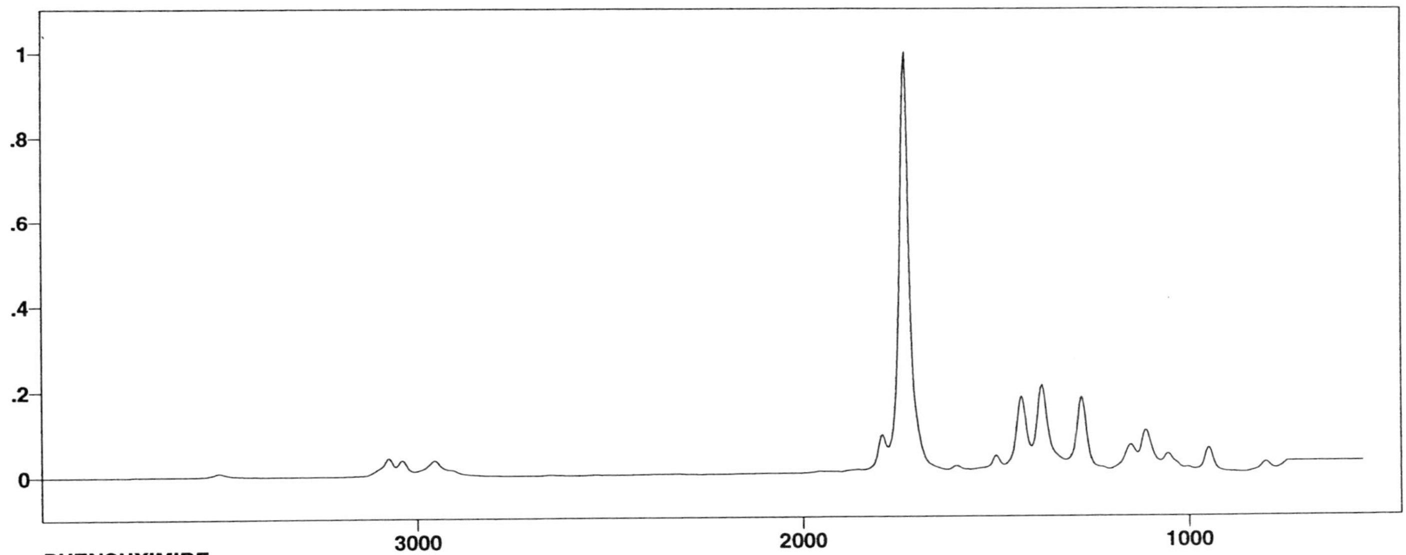

PHENSUXIMIDE

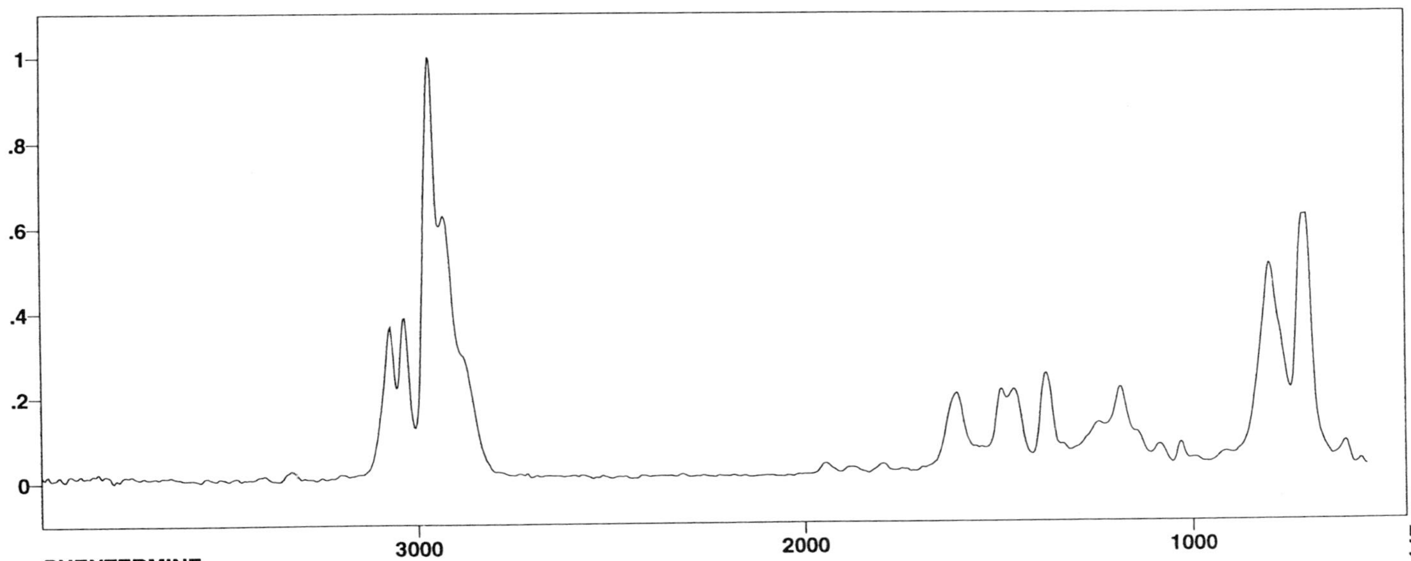

PHENTERMINE

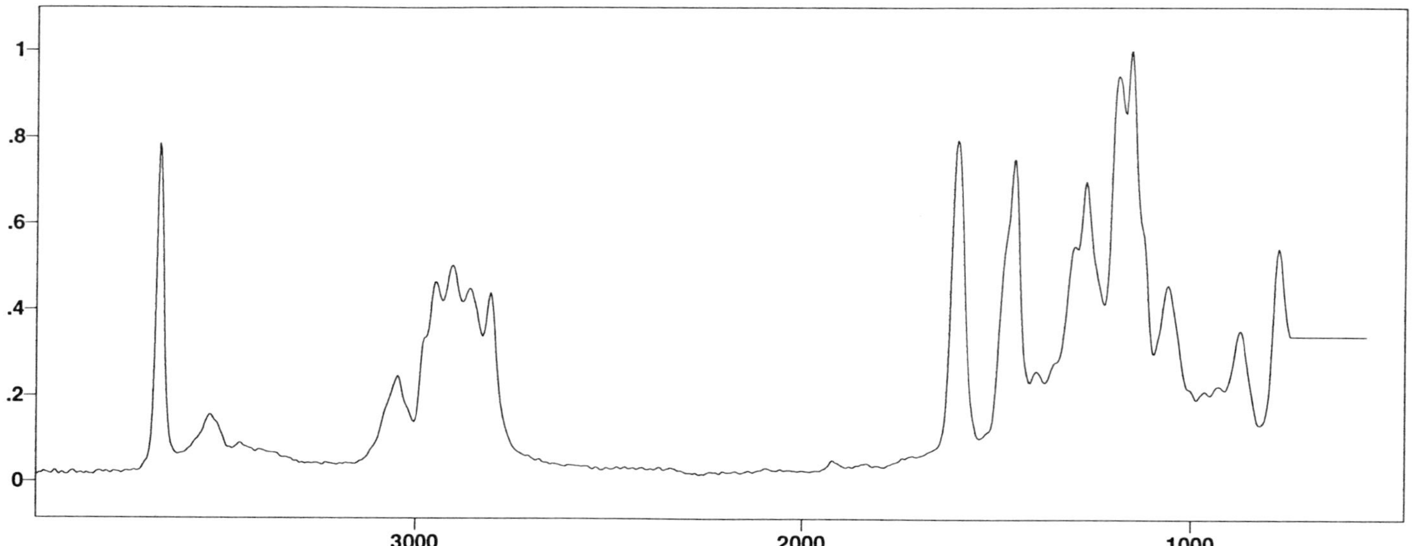

PHENYLEPHRINE

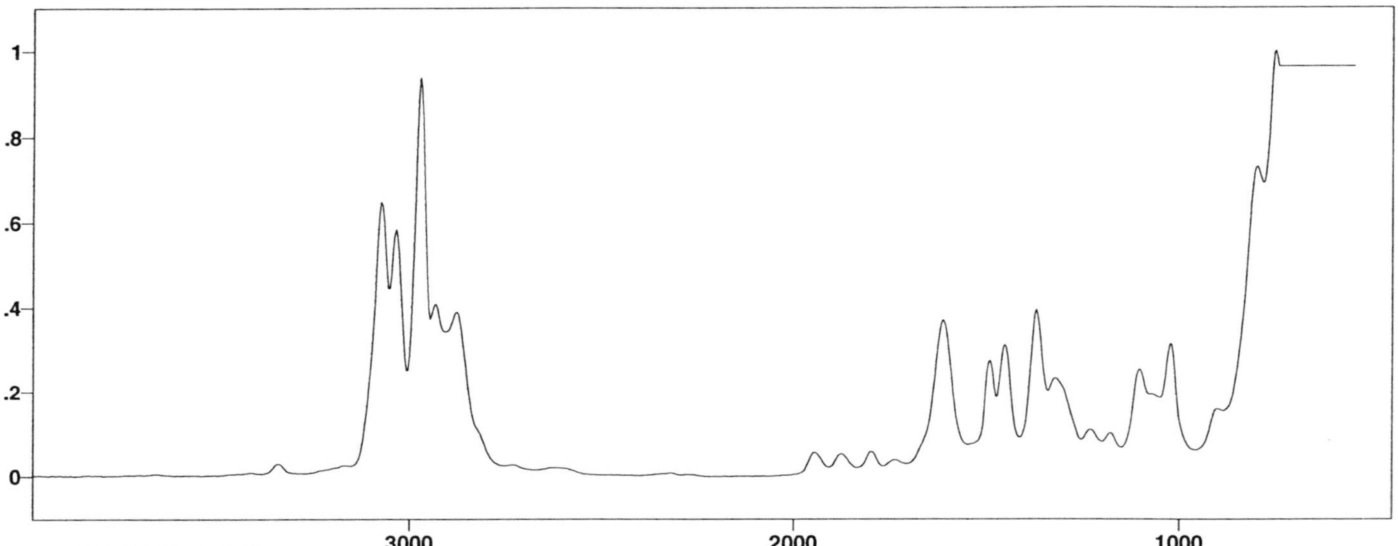

PHENYLETHYLAMINE

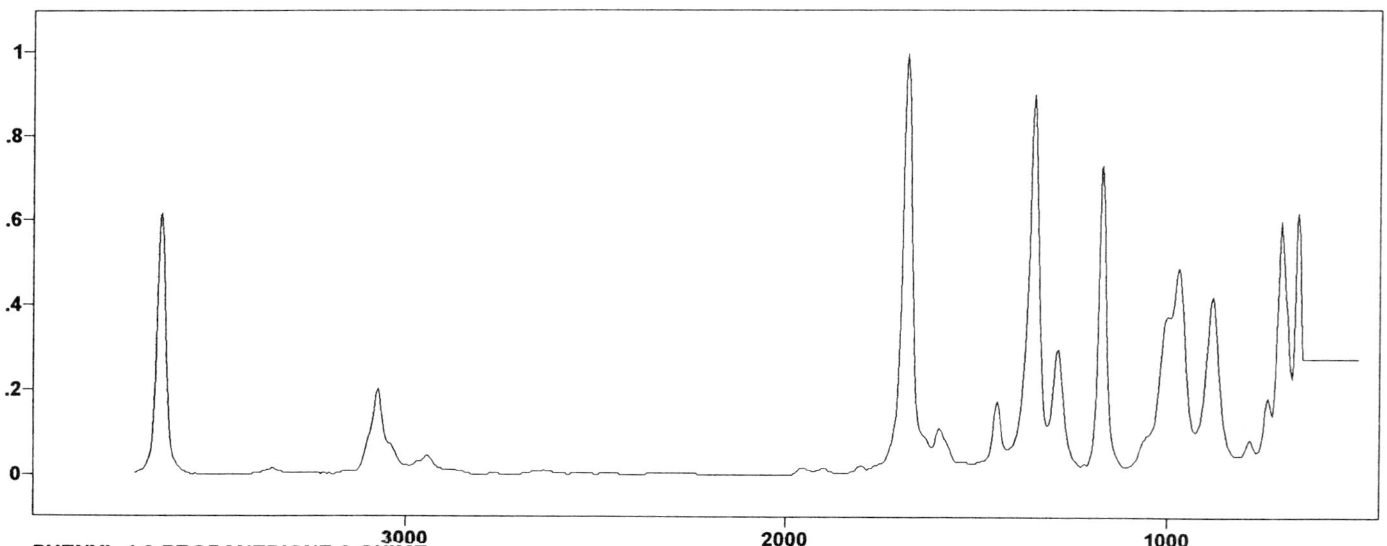

PHENYL-1,2-PROPANEDIONE-2-OXIME

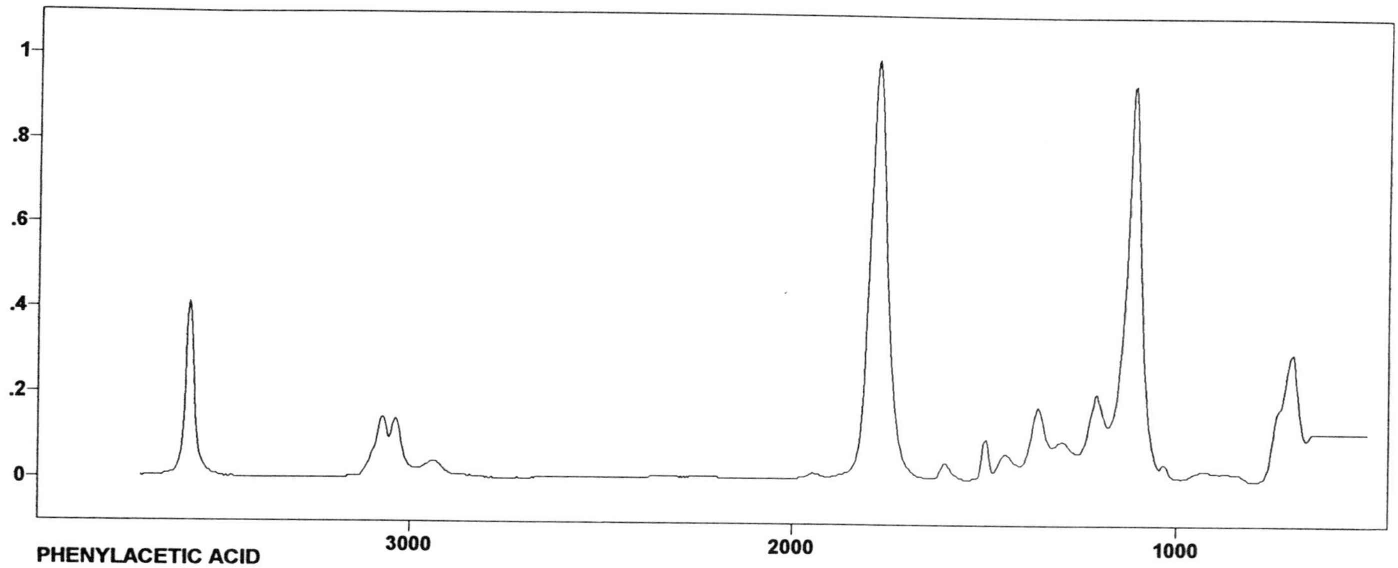

PHENYLACETIC ACID

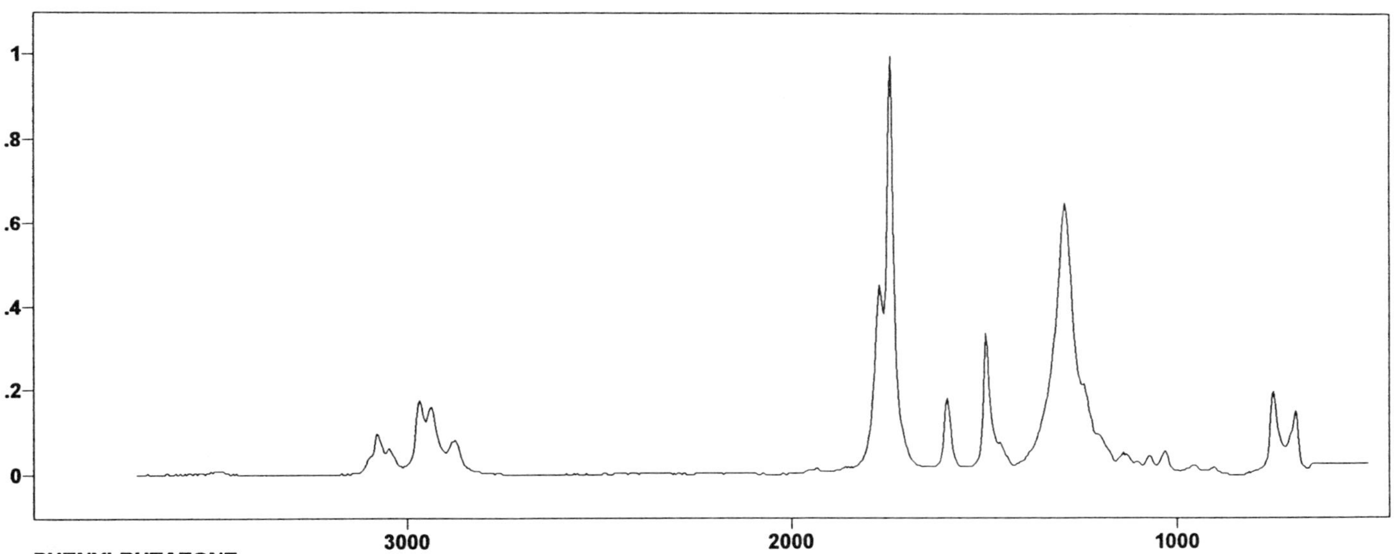

PHENYLBUTAZONE

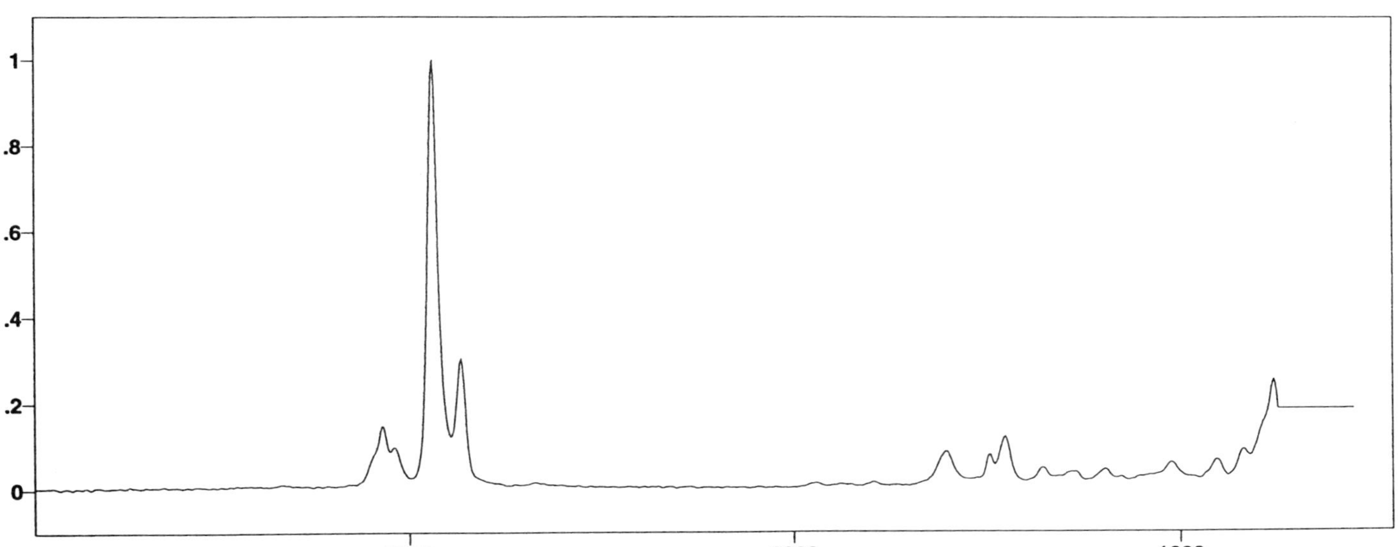

1-PHENYLCYCLOHEXYLAMINE

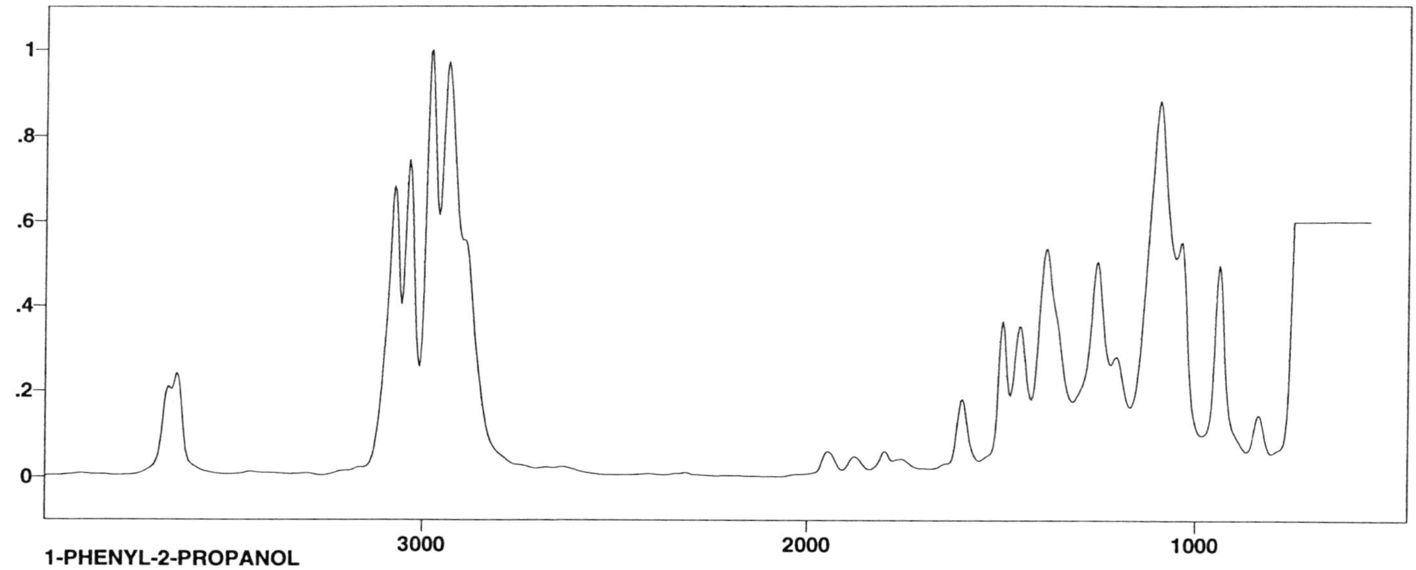

1-PHENYL-2-PROPANOL

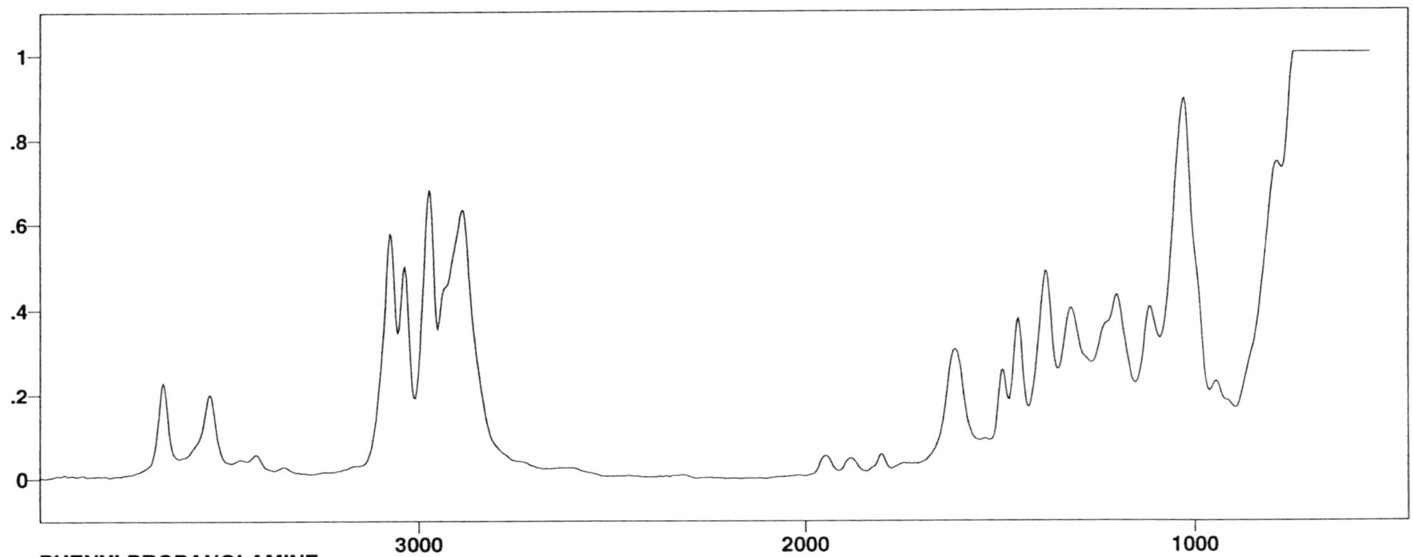

PHENYLPROPANOLAMINE

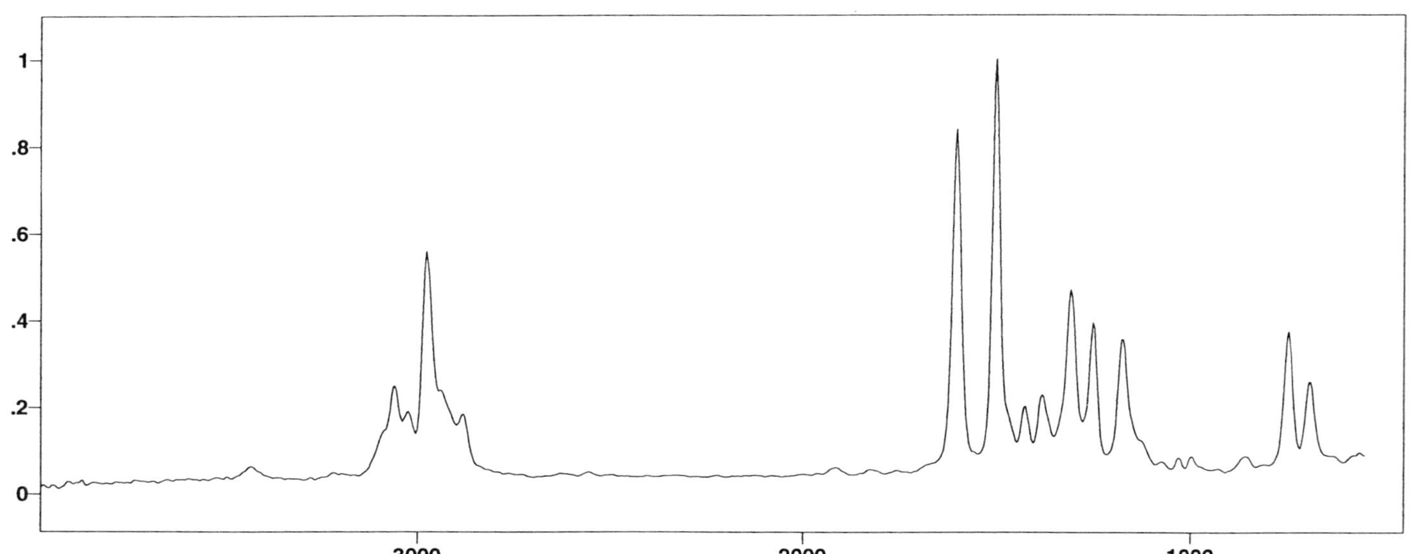

N-PHENYLisoPROPYLAMINE

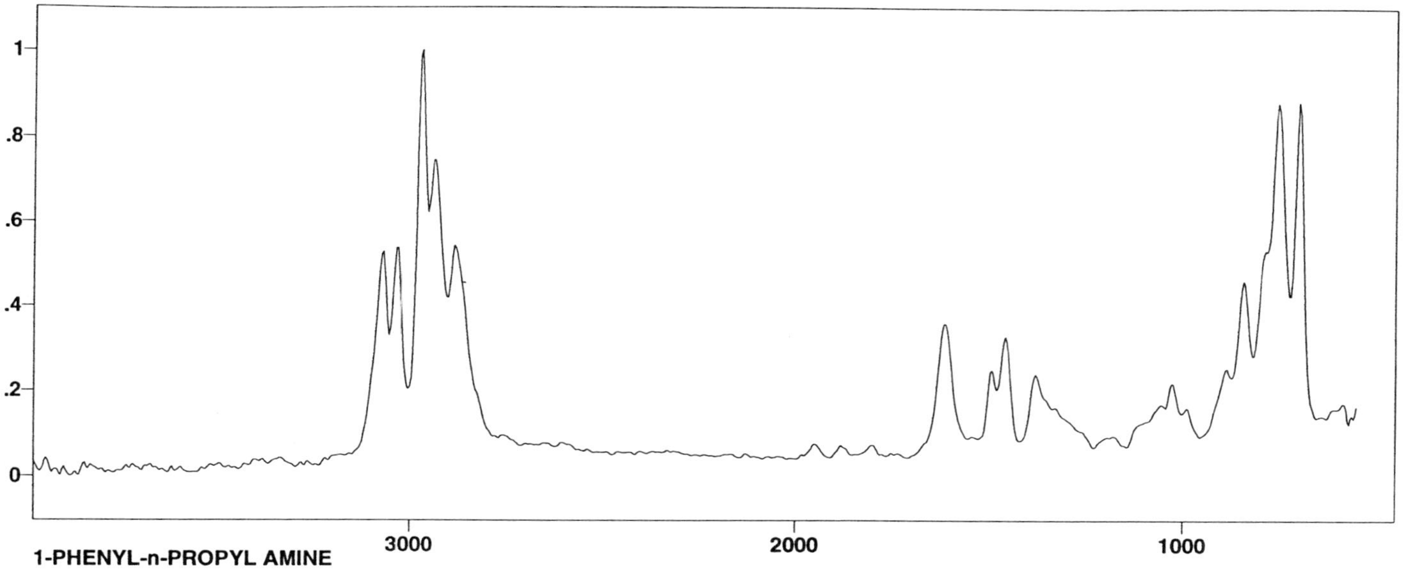

1-PHENYL-n-PROPYL AMINE

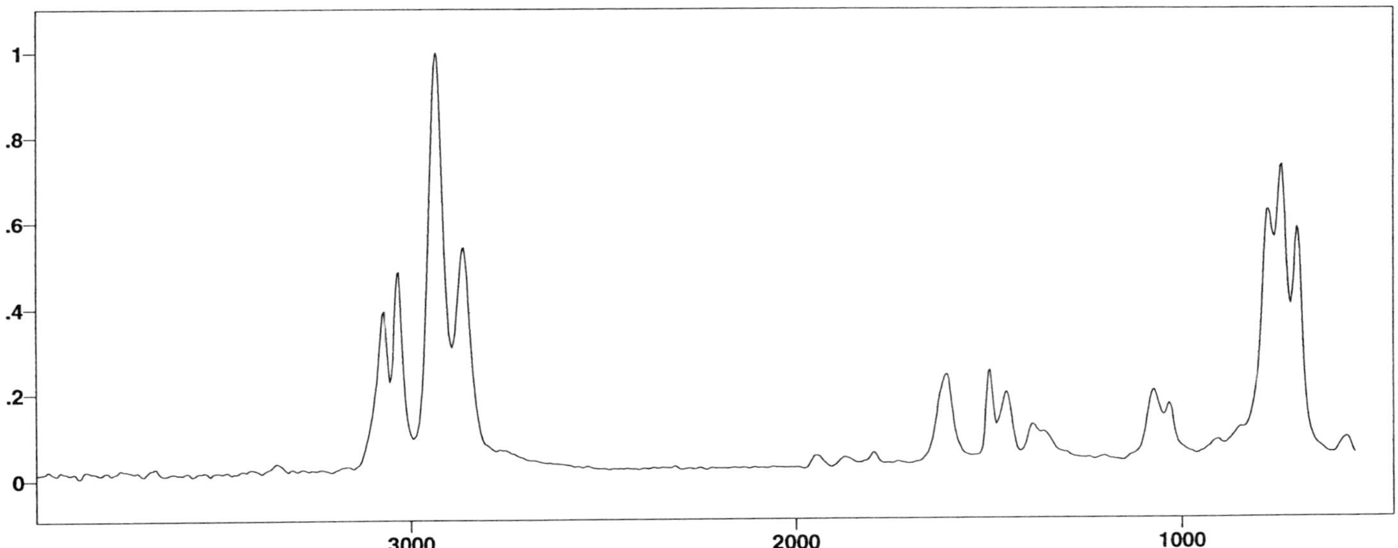

3-PHENYL-1-PROPYLAMINE

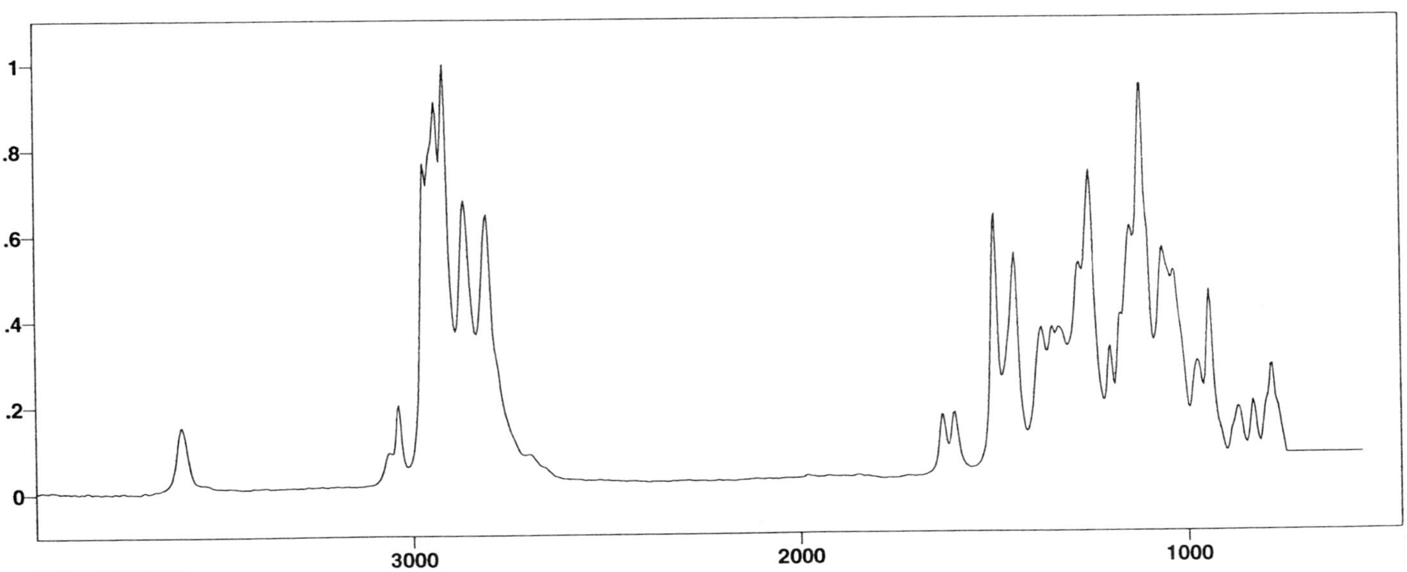

PHOLCODINE

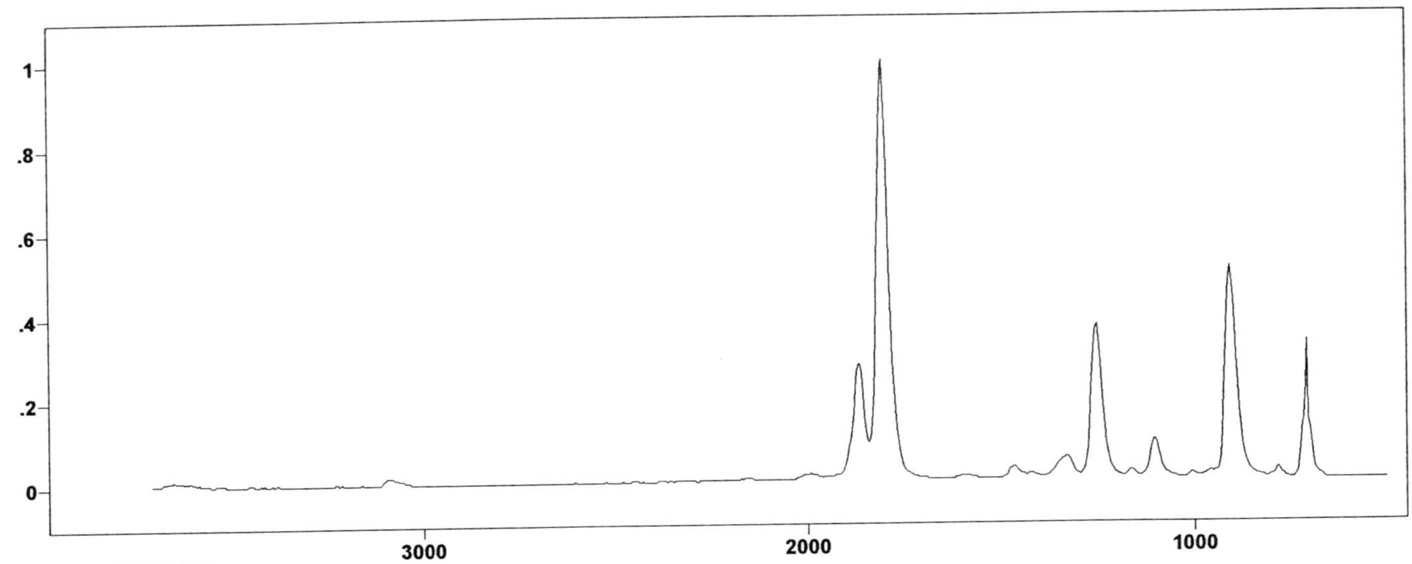

PHTHALIC ACID

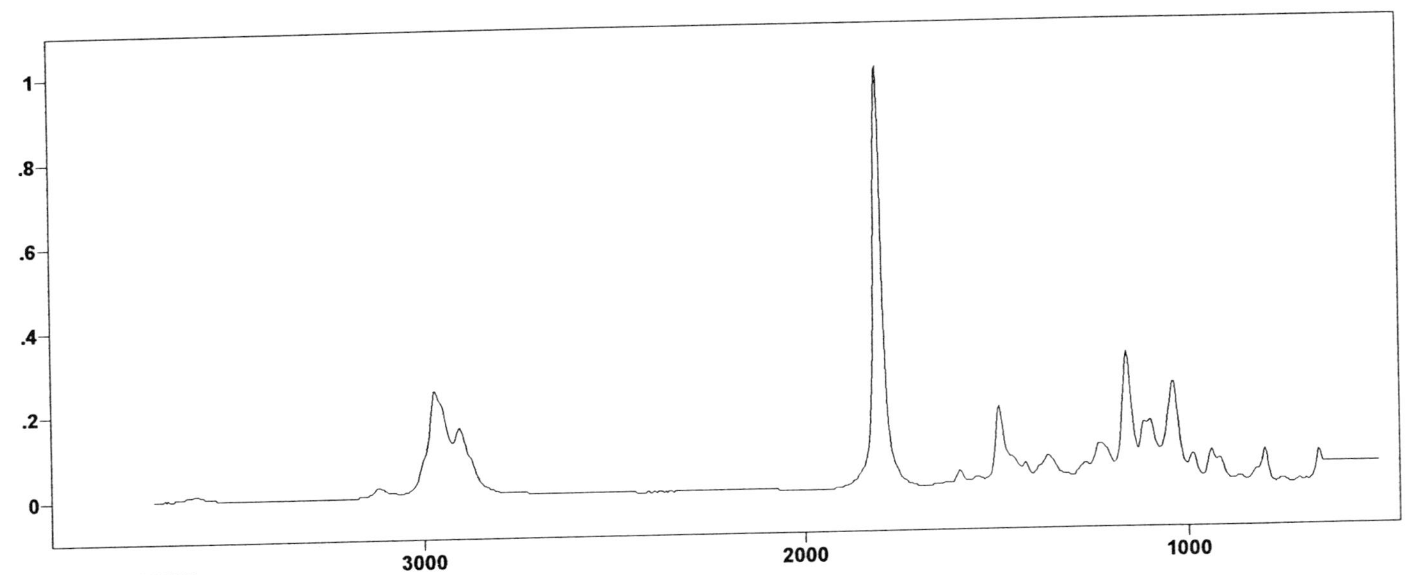

PILOCARPINE

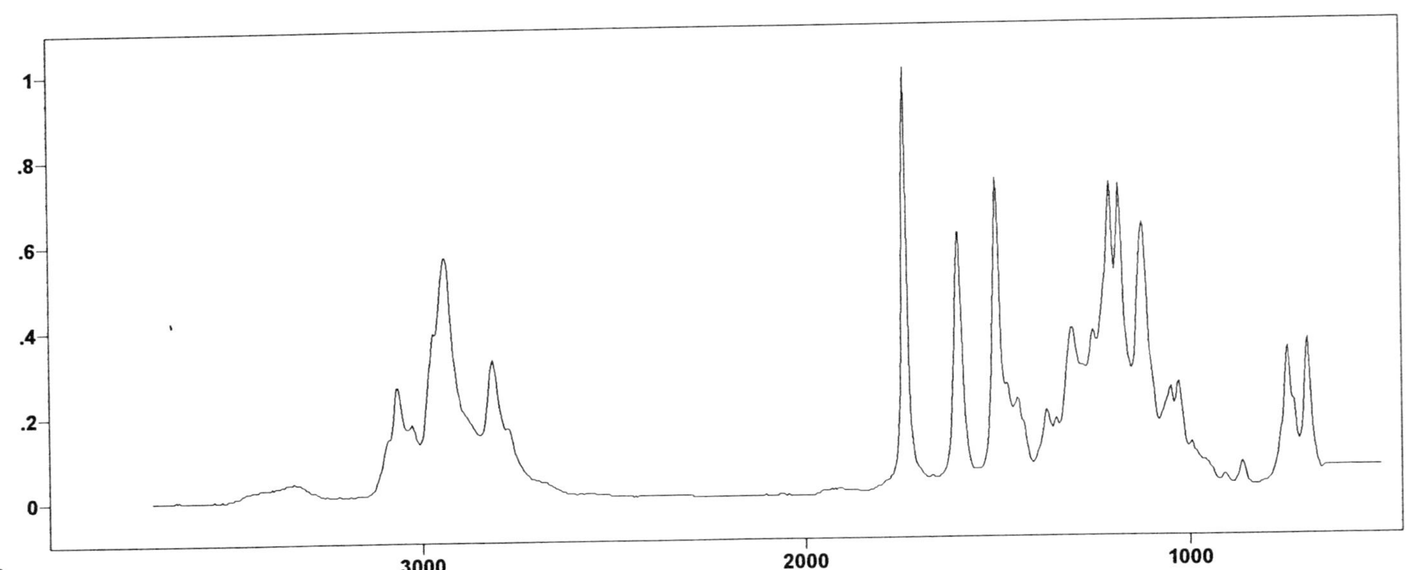

PIMINODINE

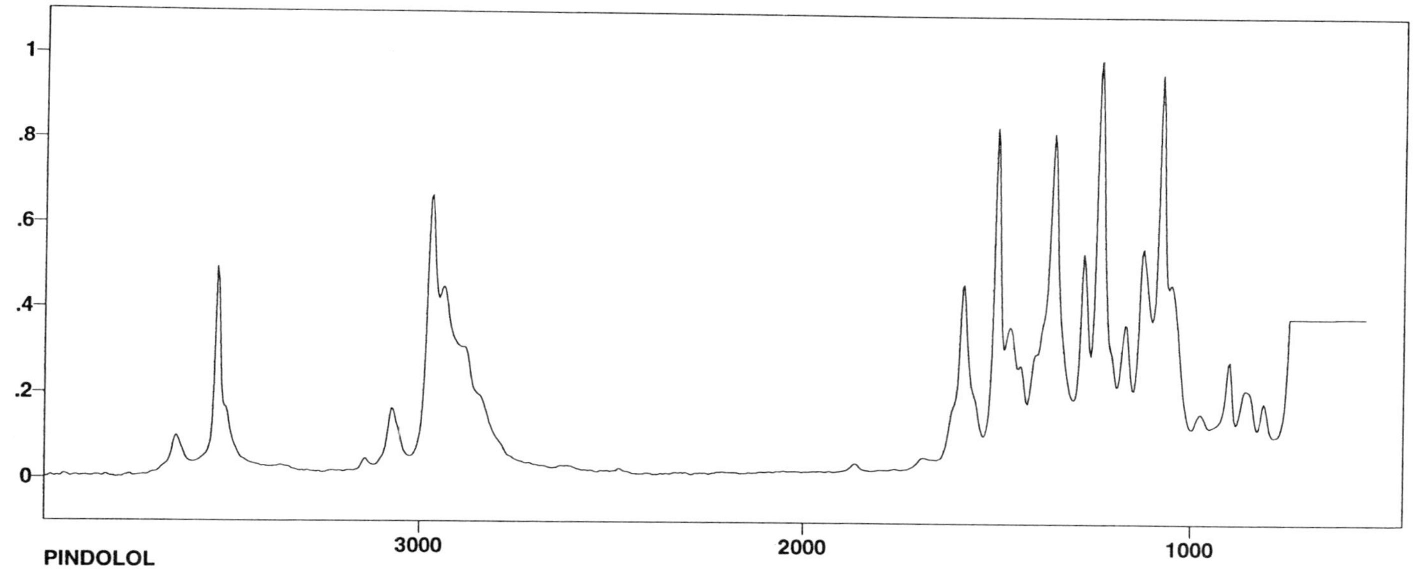

PINDOLOL

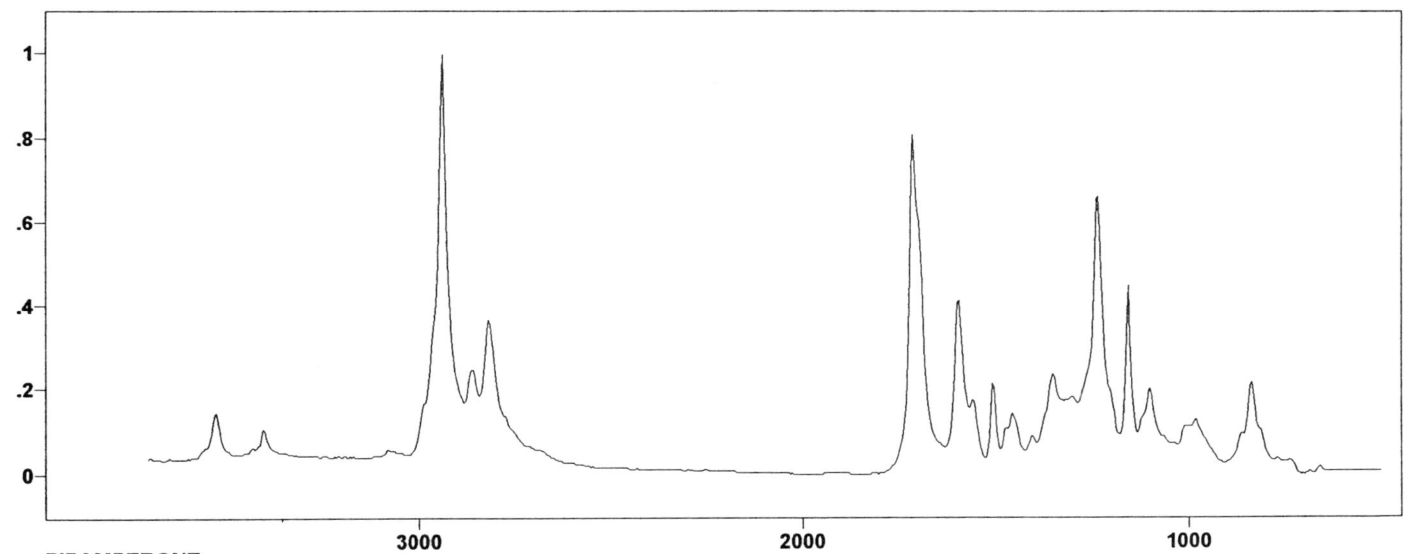

PIPAMPERONE

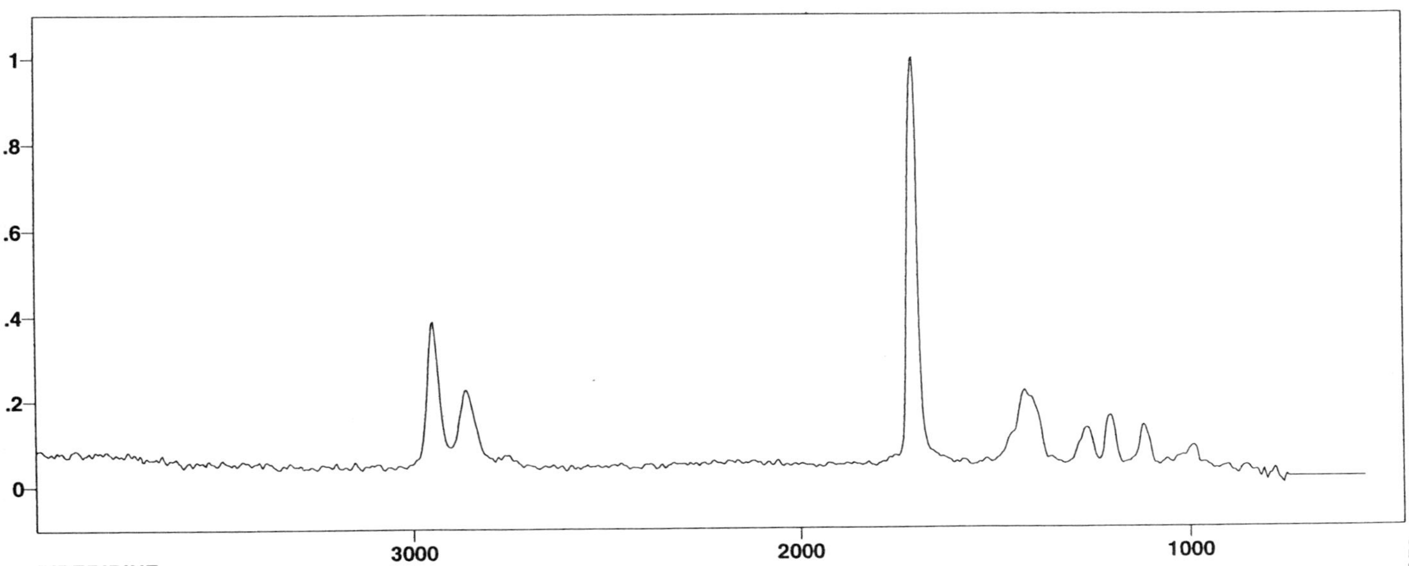

PIPERIDINE

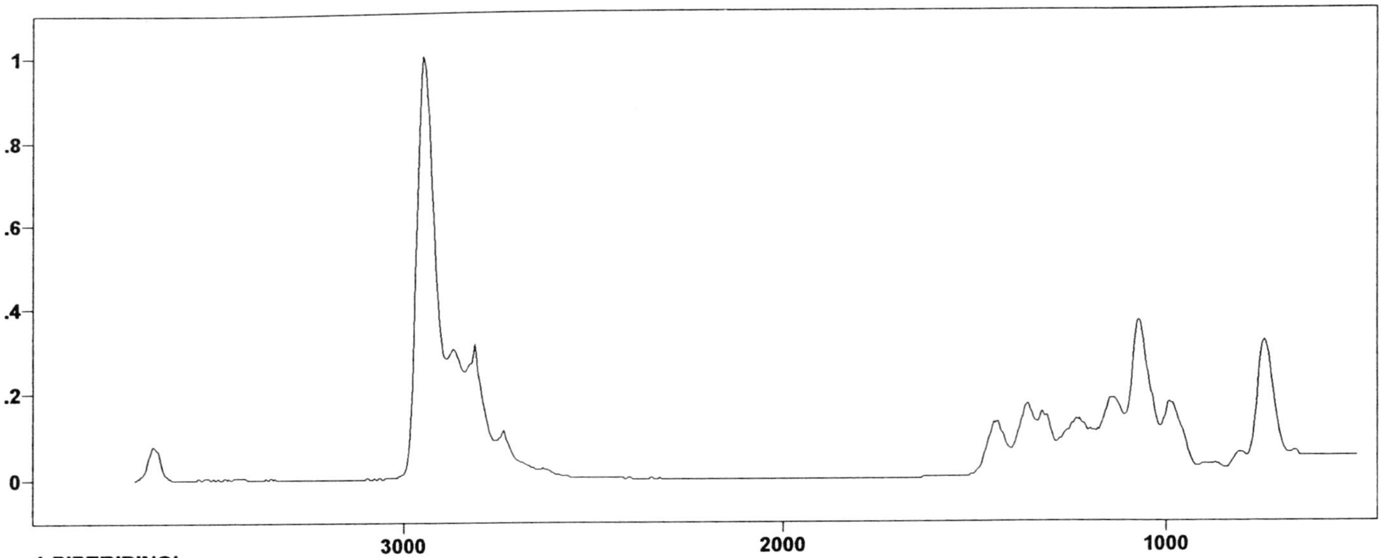

4-PIPERIDINOL

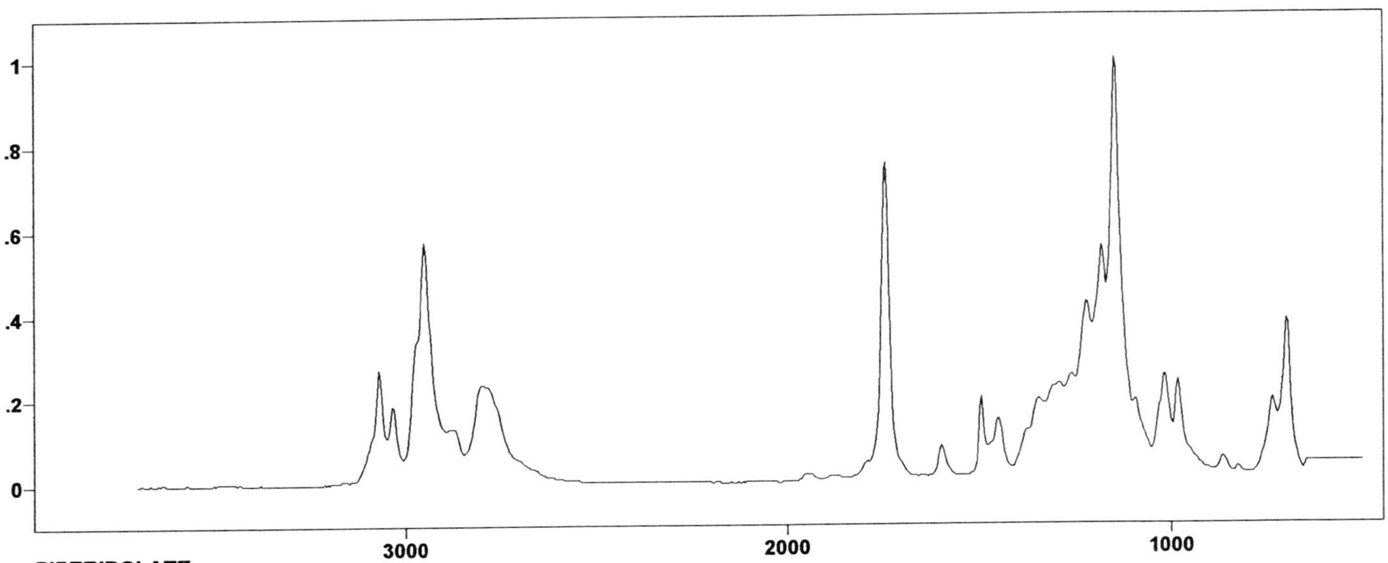

PIPERIDOLATE

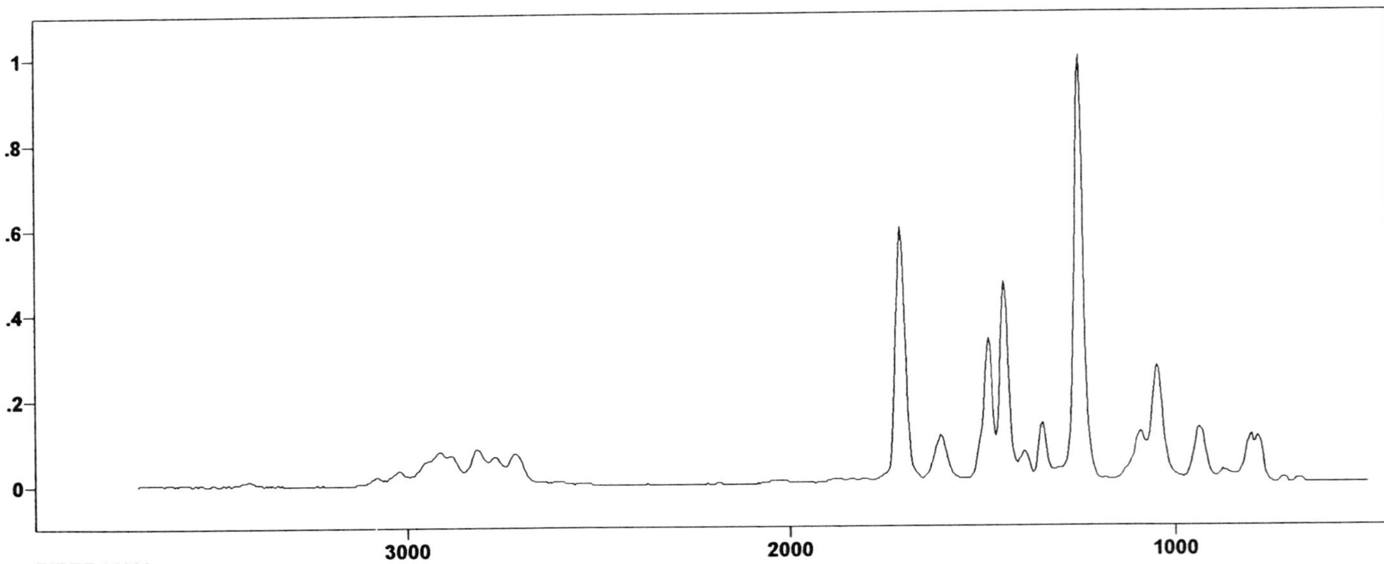

PIPERONAL

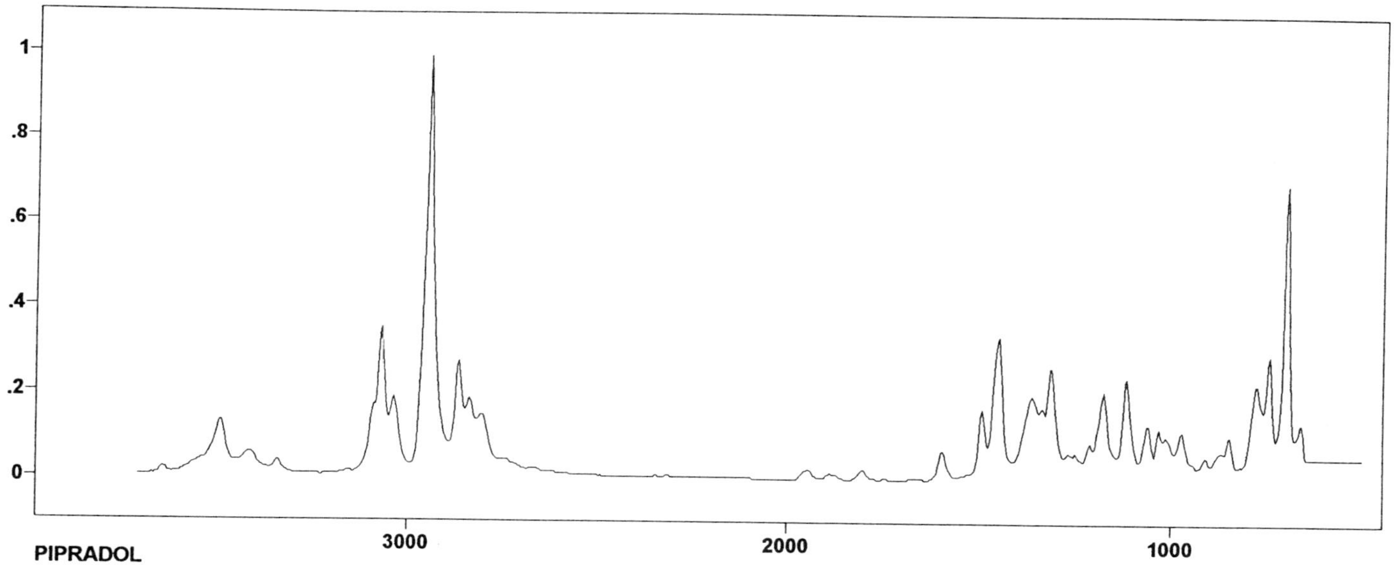

PIPRADOL

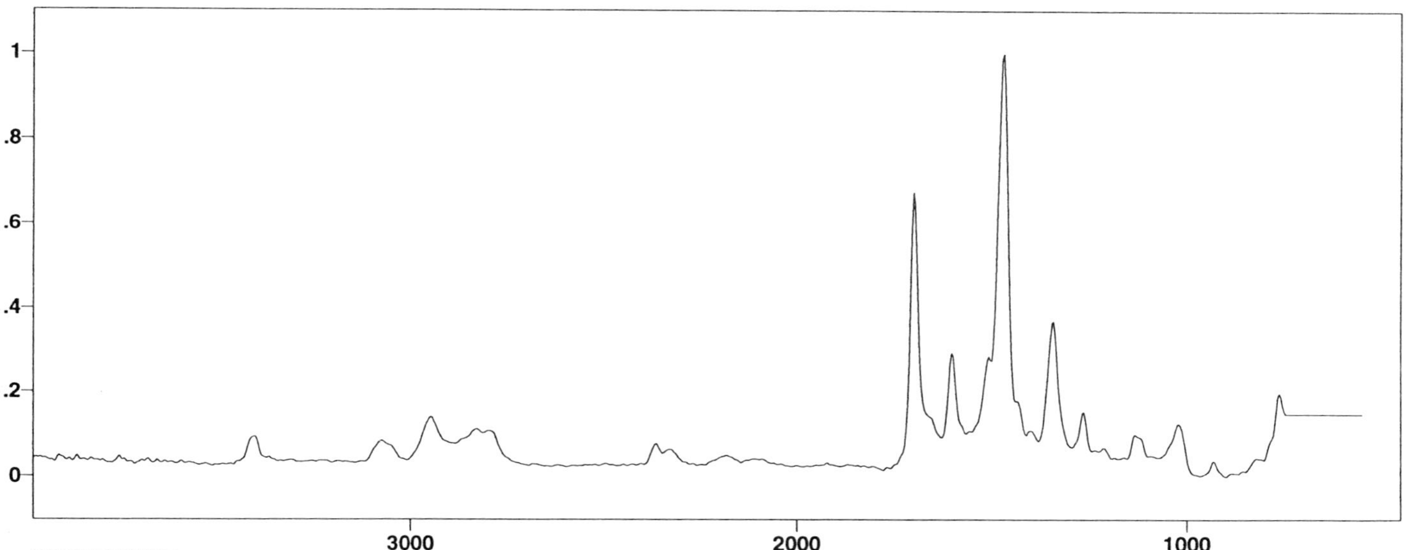

PIRENZEPINE

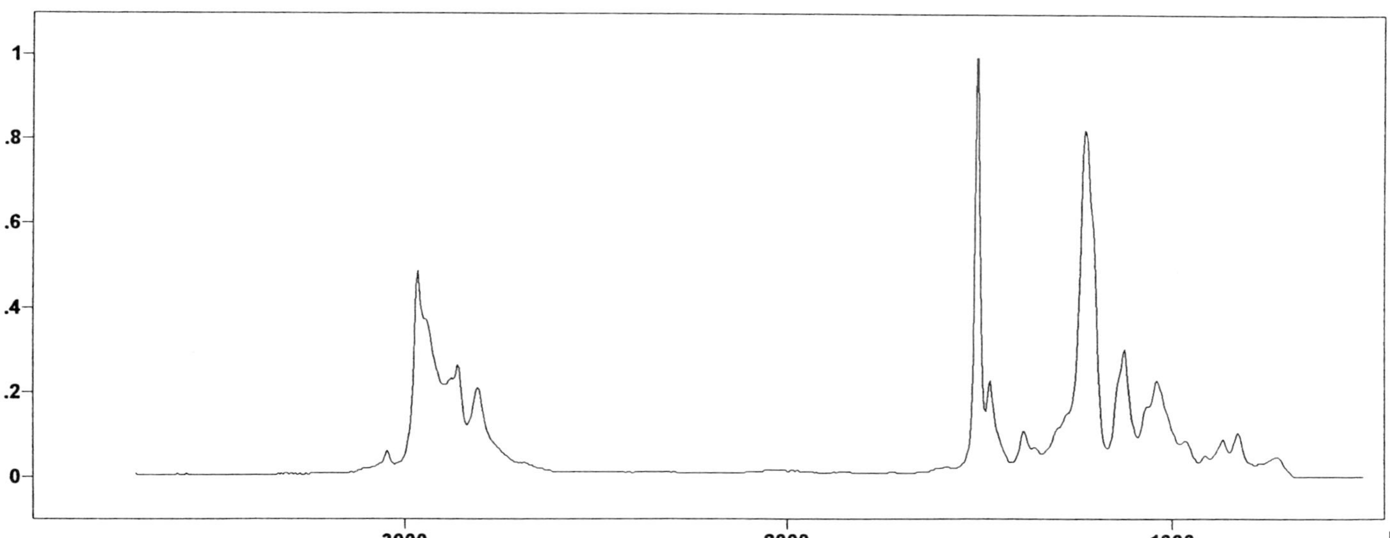

PRAMOXINE

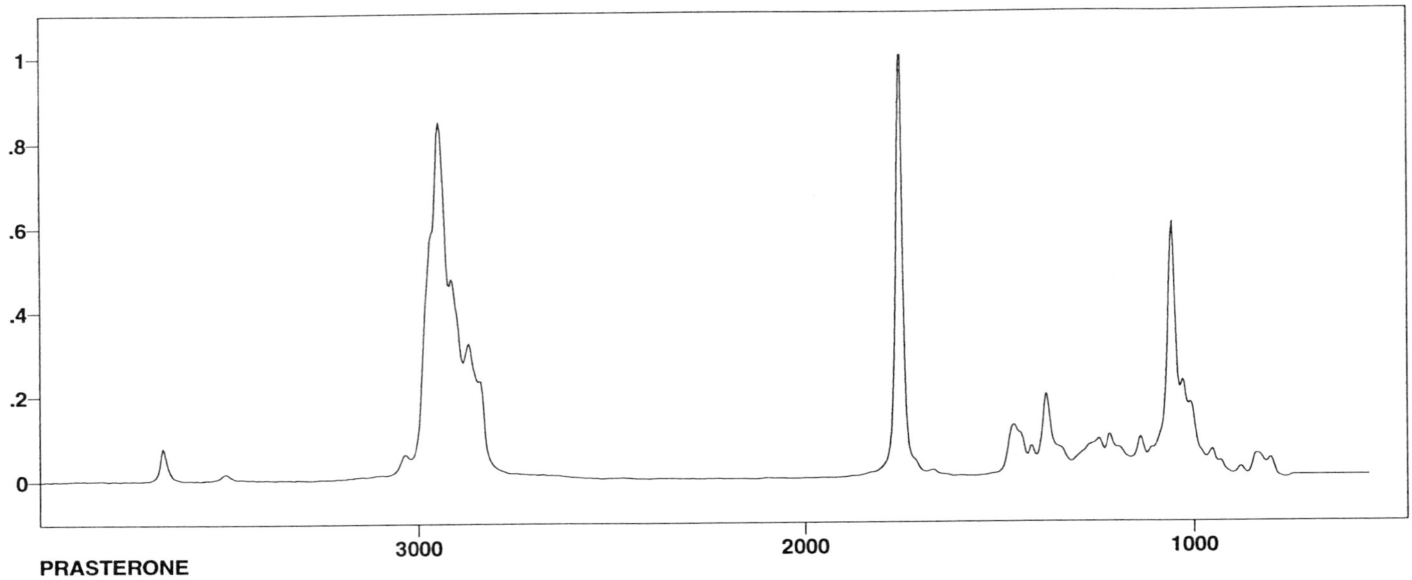

PRASTERONE

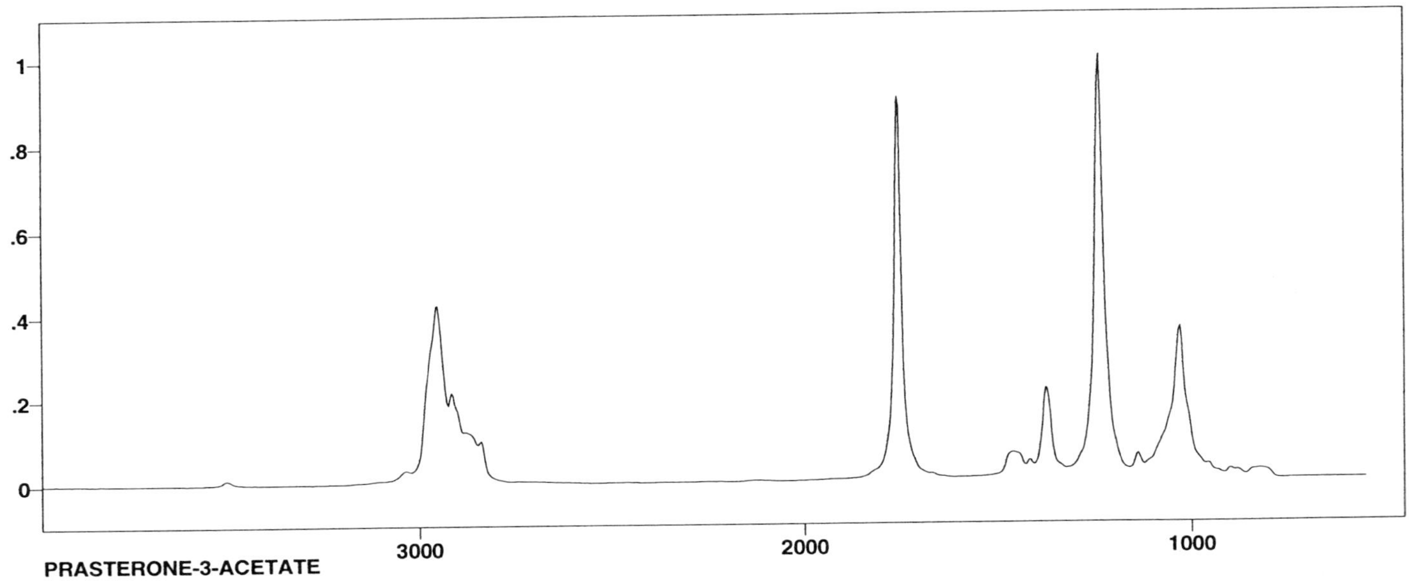

PRASTERONE-3-ACETATE

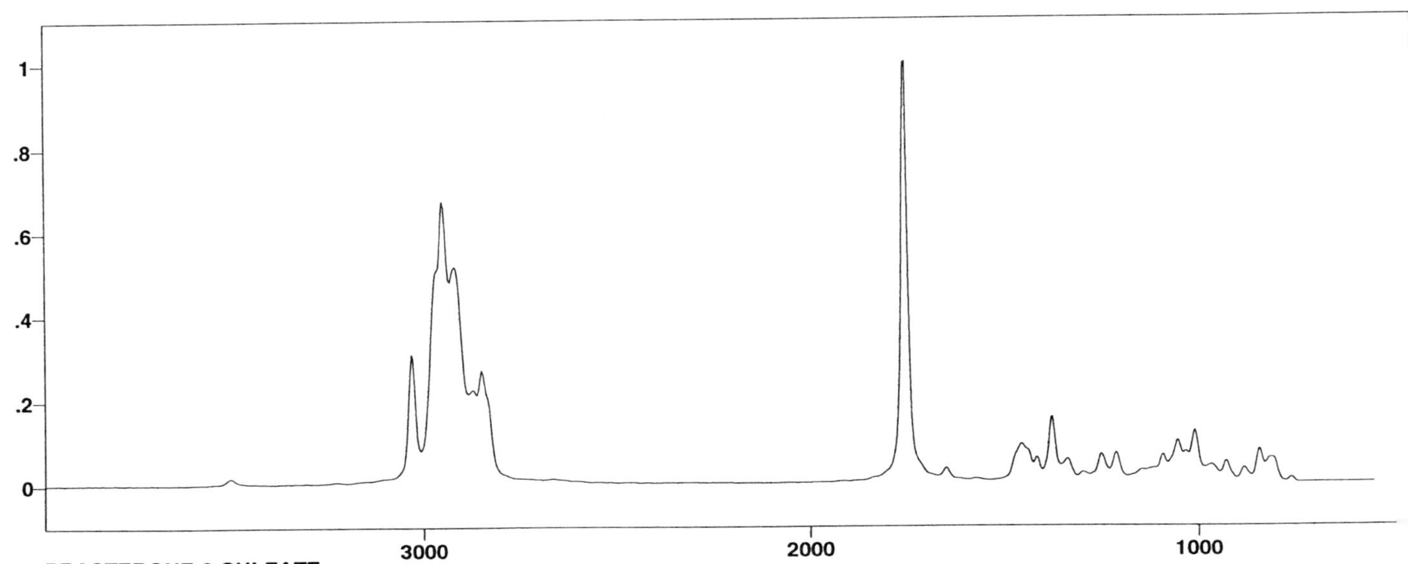

PRASTERONE-3-SULFATE

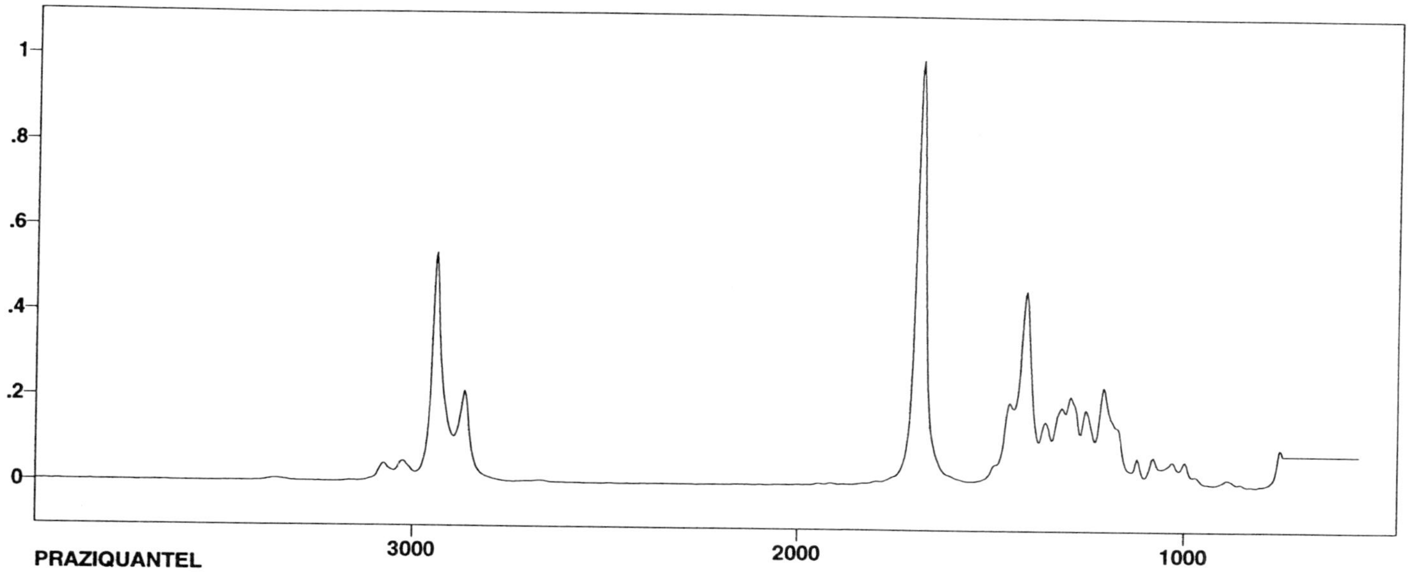

PRAZIQUANTEL

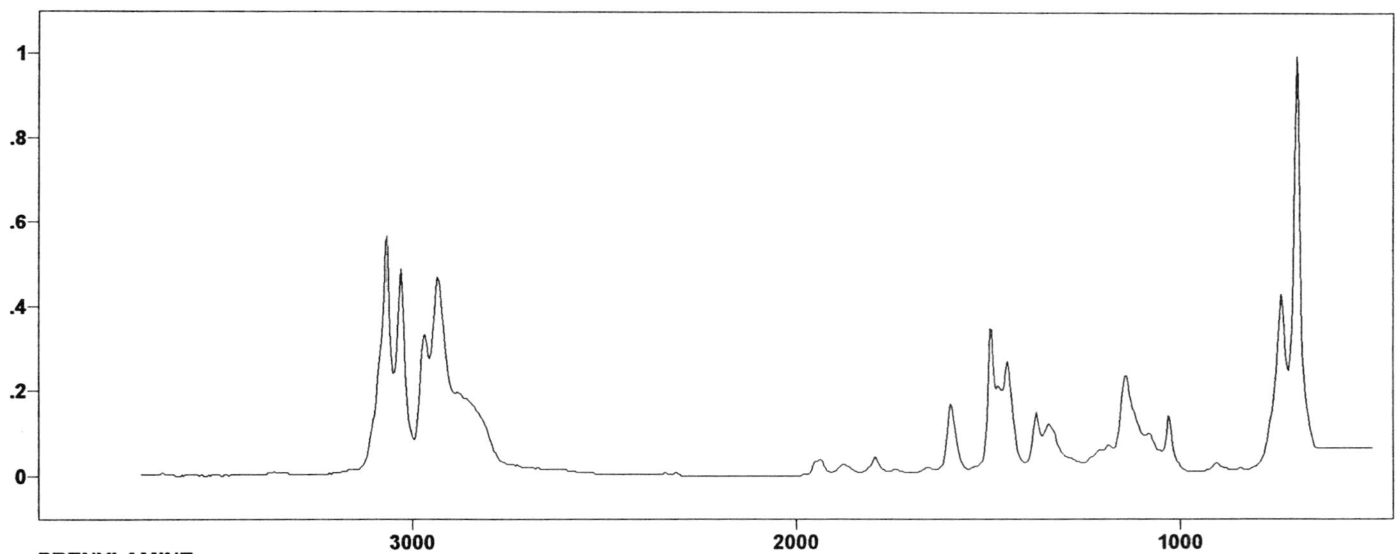

PRENYLAMINE

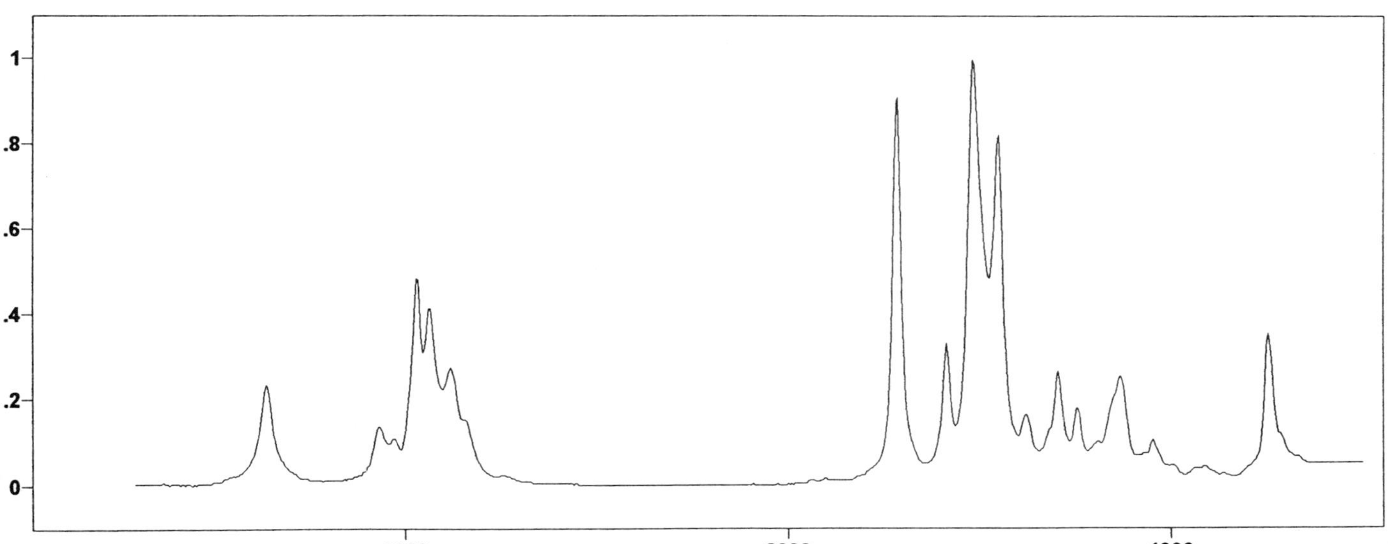

PRILOCAINE

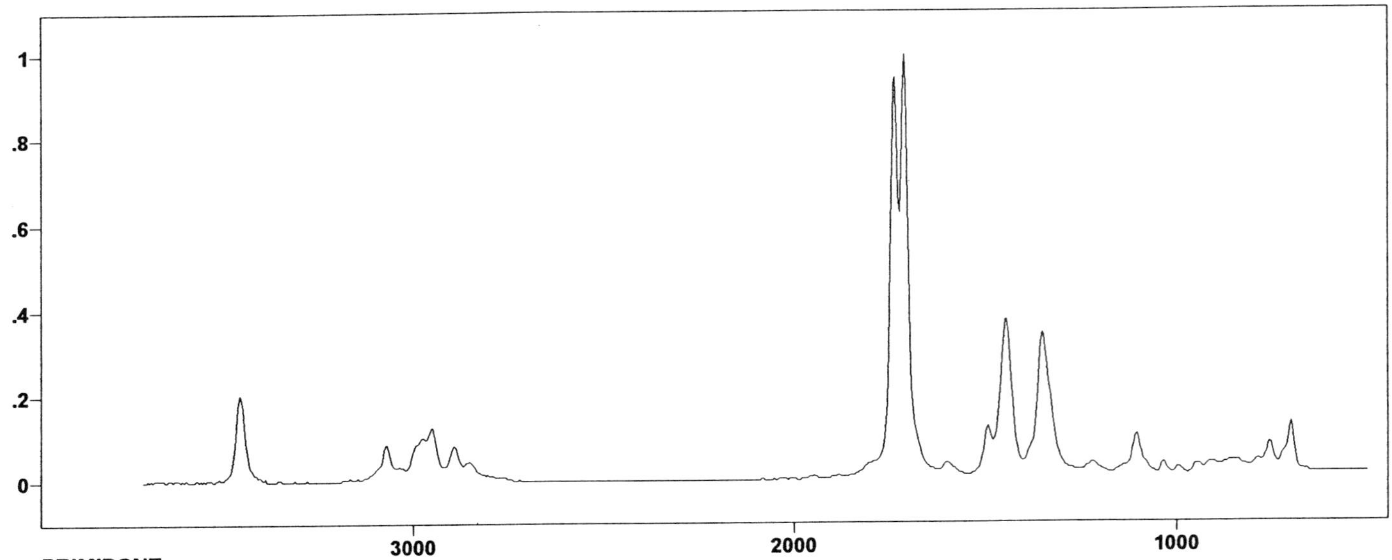

PRIMIDONE

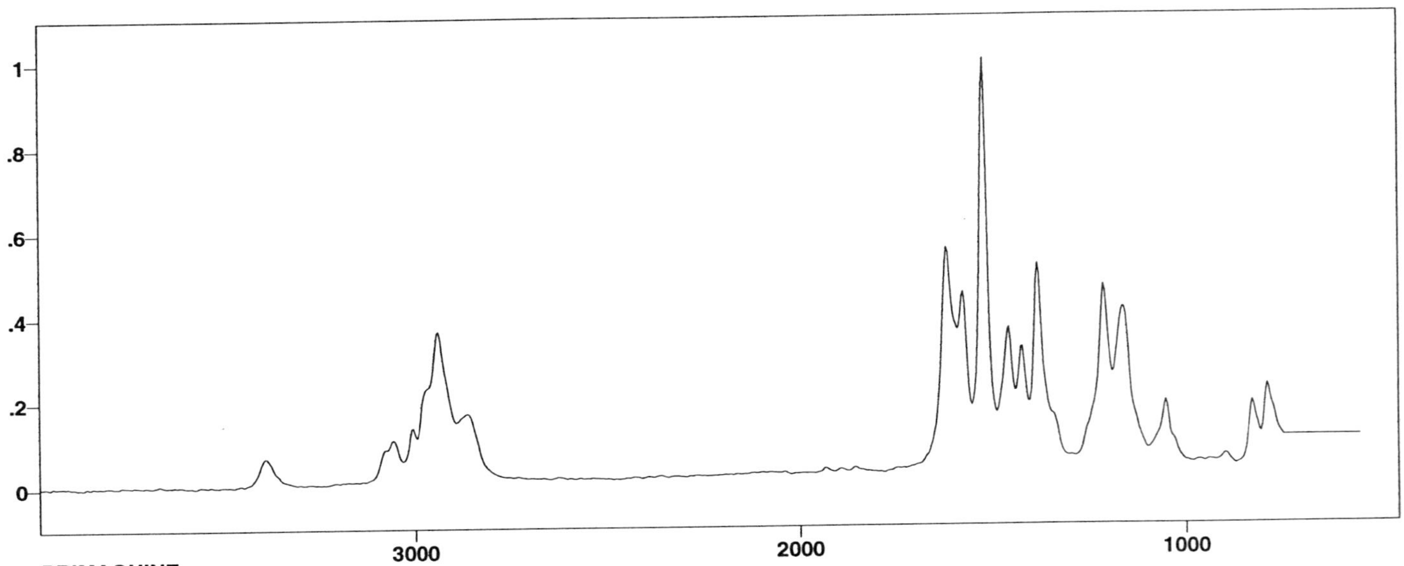

PRIMAQUINE

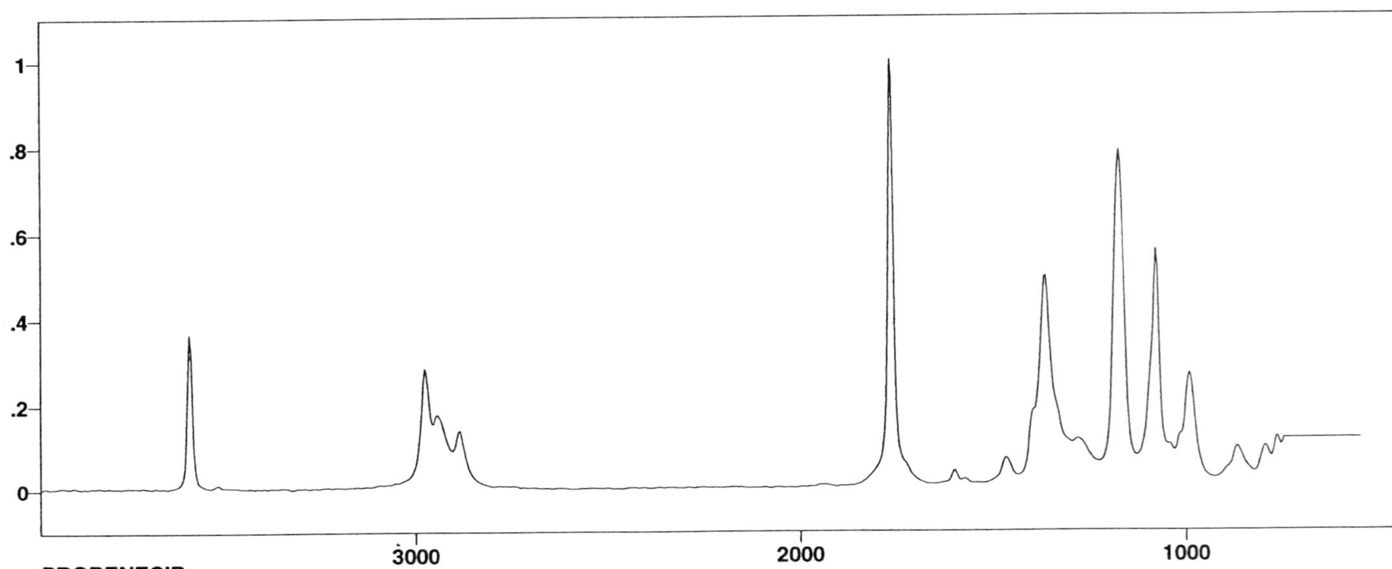

PROBENECID

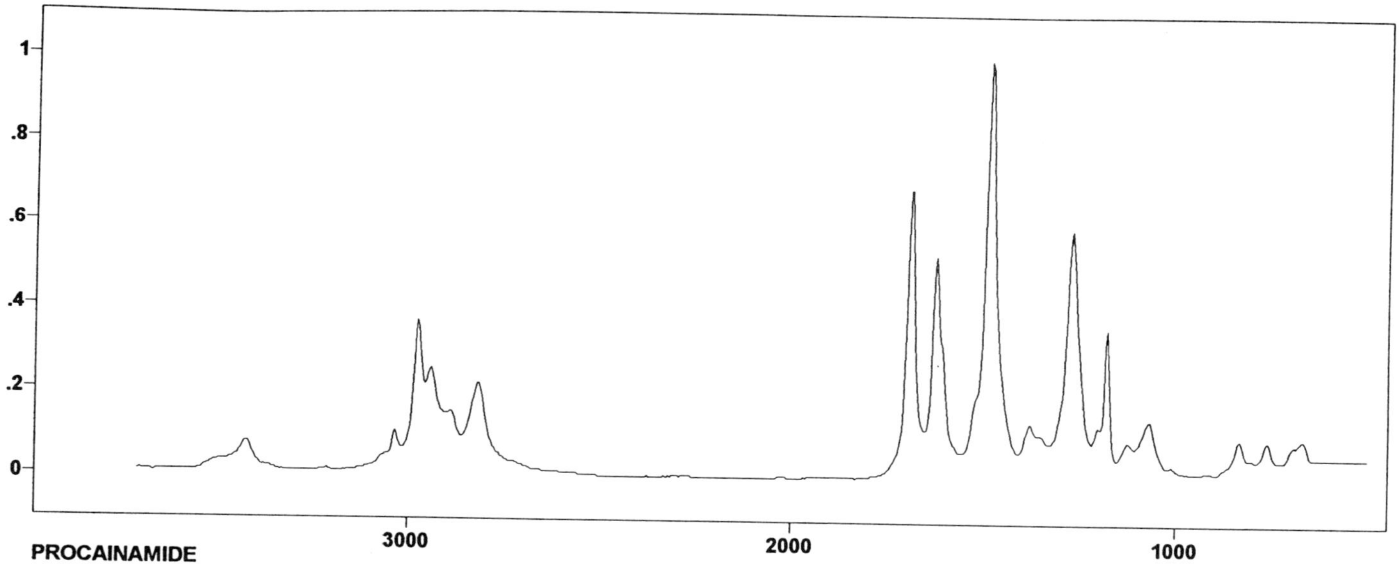

PROCAINAMIDE

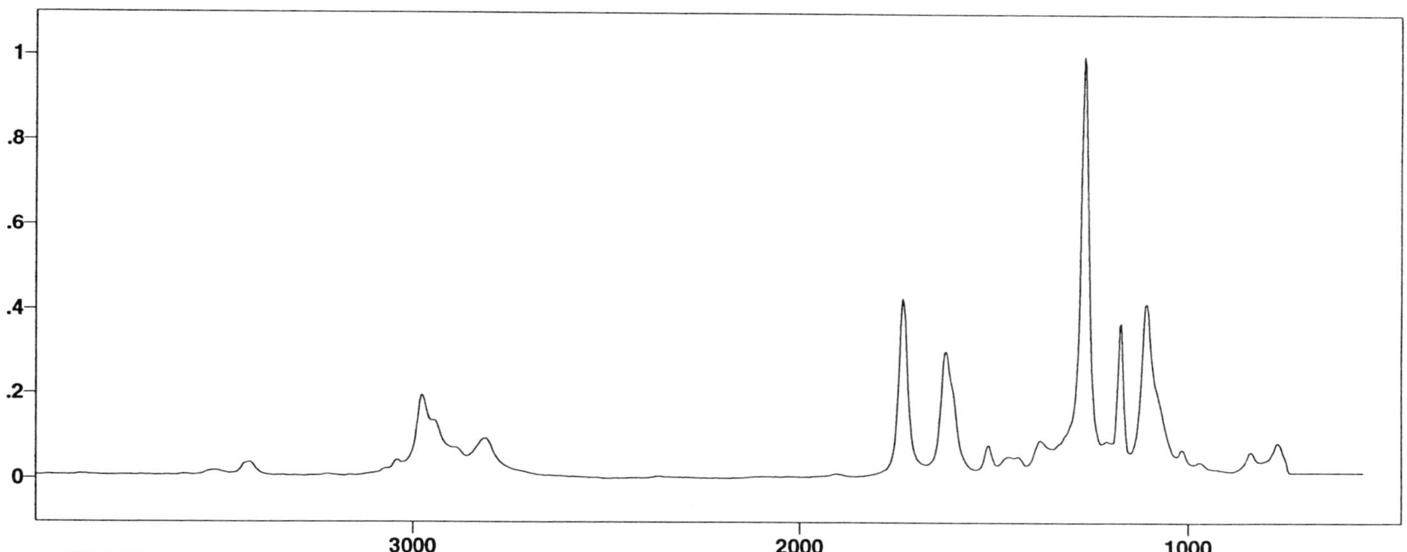

PROCAINE

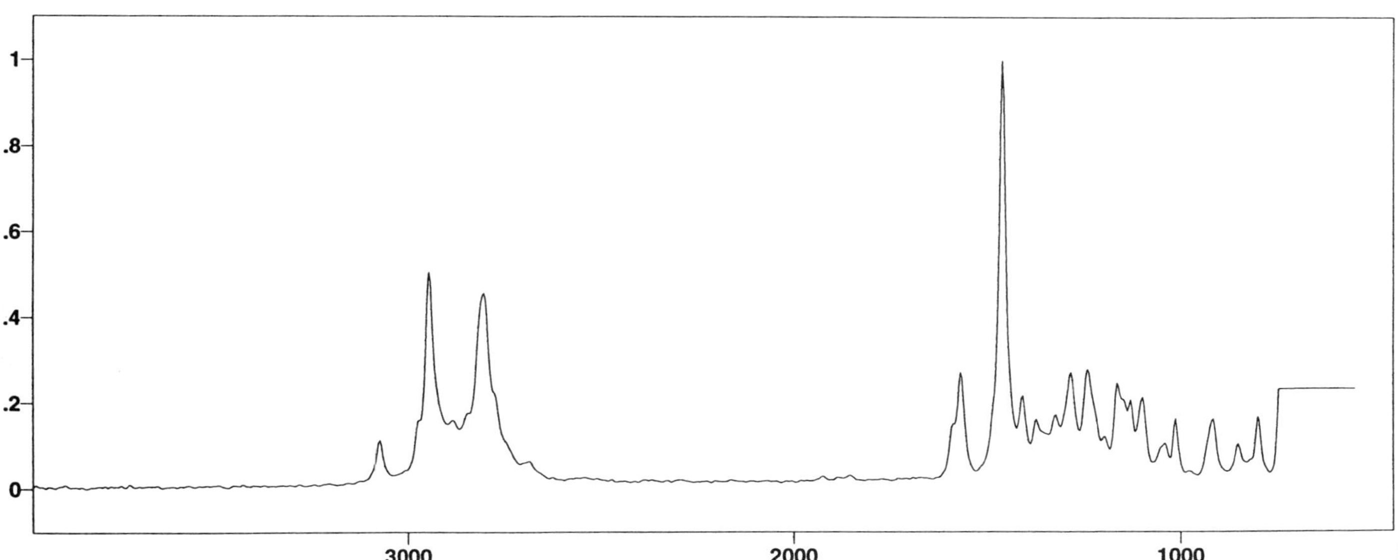

PROCHLORPERAZINE

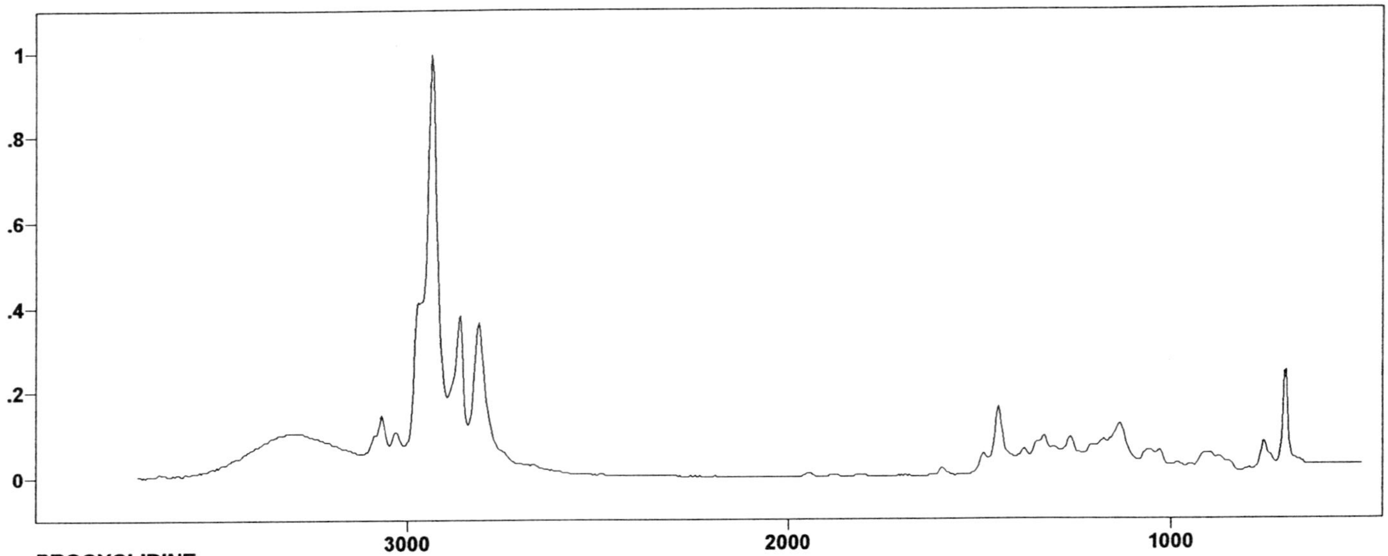

PROCYCLIDINE

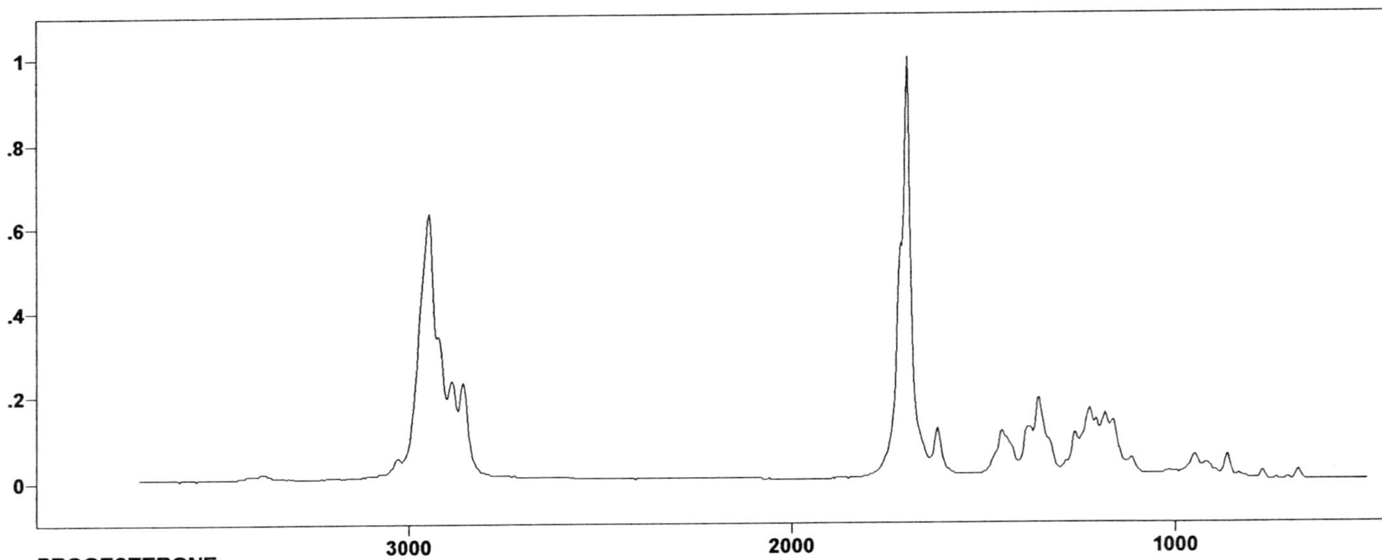

PROGESTERONE

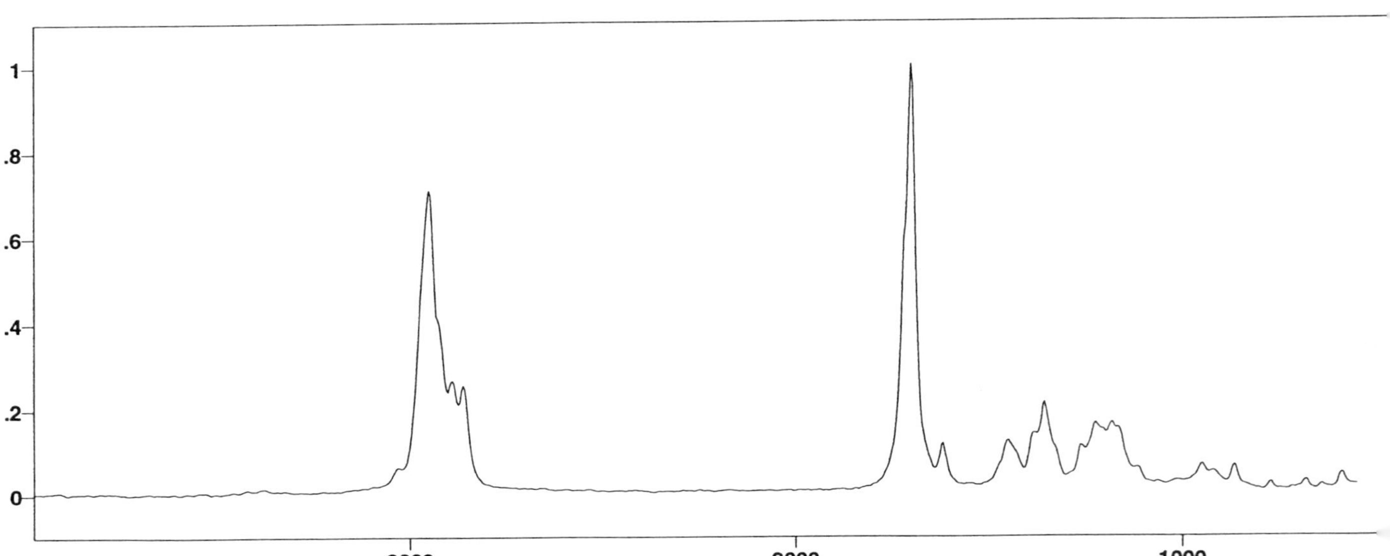

PROGESTERONE(4-PREGNENE-3,20-DIONE)

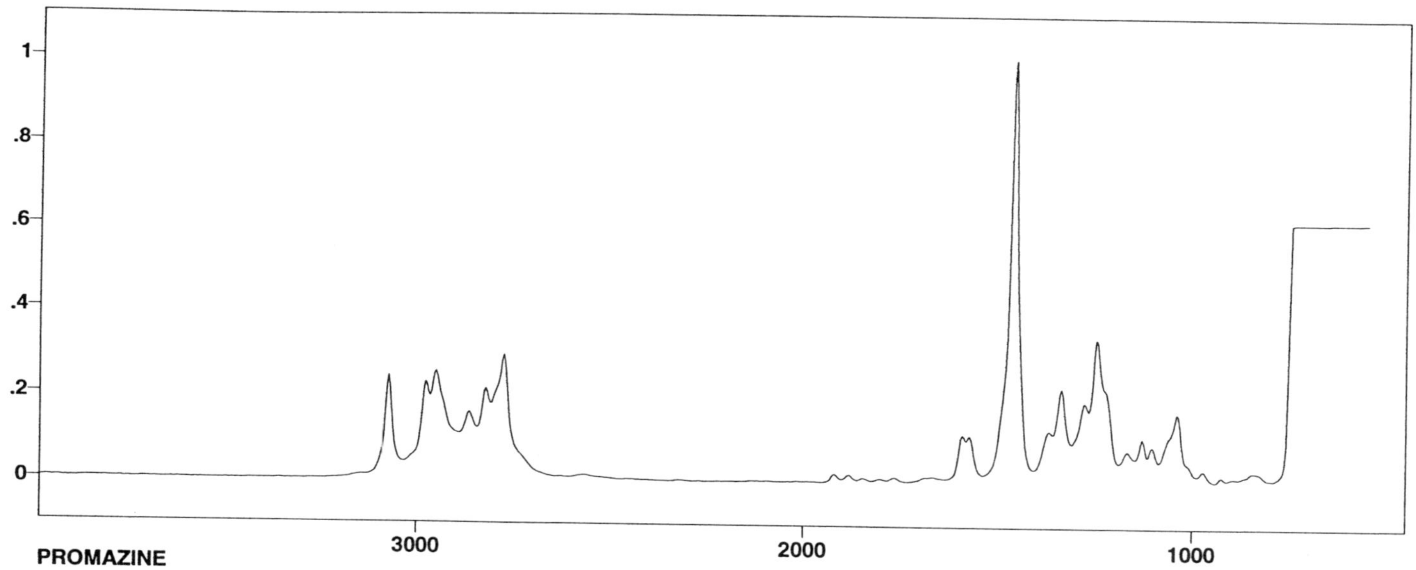

PROMAZINE

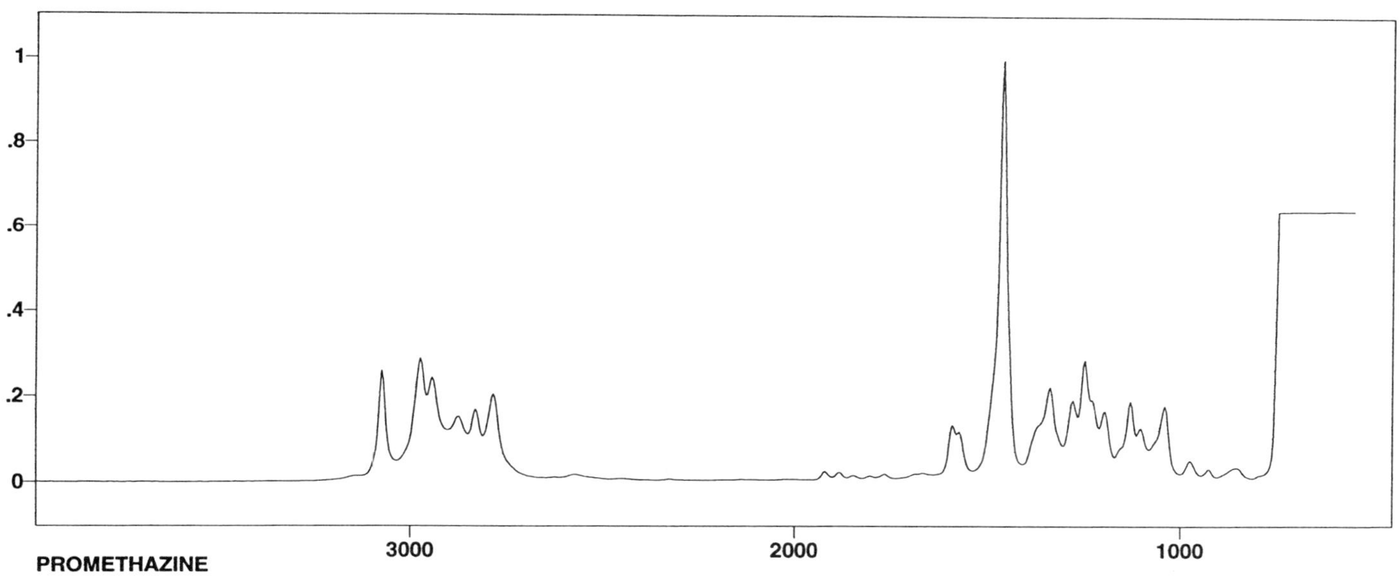

PROMETHAZINE

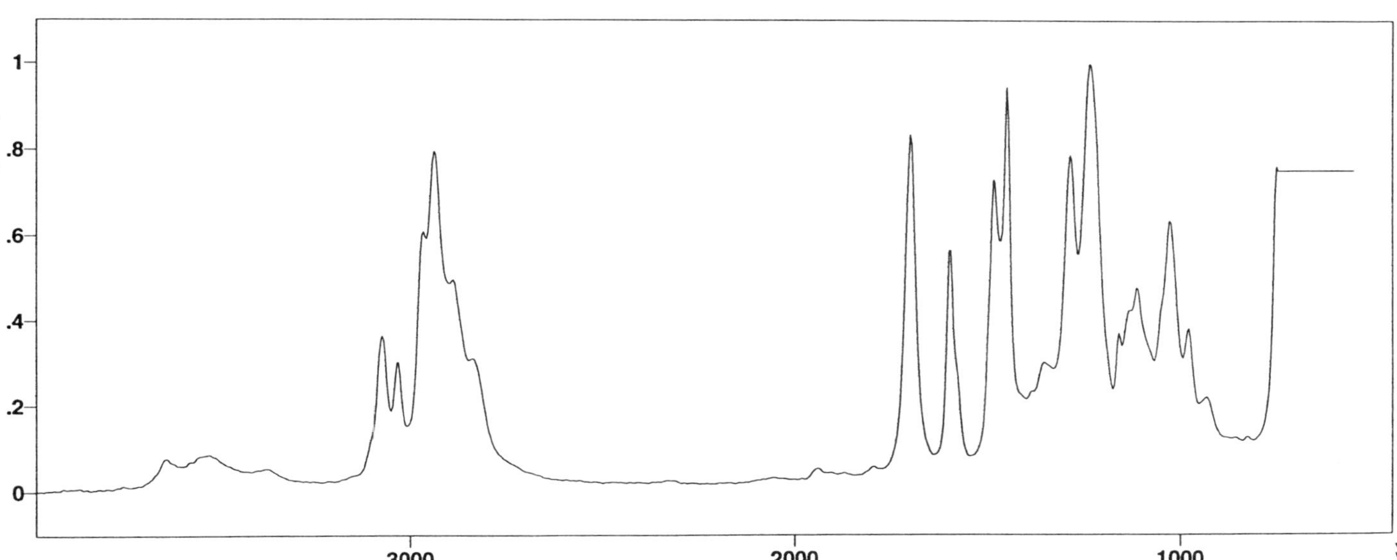

PROPAFENONE

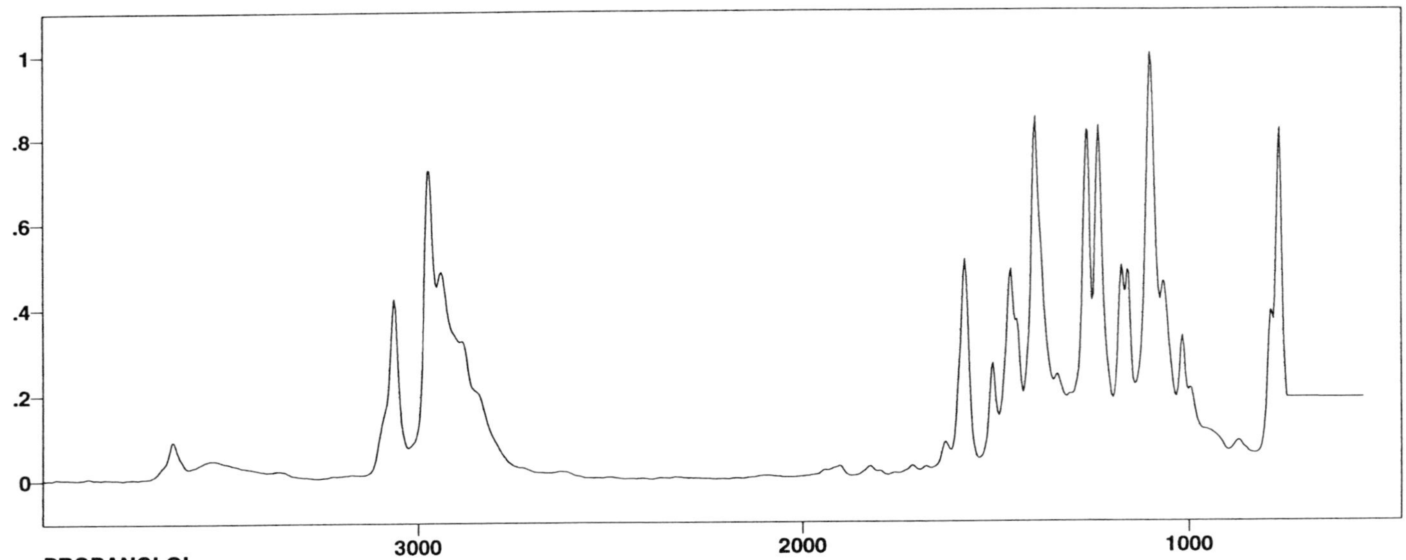

PROPANOLOL

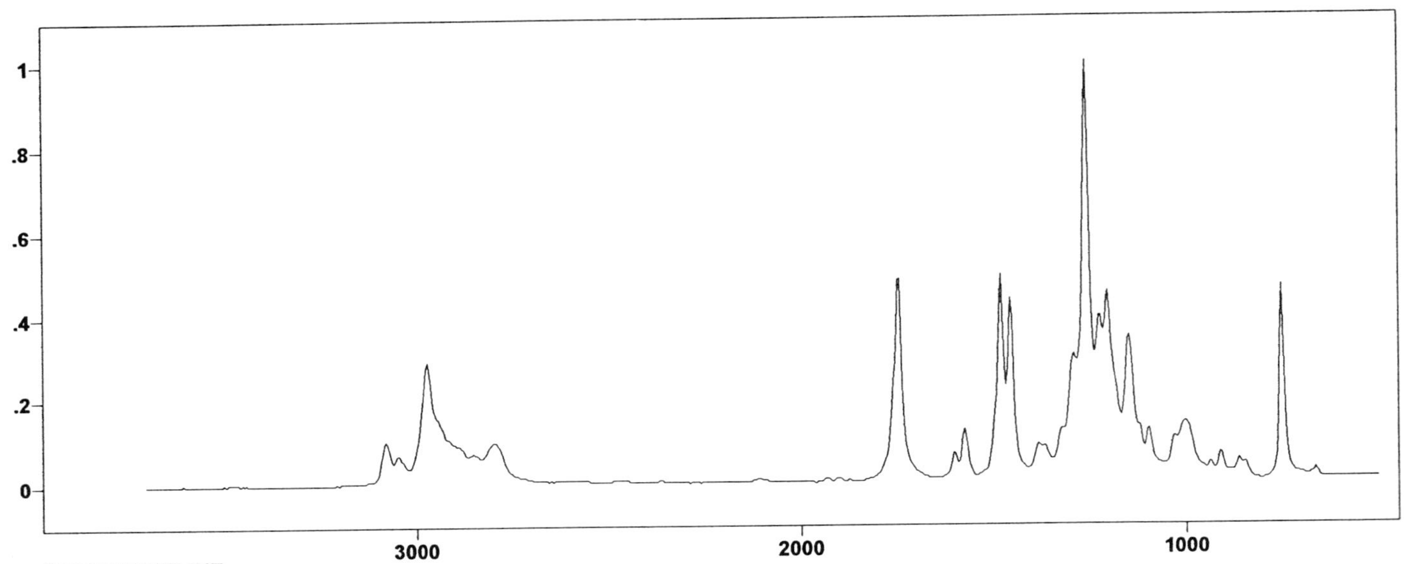

PROPANTHELINE

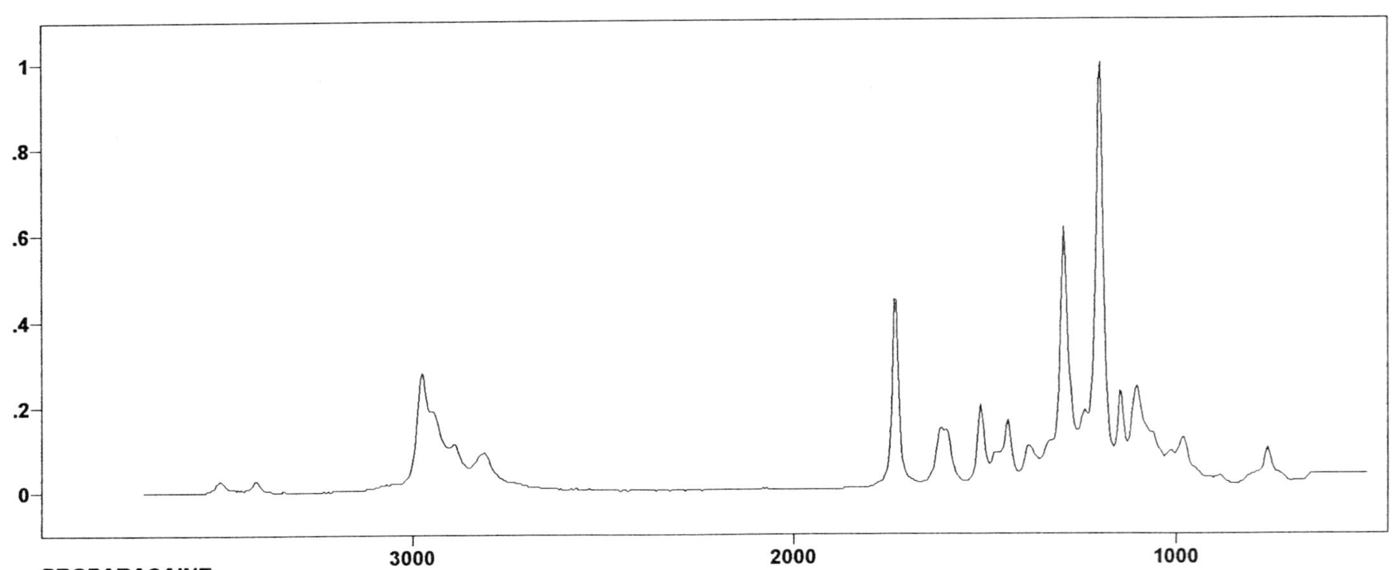

PROPARACAINE

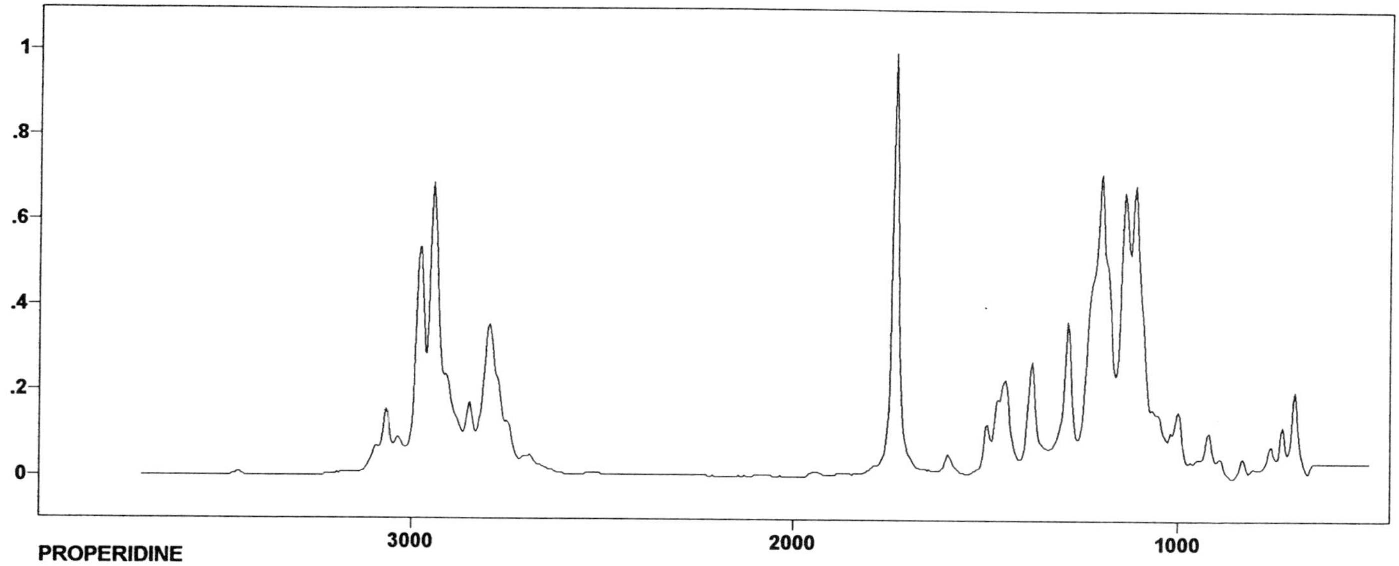

PROPERIDINE

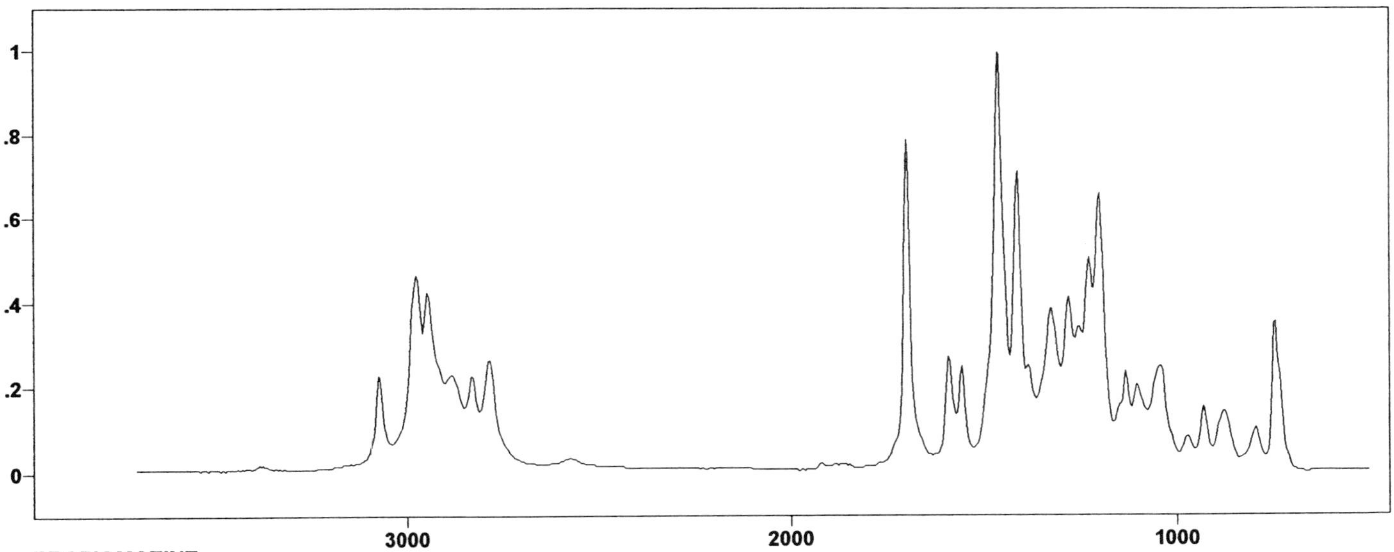

PROPIOMAZINE

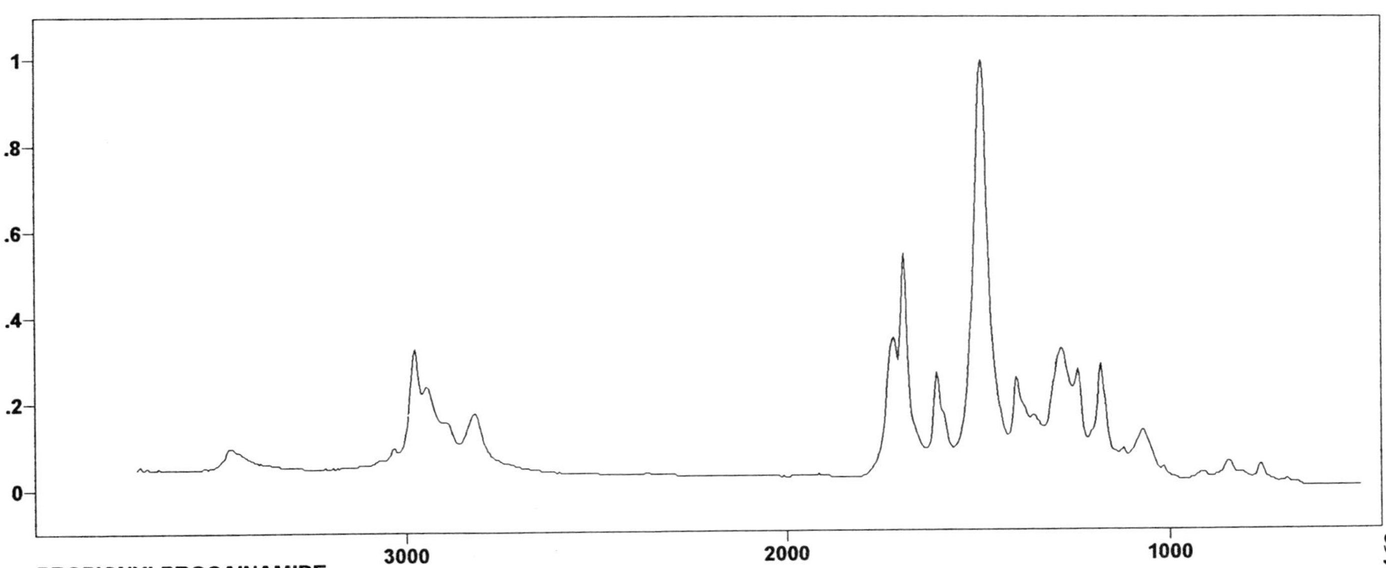

PROPIONYLPROCAINAMIDE

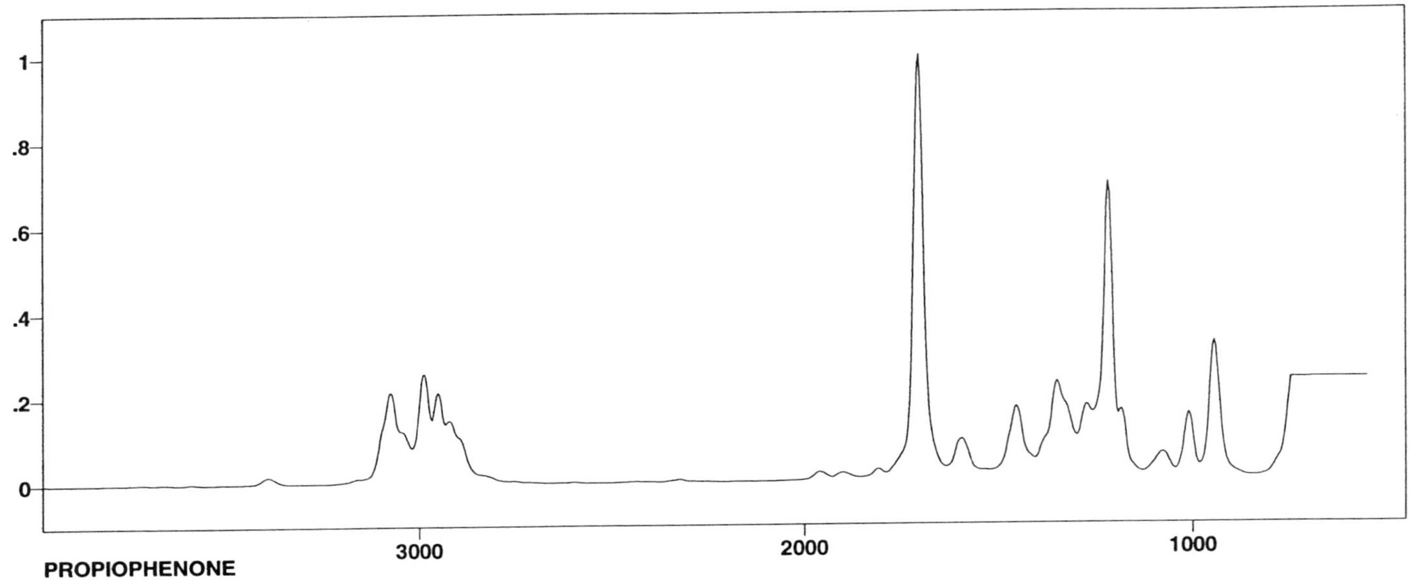

PROPIOPHENONE

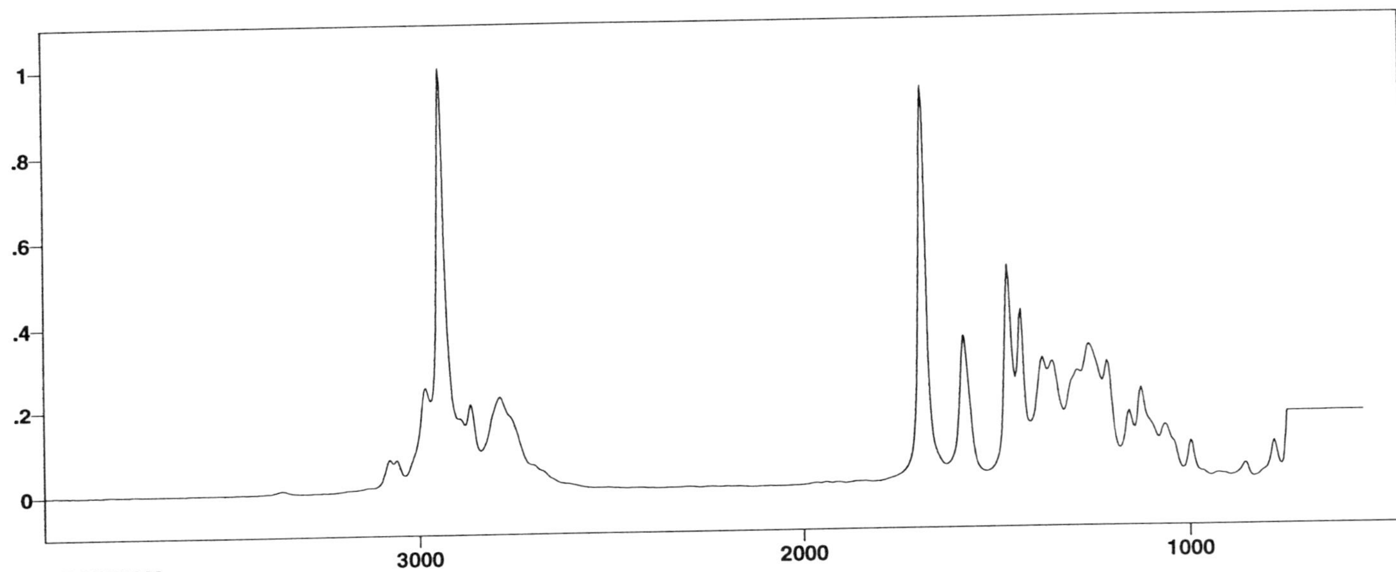

PROPIRAM

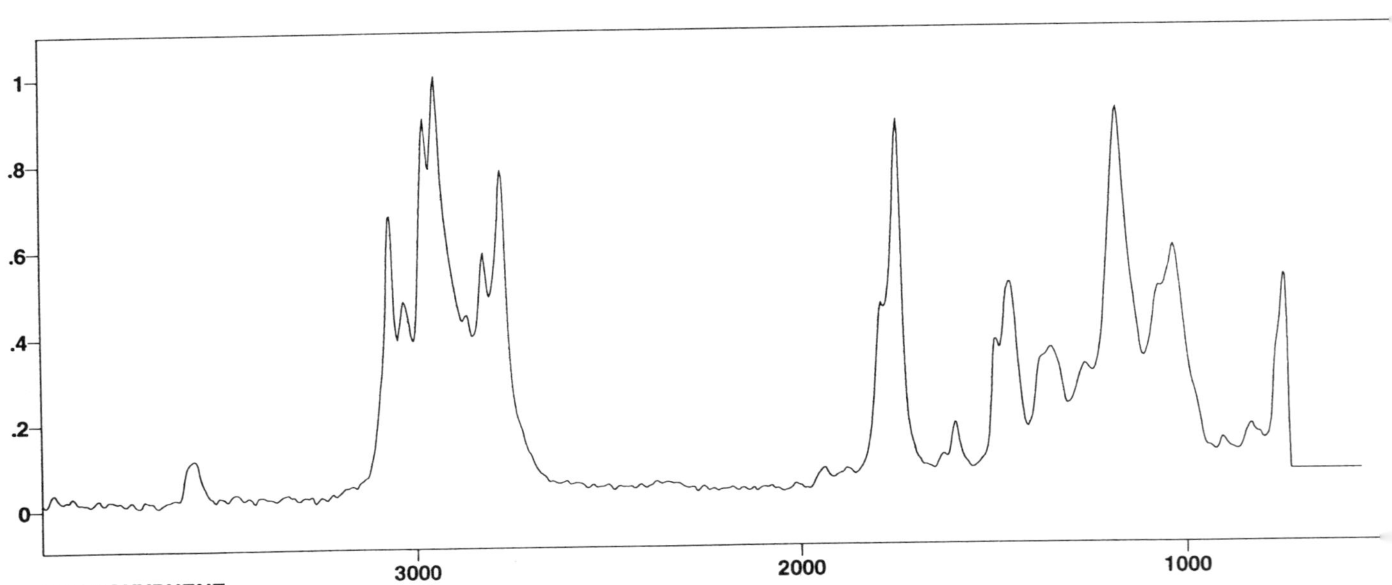

PROPOXYPHENE

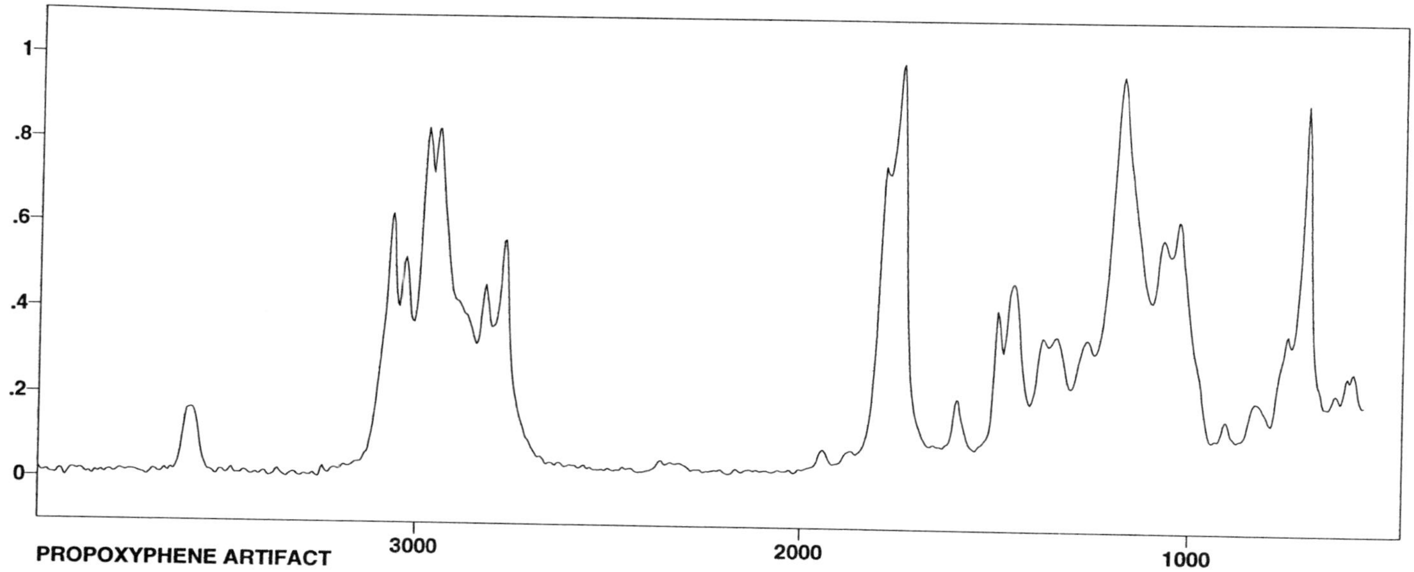

PROPOXYPHENE ARTIFACT

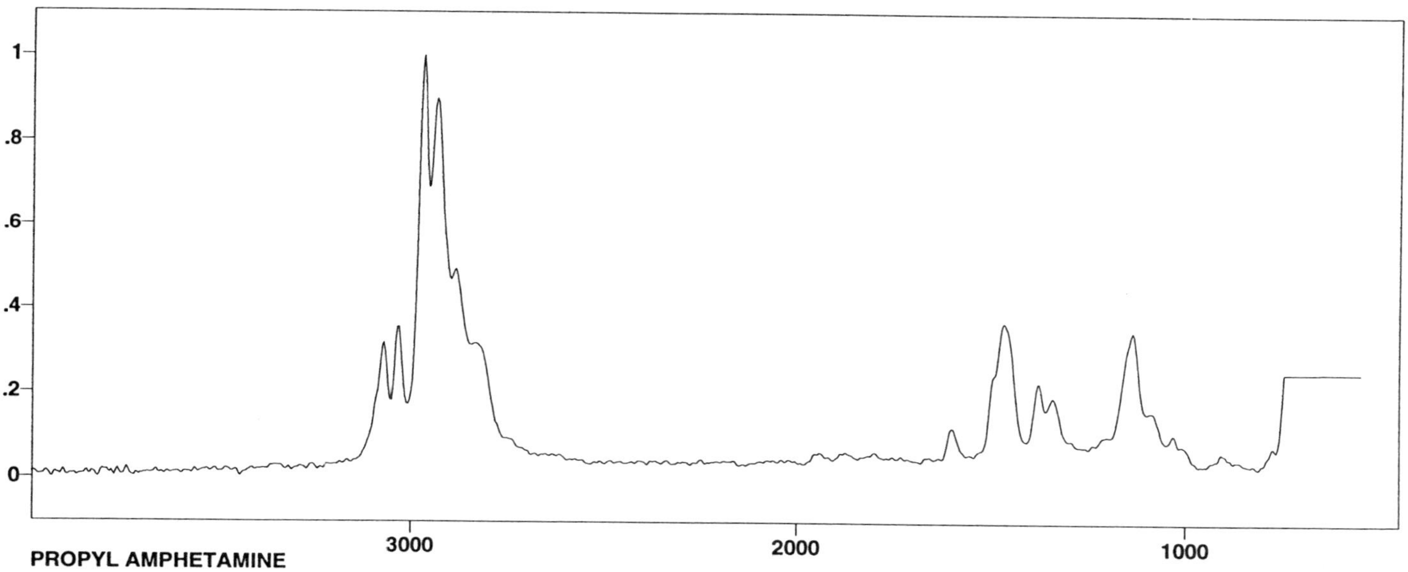

PROPYL AMPHETAMINE

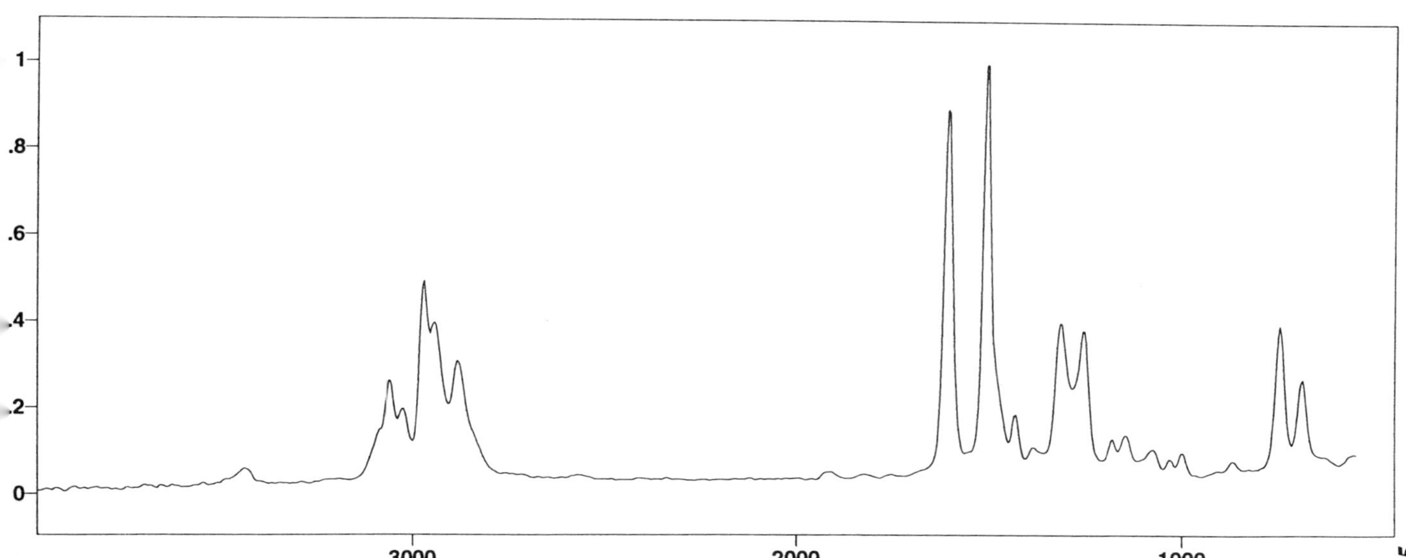

N-PROPYLANILINE

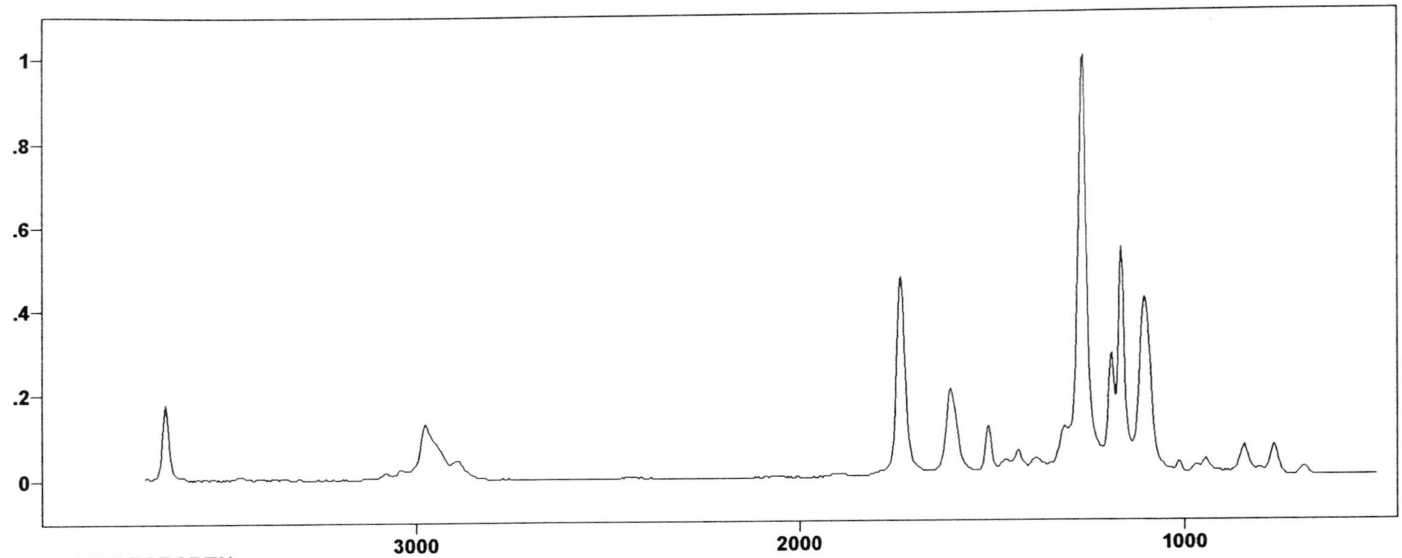

PROPYLPARABEN

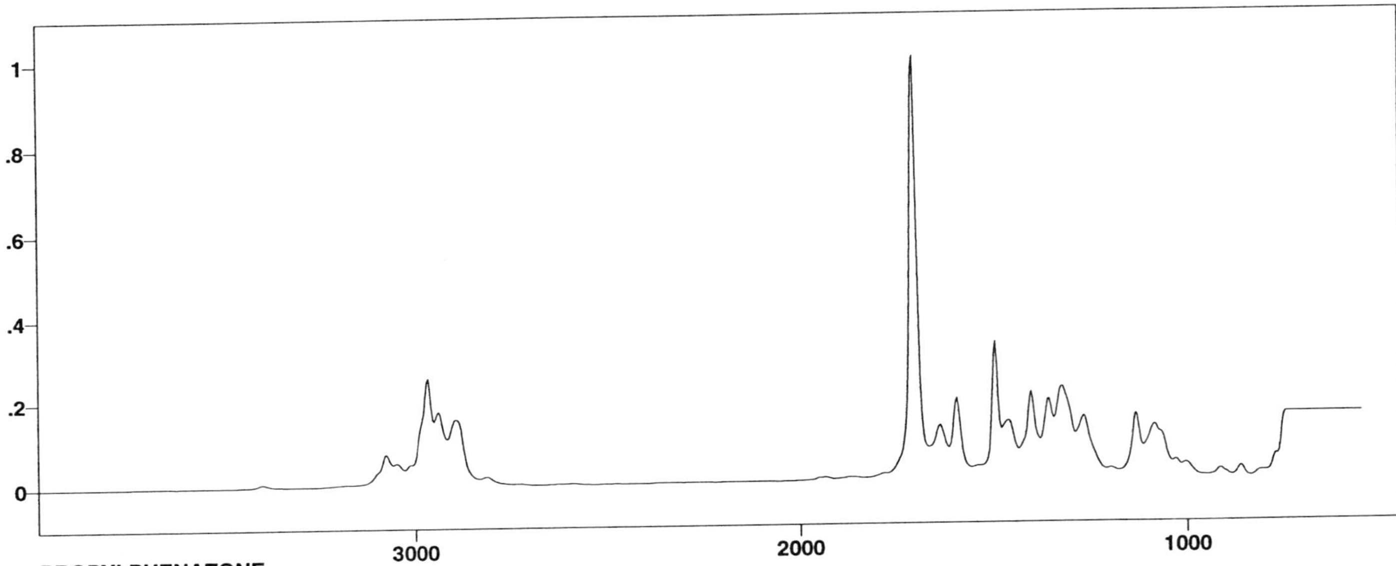

PROPYLPHENAZONE

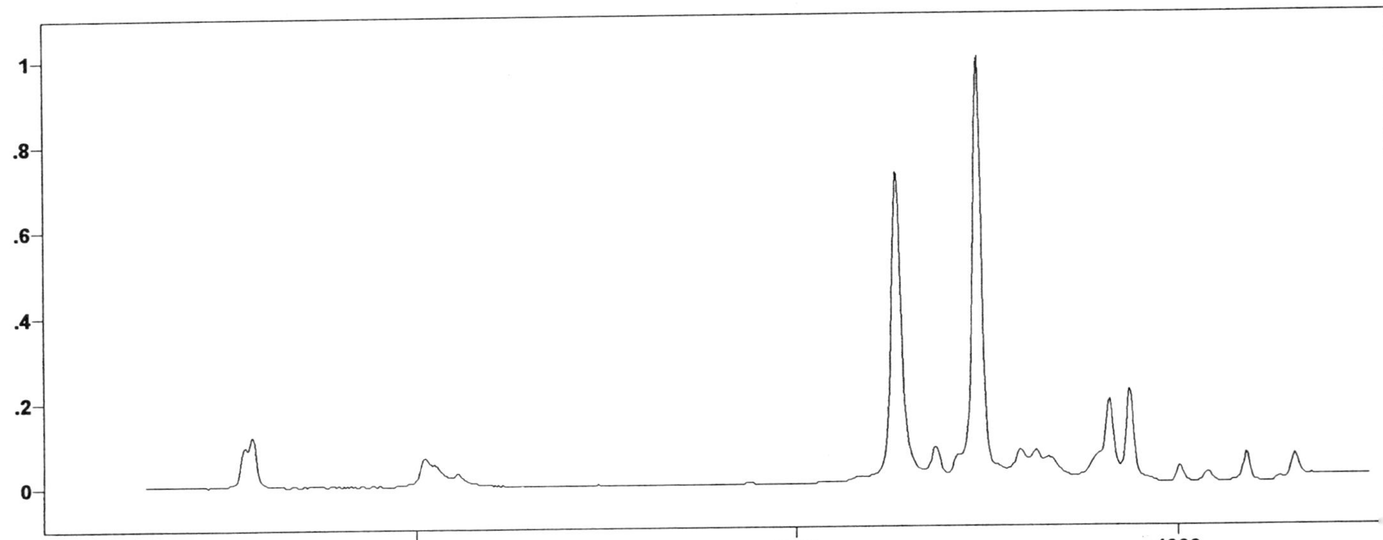

PROPYLTHIOURACIL

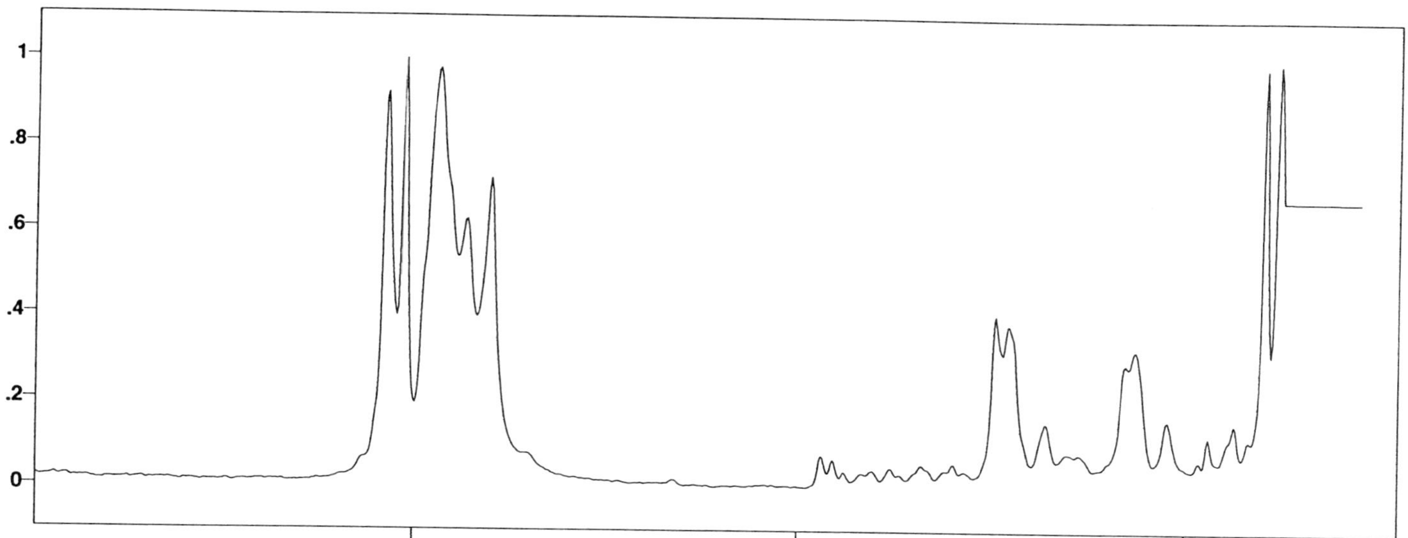

PROTRIPTYLINE

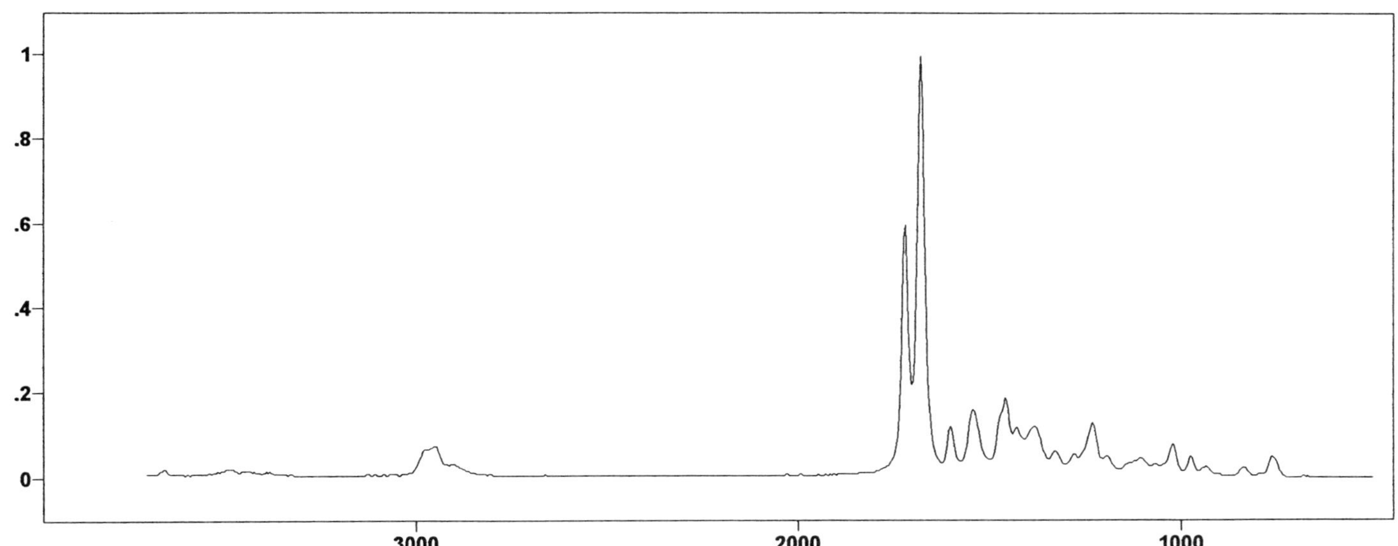

PROXYPHYLLINE

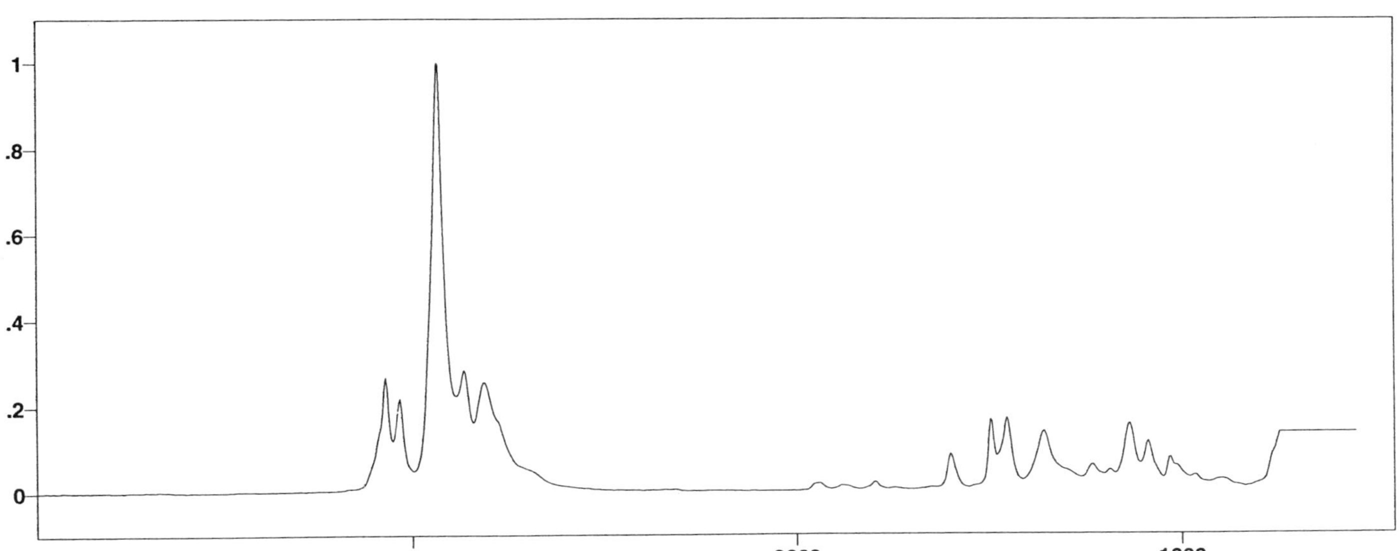

PROZAPINE

PSEUDOEPHEDRINE

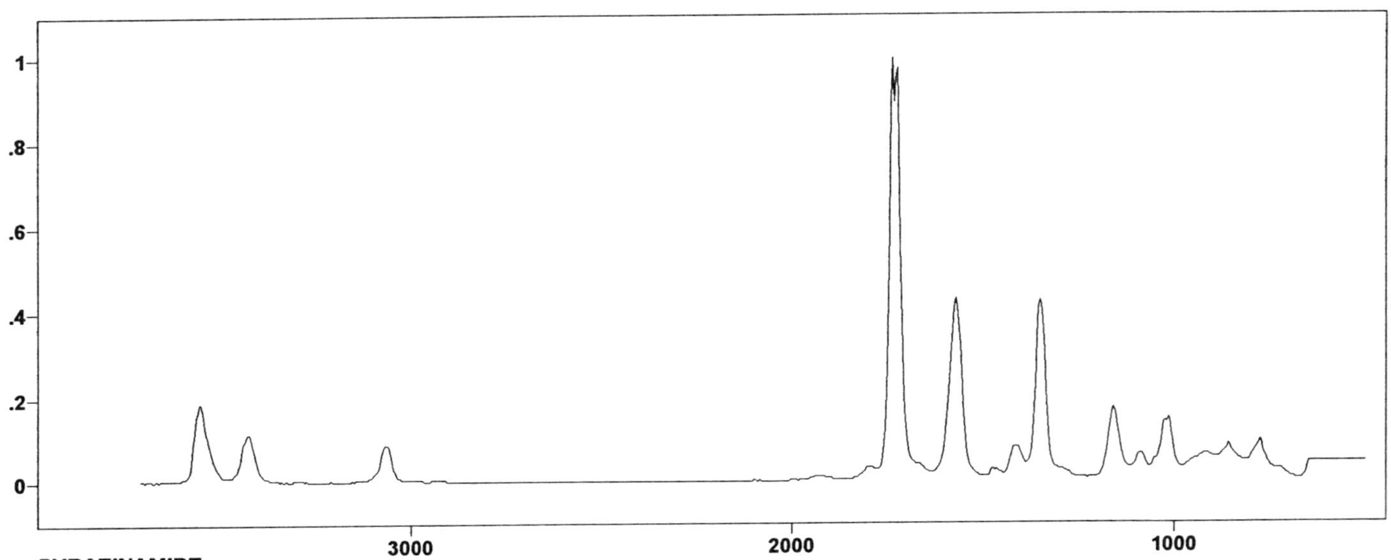

PYRAZINAMIDE

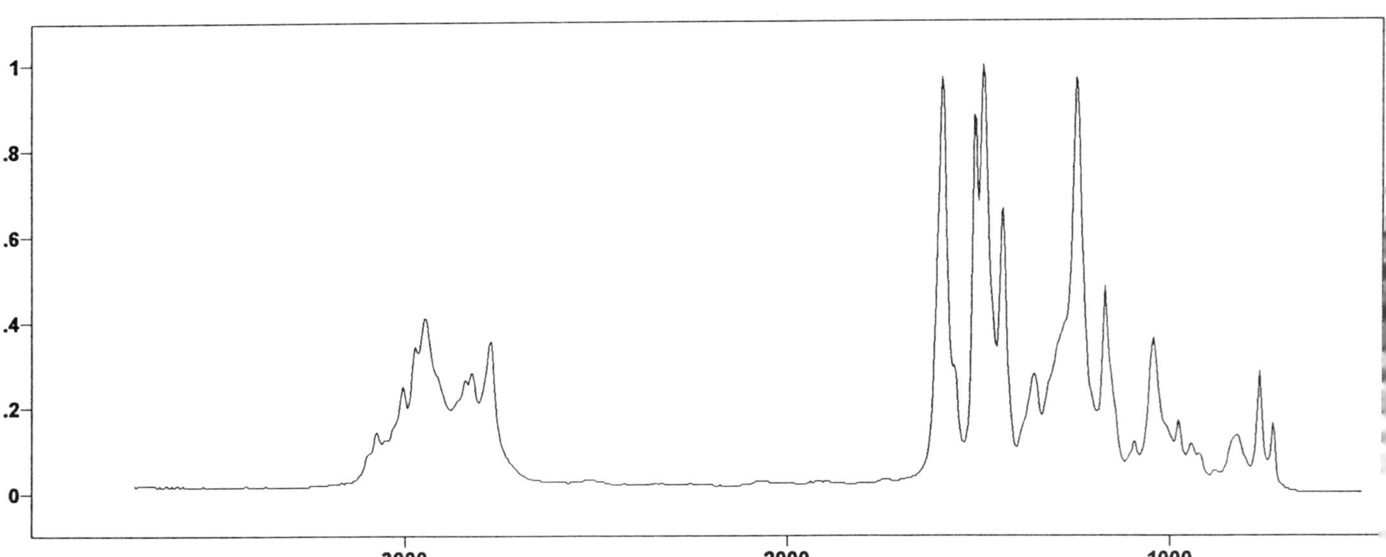

PYRILAMINE

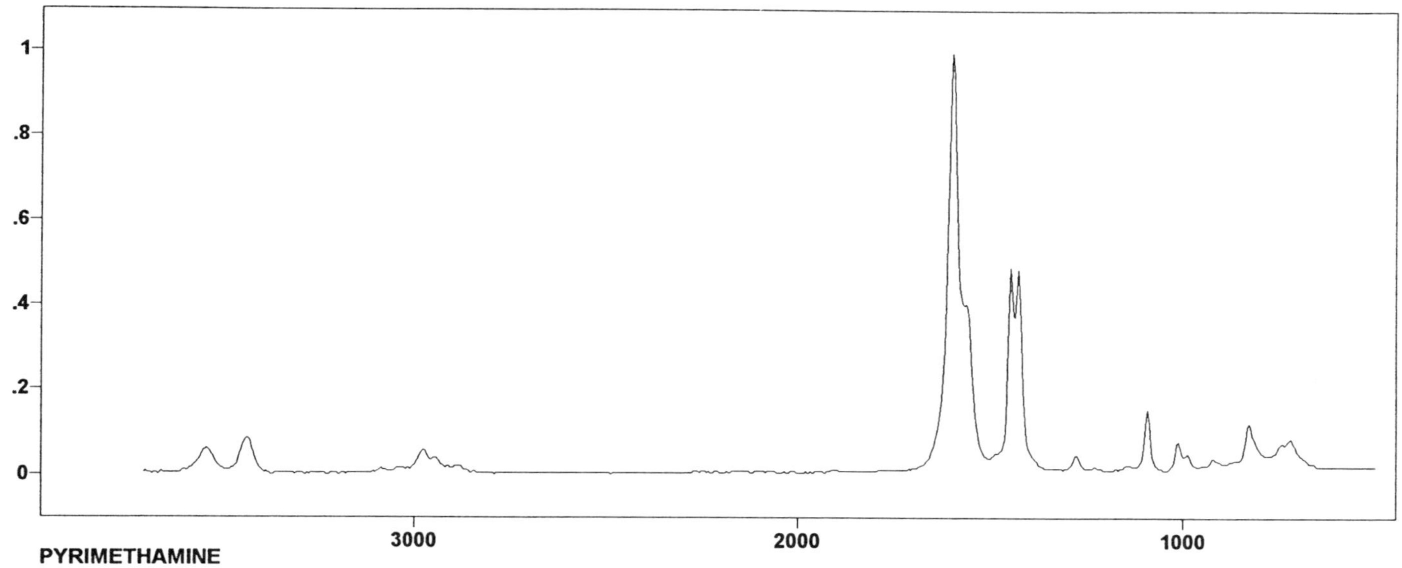

PYRIMETHAMINE

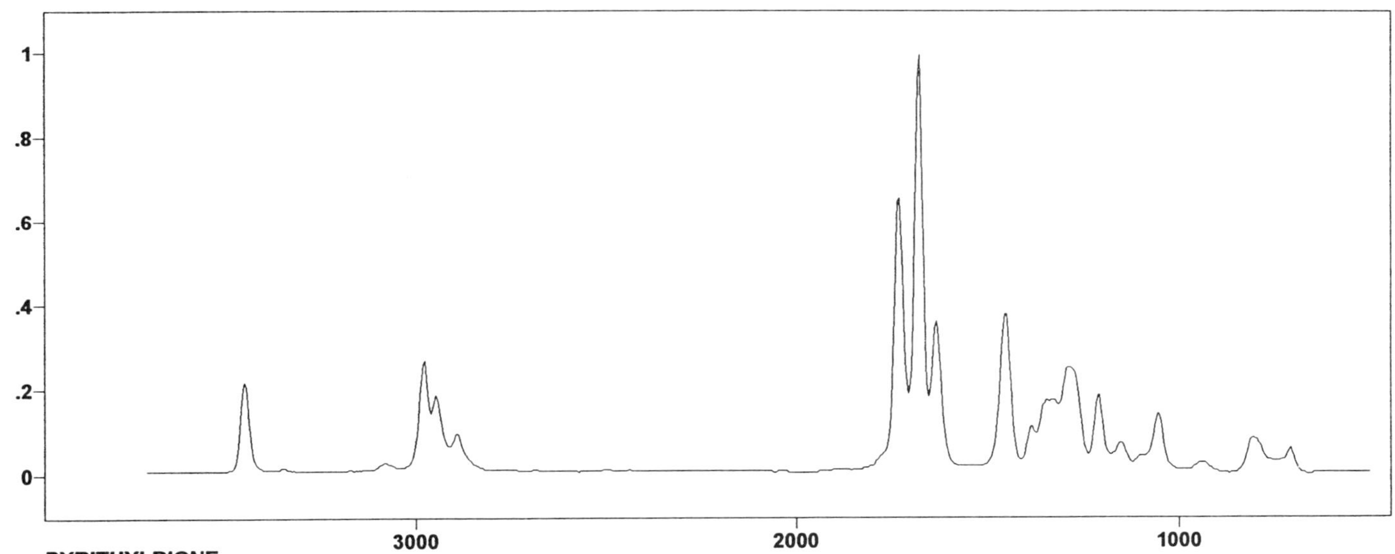

PYRITHYLDIONE

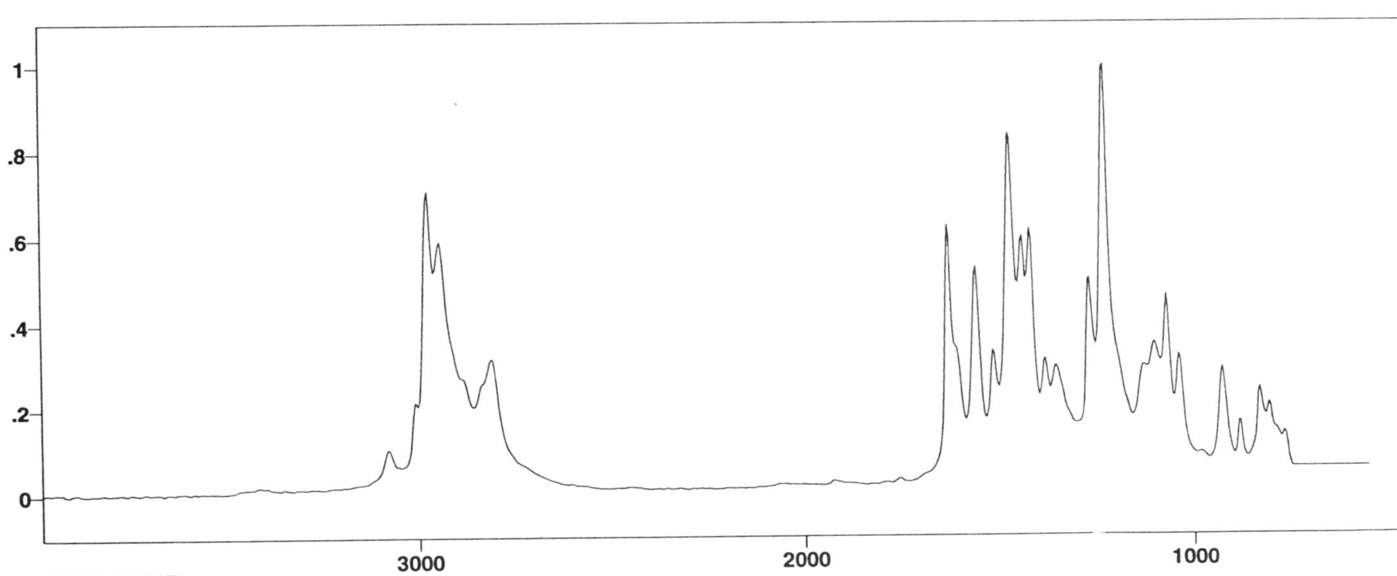

QUINACRINE

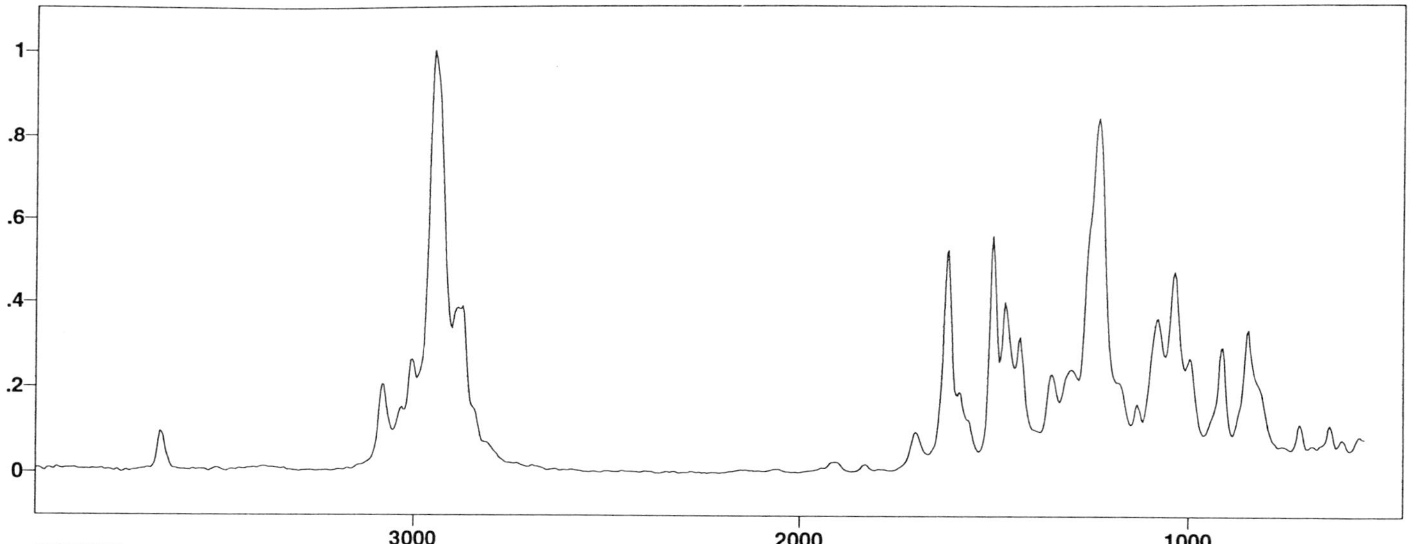

QUININE

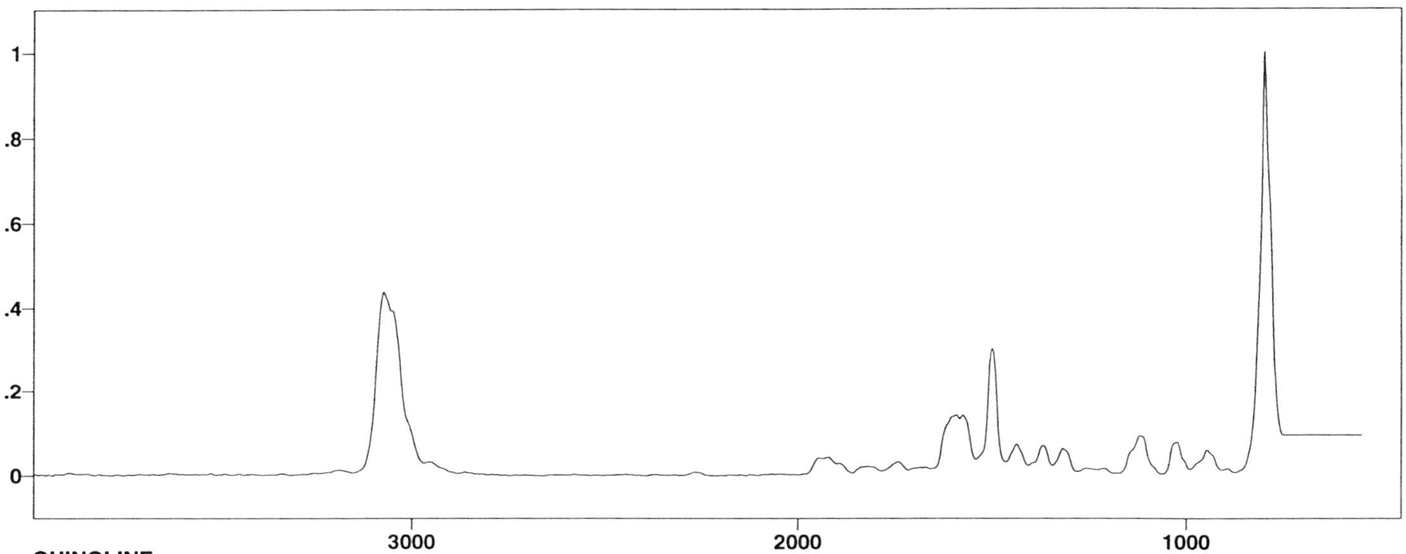

QUINOLINE

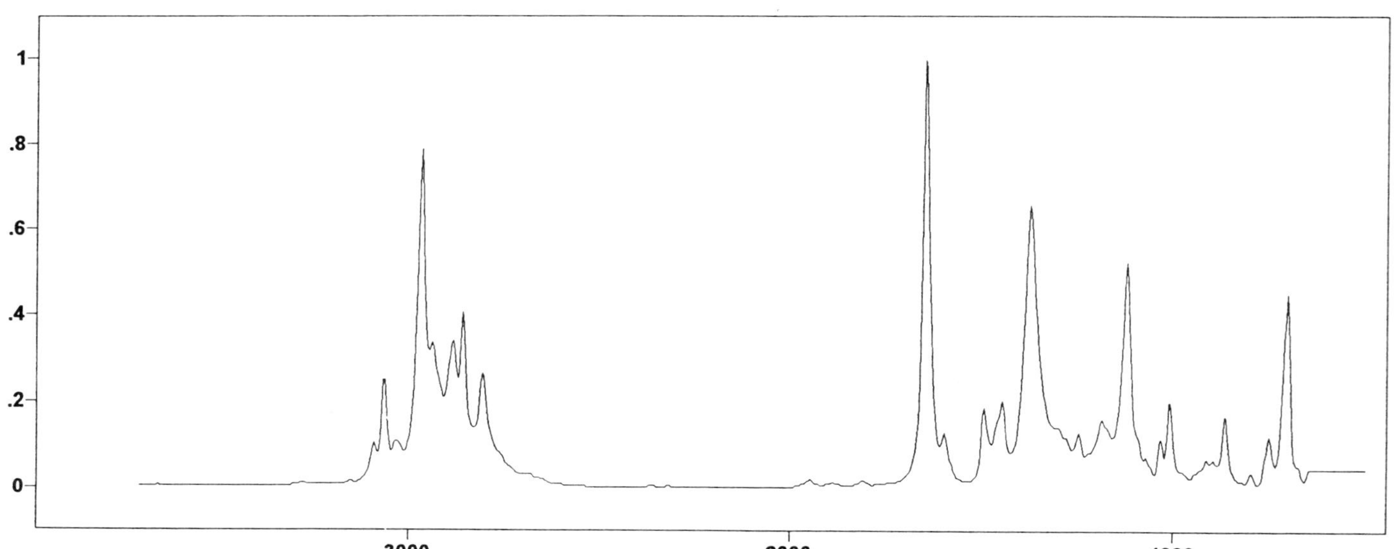

RACEMORAMIDE

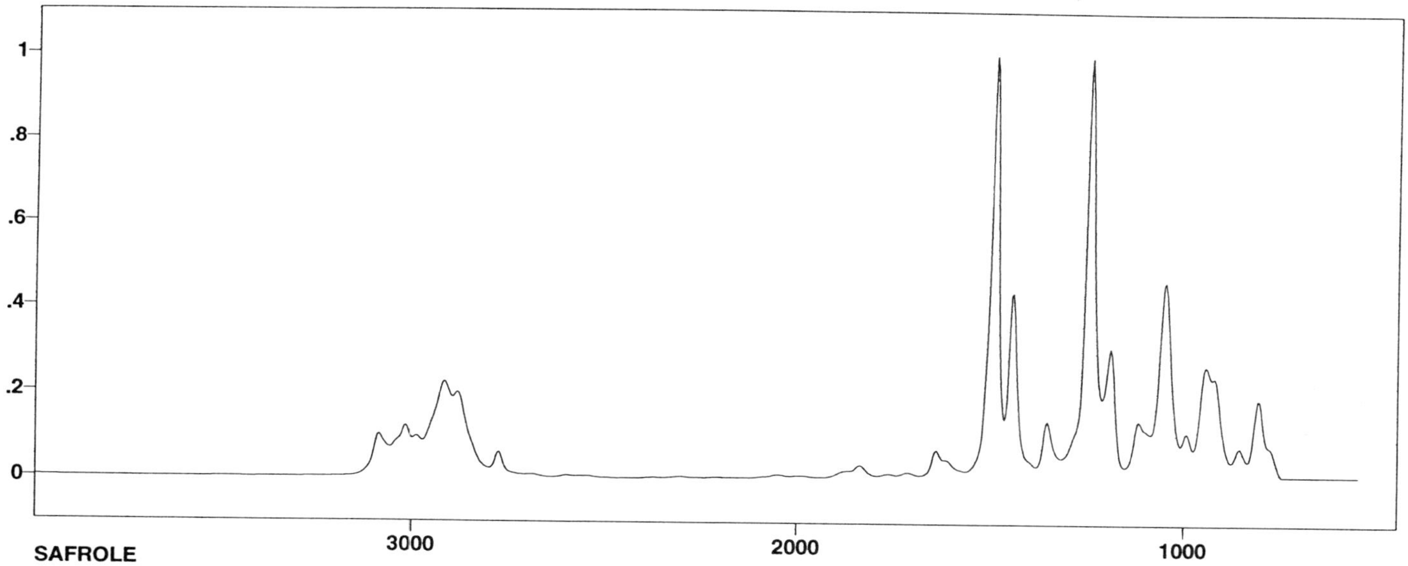

SAFROLE

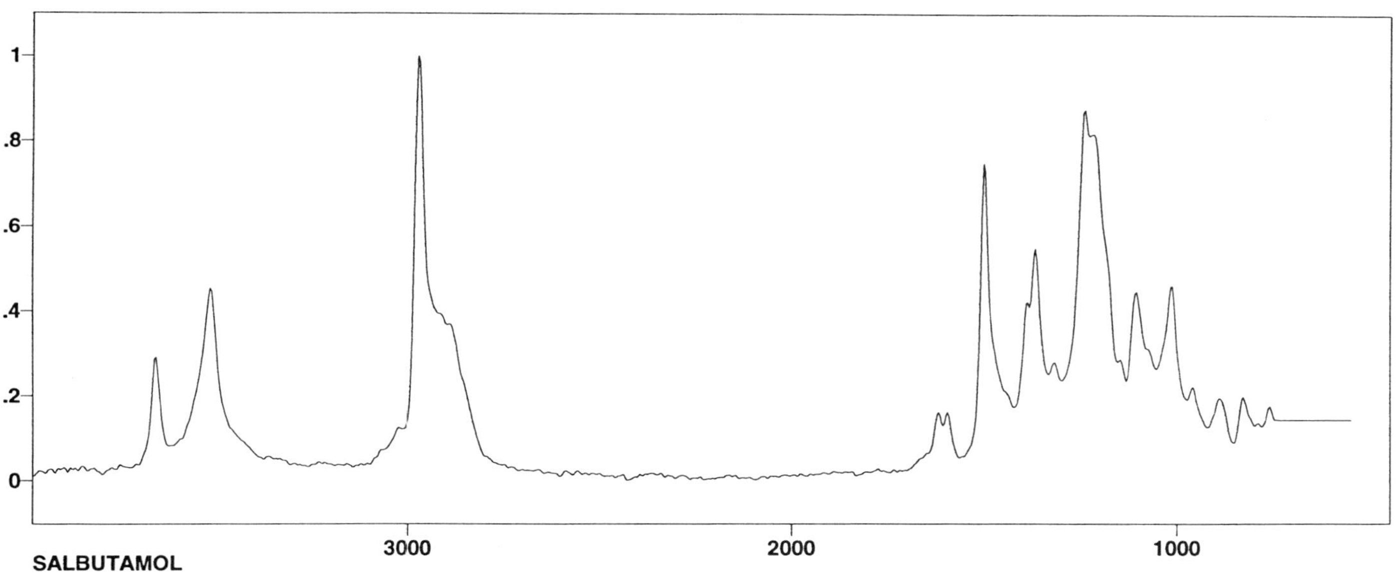

SALBUTAMOL

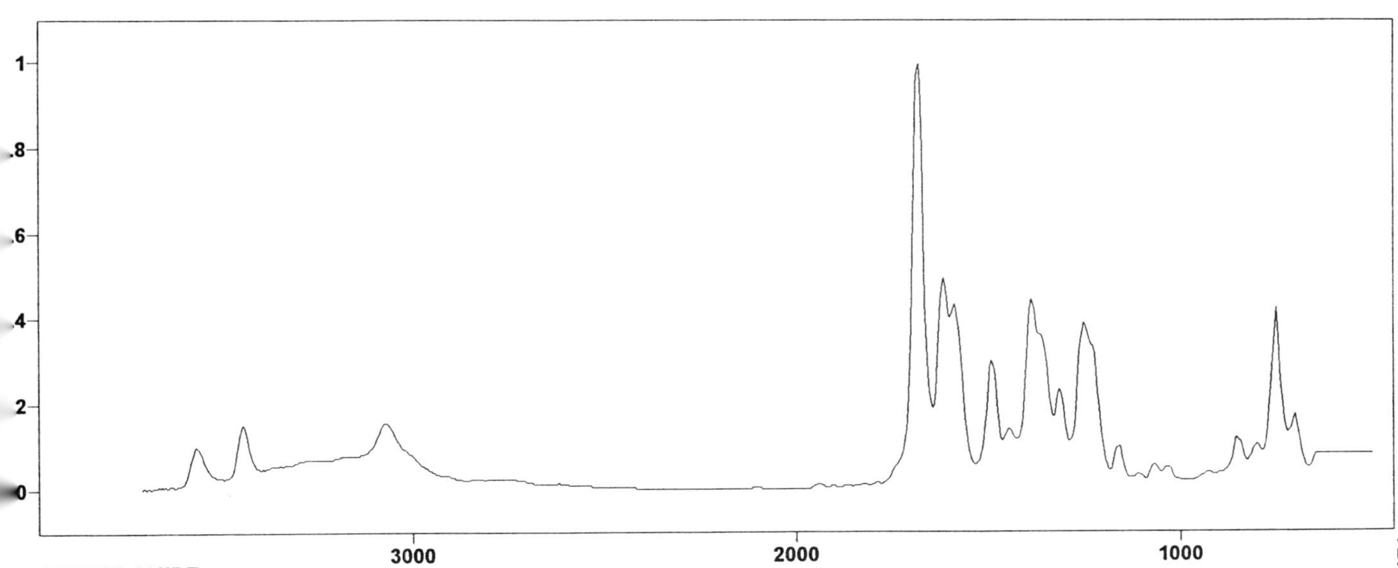

SALICYLAMIDE

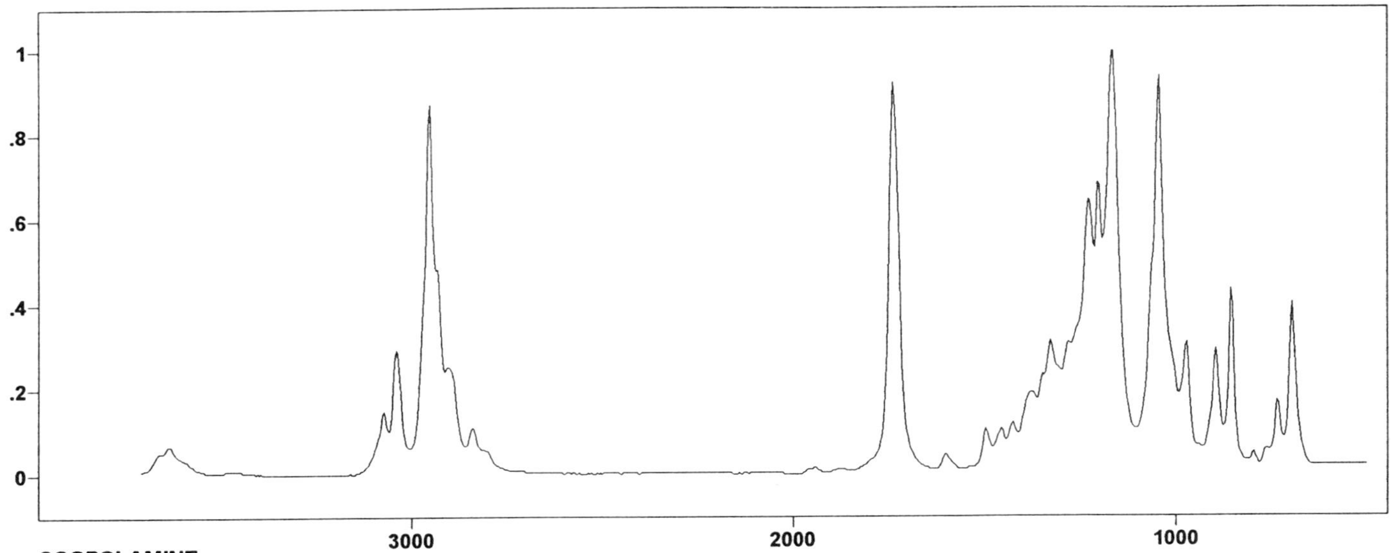

SCOPOLAMINE

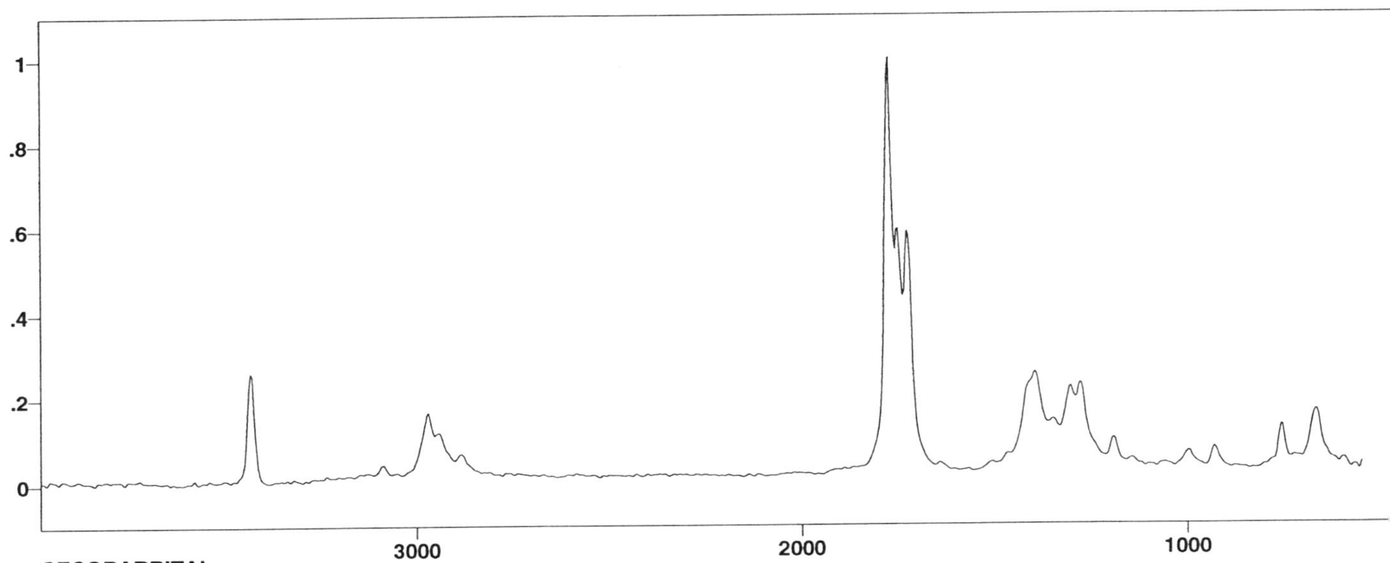

SECOBARBITAL

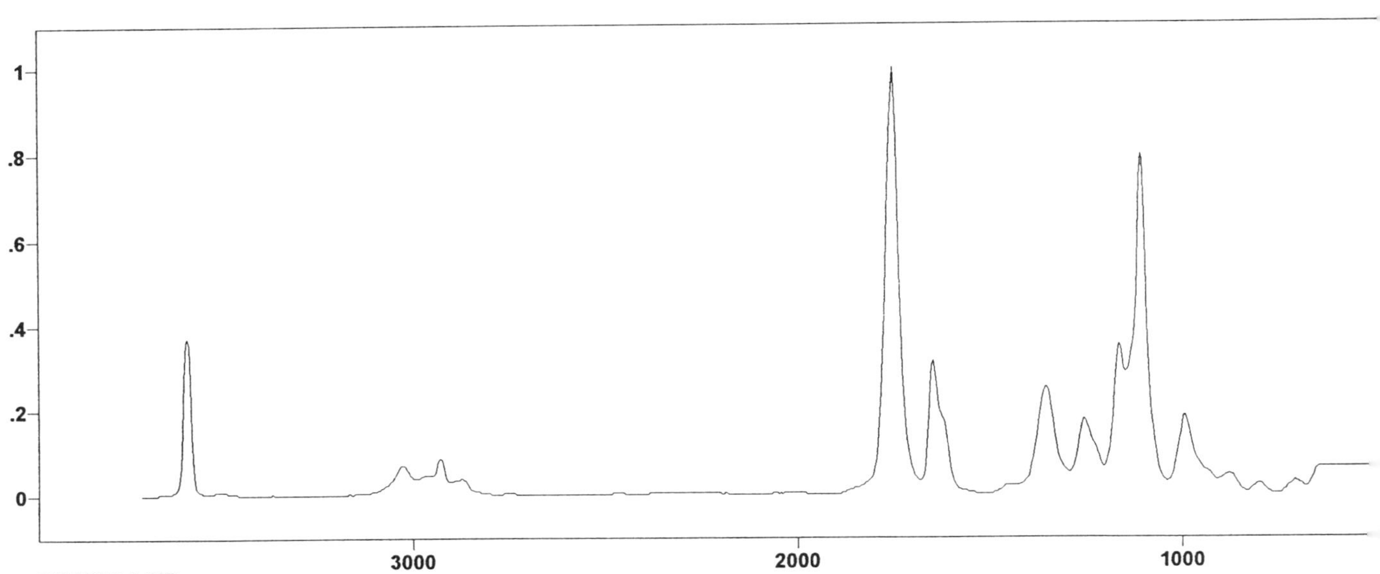

SORBIC ACID

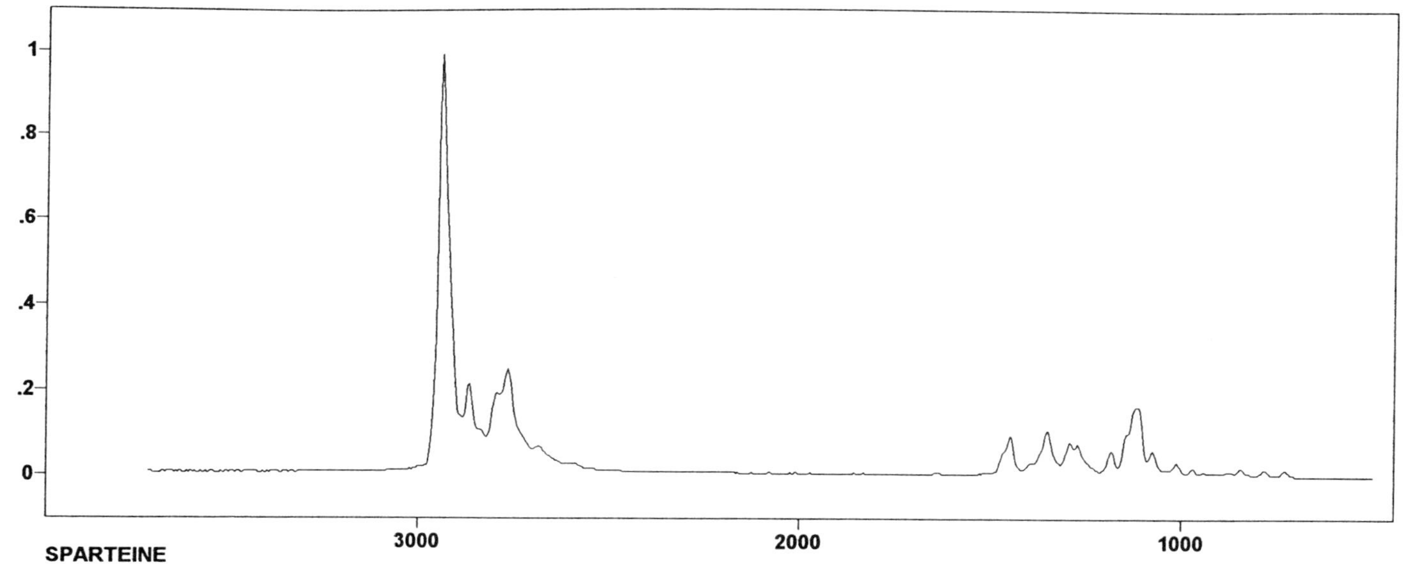

SPARTEINE

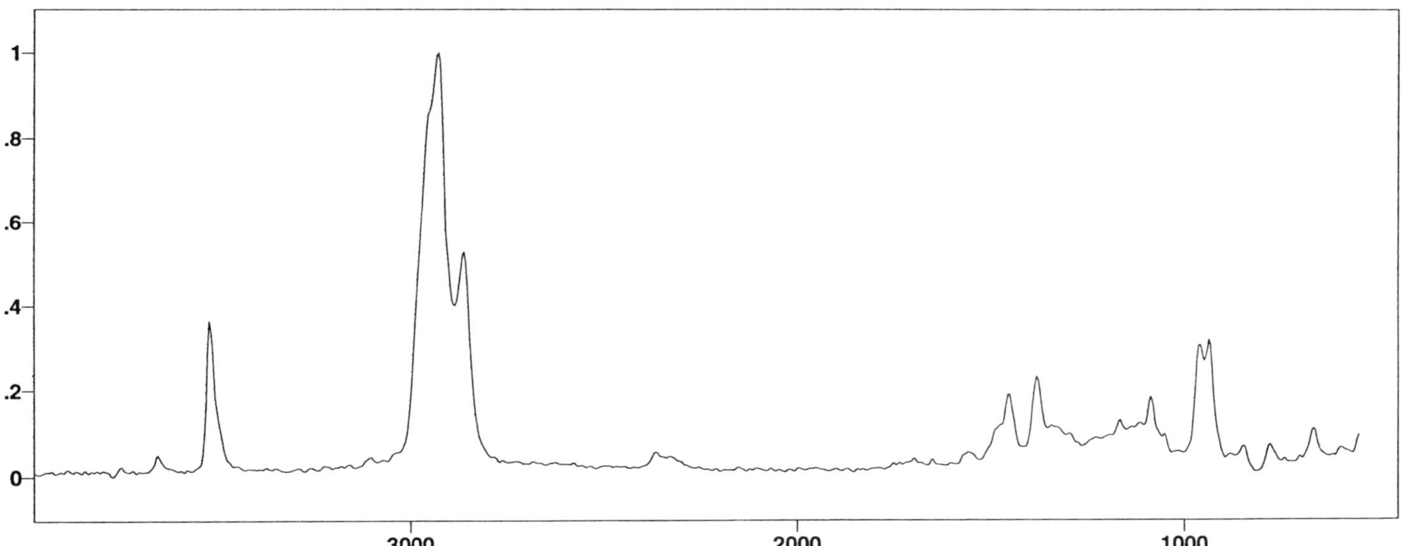

STANOZOLOL

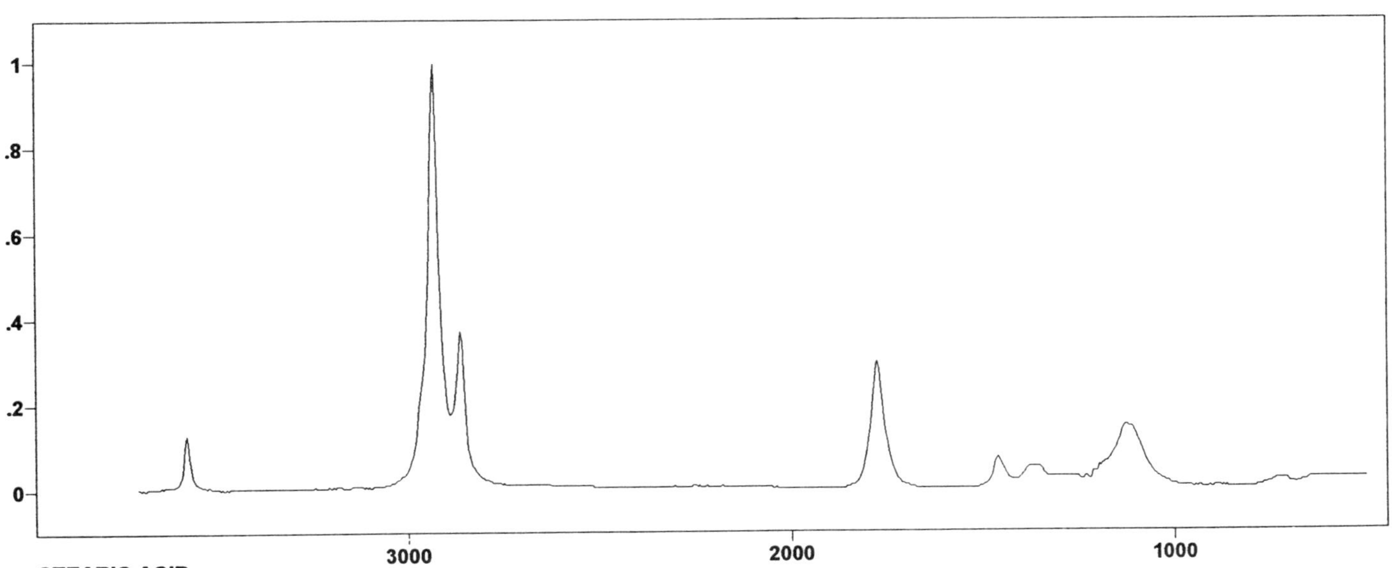

STEARIC ACID

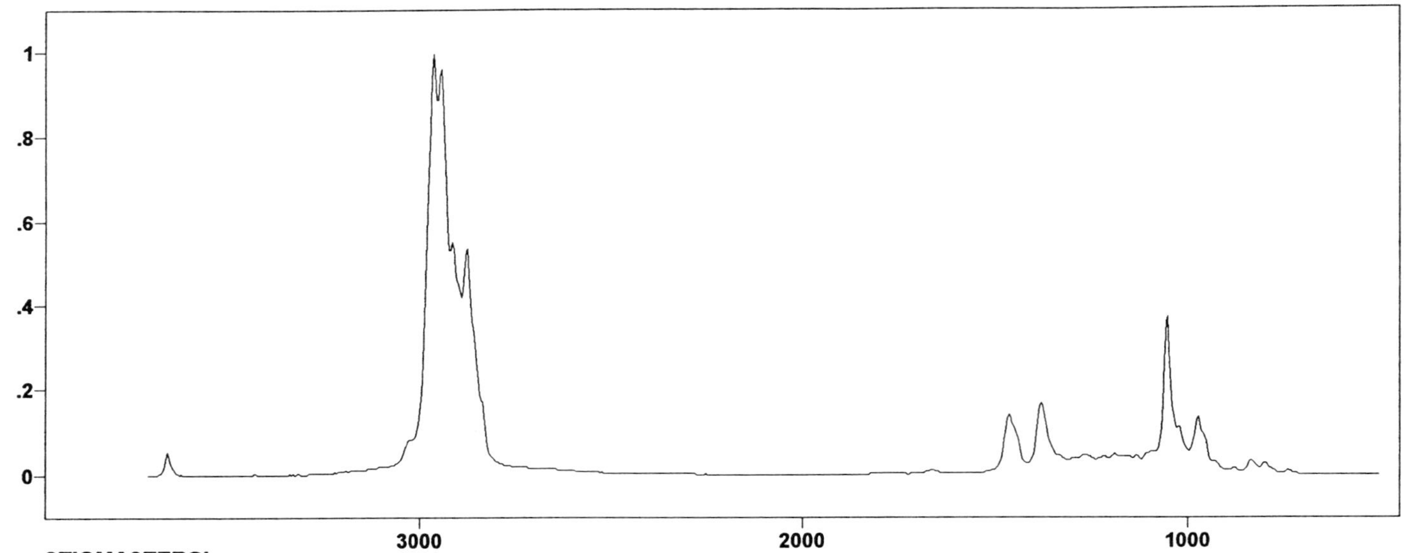

STIGMASTEROL

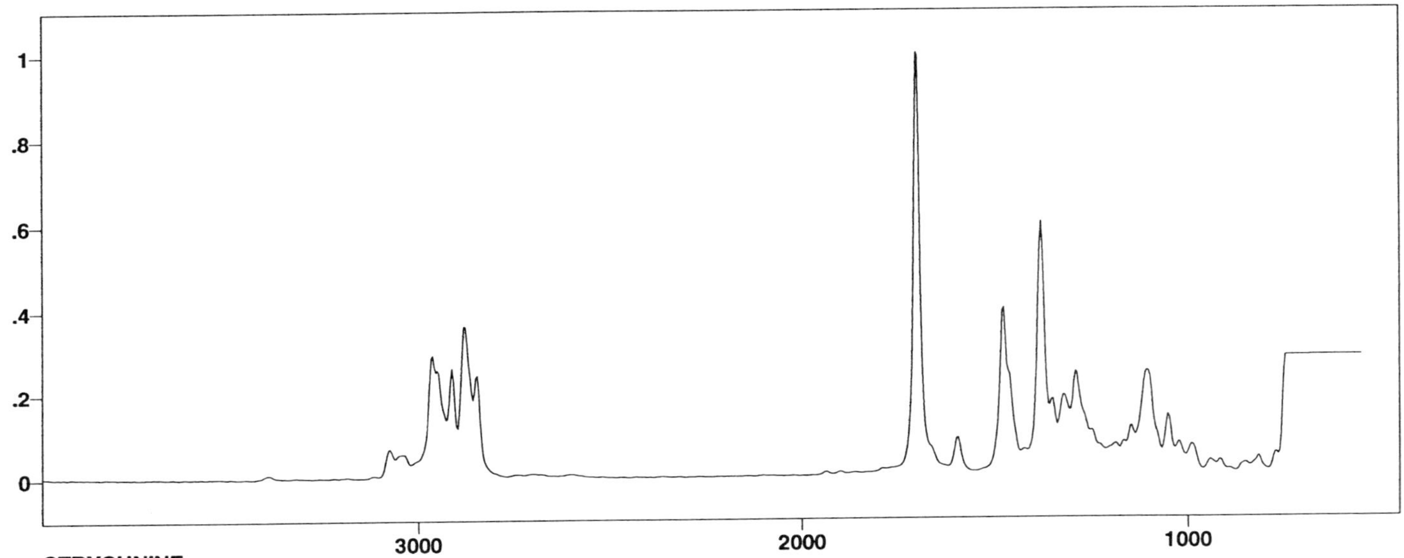

STRYCHNINE

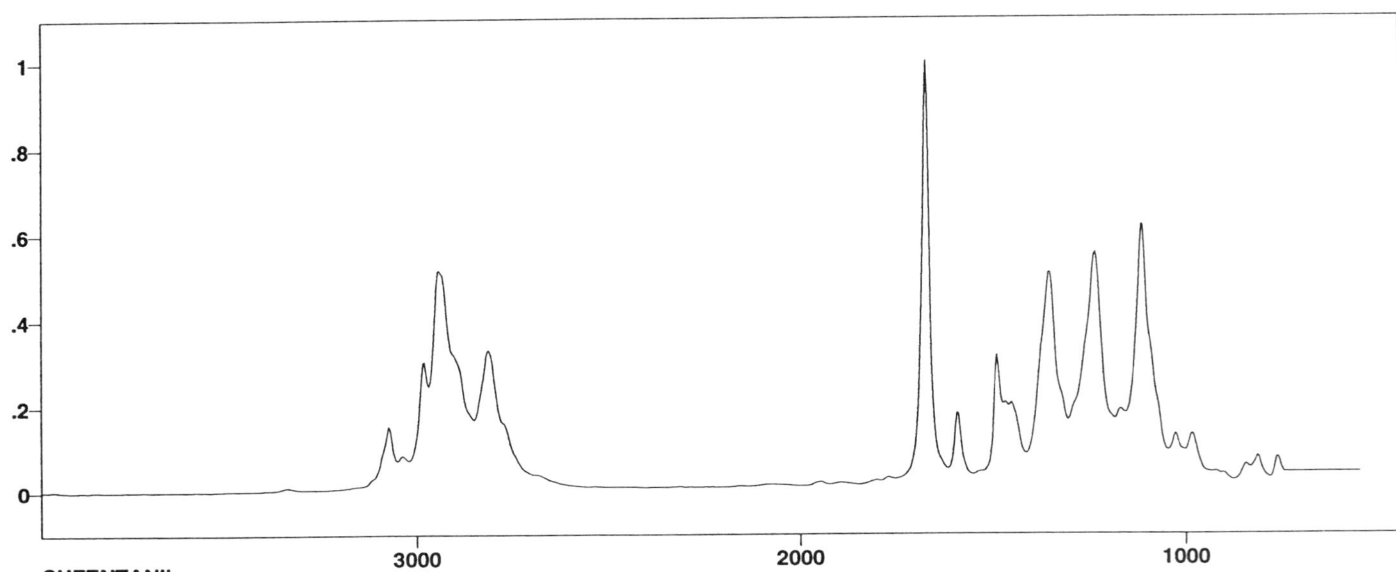

SUFENTANIL

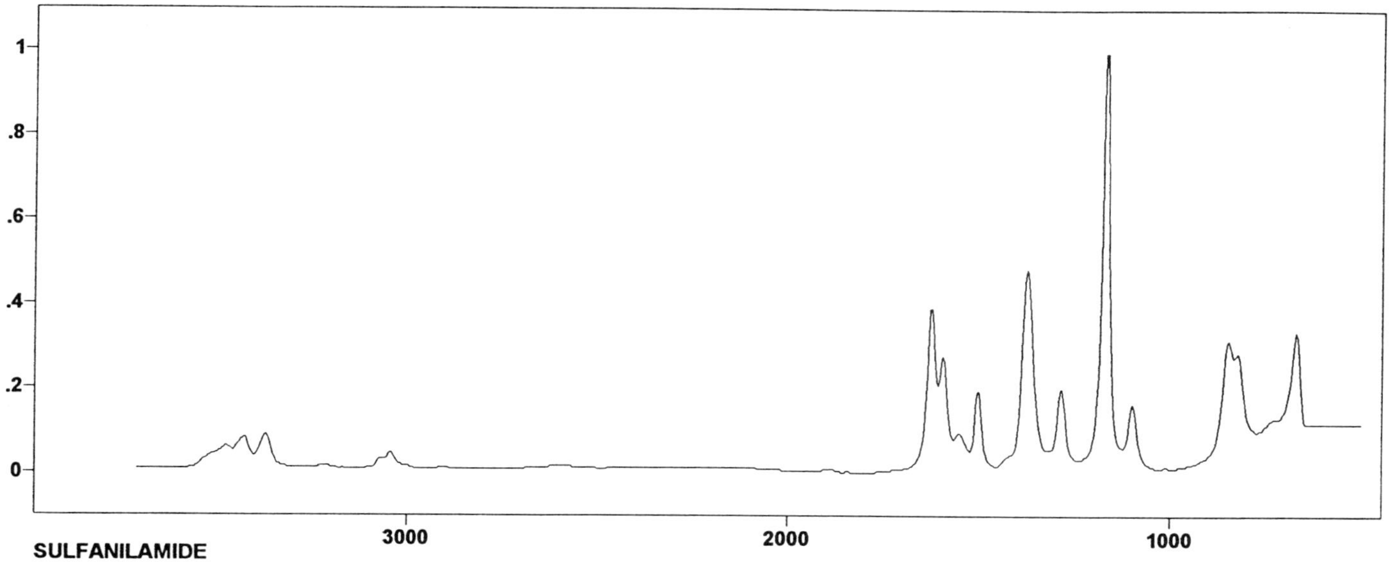

SULFANILAMIDE

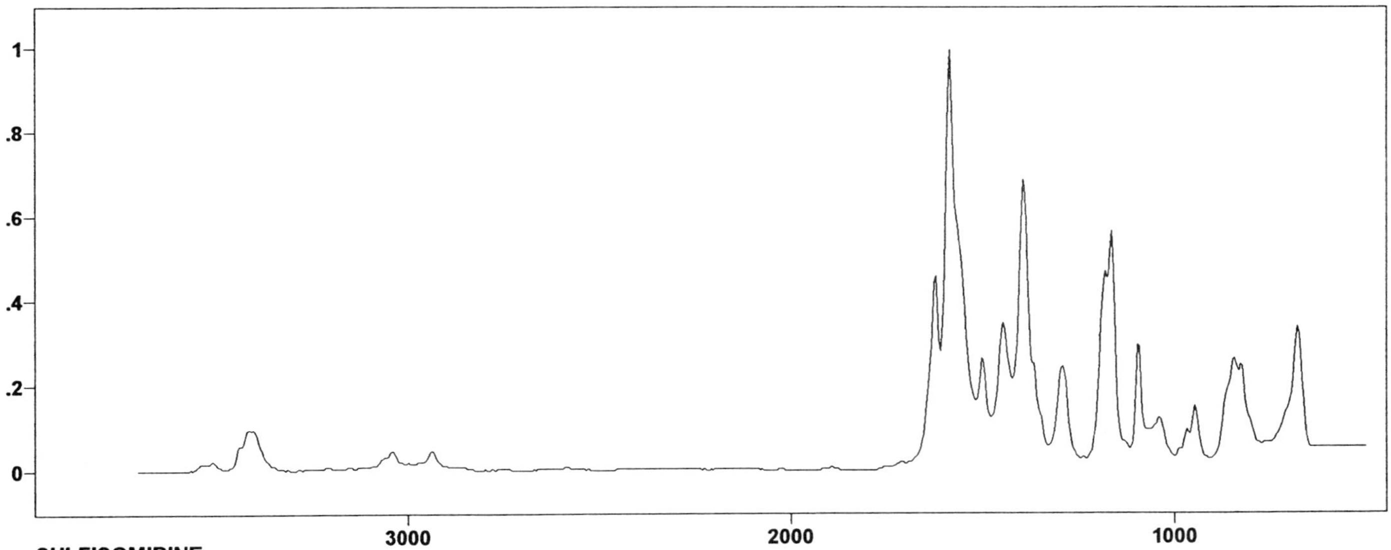

SULFISOMIDINE

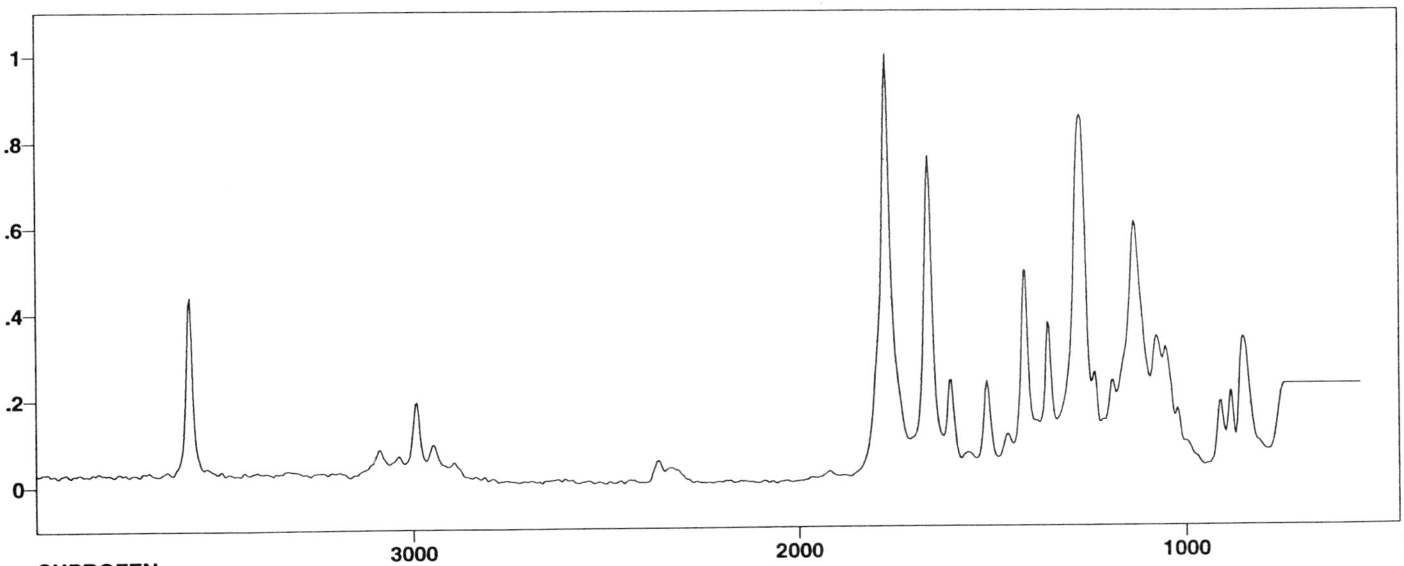

SUPROFEN

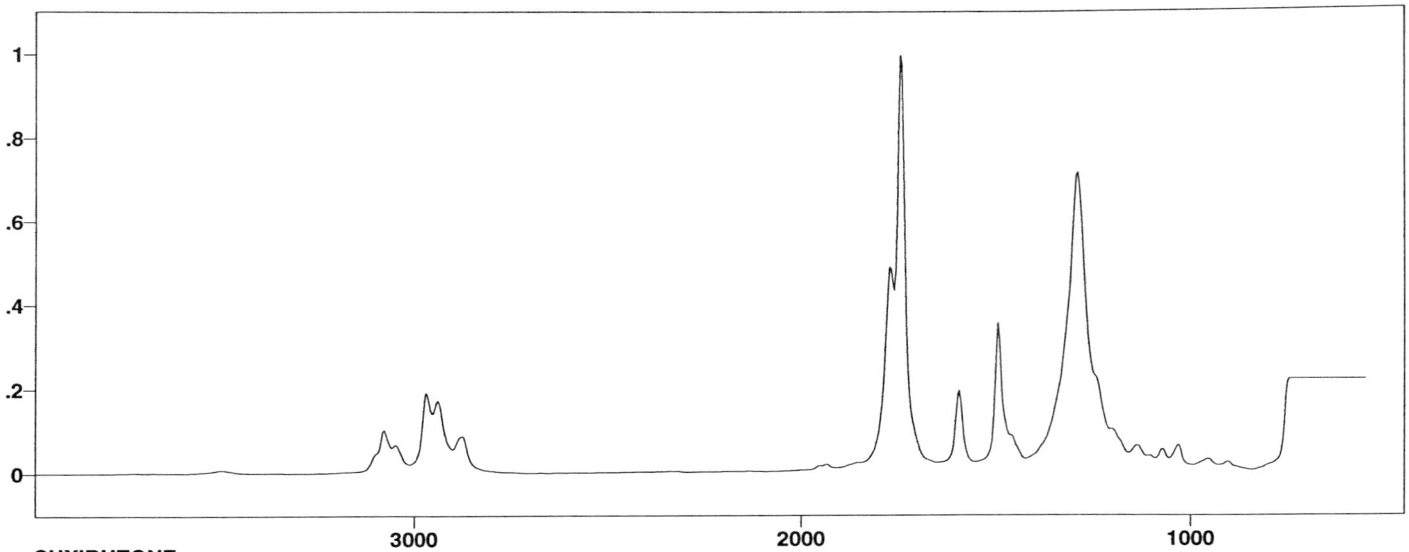

SUXIBUZONE

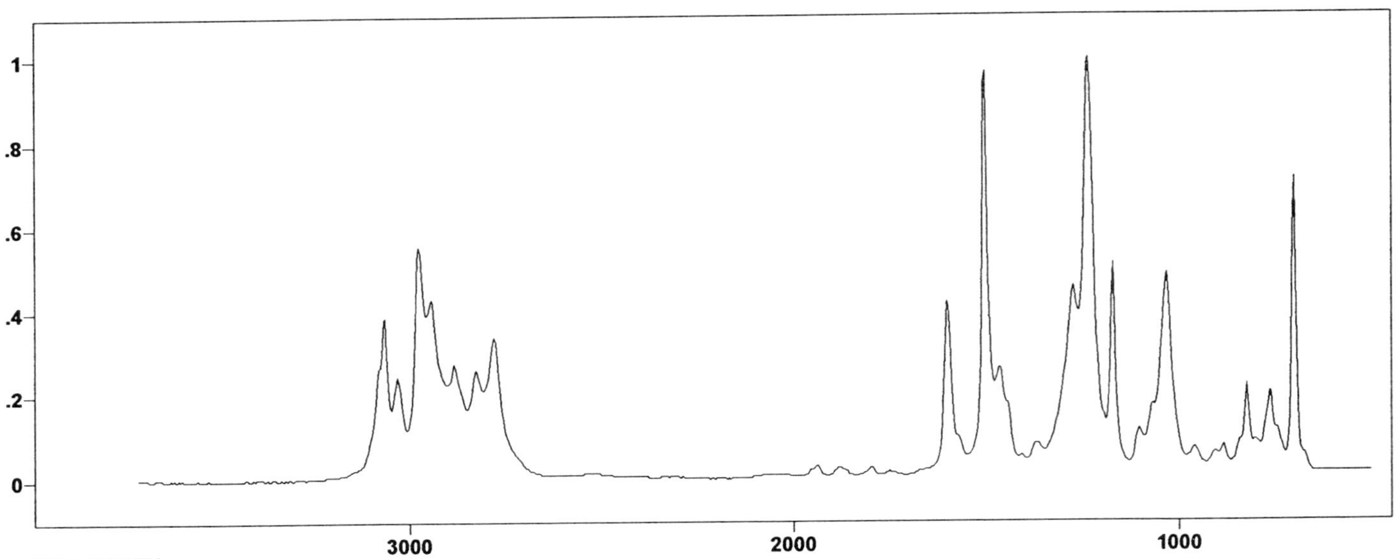

TAMOXIFEN

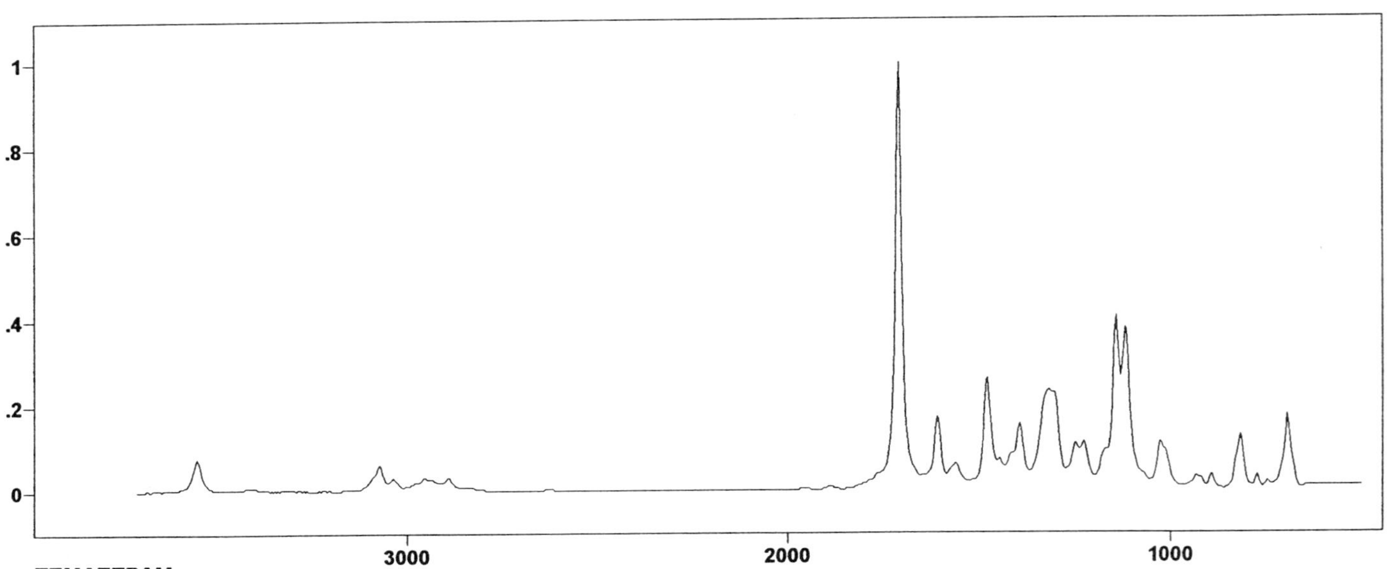

TEMAZEPAM

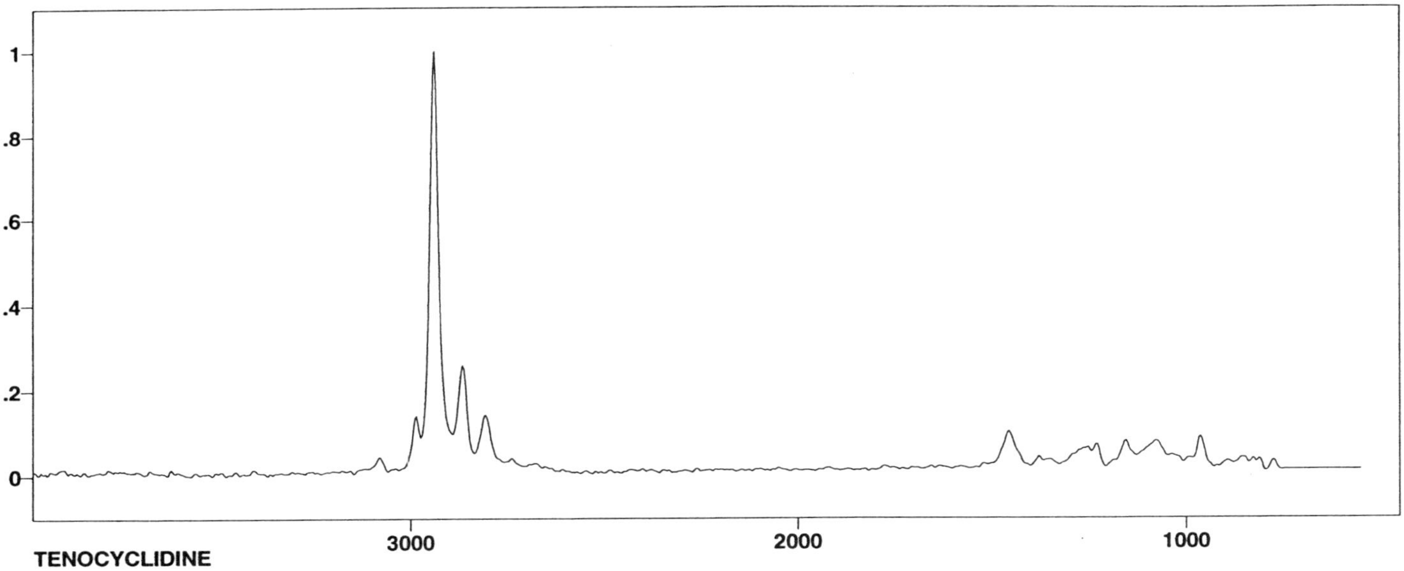

TENOCYCLIDINE

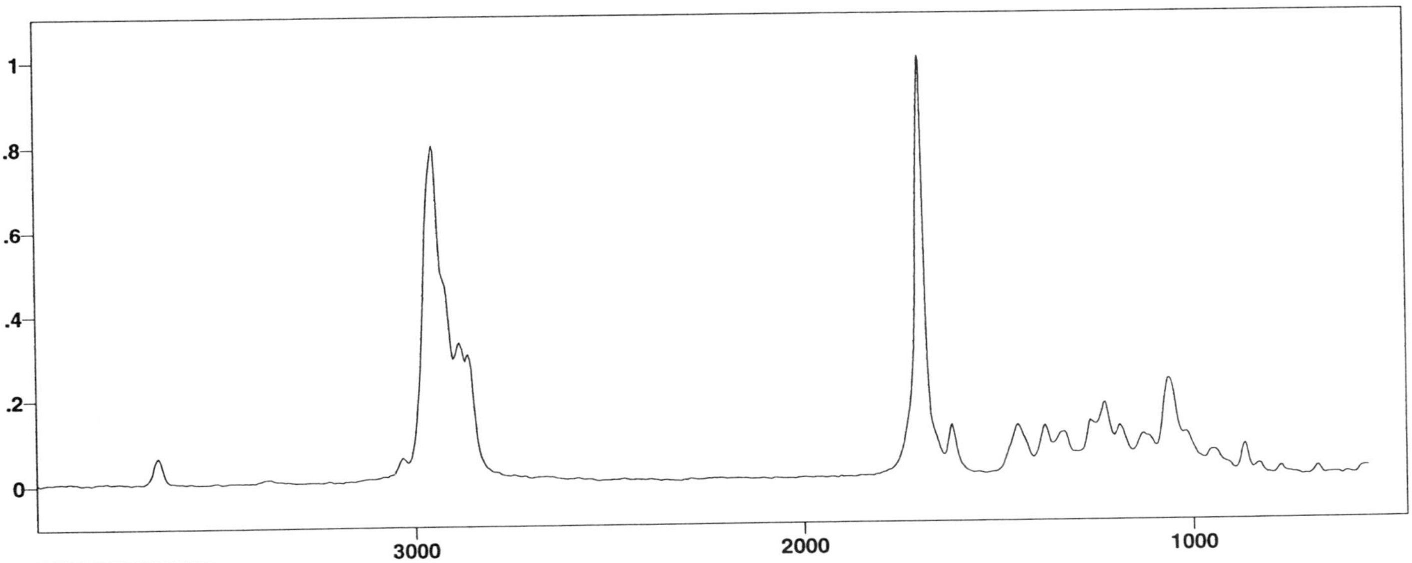

TESTOSTERONE

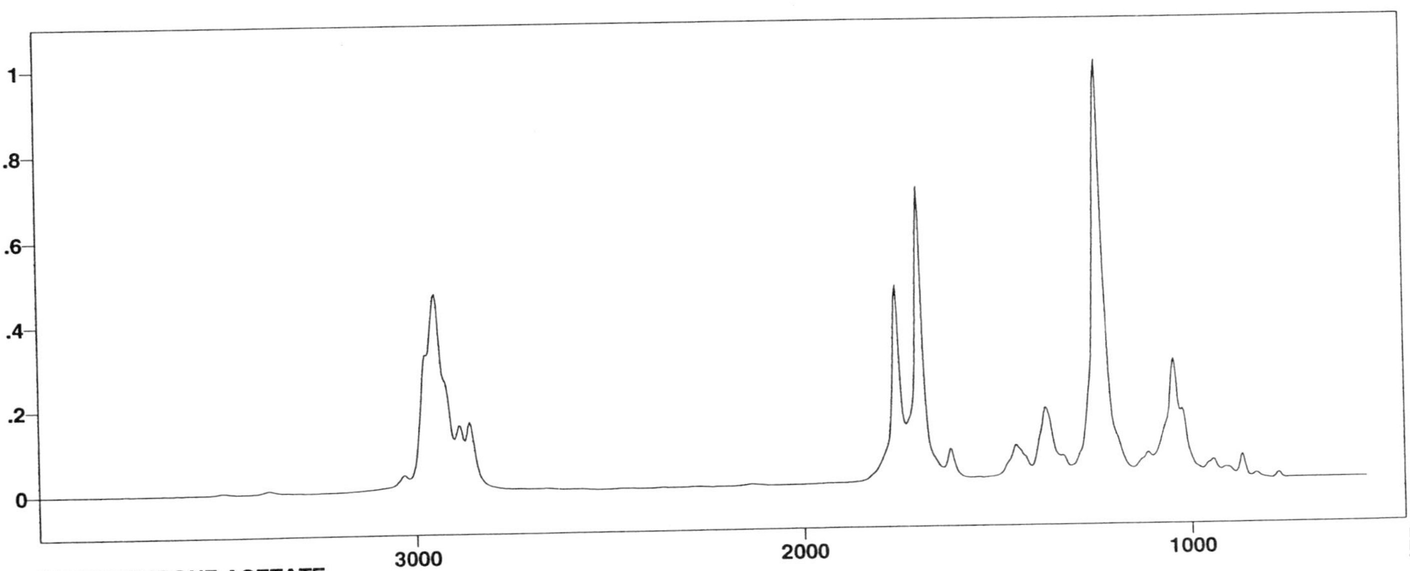

TESTOSTERONE ACETATE

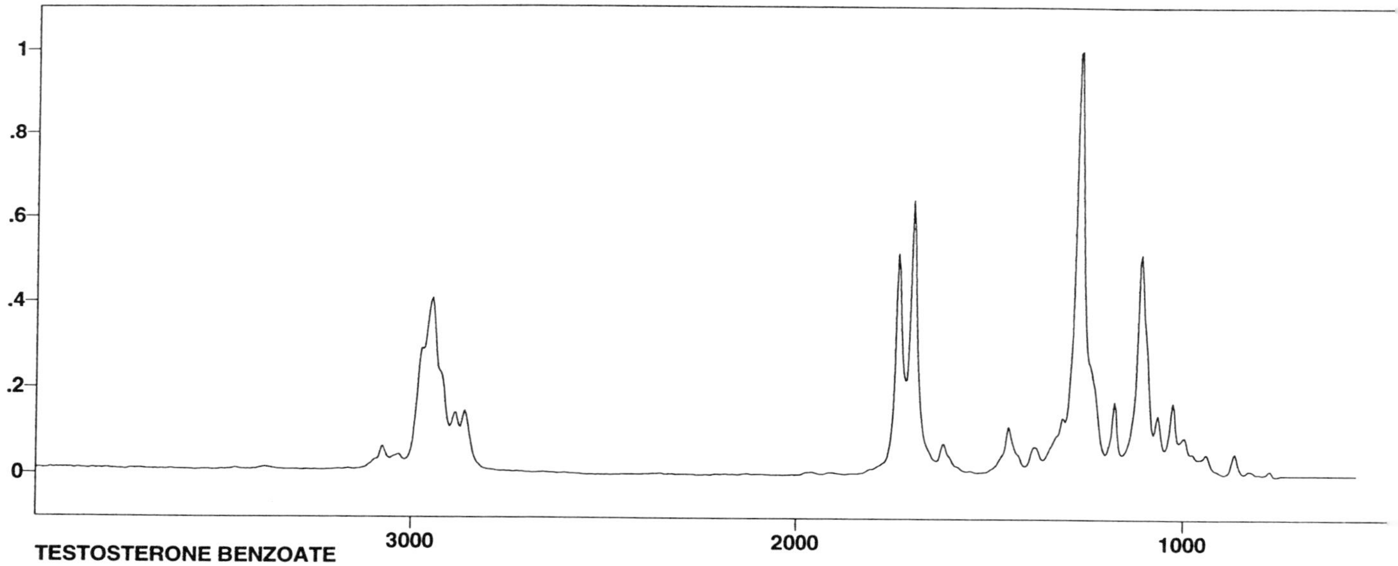

TESTOSTERONE BENZOATE

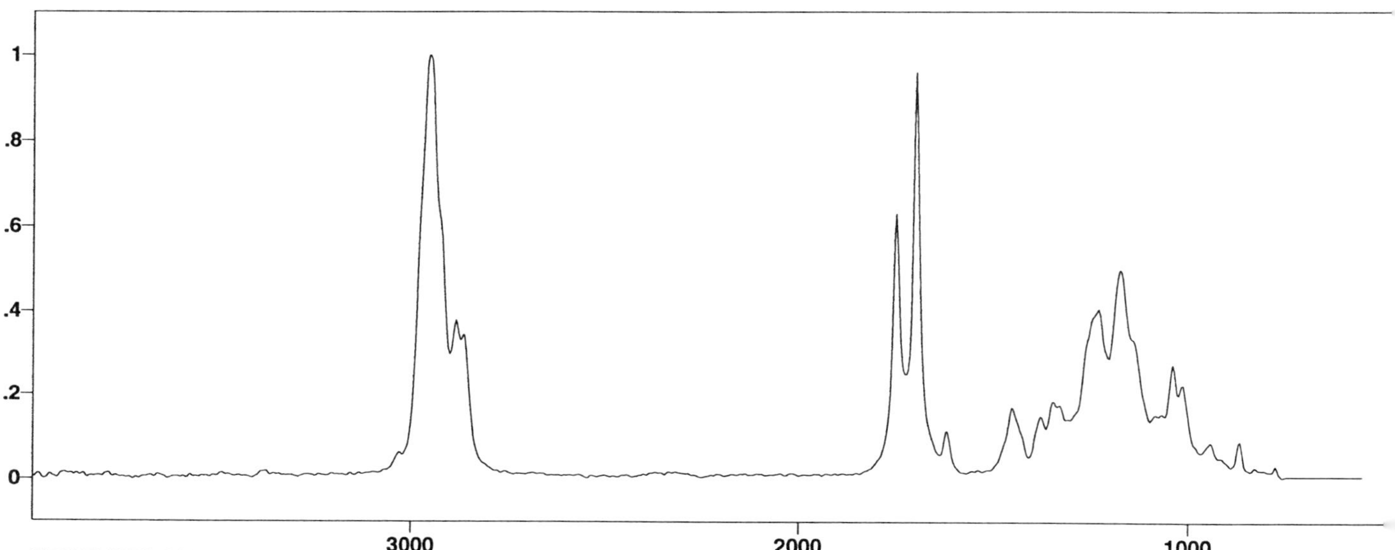

TESTOSTERONE-17-BETA-CYPIONATE

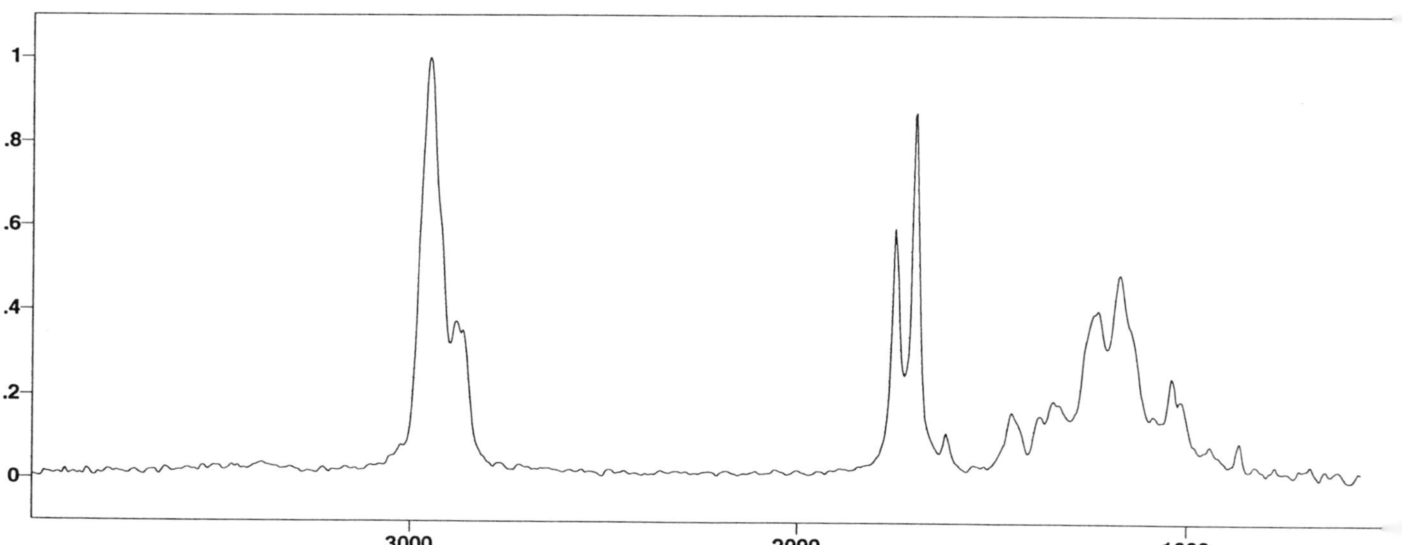

TESTOSTERONE CYPIONATE

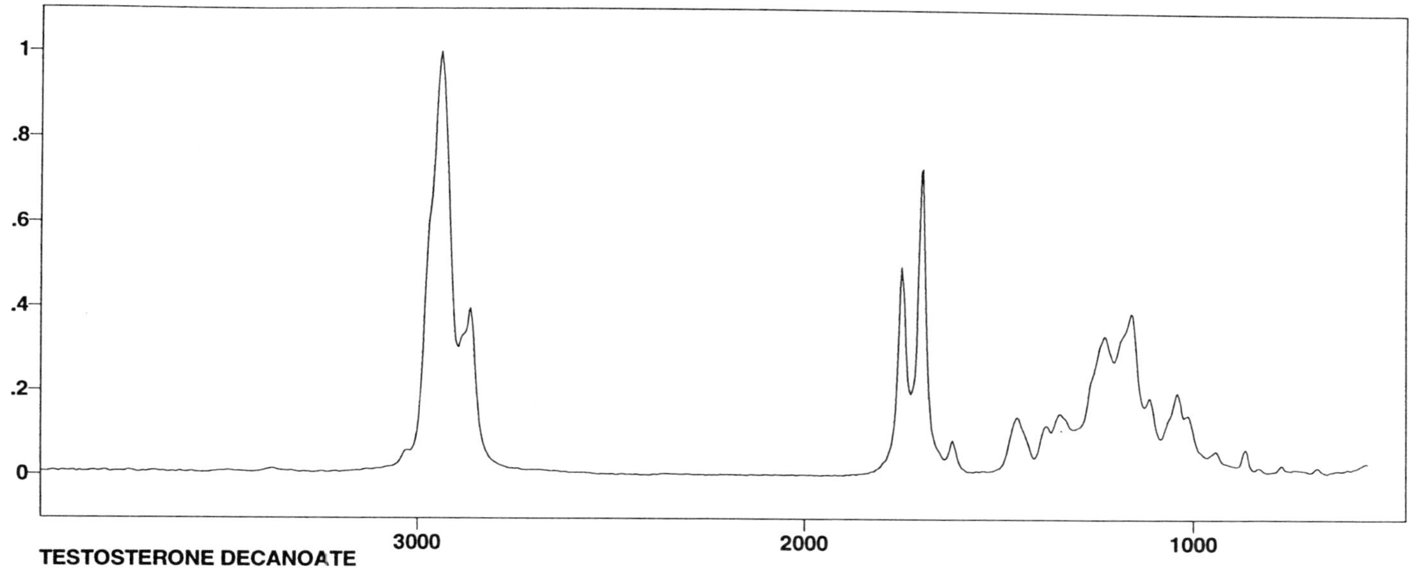

TESTOSTERONE DECANOATE

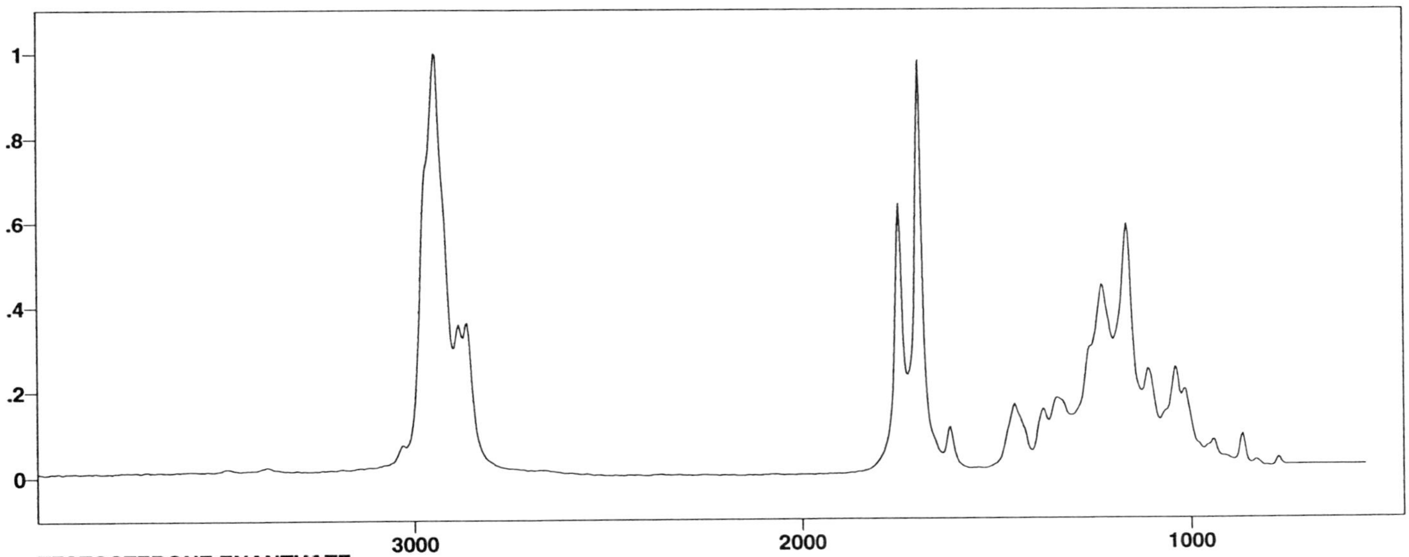

TESTOSTERONE ENANTHATE

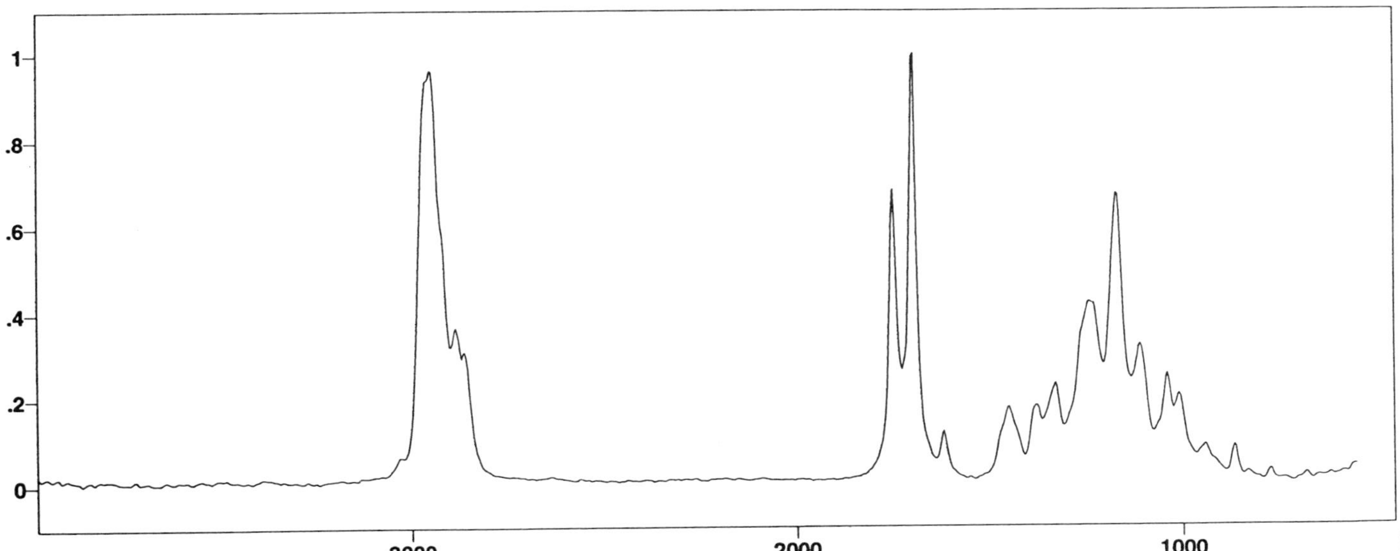

TESTOSTERONE ISOCAPROATE

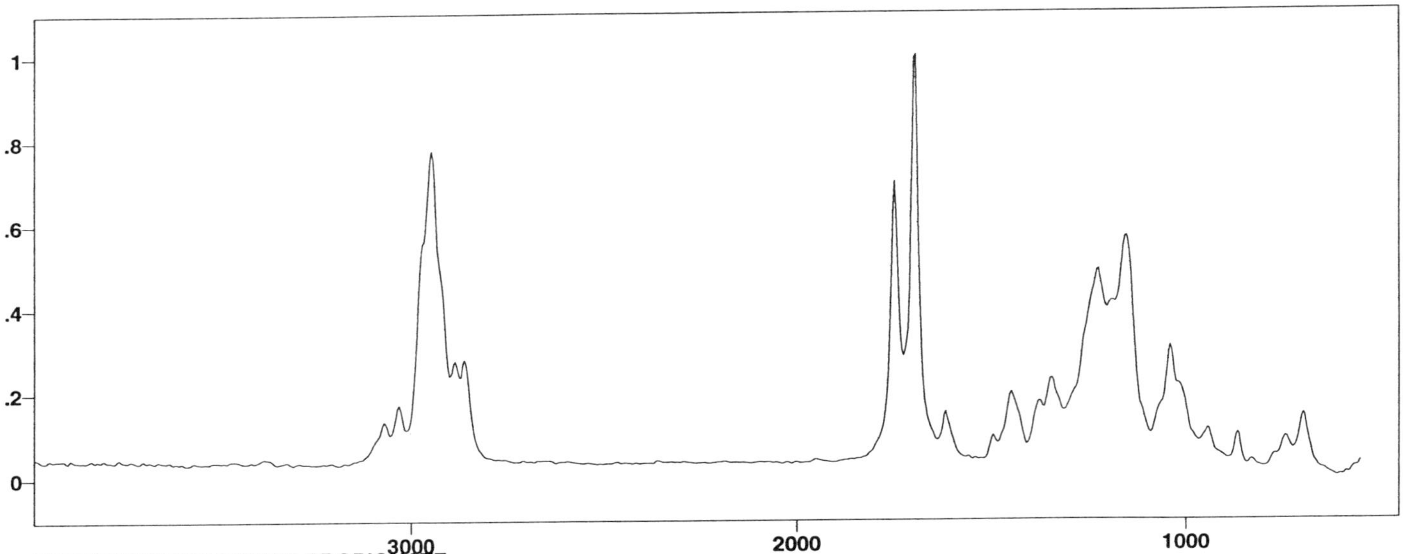

TESTOSTERONE PHENYLPROPIONATE

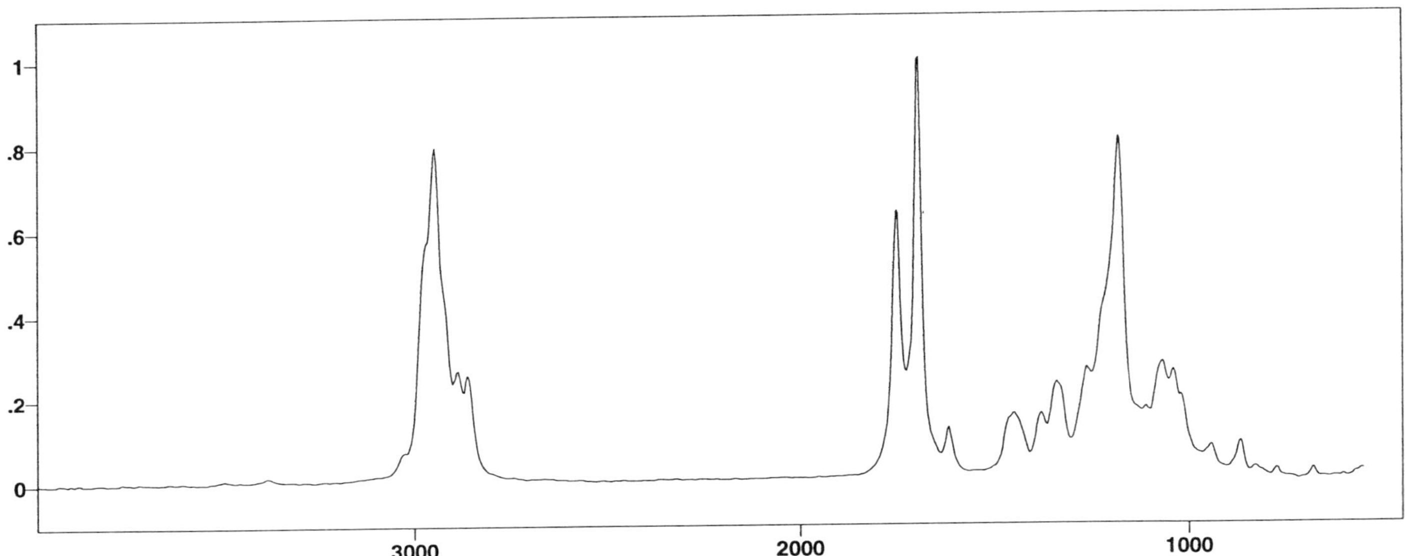

TESTOSTERONE PROPIONATE

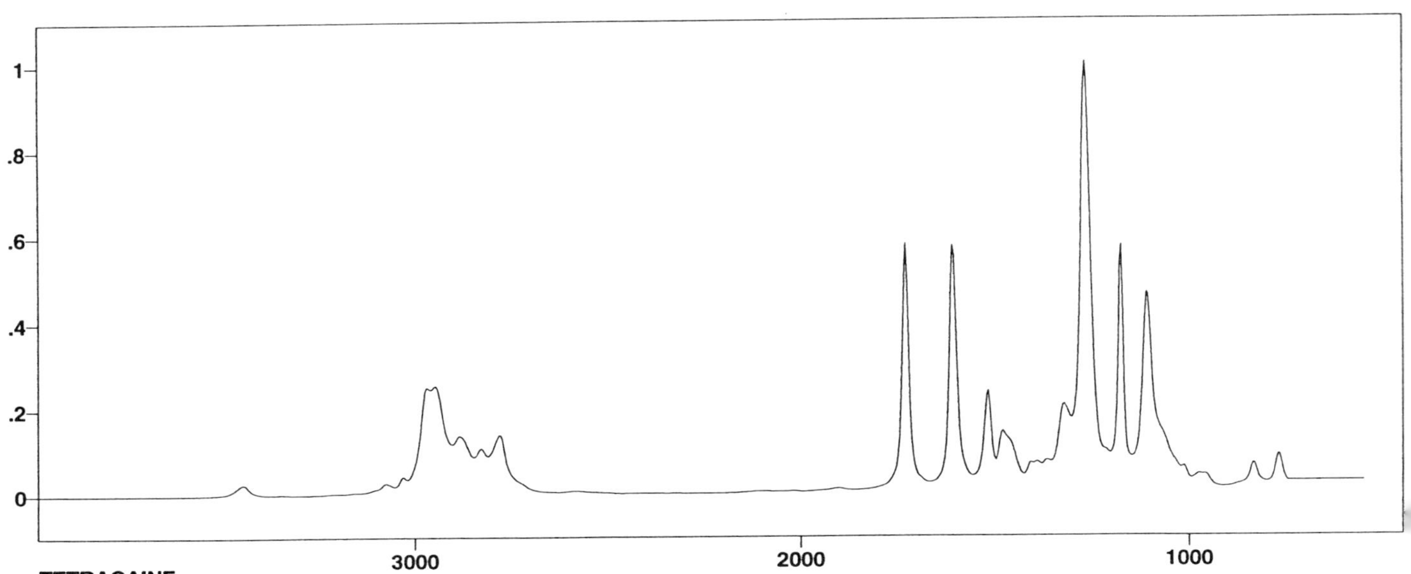

TETRACAINE

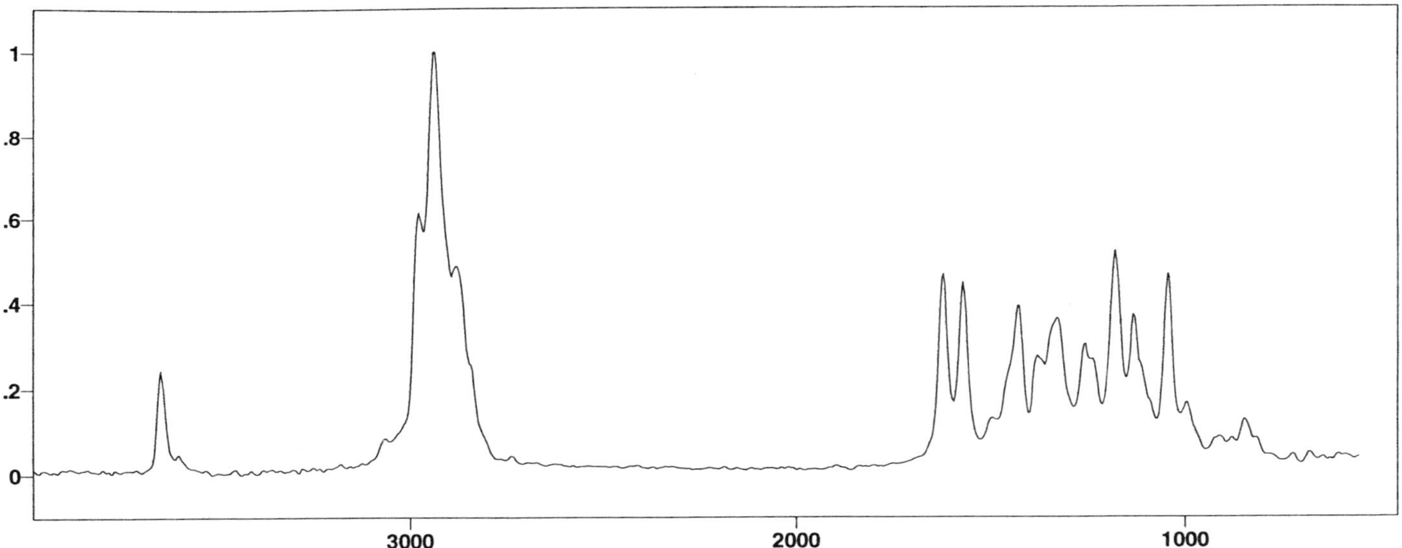

TETRAHYDROCANNABINOL

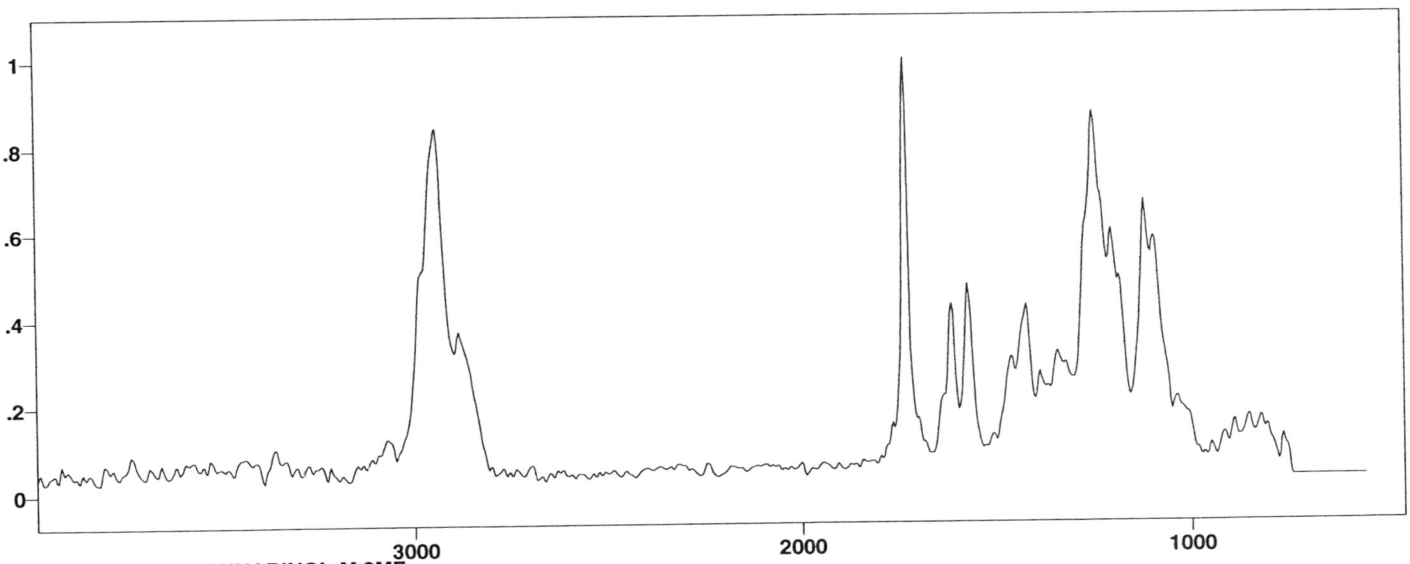

TETRAHYDROCANNABINOL-M 2ME

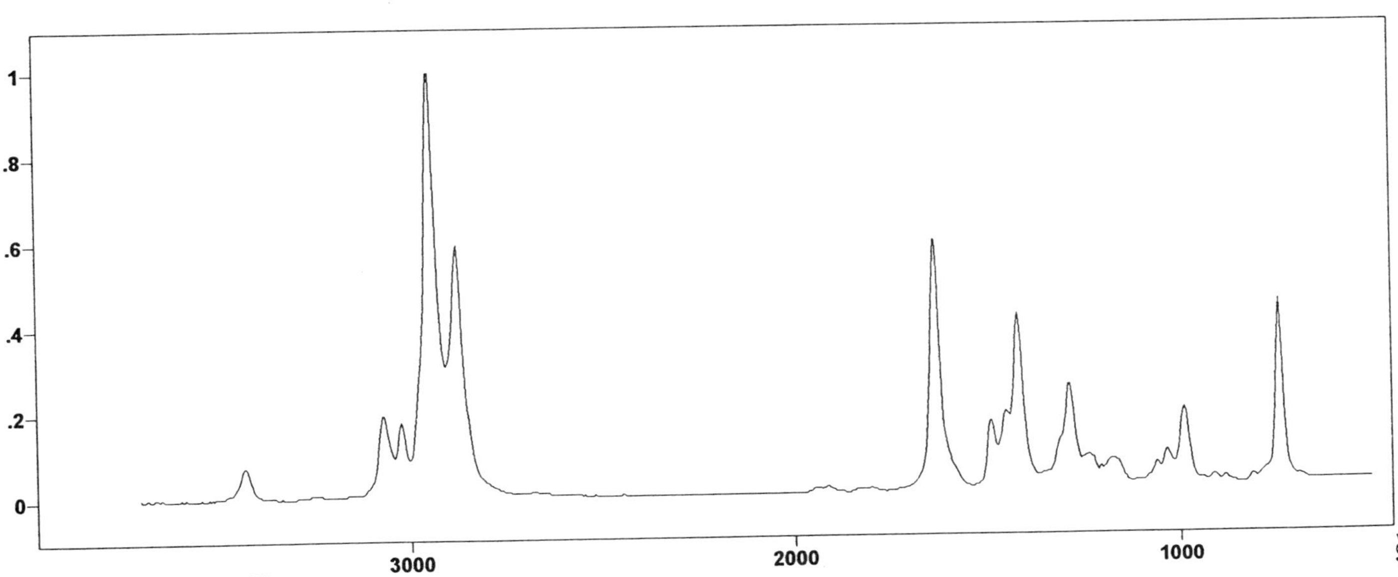

TETRAHYDROZOLINE

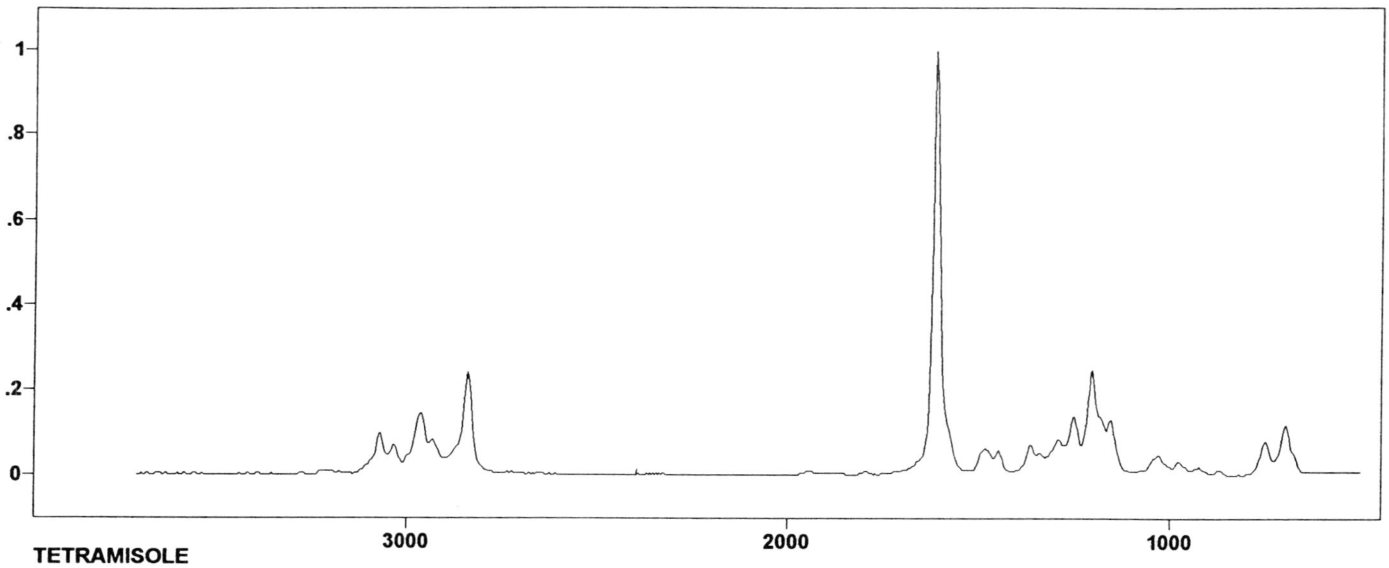

TETRAMISOLE

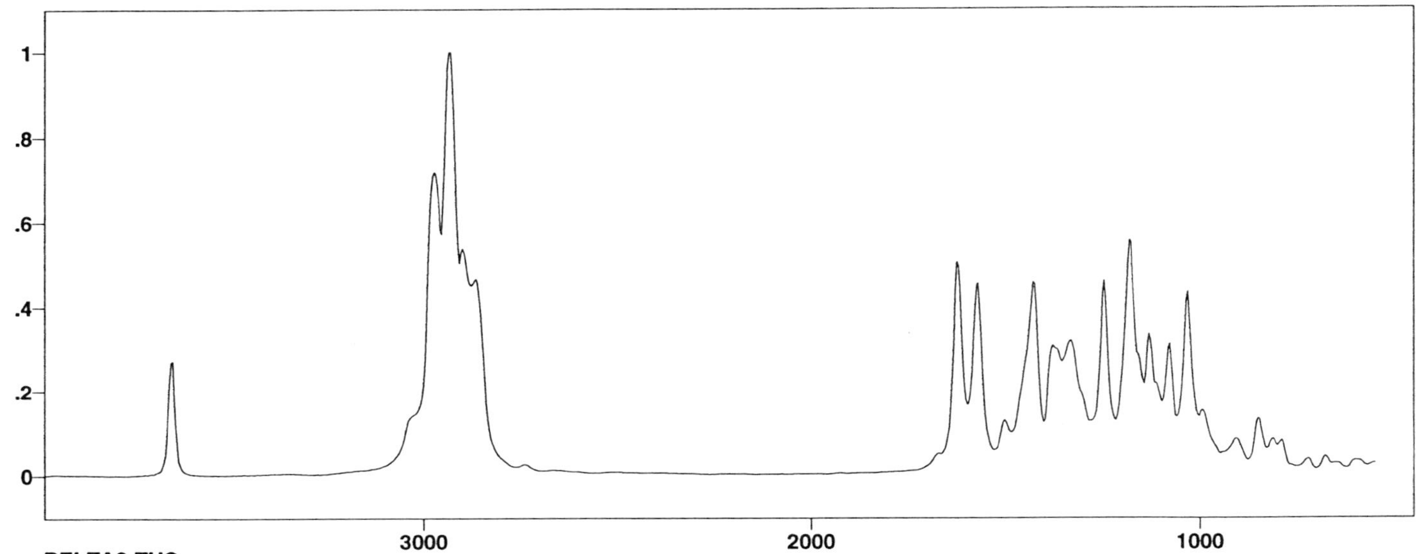

DELTA8-THC

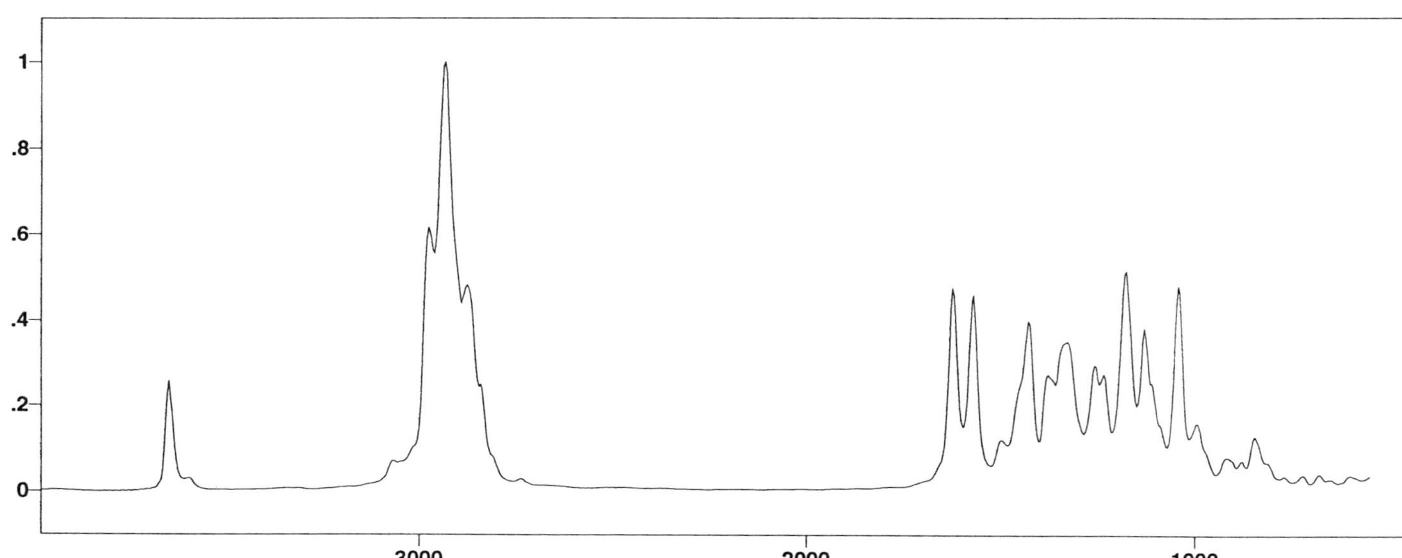

DELTA 9 THC

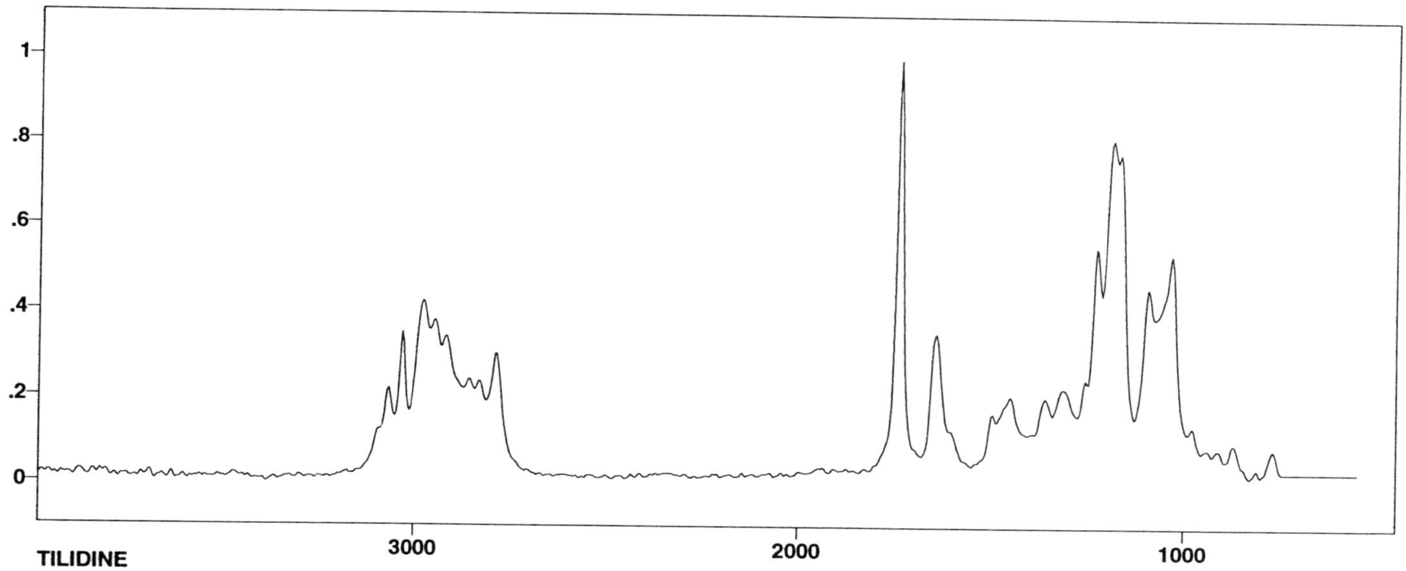

TILIDINE

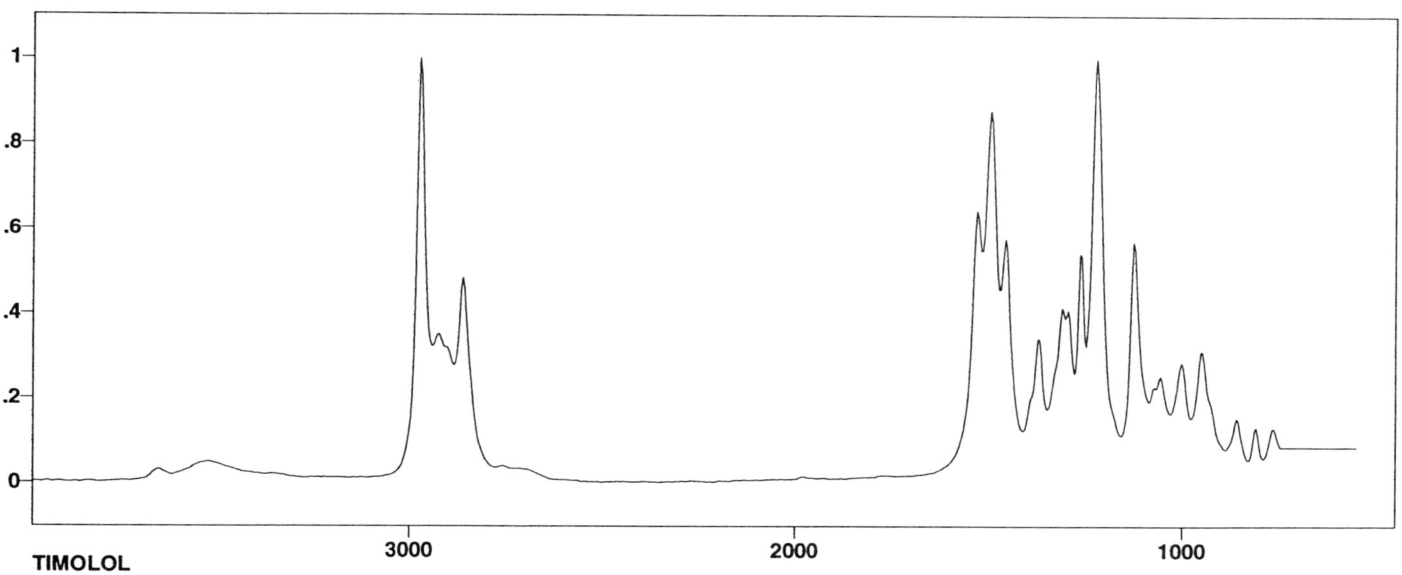

TIMOLOL

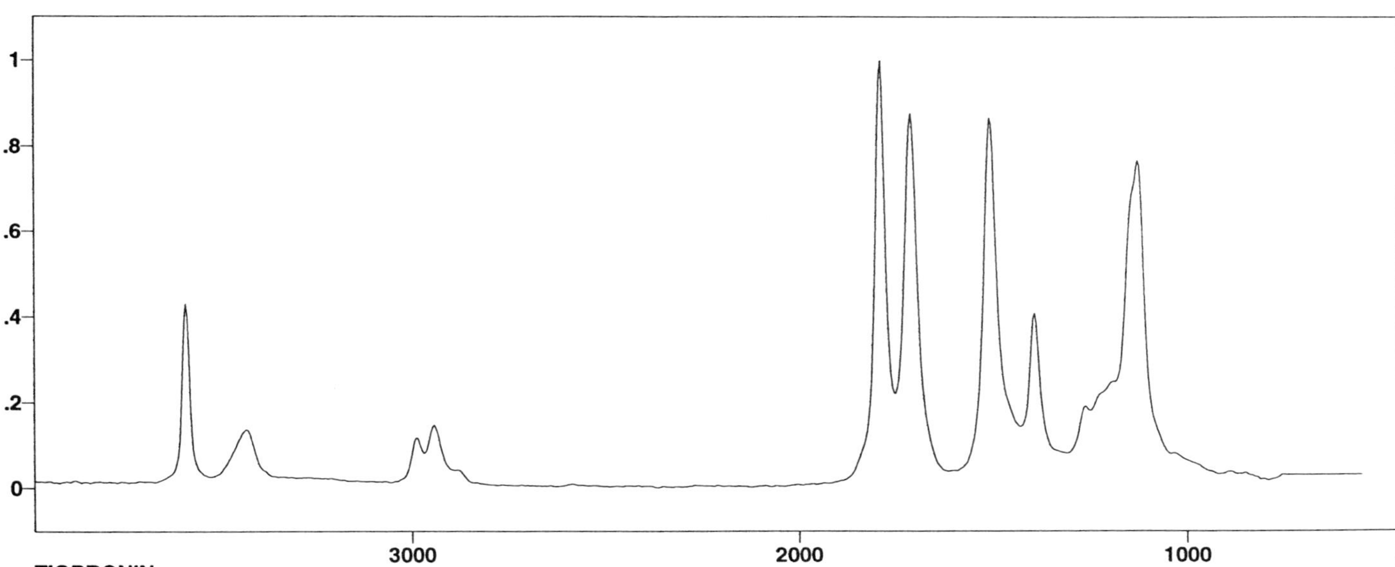

TIOPRONIN

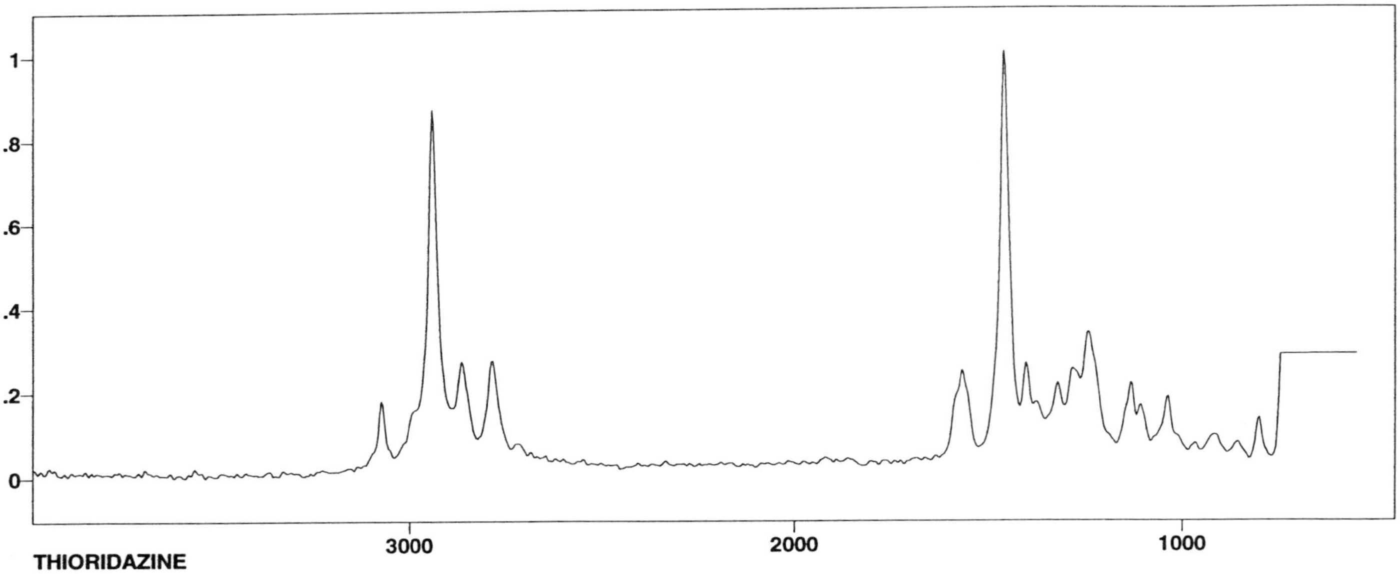

THIORIDAZINE

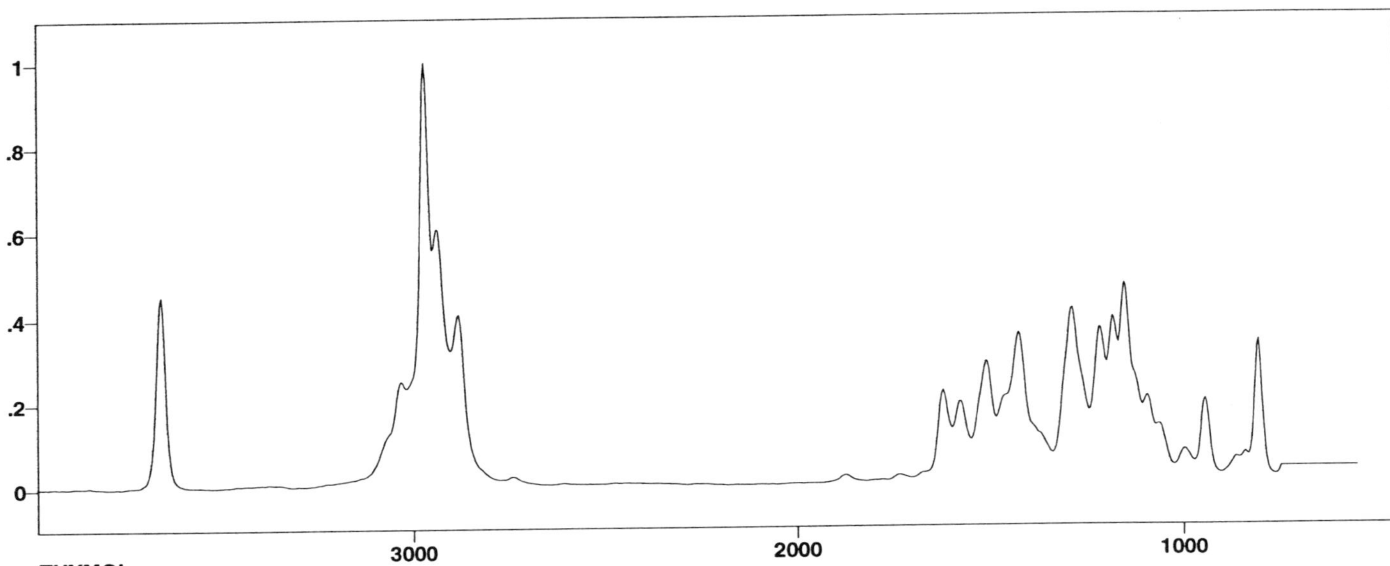

THYMOL

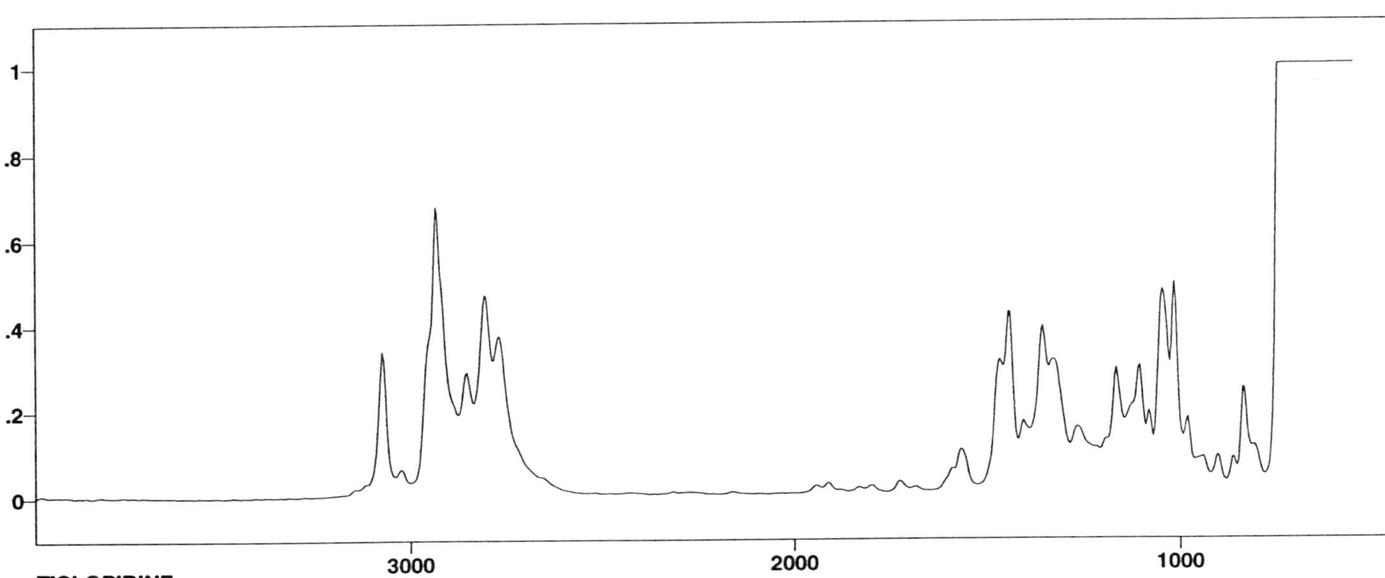

TICLOPIDINE

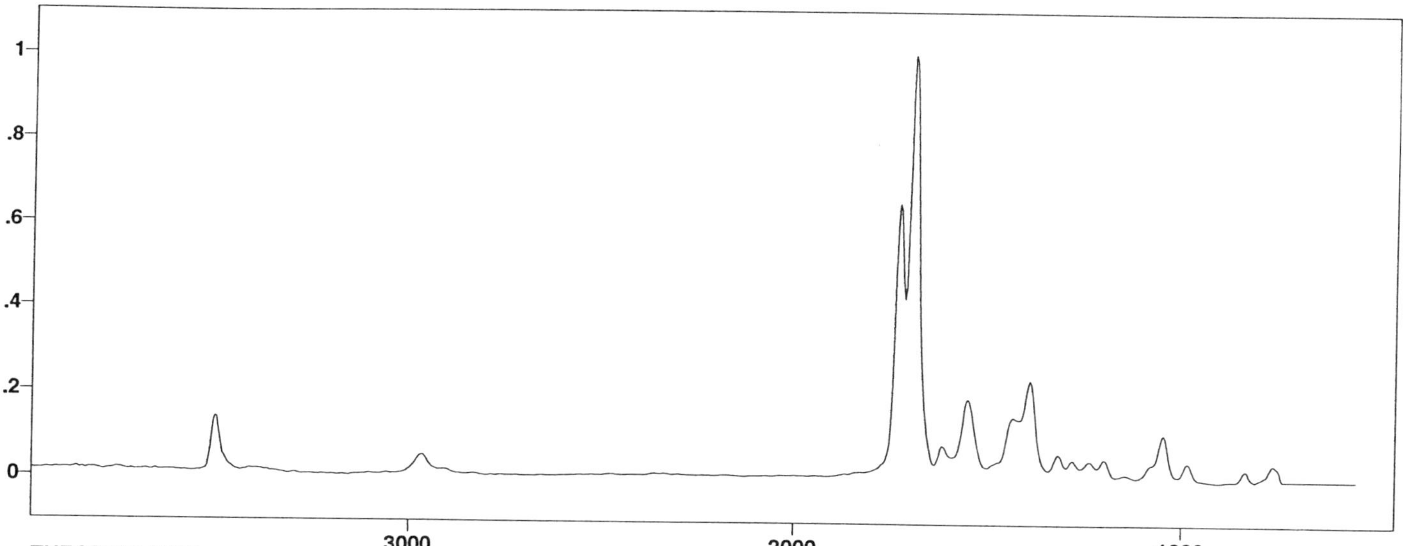

THEOPHYLLINE

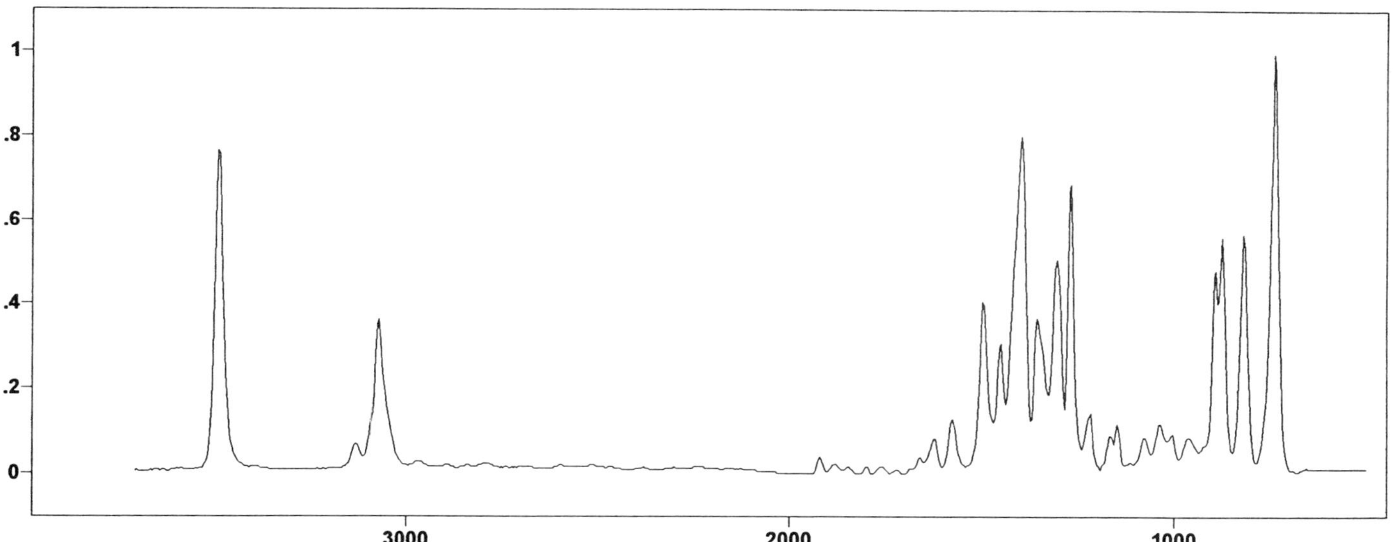

THIABENDAZOLE

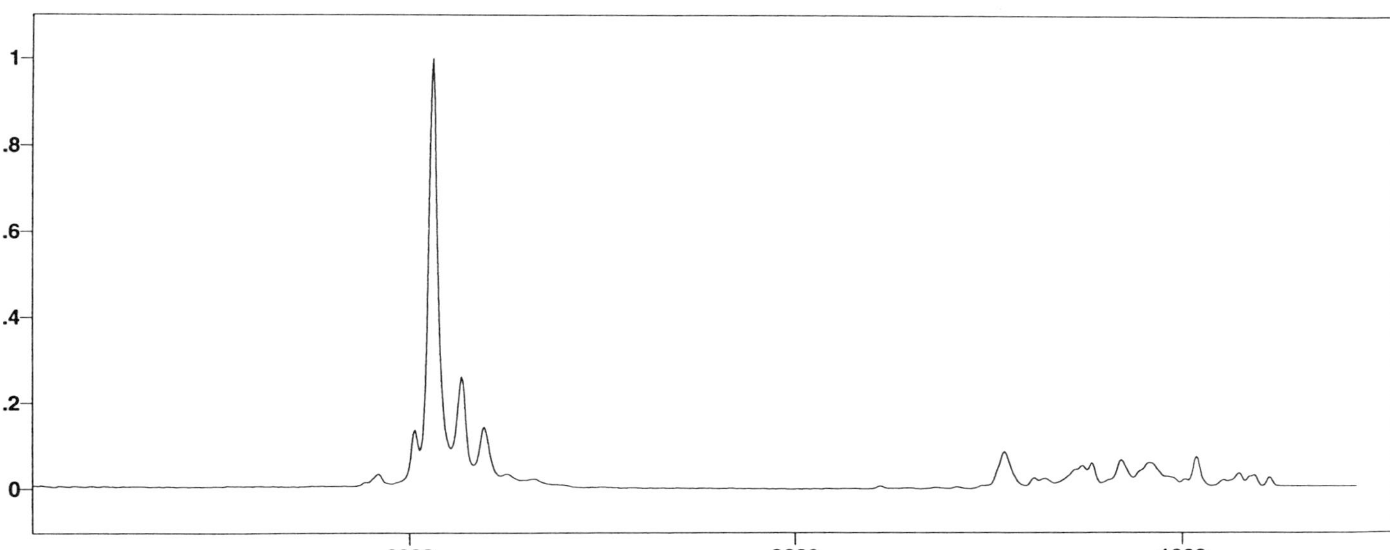

1-{1-(2-THIENYL)CYCLOHEXYL}PIPERIDINE

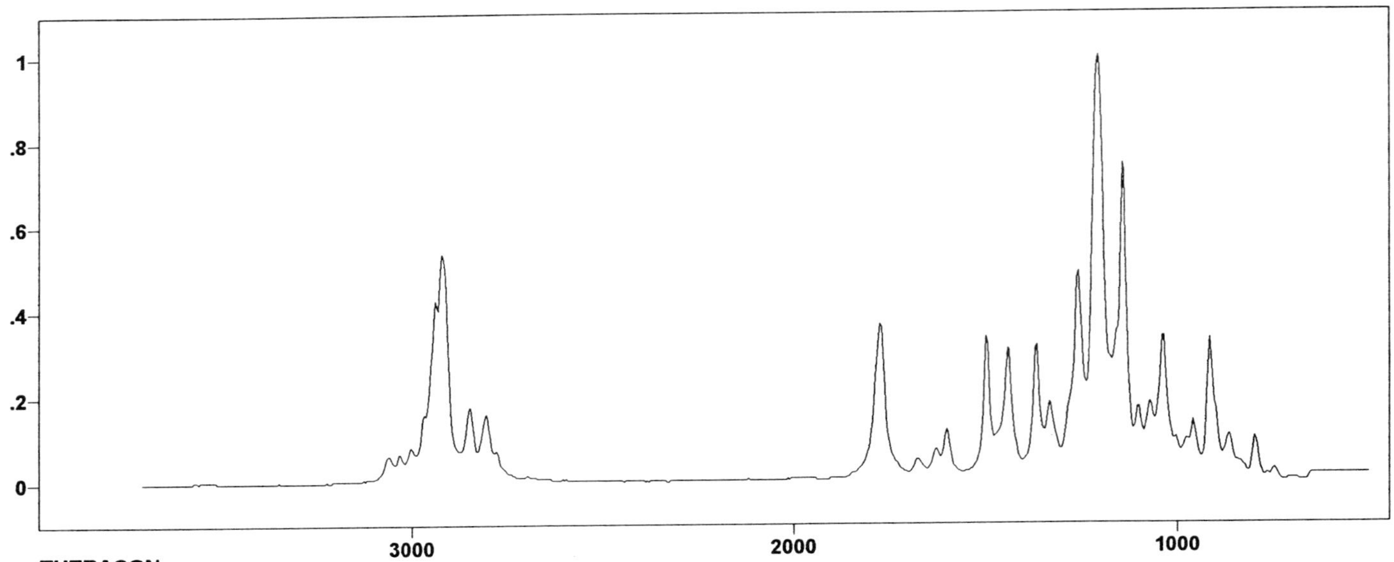

THEBACON

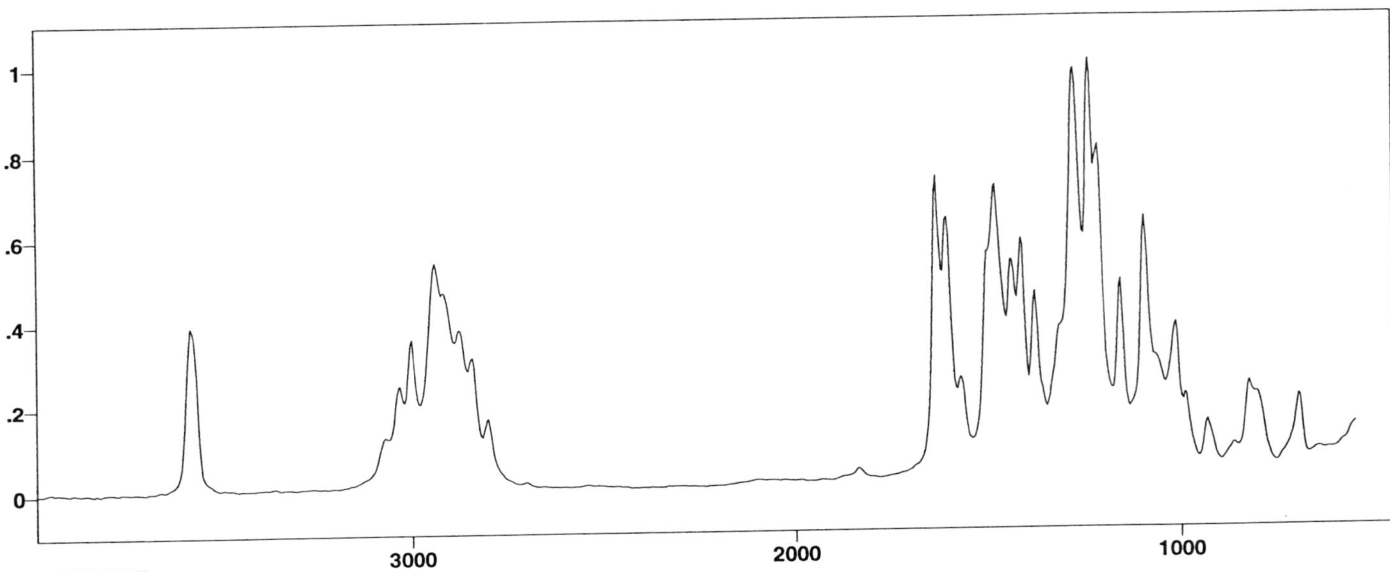

THEBAINE

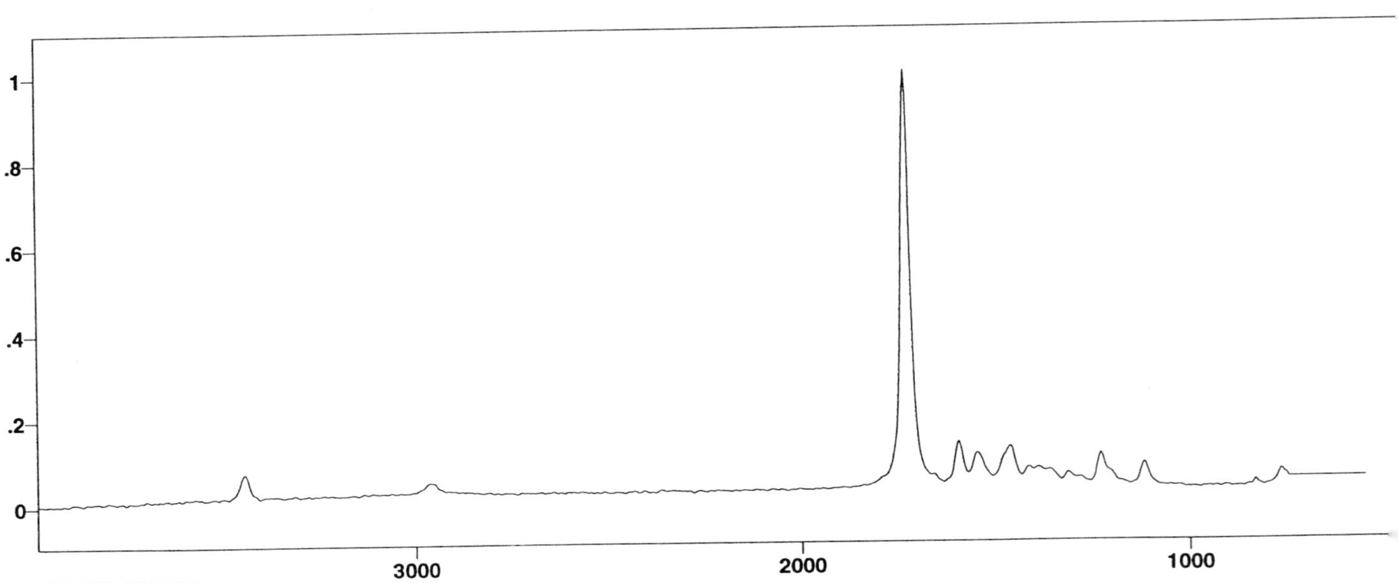

THEOBROMINE

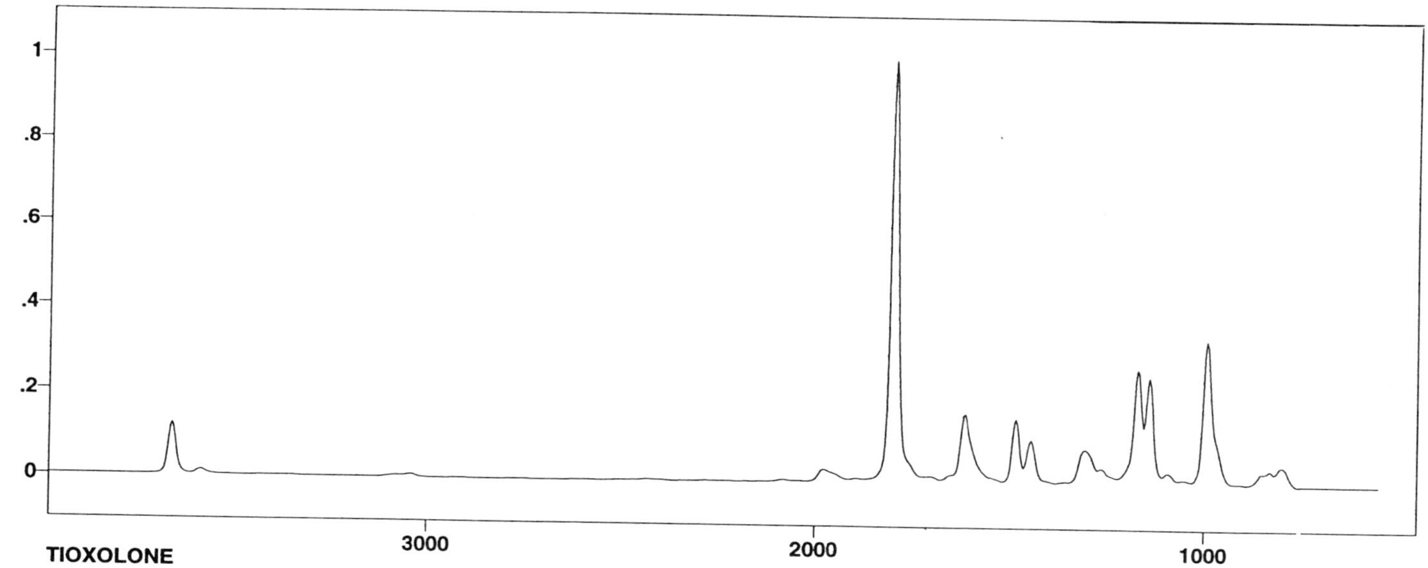

TIOXOLONE

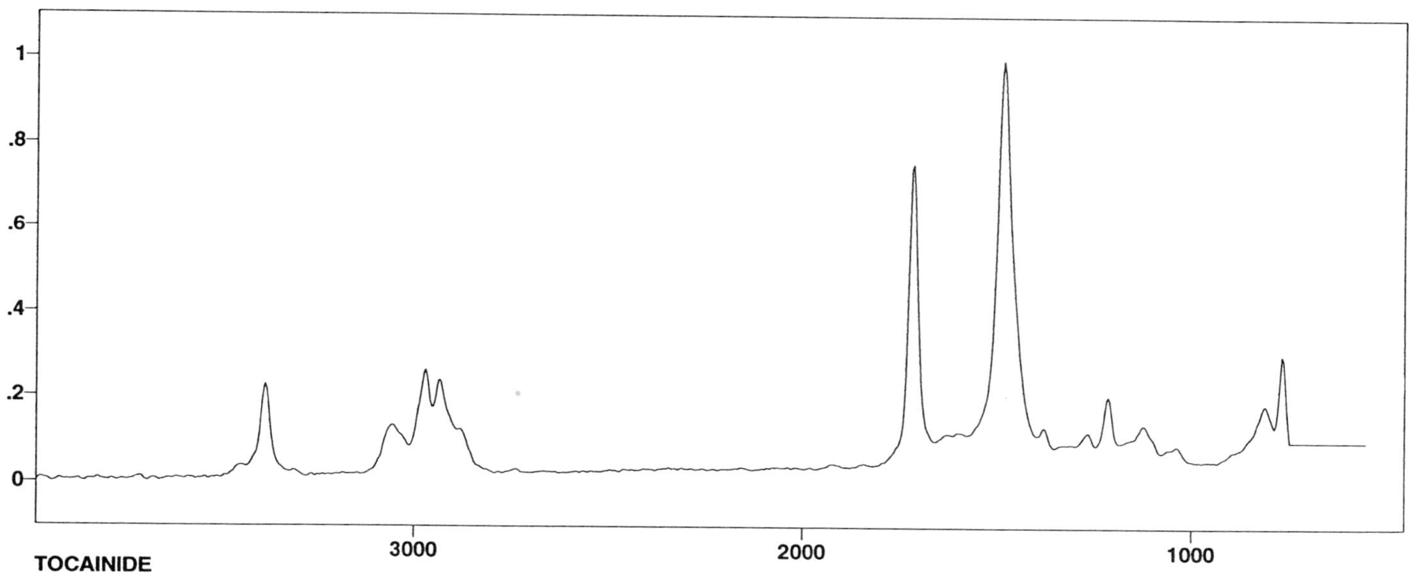

TOCAINIDE

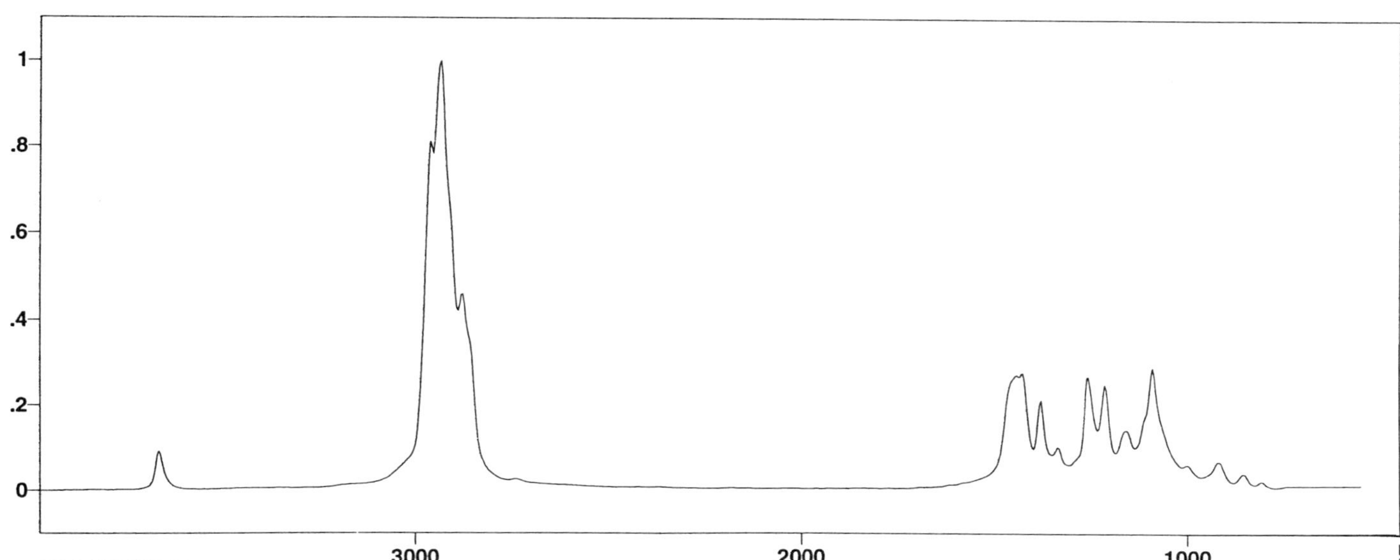

TOCOPHEROL

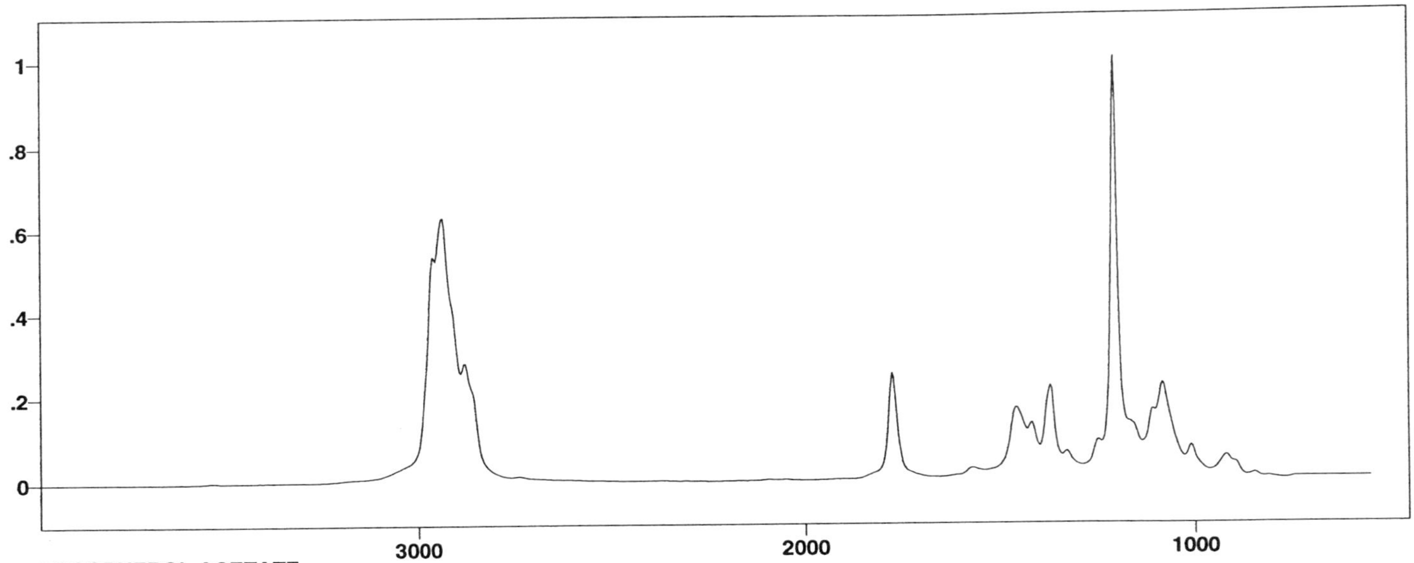

TOCOPHEROL ACETATE

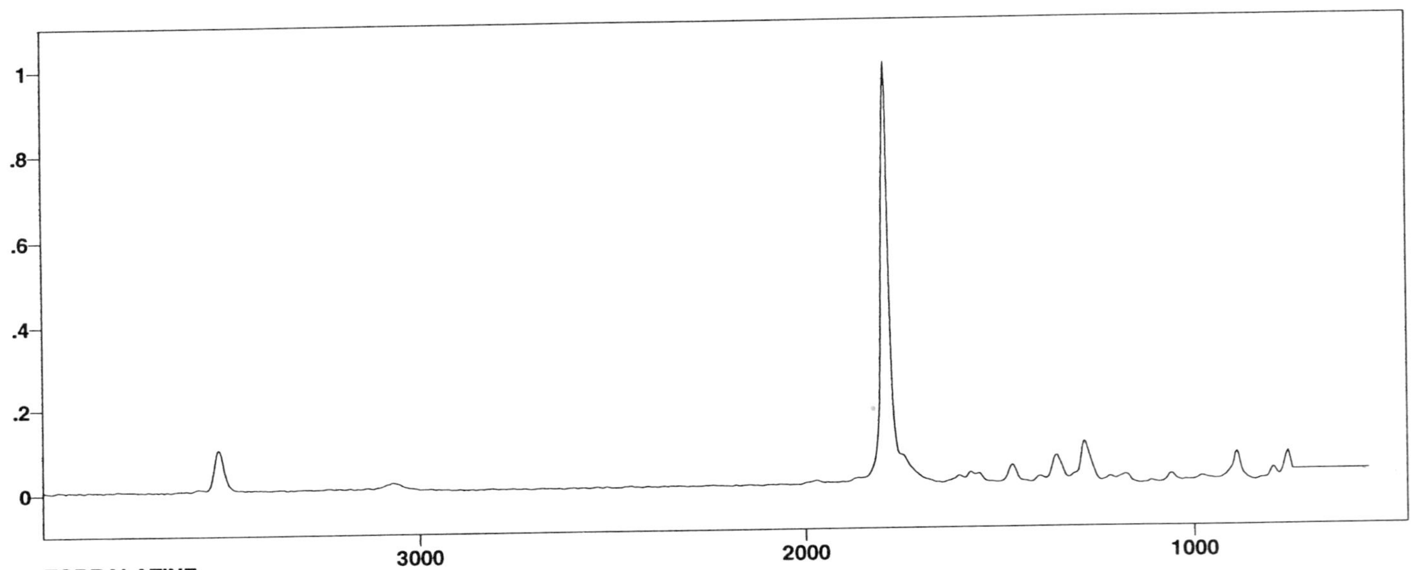

TODRALAZINE

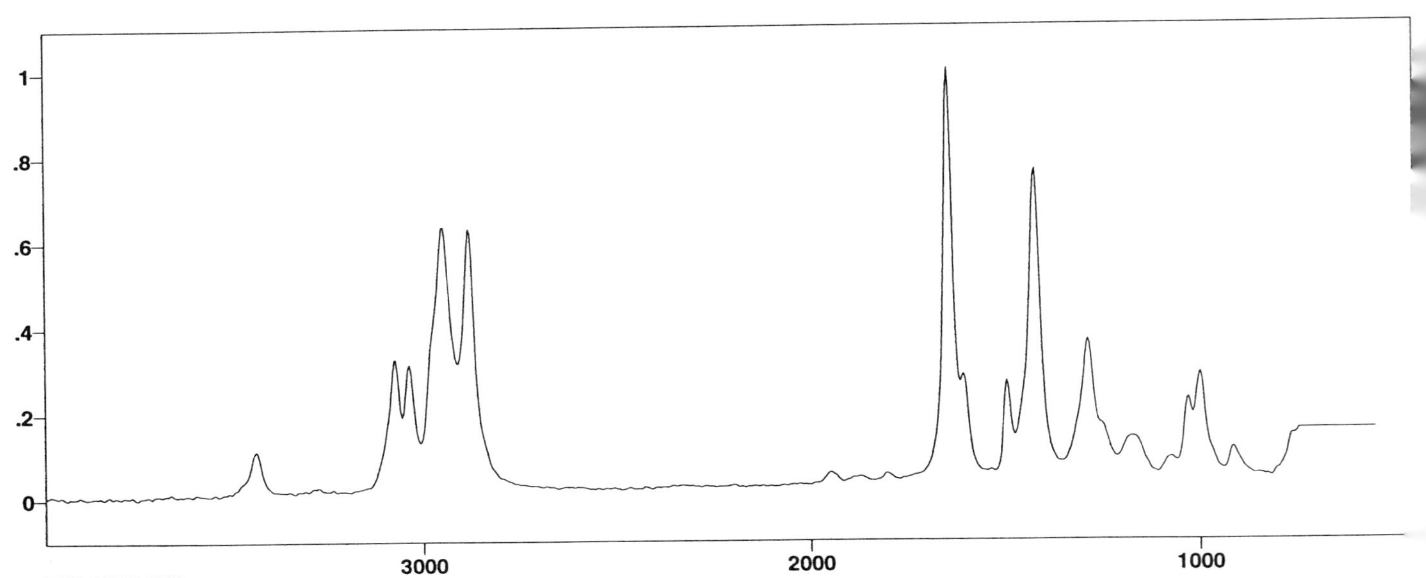

TOLAZOLINE

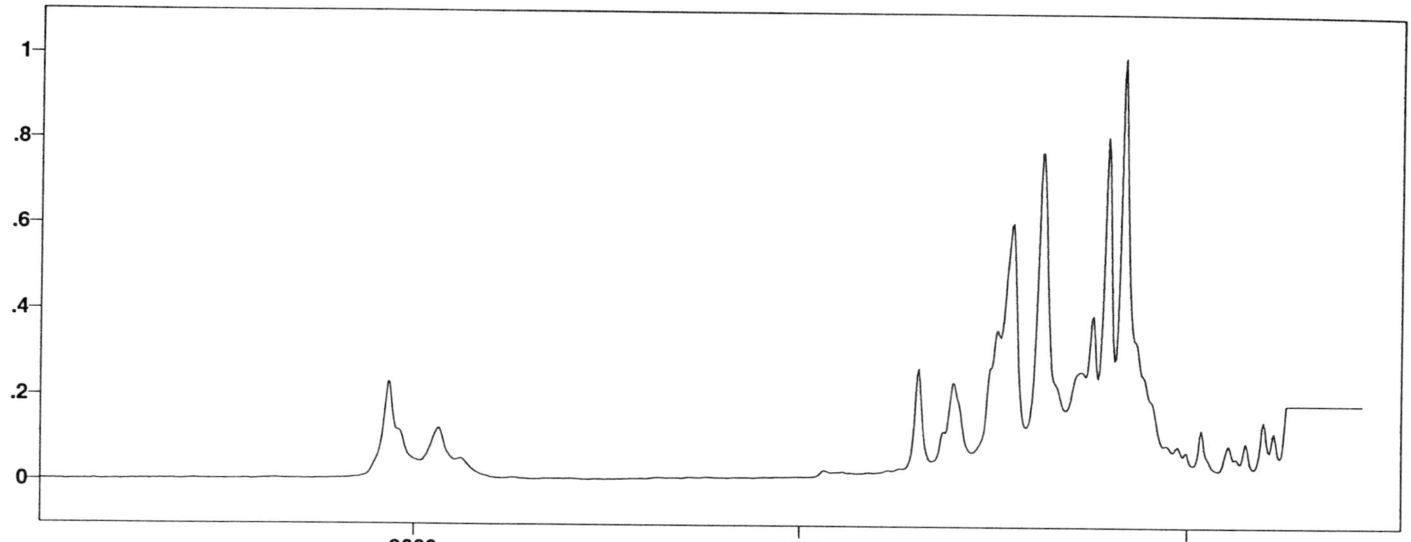

TOLNAFTATE

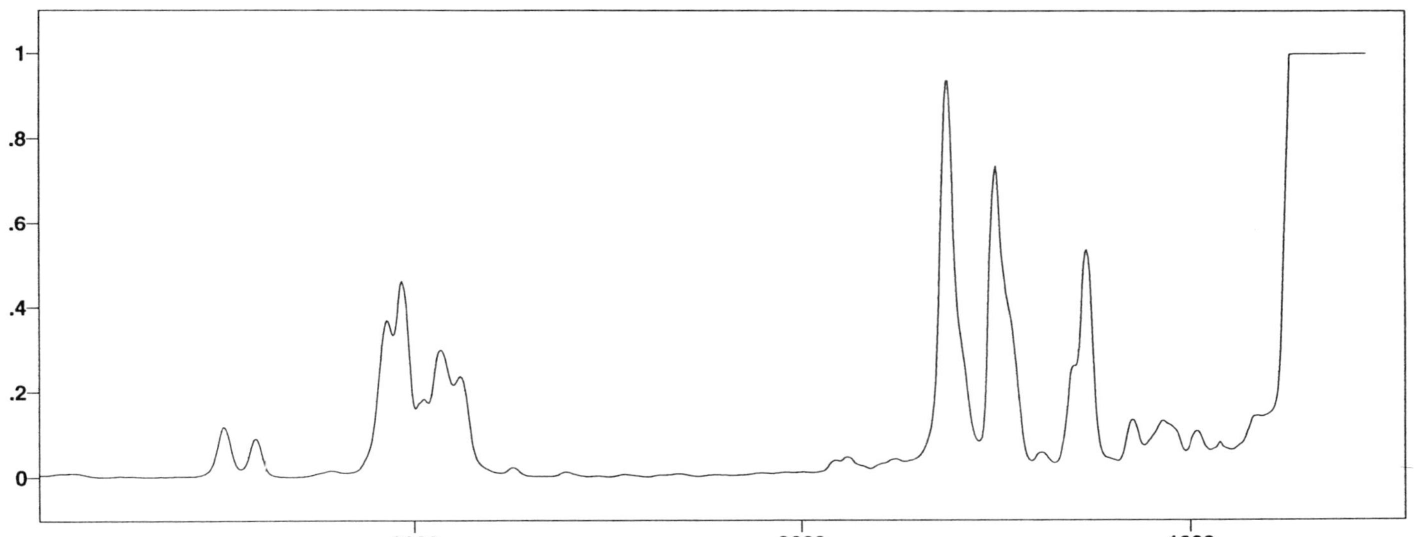

TOLUIDINE

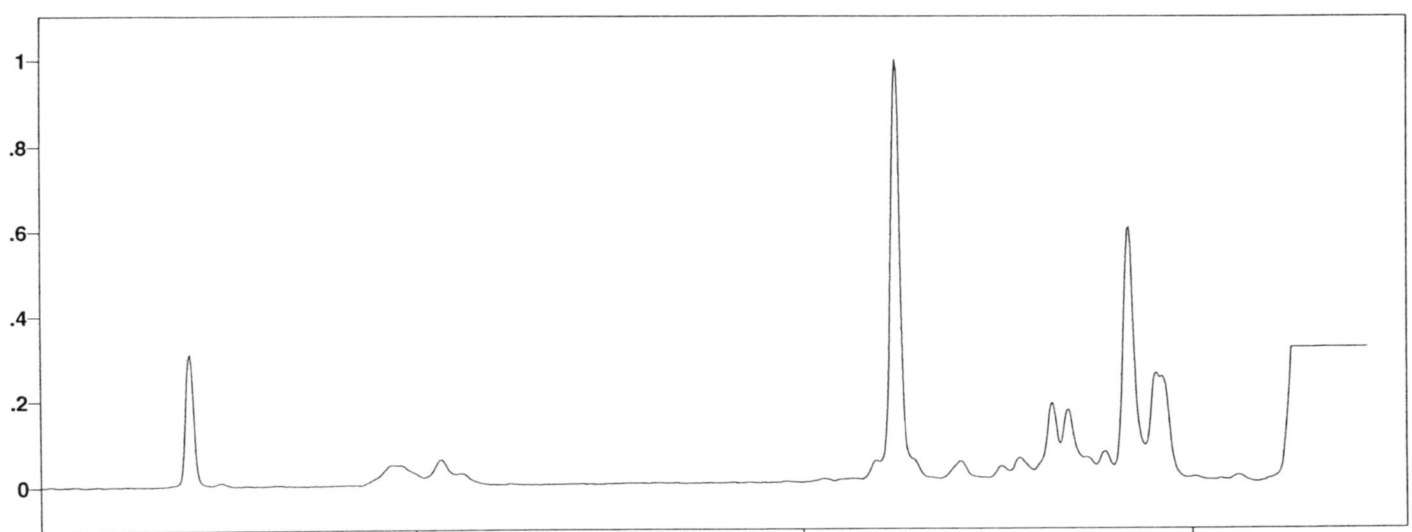

TOUIC ACID

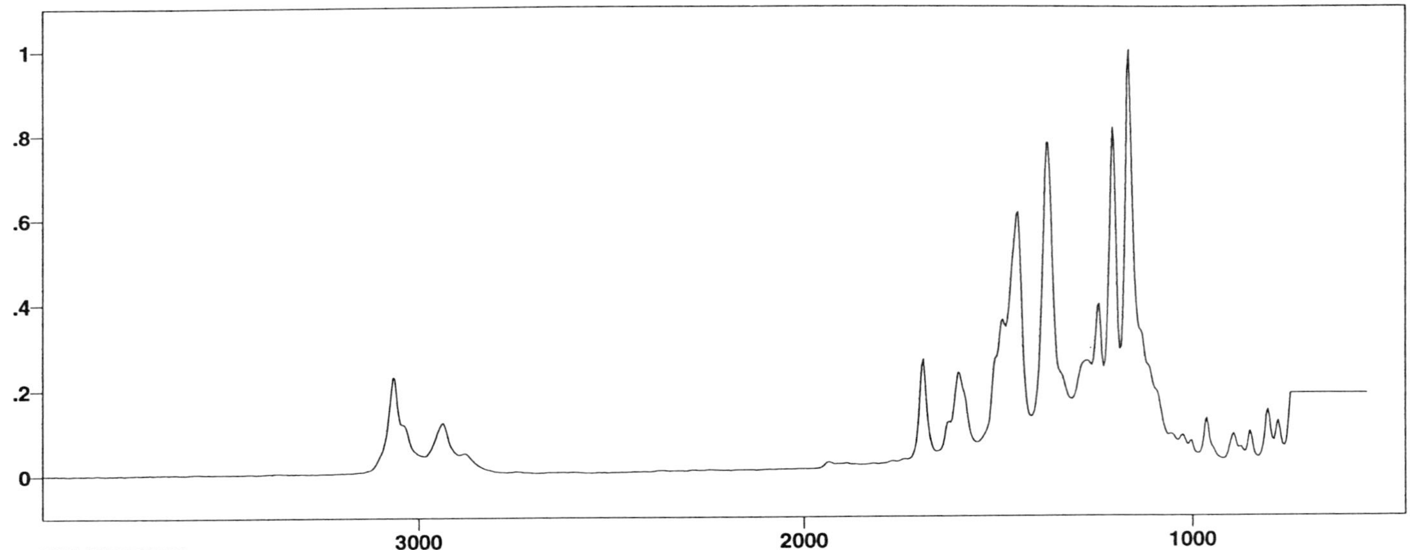

TOLNAFTATE

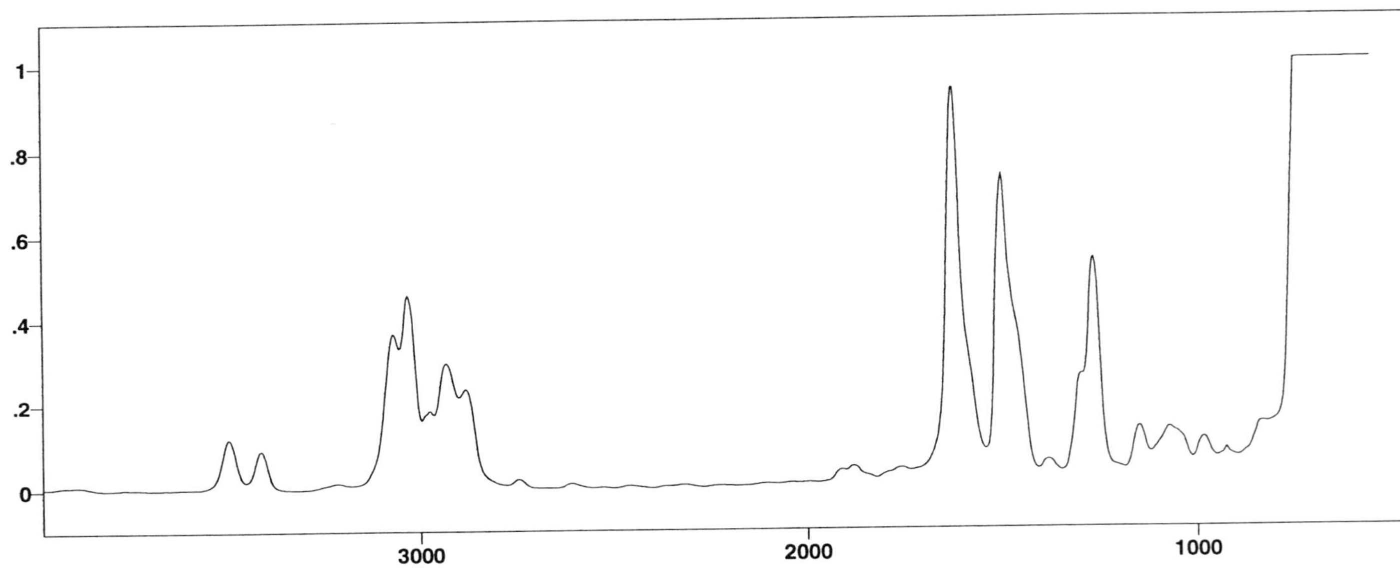

TOLUIDINE

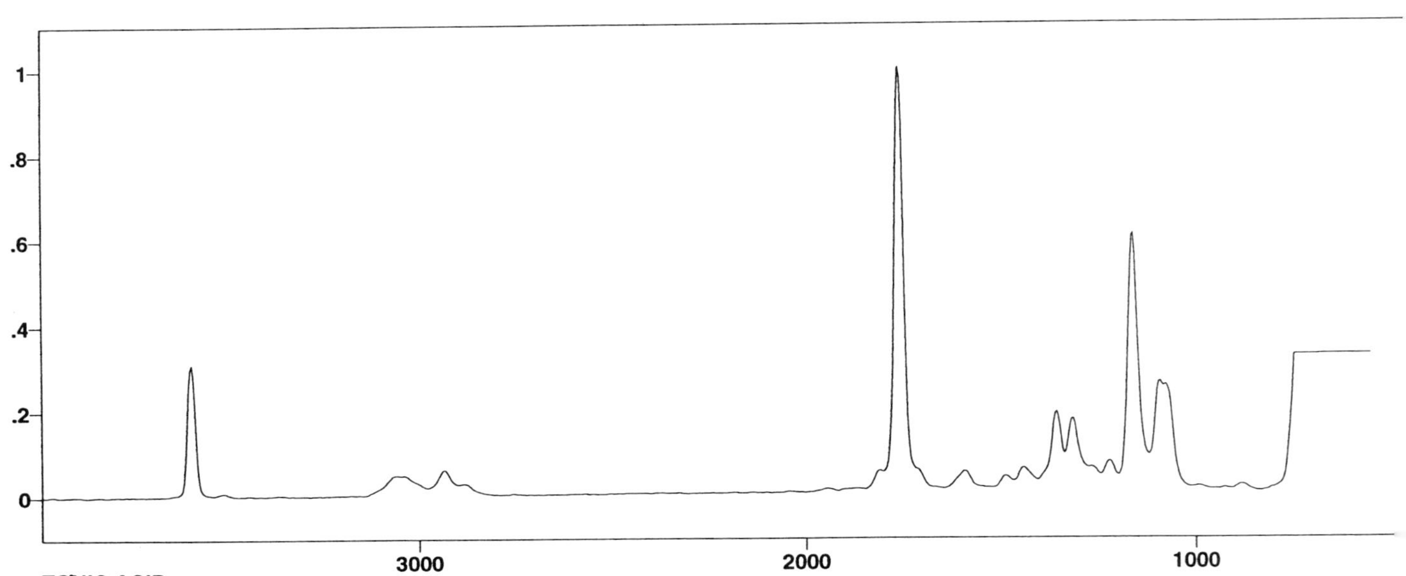

TOLUIC ACID

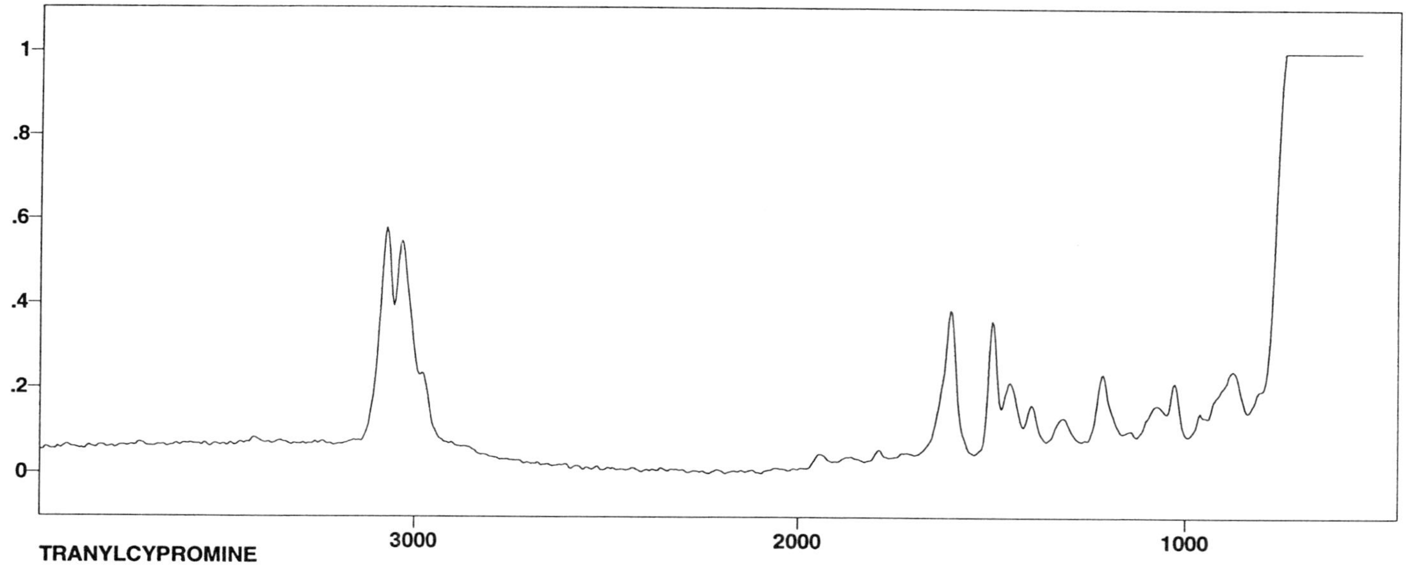

TRANYLCYPROMINE

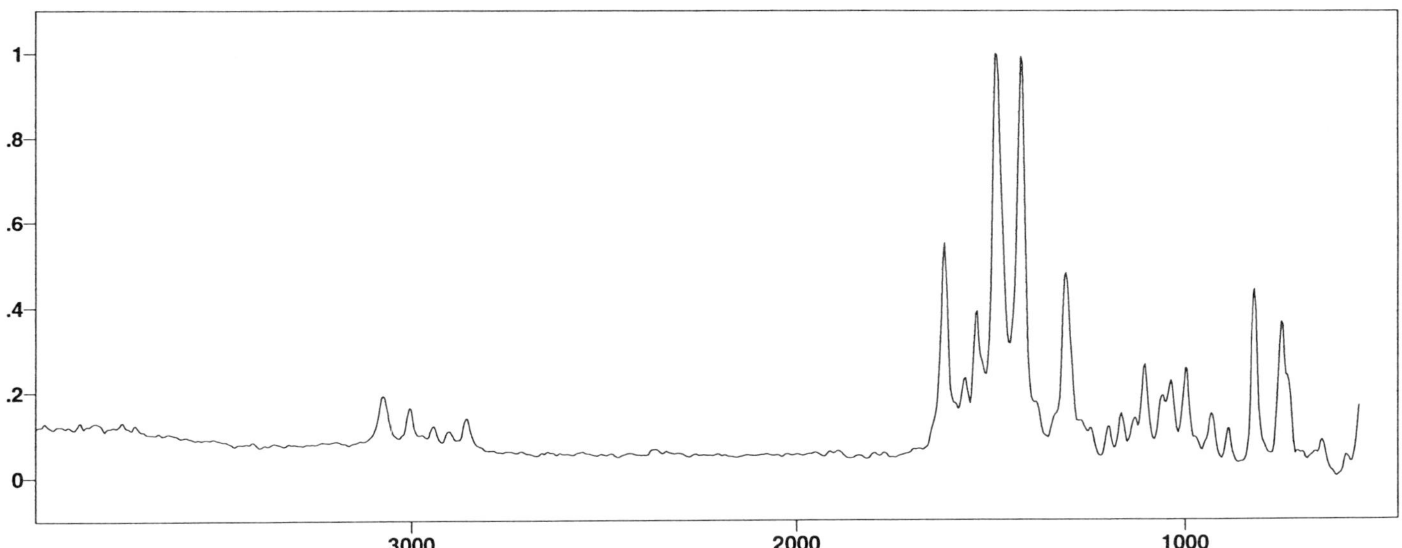

TRIAZOLAM

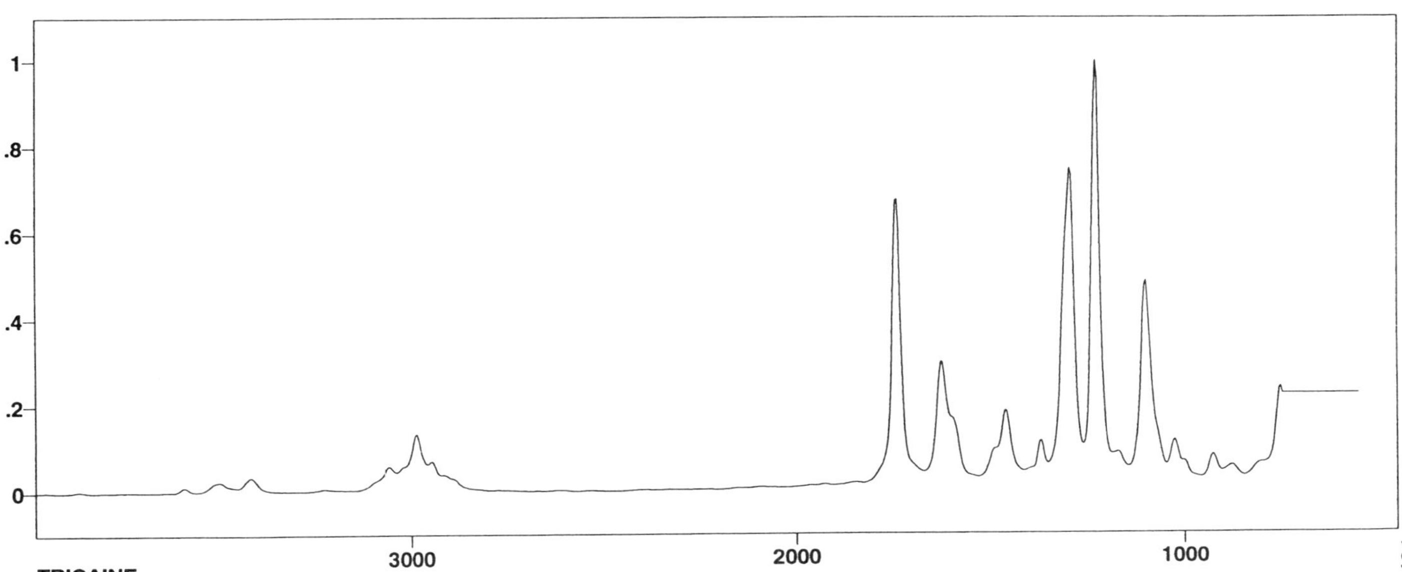

TRICAINE

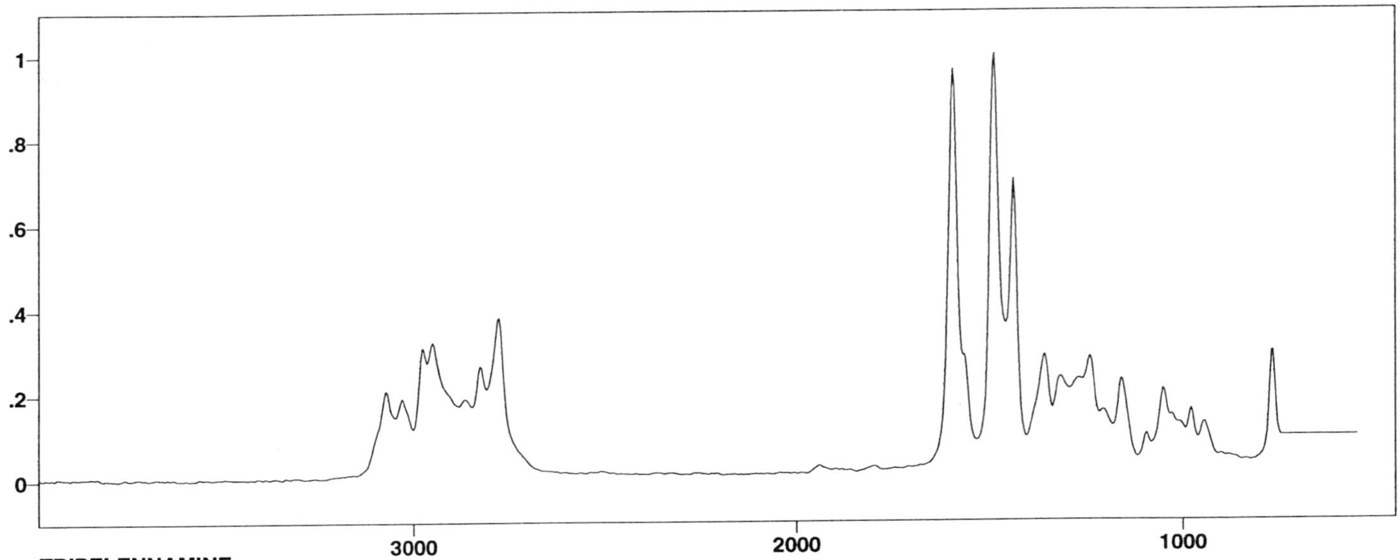

TRIPELENNAMINE

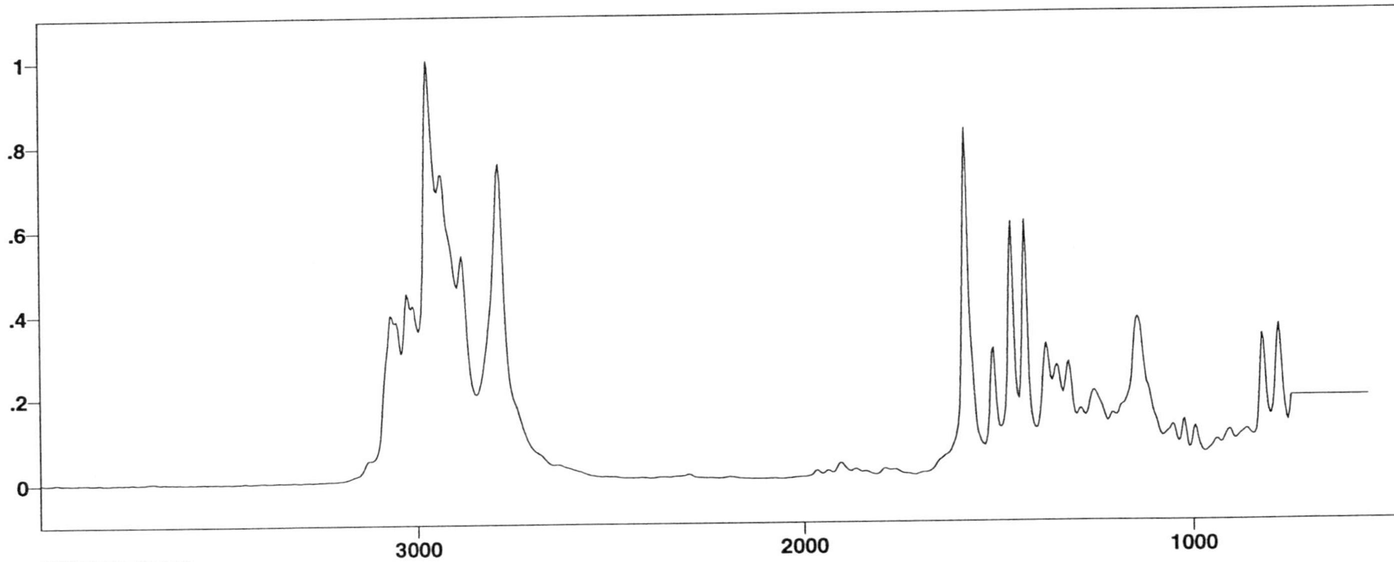

TRIPROLIDINE

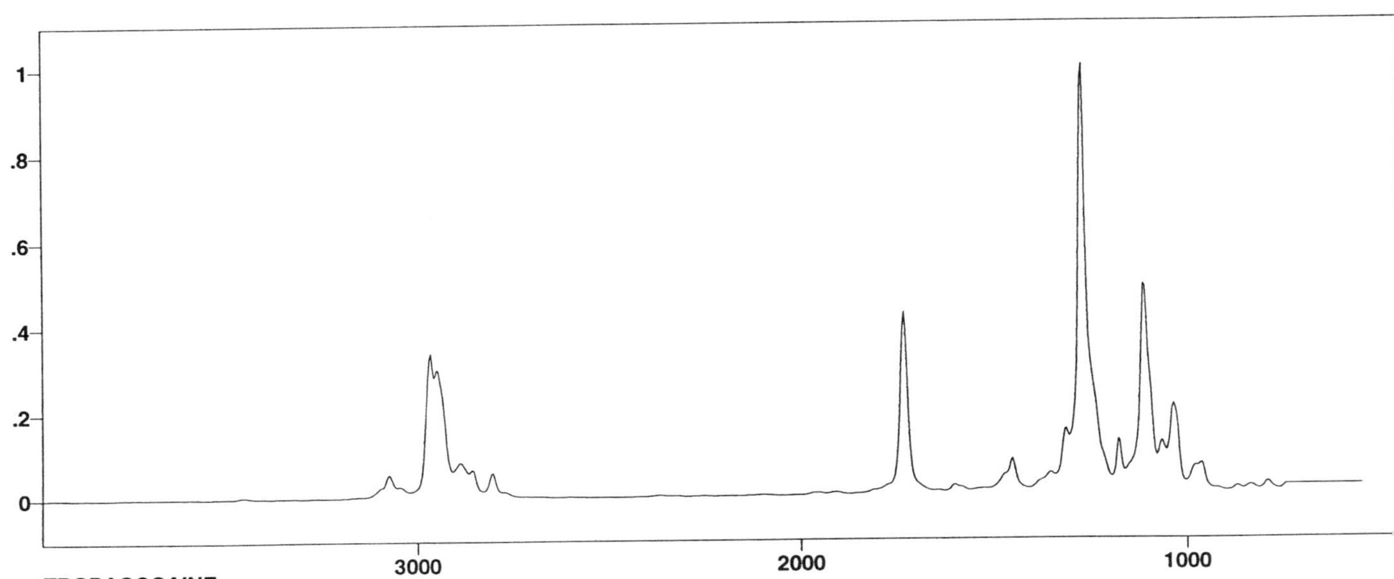

TROPACOCAINE

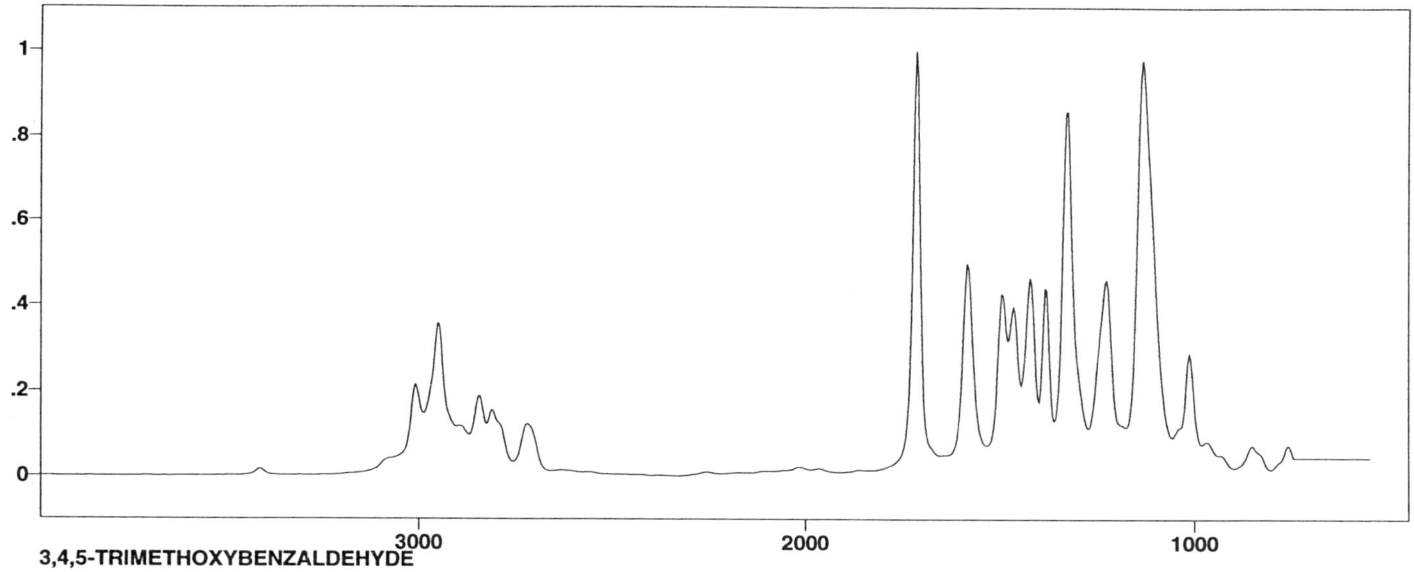

3,4,5-TRIMETHOXYBENZALDEHYDE

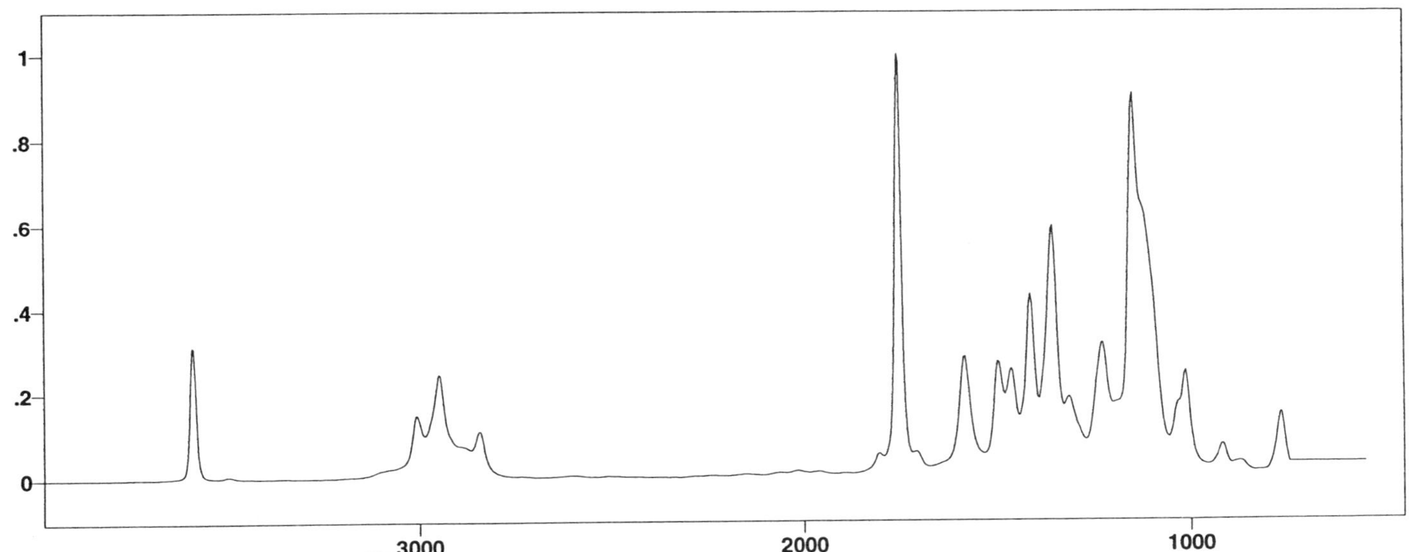

3,4,5-TRIMETHOXYBENZOIC ACID

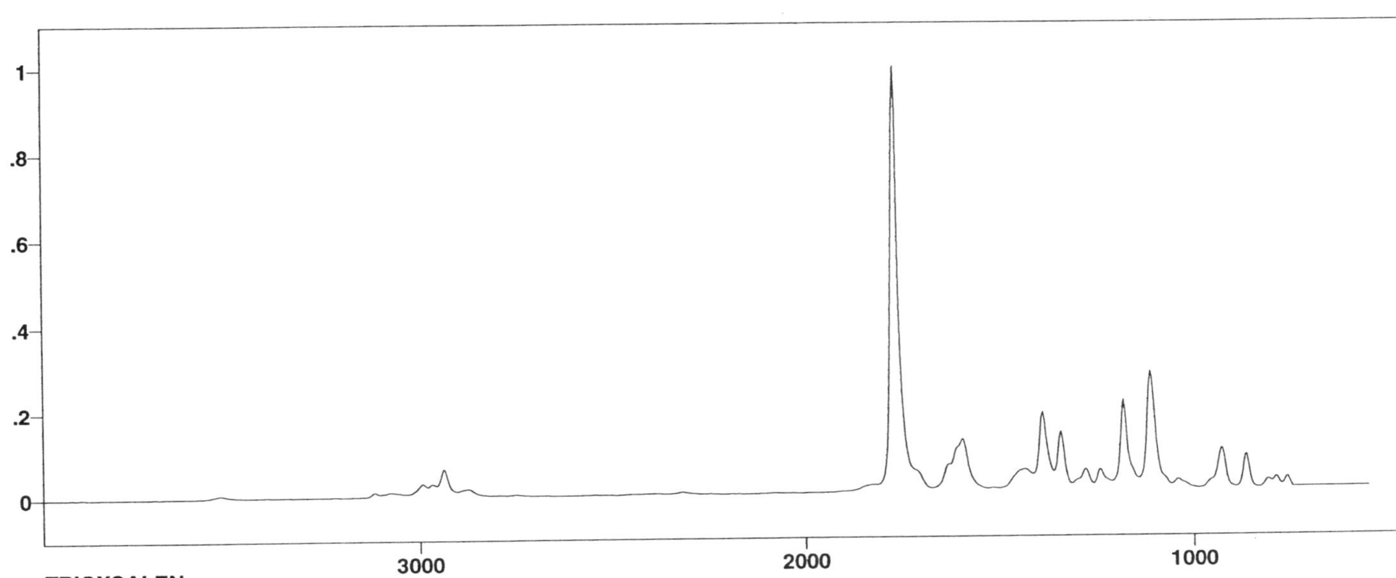

TRIOXSALEN

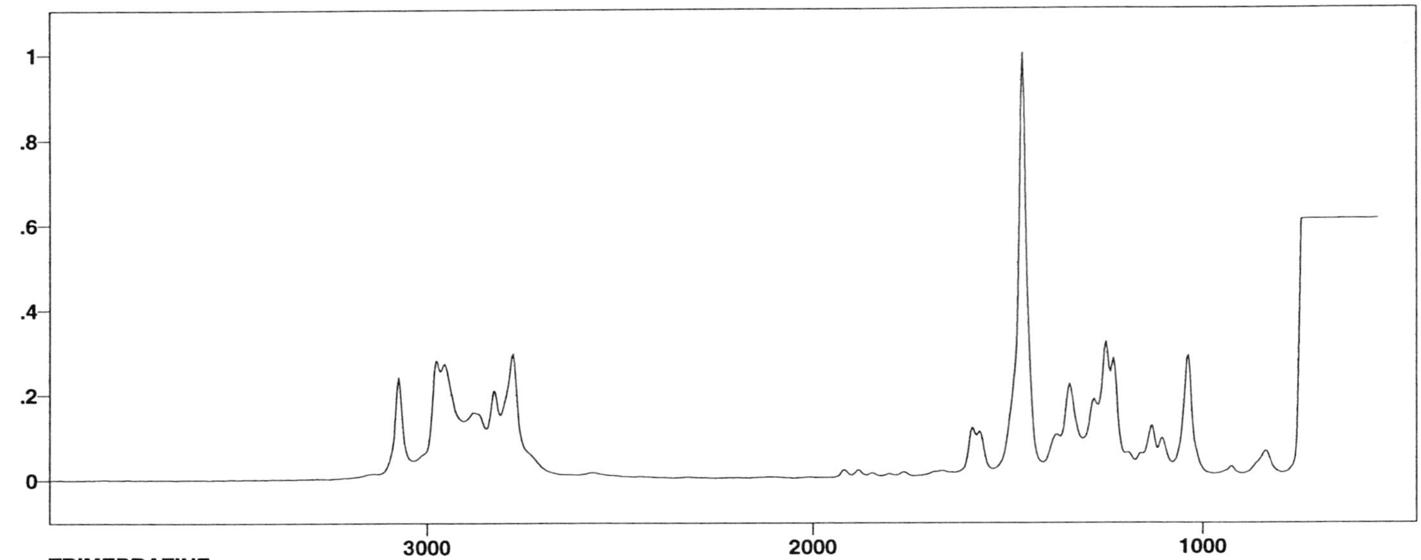

TRIMEPRAZINE

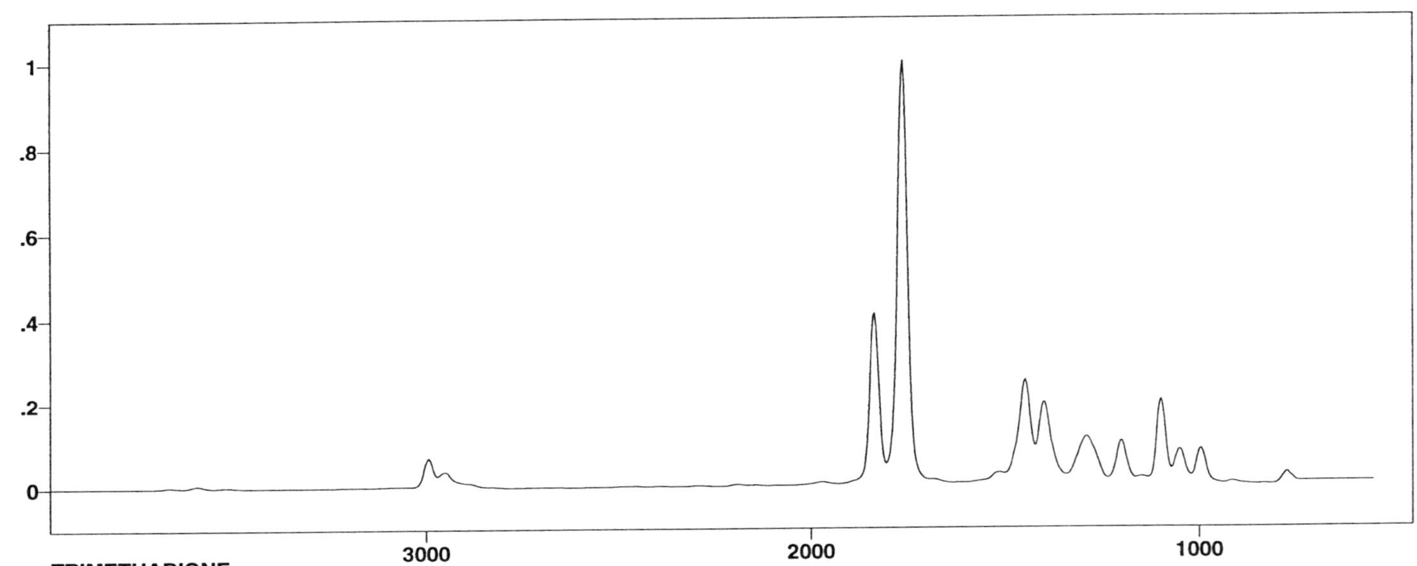

TRIMETHADIONE

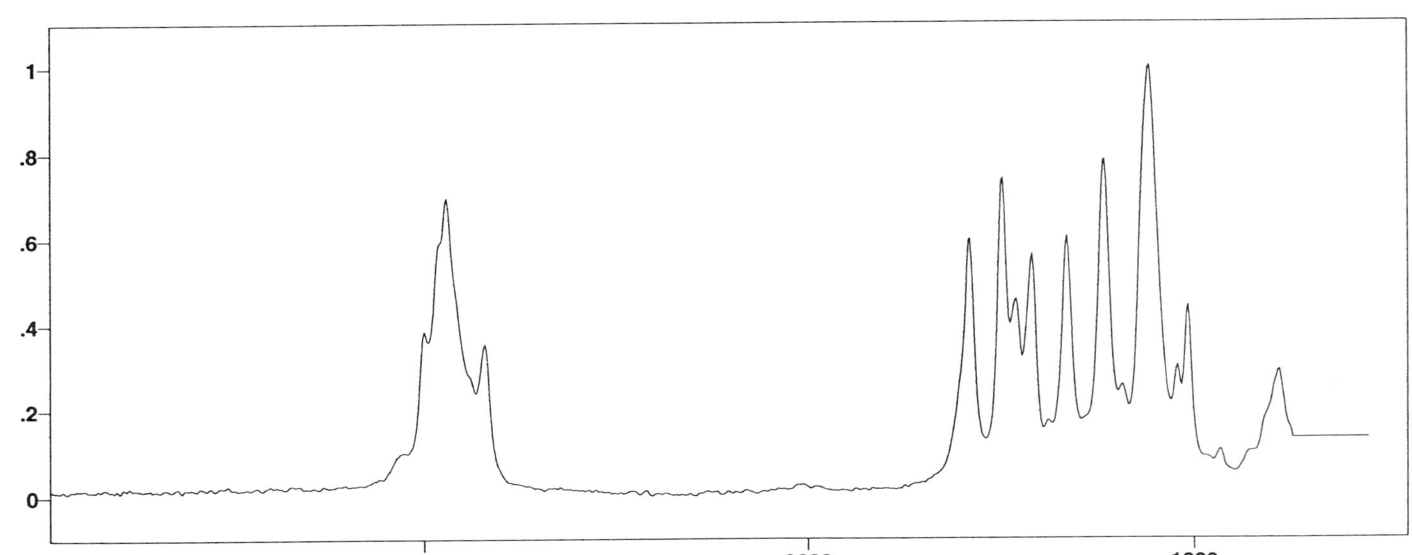

3,4,5-TRIMETHOXYAMPHETAMINE

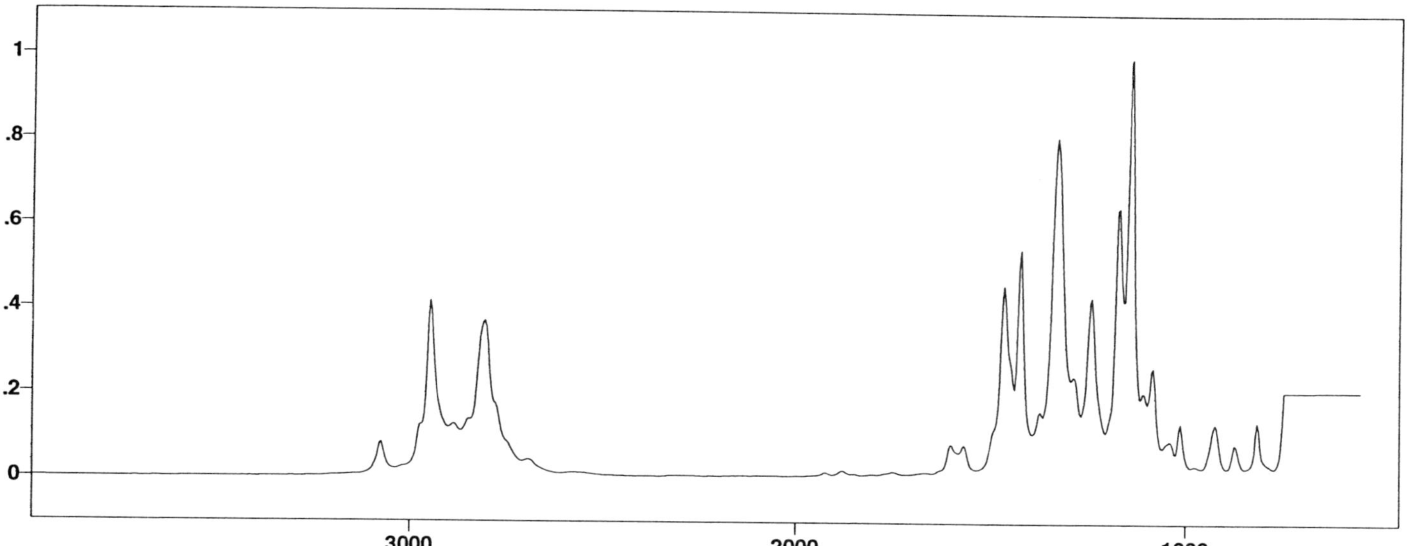

TRIFLUOPERAZINE

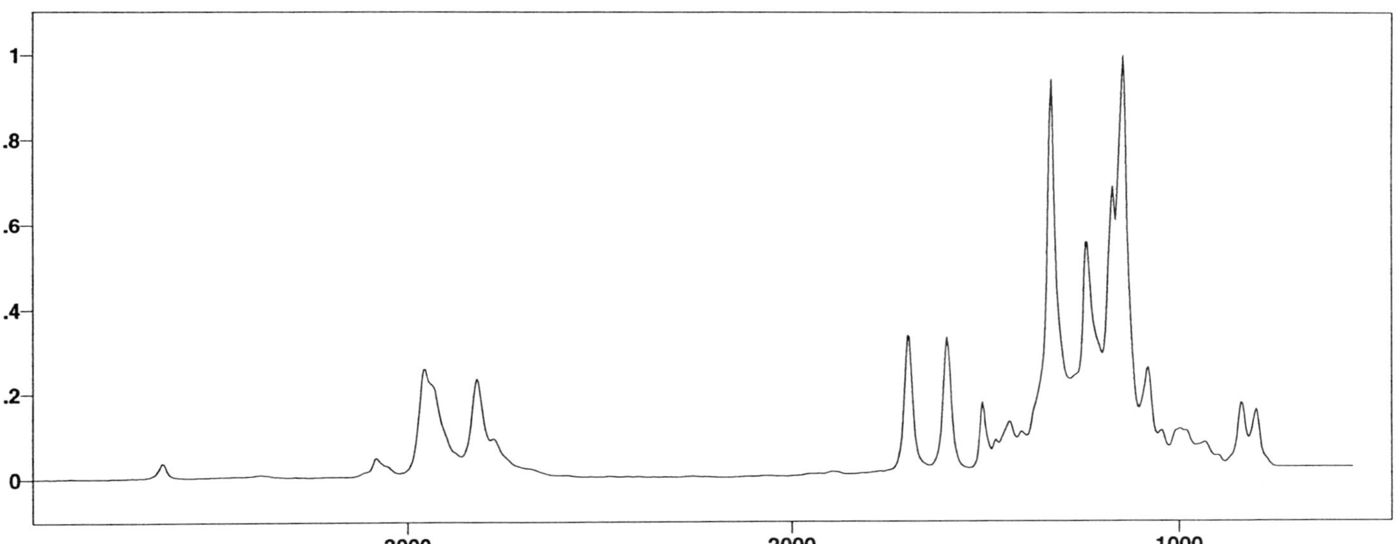

TRIFLUPERIDOL

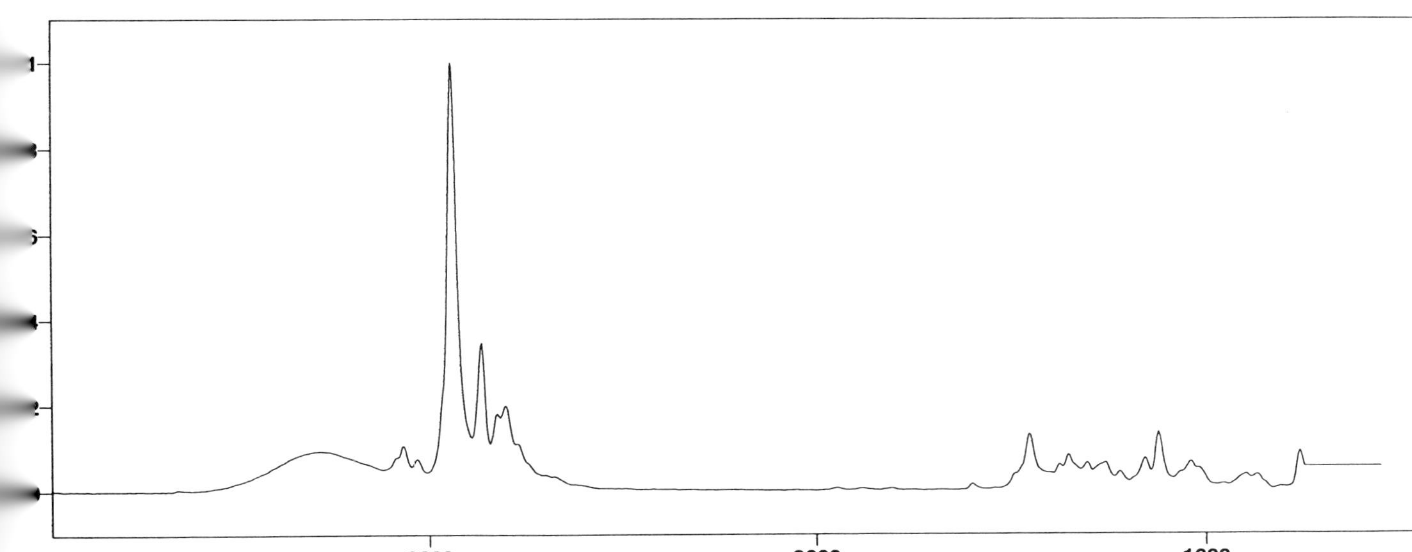

TRIHEXYPHENIDYL

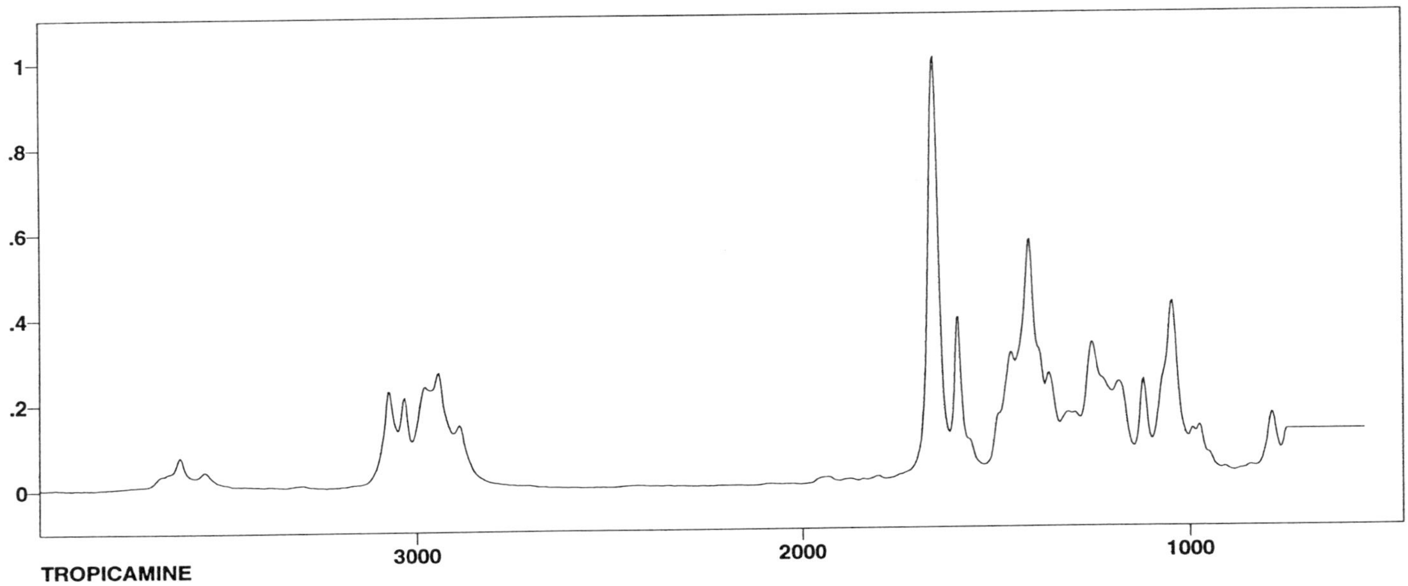
1
.8
.6
.4
.2
0
3000
2000
1000
TROPICAMINE

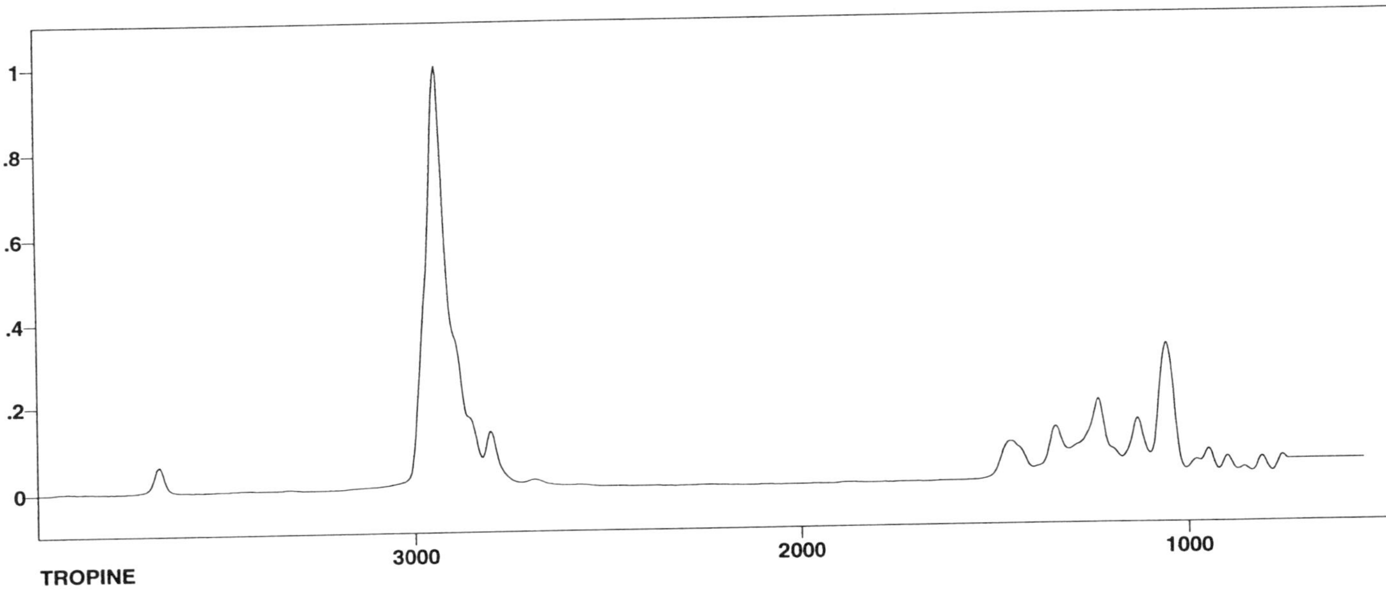
1
.8
.6
.4
.2
0
3000
2000
1000
TROPINE

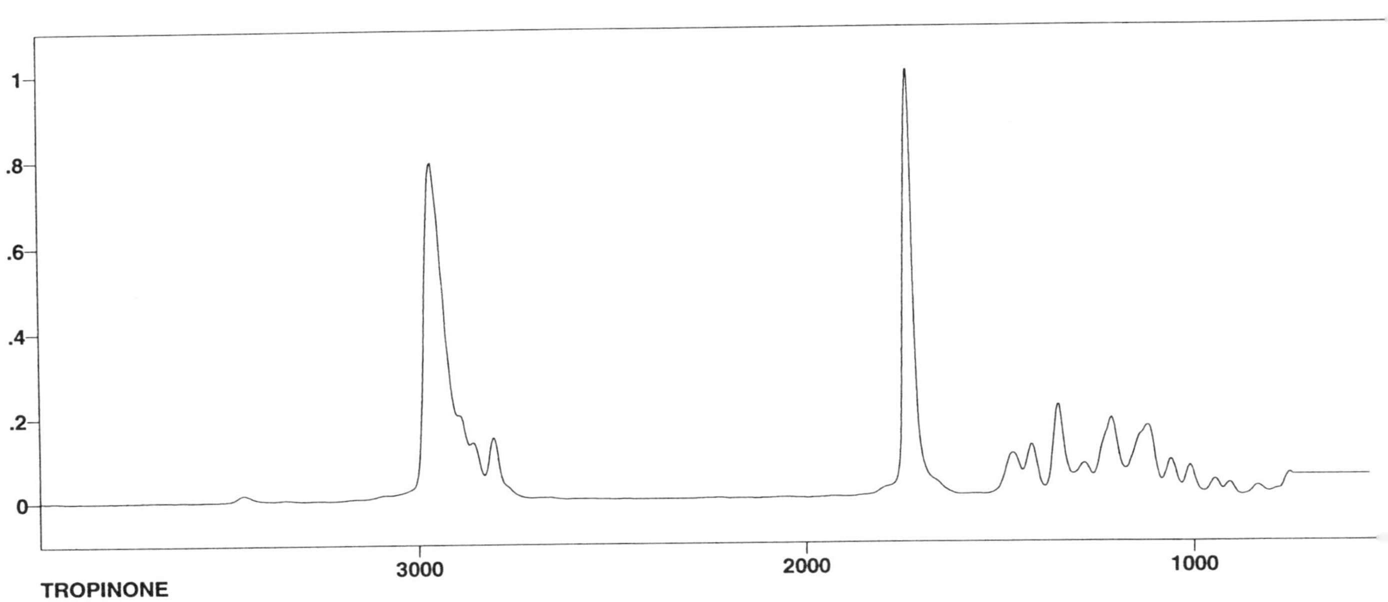
1
.8
.6
.4
.2
0
3000
2000
1000
TROPINONE

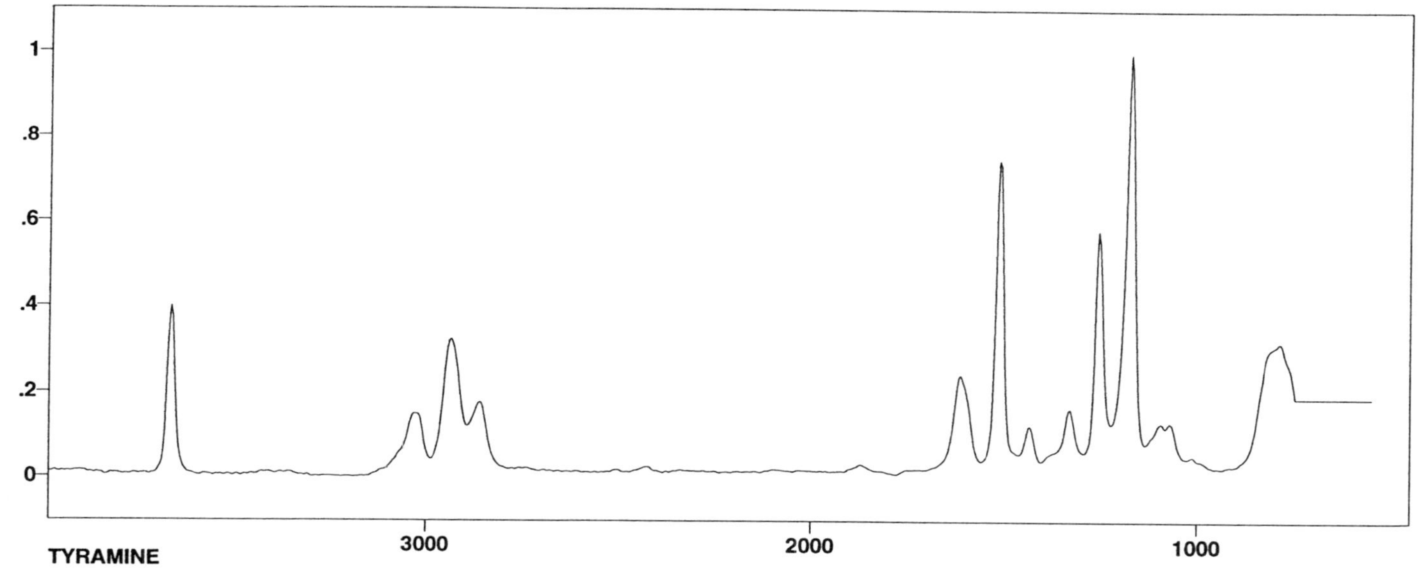

TYRAMINE

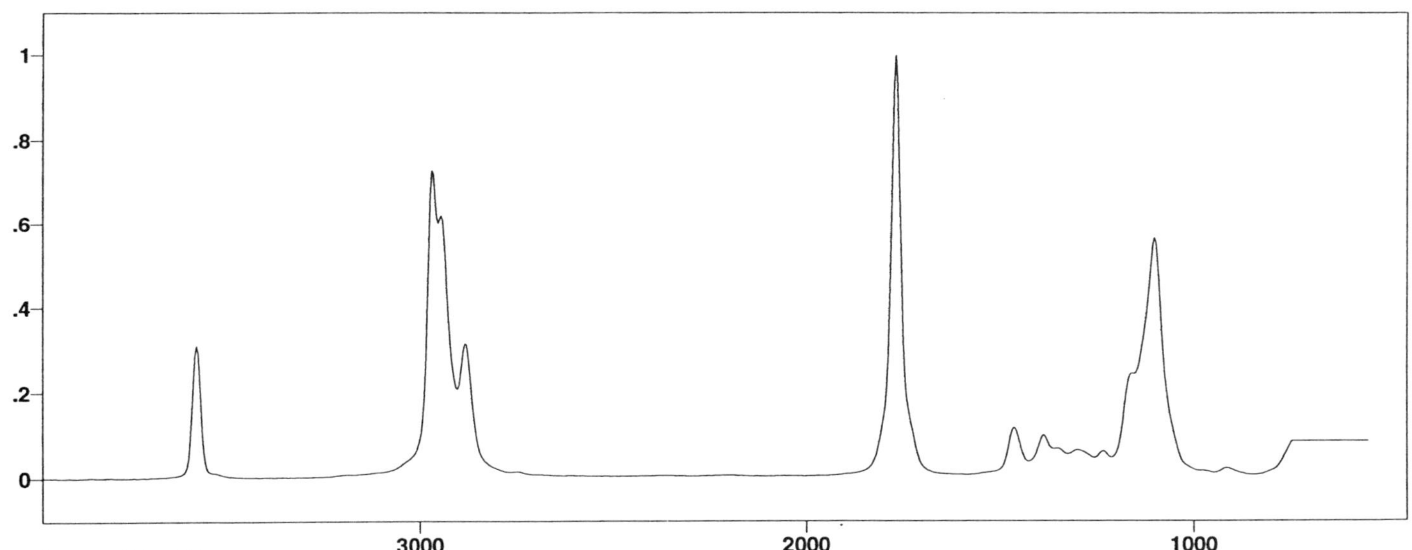

VALPROIC ACID

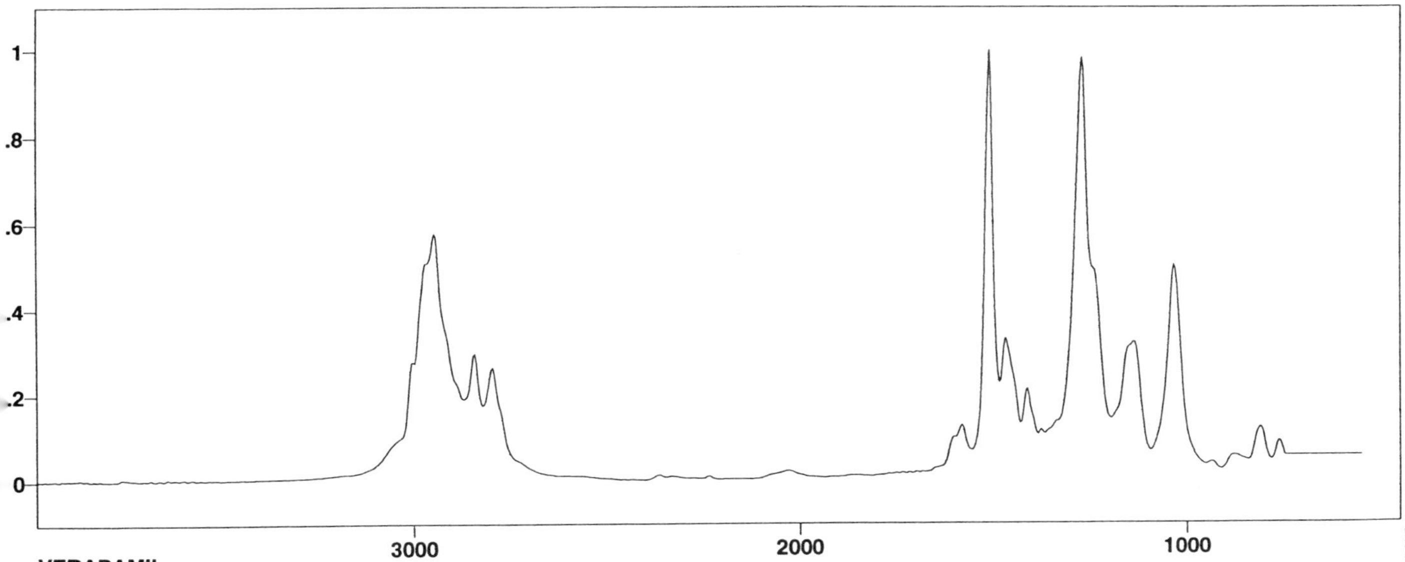

VERAPAMIL

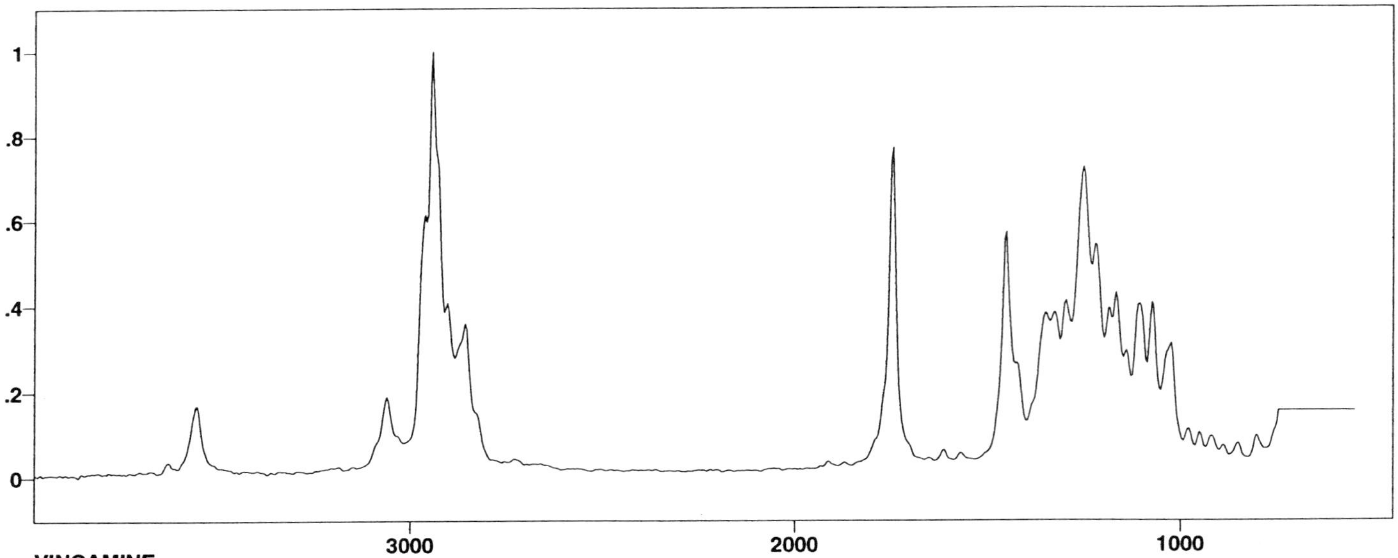

VINCAMINE

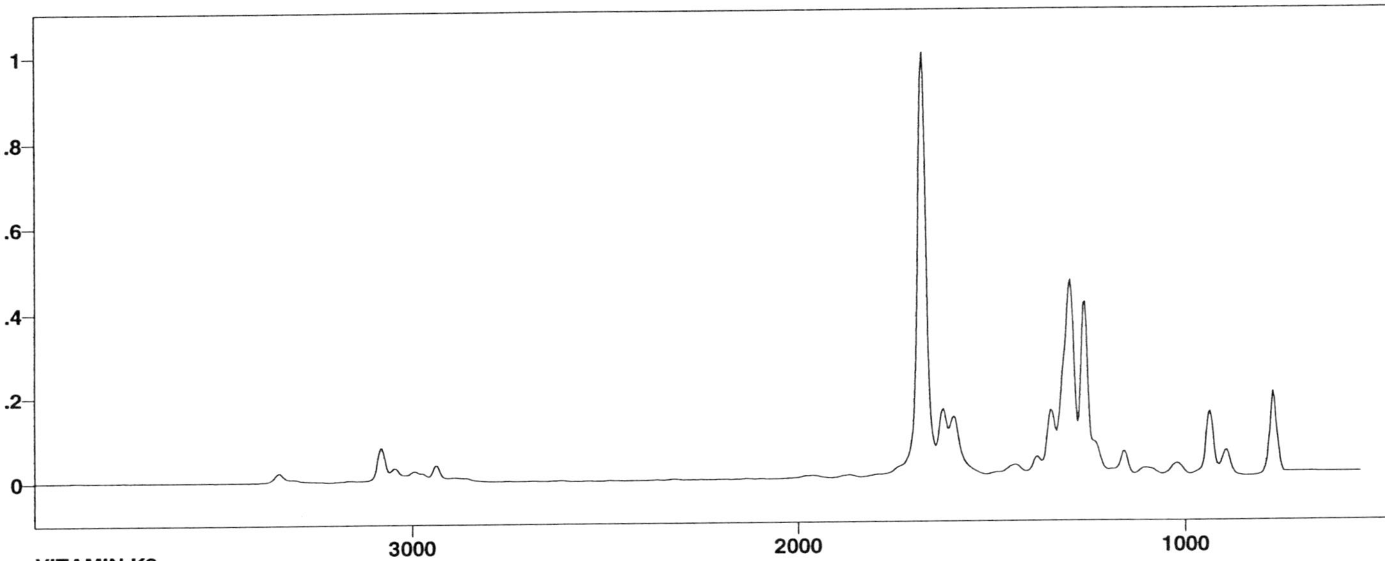

VITAMIN K3

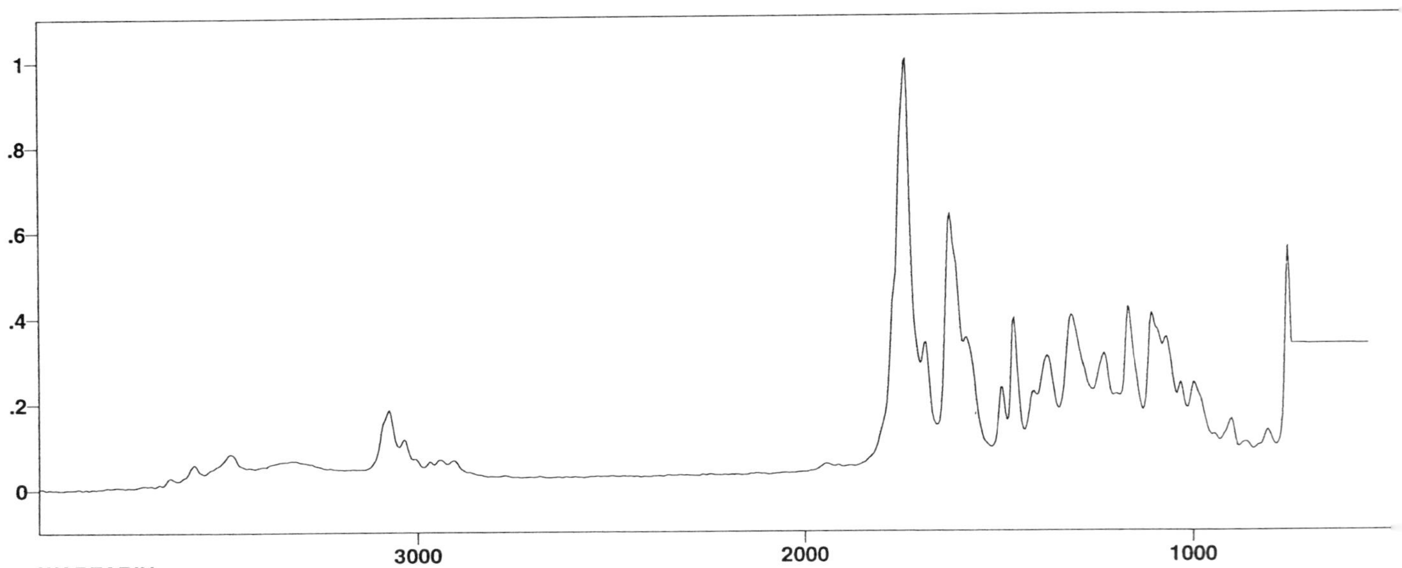

WARFARIN

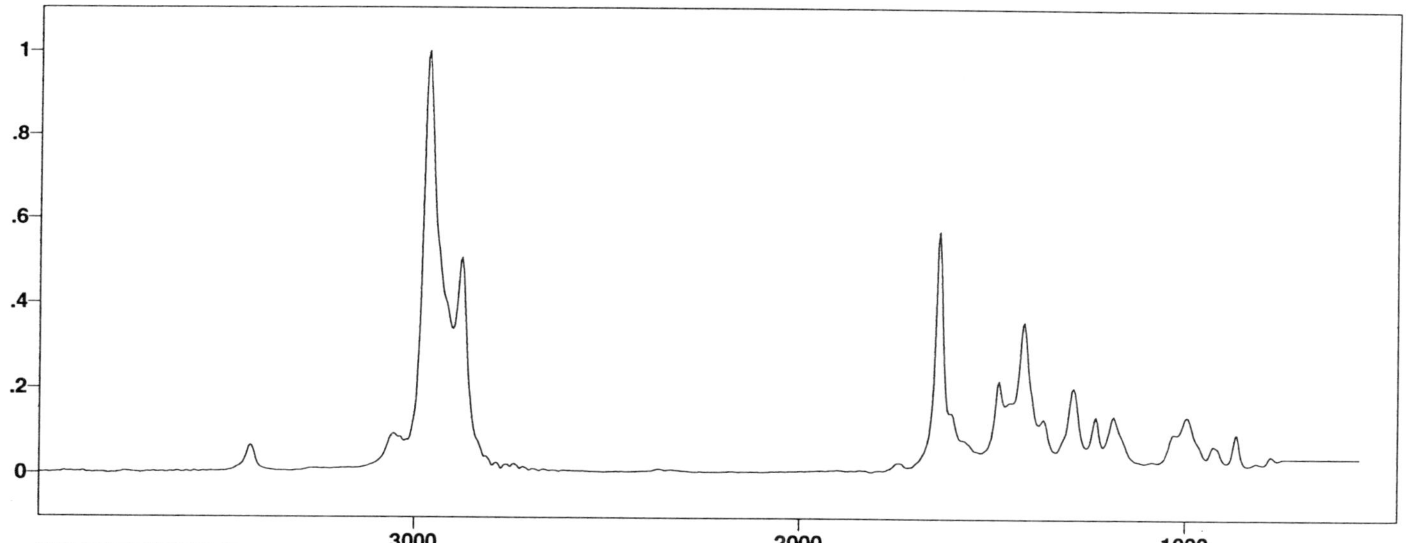

XYLOMETAZOLINE

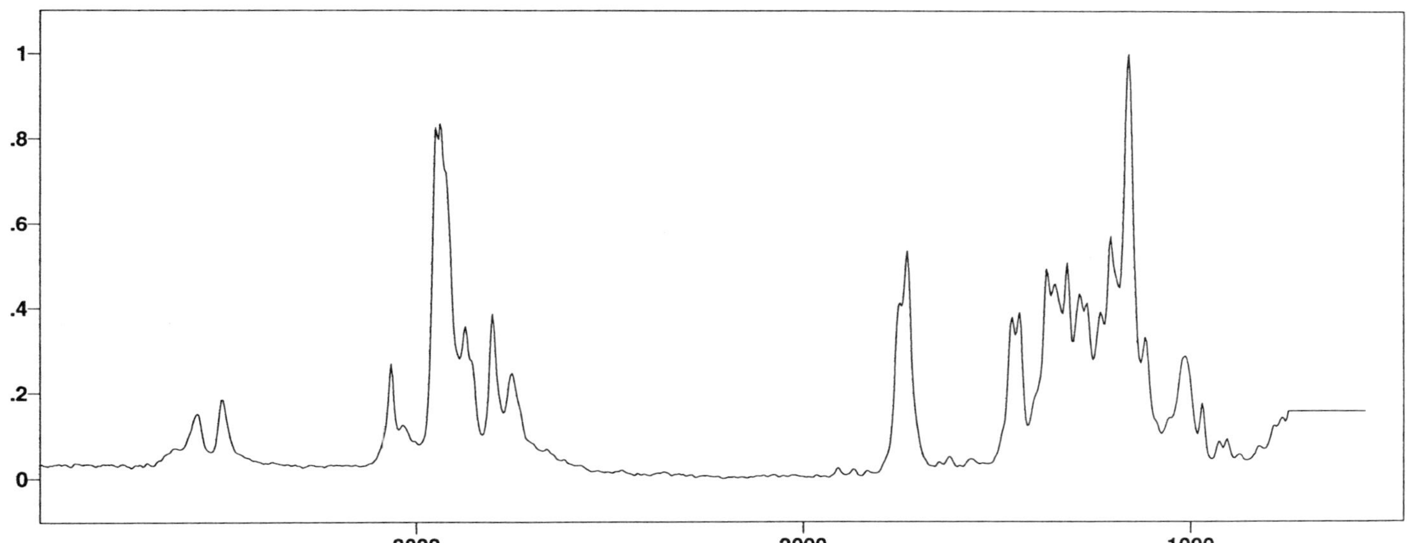

YOHIMBINE

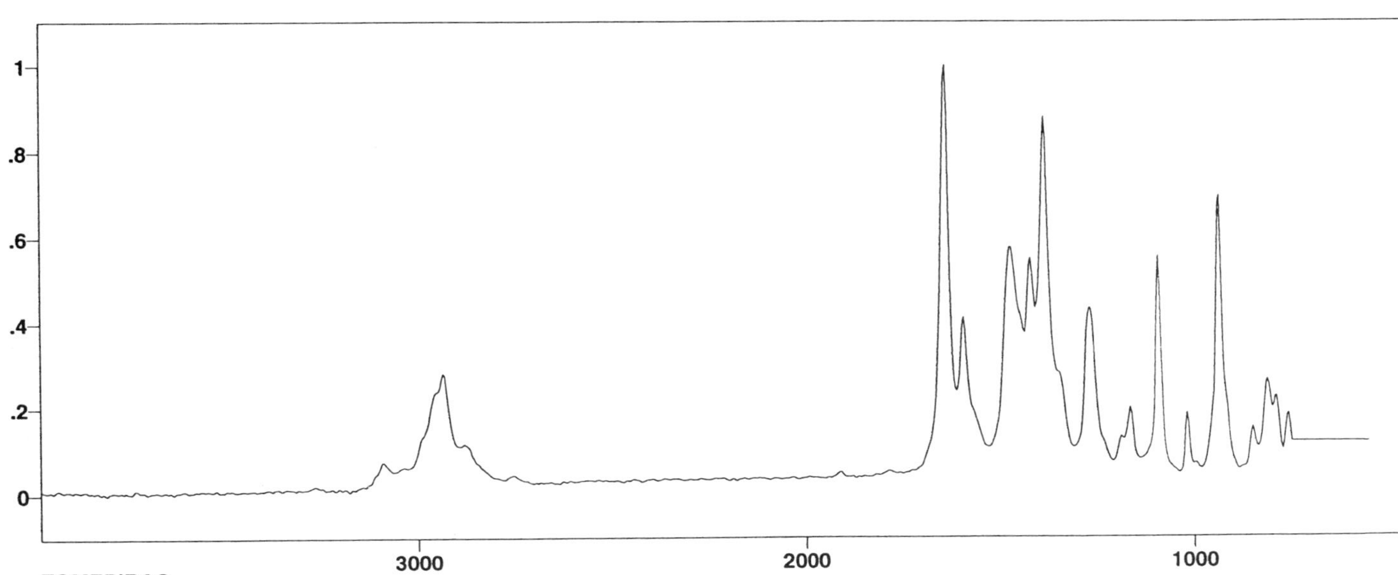

ZOMEPIRAC

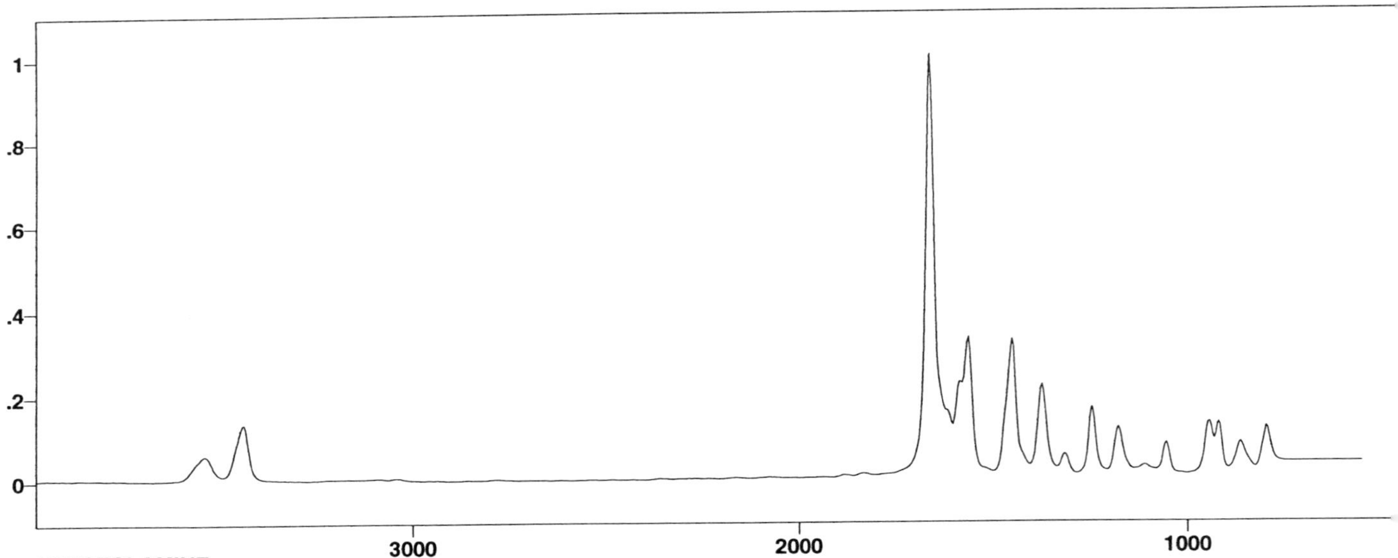

ZOXAZOLAMINE

MASS SPECTRA OF PESTICIDES

ALACHLOR

$C_{14}H_{20}NO_2Cl$

M.W. = 69.72 (269 .12)

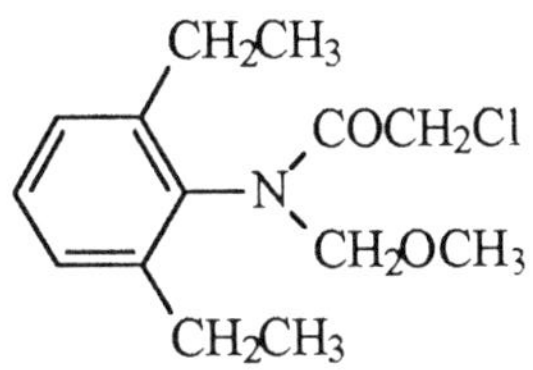

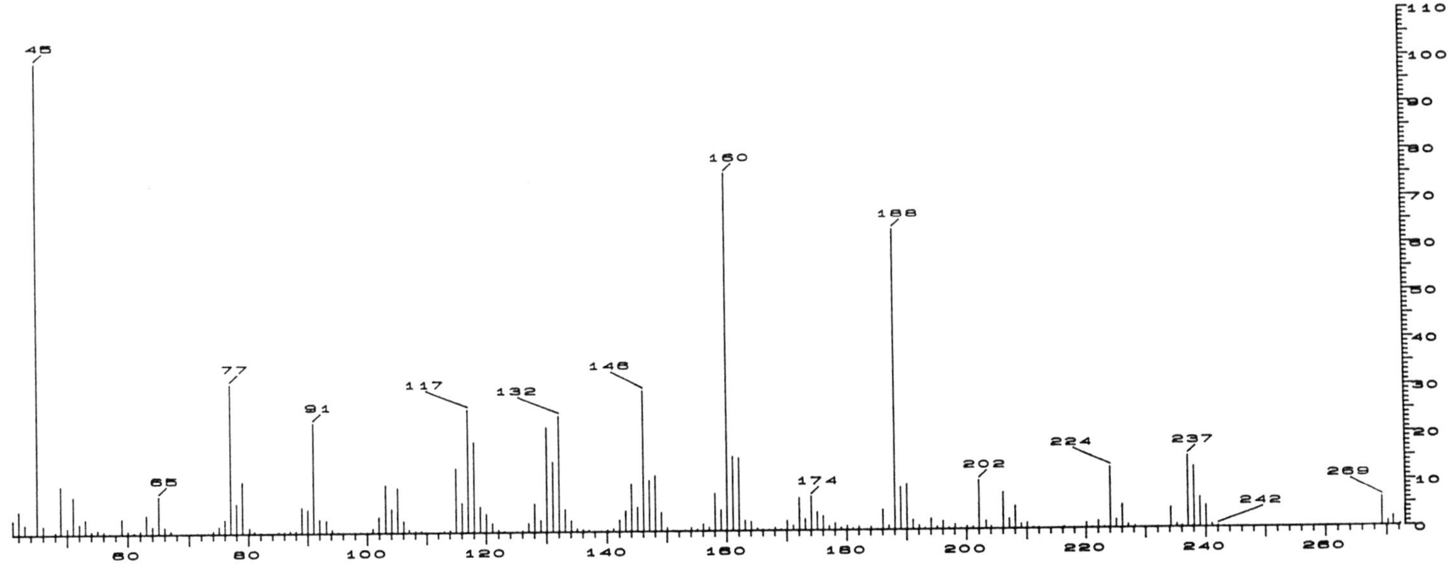

ALDRIN

$C_{12}H_8Cl_6$ M.W. = 364.91 (361.88)

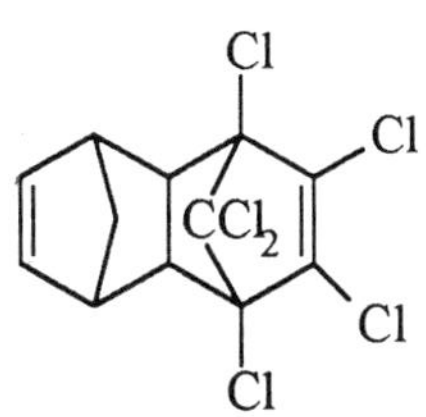

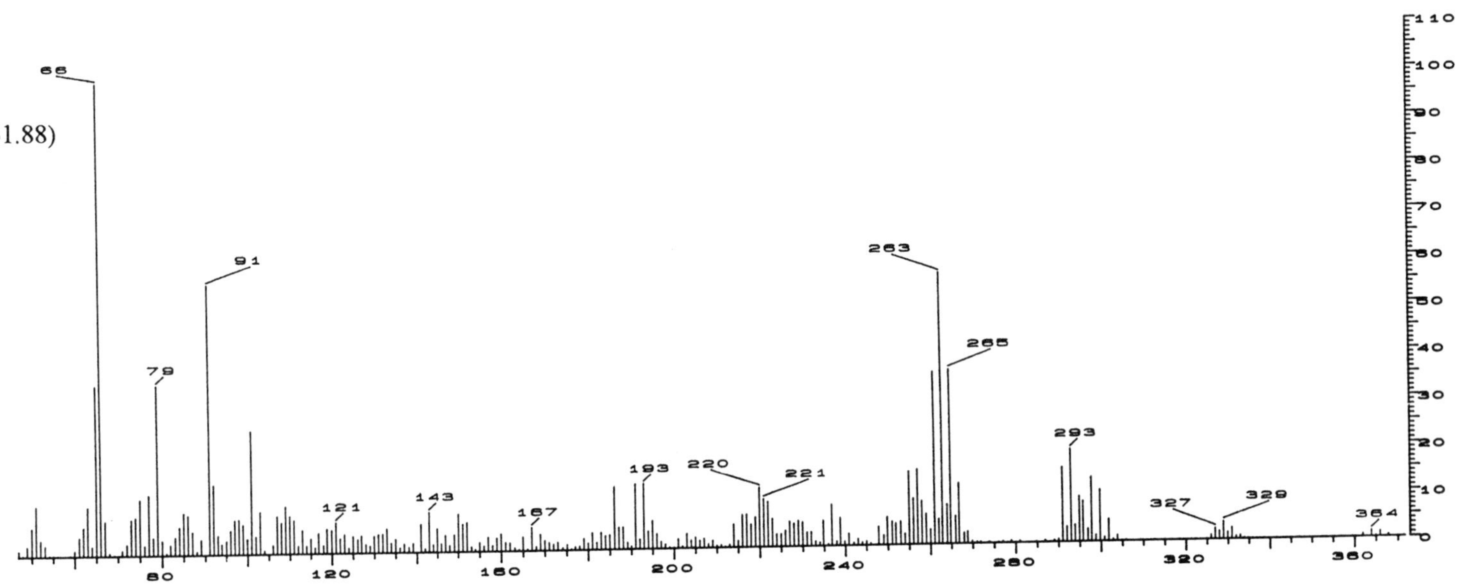

AMETRYNE

$C_9H_{17}N_5S$ M.W. = 227.32 (227.12)

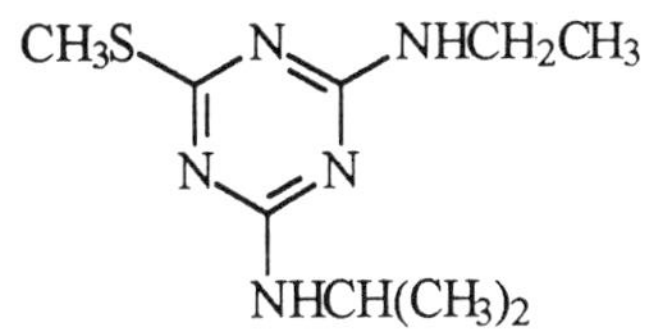

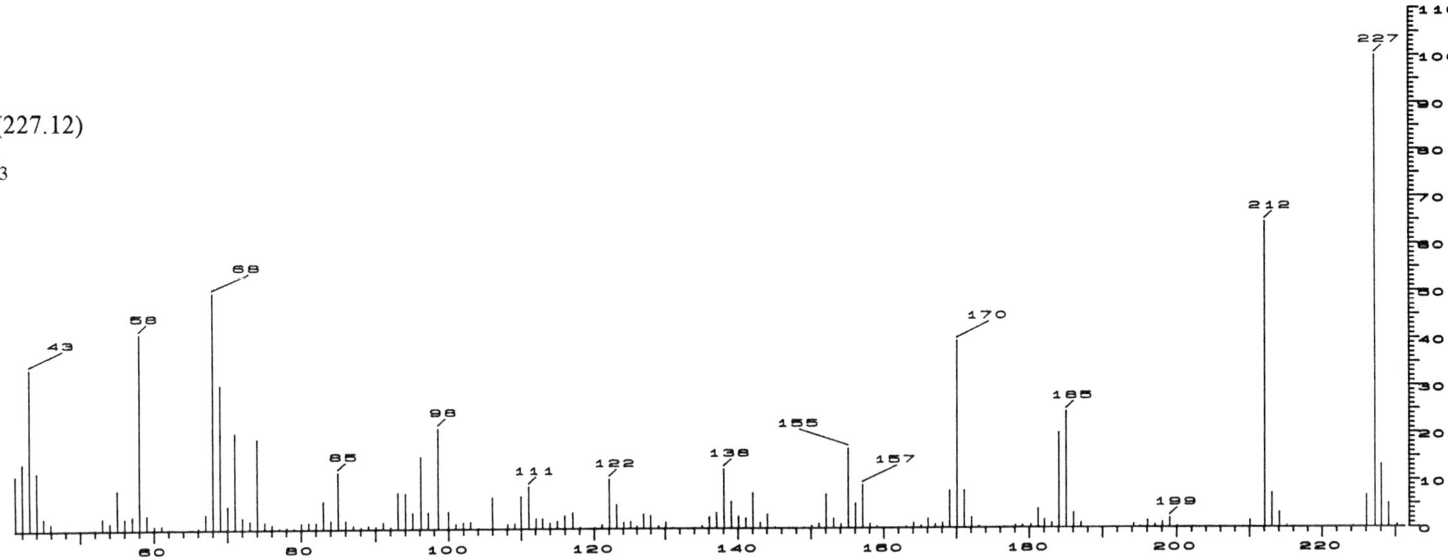

AMINOCARB

$C_{11}H_{16}N_2O_2$ M.W. = 208.26 (208.12)

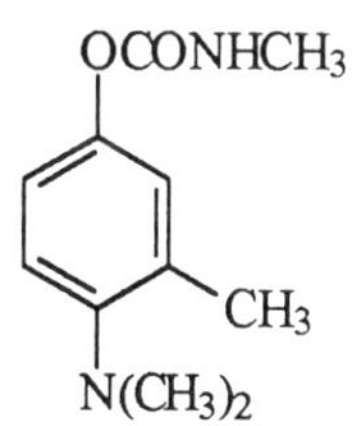

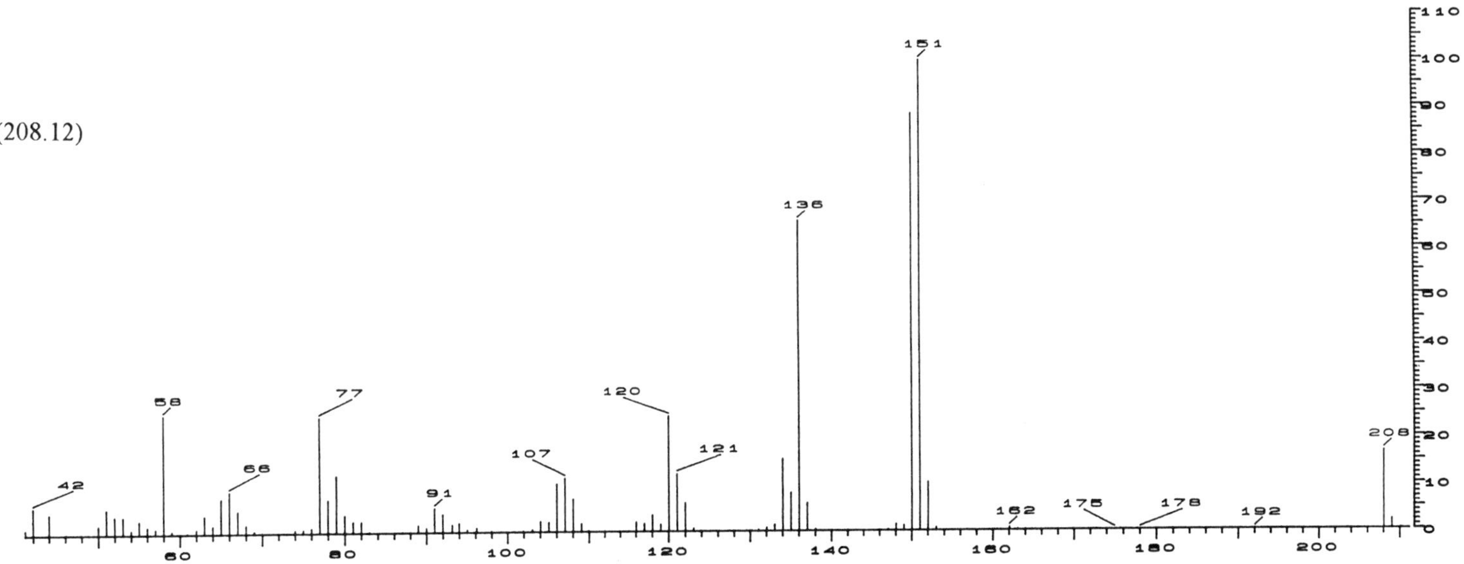

AMITRAZ

$C_{19}H_{23}N_3$ M.W. = 293.41 (293.19)

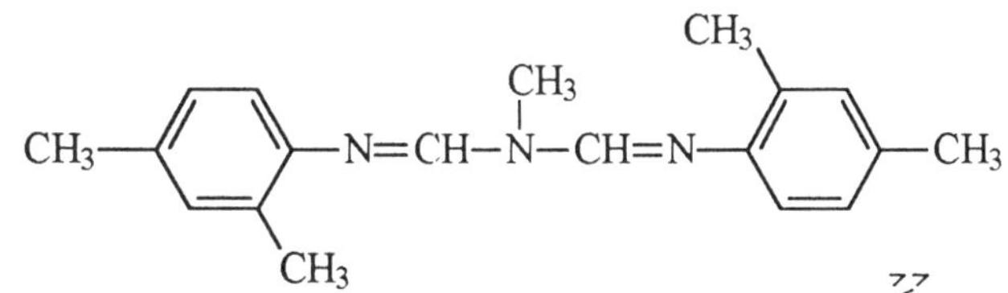

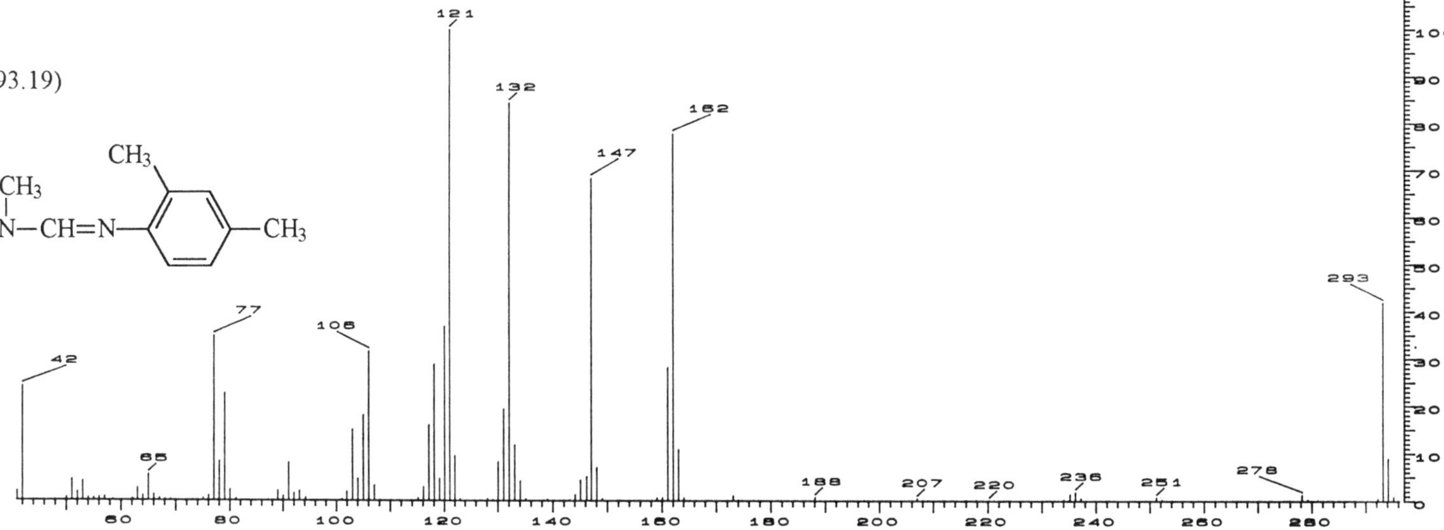

AMITROLE

$C_2H_4N_4$ M.W. = 84.08 (84.04)

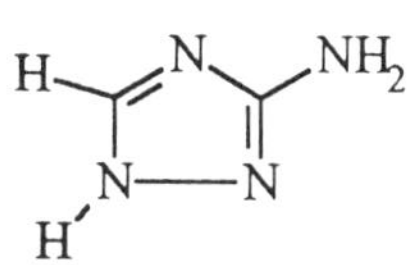

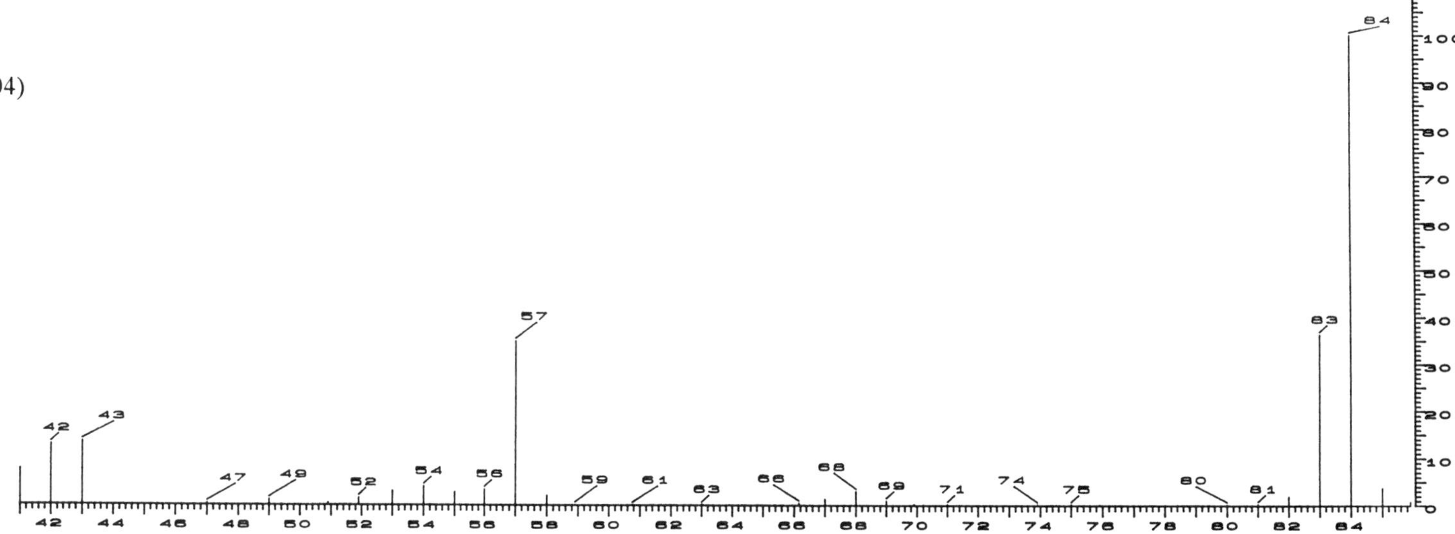

ANILAZINE

$C_9H_5N_4Cl_3$ N\M.W. = 275.52 (273.96)

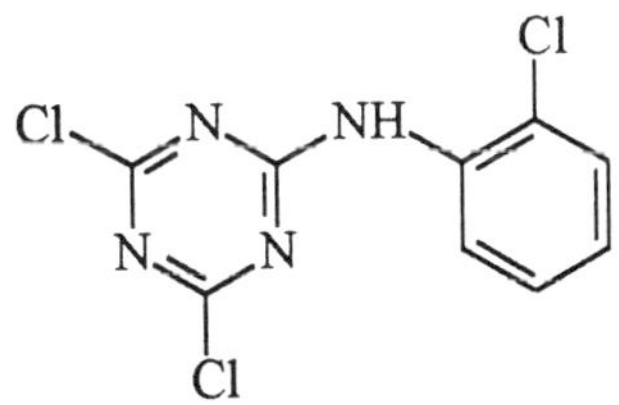

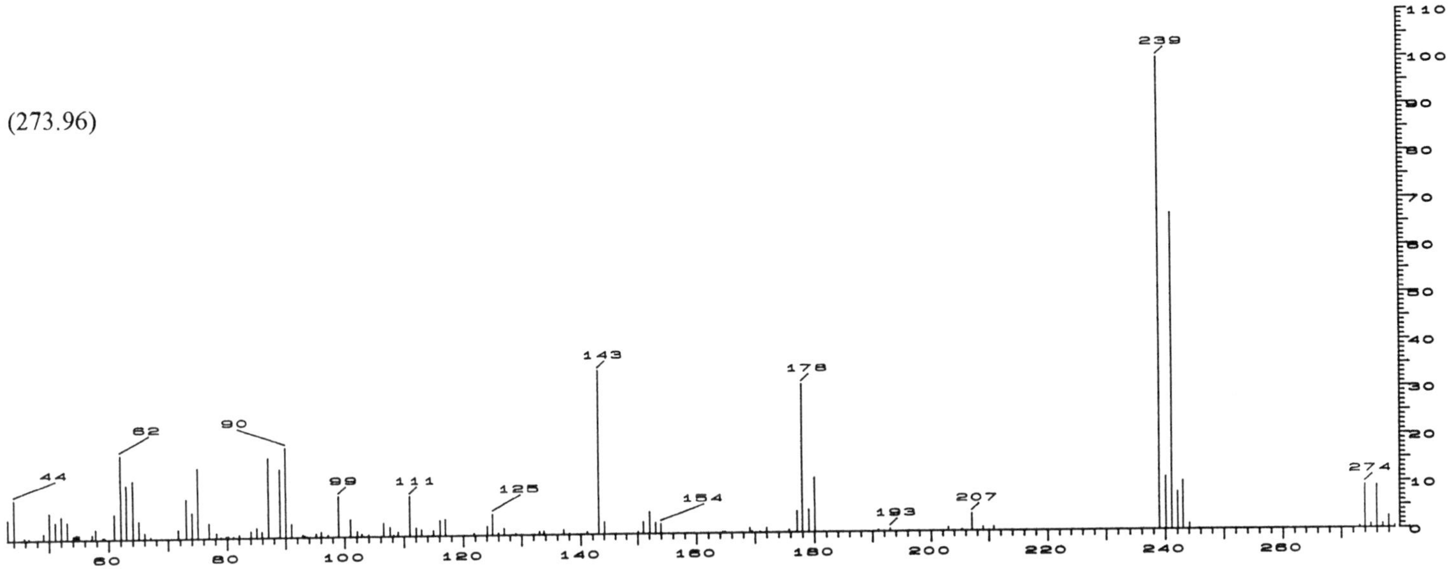

ATRATON

$C_9H_{17}N_5O$ M.W. = 211.26 (211.14)

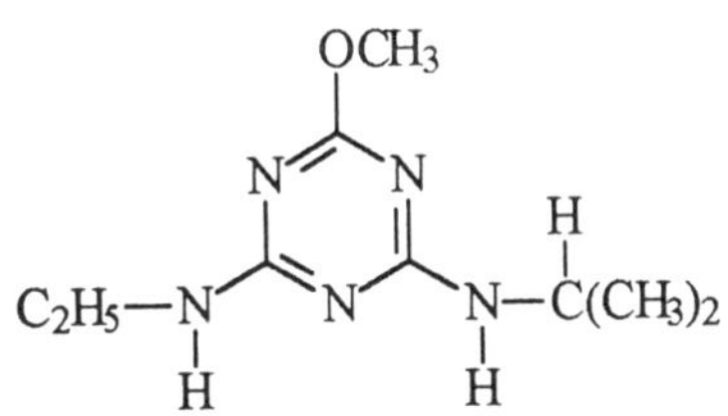

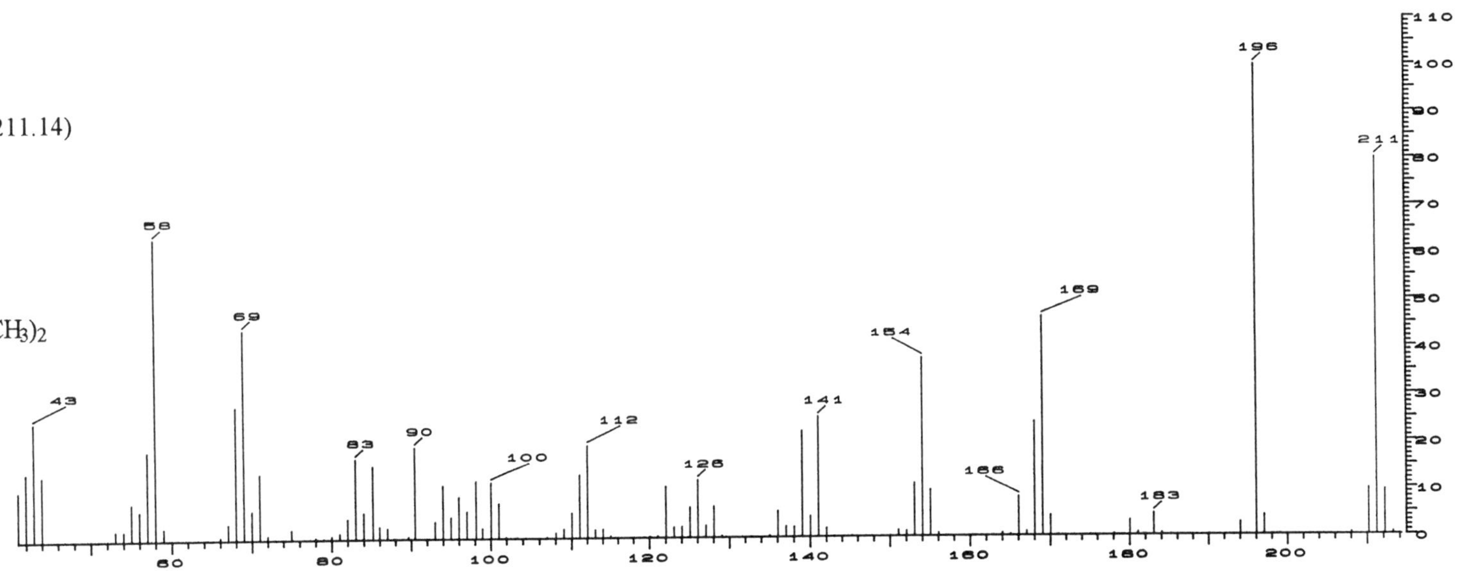

ATRAZINE

$C_8H_{14}N_5Cl$ M.W. = 215.68 (215.09)

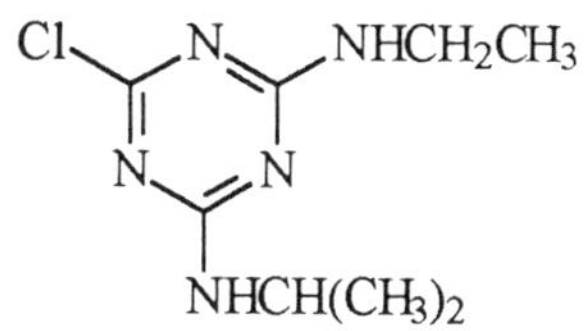

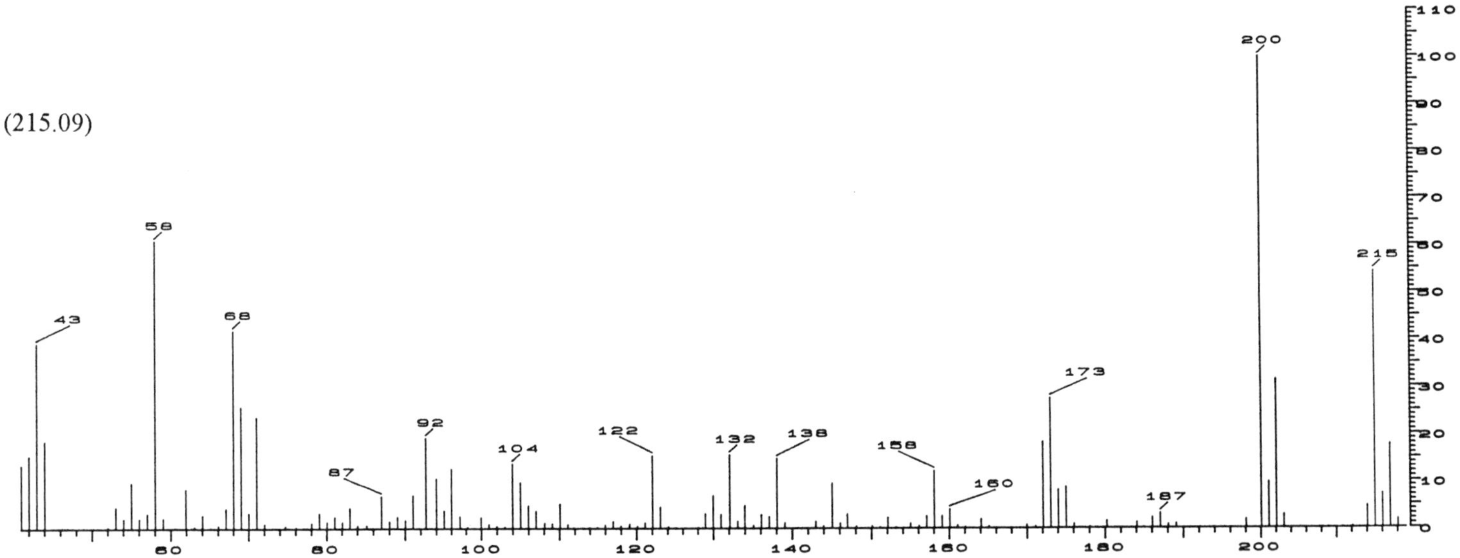

AZINPHOS-METHYL

$C_{10}H_{12}N_3O_3S_2P$ M.W. = 317.31 (317.01)

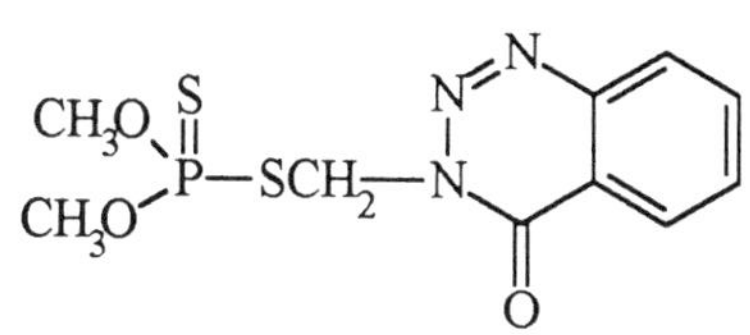

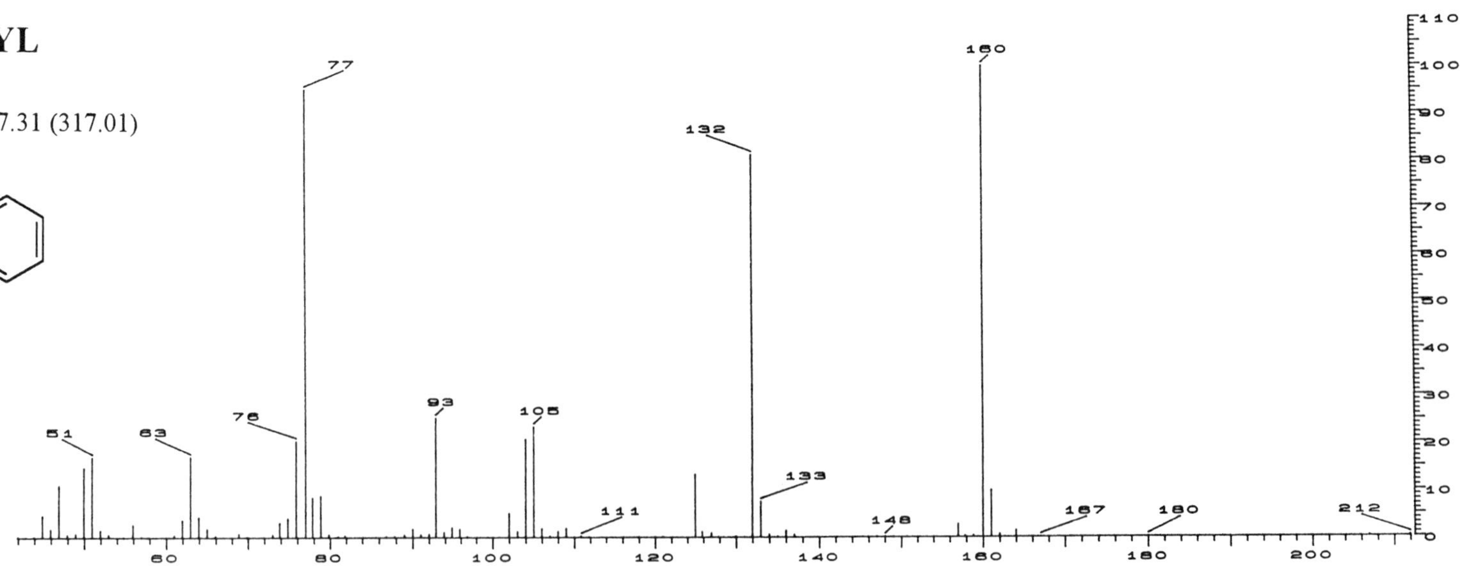

BARBAN

$C_{11}H_9NO_2Cl_2$ M.W. = 258.10 (257.00)

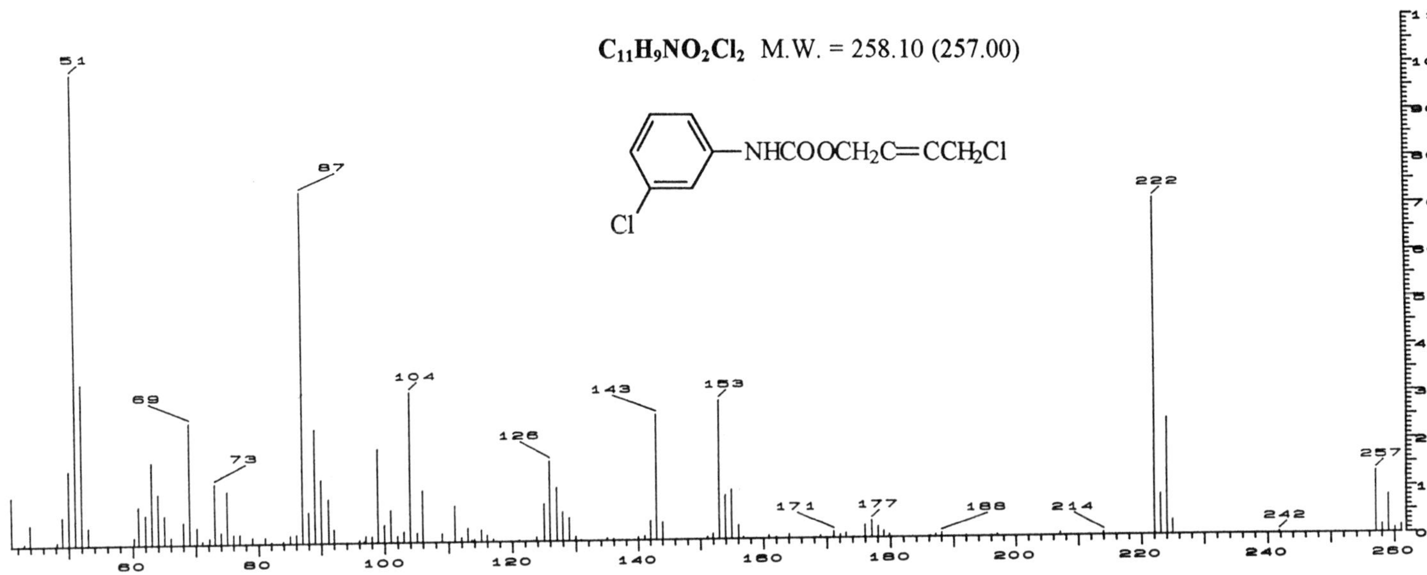

BENDIOCARB

$C_{11}H_{13}NO_4$ M.W. = 223.22 (223.08)

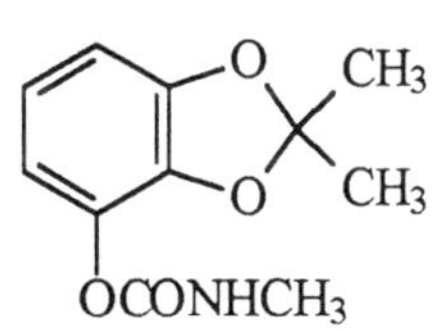

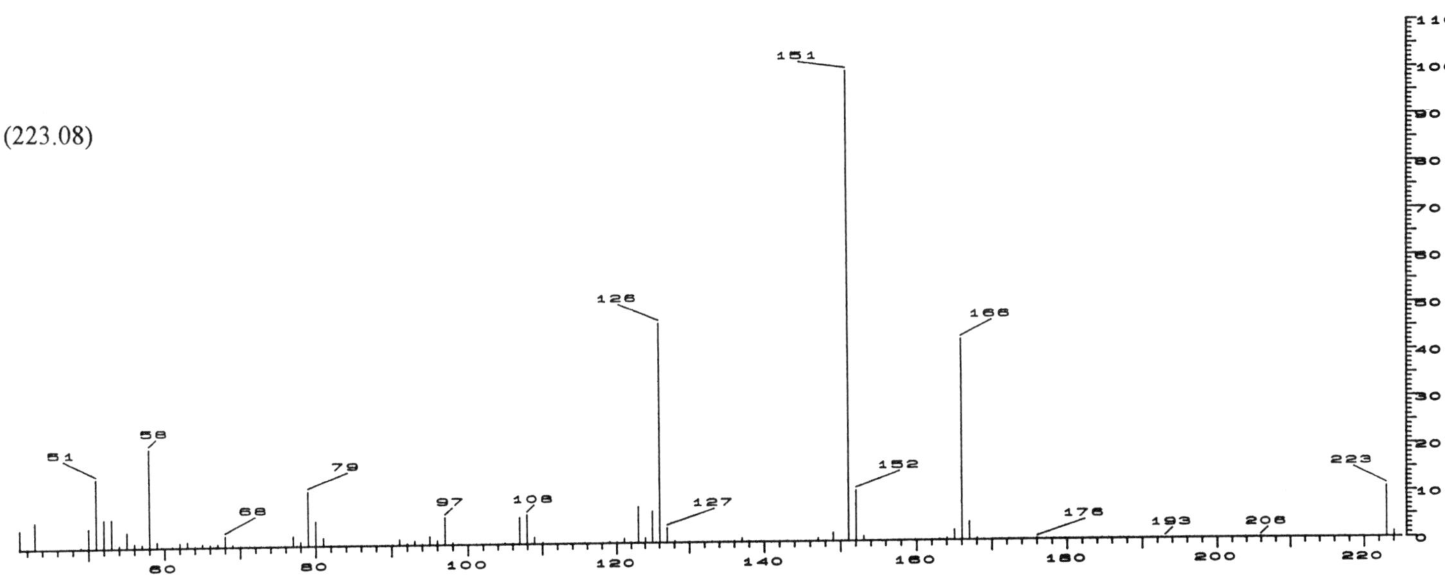

BENFLURALIN

$C_{13}H_{16}N_3O_4F_3$ M.W. = 335.28 (335.11)

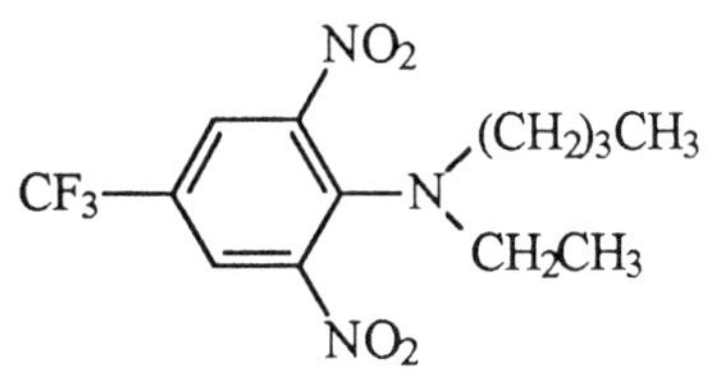

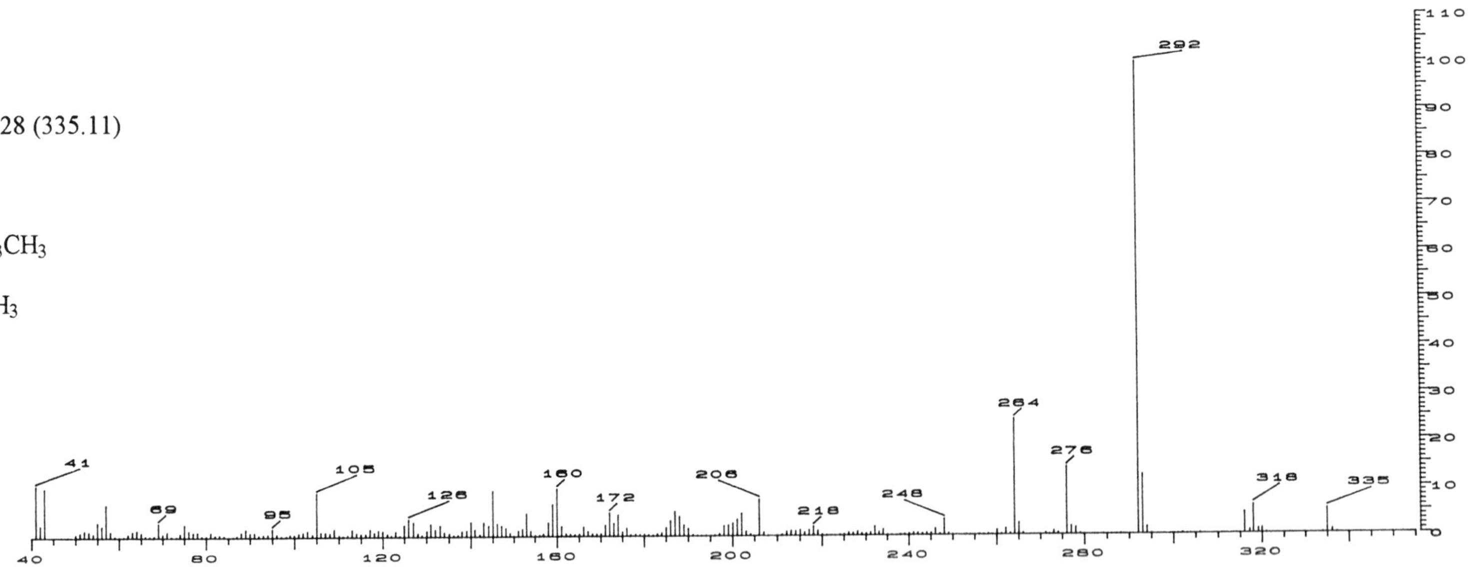

BENZOYLPROP-ETHYL

$C_{18}H_{17}NO_3Cl_2$ M.W. = 366.24 (365.06)

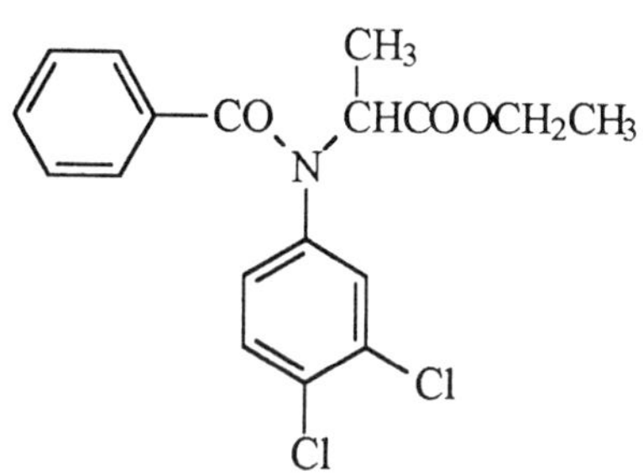

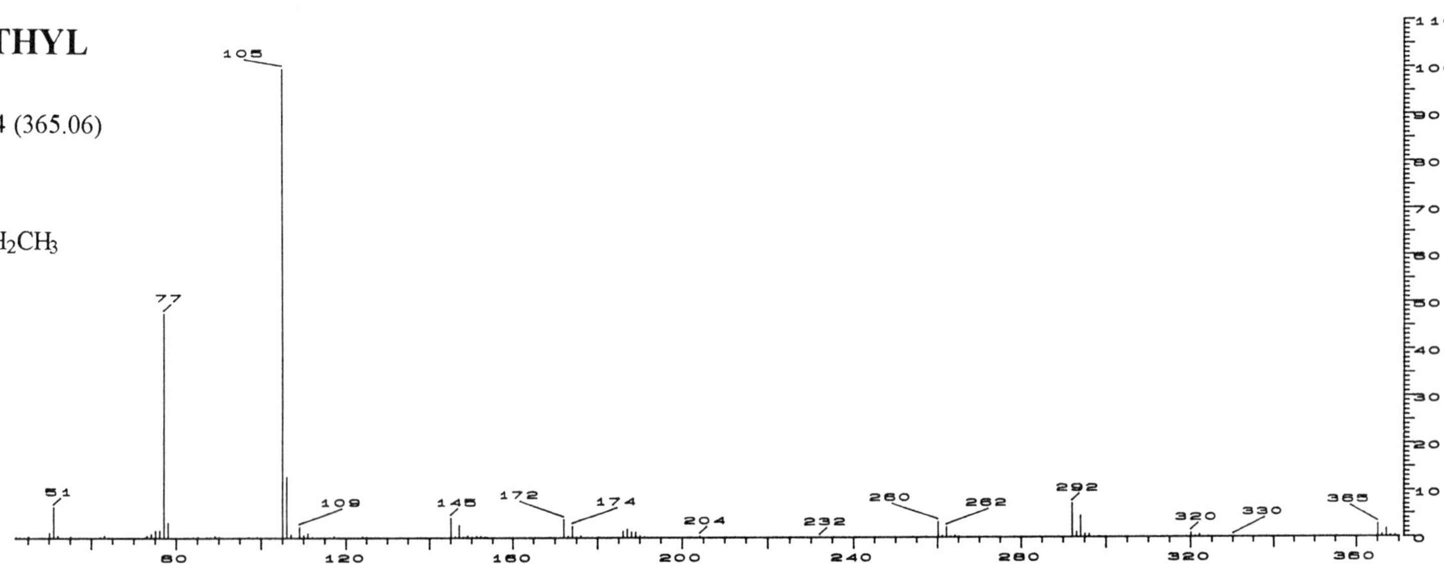

BIFENOX

$C_{14}H_9NO_5Cl_2$ M.W. = 342.13 (340.99)

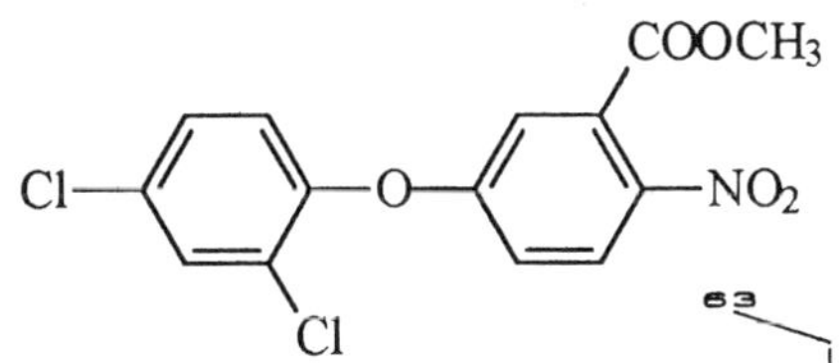

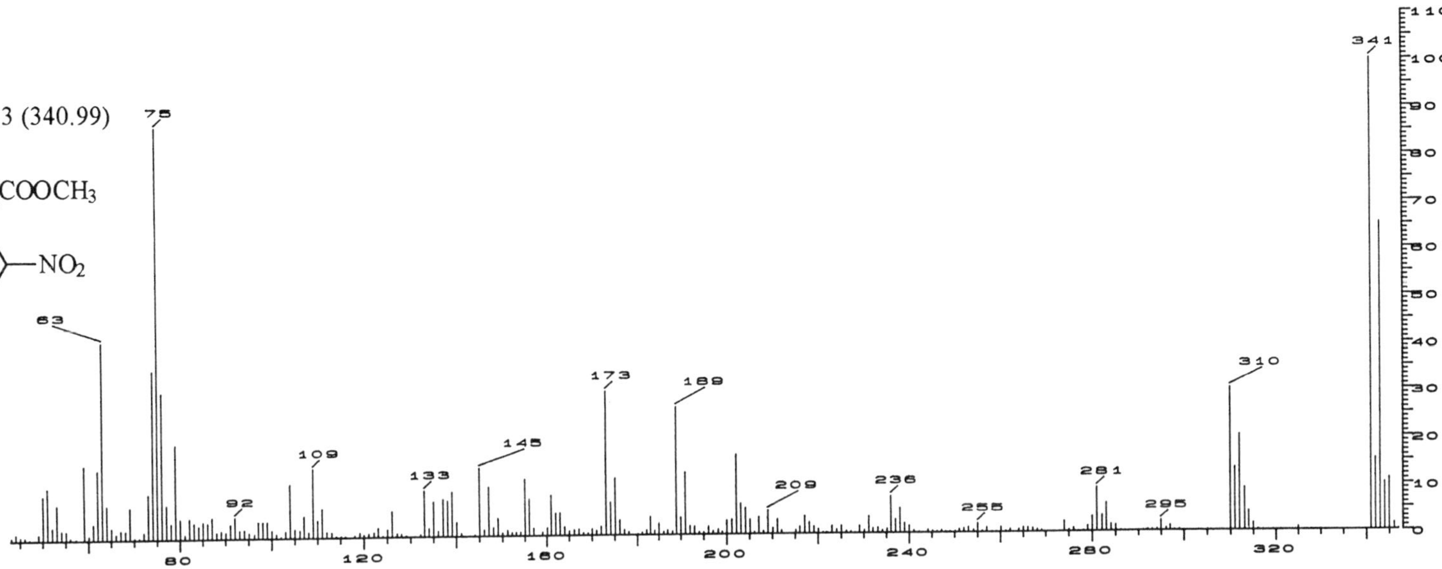

BINAPACRYL

$C_{15}H_{18}N_2O_6$ M.W. = 322.31 (322.12)

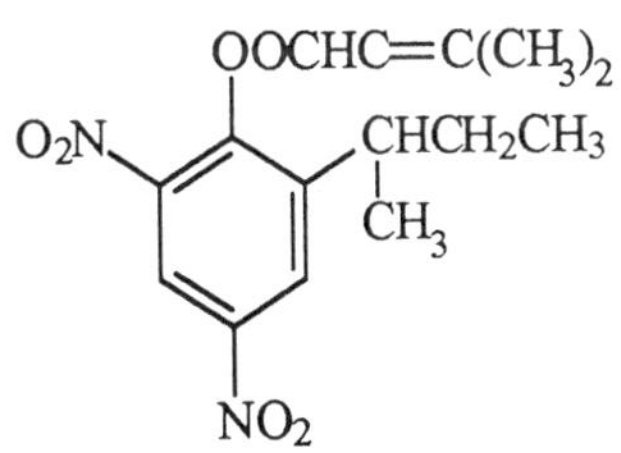

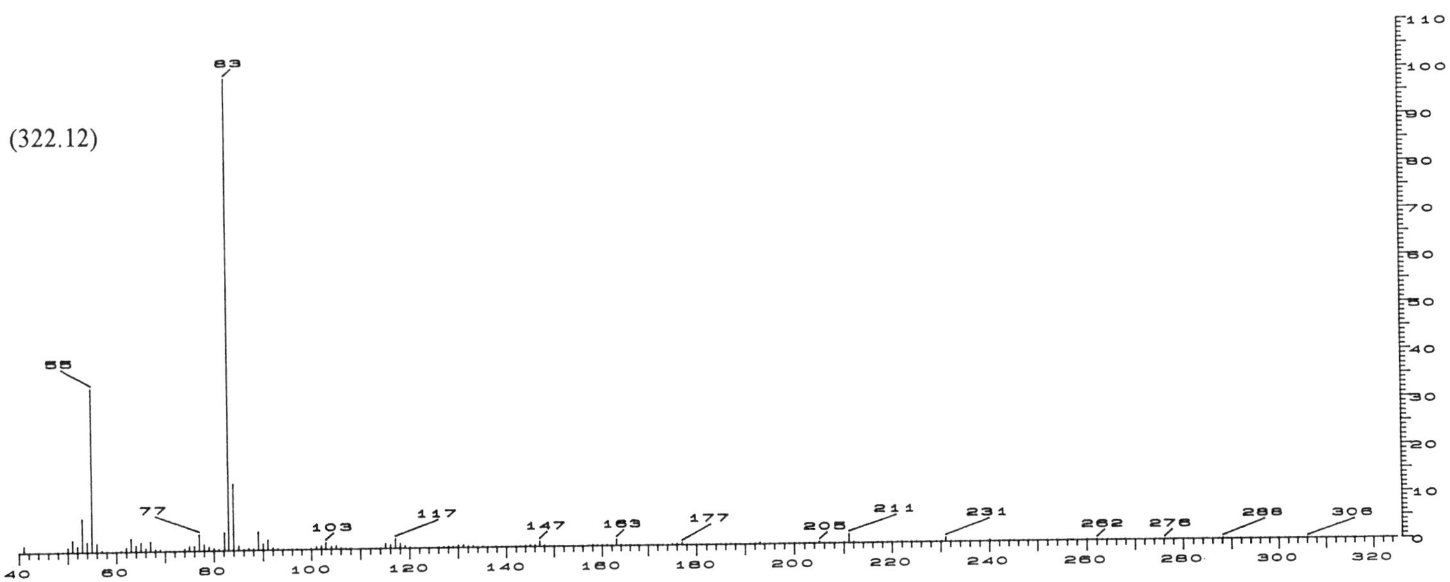

2-sec-BUTYLPHENYL METHYLCARBAMATE

$C_{12}H_{17}NO_2$ M.W. = 207.27 (207.13)

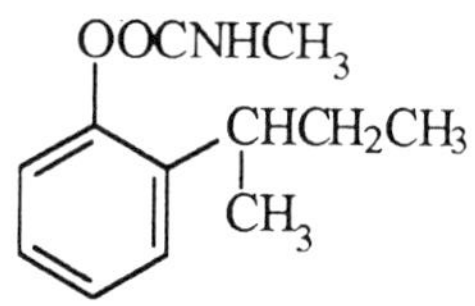

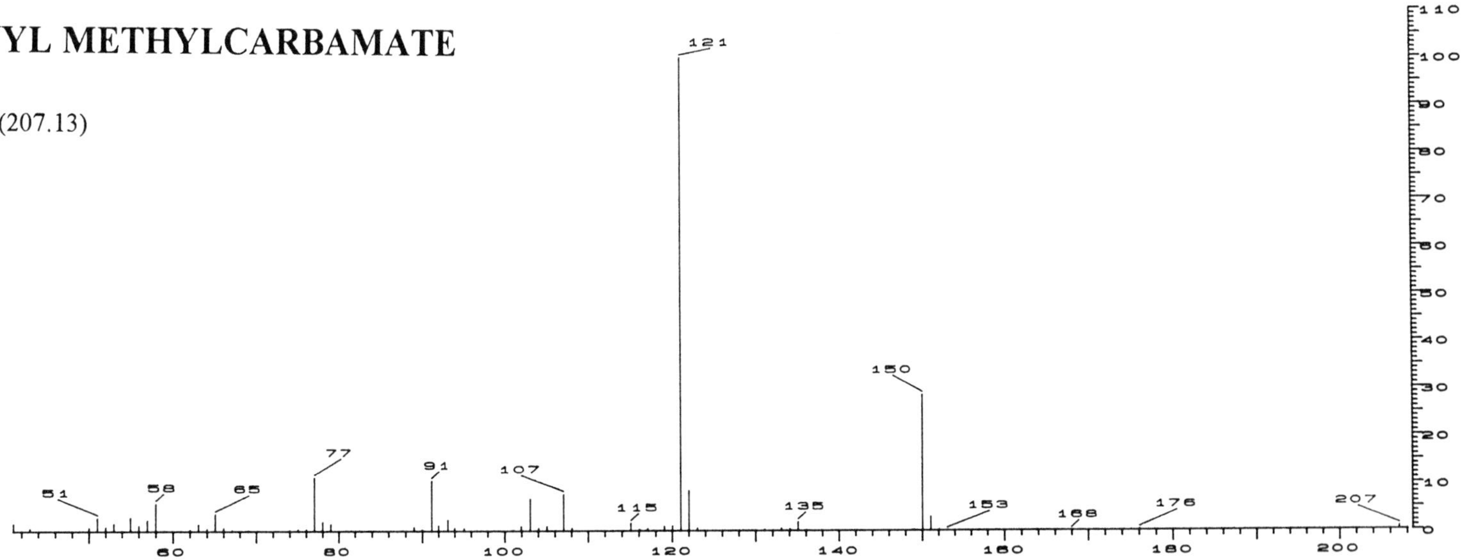

BROMACIL

$C_9H_{13}N_2O_2Br$ M.W. = 261.11 (260.02)

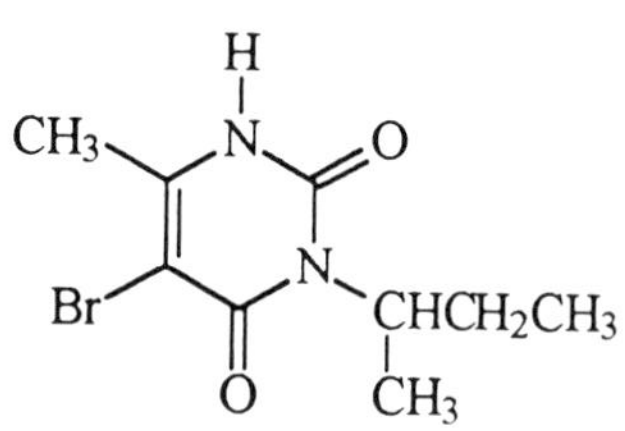

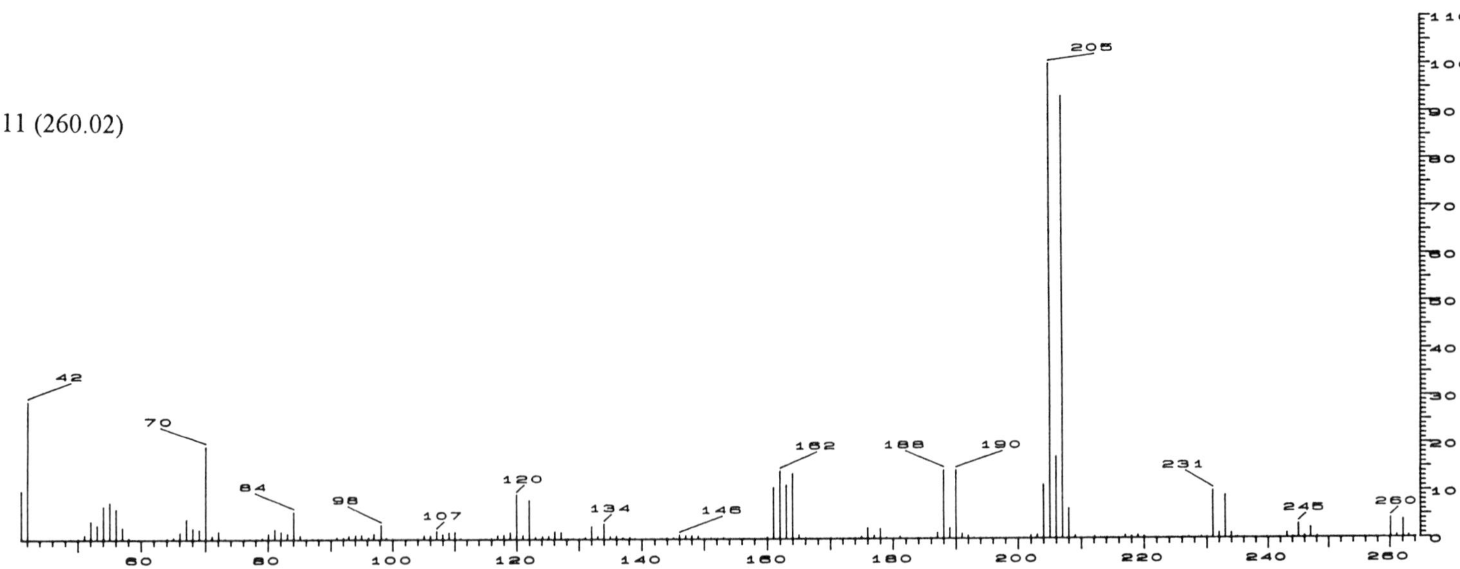

BROMOPHOS

$C_8H_8O_3SCl_2PBr$ M.W. = 365.99 (363.85)

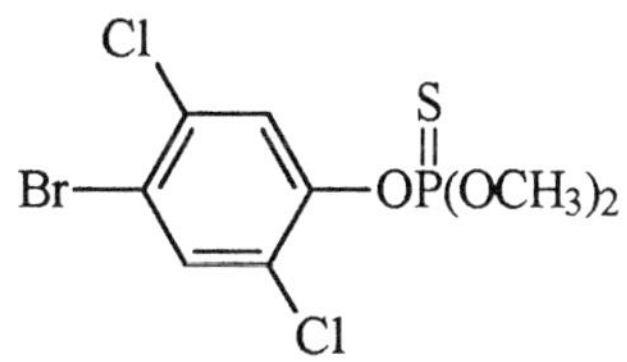

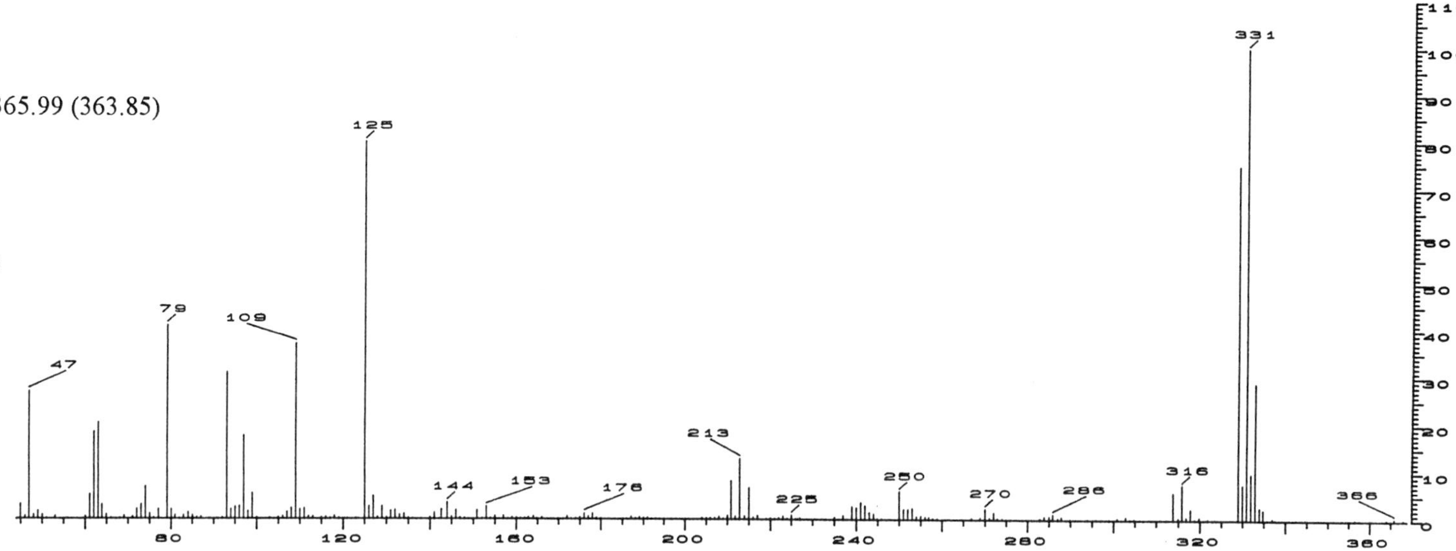

BROMOXYNIL

$C_7H_3NOBr_2$ M.W. = 276.91 (274.86)

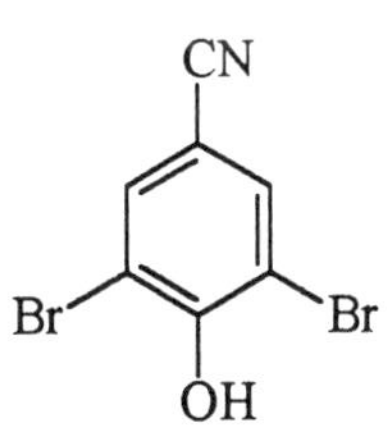

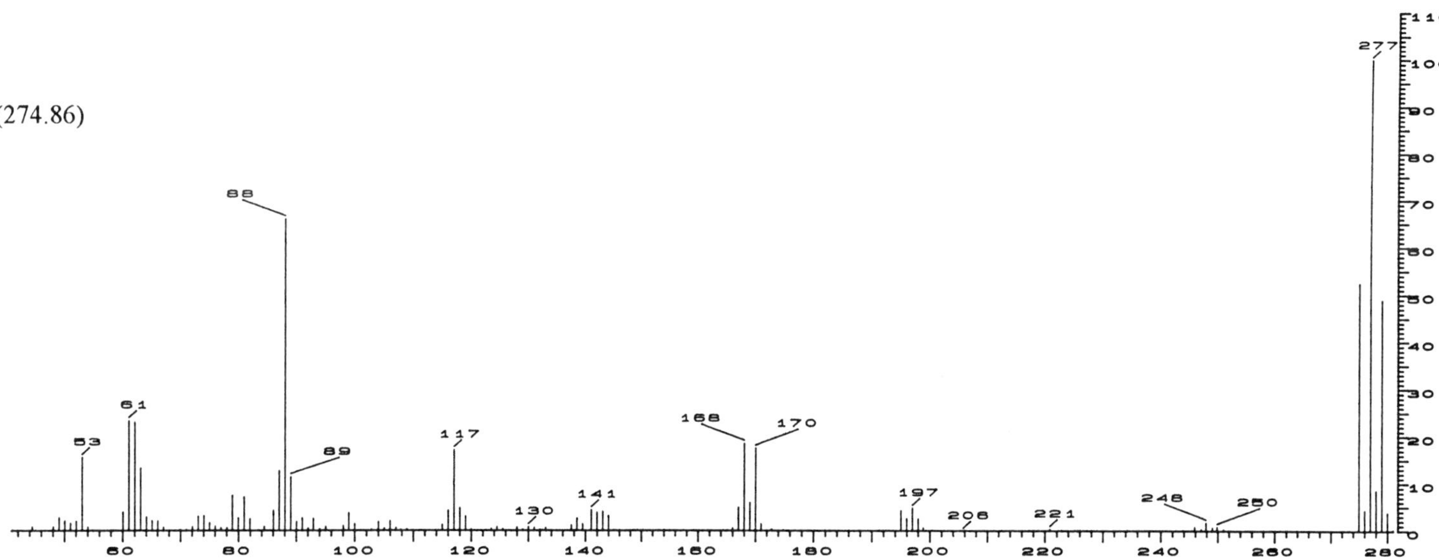

BUTACHLOR

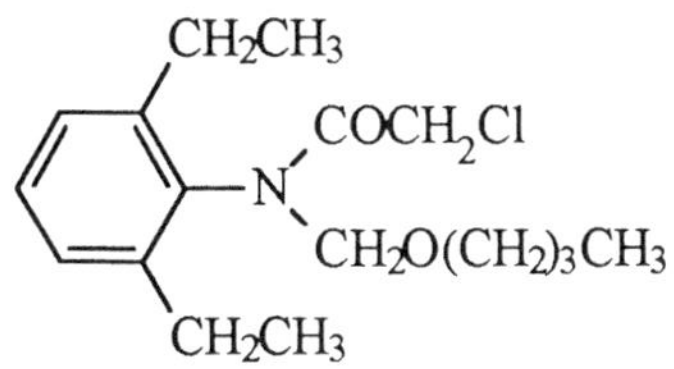

$C_{17}H_{26}NO_2Cl$ M.W. = 311.85 (311.17)

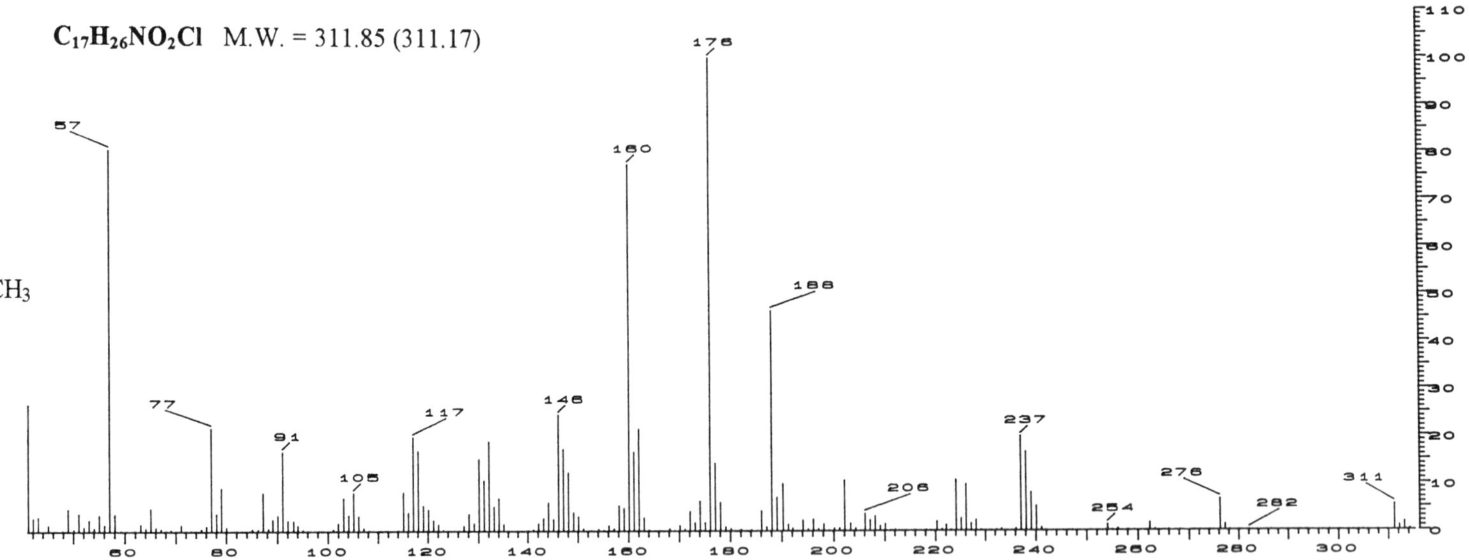

CAPTAFOL

$C_{10}H_9NO_2SCl_4$ M.W. = 349.05 (346.91)

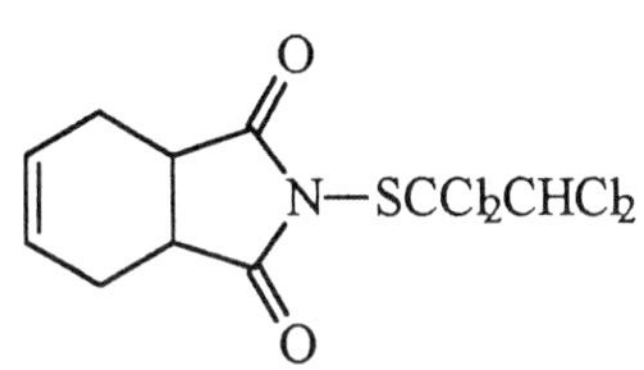

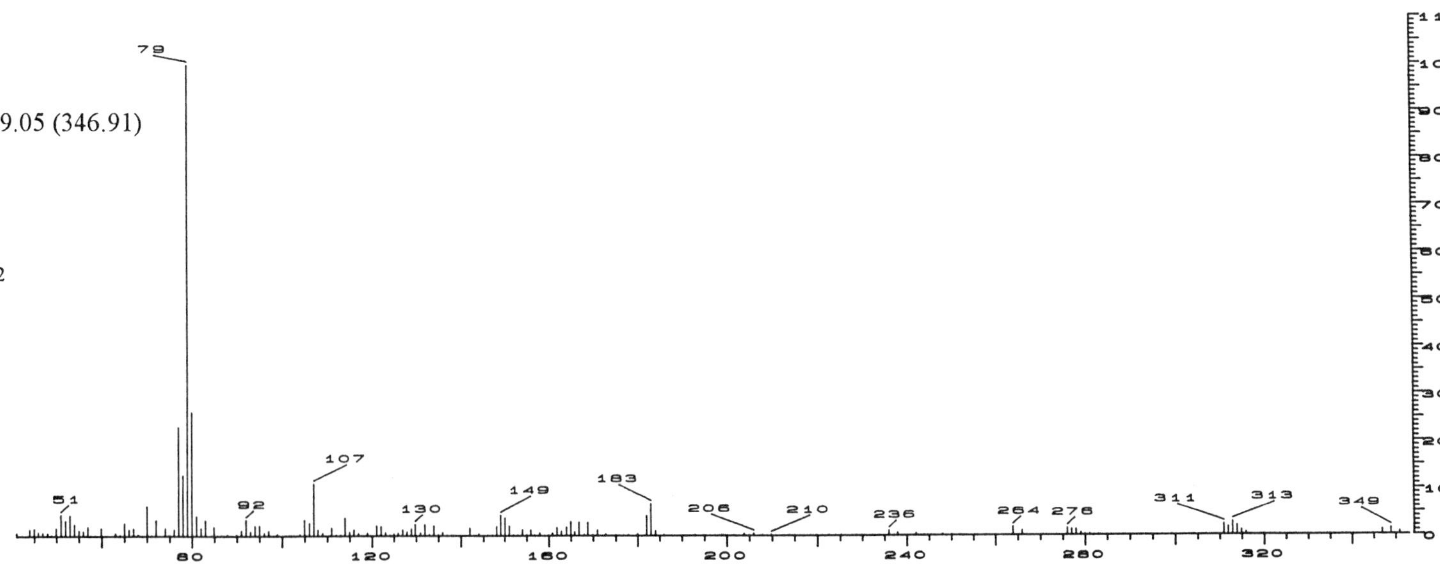

CAPTAN

$C_9H_8NO_2SCl_3$ M.W. = 300.58 (298.93)

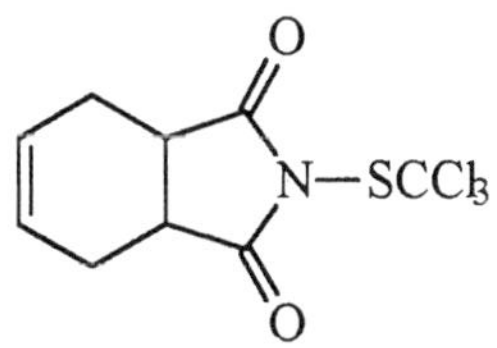

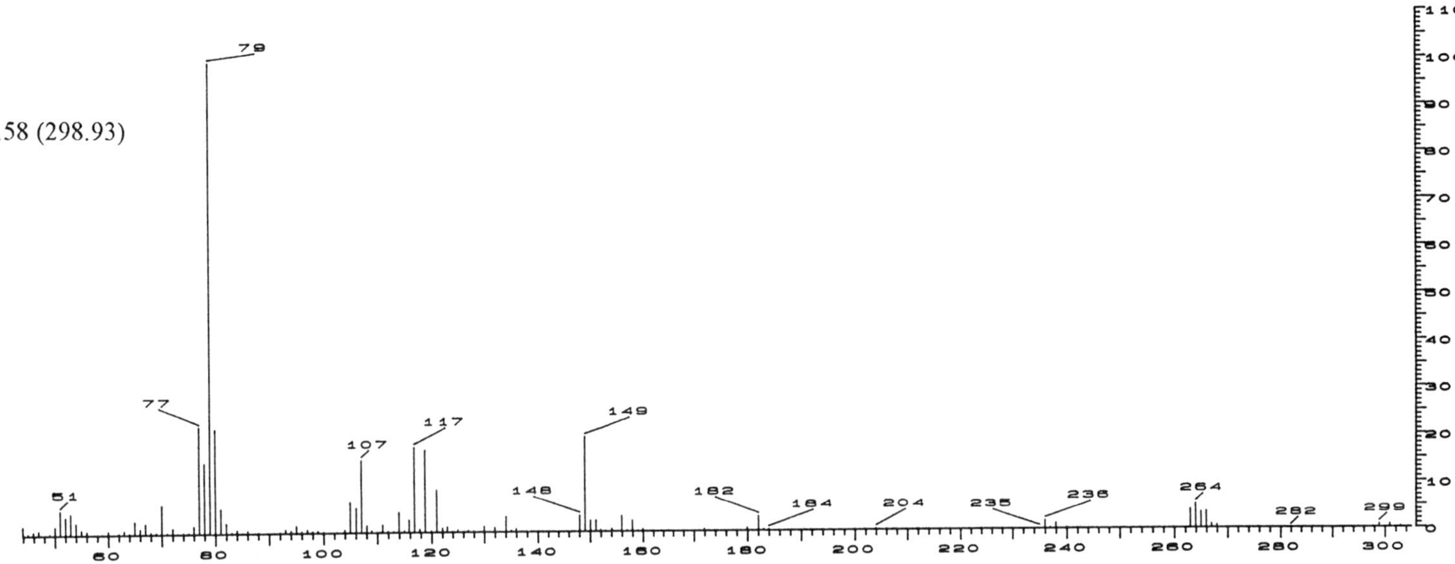

CARBARYL

$C_{12}H_{11}NO_2$ M.W. = 201.22 (201.08)

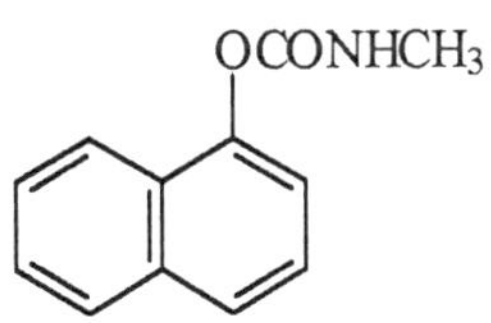

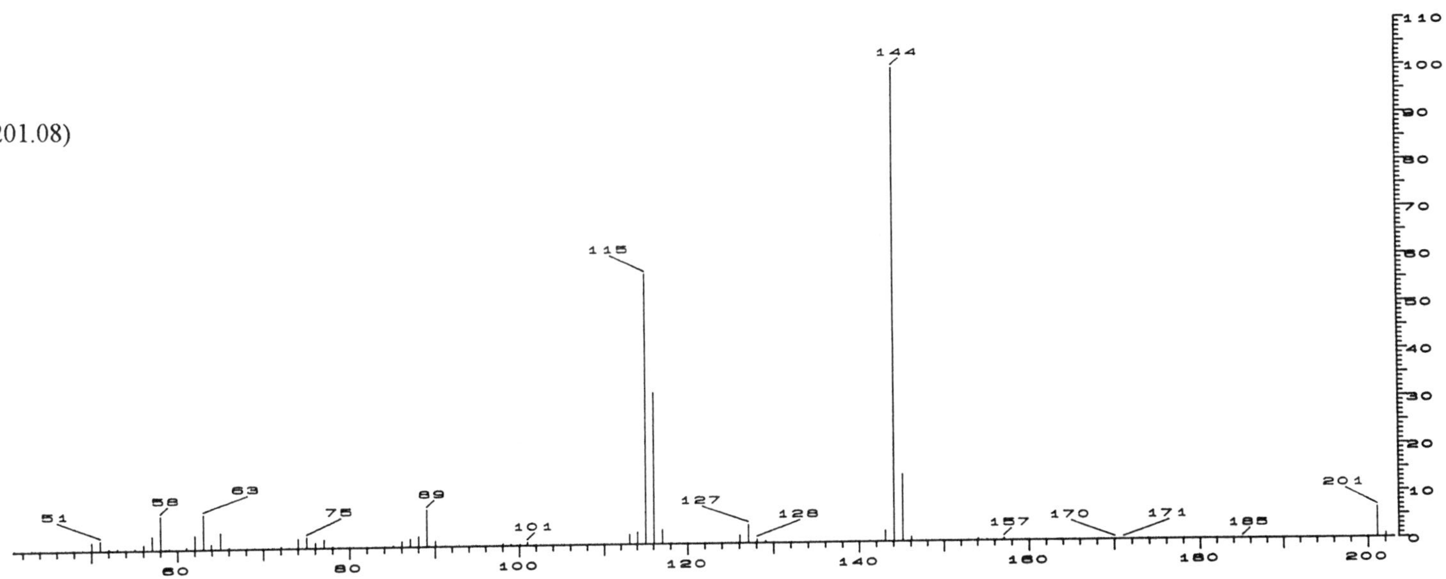

CARBOFURAN

$C_{12}H_{15}NO_3$ M.W. = 221.25 (221.11)

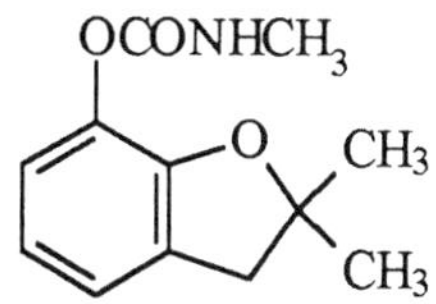

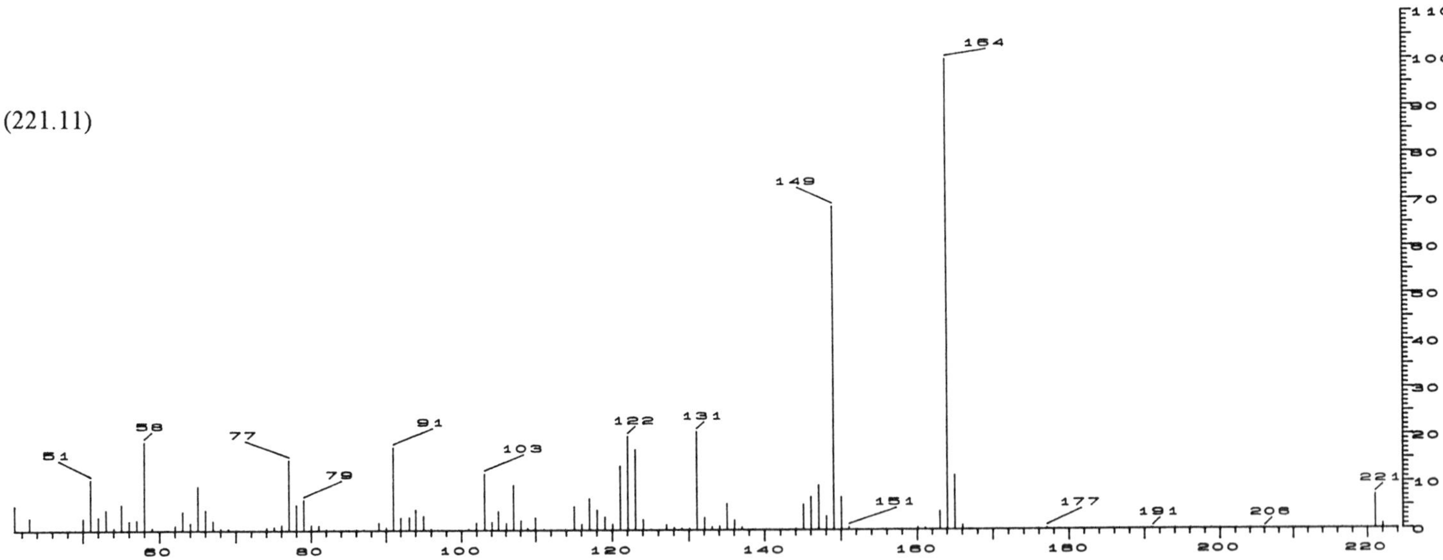

CARBOXIN

$C_{12}H_{13}NO_2S$ M.W. = 235.30 (235.07)

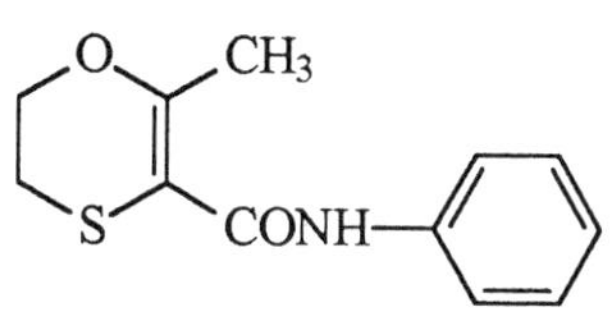

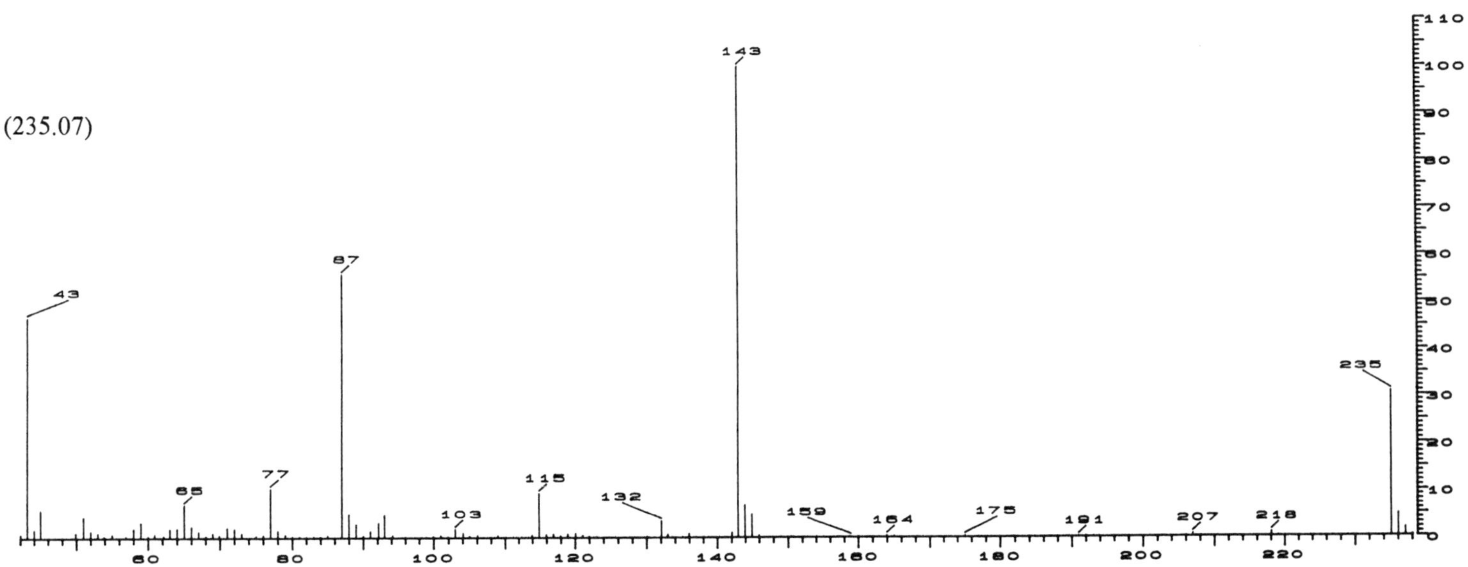

CHLORDANE-ALPHA

$C_{10}H_6Cl_8$ M.W. = 409.80 (405.80)

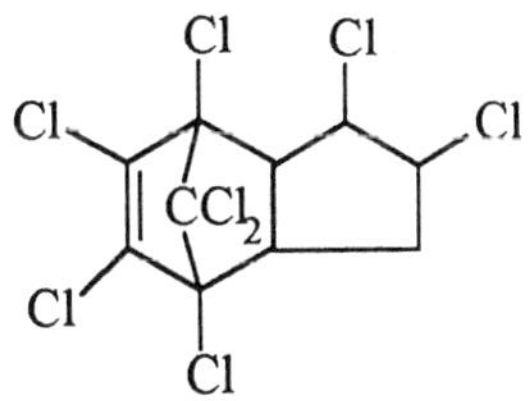

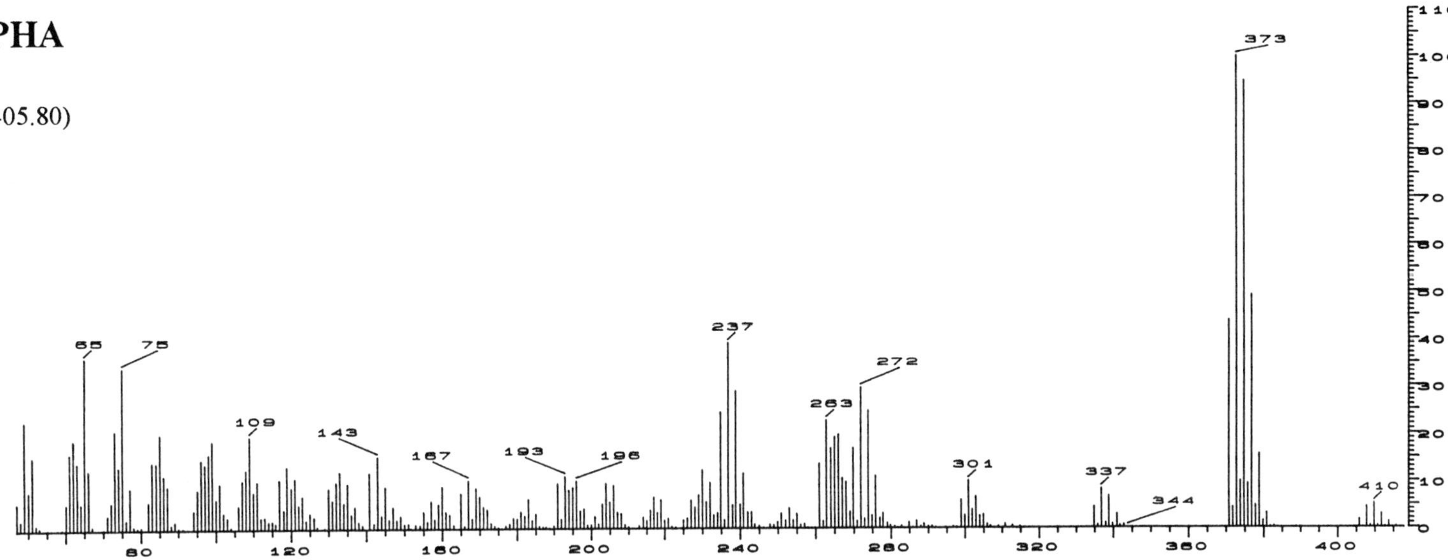

CHLORDANE-GAMMA

$C_{10}H_6Cl_8$ M.W. = 409.80 (405.80)

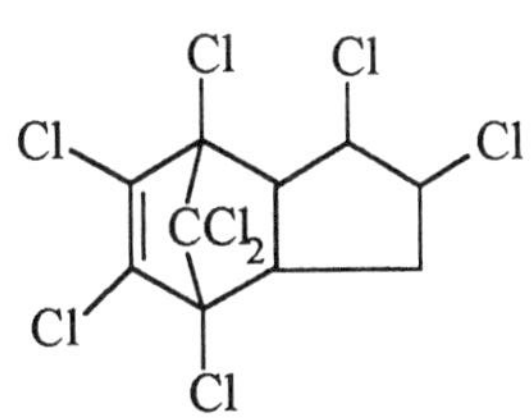

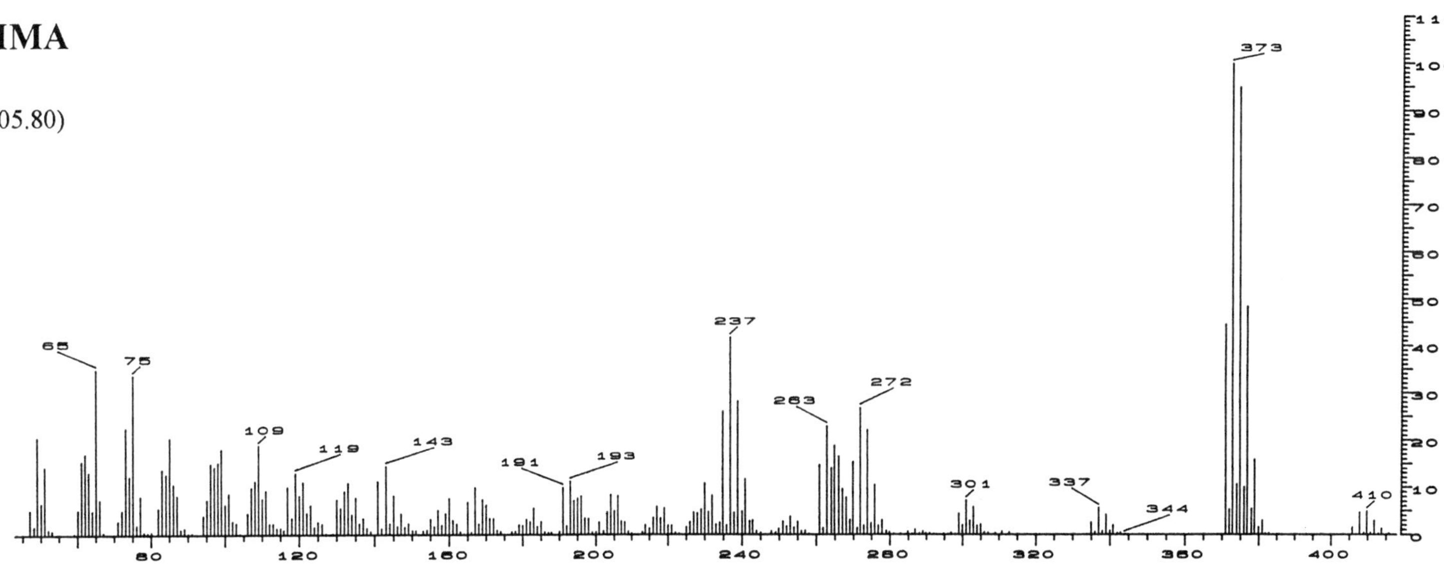

CHLORDENE-ALPHA

$C_{10}H_6Cl_6$ M.W. = 338.90 (335.86)

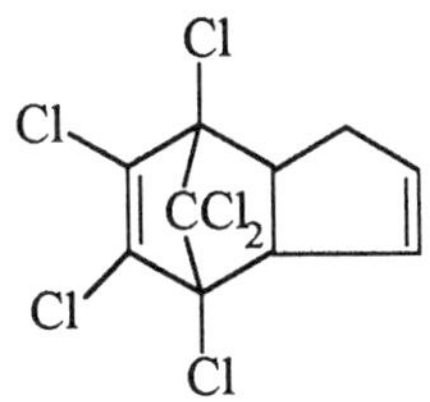

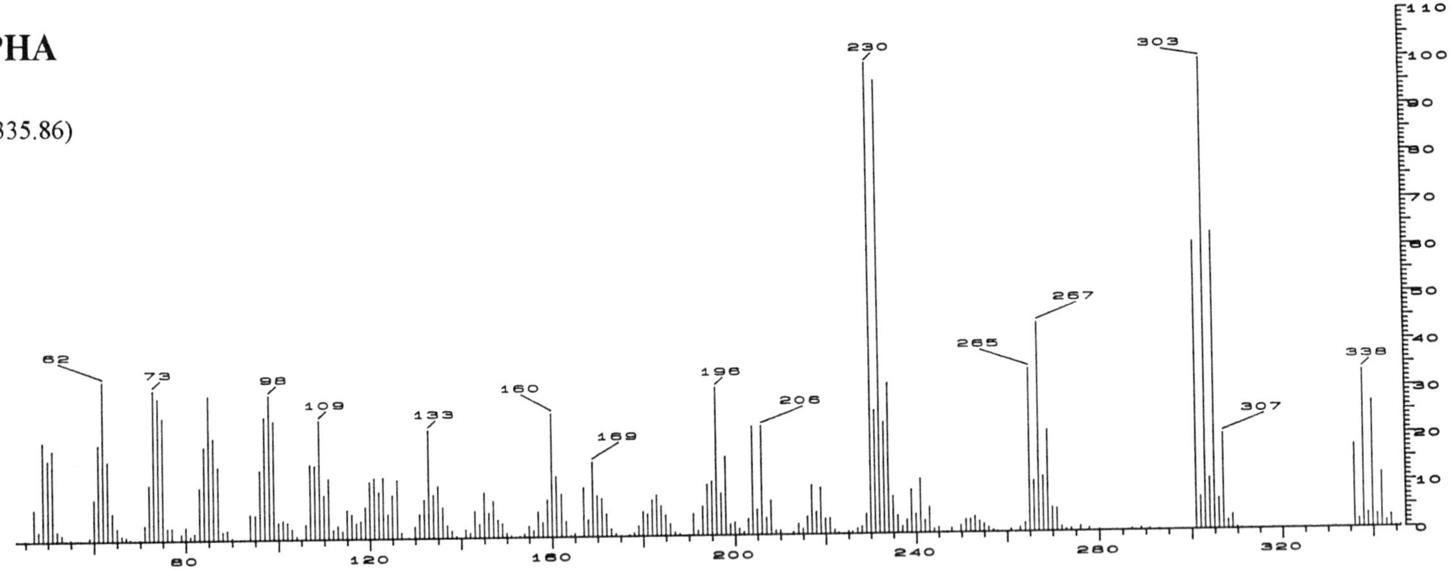

CHLORDENE-BETA

$C_{10}H_6Cl_6$ M.W. = 338.90 (335.86)

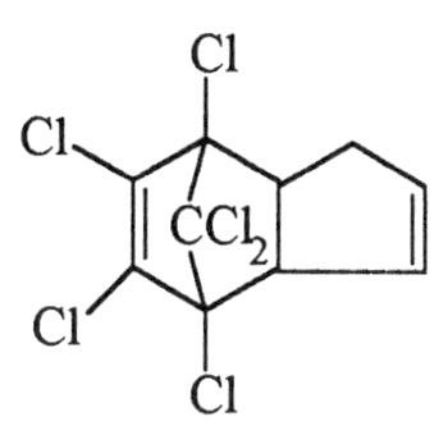

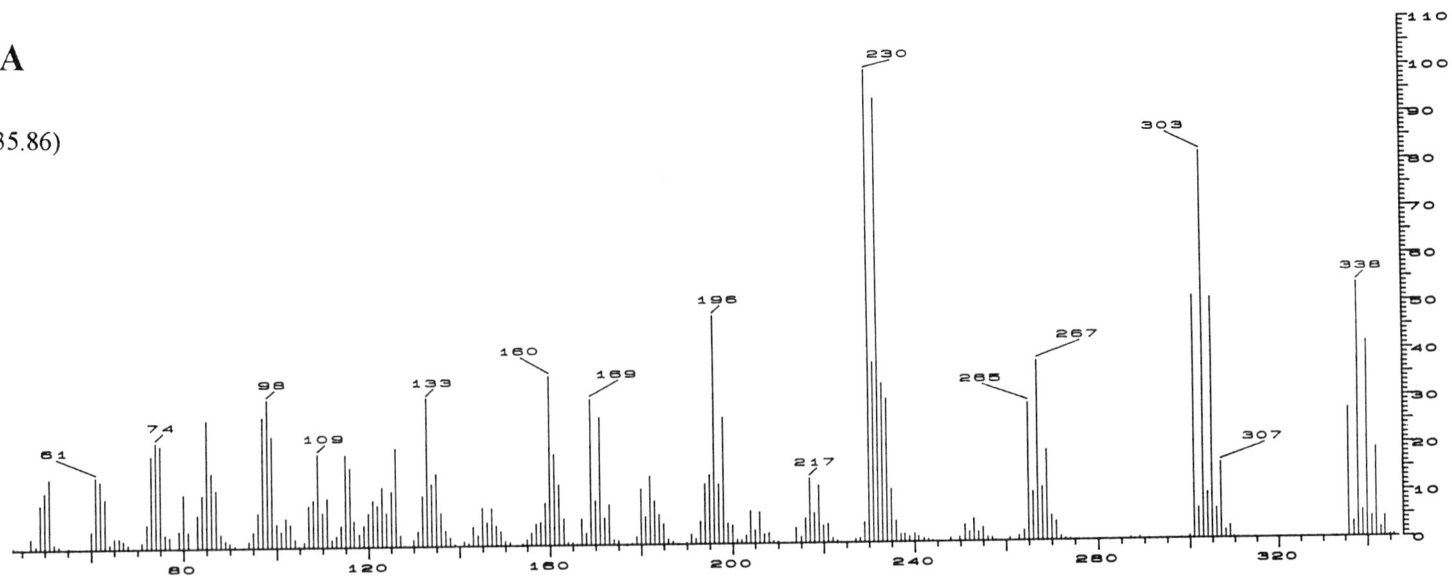

CHLORDENE-GAMMA

$C_{10}H_6Cl_6$ M.W. = 338.90 (335.86)

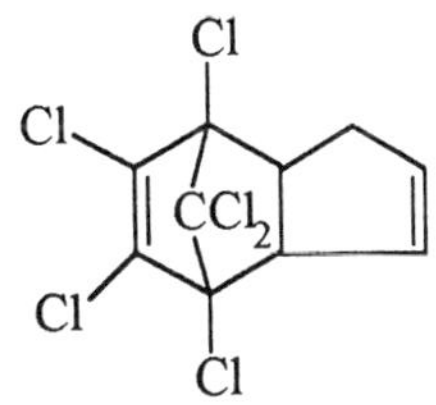

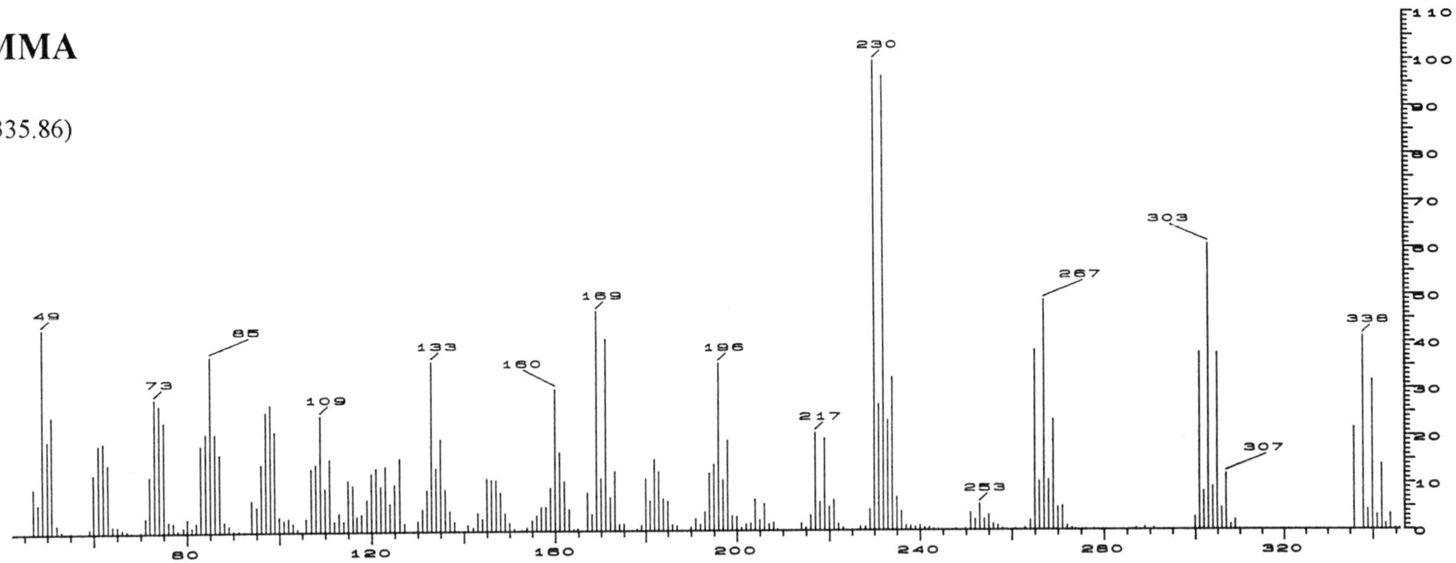

CHLORDIMEFORM

$C_{10}H_{13}N_2Cl$ M.W. = 196.67 (196.08)

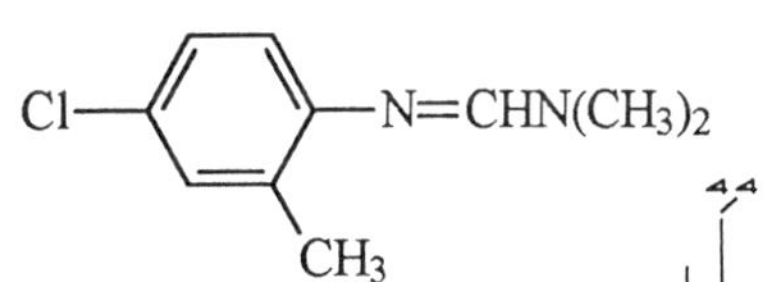

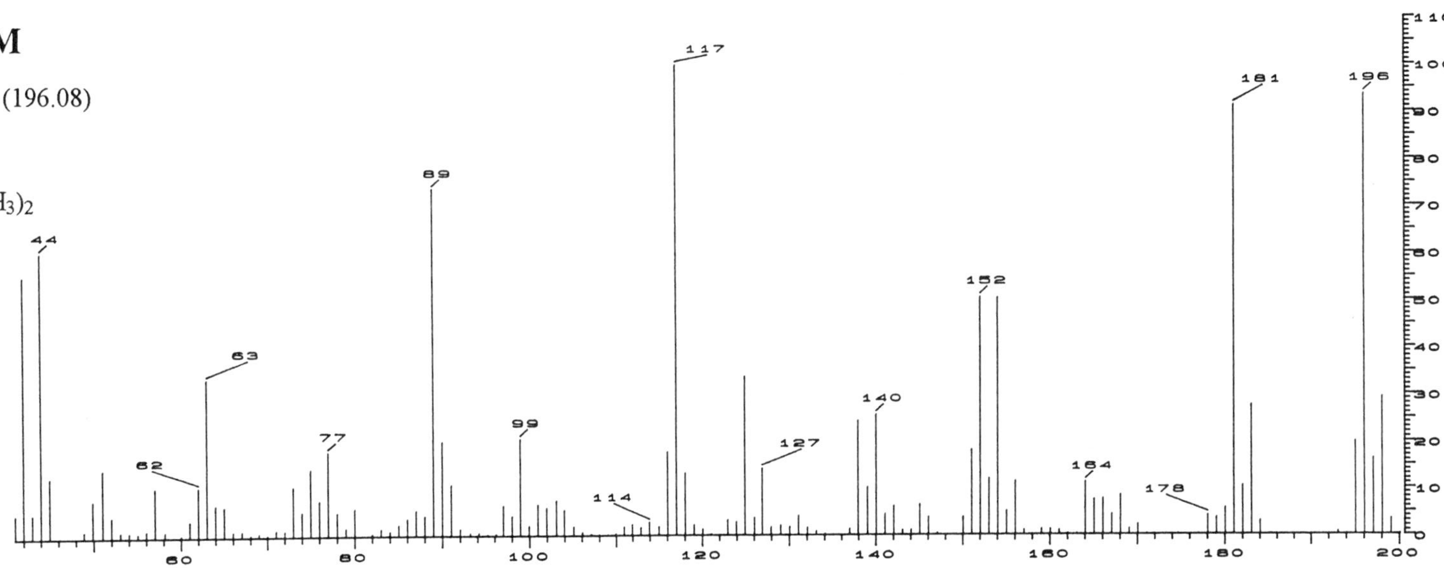

CHLORFENVINPHOS

$C_{12}H_{14}O_4PCl_3$ M.W. = 359.57 (357.97)

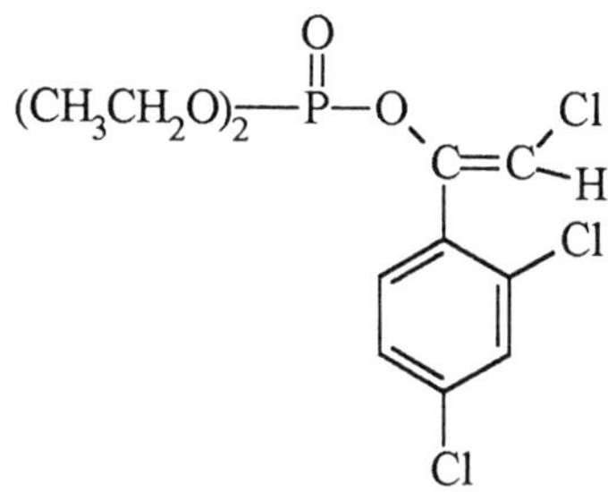

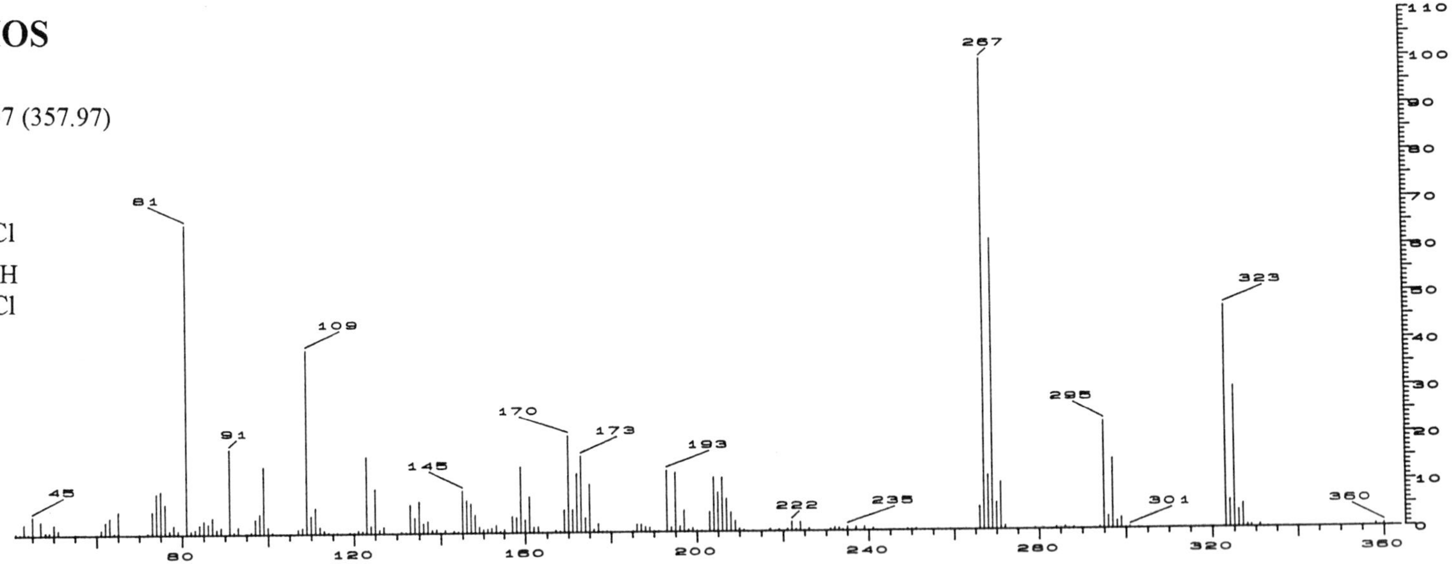

CHLORIDAZON

$C_{10}H_8N_3OCl$ M.W. = 221.64 (221.04)

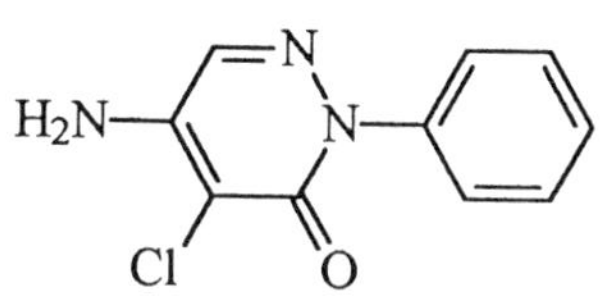

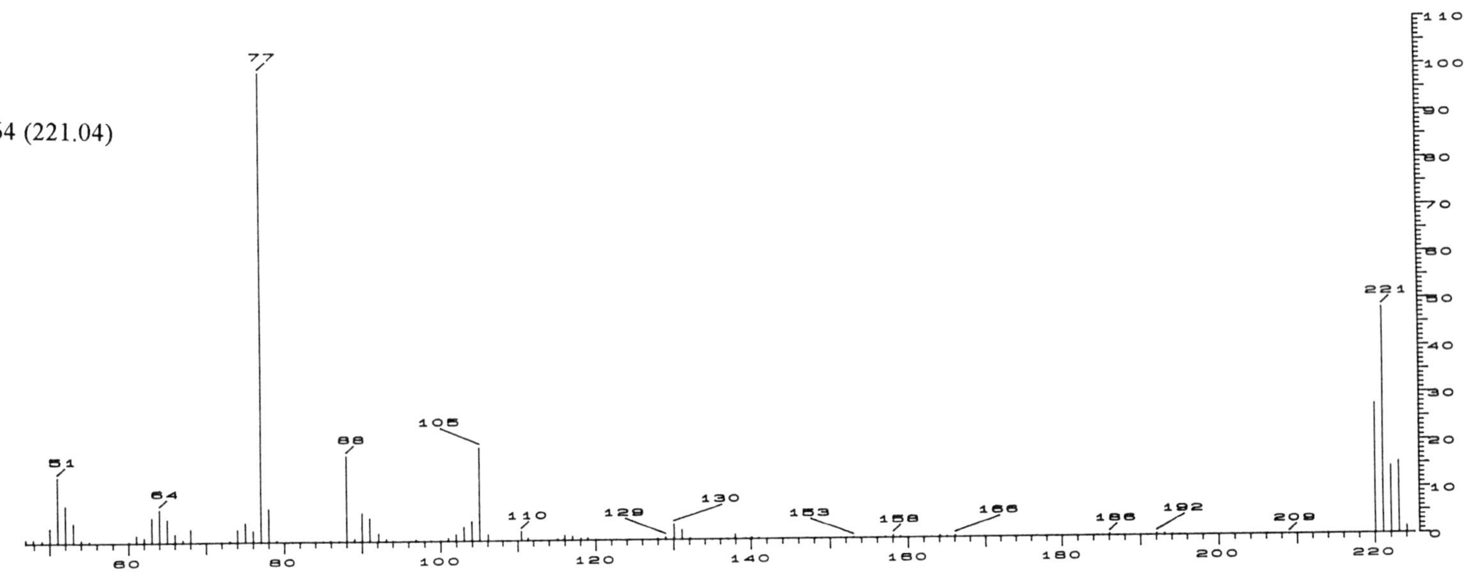

CHLOROBENZILATE

$C_{16}H_{14}O_3Cl_2$ M.W. = 325.20 (324.03)

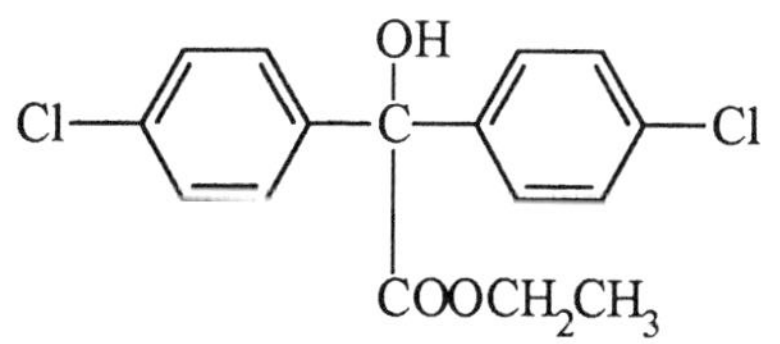

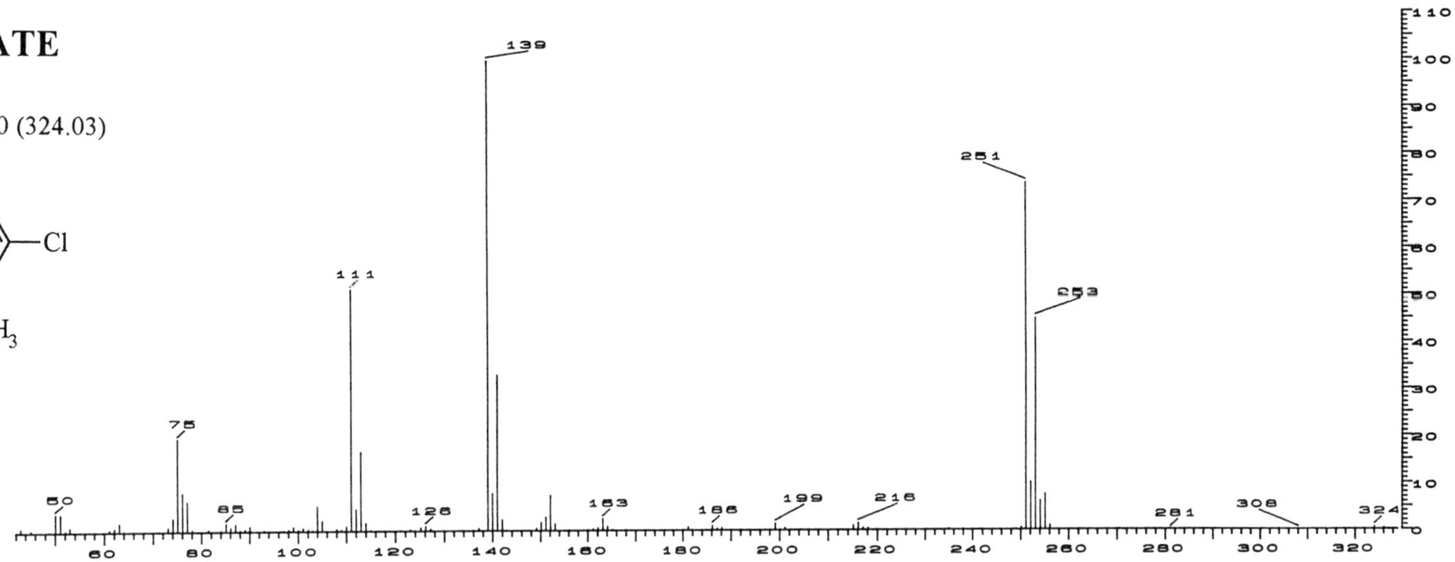

CHLORONEB

$C_8H_8O_2Cl_2$ M.W. = 207.05 (205.99)

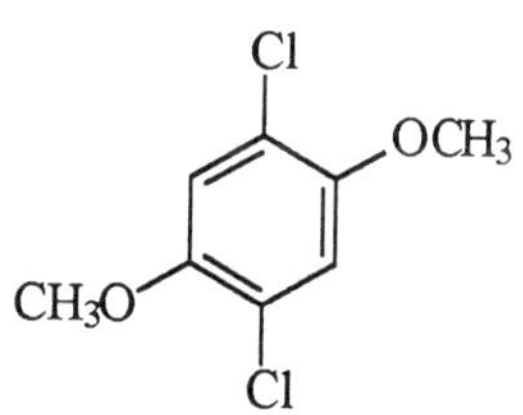

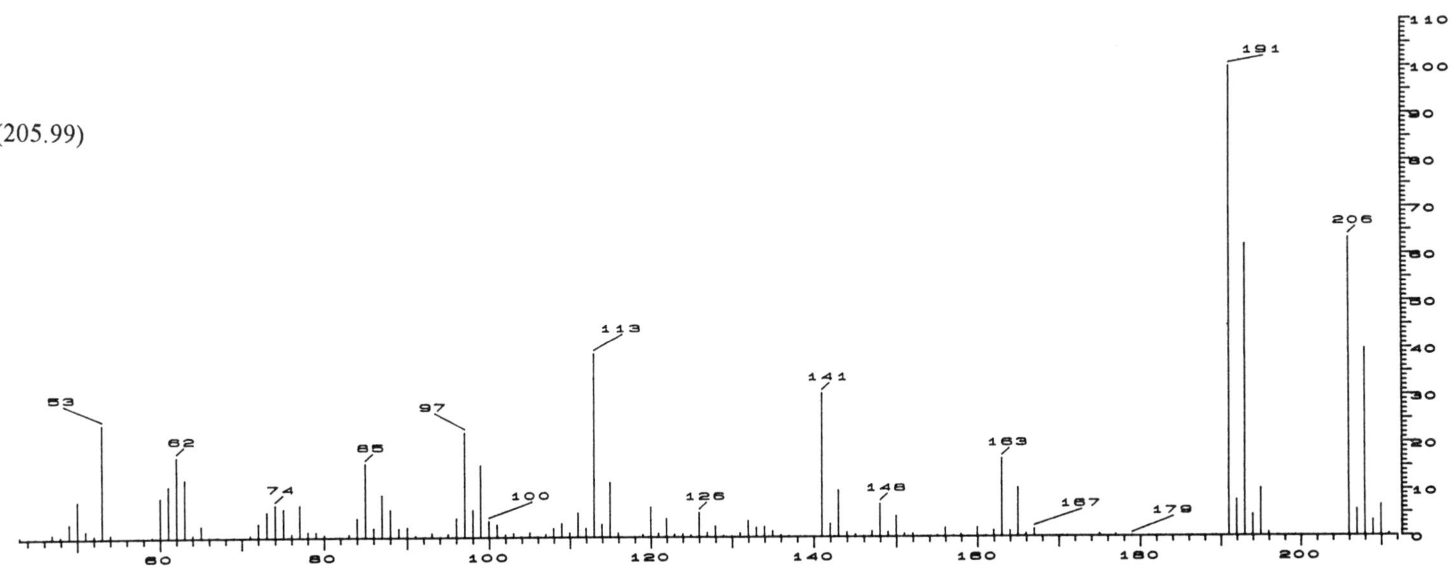

CHLOROTHALONIL

$C_8N_2Cl_4$ M.W. = 265.91 (263.88)

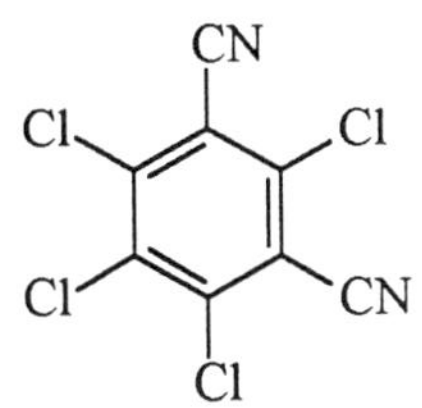

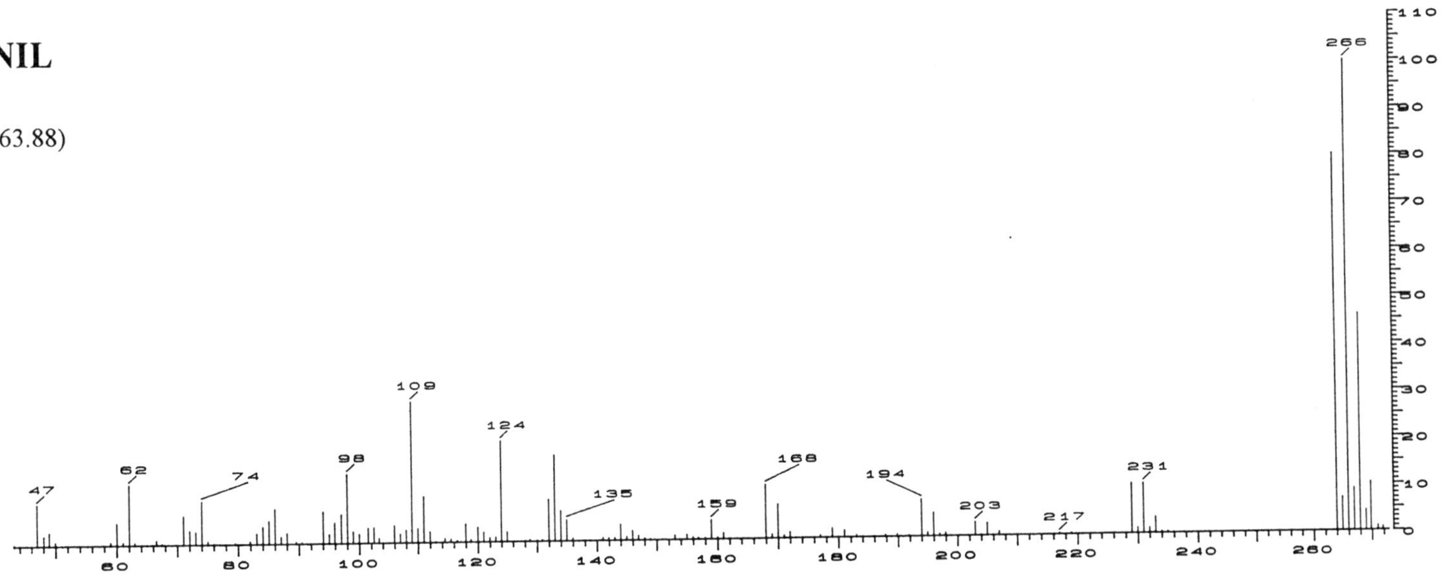

CHLORPYRIFOS

$C_9H_{11}NO_3SCl_3P$ M.W. = 350.58 (348.93)

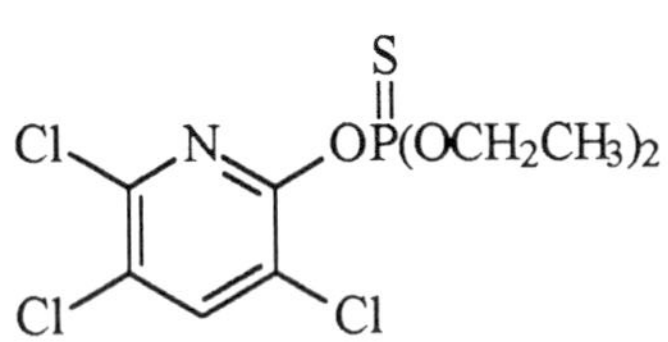

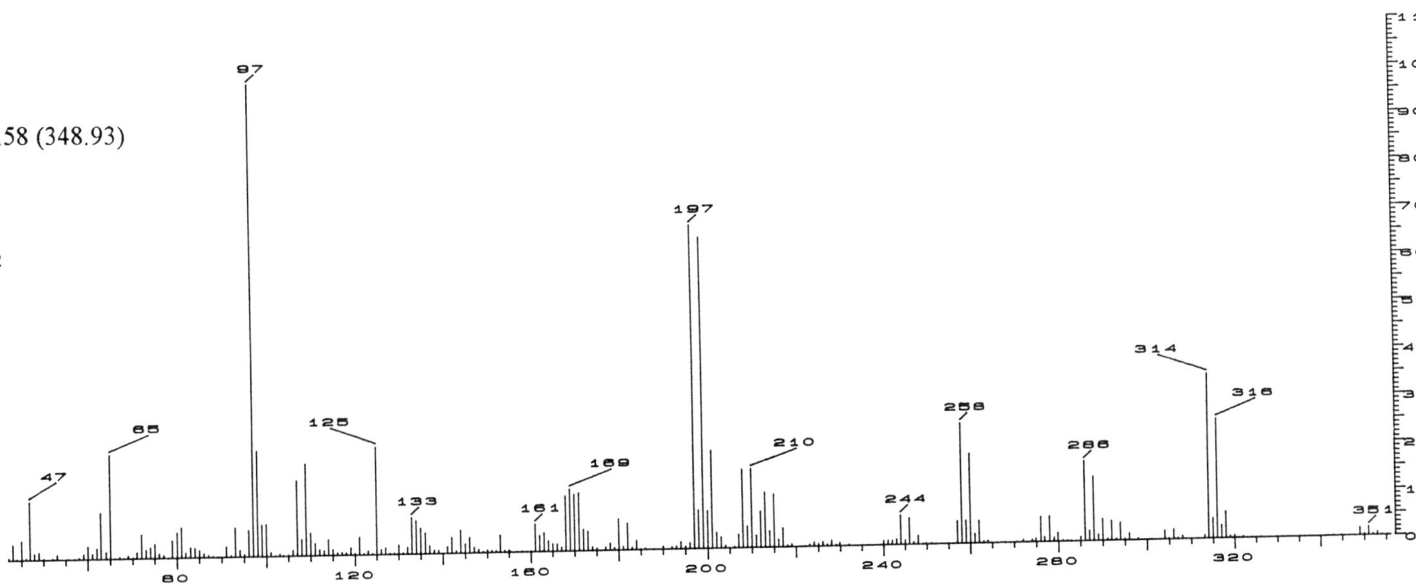

CHLORPYRIFOS-METHYL

$C_7H_7NO_3SCl_3P$ M.W. = 322.53 (320.89)

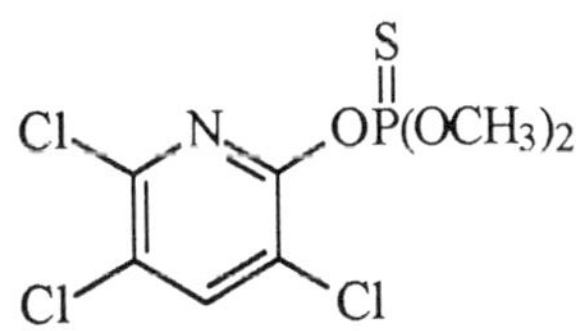

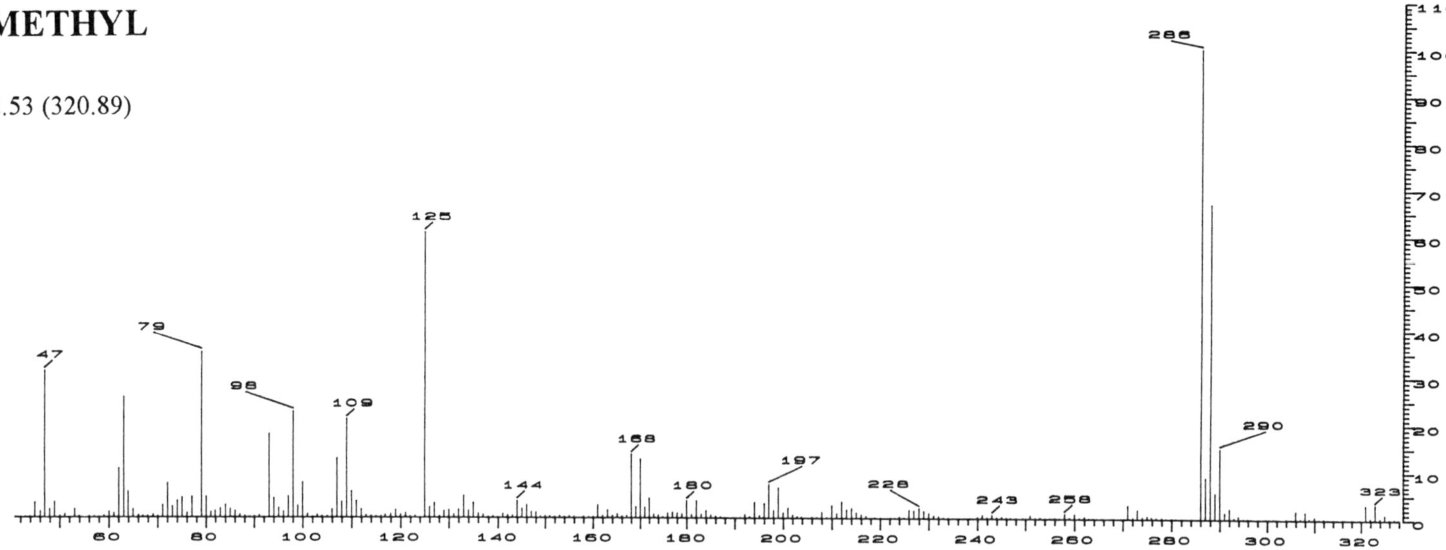

CHLORPYRIFOS-OA

$C_9H_{11}NO_4Cl_3P$ M.W. = 334.52 (332.95)

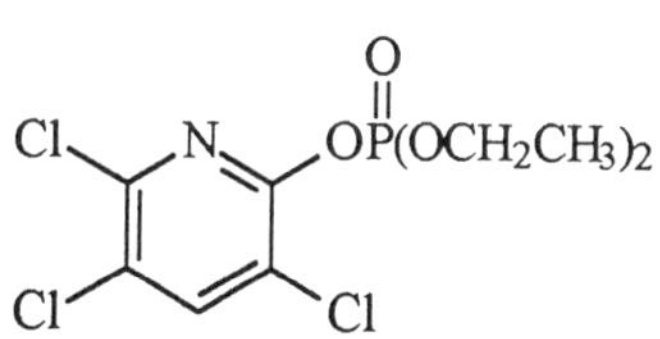

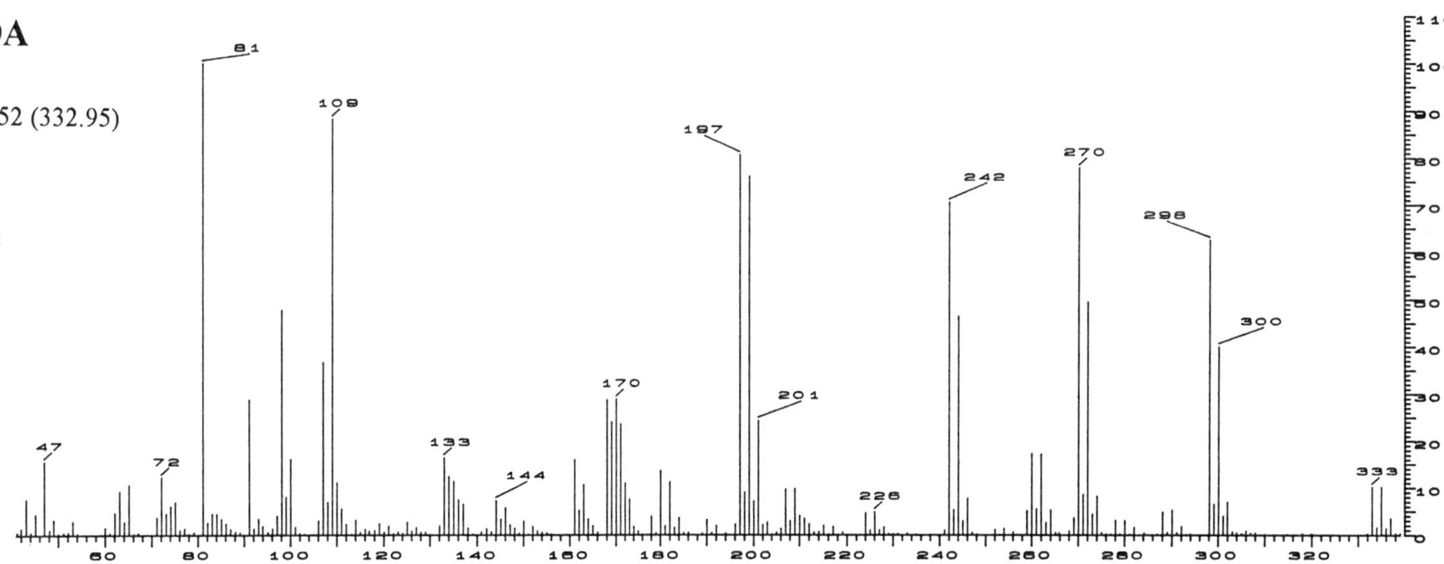

COUMAPHOS

$C_{14}H_{16}O_5ClPS$ M.W. = 362.76 (362.01)

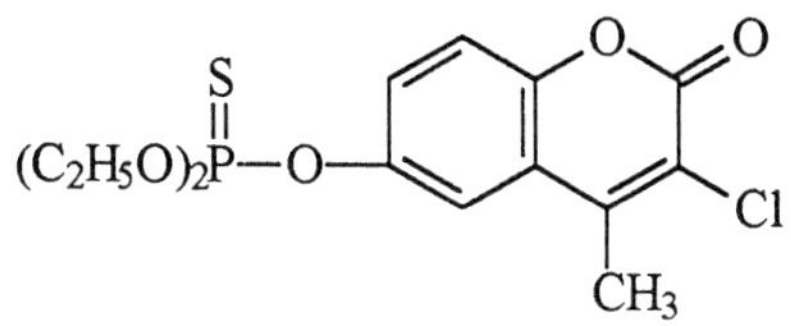

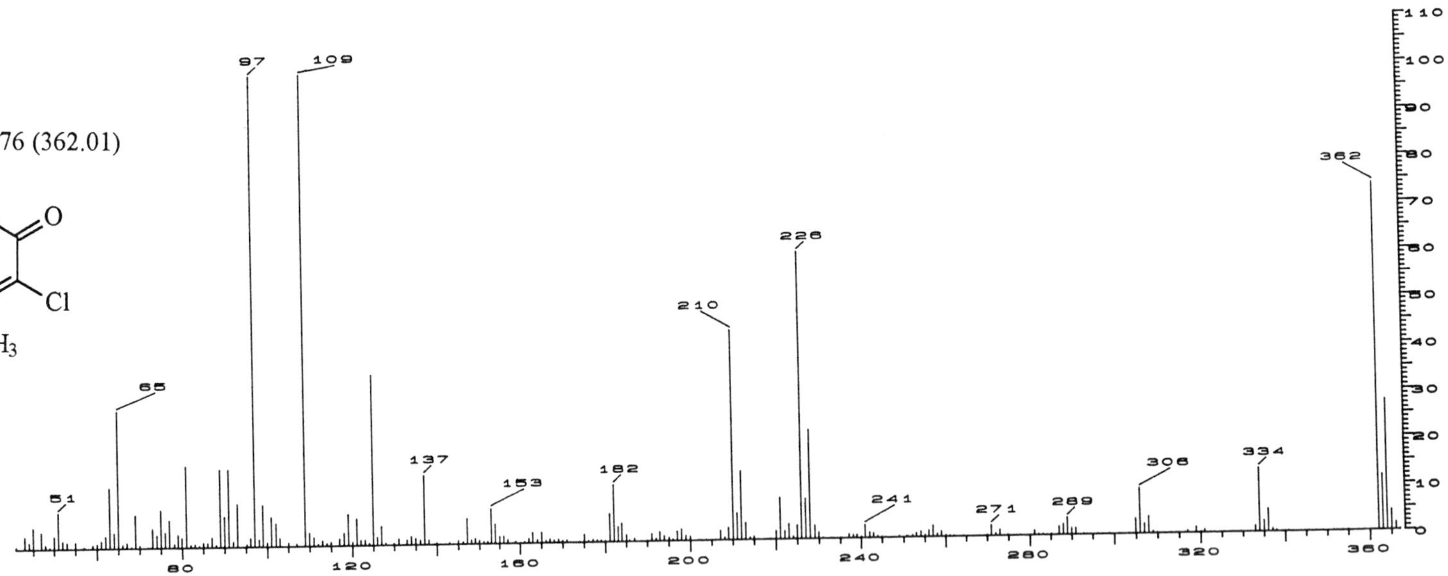

COUMAPHOS-OA

$C_{14}H_{16}O_6ClP$ M.W. = 346.70 (346.04)

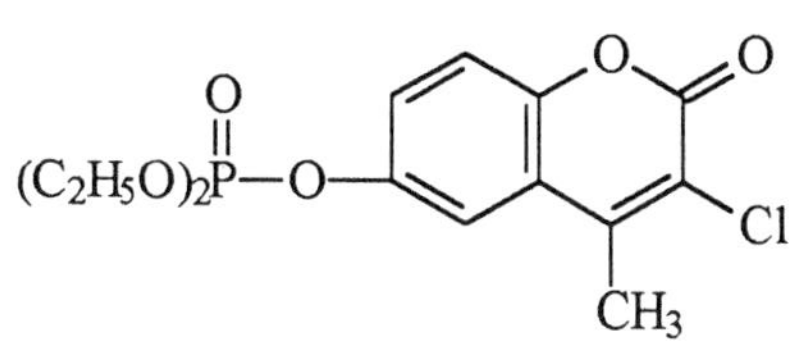

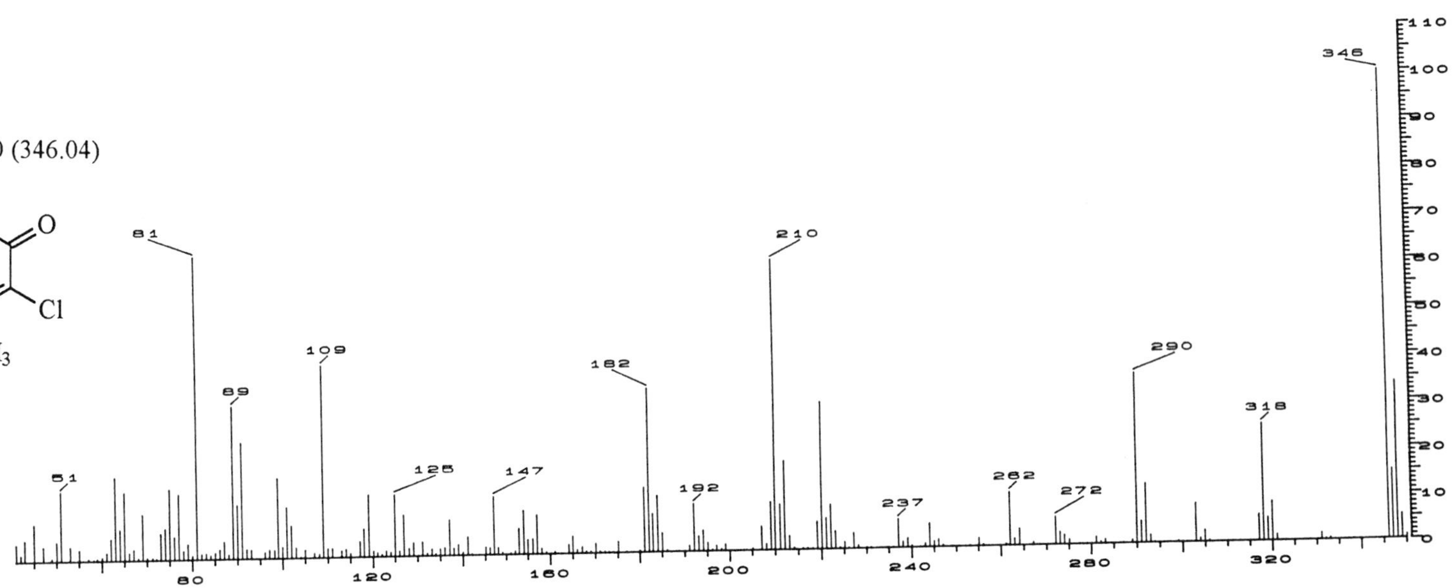

CPMC

$C_8H_8NO_2Cl$ M.W. = 185.61 (185.02)

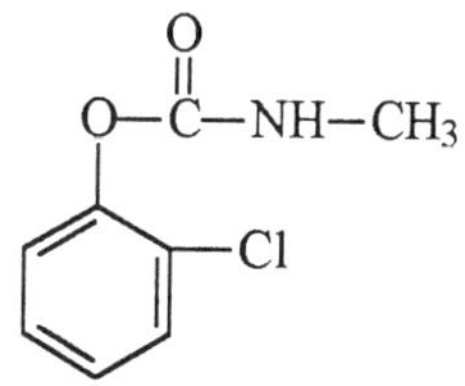

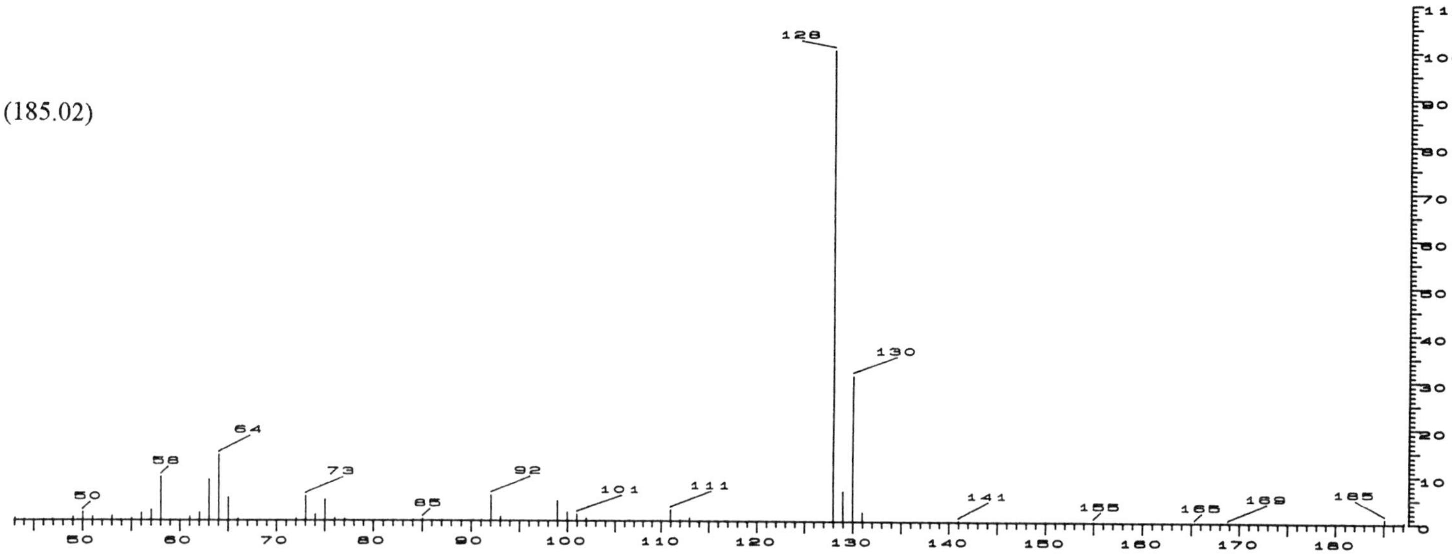

CROTOXYPHOS

$C_{14}H_{19}O_6P$ M.W. = 314.27 (314.09)

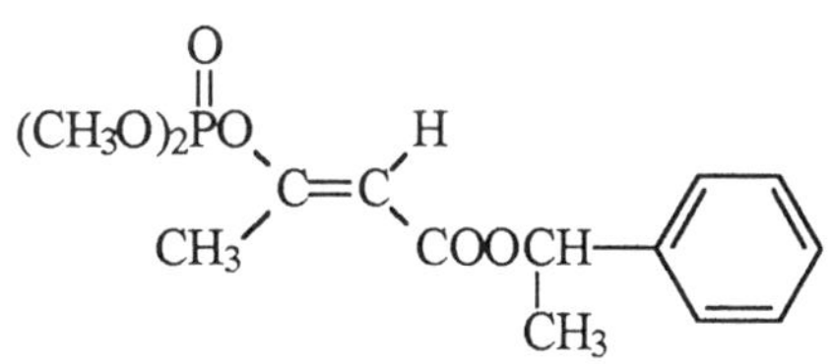

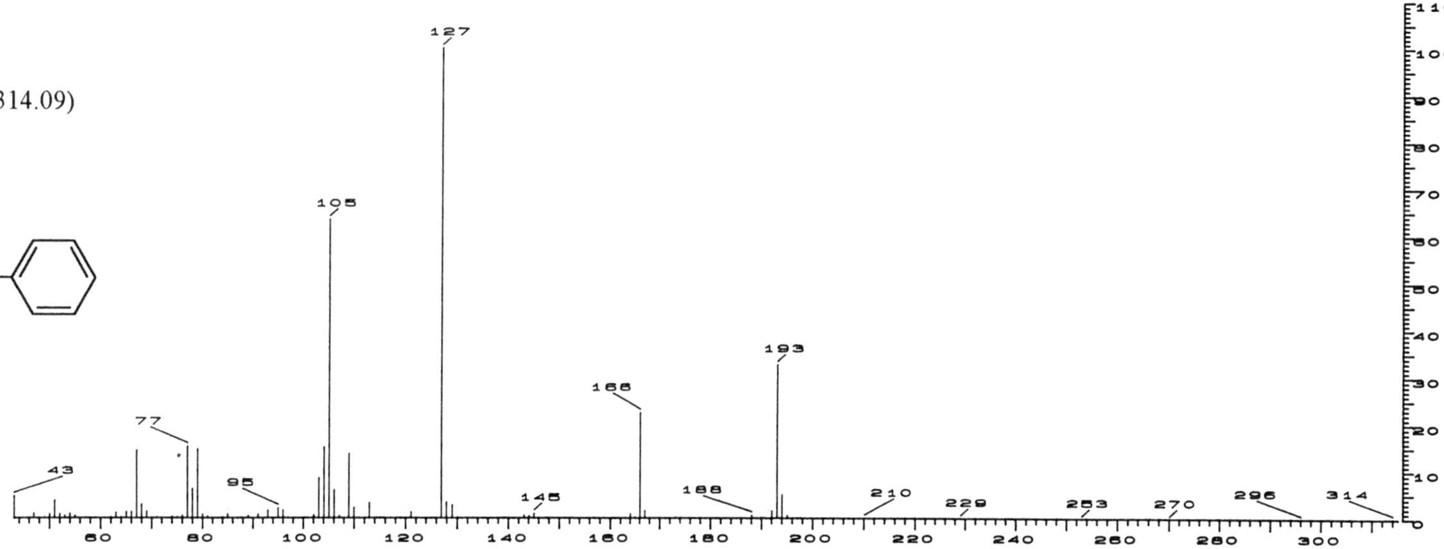

CRUFOMATE

$C_{12}H_{19}NO_3PCl$ M.W. = 291.71 (291.08)

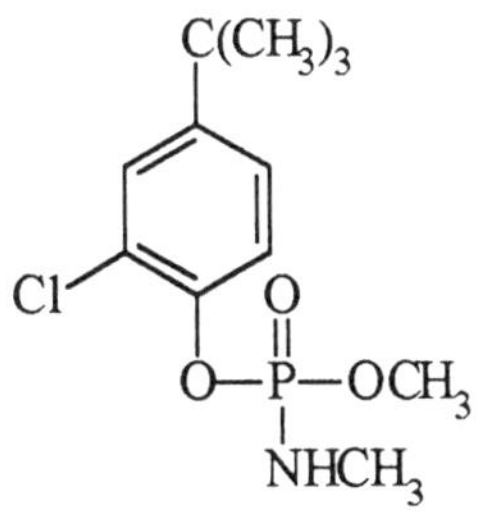

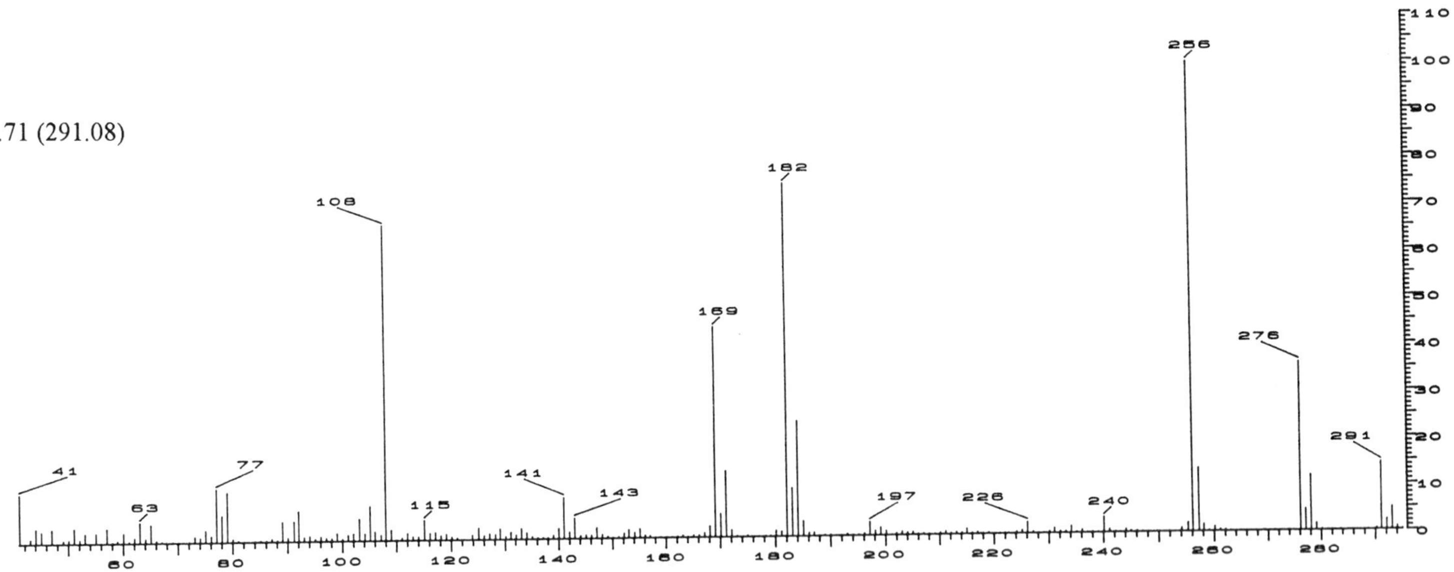

CYANAZINE

$C_9H_{13}N_6Cl$ M.W. = 240.68 (240.09)

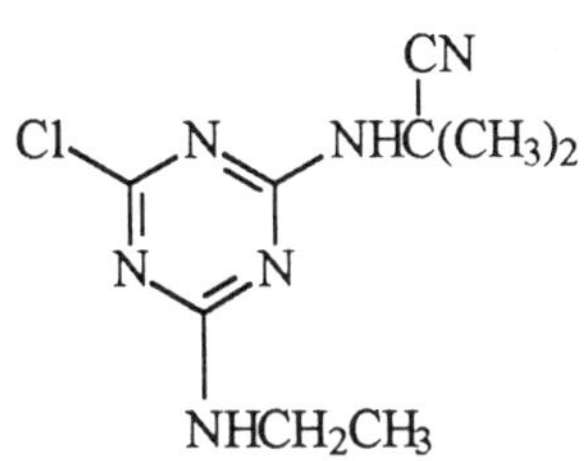

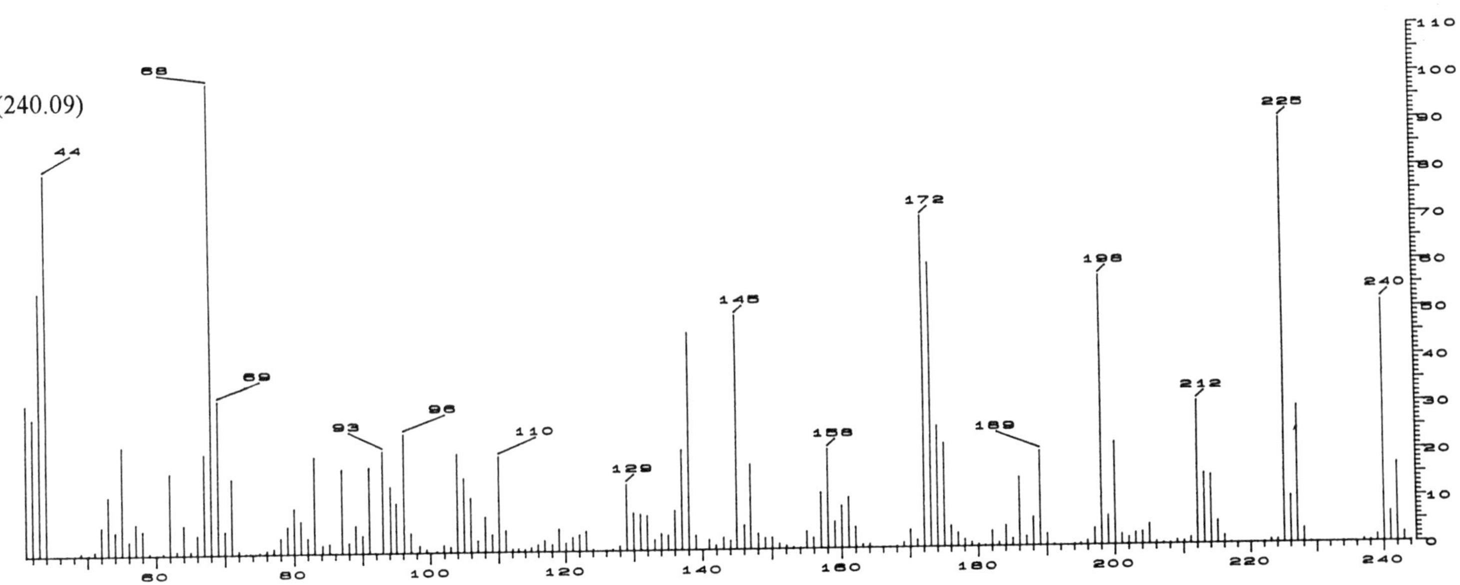

CYCLOATE

$C_{11}H_{21}NOS$ M.W. = 215.35 (215.13)

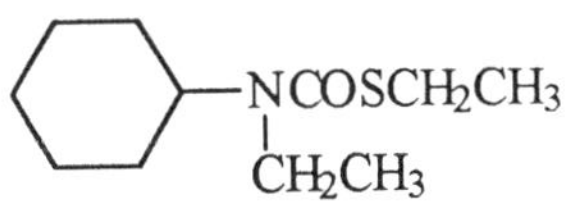

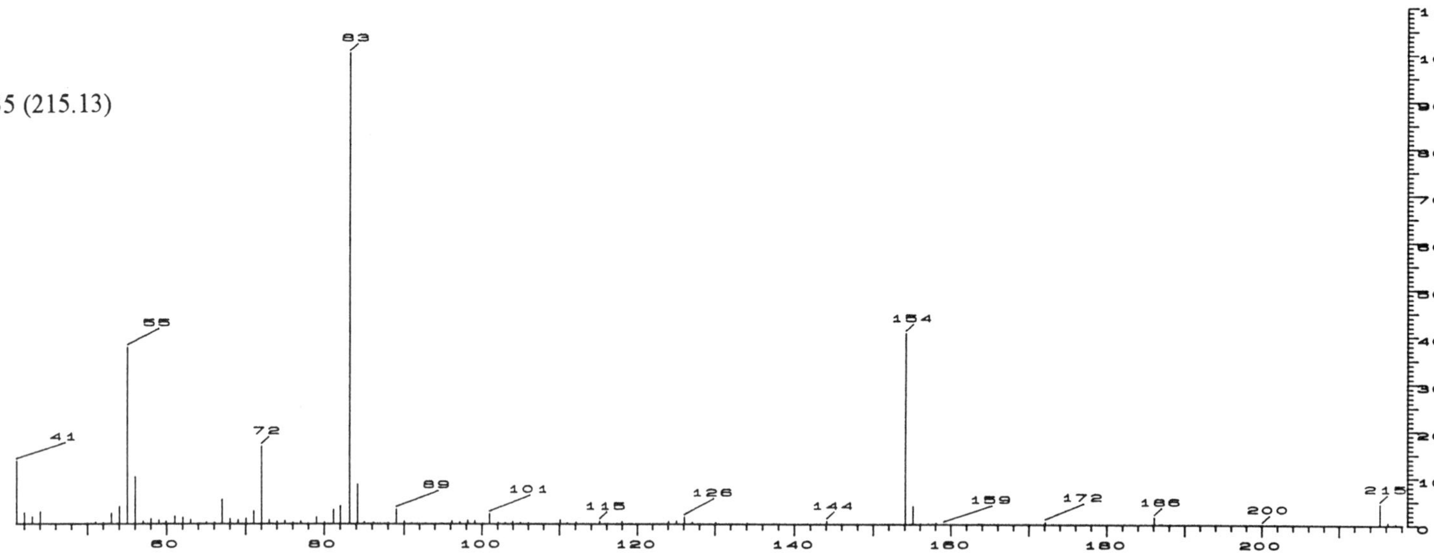

CYPERMETHRIN

$C_{22}H_{19}NO_3Cl_2$ M.W. = 416.30 (415.07)

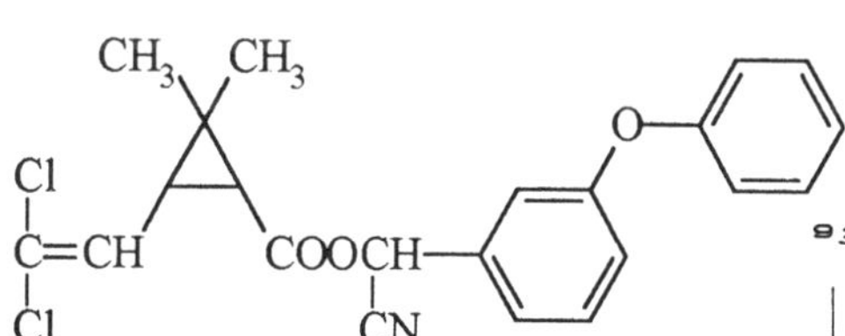

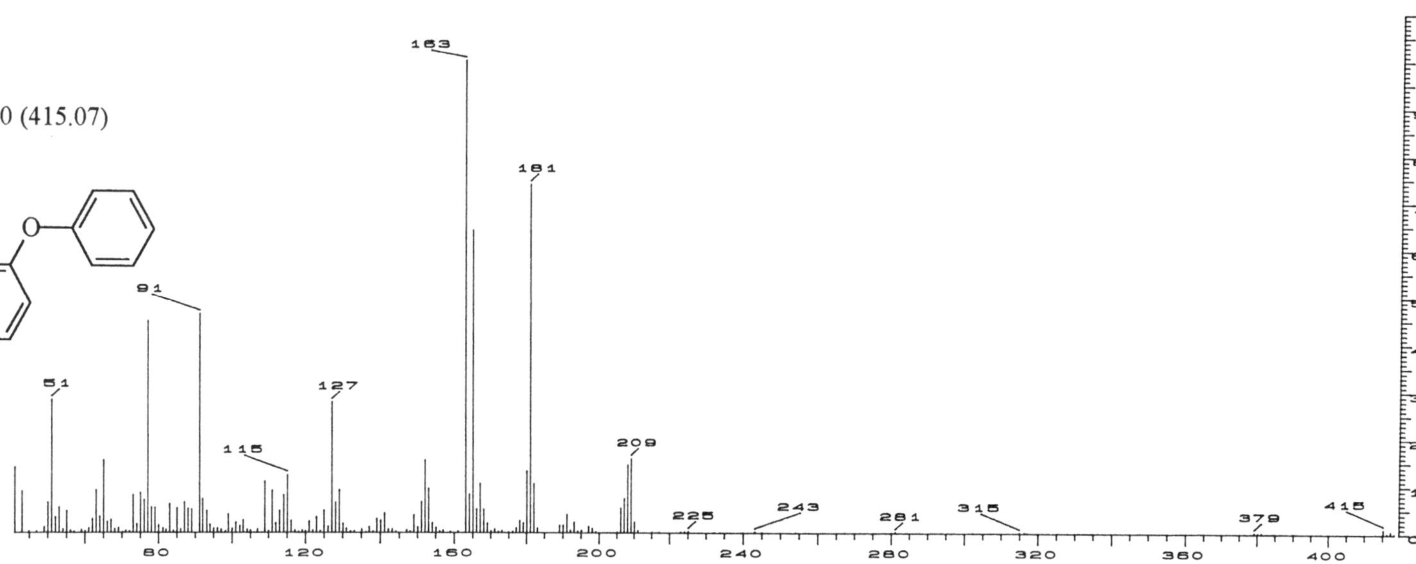

CYPRAZINE

$C_9H_{14}N_5Cl$ M.W. = 227.69 (227.09)

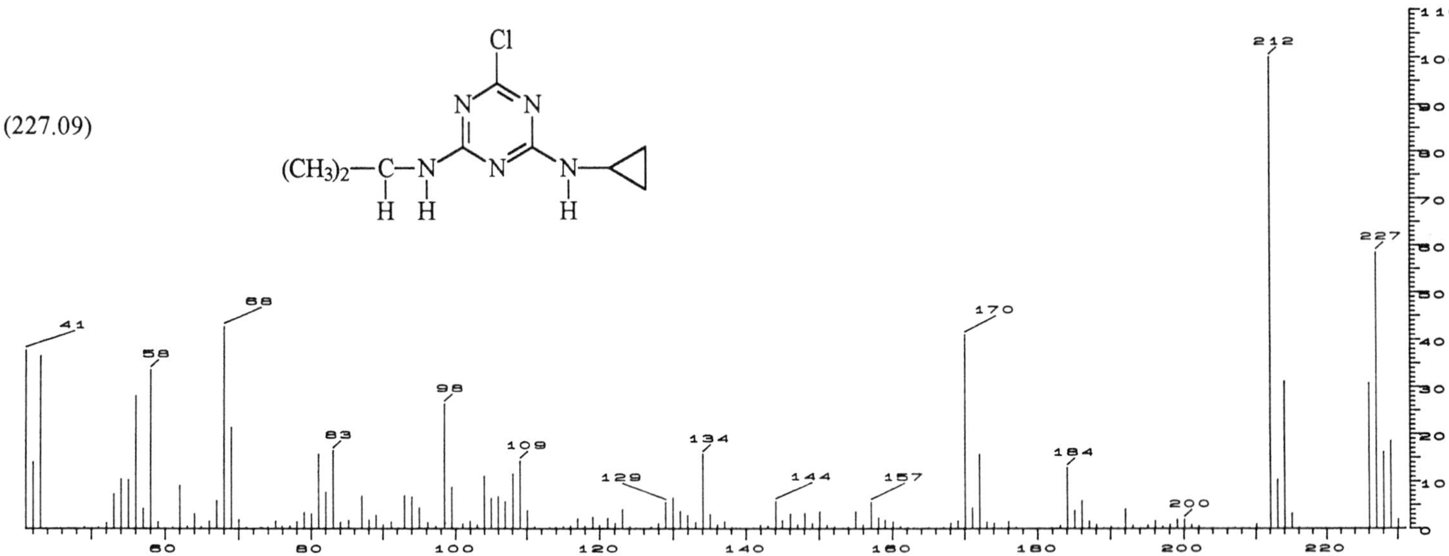

CYTHIOATE

$C_8H_{12}NO_5S_2P$ M.W. = 297.28 (296.99)

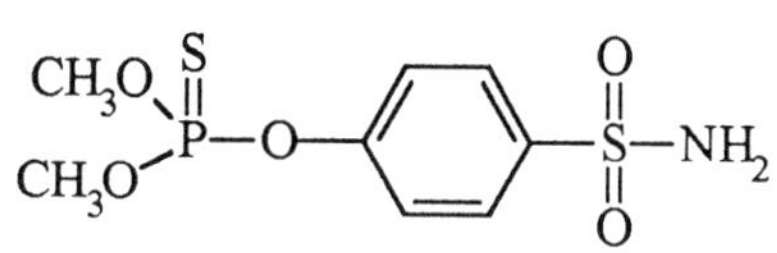

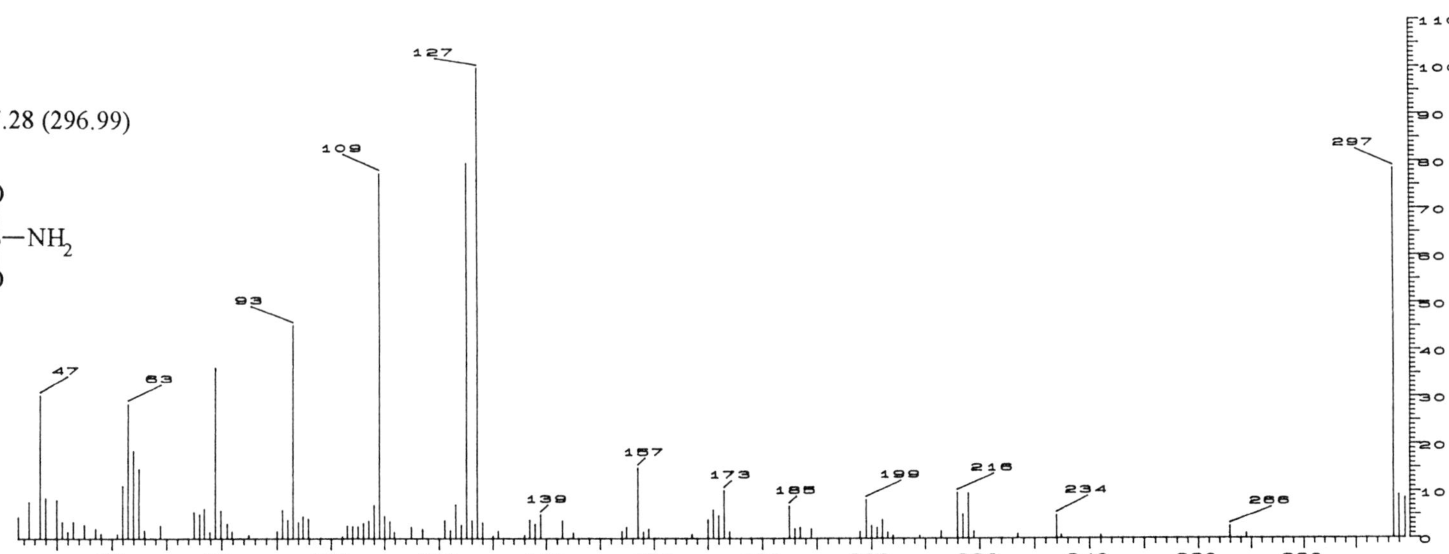

2,4-D

$C_8H_6O_3Cl_2$ M.W. = 221.03 (219.97)

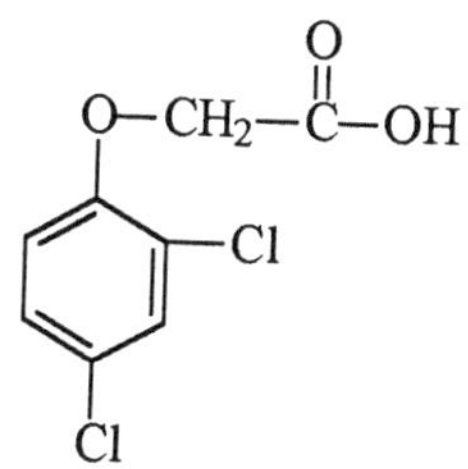

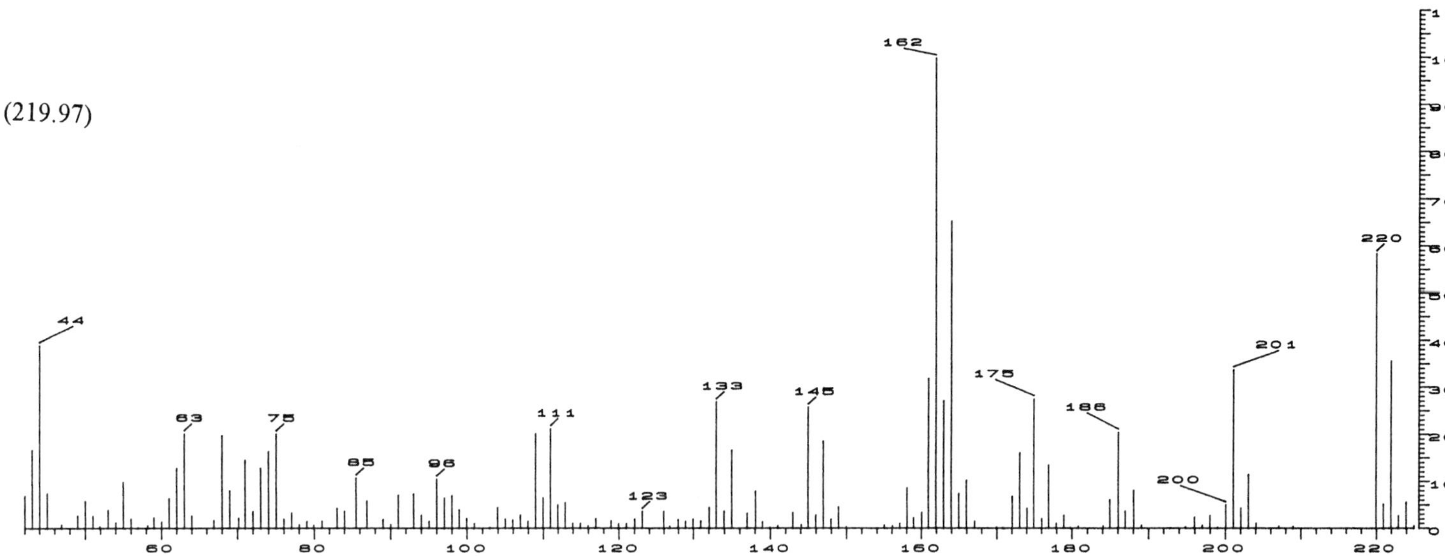

2,4-D ISOPROPYL ESTER

$C_{11}H_{12}O_3Cl_2$ M.W. = 263.12 (262.01)

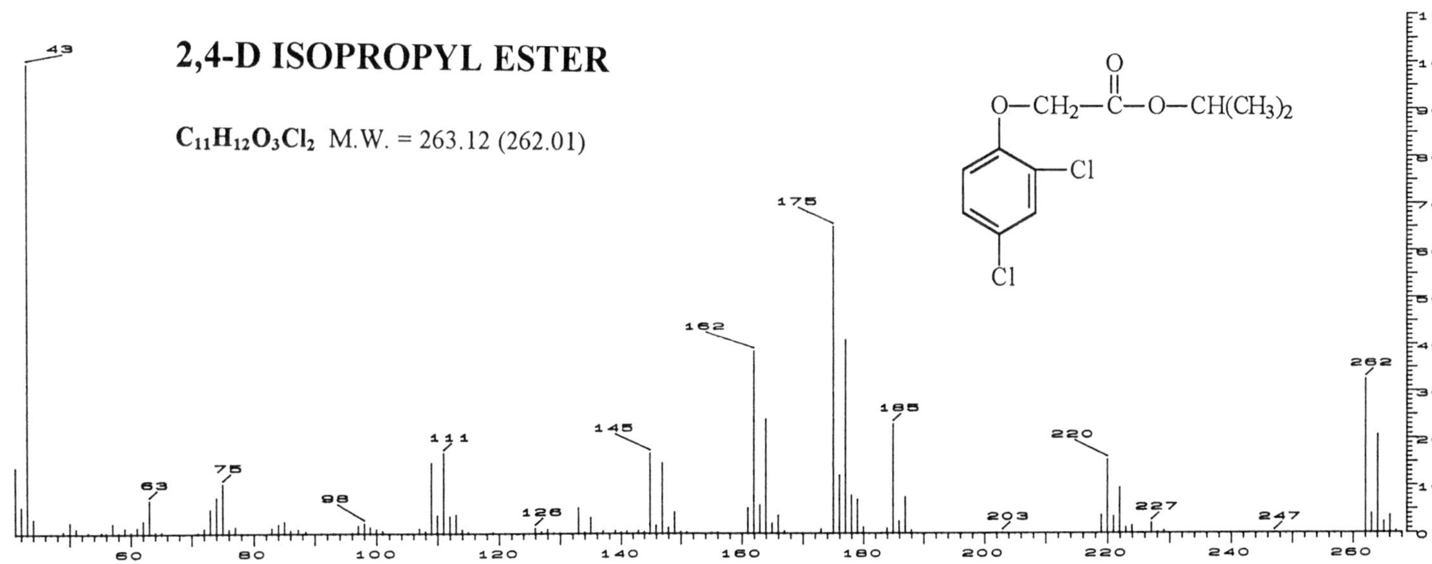

2,4-D BUTOXYETHYL ESTER

$C_{14}H_{18}O_4Cl_2$ M.W. = 321.20 (320.06)

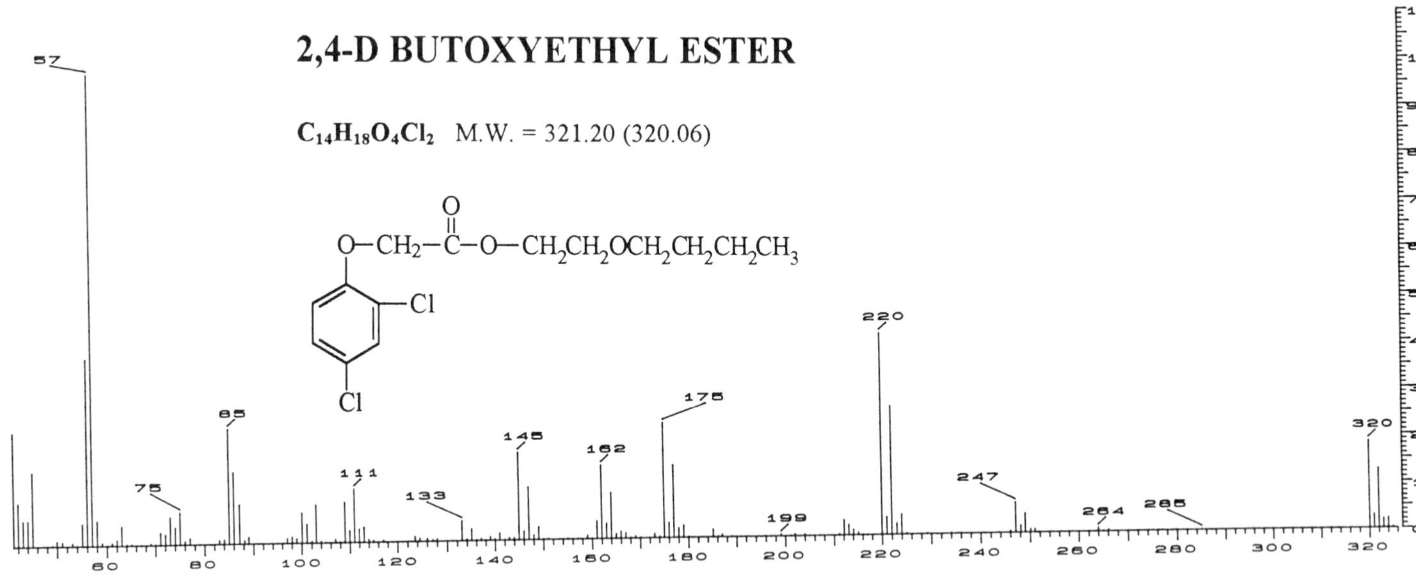

2,4-D BUTYL ESTER

$C_{12}H_{14}O_3Cl_2$ M.W. = 277.14 (276.03)

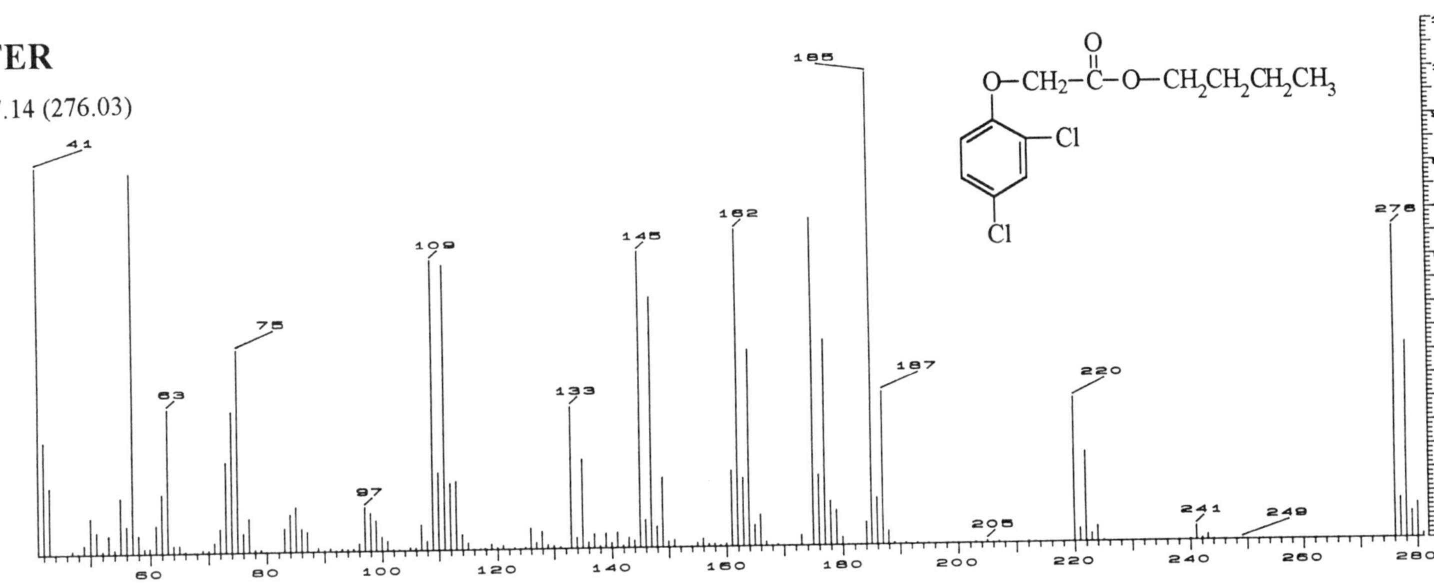

2,4-D ISOOCTYL ESTER

$C_{16}H_{22}O_3Cl_2$ M.W. = 333.25 (332.09)

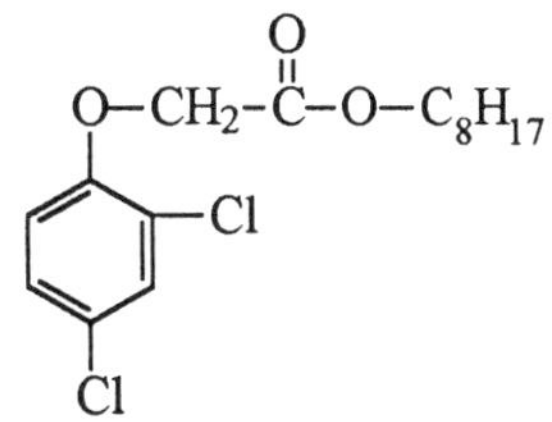

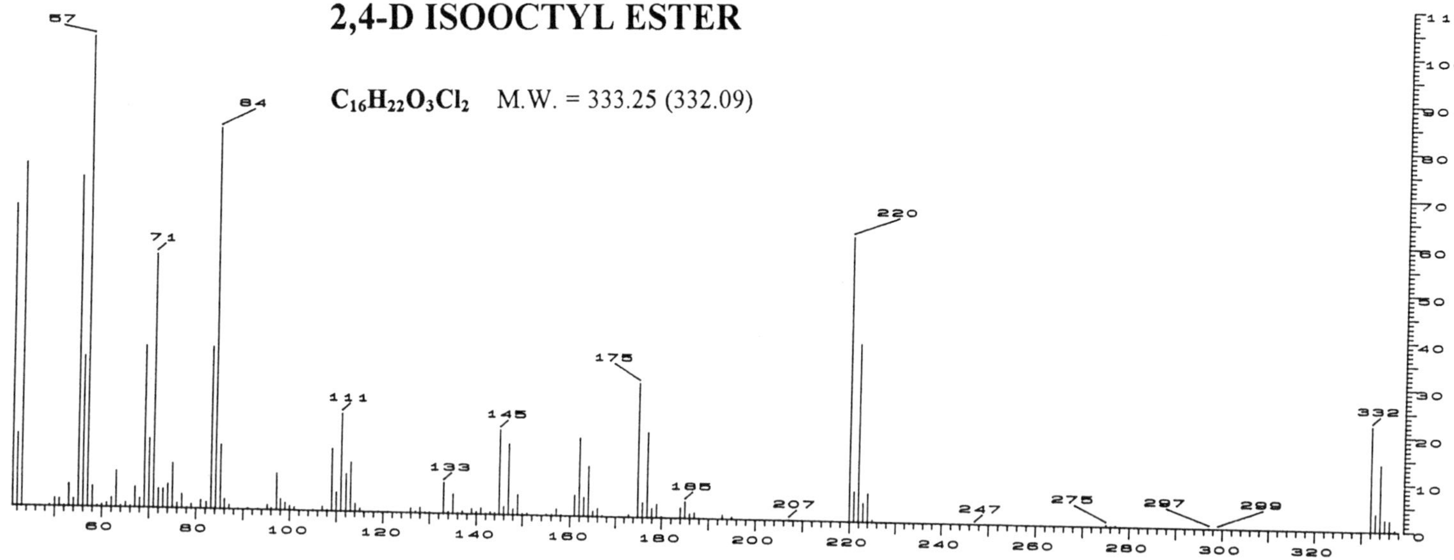

DB, (2,4-)

$C_{10}H_{10}O_3Cl_2$ M.W. = 249.09 (248.00)

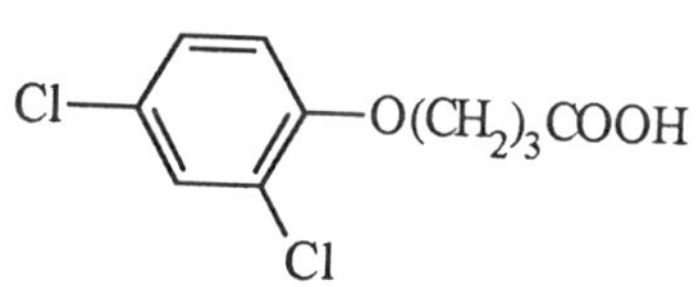

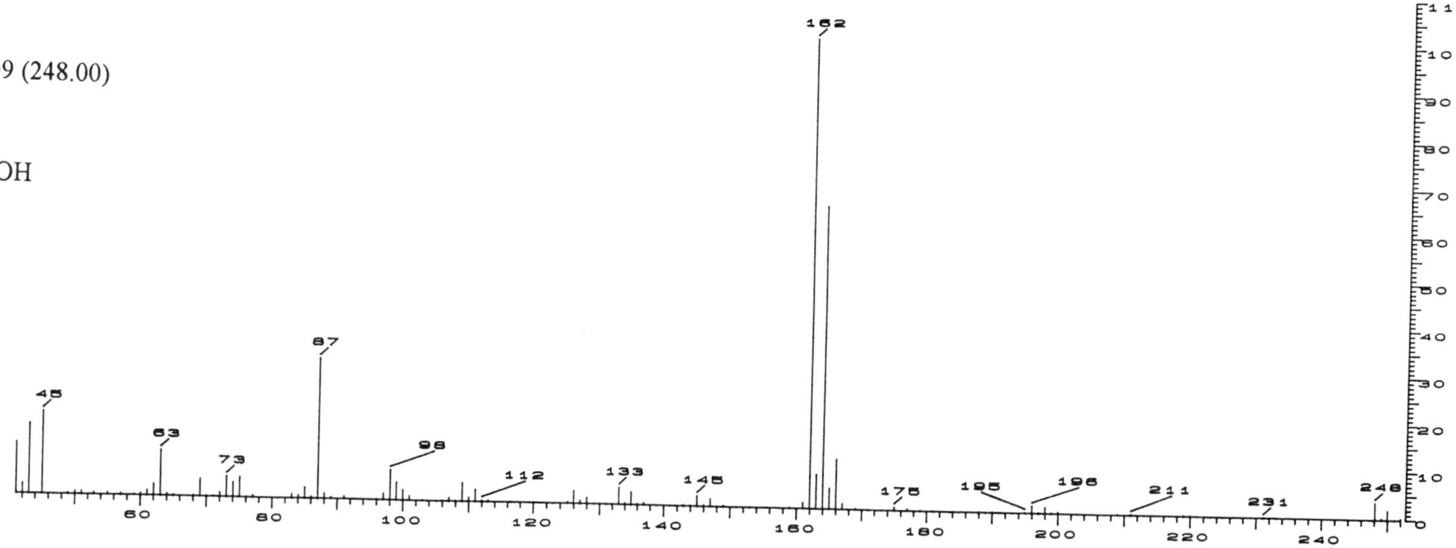

DB, (2,4-, ISOOCTYL ESTER)

$C_{18}H_{26}O_3Cl_2$ M.W. = 361.30 (360.13)

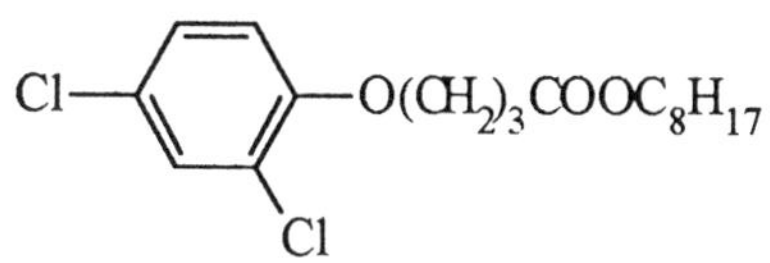

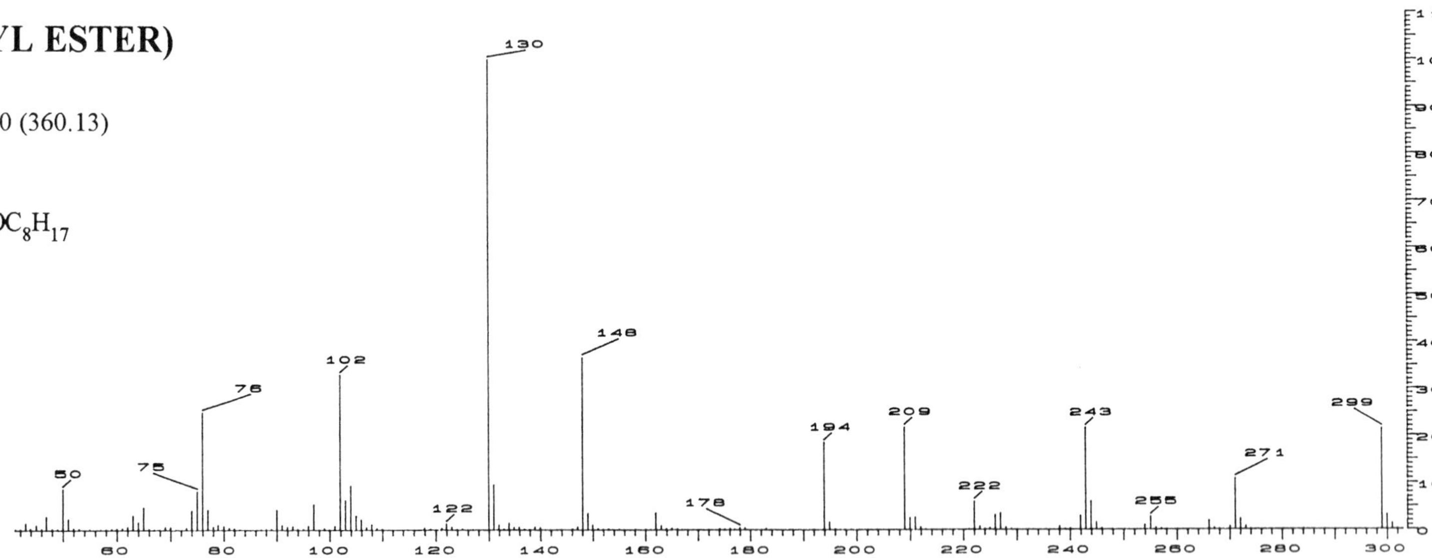

DDT-(o,p'-)

$C_{14}H_9Cl_5$ M.W. = 354.49 (351.91)

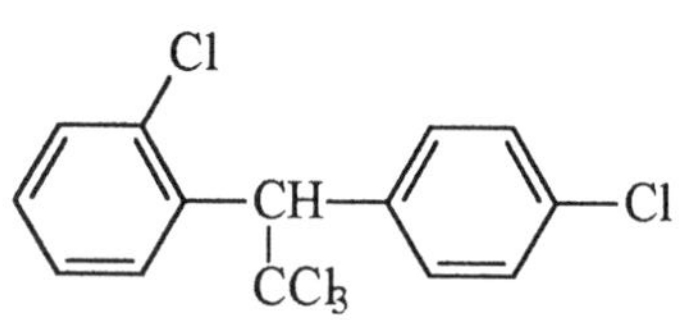

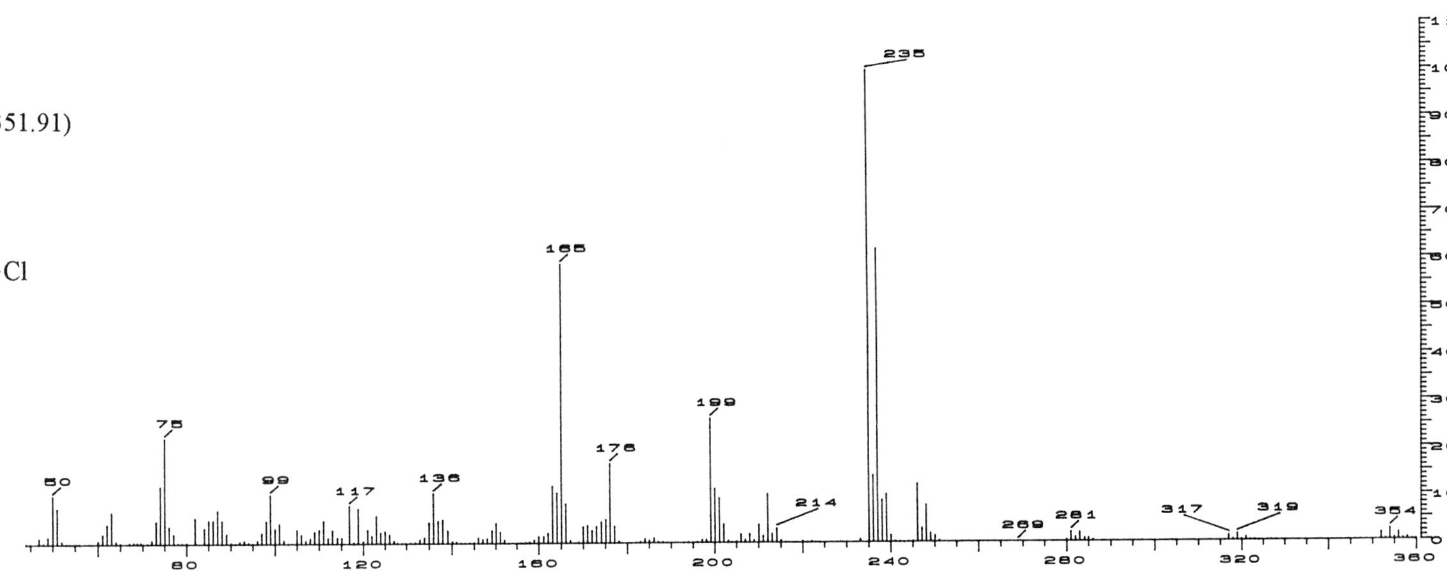

DDT- (p,p'-)

$C_{14}H_9Cl_5$ M.W. = 354.49 (351.91)

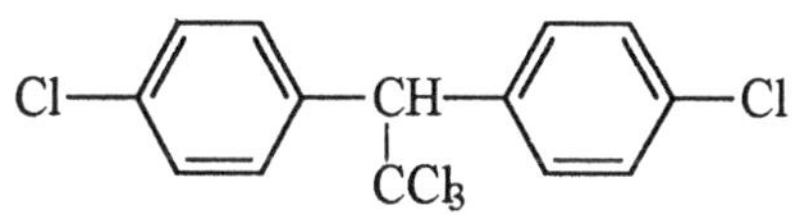

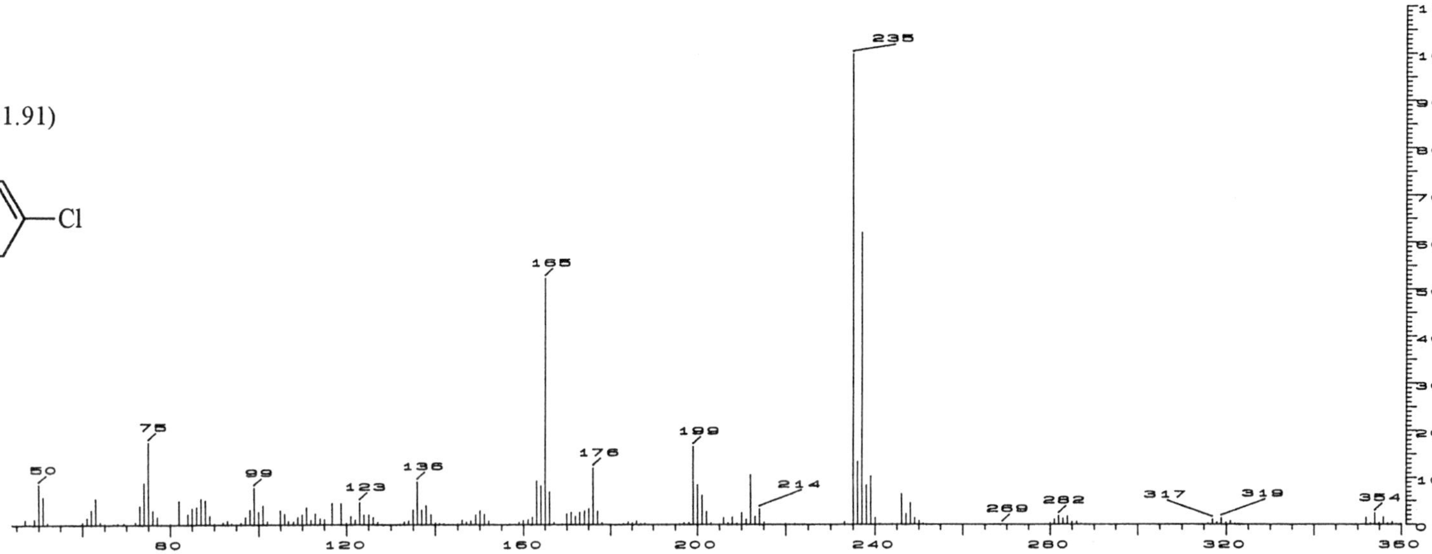

DEMETON-O

$C_8H_{19}O_3S_2P$ M.W. = 258.33 (258.05)

$CH_3CH_2SCH_2CH_2OP(=S)(OCH_2CH_3)_2$

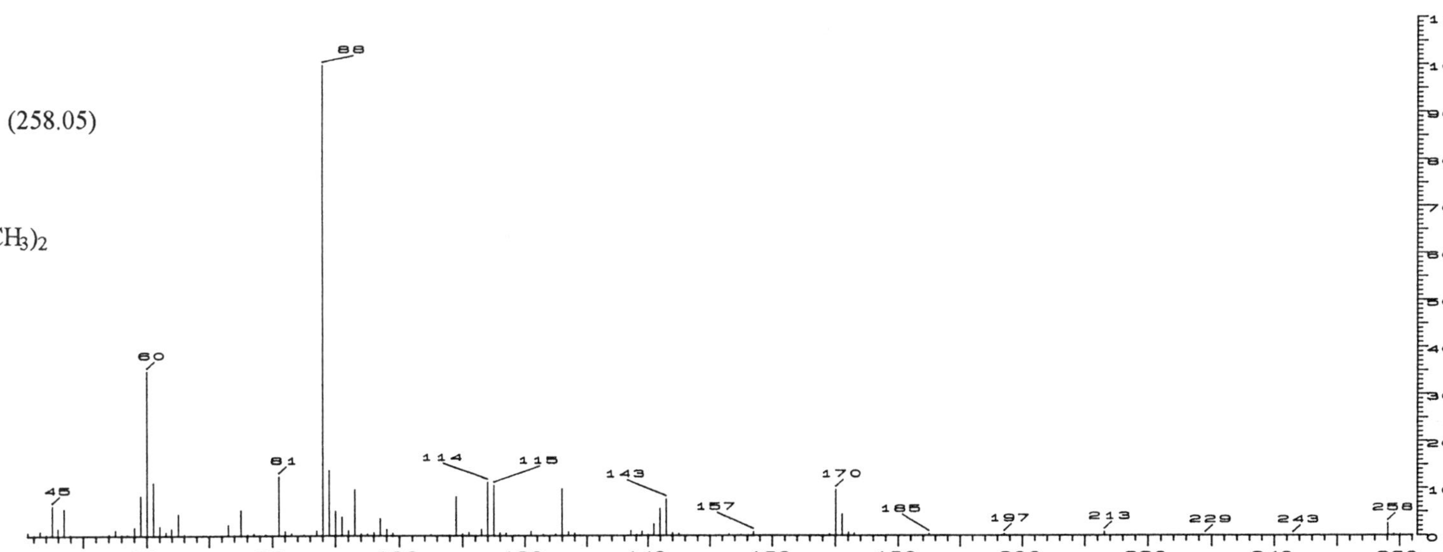

DEMETON-S

$C_8H_{19}O_3S_2P$ M.W. = 258.33 (258.05)

$CH_3CH_2SCH_2CH_2SP(=O)(OCH_2CH_3)_2$

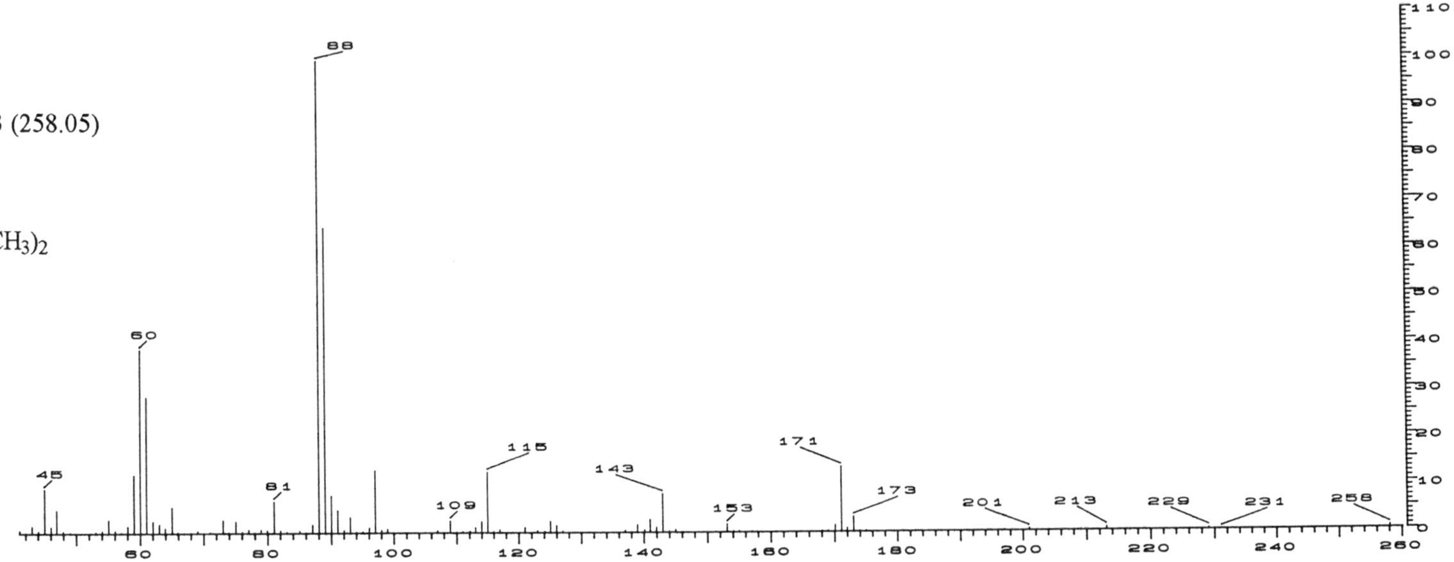

DEMETON-S SULFONE

$C_8H_{19}O_4S_2P$ M.W. = 274.33 (274.06)

$CH_3CH_2S(\rightarrow O)CH_2CH_2SP(=O)(OCH_2CH_3)_2$

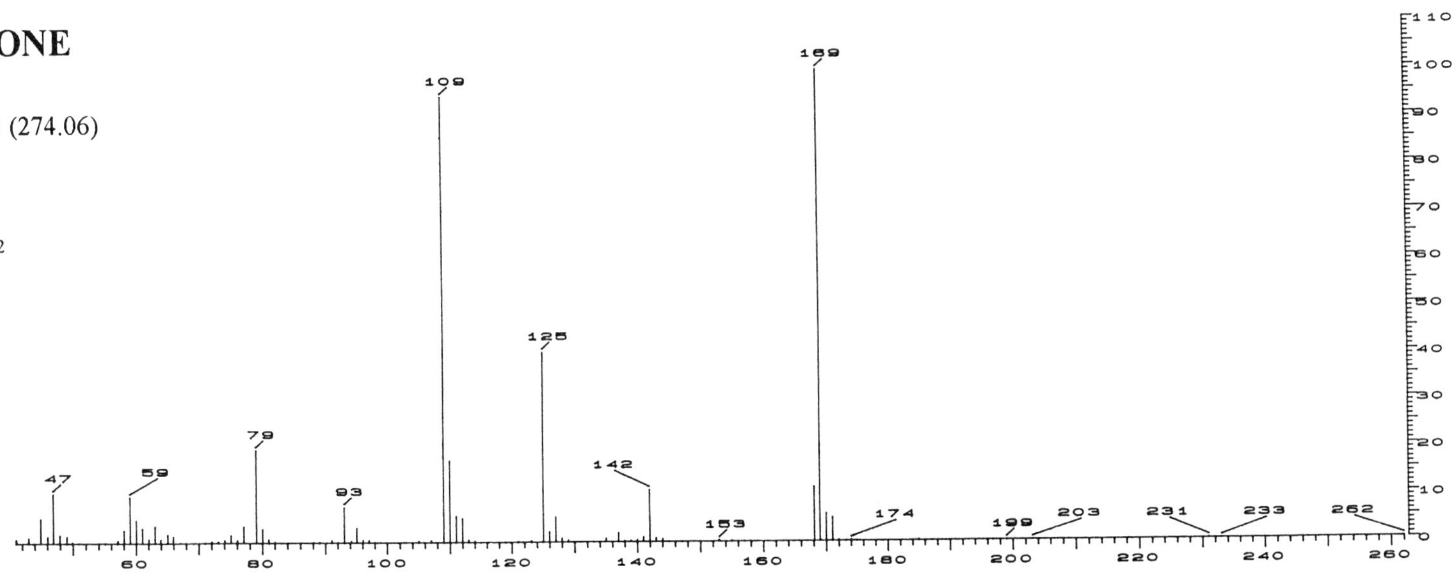

DIAZINON

$C_{12}H_{21}N_2O_3SP$ M.W. = 304.34 (304.10)

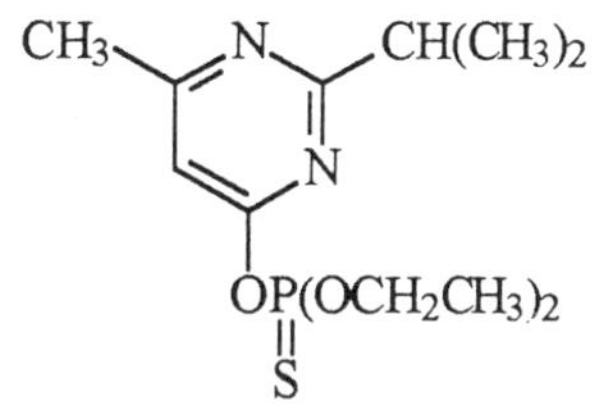

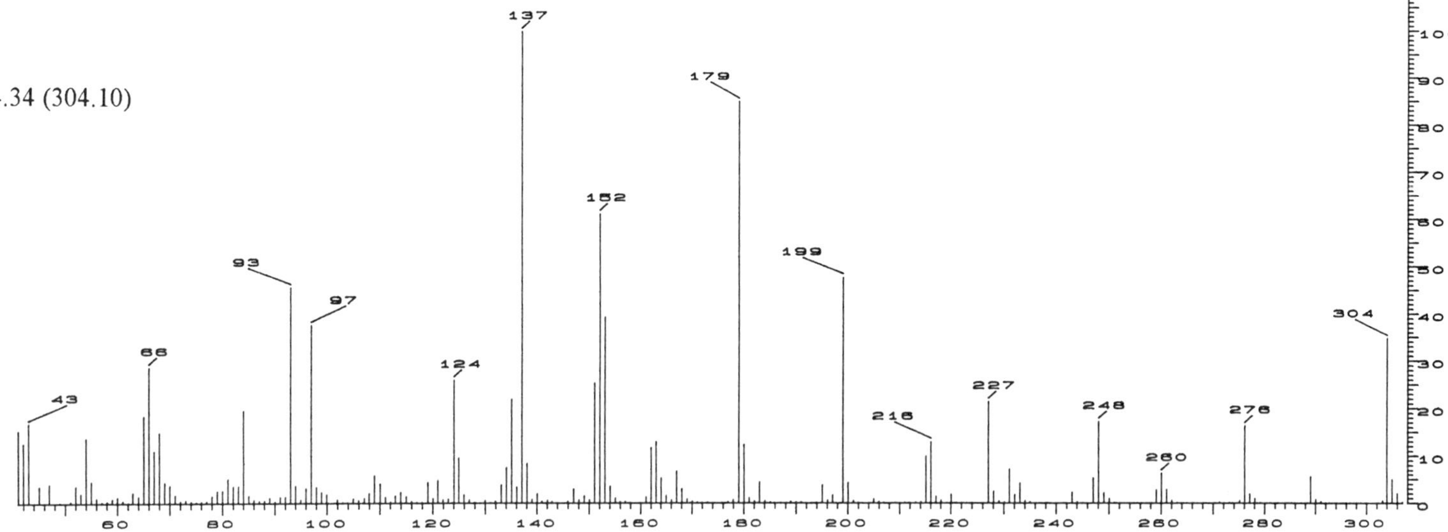

DIBUTYLCHLORENDATE

$C_{17}H_{20}O_4Cl_6$ M.W. = 501.06 (497.95)

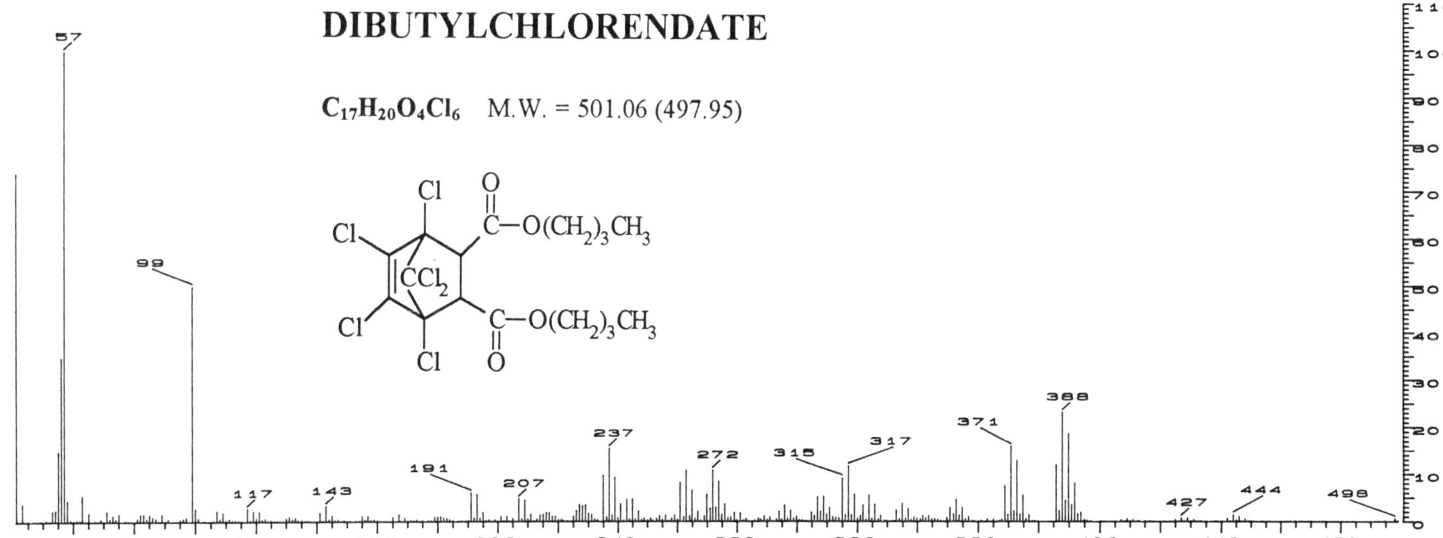

DICAMBA

$C_8H_6O_3Cl_2$ M.W. = 221.04 (219.95)

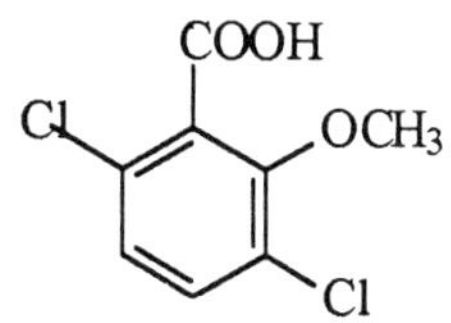

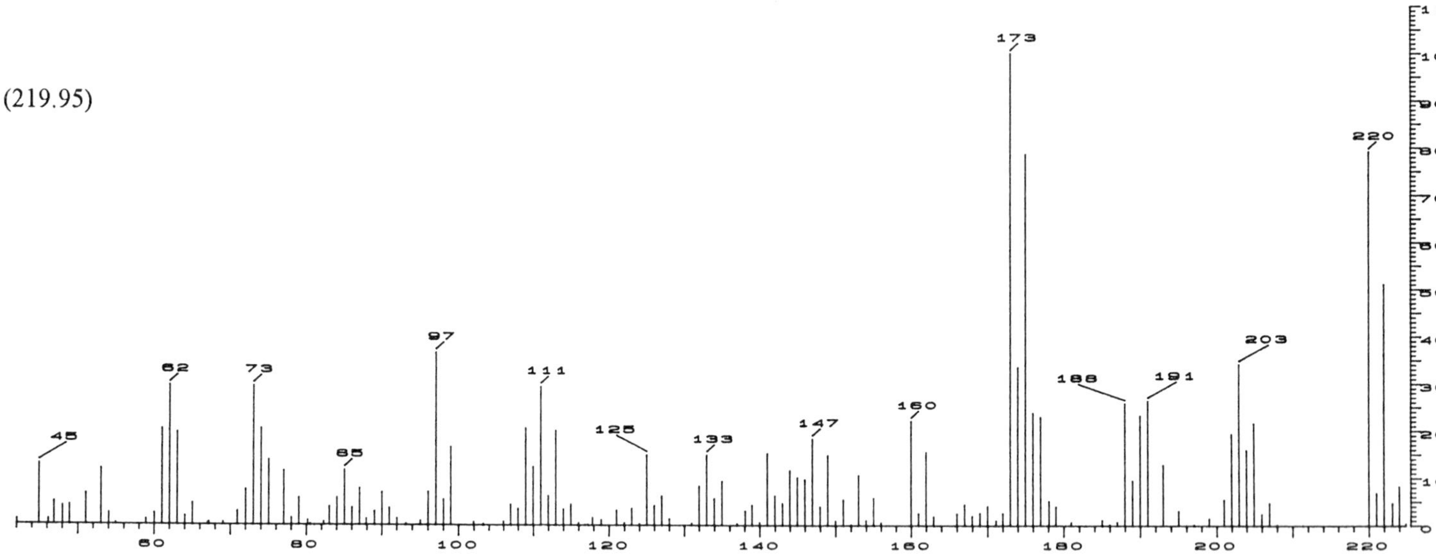

DICHLOFENTHION

$C_{10}H_{13}O_3SPCl_2$ M.W. = 315.15 (313.97)

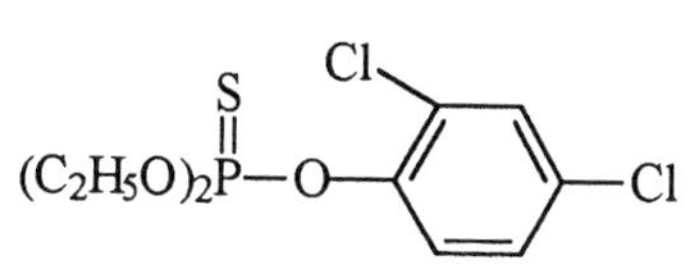

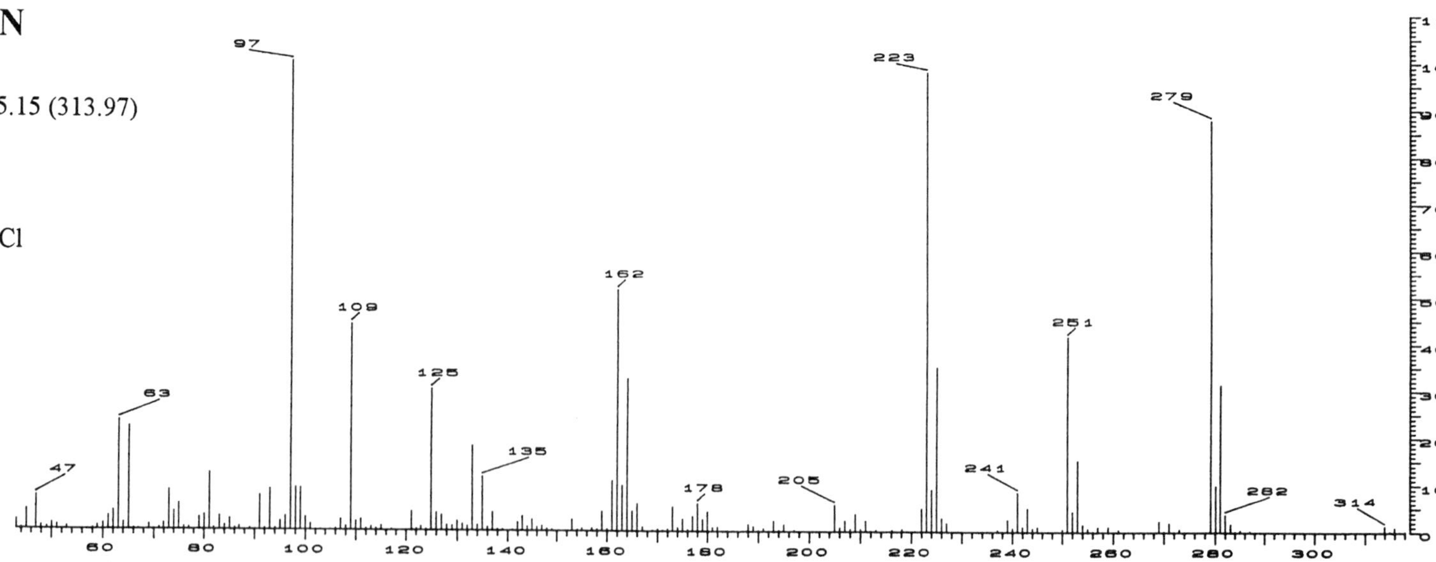

DICHLONE

$C_{10}H_4O_2Cl_2$ M.W. = 227.04 (225.96)

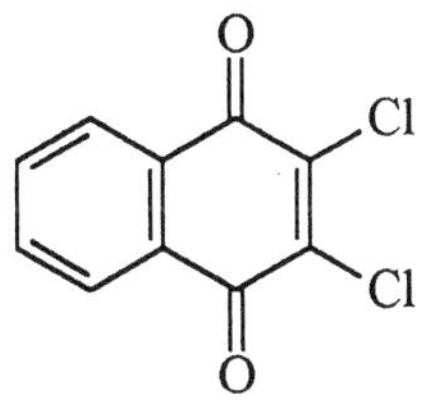

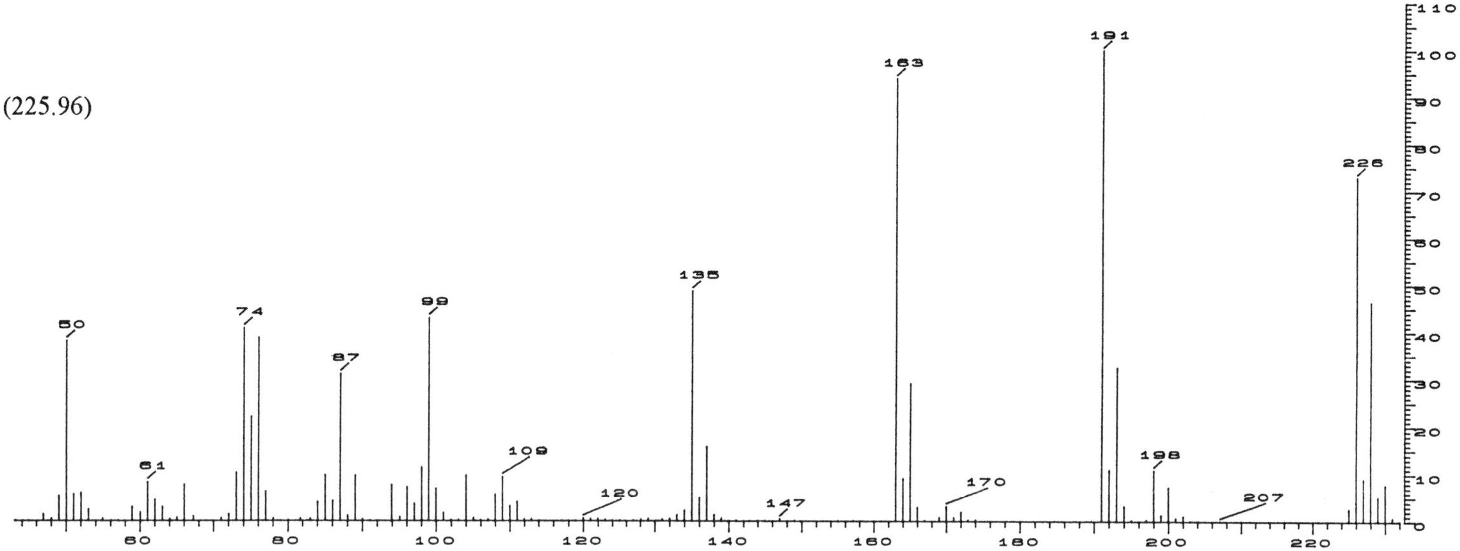

DICHLORVOS

$C_4H_7O_4PCl_2$ M.W. = 220.97 (219.95)

$Cl_2C{=}CHOP(O)(OCH_3)_2$

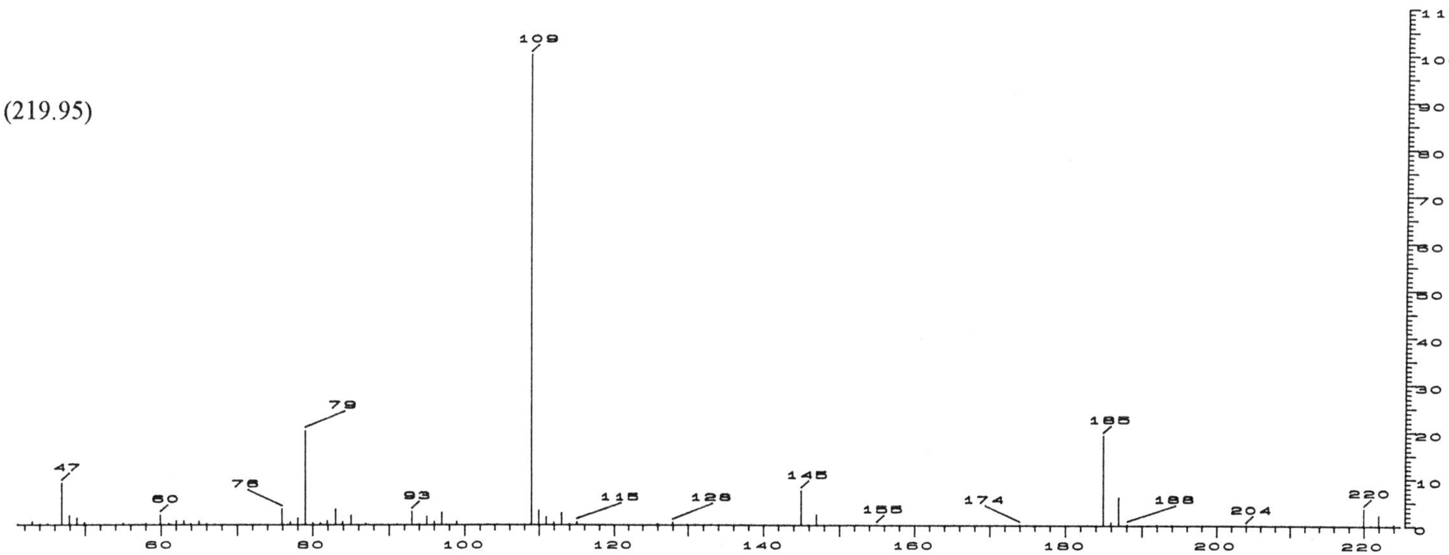

DICOFOL

$C_{14}H_9OCl_5$ M.W. = 370.48 (367.91)

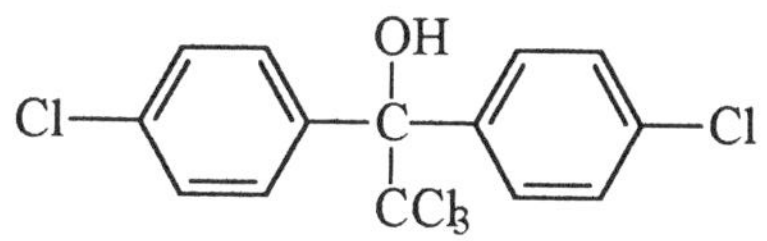

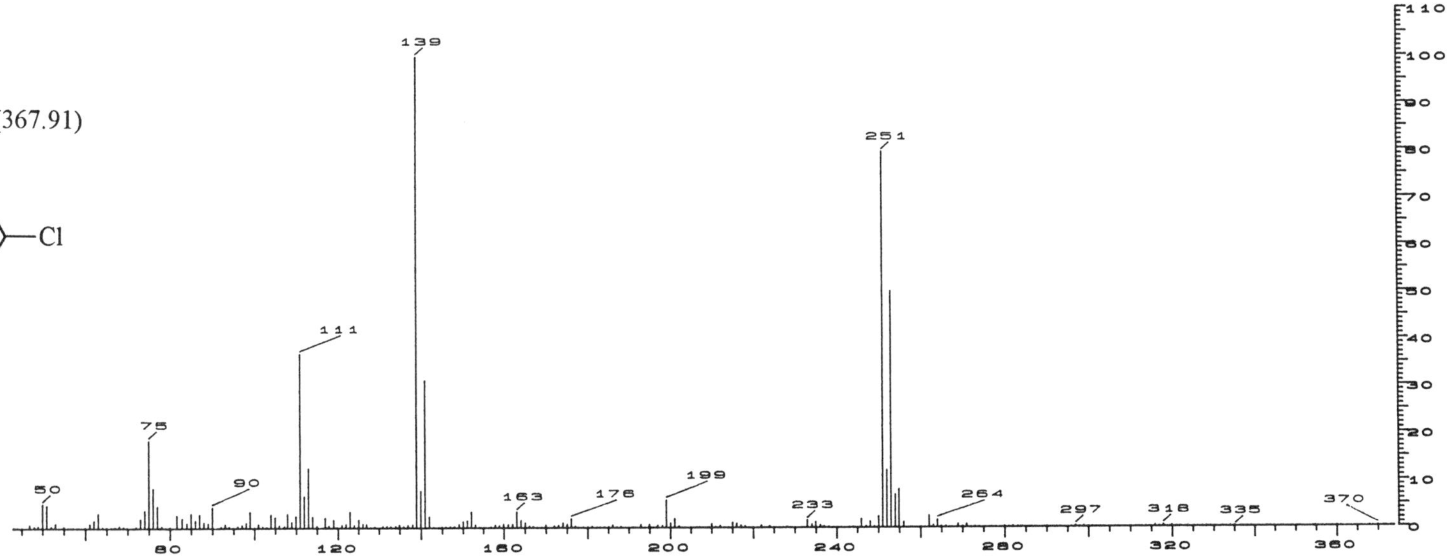

DICROTOPHOS

$C_8H_{16}NO_5P$ M.W. = 237.19 (237.08)

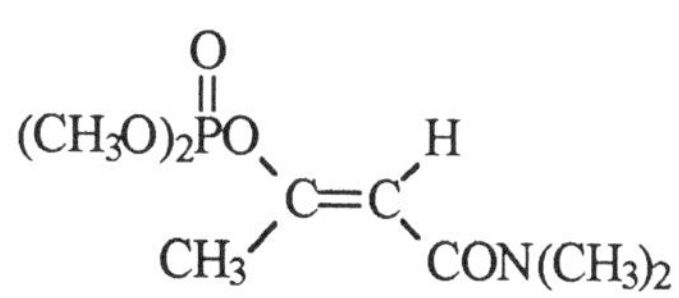

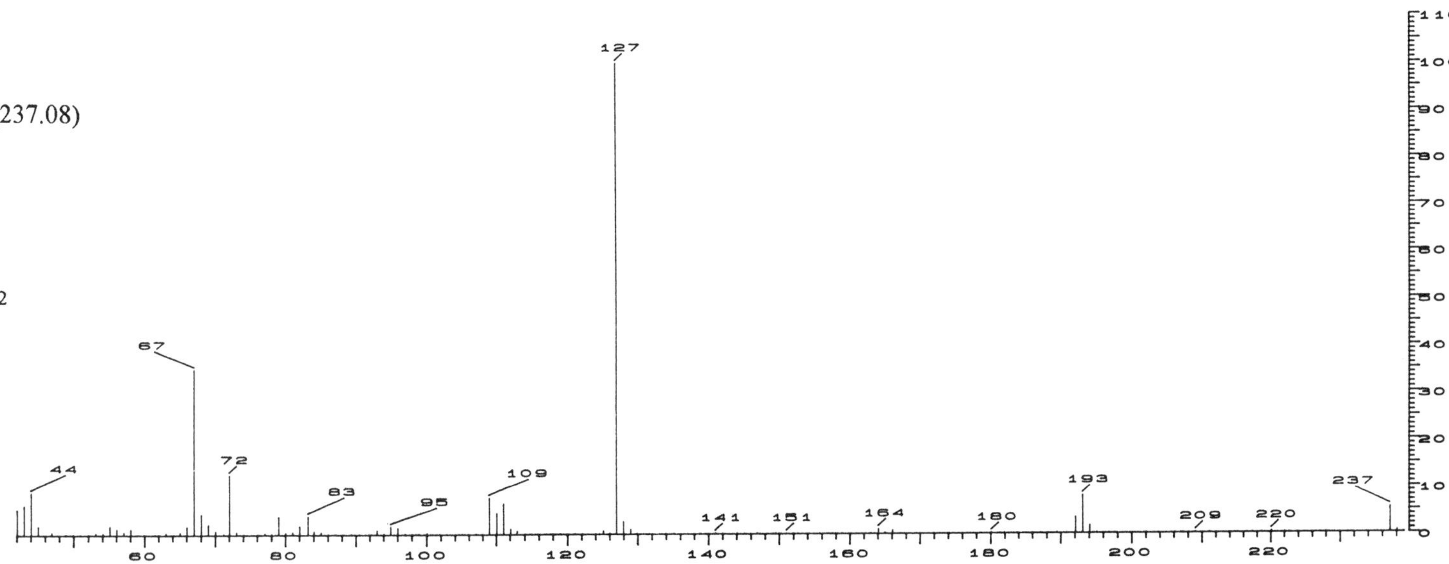

DIELDRIN

$C_{12}H_8OCl_6$ M.W. = 380.91 (377.87)

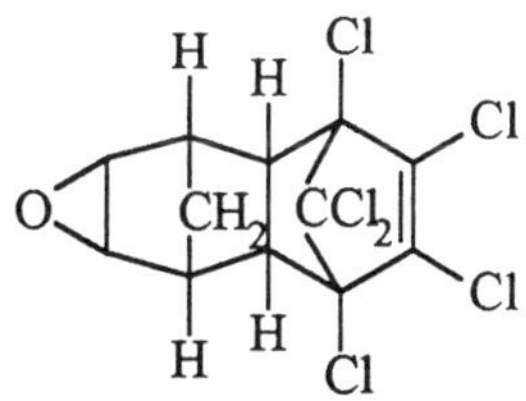

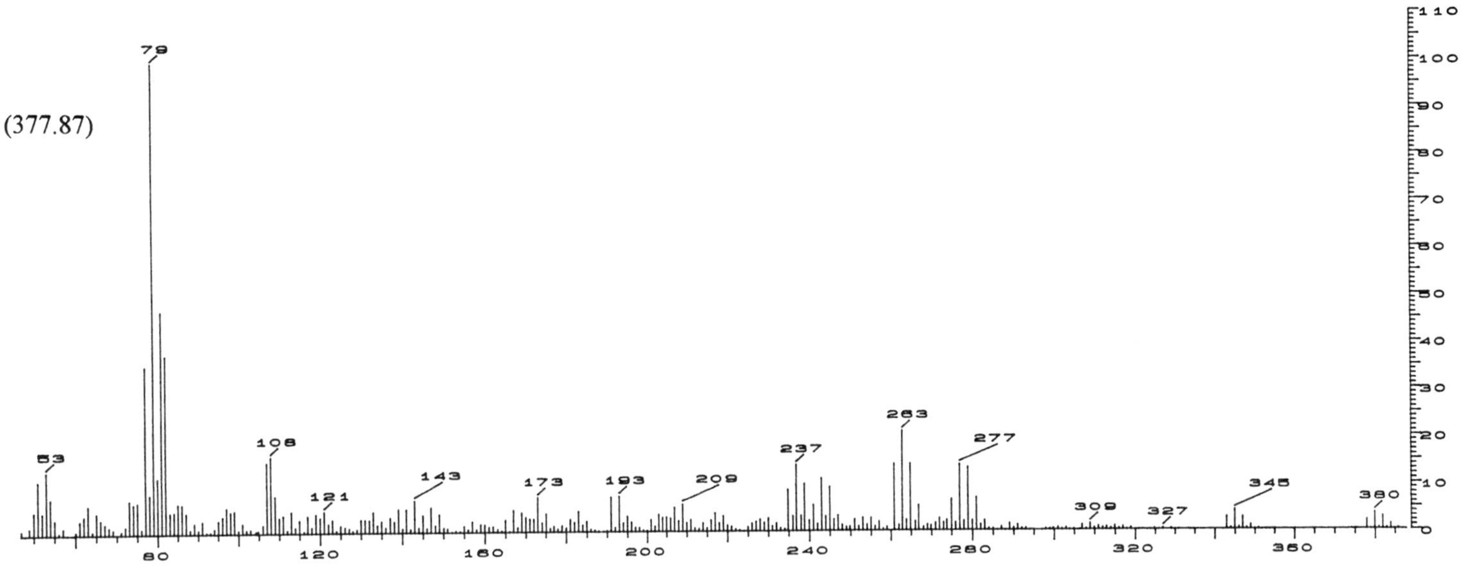

DIETHATYL-ETHYL

$C_{16}H_{22}NO_3Cl$ M.W. = 311.80 (311.13)

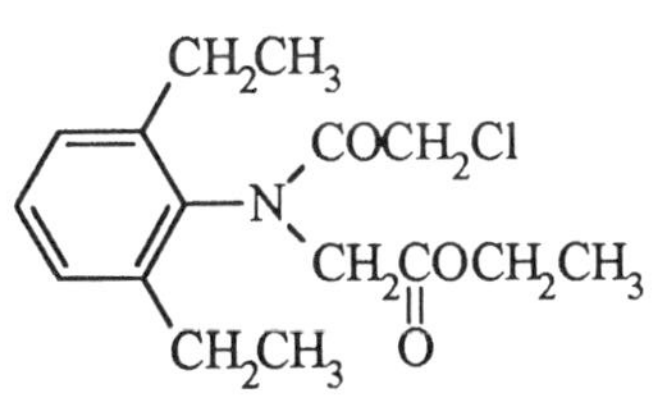

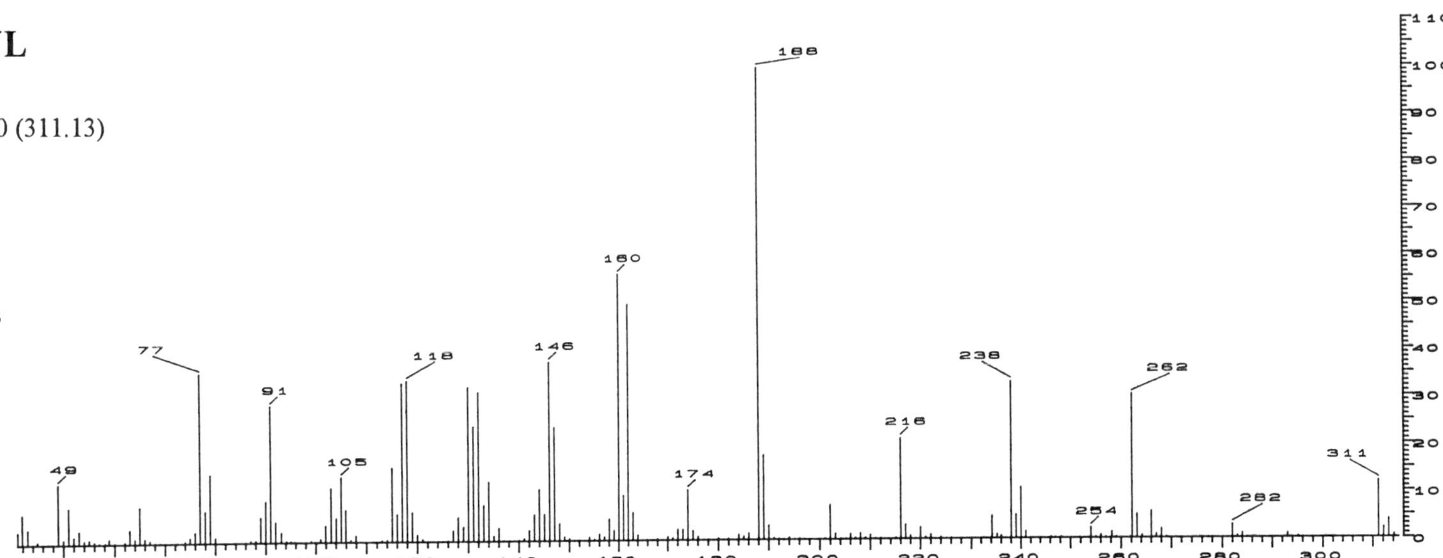

DIMETHIPIN

$C_6H_{10}O_4S_2$ M.W. = 210.26 (210.00)

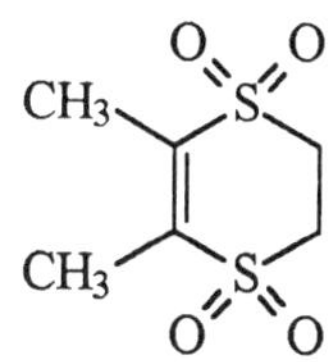

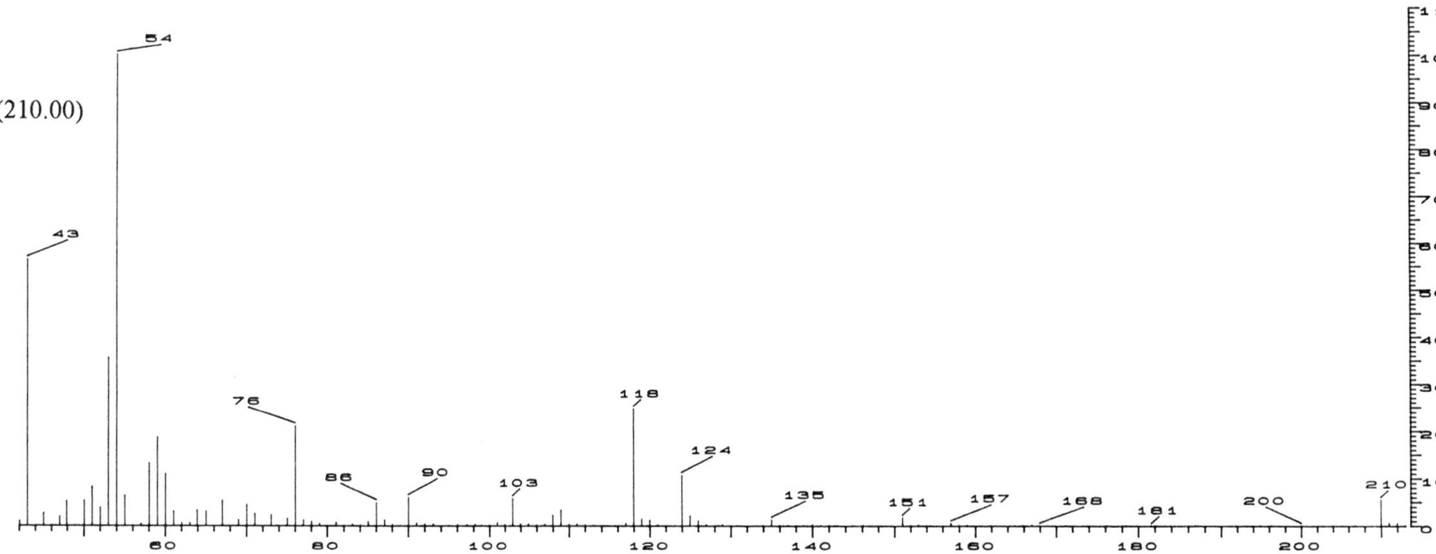

DIMETHOATE

$C_5H_{12}NO_3S_2P$ M.W. = 229.24 (229.00)

$CH_3NHCOCH_2S\overset{S}{\overset{\|}{P}}(OCH_3)_2$

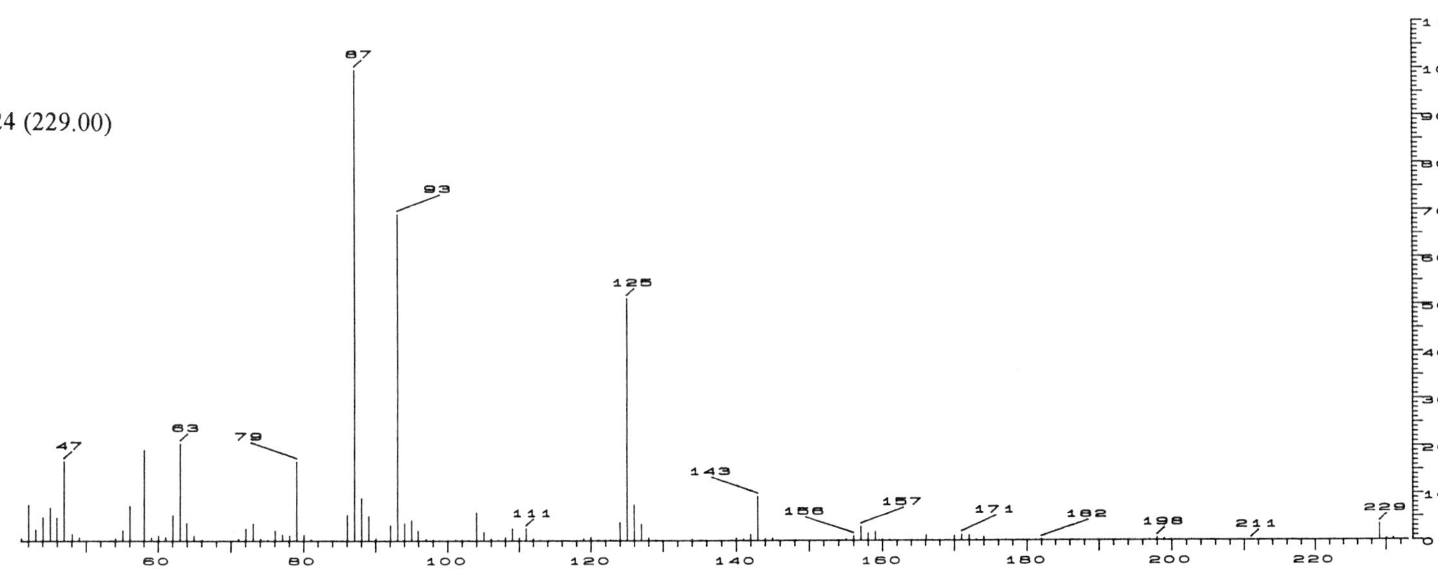

DIMETHOATE-OA

$C_5H_{12}NO_4SP$ M.W. = 213.18 (213.02)

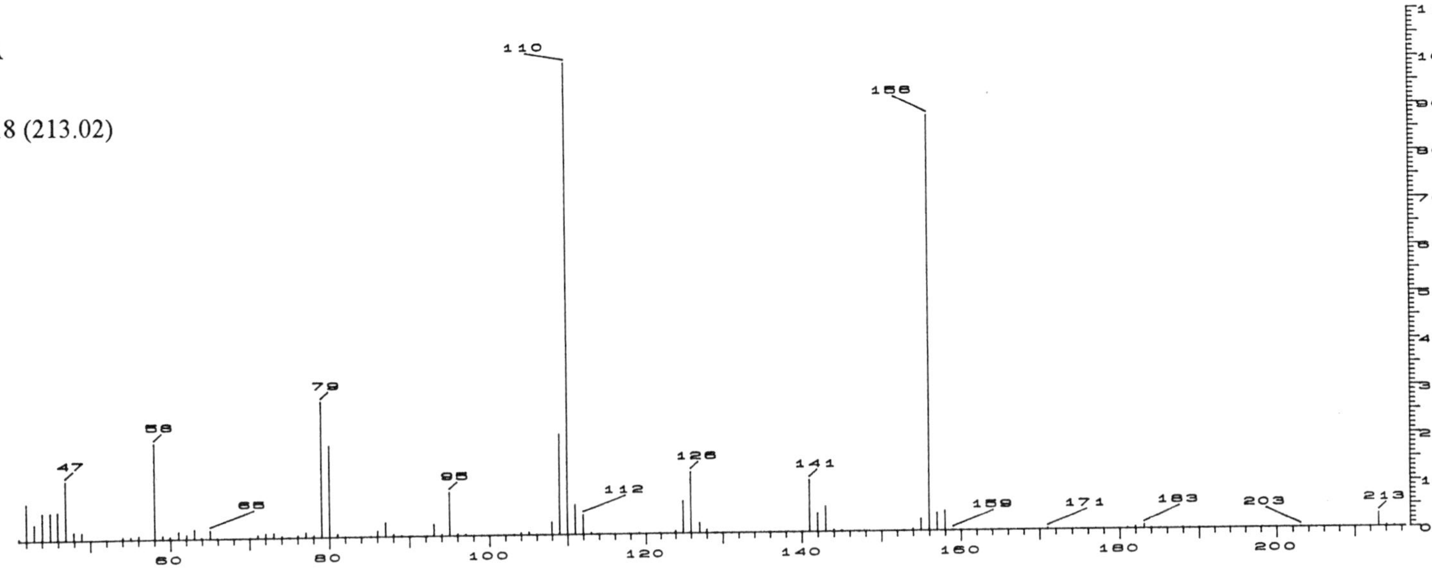

DINOSEB

$C_{10}H_{12}N_2OS$ M.W. = 240.21 (240.07)

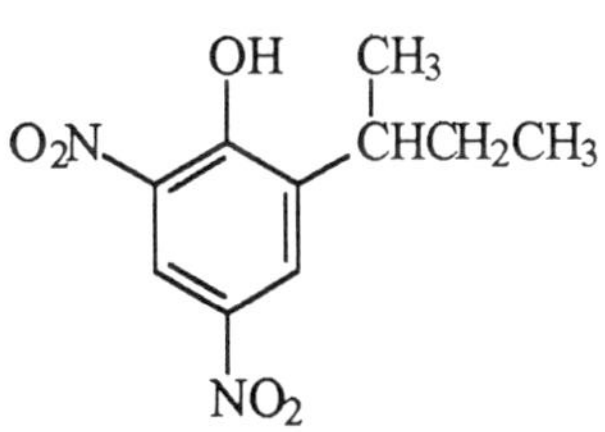

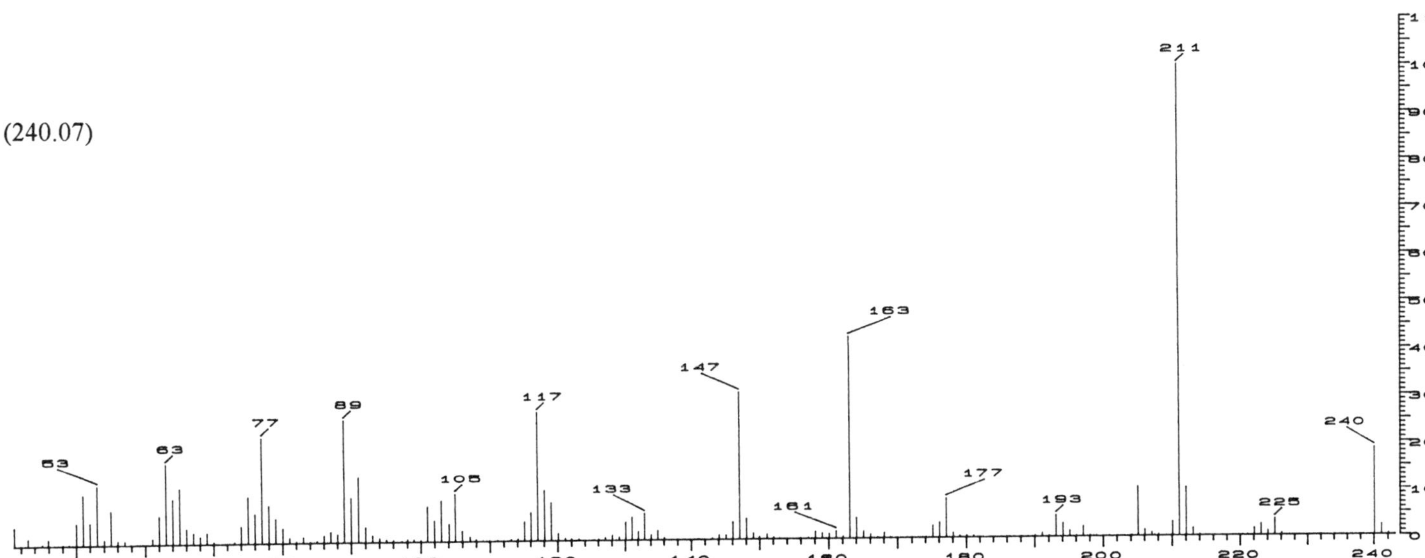

DIPROPETRYN

$C_{11}H_{21}N_5S$ M.W. = 255.38 (255.15)

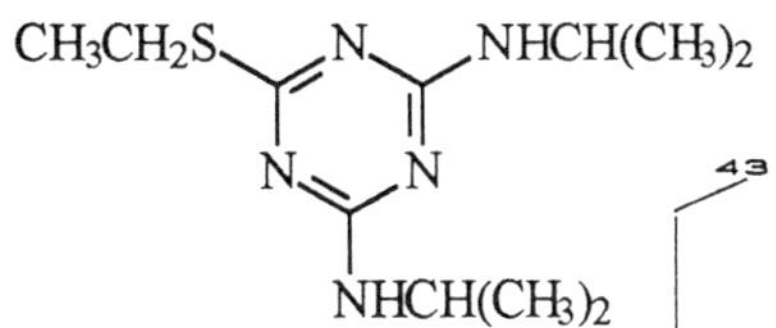

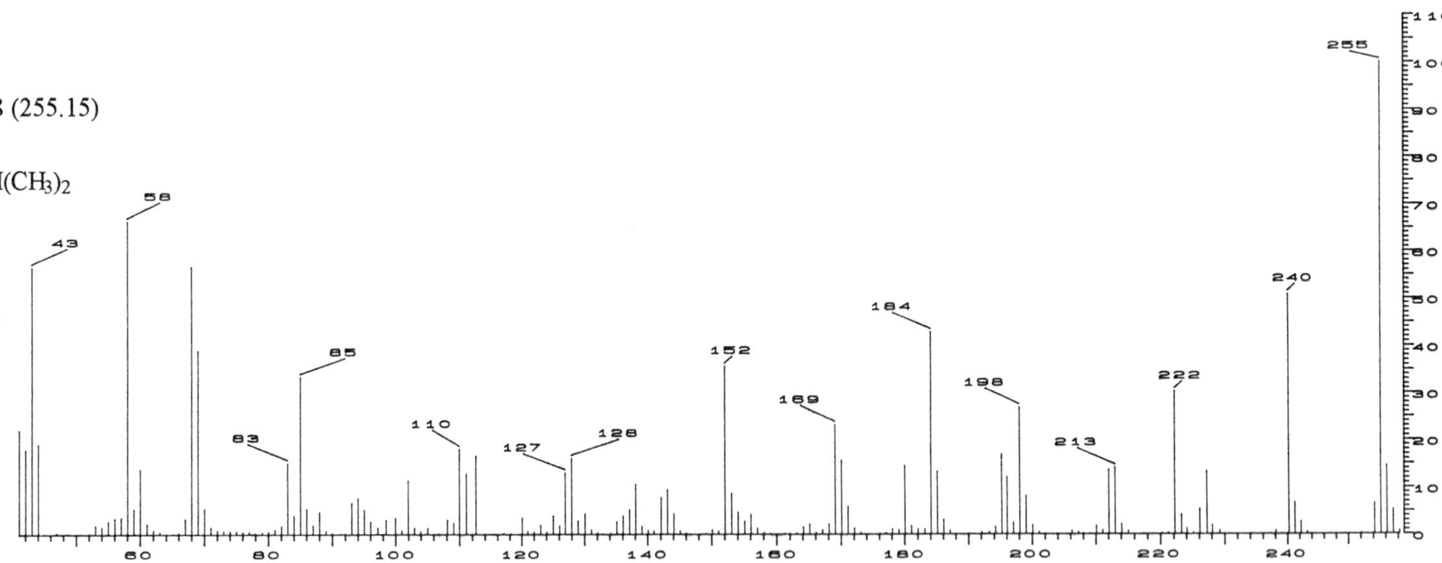

DISULFOTON

$C_8H_{19}O_2S_3P$ M.W. = 274.39 (274.03)

$CH_3CH_2SCH_2CH_2SP(=S)(OCH_2CH_3)_2$

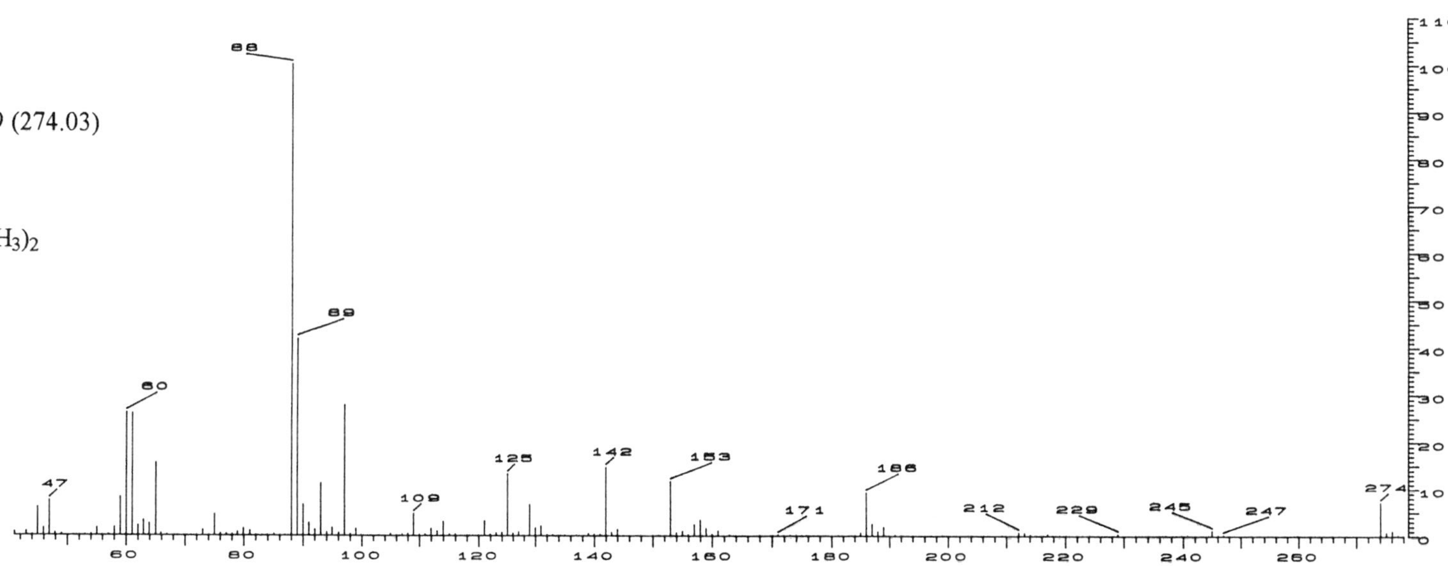

DITALIMPHOS

$C_{12}H_{14}NO_4SP$ M.W. = 299.28 (299.04)

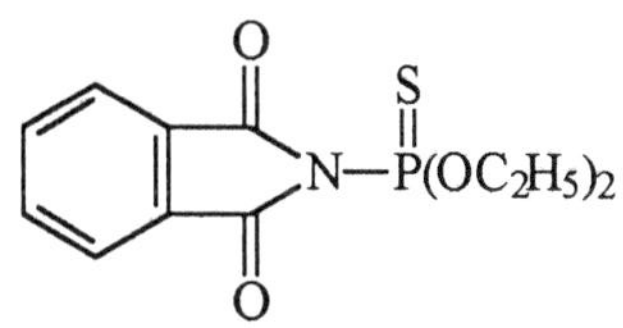

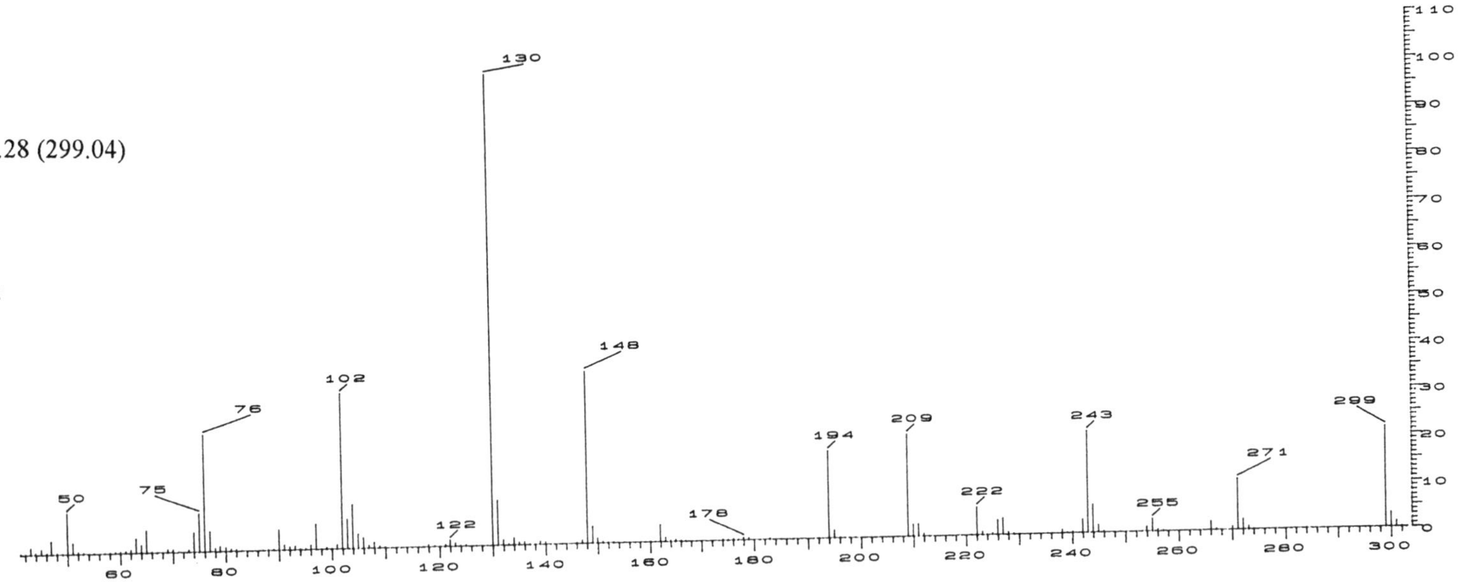

DIURON

$C_9H_{10}N_2OCl_2$ M.W. = 233.09 (232.02)

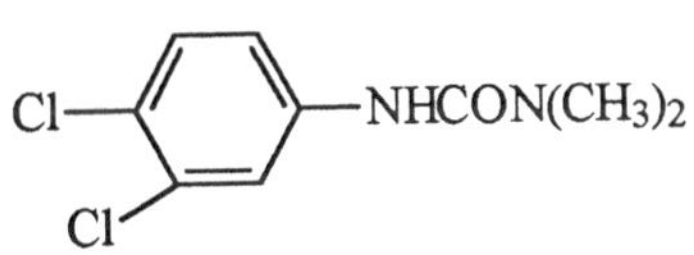

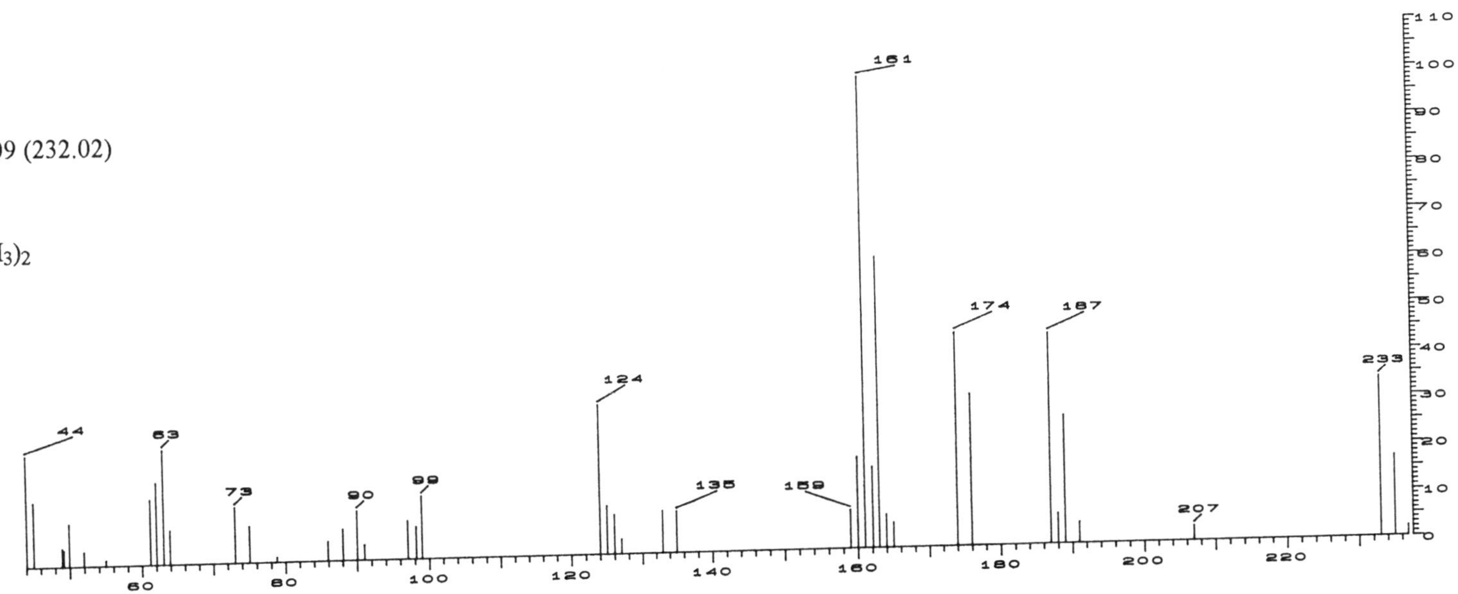

EDIFENPHOS

$C_{14}H_{15}O_2S_2P$ M.W. = 310.36 (310.23)

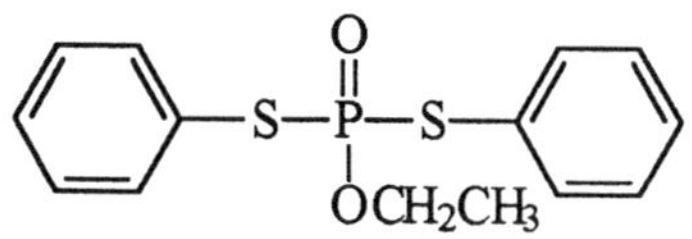

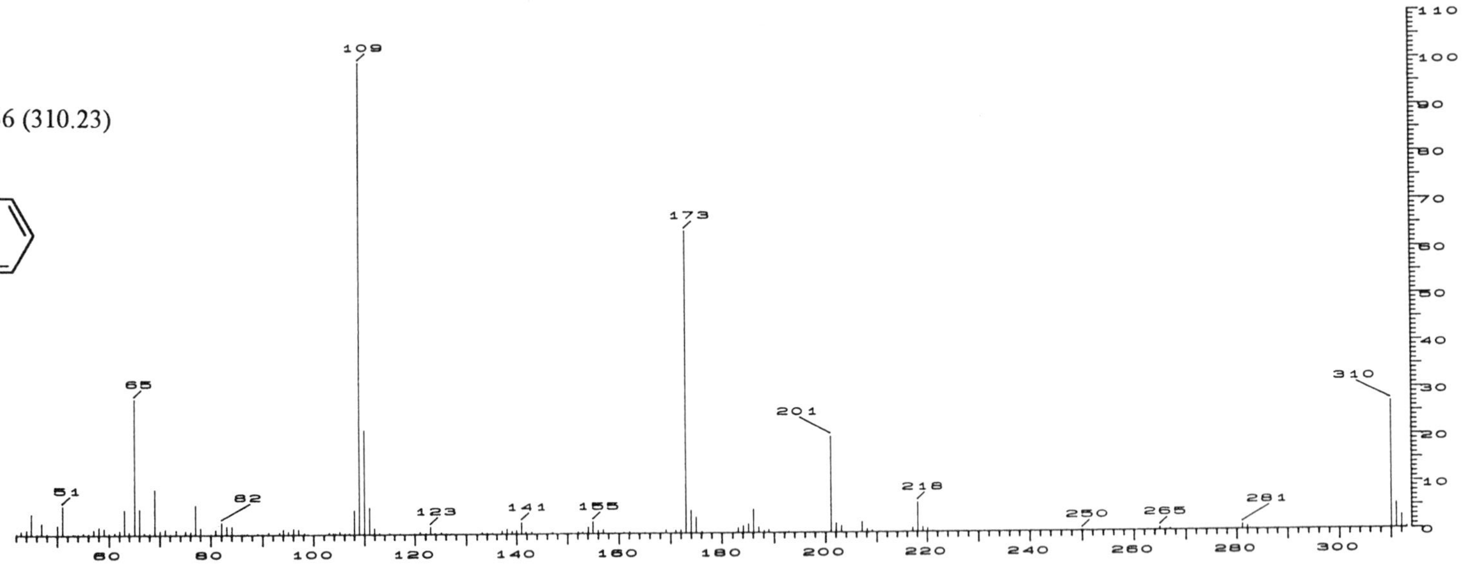

ENDOSULFAN I

$C_9H_6O_3SCl_6$ M.W. = 406.92 (403.82)

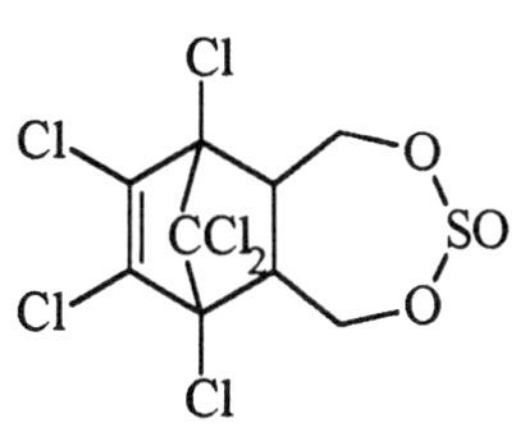

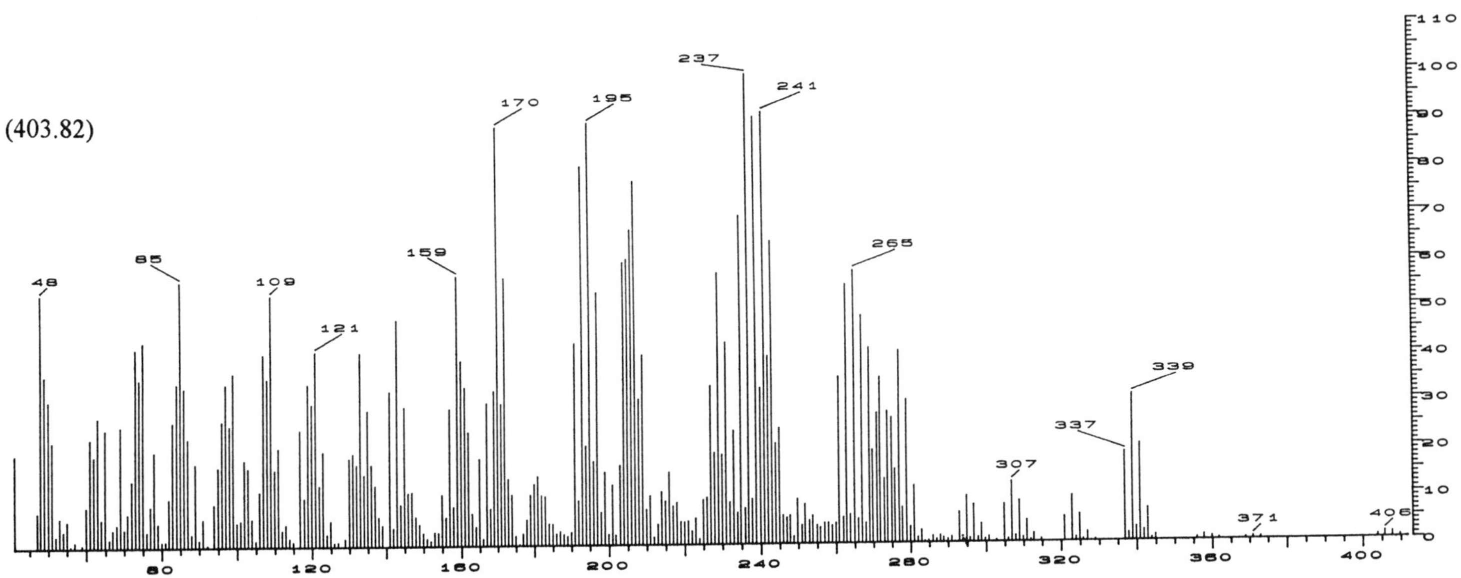

ENDOSULFAN II

$C_9H_6O_3SCl_6$ M.W. = 406.92 (403.82)

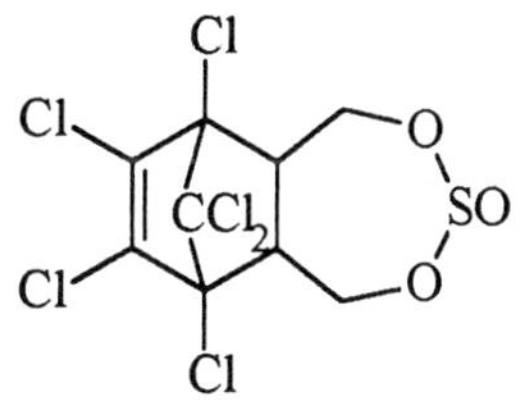

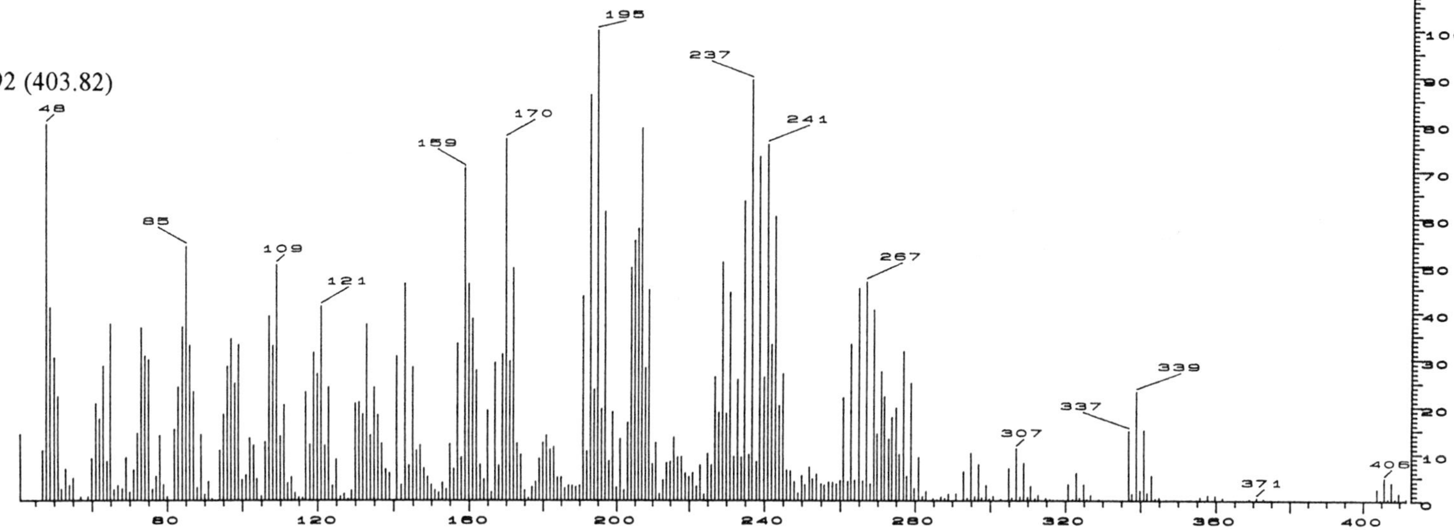

ENDRIN

$C_{12}H_8OCl_6$ M.W. = 380.91 (377.87)

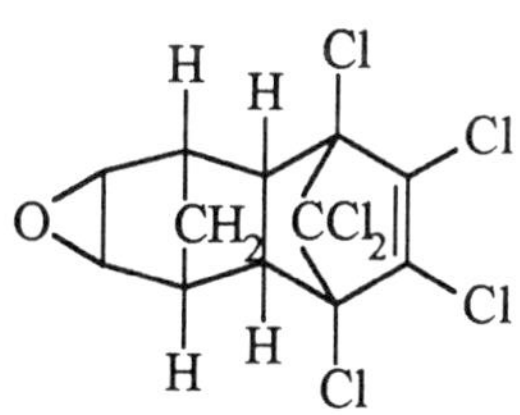

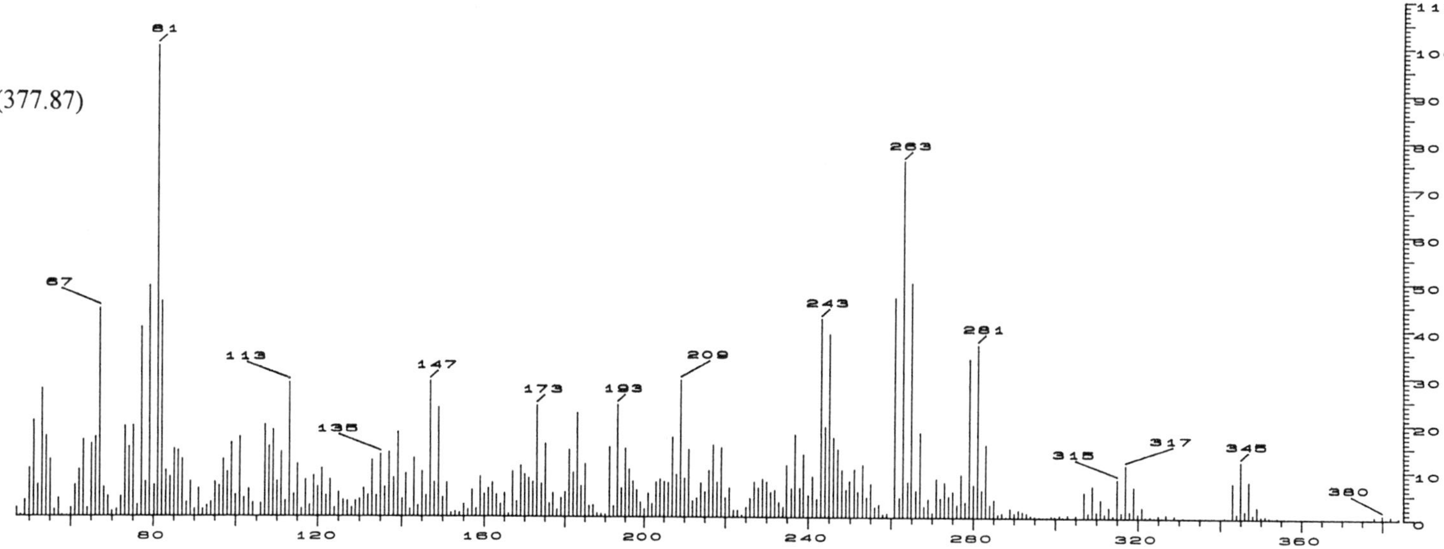

EPN

$C_{14}H_{14}NO_4SP$ M.W. = 323.30 (323.04)

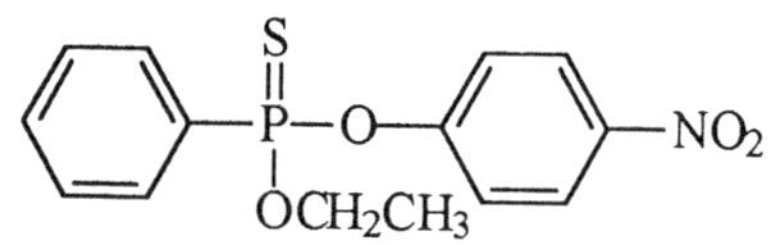

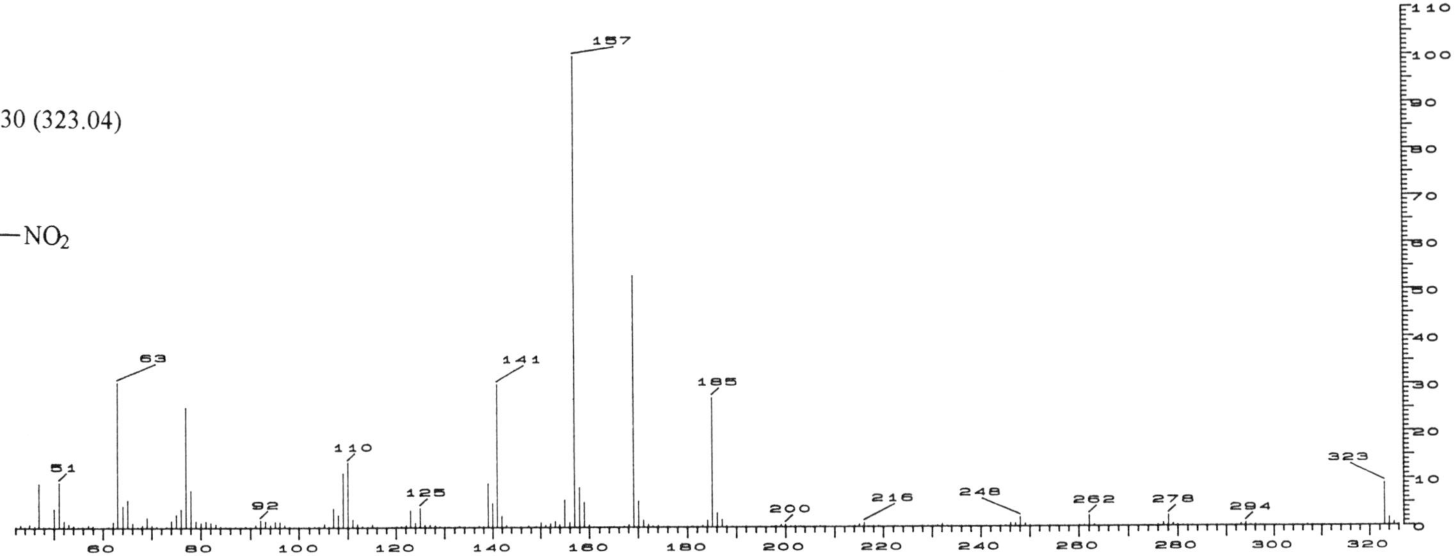

ETHION

$C_9H_{22}O_4S_4P_2$ M.W. = 384.45 (383.99)

$(CH_3CH_2O)_2P(=S)SCH_2SP(=S)(OCH_2CH_3)_2$

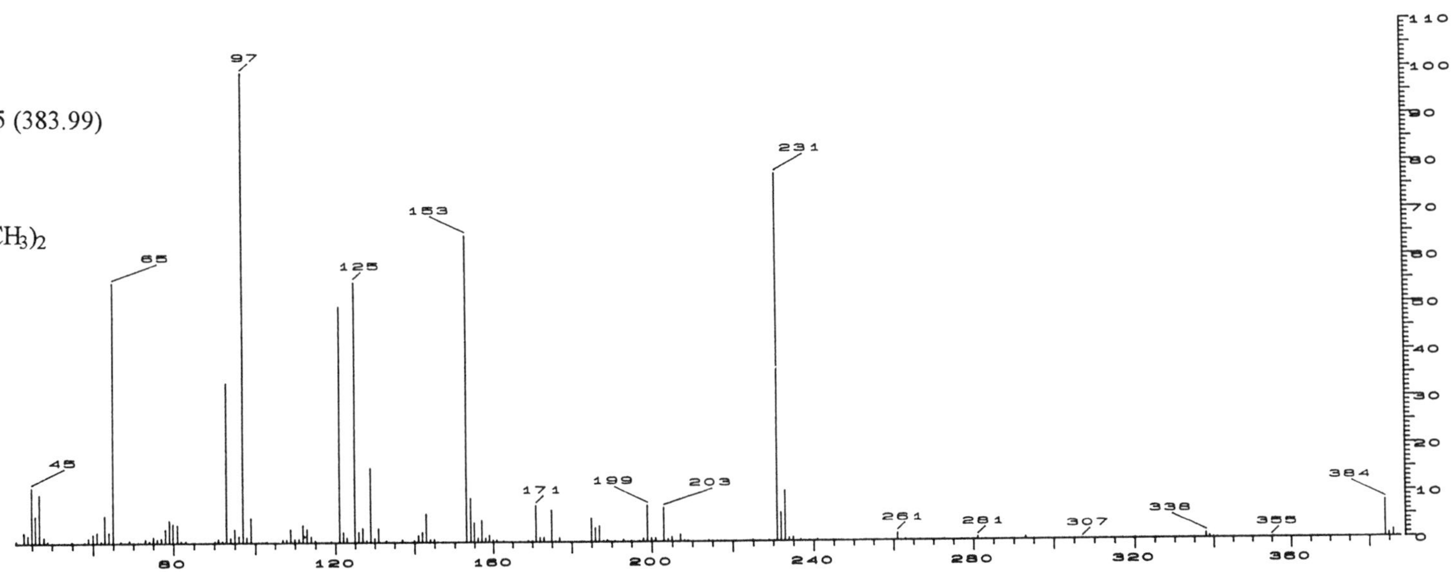

ETHIRIMOL

$C_{11}H_{19}N_3O$ M.W. = 209.29 (209.15)

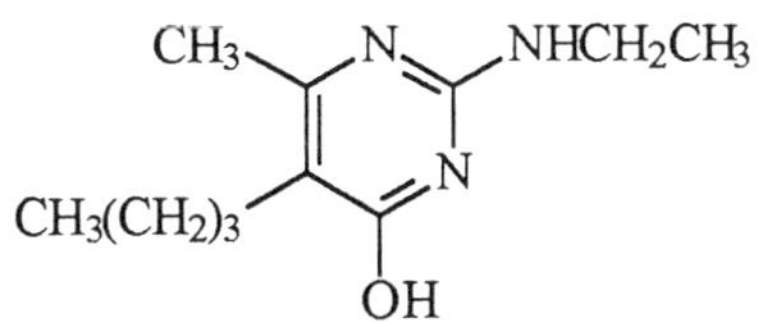

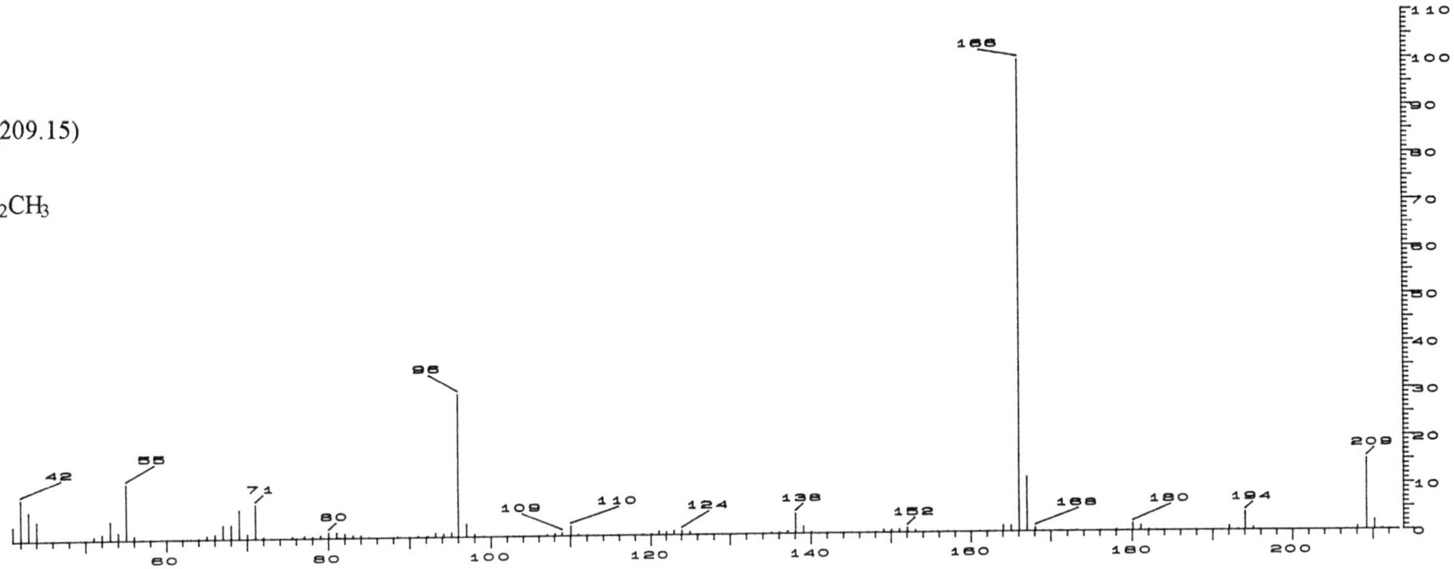

ETHOFUMESATE

$C_{13}H_{18}O_5S$ M.W. = 286.34 (286.09)

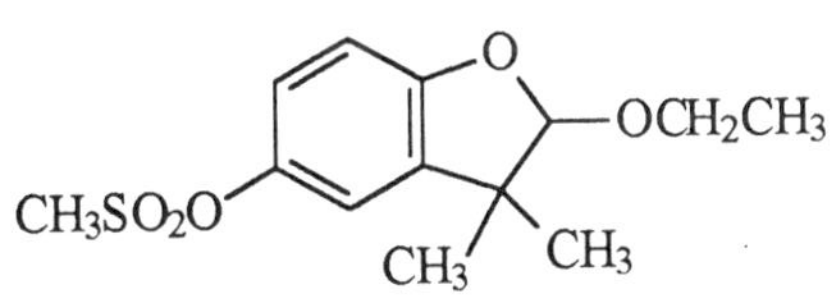

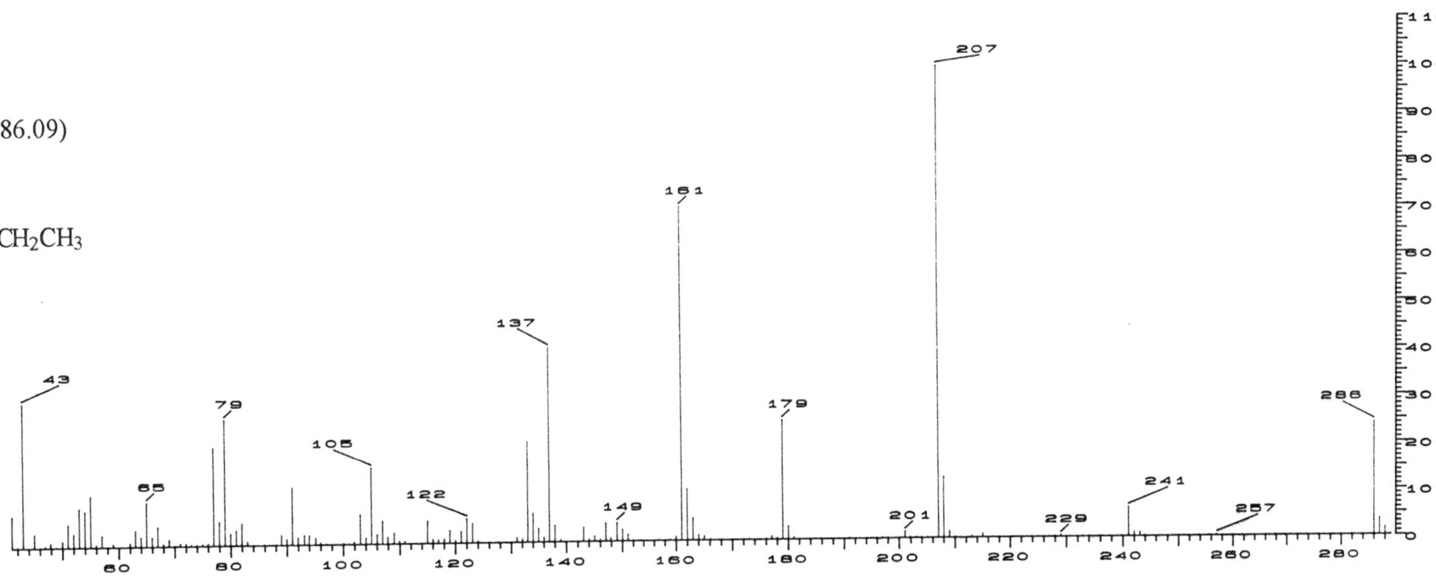

ETHOPROHOS

$C_8H_{19}O_2PS_2$ M.W. = 242.33 (242.06)

$CH_3CH_2OP(=O)(SCH_2CH_2CH_3)_2$

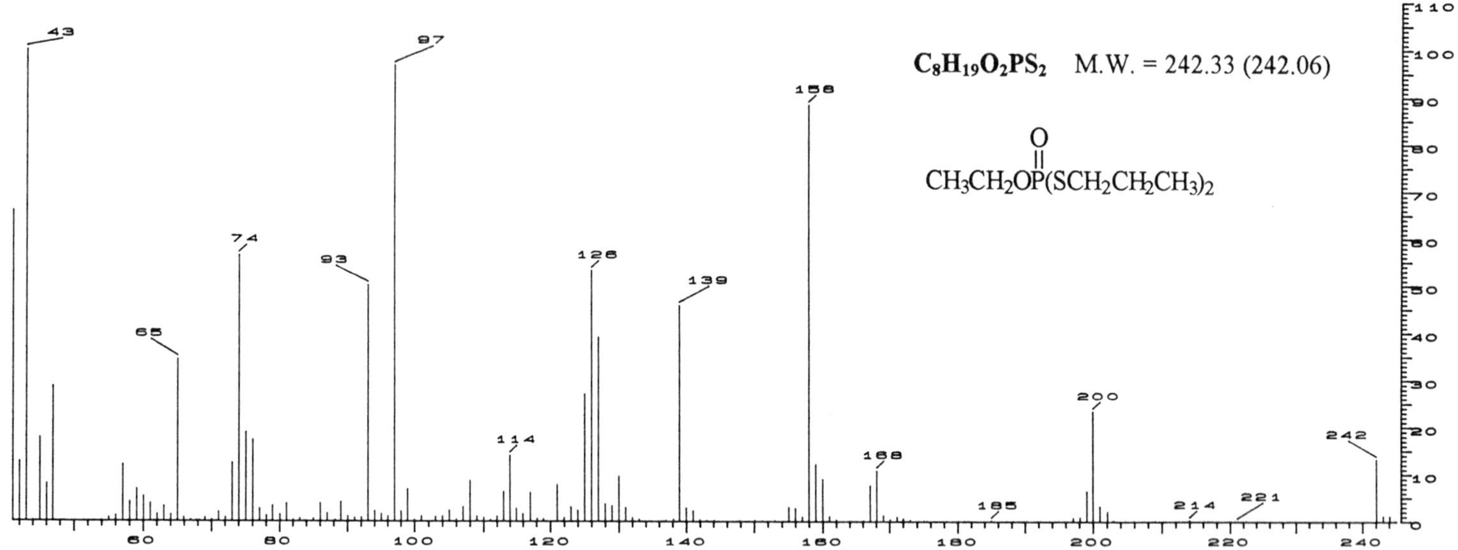

ETRIDIAZOLE

$C_5H_5N_2OCl_3S$ M.W. = 247.52 (245.92)

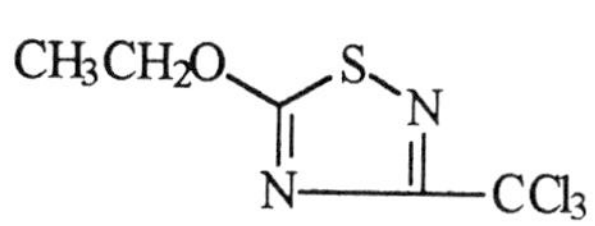

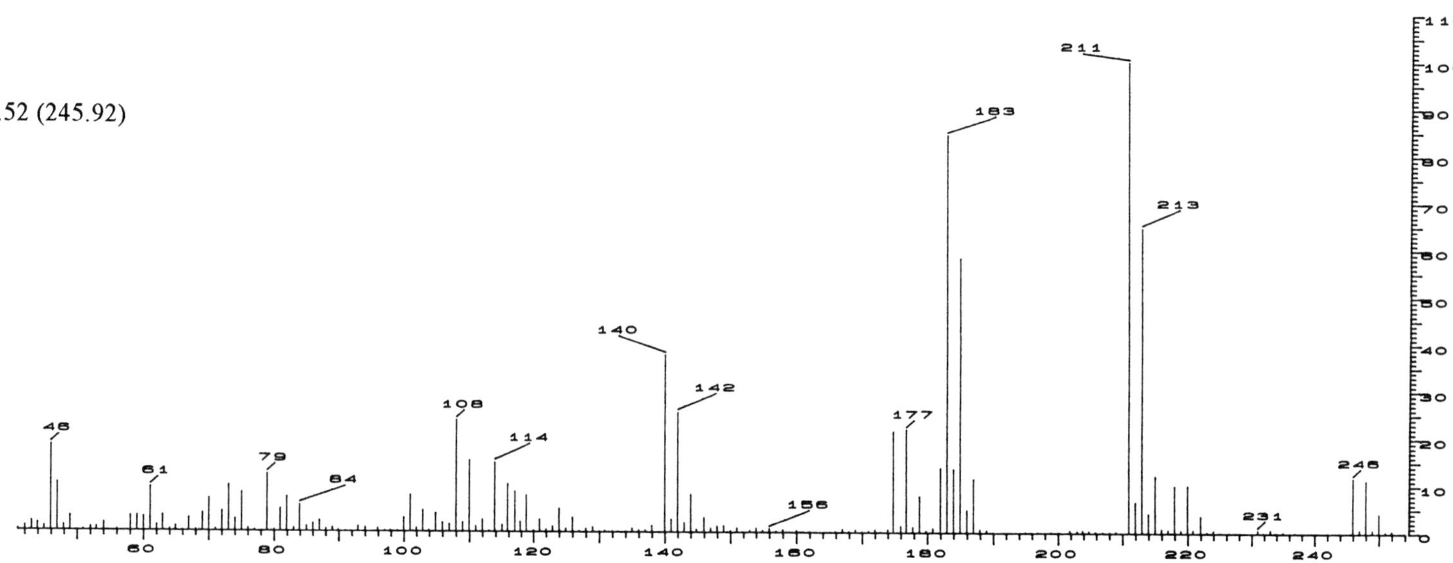

FENAMIPHOS

$C_{13}H_{22}NO_3PS$ M.W. = 303.35 (303.11)

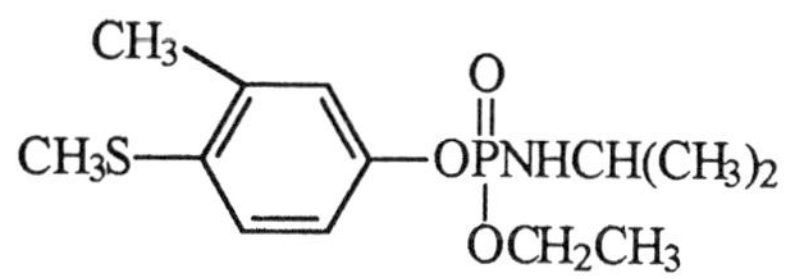

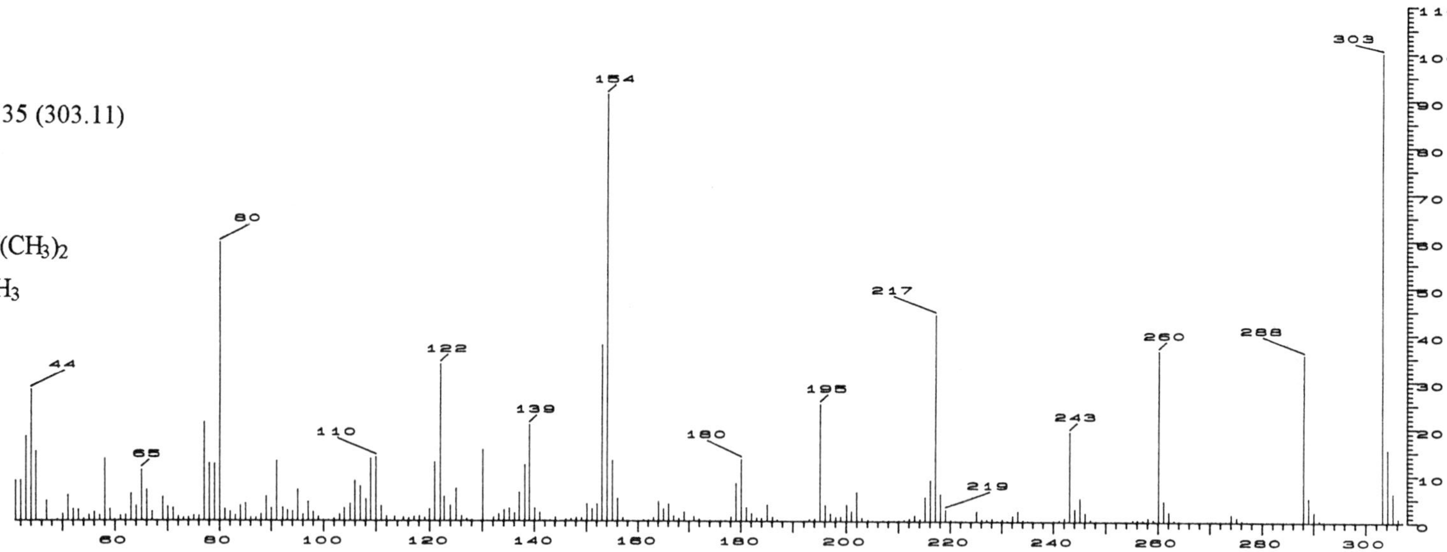

FENCHLORPHOS

$C_8H_8O_3Cl_3SP$ M.W. = 321.54 (319.90)

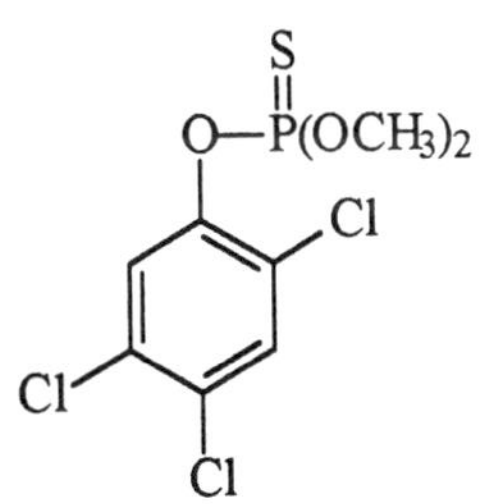

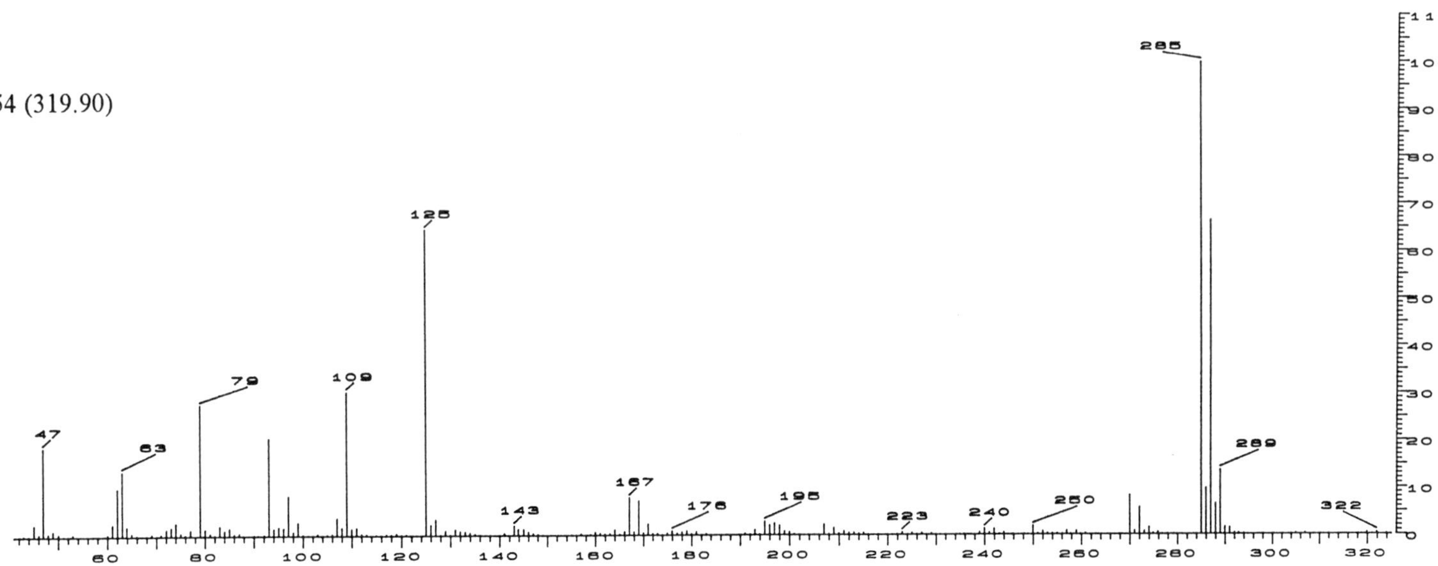

FENOPROP

$C_9H_7O_3Cl_3$ M.W. = 269.51 (267.94)

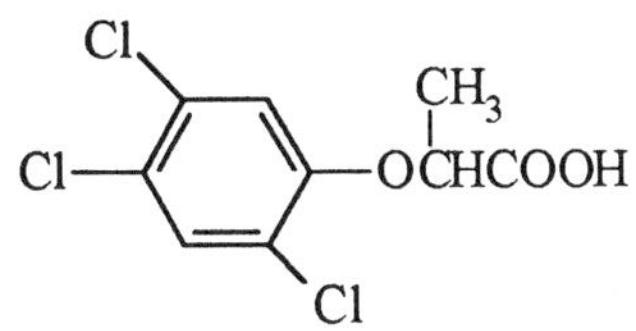

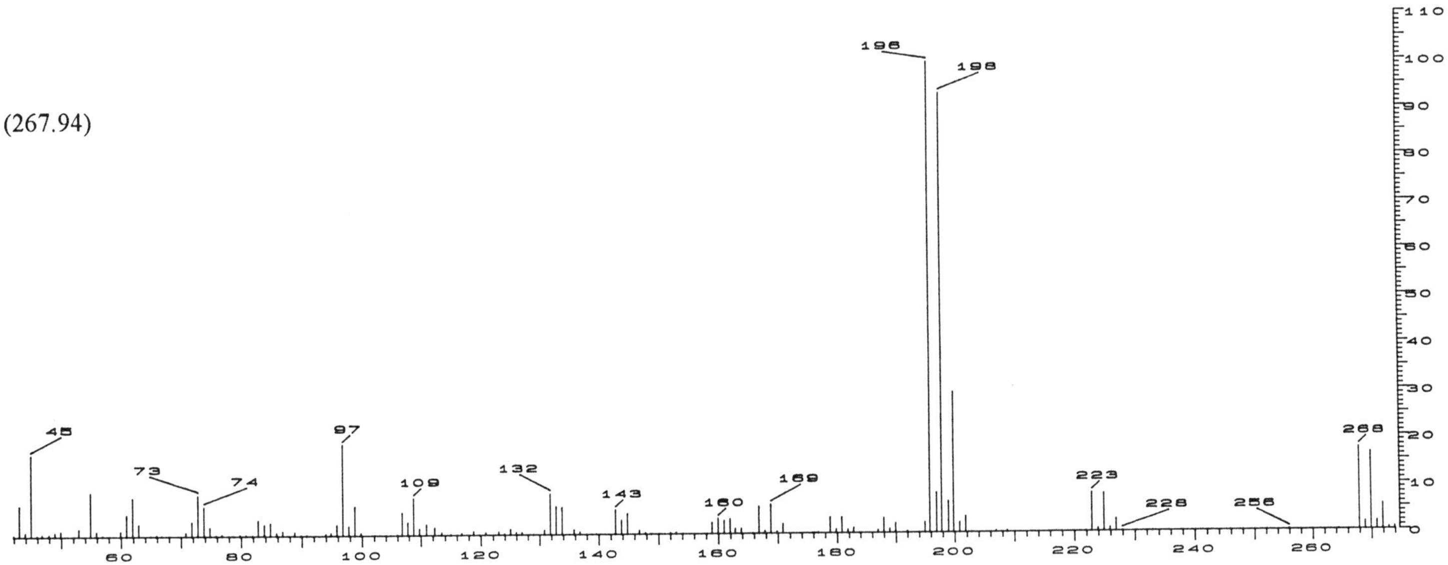

FENSULFOTHION

$C_{11}H_{17}O_4S_2P$ M.W. = 308.34 (308.03)

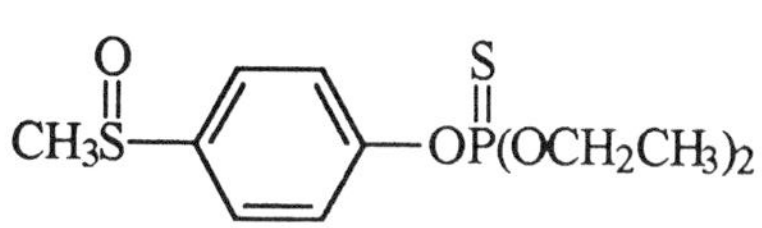

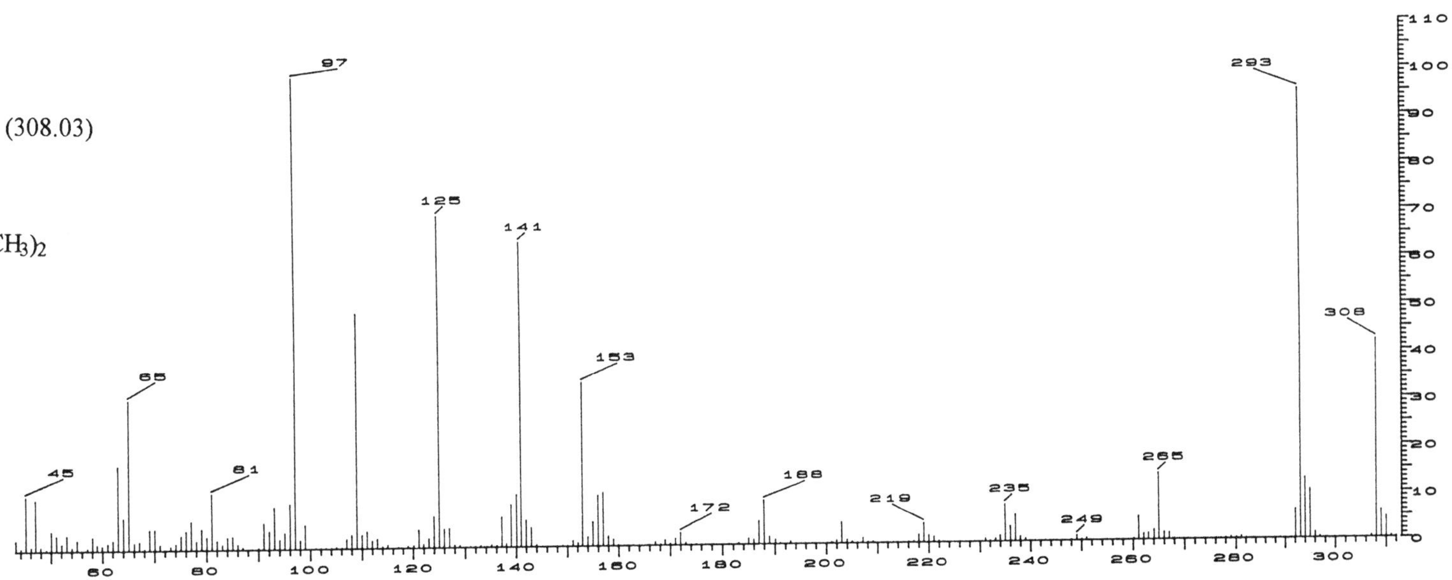

FENTHION

$C_{10}H_{15}O_3S_2P$ M.W. = 278.32 (278.02)

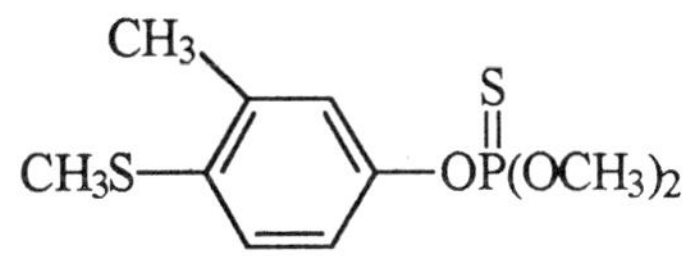

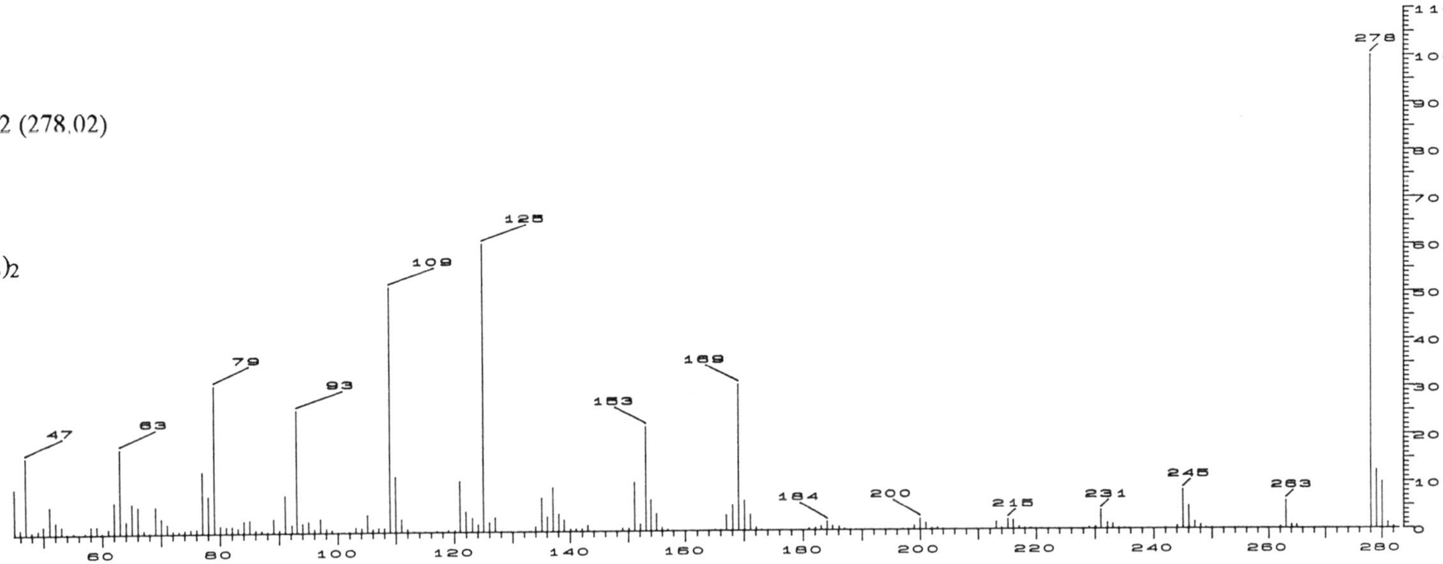

FENTHION-OA

$C_{10}H_{15}O_4PS$ M.W. = 262.26 (262.04)

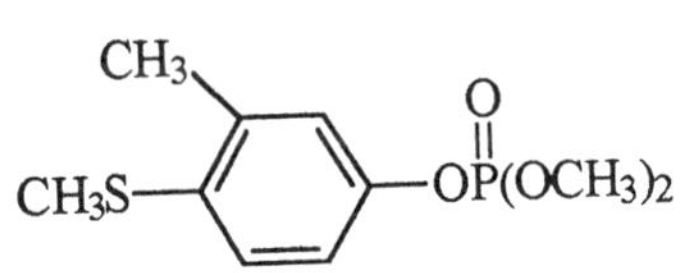

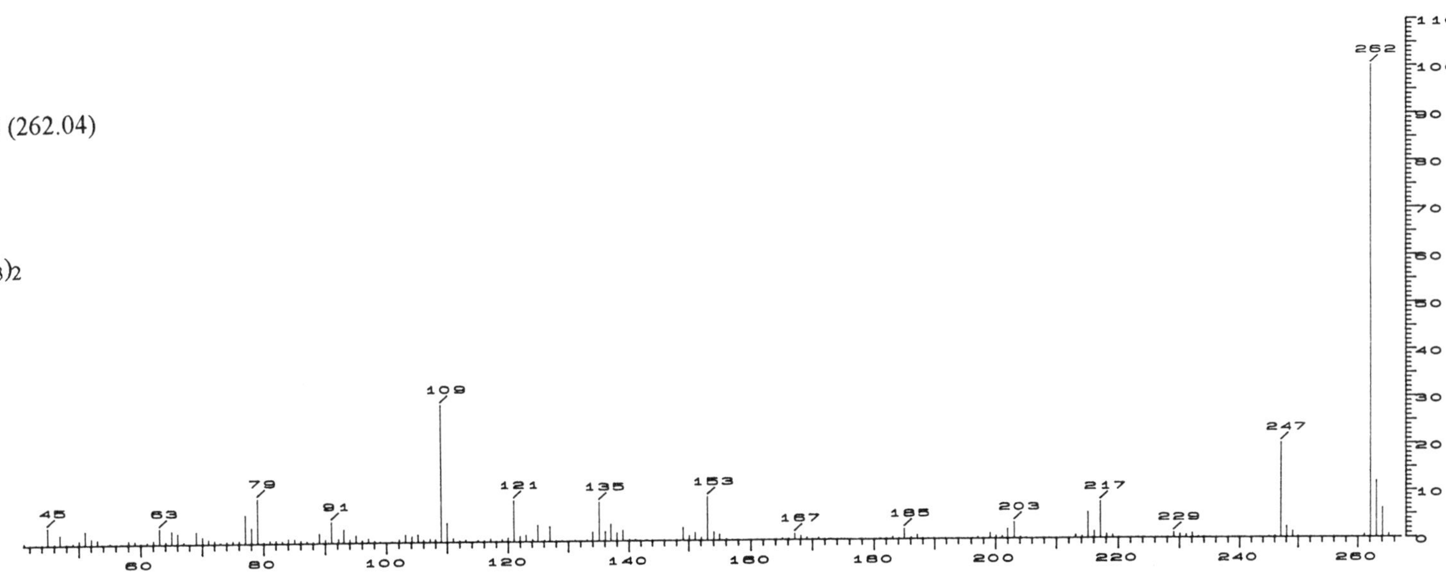

FENVALERATE

$C_{25}H_{22}NO_3Cl$ M.W. = 419.90 (419.13)

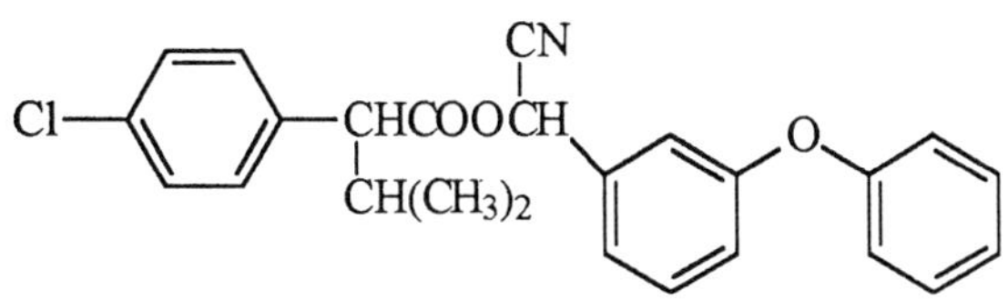

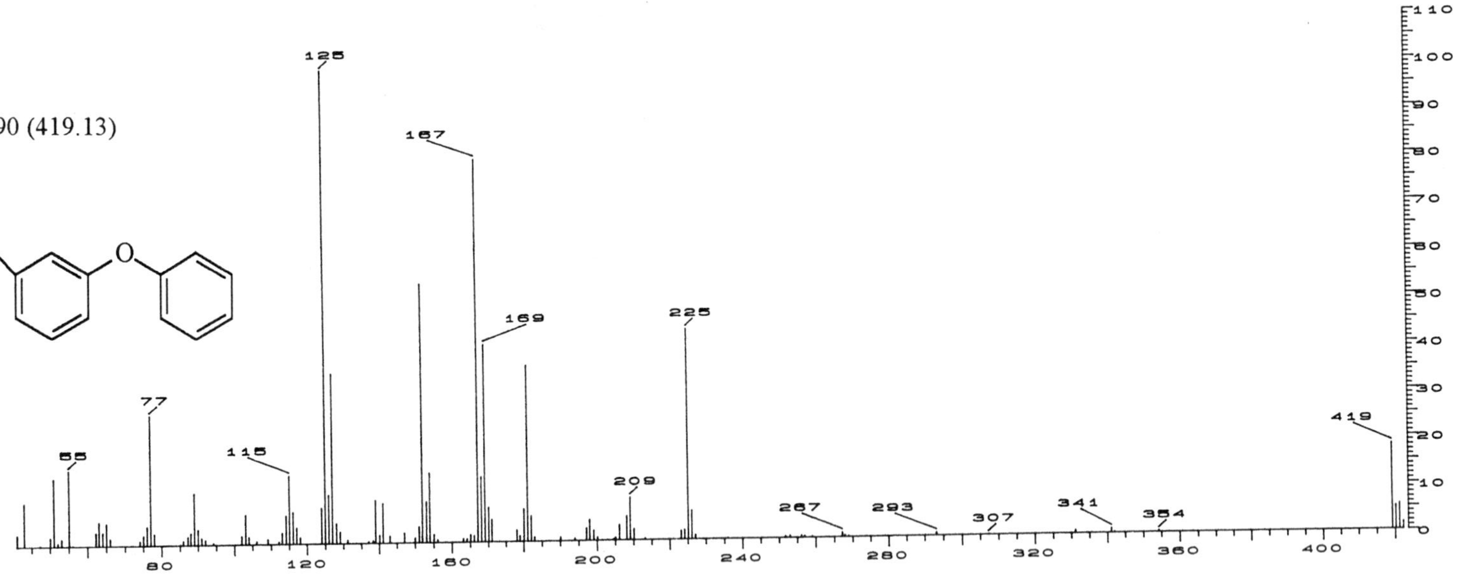

FLUCHLORALIN

$C_{12}H_{13}N_3O_4ClF_3$ M.W. = 355.70 (355.05)

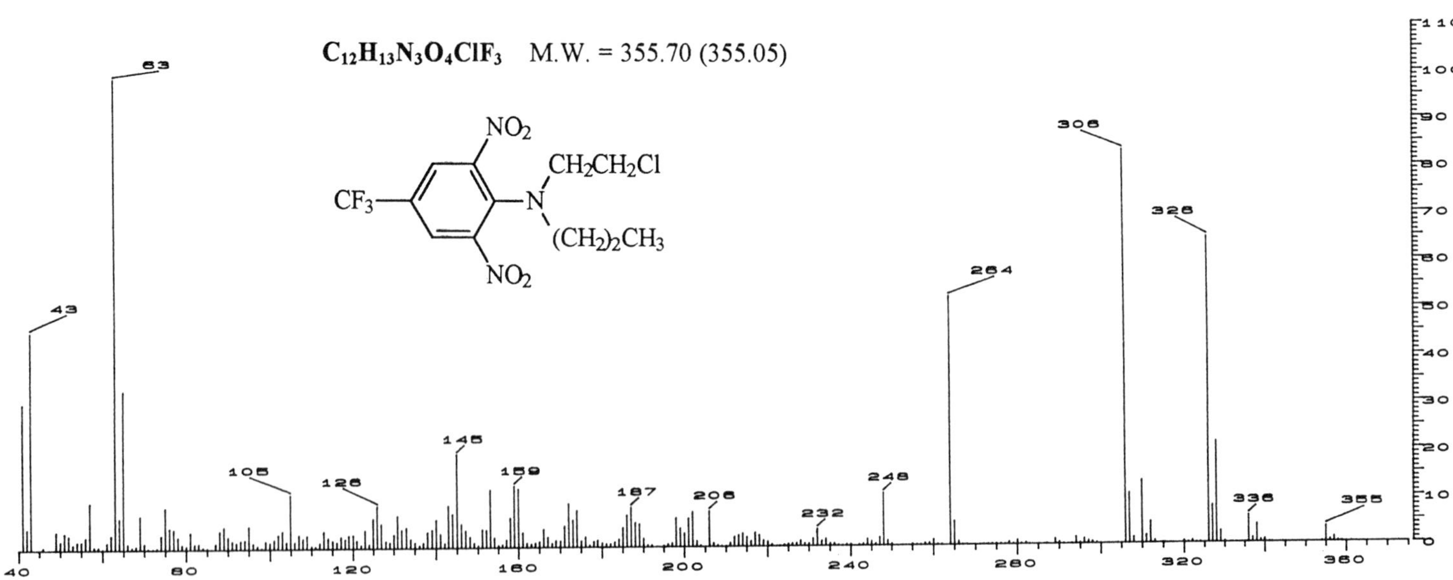

FLUCYTHRINATE

$C_{26}H_{23}NO_4F_2$ M.W. = 451.46 (451.16)

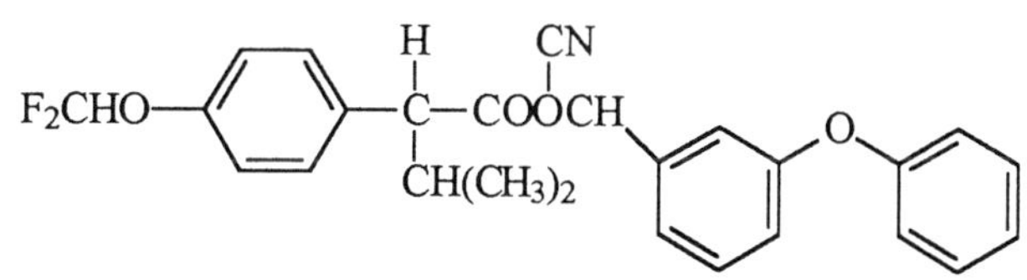

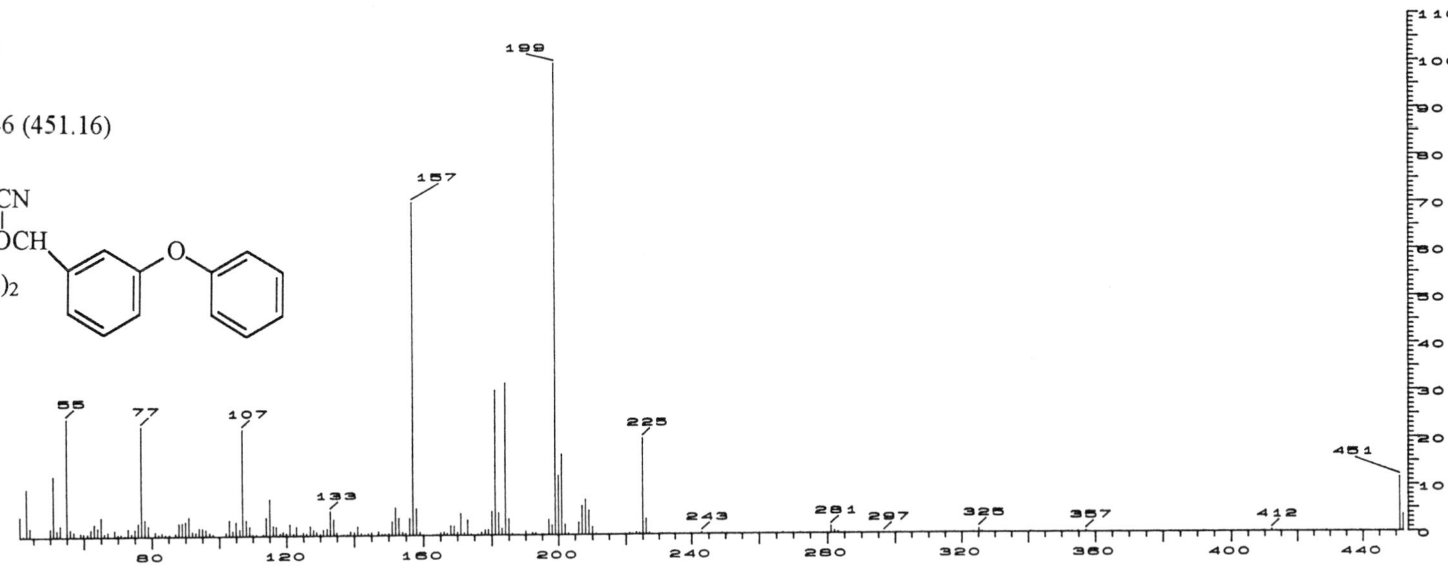

FLUORODIFEN

$C_{13}H_7N_2O_5F_3$ M.W. = 328.20 (328.03)

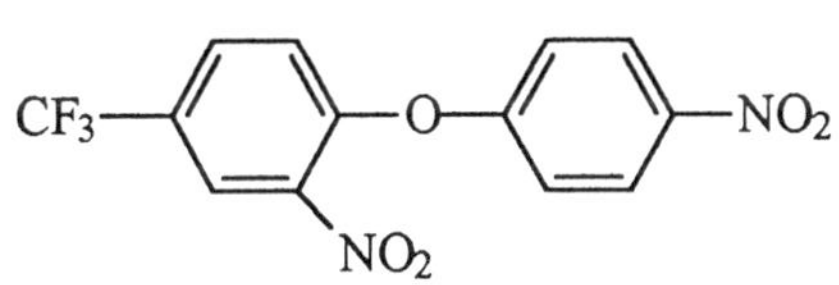

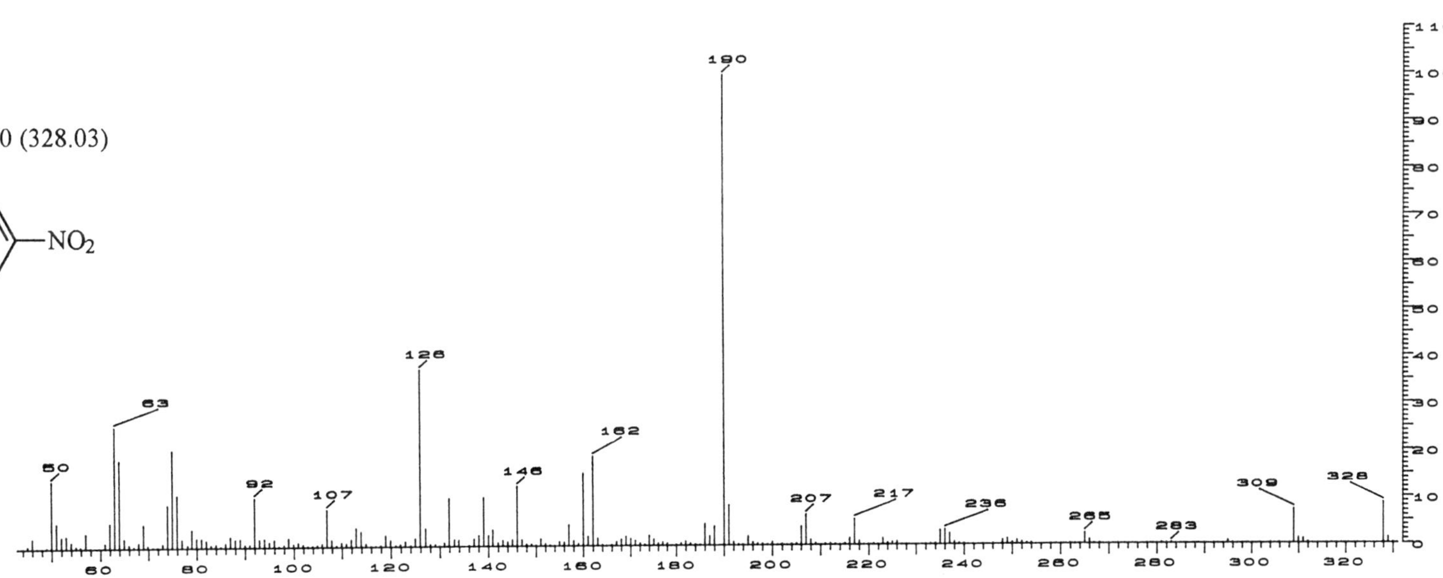

FLURECOL-BUTYL ESTER

$C_{18}H_{18}O_3$ M.W. = 282.34 (282.13)

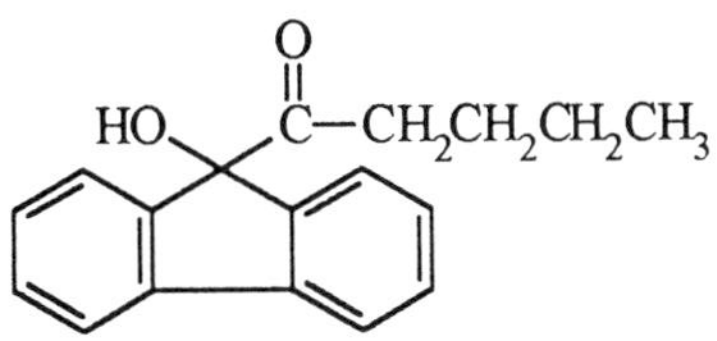

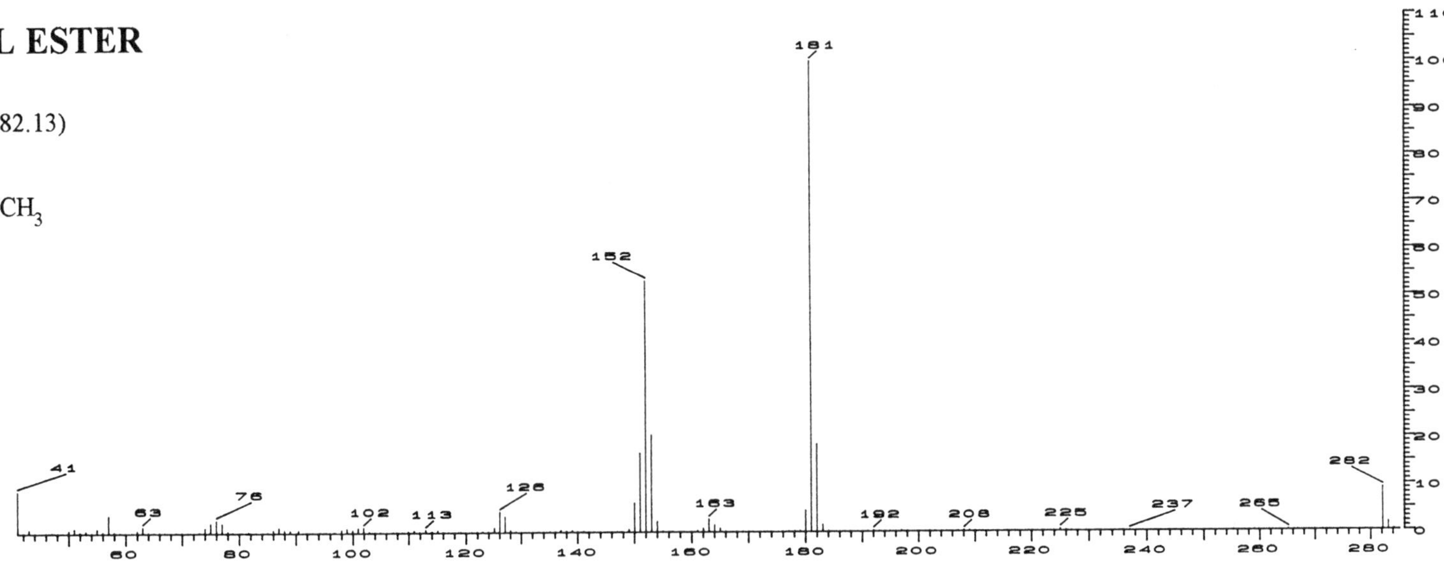

FOLPET

$C_9H_4NO_2Cl_3S$ M.W. = 296.55 (294.90)

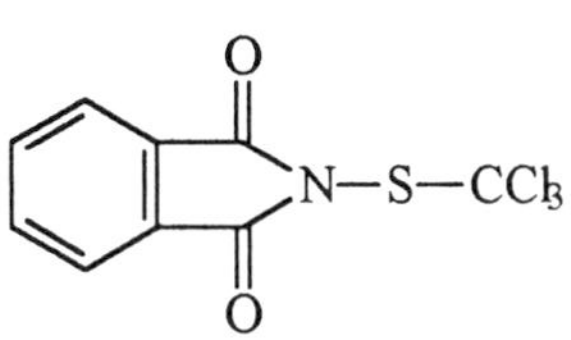

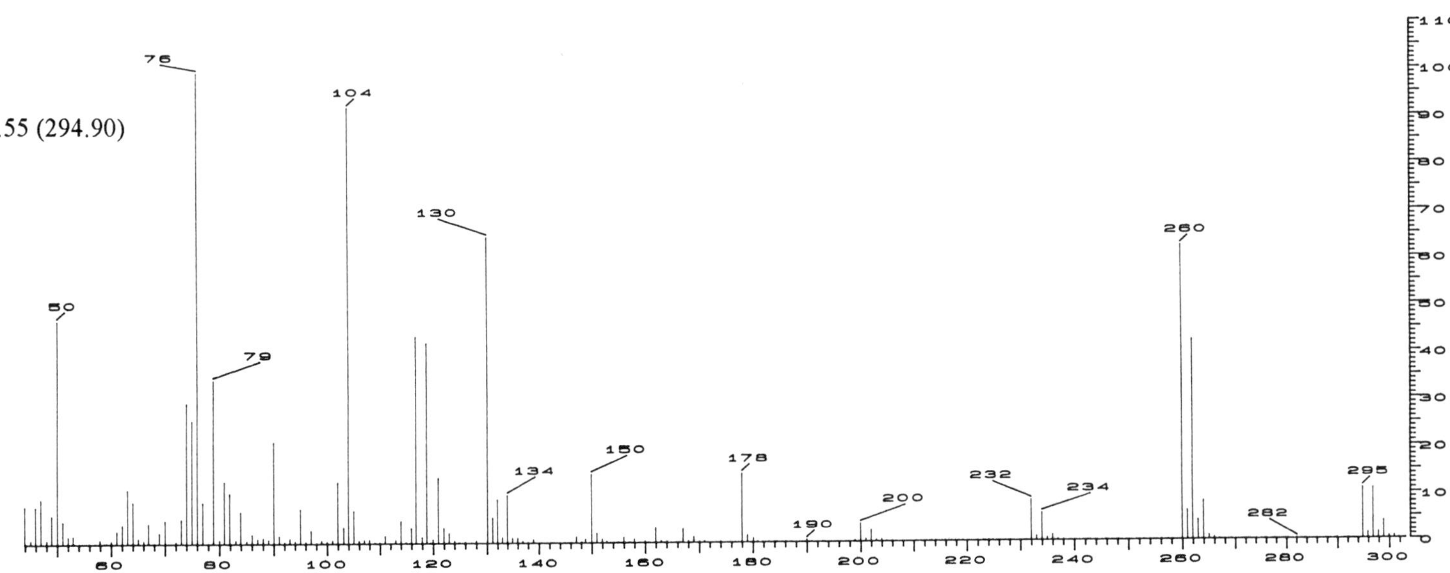

FONOFOS-OA

$C_{10}H_{15}O_2PS$ M.W. = 230.26 (230.05)

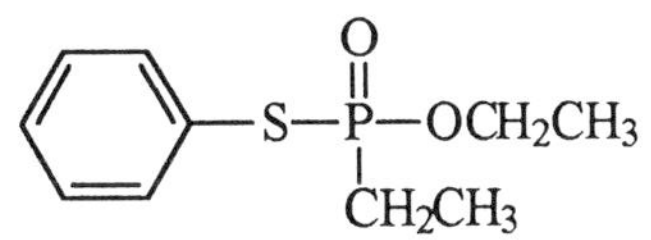

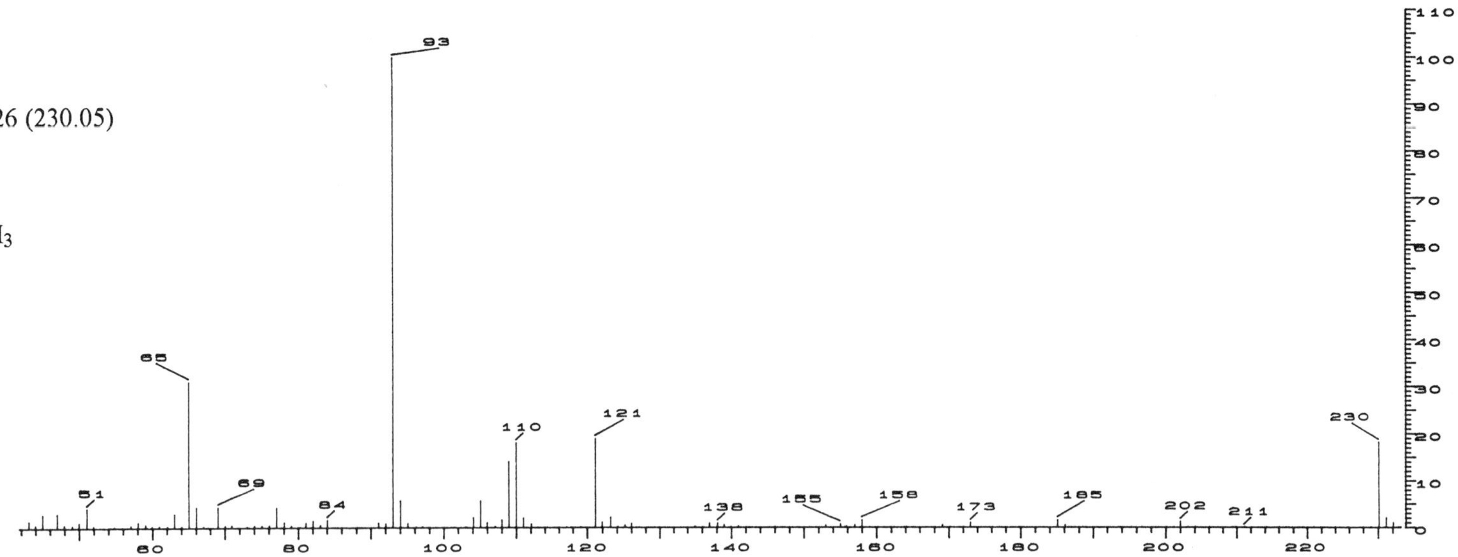

FOSTHIETAN

$C_6H_{12}NO_3PS_2$ M.W. = 241.25 (241.00)

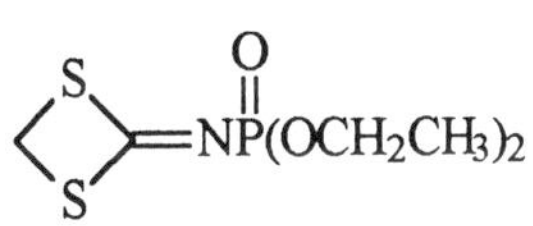

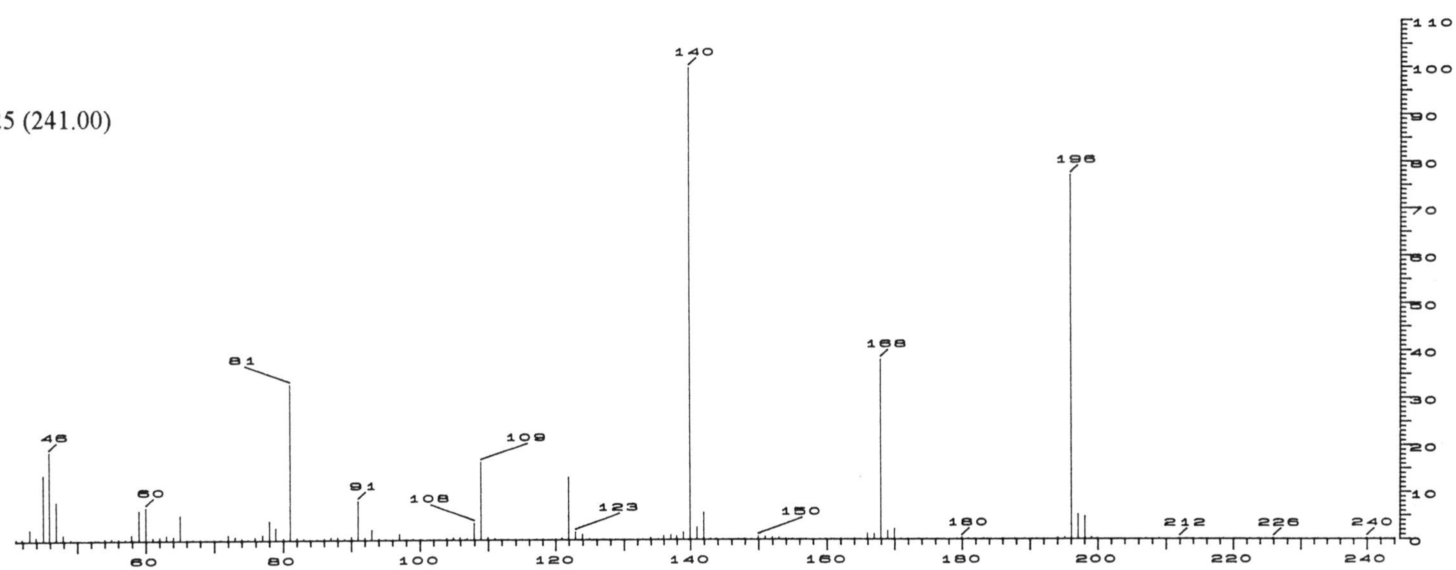

HEPTACHLOR

$C_{10}H_5Cl_7$ M.W. = 373.35 (369.82)

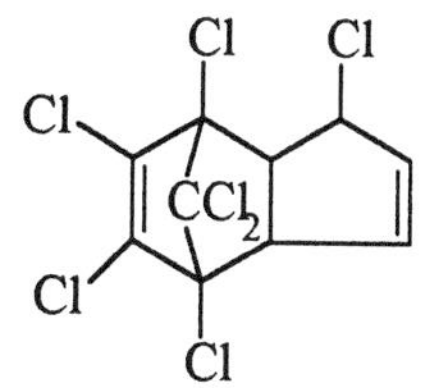

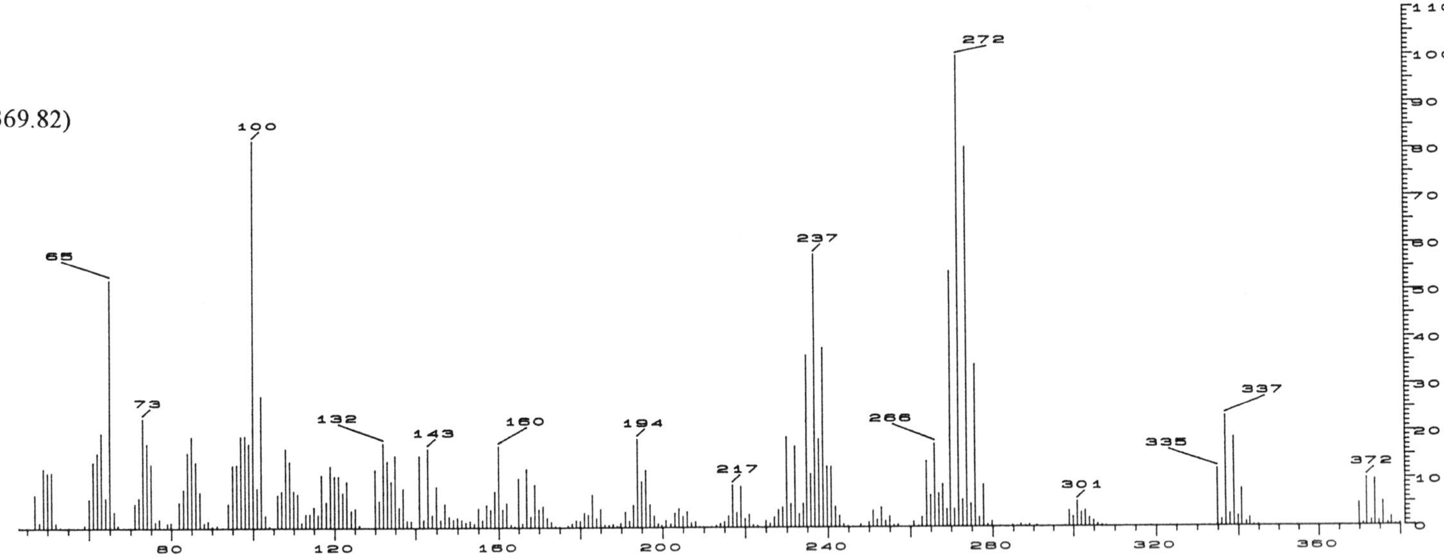

HEPTACHLOR EPOXIDE

$C_{10}H_5OCl_7$ M.W. = 389.32 (385.82)

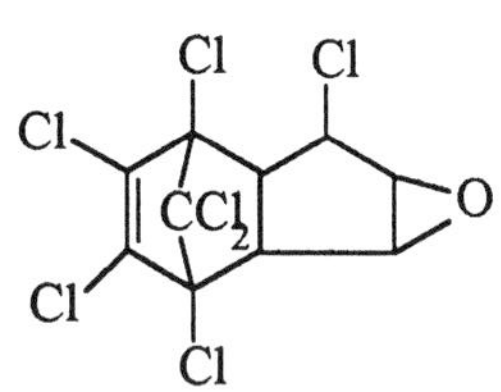

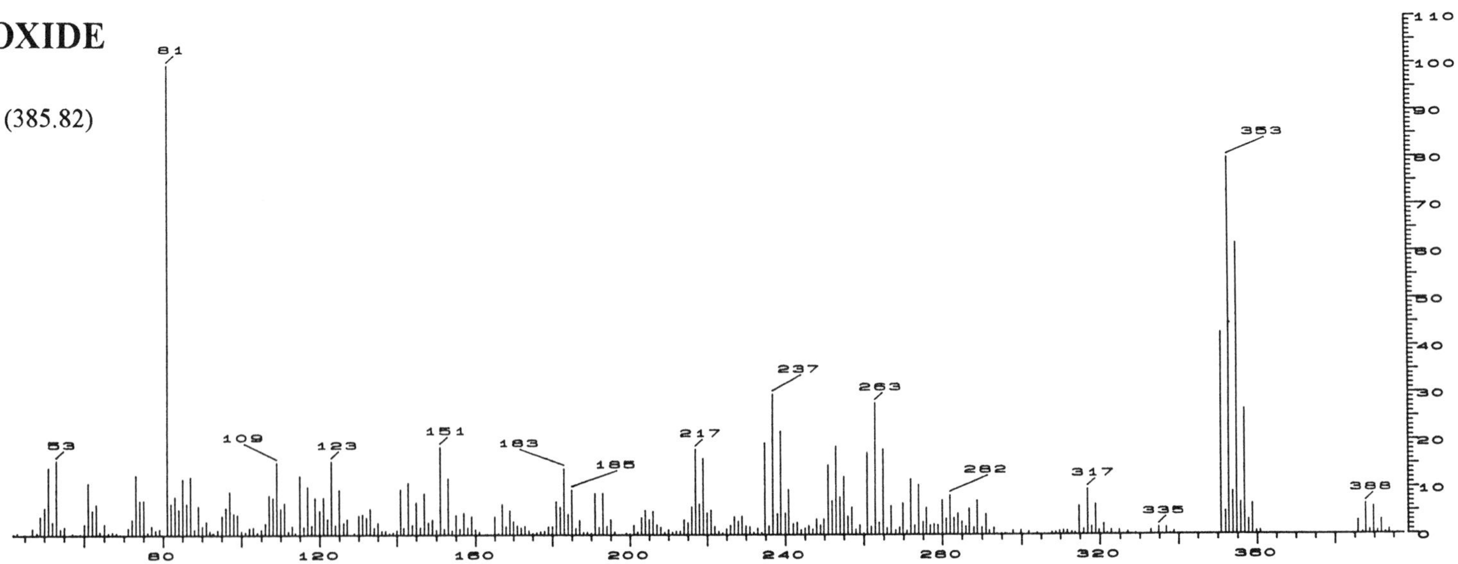

HEXACHLOROBENZENE

C_6Cl_6 M.W. = 284.78 (281.81)

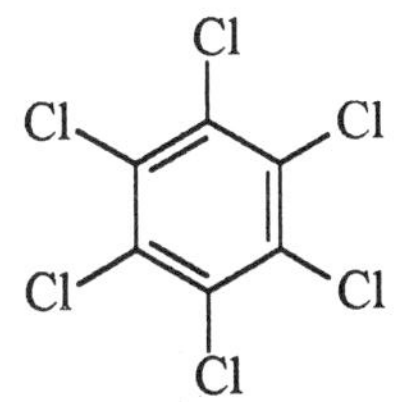

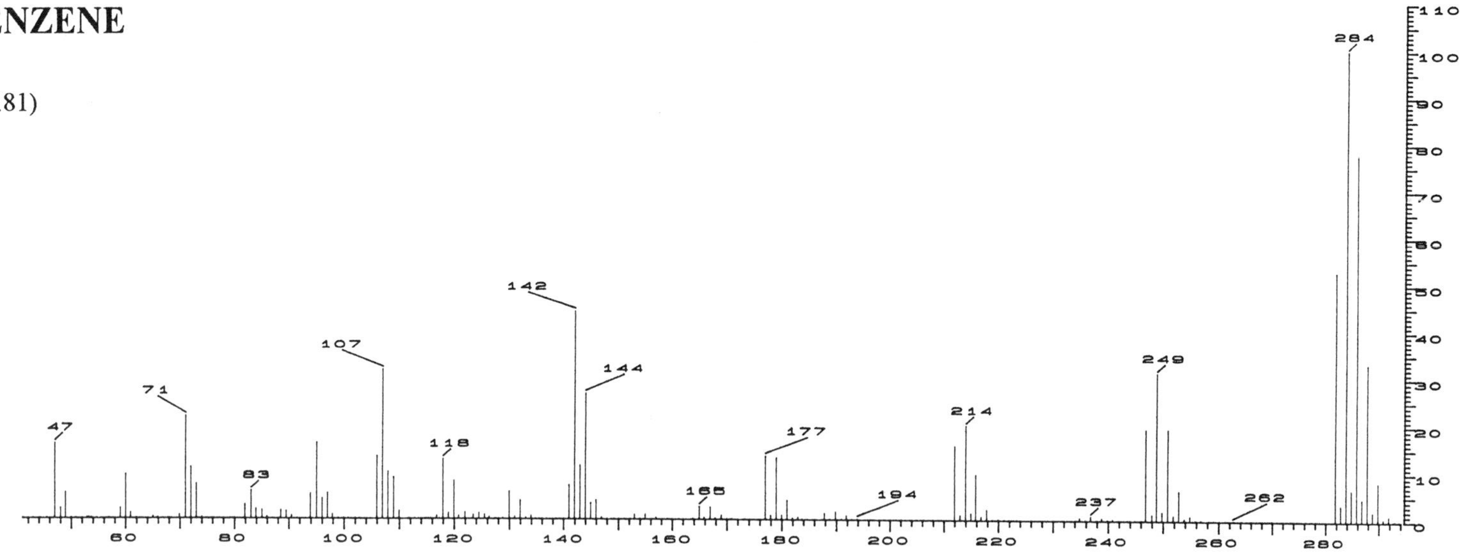

HEXAZINONE

$C_{12}H_{20}N_4O_2$ M.W. = 252.31 (252.16)

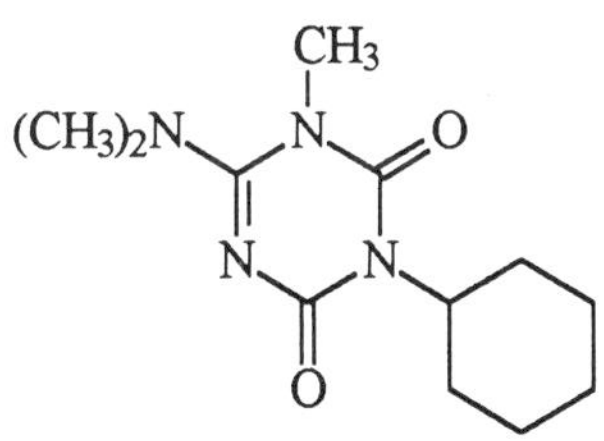

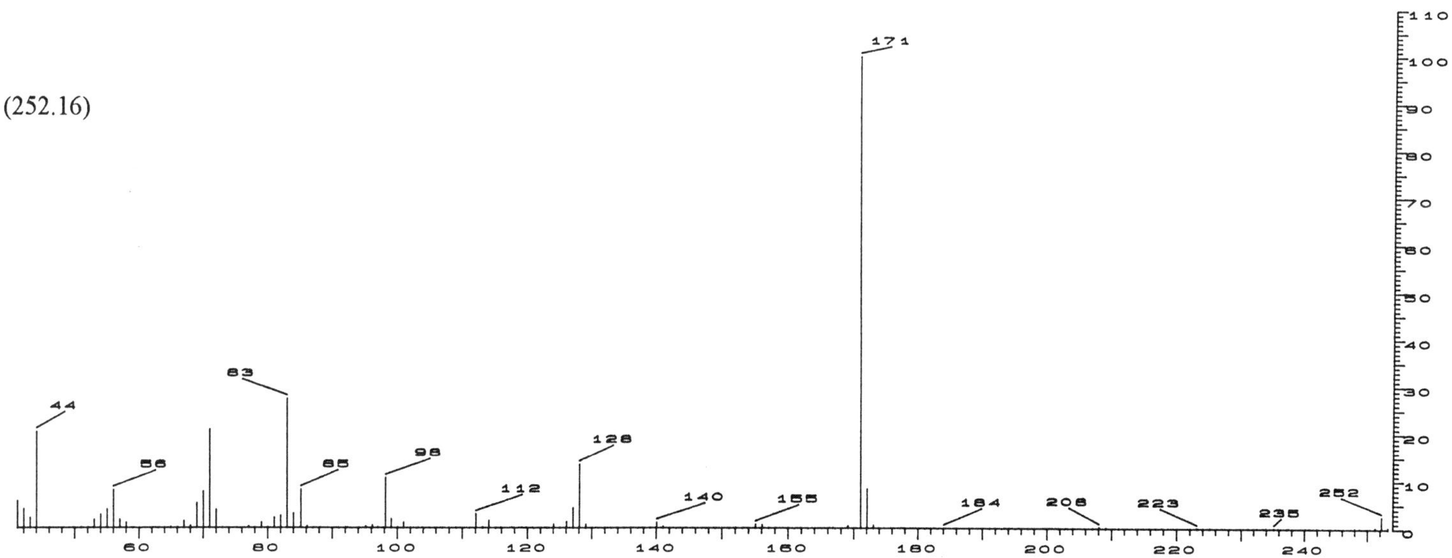

1-HYDROXYCHLORDENE

$C_{10}H_6OCl_6$ M.W. = 354.87 (351.85)

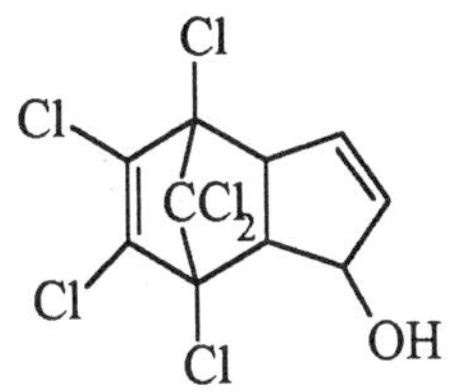

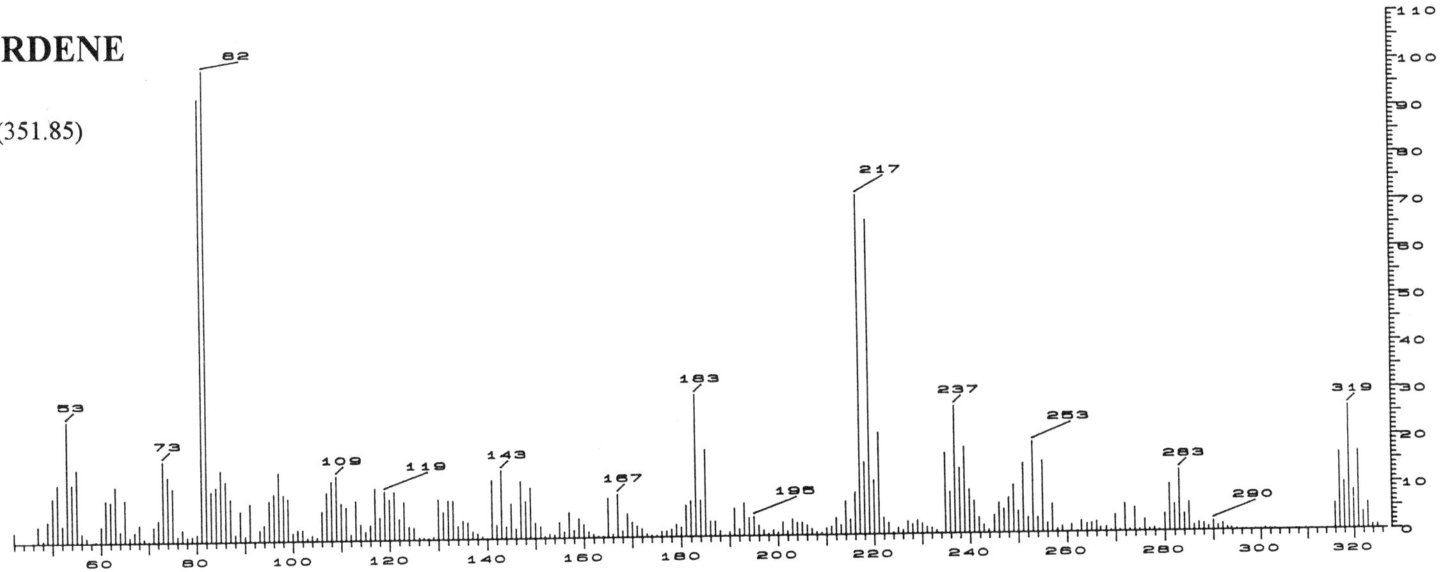

IBP

$C_{13}H_{21}O_3SP$ M.W. = 288.34 (288.09)

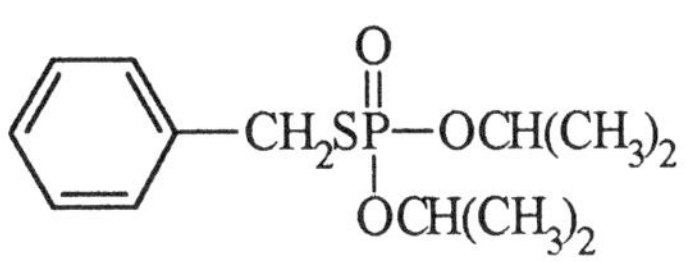

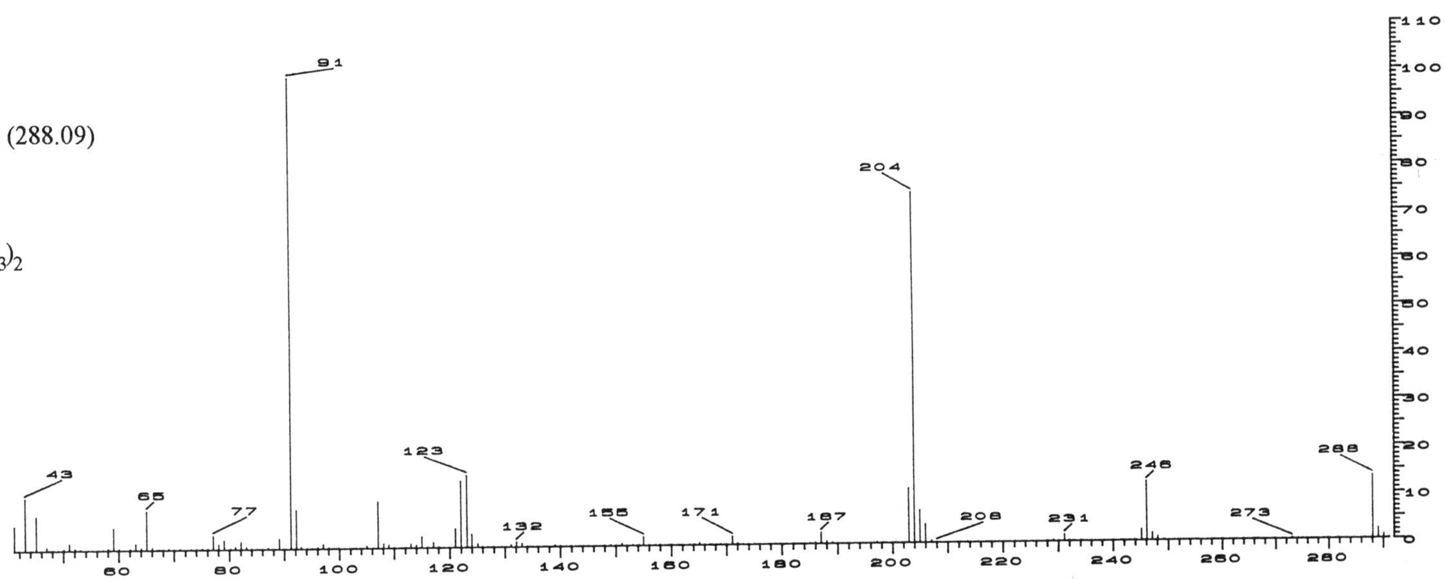

IPRODIONE

$C_{13}H_{13}N_3O_3Cl_2$ M.W. = 330.17 (329.03)

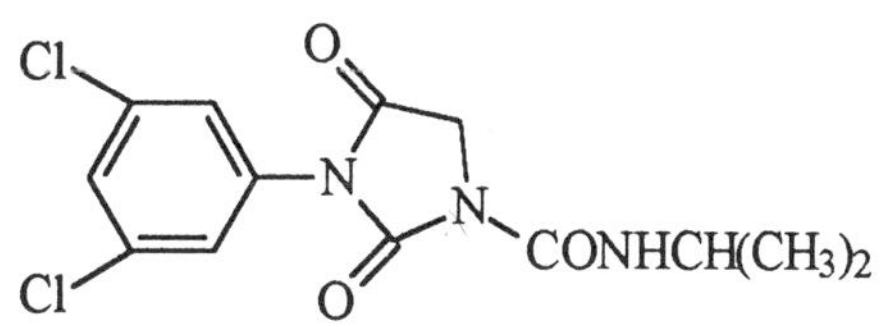

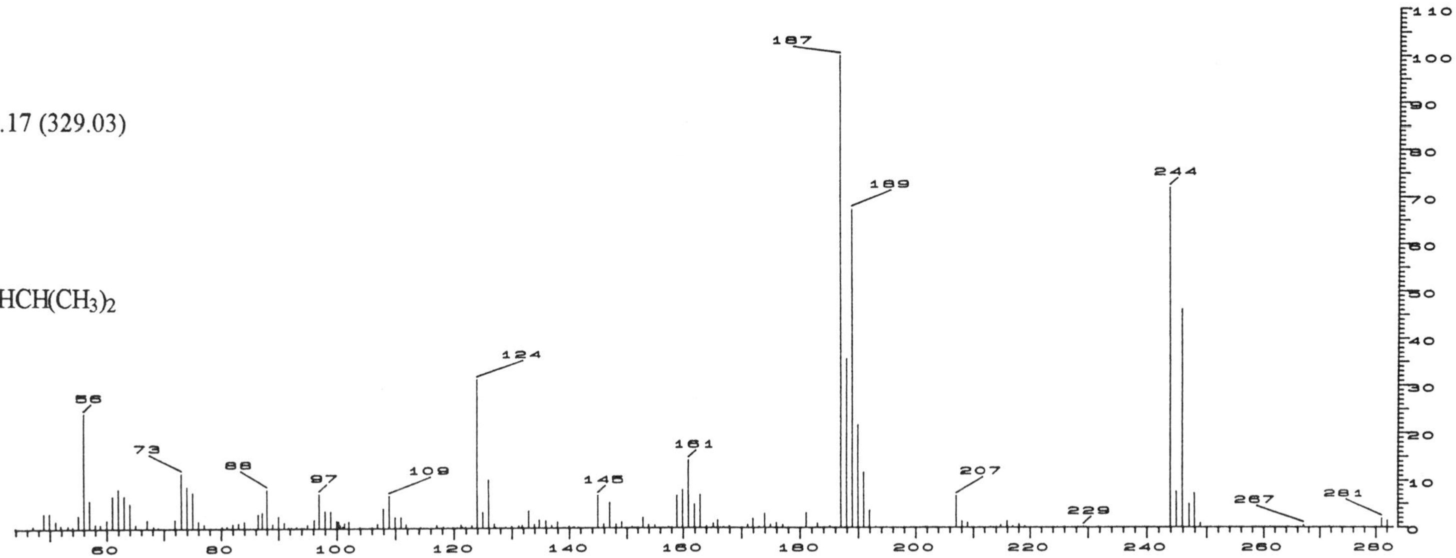

ISODRIN

$C_{12}H_8Cl_6$ M.W. = 364.91 (361.88)

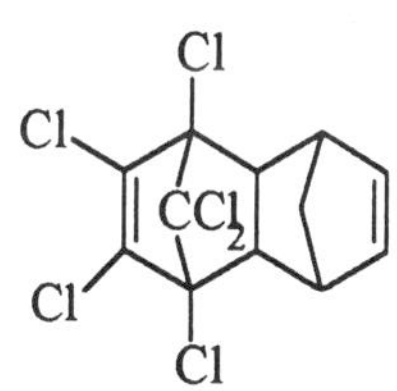

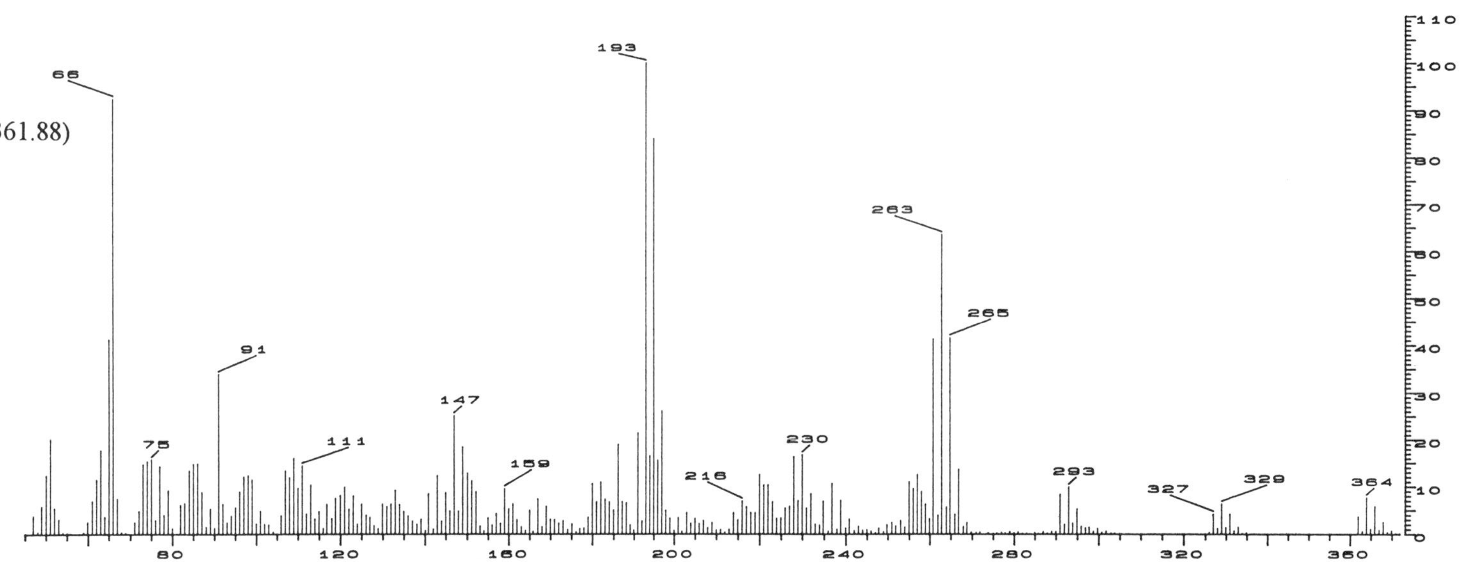

ISOFENPHOS

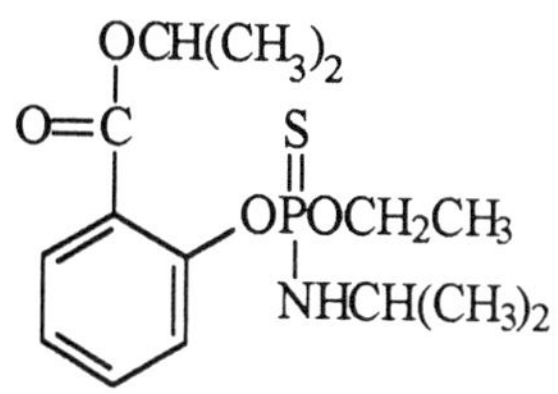

$C_{15}H_{24}NO_4SP$ M.W. = 345.39 (345.12)

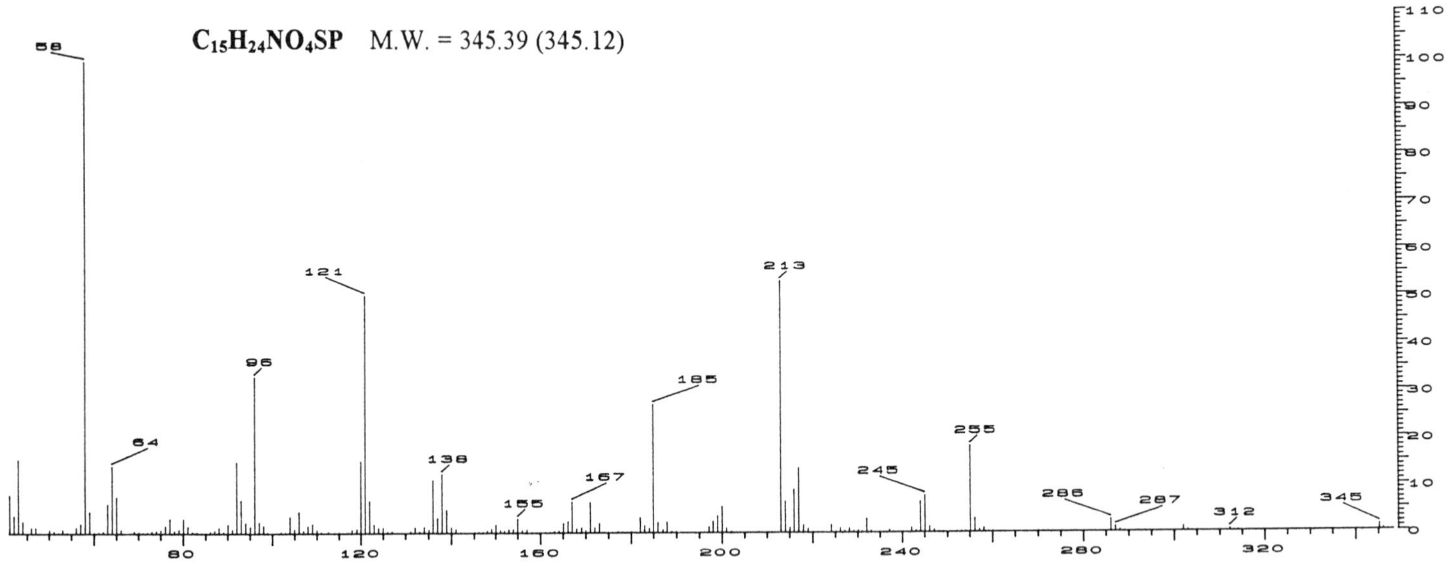

ISOPROPALIN

$C_{15}H_{23}N_3O_4$ M.W. = 309.36 (309.17)

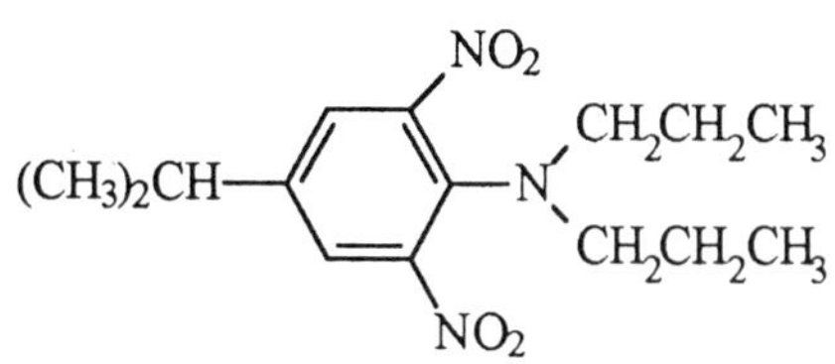

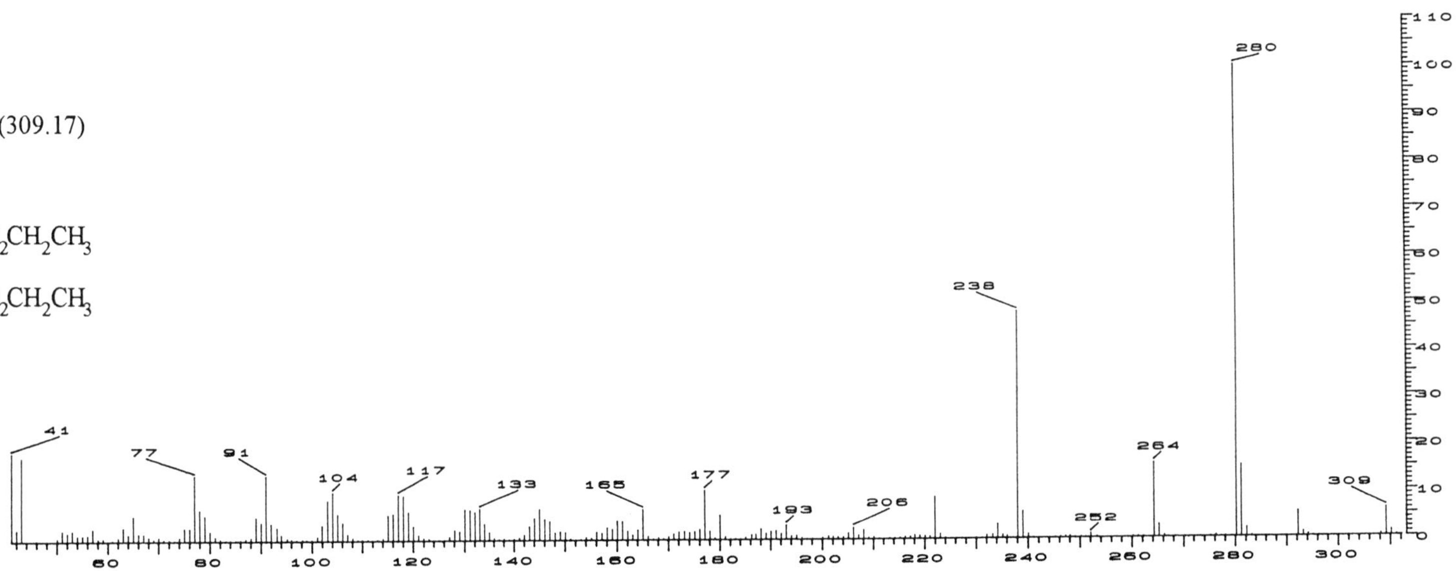

ISOPROTURON

$C_{12}H_{18}N_2O$ M.W. = 206.28 (206.14)

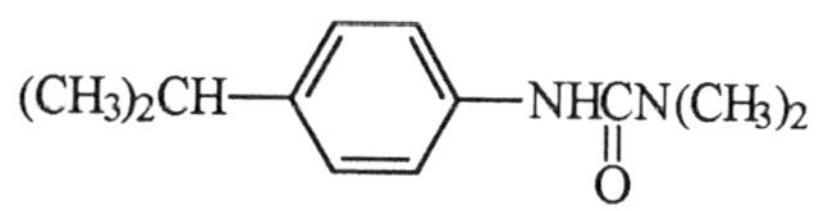

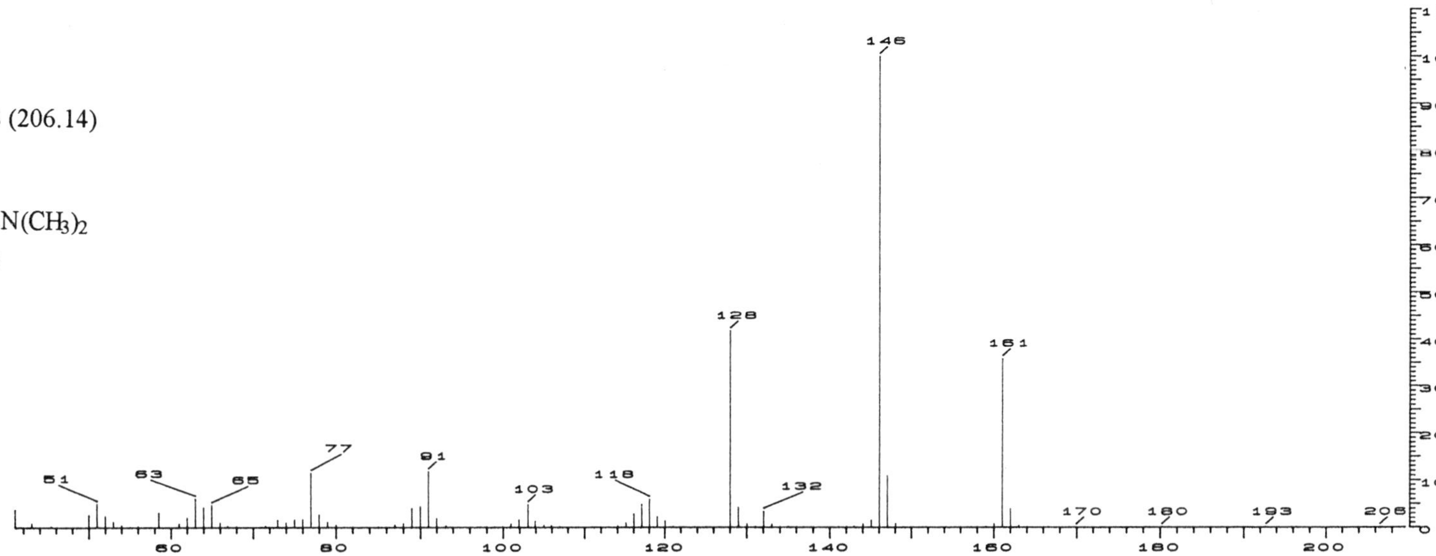

3-KETOCARBOFURANPHENOL

$C_{10}H_{10}O_3$ M.W. = 178.18 (178.06)

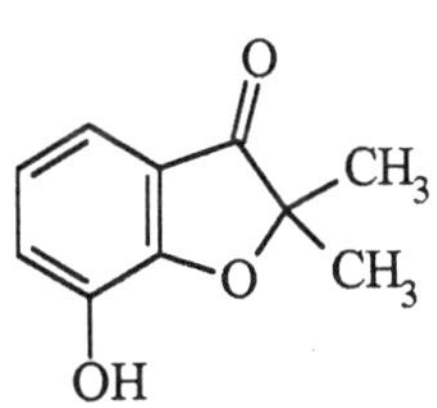

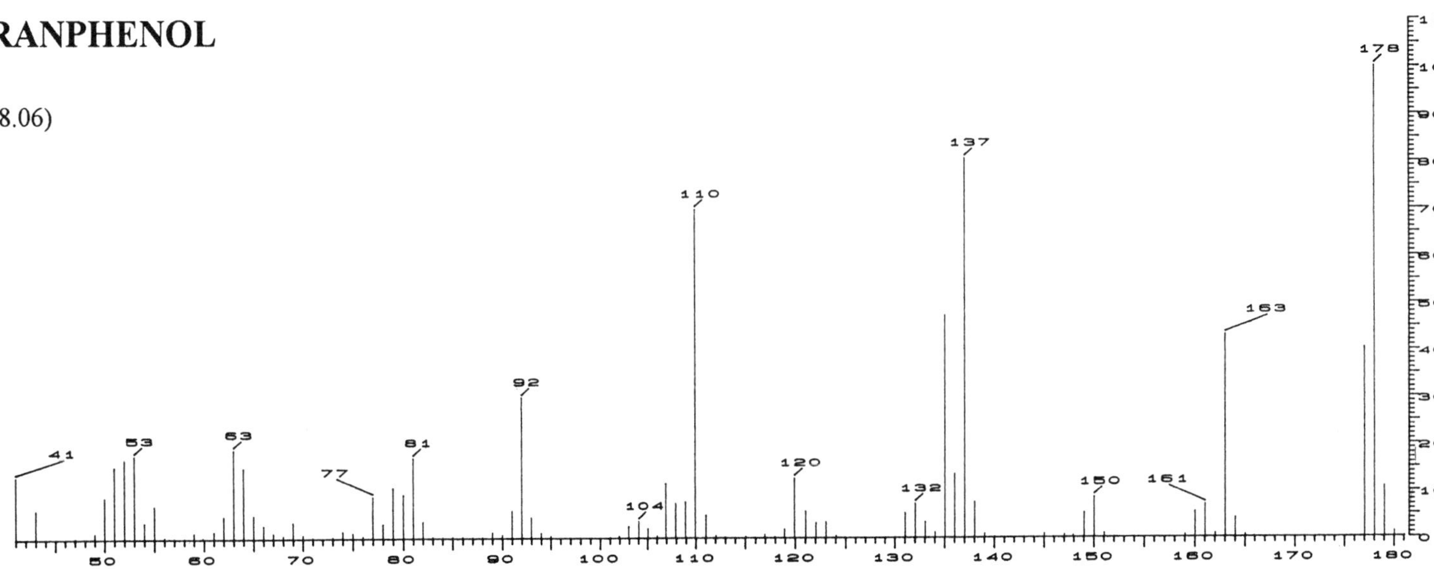

LEPTOPHOS

$C_{13}H_{10}O_2BrCl_2SP$ M.W. = 412.06 (409.87)

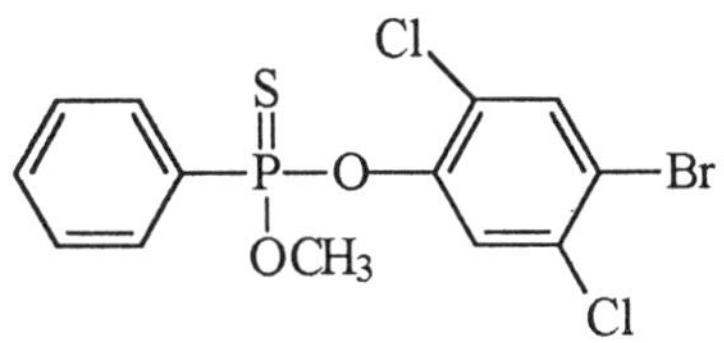

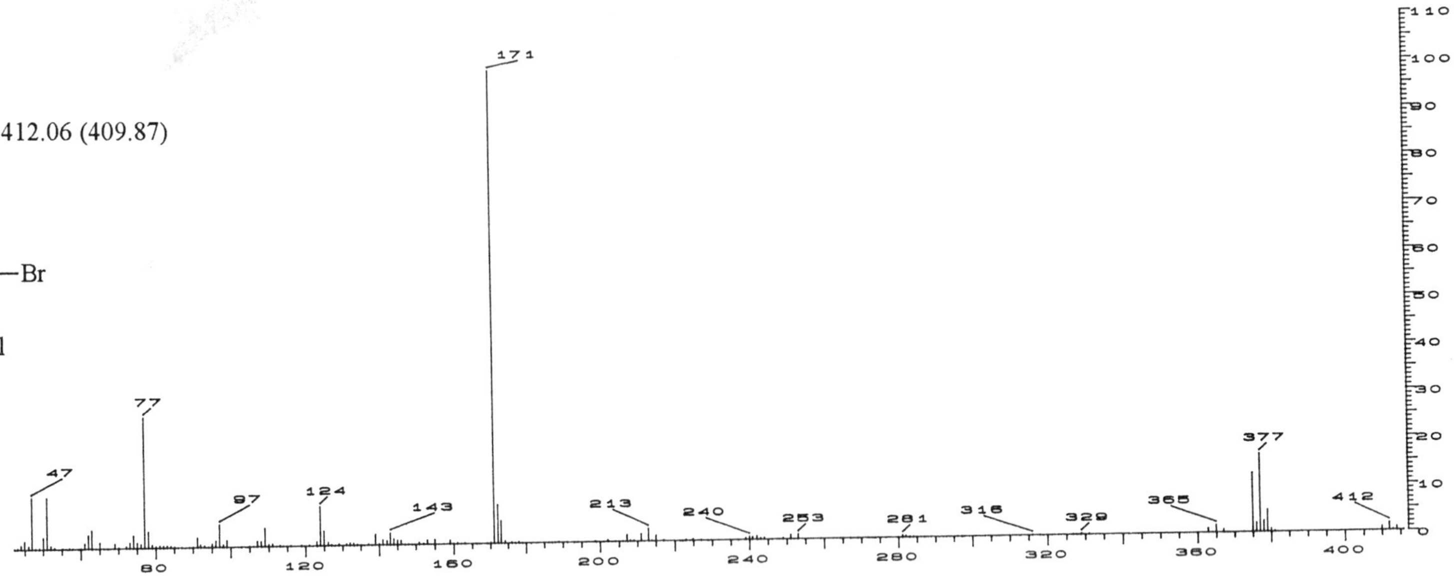

LETHANE 384

$C_9H_{17}NO_2S$ M.W. = 203.29 (203.10)

$C_4H_9O(CH_2)_2—O—(CH_2)_2—S—CN$

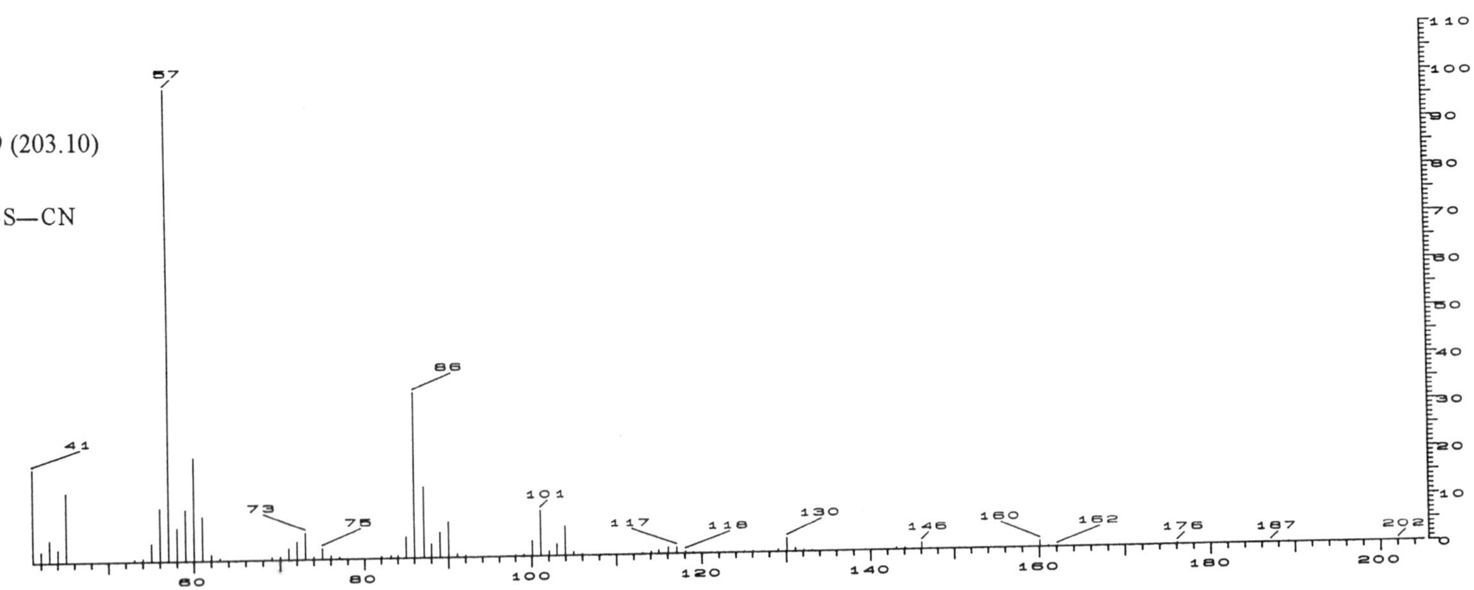

LINDANE

$C_6H_6Cl_6$ M.W. = 290.83 (287.86)

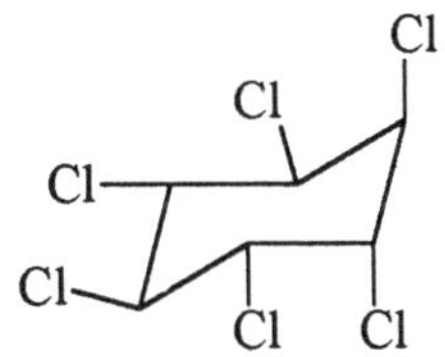

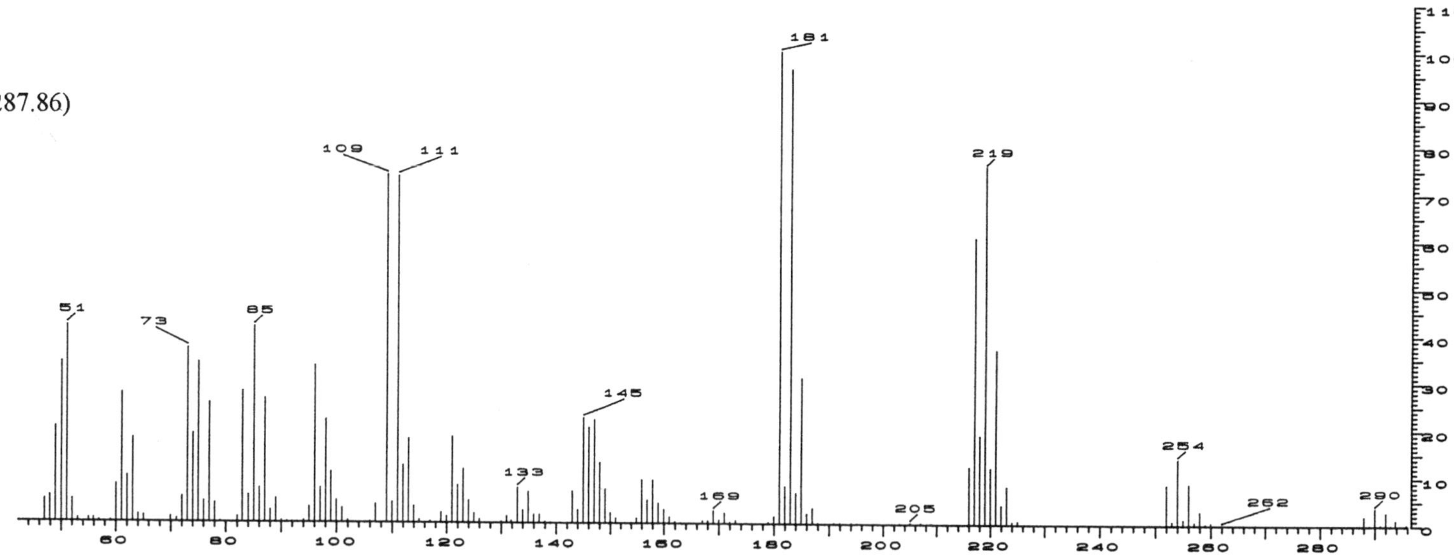

MALATHION

$C_{10}H_{19}O_6S_2P$ M.W. = 330.35 (330.04)

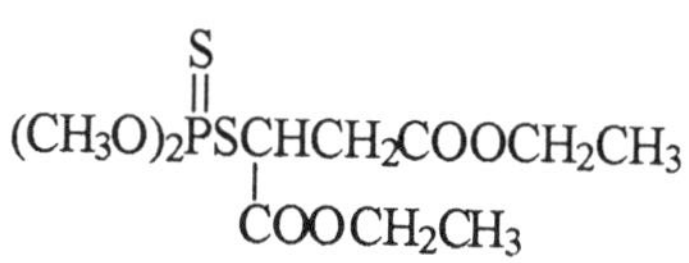

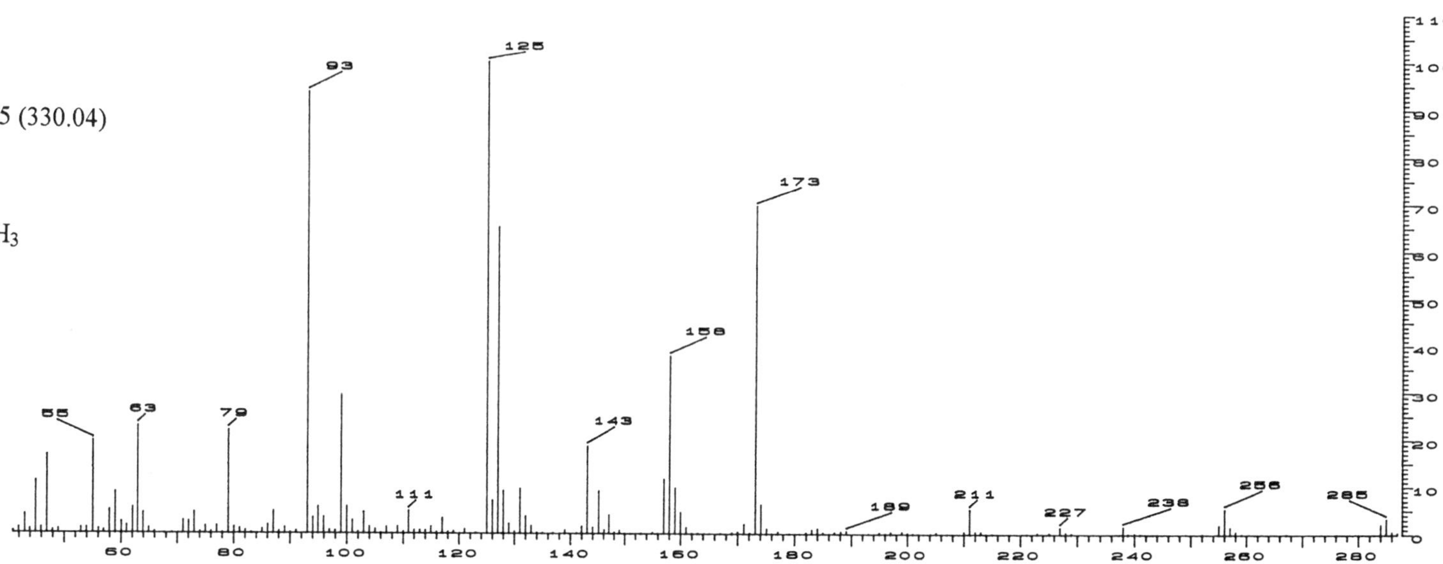

MALATHION-OA

$C_{10}H_{19}O_7PS$ M.W. = 314.29 (314.06)

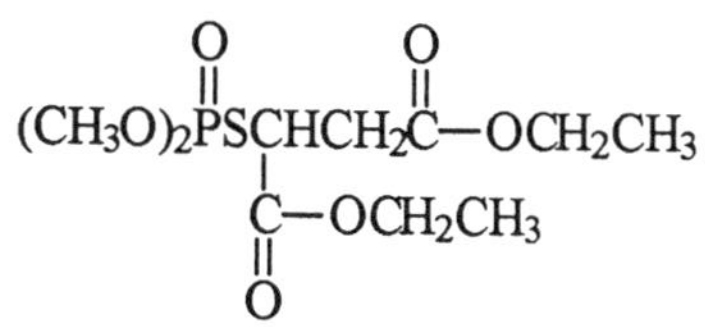

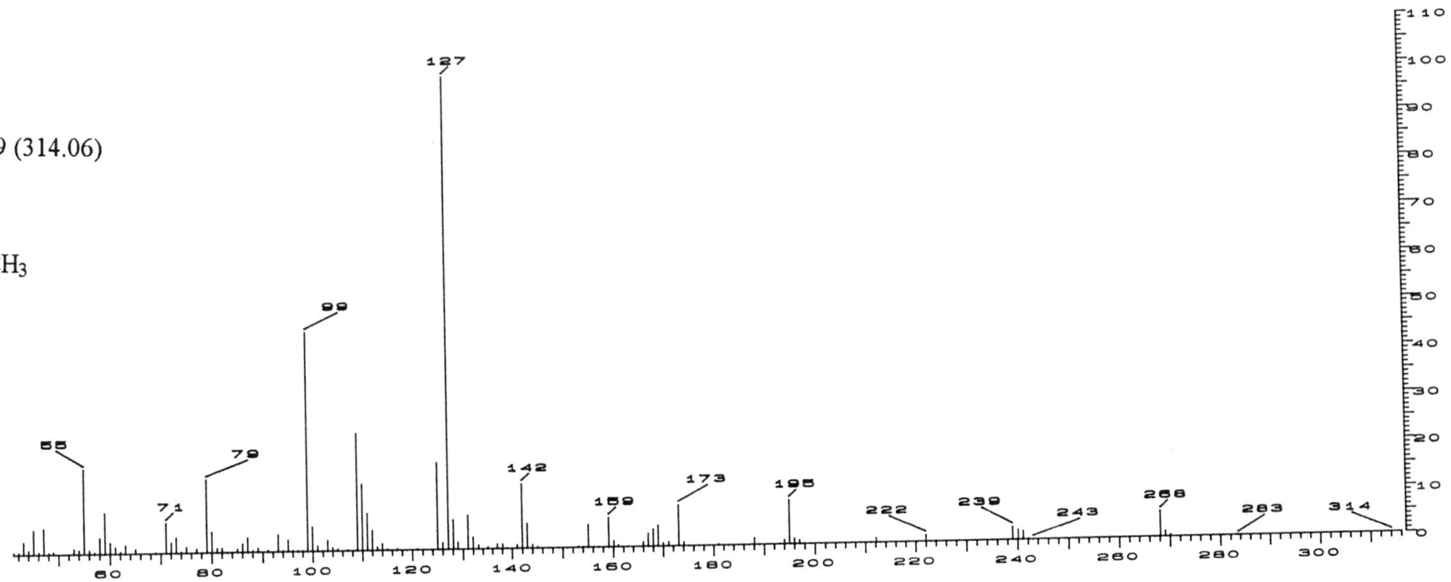

MCPA-ISOOCTYL ESTER

$C_{17}H_{25}O_3Cl$ M.W. = 312.83 (312.15)

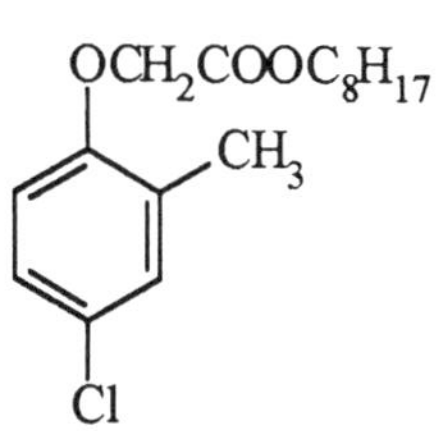

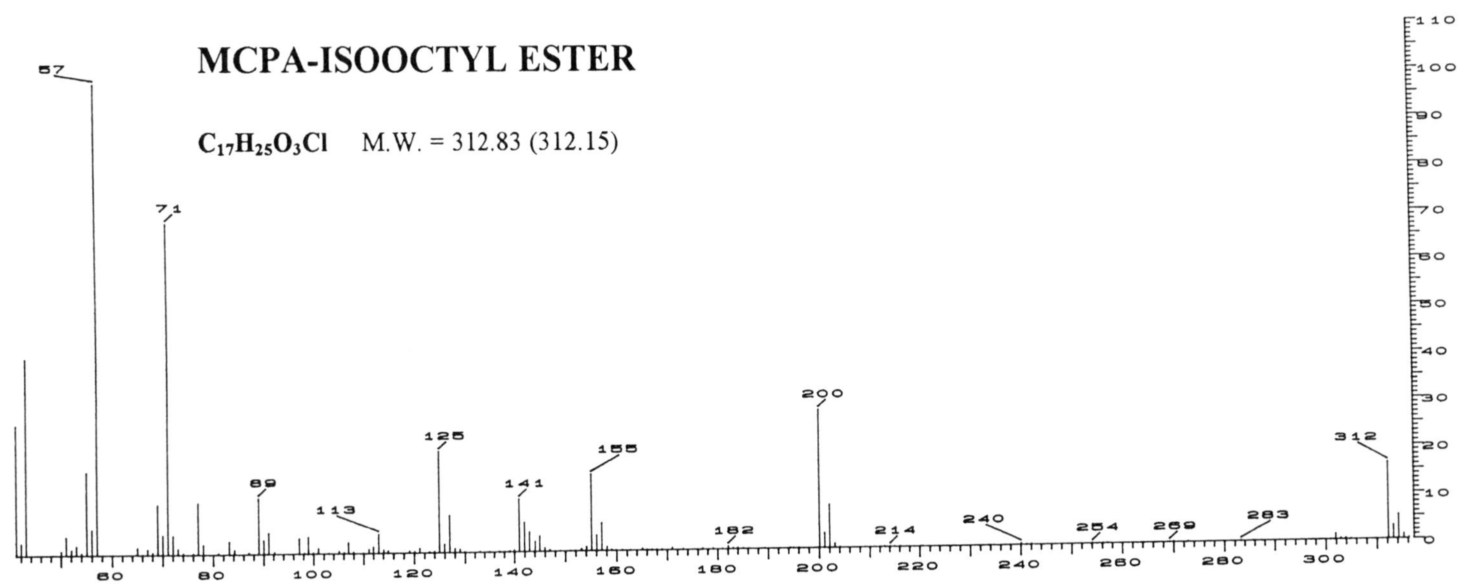

MCPP-ISOOCTYL ESTER

$C_{18}H_{27}O_3Cl$ M.W. = 326.86 (326.16)

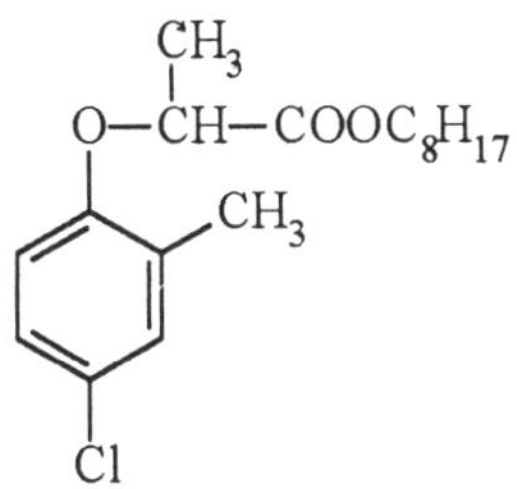

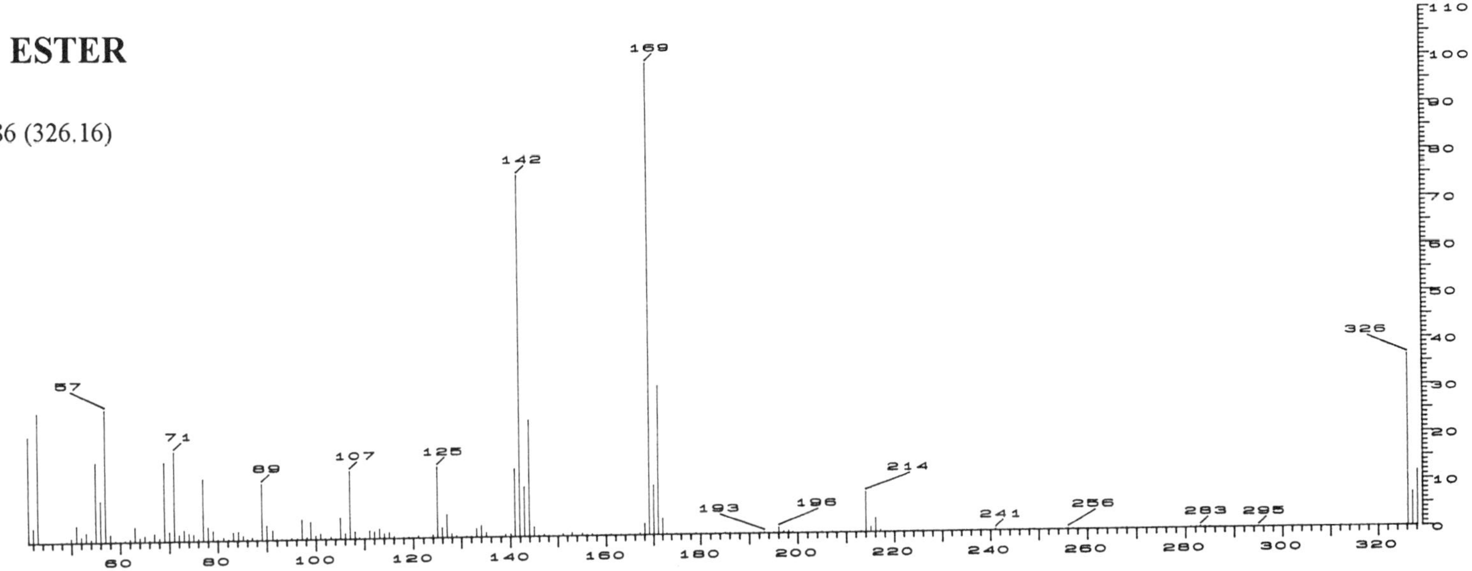

MECARBAM

$C_{10}H_{20}NO_5S_2P$ M.W. = 329.36 (329.05)

$CH_3CH_2OCON(CH_3)COCH_2SP(=S)(OCH_2CH_3)_2$

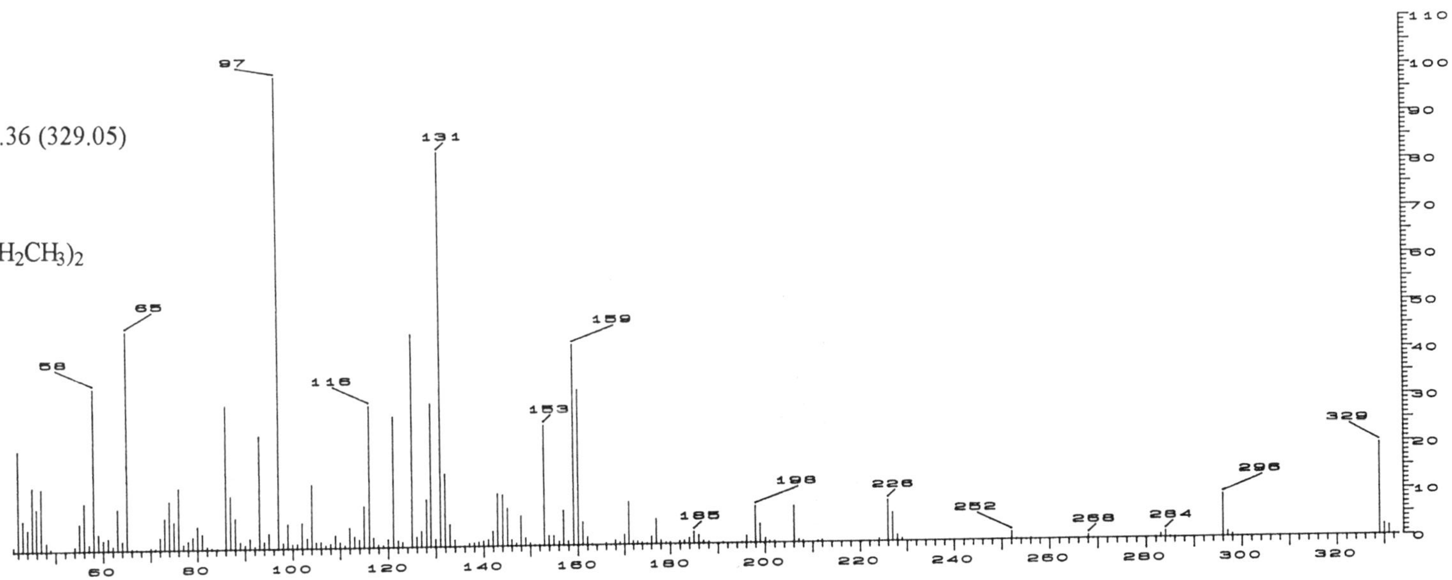

MEPHOSFOLAN

$C_8H_{16}NO_3S_2P$

M.W. = 269.31 (269.03)

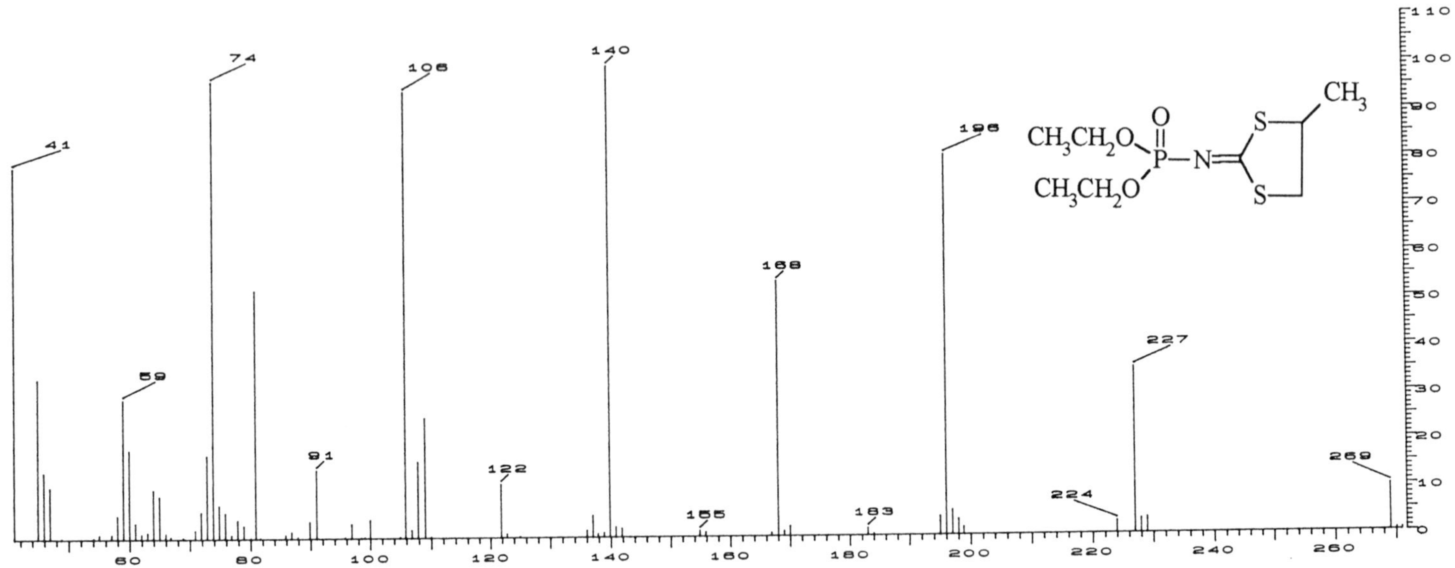

METALAXYL

$C_{15}H_{20}NO_4$

M.W. = 279.33 (279.15)

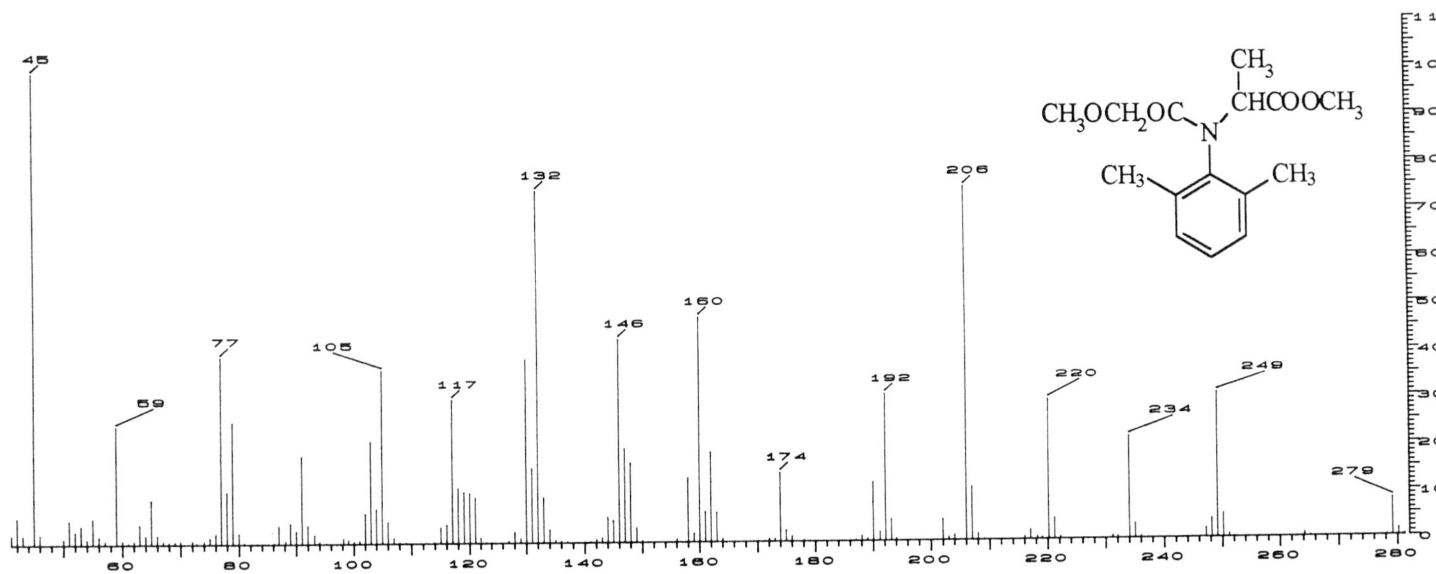

METHAMIDOPHOS

$C_2H_8NO_2PS$ M.W. = 141.12 (141.00)

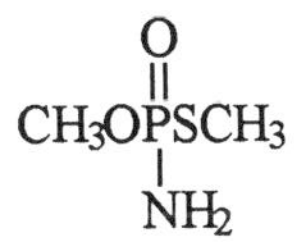

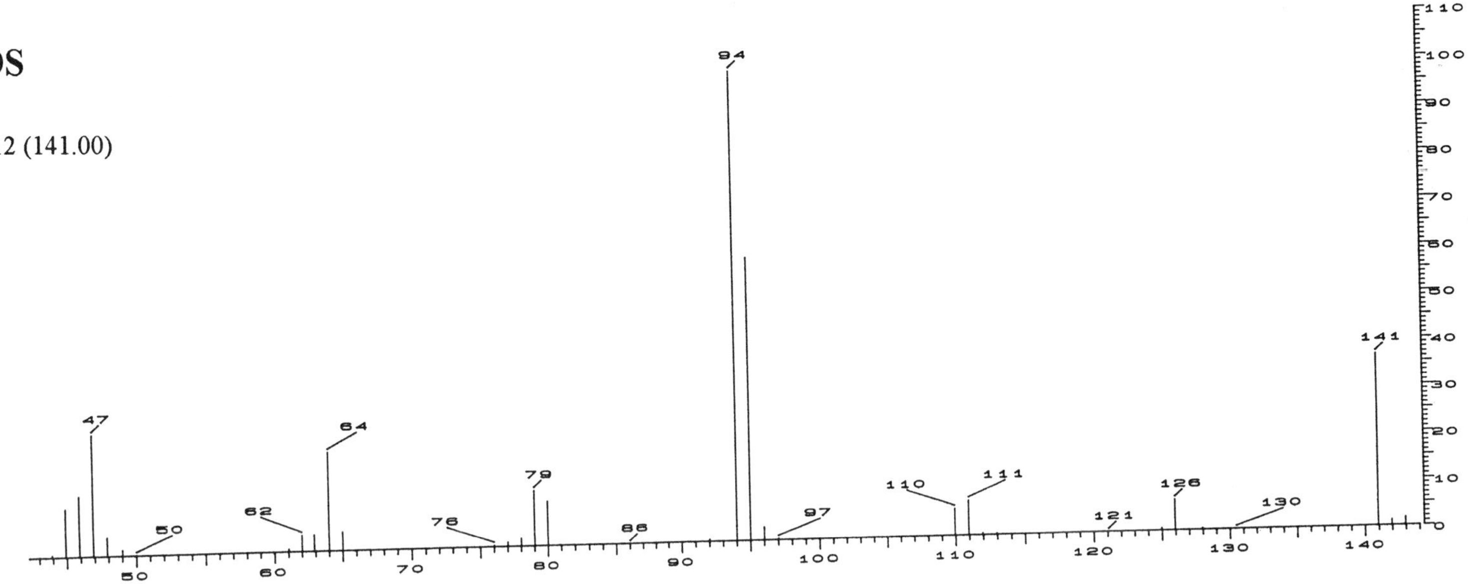

METHAZOLE

$C_9H_6N_2O_3Cl_2$ M.W. = 261.06 (259.98)

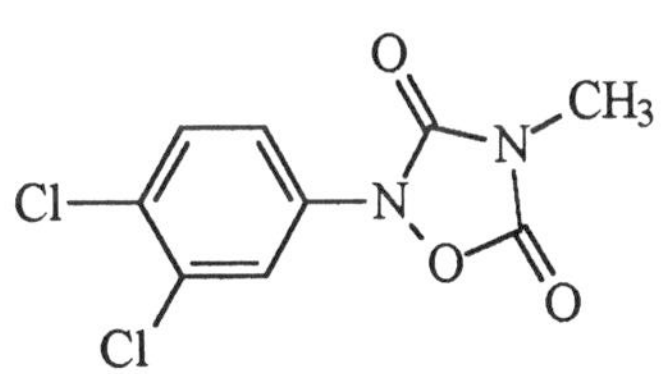

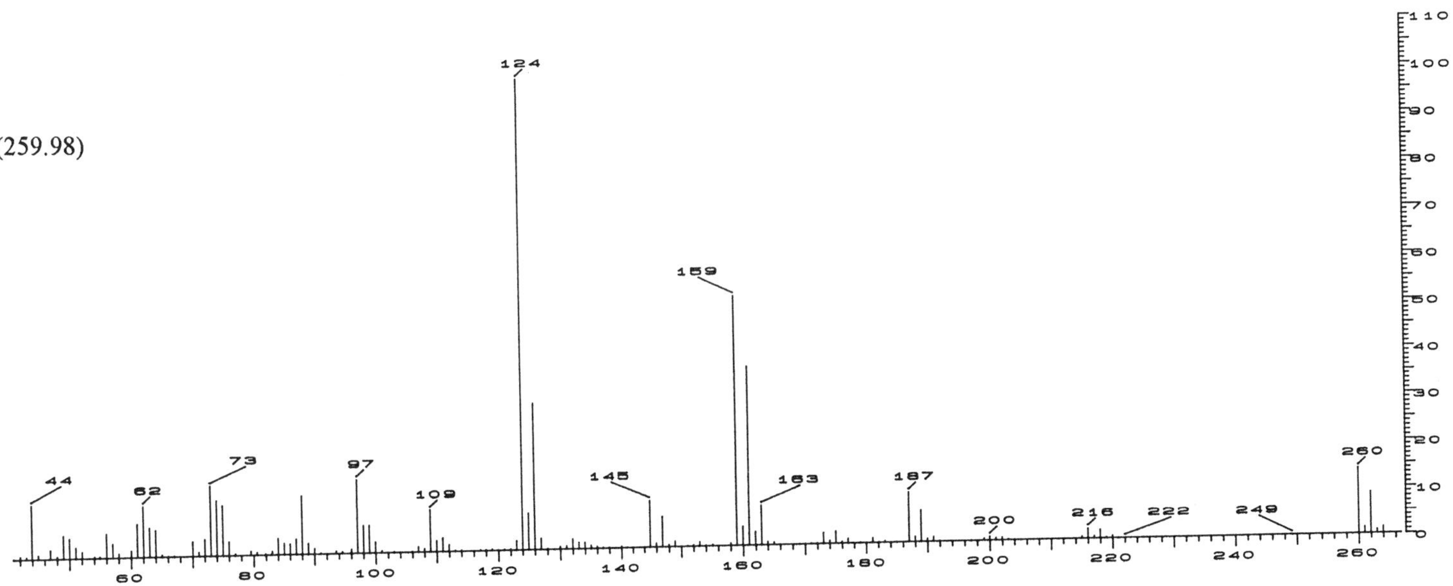

METHIDATHION

$C_6H_{11}N_2O_4S_3P$ M.W. = 302.31 (301.96)

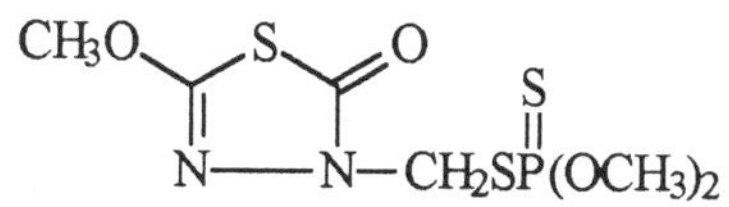

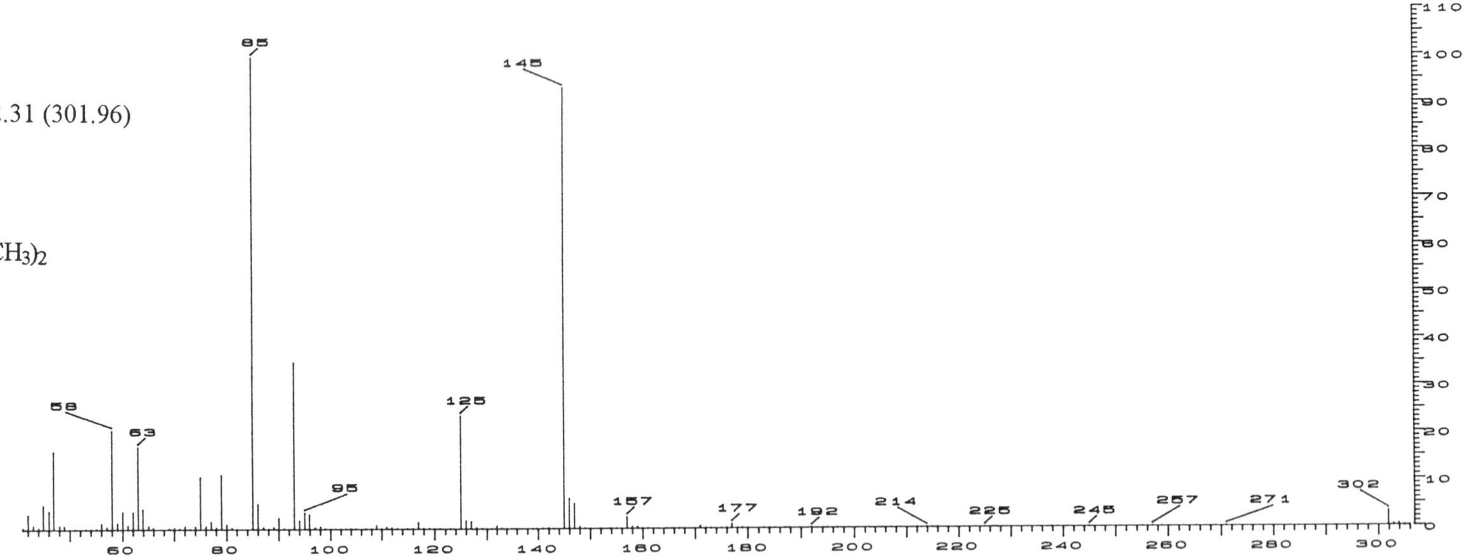

METHIOCARB

$C_{11}H_{15}NO_2S$ M.W. = 225.30 (225.08)

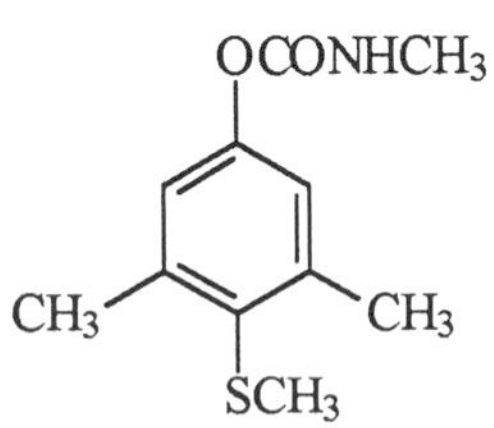

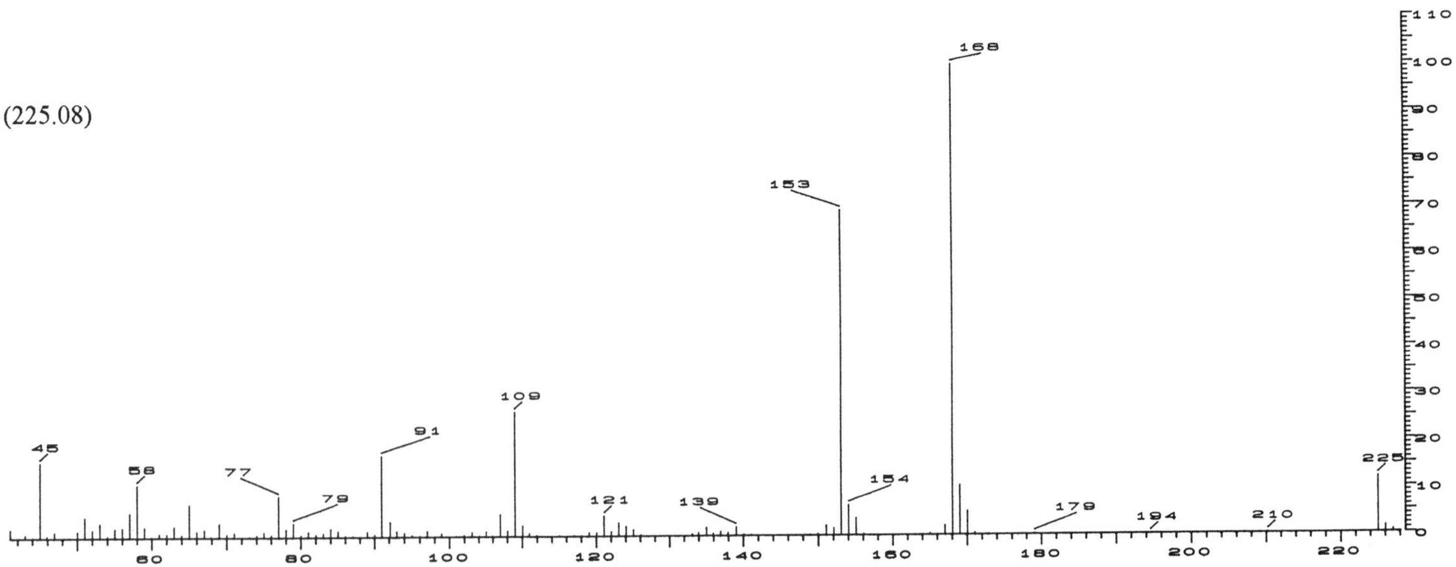

METHOPRENE

$C_{19}H_{34}O_3$ M.W. = 310.47 (310.25)

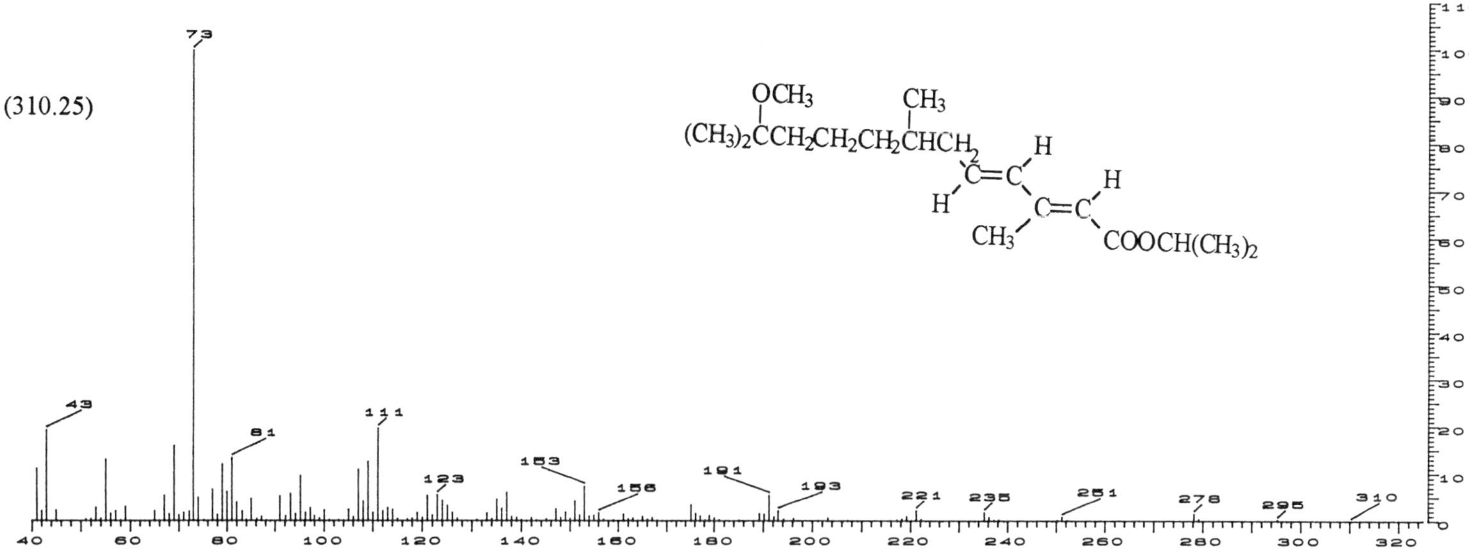

METHOXYCHLOR

$C_{16}H_{15}O_2Cl_3$ M.W. = 345.65 (344.01)

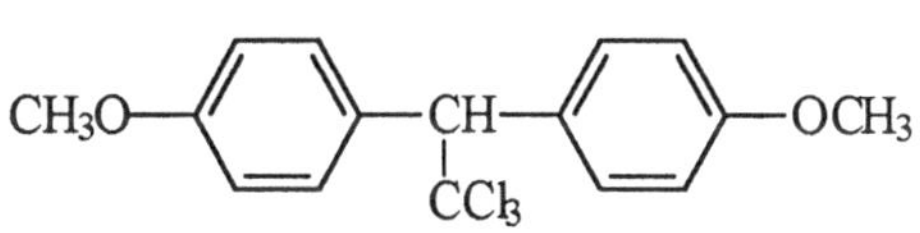

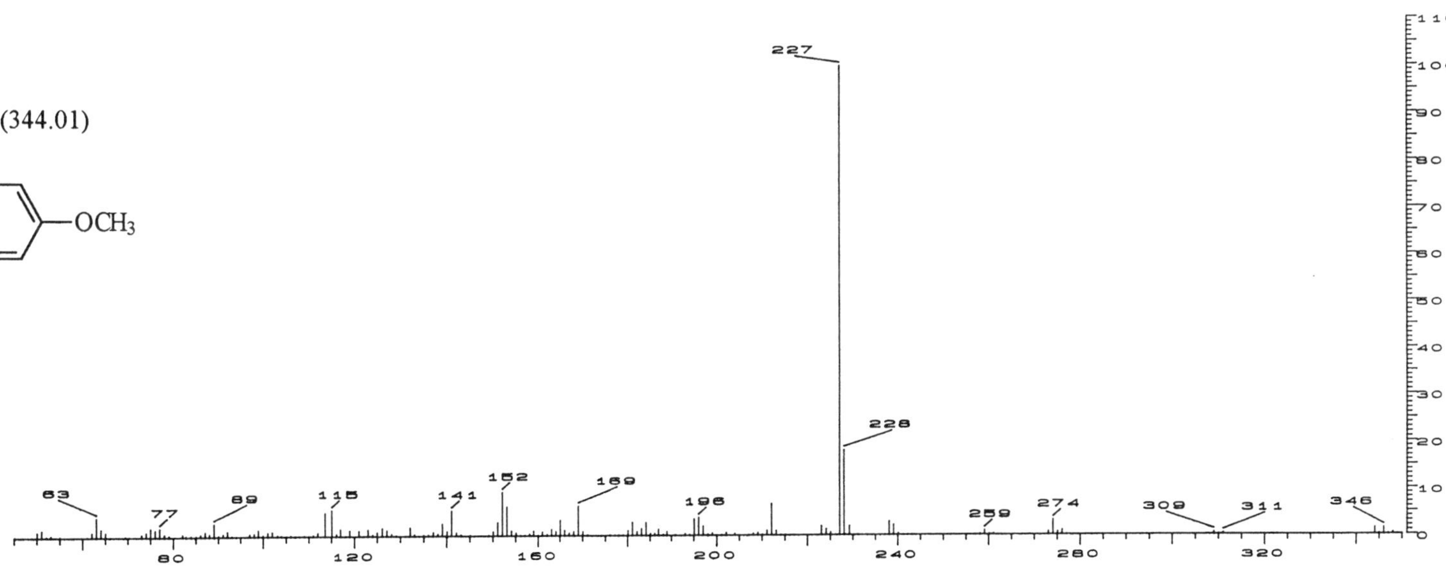

METHYLPAROXON

$C_8H_{10}NO_6P$ M.W. = 247.14 (247.02)

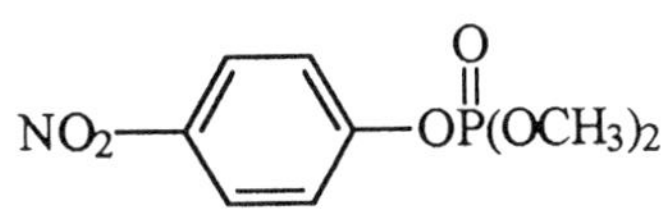

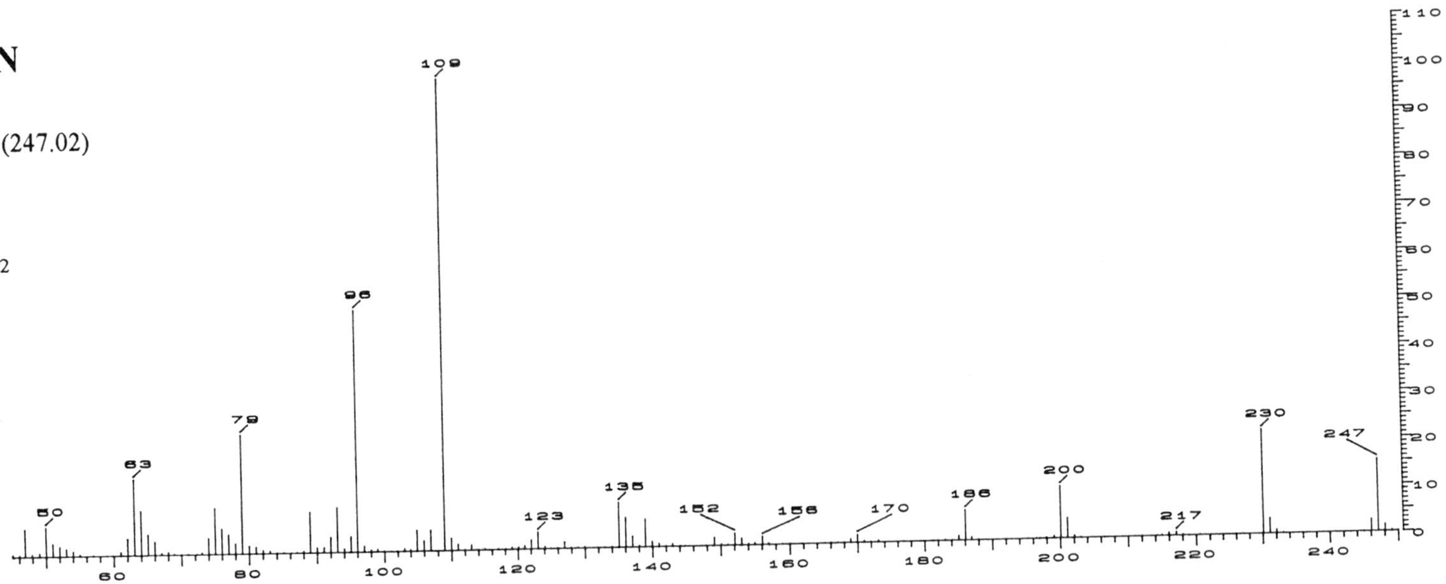

METOBROMURON

$C_9H_{11}N_2O_2Br$ M.W. = 259.10 (258.00)

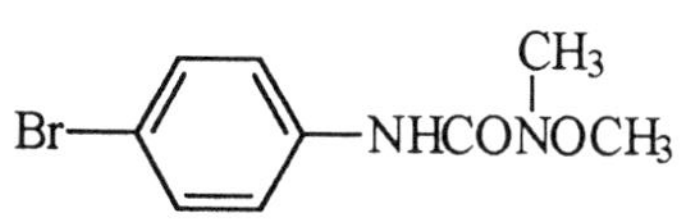

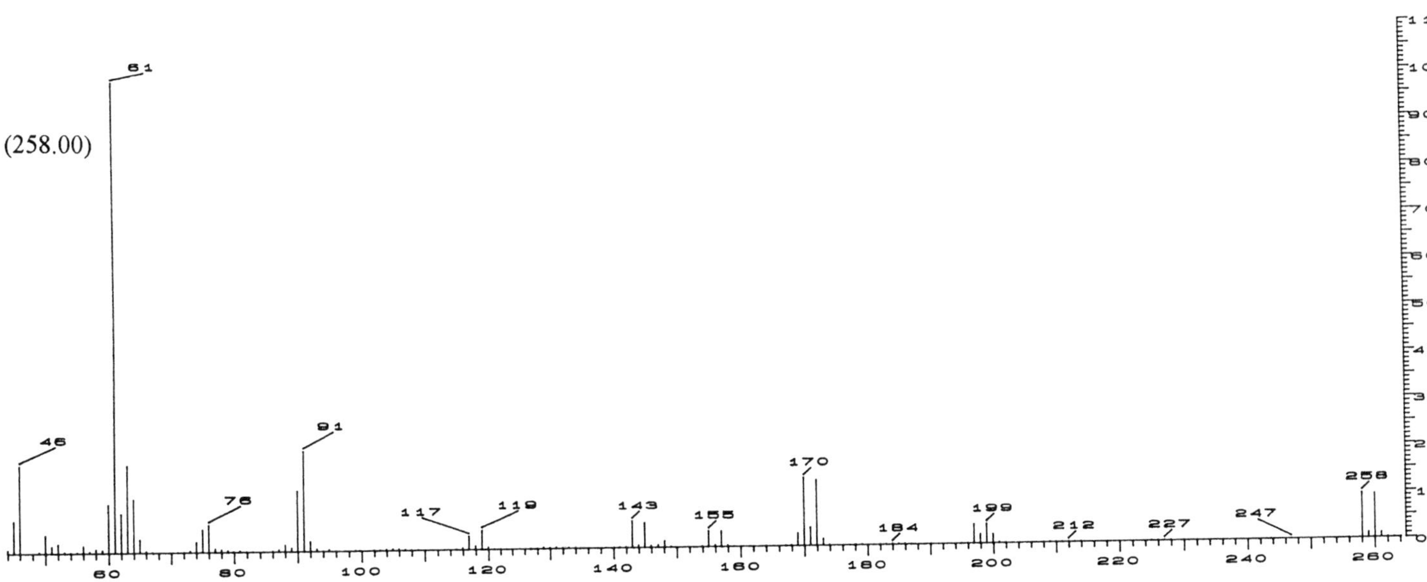

METOLACHLOR

$C_{15}H_{22}NO_2Cl$ M.W. = 283.79 (283.13)

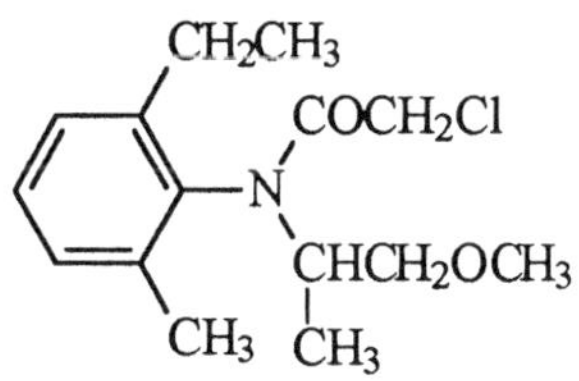

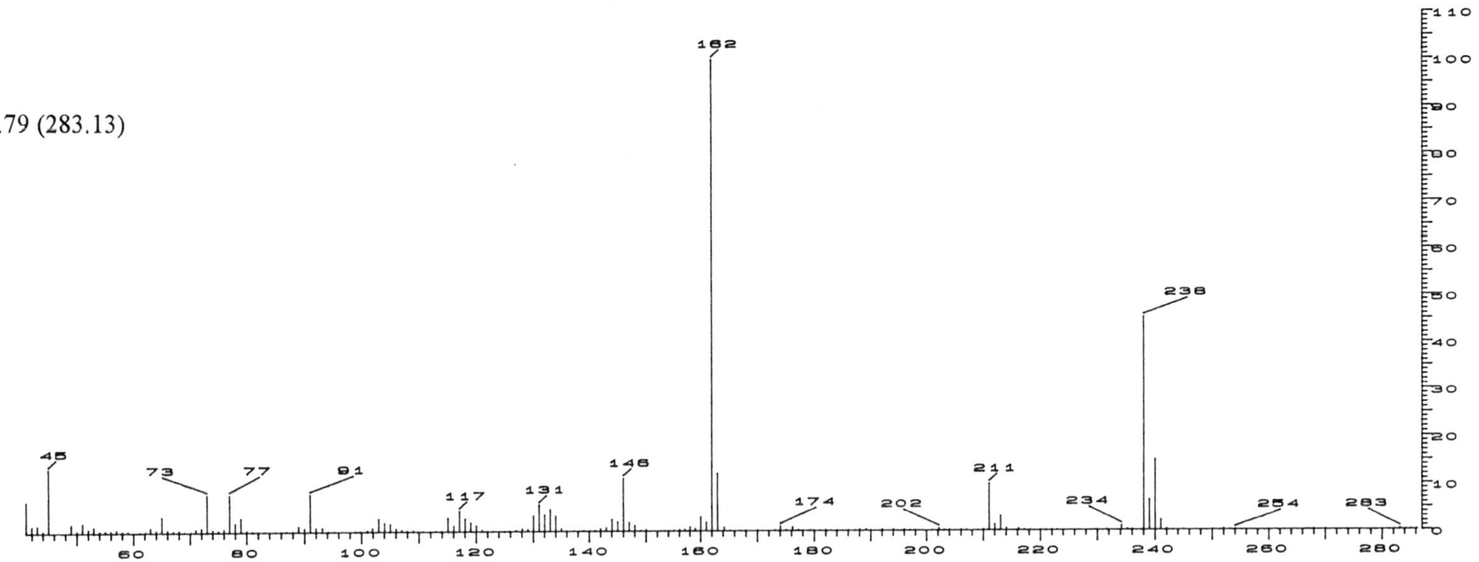

METOXURON

$C_{10}H_{13}N_2O_2Cl$ M.W. = 228.67 (228.07)

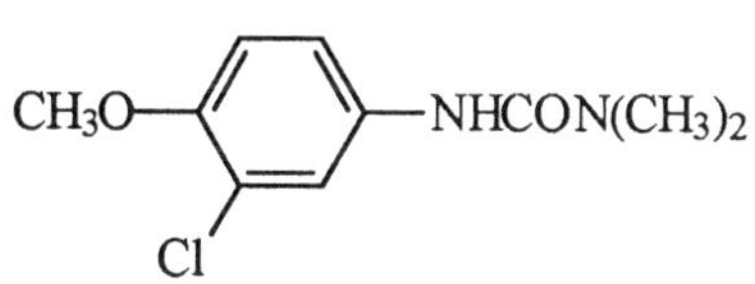

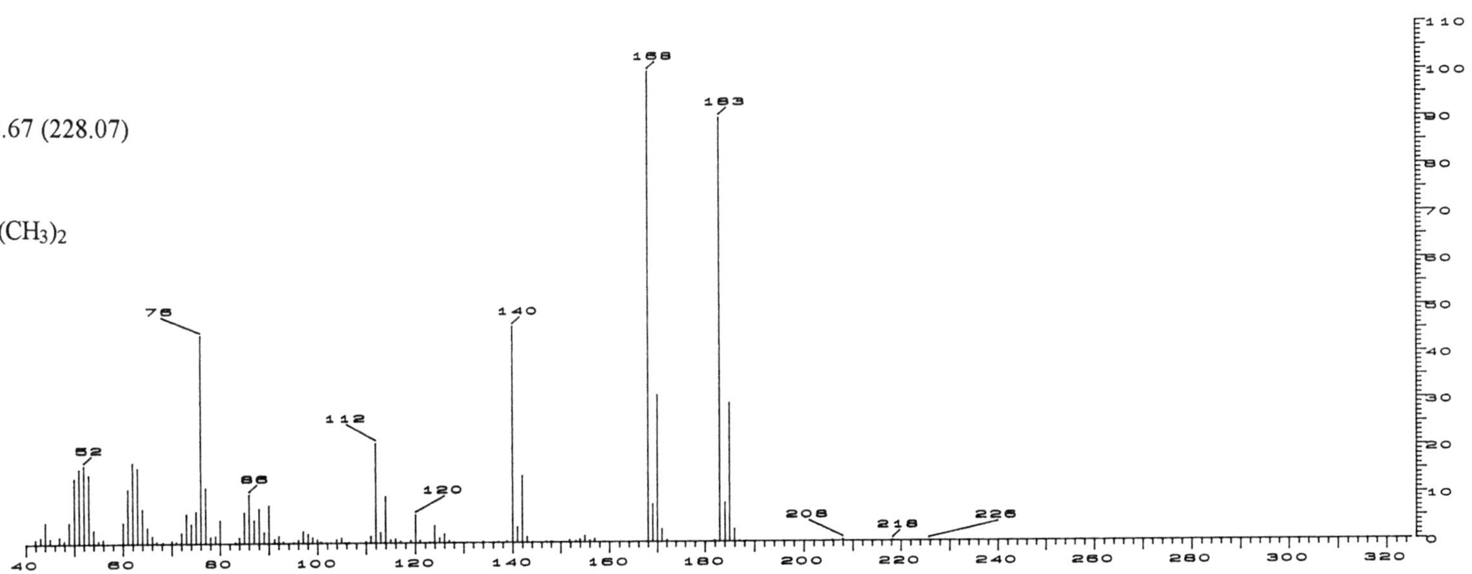

MEVINPHOS

$C_7H_{13}O_6P$ M.W. = 224.15 (224.04)

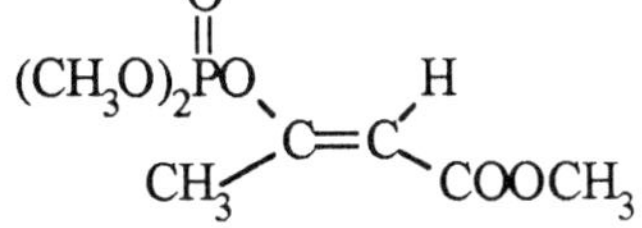

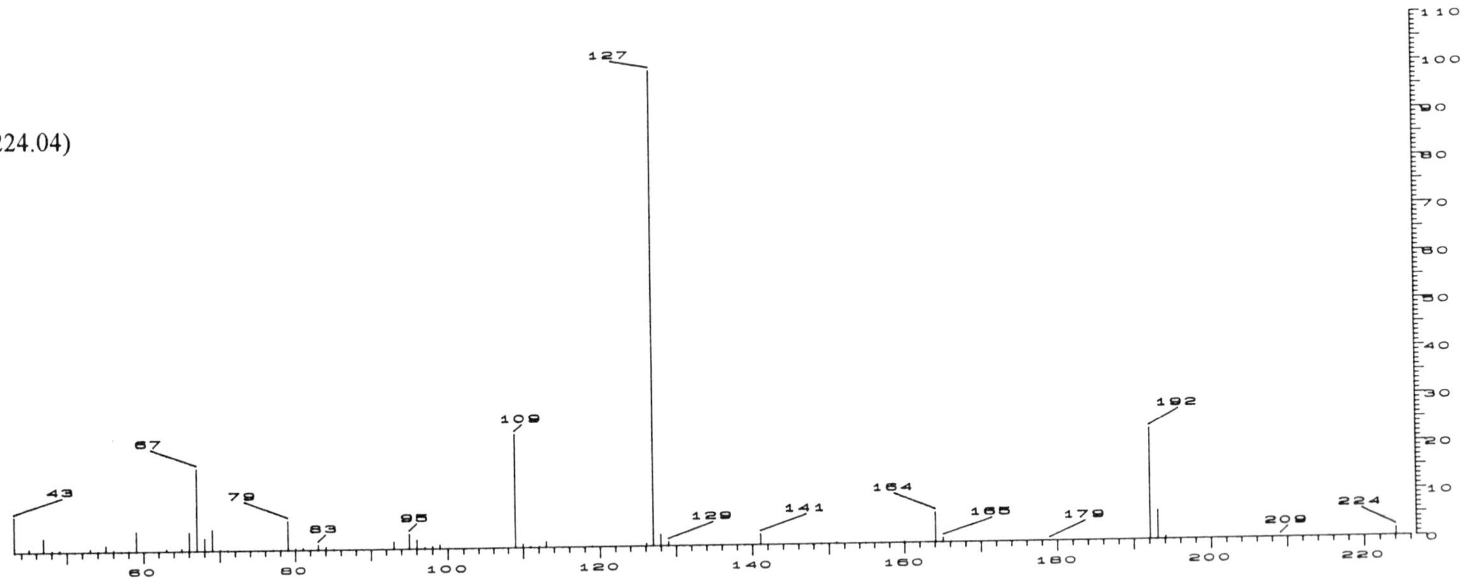

MEXACARBATE

$C_{12}H_{18}N_2O_2$ M.W. = 222.28 (222.14)

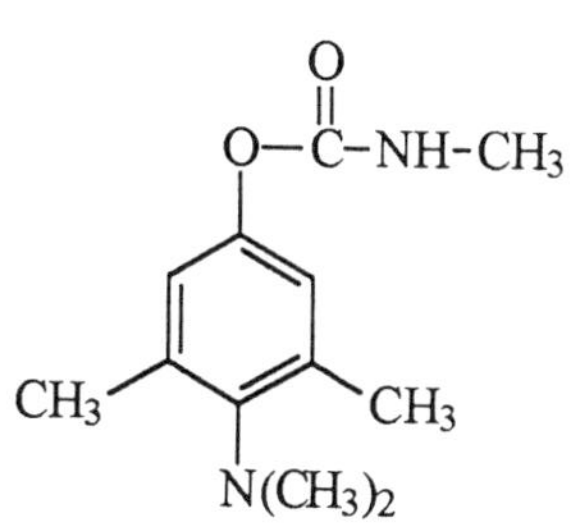

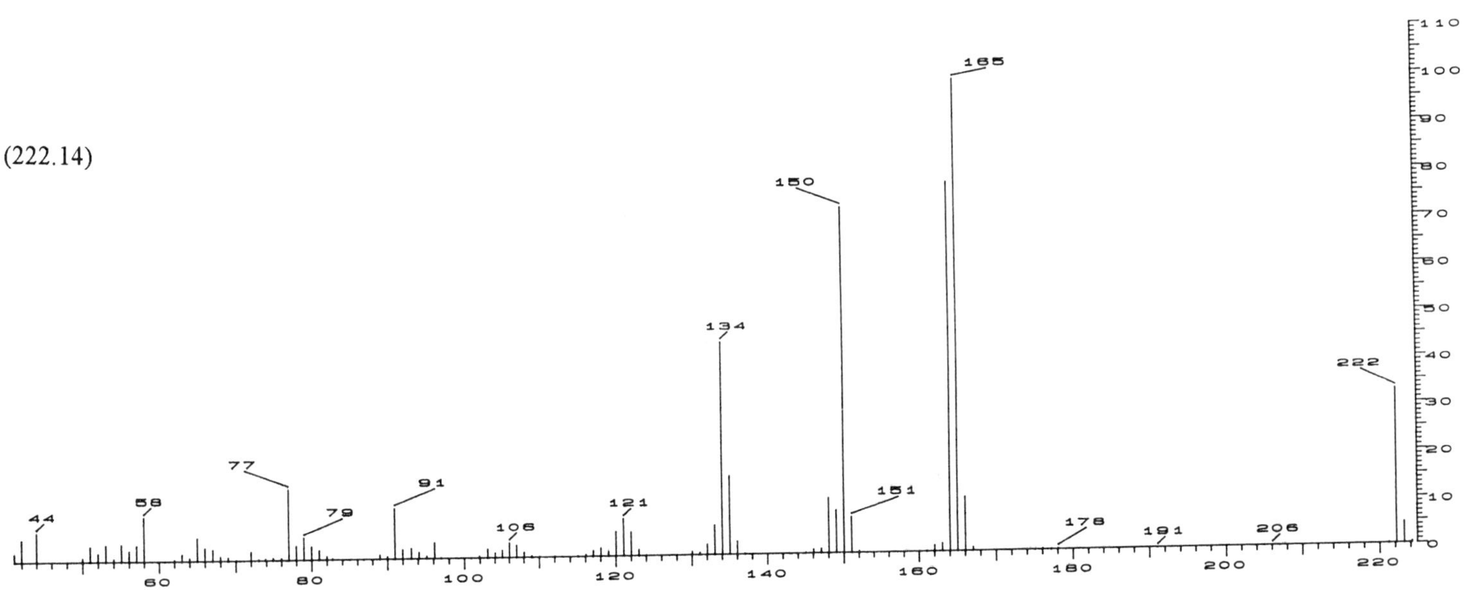

MGK-264

$C_{17}H_{25}NO_2$ M.W. = 275.39 (275.19)

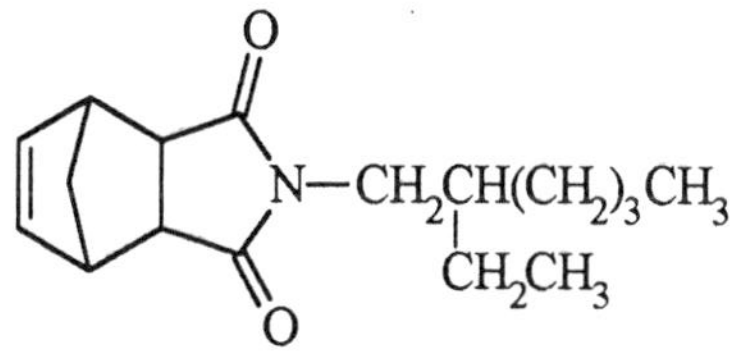

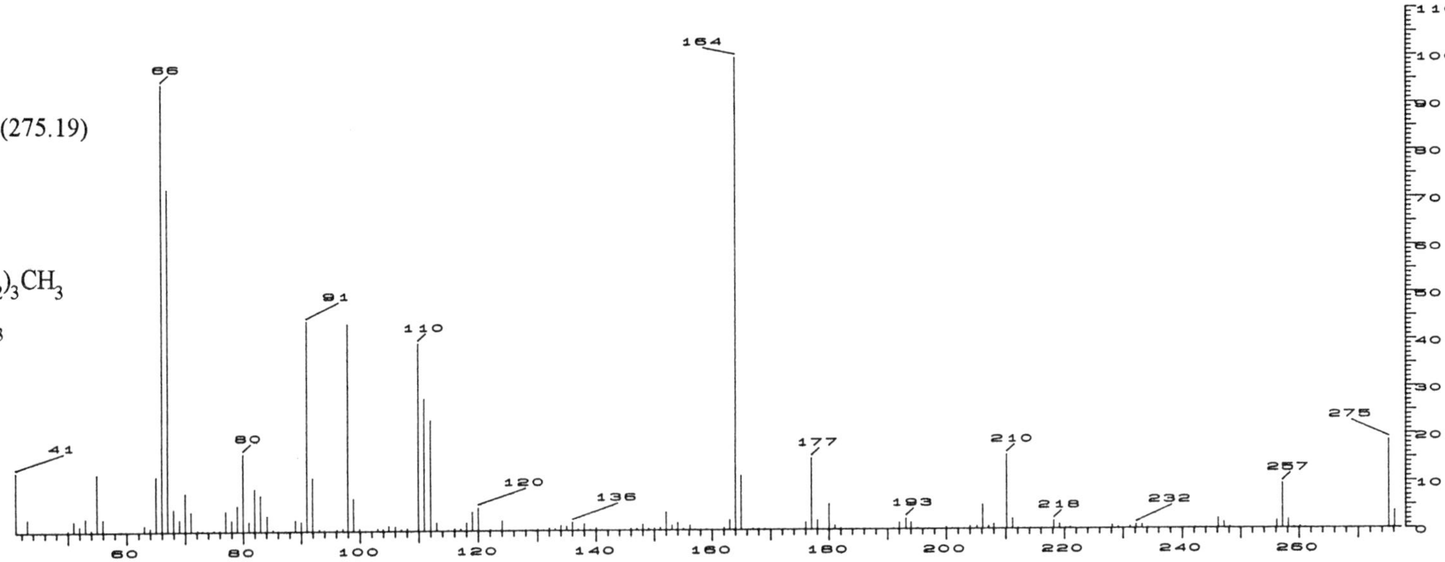

MIREX

$C_{10}Cl_{12}$ M.W. = 545.59 (539.63)

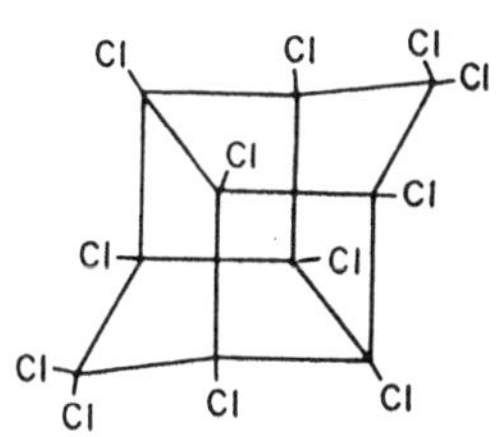

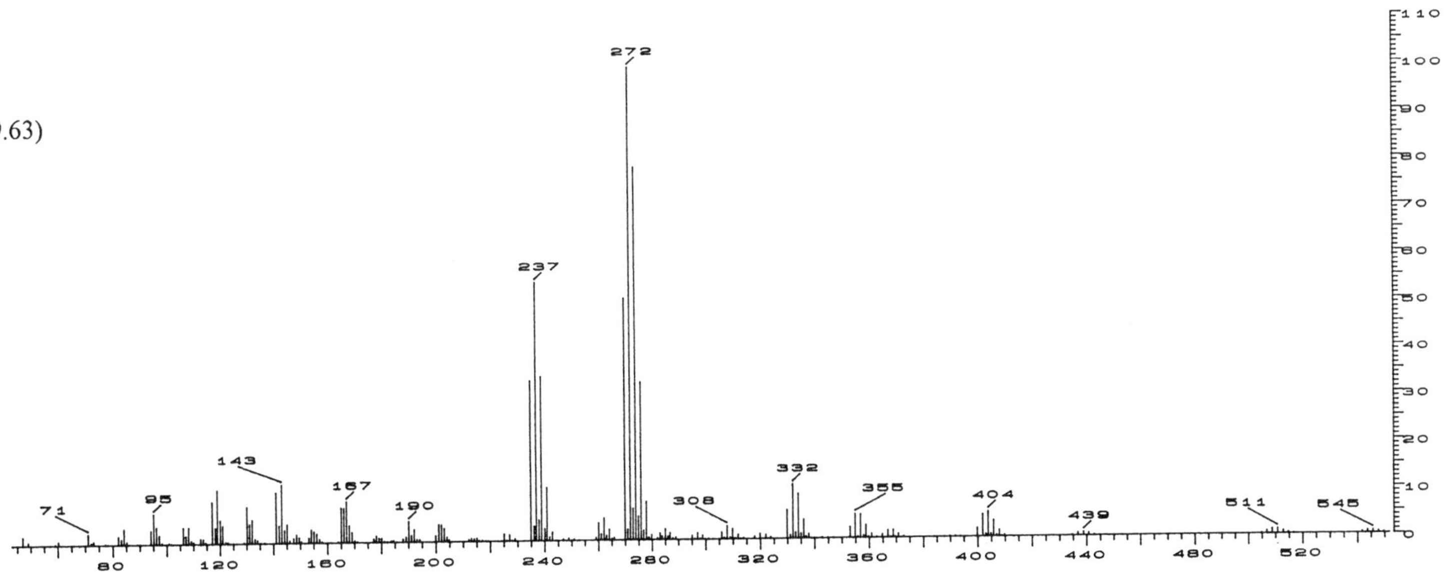

MONALIDE

$C_{13}H_{18}NOCl$ M.W. = 239.74 (239.11)

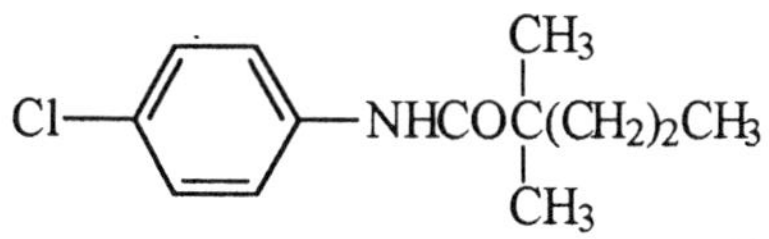

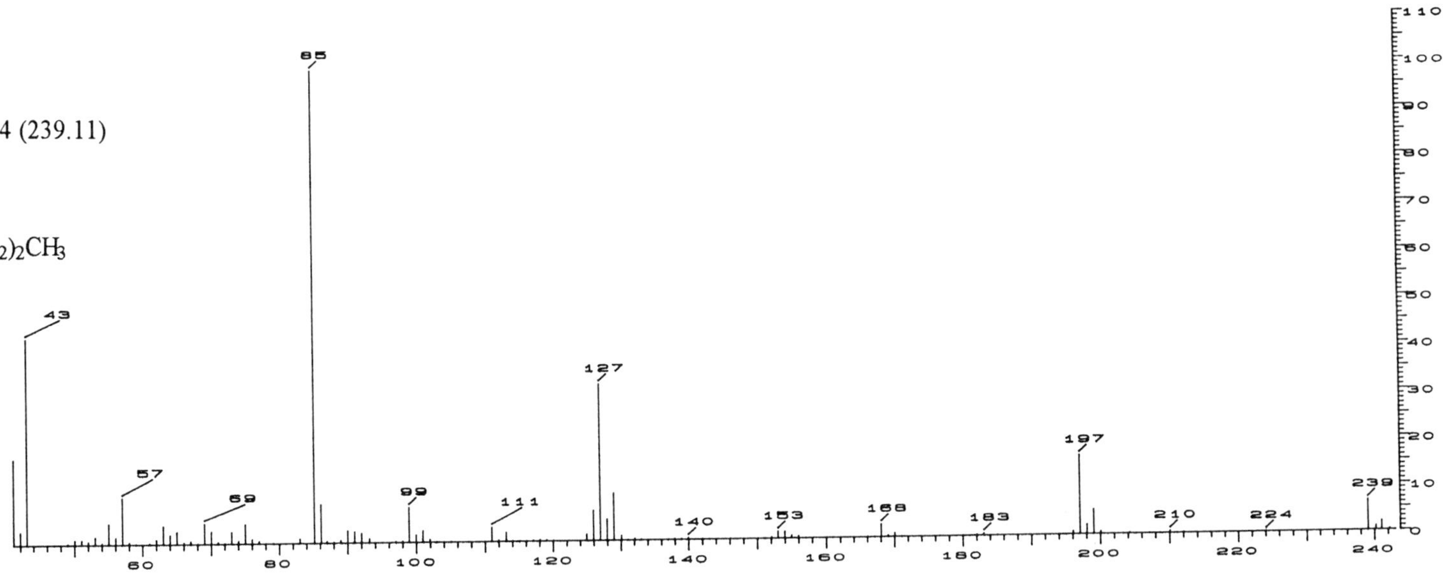

MONOCROTOPHOS

$C_7H_{14}NO_5P$ M.W. = 223.16 (223.06)

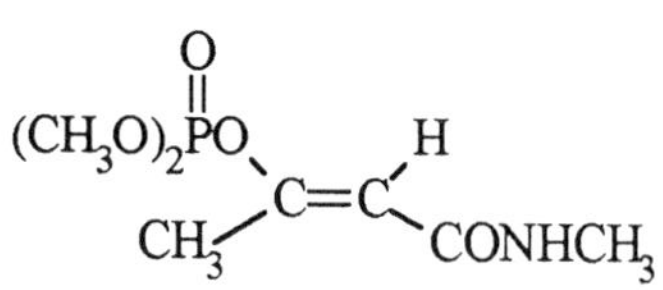

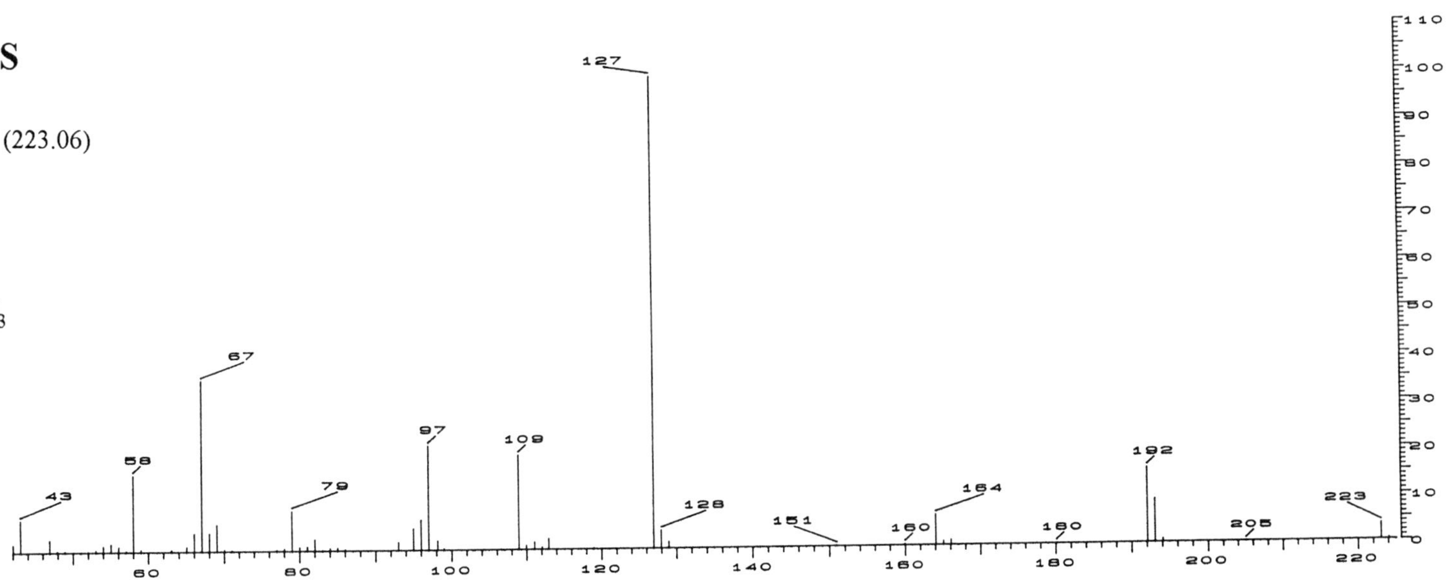

MONURON

$C_9H_{11}N_2OCl$ M.W. = 198.65 (198.06)

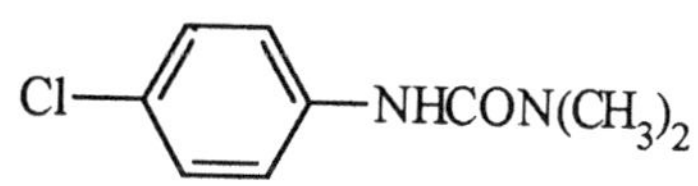

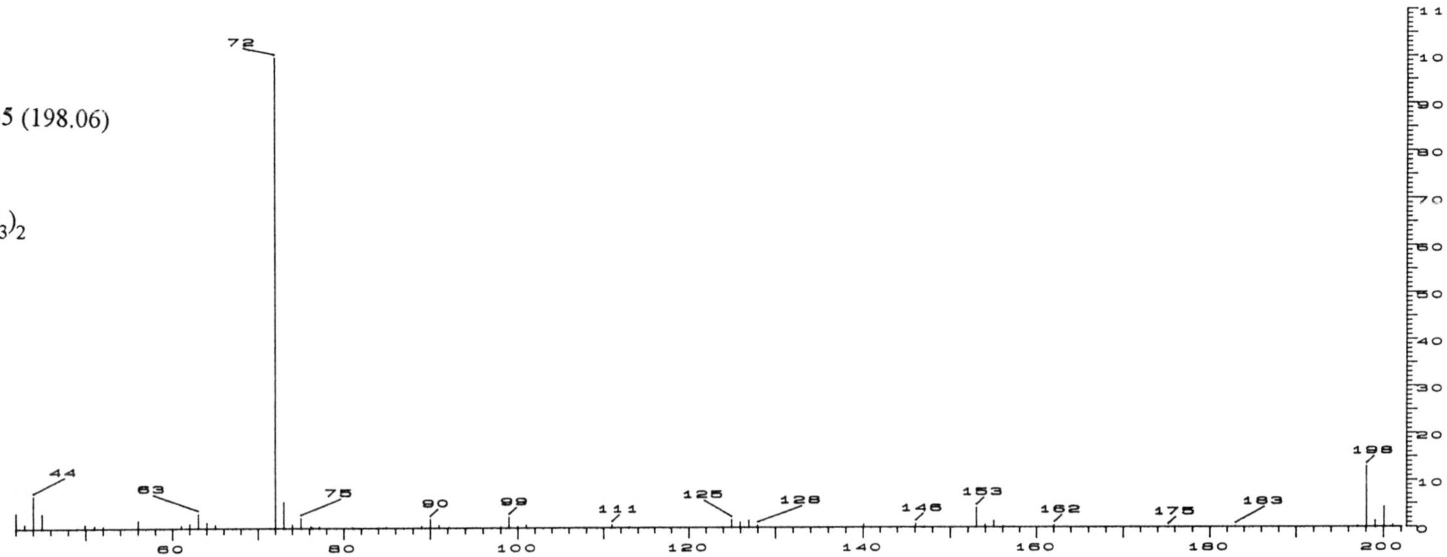

NAPHTHALENEACETAMIDE

$C_{12}H_{11}NO$ M.W. = 185.22 (185.08)

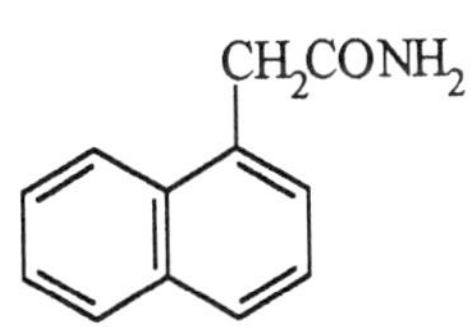

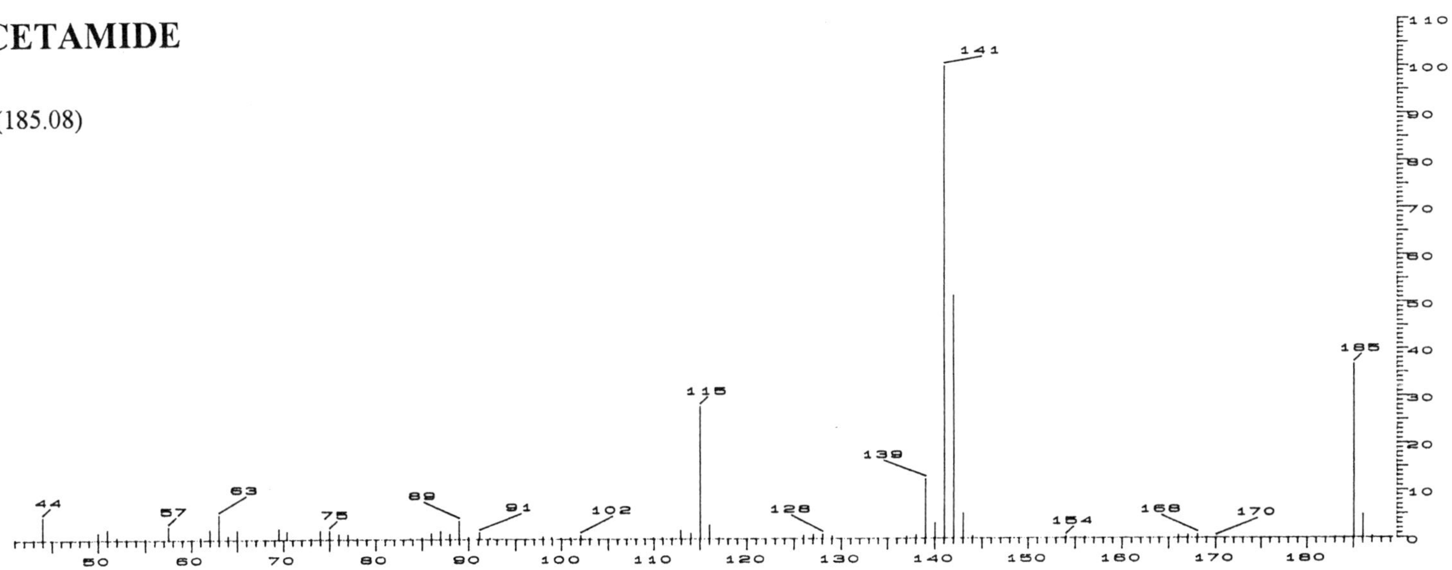

NAPROPAMIDE

$C_{17}H_{21}NO_2$ M.W. = 271.35 (271.16)

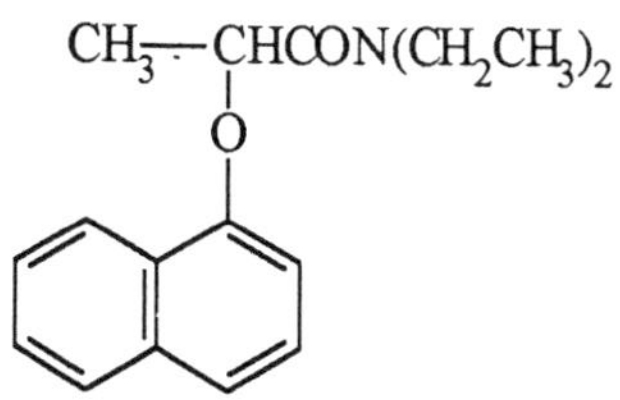

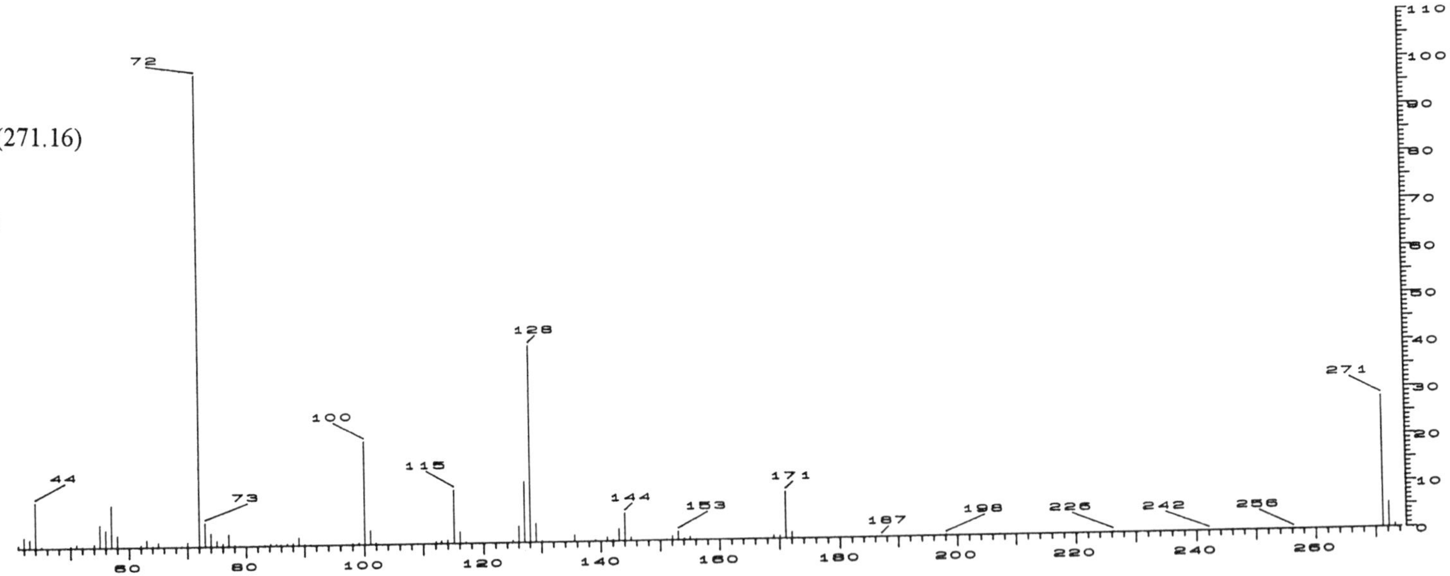

NITRAPYRIN

$C_6H_3NCl_4$ M.W. = 230.90 (228.90)

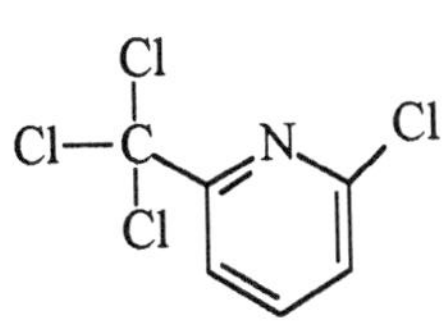

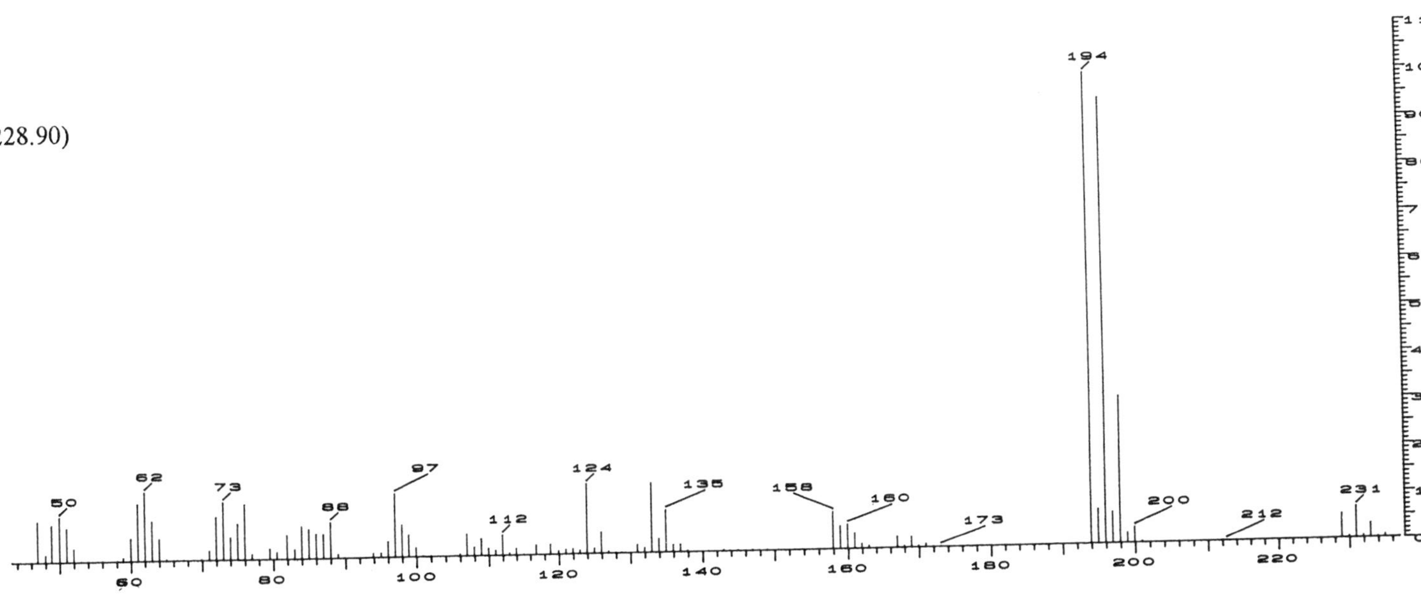

NITROFEN

$C_{12}H_7NO_3Cl_2$ M.W. = 284.09 (282.98)

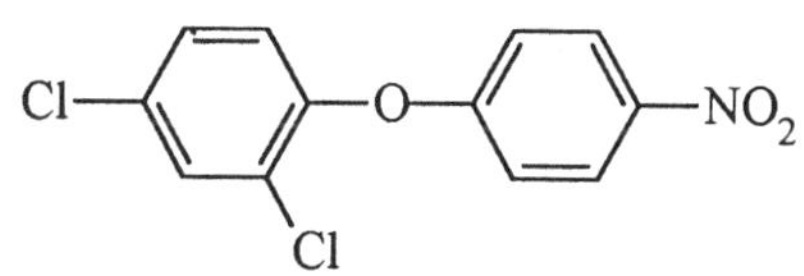

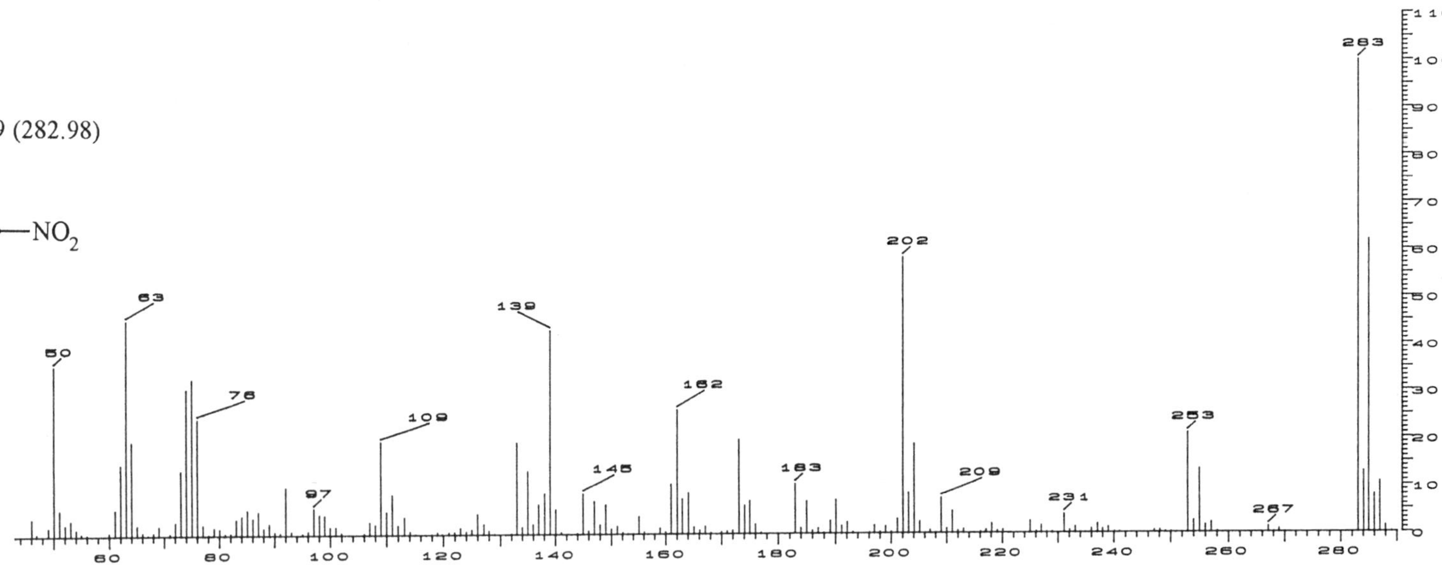

NORFLURAZON

$C_{12}H_9N_3OF_3Cl$ M.W. = 303.67 (303.04)

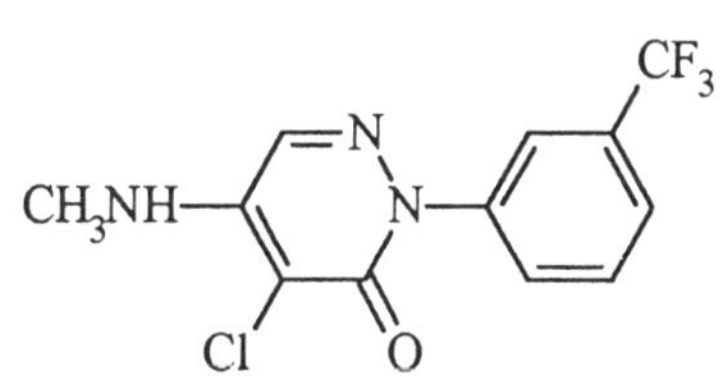

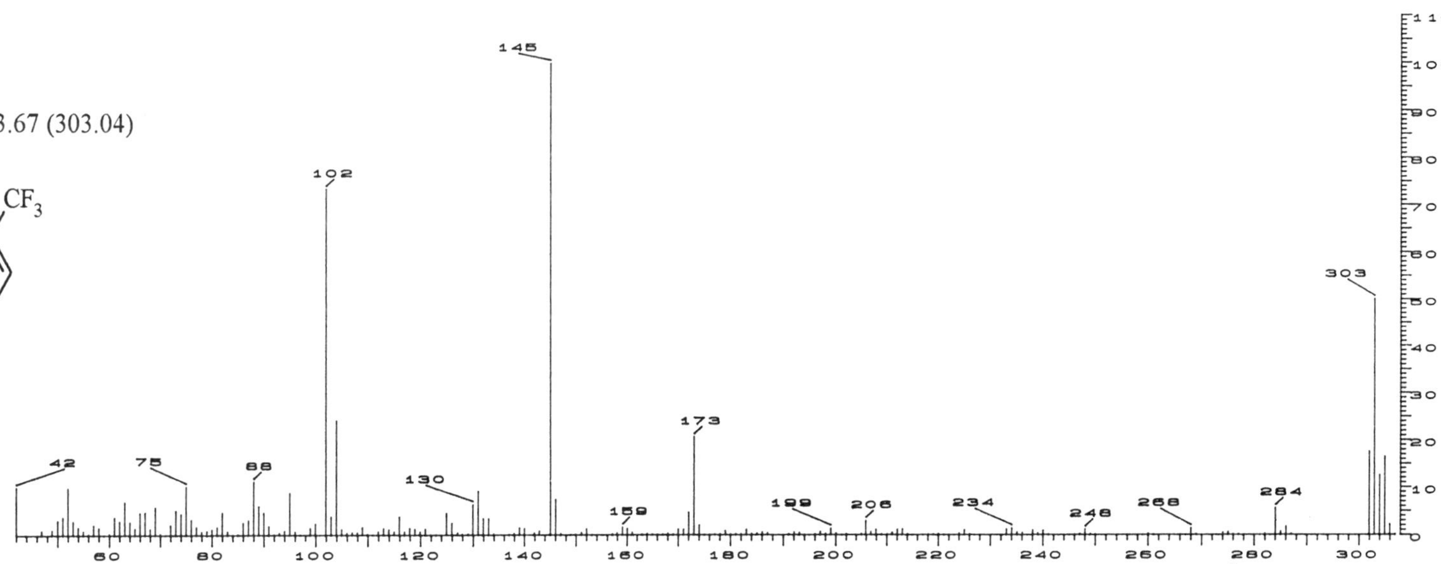

ORYZALIN

$C_{12}H_{18}N_4O_6S$ M.W. = 346.35 (346.09)

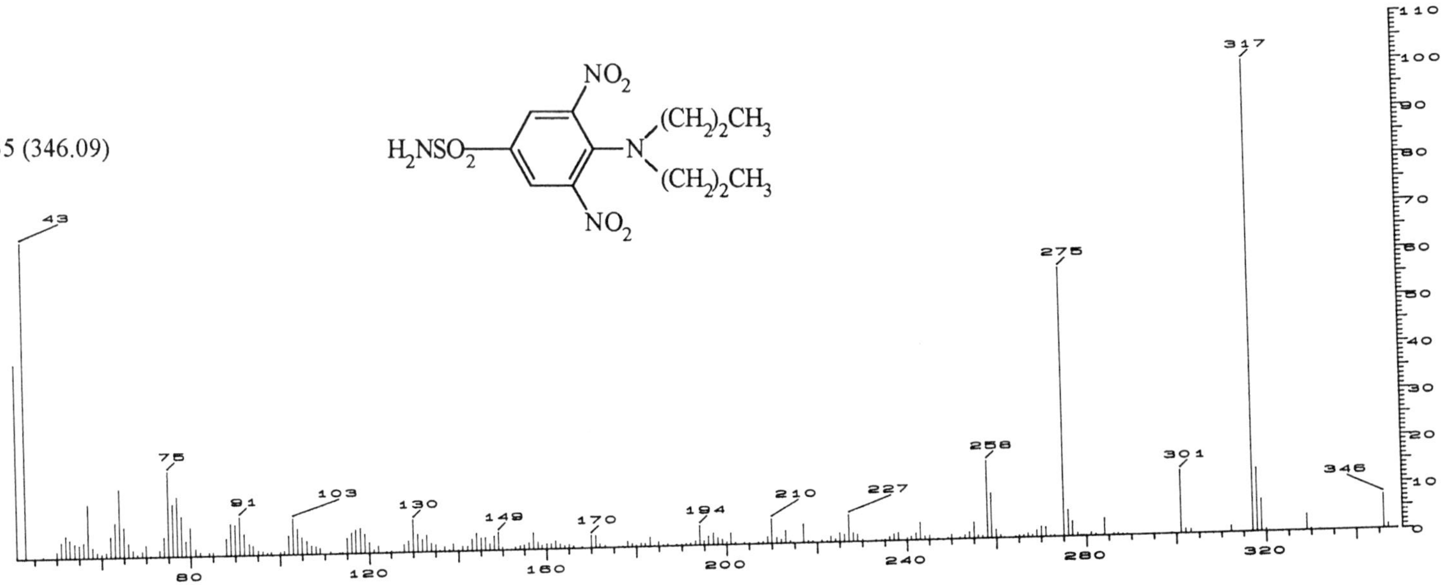

ORYZALIN-DIMETHYL

$C_{10}H_{14}N_4O_6S$ M.W. = 374.41 (374.13)

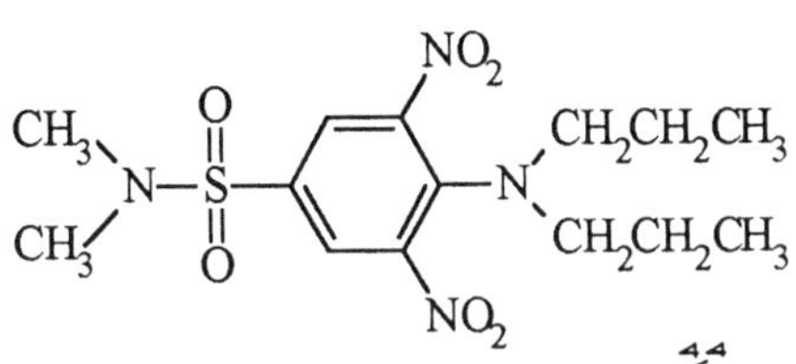

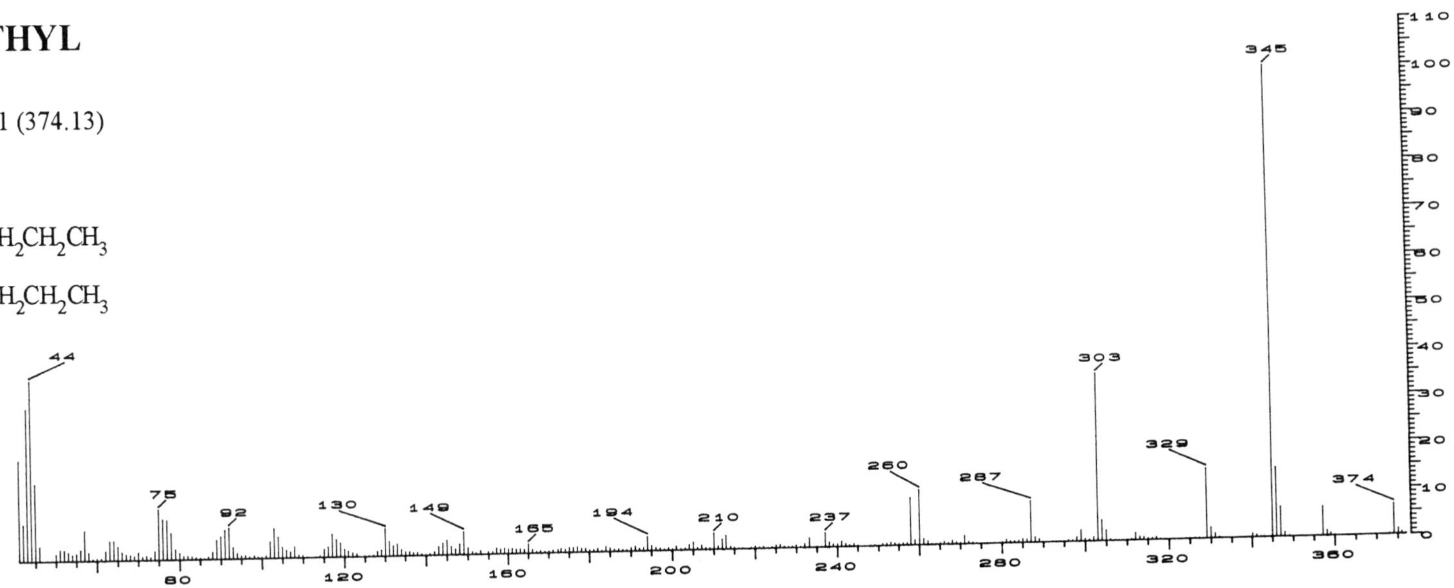

OXADIAZON

$C_{15}H_{18}N_2O_3Cl_2$ M.W. = 345.22 (344.07)

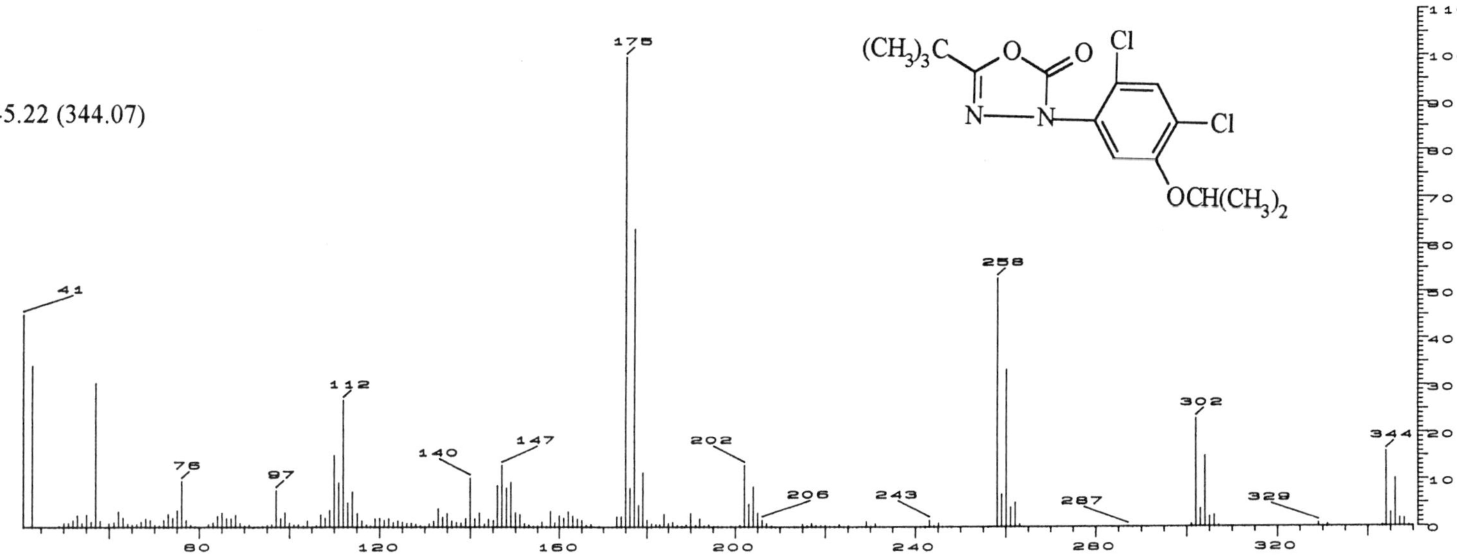

OXAMYL

$C_7H_{13}N_3O_3S$ M.W. = 219.25 (219.07)

$(CH_3)_2NCOC(SCH_3){=}NOCONHCH_3$

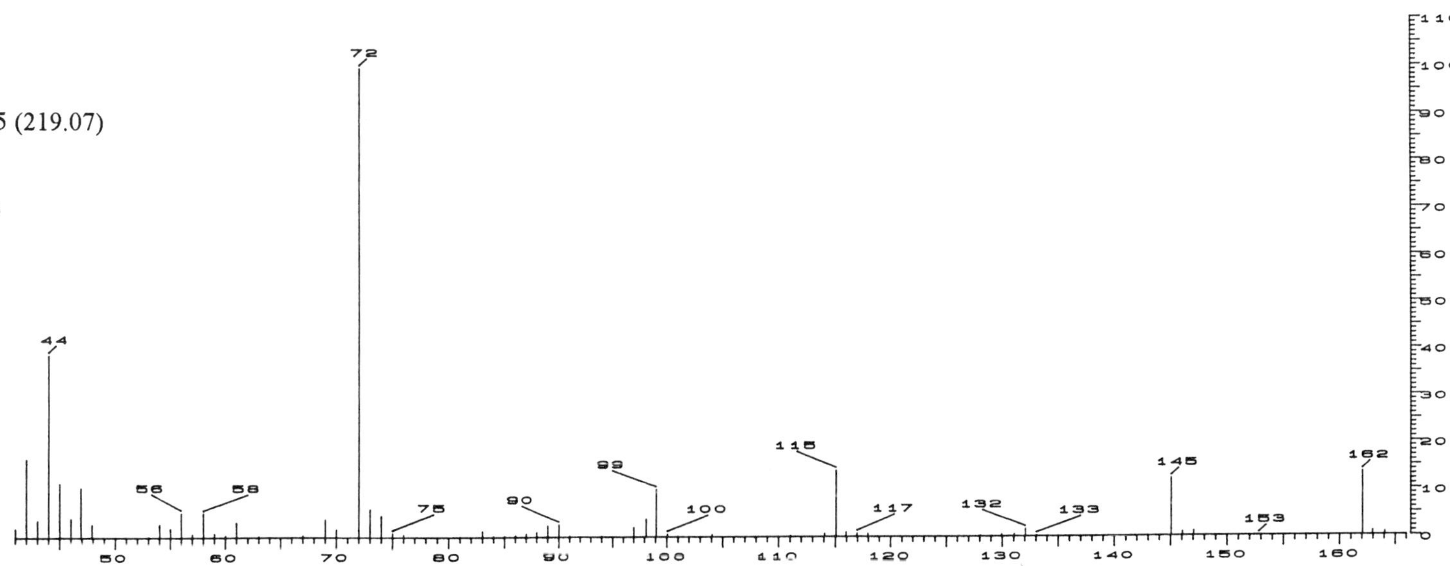

OXYCHLORDANE

$C_{10}H_4OCl_8$ M.W. = 423.76 (419.78)

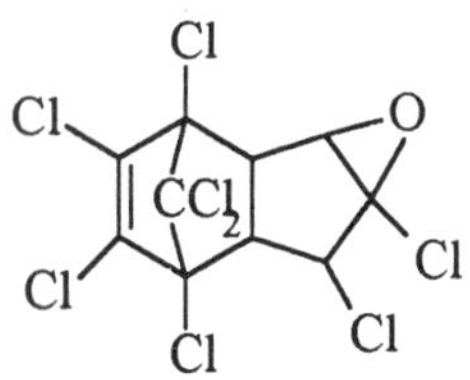

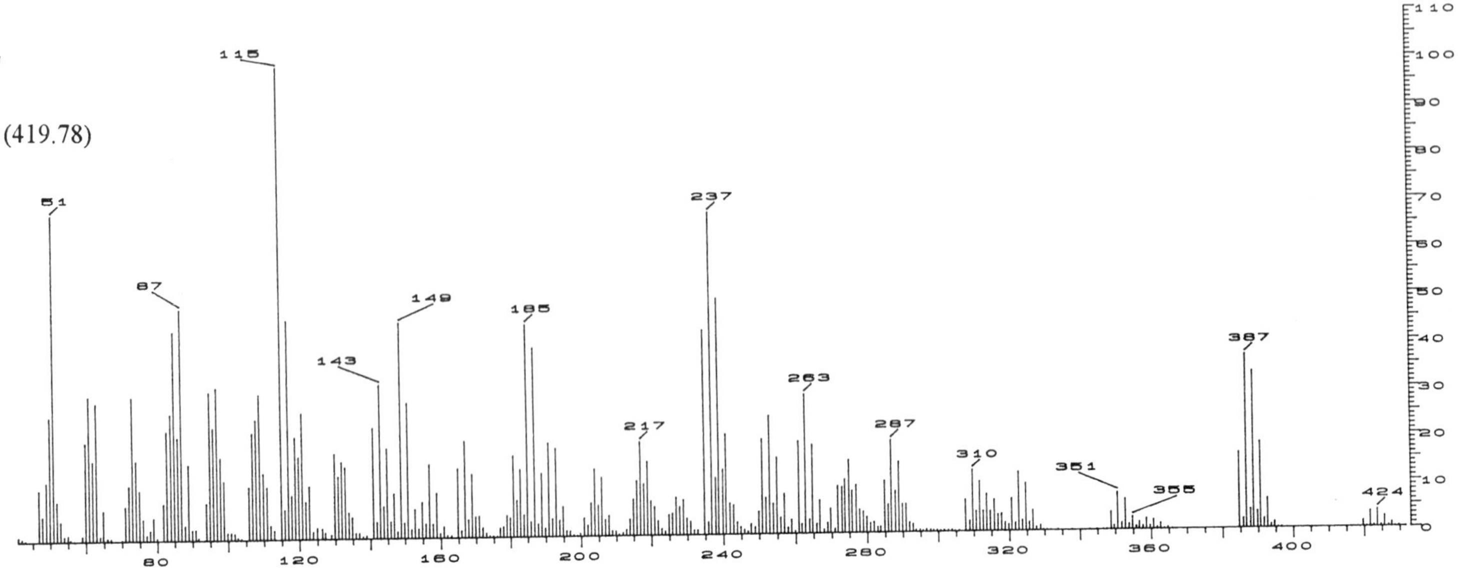

PARAOXON

$C_{10}H_{14}NO_6P$ M.W. = 275.19 (275.06)

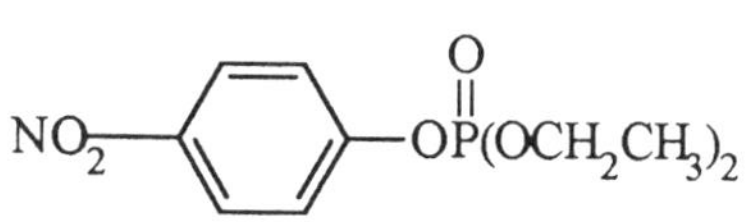

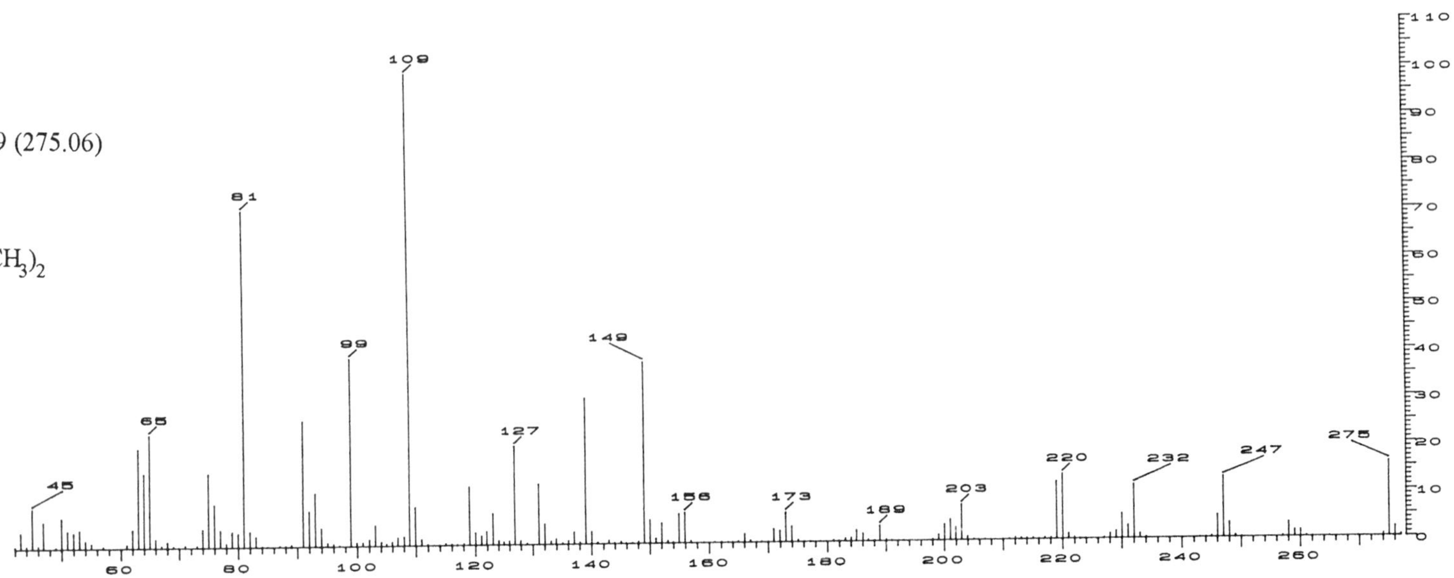

PARATHION

$C_{10}H_{14}NO_5SP$ M.W. = 291.25 (291.03)

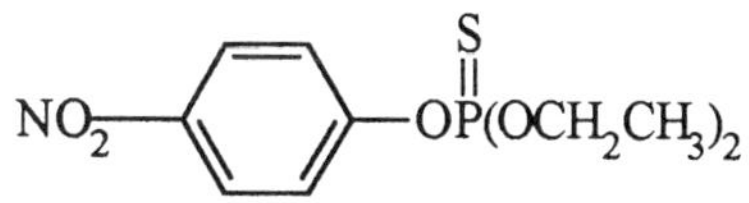

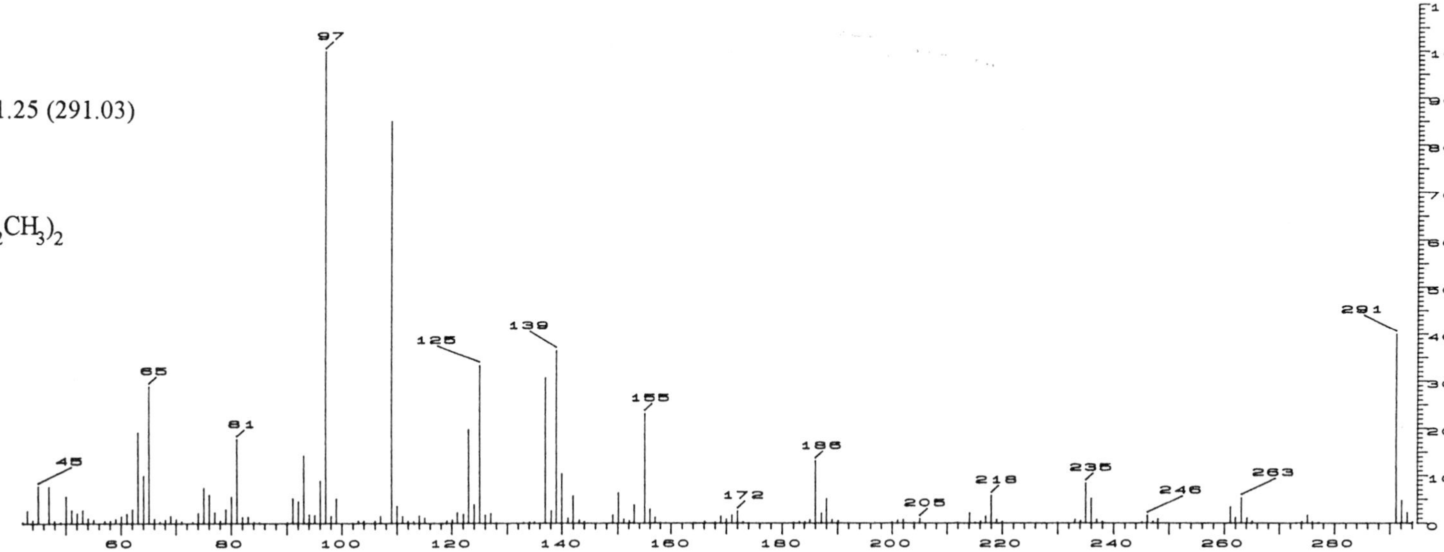

PARATHION-METHYL

$C_8H_{10}NO_5SP$ M.W. = 263.20 (263.00)

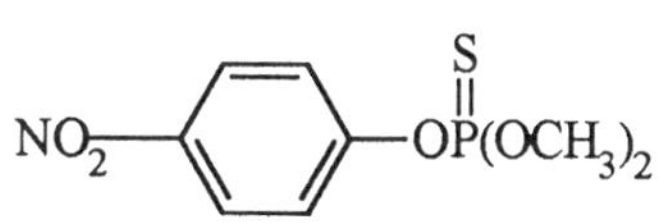

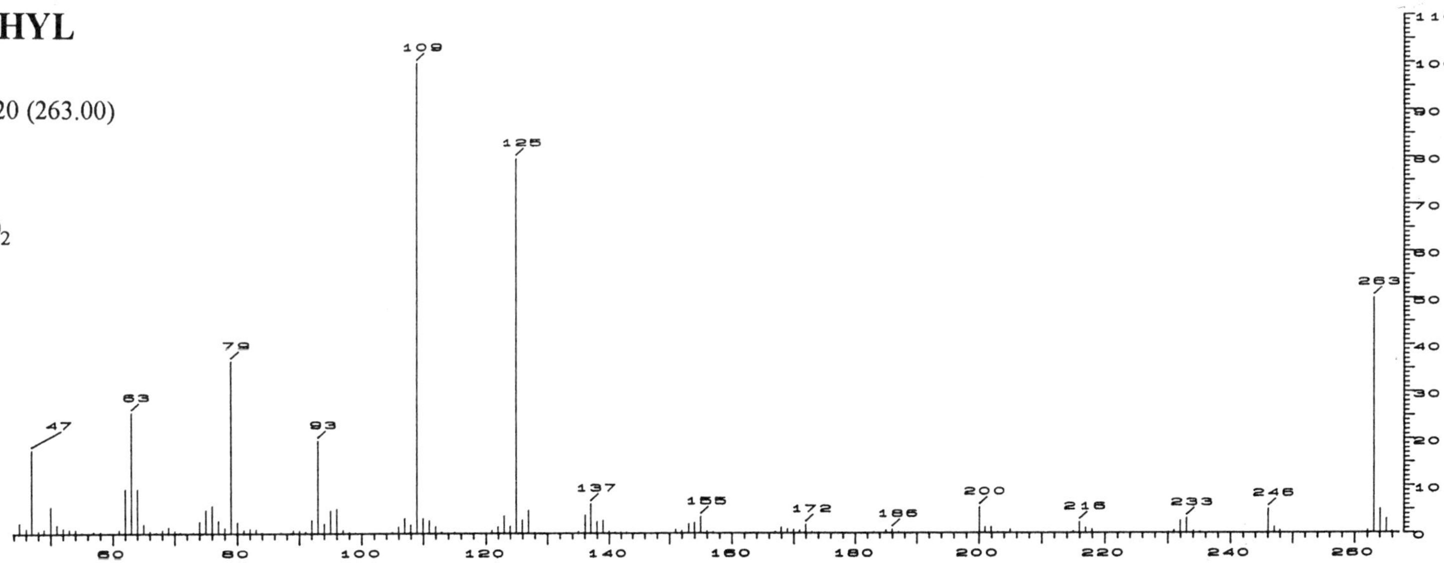

PEBULATE

$C_{10}H_{21}NOS$ M.W. = 203.34 (203.13)

$CH_3(CH_2)_3$—$NCOSCH_2CH_2CH_3$
$|$
CH_2CH_3

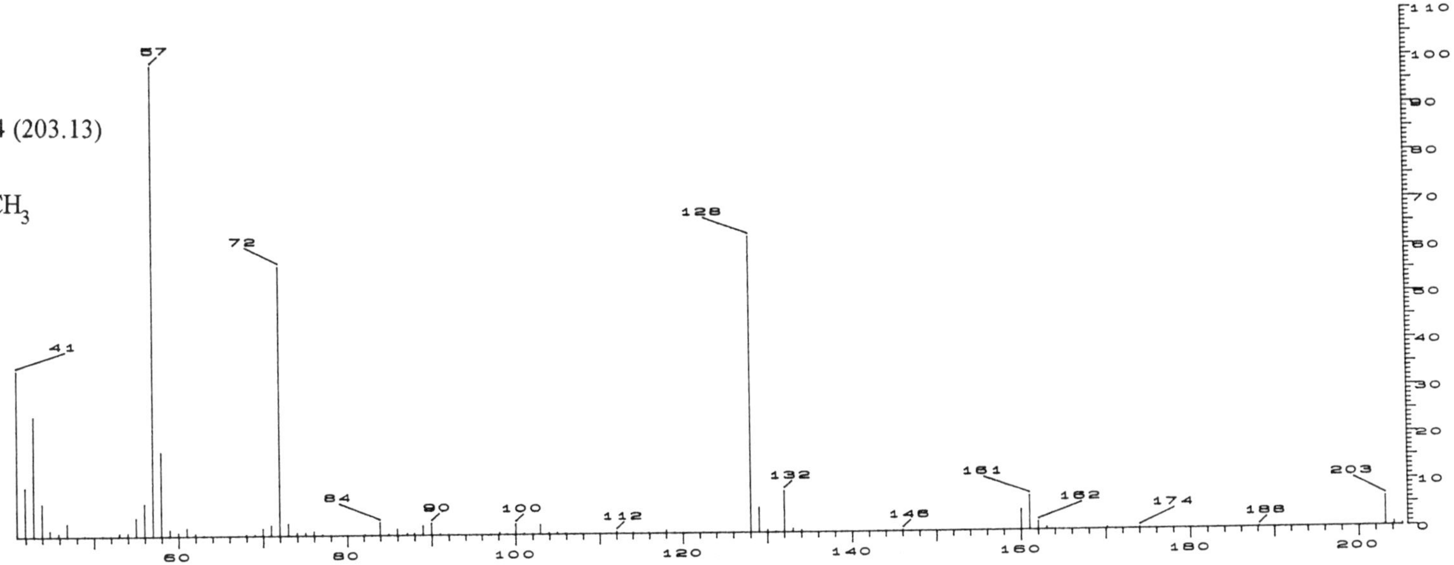

PENTACHLOROANILINE

$C_6H_2NCl_5$ M.W. = 265.35 (262.86)

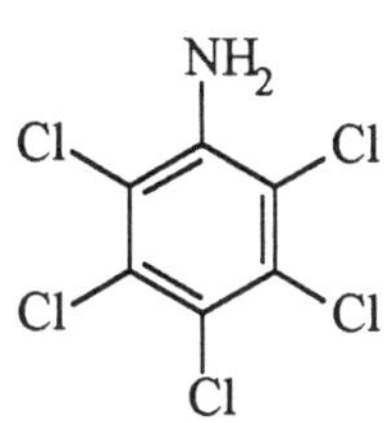

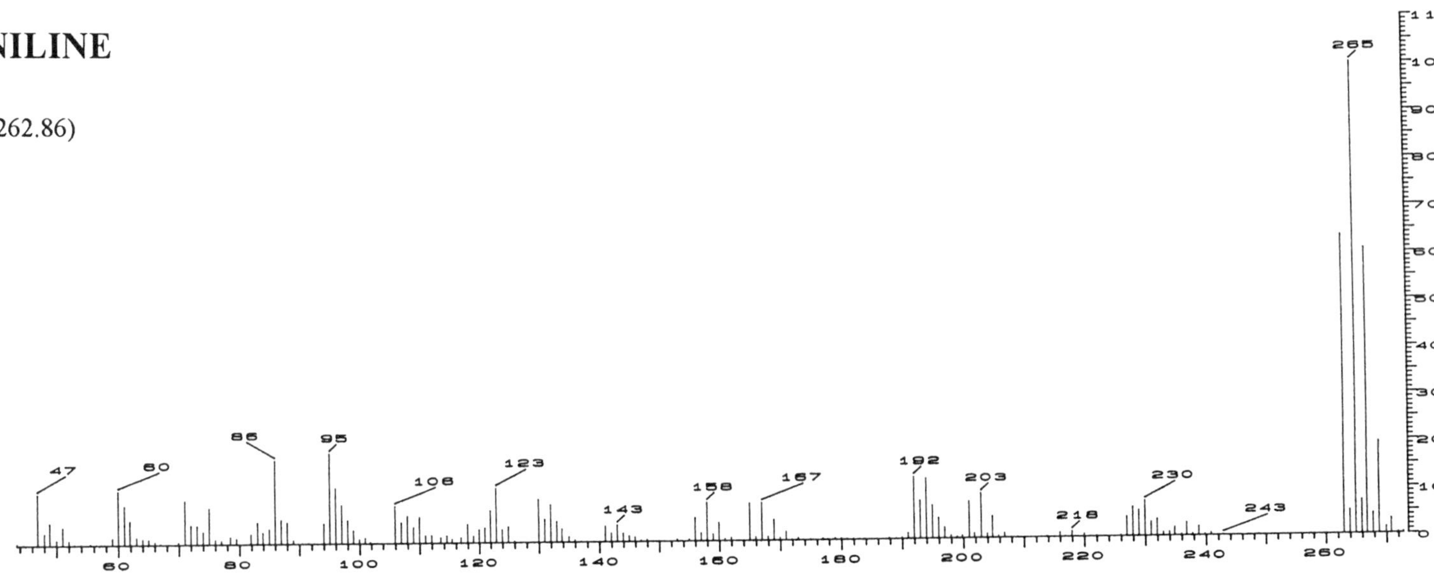

PENTACHLOROPHENOL

C_6HOCl_5 M.W. = 266.33 (263.85)

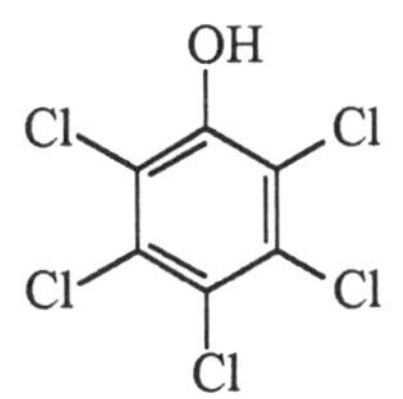

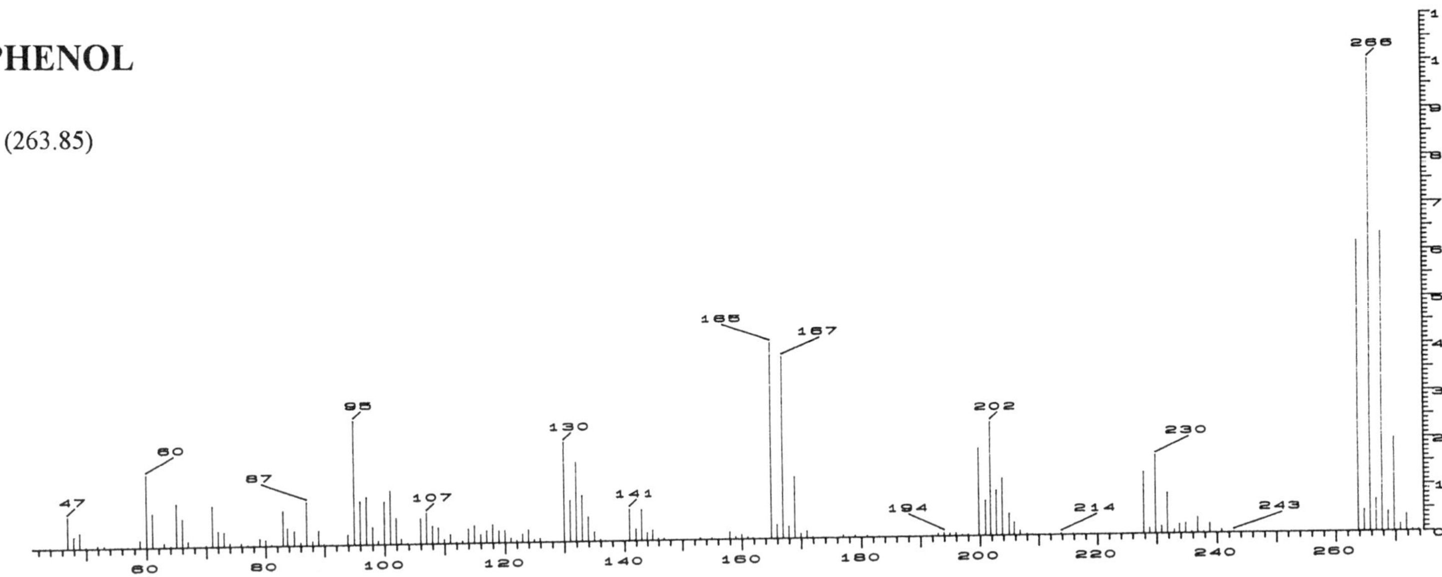

PENTANOCHLOR

$C_{13}H_{18}NOCl$ M.W. = 239.74 (239.11)

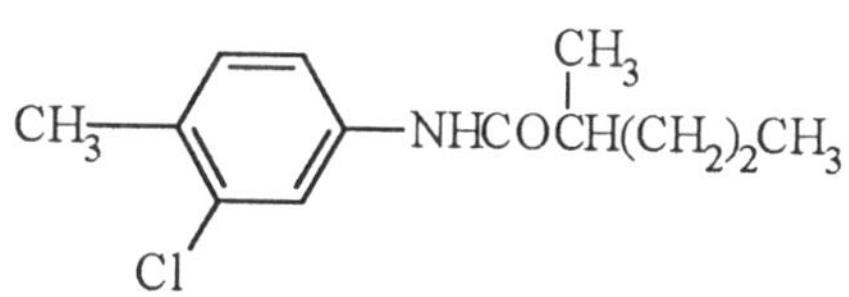

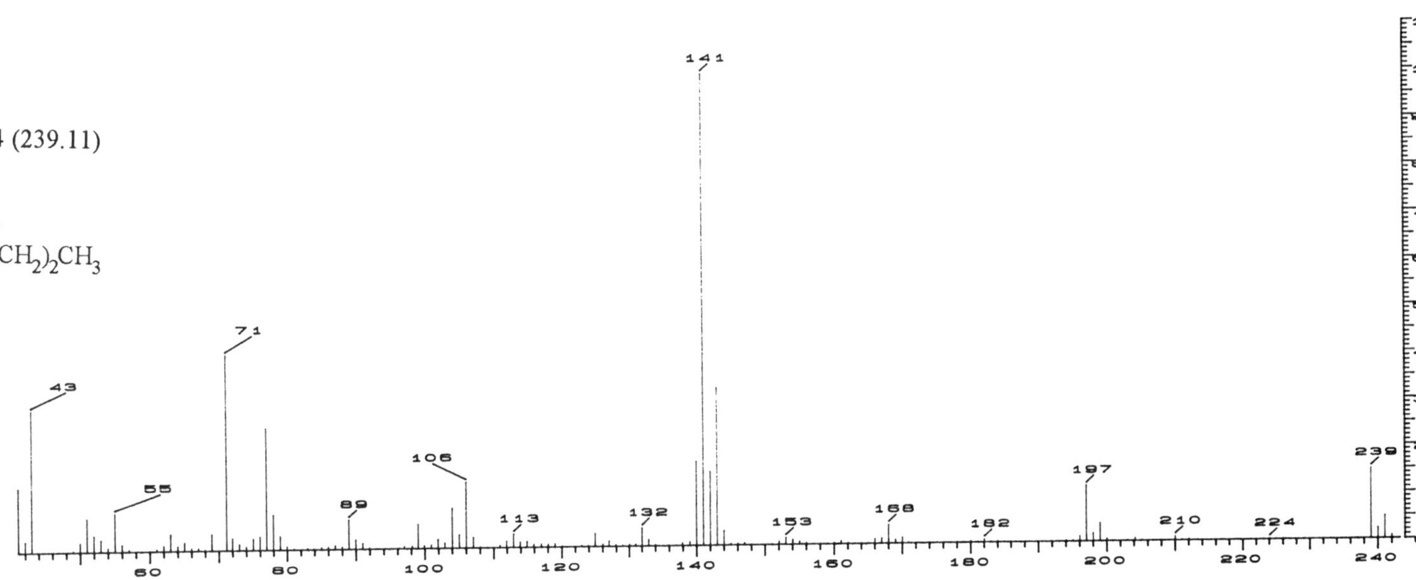

PERTHANE

$C_{18}H_{20}Cl_2$ M.W. = 307.26 (306.09)

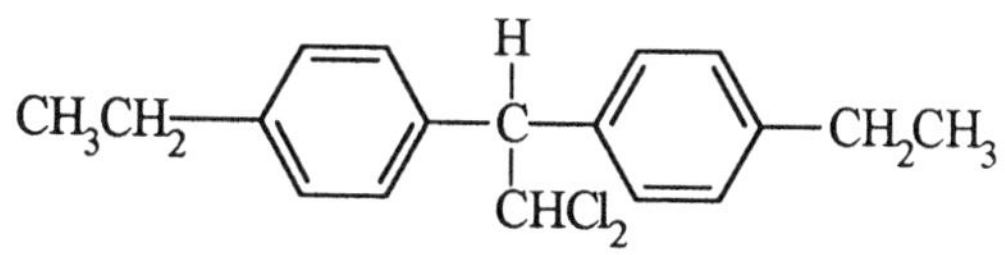

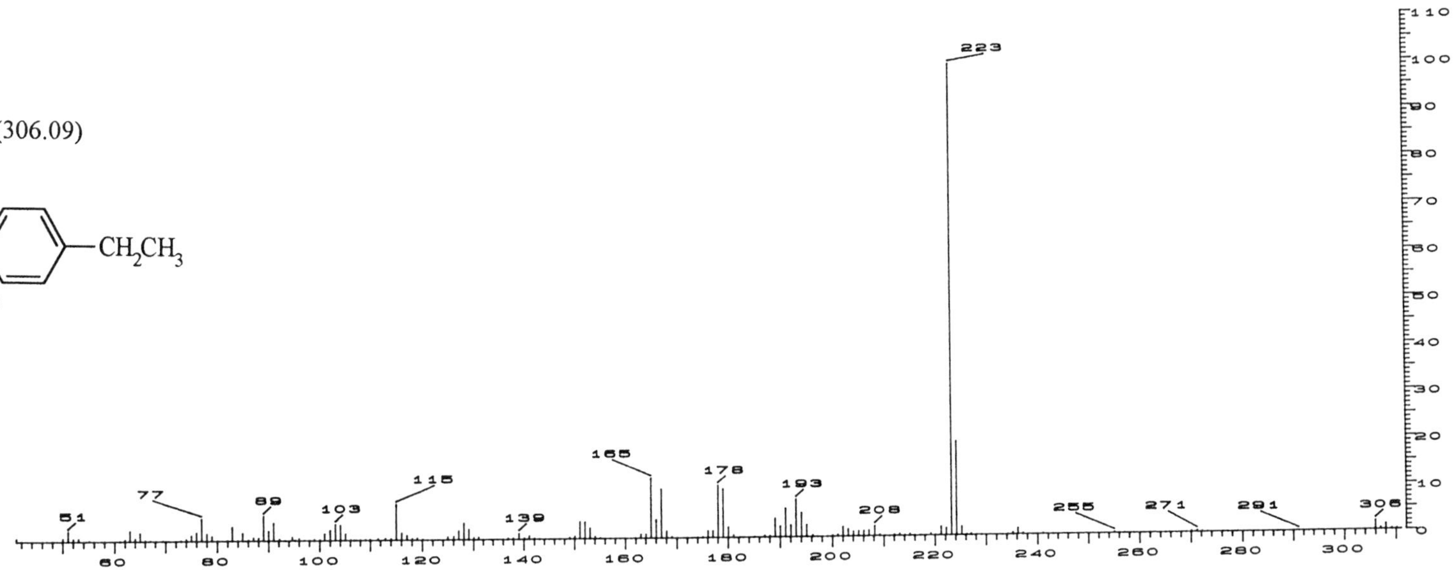

PHENOTHRIN

$C_{23}H_{26}O_3$ M.W. = 350.45 (350.19)

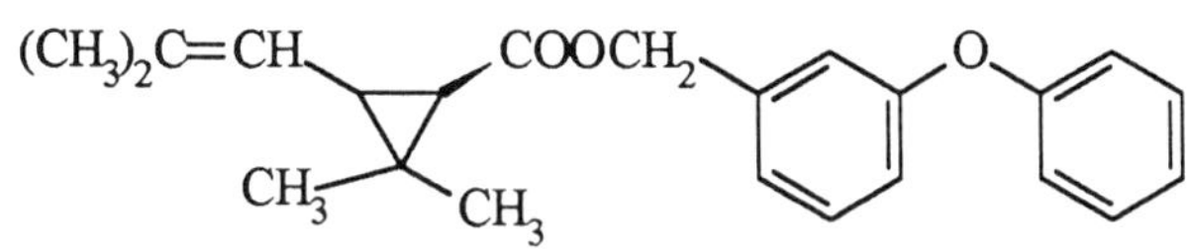

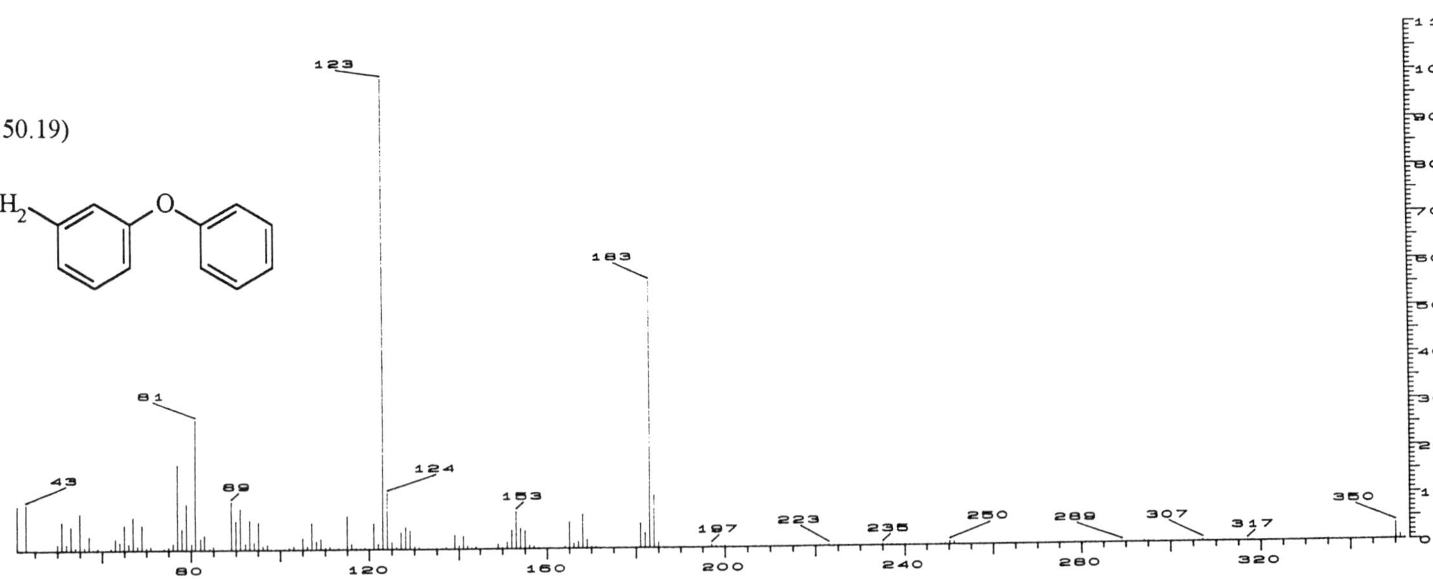

PHENTHOATE

$C_{12}H_{17}O_4S_2P$ M.W. = 320.35 (320.03)

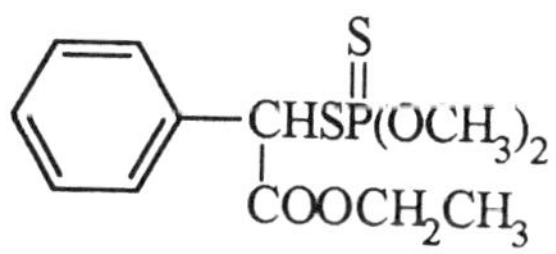

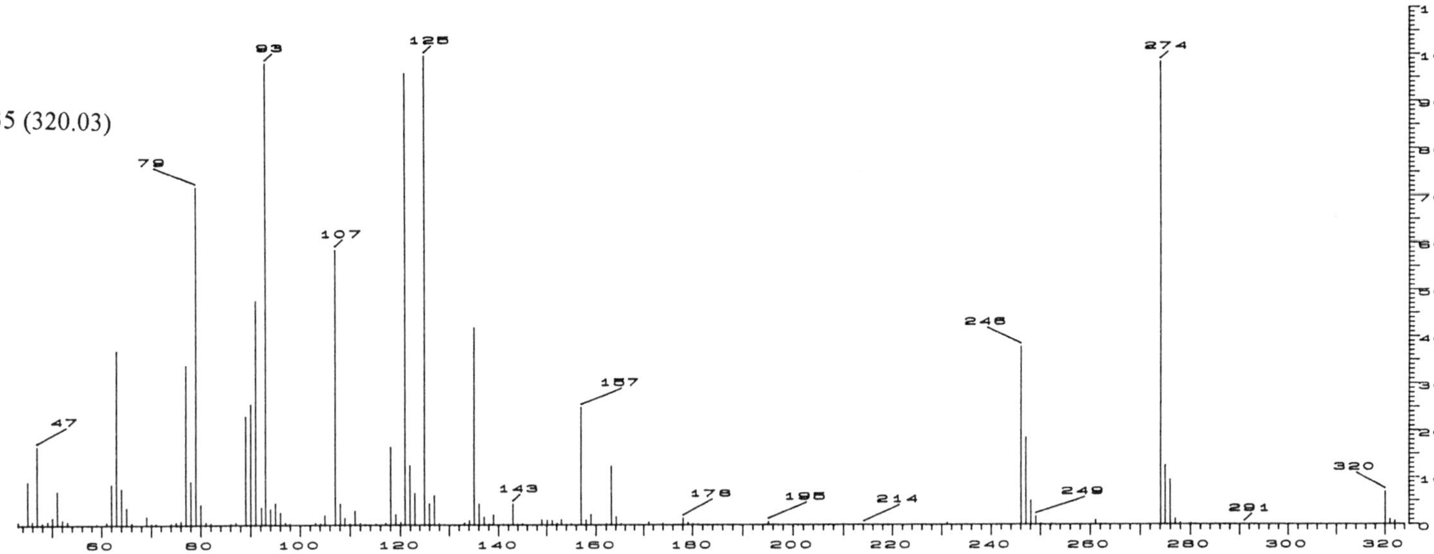

PHORATE

$C_7H_{17}O_2S_3P$ M.W. = 260.36 (260.12)

$CH_3CH_2SCH_2SP(=S)(OCH_2CH_3)_2$

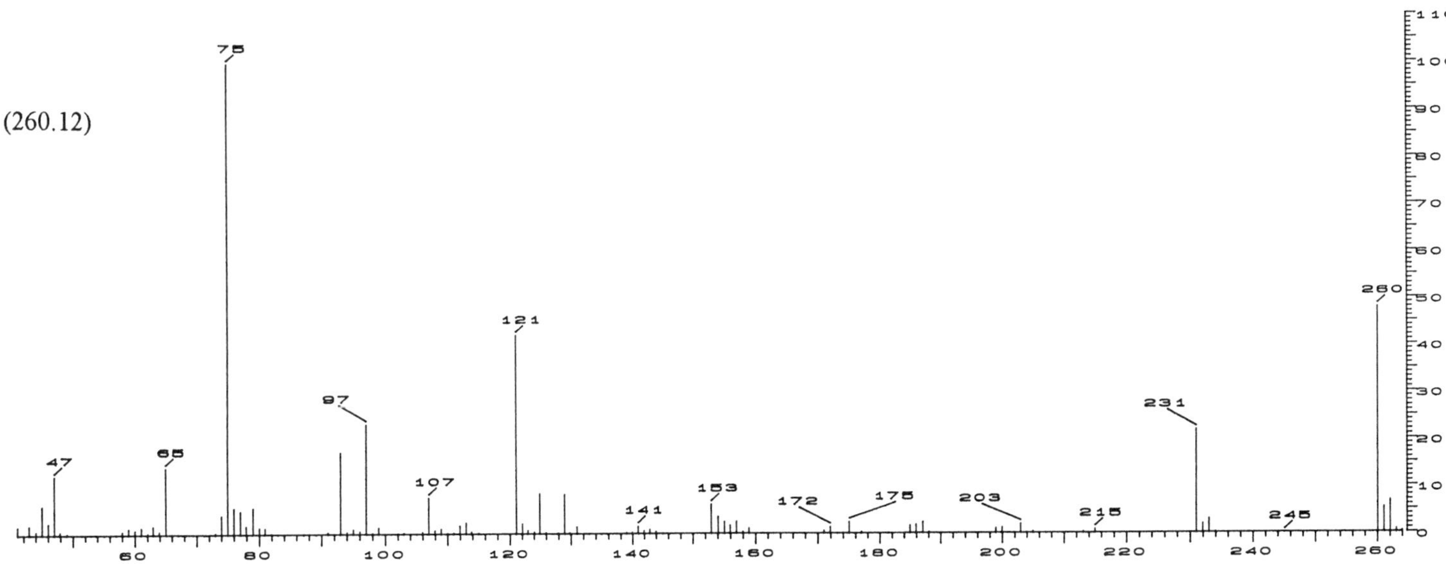

PHOSALONE-OA

$C_{12}H_{15}NO_5SPCl$ M.W. = 351.74 (351.01)

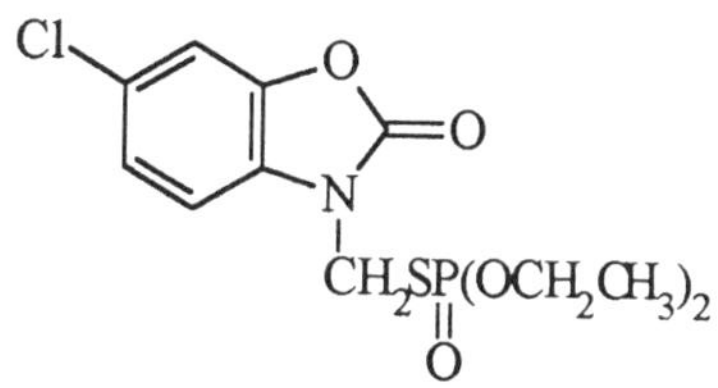

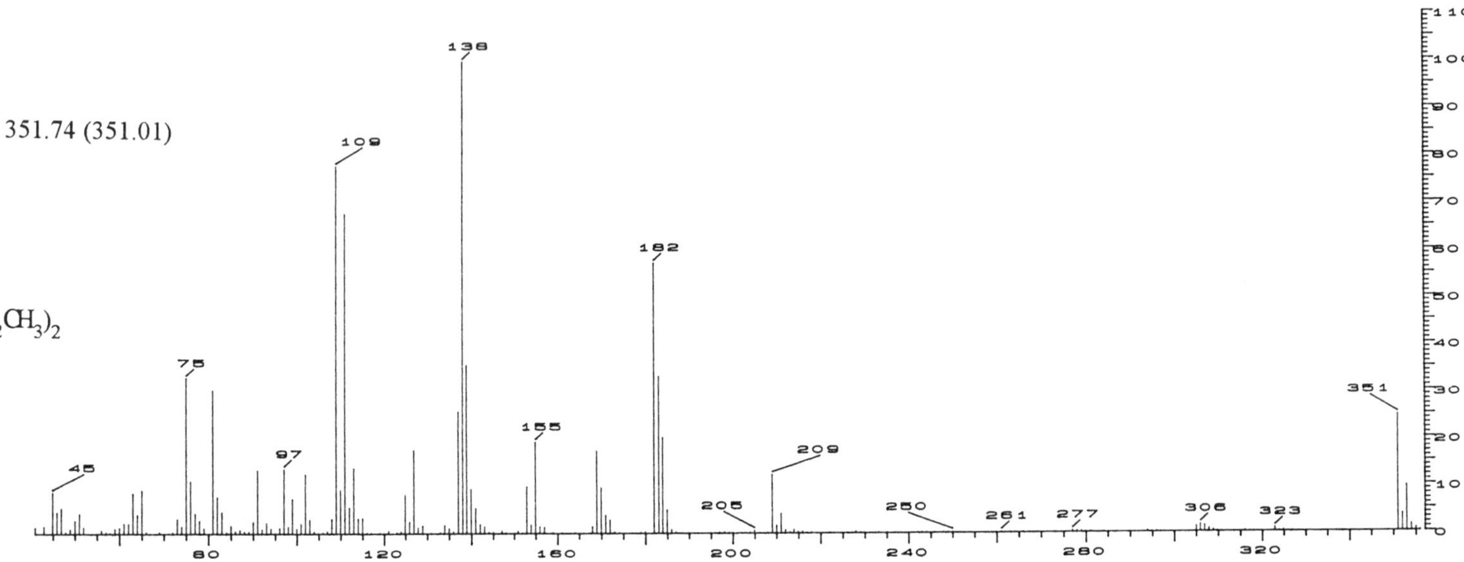

PHOSALONE

$C_{12}H_{15}NO_4S_2PCl$ M.W. = 367.80 (366.99)

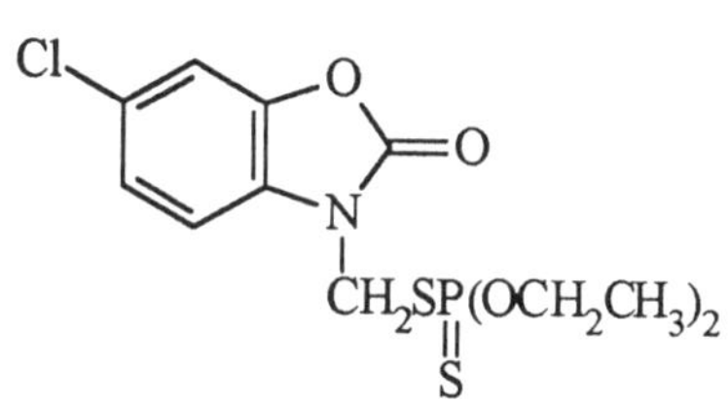

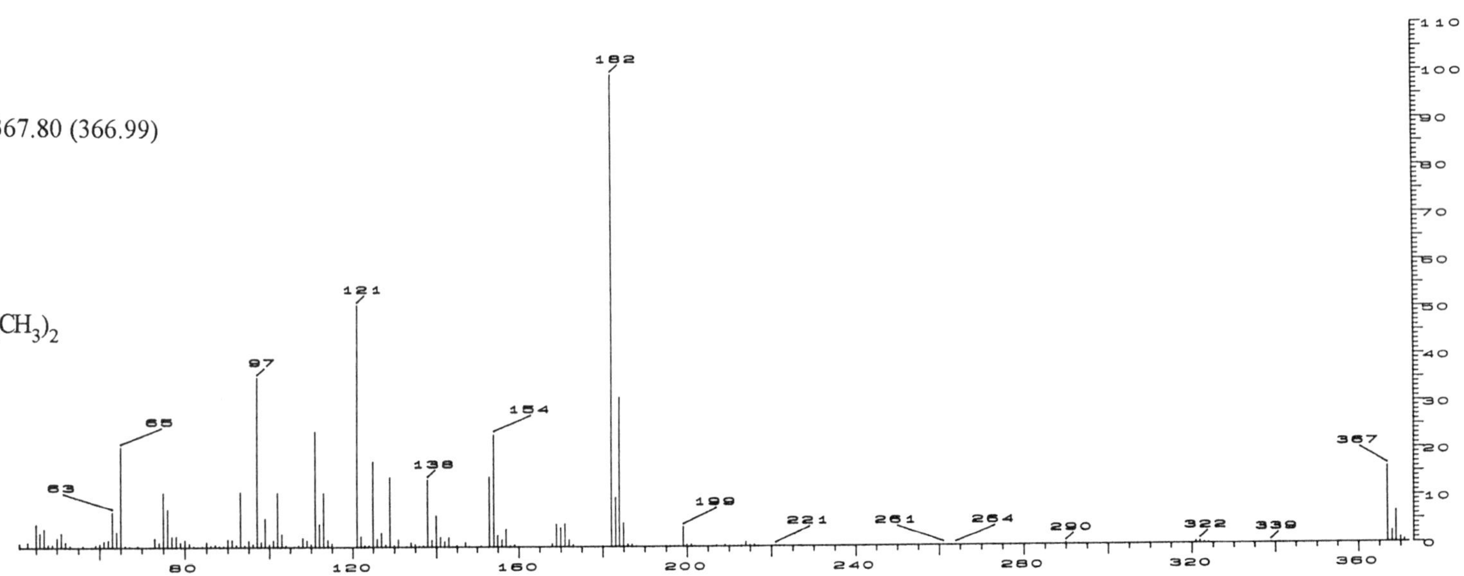

PHOSMET

$C_{11}H_{12}NO_4S_2P$ M.W. = 317.31 (316.99)

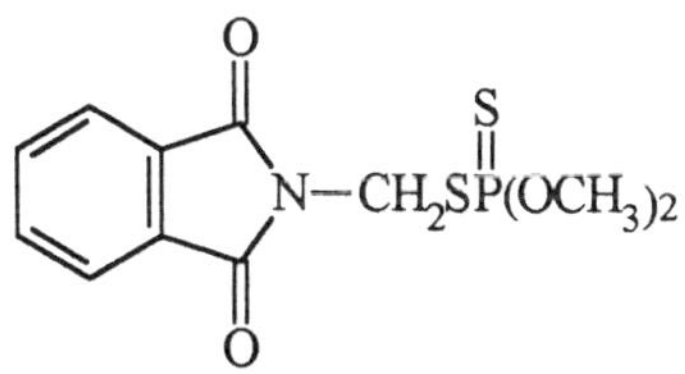

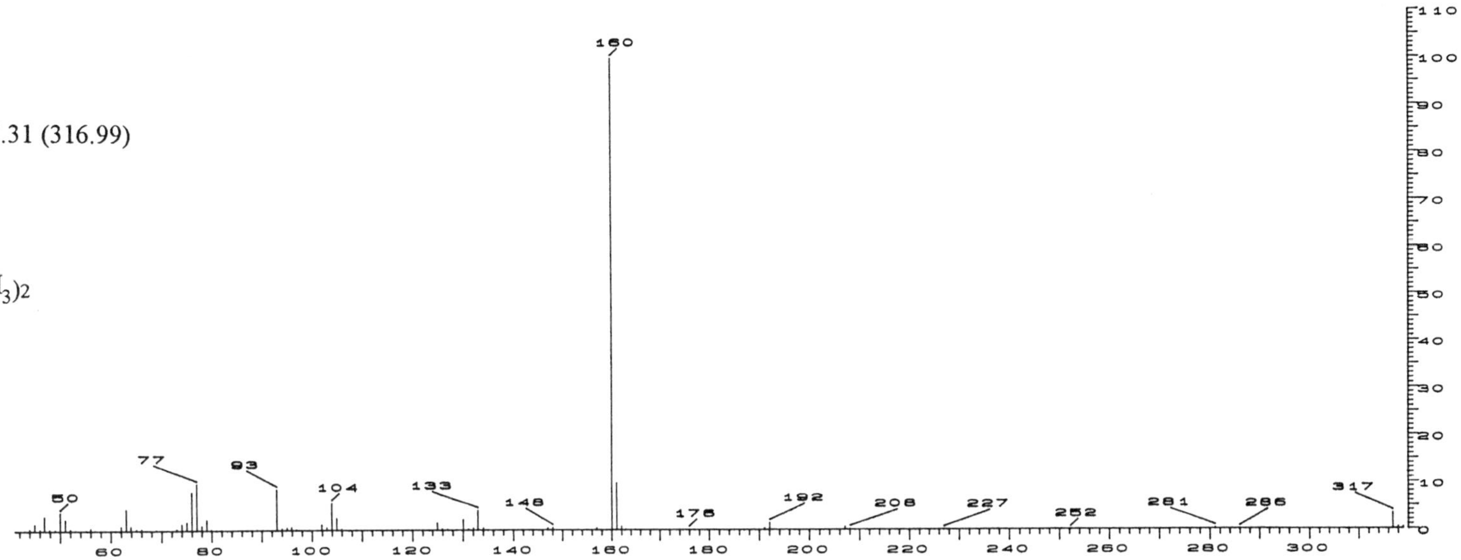

PHOSMET-OA

$C_{11}H_{12}NO_5SP$ M.W. = 301.25 (301.02)

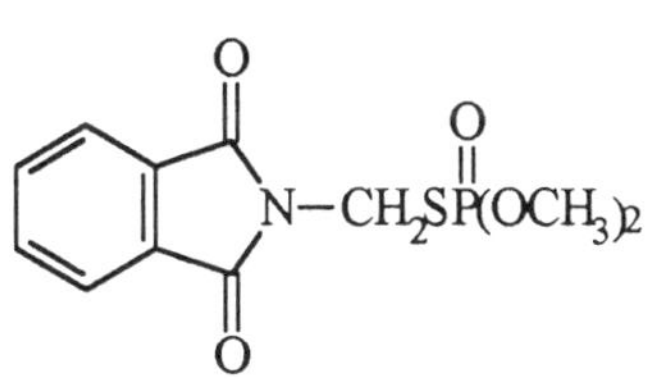

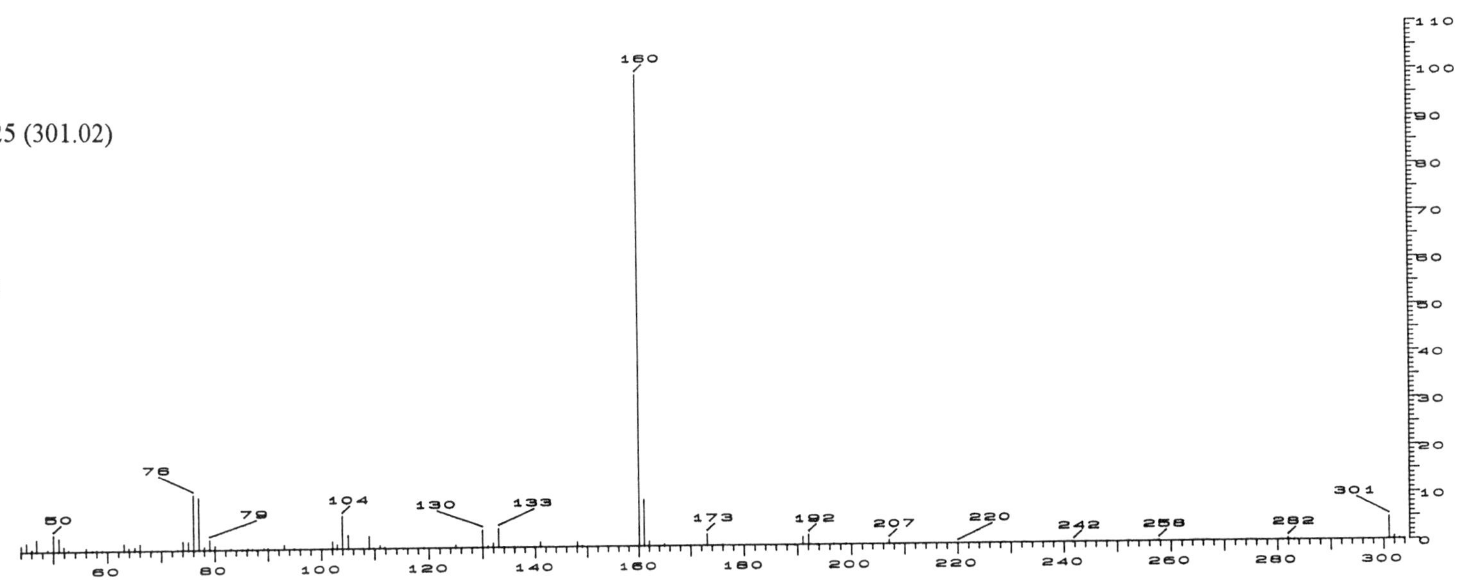

PHOSPHAMIDON

$C_{10}H_{19}NO_5PCl$ M.W. = 299.69 (299.07)

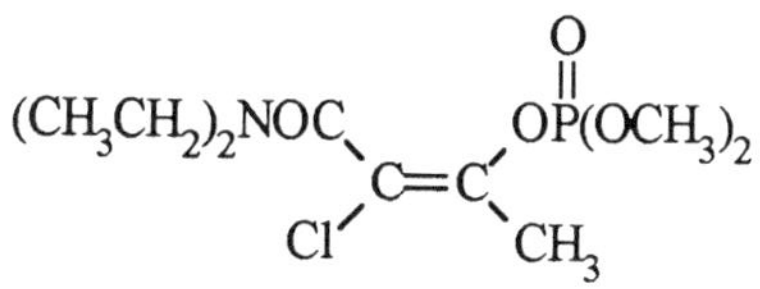

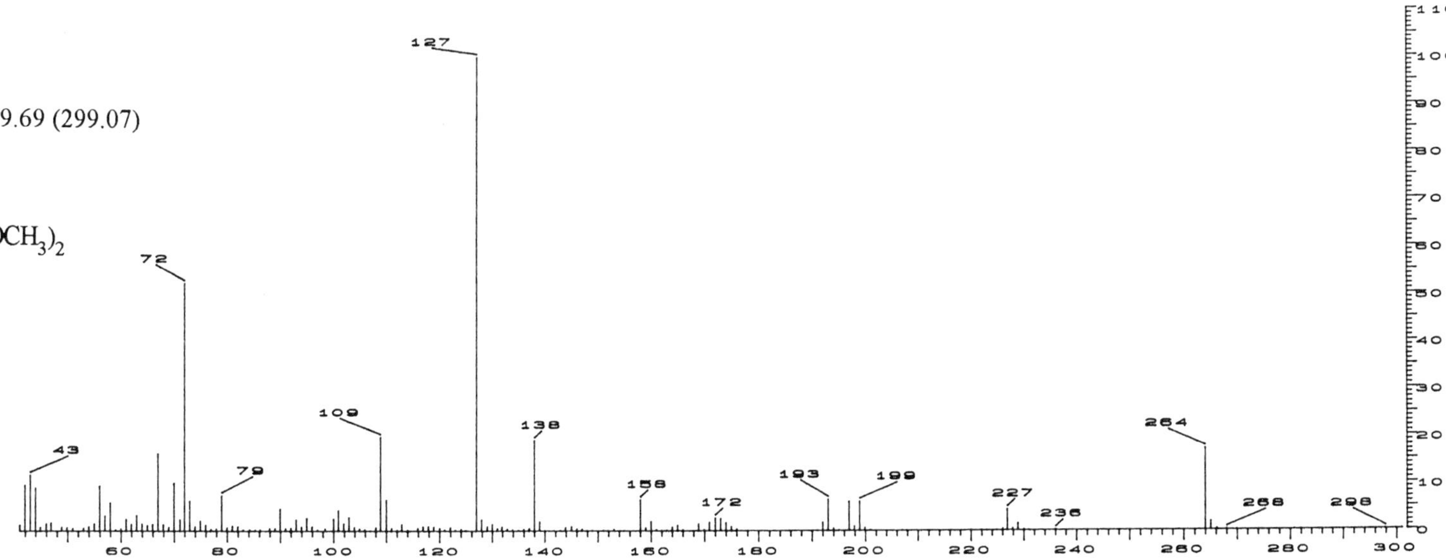

PHOXIM

$C_{12}H_{15}N_2O_3SP$ M.W. = 298.29 (298.05)

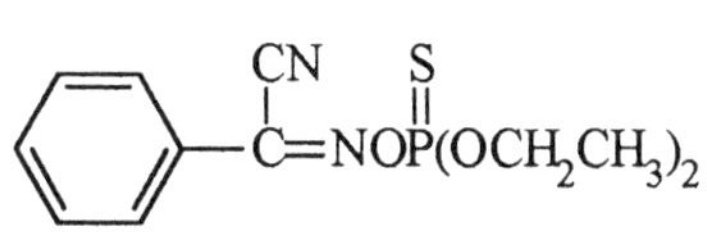

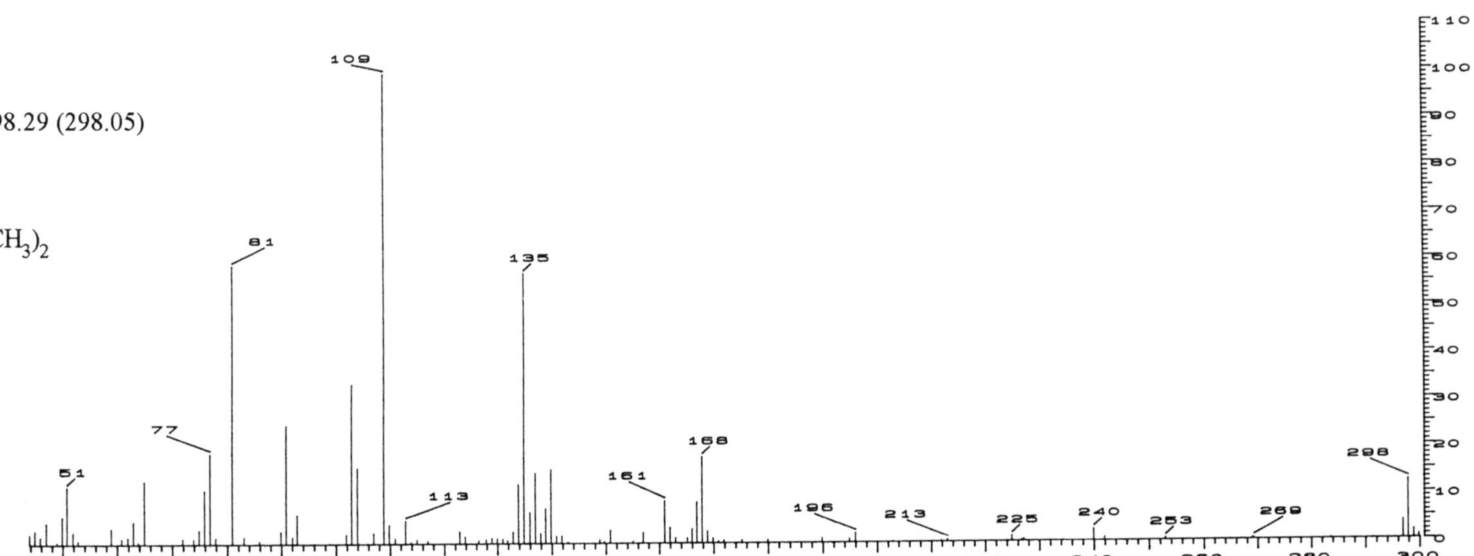

PHOXIM-OA

$C_{12}H_{15}N_2O_4P$ M.W. = 282.23 (282.08)

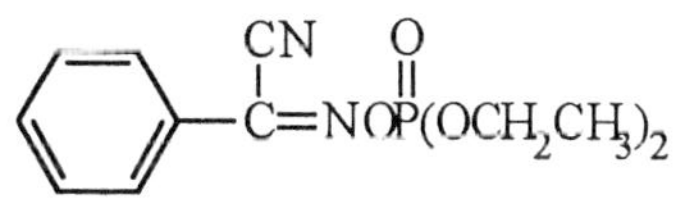

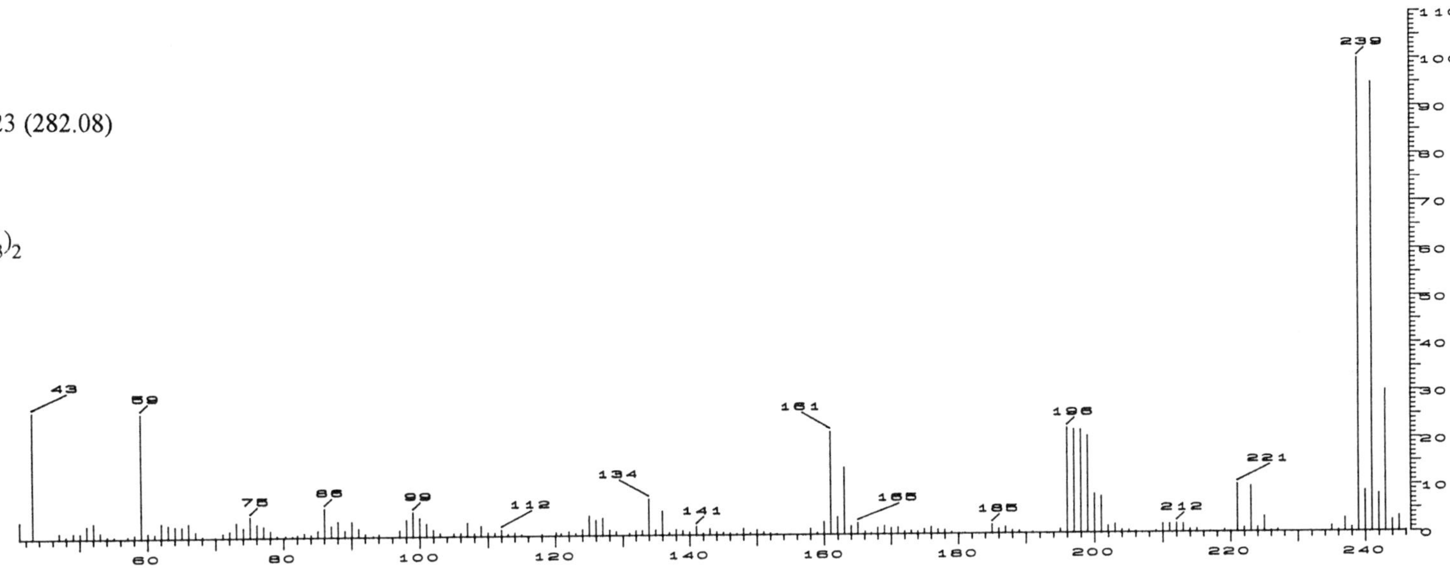

PICLORAM

$C_6H_3N_2O_2Cl_3$ M.W. = 241.46 (239.93)

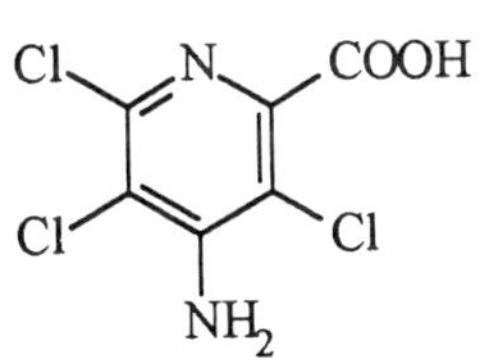

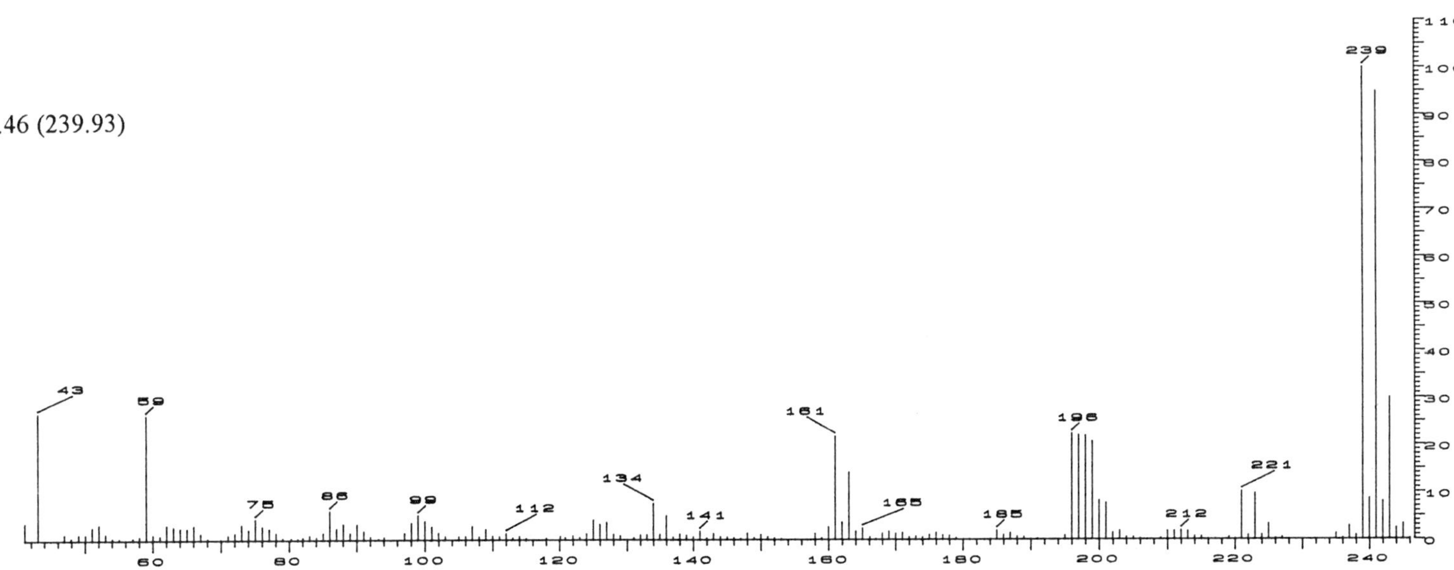

PINDONE

$C_{14}H_{14}O_3$ M.W. = 230.26 (230.09)

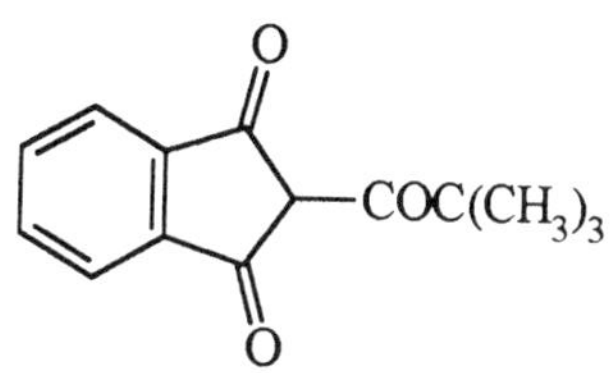

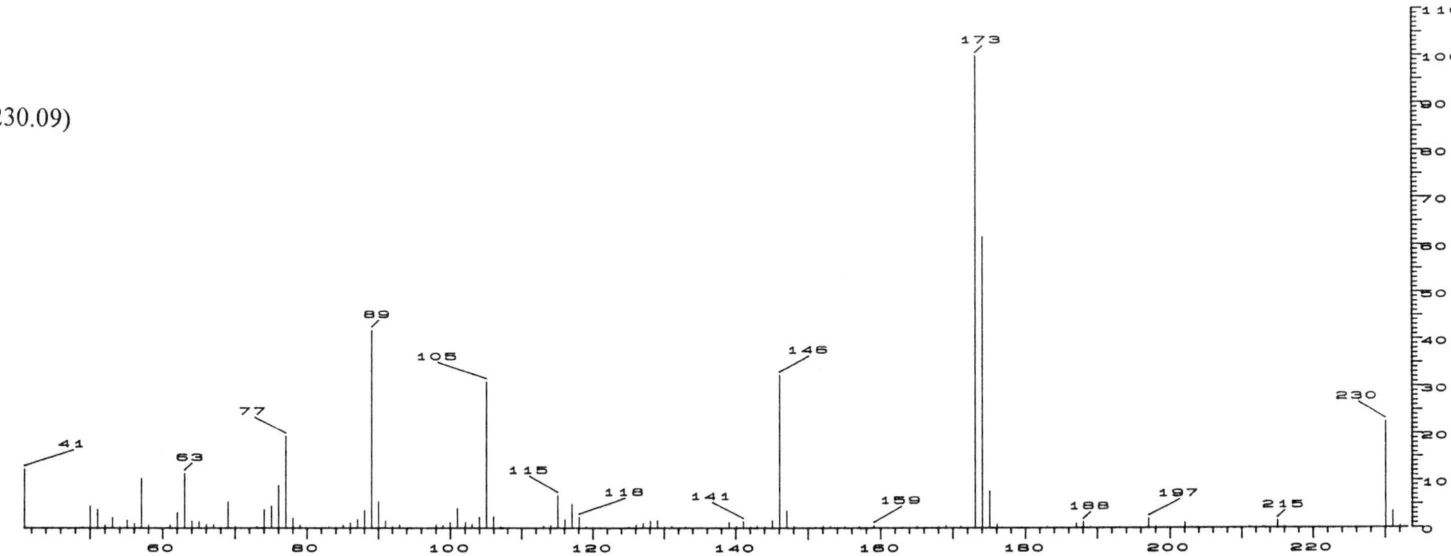

PIRIMICARB

$C_{11}H_{18}N_4O_2$ M.W. = 238.28 (238.14)

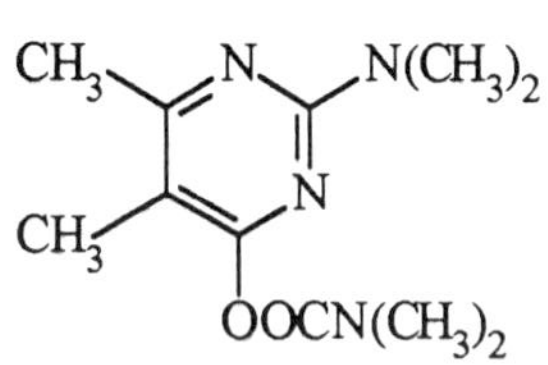

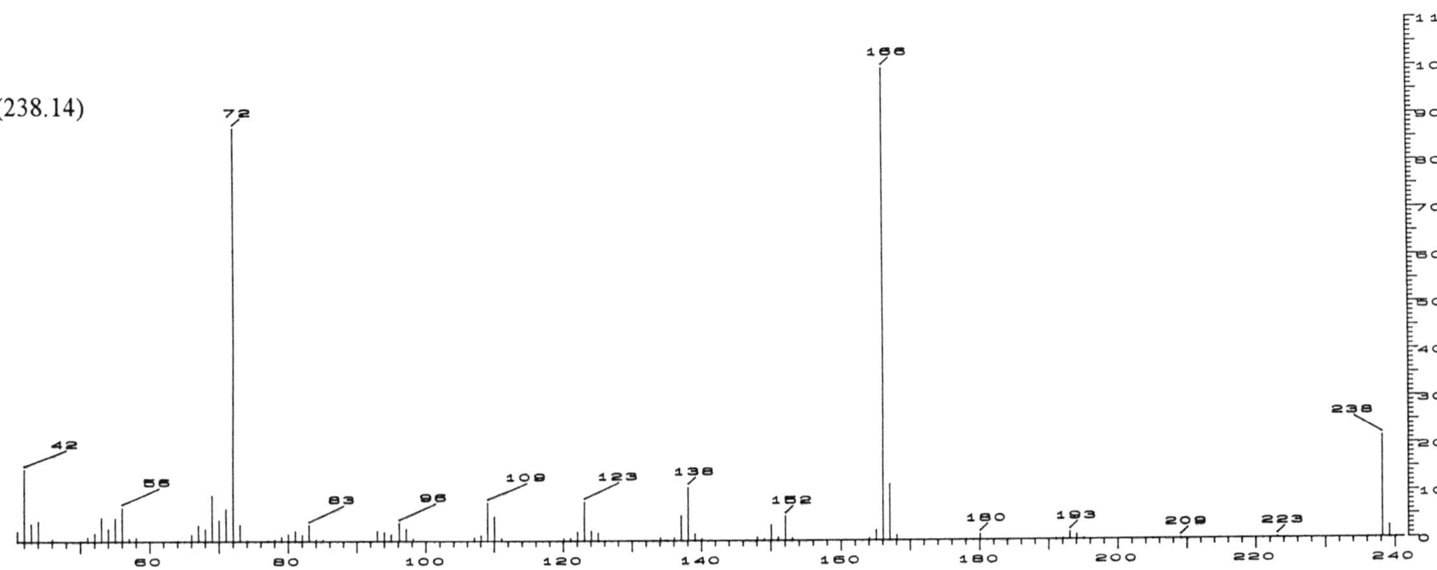

PIRIMIPHOS-ETHYL-OA

$C_{13}H_{24}N_3O_4P$ M.W. = 317.32 (317.15)

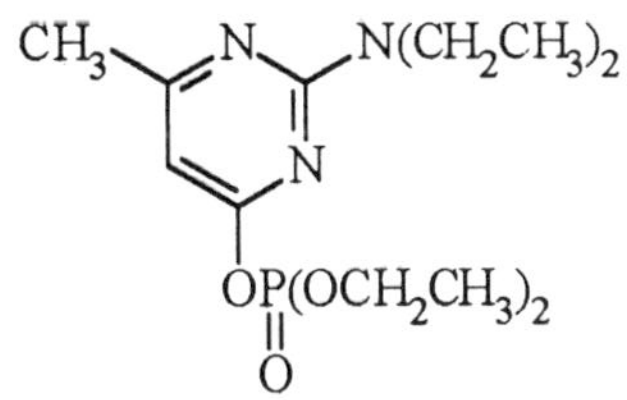

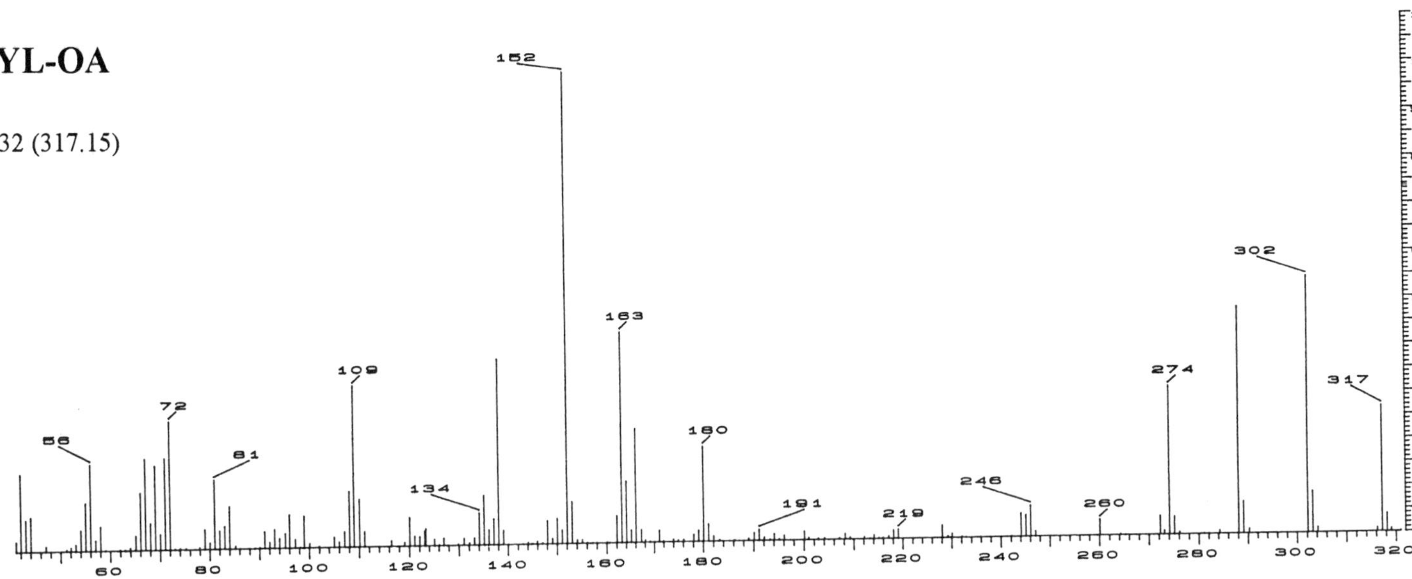

PROCYMIDONE

$C_{13}H_{14}NO_2Cl_2$ M.W. = 284.14 (283.02)

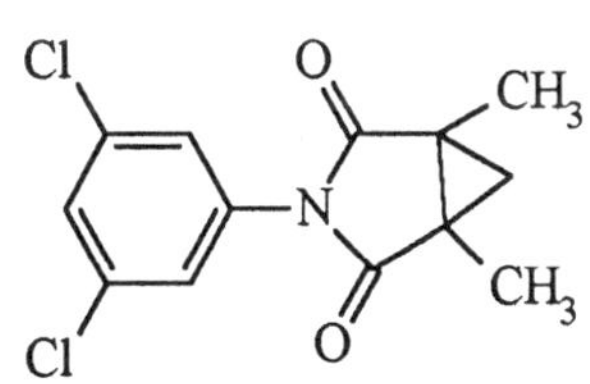

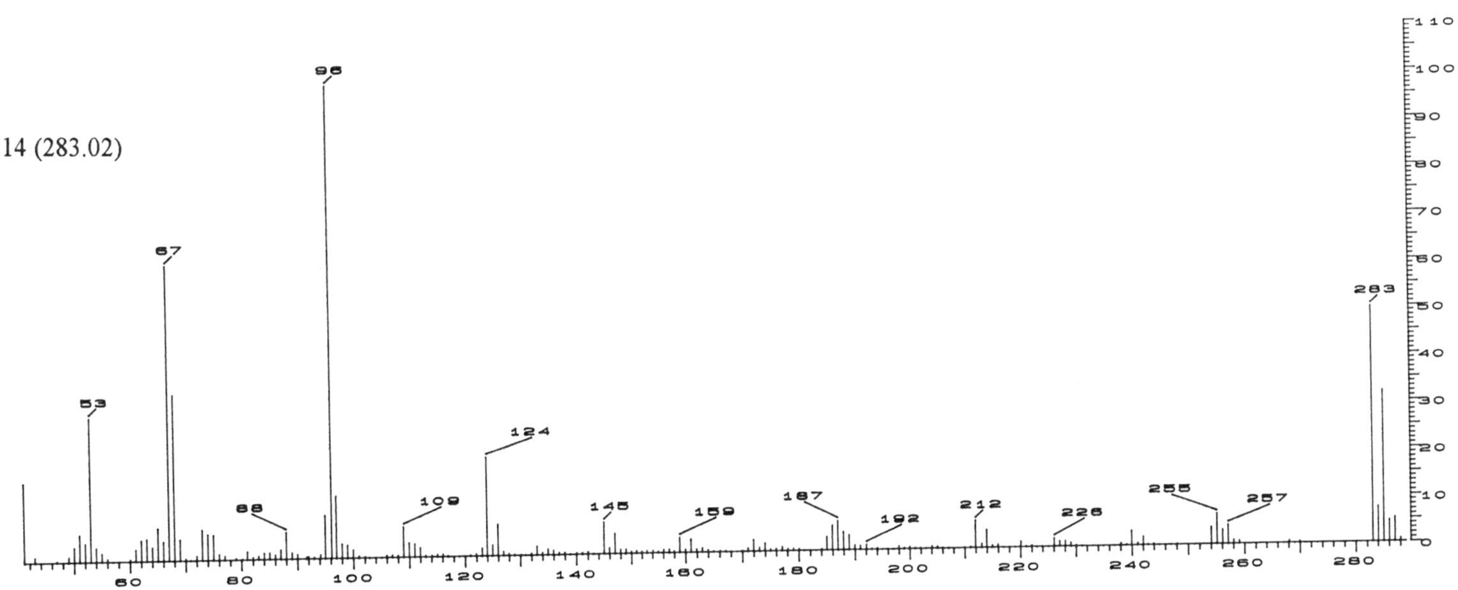

PROFENOFOS

$C_{11}H_{15}O_3SPClBr$ M.W. = 373.62 (371.94)

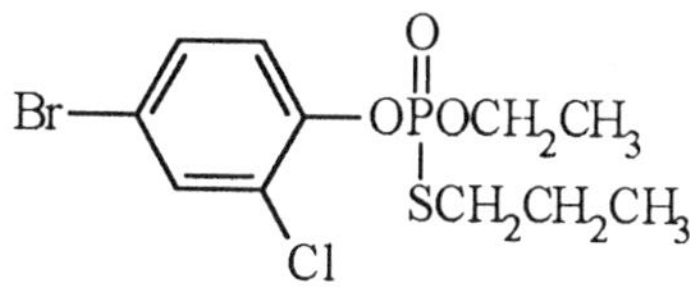

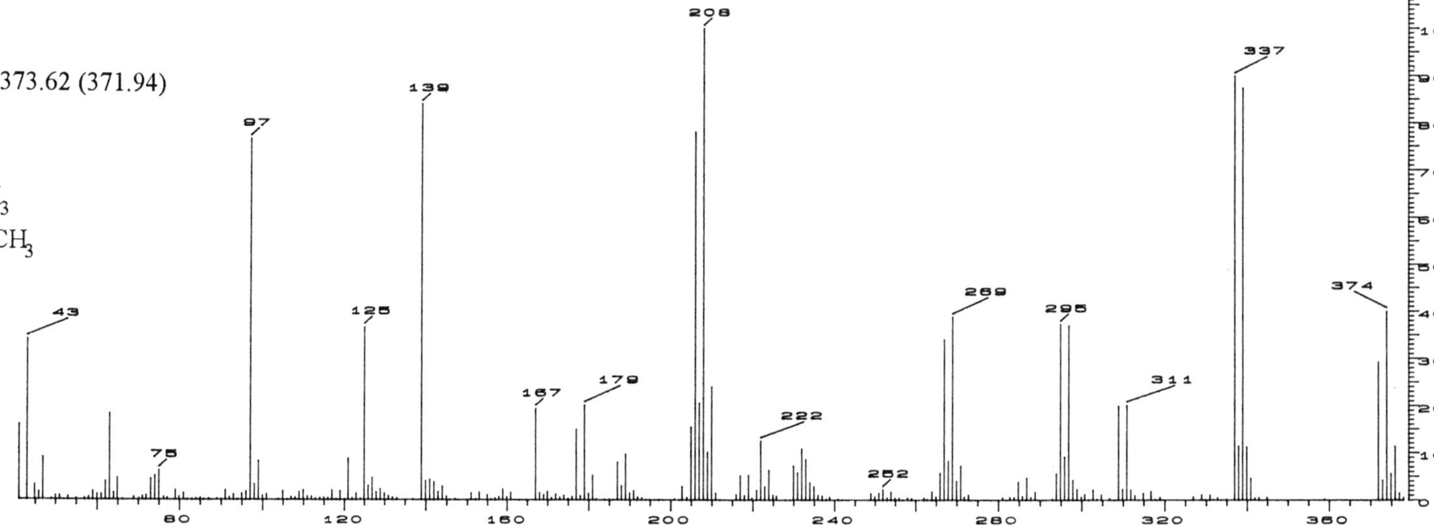

PROFLURALIN

$C_{14}H_{16}N_3O_4F_3$ M.W. = 347.29 (347.11)

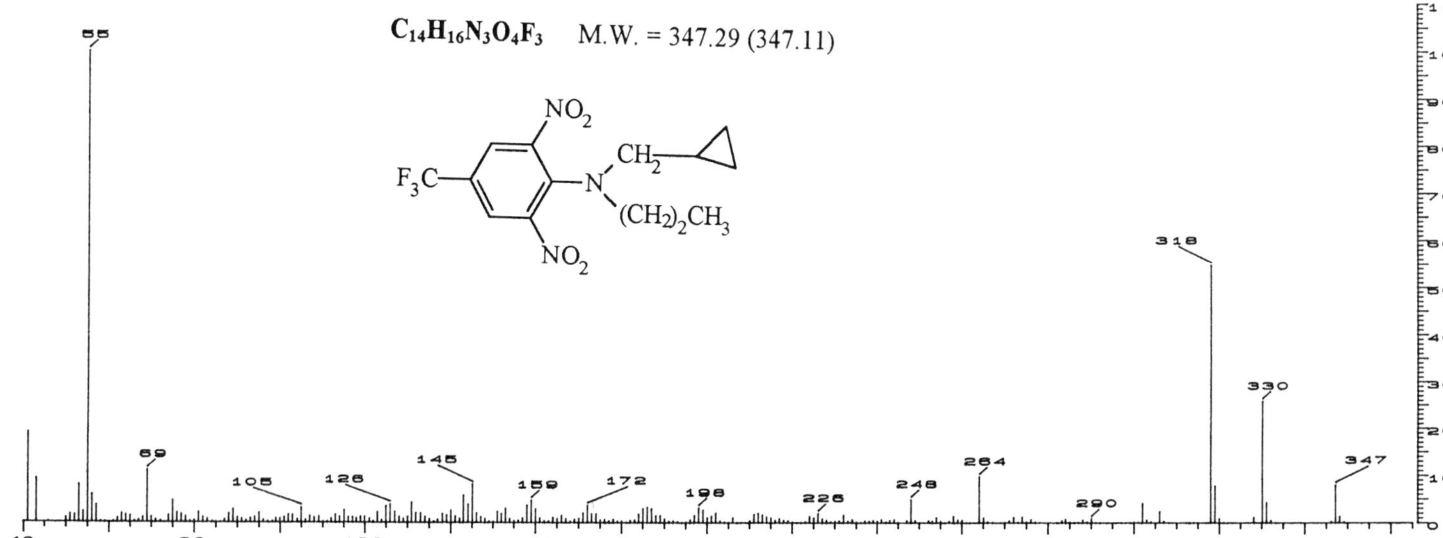

PROMETON

$C_{10}H_{19}N_5O$ M.W. = 225.29 (225.16)

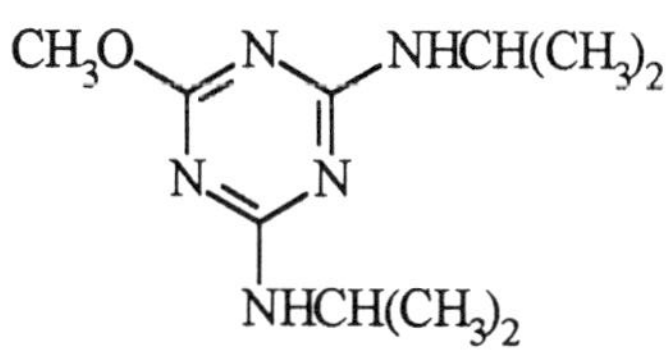

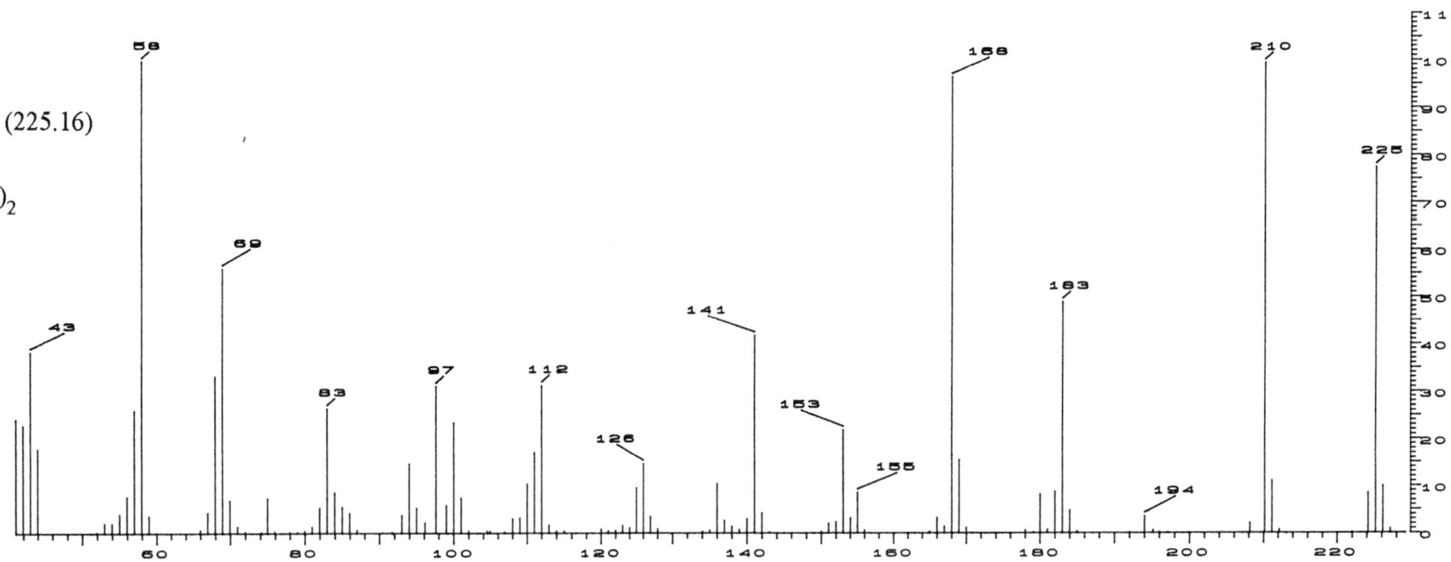

PRONAMIDE

$C_{12}H_{11}NOCl_2$ M.W. = 256.13 (255.02)

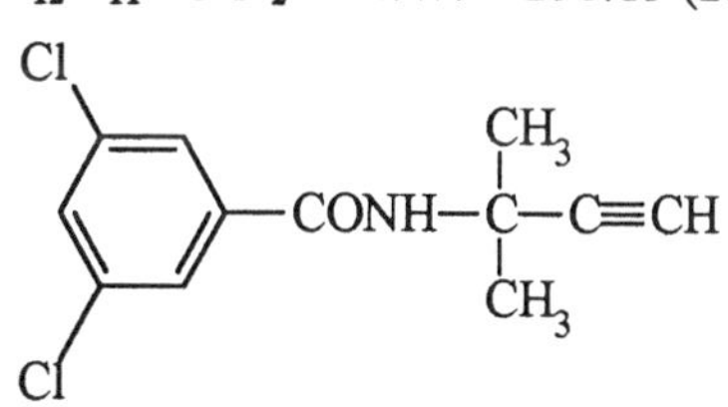

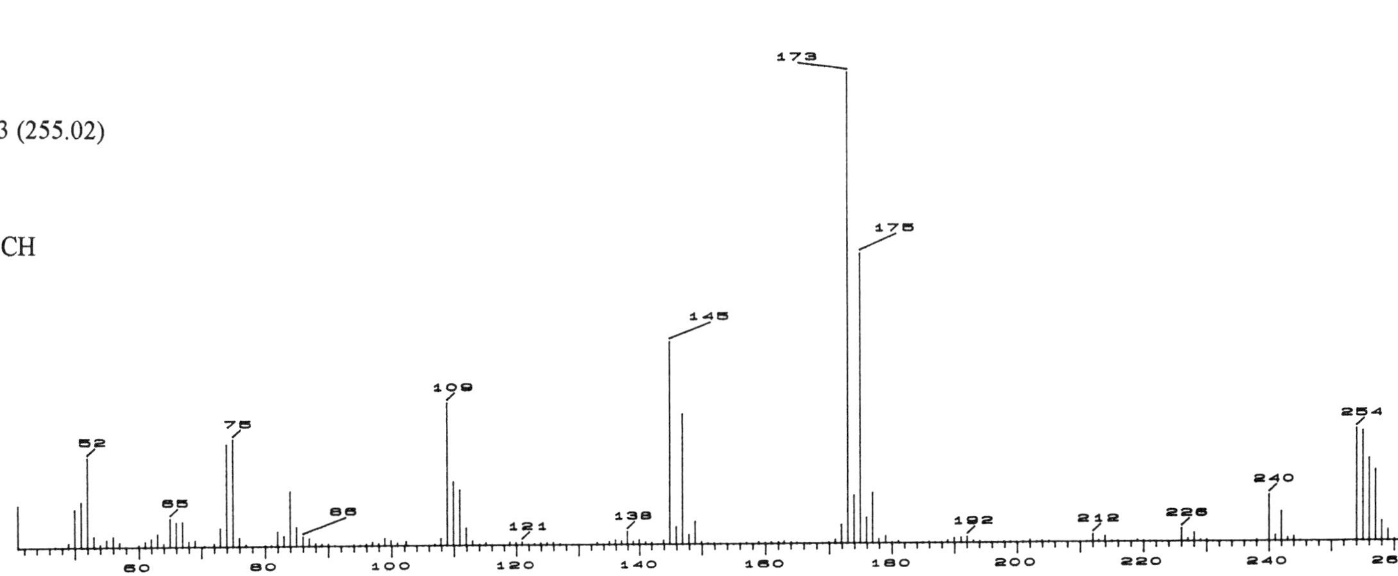

PROPANIL

$C_9H_9NOCl_2$ M.W.= 218.08 (217.01)

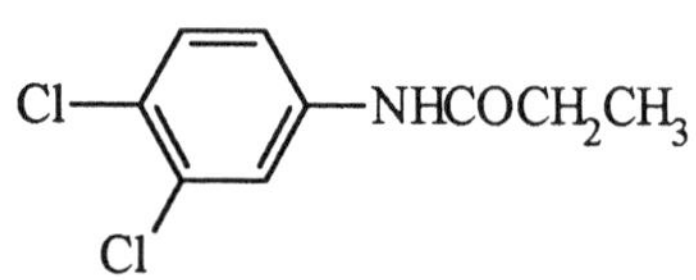

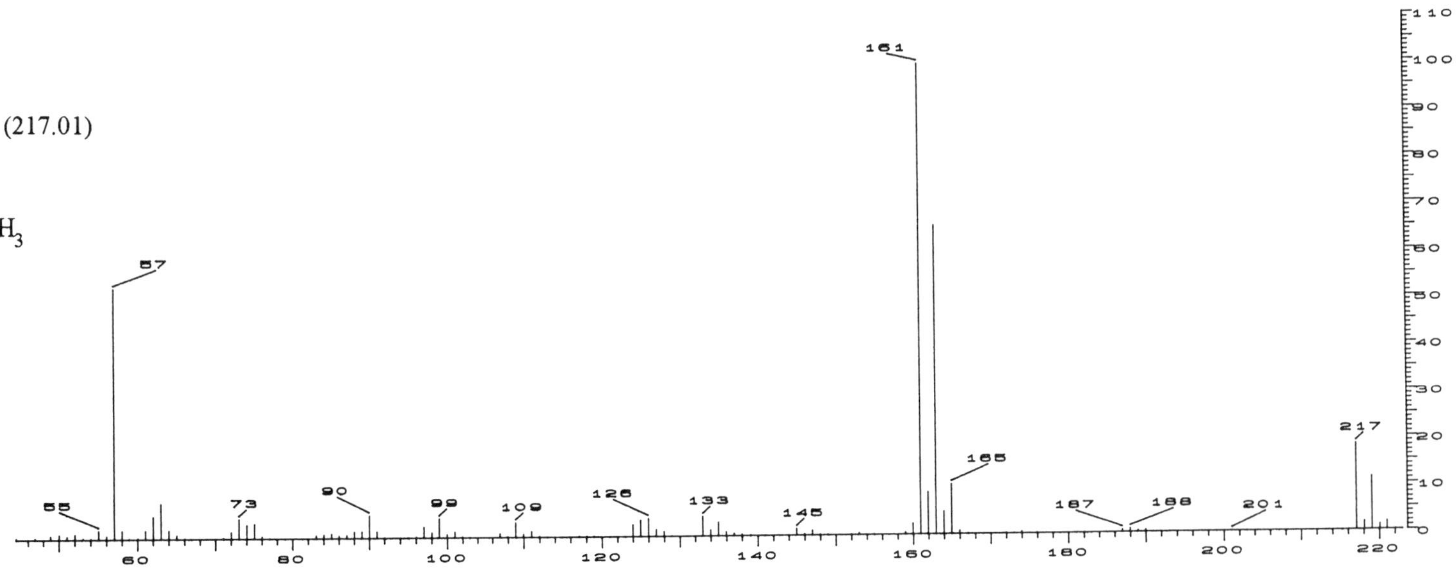

PROPAZINE

$C_9H_{16}N_5Cl$ M.W. = 229.71 (229.11)

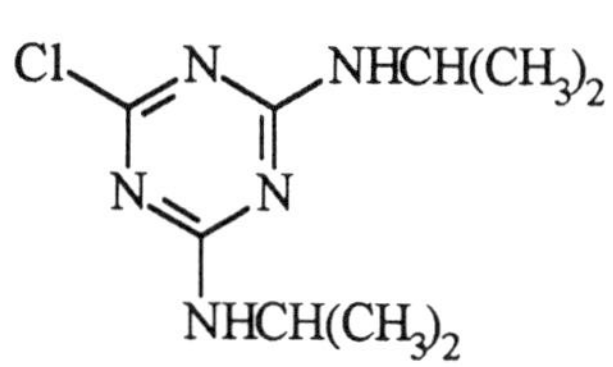

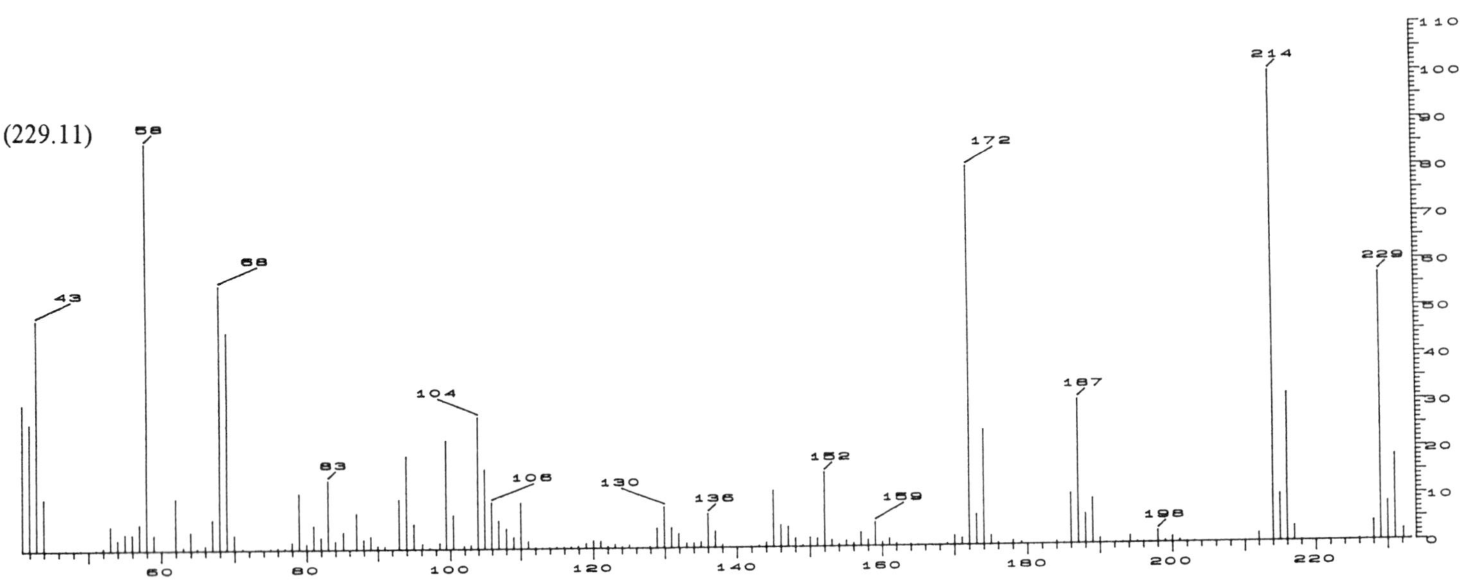

PROPETAMPHOS

$C_{10}H_{20}NO_4SP$ M.W. = 281.30 (281.09)

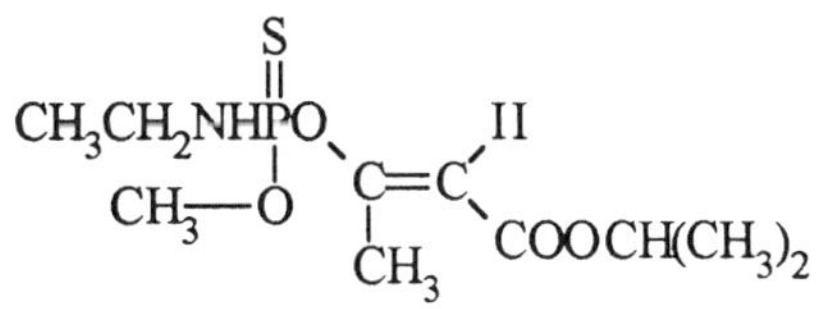

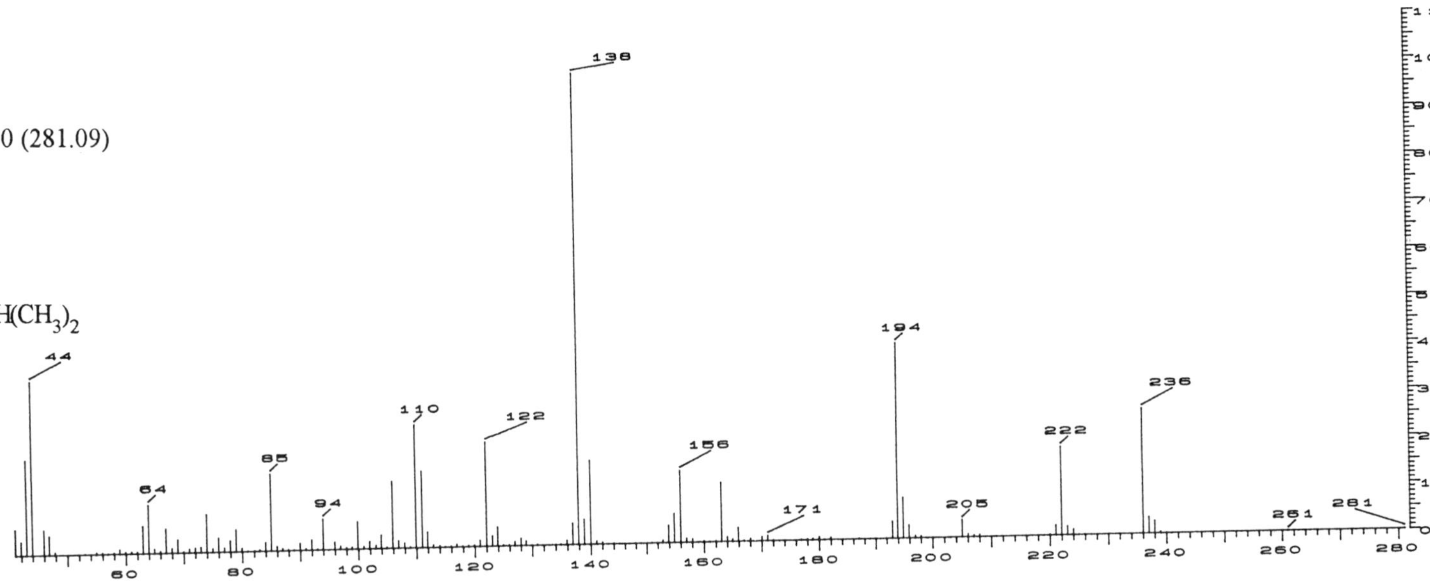

PROPOXUR

$C_{11}H_{15}NO_3$ M.W. = 209.24 (209.11)

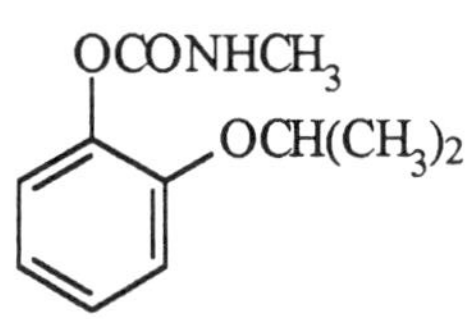

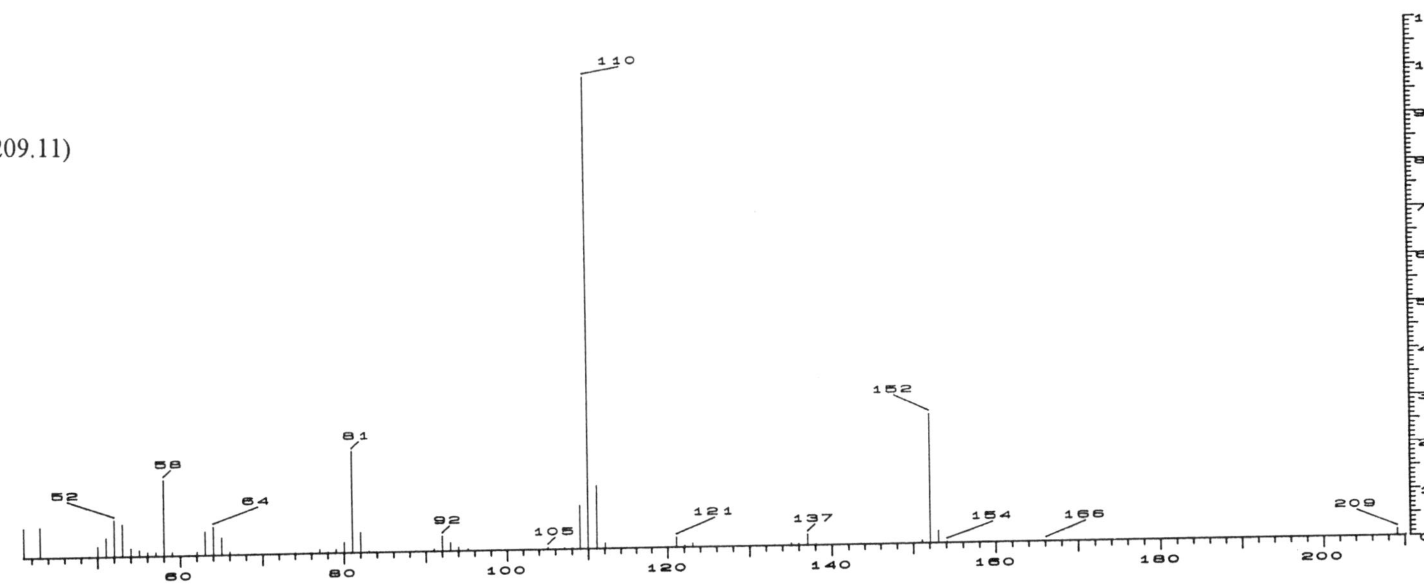

PYRAZOPHOS

$C_{14}H_{20}N_3O_5PS$ M.W. = 373.36 (373.09)

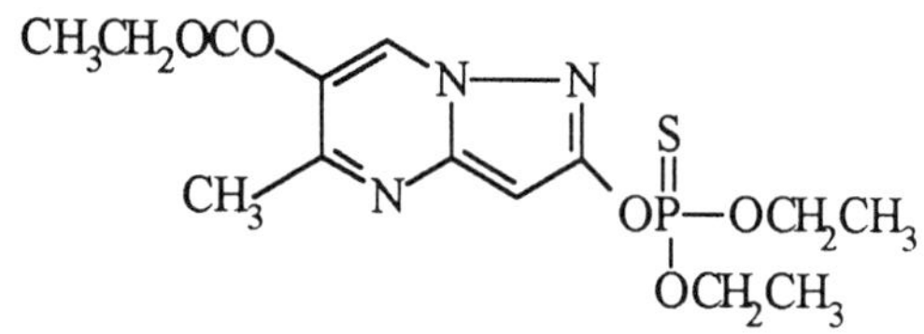

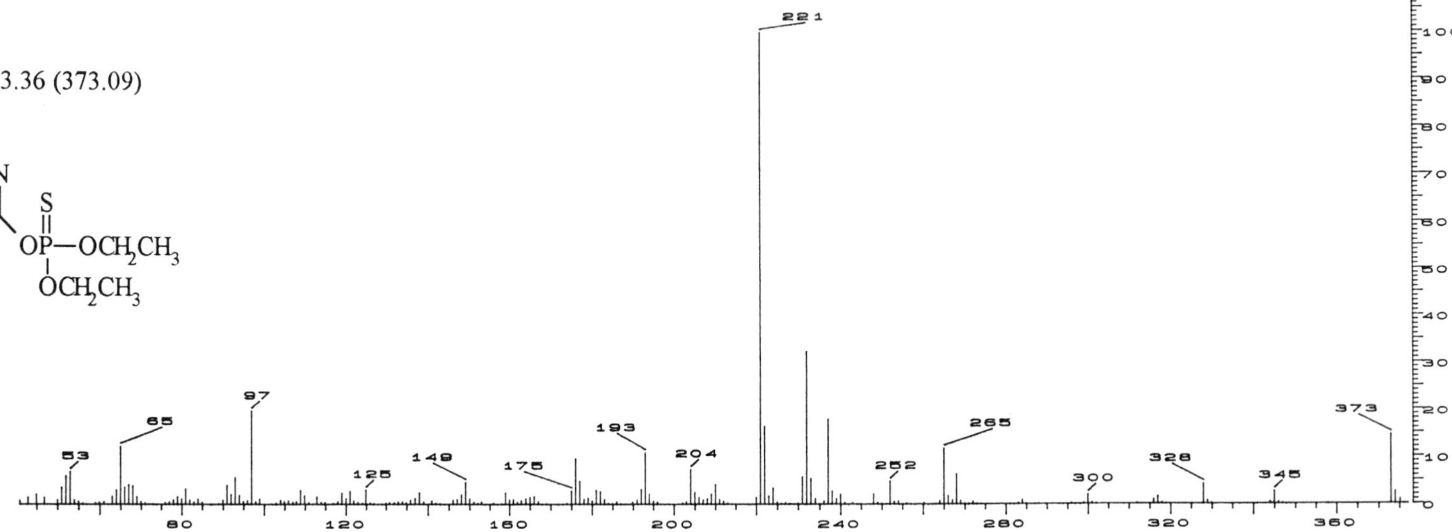

PYRIDINOL (3,5,6-TRICHLORO-2-)
Chlorpyrifos metabolite

$C_5H_2NOCl_3$ M.W. = 198.43 (196.92)

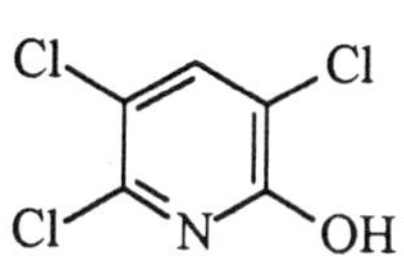

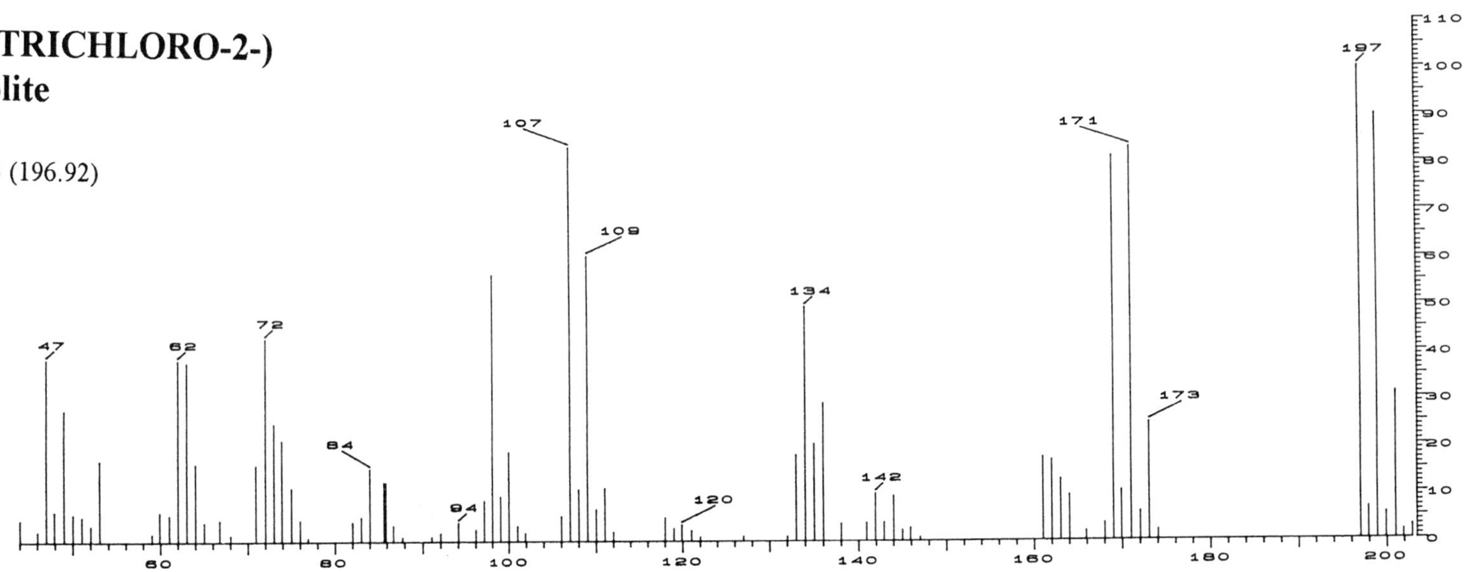

PYRIMIDINOL (2-ISOPROPYL-6-METHYL-4-)
Diazinon metabolite

$C_8H_{12}N_2O$ M.W. = 152.19 (152.09)

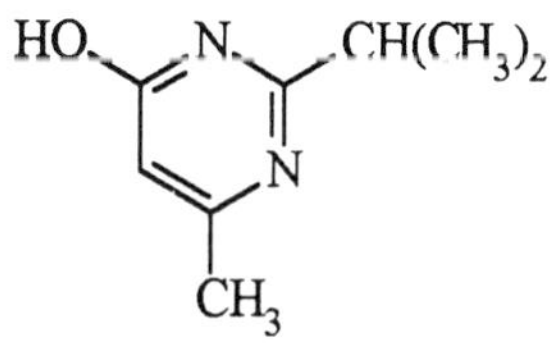

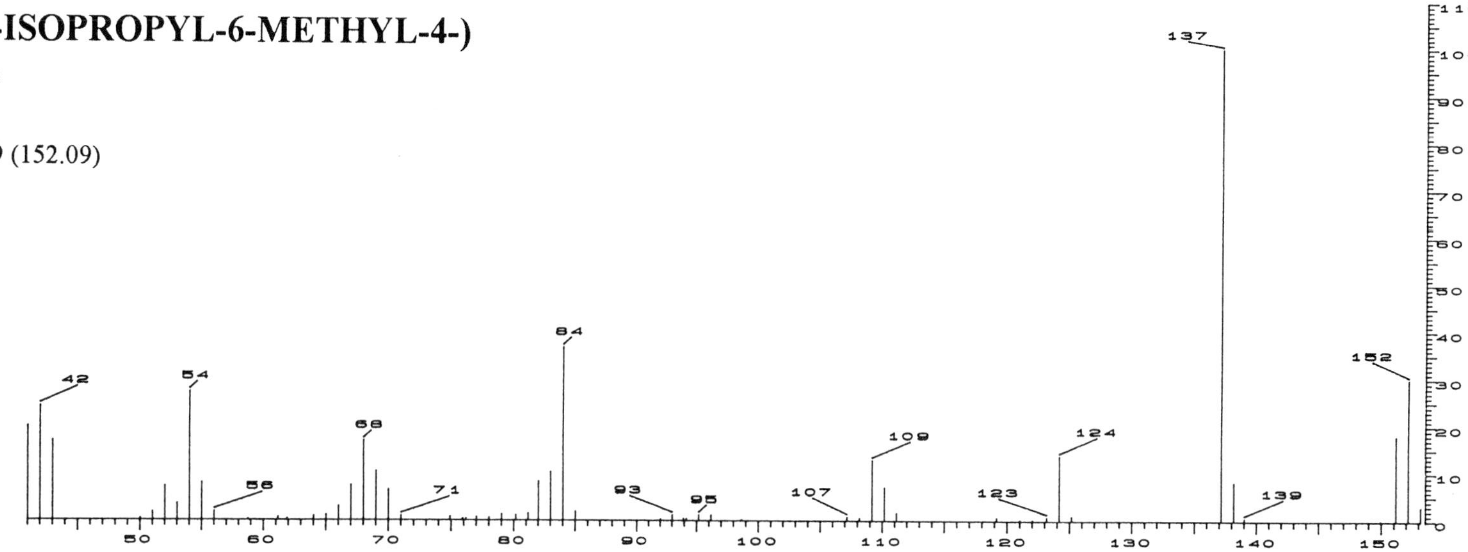

QUINTOZENE

C_6NO_2C5 M.W. = 295.33 (292.84)

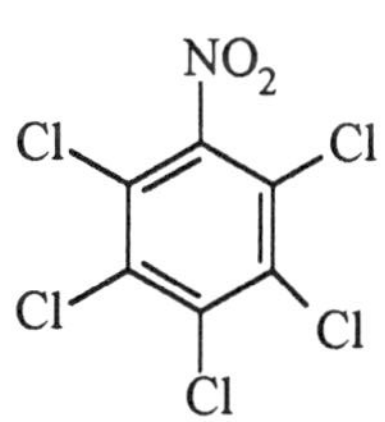

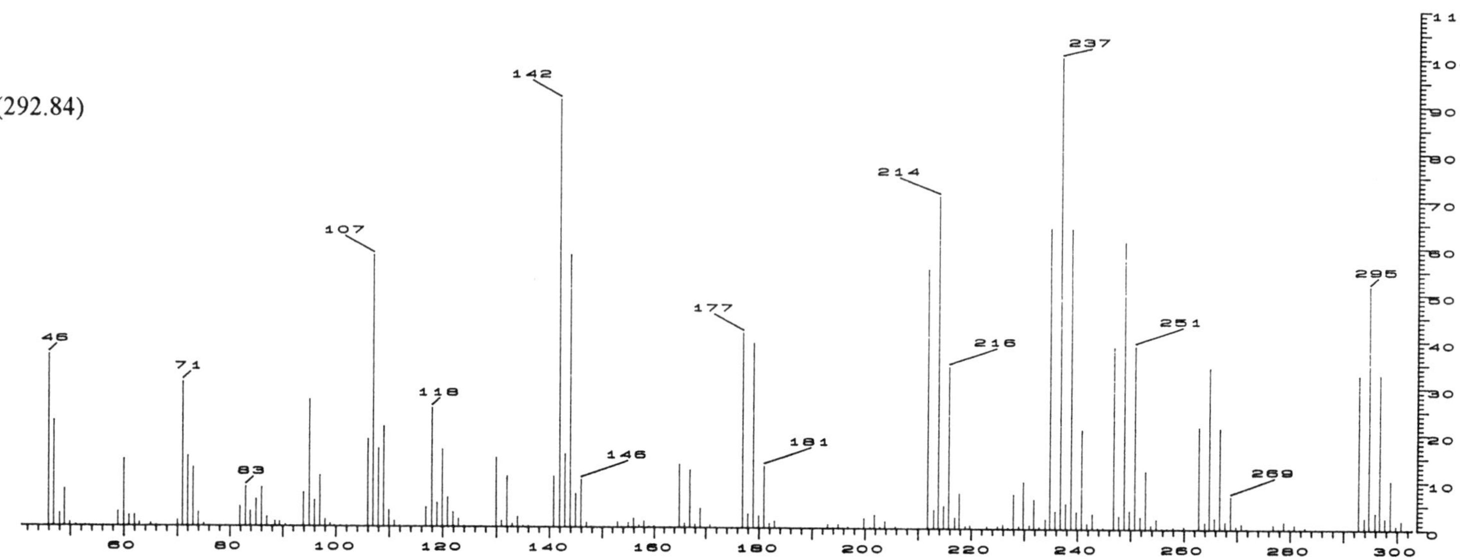

SALITHION

$C_8H_9O_3SP$ M.W. = 216.17 (216.00)

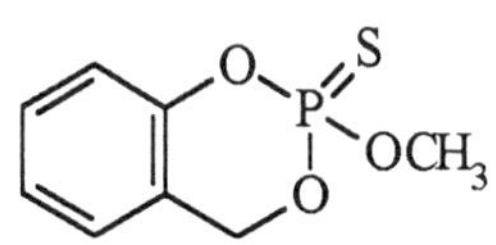

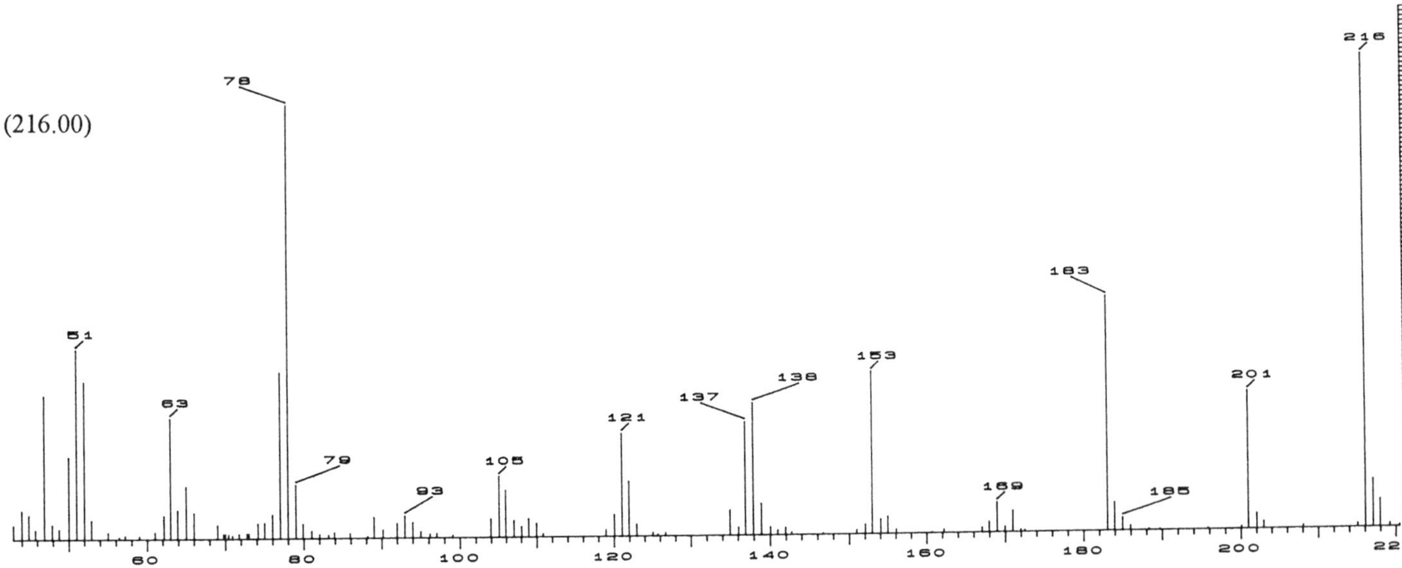

SIDURON

$C_{14}H_{20}N_2O$ M.W. = 232.32 (232.16)

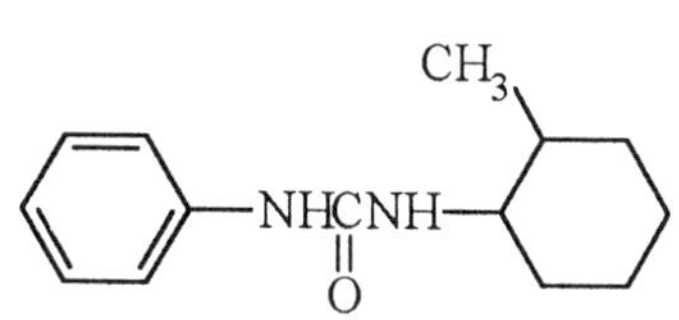

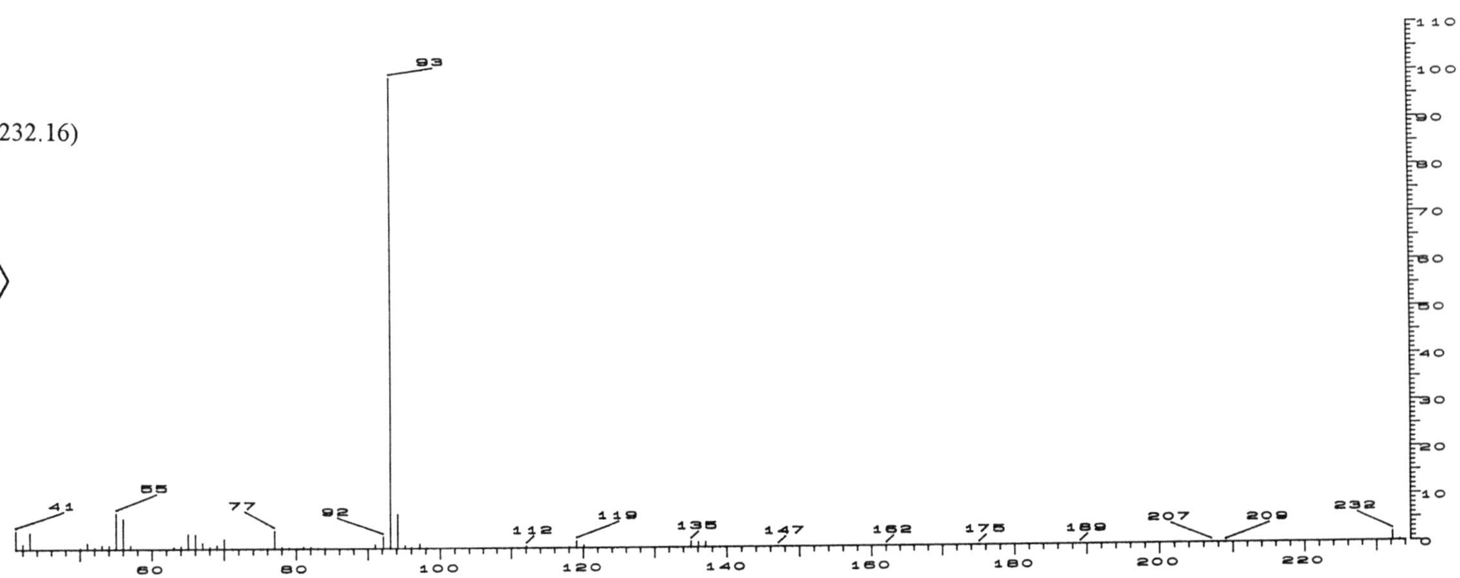

SIMAZINE

$C_7H_{12}N_5Cl$ M.W. = 201.65 (201.08)

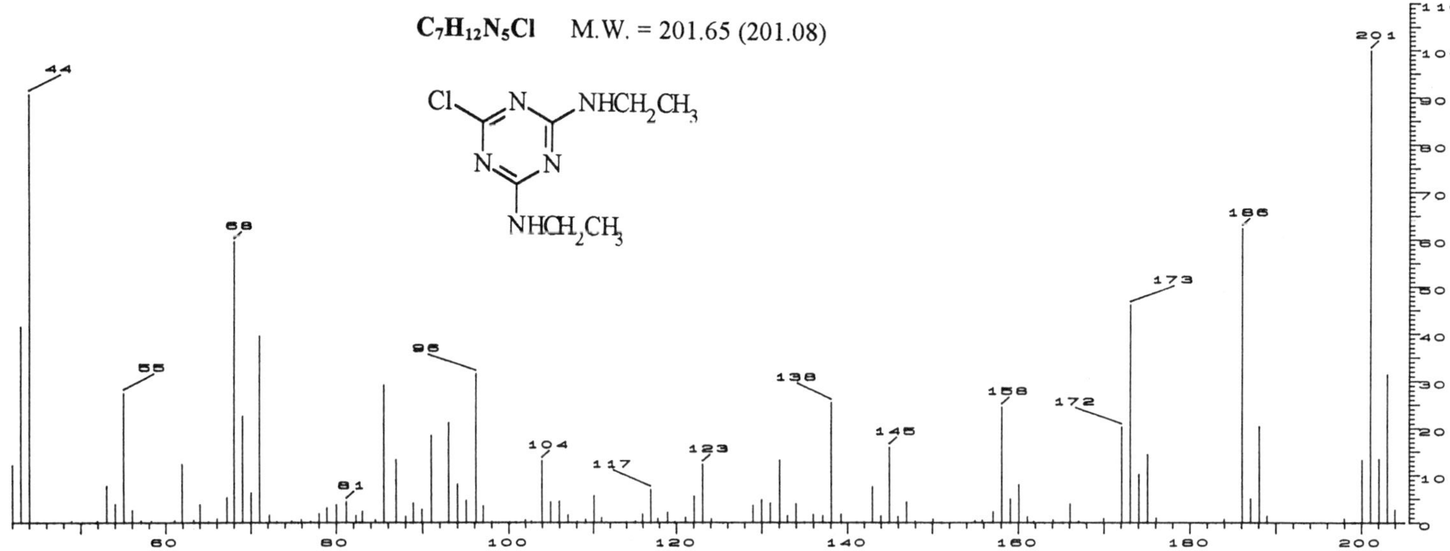

SIMAZINE METABOLITE

$C_5H_8N_5Cl$ M.W. = 173.60 (173.05)

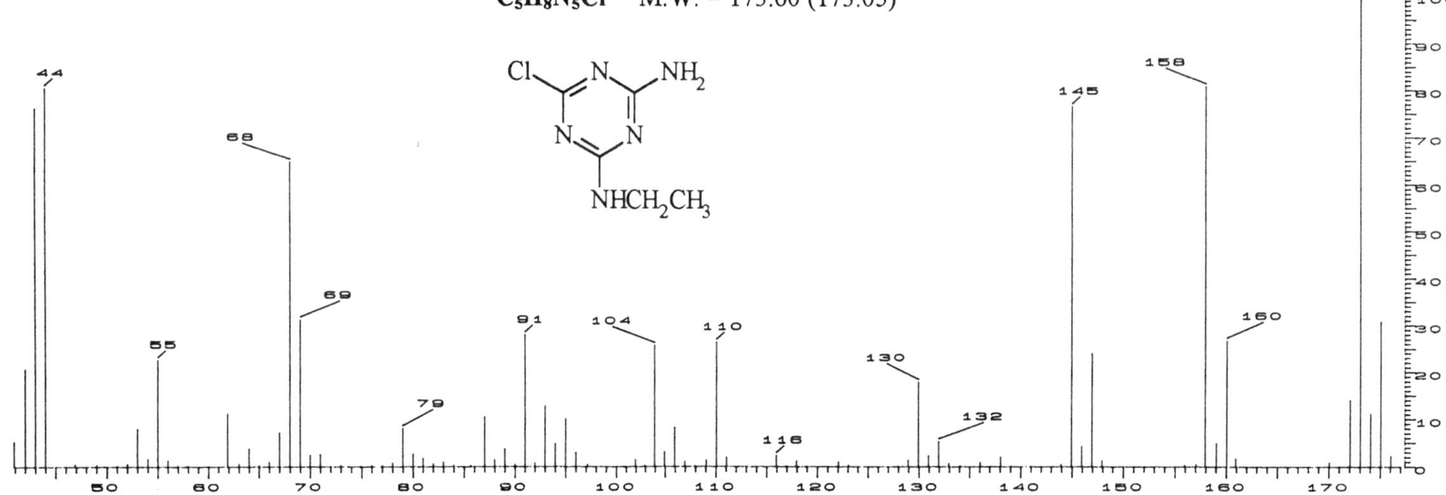

SIMETRYNE

$C_8H_{15}N_5S$ M.W. = 213.30 (213.10)

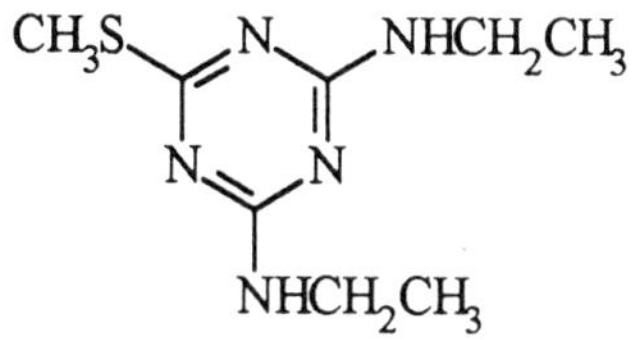

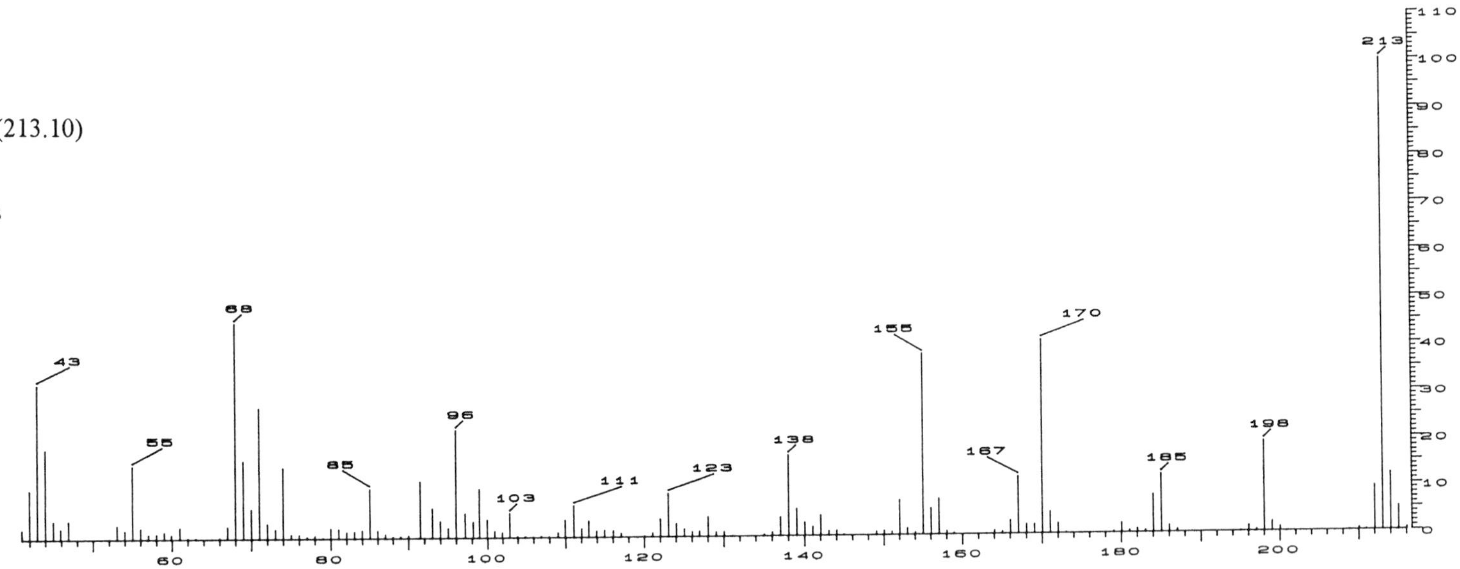

T (2,4,5-, BUTOXYPROPYL ESTER)

$C_{15}H_{19}O_4Cl_3$ M.W. = 369.67 (368.03)

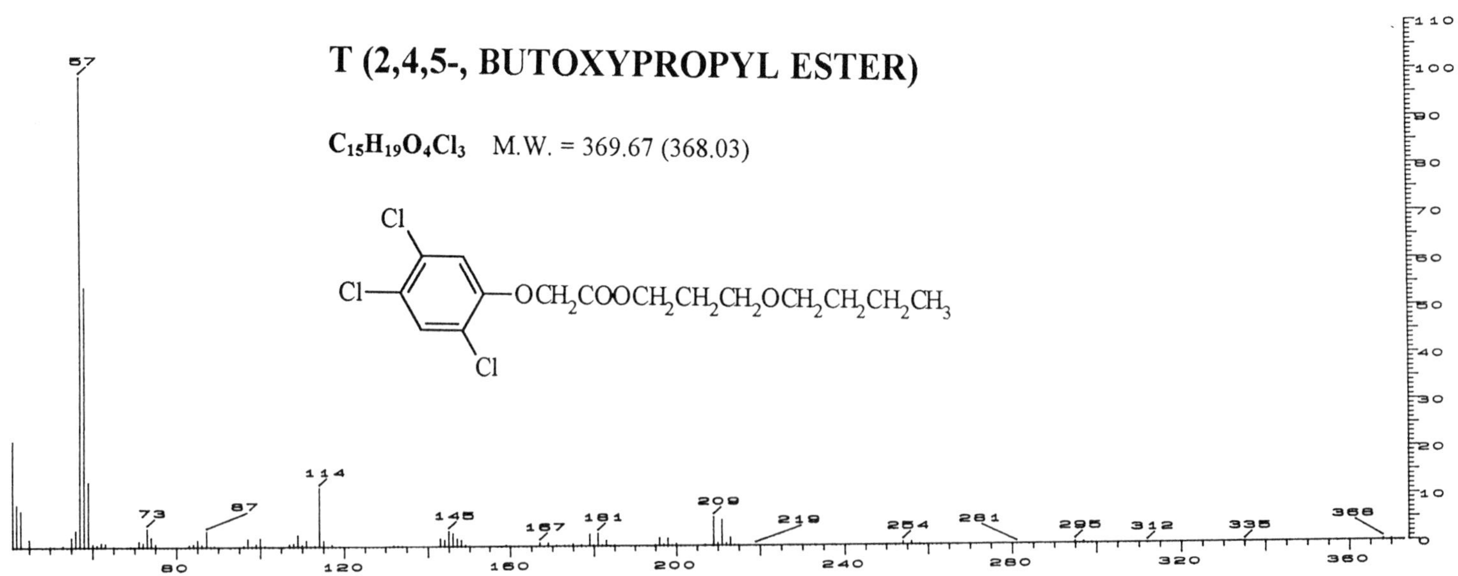

T (2,4,5-, BUTOXYPROPYLENE GLYCOL ESTER)

$C_{15}H_{19}O_4Cl_3$ M.W. = 369.67 (368.03)

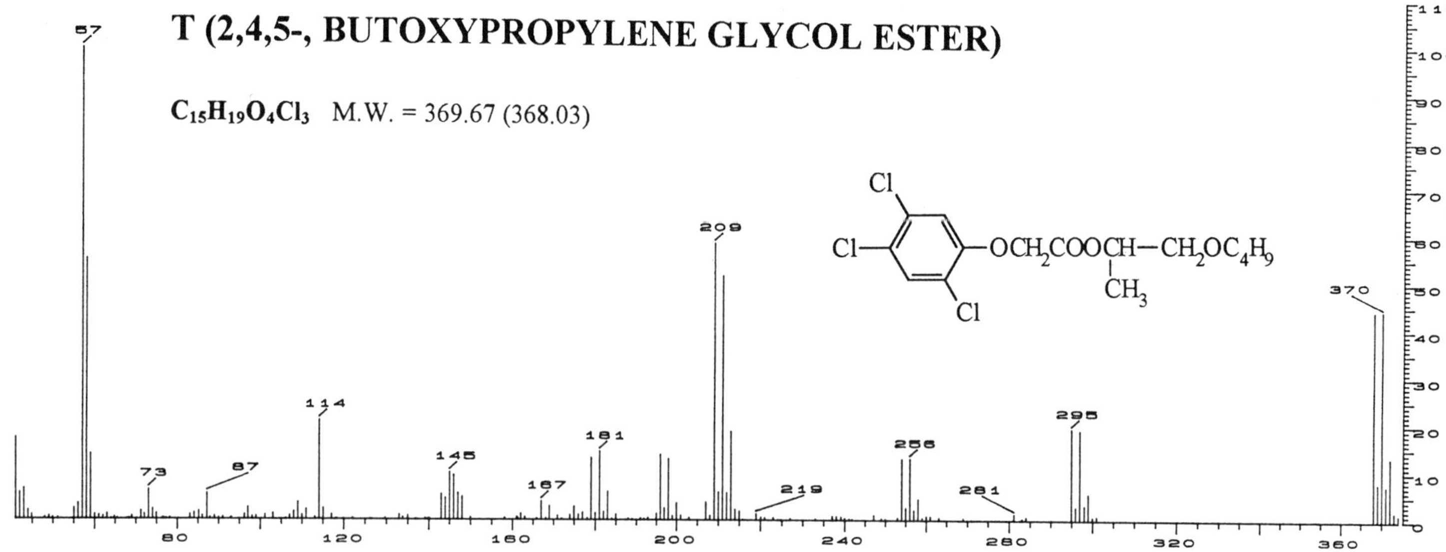

T (2,4,5-, BUTYL ESTER)

$C_{12}H_{13}O_3Cl_3$ M.W. = 311.59 (309.97)

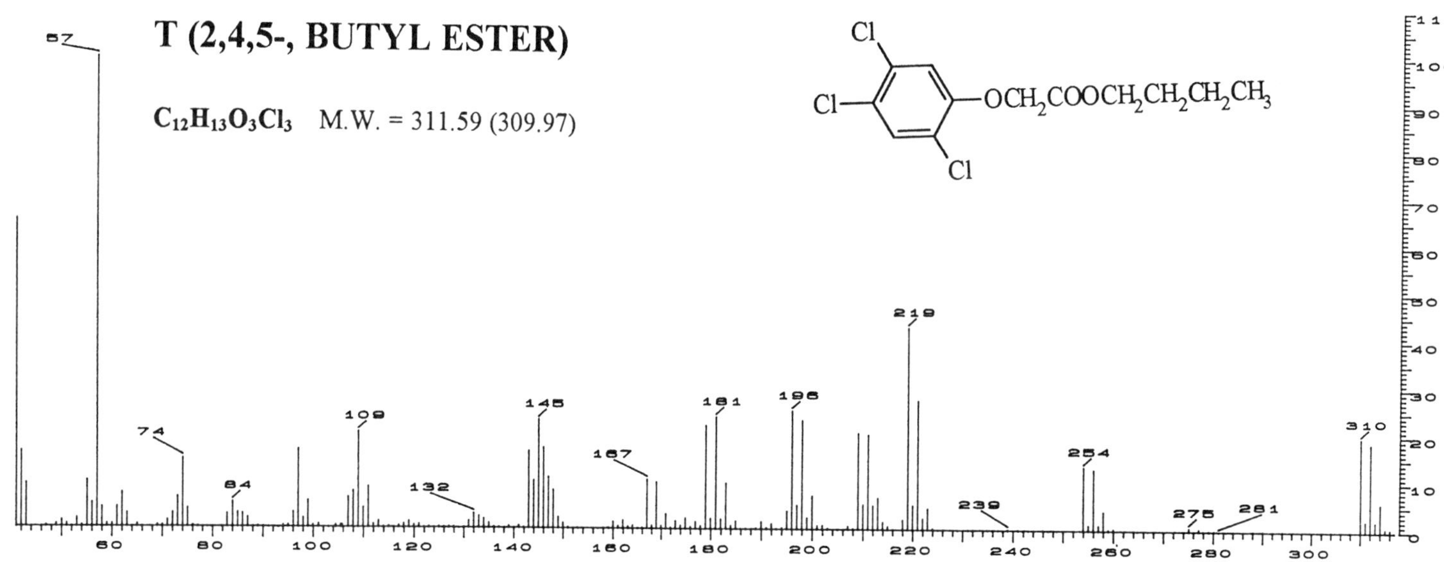

T (2,4,5-, ISOOCTYL ESTER)

$C_{16}H_{21}O_3Cl_3$ M.W. = 367.69 (366.06)

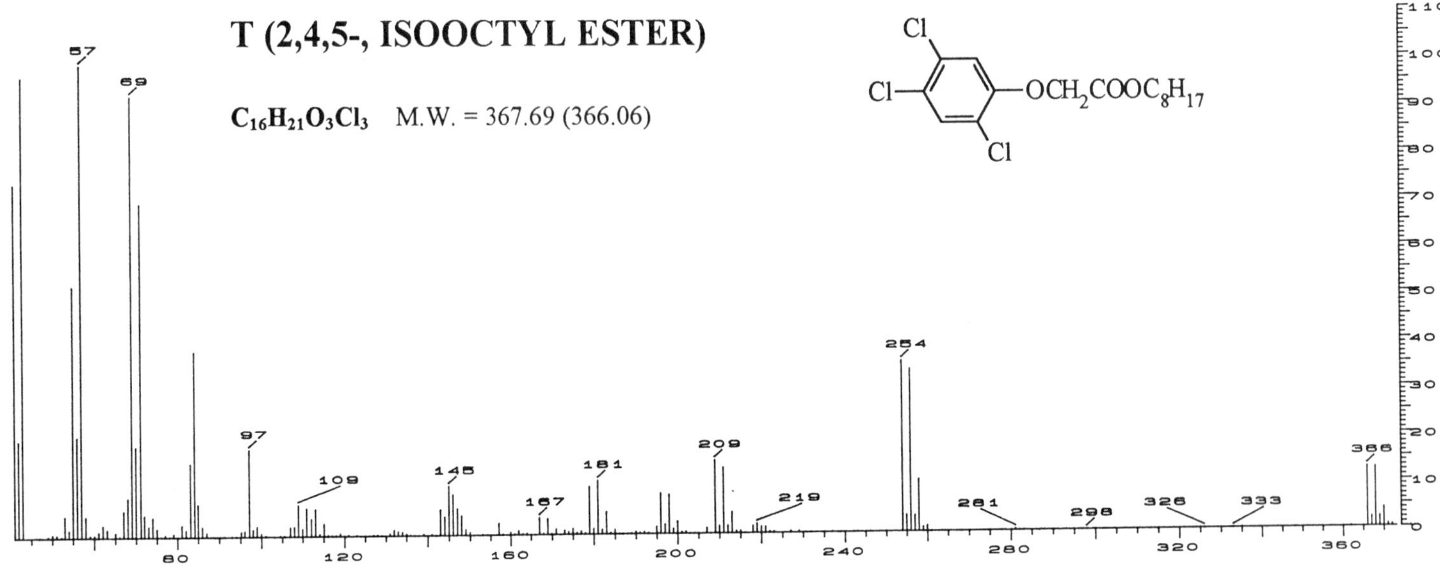

T (2,4,5-, ISOPROPYL ESTER)

$C_{11}H_{11}O_3Cl_3$ M.W. = 297.56 (295.78)

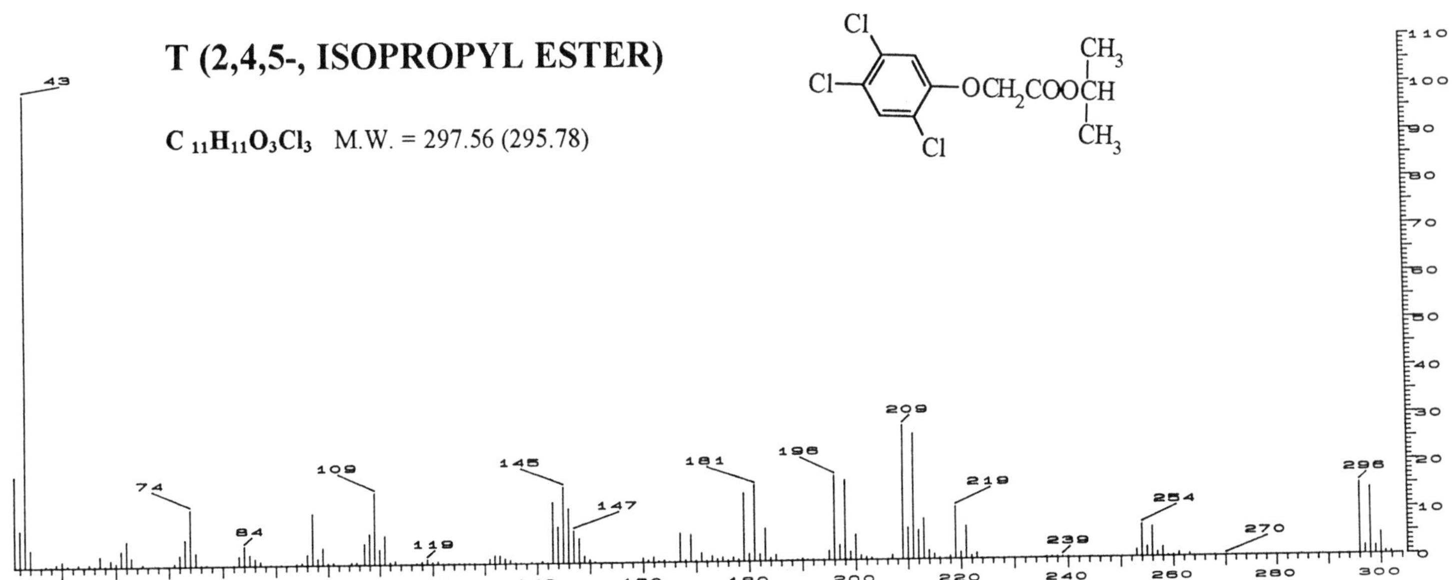

T (2,4,5-, METHYL ESTER)

$C_9H_7O_3Cl_3$ M.W. = 269.51 (267.95)

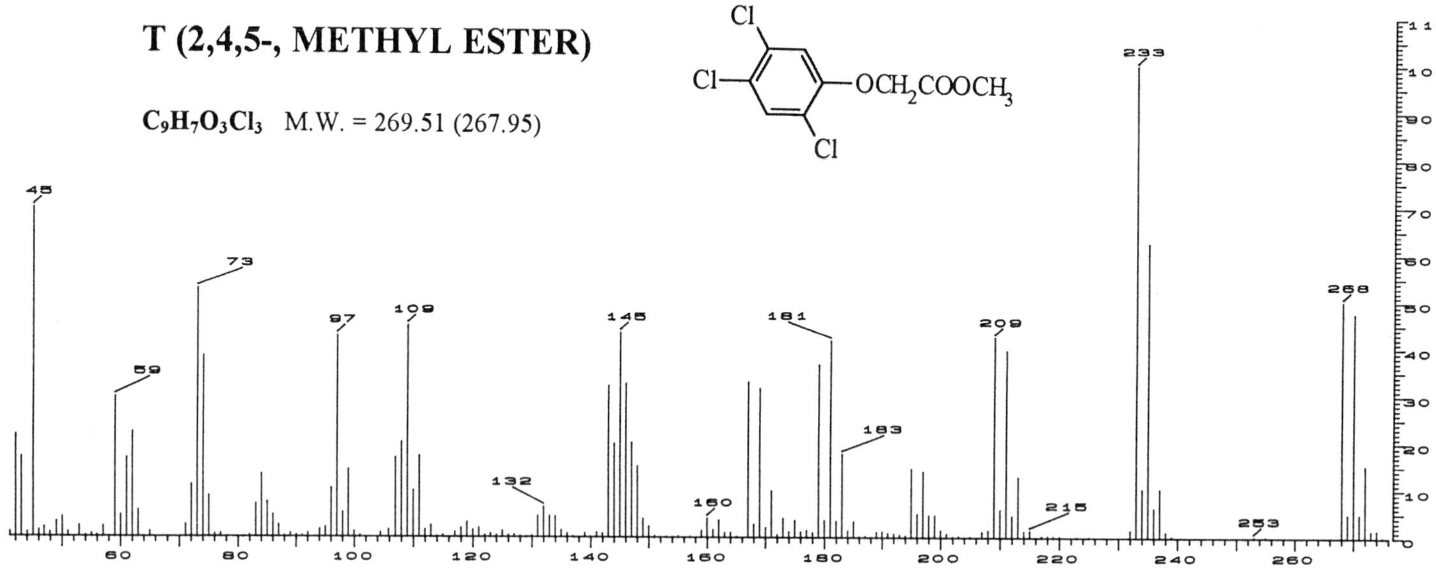

TECNAZENE

$C_6HNO_2Cl_4$ M.W. = 260.89 (258.88)

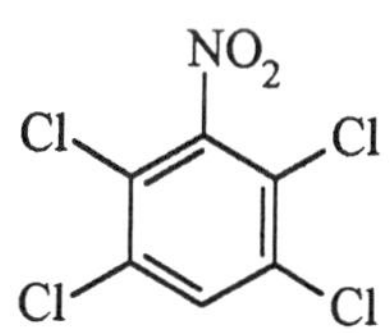

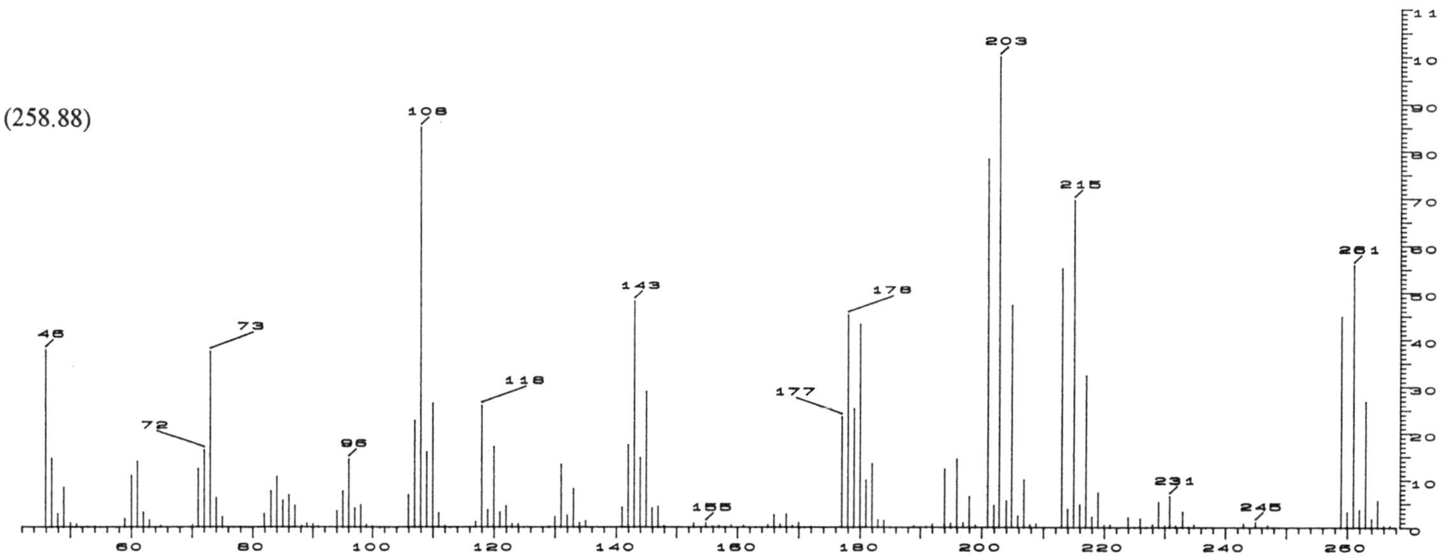

TEMEPHOS

$C_{16}H_{20}O_6S_3P_2$ M.W. = 466.45 (465.99)

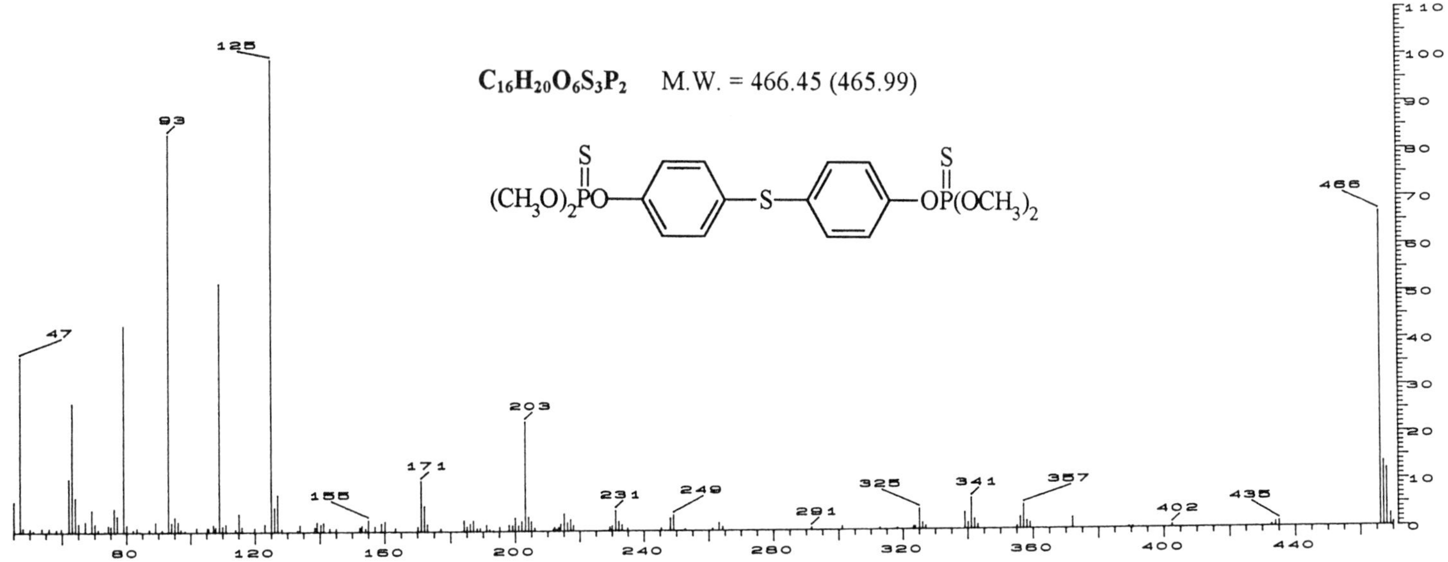

TERBACIL

$C_9H_{13}N_2O_2Cl$ M.W. = 216.66 (216.07)

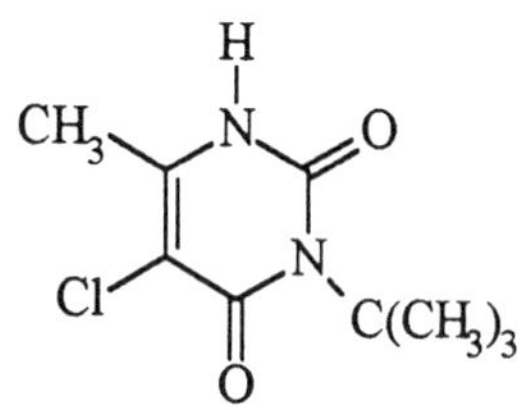

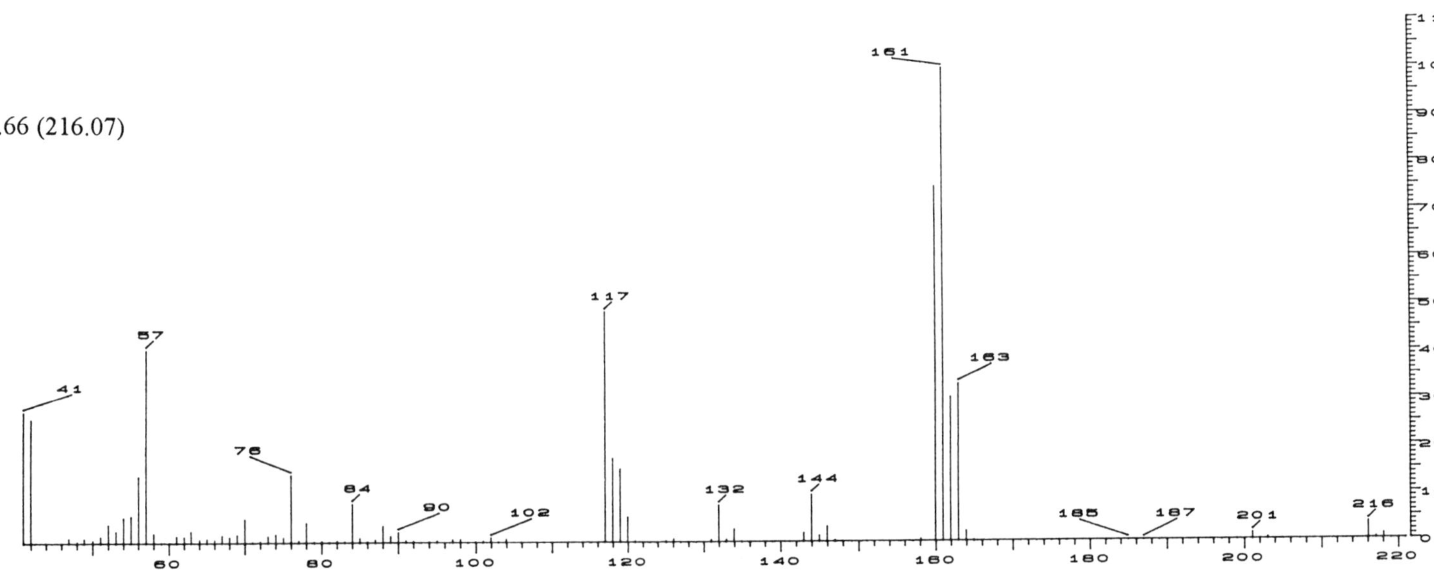

TERBUFOS

$C_9H_{21}O_2S_3P$ M.W. = 288.41 (288.04)

$(CH_3)_3CSCH_2SP(S)(OCH_2CH_3)_2$

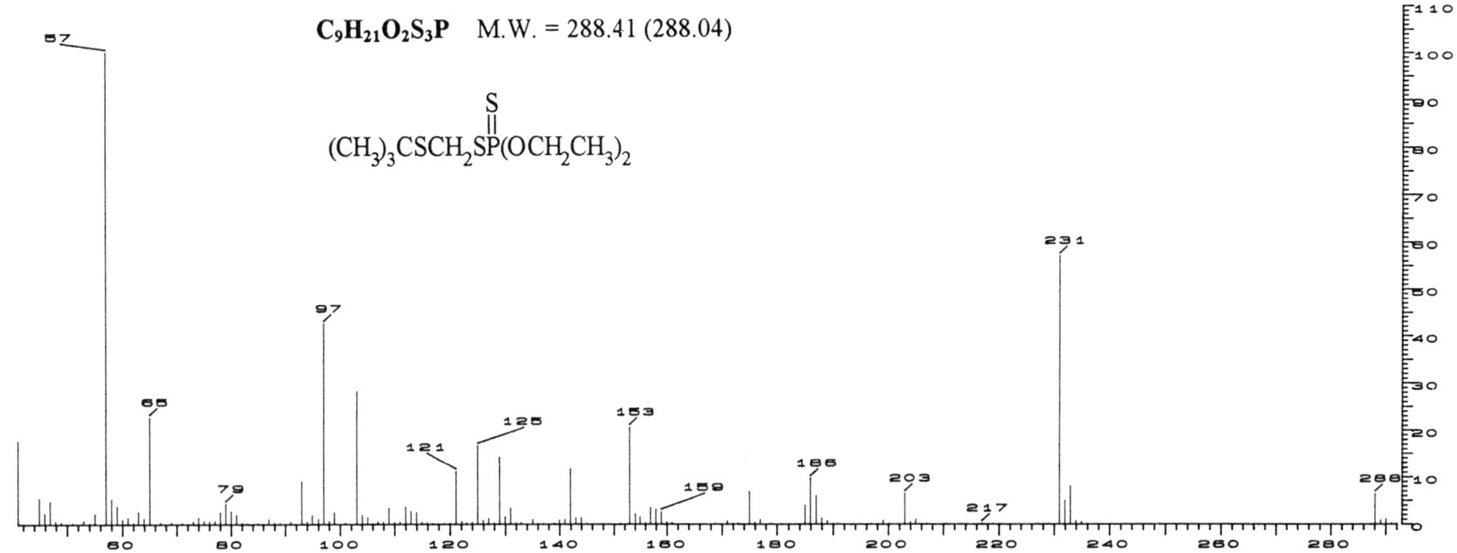

TERBUTHYLAZINE

$C_9H_{16}N_5Cl$ M.W. = 229.71 (229.11)

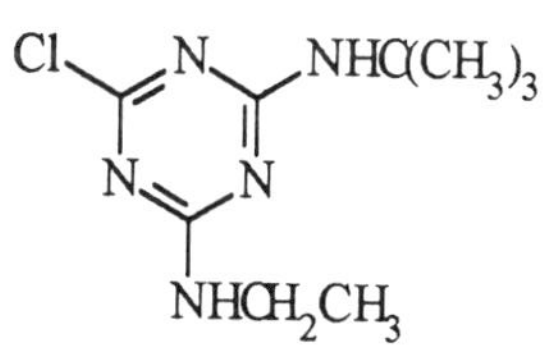

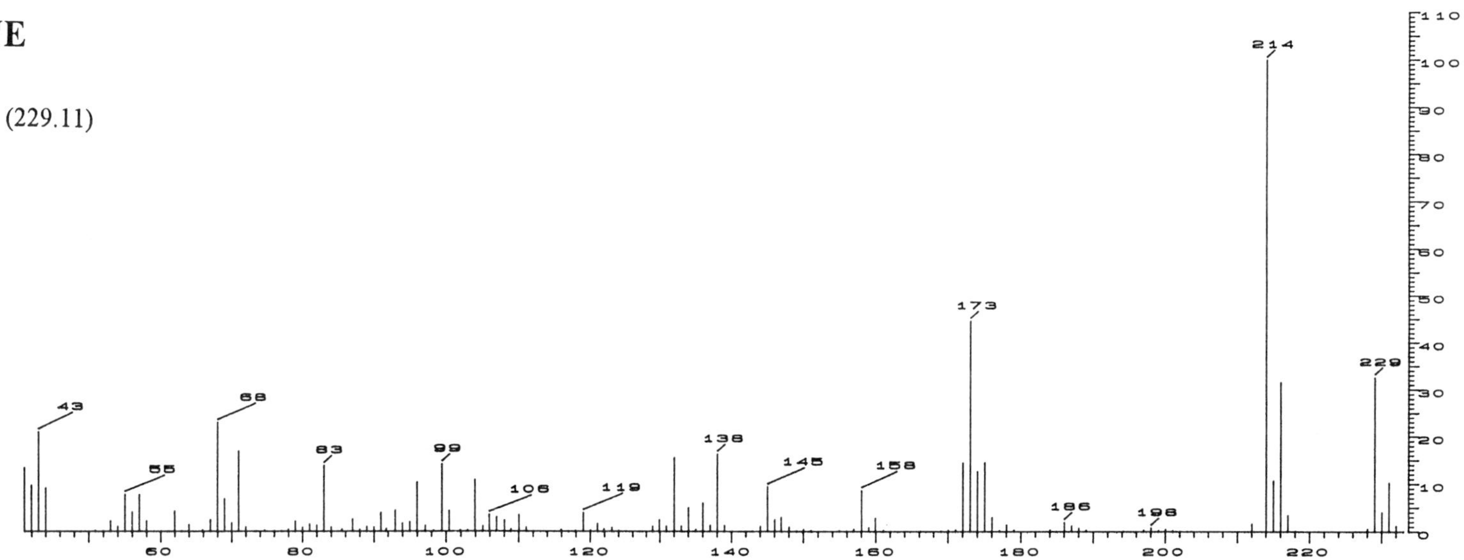

TERBUTRYNE

$C_{10}H_{19}N_5S$ M.W. = 241.35 (241.14)

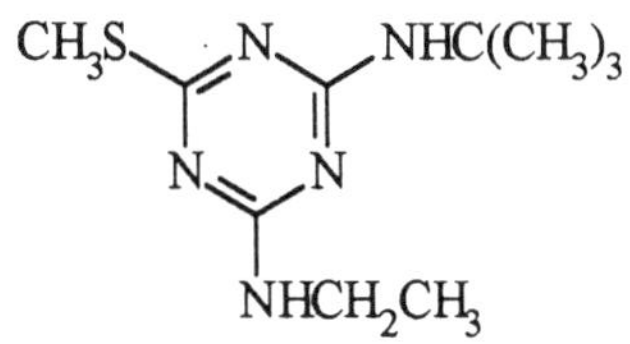

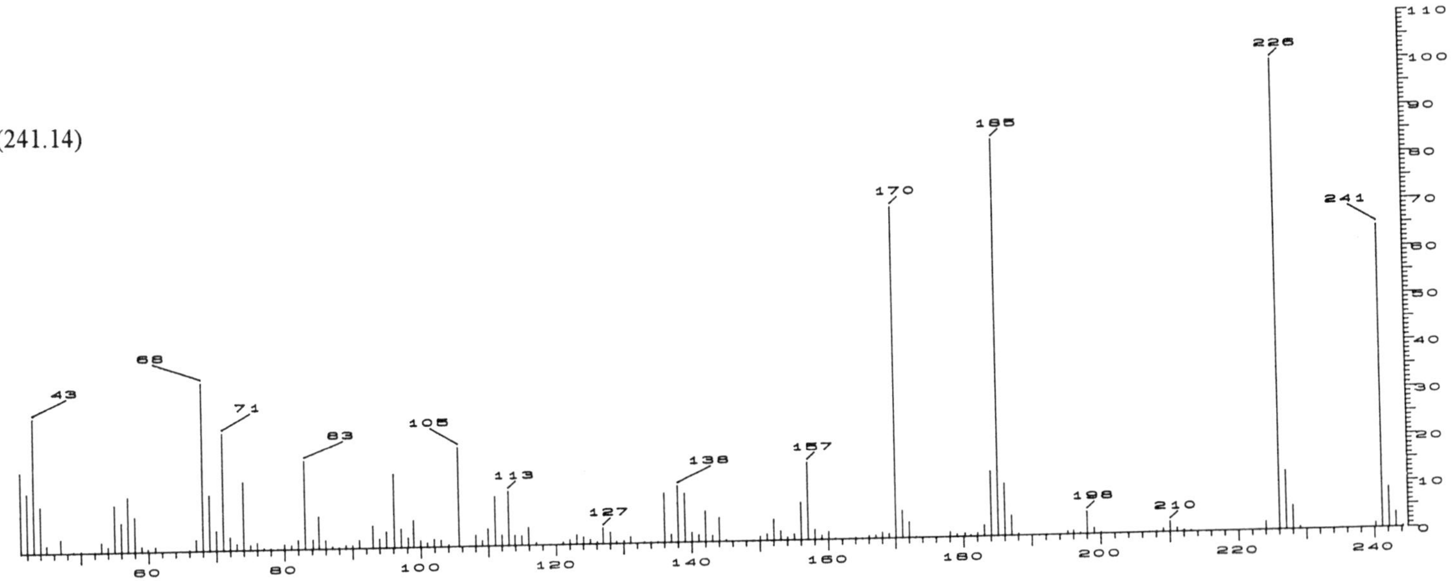

TETRACHLORVINPHOS

$C_{10}H_9O_4Cl_4P$ M.W. = 365.96 (363.90)

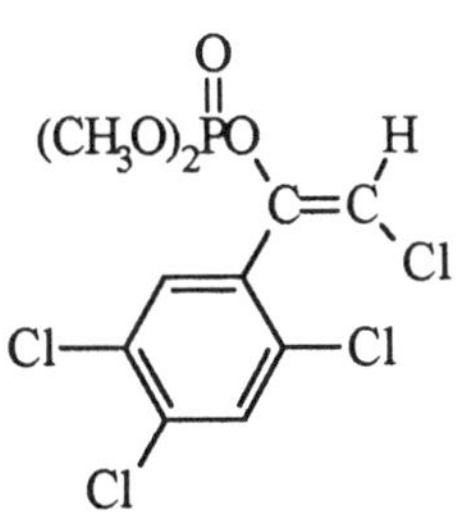

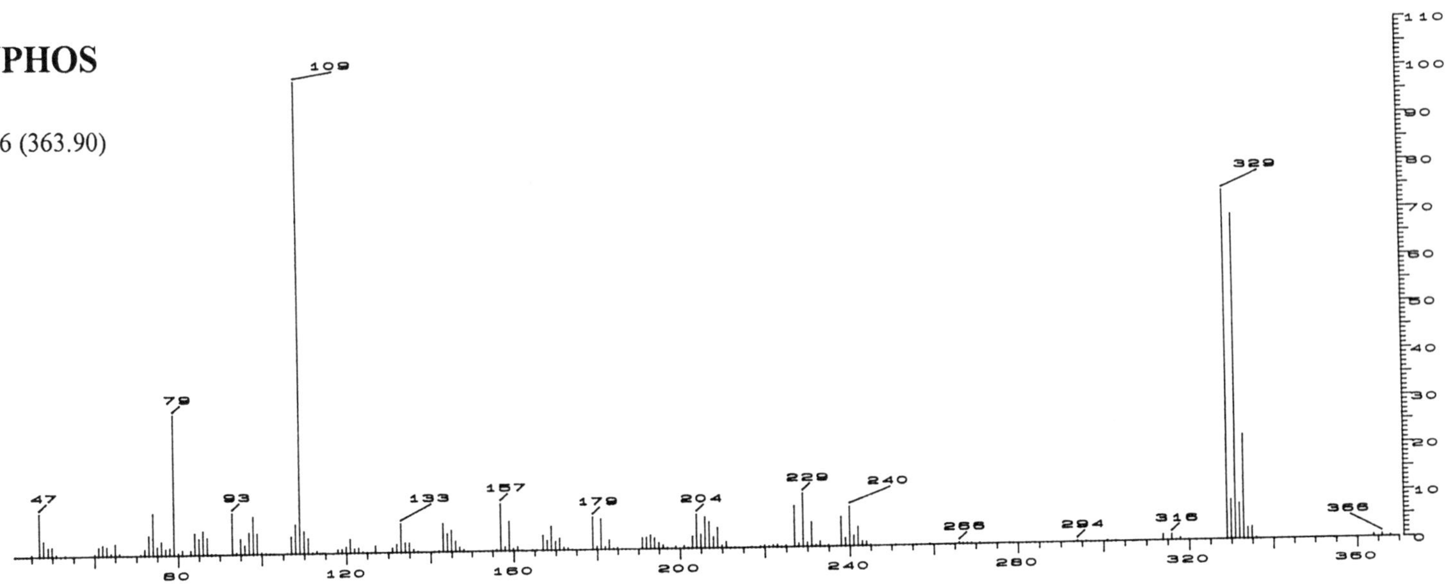

TETRADIFON

$C_{12}H_6O_2Cl_4S$ M.W. = 356.05 (353.88)

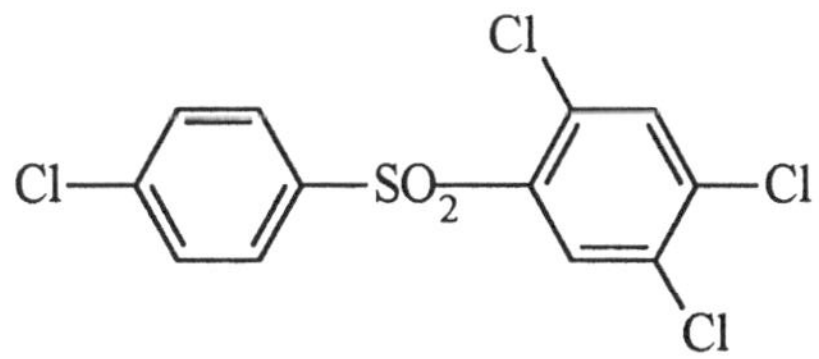

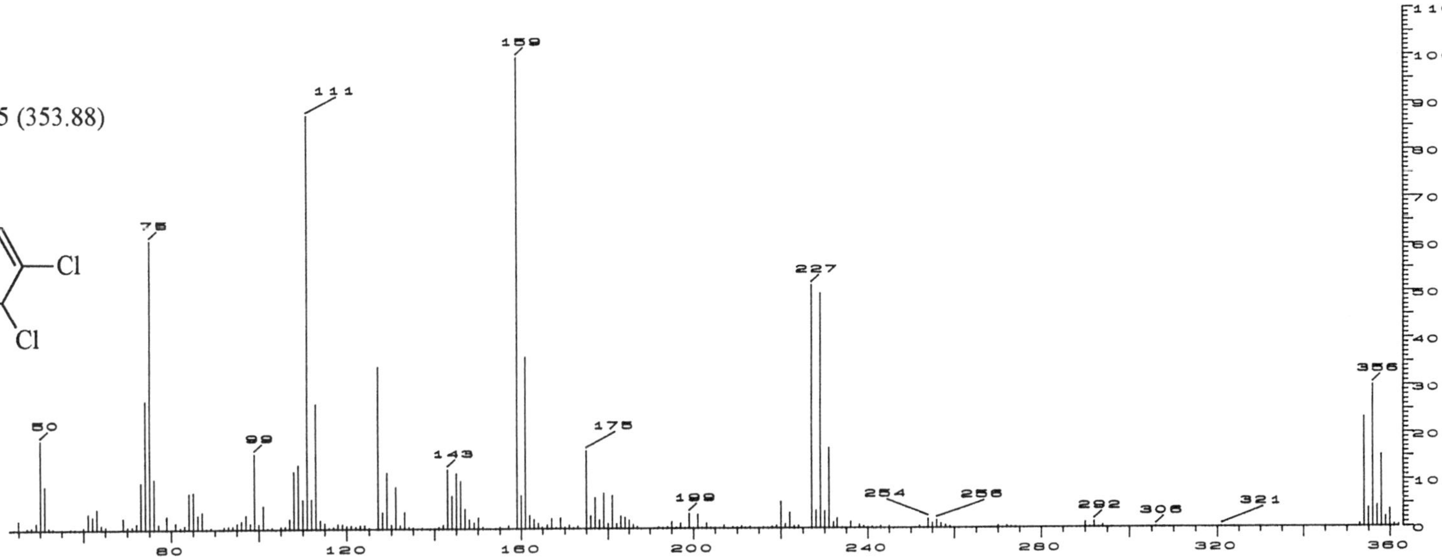

TETRAHYDROPHTHALIMIDE
Captan metabolite

$C_8H_9NO_2$ M.W. = 151.16 (151.06)

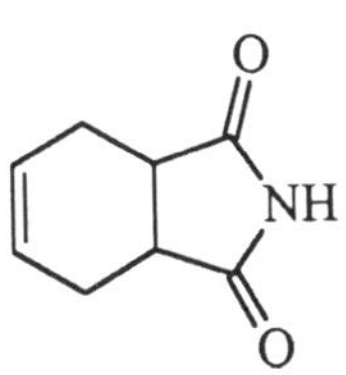

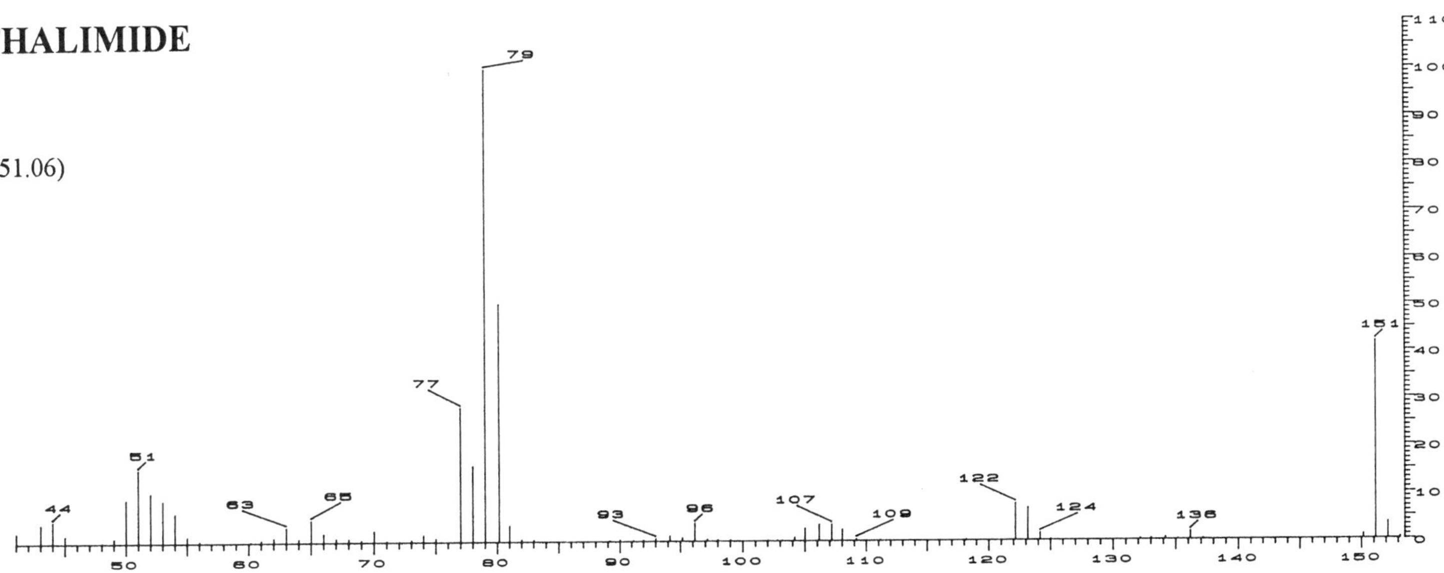

THIOMETON

$C_6H_{15}O_2S_3P$ M.W. = 246.33 (246.00)

$CH_3CH_2SCH_2CH_2SP(S)(OCH_3)_2$

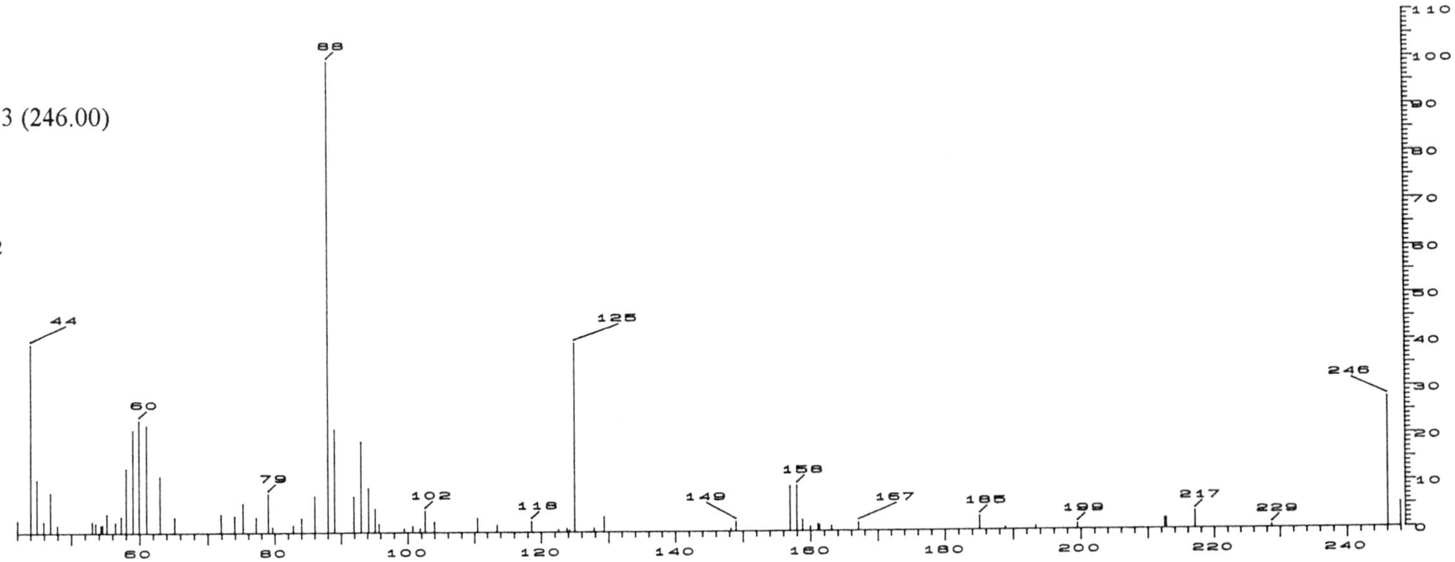

THIONAZIN

$C_8H_{13}N_2O_3SP$ M.W. = 248.23 (248.04)

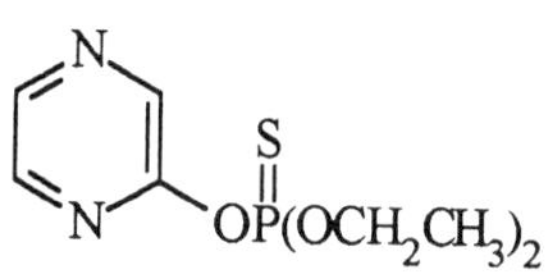

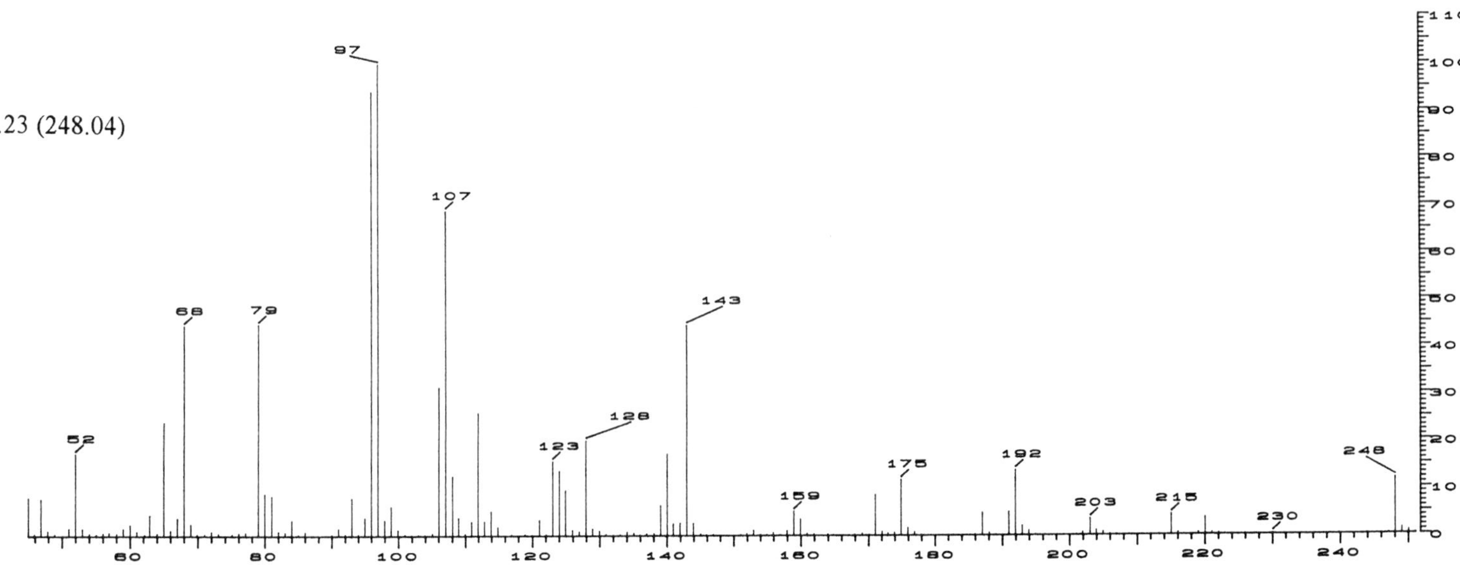

TOLYLFLUANID

$C_{10}H_{13}N_2O_2FCl_2S_2$ M.W. = 347.24 (345.98)

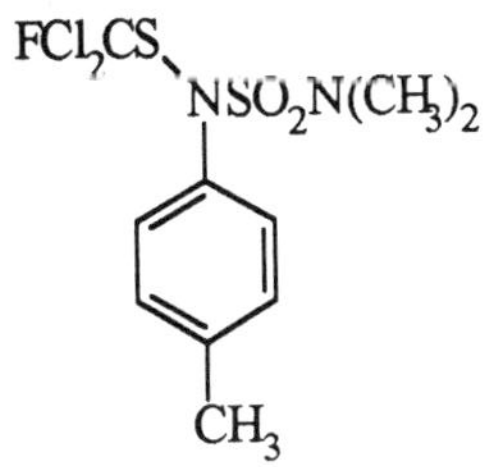

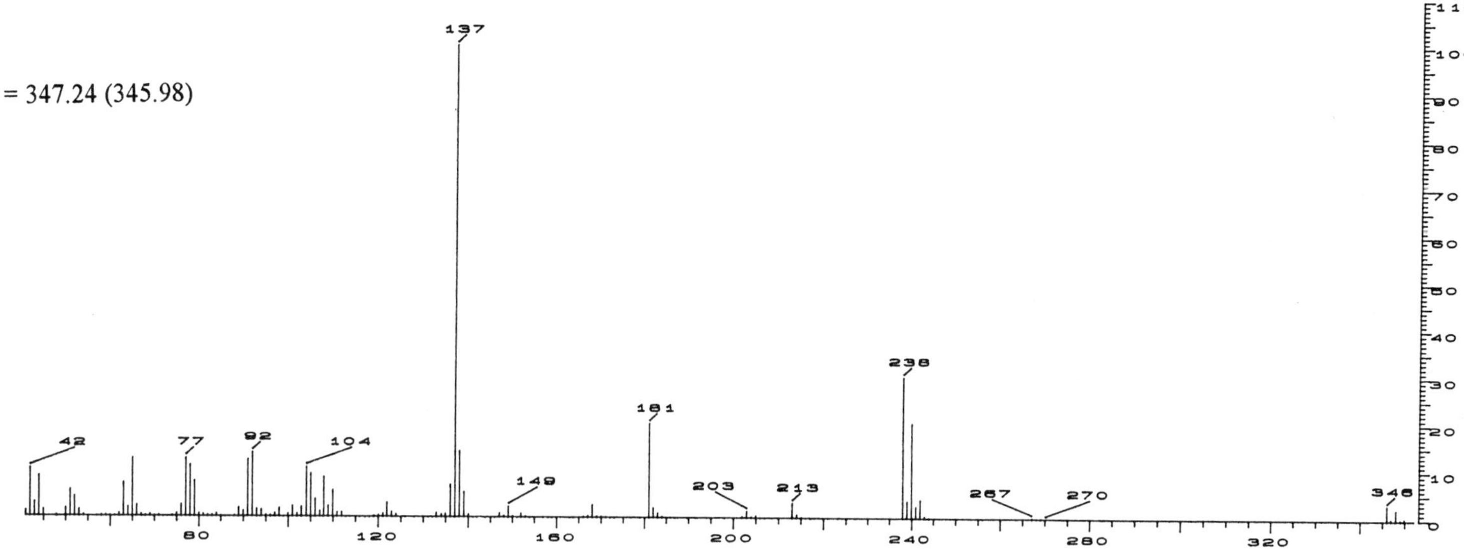

TRIALLATE

$C_{10}H_{16}NOCl_3S$ M.W. = 304.66 (303.00)

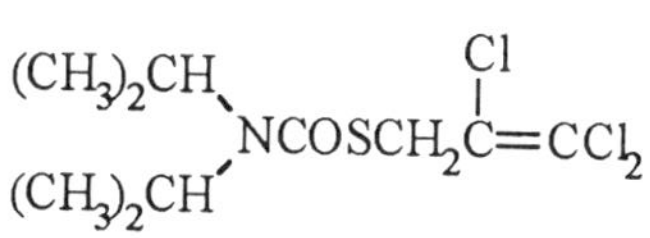

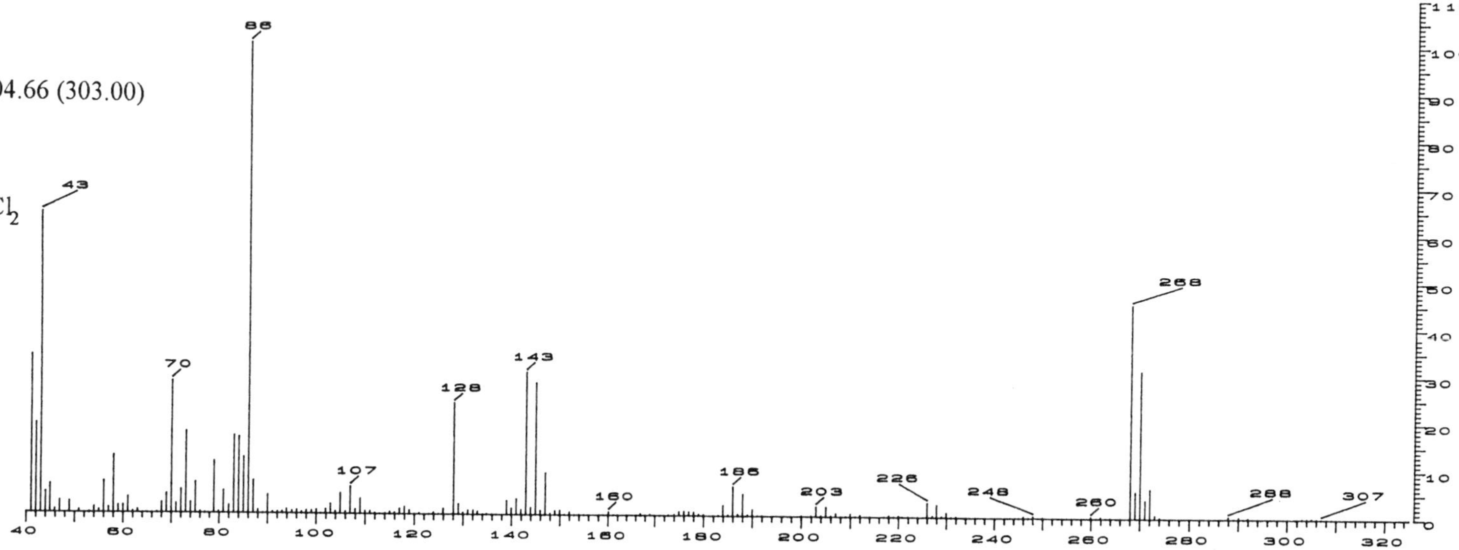

TRIAZOPHOS

$C_{12}H_{16}N_3O_3P_5$ M.W. = 313.31 (313.07)

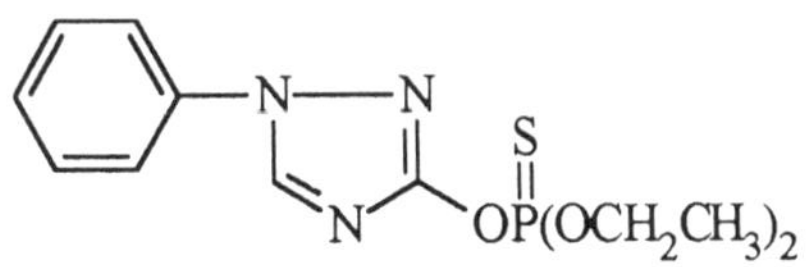

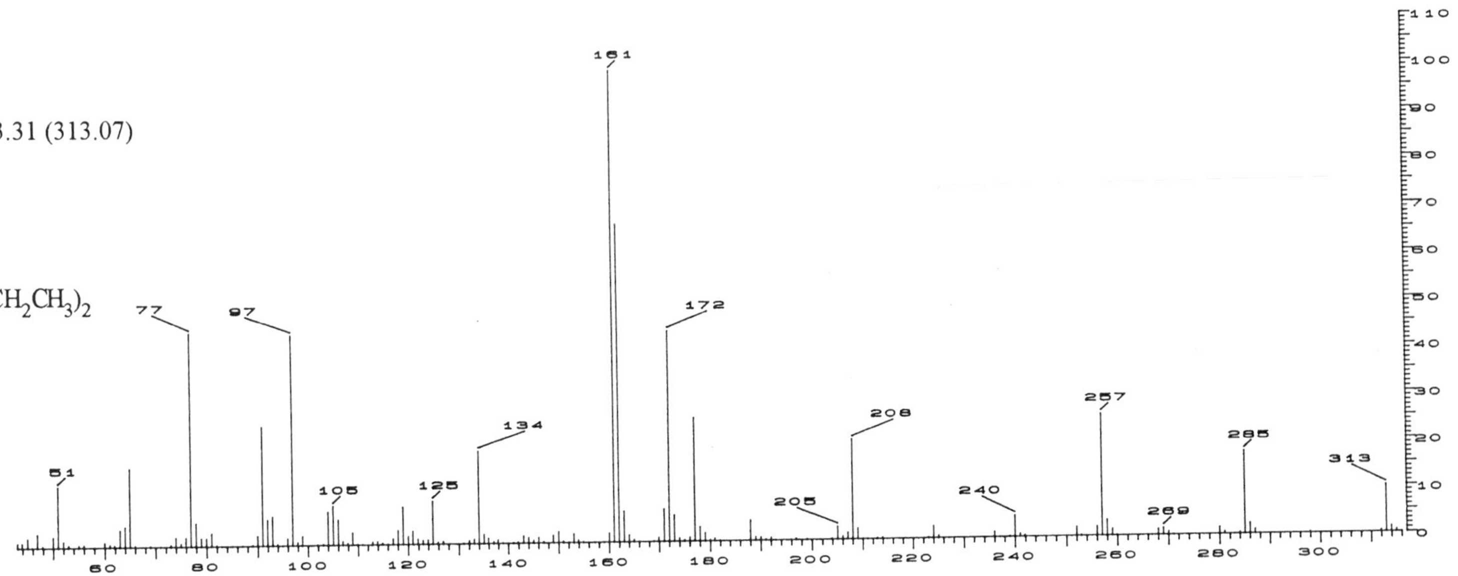

TRIETAZINE

$C_9H_{16}N_5Cl$ M.W. = 229.71 (229.11)

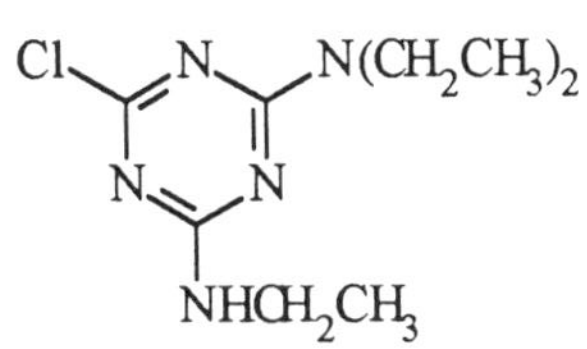

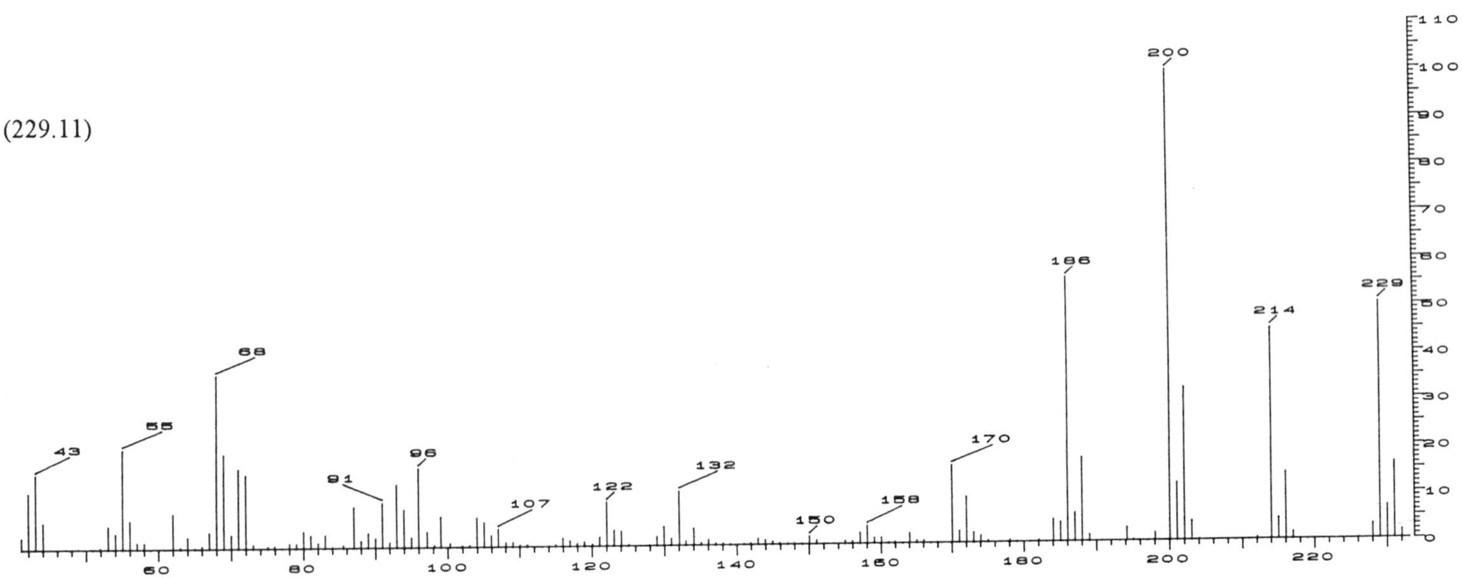

TRIFLURALIN

$C_{13}H_{16}N_3O_4F_3$ M.W. = 335.28 (335.11)

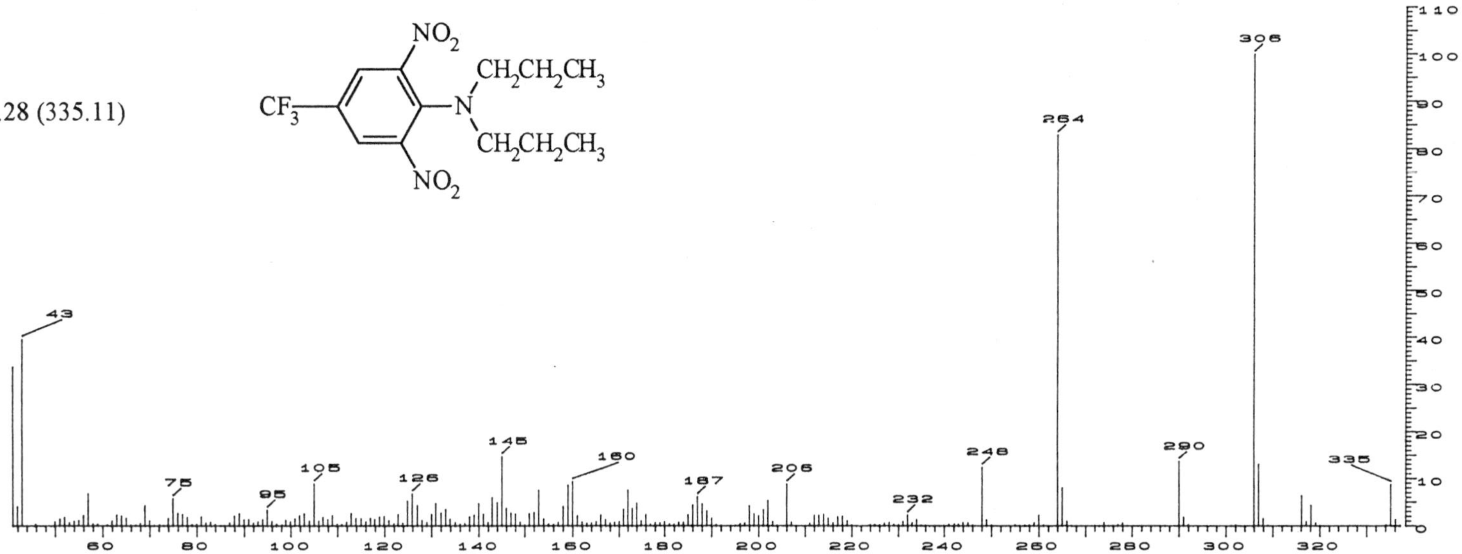

VALONE

$C_{14}H_{13}O_3$ M.W. = 230.25 (230.09)

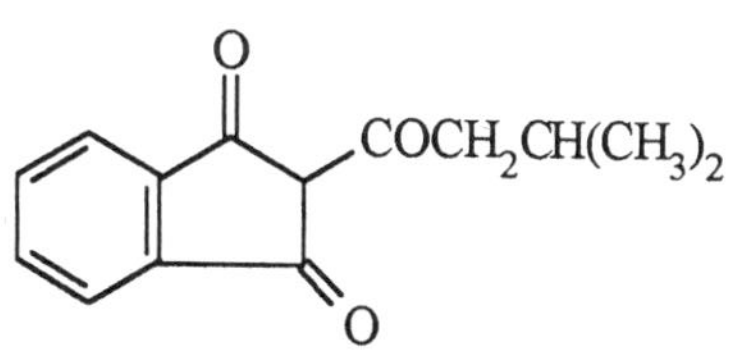

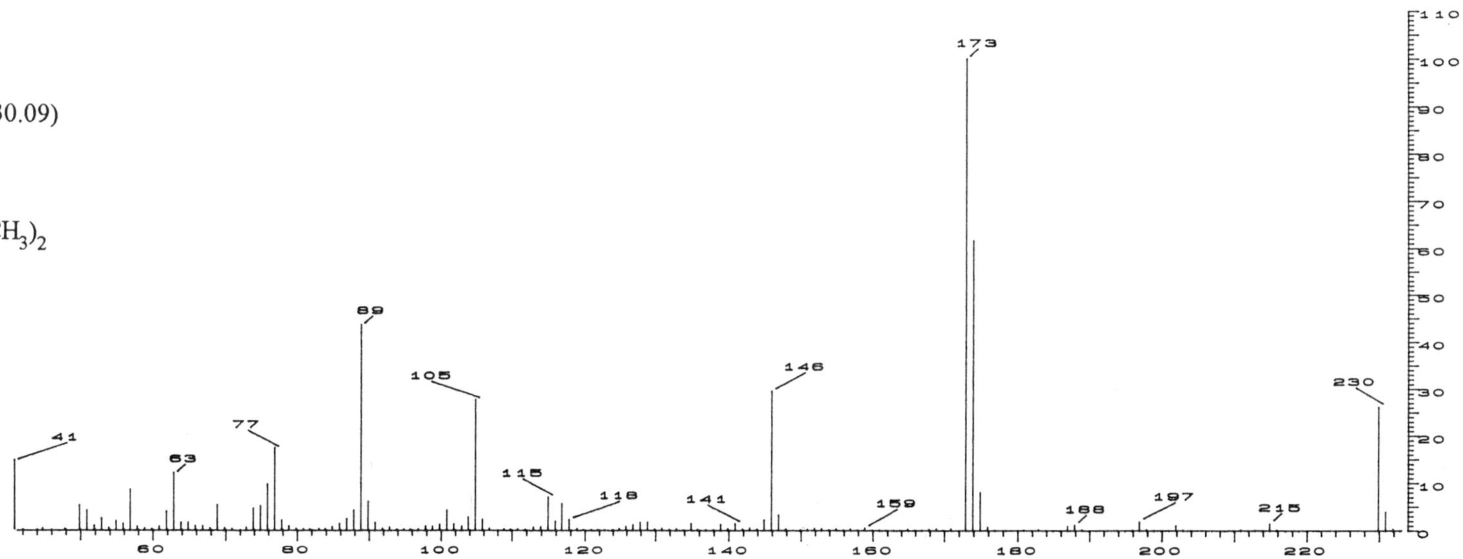

VINCLOZOLIN

$C_{12}H_9NO_3Cl_2$ M.W. = 286.11 (285.00)

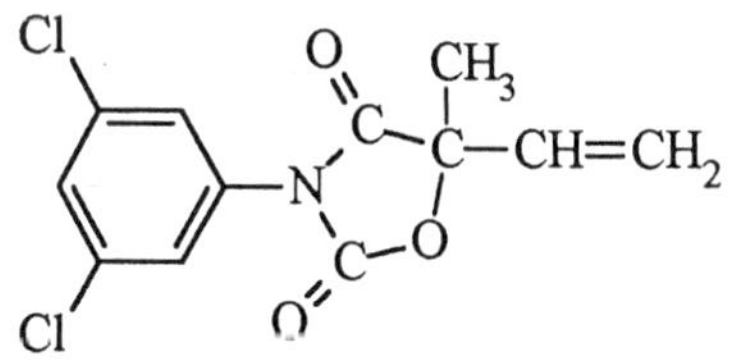

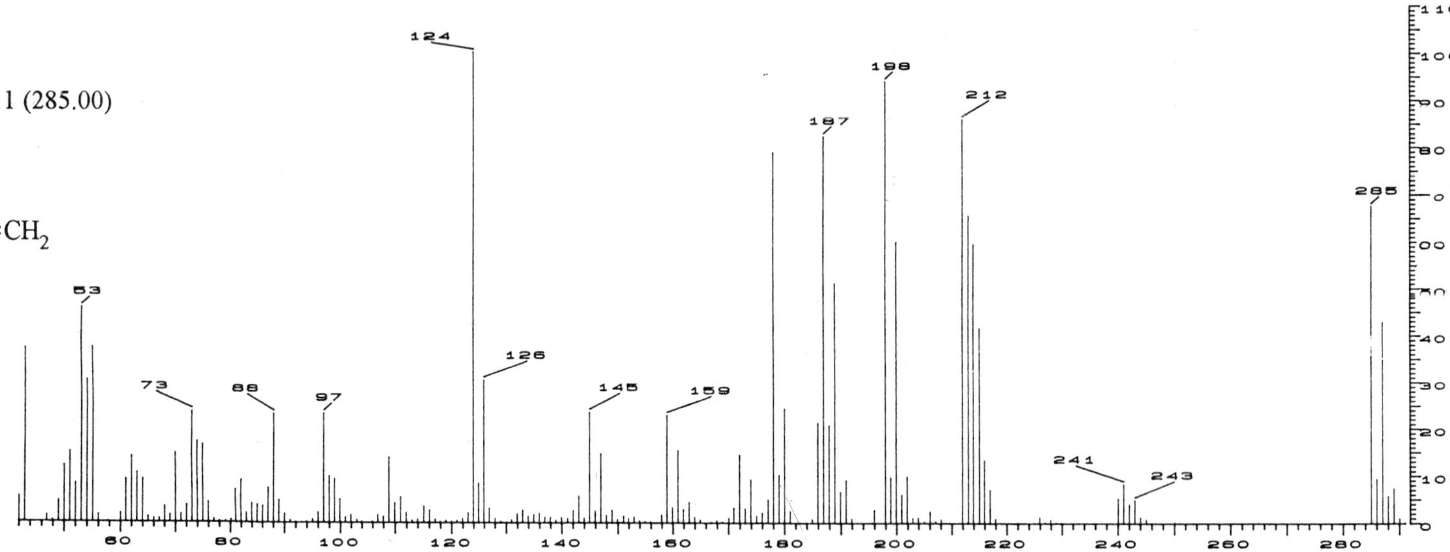